AF593560

Handbook of Industrial Chemical Additives

An International Guide by
PRODUCT, TRADE NAME FUNCTION, AND SUPPLIER

Compiled by
Michael and Irene Ash

Contains over 18,000 entries for chemical trademark products currently sold throughout the world

VCH Publishers
New York

Michael Ash
Irene Ash
Synapse Information Resources, Inc.
1247 Taft Ave.
Endicott, NY 13760

Library of Congress Cataloging-in-Publication Data Applied For.

Printed in the United States of America.

ISBN 1-56081-521-3

Printing history:

10 9 8 7 6 5 4 3 2 1

Published by:

VCH Publishers, Inc.
220 East 23rd St.
Suite 909
New York, New York 10010

Preface

This single-volume edition of the Encyclopedia of Industrial Chemical Additives contains extensive format and data revisions. It is the most current source of data on chemical trademark products available in printed form. It is the only reference that classifies additive products by function, major chemical component, and manufacturer.

This new Encyclopedia focuses on chemicals that provide functional enhancements to the properties of various end products, e.g., absorbents, antioxidants, antistats, corrosion inhibitors, enzymes, lubricants, tackifiers, thickeners, water repellents. It features expanded definitions for product entries, identification of new CAS numbers and CTFA names, and approximately 20% new product entries. Discontinued trademark products have been deleted, and product-source name and address changes are reflected.

What makes this reference unique and especially useful for anyone doing chemical research is the inclusion of cross references to chemical component, function, and manufacturer.

The book is divided into four sections:

Part I—CHEMICAL TRADENAMES DICTIONARY contains an alphabetical listing of over 18,000 additive tradename product entries that have been gathered from over 700 manufacturer sources. Each entry references its manufacturer and contains information about its chemical composition, ionic nature, applications, color, odor, form, solubility, specific gravity, pH, active ingredients, etc.

Part II— CHEMICAL COMPONENT CROSS REFERENCE contains an alphabetical listing of the major chemicals contained within the trademark products listed in Part I. Each entry is followed by the tradename product(s) that contain this chemical. Wherever possible, the synonyms, CAS (Chemical Abstract Service Registry) numbers, RD (Recognized Disclosure) numbers, and CTFA (Cosmetic, Toiletry, and Fragrance Association) adopted names for the chemical entry are included. This provides the user with the capability of finding tradename additive products that contain specific chemical compounds. For example, by locating alphabetically the chemical *dioctyl phthalate*, the user is able to retrieve an alphabetical list of all the tradename products that contain that chemical. The user can then refer back to the first section for detailed descriptions of the specific tradename products listed here.

Part III—A FUNCTIONAL CLASSIFICATION OF TRADENAMES contains an alphabetical listing of major chemical functional categories such as antimicrobial, lubricant, solvent, tackifier, uv absorber, etc. Each heading is followed by an alphabetical listing of the tradename products that perform that function.

Part IV—CHEMICAL MANUFACTURERS' DIRECTORY contains all the manufacturers of products referenced in this work. This edition includes telephone and fax numbers where possible and an alphabetical listing of all the product lines that are attributed to each manufacturer.

This book is the culmination of many years of research. We are especially grateful to Roberta Dakan for her skill and dedication in the development and maintenance of this tradename database resource. Her talent and persistence have been instrumental in the production of this reference work.

M. & I. Ash

NOTE

The information contained in this series is accurate to the best of our knowledge; however, no liability will be assumed by the publisher for the correctness or comprehensiveness of such information. The determination of the suitability of any of the products for prospective use is the responsibility of the user. It is herewith recommended that those who plan to use any of the products referenced seek the manufacturer's instructions for the handling of that particular chemical.

Unless otherwise specified, when the temperatures are not given for properties such as viscosity, density, solubility, etc., a standard temperature of 25 C is to be assumed.

OTHER BOOKS BY MICHAEL AND IRENE ASH

A Formulary of Paints and Other Coatings, Volumes I and II
A Formulary of Detergents and Other Cleaning Agents
A Formulary of Adhesives and Sealants
A Formulary of Cosmetic Preparations
The Thesaurus of Chemical Products, Volumes I and II
Encyclopedia of Plastics, Polymers and Resins, Volumes I—IV
Vol. I, What Every Chemical Technologist Wants to Know About...Emulsifiers and Wetting Agents
Vol. II, What Every Chemical Technologist Wants to Know About...Dispersants, Solvents, and Solubilizers
Vol. III, What Every Chemical Technologist Wants to Know About...Plasticizers, Stabilizers, and Thickeners
Vol. IV, What Every Chemical Technologist Wants to Know About...Conditioners, Emollients, and Lubricants
Vol. V, What Every Chemical Technologist Wants to Know About...Resins
Vol. VI, What Every Chemical Technologist Wants to Know About...Polymers and Plastics
The Condensed Encyclopedia of Surfactants
Chemical Products Desk Reference

ABBREVIATIONS

@ .... at
ABS .... acrylonitrile-butadiene-styrene
absorp. .... absorption
act. .... active
adsorp. .... adsorption
AEB .... average extent of burning
agric. .... agricultural
a.i. .... active ingredient
anhyd. .... anhydrous
APHA .... American Public Health Association
applic(s). .... application(s)
aq. .... aqueous
ASTM .... American Society for Testing and Materials
ATB .... average time of burning
aux. .... auxiliary
avail. .... available
avg. .... average
BHA .... butylated hydroxyanisole
BHT .... butylated hydroxytoluene
biodeg. .... biodegradable
bk., blk. .... black
BMC .... bulk molding compound
b.p. .... boiling point
BR .... butadiene rubbers, polybutadienes
br., brn. .... brown
brnsh. .... brownish
B/S .... butadiene-styrene
C .... degrees Centigrade
CAB .... cellulose acetate butyrate
cap .... capillary
CAS .... Chemical Abstracts Service
CC .... closed cup
cc .... cubic centimeter(s)
CCl_4 .... carbon tetrachloride
char. .... characteristic
cm .... centimeter(s)
cm^3 .... cubic centimeter(s)
CMC .... carboxymethylcellulose
COC .... Cleveland Open Cup
compd. .... compound
compr. .... compression
conc. .... concentrated, concentration
cps .... centipoise(s)
CR .... chloroprene rubber, polychloroprene
cryst. .... crystalline
cs or cSt .... centistoke(s)
CTFA .... Cosmetic, Toiletry and Fragrance Association
DEA .... diethanolamine, diethanolamide
decomp. .... decomposes
DEG .... diethylene glycol
dens. .... density

deriv. .... derivative(s)
dg .... decigram(s)
diam. .... diameter
dielec. .... dielectric
disp. .... dispersible, dispersion
dist. .... distilled
distort. .... distortion
dk. .... dark
DOT .... Department of Transportation
EDTA .... ethylene diamine tetraacetic acid
elec. .... electrical
elong. .... elongation
EO .... ethylene oxide
EP .... extreme pressure
EPDM .... ethylene-propylene-diene rubbers
EPM .... ethylene-propylene rubbers
equip. .... equipment
ESCR .... environmental stress crack resistance
esp. .... especially
EVA .... ethylene vinyl acetate
exc. .... excellent
F .... degrees Fahrenheit
FDA .... Food and Drug Administration
FFA .... free fatty acid
flam. .... flammable, flammability
flex. .... flexural
f.p. .... freezing point
ft .... foot, feet
F-T .... Fischer-Tropsch
G .... giga
g .... gram(s)
G-H .... Gardner-Holdt
gal .... gallon(s)
gr. .... gravity
gran. .... granules, granular
grn. .... green
GRP .... glass-reinforced polyester
h .... hour(s)
HAF .... high abrasion furnace carbon black
HB .... horizontal burining
HCl .... hydrochloric acid
HDPE .... high-density polyethylene
Hg .... mercury
HLB .... hydrophilic lipophilic balance
hyd. .... hydroxyl
hydrog. .... hydrogenated
i.b.p. .... initial boiling point
IIR .... isobutylene-isoprene rubber
in. .... inch(es)
incl .... including

indent. .... indentation
ingred. .... ingredient(s)
inj. .... injection
insol. .... insoluble
IPA .... isopropyl alcohol, isopropanol
IPM .... isopropyl myristate
IPP .... isopropyl palmitate
IR .... isoprene rubber (synthetic)
J .... joule
k .... kilo
kg .... kilogram(s)
l .... liter(s)
lb .... pound(s)
LDPE .... low-density polyethylene
liq. .... liquid
LLDPE .... linear low-density polyethylene
lt. .... light
M .... mega
m .... milli or meter(s)
max. .... maximum
MDA .... methylene dianiline
MDI .... methylene diphenylene isocyanate
MEA .... monoethanolamine, monoethanolamide
med. .... medium
MEK .... methyl ethyl ketone
mfg. .... manufacture
mg .... milligram(s)
MIBK .... methyl isobutyl ketone
MIL .... Military Specifications
min .... minute(s), mineral, minimum
MIPA .... monoisopropanolamine, monoisopropanolamide
misc. .... miscible
mixt. .... mixture(s)
ml .... milliliter(s)
mm .... millimeter(s)
mN .... millinewton(s)
mod. .... moderately
m.p. .... melting point
MT .... medium thermal
m.w. .... molecular weight
nat. .... natural
NBR .... nitrile-butadiene rubber
NC .... nitrocellulose
NCR .... nitrile-chloroprene rubber
NF .... National Formulary
no. .... number
nonflam. .... nonflammable
NR .... isoprene rubber (natural)
NV .... nonvolatiles
OC .... open crucible

org. .... organic
o/w .... oil-in-water
Pa .... Pascal
PAN .... polyacrylonitrile
PC .... polycarbonate
pcf .... pounds per cubic foot
PEEK .... polyetheretherketone
PEG .... polyethylene glycol
PET .... polyethylene terephthalate
petrol .... petroleum
pH .... hydrogen-ion concentration
phr .... parts per hundred of rubber or resin
pkg. .... packaging
PMCC .... Pensky-Martens closed cup
PO .... propylene oxide
POE .... polyoxyethylene
POP .... polyoxypropylene
powd. .... powder
PP .... polypropylene
PPG .... polypropylene glycol
PPS .... polyphenylene sulfide
pract. .... practically
prep. .... preparation(s)
prod. .... product(s), production
PS .... polystyrene
psi .... pounds per square inch
pt. .... point
PTFE .... polytetrafluoroethylene
PU .... polyurethane
PVAc .... polyvinyl acetate
PVAL .... polyvinyl alcohol
PVC .... polyvinyl chloride
PVDC .... polyvinylidene chloride
PVDF .... polyvinylidene fluoride
PVF .... polyvinyl fluoride
quat. .... quaternary
R&B .... Ring & Ball
RD .... Recognized Disclosure
rdsh. .... reddish
ref .... refractive
resist .... resistivity
resp. .... respectively
RIM .... reaction injection molded/molding
rpm .... revolutions per minute
R.T. .... room temperature
s .... second(s)
SAN .... styrene-acrylonitrile
sapon .... saponification
sat. .... saturated
S/B .... styrene/butadiene

SBR ... styrene-butadiene rubber
SCR ... styrene-chloroprene rubber
SE ... self-emulsifying
sec. ... secondary
sm. ... small
SMC ... sheet molding compound
soften. ... softening
sol'n. ... solution
sol. ... soluble, solubility
solid. ... solidification
solv(s) ... solvent(s)
sp. ... specific
spec. ... specification
std. ... standard
Stod. ... Stoddard solvent
str. ... strength
surf. ... surface
SUS ... Saybolt Universal Seconds
syn. ... synthetic
t ... tertiary
TCC ... Taggart closed cup
TDI ... toluene diisocyanate
TEA ... triethanolamine
tech. ... technical
temp. ... temperature
tens. ... tension or tensile
tert. ... tertiary
THF ... tetrahydrofuran
thru ... through
typ. ... typical
UL ... Underwriter's Laboratory
unsat. ... unsaturated
USP ... United States Pharmacopeia
uv ... ultraviolet
VA ... vinyl acetate
veg. ... vegetable
visc. ... viscous, viscosity
vol. ... volume
wh. ... white
w/o ... water-in-oil
wt. ... weight
yel. ... yellow
ylsh. ... yellowish
... number
% ... percent
< ... less than
> ... greater than
≈ ... approximately

TABLE OF CONTENTS

A-1. [Monsanto] N,N′-diphenylthiourea (thiocarbanilide); primary accelerator for latex and repair stocks, CR, CR latex, and EPDM sponge compds.; powd.

AA. [PMC] Anthranilic acid; antioxidant for fats, greases, lube oils, and polyamides; sludge preventative in furnace and lube oils; chelating agent and sequestrant; corrosion inhibitor; stabilizer of can lacquers, oils, and lubricants; cream cryst. or powd.; m.p. 147 C min.; 0.5 g dissolves in 100 ml water, 28 g dissolves in 100 ml methanol; 99% min. purity.

AA-81C. [Dover] Tetrachlorobutyl alcohol; reactive flame retardant for highly resilient and rigid PUF for pour-in-place applic.; Gardner 18 liq.; sp.gr. 1.51; visc. 30 cps; 65% chlorine.

AA Standard. [CasChem] Castor oil; emollient for industrial applics. where lt. color, high purity, and low acidity are desirable; sol. in alcohols, esters, ethers, ketone, and aromatic solvs.

AA USP. [CasChem] Castor oil; emollient for cosmetic and pharmaceutical purposes; sol. see AA Standard.

Aastar. [Am. Cyanamid/Ag] Phorate (15.00%) and flucythrinate (0.75%); soil and systemic insecticide for corn; restricted use.

AB. [Angus] 2-Amino-1-butanol; pigment dispersant, neutralizing amine, corrosion inhibitor, acid-salt catalyst, pH buffer, chemical and pharmaceutical intermediate, solubilizer; m.w. 89.1; water-sol.; m.p. –2 C; b.p. 178 C; flash pt. 193 F (TCC); pH 11.1 (0.1M aq. sol'n.).

Abate® 1-SG, 2-CG, 4-E, 5CG. [Am. Cyanamid/Ag] Temephos; gran. and emulsifiable conc. herbicide for control of mosquito larvae in standing water, ponds, etc.; 1, 2, 44.6, and 5% act. resp.

Abil® 10-100000. [Goldschmidt] Polydimethyl siloxanes; nonionic; conditioner for skin and hair care, sunscreens, tanning creams, or lotions; aftershave, aerosol preparations; liq.; 100% act.

Abil® AV 20-1000, AV 8853. [Goldschmidt] Phenylmethyl polysiloxanes; nonionic; conditioner for skin and hair care, sunscreen, tanning creams or lotions; aftershave, aerosol preparations; liq.; 100% act.

Abil® B 8. [Goldschmidt] Polysiloxane polyether copolymer; nonionic; surfactant, conditioner used in hair care prods., deodorants, antiperspirants, creams, and lotions; liq.; water-sol.

Abil® B 8839. [Goldschmidt] Decamethyl cyclopentasiloxane; nonionic; conditioner for hair care prods., aerosols, sticks, shaving preparations, deodorants, antiperspirants; liq.; 100% act.

Abil® B 8842, B 8843, B 8851, B 8852, B 8863, B 8873. [Goldschmidt] Dimethicone copolyol; nonionic; surfactant, conditioner used in personal care prods.; emollient for hair and skin care prods., aerosol shaving lather, perfumes and colognes; liq.; water-sol.; 100% conc.

Abil® B 9800, B 9801. [Goldschmidt] Polysiloxane polyalkylene copolymer; nonionic; emollient in cosmetic creams and lotions; spreading agent; used in pigmented prods. and nonaq. systems; liq.; oil-sol. (B 9800); 100% conc.

Abil® B 9905, B 9907, B 9908, B 9909. [Goldschmidt] Polysiloxane polydimethyl dimethylammonium acetate copolymer; cationic; conditioner used in hair and skin prods.; liq.; 30% conc.

Abil® B 9806. [Goldschmidt] Cetyl dimethicone copolyol; surfactant, emollient for creams and lotions; emulsifier for cyclomethicone.

Abil® B 9950. [Goldschmidt] Dimethicone propyl PG-betaine; amphoteric; silicone surfactant, conditioner; used in hair and skin care prods.; liq.; 30% conc.

Abil® B 88183. [Goldschmidt] Dimethicone copolyol; nonionic surfactant used as foam formers and providing lubricating and gloss properties; refatting agent for skin prods.; increases slip in shaving creams; also for shampoos, shower gels, hand cleaners, aerosols, antiperspirants; lt. yel. liq.; sol. in water, alcohol, 1,2-propylene glycol; sp.gr. 1.024; visc. 95 ± 15 $mm^2 s^{-1}$; cloud pt. 71 ± 3 C (4% aq.); ref. index 1.375; surf. tens. 34.5 $mN{\cdot}m^{-1}$ (1% aq.); 35% act.

Abil® K 4. [Goldschmidt] Octamethyl cyclotetrasiloxane; nonionic; emollient, conditioner used in hair care prods., aerosols, shaving preparations, deodorants, antiperspirants; liq.; 100% act.

Abil®-Quat 3270, 3272. [Goldschmidt] Quaternium-80; cationic; conditioner, antistat for shampoos and hair rinses; also refatting agent for skin cleansers; amber clear liqs.; sol. in water, alcohol, 1,2-propylene glycol, glycerol; sp.gr. 1.014 and 1.008 resp.; visc. 400 ± 100 and 1000 ± 200 $mm^2 s^{-1}$ resp.; flash pt. 90 C; ref. index 1.443 and 1.429 resp.; pH 7.2 ± 1 (30% aq.); surf. tens. 31.5 ± 2 and 44 ± 2 g/cm^3 resp.; 50 ± 1% act.

Abil® S201, S255. [Goldschmidt] Dimethicone thiosulfate; conditioner and setting lotion; improves gloss and sheen of shampoos and conditioners.

Abil®-Wax 2434. [Goldschmidt] Stearoxy dimethicone; nonionic; wax improving applic. and skin care properties of emulsions; spreading and emollient properties for protection against aq. media; water barrier for creams and lotions; pale yel. liq. > 40 C, waxy < 10 C; sol. in cyclomethicone, min. oil, IPM, sunflower seed oil; sp.gr. 0.88 (35 C); m.p. 25 C; pour pt. 20–30 C; flash pt. > 100 C; 100% act.

Abil®-Wax 2440. [Goldschmidt] Behenoxy dimethicone; nonionic; wax improving applic. and skin care properties of emulsions; spreading and emollient properties for protection against aq. media; reduces whitening during applic. of creams and lotions; pigment solubilizer; pale yel. waxy solid; disp. in min. oil, IPM, sunflower seed oil; sp.gr. 0.90 (50 FC); m.p. 35 C; pour pt. 30–35 C; flash pt. > 100 C; 100% act.

Abil®-Wax 9800. [Goldschmidt] Stearyl dimethicone; nonionic; wax improving color, luster, and spreadability of pigmented prods.; spreading, penetrating, and emollient properties for skin care prods.; wh. waxy liq.; sol. in min. oil, IPM, sunflower seed oil; sp.gr. 0.86 (35 C); pour pt. 20 C; flash pt. > 40 C; 100% act.

Abil®-Wax 9801. [Goldschmidt] Cetyl dimethicone; nonionic; wax providing emolliency and applic. benefits for antiperspirants; pigment solubilizer; also for skin care prods.; pale yel. liq. wax; sol. in cyclomethicone, min. oil, IPM, sunflower seed oil; sp.gr. 0.86; pour pt. 10 C; flash pt. 40 C; 100% act.

Abil®-Wax 9809. [Goldschmidt] Stearyl methicone; wax providing water barrier for night creams and protective lotions; sol. in min. oil; m.p. 40 C.

Abil®-Wax 9810. [Goldschmidt] C24–28 alkyl methicone; high m.w. wax used as thickening agent for min. oils and cosmetic esters; water barrier for night creams and protective lotions; provides gloss and smoothness for lipsticks; gels min. oil; m.p. 60 C.

Abil® ZP 2434. [Goldschmidt] Polysiloxane polyalkyl copolymer; nonionic; emollient used in cosmetic creams and lotions, pigmented prods. and nonaq. systems; liq.; 100% conc.

Abiol. [Tri-K] Imidazolidinyl urea NF; preservative.

Abitol®. [Hercules] Dihydroabietyl alcohol; resinous plasticizer and tackifier in plastics, lacquers, inks, and adhesives; chemical intermediate; colorless visc. liq.; low odor; sol. in alcohols, esters, ketones, chlorinated solvs., and in aliphatic, aromatic, or terpene hydrocarbons; insol. in water; sp.gr. 1.008; dens. 8.4 lb/gal; visc. 400 poises (40 C); acid no. 0.3; sapon. no. 15; ref. index 1.5262; flash pt. (COC) 185 C; 100% liq. or 90% solids toluene or xylene sol'ns.

Ablumine 230. [Taiwan] N-alkyl dimethyl ammonium chloride; cationic; algicide for industrial cooling towers and swimming pools; liq.; 50% act.

Ablumine 280. [Taiwan] Stearyl dimethyl benzyl ammonium chloride; antistat and hair conditioner; water-disp.

Ablumine PN. [Taiwan] Complex alkyl dimethyl benzyl chloride; cationic; retarder for acrylic fiber; prevents unlevel dyeing.

Ablunol 200ML, 400ML, 600ML. [Taiwan] PEG 200, 400, 600 laurate resp.; nonionic; emulsifiers, lubricants, dispersing and leveling agents used in cosmetic, textile, paint and other industrial uses; liq.; HLB 9.5, 13.1, 14.8; 100% conc.

Ablunol 200MO, 400MO, 600MO, 1000MO. [Taiwan] PEG 200, 400, 600, 1000 oleate resp.; nonionic; see Ablunol 200ML; liq.; HLB 7.9, 11.5, 13.5, 15.2 resp.; 100% conc.

Ablunol 200MS, 400MS, 600MS, 1000MS. [Taiwan] PEG 200, 400, 600, 1000 stearate resp.; nonionic; see Ablunol 200ML; solid; HLB 8.0, 11.6, 13.6, 15.2 resp.; 100% conc.

Ablunol 6000DS. [Taiwan] PEG 6000 distearate; thickener for cosmetic, pigment preparations; flake.

Ablunol CO5, CO10, CO15, CO30, CO45. [Taiwan] Castor oil, ethoxylated; nonionic; emulsifier, lubricant, antistat; liq.; HLB 4.0, 6.4, 8.5, 11.8, 13.6 resp.

Ablunol NP4. [Taiwan] Nonoxynol-4; nonionic; detergent and dispersant for use in petrol. oils; intermediate for mfg. of surfactants and antistatic agents; co-emulsifier for fats, oils and waxes; liq.; oil-sol.; HLB 8.8; 100% conc.

Ablunol NP6. [Taiwan] Nonoxynol-6; nonionic; emulsifier or coemulsifier; coupling agent; intermediate for mfg. of surfactants; emulsifier for silicone and min. oil; liq.; HLB 10.8.

Ablunol S-60. [Taiwan] Sorbitan stearate; nonionic; preparation of silicone defoamer emulsions; textile process lubricant; solid; HLB 4.7; 100% conc.

Abluphat LP. [Taiwan] Phosphate ester sodium salt; antistat for syn. fibers; liq. 30% act.

Abluphat LPF, LPI, LPX. [Taiwan] Phosphate ester free acid; see Abluphat LP; liq.; 100% act.

Ablupol SAE. [Taiwan] Silicone compd.; antifoamer; liq.; water-disp.

Ablusoft A. [Taiwan] Aliphatic quat. ammonium compd.; cationic; softener for textile prods.; flake; 100% conc.

Ablusoft ND. [Taiwan] Polyamide; nonionic; softener, antistatic properties in resin finishing of cotton; flake; 100% conc.

Ablusoft SN, SNC. [Taiwan] Silicone, amine-modified; cationic; softener for fabrics.

Ablusol ML. [Taiwan] Sodium naphthalene sulfonate formaldehyde condensate; plasticizer and water reducing agent used in pourable and high strength concrete; liq.

Ablusol NL. [Taiwan] Sodium naphthalene sulfonate formaldehyde condensate; anionic; dye dispersant and antiagglomerant; used in the process of polyester and acetate fabrics; liq.; 40% conc.

Abluton CMN. [Taiwan] Alkyl naphthalene; anionic; dye carrier for polyester fiber and its blends; liq.; 100% conc.

Abluton CTP. [Taiwan] Chlorinated solvs.; carrier for bleaching polyester fiber and its blends; emulsifiable.

A.B.S. 87%. [Triantaphyllou] Alkylbenzene sulfonic acid; detergent intermediate; liq.; 87% conc.

ABT-2500. [Pfizer] Talc; antiblocking agent for polyolefin films; 2.3 μ avg. particle size; dry brightness 88 ± 2 dB; sp.gr. 2.7; bulk dens. 13 lb/ft³ (loose), 37 lb/ft³ (tapped); 60% SiO_2, 33% MgO.

Accelerator 399. [Texaco] Epoxy curing promoter

for use with amine hardeners; clear pale yel. liq.; sp.gr. 1.089; visc. 880 cps; pour pt. 5 F; flash pt (PM) 180 F.

Accelerator BZ. [Akrochem] Zinc dibutyl dithiocarbamate; nondiscoloring, nonstaining accelerator for EPDM, NR and SR latexes; stabilizer and antioxidant in uncured rubber; off-wh. to cream powd. or pellets; sol. in acetone; sp.gr. 1.25; m.p. 104 C; 95% min. purity.

Accelerator D. [Anchor] Cyclo amine blended with organo metallic salt; accelerator recommended for use in lt.-colored compds., esp. shoe soling industry; wh. to beige nondusting powd.

Accelerator EZ. [Akrochem] Zinc diethyl dithiocarbamate; nondiscoloring, nonstaining accelerator for NR and SR latexes; off-wh. powd.; sp.gr. 1.5; m.p. 172 C min.; 98% min. purity.

Accelerator MF. [Akrochem] 2-Benzothiazyl-N-morpholine disulfide; accelerator in natural and syn. rubbers, e.g., tire and mech. goods; yel-br. powd.; sp.gr. 1.48; m.p. 125 C.

Accelerator MZ. [Akrochem] Zinc dimethyl dithiocarbamate; nondiscoloring, nonstaining accelerator for NR, IR, BR, SBR, IIR, and EPDM rubbers and NR latex; off-wh. powd. or pellets; sp.gr. 1.70; m.p. 240 C min.; 98% min. purity.

Accelerator P.P.D. [Akrochem] Piperidinium pentamethylene dithiocarbamate; accelerator for latex; in rubber cements, compds. containing wh. factice and for tank linings; peptizing agent; nondiscoloring and nonstaining; creamy wh. powd.; sol. in hot water, aromatic hydrocarbons; sp.gr. 1.2; m.p. 168 C min.; 98% min. purity.

Accelerator Z.B.E.D. [Akrochem] Zinc dibenzyl dithiocarbamate; accelerator; nondiscoloring and nonstaining; creamy wh. powd.; sp.gr. 1.41; m.p. 180–188 C; 95% min. purity.

Accelerator Z.P.D. [Akrochem] Zinc cyclopentamethylene dithiocarbamate; ultra-accelerator for dry, natural rubber, esp. footwear; nonpigmenting and nonstaining; off-wh. powd.; sp.gr. 1.6; m.p. 225 C min.; 98% min. purity.

Accobetaine CL. [Capital City] Complex coco betaine; detergent, wetting agent, emulsifier, high foaming agent, solubilizer, household and cosmetic uses; yel. liq.; pleasant odor; water-sol.; f.p. 0 C; 35% act.

Accomeen C2, C5, C10, C15. [Capital City] PEG-2, -5, -10, -15 cocamine resp.; nonionic; emulsifier, antistat, surfactant; Gardner 4–6 liq., amine odor; sol. in org. solv.; C5, C10 sol. in water; dens. 7.25, 8.15, 8.3, and 8.7 lb/gal resp.; sp.gr. 0.87, 0.98, 1.0, and 1.04 resp.; 99% act.

Accomeen S2, S5, S10. [Capital City] PEG-2, -5, -10 soyamine resp.; nonionic; emulsifier, antistat, surfactant; Gardner 5–10 liq.; amine odor; sol. in org. solv.; S5 water-disp., S10 water-sol.; dens. 7.6, 7.9, 8.5 lb/gal resp.; sp.gr. 0.91, 0.95, and 1.02 resp.; surf. tens. 31.3 (0.1%, S2); 99% act.

Accomeen T2, T5, T15. [Capital City] PEG-2, -5, -15 tallow amine; nonionic; emulsifier, antistat, surfactant, dispersant; Gardner 3–5 liq. to paste.; amine odor; sol. in org. solv., insol. in water; dens. 7.7 and 8.1 lb/gal (T2, T15); sp.gr. 0.92 and 0.97 (T2, T15); 99% act.

Accomid C. [Capital City] Cocamide DEA; nonionic; detergent, stabilizer, visc. improver, foam booster for shampoos and dishwashes; Gardner 4 liq.; mild odor; water-sol.; dens. 8.3 lb/gal; sp.gr. 0.99; 98% act.

Accomid PK. [Capital City] Palm kernelamide DEA (1:1); nonionic; visc. builder, foam booster/stabilizer, emulsifier for shampoos, liq. soaps, dish detergents, bubble bath prods.; Gardner 5 liq.; pH 9–10.6 (10% aq.).

Acconon 200-DL. [Capital City] PEG-4 dilaurate; nonionic surfactant used as emulsifier, dispersant, solubilizer, visc. control agent for cosmetics, pharmaceuticals, and industrial applics.

Acconon 200-MS. [Capital City] PEG-4 stearate; nonionic surfactant used as emulsifier, dispersant, solubilizer, visc. control agent for cosmetics, pharmaceuticals, and industrial applics.; Gardner 4 max. solid; HLB 8; pH 5.5–6.5.

Acconon 300-MO. [Capital City] PEG 300 oleate; nonionic; emulsifier, lubricant, chemical intermediate; for cosmetics, food, agriculture, plastics; Gardner 6 liq.; sol. in org. solv.; dens. 8.3 lb/gal resp.; sp.gr. 0.99; m.p. < –5 C; 99% act.

Acconon 400-DO. [Capital City] PEG-8 dioleate; nonionic surfactant used as emulsifier, dispersant, solubilizer, visc. control agent for cosmetics, pharmaceuticals, and industrial applics.

Acconon 400-ML. [Capital City] PEG-8 laurate; nonionic surfactant used as emulsifier, dispersant, solubilizer, visc. control agent for cosmetics, pharmaceuticals, and industrial applics.

Acconon 400-MO. [Capital City] PEG-8 oleate; nonionic; emulsifier, dispersant, lubricant, chemical intermediate, solubilizer, visc. control agent; for cosmetics, pharmaceuticals, food, agric., plastics; Gardner 4 max. liq.; sol. in org. solv.; water-disp.; dens. 8.4 lb/gal; sp.gr. 1.01; HLB 12; m.p. < 10 C; pH 5.5–6.5; 99% act.

Acconon 400-MS. [Capital City] PEG-8 stearate; nonionic surfactant used as emulsifier, dispersant, solubilizer, visc. control agent for cosmetics, pharmaceuticals, and industrial applics.; Gardner 4 max. solid; HLB 12; pH 5.5–6.5.

Acconon 1300. [Capital City] PPG-3 laureth-9; nonionic surfactant used as emulsifier, dispersant, solubilizer, visc. control agent for cosmetics, pharmaceuticals, and industrial applics.; Gardner 3 max. liq.; sp.gr. 1.016–1.019; dens. 8.35–8.45 lb/gal; HLB 11; cloud pt. 135–145 F; pH 6.0–7.0.

Acconon CA-5. [Capital City] PEG-5 castor oil; nonionic surfactant used as emulsifier, dispersant, solubilizer, visc. control agent for cosmetics, pharmaceuticals, and industrial applics.; Gardner 3 max. liq.; HLB 16; pH 6.0–7.0.

Acconon CA-8. [Capital City] PEG-8 castor oil; nonionic surfactant used as emulsifier, dispersant, solubilizer, visc. control agent for cosmetics, pharmaceuticals, and industrial applics.; Gardner 3 max. liq.; HLB 8; pH 6.0–7.0.

Acconon CA-9. [Capital City] PEG-9 castor oil; nonionic surfactant used as emulsifier, lubricant, dispersant, solubilizer, visc. control agent for cosmetics, pharmaceuticals, and industrial applics.; water-disp.

Acconon CA-15. [Capital City] PEG-15 castor oil; nonionic surfactant used as emulsifier, lubricant, dispersant, solubilizer, visc. control agent for cosmetics, pharmaceuticals, and industrial applics.; Gardner 3 max. liq.; water-disp.; HLB 16; pH 6.0–7.0.

Acconon CA-25. [Capital City] PEG-25 castor oil; lubricant, dispersant, emulsifier for cosmetics; water-disp.

Acconon CON. [Capital City] PEG-10 propylene glycol glyceryl laurate; nonionic surfactant used as emulsifier, dispersant, solubilizer, visc. control agent for cosmetics, pharmaceuticals, and industrial applics.; Gardner 2 max. liq.; HLB 10; pH 6.0–7.0.

Acconon E. [Capital City] PPG-15 stearyl ether; nonionic surfactant used as emulsifier, dispersant, solubilizer, visc. control agent for cosmetics, pharmaceuticals, and industrial applics.; Gardner 3 max. liq.; HLB 16; pH 6.0–7.0.

Acconon ETG. [Capital City] Glycereth-26; humectant; lubricant for skin care prods., creams, lotions; Gardner 3 max. liq.; sol. in water, alcohol, ketones, esters; HLB 15; pH 5.5–6.5.

Acconon GTO. [Capital City] Glyceryl trioleate; lubricant for textile processing used in cosmetics; sol. in oils, org. solvs.

Acconon TGH. [Capital City] PEG-20-PPG-10 glyceryl stearate; nonionic surfactant used as emulsifier, dispersant, solubilizer, visc. control agent, wetting and foaming agent for cosmetics, pharmaceuticals, and industrial applics.; Gardner 3 max. liq., mild odor; HLB 16; pH 6.0–7.0; 100% conc.

Acconon W-230. [Capital City] Ceteareth-20; nonionic surfactant used as emulsifier, dispersant, solubilizer, visc. control agent for cosmetics, pharmaceuticals, and industrial applics.; Gardner 3 max. solid; HLB 15; pH 6.5–7.5.

A-C® Copolymer 400A. [Allied-Signal] Low m.w. EVA copolymer; pigment dispersant; PS color concs.; powd.

A-C® Copolymer 540A. [Allied-Signal] Low m.w. ethylene-acrylic acid copolymer; plastics lubricant and processing aid, pigment dispersant; internal lubricant PVC, nylon 6, nylon color concs.; powd.

Accoquat 2C-75. [Capital City] Dicocodimonium chloride; cationic; emulsifier, coupling agent; used for car spray waxes, dust control oil, spot removal; amber liq., mild odor; dens. 7.2 lb/gal; sp.gr. 0.86; flam.; 75% act. in IPA.

Accoquat 2C-75H. [Capital City] Dicocodimonium chloride; cationic; emulsifier, coupling agent; amber liq., mild odor; 75% act. in hexylene glycol.

Accosoft 440-75. [Capital City] Methyl bis (hydrog. tallowamidoethyl) 2-hydroxyethyl ammonium methyl sulfate; fabric softener quat. for textile industry; cream colored semisolid; IPA odor; m.p. 118 F; pH 5.0–6.5 (10% IPA/water); 74.0–75.5% solids in IPA.

Accosoft 540. [Capital City] Methyl bis (modified tallowamidoethyl) 2-hydroxyethyl ammonium methyl sulfate; fabric softener quat. offering improved softening and good hand without a greasy feel; yel. opaque semisolid; mild odor; visc. 700 cps (6.5% disp.); m.p. 38.5 C; pH 5.5–6.5 (10% IPA/water); 87.0–89.0% solids in IPA.

Accosoft 550-90 HHV. [Capital City] Methyl bis (tallowamidoethyl) 2-hydroxyethyl ammonium methyl sulfate; cationic; fabric softener and antistat for laundry prods.; yel. opaque semisolid; mild odor; readily dispersible in water; visc. 50–150 cps (6.5% disp.); pH 5.5–6.5 (10% IPA/water); 89.0–90.5% solids in IPA.

Accosoft 550L-90. [Capital City] Methyl bis (tallowamidoethyl) 2-hydroxyethyl ammonium methyl sulfate; fabric softener quat. for household laundry prods.; yel. visc. liq.; mild odor; visc. 40–140 cps (6.5% disp.); pour pt. 52 F; pH 5.0–6.0 (10% IPA/water); 86.5–88.0% solids in IPA.

Accosoft 620-90. [Capital City] Methyl bis (tallowamidoethyl) 2-hydroxypropyl ammonium methyl sulfate; cationic; fabric softener and antistat for household prods.; U.S. patent #3,933,871; yel. visc. liq.; mild odor; readily dispersible in water; visc. 40–100 cps (6.5% disp.); pH 5.0–6.5 (10% IPA/water); 89.0–90.5% solids in IPA.

Accosoft 707. [Capital City] Quaternium-18; cationic surfactant used in household and industrial fabric softener, fabric antistat; Gardner 3 paste; sol. in IPA; flam.; dens. 7.3; sp.gr. 8.7; 75% act.

Accosoft 748. [Capital City] Quaternium-18 methosulfate; cationic surfactant used in household and industrial fabric softener or antistat esp. suited to dryer use; Gardner 4 paste, solid; m.p. 125–135 C; biodeg.; flam.; 90% act. in IPA.

Accosoft 750. [Capital City] Methyl bis (oleylamidoethyl) 2-hydroxyethyl ammonium methyl sulfate; fabric softener quat. for heavy duty liq. laundry detergents; amber liq.; IPA odor; pour pt. < 32 F; pH 5.0–6.5 (10% IPA/water); 89.0–90.5% solids in IPA.

Accosoft 808HT-75. [Capital City] Methyl-1-hydrog. tallowamidoethyl-2-hydrog. tallow imidazolinium-methyl sulfate; fabric softener and antistat mainly for dryer applics.; also for rinse cycle applics.; cream-colored solid @ 77 F, lt. yel. liq. @ 120 F; IPA odor; pH 4.6–7.0 (10% IPA/water); 74.0–76.0% solids in IPA.

Accosoft 870. [Capital City] Hydrog. tallow imidazolinium quat. ammonium compd. and nonionic ethoxylate blend; softener, antistat for coating substrates for the dryer addition of laundry softening compds.; cream-colored solid; m.p. 122 F; pH 4.0–8.5 (10% IPA/water); 92.0–95.0% solids in IPA; 67.5–70.0% quat. act.

Accosoft A-155. [Capital City] Tallow amido amine quat. compd.; cationic surfactant, household fabric softener base, dispersion, antistat agent; Gardner 7 gel; biodeg.; 75 or 95% act.

Accosoft M 1154. [Capital City] Tallow amine quat. deriv.; cationic; fabric softener for detergent softener systems; liq.; 98% conc.

Accosperse 20. [Capital City] Polysorbate 20; nonionic; emulsifier, solubilizer; liq.; 100% conc.

Accosperse 60. [Capital City] Polysorbate 60; nonionic; emulsifier, solubilizer; amber liq.; sp.gr. 1.1; visc. 500–600 cps; HLB 14.9–15.0; sapon. no. 45–55; 98% act.

Accosperse 80. [Capital City] Polysorbate 80; nonionic; emulsifier, solubilizer; amber liq.; sp.gr. 1.1; visc. 400–450 cps; HLB 15.0; sapon. no. 45–55; 98% act.

Acetamin 24. [KAO SA] Cocamine acetate; cationic; surface coating agent for pigments, anticaking agent for fertilizer; emulsifier, dispersant, and softening agent for textiles; min. flotation reagent; solid; 100% conc.

Acetamin 86. [KAO SA] Stearamine acetate; see Acetamine 24; flake; 100% conc..

Acetamin C. [KAO SA] n-Cocamine acetate; cationic; flotation of mins., anticaking agents, emulsifier bactericide; paste, solid; 100% conc.

Acetamin HT. [KAO SA] n-Hyd. tallow amine acetate; cationic; see Acetamin C; solid.

Acetamin T. [KAO SA] n-Tallow amine acetate; cationic; see Acetamin C; solid.

Acetol® 1706. [Henkel/Emery] Cetyl acetate, acetylated lanolin alcohols; water repellent; strongly hydrophobic emollient, penetrant, and cosolv. use in suntan preparations and baby prods.; sol. in IPM, min., castor, and olive oil; water-insol.

Acetoquat CPB. [Aceto] Cetyl pyridinium bromide; cationic; germicide, sanitizing agent; powd.; 95% act.

Acetoquat CPC. [Aceto] Cetyl pyridinium chloride; cationic; germicide, sanitizing agent; powd.; 100% act.

Acetoquat CTAB. [Aceto] Cetyl trimethyl ammonium bromide; cationic; germicide, sanitizing agent; powd.; 95% act.

Acetulan®. [Amerchol] Cetyl acetate/acetylated lanolin alcohol; binder for pressed powds.; emollient, plasticizer, cosolv., NV and sebum solv. for personal care prods.; lubricant for clay, talc, and starch; stabilizer for lanolin; solubilizer in aerosols; penetrant and spreading agent; pale yel. thin oily liq.; odorless; sol. in ethanol, min., castor and veg. oil, IPM, IPP, IPA, silicone, butyl stearate, sulfonated castor oil, ethyl acetate, and org. solv.; insol. in water; sp.gr. 0.850–0.880; visc. 10 cps; acid no. 1.0 max.; sapon. no. 180–200; hyd. 8.0 max.; pH neutral.

Acetyl Peroxide-IB25. [Atochem] Acetyl peroxide; initiator for bulk, sol'n., and suspension polymerization, high-temp. and R.T. curing of polyester resins; sol'n.; 25% act.; 3.12–3.39% act. oxygen.

Acid Thickener. [Exxon] Cationic surfactant, visc. builder, corrosion inhibitor for acid-based cleaners; amber paste; sp.gr. 0.91; dens. 7.6 lb/gal; visc. 5000 cps (9.5% HCl with 3% Acid Thickener); pour pt. 80 F; flash pt. > 200 F.

Acihib A9. [ICI Australia] Amine deriv.; controls corrosion in HCl, sulfuric, phosphoric, and other acids; acid restrainer; liq.; sol. in acidic media, org. solvs.

ACL 56. [Monsanto] Sodium dichloroisocyanurate dihydrate; bleaching compd., sanitizer, disinfectant; 56% avail. chlorine.

ACL 59. [Monsanto] Potassium dichloroisocyanurate; bleaching compd., sanitizer disinfectant, oxidizer in dishwashing compds.; 59% avail. chlorine.

ACL 60. [Monsanto] Sodium dichloroisocyanurate; bleaching compd., sanitizer, disinfectant, oxidizer in dishwashing compds.; 63% avail. chlorine.

ACL 66. [Monsanto] (Monotrichloro) tetra (monopotassium dichloro) pentaisocyanurate; bleaching compd., sanitizer, detergent; 66.5% avail. chlorine.

ACL 85. [Monsanto] Trichloroisocyanuric acid; bleaching compd., sanitizer, disinfectant, detergent; 90% avail. chlorine.

Acofor. [Reichhold] Dist. tall oil fatty acids; latex stabilizer, dispersant (as soap) for pigments and fillers; Gardner 3 color; sp.gr. 0.903; flash pt. 400 C (COC).

A-C® Polyethylene 6, 6A, 8A, 9A, 8, 9. [Allied-Signal] Polyethylene wax; processing lubricant, melt index modifier, pigment dispersant, mold release aid; external lubricant PVC, color concs., polyolefin flow modifiers; thickener for cosmetic and pharmaceutical gels; powd. (A grades), others prilled or diced; sol. in hot min. oil and fatty esters.

A-C® Polyethylene 400. [Allied-Signal] EVA copolymer; thickener for cosmetic gels; sol. in hot ternary systems; low m.w.

A-C® Polyethylene 617, 617A. [Allied-Signal] Polyethylene; see A-C Polyethylene 6; prilled/diced and powd. resp.; sol. in hot min. oil and fatty esters.

A-C® Polyethylene 629A. [Allied-Signal] Oxidized low m.w. polyethylene; processing lubricant, mold release aid; PVC lubricant; powd.

Acralen AFR. [Bayer] Acrylic acid ester, acrylonitrile; self-crosslinking binder for nonwoven fabrics, textile finishing, lamination; yel. transparent film; dens. 1.03 g/cc; pH 6.5–7.0; hardness (Shore A) 30; 49.5–50.5% solids.

Acralen ATR. [Bayer] Acrylic acid ester, styrene latex; self-crosslinking binder for nonwoven fabrics, lamination; colorless, transparent film; dens. 1.03 g/cc; pH 4.7–5.2; hardness (Shore A) 50; 47.5–48.5% solids.

Acralen BS. [Bayer] Butadiene, styrene, acrylonitrile latex; self-crosslinking binder for nonwoven fabrics, tech. fabrics; yel. transparent film; dens. 1.02 g/cc; pH 7.5–8.5; hardness (Shore A) 93; 39.5–40.5% solids.

Acrawax® C. [Lonza] N,N′-ethylene distearamide; internal and surf. lubricant in resins and plastics; processing aid; plasticizer for resin; flow improver; pigment dispersant; used in hot-melt adhesives and coatings; powd. grade used as lubricant, processing aid, detackifier, mold release, and antiblocking agent; cream hard waxy solid; 10% max. on 10 mesh (beads); 2% max. on 40 mesh (prilled); 1% max. on 100 mesh (powd.); 0.1% max. on 325 mesh (atomized); insol. in water; sp.gr. 0.97; m.p. 140–145 C; acid no. 8 max.; flash pt. 285 C.

Acrawax® C DF #1, #2. [Lonza] Ethylenedistearamide; defoamer grade lubricant; Gardner 3 prilled, powd., or atomized; m.p. 143 and 145 C resp.; acid no. 5.

Acrawax® C SG. [Lonza] Ethylenedistearamide; lubricant for thermoplastics; vinyl siding grade; Gardner 3 prilled; m.p. 143 C; acid no. 5.

Acrilev AM, AM-Special. [Finetex] Phosphate ester, potassium salt; anionic; detergent, wetter, dye leveler used in textiles; liq.; 35 and 65% conc. resp.

Acritamer 941. [RITA] Polyacrylic acid polymers; suspending and visc. agent; water-disp.

Acryloft, Acryloft Conc. [Rhone-Poulenc] Quat. ammonium compd., cationic; textile softener; liq.

Acryloid® 150 Series. [Rohm & Haas] Methacrylate polymers; lubricant; pour pt. depressant for lube oils, dewaxing aid.

Acryloid® 700 Series. [Rohm & Haas] Methacrylate polymers; visc. index improvers, pour pt. depressants for crankcase and industrial lubricants.

Acryloid® 900 Series. [Rohm & Haas] Methacrylate copolymers; visc. index improvers, pour pt. depressants, low temp. sludge dispersants in crankcase lubricants and transmission fluids.

Acryloid® 950 Series. [Rohm & Haas] Methacrylate copolymer; visc. index improvers; pour pt. depressants for multigrade motor oils and transmission fluids; lubricant.

Acryloid® 1017. [Rohm & Haas] Methacrylate copolymers; lubricant, visc. index improvers for multigrade gear lubricants.

Acryloid® 1019. [Rohm & Haas] Methacrylate copolymers; visc. index improvers for multigrade gear lubricants.

Acryloid® HF-Series. [Rohm & Haas] Methacrylate polymers; lubricant; visc. modifier for military hydraulic oils.

Acryloid® W-1600. [Rohm & Haas] Methacrylate copolymers; lubricant; visc. index improver for shock absorber fluids.

Acryloid® WR-97. [Rohm & Haas] Hydroxyl-type acrylic polymer; thermosetting sol'n. resin reducible with water or solv.; crosslinkable with urea and melamine resins; pigment dispersant; also for exterior coil coating, prod. and appliance finishing; liq.; visc. 14,000–17,000 cps; flash pt. (PMCC) 67 F; 70% solids in IPA/butyl Cellosolve (83/17).

Acrysol® 51. [Rohm & Haas] Acrylic copolymer sodium salt; thickener for rubber latexes.

Acrysol® A-1, A-3, A-5. [Rohm & Haas] Polyacrylic acid sol'n.; thickener for cleaners, binders, adhesives, emulsion paints, emulsion polymers; liq.

Acrysol® A-41. [Rohm & Haas] Methacrylic acid copolymer; scale inhibitor; 30% solids.

Acrysol® ASE-60. [Rohm & Haas] Crosslinked acrylic emulsion copolymer; used to suspend pigments and fillers in water-based paints, inks, or other coatings, and the abrasive particles in waxes or polishes; visc. modifier for emulsion and latex compds.; binder; thickener; milky liq.; sp.gr. 1.054; visc. 10 cps; pH 3.5; 28.0 ± 0.5% solids.

Acrysol® ASE-75. [Rohm & Haas] Polyacrylic acid; thickener for latexes and emulsions for paints, flocking adhesives.

Acrysol® ASE-95. [Rohm & Haas] Acrylic copolymer emulsion; thickener for fabric laminants, pigment disp. polar solvs., surfactant sol'ns., paints, inks, waxes, and polishes; alkali-sol.

Acrysol® ASE-108. [Rohm & Haas] Crosslinked swellable acrylic copolymeric emulsion; thickener/stabilizer for household and industrial all-purpose cleaners and abrasive cleaners, flocking adhesives.

Acrysol® G-110. [Rohm & Haas] Ammonium polyacrylate sol'n.; thickening and stabilizing agent for syn. latices; used in coatings, adhesives, dipped, cast, and molded goods, cements for rug backing, spraying, spreading, brushing, and extruding compds.; colorless slightly hazy liq.; sp.gr. 1.06; visc. 90–170 cps (5% aq.); pH 8.5–9.5.

Acrysol® GS. [Rohm & Haas] Sodium polyacrylate; thickener for natural and syn. latexes for paints, films, coatings, and adhesives.

Acrysol® HV-1. [Rohm & Haas] Sodium polyacrylate; all-purpose thickener for rubber rug backing, unholstery backing.

Acrysol® ICS-1. [Rohm & Haas] Acrylic polymer; cosmetic and industrial thickener; liq. polymer emulsion; alkali-sol..

Acrysol® LMW-10, LMW-20, LMW-45. [Rohm & Haas] Polyacrylic acid; scale inhibitor, detergent addtive; 48% solids.

Acrysol® LMW-10N, LMW-20N. [Rohm & Haas] Sodium salt of polyacrylic acid; scale inhibitor, detergent additive; 43% solids.

Acrysol® LMW-45N. [Rohm & Haas] Sodium salt of polyacrylic acid; scale inhibitor and detergent additive; 45% solids.

Acrysol® LMW-100N. [Rohm & Haas] Sodium salt of polyacrylic acid; detergent additive; 39–41% solids.

Acrysol® LMW-400N. [Rohm & Haas] Sodium salt of polyacrylic acid; detergent additive; 34–35% solids.

Acrysol® LMW-600N. [Rohm & Haas] Sodium salt of polyacrylic acid; detergent additive; 25% solids.

Acrysol® QR-1086. [Rohm & Haas] Acrylate copolymer; calcium phosphate stabilizer, antiscale deposition polymer for all-org. cooling water treatment; 43% solids.

Acrysol® TT-615. [Rohm & Haas] Acrylic polymer emulsion; thickener for coatings, textile printing pastes, and adhesives; liq.; alkali-sol.

Acrysol® TT-675. [Rohm & Haas] Acrylic polymer emulsion; thickener for textile dyeing and printing; liq.

Acrysol® WTP-1. [Rohm & Haas] Acrylate terpolymer; boiler sludge dispersant; 43.5% solids.

Actafoam F-2. [Olin] Activator-stabilizer for vinyl foams containing Kempore blowing agents; gas release accelerator; paste, powd.

Actafoam R-3, R-10. [Olin] Activator-stabilizer for vinyl foams containing Kempore blowing agents; gas release accelerator; liq.

Actan SP. [Troy] Tannic acid conc.; corrosion inhibitor used in surf. pretreatment for field applic.; base for applic. of coatings systems; liq.

Actiflo® 68, 70. [Central Soya] Natural lecithin; amphoteric; emulsifier, wetting agent, dispersant; used in the food industry; liq.; 66–72% act.

Activ-8, Activ-8 in Hexylene Glycol. [Vanderbilt] Sol'n. forms containing 1,10-phenanthroline; drier accelerator and stabilizer used in combination with manganese and/or cobalt in coating systems that cure by oxidative polymerization; dens. 0.95 and 1.03 mg/ m^3 resp.; flash pt. 36 and 99 C resp.; 38% act.

Activator 1102. [Anchor] Dibutyl ammonium oleate; accelerator/activator for natural and syn. rubbers; lubricant; dk. red/brn. low-visc. liq.; sp.gr. 0.87.

Activator 2013-P®. [TSE] Carbodiimide; activator; improves resistance to hydrolysis by acids, bases, and hot water in vulcanizates based on millable PU; dk. amber liq.; aromatic odor; sp.gr. 1.05; reacts slowly with water; b.p. 75 C; flash pt. 176 F (TOC).

Activator STAG. [Akrochem] Complex sec. amine, surface treated; relatively nonstaining, nondiscoloring activator for thiazole-type accelerators; primary accelerator for natural rubber; strong sec. accelerator in SBR; lt. blue nondusting powd.; sp.gr. 1.26 ± 0.03; m.p. 130 C min.

Acto 450. [Exxon] Alkylaryl sodium sulfonate; anionic; detergent, wetting agent, emulsifier, rust preventative; reddish-amber liq.; alcohol odor; m.w. 470; oil sol., slight sol. in water but readily disp.; dens. 8.24 lb/gal; sp.gr. 0.99; flam.; 44% act.

Acto 500. [Exxon] Alkylaryl sodium sulfonate; anionic; see Acto 450; amber, visc. liq.; petrol. odor; m.w. 460; oil sol., slight sol. in water but readily disp.; dens. 8.33 lb/gal; sp.gr. 1.00; 50% act.

Acto 630. [Exxon] Alkylaryl sodium sulfonate; anionic; see Acto 450; amber visc. liq.; bland, petrol. odor; flam.; m.w. 470; oil sol., slight sol. in water but readily disp.; dens. 8.41 lb/gal; sp.gr. 1.01; 63% act.

Acto 632. [Exxon] Alkylaryl sodium sulfonate; anionic; see Acto 450; reddish-amber liq.; alcohol odor; flam.; m.w. 430; oil sol., slight sol. in water but readily disp.; dens. 8.41 lb/gal; sp.gr. 1.01; 63% act.

Acto 636. [Exxon] Alkylaryl sodium sulfonate; anionic; see Acto 450; orange, visc. liq.; alcohol odor; flam.; m.w. 470; oil sol., slight sol. in water but readily disp.; dens. 8.5 lb/gal; sp.gr. 1.02; 63% act.

Acto 639. [Exxon] Alkylaryl sodium sulfonate; anionic; see Acto 450; reddish-amber liq.; alcohol odor; flam.; m.w. 520; oil sol., slight sol. in water but readily disp.; dens. 8.5 lb/gal; sp.gr. 1.02; 63% act.

Actrabase PS-470. [Climax Performance] Med. m.w. petrol. sulfonate, sodium salt; anionic; emulsifier and rust inhibitor for cutting and lube oils; dispersant for sol. oil and semi-syns. for metalworking fluids; visc. liq.; oil-sol.; 100% conc.

Actracor 129, 856. [Climax Performance] Carboxylic acid amine salt; corrosion inhibitor for syn. and semi-syn. metal working fluids; water-sol.

Actracor 401. [Climax Performance] Corrosion inhibitor for water-glycol hydraulic fluids; water-sol.

Actracor 800. [Climax Performance] Hydrocarbon; corrosion inhibitor; metal working additive; oil- and solv.-sol.

Actracor 1987. [Climax Performance] Amine carboxylate; corrosion inhibitor for metalworking; water-sol.

Actracor M. [Climax Performance] Ethanolamine-borate ester; corrosion inhibitor in cutting oils; water-sol.

Actracor T. [Climax Performance] Triethanolamine borate ester; corrosion inhibitor in cutting oils; water-sol.

Actrafoam A, B, C, S. [Climax Performance] Blend of glycols, fatty acids, and nonionic surfactants in a hydrocarbon base; general purpose defoamer (A, B); defoamer for water sewage applics. (C, S).

Actrafos 104, 109. [Climax Performance] Phosphate ester; coupler for sulfated oils in cleaner formulations; lubricant for syn, semi-syn. and water-based cutting, grinding, and drawing fluids; liq.; water-sol.

Actrafos 110, 110A. [Climax Performance] Complex aliphatic hydroxyl compd. phosphate ester; pressure additive for cutting and rolling oils; hydrotrope for cleaning compds.; lubricant emulsifier and rust inhibitor; exc. for aluminum; 110A has higher m.p.; liq.; water-sol.; 100% conc.

Actrafos 139. [Climax Performance] Complex aliphatic hydroxyl compd. phosphate ester; pressure additive for cutting and rolling oils; hydrotrope for cleaning compds.; liq.; water-sol.; 100% conc.

Actrafos 152A. [Climax Performance] Org. phosphate ester; anionic; extreme pressure lubricant for cutting oils; high phosphorus content; liq.; water-sol.; liq.; 100% conc.

Actrafos 161, 315. [Climax Performance] Phosphate ester; anionic; lubricant and emulsifier for use in cutting oils; liq.; sol. in water and oil; 100% conc.

Actrafos 186. [Climax Performance] Aliphatic alcohol phosphate ester; anionic; lubricant and emulsifier for use in cutting oils; liq.; oil and water sol.; 100% conc.

Actrafos 208. [Climax Performance] Phosphate ester; coupler for nonionics in cleaner formulations; lubricant and emulsifier for cutting fluids; stable in electrolytes; sol. in water and oil.

Actrafos 216. [Climax Performance] Phosphate ester; coupler for sulfated oils in cleaner formulations; lubricant, emulsifier for syn, semi-syn. and water-based cutting, grinding, and drawing fluids; liq.; water-sol.

Actrafos 306. [Climax Performance] Phosphate ester; coupler for cleaner formulations; lubricant, emulsifier for cutting, grinding, and drawing fluids; liq.; water-sol.

Actrafos 314. [Climax Performance] Phosphate ester; see Actrafos 208; water-sol.

Actrafos 407. [Climax Performance] Phosphate ester; coupler for sulfated oils in cleaners formulations; lubricant for cutting, grinding, and drawing fluids; liq.; water-sol.

Actrafos 800. [Climax Performance] Phosphate ester; defoamer in metal working fluids; liq.

Actrafos 822. [Climax Performance] Complex aliphatic alcohol phosphate ester; anionic; lubricant and emulsifier for use in cutting oils; liq.; oil and water sol.; 100% conc.

Actrafos S-104, S-109. [Climax Performance] Phosphate ester; anionic; used in syn. and semisynthetics; solubilizer for nonionic surfactants in alkaline cleaning systems; liq.; 100% conc.

Actrafos SA-208. [Climax Performance] Linear alcohol phosphate ester; anionic; lubricant and emulsifier in cutting oils; liq.; oil and water sol.; 100% conc.

Actrafos SA-216. [Climax Performance] Phosphate ester; anionic; lubricant and emulsifier for cutting oils; liq.; sol. in oil and water; 100% conc.

Actrafos SN-315. [Climax Performance] Phosphate ester; anionic; lubricant, emulsifier for metal working; aluminum corrosion inhibitor; liq.; oil-sol.; 100% conc.

Actrafos SP-407. [Climax Performance] Phosphate ester; low foaming lubricant and rust inhibitor for syn. metalworking fluids; water-sol.

Actrafos T. [Climax Performance] Tridecyl alcohol phosphate ester; anionic; extreme pressure lubricant and release agent for cutting oils; liq.; oil-sol.; 100% conc.

Actralube 100. [Climax Performance] Syn. ester; lubricant for water-sol. cutting fluids.

Actralube 153. [Climax Performance] Soap; corrosion inhibitor, lubricant; metal working additive; water-sol.

Actralube 310. [Climax Performance] Fatty ester; lubricity additive for cutting oil; liq.; oil-sol.

Actralube 1200. [Climax Performance] Modified triglyceride; lubricity additive; nonionic; visc. liq.; 100% act.

Actralube 7142. [Climax Performance] Blown rapeseed oil; metal lubricant additive; oil-sol.

Actralube SOS. [Climax Performance] Ester; nonionic; emulsifier, lubricant, metalworking additive; substitute for sperm oil; liq.; oil-sol.; 100% conc.

Actralube Syn-147. [Climax Performance] Complex diester; nonionic; lubricity additive for syn. cutting fluids; liq.; water-sol.; 90% conc.

Actralube Syn-153. [Climax Performance] Lt. duty syn. lubricant and rust inhibitor; water-sol.

Actralube Syn-6147. [Climax Performance] Ester, sulfurized; nonionic; lubricant and E.P. additive for metal working fluids; liq.; 100% act.

Actramide 176. [Climax Performance] Alkanolamide; cutting fluids; water-sol.

Actramide 189. [Climax Performance] Alkanolamide; nonionic; foam stabilizer for cleaners and shampoos; liq.

Actramide 202. [Climax Performance] 2:1 Tall oil fatty acid alkanolamide; nonionic; emulsifier for sol. oils, metalworking fluids and emulsion cleaners; corrosion inhibitor; liq.; water-sol.

Actramide 410. [Climax Performance] Alkanolamide; nonionic; sec. emulsifier with lubricating properties; liq.; oil-sol.; 100% conc.

Actramide 840. [Climax Performance] Alkanolamide; nonionic; foam booster and thickener; liq.; 100% conc.

Actramide 933. [Climax Performance] Alkanolamide; nonionic; visc. builder and foam stabilizer, low foaming lubricant for cutting oils; liq.; water-sol.; 100% conc.

Actramide 938. [Climax Performance] Alkanolamide; nonionic; foam stabilizer for cleaners and shampoos; liq.

Actramide 5264. [Climax Performance] Modified 2:1 tall oil fatty acid alkanolamide; emulsifier, lubricant, rust inhibitor; water-sol.

Actrasol 167A. [Climax Performance] Sulfated castor oil, sodium neutralized; anionic; acid etching additive; liq.; 80% conc.

Actrasol 6092. [Climax Performance] Sulfated rapeseed oil; lubricant, emulsifier in pigment flushing, cleaners, textiles, paper processing.

Actrasol C-50, C-75, C-85. [Climax Performance] Sulfated castor oil; anionic; pigment wetting and disp.; liq.; 50, 70, and 75% conc. resp.

Actrasol CS-75. [Climax Performance] Sulfonated soyabean oil, sodium neutralized; lubricant, emulsifier in pigment flushing, cleaners, textiles, paper processing.

Actrasol EO. [Climax Performance] Sulfated glyceryl trioleate, sodium neutralized; see Actrasol CS-75.

Actrasol KAP. [Climax Performance] Sulfated blend of oils, sodium neutralized; anionic; sperm oil substitute; fat liquor for leather, metalworking; liq.; 75% conc.

Actrasol MY-75. [Climax Performance] Sulfated methyl ester of soya fatty acid, sodium neutralized; lubricant and emulsifier in metalworking fluids, water-based drilling muds; oil field defoamer.

Actrasol OY-75. [Climax Performance] Sulfated soyabean oil, sodium neutralized; lubricant, emulsifier in pigment flushing, cleaners, textiles, paper processing.

Actrasol PSR. [Climax Performance] Sulfated ricinoleic acid, potassium neutralized; anionic; pigment wetting and disp.; liq.; 75% conc.

Actrasol SBO. [Climax Performance] Sulfated butyl oleate, sodium neutralized; see Actrasol CS-75.

Actrasol SP. [Climax Performance] Sulfonated tall oil fatty acid, sodium neutralized; wet process phosphoric acid defoamer; liq.; 50% conc.

Actrasol SR 75. [Climax Performance] Sulfated oleic acid, ammonium neutralized; anionic; mold release agent; liq.; water-disp.; 75% conc.

Actrasol SRK 75. [Climax Performance] Sulfated oleic acid, potassium neutralized; anionic; mold release agent; liq.; water-disp.; 75% conc.

Actrasol SS. [Climax Performance] Sulfonated and saponified stearic acid, plasticized with min. oil; rust preventative, pre-lube, metal polish.

Actrol 4DP. [Climax Performance] PEG diester; nonionic; emulsifier and lubricity additive for metalworking fluids; liq.; HLB < 5; 95% conc.

Actrol 4MP, 628. [Climax Performance] PEG ester; nonionic; see Actrol 4DP; liq.; HLB > 10; 95% conc.

Actrol 6M25P. [Climax Performance] PEG 600 tallate.

ACuflow AF-1. [Allied-Signal] Ethylene copolymer

and aluminum stearate mixt.; processing additive; wh. gran. solid; waxy odor; negligible sol. in water; sp.gr. ≈ 1; m.p. < 160 C; flash pt. > 260 C (CC).

Acylan. [Croda] Acetylated lanolin; lipid emollient for personal care and pharmaceutical prods.; sol. in min. oil and soft waxy hydrophobic films; Gardner 11 soft solid; bland odor; sol. in min. oil; m.p. 32–39 C; acid no. 2.0 max.; sapon. no. 100–125; 100% act.

Acylglutamate CS-11. [Ajinomoto] Sodium N-cocoyl-L-glutamate; anionic; detergent, emollient for personal care prods.; bacteriostatic effect; powd.; biodeg.; 100% act.

Acylglutamate CT-12. [Ajinomoto] TEA N-cocoyl-L-glutamate; anionic; see Acylglutamate CS-11, liq.; 20% act.

Acylglutamate GS-11. [Ajinomoto] Sodium hydrog. tallow glutamate, sodium cocoyl glutamate; see Acylglutamate CS-11; powd.; biodeg.; 100% conc.

Acylglutamate HS-11. [Ajinomoto] Sodium N-hydrog. tallowyl-L-glutamate; see Acylglutamate CS-11; powd. 100% conc.

Acylglutamate LS-11. [Ajinomoto] Sodium N-lauroyl-L-glutamate; see Acylglutamate CS-11; powd.; 100% conc.

Acylglutamate MS-11. [Ajinomoto] Sodium myristoyl glutamate; anionic; see Acylglutamate CS-11; powd.; 100% conc.

Additin 30. [Mobay] Phenyl-α-naphthylamine; staining antioxidant for rubber tech. goods and heavily stressed goods; antiflexcracking agent for NR and IR; storage stabilizer for petrol. prods.; lt. brn. to lt. violet flakes and fused; dens. 1.11 g/cm³; m.p. ≥ 58 C.

Adimoll BO. [Bayer] Benzyloctyl adipate; plasticizer used for PVC articles with resistance to low temps., calendering, coating, extrusion, inj. molding, film, expanded imitation leather; useful in VC copolymers, NC, ethyl cellulose, PS, nat., S/B, N/B, chlorinated, and butyl rubbers; Hazen < 80; dens. 1.002–1.008 g/cc; visc. 16–17 mPa.s; acid no. < 0.1; sapon. no. 305–330; ref. index 1.4805–1.4830; flash pt. 200–220 C (OC); 70:30 PVC:plasticizer suspension: tens. str. 20.5 mPa; tens. elong. 350%; hardness (Shore D) 26.

Adimoll DN. [Bayer] Diisononyl adipate; plasticizer used in PVC rigid articles which must have stability, calendering, spread coating, extrusion, inj. molding, film, tarpaulins, protective clothing, expanded imitation leather, wire insulation and cables; suitable for VC copolymers, ethyl cellulose, PS, polymethylmethacrylate, nat., S/B, N/B, chlorinated, and butyl rubbers; Hazen < 50; dens. 0.910–0.920 g/cc; visc. 22–25 mPa.s; acid no. < 0.1; sapon. no. 270–290; ref. index 1.4465–1.4490; flash pt. 200–210 (OC); 70:30 PVC:plasticizer suspension: tens. str. 22 mPa; tens. elong. 360%; hardness (Shore D) 32.

Adimoll DO. [Bayer] Dioctyl adipate; plasticizer used for PVC articles with low-temp. resistance, calendering, coating, extrusion, inj. molding, film, tarpaulins, rainwear, tubing, cable sheathing, shoes, shoe soles; used in VC copolymers, PVAc, polyvinyl butyral, NC, ethyl cellulose, PS, nat., S/B, N/B, chlorinated, and butyl rubbers; Hazen < 20; dens. 0.923–0.926 g/cc; visc. 12–14 mPa.s; acid no. < 0.1; sapon. no. 270–290; ref. index 1.4465–1.4480; flash pt. 200–210 C (OC); 70:30 PVC:plasticizer suspension: tens. str. 19.5 mPa; tens. elong. 360%; hardness (Shore D) 26.

ADM. [Climax Performance] Ammonium dimolybdate; corrosion inhibitor for vapor phase inhibitor programs.

Adma® 2. [Ethyl] Dodecyldimethylamine; cationic; intermediate for quat. ammonium compds., amine oxides, and betaines; liq.;100% conc.

Adma® 4. [Ethyl] Tetradecyldimethylamine; cationic; see Adma 2; liq.; 100% conc.

Adma® 6. [Ethyl] Hexadecyldimethylamine; cationic; see Adma 2; liq.; 100% conc.

Adma® 8. [Ethyl] Octadecyldimethylamine; cationic; see Adma 2; liq.; 100% conc.

Adma® 24. [Ethyl] C_{12}/C_{14} alkyldimethylamine; cationic; see Adma 2; liq.; 100% conc.

Adma® 46. [Ethyl] C_{14}/C_{16} alkyldimethylamine; cationic; see Adma 2; liq.; 100% conc.

Adma® 246-451, 246-621. [Ethyl] $C_{12}/C_{14}/C_{16}$ alkyldimethylamine; cationic; see Adma 2; liq.; 100% conc.

Adma® C8. [Ethyl] Octyldimethylamine; cationic; see Adma 2; liq.; 100% conc.

Adma® C10. [Ethyl] Decyldimethylamine; cationic; see Adma 2; liq.; 100% conc.

Admex 433. [Sherex] Low-visc. polyester; polymeric plasticizer for vinyl; used in plastisols and organosols, foams, calendered films, sheeting, profile extrusions, and coated fabrics; Gardner 3; sp.gr. 1.090; dens. 9.10 lb/gal; visc. 1900 cps; pour pt. 18 F; acid no. 2.0; ref. index 1.5050; flash pt. 455 F (COC); fire pt. 500 F (COC); 50 phr in PVC: tens. str. 2935 psi; tens. elong. 295%; hardness 98A.

Admex 500. [Sherex] Complex polyester; polymeric plasticizer used in coated fabrics, films, pressure sensitive tapes, wire coatings, plastisols, and foams; Gardner 2; sp.gr. 1.061; dens. 8.85 lb/gal; visc. 1200 cps; pour pt. –31 F; acid no. 1.5; ref. index 1.464; flash pt. (COC) 485 F; 50 phr in PVC: tens. str. 2885 psi; elong. 355%; 100% mod. 1590 psi; hardness 89 (A).

Admex 515. [Sherex] Low-visc. polyester; plasticizer for vinyl compds.; calendered and coated fabric, vinyl foam, plastisols, organosols, elec. insulation, and profile extrusions; excellent for pigment disp.; APHA 150; sp.gr. 1.050; dens. 8.75 lb/gal; visc. 575 cps; pour pt. 35 F; acid no. 2.5; ref. index 1.4630; flash pt. 475 F (COC); fire pt. 555 F (COC); 50 phr in PVC: tens. str. 3190 psi; tens. elong. 320%; hardness 95A.

Admex 522. [Sherex] Low-visc. polyester; plasticizer; used in profile extrusions, calendered film and sheeting, coated fabrics, flooring, and wire coatings; Gardner 3; sp.gr. 1.060; dens. 8.84 lb/gal; visc. 795 cps; pour pt. –5 F; acid no. 2.0; ref. index 1.5040; flash pt. 465 F (COC); fire pt. 500 F (COC); 50 phr in PVC: tens. str. 3095 psi; tens. elong. 275%; hardness 97A.

Admex 523. [Sherex] Med.-visc. polyester; general

purpose plasticizer used in organosols, plastisols, elec. wire insulation, coated fabrics, calendered film and sheeting, profile extrusions, vinyl foam, and vinyl-foam laminates; Gardner 3; sp.gr. 1.100 dens. 9.17 lb/gal; visc. 3960 cps; pour pt. 20 F; acid no. 2.0; ref. index 1.5140; flash pt. 450 F (COC); fire pt. 490 F (COC); 50 phr in PVC: tens. str. 3360 psi; tens. elong 210%; hardness 99A.

Admex 525. [Sherex] Low-visc. polyester; plasticizer; as grinding vehicles for color pastes; coated fabrics, paper coatings, pigment disps., plastisols, and organosols, tapes, vinyl foams, and extruded profiles; Gardner 1; sp.gr. 1.035; dens. 8.64 lb/gal; visc. 310 cps; pour pt. –30 F; acid no. 3.0; ref. index 1.4609; flash pt. 525 F (COC); fire pt. 555 F (COC); 50 phr in PVC: tens. str. 3100 psi; tens. elong. 320%; hardness 91A.

Admex 526. [Sherex] Complex polyester; polymeric plasticizer used in automotive seam sealants, coated and calendered fabrics and films, and extruded and molded profiles; Gardner 1; sp.gr. 1.038; dens. 8.66 lb/gal; visc. 1500 cps; pour pt. –4 F; acid no. 1.5; ref. index 1.462; flash pt. (COC) 505 F; @ 50 phr in PVC: tens. str. 2695 psi; elong. 320%; 100% mod. 1940 psi; hardness 89 (A).

Admex 527. [Sherex] Complex polyester; low visc. primary polymeric plasticizer for primary vinyl insulation, PVC resin; used in elec. insulation, profile extrusion, and coating applics.; Gardner 1 liq.; mild odor; sp.gr. 1.035; dens. 8.64lb/gal; visc. 310 cps; pour pt. –30 F; acid no. 3; ref. index 1.4615; flash pt. (COC) 525 F; @ 50 phr in PVC: tens. str. 3160 psi; elong. 325%; 100% mod. 2150 psi; hardness 91 (A); dissip. factor 0.140 (0.1 KHz); dielec. const. 5.4 (0.1 KHz, 23 C); vol. resist. 2.0×10^{10} ohm-cm.

Admex 529. [Sherex] Med. m.w. polyester; plasticizer; for plastisol applics., coated fabrics, calendered and extruded film and sheeting, wire coatings, floorings, vinyl tapes and foam; Gardner 2; sp.gr. 1.122; dens. 9.40 lb/gal; visc. 4150 cps; pour pt. 32 F; acid no. 3.0; ref. index 1.4695; flash pt. 536 F (COC); fire pt. 583 F (COC); @ 50 phr in PVC: tens. str. 3265 psi; tens. elong. 280%; hardness 96A.

Admex 710. [Sherex] Epoxidized soybean oil; synergistic with metallic stabilizers; plasticizer and stabilizer for PVC compds., general purpose films, food wraps, and beverage transfer tubing, hose, gaskets, flooring, NC and butyrate coatings; APHA 150; sp.gr. 0.994; dens. 8.26 lb/gal; visc. 368 cps; pour pt. 25–35 F; acid no. 0.3; ref. index 1.471; flash pt. 590 F (COC); fire pt. 600 F (COC); @ 50 phr in PVC: tens. str. 3010 psi; tens. elong. 385%; hardness 93A.

Admex 711. [Sherex] Epoxidized soybean oil; see Admex 710; APHA 150; sp.gr. 0.994; dens. 8.26 lb/gal; visc. 368 cps; pour pt. 25–35 F; acid no. 0.3; ref. index 1.471; flash pt. 590 F (COC); fire pt. 600 F (COC); @ 50 phr in PVC: tens. str. 3010 psi; tens. elong. 385%; hardness 93A.

Admex 746. [Sherex] Epoxidized tallate ester; plasticizer and stabilizer for calendered film and sheeting, PVC compds.; plastisol visc. depressant.

Admex 752. [Sherex] High m.w. ester; plasticizer; used for automotive upholstery, sheeting, and coated fabrics, rotational castings and slush moldings, dipping compds., profile extrusions, and pigment disps.; Gardner 1; sp.gr. 0.975; dens. 8.10 lb/gal; visc. 130 cps; pour pt. 10 F; acid no. 0.5; ref. index 1.4614; flash pt. 545 F (COC); fire pt. 585 F (COC); @ 50 phr in PVC: tens. str. 3000 psi; tens. elong. 310%; hardness 90A.

Admex 760. [Sherex] High m.w. polyester; plasticizer; pressure-sensitive tapes, elec. insulation, vinyl sol'n. coatings, dry-cleanable films, coating fabric and upholstery; Gardner 3; sp.gr. 1.150; dens. 9.60 lb/gal; visc. 117,600 cps; pour pt. 38 F; acid no. 2.5; ref. index 1.4700; flash pt. 560 F (COC); fire pt. 625 F (COC); @ 50 phr in PVC: tens. str. 3180 psi; tens. elong. 230%; hardness 97A.

Admex 761. [Sherex] Med. m.w. polyester; plasticizer; resists migration into PS, NC, syn. finishes, and rubber; elec. tape, wire coatings and insulation, coated fabric, calendered film and sheeting, vinyl foams, profile extrusion, pigment disps., and dipping compds.; Gardner 3; sp.gr. 1.110; dens. 9.25 lb/gal; visc. 5330 cps; pour pt. 35 F; acid no. 3.0; ref. index 1.4800; flash pt. 460 F (COC); fire pt. 590 F (COC); @ 50 phr in PVC: tens. str. 3310 psi; tens. elong. 250%; hardness 97A.

Admex 770. [Sherex] Med. m.w. polyester; plasticizer with nonmigration properties to SBR and neoprene; elec. insulation and tape, high-clarity film and sheeting, coated fabrics, pigment disps., and dipping compds.; Gardner 1; sp.gr. 1.110; dens. 9.25 lb/gal; visc. 5570 cps; pour pt. 30 F; acid no. 2.0; ref. index 1.4660; flash pt. 560 F (COC); fire pt. 600 F (COC); @ 50 phr in PVC: tens. str. 3195 psi; tens. elong 290%; hardness 96A.

Admex 775. [Sherex] Med. m.w. polyester; plasticizer; elec. tapes, coated fabric, paper coatings, refrigerator and freezer gaskets, and other profile extrusions; Gardner 1; sp.gr. 1.095; dens. 9.14 lb/gal; visc. 6130 cps; pour pt. 0 F; acid no. 1.3; ref. index 1.4670; flash pt. 550 F (COC); fire pt. 599 F (COC); @ 50 phr in PVC: tens. str. 3060 psi; tens. elong. 360%; hardness 96A.

Admex 780. [Sherex] Med. m.w. polyester; polymeric plasticizer; Gardner 1–3; sp.gr. 1.08; dens. 9.03 lb/gal; visc. 14–19 stokes; pour pt. 50 F; acid no. 3 max.; ref. index 1.465; flash pt. (COC) 540 F; @ 50 phr in PVC: tens. str. 3065 psi; elong. 350%; 100% mod. 1965 psi; hardness 94 (A).

Admex 890. [Sherex] Med. m.w. polyester; low-odor plasticizer for refrigerator gasket compds.; migration resistance against styrene and ABS polymers; used in elec. tapes, profile extrusions, calendered film and sheeting, and coated fabric; Gardner 1; sp.gr. 1.097; dens. 9.06 lb/gal; visc. 4890 cps; pour pt. 35 F; acid no. 2.0; ref. index 1.4665; flash pt. 579 F (COC); fire pt. 613 F (COC); @ 50 phr in PVC: tens. str. 3240 psi; tens. elong. 270%; hardness 95A.

Admex 910. [Sherex] Med. m.w. polyester; polymeric plasticizer for styrene and ABS polymers; used in refrigerator and freezer gaskets, calendered film and sheeting and other extruded profiles; APHA 150;

sp.gr. 1.048; dens. 8.75 lb/gal; visc. 4480 cps; pour pt. –2 F; acid no. 1.5; ref. index 1.471; flash pt. (COC) 535 F; @ 50 phr in PVC: tens. str. 2865 psi; elong. 315%; 100% mod. 2030 psi; hardness 92 (A).

Admex ELO. [Sherex] Epoxidized linseed oil; heat and lt. stabilizer, and plasticizer for food pkg. and equip., calendered film and sheeting, and profile extrusions; APHA 150; sp.gr. 1.032; dens. 8.58 lb/gal; visc. 815 cps; pour pt. 30 F; acid no. 1.0; ref. index 1.477; flash pt. 590 F (COC); fire pt. 635 F (COC); @ 50 phr in PVC: tens. str. 3220 psi; tens. elong. 255%; hardness 93A.

Admul WOL 1403. [Quest Int'l.] Polyglycerol polyricinoleate; nonionic; visc. modifier in chocolate; liq.; 100% conc.

Adogen® 58. [Sherex] Erucyl amide; antiblock, antitack, and slip agent for polyolefin film mfg.; improve dye sol. and control visc. in printing inks; Gardner 6 max.;m.p. 75–85 C; iodine no.72–82; 97% min. amide.

Adogen® 73. [Sherex] Oleyl amide; see Adogen 58; Gardner 5 max.; m.p. 72–76 C; iodine no. 80–92; 97% min. amide.

Adogen® 115 (D). [Sherex] Soya amine, distilled; corrosion inhibitor for the formulation of oil-sol. downhole corrosion inhibitors used in the drilling and prod. of oil and gas; lt. yel. liq.; 100% conc.

Adogen® 140 (D). [Sherex] Hydrog. tallow amine, distilled; corrosion inhibitor for the formulation of oil-sol. downhole corrosion inhibitors used in the drilling and prod. of oil and gas; wh. solid; 100% conc.

Adogen® 142 (D). [Sherex] Stearyl amine, distilled; corrosion inhibitor for the formulation of oil-sol. downhole corrosion inhibitors used in the drilling and prod. of oil and gas; wh. solid; 100% conc.

Adogen® 151. [Sherex] Tall oil amine; corrosion inhibitor for the formulation of oil-sol. downhole corrosion inhibitors used in the drilling and prod. of oil and gas; 100% conc.

Adogen® 160 (D). [Sherex] Cocamine, distilled; corrosion inhibitor for the formulation of oil-sol. downhole corrosion inhibitors used in the drilling and prod. of oil and gas; lt. yel. liq.; 100% conc.

Adogen® 170 (D). [Sherex] Tallow amine, distilled; corrosion inhibitor for the formulation of oil-sol. downhole corrosion inhibitors used in the drilling and prod. of oil and gas; wh. paste; 100% conc.

Adogen® 172 (D). [Sherex] Oleyl amine, distilled; corrosion inhibitor for the formulation of oil-sol. downhole corrosion inhibitors used in the drilling and prod. of oil and gas; amber liq.; 100% conc.

Adogen® 185. [Sherex] C_{12-15} ether amine; corrosion inhibitor; used in oilfield brine formulations; lt. yel. liq.; 100% conc.

Adogen® 188. [Sherex] C_{8-10} ether amine; corrosion inhibitor; used in oilfield brine formulations; 100% conc.

Adogen® 240. [Sherex] Dihydrog. tallow amine; corrosion inhibitor for the formulation of drilling and prod. corrosion inhibitors; wh. solid; 100% conc.

Adogen® 283. [Sherex] Ditridecyl amine; solv. extraction reagent for the extraction of vanadium, tungsten, and iron; amber liq.; sp.gr. 0.828; dens. 6.89 lb/gal; amine value 140; 80% sec. amine; 15% tert. amine.

Adogen® 342-D. [Sherex] Dimethyl stearamine, distilled; corrosion inhibitor; oil-sol. intermediate for formulation in refinery corrosion inhibitors and gasoline additive packages; 100% conc.

Adogen® 343. [Sherex] Dihydrog. tallow methyl amine; intermediate for the mfg. of quat. ammonium compds.; Gardner 3.0 max. wh. solid; m.p. 35 C; iodine no. 4.0 max.; tert. amine value 102; neutral. equiv. 520; 95.0% min. tert. amine.

Adogen® 345-D. [Sherex] Dimethyl hydrog. tallow amine, distilled; corrosion inhibitor; oil-sol. intermediate for formulation in refinery corrosion inhibitors and gasoline additive packages; 100% conc.

Adogen® 363. [Sherex] Trilauryl amine; solv. extraction reagent for the extraction of uranium and molybdenum; liq.; sp.gr. 0.820; dens. 6.81 lb/gal; amine value 106; 95.1% tert. amine; 2.9% sec. amine.

Adogen® 364. [Sherex] Tri C_8–C_{10} amine; solv. extraction reagent for the extraction of uranium, vanadium, and chromium; yel. liq.; sp.gr. 0.813; dens. 6.76 lb/gal; amine value 144; 96.2% tert. amine; 0.7% sec. amine.

Adogen® 367-D. [Sherex] Distilled coco dimethyl amine, distilled; corrosion inhibitor; oil-sol. intermediate for formulation in refinery corrosion inhibitors and gasoline additive packages; 100% conc.

Adogen® 368. [Sherex] Tri C_8–C_{10}–C_{12} amine; solv. extraction reagent for the extraction of uranium and molybdenum; liq.; sp.gr. 0.816; dens. 6.78 lb/gal; amine value 131; 95.8% tert. amine; 1.4% sec. amine.

Adogen® 381. [Sherex] Triisooctyl amine; solv. extraction reagent for the extraction of cobalt and uranium; yel. liq.; sp.gr. 0.817; dens. 6.79 lb/gal; amine value 155; 95.1% tert. amine; 1.9% sec. amine.

Adogen® 382. [Sherex] Triisodecyl amine; solv. extraction reagent for the extraction of uranium and vanadium; yel. liq.; sp.gr. 0.823; dens. 6.84 lb/gal; amine value 126; 95.1% tert. amine; 2.9% sec. amine.

Adogen® 383. [Sherex] Tritridecyl amine; solv. extraction reagent for the extraction of molybdenum, rhenium, and uranium; yel. liq.; sp.gr. 0.835; dens. 6.94 lb/gal; amine value 99; 96.0% tert. amine; 1.9% sec. amine.

Adogen® 415. [Sherex] Soya trimethyl ammonium chloride; cationic; antistat, emulsifier, dispersant; used in corrosion inhibitor formulations for oilfield brines and HCl acidizing systems; liq.; m.w. 342; flash pt. (PM) 58 F; 50% conc.

Adogen® 432. [Sherex] Dialkyl (C_{12}-C_{18}) dimethyl ammonium chloride; cationic; fabric softener conc. for home and commercial laundries; textile processing; Gardner 3 max. liq.; m.w. 521; flash pt. 70 F (PM); 67–69% quat.

Adogen® 432-ET. [Sherex] Quaternium 31 (dicetyldimonium chloride); base for hair conditioners; liq.; 67–69% solids in ethanol.

Adogen® 432 CG. [Sherex] Dicetyldimonium chloride; cationic; conditioner, creme rinse base; lt. clear to hazy liq.; disp. in water; pour pt. < 50 F; 74% conc. in aq. IPA.

Adogen® 441. [Sherex] Hydrog. tallow trimethyl ammonium chloride; specialty quat.; liq.; Gardner 5 max.; m.w. 341; flash pt. 58 F (PM); 49–52% quat.

Adogen® 442. [Sherex] Quaternium-18; see Adogen 432; Gardner 4 max. paste; m.w. 569; flash pt. 65 F (PM); 74–77% quat.

Adogen® 442-100P. [Sherex] Quaternium-18; base for hair conditioners; powd.; 100% solids.

Adogen® 442 H. [Sherex] Quaternium-18; viscosifier for Bentonite clays for oil-based muds in the petrol. industry; paste; flash pt. (CC) 65 F; 75% conc.

Adogen® 444. [Sherex] Palmityl trimethyl ammonium chloride; cationic; specialty quat.; emulsifier, dispersant, creme rinse, fermentation aid; liq.; Gardner 6 max.; m.w. 319; flash pt. (PM) 58 F; 49–52% quat.

Adogen® 461. [Sherex] Coco trimethyl ammonium chloride; cationic; emulsifier, dispersant; used in corrosion inhibitor formulations for oil-field brines and HCl acidizing systems; 50% conc.

Adogen® 462. [Sherex] Dicocodimonium chloride; cationic; antistat, emulsifier, flocculating agent, dispersant used in corrosion inhibitor formulations for oil-field chemicals; Gardner 5 max. liq.; m.w. 439; flash pt. 68 F (PM); 74–77% quat.

Adogen® 464. [Sherex] Methyl tri (C_8–C_{10}) ammonium chloride; cationic; flotation agent, emulsifier for solv. extraction of metals; in corrosion inhibitor formulations for oilfield brines and HCl acidizing systems; Gardner 5 max. liq.; m.w. 437; flash pt. 156 F (PM); 85% quat.

Adogen® 470. [Sherex] Ditallow dimethyl ammonium chloride; specialty quat.; liq.; Gardner 6 max.; m.w. 564; flash pt. 65 F (PM); 74–77% quat.

Adogen® 471. [Sherex] Tallow trimethyl ammonium chloride; cationic; dispersant, antistat, emulsifier; used in corrosion inhibitor formulations for oilfield brines and HCl acidizing systems; Gardner 6 max. liq.; m.w. 339; flash pt. 58 F (PM); 49–52% quat.

Adogen® 477. [Sherex] N-tallow pentamethyl propane diammonium dichloride; cationic; emulsifier, dispersant; used in corrosion inhibitor formulations for oilfield brines and HCl acidizing systems; liq.; Gardner 11 max.; m.w. 463; flash pt. 47 F (PM); 49–52% quat.

Adogen® 551. [Sherex] Tall oil diamine; corrosion inhibitor for the formulation of oil-sol. downhole corrosion inhibitors used in the drilling and prod. of oil and gas; 100% conc.

Adogen® 560. [Sherex] Coco diamine; cationic; emulsifier for asphalt; corrosion inhibitor for the formulation of oil-sol. downhole corrosion inhibitors used in the drilling and prod. of oil and gas; amber liq.; 100% conc.

Adogen® 570-DO. [Sherex] Tallow diamine dioleate; cationic corrosion inhibitor for use in petrol. prod. where protection against high brine environments is important; in industrial oils (cutting, drawing, rolling stocks) as a wetting agent, lubricity aid; Gardner 11 max. soft solid; m.w. 449; insol. in water; sol. to ≥ 5% in naphtha, min. oil, or IPA; amine value 125 min.

Adogen® 570-S. [Sherex] Tallow diamine; cationic; see Adogen 560; amber solid; 100% conc.

Adogen® 572. [Sherex] Oleyl diamine; cationic; see Adogen 560; amber liq.; 100% conc.

Adogen® 588. [Sherex] C_{8-10} ether amine; corrosion inhibitor; used in oilfield brine formulations; 100% conc.

Adogen® 2283. [Sherex] Ditridecyl amine; solv. extraction reagent for the extraction of vanadium and tungsten; liq.; sp.gr. 0.828; dens. 6.89 lb/gal; amine value 140; 15% tert. amine; 80% sec. amine.

Adogen® 2363. [Sherex] Trilauryl amine; solv. extraction reagent for the extraction of uranium and molybdenum; liq.; sp.gr. 0.820; dens. 6.81 lb/gal; amine value 106; 95.1% tert. amine; 2.9% sec. amine.

Adogen® 2364. [Sherex] Tri C_8–C_{10} amine; solv. extraction reagent for the extraction of uranium, vanadium, and chromium; liq.; sp.gr. 0.813; dens. 6.76 lb/gal; amine value 144; 96.2% tert. amine; 0.70% sec. amine.

Adogen® 2368. [Sherex] Tri C_8–C_{10}–C_{12} amine; solv. extraction reagent for the extraction of uranium and molybdenum; liq.; sp.gr. 0.816; dens. 6.78 lb/gal; amine value 131; 95.8% tert. amine; 1.4% sec. amine.

Adogen® 2381. [Sherex] Triisooctyl amine; solv. extraction reagent for the extraction of cobalt and uranium; liq.; sp.gr. 0.817; dens. 6.79 lb/gal; amine value 155; 95.1% tert. amine; 1.9% sec. amine.

Adogen® 2382. [Sherex] Tritridecyl amine; solv. extraction reagent for the extraction of molybdenum, rhenium, and uranium; liq.; sp.gr. 0.823; dens. 6.84 lb/gal; amine value 126; 95.1% tert. amine; 2.9% sec. amine.

Adogen® 2464. [Sherex] Tri C_8–C_{10} methyl ammonium chloride; solv. extraction reagent for the extraction of vanadium, tungsten, chromium, and uranium; liq.; 86% quat.

Adogen® R-6. [Sherex] 1:1 blend tallow trimethyl and dicoco dimethyl ammonium chloride; specialty quat.; liq.; Gardner 6 max.; m.w. 389; flash pt. 58 F (PM); 49–52% quat.

Adol® 42. [Sherex] Unsat. fatty alcohol; coemulsifier, lubricant, foam control agent, cosolvent, plasticizer, stabilizer, emollient, intermediate; used in metal rolling oils and lubricants, gas scrubbing, printing inks, auto antifreeze, textile dyeing and finishing, emulsion systems, paper pulping, cosmetics, cationic surfactants, min. processing/oil field chemicals, fabric softeners; Lovibond 5Y/0.5R color; m.w. 258; sol. in fatty alcohols, IPA, benzene, trichlorethylene, ethyl ether, acetone, turpentine, VMP naphtha, kerosene, lt. min. oil; sp.gr. 0.814 (60/25 C); cloud pt. 36 C max.; acid no. 1.5 max.; sapon. no. 3.5 max.

Adol® 52. [Sherex] Cetyl alcohol; emollient; waxy flake.

Adol® 52 NF. [Sherex] Cetyl alcohol; see Adol 42; Lovibond 5Y/0.5R max.; m.w. 247; sol. see Adol 42; sp.gr. 0.815 (60/25 C); m.p. 45–50 C; acid no. 1.0 max.; sapon. no. 3.0 max.

Adol® 54. [Sherex] Cetyl alcohol; see Adol 42; Lovibond 5Y/O.5R; m.w. 249; sol. see Adol 42; sp.gr. 0.816; solid. pt. 46–50 C; acid no. 1.0 max.; sapon no. 3.0 max.

Adol® 60. [Sherex] Behenyl alcohol; lubricant, foam control agent, cosolv., plasticizer, emollient; Lovibond 10Y/1.0R; m.w. 316; sp. gr. 0.79 (100/25 C); acid no. 1.5 max.; iodine no. 6 max.; sapon. no. 3 max.

Adol® 61 NF. [Sherex] Stearyl alcohol; used in emulsifiers, surfactants, cosmetics, wax formulations; see also Adol 42; wh. cryst. solid; odorless; sol. in IPA, acetone, naphtha, lt. min. oil; m.w. 272; sp.gr. 0.817 (60/25 C); m.p. 56–60 C; acid no. 0.5; sapon. no. 1.0 max.

Adol® 62. [Sherex] Stearyl alcohol; emollient; waxy flake.

Adol® 62 NF. [Sherex] Stearyl alcohol; emollient, glass frit binders, waxes, emulsion stabilizers, esters, tertiary amines, surfactants, polymers, chemical intermediate; cosmetic formulations; see also Adol 42; Lovibond 5Y/0.54 max; odorless; m.w. 272; sol. see Adol 61NF; sp.gr. 0.817 (60/25 C); visc. 42 SSU (210 F); m.p. 56–60 C; b.p. 337–360 C (760 mm, 90%); acid no. 1.0 max.; sapon. no. 3.0 max.

Adol® 63. [Sherex] Stearyl alcohol; prime base for detergents; used in plasticizers, tert. amines, lube oil additives, textile auxiliaries, mold lubricants, polymers, org. synthesis, chemical intermediates; see also Adol 42; Lovibond 5Y/0.5R max.; m.w. 268; sp.gr. 0.816 (60/25 C); visc. 44 SSU (210 F); solid pt. 48–53 C; b.p. 312–344 C (760 mm, 90%); acid no. 1.0 max.; sapon. no. 3.0 max.

Adol® 64. [Sherex] Stearyl alcohol; see Adol 63, Adol 42; Lovibond 5Y/0.5R max.; odorless; m.w. 269; sol. in IPA, acetone, naphtha, lt. min. oil; sp.gr. 0.817 (60/25 C); visc. 44 SSU (210 F); solid. pt. 51–54 C; acid no. 1.0 max.; sapon. no. 3.0 max.

Adol® 66. [Sherex] Isostearyl alcohol; see Adol 42; Lovibond 5Y/0.5R max.; m.w. 295; sol. see Adol 42; sp.gr. 0.861; cloud pt. 8.0 C max.; acid no. 1.0 max.; sapon. no. 2.0 max.

Adol® 80, 85 NF. [Sherex] Oleyl alcohol; see Adol 42; Lovibond 5Y/0.5R max. and 3Y/0.3R max. resp.; m.w. 263 and 267; sol. see Adol 42; sp.gr. 0.840; cloud pt. 13 and 10 C max.; acid no. 1.0 max.; sapon. no. 3.0 and 2.0 max.

Adol® 85. [Sherex] Oleyl alcohol; see Adol 60; Lovibond 3Y/0.3R; m.w. 268; sp.gr. 0.84; cloud pt. 10 C max.; acid no. 1.0; iodine no. 85–95; sapon no. 2 max.

Adol® 90. [Sherex] Oleyl alcohol; see Adol 60; APHA 50; m.w. 268; sp.gr. 0.840; cloud pt. 6 C max.; acid no. 0.8 max.; iodine no. 85.5–95.5; sapon. no. 2 max.

Adol® 90 NF. [Sherex] Oleyl alcohol, cosmetic and pharmaceutical grade; emollient; coupling agent; plasticizer for hair sprays, lubricant for aerosols; emulsion stabilizer; used in lotions, creams, bath oils; lt. clear liq.; low odor; m.w. 268; b.p. 282–349 C (760 mm, 90%); sol. in ethanol, IPA, benzene, ethyl ether, acetone, turpentine, VM&P naphtha, kerosene, lt. min. oil; sp.gr. 0.840; acid no. 0.5; iodine no. 90; sapon. no. 1.5; hyd. no. 210; cloud pt. 5 C.

Adol® 320, 330, 340. [Sherex] Oleyl alcohol; see Adol 42; Lovibond 5Y/0.5R max.; m.w. 261; sp.gr. 0.845; cloud pt. 19, 22, and 28 C max. resp.; acid no. 1.0 max.; sapon. no. 3.0 max.

Adol® 520. [Sherex] Cetyl alcohol; emollient; waxy flake.

Adol® 520 NF. [Sherex] Cetyl alcohol; see Adol 42; APHA 50 max.; m.w. 246; sol. see Adol 42; sp.gr. 0.815 (60/25 C); m.p. 45–50; solid pt. 48–53 C; acid no. 1.0 max.; sapon. no. 2.0 max.

Adol® 620. [Sherex] Stearyl alcohol; emollient; waxy flake.

Adol® 620 NF, 630. [Sherex] Sat. fatty alcohol; see Adol 42; APHA 50 and 100 max. resp.; m.w. 267 and 265; sol. see Adol 42; sp.gr. 0.817 and 0.816 (60/25 C); m.p. 55–60 and 48–53 C; acid no. 2.0 max.; sapon. no. 2.0 max.

Adol® 640. [Sherex] Cetearyl alcohol; see Adol 60; APHA 40 max.; m.w. 253; sp.gr. 0.82 (50/50 C); m.p. 43–46 C; acid no. 1.0 max.; iodine no. 1.5 max.; sapon. no. 1.0 max.

Adox 3125. [Int'l Dioxide] Sodium chlorite; antimicrobial for water and waste water treatment; 25% conc. in water.

Adsee® 775. [Witco] POE ethers and special resins; nonionic; spreader sticker and penetrant for agric. spray; surfactant for monosodium methane arsonate formulations; liq.; 100% conc.

Adsee® 799. [Witco] Alkyl POE ether; nonionic; agric. surfactant; soil penetrant; liq.; 100% conc.

Adtac® B10BHT, B25BHT. [Hercules] Aliphatic hydrocarbon resin; tackifier for nat. rubber, syn. rubbers, thermoplastic block copolymer elastomers used in adhesives, esp. pressure-sensitive systems; Gardner 5 liq. and Gardner 5.5 semisolid resp.; sol. in aliphatic and aromatic hydrocarbons, ethylhexanol, MIBK, trichloroethylene; insol. in water; dens. 0.916 and 0.923 kg/l; visc. 125 and 310 cps (99 C); soften. pt. (R&B) 10 and 25 C; acid no. < 1.

Adtac® L100BHT. [Hercules] Aliphatic hydrocarbon resin; tackifier resin in adhesive applics.; saturant and waterproofing agent in paper, textiles, coatings; reinforcing, tackifier, and binder resin in construction adhesives and other bldg. materials; used in food pkg. and processing operations; Gardner 8 max. solid, flake; sol. in aliphatic, aromatic, and chlorinated hydrocarbons, MIBK, ethyl ether, and long-chain alcohol; dens. 0.97 kg/l; visc. 100 poise (150 C); acid no. < 1; sapon. no. < 2; flash pt. 257 C (COC).

Advastab® LS-202. [Morton Int'l.] Organotin comp.; lubricating stabilizer in PVC pipe prod. on multiple screw extruders; catalyst for rigid foams and molded flexible systems made with a premix; off-wh. to lt. yel. granulated waxy solid; dens. 485 g/l; soften pt. 65-119 C; 100% solid.

Advastab® LS-203. [Morton Int'l.] Organotin compd.; lubricating stabilizer for high-output PVC pipe prod. on multiple screw extruders; off-wh. to lt. yel. gran. waxy solid; sp.gr. 0.92; bulk dens. 30.6 lb/ft^3; drop pt. 110 C.

Advastab® T-52N Conc. [Morton Int'l.] Butyltin carboxylate; PVC and PVC-PVAc heat and lt. stabilizer for sol'n. resin systems, calendering and molding; urethane catalyst for rigid foams and molded flexible systems made with a premix; Gardner 1 liq.; sp.gr. 1.21; dens. 10 lb/gal; visc. 150 cs; 20% tin.

Advastab® T-290. [Morton Int'l.] Butyltin carboxylate; heat and lt. stabilizer for rigid PVC, esp. high heat distort. temp. applics.; wh. fine powd.; sp.gr. 1.40; dens. 6.95 lb/gal.

Advastab® T-340. [Morton Int'l.] Dibutyltin maleate; heat stabilizer for PVC compds. in rigid sheeting, moldings, and extrusions; wh. cryst. solid; sol. in ethanol, benzene, acetone; insol. in water; sp.gr. 1.36–1.42; dens. 11.50–11.82 lb/gal; 34% tin.

Advastab® TM-180. [Morton Int'l.] Butyltin mercaptide; heat stabilizer esp. for prolonged/severe processing temps. of rigid PVC and PVC-PVA formulations; recommended for sol'n. vinyls containing reactive terpolymers, NSF applics., fluidized-bed powd. coatings; Gardner 4 max. clear liq.; sp.gr. 1.110–1.130; dens. 9.24–9.48 lb/gal; visc. 50 cps max.

Advastab® TM-181. [Morton Int'l.] Methyltin mercaptide; stabilizer for rigid PVC processes (extrusion, calendering, inj. and blow molding) and PVC-PVA formulations with prolonged/severe processing temps., for chlorinated PVC compds; tin catalyst in PU polymerizations; APHA 70 clear liq.; sp.gr. 1.175; dens. 9.8 lb/gal; visc. 50 cs.

Advastab® TM-181-FS. [Morton Int'l.] Methyltin mercaptide; heat stabilizer for rigid PVC and rigid PVC copolymers for food pkg., bottles, profile, and inj. moldings; Gardner 1 max. clear liq.; sp.gr. 1.17; dens. 9.7 lb/gal; visc. 50 cs.

Advastab® TM-181 S. [Morton Int'l.] Methyltin; heat stabilizer for rigid PVC extrusions, calendering, inj. and blow molding; also for CPVC and PVCA formulations; Gardner 2 or less clear liq.; sp.gr. 1.17; dens. 9.7 lb/gal; visc. 50 cs.

Advastab® TM-185. [Morton Int'l.] Methyltin mercaptide; PVC stabilizer for critical clear rigid PVC bottle and sheet processing, inj. molding, extruded profile, expanded rigid foams, siding, fluid-bed powd. coatings, sol'n. vinyl coil coatings; Gardner 2–3 clear liq.; sp.gr. 1.193; dens. 9.93 lb/gal; visc. 101.8 cps.

Advastab® TM-692. [Morton Int'l.] Methyltin mercaptide; stabilizer for most rigid PVC applics. for potable water, sewer, irrigation pipe, conduit, and duct, and extrusion applics., profile, foamed profile; liq.; sp.gr. 0.99; dens. 8.3 lb/gal; visc. 60 cs; cloud pt. –7 C; pour pt. –18 C.

Advastab® TM-694. [Morton Int'l.] Tin; stabilizer for rigid PVC formulations; esp. for extrusion of potable water, sewer and irrigation pipe, elec. conduit and telephone duct, profiles and tubing; liq.; sp.gr. 0.98; dens. 8.2 lb/gal; visc. 59.2 cs; pour pt. –12.0 C; cloud pt. –1.0 C.

Advastab® TM-697. [Morton Int'l.] Tin stabilizer for rigid PVC applics., esp. for large diameter pipe; also for profile and inj. moldings; liq.; sp.gr. 1.03; dens. 8.6 lb/gal; visc. 56 cs; cloud pt. –2 C; pour pt. –9 C.

Advastab® TM-2082. [Morton Int'l.] Methyltin mercaptide; heat stabilizer for rigid PVC and rigid PVC copolymers for food pkg., bottles; Gardner 1 max. clear liq.; sp.gr. 1.17; dens. 9.7 lb/gal; visc. 50 cs.

Advawax 140. [Morton Int'l.] Ester; syn. lubricant; wh. small flake.

Advawax 240. [Morton Int'l.] N,N′-ethylene bis-oleamide; syn. wax used as plastics processing lubricant and release agent, antistat, m.p. modifier for waxes, industrial asphalts and tar, pigment dispersing agent for resin systems; polyamide-paraffin coupling agent; used in adhesive tapes, coatings, food pkg. materials; color 3; sm. beads; sol. in ethanol, Cellosolve, MEK, MIBK, benzene, xylene, kerosene, heptane, CCl_4, naphtha; dens. 5 lb/gal; m.p. 113-118 C; flash pt. 270 C (COC); acid no. 10.

Advawax 275, 280, 290. [Morton Int'l.] N,N′-ethylene bisstearamide; syn. wax used as plastics processing lubricant and release agent; m.p. modifier for waxes and resin blends and industrial asphalt and tar; pigment dispersing agent for resin systems; papermaking defoamer; used in adhesvie tapes, coatings, PP; 290 grade also for food pkg. materials; sm. beads; sol. in Cellosolve, MIBK, benzene, xylene, kerosene, heptane, naphtha; dens. 5.3, 4.9, and 4.75 lb/gal resp.; m.p. 134-137, 139–143, and 143–146 C; flash pt. 280, 280, and 290 C (COC); acid no. 8.

AE-1214/6. [Procter & Gamble] Ethoxylated linear alcohols; nonionic; lubricant, wetting agent, used in textiles; solid; 100% act.

AEPD®. [Angus] 2-Amino-2-ethyl-1,3-propanediol; pigment dispersant, neutralizing amine, corrosion inhibitor, acid-salt catalyst, pH buffer, chemical and pharmaceutical intermediate, solubilizer; m.w. 119.2; water-sol.; m.p. 37.5 C; b.p. 152 C; flash pt. > 200 F (TCC); pH 10.8 (0.1M aq. sol'n.).

Aerosil® 200. [Degussa] Microfine silica; anticaking and free-flow agent with high absorp. capacity; wh. fluffy powd.; dens. 50 g/l; surf. area 200 m²/g; pH 3.6–4.3 (4% aq. disp.); > 99.8% SiO_2.

Aerosil® Colloidal Silicas. [Degussa] Colloidal silicon dioxide; anticaking agent, thixotropic agent, suspending agent; wh. powd.

Aerosil® R 972. [Degussa] Microfine silica; anticaking and free-flow agent; wh. fluffy powd.; dens. 50 g/l; surf. area 110 m²/g; pH 3.6–4.3 (4% aq. disp.); > 99.8% SiO_2.

Aerosil® R-974, R-976. [Degussa] Colloidal silicon dioxide; hydrophobic anticaking agent with suspending char.

Aerosol® 18. [Am. Cyanamid] Disodium stearyl sulfosuccinamate; anionic; emulsifier, dispersant, foamer, detergent, solubilizer for soaps and surfactants; alkaline cleaner formulations, brick and tile cleaners, emulsion polymerization of vinyl chloride and SBRs; emulsifying oils and waxes, household detergents, cleaning paper mill felts; foamer for foamed latexes and plastics; Gardner 10 max. semiliq., creamy paste; m.w. 493; water-disp.; sp.gr.

1.16; acid no. 4.0 max.; surf. tens. 41 dynes/cm; biodeg.; 35 ± 1.5% solids.

Aerosol® 22. [Am. Cyanamid] Tetrasodium dicarboxyethyl stearyl sulfosuccinamate; anionic; emulsifier, dispersant, solubilizer, surfactant; emulsion polymerization of vinyl monomers; polishing waxes; surf. tension depressant for writing and drawing inks; demulsifier for w/o emulsions; cleaning of paper mill felts; lt. tan clear to cloudy liq.; water-sol.; sp.gr. 1.12; dens. 9.4 lb/gal; visc. 53 cps; m.p. > 200 C; surf. tens. 41 dynes/cm; biodeg.; 35% act.

Aerosol® 22N. [Am. Cyanamid] Tetrasodium N-(1,2-dicarboxy ethyl)-N-octadecyl sulfosuccinamate; anionic; dispersant for pigments, silicones, and UV absorbers; solubilizer with salt tolerance; emulsion polymerization, latex foam stabilization; antigelling agent; industrial, household, cosmetic, and metal cleaners; liq.; sol. in water; 35% conc.

Aerosol® 30. [Am. Cyanamid] Cocamidopropyl betaine; amphoteric; visc. builder, surfactant, foam booster; used in personal care prods.; disinfectants; yel. clear liq.; f.p. 0 C; water-sol.; sp.gr. 1.04; surf. tens. 35 dynes/cm; partially biodeg.; 35% act.

Aerosol® 200. [Am. Cyanamid] Disodium (ethoxylated alkylamide) sulfosuccinate; emulsifier, foaming agent, dispersant; Gardner 7 max. clear yel. liq.; mild odor; water sol.; dens. 8.93 lb/gal; sp.gr. 1.07; visc. 200 cps max.; cloud pt. < 0 C; surf. tens. 35 dynes/cm; biodeg.; 30% act.

Aerosol® 413. [Am. Cyanamid] Disodium ethanol alkyl (C18) amide sulfosuccinate; dispersant, emulsifier, foamer, wetting agent; used for vinyl acetate and acrylate emulsions; shampoos; foaming applics.; yel. clear liq.; water-sol.; dens. 8.95 lb/gal; visc. 70–90 cps; surf. tens. 38 dynes/cm; biodeg.; 35% act.

Aerosol® 501. [Am. Cyanamid] Disodium alkyl sulfosuccinate; anionic-nonionic; dispersant, emulsifier, wetting agent, dispersant, foaming agent; used for acrylic and vinyl acetate emulsions; self-crosslinking latexes; textile wetting and foaming applics.; APHA 60 max. clear liq.; water-sol.; sp.gr. 1.16; dens. 9.66 lb/gal; visc. 260 cps; f.p. –3 C; surf. tens. 28 dynes/cm; 50% act.

Aerosol® A-102. [Am. Cyanamid] Disodium deceth-6 sulfosuccinate; anionic/nonionic; emulsifier, solubilizer, foamer, dispersant, surfactant; used in emulsion polymerization of PVAc/acrylics; textile industry in emulsions, wetting agent, and pad-bath additive; colorless to lt. yel. clear liq.; water-sol.; sp.gr. 1.08; dens. 9.01 lb/gal; visc. 40 cps; f.p. –4 C; surf. tens. 29 dynes/cm; 30% act.

Aerosol® A-103. [Am. Cyanamid] Disodium nonoxynol-10 sulfosuccinate; anionic/nonionic; emulsifier, solubilizer, wetting agent, surfactant, surf. tens. depressant; used in PVAc acrylic emulsions; textile emulsions, pad-bath additive, textile wetting; colorless to lt. yel. clear liq.; water-sol.; sp.gr. 1.09; dens. 9.1 lb/gal; visc. 170–190 cps; f.p. –9 C; surf. tens. 34 dynes/cm; 35% act.

Aerosol® A-196, A-196-40. [Am. Cyanamid] Dicyclohexyl sodium sulfosuccinate; anionic; dispersant, surfactant; sole emulsifier for modified S/B; post additive to stabilize latex and promote adhesion; wh. pellets, clear to cloudy liq.; surf. tens. 39 dynes/cm; biodeg.; 85% act. pellets, 40% act. aq. sol'n. resp.

Aerosol® A-268. [Am. Cyanamid] Disodium isodecyl sulfosuccinate; anionic surfactant, sole emulsifier for PVC latexes-vinyl, vinylidene chloride, acrylics, surf. tens. depressant, solubilizer; clear liq.; water sol.; dens. 9.96 lb/gal; sp.gr. 1.19; visc. 150 cps; f.p. –5 C; surf. tens. 28 dynes/cm; 50% act.

Aerosol® AY. [Am. Cyanamid] Sodium diamyl sulfosuccinate; anionic; emulsifier in emulsion polymerization; antigelling agent; used for crystal growth control, electroplating, filtration, leaching ores and slags, wetting agent and dispersant for paint pigments; paper processing, soldering fluxes; APHA 100 max. hard waxy solid; m.w. 360; sol. in polar org. solvs., 39.2 g/100 ml water; sp.gr. 1.2; m.p. 222–227 C; acid no. 2.5 max.; surf. tens. 50.2 dynes/cm (0.01% aq.); 100% solids.

Aerosol® AY-65, AY-100. [Am. Cyanamid] Diamyl sodium sulfosuccinate; anionic; wetting agent, dispersant, surfactant; used in agriculture, emulsion polymerization, electroplating, ore leaching, cleaning of porcelain, tile, brick, cement; clear liq.; sol. in water/ethanol (AY-65), sol. in water and org. solvs. (AY-100); sp.gr. 1.081 and 1.2 resp.; dens. 9.0 and 10.0 lb/gal; f.p. < 18 C; surf. tens. 29 dynes/cm; biodeg.; 65 and 100% act.

Aerosol® C-61. [Am. Cyanamid] Ethoxylated alkyl guanidine-amine complex; cationic; antistat; wetting agent in acid digestion of ores; settling agent; alkaline, cement, brick, and tile cleaner formulations for crystal growth control, emulsion breaking, alkaline metal and paint brush cleaners, paint removers, dispersant for pigments in aq. systems; textile softener; demulsifying agent; foaming agent; lt. tan creamy paste; strong ammoniacal odor; sp.gr. 1.00; dens. 8 lb/gal; flash pt. 80 C; surf. tens. 34 dynes/cm; partially biodeg.; 70.4% act.

Aerosol® IB-45. [Am. Cyanamid] Sodium diisobutyl sulfosuccinate; anionic; emulsifier, wetting agent; emulsion polymerization of styrene, butadiene and copolymers; dye and pigment dispersant; for leaching, electroplating; clear liq.; water-sol.; extremely hydrophilic; f.p. 20–21 C; surf. tens. 49 dynes/cm; biodeg.; biodeg.; 45% act.

Aerosol® MA-80. [Am. Cyanamid] Dihexyl sodium sulfosuccinate; anionic; dispersant, textile wetting agent, emulsifier, solubilizer; used for emulsion polymerization, battery separators, electroplating, ore leaching; germicidal act.; APHA 50 max. clear slt.ly visc. liq.; sol. in water, alcohol, and org. solvs.; sp.gr. 1.13; dens. 9.4 lb/gal; f.p. –28 C; m.p. 199–292 C; surf. tens. 28 dynes/cm; slowly biodeg.; 80% act.

Aerosol® OS. [Am. Cyanamid] Sodium isopropylnaphthalene sulfonate; anionic; emulsifier, dispersant and wetting agent; used in alkaline cleaning formulations; antigelling agents, automotive radiator cleaners, metal, cement, brick, and tile cleaners for crystal growth control, electroplating, filtration, glass cleaning, household detergents, leaching ores and slags, pigment disps.; soap additive; wh. to lt.

cream powd.; sol. in polar org. solvs., 50% in water; sp.gr. 1.43; dens.. 11.93 lb/gal; m.p. 205 C; surf. tens. 37 dynes/cm; pH 9 ± 1.0 (5% aq.); slowly biodeg.; 75 act., 25% sodium sulfate.

Aerosol® OT-75%. [Am. Cyanamid] Dioctyl sodium sulfosuccinate; anionic; wetting agent and surf. tens. depressant used in textile, rubber, petrol., paper, metal, paint, plastic, and agric. industries; antistat for cosmetics, dry cleaning detergents, emulsion, plastic, pipelines, and suspension polymerization; emulsifier wax for polish, firefighting, germicide, metal cleaner, mold release agent; dispersant in paints and inks, paper, photography, process aid, rust preventative, soldering flux, wallpaper removal; APHA 100 max. clear visc. liq.; m.w. 444; sol. org. solv.; limited water sol.; sp.gr. 1.09; visc. 200 cps; flash pt. 85 C (OC); acid no. 2.5 max.; surf. tens. 28.7 dynes/cm (0.1% aq.); biodeg.; 75 ± 2% solids in water/alcohol.

Aerosol® OT-100%. [Am. Cyanamid] Dioctyl sodium sulfosuccinate; anionic; see Aerosol OT-75%; also mold release agent; APHA 100 max. waxy solid; m.w. 444; sol. in polar and nonpolar solv.; sol. in oil, fat, and wax @ 75 C; disp. in water; sp.gr. 1.1; m.p. 153–157 C; acid no. 2.5 max.; surf. tens. 28.7 dynes/cm (0.1% aq.); 100% conc.

Aerosol® OT-B. [Am. Cyanamid] Dioctyl sodium sulfosuccinate, sodium benzoate; anionic; see Aerosol OT-75%; wh. powd.; bulk particle size 15–150 μ; m.w. 444; water-disp.; sp.gr. 1.1; m.p. < 300 C; acid no. 2.5 max.; surf. tens. 28.7 dynes/cm (0.1% aq.); 85% Aerosol OT; 15% sodium benzoate.

Aerosol® TR-70. [Am. Cyanamid] Ditridecyl sodium sulfosuccinate; anionic; emulsifier, surfactant; used in emulsion polymerization of vinyl chloride and vinyl acetate, suspension polymerization of vinyl chloride; dispersant for resins, pigments, polymers, and dyes in org. systems; pigment dispersant in printing inks; rust preventative; clear sol'n.; sol. in org. media; limited water sol.; sp.gr. 0.995; dens. 8.3 lb/gal; visc. 110 cps; f.p. –40 C; surf. tens. 26 dynes/cm; biodeg.; 70% act. in water/alcohol.

Aerothene MM Solvent. [Dow] Inhibited methylene chloride; solv. for various aerosol prods. such as hair sprays, lubricants, insecticides, and coatings; other tech. grades used in pharmaceuticals, extraction processes, electronics, adhesives; sol. in most org. solvs.; low water sol.

Aerothene TT Solvent. [Dow] Inhibited 1, 1, 1-trichloroethane; solv. for various aerosol prods.; vapor depressant; other tech. grades used in electronics, coatings, and degreasers; sol. see Aerothene MM Solvent.

Aethoxal B. [Henkel] Fatty alcohol polyalkylene glycol ether; nonionic; superfatting agent, emollient for bath oils, skin and personal care prods.; liq.; forms spontaneous emulsion in warm water; oil-sol.; 99–100% conc.

AF 10 FG. [Harcros] Silicone antifoam; nonionic; antifoam agent used for general food, poultry, and meat processing applics.; wh. liq.; disp. in water; sp.gr. 1.00; dens. 8.3 lb/gal; flash pt. > 212 F (PMCC); 10% act.

AF 10 IND. [Harcros] Silicone antifoam; nonionic; antifoam for agric., cutting oils, drilling muds, effluent, inks, chemicals, detergents and textiles; wh. liq.; water-disp.; sp.gr. 1.00; dens. 8.3 lb/gal; flash pt. > 212 F (PMCC); pH 4–5 (1% aq.); 10% act.

AF 10 TF. [Harcros] Silicone antifoam; nonionic; antifoam for drilling muds, effluent, cutting oils, agric. formulations; wh. liq.; water-disp.; sp.gr. 1.00; dens. 8.3 lb/gal; flash pt. > 212 F (PMCC); 10% act.

AF 30 FG. [Harcros] Silicone antifoam; nonionic; antifoam for general food, meat, and poultry processing; wh. liq.; water-disp.; sp.gr. 1.01; dens. 8.4 lb/gal; flash pt. > 212 F (PMCC); pH 4–5 (1% aq.); 30% act.

AF 30 IND. [Harcros] Silicone antifoam; nonionic; antifoam for effluent, agric. formulations, antifreeze, detergents; wh. liq.; water-disp.; sp.gr. 1.01; dens. 8.4 lb/gal; flash pt. > 212 F (PMCC); pH 4–5 (1% aq.); 30% act.

AF 60. [GE] Dimethyl polysiloxane nonionic emulsion in water; defoaming agent used in adhesive, ink, latex, soap, starch, and paint mfg. and other aq. industrial systems; wh. fluid, sorbic acid odor; sol. in water; sp.gr. 1.01; dens. 8.4 lb/gal; visc. 300 max. cps; 30% silicone content; 44.5 ± 1.5% total solids.

AF 66. [GE] Filled polydimethylsiloxane; antifoamer in nonaq. industrial applic.; adhesive, ink, and paint mfg., resin polymerization, petrol. processing; grayish wh. fluid; odorless; sol. in aliphatic, aromatic, chlorinated solvs.; negligible sol. in water; sp.gr. 1.01; dens. 8.4 lb/gal; visc. 500 cSt max.; flash pt. 600 F min. (OC); 100% act.

AF 70. [GE] 100% silicone compd.; defoamer in petrol. refining, cutting oils, chem. processing, antifoam formulating; food additive for food processing; sol. in aliphatic, aromatic, and chlorinated hydrocarbons.

AF 72. [GE] PEG-40 stearate, sorbitan stearate, and silica; nonionic; food-grade antifoam agent, surfactant; also for industrial applics. such as textile dyeing and finishing, leather finishing, latex processing, soap and detergent mfg., adhesive mfg., and as a boiler feed water defoamer; wh. fluid; slight sorbic acid odor; sol. in water; sp.gr. 1.01; dens. 8.4 lb/gal; visc. 1000 cps; 30% silicone content; 44.2 ± 1% total solid.

AF 75. [GE] PEG-40 stearate, sorbitan stearate, and silica; nonionic; food-grade antifoam emulsion, surfactant; used as direct and indirect food additive, in the mfg. of food-pkg. materials, chemical processing (adhesive mfg., water-based ink mfg., latex processing, soap mfg., starch processing, resin polymerization), textiles, paper, leather finishing, metalworking; wh. emulsion; slight sorbic acid odor; water-sol.; sp.gr. 1.02; dens. 8.4 lb/gal; visc. 2500 cps max.; 10% silicone content.

AF 93. [GE] 30% silicone emulsion; defoamer used in textiles, soap mfg., resin prod., for extreme acid and base pH ranges; water-disp.

AF 100 FG. [Harcros] Silicone antifoam; antifoam for food, edible oils, meat and poultry processing; translucent liq.; insol. in water; sp.gr. 1.01; dens. 8.4 lb/gal; pour pt. < 0 F; flash pt. > 600 F (PMCC); 100%

conc.

AF 100 IND. [Harcros] Silicone antifoam; antifoam for agric. formulations, drilling muds, vacuum distillations; translucent liq.; insol. in water; sp.gr. 1.01; dens. 8.4 lb/gal; pour pt. < 0 F; flash pt. > 600 F (PMCC); 100% conc.

AF 112. [Harcros] Nonsilicone antifoam; used in solvs., adhesives, latex paints, inks; creamy yel. liq.; water-insol.; sp.gr. 0.92; dens. 7.7 lb/gal; pour pt. 40 F; flash pt. > 200 F (PMCC); 100% act.

AF 6035. [Harcros] Silicone antifoam; used in delayed coker units; clear liq.; insol. in water; sp.gr. 0.84; dens. 7.0 lb/gal; pour pt. < 0 F; flash pt. 108 F (PMCC); 35% act.

AF 6050. [Harcros] Silicone antifoam for delayed coker units; sp.gr. 0.86.

AF 8805 FG, 8810 FG, 8830 FG. [Harcros] Silicone antifoam for general food, poultry, and meat processing, agric.; Kosher; sp.gr. 1.00.

AF 8805, 8810. [Harcros] Silicone antifoam for agric., cutting oils, drilling muds, effluent, inks, chemicals, detergents, textiles; sp.gr. 1.00.

AF 8820 FG. [Harcros] Silicone antifoam; general food, meat, and poultry processing, agric.; Kosher; dilutable for spa and hot tub applics.; sp.gr. 1.00.

AF 8820, 8830. [Harcros] Silicone antifoam; for effluent, agric., antifreeze, detergent applics.; dilutable; sp.gr. 1.00.

AF 9000. [GE] Silicone compd.; defoamer in many nonaq. direct and indirect food additive and industrial applics. incl. chemical processing (adhesive mfg., ink mfg., paint mfg., resin polymerization, formulating antifoam emulsions, insecticides/herbicides, wool fats), petrol. processing; sol. in aliphatic, aromatic, and chlorinated solvs.; sp.gr. 1.01; dens. 8.4 lb/gal; 100% silicone content.

AF 9020. [GE] Polydimethylsiloxane aq. (nonionic) emulsion; defoamer for industrial and food-processing systems incl. chemical processing (adhesive mfg., water-based ink mfg., latex processing, soap mfg., starch processing, paint additive, alcohol fermentation), waste treatment, petrochemical (resin polymerization, glycol dehydrators, ethylene oxide prod., urea prod.); wh. emulsion; disp. in warm or cold water with mild agitation; sp.gr. 1.01; dens. 8.4 lb/gal; visc. 2500 cps max.; 20% silicone content.

AF CM Conc. [Harcros] Silicone antifoam; nonionic; antifoam for carpet extraction machines; wh. visc. liq.; water-disp.; sp.gr. 0.97; dens. 8.1 lb/gal; flash pt. > 212 F (PMCC); pH 6–8 (1% aq.); 20% act.

AF GN-11-P. [Harcros] Nonsilicone antifoam; used for fermentation, drilling muds, effluent, adhesives, gas treating; clear liq.; water-disp; sp.gr. 1.01; dens. 8.4 lb/gal; pour pt. –50 F; flash pt. > 200 F (PMCC); pH 5–7 (1% aq.); 100% act.

AF GN-23. [Harcros] Nonsilicone antifoam for pulp and paper, effluent, wastewater treatment; sp.gr. 0.94.

AF GN-46. [Harcros] Nonsilicone antifoam; anionic; antifoam for solv. recovery, alcohol extraction; dk. liq.; oil-sol.; sp.gr. 0.90; dens. 7.5 lb/gal.; pour pt. 30 F; flash pt. > 250 F (PMCC); pH 4–6 (1% in 15% IPA/water); 100% act.

AF HL-21. [Harcros] Nonsilicone antifoam; nonionic; for latex paints, adhesives, inks, solvs.; opaque liq.; water-insol.; sp.gr. 0.83; dens. 6.9 lb/gal; flash pt. (PMCC) 270 F; 100% act.

AF HL-23. [Harcros] Nonsilicone antifoam; for cutting oil, boiler water additives, detergents, cleaning sol'ns, syn. lubricants; clear liq.; water-insol.; sp.gr. 1.00; dens. 8.3 lb/gal; pour pt. –20 F; flash pt. > 300 F (PMCC); pH 4–6 (1% in 15% IPA/water); 100% conc.

AF HL-26. [Harcros] Nonsilicone antifoam; nonionic; antifoam for waste water treatment, electroplating, coatings, cleaning aids, dyes, inks; yel. clear liq.; sol. in oil; sp.gr. 1.00; dens. 8.3 lb/gal; pour pt. < 30 F; flash pt. > 220 F (PMCC); 50% act.

AF HL-27. [Harcros] Nonsilicone antifoam; for urea and phenolic resins, latex paints, inks, adhesives; yel. liq.; water-insol.; sp.gr. 0.95; dens. 7.9 lb/gal; pour pt. 32 F; flash pt. > 300 F (PMCC); 100% act.

AF HL-28. [Harcros] Nonsilicone antifoam; for cutting oils, effluents, detergents; clear liq.; water-insol.; sp.gr. 1.00; dens. 8.3 lb/gal; pour pt. –22 F; flash pt. > 400 F (PMCC); 100% act.

AF HL-36. [Harcros] Nonsilicone antifoam for fermentation, processing beet sugar and yeast, distillation; Kosher; sp.gr. 1.00.

AF HL-40. [Harcros] Nonsilicone antifoam for latex paints, adhesives, inks, solvs., chemical processing, textiles, paper, paper coatings, electroplating; sp.gr. 0.91.

AF HL-52. [Harcros] Nonsilicone antifoam for solvs., latex paints, inks, chemical processing, adhesives, paper, paper coatings; sp.gr. 0.87.

Aflaban DF. [Eastman] Sorbic acid; animal feed preservative; inhibits broad spectrum of molds, yeasts, bacteria; wh. powd., < 15% retained on 100 mesh screen; mild and char. odor; m.w. 112.13; sol. 0.15% in water, 12.90% in 100% ethanol, 11.50% in glacial acetic acid, 9.20% in acetone, 5.50% in propylene glycol; dens. 41.00 lb/ft^3; m.p. 132–135 C; flash pt. 260 F (COC).

Aflaban KSG. [Eastman] Potassium sorbate; see Aflaban DF; wh. granular solid pellets, 98% min. retained on 100 mesh screen; mild and char. odor; m.w. 150.22; sol. 58.2% in water, 54.60% in 20% ethanol, 45.30% in 50% ethanol, 19.0% in propylene glycol; dens. 42.00 lb/ft^3; m.p. decomposes above 270 C.

Aflux 32. [Rhein Chemie] Thiodipropionic dilauryl ester; dispersant and internal lubricant for tech. molded and extruded rubber articles; wh. flakes; sp.gr. 1.02.

Aflux 42. [Rhein Chemie] Blend of polar compds. and surfactants; dispersant and internal lubricant for all rubbers, tech. molded and extruded articles, footwear; lt. brn. to brn. pellets; sp.gr. 0.9.

Aflux 54. [Rhein Chemie] Pentaerythrityl stearate; dispersant and internal lubricant for tech. molded and extruded rubber articles; decreases visc.; wh. flakes or powd.; sp.gr. 1.0.

Aflux R. [Rhein Chemie] Blend of polar compds. and

surfactants based on metal soaps, bound to highly reinforcing silica; dispersant and internal lubricant for all rubbers, rubber compds. containing carbon blk., hard rubber, tech. molded and extruded articles, tires, highly loaded compds.; lt. brn. fine powd.; sp.gr. 1.2.

Aflux S. [Rhein Chemie] Blend of polar compds. and surfactants bound to highly reinforcing silica; dispersant and internal lubricant for NR and SR rubbers, molded and extruded goods, footwear; ylsh. wh. fine powd.; sp.gr. 1.14.

Agent 613-95. [Stepan] Sodium alkyl sulfate; dispersant or flocculant in pesticide suspensions; pale yel. clear liq.; sp.gr. 1.073; dens. 8.9 lb/gal; pH 7.0–8.5; 38% act. in water.

Agent 765. [Witco SA] Fatty amine salt; corrosion inhibitor, pigments grinding and disp.; antistripping agent used in paint industry; liq.

Agent X-601-52N. [Stepan] Alkoxylated alkylphenol; wetting and dispersing agent for pesticides; pale-lt. brn. solid; sol. in water; sp.gr. 1.036; dens. 8.64 lb/gal; m.p. 40–50 C; flash pt. > 200 F (Seta); pH 5–7.

Agent X-646-12. [Stepan] Alkoxylated alkylphenol; wetting and dispersing agent for pesticide suspensions; straw liq.; sp.gr. 1.028; dens. 8.57 lb/gal; visc. 40 cps; pH 6.0; 30% conc. in water.

Agerite® DPPD. [Vanderbilt] Diphenyl-p-phenylene diamine; antioxidant used in rubber and mech. prods.; dk. blue-blksh. powd., 98% thru 100 mesh; m.w. 260.34; sol. in acetone, toluene, chloroform; insol. in water; dens. 1.28 ± 0.03 mg/m³; m.p. 144–153 C.

Agerite® Gel. [Vanderbilt] Agerite Stalite/petrol. wax, 75:25 ratio; see Agerite DPPD; lt. tan to brn. soft waxy solid; sol. in acetone, gasoline, chloroform, toluene; insol. in water; dens. 0.94 ± 0.03 mg/m³; m.p. 40–50 C.

Agerite® Geltrol. [Vanderbilt] Alkylated-arylated bisphenolic phosphite; antioxidant and polymer stabilizer for rubbers; gel inhibitor during polymerization; amber visc. liq.; sol. in toluene, chloroform, gasoline; insol. in water; dens. 0.94 ± 0.02 mg/m³ (60 C).

Agerite® GT. [Vanderbilt] Trifunctional hindered phenolic compd.; antioxidant for NR, syn. rubber, TPR, hot melts, and vulcanizates; wh. cryst. powd.; sol. in toluene, acetone, chloroform, dimethylformamide; dens. 1.03 ± 0.03 mg/m³; m.p. 217–225 C.

Agerite® Hipar T. [Vanderbilt] Dioctylated diphenylamine/diphenyl-p-phenylenediamine/diphenylamine-acetone reaction prod., 50:25:25 ratio; antioxidant for NR and syn. rubbers; gray to blksh. powd.; sol. see Agerite DPPD; dens. 1.15 ± 0.03 mg/m³; m.p. 70 C min.

Agerite® HP-S. [Vanderbilt] Dioctylated diphenylamine/diphenyl-p-phenylenediamine, 65:35 ratio; see Agerite Hipar T; also used in CR compds. for outdoor service; gray to blksh. rods, powd., 80% min. thru 100 mesh; sol. see Agerite DPPD; dens. 1.11 ± 0.03 mg/m³; m.p. 80–100 C.

Agerite® MA. [Vanderbilt] Polymerized 1,2-dihydro-2,2,4-trimethyl quinoline; antioxidant; aging protection to XLPE; amber pellets, cream-tan powd., 99.5% min. thru 200 mesh; m.w. 173.26; sol. see Agerite DPPD; dens. 1.06 ± 0.03 mg/m³; m.p. 105 C min.

Agerite® NEPA. [Vanderbilt] Alkylated diphenylamines; antioxidant for elastomers; dk. amber liq.; sol. in toluene, gasoline, alcohol; insol. in water; dens. 0.935 ± 0.02 mg/m³.

Agerite® Resin D. [Vanderbilt] Polymerized 1,2-dihydro-2,2,4-trimethylquinoline; see Agerite Hipar T; amber small pellets, cream-tan powd., 98% min. thru 100 mesh; m.w. 173.26; sol. see Agerite DPPD; dens. 1.06 ± 0.03 mg/m³; soften. pt. 74 C min.

Agerite® Stalite. [Vanderbilt] Octylated diphenylamines; antioxidant for elastomers; scorch retarder for CR; red. brn. visc. liq.; m.w. 281.45; sol. in alcohol, toluene, gasoline; insol. in water; dens. 0.99 ± 0.02 mg/m³.

Agerite® Stalite S. [Vanderbilt] Octylated diphenylamines; see Agerite Stalite; lt. tan powd.; m.w. 393.66; sol. see Agerite Stalite; dens. 1.02 ± 0.03 mg/m³; m.p. 89–103 C.

Agerite® Superflex, Superflex Solid G. [Vanderbilt] Diphenylamine-acetone; see Agerite DPPD; dk. brn. liq. and solid resp.; sol. see Agerite DPPD; dens. 1.10 ± 0.02 and 1.33 ± 0.03 mg/m³; 75% act. (Solid G).

Agerite® Superlite, Superlite Emulcon, Superlite Solid. [Vanderbilt] Polybutylated bisphenol A; antioxidant for latex, automotive and appliance prods.; amber liq., amber liq., and gray-tan solid resp.; sol. see Agerite Geltrol; dens. 0.945–0.965, 0.99 ± 0.02, and 1.26 ± 0.03 mg/m³; 95 and 73% act. (Superlite Emulcon, Superlite Solid).

Agri-safe. [Petrolite] Micronized polymer; binder used to recover fines in mfg. of pesticides, etc.; water-insol.

Agrilan F491. [Harcros UK] Complex phosphate ester; anionic; wetter/dispersant for agric. flowables; liq.

Agrilan F502. [Harcros UK] Alkylene oxide copolymer; nonionic; wetting and dispersing agent for agric. flowable systems; wh. soft solid; 100% conc.

Agrilan F513. [Harcros UK] Polyaromatic ethoxylate phosphate ester; anionic; wetting and dispersing agent for agric. toxicants; liq.; 100% conc.

Agrilan FS101, FS112. [Harcros UK] Epoxidized veg. oils; nonionic; stabilizers for emulsifiable concs. of agric. toxicants; liq.; 100% conc.

Agrilan WP178. [Harcros UK] Polyaromatic sulfonate; anionic; dispersant for wettable powds. and agrochemical suspension concs.; powd.; 96% act.

Agrisol PX401, 413. [Harcros UK] Aromatic alkoxylates; nonionic; cosolvs. for agrochemical toxicants; liq.; 100% conc.

Agriwet CA, C91, C92. [Henkel] Phosphate ester; anionic; wetting agents, detergents, emulsifiers and lubricants; compatibility and resuspension aid for pesticides; coupling agent; liq.; 100% conc.

Agriwet FOA. [Henkel] Phosphate ester; anionic; resuspension aid in pesticide formulations; liq.; 100% conc.

Ahco 759. [ICI Am.] Sorbitan laurate; nonionic; emulsifier, solubilizer for textile use; liq.; HLB 8.6; 100% conc.

Ahco 832. [ICI Am.] Sorbitan oleate; nonionic; emulsifier, solubilizer for textile use; liq.; 100% conc.

Ahco 909. [ICI Am.] Sorbitan stearate; nonionic; emulsifier, solubilizer for textile use; solid; HLB 4.7; 100% conc.

Ahco 944. [ICI Am.] Sorbitan oleate; nonionic; emulsifier, solubilizer for textile use; liq.; 100% conc.

Ahco 3998. [ICI Am.] Oleth-20; nonionic; dye leveling agent and emulsifier; liq.; 100% conc.

Ahco 7166T. [ICI Am.] Polysorbate 65; nonionic; dye leveling agent and emulsifier; solid; 100% conc.

Ahco 7596D. [ICI Am.] Polysorbate 21; nonionic; dye leveling agent and emulsifier; liq.; 100% conc.

Ahco 7596T. [ICI Am.] Polysorbate 20; nonionic; dye leveling agent and emulsifier; liq.; 100% conc.

Ahco A-117. [ICI Am.] Alkylaryl sulfonate; anionic; emulsifier, surfactant, used as dispersant, detergent, wetting agent; amber liq.; faint odor; sol. IPA; sp.gr. 1.0; visc. 7000 cps; 100% act.

Ahco AJ-110. [ICI Am.] Sulfated castor oil; anionic; wetting agent, lubricant, penetrant, dispersant, emulsifier, detergent for textile and leather industries; dyeing assistant for textiles; amber liq.; castor oil odor; water sol.; sp.gr. 1.05; HLB 11.0; 50% act. in water.

Ahco AY 2200. [ICI Am.] Phosphate ester; anionic; low-soiling textile antistat, lubricant; liq.; 45% conc.

Ahco C330. [ICI Am.] Quat. ammonium compd.; cationic; heat-stable textile fiber antistat; liq.; 100% conc.

Ahco DFO-100. [ICI Am.] Polysorbate 81; nonionic; emulsifier, solubilizer for textile use; liq.; HLB 10.0; 100% conc.

Ahco DFO-110. [ICI Am.] Polysorbate 85; nonionic; emulsifier, solubilizer for textile use; liq.; HLB 11.0; 100% conc.

Ahco DFO-150. [ICI Am.] Polysorbate 80; nonionic; emulsifier, solubilizer for textile use; liq.; HLB 15.0; 100% conc.

Ahco DFP-156. [ICI Am.] Polysorbate 40; nonionic; emulsifier, solubilizer for textile use; liq.; 100% conc.

Ahco DFS-96. [ICI Am.] Polysorbate 61; nonionic; emulsifier, solubilizer for textile use; solid; HLB 9.6; 100% conc.

Ahco DFS-149. [ICI Am.] Polysorbate 60; nonionic; emulsifier, solubilizer for textile use; solid; HLB 14.9; 100% conc.

Ahco DHS-111. [ICI Am.] POE stearate; nonionic; textile lubricant and all-purpose emulsifier; solid; HLB 11.1; 100% conc.

Ahco EO-102. [ICI Am.] POE sorbitol polyoleate; nonionic; emulsifier, solubilizer for textile use; liq.; HLB 10.2; 100% conc.

Ahco EO-114. [ICI Am.] POE sorbital polyoleate; nonionic; emulsifier, solubilizer for textile use; liq.; HLB 11.4; 100% conc.

Ahco FO-18. [ICI Am.] Sorbitan trioleate; nonionic; emulsifier, solubilizer for textile use; liq.; 100% conc.

Ahco FP-67. [ICI Am.] Sorbitan monopalmitate; nonionic; emulsifier, solubilizer for textile use; solid; 100% conc.

Ahco FS-21. [ICI Am.] Sorbitan tristearate; nonionic; emulsifier, solubilizer for textile use; solid; HLB 2.1; 100% conc.

Ahcovel Base 500. [ICI Am.] Fatty carbamide; cationic; textile softener; hard wax; 100% conc.

Ahcovel Base 700. [ICI Am.] Fatty amide acetate salt; cationic; textile softener; solid; 100% conc.

Ahcovel Base N-15. [ICI Am.] Glycerol monostearate, emulsifiable; nonionic; textile softener; solid; 100% conc.

Ahcovel Base N-62. [ICI Am.] Ester blend; nonionic; textile softener; liq.; 100% conc.

Ahcovel Base N-64. [ICI Am.] Ester blend; nonionic; knitting lubricant, softener, cellulosic and acrylic fiber finishes, antistat; solid; 100% conc.

Ahcovel OB. [ICI Am.] Fatty carbamide, modified; cationic; base for textile softeners; waxy solid; 100% act.

Ahcowet DQ-114. [ICI Am.] Trideceth-6; nonionic; wetting agent, dispersant, detergent in textiles; liq.; HLB 11.4; 100% conc.

Ahcowet DQ-145. [ICI Am.] Trideceth-12; nonionic; scour and dye leveling agent in textiles; liq.; HLB 14.5; 100% conc.

Air-Plas® 200H, 200L. [Westvaco] Amine tallate; air entraining agent with max. waterproofing and improved grinding efficiency for masonry cement applics.; brn. liq.; water-immiscible; sp.gr. 1.035 and 0.945 resp.; visc. 1200 and 300 cps; 100% act.

Airrol CT-1. [Toho] Dialkyl sulfosuccinate; anionic; wetting agent and dyeing assistant; liq.; 70% conc.

Ajicoat SPG. [Ajinomoto] Sodium-poly-L-glutamate; anionic; surface modifier, coemulsifier, codispersant; solid; 100% conc.

Ajidew A-100. [Ajinomoto] PCA; nat. humectant used in cosmetics and soaps; wh. cryst.

Ajidew N-50. [Ajinomoto] Sodium PCA; see Ajidew A-100; 50% aq. sol'n.

Ajidew SP-100. [Ajinomoto] PCA 1/2Na; see Ajidew A-100; wh. cryst.

Akolidine 10,11,12. [Lonza] Alkyl substituted pyridine mixture; corrosion inhibitor for metals, acid pickling, industrial acid cleaning, oilfield processes and treatment of oil refinery equip.; liq.

Akro-Gel. [Akrochem] Water gel of a sodium salt of a methyl tauride; mold release agent and lubricant; amber clear gel; mild odor; sol. in water at 54 C; pH 6.5–8.0 (10%).

Akrochem® 9930 Zinc Oxide Transparent. [Akrochem] Precipitated basic zinc carbonate; accelerator-activator for transparent nat. and syn. rubber goods, adhesives; wh. powd.; sp.gr. 3.5; m.p. 168 C min.; 70–72% zinc oxide.

Akrochem® Accelerator 40. [Akrochem] Aldehyde amine; accelerator; amber to orange liq., aromatic odor; sp.gr. 0.95; visc. 135 cs; flash pt. < 200 F; 40 ±

3% dihydropyridine (act.).

Akrochem® Accelerator CZ-1. [Akrochem] N,N dimethyl cyclohexyl ammonium dibutyl dithiocarbamate; accelerator for NR, SBR, or latexes; pale yel. to lt. tan crystals and lumps, strong amine odor; sp.gr. 1.03; m.p. 52.5 C; 98% purity.

Akrochem® Accelerator R. [Akrochem] 4,4′-Dithio dimorpholine; accelerator for uses where a nonblooming or nonstaining sulfur donor is required; used for EV or semi-EV compds., in syn. and nat. rubbers; wh. to cream powd., very slight odor; sol. in acetone, benzene, ethyl acetate; sp.gr. 1.36; m.p. 121 ± 5 C.

Akrochem® Accelerator Thio No. 1. [Akrochem] N,N′ diphenyl thiourea; accelerator for CR latex, NR latex, and cements, EPDM sponge compds.; activates thiazole accelerators; essentially nondiscoloring; off-wh. powd.; sol. in acetone; slightly sol. in alcohol, ether, toluene, naphtha, CCl_4; sp.gr. 1.30; m.p. 148 C; 98% purity.

Akrochem® Accelerator VS. [Akrochem] Zinc salt of dibutyl dithiophosphoric acid on silica carrier; nonblooming and nonstaining accelerator for EPDM compds. and other sulfur-curable elastomers, esp. hose and belt applics.; lt. gray powd.; sol. in naphtha, benzene, min. oils, and ethanol; sp.gr. 1.42; 62% act. on silica.

Akrochem® Antifoam A. [Akrochem] Aq. system using food-grade emulsifiers and silicone compds.; high-strength aq. system for systems requiring persistent antifoam properties; wh. milky fluid; dens. 8.3 lb/gal; visc. 15,000 cps; flash pt. (COC) none; 30% act.

Akrochem® Antioxidant 12. [Akrochem] Butylated reaction prod. of p-cresol and dicyclopentadiene; nonstaining antioxidant, stabilizer, and antiozonant in polymers, incl. nat. and syn. polyisoprene, neoprene, nitrile, SBR rubber, and latex; amber flakes or wh. powd.; sp.gr. 1.10; m.p. 115 C.

Akrochem® Antioxidant 16. [Akrochem] Alkylated phenol; nonstaining antioxidant used in SBR latex compding. urethanes, ABS, and EPDM; liq.; sp.gr. 1.08 ± 0.02; visc. (Saybolt) 60–100 s.

Akrochem® Antioxidant 33. [Akrochem] Phenolic; nonstaining antioxidant for elastomers and latex; wh. fine powd.; odorless; sol. in aliphatics, aromatics, esters, and ketones; sp.gr. 1.04; m.p. 100 C min.

Akrochem® Antioxidant 36. [Akrochem] Polymeric hindered phenolic; nonstaining antioxidant for rubber compds., SBR, butyl, CR, polyisoprene, nitrile, chlorinated rubber, polyolefins, ethyl cellulose, Elvac, PVC, epoxy, and waxes; wh. fine powd.; sol. in naphtha, toluene, and IPA; sp.gr. 1.05; m.p. (Ball & Ring) 148 C.

Akrochem® Antioxidant 58. [Akrochem] 2-Mercapto-4(5)-methyl benzimidazole, zinc salt; nonstaining antioxidant for use in rubber compds. to improve heat resistance; wh. powd.; sp.gr. 1.75; m.p. decomp. > 300 C.

Akrochem® Antioxidant DQ. [Akrochem] 2,2,4-Trimethyl-1,2-dihydroquinoline, polymerized; antioxidant in rubber goods requiring resistance to high temps.; copper and manganese inhibitor; antiozonant; yel. to amber; very weak, char. odor; sol. in acetone, ethyl acetate, alcohol, benzene, methylene chloride, and CCl_4; insol. in water; sp.gr. 1.095 ± 0.02; m.p. 75 C.

Akrochem® Antioxidant PANA. [Akrochem] Phenyl-α-naphthylamine; antioxidant for rubber prods.; antiflexcracking under dynamic stress; brn. to dk. violet cryst. lumps or flakes; sol. in benzene, acetone, CCl_4, ethyl acetate, methylene chloride, and ethanol; insol. in water; sp.gr. 1.2; m.p. 50 C min.

Akrochem® Antioxidant S. [Akrochem] Octylated diphenylamine; antioxidant protecting dynamically stressed rubber prods.; antiozonant; antiflexcracking; tan spaghetti-like gran./pellets; sol. in acetone, benzene, methylene chloride, and naphtha; insol. in water; sp.gr. 1.00; m.p. 88 C min.

Akrochem® Antiozonant MPD-100. [Akrochem] Mixed diaryl p-phenylene diamine; antiozonant, antidegradant, antioxidant, and antiflexcracking agent for most diene polymers; brn. flakes; sp.gr. 1.20; m.p. 90–105 C.

Akrochem® Antiozonant PD-2. [Akrochem] N-(1,3-dimethyl butyl)-N′-phenyl-p-phenylene diamine; antiozonant protecting rubber polymers against heat, oxidation, and flex cracking; copper, manganese inhibitor and SBR stabilizer; brn. to violet-brn. mass; aromatic odor; sol. in benzene, CCl_4, methylene chloride, acetone, ethyl acetate, and ethyl alcohol; insol. in water; sp.gr. 1.10; m.p. 45 C min.

Akrochem® Calcium Stearate. [Akrochem] Mold release agent; wh. fine powd.; 98.5% thru 325 mesh; apparent dens. 1.9 lb/gal; m.p. 145–160 C.

Akrochem® Cu.D.D. [Akrochem] Copper dimethyl dithiocarbamate; accelerator for butyl rubber, EPDM rubbers; brn. powd.; partly sol. in acetone and toluene; sp.gr. 1.75; m.p. > 300 C; 21% copper.

Akrochem® Cu.D.D.-PM. [Akrochem] Copper dimethyl dithiocarbamate disp. in a polymeric binder; dispersion form of Akrochem Cu.D.D.; brn. pellets; sp.gr. 1.4; 75% disp.

Akrochem® DCP-40C, DCP-40K. [Akrochem] Dicumyl peroxide absorbed on calcium carbonate and kaolin clay resp.; accelerator for syn. elastomers; wh. powd. or pellets; sp.gr. 1.60 and 1.56 resp.; 40 ± 1% peroxide.

Akrochem® D.E.T.U. Accelerator. [Akrochem] N,N′-diethylthiourea; accelerator for high-temp.-curing of mercaptan-modified CR (W-type neoprenes) for continuous prod. of solid and cellular extruded goods; nondiscoloring, nonstaining; wh. to off-wh. flake; m.w. 132.2; sol. in water and aromatic hydrocarbons; sp.gr. 1.1; m.p. 74 C min.

Akrochem® DOTG. [Akrochem] Diorthotolyl guanidine; accelerator; provides activation for MBTS, MBT, and other thiazole accelerators in nat. rubber, SBR, NBR, and CR; wh. nondusting powd., weak odor, bitter taste; sol. in acetone, ethyl alcohol, ethyl acetate, and methylene chloride; sp.gr. 1.18; m.p. 167 C min.

Akrochem® DPG. [Akrochem] Diphenyl guanidine; accelerator-activator for nat. rubber, SBR, and

NBR; wh. nondusting powd. or pellets, weak odor, bitter taste; sol. in acetone, methylene chloride, ethyl acetate, ethyl alcohol, and benzene; sp.gr. 1.19; m.p. 142 C min.

Akrochem® ETU-22 PM. [Akrochem] Ethylene thiourea on compatible polymeric binder carrier; accelerator imparting a high state of cure to neoprene compds.; nonstaining and nondiscoloring; off-wh. pellet, slight, typ. odor; disperses in dry polymers; sp.gr. 1.2; m.p. 200 C min. for act.; 75% ethylene thiourea (act.).

Akrochem® Hydrated Alumina. [Akrochem] Alumina trihydrate; inert filler and flame-retardant smoke suppressant for latex foam, rubber, carpet backing, epoxies, reinforced polyesters, phenolics, and urethane foam; wh. powd.; avail. in 3.0, 4.0, and 8.0 grades with median particle sizes of 3.0 μ, 4.0 μ, and 8.0 μ resp.; avail. in extra wh. grades noted with suffix EW; sp.gr. 2.42; GE brightness 92–93 (reg. grade), 95–97 (EW); 65% alumina.

Akrochem® P 37. [Akrochem] Cashew nut oil-modified phenol-formaldehyde two-step thermoplastic resin; tackifier and plasticizer for nitrile rubber; dk. brn. visc. liq.; sol. in ketones and blends of alcohols and aromatics; sp.gr. 1.10; visc. ≈ 3,500,000.

Akrochem® P 40. [Akrochem] Thermosetting two-step phenolic resin; tackifier; hardening and reinforcing agent for nitrile, nitrile-SBR blends, Hypalon, and BR polymers; finely pulverized; 98% min. thru 200 mesh; sol. in alcohols and ketones; sp.gr. 1.18 (cured); m.p. 82–93 C.

Akrochem® P 49. [Akrochem] Thermosetting two-step phenolic resin, oil-modified, and supplied in the thermoplastic form; tackifier recommended for use in SBR rubber; plasticizing action during processing; amber flake; sol. in ketones and blends of alcohols and aromatics; 98% min. thru 200 mesh; sp.gr. 1.16; m.p. (B&R) 95 C.

Akrochem® P 55. [Akrochem] Thermosetting two-step phenolic resin, oil-modified; tackifier for use in SBR rubber; plasticizer during processing; reinforcing agent after cure; off-wh. finely pulverized; 98% thru 200 mesh; sol. in ketones and blends of alcohol and aromatics; sp.gr. 1.17; m.p. (Cap. Tube) 73 C.

Akrochem® P 82. [Akrochem] Thermosetting two-step phenolic resin; plasticizer; hardening and reinforcing agent for nitrile, nitrile-SBR blends, Hypalon, and BR polymers; finely pulverized; 98% min. thru 200 mesh; sol. in alcohols and ketones; sp.gr. 1.18 (cured); m.p. 82–93 C.

Akrochem® P 86. [Akrochem] Thermosetting two-step phenolic resin; thermoplastic as received; tackifier; offers plasticizer action during processing; brn. flakes; sol. in ketones and blends of alcohols and aromatics; sp.gr. 1.16; m.p. (B&R) 95 C.

Akrochem® P 87. [Akrochem] Cashew nut oil-modified phenol-formaldehyde thermosetting two-step resin; contains the necessary hexa to make it thermosetting; tackifier; modifier for nitrile rubber compds. and solv. cements; plasticizer during processing; brn. finely pulverized powd.; 2% max. retained on 200 mesh; sol. in ketones and blends of alcohols and aromatics; sp.gr. 1.17; m.p. (Cap. Tube) 65 C.

Akrochem® P-90. [Akrochem] Thermoplastic alkyl phenol formaldehyde novalak resin; tackifier compat. with most elastomers; widely used in plied-up constructions; flake; 95% thru $^3/_8$ in. screen; sp.gr. 1.00; m.p. (B&R) 90 C; acid no. 33.

Akrochem® P-124. [Akrochem] Heat-reactive resin; curing agent for butyl or halogenated butyl polymers; red lumps; faint halogen odor; sol. in aliphatic and aromatic solvs., incl. benzene, xylene, toluene, ketones, and various naphthas; sp.gr. 1.05; m.p. (Cap. Tube) 57 C; 99.5+% act.

Akrochem® P-126. [Akrochem] Heat-reactive alkyl phenolic resin; used in Neoprene AF-based contact adhesives to impart low initial visc. and exc. visc. stability; also used with the adhesive grades of chloroprene; lumps; color X or lighter; sol. in aromatic solvs. incl. toluene, benzene, and xylene, ketones, and chlorinated hydrocarbons; partially sol. in aliphatic hydrocarbons; sp.gr. 1.10; m.p. (Cap. Tube) 185 F.

Akrochem® P-133. [Akrochem] Thermoplastic, alkyl phenolic resin; tackifier resin for rubber compds. used in tire construction and mechanical goods; flakes; sol. in esters, ketones, aromatic and aliphatic chlorinated hydrocarbons; insol. in alcohols; sp.gr. 1.04; visc. (Gardner-Holdt) 0 (toluene-64% resin); m.p. (B&R) 97 C; acid no. 35.

Akrochem® P-420. [Akrochem] Nonreactive alkyl phenolic resin; tackifying resin for rubber compding. and adhesives; rdsh.-brn. lumps; sol. in aromatic and aliphatic hydrocarbons, alcohols, esters, and ketones; sp.gr. 1.025; soften. pt. (B&R) 285 F.

Akrochem® P-478. [Akrochem] Thermosetting one-step phenolic resin; plasticizer for rubber compding., adhesives; powd.; sol. in alcohol, ketones, and esters; sp.gr. 1.25; m.p. (Cap. Tube) 160 F.

Akrochem® P-486. [Akrochem] Thermoplastic cashew-modified phenolic resin without curing agent; plasticizer and reinforcing agent in nitrile rubber compds.; brn. flake; sol. in higher alcohols, ketones, and esters; limited tolerance for aromatic hydrocarbons; soften. pt. (B&R) 90–100 C.

Akrochem® P-487. [Akrochem] Cashew nut oil-modified phenol-formaldehyde two-step thermosetting resin; contains hexa necessary to make it thermosetting; modifier for nitrile rubber compds. and solv. cements; plasticizes the stock during processing; brn. powd.; 2% max. retained on 200 mesh; sol. in ketones, higher alcohols, esters; limited tolerance for aromatic hydrocarbons; sp.gr. 1.17; soften. pt. (B&R) 65 C.

Akrochem® Plasticizer LN. [Akrochem] Naphthenic rubber process oil; low-visc. plasticizer; m.w. 320; sp.gr. 0.916 (60 F); dens. 7.6 lb/gal; visc. 154 SUS (100 F); pour pt. –40 F; ref. index 1.5076; flash pt. (COC) 350 F.

Akrochem® SWS-201. [Akrochem] Dimethyl polysiloxane compd.; defoamer for controlling foam and deaeration of nonaq. systems; gray translucent fluid;

solv.-disp.; dens. 8.4 lb/gal; visc. 230 cps; flash pt. (COC) 600 F; 100% act.

Akrochem® TDEC. [Akrochem] Tellurium diethyl dithiocarbamate; fast-curing primary or sec. accelerator for use in natural rubber, SBR, NBR, EPDM, and butyl; orange to yel. pellet, char. odor; sol. in benzene, chloroform, carbon disulfide; sp.gr. 1.42; m.p. 108 C min.; 80% act.

Akrochem® TMTD. [Akrochem] Tetramethylthiuram disulfide; very act., sulfur-bearing, nondiscoloring org. accelerator and activator; for curing systems requiring very low or no sulfur and for butyl and EPDM compds.; wh. to lt. gray powd.; sp.gr. 1.42; m.p. 148 C min.

Akrochem® TMTD Pellet. [Akrochem] Tetramethylthiuram disulfide; pellet form of Akrochem TMTD Powder; wh. to lt. gray pellet; sp.gr. 1.40; m.p. 148 C.

Akrochem® VC-40C, VC-40K. [Akrochem] Bis (t-butylperoxy-isopropyl) benzene absorbed on calcium carbonate and kaolin clay resp.; accelerator for syn. elastomers; wh. powd. or pellets, low odor; sp.gr. 1.53 and 1.51 resp.; 40 ± 1% peroxide (act.).

Akrochem® Zinc Stearate. [Akrochem] Mold release agent; wh. fine powd.; 100% thru 325 mesh; apparent dens. 2.8 lb/gal; m.p. 120 C.

Akrochem® ZIPPAC. [Akrochem] Zinc-amine dithiophosphate complex coated with min. oil; accelerator; used with thiazoles and thiurams for fast cure rates in ENB-type EPDM's; nondiscoloring, nonstaining; lt. gray powd.; sp.gr. 1.05; m.p. 100 C min.

Akrofax 9844 Black. [Akrochem] Dark, soft to med.-hard, vulcanized veg. oil; nonstaining extender, processing aid, and softener in nat. and syn. rubbers; crushed lumps; sp.gr. 1.02; 32–38% acetone extract.

Akrofax 9851 Light. [Akrochem] Med.-hard, sulfur-type vulcanized veg. oil; extender, processing aid, and softener in colored rubber prods., esp. tubing and calendered prods.; lt. yel. fine crumb; sp.gr. 1.03; 14–18% acetone extract.

Akrofax 9906. [Akrochem] Extra soft, sulfur-type vulcanized veg. oil; extender, processing aid, and softener in dark-colored nat. and syn. rubber prods.; brn. crushed lumps; sp.gr. 0.98; 43–49% acetone extract.

Akroform® DCP-40 EPMB. [Akrochem] Dicumyl peroxide in EPR rubber; accelerator in masterbatch form; used with syn. elastomers; $^1/_4$ in. diced pellets; sp.gr. 0.92; 40.5–41.0% peroxide (act.)

Akroform® VC-40 EPMB. [Akrochem] Bis (t-butylperoxy-isopropyl) benzene in EPR rubber; accelerator in masterbatch form; used with syn. elastomers; $^1/_4$ in. diced pellets, low odor; sp.gr. 0.89; 40.5–41.0% peroxide (act.).

Akrolease® E-9410. [Akrochem] Silicone polymer; release agent for PU, polyester, or epoxy materials from plastic or metal molds; corrosion inhibitor for metal; water-wh. clear oil; sol. in hydrocarbon and chlorinated solvs.; sp.gr. 0.97; dens. 8.1 lb/gal; visc. 225 cps; flash pt. (PM) 280 F; 100% act.

Akrolease® E-9491. [Akrochem] Silicone emulsion; nonionic; release agent for plastics in molding operations, esp. epoxy, polyester, and PU from metal or plastic molds; improves release and lubricity; corrosion inhibitor for metals; wh., cream; visc. 225 cSt (of silicone); 35% solids in water.

Akrosperse® D-177. [Akrochem] Tetramethylthiuram disulfide; 5% oil-treated powd. version of Akrochem TMTD; wh. to lt. gray dedusted powd.; sp.gr. 1.39; m.p. 148 C.

Akrosperse® DCP-40 EPMB. [Akrochem] Dicumyl peroxide in EPR rubber; accelerator in masterbatch form; used with syn. elastomers; off-wh. thin sheets; sp.gr. 0.92; 40.5–41.0% peroxide (act.).

Akrosperse® VC-40 EPMB. [Akrochem] Bis (t-butylperoxy-isopropyl) benzene in EPR rubber; accelerator in masterbatch form; used with syn. elastomers; off-wh. thin sheets, low odor; sp.gr. 0.89; 40.5–41.0% peroxide (act.).

Akrowax 130, 145. [Akrochem] Fully refined paraffin wax derived from petrol.; contains an oxidation inhibitor; wax used in rubber to give ozone resistance; inhibits cracking and frosting in NR, SBR, CR, and NBR; also for food pkg. applics.; wh. hard cryst. material; low odor; sp.gr. 0.90 and 0.82 (100 F) resp.; visc. 39 and 41 SUS (210 F); m.p. 130 and 141 F; flash pt. 430 and 440–450 F.

Akrowax 5050. [Akrochem] Wax blend for antichecking and antiozonant protection of rubber; nonstaining, nondiscoloring; Gardner 1–2 prills; visc. 13 cps (210 F); congeal. pt. 157 F.

Akrowax Micro 23. [Akrochem] Microcryst. wax composed of mixt. of hydrocarbons incl. normal paraffins, branched paraffins, monocyclic compds., and polycyclic compds.; wax used in rubber to give ozone resistance; with Akrowax 130 to inhibit cracking in NR, SBR, NBR, and CR; ASTM 2 wax; sp.gr. 0.92; visc. 86 SUS (@ 210 F); m.p. 176 F; flash pt. (COC) 540 F.

Akrowax PE. [Akrochem] Low m.w. polyethylene; process aid in natural and syn. rubbers; wh. powd.; m.w. 5600–6400; sp.gr. 0.92; m.p. 98–108 C; acid no. 0; sapon. no. 0.

Aktiplast. [Rhein Chemie] Zinc salts of higher molecular unsat. fatty acids; peptizing and dispersing agents for molded, extruded, expanded rubber, hard rubber; vulcanization accelerator; lt. brn. pellets; sp.gr. 1.08.

Aktiplast 6 N. [Rhein Chemie] Dixylyldisulfides; reclaiming agent for scrap rubber; dk. brn. liq.; sp.gr. 1.12.

Aktiplast 6 R. [Rhein Chemie] Diaryldisulfides; reclaiming agent for scrap rubber; brn. liq.; sp.gr. 1.12.

Aktiplast AS. [Rhein Chemie] N,N-bis-stearoylethylenediamide; peptizing and dispersing agents for tech. NR articles; ylsh. pellets; sp.gr. 1.00.

Aktiplast F. [Rhein Chemie] Activated zinc salts of higher molecular unsat. fatty acids; peptizing and dispersing agents for molded and extruded rubber, conveyor belts, tire compds.; vulcanization accelerator; brn. pellets; sp.gr. 1.08.

Aktiplast PP. [Rhein Chemie] Zinc salts of higher molecular fatty acids; peptizing and dispersing agents for tire compds., molded, extruded, and hard rubber; vulcanization accelerator; lt. brn. pellets; sp.gr. 1.08.

Aktiplast T. [Rhein Chemie] Zinc salts of higher molecular unsat. fatty acids; see Aktiplast; lt. yel.-brn. pellets; sp.gr. 1.05.

Akypo RLM 25. [ChemY GmbH] Fatty alcohol polyglycol ether carboxylic acid; nonionic; emulsifier, dispersant for emulsion and dispersion use; liq.; 92% conc.

Akypo RLM 45. [ChemY GmbH] Laureth-5 carboxylic acid; nonionic; emulsifier, dispersant for emulsion and detergent use; liq.; HLB 11.0; 90% conc.

Akypo RLM 100. [ChemY GmbH] Laureth-10 carboxylic acid; nonionic; see Akypo RLM-Q38; liq.; HLB 14.8; 90% conc.

Akypo RLM 130. [ChemY GmbH] Fatty alcohol polyglycol ether carboxylic acid; nonionic; see Akypo RLM-Q38; liq.; HLB 15.9; 90% conc.

Akypo RLM Q38. [ChemY GmbH] Fatty alcohol polyglycol ether carboxylic acid; nonionic; emulsifier, dispersant, additive for personal care prods., household and industrial formulas; primary emulsifier for syn. latex; liq.; HLB 10.2; 90% conc.

Akypo RLML 160. [ChemY GmbH] Fatty alcohol polyglycol ether carboxylic acid; nonionic; see Akypo RLM-Q38; liq.; HLB 16.6; 90% conc.

Akypogene Jod MB 1918. [ChemY GmbH] Iodine and fatty alcohol polyglycol-ether carboxylic acid; anionic; disinfectant; liq.; 70% act.

Akypomine AT. [ChemY GmbH] Alkylphenol polyglycol ether carboxylic acid; anionic; min. flotation; liq.; 90% conc.

Akypomine BC 50, 100. [ChemY GmbH] Fatty alcohol ether sulfate; anionic; min. flotation; paste; 50 and 95% conc. resp.

Akypopress DB. [ChemY GmbH] Syn. polymer; anionic; depressant for metal ions; liq.; 30% conc.

Akypoquat 40. [ChemY GmbH] Quat. ester; cationic; softener; antistat; liq.; 83% act.

Akypoquat 129. [ChemY GmbH] Quat. ester; cationic; raw material for mfg. of laundry softeners; molecule is fully biodeg.; liq.; 80% act.

Akypoquat 131. [ChemY GmbH] Behenoyl PG trimonium chloride; cationic; raw material for cosmetic hair prods.; molecule is fully biodeg.; liq.; 70% act.

Akypoquat 132. [ChemY GmbH] Quat. ester; cationic; see Akypoquat 131; liq.; 70% act.

Akyporox CO 400. [ChemY GmbH] Alkyl phenol polyglycol ether; nonionic; solubilizer for cosmetic prods.; liq.; 100% act.

Akyporox CO 600. [ChemY GmbH] Alkyl phenol polyglycol ether; nonionic; see Akyporox CO 400; liq.; 100% act.

Akyporox RC 200. [ChemY GmbH] Alkyl phenol polyglycol ether; nonionic; solubilizer for cosmetic prods.; solid.

Al-0104 T 3/16´´, Al-1404 T 3/16´´. [M&T Harshaw] Alumina; catalyst support, drying agent, and for dehydration reactions; tablet, 3/16´´ diam.; dens. 52 and 58 lb/ft³ resp.; 99 and 97% conc.

Al-0109 P. [M&T Harshaw] Alumina; used as a catalyst for support and for chromatographic adsorp.; wh. powd., 50% on 200 mesh, 29% on 325 mesh; dens. 55 lb/ft³; 92% conc.

Al-1401 P(MS). [M&T Harshaw] Microspheroidal alumina; catalyst for fluid bed operations; powd.; dens. 53 lb/ft³; 97% conc.

Al-1602 T 1/8´´, Al-1602 P. [M&T Harshaw] Silicated alumina; catalyst support; 1/8´´ diam. tablet and powd. resp.; dens. 52 and 63 lb/ft³; 91% conc.

Al-1609 P. [M&T Harshaw] Acidic alumina; catalyst; powd., 85% < 80μ particle size; dens. 53 lb/ft³; 91% conc.

Al-3438 T 1/8´´. [M&T Harshaw] Alumina; catalyst; tablet, 1/8´´ diam.; dens. 44 lb/ft³; 98% conc.

Al-3916 P. [M&T Harshaw] Alumina; see Al-1401 P(MS); powd., 90% < 80μ particle size; dens. 49 lb/ft³.

Al-3945 E 1/16´´. [M&T Harshaw] Alumina; catalyst; extruded, 1/16´´ diam.; dens. 40 lb/ft³; 99% conc.

Al-3970 P. [M&T Harshaw] Alumina; catalyst support; powd., 83% < 80μ particle size; dens. 50 lb/ft³.

Al-3980 T 5/32´´. [M&T Harshaw] Alpha alumina; catalyst; inert support; tablet, 5/32 diam.; dens. 75 lb/ft³.

Al-3996 R 3.5 X 1.5 mm. [M&T Harshaw] Alumina; catalyst used as drying agent and for dehydration reactions; ring, 3.5 and 1.5 mm internal and external diam.; dens. 34 lb/ft³; 99% conc.

Al-4028 T 3/16´´. [M&T Harshaw] Alumina; catalyst support; tablet, 3/16´´ diam.; dens. 34 lb/ft³.

Al-4126 E 1/16´´. [M&T Harshaw] Alumina; catalyst; extruded, 1/16´´ diam.; dens. 28 lb/ft³; 99% conc.

Alacsan 7LUF. [Rhone-Poulenc Surf.] Olealkonium chloride; cationic; conditioner for hair rinse formulations; useful for clear prod. where max. manageability is desired; detergent; Gardner 5 max. visc. liq; 55% act. Discontinued.

Albalan. [Westbrook Lanolin] De-oiled lanolin; nonionic; emollient, emulsifier; forms stable w/o emulsions; soft wax; HLB 5.0; 100% conc.

Albrite® PA-75. [Albright & Wilson] Org. acid phosphate; acid catalyst in resin curing.

Albrite® T2CEP. [Albright & Wilson] Tris 2-chloroethyl phosphite; sec. stabilizer for plastics, esp. acrylics; mobile colorless liq.; 11.5% P.

Albrite® TBPO. [Albright & Wilson] Tributyl phosphate; flame retardant plasticizer in plastics and coatings; solv. extractant; useful in hydraulic fluids; 11.6% P.

Albrite® TIOP. [Albright & Wilson] Triisooctyl phosphite; sec. stabilizer and mold release agent for cast acrylics; colorless mobile liq.; 7.4% P.

Albrite® TPP. [Albright & Wilson] Triphenyl phosphite; costabilizer for PVC; reactive diluent for resin systems; colorless mobile liq.; 10% P.

Alcalase® 0.6 L. [Novo] Proteolytic enzyme prepared from Bacillus licheniformis; major component: Subtilisin Carlsberg; food-grade enzyme for hydrolysis of proteins; used for improvement of functionality of veg. and animal proteins; endopeptidase action on serine sites; brn. clear liq.; m.w. 27,3000; water-sol.; dens. 1.25 g/cm³ (undiluted).

Alcalase® 2.0 T. [Novo] Proteinase; enzyme for laundry powd. detergent; off-wh. dustfree granulate.

Alcalase® 2.5 L. [Novo] Proteinase; enzyme for non-built liq. detergent and prespotters; lt. brn. liq.

Alcamate FA 82. [Alcolac Ltd.] Disodium N-alkyl sulfosuccinamate; anionic; foam booster and stabilizer for carpet backing; paste; 35% conc.

Alcamizer I, II, III. [Kyowa] Heat stabilizer for PVC that reacts in a unique way with HCl in PVC; offers high heat stability, nontoxicity, high transparency; wh. powd.; odorless; sp.gr. 2.1; dens. 0.42, 0.45, and 0.23 g/ml resp.

Alcogum. [Alco] Sodium polyacrylate; thickener; control of visc. in aq. systems.

Alcogum 9639. [Alco] Ammonium polyacrylate; thickener for natural and syn. latexes; used in dipped molded or cast goods; water sol'n.

Alcogum L-15, L-26, L-28, L-31, L-35, L-36. [Alco] Polyacrylic acid; anionic; thickener for adhesives, paints, paper coatings, natural and syn. latexes, alkali reactive emulsion; FDA approved; milky wh. liq.; sp.gr. 1.05, 1.05, 1.04, 1.05, 1.04, and 1.04 resp.; visc. 15, 500, 20, 20, 15, 35 cps; pH 3.0, 3.5, 3.1, 3.1, 3.5, 3.0; 30, 20, 20, 40, 20, 20% solids resp.

Alcolan®. [Amerchol] Petrolatum, sorbitan sesquiolate, and lanolin alcohol; nonionic; w/o emulsifier, emollient base for cosmetic and pharmaceutical emulsions; paste; 100% conc.

Alcolan® 36W. [Amerchol] Lanolin, petrolatum, and sorbitan sesquioleate; nonionic; see Alcolan; paste; 100% conc.

Alcolan® 40. [Amerchol] Petrolatum, lanolin, beeswax, sorbitan sesquiolate, and polysorbate 81; see Alcolan; paste; 100% conc.

Alconate® 2CSA. [Alcolac] Tetrasodium dicarboxyethyl stearyl sulfosuccinamate; anionic; surfactant in caustic industrial and household cleaners; emulsifier for emulsion polymers, pesticides; dispersant for pigments, UV absorbers, and silicones; liq.; sol. in calcium, alkaline, and electrolyte sol'ns., water; 35% conc.

Alconate® D-6. [Alcolac] Disodium deceth-6 sulfosuccinate; anionic; solubilizer, dispersant, emulsifier used in emulsion polymers; wetting agent for dry powds.; liq.; 31% act.

Alconate® LEA. [Alcolac] Disodium lauramido MEA sulfosuccinate; anionic; detergent, dispersant for rug, upholstery, and fabric cleaners; Gardner 3 opaque liq.; dens. 9.0 lb/gal; cloud pt. 20 C; pH 6.5 (5%); 40% solids.

Alconate® SBDO. [Alcolac] Dioctyl sodium sulfosuccinate; wetting agent, surfactant, dispersant, and penetrant for industrial and mining applics.; liq.; 70% act.

Alconate® SBG-280. [Alcolac] Disodium oleamido MEA-sulfosuccinate; thickener, conditioner for hair and skin prods.; liq.; 30% act.

Alconate® SBN-862. [Alcolac] Disodium nonoxynol-10 sulfosuccinate; emulsifier, dispersant for emulsion polymerization of vinyl acetate and acrylates; liq.; 32% act.

Alconate® SBTA-269. [Alcolac] Disodium alkyl (C18) sulfosuccinamate; foam generator for latex systems; liq.; 36% act.

Alconate® SBU-185. [Alcolac] Disodium undecylenamido MEA-sulfosuccinate; antimicrobial for dandruff shampoos and medicated treatments; liq.; 45% act.

Alconox. [Alconox] Blend of alkylaryl sulfonates, lauryl alcohol sulfates, phosphates, carbonates; anionic; detergent with wetting, sequestering and synergistic agents; for manual cleaning of laboratory and hospital glassware and instruments; colorless powd.; odorless; sol. in water; biodeg.

Alcophor AC. [Henkel] Tannin compds., modified; corrosion inhibitor used in petrol.-based paint systems.

Alcosperse 107, 124, 149, 157. [Alco] Sodium polyacrylate; dispersant for pigments, high solids slurries, paper coating paint, textile, mining, and ceramic applics.; liq., dry form; water-sol.

Alcosperse 160. [Alco] Polymer; anionic; see Alcosperse 107; liq.; water-sol.

Alcosperse 175. [Alco] Specialty copolymer; see Alcosperse 107; liq.; water-sol.

Alcosperse 249. [Alco] Ammonium polyacrylate; see Alcosperse 107; liq.; water-sol.

Alcosperse 404. [Alco] Polyacrylic acid sol'n.; see Alcosperse 107; water-sol.

Alcosperse 602-N. [Alco] Sodium polyacrylate; see Alcosperse 107; liq.; water-sol.

Alcotab. [Alconox] Blend of alkylaryl sulfonates, lauryl alcohol sulfates, phosphates, carbonates; anionic; detergent with wetting, sequestering and synergistic agents; tablet; biodeg.

Aldo® DC. [Lonza] Propylene glycol dicaprylate/dicaprate; coupling agent for mixed solv. systems; emulsifier for food, cosmetic, industrial use; Gardner 2 max. liq.; sol. in ethanol, min. and veg. oil; insol. in water; sp.gr. 0.92; HLB 2 ± 1; iodine no. 0.5 max.; sapon. no. 315–335.

Aldo® HMS. [Lonza] Glyceryl stearate; nonionic; emulsifier for w/o systems; stabilizer, consistency builder, emollient in personal care prods.; wh. beads; sol. in ethanol, min. and veg. oil; disp. in water; HLB 2.8; m.p. 61–68 C; sapon. no. 165–175; 52–56% mono content.

Aldo® ML. [Lonza] Glyceryl monodilaurate; nonionic; see Aldo DC; liq.; HLB 5.2; 100% conc.

Aldo® MLD. [Lonza] Glyceryl laurate, disp.; nonionic; emulsifier; see Aldo DC; cream soft solid; sol. in ethanol, ethyl acetate, toluol, naphtha, min. and veg. oils, disp. in water; sp.gr. 0.97; m.p. 21–26 C; HLB 6.8; sapon. no. 185–195; pH 7.5–8.5 (5% aq.); 100% conc.

Aldo® MO FG. [Lonza] Glyceryl mono- and dioleate; nonionic; liq. emulsifier and antifoam for foods; emulsifier, solubilizer for cosmetic, pharmaceutical, and industrial applics.; liq.; HLB 3.0; 100% conc.

Aldo® MOD FG. [Lonza] Glyceryl monodioleate, disp.; nonionic; liq. emulsifier and antifoam for foods; emulsifier, solubilizer for cosmetic, pharmaceutical, and industrial applics.; liq.; HLB 4.0; 100% conc.

Aldo® MS-20 FG. [Lonza] PEG-20 glycerol stearate; nonionic; emulsifier, antifoam for foods; emulsifier, solubilizer for cosmetic, pharmaceutical, and industrial applics.; solid; HLB 13.0; 100% conc.

Aldo® MS LG. [Lonza] Glyceryl stearate; nonionic; see Aldo HMS; wh. beads; sol. in ethanol, min. and veg. oil; disp. in water; HLB 3.3; m.p. 58–62 C; sapon. no. 160–170; 40–45% mono content.

Aldo® PMS. [Lonza] Propylene glycol stearate; nonionic; emulsifier, stabilizer, solubilizer, emollient, lubricant, thickener, plasticizing agent, antiirritant, and conditioner in pharmaceutical, cosmetic, and toiletry preparations; foam modifier for detergent systems; wh. flakes; sol. in ethanol, min. and veg. oils; insol. in water; HLB 3 ± 1; iodine no. 1.0 max.; sapon. no. 170–185.

Aldomax GA-100. [UOP] Immobilized amyloglucosidase enzyme; for saccharification of low visc., higher DE (50-95) starch containing streams; tan granular solid.

Aldosperse® ML 23. [Lonza] PEG-23 glyceryl laurate; nonionic; emulsifier, solubilizer, suspending and dispersing agent used in personal care prods.; straw liq.; sol. in ethanol, veg. oil; sp.gr. 1.09; HLB 17 ± 1; sapon. no. 42–50.

Aldosperse® ML 230. [Lonza] PEG-23 glyceryl laurate; conditioner for textiles; Gardner 21 liq.; HLB 17 ± 1; acid no. 4; hyd. no. 84.

Aldosperse® MS 20. [Lonza] PEG-20 glyceryl stearate; nonionic; emulsifier, solubilizer, suspending and dispersing agent used in personal care prods., food industry, textiles; wh. soft solid; sol. in ethanol; disp. in water; HLB 13 ± 1; m.p. 32–38 C; acid no. 2; sapon. no. 65–75; hyd. no. 73.

Aldosperse® O 20 FG. [Lonza] 80% Glyceryl stearate, 20% polysorbate 80; nonionic; emulsifier, solubilizer for frozen desserts, cosmetic, pharmaceutical, and industrial applics.; bead; HLB 5.0; 100% conc.

Alfol® 4. [Vista] n-Butyl alcohol; intermediate for amines, plasticizers, fatty acid esters, detergents; colorless liq.; char. penetrating odor; sol. in water; dens. 6.76; sp.gr. 0.809; visc. 3.4; m.p. –128 C; b.p. 117.7; 99.3% act.

Alfol® 6. [Vista] Hexyl alcohol; detergent intermediate; colorless liq.; dens. 6.83; sp.gr. 0.823; visc. 5.5; m.p. –49 C; b.p. 315 C; 98.5% act.

Alfol® 8. [Vista] Caprylic alcohol; wetting agent, emulsifier, surfactant intermediate, raw material; colorless liq.; mild aromatic odor; dens. 6.86; sp.gr. 0.823; m.p. 5 C; b.p. 383 C; 99.2% act.

Alfol® 10. [Vista] Decyl alcohol; emulsifier, surfactant intermediate; colorless liq., wh. waxy solid; sweet, fat-like odor; sol. in alcohol, ether, benzene, glacial acetic acid; water insol.; dens. 6.90; sp.gr. 0.828; m.p. 44 C; b.p. 448 C; 99% act.

Alfol® 12. [Vista] Lauryl alcohol; detergent intermediate; colorless liq.; sp.gr. 0.831; m.p. 75 C; b.p. 498 C; 99.1% act.

Alfol® 14. [Vista] Myristyl alcohol; detergent intermediate; wh. solid, cryst.; typ. fatty alcohol odor; sp.gr. 0.824; m.p. 99 C; b.p. 552 C; 99% act.

Alfol® 16. [Vista] Cetyl alcohol; detergent intermediate; emollient used in cosmetics; plastics additive; wh. waxy solid; typ. fatty alcohol odor; sol. in alcohol, acetone, ether; water-insol.; dens. 6.77; sp.gr. 0.813; visc. 6.77; m.p. 117 C; b.p. 604 C; 98.9% act.

Alfol® 18. [Vista] Stearyl alcohol; surfactant intermediate; emollient used in cosmetics; plastics additive; wh. waxy solid; typ. fatty alcohol odor; sol. in alcohol, acetone, ether; water-insol.; dens. 6.71; sp.gr. 0.8075; visc. 13.5; m.p. 135 C; b.p. 640 C; 98.7% act.

Alfol® 20. [Vista] Blend of C_{20} and higher even-carbon-number primary linear alcohols; lubricant, defoamer; emollient; used in fuel oil; waxes and polishers used in paper pulp defoamers; off-wh. solid; sol. in alcohol, acetone, ether; water-insol.; sol. in alcohol, acetone, ether; water-insol.; dens. 6.80; sp.gr. 0.817; m.p. 131.

Alfol® 22. [Vista] C_{22} and higher linear alcohols; see Alfol® 20; sol. in alcohol, acetone, ether; water-insol.

Alfol® 610. [Vista] C_6-C_{10} linear primary alcohol; intermediate for surfactants, plasticizers; colorless liq.; mild, sweet odor; sol. in alcohol, ether, benzene, insol. in water; dens. 6.92; sp.gr. 0.829; visc. 11.0; m.p. –8 C; b.p. 350 C; 99% act.

Alfol® 610ADE. [Vista] C_6, C_8, C_{10} alcohol blend; chemical intermediate; also for lube oil additives, plasticizers, surfactant feedstocks; clear colorless liq.; m.w. 140.

Alfol® 810. [Vista] C_8–C_{10} linear primary alcohol; intermediate for plasticizers and surfactants; colorless liq.; mild, sweet odor; sol. in alcohol, ether, benzene; dens. 6.93; sp.gr. 0.831; visc. 8.9; m.p. 7 C; b.p. 400 C; 99% act.

Alfol® 810FD. [Vista] C_8, C_{10} alcohol blend; chemical intermediate; also for lube oil additives, plasticizers, surfactant feedstocks; clear colorless liq.; m.w. 141.

Alfol® 1012. [Vista] C_{10}–C_{12} linear primary alcohol; intermediate for surfactants; lubricant and metal rolling rolls; colorless liq.; sweet odor; sol. in alcohol, acetone, ether; water-insol.; dens. 7.0; sp.gr. 0.834; visc. 15; m.p. 35–40 C; b.p. 427 C; 99% act.

Alfol® 1214. [Vista] C_{12}–C_{14} linear primary alcohol; see Alfol 1012; colorless liq.; sweet, typ. fatty alcohol odor; sol. see Alfol 20; dens. 7.0; sp.gr. 0.838; visc. 14.3; m.p. 73–75 C; b.p. 518 C; 99% act.

Alfol® 1214-GC. [Vista] C_{12} and C_{14} alcohol blend; chemical intermediate; also for lube oil additives, plasticizers, surfactant feedstocks; clear colorless liq.; m.w. 195.

Alfol® 1216. [Vista] C_{12}-C_{16} linear alcohols; intermediate for surfactants, plastics; lubricant and metal rolling rolls; clear, colorless liq.; sweet, typ. fatty alcohol odor; sol. see Alfol 20; dens. 7.0; sp.gr. 0.830; visc. 15.5; m.p. 64–70 C; b.p. 514 C; 98.5% act.

Alfol® 1216-CO. [Vista] C_{12} and C_{14} alcohol blend; chemical intermediate; also for lube oil additives, plasticizers, surfactant feedstocks; clear colorless liq.; m.w. 198.

Alfol® 1218. [Vista] C_{12}–C_{18} linear primary alcohol; intermediate for surfactants; wh. solid; dens. 6.85; sp.gr. 0.825; visc. 15.5; m.p. 99–101 C; b.p. 524 C; 100% act.

Alfol® 1412. [Vista] Myristyl alcohol; intermediate for surfactants, plastics; wh., soft waxy solid; typ. fatty alcohol odor; dens. 7.0; sp.gr. 0.838; visc. 16.7; m.p. 72–75 C; b.p. 560 C; 99.4% act.

Alfol® 1412. [Vista] C_{12}, C_{14}, C_{16} alcohol blend; chemical intermediate; also for lube oil additives, plasticizers, surfactant feedstocks; clear colorless liq.; m.w. 205.

Alfol® 1418-DDB. [Vista] C_{14}, C_{16}, C_{18} alcohol blend; chemical intermediate; also for lube oil additives, plasticizers, surfactant feedstocks; wh. solid; m.w. 243.

Alfol® 1618. [Vista] C_{16}–C_{18} linear primary alcohol; intermediate for surfactants; wh. waxy solid; typ. fatty alcohol odor; sol. in water, alcohol, chloroform, ether, benzene, glacial acetic acid; dens. 6.81; sp.gr. 0.820; visc. 22; m.p. 110–115 C; b.p. 609 C; 98.7% act.

Alfol® 1618-CG. [Vista] C_{16}, C_{18}, C_{20+} alcohol blend; chemical intermediate; also for lube oil additives, plasticizers, surfactant feedstocks; wh. solid; m.w. 261.

Alfol® 1620. [Vista] C_{16}–C_{20} linear primary alcohol; intermediate for surfactants; wh. waxy solid; typ. fatty alcohol odor; dens. 6.81; sp.gr. 0.817; visc. 13.5; m.p. 110–113 C; b.p. 605 C; 98.5% act.

Alfol® C22. [Vista] Blend of C_{22} and higher even-carbon-number primary linear alcohols; lubricant, defoamer; off-wh. solid; dens. 6.81; sp.gr. 0.817; m.p. 132.

Alfonic® 610-50R. [Vista] Ethoxylated alcohol (3.1 EO); surfactant intermediate; clear liq.; HLB 10.0.

Alfonic® 810-40. [Vista] Ethoxylated alcohol (2.2 EO); surfactant intermediate; clear liq.; HLB 8.0.

Alfonic® 810-60. [Vista] Ethoxylated alcohol (4.8 EO); surfactant intermediate; clear liq.; HLB 12.0.

Alfonic® 1012-40. [Vista] Ethoxylated alcohol (2.5 EO); surfactant intermediate; clear liq.; HLB 8.0.

Alfonic® 1012-60. [Vista] Ethoxylated linear alcohols (60% E.O.); nonionic; detergent, wetting agent, emulsifier, foaming agent, intermediate, dispersant; APHA 20 liq.; sweet odor; dens. 8.2; sp.gr. 0.98; visc. 20.5 cps; m.p. 36; biodeg.; 99.8% act.

Alfonic® 1214-GC-30. [Vista] Ethoxylated alcohol (2.0 EO); surfactant intermediate; clear liq.; HLB 6.2.

Alfonic® 1214-GC-40. [Vista] Ethoxylated alcohol (2.8 EO); surfactant intermediate; clear liq.; HLB 7.8.

Alfonic® 1216-22. [Vista] Ethoxylated alcohol (1.3 EO); surfactant intermediate; clear liq.; HLB 4.4.

Alfonic® 1412-40. [Vista] Ethoxylated linear alcohols (40% E.O.); nonionic; detergent intermediate, emulsifier, dispersant; clear liq.; slight odor; oil-sol.; dens. 7.64; sp.gr. 0.91; visc. 52 cps; m.p. 41; biodeg.; 100% act.

Algaesil. [Mason] Colloidal silver; algicide.

Algaetrol 76. [Harcros] 7% Copper TEA compd.; dual action algicide used to prevent and kill algae in swimming pools.

Algonina AC. [Tessilchimica] Cyclic and amino amide compds.; cationic; laundry softener for natural and syn. fibers; flakes; 100% act.

Alipal® CD-128. [Rhone-Poulenc Surf.] Ammonium alcohol ethoxylate sulfate; anionic; surfactant for aq. systems; air-entraining agent, foamer, frother for concrete, cements, gypsum wallboard; solubilizer and wetting agent; straw liq.; good sol. in electrolyte systems; sp.gr. 1.056; pour pt. –5 C; 56% act.

Alipal® CO-433. [Rhone-Poulenc Surf.] Sodium nonoxynol-4 sulfate; anionic; detergent, emulsifier, lime-soap dispersant, wetting agent, foamer; dishwashing formulations, scrub soaps, car washes, rug and hair shampoos, emulsion polymerization, petrol. waxes, textile wet processing, cosmetics, pesticides; antistat for plastics and syn. fibers; Varnish 4 max. clear liq.; mild aromatic odor; water-sol.; sp.gr. 1.065; dens. 8.9 lb/gal; visc. 2500 cps; surf. tens. 32 dynes/cm (1%); 28% act.

Alipal® CO-436. [Rhone-Poulenc Surf.] Ammonium nonoxynol-4 sulfate; anionic; see Alipal CO-433; Varnish 5 max clear liq.; alcoholic odor; water-sol.; sp.gr. 1.065; dens. 8.9 lb/gal; visc. 100 cps; surf. tens. 34 dynes/cm (1%); 58% act.

Alipal® CO-526. [Rhone-Poulenc Surf.] Sodium salt of a sulfated alkylphenoxypoly (ethyleneoxy)-ethanol; anionic; detergent, wetting, dispersing, emulsifying and foaming agent; Varnish 10 max. color, hazy liq.; alcoholic odor; water sol.; dens. 9.3 lb/gal; sp.gr. 1.115; visc. 400 cps; surf. tens. 35 dynes/cm (1%); 58% act.

Alipal® EO-526. [Rhone-Poulenc Surf.] Sodium nonoxynol-4 sulfate; anionic; see Alipal CO-433; Varnish 10 max. hazy liq.; alcoholic odor; water-sol.; sp.gr. 1.115; dens. 9.3 lb/gal; visc. 400 cps; 58% act.

Alipal® EP-110. [Rhone-Poulenc Surf.] Ammonium salt of sulfated nonylphenoxypoly ethanol; anionic; primary emulsifier and stabilizing agent for the preparation of vinyl acetate, vinyl acetate/acrylic, all acrylic, styrene/acrylic, and S/B emulsion copolymers; pale yel. liq.; sp.gr. 1.04; visc. 91.0 cks; pour pt. 0 C; surf. tens. 38.3 dynes/cm (1%); 30% act.

Aliquat 336-PTC. [Henkel] Tricaprylyl methyl ammonium chloride; phase tranfer agent for catalyzing reactions in heterogeneous media between ionic salt and org. substrates; amber visc. liq.; m.w. 442; sol. in acetonitrile, chloroform, diethyl ether, hexane, IPA, THF, and toluene; sp.gr. 0.884; visc. 1450 cps (30 C); pour pt. –15 C; acid no. 0.0–1.0; flash pt. 132 C (CC).

Alkafloc A Series. [Alkaril] Med. to high m.w. polyelectrolytes; flocculants in solid-liq. separation processes, e.g., for municipal sewage, steel mill scale, etc., min. and coal beneficiation processes, treating waste effluents, paper industry; emulsion form; 30–50% solids.

Alkafloc CE Series. [Alkaril] High m.w. cationic polyelectrolytes; flocculant for industrial and municipal sludges, paper industry, solid-liq. separation processes in industrial waste treatment, demulsification of oils; emulsion form; 20–50% conc.

Alkafloc CS Series. [Alkaril] Low to med. m.w. cationic polyelectrolytes; see Alkafloc CE Series; aq. sol'n.; 20–50% conc.

Alkafloc CSM Series. [Alkaril] Cationic polymers of higher m.w. than CS series; see Alkafloc CE Series; aq. sol'n.; 3–8% solids.

Alkafloc N Series. [Alkaril] High m.w. nonionic flocculants; flocculant in water clarification, filtration aids in dewatering industrial sludges, in ore processing in solid-liq. separation processes, retention and drainage aids in paper mills; emulsion form; 30–50% solids.

Alkali Surfactant. [Exxon] Amphoteric surfactant used in alkaline formulations for hard surf. cleaning; solubilizer for nonionics into high electrolyte cleaners; lt. amber liq.; sol. 5% in water, IPA, 5% sodium lauryl sulfate, 5% benzalkonium chloride, 10% sodium hydroxide; sp.gr. 1.04; pH 5–9 (5% sol'n.); 35% act. in water.

Alkalube DTDA. [Alkaril] Diester of adipic acid and tridecyl alcohol; syn. lubricant used in jet and engine lubes, and rotary and reciprocating compressor lubricants; clear liq.; dens. 0.9 g/ml; pour pt. -60 C max.; acid no. 1.0 max.; flash pt. 332 C (COC).

Alkamide 1182. [Alkaril] Lauramide DEA, modified; nonionic; foam stabilizer and booster, visc. builder in personal care preparations; liq.; 100% conc.

Alkamide 1423. [Alkaril] Cocamide DEA; nonionic/anionic; detergent, wetting, thickening, foaming and foam stabilizing agent in liq. soap formulations; liq.; 100% conc.

Alkamide 1509. [Alkaril] Capramide DEA; nonionic/anionic; foam booster/stabilizer for personal care products; liq.; 100% conc.

Alkamide 2104. [Alkaril] 2:1 Cocamide DEA; nonionic-anionic; detergent, emulsifier, foam booster; base for floor and general purpose cleaners; lubricant for syn. grinding and cutting fluids; amber visc. liq.; low odor; sol. in aromatic, aliphatic, and chlorinated solvs., water; 100% act.

Alkamide 2110. [Alkaril] 2:1 Coconut alkanolamide; nonionic; detergent base to solubilize high silicate concs. in liq. steam cleaners and wax strippers; amber liq.; sol. in water, aromatic solv., perchloroethylene; dens. 1.0 g/ml; pH 8–11 (1% DW); 100% act.

Alkamide 2112. [Alkaril] 2:1 Fatty acid alkanolamide, sulfonate blend; nonionic-anionic; detergent base for floor cleaners, alkaline strippers, and degreasers; detergent, wetting, lubricant, dispersing and air entertaining agent for construction materials; brn. visc. liq.; low odor; sol. in water, aromatic solv, perchloroethylene; dens. 1.0 g/ml; cloud pt. 45 F max.; pH 8–11; 100% act.

Alkamide 2124. [Alkaril] 2:1 Lauramide DEA; nonionic-anionic; detergent, wetting agent, emulsifier, foaming agent, foam stabilizer, thickener for shampoos, bubble baths, liq. detergents; amber visc. liq.; sol. in aromatic hydrocarbons and chlorinated solvs., water; sp.gr. 1.0–1.05; 100% act.

Alkamide 2204. [Alkaril] 2:1 Cocamide DEA; nonionic-anionic; detergent, rust inhibitor; base for hand soap, floor cleaners; amber visc. liq.; low odor; water sol.; 100% act.

Alkamide CDE. [Alkaril] Cocamide DEA; nonionic; detergent, emulsifier, stabilizer, thickener, foam stabilizer for personal care and detergent prods.; lt. amber liq.; low odor; sol. in detergent systems, aromatic and chlorinated aliphatic solv., petrol. solv., min. oils; water-disp.; sp.gr. 1.04; dens. 1.0 g/ml; pH 8–11 (1% DW); 90% act.

Alkamide CDE-Extra. [Alkaril] Cocamide DEA (1:1); nonionic; emulsifier, stabilizer, thickener, and foam stabilizer for personal care and detergent prods.; liq.; 100% conc.

Alkamide CDO. [Alkaril] Cocamide DEA; nonionic; detergent, emulsifier, foam stabilizer, thickener for personal care and detergent prods.; lt. amber liq.; low odor; sol. in aromatic and chlorinated aliphatic solv., petrol. solv.; min. oils, water; sp.gr. 1.04; dens. 1.0 g/ml; pH 8–11 (1% DW); 82% act.

Alkamide CL63. [Alkaril] Coconut-lauric DEA; nonionic; detergent, thickener, emulsifier, foam stabilizer in toiletry and cleaning preparations; amber clear to cloudy liq.; sol. in detergent systems, aromatic and chlorinated aliphatic solv., petrol. solv., min. oils; water-disp.; sp.gr. 1.03–1.05; dens. 1.0 g/ml; pH 8–11 (1% DW); 92% act.

Alkamide CME. [Alkaril] Cocamide MEA; nonionic; detergent, foam boosters, visc. builder, opacifier; wh. waxy solid; sol. in detergent systems, various solv., min. oils; m.p. 56 C; 95% act.

Alkamide CMO. [Alkaril] Cocamide MEA; detergent, foam boosters, visc. builder, opacifier; cream-colored, waxy solid; sol. in detergent systems, various solvs., min. oils; m.p. 50 C; 90% act.

Alkamide HTDE. [Alkaril] Stearamide DEA; nonionic; detergent, thickener, visc. builder, emulsifier for kerosene, veg. and min. oil, microcryst. wax; cream-colored waxy solid; low odor; sol. in min. spirits, aromatic solvs., perchloroethylene; water-disp.; m.p. 50–55 C; pH 8–11 (1%); 90% act.

Alkamide HTME. [Alkaril] Stearamide MEA; nonionic; detergent, opacifier, thickener for cosmetics; cream waxy solid; low odor; insol. in water; m.p. 85–90 C; pH 8–11 (1%); 90% act.

Alkamide L7DE. [Alkaril] Lauric-myristic DEA; nonionic; detergent, foam booster/stabilizer, superfatting and thickening agent for toiletry and cleaning formulations; fortifier for perfumes in soaps; wh. paste, liq. > 25 C; low odor; sol. in min. spirits, aromatic solv., perchloroethylene; water-disp.; sp.gr. 1.03–1.05; m.p. 15 C; pH 8–11 (1% DW); 92% act.

Alkamide L7DE-BT. [Alkaril] Lauric-myristic DEA, modified; nonionic; detergent, foam stabilizer, visc. builder for cosmetic preparations; straw liq.; bland odor; sol. in water, aromatic solv., perchloroethylene; sp.gr. 0.99; dens. 1.0 g/ml; visc. 275 cps; f.p. 0 C; cloud pt. 0 C; pH 8–11 (1% DW); 84% act.

Alkamide L7DE-PG. [Alkaril] Lauric-myristic alkanolamide, modified; nonionic; liq. version of Alkamide L7DE; liq.

Alkamide L7ME. [Alkaril] Lauric-myristic MEA; nonionic; foam booster and visc. for liq. and powd. detergents; solid; 100% conc.

Alkamide L9DE. [Alkaril] Lauramide DEA, high purity; nonionic; detergent, emulsifier, foam stabilizer and booster, thickener for toiletry and cleaning applics., industrial and household detergents; wh. waxy solid, liq. > 25 C; low odor; sol. in aromatic and chlorinated solvs., min. oils; sp.gr. 1.04; f.p. 25 C; 92% act.

Alkamide L9ME. [Alkaril] Lauramide MEA; nonionic; detergent, foam stabilizer and booster, visc. builder; wh., waxy solid; low odor; sol. in detergent systems, various solvs., min. oils; m.p. 60 C; 95% act.

Alkamide LIPA. [Alkaril] Lauramide MIPA; nonionic; foam booster/stabilizer in detergents and laundry compds.; visc. builder in liq. formulations; mild to skin; lt. yel. solid; low odor; disp. in water; m.p. 63–68 C; pH 8–11 (1% DW); 95% amide.

Alkamide RDO. [Alkaril] Ricinolamide DEA; nonionic; emulsifier, lubricant, antiblocking and mold release agent used in the making of cutting and sol. oils, cleaners; brn. clear liq.; low odor, water-disp.; dens. 1.0 g/ml; pH 8–11; 100% act.

Alkamide SDO. [Alkaril] Soyamide DEA; nonionic; foam stabilizer, visc. builder, superfatting agent for toiletries, cutting and sol. oils, textiles, household and industrial cleaners, corrosion inhibitor; amber, visc. liq.; low odor; sol. in min. oil and spirits, perchloroethylene; water-disp.; sp.gr. 1.04; dens. 1.0 g/ml; pH 8–11 (1%); 80% act.

Alkamide STEDA. [Alkaril] Ethylene bisstearamide; nonionic; defoamer in pulp and paper industry; lubricant, mold release agent, plasticizer, antistat and pigment dispersant in resins, plastics, wire drawing; lt. tan solid; insol. in water; m.p. 141–145 C; 100% conc.

Alkamidox C-2. [Alkaril] Coconut MEA ethoxylate; nonionic; thickener, foam stabilizer, and emulsifier for formulated detergents and cosmetics; tan paste.; sol. in aromatic solv., perchloroethylene; disp. in water; congeal pt. 19–21 C; pH 10.0–11.5 (5% DW).

Alkamidox C-5. [Alkaril] Coconut MEA ethoxylate; nonionic; emulsifier, foam stabilizer, and thickener used in alkaline industrial cleaners; carwash formulas; liq. dish and laundry detergents; cosmetics; yel. liq.; sol. in water, aromatic solv., perchloroethylene; dens. 1.0 g/ml; congeal pt. 13–15 C; pH 10.0–11.5 (5% DW).

Alkamidox CME-2. [Alkaril] PEG-3 cocamide; nonionic; detergent, foam stabilizer; cream-colored solid; sol. in ethanol, xylene, insol. in kerosene, min. oil; 100% act.

Alkamidox CME-5. [Alkaril] PEG-6 cocamide; nonionic; detergent, foam stabilizer; straw liq.; sol. in ethanol, xylene, insol. in kerosene, min. oil; 100% act.

Alkamidox L-2. [Alkaril] Lauric MEA ethoxylate; nonionic; foam stabilizer used in personal care prods.; wh. solid; sol. in aromatic solv., perchloroethylene; disp. in water; congeal pt. 30–35 C; pH 10.0–11.5 (5% DW).

Alkamidox L-5. [Alkaril] Lauric MEA ethoxylate; nonionic; see Alkamidox C-5; wh. solid; sol. in water, aromatic solv., perchloroethylene; congeal pt. 15–20 C; pH 10.0–11.5 (5% DW).

Alkaminox C-2. [Alkaril] PEG-2 cocamine; cationic; textile scouring, dyeing assistant, softening, antistatic agent; used as corrosion inhibitor in steam generating and circulating systems; coemulsifier; clear amber liq.; sol. in most org. solv., min. acid, insol. in water; sp.gr. 0.87; 100% conc.

Alkaminox C-5. [Alkaril] PEG-5 cocamine; see Alkaminox C-2; Gardner 12 liq.; sol. in aromatic solv., perchloroethylene; water-disp.; dens. 0.98 g/ml; HLB 14; 98% min. tert. amine.

Alkaminox C-10. [Alkaril] PEG-10 cocamine; cationic; used in cosmetics; Gardner 12 liq.; sol. in water, aromatic solv., perchloroethylene; dens. 1.02 g/ml; HLB 16; 98% min. tert. amine.

Alkaminox HT-12, -25, -30. [Alkaril] Hydrog. primary tallow amine ethoxylate; cationic; emulsifier, dispersant; textile scouring and desizing assistants; softening and antistatic agent; solid; 100% conc.

Alkaminox SO-5. [Alkaril] PEG-5 soyamine; cationic; imparts emulsifying, corrosion resistance, and nonfoaming properties in metal cleaning and formulation of cleaners; Gardner 14 liq.; sol. in perchloroethylene; disp. in water; dens. 0.98 g/ml; HLB 14; 98% min. tert. amine.

Alkaminox T-2. [Alkaril] PEG-2 tallow amine; cationic; emulsifier, dispersant, scouring, desizing aid, internal antistat, softener for textiles, polyethylene; Gardner 8 max. paste; sol. in most org. solv., min. acid, insol. in water; sp.gr. 0.92; HLB 4; 100% conc.

Alkaminox T-5. [Alkaril] PEG-5 tallow amine; cationic; emulsifier, dispersant, used in textile industry, coatings; corrosion inhibitor in steam generating and circulating systems, emulsifier for wax emulsions; Gardner 8 max. clear liq.; sol. in most org. solv., min. acid, insol. in water; sp.gr. 0.97; HLB 9; 98% min. tert. amine.

Alkaminox T-10. [Alkaril] PEG-10 tallow amine; cationic; emulsifier, dispersant, leveling agent for dyeing of nylon with anionic dyes; Gardner 14 liq., paste; sol. in water, perchloroethylene; dens. 1.01 g/ml; 98% min. tert. amine.

Alkaminox T-12. [Alkaril] PEG-12 tallow amine; cationic; see Alkaminox T-10; Gardner 14 liq., paste; sol. in water; dens. 1.03 g/ml; HLB 13.5; 98% min. tert. amine.

Alkaminox T-15. [Alkaril] PEG-15 tallow amine; cationic; see Alkaminox T-5; Gardner 12 paste; water-sol.; sp.gr. 1.03; HLB 14.2; 98% min. tert. amine.

Alkaminox T-25, T-30, T-50. [Alkaril] PEG-25, -30, -50 tallow amine; cationic; coemulsifier, leveling agent for differential dyeing of nylon; Gardner 12, 12, and 23 resp. solids; water-sol.; HLB 16, 16.6, 17.5; 98% min. tert. amine.

Alkamox C2-0. [Alkaril] Bis (2-hydroxyethyl) cocamine oxide; nonionic; foaming agent, stabilizer, thickener, and emollient for personal care prods. and detergents; corrosion inhibitor for nonferrous metals; dispersant in electroplating systems; clear liq.; sol. in water; dens. 1.0 g/ml; 29–31% amine oxide.

Alkamox CAPO. [Alkaril] Cocamidopropylamine oxide; nonionic; detergent, wetting agent, foaming agent and foam stabilizer for rug shampoos, laundry detergents, dishwashing, shampoos, cleaners, antistatic softeners; foam stabilizer in foam rubber, electroplating, paper coatings; colorless clear liq.; faint odor; water sol.; disp. in min. oil; dens. 1.0 g/ml; f.p. < 0 C; 29–31% amine oxide.

Alkamox LO. [Alkaril] Lauramine oxide; nonionic-cationic; see Alkamox CAPO; colorless clear liq.; faint odor; water sol.; disp. in min. oil; dens. 1.0 g/ml; f.p. < 0 C; 29–31% amine oxide.

Alkamox ODM. [Alkaril] Oleyl dimethylamine oxide; nonionic/cationic; foaming, wetting and conditioning agent; liq.; 50% conc.

Alkamuls 200-DL. [Alkaril] PEG-4 dilaurate; nonionic; w/o emulsifier, dispersant, defoamer, coemulsifier, and lubricant in industrial and textile oils, cosmetics; as paper softener; yel. liq. to paste; sol. in min. oil and spirits, aromatic solv., perchloroethylene; disp. in water; dens. 0.96 g/ml; HLB 7.4; sapon. no. 190–200; 100% conc.

Alkamuls 200-DO. [Alkaril] PEG-4 dioleate; nonionic; see Alkamuls 200-DL; amber liq.; sol. in min. oil, min. spirits, aromatic solv., perchlorethylene; disp. in water; dens. 0.95 g/ml; HLB 6.0; sapon. no. 147–157; 100% conc.

Alkamuls 200-DS. [Alkaril] PEG-4 distearate; nonionic; lubricant and softener in textile applics.; opacifier and emulsifier in cosmetic preparations; cream solid; sol. in min. oil and spirits, aromatic solv., perchloroethylene; disp. in water; HLB 5.0; sapon. no. 160–170; 100% conc.

Alkamuls 200-ML. [Alkaril] PEG-4 laurate; nonionic; emulsifier, coupling agent, and solubilizer in metal working fluids, industrial and textile lubricants; yel. liq.; disp. in water; dens. 1.00 g/ml; HLB 9.8; sapon. no. 142–152.

Alkamuls 200-MO. [Alkaril] PEG-4 oleate; nonionic; w/o emulsifier for min. oils, fatty acids, and solv.; used in industrial applics. such as cleaners, degreasers and pesticide formulations; lubricant-softener in tanning and textile processes; yel. liq.; disp. in water; dens. 0.99 g/ml; HLB 8.0; sapon. no. 111–121.

Alkamuls 200-MS. [Alkaril] PEG-4 stearate; nonionic; emulsifier for fats and oils; softener and lubricant for textiles and leather; wh. solid; disp. in water; HLB 8.0; sapon. no. 120–130.

Alkamuls 400-DL. [Alkaril] PEG-8 dilaurate; nonionic; lipophilic emulsifier, solubilizer, dispersing agent used in personal care prods. and industrial applics., agric. chemical sprays, industrial and textile lubricants; yel. liq. to paste; sol. in min. oil and spirits, aromatic solv., perchloroethylene; disp. in water; dens. 1.0 g/ml; HLB 10.0; sapon. no. 132–142; 100% act.

Alkamuls 400-DO. [Alkaril] PEG-8 dioleate; nonionic; emulsifier, solubilizer, lubricant, wetting agent for cosmetic, textile, metalworking, and agric. uses; clear, amber liq.; sol. in min. spirits and oil, aromatic solv, perchloroethylene; water-disp.; sp.gr. 0.97–0.99; HLB 7.2; sapon. no. 105–115; 100% act.

Alkamuls 400-DS. [Alkaril] PEG-8 distearate; nonionic; emulsifier and thickener for cosmetic and industrial emulsions; lubricant and softener in textile applic., opacifier and emulsifier in cosmetic preparations; wh. waxy flakes; sol. in min. oil and spirits, aromatic solv., perchloroethylene; water-disp.; HLB 7.8; sapon. no. 120–130; 100% act.

Alkamuls 400-ML. [Alkaril] PEG-8 laurate; nonionic; defoamer, leveling agent for latex paints; dispersant for pigments, dyes; emulsifier for cosmetics and toiletries; clear amber liq.; sol. (@ 10%) in aromatic solv., perchlorethylene; disp. in water, min. oil, min. spirits; sp.gr. 1.02–1.04; cloud pt. 50 F; HLB 12.8; sapon. no. 191–201; 100% act.

Alkamuls 400-MO. [Alkaril] PEG-9 oleate; nonionic; emulsifier, wetting agent, dispersant, lubricant used in dairy industry, cosmetic and industrial applics.; amber clear liquid; sol. in min. oil; disp. in water; sp.gr. 1.01–1.03; HLB 11.0; 100% act.

Alkamuls 400-MS. [Alkaril] PEG-8 stearate; nonionic; see Alkamuls 400-DL; also textile lubricant, softener; wh. paste; sol. in aromatic solv., perchloroethylene; water-disp.; HLB 11.2; sapon. no. 80–90; 100% act.

Alkamuls 600-DL. [Alkaril] PEG-12 dilaurate; nonionic; see Alkamuls 400-DL; yel. liq., paste; sol. in water, aromatic solv., perchloroethylene; dens. 1.03 g/ml; HLB 11.5; sapon. no. 106–116; 100% conc.

Alkamuls 600-DO. [Alkaril] PEG-12 dioleate; nonionic; see Alkamuls 400-DL; amber liq.; sol. in min. spirits, aromatic solvs., perchloroethylene; disp. in min. oil; sp.gr. 1.01–1.03; HLB 10.0; sapon. no. 99–104; 100% act.

Alkamuls 600-DS. [Alkaril] PEG-12 distearate; nonionic; see Alkamuls 200-DS; cream solid; sol. in min. oil and spirits, aromatic solv, perchloroethylene; disp. in water; HLB 10.6; sapon. no. 96–106; 100% conc.

Alkamuls 600-ML. [Alkaril] PEG-12 laurate; nonionic; see Alkamuls 400-ML; yel. liq. to paste; sol. (@ 10%) in water; disp. in min. oil, min. spirits, aromatic solv., perchlorethylene; dens. 1.05 g/ml; HLB 14.6; sapon. no. 67–77; 100% act.

Alkamuls 600-MO. [Alkaril] PEG-12 oleate; nonionic; dispersant, o/w emulsifier for cosmetic and industrial applics.; liq.; HLB 13.0; 100% act.

Alkamuls 600-MS. [Alkaril] PEG-12 stearate; nonionic; defoamer, leveling agent for latex paints; dispersant and dye leveling agent for textiles; wh. solid; sol. @ 10% in water, perchlorethylene; disp. in min. oil, min. spirits; HLB 13.2; sapon. no. 60–70; 100% act.

Alkamuls AG-900, -902, -906. [Alkaril] Nonionic ethoxylate blend; spreading and wetting agents for aq. pesticide systems; clear liq.; HLB 10.0–14.0; pH 7.0–7.5 (5% DW); 92% act.

Alkamuls AG-903. [Alkaril] Anionic-nonionic

blend; foaming adjuvant with wetting, spreading, and sticking properties for agric. applics.; clear liq.; pH 6.0–8.0 (5% DW); 38% act.

Alkamuls AG-904, -905, -907. [Alkaril] Anionic-nonionic blend; agric. spreading and wetting agent; clear liq.; pH 6.0–8.0 (5% DW); 84, 75, and 83% act. resp.

Alkamuls DEG-DO. [Alkaril] DEG dioleate; nonionic; w/o emulsifier, dispersant, defoamer, coemulsifier, and lubricant in industrial and textile oils; as paper softener; amber liq.; sol. in min. oil, min. spirits, aromatic solv., perchlorethylene; disp. in water; dens. 0.93 g/ml; HLB 2.9; sapon. no. 174–184; 100% conc.

Alkamuls DEG-DT. [Alkaril] DEG ditallate; nonionic; dispersing agent, emulsifier; br. liq.; faint odor; 100% act.

Alkamuls DEG-ML. [Alkaril] DEG laurate; nonionic; defoamer, dispersant, emulsifier, lubricant for industrial and textile oils; coemulsifier; yel. liq. to paste; disp. (@ 10%) in min. oil, min. spirits, aromatic solv., perchlorethylene; insol. water; dens. 0.96 g/ml; HLB 6.0; sapon. no. 190–200; 100% act.

Alkamuls DEG-MO. [Alkaril] DEG oleate; nonionic; coemulsifier for w/o and o/w emulsions; lubricant component in textile processing and leather tanning; yel. liq.; sol. (10%) in min. spirits; disp. in min. oil, aromatic solv., perchloroethylene; insol. water; dens. 0.95 g/ml; HLB 3.5; sapon. no. 140–150; 100% act.

Alkamuls DEG-MS. [Alkaril] DEG stearate; nonionic; opacifier, pearlescent, emulsifier, plasticizer for liq. cosmetics and detergent prods.; brn. liq.; faint odor; HLB 4.7; sapon. no. 154–164; 100% act.

Alkamuls DTDA. [Alkaril] Diester of adipic acid and tridecyl alcohol; nonionic; syn. lubricant characterized by wide operating temp. range and high visc. index; liq.; 100% act.

Alkamuls EGMS. [Alkaril] PEG stearate; nonionic; opacifying and pearlescing agent for liq. cosmetic and detergent compds.; wh. solid; insol. in water; HLB 2.9; sapon. no. 174–184.

Alkamuls GML-45. [Alkaril] Glyceryl laurate; nonionic; emulsifier, lubricant, mold release agent, antistat in cosmetic and industrial applics.; liq. to paste; insol. in water; dens. 0.97 g/ml; HLB 3; 40% min. monoglyceride.

Alkamuls GMO-45. [Alkaril] Glyceryl oleate; nonionic; mold release agent and rust preventive additive for compded. oils; lubricant in syn. fiber spin finishes; processing aid for PVC film; emulsifier, antistat in cosmetic and industrial applics.; amber liq. to paste; sol. in aromatic solv., perchloroethylene; dens. 0.93–0.98 g/ml; HLB 3; 40% min. monoglyceride.

Alkamuls GMO-45LG. [Alkaril] Glyceryl oleate; see Alkamuls GMO-45; amber liq. to paste; sol. in aromatic solv., perchloroethylene; insol. in water; dens. 0.93 g/ml; HLB 3; 40% min. monoglyceride.

Alkamuls GMO-55LG. [Alkaril] Glyceryl oleate; processing aid for PVC film functioning as an internal lubricant, antistat, and antifogging agent; emulsifier for cosmetic and industrial applics.; amber liq., paste; sol. in aromatic solv., perchloroethylene; dens. 0.96 gl/ml; HLB 3; 50% min. monoglyceride.

Alkamuls GMR. [Alkaril] Glyceryl ricinoleate; nonionic; emulsifier, stabilizer, opacifier for textiles, leather, foods; liq.; 100% act.

Alkamuls GMS-45. [Alkaril] Glyceryl stearate; nonionic; emulsifier, lubricant, antistat for personal care and industrial prods.; textile lubricant-softener; off-wh. flakes; insol. in water; HLB 3; 40% min. monoglyceride.

Alkamuls GMS. [Alkaril] Glyceryl stearate; nonionic; emulsifier, wetting agent for cosmetic, agric., textile industries; coupler used to bind waxes together; emollient and thickener in cosmetic creams; solid; water-disp.; HLB 3.4; 100% conc.

Alkamuls GTO. [Alkaril] Glyceryl trioleate; nonionic; lubricant for textiles, leather, and metals; amber liq.; sol. in min. oil and spirits, aromatic solv., perchloroethylene; insol. in water; dens. 0.95 g/ml; HLB 0.8; 100% conc.

Alkamuls PEMO. [Alkaril] Pentaerythritol oleate; corrosion inhibitor; deposits residual rust preventive film to protect automotive and other steel parts; liq.

Alkamuls PSML-4. [Alkaril] Polysorbate 21; nonionic; emulsifier and coupling agent; PVC emulsion polymerization; yel. liq.; disp. in water; dens. 1.0 g/ml; HLB 13.3; sapon. no. 100–115; hyd. no. 215–255; 100% conc.

Alkamuls PSML-20. [Alkaril] Polysorbate 20; nonionic; emulsifier, solubilizer, antistat, lubricant for textiles, cosmetics, pharmaceuticals; yel. liq.; sol. in water, aromatic solv.; dens. 1.1 g/ml; HLB 16.7; sapon. no. 40–50; 97% act.

Alkamuls PSMO-5. [Alkaril] Polysorbate 81; nonionic; emulsifier, solubilizer, antistat, lubricant for paint, food, cosmetic, insecticides, herbicides, fungicides, textiles, cutting oils; amber liq. to paste; dens. 1.0 g/ml; HLB 10; sapon. no. 96–104; 98.5% act.

Alkamuls PSMS-4. [Alkaril] Polysorbate 61; nonionic; emulsifier; fiber-to-metal lubricant for fibers and yarns; used in suppositories in pharmaceutical industry; tan solid; typ. odor; water-disp.; dens. 1.1 g/ml; HLB 9.6; sapon. no. 98–113; 97% act.

Alkamuls PSMS-20. [Alkaril] Polysorbate 60; nonionic; wetting agent, emulsifier for cosmetic and food applics., textiles, paper coatings; fiber-to-metal lubricant for fibers and yarns; yel. visc. liq., gel on standing; typ. odor; water disp.; dens. 1.1 g/ml; HLB 14.9; sapon. no. 45–55; 97% act.

Alkamuls PSTO-20. [Alkaril] Polysorbate 85; nonionic; emulsifier for cosmetic and food applics.; textile and leather lubricant; amber liq.; typ. odor; water disp.; dens. 1.0 g/ml; HLB 11; sapon. no. 80–95; 97% act.

Alkamuls PSTS-20. [Alkaril] Polysorbate 65; nonionic; wetting agent, emulsifier for cosmetic and food formulations; lubricant, softener for textiles; tan solid; typ. odor; sol. in min. oils, min. spirits; water disp.; dens. 1.0 g/ml; HLB 10.5; sapon. no. 88–98; 97% act.

Alkamuls SML. [Alkaril] Sorbitan laurate; nonionic; emulsifier for oils and fats in cosmetic and industrial oil prods.; corrosion inhibitor; antistat for PVC; amber liq.; typ. odor; disp. in water; moderately sol. most alcohols, veg. and min. oils; sp.gr. 1.05 (60 F); HLB 8.6; sapon. no. 160–170; hyd. no. 320–350; 97% act.

Alkamuls SMO. [Alkaril] Sorbitan oleate; nonionic; emulsifier, coupling agent, wetting agent for medicants, petrol. oils, fats, and waxes in the industrial, textile, and cosmetic industries; textile and leather lubricant and softener; corrosion inhibitor; amber liq.; sol. in most veg., min. oils, aromatic solv., perchloroethylene; insol. in water; sp.gr. 1.0; HLB 4.3; sapon. no. 145–160; hyd. no. 193–210; 97% act.

Alkamuls SMS. [Alkaril] Sorbitan stearate; nonionic; emulsifier and coupling agent; used to prepare silicone defoamer emulsions for industrial applics., paraffin wax emulsions for processing paper coatings; textile process lubricant; internal PVC film lubricant; cosmetics; foods; cream flakes; sol. (10%) in aromatic solv., perchloroethylene; dens. 1.0 g/ml; HLB 4.7; sapon. no 147–157; hyd. no. 235–260; 98.5% act.

Alkamuls STO. [Alkaril] Sorbitan trioleate; nonionic; emulsifier and coupling agent; used to compd. textile and leather softener finishes; amber liq.; typ. odor; sol. in min. oil and spirits, aromatic solv., perchloroethylene; insol. in water; dens. 1.0 g/ml; HLB 1.8; sapon. no. 170–190; hyd. no. 55–70; 100% act.

Alkamuls STS. [Alkaril] Sorbitan tristearate; hydrophobic emulsifier for use as a fiber-to-metal lubricant for syn. and cotton fibers; cosmetics, foods; cream flakes; typ. odor; sol. (10%) in aromatic solv., perchloroethylene; insol. in water; dens. 1.0 g/ml; HLB 2.1; sapon. no. 176–188; hyd. no. 66–80; 100% act.

Alkanol 189-S. [DuPont] Sodium alkyl sulfonate; anionic; wetting agent, detergent, penetrant; reddish-br. liq.; visc. 30 cps; surf. tens. 31.2 dynes/cm (1%, 90F); 31.5% act.

Alkanol A-CN. [DuPont] EO condensate; antiprecipitant, dyeing assistant, dye solubilizer, contrast agent, nonionic surfactant; lt. amber liq.; 50% sol. in water; dens. 8.2 lb/gal; surf. tens. 38 dynes/cm (1% conc.).

Alkanol CNR. [DuPont] Amphoteric surfactant; dye assistant; amber-colored clear liq.; misc. with water; dens. 8.61 lb/gal; sp.gr. 1.033.

Alkanol ND. [DuPont] Alkyl diaryl sulfonate; anionic; dyeing assistant, surfactant; yel. liq.; terpene odor; misc. with water; dens. 9.6 lb/gal; visc. slight; 45% act.

Alkanol S. [DuPont] Sodium tetrahydronaphthalene sulfonate; anionic; dispersant, stabilizer, solubilizing agent; lt. cream-colored, small flakes; water sol. 20%; dens. 2.6 lb/gal; 98% act.

Alkanol WXN. [DuPont] Sodium alkylaryl sulfonate; anionic; wetting, rewetting agent, dyeing assistant; lt. yel. liq.; alcoholic odor; misc. in water; dens. 8.6 lb/gal; cloud pt. 0 C max.; surf. tens. 26 dynes/cm (1%); 30% act.

Alkanol XC. [DuPont] Sodium alkylnaphthalene sulfonate; anionic; wetting agent, dispersant, penetrant; used in bleaching and dyeing of textiles, leather, paper; paper and paper mill felts; reduces shrinkage in ceramics mfg.; dry colors mfg.; lt. cream powd.; sol. in ethyl alcohol, acetone, benzene; 8–10% in water; sp.gr. 0.55.

Alkaphos 3. [Alkaril] Aliphatic phosphate ester, acid form; anionic; emulsifier, wetting agent, antistat, solubilizer, detergent in dry cleaning soap; lubricant for industrial coolant formulation and textile yarn; pale yel. visc. liq.; low odor; sol. in min. oil and spirits, aromatic solv., perchloroethylene; disp. in water; dens. 1.03 g/ml; visc. 863 cps; 100% act.

Alkaphos 6. [Alkaril] Mixed base phosphate ester, acid form; anionic; emulsifier, wetting agent, antistat used in textile operations, dry cleaning charge soap formulation; solv. degreaser; lubricant in industrial coolant formulations; yel. liq.; sol. in min. spirits, aromatic solv., perchloroethylene; disp. in water; dens. 1.06 g/ml; visc. 1045 cps; 100% act.

Alkaphos 10. [Alkaril] Free acid of complex org. phosphate esters; anionic; see Alkaphos 3; sp.gr. 1.085; visc. 2100 cps; 100% act.

Alkaphos B6-56A. [Alkaril] Complex phosphate ester free acid; detergent with antisoil redeposition and antistat properties; for dry cleaning formulations; antistatic lubricant for textiles; sol. in hydrocarbon and perchlorethylene solvs.; water sol./disp.

Alkaphos K380. [Alkaril] Aliphatic phosphate ester, potassium salt; anionic; solubilizer, hydrotropic surfactant for industrial cleaners; clear yel. liq.; water-sol.; dens. 1.04 g/ml; cloud pt. < 0 C; 44% act.

Alkaphos L3-64A. [Alkaril] Complex phosphate ester free acid; see Alkaphos B6-56A; sol. in hydrocarbon and perchloroethylene solvs.

Alkaphos L6-36S. [Alkaril] Complex phosphate ester sodium salt; anionic; solubilizer for surfactants and alkaline salts in cleaners for consumer and industrial applics.; liq.; 41% conc.

Alkaphos QS. [Alkaril] Aliphatic phosphate ester; acid form; anionic; detergent, emulsifier, wetting agent, and solubilizer for compded. highly alkaline and high electrolyte content liq. cleaning preparations; yel. liq.; bland odor; sol. in water, caustic soda, caustic potash; dens. 1.10 g/ml; cloud pt. 68 C (1% in 15% NaOH); 80% act.

Alkaphos R9-47A. [Alkaril] Complex phosphate ester free acid; coupler for nonionics in liq. alkaline systems; wetting agent for flake or granular caustic soda; lubricant for water based systems; water-sol.

Alkapol PEG 200. [Alkaril] PEG 200; intermediate for processing nonionic surfactants; binders and lubricants for compressed tablets; plasticizers in starch pastes and polyethylene sheet; paper softener; humectant and solv. for dyes in stamp pad inks; solv. in cosmetics and industrial applics.; antistat, lubricant in textile and plastic processing; liq., solid; m.w. 190-210; sol. in water @ 10%; dens. 1.13 g/ml; pH 5-8 (5% DW); 100% conc.

Alkapol PEG 300. [Alkaril] PEG 300; see Alkapol PEG 200; liq.; m.w. 285-315; water-sol.; dens. 1.13 g/ml; pH 5-8 (5% DW); 100% conc.

Alkapol PEG 400. [Alkaril] PEG-9; intermediate,

plasticizer, solv., lubricant, coupler in cosmetic lotions; wh. liq.; water sol.; m.w. 380–420; sp.gr. 1.13; f.p. 4-8 C; pH 5–8; 100% conc.

Alkapol PEG 600. [Alkaril] PEG-14; plasticizer, solv.; lubricant, binder for pharmaceuticals; color stabilizer for fuel oils; intermediate for processing surfactants; wh. liq.; m.w. 570–630; water sol.; sp.gr. 1.13 (30/15.5 C); f.p. 20-25 C; pH 5–8; 100% conc.

Alkapol PEG 1000. [Alkaril] PEG 1000; solv. or cosolv. in industrial applics.; humectant used in food, pharmaceutical and cosmetic applics.; emollient, lubricant used to make suppositories; solid; m.w. 950-1050; sol. in water @ 10%; pH 5-8 (5% DW); 100% conc.

Alkapol PEG 1500. [Alkaril] PEG 1500; rubber industry lubricant for calendering the latex onto a textile substrate; lubricant for rayon hosiery knitting; solv. or cosolv. in industrial applics.; solid; m.w. 1400-1600; sol. in water @ 10%; pH 5-8 (5% DW); 100% conc.

Alkapol PEG 3350. [Alkaril] PEG 3350; mold release agent, antistat for rubber prods.; binder for tablets in pharmaceuticals; waxy flake; m.w. 3000-3700; sol. in water @ 10%; pH 5-8 (5% DW); 100% conc.

Alkapol PEG 6000. [Alkaril] PEG 6000; binding agent in abrasive formulations; mold release agent for rubber articles; chemical intermediate; waxy flake; m.w. 5400-6600; sol. in water @ 10%; pH 5-8 (5% DW); 100% conc.

Alkapol PEG 8000. [Alkaril] PEG 8000; base for water-sol. sticks; spreader and binder in aq. shoe polish formulations, lubricant and binder aid for ceramic extrusion; waxy flake; m.w. 7000-9000; sol. in water @ 10%; pH 5-8 (5% DW); 100% conc.

Alkapol PPG-425. [Alkaril] PPG-425; intermediate for nonionic surfactants; lubricant, carrier; solv. or cosolv. for cosmetic, pharmaceutical, and industrial formulations; release agent for plastisols; liq.; m.w. 400-450; sol. in perchloroethylene and water; dens. 1.01 g/ml; pH 6-8; 100% conc.

Alkapol PPG-1000. [Alkaril] PPG-1000; see Alkapol PPG-425; liq.; 100% conc.

Alkapol PPG-1200. [Alkaril] PPG-1200; intermediate for nonionic surfactants; antifoam agent in emulsions; component in hydraulic fluids and greases; liq.; m.w. 1150-1250; sol. in perchloroethylene; dens. 1.01 g/ml; pH 6-8; 100% conc.

Alkapol PPG-2000. [Alkaril] PPG-2000; antifoam in industrial applics.; additive for tire lubricants; carrier or cosolvent in cosmetic preparations; liq.; m.w. 1900-2100; sol. in perchloroethylene, aromatic solv.; dens. 1.01 g/ml; pH 6-8; 100% conc.

Alkapol PPG-4000. [Alkaril] PPG-4000; see Alkapol PPG-2000; liq.; m.w. 3900-4100; sol. in perchloroethylene; dens. 1.01 g/ml; pH 6-8; 100% conc.

Alkaquat 1068. [Alkaril] Blended quat. compd.; tallow deriv.; softener conc.; pale yel. soft paste; sol. in min. oil and spirits, aromatic solv., perchloroethylene; disp. in water; dens. 0.94 g/ml; pH 5.0–7.5; 80.0–83.0% act.

Alkaquat C. [Alkaril] Coconut imidazoline quat.; cationic; antistat, fabric softener, dye retardant for textile industry, laundry detergents; amber liq.; sol. in aromatic solv., perchlorethylene; disp. in water, min. oil, min. spirits; dens. 0.99 g/ml; f.p. < 0 C; pH 6.0–6.8; 74–76% act.

Alkaquat DAET. [Alkaril] Ditallow quat. compd.; cationic; substantive antistat, softener for home and commercial laundering; yel. paste; sol. see Alkaquat 1068; dens. 0.99 g/ml; pH 4.0–7.0; 89.0–91.0% act.

Alkaquat DAPT. [Alkaril] Ditallow quat. compd.; antistat; high conc. household and commerical laundry softeners; amber liq.; sol. see Alkaquat 1068; dens. 0.96 g/ml; pH 4.0–7.0; 89.0–91.0% act.

Alkaquat DHTS. [Alkaril] Dialkyl dimethyl methosulfate quat. compd.; hydrog. tallow deriv.; conc. used in mfg. of dryer sheet softeners; almost wh. solid; sol. in aromatic solv., perchloroethylene; water-disp.; pH 5.5–7.5; 88.0–90.0% act.

Alkaquat DMB-451. [Alkaril] N-alkyl (C_{12}-C_{14}-C_{16}) methyl benzyl ammonium chloride; cationic; wetting agent, emulsifier, biocide, disinfectant for use in beverage industry, dairy industry, food processing, water treatment, paper industry, pest control, preservatives, antidandruff rinses, general disinfection and sanitization for hospitals, laundries; pale-yel. liq.; sol. in water, ethanol, acetone, aliphatic solv.; sp.gr. 0.96; surf. tens. 33 dynes/cm (1%); 50% min. act.in ethanol (10%), water (40%).

Alkaquat DMB-ST, 25%. [Alkaril] Alkyl benzyl dimethyl quat., stearic deriv.; strongly cationic; antistat, conditioner, softener for fibers, hair, paper prods.; wh. paste; sol. in water; dens. 0.95 g/ml; pH 3.0–5.0 (0.5% aq.); 24–26% act.

Alkaquat O. [Alkaril] Oleic imidazoline quat.; cationic; antistat, fabric softener for commercial and institutional use; laundry detergent; amber liq.; solv. odor; sol. see Alkaquat 1068; sp.gr. 0.95; f.p. < 0 C; pH 6.0–6.8; 75% act.

Alkaquat S. [Alkaril] Soya imidazoline quat.; antistat, fabric softener for textile applics.; amber liq.; sol. see Alkaquat 1068; dens. 0.98 g/ml; pH 6.0-6.8; 74–76% act.

Alkaquat ST. [Alkaril] Modified dialkyl imidazolinium quat., stearyl deriv.; fabric softener and antistat for textile fibers; cream solid; IPA solv. odor; sol. in min. oil, aromatic solv., perchlorethylene; disp. in water, min. spirits; m.p. 45 C; pH 6.0–6.8 (10% in IPA); 74–76% act.

Alkaquat T. [Alkaril] Ditallow imidazolinium quat.; cationic; textile surfactant, softening agent, antistat in home and commercial laundries; pale yel. slightly visc. fluid; typ. odor; sol. in min. oil, perchlorethylene; disp. in water, min. spirits, aromatic solv.; sp.gr. 0.95; pH 6.0–6.8; 74-76% act.

Alkaquest EDTA. [Alkaril] Tetrasodium EDTA; chelating agent for general purpose applics.; straw liq.; m.w. 380; sp.gr. 1.280; 37.5% act.

Alkaril SRP-C. [Alkaril] POE terephthalate polymer; cationic; textile antisoil, oil scavanger plus substantive hand and softening; milky emulsion.

Alkasperse A-2H, A-5H, A-10H. [Alkaril] Polyacrylic acid; dispersant for scale control; boiler water compds.; A-5H also mud dispersant; liq.; water-sol.

Alkasperse A-20. [Alkaril] Polyacrylate; dispersant for boiler water compds.; water-sol.

Alkasperse DM-5, MM-2. [Alkaril] Polycarboxylate; paint dispersant; water-sol.

Alkasurf 075-5, 075-7, 075-9. [Alkaril] Oleic acid ethoxylated; nonionic; emulsifier, lubricant, antistat for textile fiber additives; liq.; water-disp. (075-7, 075-9); HLB 8.0, 10.0, and 11.0 resp.; 100% conc.

Alkasurf BA-PE 70, BA-PE 80. [Alkaril] Primary aliphatic alcohol alkoxylate, straight chain; nonionic; surfactant for emulsification and dispersion applications in insecticides/herbicides, latex polymerization and in leather finishing; soft solid (BA-PE 70); HLB 16.1, 26.1 resp.; 100% conc.

Alkasurf CA. [Alkaril] Calcium dodecylbenzene sulfonic acid; anionic; industrial o/w emulsifier; dispersant in polyester yarn dyeing; used in agric. chemicals; amber liq.; sol. in min. oil, min. spirits, aromatic solv., perchlorethylene; disp. in water; dens. 1.00 g/ml; HLB 10.5; pH 6.0–8.0 (5% DW); 59–61% act.

Alkasurf CO-5. [Alkaril] PEG-5 castor oil; nonionic; emulsifier and dispersant with antifoaming properties for chlorinated solv.; pigment dispersant in latex paints; emulsifier in textile processing; Gardner 4 liq.; disp. in water; dens. 0.98 g/ml; HLB 3.8; sapon. no. 145-150.

Alkasurf CO-10. [Alkaril] PEG-10 castor oil; nonionic; see Alkasurf CO-5; Gardner 4 liq.; disp. in water; dens. 1.00 g/ml; HLB 7.2; cloud pt. 45-49 C; sapon. no. 121-126; 100% conc.

Alkasurf CO-15. [Alkaril] PEG-15 castor oil; nonionic; dispersant, detergent, wetting agent, defoamer, antistat; emulsifier in syn. fiber lubricants; vat dyeing assistant in the textile industry; in cutting oils and hydraulic fluids; Gardner 4 liq.; disp. in water; dens. 1.02 g/ml; HLB 8.6; cloud pt. 56-60 C; sapon. no. 104-109; 100% conc.

Alkasurf CO-20. [Alkaril] PEG-20 castor oil; nonionic; emulsifier, lubricant, softener, coemulsifier, antistat, degreaser for textiles, leather processing; pigment dispersants; Gardner 4 liq.; sol. in water, aromatic solv., perchloroethylene; dens. 1.02 g/ml; HLB 10.3; cloud pt. 65-69 C; sapon.no. 91-96; 100% conc.

Alkasurf CO-25. [Alkaril] PEG-25 castor oil; nonionic; detergent, wetting agent, emulsifier, lubricant for textiles, cosmetics; yel. liq., typ. odor; sol. see Alkasurf CO-20; dens. 1.03 g/ml; HLB 11.0; sapon. no. 81–86; cloud pt. 46–50 C; 100% act.

Alkasurf CO-30. [Alkaril] PEG-30 castor oil; nonionic; emulsifier, lubricant for textiles; Gardner 4 liq.; sol. see Alkasurf CO-20; dens. 1.04 g/ml; HLB 11.7; cloud pt. 48-61 C; sapon. no. 73-78; 100% conc.

Alkasurf CO-40. [Alkaril] PEG-40 castor oil; nonionic; emulsifier for oils and waxes; coemulsifier for dye carriers in textiles; lubricant for textile/leather processing; liq.; water-sol.; HLB 13.0; 100% conc.

Alkasurf CO-40M. [Alkaril] PEG-40 castor oil; nonionic; lubricant and softener for textiles; coemulsifier for dye carrier systems; Gardner 4 liq.; sol. in water, aromatic solv., perchloroethylene; dens. 1.04 g/ml; HLB 13.0; cloud pt. 65-71 C; sapon.no. 58-63; 90% conc.

Alkasurf CO-200. [Alkaril] PEG-200 castor oil; nonionic; emulsifier used as lubricant and antistatic humectant for syn. fiber processing in the textile industry; Gardner 4 solid; water-sol.; HLB 18.1; cloud pt. 78-84 C; sapon. no. 12-18; 50% conc.

Alkasurf DA-3, -4 -6. [Alkaril] Deceth-3, -4, -6; nonionic; wetting agent and penetrant used in pressure dyeing of fabrics and textile or industrial applics.; hazy liq.; water-disp.; HLB 9.0, 10.5, and 12.5 resp.; 100% conc.

Alkasurf L-2. [Alkaril] PEG-2 laurate; nonionic; coemulsifier, dispersing agent, defoamer, emulsifier, and lubricant for industrial and textile oils; yel. liq. to paste; insol. in water; dens. 0.96 g/ml; HLB 6.0; sapon. no. 190-200.

Alkasurf L-5. [Alkaril] PEG-5 laurate; nonionic; emulsifier, coupling agent, solubilizer in metal working fluids, industrial and textile lubricants; yel. liq.; disp. in water; dens. 1.00 g/ml; HLB 9.8; sapon.no. 145-152.

Alkasurf L-9. [Alkaril] PEG-9 laurate; nonionic; emulsifier and coemulsifier for cosmetic and toiletry preparations; defoamer and leveling agent for latex paints; dispersant for pigments and dyes; yel. liq. to paste; sol. in aromatic solv., perchloroethylene; disp. in water; dens. 1.03 g/ml; HLB 12.8; sapon. no. 91-101.

Alkasurf L-14. [Alkaril] PEG-14 laurate; see Alkasurf L-9; yel. liq.to paste; sol. in water; dens. 1.05 g/ml; HLB 14.6; sapon. no. 67-77.

Alkasurf LA Acid. [Alkaril] Dodecylbenzene sulfonic acid; anionic; detergent, wetter, emulsifier, penetrant, foamer, intermediate used in liq. dishwashing detergents, cleaners, personal care prods., and degreasers; br. visc. liq.; sol. in water, min. oil, perchloroethylene; dens. 1.06 g/ml; biodeg.; 96% min. act.

Alkasurf LA-1. [Alkaril] Fatty alcohol (C_{12}-C_{15}) ethoxylate; nonionic; detergent, intermediate, emulsifier, dispersant, wetting agent for mfg. of shampoos; dye retardant in hair color preparations; sol. in min. oil and spirits, aromatic solv., perchloroethylene; insol. in water; dens. 0.88 g/ml; HLB 3.6; 100% conc.

Alkasurf LA-3. [Alkaril] C12–15 pareth-3; nonionic; emulsifier, detergent base; for dishwashing, personal care prods., industrial applics.; textile lubricant; translucent to opaque liq.; sol. in min. oil, aromatic solv, perchloroethylene; sp.gr. 0.93; visc. 19 cps (100 F); HLB 7.8; cloud pt. 58-62 C; 100% act.

Alkasurf LA-7. [Alkaril] C12–15 pareth-7; nonionic; detergent, emulsifier, wetting agent, dispersant for sanitizers, metal cleaners, hard surface cleaners; lt. liq.; water sol.; sp.gr. 0.96; visc. 32 cps (100 F); m.p. 11-15 C; HLB 12; cloud pt. 44-48 C; flash pt. 440 F (COC); pour pt. 70 F; biodeg.; 100% act.

Alkasurf LA-EP15. [Alkaril] Oxyethylated straight chain alcohol, modified; nonionic; detergent, dispersant, wetting agent, emulsifier used in machine dishwasher formulations, metal cleaners, household and industrial formulations; liq.; sol. in water, perchloroethylene; HLB 11.9; cloud pt. 25-30 C (1% DW); pH 5-8 (5%); biodeg.

Alkasurf LA-EP16. [Alkaril] Oxyethylated straight chain alcohol, modified; nonionic; detergent, wetting agent, emulsifier and dispersant for industrial and household cleaners; liq.; sol. in water, aromatic solv.; HLB 13.0; cloud pt. 56-63 C (1% DW); pH 5-8 (5%); 100% conc.

Alkasurf LA23-3. [Alkaril] Fatty alcohol (C_{12}-C_{13}) ethoxylate; nonionic; detergent, emulsifier, intermediate, dispersant for mfg. of dishwashing, shampoo, bubble bath, and specialty industrial preparations; liq.; sol. in min. oil, aromatic solv., perchloroethylene; disp. in water; dens. 0.93 g/ml; HLB 8.1; 100% conc.

Alkasurf LAN-1. [Alkaril] Laureth-1; nonionic; detergent-intermediate for mfg. of shampoo ethoxylates; liq.; sol. in min. oil and spirits, aromatic solv., perchloroethylene; insol. in water; dens. 0.88 g/ml; HLB 3.7; cloud pt. 37-39 C; 100% conc.

Alkasurf LAN-23. [Alkaril] Laureth-23; nonionic; emulsifier, stabilizer, solubilizer, surfactant for cosmetics; post ad stabilizer for syn. latexes; almost colorless solid; char. odor; sol. in water, aromatic solv., perchloroethylene; HLB 17; cloud pt. 90.5–93.5 C; biodeg.; 100% act.

Alkasurf NP-1. [Alkaril] Nonoxynol-1; nonionic; emulsifier and dispersing agent for petroleum oils; coemulsifier and retardant in hair care formulations; defoamer; liq.; sol. see Alkasurf LAN-1; dens. 0.99; HLB 4.6; pH 5-8 (5% DW); 99.0% min. act.

Alkasurf NP-4. [Alkaril] Nonoxynol-4; nonionic; emulsifier, detergent, dispersant, intermediate, stabilizer; plasticizer, antistat for plastics, surfactants, household, industrial, and cosmetic use, fat liquoring, cutting and sol. oils; lt. liq.; low odor; water insol.; sp.gr. 1.02; HLB 9; 100% act.

Alkasurf NP-5. [Alkaril] Nonoxynol-5; nonionic; emulsifier, coemulsifier, dispersant; lt. liq.; low odor; oil-sol., water insol.; sp.gr. 1.03; HLB 10; 100% act.

Alkasurf NP-6. [Alkaril] Nonoxynol-6; nonionic; emulsifier, coemulsifier, oil-sol. dispersant; used for household and industrial cleaners; intermediate; plasticizer and antistat for plastics; emulsifier for min. oils; insecticides, fungicides, herbicides; fat liquoring; making of cutting and sol. oils; dispersing waxes, pigments, resins; printing; preparation of emulsified paint; lt. liq.; low odor; oil-sol., water insol.; sp.gr. 1.04; HLB 11; 100% act.

Alkasurf NP-9. [Alkaril] Nonoxynol-9; nonionic; detergent, wetting agent, emulsifier, dispersant for household and industrial cleaners, textile processing, laundry detergent, pesticides; lt. liq.; low odor; water sol.; sp.gr. 1.06; HLB 13.4; 100% act.

Alkasurf NP-10. [Alkaril] Nonoxynol-10; nonionic; detergent, wetting agent, emulsifier, solubilizer, dispersant for textiles, household and industrial cleaners, antimicrobials; lt. liq.; low odor; sol. in water, aromatic solv, perchloroethylene; sp.gr. 1.06; HLB 13.5; cloud pt. 62–66 C; pH 5–8; 100% act.

Alkasurf NP-20, NP-20 70%. [Alkaril] Nonoxynol-20; nonionic; wetting agent, stabilizer, dispersant, penetrant; textile scouring agent; coemulsifier for oils, fats, waxes, and solv.; demulsifier for petroleum oil emulsions; wh. solid; sol. in water; dens. 1.08 g/ml; HLB 16.0; cloud pt. 71 C; pH 5-8 (5% DW); 99 and 70% min. act. resp.

Alkasurf NP-30 70%. [Alkaril] Nonylphenol ethoxylate; nonionic; solubilizer and coemulsifier for highly polar substances; liq.; sol. in water; dens. 1.09 g/ml; HLB 17.1; cloud pt. 74 C; pH 5-8 (5% DW); 70% min. act.

Alkasurf NP-35 70%. [Alkaril] Nonylphenol ethoxylate; nonionic; emulsifier and stabilizer for emulsion polymerization for latex paint, floor finishes, paper coatings, and textile finishes; liq.; sol. in water; dens. 1.09 g/ml; HLB 17.4; cloud pt. 76 C; pH 5-8 (5% DW); 70% min. act.

Alkasurf NP-40. [Alkaril] Nonoxynol-40; wetting agent, stabilizer, penetrant, emulsifier, dispersant, nonionic; lt. liq.; low odor; water sol.; f.p. 50-60F; 70% act.

Alkasurf NP-40 70%. [Alkaril] Nonoxynol-40; nonionic; emulsifier, stabilizer, surfactant; coemulsifier and vinyl/acrylic latex stabilizer; liq.; water-sol.; dens. 1.09 g/ml; HLB 17.6; cloud pt. 76 C (1% in 10% NaCl); pH 5–8 (5% DW); 70% conc.

Alkasurf NP-50 70%. [Alkaril] Nonoxynol-50; coemulsifier and vinyl/acrylic latex stabilizer; liq.; sol. in water; dens. 1.09 g/ml; HLB 18.0; cloud pt. 76 C; pH 5-8 (5% DW); 70% conc.

Alkasurf O-2. [Alkaril] PEG-2 oleate; nonionic; coemulsifier for w/o and o/w emulsions; lubricant component in textile processing and leather tanning; cutting oils, solv. cleaners, degreasers, pesticides; yel. liq.; sol. in min. spirits; insol. in water; dens. 0.95 g/ml; HLB 3.5; sapon. no. 140-150; 100% conc.

Alkasurf O-5. [Alkaril] PEG-5 oleate; nonionic; w/o emulsifier for min. oils, fatty acids and solv.; used in cutting oils, solv. cleaners, degreasers, pesticide formulations, as a lubricant softener in the tanning process and in lubricants and softeners for the textile industry; yel. liq.; disp. in water; dens. 0.99 g/ml; HLB 8.0; sapon. no. 111-121; 100% conc.

Alkasurf O-7. [Alkaril] PEG-7 oleate; nonionic; see Alkasurf O-2; liq.; HLB 10.0; 100% conc.

Alkasurf O-9. [Alkaril] PEG-9 oleate; nonionic; surfactant used as a dyeing assistant in the textile industry and as an emulsifier for neats-foot oil fat liquors in leather processing; yel. liq.; sol. in aromatic solv., perchloroethylene; disp. in water; dens. 1.02 g/ml; HLB 11.0; sapon. no. 75-85; 100% conc.

Alkasurf OA-2. [Alkaril] Oleth-2; nonionic; emulsifier and solubilizer for topical cosmetics; liq.; sol. in min. oil, aromatic solv., perchloroethylene; insol. in water; dens. 0.90 g/ml; HLB 4.9; cloud pt. 47-50 C; 100% conc.

Alkasurf OA-10. [Alkaril] Oleth-10; see Alkasurf OA-2; paste; sol. in water, aromatic solv.; HLB 12.4; cloud pt. 53-60 C (1% DW); 100% conc.

Alkasurf OA-20. [Alkaril] Oleth-20; nonionic; emulsifier, dispersant, solubilizer and detergent; dyeing assistant for wool/acrylic blends; wh. solid; sol. in water, aromatic solv.; HLB 15.3; cloud pt. 73-77 C; 100% conc.

Alkasurf OP-1. [Alkaril] Octoxynol-1; nonionic; emulsifier and dispersant for petroleum oils; coemulsifier and retardant in hair color preparations; coupling agent; liq.; sol. in min. oil and spirits, aromatic solv., perchloroethylene; dens. 0.99 g/ml; HLB 3.6; pH 5-8 (5% DW); 100% act.

Alkasurf OP-5. [Alkaril] Octoxynol-5; nonionic; emulsifier, dispersant, surfactant, dry cleaning detergents, insecticides and wax emulsions; lt.-colored liq.; char. odor; sol. in min. spirits, aromatic solv., perchloroethylene; insol. in water; sp.gr. 1.05 (15/15 C); HLB 10.4; cloud pt. 63–66 C; pH 5–8; 100% act.

Alkasurf OP-10. [Alkaril] Octoxynol-10; nonionic; detergent, emulsifier, wetting agent, dispersant for household and industrial cleaners, textile processing, wool scouring, metal cleaning; lt.-colored liq.; low odor; water sol.; sp.gr. 1.1 (15/15 C); HLB 13.5; 100% act.

Alkasurf OP-12. [Alkaril] Octoxynol-12 (12–13 EO); nonionic; detergent, emulsifier, solubilizer-coupler-hydrotrope, wetting agent, foam stabilizer for metal cleaning, industrial and household liq. detergents; lt.-colored liq.; low odor; sol. in water, aromatic solv., perchloroethylene; sp.gr. 1.07; HLB 14.6; cloud pt. 85–92 C; pH 5–8; 100% act.

Alkasurf OP-16. [Alkaril] Octoxynol-16; nonionic; detergent, solubilizer, coupler, hydrotrope, and foam stabilizer; bottle washing compds.; paste; sol. in water; dens. 1.08 g/ml; HLB 15.8; cloud pt. 67 C; pH 5-8; 100% act.

Alkasurf OP-30 70%. [Alkaril] Octoxynol-30; nonionic; detergent, emulsifier, wetting agent; coemulsifier and vinyl/acrylic latex stabilizer; lt.-colored liq.; low odor; water sol.; sp.gr. 1.1; HLB 17.3; 70% act.

Alkasurf OP-40 70%. [Alkaril] Octoxynol-40; nonionic; see Alkasurf OP-30, 70%; liq.; HLB 18.0; 70% conc.

Alkasurf S-1. [Alkaril] PEG-1 stearate; nonionic; coemulsifier, opacifying and pearlescing agents for liq. cosmetic and detergent prods.; wh. solid; insol. in water; HLB 2.9; sapon. no. 174-184; 100% conc.

Alkasurf S-2. [Alkaril] PEG-2 stearate; nonionic; see Alkasurf S-1; wh. solid; insol. in water; HLB 3.7; sapon. no. 154-164; 100% conc.

Alkasurf S-5. [Alkaril] PEG-5 stearate; nonionic; emulsifier for fats and oils; softener and lubricant for textiles and leather; wh. solid; disp. in water; HLB 8.0; sapon. no. 120-130; 100% conc.

Alkasurf S-8. [Alkaril] PEG-8 stearate; lubricant and softener in textile compositions for syn. fibers; wh. solid; sol. in aromatic solv., perchloroethylene; disp. in water; HLB 11.2; sapon. no. 80-90.

Alkasurf S-14. [Alkaril] PEG-14 stearate; nonionic; defoamer and leveling agent for latex paints; pigment dispersant and dye leveling agent for textiles; wh. solid; sol. in water, perchloroethylene, aromatic solv.; HLB 13.2; sapon. no. 60-70; 100% conc.

Alkasurf S-40. [Alkaril] PEG-40 stearate; nonionic; lubricant and emulsifier in production of textile lubricants and softeners; cosmetics, pharmaceuticals, silicone preparations; wh. flake; sol. in water @ 10%, in aromatic solv. @ 1%; HLB 17.0; sapon. no. 25-35; 100% conc.

Alkasurf S65-8, S65-40. [Alkaril] Stearic acid, ethoxylated; nonionic; see Alkasurf S-8 and S-40; solid and flake resp.; HLB 11.2 and 17.0; 100% conc.

Alkasurf SA-2. [Alkaril] Steareth-2; nonionic; emulsifier, solubilizer in topical cosmetic applics.; wh. solid; sol. in aromatic solv., perchloroethylene; HLB 4.9; cloud pt. 57-61 C; 100% conc.

Alkasurf SA-10. [Alkaril] Steareth-10; nonionic; see Alkasurf SA-2; wh. solid; sol. in water; HLB 12.4; cloud pt. 60-64 C; 100% conc.

Alkasurf SA-20. [Alkaril] Steareth-20; nonionic; detergent and lubricant for fiber and fabric scouring; emulsifier for topical cosmetics; aids disp. and suspension in roll-on deodorants; wh. solid; sol. in water @ 10%, in aromatic solv. and perchloroethylene @ 1%; HLB 15.3; cloud pt. 73-77 C; 100% conc.

Alkasurf SS-CDE. [Alkaril] Sodium cocamido DEA sulfosuccinate; anionic; foamer, foam stabilizer used in personal care prods. and liq. detergent preparations; amber liq.; sol. in water; dens. 1.1 g/ml; pH 5.0-6.0; 39-41% conc.

Alkasurf SS-DA-3. [Alkaril] Sodium deceth-3 sulfosuccinate; anionic; emulsifier, dispersant used in emulsion polymerization of acrylates; liq. 35% conc.

Alkasurf SS-DA-6. [Alkaril] Sodium deceth-6 sulfosuccinate; anionic; detergent, foaming and wetting agent, emulsifier, solubilizer, sequestrant; pale yel. liq.; bland odor; water sol.; sp.gr. 1.1; f.p. < 0 C; 30% act.

Alkasurf SS-DA4-HE. [Alkaril] Disodium deceth-4 sulfosuccinate; anionic; emulsifier, dispersant used in emulsion polymerization; surfactant in personal care prods.; liq.; 30% conc.

Alkasurf SS-L7DE. [Alkaril] Sodium lauramido DEA-sulfosuccinate; anionic; see Alkasurf SS-CDE; pale yel. liq.; sol. in water; dens. 1.1 g/ml; pH 5.0-7.5; 39-41% conc.

Alkasurf SS-NO. [Alkaril] Tetrasodium N-alkyl sulfosuccinamate; anionic; emulsifier, solubilizer, wetting agent; liq.; 35% conc.

Alkasurf SS-O-75. [Alkaril] Sodium dioctyl sulfosuccinate; anionic; wetting agent, emulsifier for industrial and textile industries, resin treatments, dry cleaning systems; paint pigment dispersant; clear colorless liq.; typ. odor; 1% water sol., sol. in most org. liq.; sp.gr. 1.09; f.p. < –20 C; surf. tens. 29 dynes/cm (0.1% aq.); pH 5–6; 75% act.

Alkasurf SS-TA. [Alkaril] Disodium N-octadecyl sulfosuccinamate; anionic; detergent, emulsifier, dispersant, solubilizer; blowing agent, foamer for SBR latexes, carpet applics.; cream-colored paste; low odor; disp. in water; dens. 1.0 g/ml; pH 8–10 (5% DW); 35% act.

Alkasurf SS-TA 125. [Alkaril] Sodium N-octadecyl sulfosuccinamate; anionic; see Alkasurf SS-TA; liq.; 35% conc.

Alkasurf TA-20. [Alkaril] Talloweth-20; nonionic;

detergent, lubricant and dyeing assistant for fiber blends in textile industry; coemulsifier in topical cosmetrics; wh. solid; sol. in water, aromatic solv., perchloroethylene; HLB 15.4; cloud pt. 73–77 C; 100% conc.

Alkasurf TA-30. [Alkaril] Talloweth-30; nonionic; see Alkasurf TA-20; wh. solid; sol. in water; HLB 16.7; cloud pt. 74-79 C; 100% conc.

Alkasurf TA-40. [Alkaril] Talloweth-40; nonionic; see Alkasurf TA-20; wh. solid; sol. in water; HLB 17.4; cloud pt. 75-79 C; 100% conc.

Alkasurf TA-50. [Alkaril] Talloweth-50; nonionic; see Alkasurf TA-20; wh. solid; sol. in water; HLB 17.8; cloud pt. 75-79 C; 100% conc.

Alkasurf TDA-5. [Alkaril] Trideceth-5; nonionic; wetting, emulsifying, and dispersing agent for textile scouring, cutting oils; intermediate for phosphate esters; liq.; sol. in min. oil and spirits, aromatic solv., perchloroethylene; disp. in water; dens. 0.96 g/ml; HLB 10.5; cloud pt. 65-68 C; 100% conc.

Alkasurf TDA-6. [Alkaril] Trideceth-6; nonionic; emulsifier, wetting, dispersant; dye leveling agent in textiles; solv. degreasers and industrial cleaning compds.; liq.; HLB 11.4; 100% conc.

Alkasurf TDA-7. [Alkaril] Trideceth-7; nonionic; emulsifier, rewetting agent for paper towels; solubilizer and foam stabilizer in liq. detergents; intermediate; liq.; sol. in water, min. oil and spirits, aromatic solv., perchloroethylene; dens. 0.98 g/ml; HLB 12.1; cloud pt. 72-75 C; 100% conc.

Alkasurf TDA-7.5. [Alkaril] Trideceth-7.5; nonionic; emulsifier, wetting agent for industrial applics.; foam builder, solubilizer; chemical specialty mfg.; almost water-wh. opaque liq.; mild odor; water sol.; sp.gr. 1.00; 100% act.

Alkasurf TDA-8.5. [Alkaril] Trideceth-8.5; emulsifier, rewetting agent for paper towels; solubilizer and foam stabilizer in liq. detergents; intermediate; liq.; sol. see Alkasurf TDA-7; dens. 0.98 g/ml; HLB 12.5; cloud pt. 52-56 C.

Alkasurf TDA-12. [Alkaril] Trideceth-12; nonionic; emulsifier, high-temp. detergent; leveling agent, solubilizer; wh. solid; mild odor; water sol.; m.p. 27 C; 100% act.

Alkasurf TDA-15. [Alkaril] Trideceth-15; nonionic; emulsifier, high-temp. detergent, leveling agent, solubilizer; wh. solid; mild odor; water sol.; m.p. 30 C; 100% act.

Alkaterge®-E. [Angus] Ethyl hydroxymethyl oleyl oxazoline; amphoteric; detergent, emulsifier, wetting agent, antifoamer, antioxidant; used in salt, soap, paper, textiles, and metal cleaners; emulsion stabilizer; acid acceptor; pigment grinding and disp.; Gardner 15 max. clear liq.; sol. in most org. liq., slight sol. in water; sp.gr. 0.9; dens. 7.74 lb/gal; visc. 155 cp; f.p. –31 C; flash pt. > 200 F; surf. tens. 40 dynes/cm; 79% act.

Alkaterge®-T. [Angus] Oxazoline-type compd.; amphoteric; detergent, emulsifier, dispersant, corrosion inhibitor; buff to br., waxy solid; high sol. in aromatic hydrocarbons; flash pt. > 200 F; surf. tens. 30.4 dynes/cm.

Alkaterge®-T-IV. [Angus] Oxazoline deriv., ethoxylated; nonionic; acid scavenger, offers filming protection to metal surfs.; corrosion inhibitor; emulsifier; dispersant in aq. and nonaq. systems; liq.; oil-sol.; water-disp.; HLB 8.0; 60% conc.

Alkateric 2CIB. [Alkaril] Cocoamphodiacetate; amphoteric; foaming agent, wetting agent, emulsifier for industrial and liq. detergent formulations; used in personal care prods.; solubilizer, coupling agent; amber clear liq.; sol. in water; dens. 1.18 g/ml; pH 8-9.5; 49.5-50.5% solids.

Alkateric BC [Alkaril] Coco betaine; amphoteric; detergent and base for shampoos; hair conditioner additive; coupler for shampoos; dyeing assistant in textiles, household, cosmetics; pale straw liquid; low odor; water sol; f.p. < 0 C; pH 7.5-8.5; 31% act.

Alkateric CAB. [Alkaril] Cocamido betaine; amphoteric; detergent, foaming agent, conditioner, coupler, solubilizer, visc. builder for shampoos, household use, cosmetics; amber liq.; bland odor; water sol.; 32% act.

Alkateric EHCIB. [Alkaril] Dicarboxylic caprylic deriv., disodium salt; amphoteric; coupler and solubilizer with high electrolyte tolerance; wetting agent and emulsifier for alkaline cleaners; bottle washing compds., wax strippers, degreasers; clear liq.; water-sol.; 34% conc.

Alkateric STB. [Alkaril] Stearyl betaine; amphoteric; detergent, surfactant/softener for fabric and hair, household and cosmetic applics.; pale yel. liq.; 50% act.

Alkateric TB. [Alkaril] Tallow betaine; amphoteric; detergent, surfactant/softener for fabric and hair; amber liq.; 70% act.

Alkatronic EDP 8-4, 28-1, 28-2, 28-7, 38-1, 38-4, 38-8. [Alkaril] Ethylene diamine POP compd., ethoxylated; nonionic; detergent, dispersant, defoamer, wetting and gelling agent used in emulsion polymerization and dye leveling processes; antistat and acid release agents; liqs. except 38-4 (paste), 38-8 (flake); HLB 16.0, 3.0, 7.0, 27.0, 2.5, 14.5, and 30.5 resp.; 100% conc.

Alkatronic EGE 17-1, 17-2, 17-4, 17-8, 25-1, 25-2, 25-5, 25-8, 25-4. [Alkaril] Propoxylated POE glycol; nonionic; surfactants used in defoaming and foam control agents in pulp and paper industries; dye leveling and wetting agent in textile processing; liq.; 100% conc.

Alkatronic PGP 10-1. [Alkaril] POP glycol ethoxylate; nonionic; surfactant used as oil demulsifier, visc. control agent, wetting agent; provides foam control in industrial applic.; liq.; sol. in water @ 1%, in perchloroethylene @ 10%; dens. 1.02 g/ml; HLB 4.5; cloud pt. 40 C (1%); pH 5.0-7.5 (5% DW); 100% conc.

Alkatronic PGP 12-2. [Alkaril] POP glycol ethoxylate; nonionic; surfactant used in dishwasher rinse aids; oil demulsifier; liq.; sol. in water; dens. 1.03 g/ml; HLB 8.0; cloud pt. 38 C (1%); pH 5.0-7.5 (5% DW); 100% conc.

Alkatronic PGP 18-1. [Alkaril] POP glycol ethoxylate; nonionic; surfactant with foam control for industrial applics.; liq.; sol. in perchloroethylene and water; dens. 1.01 g/ml; HLB 3; cloud pt. 24 C (1%); pH 5.0-7.5 (5% DW); 100% conc.

Alkatronic PGP 18-4. [Alkaril] POP glycol ethoxylate; nonionic; surfactant used in detergent formulations; emulsifier and dispersing agent; paste; sol. in water @ 10%, in perchloroethylene @ 1%; dens. 1.05 g/ml; HLB 15.0; cloud pt 62 C (1%); pH 5.0-7.5 (5% DW); 100% conc.

Alkatronic PGP 18-8. [Alkaril] POP glycol ethoxylate; nonionic; surfactant used as lime soap dispersant, wetting agent, latex stabilizer; flake; sol. in water; dens. 1.06 g/ml; HLB 29; cloud pt. > 100 C (1%); pH 5.0-7.5 (5% DW); 100% conc.

Alkatronic PGP 18-8LF. [Alkaril] POP glycol ethoxylate; nonionic; see Alkatronic PGP 18-8; flake; sol. in water; dens. 1.06 g/ml; cloud pt. > 100 C (1%); pH 5.0-7.5 (5% DW).

Alkatronic PGP 23-7. [Alkaril] POP glycol ethoxylate; nonionic; surfactant used as dispersant, gelling and visc. control agent; flake; sol. in water; dens. 1.04 g/ml; HLB 24.0; cloud pt. > 100 C (1%); pH 5.0-7.5 (5% DW); 100% conc.

Alkatronic PGP 23-8. [Alkaril] POP glycol ethoxylate; nonionic; see Alkatronic PGP 23-7; flake; sol. in water; dens. 1.06 g/ml; HLB 28.0; cloud pt. > 100 C (1%); pH 5.0-7.5 (5% DW); 100% conc.

Alkatronic PGP 33-1. [Alkaril] POP glycol ethoxylate; nonionic; surfactant, foam control agent in industrial prods. and processes such as antifreeze and dry cleaning detergents; liq.; sol. in perchloroethylene @ 10%; insol. in water; dens. 1.02 g/ml; HLB 1.0; cloud pt. 16 C (1%); pH 5.0-7.5 (5% DW); 100% conc.

Alkatronic PGP 33-8. [Alkaril] POP glycol ethoxylate; nonionic; surfactant used as dispersant, gelling agent and visc. control agent; flake; sol. in water; dens. 1.06 g/ml; HLB 27.0; cloud pt. > 100 (1%); pH 5.0-7.5 (5% DW) 100% conc.

Alkatronic PGP 40-7. [Alkaril] POP glycol ethoxylate; nonionic; surfactant used as dispersant, gelling agent and visc. control agent; flake; sol. in water; dens. 1.05 g/ml; HLB 22.0; cloud pt. > 100 C (1%); pH 5.0-7.5 (5% DW); 100% conc.

Alkatrope SX-40. [Alkaril] Sodium xylene sulfonate; hydrotrope, solubilizer and coupling agent; lt.-colored, clear liq.; 41% act.

Alkawet® NP-6. [Lonza] Mono and dialkylarylphenoxy POE acid phosphates; anionic; emulsifier; intermediate surfactant for cleaners; liq.; 100% conc.

Alkazine C. [Alkaril] Coco hydroxyethyl imidazoline; cationic; emulsifier, antistat, corrosion inhibitor, softener for textiles, plastics, asphalt, tar emulsion breakers; water repellent treatment of cement, concrete, and plaster; used in cutting oils and syn. coolants; petrol. industry; antifungal agent for wood; slime control additive in paperboard; brn. paste; amine odor; sol. in aq. acid; disp. in water; dens. 0.96 g/ml; f.p. 20 C; 85% act.

Alkazine O. [Alkaril] Oleic hydroxyethyl imidazoline; cationic; emulsifier, corrosion inhibitor, antistat agent, wetting and flocculating agent, lubricant used in car wax emulsions, cleaners, paint mfg., agric. applics., syn. coolants; dispersant for clay and pigments in solv. systems; brn. liq.; amine odor; sol. in min. oil and spirits, aromatic solv., perchloroethylene; water-disp.; dens. 0.94 g/ml; f.p. 5 C; 85% act.

Alkazine ST. [Alkaril] Stearic hydroxyethyl imidazoline; cationic; emulsifier, dispersant, corrosion inhibitor used in waxes, pigments, resins, household formulations; tan, waxy solid; amine odor; solid in acid and org. systems; pH alkaline; 85% act.

Alkazine TO. [Alkaril] Tall oil hydroxyethyl imidazoline; cationic; emulsifier, corrosion inhibitor in oil burning systems, pickling bath operations, asphalt emulsions; dispersant for clay and pigments in solv. systems; brn. liq.; amine odor; sol. in acid systems; pH alkaline; 80% act.

Alkylate 215. [Monsanto] Linear alkylbenzene; detergent, intermediate used in lt.-duty detergents; liq.; 100% conc.

Alkylate 225. [Monsanto] Linear alkylbenzene; detergent, intermediate used in liq. and powd. heavy duty detergent; liq.; 100% conc.

Alkylate 230. [Monsanto] Linear alkylbenzene; detergent, intermediate used in lt. and heavy duty detergents; liq.; 100% conc.

Alkylate A215. [Monsanto] Alkylbenzene; intermediate, detergent; colorless liq.; m.w. 231-241; dens. 7.2 lb/gal; sp.gr. 0.855-0.870; biodeg.

Alkylate A225, A230. [Monsanto] Alkylbenzene; intermediate; colorless liq.; m.w. 238-248 and 231–266 resp.; dens. 7.2 lb/gal; sp.gr. 0.855-0.870; biodeg.

Alkylate H230H, H230L. [Monsanto] By-prod. of alkylbenzene mfg.; anionic; intermediate, surfactant; yel. liq.; m.w. 360 and 335 resp.; sp.gr. 0.890-0.910; visc. 140-220 and 100–160 SUS (100 F); i.b.p. 290 C min.

AlNel A-100. [Advanced Refractory Tech.] Aluminum nitride; filler for elastomeric, epoxy, polymeric, and other filled systems; provides high thermal conductivity, good dielec. strength, corrosion resistance; gray to wh. powd.; 3.0–4.0 μ avg. particle diam.; dens. 3.26 g/cc; sublimes at 2450 C; conduct. 160–220 W/mK; dielec. const. 8.2–9.0.

AlNel A-200. [Advanced Refractory Tech.] Aluminum nitride; filler for elastomeric, epoxy, polymeric, and other filled systems; provides high thermal conductivity, good dielec. strength, corrosion resistance; gray to wh. powd.; 3.0–4.0 μ avg. particle diam.; dens. 3.26 g/cc; sublimes at 2450 C; conduct. 100–150 W/mK; dielec. const. 8.0–9.0.

AlNel AG. [Advanced Refractory Tech.] Aluminum nitride; filler for elastomeric, epoxy, polymeric, and other filled systems; provides high thermal conductivity, good dielec. strength, corrosion resistance; fine to coarse grains.

Aloe-Moist. [Terry Labs] Aloe vera gel, polyglyceryl methacrylate, and propylene glycol; moisturizer with soothing, cooling effect used for first aid preps., acne and facial preps., moisturizers, creams, shaving prods., depilatories, shampoos, conditioners, suncare prods.; clear to slightly hazy colorless visc. gel; sol. in water; sp.gr. 1.1; visc. 75,000–150,000 cps;

pH 5.0–5.6.

Aloe-Moist/A 0.05% Stabilized Beta Carotene. [Terry Labs] Aloe vera prod.; see Aloe-Moist; colorless, clear to slightly hazy visc. gel; sol. in water; sp.gr. 1.1; visc. 75,000–150,000 cps; pH 5.0–5.6.

Aloe-Moist/A 0.10% Stabilized Beta Carotene. [Terry Labs] Aloe vera prod.; see Aloe-Moist; colorless, clear to slightly hazy visc. gel; sol. in water; sp.gr. 1.1; visc. 75,000–150,000 cps; pH 5.0–5.6.

Alox® 100. [Alox] Org. acids, oxy-acids, esters, lactones, and some unsaponifiable matter; film forming rust preventive by conversion to metallic soaps; oil-sol. lubricity agent by conversion of free acid to esters; emulsifier or de-emulsifier by conversion to amine or alkanolamine soaps; corrosion inhibitor by blending with other surf. act. agents; solid; m.w. 350; sp.gr. 0.917; dens. 7.7 lb/gal; m.p. 109 F; i.b.p. 149 F; flash pt. (COC) 305 F; fire pt. (COC) 340 F; acid no. 80; sapon. no. 148.

Alox® 100D. [Alox] Org. acids, oxy-acids, esters, lactones, and some unsaponifiable matter; see Alox 100; solid; m.w. 422; sp.gr. 0.864; dens. 7.2 lb/gal; m.p. 122 F; i.b.p. 185 F; flash pt. (COC) 343 F; fire pt. (COC) 415 F; acid no. 16; sapon. no. 32.

Alox® 102. [Alox] Org. acids, oxy-acids, esters, lactones, and some unsaponifiable matter; see Alox 100; solid; m.w. 291; sp.gr. 0.965; dens. 8.1 lb/gal; m.p. 102 F; i.b.p. 131 F; flash pt. (COC) 348 F; fire pt. (COC) 295 F; acid no. 128; sapon. no. 225.

Alox® 111. [Alox] Mold release or parting compd. for cast concrete prods.; rust preventive protecting metal forms; sp.gr. 0.92–0.93; dens. 7.6–7.7 lb/gal; visc. 40–50 cs (100 F); flash pt. (OC) 154 C min.; pour pt. 18 C max.; sapon. no. 30–40.

Alox® 152. [Alox] Lubricity agent in min. oil sol'ns., cutting and metalworking formulations, quench oil additives, for automotive and industrial lubricants; brn. soft consistency; dens. 7.3 lb/gal; flash pt. 225 F min.; acid no. 17–25; sapon. no. 125–140.

Alox® 162. [Alox] Lubricity agent in min. oil sol'ns., for railroad journal oils, for automotive and industrial lubricants; brn. soft consistency; sol. in naphthenic oils, some paraffinic oils; sp.gr. 0.86–0.89; dens. 7.3 lb/gal; m.p. 32–35 C; flash pt. 104 C min.; acid no. 12–23; sapon. no. 94–120.

Alox® 164. [Alox] Lubricity agent for crankcase oils to improve antiwear, detergency, solvency, and antirust properties; visc. 35–45 SUS (40 C); flash pt. 38 C min.; pour pt. 15.6 C; acid no. 10–15; sapon. no. 45–55.

Alox® 165. [Alox] Nonemulsifiable, nonstaining, water-displacing rust preventive, lubricant; forms thin films for protection of ferrous and nonferrous metals from atmospheric corrosion, even in high humidity; soft, waxy brn. solid; mild odor; sp.gr. 0.91–0.95; dens. 7.8 lb/gal; m.p. 35 C; flash pt. 121 C min.; acid no. 5–11; sapon. no. 85–105.

Alox® 318F. [Alox] Corrosion inhibitor for use on ferrous and nonferrous metals, esp. for bright or highly polished steel surfs.; brn. solid; sol. in solv. and oil; dens. 7.7 lb/gal (15.6 C); m.p. 39 C min.; flash pt. 177 C min.; acid no. 13 ± 2; sapon. no. 32 ± 3.

Alox® 319F. [Alox] Additive for compding. corrosion preventives, preservative lubricants in oil or solv. blends; protects bright or highly polished steel surfs.; exc. in acid atmospheres; dk. solid; dens. 7.7 lb/gal (15.6 C); m.p. 37 C min.; flash pt. (OC) 177 C min.; fire pt. (OC) 204 C min.; acid no. 42–46; sapon. no. 50–60.

Alox® 350. [Alox] Oxidized petrol. fraction esters; corrosion inhibitor and lubricant for making cutting and sol. oils; brn. paste; mild odor; sol. in nonpolar solv.; sol. enhanced by use of lecithin; sp.gr. 0.91; dens. 7.7 lb/gal; m.p. 100 F; flash pt. 121 C min.; acid no. 33 max.; sapon. no. 125–140; pH acid.

Alox® 436A. [Alox] Mold release, lubricant, corrosion inhibitor for concrete molds; brn. liq.; dens. 7.1 lb/gal; flash pt. 102 C.

Alox® 488. [Alox] Oxidized petrol. fractions; upper cylinder lubricant for internal combustion engines; cleans and prevents formation of carbon and lacquer deposits in engines; dens. 7.3–7.4 lb/gal (15.6 C); visc. 140–190 SUS (40 C); flash pt. > 110 C; fire pt. > 121 C; acid no. 6–10; sapon. no. 16–26.

Alox® 502. [Alox] Oxygenated hydrocarbon partially neutralized with barium; corrosion inhibitor; base; preservative lubricants; dk.; mild odor; sol. see Alox 350; sp.gr. 0.97; pH neutral; flash pt. 390 F; 100% act.

Alox® 502A. [Alox] Corrosion inhibitor, preservative lubricant meeting MIL-L-21260 specs.; sp.gr. 0.97 (60 F); dens. 8.1 lb/gal; flash pt. 380 F; fire pt. 400 F; acid no. nil; sapon. no. nil.

Alox® 575. [Alox] Oxygenated hydrocarbon with barium and a sodium petrol. sulfonate; emulsifier, corrosion inhibitor for cutting and sol. oils; tan liq., gel; mild odor; oil sol.; sp.gr. 1.02 (15.6 C); dens. 8.5 lb/gal; flash pt. 400 F; pour pt. 80 F; acid no. 9–11; sapon. no. 28–31; 100% act.

Alox® 600. [Alox] Mixt. of high m.w. org. acids, lactones, and esters derived from petrol.; corrosion inhibitor and chemical intermediate for metalworking compds., automobile undercoating compds., slushing oils; sp.gr. 0.95; dens. 7.9 lb/gal; m.p. 46 C; flash pt. 174 FC; acid no. 45–60; sapon. no. 100–145.

Alox® 601. [Alox] Org. acids, oxy-acids, esters, lactones, and some unsaponifiable matter; see Alox 100; solid; m.w. 807; sp.gr. 0.927; dens. 7.7 lb/gal; m.p. 135 F; i.b.p. 149 F; flash pt. (COC) 375 F; fire pt. (COC) 495 F; acid no. 25; sapon. no. 60.

Alox® 604. [Alox] Oxygenated hydrocarbon partially neutralized with calcium; corrosion inhibitor; base for soft film cut back rust preventives; dk. paste; mild odor; sol. see Alox 350; sp.gr. 0.96; dens. 7.99 lb/gal; m.p. 220 F; pour pt. 150 F; flash pt. 400 F; 100% act.

Alox® 606, 606-55, 606-70. [Alox] Heavy-duty rust preventive, protective coating for metal surfs., esp. for automotive rustproofing; 606-55 is a total solv. cutback grade with 55% Alox 606 and 45% min. spirits; 606-70 is a partial solv. cutback grade with 70% Alox 606 and 30% min. spirits; sp.gr. 0.98, 0.895, and 0.92 resp.; dens. 8.2, 7.5, and 7.7 lb/gal resp.; m.p. 138, liq., and 60 C min.; flash pt. 232, 42, and 42 C min.

Alox® 707. [Alox] Org. acids, lactones, and esters;

corrosion inhibitor; solid; dens. 8.8 lb/gal; m.p. 185 F min.; flash pt. 410 F (OC).

Alox® 904. [Alox] Base for formulation of fingerprint removers or rifle bore cleaners; corrosion preventive; dk. colored visc. liq.; sp.gr. 0.980; dens. 8.16 lb/gal; visc. 900–5000 cs (100 F); flash pt. (COC) 146 C; pour pt. 18 C; acid no. 6–10; sapon. no. 16–25.

Alox® 940. [Alox] Oxygenated hydrocarbon partially neutralized with calcium and a sodium petrol. sulfonate; corrosion inhibitor; oil and solv. rust preventive base; dk. visc. liq.; mild odor; sp.gr. 0.97; dens. 8.0 lb/gal; pH slightly acid; flash pt. 300 F min. (OC); 100% act.

Alox® 940A. [Alox] Oxygenated hydrocarbon, calcium soap, with sodium petrol. sulfonate; rust preventive additive for use with either solv. or oil; oil sol'ns. are lubricants; visc. liq.; sp.gr. 0.94–0.97 (15.6 C); flash pt. (OC) 149 C; acid no. 9 max.; sapon. no. 31 max.

Alox® 1680. [Alox] Oxygenated hydrocarbon containing minor amt. of phosphatide; corrosion inhibitor; penetrating oil additive; dk. liq.; mild odor; see Alox 350; sp.gr. 0.92 (15.6 C); dens. 7.6 lb/gal (15.6 C); visc. 220 SUS (100 F); flash pt. 350 F; 100% act.

Alox® 1689. [Alox] Ester of an oxygenated hydrocarbon; nonionic; emulsifier, lubricity agent for cutting and sol. oils, marine engine lubricants; rolling oil additive for stainless steel sheets and titanium sheets; dk. brn. solid; mild odor; sol. in nonpolar solv.; sp.gr. 0.93; dens. 7.7 lb/gal; m.p. 115 F; flash pt. (OC) 400 F; acid no. 16 max.; sapon. no. 75 max.; 100% act.

Alox® 1727D. [Alox] Nonstaining rust preventive for slushing oils; protects sheet steel; brn. visc. liq.; sol. in naphthenic min. oils, kerosene, many paraffin oils; sp.gr. 0.94–0.97 (15.6 C); flash pt. 350 F min.; acid no. 2–6; sapon. no. 20–40.

Alox® 1843. [Alox] Org. acid amine salt; corrosion inhibitor; surfactant; pale yel. liq.; mild odor; sol. in water, alcohols, glycols; sp.gr. 1.12; dens. 9.3 lb/gal; visc. 138 cs (100 F); pour pt. 15 F; pH neutral; biodeg. 80%; 75% act. in water.

Alox® 1846. [Alox] Corrosion inhibitor for jet fuels and lt. distillates; sol. in petrol. hydrocarbons; insol. in water; dens. 7.39 lb/gal; flash pt. (COC) 79 C; pour pt. –7 C.

Alox® 2000. [Alox] Corrosion inhibitor and penetrant for metal conditioning compds.; dk. colored, very visc. liq.; sp.gr. 0.954; dens. 7.96 lb/gal; flash pt. (OC) 193 C; pour pt. 4 C; acid no. 17.5; sapon. no. 48.

Alox® 2019A. [Alox] Oxygenated hydrocarbon, amine and calcium-neutralized; base for fingerprint removers; rust preventive protecting metal parts during long-term indoor storage; sp.gr. 0.94; dens. 7.9 lb/gal; visc. 78.8 cs (37.8 C); flash pt. (CC) 58 C; acid no. 98.

Alox® 2028. [Alox] Oxygenated hydrocarbon, barium soap; nonemulsifiable, nonstaining, water-displacing rust preventive; can be used in solv. or min. oil sol'ns.; sp.gr. 0.95; dens. 7.9 lb/gal; m.p. 35 C; flash pt. 121 C min.; acid no. 3–5; sapon. no. 61–69.

Alox® 2028L. [Alox] Oxygenated hydrocarbon, calcium soap; nonemulsifiable, nonstaining, water-displacing rust preventive, lubricant; visc. brn. liq.; mild odor; sp.gr. 0.93–0.95; dens. 7.8 lb/gal; flash pt. 168 C min.; pour pt. –18 C max.; acid no. 2–10; sapon. no. 11–19.

Alox® 2116. [Alox] Oxygenated hydrocarbon; corrosion inhibitor for greases.

Alox® 2116A. [Alox] Oxygenated hydrocarbon, partial barium soap; corrosion inhibitor used for calcium complex and lithium hydroxy stearate greases; dk. paste; mild odor; see Alox 350; sp.gr. 0.97 (15.6 C); m.p. 30 C; flash pt. (OC) 180 C; fire pt. 190 C; pH acid; 100% act.; 1.6% barium.

Alox® 2138F. [Alox] Bullet lubricant; solid; sp.gr. 0.95; dens. 7.9 lb/gal; m.p. 71 C; flash pt. 177 C; fire pt. 193 C; acid no. 29.5; sapon. no. 122.8.

Alox® 2162. [Alox] Oxygenated hydrocarbon, barium soap; corrosion inhibitor; thin film rust preventive for protection against limited salt spray and acid atmospheres; meets MIL-C-23411; dk. paste; mild odor; sol. see Alox 350; sp.gr. 0.96; dens. 8.0 lb/gal (15.6 C); m.p. 35 C; flash pt. 121 C; acid no. 12–16; sapon. no. 70–80pH slightly acid; 100% act.

Alox® 2200. [Alox] Corrosion inhibitor, preservative lubricant meeting MIL-L-21260A for combustion engines; also for preservative lubricants; brn. visc. liq.; sp.gr. 1.01 (15.6 C); dens. 8.39 lb/gal; flash pt. > 190 C; 4.7% barium.

Alox® 2201, 2201E. [Alox] Corrosion inhibitor for general purpose, water-displacing lubricating oils; dens. 7.9–8.1 and 7.7–7.9 lb/gal resp.; flash pt. (COC) 177 C min.; acid no. 1.5–4.5 and 1.5–7.0; 3% barium (2201).

Alox® 2211Y. [Alox] Thixotropic rust preventive which deposits film providing long-term protection against humidity and salt fog; can be cutback in solv. or blended 50/50 with naphthenic oil; sp.gr. 0.89; dens. 7.4 lb/gal; m.p. 82–104 C; flash pt. 177 C min.

Alox® 2213A. [Alox] Oil and solv. sol. rust preventive producing water-displacing thin, soft films with resistance to salt spray and acid atmospheres; lubricant, penetrant; sp.gr. 0.93–0.96 (15 C); dens. 7.75–8.08 lb/gal; m.p. 35–46 C; flash pt. 149 C; acid no. 25–34; sapon. no. 85–97.

Alox® 2213C. [Alox] Oil and solv. sol. rust preventive producing water-displacing thin, soft films with superior resistance to salt spray; penetrant, lubricant; sp.gr. 0.96–0.98; dens. 8.1 lb/gal; m.p. 37 ± 5 C; flash pt. 177 C min.; acid no. 15–30; sapon. no. 55–75.

Alox® 2248C, 2248C-50. [Alox] Corrosion inhibitor; additive improving the low temp. flexibility and adhesion of blown asphalts making them effective for protecting steel; 2248C-50 is a 50% cutback soft paste; solid and paste resp.; sp.gr. 0.93 and 0.86; dens. 7.8 and 7.2 lb/gal; m.p. 66–71 C (2248C); flash pt. 232 and 42 C min.; pour pt. 40 C (2248C-50); 100 and 50% act.

Alox® 2263. [Alox] Emulsifiable corrosion inhibitor for protection against salt spray and high humidity conditions on steel and phosphated steel parts; provides thin, oily protective film; sp.gr. 0.98 (15.6 C); dens. 8.2 lb/gal; flash pt. 204 C min.; pour pt. 79 C; acid no. 7–9; sapon. no. 9–11.

Alox® 2276A. [Alox] Solventless, hot melt or spray rustproofing compd. providing long-term protection of metal surfs.; deposits brn. resilient film for auto frames, chassis components, and metal surfs. exposed to severe conditions such as road de-icing salts; sp.gr. 0.96; dens. 8 lb/gal; m.p. (drop) 100 C; flash pt. 218 C min.; fire pt. 218 C min. 99% solids min.

Alox® 2278. [Alox] Nonstaining rust preventive for protecting coiled or stacked steel from high humidity conditions and acid fumes; brn. soft waxy solid; mild odor; dens. 8.0 lb/gal; m.p. 35 ± 3 C; flash pt. 121 C min.; fire pt. 121 C min.; acid no. 4–8; sapon. no. 60–75.

Alox® 2283. [Alox] Lubricant, rust and corrosion inhibitor for wire rope mfg.; dk. visc. liq.; sp.gr. 0.93; dens. 7.8 lb/gal; visc. 100 SUS min. (99 C); flash pt. 191 C min.; pour pt. –1 C.

Alox® 2284A, 2284AB. [Alox] Mixts. of org. acids, lactones, and esters; used in corrosion inhibitors for sour crude oil wells; lubricity agent for water pump lubricants and preservatives; sp.gr. 0.95 (15.6 C); dens. 7.9 lb/gal; visc. 2 and 4 cs resp. (99 C); acid no. 160; sapon. no. 276 and 265 resp.

Alox® 2287. [Alox] Rust inhibitor, protective coating for ballast tank flotation; biodeg.; sp.gr. 0.99; dens. 8.3 lb/gal; visc. 400–600 SUS (99 C); m.p. (drop) 52–57 C; flash pt. 205 C; acid no. 10; sapon. no. 75–100.

Alox® 2289. [Alox] Corrosion preventive in oil, solv., or oil/solv. blends effective for protection of ferrous metals in acid fume environments; brn. waxy solid; sp.gr. 0.90–0.94; dens. 7.7 lb/gal; m.p. (drop) 43–49 C; flash pt. (OC) 177 C min.; fire pt. (OC) 215 C min.; acid no. 16–21; sapon. no. 37–45.

Alox® 2290, 2290A. [Alox] Corrosion inhibitors for slushing oils, preservative lubricants, min. and sol. metalworking oils, water-based rust preventives; slight and no odor resp.; sp.gr. 1.00–1.04 and 1.01–1.05 resp. (15.6 C); dens. 8.3–8.7 and 8.4–8.8 lb/gal; flash pt. 204 C min.; acid no. 11–17 and 10–15; sapon. no. 30–60 and 50–75.

Alox® 2291. [Alox] Barium sulfonate; rust preventive; visc. liq.

Alox® 2292. [Alox] Calcium sulfonate; rust preventive; visc. liq.

Alperox-F. [Atochem] Lauroyl peroxide; initiator for bulk, sol'n., and suspension polymerization, high-temp. curing of polyester resins, and cure of acrylic syrup; flakes; 98% act.; 3.93% act. oxygen.

Alphaflex 101®. [Alphaflex] PTFE and MoS_2; reinforcing agent for rubber; added at 2–12 phr; gray soft gran.; 0.1–1.0 mm particle size; insol. in water; sp.gr. 2.4.

Alphaflex 202®. [Alphaflex] PTFE and MoS_2; friction-reducing additive for rubber; added at 20–30 phr; gray hard particles; 2–20µ particle size; insol. in water; sp.gr. 2.5.

Alphenate PE Extra. [Henkel-Nopco] Alkyl polyglycol ester and sulfate semiester of alkylaryl polyglycol ether blend; anionic/nonionic; dyeing assistant for polyester and acrylic fibers; liq.; 50% conc.

Alphenate TFC-76. [Henkel-Nopco] Sodium alkyl sulfate, modified; anionic; dye leveling agent; liq.; 50% conc.

Alphoxat MS 130. [Zschimmer & Schwarz] Blended fatty acid with 30 moles EO; dispersant for chemotech. prods.; wax; 100% act.

Alphoxat O 110, O 115. [Zschimmer & Schwarz] Fatty acid polyglycol ester; defoamer; aux. agent for textile, ceramic, and chemical tech. industries; liq. to waxy.

Alphoxat S 110, 120. [Zschimmer & Schwarz] Fatty acid polyglycol ester; nonionic; emulsifier and dispersant, greasing agent for textile and chemical technical industries; waxy; 100% conc.

Alrosol Conc. [Ciba-Geigy] Fatty acid amide; foam stabilizer, dispersing, emulsifying, wetting agent for SR and latexes; clear yel., visc. liq., mild odor; sp.gr. 0.99.

Alrosperse 11P Flake. [Ciba-Geigy] Fatty acid amide; nonionic; emulsifier, lubricant used in hair rinses, hand modifiers for textiles, spreading agent in paste waxes and polishes; off-wh., waxy flake.

Alrowet® D-65. [Ciba-Geigy] Dioctyl sodium sulfosuccinate; anionic; wetting and rewetting agent for textile applics.; dispersant, antistat, wetting agent for NR and SR latexes; clear liq., slight odor; sp.gr. 1.10; 65% conc.

Altax®. [Vanderbilt] Benzothiazyl disulfide; accelerator for nat. and syn. rubbers; primary accelerator and scorch modifying sec. accelerator in NR and SBR copolymers; retarder-plasticizer in neoprene (G types); cure modifier in W types; cream to lt. yel. powd.; also avail. in rods, dustless and thread-grade; m.w. 332.48; dens. 1.51 mg/m^3; m.p. 159–170 C.

Altowhite LL. [ECC Int'l.] Surf.-modified calcined aluminum silicate; extender pigment; exhibits improved optical properties in paint systems; powd., median particle size 1.8 µ; sp.gr. 2.62; dens. 22.5 lb/solid gal; ref. index 1.56; pH 6.0–7.0 (20% aq. slurry).

Alubrasoft SJ-2, SJ-Wax. [PPG-Mazer] Cationic fatty polyamide; softener used as fiber finish for nat. and producer-dyed synthetics; liq. and solid resp.; 10 and 100% act.

Alubrasoft Super 15, 25, 100. [PPG-Mazer] Cationic fatty polyamide; softener, antistat, conditioner, lubricant; imparts soft hand to syn. and nat. fibers; liq.; 15, 25, and 100% act. resp.

Alubrasol FE. [PPG-Mazer] Polyamide salt; cationic; antistat with thermal stability used on carpets; paste.

Alubraspin HS-100. [PPG-Mazer] Fatty alcohol alkoxylate; antistatic lubricant for nylon BCF; liq.; 100% act.

Aluminum Oxide C. [Degussa] Aluminum oxide; free-flow and anticaking agent; aids in reducing electrostatic charges of powder substances; wh. fluffy powd.; dens. 60 g.; surf. area 100 m^2/g; pH 4–5 (4% aq. disp.); > 99.6 Al_2O_3.

Aluminum Stearate EA. [Witco] Aluminum distearate; binder, emulsifier, anticaking agent for food applics.; wh. powd., 98% thru 200 mesh; sol. in turpentine, benzene, toluene, xylene, CCl_4, veg. and min. oil, oleic acid, waxes; insol. in water; sp.gr. 1.01; soften. pt. 160 C.

Amax®. [Vanderbilt] N-oxydiethylene benzothiazole-2-sulfenamide; accelerator for SBR, NR, IR, BR; delayed action primary accelerator and scorch modifying sec. accelerator; tan flake; m.w. 252.35; dens. 1.37 mg/m³; m.p. 70–90 C.

Amax® No. 1. [Vanderbilt] N-oxydiethylene benzothiazole-2-sulfenamide plus benzothiazyl disulfide; accelerator for SBR, NR, IR, BR; delayed action primary accelerator and scorch modifying sec. accelerator; lt. tan fl.; dens. 1.40 mg/m³; m.p. 70–90 C.

Ambergum® 721. [Aqualon] Cellulose deriv.; replaces gum arabic in lithographic printing processes; visc. 25–60 mPas @ 6%; 98% min. purity.

Ambergum® 1221. [Aqualon] Anionic polysaccharide based on cellulose deriv.; replaces gum arabic in lithographic printing processes; gran. powd.; visc. 60–200 mPas @ 12%; 98% min. purity.

Ambergum® 3021. [Aqualon] Low m.w. cellulosic polymer, aq. sol'n.; anionic; pigment dispersant, binder, stabilizer, flow control agent, liq. thickener, and protective colloid; FDA approved; pale amber liq.; dens. 9.5 lb/gal; visc. 1200 cps; pH 4–6; 30 ± 0.5% solids.

Amberlig. [Daishowa] Calcium lignosulfonate; anionic; fly ash retarder, handling and processing aid; liq.; 50–58% conc.

Amberol ST-149. [Rohm & Haas] Nonheat-hardening phenolic resin; used in prod. of tung oil varnishes; as tackifier for butyl rubber caulks and sealants; solid or crushed; sol. see Amberol ST-137; sp.gr. 1.04; visc. (G-H) F-L (64% in toluene); m.p. (R&B) 95 C; acid no. 20–40.

Ambroxan. [Henkel] 8-Alpha-12-oxido-13,14,15,16 tetra-norlabdane; fragrance raw material-ambergris type.

Amdro®. [Am. Cyanamid/Ag] Hydramethylnon; insecticide for control of fire ants in pastures, lawn, turf, nonagric. lands; 0.88% act.

Amerchol® 400. [Amerchol] Petrolatum, lanolin alcohol, and cetyl alcohol; nonionic; aux. emulsifier, emulsion stabilizer for o/w and w/o systems, incl. makeup, pharmaceuticals; emollient, lubricant; lt. cream soft solid, pract. odorless; HLB 9.0; m.p. 49–50 C; acid no. 1.5; sapon. no. 8; 100% conc.

Amerchol® BL. [Amerchol] Lanolin, min. oil, and lanolin alcohol; nonionic; absorp. base, aux. emulsifier for o/w systems, conditioner, emollient, moisturizer, stabilizer for cosmetics and pharmaceuticals; yel.-amber semisolid, slight char. sterol odor; oil sol.; HLB 8.0; acid no. 2.0 max.; sapon. no. 60–70; 100% conc.

Amerchol® C. [Amerchol] Petrolatum, lanolin, lanolin alcohol; nonionic; see Amerchol BL; pale yel.-cream soft solid, slight, char. sterol odor; oil sol.; HLB 9.5; m.p. 40–46 C; acid no. 1.0 max.; sapon. no. 10–20; 100% conc.

Amerchol® CAB. [Amerchol] Petrolatum, lanolin alcohol; nonionic; emollient, emulsifier for therapeutic ointments, burn preparations, dermatological prods., hypoallergenic preparations; stabilizer, plasticizer; pale cream soft solid, faint, char. sterol odor; oil sol.; HLB 9.0; m.p. 40–46 C; acid no. 1 max.; sapon. no. 1.0 max.; 100% conc.

Amerchol® H-9. [Amerchol] Petrolatum, lanolin, lanolin alcohol; nonionic; emollient, emulsifier, penetrant, stabilizer, absorp. base for cosmetics, pharmaceutical ointments, burn ointments; plasticizer; pale yel. soft solid, slight, char. sterol odor; oil sol.; HLB 9; m.p. 55–62 C; acid no. 1 max.; sapon. no. 15–27; 100% conc.

Amerchol® L-99. [Amerchol] Min. oil, lanolin alcohol; nonionic; emulsifier, stabilizer, conditioner, emollient, moisturizer for creams and lotions, hair and skin preparations, dermatological specialties; lt. yel. oily liq., faint char. sterol odor; oil misc.; sp.gr. 0.855–0.875; HLB 8; acid no. 2 max.; sapon. no. 2 max.; 100% conc.

Amerchol® L-101. [Amerchol] Min. oil, lanolin alcohol; nonionic; emollient, penetrant, emulsifier, moisturizer, softener, stabilizer for cosmetics, creams, makeup, hair dressing, pharmaceuticals, aerosols, baby prods., textile finishes; plasticizer for hair sprays; pale yel. oily liq., faint char. sterol odor; oil sol.; sp.gr. 0.840–0.860; visc. 20–30 cps; HLB 8; acid no. 1 max.; sapon. no. 1 max.; 100% conc.

Amerchol® L-500. [Amerchol] Min. oil, lanolin alcohol, octyldodecanol; nonionic; emulsifier, stabilizer, emollient, moisturizer, conditioner for hair and skin prods., creams, lotions, makeup, aerosols, pharmaceutical vehicles, baby toiletries; cloudy liq. at R.T.; oil-misc.; HLB 6; acid no. 2 max.; sapon. no. 5 max.; 100% conc.

Amerchol® Polysorbate 80. [Amerchol] Polysorbate 80; nonionic; emulsifier, solubilizer used in cosmetic, toiletry, and pharmaceutical prods.; liq.; sol. in water and alcohol; HLB 15.0; 100% act.

Amerchol® RC. [Amerchol] Petrolatum, lanolin alcohol, stearyl alcohol, and stearone; aux. emulsifier, stabilizer, emollient, conditioner for creams, lotions, makeup; aids pigment disp., color definition; pale ivory soft solid, slight char. sterol odor; misc. with common oil phase ingred.; m.p. 55–64 C; acid no. 2 max.; sapon. no. 10 max.

Amerfloc Plus 5270. [Drew] High m.w., highly anionic fluid polymer; flocculant and coagulant aid for sludge conditioning; for dewatering industrial slurries and water clarification applics. in paper industry, food processing, nonpotable water clarification, oily waste treatment; wh. visc. liq.; sp.gr. 1.1; visc. 1500 cps; pour pt. < 30 F; flash pt. 52 C (PMCC).

Amerfloc Plus 5495. [Drew] High m.w., cationic polymer; flocculant and coagulant aid for sludge conditioning, dewatering industrial and municipal slurries, in water clarification for paper mill, food processing, nonpotable water, oily waste treatment; wh. visc. liq.; f.p. –1 C; sp.gr. 1.04; visc. 1800 cps; flash pt. 55 C (CC).

Amerlate® LFA. [Amerchol] Lanolin acid; anionic; emulsifier, stabilizer, emollient for fatty acid systems, aerosol shave creams, cream shampoos, wax systems, household prods.; pigment dispersant; increases tack and plasticity of wax films; ylsh.-tan firm, waxy solid; mild, waxy odor; m.p. 55–62 C acid no. 130–150; sapon. no. 170–190; 100% conc.

Amerlate® P. [Amerchol] Isopropyl lanolate; nonionic; conditioner, penetrant, lubricant, moisturizer, emollient, w/o emulsifier, stabilizer, opacifier for cosmetics and pharmaceuticals; pigment dispersant; wetting agent and dispersant for solids; plasticizer for wax and pigment systems; yel. buttery solid, faint char. odor; HLB 9; acid no. 18 max.; sapon. no. 130–155; 100% act.

Amerlate® W. [Amerchol] Isopropyl lanolate; nonionic; dispersant/wetting agent for pigments in personal care prods., pharmaceuticals; emulsifier, softener, lubricant, emollient; yel. buttery solid; faint char. odor; insol. in water; HLB 9; acid no. 18 max.; sapon. no. 135–165; hyd. no. 35–55; 100% act.

Amerlate® WFA. [Amerchol] Lanolin acid; anionic; see Amerlate® LFA; also wets and disperses pigments in makeups; ylsh.-tan waxy solid, mild waxy odor; misc. with warm min. oil, IPP, IPM, castor oil; m.p. 60 C; acid no. 120; sapon. no. 165; 100% conc.

Ameroxol® LE-4, LE-23. [Amerchol] Laureth-4 and -23 resp.; nonionic; o/w emulsifier and solubilizer; liq.; HLB 9.7 and 17.0; 100% conc.

Ameroxol® OE-2. [Amerchol] Oleth-2; nonionic; solubilizer, emulsifier, dispersant, stabilizer, lipophilic cosolv. for creams and lotions, shampoos, and detergents, fluid and gelled transparent emulsions, fragrance prods., and aerosols; pale straw-colored clear liq.; bland odor; sol. in min. oil, isopropyl esters, anhyd. ethanol; HLB 5.0; acid no. 0.2 max.; sapon. no. 2 max.; pH 4.5–7.0; 100% conc.

Ameroxol® OE-10, OE-20. [Amerchol] Oleth-10 and -20 resp.; nonionic; emulsifier, stabilizer, solubilizer for cosmetics and toiletries; wh. semisolid and waxy solid resp.; bland odor; sol. in ethanol, water, hydroalcoholics, glycols; HLB 12 and 15; cloud pt. 47–55 and 87–93 C; acid no. 0.5 max.; sapon. no. 2 max.; pH 4.5–7.0 and 4.5–7.0 (10% aq.); 100% conc.

Amerscreen® P, P 80/20. [Amerchol] 2 mole propoxylate of ethyl p-aminobenzoate; uv absorber, sunscreen for cosmetics, pharmaceutical, dermatological applics.; faint yel. gel and liq. resp.; pract. odorless; sol. in ethanol, IPA, propylene glycol, castor oil; acid no. 1 max.; sapon. no. 195–215 and 160–180; 100% act. and 80% act. in propylene glycol.

Amersette®. [Amerchol] Methacryloyl ethyl betaine/methacrylates copolymer; hair fixative; forms flexible clear film which gives natural hold and look; compatible with various propellants and additives; does not require neutralization; 50% alcoholic sol'n.

Amerstat® 233. [Drew] 3,5-Dimethyl tetra-hydro-2-H,1,3,5-thiadiazone-2-thione; antimicrobial in industrial water systems, preservative in aq. systems; water-misc.

Amerstat® 250. [Drew] 2-Methyl-4-isothiazolin 3-one and 5-chloro-2 methyl-4-isothiazolin-3-one; paper mill slimicide; water-misc.

Amerstat® 251. [Drew] 5-Chloro-2-methyl-4-isothiazolin-3-one and 2-methyl-4-isothiazolin-3-one; antimicrobial, preservative in latex; water-sol.

Amerstat® 252. [Drew] 5-Chloro-2-methyl-4-isothiazolin-3-one and 2-methyl-4-isothiazolin-3-one; antimicrobial, preservative for adhesives and min. slurries; water-sol.

Amerstat® 272. [Drew] Sodium dimethyl dithiocarbamate and disodium methylene bisdithiocarbamate; paper and sugar mill slimicide; water-misc.

Amerstat® 274. [Drew] Sodium dimethyl dithiocarbamate and disodium ethylene bisdithiocarbamate ethylene diamine; beet sugar mill antimicrobial agent; water-misc.

Amerstat® 282. [Drew] Methylene bis (thiocyanate); antimicrobial in industrial water systems; preservative in water-containing systems; water-disp.

Amerstat® 294. [Drew] Bis (trichloromethyl) sulfone; paper mill slimicide; water-disp.

Amerstat® 300. [Drew] 2,2, Dibromo-3-nitrilo propionamide; paper mill slimicide, antimicrobial agent for enhanced oil recovery systems; preservative for metal working fluids containing water; water-misc.

Amerzine 35. [Drew] Hydrazine sol'n. containing an org. catalyst; corrosion inhibitor for removing oxygen in feed and boiler water, removes condensate, passivates iron, and copper surfs. in steam generating systems; yel. clear liq.; mildly ammoniacal odor; f.p. –82 F; dens. 8.54 lb/gal; 35% sol'n.

Amfotex FV-15. [Pulcra SA] Sulfonate coconut deriv.; foamer, corrosion inhibitor used in metal cleaning processes; liq.; 38% conc.

Amfotex FV-28. [Pulcra SA] Cocamido betaine; detergent, foaming agent, solubilizer, thickener, conditioner used in personal care prods.; liq.; water-sol.; 30% conc.

AMG 200 L. [Novo] Amyloglucosidase derived from *Aspergillus niger*; enzyme used in starch industry; catalyzes the reverse reaction; dk. brn. liq.; slight smell typ. of fermentation prods.; water-sol.; dens. 1.17 g/ml; pH 3–5.

Amgard® CHT. [Albright & Wilson] Stabilized encapsulated red phosphorus; flame retardant for thermoplastic and thermoset resins; 95% P.

Amgard® CPC 102. [Albright & Wilson] Phosphorus-based; flame retardant in masterbatch form for LDPE and crosslinked LDPE incl. film, sheet, inj. molding, extrusion, wire and cable, and jacketing applics.; red pellets; sp.gr. 1.4; 50% P in LDPE carrier.

Amgard® CPC 452. [Albright & Wilson] Phosphorus-based; flame retardant in masterbatch form for nylon resins for inj. molding, glass-filled resin, extrusion, and elec. applics.; red pellets; sp.gr. 1.3; bulk dens. 0.6 g/cc; 50% P in nylon 6/6 carrier.

Amgard® CPC 702. [Albright & Wilson] Phosphorus-based; flame retardant for thermoset epoxy resins; used for dielec./conductive adhesives, potting compds., coatings, encapsulants, molding compds., laminates; purple-red paste; 20–30 μ mean particle size solids; sp.gr. 1.6; 50% P in liq. bisphenol-A epoxy resin carrier.

Amgard® MC. [Albright & Wilson] Ammonium polyphosphate; flame retardant for intumescent coatings, plastics, and rubber; low water sol.; 29.5% P.

Amgard® ND. [Albright & Wilson] Dimelamine phosphate; flame retardant for intumescent coatings, plastics, and paper; very low water sol.; 45.8% N, 8.3% P.

Amgard® NH. [Albright & Wilson] Melamine phosphate; flame retardant for intumescent coatings, plastics, and paper; 38.6% N, 13.4% P.

Amgard® PI. [Albright & Wilson] Ammonium polyphosphate; flame retardant for nat. rubber and particle board; limited water sol.; 25–26% P.

Amgard® TBEP. [Albright & Wilson] Tributoxyethyl phosphate; plasticizer and defoamer, leveling agent for acrylic and styrenic floor polishes; clear mobile liq.

Amgard® TOF. [Albright & Wilson] Trioctyl phosphate; plasticizer for vinyl resins, syn. rubbers, NC; aids low flam. chars.; clear mobile liq.

Amical® 48. [Angus] Diiodomethyl p-tolyl sulfone; mildewcide, fungicide for adhesives and sealants, and in lumber, construction, home improvement, textile, and automotive industries; avail. as fine tan powd. and visc. lt. tan liq. disp.; insol. in water; sp.gr. 2.20 g/cc (powd.); m.p. 147–150 C (powd.); 95% (powd.) and 44–50% act. (liq.).

Amical® 50. [Angus] Diiodomethyl p-tolyl sulfone with color suppressant; see Amical 48; also preservative for latex paints; for color-critical applics.; tan fine powd. and visc. lt. tan liq. disp.; sol. (mg/ml): 1000 mg in dimethyl formamide, 350 mg in acetone, 263 mg in n-propyl acetate, 220 mg in tributyl phosphate, 182 mg in methyl Cellosolve, 114 mg in Carbitol acetate, 80 mg in benzene, 75 mg in Cellosolve acetate, 58 mg in dibutyl phthalate, 43 mg in toluene, 33 mg. in xylene, 0.1 mg. in water; sp.gr. 1.96 g/cc (powd.); 75% (powd.) and 37–43% act. (liq.).

Amical® Flowable. [Angus] Diiodomethyl-p-tolyl sulfone aq. suspension; preservative, mildewcide, algicide for polymeric systems, esp. latex paints, caulks, adhesives, leather; lt. gray finely divided suspension; sol. 0.1 mg/l in water; sp.gr. 1.32–1.33; dens. 11.05 lb/gal; visc. 600–1000 cps; b.p. 100 C; f.p. 0 C; pH 7.0–8.5; 40% act. in water.

Amical® WP. [Angus] Wettable powd. mildewcide for paints, caulks, adhesives, leather preservation; fine tan powd.; nil sol. in water; pH 9.0.

Amiet 102, 105, 110, 115, 202, 205, 210, 215, 302, 305, 310, 315, 402, 405, 410, 415, 502, 505, 510, 515. [KAO SA] Tert. amines EO condensation prods. of coco (100 series), soya (200), hydrog. tallow (300), oleyl (400), tallow (500); cationic/nonionic; antistat, textile dyeing assistant, softening and dispersing agent; liq., paste; 100% conc.

Amigase®. [Int'l. Bio-Synthetics] Amyloglucosidase; enzyme for hydrolyzing dextrins to dextrose; liq.

Amihope LL-11. [Ajinomoto] Lauroyl-1-lysine; amphoteric; surface modifier, coemulsifier, codispersant; solid; 100% conc.

Amiladin. [Dai-ichi Kogyo] PEG alkyl amine ether; nonionic; leveling agent for dyestuffs; paste; 50% conc.

Amiladin C-1802. [Dai-ichi Kogyo] PEG alkyl amine ether; nonionic; dispersant, washing agent; liq.; 100% conc.

Amine 2M12. [Berol Nobel] Dimethyldodecylamine; cationic; intermediate for detergent; liq.; 97% conc.

Amine 2M14. [Berol Nobel] Dimethyltetradecylamine; cationic; intermediates for detergent; liq.; 97% conc.

Amine 2M16. [Berol Nobel] Dimethylhexadecylamine; cationic; intermediate for detergents; liq.; 97% conc.

Amine 2MHBG. [Berol Nobel] Dimethylhydrogenated tallow amine; cationic; intermediate for detergent; liq.; 97% conc.

Amine 2MKK. [Berol Nobel] Dimethylcocoamine; cationic; intermediate for detergents; liq.; 97% conc.

Amine BG. [Berol Nobel] Tallow amine; cationic; emulsifier, corrosion inhibitor; paste; 98% conc.

Amine C. [Ciba-Geigy] Imidazoline deriv.; cationic; emulsifier, dispersant, detergent; used in acid cleaners; antistat for textiles, plastics; paints; corrosion inhibitor; cutting and rust preventive oils; yel. liq.; sol. in polar org. solv., hydrocarbons, relatively insol. water; 100% act.

Amine CS-1135. [Angus] Oxazolidine; corrosion inhibitor, alkaline prod. for metalworking fluids and aq. systems; 78% aq. sol'n. of water and hydrocarbon sol. act. materials.

Amine D®. [Hercules] 92% Dehydro-abietylamine derived from rosin; used as asphalt additive; flotation reagent, and for prod. of algicides, preservatives and corrosion inhibitors; sol. in solv.; water-disp.

Amine D® Acetate. [Hercules] 92-95% Dehydroabietylamine acetate derived from rosin; see Amine D; water-sol.

Amine HBG. [Berol Nobel] Tallow amine, hydrog.; emulsifier, corrosion inhibitor; solid; 98% conc.

Amine KK. [Berol Nobel] Cocoamine; cationic; emulsifier, corrosion inhibitor; liq.; 98% conc.

Amine O. [Ciba-Geigy] Oleyl imidazoline; cationic; emulsifier, dispersant, detergent, antistat used in acid cleaners and corrosion inhibitor formulations; amber liq.; sol. in polar org. solv., hydrocarbons, relatively insol. water; 100% act.

Amine OL. [Berol Nobel] Oleylamine; cationic; emulsifier, corrosion inhibitor; liq.; 98% conc.

Amine PMT®. [Angus] Aliphatic ditert. amine; alkaline material, corrosion inhibitor in aq. systems; liq.

Amine S. [Ciba-Geigy] Stearyl imidazoline; cationic; emulsifier, dispersant, detergent, antistat; tan wax; sol. in polar org. solv., hydrocarbons, relatively insol. water; 100% act.

Amine T. [Ciba-Geigy] Tall oil fatty acid imidazoline; see Amine O; dk. amber liq.; sol. in polar org. solv., hydrocarbons, relatively insol. water; 100% act.

Aminodermin CLR. [Henkel] Sulfur rich amino acid conc.; conditioner for structurally damaged hair; wh. powd.

Aminofoam C. [Croda] TEA-lauroyl animal collagen amino acids; anionic; detergent, conditioner for skin and hair care cleansing systems; liq.; water-sol.; m.w. 550; 40% conc.

Aminofoam K. [Croda] TEA-lauroyl animal keratin amino acids; foaming agent for shampoos, conditioners, facial cleansers; lt. amber liq.; m.w. 550; water-sol.; 40% act.

Amino Gluten MG. [Croda] Maize gluten amino acids, sodium chloride; conditioner for skin creams and lotions and hair conditioners; humectant for cosmetics and pharmaceuticals; amber liq.; m.w. 150; water-sol.; 15% act. in water.

Aminol CA-2. [Finetex] Ricinoleamide DEA; nonionic; softening agent, emulsifier, cosmetic emulsions; liq.; 100% conc.

Aminol CM, CM Flakes, CM-C Flakes, CM-D Flakes. [Finetex] Cocamide MEA; nonionic; soap additive and foam stabilizer, thickener, and foam booster; hair shampoos; solid, flakes; 100% conc.

Aminol COR-4, -4C. [Finetex] Lauramide DEA; nonionic; foam booster and thickener; detergent in drycleaning applications; liq.; 100% conc.

Aminol HCA. [Finetex] Cocamide DEA; nonionic; foam stabilizer in shampoos and household detergents; liq.; 100% conc.

Aminol KDE. [Finetex] Cocamide DEA; nonionic; foam stabilizer for personal care prods.; liq.; 100% conc.

Aminol LM-30C Special. [Finetex] Lauramide DEA; nonionic; foam booster or stabilizer in personal care prods.; liq.; 100% conc.

Aminol N-1918. [Finetex] Stearamide DEA and diethanolamine; nonionic; softener for textile; solid; 100% conc.

Aminol OF. [Finetex] Oleamide DEA; nonionic; thickening agent for hair colors; liq.; 100% conc.

Amino Silk. [Amerchol] Silk amino acids; silk protein deriv. used for conditioning in hair and skin care prods.; water sol'n.

Aminox Series. [Arjay] Oxyalkylated amines; corrosion inhibitors, solid wetters, demulsifier additives; liq.; oil-sol.; water-disp.

Aminoxid WS 35. [Goldschmidt] Cocamidopropylamine oxide; nonionic; detergent, emulsifier, wetting agent, softener, foam stabilizer for detergent preparations, cosmetic and pharmaceutical emulsions; liq.; 35% conc.

Amiter LGS-5. [Ajinomoto] N-lauroyl-L-glutamic acid, POE (5) stearyl ether diester; nonionic; emulsifier and emulsion stabilizer in cosmetics; wh. soft wax; HLB 5.4; 100% conc.

Amizyme. [Premier] Alpha-amylase (bacterial); enzyme with starch liquefying and heat stability properties; preparation of paper coatings and sizes; used in paper industry; tablets, liq., powd.; dissolves in water.

Ammonyx 4, 4B, 485, 4002. [Stepan] Stearalkonium chloride; cationic; emulsifier, conditioner, softener, emollient for cosmetics; paste except 4002 (powd.); sp.gr. 0.99, 0.99, 0.45, 0.47, 0.52 resp.; flash pt. > 200, > 200, 122, 130, and 170 F; 24–28, 24–26, 87 min., 92 min., and 96% min. solids.

Ammonyx 4-IPA. [Stepan] Stearalkonium chloride; conditioner, softener, emollient for hair rinses, skin prods.; cationic emulsifier. Discontinued.

Ammonyx 856. [Stepan] Tallowalkonium chloride; cationic; industrial grade pigment dispersant and suspending agent, demulsifier for solv. extraction processes; paste; sp.gr. 0.90. Discontinued.

Ammonyx 4080. [Stepan] 1-Methyl-1-tallow amidoethyl-2-heptadecyl imidazolinium methosulfate; cationic; fabric softener for home and industrial applic.; liq.; sp.gr. 0.97; flash pt. 68 F; 74–76% quat.; 81% max. solids. Discontinued.

Ammonyx CA-Special. [Stepan] Stearalkonium chloride; conditioner, softener, emollient for hair rinses, skin prods.; cationic emulsifier; paste; 20–22.5% act.

Ammonyx CDO. [Stepan] Cocamidopropyl amine oxide; nonionic; wetting, foaming agent, foam stabilizer, conditioner for bubble baths, bath oils, dishwashing, hair color systems, softeners, cleansers; liq.; sp.gr. 1.02; flash pt. > 200 F. Discontinued.

Ammonyx CETAC, CETAC-30. [Stepan] Cetrimonium chloride; cationic; emulsifier, conditioner, softener, emollient for cosmetics; liq.; 25 and 30% min. solids resp. Discontinued.

Ammonyx CO. [Stepan] Palmitamine oxide; nonionic; conditioner, detergent, foam stabilizer, visc. builder used in cosmetic, household, and janitorial prods.; wetting agent in conc. electrolyte sol'ns.; liq.; sp.gr. 0.96; flash pt. > 200 F; 29.0–31.0% amine oxide. Discontinued.

Ammonyx DMCD-40. [Stepan] Lauramine oxide; nonionic; wetting, foaming agent, foam stabilizer for cosmetic, home and janitorial prods.; liq.; sp.gr. 0.91; flash pt. 86 F; 40–42% amine oxide. Discontinued.

Ammonyx KP. [Stepan] Olealkonium chloride; cationic; conditioner, antistat in clear hair rinses; liq.; sp.gr. 0.98; flash pt. > 200 F; 52% min. solids.

Ammonyx LO. [Stepan] Lauramine oxide; nonionic; see Ammonyx CO; also wetting, foaming agent, foam stabilizer, grease emulsifier for personal care prods.; liq., sp.gr. 0.96; flash pt. > 200 F; 29–31% amine oxide.

Ammonyx MCO, MO. [Stepan] Myristamine oxide; nonionic; wetting and foaming agent, foam stabilizer for cosmetics, home and janitorial prods.; liq.; sp.gr. 0.96; 29–31% amine oxide. Discontinued.

Ammonyx SO. [Stepan] Stearamine oxide; nonionic; see Ammonyx CO; also conditioner, emulsifier; paste; sp.gr. 0.99; flash pt. > 200 F; 24.5–26.5% amine oxide. Discontinued.

Ammonyx SQ. [Stepan] Stearamine oxide; nonionic; conditioner, emulsifier used in personal care prods.; paste; sp.gr. 0.99; 25% amine oxide. Discontinued.

Ammonyx T. [Stepan] Cetalkonium chloride; cationic; emulsifier, wetting agent, germicide; latex dispersions; sp.gr. 0.99; 25% conc. Discontinued.

Ammonyx TDO. [Stepan] Tallow amidopropylamine oxide; nonionic; wetting, foaming agent, foam stabilizer, conditioner for bubble baths, bath oils, dishwashing, hair color systems, softeners, cleansers; liq.; 30% conc. Discontinued.

Amoco® Resin 18-210, 18-240, 18-290. [Amoco] Poly-α-methylstyrene; extrusion and molding process aid in ABS, PVC, CPVC, and semirigid vinyl, thermoplastic urethanes, molded rubbers, and thermoplastic elastomers; modifier and reinforcer in adhesives, thermoplastic powd. coatings, hot-melt coatings, solv.-reduced coatings, and overprint varnishes; also for vinyl house siding, shoe soles, lead-stabilized PVC pipe and fittings; Gardner 1 color; m.w. 685, 790, and 960 resp. (60 F); sp.gr. 1.075 (60/60 F); visc. (G-H) J–L, U–V, and Z-Z_1 (60% in toluene); ref. index 1.61; cloud pt. 170, 212, and 251 F; soften. pt. 210, 245, and 286 F.

Amonyl BR 1244. [Seppic] Lauralkonium bromide; germicide; liq.

AMP. [Angus] 2-Amino-2-methyl-1-propanol; anionic; emulsifier, catalyst; dispersant for pigments and latex paints; corrosion inhibitor; stabilizer; resin solubilizer; APHA 20 solid; m.w. 89.14; sp.gr. 0.928; dens. 7.78; visc. 102 cp (30 C); m.p. 30 C; b.p. 165 C; flash pt. 172 F; 100% act.

AMP-95. [Angus] 2-Amino-2-methyl-1-propanol; anionic; see AMP; colorless liq.; m.w. 89.14; dens. 7.85; sp.gr. 0.942; visc. 147 cp; f.p. –2 C; flash pt. 182 F (TCC); surf. tens. 36–38 dynes/cm; 95% act.

AMPD. [Angus] 2-Amino-2-methyl-1,3-propanediol; pigment dispersant, neutralizing amine, corrosion inhibitor, acid-salt catalyst, pH buffer, chemical and pharmaceutical intermediate, solubilizer; m..w. 105.1; sol. 250 g/100 ml water; m.p. 109 C; b.p. 151 C; pH 10.8 (0.1M aq. sol'n.).

Ampho B11-34. [Capital City] Complex cocobetaine derived from dimethyl amine propyl amine; amphoteric; high foaming surfactant with foam stability, wetting agent, solubilizer used in cosmetic formulations; lt. yel., clear liq.; mild odor; water sol.; dens. 8 lb/gal; sp.gr. 0.96; biodeg.; 35% act.

Amphocerin K. [Henkel] Cetearyl alcohol and emollients; cream base for mfg. of lt. and smooth creams and ointments of the w/o type; translucent soft solid.

Ampholan U 203. [Harcros UK] Amphoteric surfactant with carboxyl and amino groups; amphoteric; detergent, foaming and stabilizing agent, dispersant; pale amber liq.; char. odor; water sol.; sp.gr. 1.035; visc. 132 cs; flash pt. > 200 F (COC); pour pt. < 0 C; pH 6.0–8.0 (1% aq); 30% act.

Amphomer®. [Nat'l. Starch] Methacrylate copolymer; hair fixative resin for maximum curl; hair sprays, setting lotions, conditioners; sol. in ethanol and hydrocarbon/ethanol blends; when neutralized, sol. in water and water/ethanol blends.

Amphoram CB A30. [Ceca SA] Coco-alkyl betaine; amphoteric; detergent, bactericidal, emulsifier; liq.; 30% conc.

Amphoram CT 30. [Ceca SA] Coco-alkyl taurine; amphoteric; detergent, bactericidal, emulsifier; liq.; 30% conc.

Amphosol CA. [Stepan] Cocamidopropyl betaine; amphoteric; visc. builder, foam enhancer, base for cosmetics and liq. detergents; straw clear liq.; pH 5.0 (10%); 30% act. Discontinued.

Amphosol CG. [Stepan] Cocamidopropyl betaine; foam booster, visc. builder, and lime soap dispersant; used in personal care prods.; amber clear liq.; pH 4.5–6.5; 29–31% act.

Amphoterge® S. [Lonza] Stearoamphoacetate; amphoteric; textile softener; used in creme rinses; paste; 25% conc.

Amphoteric 400. [Exxon] Iminopropionate, partial sodium salt; amphoteric; detergent, coupler for detergents, laundry detergents; defoamer in latex paints; corrosion inhibitor in metalworking lubricants; leather lubricant; yel. liq.; sol. in glycols, water, alcohols; sp.gr. 1.09; pH 6 (5%); 50% solids.

Amphoteric C. [Exxon] Cocoamphopropionate; amphoteric; high foam wetting agent, detergent, coupler for hard. surf. cleaners, shampoos, etc.; corrosion inhibitor in metalworking lubricants; visc. builder; fabric softener; emulsion polymerization; fire fighting foams; clear lt. yel. liq.; sol. in glycols, water, alcohols; sp.gr. 1.04; pH 6 (5%); 35% solids.

Amphoteric L. [Exxon] Coco amphoteric; foam stabilizer/booster, surfactant for liq. detergents; lt. amber liq.; sp.gr. 1.04; pour pt. 35 F; pH 5–8 (5%); 35% min. act.

Amphoteric N. [Exxon] Sodium C12–15 alkoxypropyl iminodipropionate; amphoteric; high foam wetting agent, coupler for shampoos, detergents; corrosion inhibitor in metalworking lubricants; visc. builder; fire fighting foams; clear lt. amber liq.; sol. in glycols, water, alcohol; sp.gr. 1.04; pH 6 (5%); 35% solids in water.

Amsco 140 Solvent 66/3. [Unical] Aliphatic solv. consisting primarily of C_{10-12} sat. hydrocarbons; med.-boiling, combustible liq. solv.; specially treated to reduce aromatics and olefins; meets Los Angeles Rule 66/San Francisco Reg. 3; colorless clear liq.; mild aliphatic odor; b.p. 366 F; misc. with most common org. solvs.; sp.gr. 0.784; dens. 6.53 lb/gal; visc. 1.72 cSt; vapor pressure < 1 mm Hg; ref. index 1.4319; flash pt. (TCC) 141 F; combustible; 51.3 vol.% cycloparaffins; 47.4 vol.% paraffins.

Amsco Benzene. [Unical] Aromatic hydrocarbon; low-boiling, flam. liq. solv.; nitration grade quality; colorless clear liq.; lt. aromatic odor; b.p. 79.7 C; misc. with most common org. solvs.; sp.gr. 0.884; dens. 7.36 lb/gal; visc. 0.75 cSt; vapor pressure 76.2 mm Hg; solid. pt. 5.49 C; ref. index 1.4987; flash pt. (TCC) 12 F; 99.9 vol.% benzene.

Amsco Cyclohexane. [Unical] C_6 sat. cycloparaffin; low-boiling, flam. liq. solv.; high purity; colorless clear liq.; mild char. odor; b.p. 80.6 C; misc. with most common org. solvs.; sp.gr. 0.783; dens. 6.52 lb/gal; visc. 1.25 cSt; vapor pressure 80.0 mm Hg; solid. pt. 6.39 C; ref. index 1.4229; flash pt. (TCC) –1 F; 99.9+ vol.% cyclohexane.

Amsco Heptane. [Unical] Aliphatic consisting primarily of C_7 sat. hydrocarbons; low-boiling, flam. liq. solv.; specially treated to reduce aromatics and olefins and meets the air quality requirements of Los Angeles Rule 66/San Francisco Reg. 3; colorless clear liq.; lt. aliphatic odor; b.p. 92.6 C; misc. with most common org. solvs.; sp.gr. 0.699; dens. 5.82 lb/gal; visc. 0.61 cSt; vapor pressure 45.3 mm Hg; ref. index 1.3906; flash pt. (TCC) < 20 F; 89.0 vol.% paraffins, 26.5 vol.% n-heptane, 11.0 vol.% cycloparaffins.

Amsco Hexane. [Unical] Aliphatic consisting primarily of C_6 sat. hydrocarbon; low-boiling, flam. liq. solv.; specially treated to reduce aromatics and olefins; meets the air quality requirements of Los Angeles Rule 66/San Francisco Reg. 3; colorless clear liq.; lt. aliphatic odor; b.p. 65.3 C; misc. with most common org. solvs.; sp.gr. 0.674 dens. 5.61 lb/gal; visc. 0.50 cSt; vapor pressure 139.9 mm Hg; ref. index 1.3778; flash pt. (TCC) < 0 F; 89.0 vol.% paraffins, 48.7 vol.% *n*-hexane, 11.0 vol.% cycloparaffins.

Amsco Lactol Spirits. [Unical] Aliphatic consisting primarily of C_{6-8} hydrocarbons; low-boiling, flam. liq. solv.; meets air quality requirements of Los Angeles Rule 66/San Francisco Reg. 3; colorless clear liq.; lt. aliphatic odor; b.p. 202 F; misc. with most common org. solvs.; sp.gr. 0.746; dens. 6.21 lb/gal; visc. 0.70 cSt; vapor pressure 42.0 mm Hg; ref. index 1.4117; flash pt. (TCC) < 20 F; 46.4 vol.% paraffins, 41.5 vol.% cycloparaffins, 12.1 vol.% aromatics.

Amsco Min. Spirits 66/3. [Unical] Low aromatic Stod., aliphatic consisting primarily of C_{9-12} sat. hydrocarbons with < 2% aromatics; med.-boiling, combustible liq. solv.; specially treated to reduce aromatics and olefins; meets air quality requirements of Los Angeles Rule 66/San Francisco Reg. 3; colorless clear liq.; mild aliphatic odor; b.p. 316 F; misc. with most common org. solvs.; sp.gr. 0.773; dens. 6.44 lb/gal; visc. 1.36 cSt; vapor pressure 2.9 mm Hg; ref. index 1.4262; flash pt. (TCC) 107 F; 50.3 vol.% cycloparaffins, 48.6 vol.% paraffins.

Amsco Min. Spirits 75. [Unical] Low aromatic Stod.; aliphatic solv. consisting primarily of C_{9-12} hydrocarbons with < 8% aromatics; med.-boiling, combustible liq. solv.; specially treated to reduce aromatics and olefins; meets air quality requirements of Los Angeles Rule 66/San Francisco Reg. 3; colorless clear liq.; mild aliphatic odor; b.p. 315 F; misc. with most common org. solvs.; sp.gr. 0.779; dens. 6.48 lb/gal; visc. 1.330 cSt; vapor pressure 3.0 mm Hg; ref. index 1.4308; flash pt. (TCC) 107 F; 48.2 vol.% paraffins, 44.3 vol.% cycloparaffins, 7.5 vol.% aromatics.

Amsco Naphthol Spirits 66/3. [Unical] Low aromatic rapid-drying Stod.; aliphatic consisting primarily of C_{9-11} sat. hydrocarbons; med.-boiling, combustible liq. solv.; specially treated to reduce aromatics and olefins; meets air quality requirements of Los Angeles Rule 66/San Francisco Reg. 3; colorless clear liq.; mild aliphatic odor; b.p. 316 F; misc. with most common org. solvs.; sp.gr. 0.769; dens. 6.40 lb/gal; visc. 1.25 cSt; vapor pressure 2.7 mm Hg; ref. index 1.4257; flash pt. (TCC) 106 F; 50.1 vol.% paraffins, 49.1 vol.% cycloparaffins.

Amsco Nonene. [Unical] Propylene trimer; olefinic solv. consisting primarily of branched C_9 unsat. hydrocarbons; volatile, low-boiling, flam. liq. solv.; colorless clear liq.; mild olefinic odor; b.p. 137.9 C; misc. with most common org. solvs.; sp.gr. 0.745; dens. 6.20 lb/gal; visc. 0.84 cSt; vapor pressure 6.7 mm Hg; ref. index 1.4221; flash pt. (TCC) 75 F; 99.9+ vol.% total olefins, 99.6+ vol.% monoolefins.

Amsco Odorless Min. Spirits. [Unical] Aliphatic solv. consisting primarily of C_{8-12} sat. hydrocarbons; med.-boiling, combustible liq. solv.; specially treated to reduce odor, aromatics, and olefins; meets requirements of Los Angeles Rule 66/San Francisco Reg. 3; colorless clear liq.; odorless; b.p. 347 F; misc. with most common org. solvs.; sp.gr. 0.759; dens. 6.32 lb/gal; visc. 1.91 cSt; vapor pressure 1.2 mm Hg; ref. index 1.4218; flash pt. (TCC) 125 F; 85.7 vol.% paraffins, 14.3 vol.% cycloparaffins.

Amsco Regular Min. Spirits. [Unical] Stod.; aliphatic solv. consisting primarily of C_{9-12} hydrocarbons; med.-boiling, combustible liq. solv.; colorless clear liq.; mild aliphatic odor; b.p. 315 F; misc. with most common org. solvs.; sp.gr. 0.788; dens. 6.56 lb/gal; visc. 1.29 cSt; vapor pressure 3.1 mm Hg; ref. index 1.4378; flash pt. (TCC) 108 F; 46.1 vol.% paraffins, 37.9 vol.% cycloparaffins, 16.0 vol.% aromatics.

Amsco Rubber Solvent. [Unical] Aliphatic solv. consisting primarily of C_{4-9} hydrocarbons; low-boiling, flam. liq. solv.; meets air quality requirements of Los Angeles Rule 66/San Francisco Reg. 3; colorless clear liq.; lt. aliphatic odor; b.p. 121 F; misc. with most common org. solvs.; sp.gr. 0.697; dens. 5.80 lb/gal; visc. 0.57 cSt; vapor pressure 180 mm Hg; ref. index 1.3934; flash pt. (TCC) < 0 F; 73.0 vol.% paraffins, 19.4 vol.% cycloparaffins, 7.6 vol.% aromatics.

Amsco Solv D. [Unical] Aromatic hydrocarbon; rule 66 solv.; colorless clear liq.; odorless; b.p. 162.2 C; sp.gr. 0.859 (15.56 C); dens. 7.15 lb/gal (14.45 C); vapor pressure 2.1 mm Hg; flash pt. (TCC) 119 F; 77% aromatics; 13% cycloparaffins, 10% paraffins.

Amsco Solv F. [Unical] Aromatic hydrocarbon; rule 66 solv.; b.p. 170.6 C; sp.gr. 0.876 (15.56 C); dens. 7.30 lb/gal (14.45 C); vapor pressure 1.3 mm Hg; flash pt. (TCC) 120 F; 70% aromatics; 17% paraffins; 13% cycloparaffins.

Amsco Solv G. [Unical] Aromatic solv. consisting primarily of C_{9-11} alkyl benzenes; med.-boiling, combustible liq. solv.; colorless clear liq.; mild aromatic odor; b.p. 363 F; misc. with most common org. solvs.; sp.gr. 0.900; dens. 7.49 lb/gal; visc. 1.36 cSt; vapor pressure 0.7 mm Hg; ref. index 1.5140; flash pt. (TCC) 149 F; 99.9+ vol.% total aromatics.

Amsco Special Naphtholite 66/3. [Unical] Low aromatic VM&P naphtha; aliphatic solv. consisting primarily of C_{8-9} sat. hydrocarbons; med.-boiling, flam. liq. solv.; specially treated to reduce aromatics and olefins; meets air quality requirements of Los Angeles Rule 66/San Francisco Reg. 3; colorless clear liq.; lt. aliphatic odor; b.p. 249 F; misc. with most common org. solvs.; sp.gr. 0.750; dens. 6.24 lb/gal; visc. 0.93 cSt; vapor pressure 13.5 mm Hg; ref. index 1.4156; flash pt. (TCC) 54 F; 61.9 vol.% cycloparaffins, 37.4 vol.% paraffins.

Amsco Super High Flash Naphtha. [Unical] Aromatic solv. consisting primarily of C_{8-10} alkyl benzenes; med.-boiling, combustible liq. solv.; specially treated to reduce odor, aromatics, and olefins; meets requirements of Los Angeles Rule 66/San Francisco

Reg. 3; colorless clear liq.; mild aromatic odor; b.p. 318 F; misc. with most common org. solvs.; sp.gr. 0.873; dens. 7.27 lb/gal; visc. 0.95 cSt; vapor pressure 2.7 mm Hg; ref. index 1.4969; flash pt. (TCC) 114 F; 99.9+ vol.% total aromatics.

Amsco Tetramer. [Unical] Propylene tetramer, dodecene; olefinic solv. consisting primarily of branched C_{11-12} unsat. hydrocarbons; med.-boiling, combustible liq. solv.; colorless clear liq.; mild olefinic odor; b.p. 183.9 C; misc. with most common org. solvs.; sp.gr. 0.772; dens. 6.43 lb/gal; visc. 1.57 cSt; vapor pressure 0.8 mm Hg; bromine no. 112.1; ref. index 1.4341; flash pt. (TCC) 137 F; 99.8+ vol.% total olefins.

Amsco Textile Spirits. [Unical] Aliphatic solv. consisting primarily of C_{6-7} sat. hydrocarbons; low-boiling, flam. liq. solv.; specially treated to reduce aromatics and olefins; meets air quality requirements of Los Angeles Rule 66/San Francisco Reg. 3; colorelss clear liq.; lt. aliphatic odor; b.p. 162 F; misc. with most common org. solvs.; sp.gr. 0.688; dens. 5.73 lb/gal; visc. 0.53 cSt; vapor pressure 105.5 mm Hg; ref. index 1.3855; flash pt. (TCC) < 0 F; 81.2 vol.% paraffins, 18.8 vol.% cycloparaffins.

Amsco Toluene. [Unical] C_7 aromatic hydrocarbon; low-boiling, flam. liq. solv.; nitration grade quality; colorless clear liq.; lt. aromatic odor; b.p. 110.3 C; misc. with most common org. solvs.; sp.gr. 0.871; dens. 7.25 lb/gal; visc. 0.69 cSt; vapor pressure 23.8 mm Hg; ref. index 1.4933; flash pt. (TCC) 45 F; 99.9+ vol.% toluene.

Amsco Xylene. [Unical] Aromatic solv. consisting primarily of C_8 mixed isomers; low-boiling, flam. liq. solv.; five degree quality; colorless clear liq.; lt. aromatic odor; b.p. 137.7 C; misc. with most common org. solvs.; sp.gr. 0.871; dens. 7.25 lb/gal; visc. 0.77 cSt; vapor pressure 6.6 mm Hg; ref. index 1.4949; flash pt. (TCC) 81 F; 97.8 vol.% total C_8 aromatics; 99.6 vol.% total aromatics.

Amyl Cadmate®. [Vanderbilt] Cadmium diamyldithiocarbamate; ultra accelerator for natural and polyisoprene rubbers; lt. amber liq.; m.w. 577.25; dens. 1.08 mg/m³; flash pt. 157 C; 50% conc.

Amyl Ledate®. [Vanderbilt] Lead diamyldithiocarbamate; ultra accelerator for nat. and polyisoprene rubber; lt. amber liq.; m.w. 672.05; dens. 1.10 mg/m³; 50% conc. in oil.

Amyl Zimate®. [Vanderbilt] Zinc diamyldithiocarbamate; ultra accelerator for nat. and syn. rubbers; lt. amber liq.; m.w. 530.22; dens. 0.99 mg/m³; 50% conc.

Amyx A-25S. [Clough] Stearyl dimethyl benzyl ammonium chloride; cationic; conditioner, softener, and emollient for hair rinses, skin creams, and lotions; emulsifier; paste; 25% min. act.

Ancamine® 2049. [Air Prods.] Modified cycloaliphatic diamine crosslinker.

Anchoid. [Anchor] Sodium naphthalene formaldehyde sulfonate; dispersant for latex applics., cement, concrete, ceramics; buff powd.; sol. in cold water.

Anchor DBD. [Anchor] Activated dithiocarbamate; nonstaining, very fast accelerator used with dry nat. rubber, SBR stocks, and nat. rubber latex; recommended for cements, adhesives, and calendered and extruded compds.; brn. visc. liquid; sol. in hydrocarbons, ketones, alcohols, carbon disulfide, and chlorinated solvs.; sp.gr. 1.10.

Anchor DNPD. [Anchor] N,N′-di-beta-naphthyl-p-phenylene diamine; nonstaining antioxidant for rubber, SBR, NBR, BR, and polyethylene, latex foam, thread, and dipped goods, in adhesives, cable insulation and sheathing; brnsh.-gray fine powd.; sol. in acetone, chloroform and benzene; insol. in petrol. solvents, ethanol, CCl_4, water and aq. alkalis; sp.gr. 1.28; m.p. 230 C.

Anchor DOTG. [Anchor] Diortho-tolylguanidine; med.-speed accelerator; best as sec. accelerator in CR; off-wh. powd.; sol. in chloroform, methanol; sp.gr. 1.10; m.p. 175 C.

Anchor DPG. [Anchor] Diphenyl guanidine; med.-speed accelerator used to activate thiazoles and sulfenamides; for vulcanization of thick articles, footwear applics.; sec. gelling agent in latex foam; peptizer for sulfur-modified CR; wh. powd.; sol. in benzene hydrocarbons, alcohol, acetone, chloroform; sp.gr. 1.19; m.p. 147 C.

Anchor HDPA. [Anchor] Alkylated diphenyl amines, mainly heptylated; antioxidant for use in foamed latex carpet backings; dk. brn. visc. liq.; sol. in acetone, CCl_4, most esters and min. oils; insol. in water; sp.gr. 0.96.

Anchor HDPA/SE. [Anchor] Alkylated diphenylamines (principally heptylated diphenylamine) with selected surface-active agents; antioxidant for the compding. of nat. and syn. rubber latices; for latex carpet backings, latex-based adhesives; dk. brn. visc. liq.; mild char. odor; sol. in acetone, CCl_4, most esters and min. oils; insol. in water; readily emulsified in cold water; sp.gr. 0.98; visc. 20 poise.

Anchor ODPA. [Anchor] Octylated diphenylamine; low staining antioxidant for use in nat. and syn. rubbers, for footwear, cable insulation, mechanical rubber goods; lt. brn. flakes; sol. in petrol, benzene, ethylene dichloride, and acetone; insol. in water; sp.gr. 0.99; m.p. 82 C min.

Anchor PBN. [Anchor] Phenyl-beta-naphthylamine; staining antioxidant for use in rubber goods, e.g., tires and belting; stabilizer in mfg. of staining grades of syn. rubber; fawn to dk. gray flakes; sol. in acetone, chloroform, toluene, ethyl acetate; slightly sol. in alcohol and petrol. spirit; insol. in water; sp.gr. 1.23; m.p. 104 C.

Anchor ZDBC. [Anchor] Zinc diethyl dithiocarbamate; ultra fast, nonstaining accelerator used in latex and dry rubber compding; antioxidant in pressure-sensitive and hot-melt adhesives; wh. to cream powd.; sol. in petrol. and benzene hydrocarbons, carbon disulfide and chlorinated solvs.; sp.gr. 1.24; m.p. 106 C.

Anchor ZDEC. [Anchor] Zinc diethyl dithiocarbamate; ultra fast, nonstaining accelerator used in nat. and syn. rubbers incl. SBR, NBR, polyisoprene, butyl, chlorobutyl, and EPDM for foam, cable insulation, molded goods, etc.; wh. to ylsh. powd.; sol. in benzene hydrocarbons, carbon disulfide, chlorinated solvs., and dilute alkalis; sp.gr. 1.48; m.p. 178 C.

Ancor 120. [Air Prod.] Amine/acetylenic alcohol; corrosion inhibitor for ferrous and ferric environments; steel pickling; electroplating brightener.

Angamol 75. [Lubrizol] Zinc dialkyl dithiophosphate in compd. formula; antiwear agent used to prepare hydraulic fluids; lubricant; liq.; oil-sol.

Anstex AK-25. [Toho] Special phosphate; antistat for syn. fibers; liq.; 55% act.

Antara® HR-719. [Rhone-Poulenc Surf.] Nonfoaming phosphate ester lubricant used in syn., semi-syn., and aq.-based cutting, grinding, drawing fluids; liq.

Antara® LB-400. [Rhone-Poulenc Surf.] Fatty alcohol polyethyleneoxy phosphate ester acid; anionic; lubricant, EP additive, rust inhibitor, wetting agent, emulsifier, with detergency, dispersancy properties; chain lubricants, circulating, diesel, rolling, turbine, gear, and steel-mill slushing oils, cutting fluids, drawing compds., greases, hydraulic fluids; glass cutting and polishing lubricants; amber opaque visc. liq.; sol. in oil, water, and aliphatic and aromatic solvs.; sp.gr. 1.04; 99% act.

Antara® LE-500. [Rhone-Poulenc Surf.] Phosphate acid ester; aromatic hydrophobic base; anionic; lubricant, emulsifier used in oil- and water-based cutting fluids, hydraulic fluids, and rolling oils; yel. clear visc. liq.; sol. in oil, water, aliphatic and aromatic solv.; sp.gr. 1.11; 100% act.

Antara® LE-700. [Rhone-Poulenc Surf.] Phosphate acid ester; aromatic hydrophobic base; anionic; lubricant used in water-based cutting fluids; slight cloudy, VCS 8 visc. liq.; water-sol.; 100% act.

Antara® LF-200. [Rhone-Poulenc Surf.] Complex org. phosphate ester free acid; anionic; lubricant; VCS 8, slight cloudy visc. liquid; 100% act.

Antara® LK-500. [Rhone-Poulenc Surf.] Complex phosphate ester of an ethoxylated linear alcohol; low-foaming lubricant for metalworking fluids, with EP, antiwear, and rust inhibition properties; pale yel., slightly visc. clear liq.; sol. in xylene, ethanol, ethylene glycol, CCl_4, butyl Cellosolve, water; sp.gr. 1.11; visc. 21 cSt (100 C); biodeg.; 100% act.

Antara® LM-400. [Rhone-Poulenc Surf.] Phosphate acid ester; aromatic hydrophobic base; anionic; lubricant for rust inhibitor in formulating aq. cutting fluids; used in rolling and cutting oils, hydraulic fluids; yel. clear liq.; sol. in oil, aliphatic and aromatic solv.; water-disp.; sp.gr. 1.05; 100% act.

Antara® LM-600. [Rhone-Poulenc Surf.] Phosphate acid ester; aromatic hydrophobic base; anionic; lubricant and emulsifier for use in oil- and water-based cutting and hydraulic fluids, and rolling oils; yel. clear visc. liq.; sol. in oil, water, aliphatic and aromatic solv.; sp.gr. 1.06; 100% act.

Antara® LP-700. [Rhone-Poulenc Surf.] Free acid of a complex org. phosphate ester; anionic; lubricant; yel., clear visc. liq.; sol. in water, aliphatic and aromatic solv.; sp.gr. 1.20; 100% act.

Antara® LS-500. [Rhone-Poulenc Surf.] Phosphate acid ester; aliphatic base; anionic; see Antara LM-600; liq.; sol. see Antara LM-600; 100% act.

Antarol W 160. [Aquatec Quimica] Min. oil and fatty acid ester; antifoamer used in mfg. of acrylonitrile monomer; water-disp.

Antaron FC-34. [Rhone-Poulenc Surf.] Monocarboxyl coco imidazoline compd.; amphoteric; detergent, wetting agent, emulsifier, dispersant, emollient, surfactant; fulling agent for woolen/worsted fabrics; emulsifier for leather processing; in bubble baths, hair, upholstery, rug shampoos, liq. dishwashing and hard-surf. detergents; amber clear semivisc. liq.; sol. in water and high electrolyte sol'ns.; surf. tens. 32.0 dynes/cm (0.0155%); > 38% conc.

Antaron MC-44. [Rhone-Poulenc Surf.] Dicarboxylic coco imidazoline, sodium salt; amphoteric; emulsifier, solubilizer, coupling agent for nonirritating shampoos, skin cleaners, cosmetics, industrial and household cleaners; amber visc. liq.; 38% act.

Antaron PC-37. [Rhone-Poulenc Surf.] Ethyl PEG-15 cocamine sulfate; amphoteric; surfactant; solubilizer in ultra-mild baby and adult conditioning shampoos; amber clear slightly visc. liq.; misc. in water, ethanol; sp.gr. 1.15; visc. 1100–1200 cps; surf. tens. 39 dynes/cm (0.1%); biodeg.; 75% act.

Antarox® G-200. [Rhone-Poulenc Surf.] Alkylpoly(ethyleneoxy) glycol amide; nonionic; dispersant for ferrous metals; spin bath additive used in mfg. of tire cords and textile filaments; mfg. of cellophane and steelmaking; brn. opaque visc. liq.; ref. index 1.48; cloud pt. 64–68 C; 100% act.

Anthium Dioxide. [Int'l Dioxide] Chlorine dioxide, stabilized; broad spectrum biocide, preservative; water-misc.

Anthoxan. [Henkel] 4-Isopropyl-5, 5-dimethyl-1, 3-dioxane; raw material for herbal fragrances.

Antiblaze® 19. [Albright & Wilson] Neutral cyclic diphosphonate ester; flame retardant for textiles and plastics incl. rigid RIM; liq.; 21% P.

Antiblaze® 78. [Albright & Wilson] Chlorinated phosphonate ester; flame retardant for rigid and flexible urethanes, bonded foam, semidurable textile applics., and phenolic-based laminates; liq. 12% P, 34% Cl.

Antiblaze® 80. [Albright & Wilson] Tris (2-chloropropyl) phosphate; flame retardant for rigid urethane systems; enhances flourocarbon compatibility; liq.; 9.3% P, 33% Cl.

Antiblaze® 100. [Albright & Wilson] Chlorinated diphosphate ester; flame retardant for virgin and bonded flexible urethane foam; used in textiles, paper, adhesives, and epoxy and phenolic-based laminates; low volatility; liq.; 10.6% P, 36% Cl.

Antiblaze® 125. [Albright & Wilson] Chlorinated phosphorus ester; low visc. flame retardant for bonded flexible urethane foams and fabric backcoatings; liq.; 11.2% P, 34% Cl.

Antiblaze® 150. [Albright &Wilson] Chlorinated phosphorus ester; flame retardant for flexible PU foams; liq.; 11.3% P, 35% Cl.

Antiblaze® 175. [Albright & Wilson] Chlorinated phosphorus ester; low visc. flame retardant for molded flexible foams, pkg. components; liq.; 10.1% P, 35% Cl.

Antiblaze® 195. [Albright & Wilson] Tris (dichloropropyl) phosphate; flame retardant for flexible PU foams; exc. hydrolytic stability; liq.; 7.2% P, 49.1% Cl.

Antiblaze® 1045. [Albright & Wilson] Neutral cyclic diphosphonate ester; cost-effective flame retardant for thermoplastics; liq.; 20% P.

Antiblaze® DMMP. [Albright & Wilson] Dimethyl methylphosphonate; flame retardant for rigid urethane foams, unsat. polyester resins, and water-borne latexes; also used as a solv.; liq.; 25% P.

Antiblaze® TDCP/LV. [Albright & Wilson] Low visc. version of Antiblaze 195; flame retardant for flexible PU foams; liq.; 7.4% P, 47% Cl.

Anticake 17. [Exxon] Amine; anticaking agent for fertilizers; water-insol.

Antifoam 124. [Continental] Nonsilicone; all purpose defoamer; water-disp.

Antifoam Base 263. [Soluol] Silicone containing compd.; defoamer; water-sol.

Antifoam C-196. [Rhone-Poulenc Surf.] Silicon emulsion; nonionic; foam control in aq. systems; liq.

Antifoam CM Conc. [Harcros] Blend of antifoaming and defoaming agents with silicone synergist, o/w emulsion, nonionic emulsifier; used in industrial processing; carpet cleaning, wood pulping, paper coating, and textile mfg.; wh. liq. lt. cream; disp. in water; sp.gr. 0.972; pH 6.5–7.5 (1% aq.).

Antifoam Compd. SWS-201. [Wacker Silicones] Dimethylpolysiloxane compd. with silica fillers; antifoam for industrial applic.; gray translucent liq.; odorless; negligible sol. in water; sp.gr. 1.01; dens. 8.4 lb/gal; visc. 230 cps; flash pt. 600 F (COC); 100% act.

Antifoam Compd. SWS-202. [Wacker Silicones] Dimethylpolysiloxane compd.; antifoam for aq. systems, industrial applics., indirect food contact; cream visc. liq.; odorless; disp. in water; sp.gr. 1.01; dens. 8.4 lb/gal; visc. 10,000 cps; 100% act.

Antifoam Compd. SWS-203. [Wacker Silicones] Dimethylpolysiloxane compd. with filler; antifoam agent for industrial and indirect food applics.; gray translucent liq.; odorless; neg. sol. in water; sp.gr. 1.01; dens. 8.4 lb/gal; visc. 2400 cps; flash pt. 600 F (COC); 100% act.

Antifoam CTN. [Continental] Silicone type; all purpose defoamer incl. effluents; water-disp.

Antifoam E-20. [Kao] Emulsified denatured silicone; defoamer for tech. applic.; liq.

Antifoam Emulsion Q-75. [Wacker Silicones] Silicone fluid aq. sol'n.; antifoam agent for food and industrial applics. in aq. systems; wh. milky emulsion, liq.; mild odor; b.p. 212 F; sol. in water; sp.gr. 0.991–1.01; dens. 8.3 lb/gal; visc. 3,000–15,000 cps; 30% act.

Antifoam Emulsion SWS-211, -213. [Wacker Silicones] Silicone fluid aq. emulsion; antifoam agent for food applics.; wh. liq.; odorless; b.p. 212 F; sol. in water; sp.gr. 1.01; dens. 8.3 lb/gal; visc. 5000 and 15,000 cps resp.; 10 and 30% act.

Antifoam Emulsion SWS-214. [Wacker Silicones] Silicone fluid aq. emulsion; antifoam agent for industrial applics.; wh. liq.; odorless; b.p. 212 F; sol. in water; sp.gr. 0.991; dens. 8.3 lb/gal; visc. 3000 cps; 10% act.

Antifoam-G. [Soluol] Silicone emulsion; defoamer; water-sol.

Antifoam Q-41. [Soluol] Silicone compd.; defoamer; sol. in solvs.

Antil® 141. [Goldschmidt] PEG-55 propylene glycol oleate; thickener for surfactant systems, e.g., shampoos, shower gels; solubilizer for essential oils, fragrances, and perfumes; 40% act. liq. and 100% act. waxy solid.

Antilux 110. [Rhein Chemie] Paraffins and micro waxes; sun-checking, anti-weathering and ozone protective wax for tech. molded and extruded rubber goods, cellular rubber; pale blue flakes; sp.gr. 0.92; solid. pt. 60–64 C.

Antilux 111. [Rhein Chemie] Paraffins and micro waxes; sun-checking, anti-weathering and ozone protective wax for tires, conveyor belts, tech. rubber goods; pale grn. flakes; sp.gr. 0.92; solid. pt. 64–68 C.

Antilux 500. [Rhein Chemie] Paraffins and micro waxes; sun-checking, anti-weathering and ozone protective wax for tech. molded and extruded rubber goods, cellular rubber; wh. to ylsh. flakes; sp.gr. 0.91; solid. pt. 53–57 C.

Antilux 550. [Rhein Chemie] Paraffins and micro waxes; sun-checking, anti-weathering and ozone protective wax for tech. molded and extruded rubber goods, cellular rubber; pale yel. flakes; sp.gr. 0.91; solid. pt. 56–60 C.

Antilux 600. [Rhein Chemie] Paraffins and micro waxes; sun-checking, anti-weathering and ozone protective wax for tech. molded and extruded articles, cellular rubber; pale yel. flakes; sp.gr. 0.91; solid. pt. 58–62 C.

Antilux 620. [Rhein Chemie] Paraffins and micro waxes; sun-checking, anti-weathering and ozone protective wax for tech. molded and extruded rubber goods, cellular rubber, tires; pale yel. flakes; sp.gr. 0.92; solid. pt. 58–62 C.

Antilux 654. [Rhein Chemie] Paraffins and micro waxes; sun-checking, anti-weathering and ozone protective wax for tech. molded and extruded rubber goods, tires, conveyor belts; wh. to pale yel. flakes; sp.gr. 0.92; solid. pt. 63–67 C.

Antilux 660. [Rhein Chemie] Paraffins and micro waxes; sun-checking, anti-weathering and ozone protective wax for tires, conveyor belts, tech. rubber goods; blue flakes; sp.gr. 0.92; solid. pt. 63–67 C.

Antilux 750. [Rhein Chemie] Micro waxes; sun-checking, anti-weathering and ozone protective wax for inj. molded and extruded rubber goods; lubricant for cables; med. brn. flakes; sp.gr. 0.9; solid. pt. 75 C.

Antilux L. [Rhein Chemie] Paraffins and micro waxes; sun-checking, anti-weathering and ozone protective wax for rubber articles in contact with foodstuffs, surgical and pharmaceutical rubber articles; wh. flakes; sp.gr. 0.91; solid. pt. 53–57 C.

Antioxidant 235. [Akrochem] 2,2´-Methylene-bis (4 methyl-6 tert. butylphenol); nonstaining antioxidant in elastomers for medical equipment, rubber thread, and latex compds.; off-wh. cryst. powd.; sol. in ethanol, acetone, methylene chloride, and benzene; insol. in water; sp.gr. 1.04; m.p. > 124 C.

Antioxidant 425. [Am. Cyanamid] 2,2′-Methylene-bis (4-ethyl-6-tert-butylphenol); antioxidant of impact molding resins from oxidation degradation; stabilizer for acrylics and ABS; used in molded prods.; cream to wh. powd.; m.w. 368.5; sol. in acetone, dioxane, ethyl acetate, chloroform, ethanol, benzene, n-heptane, water; sp.gr. 1.10; m.p. 117–123 C.

Antiozonant AFS/LG. [Mobay] Cyclic acetal; antiozonant for diene rubber goods; gray to beige tablets; dens. 1.06 g/cm3; m.p. 90 C.

Antistat 680. [Laurel] Quat. ammonium compd.; cationic; antistat with compatibility and heat stability; lt. amber liq.

Antistatic Agent 106G-90%. [Hexcel] Bis (2-hydroxy ethyl) octyl methyl ammonium para-toluene sulfonate; internal antistat in thermoset and engineering resins; used at high temps.; sol. in water and alcohol.

Antistatic Agent 273-C. [Hexcel] Bis-2-hydroxyethyl cocamine; internal antistat LDPE, HDPE, PP processed at lower temp. applics.; liq.

Antistatic Agent 273-E. [Hexcel] Bis-2-hydroxyethyl stearylamine; see Antistatic Agent 273-C; m.p. 45–50 C.

Antistatic Plasticizer KA. [Bayer] Polyglycol ether; plasticizer to improve low-temp. flexibility and elastic behavior of vulcanizates; slightly discolors lt.-colored goods; brn. visc. liq.; dens. 1.12–1.14 g/cc.

Antistaticum RC 100. [Rhein Chemie] Fatty alkyl ether of polyethylene glycol; antistatic plasticizer for NR and syn. rubber and PVC plastics; whitish visc. liq.; sp.gr. 0.94.

Antisun®. [Frank B. Ross] Antiozonant, anticracking and sunchecking wax for the rubber industry; pale yel.; sp.gr. 0.920–0.930; m.p. 70–74 C; acid no. nil; sapon. no. 1.0 max.

Anti UVA. [Aceto] 2 Hydroxy-4-methoxy-benzophenone; sunscreen agent; powd.

Antozite® 1. [Vanderbilt] N,N′-di(2-octyl)-p-phenylenediamine; antiozonant for NR, IR, BR, and SBR; dk. redsh. brn. liq.; m.w. 332.57; sol. in chloroform, toluene, petrol. ether, ethyl alcohol; insol. in water; dens. 0.90 ± 0.02 mg/m³; visc. 110–145 SUS (40 C).

Antozite® 2. [Vanderbilt] N,N′-di-3(5-methylheptyl)-p-phenylenediamine; see Antozite 1; dk. redsh. brn. liq.; m.w. 332.57; sol. see Antozite 1; dens. 0.90 ± 0.02 mg/m³; visc. 180–270 SUS (40 C).

Antozite® 67, 67F. [Vanderbilt] N-(1,3-dimethylbutyl)-N′-phenyl-p-phenylenediamine; see Antozite 1; gray-blk. semisolid and dk. purple flakes resp.; m.w. 268.40; sol. in toluene, acetone, ethyl acetate, ethylene dichloride; insol. in water; dens. 0.986–1.00 mg/m³; cryst. pt. 40–44 and 44–49 C.

Antracol®. [Bayer] Propineb; fungicide for use on grapes, hops, tobacco, potatoes, vegetables, rice, fruit, etc.; wh. powd.; m.p. 150 C (decomposes).

AO-14-2. [Exxon] Bishydroxyethylisodecyloxypropylamine oxide; foam stabilizers/boosters in liq. detergents; Gardner 2 clear liq.; sp.gr. 0.956 (15 C); pour pt. < 20 F; 50% act.

AO 728 Special. [Exxon] Amine oxide; foam booster and stabilizer for industrial and household detergents; liq.; 50% conc.

Aquagard. [Soluol] Silicone emulsion; water repellent; specific aquaguard system avail. for particular fiber and applic. requirement.; water-sol.

Aquagard 170. [Soluol] Organo-polysiloxane; nonionic; curing water repellent emulsion for fibers and fabrics; wh. emulsion; 30% silicone content.

Aqualease 6100. [George Mann] Water-based, environmentally safe mold release agent for urethane elastomers and foams.

Aqualease 6102. [George Mann] Water-based environmentally safe mold release agent for thermoset resins incl. urethane elastomers, epoxies, polyester, and vinyl ester polymers; offers reduced corrosion on steel molds over Aqualease 6100.

Aqualease 6201, 6301. [George Mann] Water-based mold release agent for thermosetting resins incl. urethane elastomers, epoxies, filled polyester, and vinyl ester polymers.

Aqualon® C, R. [Aqualon] Internally cross-linked form of sodium CMC; absorbent fibers used in feminine-care prods., disposable diapers, absorbent pads, surgical dressings; as binder fibers for wet-laid nonwovens; antistatic agents for nonwovens; fibers of 3 mm avg. length; pH 5–7 (1% aq. disp.).

Aqualon® Cellulose Gum. [Aqualon] Sodium CMC; suspending agent for abrasive and polishing agents, and prevents syneresis in toothpaste; rheology control agent in creams and lotions; adhesive and cohesive agent used in denture adhesive and ostomy adhesive prods.; water-sol.

Aqualon® CMC-T. [Aqualon] Sodium CMC, tech. grades; binder, thickener, stabilizer, suspending agent, film-former, rheology control aid, water-retention aid used for adhesives, aerial-drop fluids, ceramics, coatings, detergents, lithography, paper, textiles, and tobacco; water sol.; sp.gr. 1.0068 (2%); bulk dens. 0.75 g/ml; visc. 15–30 cps (CMC-7L3T, 2%), 25–50 (CMC-7LT, 2%, CMC-7L1T, 4%), 80–200 cps (CMC-7L2T, 4%), 300–600 cps (CMC-7MT, 2%), 1600–3100 cps (CMC-7M31T, 2%); surf. tens. 71 dynes/cm (1%); ref. index 1.3355 (2%); pH 6.5–8.5 (1%).

Aqualon® CMC-Warp Size. [Aqualon] Sodium CMC; used as a size for fabrics made from polyester and cotton blends, cotton, syn. fibers, cellulose yarns and other blends of cotton or syn. fibers; water-sol.

Aqualose L30. [Westbrook Lanolin] PEG-30 lanolin; nonionic; emollient, emulsifier, plasticizer, solubilizer; wax.; HLB 14.0; 100% conc.

Aqualose L75, L75/50. [Westbrook Lanolin] PEG-75 lanolin USP; nonionic; emollient, emulsifier for personal care prods.; plasticizer in aerosol hair sprays; solubilizer for perfume and germicidal agents; conditioner for shampoos; superfatting agent for soap; wax and gel resp.; HLB 14.0–16.0 (L75); 100 and 50% conc.

Aqualose LL100. [Westbrook Lanolin] PPG-40-

PEG-60 lanolin oil; nonionic; emollient, emulsifier, plasticizer, solubilizer used in aq./alcoholic preparations of alcohol content; liq.; HLB 13.0; 100% conc.

Aqualose SLT. [Westbrook Lanolin] Lanolin deriv.; emollient; paste.

Aqualose SLW. [Westbrook Lanolin] Lanolin deriv.; emollient used in shampoos and other aq. or dilute alcoholic preparations; paste.

Aqualose W20, W20/50. [Westbrook Lanolin] Laneth-20; nonionic; plasticizer and solubilizer for hydrophobic substances; emollient, emulsifier; wax and gel resp.; 50% conc. (W20/50).

Aqualox® 225-100, 225A-100. [Alox] Amine salt; surfactant, corrosion inhibitor, lubricant, antiwear additive effective in inhibiting the attack of ferrous metals by aq. sol'ns.; for metalworking formulations; use 225-100 for EP/antiwear use; 225A-100 grade contains no phosphorous; pale yel.; mild odor; sol. in water @ < 2% and > 50%; sp.gr. 1.06 (15.6 C); dens. 8.8 lb/gal (15.6 C); visc. < 550 cs (40 C); flash pt. 150 C; pour pt. –25.5 C; pH 9.7 ± 0.3; biodeg. (225A-100).

Aqualox® 232. [Alox] Amine salts of org. acids; corrosion inhibitor, low foaming surfactant for syn. metalworking formulations, esp. in aq. sol'n.; pale amber liq.; mild odor; sol. in water, methanol, ethanol, IPA, glycol, water/alcohol mixts.; sp.gr. 1.15 (15.6 C); dens. 9.6 lb/gal (15.6 C); visc. 75 ± 15 cs (100 F); pour pt. –29 C max.; pH 7.6 ± 0.3; biodeg.; 75% act.

Aqualox® 262, 263. [Alox] Alkanolamide, 2:1; lubricant; metal working fluid additive providing increased corrosion resistance and lubricity; oil-sol.; water-disp.

Aqualox® 2268. [Alox] Low foaming corrosion inhibitor, lubricity and antiwear agent protecting ferrous metals from aq. sol'ns. and steel against high humidity conditions during extended indoor storage; emulsions deposit very thin waxy films; brn. waxy soft solid; mild odor; sp.gr. 0.92 (15.6 C); dens. 7.6 lb/gal; m.p. 57 ± 3 C; flash pt. 135 C min.; pH 9–10 (1% aq.).

Aqualox® 2295. [Alox] Heavy-duty emulsifiable rust preventive conc.; provides residual film for long-term protection of metal surfs. against salt water and spray; used for automotive rustproofing applics.; tan paste; sp.gr. 0.98 ± 0.02; dens. 7.8–8.0 lb/gal; pour pt. 43 C; 42 ± 3% NV; 55 ± 1% water.

Aquamolin Brands. [Hoechst-Celanese] EDTA, tetrasodium salt; sequestering agent.

Aquasan 124. [Laurel] Methylated amide; cationic; water and oil repellent, fluorochemical extender; milky wh. liq.

Aquasan 542. [Laurel] Fluorinated hydrocarbon finish; cationic; replacement for silicone water repellents; for water repellency after drycleaning, durable oil repellency; wh. creamy liq.

Aquasorb A250. [Aqualon] Carboxymethylcellulose; absorbent for urine, blood, and other body fluids; used in feminine hygiene prods., medical disposables, disposable diapers; wh. to lt. tan powd.; pH 6.5–8.5; 99.5% min. purity.

Aquatreat AR4, AR6. [Alco] Polyacrylic acid; dispersant and sludge conditioner for water treatment; m.w. 60,000 and 100,000 resp.; pH 2.2–3.5; 24–26% solids.

Aquatreat AR602. [Alco] Sodium polyacrylate; general purpose boiler water dispersant and antiscalant; m.w. 4500; pH 7–8; 44–46% solids.

Aquatreat AR231, AR232. [Alco] Sodium polymethacrylate; boiler water dispersant and sludge conditioner; m.w. 6500 and 9500 resp.; pH 8–9; 29–31% solids.

Aquatreat DNM-9, DNM-25, DNM-30, DMN-360. [Alco] Dithiocarbamate salts; biocide, fungicide, and algicide used in water treatment, paper, sugar, and petrol. applics.; 9, 25, 30, 36% act. in water resp.

Aquatreat KM. [Alco] Potassium dimethyldithiocarbamate; see Aquatreat DNM-9; 50% act. in water.

Aquatreat SDM. [Alco] Sodium dimethyldithiocarbamate; see Aquatreat DNM-9; 40% act. in water.

Aquazym® 120L. [Novo] Bacterial diastase; enzyme in textile industry used for starch desizing; liq.

Aqucar Microbiocides. [Union Carbide] Glutaraldehyde; water treatment biocide for recirculating cooling towers; aq. sol'ns.

Aramide® CDM-4. [Arjay] Alkanolamine condensate; nonionic; thickener, foam stabilizer, emulsifier, lubricant, and visc. builder for metal working fluids, all purpose cleaners, liq.; sol. in aromatic hydrocarbon and water; 100% conc.

Aramide® CDR. [Arjay] Alkanolamine condensate; nonionic; thickener for dishwash detergents and cosmetic use; liq.; 100% conc.

Aramide® CDS. [Arjay] Super coconut alkanolamide; nonionic; emulsifier, foam stabilizer, thickener for liq. detergents; liq.; 100% conc.

Aramide® CDX. [Arjay] Coconut alkanolamide; nonionic; solubilizer, foam stabilizer, detergent and visc. builder; liq.; 100% conc.

Aramide® LDS. [Arjay] Super lauric DEA; nonionic; foam stabilizer, lubricant and visc. builder; liq.; 100% conc.

Aramide® LMDS. [Arjay] Lauric/myristic alkanolamide; nonionic; detergent, foam stabilizer in detergents and personal care prods.; solid; 100% conc.

Arbyl 18/50. [Grünau] Fatty alcohol polyglycol ether; nonionic; precleaning and leveling agent; liq.; 50% conc.

Arconate® 1000. [Arco] Propylene carbonate; high boiling, Rule 66 exempt solv.; used in paint strippers, resin/polymer cleaning/flushing operations, paints, high-bake coatings, cosmetics, etc.; liq., Pt-Co color 300–500, mild odor; sp.gr. 1.203–1.210; dens. 10.1 lb/gal; visc. 4 cs; b.p. 242 C; f.p. –50 C; flash pt. 132 C (TOC); ref. index 1.419; pH 4.5–6.5 (10% aq.); 99.0% min. assay.

Arconate® HP. [Arco] Propylene carbonate; high boiling, Rule 66 exempt solv.; used in paint strippers, resin/polymer cleaning/flushing operations, paints, high-bake coatings, cosmetics, etc.; liq., Pt-Co color 40 max., mild odor; sp.gr. 1.203–1.210; dens. 10.1 lb/gal; visc. 4 cs; b.p. 242 C; f.p. –50 C; flash pt. 132 C (TOC); ref. index 1.419; pH 6.5–7.5 (10% aq.);

99.6% min. assay.

Arcor Series. [Arjay] Org. amine salt concs.; corrosion inhibitor concs.; liq.; oil-sol.; water-disp.

Arcosolv® DPM. [Arco] Dipropylene glycol monomethyl ether; solv. for coatings, cleaners, inks, agric. prods., cosmetics, chemical intermediate applics.; APHA 15 max. liq.; mild, pleasant char. odor; m.w. 148.2; f.p. –80 C; b.p. 188.3 (760 mm Hg); sol. in water; misc. with a number of org. solvs.; sp.gr. 0.950–0.953; dens. 7.91 lb/gal; visc. 3.6 cstk; ref. index 1.422; flash pt. (TCC) 75 C; surf. tens. 28.2 dynes/cm; sp. heat 0.54 cal/g/°C; combustible.

Arcosolv® DPM Acetate. [Arco] Dipropylene glycol monomethyl ether acetate; solv. where a slow evaporating nonhydroxylic solv. is required; effective in coatings; as a coalescent in waterborne emulsion systems; APHA 15 max. color; m.w. 190.2; b.p. 205 C (760 mm Hg); 12.3% sol. in water; sp.gr. 0.972; dens. 8.14 lb/gal; visc. 2.2 cstk; ref. index 1.4142; flash pt. (Seta) 186 F; surf. tens. 28.3 dynes/cm.

Arcosolv® PM. [Arco] Propylene glycol monomethyl ether; solv. for coatings, cleaners, inks, agric. prods., cosmetics, chemical intermediate applics.; APHA 10 max. liq.; mild, pleasant char. odor; m.w. 90.1; f.p. –95 C; b.p. 120.1 C (760 mm Hg); sol. in water; misc. with a number of org. solvs.; sp.gr. 0.918–0.921; dens. 7.65 lb/gal; visc. 1.8 cstk; ref. index 1.404; flash pt. (TCC) 32 C; surf. tens. 26.5 dynes/cm; sp. heat 0.57 cal/g/°C; flam.

Arcosolv® PM Acetate. [Arco] Propylene glycol monomethyl ether acetate; slow-evaporating solv. with good solvency for many commonly used coating resins, e.g., acrylics, NC, and urethanes; used in lacquers, water-based paints; APHA 10 max. liq.; mild, ester-like odor; m.w. 132.2; b.p. 146.0 C (760 mm Hg); 18.5% sol. in water; sp.gr. 0.963–0.966; dens. 8.02 lb/gal; visc. 1.1 cstk; ref. index 1.400; flash pt. (TCC) 47 C; surf. tens. 27.4 dynes/cm; combustible.

Arcosolv® PTB. [Arco] Propylene glycol mono-t-butyl ether; solv. offering a blend of hydrophobicity and hydrophilicity for coatings, cleaners, electronic, and ink applics.; strong coupling ability; used in lt.-duty and hard-surf. cleaners, water-reducible polyester and alkyd resin prod.; chemical intermediate in the synthesis of monomeric and polymeric prods.; cosolv.; APHA 15 max. liq.; m.w. 132.2; f.p. –56 C; b.p. 151 C (760 mm Hg); partial water sol. at ambient temps.; sp.gr. 0.870–0.874; visc. 3.8 cstk; ref. index 1.4116; flash pt. (TCC) 45 C; surf. tens. 24.2 dynes/cm.

Arcosolv® TPM. [Arco] Tripropylene glycol monomethyl ether; solv. for coatings, cleaners, inks, agric. prods., cosmetics, chemical intermediate applics.; slow evaporating; APHA 15 max. liq.; mild, pleasant char. odor; m.w. 206.3; f.p. –79 C; b.p. 242.4 C (760 mm Hg); sol. in water; misc. with a number of org. solvs.; sp.gr. 0.962–0.965; dens. 8.03 lb/gal; visc. 5.8 cstk; ref. index 1.430; flash pt. (PMCC) 114 C; surf. tens. 29.0 dynes/cm; sp. heat 0.51 cal/g/°C.

Ardril FCA 154. [Aquaness] Amine lignite; suspending agent; high temp. filtration control agent for invert mud systems; solid; oil-disp.

Ardril FCA 2924. [Aquaness] Modified polymeric hydrocarbon resin; suspending agent; high temp. filtration control agent for invert mud systems; used at temp. 475 F; solid; water-insol.

Aremsan C40. [Ronsheim & Moore] Cocamidopropyl dimethyl benzyl ammonium chloride; general bactericide; for preparation of hair conditioners; liq.

Argobase 125. [Westbrook Lanolin] Lanolin alcohols extract; nonionic; emollient, w/o emulsifier used in personal care prods.; liq.; oil-sol.; HLB 3.0; 15% conc.

Argobase EU, LI, SI. [Westbrook Lanolin] Sterols and sterol esters lanolin extracts; nonionic; emollient, emulsifier, stabilizer; absorp. base; paste, liq., paste resp.; HLB 4.0; 6, 24, and 92% conc.

Argobase MS-5. [Westbrook Lanolin] Sterols and sterol esters lanolin extracts; nonionic; w/o emulsifier, emollient, stabilizer; paste; HLB 5.0; 25% conc.

Argonol 40. [Westbrook Lanolin] Isobutylated lanolin oil; nonionic; w/o emulsifier, emollient used in personal care prods.; fluid; 100% conc.

Argonol 50 Pharmaceutical. [Westbrook Lanolin] Lanolin oil; emollient, lubricant; liq.; odorless; sol. in min. oil.

Argonol 50 Super, 60. [Westbrook Lanolin] Lanolin oil; emollient, lubricant; liq.

Argonol 60. [Westbrook Lanolin] Lanolin oil; emollient, lubricant; liq.

Argonol ACE5. [Westbrook Lanolin] Lano-cetyl acetate; emollient; liq.

Argowax Dist. [Westbrook Lanolin] Lanolin alcohol; nonionic; gelling agent, emulsifier; lt. wax; slight odor; HLB 2.6; 100% conc.

Argowax Std. [Westbrook Lanolin] Lanolin alcohol; nonionic; see Argowax Dist.; wax; 100% conc.

Aristol A. [Pilot] 60% mono substituted C_{20-24} benzene; lubricant; lube oil additive; chemical feedstock for sulfonation to produce emulsifiers and corrosion preventatives; m.w. 500; pour pt. 65 F.

Aristol B. [Pilot] 80% mono substituted C_{16-18} benzene; sulfonatable oils for lube additive mfg., and in elec., refrigeration, textile, and special lubrication end uses; m.w. 360; pour pt. 15 F.

Aristol D. [Pilot] 50% mono substituted C_{16-18} benzene; see Aristol B; m.w. 450; pour pt. 15 F.

Aristol E. [Pilot] 90% disubstituted C_{12} benzene; see Aristol B; m.w. 420; pour pt. –60 F.

Aristonate 430. [Pilot] Sodium alkylbenzene sulfonate; emulsifier and dispersant for liqs.; primary ingred. in sol., textile, leather processing, and neat oils, metal degreasers, dry cleaning soaps, and rust preventives; Gardner 10 thick liq.; mild odor; oil-sol.; dens. 8.5 lb/gal; visc. 690 SSU (210 F); pH 9.5; 60–62% act. sulfonate.

Aristonate 460. [Pilot] Sodium alkylbenzene sulfonate; emulsifier and dispersant for liqs., dispersant/wetting agent for solids and liq. and solid systems; rust and corrosion inhibitor; primary ingred. in sol., neat, and textile oils, high water base hydraulic fluids; thick liq.; mild odor; sol. see Aristonate 430;

dens. 8.5 lb/gal; visc. 121 cSt (210 F).

Aristonate 500. [Pilot] Sodium alkylbenzene sulfonate; emulsifier and dispersant for liqs., dispersant/ wetting agent for solids and liq./solid systems; rust and corrosion inhibitor; used as primary ingred. in sol. and neat oils, metalworking fluids, ore flotation reagents; thick liq.; mild odor; sol. see Aristonate 430; dens. 8.5 lb/gal; visc. 134 cSt (210 F); pH 9.0.

Arklear Series. [Arjay] Polymers, polyamines; oil production for reverse o/w emulsions; water clarifier; flocculant.

Arkomon A Conc. [Hoechst-Celanese] Fatty acid sarcoside, sodium salt; anionic; basic material for detergents and cleaning agents; corrosion inhibitor; paste; 60% conc.

Arkomon SO. [Hoechst-Celanese] Acylamino carboxylic acid; corrosion inhibitor for metal working fluids.; liq.

Arlacel® 989. [ICI Am.] PEG-7 hydrog. castor oil; nonionic; emulsifier and softener for w/o lotions, cosmetic uses; waxy, solid; 100% conc.

Arlacide A. [ICI Am.] Chlorhexidine acetate; cationic; preservative/bactericide for liq. and powd. preparations; powd.; water-sol.

Arlacide G. [ICI Am.] Chlorhexidine gluconate; cationic; see Arlacide A; 20% aq. sol'n.

Arlacide H, HM. [ICI Am.] Chlorhexidine hydrochloride; cationic; see Arlacide A; powd. and micronized powd. resp.

Arlamol® E. [ICI Am.] PPG-15 stearyl ether; nonionic; emollient, solv. for personal care prods.; oily liq.; sol. in min. oil, isopropyl esters, cottonseed oil, ethanol, IPA, hexadecyl alcohol; sp.gr. 0.95; visc. 90 cps; pour pt. typ. < 0 C; 100% act.

Arlamol® GM. [ICI Am.] Ethoxylated glyercide; nonionic; emollient/refattening agent for bath and shower prods.; liq.; 100% act.

Arlamol® S3, S7. [ICI Am.] PPG-(15)-stearyl ether and cyclomethicome; emollient used in cosmetic prods.; liq.

Arlasolve 200. [ICI Am.] Isoceteth-20; nonionic; surfactant, emulsifier, solubilizer for cosmetics; wh. soft, waxy solid; sol. in water, propylene glycol, ethanol; sp.gr. 1.0; HLB 15.7; flash pt. > 230 F; 100% act.

Arlatone 285. [ICI Am.] POE castor oil; nonionic; cosmetic grade surfactant, coupling agent, solubilizer for perfumes, fragrances; semifluid; HLB 14.4; 100% conc.

Arlatone 289. [ICI Am.] POE hydrog. castor oil; nonionic; coupling agent, solubilizer, emulsifier for o/w creams and lotions; waxy; HLB 14.4; 100% conc.

Arlatone 507. [ICI Am.] Octyl dimethyl PABA; uv lt. absorber which absorbs radiation in the erythemal range and transmits radiation in the tanning range; sunscreen agent for emulsions, hydro-alcoholic sol'ns., and anhyd. systems; pale yel. clear, low-visc. liq.; faint, pleasant odor; sol. (1.5%) in IPA, ethanol, min. oil, peanut oil, PPG-15 stearyl ether, propylene glycol; insol. in glycerin and water; visc. 100 cps; HLB 10; flash pt. (Seta OC) 204 C.

Arlatone 650. [ICI Am.] Ethoxylated castor oil; vitamin solubilizer; liq.

Arlatone 827. [ICI Am.] Ethoxylated castor oil; nonionic; vitamin solubilizer; liq.; 100% act.

Arlatone 970. [ICI Am.] POE sorbitan fatty acid ester; nonionic; coupling agent, solubilizer; liq.; 100% conc.

Arlatone 975. [ICI Am.] Ethoxylated hydrog. castor oil; solubilizer for perfumes and essential oils; liq.

Arlatone 980. [ICI Am.] Ethoxylated stearyl stearate; see Arlatone 975; liq.

Arlatone 985. [ICI Am.] Ethoxylated stearyl stearate; visc. stabilizer for o/w milks; solid.

Arlatone G. [ICI Am.] PEG-25 hydrog. castor oil; nonionic; surfactant, solubilizer; formulates clear gels; yel. visc. liq. to soft paste; sol. water, ethanol and IPA; sp.gr. 1.0; HLB 10.8; flash pt. > 300 F; 100% act.

Arlatone T. [ICI Am.] PEG-40 sorbitan peroleate; nonionic; emulsifier, solubilizer, antistat, lubricant; for bath oils, textile industry; yel. liq.; sol. in veg., min. oils, IPM, IPP; water disp.; sp.gr. 1; visc. 175 cps; HLB 9; flash pt. > 300F; 100% act.

Armac® 18D. [Akzo] Octadecylamine acetate, distilled; cationic; wetting agent, emulsifier, corrosion inhibitor, flotation agent for pigment flushing, froth flotation of mins., flocculation; wh. flake; water sol.; sp.gr. 0.827 (80 C); m.p. 85 C; flash pt. 320 F (COC); 100% act.

Armac® C. [Akzo] Cocamine acetate; cationic; wetting agent, emulsifier, corrosion inhibitor, flotation reagent, stripping agent; yel. solid; m.w. 260; water sol.; sp.gr. 0.852 (80 C); HLB 11.2; m.p. 50 C; flash pt. 285 F (COC); 95% conc.

Armac® HT. [Akzo] (Hydrog. tallow) amine acetate; cationic; wetting agent, emulsifier, corrosion inhibitor, flotation reagent; wh. solid; water-sol.; sp.gr. 0.820 (80 C); HLB 10.7; m.p. 60 C; flash pt. 325 F (COC); 95% act.

Armac® OD. [Akzo] Oleylamine acetate; corrosion inhibitor; min. flotation agent in mining industry; Gardner 13 max. solid; m.p. 55 C; flash pt. 170 C (COC); 95% min. act.

Armac® T. [Akzo] Tallowamine acetate; cationic; flotation agent, corrosion inhibitor, lubricant, emulsifier; metal processing; Gardner 10 max. solid; water-sol.; sp.gr. 0.87 (60 C); HLB 10.8; m.p. 45–61 C; flash pt. 168 C (COC); 94% min act.

Armeen® 2C. [Akzo] Dicocamine (sec. amine); cationic; emulsifier, flotation agent, corrosion inhibitor; solid; sol. in chloroform, slt. sol. in IPA, toluene, CCl_4, kerosene; sp.gr. 0.793 (60/40 C); visc. 49.1 SSU (60 C); m.p. 104–117 F; flash pt. 350 F; pour pt. 80 F; 100% conc.

Armeen® 2HT. [Akzo] Di(hydrog. tallow) amine (sec. amine); cationic; emulsifier, flotation agent, corrosion inhibitor, pour pt. depressant for petrol. refined prods.; Gardner 2 max. solid; sol. in chloroform, slt.ly sol. in toluene, CCl_4, kerosene; sp.gr. 0.803 (60/4 C); visc. 63.9 SSU (70 C); m.p. 140–149 F; pour pt. 115 F; flash pt. (COC) 435 F; pour pt. 115 F; 100% conc.

Armeen® 12D. [Akzo] Lauramine (primary amine); cationic; emulsifier, flotation agent, corrosion inhibitor; lubricant for metal treatment; Gardner 2 max. liq.; sol. in methanol, ethanol, acetone, IPA, chloroform, toluene, CCl_4, kerosene; sp.gr. 0.801; visc. 42.2 SSU; m.p. 79–86 F; flash pt. (COC) 235 F; pour pt. 80 F; 98% conc.

Armeen® 16D. [Akzo] Palmitamine (primary amine); cationic; emulsifier, flotation agent, corrosion inhibitor; solid; sol. in methanol, ethanol, IPA, chloroform, toluene, CCl_4, slt. sol. in acetone, kerosene; sp.gr. 0.789 (60/4 C); visc. 37.5 SSU (55 C); m.p. 100–118 F; flash pt. 285 F; pour pt. 100 F; 100% conc.

Armeen® 18. [Akzo] Stearamine (primary amine); cationic; emulsifier, flotation agent, corrosion inhibitor, anticaking agent; hard rubber mold release agent; solid; sol. in ethanol, IPA, chloroform, toluene, CCl_4, slt. sol. in acetone, kerosene; sp.gr. 0.792 (60/4 C); visc. 45.6 SSU; m.p. 122–133 F; flash pt. 320 F; pour pt. 115 F; 100% conc.

Armeen® 18D. [Akzo] Stearamine (primary amine); cationic; emulsifier, flotation agent, corrosion inhibitor, anticaking agent; solid; sol. in ethanol, IPA, chloroform, toluene, CCl_4, slt. sol. in methanol, kerosene; sp.gr. 0.791–0.792 (60/4 C); visc. 43.7 SSU; m.p. 122–133 F; flash pt. 300 F; pour pt. 110 F; 97% conc.

Armeen® C, CD. [Akzo] Cocamine (primary amine); cationic; emulsifier, flotation agent, corrosion inhibitor, stripping agent for paints; liq.; sol. in methanol, ethanol, acetone, IPA, chloroform, toluene, CCl_4, kerosene; sp.gr. 0.805 and 0.804 resp.; visc. 44.2 and 43 SSU (35 C); m.p. 54–59 and 57–63 F; flash pt. 240 and 230 F; pour pt. 45 and 55 F; 95 and 98% conc.

Armeen® DM12D. [Akzo] Dimethyl lauramine; cationic; surfactant intermediate; yel. liq.; amine odor; water insol.; sp.gr. 0.78; visc. 3.30 cs; f.p. –15 C; b.p. 80–115 C (3 mm Hg); flash pt. 122 C (COC); 98% act.

Armeen® DM14D. [Akzo] Dimethyl myristamine; cationic; surfactant intermediate; yel. liq.; amine odor; water insol.; sp.gr. 0.79; visc. 4.7 cs; f.p. –8 C; b.p. 100–125 C (3 mm Hg); flash pt. 28 C (COC); 98% act.

Armeen® DM16D. [Akzo] Dimethyl palmitamine; cationic; chemical intermediate, raw material for surfactants; yel. liq.; amine odor; water insol.; sp.gr. 0.80; visc. 6.75 cs; f.p. 8 C; b.p. 100–136 C (3 mm Hg); flash pt. 150 C (COC); 98% act.

Armeen® DM18D. [Akzo] Dimethyl stearamine; cationic; chemical intermediate, raw material for surfactants; yel. liq., amine odor; water insol.; sp.gr. 0.79; f.p. 20 C; b.p. 145–160 C (3 mm Hg); flash pt. 178 C (COC); 98% act.

Armeen® DMAD. [Akzo] Alkyl (C_{10}–C_{18}) dimethyl amine; chemical intermediate, raw material for surfactants; yel. liq.; amine odor; water insol.; sp.gr. 0.79; visc. 4.44 cs; b.p. 45–150 C (3 mm Hg); flash pt. 130 C (COC); 98% act.

Armeen® DMCD. [Akzo] Dimethyl cocamine; cationic; chemical intermediate, raw material for surfactants; yel. liq.; amine odor; water insol.; sp.gr. 0.79; visc. 4.20 cs; b.p. 42–150 C (3 mm Hg); flash pt. 108 C (COC); 98% act.

Armeen® DMHTD. [Akzo] Dimethyl hydrog. tallow amine; cationic; chemical intermediate, raw material for surfactants; yel. liq.; amine odor; water insol.; sp.gr. 0.80 ; visc. 8.75 cs; f.p. 18 C; b.p. 100–155 C; flash pt. 162 C (COC); 98% act.

Armeen® DMMCD. [Akzo] Dimethyl cocamine; nonionic; chemical intermediate, raw material for surfactant; yel. liq.; amine odor; water insol.; sp.gr. 0.79; visc. 3.95 cs; f.p. –19 C; b.p. 90–125 C (3 mm Hg); flash pt. 125 C (COC); 98% act.

Armeen® DMOD. [Akzo] Oleyl dimethylamine; cationic; surfactant intermediate; liq.; 100% conc.

Armeen® DMSD. [Akzo] Soyaalkyl dimethylamine; cationic; surfactant intermediates; liq.; 100% conc.

Armeen® DMTD. [Akzo] Tallowalkyl dimethylamine; cationic; surfactant intermediates; liq.; 100% conc.

Armeen® HT, HTD. [Akzo] (Hydrog. tallow) amine (primary amine); cationic; emulsifier, flotation agent, corrosion inhibitor, chemical intermediate, anticaking agent; Gardner 9 solid; sol. in methanol, ethanol, IPA, chloroform, toluene, CCl_4, sp.gr. 0.795 and 0.794 resp. (60/4 C); visc. 47.5 and 44.1 SSU (55 C); m.p. 79–136 and 70–120 F; cloud pt. 115 and 110 F; flash pt. 320 and 315 F; pour pt. 110 and 100 F; 95 and 100% conc.

Armeen® N-CMD. [Akzo] N-coco morpholine; catalyst in PU foams; Gardner 3 liq.; sol. in acetone, IPA; sp.gr. 0.87; visc. 50 SSU (55 C); m.p. –29/–10 C; pour pt. –10 C; flash pt. 155 C (COC); 97% min. act.

Armeen® O. [Akzo] Oleamine (primary amine); cationic; emulsifier, wetting agent, corrosion inhibitor, dispersant, chemical intermediate, oil additive; cosmetics; Gardner 8 paste; sol. in acetone, methanol, ethanol, IPA, chloroform; toluene, CCl_4, kerosene, wh. min. oil; sp.gr. 0.820 (38/4 C); visc. 57.0 SSU; m.p. 50–68 F; flash pt. 320 F; 100% conc.

Armeen® OD. [Akzo] Oleamine (primary amine); cationic; wetting agent, lube oil additive, emulsifier, corrosion inhibitor, cosmetic industry dispersant, chemical intermediate; Gardner 2 paste; water insol.; sp.gr. 0.79 (60 C); visc. 56.6 SSU; m.p. 8–18 C; flash pt. 154 C (COC); 98% act.

Armeen® SD. [Akzo] Soyamine (primary amine); cationic; emulsifier, flotation agent, corrosion inhibitor; Gardner 3 paste; sol. in methanol, ethanol, acetone, IPA, chloroform, toluene, CCl_4; sp.gr. 0.81 (38/4 C); visc. 46.2 SSU (35 C); m.p. 81–86 F; cloud pt. 85 F; flash pt. 305 F; pour pt. 70 F; 100% conc.

Armeen® SZ. [Akzo] Sodium N-coco beta-amino butyrate; detergent, wetting agent, corrosion inhibitor, stabilizer, dispersion and softening aid, leveling agent, fungicide for paints, pigments, dyeing; yel. liq.; water sol.; sp.gr. 0.96; visc. 19 cps (55 C); flash pt. 40 C; pH 9.9–10.8; 39–41% act. in 18% IPA, 40% water.

Armeen® TD. [Akzo] Tallowamine (primary

amine); cationic; emulsifier, flotation agent, corrosion inhibitor; Gardner 2 paste; sol. see Armeen T; sp.gr. 0.812 (38/4 C); visc. 45.2 SSU (35 C); m.p. 64–118 F; cloud pt. 102 F; flash pt. 305 F; pour pt. 70 F; 98% act.

Armeen® Z. [Akzo] Cocaminobutyric acid; amphoteric; pigment softening, dispersing agent; antifogging agent, foam booster, stabilizer, wetting agent in alkaline paint strippers, latex emulsions, latex rubber reclamation, inks, plastic films, cosmetics; cooling tower corrosion inhibitor; Gardner 8 pumpable slurry; sol. in water, IPA, ethyl acetate; sp.gr. 0.98; visc. 247 SSU; flash pt. (TCC) 175 F; pour pt. 65 F; pH 6.5–7.5 (10% aq.); 100% conc.

Armeen® Z-9. [Akzo] N-coco beta-aminobutyric acid; emulsifier, pigment dispersant aid; yel. liq.; water sol.; sp.gr. 0.90; flash pt. 33 C; 46–50% act. in 40% water/10% IPA.

Armid® 18. [Akzo] Stearamide, antiblock agent; internal lubricant and slip agent for processed plastics, coatings, and films; builder, foam visc. stabilizer, and foam booster in syn. detergent formulations; water repellent for textiles; improves dye solubility in printing inks, dyes, carbon paper coatings, and fusible coatings for glassware and ceramics; intermediate for syn. waxes; pigment dispersant; thickener for paint; Gardner 7 flake; water-insol.; sp.gr. 0.52 (100 C); m.p. 99–109 C; flash pt. 225 C; 90% act.

Armid® C. [Akzo] Cocoamide; see Armid 18; Gardner 10 flake; insol. in water; sp.gr. 0.845 (100 C); m.p. 85 C; flash pt. 174 C; 90% act.

Armid® HT. [Akzo] (Hydrog. tallow) amide; see Armid 18; also antifoam in steam generator systems, lubricant additive; Gardner 7 flakes, powd. 99% -60 mesh; m.w. 277; insol. in water; sp.gr. 0.851 (100 C); visc. 16 cps; m.p. 98–103 C; flash pt. 225 C; 90% act.

Armid® O. [Akzo] Oleamide; see Armid 18; also release agent in cosmetics, penetrant in paper manufacture; Gardner 7 flake, solid; bland odor; m.w. 279; sol (g/100 ml solv. with heating) 59 g in 95% IPA; 30 g in 95% ethanol; 15 g in acetone and trichloroethylene; 11 g in ethyl acetate and MIBK; insol. in water; sp.gr. 0.830 (100 C); visc. 25 cps; m.p. 68 C; flash pt. 207 C; 90% act.

Armocure® 100. [Akzo] Aliphatic polyamine; curing agent for flexible epoxy resins used in industrial applic.; adhesives and preparation of laminating resins; Gardner 3 max.; sp.gr. 0.831 (40 C); visc. 19 cP; cloud pt. 15 C max.; 89% min. act.

Armoflo® 48, 65. [Akzo] Anticaking and antidusting agents for hygroscopic materials such as ammonium sulfate, ammonium nitrate, diammonium phosphate, and urea; also for treating thermosetting plastics, detergents, and insecticides; liq.; sp.gr. 0.854 (60 F) and 0.824 (24 C) resp.; m.p. -4 C and 5 F resp.

Armoflo® 66. [Akzo] Cationic conditioning agent providing dust-free anticaking to fertilizers; inhibits corrosion of iron and steel equipment; liq.; oil-sol.; sp.gr. 0.826 (140 F); visc. 59 SSU (140 F); m.p. 86 F; flash pt. (COC) 375 F.

Armogard® F-201. [Akzo] Organic amine-based formulation; multifunctional fuel oil additive for use by refiners, fuel oil distributors, and lg. scale oil consumers; used as corrosion and sludge inhibitor and dispersant; clear amber liq.; sol. in hydrocarbon fuel oils; sp.gr. 0.894 (60 F); visc. 35 cP; pour pt. 30 F; flash pt. 145 F (PM).

Armogard® G-101. [Akzo] Aliphatic amidoamine; multifunctional gasoline additive for use as corrosion inhibitor, carburetor detergent, and gasoline deicer; clear amber liq.; sol. in gasoline; sp.gr. 0.855 (60 F); dens. 7.12 lb/gal; visc. 6.5 cP; pour pt. 15 F; flash pt. 66 F (PM).

Armohib® 28, 31. [Akzo] Aliphatic nitrogen inhibitor; acid corrosion inhibitor used in metal cleaning and pickling; used in HCl and HCl/HF mixts.; amber liq.; sol. in acids; sp.gr. 0.95 and 1.042 resp.; dens. 7.69 and 8.66 lb/gal; pour pt. 52 and 24 F; flash pt. 90 and 340 F; 75 and 100% act.

Armohib® F-401. [Akzo] Amine salt; cationic; chemisorbed film-forming type corrosion inhibitor used in fuel oil storage and distribution systems; amber clear liq.; sol. in fuel oils; sp.gr. 0.848; dens. 7.04 lb/gal; visc. 17 cP; pour pt. < -55; flash pt. 106 F (CC); cloud pt. < -55 F.

Armoslip® 18. [Akzo] Stearamide; internal lubricant, mold release agent, antiblock and slip agent for PE, PP, cellophane; flakes, pellets.

Armoslip® CPM. [Akzo] Oleamide; internal lubricant, mold release agent, slip agent in polyolefins; flakes, pellets.

Armoslip® EXP. [Akzo] Erucamide; see Armoslip CPM; flakes, pellets.

Armoslip® HT. [Akzo] (Hydrog. tallow) amide; see Armoslip CPM; flakes, pellets.

Armosoft® L. [Akzo] Ditallow dimonium chloride; cationic; lubricant, fabric softener for household and commercial use; Gardner 6 hazy to clear liq.; sp.gr. 0.85 (100 F); dens. 7.09 lb/gal; visc. 45.5 cps (120 F); pour pt. 70–75 F; pH 6–9; 75% act. in IPA.

Armosoft® WA-201, WA-203, WA-204. [Akzo] Detergent, softener, antistat for home laundry formulations; Gardner 3 max. clear liq.; sol. in water; sp.gr. 1.038, 1.010, and 1.035 resp.; visc. 132, 110, and 230 cps (120 F); cloud pt. 45 F (30%), 50 F (35%), and 50 F (35%); clear pt. 95 F; flash pt. 145, 105, and 115 F (PM); pour pt. 40 and 0 F (WA-203 and WA-204); 71–74, 83–87, and 83–87% act. in aq. ethanol.

Armostat® 310. [Akzo] N-Tallow alkyl-2,2′-iminobisethanol; permanent internal antistat for HDPE, LDPE; yel. paste; typ. odor; m.w. 350; sp.gr. 0.916; i.b.p. > 300 C (760 mm Hg); f.p. 32 C; flash pt. (PMCC) > 204 C; 97% act.

Armostat® 350. [Akzo] Ethoxylated amine; internal antistat for LDPE; pellets.

Armostat® 375. [Akzo] Bis (2-hydroxyethyl) tallow amine (73%) and polyethylene(25%); internal antistat for high and low dens. polyethylene; FDA approved; off-wh. pellets; very slight amine odor.

Armostat® 410. [Akzo] N-Coco alkyl-2,2′-iminobisethanol; permanent internal antistat for PP, PS, ABS, SAN resins; FDA approved; clear yel. liq.; typ. odor; m.w. 285; sp.gr. 0.874; f.p. 12 C; i.b.p. 170 C;

flash pt. (PMCC) > 204 C; 97% min. act.

Armostat® 450. [Akzo] Ethoxylated amine; internal antistat for PS; pellets; 50% conc.

Armostat® 475. [Akzo] Bis (2-hydroxyethyl) coco amine (73%) and polypropylene (25%); internal antistat for PP, HDPE; off-wh. pellets; slight amine odor; sp.gr. 0.95; m.p. > 180 C; flash pt. (PMCC) > 200 C.

Armostat® 550, 575. [Akzo] Ethoxylated amine; internal antistat for SAN, ABS; pellets and gran. resp.

Armostat® 801. [Akzo] Glycerol stearate; internal antistat for polyethylene, PP, and PVC; GRAS for food additive use; flakes.

Armostat® 810. [Akzo] Glycerol oleate; internal antistat for polyethylene, PP, and PVC; liq.

Armotan® PML 20. [Akzo] Polysorbate 20; nonionic; o/w emulsifier, solubilizer for bath oils; liq.; 100% conc.

Armotan® PMS 20. [Akzo] PEG-20 sorbitan stearate; solubilizer for perfume and bath oils; solid.

Arneel® 18 D. [Akzo] Stearyl nitrile; detergent, wetting agent, rust inhibitor; metal wetting with min. oils; plasticizer for syn. rubbers and plastics; Gardner 6; f.p. 39 C; b.p. 330–360 C.

Arneel® C. [Akzo] Coco nitrile; detergent, wetting agent, rust inhibitor; Gardner 6.

Arneel® OD. [Akzo] Oleyl nitrile; detergent, wetting agent, rust inhibitor; Gardner 3; sp.gr. 0.83; f.p. 5 C; b.p. > 325 C.

Arneel® T. [Akzo] Tallow nitrile; detergent, wetting agent, rust inhibitor; f.p. 4 C; b.p. 330–360 C.

Arnica Oil CLR. [Henkel] Arnica extract, soybean oil, tocopherol; emollient, conditioner; protective skin and hair care prods.; yel. oil; herbal odor.

Arnnate Series. [Arjay] Alkylamine sufonate; anionic; additives for emulsion breakers, corrosion inhibitors and emulsifiers; liq.; oil-sol.

Arnox 500, 515. [DeSoto] Primary amine oxide adducts; nonionic; surface active agents used as corrosion inhibitors for industrial acid cleaning and refinery applics.; liq.; 100% conc.

Arnox 906. [DeSoto] Alkyl phenol PEG ether; nonionic; emulsifier and detergent for styrene-butadine latices; used in leather industry for degreasing and penetration for dyes and finishes; liq.; HLB 10.9; 100% conc.

Arnox 908. [DeSoto] Alkyl phenol PEG ether; nonionic; emulsifier, detergent, solubilizer for dry cleaning mixture; wetting agent; liq.; oil-sol., disp. in water; HLB 12.3; 100% conc.

Arnox 910. [DeSoto] Alkyl phenol PEG ether; nonionic; wetting agent for wettable powds., clays, dust control modifier; lime soap dispersing agent; detergent and visc. builders; used in the pulp and paper industry; liq.; HLB 12.9; 100% conc.

Arnox 1003. [DeSoto] Fatty alcohols straight chain PEG ether; nonionic; emulsifier, detergent, and defoamer intermediate; liq.; HLB 8.1; 100% conc.

Arnox 1007, 1009. [DeSoto] Fatty alcohols straight chain PEG ester; nonionic; detergent intermediate, wetting agent; variety of uses for paint, oilfield, and textile industries; liq.; HLB 13.1 and 13.7 resp.; 100% conc.

Arnox 1100, 1110. [DeSoto] Primary amine oxide adducts; nonionic; surface active agents used as corrosion inhibitors for industrial acid cleaning and refinery applics.; liq.; 100% conc.

Arnox TGE-05, 16, 36, 40, 80. [DeSoto] Castor oil ethoxylates; nonionic; emulsifier, dispersing, plasticizing, and lubricating agents; used in metal, leather, textile and coating industries; liq., solid; 100% conc.

Aromatic Oil 745. [Neville] Aromatic plasticizer used in adhesives, rubber (cements, mech. and molded goods, tires), caulking compds.; dk. brn. liq.; m.w. 200; sol. in ethers and chlorinated, aromatic, naphthenic, and terpene hydrocarbons; sp.gr. 1.050; flash pt. (COC) 265 F.

Aromox® 18/12. [Akzo] Bis(2-hydroxyethyl) stearamine oxide; cationic; wetting agent, emulsifier, stabilizer, antistat, foaming agent; Gardner 2 soft paste; sp.gr. 0.927; visc. 54 cps; flash pt. 130 F; pour pt. > 120 F; 50% act. in aq. IPA.

Aromox® C/12. [Akzo] Dihydroxyethyl cocamine oxide, IPA; cationic; wetting agent, emulsifier, stabilizer, antistat, foaming agent for detergents, shampoos, cosmetics, textiles, metal plating, petrol. additives, paper, plastics, rubber; Gardner 2 clear liq.; sp.gr. 0.949; visc. 52 cp; cloud pt. 18 F; flash pt. 82 F; pour pt. 0 F; surf. tens. 33 dynes/cm; biodeg.; 50% act. in aq. IPA.

Aromox® C/12-W. [Akzo] Dihydroxyethyl cocamine oxide; cationic; see Aromox C/12; Gardner 2 clear liq.; sp.gr. 0.997; visc. 2097 cp; HLB 18.4; flash pt. 212 F; pour pt. 35 F; surf. tens. 30.8 dynes/cm; biodeg.; 40% act. in water.

Aromox® CD/12. [Akzo] Dihydroxyethyl cocamine oxide; cationic; see Aromox C/12; Gardner 1 clear liq.; sp.gr. 0.949; visc. 52 cp; cloud pt. 18 F; flash pt. < 80 F; pour pt. 0 F; biodeg.; 50% act. in aq. IPA.

Aromox® DM16. [Akzo] Palmitamine oxide, IPA; cationic; see Aromox C/12; Gardner 1 clear liq.; sp.gr. 0.885; visc. 19 cp; cloud pt. 44 F; flash pt. 80 F; pour pt. 0 F; surf. tens. 31.6 dynes/cm; biodeg.; 40% act. in aq. IPA.

Aromox® DM16D-W. [Akzo] Hexadecyldimethylamine oxide; detergent, emollient in cosmetics; Gardner 1 max. liq., paste; sp.gr. 0.94 (60 C); m.p. 33 C; nonflam.; pH 7–8; 25% min act. in water.

Aromox® DM18D-W. [Akzo] Octadecyldimethylamine oxide; detergent, emollient in cosmetics; paste; sol. in water, min. oil, isopropyl esters, cottonseed oil, ethanol, IPA; sp.gr. 0.95; visc. 90 cps; pour pt. < 0 C; pH 7–8; 25% min act. in water.

Aromox® DMC. [Akzo] Cocamine oxide, IPA; cationic; see Aromox C/12; Gardner 1 clear liq.; sp.gr. 0.89; visc. 18 cp; cloud pt. 6 F; flash pt. < 80 F; pour pt. < 25 F; surf. tens. 33.1 dynes/cm; biodeg.; 40% act. in aq. IPA.

Aromox® DMCD. [Akzo] Cocamine oxide, IPA; detergent, thickener for household and cosmetic prods.; Gardner 1 max. liq.; sp.gr. 0.89; HLB 18.6; flash pt. (APCC) 21 C; pH 6–9; 39% min. act.

Aromox® DMC-W. [Akzo] Cocamine oxide; cationic; see Aromox C/12; Gardner 1 clear liq.; sp.gr.

0.971; visc. 17 cp; cloud pt. 34 F; flash pt. > 212 F; pour pt. 35 F; surf. tens. 32.5 dynes/cm; biodeg.; 30% act. in water.

Aromox® DMHT. [Akzo] Hydrog. tallow amine oxide, IPA; cationic; see Aromox C/12; Gardner 1 liq.; sp.gr. 0.886; visc. 32 cp; cloud pt. 64 F; flash pt. < 85 F; pour pt. 35 F; surf. tens. 36.2 dynes/cm (0.1%); biodeg.; 40% act. in aq. IPA.

Aromox® DMMC-W. [Akzo] Lauramine oxide; cationic; wetting agent, emulsifier, foam stabilizer, antistat, foaming agent for detergents, shampoos; Gardner 1 clear liq.; sp.gr. 0.96; visc. 90 cp; cloud pt. 22 F; flash pt. > 212 F; pour pt. 36 F; surf. tens. 31.2 dynes/cm (0.1%); 30% act. in water.

Aromox® DMMCD-W. [Akzo] Amine oxide; see Aromox C/12; liq.; 29% conc.

Aromox® T/12. [Akzo] Dihydroxyethyl tallow amine oxide, IPA; cationic; see Aromox C/12; Gardner 4 clear liq.; sp.gr. 0.94; visc. 77 cp; cloud pt. 60 F; flash pt. 90 F; pour pt. 55 F; surf. tens. 33.0 dynes/cm (0.1%); 50% act. in aq. IPA.

Arosurf® 66-E2. [Sherex] Isosteareth-2; nonionic; emulsifier, emollient for personal care prods.; o/w and w/o systems; coupling agent, emulsion stabilizer; Gardner 1 liq.; HLB 4.6; m.p. –5 C; pH 7 (1% DW); 100% conc.

Arosurf® 66-E10. [Sherex] Isosteareth-10; nonionic; see Arosurf 66-E2; also used as a detergent; Gardner 1 semisolid, HLB 12.0; m.p. 22 C; pH 7 (1% DW); 100% conc.

Arosurf® 66-E20. [Sherex] Isosteareth-20; nonionic; see Arosurf 66-E2; Gardner 1 soft solid; HLB 15.0; m.p. 35 C; pH 7 (1% DW); 100% conc.

Arosurf® 66-PE12. [Sherex] PPG-3-isosteareth-9; nonionic; see Arosurf 66-E2; Gardner 1 liq.; HLB 12.2; m.p. –10 C; pH 7 (1% DW); 100% conc.

Arosurf® MG-70. [Sherex] Primary ether amine; flotation reagent for reverse silica flotation from iron ores; used in phosphate and glass sand industries; amber liq.; dens. 7.08 lb/gal; visc. 6 cps; pour pt. < –30 C; amine value 238; flash pt. (Seta) > 205 F.

Arosurf® MG-70A3. [Sherex] Isodecyl ether amine acetate; see Arosurf MG-70; amber liq.; dens. 7.34 lb/gal; visc. 34 cps; pour pt. < –30 C; acid no. 70; amine value 225; flash pt. (Seta) > 230 F.

Arosurf® MG-70A5. [Sherex] C_{10} ether amine acetate; see Arosurf MG-70; amber liq.; dens. 7.43 lb/gal; visc. 70 cps; pour pt. < –30 C; acid no. 108; amine value 220; flash pt. (Seta) > 230 F.

Arosurf® MG-83A. [Sherex] Tridecyl ether diamine acetate; see Arosurf MG-70; MG-83A is a partially neutralized version; amber liq.; dens. 7.65 lb/gal; visc. 920 cps; pour pt. 0 C; acid no. 112; amine value 288; flash pt. (Seta) > 210 F.

Arosurf® MG-84A3. [Sherex] C_{14} ether diamine acetate; see Arosurf MG-70; MG-84A3 is extremely selective for certain magnetic concs.; wh. waxy solid; dens. 7.31 lb/gal (60 C); pour pt. 30 C; acid no. 87; amine value 270; flash pt. (Seta) > 210 F.

Arosurf® MG-91. [Sherex] C_9–C_{11} ether amine; see Arosurf MG-70; amber liq.; dens. 7.04 lb/gal; visc. 10 cps; pour pt. –13 C; amine value 245; flash pt. (Seta) 106 F.

Arosurf® MG-91A3. [Sherex] C_9–C_{11} ether amine acetate; see Arosurf MG-70; amber liq.; dens. 7.32 lb/gal; visc. 28 cps; pour pt. –15 C; acid no. 70; amine value 230; flash pt. (Seta) 167 F.

Arosurf® MG-91A5. [Sherex] Primary ether amine; see Arosurf MG-70; amber liq.; dens. 7.42 lb/gal; visc. 60 cps; pour pt. 9 C; acid no. 108; amine value 220; flash pt. (Seta) 155 F.

Arosurf® MG-98. [Sherex] Primary ether amine; see Arosurf MG-70; MG-98 series are the most surface active and have high inherent frothing capability; amber liq.; dens. 7.03 lb/gal; visc. 7 cps; pour pt. –25 C; amine value 260; flash pt. (Seta) > 210 F.

Arosurf® MG-98A. [Sherex] C_8–C_{10} ether amine acetate; see Arosurf MG-70, MG-98; amber liq.; dens. 7.61 and 7.37 lb/gal resp.; visc. 225 and 31 cps; pour pt. –15 and –25 C; acid no. 168 and 77; amine value 220 and 237; flash pt. (Seta) 250 and 165 F.

Arosurf® MG-101D. [Sherex] Arachidyl-behenyl amine; flotation reagent for potash flotation; flakes.

Arosurf® MG-102. [Sherex] Primary hydrogenated C_{16-22} amine; ore flotation agent for sylvite (KCl) ores; Gardner 3 solid and flake; m.w. 286; m.p. 45 C; iodine no. 3.0; 95.0% primary amine.

Arosurf® MG-140. [Sherex] Hydrog. tallow amine; flotation reagent for potash flotation and anticaking applics.; wh. solid.

Arosurf® MG-148. [Sherex] 14-18 IV tallow amine; flotation reagent for potash flotation; wh. solid.

Arosurf® MG-160. [Sherex] Coconut amine; flotation reagent for potash, silica, and mica flotation; liq.

Arosurf® MG-170. [Sherex] Tallow amine; flotation reagent for mica, feldspar, and silica flotation; paste.

Arosurf® MG-172. [Sherex] Oleyl amine; flotation reagent for mica, feldspar flotation and potash anticaking applics.; liq.

Arosurf® MG-570. [Sherex] Tallow diamine; flotation reagent for silica, mica, pyrite, and pyrochlore flotation; paste.

Arosurf® MG-583. [Sherex] Diamine; see Arosurf MG-70; amber liq.; dens. 7.30 lb/gal; visc. 26 cps; pour pt. –15 C; amine value 330; flash pt. (Seta) 125 F.

Arosurf® MG-584. [Sherex] Diamine; see Arosurf MG-70; MG-584 is extremely selective for certain magnetic concs.; amber liq.; dens. 7.20 lb/gal; visc. 5 cps; pour pt. 21 C; amine value 310; flash pt. (Seta) > 210 F.

Arosurf® MG-609, MG-612, MG-613, MG-617. [Sherex] Surf. act. reagent for the flotation of coarse particles (+20 mesh) from a variety of ores such as phosphate, potash, silica; liq.; partially sol. in water; dens. 7.32, 7.38, 7.44, and 7.49 lb/gal resp.; visc. 9.5, 12.0, 13.0, and 17.0 cps; pour pt. < –38, < –38, < –33, and < –19 C; flash pt. (Seta) 102, 102, 104, 111 F.

Arosurf® MG-2057. [Sherex] Mixt. of higher m.w. alcohols; frother for use in froth flotation circuits; recommended for the flotation of potash and iron ore; clear liq.; b.p. 125 C min.; 3–5% sol. in water; sp.gr. 0.82; visc. 5–10 cps; flash pt. (TCC) 100 F min.

Arosurf® MG-3014. [Sherex] Cationic amine; surface active reagent for the flotation of silica from phosphate ores; yel. liq.; sp.gr. 0.8786; dens. 7.32 lb/gal; visc. 24 cps; pour pt. 10 C; flash pt. 130 F.

Arosurf® MG-3609, MG-3612, MG-3613, MG-3617. [Sherex] Surf. act. reagent for the flotation of coarse particles (+20 mesh) from a variety of ores such as phosphate, potash, silica; liq.; partially sol. in water; dens. 7.32, 7.38, 7.44, and 7.49 lb/gal resp.; visc. 9.5, 12, 13, 17 cps; pour pt. < –38, < –38, < –33, and < –19 C; flash pt. (Seta) 102, 102, 104, and 111 F.

Arosurf® MSF (Monomolecular Surface Film). [Sherex] Poly (oxy-1,2-ethanediyl), α-isooctadecyl-ω-hydroxy; antimicrobial; effective in controlling the larval and pupal stages of most species of mosquitoes by physical action; lt. amber clear liq.; essentially odorless; b.p. > 320 C; essentially insol. in water; sp.gr. 0.91; dens. 7.59 lb/gal; visc. 64–70 cps (22 C); m.p. –7 to –3 C; flash Pt. 202 C; surf.tens. 28.2 dynes/cm; 100% act.

Arosurf® PT Series. [Sherex] Quat. ammonium salts; phase transfer catalysts for various applics., incl. monomer and polymer synthesis, pharmaceuticals, agrochemicals, mining, and general org. synthesis; sol. in org. media; slightly sol. or disp. in aq. systems.

Arosurf® PT 20. [Sherex] Benzyl trimethyl ammonium chloride; see Arosurf PT Series; liq.; m.w. 185; sol. see Arosurf PT Series; 60% quat.

Arosurf® PT 40. [Sherex] Methyl trialkyl ammonium chloride; see Arosurf PT Series; paste; m.w. 805; sol. see Arosurf PT Series; 50% quat.

Arosurf® PT 41. [Sherex] Alkyl trimethyl ammonium chloride; see Arosurf PT Series; liq.; m.w. 341; sol. see Arosurf PT Series; 46% quat.

Arosurf® PT 50. [Sherex] Dimethyl alkyl benzyl ammonium chloride; see Arosurf PT Series; liq.; m.w. 358; sol. see Arosurf PT Series; 50% quat.

Arosurf® PT 61. [Sherex] Alkyl trimethyl ammonium chloride; see Arosurf PT Series; liq.; m.w. 279; sol. see Arosurf PT Series; 50% quat.

Arosurf® PT 62. [Sherex] Dialkyl dimethyl ammonium chloride; see Arosurf PT Series; liq.; m.w. 322; sol. see Arosurf PT Series; 75% quat.

Arosurf® PT 64. [Sherex] Methyl tri (C_{8-10}) ammonium chloride; see Arosurf PT Series; liq.; m.w. 439; sol. see Arosurf PT Series; 85% quat.

Arosurf® PT 71. [Sherex] Alkyl trimethyl ammonium chloride; see Arosurf PT Series; liq.; m.w. 339; sol. see Arosurf PT Series; 50% quat.

Arosurf® PT Bu3. [Sherex] Methyl tributyl ammonium chloride; see Arosurf PT Series; liq.; m.w. 235; sol. see Arosurf PT Series; 75% quat.

Arosurf® PT Bu4. [Sherex] Tetrabutyl ammonium bromide; see Arosurf PT Series; liq.; m.w. 322; sol. see Arosurf PT Series; 75% quat.

Arosurf® TA-100. [Sherex] Distearyl dimonium chloride; cationic; fabric softener conc., conditioner for home and commerical laundry and textile processing; Gardner 3 max. powd.; m.w. 583; water-disp.; flash pt.> 200 F (PM); 93% quat. min.

Arosurf® TA-101. [Sherex] Distearyl dimonium chloride, modified; cationic; fabric softener for commercial and institutional laundries; imparts static control; Gardner 4 max. powd.; disp. in water; bulk dens. 24 lb/ft³; flash pt. > 200 F (PM); 100% solids.

Arphos DI-403-S. [Arjay] Phosphate ester salt; anionic; coupling agent, detergent, wetting agent; liq.; 100% conc.

Arphos DI-404. [Arjay] Complex phosphate ester, free acid form; anionic; wetting and coupling agent for heavy-duty alkaline cleaners; liq.; 100% conc.

Arphos MD-200. [Arjay] Complex phosphate ester, free acid form; anionic; EP lubricant, coupling agent, low-foam detergent; liq.; 100% conc.

Arphos MH-202. [Arjay] Complex phosphate ester, free acid form; anionic; coupling agent, detergent for heavy-duty detergents; liq.; 100% conc.

Arphos MK-203. [Arjay] Complex phosphate ester, free acid form; anionic; solubilizer, coupling agent for nonionic surfactants in presence of alkaline builders; liq.; 100% conc.

Arquad® 2C. [Akzo] Dialkyl dimethyl ammonium chloride in aq.-IPA sol'n.; microbicide for formulating disinfectants, sanitizers, and algicides; used in mold growth control, swimming pool algicides, air conditioning cooling towers, sec. oil recovery (corrosion inhibition); semiliq.; m.w. 447 (of act.); sol. in water; sp.gr. 0.89; pour pt. 10 F; pH 6–9 (10%); flash pt. (PM) 74 F; 75% act.

Arquad® 2C-75. [Akzo] Dicocodimonium chloride, IPA; cationic; emulsifier, foaming, wetting, dispersing agents, corrosion inhibitor, softener, dyeing aid, antistat for textiles, paper, cosmetics; industrial, agriculture, plastics, petrol. industry, acid pickling baths; bactericide, algicide; Gardner 7 semiliq.; m.w. 447; sp.gr.0.89; HLB 11.4; flash pt. < 80 F; pour pt. 10 F; surf. tens. 30 dynes/cm (0.1%); pH 9; flam.; 75% act. in aq. IPA.

Arquad® 2HT. [Akzo] Dimethyl/dihydrog. tallow ammonium chloride; cationic; fabric softener; paste; 75% act.

Arquad® 2HT-75. [Akzo] Quaternium-18, IPA; cationic; emulsifier, foaming, wetting, dispersing agents, corrosion inhibitor, antistat, bacteriostat for paper softening, household laundry, hair conditioning; soft wh. paste; m.w. 573; water-disp.; sp.gr. 0.87; dens. 7.22 lb/gal; visc. 47.5 cps (120 F); f.p. 95 F; HLB 9.7; flash pt. 112 F; pour pt. 90–100 F; surf. tens. 37 dynes/cm (0.1%); flam.; 75% act. in aq. IPA.

Arquad® 2S-75. [Akzo] Disoyadimonium chloride; see Arquad 2C-75; Gardner 16 max. paste; m.w. 605; water sol.; sp.gr. 0.889; HLB 10.9; pH 5–8; flash pt. (PM) 84 F; 74–77% act. in aq. IPA.

Arquad® 12-33. [Akzo] Laurtrimonium chloride, IPA; cationic; see Arquad 2C-75; Gardner 7 liq.; m.w. (act.) 263; sp.gr. 0.98; f.p. 5 F; HLB 17.1; flash pt. 140 F; pH 5–8 (10% aq.); biodeg.; 33% act. in aq. IPA.

Arquad® 12-50. [Akzo] Laurtrimonium chloride, IPA; cationic; see Arquad 2C-75; Gardner 1 liq.; m.w. (act.) 263; sp.gr. 0.89; f.p. 13 F; HLB 17.1; flash pt. < 80 F; surf. tens. 33 dynes/cm; pH 5–8 (10% aq.);

biodeg.; 50% act. in aq. IPA.

Arquad® 16-29. [Akzo] Cetrimonium chloride; cationic; see Arquad 2C-75; Gardner 6 liq.; m.w. (act.) 319; sp.gr. 0.96; f.p. 61 F; HLB 15.8; flash pt. > 212 F; pH 5–8 (10% aq.); biodeg.; 29% act. in water.

Arquad® 16-50. [Akzo] Cetrimonium chloride, IPA; cationic; see Arquad 2C-75; Gardner 6 liq.; m.w. (act.) 319; sp.gr. 0.88; f.p. 61 F; HLB 15.8; flash pt. < 80 F; surf. tens. 34 dynes/cm; pH 5–8 (10% aq.); biodeg.; 50% act. in aq. IPA.

Arquad® 18-50. [Akzo] Steartrimonium chloride, IPA; cationic; see Arquad 2C-75; also as dye leveling agent, visc. stabilizer, in lubricant compding.; Gardner 7 liq.; m.w. (act.) 347; sp.gr. 0.88; HLB 15.7; flash pt. < 80 F; surf. tens. 34 dynes/cm; pH 5–8 (10% aq.); biodeg.; 50% act. in aq. IPA.

Arquad® 88. [Akzo] Dimethyl/dihydrog. tallow ammonium chloride; cationic; fabric softener; solid; 88% act.

Arquad® 210/50. [Akzo] Didecyl dimethyl ammonium chloride; cationic; bactericide, algicide; liq.; 49-52% act.

Arquad® B-50 USP. [Akzo] Alkyl (C_8–C_{18}) dimethylbenzyl ammonium chloride USP; cationic; disinfectant, germicide, bactericide, algicide; Gardner 2 max. liq., visc. mass; sp.gr. 0.98; 50% act.

Arquad® B-90 USP. [Akzo] Alkyl (C_8–C_{18}) dimethylbenzyl ammonium chloride USP; cationic; disinfectant, germicide; Gardner 6 max. liq., visc. mass; 90% act.

Arquad® B-100. [Akzo] Benzalkonium chloride; cationic; antimicrobial for industrial applics., sec. oil recovery, textiles, cosmetics, pharmaceuticals, sanitizers; Gardner 2 liq.; m.w. 380; sol. in acetone, alcohol, most polar solvs., water; sp.gr. 0.967; pour pt. 0 F; flash pt. < 80 F; pH 7–8; 50% act. in aq. IPA.

Arquad® C-33. [Akzo] Cocotrimonium chloride; cationic; see Arquad 2C-75; Gardner 7 liquid; m.w. (act.) 278; water-sol.; sp.gr. 0.98; f.p. 25 F; HLB 16.5; flash pt. < 80–140 F; pH 6–9 (10% aq.); biodeg.; 33% act. in aq. IPA.

Arquad® C-33-W. [Akzo] Cocotrimonium chloride; cationic; emulsion-break retardant in cosmetics; liq.; sp.gr. 0.96; f.p. –3 C; nonflam.; pH 5–8; 32–35% act.

Arquad® C-50. [Akzo] Cocotrimonium chloride, IPA; cationic; see Arquad 2C-75; Gardner 7 liq.; m.w. (active) 278; water-sol.; sp.gr. 0.89; f.p. 5 F; HLB 16.5; flash pt. < 80 F; surf. tens. 31 dynes/cm (0.1%); pH 5–8 (10% aq.); biodeg.; 50% act. in aq. IPA.

Arquad® DM14B-90. [Akzo] Myristyl dimethylbenzyl ammonium chloride dihydrate; cationic; bactericide, fungicide, germicide, disinfectant; wh. powd.; 90% act.

Arquad® DM18B-90. [Akzo] Octadecyl-dimethylbenzyl ammonium chloride; conditioner, emulsifier, softener for cosmetics, pharmaceuticals; wh. powder; water sol.; pH 5–8 (1%); 90% min act.

Arquad® DMCB. [Akzo] Alkyl dimethyl benzyl ammonium chloride; microbicide for disinfectants, sanitizers, algicides for use in swimming pools, air conditioning cooling towers, bathroom cleaners, petrol. recovery; liq., Gardner 4 max.; m.w. 354; sol. in water and most common org. solv.; sp.gr. 0.935; flash pt. 80 F (PM); 79% act. in aq. IPA.

Arquad® DMHTB-75. [Akzo] Hydrog. tallow dimethylbenzyl ammonium chloride; bactericide, disinfectant, softening agent for textiles; Gardner 4 max. soft paste; f.p. 60 C; 75% act.

Arquad® DMMCB-50. [Akzo] Alkyl (C_{12},C_{14},C_{16}) dimethyl benzyl ammonium chloride; cationic; antistat, flocculant, emulsifier, softener, corrosion inhibitor used in cosmetics, soil stabilization, textiles, fabric softener, fungicide, bactericide; liq.; 50% act.

Arquad® DMMCB-75. [Akzo] Alkyl (C_{12}–C_{14}, C_{16}) dimethyl benzyl ammonium chloride; disinfectant, bactericide, germicide, fungicide; liq.; cloud pt. –36 F; flash pt. < 80 F; biodeg.; flam.; 50% act.

Arquad® NF-50. [Akzo] Dialkyldimethyl ammonium chloride; bactericide, algicide; liq.; 50% act.

Arquad® S-2C-50. [Akzo] 1:1 mixt. of oleyltrimethyl ammonium chloride and dicocodimethyl ammonium chloride; bactericide, wetting agent, corrosion inhibitor; oil recovery; Gardner 10 max. liq.; sp.gr. 9.87; HLB 13.5; f.p. –15 C; pH 5–8; 50–55% act.

Arquad® S. [Akzo] Alkyl trimethyl ammonium chloride; microbicide for formulating disinfectants, sanitizers, and algicides; used in mold growth control, swimming pool conditioning, air conditioning systems, industrial water, sec. oil recovery; liq.; m.w. 342 (of act.); sol. in water; sp.gr. 0.89; flash pt. (PM) 60 F; pH 5–8 (10%); 49% act.

Arquad® S-50. [Akzo] Soytrimonium chloride, IPA; cationic; see Arquad 2C-75; also used in bitumen emulsions; slime control agent in water systems; Gardner 8 max. liq.; m.w. 343; water-sol.; sp.gr. 0.89; HLB 15.6; f.p. 20 C; flash pt. (PM) < 80 F; pH 5–8 (10% aq.); 49–52% act. in IPA.

Arquad® T-2C-50. [Akzo] 1:1 mixt. of tallow trimonium chloride and dicoco dimonium chloride; cationic; see Arquad 2C-75; Gardner 8 liq.; m.w. (act.) 394; sp.gr. 0.87; HLB 13.0; flash pt. < 80 F; flam.; 50% act. in aq. IPA.

Arquad® T-27W. [Akzo] Tallow trimonium chloride; see Arquad 2C-75; Gardner 3 max. liq.; m.w. 343; water sol.; HLB 14.2; pH 5–8 (10% aq.); 26–29% act. in aq. IPA.

Arquad® T-50. [Akzo] Tallowtrimonium chloride, IPA; cationic; see Arquad 2C-75; also used in mfg. of antibiotics; Gardner 8 max. liq.; m.w. 340; water-sol.; sp.gr. 0.881; HLB 14.2; flash pt. < 80 F (PM); pour pt. 15–48 F; pH 5–8 (10% aq.); biodeg.; 50% act. in aq. IPA.

Arquest 709. [Arjay] Amino tris(methylene phosphonic acid); scale and corrosion inhibitor in aq. systems; aq. process additive for paper, textiles, and metals; industrial and commercial formulations for cleaning various substrates; straw-colored aq. sol'n.; 50% act.

Arquest 710. [Arjay] Hydroxyethylidene diphosphonic acid; scale inhibitor for aq. systems; liq.; water-sol.

Arsenal®. [Am. Cyanamid/Ag] 2-[4,5-Dihydro-4-

methyl-4-(1-methylethyl)-5-oxo-1H-imidazol-2-yl]-3-pyridinecarboxylic acid with 2-propanamine (1:1) salt; nonselective broad-spectrum herbicide for weed control on railroad, highway right-of-ways, airports, other industrial uses; 27.6% act.

Arsil 486-S. [Arjay] Silicone sol'n.; nonionic; pesticide formulations, latex processing, glycol formulations, ink processing, soap and detergent mfg., wax and polish emulsions; liq.; 10% conc.

Arsperse Series. [Arjay] Oxyalkylated fatty org. acids; dispersant in oil-sol. corrosion inhibitors to make them disp. in water; coupler; sol. in water, aromatic hydrocarbons, solid.

Arsul DDB. [Arjay] Branched chain dodecylbenzene sulfonic acid; anionic; chemical intermediate for emulsifiers, dispersants, corrosion inhibitors, emulsion breakers, cleaning compds.; liq.; 90% conc.

Arsul LAS. [Arjay] Linear dodecylbenzene sulfonic acid; anionic; see Arsul DDB; liq.; biodeg.; 98% conc.

Arubren CP. [Bayer] Chlorinated paraffin hydrocarbon; plasticizer used to improve the flame retardant properties of vulcanizates; FDA regulated for adhesives applic.; emulsion form is used for latex goods; yel. transparent, high-visc. liq.; dens. 1.46–1.50 g/cc; 72% chlorine content.

Arylan HE Acid. [Harcros UK] Alkylbenzene sulfonic acid, branched chain; anionic; intermediate for detergent mfg. and prod. of wetting agents and emulsifiers; liq.; 95% conc.

Arylan S Acid. [Harcros UK] Dodecylbenzene sulfonic acid (branched chain); anionic; intermediary; liq.; 96% conc.

Arylan SBC Acid. [Harcros UK] Dodecylbenzene sulfonic acid; anionic; detergent intermediate; biodeg.; liq.; 97% conc.

Arylan SC Acid. [Harcros UK] Dodecylbenzene sulfonic acid; detergent intermediate; br. liq.; visc. 1800 cs; 97% act.

Arylan SNS. [Harcros UK] Sodium naphthalene sulfonic acid formaldehyde conc.; dispersant; aq. suspensions; buff powd.; pH neutral; 90% act.

Arylan SP Acid. [Harcros UK] Dodecylbenzene sulfonic acid; anionic; detergent intermediate; liq.; 97% conc.

Arzoline Series. [Arjay] Alkyl cyclic amido imidazoline amines; corrosion inhibitor base; water sol. inhibitors; sol. in aromatic naptha; water-insol.

ASA. [Ethyl] Alkenyl succinic anhydride; intermediate for defoamers, demulsifiers, emulsifiers, foam boosters, wetting agents, detergents, dispersants; liq.

Asol. [Lucas Meyer] Lecithin fraction; nonionic; antispatter agent, release agent, emulsifier for food, cosmetics, pharmaceuticals; liq.; 40–100% conc.

AST-1001. [Merix] Light stabilizer for plastics; nonflam.; faint pleasant odor; neutral pH.

Aston 123. [Lyndal] Thermosetting polyamine; durable antistat with high resistance to laundering and dry cleaning on all substrates; water-emulsifiable.

Aston AP Conc. [Lyndal] Polyamine, cationic; antistat and softener used in textile and industrial applic.; liq.; sp.gr. 1.04; flash pt. > 200 F; 100% solids.

Aston OI. [Lyndal] Fatty imidazoline; antistat for PP and polyester; water-sol.

Astramyl SB. [Henkel-Nopco] Complex fatty amidoamines; corrosion inhibitor and asphalt additive; semi-liq.

Atlas G-3780A. [ICI Am.] PEG-20 tallow amine; nonionic; surfactant, antistat for textiles; amber liq. (may become hazy); sol. in water, lower alcohols, acetone, ethyl acetate, ethylene glycol; sp.gr. 1.04; visc. 250 cps; HLB 15.5; flash pt. > 300 F; 100% conc.

Atlox 1096. [ICI Am.] POE sorbitol hexaoleate; nonionic; emulsifier and coupling agent used in textile industry; liq.; HLB 11.4; 100% conc.

Atlox 1285. [ICI Am.] POE triglyceride; nonionic; emulsifier and coupling agent used in textile industry; semisolid; HLB 14.4; 100% conc.

Atlox 4853 B. [ICI Am.] Blend of ionic and nonionic surfactants; emulsifier, wetter, dispersant for pesticides, herbicides; liq.; HLB 11.9; biodeg.

Atlox 4862. [ICI Am.] Anionic suspending agent; wetting/suspending agent, anionic; powd.; 100% conc.

Atlox 4868B. [ICI Am.] POE triglyceride/alkylaryl sulfonate; wetting/suspending agent; pesticide emulsifier; liq.; HLB 11.9; biodeg.

Atlox 4870B. [ICI Am.] Polyethylene sorbitan esters of fatty and resin acids, alkylaryl sulfonate blend; wetting/suspending agent; liq.; HLB 13.0; biodeg.

Atlox 4875. [ICI Am.] POE ester; wetting and dispersing agent for pesticides and herbicides; liq.; HLB 18.2.

Atlox 4995. [ICI Am.] POE alkyl ether; nonionic; wetting/suspending agent for pesticides, herbicides; powd.; HLB 13.6; 100% conc.

Atlox 4996 B. [ICI Am.] Nonionic; wetting/suspending agent; powd.

Atmer 100. [ICI Am.] Sorbitan monolaurate; antifog agent for PE and EVA food-wrapping films; antistat; cling additive; red-brn. liq.; visc. 3900–5000 mPa•s.

Atmer 101. [ICI Am.] Sorbitan ester; antifog, antistat; amber liq.; visc. 950–1100 mPa•s.

Atmer 102. [ICI Am.] Sorbitan ester; antifog; tan solid; m.p. 44–47 C.

Atmer 103. [ICI Am.] Sorbitan ester; antifog agent for long-lasting properties in LDPE, EVA, and PVC agric. film; pale cream solid; m.p. 50–53 C.

Atmer 104. [ICI Am.] Sorbitan ester; antifog; cream solid; m.p. 48–53 C.

Atmer 105. [ICI Am.] Sorbitan monooleate; antifog, antistat, cling additive for low-temp. LDPE film applics.; amber liq.; visc. 950–1100 mPa•s.

Atmer 106. [ICI Am.] Sorbitan trioleate; antifog, cling additive for LDPE film; amber liq.; visc. 170–230 mPa•s.

Atmer 107. [ICI Am.] Sorbitan ester; antifog, antistat; yel.-amber liq.; visc. 3500–5500 mPa•s.

Atmer 110. [ICI Am.] POE sorbitan laurate; antifog, antistat for thermoplastics incl. PE, PP, PVC, styrenic copolymers; yel. liq.; visc. 250–450 mPa s.

Atmer 111. [ICI Am.] POE sorbitan ester; antifog; yel. liq.; visc. 650 mPa•s.

Atmer 112. [ICI Am.] POE sorbitan ester; antifog; pale yel. liq.; visc. 400–650 mPa•s.

Atmer 113. [ICI Am.] POE sorbitan stearate; antifog, antistat for plastics; pale yel. semisolid; visc. 75–175 mPa•s (50 C).

Atmer 114. [ICI Am.] POE sorbitan ester; antifog agent; ivory solid; m.p. 36–40 C.

Atmer 115. [ICI Am.] POE sorbitan ester; antifog; pale yel. solid; m.p. 30–35 C.

Atmer 116. [ICI Am.] POE sorbitan ester; antifog agent for flexible PVC food-wrapping film; yel.-brn. liq.; visc. 375–480 mPa•s.

Atmer 117. [ICI Am.] POE sorbitan ester; antifog; yel.-brn. liq.; visc. 400–500 mPa•s.

Atmer 118. [ICI Am.] POE sorbitan ester; antifog; yel.-brn. liq.; visc. 250–450 mPa•s.

Atmer 121. [ICI Am.] Glycerol oleate; antifog agent and cling additive for PVC food-wrapping film; pale yel. liq.; visc. 140 mPa•s.

Atmer 122. [ICI Am.] Glycerol stearate; antistat for LDPE, PP, flex, PVC; antifog; wh. powd.; m.p. 63 C.

Atmer 123. [ICI Am.] Glycerol stearate; antifog, antistat for polyolefins and expandable PS; wh. powd.; m.p. 63 C.

Atmer 124. [ICI Am.] Glycerol ester; antifog, antistat for expandable polymers; wh. powd.; m.p. 63 C.

Atmer 125. [ICI Am.] Glycerol stearate; antifog, antistat for polyolefins esp. PP; wh. powd.; m.p. 63 C.

Atmer 126. [ICI Am.] Glycerol stearate; antifog, antistat for expandable polymers; wh. powd.; m.p. 63 C.

Atmer 127. [ICI Am.] Glycerol ester; antifog, antistat for expandable polymers; wh. powd.; m.p. 67 C.

Atmer 128. [ICI Am.] Glycerol stearate; antifog, antistat for expandable polymers; wh. powd.; m.p. 69 C.

Atmer 129. [ICI Am.] Glycerol stearate; antifog, antistat for LDPE, PP, plasticized PVC; can be used in combination with Atmer 163; wh. powd.; microbeads; m.p. 69 C.

Atmer 132. [ICI Am.] Ethoxylated nonyl phenol; antifog agent for flexible PVC food-wrapping film; colorless liq.; visc. 360 mPa•s (20 C).

Atmer 138, 139. [ICI Am.] POE ester; antistat, expandable polymer; wh. cream solid; m.p. 38–40 and 46 C resp.

Atmer 151. [ICI Am.] Alkoxylated alcohol; antistat for flexible PVC, visc. depressant; colorless liq.; visc. 72 mPa s (30 C).

Atmer 152. [ICI Am.] Blend; foam stabilizer; brn. liq.; visc. 1650 mPa•s.

Atmer 153. [ICI Am.] Alkoxylated fatty derivs.; plasticizer; PVC plastisol visc. depressant; pale yel. liq.; visc. 150 mPa s.

Atmer 154. [ICI Am.] POE fatty acid ester; visc. depressant; antistat; pale yel. liq.; visc. 90 mPa s.

Atmer 160. [ICI Am.] Quat. ammonium compd.; antistat for flexible PVC; yel. liq.; visc. 1010 mPa s (20 C).

Atmer 163. [ICI Am.] Ethoxylated fatty amine; antistat for polyolefins and styrenics (ABS, HIPS); pale yel. liq.; visc. 150 mPa s.

Atmer 164. [ICI Am.] Syn. quat. ammonium compd.; internal antistat for PU; yel. liq.; visc. 150 mPa•s (40 C).

Atmer 165. [ICI Am.] Aryl polyol diacetal; nucleating agent; wh. powd.; m.p. 205–230 C.

Atmer 166. [ICI Am.] Blend; nucleating agent; wh. powd.; m.p. 205–230 C.

Atmer 171. [ICI Am.] POE triglyceride; cell regulator; pale yel. liq.; visc. 1000–1300 mPa•s.

Atmer 172. [ICI Am.] Proprietary blend; internal antistat for polyolefins esp. PP; wh. powd.

Atmer 175. [ICI Am.] Alkylaryl sulfonate; antistat; red-brn. liq.; visc. 4000–9000 mPa•s.

Atmer 176. [ICI Am.] POE sorbitol ester; antistat; dk. amber liq.; visc. 550 mPa•s.

Atmer 177. [ICI Am.] POE fatty alcohol; antistat; colorless liq.; visc. 35 mPa•s.

Atmer 178. [ICI Am.] POE fatty alcohol; antistat; pale yel. liq.; visc. 40 mPa•s.

Atmer 180. [ICI Am.] Glycerol fatty acid ester; lubricant; yel.-brn. liq.; visc. 3000 mPa•s.

Atmer 184. [ICI Am.] Glycerol fatty acid ester; antifog agent for long-lasting properties in EVA and LDPE agric. film; amber powd.; m.p. 40–53 C.

Atmer 190. [ICI Am.] Alkyl sulfonate; antistat for PVC and HIPS; pale yel. solid; m.p. 140 C.

Atmer 191. [ICI Am.] Alkyl sulfonate; see Atmer 190; solid; fine powd.

Atmer 502. [ICI Am.] POE alkyl ether; antifog; wh. semisolid.

Atmer 507. [ICI Am.] Aliphatic polyether; nonionic; visc. depressant, surfactant; amber liq.; disp. in water; sol. in acetone, ethylene acetate, dioxane, cellosolve, lower alcohols; sp.gr. 1.1; visc. 300 mPa•s; flash pt. > 300 F.

Atmer 508. [ICI Am.] POE alkylaryl ether; nonionic; surfactant; plasticizer; PVC plastisol visc. depressant; pale yel. liq.; sol. in lower alcohols, cellosolve, many petrol. solv.; sp.gr. 1.0; visc. 220 cps; flash pt. > 300 F.

Atmer 645, 646, 647, 649, 650, 654. [ICI Am.] Blend; nonionic; antifog agents for PE, EVA, and PVC food-wrapping films; yel. liq. except 654 (yel. paste); visc. 550 mPa•s (grade 645).

Atmer 655, 656, 657, 658. [ICI Am.] Blend; nonionic; antifog; yel. liq. (655); visc. 650 mPa•s.

Atmer 685. [ICI Am.] Blend; nonionic; antifog agents for PE, EVA, and PVC food-wrapping films; pale yel. liq.; visc. 170 mPa•s.

Atmer 1001. [ICI Am.] Sorbitol ester; antistat; orange-red liq.; visc. 250 mPa•s.

Atmer 1002. [ICI Am.] Quat. ammonium compd.; internal antistat for PU; yel. liq.

Atmer 1004, 1005, 1006. [ICI Am.] Quat. ammonium compd.; external antistat for all polymers; amber, yel. and red-brn. liq. resp.; visc. 100, 500, and 1000 mPa•s resp.

Atmer 1007. [ICI Am.] Glycerol oleate; antifog, cling agent for LDPE and PVC films; pale yel. paste; visc. 530 mPa•s.

Atmer 1008, 1009. [ICI Am.] Glycerol ester; antistat; wh. powd.; m.p. 63 C.

Atmer 1010. [ICI Am.] Glycerol oleate; antifog agent and cling additive for PVC food-wrapping film, LDPE; pale yel. liq.; bland odor; sol. in lower alcohols, min. oils, veg. oils; insol. in water; sp.gr. 0.94; HLB 3.0; pour pt. 11 C; b.p. > 100 C.; flash pt. > 300 F (COC)
Atmer 1012. [ICI Am.] Glycerol ester; antifog, cling agent; pale yel. liq.; visc. 140 mPa•s.
Atmer 1020. [ICI Am.] Glycerol ester; antifog, cling agent; pale yel. liq.
Atmer 1021. [ICI Am.] POE glyceride ester; antistat, lubricant; tan solid.
Atmer 1022, 1023. [ICI Am.] POE sorbitol ester; cling agent; pale yel. liq.; visc. 200 mPa•s.
Atmer 1024. [ICI Am.] Alcohol phosphate ester; antistat; colorless liq.; visc. 25 mPa•s.
Atmer 1025. [ICI Am.] POE oleate; lubricant, internal antistat for flexible PVC; amber liq.; visc. 90 mPa•s.
Atmer 1026. [ICI Am.] POE ester; lubricant; lt. yel. liq.
Atmer 1027. [ICI Am.] Quat. ammonium compd.; internal antistat for flexible PVC; amber-brn. liq.; visc. 2500 mPa•s.
Atmer 1030. [ICI Am.] Hydrog. oil; lubricant.
Atmer 1040. [ICI Am.] EO-PO block copolymer; lubricant; wh. flake; visc. 2800 mPa•s (77 C).
Atmer 1041, 1042. [ICI Am.] EO-PO block copolymer; lubricant; colorless liq. and paste resp.; visc. 800 mPa•s and 305 mPa•s (60 C) resp.
Atmer 7000. [ICI Am.] 50% Atmer 163; antistat for inj. molding PP; pellets/powd.
Atmer 7001. [ICI Am.] 50% Atmer 129/163 (2:1); antistat for inj. molding PP; pellets/powd.
Atmer 7002. [ICI Am.] 50% Atmer 129; antistat for inj. molding PP; pellets/powd.
Atmer 7003. [ICI Am.] 50% Atmer 163; antistat for PP film; pellets/powd.
Atmer 7004. [ICI Am.] 50% Atmer 129; antistat for PP film; pellets/powd.
Atmer 7005. [ICI Am.] 50% Atmer 129/163 (73:27); antistat for inj. molding PP; pellets/powd.
Atmer 7006. [ICI Am.] 50% Atmer 129/163 (2:1); antistat for PP film; pellets/powd.
Atmer 7007. [ICI Am.] 50% Atmer 129/163 (4:1); antistat for inj. molding PP; pellets/powd.
Atmer 7008. [ICI Am.] 60% Atmer 163; antistat for inj. molding PP; pellets/powd.
Atmer 7009. [ICI Am.] 60% Atmer 129; antistat for inj. molding PP; pellets/powd.
Atmer 7010. [ICI Am.] 50% Atmer 138; antistat for PP film; pellets/powd.
Atmer 7100. [ICI Am.] 50% Atmer 103; antifog for PE film; pellets/powd.
Atmer 7101. [ICI Am.] 50% Atmer 129; antifog, antistat for PE film; pellets/powd.
Atmer 7102. [ICI Am.] 50% Atmer 129; antistat for inj. molding PE; pellets/powd.
Atmer 7103. [ICI Am.] 50% Atmer 129/163 (2:1); antistat for inj. molding PE; pellets/powd.
Atmer 7104. [ICI Am.] 50% Atmer 163; antistat for inj. molding PE; pellets/powd.
Atmer 7105. [ICI Am.] 50% Atmer 163; antistat for PE film; pellets/powd.
Atmer 7106. [ICI Am.] 50% Atmer 129/163 (2:1); antistat for PE film; pellets/powd.
Atmer 7107. [ICI Am.] 50% Atmer 184; antifog for PE film; pellets/powd.
Atmer 7108. [ICI Am.] 50% Atmer 685; antifog for LLDPE film; pellets/powd.
Atmer 7109. [ICI Am.] 50% Atmer 129/163 (4:1); antistat for inj. molding PE; pellets/powd.
Atmer 7110. [ICI Am.] 50% Atmer 1007; antistat for PE film; pellets/powd.
Atmer 7111. [ICI Am.] 50% Atmer 129/163 (4:7); antistat for PE film; pellets/powd.
Atmer 7112. [ICI Am.] 50% Atmer 122; antistat for inj. molding PE; pellets/powd.
Atmer 7113. [ICI Am.] 50% Silicone F111-300; mold release for inj. molding PE; pellets/powd.
Atmer 7114. [ICI Am.] 50% Atmer 121; antifog for PE film; pellets/powd.
Atmer 7115. [ICI Am.] 50% Atmer 184/114 (3:1); antifog for PE film; pellets/powd.
Atmer 7200. [ICI Am.] 50% Atmer 163; antistat for inj. molding PS/HIPS; pellets/powd.
Atmer 7201. [ICI Am.] 50% Atmer 163; antistat for inj. molding ABS; pellets/powd.
Atmer 7202. [ICI Am.] 50% Atmer 190; antistat for inj. molding HIPS; pellets/powd.
Atmer 7203. [ICI Am.] 50% Atmer 190; antistat for inj. molding ABS; pellets/powd.
Atmul 700H. [ICI Am.] Hydrated mono- and diglycerides with polysorbate 60 (1%); food emulsifier, softener for yeast-raised bakery prods.; creamy wh., soft plastic; water disp.; HLB 4.0; flash pt. > 300 F.
Atolene RW. [Dexter] Oleic acid sulfated ester; anionic; lubricant, wetting, dye leveling agent used in textile processing; liq.
Atolex LDA/40. [Standard Chem. UK] Dyebath aux., nonionic; leveling and disp. agent.
Atpet 80. [ICI Am.] Sorbitan oleate; nonionic; rust, corrosion inhibitor; amber liq.; disp. in water, sol. in most petrol. oils, veg. oils, lower alcohols, chlorinated solv.; sp.gr. 1; visc. 1200 cps; HLB 4.3; flash pt. > 300 F; pour pt. 0 F; 100% act.
Atpet 100. [ICI Am.] Sorbitan oleate; nonionic; rust inhibitor; red-brn. visc. liq.; sol. in min. and veg. oils, lower alcohols, chlorinated solv.; water-disp.; sp.gr. 1.0; visc. 900 cps; HLB 4.3; flash pt. > 300 F; fire pt. > 300 F.
Atpet 200. [ICI Am.] Sorbitan oleate; nonionic; rust inhibitor; rdsh.-brn. liq.; sol. in min. and veg. oils, lower alcohols, chlorinated solv.; water-disp.; sp.gr. 1.0; visc. 1200 cps; HLB 4.3; pour pt. 0 F; flash pt. > 300 F; fire pt. > 300 F.
Atplus 535. [ICI Am.] Anionic; agric. surfactant; stabilizer for liq. fertilizer-pesticide mixtures, for repackaging/resale; lt. amber liq.; complete sol. in liq. fertilizers; dens. 10.5 lb/gal; sp.gr. 1.25; visc. 40 cps; flash pt. > 200 F.
Atplus 540. [ICI Am.] Alkoxylated alcohol ether; agric. foaming agent, spray tank additive; amber liq.; dens. 8.1 lb/gal; sp.gr. 0.97; visc. 30 cps; flash pt. 65

F; pour pt. 20 F; 75% conc.

Atplus 1034. [ICI Am.] Aromatic sulfates and polyhydroxy compds.; agric. foaming agent, spray tank additive; amber liq.; dens. 10 lb/gal; sp.gr. 1.2; visc. 140 cps; flash pt. > 200 F; pour pt. –35 F.

Atrabute+. [Griffin] Atrazine and butylate; flowable herbicide to control wide range of weeds on corn; 56.80% butylate; 13.90% atrazine.

Atrabute+ II. [Griffin] Atrazine and butylate; see Atrabute+; 51.42% butylate, 19.67% atrazine.

Attacote. [Engelhard] Attapulgite clay; anticaking, anti-agglomerating agent for conditioning ammonium nitrate and sulfate crystals, urea, gran. fertilizers, etc.; fine cream powd.; sp.gr. 2.47; dens. 15–18 lb/ft^3; bulking value 0.0486 gal/lb; pH 7.5–9.5.

Attaflow. [Engelhard] Attapulgite clay; replacement for dry gelling clay; economy and ease of handling for suspension formulators; gray liq., 0.4% 325-mesh residue; dens. 10 lb/gal; pH 8.5.

Attagel 40. [Engelhard] Colloidal attapulgites; thickener, stabilizer, suspending and flatting agent in paints; suspending, bodying, and sag control agent in adhesives, sealants, and mastics; lt. cream powd., particle size 0.14 μ, 0.3 325-mesh residue; sp.gr. 2.36; dens. 19–22 lb/ft^3; pH 7.5–9.5; 11% conc.

Attagel 50. [Englehard] Colloidal attapulgites; see Attagel 40; lt. cream powd., particle size 0.14 μ, 0.01% 325-mesh residue; sp.gr. 2.36; dens. 19–22 lb/ft^3; pH 7.5–9.5; 11% conc.

Attagel 150. [Engelhard] Attapulgite clay; thickener, geller, stabilizer, thixotropic agent; used for drilling muds, forming stable suspensions when drilling oil wells; lt. cream powd., particle size 0.12 μ, 8.0% 325-mesh residue; sp.gr. 2.36; dens. 38–45 lb/ft^3; pH 7.5–9.5; 11% conc.

Attagel 350. [Englehard] Attapulgite clay; thickener, geller, stabilizer, thixotropic agent; used in liq. fertilizer and feed for suspending solids, plant, or animal nutrients; lt. cream powd., particle size 0.12 μ, 8.0% 325-mesh residue; sp.gr. 2.36; dens. 38–45 lb/ft^3; pH 7.5–9.5; 11% conc.

Autopoon NI. [Zschimmer & Schwarz] Cationic surfactants, solvs., and solubilizers; cationic; water-repellent conc. for preps. for lacquered surfaces; clear red-brn. liq.; sol. in water, IPA; sp.gr. 0.95; cloud pt. 0 C; pH 4.5–5.5 (1%); 55% act.

Avenge®. [Am. Cyanamid/Ag] Difenzoquat methyl sulfate; wild oat herbicide for barley and wheat crops; 31.2% act.

Avamid 150. [Mona] Avocadamide DEA, avocado oil; SE foam stabilizer, visc. builder, lubricant for conditioning shampoos, hair rinses, creams and lotions; imparts smooth, silky feel to skin and hair; clear amber liq.; pH 10.5 (10%); 100% act.

Avicel PH-101. [FMC] Microcryst. cellulose NF; binder, disintegrant, flow aid, and filler for pharmaceuticals and animal health prods.; absorbent; peptizing agent; anticaking agent for oils to make sticky substances free flowing; wh. powd., < 30% +200 mesh; odorless; insol. but disp. in water; pH 5.5–7.0.

Avicel PH-102. [FMC] Microcryst. cellulose NF; see Avicel PH-101; wh. powd., > 45% +200 mesh; odorless; sol. see Avicel PH-101; pH 5.5–7.0.

Avicel PH-103. [FMC] Microcryst. cellulose NF; see Avicel PH-101; wh. powd., < 30% +200 mesh; odorless; sol. see Avicel PH-101; pH 5.5–7.0.

Avicel PH-105. [FMC] Microcryst. cellulose NF; see Avicel PH-101; wh. fine powd., < 1% +400 mesh; insol. in water and org. solvs.; pH 5.0–7.0.

Avicel RC-591. [FMC] Microcryst. cellulose, cellulose gum; cosmetic base; lg. surf. area for absorbing ingred. matrix onto RC.; opacifier for wh. lotions and creams; adsorbent for dry cream bases and sachets; water-insol.

Avirol 200. [Henkel] Ammonium lauryl sulfate; foaming agent and suspension aid in rock drilling, fire fighting, dispersant for dry wall mfg., dyes and pigments; lt. amber liq.; typ. odor; visc. 1000 cps; cloud pt. 10 C max; 28% fatty alcohol sulfate.

Avirol SA-4106. [Henkel] Sodium 2-ethylhexyl sulfate; anionic; wetting agent, emulsifier, stabilizer for plastics, rubber, adhesives, food contact paper; amber liq.; sp.gr. 1.1; dens. 9.2 lb/gal; visc. 200 cps; cloud pt. 5 C max.; surf. tens. 38 dynes/cm; pH 7–10 (10%); biodeg.; 43–46% solids.

Avirol SL-2010. [Henkel] Sodium lauryl sulfate; anionic; dispersing agent and emulsifier for acrylates, styrene acrylic, vinyl chloride, vinylidene chloride and vinyl acetate copolymers; liq.; 30% conc.

Avirol SO-70P. [Henkel] Sodium dioctyl sulfosuccinate; anionic; wetting agent, emulsifier for antifog compositions, cleaners for automobiles, dry cleaning formulations, and glass surfaces; pigment dispersion in paints and inks; latex paint stabilization; used in pesticides; APHA 100 clear liq.; sol. in polar and nonpolar solv.; sp.gr. 1.06; dens. 8.8 lb/gal; flash pt. 280 F (PMCC); cloud pt. –5 C; 70% solids.

Avitex NA. [DuPont] Complex higher alkylamine; cationic; softener, antistat for textile fibers; lt. cream to wh. soft, homogeneous paste; mild odor; water-disp. @ 160–180 F; dens. 8.42 lb/gal; pH 6.5–7.5 (1% aq.); flash pt. > 93 C (TCC).

Avitex R. [DuPont] Complex higher alkylamine; cationic; softener and antistat for fabrics; modifier with hand builders and resins; lt. tan soft, homogenous paste; mild odor; water-disp. @ 160–180 F; sp.gr. 0.98; dens. 8.18 lb/gal; pH 6.5–7.5 (1% aq.).

Avocado Oil CLR. [Henkel] Avocado fatty oil; emollient; conditioner for skin preparations; ylsh.-grn. to brnsh.-grn. oil; faint char. odor.

Axel P 100. [Aquatec Quimica] Fatty alcohol and fatty ester; antifoamer for paper machines; paste.

Axol®. [Goldschmidt] Citric acid esters; lubricants for rubber and plastics applics.

Axol® E 61. [Goldschmidt] Acetylated monoglyceride; nonionic; food emulsifier, lubricant, solv., plasticizer and coating material for foodstuffs and cosmetics; waxy solid; HLB 2–3; 100% conc.

Axol® E 66. [Goldschmidt] Acetic acid ester of mono/diglycerides; nonionic; lubricant, solv. and plasticizer for foodstuffs and cosmetics; liq.; HLB 2–3; 100% conc.

AXS. [Vista] Ammonium xylene sulfonate; hydrotrope for detergent formulations; near colorless liq.; slight ammonia odor; dens. 9.5; sp.gr. 1.141; visc. 4.8; m.p. 14; b.p. 212; 42% act.

Az-Cup MC. [Hercules] Azidosilane in methylene chloride sol'n.; difunctional coupling agent for bonding most siliceous reinforcing fillers to most polymers; lt. tan liq.; pleasant aromatic odor; f.p. –97 C (of solv.); b.p. 40 C (of solv.); slightly sol. in water; sp.gr. 1.1; dens. 9.1 lb/gal; flash pt. (Seta) > 93.3 C; 50% sol'n. in methylene chloride.

AZO-33. [Asarco] Zinc oxide; antimicrobial, fungistat for paints; improves weathering and resistance to uv rays; accelerator-activator and reinforcing pigment in rubbers; sp.gr. 5.6; dens. 46.65 lb/gal; 99.20% ZnO.

AZO-55. [Asarco] Zinc oxide; activator and reinforcing agent for rubber compding.; also used in paints; small nodulars; sp.gr. 5.6; dens. 32 lb/ft^3; 99.2% conc.

AZO-55TT. [Asarco] Surface-treated zinc oxide; activator and reinforcing agent for rubber compding., esp. neoprene stocks; small nodulars; sp.gr. 5.6; dens. 42 lb/ft^3; 99.2% conc.

AZO-66. [Asarco] Zinc oxide; activator for rubber and high purity chemicals; used in wire and cable applics., latex, surgical tapes, architectural enamels, ceramics; small nodular particles; sp.gr. 5.6; dens. 30 lb/ft^3; 99.2% conc.

AZO-66TT. [Asarco] Surface-treated zinc oxide; activator; dens. 36 lb/ft^3.

AZO-77. [Asarco] Zinc oxide; activator-accelerator for rubber compding.; reinforcing pigment; stabilizes certain exterior plastics and protects against uv rays; very small nodulars; sp.gr. 5.6; dens. 28 lb/ft^3; 99.7% conc.

AZO-77TT. [Asarco] Surface-treated zinc oxide; activator-accelerator; dens. 32 lb/ft^3.

Azodox-55. [Asarco] Zinc oxide; activator; deaerated form of AZO-55 for faster incorporation; dens. 48 lb/ft^3.

Azodox-55TT. [Asarco] Zinc oxide; activator; deaerated form of AZO-55TT for faster incorporation; dens. 58 lb/ft^3.

B

B2500. [Huls] Bis (trimethylsilyl) acetamide; blocking agent; silylating reagent for amino-acids, amides, phenols, carboxylic acids, ureas, enols, imides, alcohols.

B2539. [Huls] N,O-Bis (trimethylsilyl) carbamate; blocking agent for alcohols, phenols, carboxylic acids, and other substrates.

B2559.5. [Huls] Bis (trimethylsilyl) sulfate; blocking agent; suitable for silylation of ketones.

B2570. [Huls] Bis (trimethylsilyl) trifluoroacetamide; blocking agent; suitable for carboxylic acids, alcohols, phenols, amides, and ureas.

B2595. [Huls] Bis (trimethylsilyl) urea; blocking agent; silylates alcohols and carboxylic acids; used in synthesis of penicillins and cephalosporins.

B2790. [Huls] t-Butyldimethylchlorosilane; sterically hindered blocking agent; highly selective protecting reagent for hydroxyl groups based on steric hindrance.

B2792. [Huls] 2-(t-Butyldimethylsiloxy) pent-2-en-4-one; sterically hindered blocking agent.

B2794. [Huls] t-Butyldimethylsilylimidazole; sterically hindered blocking agent; used in nucleoside synthesis.

B2796. [Huls] t-Butyldimethylsilyl trifluoromethanesulfonate; sterically hindered blocking agent.

B2805. [Huls] t-Butyldiphenylchlorosilane; sterically hindered blocking agent.

B2809. [Huls] (p-t-Butylphenethyl) dimethylchlorosilane; sterically hindered blocking agent.

BA-43. [Great Lakes] Bisacrylate of ethoxylated tetrabromobisphenol A; functionally reactive flame retardant used in chemical, irradiation, or uv curing systems; enhances FR and crosslinking performance of elastomers; wh. powd; m.w. 740.1; sol. 0.2 g/100 g solv. in toluene; sol. < 0.1 g/100 g in water, methanol, methylene chloride, MEK; m.p. 115–120 C; 43% bromine.

BA-50. [Great Lakes] Bis (2-hydroxyethyl ether) of tetrabromobisphenol A; difunctional alcohol providing flame retardance; for unsat. polyester and epoxy thermoset resins; used for laminates for electronic circuit boards; corrosion resistance of systems useful in materials for tanks, ducts, and hoods; off-wh. powd.; sol. g/100 g solv. 80 g MEK, 62 g methanol, 48 g methylene chloride, 22 g triethylene glycol, 21 g acetone, 17 g acetate, < 0.1 g in water; m.p. 105–119 C; acid no. < 3.5; 50.6% bromine.

BA-50P. [Great Lakes] Bis (2-hydroxyethyl ether) of tetrabromobisphenol A; reactive flame retardant for thermoplastic polyesters and urethanes; wh. powd.; m.w. 632; sol. (g/100 g solv.); 80 g MEK, 62 g methanol, 48 g methylene chloride, 22 g triethylene glycol, 21 g acetone, 17 g ethyl acetate, < 0.1 g in water; m.p. 113–119 C; acid no. < 0.5; 50.6% bromine.

BA-59. [Great Lakes] Tetrabromobisphenol A; aromatic-based flame retardant for thermoplastic and thermoset resin systems, epoxy systems; wh. powd.; m.w. 543.7; f.p. 180 C; sol. (g/100 g solv.) 225 g acetone, 168 g MEK, 80 g methanol, 0.1 g water; 58.8% bromine.

BA-59P. [Great Lakes] Tetrabromobisphenol A; see BA-59; also used as reactive flame retardant for PC and additive for styrenic thermoplastics; wh. powd.; m.w. 543.7; f.p. 180 C; sol. see BA-59; 58.8% bromine.

Babinar-715. [Marubishi] Fatty acid polyethylene polyamine condensate deriv.; cationic; textile softener for polyacrylonitrile fiber; liq.

Babinar-801C. [Marubishi] Fatty acid polyethylene polyamine condensate deriv.; cationic; antistat, softener; paste.

Baerostab® UBZ 630. [Bärlocher; Vanderbilt] Barium zinc complex; stabilizer for PVC plastisols, transparent topcoats; Gardner 2 max. liq.; dens. 1.064 ± 0.01 mg/m^3.

Baerostab® UBZ 632. [Bärlocher; Vanderbilt] Barium zinc complex; stabilizer for PVC plastisols, esp. topcoats for lt. pigmented cushion vinyl flooring; Gardner 3 max. liq.; dens. 1.01–1.04 mg/m^3.

Baerostab® UBZ 791. [Bärlocher; Vanderbilt] Barium zinc; stabilizer for flexible and semirigid PVC, esp. sheet and film, and where color hold under dynamic stress is important; Gardner 4–5 liq.; dens. 1.05 ± 0.01 mg/m^3; ref. index 1.490 ± 0.002.

Baerostab® UBZ 793. [Bärlocher; Vanderbilt] Barium zinc; general purpose stabilizer for transparent and pigmented, flexible and semirigid PVC and esp. suspension PVC; Gardner 10 max. liq.; dens. 1.08 ± 0.02 mg/m^3.

BAN 120 L. [Novo] Alpha-amylase derived from Bacillus subtilis; enzyme used for partial breakdown of gelatinized starch into dextrins; used in starch industry for batchwise liquefaction of starch in prod. of dextrins, dextrose, and glucose syrups; used in alcohol industry to break down the starch contained in mash; paper industry for prod. of starch pastes of various viscs. for sizing and coating preparations; brn. liq.; slight smell typ. of fermentation prods.; water-sol.; dens. 1.2 g/ml; pH 6–8.

BAN 240 L. [Novo] Alpha-amylase derived from

Bacillus subtilis; see BAN 120 L; brn. liq.; slight smell typ. of fermentation prods.; water-sol.; dens. 1.2 g/ml; pH 6–8.

BAN 360 S. [Novo] Alpha-amylase derived from Bacillus subtilis; see BAN 120 L; lt. brn. powd.; slight smell typ. of fermentation prods.; water-sol.; dens. 1.2 g/ml; pH 6–8.

BAN 1000 S. [Novo] Alpha-amylase derived from Bacillus subtilis; see BAN 120 L; lt. brn. powd.; slight smell typ. of fermentation prods.; water-sol.; dens. 1.2 g/ml; pH 6–8.

Baragel®. [Rheox] Tetra-alkyl or trialkyl aryl ammonium smectites; lubricant; rheological visc. modifying agent for bodying syn. and other org. fluids of varying polarity; off wh. powd.; sp.gr. 1.6.

Baragel® 24. [Rheox] Organophilic clay; rheological additive; lt. cream powd.; sp.gr. 1.7.

Baragel® 3000. [Rheox] Organically modified montmorillonite clay; rheological additive; self-activating grease additive; lubricant; off-wh. powd.; sp.gr. 1.7.

Barco Bond MB-14X. [Astro] Two-part epoxy adhesive; general purpose bonding agent for factor use; paste; clear after mixing green tube (resin) and red tube (curing agent).

Barco Bond MB-127. [Astro] Two-part epoxy adhesive; general purpose, thixotropic, resilient bonding agent for factory use; part A—resin, part B—curing agent; translucent after cure; sp.gr. 1.15–1.25 (A), 0.98–0.99 (B); visc. 300,000 cps (A); distort. temp. 185 F; dielec. str. 1500 V/mil; vol. resist. 10^{12} ohm/ cm^2.

Barco Bond MB-165, MB-175, MB-185. [Astro] Epoxy adhesive; used for elec./electronic parts, as sealing insulation, general purpose repair material for marine and household use; clear after cure.

Bardac® 205M. [Lonza] Alkyl dimethyl benzyl ammonium chloride, octyl decyl dimethyl ammonium chloride, didecyl dimethyl ammonium chloride; disinfectant, sanitizer, bacteriostat; liq.

Bardac® 2050. [Lonza] Quaternium-24; disinfectant, sanitizer, bacteriostat cleared by FDA at only 150 ppm for use in "no rinse" sanitizing sol'ns.; 50% liq.

Bardac® 2250. [Lonza] Didecyldimonium chloride; disinfectant, sanitizer; 50% liq.

Bardac® LF, LF80. [Lonza] Dioctyl dimethyl ammonium chloride; low-foaming microbicide for recirculating cooling water systems, swimming pool algicides; liq.; m.w. 312; sp.gr. 0.93 and 0.897 resp.; dens. 7.72 and 7.48 lb/gal resp.; 50 and 80% act. resp. in aq. ethanol.

Bar-Dust® 105, 110. [Petrolite] Polymer; anticaking aid for urea and NPK fertilizers; m.p. 152 and 145 F resp.

Barisol Super BRM. [Dexter] Complex multicarbon alcohol phosphate potassium salt; anionic; textile wetting agent for dye leveling, pectin removal from cottons; dispersant; liq.; 30% conc.

Barium Petronate 50-S Neutral. [Witco] Barium sulfonate neutral; anionic; emulsifier; fuel oil additive; rust preventive; liq.; 50% conc.

Barium Petronate Basic. [Witco] Barium carbonate salt of hydrocarbon sulfonic acid; anionic; emulsifier; lubricant additive for lube and industrial oils and fuels, specialty oils and greases; 40–45% act.

Barlox® 10S. [Lonza] Decylamine oxide; nonionic; detergent; visc. builder, emollient; liq.; 30% conc.

Barlox® 12. [Lonza] Cocamine oxide; nonionic; detergent; visc. builder, emollient; liq.; 30% conc.

Barlox® 14. [Lonza] Myristamine oxide; nonionic; detergent; visc. builder, emollient; liq.; 30% conc.

Barlox® C. [Lonza] Cocamidopropylamine oxide; foam stabilizer and visc. builder for shampoos; 40% conc.

Barquat® 1552. [Lonza] 42% Benzalkonium chloride and 8% n-dialkyl methyl benzyl ammonium chloride; algicide for swimming pool and recirculating cooling towers; Gardner < 2 clear liq.; m.w. 399; sp.gr. 0.955; dens. 7.96 lb/gal; 50% act.

Barquat® 4250. [Lonza] Alkyl dimethyl benzyl ammonium chloride/alkyl dimethyl ethyl benzyl ammonium chloride; germicide, disinfectant sanitizer; 50% liq.

Barquat® 4250Z. [Lonza] Alkyl dimethyl benzyl ammonium chloride/alkyl dimethyl ethyl benzyl ammonium chloride; disinfectant, sanitizer (no potable water rinse required for sanitizing applics.); 50% liq.

Barquat® 4280. [Lonza] Alkyl dimethyl benzyl ammonium chloride/alkyl dimethyl ethyl benzyl ammonium chloride; see Barquat 4250 but used where a minimum of water is desirable in formulating; 80% liq.

Barquat® 4280Z. [Lonza] Alkyl dimethyl benzyl ammonium chloride/alkyl dimethyl ethyl benzyl ammonium chloride; see Barquat 4250Z but used where a minimum of water is desirable in formulating; 80% liq.

Barquat® MB-50. [Lonza] Benzalkonium chloride; germicide, disinfectant, sanitizer; 50% liq.

Barquat® MB-80. [Lonza] Benzalkonium chloride; see Barquat MB-50 but used where a minimum of water is desirable in formulating; 80% liq.

Barquat® MS-100. [Lonza] Myristalkonium chloride; antimicrobial; used in powd. and tableted formulations where solubility and high potency are required; 100% powd.

Barquat® MX-50. [Lonza] Myristalkonium chloride; see Barquat MB-50; 50% liq.

Barquat® MX-80. [Lonza] Myristalkonium chloride; see Barquat MX-50 but used where a minimum of water is desirable in formulating; 80% liq.

Barquat® OJ-50. [Lonza] Alkyl dimethyl benzyl ammonium chloride; for algae control in pools and cooling towers; 50% liq.

Barre® Common Degras. [RITA] Wool grease deriv.; nonionic; emollient; leather softener; w/o emulsion for corrosion prevention; dk. paste; oil-sol.; 100% conc.

Bartex® 80. [Hitox] Barium sulfate; extender pigment for gloss enamels, powd. coatings, semigloss latexes, house paints, primers; filler in many plastics applics., in rubber goods, and in ceramics; wh.

powd.; 7–8 μ avg. particle size; trace retention on 325 mesh; Hegman grind 5–6; sp.gr. 4.36; oil absorp. 11; dry brightness 89+; pH 7; 98% $BaSO_4$.

Base 3059 E. [Henkel-Nopco] Oxyethylenated alkylamidoamine; corrosion inhibitor and asphalt additive; antistripping agent; semi-liq.

BASF Wax A. [BASF] LDPE-wax; plastics and rubber processing, printing inks, pigment master batches, hot melts, and polishes; wh. powd.

BASF Wax AF. [BASF] LDPE-wax; printing inks; wh. micronized powd.

BASF Wax AF 30. [BASF] HDPE-wax; printing inks; wh. micronized powd.

BASF Wax AF 31. [BASF] HDPE-wax; printing inks; wh. micronized powd.

BASF Wax AF 32. [BASF] HDPE-wax; printing inks; wh. micronized powd.

BASF Wax AH 3. [BASF] HDPE-wax; matting agent for paints; lubricant for plastic processing; wh. powd.

BASF Wax AH 6. [BASF] HDPE-wax; printing inks, hot melts, polishes; wh. powd.

BASF Wax AL 3. [BASF] LDPE-wax; master batches; lubricant for plastic processing; wh. powd.

BASF Wax AL 61. [BASF] Polyethylene wax; polishes, plastics and rubber processing, printing inks, pigment master batches, hot melts; wh. powd.

BASF Wax AM 3. [BASF] LDPE-wax; matting agent for paints; wh. powd.

BASF Wax AM 6. [BASF] LDPE-wax; see BASF Wax AM 3; wh. powd.

BASF Wax DG. [BASF] Montan ester wax; cleaning agents and polishes, emulsions for surf. treatment for paper, wood, and metal; pale yel. flakes.

BASF Wax E. [BASF] Montan ester wax; lubricant for plastic processing, emulsions for surf. treatment; pale yel. flakes.

BASF Wax ES. [BASF] Montan ester wax; cleaning agent and polishes; pale yel. flakes.

BASF Wax EVA 1. [BASF] EVA copolymer; metallic paints, master batches, hot melts; wh. granulate.

BASF Wax FB. [BASF] Modified polyethylene wax; liq. polishes, protective and rustproofing wax disps.; off-wh. flakes.

BASF Wax L. [BASF] Montan acid wax; component in emulsifying systems for wax emulsions, dissolving agent for color bases; pale yel. flakes.

BASF Wax LCP. [BASF] Montan ester wax; carbon and pencil carbon papers; color compositions and backings; lt. brn. flakes.

BASF Wax LCP 2. [BASF] Montan ester wax; see BASF Wax LCP; lt. brn. flakes.

BASF Wax LG. [BASF] Montan ester wax; cleaning agent and polishes, emulsions for surf. treatment for paper, wood, and metal; pale yel. flakes.

BASF Wax LGE. [BASF] Montan ester wax with emulsifier; cleaning agents and dry-bright polish emulsions; pale yel. flakes.

BASF Wax LS. [BASF] Montan acid wax; emulsifying component for paraffin wax emulsions; pale yel. flakes.

BASF Wax MCE. [BASF] Carnauba wax deriv.; cleaning agent and dry-bright polish emulsions; lt. brn. flakes.

BASF Wax OA. [BASF] Oxidized LDPE-wax; impregnating agent for textile and paper, dry-bright polishes; wh. pastills.

BASF Wax OA 2. [BASF] Oxidized HDPE-wax; dry-bright polishes, impregnating agents for paper, protection coatings, lubricant for PVC processing; wh. pastills.

BASF Wax OA 3. [BASF] Oxidized HDPE-wax; dry-bright polishes; wh. powd.

BASF Wax OP. [BASF] Partially saponified montan ester wax; solv.-based polishes and as lubricant for plastic processing; pale yel. flakes.

BASF Wax S. [BASF] Montan acid wax; component in emulsifying systems for wax emulsions, dissolving agents for color bases; pale yel. flakes.

BASF Wax SG. [BASF] Montan ester wax; see BASF Wax LG; pale yel. flakes.

BASF Wax V. [BASF] Vinylether polymer; polishes for stone floors; wh. flakes.

Basicop. [Griffin Ag] Basic copper sulfate; wettable powd. agric. fungicide for wide range of crops; 53% act.

Basopon LN. [BASF] Alkylaryl polyglycol ether; nonionic; detergent with dispersant for wool and syn. fibers; liq.

Baycor®. [Bayer] Bitertanol; broad spectrum fungicide; colorless crystals; m.w. 337.4; sol. (g/1000 ml): 100–200 g methylene chloride, 30–100 g 2-propanol, 10–30 g toluene, 0.005 g water.

Bayer 5072. [Bayer] p-Dimethylamino-benzenediazo sodium sulfonate; fungicide for the prevention of crop damage by soil fungi; ylsh.-brn. powd.; m.w. 251.2; sol. in dimethylformamide; sol. 2–3% in water.

Bayfidan®. [Bayer] Triadimenol; systemic fungicide; colorless crystals; m.w. 295.8; sol. (g/1000 ml): 100–200 g in dichloromethane and 2-propanol, 20–50 g in toluene, 0.095 g in water; m.p. 118–130 C.

Baygon®. [Bayer] Propoxur; broad spectrum insecticide for the control of household and hygiene pests; colorless crystals; m.w. 209.2; sol. (g/1000 ml): > 1000 g in methylene chloride, 100–1000 g in IPA, toluene, 1.9 g in water; m.p. 90.7 C.

Bayhibit. [Mobay] 2-Phosphono-butane-tricarboxylic acid-1,2,4; corrosion and scale inhibitor in cooling systems; deflocculation of ceramic slips and oil drilling sludges, and stabilization of pigment suspensions; formulation of cleaning agents; pickling and cleaning agent for oxidized metal surfs.; sequestrant; dispersant; colorless to straw liq.; odorless; m.w. 270; water-misc.; dens. 1.28 g/cm^3; visc. 20 mPa; pH 2 (1% aq.); 45–50 aq.

Bayhibit-AM. [Mobay] 2-Phosphono-butane-tricarboxylic acid-1,2,4; corrosion and scale inhibitor in cooling systems; sea water evaporation units, water used for flooding in oil drilling; deflocculation of ceramic slips and oil drilling sludges, stabilization of pigment suspensions; formulation of cleaning agents; pickling and cleaning agent for oxidized metal surfaces; sequestrant; dispersant; colorless-straw liq.; m.w. 270; misc. with water; dens. approx.

1.28 g/cm³; visc. approx. 20 mPa; pH approx. 2; 45-50% aq. sol'n.

Bayleton®. [Bayer] Triadimefon; systemic fungicide; colorless crystals; m.w. 293.7; sol. (g/1000 ml): > 200 g in toluene and dichloromethane, 100–200 g in 2-propanol, 0.07 g in water; m.p. 82.3 C.

Bayluscide®. [Bayer] Clonitralid; molluscicide for the control of water snails; deep-yel. to grayish-yel.; m.w. 388.2; sol. 230 ppm ± 50 ppm in water; m.p. 230 C.

Bayrusil®. [Bayer] Diethchinalphion; insecticide for the control of biting and sucking pests; wh. cryst.; m.w. 298; sol. in most org. solvs.; insol. in water; sp.gr. 1.230 (20/4 C); m.p. 35–36 C; b.p. decomposes @ 120 C; ref. index 1.5624.

Baytan®. [Bayer] Triadimenol; systemic fungicidal seed dressing for cereals; colorless cryst.; m.w. 295.8; sol. (g/1000 ml); 100–200 g in methylene chloride, 10–100 g in 2-propanol, 10–30 g in toluene, 0.095 g in water.

Baytex®. [Bayer] Fenthion; insecticide for control of hygiene pests; colorless oily liq. to ylsh.-brn. liq.; m.w. 278.3; sol. (g/1000 ml): > 1000 g in methylene chloride, toluene; 30–100 g in n-hexane, 0.0063 g in water; sp.gr. 1.246; b.p. 87 C (0.01 mbar).

Baythion®. [Bayer] Phoxim; insecticide for the control of stored-product pests; lt. yel. to reddish brn.; m.w. 298.3; sol. in alcohols, ketones, aromatic hydrocarbons, chlorinated aliphatic hydrocarbons; sol. 0.7 mg/100 g in water; sp.gr. 1.176 (20/4 C); m.p. 5–6 C.

Baythroid®. [Bayer] Cyfluthrin; synthetic pyrethroid with good insecticidal activity; ylsh.-brn. mass of oily to pasty consistency; clear ylsh.-brn. liq. > 60 C; m.w. 434.3; sol. in dichloromethane, toluene; sp.gr. 1.27–1.28.

BBTS. [Akrochem] N-t-butyl-2-benzothioazole sulfenamide; delayed-action accelerator for natural and syn. rubbers; lt. tan powd.; sp.gr. 1.29 ± 3; m.p. 104 C min.

BE-51. [Great Lakes] Bis (allyl ether) of tetrabromobisphenol A; flame retardant used in EPS and foamed PS; wh. solid; m.w. 624.0; sol. (g/100 g solv.); 47 g methylene chloride, 42 g toluene, 33 g styrene, 12 g MEK, 0.1 g in water and methanol; m.p. 115–120 C; 51% bromine.

Bearflex® LAO. [Witco] Extender oil offering lt. color and low aniline pts.; for elastomer compounding, esp. nitrile, neoprene, SBR, and natural rubber; m.w. 238; sp.gr. 0.9725 (15.6 C); dens. 8.1 lb/gal; visc. 46 cSt (40 C); flash pt. (COC) 174 C; pour pt. –21 C; ref. index 1.5377.

Belclene® 200. [Ciba-Geigy] Polymaleic acid; scale and deposit control agent for industrial cooling and boiler water systems; 50% neutral aq. sol'n.

Belclene® 201. [Ciba-Geigy] Polymaleic acid sodium salt; see Belclene® 200; 50% neutral aq. sol'n.

Belclene® 283. [Ciba-Geigy] Polymaleic copolymer; see Belclene® 200; 50% aq. sol'n.

Belclene® 313. [Ciba-Geigy] Simazine; algicide for utility cooling and waste water treatment ponds; powd.

Belclene® 322. [Ciba-Geigy] N-(1,2-dimethylpropyl)-N′-ethyl-6-(methylthio)-1,3,5-triazine-2,4-diamine; control of algae in recirculating water in cooling towers and decorative/ornamental fountains; liq.

Belclene® 329. [Ciba-Geigy] Terbuthylazine; control of algae in recirculating water in cooling towers; aq. disp.

Belclene® 400. [Ciba-Geigy] Polymeric additive; scale inhibitor and deposit control additive, particularly calcium phosphate, for industrial water systems; 50% aq. sol'n.

Belclene® 500. [Ciba-Geigy] Polyanic phosphinocarboxylic acid deriv.; corrosion inhibitor enhancer for industrial cooling and boiler water systems; 35% aq. sol'n.

Belcor 575. [Ciba-Geigy] Hydroxy phosphinocarboxylic acid; org. corrosion inhibitor for ferrous metals; 50% aq. sol'n.

Belsperse 161. [Ciba-Geigy] Phosphinocarboxylic acid; universal dispersant for industrial water and boiler water systems; aq. sol'n.; 50% act.

Benathix. [Rheox] Organically modified montmorillonite clay; rheological additive for unsat. polyester; off-wh. powd.; sp.gr. 1.74.

Benecel CM. [Aqualon] Carboxymethylmethylcellulose; thickener, stabilizer, rheology control agent, film-former, suspending agent, water-retention aid, binder for food, pharmaceutical, and cosmetic industries; wh. to creamy powd. or gran.; sol. in cold water to form a colloidal sol'n., sol. in polar solv., e.g., IPA, mixts. of methanol and methylene chloride; dens. 1.0032 g/ml (2% aq.); bulk dens. 250–500 g/l; visc. 2000 cps (2%); f.p. 0 C (2% aq.); surf. tens. 45–55 mN/m (0.1%); pH 5.5–8.0 (1%).

Benecel M. [Aqualon] Methylcellulose; see Benecel CM for applics. and props.; visc. 10–12,000 cps (2%).

Benecel ME. [Aqualon] Methylhydroxyethylcellulose; see Benecel CM for applics. and props.; visc. 100–40,000 cps (2%).

Benecel MP. [Aqualon] Methylhydroxypropylcellulose; see Benecel CM for applics. and props.; visc. 3–40,000cps (2%).

Benol. [Witco] Wh. min. oil NF; lubricant used in food, drug and cosmetic industry; water wh.; visc. (Saybolt) 95–105 (100 F); pour pt. 20 F.

Bentone® 14. [Rheox] Smectite clay org. deriv.; antisettling agent for solv.-based coatings; for trade sales and industrials; lt. tan powd.; dens. 2.30 g/cm³; 100% NV.

Bentone® 27. [Rheox] Stearalkonium hectorite, anhyd.; thixotrope, gellant, thickener for solv.-based coatings; lip and eye care prods.; used for intermediate to high polarity org. systems; creamy wh. powd.; dens. 1.80 g/cm³; 100% NV.

Bentone® 34. [Rheox] Quaternium-18 bentonite; see Bentone 27; also for antiperspirant creams and lotions; very lt. cream powd.; dens. 1.70 g/cm³; 100% NV.

Bentone® 38. [Rheox] Quaternium-18 hectorite; see Bentone 27; creamy wh. fine powd.; dens. 1.70 g/

cm^3; 100% NV.

Bentone® 38 CG. [Rheox] Organophilic clay; rheological additive; lt. cream powd.; sp.gr. 1.7.

Bentone® 128. [Rheox] Organically modified montmorillonite clay; rheological additive for printing inks; thickener; off-wh. powd.; sp.gr. 1.60.

Bentone® 500. [Rheox] Org. deriv. of a montmorillonite clay; thickener and rheological modifier offering ease of disp. with only moderate shear for use in inks, paints, sealants, cosmetics; thickener for lithographic, letterpress, metal deco and uv-cured inks, paint systems; lt. cream powd.; sp.gr. 1.7.

Bentone® EW. [Rheox] Chemically modified magnesium aluminum silicate; suspending agent, rheological additive, thickener, gellant for water-reducible industrial finishes and inks; milky wh. powd.; water-disp.; dens. 2.6 g/cm^3; 100% NV.

Bentone® Gel CAO. [Rheox] Castor oil, stearalkonium hectorite; suspending agent for anhyd. formulations; rheological additive; lt. buff opaque gel; dens. 8.33 lb/gal; flash pt. (Seta) > 230 F.

Bentone® Gel IPM. [Rheox] IPM, stearalkonium hectorite, propylene carbonate; suspending agent for anhyd. formulations and emulsions; rheological additive; lt. buff opaque gel; dens. 7.43 lb/gal; flash pt. (Seta) > 230 F.

Bentone® Gel ISD. [Rheox] 87% Isododecane; rheological additive; lt. buff gel; dens. 6.47 lb/gal; flash pt. (Seta) 112 F.

Bentone® Gel LOI. [Rheox] Organically modified montmorillonite clay in liq. lanolin/isopropyl palmitate; suspending agent for anhyd. formulations; disp. with many cosmetic oils; rheological additive; gold opaque gel; dens. 7.6 lb/gal; flash pt. (Seta) > 230 F.

Bentone® Gel M-20. [Rheox] Rheological additive; lt. buff gel; sp.gr. 1.00; flash pt. (PMCC) > 170 F); 1.8% propylene carbonate.

Bentone® Gel MIO. [Rheox] Min. oil, quaternium-18 hectorite, propylene carbonate; see Bentone GEL IPM; lt. buff opaque gel; dens. 7.29 lb/gal; flash pt. (Seta) > 230 F.

Bentone® Gel MIO A-40. [Rheox] Organically modified montmorillonite clay in min. oil and SDA 40; see Bentone GEL LOI; lt. buff opaque gel; sp.gr. 0.88; dens. 7.31 lb/gal; flash pt. (PMCC) 130 F.

Bentone® Gel OMS. [Rheox] Organo-clay dispersed in Rule 66 min. spirits; thixotrope; gellant for solv.-based coatings; post additive; tan paste; dens. 0.85 g/cm^3; 15% Bentone 38.

Bentone® Gel S-130. [Rheox] Organically modified montmorillonite clay dispersed in odorless min. spirits; suspending agent for anhyd. formulations; eye makeup; opaque gel.

Bentone® Gel SS-71. [Rheox] Petrol. distillate, quaternium-18 hectorite, propylene carbonate; see Bentone Gel S-130; lt. buff opaque gel; dens. 7.53 lb/gal; flash pt. (Seta) 135 F.

Bentone® Gel SS-71LR. [Rheox] Rheological additive in min. spirits; lt. buff gel; dens. 7.53 lb/gal; flash pt. (Seta) 135 F.

Bentone® Gel TN. [Rheox] Rheological additive; lt. buff gel; sp.gr. 0.96; flash pt. (TCC) > 300 F.

Bentone® Gel VS-5. [Rheox] Organically modified montmorillonite clay in cyclomethicone; suspending agent for anhyd. formulations dispersible with cosmetic oils; rheological additive; lt. beige opaque gel; sp.gr. 1.01; dens. 8.43 lb/gal; flash pt. (Seta) 80 F.

Bentone® Gel VS-5 PC. [Rheox] Rheological additive in propylene carbonate; lt. beige gel; sp.gr. 1.00; flash pt. (PMCC) > 170 F.

Bentone® LT. [Rheox] Hectorite, hydroxyethyl-cellulose; thickener, gellant for water-based coatings; latex paint systems; excellent leveling and flow control; easy dispersing; milky wh. soft powd.; dens. 1.9 g/cm^3; 100% NV.

Bentone® SA-38. [Rheox] Smectite clay org. deriv.; see Bentone 38; creamy wh. fine powd.; dens. 1.50 g/cm^3; 100% NV.

Bentone® SD-1. [Rheox] Organically modified montmorillonite clay; suspending agent; dispersing thickener, rheogolical and suspending additive for aliphatic systems; lt. cream powd.; sp.gr. 1.47.

Bentone® SD-2. [Rheox] Organically modified montmorillonite clay; suspending agent; dispersing thickener and suspending additive for moderate to highly polar systems; lt. cream powd.; sp.gr. 1.62.

Bentone® SD-3. [Rheox] Organically modified montmorillonite clay; thickener and suspension agent for aromatic systems; lt. cream powd.; sp.gr. 1.60.

Benzoflex® 2-45. [Velsicol] Diethylene glycol dibenzoate; plasticizer compatible with PVAc homopolymer and copolymer emulsions; produces adhesives with quick grab and set times; solvator and compatible with PVC; APHA 100 max.; mild ester odor; m.w. 314.3; f.p. 16 and 28 C; b.p. 240 C; sol. in aliphatic and aromatic hydrocarbons; sol. < 0.01% in water; sp.gr. 1.178; dens. 9.8 lb/gal; ref. index 1.5424; flash pt. 199 C (COC); 98.0% min. assay.

Benzoflex® 9-88. [Velsicol] Dipropylene glycol dibenzoate; high solvating monomeric plasticizer; solvator for PVC; used in vinyl floor coverings; latex adhesives; enhances film formation and surf. wetting in PVA homopolymer emulsion adhesives; plasticizer in elastomers; APHA 100 max. clear oily liq.; mild ester odor; m.w. 342.3; f.p. –40 C; b.p. 232 C; sol. see Benzoflex 2-45; sp.gr. 1.120; dens. 9.346 lb/gal; pour pt. –19 C; ref. index 1.5282; flash pt. 199 C (COC); vol. resist. 10 10 ohm cm; 98.0% min. assay.

Benzoflex® 9-88 SG. [Velsicol] Dipropylene glycol dibenzoate; plasticizer for cast urethane applics., graphic art printing rolls, PU uses; APHA 150 max. clear oily liq.; mild ester odor; m.w. 342.3; f.p. –40 C; b.p. 232 C; sol. see Benzoflex 2-45; sp.gr. 1.120; dens. 9.346 lb/gal; pour pt. –19 C; ref. index 1.5282; flash pt. 199 C (COC); vol. resist. 10 10 phm cm; 99.0% min. assay.

Benzoflex® 50. [Velsicol] Diethylene glycol dibenzoate/dipropylene glycol dibenzoate; monomeric plasticizer used in PVAc adhesives, acrylic latex caulk formulations, plastisol, and dry-blended vinyl formulations; APHA 100 max.; mild ester odor; m.w. 328.3; f.p. < 0 C; b.p. 240 C; sol. see Benzoflex 2-45; sp.gr. 1.154; dens. 9.6 lb/gal; pour pt. –21 C; ref.

index 1.535; flash pt. 199 C min. (COC); 98.0% min. assay.

Benzoflex® 131. [Velsicol] Isodecyl benzoate; plasticizer for plastisols, adhesives, sealants, caulks.

Benzoflex® 284. [Velsicol] Propylene glycol dibenzoate; solvating plasticizer for PVC applic., latex caulk formulations, PVAc adhesives, and castable PU; used in latex paints as coalescing agent; APHA 100 max.; mild ester; m.w. 284.3; f.p. –3 C; b.p. 233 C; sol. in aliphatic and aromatic hydrocarbons; sol. 0.01% in water; sp.gr. 1.146; dens. 9.55 lb/gal; pour pt. –16 C; ref. index 1.544; flash pt. 199 C min. (COC); 98.0% min. assay.

Benzoflex® 352. [Velsicol] 1,4-Cyclohexane dimethanol dibenzoate; plasticizer, modifier for hot-melt adhesives (EVA, block copolymer, polyester, polyamide, PU), delayed tack latex and hot-melts; m.p. 118 C.

Benzoflex® 400. [Velsicol] Polypropylene glycol dibenzoate; plasticizer for PU and polysulfide sealants, acrylic coatings, caulks, PVC, adhesives.

Benzoflex® P-200. [Velsicol] PEG 200 dibenzoate; plasticizer for PVAc adhesive formulations, phenol-formaldehyde resins, alkyd-modified phenol-formaldehyde varnishes; limited compat. with PVC; APHA 100 max.; m.w. 408.2; f.p. –30 C; b.p. 217 C; dens. 9.7 lb/gal; ref. index 1.5252; flash pt. 478 F (TOC); 98.0% min. assay.

Benzoflex® S-312. [Velsicol] Neopentyl glycol dibenzoate; process aid, modifier, plasticizer for thermoplastics, hot-melt adhesives, coatings.

Benzoflex® S-404. [Velsicol] Glyceryl tribenzoate; process aid, modifier for thermoplastics, hot-melt adhesives.

Benzoflex® S-552. [Velsicol] Pentaerythritol tetrabenzoate; plasticizer/extender for coatings, modifier for hot-melt adhesives, aq. adhesives, delayed tack adhesives, process aid for thermoplastics.

Berol 07. [Berol Nobel] Fatty alcohol ethoxylate; nonionic; detergent, equalizer, and dispersant for washing machine detergents; textile industry washing; dyes; wh. wax; sol. in water, ethanol, xylene, trichloroethylene; dens. 1.09 g/cm³; visc. 50 cps (50 C); HLB 15.0; m.p. 36–38 C; pH 6–7 (1% aq.); flash pt. > 100 C; cloud pt. 74–75 C; biodeg.; 100% act.

Berol 08 Powd. [Berol Nobel] Fatty alcohol ethoxylate; nonionic; detergent, dispersant; wh. powd.; sol. in water, ethanol, trichlorethylene, disp. in xylene; sp.gr. 0.55–0.65; visc. 200 cps (70 C); m.p. 52–54 C; HLB 18.7; cloud pt. 76–77 C; flash pt. > 100 C; surf. tens. 40–48 dynes/cm; pH 6–7 (1% aq.); biodeg.; 100% act.

Berol 09. [Berol Nobel] Nonylphenol ethoxylate; nonionic; detergent, wetting agent, emulsifier, dispersant; Hazen < 150 clear liq.; sol. in water, ethanol, xylene, trichlorethylene; disp. in paraffin oil, wh. spirit, lt. fuel oil; sp.gr. 1.05; visc. 350 cps; HLB 13.3; cloud pt. 52–58 C; flash pt. > 100 C; pour pt. 5 C; surf. tens. 33 dynes/cm; pH 6–7 (1% aq.); biodeg.; 100% act.

Berol 17. [Berol Nobel] Resin acid EO adduct; nonionic; lubricant, emulsifier for textile industry; Gardner 15, clear liq.; sol. in water, ethanol, trichloroethylene, xylene, disp. in acetone, propylene glycol; sp.gr. 1.11; visc. 50 cps; HLB 14; cloud pt. 45 C; flash pt. > 100 C; pour pt. 15 C; surf. tens. 38 dynes/cm; pH 7.5–8.5 (1% aq.); 100% act.

Berol 081 Flakes. [Berol Nobel] Fatty alcohol ethoxylate; nonionic; detergent, emulsifier, dispersant, stabilizer; wh. flakes; sol. in water, ethanol, trichloroethylene, disp. in xylene; sp.gr. 0.55–0.65; m.p. 46–50 C; HLB 16.7; cloud pt. 75–76 C (1% in 10% NaCl); flash pt. > 100 C; surf. tens. 43 dynes.cm; pH 6–7 (1% aq.); biodeg.; 100% act.

Berol 081 Powd. [Berol Nobel] Fatty alcohol ethoxylate; nonionic; detergent, emulsifier, dispersant, stabilizer; wh. powd.; sol. in water, ethanol, trichloroethylene, disp. in xylene; sp.gr. 0.55–0.65; m.p. 46–50 C; HLB 16.7; cloud pt. 75–76 C (1% in 10% NaCl); flash pt. > 100 C; surf. tens. 43 dynes.cm; pH 6–7 (1% aq.); biodeg.; 100% act.

Berol 151. [Berol Nobel] Lauric/myristic acid diethanolamide; nonionic; foam stabilizer, booster for shampoos, bubble bath, modifier, dispersant; Hazen ≤ 200 clear liq.; sol. in ethanol, propylene glycol, xylene; disp. in water, wh. spirit, paraffin oil; sp.gr. 0.98; visc. 1100 cps; HLB 12; flash pt. >100 C; pour pt. –5 C; surf. tens. 29.5 dynes/cm; pH 9.5 (1% aq.); biodeg.; 100% act.

Berol 196. [Berol Nobel] Alkyl glyceride EO adduct; nonionic; emulsifier for foodstuffs, superfatting agent in bath oils, hand cleaners; Gardner ≤ 5, clear liq.; sol. in ethanol, propylene glycol, trichloroethylene, wh. spirit, opaquely sol. in water; sp.gr. 1.020; visc. 500 cps; HLB 9; cloud pt. 58 C; flash pt. > 100 C; pour pt. –15 C; pH 5–7 (1% aq.); biodeg.; 100% act.

Berol 259. [Berol Nobel] Nonylphenol ethoxylate; nonionic; stabilizer, foam depressor; liq. cleaner, wetting agent; Hazen ≤ 500 clear liq.; sol. in ethanol, xylene, trichloroethylene, paraffin oil, wh. spirit, lt. fuel oil; sp.gr. 1.01; visc. 550 cps; HLB 5.7; cloud pt. 50–54 C; flash pt. > 100 C; pour pt. < 0 C; pH 6–7 (1% aq.); biodeg.; 100% act.

Berol 263. [Berol Nobel] Alkyl phenol EO adduct; nonionic; detergent; dispersant for pigments; Hazen < 200 soft wax; sol. in water, ethanol, xylene, trichloroethylene, disp. in paraffin oil, wh. spirit; sp.gr. 1.06; visc. 325 cps; HLB 14.7; cloud pt. 92–94 C; flash pt. > 100 C; pour pt. 20 C; surf. tens. 36 dynes/cm; pH 6–7 (1% aq.); biodeg.; 100% act.

Berol 305. [Berol Nobel] Amine oxide; nonionic; foam controller; pale yel. clear liq.; sol. in water, ethanol, acetone; sp.gr. 0.96; cloud pt. 95 C (9% NaOH); pour pt. 0 C; surf. tens. 31 dynes/cm; 28% act.

Berol 371. [Berol Nobel] EO/PO block polymer; nonionic; detergent, antistat, emollient, solv.; textile industry, cosmetics; Hazen < 150 soft paste; m.w. 2900; sol. in water, ethanol, xylene, trichloroethylene; sp.gr. 1.05; visc. 700 cps; cloud pt. 62–66 C (1% aq.); flash pt. > 100 C; pour pt. 28 C; surf. tens. 38 dynes/cm (0.1%); pH 5–7 (1% aq.); 100% act.

Berol 372 Flakes. [Berol Nobel] EO/PO block polymer; nonionic; detergent, rinse aid additive, antistat, emollient, solv., emulsifier, foam depressant/detergent; wh. flakes; m.w. 9000; sol. in water, ethanol, trichloroethylene; sp.gr. 0.55–0.65; cloud pt. 70 C (1% in 10% NaCl); flash pt. > 100 C; surf. tens. 42 dynes/cm (0.1%); pH 5–7 (1% aq.); 100% act.

Berol 374. [Berol Nobel] EO/PO block polymer; nonionic; emollient, emulsifier; emulsion polymerization; wh. flakes; m.w. 2200; sol. in ethanol, xylene, trichloroethylene, wh. spirit; disp. water; sp.gr. 1.05; visc. 450 cps; cloud pt. 24–26 C (1% aq.); flash pt. > 100 C; pour pt. < –10 C; surf. tens. 40 dynes/cm (0.1%); pH 5–7 (1% aq.); 100% act.

Berol 472. [Berol Nobel] Sodium cetyl sulfate; anionic; emulsifier, dispersant; used for emulsion polymerization of vinyl chloride, vinyl acetate, S/B, acrylics; Hazen < 150 cloudy liq.; sol. in propylene glycol; disp. in water, ethanol; sp.gr. 1.055; visc. 20 cps; flash pt. > 100 C; surf. tens. 37 dynes/cm (0.1%); pH 7–8 (1% aq.); biodeg.; 32 ±1% act.

Berol 475. [Berol Nobel] Sodium alkyl ether sulfate; anionic; degreaser, dispersant for water-based pigment pastes, cleaner; personal care and textile use; Hazen < 300 soft paste; sol. in propylene glycol and water, disp. in ethanol, trichloroethylene, xylene; sp.gr. 0.97; flash pt. > 100 C; surf. tens. 37 dynes/cm (0.1%); pH 6.5–8.0 (1% aq.); biodeg.; 39–41% act.

Berol 496. [Berol Nobel] Linear sodium alkylaryl sulfonate; anionic; detergent, foaming, wetting, penetrant, emulsifier for textile industry; Hazen 150 soft paste; sol. in water, ethanol, propylene glycol, acetone; disp. in trichloroethylene, spindle oil, wh. spirit; sp.gr. 1.060; clear pt. 0 C; flash pt. > 100 C; surf. tens. 35 dynes/cm; pH 6–8; biodeg.; 80% act.

Berol 516. [Berol Nobel] Mixt. of an alkyl ether phosphate ester and an n-primary alcohol EO adduct; anionic/nonionic; detergent, corrosion inhibitor; Gardner < 5 paste; sol. in ethanol, acetone, xylene, propylene glycol, white spirit, disp. in water; sp.gr. 1.050 (30 C); visc. 330 cps (30 C); clear pt. 40 C; flash pt. > 100 C; pour pt. 30 C; surf. tens. 34.4 dynes/cm (0.1%); pH 2.3 (1% aq.); biodeg.; 100% act.

Berol 518. [Berol Nobel] Mixt. of an alkyl ether phosphate ester and an n-primary alcohol EO adduct; anionic/nonionic; detergent, foam controller; Gardner < 5 soft wax; sol. in ethanol, acetone, xylene, white spirit, trichloroethylene, lt. fuel oil; disp. in water; sp.gr. 1.01 (50 C); visc. 1500 cps (35 C); clear pt. 35 C; flash pt. > 100 C; pour pt. 35 C; surf. tens. 38 dynes/cm (0.1%); pH 2.3 (1% aq.); biodeg.; 100% act.

Berol 521. [Berol Nobel] Potassium alkyl phosphate esters; anionic; corrosion inhibitor, solubilizer of nonionic surfactants in high electrolyte conc.; Hazen < 300 clear liq.; sol. in water, propylene glycol; sp.gr. 1.145; visc. 315 cps; clear pt. 25 C; flash pt. > 100 C; pour pt. –10 C; pH 7–8 (1% aq.); biodeg.; 40% act. in water.

Berol 522. [Berol Nobel] Potassium alkyl phosphate esters; anionic; solubilizer of nonionic surfactants in high electrolyte conc.; Hazen < 300 clear liq.; sol. in water, propylene glycol; sp.gr. 1.245; visc. < 125 cps; clear pt. < 0 C; flash pt. > 100 C; pour pt. < –10 C; pH 8–9 (1% aq.); biodeg.; 40–42% act.

Berol 525. [Berol Nobel] Alkyl ether phosphate ester, straight-chain primary alcohol EO adduct; anionic/nonionic; detergent, solubilizer; liq. alkaline industrial cleaners; household and institutional slightly alkaline detergents; Hazen < 150 paste; sol. in water, ethanol, acetone, propylene glycol, trichloroethylene, xylene, wh. spirit; dens. 1.060 g/cm³; visc. 1500–2000 cps; pour pt. 23 C; pH 2.3 (1% aq.); flash pt. > 100 C; clear pt. 30 C; biodeg.; 100% act.

Berol 594. [Berol Nobel] Hydroxyethyl-2-alkylimidazoline; strong cationic activity in acid sol'ns., oils, solv.; adhesion aid in solv.-based paints and varnishes, dispersing agent, wetting and foaming agent, corrosion inhibitor; Gardner ≤ 12 clear liq.; m.w. 360; sol. in ethanol, xylene, trichloroethylene, wh. spirit, paraffin oil, and dilute sulfuric, HCl, phosphoric and acetic acids; disp. in water; sp.gr. 0.950; visc. 300 cps; clear pt. < –10 C; flash pt. > 100 C; pour pt. < 0 C; surf. tens. 32 dynes/cm (0.1%); pH 10–11 (1% aq.); 100% act.

Berol 730. [Berol Nobel] PP glycols; antifoaming agent, lubricant, release agent, hydraulic fluid lubricant, substitute for different oils, dust adhesive in air filters; Hazen < 100 clear liq.; m.w. 1200; sol. in min. spirit, acetone, xylene, methanol, CCl_4 (5%), water (< 2%); sp.gr. 1.0; visc. 160 cps; flash pt. > 200 C; pour pt. –40 C; 100% act.

Berol 733. [Berol Nobel] Potassium salt of phosphated alkylphenol EO adduct; anionic; detergent, solubilizes nonionic surfactants in presence of inorg. salts, liq. alkaline industrial cleaners; sol. in water, 6% NaOH, 35% tetrapotassium pyrophosphate, 50% sodium metasilicate, propylene glycol; sp.gr. 1.050; visc. 400 cps; clear pt. –5 C; flash pt. > 100 C; pour pt. –7 C; surf. tens. 41 dynes/cm; pH 9–10 (1% aq.); biodeg.; 39–41% act.

Berol 740. [Berol Nobel] PP glycols; antifoaming agent, lubricant, release agent, hydraulic fluid lubricant, substitute for different oils, dust adhesive fluid in air filters; Hazen < 100, clear liq.; m.w. 1800; sol. in min. spirit, acetone, xylene, methanol, CCl_4 (5%), water (< 0.1%); sp.gr. 1.0; visc. 200 cps; flash pt. > 200 C; pour pt. –35 C; 100% act.

Berol 752. [Berol Nobel] Anionic; surfactant, foam stabilizer; Gardner < 7 visc. liq.; disp. in water, propylene glycol, dioctylphthalate, tributyl phosphate; sp.gr. 1.070; visc. 10,000 cps; flash pt. > 100 C; pour pt. 25 C; pH 6–8 (1% in 1:1 water/ethanol); 95–100% act.

Berol 777. [Berol Nobel] Linear alkylaryl sulfonate and alkyl phenol EO adduct; anionic/nonionic; detergent used in textile industry, pigment dispersant; Hazen ≤ 250 clear liq.; sol. in water, propylene glycol; disp. in ethanol, acetone, trichloroethylene, xylene; sp.gr. 1.055; visc. 425 cps; clear pt. 1 C; flash pt. > 100 C; pour pt. 0 C; surf. tens. 38 dynes/cm; pH 6–7 (1% aq.); 34% act.

Berol 786. [Berol Nobel] O/w emulsion; emulsifier, corrosion inhibitor; yel.-wh. liq.; sp.gr. 0.994; visc. 100 cps; flash pt. > 100 C; pour pt. 0 C.

Berol 808. [Berol Nobel] Anionic/cationic/nonionic; detergent, textile softener, antistat agent, post additive to powd. formulations; clear, yel. liquid; sol. in ethanol, propylene glycol, wh. spirit, lt. fuel oil; opaquely sol. in water; sp.gr. 1.013; visc. 105 cps; cloud pt. > 90 C (1% aq.); flash pt. 60 C; surf. tens. 32 dynes/cm (0.1%); pH 5.5 (1% aq.); biodeg. under normal sewage treatment cond.; 80% act.

Berol 809. [Berol Nobel] Combined anionic/cationic/ nonionic surfactant; detergent, textile softener, antistat agent; lt. yel., clear liq.; sol. in water, ethanol, wh. spirit; sp.gr. 1.04; visc. 170 cps; cloud pt. > 55 C (opal 1% aq.); flash pt. > 100 C; pour pt. –4 C; surf. tens. 31.5 dynes/cm (0.1%); pH 6.3 (1% aq.); biodeg. under normal sewage treatment cond.; 80% act.

Berol Dinonylphenol. [Berol Nobel] Dinonylphenol; chemical intermediate, stabilizer, plasticizer, hydroprobe for nonionic surfactants, used as component in lubricating oils, corrosion inhibitors; yel., visc. liq.; phenol-like odor; m.w. 346; sol. in ethanol, acetone, trichloroethylene, butyl glycol, wh. spirit, min. oil; sp.gr. 0.911; visc. 400 cps (40 C); b.p. 326–334 C (5%); flash pt. > 500 C; pour pt. 0 C; 100% act.

Berol EGA 07. [Berol Nobel] Fatty alcohol ethoxylate; nonionic; textile dye leveling and disp. agent, protective colloid; Hazen < 150 clear liq.; sol. in water, ethanol, xylene, trichloroethylene; disp. in paraffin oil, wh. spirit, lt. fuel oil; sp.gr. 1.01; visc. 10 cps; HLB 15.0; cloud pt. 74–76 C (5% in 10% NaCl); flash pt. > 100 C; pour pt. < 0 C; surf. tens. 40 dynes/ cm (0.1%); pH 6–7 (5% aq.); biodeg.; 20% act.

Berol EGA 192. [Berol Nobel] Alkyl glyceride EO adduct; nonionic; textile dye equalizing and dispersing agent; Gardner < 3 clear liq.; sol. in water, ethanol, acetone; disp. in propylene glycol, xylene; sp.gr. 1.080; visc. 120 cps; HLB 18; cloud pt. 62 C (1% in 10% NaCl); flash pt. > 100 C; pour pt. 13 C; surf. tens. 42 dynes/cm; pH 6–7 (1% aq.); biodeg.; 85% act.

Berol Fintex 10. [Berol Nobel] Fatty acid EO adduct; nonionic; textile softener and antistat, equalizing agent, emulsifier for greases and oils; wh. wax; sol. in ethanol, trichloroethylene, xylene; disp. in water, propylene glycol, acetone, wh. spirit; sp.gr. 1.010 (36 C); visc. 80 cps (36 C); HLB 12; cloud pt. 75–80 C; flash pt. > 100 C; pour pt. 35 C; surf. tens. 42 dynes/ cm (0.1%); pH 6–7 (1% aq.); biodeg.; 100% act.

Berol Fintex 42. [Berol Nobel] Weakly cationic surfactant; textile softener for natural and syn. fibers, antistat, used in anticrease treatment with resins, adhesive bath additive; lt. yel. liq. disp.; sp.gr. 1.0; visc. 500 cps.; flash pt. > 100 C; pour pt. 0 C; pH 9 (1% aq.); 25% act.

Berol Fintex 572. [Berol Nobel] Quat. ammonium compd.; cationic; textile softener, antistat; liq.; visc. 1000 mPas; 67% conc.

Berol Fintex 573. [Berol Nobel] Alkyl polyglycol ether ammonium ethyl sulfate and nonionic wetting agent; cationic; surfactant, antistat for syn. textiles and plastics, bactericide; Gardner ≤ 7 clear liq.; sol. in water, ethanol, propylene glycol, acetone, trichloroethylene, xylene, white spirit; sp.gr. 1.050; visc. 750 cps; cloud pt. 70 C (1% in 10% NaCl); flash pt. > 100 C; pour pt. 2 C; surf. tens. 35 dynes/cm (0.1%); pH 7 ± 0.5 (1% aq.); 100% act.

Berol Fintex 577. [Berol Nobel] Quat. ammonium compd.; cationic; softening, antistatic and rewetting agent for natural and syn. textiles, hair cream rinses; antibacterial properties; Hazen < 300, clear liq.; sol. in ethanol, xylene, propylene glycol, wh. spirit; disp. in hot or cold water; sp.gr. 0.94; visc. 100 cps; clear pt. < 15 C; flash pt. 25 C; pour pt. 5 C; surf. tens. 32 dynes/cm; pH 6–7 (1% aq.); 70% act.

Berol Fintex 755. [Berol Nobel] Mixt. of linear alkylaryl sulfonate n-primary alcohol; anionic/nonionic; textile wetting, cleaning, dispersing agents, leveling agent in dyeing, detergent; Gardner ≤ 5 clear liq.; sol. in water, ethanol, IPA, trichloroethylene (opal), xylene; sp.gr. 1.050; visc. 150 cps; clear pt. 0 C; flash pt. > 100 C; pour pt. –10 C; surf. tens. 36 dynes/cm (0.1%); pH 6–7 (1% aq.); biodeg.; 70% act.

Berol Lanco. [Berol Nobel] Alkyl phenol EO adduct; nonionic; textile wetting and cleaning agent for natural and syn. fibers, emulsifier for waxes, oils, solv., and paint resins, emulsion polymerization, dispersant for paint industry; Hazen < 150 clear liq.; sol. in water, ethanol, xylene, trichloroethylene; sp.gr. 1.05; visc. 350 cps; HLB 13.3; cloud pt. 52–58 C (1% aq.); flash pt. > 100 C; pour pt. 5 C; surf. tens. 33 dynes/cm; pH 6–7 (1% aq.); biodeg.; 100% act.

Berol PEG 400. [Berol Nobel] Calcium alkylaryl sulfonate; anionic; stabilizer and conservation of wood; liq.; 100% conc.

Berol PEG 4000. [Berol Nobel] Calcium alkylaryl sulfonate; anionic; stabilizer and conservation of wood; flakes; 100% conc.

Berol WASC. [Berol Nobel] Nonylphenol ethoxylate; nonionic; household and industrial detergent, rewetting agent, textile washing agent, dispersant for paint and varnish industry; Hazen < 150, clear liq.; sol. in water, ethanol; disp. in xylene, trichloroethylene, paraffin oil, wh. spirit; sp.gr. 1.06; visc. 375 cps; HLB 14.1; cloud pt. 75–80 C (1% aq.); flash pt. > 100 C; pour pt. 0 C; surf. tens. 34 dynes/cm (0.1%); pH 6–7 (1% aq.); biodeg.; 95% act.

Be Square® 175. [Petrolite] Microcryst. wax; plastic wax offering high ductility, flexibility at very low temps., low surf. tens., solvency for amorphous hydrocarbon resins and syn. elastomers; provides protective barrier properties against moisture vapor and gases; uses incl. hot-melt laminating adhesives for papers, films, and foils; hot-melt coatings, in antisunchecking agents in rubber goods, elec. insulating agents, leather treating agents, water repellents for textiles, solv.-based rustproof coatings, cosmetic ingreds., and as plasticizer for other petrol. waxes used in crayons, dental compds., chewing gum base, and candles; color 1.5 (D1500); also avail. in black; dens. 0.93 g/cc; visc. 11 cps (98.8 C); m.p. 83.3 C; flash pt. 293.3 C.

Be Square® 185. [Petrolite] Hard microcryst. wax consisting of n-paraffinic, branched paraffinic, and naphthenic hydrocarbons; wax used in hot-melt coatings and adhesives, cup and paper coatings, printing inks, plastic modification (as lubricant and processing aid), lacquers, paints, and varnishes, as binder in ceramics, for potting, filling, and impregnant in elec./electronic components, in investment casting, rubber and elastomers (plasticizer, antisunchecking, antiozonant), as emulsion wax size in papermaking, as fabric softener ingred. in permanent-press fabrics, in emulsion and latex coatings, and in cosmetic hand creams and lipsticks; color 1.0 (D1500); very low sol. in org. solvs.; sp.gr. 0.92; visc. 11.3 cps (98.9 C); m.p. 87.8 C.

Be Square® 195. [Petrolite] Hard microcryst. wax consisting of n-paraffinic, branched paraffinic, and naphthenic hydrocarbons; see Be Square 185; color 0.5 (D1500) wax; also avail. in wh. and brn.; very low sol. in org. solvs.; sp.gr. 0.93; visc. 12.5 cps (98.9 C); m.p. 91.1 C.

BIBBS. [Akrochem] N,N-diisopropyl benzothiazole-2-sulfenamide; delayed-action sulfenamide accelerator; off-wh. to lt. tan powd.; sol. in benzene, acetone, ethanol, methanol, and CCl_4; sp.gr. 1.21; m.p. 57 C.

Biju. [Mearl] Bismuth oxychloride; colorant and pearlescent for frosted cosmetics; pearl nail enamel because of brilliance and smoothness; insol.

Bi-Lite®. [Van Dyk] Bismuth oxychloride/mica; pearlescent pigments for cosmetic eye, face, lip, and body make-up; wh.

Bilt-Cote H-1, H-5. [Vanderbilt] Air-floated kaolin clay; treated kaolin agric. conditioner designed for coating ammonium nitrate prills as an anticaking agent; chemically modified to facilitate change in surf. tens. of moisture preventing water bridging between gran.; cream particulate, particle size 0.2μ, 1–4% +325-mesh residue; dens. 2.60 mg/m³; 44.9% SiO_2; 37.8% Al_2O_3; 1 and 5% additive modification resp.

Bilt-Cote S-1, S-5. [Vanderbilt] Air-floated kaolin clay; see Bilt-Cote H-1, H-5; cream particulate, particle size 15 μ, 1–4% +325-mesh residue; dens. 2.60 mg/m³; 44.3% SiO_2; 39.3% Al_2O_3; 1 and 5% additive modification resp.

Bina QAT-43. [Ciba-Geigy] Lauramidopropyl acetamidodimonium chloride; conditioner and emulsifier for personal care prods..

Bioban® BNPD. [Angus] Biocide for metalworking fluids; powd.; sol. in water, glycols, and other polyols; flash pt. > 200 F (CC); pH 4.15 (20% sol'n.).

Bioban® CS-1135. [Angus] Oxazolidine; preservative, antibacterial agent for water-based paints, latexes, emulsions, metalworking fluids; for oilfield water-flooding operations; aids corrosion protection; liq.; sol. in water, polar and nonpolar solvs.; 78% act.

Bioban® CS-1246. [Angus] Biocide for metalworking fluids, latex paints, caulks, adhesives, oilfield water systems, drilling muds; liq.; sol. in water, alcohol, glycols, aromatic solvs., min. oils; sp.gr. 1.09; flash pt. 175 F (CC); pH 8.9.

Bioban® CT. [Angus] Biocide, in-can paint preservative; also for emulsions, caulks, adhesives; liq. sol'n.; sol. in water; insol. in hydrocarbons; sp.gr. 1.22; flash pt. > 200 F (CC); pH 5.0 (0.1 M sol'n.).

Bioban® GK. [Angus] Biocide for metalworking fluids, latex paints, emulsions, caulks, and adhesives; liq.; sol. in water; insol. in hydrocarbons; sp.gr. 1.16; flash pt. 167 F (CC); pH 10.0–11.0.

Bioban® N-95. [Angus] Oxazolidine; preservative for aq. metalworking fluids; liq.; water-misc.; 50% act. aq. sol'n.

Bioban® P-1487. [Angus] Morpholine deriv.; preservative, antibacterial and antifungal agent used in metalworking fluids, lubricants and cutting oils; liq.; oil-sol.; mod. water-sol.; 90% act.

Biobrom C-103L. [Dead Sea Bromine] 2,2-Dibromo-3-nitrilopropionamide; biocide for industrial water systems, incl. cooling towers, pulp and paper-mill effluents, oil-recovery systems, metal-cutting coolants, air conditioning systems; clear to amber liq., mild antiseptic odor; misc. with water; sp.gr. 1.18–1.22 (20/4 C); f.p. –20 to –25 C; flash pt. 159–163 C; 20% min. act.

Biobrom C-103 Tech. [Dead Sea Bromine] 2,2-Dibromo-3-nitrilopropionamide; biocide for industrial water systems, incl. cooling towers, pulp and paper-mill effluents, oil-recovery systems, metal-cutting coolants, air conditioning systems; wh. to off-wh. powd. or gran., mild antiseptic odor; m.w. 241.84; sol. (g/100 g): 120 g DMF, 35 g acetone, 25 g ethanol, 1.5 g water; sp.gr. 2.375; m.p. 124–126 C; 98% min. act., 66% Br.

BioCare Polymer HA-24. [Amerchol] Polyquaternium-24, hyaluronic acid; emollient, humectant, conditioner, softener, moisturizer, lubricant for hair and skin; substantive to protein substrates; opalescent visc. liq.

Bio-Dac 50-20. [Lonza] Dialkyl (C_8–C_{10}) dimethyl benzyl ammonium chloride; cationic; antimicrobial, disinfectant, sanitizer, bacteriostat; pale yel. to water-wh. clear liq.; sp.gr. 0.925; dens. 7.7 lb/gal; pH 6–9 (10%); flash pt. (Seta) 110 F; 50% act.

Bio-Dac 50-22. [Lonza] Didecyl dimonium chloride; cationic; antimicrobial, disinfectant; sanitizing applics.; pale yel. to water-wh. clear liq.; sp.gr. 0.927; dens. 7.73 lb/gal; pH 6–9 (10%); flash pt. (Seta) 109 F; 50% act.

Bio-Dac 205. [Lonza] Benzalkonium chloride and dialkyl (C_8–C_{10}) dimethyl benzyl ammonium chloride; cationic; antimicrobial, disinfectant, sanitizer; pale yel. to water-wh. clear liq.; sp.gr. 0.946; dens. 7.89 lb/gal; pH 6–9 (10%); flash pt. (Seta) 116 F; 50% act.

Biolase Brands. [Hoechst-Celanese] Bacterial amylases; enzymatic desizing agent; powd., liq.

Biopal® NR-20. [GAF] Nonoxynol-12 iodine; nonionic; used in formulating no rinse sanitizing sol'ns.; dk., reddish brn., visc. liq.

Biopal® NR-20-W. [GAF] Nonylphenoxypoly (ethyleneoxy) ethanol-iodine complex; anionic; formulates "no rinse" sanitizing sol'n.; liq.; 20% conc.

Biopal® VRO-20. [GAF] Nonoxynol-9 iodine; nonionic; germicide; used for cleaning, sanitizing, and disinfecting hospital, biological laboratory, and dairy equipment; breweries; multiple-use eating and drinking utensils in bars, restaurants; dk. brn. pourable liq.;

water sol.; sp.gr. 1.34; visc. 1700 cps; pour pt. 10.4 C.

Biopen 302. [Aquatec Quimica] Thiochloride compd.; microbiocide for adhesives and latex; liq.; water-sol.

Biopen 315. [Aquatec Quimica] Isothiazolines blend; microbiocide for water-sol. paints and syn. adhesives; liq.; water-sol.

Biopen 319. [Aquatec Quimica] Organonitrogenous compd.; microbiocide for metal work fluids and syn. adhesives; liq.; water-sol.

Biopen 350. [Aquatec Quimica] Org. tin compd.; microbiocide and fungicide for leather preservation; liq.; water-sol.

Bio-Pruf®. [Morton Int'l.] Antimicrobials.

Bio-Quat 50-24. [Lonza] Benzalkonium chloride USP; cationic; antimicrobial, germicide, disinfectant, sanitizer, fungicide, deodorant; algicide and biocide for swimming pools and cooling waters; pale yel. to water-wh. clear liq.; sp.gr. 0.988; dens. 8.23 lb/gal; visc. 60 cps; pH 7.2–8.0 (10%); 50% act.

Bio-Quat 50-25. [Lonza] Benzalkonium chloride; antimicrobial for formulating germicides and algicides; pale yel. to water-wh. clear liq.; sp.gr. 0.984; dens. 8.20 lb/gal; pH 7.2–8.0 (10%); 50% act.

Bio-Quat 50-28. [Lonza] Benzalkonium chloride; see Bio-Quat 50-24; pale yel. to water-wh. clear liq.; sp.gr. 0.979; dens. 8.16 lb/gal; visc. 105 cps; pH 7.2–8.0 (10%); 50% act. in IPA.

Bio-Quat 50-30. [Lonza] Benzalkonium chloride; cationic; see Bio-Quat 50-25; pale yel. to water-wh. clear liq.; sp.gr. 0.985; dens. 8.21 lb/gal; pH 7.2–8.0 (10%); flash pt. (PMCT) 134 F; 50% act.

Bio-Quat 50-35. [Lonza] Benzalkonium chloride and alkyl dimethyl ethyl benzyl ammonium chloride; cationic; see Bio-Quat 50-24; pale yel. to water-wh. clear liq.; sp.gr. 0.986; dens. 8.13 lb/gal; visc. 68 cps; pH 7.2–8.0 (10%); 50% act.

Bio-Quat 50-36. [Lonza] Benzalkonium chloride and alkyl dimethyl ethyl ammonium chloride; cationic; antimicrobial, germicide, disinfectant, sanitizer, fungicide, deodorant; pale yel. to water-wh. clear liq.; sp.gr. 0.977; dens. 8.13 lb/gal; visc. 68 cps; pH 7.2–8.0 (10%); 50% act.

Bio-Quat 50-40. [Lonza] Benzalkonium chloride USP; cationic; see Bio-Quat 50-24; pale yel. to water-wh. clear liq.; sp.gr. 0.979; dens. 8.16 lb/gal; visc. 425 cps; pH 7.2–8.0 (10%); 50% act.

Bio-Quat 50-42. [Lonza] Benzalkonium chloride; see Bio-Quat 50-25; pale yel. to water-wh. clear liq.; sp.gr. 0.98–0.99; dens. 8.24 lb/gal; pH 7.2–8.0 (10%); flash pt. (PMCT) 135 F; 50% act.

Bio-Quat 50-60. [Lonza] Benzalkonium chloride; cationic; antimicrobial for germicides and algicides; pale yel. to water-wh. clear liq.; sp.gr. 0.961; dens. 8.01 lb/gal; visc. 105 cps; pH 7.2–8.0 (10%); flash pt. 116.6 F; 50% act.

Bio-Quat 50-65. [Lonza] Benzalkonium chloride; see Bio-Quat 50-25; pale yel. to water-wh. clear liq.; sp.gr. 0.987; dens. 8.23 lb/gal; pH 7.2–8.0 (10%); flash pt. (PMCT) 135 F; 50% act.

Bio-Quat 50-MAB. [Lonza] Alkyl dimethyl ethyl ammonium bromide; cationic; see Bio-Quat 50-25; pale yel. to water-wh. clear liq.; sp.gr. 0.950; dens. 7.92 lb/gal; pH 7.2–8.0 (10%); 50% act.

Bio-Quat 50-MAC. [Lonza] Alkyl dimethyl dichlorobenzyl ammonium chloride; cationic; antimicrobial, algicide, slimicide for industrial water treatment; pale yel. to water-wh. clear liq.; sp.gr. 1.038; dens. 8.65 lb/gal; visc. 75 cps; pH 7.2–8.0 (10%); 50% act.

Bio-Quat 80-24. [Lonza] Benzalkonium chloride USP; cationic; see Bio-Quat 50-24; pale yel. to water-wh. clear liq.; sp.gr. 0.916; dens. 7.63 lb/gal; visc. 350 cps; pH 7.2–8.0 (10%); flash pt. (PMCT) 75 F; 80% act.

Bio-Quat 80-28. [Lonza] Benzalkonium chloride USP; cationic; see Bio-Quat 50-24; pale yel. to water-wh. clear liq. above gel pt.; sp.gr. 0.922; dens. 7.73 lb/gal; visc. 405 cps; gel pt. 4–8 C; pH 7.2–8.0 (10%); flash pt. (PMCT) 75 F; 80% act.

Bio-Quat 80-35. [Lonza] Benzalkonium chloride and alkyl ethylbenzyl ammonium chloride; cationic; antimicrobial; see Bio-Quat 50-24; pale yellow to water-white clear liq.; sp.gr. 0.927; dens. 7.72 lb/gal; visc. 330 cps; pH 7.2–8.0 (10%); flash pt. (PMCT) 75 F; 80% act.

Bio-Quat 80-36. [Lonza] Benzalkonium chloride and alkyl dimethyl ethyl benzyl ammonium chloride; cationic; antimicrobial, germicide, disinfectant, sanitizer, fungicide, deodorant, algicide, biocide; pale yel. to water-wh. clear liq.; sp.gr. 0.922; dens. 7.72 lb/gal; visc. 330 cps; pH 7.2–8.0 (10%); flash pt. (PMCT) 82 F; 80% act.

Bio-Quat 80-40. [Lonza] Benzalkonium chloride USP; see Bio-Quat 50-24; pale yel. to water-wh. clear liq.; sp.gr. 0.922; dens. 7.73 lb/gal; pH 7.2–8.0 (10%); flash pt. (PMCT) 75 F; 80% act.

Bio-Quat 80-42. [Lonza] Benzalkonium chloride; see Bio-Quat 50-25; pale yel. to water-wh. clear liq.; sp.gr. 0.95; dens. 8.21 lb/gal; pH 7.2–8.0 (10%); flash pt. (PMCT) 75 F; 80% act.

Bio-Quat ASH-29. [Lonza] Cetyl trimethyl ammonium chloride; cationic; antistat used on fabrics and with plastics; aq.; 29% act.

Bio-Quat IM-50. [Lonza] 1-Hydroxyethyl-1 benzyl-2-coco imidazolinium chloride; cationic; algicide, bactericide, fungicide, slimicide for recirculating cooling systems, oil field discovery, and salt water disposal systems; liq.; 50% act.

Bio-Quat T-501. [Lonza] Benzalkonium chloride and alkyl ammonium bromide; cationic; see Bio-Quat 50-24; pale yel. to water-wh. clear liq.; sp.gr. 0.972; dens. 8.11 lb/gal; visc. 68 cps; pH 7.2–8.0 (10%); 50% act.

Bio-Quat T-502. [Lonza] Alkyl dimethyl dichlorobenzyl ammonium chloride, alkyl dimethyl ammonium chloride, and alkyl ethyl ammonium bromide; cationic; see Bio-Quat 50-24; pale yel. to water-wh. clear liq.; sp.gr. 0.972; dens. 8.31 lb/gal; visc. 68 cps; pH 7.2–8.0 (10%); 50% act.

Bio Soft S-100. [Stepan] Linear dodecylbenzene sulfonic acid; anionic; detergent, emulsifier, intermediate for formulation of built detergents; dk. visc. liq.; water sol.; sp.gr. 1.04; biodeg.; 97% act.

Biosperse 240. [Drew] 2,2-Dibromo-3-nitrilopropionamide and solubilizing agents; antimicrobial for control of bacteria and algae in recirculating cooling water systems, air washer systems, evaporative condensers; yel. to pale grn. clear liq.; little or no odor; sol. in water; sp.gr. 1.15; dens. 9.6 lb/gal; f.p. –32.5 F; pH 1.2; flash pt. 225 F.

Biosperse 250. [Drew] 5-Chloro-2-methyl-4-isothiazolin-3-one and 2-methyl-4-isothiazolin-3-one; broad-spectrum antimicrobial for control of bacteria, fungi, and algae in industrial recirculating cooling water systems, oil field aq. systems; preservative in aq. metalworking fluids; pale grn. to blue liq.; little or no odor; sol. in water; sp.gr. 1.02; dens. 8.6 lb/gal; pH 2.4–4.4.

Biosulphur Fluid. [Henkel] Hydro alcohol solubilized sulphur; conditioner; applic. to skin with excessive sebum secretion; dk. brn. visc. prod.

Biosulphur Powder. [Henkel] Micro grained act. sulfur with protective colloid; see Biosulphur Fluid; beige micro-powd.

Bio-Surf DC-730. [Lonza] Quat. ammonium compd.; cationic; emulsifier and dispersant; liq.; oil-sol.; 75% conc.

Bio-Surf I-20. [Lonza] Nonylphenoxypoly (ethyleneoxy) ethanol-iodine complex, iodophor conc.; nonionic; antimicrobial, germicide, disinfectant, sanitizer for cleaning, sanitizing, disinfecting in hospital, food and beverage plants, breweries, restaurants; very dk. brn. visc. liq.; mild, pleasantly clean halogen odor; sol. in IPA, monobutyl glycol ether, propylene glycol; sp.gr. 1.34; dens. 11.2 lb/gal; visc. 580 cps; f.p. < 0 C; pH 1.68 (10% aq.); flash pt. (PMCC) > 93 C; 92% act. in water.

Bio-Surf I-21LF. [Lonza] Polyethoxypoly propoxypoly ethoxyethanol-iodine complex iodophor conc.; nonionic; antimicrobial to mfg. sanitizers and disinfectants for food and beverage plants; very dk. red brn. visc. liq.; mild, pleasant clean halogen odor; sol. see Bio-Surf I-20; sp.gr. 1.39; dens. 11.5 lb/gal; f.p. < 0 C; pH 1.68 (10% aq.); flash pt. (PMCC) > 93 C; 91.3% act. in water.

Bio-Surf PBC-460. [Lonza] Lauryl myristyl dimethyl amine oxides; nonionic/cationic; detergent aid for stabilization of foam; emollient; liq.; 30% conc.

Bio Terge PAS 8. [Stepan] Sodium octyl sulfonate; solubilizer and wetting agent for acid, alkaline, and oxidizing systems; liq.

Biozan. [Hercules] Xanthan gum; suspending agent, thickener, emulsifier in slurry explosives, foundry coatings, acid and caustic cleaning compds., cosmetics, pharmaceuticals, oil field chemicals, aq. systems; food grade avail.; water-sol.

Biozan SPX 5423. [Hercules] Xanthan gum; industrial grade thickener, suspending agent; tan powd.; visc. 1300–1600 (1%).

Bismate®. [Vanderbilt] Bismuth dimethyldithiocarbamate; ultra accelerator for NR, IR, BR, and SBR; for high temp., high speed vulcanization; lemon yel. powd., fineness 99.9% thru 100 mesh; rods; m.w. 569.66; sol. in chloroform; practically insol. in water; dens. 2.04 ± 0.03 mg/m³; m.p. > 230 C with decomposition (powd.); > 228 C with decomposition (rods); 35.0–38.0% bismuth content (powd.); 32.0–34.0% bismuth content (rods).

Bismet. [Akrochem] Bismuth dimethyldithiocarbamate; accelerator for SBR, NR, IR, and BR compds. that are high-temp. cured; nonstaining; yel. powd.; sp.gr. 1.9; m.p. > 230 F.

Bitrex. [Macfarlan Smith] Denatonium benzoate NF; aversive (bitter) agent used to minimize danger of prod. ingestion; denaturant for ethanol; wh. gran., odorless; m.w. 446.5; sol. in water, chloroform, IPA; highly sol. in methanol, ethanol; m.p. 163–170 C; pH 6.5–7.5; 99.5–101.0% assay.

BL 3. [Releasomers] Semipermanent release agent for applic. to flexible molds and metal molds; effective for halogenated or peroxide-cured elastomers.

Bladafum®. [Bayer] Sulfotepp; insecticide; fumigant for the control of greenhouse pests; pale yel. liq.; m.w. 322.3; sol. 25 ppm in water; misc. with most org. solvs.; sp.gr. 1.196 (25/4 C); b.p. 92 C (0.1 mm Hg).

Blanc fixe F, N, micro®. [Sachtleben] Syn. precipitated barium sulfate (micronized grade, micro®); wh. inert filler resistant to weathering; improves hardness and stiffness of plastics and can render plastics opaque to x-rays; particle size 1, 3, and 0.7 µm resp.; dens. 4.4 g/ml; pH 9.

Blancol®. [GAF] Sodium naphthalenesulfonate-formaldehyde condensate; anionic; thinning, dispersant, peptizing agent, dye-leveling agent used in paper industry for slime control, improved retention of fillers or fines, improved sizing; tan to brn. free-flowing coarse gran.; odorless; also avail. liq. form; sol. in warm or cold water; dens. 0.77–0.82 g/ml; 88–90% act.

Blancol® N. [GAF] Sodium naphthalenesulfonate-formaldehyde condensate; see Blancol; tan to brn. gran. powd.; odorless; sol. in water; dens. 0.65–0.75 g/ml; 88–90% act.

Blandol. [Witco] Wh. min. oil N.F.; lubricant used in food, drug, and cosmetic industry; saybolt visc. 100 F-80/90; pour pt. 20 F.

Blanose Cellulose Gum. [Aqualon] Sodium CMC; thickener, binder, stabilizer, suspending agent, water-retention aid, and crystal growth inhibitor for pharmaceutical, cosmetics, toothpaste, and food applics.; food grades designated by F; powd.; visc. @ 2%: 40 MPas (7LF), 500 mPas (7MF), 600 mPas (7M8SF, 12M8P), 2000 mPas (7M31F, 9M31F, 12M31P), 5000 mPas (7M65F, 9M65F), 20,000 mPas (7HF, 7H3SF, 7HOF, 9HF), 40,000 mPas (7H4F, 9H4F); pH 6.5–8.5; 8% max. moisture, 99.5% min. purity.

Blanose Refined CMC. [Aqualon] Sodium CMC, industrial grades; thickener, binder, stabilizer, suspending agent, film-former, antiredeposition aid for paper industry, textile sizing, welding rods, drilling muds, paints, household prods., water-based adhesives; avail. in fibrous form, and in coarse, med., fine, and very fine particle sizes; visc. @ 2%: 1.5 mPas (7UL), 2 mPas (7EL), 3 mPas (7L1), 4 mPas (12UL),

5 mPas (7L2), 25 mPas (7L3), 40 mPas (7L, 9L), 75 mPas (7M1, 9M1), 150 mPas (7M2), 450 mPas (7M, 9M), 900 mPas (7M12), 2200 mPas (7M31), 5000 mPas (7M65, 9M65), 20,000 mPas (7H, 9H), 40,000 mPas (7H4), 80,000 mPas (7H9); pH 6.5–8.5; 8% max. moisture, 98% min. purity.

Blendex® 101. [GE Specialty] ABS; modifier resin providing good impact str. and improved processing to PVC compds.; used in calendered films, semiflexible PVC film; sp.gr. 1.01; bulk dens. 21 lb/ft³; tens. str. 5275 psi; Izod impact str. 6.9 ft lb/in. (1/4 in.); hardness Rockwell R94; distort. temp. 217 F (@ 264 psi).

Blendex® 131. [GE Specialty] ABS; modifier resin for calendered films requiring good thermoforming properties, semiflexible PVC film applics.; base resin in calendered ABS sheet; wh. powd.; sp.gr. 1.01; bulk dens. 17.5 lb/ft³; tens. str. 4200 psi; Izod impact str. 6.9 ft lb/in. (1/4 in.); hardness Rockwell R81; distort. temp. 215 F (264 psi).

Blendex® 201. [GE Specialty] ABS; modifier resin for improving dimensional stability and processing of PVC, plasticized PVC, rigid PVC sheet applics.; sp.gr. 1.04; bulk dens. 21.3 lb/ft³; tens. str. 6750 psi; Izod impact str. 5.8 ft lb/in. (1/4 in.); hardness Rockwell R108; distort. temp. 218 F (264 psi).

Blendex® 310. [GE Specialty] ABS; impact modifier for rigid PVC, epoxies, PU, polyesters; sp.gr. 0.98; bulk dens. 18.0 lb/ft³; tens. str. 2475 psi; Izod impact str. 5.8 ft lb/in. (1/4 in.); hardness Shore D65; distort. temp. 189 F (264 psi).

Blendex® 336. [GE Specialty] ABS; impact modifier for clear and opaque PVC, incl. pipe and conduit, calendered or extruded film and sheet, inj. molded prods.; wh. powd.; sp.gr. 0.98; bulk dens. 14.0 lb/ft³; tens. str. 600 psi; Izod impact str. 5.5 ft lb/in. (1/4 in.); hardness Shore D42; distort. temp. 114 F (264 psi).

Blendex® 338. [GE Specialty] Impact modifier providing toughness to opaque sheet, profile, and inj. molding PVC applics.; also for PC and polyesters; sp.gr. 0.95; bulk dens. 17.0 lb/ft³; tens. str. 530 psi; Izod impact str. 5.2 ft lb/in. (1/4 in.); hardness Shore D40; distort. temp. 127 F (264 psi).

Blendex® 405. [GE Specialty] ABS; clear modifier resin providing good low-temp. impact resistance to clear calendered and extruded film, rigid PVC; sp.gr. 0.99; bulk dens. 17.5 lb/ft³; tens. str. 2875 psi; Izod impact str. 1.2 ft lb/in. (1/4 in.); hardness Shore D63; distort. temp. 174 F (264 psi).

Blendex® 420. [GE Specialty] Impact modifier for clear, rigid PVC applics.; sp.gr. 1.00; bulk dens. 16.9 lb/ft³; tens. str. 980 psi; Izod impact str. 5.1 ft lb/in. (1/4 in.); hardness Shore D52; distort. temp. 120 F (264 psi).

Blendex® 424. [GE Specialty] Impact modifier for high transparency, rigid PVC applics, incl. extruded and calendered sheet requiring clarity and toughness with good vacuum formability; particulate 99% < 8 mesh; sp.gr. 0.97; bulk dens. 15.8 lb/ft³; ref. index 1.539; tens. str. 940 psi; hardness Shore D57.

Blendex® 467. [GE Specialty] ABS; modifier resin for transparent vinyl sheet and bottle applics.; sp.gr. 0.96; bulk dens. 15.6 lb/ft$_3$; tens. str. 1425 psi; Izod impact str. 6.5 ft lb/in. (1/4 in.); hardness Shore D50; distort. temp. 145 F (264 psi).

Blendex® 586. [GE Specialty] Poly (α-methylstyrene-styrene acrylonitrile); high heat modifier resin used to upgrade PVC compds.; sp.gr. 1.09; bulk dens. 20.0 lb/ft³; tens. str. 9000 psi; Izod impact str. 0.5 ft lb/in. (1/8 in.); hardness Rockwell R120; distort. temp. 255 F (264 psi).

Blendex® 590. [GE Specialty] High m.w. methyl methacrylate styrene acrylonitrile copolymer; modifier, process aid for polymers/applics. requiring higher molten elasticity; used in rigid and semirigid PVC, low-dens. foamed sheet and profiles, clear compds. for blow molding and thermoforming; wh. free-flowing powd.; sp.gr. 1.14; ref. index 1.538.

Blendex® 702. [GE Specialty] Poly (α-methylstyrene-styrene acrylonitrile)/ABS; see Blendex 586; sp.gr. 1.06; bulk dens. 16.7 lb/ft³; tens. str. 7375 psi; Izod impact str. 2.0 ft lb/in. (1/8 in.); hardness Rockwell R116; distort. temp. 241 F (264 psi).

Blendex® 703. [GE Specialty] Modifier to raise the heat distort. temp. and impact resistance of PVC compds.; sp.gr. 1.05; bulk dens. 17.0 lb/ft³; tens. str. 6800 psi; Izod impact str. 3.3 ft lb/in. (1/8 in.); hardness Rockwell R114; distort. temp. 242 F (264 psi).

Blendex® 27920. [GE Specialty] Impact modifier for clear, rigid PVC applics. incl. extruded and calendered sheet requiring clarity, toughness, and good vacuum formability; particulate 99% < 8 mesh; sp.gr. 1.0; tens. str. 1100 psi; hardness Shore D54.

BLO®. [GAF] γ-Butyrolactone; solv. for PAN, PS, fluorinated hydrocarbons, cellulose triacetate, shellac; used in paint removers, petrol. processing, hectograph process, speciality inks; intermediate for aliphatic and cyclic compds.; reaction and diluent solv. for pesticides; used in dyeing of acetate; wetting agent for cellulose acetate films, fibers, solv. welding of plastic films in adhesive applics.; liq.; f.p. –44 C; b.p. 204 C; flash pt. 98 C (OC).

Bohrmittel Hoechst. [Hoechst-Celanese] Alkylsulfamido carboxylic acid, sodium salt; anionic; o/w corrosion inhibitor, lubricant, and emulsifier for metalworking fluids; liq.

Boisambrene. [Henkel] Formaldehyde methylcyclododecylacetal; raw material for fragrances with woody note.

Boisambrene Forte. [Henkel] Formaldehyde ethyl cyclododecylacetal; see Boisambrene.

Bonding Agent 2001, 2005. [Bayer] Reactive aromatic polyisocyanate; one part bonding agent used in coating of syn. fiber fabrics with PVC pastes; 2001 is solventless; 2005 contains solv.; liq.

Bonding Agent TN. [Mobay] Polyester mixt. of ethyl acetate and methylene chloride (1:1) ratio; 2-part bonding agent in textile processing; sol'n.; dens. 1.095 g/cm³; visc. 180 ± 30 mPas.; pour pt. < 7 C; acid no. 4; flash pt. 8 C; cloud pt. < 15 C; 70% sol'n.

Bonding Agent TN/S50. [Mobay] Polyester in benzyl butyl phthalate; see Bonding Agent TN; sol'n.; dens. 1.110 g/cm³; visc. 1050 ± 150 mPa•s; pour pt.

< 12 C; acid no. 4; flash pt. 200 C; cloud pt. < 18 C; 50% sol'n.

Borax. [U.S. Borax] Sodium tetraborate decahydrate; dispersant, wetting agent for NR, SR latexes; mold lubricant for general dry rubber molding; wh. powd., odorless; sol. in water, glycerin; sp.gr. 1.73.

Bovinol 30. [RITA] Serum albumin; whole protein skin, hair, and nail conditioner; 30% sol'n.

Bozemine N 60. [Hoechst-Celanese] Polyethylene base; plasticizer for textiles from cellulosic and syn. fiber and wool; beige emulsion.

Bozemine N 609. [Hoechst-Celanese] Silicone base; plasticizer for textiles from wool, cellulosic and syn. fibers; wh. liq. disp.

Bravo W-75. [Henkel/Process] Chlorothalonil; broad-spectrum agric. fungicide; 75% act.

Breakerase G. [Int'l. Bio-Synthetics] Mannan depolymerase; enzyme for controlled hydrolysis of guar gum and related polymers; liq.

Bretol. [Hexcel] Cetethyldimonium bromide; antibacterial, disinfectant for veterinary, pharmaceutical, dental applics.; formulation of solder flux; wh. powd.; char. odor; m.w. 378.47; sol. in water, alcohol and chloroform; m.p. 164-190 C; pH 6-8; 99% min. assay.cacs

Brew(n)zyme®. [Miles Lab] Amylase/protease betaglucanase; enzyme for brewing with barley; liq., powd.

Brij® 56. [ICI Am.] Ceteth-10 with antioxidants; nonionic surfactant, emulsifier; solubilizer for fragrances; wh. waxy solid; sol. in alcohol; HLB 12.9; flash pt. > 300F; pour pt. 31 C.

Brij® 58. [ICI Am.] Ceteth-20 with preservatives; nonionic surfactant, emulsifier; solubilizer for fragrances; wh., waxy solid; sol. in water, alcohol; HLB 15.7; flash pt. > 300 F; pour pt. 38 C; 100% conc.

Brij® 76. [ICI Am.] Steareth-10 with preservatives; nonionic surfactant, o/w emulsifier, solubilizer for fragrances, topical cosmetics; wh. waxy solid; sol. in propylene glycol, ethanol; HLB 12.4; flash pt. >300 F; pour pt. 38 C; 100% conc.

Brij® 78. [ICI Am.] Steareth-20 with preservatives; nonionic surfactant, emulsifier; solubilizer for fragrances; wh. waxy solid; sol. in alcohol; HLB 15.3; flash pt. >300 F; pour pt. 38 C; 100% conc.

Brij® 96. [ICI Am.] Oleth-10 with antioxidants; nonionic; o/w emulsifier, solubilizer for fragrances, topical cosmetics; yel. liq. with some solids; sol. in water and alcohol; visc. 100 cps; HLB 12.4; flash pt. > 300 F; pour pt. 16 C; 100% conc.

Brij® 97. [ICI Am.] Oleth-10 with preservatives; nonionic; see Brij® 96; pale yel. liq.; sol. in water and alcohol; visc. 100 cps; HLB 12.4; flash pt. > 300 F; pour pt. 16 C; 100% conc.

Brij® 98, 99. [ICI Am.] Oleth-20 with antioxidants; nonionic surfactant, emulsifier; solubilizer for fragrances; cream-colored soft waxy solid; sol. in water, alcohol, propylene glycol; HLB 15.3; flash pt. > 300 F; pour pt. 30 and 33 C respectively; 100% conc.

Brij® 700. [ICI Am.] Steareth-100; nonionic; emulsifier for cosmetics and pharmaceuticals; oil solubilization; wh. solid; HLB 18.8; pour pt. 55 C; 100% conc.

Brij® 721. [ICI Am.] Steareth-21; nonionic; cosmetic emulsifier; solubilizer for fragrances; colorless waxy solid; waxy odor; HLB 15.5; pour pt. approx. 45 C; flash pt. > 230 F (PMCC); 100% conc.

Britesorb®. [PQ] Silica hydrogel based composition; preservative, stabilizer for beer; wh. powd.; water-insol.

Britesorb® A 100. [PQ] Silica hydrogel; preservative, stabilizer for beer; wh. powd.; water-insol.

Britol®. [Witco] Min. oil USP; white oil functioning as binder, carrier, conditioner, defoamer, dispersant, extender, heat transfer agent, lubricant, moisture barrier, plasticizer, protective agent, and/or softener in adhesives, agric., chemicals, cleaning, cosmetics, food, pkg., plastics, and textiles industries; sp.gr. 0.869–0.885; visc. 57–60 cSt (40 C); flash pt. 199 C; pour pt. –15 C.

Bromat. [Hexcel] Cetrimonium bromide; cationic; surfactant, emulsifier, germicide for cosmetic and topical preparations; conditioner, emulsifier and antistat in hair cream rinse formulations; wh. powd.; char. odor; m.w. 364.44; sol. in alcohol, chloroform; pH 5-8; surf. tens. 40.0 dynes; 98% min. assay.

Bromoklor 50. [Ferro] Halogenated aliphatic liqs. containing bromine and chlorine; flame retardant used in plasticized PVC; uses incl. disp. molded or coated automotive parts, interior trim, pkg. closures, boots, hand grips, flooring, upholstery, carpet backing, furniture and wall covering, and pkg. film; plasticizing properties in PVC (not primary); Gardner 3 liq.; sp.gr. 1.38; visc. 270 cps; 30% bromine; 20% chlorine.

Bromoklor 70. [Ferro] Halogenated aliphatic liq. containing bromine and chlorine; see Bromoklor 50; Gardner 5 liq.; sp.gr. 1.65; visc. 5000 cps; 35% bromine; 35% chlorine.

Bronidox L, L 5. [Henkel] 5-Bromo-5-nitro-1,3-dioxane dissolved in 1,2 propylene glycol; preservative for shampoos, foam bath and other surfactant preparations; sol. in aq. systems; liq.

Bronopol. [Angus] 2-Bromo-2-nitropropane-1,3-diol; broad spectrum antimicrobial agent; cryst. powd.; sol. in water, alcohol, and glycol.

Bronopol-Boots. [Angus] 2-Bromo-2-nitropropane-1,3-diol; antibacterial agent, stable in presence of surfactants, incl. nonionics; cryst.; water-sol.

BSWL 202. [Eagle Picher] Basic silicate wh. lead; wh. pigment acting as heat stabilizer for chlorinated polyethylene, chlorosulfonated polyethylene, PVC, and polyepichlorohydrin; rust-inhibitive pigment in the automobile industry; used in industrial or maintenance paints; 99.5% –325 mesh; sp.gr. 6.40; dens. 53.31 lb/solid gal; apparent dens. 13 g/in.^{3}; bulking value 0.0188 gal/lb; oil absorp. 16.0; brightness in oil 86.0% (Beckman); 78.6% Pb; 15.4% SiO_2.

BTC 50, 50 USP. [Stepan] Benzalkonium chloride; cationic; antimicrobial for hard surf. disinfection, sanitization, deodorization; liq.; sp.gr. 0.96; flash pt. 126 F; 50% min. quat.

BTC 65, 65 USP. [Stepan] Benzalkonium chloride; cationic; see BTC 50; liq.; sp.gr. 0.97; flash pt. 132 F;

50% min. quat.

BTC 99. [Stepan] Didecyl dimonium chloride; low foaming algicide and slimicide for swimming pool and industrial water treatment; liq.; 50% act.

BTC 100. [Stepan] Benzalkonium chloride; cationic; see BTC 50; solid; sp.gr. 0.95; flash pt.> 200 F; 95% min. quat.

BTC 776. [Stepan] Benzalkonium chloride and dialkyl methyl benzyl ammonium chloride; cationic; algicide and slimicide for swimming pool and industrial water treatment; liq.; sp.gr. 0.96; flash pt. 102 F; 50% min. quat.

BTC 812. [Stepan] Octyl dodecyl dimethyl ammonium chloride; cationic; antimicrobial for hard surf. disinfection and sanitization; fungicide; liq.; water-sol.; sp.gr. 0.93; flash pt. 126 F; 50% min. quat.

BTC 818. [Stepan] Octyldecyl dimethyl, dioctyl dimethyl, and didecyl dimethyl ammonium chlorides, cationic; disinfectant, sanitizer, and fungicide for hard surfaces; liq.; water-sol.; sp.gr. 0.93; flash pt. 86 F; 50.0% min. quat.

BTC 824. [Stepan] Myristalkonium chloride; cationic; antimicrobial for hard surf. disinfection and sanitization; algicide for swimming pools and industrial water treatment; liq.; sp.gr. 0.96; flash pt. 120 F; 50% min. quat.

BTC 824 P-100. [Stepan] Tetradecyl dimethyl benzyl ammonium chloride monohydrate; cationic; antimicrobial for tablet mfg. of disinfectants, sanitizers, deodorizers; powd.; water-sol.; sp.gr. 0.40; flash pt. > 200 F; 95% min. quat.

BTC 835. [Stepan] Benzalkonium chloride; cationic; see BTC 824; liq.; sp.gr. 0.97; flash pt. 130 F; 50% min. quat.

BTC 885. [Stepan] n-Alkyl (50% C_{14}, 40% C_{12}, 10% C_{16}) dimethyl benzyl, octyldecyl dimethyl, dioctyl dimethyl, and didecyl dimethyl ammonium chlorides; cationic; germicide for formulation of disinfectant, sanitizer, and fungicidal prods. used in hospitals, nursing homes, and public institutions; liq.; sp.gr. 0.95; flash pt. 116 F; 50.0% min. quat.

BTC 1010. [Stepan] Didecyl dimonium chloride; cationic; see BTC 812; liq.; water-sol.; sp.gr. 0.89; flash pt. 86 F; 50% min. quat.

BTC 1010-80. [Stepan] Quaternium-12; fungicide for hard-surf. disinfection and sanitization; liq.; 80% act.

BTC 1100. [Stepan] N-tetradecyl dimethyl-naphthylmethyl ammonium chlorides; cationic; nondusting powd. used as disinfectant, sanitizer, algicide, and deodorizer; sp.gr. 0.50; flash pt. > 200 F; 99% min. quat.

BTC 1326. [Stepan] n-Alkyl (60% C_{14}, 30% C_{16}, 5% C_{12}, 5% C_{18}) dimethyl benzyl and n-alkyl (68% C_{12}, 32% C_{14}) dimethyl ethylbenzyl ammonium chlorides; cationic; hard surface disinfectant/surfactant for sanitizers for hospitals, nursing homes, public institutions, and industry; powd.; sp.gr. 0.75; flash pt. > 200 F; 13% min. quat.

BTC 2125, 2125 80%, 2125 P-40. [Stepan] Benzalkonium chloride, alkyl dimethyl ethylbenzyl ammonium chloride; cationic; antimicrobial, hard surf. disinfectant, sanitizer, fungicide for hospitals, public institutions; algicide in swimming pool and industrial water treatment; deodorizer; liqs. except 2125 P-40 (powd.); sp.gr. 0.97, 0.94, and 0.55 resp.; flash pt. > 200 F except 2125 80% (130 F); 50, 80, and 40% min. quat. resp.

BTC 2125 M. [Stepan] Myristalkonium chloride and quaternium-14; cationic; disinfection and sanitization; liq.; sp.gr. 0.97; flash pt. > 200 F; 50% min. quat.

BTC 2125M 80%, 2125MP-40. [Stepan] Benzalkonium chloride and alkyl dimethyl ethylbenzyl ammonium chloride; cationic; broad-spectrum bacteriological control agent used in disinfectant and sanitizer formulations for hospitals, public institutions, industry; liq. and powd. resp.; sp.gr. 0.94 and 0.55 resp.; flash pt. > 130 and 200 F resp.; 80 and 40% min. quat. resp.

BTC 2565. [Stepan] Benzalkonium chloride; cationic; see BTC 776; liq.; sp.gr. 0.96; flash pt. 110 F; 50% min. quat.

BTC 2568. [Stepan] Benzalkonium chloride; see BTC 99; liq.; 50% act.

BTC 8248, 8249. [Stepan] Benzalkonium chloride; cationic; see BTC 824; liq.; sp.gr. 0.95 and 0.94 resp.; flash pt. 110 and 118 F resp.; 80 and 90% min. quat. resp.

BTC E-8358. [Stepan] Benzalkonium chloride; cationic; see BTC 824; liq.; sp.gr. 0.93; flash pt. 100 F; 80% min. quat.

BTCO 1010. [Stepan] Didecyl dimonium chloride; cationic; disinfectant, fungicide for hard surf. disinfection and sanitization; liq.; sp.gr. 0.89; flash pt. 86 F; 50% min. quat.

Bubble Breaker® 259, 260, 613-M, 622, 730, 737, 746, 748, 900, 913, 917. [Witco] Blend of org., nonsilicone compds.; defoamer used in the mfg. of water-based paper coatings, textile processing formulations, agric. chemical prods., paints, adhesives, and inks.

Bubble Breaker® 776, 3017-A. [Witco] Disp. of reacted silica in hydrocarbon solv.; defoamer used in latex mfg. operations, formulation of water-based paints and adhesives.

Bubble Breaker® 1840X. [Witco] Sodium salt of sulfonated fatty acid; defoamer for wet process prod. of phosphoric acid; corrosion inhibitor, detergent; sol. in water, alcohol.

Bubble Breaker® 3009-F. [Witco] Blend of org., nonsilicone compds.; defoamer for wash water of sliced potatoes.

Bubble Breaker® 3017-A. [Witco] Disp. of reacted silica in hydrocarbons; defoamer; monomer stripping and water based flexographic inks and water based coatings; water-disp.

Bubble Breaker® 3056-A. [Witco] Disp. of reacted silica in hydrocarbon solv.; defoamer used in latex mfg. operations, formulation of water-based paints and adhesives, effluent water, asphalt emulsions, PVC monomer stripping; water-disp.

Bubble Breaker® 3073-7, D. [Witco] Blend of org., nonsilicone compds.; defoamer for water-based drilling for the petrol. industry.

Bubble Breaker® PR. [Witco] Sodium salt of sulfonated fatty acid; defoamer for wet process prod. of phosphoric acid; sol. in water, IPA.

Bumyr. [Amerchol] Butyl myristate; emollient; cosmetic ingred.; oil-sol.

Butac®. [Whitney & Oettler] Resin acids-amine resin soaps blend; tackifier for IIR, NR, SBR, molding aid for NBR, SBR; improves pigment disps.; activates cure slightly; rdsh. brn. solid; shatters at R.T.; sp.gr. 1.075–1.085; soften. pt. (R&B) 135–152 F.

Butasan®. [Monsanto] Zinc dibutyldithiocarbamate; accelerator for nonblooming EPDM cures; antioxidant for noncuring applics., e.g., adhesives; IIR stabilizer; 7 mm pellets or powd.

Butoxyne 497. [GAF] Butynediol hydroxyethyl ethers; corrosion inhibitor for specialty applic.; pickling inhibitor for plating copper; nickel brightener in electroplating; liq.; 100% act.

Butyl Diglyme. [Ferro] Diethylene glycol dibutyl ether; solv. which tends to solvate cations; used in electrochemistry, polymer and boron chemistry; physical processes such as gas absorption, extraction, stabilization; used in industrial prods. such as fuels, lubricants, textiles, pharmaceuticals, pesticides; colorless clear liq.; mild nonresidual odor; m.w. 218.34; f.p. –60.2 C; b.p. 256 C (760 mm Hg); sol. 0.3% in water; sp.gr. 0.8814; dens. 7.36 lb/gal; visc. 2.4 cP; ref. index 1.4235; pH neutral; flash pt. 118 C (CC); 98.5% min. purity.

Butyl Dioxitol. [Shell] Diethylene glycol monobutyl ether; solv. for nitrocellulose, phenolics, Epon resins, alkyds, and acrylics, oils, and dyes; used in lacquers and inks which require slow evaporating solv.; lacquer and enamel formulations requiring improved flow-out and gloss; coupling solv. in preparation of specialized cleaning sol'ns. and cutting oils; coalescing agent in latex paints; component of brake fluids in automotive industry; colorless liq.; mild odor; water-misc.; sp.gr. 0.949–0.952; 0.10% max. water.

Butyl Eight. [Vanderbilt] Activated dithiocarbamate; ultra accelerator; for accelerating vulcanization of spread solv. cements, calendered and extruded stocks at room or slightly elevated temps.; accelerated stocks should be used within 8-12 h to prevent precure; redsh. brn. liq.; sol. in acetone, toluene, chloroform, alcohol, carbon disulfide; dens. 1.01±0.02 mg/m^3; flash pt. 40 C.

Butyl Namate®. [Vanderbilt] Sodium di-n-butyldithiocarbamate aq. sol'n.; accelerator for latexes of all elastomers; pale amber liq.; m.w. 227.36; misc. with water; dens. 1.09±0.02 mg/m^3; 47% min. assay.

Butyl Oxitol. [Shell] Ethylene glycol monobutyl ether; solv. for nitrocellulose, alkyds, phenolics, acrylics, Epon resins, nat. resins, dyes, waxes, oils, and org. materials; used in surf. coating formulations such as lacquers and enamels where it imparts gloss and improved leveling chars; coupling agent in cleaners and cutting oils; colorless liq.; mild odor; water-misc.; sp.gr. 0.898–0.901; 0.20% max. water.

Butyl Tuads®. [Vanderbilt] Tetrabutylthiuram disulfide; ultra accelerator; sulfur donor accelerator for sol. cure systems in natural and polyisoprene rubbers; used in conjunction with Amyl Ledate in EPDM; dk. amber liq.; m.w. 408.76; sol. in toluene, chloroform, carbon disulfide; practically insol. in water; dens. 1.06 ± 0.02 mg/m^3; 7.8% avail. sulfur.

Butyl Zimate®. [Vanderbilt] Zinc di-n-butyldithiocarbamate; ultra accelerator for EPDM and natural and syn. latexes; provides fast, flat cures in SBR, nitrile, and neoprene latexes; functions as nondiscoloring antioxidant in noncuring applics. and stabilizer in IIR; antioxidant in thermoplastic rubbers and hot melts; wh.-cream powd., fineness 99.9% thru 100 mesh; also avail. as 50% act. slurry; m.w. 474.13; sol. in toluene, carbon disulfide, chloroform, gasoline; dens. 1.21 ± 0.03 mg/m^3; m.p. 104–112 C; 13.0–15.0% zinc content.

Butynediol. [Rhone-Poulenc Surf.] 2-Butyne-1,4-diol; corrosion inhibitor in acid pickling baths; brightener in nickel electroplating; defoliator for agric. processing; cosolv. in paint removers; retards evaporation; polymerization accelerator, crosslinking agent, chain extender; starting material for inhibitors, pesticides, pharmaceuticals, textile prods., and propellants; flake, sol'n.; b.p. 140 C; sp.gr. 1.04; m.p. > 52 C; flash pt. 152 C; 96% min. flake; 35% aq.

C

C2® Sodium Chlorite. [Olin] Sodium chlorite; bacterial slimicide for wh. water systems in paper mills; wh. flake; dens. 53 lb/ft³ (loose); 69 lb/ft³ (packed); 79% min. sodium chlorite.

CA-57. [Great Lakes] Tech. grade of chlorendic anhydride; reactive flame retardant used in polymer systems (alkyds, urethanes, unsat. polyester resins, and epoxy resins); uv stability and corrosion resistance; wh. cryst. solid; m.w. 371; sol. (g/100 g solv.); 127.0 g acetone, 40.4 g benzene, 19.3 g raw linseed oil; insol. in water; m.p. 235–250 C; 54.5% chlorine.

CA0567. [Hüls] Allyltrimethoxy silane; coupling agent, chem. intermediate, blocking agent, release agent, lubricant, primer, reducing agent; liq.; m.w. 162.3; sp.gr. 0.95; b.p. 146–148 C; flash pt. 46 C; 98% purity.

CA0570. [Hüls] Allyltrimethyl silane; see CA0567; liq.; m.w. 114.2; sp.gr. 0.719; b.p. 85–86 C; flash pt. 7 C; ref. index 1.4074; 97% purity.

CA0699. [Hüls] N-(2-Aminoethyl)-3-aminopropylmethyldimethoxy silane; see CA0567; liq.; m.w. 206.4; sp.gr. 0.975; b.p. 129–130 C (10 mm); ref. index 1.4447; 95% purity.

CA0742. [Hüls] 3-Aminopropylmethyldiethoxy silane; see CA0567; liq.; m.w. 191.4; sp.gr. 0.92; b.p. 85–88 C (8 mm); flash pt. 125 C; ref. index 1.427; 97% purity.

CA0900. [Hüls] Amyltrichlorosilane; see CA0567; liq.; m.w. 205.6; sp.gr. 1.13; b.p. 171–172 C; flash pt. 30 C; ref. index 1.438; 97% purity.

Cab-O-Sil®. [Cabot] Fumed silica; dispersant, anticaking agent for foods, agric. prods., and powds. for cosmetics and coatings industries.

Cab-O-Sperse®. [Cabot] Fumed silica; thickener; rheological control in water-based systems.

Cachalot C-50. [M. Michel] Cetyl alcohol; emollient used in cosmetics; sol. in acetone, alcohol, aromatic hydrocarbons, carbon disulfide, chloroform, glycol and diglycol ethers.

Cachalot C-51. [M. Michel] Cetyl alcohol; conditioner, lubricant used in cosmetics; sol. in acetone, alcohol, aromatic hydrocarbons, chloroform, glycol and diglycol ethers.

Cachalot M-43. [M. Michel] Myristyl alcohol; see Cachalot C-50; sol. see Cachalot C-50.

Cachalot O-15. [M. Michel] Oleyl alcohol; conditioner, lubricant for cosmetics; corrosion inhibitor additive to lube oils; sol. see Cachalot C-51.

Cachalot S-53T Series. [M. Michel] Straight chain aliphatics; mold release and processing aid for vinyl resins and rubber; sol. in aromatic hydrocarbons and alcohol.

Cachalot S-54. [M. Michel] Stearyl alcohol; see Cachalot C-51; sol. see Cachalot C-51.

Cachalot S-56. [M. Michel] Stearyl alcohol; see Cachalot C-50; sol. see Cachalot C-50.

Cadet BPO-70. [Akzo] Benzoyl peroxide; initiator for elevated temp. curing of unsat. polyester resins in applics. for matched metal die molding; pultrusion, vacuum bag molding, continuous laminating, hot-cure casting, inj. molding; gran.; 70% conc.

Cadet BPO-70W. [Akzo] Benzoyl peroxide; initiator for elevated-temp. curing of unsat. polyester resins in applics. such as matched metal die molding, pultrusion, vacuum bag molding, continuous laminating, hot-cure casting, inj. molding; also for PS, specialized PVC resin, acrylic polymers; gran.; 70% BPO with 30% water.

Cadet BPO-78. [Akzo] Benzoyl peroxide; see Cadet BPO-70; gran.; 78% conc.

Cadet BPO-78W. [Akzo] Benzoyl peroxide; see Cadet BPO-70W; gran.; 78% BPO with 22% water.

Cadox 40E. [Akzo] Benzoyl peroxide with plasticizer; initiator for ambient-temp. polyester cures; emulsion; 40% conc.

Cadox BCP. [Akzo] Benzoyl peroxide with di- and tri-calcium phosphate; initiator for ambient-temp. polyester cures; used in adhesives and repkg. applic.; powd.; 35% conc.

Cadox BEP-50. [Akzo] Benzoyl peroxide with phlegmatizer; initiator for ambient-temp. polyester cures; used in mine roof bolts and autobody repair putties; red, wh., and blue paste; 50% conc.

Cadox BFF-50. [Akzo] Benzoyl peroxide with dicyclohexyl phthalate; see Cadet BPO-70; also has excellent sol. chars. for sol'n. polymerization applic.; dry gran.; 50% conc.

Cadox BFF-60W. [Akzo] Benzoyl peroxide with dicyclohexyl phthalate; see Cadox BFF-50; gran.; 50% conc.

Cadox BP-55. [Akzo] Benzoyl peroxide with phlegmatizer; initiator for elevated temp. curing of unsat. polyester resins in applic. for matched metal die molding, pultrusion, vacuum bag molding, continuous laminating, hot-cure casting, inj. molding; ambient-temp. polyester cures; disperses rapidly in polyester formulations; paste; 55% conc.

Cadox BS. [Akzo] Benzoyl peroxide; cross-linking agent used for curing silicone rubbers; paste; 50% act. with silicone fluid.

Cadox BT-50. [Akzo] Benzoyl peroxide; initiator for ambient-temp. polyester cures; esp. for automotive body putty pastes; paste; 50% act. in plasticizer.

Cadox BTA. [Akzo] Benzoyl peroxide; initiator for ambient-temp. polyester cures; powd.; 35% act. on calcium sulfate.

Cadox F-85. [Akzo] Ketone peroxide; initiator for ambient-temp. polyester cures; liq.; 8.5% act. oxygen.

Cadox L-50. [Akzo] MEK peroxide; initiator for ambient-temp. polyester cures; liq.; 9% act. oxygen.

Cadox M-30. [Akzo] MEK peroxide in DMP/DAP; initiator for ambient-temp. polyester cures; used for spray-up applics.; liq.; 5.3% act. oxygen.

Cadox M-50. [Akzo] MEK peroxide in DMP; initiator for ambient-temp. polyester cures; excellent batch-to-batch consistency; versatile ambient-temp. curing catalyst; liq.; 9.0% act. oxygen.

Cadox M-105. [Akzo] MEK peroxide; see Cadox M-50; liq.

Cadox MDA-30. [Akzo] MEK peroxide in DAP; initiator for ambient-temp. polyester cures; low visc.; used for airless spray-gun applic.; liq.; 5.3% act. oxygen.

Cadox OS. [Akzo] Bis(o-chlorobenzoyl)-peroxide; see Cadox BS; paste; 50% act. BPO with silicone fluid.

Cadox PS. [Akzo] Bis(p-chlorobenzoyl)-peroxide; see Cadox BS; paste; 50% act. with silicoine fluid.

Cadox TDP. [Akzo] 2,4-Dichlorobenzoyl peroxide in plasticizer; highly reactive initiator used with amine-accelerated polyester resins; paste; 50% act.

Cadox TS-50. [Akzo] Bis(2,4-dichlorobenzoyl) peroxide; see Cadox BS; paste; 50% act.

CAE. [Ajinomoto] N^2 cocoyl-L-arginine ethyl ester DL-pyrrolidone carboxylic acid salt; cationic; surfactant, foamer, and antistat; preservative for cosmetics; powd.; 100% conc.

Calamide C. [Pilot] Cocamide DEA; nonionic; emollient, foam stabilizer, visc. builder; liq. detergents and personal care prods.; emulsifier for degreasing; lubricant, rust preventive for metal cleaning and cooling systems; detergent and solubilizer for washing floors, walls, metals; amber clear liq.; water-sol.; pH 9.5; 100% conc.

Calamide CW-100. [Pilot] Cocamide DEA and diethanolamine; nonionic; emollient, lubricant, foam stabilizer, visc. builder, detergent, solubilizer used in bubble baths, shampoos, liq. detergents; heavy-duty cleaners; emulsion systems; yel. clear liq.; water-sol.; dens. 8.5 lb/gal; pH 10.5; 100% conc.

Calamide O. [Pilot] Oleamide DEA; nonionic; emollient, lubricant, foam stabilizer, visc. builder, emulsifier; used in personal care prods., industrial applics.; amber clear liq.; disp. water and oil; pH 10; 100% conc.

Calcium Petronate. [Witco] Calcium petrol. sulfonate; anionic; detergent and rust inhibitor in lube oil additives; dk. br. visc. liq.; sol. in org. solv.; sp.gr. 0.98; 44% act.

Calcium Stearate, Regular. [Witco] Calcium stearate; lubricant for ceramics, coated foundry sands, plastics; plasticizer/lubricant for paper coatings; aux. emulsifier; water repellent for cosmetics; flatting, pigment suspending, and thickener for solv.-based paints and varnishes; wh. powd.; sp.gr. 1.03.

Calendula Oil CLR. [Henkel] Soybean oil, calendula extract, tocopherol; emollient, conditioner; emulsified and oily preparation for skin care; rdsh.-yel. oil; herbal odor.

Calfax 10L-45. [Pilot] Sodium linear decyl diphenyl oxide disulfonate; surfactant, solubilizer, dispersant for dye bath leveling, pigment dispersion, heavy-duty cleaners, latex emulsification, agric. chemicals; lt. amber aq. sol'n.; sol. in water, org. solvs.; pH 8.5; biodeg.; 45% act.

Calfoam ES-30. [Pilot] Sodium laureth sulfate; anionic; foam stabilizer, flash foamer, wetter for detergent systems, personal care prods.; emulsion polymerization; yel. clear liq.; mild odor; dens. 8.8 lb/gal; pH 8.0; 30% solids.

Calfoam NEL-60. [Pilot] Ammonium alcohol ether sulfate; anionic; flash foamer, foam stabilizer, detergent, emulsifier, wetter for liq. detergents, bubble baths, shampoos, car washing; lime soap dispersant; clear liq.; faint alcohol odor; sol. in aq. systems; dens. 8.58 lb/gal; pH 7.5; 57.5% act.

Califlux® 90. [Witco] Aromatic oil; extender oil offering low aniline pts.; for elastomer compounding, esp. nitrile, neoprene, SBR, and natural rubber; m.w. 325; sp.gr. 0.9840 (15.6 C); dens. 8.2 lb/gal; visc. 519 cSt (40 C); flash pt. (COC) 208 C; pour pt. –6 C; ref. index 1.5534.

Califlux® 510. [Witco] Aromatic oil; see Califlux 90; m.w. 275; sp.gr. 0.9895 (15.6 C); dens. 8.2 lb/gal; visc. 138 cSt (40 C); flash pt. (COC) 168 C; pour pt. –3 C; ref. index 1.5516.

Califlux® GP. [Witco] Aromatic oil; see Califlux 90; m.w. 311; sp.gr. 1.0035 (15.6 C); dens. 8.4 lb/gal; visc. 1050 cSt (40 C); flash pt. (COC) 212 C; pour pt. 6 C; ref. index 1.5707.

Califlux® LP. [Witco] Aromatic oil; see Califlux 90; m.w. 246; sp.gr. 0.9826 (15.6 C); dens. 8.2 lb/gal; visc. 36 cSt (40 C); flash pt. (COC) 160 C; pour pt. –24 C; ref. index 1.5503.

Califlux® LV. [Witco] Aromatic oil; see Califlux 90; m.w. 300; sp.gr. 1.000 (15.6 C); dens. 8.3 lb/gal; visc. 500 cSt (40 C); flash pt. (COC) 198 C; pour pt. 0 C; ref. index 1.5651.

Califlux® SP. [Witco] Aromatic oil; see Califlux 90; m.w. 318; sp.gr. 1.0107 (15.6 C); dens. 8.4 lb/gal; visc. 2750 cSt (40 C); flash pt. (COC) 216 C; pour pt. 12 C; ref. index 1.5750.

Califlux® TT. [Witco] Aromatic oil; see Califlux 90; m.w. 325; sp.gr. 1.0180 (15.6 C); dens. 8.5 lb/gal; visc. 6000 cSt (40 C); flash pt. (COC) 216 C; pour pt. 18 C; ref. index 1.5829.

Calimulse PRS. [Pilot] Isopropylamine dodecylbenzene sulfonate; anionic; emulsifier, solubilizer; dry cleaning; degreasers; latex emulsifier; pigment dispersant; agric. sprays, oil slick emulsifiers; clear amber liq.; dens. 8.5 lb/gal; pH 4.8; 90% amine sulfonate.

Calnox Polymers. [Arjay] Polyacrylic polymers; scale inhibitor conc. bases for inhibiting calcium carbonate, calcium sulfate, barium sulfate and strontium sulfate scale; liq.; water-sol.

Calsoft L-40. [Pilot] Sodium dodecylbenzene sulfonate; anionic; emulsion stabilizer; wetting and foaming agent, detergent, emulsifier for household and industrial detergents, agric. emulsions, dye bath leveling, rug cleaners, bubble baths, ore flotation, and air entrainment in concrete and gypsum board; washing fruits and vegetables; lt. yel. visc. liq.; water-sol.; sp.gr. 1.07; pH 7.5; 42% solids.

Calsoft L-60. [Pilot] Sodium dodecylbenzene sulfonate; anionic; see nCalsoft L-40; water-wh. pasty liq.; odorless; dens. 8.7 lb/gal; pH 7.4; 60% solids.

Calsoft LAS-99. [Pilot] Dodecylbenzene sulfonic acid, linear; detergent, emulsifier, intermediate for liq. and dry detergents; Klett 50 syrupy liq.; water-sol.; sp.gr. 1.06; dens. 8.83 lb/gal; visc. 1100 cps; 97.5% act.

Calsuds CD-6. [Pilot] Cocamide DEA; nonionic/anionic; foamer, foam builder, wetting agent, visc. modifier, and emulsifier; used in liq. detergents, shampoos, wool-washing compds., hand, felt, and janitorial cleaners, textile scours, agric. sprays; clear visc. liq.; dens. 8.6 lb/gal; visc. 2100 cps; pH 9.5 (10%); 100% conc.

Calwhite®. [Georgia Marble] Ground calcium carbonate; general-use filler; 7.8 μ median particle size; 0.008% retained on #325 wet screen; 5 Hegman grind; dry brightness 94%; pH 9.0–9.5.

Calwhite® II. [Georgia Marble] Calcium carbonate; economical extender filler for liq. polyester systems, BMC-SMC, and similar applic. using polyester resins; particulate; particle size 1.9–2.2 SSD; oil absorp. 12–14%.

Camel-CAL®. [Genstar] Calcium carbonate; filler for water-based coatings and inks, paper, and PVC pipe; 0.7 μ avg. particle dia.; 100% finer than 7 μ; 65% finer than 1 μ; 0.14% sol. in water; sp.gr. 2.70–2.71; dens. 22.57 lb/gal solid; bulk dens. 35 lb/ft³ (loose); oil absorp. 28 cc/100 g; ref. index 1.6; pH 9.5 (sat. sol'n.); hardness (Moh) 3.0.

Camel-CAL® Slurry. [Genstar] Calcium carbonate slurry; see Camel-CAL; 0.7 μ avg. particle dia.; 100% finer than 7 μ; 90% finer than 2 μ; 0.14% sol. in water; sp.gr. 2.70–2.71; visc. 90–150; dens. 15.87 lb/gal; ref. index 1.6; pH 9.5 (sat. sol'n.); hardness (Moh) 3.0; 75 ± 0.5% solids.

Camel-CAL® ST. [Genstar] Stearate-coated calcium carbonate; see Camel-CAL; 0.7 μ avg. particle dia.; 100% finer than 7 μ; 65% finer than 1 μ; 0.14% sol. in water; sp.gr. 2.70–2.71; dens. 22.57 lb/gal solid; bulk dens. 25 lb/ft³ (loose); oil absorp. 28 cc/100 g; ref. index 1.6; pH 9.5 (sat. sol'n.); hardness (Moh) 3.0.

Camel-CARB®. [Genstar] Calcium carbonate; filler/extender used in interior flat paint and exterior house paints, rubber compds., putty and caulk, ceramics, adhesives, linoleum, floor tile, and textile coatings; 7.0 μ avg. particle dia.; 99.5% finer than 44 μ; 50.0% finer than 7 μ; 0.03% sol. in water; sp.gr. 2.70–2.71; dens. 22.57 lb/gal solid; bulk dens. 58 lb/ft³ (loose); oil absorp. 13 cc/100 g; ref. index 1.6; pH 9.5 (sat. sol'n.); hardness (Moh) 3.0.

Camel-FIL. [Genstar] Calcium carbonate; filler designed for high filler loading in glass-reinforced polyester; also for PVC (rigid and flexible), PP, rubber automotive goods and floor tiles, caulks, sealants, and adhesives; 5.5–6.0 μ avg. particle dia.; 97% finer than 25 μ; 46% finer than 6 μ; 0.06% sol. in water; sp.gr. 2.70–2.71; dens. 22.57 lb/gal solid; bulk dens. 55 lb/ft³ (loose); oil absorp. 17 cc/100 g; ref. index 1.6; pH 9.5 (sat. sol'n.); hardness (Moh) 3.0.

Camel-TEX®. [Genstar] Calcium carbonate; fine-ground, general-purpose filler used in interior flat paints, primers, and sealers, polyester fiberglass premixes, preforms, and hand lay-up gel coats, rubber automotive prods., household prods., tubing, medical prods., and closures, putty, caulk, bath tub sealers, body deadeners, and adhesives; 5.0 μ avg. particle dia.; 99.7% finer than 25 μ; 4–5 Hegman; 0.04% sol. in water; sp.gr. 2.70–2.71; dens. 22.57 lb/gal solid; bulk dens. 50 lb/ft³ (loose); oil absorp. 14 cc/100 g; ref. index 1.6; pH 9.5 (sat. sol'n.); hardness (Moh) 3.0.

Camel-WITE®. [Genstar] Calcium carbonate; filler for paint, paper, paper coating, PVC, rubber (automotive goods, footwear, medical supplies), thermoplastics, thermosets, and in caulks, glazing compds., ceramics, adhesives, food processing; very wh. dry powd.; 3.0 μ avg. particle dia.; 99.9% finer than 12 μ; 50% finer than 3 μ; 0.08% sol. in water; sp.gr. 2.70–2.71; dens. 22.57 lb/gal solid; bulk dens. 40 lb/ft³ (loose); oil absorp. 15 cc/100 g; ref. index 1.6; pH 9.5 (sat. sol'n.); hardness (Moh) 3.0.

Camel-WITE® Slurry. [Genstar] Calcium carbonate aq. slurry; see Camel-WITE; very wh. slurry; 3.0 μ avg. particle dia.; 99.9% finer than 12 μ; 50% finer than 3 μ; 0.08% sol. in water; sp.gr. 2.70–2.71; dens. 15.3 lb/gal; bulk dens. 40 lb/ft³ (loose); Visc. 100–200; ref. index 1.6; pH 9.5 (sat. sol'n.); hardness (Moh) 3.0; 72.0 ± 0.5% solids.

Camel-WITE® ST. [Genstar] Surface-treated calcium carbonate; see Camel-WITE; very wh. dry powd.; 3.0 μ avg. particle dia.; 99% finer than 12 μ; 50% finer than 3 μ; 0.08% sol. in water; sp.gr. 2.70–2.71; dens. 22.57 lb/gal solid; bulk dens. 35 lb/ft³ (loose); oil absorp. 15 cc/100 g; ref. index 1.6; pH 9.5 (sat. sol'n.); hardness (Moh) 3.0.

CAO®-1. [PMC] 2,6-Ditert.-butyl-p-cresol (BHT); antioxidant for rubber elastomers and polymerics; additive for animal, veg., and min. oil, fats, and greases; stabilizer for petrol. prods., waxes, insecticides.

CAO®-3. [PMC] 2,6-Ditert.-butyl-p-cresol (BHT); see CAO-1; also used in food industry.

CAO®-3 Blend 29. [PMC] 2,6-Ditert.-butyl-p-cresol (BHT); antioxidant; additive in direct animal feeds.

CAO®-5. [PMC] 2,2′-Methylene-bis (4-methyl-6-tert.-butyl-phenol); antioxidant used in adhesives and rubber.

CAO®-14. [PMC] 2,2′-Methylene-bis (4-methyl-6-tert.-butyl-phenol); see CAO-5.

CAO®-41. [PMC] Hindered phenolic; antioxidant for non-food and non-aq. prods.

CAO®-42. [PMC] Hindered phenolic; antioxidant/ stabilizer for petrol. prods.; liq.; 100% act.
Capital® 5330. [Capital City] Hydrog. cottonseed oil; Lovibond R2.5 max.; m.p. 142–146 F; iodine no. 5 max.; sapon. no. 189–198.
Capital® 5370. [Capital City] Hydrog. soya flakes; Lovibond R2.0 max.; m.p. 153–157 F; iodine no. 5 max.; sapon. no. 189–197.
Capital® 5380. [Capital City] Flaked stearine; Lovibond R4.0 max.; m.p. 138–141 F; iodine no. 5 max.; sapon. no. 196–202.
Caplube 8385. [Capital City] Castor oil, ethoxylated; nonionic; emulsifier, dispersant, lubricant used in textile compding.; liq.; HLB 14.5; 100% conc.
Caplube 8410. [Capital City] Triglycerol oleate; nonionic; emulsifier, lubricant for textiles; liq.; HLB 8.8; 100% conc.
Caplube 8440. [Capital City] Decaglycerol tetraoleate; nonionic; emulsifier, antistat, lubricant, and coupler for textile finishes; liq. HLB 6.7; 100% conc.
Caplube 8442. [Capital City] Decaglycerol decaoleate; nonionic; emulsifier, lubricant; HLB 3.2; 100% conc.
Caplube 8445. [Capital City] Decaglycerol tetracocate; nonionic; emulsifier, lubricant, antistat for textile applic.; liq.; HLB 8.8; 100% conc.
Caplube 8448. [Capital City] Decaglycerol decastearate; nonionic; lubricant for thread finish compding.; opacifier; solid; HLB 3.5; 100% conc.
Caplube 8508. [Capital City] Coconut ethoxylate; nonionic; emulsifier, lubricant for textile applics.; liq.; 100% conc.
Capmul® GMO. [Capital City] Glyceryl oleate; nonionic; food emulsifier, wetting control agent; dispersant for pigments, solids; defoamer; liq.; sol. in org. solvs. and oils; HLB 3.4; 100% conc.
Capmul® GMS. [Capital City] Glyceryl stearate; nonionic; stabilizer; internal lubricant for cosmetics; food emulsifier used in margarine, yeast-raised baked goods; beads, flakes; HLB 3.2; 100% conc.
Capmul® MCM. [Capital City] Glyceryl caprate/ caprylate; co-solv. and coupler for org. compounds; w/o emulsifier; sol. in oil and alcohol.
Capmul® O. [Capital City] Sorbitan oleate; nonionic; food emulsifier and dispersant; plastic, solid; HLB 4.3; 100% conc.
Capmul® POE-L. [Capital City] Polysorbate 20; food emulsifier and solubilizer for flavors; liq.; HLB 16.7; 100% conc.
Capmul® POE-O. [Capital City] Polysorbate 80; nonionic; food emulsifier for frozen desserts; solubilizer for oils into water systems; used for flavors, fragrances, vitamins, pharmaceuticals, solv.; liq.; water-sol.; HLB 15.0; 100% conc.
Capmul® POE-S. [Capital City] Polysorbate 60; nonionic; food emulsifier; solubilizer for oils into water systems; liq.; water-sol.; HLB 14.9; 100% conc.
Caprol® 2G4S. [Capital City] Polyglyceryl-2 tetrastearate; nonionic; food emulsifier; opacifier; wax modifier; thickener; solid; sol. in oils, waxes; HLB 2.5; 100% conc.
Caprol® 3GO. [Capital City] Polyglyceryl-3 oleate; nonionic; food emulsifier for frozen desserts, veg. dairy prods., diet spreads; wetting agent for dyes and pigments in cosmetics; defoamer; liq.; sol. in org. solvs. and oils; HLB 6.2; 100% conc.
Caprol® 3GS. [Capital City] Polyglyceryl-3 stearate; nonionic; food emulsifier, stabilizer and whipping agent used in frozen desserts and fat reduction; solid; HLB 6.2; 100% conc.
Caprol® 10G2O. [Capital City] Polyglyceryl-10 dioleate; o/w emulsifier, humectant, lubricant; sol. in alcohol; water-disp.
Caprol® 10G4O. [Capital City] Polyglyceryl-10 tetraoleate; nonionic; food emulsifier, visc. control, stabilizer; liq.; HLB 6.2; 100% conc.
Caprol® 10G10O. [Capital City] Polyglyceryl-10 decaoleate; nonionic; food emulsifier, solubilizer, lubricant, and dispersant; liq.; sol. in oils and org. solvs.; HLB 2.5; 100% conc.
Caprol® 10G10S. [Capital City] Polyglyceryl-10 decastearate; lubricant for thread finishes, wax additive, crystal modifier; thickener; opacifier; sol. in oils, org. solvs., and waxes.
Capstat. [Capital City] Alkyl ammonium ethyl sulfate quat.; antistat for fabrics; water-disp.
Captax®. [Vanderbilt] 2-Mercaptobenzothiazole; thiazole accelerator; primary accelerator for natural and syn. rubbers; nonstaining and nondiscoloring; lt. yel. to buff powd., 99.8% thru 100 mesh; m.w. 167.25; dens. 1.50 ± 0.03 mg/m³; sol. in dil. caustic, toluene, carbon disulfide, chloroform, alcohol; m.p. 164–175 C.
Captax®-Tuads Blend. [Vanderbilt] 2-Mercaptobenzothiazole, tetramethylthiuram disulfide, ratio 1:2; thiuram ultra accelerator for good scorch and a moderate cure rate in butyl rubber (IIR); cream to yel. tan powd., 99.9% thru 100 mesh; rods; sol. in toluene, chloroform, carbon disulfide; practically insol. in water; dens. 1.42 ± 0.03 mg/m³; m.p. 98–126 C (powd.); 96 C min. (rods).
Captex® 200. [Capital City] Propylene glycol dicaprylate/dicaprate; carrier, coupler, solv. for flavors, fragrance oil, sol. colorants, vitamins, medicinals, cosmetics; emollient for creams, lotions, makeup; sol. in alcohol, oils, hydrocarbons, ketones.
Captex® 300. [Capital City] Caprylic/capric triglyceride; solv. for colors and perfumes; emollient, moisturizer in cosmetics, toiletries, pharmaceuticals; plasticizer; sol. see Captex 200.
Captex® 350. [Capital City] Caprylic/capric/lauric triglyceride; emollient, solv., carrier, fixative, and extender for pharmaceutical, nutritional, and cosmetic applics.; Lovibond R1.0 max.; visc. 33 cps; cloud pt. 0 C max.; acid no. 0.1; iodine no. 1.5.
Captex® 355. [Capital City] Caprylic/capric triglyceride; lubricity vehicle for cosmetics and pharmaceuticals; carrier for essential oils; Lovibond R1.0 max. clear liq.; neutral odor; bland flavor; misc. with most org. solvs. incl. 95% ethanol; sp.gr. 0.92–0.96; visc. 20–25 cps; cloud pt. –2 C or less; acid no. 0.1 max.; iodine no. 0.5 max.; sapon. no. 330–350.
Captex® 800. [Capital City] Propylene glycol dioc-

tanoate; nonoily lubricant imparting rich feel to skin in cosmetics and pharmaceuticals; carrier for essential oils, flavors; vehicle for vitamins, medicinals, nutritional prods.; APHA 100 max. clear liq.; neutral odor; bland flavor; misc. with most org. solvs. incl. 95% ethanol; visc. 5–13 cps; cloud pt. –20 C or less; acid no. 1.0 max.; iodine no. 1.0 max.

Captex® 810B. [Capital City] Coconut fatty acid ester; emollient, solv., carrier, fixative, and extender in pharmaceuticals, nutritional, and cosmetic applics.; Lovibond R1.0 max.; visc. 30 cps; cloud pt. 10 C; acid no. 0.1 max.; iodine no. 51.0 max.

Captex® 8000. [Capital City] Caprylic triglyceride; nonoily lubricant imparting rich feel to the skin; for cosmetics and pharmaceuticals; carrier for essential oils, flavors; vehicle for vitamins, medicinals, nutritional prods.; APHA 150 max. clear liq.; neutral odor; bland flavor; misc. with most org. solvs. incl. 95% ethanol; visc. 20 cps; acid no. 1.0 max.; iodine no. 1.0 max.; sapon. no. 345–360.

Carbonox. [Baroid] Altered lignite; anionic; thinner, emulsifier, and dispersant for oil well drilling muds; powd.; 100% conc.

Carbopol® 615, 616, 617. [Goodrich] High m.w. polyacrylic acid, crosslinked with polyalkenyl polyether; anionic; emulsifier, thickener, coupler, and insol. particle suspending agent used in liq. detergents; prevents soil redeposition; wh. powd.; slightly hygroscopic water sol./disp.; odorless.

Carbopol® 907. [Goodrich] Polyacrylic acid; anionic; emulsifier, thickener, stabilizer, suspending agent; used for drilling muds, photosensitive emulsions, water tratment; solid; sol. in water, polar solvs., many nonpolar solvs. blends; 100% conc.

Carbopol® 910. [Goodrich] Carbomer 910; anionic; emulsifier, thickener, stabilizer, suspending agent; used for flocking, dip coting, textile back coating; powd.; sol. see Carbopol 907; 100% conc.

Carbopol® 934. [Goodrich] Carbomer 934; anionic; emulsifier, thickener, stabilizer, suspending agent; for lubricating, quenching and silicone emulsions; graphite, polyethylene, fiber and paper suspensions; sol. see Carbopol 907; 100% conc.

Carbopol® 934P. [Goodrich] Carbomer 934P; suspending agent; high purity grade for pharmaceutical industry; for thickening, suspending, and emulsifying; for topical lotions and sustained release tablets; sol. see Carbopol 907.

Carbopol® 940. [Goodrich] Carbomer 940; anionic; emulsifier, thickener, stabilizer, suspending agent used in cosmetic applics., for die-casting and forging lubricants, thixotropic paints; solv. thickening with or without neutralizing; powd.; sol. see Carbopol 907; 100% conc.

Carbopol® 941. [Goodrich] Carbomer 941; anionic; emulsifier, thickener; emulsion stabilization of shampoos, lotions, and thin gels with good clarity; powd.; sol. see Carbopol 907; 100% conc.

Carbopol® 1342. [Goodrich] Crosslinked polyacrylic acid copolymerized with a hydrophobic comonomer; emulsifier, stabilizer, moisturizer, thickener, gellant for cosmetic emulsions, skin care prods.; wh. powd.; slightly acetic odor; visc. 10,500–26,900 cP (1%).

Carbopol® 1706. [Goodrich] High m.w. polyacrylic acids, crosslinked with polyalkenyl polyether; anionic; emollient; additive to toilet bar soap prods. for plasticity, bar texture, bar feel, moisture/perfume retention, pigment disp., skin-mildness, lather volume and lather creaminess; prevents cracking of soap bar during use; wh. powd.; slightly hygroscopic water sol./disp.; odorless.

Carbopol® 1720. [Goodrich] High m.w. polyacrylic acids, crosslinked with polyalkenyl polyether; anionic; see Carbopol 1706; wh. powd.; sol. see Carbopol 1706; odorless.

Carbopol® 1731. [Goodrich] High m.w. polyacrylic acids, crosslinked with polyalkenyl polyether; anionic; see Carbopol 1706; wh. powd.; sol. see Carbopol 1706; odorless.

Carbopol® 1754. [Goodrich] High m.w. polyacrylic acids, crosslinked with polyalkenyl polyether; anionic; see Carbopol 1706; wh. powd.; sol. see Carbopol 1706; odorless.

Carbowax® Compound 20M. [Union Carbide] PEG-350; binder, lubricant for ceramics, powd. metallurgy, toilet bowl cleaners; flake; m.w. 15,000–20,000; sol. 65% in water; sp.gr. 1.0540 (120 C); dens. 8.67 lb/gal (130 C); visc. 18,650 cSt (210 F); m.p. 61–64 C; flash pt. (CCC) > 350 F; surf. tens. 52.0 dynes/cm (50% aq.); pH 6.5–8.0

Carbowax® MPEG 350. [Union Carbide] PEG-6 methyl ether; intermediate, lubricant for adhesives, inks, mining, soaps and detergents; liq.; m.w. 335–365; sol. in water; sp.gr. 1.0891; dens. 9.13 lb/gal; visc. 3.9 cSt (210 F); f.p. –5 to 10 C; surf. tens. 40 dynes/cm; ref. index 1.455; pH 4.0–7.5 (5% aq.)

Carbowax® MPEG 550. [Union Carbide] PEG-10 methyl ether; see Carbowax MPEG 350; liq.; m.w. 525–575; sol. in water; sp.gr. 1.1039; dens. 8.97 lb/gal (55 C); visc. 6.6 cSt (210 F); f.p. 15–25 C; flash pt. (CCC) 360 F; surf. tens. 37.5 dynes/cm (40 C); ref. index 1.455 (40 C); flash pt. (CCC) > 350 F; pH 4.0–7.5 (5% aq.).

Carbowax® MPEG 750. [Union Carbide] PEG-16 methyl ether; chemical intermediate, lubricant for adhesives; soft solid; m.w. 715–785; sol. in water; sp.gr. 1.0760 (60 C); dens. 9.02 lb/gal (55 C); visc. 10.3 cSt (210 F); f.p. 27–32 C; surf. tens. 40.7 dynes/cm (40 C); ref. index 1.459 (40 C); flash pt. (CCC) 415 F; pH 4.0–7.5 (5% aq.).

Carbowax® MPEG 2000. [Union Carbide] PEG-40 methyl ether; chemical intermediate, lubricant for adhesives, toilet bowl cleaners; flake; m.w. 1900–2100; sol. 68% in water; sp.gr. 1.0871 (60 C); dens. 9.18 lb/gal (55 C); visc. 45.5 cSt (210 F); m.p. 49–54 C; flash pt. (Seta CC) 355 F; pH 6.0–8.0 (5% aq.).

Carbowax® MPEG 5000. [Union Carbide] PEG-100 methyl ether; see Carbowax MPEG 2000; flake; m.w. 4750–5250; sol. 64% in water; sp.gr. 1.0907 (60 C); dens. 8.95 lb/gal (80 C); visc. 320 cSt (210 F); m.p. 57–63 C; flash pt. (CCC) 415 F; pH 6.0–8.0 (5% aq.)

Carbowax® PEG 200. [Union Carbide] PEG-4; antistat, intermediate, humectant, lubricant, release agent, plasticizer for adhesives, inks, lubricants; liq.; m.w. 190–210; sol. in water; sp.gr. 1.1239; dens. 9.38 lb/gal; visc. 4.3 cSt (210 F); flash pt. (CCC) > 300 F; surf. tens. 44.5 dynes/cm; ref. index 1.459; pH 4.5–7.5 (5% aq.)

Carbowax® PEG 300. [Union Carbide] PEG-6; antistat, intermediate, dye carrier, humectant, lubricant, release agent, plasticizer for adhesives, capsules, creams and lotions, deodorant sticks, inks, lipsticks; also as glycerin replacement; liq.; m.w. 285–315; sol. in water; sp.gr. 1.1250; dens. 9.38 lb/gal; visc. 5.8 cSt (210 F); f.p. –15 to –8 C; flash pt. (CCC) > 350 F; surf. tens. 44.5 dynes/cm; ref. index 1.463; pH 4.5–7.5 (5% aq.).

Carbowax® PEG 400. [Union Carbide] PEG-8; antistat, intermediate, dye carrier, humectant, lubricant, release agent, plasticizer for adhesives, capsules, ceramic glazes, creams and lotions, deodorant sticksw, inks, lipsticks; liq.; m.w. 380–420; sol. in water, methanol, ethanol, acetone, trichloroethylene, Cellosolve®, Carbitol®, dibutyl phthalate, toluene; sp.gr. 1.1254; dens. 9.39 lb/gal; visc. 7.3 cSt (210 F); f.p. 4–8 C; flash pt. (CCC) > 350 F; surf. tens. 44.5 dynes/cm; ref. index 1.465; pH 4.5–7.5 (5% aq.).

Carbowax® PEG 540 Blend. [Union Carbide] PEG-6 and PEG-32 (41:59); base for ointments suppositories; also for adhesives, creams and lotions, deodorant sticks; soft solid; m.w. 500–600; sol. in methylene chloride, 73% in water, 50% in trichloroethylene, 48% in methanol; sp.gr. 1.0930 (60 C); dens. 9.17 lb/gal (55 C); visc. 15.1 cSt (210 F); f.p. 38–41; flash pt. (CCC) > 350 F; pH 4.5–7.5 (5% aq.).

Carbowax® PEG 600. [Union Carbide] PEG-12; antistat, intermediate, humectant, lubricant, release agent, plasticizer for adhesives, capsules, ceramic glaze, creams and lotions, dentifrices, deodorant sticks, inks, lipsticks, wood treatment; liq.; m.w. 570–630; sol. in water; sp.gr. 1.1257; dens. 9.40 lb/gal; visc. 10.8 cSt (210 F); f.p. 20–25 C; flash pt. (CCC) > 350 F; surf. tens. 44.5 dynes/cm; ref. index 1.46; pH 4.5–7.5 (5% aq.).

Carbowax® PEG 900. [Union Carbide] PEG-20; antistat, intermediate, lubricant, release agent, ointment and suppository base, plasticizer for adhesives, ceramic glaze, creams and lotions, dentifrices, deodorant sticks, wood treatment; soft solid; m.w. 855-900; sol. 86% in water; sp.gr. 1.0927 (60 C); dens. 9.16 lb/gal (55 C); visc. 15.3 cSt (210 F); f.p. 32–36 ; flash pt. (CCC) > 350 F; pH 4.5–7.5 (5% aq.).

Carbowax® PEG 1000. [Union Carbide] PEG-20; see Carbowax PEG 900; soft solid; m.w. 950–1050; sol. 80% in water; sp.gr. 1.0926 (60 C); dens. 9.16 lb/gal (55 C); visc. 17.2 cSt (210 F); f.p. 37–40 ; flash pt. (CCC) > 350 F; pH 4.5–7.5 (5% aq.).

Carbowax® PEG 1450. [Union Carbide] PEG-32; antistat, intermediate, lubricant, release agent for adhesives, ceramic glaze, creams and lotions, dentifrices, deodorant sticks, wood treatment; soft solid or flake; m.w. 1300–1600; sol. 72% in water; sp.gr. 1.0919 (60 C); dens. 9.17 lb/gal (55 C); visc. 26.5 cSt (210 F); f.p. 43–46 ; flash pt. (CCC) > 350 F; pH 4.5–7.5 (5% aq.);

Carbowax® PEG 3350. [Union Carbide] PEG-75; antistat, intermediate, dye carrier, lubricant, release agent, tablet binder for adhesives, ceramic glaze, creams and lotions, dentifrices, mining, soaps and detergents, tablet coating, toilet bowl cleaners; flake or powd.; m.w. 3000–3700; sol. 67% in water; sp.gr. 1.0926 (60 C); dens. 8.94 lb/gal (80 C); visc. 90.8 cSt (210 F); f.p. 54–58 ; flash pt. (CCC) > 350 F; pH 4.5–7.5 (5% aq.).

Carbowax® PEG 4600. [Union Carbide] PEG-100; antistat, intermediate, dye carrier, lubricant, release agent, tablet binder for adhesives, ceramic glaze, creams and lotions, mining, soaps and detergents, tablet coating, toilet bowl cleaners; flake or powd.; m.w. 4400–4800; sol. 65% in water; sp.gr. 1.0926 (60 C); dens. 8.95 lb/gal (80 C); visc. 184 cSt (210 F); f.p. 57–61 ; flash pt. (CCC) > 350 F; pH 4.5–7.5 (5% aq.).

Carbowax® PEG 8000. [Union Carbide] PEG-150; antistat, ceramic binder, intermediate, dye carrier, lubricant, release agent, tablet binder for adhesives, creams and lotions, mining, powd. metallurgy, soaps and detergents, tablet coating, toilet bowl cleaners; flake or powd.; m.w. 7000–9000; sol. 63% in water; sp.gr. 1.0845 (70 C); dens. 8.96 lb/gal (80 C); visc. 822 cSt (210 F); m.p. 60–63 ; flash pt. (CCC) > 350 F; pH 4.5–7.5 (5% aq.)

Carbowax® Sentry. [Union Carbide] PEGs 200, 300, 400, 540, 600, and 1000; emollient base for creams, lotions, pharmaceuticals and makeup; carrier for medicinals in cosmetics and pharmaceuticals; numerical value approximates m.w.; water-sol.

Cardipol® LP. [Petrolite] Oxidized polyethylene wax for formulating hot-melt adhesives; color 1.0 (D1500) wax; m.p. 115.5 C; acid no. 4; sapon. no. 9.

Cardipol® LP 0-25. [Petrolite] Oxidized polyethylene; see Cardipol LP; color 1.5 (D1500) wax; m.p. 104.4 C; acid no. 32; sapon. no. 72.

Cardis® 314. [Petrolite] Oxidized microcryst. wax; used in the formulation of emulsions, polishes, and coatings; modifier in solv. polish systems; color 2.0 (D1500) wax; m.p. 87.8 C; acid no. 18; sapon. no. 47.

Cardis® 319. [Petrolite] Oxidized microcryst. wax; used in the formulation of emulsions, polishes, and coatings; color 1.5 (D1500) wax; m.p. 93.3 C; acid no. 28; sapon. no. 65.

Cardis® 320. [Petrolite] Oxidized microcryst. wax; used in the formulation of emulsions, polishes, and coatings; color 2.0 (D1500) wax; m.p. 89.4 C; acid no. 35; sapon. no. 76.

Cardolite® NC-507. [Cardolite] 3-(n-Pentadecyl) phenol; starting raw material for surfactants, antioxidants, anticorrosives; lubricant additive; cosolv. for insecticides, germicides; resin modifier; tan waxy solid; m.w. 304; sol. in org. solvs. incl. aliphatic hydrocarbons; sp.gr. > 1.0; m.p. 44–48 C (760 mm Hg); b.p. 243–249 C (10 mm Hg); ref. index 1.494.

Cardolite® NC-510. [Cardolite] 3-(n-Pentadecyl) phenol; starting raw material for surfactants; coupling agent for pigments and dyes; cosolv. for insecticides, germicides; resin modifier for phenolic-aldehyde, polyester, PC polymers; also in photographic industry; slight pink to wh. waxy solid; m.w. 304; sol. in org. solvs. incl. aliphatic hydrocarbons; sp.gr. > 1.0; m.p. 49–51 C (760 mm Hg); b.p. 190–195 C (1 mm Hg); ref. index 1.4750.

Cardolite® NC-511. [Cardolite] 3-(n-Penta-8′-decenyl) phenol; starting raw material for surfactants, antioxidants, anticorrosives; lubricant additive; cosolv. for insecticides, germicides; resin modifier; also in drilling muds; brn. to amber oily liq.; m.w. 302; sol. in org. solvs. incl. aliphatic hydrocarbons; sp.gr. 0.930; b.p. 223–227 C (10 mm Hg); ref. index 1.5112.

Cardolite® NC-512. [Cardolite] 1,8-Bis (hydroxyphenyl) pentadecane; brn. visc. liq.; m.w. ≈ 410; sol. in alcohols, dilute NaOH, aromatic hydrocarbons, ketones; sp.gr. 1.0; visc. 5000 poises.

Cardolite® NC-513. [Cardolite] Reactive flexibilizer, reactive intermediate in polyester, alkyl, and styrene maleic resins; diluent; used in elec./electronic applics.

Cardolite® NC-540, NC-541, NC-546, NC-549, NC-550. [Cardolite] Phenalkamine; curing agents for coatings, castings, adhesives, potting and encapsulation, laminates.

Cardolite® NC-548. [Cardolite] Acetate ester of cardanol; diluent/accelerator for epoxy mfg.; used to prepare adducts, in elec. applics.

Cardolite® NC-552. [Cardolite] Low m.w. furan resin; visc. reducer improving thermal and elec. properties and chem. resistance in epoxy systems.

Cardolite® NC-700. [Cardolite] Cardanol; accelerator for epoxy systems; diluent; flexibilizer for phenolic systems.

Cardolite® NC-1307. [Cardolite] Terpene-based extender/flexibilizer/diluent for epoxy mfg.; also for concrete coatings.

Carnapol 77X. [Reilly-Whiteman] Fatty diamide; nonionic; high m.p. wax for release agent and wax finishes; solid; 100% act.

Carnation. [Witco/Sonneborn] Wh. min. oil NF; emollient and lubricant for cosmetic formulations and baby oil; saybolt visc. 100 F-65/75; pour pt. 20 F.

Carrot Oil CLR. [Henkel] Soybean oil, carrot oil, carrot extract, beta-carotene, tocopherol; nonionic; emollient, conditioner, superfatting agent; emulsified and oily preparations for care of skin and hair; deep red oil; faint char. odor.

Carrybon B. [Sanyo] Polycarboxylic acid type high m.w. surfactant; dispersant for pigments; used for emulsion paint.

Carrybon L-400. [Sanyo] Polycarboxylic acid type high m.w. surfactant; pigment dispersant; liq.

Carryol MD-112. [Sanyo] Fatty acid derivs.; cold flow improver and dispersant for fuel oils; paste.

Carsamide® 7644. [Lonza] Modified cocamide DEA; nonionic/anionic; emulsifier, thickening agent, controlled suds detergent, textile fulling and scouring agent; Gardner 6 liq.; sol. in water, alcohol, chlorinated and aromatic hydrocarbons; disp. in min. oil, kerosene, natural fats and oils, min. spirits; dens. 8.3 lb/gal.

Carsamide® C-3. [Lonza] Cocamide DEA; intermediate for liq. detergent formulation, emulsifier for aromatic solv., improves the foaming, detergency and hard water resistance of soaps.

Carsamide® CA. [Lonza] Cocamide DEA; nonionic; detergent, dispersant, emulsifier, wetting agent for industrial, cosmetic, and household cleaners; sol. in water, alcohol, chlorinated and aromatic hydrocarbons, Polysolve, Cellosolves, Carbitols, and natural fats and oils; sp.gr. 1.0; dens. 8.3 lb/gal; biodeg.; 100% act.

Carsamide® CMEA. [Lonza] Cocamide MEA; nonionic; foam booster, stabilizer, visc. builder for shampoos and detergents; flake; 100% conc.

Carsamide® SAC. [Lonza] Cocamide DEA; nonionic; detergent for industrial, institutional, cosmetic, and household cleaners; dispersant, emulsifier, wetting agent, visc. builder; foam stabilizer for shampoos, bubble baths, detergents; pale amber visc. liq.; mild fatty odor; sol. see Carsamide CA; dens. 8.2 lb/gal; 100% act.

Carsamide® SAL-7. [Lonza] Lauramide DEA; nonionic; detergent, emulsifier, foaming agent, foam stabilizer, thickener for household, institutional and industrial uses; Gardner 2 solid; mild, fatty odor; sol. in alcohol, chlorinated and aromatic hydrocarbons; dens. 8.2 lb/gal; m.p. 104–114 F; biodeg.; 100% conc.

Carsamide® SAL-9. [Lonza] Lauramide DEA; nonionic; see Carsamide SAL–7; solid; dens. 8.1 lb/gal; biodeg.; 100% conc.

Carsamine Oxide LM. [Lonza] Lauryl/myristyl dimethylamine oxide; mildly cationic under acidic conditions; foaming and conditioning agent for personal, household and industrial uses, foam stabilizer, wetting agent in conc. electrolyte sol'ns., Gardner 2 clear liq., 29% act.

Carsemol A-500. [Lonza] Ethyl palmitate, ethyl myristate, ethyl stearate and ethyl pelargonate; used for emollient creams and lotions, cosmetics; sp.gr. 0.865; cloud pt. 11 C.

Carsodet TD-7C. [Lonza] Trideceth-7 carboxylic acid; anionic; acidulant, emollient, surfactant; neutralized with other materials to produce mild anionics, topical cleansing prods., shampoos; colorless liq.; dens. 8.5 lb/gal; pH 1; 90% conc.

Carsofos 4326 [Lonza] Anionic/nonionic blend; detergent, wetting agent, emulsifier, penetrant, for all-purpose cleaners; pale amber liq.; mild, fatty odor; dens. 8.25 lb/gal; sp.gr. 0.989; pH 7.5–8.5; 92% solids.

Carsonam 3. [Lonza] Cocamidopropyl betaine; amphoteric; detergent, wetting and foaming agent, emulsifier, and coupler; used in shampoos; germicides; liq. hand cleaners, dishwashing compds.; Gardner 5 liq.; dens. 8.45 lb/gal; pH 5.5–7.5 (10%); biodeg.; 30% act.; 64% water.

Carsonam 33-S. [Lonza] Cocamidopropyl betaine; amphoteric; see Carsonam 3; Gardner 6 max. liq.; dens. 8.45 lb/gal; pH 5.5–7.5 (10%); 30% act.; 63% water.

Carsonam 3147. [Lonza] Cocamidopropyl betaine;

amphoteric; see Carsonam 3; Gardner 5 max. liq.; dens. 8.45 lb/gal; biodeg.; 33% act.; 64% water.

Carsonam BCW. [Lonza] Coco fatty betaine; foaming and conditioning agent used in household and industrial detergents, shampoos, bubble baths; Gardner 2 liq.; sp.gr. 1.044; dens. 8.5 lb/gal; pH 8.0; biodeg.; 44% solids, 56% water, 8% sodium.

Carsonam C. [Lonza] Coco imidazoline betaine; amphoteric; emulsifier, detergent, sequestrant, mild germistat, and fungistat; used in human, rug, and upholstery shampoos; heavy-duty liq. cleaners, dishwashing compds.; Gardner 5 liq.; pH 9.7; biodeg.; 37% solids.

Carsonam C-SF. [Lonza] Coco imidazoline betaine (salt free); see Carsonam C; Gardner 5 liq.; pH 9.7; 37% solids.

Carsonam C-SPCL. [Lonza] Coco imidazoline betaine; amphoteric; see Carsonam C; Gardner 7 liq.; pH 6.0; biodeg.; 48% solids.

Carsonam DC. [Lonza] Coco imidazoline betaine dicarboxylate; amphoteric; see Carsonam C; Gardner 9 gel; pH 8.5 (5%); biodeg.; 50% solids.

Carsonam DC-70%-SF. [Lonza] Coco imidazoline betaine dicarboxylate (salt free); amphoteric; see Carsonam C; Gardner 11 gel; pH 9.6 (10%); biodeg.; 70% solids.

Carsonam DC-SF. [Lonza] Coco imidazoline betaine dicarboxylate (salt free); amphoteric; see Carsonam C; Gardner 8 gel; pH 9.6 (10%); biodeg.; 38% solids.

Carsonam L. [Lonza] Lauryl imidazoline betaine; amphoteric; see Carsonam C; Gardner 6 liq.; sp.gr. 1.1; pH 9.3; biodeg.; 43% solids.

Carsonol® SHS. [Lonza] Sodium 2-ethylhexyl sulfate; anionic; low foaming detergent, wetting agent, penetrant, emulsifier used in caustic sol'ns. for peeling of fruits and vegetables; Gardner 3 clear liq.; dens. 9.6 lb/gal; sp.gr. 1.15; visc. 35 cps; pH 10.3 (10%); 40% act.

Carsonon® L-985. [Lonza] Laureth-9 (9-10 EO); nonionic; wetting agent, detergent, emulsifier, dispersant used for maintenance and institutional cleaners, in textile, paper, and paint industries; colorless liq.; sol. in acid and alkaline sol'ns.; dens. 8.32 lb/gal; cloud pt. 137 F; pour pt. 39–43 F; pH neutral (3%); biodeg.; 85% act.

Carsonon® N-4. [Lonza] Nonoxynol-4; nonionic; emulsifier, detergent, wetting agent, dispersant for household and industrial uses; intermediate; drycleaning detergent; pale liq.; sol. in kerosene, alcohols, aromatic and chlorinated solv., Stod.; sp.gr. 1.02; dens. 8.6 lb/gal; visc. 175–250 cps; flash pt. 500–600 F; pour pt. -5 ± 2 F; solid. pt. –20 ± 2 F; 100% act.

Carsonon® N-8. [Lonza] Nonoxynol-8 (8–8.5 mole); nonionic; emulsifier, detergent, wetting agent, dispersant for household and industrial uses; pale-colored liq.; mild, pleasant odor; sol. in water, alcohol, aromatic, chlorinated and aliphatic solv.; dens. 8.75 lb/gal; visc. 200–300 cps; cloud pt. 82–86 F; flash pt. 435 F; pour pt. 45 F; pH neutral (1%); 100% act.

Carsonon® N-9. [Lonza] Nonoxynol-9 (9–10 mole); nonionic; emulsifier, detergent, wetting agent, dispersant for household and industrial uses; pale-colored liq.; mild, pleasant odor; sol. see Carsonon N-8; dens. 8.83 lb/gal; visc. 180–250 cps; flash pt. 540 F; pour pt. 35–39 F; pH neutral (1%); 100% act.

Carsonon® N-10. [Lonza] Nonoxynol-10 (10–11 mole); nonionic; emulsifier, detergent, wetting agent, dispersant for household and industrial uses; pale-colored liq.; mild, pleasant odor; sol. see Carsonon N-8; dens. 8.84 lb/gal; visc. 180–250 cps; cloud pt. 143–146 F; flash pt. 540 F; pour pt. 35–39 F; pH neutral (1%); 100% act.

Carsonon® N-30, 70%. [Lonza] Nonoxynol-30; emulsifier, stabilizer; pale yel. liq. to semisolid; sp.gr. 1.09; cloud pt. 167–171 F (1%); pH 6.0–7.0 (3%); 70% act., 30% water.

Carsonon® N-40, 70%. [Lonza] Nonoxynol-40; emulsifier, stabilizer; wh. to yel. semisolid; cloud pt. 212 F (1%); pH 6.0–7.0 (3%); 70% act., 30% water.

Carsonon® N-50, 70%. [Lonza] Nonoxynol-50; emulsifier, stabilizer; wh. to yel semisolid; cloud pt. 212 F (1%); pH 6.0–7.0 (3%); 70% act., 30% water.

Carsonon® N-100, 70%. [Lonza] Nonoxynol-100; emulsifier, stabilizer; wh. to off-wh. semisolid; water sol.; cloud pt. 212 F (1%); pH 6.0–7.0 (3%); 70% act., 30% water.

Carsonon® TD-10. [Lonza] Trideceth-10; nonionic; emulsifier, detergent, wetting agent, foam builder, solubilizer used for household, industrial cleaners, textile processing, paper industry; Gardner 1 liq.; water sol.; dens. 8.5 lb/gal; HLB 13.7; cloud pt. 162–166 F (1% aq.); flash pt. > 300 F (COC); pour pt. 67 F; pH 5.0–7.0 (3% aq.); 100% act.

Carsoquat® 621. [Lonza] Benzalkonium chloride; cationic; corrosion inhibitor, surfactant; almost water-wh. to pale yel. liq.; mild odor; misc. with water, lower alcohols, most polar solv.; sp.gr. 0.96; dens. 8 lb/gal; 50% act.

Carsoquat® 621 (80%). [Lonza] Benzalkonium chloride; cationic; corrosion inhibitor, surfactant; straw-colored clear liq.; m.w. 361; pH 7–8 (1%); 80% act.

Carsoquat® CT 429. [Lonza] Cetyl trimethyl ammonium chloride; surfactant; conditioner in hair and skin care preparations; antistat; water-sol.

Carsoquat® CTM-29. [Lonza] Cetyl trimethyl ammonium chloride; cationic surfactant, conditioners in specialty hair preparations, coagulating agent in the mfg. of antibiotics and other pharmaceuticals; almost clear to pale yel. liq.; m.w. 319; dens. 8.1 lb/gal; sp.gr. 0.968; pH 7.0–9.0 (10% act.); 29% act.

Carsoquat® CTM-429. [Lonza] Cetyl trimethyl ammonium chloride; cationic; see Carsoquat CTM-29; clear to pale yel. liq.; m.w. 319; sp.gr. 0.968; dens. 8.1 lb/gal; pH 3.5–4.0 (2%); 29% act.

Carsoquat® SDQ 25. [Lonza] Stearalkonium chloride; cationic; conditioner, softener, creme hair rinse and conditioner for cosmetics and natural fibers; wh. thick creamy paste; mild, sweet odor; pH 3.0–4.0 (1% disp.); 20–21% act.

Carsoquat® SDQ 85. [Lonza] Stearalkonium chloride; cationic; softener, hair conditioner; wh. waxy flake; mild, sweet odor; water-disp.; m.p. 140 F; pH 5.0–7.0 (0.5%); 85% act.

Carsosoft® S-75. [Lonza] Quaternium-27; cationic; fabric softener base, exc. rewet properties and non-yellowing; Gardner 4 liq.; m.w. 720; pH 5–7 (5%); 75% act. in IPA.

Carsosoft® S-90. [Lonza] Quaternium-27; cationic; see Carsosoft S-75; yel. semisolid; m.w. 723; pH 5–7 (5% aq.); 85–88% act.

Carsosoft® S-90-M. [Lonza] Modified alkylamidoethyl alkyl imidazolinium methyl methosulfate; cationic; see Carsosoft S-75; Gardner 4 liq.; water-disp.; pH 5.0–7.0 (5%); 88–91% act. in IPA.

Carsosoft® T-75. [Lonza] Proprietary difatty quat.; cationic; fabric softener base; substantive quat. for home and industrial laundry softeners; liq.; 70% act.

Carsosoft® T-90. [Lonza] Quat. ditallow diamido; cationic; fabric softener, rewetting agent, antistat, and antilint for home and commercial laundries; yel. liq.; pH 5–7 (5%); 88–90% act. in IPA.

Carsosoft® V-90. [Lonza] Quaternium-18 methosulfate and IPA; wh. (Gardner 4) paste; pH 5.5 (10% in 1:1 IPA/aq.); 91% solids.

Carsosoft® V-100. [Lonza] Quaternium-18 methosulfate; wh. flake or powd.; pH 5.5–6.5 (10% in 1:1 IPA/aq.); 95% solids.

Carsosulf SXS-Liq. [Lonza] Sodium xylene sulfonate; anionic; coupler, solubilizer for liq. detergent systems and cleaning compd.; liq.; sp.gr. 1.18–1.22; pH 7.5–10.5; nonflam.; biodeg.; 40% act.

Carsosulf UL-100 Acid. [Lonza] Linear dodecylbenzene sulfonic acid; m.w. 316; dens. 8.84 lb/gal; sp.gr. 1.06; biodeg.; 97 ± 1% act.

Carstab 700. [Morton Int'l.] 2-Hydroxy-4-n-octoxybenzophenone; uv lt. stabilizer for PP, HDPE, LDPE, EVA, flexible and rigid PVC, PC, acetals, epoxies, some cellulosics; suitable for food pkg. use; off-wh. cryst. powd.; m.w. 326; sol. (g/100 g solv.): 74.3 g acetone, 72.7 g benzene, 180 g ethyl acetate, 65 g MEK; water-insol.; sp.gr. 1.16 g/cc; f.p. 47.4 C; 98.5% pure.

Carstab 701. [Morton Int'l.] 2-Hydroxy-4-isooctoxybenzophenone; see Carstab 700; pale yel. liq.; m.w. 326; sol. (g/100 g solv.): 14 g methanol, 20 g 95% ethanol; misc. with acetone, benzene, hexane, ethyl acetate, MEK, perchloroethylene; water-insol.; sp.gr. 1.066 g/cc; pour pt. –15 C.

Carstab 702. [Morton Int'l.] 2-Hydroxy-4-isooctoxybenzophenone; see Carstab 700; pale yel. liq.; m.w. 326; sol. 20 g/100 g 95% ethanol; misc. with acetone, heptane, MEK, perchloroethylene, toluene; water-insol.; sp.gr. 1.504; pour pt. –20 C.

Carstab 705. [Morton Int'l.] 2-Hydroxy-4-n-octoxybenzophenone; uv lt. stabilizer for PP, PE, EVA, cellulosics, epoxies, in cosmetics, protective coatings; off-wh. cryst. powd.; sol. in most common org. solvs. incl. aliphatic and aromatic hyrocarbons, ketones, esters; dens. 5.9 lb/gal; m.p. 47 C.

Carstab DLTDP. [Morton Int'l.] Dilauryl thiodipropionate; heat aging stabilizer in conjunction with primary antioxidants; for PP, HDPE; wh. flake; m.w. 514; sol. 30 g/100 g benzene, 7 g/100 g ethanol; sp.gr. 0.896 (80 C); f.p. 41 C; acid no. 0.5; 99T purity.

Carstab DMTDP. [Morton Int'l.] Dimyristyl thiodipropionate; antioxidant for polyolefins; most effective in synergistic combination with hindered phenolics; wh. sm. flake; good odor; m.w. 570; sol. in benzene, chloroform, toluene; sp.gr. 0.885 (80 C); dens. 3.99 lb/gal; m.p. 51 C; acid no. 0.4; 99% purity.

Carstab DSTDP. [Morton Int'l.] Distearyl thiodipropionate; long-term heat aging stabilizer in conjunction with primary antioxidants; for PP, HDPE; wh. cryst. flake or powd.; m.w. 682; sol. 29 g/100 g benzene; water-insol.; sp.gr. 0.858 (80 C); f.p. 65 C; acid no. 1.0; 98% pure.

Cartaretin® F-4. [Sandoz] Adipic acid/dimethylaminohydroxypropyl diethylenetriamine copolymer; cationic; conditioner for hair and skin; wetting agent for sizing; lt. yel. liq.; natural odor; sol. in water and alcohol; sp.gr. 1.07; visc. 400 cps; f.p. –4 C; b.p. 220 F; 30% act.

Cartaretin® F-8. [Sandoz] Adipic acid/dimethylaminohydroxypropyl diethylenetriamine copolymer; see Cartaretin F-4; amber liq.; natural odor; sol. see Cartaretin F-4; sp.gr. 1.07; visc. 800 cps; f.p. –4 C; b.p. 220 F; 30% act.

Cartaretin® F-23. [Sandoz] Polymer; cationic; see Cartaretin F-4; sol. see Cartaretin F-4.

Cassastat WF. [Hoechst-Celanese] Quat. ammonium salt; antistat for syn. fibers; lt. yel. liq.

Castor Oil Sulfated 50%. [Clough] Sulfated castor oil; anionic; emulsifier, dyeing assistant, dye dispersant; liq.; 50% min. conc.

Castor Oil Sulfated 68%. [Clough] Sulfated castor oil; anionic; emulsifier, dyeing assistant, dye dispersant; liq.; 69.5% min. conc.

Castorwax® MP-70. [CasChem] Hydrog. castor oil; wax for anhyd. prods. requiring a soft creamy texture; sol. see Castorwax.

Castorwax® MP-80. [CasChem] Hydrog. castor oil; release agent; wax used for formulating antiperspirant sticks, suspending aid for aluminum chlorhydrate; wh. flakes; m.p. 80 C; 100% act.

Cata-Chek 820. [Ferro] Dibutyl tin dilaurate; heat and lt. stabilizer in flexible vinyl formulations requiring a high degree of clarity; nonsulfur staining; furnishes good lubrication and provides min. water blushing; Gardner 3 clear liq.; sp.gr. 1.045; visc. 34 cps.

Catalpo®. [Engelhard] Hydrous aluminum silicate; reinforcing extender for rubber and polymer systems; wh. highly pulverized powd.; 0.6 μm avg. particle size; 0.01% max. residue +325 mesh; sp.gr. 2.58 g/cc; bulk dens. 18 lb/ft³ (loose), 33 lb/ft³ (tamped); bulking value 0.047 gal/lb; oil absorp. 40–45 lb/100 lb; ref. index 1.56; pH 6–8; brightness (GE) 85–87; 45.4% SiO_2; 38.5% Al_2O_3.

Catalpo® X-1. [Engelhard] Hydrous aluminum silicate; see Catalpo; wh. spray-dried beads; 0.6 μm avg. particle size; 0.01% max. residue +325 mesh; sp.gr. 2.58 g/cc; bulk dens. 48 lb/ft³ (loose), 55 lb/ft³ (tamped); bulking value 0.047 gal/lb; oil absorp. 40–45 lb/100 lb; ref. index 1.56; pH 6–8; brightness

(GE) 85–87; 45.4% SiO_2; 38.5% Al_2O_3.

Catapal® Alumina. [Vista] High-purity alumina; catalyst binders and catalyst supports.

Catigene 4513, 4513/80. [Stepan] Alkyl dimethyl benzyl ammonium and ethyl benzylammonium chloride; germicidal, algicidal, fungicidal, deodorizing, antistat; liq.

Catigene BR/80B. [Stepan] Alkyl dimethyl benzyl ammonium bromide; bactericide, fungicide, algicide, disinfectant cleaners, germicide; liq.

Catigene CA 30. [Stepan] Alkyl amino propyl trimethyl ammonium methoxysulfate; conditioner; antistat for hair prods.; liq.

Catigene CLP 50. [Stepan] Lauryl pyridinium chloride; cationic; hair conditioner, antistat; shampoo base; liq.; 50% act.

Catigene CS 40. [Stepan] Alkyl benzyldimethyl ammonium chloride; hair rinses, textile softeners, antistat; liq., paste.

Catigene CT 30. [Stepan] Cetyl trimethyl ammonium methosulfate; see Catigene CS 40; paste; 30% act.

Catigene CT 30/70. [Stepan] Alkyl trimethyl ammonium methoxysulfate; see Catigene CA 30; liq., solid.

Catigene DC 100. [Stepan] Myristalkonium chloride hydrate; germicidal, algicidal, fungicial, deodorizing, antistat; wh. powd.

Catigene DE 80. [Stepan] Didecyldimethyl ammonium methoxysulfate; germicidal, algicidal, fungicidal, deodorizing, antistat; liq.;

Catigene LT 45. [Stepan] Alkyl trimethyl ammonium methoxysulfate; antistat, conditioner, bactericidal agent for personal care prods.; liq.

Catigene ST 30. [Stepan] Stearyl trimethyl ammonium methosulfate; hair rinse, textile softener and antistat; paste; 30% act.

Catigene ST 30/70. [Stepan] Alkyl trimethyl ammonium methoxysulfate; see Catigene CA 30; liq., solid.

Catigene T 50, T 80, T 80 F. [Stepan] Alkyl dimethyl benzyl ammonium chlorides; germicidal, algicidal, fungicidal, deodorizing, antistat; liq.

Catinal CB-50. [Toho] Lauralkonium chloride; disinfectant, germicide, antistat; liq.

Catinal HTB. [Toho] Alkyl trimethyl ammonium bromide; cationic; dyeing assistant for PAN fiber dyeing; liq.; 30% conc.

Catinal HTB-70. [Toho] Alkyl trimethyl ammonium bromide; base material for hair rinse, antistat, germicide; waxy.

Catinal OB-80E. [Toho] Stearyl dimethyl benzyl ammonium chloride; base material for hair rinse; waxy.

Catinex KB-10. [Pulcra SA] Nonylphenol, ethoxylated; nonionic; antifoaming and intermediate emulsifier; liq.; HLB 6.6; 100% conc.

Catinex KB-11. [Pulcra SA] Nonylphenol, ethoxylated; nonionic; detergent intermediate emulsifier; liq. surface act. agent used in petrol. prods.; liq.; HLB 8.9; 100% conc.

Catinex KB-15. [Pulcra SA] Nonylphenol, ethoxylated; nonionic; detergent, dispersant, and wetting agent; solv. cleaner formulations; emulsifier and stabilizer for paint additives; de-icing fluid and sludge dispersing additive in petrol. prods.; liq.; HLB 10.9; 100% conc.

Catinex KB-16. [Pulcra SA] Nonylphenol, ethoxylated; nonionic; detergent, dispersant, emulsifier and wetting agent; surface act. agent; liq.; water-sol.; HLB 12.3; 100% conc.

Catinex KB-18, -19, -20, -22. [Pulcra SA] Nonylphenol, ethoxylated; dispersant; solubilizer, detergent, emulsifier, wetting agent for anionic detergents; water-sol.; HLB 12.8, 13.1, 13.3, 14.1 resp.; 100% conc.

Catinex KB-23, -24, -25, -26. [Pulcra SA] Nonylphenol, ethoxylated; nonionic; see Catinex KB-16; liq., paste; HLB 14.7, 16, 17.1, 16.7 resp.; 100% conc.

Catinex KB-27. [Pulcra SA] Nonylphenol, ethoxylated; nonionic; detergent, dispersant, and wetting agent; emulsifier for oils, fats, and waxes; metal cleaner and degreaser; solid; HLB 15.0; 100% conc.

Catinex KB-31. [Pulcra SA] Nonylphenol, ethoxylated; nonionic; surfactant for foam control; coemulsifier; dispersant for petrol. oils; liq.; HLB 4.6; 100% conc.

Catinex KB-32. [Pulcra SA] Nonylphenol, ethoxylated; nonionic; see Catinex KB-15; liq.; HLB 11.6; 100% conc.

Catinex KB-40. [Pulcra SA] Octylphenol, ethoxylated; nonionic; detergent, emulsifier, dispersant, surface act. agent used in washing formulations and emulsion cleaners; liq.; HLB 11.2; 100% conc.

Catinex KB-42. [Pulcra SA] Octylphenol, ethoxylated; nonionic; wetting agent, detergent, dispersant, emulsifier for household and industrial cleaners, textile processes, wool scouring; solubilizer for perfumes; emulsifier for insecticides and herbicides; liq.; water-sol.; HLB 13.3; 100% conc.

Catinex KB-43. [Pulcra SA] Octylphenol, ethoxylated; nonionic; solubilizer for anionic detergents; used in hot cleaning systems, metal pickling operations and electrolytic cleaning; paste; water-sol.; HLB 14.7; 100% conc.

Catinex KB-44. [Pulcra SA] Octylphenol, ethoxylated; nonionic; detergent, wetting agent, coemulsifier and stabilizer for vinyl-acrylic latex; emulsifier for fats and waxes; solid; HLB 17.3; 100% conc.

Catinex KB-45. [Pulcra SA] Octylphenol, ethoxylated; nonionic; primary emulsifier for vinyl acetate and acrylate polymerization; post stabilizer for syn. latex; dyeing assistant and emulsifier for fats and waxes; solid; HLB 17.9; 100% conc.

Catinex KB-48. [Pulcra SA] Octylphenol, ethoxylated; nonionic; see Catinex KB-43; solid; HLB 15.7; 100% conc.

Catinex KB-49. [Pulcra SA] Octylphenol, ethoxylated; nonionic; emulsifier, detergent, dispersant, surface act. agent used in washing formulations and emulsion cleaners; liq.; HLB 10.2; 100% conc.

Catinex KB-50. [Pulcra SA] Octylphenol, ethoxylated; nonionic; emulsifier, detergent, wetting agent; coemulsifier and stabilizer for vinyl-acrylic latex; emulsifier for fats and waxes; solid; HLB 16.2; 100% conc.

Catinex KB-51. [Pulcra SA] Octylphenol, ethoxylated; nonionic; emulsifier, detergent, dispersant, surface act. agent used in wax based washing formulations and emulsion cleaners; liq.; HLB 9.2; 100% conc.

Catinex LB-12. [Pulcra SA] Octylphenoxy polyethoxyethanol; coupling agent in naphthol dyeing; 70% act.

Cation DS. [Sanyo] Distearyl dimethyl ammonium chloride; cationic; fabric softener; solid; 90% act.

Cation SF-10. [Sanyo] Alkyl imidazoline type quat. compd.; cationic; conc. softener; easily dilutable with water; paste; 75% act.

Cation Softener X. [Onyx] Alkyl imidazoline deriv.; conditioner, softener, lubricant, glass fiber mordant; paste; sp.gr. 0.92; flash pt. 60 F; 60–66% solids.

Cationic Guar C-261. [Henkel] Guar hydroxypropyl trimonium chloride; cationic; thickener, emulsion stabilizer and additive for shampoos; solid; 100% conc.

Catisol AO 100. [Stepan] Oleylamine acetate; cationic; emulsifier, wetting agent, antistat, anticorrosion agent, lubricant, flotation, flocculation; wax; 100% conc.

CB-4-34. [Neville] Aromatic plasticizer used in adhesives, rubber (cements, mech. and molded goods, tires), caulking compds.; Gardner 14 (50% in toluene) liq.; m.w. 515; sol. in ethers and chlorinated, aromatic, naphthenic, and terpene hydrocarbons; sp.gr. 1.003; soften. pt. (R&B) 28 C; flash pt. (COC) 390 F.

CB2405. [Hüls] Bis (glycidoxypropyl) tetramethyldisiloxane; coupling agent, chem. intermediate, blocking agent, release agent, lubricant, primer, reducing agent; amber liq.; m.w. 362.6; sp.gr. 0.99; b.p. 184–187 C (2 mm); flash pt. 150 C; ref. index 1.446; 97% purity.

CB2408. [Hüls] Bis (hydroxyethyl) aminopropyltriethoxy silane; see CB2405; liq.; m.w. 309.5; sp.gr. 0.92; flash pt. 24 C; ref. index 1.409; 62% in ethanol.

CB2409.5. [Hüls] Bis-(N-methylbenzamide) ethoxymethyl silane; see CB2405; amber visc. liq.; m.w. 356.5; sp.gr. 1.13; b.p. 210 C (1 mm); flash pt. > 110 C; 90% purity.

CB2493. [Hüls] Bis (trimethoxysilylpropyl) ethylene diamine; see CB2405; amber liq.; m.w. 468.8; flash pt. 15 C; 62% in methanol.

CB2494. [Hüls] Bis [3-(triethoxysilyl) propyl] tetrasulfide; see CB2405; amber liq.; m.w. 537.4; sp.gr. 1.07; b.p. 250 C; flash pt. 91 C; ref. index 1.074; 95% purity.

CB2595. [Hüls] Bis (trimethysilyl) urea; see CB2405; solid; m.w. 204.1; m.p. 222 C; 97% purity.

CB2790. [Hüls] t-Butyldimethylchloro silane; see CB2405; solid; m.w. 150.7; sp.gr. 0.81; b.p. 124–126 C; flash pt. 3 C; 98% purity.

CB2805. [Hüls] t-Butyldiphenylchloro silane; see CB2405; liq.; m.w. 274.9; sp.gr. 1.07; b.p. 90 C (0.015 mm); flash pt. 112 C; ref. index 1.568; 97% purity.

CB2809. [Hüls] (p-t-Butylphenethyl) dimethylchloro silane; see CB2405; liq.; m.w. 254.9; b.p. 112 C (2 mm); flash pt. 60 C; 97% purity.

CBTS. [Akrochem] N cyclohexyl-2-benzothiazole sulfenamide; delayed-action sulfenamide accelerator for natural, reclaim, and syn. rubbers; nondiscoloring; wh. to lt. gray powd. and pellets; disp. readily at temps. as low as 160 F; sp.gr. 1.28; m.p. 94–102 C.

CC-16. [Baroid] Sodium salt of a humic acid material; anionic; drilling fluid additive, thinner and emulsifier for drilling muds, gel dispersing agent; powd.; sol. in water-based drilling fluids; 100% act.

CC-603. [Stepan/PVO] Blend of carrageenan, guar gum and dextrose; food grade stabilizer for use in whipped toppings, fillings, coffee whiteners.

CC3005. [Hüls] 2-Chloroethylmethyldichloro silane; see CB2405; liq.; m.w. 177.6; sp.gr. 1.26; b.p. 157 C; flash pt. 32 C; ref. index 1.440; 95% purity.

CC3270. [Hüls] Chloromethyldimethylchloro silane; see CB2405; liq.; m.w. 143.1; sp.gr. 1.09; b.p. 115–116 C; flash pt. –5 C; ref. index 1.436; 98% purity.

CC3275. [Hüls] Chloromethylmethyldichloro silane; see CB2405; liq.; m.w. 163.5; sp.gr. 1.29; b.p. 121–122 C; flash pt. 21 C; ref. index 1.450; 95% purity.

CC3285. [Hüls] Chloromethyltrimethyl silane; see CB2405; liq.; m.w. 122.7; sp.gr. 0.88; b.p. 97–98 C; flash pt. –10 C; ref. index 1.418; 98% purity.

CC3290. [Hüls] 3-Chloropropylmethyldimethoxy silane; see CB2405; liq.; m.w. 182.7; sp.gr. 1.019; b.p. 70–72 C (11 mm); flash pt. 45 C; ref. index 1.4242; 97% purity.

CC3291. [Hüls] Chloropropyltrichloro silane; see CB2405; liq.; m.w. 212.0; sp.gr. 1.36; b.p. 81–82 C (27 mm); flash pt. 66 C; ref. index 1.467; 96% purity.

CC3433. [Hüls] 2-Cyanoethyltriethoxy silane; see CB2405; liq.; m.w. 217.3; sp.gr. 0.980; b.p. 224–225 C; flash pt. 86 C; ref. index 1.414; 97% purity.

CC3555. [Hüls] 3-Cyanopropyltrichloro silane; see CB2405; liq.; m.w. 202.6; sp.gr. 1.28; b.p. 93–94 C (8 mm); flash pt. 84 C; ref. index 1.465; 97% purity.

C-Cure 111. [R.H. Carlson] Adduct of an aliphatic polyamine and an epoxy resin; curing agent for R.T. curing sol'n. epoxy coatings used over substrates; Gardner 4 max. sol'n.; Gardner visc. Q-U; 50% sol'n. in MIBK, cellosolve, and xylene.

CD-75P. [Great Lakes] Hexabromocyclododecane; flame retardant used in thermoplastic and thermosetting polymers; textile treatments, latex binders, adhesives, unsat. polyesters, and coatings; wh. powd.; m.w. 641.7; sol. (g/100 g solv.); 18 g MEK, 12 g toluene, 10 g styrene, 0.1 g water; m.p. 185–195 C; 90% assay; 75% bromine.

CD3770. [Hüls] Decamethylcyclopentasiloxane; see CB2405; liq.; m.w. 310.7; sp.gr. 0.96; flash pt. 55 C; ref. index 1.398; 98% purity.

CD3780. [Hüls] Decamethyltetrasiloxane; see CB2405; liq.; m.w. 310.7; sp.gr. 0.85; b.p. 194–195 C; flash pt. 63 C; ref. index 1.388; 95% purity.

CD4450. [Hüls] N,N-Diethylaminotrimethyl silane; see CB2405; liq.; m.w. 145.3; sp.gr. 0.76; b.p. 126–127 C; flash pt. 10 C; ref. index 1.411; 97% purity.

CD5400. [Hüls] Dimethylaminotrimethyl silane; see CB2405; liq.; m.w. 117.3; sp.gr. 0.75; b.p. 86–87 C; flash pt. –10 C; ref. index 1.438; 98% purity.

CD5430. [Hüls] 3,3-Dimethylbutyldimethylchloro silane; see CB2405; liq.; m.w. 178.7; sp.gr. 0.89; b.p. 164–166 C; flash pt. 33 C; 97% purity.

CD5470. [Hüls] Dimethylchloro silane; see CB2405; liq.; m.w. 94.6; sp.gr. 0.85; b.p. 36 C; flash pt. –22 C; 96% purity.

CD5487. [Hüls] Dimethylcyclosiloxanes; see CB2405; liq.; sp.gr. 0.95; flash pt. 55 C; 99% purity.

CD5550. [Hüls] Dimethyldichloro silane; see CB2405; liq.; m.w. 129.0; sp.gr. 1.074; b.p. 70–71 C; flash pt. –9 C; ref. index 1.402; 99% purity.

CD5600. [Hüls] Dimethyldiethoxy silane; see CB2405; liq.; m.w. 148.3; sp.gr. 0.84; b.p. 114–115 C; flash pt. 29 C; ref. index 1.381; 95% purity.

CD5610. [Hüls] (2,3-Dimethylpropyl) dimethylchloro silane; see CB2405; liq.; m.w. 157.4; sp.gr. 0.9; b.p. 153 C; ref. index 1.4360; 97% purity.

CD5636. [Hüls] Dimethyloctadecylchloro silane; see CB2405; solid; m.w. 347.1; b.p. 185 C (0.2 mm) m.p. 27–28 C; flash pt. 201 C; 97% purity.

CD5950. [Hüls] Diphenyldichloro silane; see CB2405; liq.; m.w. 253.2; sp.gr. 1.22; b.p. 309–310 C; flash pt. 142 C; ref. index 1.582; 97% purity.

CD6000. [Hüls] Diphenyldiethoxy silane; see CB2405; liq.; m.w. 272.4; sp.gr. 1.033; b.p. 107 C (15 mm); flash pt. 175 C; ref. index 1.527; 97% purity.

CD6010. [Hüls] Diphenyldimethoxy silane; see CB2405; liq.; m.w. 244.4; sp.gr. 1.08; b.p. 161 C (2 mm); flash pt. 120 C; ref. index 1.545; 97% purity.

CD6150. [Hüls] Diphenylsilanediol; see CB2405; solid; m.w. 216.3; 95% purity.

CD6210. [Hüls] Divinyltetramethyldisiloxane; see CB2405; liq.; m.w. 186.4; sp.gr. 0.82; b.p. 139 C; flash pt. 30 C; ref. index 1.412; 98% purity.

CD6220. [Hüls] Dodecyltrichloro silane; see CB2405; liq.; m.w. 332.6; sp.gr. 1.03; b.p. 152–153 C (3 mm); flash pt. 130 C; ref. index 1.472; 95% purity.

CD6250. [Hüls] 2-(3,4-Epoxycyclohexyl) ethyltrimethoxy silane; see CB2405; liq.; m.w. 246.3; sp.gr. 1.07; b.p. 310 C; flash pt. 146 C; ref. index 1.449; 97% purity.

CE6250. [Hüls] 2-(3,4-Epoxycyclohexyl) ethyltrimethoxysilane; coupling agent, chem. intermediate, blocking agent, release agent, lubricant, primer, reducing agent; liq.; m.w. 246.3; sp.gr. 1.07; b.p. 310 C; flash pt. 146 C; ref. index: 1.449; 97% purity.

CE6345. [Hüls] Ethyltriacetoxysilane; see CE6250; liq.; m.w. 234.3; sp.gr. 1.14; b.p. 107 (8 mm); flash pt. 8 C; ref. index 1.412; 95% purity.

CE6350. [Hüls] Ethyltrichloro silane; coupling agent, chem. intermediate, blocking agent, release agent, lubricant, primer, reducing agent; liq.; m.w. 163.5; sp.gr. 1.24; b.p. 99–100 C; flash pt. 16 C; ref. index 1.426; 98% purity.

Cedepal CA-520. [Stepan Canada] Octoxynol-5; nonionic; emulsifier, detergent, dispersant, surfactant; liq.; oil-sol.; HLB 10.0; 100% conc.

Cedepal CA-897. [Stepan Canada] Octoxynol-40; nonionic; emulsifier, stabilizer; liq.; HLB 18.0; 70% conc.

Cedepal CO-210. [Stepan Canada] Nonoxynol-1 (1.5 EO); nonionic; defoaming, dispersant, detergent, emulsifier, stabilizer intermediate; surfactant and chemical intermediate; liq.; oil-sol.; HLB 4.6; 100% conc.

Cedepal CO-430. [Stepan Canada] Nonoxynol-4; nonionic; detergent, emulsifier, dispersant, stabilizer intermediate; surfactant and chemical intermediate; liq.; HLB 8.8; 100% conc.

Cedepal CO-436. [Stepan Canada] Ammonium alkylaryl ether sulfate; anionic; detergent, wetting agent, lime soap dispersant; emulsifier in polymers; liq.; 60% conc.

Cedepal CO-500. [Stepan Canada] Nonylphenol ethoxylate; nonionic; emulsifier, detergent, dispersant, stabilizer, intermediate; surfactant; oil-sol.; liq.; HLB 10.0; 100% conc.

Cedepal CO-530. [Stepan Canada] Nonoxynol-6; nonionic; emulsifier, detergent, dispersant; surfactant and intermediate; liq.; oil-sol.; HLB 10.8; 100% conc.

Cedepal CO-890. [Stepan Canada] Nonylphenol ethoxylate; nonionic; emulsifier, stabilizer; wax; HLB 17.8; 100% conc.

Cedepal CO-977. [Stepan Canada] Nonylphenol ethoxylate; nonionic; emulsifier, stabilizer; liq.; HLB 18.2; 70% conc.

Cedepal CO-990. [Stepan Canada] Nonylphenol ethoxylate; nonionic; emulsifier, stabilizer, surfactant; wax; water-sol.; HLB 19.0; 100% conc.

Cedepal CO-997. [Stepan Canada] Nonylphenol ethoxylate; nonionic; emulsifier, stabilizer; liq.; HLB 19.0; 70% conc.

Cedepal FA-406. [Stepan Canada] Primary alcohol ethoxylate; nonionic; emulsifier, detergent, wetting and foam stabilizing; liq.; HLB 12.8; 100% conc.

Cegesoft C 17. [Henkel KGaA] Myristyl lactate; emollient; lipid with low distinct fatty char., sol. of alcohol; solubilizer for solid and oil-sol. additives; solid.

Cegesoft C 19. [Henkel KGaA] Cetyl lactate; see Cegesoft C 17; solid.

Cegesoft C 25. [Henkel HGaA] 2-Ethyl hexyl palmitate; emollient; lipid with low distinct fatty char. and spreadability; solubilizer for alcohol-sol. systems and cosmetic emulsions; solid.

Celite® 21-A. [Manville] Diatomaceous silica; filler used as inert, porous catalyst carrier for nickel in hydrogenation, vanadium catalyst used in mfg. of sulfuric acid, and phosphoric acid catalyst of petrol. industry; 1/4″ pellet; dens. 36 pcf; surf. area 77 m^2/g; hardness 8 kg (Monsanto); 58.56% SiO_2, 16.75% CaO, 7.38% Al_2O_3.

Celite® 110. [Manville] Diatomaceous silica; extender pigment and flatting agent for solv.- and water-thinned paints; increases toughness and durability; add "tooth" for adhesion of subsequent coats, improves sanding properties, control vapor permeability, reduces blistering and peeling tendencies; aids faster drying with more rapid solv. release, and in traffic paints, improves night reflectivity; wh. particulate; particle size 16.0 μ; dens. 19.2 lb/solid gal; oil absorp. (Spatula) 130; 90% SiO_2, 4% Al_2O_3.

Celite® 209. [Manville] Diatomaceous silica; functional filler used for agric. chemicals; grinding aid, and conditioner; low conc. toxicants (up to 50%) where greater inertness is required; carrier for liq. seed inoculants; improves flow properties in fertilizers; buff particulate, 0.5% retained on 325 mesh screen; sp.gr. 2.10; dens. 8.0 lb/ft^3; oil absorp. 175%; ref. index 1.43; pH 7.0 max.; 86.7% SiO_2.

Celite® 219. [Manville] Diatomaceous silica; see Celite 21-A; powd.; dens. 8.5 pcf; surf. area 3 m^2/g; 92% SiO_2, 0.5% CaO, 3.3% Al_2O_3.

Celite® 263. [Manville] Diatomaceous silica; functional filler used in prod. of various specialty papers and "wet laid" prods.; high porosity, low loose wt. dens., high brightness, lg. surf. area, and inertness; lt. wgt. bulking agent, drainage aid for slow draining stock; opacity builder, friction material in clutch papers, aid in uniform disp. and suspension of fibers; wh. particulate; particle size 1–4 μ, 0.5% 325-mesh residue; sp.gr. 2.30; dens. 32.0 lb/ft^3; oil absorp. 130; ref. index 1.45–1.49.

Celite® 266. [Manville] Diatomaceous silica; see Celite 110; buff particulate; particle size 3.3 μ, 0.5% retained on 325-mesh screen; dens. 17.5 lb/solid gal; oil absorp. (Spatula) 135; ref. index 1.42; 89% SiO_2; 3% Al_2O_3.

Celite® 270. [Manville] Diatomaceous silica; functional filler for rubber industry; processing aid in rubber compds.; semireinforcing filler in mechanical rubber goods; pinkish particulate; particle size 5.1 μ, 0.7% 325-mesh residue; sp.gr. 2.15; dens. 8.6 lb/ft^3; oil absorp. 150%; 92.8% SiO_2, 3.3% Al_2O_3.

Celite® 275. [Manville] Diatomaceous silica; wax; polish mild abrasive silver/auto polish; inert.

Celite® 281. [Manville] Diatomaceous earth; see Celite 110; wh. particulate; particle size 7.8 μ, 1.5% retained on 325-mesh screen; dens. 19.2 lb/solid gal; oil absorp. (Spatula) 110; ref. index 1.46; 93% SiO_2, 2% Al_2O_3.

Celite® 289. [Manville] Diatomaceous silica; see Celite 110; grey particulate; median particle size 7.5 μ, 4% retained on 325-mesh screen; dens. 17.5 lb/solid gal; oil absorp. (Spatula) 155; ref. index 1.42; 87% SiO_2, 3% Al_2O_3.

Celite® 292. [Manville] Diatomaceous silica; see Celite 270; buff particulate; median particle size 7.5 μ, 3.7% 325-mesh residue; sp.gr. 2.10; dens. 6.4 lb/ft^3; oil absorp. 215%; ref. index 1.43; pH 7.0 max.; 86.0% SiO_2, 3% Al_2O_3.

Celite® 305. [Manville] Diatomaceous silica; see Celite 263; lt. gray particulate; particle size 1–4 μ, 1.0% 325-mesh residue; sp.gr. 2.10; dens. 24.0 lb/ft^3; oil absorp. 175; ref. index 1.40–1.46.

Celite® 315. [Manville] Diatomaceous silica; functional filler used in polishes and cleaners; intermediate grade used in formulations where greater degree of abrasion is required; pinkish particulate; median particle size 5.1 μ, 0.5% 325-mesh residue; sp.gr. 2.15; dens. 8.7 lb/ft^3; oil absorp. 150%; ref. index 1.45; pH 7.0 max.

Celite® 321. [Manville] Diatomaceous silica; see Celite 263; lt. gray particulate; particle size 1–8 μ, 8.0% 325-mesh residue; sp.gr. 2.10; dens. 17.0 lb/ft^3; oil absorp. 210; ref. index 1.40–1.46.

Celite® 321A. [Manville] Diatomaceous silica; see Celite 263; lt. gray particulate; particle size 1–8 μ, 8.0% 325-mesh residue; sp.gr. 2.10; dens. 20.0 lb/ft^3; oil absorp. 210; ref. index 1.40–1.46.

Celite® 350. [Manville] Diatomaceous silica; see Celite 270; pink particulate; median particle size 5.0 μ, 0.5% 325-mesh residue; sp.gr. 2.15; dens. 8.0 lb/ft^3; oil absorp. 165%; ref. index 1.45; pH 7.0 max.; 92.8% SiO_2, 3.3% Al_2O_3.

Celite® 388. [Manville] Diatomaceous silica; see Celite 263; wh. particulate; particle size 2–10 μ, 1.5% 325-mesh residue; sp.gr. 2.30; dens. 24.0 lb/ft^3; oil absorp. 130; ref. index 1.45–1.49.

Celite® 400. [Manville] Diatomaceous silica; see Celite 209; buff particulate, 5.0% retained on 325 mesh screen; sp.gr. 2.10; dens. 7.0 lb/ft^3; oil absorp. 210%; ref. index 1.43; pH 7.0 max.; 85.8% SiO_2.

Celite® 408. [Manville] Diatomaceous silica; see Celite 21-A; 5/32″ pellets; dens. 39 pcf; surf. area 17 m^2/g; hardness 12 kg (Monsanto); 84.0% SiO_2, 8.20% Al_2O_3.

Celite® 410. [Manville] Diatomaceous silica; see Celite 21-A; 1/4″ pellets; dens. 37 pcf; surf. area 20 m^2/g; hardness 13 kg (Monsanto); 83.06% SiO_2, 8.57% Al_2O_3.

Celite® 499. [Manville] Diatomaceous silica; see Celite 110; wh. particulate; median particle size 6.8 μ, trace retained on 325-mesh screen; dens. 19.2 lb/solid gal; oil absorp. (Spatula) 105; ref. index 1.46; 93% SiO_2, 2% Al_2O_3.

Celite® 1200. [Manville] Diatomaceous silica; see Celite 21-A; 1/4″ pellets; dens. 50 pcf; surf. area 18 m^2/g; hardness 8 kg (Monsanto); 84.51% SiO_2, 7.54% Al_2O_3.

Celite® CAS-30KR. [Manville] Diatomaceous silica; see Celite 21-A; 1/8″ pellets; dens. 38 pcf; surf. area 122 m^2/g; hardness 3 kg (Monsanto); 53.79% SiO_2, 36.55% Al_2O_3.

Celite® CS-22R. [Manville] Diatomaceous silica; see Celite 21-A; 1/8″ pellets; dens. 35 pcf; surf. area 19 m^2/g; hardness 5 kg (Monsanto); 82.68% SiO_2, 9.58% Al_2O_3.

Celite® F.C. [Manville] Diatomaceous silica; see Celite 21-A; powd.; dens.7.0 pcf; surf. ara 20 m^2/g; 86% SiO_2, 3.8% Al_2O_3.

Celite® HSC. [Manville] Diatomaceous silica; functional filler used in polishes and cleaners; suitable for fast-cutting buffing compds.; wh. particulate; median particle size 17.5 μ, 17.6% 325-mesh residue; sp.gr. 2.30; dens. 10.7 lb/ft^3; oil absorp. 185%; ref. index 1.47; pH 9.5 max.

Celite® R-625. [Manville] Diatomite; filler for use as inert, porous support/carrier for catalysts in industrial processes; 1/8 in. pellet; 5.5 μm mean pore diam.; dens. 32 lb/ft^3; surf. area 12.4 m^2/g; hardness 1.6 kg; 84.8% SiO_2.

Celite® R-633. [Manville] Diatomite; see Celite R-

625; 30/50 mesh sphere; 6.5 µm mean pore diam.; dens. 22 lb/ft³; surf. area 1.3 m²/g; hardness 27–28%; 88.2% SiO_2.

Celite® R-643. [Manville] Diatomite; see Celite R-625; 30/50 mesh crushed pellet; 0.03 µm mean pore diam.; dens. 24 lb/ft³; surf. area 67 m²/g; hardness 9–14%; 87.0% SiO_2.

Celite® R-685. [Manville] Diatomite; see Celite R-625; powd.; 0.9 µm mean pore diam.; dens. 17 lb/ft³; surf. area 140 m²/g; 63.9% SiO_2.

Celite® R-690. [Manville] Diatomite; see Celite R-625; pellet; 0.19 µm mean pore diam.; dens. 38 lb/ft³; surf. area 122 m²/g; hardness 3 kg; 54.0% SiO_2.

Celite® Snow Floss. [Manville] Diatomaceous silica; see Celite HSC; buff particulate; median particle size 3.0 µ, trace 325-mesh residue; sp.gr. 2.10; dens. 8.0 lb/ft³; oil absorp. 175%; ref. index 1.43; pH 7.0 max.

Celite® Super Fine Super Floss. [Manville] Diatomaceous silica; see Celite 110; used in polishes and cleaners; antiblocking agents in LDPE film; wh. particulate; median particle size 3.5 µ, trace 325-mesh residue; sp.gr. 2.30; dens. 7.5 lb/ft³; dens. 19.2 lb/solid gal; oil absorp. (Spatula) 100; ref. index 1.46; pH 9.4 max.; 93% SiO_2, 2% Al_2O_3.

Celite® Super Floss. [Manville] Diatomaceous silica; see Celite 110; also used in polishes and cleaners; processing aid for rubber goods where added to oil as an absorbent, in silicone rubbers; antiblocking agent in LDPE film; particulate; median particle size 5.5 µ, 0.1% retained on 325-mesh screen; dens. 19.2 lb/solid gal; oil absorp. (Spatula) 105; ref. index 1.46; 93% SiO_2, 2% Al_2O_3.

Celite® White Mist. [Manville] Diatomaceous silica; see Celite 110; see Celite Super Fine Super Floss; wh. particulate; median particle size 5.5 µ, 0.05% 325-mesh residue; sp.gr. 2.30; dens. 7.5 lb/ft³; dens. 19.2 lb/solid gal; oil absorp. (Spatula) 105; ref. index 1.46; pH 9.3 max.; 93% SiO_2, 2% Al_2O_3.

Cellolyn® 95-80T. [Hercules] Balsamic alkyd resin derived from tech. hydroabietyl alcohol; resin sol'n. used in mfg. of wood lacquers, wood finishes, in grinding vehicles for lacquers; extender for oil alkyds in pigmented lacquers, paper lacquers, printing inks; Gardner 2 liq.; sol. in lacquer solvs.; sp.gr. 1.02; dens. 8.5 lb/gal; visc. (G-H) S; acid no. 9; flash pt. (COC) 27 C; 80% solids in toluene/IPA (16.2/3.8).

Cellolyn® 98-80T. [Hercules] Alkyd resin sol'n.; plasticizing resin used in mfg. of printing inks for flexible substrates, in lacquer finishes, and general applics.; Gardner 2 liq.; sol. in lacquer solvs.; sp.gr. 1.06; dens. 8.8 lb/gal; visc. (G-H) Z; acid no. 18; flash pt. (COC) 27 C; 80% solids in toluene/IPA (15.8/4.2).

Cellopal 100. [Sybron] Polyethoxy alkylphenol sulfonate, TEA salt; anionic; emulsifier, detergent, dispersant, wetting, and foaming agent; liq.; 100% conc.

Cellosize® Hydroxyethyl. [Union Carbide] Ethoxylated cellulose; thickener; rapid dispersing material; sol. in water.

Celluclast 2.0 L Type X. [Novo] Cellulase derived from the fungus Trichoderma reesei; enzyme for breakdown of cellulosic matter for prod. of fermentable sugar, reduction of visc., or increase yield of valuable prods. of plant origin; brn. liq.; water-sol.; dens. 1.2 g/ml; pH 6–8.

Celluclast 200L. [Novo] Cellulase; enzyme for breakdown of cellulosic material for improved processing and yields; liq.

Celluclast 200 L Type N. [Novo] Cellulase derived from the fungus Trichoderma reesei; see Celluclast 2.0 L Type X; brn. liq.; water-sol.; dens. 1.2 g/ml; pH 6–8.

Celluferm. [Finnsugar Biochemics] Cellulase (fungal); enzyme for fruit and veg. processing; tan powd.

Celluzyme 2400 T. [Novo] Cellulase; enzyme for laundry powd. detergents and laundry additives; off-wh. dustfree granulate.

Celogen® AZ. [Uniroyal] Azodicarbonamide; chemical blowing agent for thermoset and thermoplastic polymers; for inj.-molding structural foam, extrusion of profiles, sheet, pipe, and wire coatings, and vinyl plastisol, coating, and calendering; yel. fine powd.; avail. in particle size grades AZ 120 (2 µ avg. particle size), AZ 130 (3 µ), AZ 150 (5 µ), AZ 180 (8 µ), AZ 199 (12 µ); somewhat sol. in polyalkylene glycols and dimethylformamide; sp.gr. 1.65; decomp. pt. 205–215 C; gas yield 220 cc/g STP; gas: 65% N_2, 24% CO, 5% CO_2, 5% NH_3.

Celogen® AZ 3990. [Uniroyal] Modified azodicarbonamide; chemical blowing agent for cellular vinyl; also used in inj.-molding structural foam and extrusion; yel. fine powd.; avail. in particle size grades AZ 2990 (2 µ avg. particle size), AZ 3990 (3 µ), AZ 5990 (5 µ); somewhat sol. in polyalkylene glycols and dimethylformamide; sp.gr. 1.65; decomp. pt. 205–215 C; gas yield 220 cc/g STP; gas: 65% N_2, 24% CO, 5% CO_2, 5% NH_3.

Celquat. [Nat'l. Starch] Cellulose; cationic; conditioner, hair fixative for setting lotions, conditioners, shampoo additive and skin prods.; water-sol.

Cenegen 7. [Crompton & Knowles] Alkylaryl sulfonate; anionic; retarding and leveling agent for nylon dyeing; good barre coverage; liq.

Cenegen NWA. [Crompton & Knowles] Aryl polyethoxy condensate; dyeing assistant for premetallized dyes on nylon and wool; liq.

Cenekol® Liq., NCS Liq. [Crompton & Knowles] Sulfonated phenolic condensate; acid dye fixative for nylon and stable to electrolytes; liq.

Centrocap® 162SS, 162US. [Central Soya] Special grade lecithin; designed for encapsulation where clarity and brilliance are required; amber fluid; visc. 3500 cP; acid no. 28.

Centrocap® 273SS, 273US. [Central Soya] Special grade lecitin; see Centrocap 162SS; amber fluid; visc. 1500 cP; acid no. 27.

Centrol® Series. [Central Soya] Natural soybean lecithin; amphoteric; wetting agent, emulsifier, emollient, softener, lubricant, solubilizer; cosmetics; food, glass, metal, and ceramic processing; pharmaceuticals; coatings mfg.; plastic, rubber, paper, and printing industry; masonry and asphalt prods.; petrol. industry as specialty oils and greases; pesticides; textiles and leather; liq.; 100% act.

Centrol® 2FSB, 2FUB, 3FSB, 3FUB. [Central Soya] Std. grade lecithin; amber fluid; visc. 6000 cP; acid no. 27.

Centrol® CA. [Central Soya] Special grade lecithin; amber fluid; visc. 4000 cP; acid no. 20.

Centrolene® Series. [Central Soya] Lecithin; amphoteric; wetting agent, emulsifier, release agent, dispersant, lubricant, anticorrosive agent, emollient; food and metal processing; cosmetics; pharmaceuticals; coatings mfg.; plastic and rubber industries; glass and ceramics; paper and printing; masonry and asphalt prods.; petrol. industry as specialty greases and oils; textiles and leather; liq.; 100% act.

Centrolene® A, S. [Central Soya] Hydroxylated lecithin; Gardner 11 and 12 resp. heavy-bodied fluid; acid no. 27.

Centrolex® Series. [Central Soya] Lecithin; amphoteric; wetting agent, emulsifier, dietary supplement, dispersant, emollient, used in cosmetics, food processing, pharmaceuticals, industrial uses; gran.; 100% act.

Centrolex® C. [Central Soya] Special grade lecithin; yel. gran.; bulk dens. 0.44 g/cc; acid no. 27.

Centrolex® D. [Central Soya] Special grade lecithin; fine yel. powd.; bulk dens. 0.46 g/cc; acid no. 27.

Centrolex® F. [Central Soya] Special grade lecithin; yel. powd.; bulk dens. 0.45 g/cc; acid no. 27.

Centrolex® G. [Central Soya] Special grade lecithin; yel. powd.; bulk dens. 0.46 g/cc; acid no. 27.

Centrolex® P. [Central Soya] Special grade lecithin; yel. gran.; bulk dens. 0.38 g/cc; acid no. 27.

Centrolex® R. [Central Soya] Special grade lecithin; yel. gran.; bulk dens. 0.38 g/cc; acid no. 27.

Centromix® Series. [Central Soya] Lecithin; emulsifier, wetting agent, dispersant, penetrant, used in cosmetics, food processing, pharmaceuticals, industrial uses; liq., paste; 100% act.

Centromix® CPS. [Central Soya] Special grade lecithin; amber fluid; visc. 8500 cP; acid no. 23.

Centromix® E. [Central Soya] Special grade lecithin; amber fluid; visc. 6500 cP; acid no. 17.

Centrophase® 152. [Central Soya] Special grade lecithin; amber fluid; visc. 1200 cP; acid no. 22.

Centrophase® C. [Central Soya] Special grade lecithin; amber fluid; visc. 1000 cP; acid no. 27.

Centrophase® HR. [Central Soya] Lecithin, heat resistant; multifunctional ingredient; food substance; lubricant and release agent for heated surfaces; Gardner 13 amber fluid; sol. in liq. veg. and min oil, aliphatic, aromatic and halogenated hydrocarbons, petroleum ether; dens. 1.01 g/cc; visc. 2000 cP; pour pt. 32 F; 52% acetone insol.

Centrophase® HR2B, HR2U. [Central Soya] Special grade lecithin; amber fluid; visc. 5000 cP; acid no. 22.

Centrophase® HR4B, HR4U. [Central Soya] Special grade lecithin; heavy-bodied fluid; visc. 3300 cP; acid no. 24.

Centrophase® HR6B. [Central Soya] Special grade lecithin; amber fluid; visc. 2500 cP; acid no. 23.

Centrophase® NV. [Central Soya] Special grade lecithin; amber fluid; visc. 215 cP; acid no. 21.

Centrophil® Series. [Central Soya] Lecithin; wetting agent, emulsifier, dispersant, release agent, dietary supplement, used for cosmetics, food processing, pharmaceuticals, industrial uses; most are food approved, some for industrial use only; liq., solid, other; 100% act.

Centrophil® K. [Central Soya] Special grade lecithin; amber plastic; acid no. 20.

Centrophil® M, W. [Central Soya] Special grade lecithin; amber fluid; visc. 150 cP; acid no. 12.

Cenwax G. [Union Camp] Hydrog. castor oil; lubricant, wax modifier; used in coatings; Lovibond 2Y/0.7R (1´´) color; m.w. 87 C; acid no. 3.0; sapon. no. 180; flash pt. 320 C (COC); fire pt. 338 C.

Cenwax ME. [Union Camp] Methyl hydroxystearate; lubricant; Lovibond 2Y/0.8R (1´´); m.p. 52 C; acid no. 6; sapon. no. 178; flash pt. 216 C; fire pt. 229 C (COC).

Ceralan®. [Amerchol] Lanolin alcohols; nonionic; emollient; w/o emulsifier; wax; 100% conc.

Ceraloid 356. [Rhone-Poulenc] Paraffin free block; wax for yarn and thread lubricant; emulsifiable.

Ceramid. [Lonza] Amide wax; syn. wax; yel. solid; m.p. 80 C; flash pt. 245 C.

Ceramitalc 1. [Vanderbilt] Talc; used in ceramic applic.; used as aux. flux in elec. porcelain, hotel china, and sanitaryware bodies; improved resistance to crazing in glazed ware; particle size 11 μ; dens. 2.85 mg/m³; 57.30% SiO_2, 30.7% MgO, 6.17% CaO.

Ceramitalc 10A. [Vanderbilt] Talc; see Ceramitalc No. 1; particle size 16 μ; dens. 2.85 mg/m³; 54.80% SiO_2, 30.7% MgO, 7.80% CaO.

Ceramitalc 10AC. [Vanderbilt] Magnesium silicate; filler used in art pottery and hobby ceramic casting slips where control of the specific resistance is essential in high talc bodies; powd.; particle size 31 μ, 2.5–3.2% +200 mesh residue; dens. 2.85 mg/m³; 52.20% SiO_2, 32.4% MgO, 7.93% CaO.

Ceramitalc HDT. [Vanderbilt] Magnesium silicate; talc used in ceramic wall tile and artware; eliminates laminations, improves prod. dens., and permits faster pressing; prevents crazing in bodies that can be safely fired at rapid cycles; powd. particle size 37 μ, 2.3–4.0% +200 mesh residue; dens. 2.85 mg/m³; 56.00% SiO_2, 29.6% MgO, 8.71% CaO.

Ceranine AT. [Sandoz] Fatty acid deriv.; cationic; softener and antistatic agent for syn. fibers; paste.

Ceranine AW. [Sandoz] Fatty acid deriv.; cationic; softener and antistatic agent for syn. fibers; solid.

Ceranine HC. [Sandoz] Fatty acid deriv.; cationic; softener and antistatic agent for all fibers; paste.

Ceranine HCA. [Sandoz] Fatty amide deriv.; cationic; emulsifier for skin care prods., softener for textiles; cream-colored gran.; slight odor; disp. in water; m.p. 71–81 C; 100% act.

Ceranine HCA Gran. [Sandoz] Fatty amide complex; antisoil additive for rug care prods.; fabric conditioner and softener additive; water-disp.

Ceranine HCL, HCS. [Sandoz] Fatty acid deriv.; cationic; softener for all fabrics; liq., flake.

Ceranine L. [Sandoz] Polyethylene disp.; nonionic; softener for cellulosic fibers and their blends; liq.;

60% conc.

Ceranine PAP. [Sandoz] Polymerization prod.; cationic; permanent softener for polyamide fibers; liq.

Ceranine PNA. [Sandoz] Fatty acid deriv.; cationic; softener for syn. fibers; paste/powd.

Ceranine PNL. [Sandoz] Substituted fatty acid amide; cationic; softener for syn. fibers; liq., flakes.

Ceranine PNS. [Sandoz] Fatty acid deriv.; cationic; softener for all fibers; flake/paste.

Ceranine PNS Gran. [Sandoz] Fatty amide complex; see Ceranine HCA Gran.; water-disp.

Ceranine RW. [Sandoz] Fatty acid deriv.; cationic; softener for all fibers; liq./paste.

Ceranine RWN. [Sandoz] Fatty acid deriv.; softener for all fibers, cationic; liq.

Ceranine SG. [Sandoz] Fatty acid ester; nonionic/anionic; softener for all fibers; paste/solid.

Ceranine SIT. [Sandoz] Fatty acid deriv.; cationic; softener for all fibers; paste/solid.

Ceranine SWP. [Sandoz] Fatty acid deriv.; softener for all fibers, cationic; liq.

Ceranine VR. [Sandoz] Fatty alcohol sulfates; anionic; softener for all fibers; paste.

Ceranine VS. [Sandoz] Fatty acid condensation prod. in aq. disp.; cationic; softener with hydrophobic properties for cellulosic fibers and their blends with syn. fibers; liq.

Ceranine WN. [Sandoz] Fatty acid ester; nonionic; softener for cellulosic and syn. fibers; liq., flakes.

Ceraphyl 28. [Van Dyk] Cetyl lactate; emollient binder for pressed powds., lipsticks, hair prods.; sol. in min. oil, IPM, and 95% ethanol.

Ceraphyl 31. [Van Dyk] Lauryl lactate; see Ceraphyl 28; also antitack agent in antiperspirants; sol. in min. oil, 95% ethanol, propylene glycol, and IPM.

Ceraphyl 41. [Van Dyk] C_{12-15} alcohols lactate; emollient; for sheen on hair; antitack in antiperspirants; sol. in aerosol, min. oil, 60% ethanol, propylene glycol and IPM.

Ceraphyl 45. [Van Dyk] Dioctyl maleate; binder, emollient for cosmetic prod. applics.; clear liq.; sol. in most org. solvs.; water-insol.

Ceraphyl 50. [Van Dyk] Myristyl lactate; lubricant, emollient for skin prods., alcoholic preps., shaving lotions, colognes, makeup, medicated prods.; wh. to yel. liq. to soft solid; sol. in peanut oil, min. oil, ethanol, propylene glycol, IPM, oleyl alcohol; sp.gr. 0.892–0.904; acid no. 2 max.; sapon. no. 166–181.

Ceraphyl 50-S. [Van Dyk] Myristyl lactate; emollient binder for pressed powds.; sol. in min. oil, 95% ethanol, propylene glycol, and IPM; water-insol.

Ceraphyl 55. [Van Dyk] Tridecyl neopentanoate; emollient for creams and lotions, binder for pressed powds.; clear liq.; sol. see Ceraphyl 45.

Ceraphyl 60. [Van Dyk] Quaternium-22; conditioner, emollient, moisturizer, humectant, antistat; highly substantive to skin and hair; amber liq.; water-sol.

Ceraphyl 65. [Van Dyk] Quaternium-26; emollient; hair conditioner; mild foaming aux. emulsifier with antistatic, antitangle properties for shampoos, rinses and other hair prods.; sol. in water, 70% and 99% ethanol.

Ceraphyl 70. [Van Dyk] Quaternium-70, propylene glycol; antitangle, antistatic ingred. used in all types of hair conditioners; aux. emulsifier and emollient for skin creams and lotions; water-disp.

Ceraphyl 85. [Van Dyk] Stearamidopropyl cetearyl dimonium tosylate, propylene glycol; see Ceraphyl 65; waxy solid; 60% act.

Ceraphyl 140. [Van Dyk] Decyl oleate; nonionic; emollient; binder for pressed powds.; pigment dispersant; co-solv.; liq.; 100% act.

Ceraphyl 140-A. [Van Dyk] Isodecyl oleate; nonionic; emollient binder for pressed powd.; make-up solubilizer; wetting agent for iron oxides; cleansing agent for emulsions; liq.; 100% act.

Ceraphyl 230. [Van Dyk] Diisopropyl adipate; emollient; coupler; increased spread of bath oils; reduces oiliness of min. oil; sol. in min. oil, 50% ethanol, propylene glycol, and IPM; water-insol.

Ceraphyl 368. [Van Dyk] Octyl palmitate; emollient binder for pressed powds., blushers; gloss agent in lipsticks; antitack for antiperspirants; liq.; sol. in most org. solvs., IPM; water-insol.

Ceraphyl 375. [Van Dyk] Isostearyl neopentanoate; see Ceraphyl 368; liq.; sol. see Ceraphyl 368.

Ceraphyl 424. [Van Dyk] Myristyl myristate; emollient; increases visc. of creams and lotions at low concs.; sol. in min. oil and IPM; water-insol.

Ceraphyl 494. [Van Dyk] Isocetyl stearate; emollient for skin and hair prods.; liq.; sol. in most org. solvs.; water-insol.

Ceraphyl 847. [Van Dyk] Octyldodecyl stearoyl stearate; emollient binder for pigmented sticks and emulsions with dispersing and mold release properties; liq.; sol. see Ceraphyl 494.

Ceraphyl GA. [Van Dyk] Maleated soybean oil; emollient for creams and lotions; hair and skin conditioner; clear liq.; sol. in most org. solvs.; water-insol.

Ceraphyl ICA. [Van Dyk] Isocetyl alcohol; emollient for creams and lotions, binder for pressed powds., hair and skin conditioner; clear liq.; sol. see Ceraphyl 45.

Ceraphyl IPL. [Van Dyk] Isopropyl linoleate; emollient; skin and hair conditioner and lubricant; superfatting agent in detergent systems; sol. in min. oil and IPM; water-insol.

Cerasynt 303. [VanDyk] Diethylaminoethyl stearate; nonionic; visc. builder in hair dyes; pharmaceutical emulsifier; dispersant, wetting agent; liq.; sol. in 95% ethanol and IPM; 100% conc.

Cerasynt 945. [VanDyk Glyceryl stearate, laureth-23; nonionic; acid-stable emulsifier and gellant; thickener; partly sol. in water; 100% conc.

Cerasynt D. [Van Dyk] Stearamide MEA-stearate; nonionic; opacifier, thickener for liq. cream shampoos; aux. emulsifier in hydrocarbon aerosol systems such as shave creams; flake; water-insol.; 100% conc.

Cerasynt GMS. [Van Dyk] Glyceryl stearate; nonionic; sec. o/w emulsifier for creams and lotions; visc. builder for emulsions; wh. to cream flakes;

forms visc. disp. in peanut oil, min. oil; m.p. 56–59 C; acid no. 3 max.; iodine no. 2 max.; sapon. no. 162–175.

Cerasynt IP. [Van Dyk] Glycol stearate and other ingreds.; nonionic; opacifier and pearling agent for lotion shampoos and liq. soaps; flake; water-insol.; 100% conc.

Cerasynt LP. [Van Dyk] Glycol stearate, sodium laureth sulfate, hexylene glycol; opacifier, pearlescent for shampoos; wh. opaque pourable liq.; disp. in water; sp.gr. 1.0±0.4; visc. < 5000 cps; pH 6.0±0.5; 67 ± 3% water.

Cerasynt M. [Van Dyk] Glycol stearate; opacifier, thickener, pearlescent for liq. and cream shampoos; sec. emulsifier for cosmetics and pharmaceuticals; flake; water-insol.; 100% conc.

Cerasynt MN. [Van Dyk] Glycol stearate SE; nonionic; opacifier for liq. and cream shampoos; primary emulsifier in cosmetics and pharmaceuticals; flake, liq. disp. in water; 100% conc.

Cerasynt PA. [Van Dyk] Propylene glycol stearate; nonionic; opacifier for liq. and cream shampoos; sec. emulsifier; flake; water-insol.; 100% conc.

Cerasynt SD. [Van Dyk] Glyceryl stearate; nonionic; opacifier/thickener for liq. and cream shampoos; aux. emulsifier in cosmetics, pharmaceuticals; flake; water-insol.; 100% conc.

Cereclor 42. [ICI Am.] Chlorinated paraffin; plasticizer extender for PVC; EP additive in gear oils and metalworking lubricants; plasticizer for paints; flame retardant additive for plastics, rubbers, and textiles; 42 grade esp. for adhesives, sealants, paints plasticizers, and in neoprene, flame retardant applics.; Gardner 1 clear visc. liq.; misc. with benzene, esters, ketones, ethers, cyclohexanol, petrol. ether, trichloroethylene, linseed and castor oils; dens. 1.17 g/ml; visc. 24 poise; 42% chlorine content.

Cereclor 42P. [ICI Am.] Chlorinated paraffin; see Cereclor 42; 42P esp. as PVC plasticizer and in flame retardant applics.; Gardner 4 clear visc. liq.; sol. see Cereclor 42; dens. 1.16 g/ml; visc. 28 poise; 42% chlorine content.

Cereclor 50LV. [ICI Am.] Chlorinated paraffin; see Cereclor 42; 50LV esp. as lubricant additive; Gardner 1 clear visc liq.; sol. see Cereclor 42; dens. 1.19 g/ml; visc. 1 poise; 50% chlorine content.

Cereclor 51L. [ICI Am.] Chlorinated paraffin; see Cereclor 42; 51L esp. as lubricant additive; Gardner 6 clear visc. liq.; sol. see Cereclor 42; dens. 1.25 g/ml; visc. 13 poise; 51% chlorine content.

Cereclor 52P. [ICI Am.] Chlorinated paraffin; see Cereclor 42; 52P esp. as adhesives, sealants, and paints plasticizer, and in flame retardant applics.; Gardner 4 clear visc. liq.; sol. see Cereclor 42; dens. 1.25 g/ml; visc. 12 poise; 52% chlorine content.

Cereclor 70L. [ICI Am.] Chlorinated paraffin; see Cereclor 42; 70L esp. as lubricant additive and in flame retardant applics.; Gardner 11 clear visc. liq.; sol. see Cereclor 42; dens. 1.55 g/ml; visc. 20,000 poise; 70% chlorine content.

Cereclor AP45. [ICI Am.] Chlorinated paraffin; see Cereclor 42; AP45 esp. in flame retardant and specialty applics.; Gardner < 1 clear visc. liq.; sol. see Cereclor 42; dens. 1.16 g/ml; visc. 1 poise; 45% chlorine content.

Cereclor AP52. [ICI Am.] Chlorinated paraffin; see Cereclor 42; AP52 esp. in flame retardant and specialty applics.; Gardner < 1 clear visc. liq.; sol. see Cereclor 42; dens. 1.25 g/ml; visc. 11 poise; 52% chlorine content.

Cereclor LP4446. [ICI Am.] Chlorinated paraffin; see Cereclor 42; LP4446 esp. as lubricant additive and in flame retardant applics.; Gardner 4 clear visc. liq.; sol. see Cereclor 42; dens. 1.19 g/ml; visc. 45 poise; 44% chlorine content.

Cereclor LP4985. [ICI Am.] Chlorinated paraffin; see Cereclor 42; LP4985 esp. as adhesives, sealants, and paints plasticizer, lubricant additive, in flame retardant applics.; Gardner 8 clear visc. liq.; sol. see Cereclor 42; dens. 1.28 g/ml; visc. 169 poise; 50% chlorine content.

Cereclor S45. [ICI Am.] Chlorinated paraffin; see Cereclor 42; S45 esp. as adhesives, sealants, paints, PVC, and rubber plasticizers, and in flame retardant applics.; Gardner 1 clear visc. liq.; sol. see Cereclor 42; dens. 1.17 g/ml; visc. 2 poise; 45% chlorine content.

Cereclor S52. [ICI Am.] Chlorinated paraffin; see Cereclor 42; S52 esp. as adhesives, sealants, paints, PVC, and rubber plasticizers, and in flame retardant applics.; Gardner 1 clear visc. liq.; sol. see Cereclor 42; dens. 1.25 g/ml; visc. 12 poise; 52% chlorine content.

Cereflo® 200 L. [Novo] Bacterial beta-glucanase derived from Bacillus subtilis; enzyme used as supplementary glucanase preparation when mashing malt or mixts. of malt and barley; liq.; water-sol.; dens. 1.2 g/ml.

Cerol A Liq. [Sandoz] Cationic zirconium/wax emulsion; water repellent for use on cotton and polyester/cotton blends.

Cerol C. [Sandoz] Fatty acid chrom-complex; cationic; waterproofing agent, fast to washing and dry cleaning; liq.

Cerol EWL. [Sandoz] Wax emulsions; cationic; water repellent for textiles and non-wovens; liq.; water-disp.

Cerol M. [Sandoz] Fatty acid chrome complex; cationic; waterproofing agent, fast to washing and dry cleanings; liq

Cerol ZN. [Sandoz] Paraffin disp. with zirconium; cationic; waterproofing agent for textiles; liq.

Ceroxin Special. [Henkel] Hydrog. castor oil; thickener used in petrol. solv. paint systems.

Cetal. [Amerchol] Cetyl alcohol NF; emollient used in emulsions, oils, and makeup; visc. control in emulsions; aux. emulsifier; wh. waxy solid; mild char. odor; m.p. 45–51 C; acid no. 1.0 max.; sapon. no. 2.0.

Cetats. [Hexcel] Cetrimonium tosylate; cationic; additive in toothpaste formulations, cooling tower algicide; wh. powd.; char. odor; m.w. 455.72; sol. in alcohol, slightly sol. in water; pH 5–8; 99% min. assay.

Cetax 16. [Aquatec Quimica] Cetyl alcohol; emollient, consistency agent for creams and lotions; superfatting agent for hair prods.; flakes.

Cetax 18. [Aquatec Quimica] Stearyl alcohol; see Cetax 16; flakes.

Cetax 50. [Aquatec Quimica] Cetyl/stearyl alcohol; see Cetax 16; flakes.

Cetax DR. [Aquatec Quimica] Ester mixt.; nonionic; see Cetax 16; flakes.

Cetina. [Robeco] Cetyl esters and stearamide DEA; nonionic; emulsifier, lubricant, emollient for cosmetics; flake; 100% conc.

Cetina TE. [Robeco] Cetyl esters, stearamide DEA; emulsifiable fraction of cetyl esters, wax for satiny feel.

Cetiol. [Henkel] Oleyl oleate; emollient; oily component of strong greasy char., for pharmaceutical prods.; liq.

Cetiol 868. [Henkel] Octyl stearate; emollient; superfatting oil; liq.; 100% act.

Cetiol 1414E. [Henkel] Myreth-3-myristate; emollient; cosmetic preparations such as creams, lotions, lipsticks, and blooming bath oils; sol. in alcohols, cosmetic oils, glycols, esters, ketones, and aromatics; water-disp.

Cetiol A. [Henkel] Hexyl laurate; vehicle for topical act. ingreds. used in skin lubricants and personal care prods.; emollient; clear oily liq.; odorless; sol. @ 10% in min. and castor oil, IPM, oleyl alcohol, ethyl alcohol-SD 40 (95%), silicone fluid; dens. 0.847 g/ml; HLB 12; solid. pt. < 0 C; flash pt. > 165 C; cloud pt. < 5 C; sapon. no. 190-205; ref. index 1.435; 100% act.

Cetiol B. [Henkel] Dibutyl adipate; emollient; oily component for day creams; liq.; 100% act.

Cetiol G16S. [Henkel] Isocetyl stearate; emollient, lubricant for cosmetic preparations; lipsticks; clear oily liq.; sol. in alcohol esters, min. oil; water-insol.

Cetiol G20S. [Henkel] Octadodecanol stearate; see Cetiol G16S; clear oily liq.; sol. see Cetiol G16S.

Cetiol HE. [Henkel] PEG-7 glyceryl cocoate; nonionic; emollient oil, superfatting agent for aq. formulations in personal care prods.; dispersant for biologically act. ingreds.; clear oil; sol. @ 10%: sol. in water, castor oil, oleyl and ethyl alcohol-SD 40 (95%), ethyl alcohol 3A (70% aq.); dens. 1.050 g/ml; solid. pt. 0 C; cloud pt. < 0 C; sapon. no. 90–100; ref. index 1.460

Cetiol J 600. [Henkel] Oleyl erucate; emollient; fatty component for cosmetic preparations, jojoba oil substitute; liq.; 100% act.

Cetiol LC. [Henkel] Coco caprylate/caprate; emollient oil used in personal care prods.; clear liq.; odorless; sol. @ 10% in min. and castor oil, IPM, oleyl alcohol, silicone fluid; insol. in water; dens. 0.843 g/ml; HLB 9; solid. pt. < 15 C; flash pt. > 180 C; cloud pt. 15 C max.; sapon. no. 165-173; ref. index 1.440

Cetiol MM. [Henkel] Myristyl myristate; emollient; wax ester with superfatting properties; wax-like substance.

Cetiol R. [Henkel] Trihydroxy methoxystearin; emollient; fatty oil for makeup preparations; castor oil substitute; clear to turbid liq.; 100% act.

Cetiol S. [Henkel] Dioctylcyclohexane; emollient, superfatting agent; used in cosmetic and pharmaceutical creams and emulsions; liq.; 100% act.

Cetiol SB 45. [Henkel] Shea butter; emollient, consistency giving agent for o/w and w/o creams and emulsions; native fatting agent for creams, lotions, anhyd. creams; soft wax.

Cetiol SN. [Henkel] Cetearyl isononanoate; emollient for applic. in skin care, massage and sun protection preparations; oily component with expressed hydrophobic effect.

Cetiol V. [Henkel] Decyl oleate; emollient, carrier for lipid sol. substances used in personal care prods. and pharmaceutical topical applics.; yel. low visc. liq.; sol. @ 10%: sol. in min. and castor oil, IPM, oleyl alcohol; insol. in water; dens. 0.86 g/ml; HLB 9; solid. pt. <0 C; flash pt. 240-260 C; cloud pt. < 10 C; sapon. no. 130-140; ref. index 1.450

Cetodan. [Grindsted] Acetylated monoglycerides; nonionic; food emulsifier, aerating agent for shortenings, toppings, cakes; edible coating; plasticizer for chewing gum base; antifoam agent, lubricant; solid, liq.; HLB 1.5; 100% conc.

Cetodan 50-00A. [Grindsted] Monoglyceride acetic acid ester; food emulsifier; aerating and foam stabilizing agent in food prods.; block; m.p. 40 C; sapon. no. 280

Cetodan 70-00A. [Grindsted] Monoglyceride acetic acid ester; see Cetodan 50-00A; m.p. 35 C; sapon. no. 315

Cetodan 90-40. [Grindsted] Monoglyceride acetic acid ester; see Cetodan 50-00A; m.p. 10 C; sapon. no. 370

CF-42,500-T, Medium. [Custom Fibers] Cellulose fibers; asbestos replacement fibers providing increased visc. and sag resistance, dispersion, and fiber reinforcement for asphalt plastic roof cement, caulks, putties, aluminum roof coating, adhesives; gray finely divided material; sp.gr. 1.58; dens. 13.2 lb/gal; pH 6.9; oil absorp. 500–600%.

CF-70,000-WDK, Ex. Superfine. [Custom Fibers] Cellulosic fibers containing an anionic wetting agent; asbestos replacement fibers providing increased visc. and sag resistance, dispersion, and fiber reinforcement for textured coatings, adhesives, roof coatings, caulks and sealants, paints, flocking material; wh. finely divided material; sp.gr. 1.54; bulk dens. 13 lb/ft^3; flash pt. none; pH 6.5 ± 0.5; water absorp. 360%; oil absorp. 270%.

CF-72,500-T Superfine. [Custom Fibers] Cellulose fibers; asbestos replacement fibers providing increased visc. and sag resistance, dispersion, and fiber reinforcement for asphalt undercoatings, butyl caulks and tapes, sealants, adhesives; gray; sp.gr. 1.58; dens. 13.2 lb/gal; pH 7.4; oil absorp. 270%.

CG6710. [Hüls] (3-Glycidoxypropyl)-methyldiethoxy silane; coupling agent, chem. intermediate, blocking agent, release agent, lubricant, primer, reducing agent; liq.; m.w. 248.4; sp.gr. 0.98; b.p. 122–126 C (5 mm); flash pt. 122 C; ref. index 1.431; 95% purity.

CH7250. [Hüls] 1,1,3,3,5,5-Hexamethylcyclotrisilazane; coupling agent, chem. intermediate, blocking agent, release agent, lubricant, primer, reducing agent; liq.; m.w. 219.5; sp.gr. 0.920; b.p. 186–188 C; flash pt. 61 C; ref. index 1.445; 97% purity.
CH7260. [Hüls] Hexamethylcyclotrisiloxane; see CH7250; solid; m.w. 222.4; b.p. 134 C; 96% purity.
CH7280. [Hüls] Hexamethyldisilane; see CH7250; liq.; m.w. 146.4; sp.gr. 0.729; b.p. 112–113 C; flash pt. 40 C; ref. index 1.421; 97% purity.
CH7310. [Hüls] Hexamethyldisiloxane; see CH7250; liq.; m.w. 162.4; sp.gr. 0.76; b.p. 99–100 C; flash pt. –1 C; ref. index 1.377; 97% purity.
CH7332. [Hüls] Hexyltrichloro silane; see CH7250; liq.; m.w. 219.6; sp.gr. 1.11; b.p. 191–192 C; flash pt. 35 C; ref. index 1.444; 97% purity.
Charlab Leveler AT Special. [Catawba-Charlab] Sodium hydrocarbon sulfonate; anionic; detergent, emulsifier, textile leveling agent, softener, lubricant; amber solid; sol. in water (160 F); sp.gr. 1.0; m.p. 110–125; 80% act.
Charlab Leveler DSL. [Catawba-Charlab] Sodium hydrocarbon sulfonate; anionic; detergent, emulsifier, textile leveling agent, softener, lubricant; lt. brn. paste; sol. in water (160 F); dens. 1.03 lb/gal; 40% act.
Cheelox® B-13. [GAF] Tetrasodium EDTA, trisodium HEDTA; sequestrant for calcium, magnesium, and kier boiling cotton, soaping naphthol dyeings, and washing woolen piece goods after fulling; clarifier for liq. soaps and shampoos, industrial cleaner sol'ns., and herbicide formulations; liq.; 40% act.
Cheelox® BF-12. [GAF] Tetrasodium EDTA; sequestrants; components of soaker-alkali cleaners to prevent and remove scale formation; water softeners for sanitizers and detergent-sanitizers; rayon mfg.; liq.; 25% act.
Cheelox® BF-13. [GAF] Tetrasodium EDTA; see Cheelox BF-12; liq.; 38% act.
Cheelox® BF-78. [GAF] Tetrasodium EDTA; sequestrant used in applics. where a solid material is preferred; powd.; 89% act.
Cheelox® BF Acid. [GAF] EDTA; sequestrant; intermediate for preparation of org. and inorg. salts, metallic chelates; preparation of chelates of iron, copper, zinc, and manganese; powd.; 99.5% act.
Cheelox® DTPA-14. [GAF] DTPA, pentasodium salt; all purpose chelating agent used in pulp bleaching applics. using hydrogen peroxide; clear liq.
Cheelox® FE-12. [GAF] Mixed alkyldiaminepolyacetic acids, as sodium salts and alkanolamines; sequestrant for iron ions in aq. systems; liq.; 27% act.
Cheelox® HE-24. [GAF] Trisodium HEDTA; chelating agent used for iron; sequestrant for magnesium; liq.; 41% act.
Cheelox® NTA-14. [GAF] Nitriloacetate trisodium salt; sequestrant used for iron; 40% act.
Cheelox® NTA-Na3. [GAF] Nitriloacetate trisodium salt; see Cheelox NTA-14; 99.5% act.
Chel DM-41. [Ciba-Geigy] Trisodium HEDTA; chelating agent used in bar soaps, photographic developer baths, textiles, and min. separations; yel. clear sol'n.; 41% act.
Chel DTPA. [Ciba-Geigy] Pentetic acid; chelating agent used for stabilizing peroxides, biological preparations, cosmetic textiles, scale removal, and rare earth separations; wh. cryst. powd.; 98% act.
Chel DTPA-41. [Ciba-Geigy] Pentasodium diethylenetriaminepentaacetic acid; chelating agent for metals; yel. clear aq. sol'n.; 41% act.
Chelon 80. [Rhone-Poulenc Basic] Pentasodium diethylenetriaminepentaacetate aq. sol'n., tech. grade; chelating agent used in chemical industry for cosmetics; redox control and metal ion catalysis, agriculture, metal finishing, rare earth separation, radioactive decontamination, fermentation control, antibiotic and pharmaceutical production; peroxide bleaching of wood pulp; stabilizer in textile industry; pale straw clear aq. sol'n.; sp.gr. 1.30; dens. 10.8 lb/gal; chel. value 80 mg $CaCO_3$/g (@ pH 11); pH 11.5 (1% aq.); 40.2% min. act.
Chelon 100. [Rhone-Poulenc Basic] Tetrasodium EDTA; chelating agent for heavy metal and alkaline earth ions; pale straw clear aq. sol'n.; sp.gr. 1.31; dens. 10.9 lb/gal; chel. value 102 mg $CaCO_3$/g (@ pH 11); pH 11.5 (1% aq.); 39% min. act.
Chelon 120. [Rhone-Poulenc Basic] Trisodium n-hydroxyethylenediaminetriacetate, tech. grade; chelating agent for iron control in soap and cosmetic prods.; germicides; lt. straw clear aq. sol'n.; sp.gr. 1.29; dens. 10.7 lb/gal; chel. value 120 mg $CaCO_3$/g (@ pH 11); pH 11.5 (1% aq.); 41.3% min. act.
Chemal BP-261, -262, -262LF, -263, -264. [Chemax] Difunctional blocked polymers; nonionic; defoamer, emulsifier, demulsifier, dispersant, binder, stabilizer, wetting agent, and chemical intermediate; liq., cloud pt. 24, 32, 28, 34, and 58 C resp.; 100% act.
Chemal BP-268, -268/50. [Chemax] Difunctional blocked polymer; nonionic; see Chemal BP-261; solid and liq. resp.; cloud pt. > 100 C; 100% conc.
Chemal DA-4. [Chemax] Deceth-4; nonionic; wetting and penetrating agent for textile processing, clay soils, and fire fighting prods.; emulsifier for polyethylene emulsions; liq.; hyd. no. 165–185; 100% conc.
Chemal DA-6. [Chemax] Deceth-6; see Chemal DA-4; liq.; hyd. no. 131–145; 100% conc.
Chemal DA-9. [Chemax] Deceth-9; see Chemal DA-4; liq.; hyd. no. 95–110; 100% conc.
Chemal LA-4. [Chemax] Laureth-4; nonionic; o/w emulsifier, lubricant, detergent for cosmetic, silicone polish, and mold release prods.; liq.; hyd. no. 150–165; 100% conc.
Chemal LA-9. [Chemax] Laureth-9; nonionic; see Chemal LA-4; liq.; hyd. no. 90–110; 100% conc.
Chemal LA-12. [Chemax] Laureth-12; nonionic; see Chemal LA-4; liq.; hyd. no. 72–87.
Chemal LA-23. [Chemax] Laureth-23; nonionic; see Chemal LA-4; solid; hyd. no. 40–55.
Chemal OA-4. [Chemax] Oleth-4; nonionic; dispersant, detergent; emulsifier and solubilizer for topical cosmetic applics.; stabilizer and anticoagulant for natural and syn. latices; emulsifier for waxes used in coating citrus fruit; liq., semisolid.
Chemal OA-5. [Chemax] Oleth-5; nonionic; emulsi-

fier, lubricant, and solubilizer; liq.; 100% conc.

Chemal OA-9. [Chemax] Oleth-9; see Chemal OA-4; liq

Chemal OA-20. [Chemax] Oleth-20; see Chemal OA-4; also lubricant; semisolid

Chemal OA-20/70. [Chemax] Oleth-20; nonionic; emulsifier, lubricant, solubilizer; 70% conc.

Chemal TDA-3. [Chemax] Trideceth-3; nonionic; wetting agent, detergent, emulsifier, dispersant, foam stabilizer; solubilizer, penetrant for scouring and dye leveling in textiles, in cleaning and dishwashing compds.; liq.; 100% conc.

Chemal TDA-6. [Chemax] Trideceth-6; nonionic; see Chemal TDA-3; liq.; 100% conc.

Chemal TDA-9. [Chemax] Trideceth-9; nonionic; see Chemal TDA-3; liq.; 100% conc.

Chemal TDA-12. [Chemax] Trideceth-12; nonionic; see Chemal TDA-3; liq.; 100% conc.

Chemal TDA-15. [Chemax] Trideceth-15; nonionic; see Chemal TDA-3; solid; 100% conc.

Chemal TDA-18. [Chemax] Trideceth-18; nonionic; see Chemal TDA-3; 100% conc.

Chemax CO-5. [Chemax] PEG-5 castor oil; nonionic; emulsifier, lubricant for textiles; pigment dispersant in latex paints; essential oils solubilizer; liq.; oil-sol.; HLB 3.8; 100% conc.

Chemax CO-25. [Chemax] PEG-25 castor oil; nonionic; emulsifier; pigment dispersant in textiles; liq.; HLB 10.5

Chemax CO-28. [Chemax] PEG-28 castor oil; nonionic; see Chemax CO-25; HLB 11.1

Chemax CO-30. [Chemax] PEG-30 castor oil; nonionic; see Chemax CO-25; liq.; HLB 11.7; 100% conc.

Chemax CO-36. [Chemax] PEG-36 castor oil; see Chemax CO-25; liq.; HLB 12.6

Chemax CO-40. [Chemax] PEG-40 castor oil; nonionic; see Chemax CO-25; liq.; HLB 12.9; 100% conc.

Chemax CO-80. [Chemax] PEG-80 castor oil; nonionic; see Chemax CO-25; solid; HLB 15.8; 100% conc.

Chemax CO-200/50. [Chemax] PEG-200 castor oil; nonionic; see Chemax CO-25; liq.; HLB 18.1; 50% conc.

Chemax CS-10. [Chemax] Silicone defoamer; industrial, cosmetic, and food applics.; emulsion; 10% act.

Chemax CS-30. [Chemax] Silicone defoamer; industrial, cosmetic, and food applics.; emulsion; 30% conc.

Chemax CS-100. [Chemax] Silicone compd.; see Chemax CS-10; liq.; 100% act.

Chemax DNP-8. [Chemax] Nonyl nonoxynol-8; nonionic; emulsifier for nonpolar solv. and oils; detergent for cellulosic and syn. fibers; dispersant for hard surface cleaners and laundry compds.; liq.; pour pt. 32 F; 100% conc.

Chemax DNP-15. [Chemax] Nonyl nonoxynol-15; see Chemax DNP-8; liq.; pour pt. 64 F.

Chemax DNP-18. [Chemax] Nonyl nonoxynol-18; nonionic; emulsifier, detergent, solubilizer; solid; 100% conc.

Chemax DNP-150. [Chemax] Nonyl nonoxynol-150; see Chemax DNP-8; solid.

Chemax E-200 ML. [Chemax] PEG-4 laurate; nonionic; emulsifier for min. and cutting oils; dispersant, detergent, lubricant; coemulsifier and defoamer in water-based coatings; liq.; HLB 9.8; 100% conc.

Chemax E-200 MO. [Chemax] PEG-5 oleate; nonionic; oil-sol. emulsifier for min. and fatty oils, solv.; degreaser, dispersant, detergent, lubricant; liq.; HLB 8.0; 100% conc.

Chemax E-200 MS. [Chemax] PEG-5 stearate; nonionic; emulsifier for min. oils and fats used in polishes and metal buffing compds.; dye assistant, lubricant, softener, antistat; soft solid; HLB 7.9; 100% conc.

Chemax E-400 ML. [Chemax] PEG-8 laurate; nonionic; visc. control agent in plastisol formulations; wetting agent and defoamer in latex paint; see also Chemax E-200 ML; liq.; HLB 13.1; 100% conc.

Chemax E-400 MO. [Chemax] PEG-9 oleate; nonionic; emulsifier and lubricant for solv. and oils in pesticides and metal cleaners; see also Chemax E-200 MO; liq.; HLB 11.4; 100% conc.

Chemax E-400 MS. [Chemax] PEG-9 stearate; nonionic; lubricant and softener for syn. fibers see Chemax E-200 MS; soft solid; HLB 11.6; 100% conc.

Chemax E-600 ML. [Chemax] PEG-14 laurate; nonionic; water-sol. emulsifier and dispersant; liq.; HLB 14.9; 100% conc.

Chemax E-600 MO. [Chemax] PEG-14 oleate; nonionic; surfactant used as coemulsifier and lubricant in industrial formulations; liq.; HLB 13.5; 100% conc.

Chemax E-1000 MO. [Chemax] PEG-20 oleate; nonionic; PEG-20 oleate; nonionic; see Chemax E-200 MO; solid; 100% conc.

Chemax HCO-5. [Chemax] PEG-5 hydrog. castor oil; nonionic; emulsifier, lubricant, softener, dispersant; coemulsifier for syn. esters; liq.; oil-sol.; HLB 3.8; 100% conc.

Chemax HCO-16. [Chemax] PEG-16 hydrog. castor oil; nonionic; emulsifier, lubricant, dispersant, softener; liq.; HLB 8.4; 100% conc.

Chemax HCO-25. [Chemax] PEG-25 hydrog. castor oil; nonionic; emulsifier, lubricant, and softener; liq.; HLB 10.8; 100% conc.

Chemax HCO-200/50. [Chemax] PEG-200 hydrog. castor oil; nonionic; emulsifier and lubricant; liq.; water-sol.; HLB 18.1; 50% act.

Chemax NP-1.5. [Chemax] Nonoxynol-1.5; nonionic; surfactant, emulsion stabilizer in latex emulsions; detergent and dispersant for petrol. oils; emulsifier for silicone and agric. prods.; coupling and wetting agent; liq.; pour pt. –3 F.

Chemax NP-4. [Chemax] Nonoxynol-4; nonionic; see Chemax NP-1.5; liq.; water oil sol.; pour pt. –15 F; 100% conc.

Chemax NP-6. [Chemax] Nonoxynol-6; nonionic; see Chemax NP-1.5; liq.; pour pt. –26 F; 100% conc.

Chemax NP-9. [Chemax] Nonoxynol-9; nonionic; see Chemax NP 1.5; liq.; pour pt. 31 F; 100% conc.

Chemax NP-10. [Chemax] Nonoxynol-10; nonionic; see Chemax NP-1.5; liq.; pour pt. 49 F; 100% conc.

Chemax NP-40. [Chemax] Nonoxynol-40; nonionic; polymerization emulsifier for vinyl acetate and acrylic emulsions; stabilizer for syn. latices; wetting agent in electrolyte sol'ns.; solid; pour pt. 112 F; 100% conc.

Chemax NP-40/70. [Chemax] Nonoxynol-40; see Chemax NP-40; liq.; pour pt. 46 F.

Chemax NP-50. [Chemax] Nonoxynol-50; see Chemax NP-40; solid; pour pt. 114 F.

Chemax NP-50/70. [Chemax] Nonoxynol-50; see Chemax NP-40; liq.; pour pt. 52 F.

Chemax NP-100. [Chemax] Nonoxynol-100; see Chemax NP-40; solid; pour pt. 122 F.

Chemax NP-100/70. [Chemax] Nonoxynol-100; see Chemax NP-40; liq.; pour pt. 68 F.

Chemax OP-3. [Chemax] Octoxynol-3; nonionic; emulsifier, detergent, stabilizer, dispersant, wetting agent; pesticides and floor finishes; liq.; pour pt. –9 F.

Chemax OP-5. [Chemax] Octoxynol-5; see Chemax OP-3; liq.; oil sol.; pour pt. –15 F; 100% conc.

Chemax OP-40. [Chemax] Octoxynol-40; nonionic; see Chemax OP-5; solid; 100% conc.

Chemax PEG 400 DO. [Chemax] PEG-8 dioleate; emulsifier and solubilizer for solv., fats, and min. oils; liq.; oil-sol.; HLB 8.5.

Chemax SBO. [Chemax] Sulfated butyl oleate; softener, emulsifier, wetting agent in textile and metal working industries; liq.

Chemax SCO. [Chemax] Sulfated castor oil; see Chemax SBO; liq.

Chemax TO-8. [Chemax] PEG-8 tallate; nonionic; emulsifier, lubricant, dye assistant; liq.; 100% conc.

Chemax TO-16. [Chemax] PEG-16 tallate; nonionic; foam detergent and emulsifier; lubricant, dye assistant; liq.; HLB 13.3; 100% conc.

Chemazine 18, C, O, TO. [Chemax] Imidazoline; cationic; softener, antistat, emulsifier, corrosion inhibitor, filming agents; solid (18, C), liq. (O, TO).

Chemeen 18-2. [Chemax] PEG-2 stearamine; emulsifier and antistat in textiles, metal buffing, and rubber compds.; lubricant for fiber glass; solid; m.w. 362.

Chemeen 18-5. [Chemax] PEG-5 stearamine; see Chemeen 18-2; solid; m.w. 540.

Chemeen 18-50. [Chemax] PEG-50 stearamine; see Chemeen 18-2; solid; m.w. 2380.

Chemeen C-2. [Chemax] PEG-2 cocamine; emulsifier, antistat, dye leveler, wetting agent, lubricant, dispersant; substantive to metals, fibers, and clays; liq.; m.w. 290; 100% conc.

Chemeen C-5. [Chemax] PEG-5 cocamine; see Chemeen C-2; liq.; m.w. 425; 100% conc.

Chemeen C-10. [Chemax] PEG-10 cocamine; see Chemeen C-2; liq.; m.w. 645; 100% conc.

Chemeen C-15. [Chemax] PEG-15 cocamine; see Chemeen C-2; liq.; m.w. 890; 100% conc.

Chemeen DT-3. [Chemax] PEG-3 tallow diamine; emulsifier, textile dyeing assistant, corrosion inhibitor used in preparation of asphalt and agric. chemical emulsions; liq.; m.w. 535.

Chemeen DT-15. [Chemax] PEG-15 tallow aminopropylamine; see Chemeen DT-3; liq.; m.w. 1020.

Chemeen DT-30. [Chemax] PEG-30 tallow diamine; see Chemeen DT-3; liq.; m.w. 1665.

Chemeen HT-2. [Chemax] PEG-2 hydrog. tallow amine; emulsifier, antistat, lubricant; solid; 100% conc.

Chemeen HT-5. [Chemax] PEG-5 hydrog. tallow amine; emulsifier and antistat in textiles, metal buffing and rubber compds., lubricant for fiber glass; solid; m.w. 495.

Chemeen HT-15. [Chemax] PEG-15 hydrog. tallow amine; see Chemeen HT-5; solid; m.w. 925

Chemeen HT-50. [Chemax] PEG-50 hydrog. tallow amine; see Chemeen HT-5; solid; m.w. 2470; 100% conc.

Chemeen O-30, O-30/80. [Chemax] PEG-30 oleamine; emulsifier and textile dyeing assistant; antiprecipitant in cross dyeing; solid; m.w. 1595; 100% and 80% conc.

Chemeen S-2, -5. [Chemax] PEG-2, -5 soya amine; emulsifier, antistat, lubricant; liq.; 100% conc.

Chemeen S-30. [Chemax] PEG-30 soya amine; see Chemeen S-2; solid; 100% conc.

Chemeen S-30/80. [Chemax] PEG-30 soya amine; see Chemeen S-2; liq.; 100% conc.

Chemeen T-2. [Chemax] PEG-2 tallow amine; antistat for carpet shampoos; emulsifier, lubricant, dispersant, softener, antiprecipitant, leveling and migrating agent in textile dyeing process; liq.; m.w. 350; 100% conc.

Chemeen T-5. [Chemax] PEG-5 tallow amine; see Chemeen T-2; liq.; m.w. 490; 100% conc.

Chemeen T-10. [Chemax] PEG-10 tallow amine; see Chemeen T-2; liq.; m.w. 700; 100% conc.

Chemeen T-15. [Chemax] PEG-15 tallow amine; see Chemeen T-2; liq.; m.w. 930; 100% conc.

Chemeen T-20. [Chemax] PEG-20 tallow amine; see Chemeen T-2; liq.; m.w. 1120; 100% conc.

Chemfac PA-080. [Chemax] Phosphate ester; anionic; detergent, emulsifier, wetting agent, lubricant, antistat; liq.; 100% conc.

Chemfac PB-082, -104, -106, –109, -133, -135. [Chemax] Phosphate esters; anionic; detergent, wetting, and coupling agent, antistat; emulsifier for detergents in alkali sol'ns.; liq.; 99% min. act.

Chemfac PB-184. [Chemax] Oleth-4 phosphate; anionic; see Chemfac PA-080; liq.; 100% conc.

Chemfac PB-264, -804. [Chemax] Phosphate esters; see Chemfac PA-080; liq.; 100% conc.

Chemfac PC-006. [Chemax] POE-phenol phosphate; hydrotrope, detergent, emulsifier; foaming hydrotrope; liq.; 99% min. act.

Chemfac PC-188, PD-600. [Chemax] Phosphate esters; see Chemfac PA-080; liq.; 100% conc.

Chemfac PD-990. [Chemax] Phosphate ester; see Chemfac PA-080; 90% conc.

Chemfac PX-322. [Chemax] Phosphate ester; anionic; hydrotrope for solubilizing surfactants in alkali or other electrolytes; liq.; 50% conc.

Chemical 39 Base. [Sandoz] Stearamidoethyl ethanolamine; cosmetic and toiletry base; cationic emulsifying agent for creams and lotions; conditioning additive for hair prods.; lubricant in skin prods.;

Gardner 3 color; very slight amine odor; insol. in water; disp. as the amine salt.

Chemical Base 6532. [Sandoz] Stearamidoethyl diethylamine; cationic; emulsifier, emollient, conditioning agent for skin and hair; yel. tan solid; slight odor; m.p. 35–45 C; 100% act.

Chemiflex 315C. [Sanyo] Polyamide; adhesion promotor for PVC plastisols; liq.; 100% conc.

Chemiflex 315XA(80). [Sanyo] Blocked isocyanate; adhesion promoter for PVC plastisols to ED paint; liq.; 80% conc.

Chemistat 6120. [Sanyo] Polymer; anionic; electroconductive agent for electro fax paper; antistat for paper; lt. brnsh. liq.; water-sol.

Chemistat 6300. [Sanyo] Quat. ammonium polymer; electroconductive agent for dielectric paper of facsimile and electrofax paper; antistat for paper; lt. brnsh. liq.; water-sol.; 33% act.

Chemistat 6300 H. [Sanyo] Quat. ammonium polymer; cationic; antistat, electroconductive agent for dielec. paper of facsimile and electrofax paper; yel. liq.; pH 6.5; visc. 150 cps; 33% act.

Chemlink® 2000. [Sartomer] Long chain acrylated diol; curing agent; APHA G5 clear liq.; sp.gr. 0.934; visc. 17 cps; 99% react. esters.

Chemlink® 2100. [Sartomer] Long chain methacrylated diol; curing agent; APHA G6 clear liq.; sp.gr. 0.93; visc. 17 cps; 99+% react. esters.

Chemlink® 3000. [Sartomer] Epoxy acrylate; curing agent; APHA 3–5G clear visc. liq.; sp.gr. 1.20; dens. 10 lb/gal; visc. 1×10^6 cps.

Chemlink® 5000. [Sartomer] Acrylated BR; curing agent; APHA G3 clear visc. liq.

Chemlink® 9000. [Sartomer] Polyalkoxylated diacrylate; curing agent; APHA 175 clear liq.; m.w. 800; sp.gr. 1.002; visc. 120 cps; 99% react. esters.

Chemlink® 9001. [Sartomer] Polyalkoxylated diacrylate; curing agent; APHA 250 clear liq.; m.w. 1150; sp.gr. 0.991; visc. 360 cps; 99% react. esters.

Chemlink® 9003. [Sartomer] Alkoxylated aliphatic triacrylate (monomer); low skin irritation curing agent; APHA 50 clear liq.; sp.gr. 1.010; visc. 45 cps; ref. index 1.4456; flash pt. (PMCC) 340 F; 99+% react. esters.

Chemlink® 9008. [Sartomer] Functionalized triacrylate ester (monomer); high adhesion-promoting curing agent; APHA 260 clear liq.; pungent odor; sp.gr. 1.15; visc. 85 cps; ref. index 1.4700; flash pt. (PMCC) > 150 F; 98% react. esters.

Chemlink® 9012. [Sartomer] Functionalized aliphatic triacrylate (monomer); high adhesion-promoting curing agent; APHA 250 clear liq.; mild odor; b.p. > 200 C; sp.gr. 1.15; visc. > 150 cps; ref. index 1.4731; flash pt. (PMCC) 192 F; 98% react. esters.

Chemlink® 9013. [Sartomer] Functionalized aliphatic monoacrylate (monomer); high adhesion-promoting curing agent; APHA 50 clear liq.; mild odor; m.w. 121 (10 mm Hg); b.p. > 200 C; sp.gr. 0.892; dens. 5–10 lb/gal; ref. index 1.4348; flash pt. (PMCC) > 200 F; 99% react. esters.

Chemlink® 9014. [Sartomer] Functionalized aliphatic diacrylate (monomer); high adhesion-promoting curing agent; APHA 65 clear liq.; mild odor; m.w. > 400; sp.gr. 1.028; visc. 50 cps; ref. index 1.4449; flash pt. (PMCC) 340 F; 98.5% react. esters.

Chemlink® 9015. [Sartomer] Functionalized aliphatic monoacrylate (monomer); high adhesion-promoting curing agent; APHA 50 clear liq.; mild odor; m.w. 121 (10 mm Hg); sp.gr. 0.885; dens. 5–10 lb/gal; ref. index 1.4350; flash pt. (PMCC) > 200 F; 99% react. esters.

Chemlink® 9020. [Sartomer] Alkoxylated aliphatic triacrylate (monomer); low skin irritation curing agent; APHA 50 clear liq.; sp.gr. 1.0704; visc. 15 cps; ref. index 1.4590; flash pt. (PMCC) > 150 F; 99+% react. esters.

Chemlink® 9021. [Sartomer] Alkoxylated aliphatic triacrylate (monomer); low skin irritation curing agent; APHA 50 clear liq.; sp.gr. 1.0583; visc. 80 cps; ref. index 1.4572; flash pt. (PMCC) > 200 F; 99% react. esters.

Chemlink® 9022. [Sartomer] Alkoxylated aliphatic tetraacrylate (monomer); low skin irritation curing agent; APHA 50 clear liq.; sp.gr. 1.0787; visc. 151 cps; ref. index 1.4615; flash pt. (PMCC) > 200 F; 99+% react. esters.

Chemlink® 9023. [Sartomer] Alkoxylated tetraacrylate ester (monomer); low skin irritation curing agent; APHA 100 ± 25 clear liq.; sp.gr. 1.1554; visc. 211 cps; ref. index 1.4760; flash pt. (PMCC) > 200 F; 99% react. esters.

Chemlink® 9024. [Sartomer] Alkoxylated aliphatic diacrylate ester (monomer); low skin irritation curing agent; APHA 50 clear liq.; mild acrylic odor; sp.gr. 1.0336; visc. 45 cps; ref. index 1.4650; flash pt. (PMCC) > 200 F; 99% react. esters.

Chemlink® 9025. [Sartomer] Alkoxylated aliphatic diacrylate ester (monomer); low skin irritation curing agent; APHA 75 clear liq.; sp.gr. 1.1496; visc. 15 cps; ref. index 1.4589; flash pt. (PMCC) > 200 F; 99+% react. esters.

Chemlink® 9040. [Sartomer] Functionalized aliphatic diacrylate (Pro 62) (monomer); high adhesion-promoting curing agent; APHA 200 clear liq.; mild acrylic odor; sp.gr. 1.028; dens. 10–15 lb/gal; visc. 45 cps; flash pt. (PMCC) 340 F; 99% react. esters.

Chemlink® 9503. [Sartomer] Aliphatic urethane acrylate; curing agent; APHA 100 ± 25 clear liq.; weak odor; visc. 67,000 cps.

Chemlink® 9504. [Sartomer] Aliphatic urethane acrylate; curing agent; APHA 75 ± 25 clear visc. liq.; weak odor; visc. 25,000 cps (60 C).

Chemlink® 9505. [Sartomer] Aliphatic urethane acrylate; curing agent; APHA 100 ± 25 clear visc. liq.; weak odor.

Chemlok® 205. [Lord] One-coat bonding agent for NBR and primer for cover coat adhesive; gray; dens. 7.6–8.1 lb/gal; visc. 85–165 cps; flash pt. (PMCC) 66 F; 22–26% solids in MIBK, MEK.

Chemlok® 210. [Lord] Adhesive for bonding castable and RIM urethane elastomers to metal; clear to hazy; dens. 7.3–7.6 lb/gal; visc. 250–350 cps; flash pt. (PMCC) 27 F; 22.5–25.5% solids in MEK.

Chemlok® 212. [Lord] Adhesive for bonding castable and RIM urethane to metal; blue; dens. 7.90 ± 0.15 lb/gal; visc. 350 ± 50 cps; flash pt. (Seta) 29 F; 24 ± 1.5% solids in MEK, ethyl 3-ethoxypropionate.
Chemlok® 218. [Lord] One-coat adhesive for castable urethane; clear amber; dens. 7.9–8.2 lb/gal; visc. 750–1050 cps; flash pt. (PMCC) 36 F; 18–21% solids in toluene, MIBK.
Chemlok® 220. [Lord] General purpose cover coat bonding agent; bk.; dens. 8.3–9.1 lb/gal; visc. 135–300 cps; flash pt. (PMCC) 83 F; 23–27% solids in toluene, xylene.
Chemlok® 222. [Lord] General purpose cover coat bonding agent; bk.; dens. 8.4–8.8 lb/gal; visc. 120–300 cps; flash pt. (PMCC) 89 F; 23.5–27.5% solids in toluene, xylene.
Chemlok® 233. [Lord] General purpose cover coat bonding agent; bk.; dens. 9.05–9.45 lb/gal; visc. 100–300 cps; flash pt. (PMCC) 92 F; 22–26% solids in toluene, xylene.
Chemlok® 234B. [Lord] General purpose cover coat and PV bonding agent; bk.; dens. 8.9–9.2 lb/gal; visc. 450–900 cps; flash pt. (PMCC) 83 F; 23–26.5% solids in toluene, xylene.
Chemlok® 235. [Lord] Nonconductive general purpose cover coat bonding agent; br.; dens. 8.8–9.2 lb/gal; visc. 400–800 cps; flash pt. (PMCC) 83 F; 21–25% solids in toluene, xylene.
Chemlok® 236A. [Lord] Bonding agent; general purpose cover coat for EPDM, nonpolar elastomer; bk.; dens. 8.3–8.6 lb/gal; visc. 300–700 cps; flash pt. (PMCC) 73 F; 16–19% solids in toluene, xylene.
Chemlok® 238. [Lord] Bonding agent; general purpose cover coat for EPDM, nonpolar elastomers; bk.; dens. 7.5–7.9 lb/gal; visc. 200–800 cps; flash pt. (PMCC) 92 F; 16–19% solids in toluene, xylene.
Chemlok® 246. [Lord] Bonding agent; cover coat for Parel; bk.; dens. 7.1–7.5 lb/gal; visc. 15–120 cps; flash pt. (PMCC) 30 F; 15–20% solids in MEK, toluene.
Chemlok® 250. [Lord] Bonding agent; general purpose one-coat adhesive; bk.; dens. 9.3–9.7 lb/gal; visc. 200–550 cps; flash pt. (PMCC) 94 F; 23.5–27.5% solids in toluene, xylene.
Chemlok® 252. [Lord] Bonding agent; general purpose one-coat adhesive; bk.; dens. 10.5–11.0 lb/gal; visc. 250–850 cps; flash pt. (PMCC) > 200 F; 17.5–20.5% solids in toluene, xylene.
Chemlok® 253. [Lord] Bonding agent; general purpose one-coat adhesive; bk.; dens. 10.4–10.9 lb/gal; visc. 800–1800 cps; flash pt. (PMCC) > 200 F; 22.0–25.0% solids in toluene, xylene.
Chemlok® 402. [Lord] Bonding agent; one-coat textile to rubber adhesive; bk.; dens. 9.8–10.5 lb/gal; visc. 100–350 cps; flash pt. (PMCC) > 200 F; 13.5–16.5% solids in toluene, xylene.
Chemlok® 459. [Lord] Bonding agent; primer promoting adhesion to thermoplastic elastomers and polyolefins; amber; dens. 10.4 lb/gal; visc. 70 cps; flash pt. > 200 F; 7% solids.
Chemlok® 607. [Lord] Silane-based one-coat adhesive for bonding elastomers, esp. silicone, fluroelastomers, EPDM, nitrile, epichlorohydrin, polyacrylate, hydrog. nitriles; clear to slightly yel. liq.; dens. 6.9 lb/gal; visc. 5 cSt; flash pt. (Seta) 9 C; 6.1–6.9% NV in ethanol/methanol.
Chemlok® 608. [Lord] Bonding agent; one-coat adhesive for silicone and fluorosilicone; clear to hazy yel.; dens. 7.0 lb/gal; visc. 2 cps; flash pt. (PMCC) 38 F; 17–20% solids in methanol, ethanol.
Chemlok® 610. [Lord] Silane-based one-coat adhesive for bonding elastomers, esp. fluroelastomers, nitrile, epichlorohydrin, polyacrylate, hydrog. nitriles; clear red liq.; dens. 8.4 lb/gal; visc. 5 cSt; flash pt. (Seta) 93 C; 2.5–3.0 NV in water.
Chemlok® 810. [Lord] Bonding agent; waterborne one-coat adhesive for nitrile; pink; dens. 9.3 lb/gal; visc. 50 cps; flash pt. (PMCC) > 200 F; 38% solids in water.
Chemlok® AP-133. [Lord] Silane-based one-coat adhesive for bonding elastomers, esp. silicone, fluroelastomers, EPDM, nitrile, polyacrylate, hydrog. nitriles; clear to slightly yel. liq.; dens. 6.65–6.95 lb/gal; visc. 5 cSt max.; flash pt. (Seta) 14 C; 4.8–6.2% NV in methanol/toluene/ethanol.
Chemlok® BN. [Lord] Bonding agent; one-coat for specialty elastomers; bk.; dens. 7.7–8.1 lb/gal; visc. 20–35 cps; flash pt. (PMCC) 42 F; 35–39% solids in MEK, ethanol, MIBK.
Chemlok® KP-1001. [Lord] Silane-based one-coat adhesive for bonding elastomers, esp. silicone; clear to slightly yel. liq.; dens. 6.55–6.85 lb/gal; visc. 8 cSt max.; flash pt. (Seta) 14 C; 4.8–7.2% NV in IPA/methanol/ethanol
Chemlok® Y-1520A. [Lord] Silane-based one-coat adhesive for bonding elastomers, esp. fluroelastomers, nitrile, epichlorohydrin, polyacrylate, hydrog. nitriles; clear to slightly yel. liq.; dens. 6.9 lb/gal; visc. 5 cSt max.; flash pt. (Seta) 17 C; 8% NV in ethanol/methanol.
Chemlok® Y-1530. [Lord] Silane-based one-coat adhesive for bonding elastomers, esp. silicone, fluroelastomers, EPDM, nitrile, polyacrylate, polyphosphazene; clear to slightly yel. liq.; dens. 6.8 lb/gal; visc. 5 cSt max.; flash pt. (Seta) 17 C; 7% NV in ethanol/methanol.
Chemlok® Y-1540. [Lord] Silane-based one-coat adhesive for bonding elastomers, esp. silicone, fluroelastomers, silicone/EPDM blends, hydrog. nitriles; clear to slightly red liq.; dens. 7.05 lb/gal; visc. 5 cSt; flash pt. (Seta) 12 C; 12% NV in ethanol/methanol.
Chemlok® Y-4310. [Lord] Silane-based one-coat adhesive for bonding elastomers, esp. fluroelastomers, EPDM, nitrile, polyacrylate, hydrog. nitriles; clear to slightly yel. liq.; dens. 6.55–6.85 lb/gal; visc. 5 cSt max.; flash pt. (Seta) 13 C; 4.8–6.2% NV in methanol/toluene/ethanol.
Chemlok® Y-5323. [Lord] Silane-based one-coat adhesive for bonding elastomers, esp. fluroelastomers, nitrile, epichlorohydrin, polyacrylate, hydrog. nitriles; clear to slightly yel. liq.; dens. 6.65–6.95 lb/gal; visc. 5 cSt max.; flash pt. (Seta) 13 C; 9.8–11.8% NV in ethanol/methanol.

Chemprene R-10. [Chemfax] Thermoplastic isoprenoidal polymer; used in compding. of syn., natural, and reclaim rubber; softener, tackifier, and reinforcing agent; used in calendering and extruding, hard rubber, molded goods, in camelback, tubes, and tire stocks, in rubber cements, wire and cable insulation, floor tile, shoe compds., caulks; Barrett 2 visc. liq. to semisolid; sol. in min. spirits, ketones, cyclic alchols, etc.; sp.gr. 1.02; soften. pt. 10 C.

Chemprene R-25. [Chemfax] Thermoplastic isoprenoidal polymer; see Chemprene R-10; Barrett 2 visc. liq. to semisolid; sol. see Chemprene R-10; sp.gr. 1.04; soften. pt. 25 C.

Chemprene R-50. [Chemfax] Thermoplastic isoprenoidal polymer; see Chemprene R-10; Barrett 2 visc. liq. to semisolid; sol. see Chemprene R-10; sp.gr. 1.06; soften. pt. 50 C.

Chemprene R-70. [Chemfax] Thermoplastic isoprenoidal polymer; see Chemprene R-10; Barrett 2 solid; sol. see Chemprene R-10; sp.gr. 1.08; soften. pt. 70 C.

Chemprene R-100. [Chemfax] Thermoplastic isoprenoidal polymer; see Chemprene R-10; Barrett 2 solid, flakes; sol. see Chemprene R-10; sp.gr. 1.10; soften. pt. 100 C.

Chemprene R-115. [Chemfax] Thermoplastic isoprenoidal polymer; see Chemprene R-10; solid and flakes; sol. see Chemprene R-10; sp.gr. 1.10; visc. (G-H) Z1 (70% in min. spirits); soften. pt. 115 C; acid no. 0; sapon. no. 0.

Chemquat 12-33. [Chemax] Laurtrimonium chloride; corrosion inhibitor, antistat, gel sensitizer in latex foam prod.; visc. depressant in softener formulations; 33% act.

Chemquat 12-50. [Chemax] Laurtrimonium chloride; see Chemquat 12-33; 50% act.

Chemquat 16-50. [Chemax] Cetrimonium chloride; see Chemquat 12-33; 50% act.

Chemstat 122. [Chemax] Bis (2-hydroxyethyl) alkyl amine; internal antistat for polyolefins; liq.

Chemstat 122/60DC. [Chemax] Bis (2-hydroxyethyl) alkyl amine; internal antistat for polyolefins for easier dry cleaning; powd.

Chemstat 172. [Chemax] Bis (2-hydroxyethyl) alkyl amine; see Chemstat 122; liq.

Chemstat 182. [Chemax] Bis (2-hydroxyethyl) alkyl amine; see Chemstat 122; paste.

Chemstat 182/67DC. [Chemax] Bis (2-hydroxyethyl) alkyl amine; see Chemstat 122/60DC; powd.

Chemstat 192. [Chemax] Bis (2-hydroxyethyl) alkyl amine; see Chemstat 122; solid.

Ches® 500. [CasChem] Emulsifier and select hydrocolloids blend; stabilizer; cold hot emulsion system; unique ambient temp. emulsifier which yields stable,, aesthetic o/w emulsions; powd.

Chimassorb® 944FL. [Ciba-Geigy] Polymeric hindered amine [N,N′-bis (2,2,6,6-tetramethyl-4-piperidinyl)-1,6-hexanediamine, polymer with 2,4,6-trichloro-1,3,5-triazine and 2,4,4-trimethyl-1,2-pentanamine]; lt. stabilizer; heat stabilizer for polyolefins; wh. to off-wh. powd.; m.w. > 2500; sol. > 50% in acetone, benzene, chloroform, methylene chloride; 40% in hexane; soften. pt. 100–135 C.

Chimassorb® 944LD. [Ciba-Geigy] Polymeric hindered amine (N,N′-bis (2,2,6,6-tetramethyl-4-piperidinyl)-1,6-hexane-diamine, polymer with 2,4,6-trichloro 1,3,5-triazine and 2,4,4-trimethyl-1,2-pentanamine; lt. stabilizer for use in HDPE and PP; antioxidant; wh. powd.; m.w. > 2500; very sol. in chloroform and hydrocarbons; slightly sol. in low alcohols; insol. in water; soften. pt. 115–125 C.

Chimin AX. [Tessilchimica] Cocamidopropyl betaine; amphoteric; conditioner, emulsifier, detergent for cosmetic use in personal care prods.; liq.; 30% conc.

Chimin BX. [Tessilchimica] Stearyl betaine; ampholite; dispersant, wetting, antistat; soft paste; 40% conc.

Chimin KSP. [Tessilchimica] Fatty alcohol ester; nonionic; fat liquoring of leather, substitute for sperm oil, textile lubricant; liq. 100% conc.

Chimin P10. [Tessilchimica] Alkyl polyphosphate; anionic; wetting and penetrating agent; liq.; 100% conc.

Chimin P20. [Tessilchimia] Complex phosphate ester acid form; antifoamer in industrial washing; flakes.

Chimin P45. [Tessilchimica] Sodium alkylpolyglycol ether phosphate; anionic; detergent, antistat, wetting agent for textile, paper, cleaners, machine washing, and personal care prods.; liq.; 30% conc.

Chimin P50. [Tessilchimica] Phosphoric acid complex org. ester; nonionic; emulsifier, antistat for use in leather, textile and cosmetic applics.; liq.; sol. in water and solv.; 100% conc.

Chimipal APG 400. [Tessilchimica] Polyglycol laurate; nonionic; pigment dispersant and wetter, emulsion polymerization; liq.; HLB 13.0; 100% conc.

Chimipal DCL. [Tessilchimica] Cocamide DEA; nonionic; thickener and foam stabilizer, detergent, and shampoo additive; liq.; 100% conc.

Chimipal DS 6000. [Tessilchimica] PEG-6000 distearate; thickener, conditioner for shampoos, bubble baths, and cosmetic preparations; flakes.

Chimipal FV. [Tessilchimica] Monoglyceride, ethoxylated; nonionic; superfatting agent in cosmetic preparations; liq.; water-sol.; 100% conc.

Chimipal LDA. [Tessilchimica] Lauramide DEA; nonionic; thickener and foam stabilizer, detergent and shampoo additive; 100% conc.

Chimipal MC. [Tessilchimica] Cocamide MEA; nonionic; thickener and foam stabilizer, detergent and shampoo additive; flakes; 100% conc.

Chimipal NH. [Tessilchimica] Fatty amine ethoxylate; cationic; leveling agent, emulsifier; 100% conc.

Chimipal OLD. [Tessilchimica] Oleamide DEA; nonionic; thickener for shampoos, bubble baths and liq. detergents; liq.; 100% conc.

Chimipal OS 2. [Tessilchimica] Fatty alcohol ethoxylate blend; nonionic; powd. detergent, wax emulsifier, leveling agent; solid; 100% conc.

Chimipal PE 300, 302. [Tessilchimica] Polypropylene glycol ethoxylate; nonionic; detergent, dispersant, emulsifier, foam controller; liq.; 100% conc.

Chimipal PE 400. [Tessilchimica] Straight chain alcohol EO/PO; nonionic; detergent, dispersant, wetting, emulsifier, deduster; liq.; 100% conc.

Chimipal SGE. [Tessilchimica] Ethylene glycol stearate; pearlescent agent; emulsifier; solid.

Chlorez 700. [Dover] Resinous chlorinated paraffin; flame retardant for LDPE; soften. pt. 100 C; 70% chlorine.

Chlorez 700-DF. [Dover] Chlorinated paraffin; flame retardant for paints, printing inks, plastics, foams, adhesives, paper and fabric coatings; APHA 150 max. gran.-flakes; sp.gr. 1.7; soften. pt. (B&R) 95–110 C; ref. index 1.54–1.55 (105 C); 69–72% chlorine.

Chlorez 760. [Dover] Resinous chlorinated paraffin; flame retardant for LDPE and PP; soften. pt. 160 C; 74% chlorine.

Chloropren-Faktis A. [Rhein Chemie] Sulfur factice; processing aid for tech. molded, extruded, and calendered rubber articles, CR; dk. brn. crushed lumps; sp.gr. 1.03.

Chloropren-Faktis NW. [Rhein Chemie] Sulfur chloride factice; processing aid for oil and petrol-resistant molded and extruded CR, NBR rubber goods; wh. fine powd.; sp.gr. 1.10.

Chlorothene SM Solvent. [Dow] Trichloroethane; formulation solv.; compat. with urethane prods.; cuts greases and oils readily; sol. in most org. solvs.; very low water sol.

CHP-158. [Witco/Argus] 80% sol'n. of cumene hydroperoxide; initiator for vinyl monomers and the curing of unsat. polyester resins; 8.4% act. oxygen.

CHPTA. [ChemY GmbH] 3 Chloro-2-hydroxypropyltrimethyl ammonium chloride; cationic; starch modifier for textiles; liq.; 60% conc.

Chromosol SS. [Nippon Senka] Silicon complex; fixing agent for acid dyes on wool; liq.

Chupol C. [Takemoto] Polyglycol ether; nonionic; reducing agent for concrete; liq.

Chupol EX. [Takemoto] Hydroxylated carboxylic acid; anionic; water reducing agent for concrete; liq.

CHY300EG. [Hüls] Hexamethyldisilazane; adhesion promoter for photoresists incl. positive and negative photoresists on SiO_2 substrates; highly purified for electronic applics.; APHA 8 max. liq.; ammonia odor; m.w. 161.4; b.p. 126.0 C; insol. in water, reacts slowly; sp.gr. 0.7742; visc. 0.90 cstk; ref. index 1.4080; flash pt (CC). 27 C; dielec. const. 2.27 (1000 Hz); 99% conc.

CI7840. [Hüls] Isocyanatopropyltriethoxy silane; coupling agent, chem. intermediate, blocking agent, release agent, lubricant, primer, reducing agent; liq.; m.w. 247.4; sp.gr. 0.99; b.p. 130 C (20 mm); flash pt. 98 C; ref. index 1.419; 95% purity.

Cibaphasol® AS. [Ciba-Geigy] Sulfuric acid ester; anionic; leveling and penetrating agent for continuous dyeing and printing of nylon; liq

Cirrasol AEN-XB. [ICI Specialty] Fatty alcohol ethoxylate; nonionic; fiber processing aid; antistat for PVC belting; emulsifier for paraffin waxes, oleins, castor oil into water; liq.; HLB 10.2; 100% conc.

Cirrasol AEN-XF. [ICI Specialty] Fatty alcohol ethoxylate; nonionic; see Cirrasol AEN-XB; semisolid; HLB 13.8; 100%

Cirrasol AEN-XZ. [ICI Specialty] Alkyl phenol ethoxylate; nonionic; fiber processing aid; detergent base, emulsifier, dispersant, wetting, dedusting agent, and antistat; liq.; HLB 12.3; 100% conc.

Cirrasol ALN-FP. [ICI Specialty] Fatty acid ethoxylate; nonionic; fiber processing aid; antistat, emulsifier, and gelling agent; semisolid; HLB 11.6; 100% conc.

Cirrasol ALN-GM. [ICI Specialty] Polyolester ethoxylate; nonionic; fiber processing aid; antistat for plastics and rubber; wetting agent; liq.; HLB 16.9; 100% conc.

Cirrasol ALN-WF. [ICI Specialty] Fatty alcohol ethoxylate; nonionic; fiber processing and antistat lubricant for syn. fibers; emulsifier for waxes; flakes; HLB 14.9; 100% conc.

Cirrasol ALN-WY. [ICI Specialty] Triglyceride, ethoxylated; nonionic; fiber processing aid; wool lubricant; liq.; 90% conc.

Cirrasol EN-MB. [ICI Specialty] Fatty alcohol ethoxylate; nonionic; fiber processing aid; emulsifier/stabilizer for textile lubricant compositions; liq.; 100% conc.

Cirrasol LC-HK. [ICI Specialty] Amine alkyl deriv.; cationic; fiber processing aid; lubricant and softener; paste; 33% conc.

Cirrasol LC-PQ. [ICI Specialty] Diamine fatty acid deriv.; cationic; fiber processing aid; lubricant and softener; paste; 25% conc.

Cirrasol LN-GS. [ICI Specialty] Syn. alcohol fatty acid ester; fiber processing aid; heat-stable lubricant; liq.; 100% conc.

Cithrol 2DL. [Croda Ltd.] PEG-4 dilaurate; nonionic; used in cosmetic and industrial applics.; dispersant, w/o emulsifier, wetting agent and co-solv.; paste; 97% conc.

Cithrol 2DO. [Croda Ltd.] PEG-4 dioleate; nonionic; see Cithrol 2 DL; liq.; 97% conc.

Cithrol 2DS. [Croda Ltd.] PEG-4 distearate; nonionic; antistat in textile finishing; paste; HLB 5.2; 97% conc.

Cithrol 2ML. [Croda Ltd.] PEG-4 laurate; nonionic; wetting agent, emulsifier, detergent, thickener, solubilizer, dispersant, softener, lubricant, antistat, dye asistant, penetrant for cosmetics, textiles, glass fiber, metal treatment; liq.; m.w. 200; HLB 8.8; sapon. no. 158–168; 100% conc.

Cithrol 2MO. [Croda Ltd.] PEG-4 oleate; nonionic; wetting agent, penetrant, detergent, emulsifier, solubilizer, thickening agent, dispersant, textile aux., softener, lubricant for textiles, cosmetics, metalworking; m.w. 200; HLB 6.2; sapon. no. 110–112; 100% conc.

Cithrol 2MS. [Croda Ltd.] PEG-4 stearate; nonionic; emulsifier for insecticides and cosmetics, detergent, wetting agent, solubilizer and thickening agent for perfumery, antifrothing agent, foaming agent; m.w. 200; m.p. 39–41 C; HLB 6.3; sapon. no. 110–120; 100% conc.

Cithrol 3DS. [Croda Ltd.] PEG-6 distearate; nonionic; see Cithrol 2 DS; solid; 97% conc.
Cithrol 3MS. [Croda Ltd.] PEG-6 stearate; see Cithrol 2MS; paste; m.w. 300; m.p. 30–33 C; HLB 10.7; sapon. no. 95–105; 100% conc.
Cithrol 4DL. [Croda Ltd.] PEG-8 dilaurate; nonionic; see Cithrol 2 DL; liq.; 97% conc.
Cithrol 4DO. [Croda Ltd.] PEG-8 dioleate; nonionic; see Cithrol 2 DL; liq.; 97% conc.
Cithrol 4DS. [Croda Ltd.] PEG-8 distearate; nonionic; see Cithrol 2 DS; solid; 97% conc.
Cithrol 4ML. [Croda Ltd.] PEG-8 laurate; nonionic; see Cithrol 2ML; m.w. 400; HLB 19; sapon. no. 92–98.
Cithrol 4MO. [Croda Ltd.] PEG-8 oleate; nonionic; see Cithrol 2MO; liq.; m.w. 400; HLB 11.4; sapon. no. 85–93; 100% conc.
Cithrol 4MS. [Croda Ltd.] PEG-8 stearate; nonionic; see Cithrol 2MS; paste; m.w. 400; m.p. 31–34 C; HLB 11; sapon. no. 95–105; 10% conc.
Cithrol 6DL. [Croda Ltd.] PEG-12 dilaurate; nonionic; see Cithrol 2 DL; liq.; 97% conc.
Cithrol 6DO. [Croda Ltd.] PEG-12 dioleate; nonionic; see Cithrol 2 DL; liq.; 97% conc.
Cithrol 6DS. [Croda Ltd.] PEG-12 distearate; nonionic; see Cithrol 2 DS; solid; 97% conc.
Cithrol 6ML. [Croda Ltd.] PEG-12 laurate; nonionic; see Cithrol 2ML; also for leather treatment; m.w. 600; HLB 15.8; sapon. no. 64–74.
Cithrol 6MO. [Croda Ltd.] PEG-12 oleate; nonionic; see Cithrol 2MO; liq.; m.w. 600; HLB 13.1; sapon. no. 65–75; 100% conc.
Cithrol 6MS. [Croda Ltd.] PEG-12 stearate; nonionic; see Cithrol 2MS; paste; m.w. 600; m.p. 33–35 C; HLB 14; sapon. no. 68–76; 100% conc.
Cithrol 10DL. [Croda Ltd.] PEG-20 dilaurate; nonionic; see Cithrol 2 DL; paste; 97% conc.
Cithrol 10DO. [Croda Ltd.] PEG-20 dioleate; nonionic; see Cithrol 2 DL; liq.; 97% conc.
Cithrol 10DS. [Croda Ltd.] PEG-20 distearate; nonionic; see Cithrol 2 DS; solid; 97% conc.
Cithrol 10ML. [Croda Ltd.] PEG-20 laurate; nonionic; see Cithrol 2ML; sol. in water, alcohol, polar solv.; m.w. 1000; HLB 18; sapon. no. 51–55; 100% conc.
Cithrol 10MO. [Croda Ltd.] PEG-20 oleate; nonionic; see Cithrol 2MO; liq.; m.w. 1000; HLB 16.2; sapon. no. 35–40; 100% conc.
Cithrol 10MS. [Croda Ltd.] PEG-20 stearate; nonionic; see Cithrol 2MS; solid; m.w. 1000; m.p. 34–40 C; HLB 16; sapon. no. 36–50; 100% conc.
Cithrol 15MS. [Croda Ltd.] PEG 1500 stearate; nonionic; see Cithrol 2MS; solid; m.w. 1500; m.p. 30-41 C; HLB 17; sapon. no. 30–55; 100% conc.
Cithrol 40MO. [Croda Ltd.] PEG-75 oleate; see Cithrol 2MO; m.w. 4000; HLB 18.7; sapon. no. 10–15.
Cithrol 40MS. [Croda Ltd.] PEG-75 stearate; nonionic; see Cithrol 2MS; solid; m.w. 4000; m.p. 30–41 C; HLB 18.8; sapon. no. 12–17; 100% conc.
Cithrol 60ML. [Croda Ltd.] PEG 6000 laurate; see Cithrol 2ML; m.w. 6000; HLB 19.2; sapon. no. 8–13.
Cithrol 60MO. [Croda Ltd.] PEG 6000 oleate; see Cithrol 2MO; m.w. 6000; HLB 19.0; sapon. no. 5–10.
Cithrol A. [Croda Ltd.] PEG-8 oleate; nonionic; w/o emulsifier, dispersant, antistat for textile, paper processing, cutting oils, polishes, emulsion cleaners, rubber latex, wool lubricants; amber liq.; slight agreeable odor; sol. in wh. spirit and liq. paraffin; completely sol. in xylol, ethyl acetate, methylated spirits, oleic acid; disp. in water; sp.gr. 1.013; HLB 11.4; flash pt. 215 C; sapon. no. 85–93; pH 6.5–7.5 (10%); 97% conc.
Cithrol DGDL N/E. [Croda Ltd.] Diethylene glycol dilaurate; nonionic; see Cithrol A; paste; 100% conc.
Cithrol DGDL S/E. [Croda Ltd.] Diethylene glycol dilaurate SE; anionic; see Cithrol A; paste; 100% conc.
Cithrol DGDO N/E. [Croda Ltd.] Diethylene glycol dioleate; nonionic; see Cithrol A; liq.; 100% conc.
Cithrol DGDO S/E. [Croda Ltd.] Diethylene glycol dioleate SE; anionic; see Cithrol A; liq.; 100% conc.
Cithrol DGDS N/E. [Croda Ltd.] Diethylene glycol distearate; nonionic; see Cithrol A; solid; 100% conc.
Cithrol DGDS S/E. [Croda Ltd.] Diethylene glycol distearate SE; anionic; see Cithrol A; solid; 100% conc.
Cithrol DGML N/E. [Croda Ltd.] Diethylene glycol laurate; nonionic; see Cithrol A; paste; 100% conc.
Cithrol DGML S/E. [Croda Ltd.] Diethylene glycol laurate SE; anionic; see Cithrol A; paste; 100% conc.
Cithrol DGMO N/E. [Croda Ltd.] Diethylene glycol oleate; nonionic; see Cithrol A; liq.; 100% conc.
Cithrol DGMO S/E. [Croda Ltd.] Diethylene glycol oleate SE; anionic; see Cithrol A; liq.; 100% conc.
Cithrol DGMS N/E. [Croda Ltd.] Diethylene glycol stearate; nonionic; see Cithrol A; solid; 100% conc.
Cithrol DGMS S/E. [Croda Ltd.] Diethylene glycol stearate SE; anionic; see Cithrol A; solid; 100% conc.
Cithrol DPGML N/E. [Croda Ltd.] Dipropylene glycol laurate; nonionic; see Cithrol A; liq.; 100% conc.
Cithrol DPGML S/E. [Croda Ltd.] Dipropylene glycol laurate SE; anionic; see Cithrol A; liq.; 100% conc.
Cithrol DPGMO N/E. [Croda Ltd.] Dipropylene glycol oleate; nonionic; see Cithrol A; liq.; 100% conc.
Cithrol DPGMO S/E. [Croda Ltd.] Dipropylene glycol oleate SE; anionic; see Cithrol A; liq.; 100% conc.
Cithrol DPGMS N/E. [Croda Ltd.] Dipropylene glycol stearate; nonionic; see Cithrol A; solid; 100% conc.
Cithrol DPGMS S/E. [Croda Ltd.] Dipropylene glycol stearate SE; anionic; see Cithrol A; solid; 100% conc.
Cithrol EGDL N/E. [Croda Ltd.] Ethylene glycol dilaurate; nonionic; see Cithrol A; solid; 100% conc.
Cithrol EGDL S/E. [Croda Ltd.] Ethylene glycol dilaurate SE; anionic; see Cithrol A; solid; 100% conc.
Cithrol EGDO N/E. [Croda Ltd.] Ethylene glycol dioleate; nonionic; see Cithrol A; liq.; 100% conc.
Cithrol EGDO S/E. [Croda Ltd.] Ethylene glycol

dioleate SE; anionic; see Cithrol A; liq.; 100% conc.

Cithrol EGDS N/E. [Croda Ltd.] Ethylene glycol distearate; nonionic; see Cithrol A; solid; 100% conc.

Cithrol EGDS S/E. [Croda Ltd.] Ethylene glycol dioleate SE; anionic; see Cithrol A; solid; 100% conc.

Cithrol EGML N/E. [Croda Ltd.] Ethylene glycol laurate; nonionic; see Cithrol A; liq.; 100% conc.

Cithrol EGML S/E. [Croda Ltd.] Ethylene glycol laurate SE; anionic; see Cithrol A; liq.; 100% conc.

Cithrol EGMO N/E. [Croda Ltd.] Ethylene glycol oleate; nonionic; see Cithrol A; liq.; 100% conc.

Cithrol EGMO S/E. [Croda Ltd.] Ethylene glycol oleate SE; anionic; see Cithrol A; liq.; 100% conc.

Cithrol EGMR N/E. [Croda Ltd.] Ethylene glycol ricinoleate; nonionic; see Cithrol A; liq.; 100% conc.

Cithrol EGMR S/E. [Croda Ltd.] Ethylene glycol ricinoleate SE; anionic; see Cithrol A; liq.; 100% conc.

Cithrol EGMS N/E. [Croda Ltd.] Ethylene glycol stearate; nonionic; see Cithrol A; solid; 100% conc.

Cithrol EGMS S/E. [Croda Ltd.] Ethylene glycol stearate SE; anionic; see Cithrol A; solid; 100% conc.

Cithrol G Range. [Croda Ltd.] Glyceryl esters; dispersant for cosmetic and industrial lipophilic systems; stabilizer for o/w chemicals; thickener for cosmetic and pharmaceutical preparations; wax-like solid.

Cithrol GDL N/E. [Croda Ltd.] Glyceryl dilaurate; nonionic; see Cithrol A; liq.; 100% conc.

Cithrol GDL S/E. [Croda Ltd.] Glyceryl dilaurate SE; anionic; see Cithrol A; liq.; 100% conc.

Cithrol GDO N/E. [Croda Ltd.] Glyceryl dioleate; nonionic; emulsifier, coemulsifier, stabilizer, wetting agent, lubricant, and antistat; used in cosmetic, pharmaceutical, industrial, food applics.; liq.; 100% conc.

Cithrol GDO S/E. [Croda Ltd.] Glyceryl dioleate SE; anionic; see Cithrol GDO N/E; liq.; 100% conc.

Cithrol GDS N/E. [Croda Ltd.] Glyceryl distearate; nonionic; see Cithrol GDO N/E; solid; 100% conc.

Cithrol GDS S/E. [Croda Ltd.] Glyceryl distearate SE; anionic; see Cithrol GDO N/E; solid; 100% conc.

Cithrol GML N/E. [Croda Ltd.] Glyceryl laurate; nonionic; see Cithrol GDO N/E; liq.; 100% conc.

Cithrol GML S/E. [Croda Ltd.] Glyceryl laurate SE; anionic; see Cithrol GDO N/E; liq.; 100% conc.

Cithrol GMO N/E. [Croda Ltd.] Glyceryl oleate; nonionic; see Cithrol GDO N/E; liq.; 100% conc.

Cithrol GMO S/E. [Croda Ltd.] Glyceryl oleate SE; anionic; see Cithrol GDO N/E; liq.; 100% conc.

Cithrol GMR N/E. [Croda Ltd.] Glyceryl ricinoleate; nonionic; see Cithrol GDO N/E; liq.; 100% conc.

Cithrol GMR S/E. [Croda Ltd.] Glyceryl ricinoleate SE; anionic; see Cithrol GDO N/E; liq.; 100% conc.

Cithrol GMS A/S ES 0743. [Croda Ltd.] Glyceryl stearate; emulsifier and opacifier for cosmetics; Gardner 3 color; m.p. 52 C; sapon. no. 96.

Cithrol GMS A/S ES 0743. [Croda Ltd] Glyceryl stearate; emulsifier and opacifier for cosmetics; base for pharmaceutical creams and ointments; Gardner 3; m.p. 52 C; sapon. no. 96.

Cithrol GMS Acid Stable. [Croda Ltd.] Glyceryl stearate, S/E; nonionic; see Cithrol GDO N/E; solid; 100% conc.

Cithrol GMS N/E. [Croda Ltd.] Glyceryl stearate; nonionic; see Cithrol GDO N/E; solid; 100% conc.

Cithrol GMS S/E. [Croda Ltd.] Glyceryl stearate SE; anionic; see Cithrol GDO N/E; solid; 100% conc.

Cithrol L, O Range. [Croda Ltd.] Polyglycol esters; dispersant for cosmetic and industrial hydrophilic systems; liq., solid; water-sol..

Cithrol PGML N/E. [Croda Ltd.] Propylene glycol laurate; nonionic; see Cithrol GDO N/E; liq.; 100% conc.

Cithrol PGML S/E. [Croda Ltd.] Propylene glycol laurate SE; anionic; see Cithrol GDO N/E; liq.; 100% conc.

Cithrol PGMO N/E. [Croda Ltd.] Propylene glycol oleate; nonionic; see Cithrol GDO N/E; liq.; 100% conc.

Cithrol PGMO S/E. [Croda Ltd.] Propylene glycol oleate SE; anionic; see Cithrol GDO N/E; liq.; 100% conc.

Cithrol PGMR N/E. [Croda Ltd.] Propylene glycol ricinoleate; nonionic; see Cithrol GDO N/E; liq.; 100% conc.

Cithrol PGMR S/E. [Croda Ltd.] Propylene glycol ricinoleate SE; anionic; see Cithrol GDO S/E; liq.; 100% conc.

Cithrol PGMS N/E. [Croda Ltd.] Propylene glycol stearate; nonionic; see Cithrol GDO N/E; solid; 100% conc.

Cithrol PGMS S/E. [Croda Ltd.] Propylene glycol stearate SE; anionic; see Cithrol GDO S/E; solid; 100% conc.

Cithrol S Range. [Croda Ltd.] Polyglycol esters; dispersant for cosmetic and industrial hydrophilic systems; liq./solid; water-sol.

Citroflex A-2. [Morflex] Acetyltriethyl citrate; plasticizer for cellulosics; m.w. 318.

Citroflex A-4. [Morflex] Acetyl tri-n-butyl citrate; plasticizer for PVC, PVDC, esp. food films, medical articles; APHA 30 max. clear iq., mild odor; m.w. 402.5; sol. in toluene, heptane; sp.gr. 1.045–1.055; visc. 33 cps; flash pt. (COC) 400 F; pour pt. –59 C; ref. index 1.441–1.443; 99% min. ester content.

Citroflex A-6. [Morflex] Vinyl plasticizer for medical applics. and other toxicologically sensitive areas; APHA 100 max. clear liq., mild odor; m.w. 486; sol. in toluene, heptane; sp.gr. 1.003–1.007; visc. 36 cps; flash pt. (COC) > 500 F; pour pt. –57 C; ref. index 1.445–1.449; 99% min. ester content.

Citroflex B-6. [Morflex] Vinyl plasticizer for medical applics.; APHA 100 max. clear liq., mild odor; m.w. 514; sol. in toluene, heptane; sp.gr. 0.991–0.995; visc. 28 cps; flash pt. (COC) > 500 F; pour pt. –55 C; ref. index 1.444–1.448; 99% min. ester content.

Citroflex C-2. [Morflex] Triethyl citrate; plasticizer for cellulosics; m.w. 276.

Citroflex C-4. [Morflex] Tri-n-butyl citrate; plasticizer for cellulosics; m.w. 360.

Citrosol 50. [Pfizer] Citric acid in water, 50% sol'n.; industrial grade sequestrant for di- and trivalent ions;

for copper, iron, and nickel ions.

Citrosol 50E. [Pfizer] Citric acid sol'ns., 50% industrial grade; sequestrant for copper and iron in cleaning formulations; dk. brn. sol'n.; burnt sugar odor; f.p. –25 C; water-misc.; 50–56% conc.

Citrosol 50T. [Pfizer] Citric acid sol'n., 50% industrial grade; see Citrosol 50E; yel.-grn. clear to hazy sol'n.; odorless; f.p. –12 C; water-misc.; 50–52% conc.

Citrosol 50W. [Pfizer] Citric acid sol'ns., 50% industrial grade; see Citrosol 50E; clear, colorless to yel.-grn. sol'n.; odorless; f.p. –15 C; water-misc.; 50–52% conc.

Clarase®. [Miles Lab] Fungal alpha-amylase; enzyme; hydrolysis of starch and dextrins with formation of substantial quantities of maltose; liq., powd.

Clarex® L. [Miles Lab] Pectinase; enzyme; depectinization of fruit juices, grape processing, berry processing and wine prod.; liq.

Clearate LV, WD. [Cleary] Soya lecithin deriv.; nonionic; pigment dispersant, wetting agent, and antiflocculant in paint systems; liq.; 56% conc.

Clearate Special Extra. [Cleary] Lecithin; nonionic; dispersant in inks, paints, foods; wetting agent; liq.

Clearcol. [Croda] Soluble animal collagen; protein for moisturizing and conditioning of facial systems where clarity is important; clear to slightly hazy visc. liq.; m.w. 300,000; 1% act.

Clink PR-46. [Yoshimura] Alkyl phosphate; rust preventer for steel or parts of loom after washing; liq.

Clorafin® 40. [Hercules] Chlorinated paraffin; used to flameproof and waterproof textiles; plasticizer for vinyl, resins or filmformers; additive in cutting oils and drawing compds.; Gardner 6 max. visc. liq.; misc. with common solvs. except lower aliphatic alcohols and water; sp.gr. 1.16; dens. 9.6 lb/gal; visc. 26–29 poise; 42% chloride content.

Clorafin® 50. [Hercules] Chlorinated paraffin; see Chlorafin 40; Gardner 6 max. visc. liq.; sol. see Clorafin 40; sp.gr. 1.26; dens. 10.4 lb/gal; visc. 430–620 poise; 50% chloride content.

CM-66. [Solem] Alumina trihydrate; filler used for mfg. of cultured onyx; proprietary surf. treatment to reduce matrix visc. and improve air release properties, resulting in higher filler loading.

CM-88. [George-Pacific] Amine-lignosulfonate; cationic; modifier; replaces 40–60% of a primary emulsifier; solid; dens. 1.25 g/ml; 46% solid.

CM8450. [Hüls] 3-Mercaptopropylmethyldimethoxy silane; coupling agent, chem. intermediate, blocking agent, release agent, lubricant, primer, reducing agent; liq.; m.w. 180.4; sp.gr. 0.99; b.p. 96 C (30 mm); flash pt. 97 C; ref. index 1.4502; 95% purity.

CM8750. [Hüls] Methyldichloro silane; see CM8450; liq.; m.w. 115.0; sp.gr. 1.11; b.p. 41–42 C; flash pt. –25 C; ref. index 1.422; 99% purity.

CM8930. [Hüls] Methylphenyldichloro silane; see CM8450; liq.; m.w. 191.1; sp.gr. 1.17; b.p. 205–206 C; flash pt. 83 C; ref. index 1.518; 96% purity.

CM8980. [Hüls] Methyltriacetoxy silane; see CM8450; solid; m.w. 220.3; sp.gr. 1.17; b.p. 87–88 C (3 mm); flash pt. 5 C; ref. index 1.408; 96% purity.

CM9000. [Hüls] Methyltrichloro silane; see CM8450; liq.; m.w. 149.5; sp.gr. 1.275; b.p. 66–67 C; flash pt. –10 C; ref. index 1.411; 98% purity.

CM9220. [Hüls] Methyltris (methylethylketoxime) silane; see CM8450; yel. liq.; m.w. 301.5; sp.gr. 0.98; b.p. 110–111 C (2 mm); flash pt. 3 C; 90% purity.

CMC-6CTL. [Hercules] Sodium CMC, crude grade; soil suspending agent for powd. detergents; gran. powd.; sol. in water; dens. 0.65 g/ml; visc. 25–150 cps (2%); 65% purity.

CMC-6-DG-L. [Hercules] Sodium CMC, crude grade; anionic; soil suspending agent for powd. detergents, soaps, syn. and built detergents; gran. powd.; visc. 25–150 cps; 90% purity.

CMC-T. [Hercules] Sodium CMC, tech. grade; thickener, binder, fil-former, stabilizer, and suspending aid in textile, adhesive, detergent, paper, paint, and other industries; off-wh. to tan powd., gran., or pellet; odorless; sol. in water; dens. 0.75 g/ml; visc. 20–600 cps (2%); 96% conc.

CMC Warp Size. [Hercules] Sodium CMC; warp size for cotton, syn. fibers, polyester/cotton blends; wh. to tan powd., pellets, gran. powd., fine powd.; odorless; sol. in water; dens. 43 lb/ft^3 (packed pellets); visc. 1500 cps (60%); pH 7.0–8.5 (2%).

CMHEC-37L. [Hercules] Sodium carboxymethyl hydroxyethyl cellulose; hydrophilic colloid; high water-binding capacity and good flocculating action on suspended solids; forms good films when cast from water sol'n.; lt. cream to wh. gran. powd.; sol. in water; sp.gr. 1.58 (film); dens. 0.6 g/ml; visc. 25–85 cps (2%); ref. index. 1.530 (film); tens. str. 10,000 psi; tens. elong. 5% (break).

CMTC-T. [Hercules] Sodium CMC; gum used as syn. detergent to prevent redeposition of soil on fabrics; powd.; water-sol.

Co-0127 T 3/16″. [M&T Harshaw] Cobalt, reduced and stabilized; catalyst for oxo aldehyde hydrogenations and reactions; tablet, 3/16″ diam.; dens. 57 lb/ft^3; 39% conc.

Co-0138 E 1/8″. [M&T Harshaw] Cobalt, reduced and stabilized; see Co-0127 T 3/16″; extrusion, 1/8″ diam.; dens. 33 lb/ft^3; 25% conc.

Co-0164 T 3/16″. [M&T Harshaw] Cobalt, reduced and stabilized; see Co-0127 T 3/16″; tablet; dens. 50 lb/ft^3; 25% conc.

Co-0301 E 3/16″. [M&T Harshaw] Cobalt oxide; catalyst used in synthesis of hydrogen sulfides; extrusion, 3/16″ diam.; dens. 75 lb/ft^3; 10% conc.

CO-1695. [Procter & Gamble] Cetyl alcohol; emollient.

CO-1895. [Procter & Gamble] Stearyl alcohol; emollient.

CO9745. [Hüls] Octadecyldimethyl [3-(trimethoxysilyl) propyl] ammonium chloride; coupling agent, chem. intermediate, blocking agent, release agent, lubricant, primer, reducing agent; sol'n.; m.w. 496.3; sp.gr. 0.89; flash pt. 24 C; 50% in methanol.

CO9750. [Hüls] Octadecyltrichloro silane; see CO9745; liq.; m.w. 398.0; sp.gr. 0.98; b.p. 160–162 C (3 mm); flash pt. 193 C; ref. index 1.460; 95%

purity.

CO9800. [Hüls] Octamethylcyclotetrasilazane; see CO9745; solid; m.w. 292.7; sp.gr. 0.95; b.p. 224–225 C; m.p. 97 C; flash pt. 66 C; ref. index 1.458; 98% purity.

CO9810. [Hüls] Octamethylcyclotetrasiloxane; see CO9745; liq.; m.w. 296.6; sp.gr. 0.95; b.p. 175–176 C; flash pt. 60 C; ref. index 1.393; 98% purity.

CO9816. [Hüls] Octamethyltrisiloxane; see CO9745; liq.; m.w. 236.0; sp.gr. 0.82; b.p. 152–153 C; flash pt. 38 C; ref. index 1.385; 96% purity.

CO9817. [Hüls] Octaphenylcyclotetrasiloxane; see CO9745; solid; m.w. 793.2; sp.gr. 1.18; m.p. 200–201 C; flash pt. 200 C; 99% purity.

CO9819. [Hüls] Octyldimethylchloro silane; see CO9745; liq.; m.w. 206.8; sp.gr. 0.79; b.p. 222–225 C; flash pt. 92 C; 97% purity.

CO9830. [Hüls] Octyltrichloro silane; see CO9745; liq.; m.w. 247.7; sp.gr. 1.07; b.p. 224–26 C; flash pt. 85 C; ref. index 1.447; 96% purity.

Coagulant CHA. [Bayer] Cyclohexyl amine acetate; coagulant used in dipped goods from natural latex; wh. cryst.; dens. 1.12 g/cm³; pH 7.5 (10% aq.).

Coagulant WS. [Mobay] Functional polyorganosiloxane; coagulant; heat sensitizer used in latex; ylsh. liq.; dens. 1.03 g/cm³; pH 8.

Cobee 76. [Stepan/PVO] Refined, bleached, deodorized coconut oil; emollient, superfatting agent, clouding agent, used in food, cosmetic, and pharmaceutical industries; Lovibond 20/2 soft solid; no odor or taste; m.p. 76–80 F.

Cobee 92. [Stepan/PVO] Refined, bleached, deodorized, hydrog. coconut oil; emol-lient, superfatting agent, clouding agent, used in the cosmetic and food industries; Lovibond 20/2 solid; no odor or taste; m.p. 98-114 F.

Cobee 110. [Stepan/PVO] Refined, bleached, deodorized, hydrog. coconut oil; see Cobee 92; Lovibond 20/2 solid; odorless; m.p. 110–114 F.

Cobratec® 99. [PMC] Benzotriazole; chelating agent and sesquestrant for copper ions; corrosion inhibitor for copper, brass, bronze; used in antifreeze, cleaners, coatings, detergents, functional fluids, metalworking fluids, pkg. materials, polishes, polymers, water circulating systems; photographic grade avail.; off-wh. powd. or flakes; sol. in alcohols, ketones, and glycols; slightly sol. in water; 98% min. assay.

Cobratec® CBT. [PMC] Carboxy benzotriazole (isomeric mixt.); copper and steel corrosion inhibitor; wh. to cream solid.

Cobratec® DBT-50, DBT-100. [PMC] Dodecyl benzotriazole; oil-sol. corrosion inhibitor for copper and copper alloys; 50% sol'n. in min. seal oil and dark, very visc. liq. resp.

Cobratec® PT. [PMC] Benzotriazole and other inhibitors formulated prod.; dilution in water establishes complete corrosion inhibitor baths for treating copper and copper alloys; slightly alkaline clear yel. liq.; water-misc.

Cobratec® TT-35-A. [PMC] 35% Tolyltriazole in a triethanolamine-water blend; corrosion inhibitor for copper, brass, bronze, and ferrous metals; for metalworking fluids.

Cobratec® TT-50-A. [PMC] 50% Tolyltriazole in triethanolamine; corrosion inhibitor for copper, brass, bronze, and ferrous metals; for metalworking fluids.

Cobratec® TT-50-S. [PMC] Sodium tolyltriazole; see Cobratec 99; water-misc.; 50% aq. sol'n.

Cobratec® TT-85. [PMC] Sodium tolyltriazole; see Cobratec 99; off-wh. powd.; water-sol.

Cobratec® TT-100. [PMC] Tolyltriazole; see Cobratec 99; tan gran.; sol. see Cobratec 99.

Cohedur A. [Bayer] Methylene donor; bonding agent for fabrics and cord bonded to rubber; colorless visc. liq.; dens. 1.2 g/cm³.

Cohedur A Solid. [Bayer] Methylene donor with filler; see Cohedur A; wh. powd.; dens. 1.51 g/cm³.

Cohedur AS. [Mobay] Methylene donor for resorcinol-based bonding agents; used for three-component bonding systems, comprising resorcinol, methylene donor, and reinforcing silica filler; used where fabrics and cord (incl. metals or glass) are bonded directly to rubber; colorless to lt. yel. liq.; dens. 1.2 g/cm³.

Cohedur AS Powder. [Mobay] Methylene donor absorbed on Microcel E; see Cohedur AS; beige powd.; dens. 1.55 g/cm³; 72% act.

Cohedur RK. [Bayer] Resorcinol component with filler, 1:1 ratio; bonding agent in polychloroprene; wh. powd.; dens. 1.55 g/cm³.

Cohedur RL. [Mobay] Resorcinol sol'n. in Cohedur A, 1:1 ratio; see Cohedur A; orange-brn. visc. liq.; dens. 1.2 g/cm³.

Cohedur RS. [Bayer] Homogeneous solidified melt of resorcinol and stearic acid, 2:1 ratio; see Cohedur A; orange-brn. visc. liq.; dens. 1.2 g/cm³.

Collagen Hydrolyzate Cosmetic 55. [Amerchol] Hydrolyzed animal protein; conditioner for shampoos, conditioners, hair groom aids, hair color and skin care prods.; water-sol'n.

Collagen Hydrolyzate Cosmetic N-55. [Amerchol] Enzymatic hydrolysate of animal protein; low odor and low ash; conditioner used in shampoos, conditioners, hair groom aids, hair color and skin care prods.; water-sol'n.

Collagen Hydrolyzate Cosmetic SD. [Amerchol] Hydrolyzed animal protein; see Collagen Hydrolyzate Cosmetic N-55; spray-dried powd.

Collasol. [Croda] Sol. animal collagen; humectant; hygroscopic film former; conditioner; skin care prods.; off-wh. disp.; m.w. 300,000; 3% act.

Colloid 102. [Rhone-Poulenc] Ammonium polyacrylate; dispersant for paper coatings, clay slurries, and ceramics; water-sol.

Colloid 111. [Rhone-Poulenc] Polycarboxylate, sodium salt; general purpose paint dispersant; liq.

Colloid 111M. [Rhone-Poulenc] Polycarboxylate, amine salt; paint dispersant; water spot resistance; liq.

Colloid 117/50. [Rhone-Poulenc] Polyacrylic acid; dispersant for boiler water compds.; liq.

Colloid 119/50. [Rhone-Poulenc] Polyacrylic acid sol'n.; dispersant; scale control in aq. heat transfer;

liq.

Colloid 202. [Rhone-Poulenc] Sodium polyacrylate; dispersant for antiredeposition and detergent aid in spray-dried laundry prods.; liq.

Colloid 204. [Rhone-Poulenc] Polyacrylic acid; dispersant for dishwashing detergents, industrial cleaners, chlorine stability excellent; liq.; water-sol.

Colloid 207. [Rhone-Poulenc] Sodium polyacrylate; dispersant for laundry, car wash, institutional detergents; excellent chlorine stability; soil suspension; liq.; water-sol.

Colloid 208. [Rhone-Poulenc] Sodium polyacrylate; dispersant; builder additive for consumer phosphate and non-phosphate detergents, process aid in mfg. by spray dryer or agglomerator; liq.; water-sol.

Colloid 211. [Rhone-Poulenc] Sodium polyacrylate; pigment dispersant; clay slurries; liq.

Colloid 218D. [Rhone-Poulenc] Sodium polyacrylate; dispersant for dishwashing, sequesters over wide pH and temp. range; liq.

Colloid 223. [Rhone-Poulenc] Sodium polyacrylate; dispersant for scale control; liq.

Colloid 223D. [Rhone-Poulenc] Sodium polyacrylate; see Colloid 223; dry.

Colloid 225. [Rhone-Poulenc] Polycarboxylate, sodium salt; dispersant for paints, and reacting pigments.; liq.; water-sol.

Colloid 226/35. [Rhone-Poulenc] Polycarboxylate, sodium salt; paint dispersant-ZnO and other reactive; liq.

Colloid 230. [Rhone-Poulenc] Sodium polyacrylate; calcium carbonate dispersant; liq.

Colloid 231. [Rhone-Poulenc] Polycarboxylate, sodium salt; paper coating dispersant; water-sol.

Colloid 233. [Rhone-Poulenc] Sodium polyacrylate; dispersant; builder additive for detergents, mfg. by an agglomerator/blender silicate compatible; liq.

Colloid 245D. [Rhone-Poulenc] Sodium polyacrylate; dispersant for detergent, car and truck wash, industrial cleaners, dishwash; powd.; water-sol.

Colloid 252. [Rhone-Poulenc] Polycarboxylate, ammonia salt; dispersant for ceramics, paper coatings, and clay slurries; water-sol.

Colloid 274. [Rhone-Poulenc] Polyacrylic acid; dispersant for calcium scale control; 50% act.

Colloid 350. [Rhone-Poulenc] Sodium polyacrylate; dispersant; calcium carbonate slurries; grinding aid.

Colloid 1010. [Rhone-Poulenc] Silicone emulsion; defoamer; latex cleaner; emulsifiable.

Colorin 102, 104, 202. [Sanyo] Polyether polyol; antifoamer used in fermentation of glutamic acid, yeast, antibiotics; liq.

Colorin 301, 302. [Sanyo] Polyether polyol; antifoamer used in acrylonitrile refining process; liq.

Colorol 20. [Lucas Meyer] High m.w. phosphoamino compds. dissolved in solvs. compat. with NC; amphoteric; wetting and dispersing agent, binder for NC and NC-modified lacquers; brn. visc. liq.; sp.gr. 1.02; visc. 2000 mPas; flash pt. 30 C; pH 6–7; 90% solids; contains xylene.

Colorol 46. [Lucas Meyer] Phosphoamino and surfactants; wetting agent for treated silicates; flatting agent; used for flat varnishes based on NC, NC-alkyds, polyester, acid curing resins; lt. yel. liq.; sp.gr. 0.85; flash pt. 15 C; pH 6.0; 28% solids; contains ethanol.

Colorol 70. [Lucas Meyer] Phosphoamino compds.; amphoteric; wetting and dispersing agent, binder for 2-part PU varnishes, epoxy-varnishes in systems based on PVC, PVA; ylsh. brn. liq.; sp.gr. 0.97; 50% solids; contains isopropyl acetate.

Colorol E. [Lucas Meyer] Phosphoamino compds.; amphoteric; wetting and dispersing agent for aq. systems; used for water-thinnable paints, leather finishes, alkyd systems, electrodeposition, emulsion paints; brn. visc. liq.; sp.gr. 1.038; visc. 1500 cp; pH 5.5–6.5; 90% solids.

Colorol F. [Lucas Meyer] Phosphoamino compds. modified with film-formers, dissolved in aromatic hydrocarbon; amphoteric; wetting and dispersing agent, binder for pigmented air-dry and stoving finishes, anticorrosive paints, alkyd and acrylic paints, zinc dust primers, marine paints, epoxy and bitumen paints, two-part systems, putty; brn. visc. liq.; sp.gr. 1.02; visc. 1300–1600 cp; flash pt. 74 C; pH 5.5–6.5; 90% solids; contains Shellsol AB.

Colorol Rust Binder. [Lucas Meyer] Corrosion inhibitor; penetrates and wets rust and similar metal oxides allowing water-repellent film to coat and protect; contains xylene.

Colorol Standard. [Lucas Meyer] Phosphoamino compds. modified with film-formers; amphoteric; wetting and dispersing agent, binder for pigmented air-dry and stoving finishes, anticorrosive paints, alkyd and marine paints, zinc dust paints, bituminous paints, epoxy resins, putty, solventless floor sealers, two-part finishes; visc. to plastic brn. mass; sp.gr. 1.01; pH 6.5; 99–100% solids.

Colsol. [Scher] Fatty amide; cationic; cotton and wool softener and lubricant for the textile industry; finishing agent; paste; 25% act.

Combat® HPC-40, HPF-325, HTP-40, HTP-325, SHP-40, SHP-325, SRG-40, SRG-325. [Carborundum] Boron nitride; powds. exhibiting high thermal conduct. and high dielec. str.; used as additives to silicone and epoxy resins, fluorinated hydrocarbons, silicone oils, etc. to increase thermal conduct. or decrease thermal expansion; high-temp. pressure media; friction modifiers; release agents; fine and coarse particles; bulk dens. 0.61, 0.37, 0.61, 0.37, 0.15, 0.31, 0.24, 0.31 g/cm^3 resp.

Combizell AG 070, AG 200. [Aqualon] Methylcellulose ether; thickener, rheology agent, binder, protective colloid, water-retention aid for latex-based systems incl. building materials and paints; wh. gran. powd.; visc. 4000–7000 and 15,000–20,000 mPas resp. (2%).

Combizell APR 060, APR 200. [Aqualon] Methylcellulose ether; see Combizell AG 070; wh. retarded powd.; visc. 4000–7000 and 20,000–25,000 mPas resp. (2%).

CoMo-0402 T 1/8″. [M&T Harshaw] Cobalt molybdate on silica alumina support; catalyst used for desulfurization of hydrocarbons by hydrotreating;

tablet, 1/8″ diam.; dens. 60 lb/ft^3; 15% conc.

CoMo-0603 T 1/8″. [M&T Harshaw] Cobalt molybdate on nonsilicated alumina support; see CoMo-0402 T 1/8″; tablet, 1/8″ diam.; dens. 64 lb/ft^3; 12% conc.

Comperlan COD. [Henkel] Cocamide DEA; foam and visc. increasing agent with emulsifying properties; conditioner for shampoos and other cosmetic and pharmaceutical applics.; solv.; liq.; 85% conc.

Comperlan F. [Henkel] Linoleamide DEA; nonionic; foam stabilizer and thickener for shampoos; vitamin E substitute; liq.; 100% conc.

Comperlan HS. [Henkel] Stearamide MEA; nonionic; consistency-giving factor in personal care prods.; flakes; 100% conc.

Comperlan ID. [Henkel] 1:1 Isostearamide DEA; hair conditioning agent; amber visc. liq.; 100% conc.

Comperlan KD. [Henkel] Cocamide DEA; nonionic; thickener and foamer for personal care prods.; hair conditioner; amber visc. liq.; 100% conc.

Comperlan KDO. [Henkel] 1:1 Cocamide DEA; nonionic; thickening and foaming agent for personal care prods.; amber visc. liq.; 100% conc.

Comperlan KM. [Henkel] Cocamide MEA; booster, foam stabilizer, pearly sheen improving agent for surfactant preparations; paste; 30% conc.

Comperlan LD. [Henkel] Lauramide DEA; nonionic; foam booster, stabilizer, detergency and visc. builder, emulsifier for personal care prods.; waxy solid; 100% conc.

Comperlan LD 9. [Henkel] Distearyldimonium chloride; cationic; superamide; foam and visc. builder in personal care prods.; paste; 75% conc.

Comperlan LDO, LDS. [Henkel] 1:1 Lauramide DEA; nonionic; foam booster, stabilizer, detergency and visc. builder, emulsifier for personal care prods.; amber visc. liq. 100% conc.

Comperlan LM. [Henkel] Lauramide MEA; nonionic; foam and visc. builder for personal care prods. and cleansers; solid; 100% conc.

Comperlan LMD. [Henkel] Lauramide DEA; nonionic; booster and thickener for personal care prods.; liq.; 100% conc.

Comperlan LS. [Henkel] Cocamide DEA and laureth-12; emulsifier, thickener, and foam stabilizer for personal care prods.; dissolving intermediates; liq.; 65–75% conc.

Comperlan OD. [Henkel] Oleamide DEA; foam stabilizer, thickener and superfatting agent for shampoos and bubble bath prods.; liq.; 90% conc.

Comperlan PD. [Henkel] 2:1 cocamide DEA; nonionic; thickener, foam builder/stabilizer for personal care prods., dishwashing agents; amber visc. liq.; 98% conc.

Comperlan SDO. [Henkel] Cocamide DEA; nonionic; foam and visc. builder for personal care prods.; liq.; 100% conc.

Comperlan SM. [Henkel] Cocamide MEA; nonionic; superamide, foam and visc. builder; gran.; 100% conc.

Comperlan UDM. [Henkel] Undecylenamide MEA; fungicide; antimycotic preparation for shampoos and bubble baths; solid; 85% act.

Comperlan VOD. [Henkel] Linoleamide DEA; nonionic; foam and visc. builder with superfatting effect for personal care prods.; liq.; 100% conc.

Complemix® 100. [Am. Cyanamid] Sodium dioctyl sulfosuccinate NF; anionic; emulsifier, wetting or clouding agent used in the food and beverage industry; dispersant, solubilizer, modifier; wh. waxy solid; water sol.; sp.gr. 1.1; 100% act.

Complexion CA-10. [Pulcra SA] EDTA, tetrasodium salt; chelating agent for calcium, magnesium, and other divalent and trivalent ions; liq.

Compound 8-S. [DuPont] Complex alkylaryl sulfonate; anionic; nonfoaming dispersant for dyes, carbon black; liq.; water sol.; sp.gr. 1.152; dens. 9.6 lb/gal.

Compound 9264B. [Stepan/PVO] Ethoxylated glycerol monostearate; softener, nonionic; solid.

Comsol 200. [Angus] Nitroparaffin; used in solv. blends in inks and coatings; sol. in non-aq. systems.

Conacure® AH-40. [Conap] Aromatic diamine; curing agent for urethane systems; liq.; sp.gr. 1.02; visc. 400 cps.

Conacure® AH-50. [Conap] Polyol; low-visc. curing agent for urethane systems; liq.; sp.gr. 1.15; visc. 750 cps.

Con-BACN. [Tosoh] Condensed bromoacenaphthylene; flame retardant imparting radiation resistance to rubbers and resins; used for insulated wires and cables (EPDM, CSM, crosslinked PE), protectors from x-ray, glove boxes, etc.; rdsh.-orange powd.; avg. 9 μm particle size; sol. in aromatic and halogenated hydrocarbons; insol. in water; bulk dens. 0.6 g/ml; m.p. 130–160 C; > 62% bromine content.

Conco® 2A1. [Continental] Sodium dodecyl diphenyl ether disulfonate; coupler, phenol disinfectant; emulsion polymerization of SBR, dispersant, plating compd.; liq.; water-sol.; 45% act.

Conco® 3B2. [Continental] Sodium decyl diphenyl ether disulfonate; coupler with med. foaming for herbicide and pesticide formulations; liq.; 45% act.

Conco® 4C3. [Continental] Sodium hexadecyl diphenyl ether disulfonate; coupler, dispersant; emulsion polymerization; liq.; water-sol.; 35% act.

Conco® AAS-98S. [Continental] Dodecylbenzene sulfonic acid; anionic; detergent intermediate used in mfg. of shampoos, detergent, metal cleaners; br. liq.; water sol.; sp.gr. 1.04; biodeg.; 98+% act.

Conco® ATR-98S. [Continental] Tridecyl benzene sulfonic acid; anionic; detergent intermediate, neutralized with wide range of alkalies; liq.; sp.gr. 1.04; biodeg.; 98+% act.

Conco® NI-21. [Continental] Nonoxynol-2; nonionic; dispersant, emulsifier, wetting, foaming and penetrating agent for textile wet processing, insecticide and cosmetic formulations, industrial and household detergents, latex and polymer emulsifier and post polymerization stabilizer, waxes and polishes, bottlewashing comps., polyester resin emulsifier, demulsifier for crude petrol./oil emulsions; liq.; sp.gr. 0.98–0.99; 100% act.

Conco® NI-43. [Continental] Nonoxynol-4; nonionic; see Conco NI-21; liq.; water-sol.; sp.gr. 1.02–1.04; 100% act.

Conco® NI-60. [Continental] Nonoxynol-6; nonionic; see Conco NI-21; liq.; water-sol.; sp.gr. 1.03–1.05; cloud pt. 32 F; 100% act.

Conco® NI-90. [Continental] Nonoxynol-9; nonionic; see Conco NI-21; liq.; water-sol.; sp.gr. 1.04–1.08; cloud pt. 87 F; 100% conc.

Conco® NI-100. [Continental] Nonoxynol-10; nonionic; see Conco NI-21; liq.; water-sol.; sp.gr. 1.05–1.07; cloud pt. 125–133 F; 100% conc.

Conco® NI-110. [Continental] Nonoxynol-11; nonionic; see Conco NI-21; liq.; water-sol.; sp.gr. 1.05–1.07; cloud pt. 158–165 F; 100% act.

Conco® NI-150. [Continental] Nonoxynol-15; nonionic; see Conco NI-21; liq.; water-sol.; sp.gr. 1.07–1.08; cloud pt. 203–212 F; 100% act.

Conco® NI-185. [Continental] Nonoxynol-20; nonionic; see Conco NI-21; wax; water-sol.; cloud pt. > 212 F; 100% act.

Conco® NI-187. [Continental] Nonoxynol-30; nonionic; see Conco NI-21; wax; cloud pt. > 212 F; 100% act.

Conco® NI-190. [Continental] Nonoxynol-40; nonionic; see Conco NI-21; wax; cloud pt. > 212 F; 100% act.

Conco® NI-197. [Continental] Nonoxynol-40; nonionic; see Conco NI-21; liq.; sp.gr. 1.10; cloud pt. > 212 F; 100% act.

Conco® NI-2000. [Continental] Nonoxynol-50; nonionic; see Conco NI-21; wax; cloud pt. 163 F; 100% act.

Conco® PXS. [Continental] Potassium xylene sulfonate; hydrotrope; liq.; 25% act.

Conco® SEQ. [Continental] EDTA sodium salt; chelating agent used in textile wet processing, peroxide bleaching, pigments, detergents, soap, and shampoos; liq., beads; water-sol.; sp.gr. 1.25; chel. value 80 and 200 mg $CaCO_3$/g; pH 11–12 (1%); 35% act. (liq.); 95% act. (beads).

Conco® Sulfate 2A1. [Continental] Sodium dodecyl diphenyl oxide disulfonate; anionic; detergent, emulsifier, wetting agent, dispersant, coupler for textile wet processing, hard surface and built cleaners, cosmetics, emulsion polymerization; yel./brn. liq.; typ. odor; sp.gr. 1.16; visc. 150 cps; 45% act.

Conco® Sulfate 2A1 Acid. [Continental] Dodecyl diphenyl oxide disulfonic acid; anionic; intermediate used in acidic sol'ns. or in preparation of other salts; dk. liq.; sol. in water, 15% HCl; sp.gr. 1.1; visc. 500 cps; 70% act.

Conco® Sulfate 3B2. [Continental] Sodium decyl diphenyl oxide disulfonate; anionic; wetting agent, dispersant, solubilizer, coupler for agric. chemicals, herbicides, food contact, adhesives; amber liq.; typ. odor; sol. in water, 25% NaOH, 20% HCl; surf. tens. 38 dynes/cm (0.1% aq.); 45% act.

Conco® Sulfate 3B2 Acid. [Continental] Decyl dephenyl oxide disulfonic acid; anionic; see Conco Sulfate 2A1 Acid; dk. liq.; sol. in water, 15% HCl; sp.gr. 1.1; visc. 500 cps; 70% act.

Conco® Sulfate 4C3. [Continental] Sodium hexadecyl diphenyl ether disulfonate; anionic; see Conco Sulfate 2A1; yel./br. liq.; typ. odor; sp.gr. 1.12; visc. 150 cps; 35% act

Conco® Sulfate EP. [Continental] DEA-lauryl sulfate; anionic; detergent, wetting agent, conditioner in personal care prods., household detergents; lt. yel. liq.; visc. 225 cps; cloud pt. 6 C; pH 7.5–8.5; 36–39.9% act.

Conco® Sulfate RA. [Continental] Alcohol ether sulfate; anionic; detergent, dispersant, emulsifier, penetrant, wetting agent, suspending agent, and leveling agent used in textile wet processing, heavy-duty laundry and household detergents.

Conco® Sulfate TL. [Continental] TEA-lauryl sulfate; anionic; detergent, wetting; shampoo base, pigment dispersant; lt. yel. liq.; pleasant odor; visc. 225 cps max.; cloud pt. 2.5 C; pH 6.4–7.7 (10%); 39.5–44.8% act

Conco® Sulfate WA Dry. [Continental] Sodium lauryl sulfate; anionic; shampoo base, rug shampoos, cleansing lotions and household prods.; pigment dispersant for personal care prods.; emulsion polymerization; powd.; typ. odor; pH 6.5–7.5 (10%); 92–98% act.

Conco® Sulfate WAG. [Continental] Sodium lauryl sulfate; see Conco Sulfate WA Dry; wh. needle; typ. odor; 92–98% act.

Conco® Sulfate WAS. [Continental] Sodium lauryl sulfate; see Conco Sulfate WA Dry; colorless liq.; pleasant odor; visc. 1000–5000 cps; cloud pt. 17–23 C; pH 7.4–8.6 (10%); 28.5–33.0% act

Conco® Sulfate WA Special. [Continental] Sodium lauryl sulfate; see Conco Sulfate WA Dry; wh. smooth paste; pleasant odor; visc. 19,000–31,000 cps; cloud pt. 17–21 C; pH 7.4–8.6 (10%); 28.5–33.0% act.

Conco® SXS. [Continental] Sodium xylene sulfonate; surfactant for liq. detergents, household cleaners, wood pulping, dye mfg.; solubilizer; water-wh. liq.; typ. odor; 40% min. act.

Conco® XA-C. [Continental] Palmitamine oxide; cationic, amphoteric; foamer, foam stabilizer for household and industrial detergents; post-polymerization stabilizer, electroplating brightener, cosmetic emollient, polymerization emulsifier, textile dyeing assistant; sol. in water, IPA, ethanol; biodeg.; 20% conc.

Conco® XA-L. [Continental] Lauramine oxide; cationic, amphoteric; see Conco XA-C; sol. in water, IPA, ethanol; biodeg.; 30% conc.

Conco® XA-M. [Continental] Myristamine oxide; cationic, amphoteric; see Conco XA-C; sol. in water, IPA, ethanol; biodeg.; 30% conc.

Conco® XA-MC. [Continental] Myristyl-cetyl dimethyl amine oxide; cationic, amphoteric; see Conco XA-C; sol. in water, IPA, ethanol; biodeg.; 30% conc.

Conco® XA-S. [Continental] Stearamine oxide; cationic, amphoteric; see Conco XA-C; sol. in IPA, ethanol; forms emulsions in water; biodeg.; 30% conc.

Conco® XA-T. [Continental] Tallow amine oxide;

cationic, amphoteric; see Conco XA-C; sol. in IPA, ethanol; forms emulsions in water; biodeg.; 30% conc.

Conco® XA-Y. [Continental] Cocamine oxide; cationic, amphoteric; see Conco XA-C; sol. in water, IPA, ethanol; biodeg.; 30% conc.

Concofac 277. [Continental] Phosphorylated organic ester, anionic; surfactant; detergent and wetting agent for automotive applics.; textile penetrant, scouring and wetting agent; floor, rug and kitchen cleaner, metal finishing chemical; water-wh. to lt. straw liq.; sol. in aq. sol'ns.; pH 4–6; 30–32% solids.

Concosoft 90. [Continental] Imidazoline; cationic; softener; substantive quat. for home laundry softeners; liq.; 90% act.

Condensate 640. [Continental] Alkanolamide; detergent and soil-suspending agent; used in floor and hard surface cleaners; lubricant.

Condensate PA. [Continental] Cocamide DEA; thickener, dispersant, foamer, foam stabilizer for cosmetic, personal care prods., industrial and household cleaners, textile prods., buffing and polishing compds.; yel. thick liq.; typ. odor; disp. in water; sp.gr. 0.97; alkali no. 30–50; pH 9.0–9.8; 90–95% amide content.

Condensate PC. [National Starch] Coconut-DEA condensation product; wetting and rewetting agent, detergent, high foamer, emulsifier, thickener, used in the textile industry, dishwashing; amber liq.; 100% act.

Condensate PE. [Continental] Lauramide DEA; see Condensate PA; wh. solid; typ. odor; water-disp.; sp.gr. 0.98; m.p. 97.5–99.0 F; alkali no. 30–47; pH 9.0–9.7; 90–95% amide.

Condensate PL. [Continental] Lauramide DEA and diethanolamine; nonionic; wetting agent, dispersant, detergent, thickener, foamer, foam stabilizer in personal care prods., dishwashing detergents, buffing compds., textiles, food processing compds.; lt. amber liq.; typ. odor; sp.gr. 1.02; alkali no. 175–205; pH 9–10; 100% act.

Condensate PM. [Continental] Myristamide DEA; see Condensate PA; wh. solid; typ. odor; disp. in water; sp.gr. 0.97; m.p. 100–102 F; alkali no. 30–48; pH 9.0–9.7; 90–95% amide.

Condensate PO. [Continental] Cocamide DEA; see Condensate PA; thick lt. yel. liq.; typ. odor; disp. in water; sp.gr. 0.98; alkali no. 38–50; pH 9.0–9.7; 85% min. amide.

Condensate PS. [Continental] Cocamide DEA and diethanolamine; emulsifier, wetting agent, foam builder and stabilizer; used in consumer and commercial cleaners and textile processing applics.; lt. amber liq.; char. odor; sol. in water, alcohol, CCl_4; sp.gr. 1.02; pH 8.5–9.5; 97% min. act.

Consamine CA. [Consos] Fatty alkanolamide; nonionic; detergent, emulsifier, intermediate, detergent base, thickener, fulling agent used in textiles; liq

Consamine PA. [Consos] Complex alcohol ester; anionic; wetting agent, and penetrant for textile processes; liq.

Consoft CP-50. [Consos] Quaternized imidazoline; cationic; softener for acrylics and nylon, napping lubricant and leveling agent; liq.

Consoft CPN. [Consos] Quaternized fatty amide; cationic; softener for acrylic and nylon yarn and knit goods; liq.

Consonyl FIX. [Consos] Alkyl naphthalene sulfonate; anionic; fixative for improving wash fastness of acid dyestuffs on nylon; resist for direct and naphthol dyes on nylon; liq.

Consos Castor Oil. [Consos] Sulfated castor oil; anionic; retarder and leveling agent for dyeing applic. in textiles; dispersant and lubricant; liq.

Consoscour 47. [Consos] Alkylaryl sulfonate; anionic; detergent, scouring, and leveling agent for dyeing applics.; liq.; 47% conc.

Consostat DKM. [Consos] POE alkyl amine; cationic; antistat used for wool and syn. fibers; liq.; 88% conc.

Consotard No. 895. [Consos] Quat. deriv.; cationic; retarder for dyestuffs on acrylic fiber; liq.

Continental. [Vanderbilt] Sec. kaolin clay; filler used in agric. applic.; off-wh. particle size 0.3 μ particulate, 99.5% thru 325 mesh screen; dens. 2.5 mg/m^3; dens. 21 lb/ft^3; oil absorp. 36; 44.87% SiO_2, 37.77% Al_2O_3.

Contraspum 210. [Zschimmer & Schwarz] Mixt. of fatty acids, paraffin oil, and nonionic emulsifier; nonionic; defoamer for yeast and alcohol prod.; clear yel. liq.; emulsifiable in water; sp.gr. 0.92; visc. 100 mPa•s; 100% act.

Copolymer 845. [GAF] PVP/dimethylaminoethyl methacrylate copolymer; conditioner used in personal care prods.; hair fixative; liq.; 20% aq. sol'n.

Copolymer 937. [GAF] PVP/dimethylaminoethyl methacrylate copolymer; film-forming resin providing conditioning to hair prods., skin creams and lotions; m.w. 10^6; 20% aq. sol'n.

Copolymer 958. [GAF] PVP/dimethylaminoethyl methacrylate copolymer; film-forming resin providing conditioning to hair prods., skin creams and lotions; m.w. 10^5; 50% ethanol sol'n.

Corax. [Henkel] Wax; rubber lubricant for O-rings; solid; 100% act.

Cordex DJ. [Finetex] Fatty amine deriv.; antistat for syn. fibers; water-sol.

Cordon AES-65. [Finetex] Aliphatic ester sulfate; anionic; lubricant, dye leveling, prevents chafe marks in textiles; liq.; 65% conc.

Cordon N-400. [Finetex] Complex alkyl-naphthalene sulfonate; anionic; disperse dye leveling; liq.; 22% conc.

Cordon NU 890/75. [Finetex] Sulfated castor oil; anionic; emulsifier for pine oil, plasticizer, lubricant, textile dyeing assistant; liq.; 75% conc.

Cordon PB-870. [Finetex] Sulfated tall oil; anionic; lubricant, emulsifier; liq.; 42% conc.

Corfree M1. [DuPont] Dibasic acid mixt., primarily C_{12} and C_{11}; corrosion inhibitor raw material when reacted with amines; off-wh. flaked solid; m.w. 215; soften. pt. 85–95 C; flash pt. (CC) 204 C.

Corfree M2. [DuPont] C_{12} dibasic acid (dodecanedioic acid); corrosion inhibitor raw material when reacted with amines; APHA color 10 (2 g/50 ml methanol); sol. 1.0% in acetone, 0.006% in benzene (35 C), 0.012% in water (60 C); sp.gr. 1.15; m.p. 130 ± 1 C; flash pt. (CC) 227 C98% min. act.

Corfree M3. [DuPont] Primarily C_{12} dibasic acid; corrosion inhibitor raw material when reacted with amines; off-wh. flaked solid; m.p. 128 ± 2 C; acid no. 500 ± 25; 95.7% act.

Corona Lanolin. [Croda Ltd.] Anhyd. lanolin; nonionic; conditioning emollient, moisturizing agent, cosolv., plasticizer, w/o emulsifier, superfatting agent, wetting/dispersing agent, used in cosmetics, as pharmaceutical base; paste; oil-sol.; 100% conc.

Coronet Lanolin. [Croda Ltd.] Super refined cosmetic grade lanolin BP; nonionic; conditioning emollient, moisturizing agent, co-solv., plasticizer, w/o emulsifier, superfatting agent, wetting/dispersing agent, used as pharmaceutical base; minimum odor, color, batch variation in this grade; slightly higher m.p. than USP and BP standards; paste; 100% conc.

Corrosion Inhibitor 500. [Rhone-Poulenc Surf.] Corrosion inhibitor, used where corrosion is caused by acids, etc.; clear, amber liq.; sol. in distilled water, aliphatic and aromatic solvents, various acids.

Corrosion Inhibitor 600. [Rhone-Poulenc Surf.] Corrosion inhibitor, filming type when applied to steel; sol. in distilled water, various acids; clear, amber liq., pH 6-7.

Cosmedia Guar C-261. [Henkel] Guar hydroxypropyltrimonium chloride; visc. builder; personal care prod. formulating; substantivity provides hair conditioning; stabilizer and thickener for emulsions; provides slip to finished formulations; powd.; water-disp.

Cosmedia Guar U. [Henkel] Guar gum; swelling agent for regulation of the visc. and stabilization in cosmetic emulsions; powd.

Cosmetol X. [CasChem] Castor oil; deodorized specially refined grade of castor oil containing a food grade antioxidant; high visc. veg. oil; low pour pt.; emollient.

Cosmopon BL. [Tessilchimica] Fatty alkanolamide sulfosuccinate; anionic; rug shampoo, bubble bath, conditioner; paste; 50% conc.

Cosmowax. [Croda] Stearyl alcohol, steareth-20, and steareth-10; nonionic; SE wax; o/w emulsifier, thickener, stabilizer for cosmetics, pharmaceuticals esp. antiperspirant, hair straighteners, depilatories; solid; low odor; HLB 8.0; m.p. 47–55 C; acid no. 0.5 max.; sapon. no. 1 max.; 100% conc.

Cosmowax J. [Croda] Cetearyl alcohol and ceteareth-20; nonionic; o/w emulsifier and stabilizer for personal care prods., pharmaceutical creams; creamy wh. waxy solid; low odor; HLB 8.5; m.p. 47–55 C; sapon. no. 1 max.; 100% conc.

Cosmowax K. [Croda] Stearyl alcohol and ceteareth-20; nonionic; emulsifier, stabilizer for personal care prods. and pharmaceuticals; creamy wh. waxy solid; low odor; HLB 9.0; m.p. 55–63 C; acid no. 0.5 max.; sapon. no. 1 max.; 100% conc.

Cosmowax P. [Croda] Ceteareth-20, cetearyl alcohol; nonionic; SE wax, emulsifier providing improved emulsion stability in lotions; wh. to off-wh. flakes; m.p. 46–53 C.

Cotton-Pro. [Griffin] Prometryn suspension; flowable herbicide for selective weed control in cotton and celery crops; 45.41% act.

Counter®. [Am. Cyanamid/Ag] Terbufos; systemic insecticide/nematicide for use in corn; restricted use; 15.0% act.

Coviox T-50. [Henkel] Tocopherols; antioxidant and blocking agent for cosmetic formulations; brnsh.-red visc. oil; char. odor; b.p. 190–220 C (@ 0.1 mm Hg); 500 mg/g min. total natural tocopherols; 7–9% d-alpha; 32–38% d-gamma and d-beta; 10–15% d-delta.

Coviox T-70. [Henkel] Tocopherols; see Coviox T-50; also used in food industry; clear yel.

Covitol 544. [Henkel] Tocopheryl acetate; antioxidant; pharm. grade natural Vitamin E; 400 mg/g min. assay.

Covitol 700C. [Henkel] d-Alpha tocopheryl acetate; see Covital 544; wh. to cream powd.; bland odor; 515 mg min. of d-alpha tocopheryl acetate conc. FCC/g.

Covitol 1100. [Henkel] Tocopheryl acetate; see Covitol 544; 809 mg/g min. assay.

Covitol 1185. [Henkel] Tocopheryl succinate; see Covital 544; wh. gran. powd.; bland odor; 960 mg min. of d-alpha tocopheryl acid succinate FCC/g.

Covitol 1210. [Henkel] Tocopheryl succinate; see Covitol 544; wh. to off-wh. cryst. powd.; 960 mg min. of d-alpha tocopheryl acid succinate/g.

Covitol 1360. [Henkel] Tocopheryl acetate; see Covitol 544; yel. clear visc. liq.; m.p. 25 C; 960 mg/g min. of d-alpha tocopheryl acetate.

Covitol F-350M. [Henkel] Tocopherol; see Covitol 544; cream powd.; 235 mg/g of d-alpha tocopherol conc. FCC plus 59 mg/g of d-beta, d-gamma, and d-delta tocopherol FCC.

Covitol F-600. [Henkel] d-Alpha tocopherol conc. FCC; see Covitol 544; 403 mg/g min. assay.

Covitol F-1000. [Henkel] Tocopherol; see Covitol 544; brn.-red clear visc. oil; mild odor; 671 mg/g min. assay.

CP0110. [Hüls] 2-Phenethyltrichloro silane; coupling agent, chem. intermediate, blocking agent, release agent, lubricant, primer, reducing agent; liq.; m.w. 239.6; sp.gr. 1.24; b.p. 92–96 C (3 mm); flash pt. 78 C; ref. index 1.518; 96% purity.

CP0156. [Hüls] N-Phenylaminopropyltrimethoxy silane; see CP0110; yel. liq.; m.w. 255.4; sp.gr. 1.05; b.p. 132–135 C (0.3 mm); flash pt. 146 C; ref. index 1.500; 95% purity.

CP0160. [Hüls] Phenyldimethylchloro silane; see CP0110; liq.; m.w. 170.7; sp.gr. 1.03; b.p. 192–193 C; flash pt. 61 C; ref. index 1.508; 97% purity.

CP0280. [Hüls] Phenyltrichloro silane; see CP0110; liq.; m.w. 211.6; sp.gr. 1.329; b.p. 201 C; flash pt. 91 C; ref. index 1.525; 97% purity.

CP0320. [Hüls] Phenyltriethoxy silane; see CP0110; liq.; m.w. 240.4; sp.gr. 0.99; b.p. 233–234 C; flash pt. 121 C; ref. index 1.461; 99% purity.

CP0330. [Hüls] Phenyltrimethoxy silane; see

CP0110; liq.; m.w. 198.3; sp.gr. 1.06; b.p. 211 C; flash pt. 92 C; ref. index 1.473; 97% purity.

CP0800. [Hüls] n-Propyltrichloro silane; see CP0110; liq.; m.w. 177.6; sp.gr. 1.185; b.p. 123–125 C; flash pt. 2 C; ref. index 1.429; 97% purity.

CP0810. [Hüls] n-Propyltrimethoxy silane; see CP0110; liq.; m.w. 164.3; sp.gr. 0.932; b.p. 142 C; flash pt. 35 C; ref. index 1.388; 98% purity.

CPC. [Hexcel] Cetyl pyridinium chloride USP XX; conditioner, emulsifier and antistat imparting softness in cream rinse formulations; sol. in water and alcohol.

CPF-0001. [Witco/Argus] Chlorinated paraffin; plasticizer for use in paint, coatings, caulks, sealants, rubber, and adhesives applic. where lower volatility than phthalate plasticizers and/or flame retardancy are desired; coatings, adhesives, and caulks/sealants; Gardner 3 liq.; sp.gr. 1.14; visc. 11 poise; 40% chlorine content.

CPF-0003. [Witco/Argus] Chlorinated paraffin; see CPF-0001; Gardner 6 liq.; sp.gr. 1.22; visc. 115 poise; 46% chlorine content.

CPF-0008. [Witco/Argus] Chlorinated paraffin; see CPF-0001; Gardner 4 liq.; sp.gr. 1.26; visc. 145 poise; 50% chlorine content.

CPF-0019. [Witco/Argus] Chlorinated paraffin; see CPF-0001; Gardner 5 liq.; sp.gr. 1.28; visc. 500 poise; 50% chlorine content.

CPF-0022. [Witco/Argus] Chlorinated paraffin; see CPF-0001; Gardner 3 liq.; sp.gr. 1.16; visc. 35 poise; 41% chlorine content.

CPH-27-N. [C.P. Hall] PEG-4 laurate; nonionic; solubilizer and emulsifier; Gardner 3 clear oily liq.; sol. in toluene, kerosene, ethanol, acetone; disp. in water; sp.gr. 0.985; f.p. 5 C; flash pt. 199 C; acid no. 4.0; sapon. no. 139; ref. index 1.454; 100% conc.

CPH-30-N. [C.P. Hall] PEG-8 laurate; nonionic; solubilizer and emulsifier; Gardner 2 clear oily liq.; sol. in toluene, kerosene, ethanol, acetone, water; sp.gr. 1.030; HLB 12.9; f.p. 15 C; acid no. 4.0; sapon. no. 91; ref. index 1.458.

CPH-31-N. [C.P. Hall] Glyceryl oleate; nonionic; antiblocking agent; amber liq.; m.w. 340; sol. in toluene, kerosene, ethanol, acetone, propylene glycol, min. oil; sp.gr. 0.943; f.p. 9 C; HLB 3.7; acid no. 3.0; sapon. no. 168; ref. index 1.467; 100% conc.

CPH-37-NA. [C.P. Hall] Ethylene glycol stearate; wh. flake; m.w. 300; sol. in toluene, ethanol, acetone, min. oil; m.p. 57 C; acid no. 3.0; sapon. no. 187.

CPH-39-N. [C.P. Hall] PEG-4 oleate; nonionic; spreading agent, defoamer; Gardner 2 clear oily liq.; m.w. 457; sol. in toluene, kerosene, ethanol, acetone; disp. water; sp.gr. 0.969; HLB 7.6; f.p. –5 C; acid no. 3.0; sapon. no. 125; ref. index 1.462; 100% conc.

CPH-41-N. [C.P. Hall] PEG-12 oleate; nonionic; solubilizer and emulsifier; Gardner 2 clear oily liq.; m.w. 857; sol. in toluene, kerosene, ethanol, acetone, water; sp.gr. 1.035; HLB 13.6; f.p. 25 C; acid no. 4.0; sapon. no. 67; ref. index 1.466; 100% conc.

CPH-43-N. [C.P. Hall] PEG-12 laurate; nonionic; solubilizer and dispersant; solid; sol. in toluene, kerosene, ethanol, acetone, water; sp.gr. 1.050; HLB 14.6; f.p. 25 C; flash pt. 202 C; acid no. 4.0; sapon. no. 69; ref. index 1.461; 1010% conc.

CPH-52-SE. [C.P. Hall] Propylene glycol stearate; wh. to cream solid; m.w. 333; sol. in toluene, ethanol, min. oil; disp. hot in water; m.p. 38. C; acid no. 17; sapon. no. 170.

CPH-53-N. [C.P. Hall] Glyceryl stearate; nonionic; thickener and emulsifier; wh. flake; m.w. 330; sol. in toluene, ethanol, acetone; HLB 3.7; m.p. 56 C; acid no. 2.0; sapon. no. 170; 100% conc.

CPH-79-N. [C.P. Hall] PEG-8 dilaurate; nonionic; wetting agent, pigment dispersant; Gardner 2 clear oily liq.; sol. in hexane, toluene, kerosene, ethanol, acetone; disp. in water; sp.gr. 0.990; HLB 9.7; f.p. 17 C; acid no. 9.0; sapon. no. 128; ref. index 1.457; 100% conc.

CPH-170-N. [C.P. Hall] PEG-8 dioleate; Gardner 1–2 clear oily liq.; m.w. 900; sol. in toluene, kerosene, ethanol, acetone, propylene glycol, min. oil; disp. water; sp.gr. 0.976; f.p. 5 C; acid no. 8.0; sapon. no. 118; ref. index 1.464.

CPH-205-NX. [C.P. Hall] Glyceryl oleate; nonionic; lube additive, rust preventive; amber liq.; m.w. 349; sol. in hexane, toluene, kerosene, ethanol, acetone, propylene glycol, min. oil; sp.gr. 0.954; HLB 3.0; f.p. 8 C; acid no. 3.0; sapon. no. 165; ref. index 1.470; 100% conc.

CPH-211-N. [C.P. Hall] PEG-8 dioleate; nonionic; water treatment, lubricant; Gardner 3 clear oily liq.; m.w. 914; sol. in toluene, kerosene, ethanol, acetone, propylene glycol, min. oil; disp. water; sp.gr. 0.981; HLB 8.1; f.p. 5 C; acid no. 4.0; sapon. no. 115; ref. index 1.467; 100% conc.

CPH-213-N. [C.P. Hall] PEG-12 dioleate; nonionic; lubricant emulsifier; liq.; HLB 10.5% conc.

CPH-353-SE. [C.P. Hall] Glyceryl oleate; anionic; corrosion inhibitor; amber liq.; m.w. 303; sol. in toluene, kerosene, ethanol, acetone, propylene glycol; min. oil; disp. in water; sp.gr. 0.965; f.p. –11 C; acid no. 3.0; sapon. no. 145; ref. index 1.456; 100% conc.

CPH-362-N. [C.P. Hall] Glyceryl oleate; nonionic; oil additive, rust preventive; amber liq.; m.w. 340; sol. in toluene, kerosene, ethanol, acetone, propylene glycol; disp. in min. oil; sp.gr. 0.944; HLB 4.7; f.p. –7 C; acid no. 3.0; sapon. no. 173; ref. index 1.468; 100% conc.

CPH-376-N. [C.P. Hall] POE laurate; nonionic; solubilizer and emulsifier; PVC antistat; Gardner 2 clear oily liq.; sol. in toluene, kerosene, ethanol, acetone, propylene glycol, min. oil; disp. water; sp.gr. 1.002; acid no. 4.0; sapon. no. 120; ref. index 1.455; 100% conc.

CPH-380-N. [C.P. Hall] Stearic acid ester amide; wh. solid; m.w. 600; sol. in ethanol; partly sol. in hexane, toluene, kerosene, propylene glycol; sp.gr. 0.971; m.p. 80 C; acid no. 17; iodine no. 1.0; sapon. no. 107.

CPS 076. [Hüls] (N-Trimethoxysilylpropyl)-polyethylenimine; coupling agent esp. for min.-filled and adhesive bonding applics. of high m.w. thermoplastic polyamides and polyesters; sol'n.; m.w. 1000; sp.gr. 0.91; visc. 175–250 cSt; ref. index 1.442;

flam.; 50% act. in IPA.

CPS 078.5. [Hüls] Triethoxysilyl modified poly(1,2-butadiene); coupling agent esp. for polyolefins and polyolefin elastomers, min.- and glass-filled formulations, as primer coat for metals in insert molding; sol'n.; m.w. 3500–4000; sp.gr. 0.90; visc. 35–50 cSt; flam.; 50% act. in toluene.

Cr-0211 T 5/32''. [M&T Harshaw] Chromium alumina; incineration catalyst for dehydrogenation of butane to butene; tablet, 5/32'' diam.; dens. 71 lb/ft^3; 19% conc.

Crafol AP-10. [Pulcra SA] Free fatty acid complex org. phosphate ester; lubricity additive, corrosion inhibitor and emulsifier; sol. in oil and water.

Crafol AP-11. [Pulcra SA] Alkoxylated cetyl/oleyl phosphate; conditioner for cosmetic formulations; water-disp.

Crafol AP-31, -33. [Pulcra SA] Potassium lauryl phosphate; anionic; mfg. of manmade textile fibers; antistat; spinning oil for carpets; gel, paste; 62% act.

Crafol AP-34. [Pulcra SA] Complex phosphate ester, potassium salt; anionic; see Crafol AP-31; gel, paste; 62% act.

Crafol AP-51. [Pulcra SA] Free acid of alkyl phenol ether phosphate; anionic; lubricant and corrosion inhibitor used in aq. formulations or cutting fluids; for lubricating oils, cutting fluids, hydraulic fluids; liq.; oil-sol.

Crafol AP-60. [Pulcra SA] Org. phosphate ester free acid complex; compatabilizer and coupler for alkaline formulations; lubricant compat. with conc. electrolyte sol'ns.; liq.; water-sol.

Cralane KR-13, -14. [Pulcra SA] Ethoxylated lanolin; nonionic; emollient, wetting, cleaning, and superfatting agents, o/w emulsifier, conditioner, solubilizer, and plasticizer for personal care prods. and germicides; solid; sol. in water and ethanol; 100% conc.

Cralane LR-10. [Pulcra SA] Ethoxylated lanolin (75 EO); see Cralane KR-13; visc. liq.; 50% act.

Cralane LR-11. [Pulcra SA] Ethoxylated lanolin; nonionic; see Cralane KR-13; visc. liq.; 50% conc.

Crapol AU-20. [Pulcra SA] Benzalkonium chloride; cationic; germicide, disinfectant, algicide for swimming pools, industrial waste water; liq.; 80 ± 2% conc.

Crapol AU-21. [Pulcra SA] Benzalkonium chloride, modified; cationic; retarding and leveling agent for acrylic fiber dyeing; liq.; 50 ± 1% conc.

Crapol AU-23. [Pulcra SA] Stearalkonium chloride plus additives; cationic; antistat for creams, hair conditioner; paste; water-disp.; 25 ± 1% conc.

Crapol AU-24. [Pulcra SA] Benzalkonium chloride; cationic; see Crapol AU-20; liq.; 80 ± 2% conc.

Crapol AU-25. [Pulcra SA] Quat. ammonium blend; cationic; migration and leveling agent for acrylic fiber dyeing; liq.; 40 ± 1% conc.

Crapol AU-31. [Pulcra SA] Dialkyl imidazolinium quat.; substantive antistat, and non-yel. cationic softener for textiles; water-disp.

Crapol AU-40. [Pulcra SA] Alkylamine polyglycol ether methosulfate plus additives; cationic; antistat and surface act. agent used in mfg. of heavy duty detergents; liq.; 46 ± 2% conc.

Crapol AV-10. [Pulcra SA] Coco hydroxyethyl imidazoline; cationic; corrosive inhibitor, dispersant and fluidizing agent for pigments, emulsions, and agric. prods.; acid detergent for food and dairy prods.; liq., paste; sol. in most hydrocarbon polar and chlorinated solvs., acids; water-disp.; 100% conc.

Crapol AV-11. [Pulcra SA] Lauryl hydroxyethyl imidazoline; cationic; see Crapol AV-10; liq.; 100% conc.

Crapol FU-25. [Pulcra SA] Dimethyl dihydrog. tallow ammonium chloride; antistat hair and fiber conditioner; water-disp.

Cremba. [Croda] Min. oil, petrolatum, lanolin alcohol, and lanolin; nonionic; emollient, moisturizer, and emulsifier for cosmetics and pharmaceuticals; paste; 100% conc.

Cremba B6. [Croda] Min. oil, lanolin, paraffin, lanolin alcohol, and beeswax; emollient moisturizer for surgical scrub emulsions; pale yel. soft solid.

Cremophor® EL. [BASF] PEG-36 castor oil; nonionic; solubilizer, emulsifier used for essential oils, pharmaceuticals, cosmetics, veterinary medicine; pale yel. liq.; sol. in water, ethanol, propanol, ethyl acetate, chloroform, CCl_4, benzene, toluene, and xylene; sp.gr. 1.05–1.06; visc. 700–850 cps; HLB 12–14; acid no. < 2; sapon no. 65–70; ref. index 1.471; pH 6–8 (10% aq.); 100% act.

Cremophor® MP-10. [BASF] Nonoxynol-10; solubilizer; water-sol.

Cremophor® NP 10. [BASF] Nonoxynol-10; solubilizer for perfumes, essential oils, and flavors; liq.

Cremophor® NP 14. [BASF] Nonoxynol-14; see Cremophor NP 10; liq.

Cremophor® RH 40. [BASF] PEG-40 hydrog. castor oil; nonionic; solubilizer for essential oils and perfumery, emulsifier; water-wh. paste; sol. in water, ethanol, IPA, n-propanol, ethyl acetate, benzene, toluene, and xylene; HLB 14–16; pH 6–7 (10% aq.); 100% act.

Cremophor® RH 60. [BASF] PEG-60 hydrog. castor oil; nonionic; see Cremophor RH 40; wh. paste; sol. see Cremophor RH 40; HLB 15–17; acid no. < 1; sapon. no. 40–50; pH 6–7 (10% aq.); 100% act.

Cremophor® RH 410. [BASF] PEG-40 hydrog. castor oil; nonionic; see Cremophor RH 40; liq.; sol. see Cremophor RH 40; HLB 14–16; visc. 1500–2000 cps; sapon. no. 45–55; pH 6–7 (10% aq.); 90% act.

Cremophor® RH 455. [BASF] PEG-40 hydrog. castor oil and propylene glycol; nonionic; see Cremophor RH 40; liq.; sol. see Cremophor RH 40; HLB 14–16; visc. 1000–1500 cps; acid no. < 1; sapon. no. 45–55; ref. index 1.459–1.464; pH 6–7 (10% aq.); 90% act.

Cremophor® S 9. [BASF] PEG-9 stearate; nonionic; emulsifier for o/w type, thickening agent; suspension stabilizer; lubricating and antitack effects; ylsh. wh. solid; sol. in water, alcohols, acetone, ethyl acetate, chloroform, benzene, castor oil, and oleic acid; sp.gr. 0.97 (60 C); HLB 12; acid no. 2; sapon. no. 88–98; 100% act.

Crestalans. [Croda Ltd.] Fractionated liq. lanolin and isopropyl ester deriv.; emollient, moisturizer, conditioner; personal care prods.; plasticizer for aerosol hair sprays and lacquers; superfatting agent for cosmetic solv. compositions; pale yel. clear liq.; faint sterol odor.

Crester PR. [Croda Ltd.] Polyglycerol polyricinoleate; nonionic; visc. modifier in the food industry; liq.; 100% conc.

Crill 1. [Croda] Sorbitan laurate; nonionic; emulsifier, pigment dispersant, cosolv., wetting agent, antifoam, visc. reducer, mold release, antiblock agent, corrosion inhibitor, lubricant, antistat; used for cosmetics, food and food pkg., insecticides and herbicides, leather treatment, metalworking fluids, oil slick dispersing, paints and inks, pharmaceuticals, plastics, polishes, textiles; pale yel. clear visc. liq.; sol. in ethanol, oleyl alcohol, min. oil; HLB 8.6; sapon. no. 158–170; 98% conc.

Crill 2. [Croda] Sorbitan palmitate; see Crill 1; pale tan hard waxy solid; partially sol. in propylene glycol, ethyl and oleyl alcohols, olive oil, oleic acid; m.p. 46 C; HLB 6.7; sapon. no. 140–150; 98% conc.

Crill 3. [Croda] Sorbitan stearate; emulsifier, lubricant, antistat; o/w emulsions; cosmetic and pharmaceutical creams and lotions; polishes; insecticides, herbicides; metal cleaners; buffing compds.; textile lubricants; food applics.; pale tan hard waxy solid; low odor; partially sol. in oleyl alcohol, olive oil, oleic acid; m.p. 51–54 C; HLB 4.7; sapon. no. 147–157; 98% conc.

Crill 4. [Croda] Sorbitan oleate; w/o emulsifier, wetting agent, pigment dispersant, coupler, antifoam; cosmetic and pharmaceutical applic.; aerosol polishes; insecticidal sprays; inks and surf. coatings; metal working lubricants; cutting oils; textile lubricants; dry cleaning operations; food process antifoam; oil slick dispersant; amber visc. liq.; sol. in ethyl and oleyl alcohols, min. oil, IPM, olive oil, oleic acid; HLB 4.3; sapon. no. 147–160; 98% conc.

Crill 6. [Croda] Sorbitan isostearate; w/o emulsifier, wetting agent, pigment dispersant; pale yel. visc. liq.; sol. in min. oil, olive oil, partly sol. in oleyl alcohol, IPM; HLB 4.7; sapon. no. 143–153; 98% conc.

Crill 35. [Croda] Sorbitan tristearate; emulsifier, lubricant, antistat; pale tan hard waxy solid; partly sol. in oleyl alcohol, min. and olive oil, IPM, oleic acid; m.p. 48 C; HLB 2.1; sapon. no. 176–188; 98% conc.

Crill 43. [Croda] Sorbitan sesquioleate; w/o emulsifier, wetting agent, pigment dispersant; amber visc. liq.; sol. in oleyl alcohol, min. and olive oil, oleic acid; HLB 3.7; sapon. no. 149–160; 98% conc.

Crill 45. [Croda] Sorbitan trioleate; w/o emulsifier, wetting agent, pigment dispersant; amber visc. liq.; sol. in oleyl alcohol, min. and olive oil, IPM, oleic acid; HLB 1.8; sapon. no. 172–186; 98% conc.

Crill 50. [Croda] Sorbitan oleate (tech.); see Crill 1; liq., paste; HLB 4.3; 98% conc.

Crillet 1. [Croda] Polysorbate 20; nonionic; solubilizer, emulsifier, dispersant, wetting agent; often combined with a member of the Crill range in emulsification systems; used in cosmetics, food and food pkg., household prods., insecticides, herbicides, metalworking fluids, paints and inks, pharmaceuticals, textiles; clear, yel. clear liq.; low odor; sol. in water, ethyl and oleyl alcohol, oleic acid; HLB 16.7; sapon. no. 40–51; surf. tens. 38.5 dynes/cm (0.1%); 97% conc.

Crillet 2. [Croda] Polysorbate 40; emulsifier, dispersant, penetrant, leveling agent, lubricant, antistat for textiles, cosmetics, pharmaceuticals, polishes, insecticides, herbicides; yel. amber pasty liq.; sol. in water, ethyl and oleyl alcohol, oleic acid; HLB 15.6; sapon. no. 43–49; surf. tens. 41.5 dynes/cm (0.1%); 97% conc.

Crillet 3. [Croda] Polysorbate 60; emulsifier for cosmetics and pharmaceuticals, dispersant for insecticides, herbicides, cattle dyes, penetrant, leveling agent, lubricant, antistat; yel. liq. gels to soft solid on cooling; sol. in ethyl and oleyl alcohol, oleic acid; partly sol. in water; HLB 14.9; sapon. no. 45–55; surf. tens. 42.5 dynes/cm (0.1%); 97% conc.

Crillet 4. [Croda] Polysorbate 80; emulsifier, dispersant, solubilizer, lubricant, detergent, antistat, wetting agent for cosmetics, pharmaceuticals, polishes, insecticides, leather degreasing, veterinary prods.; clear yel. amber liq.; faint char. odor; sol. in water, ethyl and oleyl alcohols, oleic acid; HLB 15.0; sapon. no. 45–55; surf. tens. 42.5 dynes/cm (0.1%); 97% conc.

Crillet 6. [Croda] PEG-20 sorbitan isostearate; see Crillet 1; clear yel. liq.; sol. in water, ethyl and oleyl alcohols, oleic acid, xylene, trichlorethylene; HLB 14.9; sapon. no. 40–50; surf. tens. 38.6 dynes/cm (0.1%); 97% conc.

Crillet 11. [Croda] Polysorbate 21; solubilizer used in cosmetics, emulsifier and dispersant for pharmaceuticals; clear yel. liq.; sol. in ethyl and oleyl alcohols, oleic acid, kerosene; partly sol. in water; HLB 13.3; sapon. no. 100–115; surf. tens. 34.7 dynes/cm (0.1%); 97% conc.

Crillet 31. [Croda] Polysorbate 61; emulsifier for cosmetics and pharmaceuticals, dispersant for insecticides, herbicides, cattle dyes, penetrant, leveling agent, lubricant, antistat; cream pale yel. waxy solid; sol. in oleyl alcohol, oleic acid; HLB 9.6; sapon. no. 98–113; surf. tens. 41.5 dynes/cm (0.1%); 97% conc.

Crillet 35. [Croda] Polysorbate 65; see Crillet 31; cream/buff waxy solid; sol. in ethyl and oleyl alcohols, oleic acid, trichlorethylene, partly sol. in water; HLB 10.5; sapon. no. 88–98; surf. tens. 42.5 dynes/cm (0.1%); 97% conc.

Crillet 41. [Croda] Polysorbate 81; emulsifier, dispersant, solubilizer, lubricant, detergent, antistat, wetting agent, for cosmetics and pharmaceuticals; clear yel. amber liq.; sol. in ethyl and oleyl alcohol, IPM, oleic acid, kerosene, butyl stearate; HLB 10.0; sapon. no. 96–104; surf. tens. 37 dynes/cm (0.1%); 97% conc.

Crillet 45. [Croda] Polysorbate 85; see Crillet 41; clear amber visc. liq.; sol. in ethyl and oleyl alcohols, IPM, oleic acid, kerosene, trichlorethylene, butyl stearate; HLB 11.0; sapon. no. 82–95; surf. tens. 41

dynes/cm (0.1%); 97% conc.

Crillon LDE. [Croda Ltd.] Lauramide DEA; nonionic; detergent and foam stabilizer; o/w emulsifiers, antistat; anticorrosive; liq.

Crillon LME. [Croda Ltd.] Lauramide MEA; nonionic; lubrication; skin protection agent; solid.

Cri-Spersion CRI-ACT-45-1/1. [Cri-Tech] 45% disp. of calcium hydroxide/magnesium oxide (1:1) on a fluorocarbon elastomer base; activator for fluoroelastomer compounding; opaque, brownish yel. chopped or slab; sol. in esters and ketones; sp.gr. 1.89–1.99.

Cri-Spersion CRI-ACT-45-LV. [Cri-Tech] 45% disp. of calcium hydroxide/magnesium oxide (2:1) on a fluorocarbon elastomer base; activator for fluoroelastomer compounding in low-visc. compounds; opaque brownish yel. chopped or slab; sol. in esters and ketones; sp.gr. 1.89–1.99.

Cri-Spersion CRI-ACT-45. [Cri-Tech] 45% disp. of calcium hydroxide/magnesium oxide (2:1) on a fluorocarbon elastomer base; activator for fluoroelastomer compounding; opaque brownish chopped or slab; sol. in esters and ketones; sp.gr. 1.85–1.95.

Croda Absorption Base. [Croda Ltd.] Lanolin, lanolin and fatty alcohols in min. oils, petrolatum; bases for cosmetic and pharmaceutical formulations; w/o emulsifier; conditioner; liq.; oil-sol.; water-insol.

Crodacel QL. [Croda] Laurdimonium hydroxyethyl cellulose; conditioner improving foaming for skin and hair care prods.; pale yel. sol'n.; m.w. 10,000; water-sol.; 20% act.

Crodacel QM. [Croda] Cocodimonium hydroxyethyl cellulose; conditioner improving foaming and imparting body to skin and hair care prods.; clear visc. sol'n.; m.w. 10,000; 20% act.

Crodacel QS. [Croda] Steardimonium hydroxyethyl cellulose; see Crodacel QM; opaque visc. sol'n.; m.w. 10,000; 20% act.

Crodacid B. [Croda] Behenic acid; gellant for stick formulations when neutralized; wh. solid.

Crodacol C. [Croda] Cetyl alcohol; thickener; bodying agent; sol. in oils and fats.

Crodacol C70. [Croda] Cetyl alcohol; sec. emulsifier, thickener, opacifier, and structural agent in anhyd. stick systems; wh. flakes.

Crodacol C95NF. [Croda] Cetyl alcohol NF; primary structural agent in antiperspirant sticks; wh. flakes.

Crodacol CS50. [Croda] Cetearyl alcohol; emollient; hair and skin lubricant; wh. flakes; oil-sol.

Crodacol S. [Croda] Stearyl alcohol; thickener; bodying agent; sol. in oils and fats.

Crodacol S70. [Croda] Stearyl alcohol; sec. emulsifier, thickener, opacifier, and structural agent in anhyd. stick systems; wh. flakes.

Crodacol S95NF. [Croda] Stearyl alcohol NF; primary structural agent in antiperspirant sticks; wh. flakes.

Crodafos Series. [Croda] Alkyl ether phosphates, avail, in acid form or neutralized with appropriate base; anionic; surfactants, detergents, wetting agents, coupling agents, corrosion inhibitors, foam stabilizers, thickeners, lubricants, antistats, for textile and other industrial uses; many and varied forms; 99% min. act.

Crodafos 25 D2 Acid. [Croda] Phosphated C_{12}–C_{15} ether (2 EO); anionic; see Crodafos Series; liq.; 99% conc.

Crodafos 25 D5 Acid. [Croda] Phosphated C_{12}–C_{15} ether (5 EO); anionic; see Crodafos Series; liq.; 99% conc.

Crodafos 25 D10 Acid. [Croda] Phosphated C_{12}–C_{15} ether (10 EO); anionic; see Crodafos Series; paste; 99% conc.

Crodafos 25D Series. [Croda] Alkyl ether phosphates based on syn. C_{12}/C_{15} alcohols; available with 2, 5, and 10 moles EO and as Crodafos 25D TEA salt; anionic; see Crodafos Series; colorless to pale yel. liq. to soft paste; pH 1.8–2.5 (3%); 99% min. act.

Crodafos CAP. [Croda] PPG-10 cetyl ether phosphate; anionic; w/o emulsifier; antistat used in personal care prods.; liq.; 100% conc.

Crodafos CDP. [Croda] Cetyl diethanolamine phosphate; anionic; emulsifier and stabilizer for o/w emulsions; powd.; water-disp.; 100% conc.

Crodafos CS Series. [Croda] Alkyl ether phosphates based on ceto stearyl alcohol; avail. with 2, 5, and 10 moles EO and as Crodafos CS TEA salt; see Crodafos Series; wh. to buff hard wax; pH 1.8–2.5 (3%); 99% min. act.

Crodafos CS2 Acid. [Croda] Phosphated ceto/stearyl ether (2 EO); anionic; emulsifier and stabilizer for o/w emulsions; solid; 99% conc.

Crodafos CS5 Acid. [Croda] Phosphated ceto/stearyl ether (5 EO); anionic; see Crodafos CS2 Acid; solid; 99% conc.

Crodafos CS10 Acid. [Croda] Phosphated ceto/stearyl ether (10 EO); anionic; see Crodafos CS2 Acid; solid; 99% conc.

Crodafos ID Series. [Croda] Alkyl ether phosphates based on isodecanol; avail. with 2, 5, and 10 moles EO and as Crodafos ID TEA salts; see Crodafos Series; colorless to pale yel. fluid to paste; pH 1.8–2.5 (3%); 99% min. act.

Crodafos N3 Acid. [Croda] Oleth-3 phosphate; anionic; surfactant, conditioner, antistat, emulsifier, gelling agent for cosmetics, pharmaceuticals, and toiletries; corrosion inhibitor and anti-gelling agent in aerosol antiperspirant systems; Gardner 11 max. acid no. 120–135; sapon. no. 125–145; pH 1–3 (2% aq.); 100% act.

Crodafos N3 Neutral. [Croda] DEA-oleth-3 phosphate; anionic; see Crodafos N3 Acid; Gardner 11 max. paste; acid no. 90–100; pH 6–7 (2% aq.); 100% conc.

Crodafos N5 Acid. [Croda] Oleth-5 phosphate; anionic; see Crodafos N3 Acid; liq.; 99% conc.

Crodafos N10 Acid. [Croda] Oleth-10 phosphate; anionic; see Crodafos N3 Acid; Gardner 11 max. paste; acid no. 70–100; sapon. no. 75–115; pH 1–3 (2% aq.); 100% act.

Crodafos N10 Neutral. [Croda] DEA-oleth-10 phosphate; anionic; see Crodafos N3 Acid; Gardner 11 max. paste; acid no. 65–85; pH 5.5–7.0 (2% aq.); 100% conc.

Crodafos O Series. [Croda] Alkyl ether phosphates based on oleyl alcohol; see Crodafos Series; pale yel. to amber slightly visc. liq. to soft paste; pH 1.8–2.5; 99% act.

Crodafos O2 Acid. [Croda] Oleth-2 phosphate; anionic; conditioner for hair and shampoo prods.; liq.; pH 1.8–2.5 (3%); 99% conc.

Crodafos O5 Acid. [Croda] Oleth-5 phosphate; anionic; see Crodafos O2 Acid; liq.; pH 1.8–2.5 (3%); 99% conc.

Crodafos O10 Acid. [Croda] Oleth-10 phosphate; anionic; see Crodafos O2 Acid; paste; pH 1.8–2.5 (3%); 99% conc.

Crodafos SG. [Croda] PPG-5 ceteth-10 phosphate; anionic; see Crodafos N3 Acid; esp. for shampoos and cream rinses; Gardner 8 max. liq.; sol. in water and surfactants; acid no. 85–105; sapon. no. 90–110; pH 1–3 (3%); 100% act.

Crodafos T2 Acid. [Croda] Phosphated tridecyl ether (2 EO); anionic; see Crodafos SG; liq.; 99% conc.

Crodafos T5 Acid. [Croda] Phosphated tridecyl ether (5 EO); anionic; see Crodafos SG; liq.; 99% conc.

Crodafos T10 Acid. [Croda] Phosphated tridecyl ether (10 EO); anionic; see Crodafos SG; paste; 99% conc.

Crodalan 0477. [Croda] Lanolin alternative; nonionic; emollient, superfatting agent, w/o emulsifier for cosmetic and pharmaceutical systems; paste; 100% conc.

Crodalan AWS. [Croda] Polysorbate 80, cetyl acetate, acetylated lanolin alcohol; nonionic; emollient, superfatting agent, conditioner, o/w emulsifier, dispersant, wetting agent, plasticizer, solubilizer used in cosmetics, pharmaceuticals, detergent systems; golden liq.; faint fatty odor; sol. in alcohol, water; sp.gr. 1.02–1.08; acid no. 3 max.; hyd. no. 55–67; pH 5–7 (10% aq.); 100% act.

Crodalan IPL. [Croda] Isopropyl lanolate; emollient, wetting agent used in bath oils, skin, hair and makeup cosmetics; pale yel. thin opalescent liq. or soft paste liquifying at skin temp.

Crodalan LA. [Croda] Cetyl acetate and acetylated lanolin alcohols; nonionic; emollient, penetrant, wetting agent, conditioner, plasticizer used in cosmetics, pharmaceuticals; pale yel. clear, thin mobile liq.; odorless; sol. in min. oil; cloud pt. 20 C; acid no. 2 max.; sapon. no. 180–200; 100% act.

Crodamet. [Croda Ltd.] POE amine; cationic; emulsifier, antistat for plastics and fibers; liq., paste; 100% conc.

Crodamide E, ER. [Croda Ltd.] Eruamide; international slip and antiblock agent for polyolefins; internal release agent for moulded thermoplastic polymers; wax.

Crodamide O, OR. [Croda Ltd.] Oleamide; see Crodamide E; wax.

Crodamide S, SR. [Croda Ltd.] Stearamide; see Crodamide E; wax.

Crodamine 1.HT. [Croda Ltd.] Hydrog. tallow amine; anticaking agent for fertilizers; see also Crodamine 1.16D; waxy solid; 100% conc.

Crodamine 1.O, 1.OD. [Croda Ltd.] Oleyl amine; cationic; see Crodamine 1.16D; solid and liq. resp.; 100% conc.

Crodamine 1.T. [Croda Ltd.] Tallow amine; cationic; see Crodamine 1.16D; liq.; 100% conc.

Crodamine 1.16D. [Croda Ltd.] Primary cetyl amine; cationic; emulsifier for herbicides, ore flotation, pigment dispersion; aux. for textiles, leather, rubber, plastics, and metal industries; liq., paste; 100% conc.

Crodamine 1.18D. [Croda Ltd.] Stearylamine; cationic; see Crodamine 1.16D; solid; 100% conc.

Crodamine 3.A16D. [Croda Ltd.] Palmityl dimethylamine; cationic; see Crodamine 1.16D; liq.; 100% conc.

Crodamine 3.A18D. [Croda Ltd.] Stearyl dimethylamine, dist.; cationic; see Crodamine 1.16D; paste; 100% conc.

Crodamine 3.ABD. [Croda Ltd.] Dimethyl behenylamine; cationic; see Crodamine 1.16D; solid; 100% conc.

Crodamine 3.AED. [Croda Ltd.] Dimethyl erucylamine; cationic; see Crodamine 1.16D; liq.; 100% conc.

Crodamine 3.AOD. [Croda Ltd.] Dimethyl oleylamine; cationic; see Crodamine 1.16D; liq.; 100% conc.

Crodamol AC. [Croda] Cetyl acetate; emollient for skin care prods.; clear colorless liq.; oil-sol.; cloud pt. 19 C.

Crodamol BE. [Croda] Behenyl erucate; thickening and opacifying agent, emollient, stabilizer, modifier for lipsticks, powd. suspensions; lt. amber waxy solid; oil-sol.; m.p. 43–46 C.

Crodamol BS. [Croda] Butyl stearate; emollient without oiliness in lotions and creams, plasticizer in hair sprays and nail varnish, wetting agent for pigments and dyes in lipsticks; water wh. liq.

Crodamol CAP. [Croda] Cetearyl octanoate; emollient simulating properties of preen gland oil; provides water repellency in skin care preps.; clear colorless liq., low odor; oil-sol.; cloud pt. 0 C.

Crodamol CL. [Croda] Cetyl lactate; emollient in personal care prods.; colorless to pale straw solid; sol. in aq. alcoholic sol'ns., various glycols.

Crodamol CP. [Croda] Cetyl palmitate; emollient for replacing spermaceti wax; wh. flakes; oil-sol.; m.p. 50–54 C.

Crodamol CSL. [Croda] Ceto stearyl lactate; emollient esp. in compositions packaged in styrene; similar to Crodamol ML with slightly higher m.p.

Crodamol CSP. [Croda] Cetearyl palmitate; improves feel and body of emulsions; replaces spermaceti wax and beeswax in personal care prods.; Gardner 3 max. wh. flakes; oil-sol.; m.p. 48–52 C; acid no. 1 max.; sapon. no. 100–115.

Crodamol CSS. [Croda] Ceto stearyl stearate; emollient used in cosmetics and pharmaceuticals; powd. wax.

Crodamol DA. [Croda] Diisopropyl adipate; emollient and latent solv. in nail polish removers; superfatting agent; detackifier in skin applic.; solv. for cosmetics; lubricant in preelectric shave lotions; coupler, cosolv., plasticizer; water wh. liq.; odorless; sol. in ethanol; sp.gr. 0.950–0.962; f.p. –1 C; acid no.

0.5 max.; sapon. no. 480–500; ref. index 1.422–1.424.

Crodamol DO. [Croda] Decyl oleate; emollient, lubricant, penetrant, for cosmetics and pharmaceuticals; pale straw liq.; mild char. odor; bland taste.

Crodamol DOA. [Croda] Dioctyl adipate; emollient used in cosmetics and pharmaceuticals; liq.

Crodamol ICS. [Croda] Isocetyl stearate; emollient, lubricant, spreading agent, controls visc. and cryst. and reduces rheological changes in emulsion systems, cosmetic uses; water wh. liq.; low visc., liq. sets at 0 C.

Crodamol IPM. [Croda] IPM, perfumery grade; spreading agent, emollient, cosolv. for cosmetic raw materials; water wh. liq.; odorless.; sp.gr. 0.847–0.853; acid no. 2 max.; ref. index 1.432–1.434; 92% act.

Crodamol IPP. [Croda] IPP; see Crodamol DOA; water wh. liq.; odorless.

Crodamol ISNP. [Croda] Isostearyl neopentanoate; see Crodamol DOA; liq.

Crodamol LL. [Croda] Lauryl lactate; emollient, promotes spreading and wetting of cosmetics, leaves dry film on skin and hair; colorless to pale straw liq.; faint, bland odor; sol. in highly polar materials such as glycols.

Crodamol ML. [Croda] Myristyl lactate; emollient, improves smoothness and gloss in lipsticks without oiliness, tack reducer in antiperspirants, hair care preparations; wh. to pale yel. liq.; sol. in aq. alcoholic sol'ns., essentially lipophilic but also sol. in a variety of glycols.

Crodamol MM. [Croda] Myristyl myristate; emollient, superfatting agent, visc. builder in emulsions; substitute for spermaceti wax and/or beeswax in cosmetic and pharmaceutical formulations; creamy wh. waxy solid.; sol. in min. oil, IPM, oleyl alcohol; insol. water; m.p. 36–39 C; acid no. 1 max.; sapon. no. 120–130.

Crodamol OHS. [Croda] Octylhydroxy stearate; see Crodamol DOA; clear liq.

Crodamol OP. [Croda] 2-Ethyl hexyl palmitate; lubricant, aids prevention of insensible perspiration and skin respiration, used in make-up, skin and hair preparations; water wh. liq.; mild odor.

Crodamol PC. [Croda] Propylene glycol dicaprylate; emollient, spreading agent, used in cosmetic systems, skin and hair preparations; liq.; slight bland odor; low visc.; low cloud pt.

Crodamol PMP. [Croda] PPG-2 myristyl ether propionate; emollient, lubricant with dry, lt. greaseless feel; coupling agent; solv. for sunscreen actives; for bath oils, creams, moisturizers, emulsions; Gardner 1 max. clear, colorless liq.; very mild char. odor; sol. (1%) in min. oil, IPM, oleyl alcohol, ethanol/water, lanolin, cetyl alcohol; cloud pt. –5 C; acid no. 0.5 max.; iodine no. 1.0 max.; sapon. no. 140–155.

Crodamol PTC. [Croda] Pentaerythritol tetracaprylate/caprate; see Crodamol DOA; lt. yel. lipophilic visc. liq.; oil-sol.; cloud pt. 10 C.

Crodamol PTIS. [Croda] Pentaerythritol tetraisostearate; see Crodamol DOA; lt. amber visc. liq.; oil-sol.; cloud pt. 0 C.

Crodamol SS. [Croda] Cetyl esters and PPG-5-ceteth-20; syn. spermaceti NF; emollient and visc. builder for cosmetic and pharmaceutical preparations; almost wh. cryst. solid; sp.gr. 0.82–0.84 (50 C); m.p. 43–47 C; acid no. 5 max.; sapon. no. 109–120.

Crodamol W. [Croda] Stearyl heptanoate; emollient and water repellent for cosmetics and toiletries, melts rapidly on applic. to skin; wh. waxy solid; mild odor; m.p. 23–27 C.

Crodapearl Liq. [Croda] Sodium laureth sulfate, hydroxyethyl stearamide MIPA; pearling agent for shampoos, bubble baths; wh. paste.

Crodapearl NI Liquid. [Croda] Hydroxyethyl stearamide-MIPA, PPG-5 ceteth-20; nonionic; pearlescent for detergent systems; wh. soft paste.

Crodapur. [Croda Ltd.] Tech. lanolin; plasticizer for tape adhesives, mastics, and putty.

Crodasinic L&O. [Croda Ltd.] Alkyl sarcosines; lubricants and metal working fluids; oil-sol.

Crodasinic LS30. [Croda Ltd.] Sodium N-lauroyl sarcosinate; anionic; foaming, wetting agent and detergent for acidic conditions; corrosion inhibitor; bacteriastat and inhibitor; used in dental care preps., pharmaceuticals, personal care prods., household and industrial applics.; clear liq.; water-sol.; 30% act.

Crodasinic LS35. [Croda Ltd.] Sodium N-lauroyl sarcosinate; anionic; foaming agent, wetting agent, detergent, lubricant, antistat, corrosion inhibitor, bacteriostat, penetrant used in dental, pharmaceutical, shampoos, depilatories, and shaving preparations, food pkg., household and industrial uses; clear liq.; water sol.; biodeg.; 35% act.

Crodasinic OS35. [Croda Ltd.] Sodium N-oleoyl sarcosinate; anionic; see Crodasinic LS35; lower foaming; liq.; 35% conc.

Crodax DP 50. [Croda Ltd.] Complex mixt. of org. acids, esters, and lactones; converted to metallic soaps and esters used in solv. or oil-based rust preventives and lubricating prods.; solid.

Crodax DP 55. [Croda Ltd.] Complex mixt. of org. acids, esters, and lactones; see Crodax DP 50; solid.

Crodax DP 58. [Croda Ltd.] Complex mixt. of org. acids, esters, and lactones; see Crodax DP 50; solid.

Crodax DP 60. [Croda Ltd.] Complex mixt. of org. acids, esters, and lactones; see Crodax DP 50; solid.

Crodax DP 100. [Croda Ltd.] Calcium soap and oxidate; used in thick film rust preventives providing protection for steel in salt environments; used for underseal, wire rope dressings, waterproofing agents; solid, liq.

Crodax DP 101. [Croda Ltd.] Calcium soap and oxidate; see Crodax DP 100; solid, liq.

Crodax DP 111. [Croda Ltd.] Calcium soap and oxidate; see Crodax DP 100; solid, liq.

Crodax DP 155. [Croda Ltd.] Calcium soap and oxidate; see Crodax DP 100; solid, liq.

Crodazoline O. [Croda Ltd.] Oleic imidazoline; lubricating metal processing aids; Gardner 6–7 liq.; m.w. 345; oil-sol.; m.p. 40–57 C; surf. tens. 26.5 (1%/0.36%); pH 11.1 (10% aq. disp.); 90% min. imidazoline.

Croderol G7000. [Croda Ltd.] Glycerin; conditioner, humectant, moisturizing agent in cosmetics and pharmaceuticals, diluent and plasticizer for many polar materials; liq..

Crodesta A-10. [Croda] Acetylated sucrose distearate; dispersant, emulsifier, wetting agent, solubilizer and detergent in cosmetics, toiletries, pharmaceuticals; yel./wh. wax; sol. in veg. and min. oils; HLB 0.5; m.p. 43–49 C; acid no. 5 max.; sapon. no. 230–290; 100% act.

Crodesta A-20. [Croda] Acetylated sucrose distearate; see Crodesta A-10; solid; HLB 1.0; 100% conc.

Crodesta F-10. [Croda] Sucrose distearate; see Crodesta A-10; creamy wh. powd.; HLB 3.0; m.p. 60–68 C; acid no. 5 max.; sapon. no. 140–200; 100% act.

Crodesta F-50. [Croda] Sucrose distearate; see Crodesta A-10; wh. powd.; HLB 6.5; m.p. 74–78 C; acid no. 5 max.; sapon. no. 93–153; 100% act.

Crodesta F-110. [Croda] Sucrose distearate and sucrose stearate; see Crodesta A-10; thickener and suspending agent; wh. powd.; HLB 12.0; m.p. 72–78 C; acid no. 5 max.; sapon. no. 85–145; 100% act.

Crodesta F-160. [Croda] Sucrose stearate; see Crodesta A-10; thickener and suspending agent; wh. powd.; HLB 14.5; m.p. 70–74 C; acid no. 5 max.; sapon. no. 75–153; 100% act.

Crodesta SL-40. [Croda] Sucrose cocoate; nonionic; see Crodesta A-10; amber liq.; HLB 15.0; acid no. 5 max.; 55% mono ester.

Crodesta SL-40PX. [Croda] Sucrose ester; solubilizer; water and glycol sol.

Crodet L4. [Croda Ltd.] PEG-4 laurate; nonionic; o/w emulsifier for cosmetics and pharmaceutical creams, lotions and ointments, wetting agent, solubilizer for perfumes or aq. alcoholic preparations; dispersant; plasticizer for hair setting sprays; pale straw liq.; sol. in ethyl, oleyl, and cetearyl alcohols, oleic acid; HLB 9.8; sapon. no. 138–150.

Crodet L8. [Croda Ltd.] PEG-8 laurate; see Crodet L4; pale straw liq.; sol. see Crodet L4; HLB 12.7; sapon. no. 95–106.

Crodet L12. [Croda Ltd.] PEG-12 laurate; see Crodet L4; solubilizer for perfumes; dispersant in aq. systems; pale straw liq.; sol. in water, ethyl, oleyl, and cetearyl alcohols, oleic acid; HLB 14.5; sapon. no. 72–82.

Crodet L24. [Croda Ltd.] PEG-24 laurate; see Crodet L4; solubilizer for perfumes; off-wh. soft paste; sol. in water, ethyl and cetostearyl alcohol; HLB 16.8; sapon. no. 42–48.

Crodet L40. [Croda Ltd.] PEG-40 laurate; see Crodet L4; off-wh. waxy solid; sol. in water, ethyl and cetostearyl alcohol; HLB 17.9; sapon. no. 26–31.

Crodet L100. [Croda Ltd.] PEG-100 laurate; see Crodet L4; pale yel. waxy solid; sol. in water, ethyl and cetostearyl alcohol; HLB 19.1; sapon. no. 11–15.

Crodet S4. [Croda Ltd.] PEG-4 stearate; nonionic; o/w emulsifier for cosmetics and pharmaceutical creams, lotions and ointments, wetting agent, solubilizer for perfumes or aq. alcoholic preparations, dispersant; off-wh. soft paste; sol. in ethyl and oleyl alcohols, oleic acid, ceto stearyl alcohol, arachis oil and isoparaffinic solv.; HLB 7.7; sapon. no. 117–129.

Crodet S8. [Croda Ltd.] PEG-8 stearate; see Crodet S4; off-wh. soft paste; sol. see Croduret S4; HLB 10.8; sapon. no. 84–94.

Crodet S12. [Croda Ltd.] PEG-12 stearate; see Crodet S4; off-wh. waxy solid; sol. in ethyl, oleyl, and cetosteary1 alcohol, water, oleic acide; HLB 13.4; sapon. no. 65–75.

Crodet S24. [Croda Ltd.] PEG-24 stearate; see Crodet S4; off-wh. waxy solid; sol. in ethyl and cetostearyl alcohol, water; HLB 15.8; sapon. no. 38–47.

Crodet S40. [Croda Ltd.] PEG-40 stearate; see Crodet S4; off-wh. waxy solid; sol. in ethyl and cetostearyl alcohol, water; HLB 16.7; sapon. no. 23–30.

Crodet S100. [Croda Ltd.] PEG-100 stearate; see Crodet S4; off-wh. waxy solid; sol. in ethyl and cetostearyl alcohol, water; HLB 18.8; sapon. no. 10–14.

Crodex A. [Croda Ltd.] Cetostearyl alcohol and sodium lauryl sulfate; anionic; emulsifying wax BP for pharmaceuticals and cosmetic uses; almost wh. waxy solid; faint char. odor; water-disp.

Crodex C. [Croda Ltd.] Cetostearyl alcohol and Cetrimide BP; cationic; emulsifying wax BPC, bacteriacides, for pharmaceuticals and cosmetics; almost wh. waxy solid; faint char. odor; water-disp.

Crodex N. [Croda Ltd.] Cetostearyl alcohol and ceteth-20; nonionic; emulsifying wax BP, wetting agent, penetrant, emulsifier for most emollient materials in cosmetics and pharmaceuticals; almost wh. waxy solid; faint char. odor; water-disp.

Crodinhib RT70, RT70S. [Croda Ltd.] Borate derivs.; corrosion inhibitor used in metal working fluids and oils; liq.; water-sol.

Croduret 10. [Croda Ltd.] PEG-10 hydrog. castor oil; nonionic; emulsifier, solubilizer, emollient, superfatting agent, detergent used for cosmetics, textiles, metalworking fluids, emulsion polymerization, insecticides, herbicides, household detergents; straw-colored liq.; sol. in oleyl alcohol, naphtha, MEK, oleic acid, trichloroethylene; HLB 6.3; cloud pt. < 20 C; sapon. no. 120–130; surf. tens. 37 dynes/cm (0.1%).

Croduret 30. [Croda Ltd.] PEG-30 hydrog. castor oil; see Croduret 10; straw colored liq.; sol. in water, ethanol, oleyl alcohol, naphtha, MEK, oleic acid, trichloroethylene; HLB 11.6; cloud pt. 48 C; sapon. no. 70–80; surf. tens. 46 dynes/cm (0.1%).

Croduret 40. [Croda Ltd.] PEG-40 hydrog. castor oil; see Croduret 10; off-wh. visc. paste; sol. in water, ethanol, naphtha, MEK, oleic acid, trichloroethylene; HLB 13.0; cloud pt. 62 C; sapon. no. 60–65; surf. tens. 46 dynes/cm (0.1%).

Croduret 50 Special. [Croda Ltd.] PEG-50 hydrog. castor oil; solubilizer and emulsifier for cosmetics and pharmaceutical applics.; soft paste; water-sol.

Croduret 60. [Croda Ltd.] PEG-60 hydrog. castor oil; nonionic; see Croduret 10; off-wh. stiff paste; sol. in water, ethanol, MEK, oleic acid, trichloroethylene; HLB 14.7; cloud pt. 71 C; sapon. no.45–50; surf. tens.

47.5 dynes/cm (0.1%).

Croduret 100. [Croda Ltd.] PEG-100 hydrog. castor oil; nonionic; see Croduret 10; off-wh. waxy solid; sol. in water, ethanol, MEK, trichloroethylene; HLB 16.5; cloud pt. 65 C; sapon. no. 25–35; surf. tens. 46.2 dynes/cm (0.1%).

Crodyne BY19. [Croda] Gelatin; protective colloid, moisturizer, conditioner for skin and hair care prods.; humectant, thickener for pharmaceutical and food applics.; buff cryst. powd.; water-sol.; m.w. 25,000; 85% act.

Crolastin. [Croda] Hydrolyzed animal elastin; moisturizer and conditioner for skin care prods. and cleansers; yel. liq.; m.w. 4000; 30% act.

Crolastin 10 Powder. [Croda] Partially hydrolyzed elastin; conditioner for skin care cosmetics; aq. sol'n.

Crolastin 30 Powder. [Croda] Partially hydrolyzed elastin; see Crolastin 10 Powder.; aq. sol'n.

Cromoist CS. [Croda] Chondroitin sulfate, hydrolyzed animal protein; moisturizer, conditioner for face and body creams and lotions; clear to slightly hazy visc. sol'n.; m.w. 115,000; 20% act.

Cromoist HYA. [Croda] Hydrolyzed animal protein, hyaluronic acid; moisturizer, conditioner for skin care prods., facial creams; amber liq.; m.w. 500,000; 15% act.

Cromul 0685. [Croda Ltd.] Complex ethoxy (6) fatty alcohol ether; emulsifier and opacifier for cosmetic and pharmaceutical creams and lotions; soft waxy solid.

Cronectin H. [Croda] Hydrolyzed fibronectin; humectant, moisturizer for skin creams and lotions; opalescent low visc. liq.; m.w. 20,000; 3% act.

Croneton®. [Bayer] Ethiofencarb; insecticide for control of aphids; colorless crystals; m.w. 225; sol. > 60 g/100 g in methylene chloride, propanol-2, toluene, 0.19 g/100 g in water; sp.gr. 1.147 (20/4 C); m.p. 33.4 C.

Cropepsol. [Croda Ltd.] Hydrolysed collagen; conditioner for skin and hair conditioner; aq. sol'n.

Cropeptone. [Croda Ltd.] Hydrolyzed animal protein; see Cropepsol; aq. sol'n.

Cropotex®. [Bayer] Flubenzimine; acaricide displaying good larvicidal activity; orange-yel. powd.; m.w. 416.1; sol. (g/1000 ml): > 200 g in dichloromethane, toluene; 0.0016 g in water; m.p. 118.7 C.

Croquat K. [Croda] Quat. deriv. of hydrolyzed keratin; conditioner and permanent conditioner for hair; aq. sol'n.

Croquat L. [Croda] Laurdimonium hydrolyzed animal protein; cationic; conditioner for clear rinses, shampoos, conditioners; clear yel. liq.; m.w. 2500; essentially water-sol.; 40% act.

Croquat M. [Croda] Cocodimonium hydrolyzed animal protein; cationic; conditioner for shampoos, perms, hair relaxers; clear yel. liq.; m.w. 2500; essentially water-sol.; 40% act.

Croquat S. [Croda] Steardimonium hydrolyzed animal protein; cationic; conditioner for cream rinses; yel. paste/gel; m.w. 2700; partly water-sol.; 40% act.

Croquat Soya. [Croda] Quaternized hydrolyzed soya protein; conditioner for skin and hair conditioner; aq. sol'n.

Croquat WKP. [Croda] Cocodimonium hydrolyzed animal keratin; cationic; permanent conditioning protein for cream rinses, shampoos, conditioners, perms, nail care prods.; clear amber liq.; m.w. 1000; water-sol.; 25% act. in water.

Crosilk 10,000. [Croda] Hydrolyzed silk; water-sol. protein conditioner providing manageability, gloss, texture in hair care prods., moisturizing and protection in skin care prods.; amber liq.; m.w. 10,000; 15% act.

Crosilk Liq. [Croda] Silk amino acids; conditioner, humectant for skin and hair care prods.; pale amber liq.; m.w. 90; water-sol.; 15% act.

Crosilk Powder. [Croda] Silk powder; protein for solid make-up where it absorbs oil, improves leveling, modifies applic. properties and provides a silky, lustrous appearance; lustrous gray/wh. powd.; m.w. > 500,000; 100% act.

Crosilkquat. [Croda] Quaternized hydrolyzed silk protein; conditioner for skin and hair conditioner; aq. sol'n.

Crossential EPO. [Croda] Evening primrose oil; contains essential fatty acids for skin care.

Crosterene. [Croda Ltd.] Stearic acid; pearlescent in cosmetic creams and lotions; binder; buffing compds., candles, and lubricants; wax-like solids.

Crotein A. [Croda] Refined collagen hydrolysates; conditioner, foam stabilizer and booster, dye leveling agent used in personal care prods.; powd.; low odor; water-sol.; 100% act.

Crotein AD Anhyd. [Croda] AMP isostearoyl hydrolyzed animal protein; conditioners for hair preparations; yel. clear liq.; sp.gr. 0.830–0.850; acid no. 35–50; pH 8.0–9.0 (10%); 27–33% solids in ethanol.

Crotein AD. [Croda] AMP isostearoyl hydrolyzed animal protein; emollient, conditioner for hair preparations; yel. clear liq.; sol. in oil and alcohol; sp.gr. 0.885–0.900; acid no. 29–39; pH 8.4–9.2 (10%); 24–26% solids in ethanol/water.

Crotein ADX. [Croda] AMP isostearoyl hydrolyzed animal protein; emollient, conditioners for hair preparations; yel. clear liq.; sol. in alcohol; sp.gr. 0.855–0.870; acid no. 55–70; pH 8.4–9.2 (10%); 36–40% solids in ethanol.

Crotein ASC. [Croda] Ethyl ester of hyrolyzed animal protein; conditioner and film modifier for alcoholic and hydroalcoholic lotions and hair care prods.; clear lt. amber liq.; m.w. 2000; sol. in water, alcohol; 20% act.

Crotein ASK. [Croda] Hydrolyzed animal keratin; conditioner, film modifier for hair setting/conditioning systems; clear amber liq.; m.w. 2000; sol. in water, alcohol; 12% act.

Crotein BTA. [Croda] Benzyltrimonium hydrolyzed animal protein; conditioner and softener for hair; detangler and bodying agent; yel. clear liq.; sol. in water; forms clear sol'n. (50% alcohol); sp.gr. 1.05–1.15; acid no. 20–35; pH 5.5–7.0 (10%); 30–35% solids.

Crotein C. [Croda] Refined collagen hydrolysates; see Crotein A; powd.; low odor; water-sol.; 100%

act.

Crotein CAA. [Croda] Animal collagen amino acids, sodium chloride; amphoteric; hair moisturizer and conditioning, aux. emulsifier; powd.; 100% act.

Crotein CAA/SF. [Croda] Animal collagen amino acids; moisturizer for skin creams and lotions; yel. liq.; m.w. 150; water-sol.; 40% act.

Crotein CCT. [Croda] TEA salt of a complex protein coconut fatty acid ester; detergent, foaming agent, used in skin cleansers, hair and body shampoos, reduces irritation of other detergents.

Crotein HKP Powd. [Croda] Hair keratin amino acids, sodium chloride; amphoteric; substantive conditioner and moisturizer for shampoos, cream rinses; lt. brn. powd.; m.w. 150; water-sol.; 50% act.

Crotein HKP/SF. [Croda] Animal keratin amino acids; conditioner for hair and nail care prods.; pale straw liq.; m.w. 150; compat. with aq. alcohol; 25% act.

Crotein HWE. [Croda] Hydrolyzed whole egg; conditioner for hair and skin care prods.; powd.; water-disp.

Crotein IP, IPX. [Croda] Isostearoyl hydrolyzed animal protein; conditioner; IP used in nail enamels, solvent-based nail polish and removers; IPX for use in anhyd. pomades, brilliantines, bath oils; brn. visc. liq.; sol. (@ 10%) in min. oil, ethanol (slight haze); immiscible with water; acid no. 180–200 and 130–150 resp.; pH 6–8 (10% aq. disp.); 100% act.

Crotein K. [Croda] Hydrolyzed animal keratin; proteinic conditioner and moisturizer for hair and nail care prods.; lt. tan powd.; m.w. 2000; 60% act.

Crotein O. [Croda] Refined collagen hydrolysates; conditioning agent, foam booster and stabilizer, dye leveling agent in hair prods., substantive to hair and skin; powd.; low color and odor; powd.; water-sol.; 100% act.

Crotein Q. [Croda] Steartrimonium hydrolyzed animal protein; conditioner for hair care preparations; off-wh. powd.; m.w. 12,500; sol. in water; sol. @ 5% in 50% ethanol sol'n.; visc. 15–25 mps (10%); pH 5.5–6.5 (10%); 95% act. min.

Crotein SPA. [Croda] Hydrolyzed animal protein; conditioner for hair care preparations; foam stabilizer/booster in shampoo; peptizing aid in shampoos; dye leveling aid in hair dyes and bleaches; also for skin treatment, in depilatories; wh. powd., bland, pleasant odor; m.w. 2000; sol. in 50% ethanol, water; visc. 20–25 mps (10%); pH 5.5–6.5 (10%).

Crotein SPA55. [Croda] Hydrolyzed animal protein; conditioner for personal care prods.; amber liq.; m.w. 4000; 55% act.

Crotein SPC. [Croda] Hydrolyzed animal protein; see Crotein SPA; wh. powd.; bland, pleasant odor; m.w. 10,000; sol. in 65% ethanol @ 1% solid protein; water-sol.; visc. 40–50 mps (10%); pH 5.5–6.5 (10%).

Crotein SPO. [Croda] Hydrolyzed animal protein; see Crotein SPA; wh. powd.; bland, pleasant odor; m.w. 1000; sol. in water, 20% ethanol @ 1% solid protein; visc. 15–20 mps (10%); pH 5.5–6.5 (10%).

Crotein WKP. [Croda] Hydrolyzed animal keratin; conditioner for hair prods.; pale yel. liq.; m.w. 600; 22% act.

Cru Fluid 350. [Crucible] Silicone fluid; defoamer for food applics.; fluid.

Cru Release 900 Series. [Crucible] Silicone emulsion; defoamer for food/pharmaceutical applics. incl. drug extraction and separation, antibiotics, beverage processing.

Cru Thix 46. [Crucible] Polyacrylate and gum; dispersant for latex compding.; thickener; pale wh. transparent, very visc. liq.; sp.gr. 1.0.

Crylcon 3000 Series. [DuPont] Methacrylate; bonding agent for use in concrete, filled and unfilled mortar; sp. dens. 2243 kg/m^3; tens. str. 8.3 MPa.

Crylcon 7000 Series. [DuPont] Methacrylate; bonding agent used in filled and unfilled mortar; sp. dens. 2243 kg/m^3; tens. str. 8.3 MPa.

Crystal® Crown. [CasChem] Castor oil; emollient; deodorized specially refined grade; sol. in alcohols, esters, ethers, ketone, and aromatic solvs.

Crystal Inhibitor #5. [Harcros] Nonionic blend; retards formation of crystals in 2,4-D amine formulation dilutions.

Crystal® O. [CasChem] Castor oil; emollient; deodorized, refined grade; sol. in alcohols, esters, ethers, ketones, and aromatic solvs.; faint mild odor.

Crystamet 1020. [Crosfield] Sodium metasilicate, pentahydrate; alkaline silicate used for formulating specialty detergents and contributing buffering capacity and corrosion inhibition of soft metals and ceramic glazes; wh. free-flowing gran.; 53% on 20 mesh; m.w. 212.14; sol. 61 g/100 g water (30 C); bulk dens. 55 lb/ft^3 (loose); m.p. 72.2 C; pH 12.4 (1%); 28.5–30.0% sodium oxide, 27.5–29.0% silica.

Crystamet 2040. [Crosfield] Sodium metasilicate pentahydrate; see Crystamet 1020; wh. free-flowing gran.; 90% on 40 mesh; m.w. 212.14; sol. 61 g/100 g water (30 C); bulk dens. 55 lb/ft^3 (loose); m.p. 72.2; pH 12.4 (1%); 28.5–30.0 sodium oxide, 27.5–29.0% silica.

Crystamet 3080. [Crosfield] Sodium metasilicate pentahydrate; see Crystamet 1020; wh. free-flowing gran.; 63% on 60 mesh; m.w. 212.14; sol. 61 g/100 g water (30 C); bulk dens. 55 lb/ft^3 (loose); m.p. 72.2; pH 12.4 (1%); 28.5–30.0 sodium oxide, 27.5–29.0% silica.

Crystic Release Agent No. 1. [Scott Bader] Cellulose acetate sol'n.; quick-drying sol'n. for use as a primary sealer on wood, plaster, other porous surfs.; colored or colorless liq.; flash pt. < 32 C.

Crystic Release Agent No. 2. [Scott Bader] PVAL sol'n.; quick-drying sol'n. for use on metal or reinforced polyester molds, or as a sec. release agent on molds treated with Release Agent No. 1; colored or colorless liq.; flash pt. < 32 C.

CS1590. [Hüls] 3-(N-Styrylmethyl-2-aminoethylamino) propyltrimethoxy silane hydrochloride; coupling agent, chem. intermediate, blocking agent, release agent, lubricant, primer, reducing agent; liq.; m.w. 375.0; sp.gr. 0.92; flash pt. 13 C; ref. index 1.395; 40% in methanol.

CT-88 Aerosol. [Chem-Trend] Fluoropolymer/organic binder disp.; dry-film release agent and lubricant for molding operations, esp. where abrasion is a problem; effective on vinyls, polyesters, epoxies, phenolics, and PP and for releasing silicone rubber from metal or plastic molds; applics. incl. casting, potting, hand layup, encapsulating, filament winding, inj. molding, compression molding; off-wh. aerosol; cloudy disp.; ethereal odor; b.p. ≈ 165 F; nil sol. in water; 97.3% volatile.

CT1800. [Hüls] Tetrachloro silane; see CS1590; liq.; m.w. 169.9; sp.gr. 1.481; b.p. 51–52 C; flash pt. none; ref. index 1.415; 99% purity.

CT2015. [Hüls] 1,1,4,4-Tetramethyldichlorodisilethylene; see CS1590; solid; m.w. 215.3; b.p. 198–200 C; m.p. 37 C; flash pt. 44 C; 97% purity.

CT2030. [Hüls] Tetramethyldisiloxane; see CS1590; liq.; m.w. 134.3; sp.gr. 0.76; b.p. 70–71 C; flash pt. –10 C; ref. index 1.370; 95% purity.

CT2050. [Hüls] Tetramethyl silane; see CS1590; liq.; m.w. 88.2; sp.gr. 0.641; b.p. 26.6–26.7 C; flash pt. –27 C; ref. index 1.359; 99.9% purity.

CT2090. [Hüls] Tetrapropoxy silane; see CS1590; liq.; m.w. 264.4; sp.gr. 0.92; b.p. 224–225 C; flash pt. 95 C; ref. index 1.401; 98% purity.

CT2500. [Hüls] Triethoxy silane; see CS1590; liq.; m.w. 164.3; sp.gr. 0.88; b.p. 131 C; flash pt. 26 C; ref. index 1.337; 95% purity.

CT2520. [Hüls] Triethylchloro silane; see CS1590; liq.; m.w. 150.7; sp.gr. 0.90; b.p. 144–145 C; flash pt. 29 C; ref. index 1.431; 97% purity.

CT2523. [Hüls] Triethyl silane; see CS1590; liq.; m.w. 116.3; sp.gr. 0.73; b.p. 107–108 C; flash pt. 0 C; ref. index 1.412; 98% purity.

CT2902. [Hüls] 1-Trimethoxysilyl-2-(chloromethyl) phenylethane; see CS1590; amber liq.; m.w. 274.8; b.p. 161 C (1.5 mm); flash pt. 130 C; 95% purity.

CT2925. [Hüls] N-Trimethoxysilylpropyl-N,N,N-trimethyl ammonium chloride; see CS1590; amber liq.; m.w. 257.8; sp.gr. 0.93; flash pt. 10 C; ref. index 1.397; 50% in methanol.

CT2950. [Hüls] Trimethylchloro silane; see CS1590; liq.; m.w. 108.7; sp.gr. 0.86; b.p. 57-58 C; flash pt. –18 C; ref. index 1.389; 99% purity.

CT2970. [Hüls] Trimethylethoxy silane; see CS1590; liq.; m.w. 118.3; sp.gr. 0.76; b.p. 75–76 C; flash pt. –5 C; ref. index 1.374; 97% purity.

CT3250. [Hüls] Trimethylsilyl acetamide; see CS1590; solid; m.w. 131.2; b.p. 103–105 C (33 mm); 98% purity.

CT3254. [Hüls] Trimethylsilylacetate; see CS1590; liq.; m.w. 132.2; sp.gr. 0.89; b.p. 103–104 C; flash pt. 4 C; ref. index 1.389; 97% purity.

CT3600. [Hüls] Trimethylsilyl imidazole; see CS1590; liq.; m.w. 140.3; sp.gr. 0.95; b.p. 99 C (19 mm); flash pt. 80 C; ref. index 1.475; 98% purity.

CT3610. [Hüls] Trimethylsilyl iodide; see CS1590; liq.; m.w. 200.1; sp.gr. 1.47; b.p. 106–107 C; flash pt. –2 C; ref. index 1.474; 95% purity.

CT3795. [Hüls] Trimethylsilyl trifluoromethane sulfonate; see CS1590; liq.; m.w. 222.3; sp.gr. 1.23; b.p. 140–141 C; flash pt. –3 C; 95% purity.

Cu-0203 T 1/8″. [M&T Harshaw] Copper chromite; catalyst for dehydrogenation of alcohols, decomposition of methanol, oxygen scavenging, and dehydrogenation of piperazines to pyrazines; tablet, 1/8″ diam.; dens. 137 lb/ft^3; 79% conc.

Cu-0211 P. [M&T Harshaw] Copper chromite; see Cu-0203 T 1/8″; powd.; dens. 32 lb/ft^3.

Cu-0223 P. [M&T Harshaw] Copper chromite; see Cu-0203 T 1/8″; powd.; dens. 52 lb/ft^3.

Cu-0233 T 1/8″. [M&T Harshaw] Copper chromite; see Cu-0203 T 1/8″; tablet, 1/8″ diam.; dens. 137 lb/ft^3.

Cu-0803 T 1/8″. [M&T Harshaw] Copper; catalyst used to remove trace amounts of oxygen from gas steams; tablet, 1/8″ diam.; dens. 67 lb/ft^3.

Cu-1106 P. [M&T Harshaw] Copper chromite stabilized with barium; catalyst for methyl ester hydrogenolysis; powd.; dens. 30 lb/ft^3.

Cu-1107 T 1/8″. [M&T Harshaw] Copper chromite stabilized with barium; catalyst for methyl ester hydrogenolysis; tablet, 1/8″ diam.; dens. 105 lb/ft^3.

Cu-1117 T 1/8″. [M&T Harshaw] Copper chromite stabilized with barium; hydrogenation catalyst; tablet, 1/8″ diam.; dens. 118 lb/ft^3.

Cu-1129 P. [M&T Harshaw] Copper chromite stabilized with barium; hydrogenation catalyst; powd.; dens. 47 lb/ft^3.

Cu-1132 T 1/8″. [M&T Harshaw] Copper chromite stabilized with barium; see Cu-1129 P; tablet, 1/8″ diam.; dens. 110 lb/ft^3.

Cu-1413 P. [M&T Harshaw] Copper chromite; catalyst for methyl ester hydrogenolysis; powd.; dens. 57 lb/ft^3.

Cu-1422 T 1/8″. [M&T Harshaw] Copper chromite; see Cu-1413 P; tablet, 1/8″ diam.; dens. 100 lb/ft^3.

Cu-1800 P. [M&T Harshaw] Copper chromite; catalyst for methyl ester hydrogenolysis and dehydrogenation; powd.; dens. 41 lb/ft^3.

Cu-1803 P. [M&T Harshaw] Copper chromite; see Cu-1800 P; powd.; dens. 20 lb/ft^3.

Cu-1808 T 1/8″. [M&T Harshaw] Copper chromite; catalyst; tablet, 1/8″ diam.; dens. 93 lb/ft^3.

Cu-1809 T 1/8″. [M&T Harshaw] Copper chromite; catalyst; tablet, 1/8″ diam.; dens. 93 lb/ft^3.

Cu-1920 P. [M&T Harshaw] Copper chromite stabilized with manganese; catalyst suggested for selective hydrogenation of edible oils; hydrogenolysis of methyl esters; powd.; dens. 37 lb/ft^3.

Cu-2501 G 4-10. [M&T Harshaw] Copper catalyst; alkali-promoted; gran.; dens. 30 lb/ft^3.

Cu-3829 P. [M&T Harshaw] Copper chromite; catalyst for the hydrogenation of nitroaromatic molecules; powd.; dens. 80 lb/ft^3.

Culminal CMMC. [Aqualon] Carboxymethyl methylcellulose; nonionic; thickener, binder, film-former, protective colloid, water-retention aid, adhesive aid for building prods., latex paints, stuccos, S-PVC prod., tobacco sheeting; avail. in gran., powd. (P), fine powd. (PF), very fine powd. (PFF), and delayed hydration grades (R); visc. 1000–2200 mPas (2%).

Culminal MC. [Aqualon] Methylcellulose; nonionic; see Culminal CMMC; visc. 10–15,000 mPas (2%).

Culminal MHEC. [Aqualon] Methylhydroxyethylcellulose; nonionic; see Culminal CMMC; visc. 100–40,000 mPas (2%).

Culminal MHPC. [Aqualon] Methylhydroxypropylcellulose; nonionic; see Culminal CMMC; visc. 20–40,000 mPas (2%).

Cumate®. [Vanderbilt] Copper dimethyldithiocarbamate; ultra accelerator for SBR, IIR for high speed vulcanization; used with thiazole modifier to control scorch rate; dk. brn. powd., 99.9% thru 100 mesh; rods; m.w. 303.98; practically insol. in water; dens. 1.75 ± 0.03 mg/m³ (powd.); m.p. > 325 C (powd.); 300 C min. (rods); 20.0–22.0% copper content (powd.); 19.0–21.0% copper content (rods).

Cunilate 2174-NO. [Morton Int'l.] Antimicrobial for solv. systems in adhesive/latex applics.; FDA approved preservatives.

Cunilate 2419, 2419-75. [Morton Int'l.] Antimicrobial for textile applics., pulp slurries, paperboard, concrete safes.

Cuniphen 2713. [Morton Int'l.] Antimicrobial for textile applics.

Cuniphen 2721, 2721-C. [Morton Int'l.] Antimicrobial for textile applics.

Cuniphen 2778-1. [Morton Int'l.] Antimicrobial for in-can preservation; FDA approved.

Curaterr®. [Bayer] Carbofuran; insecticide and nematocide; wh. cryst.; m.w. 221.3; sol. 15% in acetone, 12% in methylene chloride, 250–700 ppm in water; m.p. 150–152 C.

Cure-Rite 18. [Akrochem] Thiocarbamyl sulfenamide; nonstaining primary accelerator for EPDM, SBR, nitrile, natural and butyl rubbers; wh. powd. and pellets; m.w. 248; sol. in benzene, chloroform, CCl_4, ether, and alcohol; sp.gr. 1.35 ± 0.05; m.p. 133 C min.

Cure-rite Accelerator. [Goodrich] Thiocarbamyl sulfenamide; accelerators for SBR, EPDM, and other general purpose rubbers, and for special elastomers incl. nitriles and butyls; provides fast cure time.

Curezol® 2E4MZ. [Pacific Anchor] 2-Ethyl 4-methyl imidazole; epoxycuring agent for adhesives, composites, filament winding, solder-resistant inks, potting compds.; accelerator for anhydride curing agents; Gardner 2 solid; dens. 8.2 lb/gal; visc. 95 poise.

Curezol® 2E4MZ-CN. [Pacific Anchor] Imidazole; epoxy curing agent for fiber-reinforced composites; accelerator for anhydrides; Gardner 7 color; dens. 8.6 lb/gal; visc. 2.2 poise.

Curezol® 2MZ-AZINE. [Pacific Anchor] Imidazole; epoxy curing agent for solder-resistant inks, insulating powds., structural adhesives; accelerator for dicyandiamide and anhydrides; wh. powd.; m.p. 477-484 F.

Curezol® 2PHZ. [Pacific Anchor] Imidazole; epoxy curing agent for adhesives, casting, potting, encapsulation; accelerator for dicyandiamide and anhydrides; adhesive for surf. mounting of devices onto circuit boards; ylsh.-pink powd.; m.p. 415–437 F (decomp.).

Curezol® 2PZ. [Pacific Anchor] 2-Phenyl imidazole; epoxy curing agent for printed circuit boards, molding compds., potting; accelerator for dicyandiamide and anhydrides; pale pink powd.; m.p. 279–297 F.

Curezol® 2PZ-CNS. [Pacific Anchor] Imidazole; epoxy curing agent for transfer molding powds., elec. powd. coatings; accelerator for anhydrides, solid epoxy resins; wh. powd.; m.p. 356–360 F.

Curezol® 2PZ-OK. [Pacific Anchor] Imidazole; epoxy curing agent for molding compds., powd. coatings; accelerator for dicyandiamide; longer pot life than Curezol 2PZ; wh. powd.; m.p. 284 F (decomp.).

Curezol® C17Z. [Pacific Anchor] Imidazole; epoxy curing agent with long latency; for structural adhesives, powd. coatings, molding powds.; accelerator for anhydrides and dicyandiamides; wh. powd.; m.p. 187–196 F.

Curithane 103. [Upjohn] Polymethylene polyaniline; amine curing agent for PU; intermediate in mfg. of polyamides, polyimides, coatings, and plastics; sp.gr. 1.065; visc. 82 cps (70 C); soften. pt. 55 C; flash pt. 460 F (CC); 0.2% volatiles.

Cutina AGS. [Henkel] Glycol distearate; pearlescent and opacifier; emulsion shampoos and foam baths; wh. flakes.

Cutina BW. [Henkel] Glyceryl hydroxystearate, cetyl palmitate, microcryst. wax, trihydroxystearin; wax for use as beeswax substitute; flakes.

Cutina CBS. [Henkel] Glyceryl stearate, cetearyl alcohol, cetyl palmitate, and cocoglycerides; nonionic; cream base for mfg. of creams and lotions of the o/w type; visc. stabilizer; flakes; 100% conc.

Cutina CP. [Henkel] Cetyl palmitate; syn. spermaceti; consistency factor for creams, ointments, liq. emulsions, fatty make-ups, and sticks; wh. waxy flakes; insol. in water; HLB 9; m.p. 50 C; sapon. no. 116-126.

Cutina CP-A. [Henkel] Cetyl palmitate; emollient for cosmetics, preparations such as creams, lotions, and lipsticks; substitute for nat. spermaceti; flakes; sol. in alcohols, esters, cosmetic oils, glycols, ketones, and aromatics; water-insol.

Cutina EGMS. [Henkel] Ethylene glycol stearate; nonionic; opacifying and pearly sheen producing wax for shampoos, shower, and bath preparations; consistency factor for creams and emulsions; wh. flakes.

Cutina GMS. [Henkel] Glyceryl stearate; nonionic; cream base for o/w and w/o emulsions; visc. agent for creams, ointments, sticks; beads; 100% conc.

Cutina HR. [Henkel] Hydrog. castor oil; lubricant for tablets, high melting consistency giving factor, thickener for oils; powd.; 100% act.

Cutina LM. [Henkel] Castor oil, glyceryl ricinoleate, octyl dodecanol, carnauba, candelilla, and microcryst. waxes, cetyl alcohol, beeswax, and min. oil; wax base for cosmetic stick preparations; yel. wax; slight odor; sp.gr. 0.90; m.p. 70-74 C; sapon. no. 120-130; 100% fatty matter.

Cutina MD. [Henkel] Glyceryl stearate; stabilizer used in ointments, creams, and liq. emulsions; solid.

Cutina TS. [Henkel] Fatty acid polyglycol ester; nonionic; opacifying and pearly sheen producing wax for shampoos, shower and bath preparations.

Cutless 50. [GAF] Ampholytic polymer; softener, lubricant and finish for fibers; liq.

CV4772. [Hüls] Vinylmethyldichloro silane; coupling agent, chem. intermediate, blocking agent, release agent, lubricant, primer, reducing agent; liq.; m.w. 141.1; sp.gr. 1.09; b.p. 92–93 C; flash pt. 30 C; ref. index 1.427; 98% purity.

CV4800. [Hüls] Vinyltriacetoxy silane; see CV4772; liq.; m.w. 232.3; sp.gr. 1.167; b.p. 112–113 C (1 mm); flash pt. 88 C; ref. index 1.423; 96% purity.

CV5050. [Hüls] Vinyl tris (methylethylketoxime) silane; see CV4772; liq.; m.w. 313; sp.gr. 0.98; b.p. 310 C; flash pt. 4 C; 90% purity.

CV5100. [Hüls] Vinyl tris (trimethylsiloxy) silane; see CV4772; liq.; m.w. 274.7; sp.gr. 0.866; b.p. 49 C (20 mm); ref. index 1.3971; 97% purity.

Cyanamer A-370. [Am. Cyanamid] Modified polyacrylamide polymer; dispersant, antiprecipitant, solubilizer for adhesive tapes, agriculture, cement, ceramics, coatings, detergents, dyes, printing inks, processing, textile and leather; binder, flocculating and suspending agent and lubricant; thickener and stabilizer; off-wh. to med. tan powd.; water-sol.; sp. visc. 3.7 ± 0.5; ≤ 7% moisture.

Cyanamer P-26. [Am. Cyanamid] Modified polyacrylamide; dispersant, solubilizer, antiprecipitant, thickener, binder, flocculating, suspending, and crosslinking agent, lubricant; used in adhesives, agriculture, cosmetics, detergents, foundry molds, metal processing, plaster, printing inks, processing, textile and leather, latex mfg., cements; wh. powd.; m.w. 200,000; water-sol.; visc. 300–800 cps (10%); pH 10–12.

Cyanamer P-35. [Am. Cyanamid] Modified polyacrylamide; dispersant and antiprecipitant for use in aq. systems; reduces slime and sludge formation for pipes and processing equipment; latexes and emulsions; hydrotrope in detergents; used in adhesive tapes, agriculture, cement, ceramics, graphic arts, insulation, latex mfg., metal processing, phospher processing for TV tubes, printing ink, textile, and leather; wh. powd.; water-sol.; visc. 6.5–10.5 (15%); pH 10–12 (15%); 93% min. solids.

Cyanamer P-70. [Cyanamid BV] Modified polyacrylamide; dispersant and suspending agent in aq. systems; antiscalant for calcium carbonate, calcium, barium, and strontium sulphate; water-sol.

Cyanamer P-80. [Cyanamid BV] Sulfonated maleic anhydride copolymer; see Cyanamer P-70; yel. liq.; water-sol.

Cyanamer P-250. [Am. Cyanamid] Polyacrylamide; nonionic; thickener, binder, sizing, flocculating, suspending, crosslinking agent, filtering aid, lubricant; used in adhesives, agric., cement, coatings, cosmetics, detergents, fire fighting, graphic arts, insulation, latex mfg., plaster, printing ink, processing, textile and leather, ceramics, dyes and pigments, used for polymer recovery in adhesive tapes, paints, latex mfg. and processing; wh. powd.; m.w. 5–6 × 10^6; water-sol.; visc. 1.8–2.2 cps (0.1%); pH 6.0–6.5 (1%).

Cyanox® 425. [Am. Cyanamid] 2,2′-Methylenebis (4-ethyl-6-tert-butylphenol); antioxidant for impact molding resins; stabilizes acrylics and ABS; end-uses incl. extruded and molded prods.; cream to wh. powd; m.w. 368.5; sol. (g/100 g sol'n.) > 60.0 g in acetone and dioxane, 56.5 g in ethyl acetate, 50.0 g in chloroform, 42.8 g in 95% ethanol, 33.3 g in benzene, 9.1 g in n-heptane; < 0.1 g in water; sp.gr. 1.10; m.p. 117–123 C.

Cyanox® 711. [Am. Cyanamid] Ditridecyl thiodipropionate; sec. antioxidant for stabilizing polyolefins, ABS, petrol. lubricants, SBR latex compositions; colorless to slightly yel. clear liq.; m.w. 543; sol. (g/100 g sol'n.) 13.8 g in 95% ethanol; miscible with ethyl acetate, n-heptane, MEK, and toluene; sp.gr. 0.936; acid no. 3.0; sapon. no. 200–210.

Cyanox® 1212. [Am. Cyanamid] Mixed lauryl-stearyl thiodipropionate; sec. antioxidant for stabilizing polyolefins; applications incl. pipe, hot-melt adhesives, and molded olefin prods.; wh. waxy cryst. flakes; sol. (g/100 g sol'n.) 30.0 g in toluene, 4.2 g in n-heptane; 0.05 g in water and 95% ethanol; sp.gr. 1.018; set pt. 49 C; acid no. 1.0.

Cyanox® 1735. [Am. Cyanamid] Phenolic-phosphite composition; antioxidant esp. effective in polyolefins; specialty lubricants; plastics; pale yel. liq.; sol. in most org. solvs.; insol. in water; dens. 0.90 g/cc.

Cyanox® 1790. [Am. Cyanamid] 1,3,5-Tris (4-tert-butyl-3-hydroxy-2,6-dimethylbenzyl)-1,3,5-triazine-2,4,6-(1H,3H,5H)-trione; antioxidant for use in polyolefin pipe, film, household appliances, olefin and urethane fibers, styrenics and polyesters; off-wh. powd.; sol. (g/100 g sol'n.) >10.0 g in cyclohexane, styrene, toluene, MEK; 4.6 g in ethanol; insol. in water; m.p. 145–155 C.

Cyanox® 2246. [Am. Cyanamid] 2,2′-Methylenebis (4-methyl-6-tert-butylphenol); antioxidant preventing thermal oxidation of ABS, polyethylene, PP, and EVA; oxidation inhibitor for fats, oils, and paraffin wax; polymerization inhibitor in chemical processes; for ABS, hot-melt adhesives, latex carpet backing, and specialty olefin applics.; cream to wh. powd.; m.w. 340.5; sol. (g/100 g sol'n.) > 60.0 g in acetone, 59.9 g in dioxane, 54.4 g in ethyl acetate, 47.6 g in chloroform, 36.7 g in 95% ethanol, 35.8 g in benzene, 5.8 g in heptane, < 0.1 g in water; sp.gr. 1.08; m.p. 121–128 C.

Cyanox® 2777. [Am. Cyanamid] 1:2 blend of 1,3,5-tris (4-t-butyl-3-hydroxy-2,6-dimethylbenzyl)-1,3,5-triazine-2,4,6-(1H,3H,5H) trione and tris (2,4-di-t-butylphenyl) phosphite; antioxidant, stabilizer for polymers, esp. polyolefins in high-temp. processing; off-wh. powd.; sol. (g/100 g) 13.7 g ethyl acetate, 3.2 g acetone, 2.5 g hexane; sp.gr. 1.07; bulk dens. 39.6 lb/ft^3; m.p. 166–171 C.

Cyanox® LTDP. [Am. Cyanamid] Dilauryl thiodipropionate; sec. antioxidant in ABS, PP, and polyethylene; used in food pkg. materials, automotive, appliance, battery casing, pipe; stabilization of oils, lubricants, sealants, and adhesives; wh. waxy cryst. flakes; m.w. 514; sol. (g/100 g sol'n.) 51.2 g in MEK and acetone, 40.5 g in *n*-heptane, 39.4 g in ethyl acetate; < 0.5 g in 95% ethanol; sp.gr. 0.915; set pt. 40 C; acid no. 1.0.

Cyanox® MTDP. [Am. Cyanamid] Dimyristyl thiodipropionate; sec. antioxidant for protection of polyolefins; wh. waxy cryst. flakes; set pt. 47 C; acid no. 0.5. 97% purity.

Cyanox® STDP. [Am. Cyanamid] Distearyl thiodipropionate; sec. antioxidant used in polyolefins and other polymers; used in automotive, appliance, container film, sealant, and adhesive applics.; wh. waxy cryst. flakes; m.w. 683; sol. (g/100 g sol'n.) 10.7 g in toluene, 1.5 g in *n*-heptane; < 0.5 g in ethyl acetate, 95% ethanol, and MEK; sp.gr. 1.027; set pt. 64 C; acid no. 1.5 max.

Cyasorb® UV 9. [Am. Cyanamid] 2-Hydroxy-4-methoxybenzophenone; lt. stabilizer and uv absorber for plastics and coatings; esp. for flexible and rigid PVC, unsat. polyesters, and acrylics; used in outdoor sheeting and glazing applics., molded products, adhesives; pale cream to wh. powd.; m.w. 228; b.p. 150–160 C (5 mm Hg); sol. (g/100 g sol'n.) 59.5 g methylene chloride, 56.2 g benzene, 51.2 g styrene, 34.5 g CCl_4, 18.7 g di-2-ethylhexyl phthalate; sp.gr. 1.324; bulk dens. 3.2 lb/gal; set pt. 62 C.

Cyasorb® UV 24. [Am. Cyanamid] 2,2′-Dihydroxy-4-methoxybenzophenone; lt. stabilizer and uv absorber for coatings and plastics, e.g., alkyds, phenolics, PU coatings; stabilizer for polyester film and PVC formulations; yel. powd.; m.w. 244.2; b.p. 160–170 C (0.5–1.0 mm Hg); sol. (g/100 g sol'n.) 55.3 g MEK, 46.6 g benzene, 31.1 g di-2-ethylhexyl phthalate, 22.2 g CCl_4, 21.4 g 95% ethanol; sp.gr. 1.382; bulk dens. 3.5 lb/gal; set pt. 68 C.

Cyasorb® UV 531. [Am. Cyanamid] 2-Hydroxy-4-n-octoxybenzophenone; lt. stabilizer and uv absorber for plastics and coatings, e.g., polyethylene, PP, PVC, and EVA; uses incl. pipe, storage tanks, and auto, marine, garden prods., auto refinish and industrial coatings, adhesives and sealants; lt. yel. powd.; m.w. 326.1; sol. (g/100 g sol'n.) 74.3 g acetone, 72.7 g benzene, 69.8 g methylene chloride, 40.1 g n-hexane, 20.5 g di-2-ethylhexyl phthalate; sp.gr. 1.160; bulk dens. 3 lb/gal; set pt. 47 C.

Cyasorb® UV 1084. [Am. Cyanamid] 2,2′-Thiobis (4-t-octylphenolato]-n-butylamine nickel II; lt. and heat stabilizer for polyolefins, e.g., PP fiber, LDPE agric. films and pool liners, and molded prods.; pale gr. powd.; m.w. 571; sol. (g/100 g sol'n.) 51.2 g n-heptane, 48.8 g tetrahydrofuran, 42.8 g toluene; sp.gr. 1.13; m.p. 258–261 C.

Cyasorb® UV 2098. [Am. Cyanamid] 2-Hydroxy-4-acryloyloxyethoxy benzophenone; lt. stabilizer, uv absorber which may be chemically bonded with monomers or polymers; lt. yel. powd.; m.w. 312; sol. in common org. solvs.; insol. water; sp.gr. 1.36; m.p. 77–80 C.

Cyasorb® UV 2126. [Am. Cyanamid] Polymer of 4-(2-acryloyloxyethoxy)-2-hydroxybenzophenone; lt. stabilizer and uv absorber for films and plastics in automotive, greenhouse, home siding, and solar applics.; lt. yel. powd.; m.w. ≈ 50,000; sol. (g/100 g sol'n.) > 50 g in MEK and dimethylacethamide, 30–40 g in benzene.

Cyasorb® UV 2908. [Am. Cyanamid] 3,5-Di-t-butyl-4-hydroxybenzoic acid, n-hexadecyl ester; lt. stabilizer, free radical scavenger for polyolefins, esp. pigmented opaque formulations; antioxidant; for pipe, crates, drums, auto, marine, garden, recreational prods.; near wh. powd.; m.w. 475; sol. (g/100 g): 61.1 g chloroform, 53.5 g toluene, 35.9 g hexane; sp.gr. 1.07; bulk dens. 16 lb/ft³; m.p. 60 C.

Cyasorb® UV 3346. [Am. Cyanamid] Oligomeric hindered amine; stabilizer for polymers alone or in combination with UV absorbers; off-wh. powd., flakes.

Cyasorb® UV 3346 LD. [Am. Cyanamid] Poly [(6-morpholino-s-triazine-2,4-diyl) [2,2,6,6-tetramethyl-4-piperidyl) imino]-hexamethylene [(2,2,6,6-tetramethyl-4-piperidyl) imino]]; lt. stabilizer, free radical scavenger, thermal antioxidant for polymers, esp. polyolefins requiring weatherability; off-wh. powd.; m.w. 1600 ± 10%; sol. (g/100 g) 76 g tetrahydrofuran, 67 g acetone, 62 g ethyl acetate, 25 g toluene, 23 g ethanol; bulk dens. 32.3 lb/ft³; m.p. 110–130 C.

Cyasorb® UV 5411. [Am. Cyanamid] 2-(2-Hydroxy-5-t-octylphenyl)-benzotriazole; lt. stabilizer and uv absorber for polymeric systems incl. polyester, PVC, styrenics, acrylics, PC, and polyvinyl butyral; end-uses incl. molding, sheet, and glazing materials for window, marine, and auto applics.; also in coatings, photo prods., sealants, and elastomeric materials; near wh. powd.; m.w. 325; sol. (g/100 g sol'n.) 41.5 g in methylene chloride, 39.4 g in benzene, 33.2 g in styrene, 21.4 g in ethyl acetate; sp.gr. 1.18; m.p. 101–105 C.

Cyastab® 40, 621. [Am. Cyanamid] Basic lead carbonate; heat stabilizer for PVC; end-uses incl. wire and cable, credit cards, profile extrusions, shoe and record compds., sealants, plastisols, and specialty coatings.

Cyastab® 523, 563, 593, 800. [Am. Cyanamid] Basic lead sulfate; heat stabilizer for PVC; see Cyastab 40.

Cyastab® 712. [Am. Cyanamid] Dibasic lead phosphate; heat stabilizer for PVC; see Cyastab 40; 712 also provides outdoor weathering resistance.

Cyastab® 908, 988. [Am. Cyanamid] Dibasic lead phthalate; heat stabilizer for PVC; see Cyastab 40; 908 for high-temp. elec. insulation; 988 is a coated version.

Cyastab® 948. [Am. Cyanamid] Basic lead complex (coated phthalate-sulfate complex); see Cyastab 40.

Cyastat® 609. [Am. Cyanamid] N,N-Bis (2-hydroxyethyl)-N-(3-dodecyloxy-2-hydroxypropyl) methyl ammonium methosulfate sol'n.; antistatic agent with good heat stability; for PVC phonograph records, specialty pkg.; very sl. yel. liq.; sol. in water, acetone, alcohol, and other polar solvents of low m.w.; sp.gr. 0.96; pH 4–6 (10% sol'n.); flash pt. (CC) 14.4 C; 50% solids in IPA/water.

Cyastat® LS. [Am. Cyanamid] (3-Lauramidopropyl) trimethylammonium methyl sulfate; antistatic agent for polymeric materials, i.e., coatings, PVC, PS, polyolefins, ABS; wh. cryst. powd.; 100% thru 20-mesh screen; m.w. 410; sol. (g/100 ml) 105 g in ethanol, 75 g in water, 70 g in ethyl Cellosolve, 9.0 g in acetone; < 1.0 g in benzene, Varsol, perchloroethylene; sp.gr. 1.12; m.p. 99–103 C.

Cyastat® SN. [Am. Cyanamid] Stearamidopropyl dimethyl-B-hydroxyethyl ammonium nitrate sol'n.; antistatic agent for polymers; used for plastics, surface coatings, paper, glass, and other materials; dispersant in coatings; lt. yel. to amber liq.; m.w. 476; sol. in water, acetone, alcohol, and other polar solvents of low m.w.; sp.gr. 0.95; flash pt. (CC) 12.2 C; 50% solids in IPA/water.

Cyastat®SP. [Am. Cyanamid] Stearamidopropyl dimethyl-B-hydroxyethyl ammonium dihydrogen phosphate sol'n.; cationic; antistatic agent with surface-active properties; for plastics, waxes, textiles, and glass; emulsifier, settling, dispersing, and rewetting agent; pale yel. liq.; m.w. 509; misc. with water, acetone, alcohol, and other polar solvents of low m.w.; sp.gr. 0.94; flash pt. (CC) 14.4 C; 35% solids in IPA/water.

Cyclochem 326A. [Rhone-Poulenc Surf.] Syn. beeswax; nonionic/anionic; emulsifier for w/o and o/w emulsions incl. hair creams, cold creams, lotions and lipsticks; off-wh. flakes; m.p. 51–54 C; 100% conc.

Cyclochem CL. [Rhone-Poulenc Surf.] Cetyl lactate; penetrant to promote penetration and feel; emollient; off-wh. solid; m.p. 24 C; sapon. no. 180.

Cyclochem CP. [Rhone-Poulenc Surf.] Cetyl palmitate; lubricant, emollient; substitute for spermaceti in cold creams, lipsticks and topical cosmetics; off-wh. flakes; m.p. 50 C; sapon. 114, acid no. 3.0 max.

Cyclochem DGS. [Rhone-Poulenc Surf.] PEG-2 stearate; high melting bodying agent and opacifier.

Cyclochem EGDS. [Rhone-Poulenc Surf.] Glycol distearate pure; thickener, opacifier, pearlizing agent used in shampoos and cosmetic lotions; wh. flakes; m.p. 61 C; sapon. no. 195.

Cyclochem EGMS. [Rhone-Poulenc Surf.] Glycol stearate pure; adds visc., pearlescence and opacity to lotions and shampoos; wh. flakes; water-insol.; m.p. 56 C; sapon. no. 182.

Cyclochem EGMS/SE. [Rhone-Poulenc Surf.] Glycol stearate SE; opacifier.

Cyclochem EM 324. [Rhone-Poulenc Surf.] Glycol stearate SE; anionic; emulsifier useful in formulating o/w hair and skin conditioning creams; lubricant wax for cosmetics; off-wh. flakes; m.p. 47–50 C.

Cyclochem EM 560. [Rhone-Poulenc Surf.] Cetearyl alcohol, sodium lauryl sulfate, cetyl esters, and myristyl alcohol; anionic; emulsifier, SE, lubricant, emollient for acid o/w creams, peroxide creams and lotions; off-wh. color; m.p. 62–65 C.

Cyclochem GMO. [Rhone-Poulenc Surf.] Glyceryl oleate; nonionic; emulsifier, emollient, opacifier, provides lubricity to shave lotions and cosmetic systems; amber liq.; m.p. 9 C; sapon. no. 170; 100% conc.

Cyclochem GMS. [Rhone-Poulenc Surf.] Glyceryl monostearate pure; nonionic; strongly lipophilic emulsifier for personal care prods.; wh. flakes; m.p. 56 C; sapon. no. 170.

Cyclochem GMS 21. [Rhone-Poulenc Surf.] Glyceryl stearate SE, acid stable; anionic; emulsifier, emollient with electrolyte tolerance for creams, lotions; off-wh. flakes; m.p. 52 C; acid no. 18.0 max.; sapon. no. 157.

Cyclochem GMS 165. [Rhone-Poulenc Surf.] Glyceryl stearate and PEG-100 stearate; nonionic; acid stable, self emulsifying; electrolyte-tolerant emulsifier, emollient for creams, lotions and toiletry prods.; off-wh. flakes; m.p. 55 C; sapon. no. 93; 100% conc.

Cyclochem GMS/SE. [Rhone-Poulenc Surf.] Glyceryl stearate SE; anionic; self-emulsifying form of Cyclochem GMS; flakes; HLB 5.5; 100% conc.

Cyclochem GTIS. [Rhone-Poulenc Surf.] Triisostearin; emollient, coemulsifier, softener, provides good texture in creams, promotes softness on skin; yel. liq.; m.p. < 0 C; acid no. 8.0 max.; sapon. no. 192.

Cyclochem GTL. [Rhone-Poulenc Surf.] Trilaurin; emollient; promotes an oily feel to cream and lotions; liq./paste; m.p. 27 C; acid no. 30.0 max.; sapon. no. 245.

Cyclochem GTO. [Rhone-Poulenc Surf.] Glyceryl trioleate; nonionic; emollient; enhances feel and appearance to creams and lotions; compatible with anionic, cationic, and nonionic emulsifiers; liq.; 100% act.

Cyclochem INEO. [Rhone-Poulenc Surf.] Isostearyl neopentanoate; emollient, penetrant, plasticizer, lubricant; promotes a dry feel to the skin; off-wh. liq.; m.p. < 0 C; sapon. no. 149.

Cyclochem LIPA. [Rhone-Poulenc Surf.] Lauramide MIPA; visc. builder, foam stabilizer; off-wh. flakes; 100% act.; 95% amide content.

Cyclochem LVL. [Rhone-Poulenc Surf.] Lauryl lactate; penetrant, emollient, pigment dispersant for liq. makeups; reduces tackiness of aluminum salt antiperspirants; off-wh. liq.; m.p. 4 C; acid no. 3.0 max.; sapon. no. 199.

Cyclochem ML. [Rhone-Poulenc Surf.] Myristyl lactate; melting ester, penetrant, emollient, pigment dispersant; off-wh. liq.; m.p. 12 C; acid no. 3.0 max.; sapon. no. 172.

Cyclochem MM/M. [Rhone-Poulenc Surf.] Myristyl myristate; emollient; imparts a soft, velvety feel which persists after rinsing; off-wh. flakes; m.p. 38 C; sapon. no. 130.

Cyclochem MST. [Rhone-Poulenc Surf.] Myristyl stearate; emollient; imparts soft, velvety feel to emulsions where a stiffer emulsion with superior opacity is desired; off-wh. flakes; m.p. 45 C; sapon no. 115.

Cyclochem NI. [Rhone-Poulenc Surf.] Emulsifying wax; nonionic; visc. building agent for creams, lotions, ointments; flakes; m.p. 48–51 C.

Cyclochem PEG 200 DL. [Rhone-Poulenc Surf.] PEG-4 dilaurate; emollient; graduated hydrophilic and lipophilic properties; used alone or in combination with anionic or cationic emulsifiers.

Cyclochem PEG 200 ML. [Rhone-Poulenc Surf.]

PEG-4 laurate; see Cyclochem PEG 200 DL.

Cyclochem PEG 400 DS. [Rhone-Poulenc Surf.] PEG-8 distearate; nonionic; see Cyclochem PEG 200 DL; wh. soft solid; m.p. 33 C; sapon. 123; acid no. 10.0 max.

Cyclochem PEG 400 ML. [Rhone-Poulenc Surf.] PEG-8 laurate; see Cyclochem PEG 200 DL; lt. straw liq.; m.p. 10 C; sapon. no. 90.

Cyclochem PEG 400 MO. [Rhone-Poulenc Surf.] PEG-8 oleate; see Cyclochem PEG 200 DL; lt. straw liw.; m.p. 10 C; sapon. no. 85.

Cyclochem PETO. [Rhone-Poulenc Surf.] Pentaerythritol tetraoleate; nonionic; emollient for gloss in lipsticks; yel. liq.; m.p. < 0 C; sapon. no. 187.

Cyclochem PETS. [Rhone-Poulenc Surf.] Pentaerythritol tetrastearate; nonionic; lubricant wax providing high heat stability in lubricants; emulsifier for nat. waxes in cosmetics; off-wh. flakes; m.p. 57 C; sapon. no. 190.

Cyclochem POL [Rhone-Poulenc Surf.] Cetearyl alcohol, ceteth-20, and glycol stearate; nonionic; lubricant SE wax for creams and lotions; emulsifier; off-wh. wax; m.p. 48–52 C; 100% conc.

Cyclochem SEG. [Rhone-Poulenc Surf.] Glycol stearate; pearlizing agent in shampoos, liq. hand soaps, and liq. detergents; emulsion stabilizer and visc. builder; flakes; m.p. 55–60 C.

Cyclochem SPS. [Rhone-Poulenc Surf.] Cetyl ester (syn. spermaceti); nonionic; provides sheen and smooth feel to creams, gloss to stick makeup, off-wh. flakes; m.p. 46 C; sapon. no. 115; 100% conc.

Cyclochem SS. [Rhone-Poulenc Surf.] Stearyl stearate; emollient; provides a feel on skin and hair that is less oily than min. oil and less drying than IPM; off-wh. flakes; water-insol.; m.p. 53 C; sapon. no. 110.

Cyclol SPS. [Witco SA] Fatty esters; nonionic; spermaceti substitute for cosmetic and pharmaceutical formulations; solid.; 100% act.

Cyclolube® 62. [Witco] Naphthenic oil distillate; process oils offering lt. color, low pour pts. for general-purpose compounding of elastomers, e.g. neoprene, SBR, isoprene, EDPM, butyl, and natural rubber; m.w. 340; sp.gr. 0.9529 (15.6 C); dens. 7.9 lb/gal; visc. 200 cSt (40 C); flash pt. (COC) 196 C; pour pt. –9 C; ref. index 1.5286.

Cyclolube® 85. [Witco] Low aromatic naphthenic oil; process oil offering exc. color and heat stability for rubber compounding, esp. EPDM and butyl; also for SBR, isoprene, and natural rubber; m.w. 450; sp.gr. 0.9129 (15.6 C); dens. 7.6lb/gal; visc. 290 cSt (40 C); flash pt. (COC) 244 C; pour pt. –18 C; ref. index 1.4954.

Cyclolube® 120. [Witco] Low aromatic naphthenic oil; see Cyclolube 85; m.w. 468; sp.gr. 0.9159 (15.6 C); dens. 7.6 lb/gal; visc. 386 cSt (40 C); flash pt. (COC) 254 C; pour pt. –12 C; ref. index 1.5006.

Cyclolube® 132. [Witco] Naphthenic oil distillate; see Cyclolube 62; m.w. 359; sp.gr. 0.9593 (15.6 C); dens. 8.0 lb/gal; visc. 425 cSt (40 C); flash pt. (COC) 214 C; pour pt. –3 C; ref. index 1.5337.

Cyclolube® 210. [Witco] Low aromatic naphthenic oil; see Cyclolube 85; m.w. 305; sp.gr. 0.9042 (15.6 C); dens. 7.5 lb/gal; visc. 20 cSt (40 C); flash pt. (COC) 158 C; pour pt. –42 C; ref. index 1.4927.

Cyclolube® 213. [Witco] Low aromatic naphthenic oil; see Cyclolube 85; m.w. 320; sp.gr. 0.9100 (15.6 C); dens. 7.6 lb/gal; visc. 42 cSt (40 C); flash pt. (COC) 180 C; pour pt. –36 C; ref. index 1.4997.

Cyclolube® 270. [Witco] Low aromatic naphthenic oil; see Cyclolube 85; m.w. 368; sp.gr. 0.9218 (15.6 C); dens. 7.7 lb/gal; visc. 133 cSt (40 C); flash pt. (COC) 206 C; pour pt. –21 C; ref. index 1.4995.

Cyclolube® 413. [Witco] Low aromatic naphthenic oil; see Cyclolube 85; m.w. 327; sp.gr. 0.8956 (15.6 C); dens. 7.5 lb/gal; visc. 29 cSt (40 C); flash pt. (COC) 168 C; pour pt. –45 C; ref. index 1.4875.

Cyclolube® 2290. [Witco] Low aromatic naphthenic oil; see Cyclolube 85; m.w. 453; sp.gr. 0.9340 (15.6 C); dens. 7.8 lb/gal; visc. 515 cSt (40 C); flash pt. (COC) 238 C; pour pt. –6 C; ref. index 1.5077.

Cyclolube® 2310. [Witco] Low aromatic naphthenic oil; see Cyclolube 85; m.w. 414; sp.gr. 0.9248 (15.6 C); dens. 7.7 lb/gal; visc. 243 cSt (40 C); flash pt. (COC) 226 C; pour pt. –9 C; ref. index 1.5059.

Cyclolube® 4053. [Witco] Low aromatic naphthenic oil; see Cyclolube 85; m.w. 400; sp.gr. 0.9042 (15.6 C); dens. 7.5 lb/gal; visc. 97 cSt (40 C); flash pt. (COC) 200 C; pour pt. –30 C; ref. index 1.4902.

Cyclolube® NN-1. [Witco] Naphthenic oil distillate; see Cyclolube 62; m.w. 266; sp.gr. 0.9218 (15.6 C); dens. 7.7 lb/gal; visc. 19 cSt (40 C); flash pt. (COC) 158 C; pour pt. –42 C; ref. index 1.5063.

Cyclolube® NN-2. [Witco] Naphthenic oil distillate; see Cyclolube 62; m.w. 295; sp.gr. 0.9279 (15.6 C); dens. 7.7 lb/gal; visc. 28 cSt (40 C); flash pt. (COC) 166 C; pour pt. –36 C; ref. index 1.5114.

Cyclomide AC 28. [Rhone-Poulenc Surf.] Fatty acid condensate; anticorrosive min. oil additive for cutting fluids and sol. cutting oil compositions; liq.; 100% act.

Cyclomide C212. [Rhone-Poulenc Surf.] Cocamide MEA; nonionic; thickener, foam builder and stabilizer for soap or syn. based washing powds.; cream-colored flakes; 100% act.; 95% amide.

Cyclomide CODI. [Rhone-Poulenc Surf.] Cocamidopropyl dimethylamine; cationic; emulsifier; base for emulsions for creams, lotions, and hair rinses; conditioner, antistat, visc. builder; yel. liq.; 100% act., 98% amide.

Cyclomide DC212. [Rhone-Poulenc Surf.] Cocamide DEA and diethanolamine; nonionic; visc. booster, foaming agent, detergent, used in cosmetic and laundry prods.; amber liq.; 100% act.

Cyclomide DC212/M. [Rhone-Poulenc Surf.] 2:1 Modified cocamide DEA; nonionic; thickener, base for hard surf. cleaners, solventless degreasers, and wax strippers; high alkalki compatibility; amber liq.; 100% act.

Cyclomide DC212/S. [Rhone-Poulenc Surf.] 1:1 Cocamide DEA; nonionic; foam stabilizer and visc. booster with low cloud pt.; liq.; 100% act.; 85% amide.

Cyclomide DC212/SE. [Rhone-Poulenc Surf.] 1:1

Cocamide DEA; nonionic; detergent, foam stabilizer, thickener, foam booster in shampoos and detergent formulations; liq.; 100% act.; 95% amide.

Cyclomide DIN295. [Rhone-Poulenc Surf.] 2:1 Linoleamide DEA; visc. builder, grease cutting aid for cleaners; lubricant for metal treatment; liq.

Cyclomide DIN295/S. [Rhone-Poulenc Surf.] Linoleamide DEA and diethanolamine; nonionic; visc. builder, thickener; conditioning to hair; liq.; 100% conc.

Cyclomide DL203. [Rhone-Poulenc Surf.] 2:1 Lauramide DEA; nonionic; lather stabilizer for shampoos and detergents; off-wh. liq., wh. paste; 100% act.

Cyclomide DL203/S. [Rhone-Poulenc Surf.] 1:1 Lauramide DEA; nonionic; foam and detergency booster for cosmetic prods.; superfatting agent; off-wh. cryst. solid; 100% act; 95% amide.

Cyclomide DL207/S. [Rhone-Poulenc Surf.] 1:1 Lauramide DEA; nonionic; foam and visc. booster, wetting agent, superfatting agent; off-wh. liq., wh. paste; 100% act.; 95% amide.

Cyclomide DO280. [Rhone-Poulenc Surf.] Oleamide DEA and diethanolamine; nonionic; emulsifier for sol. oils; corrosive inhibitor; liq.; 100% conc.

Cyclomide DO280/S. [Rhone-Poulenc Surf.] 1:1 Oleamide DEA; nonionic; thickening and superfatting agent, emulsifier, for lotion shampoos; liq.; 100% act.; 85% amide.

Cyclomide DOTS. [Rhone-Poulenc Surf.] 1:1 Linoleamide DEA; visc. builder, foam stabilizer, emulsifier, pigment dispersant; amber liq.; 100% act.; 95% amide.

Cyclomide DP240/S. [Rhone-Poulenc Surf.] 1:1 Palm kernelamide DEA; nonionic; visc. builder and foam stabilizer for org. formulations; liq.; 100% act.; 85% amide.

Cyclomide DS280. [Rhone-Poulenc Surf.] 2:1 Stearamide DEA; visc. builder, grease cutting aid for cleaners; lubricant for metal treatment; solid.

Cyclomide DS280/S. [Rhone-Poulenc Surf.] 1:1 Stearamide DEA; visc. builder, thickener for gel and economy shampoos, bath preps.; conditioner for hair and skin care prods.; solid; 95% amide.

Cyclomide IODI. [Rhone-Poulenc Surf.] Isostearamidopropyl dimethylamine; base for preparing cationic emulsions, promotes substantivity and adhesion of additives to the skin; clear liq.; 100% act.; 98% amide.

Cyclomide KD. [Rhone-Poulenc Surf.] 1:1 Cocamide DEA; nonionic; visc. builder, foam stabilizer; pale yel. liq.; 100% act.; 95% amide.

Cyclomide L203. [Rhone-Poulenc Surf.] Lauramide MEA; nonionic; foam builder/stabilizer for soap and syn. washing powds.; visc. builder; off-wh. flakes; 100% act.; 95% amide.

Cyclomide LE. [Rhone-Poulenc Surf.] Lauramide DEA; nonionic; visc. booster, foam stabilizer in shampoo; liq.; 100% conc.

Cyclomide LIPA. [Rhone-Poulenc Surf.] Lauramide MIPA; nonionic; foam booster, stabilizer and visc. booster; off-wh. flake; 100% conc.

Cyclomide Pinemulse. [Rhone-Poulenc Surf.] Modified coconut polydiethanolamide; nonionic; emulsifier and coupler for preparing stable pine oil cleaners without heat; amber liq.; 100% act.

Cyclomide RODEA. [Rhone-Poulenc Surf.] 2:1 Ricinoleamide DEA; visc. builder, grease cutting aid for cleaners; lubricant for metal treatment; liq.

Cyclomide S280. [Rhone-Poulenc Surf.] Stearamide MEA; nonionic; skin protectant in toilet bars, creams, lotions, pastes; off-wh. flakes; 100% act.; 95% amide.

Cyclomide SODI. [Rhone-Poulenc Surf.] Stearamidopropyl dimethylamine; emulsifier, conditioner; produces cationic emulsions; off-wh. solid; 100% act.; 98% amide.

Cyclomox C. [Rhone-Poulenc Surf.] Coconut dimethylamine oxide; nonionic; flash foamer and viscosifier in hard water; liq.; 30% conc.

Cyclomox CO. [Rhone-Poulenc Surf.] Cocoamidopropylamine oxide; cationic; conditioner, emollient with visc. and foam building effects; liq.; 30% conc.

Cyclomox SO. [Rhone-Poulenc Surf.] Stearamidopropylamine oxide; nonionic; foamer, viscosifier, emollient, conditioner; hydrogen peroxide stabilizer in bleaching formulations, cold waves; paste; 30% conc.

Cyclonox® BT-50. [Akzo] Cyclohexanone peroxide; initiator for ambient-temp. polyester cures; for automotive body putty, hobby and automotive kits; paste; 50% ketone peroxide in plasticizer.

Cyclopol SB-5. [Rhone-Poulenc Surf.] Sodium laneth-5 sulfosuccinate; anionic; moisturizing agent for shampoos, foam baths, skin cleansers, skin creams, hair sprays, dishwashing liq.; yel. liq.; 48–50% act.

Cyclopol SBDO. [Rhone-Poulenc Surf.] Dioctyl sodium sulfosuccinate; anionic; surf. tens. depressant, solubilizer for soaps and surfactants; liq.; 68–70% act.

Cyclopol SBG280. [Rhone-Poulenc Surf.] Disodium oleamido PEG-2 sulfosuccinate; anionic; foaming agent, nonirritating surfactant, visc. control in personal care prods.; yel. liq.; 29–31% act.

Cyclopol SBU-185. [Rhone-Poulenc Surf.] Disodium undecylenamido MEA-sulfosuccinate; anionic; fungicidal and bacteriostatic additive, antidandruff agent used in shampoos, body lotions, foot preparations; yel. liq.; 49–51% act.

Cycloryl DCA. [Rhone-Poulenc Surf.] Modified coconut polydiethanolamide; anionic; conc. detergent for human and pet shampoos; builds visc. and solubilizes difficult ingredients; thickener for agric. biocidal agents; amber liq.; 98–100% act.

Cyclosheen 100. [Rhone-Poulenc Surf.] Lauramide MIPA, glycol stearate, and cocamide MEA; nonionic; emulsifier, opacifier imparting stable pearl, improved lathering and increased visc. of liq.; shampoos and cleaners; off-wh. soft wax; m.p. 58–60 C; acid no. 2.0 max.

Cyclosheen 202. [Rhone-Poulenc Surf.] Glycol stearate, emulsifiers; pearl conc. for cold blend formulations; visc. builder, foam booster, conditioner; lotion/paste.

Cycloteric BET-C30. [Rhone-Poulenc Surf.] Cocamidopropyl betaine; amphoteric; foaming, conditioning and thickening agent for use with skin and hair preparations; yel. liq.; pH 4.5–5.5; 29–31% act.

Cycloteric BET-C41. [Rhone-Poulenc Surf.] Coco betaine; high foaming conditioning agent for shampoos, foam baths, and liq. hand soaps; 41% act.

Cycloteric BET-CB. [Rhone-Poulenc Surf.] Cocamidopropyl betaine, cosmetic grade, glycerin-free; foam booster, foaming agent, thickener, conditioner for shampoos, cosmetic, and industrial applics.; irritation mollifying agent for baby shampoos; 30% act.

Cycloteric BET-I30. [Rhone-Poulenc Surf.] Isostearamidopropyl betaine; amphoteric; foaming, conditioning and thickening agent for use with skin and hair preparations; yel. liq.; pH 5.5–6.5; 29–31% act.

Cycloteric BET-L31 [Rhone-Poulenc Surf.] Lauryl betaine; high foaming conditioning agent for shampoos, foam baths, and liq. hand soaps; 31% act.

Cycloteric BET-O30. [Rhone-Poulenc Surf.] Oleamidopropyl betaine; amphoteric; foaming, foam stabilizing, conditioning and thickening agent for use with skin and hair preparations; yel. liq.; pH 5.5–6.5; 29–31% act.

Cycloteric BET-OB50. [Rhone-Poulenc Surf.] Oleyl betaine; visc. building, gelling agent used in bath gel and shampoos; 50% act.

Cycloteric BET-T2 40. [Rhone-Poulenc Surf.] Dihydroxyethyl tallow glycinate; visc. building agent for industrial applics.; conditioner for premium shampoos; 40% act.

Cycloteric BET-W. [Rhone-Poulenc Surf.] Cocamidopropyl betaine, tech.; foam booster, foaming agent, thickener, conditioner for shampoos, cosmetic, and industrial applics.; irritation mollifying agent for baby shampoos; 30% act.

Cycloteric CAPA. [Rhone-Poulenc Surf.] Cocaminopropionic acid; high foaming conditioning agent for shampoos, skin cleansers, foam baths; surfactant for hard surf. cleaners, car washes, industrial foamers; 40% act.

Cycloteric SLIP. [Rhone-Poulenc Surf.] Sodium lauriminodipropionate; high foaming conditioning agent for shampoos, skin cleaners, foam baths; surfactant for hard surf. cleaners, car washes, industrial foamers; 30% act.

Cycloton 7LUF. [Rhone-Poulenc Surf.] Oleakonium chloride; pumpable conditioner base for creme rinse systems; liq.; 50% act.

Cycloton 75C. [Rhone-Poulenc Surf.] Ditallow-based methylsulfate quat.; economy softener conc. used as textile softener, antistat, and in paper and pulp processing; liq.; 75% act.

Cycloton D256B/99. [Rhone-Poulenc Surf.] Cetethyl dimonium bromide; cationic surfactant contributing exotic feel to hair conditioners; powd.; 99% act.

Cycloton D261C/70. [Rhone-Poulenc Surf.] Ditallowalkonium chloride; conc. base for prep. of fabric softeners, antistatic treatments, hair conditioners; soft paste; 70% act.

Cycloton D261C/75. [Rhone-Poulenc Surf.] Quaternium 18; cationic; base for fabric softeners; paste; 75–77% act.

Cycloton D261S/90. [Rhone-Poulenc Surf.] Dihydrog. tallow based dimethyl methosulfate; cationic; fabric softener for commercial and household applics.; paste; 88–90% act.

Cycloton M214B/99. [Rhone-Poulenc Surf.] Myrtrimonium bromide; cationic; conditioner with superior antistatic properties and lt. feel; flake; 99% act.

Cycloton M214C/45. [Rhone-Poulenc Surf.] Myrtrimonium bromide; cationic; used in cream rinse and conditioners; liq.; 45% act.

Cycloton M214C/99. [Rhone-Poulenc Surf.] Myrtrimonium bromide; cationic; cream rinse, conditioner.

Cycloton M242B/99. [Rhone-Poulenc Surf.] Cetrimonium bromide; cationic; quat. ammonium compd. used in conditioners to increase antistatic and comb out effects; 100% act.

Cycloton M242C/29. [Rhone-Poulenc Surf.] Cetrimonium chloride; cationic; surfactant with conditioning and emolliency effect on hair; liq.; 29% act.

Cycloton M270C/85. [Rhone-Poulenc Surf.] Stearalkonium chloride; cationic; high activity surfactant base used in conditioners and cream rinses; includes detangling and antistatic properties; solid; 85% act.

Cyclovertal. [Henkel] 3,6-Dimethyl-3-cyclohexene-1-carbaldehyde; general fragrance raw material, grn., fruity notes.

Cyfluthrin. [Bayer] Cyfluthrin; syn. pyrethroid for the control of household, public health, and stored-prod. pests; ylsh.-brn. mass of oily to pasty consistency; clear ylsh.-brn. oil above 60 C; m.w. 434.3; sol. in dichloromethane, toluene; sp.gr. 1.27–1.28.

Cygon® 400. [Am. Cyanamid/Ag] Dimethoate; systemic insecticide-miticide for field, veg., and fruit crops; 43.5% act.

Cymel 303. [Am. Cyanamid] Hexamethoxymethylmelamine; crosslinking agent used in melamine resin coating systems; used in general industrial finishes, coil-coating enamels, appliance finishes; also useful with alkyd, polyester, thermosetting acrylic, epoxy, and cellulose resins; Gardner 2 max. clear, visc. liq.; limited water-sol.; dens. 10.0 lb/gal; visc. (G-H) X–Z_2; flash pt. > 180 F; 98% NV.

Cymel 370. [Am. Cyanamid] Methylated melamine resin; crosslinking agent for solv. and water-based coatings, e.g., general purpose spray coatings, coil coatings, metal decorating enamels, varnishes; Gardner 1 max. clear liq.; sol. in methanol, ethanol, isobutanol, ethylene and propylene glycols, MEK, MIBK, toluene, xylene; partly sol. in water; sp.gr. 1.18; dens. 9.8 lb/gal; visc. (G-H) Z_2–Z_4; flash pt. (CC) 72 F; 88 ± 2% solids in IPA/butanol.

Cymel 373. [Am. Cyanamid] Partially methylated melamine formaldehyde resin; crosslinking agent for water-based coatings, e.g., emulsions or sol'ns.; Gardner 1 max. clear liq.; sol. in methanol, ethanol, IPA, ethylene and propylene glycols, MEK; sp.gr. 1.26; dens. 10.5 lb/gal; visc. (G-H) Z-Z_4; 85 ± 2% solids in water.

Cymel 380. [Am. Cyanamid] Partially methylated

melamine formaldehyde resin; see Cymel 370; Gardner 1 max. clear liq.; sol. see Cymel 370; sp.gr. 1.15; dens. 9.6 lb/gal; visc. (G-H) V-Z; flash pt. (CC) 62 F; 80 ± 2% solids in IPA/butanol.

Cymel 1141. [Am. Cyanamid] Alkylated carboxyl-modified melamine formaldehyde resin; crosslinking agent for cationic resins; used for emulsions, water-sol. coatings, and cationic electroplating applics.; Gardner 3 max. clear, visc. liq.; water-disp.; sol. in common org. solvs.; sp.gr. 1.07; dens. 8.95 lb/gal; visc. (G-H) W-Y; acid no. 22 ± 3; flash pt. (Seta CC) 93 F; 85 ± 2% solids in isobutanol.

Cyncal® 80%. [Hilton-Davis] Myristalkonium chloride; cationic; antimicrobial, antistat, disinfectant, sanitizer for use in food, beverage processing, for industrial and farm use, fabrics; lt. yel. liq.; mild, pleasant odor; m.w. 359; sol. in water, lower alcohols, ketones, glycols; dens. 7.8 lb/gal; sp.gr. 0.94; 80% act.

Cyprex® 65-W. [Am. Cyanamid/Ag] Dodine; fungicide for protection of fruits; 65.0% act.

Cythion®. [Am. Cyanamid/Ag] Malathion conc.; insecticide for formulation of liq., wettable powd., and dust insecticides for agric. crops; also for forest insect, mosquito control; 91.0% act.

Cythion® Emulsifiable Conc. [Am. Cyanamid/Ag] Malathion; water emulsifiable conc. for insect control on a wide variety of crops and ornamentals; flam.; 57% act. in 30% xylene.

D

D-70. [Georgia Marble] Ground calcium carbonate; noncolor-critical filler; 67 ± 3% passing #325 wet screen.
D-400. [Olin] PPG diol; lubricant in two- and four-cycle engines; used in the processing of rubber, to inhibit foam; used in brake fluids, cosmetics, oil and grease compds., pesticides, urethane foams, urethane coatings, adhesives, elastomers, and sealants/caulks; liq.; m.w. 400; f.p. –26 C; sp.gr. 1.009; dens. 8.40 lb/gal; visc. 70 cs; ref. index 1.445; flash pt. 204 C.
D-1000. [Olin] PPG diol; see D-400; liq.; m.w. 1000; f.p. –36 C; sp.gr. 1.005; dens. 8.37 lb/gal; visc. 150 cs; ref. index 1.448; flash pt. 230 C.
D-1200. [Olin] PPG diol; see D-400; liq.; m.w. 1200; f.p. –36 C; sp.gr. 1.006; dens. 8.37 lb/gal; visc. 190 cs; ref. index 1.448; flash pt. 232 C.
D-1300. [Olin] PPG diol; see D-400; liq.; m.w. 1300; f.p. –34 C; sp.gr. 1.005; dens. 8.37 lb/gal; visc. 225 cs; ref. index 1.449; flash pt. 235 C.
D-2000. [Olin] PPG diol; see D-400; liq.; m.w. 2000; f.p. –29 C; sp.gr. 1.005; dens. 8.37 lb/gal; visc. 325 cs; ref. index 1.449; flash pt. 238 C.
D-3000. [Olin] PPG diol; see D-400; liq.; m.w. 3000; f.p. –32 C; sp.gr. 1.004; dens. 8.36 lb/gal; visc. 580 cs; ref. index 1.449; flash pt. 238 C.
D-4000. [Olin] PPG diol; see D-400; liq.; m.w. 4000; f.p. –30 C; sp.gr. 1.003; dens. 8.36 lb/gal; visc. 895 cs; ref. index 1.450; flash pt. 343 C.
D4165. [Hüls] Di-t-butyldichlorosilane; difunctional blocking agent which can protect diols.
D4368. [Hüls] 1,3-Dichlorotetraisopropyldisiloxane; difunctional blocking agent for the simultaneous protection of the 3´- and 5´-hydroxy groups in nucleosides; used in the synthesis of glycolipids.
D4450. [Hüls] N,N-diethylaminotrimethylsilane; blocking agent; silylating reagent.
D5115. [Hüls] (N,N-dimethylamino) (3,3-dimethylbutyl) dimethylsilane; sterically hindered blocking agent.
D5400. [Hüls] N,N-dimethylaminotrimethylsilane; blocking agent; employed where removal of the amine from the reaction mixt. is desired.
D5430. [Hüls] (3,3-Dimethylbutyl) dimethylchlorosilane; sterically hindered blocking agent.
D5490. [Hüls] Dimethyldiacetoxysilane; difunctional blocking agent for derivatizing corticosteroids.
D5600. [Hüls] Dimethyldiethoxysilane; difunctional blocking agent for protection of vicinal hydroxyl groups.
D6102. [Hüls] 2-(Diphenylmethylsilyl) ethanol; blocking agent.
Dabco® 33-LV. [Air Prods.] 33% Triethylenediamine in dipropylene glycol; catalyst for PU coating.
Dabco® BDO. [Air Prods.] 1,4-Butanediol; chain extender; provides reactive H-source in prepolymer prod.; used to provide hard segments in PU.
Dabco® CL-485. [Air Prods.] Tetrafunctional hydroxyl crosslinker with amine functionality; promotes crosslinking reaction.
Dabco® Crystalline. [Air Prods.] Triethylenediamine; catalyst for PU coatings.
Dabco® DC-2. [Air Prods.] Proprietary blend of delayed-activity tin and a tert. amine; catalyst for PU coatings; 45% act.
Dabco® T-1. [Air Prods.] Dibutyltin diacetate; catalyst for PU coatings.
Dabco® T-5. [Air Prods.] Dibutyltin disulfide; catalyst for PU coatings; temp.-activated.
Dabco® T-12. [Air Prods.] Dibutyltin dilaurate; catalyst for PU coatings; liq. above 65 F.
Dabco® T-120. [Air Prods.] Butyltin mercaptide; catalyst for PU coatings.
Dabco® T-125. [Air Prods.] Dibutyltin diester; catalyst for PU coatings.
Dabco® T-131. [Air Prods.] Dibutyltin (bis) mercaptide; catalyst for PU coatings; ambient temp. cure; provides strong exotherm.
Daconil 2787. [Henkel] Chlorothalonil; broad-spectrum fungicide for use on golf courses; 40.4% act. flowable or 75% wettable powd.
Dacospin 12-R. [Henkel] Ethoxylated hydrog. castor oil; emulsifier and solubilizer for esters and glycerides; water-sol.
Dacospin 1735-A. [Henkel] Ethoxylated castor oil; lubricant for PP carpet backing giving excellent fiber-to-fiber lubricity and plasticizing properties; lt. yel. liq.; dens. 1.04 g/ml; visc. 450 cP; pH 7.3 (2%); flash pt. 266 C (COC); 100% act.
Dacospin 9212. [Henkel] Alkylolamide, high-purity; antistat and wetting agent for PP yarn; yel. clear liq.; dens. 1.00 g/ml; visc. 365 cP; pH 9.5 (2%); flash pt. 185 C (COC); 100% act.
Dacospin HT-5. [Henkel] Ethoxylated alkylaryl ether; thermally stable lubricant; water-disp.
Dacospin HT-50. [Henkel] High m.w. ester; thermally stable lubricant; water-insol.
Dacospin LS 4100. [Henkel] Ethoxylated oil; see Dacospin 1735-A; yel. clear liq.; dens. 1.13 g/ml; visc. 300 cP; pH 6.5 (5%); 100% act.
Dacospin PE-146. [Henkel] Phosphated aliphatic alcohol; antistat for syn. fibers; water-disp.
Dacospin POE(25)HRG. [Henkel] Ethoxylated

hydrog. castor oil; see Dacospin 1735-A; also used as spin finish component with good emulsifying properties, high thermal stability, and good color; pale yel. liq.; dens. 1.03 g/ml; visc. 1200 cP; pH 6.4 (5%); flash pt. 266 C (COC); 99% act.

Dai Cari XBN. [Vikon] Butyl benzoate; carrier for disperse dyes on polyesters.

Dantocol® DHE, DHE (5), DHE (10), DHE (15), DHE (20). [Lonza] Di-(2-hydroxyethyl)-5,5-dimethyl hydantoin and with 5, 10, 15, and 20 moles EO resp.; intermediates for epoxies, urethane resins, and antistatic lubricants for the textile and plastics industries; Gardner 1 liq.; hyd. no. 500, 260, 170, 127, and 105 resp.

Dantoest® DHE DL. [Lonza] Dantocol DHE dilaurate; intermediate for epoxies, urethane resins, and antistatic lubricants for the textiles industry; Gardner 3 liq.; acid no. 2; hyd. no. 10.

Dantoin® DCDMH. [Lonza] 1,3-Dichloro-5,5-dimethyl hydantoin; intermediate for custom chemical synthesis, laundry bleach formulations, and automatic dishwashing compds.; wh. powd.; m.w. 197; m.p. 130 C.

Dantoin® DMDMH-55. [Lonza] DMDM hydantoin; intermediate; wh. liq.; 55%.

Dantoin® DMH. [Lonza] DM hydantoin; intermediate for textiles and other applics.; wh. crystals; m.w. 128; m.p. 176 C.

Dantoin® DMHF. [Lonza] DMHF; intermediate for cosmetics, textiles, and other applics.; wh. solid; m.p. 75 C.

Dantoin® DMHF-75. [Lonza] DMHF; intermediate for cosmetics, textiles, and other applics.; hair lacquers and wave sets; film former; clear liq.; sol. in water, alcohol; 75% conc.

Dantoin® DMHF Refined. [Lonza] Dimethyl hydantoin formaldehyde, refined grade; hair lacquers and wave sets; film former; sol. in water and alcohol.

Dantoin® MDMH. [Lonza] MDM hydantoin; intermediate for cosmetics and other applics.; wh. crystals; m.w. 158; m.p. 112 C.

Dantosperse® DHE (5) MO, (10) MO, (15) MO, (20) MO. [Lonza] Dantocol DHE (5), (10), (15), and (20) oleate resp.; intermediate for epoxies, urethane resins, and antistatic lubricants for the textiles industry; Gardner 1–2 liq.; acid no. 2, 2, 2, and 1 resp.; hyd. no. 60, 45, 40, 31 resp.

Dapral® GT Series. [Akzo] Polyoxyalkylene glycol; nonionic; thickener for surf. act. materials, mild shampoos, and acids in cleaning formulations; solid; 100% act.

Dar Chem-11. [Unichema] Single pressed stearic acid; lubricant used in mfg. of cosmetics, creams, salves, soaps, detergent systems, emulsifiers.

Dar Chem-12. [Unichema] Double pressed stearic acid; see Dar Chem-11.

Dar Chem-13. [Unichema] Triple pressed stearic acid; lubricant; component of buffing compds., candles, and specialty lubricants, textile specialties.

Dar Chem-14. [Unichema] Stearic acid; see Dar Chem-13.

Dar Chem-030. [Unichema] Refined tallows; laundry softener emollients; lubricants; sol. in common org. solvs.; water-insol.

Darex 41. [W.R. Grace] Syn. polymer; water treatment polymer for scale and sludge inhibition; disp. particulate matter; non-corrosive to metals; pure wh. powd.; sol. in water systems; dens.(bulk) 29 lb/ft^3 (untamped); pH 7.5 (1%); surf. tens. 70 dynes/cm (1%).

Darex 515L. [W.R. Grace] Copolymer latex; pigment binder used in surface coatings, e.g., interior flat wall paints and primer sealers; wide compding. latitude; particle size 0.2 μ; dens. 8.5 lb/gal; visc. 20–60 cps; surf. tens. 33–40 dynes/cm; pH 7.0–8.5; 47–49% total solids.

Darex 529L. [W.R. Grace] Carboxylated terpolymer of butadiene, styrene, and vinylidene chloride; binder-saturant conferring flexible hand, freedom from tackiness, and good abrasion resistance; improved flame resistance; particle size 0.2 μ; dens. 9.1 lb/gal; visc. 50 cps; surf. tens. 40 dynes/cm; pH 9.5; 53% total solids.

Darex 535L. [W.R. Grace] Carboxylated terpolymer of butadiene, styrene, and vinylidene chloride; similar to Darex 529L but with more vinylidene chloride and less styrene; softer and with improved flame resistance; particle size 0.2 μ; dens. 9.3 lb/gal; visc. 40 cps; surf. tens. 40 dynes/cm; pH 8.5; 52% total solids.

Darex SP-1566. [W.R. Grace] Modified vinyl latex; biodeg. latex used in land reclamation; binder for humus, stabilizing soil against wind and rain erosion; compatible with most fertilizers; does not inhibit seed germination; water wh. films with good clarity, med. gloss, and no odor, with water resistance, and lt. stability and aging charac.; emulsion; particle size 1 μ; wh., slight odor; dens. 9.93 lb/gal; visc. 900–1500 cps; pH 4.5–5.5; 54–56% total solids.

Dariloid® Series. [Kelco] Alginates; dairy stabilizers and stabilizer/emulsifiers; thickener for food prep.; milk sol. at pasteurization temps.; lt. ivory gran. particles; pH 4.5–10.2.

Darvan® 404. [Vanderbilt] Polymerized sub. benzoid alkyl sulfonic acids, calcium salts; anionic; dispersant for agric. flowables, wettable powds., and min. slurries; powd.; 100% act.

Darvan® C. [Vanderbilt] Polyelectrolyte, ammonium salt; dispersant for electronic and ceramics field; makes slurries for spray driers; amber clear to slightly opalescent fluid; f.p. –5 C; sol. in water systems; dens. 1.11 ± 0.02 mg/m^3; visc. 75 cps; pH 7.5–9.0; 25 ± 1% solids.

Darvan® L. [Vanderbilt] Ammonium salts of alkyl phosphate; dispersant; mold lubricant for elastomers; corrosion inhibitor; cream to amber paste; dens. 1.04 ± 0.03 mg/m^3; pH 7.0–8.0 (5%).

Darvan® ME. [Vanderbilt] Sodium alkyl sulfates; anionic; latex stabilizer; NR, SR mold and stock lubricant; foam modifier, wetting agent, and emulsifier for latexes; wh. to cream powd.; 99.9% min. thru 10 mesh; dens. 1.19 ± 0.03 mg/m^3; pH 9.0–11.0 (10%).

Darvan® No. 1. [Vanderbilt] Sodium polynaphthalene sulfonate; latex dispersant; water-sol.; pH 8.0–10.5 (1%); 87% min. act.

Darvan® No. 2. [Vanderbilt] Sodium lignosulfonate; dispersing and emulsifying agent for rubber industry, esp. for zinc oxide, clays, and sulfur; dk. brn. powd.; water-sol.; dens. 1.25 ± 0.03 mg/m³; pH 7.0–8.5 (1%); 84.5% min. act.

Darvan® No. 3. [Vanderbilt] Sodium salt of polymerized substituted arylalkyl sulfonic acids combined with an inert inorg. suspending agent; dispersant for latex processing; lt. gray-brn. powd.; pH 6.0–8.0 (30% aq.); 84.5% min. act.

Darvan® No. 4. [Vanderbilt] Polymerized aryl alkyl sulfonic acid monocalcium salt; dispersant for finely divided insol. materials; brn. powd.; water-sol.; pH 5.0–8.0 (1%); 92% min. act.

Darvan® No. 6. [Vanderbilt] Polymerized alkyl naphthalene sulfonic acid, sodium salt; dispersant for NR and SR latexes; water-sol.; pH 7.8–10.4 (1%); 85% min. act.

Darvan® No. 7. [Vanderbilt] Sodium polymethacrylate; dispersant for NR, SR latexes; surf. act. agent for preparing high solids disps.; water-wh. clear to slightly opalescent liq.; dens. 1.16 ± 0.02 mg/m³; visc. 75 cps max.; pH 9.5–10.5; 25.0 ± 1% total solids.

Darvan® No. 31. [Vanderbilt] Carboxylated polyelectrolyte, sodium salt; pigment dispersant for aq. systems; latex coatings; yel. liq.; sol. in water systems; dens. 1.11 ± 0.02 mg/m³; visc. 75 cps max.; pH 9–11; 25% min. total solids.

Darvan® NS. [Vanderbilt] c-Cetyl betaine and c-decyl betaine aq. sol'n.; nonionic; stabilizer for low alkaline latex systems; latex foam modifier and wetting agent; clear to lt. amber liq.; dens. 1.05 ± 0.02 mg/m³; pH 8.5–11.5 (10%); 32.0–36.0% total solids.

Darvan® SMO. [Vanderbilt] Sodium salt of sulfated methyl oleate; dispersant; with Darvan WAQ improves smoothness and gloss of dipped CR latex films; clear to lt. amber liq.; dens. 1.08 ± 0.02 mg/m³; pH 6.0–7.5 (10%); 30–35% total solids.

Darvan® WAQ. [Vanderbilt] Sodium alkyl sulfates; anionic; latex stabilizer for NR and syn. latexes; wetting agent and emulsifier for latexes; mold and internal lubricant; clear water-wh. paste (> 30 C); dens. 1.04 ± 0.03 mg/m³; pH 7.0–9.0 (10%).

Dastar. [Croda Ltd.] Cholesterol; conditioner; cosmetic skin and hair care prods.; cryst. wax.

Daxad® 11. [W.R. Grace] Low m.w. naphthalene sulfonate formaldehyde condensate, sodium salt; dispersant for pigments in aq. media; used in agric. chemicals, mastics, caulks, sealants, pigment slurries and disps.; buff powd.; dens. 35 lb/ft³; pH 8.0–10.5 (1%); surf. tens. 70 dynes/cm (1%); 87% min. act.

Daxad® 11G. [W.R. Grace] Low m.w. naphthalene sulfonate formaldehyde condensate, sodium salt; see Daxad 11; also for inks, syn. polymers, paper coating disps., paper mill slime control, pitch control, pulp digestion, and tall oil separation; buff fine gran.; dens. 42 lb/ft³; pH 8.0–10.5 (1%); surf. tens. 70–71 dynes/cm (1%); 87% min. act.

Daxad® 11KLS. [W.R. Grace] Low m.w. naphthalene sulfonate formaldehyde condensate, potassium salt; dispersant for dyes, dyestuffs, inks, latex paints, wax emulsions, wallboard coating, ore flotation; buff powd.; dens. 35 lb/ft³; pH 7.0–8.5 (1%); 85% min. act.

Daxad® 13. [W.R. Grace] Polymerized alkyl naphthalene sulfonic acid sodium salt; dispersant for use in paper making operations, pigment slurries and disp., paper coating disp., paper mill slime control, pitch control, pulp digestion, tall oil separation; amber powd.; dens. 47 lb/ft³; pH 9.5 (1%).

Daxad® 14B. [W.R. Grace] Low m.w. naphthalene sulfonate formaldehyde condensate, sodium salt sol'n.; dispersant for emulsion polymerization, dyestuffs, tanning, herbicides, pesticides, and pitch; dk. br. liq.; 45% total solids.

Daxad® 14C. [W.R. Grace] Low m.w. naphthalene sulfonate formaldehyde condensate, sodium and potassium salts in sol'n.; dispersant for emulsion polymerization, dyestuffs, tanning, herbicides, pesticides, and pitch; designed for cold weather stability; dk. br. liq.; 45% total solids.

Daxad® 15. [W.R. Grace] Low m.w. naphthalene sulfonate formaldehyde condensate, sodium salt; industrial grade, general purpose dispersant for emulsion polymerization, dyestuffs, tanning, herbicides, pesticides, and pitch; amber powd.; dens. 35 lb/ft³; pH 9.5 (1%); surf. tens. 70 dynes/cm (1%); 85% total solids

Daxad® 16. [W.R. Grace] Low m.w. naphthalene sulfonate formaldehyde condensate, sodium salt; dispersant for emulsion polymerization, concrete, dyestuffs, tanning, herbicides, pesticides, and pitch; dk. brn. liq.; dens. 10.4 lb/gal; pH 9.5 (1%); 47.5% total solids.

Daxad® 17. [W.R. Grace] Low m.w. naphthalene sulfonate formaldehyde condensate, sodium salt; industrial grade dispersant for emulsion polymerization, dyestuffs, tanning, herbicides, pesticides, and pitch; amber gran.; ≤ 3% passes through 100-mesh screen; dens. 42 lb/ft³; pH 7.8–10.4 (1%); surf. tens. 70 dynes/cm (1%); 85% act.

Daxad® 19. [W.R. Grace] High m.w. naphthalene sulfonate formaldehyde condensate, sodium salt; dispersant and fluidifier for high-solids, aq. disps. and slurries; used in cement, gypsum, lime, coal, and other slurry systems; br. powd.; very sol. in water; dens. 38 lb/ft³; pH 9.5 (1%); surf. tens. 71 dynes/cm (1%).

Daxad® 19K. [W.R. Grace] High m.w. naphthalene sulfonate formaldehyde condensate, potassium salt; version of Daxad 19 for use where the sodium salt is undesirable; br. powd.; very sol. in water; dens. 42 lb/ft³; pH 9.5 (1%).

Daxad® 19L-33. [W.R. Grace] High m.w. naphthalene sulfonate formaldehyde condensate, sodium salt, in sol'n.; see Daxad 19; dk. br. liq.; dens. 9.8 lb/gal; pH 9.5 (1%); surf. tens. 71 dynes/cm (1%); 33% total solids.

Daxad® 19L-40. [W.R. Grace] High m.w. naphthalene sulfonate formaldehyde condensate, sodium/potassium salt, in sol'n.; cold weather-stable dispersant, water-reducing agent, and visc.-reducer for high-solids slurries like concrete, cement, gypsum, lime, coal; dk. br. liq.; dens. 10 lb/gal; pH 9.5 (1%); 40% total solids.

Daxad® 21. [W.R. Grace] Calcium salt of polymerized aryl alkyl sulfonic acids; dispersant for finely divided insol. particles in water; used for wettable powds., agric. chemicals, concrete admixtures, dyes, gypsum wallboard, high-strength cements, pigment slurries and disps., and water treatment chemicals; br. powd.; dens. 30–35 lb/ft³; pH 5.0–8.0 (1%); surf. tens. 60–65 dynes/cm (1%); 92% act.

Daxad® 23. [W.R. Grace] Sodium salts of polymerized substituted benzoid alkyl sulfonic acids; dispersant for agric. chemicals, concrete admixtures, dyes, high-strength cements, linoleum pastes, pigment slurries and disps., and water treatment chemicals; dk. br. powd.; dens. 35–40 lb/ft³; pH 7.5 (1%); surf. tens. 56–57 dynes/cm (1%); 84.5% act.

Daxad® 27. [W.R. Grace] Polymerized aryl and substituted benzoid alkyl sulfonic acid sodium salt; dispersant and suspending aid; used in agric. chemicals and pigment slurries and disp.; gray brn. powd.; dens. 48 lb/ft³; pH 8 (1%).

Daxad® 30. [W.R. Grace] Sodium polymethacrylate sol'n.; dispersant esp. for pigments in aq. sol'ns.; used in paint formulations, emulsion polymerization, water treatment, agriculture, cosmetics, industrial cleaners, in large particle suspensions; water-wh. clear liq.; sol. in water systems; sp.gr. 1.15; dens. 9.6 lb/gal; visc. 75 cps max.; pH 10.0; surf. tens. 70 dynes/cm (1%); 25% solids.

Daxad® 30-30. [W.R. Grace] Sodium polymethacrylate sol'n.; dispersant esp. for pigments in aq. sol'ns.; used in paint formulations, emulsion polymerization, water treatment (as a scale control agent for boiler systems), in large particle suspensions; water-wh. clear liq.; sol. in water systems; sp.gr. 1.21; dens. 10.1 lb/gal; visc. 150 cps max.; pH 10.0; surf. tens. 70 dynes/cm (1%); 30% solids.

Daxad® 30S. [W.R. Grace] Sodium polymethacrylate polymer; dispersant for pigments in aq. sol'ns.; used in water treatment, trade sales flat paints, dry mixes, agriculture; wh. fine powd.; sol. in water systems; dens. 32–35 lb/ft³ (tamped); pH 10.0 (1%); 90% act.

Daxad® 31. [W.R. Grace] Sodium polyisobutylene maleic anhydride copolymer sol'n.; dispersant for aq. systems; used in latex paints and coatings, enamels, polymerization, leather tanning, and water treatment; pale amber clear liq.; sol. in water systems; sp.gr. 1.11; dens. 9.2 lb/gal; visc. 30 cps; pH 10.0; surf. tens. 50 dynes/cm (1%); 25% total solids.

Daxad® 31S. [W.R. Grace] Sodium polyisobutylene maleic anhydride copolymer; see Daxad 31; wh. fine powd.; sol. in water systems; dens. 32–35 lb/ft³ (tamped); pH 10.0 (1%); 90% act.

Daxad® 32. [W.R. Grace] Ammonium polymethacrylate sol'n ; dispersant for the ceramics industry and pigment disps.; pale amber clear liq.; very sol. in water systems; dens. 9.1–9.4 lb/gal; visc. 75 cps max.; pH 8.0; 25% total solids.

Daxad® 32S. [W.R. Grace] Polymethacrylic acid buffered with ammonia; dispersant for the ceramics industry and where sodium salts are undesirable; wh. fine powd.; sol. in water systems; dens. 32–35 lb/ft³ (tamped); pH 6.3 (1%); 90% act.

Daxad® 34. [W.R. Grace] Polymethacrylic acid sol'n.; dispersant for pigments and fillers in ceramics, polymerization; pale amber clear liq.; sol. in water systems; sp.gr. 1.09; dens. 9.1 lb/gal; visc. 400 cps max.; pH 3; surf. tens. 70 dynes/cm (1%); 25% total solids.

Daxad® 34A9. [W.R. Grace] Ammonium polymethacrylate sol'n.; dispersant for pigments and fillers commonly used in latex paint systems; also for polymerization and clay coating; clear liq.; sol. in water systems; sp.gr. 1.11; dens. 9.2 lb/gal; visc. 25 cps max.; pH 9; surf. tens. 76 dynes/cm (1%); 24% total solids

Daxad® 34N10. [W.R. Grace] Sodium polymethacrylate sol'n.; dispersant for latex polymerization, industrial cleaners, and for clay in paper coating colors; pale amber clear liq.; sol. in water systems; sp.gr. 1.21; dens. 10.0 lb/gal; visc. 400 cps max.; pH 10.0; surf. tens. 70 dynes/cm (1%); 30% total solids.

Daxad® 34S. [W.R. Grace] Polymethacrylic acid; dispersant for preparations that start with acidic components and are further compded. to an alkaline pH; wh. fine powd.; sol. in water systems; dens. 32–35 lb/ft³ (tamped); pH 3.2 (1%); 90% act.

Daxad® 35. [W.R. Grace] Sodium polymethacrylate sol'n.; primary dispersant for clays, zinc oxide, and other paint or coating pigments and fillers; for latex paint formulations, water treatment; pale amber clear liq.; sol. in water systems; sp.gr. 1.28; dens. 10.6 lb/gal; visc. 350 cps; pH 7.0; surf. tens. 70 dynes/cm (1%); 40% total solids.

Daxad® 37LA7. [W.R. Grace] Ammonium polyacrylate sol'n.; anionic; dispersant in aq. systems when used at very low levels; esp. for latex paints and coatings, dispersing org. and inorg. pigments; slight amber clear, slightly hazy liq.; sol. in water and glycols; sp.gr. 1.17; dens. 9.8 lb/gal; visc. 80 cps; pH 7.0; surf. tens. 66 dynes/cm (1%); 40% total solids.

Daxad® 37L Acid. [W.R. Grace] Aq. sol'n. of low m.w. polyacrylic acid; for applics. requiring an acid-stable dispersant; also for use in water treatment as an antiscale agent and in latex flat paints; straw liq.; visc. 250 cps; pH 3.0; 50% total solids.

Daxad® 37LK9. [W.R. Grace] Potassium polyacrylate sol'n.; dispersant in aq. systems when used at very low levels; esp. for latex paints and coatings, dispersing org. and inorg. pigments; slight amber clear, slightly hazy liq.; sol. in water and glycols; sp.gr. 1.24; dens. 10.3 lb/gal; visc. 17 cps; pH 9.0; surf. tens. 71 dynes/cm (1%); 30% total solids.

Daxad® 37LN7. [W.R. Grace] Sodium polyacrylate sol'n.; dispersant, esp. for clays; deflocculant and dispersant for clay slurries; used in paper coating industry, in flat latex paints; as a fluidifier in the formulation of oil well drilling muds; as an antiredeposition agent in cleaning formulas; pale amber clear, slightly hazy liq.; infinite sol. in water; sp.gr. 1.30;

dens. 10.9 lb/gal; visc. 500 cps; pH 7.0; surf. tens. 75 dynes/cm (1%); 45% total solids.

Daxad® 37LN10. [W.R. Grace] Sodium polyacrylate sol'n.; see Daxad 37LN7; pale amber clear, slightly hazy liq.; infinite sol. in water; sp.gr. 1.30; dens. 10.9 lb/gal; visc. 500 cps; pH 10.0; surf. tens. 71 dynes/cm (1%); 45% total solids.

Daxad® 37NS. [W.R. Grace] Sodium polyacrylic acid; dispersant for solids in aq. systems; esp. for clays; as a deflocculant in the prod. of high-solids, low-visc. slurries in the pigments industry; wh. fine powd.; sol. in water systems; dens. 32–35 lb/ft^3 (tamped); pH 2.7 (1%); 90% act.

Daxad® 41. [W.R. Grace] Sodium polymethacrylate polymer; dispersant designed for scale inhibition and conditioning of suspended matter in water treatment; inhibits deposition of new scale and sludge; wh. fine powd.; sol. in water systems; pH 7.5 (1%); surf. tens. 70 dynes/cm (1%); 90% act.

Daxad® CP-2. [W.R. Grace] Cationic polyelectrolyte; dispersant for fillers, pigments, other additives; precipitates wetting agents, detergent soaps, emulsifiers, and dispersants out of industrial waters; coagulates latexes; flocculates kaolin clay disps.; breaks o/w and w/o emulsions; liq.; sol. in water; dens. 9.6 lb/gal; visc. 700 cps; pH 6.0; 50% total solids.

DBE. [DuPont] Dibasic ester mixt. (66% dimethyl glutarate, 17% dimethyl adipate, 16.5% dimethyl succinate); solv. for coatings, cleaners, inks, textile lubricants, urethane prod.; plasticizer; polymer intermediate for polyester polyols for urethanes, wet-str. paper resins, polyester resins; specialty chemical intermediate; clear colorless liq., mild odor; m.w. 159; sol. in alcohols, ketones, ethers, most hydrocarbons; slightly sol. in water and higher paraffianic hydrocarbons; sp.gr. 1.092; dens. 9.09 lb/gal; visc. 2.4 cSt; b.p. 196–225 C; f.p. –20 C; flash pt. (TCC) 100 C; surf. tens. 35.6 dynes/cm.

DBE-2, -2SPG. [DuPont] Dibasic ester mixt. (76% dimethyl glutarate, 23% dimethyl adipate, 0.5% dimethyl succinate); see DPE; clear colorless liq., mild odor; m.w. 163; sol. see DBE; sp.gr. 1.081; dens. 9.00 lb/gal; visc. 2.5 cSt; b.p. 210–225 C; f.p. –13 C; flash pt. (TCC) 104 C.

DBE-3. [DuPont] Dibasic ester mixt. (89% dimethyl adipate, 10% dimethyl glutarate, 0.5% dimethyl succinate); see DPE; clear colorless liq., mild odor; m.w. 173; sol. see DBE; sp.gr. 1.068; dens. 8.89 lb/gal; visc. 2.5 cSt; b.p. 215–225 C; f.p. 8 C; flash pt. (TCC) 102 C.

DBE-4. [DuPont] Dimethyl succinate; see DPE; clear colorless liq., mild odor; m.w. 146; sol. see DBE; sp.gr. 1.121; dens. 9.33 lb/gal; visc. 2.5 cSt; b.p. 196 C; f.p. 19 C; flash pt. (TCC) 94 C; 99.5+% conc.

DBE-5, -5SPG. [DuPont] Dimethyl glutarate; see DPE; clear colorless liq., mild odor; m.w. 160; sol. see DBE; sp.gr. 1.091; dens. 9.08 lb/gal; visc. 2.5 cSt; b.p. 210–215 C; f.p. –37 C; flash pt. (TCC) 107 C; 99.5% conc.

DBE-9. [DuPont] Dibasic ester mixt. (73% dimethyl glutarate, 25% dimethyl succinate, 1.5% dimethyl adipate); see DPE; clear colorless liq., mild odor; m.w. 156; sol. see DBE; sp.gr. 1.099; dens. 9.15 lb/gal; visc. 2.4 cSt; b.p. 196–215 C; f.p. –10 C; flash pt. (TCC) 94 C.

DBP. [Great Lakes] 2,3 Dibromopropanol; low visc. brominated alcohol reacting readily with isocyanates imparting flame retardancy to urethane foams; intermediate in synthesis of org. compds.; clear visc. liq.; f.p. 6 C; b.p. 219 C; sol. in methanol, methylene chloride, toluene, MEK; sol. 5 g/100 g water; sp.gr. 2.14; visc. 26 cps; acid no. 0.05 max.; 73% bromine

DD-8126. [Dover] Brominated paraffin; flame retardant for PUF, foam-in-place for pkg., rubber, textiles; visc. reducer; Gardner 2 liq.; sp.gr. 1.19; visc. 10 cps; 44% bromine

DD-8133. [Dover] Aromatic bromine and aliphatic chlorine; flame retardant for PP, PE, SBR, other rubbers, unsat. polyesters, and coatings; wh. solid; sp.gr. 2.3; soften. pt. 130–300 C; 41% bromine; 35% chlorine.

DD-8207. [Dover] Bromochlorinated paraffin; flame retardant for flexible, rigid PUF, laminated board, board stock, pkg., textiles, carpet backing; Gardner 2 clear liq.; pleasant odor; sol. in aliphatic, aromatic, and chlorinated hydrocarbons, esters, ketones, and higher alcohols (other than methyl and ethyl); insol. in water; sp.gr. 1.45; dens. 12 lb/gal; visc. 1700 cps; acid no. < 1; 32% bromine; 27% chlorine, trace phosphorus.

DD-8307. [Dover] Bromochlorinated paraffin with phosphorus; flame retardant for rigid and flexible PUF, RIM, textiles, rubber, and PVC for general purpose applics. in construction, agric., consumer, and industrial markets where UL rated material is required; Gardner 2 clear liq.; pleasant odor; sol. see DD-8207; sp.gr. 1.4; visc. 1000 cps; 26% bromine; 22% chlorine; 5% phosphorus.

DDBS 100. [Zohar] Alkylbenzene sulfonic acid, branched; anionic; detergent intermediate; liq.; 100% conc.

DE-60F. [Great Lakes] Pentabromodiphenyl oxide/aromatic phosphate; flame retardant additive for rigid and flexible urethanes, epoxies, laminates, unsat. polyesters, and plasticized PVC compds.; amber liq.; sol. in Freon 11, polyol, styrene, MEK, triethyl phosphate, methylene chloride, toluene, dioctylphthalate; sol. 0.1 g/100 g water; sp.gr. 1.95; dens. 16.4 lb/gal; acid no. 0.25 max.; 85% DE-71; 15% aromatic phosphate; 59–61% bromine; 1.1–1.4% phosphorus.

DE-71. [Great Lakes] Pentabromodiphenyl oxide; high visc. flame retardant for thermosetting and thermoplastic resin systems; used for unsat. polyester, rigid and flexible urethane foams, epoxies, laminates, adhesives, and coatings; amber visc. liq.; m.w. 564.7; sol. in Freon 11, polyol, styrene, MEK, triethyl phosphate, methylene chloride, toluene, dioctylphthalate, sol. < 0.1 g/100 g water; sp.gr. 2.27; dens. 19.0 lb/gal; acid no. 0.25 max.; 69–72% bromine.

DE-79. [Great Lakes] Octabromodiphenyl oxide; flame retardant for ABS, nylon, polycarbonate, and polyester thermoplastic polymers; additive for unsat. polyesters and epoxy thermoset resins; off-wh.

powd.; m.w. 801; sol. (g/100 g solv.) 25 g styrene, 20 g benzene, 19 g toluene, 11 g methylene chloride, < 0.1 g water; sp.gr. 2.6; m.p. 70–150C; 79% org. bromine.

DE-83R. [Great Lakes] Decabromobiphenyl oxide; halogenated flame retardant for thermoplastic, elastomeric, and thermoset polymer systems incl. HIPS, PBT, ABS, nylons, PP, LDPE, EPDM, unsat. polyesters, and epoxy resins; wh. powd.; 3.2 μ particle size; m.w. 959.2; sol. (g/100 g solv.) 0.2 g toluene, 0.1 g acetone and benzene, < 0.1 g. in water; sp.gr. 3.2; m.p. 300–310 C; 97% assay; 83.3% bromine.

Deatron N. [Nikko] Alkyl phosphate; anionic; antistat for nylon fabrics; liq.; 50% conc.

Decalin. [DuPont] Decahydronaphthalene; solv. and stabilizer for shoe creams and floor waxes; solv. in paint and lacquers, oils, resins, rubber, and asphalt; colorless to pale yel. liq.; m.w. 138.2; f.p. –45 C; b.p. 186 C; negligible sol. in water; sp.gr. 0.876; dens. 7.39 lb/gal; flash pt. 57 C (TCC); 97.0% min. conc.

Decanox-F. [Atochem] Decanoyl peroxide; initator for bulk, sol'n., and suspension polymerization, curing elastomers, and high-temp. cure of polyester resins; flaked solid; 98.5% act.; 3.93% min. act. oxygen.

Dechlorane® Plus 25. [Occidental] Chlorine-containing cycloaliphatic compd.; 1,2,3,4,7,8,9,10,13,13,14,14-dodecachloro-1,4,4a,5,6,6a,7,10,10a,11,12,12a-dodecahydro-1,4:7,10-dimethanodibenzo (a,e) cyclooctene; flame retardant in polymer systems (thermoplastics, thermosets, and elastomers); usually combined with antimony oxide as a synergist; wh. cryst. powd.; 2–5 (Fisher Sub-Sieve Sizer) avg. particle size; odorless; m.w. 653.77; low sol. in most solvs. and polymer systems; insol. in water; dens. 1.8 g/cc; bulk dens. 38–42 lb/ft^3; m.p. 350 C (with decomp.); pH 6.0–8.0 (methanol-water extract); flash pt. none; 65% chlorine.

Dechlorane® Plus 515. [Occidental] Chlorine-containing cycloaliphatic compd.; 1,2,3,4,7,8,9,10,13,13,14,14-dodecachloro-1,4,4a,5,6, 6a,7,10,10a,11,12,12a-dodecahydro-1,4:7,10-dimethanodibenzo (a,e) cyclooctene; see Dechlorane Plus 25; wh. cryst. powd.; 5–15 (Fisher Sub-Sieve Sizer) avg. particle size; odorless; m.w. 653.77; low sol. in most solvs. and polymer systems; insol. in water; dens. 1.8 g/cc; bulk dens. 38–42 lb/ft^3; m.p. 350 C (with decomp.); pH 6.0–8.0 (methanol-water extract); flash pt. none; 65% chlorine.

Dechlorane® Plus 2520. [Occidental] Chlorine-containing cycloaliphatic compd.; 1,2,3,4,7,8,9,10,13,13,14,14-dodecachloro-1,4,4a,5,6,6a,7,10,10a,11,12,12a - dodecahydro-1, 4:7,10-dimethanodibenzo (a,e) cyclooctene; see Dechlorane Plus 25; recommended for highly sensitive applics. involving ultrathin cross sections; wh. cryst. powd.; 2–5 (Fisher Sub-Sieve Sizer) avg. particle size; odorless; m.w. 653.77; low sol. in most solvs. and polymer systems; insol. in water; dens. 1.8 g/cc; bulk dens. 38–42 lb/ft^3; m.p. 350 C (with decomp.); pH 6.0–8.0 (methanol-water extract); flash pt. none; 65% chlorine.

Dedevap®. [Bayer] Dichlorvos; insecticide and acaricide for the control of biting and mining pests; colorless liq., slight odor; m.w. 220.98; sol. 0.88 g/100 ml in water; misc. with cyclohexanone, IPA, toluene, methylene chloride; sp.gr. 1.4 (20/4 C); b.p. 74 C (1 mm Hg).

Dee-Tac. [Olin] Detackifier for uncured rubber; paste.

Defoamer 44. [Consos] Silicone emulsion; nonionic; silicone based defoamer for most textile applics.; paste; biodeg.

Defoamer 167. [A. Harrison] Silicone emulsion; defoamer used in dyebath and finishing; water-sol.

Defoamer 357. [Hercules] Hydrocarbon oil; defoamer for mfg. and end-use foam control of water-based systems; water-disp.

Defoamer 388. [Hercules] Hydrocarbon oil; defoamer for chemical processing; liq.

Defoamer 831. [Hercules] Hydrocarbon oil; liq. pH defoamer; low visc. for metering and rapid foam control; water-disp.

Defoamer A. [Exxon] Petrol.-syn. base defoamer with cationic dispersants; cationic; used in activated sludge sewage treatment plants; amber-brn. liq.; dens. 7.0–7.4 lb/gal; pH 6–8; 99% min. act.

Defoamer A 50. [ChemY GmbH] Alkyl polyglycol ether carboxylic acid; cationic; defoamer; used in phosphoric acid prod.; liq.; 50% act.

Defoamer CK-35. [Crompton & Knowles] Ethoxylated fatty acids in petrol. hydrocarbon; defoamer; eliminates entrained gas and improves drainage on paper machines; disp.

Defoamer CK-55. [Crompton & Knowles] Fatty amide derivs., hydrophobic silica, and nonionic surfactants in petrol. oil; defoamer for standard paper machines and hydropulpers; disp.

Defoamer CK-75. [Crompton & Knowles] Ethoxylated fatty acids and petrol. oil; defoamer for paper machines and hydropulpers; disp.

Defoamer MJ. [Hart Prod.] Org. esters and min. oil; nonsilicone defoamer; water emulsifiable.

Defoamer S. [Hart Prod.] Silicone emulsion; nonionic; defoamer for industrial and textile applic.; wh. visc. liq.; sp.gr. 0.988; visc. 1500 cps; pH acid.

Defoamer S-10. [Hart Prod.] Silicone; defoamer for textile printing and processing plants, water treatment, chemical specialty mfg., paper mills, etc.; wh. visc. emulsion; odorless; disperses in water; sp.gr. 1.060; pH neutral.

Defoamer S-100. [Hart Prod.] Silicone, modified; antifoamer used in paper mills, dyehouses, finishing plants, textile printing and processing; wh. visc. emulsion; odorless; disperses in water; pH neutral.

Defoamer TIP. [Hoechst-Celanese] Phosphoric acid esters; defoamer for waste water; clear liq.; 50% act.

Defoamer/Drainage Aid CK-25. [Crompton & Knowles] Fatty acids, fatty acid derivs., and surfactants in petrol. blending oil; defoamer and drainage aid for paper industry; disp.

Degacure® K 126. [Degussa] Cycloaliphatic epoxide; binder for cationic uv-curing with Degacure KI

85; for uv-curable coatings and printing inks; clear, colorless liq.; dens. 1.16–1.18 g/cm³; visc. 300–400 MPa•s; b.p. 354 C (1013 mbar); solid. pt. –20 C; flash pt. (PMCC) 197 C.

Degacure® KI 85. [Degussa] Bis [4-(diphenylsulfonio)-phenyl] sulfide-bis-hexafluorophosphate; cationic photoinitiator for uv-curing of binder systems based on Degacure K 126; pale yel. clear liq.; weak odor; m.w. 846.7; dens. 1.28–1.30 g/cm³; visc. 14–31 mPa•s; flash pt. (PMCC) 119 C; 33–40% sol'n. in propylene carbonate.

Degressal SD 20. [BASF AG] Polypropoxylate; defoamer for cleaners and detergents; liq.; 100% act.

Degressal SD 21. [BASF AG] Polyalkoxylate; see Degressal SD 20; liq.; 100% act.

Degressal SD 23. [BASF AG] Polyalkoxylate; defoamer for chemical and tech. industries; liq.; 100% act.

Degressal SD 30. [BASF AG] Org. ester; defoamer for cleaners and detergents; liq.; 100% act.

Degressal SD 40. [BASF AG] Phosphoric acid ester; defoamer for cleaners; defoamer and plasticizer for dry-bright emulsions; liq.; 100% act.

Degressal SNC. [BASF AG] Modified phosphoric acid monoester; defoamer for detergents and cleaners; liq.; 100% act.

D.E.H. 20. [Dow] Diethylenetriamine; aliphatic polyamine curing agent for epoxy resins; used for civil engineering, adhesives, grouts, casting and elec. encapsulation; D.E.H. 20 is a general purpose curing agent; liq.; misc. with polar solvs. (water, alcohols, acetone, benzene, ethyl ethers); sp.gr. 0.949; dens. 7.89 lb/gal; visc. 6 cps; flash pt. 208 F.

D.E.H. 24. [Dow] Triethylenetetramine; see D.E.H. 20; liq.; sol. see D.E.H. 20; sp.gr. 0.978; dens. 8.13 lb/gal; visc. 22 cps; flash pt. 245 F.

D.E.H. 26. [Dow] Tetraethylenepentamine; see D.E.H. 20; D.E.H. 26 is a R.T. curing agent often used in 2-pkg. protective coating systems; liq.; sol. see D.E.H. 20; sp.gr. 0.993; dens. 8.26 lb/gal; visc. 55 cps; flash pt. 310 F.

D.E.H. 29. [Dow] Aliphatic polyamine blend; see D.E.H. 20; liq.; sp.gr. 1.01; dens. 8.46 lb/gal; visc. 300 cps; flash pt. 330 F.

D.E.H. 39. [Dow] Aminoethylpiperazine; see D.E.H. 20; liq.; sp.gr. 0.982; dens. 8.17 lb/gal; visc. 10 cps; flash pt. 175 F.

D.E.H. 40. [Dow] Accelerated dicyandiamide; epoxy curing agent for powd. coating formulations; wh. powd.; sp.gr. 1.33; dens. 0.31 g/cc.

D.E.H. 52. [Dow] Diethylenetriamine adduct; adducted aliphatic polyamine curing agent for epoxy resins; liq.

D.E.H. 58. [Dow] Diethylenetriamine modified with Bisphenol A; see D.E.H. 52; liq.

Dehydag Wax 14. [Henkel] Myristyl alcohol; consistency factor for cosmetic and pharmaceutical o/w and w/o creams, ointments, emulsions, liniments, and sticks; flakes.

Dehydag Wax 16. [Henkel] Cetyl alcohol; see Dehydag Wax 14; flakes.

Dehydag Wax 18. [Henkel] Stearyl alcohol; see Dehydag Wax 14; flakes.

Dehydag Wax 22 (Lanette). [Henkel] Behenyl alcohol; see Dehydag Wax 14; fused/flakes.

Dehydag Wax E. [Henkel] Sodium cetearyl sulfate; anionic; o/w emulsifier used in personal care prods. and powd. cleaners; wh. powd.; water-sol.; dens. 0.180 g/ml; pH 6–8; 87% min. act.

Dehydag Wax N. [Henkel] Cetearyl alcohol and sodium cetearyl sulfate; anionic; mfg. of o/w creams and liq. emulsions; solid; 100% conc.

Dehydag Wax O. [Henkel] Cetearyl alcohol; nonionic; consistency modifier used in personal care prods. and pharmaceuticals; wh. gran.; dens. 0.816 g/ml (60 C); solid. pt. 48–52 C; sapon. no. 1.0 max.; 100% conc.

Dehydag Wax SX. [Henkel] Cetearyl alcohol, sodium lauryl sulfate (90:10 ratio); anionic; surfactant used in mfg. of o/w emulsions for personal care prods.; SE base; wh. gran.; solid. pt. 50–54 C; sapon. no. 2.0 max.; 99.0% min. fatty alcohol.

Dehydag Wax W. [Henkel] Cetearyl alcohol and sodium lauryl sulfate (90:10 ratio); anionic; base for mfg. of ointments, creams and liniments; gran.; 100% conc.

Dehydol D 3. [Henkel] Fatty alcohol polyglycol ether; nonionic; emulsifier, solubilizer for oils; liq.; HLB 9.0; 100% conc.

Dehydol LS 2, LS 3, LS 4. [Henkel] Laureth-2, -3, -4 resp.; nonionic; emulsifier, solubilizer for solvs., oils, bases for prod. of sulfates; raw material for dishwashing, cleansing agent and cold cleaners; liq.; 99–100% conc.

Dehydol LT 2, LT 3, LT 4, LT 5, LT 7. [Henkel] Fatty alcohol polyglycol ether; nonionic; see Dehydol LS 2; liq.; 99–100% conc.

Dehydran 150. [Henkel] Aq. silicone compd.; defoamer for systems containing surfactants, emulsifiers, and cleaning suds; liq.

Dehydran 240. [Henkel] Modified polyalkylene glycols; defoamer for industrial cleaners, automatic dishwashing agents; liq.

Dehydran 241. [Henkel] Modified polyalkylene glycols; see Dehydran 240; liq.

Dehydran 420. [Henkel] Fatty acid ester; defoamer used in inorg. acid prod.; liq.

Dehydran 610. [Henkel] Natural fatty derivs. in emulsion form; defoamer used in waste water treatment; liq.

Dehydran 630. [Henkel] Mixt. of special fatty acid esters with aliphatic hydrocarbons; see Dehydran 610; liq.

Dehymuls K. [Henkel] Petrolatum, decyl oleate, sorbitan sesquioleate, beeswax, min. oil, ceresin, aluminum stearate; SE base for mfg. of cosmetic and pharmaceutical preparations of the w/o type; soft waxy solid; 100% act.

Dehypon Conc. [Henkel] Fatty alcohol polyglycol ether blend; nonionic; wetting agent, emulsifier and dispersant; paste; 99% conc.

Dehypon LS 24, LS 36, LS 45, LS 54. [Henkel] Fatty alcohol EO/PO adducts; nonionic; surfactant, dispersant, wetting agent for industrial cleaners, liq. laundry detergents, dishwashing agents; liq.; 100% conc.

Dehypon LT 104. [Henkel] Fatty alcohol polyglycol ether; nonionic; surfactant, antifoaming agent; paste; 100% conc.

Dehyquart A. [Henkel] Cetrimonium chloride; cationic; softener, conditioner, bactericide, fungicide, and odor inhibitor in personal care prods.; antistat for hair and fibers; emulsifier; pale yel. clear liq.; 24–26% act.

Dehyquart C. [Henkel] Lauryl pyridinium chloride; cationic; surfactant, wetting agent, fungicide, bactericide, and disinfectant used in cleaning formulations and personal care prods.; sequestrant for min. oil industry; wh. paste, powd.; 80–82% act.

Dehyquart C Crystals. [Henkel] Lauryl pyridinium chloride; cationic; bactericide, fungicide, corrosion inhibitor, sequestrant, conditioner used in personal care prods.; 90–94% act.

Dehyquart D. [Henkel] Lauryl pyridinium bisulfate; germicide, wetting agent with anticorrosive effect, emulsion breaker; liq.

Dehyquart DAM. [Henkel] Distearyl dimethyl ammonium chloride; conditioning component for hair care preparations; antistat; paste; 70–80% act.

Dehyquart E. [Henkel] Hydroxyhexadecyl dimethyl hydroxyethyl ammonium chloride; antistat, conditioner for hair care preparations; compatible with anionic systems; liq.; 28% aq.

Dehyquart LDB. [Henkel] Lauralkonium chloride; cationic; bactericide and fungicide for disinfectants; liq.; 34–36% conc.

Dehyquart LT. [Henkel] Laurtrimonium chloride; cationic; wetting agent, antistat, bactericide, demulsifier, deodorant, conditioning component for hair care prods.; liq.; 34–36% conc.

Dehyquart SP. [Henkel] Quaternium-52; cationic; emulsifier, conditioning, softening and antistatic agent used in personal care prods.; metal corrosion inhibitor; lt. yel. clear visc. liq.; pH 6.8–7.2 (10%); 49.0–51.0% solids.

Dehyquart STC-25. [Henkel] Stearalkonium chloride; cationic; antistat, conditioner for hair and fibers; emulsifier in creams, lotions; wh. slurry; 25% conc.

Dehysan Z 4904. [Henkel] Mixt. of fatty alcohol polyglycol ethers and soaps; defoamer for sugar prod. (cane and beet); liq.

Dehysan Z 7225. [Henkel] Mixt. of natural fats with hydrophilic fat derivs.; defoamer for beet sugar prod.; liq.

Dehysol. [Henkel KGaA] Nonsaponifiable thickening and antisagging agent for alkyd paints; paste.

Dehysol R. [Henkel KGaA] Thickening and wetting agent for alkyd paints; liq.

Dehyton K. [Henkel] Cocamidopropyl betaine; amphoteric; raw material for mfg. of surfactant preparations, conditioning agent; liq.; 30% conc.

Delac® MOR. [Uniroyal] N-Oxydiethylene benzothiazole-2-sulfenamide; delayed action accelerator for SBR, natural rubber, polyisoprene, polybutadiene, and nitrile rubber; tan flakes or pellets; sol. in acetone, benzene, CCl_4, methanol; insol. in water; sp.gr. 1.37; m.p. 75–90 C.

Delac® NS. [Uniroyal] N-t-butyl-2-benzothiazole sulfenamide; delayed action accelerator for natural and syn. rubbers; used in tire treads, carcass, mechanicals, and wire jackets; lt. tan powd. or flake; sol. in acetone, benzol, alcohol, chloroform, ether, naphtha; insol. in water; sp.gr. 1.29; m.p. ≥ 104 C.

Delfloc 50. [Hercules] Cationic polymer sol'n.; retention aid and flocculant for the paper industry; 12.5% solids.

Delion 624. [Takemoto] Phosphate surfactant; anionic; antistat and lubricant for fibers; paste; 60% conc.

Delion 662. [Takemoto] Phosphate surfactant; anionic; antistat for syn. fibers; paste.

Delion 964. [Takemoto] Alkyl phosphate; anionic/nonionic; spin finish oil for polyester staple; paste; 80% conc.

Delion 6067. [Takemoto] Alkyl phosphate; anionic; staple finish oil for polyester staple; paste; 60% conc.

Delion A-016. [Takemoto] Sulfonated surfactant; anionic; antistat for syn. fibers; wetting agent and detergent; liq.; 40% conc.

Delion A-160. [Takemoto] Phosphated surfactant; anionic; antistat agent for fibers; liq.

Delsette. [Hercules] Polyamide resin; hair fixative; imparts improved body and hold to hair; reactive type avail. which is resistant to shampooing; water-sol.

Deltyl® Extra. [Givaudan] IPM; emollient and aux. emulsifier in cosmetics; sol. in alcohol, min., peanut, sesame, olive, and almond oils; water-insol.

Delvet 68, 70. [Henkel] Nonionic surfactant used as chlorinated paraffin flame retardant additive; fluid; 68% act.

Delvocid. [Int'l. Bio-Synthetics] Natamycin; food grade mold and yeast inhibitor; powd.

Demulfer Series. [Toho] Demulsifier for crude oil prod.; liq.

Demol AS. [Kao] Condensed aromatic sulfonic acid salt; dispersant for pigment, carbon black, clay, dyestuff; powd.

Demol C. [Kao] Condensed aromatic sulfonic acid salt; dispersant for dyes, pigments, and agric. chemicals, esp. used at high temps.; powd., liq.

Demol EP. [Kao] Sodium C_{4-12} olefin/maleic acid copolymer; colorless dispersant for dyestuff and pigment; liq., powd.

Demol MS. [Kao] Condensed aromatic sulfonic acid; dispersant for pigment, carbon black, clay, and dyestuff; powd.

Demol N. [Kao] Condensed naphthalene sulfonic acid; see Demol MS; powd.

Demol P. [Kao] Polymeric carboxylic acid salt; dispersant for dyestuff and pigment; liq.; colorless.

Demol RN. [Kao] Condensed naphthalene sulfuric acid; dispersant for pigment, carbon black, and clay; powd.

Demol SN-B. [Kao] Condensed aromatic sulfonic acid; see Demol C; powd., liq.

Demol SS Paste. [Kao] Condensed aromatic sulfuric acid; see Demol RN; paste.

Demol SSL. [Kao] Condensed aromatic sulfonic acid; see Demol C; powd., liq.

Demol VP. [Kao] Condensed aromatic sulfonic acid; see Demol C; powd., liq.

Denlube 33-B. [Graden] Ammonium stearate disp.; frothing aid and foam stabilizer in acrylic and SBR latex systems; foam rubber additive; wh. liq.; ammoniacal odor; dens. 8.0 lb/gal; pH 10.0 ± 1.0 (2%); 30% act.

Denlube 40. [Graden] High dens. polyethylene emulsion; nonionic; lubricant in coatings; improves flow and leveling chars. for smooth even applic.; reduces cracking of dry coating and provides release; amber liq.; dens. 8.4 lb/gal; pH 8.5–9.0 (1%); 22% act.

Denlube 1025-KA. [Graden] Higher m.w. sulfonated aliphatic compd.; antiblocking agent; strip-stock lubricant during rubber processing; processing aid; cream paste; water-disp.; pH 8.5–9.0 (1%); 80% act.

Denphos P-610. [Graden] Phosphate ester; anionic; detergent, emulsifier, wetting agent, hydrotrope for household and industrial cleaners; water-wh. to pale yel. liq.; mild odor; sol. in aq. sol'ns. with high conc. of alkaline builders; sp.gr. 1.155; pH 1.5–2.5 (2%); 100% act.

Densol 284. [Graden] Highly sulfated fatty acid; anionic; stabilizer for natural and syn. latex; protects from coagulation; dk. amber liq.; fatty odor; 70% conc.

Densol 1010. [Graden] Modified sodium polyacrylate; thickener for latex adhesives, suitable for mfg. of footwear; controlled penetration; stabilizer for latex compd. systems at 0.5–1.0%; lt. amber gel; dens. 8.5 lb/gal; pH 10 (2%); 10% solids.

Densol 6920. [Graden] Syn. replacement for sulfated castor oil; anionic; emulsifier for min. oil, dispersant for colors in oil systems; liq.; sp.gr. 0.972 (60 F); pH 5.4 (10%); 75% act.

Densol BP-61, -62. [Graden] Block polymer; nonionic; defoamer, emulsifier, dispersant, wetting agent, and stabilizer; liq.; 100% conc.

Densulf TA-75. [Graden] Sulfonated tallow; softener for textiles; fatliquoring agent for leathers; plasticizer for cotton finishing; defoamer for glues; ivory soft paste, mild fatty odor; sol. in water; sp.gr. 1.09; 25% moisture.

Denwet CM. [Graden] Sodium dioctyl sulfosuccinate; anionic; wetting agent, emulsifier for emulsion polymerization, battery separators, glass cleaners; aids in dispersing, emulsifying, penetrating, and solubilizing; clear liq.; dens. 9.0 lb/gal; pH 7.0 ± 0.5 (2%); 75% act.

Denwet RG-7. [Graden] Sulfonated alkyl ester; anionic; wetting agent, coupling agent, solubilizer used in drycleaning, herbicides, pesticides; aids in dispersing, emulsifying, penetrating; clear liq.; dens. 9.0 lb/gal; pH 7.0 ± 0.5 (2%); 75% act.

DEP. [Hüls] Diethyl phthalate; plasticizer for cellulose acetate; carrier for act. ingreds., flavors, fragrances; clear colorless; odorless; liq.; water-insol.

Depasol AS-27. [Pulcra SA] Sodium dioctyl sulfosuccinate; anionic; dispersant for pigments; wetting and penetrating agent; used in drycleaning detergents and emulsion polymerization; rewetting agent for textile and paper industries; liq.; sol. in water, alcohol; 70 ± 2% conc.

Depasol CM-41. [Pulcra SA] Dialkyl sodium sulfosuccinate; anionic; see Depasol AS-27; liq.; 50 ± 2% conc.

Dequest® 2000. [Monsanto] Aminotrimethylene phosphonic acid; aq. process additives for paper, textiles, and metals; industrial and commercial formulations for cleaning various substrates; scale and corrosion inhibitor in aq. systems; straw-colored aq. sol'n.; 50% act.

Dequest® 2006. [Monsanto] Pentasodium amino trimethylene phosphonate; dispersant; see Dequest 2000; straw-colored aq. sol'n.; 40% act.

Dequest® 2010. [Monsanto] Hydroxy-ethylidene diphosphonic acid; see Dequest 2000; water wh. aq. sol'n.; 60% act.

Dequest® 2051. [Monsanto] Hexamethylene diamine tetra (methylene phosphonate); see Dequest 2000; wh. powd.; 98% act.

Dequest® 2054. [Monsanto] Hexapotassium hexamethylene diamine tetra (methylene phosphonate); see Dequest 2000; straw-colored aq. sol'n.; 35% act.

Dequest® 2060. [Monsanto] Diethylene triamine penta (methylene phosphonic acid); see Dequest 2000; dk. amber aq. sol'n.; 50% act.

Dequest® 2066. [Monsanto] Octasodium diethylene triamine penta (methylene phosphonate); see Dequest 2000; dk. amber aq. sol'n.; 25% act.

Deriphat 151C. [Henkel] N-coco-beta aminopropionic acid; amphoteric; wetting agent, detergent, emulsifier, corrosion inhibitor; hard surface cleaners; Gardner 5 clear liq.; sol. in strong acids, alkalies, ionic systems; sp.gr. 1.03; pH 5.5; 34% solids.

Deriphat 154. [Henkel] Disodium N-tallow-beta iminodipropionate; amphoteric; detergent, solubilizer for hard surface cleaning, textiles, emulsion polymerization; wh. powd.; sol. in strong acids, alkalies, and ionic systems; dens. 2 lb/gal; pH 11; 98% solids.

Deriphat 160. [Henkel] Disodium N-lauryl beta-iminodipropionate; amphoteric; detergent, solubilizer, primary emulsifier used in org. and inorg. compds.; emulsion polymerization and stabilization; wh. powd.; dens. 2.0 lb/gal; 98% solids.

Deriphat 160C. [Henkel] Monosodium-N-lauryl beta-iminodipropionate; amphoteric; detergent, solubilizer, stabilizer; used in petrol. processing and emulsion polymerization; amber clear liq.; sol. in strong acid, alkali, and ionic systems; sp.gr. 1.04; dens. 8.6 lb/gal; pH 7.5; 30% solids.

Deriphat BC, BCW. [Henkel] Coco dimethyl ammonium carboxylic acid betaine; amphoteric; foam stabilizer, wetting agent, frothing agent for liq. detergents; lime soap dispersant; liq.; 59 and 43% act. resp.

Dermalcare 326A. [Rhone-Poulenc Surf.] Syn. beeswax; flake; m.p. 51–54 C.

Dermalcare C-20. [Rhone-Poulenc Surf.] Ceteth-20; cosmetic grade emulsifier and emulsion stabilizer for creams, lotions, antiperspirants, creme rinse conditioners, personal care prods.; acid and alkali stable; flake; m.p. 47–52 C.

Dermalcare EGMS/SE. [Rhone-Poulenc Surf.] Glycol stearate, SE; emollient, moisturizer, lubricant, and conditioner for hair and skin care prods.; visc. builders and gelling/stiffening agents for makeup and deodorant applics.; flake; m.p. 56–62 C.

Dermalcare GMS-165. [Rhone-Poulenc Surf.] Glyceryl stearate, PEG 100 stearate; see Dermalcare EGMS/SE; flake; mp. 53–57 C.

Dermalcare GMS. [Rhone-Poulenc Surf.] Glyceryl stearate; see Dermalcare EGMS/SE; flake; m.p. 58–63 C.

Dermalcare GMS/SE. [Rhone-Poulenc Surf.] Glyceryl stearate, SE; see Dermalcare EGMS/SE; flake; m.p. 58–63 C.

Dermalcare GTIS. [Rhone-Poulenc Surf.] Triisostearin; see Dermalcare EGMS/SE; liq.

Dermalcare HL. [Rhone-Poulenc Surf.] Hexyl laurate; emollient; good spreadability and refatting chars.; lubricant in creams, lotions, makeup bases and other personal care skin prods.; liq.

Dermalcare LVL. [Rhone-Poulenc Surf.] Lauryl lactate; see Dermalcare EGMS/SE; liq.

Dermalcare MM/M. [Rhone-Poulenc Surf.] Myristyl myristate; see Dermalcare EGMS/SE; flake; m.p. 37–39 C.

Dermalcare MST. [Rhone-Poulenc Surf.] Myristyl stearate; see Dermalcare EGMS/SE; flake; m.p. 43–45 C.

Dermalcare PGMS. [Rhone-Poulenc Surf.] Propylene glycol stearate; see Dermalcare EGMS/SE; flake; m.p. 33–38 C.

Dermalcare SDG. [Rhone-Poulenc Surf.] PEG 2 stearate; see Dermalcare EGMS/SE; solid; m.p. 43–47 C.

Dermalcare SPS. [Rhone-Poulenc Surf.] Cetyl esters; see Dermalcare C-20; flake; m.p. 46–48 C.

Dermalcare SS. [Rhone-Poulenc Surf.] Stearyl stearate; see Dermalcare EGMS/SE; flake; m.p. 53-55 C.

Dermatein GSL. [Hormel] Fluid matrix containing glycosphingolipids, phospholipids, and cholesterol; skin lipid for barrier renewal and moisturization; used for night creams, after-shave balms, lip protectants; milky cream-wh. liq., char. odor; pH 4.25–4.75; 2–3% solids.

Desmodur® KA-8267. [Mobay] Aromatic water-disp. polyisocyanate; cross-linking agent that improves resistance to heat, greases, oils, plasticizers, org. solvs., and the adhesion to many substrates; primarily used in adhesives; dk. brn. liq.; dens. 1.2 g/cm^3; visc. 500 p; 100% solids; 27% NCO content.

Desmodur® MP-225. [Mobay] Modified polyisocyanate based on 4,4′-diphenylmethane diisocyanate; R.T. crosslinking agent for adhesives based on Baycoll polyols and Desmocoll PU; pale yel. liq.; sol. in esters, ketones, aromatic hydrocarbons, chlorinated solvs.; sp.gr. 1.2; visc. 800 cps; 22.5% isocyanate content.

Desmodur® N. [Bayer] Polyfunctional aliphatic isocycnate; two-part bonding agent used in coating syn. fiber fabrics with PVC pastes; esp. for coatings that must withstand lt.; liq.

Desmodur® N-75. [Mobay] Aliphatic polyisocyanate based on hexamethylene diisocyanate in butyl acetate/xylene (1:1); R.T. crosslinking agent for adhesives based on Desmocoll PU and Baycoll polyols; does not discolor on lt. exposure; provides slower cure rates and longer pot lives; pale yel. liq.; sol. in esters, ketones, chlorinated solvs., aromatic hydrocarbons; sp.gr. 1.1; visc. 175 cps; flash pt. (Seta) 31 C; 16.5% isocyanate content; 75% solids.

Desmodur® RC. [Bayer] Aromatic polyisocyanate; crosslinking agent to improve adhesion of rubber to other materials; yields strong adhesives with high thermal stability, and resistance to oils, grease, and solvs.; pale ylsh. liq.; 30% sol'n. in ethyl acetate.

Desmodur® RE. [Mobay] Triphenyl methane 4,4′,4′′-triisocyanate in ethyl acetate; R.T. crosslinking agent for adhesives based on Desmocoll and Baycoll polymers, nat. rubber, and syn. rubbers; ylsh.-grn. to dk. violet liq.; dilutable with methylene chloride, trichloroethylene, acetone, MEK, ethyl acetate, toluene; limited compat. with aliphatic hydrocarbons; sp.gr. 1.0; visc. 3 cps; flash pt. (CC) –3 C; 9.3% NCO content; 27% solids.

Desmodur® RFE. [Mobay] Tris (p-isocyanatophenyl)-thiophosphate in ethyl acetate; R.T. crosslinking agent for adhesives based on Desmocoll and Baycoll polymers, natural and syn. rubbers; ylsh.-brn. liq.; dilutable with methylene chloride, trichloroethylene, acetone, MEK, ethyl acetate, toluene; limited compat. with aliphatic hydrocarbons; sp.gr. 1.0; visc. 3 cps; flash pt. (CC) –3 C; 7.2% NCO content; 27% solids.

Desmodur® RU. [Bayer] Aromatic polyisocyanate; see Desmodur RC; brnsh. liq.; 20% sol'n. in methylene chloride.

Desmodur® TT. [Bayer] Aromatic diisocyanate; two-part bonding agent used in coating syn. fiber fabrics with PVC pastes; highly reactive at R.T.; powd.

Desmodur® VKS-2, VKS-4, VKS-18. [Mobay] Polymethylene polyphenyl isocyanate; R.T. crosslinking agent for adhesives based on Desmocoll and Baycoll polymers, natural and syn. rubbers; dk. brn. liq.; sol. in methylene chloride, trichloroethylene, acetone, MEK, ethyl acetate, toluene; insol. in aliphatic hydrocarbons; sp.gr. 1.2; visc. 20, 40, and 180 cps resp.; 33%, 32.5%, and 31% NCO content resp.

Desmodur® Z-4370. [Mobay] Aliphatic polyisocyanate based on isophorone diisocyanate in propylene glycol methyl ether acetate/xylene (1:1); R.T. crosslinking agent for adhesives based on Baycoll polyol, Desmocoll PU, and rubber; offers exc. lt. stability, slower cure rates, and longer pot lives; pale yel. liq.; sol. in esters, ketones, chlorinated solvs., aromatic hydrocarbons; sp.gr. 1.1; visc. 2000 cps; flash pt. (Seta) 39 C; 11.5% isocyanate content; 70% solids.

Desmorapid LA. [Mobay] Cyclic nitrogen compd. in ethyl acetate; catalyst for the crosslinking reaction of polyisocyanates with polyols, PU, natural rubber, polychloroprene and nitrile rubber-based solv. adhesives; clear pale yel. liq.; dilutable with ethyl acetate, acetone, MEK, toluene, aliphatic hydrocarbons, chlorinated solvs.; sp.gr. 0.9; flash pt. (TCC) –2 C; 19% solids.

Desmorapid PP. [Mobay] Tert. amine; catalyst accelerating the reaction of polyisocyanates with polyols and PU; catalyst for 2-component adhesive systems; pale yel. liq.; sol. in aromatic hydrocarbons, esters, ketones; limited sol. in aliphatics; sp.gr. 0.9; flash pt. (TCC) 70 C.

DeSomeen TA-2. [DeSoto] PEG-2 tallow amine; emulsifier, dispersant, textile scouring, dyeing assistant, desizing assistant, softener, antistat; paste; 99% act.

DeSomeen TA-5. [DeSoto] PEG-5 tallow amine; emulsifier and dispersant; used as textile scouring agents, dyeing assistants, desizing agents, softening agents, antistats, etc.; liq.; HLB 5.3; cloud pt. 68–74 F; 100% act.

DeSomeen TA-15. [DeSoto] PEG-15 tallow amine; see DeSomeen TA-2; liq.; cloud pt. 172–179 F; 100% act.

DeSomeen TA-20. [DeSoto] PEG-20 tallow amine; see DeSomeen TA-2; liq.; cloud pt. 179–181 F (10% NaCl); 100% act.

DeSonate SA. [DeSoto] Linear dodecylbenzene sulfonic acid; detergent intermediate, emulsifier, base for liquid and powd. detergents; liq.; 97% act.

DeSonate SA-H. [DeSoto] Branched dodecylbenzene sulfonic acid; chemical intermediate, emulsifier; liq.; 97% act.

DeSonic® 1.5N. [DeSoto] Nonoxynol-1 (1.5 EO); defoamer, detergent, emulsifer; liq.; oil-sol.; 100% act.

DeSonic® 3K. [DeSoto] Linear alcohol ethoxylate (3 EO); nonionic; detergent, emulsifier, wetting agent, defoamer for industrial detergents, textile scouring; liq.; oil-sol.; 100% act.

DeSonic® 4N. [DeSoto] Nonoxynol-4; nonionic; detergent, emulsifier, defoamer for pesticide, paint, paper, and textile industries; liq.; oil-sol.; HLB 8.8; 100% act.

DeSonic® 6C. [DeSoto] PEG-6 castor oil; nonionic; emulsifier, lubricant, dye leveler, antistat, and dispersant for textile applics.; mfg. of PU foams; softening and rewetting agents for paper; liq.; HLB 4.4; 100% conc.

DeSonic® 12-3. [DeSoto] PEG-3 linear alcohol; emulsifier, dispersant, surfactant; detergent intermediate for mfg. of ethoxysulfates for specialty industrial and dishwashing applics.; liq.; oil-sol.; HLB 8.0; 100% act.

DeSonic® 13N. [DeSoto] Nonoxynol-13; nonionic; see DeSonic 4N; liq.; HLB 14.3; 100% conc.

DeSonic® 30C. [DeSoto] PEG-30 castor oil; nonionic; emulsifier, lubricant, dye levelers, antistat, dispersant for textiles; emulsifer for PU foams; softener/rewetter for wet-strength paper; liq.; HLB 11.7; 100% act.

DeSonic® 36C. [DeSoto] PEG-36 castor oil; nonionic; see DeSonic 30C; liq.; HLB 12.6; cloud pt. 122–140 F; 100% act.

DeSonic® 40C. [DeSoto] PEG-40 castor oil; nonionic; see DeSonic 30C; liq.; HLB 13.1; cloud pt. 173–179 F; 100% act.

DeSonic® 54C. [DeSoto] PEG-54 castor oil; nonionic; see DeSonic 30C; liq.; HLB 14.4; cloud pt. 136–142 F (10% NaCl); 100% act.

DeSonic® 81-2. [DeSoto] PEG-2 linear alcohol; detergent intermediate, emulsifier, dispersant, defoamer, surfactant; liq.; oil-sol.; HLB 8.0; 100% act.

DeSonic® 315-3. [DeSoto] PEG-3 linear alcohol; detergent, emulsifier, wetting, surfactant, defoamer, intermediate; liq.; oil-sol.; HLB 8.0; 100% act.

DeSonic® S-405. [DeSoto] Octoxynol-40; coemulsifier for vinyl and acrylic polymerization; dye assistant; liq.; cloud pt. 165–176 F (10% NaCl); 70% act.

DeSonic® SMO. [DeSoto] Sorbitan monooleate; nonionic; emulsifier, fiber lubricant and softener; liq.; sol. in oils and org. solvs.; insol. in water; HLB 4.3; 100% conc.

DeSonic® SMT. [DeSoto] Sorbitan monotallate; nonionic; see DeSonic SMO; liq.; insol. in water; sol. in oils and org. solvs.; HLB 4.1; 100% conc.

DeSonic® TA-2, -15. [DeSoto] Tallow amine, ethoxylated; cationic; emulsifier, dispersant, textile scouring agent, desizing assistant, softener, and antistat; 80-100% conc.

DeSonic® TA-25CWS. [DeSoto] Tallow amine, ethoxylated; cationic; see DeSonic TA-2; liq.; 80% conc.

DeSophos 4 CP. [DeSoto] Phosphate ester; softener, antistat for textile finishing; in lubricants for filament yarns, syn. fibers, wool; emulsifier for cosmetic oils and creams, polymerization of latexes; liq.; 95% act.

DeSophos 5 AP. [DeSoto] Phosphate ester; coupling agent for nonionics in liquid alkaline detergents; moderate foamer; liq.; 100% act.

DeSophos 5 BMP. [DeSoto] Phosphate ester; coupling agent for industrial uses; liq.; 100% act.

DeSophos 5 BP. [DeSoto] Phosphate ester; coupling agent for agric. and industrial uses; liq.; sol. in water, nonpolar aliphatic solv., oils; 100% act.

DeSophos 6 DNP. [DeSoto] Phosphate ester; paraffinic oil emulsifier as TEA salt; rust inhibitor; most oil-sol. in DeSophos DNP series; liq.; 100% act.

DeSophos 6 DP. [DeSoto] Phosphate ester; dry cleaning detergent, penetrant, antistat; liq.; 100% act.

DeSophos 6 MPNa. [DeSoto] Phosphate ester; hard-surface detergent, corrosion retarder, moderate foamer in detergent concentrates; dry cleaning detergent; textile antistat; liq.; 88% act.

DeSophos 6 NPNa. [DeSoto] Phosphate ester; imparts hard surf. detergency, corrosion retardation, and moderate foaming to detergents; dry cleaning detergent; antistat for textiles; liq.; 88% act.

DeSophos 7 DP. [DeSoto] Phosphate ester; alkali-stable emulsifier, detergent, wetting agent, dispersant; rewetter in textile applics.; liq.; 100% act.

DeSophos 7 OPNa. [DeSoto] Phosphate ester; dispersible surfactant in emulsification of min. oils; textile softener and antistat; in lubricants for syn. and wool fibers; liq.; sol. in aromatic solv.; 98% act.

DeSophos 30 NP. [DeSoto] Phosphate ester; emulsifer, stabilizer; in preparation of PVAc and acrylic copolymers; liq.; 100% act.

DeSotan SMO. [DeSoto] Sorbitan monooleate; lipophilic emulsifier, fiber lubricant, softener; liq.; sol. oils, org. solv.; insol. water; HLB 4.3; 100% act.

DeSotan SMT. [DeSoto] Sorbitan monotallate; lipophilic emulsifier, fiber lubricant, softener; liq.; sol. oils, org. solv.; insol. water; HLB 4.1; 100% act.

Detergent CR. [Arol] Fatty ethanolamine condensate; detergent, wetting agent, emulsifier, thickener, penetrating and leveling agent, for use in textiles, personal care, food, and household prods.; pigment dispersant; pale amber liq.; mild odor; readily sol. in water; sp.gr. 1.01; 100% act.

Detergyl S. [ICI Ltd.] 1:2 Coco DEA; nonionic; detergent; scouring assistant for wool and cotton; penetrating assistant in melange printing; liq.; 100% conc.

Dex-Lo®. [Int'l. Bio-Synthetics] Bacterial alpha amylase; enzyme; liquefaction of starch; liq.

Dextranase Novo 25 L. [Novo] Fungal dextranase derived from Penicillium lilacinum; enzyme used in sugar industry to break down dextran in raw sugar juice; brn. liq.; slight smell typ. of fermentation prods.; water-sol.; dens. 1.25 g/ml; pH 5–7.

Dextrol OC-15. [Dexter] Complex org. phosphate ester free acid; anionic; dispersant for magnetic oxide in aromatic solvs.; liq.; 100% act.

Dextrol OC-20. [Dexter] Complex org. phosphate ester free acid; anionic; see Dextrol OC-15; liq.; 100% act.

Dextrol OC-50. [Dexter] Complex org. phosphate ester, partially sodium neutralized; anionic; detergent formulations; dispersant for paint formulations; liq.; sol. in water and solvs.; 88% conc.

Dextrol OC-70. [Dexter] Complex org. phosphate ester free acid; anionic; see Dextrol OC-15; liq.; 100% act.

Dextrozyme. [Novo] Pullulanase/amloglucosidase; enzyme; in starch industry for saccharification of dextrose and iso-syrup prod.; prod. of high conversion syrups; liq.

DF 1040. [GE] Reactive silicone fluid; forms water-repellent fluids with heat or heat/catalyst; used in textiles, particle treatment, magnesium oxide, and Calrod® units; fluid; sp.gr. 1.00; visc. 20 cstk; ref. index 1.3952; flash pt. 250 F.

Diacid® 1550. [Westvaco] Acrylinoleic acid; chemical intermediate forming high solids, low visc. soaps, esters, polyamide derivs.; as surfactant, coupling agent; hydrotrope esp. in caustic systems; Gardner 9 max. liq.; dens. 8.45 lb/gal; visc. 5000 cps (100 F); acid no. 265–277; flash pt. (COC) 455 F; biodeg.

Diamiet 503, 508, 520, AB, C. [KAO S.A.] n-Propylene alkyl fatty diamine EO condensation product; cationic; emulsifier, dispersant, corrosion inhibitor, wetting agent; liq., paste, solid; 100% conc.

Diamin DO. [KAO S.A.] Diamine dioleate salts; cationic; dispersant for oil flushing agents for pigments, oil/wax emulsifier; liq./paste; 100% conc.

Diamin HT. [KAO S.A.] n-Propylene hydrog. tallow fatty diamine; cationic; asphalt emulsifier, corrosion inhibitor, antistripping agent; solid; 100% conc.

Diamin O. [KAO S.A.] n-Propylene oleyl fatty diamine; cationic; see Diamin HT; liq. or paste; 100% conc.

Diamin S. [KAO S.A.] n-Propylene soya fatty diamine; cationic; see Diamin HT; liq., paste; 100% conc.

Diamin T. [KAO S.A.] n-Propylene tallow fatty diamine; cationic; see Diamin HT; paste; 100% conc.

Diamine B11. [Berol Nobel] Alkyl-1,3-propylene diamine; cationic; emulsifier, corrosion inhibitor; solid; 95% conc.

Diamine BG. [Berol Nobel] Tallow-1,3-propylene diamine; cationic; emulsifier, corrosion inhibitor; paste; 95% conc.

Diamine KKP. [Berol Nobel] Coco-1,3-propylene diamine; cationic; emulsifier, corrosion inhibitor; liq.; 95% conc.

Diamond Quality. [CasChem] Castor oil; emollient for color and odor; high visc.; low pour pt.

Dianol. [Dai-ichi Kogyo] Fatty acid DEA; nonionic; emulsifier, dispersant for pigments, wax, solvs.; detergent, solubilizer; liq.; 100% conc.

Dianol 300. [Dai-ichi Kogyo] Fatty acid DEA; nonionic; detergent, solubilizer, emulsifier, and dispersant solvs.; liq.; 100% conc.

Diazital O Extra Conc. [Henkel] Fatty alcohol polyglycol ethers; nonionic; dyeing assistant; solid; 100% conc.

Diazopon SS-837. [GAF] POE alkylphenol; nonionic; dispersant, solubilizer, stabilizer, anticrock and soaping agent for naphthol dyeings; used in textile dyeing processes; liq.; pH neutral; 70% conc.

Diazyme. [Miles Lab] Glucoamylase; enzyme; hydrolysis of starch dextrins to glucose; liq.

DIBA. [Croda Ltd.] Diisobutyl adipate; low temp. plasticizer for plastics and syn. rubbers; liq.

Dibactol. [Hexcel] Myristalkonium chloride; biocide used in mfg., formulating, tableting, and dry blending; wh. powd.; m.w. 404.06; pH 5–8; 99–103% assay.

Dichan 100. [Olin] Dicyclohexylamine nitrite; vapor-phase corrosion inhibitor for ferrous metals; designed for items enclosed in pkg., e.g., hot water heating systems, nuclear reactor heat-exchange units, gas recovery systems, jet aircraft engine compressors, internal combustion engines, welding electrodes, double-walled pipes; wh. powd.; m.w. 228.34; sol. (g/100 g sol'n.) 23 g in methanol, 9 g in ethanol, 2 g in IPA; 4 g/100 g water; dens. 0.406 g/cm^3; m.p. 195–198 C; vapor pressure 3.3 mm Hg; 98% min. dicyclohexylamine nitrite.

Di-Cup 40C, 40KE. [Hercules] Dicumyl peroxide supported on precipitated calcium carbonate and Burgess KE clay resp.; vulcanizing agent, and high-temp. polymerization catalyst for rubber and plastics; can be used to cross-link a wide variety of polymers; can be emulsified for applics. involving aq. systems; off-wh. free-flowing powds.; m.w. 270; sp.gr. 1.57 and 1.55 resp.; 39.5–41.5% act.

Di-Cup R, T. [Hercules] Dicumyl peroxide; see Di-Cup 40C; pale yel. to wh. granular solid and pale yel., fused, semicryst. solid resp.; m.w. 270; sol. or disp. in natural and syn. rubber compds., silicone gums, and polyester resins, veg. oils; water-insol.; sp.gr. 1.018 and 1.023 resp.; m.p. slowly melt at 38 C and 30 C resp.; 98–100% and 92–95% act. resp.

Dicyanex® 200X. [Pacific Anchor] Dicyandiamide; epoxy curing agent for printed circuit board laminates; 200 mesh powd.; dens. 4.0 lb/gal; m.p. 410 F.

Dicyanex® 325. [Pacific Anchor] Dicyanamide with 1–3% inert flow control additive; epoxy curing agent for prepregs, composites; 325 mesh powd.; dens. 3.5 lb/gal; m.p. 410 F.

Dicyanex® 1200. [Pacific Anchor] Dicyandiamide with 1–3% inert flow control additive; epoxy curing agent for prepregs, composites; 10 μ powd.; dens. 2.5 lb/gal; m.p. 410 F.

Diglycolamine® Agent (DGA®). [Texaco] 2-(2-aminoethoxy) ethanol; solv. for removal of CO_2 or H_2S from gases, for recovery of aromatics from refinery streams, for prep. of foam stabilizers, wetting agents, emulsifiers, and condensation polymers; clear colorless slightly visc. liq.; mild amine odor; m.w. 105.14; misc. with water, alcohols, aromatic hydrocarbons, ethyl ether; sp.gr. 1.0572; dens. 8.7 lb/gal; b.p. 221 C; f.p. –12.5 C; flash pt. 255 F (PMCC); ref. index 1.4598; 98% min. act.

Diglyme. [Ferro] Diethylene glycol dimethyl ether; solv. which tends to solvate cations; used in electrochemistry, polymer, and boron chemistry; physical processes such as gas absorption, extraction, stabilization; used in industrial prods. such as fuels, lubricants, textiles, pharmaceuticals, pesticides; colorless clear; ethereal, nonresidual odor; m.w. 134.17; f.p. –64 C; b.p. 162 C (760 mm Hg); water-sol.; sp.gr. 0.9451; dens. 7.88 lb/gal; visc. 2.0 cP; ref. index 1.4078; pH neutral; flash pt. 57 (CC); 99.6% min. purity.

Dilasoft RS. [Sandoz Ltd.] Fatty acid deriv.; amphoteric; hydrophilic softener for cellulosic fibers; liq.; 100% conc.

Dilasoft RW. [Sandoz Ltd.] Fatty acid deriv. modified; cationic/nonionic; softening agent with wetting properties for cellulosic fibers; liq.

Dilasoft TF. [Sandoz Ltd.] Sulfonate; softening/filling agent, anionic; liq.

Diluex. [Floridin] Attapulgite clay; absorbent carrier for pesticides in dusts and wettable powds.; fine powd.

Dimilin®. [Duphar] Diflubenzuron; insecticide interfering with chitin deposition; used for forestry ornamentals, fruit, field crops, horticulture, etc.; wh. cryst. solid; m.w. 310.7; sol. 120 g/l dimethylsulfoxide, 104 g/l dimethylformamide, 24 g/l dioxane, 6.5 g/l acetone; m.p. 210–230 C (tech.), 230–232 C (pure).

Dinoram C. [Ceca SA] N-coco propylene diamine; cationic; chemical intermediate; paste.

Dinoram O. [Ceca SA] N-oleyl propylene diamine; cationic; chemical intermediate; paste.

Dinoram SH. [Ceca SA] N-hydrog. tallow propylene; cationic; chemical intermediate; paste.

Dinoramac C. [Ceca SA] N-coco propylene diamine acetate; cationic; flotation, bactericide, emulsifier, anticaking, soil stabilization, flocculation, corrosion inhibitor; paste.

Dinoramac O. [Ceca SA] N-oleyl propylene diamine acetate; cationic; see Dinoramac C; paste.

Dinoramac S. [Ceca SA] N-tallow propylene diamine acetate; cationic; see Dinoramac C; paste.

Dinoramox S 3 to 12. [Ceca SA] POE N-tallow propylene diamine; cationic; dispersant, wetting agent, emulsifier, corrosion inhibitor used in paint, agric., chemical, and textile industries; liq.; 100% conc.

Dinoramox SH 3. [Ceca SA] PEG-3 N-hydrog. tallow propylene diamine; cationic; emulsifier, dispersant, corrosion inhibitor for use in textile and detergent industries; solid; 100% conc.

Dional 11, 113. [Hoechst-Celanese] Fluorinated hydrocarbon; solv. for dry cleaning.

Dioxitol-High Gravity. [Shell] Diethylene glycol monoethyl ether; solv. for natural and syn. resins, dyes, fats used in automotive industry as component of heavy duty brake fluids; in paint industry as solv. in nonaq. strains, and nongrain raising stains for furniture and wood uses; yarn and cloth conditioner in textile industry; dye solv. where deep penetration and bright shades are obtained; component of industrial cleaners; coupling agent; mfg. of esters; intermediate in mfg. of plasticizers, solv.; colorless liq.; water-misc.; sp.gr. 1.021–1.027; 72% Dioxitol-Low Gravity, 28% ethylene glycol.

Dioxitol-Low Gravity. [Shell] Diethylene glycol monoethyl ether; see Dioxitol-High Gravity; colorless liq.; water-misc.; sp.gr. 0.986–0.990.

Dipsal. [Scher] Dipropylene glycol salicylate; uv absorbent; used in sunscreen base; suitable as inhibitor for uv degradation of polymers and dyestuffs; does not deteriorate in contact with perspiration; used in toiletries, alcohol lotions, veg. or min.-type prods., and pharmaceutical specialties; suitable for hair applics.; useful to reduce deterioration and discoloration of polymers; yel. clear liq.; mild salicylate odor; m.w. 254; sol. in most org. solvs.; water-insol.; sp.gr. 1.165; dens. 9.66 lb/gal; sapon. no. 225–240; ref. index 1.522; flash pt. > 160 C (OC); cloud pt. < –5 C.

Dipterex®. [Bayer] Trichlorfon; insecticide with stomach and contact poison action; wh. cryst. powd.; m.w. 257.5; readily sol. in strong polar org. solvs., e.g., low alcohols, acetone; sol. 15.4 g/100 cc water; sp.gr. 1.73 (20/4 C); b.p. 100 C (0.1 mm Hg); m.p. 83–84 C.

Direx® 4L. [Griffin] Diuron suspension; flowable herbicide for control of many weeds and grasses in a variety of crops; 40% act.

Disflamoll DPK. [Mobay] Diphenylcresyl phosphate; flame retardant plasticizer for plasticized PVC prods.; used in, air ducts, tarpaulins, driving and conveyor belts, imitation leather, coatings, hoses and extruded goods, cable sheathing and insulation, soles and inj. molded items; colorless liq.; b.p. 225–235 C; sol. in solvs.; insol. in water; dens. 1.20–1.21 g/cm^3; visc. 45–50 mPa•s; acid no. < 0.2; ref. index 1.562–1.564; flash pt. > 230 C (OC).

Disflamoll DPO. [Mobay] Diphenyloctyl phosphate;

flame retardant plasticizer for typ. PVC applics., dip, rotationally, extruded and inj. molded parts, mechanical foam; colorless liq.; b.p. 225 C; sol. in solvs.; insol. in water; dens. 1.08–1.09 g/cm³; visc. 21–23 mPa.s; acid no. < 0.05; ref. index 1.508–1.511; flash pt. > 190 C.

Disflamoll TCA. [Mobay] Trichloroethyl phosphate; flame retardant used to improve fire performance in PVC; weak plasticizing action; tarpaulins, air ducts, film; clear liq.; b.p. 180 C; sol. in solvs.; partially sol. in water; dens. 1.42 g/cm³; visc. 45 mPa.s; acid no. < 0.05; ref. index 1.473; flash pt. > 220 C (OC).

Disflamoll TKP. [Mobay] Tricresyl phosphate; see Disflamoll DPK; colorless to slightly ylsh. liq.; b.p. 240–250 C; sol. in org. solvs.; insol. in water; dens. 1.175–1.185 g/cm³; visc. 65–75 mPa•s; acid no. < 0.2; ref. index 1.558–1.560; flash pt. > 230 C (OC).

Disflamoll TP. [Mobay] Triphenyl phosphate; flame retardant; gelatinizing and plasticizing agent for collodion cotton; plasticizer w/o gelatinizing properties for acetyl cellulose; reduces flam. of NC and acetyl cellulose-based plastic compds. and lacquer films; mfg. of photographic film materials, surf. coatings; wh. crystals; odorless; b.p. 250 C; sol. in org. solvs.; insol. in water; dens. 1.205 g/cm³; dens. 0.65 kg/l; visc. 11 cps; acid no. 0.05 max.; ref. index 1.555 (50 C); flash pt. 235 C (OC).

Dispatex G. [Nikko] Aromatic condensated compd.; anionic; dispersing agent for polyester dyeing; liq.; 26% conc.

Dispersal 130. [Aquatec Quimica] Polyacrylate ammonium salt; dispersant for inorg. materials; liq.

Dispersal 140. [Aquatec Quimica] Sodium polyacrylate; dispersant for inorg. materials; liq.

Dispersogen ASN. [Hoechst-Celanese] Higher alcohol, ethoxylated; dispersant for polyester dyeing; liq.; 26% conc.

Dispersol T. [ICI Ltd] Methylene bis (naphthalene sulfonic acid) disodium salt; general purpose dispersant; scale modifier for evaporation; powd.

Disrol SH. [Nippon Nyukazai] Na bis(naphthalene) sulfonate; dispersant and suspending agent; flake.

Dissolvine® A 40. [Akzo] NTA-Na_3; chelating agent; liq.; 37% act.

Dissolvine® A 92. [Akzo] NTA-Na_3; chelating agent; powd.; 92% act.

Dissolvine® A Z. [Akzo] NTA-H_3; chelating agent; powd.; 98% act.

Dissolvine® AM 2. [Akzo] EDTA-$(NH_4)_2H_2$; chelating agent; liq.; water-misc.; 40% act.

Dissolvine® AM 3. [Akzo] EDTA-$(NH_4)_3H$; chelating agent; liq.; water-misc.; 40% act.

Dissolvine® AM 4. [Akzo] EDTA-$(NH_4)_4$; chelating agent; liq.; water-misc.; 50% act.

Dissolvine® AMFE. [Akzo] EDTA-NH_4Fe NH_4OH; chelating agent; liq.; water-misc.; 52% act.

Dissolvine® CA. [Akzo] EDTA-Na_2Ca-$2H_2O$; chelating agent; powd.; 98% act.

Dissolvine® D 40. [Akzo] DTPA-Na_5; chelating agent; liq.; water-misc.; 42% act.

Dissolvine® D 88. [Akzo] DTPA-Na_5; chelating agent; powd.; 88% act.

Dissolvine® D FE. [Akzo] DTPA-Na FeH; chelating agent; powd.; 92% act.

Dissolvine® DZ. [Akzo] DTPA-H_5; chelating agent; powd.; 98% act.

Dissolvine® E 39. [Akzo] EDTA-Na_1; chelating agent; liq.; water-misc.; 39% act.

Dissolvine® H 40. [Akzo] HEDTA-Na_3; chelating agent; liq.; water-misc.; 42% act.

Dissolvine® H 84. [Akzo] HEDTA-Na_3; chelating agent; powd.; 84% act.

Dissolvine® K 3. [Akzo] EDTA KH; chelating agent; liq.; water-misc.; 50% act.

Dissolvine® MG 2. [Akzo] EDTA-Mg_2-$9H_20$; chelating agent; powd.; 98% act.

Dissolvine® NA. [Akzo] EDTA-Na4; chelating agent; powd.; 86% act.

Dissolvine® NA 2. [Akzo] EDTA-Na_2H_2-$2H_2O$; chelating agent; powd.; 99% act.

Dissolvine® NAFE. [Akzo] EDTA NaFe $3H_2O$; chelating agent; powd.; 99% act.

Dissolvine® Z. [Akzo] EDTA-H_4; chelating agent; powd.; 99% act.

Dist. Lipolan. [Lipo] Hydrog. lanolin; emollient, lubricant, and conditioner used in cosmetics, toiletries, and topical pharmaceuticals; wh./off-wh. paste; mild char. odor; insol. in water; m.p. 34–43 C; acid no. 1 max.; sapon. no. 3 max.

Disyston®. [Bayer] Disulfoton; systemic insecticide and acaricide; colorless oily liq.; m.w. 274.4; sol. in most org. solvs. (acetone, alcohol, ether); 1:40,000 in water; sp.gr. 1.14 (20/4 C); b.p. 128 C (1 mm Hg).

Dixie® Clay. [Vanderbilt] Kaolin, hydrated aluminum silicate; reinforcing min. filler and extender for polyester, styrene, epoxies, phenol formaldehyde resins; better flexural modulus and improved flow chars.; wh. to cream powd., 99.8% min. thru 325 mesh; dens. 2.62 mg/m³; oil absorp. 42; 44.87% SiO_2, 33.77% Al_2O_3.

Dizene. [PPG Industries] Orthodichlorobenzene; aromatic used in industrial plants, stockyard pens, restaurants, hotels, hospitals, camps, parks, land-fill areas, cesspools, septic tanks, grease traps, dry wells, and soil; solv. for oil and grease; ylsh. clear liq.; strong pleasant odor; dens. 10.63 lb/gal; flash pt. 150 F (TCC); 57% min. conc.

DK Ester F-10, -20, -50, -70, -90, -110, -140, -160. [Dai-ichi Kogyo] Sucrose fatty acid ester; nonionic; food additive for stabilizing emulsion, dispersant, solubilizer; powd.; HLB 1.0, 2.0, 6.0, 8.0, 9.5, 11.0, 13.0, 15.0 resp.; 100% conc.

DMP. [Hüls] Dimethyl phthalate; plasticizer for film and molding grades of cellulose acetate; clear colorless liq.

DN 25L®. [Novo] Dextranase; enzyme; reduction of dextran content in sugar juices; liq.

DOA. [Monsanto] Dioctyl adipate; plasticizer for PVC and syn. rubber compositions; low temp. flexibility.

Dobanol 23. [Shell] C_{12}–C_{13} linear primary alcohol; nonionic; base material for preparation of alcohol ethoxylates, sulfates, ethoxysulfates; liq.; mild odor; dens. 0.831 kg/l; visc. 14 cs (40 C); m.p. 17–19 C;

flash pt. 132 C; > 95% biodeg.; 100% act.

Dobanol 23-2. [Shell] C_{12}–C_{13} linear primary alcohol ethoxylate; nonionic; detergent, intermediate for prod. of laundry powds.; clear liq.; dens. 0.91 kg/l; visc. 15 cs (40 C); m.p. 4–5 C; HLB 6.2; flash pt. 148 C; 100% act.

Dobanol 23-3. [Shell] C_{12}-C_{13} linear primary alcohol ethoxylate; nonionic; detergent intermediate for cosmetic, specialty and dishwashing formulations; liq.; HLB 8.1; 100% conc.

Dobanol 23-6.5. [Shell] C_{12}–C_{13} linear primary alcohol ethoxylate (6.5 EO); nonionic; see Dobanol 23-2; wh. slurry; dens. 0.96 kg/l; visc. 26 cs (40 C); m.p. 17–19 C; HLB 11.9; flash pt. 172 C; 100% act.

Dobanol 25. [Shell] C_{12}–C_{15} linear primary alcohol; base material for preparation of alcohol ethoxylates, sulfates, and ethoxysulfates; liq.; mild odor; dens. 0.831 kg/l; visc. 14 cs (40 C); m.p. 20–22 C; biodeg.; 100% act.

Dobanol 25-3. [Shell] Pareth-25-3; nonionic; detergent, intermediate for mfg. of cosmetic, specialty industrial, and dishwashing formulations; clear liq.; dens. 0.93 kg/l; visc. 18 cs (40 C); m.p. 5–6 C; HLB 7.8; flash pt. 162 C; > 95% biodeg.; 100% act.

Dobanol 45. [Shell] C_{14}–C_{15} linear primary alcohol; detergent, intermediate, base material; liq.; mild odor; dens. 0.824 kg/l (40 C); visc. 17 cs (40 C); m.p. 29–31 C; flash pt. 152 C; biodeg.; 100% act.

Dobanol 91. [Shell] C_9–C_{11} linear primary alcohol; base material for preparation of alcohol ethoxylates, sulfates, ethoxysulfates; dens. 0.832 kg/l; visc. 9 cs (40 C); m.p. –10 to –8 C; flash pt. 107 C (PMCC); biodeg.; 100% act.

Dobanol 91-2.5. [Shell] Pareth-91-3; nonionic; detergent intermediate; clear liq.; dens. 0.93 kg/l; visc. 11 cs (40 C); HLB 8.1; flash pt. 120 C (PMCC); pH 6.8 (0.5% aq.); 95% biodeg.; 100% act.

Dobanol 235. [Shell] A blend of primary alcohols; base material for preparation of alcohol ethoxylates, sulfates, ethoxysulfates; dens. 0.831 kg/l; visc. 14 cs (40 C); m.p. 18–20 C; flash pt. 134 C (PMCC); biodeg.

Dobanol 235-3. [Shell] Primary alcohol ethoxylate; chemical intermediate; clear liq.; dens. 0.92 kg/l; visc. 17 cs (40 C); m.p. 4–5 C; HLB 7.8; flash pt. 160 C (PMCC); > 95% biodeg.; 100% act.

Dodicor Brands. [Hoechst-Celanese] Fatty amine derivs.; corrosion inhibitor for oil and gas industry; liq.

Dodigen 226, 1611. [Hoechst-Celanese] Quat. ammonium salt; antimicrobial used in swimming pools; liq.; water-sol.

Dodigen Brands. [Hoechst-Celanese] Nitrogen containing compds.; disinfectant; corrosion inhibitor for oil and gas industry; liq., paste.

Dodilube Brands. [Hoechst-Celanese] Nitrogen containing condensation prod.; lubricant for water-based drilling muds.; liq.

Doittol 14. [Henkel] Sulfate fatty acid ester; anionic; wetting agent and leveling assistant for dyeing and bleaching; liq.; 45% conc.

Doittol 891. [Henkel] Sodium fatty acid soap, modified; anionic; surfactant; rubber antiblocking agent; liq.; 20% act.

Doittol APS Conc. [Henkel] Alkylaryl polyglycol ether, semiester; anionic; dyeing assistant for dyeing polyester and blends; liq.; 60% conc.

Doittol K21. [Henkel] Fatty acids and alkylpolyglycol ethers; anitblocking and mold release agent; liq.

Doittol P690. [Henkel] Complex org. phosphate ester; amphoteric; texile fiber antistat; liq.; 100% conc.

Double Soft.® [Am. Ingredients] Hydrated emulsion of high-purity, molecularly distilled monoglyceride; crumb softener in yeast-raised prods.; ivory semi-solid; sapon. no. 66–78; pH 3.6–4.2 (2%); 40% min. alpha monoester.

Dousoft BK 5078. [Clough] Imidazoline deriv.; cationic; non-yel. softener conc. for textiles, paper, and liq. fabric softeners; liq.; 75% min.

Dovanox 23H, 23M, 25N, 231. [Lion] POE syn. alcohol; nonionic; detergent, penetrant, emulsifier, dispersant; liq., wax; 100% conc.

Dowanol® DB. [Dow] Butoxydiglycol; solv.; sol. in water and many org. liqs.

Dowanol® DE. [Dow] Ethoxydiglycol; solv. sol. in water and many org. liqs.

Dowanol® DPM. [Dow] PPG-2 methyl ether; solv.; coupler for detergents, paints, and printing inks; sol. in water and many org. liqs.

Dowanol® DPnB. [Dow] Dipropyleneglycol n-butyl ether; coupler for disp. paints; liq. cleaners; liq.; b.p. 22 C.

Dowanol® EB. [Dow] Butoxyethanol; solv.; sol. in water and many org. liqs.

Dowanol® EPH. [Dow] Phenoxyethanol; solv.; sol. in water and many org. liqs.

Dowanol® PM. [Dow] Methoxpropanol; solv.; sol. in water and many org. liqs.

Dowanol® PnB. [Dow] Propylene glycol n-butyl ether; coupler in liq. cleaners, paints, and printing inks; liq.; b.p. 170.

Dowanol® TPnB. [Dow] Tripropylene glycol n-butyl ether; coupler for disp. paints; cleaners; floor waxes; liq.; b.p. 274 C.

Dow Antimicrobial 7287, 8536. [Dow] 2,2-Dibromo-3-nitrilopropionamide; antimicrobial for control of bacteria, fungi, and yeasts in paper mills, aq. metalworking fluids, enhanced oil recovery systems, industrial recirculating water cooling towers and water systems, air washer systems; slimicide in paper mfg.; clear to amber liq.; mild antiseptic odor; misc. with water; sp.gr. 1.24–1.26 and 1.14–1.17 g/ml resp.; b.p. > 120 C; pour pt. –45 C; 20 and 5% act. resp.

Dowclene® EC Solvent. [Dow] Electrical cleaning solv.

Dowclene® LS. [Dow] Trichloroethane; solv. for leather and suede cleaning; sol. in most org. solvs.; very low water sol.

Dow Corning® 1-2531 Release Coating. [Dow Corning] Silicone resin coating; provides durable release coating for bakers' pans, waffle irons; FDA approved for food contact use.

Dow Corning® 7 Compound. [Dow Corning] Silicone compd.; lubricant, release agent for mold break-in; preservative and lubricant for rubber.

Dow Corning® 24 Emulsion. [Dow Corning] Silicone emulsion; nonionic; food grade lubricant for mfg. of paper/paperboard in contact with food; FDA approved; water-dilutable.

Dow Corning® 200 Fluid. [Dow Corning] Dimethicone; foam control agent for distillation, resin mfg., asphalt, oil refining, gas-oil separation; clear fluid; visc. 1000–10,000 cSt and 12,500–60,000 cSt grades; 100% act.

Dow Corning® 200 Fluid, Food Grade. [Dow Corning] Dimethyl silicone fluid; foam control agent for food industry, inks; thin clear fluid; visc. 350 cSt; 100% act.

Dow Corning® 203 Fluid. [Dow Corning] Silicone fluid; release film providing internal release and lubrication.

Dow Corning® 233A Fluid. [Dow Corning] Silicone fluid; release agent for thermoplastics and thermosets; heat-stable, noncarbonizing, nonoily parting film.

Dow Corning® 236 Dispersion. [Dow Corning] Silicone disp.; release agent; applied to surfaces to provide a durable, rubbery coating; prevents adhesion of most sticky materials; protects plastics from weather and shields metals from corrosion.

Dow Corning® 290 Emulsion. [Dow Corning] Silicone emulsion; high-solids, paintable release agent for plastics, elastomers, and die castings; water-dilutable.

Dow Corning® 344 Fluid, 345 Fluid. [Dow Corning] Cyclomethicone; lubricant, spreading agent, detackifier for skin cleansers.

Dow Corning® 346 Emulsion. [Dow Corning] Silicone emulsion; release agent and general lubricant for plastics and rubber, e.g., polyethylene bags; water-dilutable; 60% solids.

Dow Corning® 347 Emulsion. [Dow Corning] Silicone emulsion; nonionic; release agent for plastics and elastomers, esp. effective on hot molds; low visc. fluid base; water-dilutable; high solids.

Dow Corning® 544 Antifoam. [Dow Corning] Silicone compd.; foam control agent for detergents, textile jet dyeing, cutting oils; med. off-wh. liq.; 100% act.

Dow Corning® 556 Fluid. [Dow Corning] Phenyldimethicone; lubricant, emollient for skin care prods.; sol. in org. solvs.

Dow Corning® 1500 Silicone Antifoam. [Dow Corning] Silica filled polydimethyl siloxane; defoamer to perform over wide range of process conditions; nonaq. systems; liq.

Dow Corning® 1520 Silicone Antifoam. [Dow Corning] Silica filled polydimethyl siloxane emulsion; defoamer which controls foam in aq. systems; inert, long lasting, does not interact with other system components; liq.

Dow Corning® 3225C Formulation Aid. [Dow Corning] Silicone glycol copolymer; nonionic; used in personal care prods.; liq.; HLB 4.0; 10% conc.

Dow Corning® ACH-303. [Dow Corning/Wickhen] Aluminum chlorhydrate; act. ingred. in antiperspirant and deodorant formulations; sol. in water, methanol, ethanol, propylene glycol, glycerine; 50% conc.

Dow Corning® ACH-323. [Dow Corning/Wickhen] Aluminum chlorhydrate; for any type of antiperspirant formulation requiring dry, very discrete particle size; impalpable powd.; water-sol.

Dow Corning® ACH-331. [Dow Corning/Wickhen] Aluminum chlorhydrate; used in suspensoid solid state antiperspirant sticks; super fine powd.; water-sol.

Dow Corning® ACH7-308. [Dow Corning/Wickhen] Aluminum sesquichlorhydrate beads; act. ingreds. in antiperspirant formulations; sol. in water and 190 proof ethanol.

Dow Corning® ACH7-321. [Dow Corning/Wickhen] Aluminum chlorhydrate beads; used in antiperspirant formulations where extremely fine particle size is not required; water-sol.

Dow Corning® Antifoam 1400. [Dow Corning] Silicone compd.; foam control agent for glycol scrubbing, resin mfg., oil refining, asphalts, adhesives/coatings, metalworking, pesticide/fertilizer, and detergents industries; med. off-wh. to wh. liq.; 100% act.

Dow Corning® Antifoam 1410. [Dow Corning] Silicone emulsion, nonionic; foam control agent for inks, textile starching/sizing, cutting oils, adhesives/coatings, waste water treatment, pesticide/fertilizer industries; thin wh. cream; 10% act.

Dow Corning® Antifoam 1430. [Dow Corning] Silicone emulsion, nonionic; foam control agent for detergents, distillation, glycol scrubbing, metalworking, waste water treatment, adhesives/coatings, and textile industries; med. wh. cream; 30% act.

Dow Corning® Antifoam 1500. [Dow Corning] Silicone compd.; foam control agent for foods, pesticides/fertilizers, gas-oil separation, printing inks, asphalt; med. off-wh. liq.; 100% act.

Dow Corning® Antifoam 1510-US. [Dow Corning] Silicone emulsion; nonionic; foam control agent for pesticide/fertilizer, waste water treatment, food industries; med. wh. cream; 20% act.

Dow Corning® Antifoam 1520-US. [Dow Corning] Silicone emulsion; nonionic; foam control agent for food, resin mfg., waste water treatment, adhesives/coatings, metalworking, and textile industries; thin wh. cream; 20% act.

Dow Corning® Antifoam A. [Dow Corning] Simethicone; foam control agent for distillation, resin sizes, textile latex backing; also avail. in food grade; med. off-wh. to gray liq.; 100% act.

Dow Corning® Antifoam AF. [Dow Corning] Simethicone; nonionic; foam control agent for food, pesticide/fertilizer, paper/printing, textile, adhesives/coatings industries; thick wh. cream; 30% act.

Dow Corning® Antifoam B. [Dow Corning] Silicone emulsion; nonionic; foam control agent for adhesives/glues, textile latex backing, inks, detergents, distillation, resin mfg, pesticide/fertilizer industries; thin wh. cream; 10% act.

Dow Corning® Antifoam C. [Dow Corning] Simethicone; nonionic; food grade foam control agent for food industry and paper coatings; med. wh. cream; 30% act.

Dow Corning® Antifoam FG-10. [Dow Corning] Silicone emulsion; nonionic; foam control agent for food industry and waste water treatment; thin wh. cream; 10% act.

Dow Corning® Antifoam H-10. [Dow Corning] Silicone emulsion; nonionic; foam control agent for detergents, waste water treatment, adhesives/coatings textile, metalworking, and paper/printing industries; thin wh. cream; 10% act.

Dow Corning® Antifoam Y-30. [Dow Corning] Silicone emulsion; nonionic; foam control agent for paper/printing industry for resin sizes and inks; med. wh. cream; 30% act.

Dow Corning® AZG-368. [Dow Corning/Wickhen] Aluminum/zirconium chlorohydrate; act. ingred. for all forms of topical antiperspirants; sol. in water and ethanol.

Dow Corning® AZG-369. [Dow Corning/Wickhen] Aluminum/zirconium chlorohydrate; see Dow Corning AZG-368; sol. see Dow Corning AZG-368.

Dow Corning® AZG-370. [Dow Corning/Wickhen] Aluminum/zirconium glycine; see Dow Corning AZG-368; powd.; water-sol.

Dow Corning® AZG-374. [Dow Corning/Wickhen] Aluminum zirconium tetrachlorohydrex glycine; see Dow Corning AZG-368; water-sol.; 50% conc.

Dow Corning® FF-400. [Dow Corning] Silicone and glycol copolymer; nonionic; lubricant, stabilizer, antistat, fiber finish for textiles, threads; lt. amber fluid; sol. in polar solvs.; water disp.; sp.gr. 1.022; visc. 300 cs; HLB 7.0; flash pt. 177 C (OC); 100% act.

Dow Corning® FF-412. [Dow Corning] Organopolysiloxane; see Dow Corning FF-400; lt. amber fluid; sol. in low m.w. alcohols, glycol ethers; water disp.; sp.gr. 1.040; visc. 200 cs; flash pt. 177 C (OC); 100% act.

Dow Corning® FF-414. [Dow Corning] Organopolysiloxane; see Dow Corning FF-400; lt. amber fluid; sol. in water, alcohols, glycol ether, water disp.; sp.gr. 1.070; visc. 465 cs; flash pt. 177 C (OC); 100% act.

Dow Corning® FS-1265 Fluid. [Dow Corning] Dimethyl silicone fluid; foam control agent for aromatic/chlorinated solvs., dry cleaning, metal cleaning and degreasing, oil refining; clear fluid; visc. 300 cSt and 1000–10,000 cSt grades; 100% act.

Dow Corning® HV-490 Emulsion. [Dow Corning] Silicone emulsion; anionic; release agent for plastics and elastomers; high-visc. fluid base, fine particle size; water-dilutable.

Dow Corning® Q2-7119 Release/Parting Agent. [Dow Corning] Silicone; internal release agent for paintable parts in RIM urethanes; sol. in nonpolar org. solvs.

Dow Corning® Q2-7224. [Dow Corning] Amino functional silicone; hair conditioning ingred.; emulsion.

Dow Corning® QF1-3593A. [Dow Corning] Dimethicone, trimethylsiloxysilicate; emollient with water repellency and low slip for skin care prods.

Dow Corning® X1-6100. [Dow Corning] Organic amine and inorg. methoxysilane; coupling agent providing coupling at the org./inorg. interface in composites; useful in reinforced plastics and composites, adhesives, sealants; useful with glass, min. fillers, and metals; yel. to br. liq.; sol. in alcohols and aromatic and aliphatic solvs.; sol. in acidified water; sp.gr. 1.05; visc. 2.0 cSt; ref. index 1.4660; flash pt. (CC) 32 C; 100% act.

Dow Corning® X1-6106. [Dow Corning] Coupling agent designed to promote adhesion of two dissimilar materials; in reinforced plastic systems, provides coupling of most reinforcing agents, e.g., glass, Kevlar to most polar org. polymers and engineering plastics such as epoxies, urethanes, acrylics, polysulfones, polyphenylene sulfides, melamines, polyimides, PC, and thermoplastic polyesters; adhesion promoter for plastics and metals; wh. to slightly yel. visc. liq.; sol. in polar solvs. such as methanol and Dowanol PM; sol. in water; sp.gr. 1.19; visc. 850 cSt; ref. index 1.5104; flash pt. (CC) 155 F; 100% act.

Dow Corning® Z-6020. [Dow Corning] N-(2-aminoethyl-3-aminopropyl trimethoxy silane; coupling agent used for epoxies, phenolics, melamines, nylons, PVC, acrylics, polyolefins, polyurethanes, nitrile rubbers.

Dow Corning® Z-6030. [Dow Corning] 3-Methacryloxypropyl trimethoxysilane; coupling agent used for free-radical cross-linked polyester, rubber, polyolefins, styrenics, acrylics.

Dow Corning® Z-6032. [Dow Corning] N- [2-(Vinylbenzylamino)-ethyl] -3-aminopropyl trimethoxysilane; coupling agent used for most thermoset and thermoplastic resins.

Dow Corning® Z-6040. [Dow Corning] 3-Glycidoxy-propyl trimethoxysilane; coupling agent used for epoxies, urethane, acrylic and polysulfide sealants.

Dow Corning® Z-6075. [Dow Corning] Vinyltriacetoxy silane; coupling agent used for polyesters, polyolefins, EPDM, EPM (peroxide cured).

Dow Corning® Z-6076. [Dow Corning] 3-Chloropropyl trimethoxy silane; coupling agent used for epoxy, styrenics, nylon.

Dowfax 2A0. [Dow] Dodecyl diphenyloxide disulfonic acid; anionic; intermediate used in acidic systems or preparation of salts; black liq.; sp.gr. 1.11; visc. 510 cps; surf. tens. 30 dynes/cm (0.10%); 40% act.

Dowfax 2A1. [Dow] Sodium dodecyl diphenyloxide disulfonate; anionic; detergent, emulsifier, wetting agent in electroplating, solubilizer, leveling agent for acid dyeing of nylon, dyeing assistant; amber liq.; sol. in aq. sol'ns. of acids, alkalies, electrolytes; sp.gr. 1.16; visc. 145 cps (0.10%); HLB 16.7; surf. tens. 31 dynes/cm (0.10%); 45% act.

Dowfax 2EP. [Dow] Sodium dodecyl diphenyloxide disulfonate; anionic; see Dowfax 2A1; liq.; HLB 16.7; 45% conc.

Dowfax 3B0. [Dow] N-decyl diphenyloxide disulfonate; anionic; see Dowfax 2A1; also used in formulating cleaning prods. and agric. formulations; liq.; 40% conc.

Dowfax 3B1. [Dow] Sodium n-decyl diphenyloxide disulfonate; coupler; med. foaming, stable in bleach, acids, and alkalies; liq.; water-sol.

Dowfax 3B2. [Dow] Sodium n-decyl diphenyloxide disulfonate; anionic; detergent, wetting agent, emulsifier, solubilizer, used as pesticide spray adjuvant; red-brn. liq.; sp.gr. 1.17; visc. 120 cps; surf. tens. 38 dynes/cm (0.10%); 45% act.

Dowfax 9N. [Dow] Nonyl phenol ethoxylate; nonionic; intermediate, detergent, wetting agent, dispersant; used for household and industrial cleaners, wool scouring, textile aux., general industrial usage; liq., waxy solid; 100% conc.

Dowfax 63N10. [Dow] POE POP block copolymer; nonionic; detergent for textiles, household uses, defoamer for anionic detergent slurries; colorless liq.; slt. water sol.; sp.gr. 1.02; visc. 340 cst; 100% act.

Dowfax 63N20. [Dow] POE POP glycol copolymer; nonionic; detergent in liq. dishwashing agent, defoamer; liq.; 100% conc.

Dowfax 8390. [Dow] Sodium n-hexadecyl diphenyloxide disulfonate; see Dowfax 3B1; liq.; water-sol.

Dowfax XDS 8292.00. [Dow] Sodium n-hexyl diphenyloxide disulfonate; anionic; see Dowfax 2A1; liq.; 45% conc.

Dowfrost®. [Dow] Propylene glycol inhibited; solv. for heat transfer medium; sol. in water and org. solvs.

Dowicide® 1. [Dow] o-Phenylphenol; antimicrobial; sol. in acetone, methanol, and water.

Dowicide® A. [Dow] o-Phenylphenol sodium; antimicrobial.

Dowicide® OBCP. [Dow] o-Benzyl-p-chlorophenol; antimicrobial.

Dowicil® 75. [Dow] 1-(3-chloroallyl)-3,5,7-triaza-1-azoniaadamantane chloride (act. ingred.) and sodium bicarbonate (stabilizer); preservative for aq. end prods. such as adhesives, latex emulsions, paints, metal-cutting fluids, drilling muds, biodeg. detergents, and paper coatings; antimicrobial activity; off-wh. powd., 100% thru #5 US sieve; m.w. 251.2; b.p. decomposes @ > 60 C; sol. (act. ingred. g/100 g) in water (222.0 g), anhyd. IPA (39.5 g), propylene glycol USP (20.6 g); sp.gr. 1.54 g/cc; dens. 43 lb/ft^3; 67.5% act.

Dowicil® 200. [Dow] Quaternium-15; preservative and broad-spectrum antimicrobial used in cosmetics and personal care prods.; antimicrobial activity; off-wh. powd., 100% thru #20 US sieve; m.w. 251.2; b.p. decomposes; sol. (g/100 g) in water (127.2 g), anhyd. methanol (20.8 g), propylene glycol USP (18.7 g), 99.5% glycerin (12.6 g); dens. 25.0 lb/ft^3; 94% act.

Dowper®. [Dow] Perchloroethylene; used as dry cleaning solv.; sol. in most org. solvs.; very low water sol.

DPG. [Monsanto] Diphenylguanidine; sec. accelerator for rubber industry; 2 mm pellets or dust-suppressed powd.

DPPG. [Gattefosse Ets.] Propylene glycol dipelargonate; emollient and oily rancidless additive for cosmetic and pharmaceutical preparations; oily liq.

DPTT. [Akrochem] Dipentamethylene thiuram hexasulfide; very act. sulfur-bearing accelerator imparting heat resistance to sulfurless compds.; primary accelerator for Hypalon, butyl, and EPDM; vulcanizing agent for heat-resistant latex; nondiscoloring, nonstaining; lt. gray powd.; sp.gr. 1.50; m.p. 234 F min.; 25% avail. sulfur.

DPTT-S. [Akrochem] Dipentamethylene thiuram hexasulfide; see DPTT; buff powd.; sp.gr. 1.52; m.p. 117 C min.; 25% avail. sulfur.

DRA-1500. [Toho] POE disproportionated rosin ester; nonionic; emulsifier, dispersant; liq.; HLB 14.0; 100% conc.

Drakeol 5. [Penreco] Lt. min. oil USP; emollient; sp.gr. 0.821–0.833; dens. 6.89–7.00 lb/gal; visc. 7.6–8.7 cSt (40 C); flash pt. 320 F; pour pt. 15 F; ref. index 1.4600.

Drakeol 6. [Penreco] Lt. min. oil USP; base material for hair preps., bath oils, baby oils; sp.gr. 0.827–0.836; dens. 6.94–7.02 lb/gal; visc. 9.2–10.6 cSt (40 C); flash pt. 320 F; pour pt. 15 F; ref. index 1.4613.

Drakeol 7. [Penreco] Lt. min. oil USP; carrier, base ingred. in ointments, lotions, baby oils, sun tan lotions, makeup; solv. and emollient in creams, waterless hand cleaners; protective coating for foods; pigment dispersant, lubricant for plastics; sp.gr. 0.827–0.844; dens. 6.94–7.08 lb/gal; visc. 10.8–12.7 cSt (40 C); flash pt. 355 F; pour pt. 15 F; ref. index 1.4632.

Drakeol 8. [Penreco] Lt. min. oil USP; sp.gr. 0.830–0.844; dens. 6.97–7.08 lb/gal; visc. 12.7–14.5 cSt (40 C); flash pt. 360 F; pour pt. 15 F; ref. index 1.4655.

Drakeol 9. [Penreco] Lt. min. oil USP; base material, carrier for cosmetics; emulsified lubricant for laxatives; pigment dispersant, lubricant for plastics; plasticizer for PS; textile and paper lubricant; sp.gr. 0.838–0.854; dens. 7.03–7.16 lb/gal; visc. 14.7–16.8 cSt (40 C); flash pt. 365 F; pour pt. 15 F; ref. index 1.4665.

Drakeol 10. [Penreco] Lt. min. oil USP; plasticizer, lubricant for plastics; sp.gr. 0.844–0.864; dens. 7.08–7.25 lb/gal; visc. 17.9–20.0 cSt (40 C); flash pt. 365 F; pour pt. 15 F; ref. index 1.4692.

Drakeol 10B. [Penreco] Lt. min. oil NF; FDA approved; sp.gr. 0.867–0.878; visc. 17.7–20.2 cSt (40 C); pour pt. –40 C; flash pt. 160 C.

Drakeol 13. [Penreco] Lt. min. oil USP; base ingred. in bath oils; sp.gr. 0.848–0.867; dens. 7.11–7.27 lb/gal; visc. 24.2–26.3 cSt (40 C); flash pt. 390 F; pour pt. 15 F; ref. index 1.4726.

Drakeol 15. [Penreco] Lt. min. oil USP; sp.gr. 0.850–0.870; dens. 7.13–7.30 lb/gal; visc. 28.2–30.2 cSt (40 C); flash pt. 390 F; pour pt. 15 F; ref. index 1.4740.

Drakeol 19. [Penreco] Min. oil USP; primary plasticizer for ethyl cellulose; lubricant for textile/paper; sp.gr. 0.852–0.871; dens. 7.14–7.31 lb/gal; visc. 35.0–37.0 cSt (40 C); flash pt. 410 F; pour pt. 15 F;

ref. index 1.4725.

Drakeol 21. [Penreco] Min. oil USP; primary plasticizer for ethyl cellulose; sp.gr. 0.853–0.873; dens. 7.15–7.32 lb/gal; visc. 39.7–41.7 cSt (40 C); flash pt. 415 F; pour pt. 15 F; ref. index 1.4733.

Drakeol 32. [Penreco] Min. oil USP; plasticizer; sp.gr. 0.856–0.876; dens. 7.18–7.35 lb/gal; visc. 60.0–63.3 cSt (40 C); flash pt. 430 F; pour pt. 10 F; ref. index 1.4770.

Drakeol 34. [Penreco] Min. oil USP; plasticizer for PS; sp.gr. 0.858–0.872; dens. 7.19–7.31 lb/gal; visc. 71.0–78.3 cSt (40 C); flash pt. 475 F; pour pt. 15 F; ref. index 1.4760.

Drakeol 35. [Penreco] Min. oil USP; ingred. in laxatives; plasticizer for PS, ethyl cellulose; water repellant for paper; sp.gr. 0.864–0.876; dens. 7.25–7.35 lb/gal; visc. 65.3–70.0 cSt (40 C); flash pt. 435 F; pour pt. 10 F; ref. index 1.4785.

Draketex 50. [Penreco] Lt. min. oil USP; coning and finishing oil base for nylon and rayon prod.; sp.gr. 0.817–0.830; dens. 6.86–6.96 lb/gal; visc. 6.5–7.8 cSt (40 C); flash pt. 305 F; pour pt. 20 F; ref. index 1.4570.

Drapex 3.2. [Witco/Argus] Octyl epoxy stearate; epoxy plasticizer for vinyl compds; used with Argus Ba/Cd stabilizers to synergistically improve heat and lt. stability; imparts low temp. flexibility to vinyl compds.; suitable for use in plastisols and organosols.

Drapex 4.4. [Witco/Argus] Octyl epoxy tallate; see Drapex 3.2.

Drapex 6.8. [Witco/Argus] Epoxidized soybean oil; plasticizer for vinyl compds.; compatible with primary plasticizers; used with Argus Ba/Cd stabilizers to synergistically improve heat and lt. stability; resistant to extraction and migration; food pkg. use.

Drapex 10.4. [Witco/Argus] Epoxidized linseed oil; see Drapex 6.8.

Dresinate® 81. [Hercules] Rosin sodium soap; anionic; emulsifier ingred. and/or stabilizer in solv. cleaners and sol. oils for metalworking, disinfectants, oil-well drilling muds, drawing and grinding compds.; plasticizer; pale-colored liq.; dilutable with water and aq. sol'ns. of alcohols and glycols; dens. 8.7 lb/gal; visc. 5.7 poises (140 F); acid no. 15; 87% solids.

Dresinate® 90. [Hercules] Rosin potassium soap; anionic; see Dresinate 81; pale liq.; dilutable with water and aq. sol'ns. of alcohols and glycols; dens. 9.0 lb/gal; visc. 14.7 poises (140 F); acid no. 12; 90% solids.

Dresinate® 91. [Hercules] Rosin potassium soap; anionic; see Dresinate 81; emulsifier for metalworking oils, drilling; pale liq.; dilutable with water and aq. sol'ns. of alcohols and glycols; dens. 8.8 lb/gal; visc. 5.3 poises (140 F); acid no. 15; 88% solids.

Dresinate® 92. [Hercules] Potassium soap; anionic; see Dresinate 81; dk. liq.; dilutable with water and aq. sol'ns. of alcohols and glycols; dens. 9.0 lb/gal; visc. 9.2 poises (140 F); acid no. 15; 88% solids.

Dresinate® 93. [Hercules] Potassium soap; anionic; see Dresinate 81; dk. liq.; dilutable with water and aq. sol'ns. of alcohols and glycols; dens. 9.5 lb/gal; visc. 7.3 poises (140 F); acid no. 3; 87% solids.

Dresinate® 94. [Hercules] Potassium soap; anionic; see Dresinate 81; pale liq.; dilutable with water and aq. sol'ns. of alcohols and glycols; dens. 9.3 lb/gal; visc. 12.4 poises (140 F); acid no. 5; 87% solids.

Dresinate® 95. [Hercules] Potassium soap; anionic; see Dresinate 81; emulsifier for metalworking oils, drilling; dk. liq.; dilutable with water and aq. sol'ns. of alcohols and glycols; dens. 9.1 lb/gal; visc. 4.5 poises (140 F); acid no. 15; 88% solids.

Dresinate® 214. [Hercules] Potassium soap of a pale modified rosin; anionic; emulsifier, pigment wetting agent and dispersant, foaming agent, used in mfg. of adhesives, polymer emulsion syn. latices; paste; dilutable with water and aq. sol'ns. of alcohols and glycols; dens. 9.22 lb/gal (60 C); visc. 23 poises (140 F); 80% solids.

Dresinate® 515. [Hercules] Potassium soap of a pale modified rosin; anionic; see Dresinate 214; paste; dens. 9.24 lb/gal (60 C); visc. 64 poises (140 F); 80% solids.

Dresinate® 731. [Hercules] Sodium soap of a pale modified rosin; see Dresinate 214; paste; dilutable with water and aq. sol'ns. of alcohols and glycols; dens. 9.1 lb/gal (60 C); visc. 11 poises (140 F); 70% solids.

Dresinate® DS-60W, TX-60W. [Hercules] Sodium soaps of tall oil derivatives; used for flotations beneficiation of glass sand and phosphates and as emulsifier for asphalt; liq.; 60% solids aq. sol'ns.

Dresinate® TX. [Hercules] Tall oil sodium soap; anionic; emulsifier for oils and asphalts; detergent aid; dispersant for pigments; stabilizer for syn. rubber latices, used for industrial and household cleaners; conditioner for sulfur dusts; dk. brn. powd.; sol. in water and aq. sol'ns. of alcohols and glycols; dens. 26 lb/ft^3; 96% min. solids.

Dresinate® X. [Hercules] Sodium soap of a pale rosin; anionic; see Dresinate TX; pale cream, powd.; sol. in water and aq. sol'ns. of alcohols and glycols; dens. 26 lb/ft^3; 95% min. solids.

Dresinate® XX. [Hercules] Sodium soap of a dk. rosin; see Dresinate TX; lt. br. powd.; sol. in water and aq. sol'ns. of alcohols and glycols; dens. 28 lb/ft^3; 96% min. solids.

Dresinol 40. [Hercules] Pale rosin, aq.-based disp.; solv.-free modifier for polymeric film-formers in aq. systems (adhesives, sizings); improves stiffness or reinforcement of coating films, wetting of pigments and fillers, penetration and adhesion to substrates; improves surf. tack and bond strength between substrates in adhesives systems when blended with elastomer latices; cream liq.; visc. 550 cps; soften. pt. 87 C; pH 9.1; 40% solids.

Dresinol 42. [Hercules] Partially decarboxylated rosin aq.-based disp.; see Dresinol 40; cream liq.; sol. see Dresinol 40; visc. 600 cps; soften. pt. 46 C; pH 9.4; 40% solids.

Drewfax 0007. [Drew] Sodium dioctyl sulfosuccinate; wetting, penetrating, surf. tens. reducing agent, dispersant for textile, cosmetic, paper, metal, paint, rubber, plastics, petrol. and agric. industries; colorless liq.; dens. 9.0 lb/gal; sol. in water and various solvs.; 70% act.

Drewfax 1980. [Aquatec Quimica] Glycol ester; nonionic; visc. modifier for thermoplastics; liq.; 100% conc.

Drewfax R-122. [Drew] Fatty acid derivs. and surf. act. compds.; lubricant for easy release of dry creped paper from Yankee Dryers.

Drewfloc 3. [Drew] High m.w. org. anionic polymer; flocculant for water clarification applics. incl. potable water, min. beneficiation; wh. powd.; 2% max. sol. in cold water; pH 7.0–7.5 (1%).

Drewfloc 4. [Drew] Syn. and natural org. anionic polymers; flocculant for water clarification applics. incl. potable water treatment, and min. beneficiation; wh. powd.; 2% max. sol. in cold water; visc. 20 cps; pH 7.0–7.5 (1%).

Drewmulse 3-1-O. [Stepan/PVO] Triglycerol monooleate; nonionic; emulsifier, solubilizer, dispersant for w/o and o/w emulsions, creams, lotions, for internal, cosmetic, pharmaceutical use; Gardner 8 liq.; sol. in IPA, peanut oil, min. oil, water disp.; HLB 7.0; sapon. no. 125–150.

Drewmulse 3-1-S. [Stepan/PVO] Triglycerol monostearate; nonionic; see Drewmulse 3-1-O; Gardner 10 liq.; disp. in water, IPA, peanut oil, min. oil, propylene glycol; HLB 7.0; sapon. no. 120–140.

Drewmulse 6-2-S. [Stepan/PVO] Hexaglycerol distearate; nonionic; see Drewmulse 3-1-O; Gardner 13 solid; sol. in IPA, peanut oil, min. oil; HLB 8.0; sapon. 105–125.

Drewmulse 10. [Stepan/PVO] Glycerol mono fat from soya oil; lipophilic emulsifier used in food industry as dispersing aid, antistaling agent, antistick agent; cream plastic; m.p. 118 F; HLB 2.8; sapon. no. 170–190; 40% alpha mono.

Drewmulse 10-4-O. [Stepan/PVO] Decaglycerol tetraoleate; nonionic; see Drewmulse 3-1-O; Gardner 8 liq.; sol. in IPA, peanut oil, min. oil; water disp.; HLB 6.0; sapon. no. 125–145.

Drewmulse 10-8-O. [Stepan/PVO] Decaglycerol octaoleate; nonionic; see Drewmulse 3-1-O; Gardner 8 liq.; sol. in peanut oil, min. oil; HLB 4.0; sapon. no. 155–175.

Drewmulse 10-10-O. [Stepan/PVO] Decaglycerol decaoleate; nonionic; see Drewmulse 3-1-O; Gardner 8 liq.; sol. in peanut oil, min. oil; HLB 3.0; sapon. no. 165–185.

Drewmulse 10-10-S. [Stepan/PVO] Decaglycerol decastearate; nonionic; see Drewmulse 3-1-O; 8 Gardner solid; sol. in peanut oil, min. oil; HLB 3.0; sapon. no. 160–180.

Drewmulse 15. [Stepan/PVO] Glycerol mono cottonseed oil; see Drewmulse 10; cream soft solid; m.p. 110 F; 40% alpha mono.

Drewmulse 20/200/V. [Stepan/PVO] Glycerol stearate; see Drewmulse 10; cream beads; m.p. 140 F; 40% alpha mono.

Drewmulse 21. [Stepan/PVO] Glycerol stearate from soya oil; see Drewmulse 10; cream bead, flake; HLB 8.4; sapon. no. 160–180; 30% alpha mono.

Drewmulse 30/900. [Stepan/PVO] Glycerol monostearate; see Drewmulse 10; cream beads; m.p. 138 F; 52% alpha mono.

Drewmulse 50. [Stepan/PVO] Glycerol mono fat; see Drewmulse 10; cream colored cream; 52% alpha mono.

Drewmulse 55. [Stepan/PVO] Glycerol mono shortening from tallow and soya; see Drewmulse 10; cream plastic; m.p. 130–135 F (tallow), 116–122 F (soya); HLB 3.4 (tallow), 2.4 (soya); sapon. no. 165–185; 52% alpha mono.

Drewmulse 85. [Stepan/PVO] Glycerol monooleate from tallow and soya; see Drewmulse 10; lt. yel. semiliq.; HLB 3.4; sapon. no. 160–180; 40% alpha mono.

Drewmulse 200. [Stepan/PVO] Glycerol monooleate from tallow and soya; see Drewmulse 10; cream beads; m.p. 135–145; HLB 2.8; sapon. no. 154–170 (soya), 163–177 (tallow); 40% alpha mono.

Drewmulse 300. [Stepan/PVO] Glycerol monostearate (80%) and polysorbate (20%) blend from soya and tallow oils; see Drewmulse 10; cream beads; m.p. 135 F; HLB 4.3; 30% alpha mono.

Drewmulse 365. [Aquatec Quimica] Glycerol monostearate (67%) and polysorbate 65 (33%) blend from soya and tallow oils; see Drewmulse 10; cream beads; m.p. 136 F; HLB 5.3; 25–30% alpha mono.

Drewmulse 700. [Stepan/PVO] Glycerol monostearate (80%) and polysorbate 80 (20%) blend from soya and tallow oils; see Drewmulse 10; cream beads; m.p. 135 F; HLB 5.2; 30% alpha mono.

Drewmulse 900. [Stepan/PVO] Glycerol stearate; see Drewmulse 10; cream beads; m.p. 138 F; HLB 3.8; sapon. 160–180; 52% alpha mono.

Drewmulse 1128. [Stepan/PVO] Glyceryl stearate, PEG-120 glyceryl stearate, and PEG-120 propylene glycol stearate; nonionic; emulsifier, stabilizer, and visc. builder in cosmetics and pharmaceuticals, opacifier; Gardner 2 flake; HLB 8.4; 20% mono.

Drewmulse 1129. [Stepan/PVO] Glyceryl stearate and PEG-90 stearate; see Drewmulse 1128; flake; HLB 11.0; 20% mono.

Drewmulse CNO. [Stepan/PVO] Glyceryl cocoate; see Drewmulse 1128; Gardner 4 solid; HLB 3.8; 50% mono.

Drewmulse DGMS. [Aquatec Quimica] Diethylene glycol stearate; nonionic; emulsifier for w/o and o/w systems, opacifier and pearlescence agent for shampoos, additive for pigment milling; flakes; 85% act.

Drewmulse EGDS. [Aquatec Quimica] Ethylene glycol distearate; nonionic; see Drewmulse DGMS; flakes; 48% act.

Drewmulse EGMS. [Aquatec Quimica] Ethylene glycol stearate; nonionic; see Drewmulse DGMS; flakes; 48% act.

Drewmulse GMC-8. [Stepan/PVO] Glycerol caprylate-caprate; stabilizer and visc. builder in cosmetics and pharmaceuticals, opacifier; coupler, emulsifier; for food applics.; Gardner 3 liq.; sol. in alcohol, min. oil; HLB 8.3; 75% mono.

Drewmulse GMO. [Stepan/PVO] Glyceryl oleate; nonionic; stabilizer and visc. builder in cosmetics and pharmaceuticals, opacifier; Gardner 6 liq.; HLB 3.4; 40–45% conc.

Drewmulse GMRO. [Aquatec Quimica] Glycerol ricinoleate; nonionic; see Drewmulse DGMS; liq.

Drewmulse GMS. [Aquatec Quimica] Glyceryl monostearate; nonionic; see Drewmulse DGMS; flakes; 40–45% act.

Drewmulse HM-100. [Stepan/PVO] Glyceryl stearate and PEG-40 stearate; stabilizer and visc. builder in cosmetics and pharmaceuticals, opacifier, esp. effective in antiperspirant formulations; Gardner 2 flake; HLB 8.4; 24% mono.

Drewmulse PGMS. [Aquatec Quimica] Propylene glycol monostearate; nonionic; see Drewmulse DGMS; flakes; 48% act.

Drewmulse POE-SML. [Stepan/PVO] Polysorbate 20; nonionic; w/o emulsifier for cosmetics and pharmaceuticals, solubilizer, dispersant, wetting agent, detergent, visc. control agent; liq.; HLB 16.7; sapon. no. 40–51.

Drewmulse POE-SMO. [Stepan/PVO] Polysorbate 80; nonionic; see Drewmulse POE-SML; liq.; HLB 15.0; sapon. no. 45–55.

Drewmulse POE-SMS. [Stepan/PVO] Polysorbate 60; nonionic; see Drewmulse POE-SML; solid; HLB 14.9; sapon. no. 45–55.

Drewmulse POE-STS. [Stepan/PVO] Polysorbate 65; nonionic; see Drewmulse POE-SML; solid; HLB 10.5; sapon. no. 88–98.

Drewmulse SML. [Stepan/PVO] Sorbitan monolaurate; nonionic; see Drewmulse POE-SML; liq.; HLB 8.6; sapon. no. 159–171.

Drewmulse SMO. [Stepan/PVO] Sorbitan monooleate; nonionic; see Drewmulse POE-SML; liq.; HLB 4.3; sapon. no. 144–156.

Drewmulse SMS. [Stepan/PVO] Sorbitan monostearate; nonionic; see Drewmulse POE-SML; solid; HLB 4.7; sapon. no. 147–157.

Drewmulse STS. [Stepan/PVO] Sorbitan tristearate; nonionic; see Drewmulse POE-SML; solid; HLB 2.1; sapon. no. 170–190.

Drewmulse TP. [Stepan/PVO] Glyceryl stearate; nonionic; emulsifier, stabilizer, and visc. builder in cosmetics and pharmaceuticals, opacifier; Gardner 4 flake; HLB 3.8; 40% mono.

Drewmulse V. [Stepan/PVO] Glyceryl stearate (veg.); see Drewmulse TP; Gardner 6 bead; HLB 3.8; 40% mono.

Drewmulse V-SE. [Stepan/PVO] Glyceryl stearate SE; see Drewmulse TP; Gardner 6 bead; HLB 8.4; 30% mono.

Drewpol 3-1-O. [Stepan/PVO] Triglycerol monooleate; food emulsifier and additive; flavor and color solubilizer; amber liq.; HLB 7.0; sapon. no. 120–160.

Drewpol 3-1-S. [Stepan/PVO] Triglycerol monostearate; nonionic; see Drewpol 3-1-O; amber beads; HLB 7.0; sapon. no. 120–140.

Drewpol 3-1-SH. [Stepan/PVO] Triglycerol mono shortening; see Drewpol 3-1-O; thick, amber liq.; HLB 7.4; sapon. no. 110–140.

Drewpol 6-2-S. [Stepan/PVO] Hexaglycerol distearate; see Drewpol 3-1-O; amber beads; HLB 4.0; sapon. no. 100–130.

Drewpol 10-4-O. [Stepan/PVO] Decaglycerol tetraoleate; see Drewpol 3-1-O; amber liq.; HLB 6.0; sapon. no. 125–145.

Drewpol 10-8-O. [Stepan/PVO] Decaglycerol octaoleate; see Drewpol 3-1-O; amber liq.; HLB 4.0; sapon. no. 150–180.

Drewpol 10-10-O. [Stepan/PVO] Decaglycerol decaoleate; see Drewpol 3-1-O; amber liq.; HLB 3.0; sapon. no. 160–190.

Drewpol 10-10-S. [Stepan/PVO] Decaglycerol decastearate; see Drewpol 3-1-O; tan solid; HLB 3; sapon. no. 150–190.

Drewpon 40, 40LS. [Aquatec Quimica] TEA lauryl sulfate; anionic; emulsifier, detergent, wetting agent, dispersant used for household cleaning prods., cosmetics, emulsion polymerization; liq.; 39–42 and 38–40% act. resp.

Drewsoft 100. [Stepan/PVO] Glyceryl monostearate, modified; nonionic; softener for fibers and blends; flake; 100% conc.

Drewsperse 611. [Drew] Sodium polyacrylate; pigment and paint dispersant.

Druspin 1223. [Stepan/PVO] Ester; textile lubricant to minimize breakage in worsted carding; used for ordinary filet wire in the card clothing; liq.; 100% act.

Druspin 1400. [Stepan/PVO] Ester; textile lubricant to minimize breakage in worsted carding; used for metallic wire in card clothing; liq.; 100% act.

Druspin LO-VIS. [Stepan/PVO] Blended prod.; textile lubricant, antistat, emulsifier; liq.; low visc.; 100% act.

Druspin Supreme. [Stepan/PVO] Blend of min. oil, emulsifiers, lubricants, antistats; liq.; 100% act.

Dry Flo®. [Nat'l. Starch] Aluminum starch octenyl succinate; body powds., antiperspirants, femine hygiene sprays, food powds.; water-insol.

Drymet 59. [Crosfield] Sodium metasilicate anhyd.; alkaline silicate used for formulating specialty detergents and contributing buffering capacity and corrosion inhibition of soft metals and ceramic glazes; wh. free-flowing gran.; 55% on 40 mesh; m.w. 122.06; sol. 27 g/100 g water (30 C); bulk dens. 64 lb/ft^3 (loose); m.p. 1088 C; pH 12.6 (1%); 50.4–51.9% sodium oxide, 45.5% min. silica.

Drymet Fines. [Crosfield] Sodium metasilicate anhyd.; see Drymet 59; wh. free-flowing gran.; 64% thru 100 mesh; m.w. 122.06; sol. 27 g/100 g water (30 C); bulk dens. 57 lb/ft^3 (loose); m.p. 1088 C; pH 12.6 (1%); 50.4–51.9% sodium oxide, 45.5% min. silica.

Dry Pexol 200. [Hercules] Fortified pale rosin; dry size used with alum in paper and paperboard to produce resistance to water and aq. sol'ns.; lt. tan powd.; water-sol.; dens. 28 lb/ft^3; 96% solids.

Dry Pexol 243. [Hercules] Fortified dk. rosin; dry size for lower brightness paper grades; powd.

Dry Size XL20C. [Hercules] Pale rosin, paraffin wax, ratio 80:20; dry size used with alum to impart high level of resistance to water and aq. sol'ns. in specialized applics. like molded pulp prods.; high levels can be used without brightness, foam, or other operating problems on foaming machine; food pkg. paper/

paperboard; lt. tan; dens. 28 lb/ft3; 96% total solids.

Drysperse 401. [Witco Israel] Polyoxyalkylene glycols blended with dispersing and anticaking agents; dispersant for wettable-powd. insecticides, herbicides, and fungicide formulations; powd.

Drysperse 801. [Witco Israel] Alkylaryl polyethoxyethanol with wetting and dispersing properties; see Drysperse 401; powd.

Drysperse 902. [Witco Israel] Alkylaryl polyoxyalkylated alcohol with defoaming and dispersants; see Drysperse 401; powd.

Drysperse 908H. [Witco Israel] Polyoxylalkylene glycols blended with dispersants and anticaking resins; see Drysperse 401; powd.

D Sodium Silicate. [PQ] Sodium silicate; corrosion inhibitor for water systems; used for metals in industrial plants, textile mills, laundries, office bldgs., municipalities, oil refineries; liq.; dens. 12.8 lb/gal.

DSX 1514. [Henkel] Urethane associative thickener; rheology modifier for latex paints and adhesives; maximize high sheer visc.; opaque wh. liq.; dens. 8.9 lb/gal; visc. 2500 cps; 40% act.

DSX 1550. [Henkel] Urethane associative thickener; improves applic. and flow properties of adhesives and coatings; clear to hazy wh. liq.; dens. 8.9 lb/gal; 40% act.

Dualite M6001AE, M6017AE. [Pierce & Stevens] Hollow composite microspheres (shell: PVDC/acryolonitrile copolymer; coating: calcium carbonate); low-dens. filler for use in plastics, coatings, adhesives, BMC, SMC, rubber compding., paper mfg., etc.; M6001AE for moderate exposure to nonaggressive solvs.; M6017AE for prolonged exposure to aggressive solvs.; dens. 2.71 ($CaCO_3$), 0.02 (microspheres), 0.13 ± 0.02 (composite); 85% $CaCO_3$; 15% microspheres.

Dulectin. [Duphar] De-oiled soybean lecithin; nonionic; surfactant, emulsifier, dispersant, and stabilizer for pharmaceuticals and cosmetics; powd.; 97% conc.

Duofol T. [Hart Chem. Ltd.] Sulfated ester; anionic; wetting, disp., finishing agents for textiles; amber liq.; clear sol. 5% water; sp.gr. 1.068; visc. 70 cps; 55% act.

Duolite A-7. [Henkel] Polyfunctional amine crosslined phenol-formaldehyde resin; low-cost, macroporous, weak-base anion-exchange resin with good operating capacity and high porosity; acid adsorbent, org. scavenger, sugar deionization, decolorization, removal of high m.w. colorants (e.g. in wine processing); cream gran.; sp.gr. 1.12; dens. 320–400 g/l.

Duolite A-143. [Henkel] Quat. ammonium (type 1) PS resin; microporous, strong-base resin specially formulated for adsorption of high m.w. colorants; used for color removal from sugar syrups and other food prods.; regeneration with salt brine; beads; sp.gr. 1.09.

Duolite S-37. [Henkel] Amine crosslinked phenol-formaldehyde resin; porous, very weak-base, adsorbent resin; reversible adsorption of aromatic acids; used in removal of orgs., colorants from water, sugar sol'ns., waste prods.; ecru gran. (20–50 mesh); sp.gr. 1.12; dens. 40 lb/ft^3.

Duolite S-761. [Henkel] Hydroxyl crosslinked phenol-formaldehyde resin; macroporous, very weak-acid adsorbent resin with good resistance to attrition; regeneration with alkali; used in removal of proteins and high m.w. colorants from sugar sol'ns., wines, fermentation broths; moist gran.; sp.gr. 1.11; dens. 477 g/l.

Duomac® C. [Akzo] Coco-1,3-diaminopropane diacetate; corrosion inhibitor, bactericide in drilling fluids; sol. in IPA, sparingly sol. in water; sp.gr. 0.85; m.p. 20–24 C; 89% act.

Duomac® T. [Akzo] N-tallow-1,3-propanediamine diacetate; cationic; emulsifier, corrosion inhibitor, flotation reagent; pigment flushing, flocculation; br. paste; vinegar odor; m.w. 480; sol. in water, chloroform, ethanol, IPA; sp.gr. 0.892 (90 C); HLB 10.7; m.p. 181 F; b.p. > 300 C; flash pt. 335 F (COC); 95% conc.

Duomeen® C. [Akzo] N-coco-1,3-propanediamine; cationic; chemical intermediate; corrosion inhibitor, fuel oil additive, flotation agent; used in metals, textiles, plastics, herbicides; epoxy curing agent; dk. amber. liq.; ammonia odor; m.w. 276; sol. in naphtha, min. oil, IPA; sp.gr. 0.836; visc. 10 cP; m.p. 71 F; b.p. > 300 C; flash pt. 295 F (COC); 100% conc.

Duomeen® C Special. [Akzo] N-coco-1,3-propanediamine; cationic; see Duomeen C; yel. liq.; ammonia odor; m.w. 276; sol. in naphtha, min. oil, IPA; sp.gr. 0.836; m.p. 80 F; b.p. > 300 C; flash pt. 295 F (COC); 95% amin. diamine act.

Duomeen® CD. [Akzo] Coco-1,3 diamino propane; cationic; intermediate; Gardner 3 max. liq.; 89% conc.

Duomeen® CX. [Akzo] Diamine; cationic; chemical intermediate; liq., paste; 100% conc.

Duomeen® L-11. [Akzo] N-s-alkyl-1,3-propanediamine; see Duomeen C; yel. liq.; ammonia odor; m.w. 300; sol. in min. oil, IPA; sp.gr. 0.841; m.p. –32 F; b.p. > 300 C; flash pt. 226 F (COC).

Duomeen® L-15. [Akzo] N-s-alkyl-1,3-propanediamine; see Duomeen C; yel. liq.; ammonia odor; m.w. 383; sol. in min. oil, IPA; sp.gr. 0.842; m.p. 44 F; b.p. > 300 C; flash pt. 266 F (COC).

Duomeen® LT-4. [Akzo] 3-Tallowalkyl-1,3-hexahydropyrimidine; Gardner 8 max. liq.; amine no. 281.

Duomeen® O. [Akzo] N-oleyl-1,3-propanediamine; cationic; see Duomeen C; yel. paste; ammonia odor; m.w. 350; sol. in naphtha, IPA; sp.gr. 0.841; visc. 19 cP; m.p. 59 F; b.p. > 300 C; flash pt. 370 F (COC); 89% min. act.

Duomeen® OL. [Akzo] N-oleyl-1,3-diaminopropane; Gardner 10 max. liq.; amine no. 344.

Duomeen® OTM. [Akzo] N,N,N′-trimethyl-N′-9-octadecenyl-1,3-diaminopropane; Gardner 8 max. liq.; amine no. 281.

Duomeen® OX. [Akzo] Oleyl-1,3 diaminopropane; cationic; intermediate; liq., paste; 92% conc.

Duomeen® S. [Akzo] N-soya-1,3-diaminopropane; cationic; see Duomeen C; Gardner 12 pasty liq.; m.w. 350; sp.gr. 0.840; visc. 70 cP; m.p. 48–68 F; flash pt.

350 F (COC); 89% min. diamine act.

Duomeen® T, T Special [Akzo] N-tallow-1,3-propanediamine; see Duomeen C; textile finishing agent, dispersant for inorg pigments in paints, larvacidal oil additive; bitumen emulsifier; dk. amber paste; ammonia odor; m.w. 330 and 350 resp.; sol. in naphtha, IPA; sp.gr. 0.841; visc. 880 cP; m.p. 115 F; b.p. > 300 C; flash pt. 370 F (COC); 100% conc.

Duomeen® TDO. [Akzo] N-tallow-1,3-propanediamine dioleate; cationic; see Duomeen C; also in metal treatment as film and boundary lubricant, metal drawing additive; dispersant in paint industry; dk. semisolid; ammonia odor; m.w. 924; sol. in naphtha, min. oil, IPA; insol. in water; sp.gr. 0.865 (65 C); m.p. 77 F; b.p. > 300 C; flash pt. 455 F (COC); 33% min. act.

Duomeen® TDO-IHF. [Akzo] N-tallow trimethylene diamine dioleate; pigment dispersant; liq.; 80% act.

Duomeen® TTM. [Akzo] N,N,N′-trimethyl-N′-tallow-1,3-diaminopropane; Gardner 8 liq.; amine no. 282.

Duomeen® TX. [Akzo] Tallow-1,3-diaminopropane; emulsifier for bitumen emulsions, base material; antisettling agent for paints; Gardner 7 solid; ammoniacal odor; water insol.; sp.gr. 0.84; m.p. 46–55 C; flash pt. 180 C; 92% act.

Duoquad® O-50. [Akzo] N,N,N′,N′,N′-pentamethyl-N-octadecenyl-1,3-diammonium dichloride; Gardner 6 max. liq.; flash pt. (PMCC) 15 C; pH 6–9; 50% quat.

Duoquad® T-50. [Akzo] N,N,N′,N′,N′-pentamethyl-n-tallow-1,3-propanediammonium dichloride; cationic; detergent, corrosion inhibitor, metal cleaner; emulsifier for sec. oil recovery; Gardner 13 max. liq.; m.w. 480; water-sol.; sp.gr. 0.90; HLB 14.4; f.p. –20 C; flash pt. 13 C; pH 6–8 (10% aq.); 48–52% act. in aq. IPA.

Duowet EHS, EHS Conc. [Rhone-Poulenc Surf.] Sulfonated aliphatic polyester, anionic; wetting agent used in dyeing, sizing, desizing, raw stock processing; liq.

Duponol 80. [DuPont] Sodium n-octyl sulfate, tech.; anionic; wetting agent, dispersant, penetrant, softening agent, emulsifier used for textile, metal, and leather processing; lt. yel. liq.; m.w. 232; sol. in presence of many multivalent ions; visc. 30 cps; surf. tens. 51.7 dynes/cm (0.1%); 33–35% act.

Duponol C. [DuPont] Sodium lauryl sulfate USP; anionic; emulsifier, dispersant, detergent, wetting agent for cosmetic, dental and medical preps.; wh. spray-dried beads; mild fatty odor; water sol.; dens. 3.3 lb/gal; 90–96% conc.

Duponol D Paste. [DuPont] Sodium lauryl oleyl sulfate, tech.; anionic; detergent, emulsifier, wetting agent, dispersant used for dyes, textile, metal, and leather processing; lt. yel. paste; m.w. 328; surf. tens. 30.5 dynes/cm (0.1%); 38% act.

Duponol RA. [DuPont] Fortified sodium ether-alcohol sulfate; anionic; detergent, dispersant, emulsifier, penetrating, leveling, and wetting agent; textile processing; lt. amber clear liq.; misc. with water; sp.gr. 1.04; dens. 8.7 lb/gal; visc. 178 cps (80 F); cloud pt. 68 F; flash pt. > 212 F (TOC); surf. tens. 33 dynes/cm (0.1%).

Duponol WA Dry. [DuPont] Sodium lauryl sulfate; anionic; detergent, emulsifier, wetting agent, penetrant; wh. powd.; water sol., the act. ingred. is sol. in highly polar solv.; dens. 2.6 lb/gal; cloud pt. 37 F (1%); surf. tens. 33 dynes/cm (0.1%); 44% conc.

Duponol WN. [DuPont] Sodium octyl/decyl sulfate, tech.; anionic; wetting and penetrating agent; pale yel. liq; m.w. 246; visc. 30 cps; surf. tens. 52.5 dynes/cm (0.1%); 33–35% act.

Du Pont Retarder LAN. [DuPont] Alkyl trimethylammonium halide; cationic; dye retarder and stripping agent with azoic dyes, emulsifier; dk. red-brn. liq.; water-misc.; surf. tens. 43.4 dynes/cm (0.1%); 20% act.

Duralink HTS. [Monsanto] Hexamethylene bisthiosulfate disodium salt dihydrate; post vulcanization stabilizer for sulfur cures of NR, IR, SBR, and NBR; used in tire treads, sidewalls, and general industrial prods. incl. belting and inj. molded goods; bonding promoter for rubber-based steel adhesion; nondiscoloring, nonstaining; wh. dust-suppressed powd.; sp.gr. 1.4; very sol. in water.

Durazone 37. [Uniroyal] 2,4,6-Tris-(N-1,4-dimethylpentyl-p-phenylenediamino)-1,3,5-triazine; nonstaining antiozonant/antioxidant for natural and syn. rubbers, e.g., tires, hose, footwear, mech. goods, roofing, wire and cable; buff to lt. pink cryst. powd., char. odor; m.w. 694; sol. in org. solvs.; sp.gr. 1.13; m.p. 130 C min.

Dur-Em® 117. [Van Den Bergh] Glyceryl stearate; nonionic; textile lubricant and finishing agent; emulsifier for cosmetic and pharmaceutical creams and lotions; lubricant for thermoplastics; dispersant for inorg. pigments; HLB 2.8; m.p. 62–65 C; 40% min. alpha monoglyceride.

Dur-Em® GMO. [Van Den Bergh] Glyceryl oleate; nonionic; lubricant for personal care prods., dry cleaning bases, paints, and insecticides; HLB 2.8.

Durfax® 20. [Van Den Bergh] Polysorbate 20; nonionic; used in personal care and household prods.; food emulsifier; fabric antistat; HLB 16.5; sapon. no. 39–52.

Durfax® 60. [Van Den Bergh] Polysorbate 60; nonionic; food emulsifier; personal care prods.; preshave beard lubricant and softener prods.; paste; sol. in water; HLB 14.9; sapon. no. 45–55.

Durfax® 65. [Van Den Bergh] Polysorbate 65; nonionic; food emulsifier; pesticide dispersant; solid; water-disp.; HLB 10.5; sapon. no. 88–98.

Durfax® 80. [Van Den Bergh] Polysorbate 80; nonionic; food emulsifier; personal care prods.; antifog agent in plastics and aerosol furniture polish; liq.; water-sol.; HLB 15.9; sapon. no. 45–55.

Durfax® EOM. [Van Den Bergh] PEG-20 glyceryl stearate; food emulsifier; dough conditioner; lubricant and fabric softener; HLB 13.5; sapon. no. 65–75.

Durlac® 100W. [Van Den Bergh] Glyceryl stearate lactate; emulsifier for food; starch gelling agent in industrial processes; HLB 2.4; m.p. 46–54 C.

Durotex 7603. [Morton Int'l.] Antimicrobial for latex carpet backing, dry film preservative.

Durotex GPM. [Morton Int'l.] Antimicrobial for high tech filtration papers.

Durpeg® 400MO. [Van Den Bergh] PEG-8 oleate; nonionic; food emulsifier, processing aid in animal feeds; cosmetic prods.; amber semisolid; HLB 11.1; sapon. no. 80–88; 100% act.

Durtan® 20. [Van Den Bergh] Sorbitan monolaurate; nonionic; used in cosmetic, toiletry, and household prods.; fabric antistat and lubricant; HLB 7.4; sapon. no. 158–170.

Durtan® 60. [Van Den Bergh] Sorbitan monostearate; nonionic; food emulsifier; dispersant for inorganics used in thermoplastics; beads; HLB 4.7; sapon. no. 147–157.

Durtan® 80. [Van Den Bergh] Sorbitan monooleate; nonionic; used in water softening compds., auto polishes, pesticide formulations; antifog in plastics; rust inhibitor additive; ink pigment dispersant; HLB 4.3; sapon. no. 148–161.

Dusoran MD. [Duphar; Van Schuppen] Lanolin alcohols, dist.; nonionic; w/o emulsifier, stabilizer, softener, emollient for absorption bases, cosmetics, cleansing preps.; lt. yel., waxy solid; sol, in min. oil, abs. and 95% alcohols, chloroform, ether, lt. petrol., toluene; acid no. 2.0; sapon. 8.0; 100% act.

Dusoran NG, NO. [Van Schuppen] Wool and lanolin alcohols; stabilizer, softener, emulsifier for cosmetic and personal care prods.; tan and lt. brn. resp.; solid; sol. in min. oils, abs. and 95% alcohols, chloroform, toluene; acid no. 3.0 and 2 resp.; sapon. no. 10.

Du-Ter®. [Griffin] Triphenyltin hydroxide; flowable fungicide for pecans, potatoes, sugarbeets; restricted use; 19.7% act.

DyaFac 6-S. [Henkel] PEG-600 monostearate; emulsifier and lubricant for syn. fibers; water-disp.

DyaFac 1926-B. [Henkel] POE glyceride; see DyaFac 6-S; water-insol.

DyaFac LA9. [Henkel] PEG-400 monolaurate; solubilizer and emulsifier for fats and oils; water-sol.

Dyasulf 1761-A. [Henkel] Sulfated castor oil; dispersant; wetting agent, emulsifier; water-sol.

Dyasulf 2031. [Henkel] Sulfated oleic acid sodium salt; wetting agent and penetrant in textile operations; detergent and penetrant in fabrics; amber liq.; dens. 1.09 g/ml; pH 5.3 (2% aq.); 60% act.

Dyesperse DC. [Hart Chem. Ltd.] Naphthalene formaldehyde sulfonate; anionic; dispersant for disperse and acid dyes; amber liq.; clear sol.; sp.gr. 1.138; visc. 10 cps; ref. index 1.4028; 33% act.

Dyetone®. [Olin] Aq. sol'n. of sodium bromate; dye oxidant for vat and sulfur dyes; water-wh. to lt. yel. sol'n.; sp.gr. 1.11; pH 10–12; 12.0% min. sodium bromate.

Dymel® 22. [DuPont] Chlorodifluoromethane; aerosol propellant; environmentally safe and low in toxicity; only nonflam. liquefied propellant for general aerosol usage; m.w. 86.5; sp.gr. 1.21; b.p. –41.4 F.

Dymel® 142. [DuPont] 1,1,1-Chlorodifluoroethane; aerosol propellant; environmentally safe and low in toxicity; very low flam.; m.w. 100.5; low water sol.; sp.gr. 1.12; b.p. 14.4 F.

Dymel® 152. [DuPont] 1,1-Difluoroethane; aerosol propellant; environmentally safe and low in toxicity; moderately flam.; m.w. 66.1; low water sol.; sp.gr. 0.91; b.p. –13 F.

Dymel® A. [DuPont] Dimethyl ether; aerosol propellant; environmentally safe and low in toxicity; low flam.; m.w. 46.1; high sol. in water; misc. with most conventional solvs.; sp.gr. 0.66; b.p. –12.7 F.

Dymsol® 31-P. [Henkel] Anionic surfactant; biodeg. mechanical and chemical stabilizer for latex; liq.; 55% act.

Dymsol® L. [Henkel] PEG fatty ester; nonionic; stabilizer, wetting agent, visc. controller for latex and resin emulsions; liq.; biodeg.; 100% act.

Dymsol® MS-40. [Henkel] Polyethylene emulsion; improves water resistance and antiblocking in adhesives and coatings; opaque wh. liq.; dens. 8.0 lb/gal.

Dymsol® N. [Henkel] Fatty amido condensate; stabilizer for natural and syn. latex; improves chemical resistance and visc. uniformity; liq.

Dymsol® S. [Henkel] Sulfated syn. sperm oil; anionic; emulsion breaker, solubilizer, coupling agent; fatliquors for tanning leather; greenish-br. liq.

Dynacerin 660. [Hüls] Oleyl erucate; jojoba oil substitute; emollient for cosmetic preparations; liq. wax; sol. in fat solvs.; water-insol.

Dynakoll. [Akzo] Rosin size; water repellent for paper prods.; liq.

Dynamar FC 5157. [3M] Vulcanizing agent for polyepichlorohydrin elastomers in conj. with Dynamar FX 5166; pink 1 in. sq. chips; sp.gr. 1.2; 75% act.

Dynamar FX 5166. [3M] Cure accelerator for polyepichlorohydrin elastomers in conj. with Dynamar FC 5157; wh. free-flowing powd.; sp.gr. 1.3; 70% act.

Dynasan 110. [Hüls] Tricaprin; emollient and lubricant for cosmetic applics. such as sticks, creams, lotions, powds.

Dynasan 114. [Hüls] Trimyristin; binder, lubricant for tablets and compressed confectioneries; powd.

Dynasan 116. [Hüls] Tripalmitin; lubricant in cosmetic powds., cakes, and production of tablets; microcryst. powds.; sol. in ether and benzene.

Dynasan 118. [Hüls] Tristearin; see Dynasan 116; microcryst. powds.; sol. in ether and benzene.

Dynasil A. [Hüls] Tetraethoxysilane; coupling agent, chem. intermediate, blocking agent, release agent, lubricant, primer, reducing agent; liq.; m.w. 208.3; sp.gr. 0.934; b.p. 169 C; flash pt. 60 C; ref. index 1.383; 98% purity.

Dynasil CA. [Hüls] Tetrakis (2-ethoxyethoxy) silane; see Dynasil A; liq.; m.w. 384; sp.gr. 1.02; b.p. 200 C (10 mm); flash pt. 131 C; 90% purity.

Dynasil CM. [Hüls] Tetrakis (2-methoxyethoxy) silane; see Dynasil A; liq.; m.w. 328.4; sp.gr. 1.079; b.p. 179–182 C (10 mm); ref. index 1.4219.

Dynasil M. [Hüls] Tetramethoxysilane; see Dynasil A; liq.; m.w. 152.2; sp.gr. 1.05; b.p. 121–122 C; flash pt. 45 C; ref. index 1.368; 99% purity.

Dynasperse A. [Daishowa] Sodium lignosulfonate;

anionic; primary dye dispersant, good heat stability and milling properties, low staining and very low dye reduction; powd.; 100% act.

Dynasperse B. [Daishowa] Sodium lignosulfonate; anionic; primary dye dispersant, good heat stability, and good milling properties; powd.; 100% act.

Dynasylan® AMEO. [Hüls] Aminopropyltriethoxysilane; coupling agent, chem. intermediate, blocking agent, release agent, lubricant, primer, reducing agent; liq.; m.w. 221.4; sp.gr. 0.95; b.p. 122–123 C (30 mm); flash pt. 96 C; ref. index 1.423; 98.5% purity.

Dynasylan® AMEO-40. [Hüls] Aminoalkyl silicone sol'n.; organo-silicon compds. for bonding between org. and inorg. components of a system; liq.; water-sol.

Dynasylan® AMEO-P. [Hüls] Aminopropyltriethoxy silane; see Dynasylan AMEO-40; liq.; water-sol.

Dynasylan® AMEO-T. [Hüls] Aminopropyltriethoxy silane, tech. grade; see Dynasylan AMEO-40; liq.; water-sol.

Dynasylan® AMMO. [Hüls] Aminopropyltrimethoxysilane; see Dynasylan AMEO; liq.; water-sol.; m.w. 179.2; sp.gr. 1.01; b.p. 80 C (8 mm); flash pt. 104 C; ref. index 1.420; 98% purity.

Dynasylan® BDAC. [Hüls] Di-t-butoxydiacetoxysilane; see Dynasylan AMEO; liq.; m.w. 292.4; sp.gr. 1.0196; b.p. 102 C (5 mm); flash pt. 95 C; ref. index 1.4040; 97% purity.

Dynasylan® BSA. [Hüls] Bis(trimethylsilyl)acetamide; see Dynasylan AMEO; liq.; m.w. 203.4; sp.gr. 0.83; b.p. 71–73 C (35 mm); flash pt. 17 C; ref. index 1.418; 95% purity.

Dynasylan® CPTEO. [Hüls] 3-Chloropropyl-triethoxysilane; see Dynasylan AMEO; liq.; m.w. 240.8; sp.gr. 1.009; b.p. 98–102 C (10 mm); ref. index 1.420.

Dynasylan® CPTMO. [Hüls] 3-Chloropropyltrimethoxysilane; see Dynasylan AMEO; liq.; m.w. 198.7; sp.gr. 1.081; b.p. 183 C; flash pt. 66 C; ref. index 1.464; 98% purity.

Dynasylan® DAMO. [Hüls] N-2-aminoethyl-3-aminopropyltrimethoxysilane; see Dynasylan AMEO; liq.; m.w. 222.4; sp.gr. 1.01; b.p. 140 C (15 mm); flash pt. 121 C; ref. index 1.442; 95% purity.

Dynasylan® DAMO-M. [Hüls] Aminoalkyl sol'n.; see Dynasylan AMEO-40; liq.; water-sol.

Dynasylan® DAMO-P. [Hüls] Aminoethylamino propyltrimethoxy silane; coupler for epoxy, phenolic, melamine, nylons, PVC, acrylics, polyolefins, PU and nitrile rubber; liq.; water-sol.

Dynasylan® DAMO-T. [Hüls] Aminoethylamino propyltrimethoxy silane, tech. grade; see Dynasylan DAMO-P; liq.; water-sol.

Dynasylan® GLYMO. [Hüls] 3-Glycidoxypropyltrimethoxysilane; see Dynasylan AMEO; liq.; m.w. 236.3; sp.gr. 1.070 b.p. 120 C (2 mm); flash pt. 79 C; ref. index 1.428; 98% purity.

Dynasylan® HMDS. [Hüls] Hexamethyldisilazane; see Dynasylan AMEO; liq.; m.w. 161.4; sp.gr. 0.77; b.p. 126–127 C; flash pt. 27 C; ref. index 1.408; 98% purity.

Dynasylan® IBTMO. [Hüls] Isobutyltrimethoxysilane; see Dynasylan AMEO; liq.; m.w. 178.3; sp.gr. 0.93; b.p. 154–157; flash pt. 14 C; ref. index 1.396; 98% purity.

Dynasylan® IMEO. [Hüls] N-[3-(Triethoxysilyl)-propyl] 4,5-dihydroimidazole; see Dynasylan AMEO; liq.; m.w. 274.1; sp.gr. 1.00; b.p. 134 C (2 mm); flash pt. 78 C; ref. index 1.45; 90% purity.

Dynasylan® MEMO. [Hüls] 3-Methacryloxypropyltrimethoxysilane; see Dynasylan AMEO; liq.; m.w. 248.4; sp.gr. 1.045; b.p. 190 C; flash pt. 92 C; ref. index 1.429; 97% purity.

Dynasylan® MSTFA. [Hüls] N-Methyl-N-trimethylsilyltrifluoroacetamide; see Dynasylan AMEO; liq.; m.w. 199.3; sp.gr. 1.074; b.p. 131–132 C; flash pt. 25 C; ref. index 1.380; 97% purity.

Dynasylan® MTES. [Hüls] Methyltriethoxysilane; see Dynasylan AMEO; liq.; m.w. 178.3; sp.gr. 0.90; b.p. 141–143 C; flash pt. 38 C; ref. index 1.383; 98% purity.

Dynasylan® MTMO. [Hüls] 3-Mercaptopropyltrimethoxysilane; see Dynasylan AMEO; liq.; m.w. 196.3; sp.gr. 1.05; b.p. 219–220; flash pt. 93 C; ref. index 1.440; 97% purity.

Dynasylan® MTMS. [Hüls] Methyltrimethoxysilane; see Dynasylan AMEO; liq.; m.w. 136.2; sp.gr. 0.955; b.p. 102–103 C; flash pt. 8 C; ref. index 1.3696; 98% purity.

Dynasylan® OCTEO. [Hüls] Octyltriethoxysilane; see Dynasylan AMEO; liq.; m.w. 276.5; sp.gr. 0.88; b.p. 98–99 C (2 mm); flash pt. 100 C; ref. index 1.415; 98% purity.

Dynasylan® TCS. [Hüls] Trichlorosilane; see Dynasylan AMEO; liq.; m.w. 135.5; sp.gr. 1.342; b.p. 31–32 C; flash pt. —8 C; ref. index 1.402; 99.9% purity.

Dynasylan® TRIAMO. [Hüls] Trimethoxysilylpropyldiethylene triamine; see Dynasylan AMEO; liq.; m.w. 265.4; sp.gr. 1.03; b.p. 114–168 C (2 mm); flash pt. 137 C; ref. index 1.463; 91% purity.

Dynasylan® VTC. [Hüls] Vinyltrichlorosilane; see Dynasylan AMEO; liq.; m.w. 161.5; sp.gr. 1.24; b.p. 92–93 C; flash pt. 21 C; ref. index 1.43; 97% purity.

Dynasylan® VTEO. [Hüls] Vinyl triethoxysilane; see Dynasylan AMEO; liq.; m.w. 190.3; sp.gr. 0.90; b.p. 160–161 C; flash pt. 34 C; ref. index 1.396; 98% purity.

Dynasylan® VTMO. [Hüls] Vinyltrimethoxysilane; see Dynasylan AMEO; liq.; m.w. 148.2; sp.gr. 1.13; b.p. 123 C; flash pt. 23 C; ref. index 1.393; 98% purity.

Dynasylan® VTMOEO. [Hüls] Vinyl tris (methoxyethoxy) silane; see Dynasylan AMEO; liq.; m.w. 280.4; sp.gr. 1.04; b.p. 284–286 C; flash pt. 71 C; ref. index 1.427; 97% purity.

Dynasylan® VTPMO. [Hüls] Vinyl tris (1-methoxy-2-propoxy) silane; see Dynasylan AMEO; liq.; m.w. 322; sp.gr. 0.98; b.p. 109 C (2 mm); flash pt. 147 C; ref. index 1.404; 97% purity.

Dyqex®. [Georgia-Pacific] Sodium lignosulfonate; anionic; dye disp. extender; 46% liq.; 100% powd.

Dyrene®. [Bayer] Anilazine; fungicide; colorless cryst.; m.w. 275.5; sol. (in 1000 ml solv.): 90 g in dichloromethane, 40 g in toluene, 8 g in 2-propanol, 8 mg in water.

Dytek A. [DuPont] 2-Methylpentamethylenediamine; epoxy curing agent; also used in PUs, wet strength resins, scale and corrosion inhibitors, motor oil and gasoline additives, polyamide plastics, films, adhesives, and inks; colorless liq.; weak ammonia, fish odor; m.w. 116.2; misc. with water, acetone, ethanol, cyclohexane, n-heptane, CCl_4, xylene; sp.gr. 0.86; visc. 2.9 cp; m.p. –56 to –60 C; b.p. 193 C (760 mm); f.p. < –70 C; ; flash pt. 82 C; > 98.5% act.

E-64. [Wacker Silicones] Food-grade release emulsion; water-dilutable; 35% act.

E-130. [Wacker Silicones] Dimethyl emulsion; general release agent; water-dilutable; visc. 100 cstk; 60% act.

E-131. [Wacker Silicones] Dimethyl emulsion; general release agent; water-dilutable; visc. 350 cstk; 60% act.

E-133. [Wacker Silicones] Dimethyl emulsion; general release agent; water-dilutable; visc. 1000 cstk; 60% act.

E-155. [Wacker Silicones] Silicone fluid; lubricant for metal-to-nonmetal or nonmetal applics.; urethane/epoxy release agent; sol. in alcohol, aromatic, aliphatic, and chlorinated solvs.

E-170. [Wacker Silicones] Silicone mica glycol emulsion; manderal release; water-dilutable.

Ease Release 200 Series. [George Mann] General purpose release agent for release of most types of casting and molding systems, e.g., PU elastomers, PU foam, epoxy resin, polyester, RTV silicones, rubber, and thermoplastic polymers; effective on aluminum, chrome, RTV silicone, epoxy, polyester, rubber, and steel molds; grades avail.: 200—solv. trichloromonofluoromethane, dichlorodifluoromethane; 201—solv. trichloromonofluoromethane, dens. 12 lb/gal; 202—solv. methylene chloride, dens. 10.8 lb/gal; 203—solv. Chlorothene® SM, dens. 10.8 lb/gal; 204—solv. trichlorotrifluoroethane, dens. 12.6 lb/gal; 205—solv. petrol. hydrocarbons, dens. 6 lb/gal; 206—solv. hexane, dens. 5.7 lb/gal.

Ease Release 300 Series. [George Mann] Release agent for urethane elastomers and foams, and epoxy resins; effective on steel, aluminum, chrome, epoxy, RTV urethane, and silicone rubber mold surfaces; grades avail.: 300—solv. trichloromonofluoromethane, dichlorodifluoromethane; 301—solv. trichloromonofluoromethane, dens. 12 lb/gal; 302—solv. methylene chloride, dens. 10.8 lb/gal; 303—solv. Chlorothene® SM, dens. 10.8 lb/gal; 304—solv. trichlorotrifluoroethane, dens. 12.6 lb/gal; 305—solv. petrol. hydrocarbons, dens. 6 lb/gal; 306—solv. hexane, dens. 5.7 lb/gal.

Ease Release 400 Series. [George Mann] Release agent for PU elastomers, thiokol rubbers, EPDM, and other syn. elastomers from aluminum, chrome, epoxy, polyester, rubber, and steel molds; grades avail.: 400—solv. trichloromonofluoromethane, dichlorodifluoromethane; 401—solv. trichloromonofluoromethane, dens. 12.1 lb/gal; 402—solv. methylene chloride, dens. 10.9 lb/gal; 403—solv. Chlorothene® SM, dens. 10.9 lb/gal; 404—solv. trichlorotrifluoroethane, dens. 12.8 lb/gal; 405—solv. petrol. hydrocarbons, dens. 6 lb/gal; 406—solv. hexane, dens. 5.7 lb/gal.

Ease Release 500 Series. [George Mann] General purpose release and dry lubricant containing no silicone or oils; releases most casting and molding systems, e.g., PU elastomers, PU foams, epoxy resins, polyester, rubbers, thermoplastic polymers, phenolic resins from aluminum, chrome, epoxy, polyester, rubber, and steel molds; grades avail.: 500—solv. trichloromonofluoromethane, dichlorodifluoromethane; 504—solv. trichlortrifluoroethane, dens. 13 lb/gal; 507—solv. water, dens. 8.6 lb/gal.

Ease Release 2040 Series. [George Mann] Semipermanent release coating providing long lasting release for urethane, epoxy, rubber, aluminum, and steel molds, molding presses, coating machinery; effective with urethanes, rubbers, epoxies, hot-melt adhesives, polyesters, and other polymers and gums; grades avail.: 2040; 2044—dens. 12.3 lb/gal; 2045—solv. petrol. hydrocarbons, dens. 6.2 lb/gal.

Eastman® DTBHQ. [Eastman] 2,5-Di-tert-butylhydroquinone; antioxidant for rubber, polyesters.

Eastman® HQMME. [Eastman] Hydroquinone monomethyl ether; antioxidant for monomers.

Eastman® Inhibitor DOBP. [Eastman] 4-Dodecyloxy-2-hydroxybenzophenone; industrial grade uv absorber/stabilizer for polyethylene and PP; suitable for unsat. polyesters, PS, PVC, CAB, NC, urethane, and acrylic surf. coatings; also used as a screening agent in protecting rubber, flourescent pigments, polishes, and papers from the degrading effects of uv lt.; pale yel. flakes; little or no odor; m.w. 382.5; sol. 0.1% in water; 81% benzene; 55% in acetone; 40% in bis(2-ethylhexyl)phthalate; 29% in hexane; dens. 4.1 lb/gal (20 C); flash pt. (COC) 254 C.

Eastman® Inhibitor OABH. [Eastman] Oxalyl bis(benzylidenehydrazide); stabilizer used in polyolefins in contact w/ copper or copper-containing alloys; copper deactivator; wh. crys. powd., 99% min. thru 200 mesh; m.w. 294.16 (calcd.); sol. 5540 mg in n-methyl-2-pyrrolidinone (105 C); 1990 mg in di-methylformamide (88 C); 70.4 mg in 1,4-butanediol (102 C); 59.9 g in ethylene glycol (98 C); 67.4 mg in 1,3-propanediol (100 C); 191 mg in acetic acid (81 C); dens. 0.216 kg/l; sapon. no. 144–150; 97.0% min. assay.

Eastman® Inhibitor RMB. [Eastman] Resorcinol monobenzoate; industrial grade UV absorber/stabilizer for cellulosic plastics and PVC formulations; UV absorber; wh. cryst. solid; odorless; m.w. 214.2; b.p. 140 C; sol. (g/100 g solv.); 75.2 g in acetone; 27.7 g in ethanol; > 0.1 g in benzene, n-hexane, DOP, and water; sp.gr. 0.60; dens. 0.68 kg/l; m.p. 132–135 C; flash pt. 214 C (COC).

Eastman® MTBHQ. [Eastman] Mono-tert-butylhydroquinone; antioxidant for rubber, monomers.

Eastman® TDP 2000. [Eastman] Polymeric stabilizer, sec. antioxidant for PP; sol. in polyolefins.

Eastobrite OB-1. [Eastman] 2,2′-(1,2-Ethenediyldi-4,1-phenylene)bisbenzoxazole; optical brightener; fluorescent whitening agent for use in linear polyester, PET, nylon fibers; yel. cryst. solid; water-insol.; m.w. 414.4; sp.gr. 1.39; m.p. (DTA): 359 C.

E.B. Golden Glitter, Neutral Glitter. [Eastern Color] Pearlescent material and aq. acrylic resinous binders; for decorative textile printing.

EBS Wax. [GE] N,N′-ethylenebisstearamide; wax functioning as internal and external lubricant for plastics incl. ABS, PVC, PS, polyolefins, and thermosets; mold release and antiblocking agent for plastics; antiblock and antislip agent; defoamer for latex, paper processes, and fiber finishings; dispersant for pigments and flame retardants in polymers; off-wh. beads; m.w. 580; sol. in toluene, naphtha and partly sol. in ethanol, IPA upon heating; sp.gr. 0.97; m.p. 140–145 C; flash pt. (COC) 285 C; acid no. 7.0 max.; neutral value 3 max.

Ecco MP®. [Eastern Color] Cu-8/G-4; textile fungicide; solv./water sol. G-4.

Eccobrite RB. [Eastern Color] Stilbene; whitening agent for cotton and acetates; water-sol.

Ecco Defoamer. [Eastern Color] Silicone types; foam inhibitor for dyeing, printing, and general processing; water-disp.

Ecco Defoamer S. [Eastern Color] Petrol./silicone; defoamer for mills waste water; clear liq.

Eccolene OW. [Eastern Color] Sulfated fatty ester; anionic; wetting aid, dispersant, emulsifier, penetrant and scouring agent for textile processing; liq.; 65% conc.

Eccopel. [Eastern Color] Metal salts; semidurable water repellent; water-sol.

Ecco Polyester® Optical 525. [Eastern Color] Benzoxazole; whitening agent for polyester fiber to give optical brightness; disp. in water.

Eccoro®. [Eastern Color] Cationic/anionic imidazolines; corrosion resistance of metals; water-sol.

Ecco Spersant PDQ. [Eastern Color] Ethoxylated phenoxy condensate; nonionic; dispersant for pigments at neutral pH; liq.

Eccospin PPF. [Eastern Color] Ethoxylated/propoxylated alcohol; lubricant and antistat for PP fibers; spin finish for fiber processing; clear liq.

Eccostat. [Eastern Color] Cationic/nonionic; antistat for textiles; water-sol.

Eccoterge EO. [Eastern Color] Phenol, ethoxylated; nonionic; detergent, wetting and dye leveling agent for fabrics; kier boiling assistant; liq.; 100% conc.

Eccoterge SCH. [Eastern Color] Complexed amide, ethoxylated; anionic; wetting, scouring agent, and dyeing assistant for textiles; liq.; 60% conc.

Eccotex P. Conc. [Eastern Color] Alkylaryl compd., modified; anionic; dispersant, penetrant used in textile industry; wetting of dyestuffs; gel; 60% conc.

Eccowet® LF Conc. [Eastern Color] Sodium alkyl naphthalene sulfonate; anionic; wetting agent, dispersant used in dyeing applics.; liq.; 60% conc.

Eccowet® W-50. [Eastern Color] Sodium aliphatic ester sulfonate; anionic; wetting, penetrant used in textile industry; liq.; 50% conc.

Ecco White® FW-5. [Eastern Color] Distyrylphenyl; optical brightener for nylon and blends; water-sol.

Ecco White® OP. [Eastern Color] Benzoxazole; optical brightener for polyesters and blends; disp. in water.

Edenol 302. [Henkel] Propylene glycol dicaprylate/dicaprate; emollient oil and cosolv. for personal care prods.; yel. low visc. liq.; sol. in alcohols, min. oils, ketones; insol. in water; dens. 0.92 g/ml; HLB 9; solid. pt. < 0 C; cloud pt. < 0 C; sapon. no. 315–335; ref. index 1.435.

Edenol DCHP. [Henkel] Dicyclohexyl phthalate; plasticizer for foil lacquers, cellophane lacquers; powd.

Edenol TFO. [Henkel] Tetrahydrofurfuryl oleate; release agent for calender coatings; liq.

Edenol W 300 S. [Henkel] Diisotridecyl phthalate; barely volatile plasticizer for NC, vinyl, and other coatings; liq.

Effcol TS-214. [Matsumoto] Linear hydrocarbon sulfonate; anionic; antistat agent for warp sizing; liq.; 40% conc.

EGDS-VA. [Goldschmidt] Ethylene glycol distearate; pearling agent, opacifier; flake.

EGMS-VA. [Goldschmidt] Ethylene glycol stearate; pearling agent, opacifier; flake.

EHP. [PPG Industries] Di(2-ethylhexyl) peroxydicarbonate; initiator for polymerization of unsat. monomers; reduces time of batch runs; increases prod.; makes polymerization easier to control; improves polymer chars.; more sol. in solvs. used to dilute peroxydicarbonate polymerization initiators; colorless clear sol'ns.; m.w. 346; f.p. < –78 C; dens. 0.906 and 0.828 g/ml (75 and 40% resp.); visc. 12.9 and 4.7 cSt (75 and 40% resp.); 75% and 40% sol'ns. resp.; 3.46 and 1.85% act. oxygen resp.

Ekaline F. [Sandoz] Aliphatic polyglycol ether; nonionic; agent for washing off dyeings; leveling and dispersing agent; liq., flakes.

Ektapro EEP Solvent. [Eastman] Linear ether ester; high-performance retarder solv. for formulating enamels, lacquers, topcoats, and primers; urethane-grade quality; polymerization solv.; Pt-Co 15 max. color; moderate odor; m.w. 146.19; f.p. < –50 C; b.p. 165 C (760 mm); 2.9% sol. in water; sp.gr. 0.95; dens. 7.91 lb/gal; visc. 1.0 cP; flash pt. (Seta) 58 C; surf. tens. 27.0 dynes/cm (23 C); elec. resist. 20 megohms.

Ektasolve® DM. [Eastman] Diethylene glycol monomethyl ether; evaporating solv. used in brushing lacquers and dye stains; useful in wood stains, printing inks, and dye pastes for textiles; coalescing aid for PVAc latex paints; used in stamp pad and stencil inks; diluent for hydraulic brake fluids; water-wh. liq.; agreeable odor; m.w. 120.15; f.p. –85 C; b.p. 191 C min.; water-sol.; sp.gr. 1.023; dens. 1.02 kg/l; ref. index 1.4263; flash pt. 96 C (COC); fire pt. 96 C; > 99% volatiles by vol.

Ektasolve® DP. [Eastman] Diethylene glycol monopropyl ether; evaporating, water-misc. solv. used in sol'n. and water-dilutable coatings; act. for many coating materials incl. NC, acrylic copolymers, epoxy resins, chlorinated rubber, and alkyd resins; strong coupling agent with some resin systems in water-dilutable coatings; colorless clear liq.; mild odor; m.w. 148.2; f.p. < –90 C; b.p. 202 C min.; water-sol.; sp.gr. 0.963; dens. 0.96 kg/l; ref. index 1.429; flash pt. 93 C (TCC); fire pt. 103 C; > 99% volatiles by vol.

Ektasolve® EB. [Eastman] Ethylene glycol monobutyl ether; solv. for alkyd, phenolic, maleic, and cellulose nitrate resins; excellent retarder for lacquers, improving gloss and flow-out, blush resistance, and reducing the formation of orange peel; useful in formulating hot-spray, brushing, flow-coat, and aerosol lacquers; colorless liq.; faint odor; m.w. 118.17; f.p. –75 C; b.p. 169.0 C min.; water-sol.; sp.gr. 0.902; dens. 0.90 kg/l; ref. index 1.4193; flash pt. 62 C (TCC); fire pt. 70 C; > 99% volatiles by vol.

Ektasolve® EB Acetate. [Eastman] Ethylene glycol monobutyl ether acetate; high-boiling; useful as coalescing aid for latex paints; used in multicolor lacquers, lacquer emulsions; retarder in high-low lacquer thinners, printing inks, and epoxy coatings; used as solv. in silk-screen, stamp pad, and stencil inks and component of varnish removers; colorless liq.; mild odor; m.w. 160.21; f.p. –64 C; b.p. 186 C min.; limited sol. in water; sp.gr. 0.942; dens. 0.94 kg/l; ref. index 1.4200; flash pt. 71 C (TCC); fire pt. 82 C; > 99% volatiles by vol.

Ektasolve® EE. [Eastman] Ethylene glycol monoethyl ether; solv. with low evaporation rate and complete water misc.; imparts excellent flow properties and high gloss to thermoplastic and thermosetting coating systems; for cellulosic lacquers; useful in hot-spray lacquers in combination with esters or ketones; used in dyes, alkyd baking enamels, varnish removers, metal and glass cleaners; aid in printing and dyeing of textiles and leathers; act. in epoxy/polyamide coatings; retarder in flexographic printing inks, coupling agent in water-borne coating systems; component in hydraulic fluids; colorless liq.; mild, agreeable odor; m.w. 90.12; f.p. –94 C; b.p. 134 C min.; water-sol.; sp.gr. 0.930–0.932; dens. 0.93 kg/l; ref. index 1.4080; flash pt. 43 C (TCC); fire pt. 48 C; 100% volatiles by vol.

Ektasolve® EE Acetate. [Eastman] Ethylene glycol monoethyl ether acetate; solv.; colorless liq.; mild, ester-like odor; b.p. 156 C; sol. 23.9% in water; sp.gr. 0.975; flash pt. 54 C (TCC); > 99% volatiles by vol.

Ektasolve® EM. [Eastman] Ethylene glycol monomethyl ether (2-methoxyethanol); solv. with lowest boiling range of any commerical glycol ether; for cellulose nitrate, cellulose acetate, CAB, and other film formers used in lacquer industry; for nonreactive urethane elastomers, dyes and oils used in wood-staining and printing inks; used in coating PC plastic; adhesive for cellophane and cellulose acetate plastics; used in mfg. of plasticizers and as a penetrating and leveling agent in dyeing of leather, animal, and veg. fibers; used in printing inks, nonaq. wood stains, and hydraulic brake fluids; Pt-Co 10 max. clear liq.; char. odor; m.w. 76.11; f.p. –85 C; b.p. 123.5 C min.; water-sol.; sp.gr. 0.965; dens. 0.96 kg/l; ref. index 1.4023; flash pt. 39 C (TCC); fire pt. 44 C; 100% volatiles by vol.

Ektasolve® EP. [Eastman] Ethylene glycol monopropyl ether; slow evaporating solv. used in coatings; useful in waterborne coating systems; coupling solv. for resin/water systems; controls visc. of waterborne resins; effective for NC, acrylic, epoxy, polyamide, and alkyd resins; retarder in coating systems; colorless clear liq.; mild odor; m.w. 104.15; f.p. < –90 C; b.p. 149.5 min.; water-sol.; sp.gr. 0.9125; dens. 0.91 kg/l; ref. index 1.4136; flash pt. 49 C (TCC); fire pt. 56 C; 100% volatiles by vol.

Ektasolve® PM Acetate. [Eastman] Retarder solv.; b.p. 142–147 C; 25.9% sol. in water; dens. 7.91 lb/gal; flash pt. 46 C; surf. tens. 26.4 dynes/cm; elec. resist. 6 megohms.

Elacid CLR. [Henkel Canada] Cationic hair conditioner conc.; ivory paste; faint char. odor.

Elastocarb Tech Light, Tech Heavy. [Morton Int'l.] Magnesium carbonate; inorg. filler providing flame retardancy and smoke suppression to elastomers, plastics, and thermosets incl. EPDM, PP, PE, PVC; used in wire and cable compds., conduit/tubing, film and sheet; wh. powd. (flat platelets); 2 and 3 μ avg. particle size resp.; 0.5% retained 325 mesh; bulk dens. 0.2 and 0.26 g/cc.

Elastocarb UF. [Morton Int'l.] Magnesium carbonate; precipitated ultra-fine grade filler for use in conduit, wire and cable compds., aircraft interior components, other transportation uses; reduces smoke dens. and increases physical properties; wh. fine powd.; 90% < 5 μ, 40% < 1 μ; 40% min. magnesium oxide, 34% max. carbon dioxide.

Elastomag® 100. [Akrochem] Magnesium oxide; chemical thickener for polyester resins; anticaking agent; used in syn. rubber compding., adhesives, fuel oil additives, and as acid acceptor for specialty plastics; powd.; 99.9% thru 325 mesh; sp.gr. 3.2; bulk dens. 18 lb/ft^3 (aerated); surf. area 104–141 m^2/g; 98% MgO.

Elastomag® 100R. [Akrochem] Magnesium oxide; see Elastomag 100; powd.; 99.9% thru 325 mesh; sp.gr. 3.2; bulk dens. 18 lb/ft^3 (aerated); surf. area 86–113 m^2/g; 98% MgO.

Elastomag® 170. [Akrochem] Magnesium oxide; see Elastomag 100; 170 grade has highest activity of the Elastomag MgOs; powd.; 99.9% thru 325 mesh; sp.gr. 3.2; bulk dens. 18 lb/ft^3 (aerated); surf. area 141–188 m^2/g; 98% MgO.

Elastomag® 170 Micropellet. [Akrochem] Magnesium oxide; see Elastomag 100; dust-free micropellet; 40.0% max. < 70 mesh; 0.1% max. > 20 mesh;

sp.gr. 3.2; bulk dens. 24 lb/ft^3 (aerated); surf. area 141–188 m^2/g; 98% MgO.

Elastomag® 170 Special. [Akrochem] Magnesium oxide; chemical thickener for polyester resins; used in adhesives, fuel oil additives, and as acid acceptor for specialty plastics and rubber; powd.; 99.9% thru 325 mesh; 50.0% min. < 1 μ; sp.gr. 3.2; bulk dens. 18 lb/ft^3 (aerated); surf. area 132 m^2/g; 98% MgO.

Elastosol. [Croda] Sol. animal elastin, sol. animal collagen; conditioner, moisturizer for skin care prods., eye wrinkle creams; wh. to off-wh. visc. liq.; m.w. 200,000; 2% act.

Elcema® G 250, P 100. [Degussa] Cellulose NF; anticaking agent, tabletting aid; wh. gran. and wh. powd. resp.

Elec AC. [Kao] Amphoteric surfactant; antistat for plastics; liq.

Elec QN. [Kao] Cationic surfactant; antistat for plastics; liq.

Elec RC. [Kao] Anionic surfactant; antistat for plastics; powd.

Elec TS-5, TS-6. [Kao] Nonionic surfactant; antistat for plastics; needles.

Elec-2. [Kao] Ethoxylated fatty amine; nonionic; antistat for plastics; liq./paste; 100% act.

Elfacos C26. [Akzo BV] Hydroxyoctacosanyl hydroxystearate; nonionic; consistency regulating agent for w/o emulsions, cosmetic applics.; stabilizer; waxy substance for decorative cosmetics; pellets; 100% conc.

Elfacos ST 9. [Akzo BV] PEG-45/dodecyl glycol copolymer; nonionic; stabilizer for w/o emulsions and emollient used in cosmetics; paste; HLB 7.0; 100% conc.

Elfacos ST 37. [Akzo BV] PEG-22/dodecyl glycol copolymer; nonionic; see Elfacos ST 9; also binder in cosmetic compressed powds., emollient for emulsions and decorative cosmetics, consistency regulator for w/o emulsions; liq.; 100% conc.

Elfan® L 310. [Akzo BV] Ethylene glycol distearate; anionic; pearlescent for shampoos and bubble baths; lt. yel. solid, flakes; water disp.; sp.gr. 0.96; m.p. 64–66 C; acid no. 6 max.; sapon. no. 200; 100% conc.

Elimina-254. [Matsumoto] POE fatty alcohol sulfate; anionic; detergent, dispersant, antistat, emulsifier used in detergent and shampoo formulations, latex, dyestuffs, pigments, oil and waxes; used for syn. fibers; paste; 30% conc.

Eltesol® 4009, 4018. [Albright & Wilson] Xylene sulfonic acid modified with methanol and sulfuric acid; catalysts for curing cold-setting phenol-formaldehyde and phenol-furane resins used in the foundry industry as binders for sand in the prod. of molds and cores; clear brn. liq.; sp.gr. 1.25 and 1.30 resp.; visc. 165–195 cs and 140–170 cs; flash pt. (CC) 72 and 93 C.

Eltesol® ACS 60. [Albright & Wilson] Ammonium cumenesulfonate; hydrotrope, solubilizer, coupling agent, and visc. modifier in liq. formulations; cloud pt. depressant in detergent formulations; pale pink liq.; dens. 1.10 g/cm^3; pH 7.0–8.0 (10% aq.); 60.0% min. act.

Eltesol® AX 40. [Albright & Wilson] Ammonium xylene sulfonate; anionic; hydrotrope, cloud pt. depressant used in the detergent mfg.; solubilizer, coupler; pale yel. liq.; pH 7–8.5; 41% act.

Eltesol® CA 65. [Albright & Wilson] Cumene sulfonic acid; anionic; catalyst for foundry resins; descaling agent for metal cleaning; antistress additive and plating aid in electroplating bath; curing aid in the plastics industry; raw material in the mfg. of dyes and pigments; detergents industry; dens. 1.15 g/cm^3; visc. 30 cs; 65 ± 1.0% act. in water.

Eltesol® CA 96. [Albright & Wilson] Cumene sulfonic acid conc.; anionic; see Eltesol CA65; brn. visc. liq.; dens. 1.25 g/cm^3; visc. 1000 cs; 96.0 ± 1.0% act.

Eltesol® MGX. [Albright & Wilson] Magnesium xylene sulfonate; hydrotrope for liq and spray-dried detergent formulations; powd.

Eltesol® PSA 65. [Albright & Wilson] Phenol sulfonic acid; anionic; see Eltesol CA-65; mfg. pharmaceutical chemicals and disinfectants; dens. 1.3 g/cm^3; visc. 50 cs; 65.0 ± 1.0% act. in water.

Eltesol® PT 45. [Albright & Wilson] Potassium toluene sulfonate; see Eltesol MGX; liq.

Eltesol® PT 93. [Albright & Wilson] Potassium toluene sulfonate; anionic; see Eltesol AX 40; off wh. powd.; pH 9–10.5; 93% act.

Eltesol® PX 40. [Albright & Wilson] Potassium xylene sulfonate; see Eltesol ACS 60; straw liq.; dens. 1.10 g/cm^3; pH 7.0–10.5 (10% aq.); 40.0 ± 1.0% act. in water.

Eltesol® PX 93. [Albright & Wilson] Potassium xylene sulfonate; anionic; see Eltesol AX 40; off wh. powd.; pH 9–10.5; 93% act.

Eltesol® SCS 40. [Albright & Wilson] Sodium cumene sulfonate; hydrotrope, solubilizer for spray-dried prods.

Eltesol® ST 34. [Albright & Wilson] Sodium toluene sulfonate; anionic; see Eltesol AX 40; pale yel. liq.; pH 7–10; 34% act.

Eltesol® ST 40. [Albright & Wilson] Sodium toluene sulfonate; see Eltesol ACS-60; straw liq.; dens. 1.10 g/cm^3; pH 7.0–10.5 (10% aq.); 40.0±1.0% act.

Eltesol® ST 90. [Albright & Wilson] Sodium toluene sulfonate; anionic; see Eltesol AX 40; off wh. powd., pellets; pH 9–10.5; 89% act.

Eltesol® ST Pellets. [Albright & Wilson] Sodium toluene sulfonate; see Eltesol ACS-60; wh. pellets; dens. 0.55 g/cm^3; pH 9.0–10.5 (3% aq.); 85.0% min. act.

Eltesol® SX 30. [Albright & Wilson] Sodium xylene sulfonate; anionic; see Eltesol AX 40; pale yel. liq.; pH 7–10; 30% act.

Eltesol® SX 93. [Albright & Wilson] Sodium xylene sulfonate; anionic; see Eltesol AX 40; off wh. powd., pellets; pH 9–10.5; 93% act.

Eltesol® SX Pellets. [Albright & Wilson] Sodium xylene sulfonate; see Eltesol ACS-60; wh. pellets; dens. 0.5 g/cm^3; pH 9.0–10.5 (3% aq.); 88.0% min. act.

Eltesol® TA. [Albright & Wilson] Toluene sulfonic acid; anionic; see Eltesol CA-65; hydrotrope, intermediate; used in mfg. of acrylonitrile; dens. 1.2 g/

cm^3; visc. 15 cs; 61.2% min. act. in water.

Eltesol® TA 65. [Albright & Wilson] 65% Toluene sulfonic acid and 1.4% sulfonic acid aq. sol'n.; anionic; intermediate, catalyst in foundry and chemical industries, hardening agent in plastics, activator for nicotine insecticides; lt. amber clear liq.; dens. 1.2 g/cc; visc. 9–12 cs.

Eltesol® TA 96. [Albright & Wilson] Toluene sulfonic acid; see Eltesol TA 65; brn. thick cryst. liq.; dens. 1.3 g/cm^3; 97% act.

Eltesol® TA/E. [Albright & Wilson] 52% Toluene sulfonic acid and 11.5% sulfuric acid aq. sol'n.; catalyst for curing cold-setting foundry resins; clear amber/brn. liq.; sp.gr. 1.26; visc. 10 cs.

Eltesol® TA/F. [Albright & Wilson] 63% Toluene sulfonic acid and 1.2% sulfuric acid aq. sol'n.; catalyst for curing cold-setting foundry resins; descaling agent for metal cleaning; antistress additive and plating aid in electroplating bath; curing aid in the plastics industry; raw material in the mfg. of dyes and pigments; detergents industry; hydrotrope, intermediate; clear amber/brn. liq.; sp.gr. 1.22; visc. 15 cs.

Eltesol® TA/H. [Albright & Wilson] 54% Toluene sulfonic acid and 9% sulfuric acid aq. sol'n.; catalyst for curing cold-setting foundry resins; clear amber/brn. liq.; sp.gr. 1.26; visc. 10 cs.

Eltesol® TA/K. [Albright & Wilson] 64% Toluene sulfonic acid and 4% sulfuric acid aq. sol'n.; catalyst for curing cold-setting foundry resins; clear amber/brn. liq.; sp.gr. 1.25; visc. 17 cs.

Eltesol® XA. [Albright & Wilson] Xylene sulfonic acid aq. sol'n.; see Eltesol CA-65; amber/brn. clear liq.; dens. 1.2 g/cm^3; visc. 15 cs; 61.0% min. act.

Eltesol® XA65. [Albright & Wilson] Xylene sulfonic acid; anionic; intermediate, catalyst in preparation of esters, hardening agent in plastics, activator for nicotine insecticides; lt. amber, clear to slightly hazy liq.; dens. 1.2 g/cc; visc. 9–12 cs; 65% act.

Eltesol® XA90. [Albright & Wilson] Xylene sulfonic acid conc.; anionic; see Eltesol CA–65; brn. visc. liq.; dens. 1.35 g/cm^3; visc. 3000 cs; 95.0 ± 2.0% act.

Eltesol® XA/M65. [Albright & Wilson] Xylene sulfonic acid; see Eltesol XA65.

Elvanol 71-30. [DuPont] Fully hydrolyzed PVAL; film-forming binder used in adhesives, paper, paperboard sizing and coatings, textiles, films, building prods., hoses, gaskets, emulsification in emulsions and latices, additive for concrete, cement, and food; wh. gran.; sol. in hot water, ethanol; insol. in cold water; sp.gr. 1.30; dens. 400–432 kg/m^3; visc. 28–32 cps (4% aq.); sapon. no. 3–12; ref. index 1.54; pH 5.0–7.0; tens. str. 117 MPa; tens. elong. 10% (break) to 400%; hardness > 100; biodeg.

Elvanol 75-15. [DuPont] Fully hydrolyzed PVAL; visc. stabilizer and imparts gel resistance to aq. sol'ns.; used in adhesives, films, paper applic.; binder in cement and ceramic powd.; emulsifier; food pkg. adhesive; wh. gran.; slurries easily in cold water; visc. 13–15 cps (4% aq.); pH 5.0–7.0; tens. str. 55.2–138 MPa; biodeg.

Elvanol 90-50. [DuPont] Fully hydrolyzed PVAL; provides high film strength and binding power in low visc. systems; used in paper and paperboard coating and sizing, adhesives; pigment binder; food additive applic.; visc. 13–15 cps (4% aq.); pH 5.0–7.0.

Elvanol T-25. [DuPont] PVAL; for textile warp sizing, esp. polyester/cotton blends and other spun yarns; weaving performance; wh. gran.; slurries easily in cold water without lumping and dissolves readily on heating; bulk dens. 512 kg/m^3 (packed); visc. (Hoeppler Falling Ball) 25–31 cps (4% aq.); pH 5.0–7.0 (sol'n.).

Elvanol T-66. [DuPont] PVAL; unique grade for textile warp sizing (polyester and polyester blend spun yarns); smooth slasher operation, easy desize, weaving performance; wh. gran.; slurries easily in cold water without lumping and dissolves readily on heating to 85 C or higher for 20–30 min; bulk dens. 513 kg/m^3 (packed); visc. (Hoeppler Falling Ball) 13–15 cps (4% aq.); pH 5.0–7.0 (sol'n.); biodeg.

EM-90. [Keil] Fatty ester; sperm oil replacement; lubricant; oil-sol.

EM-980. [Keil] Modified alkanolamide; solubilizer; detergency and lubricity used in floor cleaners, paint strippers, buffing compds.; water-sol.

EM-985. [Keil] Modified alkanolamide; see EM-980; oil-sol.

EM-9400. [Keil] Fatty ester; blown sperm oil replacement; lubricant; oil-sol.

Emal 20C. [Kao] Sodium POE alkyl ether sulfate; anionic; detergent, emulsifier, foam stabilizer, shampoo base; scouring agent for syn. fibers; liq.; 25% conc.

Emal E-25C, -70C. [Kao] Sodium POE alkyl ether sulfate; anionic; detergent, emulsifier, foam stabilizer, shampoo base; liq.; 70% conc.

Emal NC-35. [Kao] Sodium POE alkylaryl ether sulfate; detergent, emulsifier, foam stabilizer, shampoo base; scouring agent for syn. fibers; liq.; 35% conc.

Emanon 3199, 3299R. [Kao] PEG fatty acid esters; thickener for cosmetic, pigment preps.; flakes.

Emanon 4110. [Kao] PEG fatty acid ester; thickener for cosmetic, pigment preps.; liq.

Emargol® KL. [Witco] Sodium sulfoacetate of mono- and diglycerides; anionic; food processing defoamer and o/w emulsifier; dough modifier; paste; disp. in oil and water.

Emasol L-106. [Kao] PEG-4 sorbitan laurate; nonionic; emulsifier; solubilizer for colorants; liq.; HLB 13.3; 100% conc.

Emasol L-120. [Kao] PEG-20 sorbitan laurate; nonionic; emulsifier for pharmaceuticals and cosmetics; stabilizer for emulsion polymerization; liq.; HLB 16.7; 100% conc.

Emasol O-10. [Kao] Sorbitan oleate; nonionic; dispersant for colorants; liq.; HLB 4.3; 100% conc.

Emasol O-15 R. [Kao] Sorbitan sesquioleate; nonionic; emulsifier and lubricant; liq.; HLB 3.7; 100% conc.

Emasol O-30. [Kao] Sorbitan trioleate; nonionic; see Emasol O-10; liq.; HLB 1.8; 100% conc.

Emasol O-105 R. [Kao] POE sorbitan oleate; nonionic; emulsifier and solubilizer; liq.; HLB 10.0;

100% conc.

Emasol O-106. [Kao] PEG-5 sorbitan oleate; nonionic; see Emasol O-105 R; liq.; HLB 10.0; 100% conc.

Emasol O-120. [Kao] PEG-20 sorbitan oleate; nonionic; see Emasol O-105 R; liq.; HLB 15.0; 100% conc.

Emasol O-320. [Kao] PEG-20 sorbitan trioleate; nonionic; see Emasol O-105 R; liq.; HLB 11.0; 100% conc.

Emasol P-10. [Kao] Sorbitan palmitate; nonionic; see Emasol O-10; solid; HLB 6.7; 100% conc.

Emasol P-120. [Kao] PEG-20 sorbitan palmitate; nonionic; emulsifier for pharmaceuticals and cosmetics; solubilizer for colorants; stabilizer for emulsion polymerization; liq.; HLB 15.6; 100% conc.

Emasol S-10. [Kao] Sorbitan stearate; nonionic; see Emasol O-10; solid; HLB 4.7; 100% conc.

Emasol S-20. [Kao] Sorbitan distearate; nonionic; emulsifier and lubricant; flake; HLB 4.4; 100% conc.

Emasol S-30. [Kao] Sorbitan tristearate; nonionic; see Emasol O-10; HLB 2.1; 100% conc.

Emasol S-106. [Kao] PEG-4 sorbitan stearate; nonionic; see Emasol S-20; solid; HLB 9.6; 100% conc.

Emasol S-120. [Kao] PEG-20 sorbitan stearate; nonionic; see Emasol S-20; solid; HLB 14.9; 100% conc.

Emasol S-320. [Kao] PEG-20 sorbitan tristearate; nonionic; see Emasol S-20; solid; HLB 10.5; 100% conc.

Emcol® 14. [Witco] Polyglycerol oleate; nonionic; w/o emulsifier; corrosion inhibitor for aerosol; spreader sticker, antifoaming agent; liq.; 100% conc.

Emcol® 150. [Witco] Isopropyl alcohol and stearyl alcohol; cationic; cosmetics and toiletries surfactant functioning as antistat, conditioner, lubricant, and substantive agent; paste; disp. in water; 25% act.

Emcol® 1655. [Witco] Cocamidopropyl dimethylamine propionate; cationic; cosmetics and toiletry surfactant used as antistat, conditioner, emollient, foaming and substantive agent; liq.; water-sol.; 40% conc.

Emcol® 3780. [Witco] Stearamidopropyl dimethylamine lactate; cationic; see Emcol 1655; water-sol.; 25% act.

Emcol® 4072. [Witco] Disodium hydrog. cottonseed glyceride sulfosuccinate; anionic; cosmetics and toiletries surfactant used as conditioner, emollient, emulsifier; paste; water-sol.

Emcol® 4100M. [Witco] Disodium myristamido MEA-sulfosuccinate; anionic; dispersant, wetting agent, foaming agent, detergent, emulsifier for bubble bath, shampoo, carpet and upholstery cleaners, textiles; wh. creamy semisolid; disp. in water; sp.gr. 1.01; acid no. 4.5; surf. tens. 37.8 dynes/cm (0.05%); pH 6.5 (3% aq.); 38% solids.

Emcol® 4161L. [Witco] Disodium oleamido MIPA sulfosuccinate; anionic; dispersant, wetting, foaming, detergent, and emulsifying agent for bubble bath, shampoos, cleansers for cosmetics and toiletries; lt. yel. clear liq.; sol. in water; sp.gr. 1.10; flash pt. (PMCC) 93 C; acid no. 6.0; surf. tens. 32.6 dynes/cm (0.5%); pH 6.5 (3% aq.); 38% solids.

Emcol® 4300. [Witco] Disodium C_{12-15} sulfosuccinate; anionic; dispersant, wetting, foaming, detergent, emulsifying agent for bubble bath, shampoo, cosmetics and toiletries; lt. clear liq.; sol. in water; sp.gr. 1.09; acid no. 5.0; surf. tens. 29.3 dynes/cm (0.05%); pH 6.2 (3% aq.); 33% solids.

Emcol® 4400-1. [Witco] Disodium lauryl sulfosuccinate; anionic; cosmetics and toiletries surfactant used as emulsifier, cleansing, foaming agent, stabilizer; paste; water-sol.

Emcol® 4500. [Witco] Sodium dioctyl sulfosuccinate; dispersant, detergent, wetting, foaming, emulsifying agent; for cosmetics, toiletries, textiles, industrial processing slurries; clear, lt. visc. liq.; sol. in perchloroethylene, CCl_4, kerosene, xylene, Stod.; disp. water and alcohol; sp.gr. 1.10; flash pt. > 93 C; acid no. 3.0; surf. tens. 26.3 dynes/cm (0.05%); pH 6.5 (3% aq.); 70% solids.

Emcol® 4560. [Witco] Sodium dioctyl sulfosuccinate; dispersant, wetting, foaming, detergent, emulsifying agent, solubilizer; used in drycleaning; clear, lt. visc. liq.; sol. in perchlorethylene, CCl_4, Stod.; slightly sol. in water; disp. in alcohol; sp.gr. 1.08; acid no. 3.0; pH 5.1 (3% aq. disp.); surf. tens. 27.4 dynes/cm (0.05%); 90% solids.

Emcol® 4580PG. [Witco] Sodium diester sulfosuccinate; anionic; detergent, emulsifier, and water solubilizer in dry cleaning; textile industry detergent and scouring agent for solv. systems; liq.; sol. in perchloroethylene, Stod.; disp. in water.

Emcol® 4600. [Witco] Sodium ditridecyl sulfosuccinate; detergent, foam modifier, wetting agent, dispersant, processing aid for pigments, in drycleaning; clear, lt. visc. liq.; sol. in perchlorethylene, CCl_4, Stod., kerosene, xylene, IPA; slightly sol. in alcohol; insol. water; sp.gr. 1.00; flash pt. > 93 C; acid no. 0.3; pH 6.6 (3% aq. disp.); 70% solids.

Emcol® 6613. [Witco] Isostearamidopropyl dimethylamine lactate; cationic; conditioner for shampoo, cream rinses, clear formulations; liq.; 100% act.

Emcol® CC-9. [Witco] PPG-9 diethylmonium chloride, cationic; dispersant, particle suspension aid, antistat, wetting agent, o/w emulsifier, conditioner, penetrant, lubricant; for cosmetics, toiletries, textiles, industrial slurries; lt. amber clear liq.; sol. in water, IPA; sp.gr. 1.01; flash pt. > 93 C; pH 6.5 (10% aq.); 100% conc.

Emcol® CC-36. [Witco] PPG-25 diethylmonium chloride; cationic; dispersant, particle suspension aid, o/w emulsifier; plasticizer for hair polymers; antistat; skin cleanser for cosmetics; used in dry cleaning systems; industrial processes; lt. amber clear liq.; sol. in water, IPA; sp.gr. 1.01; flash pt. > 93 C; pH 6.7 (10% in 10:6 IPA:water); 100% conc.

Emcol® CC-37-18. [Witco] Coco-betaine; amphoteric; foaming agent and stabilizer for cosmetics and toiletries, visc. modifier; liq.

Emcol® CC-42. [Witco] PPG-40 diethylmonium chloride, cationic; pigment dispersant, particle suspension aid, emulsifier, solv., conditioner, antistat, lubricant, corrosion inhibitor for toiletries, cosmetics, germicides, syn. fibers and plastics, textiles, industrial processes; ore flotation additive; lt. amber oily liq.; sol. in IPA, acetone, MEK, min. spirits, ethanol; partly sol. in water; sp.gr. 1.01; flash pt. > 200 C (PMCC).

Emcol® CC-55. [Witco] Polypropoxy quat. ammonium acetate; cationic; antistat; conditioner for hair rinse preparations; emulsifier for cosmetics and textile flame retardants; solv. for phenolic-type germicides for cosmetics and toiletries; antistat for syn. fibers and plastics; fabric conditioner; lubricant for textile and industrial formulations; solv. cleaning and scouring agent; corrosion inhibitor in protective coatings; pigment dispersant in nonaq. media; o/w emulsifier; lt. amber oily liq.; sol. @ 25% in ethanol, IPA, acetone, MEK; water-disp.; sp.gr. 1.02; flash pt. > 93 C (PMCC); 2.0% moisture.

Emcol® CC-57. [Witco] Polypropoxy quat. ammonium phosphate; cationic; antistat, conditioner, emulsifier, solv., lubricant, solv. cleaning and scouring agent, corrosion inhibitor, and dispersant; used in syn. fibers and plastics, personal care prods., germicides, flame retardants, textile and industrial applic., protective coatings, and pigments; lt. amber oily liq.; sol. in water, ethanol, IPA, acetone, MEK; sp.gr. 1.12; flash pt. > 93 C (PMCC).

Emcol® CC-59. [Witco] Polypropoxylated quat. ammonium phosphate; dispersant for electronics industry; liq.

Emcol® CC-422. [Witco] Polypropoxy quat. ammonium chloride; cationic; detergent and dry cleaning applics.; anticaking and cleansing agent, antistat, dispersant and o/w emulsifier for aerosol formulations; liq.; sol. in halogenated hydrocarbons, perchloroethylene, and water.

Emcol® CS-136, -143, -151, -165. [Witco] Org. phosphate ethers; anionic; detergent, emulsifier, dispersant, solubilizer; liq.

Emcol® CS-1361. [Witco] Sodium salt of complex organo-phosphate ester; used for scale control; clear liq.; sol. in hydrocarbons, min. oil, kerosene, ethanol, water; sp.gr. 1.10; pour pt. 40 F; surf. tens. 31.9 dynes/cm (0.05%); pH 5 (3% aq.); 10% moisture.

Emcol® D 75-13, -33. [Witco] Alkoxylated acetylenic alcohol; nonionic; corrosion inhibitor for use in metal processing; liq.; water-sol.

Emcol® DG. [Witco] Cocamidopropyl betaine; amphoteric; detergent, emulsifier, foaming and wetting agent, foam stabilizer for detergent industry; liq.; water-sol.

Emcol® E-607L. [Witco] Lapyrium chloride; cationic; emollient, emulsifier, foamer, cleanser, substantive agent, deodorant for cosmetics, toiletries, industrial applics.; hair conditioner; wh. cryst.; sol. 37% in water and ethanol, 28% in IPA; surf. tens. 37 dynes/cm (0.1% aq.); pH 3.9 (1% aq.); 97.5% act.

Emcol® E-607S. [Witco] Steapyrium chloride; cationic; see Emcol E-607L; wh. to off-wh. powd.; sol. 2.5% in water; pH 3.4 (1% aq.); 94% act.

Emcol® ISML. [Witco] Isostearamidopropyl morpholine lactate; cationic; cosmetics and toiletries surfactant used as antistat, conditioner, emollient, foaming and substantive agent; nonirritating base for cream rinses and conditioning shampoos; liq.; water-sol.; 25% act.

Emcol® K-8300. [Witco] Alkanolamide disodium sulfosuccinate half ester; anionic; dispersant, particle suspension aid, wetting and foaming agent, detergent; emulsifier in emulsion polymerization; amber liq.; sol. in water; sp.gr. 1.10; pH 6.5 (3% aq.); surf. tens. 32.6 dynes/cm (0.05%); 38% solids.

Emcol® L. [Witco] Lauramine oxide; cationic; cosmetics and toiletries surfactant used as antistat, cleansing and substantive agent, emollient, lubricant, and visc. modifier; detergent and foam stabilizer for industrial detergents; liq.; water-sol.

Emcol® M. [Witco] Myristamine oxide; cationic; see Emcol L; liq.; water-sol.

Emcol® NA-30. [Witco] Cocamidopropyl betaine; amphoteric; cosmetics/toiletries surfactant used as antistat, cleansing, foaming, spreading agent, conditioner, foam stabilizer, solubilizer and visc. modifier; liq.; water-sol.

Emcol® P-1020 BU. [Witco] Calcium alkylaryl sulfonate; anionic; o/w emulsifier for industrial use; intermediate for agric. emulsifiers; amber visc. liq.; oil sol.; sp.gr. 1.01–1.03; pH 6–7.5 (3% in 20% IPA); 68–70% act.

Emcol® P-1045. [Witco] Amine salt of alkylaryl sulfonic acid; anionic; wetting agent in oil-based systems, sludge dispersant in fuel oil; liq.; oil-sol.; water-disp.; 90% act.

Emcol® P-1059B. [Witco] Amine salt of alkylaryl sulfonic acid; anionic surfactant, water solubilizer; liq.; 90% act.

Emcol® Q. [Witco] Polyquaternium-5; cationic; cosmetics and toiletries surfactant used as antistat, conditioner, emulsifier, and substantive agent; liq.; water-sol.; 30% act.

Emcol® TS-230. [Witco] Org. phosphate ester; dispersant and lubricant in oil drilling muds, o/w emulsifier, flow control agent; amber clear liq.; sol. in xylene, ethanol, CCl_4, water; sp.gr. 1.22; pour pt. < 50 F; acid no. 170; surf. tens. 49.6 dynes/cm (0.05%); pH 2 (3% aq.).

EmCon E. [Fanning] Egg oil; nonionic; emollient, humectant, w/o emulsifier for food, cosmetics, household, pharmaceuticals; yel. liq.; bland odor; sol. in min. and veg. oils, most org. solvs.; dens. 9.95; 100% act.

EmCon E-5, E-20. [Fanning] Lecithin; emollient, moisturizer for hair and skin care prods.; w/o emulsifier; superfatting agent, humectant, mold release agent; yel. liq.; E-5 sol. in min. and veg. oils, slightly disp. in most org. solv., insol. water; E-20 sol. in min. and veg. oils, disp. in most org. solv., water; sp.gr. 0.95; 85.4 and 62.3% triglycerides resp.

EmCon W. [Fanning] Wheat germ oil; nonionic; emollient, conditioner, source of vitamin E; used in skin and hair care creams and lotions; flavoring agent in baking industry; food supplement; liq.; 100% conc.

Emercide® 1199. [Henkel/Emery] Liquid preservative system suitable for cosmetics; Gardner 1 liq.; sol. in triolein, IPA, IPM; disp. in water, min. oil, glycerin; dens. 9.3 lb/gal; pour pt. –10 C; flash pt. 285 F; 100% act.

Emeressence® 1150. [Henkel/Emery] Ethylene brassylate; musk chemical for fragrance or odor masking applics.; Gardner 1 liq.; sol. @ 5% in min. oil, triolein, IPA, IPM; dens. 8.7 lb/gal; visc. 41 cSt (100 F); pour pt. < 3 C; flash pt. 370 F; 100% act.

Emeressence® 1151. [Henkel/Emery] Ethylene dodecanedioate; musk chemical for fragrance or odor masking applics.; liq.; sol. see Emeressence 1150; dens. 8.75 lb/gal; flash pt. 365 F; 100% act.

Emeressence® 1160 Rose Ether. [Henkel/Emery] Phenoxyethanol; cosmetic preservative; effective against gram negative microorganisms; liq.; 100% act.

Emeressence® 1166 Styrate®. [Henkel/Emery] Aromatic with odor similar to styrax or cinnamic alcohol with warm, floral (lilac, ylang) undertones; clear colorless liq.; sol. in triolein, IPA, IPM; dens. 9.47 lb/gal; flash pt. 250 F; 100% act.

Emeressence® 1170 Chamol®. [Henkel/Emery] Aromatic reminiscent of Roman chamomile oil; clear colorless to slightly yel. liq.; sol. in min. oil, triolein, IPA, IPM; dens. 8.37 lb/gal; flash pt. 257 F; 100% act.

Emeressence® 1171 Pseudojasmone. [Henkel/Emery] Aromatic with odor similar to cis-jasmone with warm, herbaceous (celery-like) undertones; suitable for floral blends; clear, colorless liq.; sol. in min. oil, triolein, IPA, IPM; dens. 8.28 lb/gal; flash pt. 250 F; 100% act.

Emeressence® 1172 Privetone®. [Henkel/Emery] Aromatic with odor reminiscent of freshly cut privet hedge; clear, colorless liq.; sol. in min. oil, triolein, IPA, IPM; dens. 8.38 lb/gal; flash pt. 250 F; 100% act.

Emeressence® 1173 Parajasmone. [Henkel/Emery] Aromatic with floral odor of strong jasmine character, coupled with touches of gardenia and tuberose; clear, colorless liq.; sol. in triolein, IPA, IPM; dens. 8.62 lb/gal; flash pt. 355 F; 100% act.

Emeressence® 1174 Fir Balsam. [Henkel/Emery] Aromatic with sweet balsamic odor reminiscent of Canadian fir; clear, colorless liq.; sol. in min. oil, triolein, IPA, IPM; dens. 1.0 lb/gal; flash pt. 266 F.

Emerest® 1723. [Henkel/Emery] Isopropyl ester of lanolic acids; emollient, pigment dispersing and wetting agent; sol. in warm oils.

Emerest® 2301. [Henkel/Emery] Methyl oleate; nonionic; base for industrial lubricants; mold release agent, defoamer, flotation agent, plasticizer for cellulosic plastics, needle lubricants; when sulfated is useful as wetting, rewetting, and dye leveling agent in textile and leather industries; Gardner < 6 liq.; sol. 5% in min. oil, toluene, IPA, xylene; insol. in water; dens. 7.3 lb/gal; visc. 5 cSt (100 F); pour pt. –16 C; flash pt. 350 F.; 100% act.

Emerest® 2302. [Henkel/Emery] Propyl oleate; nonionic; see Emerest 2301; Gardner < 6 liq.; sol. see Emerest® 2301; dens. 7.3 lb/gal; visc. 5 cSt (100 F); pour pt. –16 C; flash pt. 350 F; 100% act.

Emerest® 2308. [Henkel/Emery] Tridecyl stearate; nonionic; lubricant used in sewing thread mfg. and fiber finish applics. where high heat stability is desired; Gardner 1 liq.; sol. see Emerest 2301; dens. 7.1 lb/gal; visc. 18 cSt (100 F); pour pt. 3 C; flash pt. 440 F; 100% act.

Emerest® 2310. [Henkel/Emery] Isopropyl isostearate; low visc. emollient, lubricant for bath oils, creams, lotions, shampoos; binder for pressed powd.; sol. in castor oil, ethanol, IPA, lanolin, min. oil, peanut oil, silicone.

Emerest® 2314. [Henkel/Emery] IPM; cosmetic emollient; sewing thread lubricant; Gardner 1 liq.; sol. 5% in min. oil, toluene, IPA, xylene; water-insol.; dens. 7.1 lb/gal; visc. 4 cSt (100 F); pour pt. –5 C; flash pt. 320 F.

Emerest® 2316. [Henkel/Emery] IPP; lubricant used for syn. fibers in applics. where low friction is essential; emollient in cosmetic formulations; high purity; Gardner 1 liq.; sol. see Emerest 2301; dens. 7.1 lb/gal; visc. 6 cSt (100 F); pour pt. 14 C; flash pt. 340 F.

Emerest® 2321. [Henkel/Emery] Butyl stearate; nonionic; lubricant esp. for aluminum foil and cans; liq.; 100% act.

Emerest® 2324. [Henkel/Emery] Isobutyl stearate; nonionic; lubricant; Gardner 1 liq.; sol. @ 5% in min. oil, toluene, IPA, xylene; dens. 7.1 lb/gal; visc. 10 cSt (100 F); pour pt. 10 C; flash pt. 370 F; 100% act.

Emerest® 2325. [Henkel/Emery] Butyl stearate; nonionic; emollient in creams and lotions; dye solubilizer in lipsticks; lubricant; Gardner 4 liq.; sol. see Emerest 2314; dens. 7.1 lb/gal; visc. 7 cSt (100 F); pour pt. 18 C; flash pt. 370 F.

Emerest® 2326. [Henkel/Emery] Butyl stearate; noninic; industrial lubricant used in aluminum foil rolling, fiber and yarn processing; Gardner 1 liq.; sol. in IPA, toluol, min. oil; dens. 7.2 lb/gal; visc. 10 cs; f.p. 20 C; flash pt. 188 C; sapon. no. 172; 100% act.

Emerest® 2328. [Henkel/Emery] Butyl oleate; nonionic; lubricant; Gardner 4 liq.; sol. @ 5% in min. oil, toluene, IPA, xylene; dens. 6.9 lb/gal; visc. 6 cSt (100 F); pour pt. –25 C; flash pt. 356 F; 100% act.

Emerest® 2350. [Henkel/Emery] Glycol stearate; nonionic; opacifying and pearlescing agent used in liq. cosmetic and detergent compds.; Gardner 1 beads; sol. @ 5% in IPA, toluol, min. oil, xylene; f.p. 50 C; HLB 2.1; flash pt. 390 F; sapon. no. 185; 100% act.

Emerest® 2355. [Henkel/Emery] Glycol distearate; nonionic; opacifier used in liq. detergent and cosmetic prods.; Gardner 4 flakes; sol. 5% in toluene, xylene; disp. in min. oil; insol. in water; f.p. 62 C; HLB 1.2; flash pt. 455 F; sapon no. 195; 100% act.

Emerest® 2380. [Henkel/Emery] Propylene glycol stearate; aux. emulsifier, opacifier, pearlescent; Gardner 2 solid; disp. @ 5% in min. oil, toluene, xylene; insol. in water; dens. 7.3 lb/gal (45 C); HLB 1.8; m.p. 36 C; flash pt. 470 F.

Emerest® 2381. [Henkel/Emery] Propylene glycol stearate SE; anionic; pearlescing and spreading agent, for cosmetic and dishwashing prods., hand lotions; Gardner 2 beads; sol in IPA, disp. in water, min. oil, hot toluol; f.p. 40 C; HLB 4.0; cloud pt. < 25 C; flash pt. 380 F; sapon. no. 170; 100% act.

Emerest® 2384. [Henkel/Emery] Propylene glycol isostearate; solubiizer for fragrances in low alcohol or oil preparations; emollient; sol. in most cosmetic oils, ethanol.

Emerest® 2388. [Henkel/Emery] Propylene glycol dipelargonate; lubricant, low visc. emollient for preshaves, bath oils, creams and lotions; colorless; odorless; sol. in castor oil, ethanol, IPA, lanolin, min. oil, silicone oil.

Emerest® 2400. [Henkel/Emery] Glyceryl stearate; nonionic; emulsifier for hand creams, cosmetics, textiles, industrial lubricants; lubricant softener for textiles; Gardner 1 beads; sol. 5% in IPA, hot toluol, hot min. oil; insol. in water; f.p. 57.5; HLB 3.9; flash pt. 415 F; sapon. no. 153; 100% act.

Emerest® 2401. [Henkel/Emery] Glyceryl stearate; nonionic; tech. grade of Emerest 2400, emulsifier for hand creams, cosmetics, industrial lubricants; Gardner 2 beads; sol. in IPA, hot toluol, hot min. oil; f.p. 58; HLB 3.9; flash pt. 425 F; sapon. no. 153; 100% act.

Emerest® 2407. [Henkel/Emery] Glyceryl stearate SE; cosmetic ester; Gardner 3 beads; sp.gr. 0.920; m.p. 58 C; acid no. 20 max.; iodine no. 1.0 max.; sapon. no. 148–158.

Emerest® 2410. [Henkel/Emery] Glyceryl isostearate; emollient, lubricant, and w/o emulsifier; exc. oxidation and color stability; Gardner 2 liq.; sol. @ 5% in min. oil, IPA; insol. in water; dens. 7.8 lb/gal; visc. 260 cSt (100 F); HLB 2.9; pour pt. 5 C; flash pt. 400 F.

Emerest® 2421. [Henkel/Emery] Glyceryl oleate; nonionic; emulsifier, in mold release agents, anti-icing fuel additive, rust preventative; in textiles as a lubricant component in syn. fiber spin finishes; vehicle for agric. insecticides; Gardner 11 liq.; sol. 5% in toluol, min. oil, xylene; dens. 7.9 lb/gal; visc. 91 cs (38 C); f.p. 6 C; HLB 3.4; flash pt. 235 C; sapon no. 170; 100% act.

Emerest® 2423. [Henkel/Emery] Triolein; nonionic; lubricant for metals, leather, textiles, emulsifier, called syn. olive oil; sulfated form used as softener in leather and textile industries; Gardner 3 liq.; sol. 5% in min. oil, xylene; dens. 7.6 lb/gal; visc. 43 cs; f.p. 9 C; HLB 0.6; flash pt. 293 C; sapon no. 197; 100% act.

Emerest® 2452. [Henkel/Emery] Polyglyceryl-3 diisostearate; dye and pigment wetter; emollient; thickener; solv.; for creams, lotions, lip prods.; Gardner 4 liq.; sol. @ 5% in min. oil, triolein, IPA, IPM; disp. in glycerin; dens. 8.2 lb/gal; HLB 6.7; pour pt. 4 C; flash pt. 455 F.

Emerest® 2485. [Henkel/Emery] Pentaerythritol tetrapelargonate; nonionic; primary lubricant base or modifier in lubricant formulations used in metal working and syn. fiber processing; Gardner 1 liq.; sol. 5% in min. oil, toluene, IPA, xylene; insol. in water; dens. 8.0 lb/gal; visc. 35 cSt (100 F); pour pt. 10 C; flash pt. 555 F.

Emerest® 2486. [Henkel/Emery] Pentaerythrityl tetrapelargonate; nonionic; nongreasy emollient for skin prods.; Gardner 2 liq.; sol. @ 5% in min. oil, triolein, IPA, IPM; dens. 8.0 lb/gal; visc. 34 cSt (100 F); HLB 3.3; pour pt. 10 C; flash pt. 550 F; 100% act.

Emerest® 2610. [Henkel/Emery] PEG-20 stearate; emulsifier for glyceryl stearate in nonionic textile lubricants and softeners; thickener; antigellant in starch sol'ns.; Gardner 1 solid; sol. @ 5% in water, glycerol trioleate, xylene; HLB 15.7; m.p. 36 C; cloud pt. 86–90 C; flash pt. 430 F.

Emerest® 2617. [Henkel/Emery] PEG-150 oleate; strongly hydrophilic emulsifier, stabilizer, and lubricant; Gardner 3 solid; sol. @ 5% in water, xylene; HLB 19.2; m.p. 58 C; cloud pt. 81 C (5% saline); flash pt. 470 F.

Emerest® 2620. [Henkel/Emery] PEG-4 laurate; nonionic; emulsifier, coupling agent, defoamer in water base coatings, visc. control additive; visc. depressant in vinyl plastisols; Gardner 1 liq.; disp. in water, min. oil, butyl stearate, glycerol trioleate, perchloroethylene, Stod.; dens. 8.2 lb/gal; visc. 40 cs; HLB 9.8; cloud pt. < 25 C; 100% act.

Emerest® 2622. [Henkel/Emery] PEG-4 dilaurate; nonionic; coemulsifer and lubricant in SE textile and industrial oils, mold release agent, visc. control agent; Gardner 2 liq.; sol. 5% in min. oil, butyl stearate, glycerol trioleate, Stod., xylene; disp. in water; dens. 8.0 lb/gal; visc. 35 cs; HLB 7.6; cloud pt. < 25 C; flash pt. 455 F; 100% act.

Emerest® 2624. [Henkel/Emery] PEG-4 oleate; lubricant component in textile processing; softener/lubricant for leather during tanning; emulsifier for min. oils, fatty oils, and solvs. for cutting oils, solvs. in metal cleaners and degreasers, w/o emulsifier for consumer pesticide aerosols; Gardner 3 liq.; sol. 5% in glycerol trioleate, xylene; disp. in water; dens. 8.1 lb/gal; visc. 34 cSt (100 F); HLB 8.3; pour pt. < –15 C; flash pt. 415 F; cloud pt. < 25 C.

Emerest® 2625. [Henkel/Emery] PEG-4 isostearate; emulsifier; component in fiber lubricants, processing aids, and conc. liq. fabric softeners; Gardner 2 liq.; sol. 5% in Stod.; water-disp.; dens. 8.2 lb/gal; visc. 50 cSt (100 F); HLB 8.3; pour pt. –8 C; flash pt. 310 F; cloud pt. < 25 C.

Emerest® 2630. [Henkel/Emery] PEG-6 laurate; hydrophilic emulsifier; lubricant component and scrooping agent for textile fibers and yarns; visc. control agent for plastisols; Gardner 2 liq.; sol. 5% in xylene; water-disp.; dens. 8.4 lb/gal; visc. 37 cSt (100 F); HLB 12.1; pour pt. 9 C; flash pt. 475 F; cloud pt. < 25 C.

Emerest® 2632. [Henkel/Emery] PEG-6 oleate; emulsifier and lubricant; SE component in formulating textile softeners; Gardner 3 liq.; sol. 5% in xylene; water-disp.; dens. 8.3 lb/gal; visc. 46 cSt (100 F); HLB 10.4; pour pt. –10 C; flash pt. 425 F; cloud pt. < 25 C.

Emerest® 2636. [Henkel/Emery] PEG-6 stearate; waxy emulsifier for oils and fats in industrial lubricants; softener and lubricant in textiles and leather; Gardner 1 solid; sol. 5% in glycerol trioleate, xylene; water-disp.; dens. 8.3 lb/gal; visc. 45 cSt (100 F); HLB 10.1; m.p. 29 C; flash pt. 460 F; cloud pt. < 25 C.

Emerest® 2640. [Henkel/Emery] PEG-8 stearate;

nonionic; emulsifier for oils and fats in mfg. of industrial lubricants, consumer prods., textile lubricants and softeners; lipophilic; thickener and stabilizer for starch coatings on paper; paper size; lubricant for channeling wire through conduit; Gardner 1 soft, waxy solid; sol. see Emerest 2632; dens. 8.5 lb/gal; visc. 57 cSt (100 F); HLB 12.0; m.p. 32 C; flash pt. 425 F; cloud pt. < 25 C; 100% act.

Emerest® 2642. [Henkel/Emery] PEG-8 distearate; nonionic; lipophilic waxy surfactant used as emulsifier and thickener in cosmetic and industrial emulsions; Gardner 2 solid; sol. (5%) in min. oil, butyl stearate, glycol trioleate, Stod., xylene; water-disp.; visc. 52 cSt (100 F); HLB 7.5; m.p. 36 C; flash pt. 470 F; cloud pt. < 25 C.

Emerest® 2644. [Henkel/Emery] PEG-8 isostearate; see Emerest 2625; Gardner 1 liq.; disp. in water, min. oil, butyl stearate, xylene; dens. 8.5 lb/gal; visc. 70 cSt (100 F); HLB 11.3; pour pt. 10 C; flash pt. 450 F; cloud pt. < 25 C.

Emerest® 2646. [Henkel/Emery] PEG-8 oleate; nonionic; emulsifier for oil-based sol. cutting oils and specialty industrial lubricants; emulsifies solvs. for industrial degreasers; stabilizes visc. of vinyl plastisols; component in textile softeners and lubricant bases; drycleaning formulations; Gardner 2 liq.; sol. 5% in xylene; water-disp.; dens. 8.5 lb/gal; visc. 52 cSt (100 F); HLB 11.8; pour pt. 5 C; flash pt. 455 F; cloud pt. < 25 C.

Emerest® 2648. [Henkel/Emery] PEG-8 dioleate; lipophilic emulsifier and solubilizer for min. oils, fats, and solvs.; emulsifier for kerosene in agric. and pesticide sprays; emulsification of latex paints, metalworking fluids, solvs.; speciality and industrial lubricants; Gardner 3 liq.; sol. in min. oil, glycerol trioleate, xylene; water-disp.; dens. 8.1 lb/gal; visc. 45 cSt (100 F); HLB 8.8; pour pt. 6 C; flash pt. 515 F; cloud pt. < 25; 100% act.

Emerest® 2650. [Henkel/Emery] PEG-8 laurate; nonionic; hydrophilic surfactant functioning as leveling and wetting agent; defoamer in latex paints; dispersant in pigment grinding; solubilizer for oils and solvs.; antiblock agent in vinyls; textile processing; Gardner 1 liq.; sol. 5% in water, xylene; dens. 8.6 lb/gal; visc. 41 cSt (100 F); HLB 13.2; pour pt. 12 C; flash pt. 450 F; cloud pt. 33 C; 100% act.

Emerest® 2652. [Henkel/Emery] PEG-8 dilaurate; emulsifier, lubricant, softener, and release agent for paper; textile industry, coupler and lubricant in syn. fiber spin finishes; Gardner 2 liq.; sol. 5% in min. oil, butyl stearate, glycerol trioleate, Stod., xylene; water-disp.; dens. 8.3 lb/gal; visc. 38 cSt (100 F); HLB 10.8; pour pt. 8 C; flash pt. 420 F; cloud pt. < 25 C; 100% act.

Emerest® 2654. [Henkel/Emery] PEG-8 pelargonate; surfactant used as base lubricant for syn. fiber spin finishes, other textile processing; coemulsifier and coupler; Gardner 1 liq.; sol. 5% in water, xylene; dens. 8.7 lb/gal; visc. 34 cSt (100 F); HLB 14.3; pour pt. 5 C; flash pt. 440 F; cloud pt. 40 C (2% saline).

Emerest® 2660. [Henkel/Emery] PEG-12 oleate; dye leveling agent in textiles; emulsifier in specialty lubricants; detergent; acid washing of printed circuit boards; Gardner 2 liq.; sol. 5% in water, xylene; dens. 8.7 lb/gal; visc. 75 cSt (100 F); HLB 13.6; pour pt. 18 C; flash pt. 545 F; cloud pt. 47 C; 100% act.

Emerest® 2661. [Henkel/Emery] PEG-12 laurate; lubricant in processing syn. fibers; Gardner 2 liq.; sol. 5% in water, xylene; dens. 8.6 lb/gal; visc. 60 cSt (100 F); HLB 14.8; pour pt. 14 C; flash pt. 525 F; cloud pt. 63 C.

Emerest® 2662. [Henkel/Emery] PEG-12 stearate; emulsifier for cosmetic and textile formulations; textile lubricants and softeners; visc. modifier in creams and lotions; Gardner 1 solid; sol. 5% in xylene; water-disp.; dens. 8.5 lb/gal; HLB 13.8; m.p. 40 C; flash pt. 440 F; cloud pt. 55 C.

Emerest® 2664. [Henkel/Emery] PEG-12 isostearate; used in fiber lubricants and processing aids; emulsifier in formulations; Gardner 5 liq.; sol. 5% in glycerol trioleate, xylene; water-disp.; dens. 8.6 lb/gal; visc. 90 cSt (100 F); HLB 13.0; pour pt. 10 C; flash pt. 490 F; cloud pt. < 25 C.

Emerest® 2665. [Henkel/Emery] PEG-12 dioleate; see Emerest® 2648; more hydrophilic; Gardner 4 liq.; sol. in butyl stearate, glycerol trioleate, Stod., xylene; water-disp.; dens. 8.3 lb/gal; visc. 64 cSt (100 F); HLB 10.3; pour pt. 19 C; flash pt. 530 F; cloud pt. < 25 C.

Emerest® 2675. [Henkel/Emery] PEG-50 stearate; hydrophilic emulsifier used for preparing solubilized oils; visc. modifier, softener or plasticizer in acrylic or vinyl resin emulsions; Gardner 1 liq.; sol. @ 5% in water; dens. 8.5 lb/gal; visc. 671 cSt (100 F); HLB 17.8; pour pt. 0 C; cloud pt. 81 C (5% saline); flash pt. 540 F; 30% act. in water.

Emerest® 2704. [Henkel/Emery] PEG-4 dilaurate; emulsifier, lubricant, dispersant for bath oils; visc. control agent for creams and lotions; Gardner 2 liq.; sol. @ 5% in min. oil, triolein, IPA; disp. in water, IPM, glycerin; dens. 8.0 lb/gal; visc. 22 cSt (100 F); HLB 7.6; pour pt. 0 C; cloud pt. < 25 C; flash pt. 455 F.

Emerest® 2711. [Henkel/Emery] PEG-8 stearate; nonionic; emulsifier, thickener for o/w and w/o systems; Gardner 1 waxy solid; sol. @ 5% in glycerin; disp. in water, min. oil, triolein, IPA, IPM; dens. 8.5 lb/gal (35 C); visc. 57 cSt (100 F); HLB 11.4; m.p. 30 C; cloud pt. < 25 C; flash pt. 500 F.

Emerest® 2712. [Henkel/Emery] PEG-8 distearate; emulsifier, opacifier, thickener for firm creams and high-visc. lotions; Gardner 2 solid; sol. @ 5% in min. oil, triolein, IPA; disp. in IPM, glycerin; dens. 7.9 lb/gal (53 C); visc. 52 cSt (100 F); HLB 7.5; m.p. 36 C; cloud pt. < 25 C; flash pt. 470 F.

Emerest® 2715. [Henkel/Emery] PEG-40 stearate; emulsifier, stabilizer, antigellant, lubricant for creams, lotions, shampoos, deodorants, makeup; Gardner 1 solid; sol. @ 5% in water, IPA; dens. 8.9 lb/gal (50 C); HLB 17.0; m.p. 50 C; cloud pt. 75–81 C (5% saline); flash pt. 515 F.

Emerest® 2717. [Henkel/Emery] PEG-2 stearate; emulsifier, thickener for creams and lotions.

Emerest® 11723. [Henkel/Emery] Isopropyl ester of

lanolic acids; emollient, pigment wetting, dispersing agent, conditioner; sol. in warm oils.

Emerlube® 5919. [Henkel/Emery] Ethoxylated veg. oil; lubricant for PP yarns, carpet backings; Gardner 1 liq.; sol. @ 5% in water, xylene; disp. in butyl stearate, glyceryl trioleate; dens. 8.8 lb/gal; visc. 106 cSt (100 F); pour pt. –1 C; cloud pt. 64 C; flash pt. 610 F.

Emersol® 110, 120. [Henkel/Emery] Stearic acid; opacifier in cosmetics, soaps, emulsifiers, chemical specialties; acid no. 2205–210; iodine no. 8–12 and 5–7 resp.

Emersol® 132 NF Lily®. [Henkel/Emery] Triple pressed stearic acid; see Emersol 110; acid no. 205–210; iodine no. 0.5 max.

Emersol® 143. [Henkel/Emery] Palmitic acid; see Emersol 110; acid no. 215–223; iodine no. 1 max.

Emersol® 871. [Henkel/Emery] Isostearic acid; acid no. 175 min.; iodine no. 12 max.

Emersol® 875. [Henkel/Emery] Isostearic acid; acid no. 187–197; iodine no. 3 max.

Emersol® 6313 NF. [Henkel/Emery] Low-titer oleic acid; food grade fatty acid; acid no. 201–204; iodine no. 88–93.

Emersol® 6320. [Henkel/Emery] DP stearic acid; food grade fatty acid; acid no. 205–210; iodine no. 3.5–5.0.

Emersol® 6321 NF. [Henkel/Emery] Low-titer wh. oleic acid; food grade fatty acid; acid no. 201–204; iodine no. 87–92.

Emersol® 6332 NF. [Henkel/Emery] TP stearic acid; food grade fatty acid; acid no. 205–211; iodine no. 0.5 max.

Emersol® 6333 NF. [Henkel/Emery] Low-linoleic content oleic acid; food grade fatty acid; acid no. 200–204; iodine no. 86–91.

Emersol® 6349. [Henkel/Emery] Stearic acid; food grade fatty acid; acid no. 203–206; iodine no. 0.5 max.

Emersol® 6351. [Henkel/Emery] Stearic acid; food grade fatty acid; acid no. 196–201; iodine no. 1.0 max.

Emerstat® 6660. [Henkel/Emery] Cationic antistat for fiber lubricants; Gardner 6 liq.; sol. @ 5% in water, xylene; disp. in butyl stearate, glyceryl trioleate; dens. 8.8 lb/gal; visc. 900 cSt (100 F); pour pt. < 0 C; flash pt. 325 F.

Emerwax® 1251. [Henkel/Emery] Syn. ester wax for lipsticks, makeup, etc.; Gardner 3 solid; disp. @ 5% in min. oil, triolein, IPM; dens. 8.03 lb/gal (75 C); m.p. 75–80 C; flash pt. 400 F; 100% act.

Emerwax® 1253. [Henkel/Emery] Beeswax substitute; for cosmetic formulations incl. sticks, cold creams, makeup; Gardner < 5 solid; disp. @ 5% in min. oil, triolein, IPM; dens. 7.1 lb/gal; m.p. 60–80 C; flash pt. 490 F; 100% act.

Emerwax® 1254. [Henkel/Emery] Japan wax substitute; Gardner 3 solid; sol. @ 5% in min. oil, triolein, IPA, IPM; dens. 7.9 lb/gal (40 C); visc. 93.4 cSt (100 F); m.p. 40 C; flash pt. 455 F; 100% act.

Emerwax® 1257. [Henkel/Emery] Emulsifying wax NF; nonionic; emulsifier for cosmetic and pharmaceutical applics.; Gardner < 2 solid; sol. @ 5% in triolein, IPA; disp. in water; dens. 8.1 lb/gal; m.p. 48–52 C; flash pt. 355 F; 100% act.

Emerwax® 1266. [Henkel/Emery] Emulsifying wax BP (blend of fatty alcohols and ethoxylated fatty alcohols); nonionic; o/w emulsifier for pharmaceuticals, creams, lotions, antiperspirants, hair care prods., depilatories; Gardner < 2 solid; sol. in IPA; disp. in min. oil, triolein; dens. 8.0 lb/gal; m.p. 47–55 C; flash pt. 355 F; 100% act.

Emery® 880. [Henkel/Emery] Fatty acid; soaps used as lubricants in metals and industrial applics.; liq.

Emery® 912. [Henkel/Emery] CP/USP glycerin; skin softener, solubilizer, visc. modifier, flavor enhancer, moisturizer, solv., humectant, and solubilizer in cosmetics, drug vehicles, and food applics., glass, ceramics, and adhesives; APHA 20 max. visc. liq.; odorless; sp.gr. 1.2517 min.; 96.0% min. glycerol.

Emery® 916. [Henkel/Emery] CP/USP glycerin; see Emery 912; APHA 20 max. visc. liq.; odorless; sp.gr. 1.2607 min.; 99.5% min. glycerol.

Emery® 918. [Henkel/Emery] CP/USP glycerin; see Emery 912; APHA 8 max.; sp.gr. 1.2615; 99.8% min. glycerol.

Emery® 1650. [Henkel/Emery] Anhyd. lanolin USP; emulsifier, emollient, conditioner, lubricant for cosmetics, sun care prods.; Gardner < 9 solid; sol. @ 5% in IPM; disp. in min. oil, triolein; dens. 7.9 lb/gal; m.p. 36–42 C; flash pt. 530 F.

Emery® 1656. [Henkel/Emery] Anhyd. lanolin USP; emulsifier, emollient, conditioner, moisturizer, pigment dispersant for pharmaceutical ointments, veterinary prods., industrial hand cleansers, cosmetics; Gardner < –12 solid; sol. in IPM; disp. in min. oil, triolein; dens. 7.9 lb/gal; m.p. 36–42 C; flash pt. 530 F.

Emery® 1660. [Henkel/Emery] Anhyd. lanolin USP, cosmetic grade; emulsifier, emollient, conditioner, moisturizer, pigment dispersant for lip prods.; Gardner < 5 solid; sol. in IPM; disp. in min. oil, triolein; dens. 7.9 lb/gal; m.p. 38–44 C; flash pt. 460 F.

Emery® 1720. [Henkel/Emery] Lanolin oil/isopropyl ester blend; emollient, modifier for skin and hair care prods., makeup, bath preps.; Gardner < 7 liq.; sol. in min. oil, triolein, IPA, IPM; dens. 7.3 lb/gal; visc. 18 cSt (100 F); pour pt. 2 C; cloud pt. 7–13 C; flash pt. 300 F; 100% act.

Emery® 1730. [Henkel/Emery] Min. oil/fraction of lanolin alcohols and sterols blend; penetrant, emulsifier, emollient, stabilizer; Gardner 11 liq.; sol. in min. oil, triolein, IPA, IPM; dens. 7.4 lb/gal; visc. 39 cSt (100 F); pour pt. 18 C; flash pt. 350 F.

Emery® 1732. [Henkel/Emery] Min. oil/fraction of lanolin alcohols and sterols blend; emollient, emulsion stabilizer, moisturizer, emulsifier for creams, lotions, hairdressings, makeup, topical pharmaceutical preps.; Gardner 5 liq.; sol. in min. oil, triolein, IPA, IPM; disp. in glycerin; dens. 7.2 lb/gal; visc. 15 cSt (100 F); pour pt. < 4 C; flash pt. 360 F.

Emery® 1740. [Henkel/Emery] Lanolin alcohol, other lanolin fractions, min. oil; emulsifier for w/o systems, emollient, moisturizer for pharmaceutical ointments and hair prods.; very high water absorp.; Gardner < 7 solid; sol. in min. oil, IPM; disp. in triolein; dens. 7.4 lb/gal; m.p. 40–46 C; flash pt. 450 F.

Emery® 1747. [Henkel/Emery] Lanolin/lanolin alcohol blend in inert hydrocarbon base; primary w/o emulsifier, aux. o/w emulsifier; emollient, moisturizer for skin and hair care prods., pharmaceutical ointments; Gardner < 6 solid; sol. in min. oil, IPM; disp. in triolein; dens. 7.4 lb/gal; m.p. 40–46 C; flash pt. 425 F.

Emery® 1780. [Henkel/Emery] Lanolin alcohol; w/o emulsifier, emollient, visc. builder, stabilizer for emulsions, personal care prods.; Gardner 10 solid; sol. @ 5% in triolein; dens. 8.2 lb/gal; m.p. 48 C; flash pt. 440 F.

Emery® 1781. [Henkel/Emery] Lanolin acids; emollient superfatting agent for soaps, personal care prods.; Gardner < 8 solid; sol. @ 5% in triolein, IPA, IPM; disp. in min. oil; m.p. 45 C; flash pt. 380 F.

Emery® 1787. [Henkel/Emery] Cetyl alcohol NF; nonionic; emulsion stabilizer, thickener, opacifier, emollient, lubricant; flakes; 100% act.

Emery® 2421. [Henkel/Emery] Antifog agent for use in food-pkg. films; liq.; HLB 3.4; pour pt. 6 C; acid no. 6.0 max.; sapon. no. 165– 175; flash pt. 235 F.

Emery® 2809. [Henkel/Emery] SAE 0W-20 multipurpose arctic lubricant hydraulic fluid; sp.gr. 0.86 (60/60 F); visc. 30.1 cSt (40 C); flash pt. 446 F; pour pt. < –67 F.

Emery® 2811. [Henkel/Emery] SF-CD 5W-30 syn. engine lubricant; sp.gr. 0.85 (60/60 F); visc. 76.0 cSt (40 C); flash pt. 465 F; pour pt. –75 F.

Emery® 2831. [Henkel/Emery] Syn. gear lubricant SAE 80W-140; sp.gr. 0.92 (60/60 F); visc. 254.1 cSt (40 C); flash pt. 405 F; pour pt. < –25 F.

Emery® 2879. [Henkel/Emery] Syn. compressor lubricant; sp.gr. 0.96 (60/60 F); visc. 151.2 cSt (40 C); flash pt. 485 F; pour pt. –25 F.

Emery® 2880. [Henkel/Emery] Syn. compressor lubricant; sp.gr. 0.96 (60/60 F); visc. 98.8 cSt (40 C); flash pt. 510 F; pour pt. –20 F.

Emery® 2890. [Henkel/Emery] Syn. compressor lubricant; sp.gr. 0.97 (60/60 F); visc. 68.3 cSt (40 C); flash pt. 500 F; pour pt. –40 F.

Emery® 2900. [Henkel/Emery] Dimer ester syn. lubricant basestock for industrial applics.; sp.gr. 0.91 (60/60 C); visc. 83 cSt (40 C); flash pt. 570 F; pour pt. –45 F.

Emery® 2902. [Henkel/Emery] Monomethyl dimerate; syn. lubricant; sp.gr. 0.92 (60/60 C); visc. 306 cSt (40 C); flash pt. 465 F.

Emery® 2905. [Henkel/Emery] Dimer ester syn. lubricant basestock for industrial applics.; sp.gr. 0.91 (60/60 C); visc. 133 cSt (40 C); flash pt. 585 F; pour pt. –50 F.

Emery® 2908. [Henkel/Emery] Dimer ester syn. lubricant basestock for industrial applics.; water-sol.; sp.gr. 1.07 (60/60 C); visc. 375 cSt (40 C); flash pt. 535 F; pour pt. 10 F.

Emery® 2911. [Henkel/Emery] Syn. lubricant basestock; sp.gr. 0.86 (60/60 C); visc. 4.87 cSt (40 C); flash pt. 340 F; pour pt. –100 F.

Emery® 2913. [Henkel/Emery] Syn. lubricant basestock; sp.gr. 0.96 (60/60 C); visc. 218 cSt (40 C); flash pt. 520 F; pour pt. –25 F.

Emery® 2914. [Henkel/Emery] Dimethyl azelate; syn. lubricant; sp.gr. 1.01 (60/60 C); visc. 3.17 cSt (40 C); flash pt. 300 F; pour pt. 20 F.

Emery® 2918. [Henkel/Emery] Polyol ester syn. lubricant basestock for aircraft, engines, automotive, and industrial applics.; sp.gr. 0.96 (60/60 C); visc. 33.2 cSt (40 C); flash pt. 550 F; pour pt. –5 F.

Emery® 2924. [Henkel/Emery] Syn. transmission lubricant SAE 50; sp.gr. 0.92 (60/60 F); visc. 138.2 cSt (40 C); flash pt. 455 F; pour pt. < –40 F.

Emery® 2929. [Henkel/Emery] Polyol ester syn. lubricant basestock for aircraft, engines, automotive, and industrial applics.; sp.gr. 0.99 (60/60 C); visc. 26.5 cSt (40 C); flash pt. 485 F; pour pt. –85 F.

Emery® 2936. [Henkel/Emery] Polyol ester syn. lubricant basestock for aircraft, engines, automotive, and industrial applics.; visc. 23 cSt (40 C); flash pt. 495 F; pour pt. < –75 F.

Emery® 2939. [Henkel/Emery] Polyol ester syn. lubricant basestock for aircraft, engines, automotive, and industrial applics.; sp.gr. 0.99 (60/60 C); visc. 25.4 cSt (40 C); flash pt. 495 F; pour pt. –85 F.

Emery® 2943. [Henkel/Emery] Syn. hydraulic fluid; sp.gr. 0.83 (60/60 F); visc. 22.1 cSt (40 C); flash pt. 310 F; pour pt. –65 F.

Emery® 2957. [Henkel/Emery] Diester syn. lubricant basestock; sp.gr. 0.92 (60/60 C); visc. 12.2 cSt (40 C); flash pt. 425 F; pour pt. –95 F.

Emery® 2958. [Henkel/Emery] Diester syn. lubricant basestock; sp.gr. 0.92 (60/60 C); visc. 10.3 cSt (40 C); flash pt. 425 F; pour pt. –95 F.

Emery® 2960. [Henkel/Emery] Diester syn. lubricant basestock; sp.gr. 0.91 (60/60 C); visc. 17.5 cSt (40 C); flash pt. 450 F; pour pt. –95 F.

Emery® 2970. [Henkel/Emery] Diester syn. lubricant basestock; sp.gr. 0.92 (60/60 C); visc. 13.7 cSt (40 C); flash pt. 435 F; pour pt. –95 F.

Emery® 2971. [Henkel/Emery] Diester syn. lubricant basestock; sp.gr. 0.92 (60/60 C); visc. 26.7 cSt (40 C); flash pt. 445 F; pour pt. –95 F.

Emery® 2976. [Henkel/Emery] Diester syn. lubricant basestock; sp.gr. 0.93 (60/60 C); visc. 9.9 cSt (40 C); flash pt. 390 F; pour pt. –95 F.

Emery® 2984. [Henkel/Emery] Syn. gear lubricant SAE 75W-90; sp.gr. 0.91 (60/60 F); visc. 136.5 cSt (40 C); flash pt. 405 F; pour pt. < –50 F.

Emery® 2990. [Henkel/Emery] Syn. compressor lubricant; sp.gr. 0.99 (60/60 F); visc. 32.4 cSt (40 C); flash pt. 490 F; pour pt. –60 F.

Emery® 2995. [Henkel/Emery] Polyol ester syn. lubricant basestock for aircraft, engines, automotive, and industrial applics.; sp.gr. 0.99 (60/60 C); visc. 26 cSt (40 C); flash pt. 480 F; pour pt. –85 F.

Emery® 3002. [Henkel/Emery] Polyalphaolefin syn. lubricant; sp.gr. 0.80 (15.6/15.6 C); visc. 5.12 cSt (40 C); flash pt. 164 C; pour pt. –30 to –50 F.

Emery® 3004. [Henkel/Emery] Polyalphaolefin syn.

lubricant; sp.gr. 0.82 (15.6/15.6 C); visc. 16.9 cSt (40 C); flash pt. 225 C; pour pt. –69 F.

Emery® 3006. [Henkel/Emery] Polyalphaolefin syn. lubricant; sp.gr. 0.83 (15.6/15.6 C); visc. 30.9 cSt (40 C); flash pt. 243 C; pour pt. –64 F.

Emery® 3008. [Henkel/Emery] Polyalphaolefin syn. lubricant; sp.gr. 0.83 (15.6/15.6 C); visc. 48.5 cSt (40 C); flash pt. 264 C; pour pt. –59 F.

Emery® 3010. [Henkel/Emery] Polyalphaolefin syn. lubricant; sp.gr. 0.84 (15.6/15.6 C); visc. 68.2 cSt (40 C); flash pt. 280 C; pour pt. –55 F.

Emery® 5430 [Henkel/Emery] Cocamidopropyl betaine; conditioner and mild foaming agent; provides visc. control for shampoos; liq.; water-sol.

Emery® 5440. [Henkel/Emery] Sodium isethionate; anionic; intermediate raw material for surfactants; liq.; 56% conc.

Emery® 5700. [Henkel/Emery] Phenyl ethanolamine; intermediate for basic dyes for syn. fibers; stabilizer for colloidal disps. of metal oxides in lubricants; in photographic color developer, emulsion polymerization of diene monomers; lt. yel. to amber liq.; m.w. 137.2; sol. in water (4%), acetone, benzene, CCl_4, ethyl acetate, ethyl ether, methanol; flash pt. (COC) 305 F; 98% min. act.

Emery® 5702. [Henkel/Emery] PEG-3 aniline; cure promoter for polyester resins; amber to brn. liq.; m.w. 225; flash pt. (COC) 410 F; 50% min. act.

Emery® 5703. [Henkel/Emery] Phenyl diethanolamine; coupling agent for disperse dyes for syn. fibers; intermediate for dyes; cure promoter for polyester resins; curing agent for urethane elastomers; intermediate; also in hair dyes, heavy-duty detergents, paint strippers; wh. to off-wh. cryst. solid; m.w. 181.2; sol. in water (2.5%), acetone, methanol; m.p. 55.5 C; flash pt. (COC) 380 F; 98% min. act.

Emery® 5704. [Henkel/Emery] N,N-Bisacetoxyethyl aniline; coupling agent for disperse dyes for syn. fibers; brn. liq.; m.w. 265; flash pt. 113 F (TCC); 45% min. act.

Emery® 5705. [Henkel/Emery] Phenyl diethanolamine, tech.; coupling agent for disperse dyes for syn. fibers; additive for paint strippers; used in heavy-duty detergent formulas; brn. to blk. cryst. solid; pungent odor; m.w. 181.2; sol. in water (2.5%), acetone, methanol; m.p. 45 C; flash pt. 380 F (COC); 98% min. act.

Emery® 5706. [Henkel/Emery] Methyl phenyl ethanolamine; coupling agent for disperse dyes for syn. fibers; photosensitive chemical for paper coatings; yel. liq.; m.w. 151.2; sol. in water (1.3%), acetone, benzene, CCl_4, methanol, ethyl acetate, ethyl ether; flash pt. (COC) 275 F; 97% min. act.

Emery® 5707. [Henkel/Emery] Ethyl phenyl ethanolamine; coupling agent for dyes for syn. fibers; cure promoter for polyester resins; photosensitive chemical for paper coatings; intermediate for oil-sol. dyestuffs; lt. tan cryst. solid; m.w. 165.2; sol. in acetone, benzene, CCl_4, ethyl ether, methanol; flash pt. (COC) 280 F; 97% min. act..

Emery® 5709. [Henkel/Emery] m-Tolyl diethanolamine; coupling agent for dyes for syn. fibers; cure promoter for polyester resins; also in mfg. of nonporous PU elastomers; tan cryst. solid; m.w. 195.2; sol. in acetone, ethanol; m.p. 67.5 C; flash pt. (COC) 385 F; 98% min. act.

Emery® 5710. [Henkel/Emery] p-Tolyl diethanolamine; cure promoter for polyester resins; intermediate for plastics; brn. cryst. solid; m.w. 195.2; m.p. 53 C; flash pt. (COC) 380 F; 98% min. act.

Emery® 5711. [Henkel/Emery] o-Tolyl ethanolamine, tech.; coupling agent for disperse dyes for syn. fibers; red-brn. liq.; m.w. 151; flash pt. (COC) 310 F; 78% min. act.

Emery® 5712. [Henkel/Emery] o-Tolyl diethanolamine; additive for heavy-duty detergents; component for industrial cleaners, paint strippers; wh. to tan cryst. solid; m.w. 195.2; m.p. 30–52 C; flash pt. (COC) 365 F; 96% min. act.

Emery® 5713. [Henkel/Emery] N,N-Bisacetoxyethyl-m-toluidine; coupling agent for disperse dyes for syn. fibers; brn. liq.; m.w. 279.2; flash pt. 129 F (TCC); 60% min. act.

Emery® 5714. [Henkel/Emery] N-Ethyl-N-hydroxyethyl-m-toluidine; coupling agent for disperse dyes for syn. fibers; photosensitive chemical for paper coatings; accelerator for polyester resins; yel.-brn. liq.; m.w. 179.2; flash pt. (COC) 280 F; 96.5% min. act.

Emery® 5715. [Henkel/Emery] m-Chlorophenyl diethanolamine; coupling agent for dyes for syn. fibers; wh. cryst. presscake; m.w. 215.7; 97% min. act.

Emery® 5717. [Henkel/Emery] m-Chlorophenyl diethanolamine; coupling agent for dyes for syn. fibers; intermediate for synthesis of pharmaceuticals; lt. yel. cryst. solid; m.w. 215.7; sol. in toluene, alcohol; slightly sol. in water; m.p. 84 C; flash pt. (COC) 420 F.

Emery® 5722. [Henkel/Emery] N-Cyanoethyl-N-methyl aniline; coupling agent for basic and disperse dyes for syn. fibers; brn. liq.; m.w. 160; flash pt. 290 F (COC); 95% min. act.

Emery® 5723. [Henkel/Emery] N-Cyanoethyl-N-ethyl aniline; coupling agent for basic and disperse dyes for syn. fibers; brn. liq.; m.w. 174; flash pt. 270 F (COC); 95% min. act.

Emery® 5724. [Henkel/Emery] N-Cyanoethyl-N-hydroxyethyl aniline; coupling agent for disperse dyes for syn. fibers; brn. liq.; m.w. 190.2; flash pt. 345 F (COC); 87% min. act.

Emery® 5725. [Henkel/Emery] N-Cyanoethyl-N-acetoxyethyl aniline; coupling agent for disperse dyes for syn. fibers; brn. liq.; m.w. 232; flash pt. 345 F (COC); 88% min. act.

Emery® 5727. [Henkel/Emery] N-Cyanoethyl-N-butyl aniline; coupling agent for disperse dyes for syn. fibers; brn. liq.; m.w. 202; flash pt. 325 F (COC); 93% min. act.

Emery® 5728. [Henkel/Emery] N-Cyanoethyl-N-ethyl-m-toluidine; coupling agent for acid and disperse dyes for syn. fibers; brn. liq. (may solidify at low temps.); m.w. 188; flash pt. 275 F (COC); 94% min. act.

Emery® 5730. [Henkel/Emery] N-Cyanoethyl-N-hydroxyethyl-m-toluidine; coupling agent for disperse dyes for syn. fibers; brn. liq.; m.w. 204; flash pt. (COC) 330 F; 87% min. act.

Emery® 5731. [Henkel/Emery] N-Cyanoethyl-N-acetoxyethyl aniline sol'n.; coupling agent for disperse dyes for syn. fibers; brn. sol'n.; m.w. 232; flash pt. 128 F (TCC); 60% min. act. in acetic acid.

Emery® 5736. [Henkel/Emery] 3(N-Cyanoethyl) amino-4-methoxy acetanilide; coupling agent and intermediate for disperse dyes for syn. fibers; brn. cryst. presscake; m.w. 233; 74.8% act.

Emery® 5737. [Henkel/Emery] 3(N-Cyanoethyl-N-hydroxyethyl) amino-4-methoxy acetanilide; coupling agent for disperse dyes for syn. fibers; brn. aq. sol'n.; m.w. 277; pH 5.5–6.5; 49 ± 2.5% tert. amine.

Emery® 5743. [Henkel/Emery] p-Nitrophenoxyethanol; intermediate for acid dyes for syn. fibers; brn. cryst. chunks; m.w. 183; flash pt. (COC) 235 F; 95% min. act.

Emery® 5751. [Henkel/Emery] 3(N-Cyanoethyl-N-ethyl) amino-4-methoxy acetanilide; coupling agent for disperse dyes for syn. fibers; brn. sol'n.; m.w. 261; flash pt. 200 F (TCC); 23.9% tert. amine, 60.2% acetic acid.

Emery® 5753. [Henkel/Emery] 3(N,N-Bisacetoxyethyl) amino-4-methoxy acetanilide; coupling agent for disperse dyes for syn. fibers; brn. sol'n.; m.w. 352; flash pt. 145 F (TCC); 60% min. act. in acetic acid.

Emery® 5758. [Henkel/Emery] 3(N,N-Bisacetoxyethyl) amino acetanilide; coupling agent for disperse dyes for syn. fibers; brn. sol'n.; m.w. 322; flash pt. 128 F (TCC); 38% min. act. in acetic acid.

Emery® 5761. [Henkel/Emery] 3(N,N-Bisacetoxyethyl) amino benzanilide; coupling agent for disperse dyes for syn. fibers; brn. sol'n.; m.w. 384; flash pt. 114 F (TCC); 37% min. act. in acetic acid.

Emery® 5770. [Henkel/Emery] N-Chloroethyl-N-ethyl aniline; coupling agent for dyes for syn. fibers; lt. yel. to amber liq.; m.w. 183.5; flash pt. (COC) 265 F; 96% min. act.

Emery® 5771. [Henkel/Emery] N-Chloroethyl-N-ethyl-m-toluidine; intermediate for basic dyes for syn. fibers; amber liq.; m.w. 197.5; flash pt. 260 F (COC); acid no. 0.5; 96% min. act.

Emery® 5774. [Henkel/Emery] POE (200) aniline; intermediate for fugitive tints; wh. to off-wh. solid; flash pt. > 600 F (COC); pH 6.0–7.5 (5% aq.).

Emery® 5776. [Henkel/Emery] POE (200) m-toluidine; intermediate for fugitive tints; pH 6.0–7.5 (5% aq.); 50% act.

Emery® 5778. [Henkel/Emery] POE (200) m-toluidine; intermediate for fugitive tints; off-wh. solid; flash pt. > 600 F (COC); pH 6.0–7.5 (5% aq.).

Emery® 5789. [Henkel/Emery] Ethoxylated naphthol sulfonic acid; brightener additive in the mfg. of tinplate; liq.; sol. in water; dens. 10.6 lb/gal; pour pt. < 25 C; flash pt. 200 F.

Emery® 5794. [Henkel/Emery] PEG-3 triethanolamine; solv. neutralizer for acid dyes; dk. brn. liq.; m.w. 281; hyd. no. 590–630; amine no. 196–204; flash pt. (COC) 315 F.

Emery® 6657. [Henkel/Emery] Amidoamine condensate; hydrophobic component for formulating fabric finishes; softener, lubricant; good whitness retention; Gardner 4 solid; disp. in water; m.p. 75 C; cloud pt. < 25 C; flash pt. 530 F.

Emery® 6686. [Henkel/Emery] PEG-600; nonionic; solv., humectant, binder, visc. modifier; Gardner < 1 liq.; sol. @ 5% in water, xylene; dens. 9.4 lb/gal; visc. 63 cSt (100 F); pour pt. 22 C; flash pt. 475 F; 100% act.

Emery® 6705. [Henkel/Emery] Phenoxyethanol; solv. for NC, cellulose acetate, ethylcellulose, and vinyl, alkyl, and ester type resins; used in printing inks and special formulations for removing industrial finishes and coatings; Gardner 2 liq.; sol. 5% in Stod., xylene; water-insol.; dens. 9.1 lb/gal; visc. 10 cSt (100 F); pour pt. 13 C; flash pt. 250 F.

Emery® 6709. [Henkel/Emery] PEG-8; nonionic; see Emery 6686; Gardner < 1 clear, visc. liq.; sol. @ 5% in water, xylene; dens. 9.4 lb/gal; visc. 45 cSt (100 F); pour pt. 6 C; flash pt. > 350 F; 100% act.

Emery® 6717. [Henkel/Emery] Internal-external lubricant for mfg. of glass fibers; Gardner 10 visc. liq.; sol. @ 5% in water, disp. in min. oil; dens. 8.3 lb/gal; pour pt. 55 C; cloud pt. < 25 C; flash pt. 540 F.

Emery® 6724. [Henkel/Emery] Methoxyl PEG ester; lubricant for nylon, polyester, and PP fibers; Gardner 1 liq.; sol. @ 5% in water, butyl stearate, glyceryl trioleate, xylene; disp. in min. oil, Stod.; dens. 8.6 lb/gal; visc. 20 cSt (100 F); pour pt. 6 C; flash pt. 470 F.

Emery® 6726. [Henkel/Emery] Methoxy PEG-400; chemical intermediate for textile applics., in cleaning formulations; Gardner 1 visc. liq.; sol. @ 5% in water, xylene; dens. 9.0 lb/gal; visc. 98.1 cSt (100 F); pour pt. –2 C; flash pt. > 250 F.

Emery® 6727. [Henkel/Emery] POE (100) aniline; intermediate for fugitive tints; straw-colored liq.; acid no. 0.5 max.; pH 6.0–8.0 (5% aq.); 50% act.

Emery® 6728. [Henkel/Emery] POE (100) m-toluidine; intermediate for fugitive tints; straw-colored liq.; pH 6.0–8.0 (5% aq.); 50% act.

Emery® 6730. [Henkel/Emery] POE (100) m-toluidine; intermediate for fugitive tints; off-wh. to yel. solid; flash pt. 300 F (COC); pH 6.0–7.5 (5% aq.).

Emery® 6748. [Henkel/Emery] Cocamidopropyl betaine; amphoteric surfactant, conditioner, mild foaming agent, visc. control agent; used in hair and bath prods., skin cleaners, specialty cosmetics; Gardner 4 liq.; sol. in water; dens. 8.7 lb/gal; 30% act.

Emery® 6749. [Henkel/Emery] POE (100) aniline; intermediate for fugitive tints; wh. to off-wh. solid; flash pt. 300 F (COC); acid no. 0.5 max.; pH 6.0–8.0 (5% aq.); 100% act.

Emery® 6752. [Henkel/Emery] Monocarboxylate coconut imidazolinium deriv.; amphoteric; surfactant used in cleaning compds. and shampoos, visc. improver and foam stabilizer, detergent; Gardner 9 liq.; sol. in water; dens. 8.7 lb/gal; 38% act.

Emery® 6760. [Henkel/Emery] Sol'n. of Emery 6717; Gardner 8 liq.; sol. @ 5% in water; disp. in butyl stearate, glyceryl trioleate; dens. 9.2 lb/gal;

visc. 1200 cSt (100 F); pour pt. 5 C; 50% act. in acetic acid and water.

Emery® 6773. [Henkel/Emery] PEG-200; chemical intermediate for coatings and adhesives, lubricants, metalworking, paper mfg., petrol. prod.; Gardner < 1 clear visc. liq.; sol. @ 5% in water, xylene; dens. 9.4 lb/gal; visc. 25 cSt (100 F); pour pt. < –15 C; flash pt. > 330 F.

Emery® 9331. [Henkel/Emery] Dibromophenol; flame retardant for phenolics and epoxies; wh. powd.; m.w. 252.9; m.p. 36 C; 63.4% bromine

Emery® 9332. [Henkel/Emery] Tribromophenol; flame retardant for phenolics, PC, and epoxies; reddish-wh. powd.; m.w. 330.8; m.p. 92–94 C; 72.4% bromine

Emery® 9336. [Henkel/Emery] Dibromoneopentyl glycol; flame retardant for polyester resins; used in PU forms or elastomers; wh. powd.; m.w. 262; sol. g/100 g solv. 102 g methanol, 83 g acetone, 60 g MEK, 52 g IPA; m.p. 110–111 C; 61% bromine

Emery® 9345. [Henkel/Emery] Tetrabromoxylene; flame retardant for PS, polyester, textiles; wh. powd.; m.w. 421.75; m.p. 251–255 C; 75% bromine

Emery® 9350. [Henkel/Emery] Tetrabromobisphenol A; flame retardant for epoxy, PC, ABS; wh. powd.; m.w. 543.9; m.p. 178–181 C; 58% bromine

Emery® 9353. [Henkel/Emery] Tetrabromobisphenol A di-2-hydroxyethyl ether; flame retardant for PC, thermoset, and thermoplastic polyester, PU; wh. powd.; m.w. 632; m.p. 105–117; 48.5% bromine

Emery® Methyl Caprylate/Caprate. [Henkel/Emery] Used in the preparation of lubricants for automotive, textile, metal rolling operations; m.p. 30–36 C.

Emery® Methyl Lardate. [Henkel/Emery] Solv.-carrier for agric. spray prods.; defoaming component in metalworking, paper deinking, pharmaceutical fermentation.

Emery® Methyl Oleate. [Henkel/Emery] See Emery Methyl Lardate; also for prep. of lubricants for automotive, textile, metal rolling operations.

Emery® Methyl Stearate. [Henkel/Emery] Used in the preparation of lubricants for automotive, textile, metal rolling operations.

Emery® PAO 2947. [Henkel/Emery] Syn. compressor lubricant; sp.gr. 0.85 (60/60 F); visc. 66.4 cSt (40 C); flash pt. 505 F; pour pt. –55 F.

Emery® PAO 2948. [Henkel/Emery] Syn. compressor lubricant; sp.gr. 0.84 (60/60 F); visc. 28.2 cSt (40 C); flash pt. 510 F; pour pt. –70 F.

Emery® PAO 2950. [Henkel/Emery] Syn. compressor lubricant; sp.gr. 0.86 (60/60 F); visc. 101.6 cSt (40 C); flash pt. 510 F; pour pt. –45 F.

Emgard® 2802. [Henkel/Emery] Dexron II/Mercon automatic transmission fluid; syn. lubricant; sp.gr. 0.84 (60/60 F); visc. 30.8 cSt (40 C); flash pt. 220 F; pour pt. < –75 F.

Emid® 6500. [Henkel/Emery] Cocamide MEA; nonionic; thickener, foam stabilizer for shampoos, liq. detergents, and rug cleaners; Gardner 8 flaked solid; sol. (5%) in min. oil, butyl stearate, glycerol trioleate, Stod.; water-disp.; dens. 7.5 lb/gal (75 C); m.p. 72 C; flash pt. 405 F; cloud pt. < 25 C; 100% act.

Emid® 6510. [Henkel/Emery] Lauramide DEA; nonionic; foam booster, stabilizer and thickener in shampoos, bubble baths, detergents; Gardner 2 solid; sol. in perchloroethylene, Stod., disp. in water, min. oil, butyl stearate; m.p. 42 C; 100% act.

Emid® 6514. [Henkel/Emery] Cocamide DEA; nonionic; thickener, foam stabilizer, and detergent component for various liq. cleaning compds.; Gardner 4 liq.; sol. (5%) in water, butyl stearate, glycerol trioleate, Stod., xylene; dens. 8.2 lb/gal; visc. 336 cSt (100 F); pour pt. 10 C; flash pt. 345 F; 100% act.

Emid® 6515. [Henkel/Emery] Cocamide DEA; nonionic; foam booster and stabilizer; inhibits redeposition of soils; thickener, superfatting agent for shampoos, bubble baths, cleansers; Gardner 4 liq.; sol. (5%) in water, butyl stearate, glycerol triolate, Stod., xylene; dens. 8.3 lb/gal; visc. 390 cSt (100 F); pour pt. 0 C; flash pt. 370 F; 100% act.

Emid® 6518. [Henkel/Emery] Lauramide DEA; foam booster, stabilizer, visc. modifier for personal care prods., soaps, bath additives; Gardner < 5 liq.; sol. @ 5% in water, triolein, IPA, IPM; disp. in min. oil; dens. 8.2 lb/gal; pour pt. < –10 C; flash pt. 300 F.

Emid® 6519. [Henkel/Emery] Lauramide DEA; see Emid 6518; Gardner < 5 liq.; sol. @ 5% in water, triolein, IPA, IPM, glycerin; dens. 8.1 lb/gal (50 C); pour pt. < –10 C; flash pt. 345 F.

Emid® 6521. [Henkel/Emery] Cocamide DEA; foam booster, stabilizer, thickener in personal care prods., liq. soaps, bath additives; Gardner < 4 liq.; sol. @ 5% in triolein, IPA, IPM; disp. in water, min. oil; dens. 8.4 lb/gal; visc. 327 cSt (100 F); pour pt. –10 C; cloud pt. < 25 C; flash pt. 345 F.

Emid® 6529. [Henkel/Emery] Cocamide DEA; thickener, foam booster, stabilizer, emulsifier; Gardner 2 liq.; sol. @ 5% in water, xylene; disp. in min. oil, butyl stearate, glyceryl trioleate; dens. 8.3 lb/gal; visc. 619 cSt (100 F); pour pt. < 25 C; flash pt. 360 F.

Emid® 6531. [Henkel/Emery] Cocamide DEA and diethanolamine; thickener, detergent, emulsifier, and foam stabilizer for cosmetics, shampoos, liq. detergents, and janitorial cleaners; deduster for powd. detergents; Gardner 3 liq.; sol. (5%) in water, Stod., xylene; water-disp.; dens. 8.4 lb/gal; visc. 360 cSt (100 F); pour pt. –10 C; flash pt. 338 F.

Emid® 6533. [Henkel/Emery] Cocamide DEA, diethanolamine; emulsifier, thickener, and foamer for controlled-suds detergents; fulling and scouring agent in textile industry; Gardner 5 liq.; sol. see Emid 6531; dens. 8.7 lb/gal; visc. 345 cSt (100 F); pour pt. –15 C; flash pt. 365 F; 100% act.

Emid® 6534. [Henkel/Emery] Cocamide DEA, diethanolamine; nonionic; thickener, foam stabilizer, emulsifier; shampoos, cosmetic formulations; Gardner 4 liq.; sol. in water, perchloroethylene, Stod. solv.; dens. 8.5 lb/gal; 100% act.

Emid® 6538. [Henkel/Emery] Modified coco DEA; nonionic; thickener and foam stabilizer used in built, liq. detergents; Gardner 6 liq.; sol. see Emid 6531; dens. 8.3 lb/gal; visc. 690 cSt (100 F); pour pt. –1 C; flash pt. 345 F; 100% act.

Emid® 6541. [Henkel/Emery]] Lauramide DEA,

diethanolamine; nonionic; emulsifier, foam stabilizer and booster in shampoos and general purpose cleaners; Gardner 3 paste; sol. (5%) in water, Stod., xylene; dens. 8.4 lb/gal; visc. 275 cSt (100 F); pour pt. –15 C; flash pt. 300 F; 100% act.

Emid® 6545. [Henkel/Emery] Oleic DEA; nonionic; foam suppressant in dye carrier and solv. emulsions; emulsifier for min. oils for antistatic fiber processing aids and yarn lubricants; Gardner 7 liq.; sol. in min. oil, butyl stearate, glycerol trioleate, Stod., xylene; water-disp.; dens. 7.7 lb/gal; visc. 290 cSt (100 F); pour pt. –3 C; flash pt. 475 F; cloud pt. < 25 C; 100% act.

Emid® 6590. [Henkel/Emery] Lauramide DEA and propylene glycol; foam booster, visc. modifier for personal care prods., bath additives, liq. soaps; Gardner < 4 liq.; sol. @ 5% in triolein, IPA, glycerin; disp. in water; dens. 8.3 lb/gal; pour pt. < 0 C; flash pt. 375 F.

Emka Defoam BC, BCK, NC. [Emkay] Nonsilicone; foam and froth control for industrial wastes.

Emka Defoam CPT. [Emkay] Nonsilicone; foam and froth control for print pastes and for clears.

Emka Defoam DP. [Emkay] Silicone base; foam and froth control for industrial processing and wastes; water-disp.

Emka Defoam PKM. [Emkay] Nonsilicone; foam and froth control for high temps. and prolonged treatment.

Emka Defoam SD. [Emkay] Silicone defoamer for industrial wastes.

Emka Defoam SMM. [Emkay] Silicone; foam and froth control of industrial wastes; paste.

Emkafix RXC. [Emkay] Quat. ammonium deriv.; cationic; dispersant and improves fixation of direct dyes; liq.

Emkal BNS, BNX Powd. [Emkay] Butyl naphthalene sodium sulfonate; anionic; dispersant, detergent, dyeing assistant, scouring, and wetting agent, emulsifier used in textile industry; liq. and powd. resp.; 35 and 80% conc. resp.

Emkal BNS Acid. [Emkay] Butyl naphthalene sulfonic acid; anionic; see Emkal BNS; liq.; 90% conc.

Emkal NNS. [Emkay] Nonyl naphthalene sodium sulfonate; anionic; see Emkal BNS; liq.; 35% conc.

Emkal NNS Acid. [Emkay] Nonyl naphthalene sulfonic acid; anionic; see Emkal BNS; liq.; 90% conc.

Emkal NOBS. [Emkay] Sodium nonyl benzene sulfonate; anionic; see Emkal BNS; liq.; 40% conc.

Emkalon AVR. [Emkay] Quat. compound; cationic; softener, antistat used in textile applics.; paste.

Emkalon CL. [Emkay] Alkylamide blend; cationic; durable fabric softener; paste; 25% act.

Emkalon CNW. [Emkay] Alkylamide blend; nonionic; waxy softener, lubricant, conditioner; paste.

Emkane Acid. [Emkay] Alkylaryl sulfonic acid; anionic; detergent, dyeing assistant, scouring and wetting agent; emulsifier used in the textile industry; liq.

Emkane HAD, HAL, HAX. [Emkay] Alkylaryl sulfonate; anionic; see Emkane Acid; paste, liq., and liq. resp.

Emkapon Jel BS. [Emkay] Fatty amide sulfonate; anionic; see Emkane Acid; liq.

Emkastat PC. [Emkay] Quat. deriv.; cationic; antistat; used in soil-resistant finishes for textiles; paste.

Emkatex AES. [Emkay] Aliphatic ester sulfate; anionic; lubricant and dyeing assistant used in textile applics..

Emkatex DX. [Emkay] Phosphated alcohol; anionic; dispersant and textile dye bath stabilizer and leveling agent used in the textile industry; liq.

Emkatex LS. [Emkay] POE deriv.; nonionic; detergent, dye leveling and retarding agent used in the textile industry; paste.

Emkatex PX29. [Emkay] Sulfated amide; anionic; dye assistant, dispersant, emulsifier, and anticrock used in textiles; jelly; 33% conc.

Emkatex WNX. [Emkay] Sulfated esters; anionic; wetting and leveling agent, lubricant used in textile industry; gel.

Emphos CS-121. [Witco] Alkylaryl ethoxylate phosphate ester; anionic; detergent, o/w emulsifier, lubricant, solubilizer, and wetting agent for metal cleaning; liq.; sol. in naphthenic and paraffinic oils; disp. in water.

Emphos CS-136. [Witco] Sodium nonoxynol-9 phosphate; anionic; lubricant; antistat; emulsifier for cutting fluids, PVAc, and acrylic film formation; detergent for hard surfaces, metal cleaners, and dry cleaning systems; waterless hand cleaner component; clear visc. liq.; sol. in water, ethanol, CCl_4, perchloroethylene, heavy aromatic naphtha, kerosene; sp.gr. 1.09; pour pt. 18 C; acid no. 28; surf. tens. 31.2 dynes/cm (0.05% aq.); pH 5.0 (3% aq.).

Emphos CS-141. [Witco] Nonyl phenol ethoxylate phosphate ester, acid form; anionic; electrolyte-compatible lubricant for water-based cutting fluids; detergent for heavy-duty, all-purpose liq. formulations; emulsifier in PVAc and acrylic film formation; textile wetting agent; water-repellent fabric finishing; clear visc. liq.; sol. 25% in water, 10% sodium hydroxide, ethanol, CCl_4, perchloroethylene, xylene, heavy aromatic naphtha; sp.gr. 1.12; pour pt. > 4 C; acid no. 65 (to pH 5.5); 110 (to pH 9.5); surf. tens. 30.6 dynes/cm (0.05%); pH 2.0 (3% aq.).

Emphos CS-147. [Witco] Alkylaryl ethoxylate phosphate ester; anionic; industrial detergent, o/w emulsifier, lubricant, solubilizer, foaming and coupling agent; liq.; water-sol.

Emphos CS-151. [Witco] Org. phosphate ester; anionic; see Emphos CS-141; clear visc. liq.; sol. see Emphos CS-141; sp.gr. 1.11; pour pt. > 50 F; acid no. 95 (to pH 9.5); pH 2 (3% aq.); surf. tens. 36.8 dynes/cm (0.05%).

Emphos CS-330. [Witco] Alkylaryl ethoxylate phosphate ester; anionic; detergent, o/w emulsifier, and lubricant for oils in metal processing.

Emphos CS-733. [Witco] Alkylaryl ethoxylate phosphate ester; anionic; industrial coupler, o/w emulsifier, lubricant with electrolyte tolerance; liq.; water-sol.

Emphos CS-1361. [Witco] Sodium nonoxynol-9 phosphate; anionic; antistat; emulsifier for transparent gels; detergent for hard surfaces, metal cleaners, and drycleaning systems; particle dispersant for aq. systems; clear liq.; sol. @ 25% in water, ethanol, CCl_4, perchloroethylene, aromatic naphtha, kerosene; sp.gr. 1.10; pour pt. 4 C; pH 5.0 (3% aq.); surf. tens. 31 dynes/cm (0.05% aq.).

Emphos D70-30C. [Witco] Sodium glyceryl oleate phosphate; anionic; antistat, emollient, emulsifier, substantive agent, and moisture barrier for personal care prods.; aerosol formulation and food processing surfactant; food-grade mold lubricant; liq.; oil-sol.

Emphos D70-31. [Witco] Mono- and diglycerides phosphate ester; anionic; pigment dispersant for oil-base paints; w/o emulsifier, lubricant, release agent, and thickener in food processing; flake; oil-sol.

Emphos F27-85. [Witco] Hydrog. veg. glycerides phosphate; anionic; mold lubricant for food use; pigment dispersant in oil-based systems; moisture barrier; tan soft solid; insol. in water; sp.gr. 1.01; pour pt. 17 C; acid no. 40 (to pH 9.5); pH 6.9 (3% aq.).

Emphos PS-21A. [Witco] Alcohol ethoxylate phosphate ester; anionic; detergent base, emulsifier, foaming and wetting agent, lubricant, dispersant; used for detergent industry and industrial surfactants; clear liq.; sol. in kerosene, xylene, IPA; water-insol.; sp.gr. 1.05; pH 2.0 (10% aq.); flash pt. > 93 C.

Emphos PS-121. [Witco] Linear alcohol ethoxylate phosphate ester, acid form; anionic; emulsifier/lubricant for cutting fluids; drycleaning detergent with antistat properties; textile wetting and scouring agent; clear liq.; sol. @ 25% in ethanol, CCl_4, perchloroethylene, xylene, heavy aromatic naphtha, kerosene, min. oil; water-disp.; sp.gr. 1.06; pour pt. < 10 C; acid no. 105 (to pH 5.5); 180 (to pH 9.5); surf. tens. 28.6 dynes/cm (0.05% aq.); pH 2.0 (3% aq.).

Emphos PS-220. [Witco] Linear alcohol ethoxylate phosphate ester, acid form; anionic; lubricant; improves lubricity and load bearing of oil-based lubricants; monomer/water emulsifier; polymer particle dispersant; emulsifier for caustic sol'ns., aliphatic solvs.; textile antistat; clear liq.; sol. @ 25% in ethanol, CCl_4, perchloroethylene, xylene, heavy aromatic naphtha, kerosene, min. oil; insol. in water; sp.gr. 1.02; pour pt. 10 C; acid no. 105 (to pH 5.5); 185 (to pH 185); surf. tens. 32.3 dynes/cm; pH 2.0 (3% aq.).

Emphos PS-236. [Witco] Alcohol ethoxylate phosphate ester; anionic; detergent base, emulsifier, coupling, foaming and wetting agent, corrosion inhibitor, lubricant, antistat, penetrant, and solubilizer used in the detergent industry, industrial surfactants, textile systems; metal cleaning and emulsion polymerization surfactant; liq.; sol. in naphthenic oils and water.

Emphos PS-400. [Witco] Linear alcohol ethoxylate phosphate ester, acid form; anionic; polymer particle dispersant; monomer/water emulsifier; textile antistat; also for pesticides, detergents, emulsion polymerization; clear liq.; sol. @ 25% in ethanol, CCl_4, perchloroethylene, xylene, heavy aromatic naphtha, kerosene, min. oil; water-disp.; sp.gr. 1.00; pour pt. > 10 C; acid no. 220 (to pH 5.5); 300 (to pH 9.5); surf. tens. 30.6 dynes/cm (0.05% aq.); pH 2.0 (3% aq.).

Emphos PS-415M. [Witco] Alcohol ethoxylate phosphate ester; anionic; industrial coupling and defoaming agent, lubricant, solubilizer; textile industry detergent, dispersant, scouring and wetting agent for aq. systems; liq.; water-sol.

Emphos PS-440. [Witco] Alcohol ethoxylate phosphate ester; anionic; industrial detergent, o/w emulsifier, lubricant, penetrant, solubilizer, coupling, foaming and wetting agent; metal processing surfactant; liq.; sol. in naphthenic oils and water.

Emphos PS-810. [Witco] Linear alcohol ethoxylate phosphate ester, acid form; anionic; emulsifier/lubricant for cutting fluids; lube additive; clear liq.; disp. in water; sp.gr. 1.03; pour pt. > 10 C; acid no. 85 (to pH 5.5), 150 (to pH 9.5); pH 2.0 (3% aq.); surf. tens. 37.6 dynes/cm (0.05% aq.).

Emphos PS-900. [Witco] Alcohol ethoxylate phosphate ester; anionic; detergent, o/w emulsifier, and lubricant for use in metal processing; liq.; sol. in naphthenic and paraffinic oils.

Emphos TS-230. [Witco] Phenol ethoxylate phosphate ester, acid form; anionic; lubricant; improves lubricity and load-bearing of water-based lubricants; industrial corrosion inhibitor, defoamer, metal processing surfactant for syn. oils; low-foaming hydrotrope for alkaline cleaners; amber clear liq.; sol. @ 25% in water, 10% sodium hydroxide, ethanol; sp.gr. 1.18; pour pt. < 10 C; acid no. 87 (to pH 5.5); 155 (to pH 9.5); pH 2.0 (3% aq.).

Empicol® 0185. [Albright & Wilson] Sodium lauryl sulfate; anionic; detergent, wetting and rewetting agent, emulsifier, dispersant, disintegrator; used in personal care prods., pharmaceuticals, dental care, textile printing inks, agric. and horticultural preparations; wh. powd.; dens. 0.35 g/cm³; pH 9.5–10.5 (1% aq.); 94.0% min. act.

Empicol® 0216. [Albright & Wilson] Complex org. phosphate ester; anionic; foam stabilizer, conditioner, and antistatic agent for toiletries; detergent used in textile processing; liq.

Empicol® 0627. [Albright & Wilson] Anionic/nonionic blend; opacifier/pearling agent in the mfg. of lotion shampoos and bath prods.; foam stabilizer and superfatting agent; pearly wh. visc. liq.; sp.gr. 1.05; pH 7.0–7.5 (5% aq.); 20 ± 1% act.

Empicol® ESB3. [Albright & Wilson] Sodium laureth sulfate; anionic; base and foam stabilizer for personal care prods.; colorless-pale yel. liq.; dens. 1.0 g/cm³; visc. 2000 ± 800 cP; cloud pt. 0 C; pH 7.0 ± 0.7 (5%); 28.0 ± 1.0% act.

Empicol® ESB3GA. [Albright & Wilson] Sodium laureth sulfate; foam stabilizer for toiletries; colorless to pale yel. clear visc. liq.; visc. 2750 ± 750 cP; pH 7.0 ± 0.5 (1%); 27.2 ± 1.2% act.

Empicol® ESB30. [Albright & Wilson] Sodium laureth sulfate; see Empicol ESB3; colorless clear liq.; m.w. (of act.) 384; dens. 1.0 g/cm³; visc. 20,000 cP; cloud pt. 0 C; pH 6.5–7.5 (5% aq.); 27.5% act.

Empicol® ESB50. [Albright & Wilson] Sodium laureth sulfate; see Empicol ESB3; pale straw hazy liq.; m.w. (of act.) 384; dens. 1.1 g/cm³; pH 6.5–7.5

(5% aq.); 50.0 ± 1.0% act.

Empicol® ESB70. [Albright & Wilson] Sodium laureth sulfate; see Empicol ESB3; pale straw hazy liq.; dens. 1.1 g/cm³; pH 7.0–9.0 (2% aq.); 68.0 ± 2.0% act.

Empicol® ESB70-AU. [Albright & Wilson] Sodium laureth sulfate; anionic; see Empicol ESB3; pale yel. hazy liq.; dens. 1.1 g/cm³; pH 8.0 ± 1.0 (2% aq.); 68.0 ± 2.5% act.

Empicol® ESC70. [Albright & Wilson] Sodium laureth sulfate; anionic; see Empicol ESC3; pale straw visc. hazy liq.; dens. 1.1 g/cm³; pH 7.0–8.0 (5% aq.); 69.5 ± 2.0% act.

Empicol® LS30. [Albright & Wilson] Sodium lauryl sulfate; wetting, dispersing, emulsifying, and foaming agent for industrial processes, detergent/cleaner formulations; pale yel. clear visc. liq.; visc. 750 ± 300 cps (35 C); pH 8.5 ± 1.0 (10%); 30 ± 1% act.

Empicol® LX. [Albright & Wilson] Sodium lauryl sulfate; emulsifier in the mfg. of plastics, resins, and syn. rubbers; foaming agent for rubber foams, personal care prods. and carpet and upholstery shampoos; lubricant in mfg. of molded rubber goods; wh. powd.; dens. 0.35 g/cm³; pH 9.5–10.5 (1% aq.); 90% act.

Empicol® LX28. [Albright & Wilson] Sodium lauryl sulfate; see Empicol LX; pale yel. liq., paste; dens. 1.05 g/cm³; pH 8.0–9.5 (5% aq.); 28% act. aq. sol'n.

Empicol® LXV. [Albright & Wilson] Sodium lauryl sulfate; see Empicol LX; wh. needles; dens. 0.5 g/cm³; pH 9.5–10.5 (1% aq.); 85% act.

Empicol® LXV/D. [Albright & Wilson] Sodium lauryl sulfate USP, BP; emulsifier in the mfg. of plastics and rubbers; foaming agent for rubber foams; lubricant for mfg. of molded rubber goods; also in mfg. of carpet and upholstery shampoos, toiletries, toothpastes; wh. needles; sp.gr. 0.5; pH 9.5–10.5 (5% aq.); 90% act.

Empicol® LY28/S. [Albright & Wilson] Sodium lauryl sulfate; coprecipitant in mfg. of photographic film; emulsion polymerization in the plastic and rubber industries; foaming agent in carpet processes; pale yel. liq. to paste; dens. 1.05 g/cm³; visc. 28 cs; cloud pt. –1 C; pH 8.5–9.5 (5% aq.); 29.0 ± 1.0% act.

Empicol® LZ. [Albright & Wilson] Sodium lauryl sulfate; detergent, wetting and foaming agent in personal care prods.; emulsifier in mfg. of rubbers, plastics, and resins by emulsion polymerization; foaming agent in mfg. of foam rubber goods; lubricant used in plastic goods; wh. powd.; dens. 0.35 g/cm³; pH 9.5–10.5 (1% aq.); 89.0% min. act.

Empicol® LZ/D. [Albright & Wilson] Sodium lauryl sulfate BP; anionic; surfactant, foaming agent, emulsifier, wetting agent for dental preps., toiletries, rubber, plastics, foam rubber; lubricant in extrusion of plastic goods, e.g., PVC; wh. powd.; sp.gr. 0.35; pH 9.5–10.5 (1% aq.); 90% min. act.

Empicol® LZ/E. [Albright & Wilson] Sodium lauryl sulfate; anionic; emulsifier in mfg. of plastics, resins, and syn. rubbers by emulsion polymerization; emulsifier and dispersant in textile printing pastes; dispersant/lubricant in extrusion of PVC; wh. powd.; dens. 0.35 g/cm³; 90.0% min. act.

Empicol® LZP. [Albright & Wilson] Sodium lauryl sulfate; see Empicol LZ; wh. powd.; dens. 0.35 g/cm³; pH 9.5–10.5 (1% aq.); 89.0% min. act.

Empicol® LZV. [Albright & Wilson] Sodium lauryl sulfate; see Empicol LZ; wh. needles; dens. 0.5 g/cm³; pH 9.5–10.5 (1% aq.); 85.0% min. act.

Empicol® LZV/D. [Albright & Wilson] Sodium lauryl sulfate; see Empicol LZ/D; wh. fine needles; sp.gr. 0.5; pH 9.5–10.5 (1% aq.); 90% min. act.

Empicol® LZV/E. [Albright & Wilson] Sodium lauryl sulfate; see Empicol LZ/E; wh. needles; dens. 0.5 g/cm³; 90.0% min. act.

Empicol® TAS90. [Albright & Wilson] Sodium tallow sulfate; fabric scouring agent, sizing additive; cream powd.; pH 10; 90% act.

Empicol® TLP. [Albright & Wilson] TEA-lauryl sulfate and lauramide MIPA; detergent used for shampoos, as foam booster, stabilizer and visc. modifier; golden yel. clear liq.; misc. with water; dens. 1.025 g/cc; visc. 1000 cs; pH 6.8–7.1 (5%); 40.5% act.

Empicol® XDB. [Albright & Wilson] Detergent complex; detergent, foaming and conditioning agent for shampoos; golden clear liq.; visc. 4000 cs; cloud pt. < 0 C; pH 7.5–8.5 (5%); 30% solids.

Empicryl® 6045. [Albright & Wilson] Alkyl methacrylate polymer in hydrocarbon oil; visc. index improver for formulation of high visc. index hydraulic fluids; yel. visc. fluid; sp.gr. 0.90; visc. 1100 mm²/s; pour pt. –3 C; flash pt. > 100 C; 60% min. oil content.

Empicryl® 6047. [Albright & Wilson] Alkyl methacrylate ester; monomer for mfg. of methacrylate polymers and copolymers; internal plasticizer for methacrylate copolymers used as adhesives and waterproofing coatings; pale yel. liq.; sp.gr. 0.9; flash pt. 80 C; acid no. 0.15 max.; hyd. no. 12 max.

Empicryl® 6052. [Albright & Wilson] Alkyl methacrylate polymer in hydrocarbon oil; see Empicryl 6045; yel. visc. fluid; sp.gr. 0.91; visc. 1000 mm²/s (100 C); pour pt. –3 C; flash pt. > 100 C; 33% min. oil content.

Empicryl® 6054. [Albright & Wilson] Alkyl methacrylate polymer in hydrocarbon oil; see Empicryl 6045; yel. visc. fluid; sp.gr. 0.93; visc. 1600 mm²/s (100 C); pour pt. –12 C; flash pt. > 100 C; 20% min. oil content.

Empicryl® 6058. [Albright & Wilson] Alkyl methacrylate polymer in hydrocarbon oil; see Empicryl 6045; pale yel. visc. fluid; sp.gr. 0.92; visc. 1100 mm²/s (100 C); pour pt –3 C; flash pt. > 100 C; 37% min. oil content.

Empicryl® 6059. [Albright & Wilson] Polyalkyl methacrylate copolymer in hydrocarbon oil; see Empicryl 6045; yel. visc. fluid; sp.gr. 0.91; visc. 900 mm²/s (100 C); pour pt –3 C; flash pt. > 100 C.

Empicryl® 6070. [Albright & Wilson] Polyalkyl methacrylate copolymer in hydrocarbon oil; visc. index improver for formulation of hydraulic oils; yel. visc. fluid; sp.gr. 0.94; visc. 800 mm²/s (100 C); pour pt 15 C; flash pt. > 100 C.

Empicryl® APD. [Albright & Wilson] Maleic anhydride/diisobutylene copolymer, disodium salt, aq. sol'n.; pigment dispersant for water based paints, pigment stabilizer; pale yel. clear liq.; sp.gr. 1.12 (18 C); visc. 16 cs; f.p. –3 C; pH 12.0–12.7; 22% min. act.

Empicryl® APD90. [Albright & Wilson] Maleic anhydride/diisobutylene copolymer, sodium salt; dispersing agent for pigments and extenders in aq. systems; agric. applics.; wh. powd.; dens. 0.35 g/cm³; pH 10.0–10.5 (25% aq.); 90.0% solids.

Empicryl® APD/B. [Albright & Wilson] Maleic anhydride/diisobutylene copolymer, disodium salt, aq. sol'n.; see Empicryl APD; sp.gr. 1.12 (18 C); visc. 16 cs; f.p. –3 C; pH 10.0–10.5; 22% min. act.

Empicryl® APD/B90. [Albright & Wilson] Maleic anhydride/diisobutylene copolymer, sodium salt; see Empicryl APD90; wh. powd.; dens. 0.35 g/cm³; pH 12.0–12.7 (25% aq.); 90.0% total solids.

Empicryl® DH122. [Albright & Wilson] Polyalkyl methacrylate copolymer in hydrocarbon oil; visc. index improver, pour pt. depressant, and low temp. sludge dispersant for formulation of multigrade crankcase oils; yel. hazy visc. fluid; sp.gr. 0.90; visc. 1200 cs (100 C); pour pt. 6 C; flash pt. > 100 C.

Empicryl® DH135. [Albright & Wilson] Polyalkyl methacrylate copolymer in hydrocarbon oil; see Empicryl DH122; yel. hazy visc. fluid; sp.gr. 0.90; visc. 1100 cs (100 C); pour pt. 3 C; flash pt. 198 C.

Empicryl® DH145. [Albright & Wilson] Polyalkyl methacrylate copolymer in hydrocarbon oil; see Empicryl DH122; yel. hazy visc. fluid; sp.gr. 0.90; visc. 1200 cs (100 C); pour pt. 3 C; flash pt. > 100 C.

Empicryl® HV110. [Albright & Wilson] Alkyl methacrylate polymer in hydrocarbon oil; visc. index improver for formulation of automotive gear oils and high visc. index hydraulic fluids; esp. for shear-stable hydraulic fluids for severe low temp. conditions; yel. visc. fluid; sp.gr. 0.95; visc. 1200 mm²/s (100 C); pour pt. 6 C; flash pt. > 100 C; 22% min. oil content.

Empicryl® HV125. [Albright & Wilson] Polyalkyl methacrylate copolymer in hydrocarbon oil; visc. index improver for formulation of high visc. index hydraulic fluids; yel. visc. fluid; sp.gr. 0.93; visc. 1000 mm²/s (100 C); pour pt. 6 C; flash pt. > 100 C; 33% min. oil content.

Empicryl® HV130. [Albright & Wilson] Polyalkyl methacrylate copolymer in hydrocarbon oil; see Empicryl HV125; yel. visc. fluid; sp.gr. 0.92; visc. 1300 mm²/s (100 C); pour pt. 9 C; flash pt. > 100 C; 41% min. oil content.

Empicryl® HV226. [Albright & Wilson] Polyalkyl methacrylate copolymer in hydrocarbon oil; see Empicryl HV125; yel. visc. fluid; sp.gr. 0.93; visc. 1000 mm²/s (100 C); pour pt. 6 C; flash pt. > 100 C; 33% min. oil content.

Empicryl® PPT38. [Albright & Wilson] Alkyl methacrylate copolymer in hydrocarbon oil; pour pt. depressant for formulation of lubricants incl. multigrade crankcase oils in conj. with hydrocarbon-based polymeric visc. index improvers; yel. visc. fluid; sp.gr. 0.88; visc. 400 mm²/s (100 C); pour pt. –18 C; flash pt. > 100 C; 64% min. oil content.

Empicryl® PPT144. [Albright & Wilson] Alkyl methacrylate polymer in hydrocarbon oil; see Empicryl PPT38; yel. visc. fluid; sp.gr. 0.90; visc. 400 mm²/s (100 C); pour pt. 0 C; flash pt. > 100 C; 35% min. oil content.

Empicryl® PPT145. [Albright & Wilson] Alkyl methacrylate polymer in hydrocarbon oil; see Empicryl PPT38; yel. visc. fluid; sp.gr. 0.91; visc. 1000 mm²/s (100 C); pour pt. 3 C; flash pt. > 100 C; 15% min. oil content.

Empicryl® PPT147. [Albright & Wilson] Alkyl methacrylate polymer in hydrocarbon oil; pour pt. depressant for use in light basestocks produced by solv. refining or hydrotreating; yel. visc. fluid; sp.gr. 0.90; visc. 400 mm²/s (100 C); pour pt. 0 C; flash pt. > 100 C; 35% min. oil content.

Empicryl® PPT148. [Albright & Wilson] Alkyl methacrylate polymer in hydrocarbon oil; see Empicryl PPT38; yel. visc. fluid; sp.gr. 0.90; visc. 400 mm²/s (100 C); pour pt. 0 C; flash pt. > 100 C; 35% min. oil content.

Empicryl® PPT148/D. [Albright & Wilson] Alkyl methacrylate polymer in hydrocarbon oil; see Empicryl PPT38; diluted form of Empicryl PPT148; yel. visc. fluid; sp.gr. 0.90; visc. 130 mm²/s (100 C); pour pt. –6 C; flash pt. > 100 C; 58% min. oil content.

Empicryl® PT1334. [Albright & Wilson] Alkyl methacrylate polymer in hydrocarbon oil; visc. index improver/pour pt. depressant for formulation of multigrade crankcase oils; yel. visc. fluid; sp.gr. 0.90; visc. 800 mm²/s (100 C); pour pt. –12 C; flash pt. > 100 C; 64% min. oil content.

Empicryl® PT1345. [Albright & Wilson] Alkyl methacrylate polymer in hydrocarbon oil; see Empicryl PT1334; yel. visc. fluid; sp.gr. 0.88; visc. 900 mm²/s (100 C); pour pt. –18 C; flash pt. > 100 C; 72% min. oil content.

Empicryl® PT1397. [Albright & Wilson] Alkyl methacrylate polymer in hydrocarbon oil; visc. index improver for formulation of high visc. index hydraulic fluids; yel. visc. fluid; sp.gr. 0.89; visc. 200 mm²/s (100 C); pour pt. –12 C; flash pt. > 100 C; 58% min. oil content.

Empicryl® PT1544. [Albright & Wilson] Alkyl methacrylate polymer in hydrocarbon oil; see Empicryl PT1334; yel. visc. fluid; sp.gr. 0.90; visc. 1300 mm²/s (100 C); pour pt. 0 C; flash pt. > 100 C; 48% min. oil content.

Empicryl® PT1764/DO. [Albright & Wilson] Alkyl methacrylate polymer in hydrocarbon oil; visc. index improver/pour pt. depressant for formulation of shear-stable multigrade crankcase oils; yel. visc. fluid; sp.gr. 0.90; visc. 750 mm²/s (100 C); pour pt. 0 C; flash pt. > 100 C; 50% min. oil content.

Empigen® 5083. [Albright & Wilson] Coco dimethylamine oxide; foam booster/stabilizer and visc. modifier for shampoos, foam baths, cleaners; improves conditioning in shampoos; solubilizer for liq. bleach prods.; pale straw liq.; m.w. 244; sp.gr. 0.98; visc. 35 cs; pH 7.5 ± 0.5 (5% aq.); 30% act.

Empigen® 5089. [Albright & Wilson] Laurtrimonium chloride; see Empigen BAC50; pale yel. liq.;

sp.gr. 0.97; pH 5.5–8.5; 34 ± 2% act. in water.

Empigen® 5107. [Albright & Wilson] Alkyl dimethylamine betaine; foaming agent/stabilizer and antistat for shampoos, foam baths, latex foam compds., oil prod.; wetting and coupling agent for cleaners, traffic film removers; pale amber liq.; m.w. 286; sp.gr. 1.04; pH 5–9; 31 ± 2% act. in water.

Empigen® AB. [Albright & Wilson] Dimethyl lauramine; precursor for mfg. of derivs.; catalyst for PU foam, resin curing agent, corrosion inhibitor, and flotation aid; pale straw liq.; char. odor; m.w. 221; dens. 0.79 g/cm³; 91.0% min. tert. amine.

Empigen® AD. [Albright & Wilson] Tert. alkyl dimethylamine; cationic; intermediate for amine deriv.; resin curing agents; corrosion inhibitor; ore flotation chemicals; liq.; 97% conc.

Empigen® AF, AG. [Albright & Wilson] C_{12}/C_{16} alkyl dimethyl amine; cationic; intermediate for amine deriv.; resin curing agents; corrosion inhibitor; ore flotation chemicals; pale straw liq.; m.w. 230 and 238 resp.; sp.gr. 0.79; 97% conc.

Empigen® AH. [Albright & Wilson] C_{14} alkyl dimethyl amine, dist.; see Empigen AB; pale straw liq.; char. odor; m.w. 241; dens. 0.79 g/cm³; 91.0% min. tert. amine.

Empigen® AM. [Albright & Wilson] Alkyl dimethyl amine, dist.; cationic; see Empigen AB; pale straw waxy solid; char. odor; dens. 0.84 g/cm³; 96.0% min. tert. amine.

Empigen® AS. [Albright & Wilson] Coco amido propyl dimethyl amine; cationic; industrial corrosion inhibitor for cutting oil formulations, oil drilling, and refinery operations; chemical intermediate for surfactants; amber liq., paste; char. odor; m.w. 290; dens. 0.91 g/cm³; 88% tert. amine.

Empigen® AT. [Albright & Wilson] Alkyl amide propyl dimethyl amine; cationic; see Empigen AS; amber liq./paste; char. odor; dens. 0.95 g/cm³; 94% conc.

Empigen® AY. [Albright & Wilson] Tert. alkyl ethoxy dimethyl amine; cationic; see Empigen AB; amber liq.; dens. 0.89 g/cm³; 94.0% min. tert. amine.

Empigen® BAC50. [Albright & Wilson] Benzalkonium chloride NF; cationic; bactericide for disinfectant and sanitizer formulations for household, institutional, agric., food processing applics., antiseptic detergents in pharmaceuticals; algicide for swimming pools; wood preservatives; masonry biocides; permanent retarders in dyeing of acrylic fibers; phase transfer catalyst; pale straw liq.; misc. with water, alcohol, acetone; sp.gr. 0.99; pH 7–9 (5%); 50% act.

Empigen® BAC50/BP. [Albright & Wilson] Benzalkonium chloride BP, NF; see Empigen BAC50; pale straw liq.; sp.gr. 0.99; pH 7–9 (5%); 50% act.

Empigen® BAC80. [Albright & Wilson] Benzalkonium chloride; see Empigen BAC50; pale straw liq./paste; m.w. 354; sp.gr. 0.97; pH 7–9; 80% act. in org. solv./water.

Empigen® BAC90. [Albright & Wilson] Benzalkonium chloride; cationic; see Empigen BAC; gel; 90% conc.

Empigen® BB. [Albright & Wilson] Lauryl betaine; amphoteric; foam booster/stabilizer, wetting agent, thickening agent, used for shampoos and detergents; almost water-wh. liq.; 30% betaine.

Empigen® BB-AU. [Albright & Wilson] Alkyl betaine; amphoteric; foam booster/stabilizer, thickener, emulsifier, dispersant, wetting and foaming agent; used in personal care prods., industrial and institutional cleaners; pale yel. liq.; sp.gr. 1.04; cloud pt. 1 C max.; pH 8.5 ± 1.0 (5% aq.); 30% act. in water.

Empigen® BCB50. [Albright & Wilson] Benzalkonium chloride; cationic; see Empigen BAC; also retarder in dyeing of acrylic fibers; pale yel. liq.; dens. 0.98 g/cm³; visc. 120 cs; cloud pt. < 0 C; pH 7.0–9.5 (5% aq.); 50.0 ± 1.5% act. in water.

Empigen® BCF 80. [Albright & Wilson] Benzalkonium chloride; cationic; bactericide, germicide; raw material; industrial applics.; liq.; 80% conc.

Empigen® BCM75, BCM75/A. [Albright & Wilson] Hydrog.-tallow dimethyl benzyl ammonium chloride; cationic; see Empigen BAC; also used in mfg. of bentones; flotation aid; wh. waxy solid; m.p. 60 C; flash pt. 40 C (CC); pH 6.0–8.5 (5%); 75.0–80.0% act.

Empigen® BCP/25. [Albright & Wilson] Stearyl dimethyl benzyl ammonium chloride; cationic; hair conditioner rinses; mobile paste; 25% conc.

Empigen® BCY40. [Albright & Wilson] Alkyl ethoxy benzyl dimethyl ammonium chloride; retarder in the dyeing of acrylic fibers; pale yel. liq.; dens. 0.99 g/cm³; visc. 400 cs; cloud pt. < 0 C; pH 7.0–9.5 (5% aq.); 40.0 ± 2.0% act. in water.

Empigen® BS. [Albright & Wilson] Cocamidopropyl dimethylamine betaine; foaming agent/stabilizer and antistat for shampoos, foam baths, latex foam compds., oil prod.; wetting and coupling agent for cleaners, traffic film removers; pale sraw liq.; m.w. 350; sp.gr. 1.00; pH 4.5–8.0; 30 ± 1% act. in water.

Empigen® BS/H. [Albright & Wilson] Cocamidopropyl dimethylamine betaine; see Empigen BS; colorless/pale sraw liq.; m.w. 350; sp.gr. 1.04; pH 4.5–8.0; 30 ± 1% act. in water.

Empigen® BS/P. [Albright & Wilson] Cocamidopropyl dimethylamine betaine; see Empigen BS; pale sraw liq.; m.w. 350; sp.gr. 1.00; pH 4.0–8.0; 25.5 ± 1.5% act. in water.

Empigen® CHB. [Albright & Wilson] Myrtrimonium bromide; see Empigen BAC50; wh./pale cream powd.; m.w. 336; dens. 0.27 g/cc; pH 5–7; 96% act.

Empigen® CHB40. [Albright & Wilson] Alkyl trimethyl ammonium bromide; bactericide; liq.; 40% conc.

Empigen® CM. [Albright & Wilson] Tallow alkyl trimethyl ammonium methosulfate; cationic; antistat and conditioning agent for personal care prods.; emulsifier and antistat for industrial use; retarder for dyeing of acrylic fibers; pale straw cloudy liq.; dens. 0.96 g/cm³; flash pt. 40 C; cloud pt. 23 C; pH 6.5–8.5 (5% aq.); 30.0 ± 1.5% act.

Empigen® CMC. [Albright & Wilson] Alkyl trimethyl ammonium chloride; hair conditioner rinses; liq.

Empigen® CSC. [Albright & Wilson] Coco amidopropyl trimethyl ammonium chloride; hair conditioner rinses; liq.; 27% conc.

Empigen® FKC75K. [Albright & Wilson] Alkyl diamido amine acetate; cationic; raw materical used in fabric softener formulations; paste; 75% conc.

Empigen® FKC75L. [Albright & Wilson] Alkyl diamido amine lactate; cationic; raw materical used in fabric softener formulations; paste; 75% conc.

Empigen® FRB75S. [Albright & Wilson] Hardened tallow alkyl imidazoline methosulfate; cationic; textile conditioning agents for domestic, commercial laundry use and textile finishing; pale amber waxy solid; m.w. 723; set pt. 40 C; pH 4.0–6.5 (8% aq.); 75% act.

Empigen® FRC75S. [Albright & Wilson] Tallow alkyl imidazoline methosulfate; cationic; textile conditioning agent, antistat, and lubricant for fibers; raw material for fabric softener formulations; pale amber hazy liq.; disp. in water; dens. 0.95 g/cm³; visc. 300 cp; flash pt. 23 C (CC); pH 4.0–6.5 (8%); 75.0% act.

Empigen® FRC90S. [Albright & Wilson] Tallow alkyl imidazoline methosulfate; cationic; see Empigen FRB75S; pale amber soft paste; m.w. 723; dens. 0.95 g/cc; set pt. 20 C; pH 4.0–6.5 (8% aq.); 90% act.

Empigen® FRG75S. [Albright & Wilson] Partially hardened tallow alkyl imidazoline methosulfate; cationic; see Empigen FRB75S; pale amber waxy solid; m.w. 723; set pt. 30 C; pH 4.0–6.5 (8% aq.); 75% act.

Empigen® FRH75S. [Albright & Wilson] Oleyl alkyl imidazoline methosulfate; cationic; see Empigen FRB75S; pale amber clear or slightly hazy liq.; m.w. 739; sp.gr. 0.95; set pt. < 0 C; pH 4.0–6.5 (8% aq.); 75% act.

Empigen® OB. [Albright & Wilson] Lauramine oxide; coactive, foam booster/stabilizer and visc. modifier for personal care prods., fire fighting foam concs.; pale straw liq.; dens. 0.98 g/cm³; visc. 25 cs; pH 7.5 ± 0.5 (5% aq.); 30.0 ± 1.5% act.

Empigen® OH25. [Albright & Wilson] n-Myristyl dimethyl amine oxide; foam booster/stabilizer and visc. modifier for shampoos, foam baths, cleaners; improves conditioning in shampoos; solubilizer for liq. bleach prods.; pale straw liq.; m.w. 257; sp.gr. 0.99; visc. 800 cs; pH 7.5 ± 0.5 (5% aq.); 25% act.

Empigen® OS/A. [Albright & Wilson] Coco amidopropyl dimethyl amine oxide; foam booster/stabilizer and visc. modifier for shampoos, foam baths, cleaners; improves conditioning in shampoos; pale straw liq.; m.w. 306; sp.gr. 1.01; visc. 65 cs; pH 6.5–8.0 (5% aq.); 30.5% act.

Empigen® OY. [Albright & Wilson] PEG-3 lauramine oxide; see Empigen OB; pale straw liq.; dens. 1.0 g/cm³; visc. 23 cs; pH 6.5 ± 0.5 (5% aq.); 25.0 ± 1.0% act.

Empilan 0004. [Albright & Wilson] Castor oil EO/PO condensate; nonionic; emulsifier and lubricant for formulation of cutting oils, grinding fluids, and textile lubricants; mfg. of brake fluids; pale amber liq.; visc. 550 cs; HLB 12.0; sapon. no. 67.5 ± 3.5.

Empilan 2020. [Albright & Wilson] Lauryl alcohol ethoxylate; stabilizer for latices produced by emulsion polymerization; wh./cream solid; dens. 1.05 g/cm³; m.p. 35 C; HLB 15.7; hyd. no. 64 ± 4.

Empilan 2125-AU. [Albright & Wilson] Soya bean DEA; visc. modifier, superfatting agent, and coactive ingred. in personal care prods.; amber clear visc. liq.; sp.gr. 1.01; pH 10.5 ± 1.0 (10% aq.).

Empilan 2502. [Albright & Wilson] Coconut DEA; nonionic; foam booster/stabilizer; liq.; 100% conc.

Empilan BQ 100. [Albright & Wilson] PEG-8 oleate; nonionic; emulsifier for insecticides, spindle oils, industrial wetting agent, scouring agent, antifoam, PVC antistat; in laundry works, glue mfg., paper coating, hand cleaning jellies, turbine oil additive; dk. brn. liq; faint, typ. odor; m.w. 400 (of glycol base); disp. in water, yields clear sol'ns. with ethanol, xylene, oleic acid; sp.gr. 1.017; visc. 130 cps; m.p. –1 C; 100% act.

Empilan CDE. [Albright & Wilson] Cocamide DEA; nonionic; foam boosting/stabilizing agent, solubilizer, detergent for use in personal care and detergent prods.; antistat in plastics; pale yel. visc. liq.; dens. 1.0 g/cm³; 100% act.

Empilan CDEY. [Albright & Wilson] Coconut DEA; nonionic; foam booster and visc. modifier in hair shampoos; pale yel. clear visc. liq.; 78.0% conc.

Empilan CDX. [Albright & Wilson] Cocamide DEA; nonionic; foam booster/stabilizer, solubilizer, and visc. modifier for toiletry and detergent formulations; antistatic agent in plastics; pale yel. visc. liq.; dens. 1.0 g/cm³.

Empilan CIS. [Albright & Wilson] Cocamide MIPA; nonionic; see Empilan CME; cream waxy flake; dens. 0.4 g/cm³; m.p. 46 C; 100% conc.

Empilan CM. [Albright & Wilson] Coconut MEA; nonionic; foam booster/stabilizer in toiletry and detergent formulations; intermediate; cream wax-like solid; 80% conc.

Empilan CME. [Albright & Wilson] Cocamide MEA; nonionic; foam booster/stabilizer in detergent systems; stabilizer for hair and carpet shampoos; base for mfg. of ethoxylated alkylolamides; visc. modifier; cream waxy flake; disp. in hot water; dens. 0.4 g/cm³; m.p. 68 C; 94.2% act.

Empilan DL 40. [Albright & Wilson] Fatty alcohol ethoxylate; nonionic; emulsifier for fatty acids and min. oils for fiber lubricants, wetting and stabilizing agent for syn. rubber latices, textile dye leveling and dispersing agent in textile industry; antistat; pale yel. clear liq.; misc. with water; dens. 1.046 g/cc; cloud pt. > 100 C in dist. water; 40% act.

Empilan DL 100. [Albright & Wilson] Fatty alcohol ethoxylate; see Empilan DL 40; stiff waxy solid; warm water sol.; m.p. 36 C; cloud pt. > 100 C (1% aq.); 100% act.

Empilan EGMS. [Albright & Wilson] Glycol stearate; opacifier/pearling and emulsifying/stabilizing agent in shampoos; emollient; wh. hard, wax-like flake; oil-sol.; m.p. 56 C; 65% diester, 30% monoester.

Empilan FD. [Albright & Wilson] Coconut DEA;

nonionic; surfactant, wetting agent, emulsifier for mfg. of personal care prods., industrial and institutional liq. detergent systems; coactive and visc. modifier for laundry detergents and hard surface cleaners; foam stabilizer and booster; pale yel. clear visc. liq.; sp.gr. 1.00–1.01; flash pt. 95 C; 80% conc.

Empilan FD20. [Albright & Wilson] Coconut DEA; nonionic; act. ingred. in liq. detergents; foam booster and visc. modifier; pale yel. clear visc. liq.; 65% conc.

Empilan FE. [Albright & Wilson] Coconut DEA; nonionic; foam booster, visc. modifier used as coactive ingred. in toiletries, liq. detergents; pale yel. clear visc. liq.; sp.gr. 0.970; visc. 250 cP; flash pt. 28 C (PMCC); 90% conc.

Empilan GMS LSE40. [Albright & Wilson] Glyceryl stearate SE; emulsifier in the baking and food industry; lubricant in prod. of PVC sheets and expanded PS; wh. microbead powd.; dens. 0.5 g/cm^3; m.p. 55–65 C; 40.0% min. monoglyceride.

Empilan GMS LSE80. [Albright & Wilson] Glyceryl stearate SE; see Empilan GMS LSE40; wh. microbead powd.; dens. 0.5 g/cm^3; m.p. 64 C; 80.0% min. monoglyceride.

Empilan GMS MSE40. [Albright & Wilson] Glyceryl stearate SE; see Empilan GMS LSE40; wh. microbead powd.; dens. 0.5 g/cm^3; m.p. 55–60 C; 36.0% min. monoglyceride.

Empilan GMS NSE40. [Albright & Wilson] Glyceryl stearate SE; nonionic; emulsifier in the baking and food industry; lubricant and antistat in prod. of PVC sheets and expanded PS; stabilizer; wh. microbead powd.; dens. 0.5 g/cm^3; m.p. 55–60 C; pH 6–8 (10% aq.); 40.0% min. monoglyceride.

Empilan GMS NSE90. [Albright & Wilson] Glyceryl stearate SE; see Empilan GMS NSE40; wh. microbead powd.; dens. 0.5 g/cm^3; m.p. 65 C; 90.0% min. monoglyceride.

Empilan GMS SE40. [Albright & Wilson] Glyceryl stearate SE; see Empilan GMS LSE40; also emulsifier for cosmetic and pharmaceuticals; wh. microbead powd.; dens. 0.5 g/cm^3; m.p. 55–60 C; pH 7–9 (10% aq.); 36.0% min. monoglyceride.

Empilan GMS SE70. [Albright & Wilson] Glyceryl stearate SE; see Empilan GMS LSE40; wh. microbead powd.; dens. 0.5 g/cm^3; m.p. 61 C; 70.0% min. monoglyceride.

Empilan KA3. [Albright & Wilson] Fatty alcohol EO condensate; nonionic; scouring and wetting agent for textiles, emulsifier, dye leveling and dispersing agent, detergent, in metal processing, cutting oils, paper industry, paints, insecticides and pesticides; pale straw clear liq.; sol. in min. oils and other org. solvs.; HLB 8.9; biodeg.; 100% act.

Empilan KA5. [Albright & Wilson] Fatty alcohol EO condensate; nonionic; see Empilan KA3; pale straw clear liq.; HLB 11.5; cloud pt. 37 C; biodeg.; 100% act.

Empilan KA590. [Albright & Wilson] Fatty alcohol EO condensate; nonionic; see Empilan KA3; water wh. clear liq.; dens. 9.8 lb/gal; HLB 11.4; pour pt. 12 F; cloud pt. 95 F; biodeg.; 90% act.

Empilan KA880. [Albright & Wilson] Fatty alcohol EO condensate; see Empilan KA3; HLB 13.6; cloud pt. 81 C; biodeg.; 80% act.

Empilan KA1080. [Albright & Wilson] Fatty alcohol EO condensate; see Empilan KA3; HLB 14.6; cloud pt. 95 C; biodeg.; 80% act.

Empilan KB 2. [Albright & Wilson] Laureth-2; nonionic; emulsifier, foam booster, superfatting agent, used in detergents and emulsifying systems; almost colorless liq.; insol. in cold water; dens. 0.9; visc. 25 cs; 100% act.

Empilan KB 3. [Albright & Wilson] Laureth-3; nonionic; see Empilan KB 2; liq.; 100% conc.

Empilan KC 3. [Albright & Wilson] Laureth-3; nonionic; see Empilan KB 2; liq.; 100% conc.

Empilan KM 11. [Albright & Wilson] Steareth-11; nonionic; emulsifier, foam control agent in syn. heavy duty detergents, soap additive; colorless to pale straw paste; 100% act.

Empilan KM 20. [Albright & Wilson] Ceteareth-20; nonionic; wetting agent, detergent for industrial/domestic applics.; emulsifier, foam control agent in syn. heavy duty detergents; colorless to pale straw waxy flake/block; 100% act.

Empilan KM 50. [Albright & Wilson] Steareth-50; nonionic; see Empilan KM 20; colorless to pale straw waxy flake/block; 100% act.

Empilan LDE. [Albright & Wilson] Lauramide DEA; nonionic; foam stabilizer, solubilizer, detergent used in shampoos and liq. detergent formulations; cream solid; dens. 1.0 g/cm^3.

Empilan LDX. [Albright & Wilson] Lauramide DEA and diethanolamine; nonionic; foam booster/stabilizer, solubilizer, and visc. modifier for toiletry and detergent formulations; antistatic agent in plastics; cream solid; dens. 1.0 g/cm^3; 100% conc.

Empilan LIS. [Albright & Wilson] Lauramide MIPA; nonionic; foam booster/stabilizer for liq. and powd. detergents, visc. modifier, base for mfg. of ethoxylated deriv.; cream waxy flake; hot water disp.; dens. 0.4 g/cc; 87.5% act.

Empilan LME. [Albright & Wilson] Lauramide MEA; foam booster/stabilizer in detergent systems; stabilizer for hair and carpet shampoos; base for mfg. of ethoxylated alkylolamides; visc. modifier; waxy flake; dens. 0.4 g/cm^3; m.p. 85 C.

Empilan LP2. [Albright & Wilson] EO condensate of Empilan CME; nonionic; liq. detergent additive as foam stabilizer, detergent booster, solubilizer; cream yel. soft paste; faint char. odor; sapon. no. 14; 100% act.

Empilan MAA. [Albright & Wilson] PEG-6 cocamide; nonionic; see Empilan LP2; paste; sapon. no. 15; 100% act.

Empilan NP9. [Albright & Wilson] Nonoxynol-9; nonionic; wetting agent, detergent, emulsifier, solubilizer; pale straw soft paste/liq.; dens. 1.0 g/cc; visc. 300 cs; cloud pt. 55 C (1% aq.); 100% act.

Empimin® 3093. [Albright & Wilson] Anionic surfactant and solv.; defoaming agent for mfg. of wet process phosphoric acid and other aq., acidic systems; straw clear liq.; m.w. 444; sp.gr. 1.10; visc. 450

cs; flash pt. (CC) > 100 C; pH 6.0 ± 1.0 (5%); 72 ± 2% act.

Empimin® 3095. [Albright & Wilson] Anionic surfactant and solv.; see Empimin 3093; straw clear liq.; m.w. 444; sp.gr. 1.05; visc. 100 cs; flash pt. (CC) > 100 C; pH 6.0 ± 1.0 (5%); 44 ± 2% act.

Empimin® AQ60. [Albright & Wilson] Ammonium alkyl triethoxysulfate; anionic; detergent, foaming and wetting agent, dispersant in institutional and industrial cleaning specialties; pale yel. clear liq.; dilutable in water; sp.gr. 1.02; flash pt. 30 C (PMCC); pH 7.0 ± 0.6 (10% aq.); 58.0 ± 3.0% act.

Empimin® BMA. [Albright & Wilson] Anionic surfactant used as air entraining agent for cementitious mixes, foaming agent for concrete; pale yel. cloudy liq.; m.w. 433; sp.gr. 1.11; visc. 140 cs; pH 7–9 (5% aq.); 34 ± 1.5% act.

Empimin® MA. [Albright & Wilson] Sodium dihexyl sufosuccinate; anionic; emulsifier, dispersant, wetting agent for emulsion polymerization; straw clear liq.; sol. in water; dens. 1.1 g/cm^3; visc. 80 cs; flash pt. 30 C (CC); pH 6.0 ± 1.0 (5%); 63.0% min. act.

Empimin® OP45. [Albright & Wilson] Sodium dioctyl sulfosuccinate; anionic; emulsifier, dispersant, wetting agent used for emulsion polymerization, oil slicks, textiles; straw clear liq.; sol. in water; dens. 1.05 g/cm^3; visc. 100 cs; flash pt. > 100 C (CC); pH 6.0 ± 1.0; 44% conc.

Empimin® OP70. [Albright & Wilson] Sodium dioctyl sulfsuccinate; anionic; see Empimin OP45; straw clear liq.; sol. in water; dens. 1.10 g/cm^3; visc. 450 cs; flash pt. > 100 C (CC); pH 6.0 ± 1.0 (5%); 72.0 ± 2.0% act.

Empimin® OT. [Albright & Wilson] Sodium dioctyl sulfosuccinate; anionic; dispersant, emulsifier, wetting agent for o/w emulsions, emulsion polymerization, filler and extender dispersions; straw clear liq.; dens. 1.0 g/cm^3; visc. 55 cs; flash pt. 25 C (CC); pH 6.0 ± 1.0 (5%); 60.0% min. act.

Empimin® SDS. [Albright & Wilson] Sodium decyl sulfate; anionic; emulsifier, dispersant, detergent, and wetting agent for industrial and institutional cleansers, mfg. of pigments, alkaline cleansers; dust suppression; pale yel. clear liq.; sp.gr. 1.06; pH 7.0 ± 0.6 (5% aq.); 30.0 ± 1.0% act.

Empimin® SQ25. [Albright & Wilson] Sodium alkyl triethoxysulfate; dispersant, wetting and foaming agent in institutional, household, and industrial cleaners; water-wh. to pale yel. clear liq.; dilutable in water; sp.gr. 1.043 ± 0.005; visc. 80 ± 45 cP; pH 7.0–0.7 (5% aq.); 25.0 ± 0.5% act.

Empimin® SQ70. [Albright & Wilson] Sodium alkyl triethoxylsulfate; see Empimin SQ25; off-wh. to pale yel. hazy visc. liq.; dens. 1.1 g/cm^3; visc. 550 cP max.; cloud pt. 0 C max.; pH 8.0 ± 0.6 (2% aq.); 68.0 ± 2.0% act.

Empiphos 4KP. [Albright & Wilson] Tetrapotassium pyrophosphate; detergent builder for liq. detergents; pigment dispersant and stabilizer in emulsion paints; clarifying agent in liq. soaps; mfg. of syn. rubber; boiler water treatment; sol. in water; dens. 75 lb/ft^3; pH 10.3.

Empiphos OMP, OSP. [Albright & Wilson] Org. polyphosphates; builders and dispersants for liq. detergents; liq.

Empiphos STP, STP/D. [Albright & Wilson] Pentasodium triphosphate, tech. anhyd.; water softener, sequestrant, peptizer, deflocculating agent; wh. free-flowing powd.; 93% thru 100 mesh; poured dens. 0.8 g/ml; pH 9.8 (1%); 95% act.

Empiphos STP/1L, STP/6L, STP/6N. [Albright & Wilson] Pentasodium triphosphate, tech. hydrated; water softener, sequestrant, peptizer, deflocculating agent; wh. free-flowing gran.; 40% thru 16 on 60 mesh (STP/1L); 55% thru 16 on 60 mesh (STP/6L); 60% thru 16 on 60 mesh (STP/6N); poured dens. 0.7, 0.55, and 0.88 g/ml resp.; pH 9.6 (1%); 95% act.

Empiphos STP/AD, STP/N. [Albright & Wilson] Pentasodium triphosphate, tech. anhyd.; water softener, sequestrant, peptizer, deflocculating agent; wh. free-flowing gran.; 55% on 16 mesh (STP/AD); 60% thru 16 on 60 mesh (STP/N); poured dens. 0.70 and 0.80 g/ml resp.; pH 9.8 (1%); 95% act.

Empiphos STP/DMST, STP/E, STP/F, STP/MST. [Albright & Wilson] Pentasodium triphosphate, tech. moisturized; water softener, sequestrant, peptizer, deflocculating agent; wh. free-flowing powd.; 90% thru 100 mesh (STP/DMST, STP/MST); 85% thru 100 mesh (STP/E); 80% thru 100 mesh (STP/F); pH 9.8 (1%); poured dens. 0.8 g/ml; 94% act.

Empiphos STP/L, STP/L16, STP/M, STP/M16. [Albright & Wilson] Pentasodium triphosphate, tech. hydrated; water softener, sequestrant, peptizer, deflocculating agent; wh. free-flowing gran.; 35% thru 16 on 60 mesh (STP/L, STP/L16); 50% thru 16 on 60 mesh (STP/M, STP/M16); poured dens. 0.60, 0.55, 0.60, and 0.55 g/ml resp.; pH 9.8 (1%); 95% act.

Empiwax SK. [Albright & Wilson] Cetearyl alcohol and sodium lauryl sulfate; SE wax as o/w emulsifier for pharmaceutical and toilet preparations and ointments; wh. flake.

Empiwax SK/BP. [Albright & Wilson] Cetearyl alcohol and sodium lauryl sulfate; see Empiwax SK; complies with B.P. specs. for emulsifying wax; wh. flake.

Emrite® 6007. [Henkel/Emery] Antifog agent for use in food-pkg. films; Gardner 4 max. clear liq.; sp.gr. 0.945–0.955; visc. 210 cSt; HLB 3.4; pour pt. –7 C; acid no. 2.0 max.; iodine no. 77; sapon. no. 163–176; ref. index 1.4675–1.4690; flash pt. 245 F; 42% min. monoester.

Emrite® 6120. [Henkel/Emery] Polysorbate 80; nonionic; food emulsifier; antifog agent for use in food-pkg. films; Gardner 6 max. liq.; sp.gr. 1.00–1.10; visc. 400 cSt; HLB 15.1; pour pt. –15 C; acid no. 2.0 max.; sapon. no. 45–55; ref. index 1.4700; flash pt. 605 F; 100% conc.

Emsorb® 2500. [Henkel/Emery] Sorbitan oleate; nonionic; coupler, emulsifier, lubricant, and softener for textile fibers and leather, cosmetics, household prods.; for formulating petrol. oils and waxes, natural fats and waxes, and alkyl esters; Gardner 8 liq.; sol. in min. oil, butyl stearate, Stod., xylene; insol. in water; dens. 8.3 lb/gal; visc. 360 cSt (100 F); HLB 4.6; pour pt. < 0 C; flash pt. 475 F; 100% conc.

Emsorb® 2502. [Henkel/Emery] Sorbitan sesquioleate; nonionic; coupler, coemulsifier for o/w systems; emulsifier for w/o systems; household aerosols, cosmetics, industrial and textile oils; Gardner 6 liq.; sol. in min. oil, butyl stearate, glycerol trioleate, Stod. solv., xylene; insol. in water; dens. 8.2 lb/gal; visc. 475 cSt (100 F); HLB 4.5; pour pt. < 0 C; flash pt. 470 F; 100% conc.

Emsorb® 2503. [Henkel/Emery] Sorbitan trioleate; nonionic; see Emsorb 2500; also coemulsifier for min. oil; processing and softening of textiles and leathers; Gardner 7 liq.; sol. see Emsorb 2502; dens. 7.9 lb/gal; visc. 100 cSt (100 F); HLB 2.1; pour pt. < 0 C; flash pt. 500 F; 100% act.

Emsorb® 2505. [Henkel/Emery] Sorbitan stearate; nonionic; coupler, hydrophobic emulsifier; coemulsifier for industrial oils, household prods., and cosmetics; textile lubricant; paper and textile processing; Gardner 4 solid; disp. in butyl stearate, perchloroethylene; insol. in water; HLB 5.2; m.p. 50 C; flash pt. 480 F; 100% conc.

Emsorb® 2507. [Henkel/Emery] Sorbitan tristearate; see Emsorb 2503; also fiber-to-metal lubricant for textile fibers; Gardner 4 waxy solid; sol. in butyl stearate, perchloroethylene, glycerol trioleate, Stod. solv., xylene; insol. in water; HLB 2.2; m.p. 54 C; flash pt. 480 F; 100% conc.

Emsorb® 2510. [Henkel/Emery] Sorbitan palmitate; coupler, emulsifier for cosmetic and household prods.; fiber-to-metal lubricant; Gardner 8 waxy solid; sol. in oil, Stod. solv., xylene; insol. in water; HLB 6.5; m.p. 47 C; flash pt. 445 F; 100% conc.

Emsorb® 2515. [Henkel/Emery] Sorbitan laurate; nonionic; coupler, emulsifier; used in household specialities, industrial oils, cosmetics, and emulsion polymerization; antifoam properties; Gardner 7 liq.; sol. in min. oil, butyl stearate, glycerol trioleate, Stod. solv., xylene; water-disp.; dens. 8.8 lb/gal; visc. 1000 cSt; HLB 8.0; pour pt. 15 C; flash pt. 430 F; cloud pt. < 25 C; 100% conc.

Emsorb® 2720. [Henkel/Emery] Polysorbate 20; o/w emulsifier, solubilizer, visc. modifier; used in creams, lotions, shampoos, conditioners, liq. soaps; Gardner 6 liq.; sol. @ 5% in water, triolein, IPA; dens. 9.2 lb/gal; visc. 160 cSt (100 F); HLB 16.7; pour pt. –10 C; flash pt. 510 F.

Emsorb® 2721. [Henkel/Emery] PEG-80 sorbitan laurate; wetting agent, dispersant, mild cleanser for baby prods.; Gardner < 6 liq.; sol. @ 5% in water, IPA; dens. 9.2 lb/gal; visc. 500 cSt (100 F); HLB 19.0; pour pt. 15 C; cloud pt. 82.5 C (5% saline); flash pt. 505 F; 72% act. in water.

Emsorb® 2722. [Henkel/Emery] Polysorbate 80; emulsifier, coemulsifier for cosmetics; dispersant for pigments in makeup; solubilizer for oils, flavors, fragrances; Gardner 5 liq.; sol. @ 5% in water, triolein, IPA; dens. 9.0 lb/gal; visc. 200 cSt (100 F); HLB 15.0; pour pt. –12 C; flash pt. 505 F.

Emsorb® 2726. [Henkel/Emery] PEG-40 sorbitan diisostearate; solubilizer for flavors in mouthwashes, for perfumes, germicides, and other polar substances in aq. systems; sol. in water, alcohol, some cosmetic oils.

Emsorb® 2728. [Henkel/Emery] Polysorbate 60; o/w emulsifier for cosmetics, hair straighteners, shaving prods., sun care prods.; binder in powds.; with Emsorb 2505 to stabilize wax emulsions; Gardner 3 waxy semisolid; sol. @ 5% in water, IPA; dens. 8.9 lb/gal; visc. 250 cSt (100 F); HLB 15.2; pour pt. 25 C; flash pt. 510 F.

Emsorb® 2729. [Henkel/Emery] Polysorbate 65 NF; o/w emulsifier, lubricant, emollient in creams and lotions; Gardner 5 waxy solid; sol. @ 5% in min. oil, triolein, IPA; disp. in water, IPM; dens. 8.4 lb/gal; HLB 11.0; m.p. 34 C; flash pt. 555 F.

Emsorb® 6900. [Henkel/Emery] PEG-20 sorbitan oleate; nonionic; dispersant for pigments in coatings; solubilizer for oils and fragrances; hydrophilic emulsifier; coemulsifier for petrol. oils, fats, solvs., and waxes in cosmetics, household prods., industrial lubricants, and textile dye carriers; emulsifier for tobacco sucker control concs.; Gardner 5 liq.; sol. in water; dens. 9.0 lb/gal; visc. 200 cSt (100 F); HLB 15.0; pour pt. –12 C; flash pt. 535 F; cloud pt. 70 C (5% saline); 100% conc.

Emsorb® 6901. [Henkel/Emery] PEG-5 sorbitan oleate; nonionic; o/w emulsifier and lubricant in industrial lubricants, textile lubricants and softeners; Gardner 6 liq.; sol. in glycerol trioleate, Stod. solv.; water-disp.; dens. 8.5 lb/gal; visc. 450 cs; HLB 10.1; cloud pt. < 25 C; 100% conc.

Emsorb® 6903. [Henkel/Emery] PEG-20 sorbitan trioleate; nonionic; o/w emulsifier for petrol. oils, fats, waxes, and alkyl esters; lubricant for metals, textiles, leather; in sol. oils for metal processing and finishing; glass fiber lubricants; automotive lubricant additives; Gardner 6 liq.; sol. in butyl stearate, glycerol trioleate, Stod. solv.; water-disp.; dens. 8.6 lb/gal; visc. 300 cs; HLB 10.9; cloud pt. < 25 C; 95% act.

Emsorb® 6905. [Henkel/Emery] PEG-20 sorbitan stearate; nonionic; o/w emulsifier for min. oil, fats, and waxes; fiber-to-metal lubricant and coemulsifier with Emsorb 2505 in paraffin wax emulsions for textiles, paper and wallboard coatings; useful as emulsifier in cosmetic and household prods.; Gardner 3 waxy semisolid; sol. 5% in Stod. solv. and water; dens. 8.9 lb/gal; visc. 250 cSt (100 F); HLB 15.2; pour 25 C; flash pt. 510 F; cloud pt. 70 C (5% saline); 100% conc.

Emsorb® 6906. [Henkel/Emery] PEG-4 sorbitan stearate; nonionic; w/o emulsifier used in household formulations; fiber-to-metal lubricant for syn. and cellulosic fibers and yarns; Gardner 3 solid; sol. 5% in Stod. solv.; water-disp.; HLB 9.0; m.p. 42 C; flash pt. 495 F; 100% conc.

Emsorb® 6907. [Henkel/Emery] PEG-20 sorbitan tristearate; nonionic; o/w emulsifier for petrol. oils and natural fats; lubricant and softener for textile procesing and finishing compds.; primary emulsifier; Gardner 5 solid; sol. 5% in min. oil, butyl stearate, glycerol trioleate, Stod. solv.; water-disp.; HLB 11.1; m.p. 34 C; flash pt. 555 F; cloud pt. < 25

C; 100% conc.

Emsorb® 6908. [Henkel/Emery] PEG-16 sorbitan tristearate; o/w emulsifier, lubricant, softener; for textile processing and finishing compds.; Gardner 4 waxy solid; sol. @ 5% in min. oil, glyceryl trioleate, Stod., xylene; disp. in water, butyl stearate; HLB 10.0; m.p. 38 C; cloud pt. < 25 C; flash pt. 540 F.

Emsorb® 6910. [Henkel/Emery] PEG-20 sorbitan palmitate; nonionic; o/w emulsifier and lubricant; Gardner 6 visc. liq.; sol. 5% in water, glycerol trioleate; dens. 9.2 lb/gal; visc. 200 cSt (100 F); HLB 15.8; pour pt. 22 C; flash pt. 485 F; cloud pt. 77 C (5%); 100% conc.

Emsorb® 6915. [Henkel/Emery] PEG-20 sorbitan laurate; nonionic; o/w emulsifier and solubilizer of petrol. oils, solvs., and fats; used in cosmetic creams and lotions; visc. modifier in shampoos; emulsifier for dye carriers, antistatic scrooping agent in primary spin finishes, and fiber processing aid in textile industry; Gardner 6 liq.; sol. in water, glycerol trioleate; dens. 9.2 lb/gal; visc. 160 cSt (100 F); HLB 16.7; pour pt. –10 C; flash pt. 510 F; cloud pt. 75–85 C; 100% conc.

Emsorb® 6917. [Henkel/Emery] PEG-16 sorbitan trioleate; o/w emulsifier, lubricant for metals, textiles, leather, glass fiber, automotive additives; Gardner 5 liq.; sol. @ 5% in min. oil, butyl stearate, glyceryl trioleate, Stod., xylene; disp. in water; dens. 8.2 lb/gal; visc. 122 cSt (100 F); HLB 10.0; pour pt. < –10 C; cloud pt. < 25 C; flash pt. 530 F.

Emthox® 2730. [Henkel/Emery] PEG-75 cocoa butter NF; emollient, humectant, emulsifier; Gardner 3 solid; sol. @ 5% in water, triolein, IPA; dens. 8.7 lb/gal (50 C); visc. 210 cSt (100 F); m.p. 31–35 C; flash pt. 530 F.

Emthox® 5882. [Henkel/Emery] Laureth-4; nonionic; dispersant for bath oil; emulsifier for creams and lotions; Gardner 1 liq.; sol. @ 5% in min. oil, triolein, IPA, IPM; disp. in water, glycerin; dens. 7.9 lb/gal; visc. 20 cSt (100 F); HLB 9.4; cloud pt. < 25 C; flash pt. 355 F; 100% conc.

Emthox® 5885. [Henkel/Emery] Ceteareth-20; emulsifier, solubilizer for cosmetics, conditioners, depilatories, hair straighteners, sun care prods.; Gardner 1 solid; sol. @ 5% in water, IPA; disp. in min. oil; dens. 7.7 lb/gal; HLB 15.3; m.p. 40 C; cloud pt. 91 C (5% saline); 560 F.

Emthox® 5940. [Henkel/Emery] Trideceth-6; emulsifier, dispersant, wetting agent, penetrant, leveling agent; for hair coloring prods.; Gardner 1 liq.; sol. @ 5% in triolein, IPA, IPM; disp. in water, min. oil, glycerin; dens. 8.2 lb/gal; visc. 32 cSt (100 F); HLB 11.4; pour pt. 12 C; cloud pt. < 25 C; flash pt. 345 F.

Emthox® 5941. [Henkel/Emery] Trideceth-9; nonionic; general purpose wetting agent and detergent; dispersant in bath oils; Gardner 1 liq.; sol. @ 5% in water, triolein, IPA; disp. in min. oil, glycerin; dens. 8.3 lb/gal; visc. 43 cSt (100 F); HLB 13.0; pour pt. 20 C; cloud pt. 54 C; flash pt. 390 F.

Emthox® 5942. [Henkel/Emery] Trideceth-11; see Emthox 5941; Gardner 1 liq.; sol. @ 5% in water, triolein, IPA; disp. in min. oil, glycerin; dens. 8.4 lb/gal; visc. 47 cSt (100 F); HLB 13.8; pour pt. 17 C; cloud pt. 73 C; flash pt. 420 F.

Emthox® 5964. [Henkel/Emery] Laureth-23; emollient, thickener for shampoos; solubilizer, dispersant; Gardner 1 solid; sol. @ 5% in water, IPA; dens. 8.6 lb/gal (70 C); HLB 16.7; m.p. 40 C; cloud pt. 93 C (5% saline); flash pt. 440 F.

Emthox® 5967. [Henkel/Emery] Laureth-12; nonionic; emollient, thickener in shampoos; emulsifier in creams and lotions; solid; HLB 14.0; 100% conc.

Emthox® 5993. [Henkel/Emery] Trideceth-3; emulsifier, dispersant; Gardner 1 liq.; sol. @ 5% in triolein, IPA; disp. in water, min. oil; dens. 7.8 lb/gal; visc. 19 cSt (100 F); HLB 7.9; pour pt. –15 C; cloud pt. < 25 C; flash pt. 300 F.

Emthox® 6957. [Henkel/Emery] Nonoxynol-40; wetting agent, detergent, dispersant, coemulsifier; for waterless hand cleaners, hair care prods., liq. soaps, bath additives; Gardner 1 solid; sol. @ 5% in water, IPA; HLB 17.8; m.p. 40 C; cloud pt. 88–92 C (5% saline); flash pt. 560 F.

Emthox® 6962. [Henkel/Emery] Nonoxynol-6; dispersant, wetting agent, coemulsifier for hair coloring prods., soaps; Gardner 2 liq.; sol. @ 5% in IPA, IPM; disp. in water, min. oil, glycerin; dens. 8.5 lb/gal; visc. 100 cSt (100 F); HLB 10.9; pour pt. –10 C; cloud pt. < 25 C; flash pt. 515 F.

Emthox® 6964. [Henkel/Emery] Nonoxynol-9; detergent, penetrant, emulsifier, scouring agent, wetting agent; used in cleansers, hair colorants, bath preps., makeup, hair conditioners, pharmaceuticals; Gardner 1 liq.; sol. @ 5% in water, IPA, IPM; dens. 8.8 lb/gal; visc. 112 cSt (100 F); HLB 13.0; pour pt. 5 C; cloud pt. 54 C; flash pt. 510 F.

Emthox® 6965. [Henkel/Emery] Nonoxynol-11; nonionic; detergent, penetrant, wetting agent, emulsifier for makeup, hair colorants, bath prods., hair conditioners, soaps; Gardner 2 liq.; sol. @ 5% in water, IPA; dens. 8.7 lb/gal; visc. 116 cSt (100 F); HLB 13.5; pour pt. 5 C; cloud pt. 71 C; flash pt. 480 F.

Emthox® 6971. [Henkel/Emery] Nonoxynol-50; detergent, wetting agent, dispersant, emulsifier; used in liq. soaps, hair colorants, makeup and nail care prods.; Gardner 1 solid; sol. @ 5% in water, IPA; HLB 18.2; m.p. 44–48 C; cloud pt. 74–77 C (10% saline); flash pt. 520 F.

Emthox® 6972. [Henkel/Emery] Nonoxynol-50; detergent, wetting agent, dispersant, coemulsifier for hair colorants and curlers, waterless hand cleaners, bath additives, liq. soaps, nail care prods.; Gardner 1 liq.; sol. @ 5% in water, IPA; dens. 9.2 lb/gal; visc. 394 cSt (100 F); HLB 18.2; pour pt. < 0 C; cloud pt. 74–77 C (10% saline); flash pt. 520 F.

Emthox® 6984. [Henkel/Emery] Octoxynol-40; nonionic; dispersant, wetting agent, emulsifier for hair coloring and curling prods., shampoos, makeup; Gardner 1 liq.; sol. @ 5% in water, IPA; dens. 9.0 lb/gal; visc. 220 cSt (100 F); HLB 17.9; pour pt. 13 C; cloud pt. 74 C (10% saline); 70% act. in water.

Emulan. [Emulan] Mink oil; emollient with high spreading coefficient for skin and hair care formulations requiring good oxidative stability; sol. in IPM and similar esters, min., animal, and veg. oils, IPA; partly sol. in ethanol; HLB 9.0.

Emulan® FM. [BASF AG] TEA monooleic ester; nonionic; w/o emulsifier for impregnation, lubrication, polishing, and cleaning; o/w emulsifier for polishes; corrosion inhibitor for metals; brn. liq.; char. odor; sol. in min. oils, fatty oils and mixts.; sp.gr. 0.95; visc. 200 mPa•s; HLB 3–5 (w/o type); 9.5 (o/w type); acid no. 0–5; sapon. no. 135; pH 10 (1% aq.); flash pt. 240 C; 100% act.

Emulan® OC. [BASF AG] Fatty alcohol oxyethylate; nonionic; o/w emulsifier for dry bright emulsions; stabilizer for emulsions and disp.; wh. or pale yel. soft wax; sol. in water, fatty acids, waxes, and polar org. solv.; sp.gr. 1.03 (50 C); HLB 17; m.p. 35–40 C; acid no. almost 0; sapon. no. 0; pH 6–7.5 (1% aq.); flash pt. 210 C; 100% act.

Emulan® OG. [BASF AG] Fatty alcohol oxyethylate; nonionic; o/w emulsifier for fatty acids, waxes, polar org. solvs.; dispersant for solid substances, grinding dyes; stabilizer for hydraulic and anticorrosion oils; waxy powd. in bead form; sol. in water and polar org. media; dens. 0.6 kg/l; HLB 17; m.p. 45–50 C; pH 6–7.5 (1% aq.); flash pt. 240 C; 100% act.

Emulan® OSN. [BASF AG] Fatty alcohol oxyethylate; nonionic; o/w emulsifier for wax emulsions; dispersant, stabilizer; wh. to pale yel. soft wax; sol. in water, fatty acids, waxes, polar org. solv.; sp.gr. 1.03 (50 C); m.p. 35–40 C; HLB 17; flash pt. 230 C; sapon no. 0; pH 6–7 (1% aq.); 100% act.

Emulan® OU. [BASF AG] Fatty alcohol oxyethylate; nonionic; o/w emulsifier for fatty acids, waxes, polar org. solvs.; dispersant for solid substances, grinding processes; stabilizer of emulsions and disps.; wh. or pale yel. soft wax; sol. in water, fatty acids, molten greases and ester waxes, polar org. solvs.; sp.gr. 1.02 (50 C); HLB 17; m.p. 40 C; solid. pt. 35 C; flash pt. 210 C; sapon. no. 0; pH 6–7.5 (1% aq.); 100% act.

Emulbon LB-75. [Toho] Polyol polyborate; lubricant; liq.

Emulbon LB-78. [Toho] Six-membered borate with monoalkyl POE group; lubricant, plasticizer, solv.; liq.

Emulbon S-83. [Toho] Di(glycerol) borate sesquioleate; nonionic (semipolar); emulsifier, dispersant; liq.; 100% conc.

Emulbon S-260. [Toho] Di(glycerol) borate sesquistearate; lubricant; solid.

Emulbon T-80. [Toho] Di(glycerol) borate POE monooleate; nonionic (semipolar); emulsifier, dispersant; liq.; 100% conc.

Emuldan HA 40. [Grindsted] Tallow or lard monodiglyceride; food emulsifier and stabilizer; beads; m.p. 60 C; sapon. no. 160–180; 40% min. monoester.

Emuldan HA 52. [Grindsted] Tallow or lard monodiglyceride; see Emuldan HA 40; beads; m.p. 62 C; sapon. no. 155–175; 52% min. monoester.

Emuldan HV 40 K. [Grindsted] Veg. oil monodiglyceride; see Emuldan HA 40; beads; m.p. 62 C; sapon. no. 155–175; 40% min. monoester.

Emuldan HV 52 K. [Grindsted] See Emuldan HA 40; beads; m.p. 65 C; sapon. no. 160–180; 52% min. monoester.

Emuldan HVF 52 K. [Grindsted] Veg. oil monodiglyceride; see Emuldan HA 40; powd.; m.p. 65 C; sapon. no. 160–180; 52% min. monoester.

Emulgade CRC. [Henkel] Hair conditioner/antistat blend for hair rinses and conditoners; wh. flakes.

Emulgane E. [Henkel] POE veg. oil; nonionic; emulsifier for animal, veg. fats and oils; dispersant for pigments; liq.; water-sol.; 100% conc.

Emulgane O. [Henkel] Alkylpolyglycolether; nonionic; dispersant and emulsifier; dyeing assistant in acid media; paste; 100% conc.

Emulgen A-60, A-90, A-500. [Kao] POE aryl ethers; nonionic; low foaming dispersant for pigment and dyestuffs; liq. to solid.

Emulphogene® BC-420. [Rhone-Poulenc Surf.] Trideceth-3; nonionic; emulsifier, detergent, and dispersant for petrol. oils, intermediate in mfg. of high-foaming anionic surfactants; VCS 2 (50 C) cloudy, slightly visc. pourable liq.; mild, pleasant odor; sol. in toluene, oil; insol. in water; sp.gr. 0.92–0.94; flash pt. > 200 F (PMCC); pour pt. –7 to –1 C; pH 6.5–8.5 (10%); 100% act.

Emulphogene® BC-720. [Rhone-Poulenc Surf.] Trideceth-10; nonionic; foam builder/stabilizer, detergent booster, wetting and rewetting agent, emulsifier, and solubilizer; for textiles, in corrosion inhibitors; VCS 2 (50 C) opaque, visc. pourable liq.; mild, pleasant odor; sol. in water, alcohol, acetone, toluene, tetrachloroethylene; sp.gr. 1.00–1.03 (40 C); cloud pt. 58–62 C (1%); flash pt. > 200 F (PMCC); pour pt. 20 C; pH 6–8 (10%); 100% act.

Emulphogene® BC-840. [Rhone-Poulenc Surf.] Trideceth-15; nonionic; solubilizer, foam builder, and detergent; emulsifier and stabilizer for syn. latices; paste; sp.gr. 1.01–1.04 (40 C); cloud pt. > 95 C (1%); flash pt. > 200 F (PMCC); pour pt. 33–40 C; pH 6–8 (10%); 100% act.

Emulphogene® DA-530, DA-630. [Rhone-Poulenc Surf.] Isodeceth-4, -6; nonionic; wetting agent for textile industry; intermediates; liq.; 100% act.

Emulphogene® TB-970. [Rhone-Poulenc Surf.] Linear aliphatic ethoxylate; nonionic; detergent, dispersant, textile wet processing aid, used in cleaners and sanitizers; off-wh. waxy solid and flakes; sol. in water, ethanol, CCl_4, Cellosolve, xylene; cloud pt. > 100 C (1% aq.); surf. tens. 50.8 dynes/cm (0.1%); 100% act.

Emulphor® EL-620. [Rhone-Poulenc Surf.] PEG-30 castor oil; nonionic; emulsifier for cosmetic and pharmaceutical preparations, dispersant, antistat, lubricant, solubilizer, wetting agent, used in leather, pesticides, herbicides, paper industries; lt. brn. clear liq. sol. in water, acetone, CCl_4, alcohols, veg. oil, ethers, toluene, xylene; dens. 8.7–8.8 lb/gal; sp.gr. 1.04–1.05; visc. 600–1000 cps; cloud pt. 40–45 C (1% aq.); flash pt. 291–295 C; surf. tens. 41 dynes/cm (0.1%); 100% act.

Emulphor® EL-719. [Rhone-Poulenc Surf.] PEG-40 castor oil; nonionic; see Emulphor EL–620; yel.

clear liq.; sol. see Emulphor EL-620; dens. 8.9–9.0 lb/gal; sp.gr. 1.06–1.07; visc. 500–800 cps; cloud pt. 83–88 C (1% aq.); flash pt. 275–279 C; surf. tens. 38 dynes/cm (0.1%); 96% act.

Emulphor® EL-980. [Rhone-Poulenc Surf.] POE-200 castor oil; antistat, syn. fiber lubricant, and dyeing assistant in textiles; emulsifier for min. oil, triglycerides, and alkyl esters; solid; 100% act.

Emulphor® EL-985. [Rhone-Poulenc Surf.] POE veg. oil; nonionic; emulsifier, antistat, lubricant, spin finish additive; VCS 4 max. visc. liq. ; water sol.; 50% act.

Emulphor® ON-870. [Stepan] Oleth-20; nonionic; emulsifier, dispersant, stabilizer, solubilizer, wetting agent in metal cleaners with min. acids and corrosion inhibitors; used in rubber-based paints, adhesives, in textile, cosmetic, pharmaceutical industries; wh. solid wax; sol. in water, xylene, ethanol, ethylene glycol, butyl Cellosolve; sp.gr. 1.04; cloud pt. > 100 C; flash pt. 93 C (PMCC); pour pt. 46 C; surf. tens. 37 dynes/cm (0.1%); 100% act.

Emulphor® ON-877. [Stepan] POE oleyl alcohol; nonionic; see Empulphor ON–870; VCS 3 max. pourable visc. liq.; sol. in water, xylene, ethanol, ethylene glycol, butyl Cellosolve; sp.gr. 1.06; cloud pt. 95–100 C; flash pt. > 93 C (PMCC); pour pt. 15 C; surf. tens. 40.5 dynes/cm (0.1%); 70% act.

Emulphor® TO-530. [Rhone-Poulenc Surf.] PEG-10 tall oil; leveling agent and dye assistant for water- and solv.-based systems; low-foaming wetting agent; useful as constituent of alkaline-based spray metal cleaners; liq.; 100% act.

Emulphor® TO-639. [Rhone-Poulenc Surf.] PEG-16 tall oil; leveling agent, dye assistant; low-foam wetting agent in metal cleaners; liq.; 97% act.

Emulphor® VN-430. [Rhone-Poulenc Surf.] PEG-5 oleate; nonionic; w/o emulsifier in cosmetics, lubricant, surfactant used in textile industry; dk. amber clear liq.; disp. in water, sol. in xylene, wh. min. oil, ethanol, butyl Cellosolve; sp.gr. 1.0; cloud pt. < 10 C (1%); flash pt. > 93 C (PMCC); pour pt. 13 C; surf. tens. 32 dynes/cm (0.1%); pH 7–7.5; 100% act.

Emulphor® VT-650. [Rhone-Poulenc Surf.] PEG-9 stearate; nonionic; emulsifier for silicone and non-silicone defoamers, lubricant for dye carriers, used in textile processing; VCS 2 soft wax; slight acetic acid odor; sol. in hydrocarbons, ethanol, ether; dens. 8.3 lb/gal; sp.gr. 0.99 (50 C); visc. 9 cs (100 C); flash pt. 162 F (CC); sapon. no. 83–94; 100% act.

Emulsifier 99. [Pilot] Dodecylbenzene sulfonic acid; intermediate to produce agric. emulsifiers, emulsion polymer assistants; thick liq.; nonbiodeg.; 98% act.

Emulsifier K-700. [Evans Chemetics] Mixt. of wetting agents, conditioning oils, lanolin; clouding agent, wetting agent, conditioner for hair preparations; ylsh. clear oily liq.

Emulsifying Oil C. [Nat'l. Starch] Sulfated fatty acid; anionic; carrier for pine oil, basis for detergent emulsifier for xylene; liq.

Emulsion E-1614. [Rohm & Haas] Modified acrylic polymer latex; anionic; vehicle for sealer/finishes applied to nonresilient flooring materials such as terrazzo, quarry tile, or slate; such finishes develop high gloss; milky-wh. liq.; dens. 8.9 lb/gal; visc. 25 cps; pH 9.4; 40% solids.

Emulsogen B2M. [Hoechst Celanese] Amine salt of alkyl sulfamidocarboxylic acids with partially chlorinated hydrocarbons; anionic; emulsifier, corrosion inhibitor for metal machining; lubricant; brn. clear oily liq.; dissolves in water, min. oils; sp.gr. 0.94–0.95; flash pt. 128 C (Marcusson); ref. index 1.46; pH 7.3 ± 0.2 (1% aq.).

Emulsogen E. [Hoechst Celanese] Alkylaryl polyglycol ethers with fatty amine salts; corrosion inhibitor and wetting agent for preparation of dewatering fluids for metalworking; brn. clear visc. liq.; dissolves in water.

Emulsogen H. [Hoechst Celanese] Alkyl sulfamidocarboxylic acid with chlorinated hydrocarbons; anionic; base for corrosion inhibitors; dk. brn. visc. liq.; dissolves to clear sol'n. in most min. oils, to turbid sol'n. in water; acid no. 47–52.

Emulsogen KW. [Hoechst Celanese] Alkyl polyglycol ether with nitrogenous condensation prod.; corrosion inhibitor; pH 7.5–8.5 (1%); flash pt. 150 C.

Emulsogen S. [Hoechst Celanese] Inorganic/nonionic blend; emulsifier/corrosion inhibitor for metalworking oils; brn. clear visc. liq.; sol. in min. oils; dens. 0.97 g/cm^3; solid. pt. –9 C; flash pt. 130 C; pH 8.5–9.5 (10% aq.).

Emulsogen STH. [Hoechst Celanese] Alkylsulfamido carboxylic acid; anionic; antioxidant, emulsifier for min. oil emulsions with good rust protection; corrosion inhibitor for metalworking fluids; golden yel. clear oily liq.; sol. with turbidity in water, clear to turbid in min. oils.

Emulsynt 1055. [Van Dyk] Polyoxyalkylene oleate/laurate; fragrance solubilizer; sol. in min. oil, 95% ethanol, IPM; water-insol.

Emulsynt GDL. [Van Dyk] Glyceryl dilaurate; emulsifier, thickener, and emollient for creams and lotions; oil sol.; disp. in hot water.

Emulthin M-35. [Lucas Meyer] Lecithin; amphoteric; emulsifier, dispersant, wetting agents for all food uses esp. for baking industry; releasing agent for waffles; flour improver; spray-dried powd.; 50% conc.

Emulvin S. [Mobay] Polythioether; nonionic; emulsifier and latex stabilizer; ylsh. brn. oily liq.; dens. 1.13 g/cm^3; pH 6.

Emulvin W. [Mobay] Aromatic polyglycol ether, nonionic; emulsifier, stabilizer, wetting agent for latices; ylsh. brn. liq.; sp.gr. 1.13; pH 6.

Emulvis®. [C.P. Hall] PEG-150 distearate; nonionic; visc. builder in cosmetics, sol. retarder for gums and resins; antiblock agent for decals; thickener in pet shampoos; melt pt. control in suppositories; wh. to cream flake; bland odor; 50% sol. in 70% IPA; 3% sol. in water; m.p. 53–58 C; pH 4–6 (3% aq.); 100% act.

Epal® 6, 8, 10, 12. [Ethyl] C_6, C_8, C_{10}, and C_{12} resp. linear primary alcohols; detergent and emulsifier intermediate; liq.; 100% conc.

Epal® 12/70, 12/85. [Ethyl] C_{12}-C_{14} linear primary alcohol; see Epal 6; liq.; 100% conc.

Epal® 14, 16NF, 18NF. [Ethyl] C_{14}, C_{16}, and C_{18} resp. linear primary alcohols; see Epal 6; waxy solid; 100% conc.

Epal® 108, 610. [Ethyl] C_6-C_{10} linear primary alcohol; see Epal 6; liq.; 100% conc.

Epal® 618. [Ethyl] C_{16}-C_{18} linear primary alcohol; waxy solid; see Epal 6; 100% conc.

Epal® 810. [Ethyl] C_8-C_{10} linear primary alcohol; see Epal 6; liq.; 100% conc.

Epal® 1012. [Ethyl] C_{10}-C_{12} linear primary alcohol; see Epal 6; liq.; 100% conc.

Epal® 1214. [Ethyl] C_{12}-C_{16} linear primary alcohol; see Epal 6; sl. visc. liq.; 100% conc.

Epal® 1218. [Ethyl] C_{12}-C_{18} linear primary alcohol; see Epal 6; waxy solid; 100% conc.

Epal® 1416. [Ethyl] C_{14}-C_{16} linear primary alcohol; see Epal 6; waxy solid; 100% conc.

Epal® 1418. [Ethyl] C_{14}-C_{18} linear primary alcohol; see Epal 6; waxy solid; 100% conc.

Epal® 1618. [Ethyl] C_{16}-C_{18} linear primary alcohol; see Epal 6; waxy solid; 100% conc.

Epilink® 148, 149. [Akzo] Modified cycloaliphatic amine; solv.-free epoxy curing agent; Gardner 2 and 3 max. color resp.; visc. 0.35–0.55 and 0.5–0.7 Pa•s; amine no. 265–285 and 285–305.

Epilink® 353, 354, 355. [Akzo] Adducted polyaminoamides in org. solvs.; epoxy curing agents; Gardner 10 max. color; visc. 1–2, 1–3.5 and 4–6 Pa•s resp.; amine no. 125–140, 100–115, and 240–265 resp.

Epilink® 356. [Akzo] Adducted polyaminoamide in xylene; epoxy curing agent; Gartdner 10 max. color; visc. 1–2 Pa•s; amine no. 125–140.

Epilink® 360, 361, 370. [Akzo] Polyaminoamide-based in water; epoxy curing agents with ability to emulsify liq. epoxy resins in water; Gardner 16 max. color; visc. 30–50, 4.5–6, and 40–60 Pa•s resp.; amine no. 150–190.

Epilink® 375. [Akzo] Polyaminoamide-based in water; see Epilink 360; Gardner 16 max. color; visc. 40–70 Pa•s; amine no. 150–190.

Epilink® 380, 381. [Akzo] Polyaminoamide-based in water; see Epilink 360; Gardner 13 max. color; visc. 25–33 and 15–23 Pa•s resp.; amine no. 225–245 and 230–255.

EPIstatic® 100. [Eagle Picher] Tribasic lead sulfate; heat stabilizer for flexible and rigid PVC compds.; wh.; 99.9% 325 mesh; sp.gr. 6.7; surf. area 2.6 m^2/g; 82% Pb.

EPIstatic® 101. [Eagle Picher] EPIstatic 100 with lt. min. oil added for dust control; wh.; 99.9% 325 mesh; sp.gr. 6.0; surf. area 2.6 m^2/g; 80% Pb.

EPIstatic® 103. [Eagle Picher] Tribasic lead sulfate, org. surface treated, oiled; heat stabilizer for flexible PVC; 99.9% –325 mesh; sp.gr. 4.8; surf. area 2.7 m^2/g; 74.5% lead.

EPIstatic® 110. [Eagle Picher] Basic lead silicosulfate; heat stabilizer for use in low-temp. flexible PVC; wh.; 99.9% 325 mesh; sp.gr. 5.1; surf. area 2.0 m^2/g; 66% Pb.

EPIstatic® 111. [Eagle Picher] EPIstatic 110 with lt. min. oil for dust control; w. 99.9% 325 mesh; sp.gr. 6.0; surf. area 2.6 m^2/g; 80% Pb.

EPIstatic® 112. [Eagle Picher] Basic lead silicosulfate, org. surface treated; enhances heat stabilization in low-temp. PVC; 99.9% –325 mesh; sp.gr. 4.2; surf. area 2.1 m^2/g; 62.0% lead.

EPIstatic® 113. [Eagle Picher] EPIstatic 112 with lt. min. oil for dust control; wh.; 99.9% 325 mesh; sp.gr. 3.7; surf. area 2.1 m^2/g; 60% Pb.

EPIthal 120. [Eagle Picher] Basic lead phthalate; heat stabilizer in 90 and 105 C PVC wire and cable compds.; wh.; 99.9% 325 mesh; sp.gr. 4.5; surf. area 2.9 m^2/g; 76% Pb.

Epodil 741. [Pacific Anchor] Butyl glycidyl ether; reactive diluent for tooling, elec. applics., flooring; Gardner 1 color; dens. 7.7 lb/gal; visc. 1–3 cps; flash pt. (Seta) 137 F.

Epodil 742. [Pacific Anchor] Cresyl glycidyl ether; reactive epoxy diluent for tooling, elec. applics., coatings, flooring; Gardner 2 color; dens. 9.0 lb/gal; visc. 5–20 cps; flash pt. (Seta) > 200 F.

Epodil 743. [Pacific Anchor] Phenyl glycidyl ether; reactive epoxy diluent for elec. applics.; Gardner 2 color; dens. 9.2 lb/gal; visc. 4–7 cps; flash pt. (Seta) > 200 F.

Epodil 745. [Pacific Anchor] p-t-Butyl phenyl glycidyl ether; reactive epoxy diluent for tooling, elec. applics., flooring; Gardner 3 color; dens. 8.5 lb/gal; visc. 20–40 cps; flash pt. (Seta) > 200 F.

Epodil 746. [Pacific Anchor] 2-Ethyl hexyl glycidyl ether; reactive epoxy diluent for exposed aggregates, potting, flooring; Gardner 2 color; dens. 7.6 lb/gal; visc. 2–15 cps; flash pt. (Seta) > 200 F.

Epodil 747. [Pacific Anchor] Aliphatic glycidyl ether comparable to Epoxide 7; reactive epoxy diluent for exposed aggregates, potting, coatings, flooring; Gardner 1 color; dens. 7.5 lb/gal; visc. 5–15 cps; flash pt. (Seta) > 200 F.

Epodil 748. [Pacific Anchor] Aliphatic glycidyl ether comparable to Epoxide 8; reactive diluent for exposed aggregates, coatings, flooring; Gardner 1 color; dens. 7.4 lb/gal; visc. 5–20 cps; flash pt. (Seta) > 200 F.

Epodil 749. [Pacific Anchor] DGE of neopentyl glycol; reactive epoxy diluent for civil engineering applics.; Gardner 1 color; dens. 8.7 lb/gal; visc. 10–25 cps; flash pt. (Seta) > 200 F.

Epodil 750. [Pacific Anchor] DGE of 1,4-butanediol; reactive epoxy diluent for laminates and civil engineering applics.; Gardner 2 color; dens. 9.2 lb/gal; visc. 60–70 cps; flash pt. (Seta) > 200 F.

Epodil 757. [Pacific Anchor] Cyclohexane-dimethanol DGE; reactive epoxy diluent for civil engineering applics.; Gardner 1 color; dens. 8.7 lb/gal; visc. 10–25 cps; flash pt. (Seta) > 200 F.

Epodil 759. [Pacific Anchor] Aliphatic glycidyl ether; reactive epoxy diluent for coatings; Gardner 1 color; dens. 7.4 lb/gal; visc. 6–9 cps; flash pt. (Seta) > 200 F.

Epodil 769. [Pacific Anchor] DGE of resorcinol; reactive epoxy diluent for elec. applics.; Gardner 6 color; dens. 10.1 lb/gal; visc. 300–500 cps; flash pt. (Seta) > 200 F.

Epodil L. [Pacific Anchor] Low m.w. aromatic hydrocarbon diluent for epoxy systems; used for solv.-free coatings and flooring; Gardner 6 color; dens. 8.6 lb/gal; visc. 0.9 poise.

Epodil VFT-V6. [Pacific Anchor] Coumarone-indene resin; extender in lt.-colored epoxy systems, solv.-free coatings and flooring; Gardner 4 color; dens. 9.1 lb/gal; visc. 4 poise.

Epolene C-10. [Eastman] Low m.w. polyolefin resin; nonemulsifiable low-density wax used in hot melt adhesives and coatings for papers and packaging materials, as paraffin modifiers, in slush and cast molding, rubber compding., and in rigid and flexible vinyl compds.; high gloss coatings, low water-vapor-transmission rates, grease- and blocking-resistance; Gardner 1; m.w. 8000; dens. 0.906 g/cc; visc. (Thermosel) 7800 cps (150 C); soften. pt. (R&B) 104 C; cloud pt. 77 C (2% in 130 F paraffin); acid no. < 0.05.

Epolene C-13. [Eastman] Polyolefin resin, low m.w.; wax designed for use with Epolene waxes or blends containing lower m.w. materials; used as petrol.-wax modifiers to increase blend visc., improve grease resistance, blocking temp., scuff resistance, and gloss of paraffin; additive for inks and as compding. resins for hot melt adhesives; Gardner 1; m.w. 12,000; dens. 0.913 g/cc; cloud pt. 81 C (2%); soften. pt. 110 C; acid no. < 0.05.

Epolene C-14. [Eastman] Polyolefin resin; see Epolene C-13; Gardner 1 solid; m.w. 23,000; dens. 0.918 g/cc; cloud pt. 84 C; soften. pt. > 133 C; acid no. < 0.05.

Epolene C-15. [Eastman] Polyolefin resin; wax used in hot melt coatings for papers and pkg. materials, as paraffin modifiers, in slush and cast molding, hot melt adhesives, rubber compding., rigid and flexible vinyl compds.; additive in candle wax to provide gloss, sheen, opacity, and mold release; Gardner 1; m.w. 4000; dens. 0.906 g/cc; visc. 3900 cps (150 C); cloud pt. 75 C; soften. pt. 102 C; acid no. < 0.05.

Epolene C-16. [Eastman] Polyolefin resin; nonemulsifiable wax used as hot melt coatings for paper (glossy barrier coatings, readily heat sealable to paper prods., metal foils, and plastic films); in petrol.-wax coatings, and in wax-copolymer coatings for increased scuff resistance, gloss stabilization, and hot tack; tolerates high levels of inorganic fillers without drastic increases in melt visc.; Gardner 1; m.w. 8000; dens. 0.908 g/cc; visc. (Thermosel) 8500 cps (150 C); soften. pt. (R&B) 106 C; cloud pt. 78 C (2% in 130 F paraffin); sapon. no. 5.

Epolene C-17. [Eastman] Polyolefin resin; see Epolene C-13; Gardner 1 solid; m.w. 19,000; dens. 0.917 g/cc; soften. pt. (R&B) 133 C; cloud pt. 81 C (2% in 130 F paraffin); acid no. < 0.05.

Epolene C-18. [Eastman] Polyolefin resin; see Epolene C-16; Gardner 1; m.w. 4000; dens. 0.905 g/cc; visc. (Thermosel): 4000 cps (150 C); soften. pt. (R&B) 102 sapon no. 5.

Epolene E-10. [Eastman] Polyolefin resin; emulsifiable wax for water-emulsion for floor polishes, imparting excellent slip resistance, toughness, and durability to polish films; finishing agent for cotton and rayon fabrics; textile softener; lubricant in clay coatings for paper to reduce dusting during calendering; Gardner 2; m.w. 3000; dens. 0.942 g/cc; visc. 900 cps (125 C); soften. pt. (R&B) 106 C; acid no. 15.

Epolene E-11. [Eastman] Polyolefin resin; low-density emulsifiable wax used in emulsion floor polishes requiring slip resistance; Gardner 2; m.w. 2200; dens. 0.941 g/cc; visc. (Thermosel) 350 cps (125 C); soften. pt. (R&B) 106 C; acid no. 15.

Epolene E-12. [Eastman] Polyolefin resin; high-dens. emulsifiable wax which imparts better tensile strength, abrasion resistance, sewability, and hand in textile lubricant; softener for syn. fabrics, floor polishes, and emulsions for coating citrus fruit; Gardner 1; m.w. 2300; dens. 0.955 g/cc; visc. 250 cps; soften. pt. (R&B) 112 C; acid no. 16.

Epolene E-14. [Eastman] Polyolefin resin; low-density and low soften. pt. emulsifiable wax, imparts slip resistance to floor polish films; Gardner 2 powd.; m.w. 1800; dens. 0.939 g/cc; visc. (Thermosel) 250 cps (125 C); soften. pt. (R&B) 104 C; acid no. 16.

Epolene E-15. [Eastman] Polyolefin resin; see Epolene E-14; Gardner 2; m.w. 3400; dens. 0.925 g/cc; visc. (Thermosel) 350 cps (125 C); soften. pt. (R&B) 100 C; acid no. 16.

Epolene E-43. [Eastman] Polyolefin resin; emulsifiable wax, imparts slip resistance to floor polishes; Gardner 11 solid; m.w. 4500; dens. 0.934 g/cc; visc. (Thermosel) 400 cps (190 C; soften. pt. (R&B) 157 C; acid no. 47.

Epolene N-10. [Eastman] Polyolefin resin; nonemulsifiable wax, easily blended with waxes to improve tens. str., abrasion resistance, and adhesion to fibrous substrates; for paper-coating applics., e.g. folding cartons (increased coverage, glossy, flexible, scuff-resistant finish), and printing inks (improves resistance to scuffing and rub-off); Gardner 1; m.w. 3000; dens. 0.925 g/cc; visc. (Thermosel) 1500 cps (125 C); soften. pt. (R&B) 111 C; cloud pt. 85 C (2% in 130 F paraffin); acid no. < 0.05.

Epolene N-11. [Eastman] Polyolefin resin; similar to Epolene N-10; also used as solid lubricant in corrugated board manufacture; Gardner 1; m.w. 2200; dens. 0.921 g/cc; visc. (Thermosel) 350 cps (125 C); soften. pt. (R&B) 108 C; cloud pt. 79 C (2% in 130 F paraffin); acid no. < 0.05.

Epolene N-12. [Eastman] Polyolefin resin; high-density, high-melting point, hard polyethylene wax; raises blocking temp. of wax blends and upgrades low-melting paraffins; also as mold release additive (rubber molding); extrusion aid (vinyl processing); Gardner 1; m.w. 2300; dens. 0.938 g/cc; visc. (Thermosel) 450 cps (125 C); soften. pt. (R&B) 117 C; cloud pt. 87 C (2% in 130 F paraffin); acid no. < 0.05.

Epolene N-14. [Eastman] Polyolefin resin; similar to Epolene N-10 with lower m.w. and density; lubricant in processing vinyl plastics; Gardner 1 powd.; m.w. 1800; dens. 0.920 g/cc; visc. (Thermosel) 150 cps (125 C); soften. pt. (R&B) 106 C; cloud pt. 77 C (2% in 130 F paraffin); acid no. < 0.05.

Epolene N-15. [Eastman] PP resin; high melting pt. and hardness; modifier for petrol. waxes to increase resistance to blocking, scuffing, and abrasion, compd. resin for hot melt adhesives; Gardner 1 solid; m.w. 11,000; dens. 0.860 g/cc; visc. 600 cps (190 C); cloud pt. 104 C; soften. pt. (R&B) 163 C; acid no. < 0.05.

Epolene N-34. [Eastman] Polyolefin resin; see Epolene N-11; Gardner 1; m.w. 2900; dens. 0.910 g/cc; visc. (Thermosel) 450 cps (125 C); soften. pt. (R&B) 103 C; cloud pt. 69 C (2% in 130 F paraffin); acid no. < 0.05.

Epoxol 5-2E. [Am. Chem. Services] Epoxidized octyl tallate; stabilizing plasticizer with low volatility; produces hardness in semirigid vinyl formulations; extrusion aid; minimizes air release problems in plastisols; Gardner 1; sp.gr. 0.9244; Gardner visc. A; acid no. 0.5; sapon. no. 139.8.

Epoxol 7-4. [Am. Chem. Services] Epoxidized soybean oil; aux. plasticizer, acid scavenger, stabilizer for PVC compds.; food pkg. materials; Co-Pt 70; low odor; misc. with esters, ketones, heptane, ethyl ether, vinyl plasticizers, aromatic and chlorinated hydrocarbons; sp.gr. 0.994; dens. 8.28 lb/gal; visc. 314 cps; pour pt. 25 F; acid no. 0.10; sapon. no. 178.1; ref. index 1.4705; flash pt. 590 F (COC).

Epoxol 8-2B. [Am. Chem. Services] Epoxidized butyl esters of linseed oil fatty acids; aux. plasticizer with low visc. low volatility, heat and lt. stabilization; acid scavenger; stabilizer for PVC compds.; food pkg. materials; Gardner 1; sp.gr. 0.965; Gardner visc. A; acid no. 0.25–0.50; sapon. no. 150.

Epoxol 9-5. [Am. Chem. Services] Epoxidized linseed oils; stabilizing plasticizer; food pkg. materials; Cobalt-platinum 70; low odor; sol. see Epoxol 7-4; sp.gr. 1.03; dens. 8.58 lb/gal; visc. 619 cps; pour pt. 30 F; acid no. 0.12; sapon. no. 172; ref. index 1.4715; flash pt. 595 F (COC).

Epoxy Modifier ML. [Pacific Anchor] Diluent/plasticizer/wetting agent for epoxy systems; used for coatings, decorative flooring and exterior patching; improves gloss and substrate wetting; Gardner 8 color; dens. 7.4 lb/gal; visc. < 0.1 poise.

Ervol. [Witco] Wh. min. oil NF; emulsifier, lubricant, emollient; for cosmetics, food contact; visc. (Saybolt) 125–135 (100 F); pour pt. 20 F.

ES-1239. [CasChem] Dimethylaminopropyl ricinolamide benzyl chloride; cationic; emulsifier with antistatic, softening, bactericidal, wetting, dispersing properties for cosmetic, textile, agric. industries; Gardner 10+ clear liq.; sol. in water, alcohol, polar solv.; sp.gr. 1.022; dens. 8.48 lb/gal; visc. 15.3 stokes; flash pt. 230 F; 70% solids in propylene glycol.

Escacure® EB3. [Sartomer] Mixt. of benzoin normal butyl ethers; photoinitiator for polyester-styrene wood filler composites; optimum absorp. 250–350 nm; yel. clear liq.; m.w. 256; sp.gr. 1.060; flash pt. 158 F.

Escacure® KB1. [Sartomer] Benzyldimethyl ketal; photoinitiator for wh. coatings, inks, photopolymers, electronic photoresists, polyester-styrene wood filler composites; optimum absorp. 250–350 nm; wh. powd.; m.w. 256; sp.gr. 1.176; m.p. 145–151 F.

Escacure® KB60. [Sartomer] 60% sol'n. of benzyldimethyl ketal; see Escacure KB1; yel. clear liq.; sp.gr. 1.200; nonflam.

Escalol® 106. [Van Dyk] Glyceryl para-aminobenzoate; sunscreen for emulsion and hydroalcoholic systems; sol. in 95% ethanol, propylene glycol, glycerin; insol. in water, min. oil.

Escalol® 507. [Van Dyk] 2-Ethylhexyl p-dimethylaminobenzoate; topical sunscreen; sol. in min. and peanut oil, ethanol, IPA; insol. in water.

Escalol® 537Q. [Van Dyk] Dimethyl PABA ethyl cetearyldimonium tosylate; protects hair from sunlight damage; off-wh. powd.

Escalol® 557. [Van Dyk] Octyl methoxycinnamate; sunscreen; pale yel. liq. with slight odor; m.w. 290.4; sol. in peanut oil, min. oil, ethanol (95%, 100%), oleyl alcohol, castor oil, IPM, cyclomethicone; sp.gr. 1.005–1.013; acid no. 1 max.; sapon. no. 189 min.; ref. index 1.542–1.548.

Escalol® 567. [Van Dyk] Benzophenone-3; sunscreen; slightly ylsh., fine cryst. powd.; m.w. 228.25; sol. in peanut oil, ethanol, oleyl alcohol, castor oil; m.p. 62 C min.; 97% min. assay.

Escalol® 587. [Van Dyk] Octyl salicylate; sunscreen; colorless to pale yel. liq.; typ. bland odor; m.w. 250.34; sol. in IPA, ethanol, min. oil, dimethicone, cyclomethicone, IPM, octyl palmitate; sp.gr. 1.013–1.022; acid no. 2 max.; sapon. no. 200–230; ref. index 1.495–1.505.

ESI-Cryl 11. [Emulsion Systems] Polymer latex, aq.; opacifier for liq. dishwash, shampoos, general cosmetics; liq.; 40% act.

ESI-Cryl 12. [Emulsion Systems] Polymer latex, aq.; opacifier for acidic cleaners; liq.; 40% act.

ESI-Cryl 25. [Emulsion Systems] Polymer latex, aq.; opacifier for liq. dishwash, shampoos, general cosmetics; liq.; 40% act.

ESI-Terge 10. [Emulsion Systems] Coconut DEA; nonionic; foam stabilizer, thickener for household, cosmetic, industrial prods.; liq.; 100% conc.

ESI-Terge AH 20. [Emulsion Systems] Amine condensate, modified; nonionic/anionic; detergent base; visc. builder; liq.; 100% conc.

ESI-Terge B-15. [Emulsion Systems] 2:1 Coconut DEA; nonionic; detergent for mild all-purpose cleaners, foam builder and thickener, wetting agent; amber liq.; water sol.; sp.gr. 1.01; dens. 8.4 lb/gal; pH 9.8–10.8 (1%); biodeg.; 100% act.

ESI-Terge C-5. [Emulsion Systems] Modified coconut DEA; nonionic; see ESI-Terge B-15; amber liq.; water sol.; dens. 8.1 lb/gal; sp.gr. 0.97; pH 8.5–9.5 (1%); biodeg.; 100% act.

ESI-Terge S-10. [Emulsion Systems] 1:1 Coconut DEA; nonionic; detergent, emulsifier, foam stabilizer, thickener, for liq. dishwashing and car washing detergents; lt. straw liq.; dens. 8.15 lb/gal; sp.gr. 0.98; pH 9–10; biodeg.; 100% act.

ESI-Terge SXS. [Emulsion Systems] Sodium xylene sulfonate; anionic; solubilizer for lt.- and heavy-duty cleaners; liq.; 40% act.

ESI-Terge T-60. [Emulsion Systems] TEA dodecylbenzene sulfonate; anionic; detergent, wetting agent, foam stabilizer, for car and dishwashing detergent, syn. hand soap; amber liq.; dens. 9.1 lb/gal; sp.gr. 1.09; f.p. 320 F; pH 6.5–7.5 (1%); biodeg.; 60% act.

Esperal® 115RG. [Witco/Argus] Dicumyl peroxide; for med.-temp. applics. as a polymerization and crosslinking agent; 99% min. purity; 5.9% act. oxygen.

Esperase® 4.0T. [Novo] Proteinase; enzyme for laundry powd. detergents; off-wh. gran.

Esperase® 8.0L. [Novo] Proteinase; enzyme for nonbuilt detergents and prespotters; industrial laundry use; lt. brn. liq.

Esperase® 16.0 L. [Novo] Proteinase; enzyme; defoamer for built liq. detergents; brn. slurry.

Espercarb® 438M-60. [Witco/Argus] 60% sol'n. di-s-butyl peroxydicarbonate in odorless min. spirits; initiator, crosslinking agent for polymers; 4.10% act. oxygen.

Espercarb® 840. [Witco/Argus] Di-2-ethylhexyl peroxydicarbonate; 98% min. purity; 4.5% act. oxygen.

Espercarb® 840M-40. [Witco/Argus] 40% sol'n. of di-2-ethylhexyl peroxydicarbonate in odorless min. spirits; 1.85% act. oxygen.

Espercarb® 840M-70. [Witco/Argus] 70% sol'n. of di-2-ethylhexyl peroxydicarbonate in odorless min. spirits; 3.2% act. oxygen.

Espercarb® 840M. [Witco/Argus] 75% sol'n. of di-2-ethylhexyl peroxydicarbonate in odorless min. spirits; 3.5% act. oxygen.

Esperfoam® FR. [Witco/Argus] Ketone peroxides; initiator for rapid cures of polyester resins; DOT org. peroxide label is not required; sol'n.; 9.0% act. oxygen.

Esperox® 10. [Witco/Argus] t-Butyl peroxybenzoate; initiator for polymerization of ethylene and styrene, and for high temp. molding of polyesters; liq.; 98% min. purity; 8.1% act. oxygen.

Esperox® 10KXL. [Witco/Argus] t-Butyl peroxybenzoate on a proprietary formulated Burgess clay carrier; initiator; powd.; 40% conc.; 3.3% act. oxygen.

Esperox® 10XL. [Witco/Argus] t-Butyl peroxybenzoate on a proprietary formulated calcium carbonate carrier; initiator; crosslinking agent for chlorinated polyethylene; powd.; 40% conc.; 3.3% act. oxygen.

Esperox® 13M. [Witco/Argus] 75% sol'n. of t-butyl peroxycrotonate in odorless min. spirits; initiator for polymerization applics.; 7.6% act. oxygen.

Esperox® 28. [Witco/Argus] t-Butyl peroxy 2-ethyl hexanoate; initiator recommended for polymerization of ethylene plus use in med. temp. molding of polyester resin systems; liq.; 98% min. purity; 7.2% act. oxygen.

Esperox® 28MD. [Witco/Argus] t-Butyl peroxy 2-ethyl hexanoate in min. spirit diluent; initiator used in high pressure polymerization of ethylene; liq.; 50% conc.; 3.7% act. oxygen.

Esperox® 28PD. [Witco/Argus] t-Butyl peroxy 2-ethyl hexanoate in butyl benzyl phthalate; initiator used in high speed, heated cures of polyester resin systems; liq.; 50% conc.; 3.7% act. oxygen.

Esperox® 31M. [Witco/Argus] t-Butyl peroxypivalate in min. spirit diluent; initiator used in polymerization of ethylenically unsat. monomers; liq.; 75% conc.; 6.8% act. oxygen.

Esperox® 33M. [Witco/Argus] t-Butyl peroxyneodecanoate in min. spirit diluent; efficient and reactive initiator for polymerization of ethylenically unsat. monomers; liq.; 75% conc.; 4.9% act. oxygen.

Esperox® 41-25. [Witco/Argus] t-Butyl peroxymaleic acid; initiator for polymerization of ethylenically unsat. resins; paste; 25% conc.; 2.2% act. oxygen.

Esperox® 497M. [Witco/Argus] t-Butyl peroxy 2-methyl benzoate sol'n. in min. spirit diluent; initiator for polymerization of ethylene and styrene, high temp. molding of polyester resin systems, and vulcanization of silicon rubber; liq.; 75% conc.; 5.8% act. oxygen.

Esperox® 545M. [Witco/Argus] t-Amyl peroxyneodecanoate in min. spirit diluent; efficient and reactive initiator for polymerization of ethylenically unsat. monomer; liq.; 75% conc.; 4.6% act. oxygen.

Esperox® 551M. [Witco/Argus] t-Amyl peroxypivalate in min. spirit diluent; initiator used in polymerization of ethylenically unsat. monomers; liq.; 75% conc.; 6.4% act. oxygen.

Esperox® 570P. [Witco/Argus] 75% sol'n. of t-amyl peroxy 2-ethyl hexanoate in butyl benzyl phthalate; initiator; replacement for benzoyl peroxide, t-butyl peroctoate, and diperoctoates; liq.; 75% conc.; 5.2% act. oxygen.

Esperox® 740M. [Witco/Argus] 75% sol'n. of cumyl peroxyneoheptanoate in odorless min. spirits; initiator for polymerization of vinyl chloride.

Esperox® 747M. [Witco/Argus] 75% sol'n. of t-amyl peroxyneoheptanoate in odorless min. spirits; initiator for polymerization of ethylenically unsat. monomers.

Esperox® 750M. [Witco/Argus] 75% sol'n. of t-butyl peroxyneoheptanoate in odorless min. spirits; initiator for polymerization of ethylenically unsat. monomers;

Esperox® 939M. [Witco/Argus] Cumyl peroxyneodecanoate in min. spirit diluent; efficient and reactive initiator for polymerization of vinyl chloride; liq.; 75% conc.; 3.9% act. oxygen.

Esperox® 5100. [Witco/Argus] t-Amyl peroxybenzoate; initiator for polymerization of ethylene, styrene, acrylates, and curing of unsat. polyester resins; 95% min. purity; 7.3% act. oxygen.

Esperox® C-496. [Witco/Argus] t-Butyl peroxy-2-ethylhexyl carbonate; initiator for polymerization of vinyl monomers, styrene, acrylates, and unsat. polyester resins;

Espesilor AC Series. [Pulcra SA] Fatty acid DEAs; thickener and superfatting agents for shampoos, bubble baths, liq. detergents; perfume boosters for soaps.

Espesilor AC-43. [Pulcra SA] Linoleic alkanolamide; solubilizer, visc. builder, thickener, w/o emulsifier, hair and fiber conditioner; sol. in alcohols, glycols, and veg. oils; gels in water.

Espesilor AC-50. [Pulcra SA] Alkanolamide; foam stabilizer, solubilizer, and visc. modifier; liq.; water-sol.

Espesilor AC-52. [Pulcra SA] Alkanolamide; lubricating oil; sol. in aromatic and chlorinated solvs., disp. in water.

Espesilor EC Series. [Pulcra SA] Fatty acid MEAs; thickener for powd. detergents; soap additives.

Estabex® 138-A. [Akzo] Epoxidized soybean oil compd.; plasticizer for PVC homopolymer and copolymer resin formulations; FDA food grade applic; Gardner 2 clear liq.; sol. in esters, ethers, ketones, higher alcohols, aromatic and aliphatic hydrocarbons; sp.gr. 0.982; dens. 8.18 lb/gal; Gardner-Holdt visc. G.

Estabex® 2307. [Akzo] Epoxidized soya bean oil; plasticizer/stabilizer for plastisol applics., organosols, surf. coatings and inks, extruded prods., rigid PVC, chlorinated paraffins, halogenated rubbers, ethyl cellulose, PVC/PVA emulsions; processing aid for rigid PVC; pigment wetter; corrosion inhibitor for agrochemicals; approved for food contact and medical applics.; slightly yel. liq.; dens. 0.99 g/ml; visc. 550 mPa•s; flash pt. 295 C; acid no. 0.8; iodine no. 3; ref. index 1.473; 6.4% oxygen content.

Estabex® 2307 DEOD. [Akzo] Deodorized version of Estabex 2307 for sensitive food pkg. applics.; slightly yel. liq.; dens. 0.99 g/ml; visc. 550 mPa•s; acid no. 0.7; iodine no. 2; ref. index 1.473; 6.4% oxygen content.

Estabex® 2381. [Akzo] Monooctyl ester; plasticizer/stabilizer for plastisol applics., organosols, surf. coatings and inks, extruded prods.; sec. stabilizer for rigid and flexible PVC compds.; also for pigment wetting; yel. clear liq.; dens. 900 kg/m³; visc. 50 mPa•s; acid no. 0.8; iodine no. 3; ref. index 1.457; 3.8% oxygen content.

Estaflex® ATC. [Akzo] Acetyl tributyl citrate; primary plasticizer for plastics incl. PP, PS, PVAc, PVB, PVC, vinyl acetate copolymers, chlorinated rubber, ethyl cellulose, nitrocellulose; suitable for coatings in contact with foodstuffs; clear liq., odorless; dens. 1050 kg/m³; visc. 40 mPa•s; acid no. 0.1; sapon. no. 555; ref. index 1.441.

Ethacure® 100. [Ethyl] Diethyl toluene diamine; high-performance curing agent for epoxy resins; used in filament winding, elec. encapsulation, prepregs, tooling, potting and casting, laminating, coating, molding, and adhesive applics.; lt. red, clear, low-visc. liq.; b.p. 308 C; dens. 1.022 g/ml (20 C); visc. 326 cs (20 C); pour pt. –9 C; equiv. wt. 44.6 g/eq; flash pt. (TCC) > 135 C.

Ethanox® 330. [Ethyl] 1,3,5-Trimethyl-2,4,6-tris (3,5-di-tert-butyl-4-hydroxybenzyl) benzene; antioxidant and stabilizer for plastic, resin, rubber, and wax; food industry; wh. cryst. powd.; odorless; m.w. 775.2; sol. in methylene chloride, benzene, acetone; insol. in water; m.p. 244 C.

Ethanox® 398. [Ethyl] 2,2′-Ethylidenebis (4,6-di-t-butylphenyl) fluorophosphonite; antixoidant for PP, LLDPE; wh. cryst. powd.; m.w. 486.7; sol. (g/100 g): 18 g in xylene, 14.7 g in cyclohexane, 6.1 g in hexane; m.p. 200 C; 6.4% phosphorus.

Ethanox® 701. [Ethyl] 2,6-di-tert-Butylphenol; antioxidant; wh. or off-wh. cryst. powd.; m.w. 206.3; b.p. 253 C; m.p. 36 C.

Ethanox® 702. [Ethyl] 4,4′-Methylenebis(2,6-di-tert-butylphenol); antioxidant for rubber, plastic, resin, adhesive, petrol. oil, and wax; wh. to lt. straw cryst. powd.; m.w. 424.7; b.p. 250 C @ 10 mm; sol. in toluene, petrol. ether, ethyl alcohol; insol. in water; dens. 33 lb/ft³; m.p. 154 C; flash pt. 400 F (COC).

Ethanox® 703. [Ethyl] 2,6-Di-tert-butyl-alpha-dimethylamino-p-cresol; see Ethanox 702; lt. yel. cryst. powd.; m.w. 263.4; b.p. 179 C @ 40 mm; sol. in ethyl alcohol, toluene; insol. in water; m.p. 94 C; flash pt. > 200 F (COC).

Ethanox® 736. [Ethyl] 4,4′-Thiobis(6-tert-butyl-o-cresol); see Ethanox 702; also synergist for carbon black; wh. to yel. straw cryst. powd.; m.w. 358.5; b.p. 312 C @ 40 mm; sol. in ethyl alcohol, toluene, NaOH; insol. in water; dens. 39 lb/ft³; m.p. 124 C; flash pt. > 210 F. (TCC)

Ethanox® Antioxidant 376. [Ethyl] Med. m.w. hindered phenolic; antioxidant for food contact use in polymers; sol. 52% in hexane, acetone, 146% in benzene; insol. in water.

Ethoduomeen® T/13. [Akzo] PEG-3 tallow aminopropylamine; cationic; emulsifier used in making of bitumen emulsions, dispersing waxes; Gardner 18 max. liq.; sp.gr. 0.95; f.p. 20 C; m.p. 17 C; b.p. 150 C; flash pt. 204 C (COC); 95% min. act.

Ethoduomeen® T/25. [Akzo] PEG-15 tallow aminopropylamine; corrosion inhibitor in water treatment chemicals in sec. oil recovery; Gardner 18 min. liq.; sp.gr. 1.02; m.p. 25 C; flash pt. 238 C (COC); 95% act.

Ethoduomeen® TD/13. [Akzo] Ethoxylated diamine from tallow fatty acid; see Ethoduomeen T/13; Gardner 4 max. clear liq.; sp.gr. 0.95; surf. tens. 34.2 dynes/cm (0.1%).

Ethoduoquad® T/20. [Akzo] POE quat. ammonium salt from tallow fatty acid; cationic; antistat, emulsifier, dyeing assistant used in textile industry, as electroplating bath additive; clear liq.; pH 6–9 (10% aq.); 72% min. act.

Ethofat® 60/15. [Akzo] PEG-5 stearate; nonionic; emulsifier, detergent, wetting agent, dispersant, suspending agent, for textile, cosmetic, agric., metal and leather treating; Gardner 8 soft paste; sol. in acetone, IPA, CCl_4, benzene, disp. in water; sp.gr. 1.01; acid no. 1 max.; sapon. no. 110–120; surf. tens. 39 dynes/cm (0.1%); pH 6–7.5 (10% aq.).

Ethofat® 60/20. [Akzo] PEG-10 stearate; nonionic; see Ethofat 60/15; Gardner 8 clear liq.; sol. see Ethofat 60/15; sp.gr. 1.02; acid no. 1 max.; sapon. no. 70–80; surf. tens. 36 dynes/cm (0.1%); pH 6–7.5 (10% aq.).

Ethofat® 60/25. [Akzo] PEG-15 stearate; nonionic; see Ethofat 60/15; Gardner 11 paste; acid no. 1 max.; sapon. no. 55–65; surf. tens. 39.5 dynes/cm (0.1%); pH 6–7.5 (10% aq.).

Ethofat® 142/20. [Akzo] PEG-10 glycol tallate; see Ethofat 60/15; Gardner 11 clear liq.; sp.gr. 1.03; acid no. 1 max.; sapon no. 72–82; surf. tens. 39.5 dynes/cm (0.1%); pH 6–7.5 (10% aq.); flash pt. (PM) > 450 F.

Ethofat® 242/25. [Akzo] PEG-15 glycol tallate; nonionic; see Ethofat 60/15; Gardner 12 clear liq.; sol. in acetone, IPA, CCl_4, dioxane, benzene, water; sp.gr. 1.08; acid no. 1 max.; sapon no. 55–65; surf. tens. 42 dynes/cm (0.1%); pH 6–7.5 (10% aq.); flash pt. (PM) 525 F; 100% conc.

Ethofat® C/15. [Akzo] PEG-5 cocoate; nonionic; see Ethofat 60/15; Gardner 9 clear liq.; sol. see Ethofat 60/15; sp.gr. 1.00; HLB 10.6; acid no. 1 max.; sapon. no. 120–130; surf. tens. 33 dynes/cm (0.1%); pH 6–7.5 (10% aq.); 100% conc.

Ethofat® C/25. [Akzo] PEG-15 cocoate; nonionic; see Ethofat 60/15; Gardner 8 paste; sol. see Ethofat 60/15; acid no. 1 max.; sapon. no. 60–70; pH 6–7.5 (10% aq.); flash pt. (PM) > 400 F.

Ethofat® O/15. [Akzo] PEG-5 oleate; nonionic; see Ethofat 60/15; 9 max. Gardner clear liq.; sol. see Ethofat 60/15; sp.gr. 0.99; acid no. 1 max.; sapon. no. 110–120; surf. tens. 35 dynes/cm (0.1%); pH 6–7.5 (10% aq.); flash pt. (PM) > 400 F; 100% conc.

Ethofat® O/20. [Akzo] PEG-10 oleate; nonionic; see Ethofat 60/15; Gardner 9 max. clear liq.; sol. see Ethofat 60/15; sp.gr. 1.03; acid no. 1 max.; sapon. no. 75–85; surf. tens. 41 dynes/cm (0.1%); pH 6–7.5 (10% aq.); flash pt. (PM) 485 F.

Ethomeen® 18/12. [Akzo] PEG-2 stearamine; cationic; emulsifier, dispersant used in textile processing; Gardner 7 solid; sol. in acetone, IPA, CCl_4, benzene; sp.gr. 0.96; flash pt. (COC) 400 F; 95% tert. amine.

Ethomeen® 18/15. [Akzo] PEG-5 stearamine; cationic; see Ethomeen 18/12; Gardner 8 liq. to paste; sol. in acetone, IPA, CCl_4, benzene; sp.gr. 0.98; surf. tens. 34 dynes/cm (0.1%); flash pt. (COC) 500 F.

Ethomeen® 18/20. [Akzo] PEG-10 stearamine; see Ethomeen 18/12; Gardner 8 clear liq.; sp.gr. 1.02; flash pt. (COC) 540 F; surf. tens. 40 dynes/cm (0.1%).

Ethomeen® 18/25. [Akzo] PEG-15 stearamine; see Ethomeen 18/12; Gardner 8 clear liq.; sp.gr. 1.04; flash pt. (COC) 560 F; surf. tens. 43 dynes/cm (0.1%).

Ethomeen® 18/60. [Akzo] PEG-50 stearamine; see Ethomeen 18/12; wh. solid; sol. in acetone, IPA, CCl_4, benzene, water; sp.gr. 1.12; flash pt. (COC) 579 F; surf. tens. 49 dynes/cm (0.1%).

Ethomeen® C/12. [Akzo] PEG-2 cocamine; cationic; emulsifier, dispersant used in textile processing; Gardner 6 max. clear liq.; sol. in acetone, IPA, CCl_4, Stod., benzene; forms gel in water; sp.gr. 0.87; flash pt. (COC) 380 F; 95% tert. amine.

Ethomeen® C/15. [Akzo] PEG-5 cocamine; see Ethomeen C/12; Gardner 6 max. clear liq.; sol. in acetone, IPA, CCl_4, Stod., benzene, water (cloudy); sp.gr. 0.98; HLB 13.9; flash pt. (TOC) 460 F; surf. tens. 33 dynes/cm (0.1%).

Ethomeen® C/20. [Akzo] PEG-10 cocamine; see Ethomeen C/12; Gardner 11 max. clear liq.; sol. see Ethomeen C/15; sp.gr. 1.02; flash pt. (COC) 560 F; surf. tens. 39 dynes/cm (0.1%).

Ethomeen® C/25. [Akzo] PEG-15 cocamine; see Ethomeen C/12; Gardner 12 max. clear liq.; sol. in acetone, IPA, CCl_4, Stod., benzene, water; sp.gr. 1.04; flash pt. (COC) 500 F; surf. tens. 41 dynes/cm (0.1%).

Ethomeen® O/12. [Akzo] PEG-2 oleamine; cationic; emulsifier, dispersant used in textile processing; Gardner 8 max. clear liq.; sp.gr. 0.90; flash pt. (COC) 470 F; surf tens. 31.5 dynes/cm (0.1%).

Ethomeen® O/15. [Akzo] PEG-5 oleamine; see Ethomeen O/12; Gardner 8 max. clear liq.; sp.gr. 0.96; flash pt. (COC) 540 F; surf tens. 35.3 dynes/cm (0.1%).

Ethomeen® O/25. [Akzo] PEG-15 oleamine; see Ethomeen O/12; Gardner 8 max. paste; sp.gr. 1.04; flash pt. (PM) 380 F.

Ethomeen® S/12. [Akzo] PEG-2 soyamine; cationic; emulsifier, dispersant used in textile processing; Gardner 14 max. clear heavy liq; sol. in acetone, IPA, CCl_4, Stod., benzene; insol. water; sp.gr. 0.91; flash pt. (TOC) 440 F; surf. tens. 31.4 dynes/cm (0.1%); 95% tert. amine.

Ethomeen® S/15. [Akzo] PEG-5 soyamine; see Ethomeen S/12; Gardner 14 max. clear heavy liq; sol. in acetone, IPA, CCl_4, Stod., benzene; forms gel or disp. in water; sp.gr. 0.95; flash pt. (TOC) 460 F; surf. tens. 33 dynes/cm (0.1%).

Ethomeen® S/20. [Akzo] PEG-10 soyamine; see Ethomeen S/12; Gardner 14 max. clear heavy liq; sol. in acetone (cloudy), IPA, CCl_4, Stod., benzene, water; sp.gr. 1.02; flash pt. (COC) 540 F; surf. tens. 40 dynes/cm (0.1%).

Ethomeen® S/25. [Akzo] PEG-15 soyamine; see Ethomeen S/12; Gardner 14–18 max. clear heavy liq; sol. see Ethomeen S/20; sp.gr. 1.04; flash pt. (COC) 540 F; surf. tens. 43 dynes/cm (0.1%).

Ethomeen® T/12. [Akzo] PEG-2 tallow amine; cationic; emulsifier, dispersant used in textile processing; Gardner 8 paste; sol. in acetone, IPA, CCl_4, Stod., benzene, water; sp.gr. 0.92; flash pt. (COC) 410 F; 95% tert. amine.

Ethomeen® T/15. [Akzo] PEG-5 tallow amine; see Ethomeen T/12; Gardner 8 max. clear liq.; sol. cloudy in acetone, IPA, CCl_4, Stod., Benzene; forms gel in water; sp.gr. 0.97; flash pt. (PM) > 400 F; surf. tens. 34 dynes/cm (0.1%).

Ethomeen® T/25. [Akzo] PEG-15 tallow amine; see Ethomeen T/12; Gardner 8 max. clear liq.; sol. cloudy in acetone, IPA, CCl_4, benzene, water; sp.gr. 1.03; flash pt. (COC) > 500 F; surf. tens. 41 dynes/cm (0.1%).

Ethomeen® T/60. [Akzo] PEG-50 tallow amine; see Ethomeen 18/12; Gardner 10 max. paste to solid; m.w. 2362–2562; sp.gr. 1.115; flash pt. > 400 F (PM).

Ethomeen® TD/15. [Akzo] Ethoxylated tert. amine from tallow fatty acid; see Ethomeen T/12; Gardner 4 max. clear liq.; sp.gr. 0.97; surf. tens. 35.8 dynes/cm (0.1%).

Ethomeen® TD/25. [Akzo] Ethoxylated tert. amine from tallow fatty acid; see Ethomeen T/12; Gardner 4 max. clear liq.; sp.gr. 1.03; surf. tens. 43.5 dynes/cm

(0.1%).

Ethomid® HT/15. [Akzo] Ethoxylated amide from hyd. tallow acid; nonionic; emulsifier, dispersant, detergent; Gardner 10 max. solid; sol. in IPA; disp. in acetone, CCl_4, Stod., benzene, water; sp.gr. 1.03; surf. tens. 37 dynes/cm (0.1%); 100% conc.

Ethomid® HT/23. [Akzo] PEG-12.5 hydrog. tallow amide; nonionic; emulsifier, dispersant, detergent, dye leveling agent; for silicone finishing agents, sizing lubricants; Gardner 8 max. solid; sp.gr. 1.029; cloud pt. 130–170 F (1%); flash pt. (PM) > 400 F; surf. tens. 37 dynes/cm (0.1%).

Ethomid® HT/60. [Akzo] PEG-50 tallow amide; see Ethomid HT/15; Gardner 11 max. hard solid; sol. in IPA, CCl_4, benzene, water; sp.gr. 1.14; surf. tens. 47 dynes/cm (0.1%).

Ethomid® O/15. [Akzo] PEG-5 oleamide; nonionic; emulsifier, dispersant, detergent; lubricant for syn. and natural fibers; Gardner 12 max. liq.; sol. in IPA, CCl_4, gels in water @ 50–80 C; sp.gr. 1.0; flash pt. (PM) 225 F; surf. tens. 35 dynes/cm (0.1%).

Ethoquad® 18/12. [Akzo] PEG-2 stearmonium chloride and IPA; cationic; antistat, emulsifier, dyeing assistant, leveling agent, antifoam used in textile industry, as electroplating bath additives; Gardner 6 max. paste; sol. in acetone, IPA, benzene, water; sp.gr. 0.919; flash pt. 71 F; surf. tens. 40.2 dynes/cm (0.1%); pH 7–8 (10% aq.); 70% act.

Ethoquad® 18/25. [Akzo] PEG-15 stearmonium chloride; cationic; see Ethoquad 18/12; Gardner 11 max. clear liq.; sol. in acetone, IPA, benzene, water, CCl_4; sp.gr. 1.058; flash pt. (PM) 146 F; surf. tens. 50.1 dynes/cm (0.1%); pH 6–9 (10% aq.); 95% act.

Ethoquad® C/12. [Akzo] PEG-2 cocomonium chloride and IPA; see Ethoquad 18/12; Gardner 9 max. clear liq.; sol. see Ethoquad 18/25; sp.gr. 0.969; flash pt. (PM) < 80 F; surf. tens. 35.4 dynes/cm (0.1%); pH 7–8 (10% aq.); 75% act.

Ethoquad® C/25. [Akzo] PEG-15 cocomonium chloride; cationic; see Ethoquad 18/12; Gardner 11 max. clear liq.; sol. see Ethoquad 18/25; sp.gr. 1.071; flash pt. (PM) 196 F; surf. tens. 43.4 dynes/cm (0.1%); pH 7–8 (10% aq.); 95% act.

Ethoquad® O/12. [Akzo] PEG-2 oleamonium chloride and IPA; cationic; see Ethoquad 18/12; Gardner 9 max. clear liq.; sol. see Ethoquad 18/25; sp.gr. 0.932; flash pt. (PM) < 80 F; surf. tens. 40.3 dynes/cm (0.1%); pH 7–8 (10% aq.); 72% act.

Ethoquad® O/25. [Akzo] PEG-15 oleamonium chloride; cationic; see Ethoquad 18/12; Gardner 11 max. clear liq.; sol. see Ethoquad 18/25; sp.gr. 1.062; flash pt. (PM) 200 F; surf. tens. 40.8 dynes/cm (0.1%); pH 7–9 (10% aq.); 95% act.

Ethosperse® CA-2. [Lonza] Ceteth-2; nonionic; o/w emulsifier, thickener, stabilizer for hair care prods., antiperspirants; wh. solid; sol. in ethanol; disp. hot in water; HLB 6 ± 1; m.p. 29–33 C.

Ethosperse® G-26. [Lonza] Glycereth-26; nonionic; emulsifier, humectant for cosmetic, pharmaceutical and industrial uses; pale straw liq.; sol. in water, methanol, ethanol, acetone, toluol; sp.gr. 1.12 (38 C); visc. 150 cps (38 C).

Ethosperse® LA-23. [Lonza] Laureth-23; nonionic; see Ethosperse CA-2; wh. solid; sol. in water, hot ethanol; HLB 17 ± 1; m.p. 30–45 C.

Ethosperse® OA-2. [Lonza] Oleth-2; nonionic; see Ethosperse CA-2; also as spreading agent and emollient in bath oils; colorless liq.; sol. in ethanol, min. oil; disp. hot in water; visc. 30 cps; HLB 4 ± 1.

Ethosperse® SL-20. [Lonza] Sorbeth-20; nonionic; emulsifier, humectant for in cosmetic, pharmaceutical and industrial uses; lt. yel. liq.; sol. in water, methanol, ethanol, acetone; sp.gr. 1.16; visc. 460 cps; HLB 16.6; 100% act.

Ethoxamine C5, SF11. [Witco SA] Ethoxylated fatty amine; corrosion inhibitor, emulsifier, dispersant, antistat; liq.

Ethoxylan® 1685. [Henkel/Emery] PEG-75 lanolin; emollient, emulsifier, dispersant, foam stabilizer, resin plasticizer for cosmetic and pharmaceutical preparations; Gardner < 11 solid; sol. @ 5% in water, IPA; dens. 9.6 lb/gal; m.p. 39 C; cloud pt. 85 C; flash pt. 530 F.

Ethoxylan® 1686. [Henkel/Emery] PEG-75 lanolin; see Ethoxylan 1685; Gardner < 10 liq.; sol. @ 5% in water, IPA; dens. 8.9 lb/gal; visc. 1767 cSt (100 F); pour pt. 1 C; cloud pt. 86 C; 50% aq. sol'n.

Ethoxyol® 1707. [Henkel/Emery] Acetylated alcohols SE; nonionic; SE emollient with lubricating and penetrating properties; aux. emulsifier, solubilizer, pigment wetting agent for cosmetics, liq. soaps; Gardner < 8 liq.; sol. @ 5% in water, IPA, IPM, glycerin; dens. 8.7 lb/gal; visc. 132 cSt (100 F); pour pt. 1; cloud pt. –1 C; flash pt. 340 F.

Ethylan® 20. [Harcros UK] Nonoxynol-20; nonionic; detergent, wetting agent, stabilizer, emulsifier, solubilizer in waxes, resins, hand cleaning gels, pesticides, latexes; wh. waxy solid; negligible odor; water sol.; sp.gr. 1.065 (40 C); visc. 160 cs; HLB 16.0; cloud pt. > 100 C (1% aq.); flash pt. > 400 F (COC); pour pt. 30 C; pH 6–8 (1% aq.); 100% act.

Ethylan® A2. [Harcros UK] PEG-4 oleate; nonionic; emulsifier, antifoam agent; lt. amber liq.; sp.gr. 0.976; visc. 67 cs; HLB 7; 100% act.

Ethylan® A3. [Harcros UK] PEG-6 oleate; nonionic; emulsifier, antifoam, dispersant for industrial uses; lt. amber liq.; sp.gr. 1.0; visc. 100 cs; HLB 8.9; 100% act.

Ethylan® A4. [Harcros UK] PEG-8 oleate; nonionic; emulsifier, antifoam agent, dispersing agent for industrial uses; plastics antistat; lt. amber liq.; sp.gr. 1.017; visc. 130 cs; HLB 10.3; 100% act.

Ethylan® A6. [Harcros UK] PEG-12 oleate; nonionic; dispersing agent for industrial uses; emulsifier; antistat for plastics; lt. amber liq.; sp.gr. 1.037; visc. 160 cs; HLB 12.3; 100% act.

Ethylan® BBC31. [Harcros UK] Modified linear alcohol ethoxylate; nonionic; low foam iodophor intermediate; liq.; 100% conc.

Ethylan® BK 1130. [Harcros UK] Alkylaryl ethoxylate; dispersant for org. pigments, agrochemical toxicants, latex stabilization; liq.; 80% conc.

Ethylan® BV. [Harcros UK] Nonyl phenol ethoxylate; nonionic; foam stabilizer and booster, solubilizer, emulsifier used in pesticides, perfumes; wh. soft paste; water sol.; sp.gr. 1.083; visc. 380; HLB 14.5; cloud pt. 90 C (1% aq.); flash pt. > 400 F (COC); pour pt. 20 C; pH 6–8 (1% aq.); 100% act.

Ethylan® C 160. [Harcros UK] PEG ester of unsat. fatty acid; nonionic; emulsifier for olein and wax, dispersing aid, perfumery solubilizer; cream waxy solid; HLB 17.6; 100% act.

Ethylan® CD913. [Harcros UK] Syn. lower fraction primary alcohol EO condensate; nonionic; detergent, o/w and w/o emulsifier, wetting agent, solubilizer; colorless, clear liq.; water insol.; sp.gr. 0.936; visc. 27 cs; HLB 8.8; flash pt. 290 F (COC); pour pt. < 0 C; pH 6–8 (1% aq.); 97% act.

Ethylan® CD916. [Harcros UK] Syn. lower fraction primary alcohol EO condensate; nonionic; see Ethylan CD913; colorless clear liq.; char. fatty alcohol odor; water sol.; sp.gr. 0.993; visc. 53 cs; HLB 12.8; flash pt. 355 F (COC); pour pt. 0 C; pH 6–8 (1% aq.); 96% act.

Ethylan® CD919. [Harcros UK] Syn. lower fraction primary alcohol EO condensate; nonionic; see Ethylan CD913; colorless clear liq.; char. fatty alcohol odor; water sol.; sp.gr. 1.008; visc. 77 cs; HLB 14.4; flash pt. 385 F (COC); pour pt. 7 C; pH 6–8 (1% aq.); 97% act.

Ethylan® CD9112. [Harcros UK] Syn. lower fraction primary alcohol EO condensate; nonionic; see Ethylan CD913; wh. waxy solid; char. fatty alcohol odor; water sol.; sp.gr. 1.031 (40 C); visc. 56 cs (40 C); HLB 15.5; flash pt. > 400 F (COC); pour pt. 23 C; pH 6–8 (1% aq.); 100% act.

Ethylan® CF71. [Harcros UK] Coconut fatty ester; nonionic; fiber lubricant; emulsifier for cosmetics, pharmaceuticals, fiber lubricant; pale straw liq.; HLB 14; 100% act.

Ethylan® CH. [Harcros UK] Ethoxylated coconut fatty acid alkylolamide; nonionic; foam stabilizer in liq. detergents, coemulsifier; amber clear liq.; mild odor; water sol.; sp.gr. 1.025; visc. 270 cs; cloud pt. 80 C (1% aq.); flash pt. > 300 F (COC); pour pt. 14 C; pH 7–9 (1% aq.); 100% act.

Ethylan® CL. [Harcros UK] Alkylolamide EO condensate; nonionic; emulsifier, foam stabilizer for shampoos; liq.; cloud pt. 26 C; 100% act.

Ethylan® CRS. [Harcros UK] Ethoxylated coconut fatty acid alkylolamide; nonionic; foam stabilizer in liq. detergent, general purpose, or hard surface cleaners and shampoos, coemulsifier; amber clear liq.; mild odor; water sol.; sp.gr. 1.026; visc. 250 cs; flash pt. > 300 F (COC); pour pt. 10 C; pH 7–9 (1% aq.); 100% act.

Ethylan® D259. [Harcros UK] Syn. primary alcohol ethoxylate; nonionic; detergent, wetting agent, emulsifier, solubilizer for wax, oils; wh. solid; water sol.; sp.gr. 0.988; visc. 36 cs (40 C); cloud pt. 76 C; 100% act.

Ethylan® D2512. [Harcros UK] Syn. primary alcohol ethoxylate; nonionic; o/w emulsifier; detergent, wetting agent, solubilizer; wh. solid; faint odor; water sol.; sp.gr. 1.001 (40 C); visc. 46 cs (40 C); HLB 14.2; cloud pt. 92 C (1% aq.); pour pt. 29 C; pH 6–8 (1% aq.); 100% act.

Ethylan® DP. [Harcros UK] Nonyl phenol ethoxylate; nonionic; foam stabilizer and booster, emulsifier, solubilizer; in pesticides; slight hazy visc. liq.; negligible odor; water sol.; sp.gr. 1.068; visc. 400 cs; HLB 14.0; cloud pt. 80 C (1% aq.); flash pt. > 400 F (COC); pour pt. 13 C; pH 6–8 (1% aq.); 100% act.

Ethylan® ENTX. [Harcros UK] Alkylphenol ethoxylate; nonionic; emulsifier for min. oil and wax, mastic plasticizer; pale straw liq.; faint odor; oil sol.; sp.gr. 1.006; visc. 440 cs; HLB 9.0; flash pt. > 350 F (COC); pour pt. 2 C; pH 6–7.5 (1% aq.); 100% act.

Ethylan® GD. [Harcros UK] Special fatty acid DEA; nonionic; emulsifier for oils and metal degreasing agents, antistat, lubricant; clear amber liq.; mild odor; oil sol.; sp.gr. 0.982; visc. 1200 cs; flash pt. > 300 F (COC); pour pt. 10 C; pH 8–11 (1% aq.); 75% act.

Ethylan® GEL-2. [Harcros UK] Polysorbate 20; w/o emulsifier, solubilizer esp. with sorbitan esters; used in agric., perfumes, fiber lubricants, textile antistats, polymer additives; liq.; HLB 16.5; 100% conc.

Ethylan® GEO-8. [Harcros UK] Polysorbate 80; see Ethylan GEL-2; liq.; HLB 15.0; 100% conc.

Ethylan® GEP-4. [Harcros UK] Polysorbate 40; see Ethylan GEL-2; liq.; HLB 15.5; 100% conc.

Ethylan® GES-6. [Harcros UK] Polysorbate 60; see Ethylan GEL-2; paste; HLB 15.0; 100% conc.

Ethylan® GL-20. [Harcros UK] Sorbitan laurate; surfactant for use as antistat and textile fiber lubricant; liq.; HLB 8.0; 100% conc.

Ethylan® GLE-21. [Harcros UK] Polysorbate 21; see Ethylan GEL-2; liq.; HLB 13.3; 100% conc.

Ethylan® GO-80. [Harcros UK] Sorbitan oleate; see Ethylan GL-20; liq.; HLB 4.5; 100% conc.

Ethylan® GOE-21. [Harcros UK] Polysorbate 81; see Ethylan GEL-2; liq.; HLB 10.0; 100% conc.

Ethylan® GP-40. [Harcros UK] Sorbitan palmitate; see Ethylan GL-20; solid; HLB 6.7; 100% conc.

Ethylan® GPS-85. [Harcros UK] Polysorbate 85; see Ethylan GEL-2; liq.; HLB 11.0; 100% conc.

Ethylan® GS-60. [Harcros UK] Sorbitan stearate; see Ethylan GL-20; solid; HLB 5.0; 100% conc.

Ethylan® GT-85. [Harcros UK] Sorbitan trioleate; see Ethylan GL-20; liq.; HLB 1.5; 100% conc.

Ethylan® HA. [Harcros UK] Nonyl phenol ethoxylate; nonionic; detergent, wetting agent, stabilizer, emulsifier in waxes, resins, hand cleaning gels; wh. waxy solid; negligible odor; water sol.; sp.gr. 1.064 (60 C); visc. 120 cs (60 C); HLB 17.4; cloud pt. > 100 C (1% aq.); flash pt. > 400 F (COC); pour pt. 43 C; pH 6–8 (1% aq.); 100% act.

Ethylan® HB Series. [Harcros UK] Aromatic alkyoxylates; cosolv. with high flash pt. for use in iodophors, rinse aids, detergent sterilizers, glass cleaners, etc.; liq.

Ethylan® HB1. [Harcros UK] Short chain ethoxylate; nonionic; coalescing agent for surface coatings; liq.; 100% conc.

Ethylan® HB1-TG. [Harcros UK] 2-Phenoxyethanol; preservative for toiletries, coatings, etc.; liq.; 100% conc.

Ethylan® HB4. [Harcros UK] Aromatic ethoxylate; nonionic; penetrant, hydrotrope, cosolv., solubilizer, coupling agent; solv. for surfactants with low aq. sol.; liq.; 100% conc.

Ethylan® HB30. [Harcros UK] Aromatic ethoxylate; nonionic; penetrant for oil well drilling muds; solid; 100% conc.

Ethylan® HP. [Harcros UK] Nonyl phenol ethoxylate; nonionic; see Ethylan HA; wh. waxy solid; negligible odor; water sol.; sp.gr. 1.072 (40 C); visc. 78 cs (60 C); HLB 16.6; cloud pt. > 100 C (1% aq.); flash pt. > 400 F (COC); pour pt. 36 C; pH 6–8 (1% aq.); 100% act.

Ethylan® KEO. [Harcros UK] Nonyl phenol ethoxylate; nonionic; detergent, wetting agent, emulsifier, foam stabilizer, solubilizer, used in pesticides, perfumes; clear colorless liq.; negligible odor; water sol.; sp.gr. 1.060; visc. 340 cs; HLB 13; cloud pt. 54 C (1% aq.); flash pt. > 400 F (COC); pour pt. 3 C; pH 6–8 (1% aq.); 100% act.

Ethylan® LD. [Harcros UK] Coconut DEA; nonionic; foam stabilizer, emulsifier for cleaners, shampoos; plastics antistat; clear straw liq.; mild odor; disp. in water; sp.gr. 0.981; visc. 960 cs; flash pt. > 350 F (COC); pour pt. 15 C; pH 8.5–10.5 (1% aq.); 90% act.

Ethylan® LDG. [Harcros UK] Coconut DEA; nonionic; detergent, shampoo foam stabilizer; liq.; 82% conc.

Ethylan® LDS. [Harcros UK] Coconut DEA; nonionic; see Ethylan LD; straw clear liq.; mild odor; water-disp.; sp.gr. 0.990; visc. 810 cs; flash pt. > 350 F (COC); pour pt. 10 C; pH 8.5–10.5 (1% aq.); 87% act.

Ethylan® LM. [Harcros UK] Coconut monoalkanolamide; nonionic; emulsifier, antistat, foam stabilizer for liq. detergents; pale straw flake; mild odor; water insol.; sp.gr. 0.909 (80 C); visc. 40 cs (80 C); flash pt. > 350 F (COC); pour pt. 63 C; pH 8.5–10.5 (1% aq.); 94% act.

Ethylan® LM2. [Harcros UK] Ethoxylated coconut fatty acid alkylolamide; nonionic; foam stabilizer, emulsifier for detergent formulations; off-wh. soft paste; mild odor; water sol.; sp.gr. 0.963 (40 C); visc. 108 cs (40 C); flash pt. > 300 F (COC); pour pt. 2 C; pH 7–9 (1% aq.); 100% act.

Ethylan® MLD. [Harcros UK] Lauric DEA; nonionic; foam booster and stabilizer in toiletries and detergent formulations, antistat for plastics; pale cream waxy flake; negligible odor; disp. in water; visc. 130 cs (60 C); flash pt. > 300 F (COC); pour pt. 40 C; pH 8–9 (1% aq.); 95% act.

Ethylan® N30. [Harcros UK] Nonyl phenol ethoxylate; nonionic; detergent, wetting agent, stabilizer, emulsifier in waxes, resins, hand cleaning gels; wh. waxy solid; negligible odor; water sol.; sp.gr. 1.064 (60 C); visc. 90 cs (60 C); HLB 17.0; cloud pt. > 100 C (1% aq.); flash pt. > 400 F (COC); pour pt. 39 C; pH 6–8 (1% aq.); 100% act.

Ethylan® N50. [Harcros UK] Nonyl phenol ethoxylate; nonionic; see Ethylan N30; wh. waxy solid; negligible odor; water sol.; sp.gr. 1.073 (60 C); visc. 135 cs (60 C); HLB 18.2; cloud pt. > 100 C (1% aq.); flash pt. > 400 F (COC); pour pt. 43 C; pH 6–8 (1% aq.); 100% act.

Ethylan® N92. [Harcros UK] Nonyl phenol ethoxylate; nonionic; see Ethylan N30; wh. waxy solid; negligible odor; water sol.; sp.gr. 1.078 (60 C); visc. 340 cs (60 C); HLB 19.0; cloud pt. > 100 C (1% aq.); flash pt. > 400 F (COC); pour pt. 50 C; pH 6–8 (1% aq.); 100% act.

Ethylan® NP 1. [Harcros UK] Nonoxynol-1; nonionic; defoamer, oil emulsifier; clear, straw liq.; oil sol.; sp.gr. 0.987; visc. 650 cs; HLB 4.5; flash pt. 300 F (COC); pour pt. < 0 C; pH 6–8 (1% aq.); 100% act.

Ethylan® OE. [Harcros UK] Cetyl/oleyl alcohol ethoxylate; nonionic; emulsifier for fatty acids, alcohols, and waxes, oil and latex stabilizer, dye leveling agent; cream waxy solid; negligible odor; water sol.; sp.gr. 1.009 (40 C); visc. 67 cs (40 C); HLB 14; cloud pt. 90 C (1% aq.); flash pt. > 350 F (COC); pour pt. 31 C; pH 6–8 (1% aq.); 100% act.

Ethylan® PQ. [Harcros UK] Modified alkylphenol ethoxylate; nonionic; solubilizer used in the mfg. of iodophor; water-wh. liq (clear @ 30 C); faint odor; water sol.; sp.gr. 1.045 (40 C); HLB 14.2; cloud pt. 83 C (1% aq.); flash pt. > 350 F (COC); pour pt. 15 C; pH 6–7.5 (1% aq.); 98% act.

Ethylan® R. [Harcros UK] Cetyl/oleyl alcohol ethoxylate; see Ethylan OE; cream waxy solid; negligible odor; water sol.; sp.gr. 1.026 (40 C); visc. 150 cs (40 C); HLB 15.4; cloud pt. > 100 C (1% aq.); flash pt. > 350 F (COC); pour pt. 36 C; pH 6–8 (1% aq.); 100% act.

Ethylan® TC. [Harcros UK] Coconut fatty amine ethoxylate; nonionic; wetting agent for metal cleaning and stripping of surface coatings, fiber antistat, used in cosmetics; dk. amber liq.; mild fatty amine odor; water sol.; sp.gr. 1.040; visc. 140 cs; HLB 15; flash pt. > 300 F (COC); pour pt. < 0 C; pH 8.5–10.0 (1% aq.); 100% act.

Ethylan® TCO. [Harcros UK] Complex amine oxide; nonionic; foam and suds stabilizer in lt.- and heavy-duty detergents, cosmetic formulations, foam booster, emulsifier; straw clear to hazy liq.; mild fatty amine odor; water sol.; sp.gr. 1.008; visc. 159 cs; flash pt. > 300 F (COC); pour pt. 0 C; pH 6–8 (1% aq.); 40% act.

Ethylan® TF. [Harcros UK] Ethoxylated coconut fatty amine; nonionic; oil emulsifier with anticorrosive properties, antistat for syn. fibers with PS; lt. amber liq.; mild fatty amine odor; disp. in water; sp.gr. 0.912; visc. 140 cs; HLB 6; flash pt. > 300 F (COC); pour pt. < 0 C; pH 8.5–10.0 (1% aq.); 100% act.

Ethylan® TH-2. [Harcros UK] Hydrog. fatty amine ethoxylate; nonionic; see Ethylan TF; pale brn. solid; mild fatty amine odor; disp. in water; sp.gr. 0.878 (60 C); visc. 33 cs (60 C); HLB 5; flash pt. > 300 F (COC); pour pt. 40 C; pH 8.5–10.0 (1% aq.); 100% act.

Ethylan® TN-10. [Harcros UK] Coconut fatty amine ethoxylate; nonionic; see Ethylan TF; also for cosmetics mfg.; dk. amber liq.; mild fatty amine odor; water sol.; sp.gr. 1.015; visc. 174 cs; HLB 14; flash pt. > 300 F (COC); pour pt. < 0 C; pH 8.5–10.0 (1% aq.); 100% act.

Ethylan® TT-15. [Harcros UK] Tallow amine ethoxylate; nonionic; wetting agent for metal cleaning and stripping of surface coatings, fiber antistat, used in cosmetics; pale brn. paste; mild fatty amine odor; water sol.; sp.gr. 1.030; visc. 252 cs; HLB 14; flash pt. > 300 F (COC); pour pt. 0 C; pH 8.5–10.0 (1% aq.); 100% act.

Ethylan® VPK. [Harcros UK] PEG unsat. fatty acid ester; nonionic; textile oil emulsifier, fiber lubricant; pale clear liq.; sp.gr. 1.043; visc. 120 cs; HLB 12.4; 100% act.

Ethyl Cadmate®. [Vanderbilt] Cadmium diethyldithiocarbamate; ultra accelerator for IIR, EPDM, and SBR; primary accelerator with a thiazole; gives heat resistant, low compr. set properties to NBR and IIR; wh. to lt. gray powd., rods, or gran.; m.w. 408.94; practically insol. in water; dens. 1.48 ± 0.03 mg/m³ (powd.); 1.39 ± 0.03 mg/m³ (rods or gran.); m.p. 68 C min.; 11.5–12.9% cadmium content.

Ethyl Diglyme. [Ferro] Diethylene glycol diethyl ether; solv. which tends to solvate cations; used in electrochemistry, polymer and boron chemistry; physical processes such as gas absorption, extraction, stabilization; used in industrial prods. such as fuels, lubricants, textiles, pharmaceuticals, pesticides; colorless; mild, nonresidual odor; m.w. 162.23; f.p. –44.3 C; b.p. 189 C; water-sol.; sp.gr. 0.9082; dens. 7.56 lb/gal; visc. 1.4 cP; ref. index 1.4115; pH netural; flash pt. 90 C (CC); surf. tens. 27.2 dynes/cm; 98.0% min. purity.

Ethyl Glyme. [Ferro] Ethylene glycol diethyl ether; see Ethyl Diglyme; colorless; mild ethereal, nonresidual odor; m.w. 118.18; f.p. –74 C; b.p. 121 C; water-sol.; sp.gr. 0.8417; dens. 7.0 lb/gal; visc. 0.7 cP; ref. index 1.3922; pH neutral; flash pt. 27 C (CC); 97.0% min. purity.

Ethyl Selenac®. [Vanderbilt] Selenium diethyldithiocarbamate; ultra accelerator for NR, SBR, IIR; also vulcanizing agent; effective in low sulfur and sulfurless heat resistant compds.; nondiscoloring in lt. stocks; used with thiazoles to balance scorch and curing chars.; yel. powd., 99.5% thru 100 mesh; m.w. 672.00; sol. in toluene, carbon disulfide, chloroform; practically insol. in water; dens. 1.32 ± 0.03 mg/m³; m.p. 59–85 C; 10.5–12.7% selenium content.

Ethyl Tellurac®. [Vanderbilt] Tellurium diethyldithiocarbamate; ultra accelerator for NR, SBR, NBR, EPDM; used with thiazole modifiers; produces high modulus vulcanization; particulary act. in IIR compds.; orange-yel. powd., 100% thru 30 mesh; rods; m.w. 720.69; sol. see Ethyl Selenac; dens. 1.44 ± 0.03 mg/m³ (powd.); m.p. 108–118 C (powd.); 106 C min. (rods); 17.5–19.5% tellurium content (powd.); 14.0–16.0% tellurium content (rods).

Ethyl Tuads®. [Vanderbilt] Tetraethylthiuram disulfide; ultra accelerator; for NR and syn. rubbers; accelerator and vulcanizing agent; cure modifier for Neoprene (retards G types, accelerates W types); nondiscoloring in lt. stocks; general applics. are same as Methyl Tuads; buff to lt. gray cubes; m.w. 296.54; sol. see Ethyl Selenac; dens. 1.27 ± 0.3 mg/m³; m.p. 63–75 C; 10.8% avail. sulfur.

Ethyl Zimate®. [Vanderbilt] Zinc diethyldithiocarbamate; ultra accelerator; primary accelerator in NR and SBR; requires a thiazole modifier for safe processing and wide cure range; nondiscoloring in lt. colored stocks; stabilizer in thermoplastic rubbers and hot melts; 1:1 combination (dry wt.) with Zetax suggested for latex foam acceleration; accelerator for butyl latex vulcanization; wh. powd., 99.9% thru 100 mesh; also avail. as 50% act. slurry; m.w. 361.92; practically insol. in water; dens. 1.48 ± 0.03 mg/m³; m.p. 171–182.5 C; 15.0–20.5% zinc content.

Etilenox KM-53. [Pulcra SA] Fatty amine deriv.; antistat for syn. fibers; water-sol.

Etocas 10. [Croda Ltd.] PEG-10 castor oil; nonionic; cosmetic and essential oil solubilizer, emulsifier, lubricant, softener, leveling agent, emollient, superfatting agent, antistat, softener, detergent; used in personal care prods., fiber processing, metalworking fluids, emulsion polymerization, insecticides; pale yel. liq.; sol. in ethanol, naphtha, MEK, trichlorethylene, disp. in water; HLB 6.3; cloud pt. < 20 C; acid no. 1.0 max.; sapon. no. 120–130; pH 6–7.5; 97% act.

Etocas 30. [Croda Ltd.] PEG-30 castor oil; humectant, o/w emulsifier, skin cleanser, conditioner, emollient, solubilizer; sol. in oil; water-disp.

Etocas 35. [Croda Ltd.] PEG-35 castor oil; nonionic; see Etocas 10; pale yel. liq.; sol. see Ethocas 35; HLB 12.5; cloud pt. 35–40 C; acid no. 1.0 max.; sapon. no. 62–72; surf. tens. 41.5 dynes/cm (0.1% aq.); pH 6–7.5; 97% act.

Etocas 40. [Croda Ltd.] PEG-40 castor oil; nonionic; see Etocas 10; pale yel. liq.; sol. see Etocas 35; HLB 13; cloud pt. 50 C; acid no. 1.0 max.; sapon. no. 60–65; pH 6–7.5; 97% act.

Etocas 60. [Croda Ltd.] PEG-60 castor oil; nonionic; see Etocas 10; pale yel. soft paste; sol. see Etocas 35; HLB 14.7; cloud pt. 60 C; acid no. 1.0 max.; sapon. no. 45–50; surf. tens. 43.2 dynes/cm (0.1% aq.); pH 6–7.5.

Etocas 100. [Croda Ltd.] PEG-100 castor oil; nonionic; see Etocas 10; also humectant; pale yel. waxy solid; sol. see Etocas 35; HLB 16.5; cloud pt. 66 C; acid no. 1.0 max.; sapon. no. 25–35; surf. tens. 41.6 dynes/cm (0.1% aq.); pH 6–7.5; 97% act.

Etoxi AC-91. [Pulcra SA] Lauryl myristyl dimethyl amine oxide; thickener for shampoos; thickener/solubilizer for sodium hypochlorite-based cleaning agents; emollient, softening, and antistatic properties; liq.

Etoxi KC-10, KC-11, MC-12. [Pulcra SA] Ethoxylated MEA; solubilizer for dodecylbenzene sulfonates and inorg. salts present in detergent systems; water-sol.

Etrofolan®. [Bayer] 2-Isopropyl-phenyl-N-methylcarbamate; insecticide effective against leafhoppers and bugs; wh. cryst. powd. or flakes; m.w. 193.2; sol. in acetone, methanol; insol. in water; m.p. 88–93 C.

Eumulgin B-1. [Henkel] Ceteareth-12; wetting agent and dispersant for paint systems; waxy solid; sp.gr.

0.95; solid. pt. 50–68 F.

Eumulgin B-2. [Henkel] Ceteareth-20; see Eumulgin B-1; waxy solid; sp.gr. 0.95; solid. pt. 50–68 F.

Eumulgin B-3. [Henkel] Ceteareth-30; see Eumulgin B-1; waxy solid; sp.gr. 0.95; solid. pt. 50–68 F.

Eumulgin HRE 40, 60. [Henkel] PEG-40 and -60 hydrog. castor oil; solubilizer and emulsifier for cosmetic and pharmaceutical preps.; wax.

Eumulgin L. [Henkel] PPG-2-ceteareth-9; nonionic; emulsifier, solubilizer; liq.; 100% conc.

Eumulgin M 8. [Henkel] Oleth-10 and oleth-5; emulsifier, solubilizer for pesticides and cosmetics; paste.

Eumulgin O 5. [Henkel] Oleth-5; emulsifier, solubilizer for pesticides, cosmetics, floor polishes; dispersant improving color acceptance of emulsion paints; liq.; water-sol.

Eumulgin O 10. [Henkel] Oleth-10; nonionic; wetting agent and dispersant for paint systems; soft waxy solid; sp.gr. 0.95; solid. pt. 50–68 F.

Eumulgin RO 40. [Henkel] PEG-40 hydrog. castor oil; o/w emulsifier, solubilizer for perfume oils; visc. liq.

Eumulgin SML 20. [Henkel] PEG-20 sorbitan laurate; solubilizer and emulsifier for cosmetics and pharmaceuticals; liq.

Eumulgin SMO 20. [Henkel] PEG-20 sorbitan oleate; see Eumulgin SML 20; visc. liq. to paste.

Eumulgin SMS 20. [Henkel] PEG-20 sorbitan stearate; see Eumulgin SML 20; wax.

Euparen®. [Bayer] Dichlofluanid; fungicide with specific action against Botrytis; colorless cryst. powd.; m.w. 333.2; sol. (g/1000 ml): 100–1000 g in dichloromethane and toluene, 10–30 g in 2-propanol, 0.002 g in water; m.p. 106 C.

Euparen® M. [Bayer] Tolylfluanid; broad spectrum fungicide; effective against Botrytis; colorless powd.; m.w. 347; sol. (g/100 ml): 57 g in benzene, 23 g in m,p-xylene, 0.4 g in water; m.p. 95–97 C.

Euperlan® K 771. [Henkel] Pearlizing substances and surfactants; pearly gloss conc. for surfactant preparations; wh. visc. liq.

Euperlan® MPK 850. [Henkel] Fatty alcohol ether sulfates with pearlescents; pearly gloss conc.; liq.

Euperlan® PK 771. [Henkel] Sodium laureth sulfate, glycol distearate, and cocamide MEA; anionic/nonionic; pearlescent base for shampoos and bath prods.; liq.; 45% conc.

Euperlan® PK 776. [Henkel] Sodium laureth sulfate, glycol distearate, and cocamide MEA; anionic/nonionic; see Euperlan PK 771; liq.; 41% conc.

Euperlan® PK 789. [Henkel] Sodium laureth sulfate, glycol distearate, and cocamide MEA; anionic; pearl conc. for emulsion-type cosmetics; wh. high-visc. emulsion.

Euperlan® PK 810. [Henkel] Sodium laureth sulfate, glycol stearate, and cocamide MEA; anionic/nonionic; pearlescent base for lotion shampoos and bath prods.; liq.; 37% conc.

Euperlan® PK 900. [Henkel] Sodium laureth sulfate, PEG-3 distearate, cocamide MEA; anionic/nonionic; pearlescent base for lotion shampoos, bath prods.; liq.; 40% conc.

Eureka 23-KP-2. [Atlas Refinery] Sulfated fish oil, sulfated neatsfoot oil, sulfonated fish oil; anionic; fatliquor for soft leather, gloving and lining leather; dens. 8.13 lb/gal; pH 5.5–6.0 (10%); 68–70% act.

Eureka 82. [Atlas Refinery] Anionic; chrome-stable fatliquor; helps to disperse natural fats to produce more uniform color in the chrome tan bath; dens. 6.56 lb/gal; pH 5–6 (10%) moisture.

Eureka 82-U. [Atlas Refinery] Anionic; see Eureka 82; dens. 7.20 lb/gal; pH 5–6 (10%); 1–2% moisture.

Eureka 99. [Atlas Refinery] Sulfonated natural oil, fatty alcohol, waxes; anionic; leather fatliquor for use on fine leathers where high tens. str. is required; fiber lubricant; flowable paste; dens. 8.20 lb/gal; pH 5.5–6.0 (10%); 50% act.

Eureka 102. [Atlas Refinery] Sulfated castor oil; anionic; emulsifier, detergent; grinding aid in pigment disps.; plasticizer in finish coatings; lt. liq.; pH 7.6; 72% act.

Eureka 395-CX. [Atlas Refinery] Sulfated neatsfoot oil, sulfated fish oil; anionic; fatliquor imparting softness and fullness to leather; used on mellow to soft shoe upper and upholstery leather; dens. 8.06 lb/gal; pH 5.5–6.0 (10%); 85% act.

Eureka 400-R. [Atlas Refinery] Sulfonated fish oil; anionic; general purpose fatliquor for soft, pale leather, upholstery leather, side leather, sheepskin, gloving leather; dens. 8.31 lb/gal; pH 5.5–6.0 (10%); 87% act.

Eureka 400-RCA. [Atlas Refinery] Sulfonated fish oil; anionic; chrome- and alum-stable fatliquor; suitable for suede, full grain soft glove and garment leathers, splits, and for the prefatting of leather during chrome or alum tannage; lt. brn. visc. liq.; dens. 8.39 lb/gal; pH 5.5–6.0 (10%); 75% act.

Eureka 400-RKM. [Atlas Refinery] Sulfonated fish oil; anionic; fatliquor for fine glove and garment leathers, suede leathers, upholstery leathers, side leathers, etc.; dens. 8.66 lb/gal; pH 5.3–5.8 (10%); 87% act.

Eureka 400-RT. [Atlas Refinery] Sulfonated natural oils, syn. lubricants, and ethoxylated prods.; anionic; leather fatliquor; suitable for suede, splits, full grain leathers, fur skins; yel. visc. liq.; dens. 8.13 lb/gal; pH 5.4–5.6 (10%); 75% act.

Eureka 400-RY. [Atlas Refinery] Sulfonated fish oil and low sulfated natural and syn. oils; anionic; fatliquor for general purpose, soft leather, glove and garment leathers; dens. 8.31 lb/gal; pH 6.5–7.0 (10%); 81% act.

Eureka 537. [Atlas Refinery] Conc. alkaline fatliquor for patent and dress-type leathers; anionic; dens. 7.44 lb/gal; pH 7.5–8.0 (10%); 93% act.

Eureka 678. [Atlas Refinery] Neatsfoot oil; anionic; mayonnaise-type fatliquor for dress upper leathers, glaze kid, sheepskin garment leathers; imparts high tens. str. and fine break; wh. flowable paste; dens. 8.12 lb/gal; pH 8.5–9.0 (10%); 46.5% act.

Eureka 691-A. [Atlas Refinery] Anionic; low-sulfated fatliquor for wh. and pastel leathers, esp. for upholstery and clothing leather; dens. 7.93 lb/gal; pH

6.0–6.5 (10%); 75% act.

Eutanol G. [Henkel] Octyl dodecanol; lubricant, emollient for cosmetics and pharmaceuticals; liq.; sol. in alcohols, esters, cosmetic oils, glycols, ketones, aromatics; insol. in water.

Eutanol G16. [Henkel] Hexyl dodecanol; universal emollient; clear liq.

Evanacid®3CS. [Evans Chemetics] Carboxymethyl mercaptosuccinic acid; metal chelate; metal deactivator for the stabilization of glyceride oils; wh. powd.; m.w. 208.2; sol. (g/100 g) 147 g water; 76 g ethanol; m.p. 136–138 C; 98% min. purity.

Evanstab® 12. [Evans Chemetics] Dilauryl thiodipropionate; antioxidant for polyethylene, PP, and polyolefins, ABS; stabilizer for polyolefins, oils and fats, food applic.; plasticizer for rubber prods.; lubricating oil additive; syn. lubricant; chemical preservative in fats and oils; wh. cryst. flakes or powd.; m.w. 514; f.p. 40.0 C min.; sol. in acetone, MEK, n-heptane, toluene, ethyl acetate, ethanol; acid no. 1.0 max.; 99.0% min. assay.

Evanstab® 13. [Evans Chemetics] Ditridecyl thiodipropionate; sec. antioxidant in ABS, polyolefins, and other polymer systems; colorless to sl. yel. clear liq.; m.w. 543; b.p. 265 C (0.25 mm); sol. (g/100 g sol'n.) 14 g 95% ethanol; miscible with toluene, ethyl acetate, n-heptane, MEK; acid no. 3.0 max.; sapon. no. 200–210; 99.0% min. assay.

Evanstab® 14. [Evans Chemetics] Dimyristyl thiodipropionate; sec. antioxidant for polyolefins; wh. cryst. flakes or powd.; m.w. 570; m.p. 48–50 C; acid no. 1.0 max.; sapon. no. 280–290; 98.5% min. assay.

Evanstab® 16. [Evans Chemetics] Dicetyl thiodipropionate; sec. antioxidant in polymers used in the mfg. of articles for food-contact use; wh. cryst. flakes or powds.; m.w. 627; m.p. 59–62 C; acid no. 1.0 max.; sapon. no. 176–183; 98.% min. assay.

Evanstab® 18. [Evans Chemetics] Distearyl thiodipropionate; sec. antioxidant for use in polyolefins; used in food-pkg. materials and edible fats and oils; wh. cryst. flakes or powd.; m.w. 683; f.p. 64.0 C min.; sol. (g/100 g sol'n.) 11 g toluene, 2 g n-heptane; < 1 g ethyl acetate, 95% ethanol, MEK, water; 98.0% min. assay.

Evanstab® A. [Evans Chemetics] 3,3′-Thiodipropionic acid; antioxidant for edible fats and oils, food-pkg. materials; wh. cryst. powd.; m.w. 178; sol. (g/100 g sol'n.) 14 g propylene glycol, 4 g water, 3 g glycerine; sol. in alcohol, acetone, and hot aromatic hydrocarbons; insol. in aliphatic hydrocarbons; m.p. 128–130 C; 99.0% min. assay.

Everflex® 81L. [W.R. Grace] Vinyl acetate copolymer emulsion; paint and coating emulsion for use as binder for clay coating of paper and paperboard; factory finishes, ceiling tile, wall board, and textile treatments; wh. emulsion, particle 0.15 μ; slight char. odor; dens. 9.9 lb/gal; visc. 50–250 cps; pH 4.5–5.5; 49–51% solids.

Everflex® 515L. [W.R. Grace] Vinyl emulsion in anionic/nonionic emulsifier system; pigment binder for use in surface coatings, e.g., interior flat wall paints and primer sealers; emulsion; 0.2 μ avg. particle size; dens. 8.5 lb/gal (solids); visc. 20–60 cps; pH 7.0–8.5; surf. tens. 33–40 dynes/cm²; 47–49% total solids.

Everflex® Solvent 80. [W.R. Grace] 80% methyl acetate, 20% methanol; solv. for use in paint remover compds. and lacquer solv.; clear water-wh.; sp.gr. 0.900–0.910; 0.005% NV.

Everflex® T. [W.R. Grace] Vinyl acrylic emulsion; high solids binder for prod. of gloss and flat interior/exterior coatings prepared by pigment slurry techniques; used in latex paints, primer sealers, exterior masonry and wood, factory finishes, ceiling tile, wall board, stains, roof coatings; wh. emulsion; particle size 0.25 μ; slight, char. odor; films of the polymer are clear, transparent, with no odor; dens. 9.7 lb/gal; visc. 3000–4000 cps; pH 4.5–5.5; 64–66% total solids.

Everflex® TMF. [W.R. Grace] Vinyl acrylic terpolymer emulsion; binder for coatings; wh. emulsion, particle 0.2 μ; slight char. odor; dens. 9.7 lb/gal; visc. 300–700 cps; pH 4.5–5.5; 54–56% solids.

Examide-DA. [Soluol] EO condensate; nonionic; detergent, wetting, emulsifier; surfactant used as scouring and leveling agent, dyeing assistant in textile industry; liq.; 50% conc.

Excel 300. [Kao] Monodiglyceride propylene glycol blend; nonionic; defoaming agent for beverages; liq.; HLB 2.8; 100% conc.

Exceparl HO. [Kao] n-Hexadecyl 2-ethylhexanoate; emollient for cosmetics; liq.

Exceparl IPM. [Kao] IPM; emollient for cosmetics use; liq.

Exceparl IPP. [Kao] IPP; emollient for cosmetics use; liq.

Exceparl OD-M. [Kao] Octyldodecyl myristate; emollient for use in cosmetics; liq.

Exceparl OD-OL. [Kao] Octyldodecyl oleate; emollient for use in cosmetics; liq.

Exceparl TGO. [Kao] 2-Ethylhexyl triglyceride; emollient for use in cosmetics; liq.

Expandex® 5PT. [Olin] 5-Phenyltetrazole; blowing agent for foaming plastics and elastomers at elevated temps.; wh. needle-like material; particle size 18 μ; sol. in ethanol and other common org. solvs.; sp.gr. 1.42; m.p. 212 C; gas yield 200 ml/g.

Exsize HA, Regular Exsize, Special Exsize, Super Exsize. [Premier Malt] Predominantly alpha-amylase with some protease; enzyme for use as desizing agents in the textile industry; solubilizes starch; liqs.; sol. in water; dens. 10.05, 9.32, 9.35, and 9.5 lb/gal resp.

Extractase L5X, P15X. [Finnsugar] Pectinase (fungal); broad act. enzyme for fruit processing; amber liq. and tan powd. resp.

EZ Mold Lubricant. [TSE] Glycol surfactant; mold release lubricant for natural and syn. rubber compds; milky wh.; odorless; sp.gr. 1.01; b.p. 100 C; flash pt. none.

EZA®. [Ethyl] Zeolite A; solv.; anticaking agent for detergents and desiccants; powd.; 4Å diam.; 100% act.

F 50. [GE] Phenyl silicone; silicone fluid providing higher temp. resistance and improved oxidative properties, increased radiation resistance, improved compat. with org. materials, improved lubricity; suggested for hydraulic fluid and lubrication in mechanical/elec. applics.; fluid; sp.gr. 1.050; visc. 70 cstk; pour pt. –100 F; ref. index 1.4280; flash pt. 430 F; surf. tens. 21.0 dynes/cm; dielec. str. 29.0 kV; vol. resist. 8×10^{12} ohm-cm.

F-221. [Wacker Silicones] Low m.w. polysiloxane; emollient; for cosmetics or applics. where it is desirable to have the properties of a silicone temporarily; sol. in alcohols, aromatic, aliphatic, and chlorinated solvs.; 100% act.

F-222. [Wacker Silicones] Siloxane; see F-221; sol. see F-221; 100% act.

F-251. [Wacker Silicones] Siloxane; emollient for personal care and cosmetic prods.; sol. in alcohol, aromatic, aliphatic, and chlorinated solvs.

F-755. [Wacker Silicones] Silicone copolymer; emollient in skin care cosmetics; fiber lubricant; cutting tool lubricant; sol. in aromatic, aliphatic, and chlorinated hydrocarbons; some sol. in alcohols; water-disp.

F-789. [Wacker Silicones] Silicone wax; lubricant; sol. in alcohol, aromatic, aliphatic, and chlorinated solvs.

Faktis Asolvan, Asolvan T. [Rhein Chemie] Sulfur factice; processing aid for NBR, CR, and other special rubbers, petrol-resistant rubbers; brn. ground; sp.gr. 1.06.

Faktis Badenia C. [Rhein Chemie] Sulfur factice; processing aid for colored molded and extruded rubber articles; lt. yel. fine powd.; sp.gr. 1.03.

Faktis Badenia T. [Rhein Chemie] Sulfur factice, min. oil extended; processing aid for lt. colored tech. molded and extruded rubber goods; yel. fine powd.; sp.gr. 1.01.

Faktis DK 10. [Rhein Chemie] Sulfur factice; processing aid for rubber hoses and profiles, surgical and foodstuff goods; brn. crushed lumps; sp.gr. 1.00.

Faktis DK 14, DK 17. [Rhein Chemie] Sulfur factice; processing aid for rubber hoses and profiles, foodstuff goods, rubberized fabrics, etc.; brn. crushed lumps; sp.gr. 1.02 and 1.04 resp.

Faktis HF Braun. [Rhein Chemie] Sulfur factice; processing aid for blended rubber compds., exp. with EPDM; brn. elastic, lumps, semisolid, powd.; sp.gr. 1.00.

Faktis Para extra weich. [Rhein Chemie] Sulfur factice, min. oil extended; processing aid for soft molded and extruded rubber goods of SBR and SBR/NR blends; brn. lumps; sp.gr. 0.98.

Faktis R Spezial, Weib MB. [Rhein Chemie] Sulfur chloride factice; processing aid for cold curing of rubber compds.; wh. fine powd.; sp.gr. 1.05 and 1.14 resp.

Faktis RC 110, RC 111, RC 140, RC 141, RC 144. [Rhein Chemie] Sulfur factice; processing aid for hoses and profiles, rubberized fabrics, tech. goods; dk. crushed lumps; sp.gr. 1.05, 1.02, 1.05, 1.05, and 1.05 resp.

Faktis Rheinau H, W. [Rhein Chemie] Sulfur chloride factice; processing aid for hot curing of wh. and bright colored rubber goods; wh. fine powd.; sp.gr. 1.06 and 1.03 resp.

Faktis T-hart. [Rhein Chemie] Sulfur factice, min. oil extended; processing aid for tech. molded and extruded SBR rubbers and blends; brn. crushed lumps; sp.gr. 1.02.

Faktis ZD. [Rhein Chemie] Sulfur factice, min. oil extended; processing aid for tech. molded and extruded rubber goods; dk. brn. lumps; sp.gr. 1.02.

Fanchem HL. [Fanning] Hydrog. lanolin; emollient for cosmetics, pharmaceuticals; sol. in alcohol and most hydrocarbon oils.

Fancol ALA. [Fanning] Acetylated lanolin alcohol; cosmetic, toiletry conditioner, emollient; oily liq.

Fancol EOA. [Fanning] Ester of oleyl alcohol; conditioner; water-sol.

Fancol HL. [Fanning] Hydrog. lanolin; emollient, moisturizer, lubricant, plasticizer, chemical intermediate, humectant, mold release agent for pharmaceuticals, cosmetics, industrial applics.; wh. solid, trace odor; sol. in IPM, min. oil, castor oil, ethyl acetate, acetone (@ 75 C); insol. water; sp.gr. 0.855–0.865; m.p. 48–53 C; sapon. no. 6 max.; ref. index 1.460–1.469 (50 C).

Fancol LA. [Fanning] Lanolin alcohol; nonionic; emollient, thickener, emulsifier, stabilizer, plasticizer, superfatting agent, dye dispersant, chemical intermediate, lubricant, humectant, mold release agent, conditioner for cosmetics, pharmaceuticals, soaps, industrial applics.; brn. solid wax-like; sol. in CCl_4, chloroform, IPM (@ 75 C), min. oil (@ 75 C), oleyl alcohol; insol. water; m.p. 56 C; acid no. 2 max.; sapon. no. 12 max.; 100% act.

Fancol LAO. [Fanning] Min. oil and lanolin alcohol; nonionic; conditioner, surfactant, stabilizer, moisturizer, humectant, penetrant, emollient, plasticizer, and primary emulsifier for use in cosmetics and pharmaceuticals; plasticizer in aerosol formulas; yel. clear oily liq.; odorless; sol. in oils; insol. in water; sp.gr. 0.84–0.86; sapon. no. 3.0 max.

Fancol OA 50, 70, 80, 90, 95. [Fanning] Oleyl alcohol; nonionic; industrial plasticizer, emulsion stabilizer, antifoam and coupling agent, aerosol lubricant, petrol. additive, pigment dispersant; rust preventive; detergent, release agent, cosolvent, softener, tackifier, spreading agent used for metalworking, petrochemicals, pulp and paper, paints and coatings, plastics and polymers, food applics., pharmaceuticals, cosmetics; chemical intermediate; liq.; sapon no. 1 max.; cloud pt. 30–40 C, 18–25 C, 12–18 C, 6 C, and 5 C max. resp.

Fancor Lanolin. [Fanning] Lanolin; plasticizer for rubber; sol. in hydrocarbon oils, most solvs.

Fancor Lanwax. [Fanning] Natural lanolin wax ester; plasticizer in wax crayons; water repellent, humectant, conditioner, corrosion inhibitor, emollient, lubricant for cosmetics, toiletries, pharmaceuticals, industrial applics.; extender and crystallization inhibitor for natural waxes in industrial applics.; emulsifier; semisolid; m.p. 48–52 C; sapon. no. 90–110.

Fancor LFA. [Fanning] Lanolin fatty acids; emollient, stabilizer, emulsifier, corrosion inhibitor for personal care and pharmaceutical prods.; used in industrial leather treating, coatings, polishes, corrosion inhibitors, lubricants; wax-like solid; m.p. 57–65 C; acid no. 135–170; sapon. no. 158–175.

Fancorp Lanolin. [Fanning] Lanolin; lubricant for txtile, metalworking compds.; sol. in hydrocarbon oils, most solvs.

Farmin 2C. [KAO S.A.] Sec. fatty di-n-alkyl amine; intermediate for textile finishing; softener; antistat; solid; 100% conc.

Farmin 20, 60, 68, 80, 86, AB, C. [KAO S.A.] Primary fatty amines; emulsifier, anticaking agent, textile additive, corrosion inhibitor, flotation reagent; liq./solid; 100% conc.

Farmin D86. [KAO S.A.] Sec. fatty di-n-alkyl amine; see Farmin 2C; solid; 100% conc.

Farmin DM20, 40, 60, 80, 86, DMC. [KAO S.A.] Tert. fatty n-alkyl dimethylamine; intermediate for corrosion inhibitors; quat. synthesis, benzalkonium chloride; liq.; 100% conc.

Farmin HT, O, S, T. [KAO S.A.] Primary fatty amines; see Farmin 20; liq.; 100% conc.

Farmin M2C, M2R86. [KAO S.A.] Tert. fatty di-n-alkyl methyl amine; see Farmin DM20; paste, solid; 100% conc.

Farmin R 24H, R 86H. [KAO S.A.] Fatty amine hydrochloride; cationic; anticaking agent for fertilizers; corrosion inhibitors; liq./solid.

FC-24. [3M] Trifluoromethanesulfonic acid; catalyst or reactant; polymerization of epoxies, styrenes, and THF, alkylation and some acylation reactions; pharmaceuticals, explosives, dyes, and intermediates; electrolytes; formation of biaryls; polymerization reactions; dehydrating agent; inhibits formation of metal oxides in welding and soldering fluxes; colorless liq.; sol. in polar org. solvs; misc. in water; m.w. 150.02; f.p. –40 C; b.p. 54 C (8 mm Hg); dens. 1.696 g/cc; visc. 2.87 cP; ref. index 1.325.

FC-60% SS. [ICI Advanced Materials] Stainless steel-filled PTFE; lubricant; sp.gr. 3.78; tens. str. 2900 psi; tens. elong 210%; 60% conc.

FC-446CS. [ICI Advanced Materials] Color stabilized bronze-filled PTFE; lubricant; pelletized; sp.gr. 3.90; tens. str. 2500 psi; tens. elong 130%; 60% conc.

FC-477 SM. [ICI Advanced Materials] Inorg. filled PTFE; lubricant; sp.gr. 2.24; tens. str. 2400 psi; tens. elong. 257%; 79.5% conc.

FC-520. [3M] Trifluoromethanesulfonic acid amine salt and diethylene glycol monoethyl ether; catalyst for condensation and ring-opening polymerizations; coatings utilizing such resins as epoxides, aminoplasts, phenolics, and acrylics; coatings for wood, paper, and plastic; lemon-yel. clear liq.; f.p. cryst. at 40 F; misc. in water; sp.gr. 1.2; dens. 10.0 lb/gal; visc. 13 cps; pH 4.5–6.0; flash pt. 230 F (PMCC); 60% sol'n.

Fekta® RT. [Goldschmidt] Algicide for swimming pools.

Fermalpha. [Finnsugar] Alpha-amylase (fungal); enzyme for food processing; used in maltose syrups and baking; optimum pH 5.0–6.0; optimum temp. 45–55 C; activity 30,000 and 52,500 SKG/g; tan powd.

Fermcolase®. [Finnsugar] Catalase (fungal); enzyme for food processing, peroxide decomposition; wider pH, temp., and peroxide range than animal catalase; activity 1000 Baker units/mL; red-amber liq.

Fermcozyme® 1307, BG, BGXX, CBB, CBBXX, M. [Finnsugar] Glucose oxidase (fungal); enzyme for food processing; antioxidant; maintains freshness by preventing oxygen or glucose deterioration; desugarization of egg prods.; optimum pH 4.5–7.5; special preparations avail. for use at pH down to 2.5; act. 750 U/mL (1307, BG grades, CBB, M); 1500 U/mL (BGXX, CBBXX); amber clear liqs.

Fermlipase. [Finnsugar] Lipase (pancreatic); enzyme for food processing; hydrolysis of yolk fat in egg albumin; digestive aid; additive to animal feed; low (reduced) proteolytic activity; activity 30X NF/mg; lt. tan powd.

Fermvertase, 10X, XX. [Finnsugar] Invertase (yeast); enzyme for food processing; sucrose inversion for confections; optimum pH 4.0–5.5; optimum temp. 55–65 C; activity 3000, 30,000, and 6000 SU/mL resp.; clear liqs.

Ferro 1288. [Ferro] Barium, cadmium, zinc stabilizer; heat stabilizer used in flexible vinyls processed by calendering, extrusion, inj. molding, or plastisol applics.; long-term stability; Gardner 8 liq.; dens 8.7 lb/gal; visc. Gardner A.

Ferro Permyl B 100. [Ferro] Proprietary; uv absorber for polyesters, coatings; sol. in highly polar solvs.

Ferrosil 14. [Kaopolite] Ferroaluminum silicate; filler for abrasion-resistant plastic systems; extender pigment for primers and other coatings; lt. buff color; 98% finer than 14 μ; sp.gr. 4.0; dens. 33.32 lb/gal solid; bulking value 0.030 gal/lb; oil absorp. (Gardner-Coleman) 18.0; ref. index 1.83; hardness (Moh)

8–9; 41.24% silicon dioxide; 22.23% iron oxide; 20.36% aluminum oxide; 12.35% magnesium oxide; 2.97% calcium oxide.

Ferro UV-Chek AM-101, AM-105, AM-205. [Ferro] Nickel-org. uv absorber for polyolefin; sol. in most org. solvs.

Ferro UV-Chek AM-300. [Ferro] 2-Hydroxy-4-n-octoxybenzophenone; uv absorber for plastics; sol. in most org. solvs.

Ferro UV-Chek AM-340. [Ferro] Hydroxybenzoate type; uv absorber for polyolefin plastics; sol. in most org. solvs.

Ferro UV-Chek AM-595. [Ferro] Mixed metal organophosphate; uv absorber for PVC, PVDC, CPE, flame-retardant plastics.

FF 680. [Great Lakes] Bis (tribromophenoxy) ethane; flame retardant for applic. where thermal stability at high processing temps. is important; for thermoplastic and thermoset systems, lt.-stable applics.; wh. cryst. powd.; sol. in boiling dichlorobenzene, p-xylene perchloroethylene; insol. at ambient temps. in water; dens. 2.58 g/ml; dens. 46.7 lb/ft^3; m.p. 223–225 C; 70.0% bromine

Ficel® AC2. [Sherex] Azodicarbonamide; chem. blowing agent for sponge rubber, cushion vinyl floor covering, expanded vinyl coated fabrics, profiles/pipe, sealants; yel. powd.; negligible sol. in water; 100% act.

Ficel® AC3, AC3F, AC4. [Sherex] Azodicarbonamide; see Ficel AC2.

Ficel® ACSP4. [Sherex] Coated azodicarbonamide; disp. in vinyl plastisols; blowing agent for natural and syn. rubbers; sol. in dimethyl sulfoxide; powd., particle size 5μ.

Ficel® AF100. [Sherex] Formulated zinc compd.; chem. blowing agent for ammonia-sensitive polymers, stain-free moldings, taint-free moldings in food applics.; wh. powd.; insol. in water; 44–51% act.

Ficel® AZDN-LF. [Sherex] 2,2-Azodiisobutyronitrile; polymerization initiator for a wide range of monomers; off-wh. powd., char. odor; insol. in water; sp.gr. 0.4; m.p. 103 C (with decomp.).

Ficel® AZDN-LMC. [Sherex] 2,2-Azodiisobutyronitrile; see Ficel AZDN-LF; LMC has low metal content.

Ficel® AZM. [Sherex] 2,2-Azodi (2-methylbutyronitrile); polymerization initiator for a wide range of monomers; high sol., lower toxicity.

Ficel® EPA, EPB, EPC, EPD, EPE. [Sherex] Formulated azodicarbonamide; chem. blowing agent for structural foam molding, wire and cable insulation, profiles, pipe, and film; nonplating, self-nucleating.

Ficel® LE. [Sherex] Modified azodicarbonamide; blowing agent one-pkg. system used in PVC plastisol and sponge rubber applics.; yel. powd.

Ficel® SCE. [Sherex] Modified azodicarbonamide; chemical blowing agent one-pkg. system used in inhibition process of chemical embossing; plastisol and sponge rubber applics.; food contact applics.; yel. powd.

Filmex®. [Quantum/USI] Ethanol; solv. and thinner; general purpose industrial applics.; sol. in conc. sulfuric acid; misc. in water, alcohol, chlorine, benzene.

Finazoline Q-4. [Finetex] Modified fatty acid amide; debonding and release agent for paper and cellulose fibers; disp. in water.

Finazoline Series. [Finetex] High m.w. imidazolines; corrosion inhibitor for metalworking and fuel oil additives; sol. in alcohols, oils, min. spirits; disp. in water.

Findet Series. [Finetex] Phosphate esters; corrosion inhibitor for metalworking and cleaning, lubricant additives; compatibilizers and coupling agents for alkaline formulations.

Finizym 200L. [Novo] Beta-glucanase; enzyme for beer filtration; liq.

Finquat CT. [Finetex] Quaternium-75; used in thioglycolate permanent waving sol'ns. to reduce processing time and condition hair; also in depilatories; liq.; water-sol.

Finsoft HCM-100. [Finetex] Quat.; hair conditioner, non-eye irritating cream rinses; solid.

Finsolv® SB. [Finetex] Isostearyl benzoate; non-greasy emollient, noncomedogenic; for cosmetics, sunscreen, antiperspirants, deodorants; lubricant; plasticizer for hair resins, in anhyd. systems; perfume solubilizer; liq.; sol. in org. solvs. and oils.

Finsolv® TN. [Finetex] C_{12-15} alcohol benzoate; emollient, solubilizer, sunscreen vehicle, deoiler for creams and oils; binder in pressed powds.; liq.; 100% conc.

Fire Retardant RCA. [GAF] Complex inorg. salts; anionic; nondurable water-sol. fire retardant textile finish; granular powd.

Firebrake® ZB. [U.S. Borax] Zinc borate; flame retardant synergistic with antimony oxide or alumina trihydrate in most halogenated polyesters and vinyl esters; delays onset of oxidation; used in chlorinated polyesters, PVC plastisols, and epoxy systems; smoke suppressant; char promoter; improved elec. properties in polyester and nylon; promotes adhesion between plastics and metals; translucent, low toxicity, and nonhygroscopic; wh. cryst. powd.; water-insol. below 100 F.

Firemaster 680. [Velsicol] Bis (tribromophenoxy) ethane; flame retardant for ABS; powd.

Flamtard H. [Alcan] Flame retardant/smoke suppressant for PVC, halogen-containing unsat. polyesters, and rubbers; reduces carbon monoxide evolution; used at 1–5 phr in halogenated polymers; wh.; 2.5 μm median particle size; bulk dens. 1.4 g/cm (untamped); surf. area 4.5 m^2/g; oil absorp. 18 g/100 g; decomp. temp. > 180 C.

Flamtard S. [Alcan] Flame retardant for plastics and rubbers incl. PVC, polychloroprene, chlorosulfonated polyethylene, halogenated polyesters; reduces smoke emission and carbon monoxide evolution on combustion; no practical temp. restrictions; wh.; used at 1–5 phr with halogenated polymers; 1.7 μm median particle size; bulk dens. 0.7 g/cc (untamped); surf. area 33 m^2/g; oil absorp. 25 g/100 g.

Flectol® H. [Monsanto] Polymerized 1,2-dihydro-2,2,4-trimethylquinoline; high activity antioxidant;

metal deactivator; staining; powd. or flakes.

Flectol® ODP. [Monsanto] Octylated diphenylamine; staining antioxidant for CR; powd.

Flectol® Pastilles. [Monsanto] Polymerized 1,2-dihydro-2,2,4-trimethylquinoline; see Flectol H; 12 mm pastilles.

Flexan® 130. [Nat'l. Starch] Sulfonated PS, sodium salt; hair fixative for setting lotions, conditioners, blow drying aid; gloss and antitstatic chars.; sol. in water and ethanol/water blends.

Flexchlor 0001. [Witco/Argus] Chlorinated paraffin; plasticizer for use in paint, coatings, caulks, sealants, rubber, and adhesives applic. where lower volatility than phthalate plasticizers and/or flame retardancy are desired; used in coatings, adhesives, caulks/sealants; Gardner 1 liq.; sp.gr. 1.10; visc. 0.5 poise; 41% conc.

Flexchlor 0002. [Witco/Argus] Chlorinated paraffin; see Flexchlor 0001; Gardner 1 liq.; sp.gr. 1.22; visc. 3 poise; 50% conc.

Flexchlor 0008. [Witco/Argus] Chlorinated paraffin; see Flexchlor 0001; Gardner 3 liq.; sp.gr. 1.27; visc. 15 poise; 52% conc.

Flexchlor 0009. [Witco/Argus] Chlorinated paraffin; see Flexchlor 0001; Gardner 3 liq.; sp.gr. 1.41; visc. 800 poise; 57% conc.

Flexchlor 0010. [Witco/Argus] Chlorinated paraffin; see Flexchlor 0001; esp. for sec. plasticizing of vinyl; Gardner 2 liq.; sp.gr. 1.30; visc. 25 poise; 55% conc.

Flexchlor 0011. [Witco/Argus] Chlorinated paraffin; see Flexchlor 0001; Gardner 8 liq.; sp.gr. 1.35; visc. 15 poise; 56% conc.

Flexchlor 0012. [Witco/Argus] Chlorinated paraffin; see Flexchlor 0001; also used for caulks/sealants applic.; Gardner 1 liq.; sp.gr. 1.35; visc. 20 poise; 58% conc.

Flexchlor 0018. [Witco/Argus] Chlorinated paraffin; see Flexchlor 0001; also demonstrates uncommon heat stability over general purpose plasticizers; Gardner 2 liq.; sp.gr. 1.35; visc. 500 poises; 55% conc.

Flexchlor 0023. [Witco/Argus] Chlorinated paraffin; see Flexchlor 0001; low volatility at elevated temps.; flame retardant plasticizer; Gardner 3 liq.; sp.gr. 1.33; visc. 55 poise; 55% conc.

Flexichem CS. [Guelph] Proprietary; waterproofing of concrete; liq. and powd.; disp. in water.

Flexol® Plasticizer 4GO. [Union Carbide] PEG di-2-ethylhexoate; plasticizer for rubber.

Flexol® Plasticizer EP-8. [Union Carbide] Octylepoxy tallate; general purpose epoxy plasticizer.

Flexol® Plasticizer EPO. [Union Carbide] Epoxidized soybean oil; general purpose epoxy plasticizer.

Flexowax C. [Lonza] Microcryst. wax; syn. wax for textiles and other applics.; tan solid; m.p. 70 C; flash pt. 255C.

Flexowax C Light. [Lonza] Microcryst. wax; syn. wax for cosmetics and other applics.; cream solid; m.p. 61 C; flash pt. 230 C.

Flexricin® 9. [CasChem] Propylene glycol ricinoleate; nonionic; wetting agent, dye solv., wax plasticizer, stabilizer for textile, household, and cosmetic applics., rewetting dried skins; Gardner 2+ liq.; sol. in toluene, butyl acetate, MEK, ethanol; sp.gr. 0.96; visc. 3 stokes; 100% act.

Flexricin® 13. [CasChem] Glyceryl ricinoleate; nonionic; wetting agent, wax plasticizer, and mold release agent for rubber polymers, antifoam agent, household and cosmetic applics., rewetting dried skins; Gardner 2 min. liq.; sol. in toluene, butyl acetate, MEK, ethanol; sp.gr. 0.985; visc. 8.8 stokes; 100% act.

Flexricin® 15. [CasChem] Glycol ricinoleate; nonionic; wetting agent, plasticizer, textile, household, and cosmetic applics., rewetting dried skins; chemical intermediate; Gardner 4 liq.; sol. in toluene, butyl acetate, MEK, ethanol; sp.gr. 0.965; visc. 3.9 stokes; 100% act.

Flexricin® 17. [CasChem] Pentaerythritol ricinoleate; plasticizer; liq.; 100% act.

Flexricin® 100. [CasChem] Ricinoleic acid deriv.; lubricant for textile, metalworking compds.; corrosion inhibitor intermediate; sol. in alcohols, ethers, ketones, aromatic hydrocarbons.

Flexricin® P-1. [CasChem] Methyl ricinoleate; low temp. lubricant plasticier for rubber, phenolic and epoxy resins; liq.; 100% act.

Flexricin® P-3. [CasChem] Butyl ricinoleate; general purpose plasticizer, lubricant for NC; liq.; 100% act.

Flexricin® P-4. [CasChem] Methyl acetyl ricinoleate; all purpose plasticizer, lubricant for vinyls and lacquers; liq.; 100% act.

Flexricin® P-6. [CasChem] Butyl acetyl ricinoleate; lubricity additive for textile finishes; plasticizer; liq.; 100% act.

Flexricin® P-8. [CasChem] Glyceryl (triacetyl ricinoleate); plasticizer for vinyl wire jacketing and semirigid vinyls; emollient; stabilizer for anhyd. pigmented systems; liq.; 100% act.

Flo-Gard AG 110, AG 130, AG 150. [PPG Industries] Amorphous silica; absorbent used as carriers for liqs., grinding and suspension aids, or anticaking agents in agric. chemical formulations (e.g. pesticides and animal feed additives); permits extended mixing cycles; useful as assay adjuster if purity of tech. grade pesticide varies considerably; provides thixotropic action in water disps. permitting formulation of flowable pesticides; wh. ultrafine powd., 70, 15, and 3 μm median agglomerate size resp.; sp.gr. 2.1; ref. index 1.46; pH 7.0; MOH Scale hardness 0; 97.5% silica content, anhyd. basis.

Flo-Gard CC 120, CC 140, CC 160. [PPG Industries] Amorphous silica; absorbent used as chemical carriers for liqs., grinding and suspension aids, or anticaking and flow-control agents in chemical formulations and processes; serves as highly reinforcing fillers in finished compds. such as rubber; permits extended mixing cycles; provides thixotropic action in water disps. permitting formulation of prods. that resist settling in storage and enhances stability of aq. disps. after dilution in the field; wh. ultrafine powd., 70, 15, and 3 μm median agglomerate size resp.; sp.gr. 2.1; ref. index 1.46; pH 7.0; MOH Scale hardness 0; 97.5% silica content, anhyd. basis.

Flo-Gard FF 310, FF 320, FF 330, FF 350, FF 370, FF 390. [PPG Industries] Amorphous silica; absorbent; used as flow control and anticaking agents, grinding and suspension aids, drying agents, and absorbent carriers in chemical formulations and processes; offers superior ability to resist overmixing in comparison with calcium silicates; wh. ultrafine powd., 15, 9, 70, 15, 3, and 3 μm median agglomerate size resp.; sp.gr. 2.1; ref. index 1.46; pH 7.0; MOH Scale hardness 0; 97.5% silica content, anhyd. basis.

Floramat. [Henkel] Ethyl-2-t-butylcyclohexylcarbonate; fragrance raw material, for floral notes.

Florco. [Floridin] Montmorillonite clay; floor absorbent for oil, grease, water, and other liqs.; offers lighter dens.; antislip agent on floors; used by automotive, steel, transportation, commercial, pet, food and beverage, institutional, and amusements industries, and for winter and home uses; gran., 8/30 mesh.

Florco X. [Floridin] Attapulgite clay; 8/30 mesh; absorbs oil and water spills; pet absorbent.

Florex Granular Grades. [Floridin] RVM and LVM attapulgites; percolation adsorbents for jet fuel treating, catalyst stripping, transformer oil purification, oil reclamation, min. and veg. oil purification; absorbent carrier for pesticides; gran.

Flotigam Brands. [Hoechst-Celanese] Fatty amines, acetates, and preps.; anticaking agents for fertilizers; wax-like, pasty, and powds.

Flow Agent WR-100. [Henkel] Acrylic copolymer resin; anticaking agent; flow agent with surf. act. properties; improves flow properties of water reducible coatings; reduces pinholing and createring in air dry and bake systems; suitable for coatings designed for spray, dip, flow coat, and electrodeposition; clear to hazy liq.; dens. 8.5 lb/gal; flash pt. 24 C; 50% act.

Fluidiram. [Ceca] Fatty amines and mixts.; anticaking agent for fertilizers; flake or waxy.

Fluilan. [Croda] Lanolin oil; nonionic; w/o emulsifier; dispersant for pigments; conditioning agent; emollient, penetrant, superfatting agent for lipsticks, baby oils, brilliantines, cleansing lotions; plasticizer for hair spray resins; moisturizer in w/o emulsions; also for soaps, shampoos, dishwashing liqs., germicidal skin cleansers; pale yel. visc. liq.; pleasant, char. odor; sol. in min. oils, IPA, fatty alcohol, hydrocarbons, and aerosol propellents; cloud pt. 18 C max.; pour pt. 8 C max.; acid no. 2 max.; sapon. no. 80–100; 100% act.

Fluilan AWS. [Croda] PPG-12 PEG-65 lanolin oil; emollient, solubilizer; plasticizer and film modifier for hair sprays; amber visc. liq., nearly odorless; sol. in water and alcohol; acid no. 3 max.; iodine no. 7–15; sapon. no. 10–25; pH 5–7 (10%).

Fluilanol. [Croda Ltd.] Lanolin oil and oleth-3; nonionic; emulsifier, emollient, dispersant, wetting agent, superfatting agent used in cosmetics, soaps, detergents; pale liq.; faint sterol odor; sol. in min. and veg. oils, higher aliphatic alcohols, hydrocarbons, aerosol propellants; partly sol. in glycols; disp. in water; 100% act.

Fluorad FC-120. [3M] Ammonium perfluoroalkyl sulfonate; anionic; wetting, leveling agent, surfactant in aq. coating systems; low visc. liq.; sp.gr. 1.06; pH 8.5–9.5; 25% act.

Fluorad FC-128. [3M] Potassium fluorinated alkyl carboxylate; anionic; wetting, spreading and leveling agent used in alkaline cleaners, leak detection sol'ns., water-based coatings; powd.; forms gel in water; surf. tens. 17 dynes/cm (0.1–0.2% aq.); pH 7–8; 100% act.

Fluorad FC-129. [3M] Potassium fluorinated alkyl carboxylate; anionic; wetting and leveling agent, used in alkaline cleaners, floor polishes; amber liq.; limited sol. in aq. acids and bases; sp.gr. 1.3; surf. tens. 17 dynes/cm (0.1–0.2% aq.); flash pt. (PMCC) 115 F; pH 7–8; 50% act., 28% water, 18% butyl Cellosolve, 4% ethanol.

Fluorad FC-134. [3M] Fluorinated alkyl quat. ammonium iodides; cationic; wetting, spreading and leveling agent, used in cleaning and polishing prods., photographic development, leak detector sol'ns., corrosion inhibitors; powd.; forms gel in water; surf. tens. 17 dynes/cm (0.1% aq.); pH 3–4; 100% act.

Fluorad FC-135. [3M] Fluorinated alkyl quat. ammonium iodides; cationic; wetting, spreading and leveling agent, used in cleaning and polishing prods.; dk. amber liq.; sol. in nonpolar org. solvs.; sp.gr. 1.2; surf. tens. 17 dynes/cm (0.2% aq.); flash pt. (PMCC) 69 F; pH 3–4; 50% act., 34% IPA, 16% water.

Fluorad FC-170-C. [3M] Fluorinated alkyl POE ethanol; nonionic; wetting, spreading and leveling agent for floor polish formulations; amber liq.; sol. in HCl, alcohols; sp.gr. 1.32; cloud pt. < 0 C; flash pt. > 220 F (PMCC); surf. tens. 20 dynes/cm (0.1% aq.); pH 7–8 (1% aq.); 95% act.

Fluorad FC-171. [3M] Fluorinated alkyl alkoxylate; nonionic; wetting, spreading and leveling agent for floor polish formulations, coatings; amber liq.; sol. in butyl Cellosolve, toluene, MEK, IPA, dibutyl phthalate, water; sp.gr. 1.4; flash pt. > 220 F (PMCC); pour pt. 20 F; surf. tens. 20 dynes/cm (0.1% aq.); 100% act.

Fluorad FC-430. [3M] Fluorinated alkyl ester; nonionic; wetting, leveling, spreading and flow control agents for water- and solv.-based coatings and polymer systems, soldering systems; straw to amber liq.; sol. > 20 g/100 ml in water, Cellosolve acetate, MEK, toluene; sp.gr. 1.15; visc. 15,000 cp; flash pt. > 300 F (TOC); surf. tens. 30.3 dynes/cm (0.1 g/100 ml aq.); 100% act.

Fluorad FC-431. [3M] Fluorinated alkyl ester; nonionic; see Fluorad FC-430; straw to amber thin liq.; sol. > 20 g/100 ml in butyl Cellosolve, Cellosolve acetate, MEK, toluene, dibutyl phthalate; sp.gr. 1.05; visc. 200 cp; flash pt. 40 F (TOC); surf. tens. 36.6 dynes/cm (0.1 g/100 ml aq.); ref. index 1.406; 50% act. in ethyl acetate.

Fluorad FC-750. [3M] Fluorinated alkyl quat. ammonium iodides; cationic; well stimulation additive, offers lowest surf. tens. for aq. acids, enhances activity of org. corrosion inhibitors; dk. liq.; sp.gr. 1.2; flash pt. 53 F (PMCC); pour pt. 32–40 F; surf. tens. 17.5 dynes/cm (100 ppm aq.); 50% solids in 2/1 IPA/water.

Fluorad FC-760. [3M] Fluorinated alkyl alkoxylates; nonionic; well stimulation additive, useful at low levels to reduce surf. tens. of fluids used for acidizing, fracturing and water flooding; amber liq.; sp.gr. 1.4; flash pt. > 220 F (PMCC); pour pt. 20 F; surf. tens. 19 dynes/cm (100 ppm aq.).

FMB 65-15 Quat, 65-28 Quat. [Huntington] n-Alkyl dimethyl benzyl ammonium chloride; for formulation of mildewcides and swimming pool algicides; 50% and 80% act. resp.

FMB 210-8 Quat, 210-15 Quat. [Huntington] Didecyl dimethyl ammonium chloride; for formulation of disinfectants, sanitizers, fungicides, water treatment microbicides, swimming pool algicides, mildewcides; 80 and 50% act. resp.

FMB 302-8 Quat. [Huntington] Octyl decyl dimethyl ammonium chloride, dioctyl dimethyl ammonium chloride, didecyl dimethyl ammonium chloride; for formulation of disinfectants, sanitizers, and fungicides; 80% act.

FMB 451-5 Quat, 451-8 Quat. [Huntington] n-Alkyl dimethyl benzyl ammonium chloride; for formulation of disinfectants, sanitizers, fungicides, water treatment microbicides, swimming pool algicides, mildewcides; 50 and 80% act. resp.

FMB 500-15 Quat U.S.P. [Huntington] n-Alkyl dimethyl benzyl ammonium chloride; preservative in OTC drug prods.; 50% act.

FMB 504-5 Quat. [Huntington] n-Alkyl dimethyl benzyl ammonium chloride, octyl decyl dimethyl ammonium chloride, dioctyl dimethyl ammonium chloride, didecyl dimethyl ammonium chloride; for formulation of disinfectants, sanitizers, fungicides, water treatment microbicides, mildewcides; 50% act.

FMB 1210-5 Quat, 1210-8 Quat. [Huntington] n-Alkyl dimethyl benzyl ammonium chlorides and didecyl dimethyl ammonium chloride; for formulation of disinfectants, sanitizers, fungicides, water treatment microbicides, swimming pool algicides; 50 and 80% act. resp.

FMB 3328-5 Quat, 3328-8 Quat. [Huntington] n-Alkyl dimethyl benzyl ammonium chloride and n-alkyl dimethyl ethylbenzyl ammonium chloride; for formulation of disinfectants, sanitizers, fungicides, water treatment microbicides, swimming pool algicides; 50 and 80% act. resp.

FMB 4500-5 Quat, 4500-8 Quat. [Huntington] n-Alkyl dimethyl benzyl ammonium chloride; for formulation of disinfectants, sanitizers, fungicides, water treatment microbicides, swimming pool algicides, mildewcides; 50 and 80% act.

FMB 6075-5 Quat, 6075-8 Quat. [Huntington] n-Alkyl dimethyl benzyl ammonium chloride and n-alkyl dimethyl ethylbenzyl ammonium chloride; for formulation of disinfectants, sanitizers, fungicides, water treatment microbicides, and swimming pool algicides; 50 and 80% act.

Foamaster® 111. [Henkel] Silicone-free defoamer for adhesives, coatings, printing inks, latex; opaque yel. liq.; disp. in aq. systems; dens. 7.5 lb/gal; 100% act.

Foamaster® 206-A, 267-A, 335-A. [Henkel] Non-silicone type based on chemically modified oils and fats; nonionic; defoamer for textile industry; liq.; water disp.; 100% act.

Foamaster® 333. [Henkel] Silicone-free defoamer for adhesives, coatings, printing inks; opaque yel. liq.; insol. in water; disp. in aq. systems containing surfactants; dens. 7.5 lb/gal; 100% act.

Foamaster® 1407-50. [Henkel] Nonionic; defoamer for fluidized paste; water disp.

Foamaster® 1719-A. [Henkel] Nonionic; defoamer for use in gelatine; liq.; 50% act.

Foamaster® 8034. [Henkel] Nonionic; defoamer for paints, adhesives, rubber, and latex; liq.; 100% act.

Foamaster® A. [Henkel] Non-min. oil defoamer for coatings, printing inks; lt. yel. visc. iq.; insol. in water; disp. in surfactant systems; dens. 8.2 lb/gal; 100% act.

Foamaster® AP. [Henkel] Defoamer for adhesives, coatings, latex; esp. compatible in associative thickener-modified paints and in high gloss systems; opaque pale yel. liq.; disp. in water; dens. 7.7 lb/gal; 100% act.

Foamaster® B. [Henkel] Nonionic; predispersible defoamer for latex systems; liq.; 100% act.

Foamaster® DD-72. [Henkel] Nonionic; defoamer for emulsion paints, wallpaper coatings, inks; liq.; 100% act.

Foamaster® DF-122NS. [Henkel] Fatty soaps, fatty esters in hydrocarbon, silicone-free; nonionic; defoamer for water-based coatings; water-disp.

Foamaster® DF-177-F, -178. [Henkel] Nonionic; foam suppressor for coating and adhesive formulations; liq.; 100% act.

Foamaster® DF-198-L. [Henkel] Defoamer for paints based on fine particle size acrylic emulsions; liq.; 100% act.

Foamaster® DNH-1. [Henkel] Nonionic; defoamer for emulsion paints and polymers; liq.; 100% act.

Foamaster® DR-187. [Henkel] Fatty acids, alcohols, esters in hydrocarbon; nonionic; defoamer for mfg. of paper and paperboard; water-disp.

Foamaster® DRY. [Henkel] Silicone; dry defoamer for wettable powds. and dry flowable pesticide formulation; powd.

Foamaster® DS. [Henkel] Silicone type; defoamer; opaque yel. liq.; water-disp.; dens. 7.6 lb/gal; 100% act.

Foamaster® FGA. [Henkel] Nonsilicone emulsion; food grade defoamer; liq.

Foamaster® FLD. [Henkel] Silicone emulsion; food grade defoamer; liq.

Foamaster® G. [Henkel] Disp. latex stripping defoamer; liq.; dens. 7.5 lb/gal; 100% act.

Foamaster® JMY. [Henkel] Nonionic; foam control agent for adhesives, rubber, and latex; lt. amber liq.; very water disp.; dens. 7.3 lb/gal; 100% act.

Foamaster® KF-99. [Henkel] Nonsilicone type based on chemically modified oils and fats; defoamer for textile industry; liq.; disp. in water; 100% act.

Foamaster® NDW. [Henkel] Nonionic; defoamer for water-based paints, adhesive and cement screeds, rubber and latex; liq.; 100% act.

Foamaster® NS-20. [Henkel] Nonionic; silicone-free defoamer for coatings or paper stock, air knife coatings; liq.; 100% act.

Foamaster® NXZ. [Henkel] Defoamer for syn. latices, paints, adhesives, blade and roll coatings, emulsifiable latex stripping; liq.; 100% act.

Foamaster® O. [Henkel] Defoamer for adhesives, coatings, printing inks; opaque off-wh. liq.; disp. in aq. systems; dens. 7.4 lb/gal.

Foamaster® P. [Henkel] Nonionic; foam controller in PVAL systems in adhesive industry; liq.; 35% act.

Foamaster® PD-1. [Henkel] Nonionic; defoamer for dry mixes, adhesives and cement screeds, joint cements, spackles; powd.; 67% act.

Foamaster® PD-1-D. [Henkel] Defoamer for coatings; off-wh. free-flowing powd.; 65% act.

Foamaster® R. [Henkel] Silica-based defoamer used for coatings; opaque off-wh. liq.; insol. in water; disp. in surfactant systems; dens. 7.5 lb/gal; 100% act.

Foamaster® S. [Henkel] Silica-based defoamer with high compatibility; used for adhesives, coatings, latex; opaque lt. tan liq.; disp. in water; dens. 7.5 lb/gal.

Foamaster® SA-3. [Henkel] Multi-hydrophobe based defoamer for coatings; opaque off-wh. liq.; insol. in water; disp. in surfactant systems; dens. 7.1 lb/gal; 100% act.

Foamaster® Soap L. [Henkel] Tallow soap; dry defoamer for wettable powd. and dry flowable pesticide formulations; flake.

Foamaster® TBD-1. [Henkel] Defoamer for water-sol. resins; liq.; 100% act.

Foamaster® TDB. [Henkel] Nonionic; foam control agent in dry adhesives, joint cements, spackle, rubber and latex; liq.; 100% act.

Foamaster® V. [Henkel] Nonionic; disp. latex stripping defoamer; liq.; 100% act.

Foamaster® VL. [Henkel] Nonionic; emulsifiable latex monomer stripping defoamer, used in latex and emulsion paints; hazy amber liq.; disp. in water; dens. 7.7 lb/gal; 100% act.

Foam Burst 2. [Ross] Silicone compd.; oil refining, cutting oils, chemical processing; base for antifoam formulations; solv.-disp.; can be emulsified in water; 100% act.

Foam Burst 5. [Ross] Silicone compd.; defoamer for food and industrial formulations; solv.-disp.; can be emulsified in water; 100% act.

Foam Burst 10. [Ross] Silicone compd.; food grade defoamer; solv.-disp.; can be emulsified in water; 100% act.

Foam Burst 87. [Ross] Silicone compd.; nonfood antifoam; emulsifiable in water.

Foam Burst 100. [Ross] Silicone emulsion; food grade defoamer; disp. in water.

Foam Burst 105. [Ross] Silicone emulsion; general purpose defoamer; disp. in water.

Foam Burst 150. [Ross] Silicone emulsion; food grade defoamer; disp. in water.

Foam Burst 151. [Ross] Silicone emulsion; defoamer; disp. in water.

Foamer. [Hart Chem. Ltd.] Fatty acid amide sulfosuccinate ester; anionic; stabilizer for shampoos, detergents, and bubble baths; yel. liq.; 40% act.

Foamex AD-50. [Lyndal] Silicone emulsion; nonionic; defoamer and antifoam.

Foamex J-10. [Lyndal] Silicone emulsion; antifoam for pressure dyeing; water-disp.

Foamgard 73. [Lyndal] Alcohol ether; defoamer for print pastes where silicone type undesirable.

Foamgard 161. [Lyndal] Nonsilicone emulsified hydrocarbon; defoamer for carpet dyeing.

Foamgard 200. [Lyndal] Nonsilicone; defoamer for dye baths, scouring baths, print pastes and emulsions, paper and pulp processing, waste treatment; water-disp.

Foamkill 8BA. [Crucible] Silicone compd.; defoamer for food applics.

Foamkill 8G. [Crucible] Silicone compd.; defoamer for food applics. incl. general nonaq. systems, drug extraction and separation, drug processing, starch extractions and processing, anaerobic fermentations, vitamins, etc.

Foamkill 8J Series. [Crucible] Org. and organo-silicone conc.; defoamer for food/pharmaceutical applics.

Foamkill 30 Series. [Crucible] Org. and organo-silicone conc.; defoamer for food/pharmaceutical applics.

Foamkill 30C, 30-HP. [Crucible] Organo-silicone emulsion concs.; defoamer for pulp/paper applics. incl. adhesives, casein, neoprene, and natural latices.

Foamkill 80J Series. [Crucible] Silicone compd.; defoamer for food/pharmaceutical applics.

Foamkill 400A. [Crucible] Organo-silicone emulsion; defoamer for pulp/paper applics. incl. adhesives backings, latex paints and coatings, water-reducible inks, floor and ceiling coatings.

Foamkill 608 Series. [Crucible] Org. and organo-silicone conc.; defoamer for paper coatings and adhesives in contact with food.

Foamkill 608, 608M. [Crucible] Org.; general-purpose defoamer for pulp/paper applications incl. acrylics, vinyl acrylics, terpolymers, copolymers, PVA, PVC, PVAC and other coatings, screen area and wet end areas.

Foamkill 608G. [Crucible] Org.; defoamer for pulp/paper applics.; predispersible with water.

Foamkill 618 Series. [Crucible] Org. and organo-silicone conc.; defoamer for food/pharmaceutical applics. incl. paper coatings and adhesives in contact with food.

Foamkill 618, 618C. [Crucible] Org. see Foamkill 608G.

Foamkill 618F, 618J. [Crucible] Org.; defoamer for pulp/paper applics. incl. adhesive backings, water cooling towers, evaporators, screen areas, wet end machine areas, deinking operations, water treatment ponds.

Foamkill 627. [Crucible] Org. conc.; defoamer for pulp/paper applics., water cooling towers, water treatment ponds; lubricant for fine papers.

Foamkill 628A. [Crucible] Org. defoamer for pulp/paper applics.

Foamkill 634 Series. [Crucible] Org. and organo-silicone conc.; defoamer for food/pharmaceutical applics.

Foamkill 634B-HP. [Crucible] Org. and organo-silicone conc.; defoamer for food/pharmaceutical applics.

Foamkill 634C. [Crucible] Org. and organo-silicone conc.; defoamer for food/pharmaceutical applics.

Foamkill 634D-HP. [Crucible] Org. and organo-silicone conc.; defoamer for food/pharmaceutical applics.

Foamkill 634F-HP. [Crucible] Org. and organo-silicone conc.; defoamer for food/pharmaceutical applics.

Foamkill 639 Series. [Crucible] Org. and organo-silicone conc.; defoamer for food/pharmaceutical applics. incl. paper coatings and adhesives in contact with food, pulp/paper applics.

Foamkill 639 Conc. [Crucible] Org. conc.; defoamer for pulp/paper applics., acrylic, PVC, PVA, PVPC, and other coatings,antifoam formulating.

Foamkill 639J. [Crucible] Org. and organo-silicone conc.; defoamer for food/pharmaceutical applics. incl. paper coatings and adhesives in contact with food, cleaning and sanitizing sol'ns., pulp/paper applics., acrylic, PVC, PVA, PVPC, and other coatings, antifoam formulating.

Foamkill 639J-F. [Crucible] Org. and organo-silicone conc.; defoamer for food/pharmaceutical applics. incl. paper coatings and adhesives in contact with food.

Foamkill 639L. [Crucible] Org. defoamer for pulp/paper applics. incl. adhesive backings, acrylic coatings, vinyl-acrylic coatings, PVC, PVA, PVAC coatings, paper formation.

Foamkill 639P. [Crucible] General-purpose defoamer for pulp/paper applics., acrylic/PVC/ PVA/ PVAC coatings.

Foamkill 644 Series. [Crucible] Org. and organo-silicone conc.; defoamer for food/pharmaceutical applics. incl. paper coatings and adhesives in contact with food, cosmetics.

Foamkill 644E, 645. [Crucible] Org.; see Foamkill 608G.

Foamkill 649 Series. [Crucible] Org. and organo-silicone conc.; defoamer for food/pharmaceutical applics. incl. paper coatings and adhesives in contact with food, pulp/paper applics.; lubricant.

Foamkill 649N, 649P. [Crucible] Org. defoamer for pulp/paper applics.; lubricant.

Foamkill 652 Series. [Crucible] Org. and organo-silicone conc.; defoamer for paper coatings and adhesives in contact with food.

Foamkill 652H. [Crucible] Org. and organo-silicone conc.; defoamer for food/pharmaceutical applics.

Foamkill 652-HF. [Crucible] Org. and organo-silicone conc.; defoamer for food/pharmaceutical applics. incl. general nonaq. systems.

Foamkill 652L. [Crucible] Conc. multi-ingred. org.; defoamer for pulp/paper applics.

Foamkill 660, 660F. [Crucible] Org. defoamer for pulp/paper applics., water cooling towers, evaporators.

Foamkill 660-HP. [Crucible] Org.; see Foamkill 618F.

Foamkill 663J. [Crucible] Org. and organo-silicone conc.; defoamer for food/pharmaceutical applics. incl. paper coatings and adhesives in contact with food, pulp/paper applics., water cooling towers; lubricant for fine papers; water-disp.

Foamkill 679. [Crucible] Organo-silicone emulsion; defoamer for pulp/paper applics. incl. adhesives backings, latex paints and coatings, water-reducible inks, floor and ceiling coatings.

Foamkill 684 Series. [Crucible] Org. and organo-silicone conc.; defoamer for food/pharmaceutical applics. incl. paper coatings and adhesives in contact with food, pulp/paper applics., water treatment ponds.

Foamkill 684A. [Crucible] Org.; defoamer for pulp/paper applics., water cooling towers, evaporators.

Foamkill 684P. [Crucible] Foamkill 700 type in min. spirits; defoamer for pulp/paper applics., waste treatment, water cooling towers; lubricant.

Foamkill 700, 700 Conc. [Crucible] All org. liq. versions of paste or brick types; defoamer for pulp/paper applics., waste treatment, water cooling towers; liq.; 60% solids (700); ≈ 100% solids (700 Conc.).

Foamkill 810F. [Crucible] Dimethicone; defoamer for food/pharmaceutical applics. incl. general aq. systems, paper coatings and adhesives in contact with food, cleaning/sanitizing sol'ns., cosmetics, pulp/paper applics.; 10% silicone emulsion.

Foamkill 830. [Crucible] Organo-silicone emulsion; defoamer for food/pharmaceutical applics. incl. paper coatings and adhesives in contact with food, cosmetics, pulp/paper applics.; emulsion; very small particle size.

Foamkill 830F. [Crucible] Dimethicone; defoamer for food/pharmaceutical applics. incl. general aq. systems, paper coatings and adhesives in contact with food, cleaning/sanitizing sol'ns., cosmetics.

Foamkill 836A. [Crucible] Silicone emulsions; defoamer for food/pharmaceutical applics. incl. paper coatings and adhesives in contact with food, cosmetics.

Foamkill 852. [Crucible] Silicone emulsions; defoamer for pulp/paper applics.

Foamkill 1001 Series. [Crucible] Org. and organo-silicone conc.; defoamer for food/pharmaceutical applics. incl. paper coatings and adhesives in contact with food, pulp/paper applics.

Foamkill FPF. [Crucible] Organo-silicone emulsion; defoamer for pulp/paper applics. incl. adhesives backings, latex paints and coatings, water-reducible inks, floor and ceiling coatings.

Foamkill GCP Series. [Crucible] Silicone fluid; defoamer for food/pharmaceutical applics., cosmetics.

Foamkill MS-1. [Crucible] Organo-silicone emulsion; defoamer for pulp/paper applics. incl. adhesives backings, latex paints and coatings, water-reducible inks, floor and ceiling coatings; flow control agent.

Foamkill MS Conc. [Crucible] Silicone emulsion; defoamer for food/pharmaceutical applics. incl. paper coatings and adhesives in contact with food, cleaning/sanitizing sol'ns., pulp/paper applics. incl. adhesives, casein, neoprene, and natural latices, antifoam formulating.

Foamkill MSC Series. [Crucible] Org. and organosilicone conc.; defoamer for food applics.

Foamkill NSP-1, NSP-3, NSP-4, NSP-5. [Crucible] Org. defoamer for pulp/paper applics.

Foamkill RP. [Crucible] Org. and organo-silicone conc.; defoamer for food/pharmaceutical applics.

Foamole A. [Van Dyk] Linoleic alkanolamide (1:1); hair conditioner for shampoos, hair dyes; visc.booster for shampoos, liq. soaps; sol. in 70% ethanol, propylene glycol, peanut oil; gels in water.

Foamole B. [Van Dyk] Minkamidopropyl dimethylamine; superfatting agent, conditioner for hair care prods.; sol. in propylene glycol; disp. in water.

Foamole M. [Van Dyk] Cocamide MEA (1:1); foam stabilizer, shampoo thickener; sol. in 70% ethanol, propylene glycol.

Folidol®-E605. [Bayer] Parathion; insecticide and acaricide; colorless liq., odor reminiscent of leeks; m.w. 291.3; sol. in most org. solvs.; low sol. in petrol. and min. oils; sp.gr. 1.265 (25/4 C); b.p. 160 C (1 mm Hg).

Folidol® M. [Bayer] Parathion-methyl; insecticide with contact, stomach, and breathing poison action; wh. cryst.; m.w. 263.2; sol. in most org. solvs.; low sol. in petrol. ether and min. oils; sp.gr. 1.36 (20/4 C); b.p. 154 C (1 mm Hg); m.p. 35–36 C.

Folimat®. [Bayer] Omethoate; systemic acaricide and insecticide; colorless to slightly ylsh. oily liq., odor reminiscent of leeks; m.w. 213.2; readily sol. in water, alcohol, acetone; less sol. in ethyl ether; practically insol. in petrol. ether; sp.gr. 1.32 (20/4 C); decomp. pt. 135 C.

Folithion®. [Bayer] Fenitrothion; broad spectrum insecticide for controlling biting and sucking insect pests; ylsh.-brn. liq.; m.w. 277.24; sol. in acetone, alcohol, chlorinated hydrocarbons; insol. in water; sp.gr. 1.308 (20/4 C); b.p. 164 C (1 mm Hg).

Fomrez B-306. [Witco] Organo-silicone; coupler used in mfg. of flexible polyester urethane foams; clear liq.; sp.gr. 1.04; acid no. 1.0; pH 5.0 (3.5% aq.).

Fomrez B-308. [Witco] Organo-silicone; see Fomrez B-306; clear liq.; sp.gr. 1.04; acid no. 1.0; pH 5.0 (3.5% aq.).

Fomrez B-320. [Witco] Organo-silicone; see Fomrez B-306; also used in mfg. of polyester microcellular urethane foams; clear liq.; sp.gr. 1.04; acid no. 1.0; pH 5.0 (3.5% aq.).

Fomrez C-2. [Witco] Stannous octoate, stabilized; catalyst for mfg. of one-shot polyether urethane foams; yel. clear liq.; sp.gr. 1.27; dens. 10.6 lb/gal; flash pt. > 121 C (COC); 28.1% total tin.

Fomrez C-4. [Witco] Stannous octoate, stabilized and anhyd. dioctyl phthalate; see Fomrez C-2; also used for foam producers; yel. clear liq.; sp.gr. 1.10; dens. 9.2 lb/gal; flash pt. > 121 C (COC); 14.1% total tin.

Fomrez SUL-3. [Witco] Dibutyltin diacetate; catalyst used in rigid urethane spray foam formulations and in R.T.-vulcanized silicone polymers; colorless to pale yel. liq.; acetic acid odor; m.w. 351; f.p. 5 C; sp.gr. 1.31; ref. index 1.4780; flash pt. 143 C (COC); 33.8% tin.

Fomrez SUL-4. [Witco] Dibutyltin dilaurate; catalyst used in mfg. of polyether-based, rigid urethane foams and R.T.-vulcanized silicone polymers; Gardner 1; m.w. 631; f.p. 8 C; sp.gr. 1.05; ref. index 1.4700; flash pt. 232 C (COC); 18.9% tin.

Fomrez UL-1. [Witco] Tin catalyst; catalyst for prep. of one-shot or prepolymer-type polyether-based, rigid urethane foams for spray or pour-in-place applic.; liq.; water wh. to pale yel.; f.p. –15 C; sp.gr. 1.01.

Fomrez UL-2. [Witco] Proprietary organotin carboxylate; catalyst used as replacement for dibutyltin dilaurate; pale yel. liq.; sp.gr. 1.14; pour pt. –25 C.

Fomrez UL-6. [Witco] Organotin; catalyst used in elastomer formulations and polyether-based urethane foam systems; water-wh. to pale yel. liq.; sp.gr. 1.13; pour pt. < –25 C.

Fomrez UL-8. [Witco] Organotin; catalyst used in fluorocarbon-blown urethane spray foams; cocatalyst in isocyanurate foam formulations; pale yel. liq.; sp.gr. 1.34; pour pt. –14 C; ref. index 1.50; flash pt. 135 C (COC).

Fomrez UL-22. [Witco] Organotin; catalyst for water/fluorocarbon-blown urethane foams; pale yel. liq.; f.p. –10 C; sp.gr. 1.03; ref. index 1.50; flash pt. 185 C (COC).

Fomrez UL-28. [Witco] Organotin; see Fomrez SUL-4; yel. liq.; sp.gr. 1.14; pour pt. –6 C; ref. index 1.47; flash pt. 153 C (COC).

Fomrez UL-29. [Witco] Organotin; see Fomrez UL-6; pale yel. liq.; sp.gr. 1.28; pour pt. –40 C; ref. index 1.50; flash pt. 185 C (COC).

Fomrez UL-32. [Witco] Organotin; see Fomrez UL-1; pale yel. liq.; f.p. –8 C; sp.gr. 0.98; ref. index 1.49; flash pt. 210 C (COC).

Fonoline® White, Yellow. [Witco] Petrolatum USP; soft, low m.p. for consumer use as petrol. jelly, ointments, industrial applics.; as emollient, protective coating, binder, carrier, lubricant, moisture barrier, plasticizer, protective agent, softener; Lovibond color 1.7Y and 30Y2.5R resp.; visc. 9–14 cSt (100 C); m.p. 53–58; pour pt. 20 F.

Foral® 85. [Hercules] Glycerol ester of highly hydrog. rosin; thermoplastic syn. resin used as tackifier resin in adhesives, and in protective and barrier-type coatings that require exceptional degree of color retention and oxidation resistance; pale solid, flakes; low odor; sol. in esters, ketones, higher alcohols, glycol ethers, and aliphatic, aromatic, and chlorinated hydrocarbons; water-insol.; sp.gr. 1.07; dens. 1.07 kg/l; soften. pt. (drop) 82 c; acid no. 9.

Foral® 105. [Hercules] Pentaerythrityl tetraabietate; see Foral 85; pale solid, flakes; low odor; sol. see Foral 85; sp.gr. 1.07; dens. 1.07 kg/l; visc. (G-H) C–D (60% in min. spirits); soften. pt. (drop) 104 C; acid no. 12.

Foral® AX. [Hercules] Highly hydrog. wood rosin; thermoplastic; acidic resin with exc. resistance to oxidation and exceptionally pale color; used as tackifier or modifying resin in adhesives and hot-melt-applied decorative, pressure sensitive, and heat-sealable coatings; very pale solid; low odor; sol. in alcohols, esters, ketones, hydrocarbons, chlorinated solvs. and min. oils; water-insol.; soften. pt. (drop) 75 C; acid no. 160; ref. index 1.4955 (100 C).

Forbest 30. [Lucas Meyer] Modified aryl-alkyl silicones; antifloating and antisilking agent for all paint and varnish systems except water-based systems; lt. blue mobile liq.; sp.gr. 0.818; flash pt. 28 C; ref. index 1.460; 50% act. in xylene and hydrocarbons.

Forbest 50. [Lucas Meyer] Modified aryl-alkyl silicones; see Forbest 30; mobile liq.; sp.gr. 0.818; flash pt. 28 C; 50% act. in xylene.

Forbest 62B. [Lucas Meyer] Based on hydrocarbon polymers; additive to improve smoothness and abrasion resistance; recommended for pigmented air-dry and stoving enamels, clear coatings; ylsh. transparent gel; sp.gr. 0.905; flash pt. 46 C; 75% solids.

Forbest 70. [Lucas Meyer] Slip agent improving abrasion resistance; used for solvent-based paints and coatings, printing inks; FDA approved; lt. beige fine powd.; 5 μ max. particle size; sp.gr. 1.9 (wetted).

Forbest 150. [Lucas Meyer] Polyaryl ester; leveling agent for air-dry paints with high pigment loading; sp.gr. 1.085; flash pt. 88 C; 97% solids.

Forbest 172. [Lucas Meyer] Org. polymers; visc. control agent for air-dry and stoving solvent systems; golden brn. liq.; sp.gr. 0.8640; visc. 42 s; flash pt. 16 C; 45% solids; contains propanol and xylene.

Forbest 260. [Lucas Meyer] Hydrocarbon polymers; slip additive improving scratch and abrasion resistance in pigmented air-dry and stoving enamels, clear coatings; brn. transparent liq.; sp.gr. 0.811; flash pt. 38 C; 40% solids in min. spirits.

Forbest 410. [Lucas Meyer] Polyethylene compd.; leafing and stabilizing agent for leafing aluminum pigments; recommended for air-dry and stoving paints; ylsh. clear liq.; sp.gr. 0.86; flash pt. 28 C; pH 10; 25% solids; contains xylene, alkylamine, butanol.

Forbest 600. [Lucas Meyer] Corrosion inhibitor preventing can corrosion for all emulsions and disp., other aq. systems; ylsh. clear liq.; sp.gr. 1.14; pH 9–9.5; contains alkylamine.

Forbest 610. [Lucas Meyer] Carboxylic acid diamine compd.; cationic; wetting and disp. agent for aq. systems; recommended for air-dry and stoving water-thinnable systems based on emulsions, alkyd, and acrylic resins; brn. clear liq.; sp.gr. 1.01; flash pt. 10 C; pH 8.6; 60% act. in ethanol.

Forbest 620. [Lucas Meyer] Based on polyoxy alkylic siloxanes and cyclic amide; leveling agent for water-thinnable systems incl. varnishes, roller paints, coil coatings, emulsion paints; lt. ylsh. grn. liq.; sp.gr. 1.0; pH 7.5.

Forbest 680. [Lucas Meyer] Alkyaryl-silicone compd.; smoothness and leveling agent for aq. systems; recommended for water-thinnable varnishes, coil coating systems; water clear liq.; sp.gr. 1.03; visc. 1600 mPas; flash pt. 66 C; pH 7.0; 50% act. in ethanol.

Forbest 780. [Lucas Meyer] Carboxylic reaction prod.; catalyst for gel formation in offset and heat-set printing inks, hydrocarbon-based structural varnishes; yel.-brn. liq.; sp.gr. 0.91; visc. 60 s; pH 9.6; 100% solids.

Forbest 850. [Lucas Meyer] Cationic; wetting agent with anticorrosion act.; for air-dry anticorrosive paints and primers, asphalt paints, marine paints, road-marking paints, caulking compds.; brn. paste; m.p. 30 C; flash pt. 150 C; 100% solids.

Forbest 1000B. [Lucas Meyer] Nonionic fatty acid blend; defoamer for solv. and aq. systems; used for solvent-based air-dry and stoving finishes, aq. systems, dip-coatings, emulsion paints, varnishes; ylsh. turbid liq.; sp.gr. 0.87; visc. 300–450 mPas; flash pt. 172 C; 100% act.

Forbest 1500W. [Lucas Meyer] Nonionic hydrocarbon-fatty acid oil-based blend; defoamer for air-dry and stoving water-reducible systems based on emulsions, alkyds, acrylics; ylsh. turbid liq.; sp.gr. 0.84; flash pt. 170 C; 100% act.

Forbest 2000C. [Lucas Meyer] Min. oil/fatty acid partial ester; defoamer for emulsion paints and varnishes, roughcasts, water-disp. alkyd emulsions, styrene-acrylates; lt. brn. cloudy, visc. liq.; sp.gr. 1.0; flash pt. > 160 C; 100% act.

Forbest 5724. [Lucas Meyer] Pretreated aluminum-2-ethyl hexanate; gelling agent for offset and heat-set printing inks, hydrocarbon-based structural varnishes; wh. powd., nonhygroscopic; m.p. 275 C.

Forbest 8209. [Lucas Meyer] Wetting, dispersing agent with antistatic properties; used for air-dry and stoving finishes and coatings; ylsh. slightly turbid visc. liq.; sp.gr. 1.08; pH 6–8; 97% solids.

Forbest CP. [Lucas Meyer] Micronized cryst. polyolefin; slip agent improving abrasion resistance in paints, coatings, printing inks; FDA approved; lt. fine powd.; 8 μ particle size; m.w. 1000–1500; m.p. 118 C.

Forbest G23. [Lucas Meyer] Organo-modified polysiloxanes and leveling components in high-boiling solvs.; smoothness and leveling agent for all air-dry and stoving solv. coatings; lt. yel. oily liq.; sp.gr. 1.0; visc. 21 s; flash pt. 43 C; ref. index 1.429–1.431; 50% solids in ethylglycol.

Forbest GLW. [Lucas Meyer] High melting, modified metallic soap; hydrophilic slip agent for aq. air-dry and stoving systems, water-based flexo printing inks; wh. fine powd.; 6–12 μ particle size; sp.gr. 1.05 (wetted); m.p. > 130 C; 3% max. water.

Forbest MF II. [Lucas Meyer] Ultra fine micronized syn. hard wax; increases smoothness and abrasion resistance for solv.-based paints and coatings, printing inks; FDA approved; lt. fine powd.; 4–7 μ particle size; m.p. 105 C; acid no. 0; sapon. no. 0.

Forbest MW 23. [Lucas Meyer] Microcryst. wax finely disp. in C_{17} fatty acid; increases slip and abrasion resistance in clear and pigmented stoving enamels, e.g., industrial paints, coil-coating finishes, printing inks (flexo, letterpress); ylsh. paste; < 10 μ particle size; sp.gr. 0.955; 35% micro wax; 99% solids.

Forbest P 22. [Lucas Meyer] Polyethylene/colophony resin disp. in butyl diglycol/naphtha solv.; increases slip and abrasion resistance in clear and pigmented stoving enamels, industrial paints, coil-coating finishes; ylsh. soft paste; < 10 μ particle size; sp.gr. 0.95; 27% polyethylene wax, 75% solids.

Forbest PAM. [Lucas Meyer] Micronized polyamide wax; improves slip and abrasion resistance in solv.-based paints, coatings, and printing inks; FDA approved; lt. fine powd.; 8 μ max. particle size; m.p. 138–142 C.

Forbest VP 13. [Lucas Meyer] Polyester-fatty alcohol compd.; wetting and disp. agent , binder for all water-thinnable paints incl. leather finishes, air-dry and stoving alkyd systems, electrodeposition, emulsion paints, pigment slurries for paper coating; ylsh. clear liq.; sp.gr. 0.90; visc. 50 s; flash pt. 40 C; pH 10.5; 50% solids.

Forbest VP 18. [Lucas Meyer] Aliphatic ester; anionic; wetting and disp. agent for pigmented air-dry and stoving paints and lacquers, e.g., anticorrosive paints, alkyd enamels, zinc paints, marine paints, bituminous paints, epoxy paints, putty, printing inks, pigment concs.; brn. clear visc. liq.; sp.gr. 0.97; visc. 24 Pas; flash pt. > 250 C; 100% solids.

Forbest VP 20. [Lucas Meyer] Compd. based on neutralized chelating agents; wetting and disp. agent for water-thinnable paints, e.g., leather finishes, air-dry and stoving alkyds, electrodeposition, emulsion paints, pigment concs.; nearly colorless thin liq.; sp.gr. 1.19; pH 6; 50% solids.

Forbest VP 33. [Lucas Meyer] Modified polyalkoxyether; wetting and disp. agent for water-thinnable paints, air-dry and stoving alkyds, emulsion paints; brn. clear visc. liq.; sp.gr. 1.05; visc. 1 Pas; pH 7; 40% solids.

Forbest VP S7. [Lucas Meyer] Acrylic polymers dissolved in Shellsol A; amphoteric; wetting and disp. agent, binder, antisettling agent for pigment pastes; brn. clear liq.; sp.gr. 0.9; visc. 18–22 Pas; 55% solids in Shellsol A.

Forbest WP. [Lucas Meyer] Wetting, disp., and antisettling agent for all air-dry and stoving solvent-based paints and systems with polar solvs.; brn. liq., mildly fragrant; sp.gr. 0.894; flash pt. 185 C; pH 9.5; 95% solids.

Fore®. [Rohm & Haas] Maneb and zinc ion coordinate; fungicide for turf, ornamentals.

Fore® WD. [Rohm & Haas] Maneb/sineb coordinate; fungicide for turf, ornamentals; solid.

Forlan. [RITA] Lanolin and derivs.; absorp. base, emollient.

Forlan C-24. [RITA] Ethoxylated sterol; emulsion stabilizer, emollient, moisturizer, solubilzer, dispersant, plasticizer for personal care prods.; sol. in alcohol, water.

Forlan L. [RITA] Lanolin replacement; emulsion stabilizer, emollient; sparingly water-sol.

Forlanit P. [Henkel KGaA] Sodium lauryl ether phosphate; anionic; wetting agent, flotation aux. for hot-baths containing cyanides; liq.; 30% conc.

Forlanon. [Henkel KGaA] Sodium lauryl ether phosphate; anionic; see Forlanit P; liq.; 10% conc.

Formasil® Silicone Emulsion 45. [Union Carbide] Dimethylsiloxane emulsion; release agent in mfg. of medical devices and appliances, food pkg. materials; wate-disp.

Formula #633. [Polymer Research] Coumarin deriv.; cationic; economical fluorescent whitener which produces a brilliant wh. on acetate, wool, silk, and nylon; can be applied in scouring and peroxide or hydrosulfite bleach baths; liq.; readily sol. in warm water.

Formula #733. [Polymer Research] Oxazole derivs.; nonionic; effective fluorescent whitener yielding intense bluish whites with outstanding lt. and wash fastness; used primarily for polyester and polyester blends with cotton and wool; suitable for whitening acetate, PP, and polyvinyl fibers; applicable by exhaust or pad thermosol procedures at all stages of processing; cream fine aq. disp.; water-misc.; pH 6.0–6.5.

Fortex®. [Petrolite] Hard microcryst. wax consisting of n-paraffinic, branched paraffinic, and naphthenic hydrocarbons; used in hot-melt coatings and adhesives, cup and paper coatings, printing inks, plastic modification (as lubricant and processing aid), lacquers, paints, and varnishes, as binder in ceramics, for potting, filling, and impregnant in elec./electronic components, in investment casting, rubber and elastomers (plasticizer, antisunchecking, antiozonant), as emulsion wax size in papermaking, as fabric softener ingred. in permanent-press fabrics, in emulsion and latex coatings, and in cosmetic hand creams and lipsticks; color 1.5 (D1500) wax; very low sol. in org. solvs.; sp.gr. 0.93; visc. 23.5 cps (98.9 C); m.p. 96.1 C.

Fosterge BA-14 Acid. [Henkel] Mono and dialkyl phenoxy POE acid phosphate; anionic; detergent intermediate for cleaners and emulsifier for min. oils and pesticides; liq.; dens. 8.50 lb/gal; 98–100% act.

Fosterge LF Acid. [Henkel] Alkylphenoxy POE acid phosphate; anionic; intermediate for foaming cleaners and solv. emulsion cleaners; emulsifier for pesticides; corrosion inhibitor; liq.; dens. 9.17 lb/gal; 98–100% act.

Fosterge LFS. [Henkel] Phosphate ester; emulsifier, dispersant, detergent.

Fosterge R Acid. [Henkel] Mono and dialkyl phosphoric acid; anionic; intermediate for emulsifying, penetrating, and anticorrosion compds.; dens. 8.5 lb/gal; liq.; 100% act.

Fosterge RD. [Henkel] Fosterge R Acid DEA salt; anionic; emulsifier, penetrant, and anticorrosion agent used in industrial cleaners; liq.; dens. 9.15 lb/gal; 86–88% act.

Fosterge W Acid. [Henkel] Alkyl phenoxy POE phosphoric acid; anionic; detergent, emulsifier, coupling agent and corrosion inhibitor; dedusting agent for alkaline powds.; liq.; dens. 9.29 lb/gal; 100% act.

Fostex AMP. [Henkel] Amino trimethylene phosphonic acid; anionic; scale inhibitor and sequestrant in water treatment; liq.; 50% act. in water.

Fostex E. [Henkel] Polymeric alkyl phosphonic acid; scale inhibitor and sequestrant; aq. sol'n.

Fostex P. [Henkel] Etidronic acid; anionic; scale inhibitor and sequestrant for water treatment; liq.; misc. in water; 59% act.

Fostex S. [Henkel] Polymeric alkyl phosphonate; anionic; scale inhibitor and dispersant for water treating applics.; liq.; 48% act. in water.

Fostex SN. [Henkel] Polymeric alkyl phosphonate sodium salt; anionic; see Fostex S; liq.; 43% act.

Fostex T. [Henkel] Hydroxyamine phosphate ester; anionic; acid intermediate for scale inhibitor; corrosion inhibitor; liq.; 74% act.

Fostex TN. [Henkel] Hydroxyamine phosphate ester sodium salt; see Fostex S; liq.; biodeg.; 70% act.

Fostex TX. [Henkel] Hydroxyamine phosphate ester amine salt; scale and corrosion inhibitor; dispersant for water treatment; liq.; 71% act.

Fostex U. [Henkel] Organophosphonic acid; anionic; scale inhibitor and sequestrant for water treatment; liq.; 53% act. in water.

FR-20. [Dead Sea Bromine] Magnesium hydroxide; flame retardant and smoke suppressant for ABS, PP, PS, rubbers.

FR-300-BA. [Dow] Decabromodiphenyl oxide; flame retardant combining high bromine content with excellent thermal stability and inertness in variety of substrates; used in PS, ABS, nylon, PPO, polyester resins and fibers, epoxy resins, phenolics, PC; polyolefins, PVC, flexible and rigid urethanes, rubber, and textile coatings; wh. powd.; m.w. 960; sol. 20 ppb in water; m.p. 285 C min.; 81–83% bromine.

FR-513. [Dead Sea Bromine] Tribromoneopentyl alcohol; flame retardant for flexible and rigid PU; flame retardant intermediate.

FR-521. [Dead Sea Bromine] Dibromoneopentyl glycol; flame retardant for rigid PU; flame retardant intermediate; liq.

FR-522. [Dead Sea Bromine] Dibromoneopentyl glycol; flame retardant for unsat. polyesers, rigid PU and foams; flame retardant intermediate; wh. cryst. powd.; m.w. 261.94; sol. (g/100 g solv.) 83 g in acetone; 52 g in IPA; 2 g in water; 0.5 g in CCl_4 and xylene; m.p. 109.0 C min.; 60.0% min. bromine.

FR-612. [Dead Sea Bromine] Dibromophenol; flame retardant for epoxy resins, phenolic resins, polyester resins; flame retardant intermediate; pink solid; m.w. 251.92; highly sol. in acetone, benzene, CCl_4, ether, n-heptane, methanol, etc.; < 0.1 g/100 g in water; m.p. 36 C min.; 63–64% bromine.

FR-613. [Dead Sea Bromine] 2,4,6-Tribromophenol; reactive flame retardant used mainly as an intermediate for polymeric flame retardants; off-wh. flakes; m.w. 330.8; 72.47% bromine.

FR-651-A. [Dow] Pentabromochlorocyclohexane; flame retardant additive used in polymer systems requiring low temp. processing; for foamed PS beads, board stock; and in epoxy and polyester resins, polyolefins, and rubber prods.; wh. cryst. solid; m.w. 513.09; sol. g/100 g solv. 7 g benzene, 5 g acetone, 4 g ether, 3 g n-heptane, 3 g methanol, < 0.1 g water; 77.87% bromine; 6.91% chloride.

FR-705. [Dead Sea Bromine] Pentabromotoluene; flame retardant for unsat. polyesters, polyethylene, PP, PS, SBR latex, textiles, rubbers.

FR-910. [Dead Sea Bromine] Nonhalogenic flame retardant for PP.

FR-913. [Dead Sea Bromine] Tribromophenyl allyl ether; aromatic flame retardant for expandable PS; synergist with hexabromocyclododecane; wh. cryst. free-flowing powd.; m.w. 371; sol. (g/100 g): 130 g in methylene chloride, 125 g in benzene, 77 g in dichloroethane, 67 g in CCl_4, 32 g in acetone; sp.gr. 2.2; m.p. 75–76.5; 64.7% Br.

FR-1025. [Dead Sea Bromine] Poly (pentabromobenzyl) acrylate; polymeric flame retardant for engineering thermoplastics, PET, PBT, nylon, PP, and PS; wh. powd.; m.w. 80,000; insol. in the common org. solvs. (aliphatic, aromatic, and chlorinated hydrocarbons, noncyclic and cyclic ethers, and ketones, DMF, and DMSO); dens. ≈ 2.05 g/cm³; m.p. 205–215 C; 70–71% bromine.

FR-1138. [Dow] Dibromoneopentyl glycol; flame retardant intermediate for unsat. polyesters, cellulose acetate, epoxy resins, PC, flexible and rigid urethanes, urethane and textile coatings, and reactive FR intermediate; wh. flakes; m.w. 262; sol. g/100 g solv. 82.5 g acetone, 60.3 g MEK, 52.0 g 2-propanol, 100 g water (100 C); 102.4 g methanol; m.p. 90–110 C; 61% bromine.

FR-1205. [Dead Sea Bromine] Pentabromodiphenyl oxide; flame retardant for use in laminates (both epoxy and phenolic), unsat. polyesters, syn. fibers, and flexible PU foams; suitable for textiles; liq.; m.w. 565; sol. in CCl_4, methylene chloride, benzene, acetone, freon, and most polyols; insol. in water; slightly sol. in methanol; dens. 2.25 g/nil; visc. 2650 cps (40 C); 69% bromine.

FR-1206. [Dead Sea Bromine] Hexabromocyclododecane; fire retardant for wide range of plastics, textiles, adhesives, and coatings; esp. for styrene-based systems; wh. to off-wh. cryst. free-flowing powd.; m.w. 641.7; sol. (g/100 ml): 10 g in styrene, 7 g in acetone, 6.5 g in toluene, < 0.1 g in water; sp.gr. 2.18; m.p. 180–190 C; 74.7% Br.

FR-1208. [Dead Sea Bromine] Octabromodiphenyl oxide; flame retardant for thermoplastics, e.g., ABS, HIPS, LDPE, PP random copolymer; recommended for inj. moldings; off-wh. cryst. powd.; m.w. 802; m.p. 125–165 C; 79.8% bromine.

FR-1210. [Dead Sea Bromine] Decabromodiphenyl oxide; flame retardant used in thermoplastics and fibers, incl. HIPS, glass-reinforced thermoplastic polyester molding resins, LDPE extrusion coatings, PP (homo and copolymers), ABS, nylon, PBT, PET, PU, SBR latex, textiles, rubber; wh. to off-wh. fine cryst. powd.; m.w. 960; sol. (g/100 g solv.) < 1 g in acetone, benzene, chlorobenzene, methylene bromide, methylene chloride, *o*-xylene; m.p. 300–305 C; 82% min. bromine.

FR-1215. [Dead Sea Bromine] Flame retardant for

flexible PU.

FR-1524. [Dead Sea Bromine] Tetrabromobisphenol-A; reactive flame retardant used in the mfg. of epoxy, PC, ABS, phenolic, PS, and polyester resins, rubber; flame retardant intermediate; wh. cryst. powd.; m.w. 543.88; sol. (g/100 g solv.) 240 g in acetone; 92 g in methanol; 0.3 g in benzene, 0.01 g in water; m.p. 181–182 C; 58.77% bromine.

FR-1525. [Dead Sea Bromine] Tetrabromobisphenol-A ethoxylate; flame retardant for nylon, PBT, PET, unsat. polyesters; flame retardant intermediate.

FR-1540. [Dow] Brominated polyester conc. in styrene; reactive additive flame retardant for unsat. polyester resins; amber visc. liq.; sp.gr. 1.46; dens. 12.2 lb/gal; visc. 5000–8000 cps; 33% bromine; 25% styrene.

FR-2124. [Dead Sea Bromine] Tetrabromobisphenol-A allyl ether; flame retardant for expandable PS.

FR-D. [FMC] Phosphorus-based diol; reactive flame retardant; effective for rigid PU foam; suitable for pour-in-place, slabstock, and spray formulations; flame retardant/friability modifier in PUR-modified isocyanurate foams; does not plasticize or affect catalyst activity; clear liq.; b.p. 235 C; dens. 1.10 g/ml; visc. 12,000 cps; 14% phosphorus.

Franklin T-11. [Franklin Limestone] Calcium limestone; filler for the plastics and rubber industries; natural beige; 4 μ mean particle size; 5–6 Hegman grind; 0.005% max. retained on US # 325; sp.gr. 2.74; pH 9.0–9.5; hardness 3.

Franklin T-13. [Franklin Limestone] Calcium limestone; see Franklin T-11; neutral beige; 6 μ mean particle size; 3–3$^1/_2$ Hegman grind; 0.15% max. retained on US # 325; sp.gr. 2.74; pH 9.0–9.5; hardness 3.

Franklin T-14. [Franklin Limestone] Calcium limestone; see Franklin T-11; neutral beige; 7 μ mean particle size; 2$^1/_2$–3 Hegman grind; 0.5% max. retained on US # 325; sp.gr. 2.74; pH 9.0–9.5; hardness 3.

Franklin T-325. [Franklin Limestone] Calcium limestone; see Franklin T-11; neutral beige; 10 μ mean particle size; 1$^1/_2$–2$^1/_2$ Hegman grind; 5.0% max. retained on US # 325; 0.08% max. retained on US # 200; sp.gr. 2.74; pH 9.0–9.5; hardness 3.

FRE. [Solem] Alumina trihydrate; filler providing low cost flame and smoke suppression in fiberglass-reinforced polyester applic; particulate; particle size 12–15 μ; sp.gr. 2.5; dens. 0.8g/cc; oil absorp. 25 ml/100 g.

Frigen 113 TR-N. [Hoechst-Celanese AG] Fluorinated hydrocarbon; solv. for removing org. and inorg. pollution.

Frigid-Go® 2815, 2816. [Henkel/Emery] Moly low-temp. multipurpose syn. grease; visc. 24.1 cSt (40 C); flash pt. 455 F; pour pt. –80 F.

Frimulsion 6G. [Hercules/PFW] Pectin, dextrose, carrageenan, sucrose, potassium chloride, locust bean gum, potassium citrate, blend; stabilizer and gelling agent for food systems; effective jellification at 0.4–1.0% by wt.; off-wh. powd., < 1% retained on 0.25-mm mesh; pH 6–7 (1%); 13% max. moisture.

Frimulsion 10. [Hercules/PFW] Refined and standardized locust bean gum and guar gum, blend; stabilizer and water-binding agent in food systems; good disp. properties; off-wh. powd., < 1% retained on 0.1-mm mesh; visc. 3000 ± 300 cps (1% aq.); pH 5–6; 13% max. moisture.

Frimulsion Q8. [Hercules/PFW] Modified food starch, gelatin, locust bean gum, guar gum, pectin, and dextrose, blend; stabilizer, mainly in heat-processed, low-fat skim milk cheese; particulate, < 1% retained on 0.25-mm mesh; pH 5–6; 13% max. moisture.

Frimulsion RA. [Hercules/PFW] Standardized blend of starch, gelatin, pectin, and dextrose; stabilizer in Swiss-style yogurt with act. cultures; added to milk before pasturization and incubation; particulate, < 1% retained on 25-mm mesh; pH 5.5–6.5; 13% max. moisture.

Frimulsion RF. [Hercules/PFW] Standardized blend of native starch and gelatin; stabilizer and texturizing agent for Swiss-style yogurt at levels of 1–2%; added to milk before pasturization and incubation; particulate, < 1% retained on 0.25-mm mesh; pH 5–6; 12% max. moisture.

Frimulsion X5. [Hercules/PFW] Standardized blend of refined guar gum and locust bean gum; stabilizer and thickener for food systems; off-wh. powd., < 1% retained on 0.18-mm mesh; visc. 3000 ± 300 cps; pH 5–6 (1% aq.); 13% max. moisture.

Fruvit®. [Bayer] Propineb; fungicide for control of downy mildews; wh. powd.; m.p. 150 C (decomp.).

Fuelsaver. [Angus] Morpholine derivs.; preservative for diesel and other hydrocarbon fuels; liq.; oil-sol.; mod. water-sol.

Fungamyl®. [Novo] Fungal alpha-amylase derived from Aspergillus oryzae; enzyme used in starch industry for prod. of high maltose syrups and high DE syrups; in brewing industry to increase fermentability of the wort; in alcohol industry for liquefaction of starch in a distillery mash; and in baking industry for supplementation of the alpha-amylase in flour for yeast-leavened doughs in order to increase the formation of fermentable sugars; activity 180–1600 FAU/g; brn. liq., lt. amber and ylsh. powd.; odor typ. of fermentation prods.; water-sol.; dens. 1.25 g/ml; pH 6–7.

Fungamyl® 800L. [Novo] Fungal alpha-amylase; enzyme for starch industry, brewing; liq.

Fungamyl® 1600S. [Novo] Fungal alpha-amylase; enzyme for baking; powd.

Fungicide ZV. [Sandoz] Zinc salts of dimethydithiocarbamic acid and 2-mercaptobenzothiazole; nonionic; mildewcide, fungicide, bacterial inhibitor for textiles, paper; wh. soft paste; disp. in water; pH 7.0–7.5.

Fungitex R. [Ciba-Geigy] Dodecyldimethyl (2-phenoxyethyl) ammonium bromide; antimicrobial, fungicide for use in mouthwashes, antiseptics, cold sterilization; water-sol.

Fungitrol 11. [Servo] Org. compd.; fungicide, bactericide for paint industry; powd.

Fyarestor 100. [Witco/Argus] Bromochlorinated

paraffin; flame retardant which has replaced chlorinated paraffins used in treating outdoor fabrics; water resistance, and some plasticizing; excellent weatherability, compatible in water- and solv.-based systems.

G

G-4 Pure. [Givaudan] Dichlorophene; antimicrobial agent in veterinary applics.; sol. in alcohol; water-insol.

G-4 Tech. [Givaudan] Dichlorophene, tech.; industrial fungicide, bactericide for textiles; lt. tan powd.; weak phenolic odor; sol. (g/100 ml solv.) 80 g acetone, 75 g MEK, 60 g t-butyl alcohol, 54 g IPA, 53 g ethanol, 45 g propylene glycol, 43 g n-butyl alcohol; m.p. 164 C min.

G-431. [Solem] Alumina trihydrate; filler featuring flame retarding and smoke suppressing properties; resin extender in polyester, vinyls, PU, latex, neoprene foam systems, wire and cable insulation, vinyl wall and floor coverings; fair processing chars.; builds visc. rapidly; suspension properties; used in epoxy elec. laminates, epoxy encapsulating/castings, FRP elec. laminates, SMC/BMC, phenolic compr. molding; particulate; median particle diam. 8–10 μ, 99.5% thru 325 mesh; sp.gr. 2.42; dens. 0.6–0.7 g/cm^3; oil absorp. 27–29; ref. index 1.57; 65.0% Al_2O_3.

Gafac® GB-520. [Rhone-Poulenc Surf.] Partial sodium salt of a complex org. phosphate ester; anionic; textile antistat, lubricant, softener, emulsifier for fiber finishes; hazy visc. liq.; sol. in xylene, butyl Cellosolve, perchloroethylene, ethanol; disp. water; dens. 8.6 lb/gal; sp.gr. 1.03–1.04; pour pt. 9 C; pH 2.5–4.5 (10%); 98% act.

Gafac® LO-529. [Rhone-Poulenc Surf.] Sodium nonoxynol-6 phosphate; anionic; detergent for household and industrial formulations; rust and corrosion retarder for detergent conc.; wax and floor finishes; antistat for acrylic carpet yarn and drycleaning detergents; clear visc. liq.; sol. in kerosene, Stod., xylene, butyl Cellosolve, perchloroethylene, water; dens. 9.1 lb/gal; sp.gr. 1.05–1.15; pour pt. 6 C; pH 5–6 (10%); 88% act.

Gafac® MC-470. [Rhone-Poulenc Surf.] Partial sodium salt of a complex org. phosphate ester; anionic; household and industrial detergent, textile antistat, lubricant, emulsifier; clear visc. liq.; sol. in min. oil, kerosene, xylene, perchloroethylene, disp. in water; dens. 8.6 lb/gal; sp.gr. 1.02–1.04; pour pt. 0 C; pH 5–6.5 (10%); 95% act.

Gafac® RA-600. [Rhone-Poulenc Surf.] Deceth-4 phosphate; anionic; detergent, foamer, dispersant, wetting agent for household, industrial, textile wet processing; coupler used in liq. alkali detergents; clear to slightly hazy visc. liq.; sol. in kerosene, xylene, butyl Cellosolve, perchloroethylene, ethanol, water; dens. 8.9 lb/gal; sp.gr. 1.06–1.08; pour pt. < 0 C; pH < 2.5 (10%); 98.50% act.

Gafac® RB-400. [Rhone-Poulenc Surf.] Free acid of complex org. phosphate ester; anionic; lubricant, antistat, emulsifier for fibers and metal; opaque visc. liq.; sol. in min. oil, kerosene, Stod., xylene, butyl Cellosolve, perchloroethylene, ethanol; disp. in water; dens. 8.6 lb/gal; sp.gr. 1.03–1.04; pour pt. 18 C; pH < 2.5 (10%); 98% act.

Gafac® RD-510. [Rhone-Poulenc Surf.] Free acid of complex org. phosphate ester; anionic; emulsifier, antistat, lubricant, solubilizer for fibers, metals, cosmetics; clear visc. liq.; sol. in min. oil, kerosene, xylene, butyl Cellosolve, perchloroethylene, ethanol; disp. in water; dens. 8.8 lb/gal; sp.gr. 1.05–1.06; pour pt. 13 C; pH < 2.5 (10%); 98% act.

Gafac® RE-410. [Rhone-Poulenc Surf.] Free acid of complex org. phosphate ester; anionic; detergent, emulsifier, wetting agent, dispersant, antistat, lubricant for pesticides, drycleaning, textile wet processing, metals; clear visc. liq.; sol. in kerosene, Stod., xylene, butyl Cellosolve, perchloroethylene, ethanol; disp. in water; dens. 9.1 lb/gal; sp.gr. 1.00–1.20; pour pt. 18 C; pH < 2.5 (10%); 100% act.

Gafac® RE-610. [Rhone-Poulenc Surf.] Nonoxynol-9 phosphate; anionic; see Gafac RE-410; also for household and industrial detergents, fabric finishes, emulsion polymerization; slightly hazy visc. liq.; sol. in xylene, butyl Cellosolve, perchloroethylene, ethanol, water; dens. 9.2 lb/gal; sp.gr. 1.1–1.12; pour pt. < 0 C; 100% act.

Gafac® RE-870. [Rhone-Poulenc Surf.] Free acid of complex org. phosphate ester; anionic; see Gafac RE-410; also for cosmetics; soft waxy paste; sol. in xylene, butyl Cellosolve, perchloroethylene, ethanol, water; dens. 9.6 lb/gal; sp.gr. 1.1–1.2 (50 C); pour pt. 29 C; pH < 2.5 (10%); 99% act.

Gafac® RE-877. [Rhone-Poulenc Surf.] Phosphate ester; emulsifier, solubilizer, detergent, dispersant, wetting agent, antistat, lubricant for pesticides, textile wet processing, fiber and metal lubricants, emulsion polymerization, cosmetics; clear visc. liq.; sol. in xylene, butyl Cellosolve, ethanol, water; sol. hazy in perchloroethylene; sp.gr. 1.155; dens. 9.6 lb/gal; pour pt. 2 C; acid no. 60–74; pH < 2.5 (10%); 75% act.

Gafac® RE-960. [Rhone-Poulenc Surf.] Free acid of complex org. phosphate ester; anionic; see Gafac RE-410; also for emulsion polymerization; soft waxy paste; sol. in butyl Cellosolve, ethanol, water; dens. 9.8 lb/gal; sp.gr. 1.17–1.18 (50 C); pour pt. 20 C; pH < 2.5 (10%); 90% act.

Gafac® RK-500. [Rhone-Poulenc Surf.] Free acid of complex org. phosphate ester; anionic; lubricant, antistat, emulsifier for metal treatment; clear, visc. liq.; sol. in xylene, butyl Cellosolve, perchloroethylene, ethanol, water; dens. 9.1 lb/gal; sp.gr. 1.05–1.15; pour pt. < 0 C; pH < 2.5 (10%); 100% act.

Gafac® RM-410. [Rhone-Poulenc Surf.] Nonyl nonoxynol-7 phosphate; anionic; detergent, emulsifier for drycleaning, pesticides; rust inhibitor; slightly hazy, visc. liq.; sol. in min. oil, kerosene, Stod., xylene, butyl Cellosolve, perchloroethylene, ethanol; disp. in water; dens. 8.8 lb/gal; sp.gr. 1.05–1.07; pour pt. 19 C; pH < 2.5 (10%); 100% act.

Gafac® RM-510. [Rhone-Poulenc Surf.] Nonyl nonoxynol-10 phosphate; anionic; emulsifier, solubilizer for pesticides, cosmetics; hazy, visc. liq.; sol. in kerosene, Stod., xylene, butyl Cellosolve, perchloroethylene, ethanol; disp. in water; dens. 8.8 lb/gal; sp.gr. 1.05–1.07; pour pt. 5 C; pH < 2.5 (10%); 100% act.

Gafac® RM-710. [Rhone-Poulenc Surf.] Nonyl nonoxynol-15 phosphate; anionic; emulsifier, solubilizer for cosmetics; hazy, visc. liq.; sol. in xylene, butyl Cellosolve, perchloroethylene, ethanol, water; dens. 8.9 lb/gal; sp.gr. 1.06–1.08; pour pt. 15 C; acid no. 33–41; pH < 3 (10%); 100% act.

Gafac® RS-410. [Rhone-Poulenc Surf.] Free acid of complex org. phosphate ester; anionic; pesticide emulsifier, drycleaning detergent, textile wetting agent, antistat, lubricant for fiber and metal treatment; hazy visc. liq.; sol. in min. oil, kerosene, Stod., xylene, butyl Cellosolve, perchloroethylene, ethanol, disp. in water; dens. 8.6 lb/gal; sp.gr. 1.03–1.04; pour pt. –15 C; pH < 2.5 (10%); 100% act.

Gafac® RS-610. [Rhone-Poulenc Surf.] Trideceth-6 phosphate; anionic; see Gafac RS-410; also penetrant; haxy, visc. liq.; sol. in kerosene, Stod., xylene, butyl Cellosolve, perchloroethylene, ethanol; disp. in water; dens. 8.7 lb/gal; sp.gr. 1.04–1.06; pour pt. < 0 C; pH < 2.5 (10%); 100% act.

Gafac® RS-710. [Rhone-Poulenc Surf.] Free acid of complex org. phosphate ester; anionic; see Gafac RS-410; opaque visc. liq.; sol. in Stod., xylene, butyl Cellosolve, perchloroethylene, ethanol, water; dens. 8.7 lb/gal; sp.gr. 1.04–1.06; pour pt. 18 C; pH < 2.5 (10%); 100% act.

Gafamide® CDD-518. [Rhone-Poulenc Surf.] Cocamide DEA; anionic; foam stabilizer, visc. builder, detergent, wetting agent used in shampoos, bubble baths, laundry, industrial cleaners; straw clear visc. liq.; little odor; sol. in water, ethanol, perchloroethylene, butyl Cellosolve; sp.gr. 1.0; visc. 1050 cst (Ostwald-Fenske); pH 10 (1%); 84% act.

Gaffix® VC-713. [Rhone-Poulenc Surf.] Vinylcaprolactam/PVP/dimethylaminoethyl methacrylate copolymer; film-forming, fixative resin for use in mousses, gels, glazes, lotions, and hairsprays; APHA 100 max. clear visc. liq., ethanolic odor; sol. in water and alcohol; b.p. 173 F; cloud pt. 44 C (1% aq.); flash pt. (CC) 55 F; 37% solids in ethanol.

GAF Latex 6000 Series. [GAF] Carboxylated SBR copolymers/S/B/acrylonitrile terpolymers; binders for fabric and paper prods.; dens. 8.4 lb/gal; visc. 84 cps; pH 8.0–9.0; 48–50% solids.

GAF Latex 7000 Series. [GAF] S/B latices; binders for padding applic.; dens. 8.47 lb/gal; visc. 200 cps; pH 9.0–9.5; 47–49% solids.

GAF Latex 8000 Series. [GAF] S/B latices; binders in pigment printing applic.; dens. 8.37 lb/gal; visc. 50 cps; pH 9.0–9.5; 37–40% solids.

Gafquat® 734. [Rhone-Poulenc Surf.] Polyquaternium-11; film-forming substantive polymer, conditioner for formulation of hair conditioners, rinses, sprays, shampoos, dyes, semipermanents, deodorants, antiperspirants, shaving preparations, antiseptics, toilet soaps, skin creams, sunburn remedies; 50% alcoholic sol'n.

Gafquat® 755. [Rhone-Poulenc Surf.] Polyquaternium-11; high m.w. film-forming polymer for superior hair and skin care prods., deodorants, antipersipirants, shaving preparations, antiseptics, toilet soaps, skin creams, sunburn remedies; sol'n.; 20% aq. sol'n.

Gafquat® 755N. [Rhone-Poulenc Surf.] Polyquaternium-11; high m.w.; see Gafquat 734; liq.; 20% aq. sol'n.

Gafsoft 325. [Rhone-Poulenc Surf.] Amine condensate; softener for textile finishing systems; solid.

Gafstat S. [Rhone-Poulenc Surf.] Nitrogen compd., modified; cationic; textile finishing agent; antistat for fabrics and fibers; amber clear liq.; water-sol.; pH 6–7; 24% min. solids.

Gafstat S-100. [Rhone-Poulenc Surf.] Nitrogen compd., modified; cationic; antistat for textiles; liq.

Gafterge AW-123. [Rhone-Poulenc Surf.] Phosphate ester; anionic; surfactant, wetting agent, penetrant, and dispersant used in textile applics., dyeing; liq.; biodeg.

Gaftex 288. [GAF] Sulfated alkyl ester; anionic; penetrant, wetting and leveling agent with softening properties for textile dyeing; liq.; sol. in water.

Gaftex CD-169. [GAF] Phosphate ester deriv. complex; anionic; surfactant, penetrant, dispersant, and wetting agent used in textile dyeing; liq.; biodeg.

Gaftex CK-160 Conc. [GAF] Polyether; nonionic; leveling agent, antiprecipitant; differential dyeing assistant for continuous dyeing of carpets on Kuster machines; excellent contrast; applic. in beck and hosiery dyeing; liq.

Gaftex COM-154, -154 Conc. [GAF] Polyether; nonionic; wetter and dispersant for cationic and disperse dyes on acrylics; compatibilizes cationic and disperse dyes; provides clear, bright shades in cross dyeings; emulsifies residual oil and lubricant in dyebath; liq.

Gaftex DN-159 Conc. [GAF] Polyether; nonionic; leveling agent for disperse dyes on nylon; prevents shaded dyeings due to temp. variations; promotes uniform dye transfer and full color development; low foaming biodeg. lubricant; liq.

Gaftex KF. [GAF] Sulfated alkyl ester; anionic; dye assistant and leveling agent for continuous dyeing of carpets and yarns; optimum foam levels in continuous dyeing of solid shades of carpets; leveling agent in thermosol dyeing of polyester; effective penetration of yarn loops on carpet back; total rinsing even in cold water; liq.

Gaftex LN-110. [GAF] Polyamide polyether; modified nonionic; leveling-migrating agent in atmospheric and pressure dyeing of nylon yarn, piece goods, and carpeting; compatibilizes anionic/cationic dyes in a single bath; also in repair dyeing; liq.
Gaftex PC-150. [GAF] Sulfate ester; anionic; leveling agent for disperse dyes on polyester; minimizes trimer deposits on polyester pkgs.; reduces disperse dye staining on cotton in dyeing of blended fibers; liq.
Gaftex PT. [GAF] Modified copolymer of methyl vinyl ether and maleic anhydride; thickener in aq. media; effective in textile print paste systems; powd.
Gaftex WSP. [GAF] Sulfonated org. ester; anionic; wetting agent and penetrant used in dyeing; liq.
Gaftex ZP. [GAF] Modified polyethoxylated complex; nonionic; leveling agent for printing nylon carpeting; liq.
Galactasol Series. [Henkel] Guar gums and derivs.; gums used to increase visc. in water and most brines; high visc. grades also as flocculants; some in series as stabilizers, suspending aids.
Gamaco®. [Georgia Marble] Calcium carbonate slurry; extender pigment; may be used in food contact applics.; sp.gr. 1.79 (slurry), 2.71 (solids); dens. 1.49 lb/gal; visc. 1000–2000 cps; 70 ± 1% solids by wt.
Gamaco® II. [Georgia Marble] Calcium carbonate; extender fillers used in liq. polyester systems, BMC/SMC and similar applic.; particulate; 3 μ median particle size; 0.005% retained on #325 wet screen; oil absorp. 14–16%.
Gama-Fil 55. [Georgia Marble] Calcium carbonate; filler; very wh.; 0.8 μ median particle size; visc. 300–500 cps; GE brightness 95.
Gama-Fil 90. [Georgia Marble] Calcium carbonate; filler; very wh. 1.8 μ median particle size; visc. 300–500 cps; GE brightness 95.
Gama-Fil D-1. [Georgia Marble] Calcium carbonate, dry; ultra fine filler; 1.0 μ median particle size; GE brightness 95.
Gama-Fil D-2. [Georgia Marble] Calcium carbonate, dry, wet ground, nondispersed; ultra fine filler for use in plasrtics, BMC/SMC, paint, caulks, sealants, adhesives, paper, foam urethane, modified acrylics, filled thermosets/thermoplastics, and rubber; NSF approved for potable water materials; 2.0 μ median particle size; Hegman grind 6; oil absorp. 16 ± 1; dry brightness 95; 96% min. conc.
Gamanase 1.5L. [Novo] Galactomannase; enzyme for visc. reduction in coffee extracts, other visc. reduction applics.; liq.
Gama-Plas®. [Georgia Marble] Dry ground calcium carbonate; filler for use in glass fiber-reinforced polyester compds. (BMC, SMC, TMC, and wet mat molding); wh. powd.; 7.0 μ median particle size; 0.20% retained on #325 wet screen; visc. 36,000 cps; dry brightness 94.
Gama-Sperse® 5. [Georgia Marble] Calcium carbonate; filler for paint, plastics, caulks, sealants, adhesives, foam urethane, filled thermosets/thermoplastics, BMC, and rubber applics.; 5.0 μ median particle size; 0.005% retained on #325 wet screen; Hegman grind 5.5; oil absorp. 14 ± 1; dry brightness 94; 96% conc.
Gama-Sperse® 80. [Georgia Marble] Natural ground calcium carbonate; general-use filler for paints, plastics, paper coatings, and rubber prods.; 2.3 μ median particle size; 0.005% retained on #325 wet screen; $6^3/_4$–7 Hegman grind; sp.gr. 2.71; bulking value 0.443 gal/lb; oil absorp. 18–20 lb oil/100 lb pigment; dry brightness 96%; pH 9.0–9.5; hardness (Moh) 3; 95.0% min. total carbonates.
Gama-Sperse® 140. [Georgia Marble] Natural ground calcium carbonate; see Gama-Sperse 80; 0.005% retained on #325 wet screen; $6^1/_4$–$6^1/_2$ Hegman grind; sp.gr. 2.71; bulking value 0.443 gal/lb; oil absorp. 15–17 lb oil/100 lb pigment; dry brightness 96%; pH 9.0–9.5; hardness (Moh) 3; 95% min. total carbonates.
Gama-Sperse® 255. [Georgia Marble] Natural ground calcium carbonate; see Gama-Sperse 80; 12.0 μ median particle size; 0.02% retained on #325 wet screen; $3^3/_4$–$4^1/_4$ Hegman grind; sp.gr. 2.71; bulking value 0.443 gal/lb; oil absorp. 10–12 lb oil/100 lb pigment; dry brightness 94%; pH 9.0–9.5; hardness (Moh) 3; 95.0% min. total carbonates.
Gama-Sperse® 6451. [Georgia Marble] Natural ground calcium carbonate; see Gama-Sperse 80; 7.8 μ median particle size; 0.008% retained on #325 wet screen; 5–$5^1/_2$ Hegman grind; sp.gr. 2.71; bulking value 0.443 gal/lb; oil absorp. 12–14 lb oil/100 lb pigment; dry brightness 96%; pH 9.0–9.5; hardness (Moh) 3; 95.0% min. total carbonates.
Gama-Sperse® 6532. [Georgia Marble] Natural ground calcium carbonate; see Gama-Sperse 80; 4.3 μ median particle size; 0.005% retained on #325 wet screen; 6–$6^1/_4$ Hegman grind; sp.gr. 2.71; bulking value 0.443 gal/lb; oil absorp. 15–17 lb oil/100 lb pigment; dry brightness 96%; pH 9.0–9.5; hardness (Moh) 3; 95.0% min. total carbonates.
Gama-Sperse® 6532 NSF. [Georgia Marble] Calcium carbonate; filler for plastics; 3.0 μ median particle size; 0.005% retained on #325 wet screen; dry brightness 94.
Gama-Sperse® CS-11. [Georgia Marble] Calcium stearate surface-modified calcium carbonate; filler for plastics and other compds., for use in food contact applics.; very wh. powd.; 4.0 μ mean avg. particle size; 0.005% max. retained on # 325 wet screen; sp.gr. 2.7; bulking value 0.443 gal/lb; hardness (Moh) 3.0.
Ganex® P. [GAF] Alkylated vinylpyrrolidone polymer; used in cosmetics and toiletries, as pigment dispersants and additives in film-forming resins; dispersants for pigments; solubilizers for dyes; petrol. additive, sludge and detergent dispersants, pour pt. depressants; protective colloids in coatings; suspending aids in polymerization; syn. sizing additive; dye assistant; antiredepositon agent in drycleaning solvs.; liq. to waxy solid; sol. in min. oil, org. solvs., other polymers.
Ganex® P904. [GAF] Butylated PVP polymer; used in cosmetics and toiletries as moisture barrier, adhesive, protective colloid, and microencapsulating resin; as dispersant for pigments; as solubilizer for

dyes; in petroleum industry as sludge and detergent dispersant and pour pt. depressant; protective colloid in coatings; suspending aid in polymerization; dyeing assistant; antiredeposition agent in drycleaning; additive for radiation curable inks and coatings; P-904 is the water-sol. member of series; esp. as dispersant in aq. agric. chemical formulations or pigmented skin care prods.; sol'n.; m.w.. 16,000; sol. in water; 45% solids in IPA, M-Pyrol, or water.

Ganex® V. [GAF] Alkylated vinylpyrrolidone copolymer; see Ganex P; liq. to waxy and gran. solids.; sol. in min. oil, org. solvs, other polymers.

Ganex® V216. [GAF] PVP/hexadecene copolymer; see Ganex P904; visc. liq.; m.w. 7300; sol. in min. oil, org. solvs., and other polymers; 100% act.

Ganex® V220. [GAF] PVP/eicosene copolymer; see Ganex P904; waxy solid; m.w. 8600; sol. in min. oil, org. solvs., and other polymers; 100% act.

Ganex® V516. [GAF] PVP/hexadecene copolymer; see Ganex P904; sol'n.; m.w. 9500; sol. in min. oil, org. solvs., and other polymers; 55% solids in IPA.

Gantrez® AN. [GAF] PVM/MA copolymer; anionic; dispersant, coupling, stabilizing agent, thickener, emulsifier, solubilizer, corrosion inhibitor, film former, antistat, used in agric., paper and textile industries, chemical processing, industrial prods., detergents, cosmetics; wh. fluffy powd.; sol. in water, acid, and several org. solvs.; dens. 0.32 g/cc; sp.gr. 1.37; soften. pt. 200–225 C; pH 2 (free acid, 5% aq.); 100% conc.

Gantrez® AN-119, -139, -149, –169, –179. [GAF] PVM/MA copolymer; thickener, dispersant and stabilizer used in emulsion polymerization; gelling agent, coupler, and suspending aid; produces clear films of high tens. and cohesive str.; used in adhesives, detergents; AN-119 is a low-visc. type; AN-139 and AN-149 are med.-visc.; AN-169 and AN-179 are high-visc.; powd.; sol. in water, acid, caustic, and org. solv.

Gantrez® B-773. [GAF] Poly(vinyl isobutyl ether); tacky polymer with excellent adhesion to plastic, metal, and coated surfs.; plasticizer and leveling agent for surf. coatings; sol. in aromatic, aliphatic, and chlorinated hydrocarbons; 70% sol'n. in hexane.

Gantrez® ES-225. [GAF] Ethyl ester of PVM/MA copolymer; anionic; copolymer forming clear, glossy films with substantivity and moisture resistance; used in hairsprays, mousses, gels and lotions, coatings, polishes; emulsion stabilizer in creams and lotions; clear, visc. sol'n.; sol. [1 g resin (100% solids) in 9 g solv.] in ethanol, IPA, diethylene glycol, THF, ethylene glycol monomethyl ether, butyl Carbitol, acetone, cyclohexanone, dioxane, water; sp.gr. 0.983; dens. 8.18 lb/gal; acid no. 275–300 (100% solids); 50% in ethanol.

Gantrez® ES-335. [GAF] Isopropyl ester of PVM/MA copolymer; anionic; see Gantrez ES-225; clear, visc. sol'n.; sol. in alcohols, alkali, esters, ketones, and glycol ethers; sp.gr. 0.957; dens. 7.98 lg/gal; acid no. 250–280 (100% solids); 50% in IPA.

Gantrez® ES-425. [GAF] n-Butyl ester of PVM/MA copolymer; anionic; see Gantrez ES-225; clear, visc. sol'n.; sol. [1 g resin (100% solids) in 9 g solv.] in ethanol, IPA, diethylene glycol, THF, ethylene glycol monomethyl ether, butyl Carbitol, ethyl acetate, acetone, cyclohexanone, dioxane, water; sp.gr. 0.977; dens. 8.13 lb/gal; acid no. 235–265 (100% solids); 50% solids in ethanol.

Gantrez® ES-435. [GAF] n-Butyl ester of PVM/MA copolymer; anionic; see Gantrez ES-225; clear, visc. sol'n.; sol. see Gantrez® ES-335; sp.gr. 0.962; dens. 8.02 lb/gal; acid no. 245–275 (100% solids); 50% in IPA.

Gantrez® M-154, M-555, M-556, M-574. [GAF] Poly(methyl vinyl ether); polymer functioning as tackifier, binder, and plasticizer; used in printing inks, textile sizes and finishes, latex modification; liqs.; sol. in water and diverse organic solvs.; thermally reversible solubility in aq. systems; 50% aq. sol'n. (M-154); 50% toluene sol'n. (M-555, M-556); 70% toluene sol'n. (M-574).

Gantrez® S-95. [GAF] Poly(methyl vinyl ether/maleic acid); hydrolyzed low m.w. polymer; water-sol. polyelectrolyte similar to Gantrez AN series; chelating agent; rapid cold-water solubility over entire pH range; powd.; pH 2 (5% aq.).

Gantrez® S-97. [GAF] Poly(methyl vinyl ether/maleic acid); see Gantrez S-95; powd.; water-sol.; pH 2 (5% aq.).

Gantrez® SP-215. [GAF] Ethyl ester of PVM/MA copolymer; hair fixative for stiffer, harder holding prods.; up to 10% can be formulated in pump hair sprays; lt. yel. to straw clear visc. liq.; 48–52% solids.

Garalease 915. [Ram] Cellulosic lacquer; film-forming release agent effective with epoxy, polyester, and urethane resins; blue; dens. 6.8 lb/gal; flash pt. 2 F (Seta CC); 96% volatiles.

Gardilene IPA. [Albright & Wilson] Isopropylamine alkylbenzene sulfonate; anionic; emulsifier, drycleaning assistant, dispersant; amber clear visc. liq.; pH 6.5; flam.; 85% act. in solv.

Gardilene S25L. [Albright & Wilson] Sodium alkylbenzene sulfonate; anionic; wetting, dispersing, and foaming agent, emulsifier in emulsion polymerization; pale yel. clear visc. liq.; visc. 3300 cps; clear pt. < 6 C; pH 7.2 (10%); 25 % act. in water.

Gardinol CX. [Albright & Wilson] Sodium cetyl/oleyl alcohol sulfate; anionic; detergent, wetting agent, dispersant, dyeing assistant and emulsifier used in fiber applics., hand cleaners, laundry detergents; off-wh. to cream soft paste, visc. liq.> 30 C; sp.gr. 1.05 (35 C); pH 7.0 ± 0.5 (10% aq.); > 80% biodeg.; 30.0 ± 1.0% act.

Gardiquat 12H. [Albright & Wilson] Benzalkonium chloride; bactericide in disinfectants and antiseptics; used in hospitals, food processing industries, and oil drilling muds; colorless to pale yel. clear liq.; sp.gr. 0.98; cloud pt. 0 C; pH 7.0±0.5 (1% aq.); 50.0± 1.0% act. in water.

Gardiquat 1450. [Albright & Wilson] Benzalkonium chloride USP; germicide, deodorant, algicide, slimicide; almost water wh. clear liq.; cloud pt. 1 C; pH 7.0 ± 0.5 (1%); flam.; 50% act.

Gardiquat 1480. [Albright & Wilson] Benzalkonium

chloride USP; germicide, deodorant, algicide, slimicide; almost water wh. clear liq.; sp.gr. 0.945; cloud pt. 1 C max.; flash pt. 32 C (PMCC); pH 7 ± 0.5 (1%); 80% act.

Gardiquat CQ. [Albright & Wilson] Base for disinfectants that will cloud on dilution with water; liq.

Gardiquat SV 480. [Albright & Wilson] Benzalkonium chloride; germicide, deodorant, algicide, slimicide; water wh. clear liq.; sp.gr. 0.929; visc. 106 s (#4 cup); cloud pt. < 0 C; flash pt. 29 C (PMCC); pH 7.2 (1%); 80% act.

Gardisperse AC. [Albright & Wilson] Sodium alkyl phenoxy POE sulfate; anionic; emulsifier, dispersant, wetting agent; amber clear liq.; visc. 120 cP max.; cloud pt. 2 C max.; pH 8.5 ± 1.0 (2%); 30.0 ± 1.0% act.

GC-44-14. [Stepan/PVO] Dipentaerythritol ester; lubricant, functional fluids for aviation, automotive, and military formulated lubricants and gas turbine engine oils; liq.; sp.gr. 1.01; visc. 56 cs (100 F); pour pt. < -80 F; flash pt. 565 F; fire pt. 625 F.

Gelatinase No. 53. [Miles Biotech] Protease; enzyme for film stripping; powd.

Gelcharg HP4. [Hercules] Hydroxypropyl guar; nonionic; high-visc., water-sol. thickener offering improved compatability with electrolytes and polar org. solvs. in dilution used in stabilizing o/w emulsions as protective colloid; may be reacted with borax or cross-linking agent to form tough, water-resistant gels; high-visc. liq.; visc. 15,000 cps (2%); pH 7.0; 10.0% moisture.

Gelot 64®. [Gattefosse] Glyceryl stearate and PEG-75 stearate; nonionic; SE base for o/w cosmetic and pharmaceutical emulsions; Gardner < 5 waxy solid; weak odor; HLB 10; m.p. 59–65 C; acid no. < 6; iodine no. < 3; sapon. no. 105–125.

Gelrite. [Canada Packers] Gelatin; gels and stabilizes foam for food use; thickener; hot water-sol.

Gemtex 18, 19, 162. [Finetex] EDTA type; chelating agent; metal ion control in textile, metal, and cosmetic use; water-sol.

Gemtex PA-75. [Finetex] Dioctyl sodium sulfosuccinate; wetting agent for textile uses and emulsion polymerization; dye leveler; bubble baths, bath oils; liq.; 75% conc.

Gemtex PA-85P, PAX-60. [Finetex] Dioctyl sodium sulfosuccinate, anhyd.; see Gemtex PA-75; liq.; 85 and 60% conc. resp.

Gemtex SC-40, -70. [Finetex] Dioctyl sodium sulfosuccinate; wetting/rewetting agent, dye leveler for textiles; liq.; 40, 70% conc. resp.

Gemtex SC-75. [Finetex] Dioctyl sodium sulfosuccinate; dispersant, wetting agent, pasting aid for textile dyeing applics.; antifog for glass cleaners, windshield washers; water and alcohol sol.; 75% conc.

Gemtex SC-75E, SC Powd. [Finetex] Dioctyl sodium sulfosuccinate; see Gemtex PA-75; liq. and powd. resp.; 75 and 83% conc. resp.

Genagen CD. [Hoechst-Celanese] Fatty acid ester; nonionic; lubricant for fiber preparation; liq.

Genamin CC, CS, OL, SH, TA Brands. [Hoechst-Celanese] Fatty amine and diamine, primary, sec., and tert.; cationic; basic material for paints and varnishes, road construction, anticaking agent; 100% conc.

Genamin CTAC. [Hoechst-Celanese] Cetrimonium chloride; cosmetic raw material for hair treatment preps.; liq.

Genamin DSAC. [Hoechst-Celanese] Distearyldimonium chloride; see Genamin CTAC; powd.

Genamin KDM. [Hoechst-Celanese] Behentrimonium chloride; cationic; conditioner and antistat in hair care prods.; yel. paste; water-sol.; pH 6–7 (1%); 80% act.

Genamin KDM-F. [Hoechst-Celanese] Alkyl trimethyl ammonium chloride; cationic; base, antistat, emulsifier for preparation of hair care prods.; conditioner; wh. waxy flake; sol. in water with use of alcohol; 80% act.

Genamin KS 5. [Hoechst-Celanese] PEG-5 stearyl ammonium chloride; cationic; cosmetic raw material, conditioner for hair care prods.; yel.-brn. aq. sol'n.; pleasant odor; water-misc.; pH 6.5 ± 0.3 (1%); 20% act.

Genamin KSE. [Hoechst-Celanese] Quat. ammonium compd.; SE base material for hair treatment preps.; powd.

Genamin O, S, T Brands. [Hoechst-Celanese] Fatty amine ethoxylate (2–25 EO); cationic; surface act. basic material for industrial fields; 100% conc.

Genamine C-020. [Hoechst-Celanese] Coconut fatty amine oxethylate; cationic; raw material for min. oil additives, insecticides, pesticides, cosmetic bases, adhesives; clear liq.; sol. in min. oil, turbid in water (10 g/l); sp.gr. 0.895 (50 C); visc. 28.8 cps (50 C); flash pt. 188 C (Marcusson); surf. tens. 26 dynes/cm; pH 9–10; 100% act.

Genamine C-050. [Hoechst-Celanese] Coconut fatty amine oxethylate; cationic; see Genamine C-020; clear liq.; sol. in water, min. oil; sp.gr. 0.95 (50 C); visc. 30.4 cps (50 C); flash pt. 224 C (Marcusson); surf. tens. 32 dynes/cm; pH 9–10; 100% act.

Genamine C-080. [Hoechst-Celanese] Coconut fatty amine oxethylate; cationic; see Genamine C-020; clear liq.; sol. in water, turbid in min. oil; sp.gr. 0.988 (50 C); visc. 35.2 cps (50 C); flash pt. 249 C (Marcusson); surf. tens. 37 dynes/cm; pH 9–10; 100% act.

Genamine C-100. [Hoechst-Celanese] Coconut fatty amine oxethylate; cationic; see Genamine C-020; clear liq.; sol. in water, turbid in min. oil; sp.gr. 1.0 (50 C); visc. 37 cps (50 C); flash pt. 273 C (Marcusson); surf. tens. 40 dynes/cm; pH 9–10; 100% act.

Genamine C-150. [Hoechst-Celanese] Coconut fatty amine oxethylate; cationic; see Genamine C-020; clear liq.; sol. in water; sp.gr. 1.027 (50 C); visc. 45.5 cps (50 C); flash pt. 287 C (Marcusson); surf. tens. 44 dynes/cm; pH 9–10; 100% act.

Genamine C-200. [Hoechst-Celanese] Coconut fatty amine oxethylate; cationic; see Genamine C-020; clear liq.; sol. in water, min. oil; sp.gr. 1.043 (50 C); visc. 55 cps (50 C); flash pt. 295 C (Marcusson); surf. tens. 45 dynes/cm; pH 9–10; 100% act.

Genamine C-250. [Hoechst-Celanese] Coconut fatty amine oxethylate; cationic; see Genamine C-020;

turbid liq.; sol. in water; sp.gr. 1.059 (50 C); visc. 65.2 cps (50 C); flash pt. 298 C (Marcusson); surf. tens. 47 dynes/cm; pH 9–10; 100% act.

Genamine O-020. [Hoechst-Celanese] Oleylamine oxethylate; cationic; see Genamine C-020; turbid liq.; sol. in min. oil; sp.gr. 0.883 (50 C); visc. 37.4 cps (50 C); flash pt. 220 C (Marcusson); pH 9–10; 100% act.

Genamine O-050. [Hoechst-Celanese] Oleylamine oxethylate; cationic; see Genamine C-020; clear liq.; sol. in min. oil; sp.gr. 0.956 (50 C); visc. 45.3 cps (50 C); flash pt. 269 C (Marcusson); surf. tens. 42 dynes/cm; pH 9–10; 100% act.

Genamine O-080. [Hoechst-Celanese] Oleylamine oxethylate; cationic; see Genamine C-020; clear liq.; sol. in water min. oil; sp.gr. 0.983 (50 C); visc. 49.6 cps (50 C); flash pt. 282 C (Marcusson); surf. tens. 39 dynes/cm; pH 9–10; 100% act.

Genamine O-100. [Hoechst-Celanese] Oleylamine oxethylate; cationic; see Genamine C-020; turbid liq.; sol. in water; sp.gr. 0.995 (50 C); visc. 52.1 cps (50 C); flash pt. 288 C (Marcusson); surf. tens. 41 dynes/cm; pH 9–10; 100% act.

Genamine O-150. [Hoechst-Celanese] Oleylamine oxethylate; cationic; see Genamine C-020; turbid liq.; sol. in water; sp.gr. 1.024 (50 C); visc. 65.1 cps (50 C); flash pt. 307 C (Marcusson); surf. tens. 45 dynes/cm; pH 9–10; 100% act.

Genamine O-200. [Hoechst-Celanese] Oleylamine oxethylate; cationic; see Genamine C-020; paste; sol. in water; sp.gr. 1.063 (50 C); visc. 75.9 cps (50 C); flash pt. 314 C (Marcusson); surf. tens. 48 dynes/cm; pH 9–10; 100% act.

Genamine O-250. [Hoechst-Celanese] Oleylamine oxethylate; cationic; see Genamine C-020; wax; sol. in water; sp.gr. 1.08 (50 C); visc. 89.2 cps (50 C); flash pt. 317 C (Marcusson); surf. tens. 51 dynes/cm; pH 9–10; 100% act.

Genamine S-020. [Hoechst-Celanese] Stearylamine oxethylate; cationic; see Genamine C-020; wax; sol. in min. oil, turbid in water; sp.gr. 0.889 (50 C); visc. 40.6 cps (50 C); flash pt. 219 C (Marcusson); pH 9–10; 100% act.

Genamine S-050. [Hoechst-Celanese] Stearylamine oxethylate; cationic; see Genamine C-020; clear liq.; sol. in min. oil, turbid in water; sp.gr. 0.943 (50 C); visc. 39 cps (50 C); flash pt. 258 C (Marcusson); surf. tens. 35 dynes/cm; pH 9–10; 100% act.

Genamine S-080. [Hoechst-Celanese] Stearylamine oxethylate; cationic; see Genamine C-020; turbid liq.; sol. in water, min. oil; sp.gr. 0.974 (50 C); visc. 43.3 cps (50 C); flash pt. 273 C (Marcusson); surf. tens. 39 dynes/cm; pH 9–10; 100% act.

Genamine S-100. [Hoechst-Celanese] Stearylamine oxethylate; cationic; see Genamine C-020; turbid liq.; sol. in water, min. oil; sp.gr. 0.987 (50 C); visc. 46.1 cps (50 C); flash pt. 277 C (Marcusson); surf. tens. 42 dynes/cm; pH 9–10; 100% act.

Genamine S-150. [Hoechst-Celanese] Stearylamine oxethylate; cationic; see Genamine C-020; turbid liq.; sol. in water; sp.gr. 1.011 (50 C); visc. 55.6 cps (50 C); flash pt. 280 C (Marcusson); surf. tens. 44 dynes/cm; pH 9–10; 100% act.

Genamine S-200. [Hoechst-Celanese] Stearylamine oxethylate; cationic; see Genamine C-020; turbid liq.; sol. in water, turbid in min. oil; sp.gr. 1.033 (50 C); visc. 67.8 cps (50 C); flash pt. 281 C (Marcusson); surf. tens. 47 dynes/cm; pH 9–10; 100% act.

Genamine S-250. [Hoechst-Celanese] Stearylamine oxethylate; cationic; see Genamine C-020; paste; sol. in water; sp.gr. 1.050 (50 C); visc. 81.8 cps (50 C); flash pt. 282 C (Marcusson); surf. tens. 49 dynes/cm; pH 9–10; 100% act.

Genaminox CS. [Hoechst-Celanese] Coco dimethyl amine oxide; nonionic; foaming agent and stabilizer, thickener for personal care prods.; liq.; 30% conc.

Genaminox KC. [Hoechst-Celanese] Cocamine oxide; nonionic; see Genaminox CS; liq.; 30% conc.

Genapol AMS. [Hoechst-Celanese] TEA-PEG-3 cocamide sulfate; anionic; detergent, foaming agent used in top-grade cosmetics cleansers; lime soap dispersant; clear yel. low-visc. liq.; weak odor; dens. 1.0 g/cm^3; visc. 200 cps max.; pH 6.5–8.0 (1%); biodeg.; 40% act.

Genapol B. [Hoechst-Celanese] EO/PO crosslinked blocked polymer; low-foaming lubricity additive; water-sol.

Genapol C-020. [Hoechst-Celanese] Coconut fatty alcohol polyglycol ether; nonionic; raw material for mfg. of textile, leather, paper auxs., detergents, emulsifiers, cosmetics; turbid liq.; sol. in min. oil, benzene, turbid in water; sp. gr. 0.892 (50 C); visc. 10.6 cps (50 C); flash pt. 170 C (Marcusson); surf. tens. 34 dynes/cm; pH 9–10; 100% act.

Genapol C-050. [Hoechst-Celanese] Coconut fatty alcohol polyglycol ether; nonionic; see Genapol C-020; turbid liq.; sol. in min. oil, benzene, turbid in water; sp. gr. 0.952 (50 C); visc. 17.6 cps (50 C); flash pt. 201 C (Marcusson); surf. tens. 32 dynes/cm; 100% act.

Genapol C-080. [Hoechst-Celanese] Coconut fatty alcohol polyglycol ether; nonionic; see Genapol C-020; paste; sol. in water, benzene, turbid in min. oil; sp. gr. 0.979 (50 C); visc. 25.4 cps (50 C); flash pt. 246 C (Marcusson); surf. tens. 36 dynes/cm; 100% act.

Genapol C-100. [Hoechst-Celanese] Coconut fatty alcohol polyglycol ether; nonionic; see Genapol C-020; paste; sol. in water, benzene, turbid in min. oil; sp. gr. 0.990 (50 C); visc. 30.4 cps (50 C); flash pt. 251 C (Marcusson); surf. tens. 38 dynes/cm; 100% act.

Genapol C-150. [Hoechst-Celanese] Coconut fatty alcohol polyglycol ether; nonionic; see Genapol C-020; wax; sol. in water, benzene; sp. gr. 1.027 (50 C); visc. 45.8 cps (50 C); flash pt. 260 C (Marcusson); surf. tens. 43 dynes/cm; 100% act.

Genapol C-200. [Hoechst-Celanese] Coconut fatty alcohol polyglycol ether; nonionic; see Genapol C-020; wax; sol. in water, benzene; sp. gr. 1.032 (50 C); visc. 61.5 cps (50 C); flash pt. 264 C (Marcusson); surf. tens. 44 dynes/cm; 100% act.

Genapol C-250. [Hoechst-Celanese] Coconut fatty alcohol polyglycol ether; nonionic; see Genapol C-020; wax; sol. in water, benzene; sp. gr. 1.055 (50 C); visc. 80.6 cps (50 C); flash pt. 267 C (Marcusson);

surf. tens. 47 dynes/cm; 100% act.

Genapol GEV. [Hoechst-Celanese] Fatty alcohol, ethoxylated; nonionic; lubricant for fiber preparation; liq.

Genapol LRO Liq., Paste. [Hoechst-Celanese] Sodium laureth sulfate; anionic; detergent, foaming agent used in cosmetic prods., personal care prods., lime soap dispersant; pale yel. clear liq., mobile paste resp.; slight odor; misc. with water; dens. 1.05 and 1.0 g/cm^3 resp.; visc. 100 cps max. (liq.); pH 6.5–8.0 (1% aq., liq.), 7.2 ± 0.6 (1% aq., paste); biodeg.; 27 and 68% conc.

Genapol O-020. [Hoechst-Celanese] Oleyl alcohol polyglycol ether; nonionic; raw material for mfg. of textile, leather, paper auxs., detergents, emulsifiers, cosmetics; clear liq.; sol. in min. oil, benzene, turbid in water; sp.gr. 0.894 (50 C); visc. 12.4 cps (50 C); flash pt. 186 C (Marcusson); 100% act.

Genapol O-050. [Hoechst-Celanese] Oleyl alcohol polyglycol ether; nonionic; see Genapol O-020; turbid liq.; sol. in benzene, turbid in water, min. oil; sp.gr. 0.936 (50 C); visc. 18.4 cps (50 C); flash pt. 225 C (Marcusson); surf. tens. 54 dynes/cm; 100% act.

Genapol O-080. [Hoechst-Celanese] Oleyl alcohol polyglycol ether; nonionic; see Genapol O-020; turbid liq.; sol. in water, benzene, turbid in min. oil; sp.gr. 0.960 (50 C); visc. 25.1 cps (50 C); flash pt. 246 C (Marcusson); surf. tens. 44 dynes/cm; 100% act.

Genapol O-100. [Hoechst-Celanese] Oleyl alcohol polyglycol ether; nonionic; see Genapol O-020; paste; sol. in water, benzene, turbid in min. oil; sp.gr. 0.989 (50 C); visc. 33 cps (50 C); flash pt. 260 C (Marcusson); surf. tens. 41 dynes/cm; 100% act.

Genapol O-120. [Hoechst-Celanese] Oleyl alcohol polyglycol ether; nonionic; see Genapol O-020; paste; sol. in water, benzene, turbid in min. oil; sp.gr. 1.0 (50 C); visc. 42.5 cps (50 C); flash pt. 265 C (Marcusson); surf. tens. 42 dynes/cm; 100% act.

Genapol O-150. [Hoechst-Celanese] Oleyl alcohol polyglycol ether; nonionic; see Genapol O-020; wax; sol. in water, benzene, turbid in min. oil; sp.gr. 1.02 (50 C); visc. 49.1 cps (50 C); flash pt. 271 C (Marcusson); surf. tens. 45 dynes/cm; 100% act.

Genapol O-200. [Hoechst-Celanese] Oleyl alcohol polyglycol ether; nonionic; see Genapol O-020; wax; sol. in water, benzene, turbid in min. oil; sp.gr. 1.037 (50 C); visc. 65.9 cps (50 C); flash pt. 278 C (Marcusson); surf. tens. 47 dynes/cm; 100% act.

Genapol O-230. [Hoechst-Celanese] Oleyl alcohol polyglycol ether; nonionic; see Genapol O-020; wax; sol. in water, benzene, turbid in min. oil; sp.gr. 1.042 (50 C); visc. 79.5 cps (50 C); flash pt. 279 C (Marcusson); surf. tens. 47 dynes/cm; 100% act.

Genapol O-250. [Hoechst-Celanese] Oleyl alcohol polyglycol ether; nonionic; see Genapol O-020; wax; sol. in water, benzene, turbid in min. oil; sp.gr. 1.042 (50 C); visc. 88.1 cps (50 C); flash pt. 280 C (Marcusson); surf. tens. 48 dynes/cm; 100% act.

Genapol PAF, PF, PL, PN Brands. [Hoechst-Celanese] Propylene and ethylene oxide polymerization prods.; nonionic; defoamer, component for detergent, cleaning and antistat; basic material for wide variety of aux.; liq., powd.; 100% conc.

Genapol PGM. [Hoechst-Celanese] Alkyl ether sulfate, pearlescent, and foam stabilizer; pearlescent for shampoos, bubble baths, and shower preps.; wh. highly visc. disp.

Genapol PGM Conc. [Hoechst-Celanese] Sodium laureth sulfate, glycol distearate, and cocamide MEA; anionic; pearl-luster conc. used in shampoos, bubble baths, cosmetics; wh. fluid paste; slight odor; misc. with water; dens. 1.0 g/cm^3; pH 7.2 ± 0.8 (1% act.); biodeg.; 40.0 ± 1.0% act.

Genapol PMS. [Hoechst-Celanese] Glycol distearate; nonionic; pearlescent agent for cosmetic washing agents and shampoo; powd.; 100% conc.

Genapol PN 30. [Hoechst-Celanese] EO/PO cross linked block polymer; nonionic; lubricant for formulating water sol. metalworking fluids; controlled foam detergents, wetting agents; yel. liq.; cloudy sol. in water; sp.gr. 1.01 (50 C); solid. pt. < –10 C; cloud pt. 29–31 C (5 g in 25 ml of 24% butyl diglycol); ref. index 1.448 (50 C); pH 10–11 (1%); 100% act.

Genapol PS. [Hoechst-Celanese] High m.w. polyether; lubricant; improves lubricating properties of min. oil-free, water-sol. metalworking fluids; ylsh. visc. liq.; sol. in water, glycols, benzene, xylene, acetone, ethyl acetate; sp.gr. 1.1; pH 6.5 ± 0.3 (5%); cloud pt. 160–170 F (1% aq.); 100% act.

Genapol S-020. [Hoechst-Celanese] Stearyl alcohol polyglycol ether; nonionic; raw material for mfg. of textile, leather, paper auxs., detergents, emulsifiers, cosmetics; wax; sol. in min. oil, benzene, turbid in water; sp.gr. 0.890 (50 C); visc. 15.2 cps (50 C); flash pt. 198 C (Marcusson); surf. tens. 50 dynes/cm; 100% act.

Genapol S-050. [Hoechst-Celanese]. Stearyl alcohol polyglycol ether; nonionic; see Genapol S-020; wax; sol. in min. oil, benzene, turbid in water; sp.gr. 0.934 (50 C); visc. 21.2 cps (50 C); flash pt. 224 C (Marcusson); surf. tens. 50 dynes/cm; 100% act.

Genapol S-080. [Hoechst-Celanese]. Stearyl alcohol polyglycol ether; nonionic; see Genapol S-020; paste; sol. in benzene, turbid in water, min. oil; sp.gr. 0.971 (50 C); visc. 28.4 cps (50 C); flash pt. 228 C (Marcusson); surf. tens. 44 dynes/cm; 100% act.

Genapol S-100. [Hoechst-Celanese]. Stearyl alcohol polyglycol ether; nonionic; see Genapol S-020; paste; sol. in benzene, turbid in water, min. oil; sp.gr. 0.978 (50 C); visc. 34 cps (50 C); flash pt. 239 C (Marcusson); surf. tens. 45 dynes/cm; 100% act.

Genapol S-150. [Hoechst-Celanese]. Stearyl alcohol polyglycol ether; nonionic; see Genapol S-020; wax; sol. in water, benzene, turbid in min. oil; sp.gr. 1.009 (50 C); visc. 50.3 cps (50 C); flash pt. 266 C (Marcusson); surf. tens. 47 dynes/cm; 100% act.

Genapol S-200. [Hoechst-Celanese]. Stearyl alcohol polyglycol ether; nonionic; see Genapol S-020; wax; sol. in water, benzene, turbid in min. oil; sp.gr. 1.031 (50 C); visc. 67.4 cps (50 C); flash pt. 267 C (Marcusson); surf. tens. 48 dynes/cm; 100% act.

Genapol S-250. [Hoechst-Celanese]. Stearyl alcohol polyglycol ether; nonionic; see Genapol S-020; wax; sol. in water, benzene, turbid in min. oil; sp.gr. 1.049

(50 C); visc. 106 cps (50 C); flash pt. 273 C (Marcusson); surf. tens. 49 dynes/cm; 100% act.

Genapol TSM. [Hoechst-Celanese] Alkyl ether sulfate plus silk lustering agent; opacifier for shampoos, bubble baths, shower preps.; wh. med.-visc. disp.

Genapol TS Powd. [Hoechst-Celanese] PEG-3 distearate; pearlescent, opacifier for shampoos, bubble baths, shower preps.; wh. powd.

Genapol X Brands. [Hoechst-Celanese] Isotridecanol polyglycol ether (3–15 EO); nonionic; basic material for mfg. of detergents, cleaning and rinsing agents; mfg. of leather and paper aux.; 100% conc.

Genapol ZRO Brands. [Hoechst-Celanese] Sodium laureth sulfate; anionic; see Genapol LRO; 70% conc.

Genencor® Cellulase 150L, 300P. [Genencor] Cellulase; enzyme for hydrolysis of cellulose; liq. and powd. resp.

Generol® 122. [Henkel] Soya sterol; nonionic; emollient, aux. or primary emulsifier, emulsion stabilizer, visc. modifier, solubilizer for cosmetics; wh. waxy flakes; sol. in abs. alcohol, min. and veg. oils; insol. in water; m.p. 135 C; acid no. 0.1; sapon. no. 2.0; HLB 3.0; 90% conc.

Generol® 122E5. [Henkel] PEG-5 soya sterol; nonionic; emollient, primary and sec. emulsifier, conditioner, stabilizer and consistency modifier in o/w emulsions; substantive to hair and in shampoo; Gardner 6 soft, waxy, amorphous solid; faint odor; sol. in ethanol, isopropyl esters; water-disp.; HLB 5; m.p. 75 C; pH 7 (1% aq.); 100% conc.

Generol® 122E10. [Henkel] PEG-10 soya sterol; nonionic; emollient, aux. emulsifier, appearance modifier, gloss enhancer, solubilizer; less substantivity in hair care formulations; emulsion stabilizer; modifies consistency of a firm cream to a pourable lotion in high solids o/w emulsions; lt. amber soft wax; faint odor; sol. in ethanol, isopropyl esters; water-disp. (translucent); HLB 12; m.p. 55 C; pH 7 (1% aq.); 100% conc.

Generol® 122E16. [Henkel] PEG-16 soya sterol; nonionic; emollient, primary or aux. emulsifier, solubilizer, stabilizer and consistency modifier in o/w emulsions, makeup formulations; improves pigment disp.; ivory hard wax; faint odor; sol. in water, isopropyl esters, veg. oils, and ethanol; HLB 15; m.p. 50 C; pH 7 (1% aq.); 100% conc.

Generol® 122E25. [Henkel] PEG-25 soya sterol; nonionic; emulsifier, solubilizer, emollient, pigment dispersant and wetter, deflocculating agent in cosmetics; ivory hard wax; faint odor; sol. in water and ethanol; at 80 C, sol. in isopropyl esters and veg. oils; HLB 17; m.p. 45 C; pH 7 (1% aq.); 100% conc.

Genesolv®. [Allied-Signal] Solvs. for specific cleaning, degreasing or dewatering tasks, flux removal.

Genu 04CG or 04CB. [Hercules] Pectin, low-methoxyl; gellant for Turkish Delight-type jelly prods. in confectionary industry; pH 4.8–5.2 (2%); 35–45% conc.

Genu 12CG or 12CB, 85–100 Grade. [Hercules] Pectin, low-methoxyl; gellant, thickener for fruit preparation used in fruit yogurt; thickener for toppings and variegated syrups in bakery industry; pH 3.8–4.4 (1%); 27–35% methoxylation; 65-73% free acid.

Genu 18CG or 18CB, 70-90 Grade. [Hercules] Pectin, low-methoxyl; gellant/thickener for fruit preparation used in fruit yogurt; gellant for canned fruit preparation used for fruit-flavored milk dessert; pH 3.2–4.0 (1%); 35–45% methoxylation; 55-65% free acid.

Genu 20AS or 20AB, 100 Grade. [Hercules] Pectin, low-methoxyl; gellant for jams and jellies, milk prods., and canned fruit preparations used for fruit-flavored milk dessert; thickener for topping and variegated syrups in bakery industry; pH 3.8–4.4 (1%); 33–40% methoxylation; 15-20% amidation.

Genu 21AS or 21AB. [Hercules] Pectin, low-methoxyl; gellant for canned fruit preparation used for fruit-flavored milk dessert, and milk prods.; pH 3.8–4.4 (1%); 33–40% methoxylation; 15-20% amidation.

Genu 102AS Buffered, 100 Grade. [Hercules] Pectin, low-methoxyl; gellant for firm gels in confectionary industry.

Genu 104AS or 15AB, 100-120 Grade. [Hercules] Pectin, low-methoxyl; gellant agent for jams and jellies, milk prods.; thickener for toppings and variegated syrups in bakery industry; pH 3.8–4.4 (1%); 27–33% methoxylation; 20-25% amidation.

Genu AA Medium-Rapid Set, 150 Grade. [Hercules] Pectin, high-methoxyl; gellant for jams and preserves in jars; pH 3.4–4.2 (4%); 67–70% methoxylation.

Genu BA-KING, 150 Grade. [Hercules] Pectin, high-methoxyl; gellant for oven-resistant jams and jellies in fruit preserving and bakery industries; pH 2.7–3.2 (4%); 68–71% methoxylation.

Genu BB Rapid Set, 150 Grade. [Hercules] Pectin, high-methoxyl; gellant for jams and preserves in jars, jellies in bakery industry; stabilizer for pulp suspension and oil emulsion in citrus drink concs.; pH 3.4–4.2 (4%); 71–75% methoxylation.

Genu DD Extra-Slow Set, 150 Grade. [Hercules] Pectin, high-methoxyl; gellant for jams and preserves in lg. containers; pH 3.4–4.2 (4%); 61–64% methoxylation.

Genu DD Extra-Slow Set C, 150 Grade. [Hercules] Pectin, high-methoxyl; gellant for jellies; texturizer, and foam stabilizer for aerated confectionary prods.; pH 3.4–4.2 (4%); 60–63% methoxylation.

Genu DD Slow Set, 150 Grade. [Hercules] Pectin, high-methoxyl; gellant agent for jams and preserves in lg. containers; pH 3.4–4.2 (4%); 63–66% methoxylation.

Genu JM. [Hercules] Pectin, high-methoxyl type; stabilizer for milk proteins in pasturized cultured milk drinks and milk/fruit juice drinks having a long shelf life; provides protective colloid effect; pH 2.7–3.2 (4%); 70–75% methoxylation.

Genu Carrageenan. [Hercules] Refined carrageenan; emulsifier, stabilizer, thickener, or gelling agent in foods, pharmaceuticals, and cosmetics; powd.

Genugel. [Hercules] Carrageenan; gellant, thickener, stabilizer, and suspender used in foods, pharmaceuticals, and cosmetics; water binder; imparts desirable body and mouthfeel.

Genulacta Series. [Hercules] Carrageenan; gellant, thickener, stabilizer, and suspender used in foods, pharmaceuticals, and cosmetics; water binder; imparts desirable body and mouthfeel.

Genu Pectins. [Hercules] High-methoxy and low-methoxy purified natural hydrocolloid derived from citrus peels; consists chiefly of partially methoxylated polygalacturonic acid; used as gelling agent for jellies, and as a visc. builder, protective colloid, and stabilizer for food systems, pharmaceutical, and cosmetic industries; lt. cream to grayish powd.; no odor and flavor.

Genuvisco. [Hercules] Carageenan; gellant, thickener, stabilizer, and suspender used in foods, pharmaceuticals, and cosmetics; water binder; imparts desirable body and mouthfeel.

Genuzan 1038A. [Hercules] Food-grade guar gum, xanthan gum; synergistic blend of water-sol. gums used in liq. animal feed; stabilizes prod. by maintaining a uniform suspension of insol. matter; additional visc. to the finished feed; improves cling and appearance; thickener for standard feed, extended-molasses feeds.

Genuzan 1063X. [Hercules] Food-grade guar gum, xanthan gum; see Genuzan 1038A.

Germaben® II. [Sutton] Diazolidinyl urea (30%), propylene glycol (56%), methylparaben (11%), and propylparaben (3%); broad-spectrum antimicrobial preservative for cosmetic prods.; pale to lt. yel. clear visc. liq.; char. mild odor; sol. @ 1% in aq. sol'n. and oil-water emulsions; sp.gr. 1.1731–1.1839; b.p. 369 F; flash pt. (TCC) > 200 F; 42.5–45.5% total solids; 5.8–6.4% N.

Germaben® II-E. [Sutton] Diazolidinyl urea (20%), propylene glycol (60%), methylparaben (10%), and propylparaben (10%); antimicrobial preservative system for cosmetics; clear pale yel. visc. liq.; char. mild odor; sol. 0.5 g/100 g water; sp.gr. 1.1353–1.1438; b.p. 369 F; flash pt. (TCC) > 200 F; 38.5–41.5% total solids; 3.8–4.4% N.

Germall® 115. [Sutton] Imidazolidinyl urea; antimicrobial preservative for cosmetics; wh. free-flowing powd.; none or char. mild odor; m.w. 406.33; sol. (g/100 g): 200 g in water, 50 g in propylene glycol, 16 g in glycerin; pH 6.0–7.5 (1% aq.); 26–28% N.

Germall® II. [Sutton] Diazolidinyl urea; broad-spectrum antimicrobial preservative for cosmetics and toiletries; wh. fine powd.; char. mild odor or odorless; m.w. 278.23; sol. in water, propylene glycol, glycerin; 19–21% N.

Geronol ACR/4. [Rhone-Poulenc SpA] Disodium laurylethoxy sulfosuccinate; anionic/nonionic; emulsifier for emulsion polymerization, detergent base, foamer, foam stabilizer, dispersant; liq.; 30% conc.

Geronol ACR/9. [Rhone-Poulenc SpA] Sodium salt alkylarylethoxy half ester sulfosuccinic acid; anionic; emulsifier for emulsion polymerization, detergent base, foamer, foam stabilizer, dispersant; liq.; 30% conc.

Geronol Aminox/3. [Rhone-Poulenc SpA] C_{16}-C_{18} alkanolamide; cationic; adjuvant for cutting oils for metals and rust protector; 100% conc.

Geronol TZ/14. [Rhone-Poulenc SpA] POE arylphenol phosphate amine salt; anionic; dispersant/suspending agent for pesticides; visc. liq.; 100% conc.

Geronol TZ/A. [Rhone-Poulenc SpA] Org. sulfonate, straight chain and POE/POP block polymer; anionic/nonionic; see Geronol TZ/14; liq.; 90% conc.

Geronol TZ/B. [Rhone-Poulenc SpA] Alkylphenol, ethoxylated; nonionic; see Geronol TZ/14; thick liq.; 100% conc.

Geropon CET/50/P. [Rhone-Poulenc SpA] Fatty acids, ethoxylated; nonionic; wetting/suspending agent for pesticides; waxy powd.; 100% conc.

Geropon FMS. [Rhone-Poulenc SpA] Sodium diphenylmethane sulfonate; anionic; dispersant and suspending agent for pesticide formulations; powd.; 70% conc.

Geropon IN. [Rhone-Poulenc SpA] Sodium diisopropyl naphthalene sulfonate; anionic; wetting agent for pesticide wettable powds.; dispersant; latex stabilizer; prevents coagulation in SBR and other elastomers; rewetting agent; humectant; powd.; 65% conc.

Geropon K/65. [Rhone-Poulenc SpA] Fatty acids with ethoxylated alcohols, modified; anionic/nonionic; wetting and suspending agent for Dodine wettable powds.; powd.; 85% conc.

Geropon K/202. [Rhone-Poulenc SpA] POE fatty alcohol; nonionic; wetting and suspending agent for Dodine wettable powds.; powd.; 50% conc.

Geropon PL/68, PL/87. [Rhone-Poulenc SpA] POE/POP block copolymers; nonionic; wetting/antistat for agric. powds. and general purpose; waxy powd.; 100% conc.

Geropon RM/77. [Rhone-Poulenc SpA] Sodium dinaphthyl methane sulfonate; anionic; dispersant/suspending agent for pesticides and general purpose; powd.; 80% conc.

Geropon RM/77-D. [Rhone-Poulenc SpA] Dinaphthalene-methane sulfonate sodium salt; anionic; dispersant, protective colloid, stabilizer of nat. and syn. elastomers; pigment grinding aid for paints; powd.; 95% conc.

Geropon RM/210. [Rhone-Poulenc SpA] Polynaphthalene-methane sulfonate, sodium salt; anionic; fluidizing and plasticizer agent for concrete and mortar; powd.; 92% conc.

Geropon SC/211. [Rhone-Poulenc SpA] Sodium salt of carboxylic copolymer with an anionic dispersant; anionic; dispersing/suspending/compatibility agent for pesticide wettable powds. and water-disp. gran.; powd.; 93% conc.

Geropon SC/213. [Rhone-Poulenc SpA] Potassium salt of carboxylic copolymer with anionic dispersant; anionic; dispersing/suspending/compatibility agent for pesticide wettable powds. and water-disp. gran.; powd.; 93% conc.

Geropon SDS. [Rhone-Poulenc SpA] Sodium dioctyl

sulfosuccinate; anionic; dispersant and wetting agent for pigments and dyes in plastics, pesticide wettable powds.; powd.; dissolves in water, partly sol. in org. solvs.; 100% conc.

Geropon T/36-DF. [Rhone-Poulenc SpA] Sodium salt of polycarboxylic acid; anionic; dispersant for final solv. stripping phase in polymerization of BR-SBR; liq.; 25% conc.

Geropon TA/72. [Rhone-Poulenc SpA] Sodium salt of polycarboxylic acid with wetting agent; anionic; dispersing/suspending/compatibility agent for pesticide wettable powds. and water-disp. gran.; powd.; 93% conc.

Geropon TA/72/S. [Rhone-Poulenc SpA] Sodium salt of polycarboxylic acid; anionic; see Geropon RM/77; powd.; water-sol.; 93% conc.

Geropon TA/764. [Rhone-Poulenc SpA] Sodium salt of polycarboxylic acid with wetting agent; anionic; dispersing/suspending/compatibility agent for pesticide wettable powds. and water-disp. gran.; powd.; 90% conc.

Geropon TA/K. [Rhone-Poulenc SpA] Potassium salt of polycarboxylic acid; anionic; dispersant, antisettling, general purpose; also for emulsion paints, dirt antiresetting in detergent compds.; sequestering agent for Ca and Mg salts; powd.; 90% conc.

Geropon TX/99. [Rhone-Poulenc SpA] Sodium-ammonium alkyl sulfosuccinate; anionic; thixotropic dispersant, antisetting agent for solv.-based paint systems; emulsion PVC plastisols; liq.; 38% conc.

Getren®. [Goldschmidt] Silicone-free release agent for foundry and steel industry, plastics incl. unsat. polyester, epoxy, and PU resins; lubricant for tire prod.

GFS®. [Kelco] Xanthan gum/galactomannan blend; gum for use in food preparations; stabilizer, suspending agent, thickener; provides rheological control to water- and milk-based foods; partly sol. in water.

Giv-Gard DXN. [Givaudan] 6-Acetoxy-2,4-dimethyl-m-dioxane; antimicrobial; used to preserve industrial emulsions; bactericide, fungicide; sol. in water, org. solvs.

Givsorb® UV-2. [Givaudan] N-(p-ethoxycarbonylphenyl)-N′-ethyl-N′-phenylformamidine; industrial UV absorber, lt. stabilizer, and antioxidant for PU, PVC, polyolefins, ABS, nylon, and acetal resins; photostable, broad spectrum screening agent for protection against the adverse effects of both UV-B and UV-A radiation; used in surf. coatings, polishes, dyestuffs, carpet treatments; wh. to pale yel. fine powd.; m.w. 296.4; sol. (g/100 g solv.): > 200 g in diethyl phthalate, Carbitol acetate, 2-methoxyethyl acetate, 2-ethoxyethyl acetate; > 100 g in toluene, acetone; > 20 g in IPP; > 10 g in IPA; dens. 1.077 g/cc; m.p. 49–52 C and 62–65 C (two allotropic cryst. forms); b.p. 225 C (1mm); m.p. 49–65 C; flash pt. (TCC): 200F.

Glissofluid® A 10, A 13. [BASF AG] Aliphatic dicarboxylic acid ester; component for syn. lubricants; liq.; visc. 3.5 and 5.3 mm^2/s resp.; 100% conc.

Glissolube® Grades. [BASF AG] Polyalkyleneoxide derivs.; syn. components for high-performance lubricants; liq.; 100% conc.

Glissoviscal B. [BASF AG] Polyisobutylenes; thickener (visc. improver) for lubricating oils; solid (block); 100% conc.

Glissoviscal SB, SG. [BASF AG] Diene-styrene copolymer; visc. index improver for lubricating oil; block and gran. resp.; 100% conc.

Glitter Adhesive #5048. [Polymer Research] Aq. nonionic emulsion; thickened emulsion used for adhering glitter to fabrics (100% cotton and cotton/polyester blends); wh. thickened pasty emulsion; gen'l. prop. 9.0 lb/gal; 43.2% solids.

Gloria. [Witco] Wh. min. oil USP; emollient, lubricant for food, drug, and cosmetic industries; water-wh.; visc. (Saybolt) 200–210 (100 F); pour pt. 15 F.

Glucam® E-10. [Amerchol] Methyl gluceth-10; nonionic; humectant for personal care prods.; freezing pt. depressant; emollient in aq. and hydroalcoholic prods.; moisturizer; foam modifier in detergent and shampoo systems; solv. and solubilizer for topical pharmaceuticals; used in emulsions, toilet articles; adds gloss, conditioning; pale yel. med. visc. syrup; odorless; sol. in water, alcohol, hydroalcoholic systems; acid no. 1.5 max.; sapon. no. 1.5 max; 100% conc.

Glucam® E-20. [Amerchol] Methyl gluceth-20; see Glucam E-10; pale yel. thin syrup; odorless; sol. see Glucam E-10; acid no. 1.0 max.; sapon. no. 1.0 max.; 100% conc.

Glucam® E-20 Distearate. [Amerchol] Methyl gluceth-20 distearate; nonionic; aux. o/w emulsifier, moisturizer, emollient and lubricant for cosmetics and pharmaceuticals; conditioner; soft solid; HLB 12.5; 100% conc.

Glucam® P-10. [Amerchol] PPG-10 methyl glucose ether; nonionic; see Glucam E-10; pale yel. heavy visc. syrups; odorless; sol. in water, alcohol, and hydroalcoholic systems, castor oil, IPM, IPP; visc. 8500 cps; acid no. 1.0 max.; sapon. no. 1.0 max.; 100% conc.

Glucam® P-20. [Amerchol] PPG-20 methyl glucose ether; nonionic; see Glucam E-10; pale yel. med. visc. syrup; odorless; sol. see Glucam P-10; visc. 1700 cps; acid no. 1.0 max.; sapon. no. 1.0 max.; 100% conc.

Glucam® P-20 Distearate. [Amerchol] PPG-20 methyl glucoside distearate; skin moisturizer, conditioner, and emollient for cosmetics and pharmaceuticals; binder and plasticizer for pressed powds.; liq.

Glucamate® DOE-120. [Amerchol] PEG-120 methyl glucoside dioleate; nonionic; thickener, emulsifier, solubilizer for shampoos; waxy solid; water-sol.; 100% conc.

Glucamate® SSE-20. [Amerchol] PEG-20 methyl glucose sesquistearate; nonionic; o/w emulsifier, solubilizer used with Glucate SS; pale yel. soft solid; sol. in water, IPA, ethanol, castor oil, corn oil; HLB 15.3; cloud pt. 74 C (1% in 5% NaCl); flash pt. 570 F (OC); sapon. no. 47; pH 6.5 (10% aq.); 100% conc.

Glucate® DO. [Amerchol] Methyl glucoside dioleate; nonionic; w/o emulsifier, aux. emulsifier for o/w systems; conditioner, emollient, lubricant, plasticizer, and pigment dispersant; liq.; HLB 5.0; 100% conc.

Glucate® SS. [Amerchol] Methyl glucose sesquistearate; nonionic; w/o emulsifier used with Glucamate SSE-20 to provide visc. stability, mildness; off wh. flakes; sol. in IPA, misc. with common oil phase ingred., water insol.; m.p. 51 C; HLB 6.4; flash pt. 530 F (OC); sapon. no. 136; pH 5.5 (10% in 1:1 IPA:water); 100% conc.

Glybrom Reagent Grade. [Lonza] 1,3-Dibromo-5,5-dimethyl hydantoin; intermediate for custom chemical synthesis, laundry bleach formulations, and automatic dishwashing compds.; wh. powd.; m.w. 286; m.p. 194 C.

Glycerox. [Croda Ltd.] POE glyceryl esters; emollient for aq./alcoholic preps.; liq.; water-sol.

Glycerox L. [Croda Ltd.] POE glycerol esters; solubilizer for oils and perfumes; liq.; water-sol.

Glycerox L15. [Croda Ltd.] PEG-15 glyceryl laurate; nonionic; solubilizer and emulsifier for cosmetics; liq.; HLB 14.0; 100% conc.

Glychlor® Reagent Grade. [Lonza] 1,3-Dichloro-5,5-dimethyl hydantoin; intermediate for custom chemical synthesis, laundry bleach formulations, and automatic dishwashing compds.; wh. powd.; m.w. 197; m.p. 133 C.

Glycolube® 810. [Lonza] Proprietary nonionic; antifog agent for PVC film; liq.; 100% conc.

Glycolube® P. [Lonza] Ester wax; specialty plastic lubricant; tan bead; m.p. 59 C; flash pt. 271 C.

Glycolube® PG. [Lonza] Specialty plastic lubricant; yel. liq.; flash pt. 204 C.

Glycolube® VL. [Lonza] Amide wax; specialty plastics lubricant and release agent; also used for textiles; tan bead; m.p. 114 C; flash pt. 270 C.

Glyconol®. [Lonza] Amide wax; syn. wax for textiles and other applics.; amber solid; m.p. 104 C; flash pt. 215 C.

Glycowax® S 932. [Lonza] Triester wax; syn. wax for cosmetics and other applics.; wh. flake; m.p. 62 C; flash pt. 310 C.

Glycox® PEMS. [Lonza] Ester wax; specialty plastic lubricant; tan flake; m.p. 51 C; flash pt. 218 C.

Glydant®. [Lonza] DMDM hydantoin in water; preservative, broad spectrum antimicrobial for cosmetics and toiletries; effective against Gram-positive and Gram-negative bacteria, fungi, and yeast; APHA 10 max. clear sol'n.; mild odor; f.p. –11 ± 0.6 C; sol. in water and ethanol; sp.gr. 1.1579 ± 0.0026; pH 6.5–7.5.

Glyecine A. [GAF] Thiodiethylene glycol; hygroscopic agent in textile printing; solv. for basic colors; dissolving aid for acid, direct, mordant, vat, and vat-ester dyes; liq.

Glytex® EL 176. [Lonza] Used in textile spinning process as lubricants, emulsifiers, and antistats for polyester and nylon fibers; Gardner 1; smoke pt. 140 C; acid no. 1; sapon. no. 93; hyd. no. 86.

Glytex® EL 882. [Lonza] See Glytex EL 176; Gardner 3+; smoke pt. 134 C; acid no. 9; sapon. no. 105; hyd. no. 32.

Glytex® EL 905. [Lonza] See Glytex EL 176; APHA 20; smoke pt. 146 C; acid no. 1; sapon. no. 46; hyd. no. 83.

Glytex® L 154. [Lonza] Used in textile spinning process as lubricants and antistats for polyester and nylon fibers; Gardner 2; smoke pt. 127 C; acid no. 0.5; sapon. no. 197; hyd. no. 5.

Glytex® L 203. [Lonza] Used in textile spinning process as lubricants and antistats for polyester and nylon fibers; APHA 50; smoke pt. 138 C; acid no. 0.5; sapon. no. 327; hyd. no. 5.

Golden Dawn Grade 1, 2. [Westbrook Lanolin] Anhyd. lanolin; nonionic; emollient, w/o emulsifier, ointment base, hair conditioner, wax crystal inhibitor, lipstick binder; soft wax, Grade 2 slightly darker color.

Golden Dawn Lanolin. [Westbrook Lanolin] Anhyd. lanolin, pharmaceutical cosmetic grade; emollient, w/o emulsifier; soft wax.

Golden Dawn Superfine. [Westbrook Lanolin] Anhyd. lanolin; nonionic; see Golden Dawn Grade 1; high quality; pale soft wax; HLB 4.0; 100% conc.

Golden Fleece DF. [Westbrook Lanolin] Extra refined anhyd. lanolin (detergent-free); nonionic; see Golden Dawn Grade 1; soft solid; HLB 4.5; 100% conc.

Golden Fleece Lanolin. [Westbrook Lanolin] Anhyd. lanolin, purified; emollient; hypoallergenic, low in trace pesticides; soft wax.

Golden Fleece P-80. [Westbrook Lanolin] Anhyd. lanolin (trace pesticides reduced by 80%); nonionic; see Golden Dawn Grade 1; pale soft wax.

Golden Fleece P-95. [Westbrook Lanolin] Anhyd. lanolin (trace pesticides reduced by 95%, low in natural free alcohols); nonionic; see Golden Dawn Grade 1; pale soft wax.

Golden Fleece RA. [Westbrook Lanolin] Anhyd. lanolin (detergent-free and low in natural free alcohols); nonionic; see Golden Dawn Grade 1; soft solid; 100% conc.

Goltix®. [Bayer] Metamitron; selective herbicide for use in fodder and sugar beet; colorless to lt. yel. cryst.; m.w. 202; sol. (g/1000 ml): 10–30 g in methylene chloride, 1–10 g in 2-propanol and toluene, 1.82 g in water; m.p. 166.6 C.

Good-rite® 3114. [Goodrich] Tris (3,5-di-t-butyl-4-hydroxybenzyl) isocyanurate; nonstaining, nondiscoloring antioxidant, stabilizer for PP film, polyethylene, EPDM, adhesives, fiber applic., and talc; processing and end-use applic.; food pkg.; wh. fine powd.; m.w. 784; sol. in dimethyl formamide, acetone, chloroform; sp.gr. 1.03; dens. 522.2 kg/m^3; m.p. 220 C; flash pt. 289 C. (COC); f.p. 302 (COC).

Good-rite® 3125. [Goodrich] 3,5-Di-t-butyl-4-hydroxyhydrocinnamic acid triester with 1,3,5-tris (2-hydroxyethyl)-s-triazine-2,4,6 (1H,3H,5H)-trione; stabilizer/antioxidant for PP, polyethylene articles for food contact applic.; wh. cryst. powd.; sol. in most org. solvs.; insol. in water.

Good-rite® Antioxidants. [Goodrich] Polymer chemical; stabilizer for PP and polyethylene, EPDM, nitrile, epichlorohydrin, and PUs; ultraviolet protection and lt. stability, high-temp. protection; resists water extraction; prolonged color retention; thermal stability in processing.

Good-rite® K-722. [Goodrich] Polyacrylic acid; anionic; dye chelating agent, sequestrant; dispersant for pigments, fillers, clay, silt, etc.; liq.

Good-rite® K-732, -752. [Goodrich] Polyacrylic acid; anionic; dispersant, sequestrant; resin used to disperse pigments, fillers, clay, silt, other suspended matter in water; liq.; 50 and 65% act. resp.

Good-rite® K-7028. [Goodrich] Low-med. m.w. linear polyacrylic acids and their sodium salts; anionic; co-builder in laundry, auto-dish, and misc. cleaners; provides detergency boosting, antisoil redeposition, antiscaling, antifilming/spotting; process and granulating aid in spray-dried detergents; dispersant; water sol'n. or dry powd.; colorless and odorless.

Good-rite® K-7058. [Goodrich] Low to med. m.w. linear polyacrylic acids and their sodium salts; anionic; co-builder in laundry, auto-dish, and misc. cleaners; provides detergency boosting, antisoil redeposition, antiscaling, antifilming/spotting; process and granulating aid in spray-dried detergents; chelating agent; water sol'n. and dry powd.; colorless, odorless.

Good-rite® K-7200. [Goodrich] Low to med. m.w. linear polyacrylic acids and their sodium salts; anionic; co-builder in laundry, auto-dish, and misc. cleaners; provides detergency boosting, antisoil redeposition, antiscaling, antifilming/spotting; process and granulating aid in spray-dried detergents; water sol'n. or dry powd.; colorless; odorless.

Good-rite® K-7600. [Goodrich] Low to med. m.w. linear polyacrylic acids and their sodium salts; anionic; co-builder in laundry, auto-dish and misc. cleaners; provides detergency boosting, antisoil redeposition, antiscaling, antifilming/spotting; anticaking agent; process and granulating aid in spray-dried detergents; water sol'n. or dry powd. forms; colorless; odorless.

Good-rite® Polyacrylates. [Goodrich] Polyacrylic resin; used as dispersant for pigments and fillers; prevents scale and nonbiological fouling in cooling water; sludge conditioner in boiler water treatment; latex thickener; used in drilling muds, pigment disps. and detergents; infinitely dilutable in water.

Good-rite® uv3034. [Goodrich] 1,1-(1,2 Ethanediyl)bis(3,3,5,5-tetramethylpiperazinone); uv lt. stabilizer in outdoor weathered PP, performance at low levels in PP stretched tape, water carryover performance in PP film, resistance to discoloration in PP, performance at low levels in HDPE; synergistic with benzophenone and benzotriazole compds.; performance in titanium dioxide pigmented systems; wh. cryst. powd.; m.w. 338; sol. (g/l sol'n.) 438 g in methylene chloride; 342 g in ethanol; 36 g in toluene; 0.67 g in heptane; dens. (apparent bulk) 626.3 kg/m^3; m.p. 136 C.

GP-0098. [Georgia-Pacific] Alcohol/fatty acid; release agent for use on cauls, presses, and glue applic. equipment in the plywood mill; clear to transparent liq.; f.p. 13 C; b.p. 80 C; dens. 6.63 lb/gal; flash pt. (PMCC) 6 C; 11.0 ± 1.0% act.

GP-2925. [Georgia-Pacific] Cationic polyamide; sizing agent for glass and min. fibers used in the mfg. of mats for roofing and other prods.; amber liq.; b.p. 100 C; sp.gr. 1.040–1045; visc. 140–200 cSt; dens. 8.7 lb/gal; pH 6.9–7.3; flash pt. none to boiling; 20.0–20.5% solids.

Graden Butyl Oleate. [Graden] Butyl oleate; low temp. plasticizer for natural and syn. elastomers; liq.; water-insol.

Gradol 250. [Graden] Sodium salt of a carboxylated polyelectrolyte; anionic; dispersant for pigments in latex paint, visc. depressant for latex, clay slurries; APHA 40 max. clear liq.; sol. in water systems; dens. 9.6 lb.gal; f.p. < –5 C; surf. tens. 69–71 dynes/cm (1%); pH 9.5–10.5 (1%); 25% act.

Gradol 250A. [Graden] Ammonium salt of a carboxylated polyelectrolyte; dispersant in ceramics industry where sodium or potassium salts undesirable; amber clear liq.; sol. in water systems; dens. 9.1–9.4 lb/gal; pH 7.5–9.0; 25% act.

Gradol 300. [Graden] Sodium salt of carboxylated polyelectrolyte; anionic; nonfoaming dispersant for pigments, fillers and other insol. fine particles in water; lt. liq.; sol. in aq. systems; sp.gr. 1.21; pH 10 (5% sol'n.); 30% solids.

Gradonic 400-ML. [Graden] PEG 400 laurate; nonionic; emulsifier for oils and solvs.; plasticizer, visc. reducer for starches; stabilizer, wetting agent for latex paints; liq.; 100% conc.

Gradonic N-95. [Graden] Nonoxynol-9.5; nonionic; detergent, wetting agent, dispersant, emulsifier; almost colorless liq.; mild odor; sol. in water, dilute inorg. salt and caustic sol'ns., min. acid sol'ns.; cloud pt. 130 F (5%); pH 7 (2%); 100% act.

grapHsize®. [Akzo] PU emulsion; surface size and treatment for paper; water repellent.

Grillocin HY-77. [RITA] Zinc ricinoleate; absorbs malodors from sol'ns. and surfs.; liq.

Grilloten LSE87. [RITA] Sucrose laurate; o/w emulsifier, solubilizer; liq.; HLB 12.5; 100% conc.

Grilloten LSE87K. [RITA] Sucrose cocoate; see Grilloten LSE87; liq.; HLB 12.5; 100% conc.

Grilloten PSE141G. [RITA] Sucrose stearate; see Grilloten LSE87; liq.; HLB 15.0; 100% conc.

Grilloten ZT12, ZT40, ZT80. [RITA] Sucrose ricinoleate; see Grilloten LSE87; liq.; HLB 9.0, 8.5, and 9.5 resp.; 100% conc.

Grindtek AML 60. [Grindsted] Acetylated palm kernel glycerides; lubricant and plasticizer for plastics and coatings; cosolv. for polar additives in low polarity systems; lt. tan liq.; sol. in ethanol, peanut oil, toluene, wh. spirit, paraffin oil; HLB 2.1.

Grindtek AMOS 90. [Grindsted] Acetylated lard glyceride; see Grindtek AML 60; yel. liq.; sol. in ethanol, peanut oil, toluene, wh. spirit, paraffin oil; HLB 1.8.

Grindtek ML 90. [Grindsted] Glyceryl laurate; component in w/o and o/w creams, lubricant, antistat, antifogging agent in plastics; antimicrobial effects reported; wh. block; sol. in ethanol, warm in propylene glycol, toluene, wh. spirit; HLB 5.3.

Grindtek MM 90. [Grindsted] Glyceryl myristate; see Grindtek ML 90; wh. block; sol. warm in ethanol, propylene glycol, toluene; HLB 5.0.

Grindtek MOL 90. [Grindsted] Glyceryl linoleate; see Grindtek ML 90; wh. semisolid; sol. in ethanol, toluene, warm in propylene glycol, peanut oil, wh. spirit, paraffin oil; HLB 4.3.

Grindtek MOP 90. [Grindsted] Lard glyceride; see Grindtek ML 90; wh. block; sol. warm in ethanol, propylene glycol, tolene, wh. spirit; HLB 4.3.

Grindtek MSP 32-6. [Grindsted] Glyceryl stearate SE; see Grindtek ML 90; wh. powd.; water-disp.; HLB 3.7.

Grindtek MSP 40. [Grindsted] Glyceryl stearate; see Grindtek ML 90; wh. powd.; sol. in toluene; HLB 2.8.

Grindtek MSP 40F. [Grindsted] Glyceryl stearate; see Grindtek ML 90; wh. fine powd.; sol. warm in toluene; HLB 2.8.

Grindtek MSP 52. [Grindsted] Glyceryl stearate; see Grindtek ML 90; wh. powd.; sol. warm in toluene; HLB 3.8.

Grindtek MSP 90. [Grindsted] Glyceryl stearate; see Grindtek ML 90; wh. powd.; sol. in ethanol, toluene; HLB 4.3.

Grindtek PGE 25. [Grindsted] Polyglyceryl-3 oleate; o/w emulsifier, antifogging agent for plastics; yel.-br. liq.; sol. in ethanol, warm in peanut oil, toluene, paraffin oil, disp. in water, propylene glycol; HLB 5.5.

Grindtek PGE 55, 55-6. [Grindsted] Polyglyceryl-3 stearate (55-6: SE grade); see Grindtek PGE 25; wh. powd., tan block; sol. warm in toluene (55), ethanol (55-6); HLB 6.8, 7.4.

Grindtek PGE-DSO. [Grindsted] Oleic/linoleic fatty acid ester of polyglycerol; o/w emulsifier, antifogging agent for plastics, effective for w/o emulsifications with veg. oils; yel.-br. visc. liq.; sol. in toluene, warm in peanut oil, wh. spirit, paraffin oil; HLB 2.8.

Grindtek PK 60. [Grindsted] Palm kernel glycerides; component in w/o and o/w creams, lubricant, antistat, antifogging agent for plastics; wh. block; sol. in ethanol, warm in peanut oil, toluene, paraffin oil, propylene glycol, tolene; HLB 4.1.

Grit-O'Cobs®. [Andersons] Corn cob meal; chemically inert plastic extender and filler; replaces wood flour in wood particle molding with phenolic resins, in profile and sheet stock prod.; filler in glue, asphalt, caulking compds., and rubber; also used in industrial abrasives, as industrial absorbent, as agric. chemical carriers, livestock feed roughage; tan gran.; essentially odorless; 20.9% sol. in 1% sodium hydroxide, 9.5% in hot water, 5.6% in alcohol, 2.5% in acetone and in 10% sulfuric acid; sp.gr. 1.2; bulk dens. 20–30 lb/ft^3; oil absorp. 100%; water absorp. 133.0%; pH 7.4 (surface); flash pt. (OC) 350 F; hardness (Mohs) 4.5; 47.1% cellulose.

Groutcide 75. [Henkel] Chlorothalanil; fungicide for Portland cement grout; powd.; 75% act.

GSD 550. [Lonza] Bromochloro-5,5-dimethyl hydantoin; intermediate for custom chemical synthesis, laundry bleach formulations, and automatic dishwashing compds.; wh. powd.; m.w. 241; m.p. 160 C.

GSP-40. [Genstar] Calcium carbonate; filler; 99.5% finer than 44 μ; 0.03% sol. in water; sp.gr. 2.7–2.71; bulk dens. 65 lb/ft^3 (loose); oil absorp. 11 cc/100 g; ref. index 1.6; pH 9.5 (sat. sol'n.); hardness (Moh) 3.0; .

GSP-80. [Genstar] Calcium carbonate; filler; Hegman grind 4+; 100% finer than 25 μ; 0.04% sol. in water; sp.gr. 2.7–2.71; bulk dens. 50 lb/ft^3 (loose); oil absorp. 14 cc/110 g; ref. index 1.6; pH 9.5 (sat. sol'n.); hardness (Moh) 3.0.

Guartec CM. [Henkel] Guar-based product; nonionic; flotation reagent used as selective depressant for silicate minerals in sulfide ores; talc depressant; lt. cream to lt. yel. powd.; sol. in water; dens. 46 lb/ft^3; visc. 15 cps; biodeg.; 13% max. moisture.

Guartec FM. [Henkel] Guar, industrial grade; nonionic; depressant for clay and siliceous gangue in flotation of metallic and nonmetallic minerals; flocculant for liq.-solid separations involving filtration; potash flotation; used as talc and mica depressant in sulfide mineral flotation processes; lt. tan powd.; sol. in water and brine; dens. 43 lb/ft^3; visc. 180 cps; pH 6.0–7.0 (1%); biodeg.; 13% max. moisture.

Guartec SJM. [Henkel] Guar-based; nonionic; flocculant; effectiveness and ease of handling; used in mineral processing, flotation depressant for clays and talc in potash and sulfide flotation; lt. tan fine powd.; free-flowing; dissolves in water; dens. 45 lb/ft^3; pH 6.5–7.5 (1%); biodeg.; 12% max. moisture.

Gusathion®. [Bayer] Azinphos-methyl; insecticide and acaricide with broad spectrum of activity; whitish cryst.; m.w. 317.33; sol. in most org. solvs.; sol. 33 ppm in water; sp.gr. 1.44 (20/4 C); m.p. 73–74 C; ref. index 1.61.

Gusathion® A. [Bayer] Azinphos-ethyl; broad spectrum insecticide for the control of biting and sucking pests; colorless cryst.; m.w. 345.4; sol. in most org. solvs.; practically insol. in water; sp.gr. 1.284 (20/4 C); b.p. 147 C (0.01 mm Hg); m.p. 53 C; ref. index 1.592.

G-White. [Huber] Calcium carbonate; functional filler extender with easy disp. and low binder demand; used in coatings, plastic and rubber fillers, building prods., ceramics flux, paper fillers, adhesives, cleaning compds., and polishing agents; wh. irreg., uniaxial particles; particle size 5.5 μ, 99% thru 500 mesh; water-sol.; sp.gr. 2.71; dens. 22.6 lb/solid gal, 55 lb/ft^3; oil absorp. 13.5 lb oil/100 lb; ref. index 1.6; hardness 3 Moh; 97.7% $CaCO_3$.

H-36. [Solem] Alumina trihydrate; filler used to suppress flame and smoke in latex, vinyl, and urethane carpet-backing compds.; particulate; median particle size 25 μ; sp.gr. 2.42; dens. 0.8g/cc; 64.9% Al_2O_3.

H7250. [Hüls] Hexamethylcyclotrisilazane; difunctional blocking agent; reagent for cyclosilylation.

H7300. [Hüls] Hexamethyldisilazane; blocking agent; silylating reagent applicable to alcohols, acids, amines, amides, and thiols; widely used in pharmaceutical mfg. processes.

Hallco® Antistat C-1047. [C.P. Hall] Laurate ester; antistat for PVC.

Hallco® C-7065. [C.P. Hall] POE laurate SE; PVC antistat; Gardner 2 clear oily liq.; sol. in kerosene, ethanol, acetone, propylene glycol, min. oil; disp. water; sp.gr. 0.998; f.p. –1 C; flash pt. 202 C; acid no. 2.5; sapon. no. 141; ref. index 1.458.

Hallco® DBS. [C.P. Hall] Dibutyl sebacate; plasticizer for adhesives; APHA 50 clear liq.; m.w. 316; sol. in hexane, toluene, kerosene, ethanol, acetone, min. oil; sp.gr. 0.934; visc. 14 cps; f.p. –11 C; flash pt. 185 C; acid no. 0.2; sapon. no. 355; ref. index 1.440.

Hallcomid® M-6. [C.P. Hall] N,N-Dimethyl caproamide; detergent, demulsifier, penetrant, solubilizer, plasticizer, softening agent, solv. for agriculture, automotive, coatings, cosmetics, drilling muds, textiles, germicides; colorless to pale yel. liq.; low odor; m.w. 143; sol. in common org. solvs., hydrocarbons, veg. oils, esters, water; visc. 7.0 cps; f.p. –40 C; b.p. 83–89 C (3 mm Hg); flash pt. 190 F; 95% conc.

Hallcomid® M-8-10. [C.P. Hall] N,N-Dimethyl caprylamide-capramide; nonionic; mutual solv. for polar and nonpolar ingred. in cosmetics, cleaners, corrosion inhibitor; colorless to pale yel. liq.; m.w. 171–199; sol. 0.17 g/100 g water; sol. in most common organic solvs., hydrocarbons, plasticizers, fatty acids, and veg. oils; sp.gr. 0.88; visc. 10.5 cps; f.p. –21 C; b.p. 115–170 C (3 mm Hg); flash pt. 245 F; 95% conc.

Hallcomid® M-12. [C.P. Hall] N,N-Dimethyl lauramide; nonionic; hydrotrope, solubilizer for polar and nonpolar emulsifiers, dyes, perfumes, corrosion inhibitor; colorless to pale yel. liq.; m.w. 227; sol. < 0.2 g/100 g water, most common org. solvs., hydrocarbons, plasticizers, fatty acids, veg. oils; sp.gr. 0.872; visc. 14.5 cps; f.p. 14–16 C; b.p. 147–163 C (3 mm Hg); HLB 8.0; flash pt. 292 F; 95% conc.

Hallcomid® M-18. [C.P. Hall] N,N-dimethyl stearamide; nonionic; solubilizer for shampoos, corrosion inhibitor; colorless to pale yel. solid; m.w. 312; sol. 0.1 g/100 g water, most common org. solvs., hydrocarbons, plasticizers, fatty acids, veg. oils; sp.gr. 0.851; visc. 13.5 cps (40 C); f.p. 38–40 C; b.p. 194–214 C (3 mm Hg); HLB 7.0; flash pt. 399 F; 95% conc.

Hallcomid® M-18-OL. [C.P. Hall] N,N-dimethyl oleamide; nonionic; solubilizer, solv., dispersant, wetting agent; slightly yel. liq.; m.w. 310; sol. see Hallcomid M-18; sp.gr. 0.875; visc. 18.5 cps; f.p. –8 C; b.p. 196–211 C (3 mm Hg); HLB 7.0; flash pt. 410 F; 95% conc.

Halobrom. [Dead Sea Bromine] 1-Bromo-3-chloro-5,5-dimethylhydantoin; broad spectrum biocide for control of algae, bacterial and fungal slimes in swimming pools and industrial water systems; nonflam.; wh./off-wh. free-flowing gran. or tablets; m.w. 241.5; sol. in many org. solvs., 0.2 g/100 g in water; 94–96% act., 33.1% Br.

Halofree 22. [Solem] Halogen-free flame-retardant additive with improved smoke-suppressing properties over ATH; used in wire and cable, inj.-molded parts, extrusions, coatings, and adhesives; powd.; 1.1 μ avg. particle size; 0.01% retained on 325 mesh; sp.gr. 2.4 g/cc; bulk dens. 0.33 g/cc (loose), 0.73 g/cc (packed); surf. area 13 m^2/g; oil absorp. 42 cc/100 g.

Haltex 300. [Hitox] Alumina trihydrate; flame retardant; fine wh. powd.; 94% on 325 mesh; sp.gr. 2.42; bulk dens. 71 lb/ft^3 (loose); pH 8; 65.30% Al_2O_3.

Haltex 310, 313, 320. [Hitox] Alumina trihydrate; flame retardant; fine wh. powd.; 10, 13, and 20 μ median particle diam. resp.; 0.30%, 2%, and 13.5% on 325 mesh; sp.gr. 2.42; bulk dens. 57, 64, and 68 lb/ft^3 (loose), 71, 82, 92 lb/ft^3 (packed); brightness 91; oil absorp. 19, 19, and 18.

Hamp-Ene® 100. [W.R. Grace] Tetrasodium EDTA; general purpose chelating agent; pale straw clear liq.; m.w. 380.2; water-misc.; sp.gr. 1.26–1.28; dens. 10.6 lb/gal; chel. value 100 mg $CaCO_3$/g min. (@ pH 11); pH 11–12 (1%); 38% min. act.

Hamp-Ene® 100S. [W.R. Grace] Tetrasodium EDTA sol'n.; high-purity form of Hamp-ene 100; pale straw clear liq.; m.w. 380.2 (anhyd.); sp.gr. 1.26–1.28; dens. 10.6 lb/gal; misc. with water; pH 11–12 (1%); 38% min. act.

Hamp-Ene® 220. [W.R. Grace] Tetrasodium EDTA, tech.; chelating agent; wh. cryst. powd.; m.w. 452.3; sol. 50% in water; dens. 6.5 lb/gal; chel. value 219 mg Ca CO_3/g min (@ pH 11); pH 10.5–11.5 (1%); 99% min. act.

Hamp-Ene® Acid. [W.R. Grace] EDTA; chelating agent; used where sodium ion is undesirable; wh. cryst. powd.; m.w. 292.3; dens. 6.0 lb/gal; chel. value

340 mg $CaCO_3$/g min.; pH 2.6–3.1; 99.3% min. act.

Hamp-Ene® Diammonium EDTA. [W.R. Grace] Diammonium EDTA; chelating agent; photographic developer; pale straw clear liq.; m.w. 326.3; water-misc.; sp.gr. 1.19–1.22; dens. 10.0 lb/gal; 44% min. act.

Hamp-Ene® Fe-62. [W.R. Grace] Chelator for complexing calcium and iron in mod. alkaline sol'ns.; misc. with water.

Hamp-Ene® Na_2. [W.R. Grace] Disodium EDTA dihydrate; chelating agent; wh. cryst. powd.; m.w. 372.3; sol. 15% in water; dens. 6.0 lb/gal; chel. value 267 mg $CaCO_3$/g min.; pH 4.8–5.8 (1%); 99.3% min. act.

Hamp-Ene® Na_3 Liq. [W.R. Grace] Trisodium EDTA; chelating agent; pale straw clear liq.; m.w. 358.2; water-misc.; sp.gr. 1.25–1.28; dens. 10.5 lb/gal; chel. value 100 mg $CaCO_3$/g min. (@ pH 11); pH 9.0–9.4 (1%); 35.8% min. act.

Hamp-Ene® Na_3T. [W.R. Grace] Trisodium EDTA trihydrate; chelating agent; wh. cryst. powd.; m.w. 412.3; sol. 40% in water; dens. 6.0 lb/gal; chel. value 242 mg $CaCO_3$/ g min.; pH 8.3–8.7 (1%); 99% min. act.

Hamp-Ene® Na_4. [W.R. Grace] Tetrasodium EDTA dihydrate; chelating agent; wh. cryst. powd.; m.w. 416.2; sol. 48% in water; dens. 6.5 lb/gal; chel. value 240 mg $CaCO_3$/g min. (@ pH 11); pH 10.7–11.7 (1%); 99.5% min. act.

Hamp-Ene® NaFe Purified Grade. [W.R. Grace] Sodium ferric ethylenediaminetetraacetate, trihydrate; chelating agent used in photographic developers, oxidation-reduction systems, and emulsion polymerization; m.w. 421.1; dens. 5.0 lb/gal; pH 4–6 (1%); 98% min. act.

Hamp-Ene® OH-1. [W.R. Grace] Chelating agent to complex iron in caustic and to prevent iron precipitation on dilution, neutralizing, or acidification; misc. in water.

Hamp-Ene® OH Powd. [W.R. Grace] TEA hydrochloride, disodium EDTA; chelating agent for iron in alkaline conditions; off-wh. to tan cryst. powd.; sol. 30% in water; dens. 6.4 lb/gal; chel. value 115 mg $CaCO_3$/g min.; pH 4.3–5.1 (15%); 100% act.

Hamp-Ene® Photoiron. [W.R. Grace] Ammonium ferric EDTA; chelating agent used in photographic developers and oxidation-reduction systems; dk. brn.-red liq.; m.w. 362.1; water-misc.; sp.gr. 1.27–1.35; dens. 10.8 lb/gal; pH 7.0–8.0; 43.8% min. act.

Hamp-Ex® 80. [W.R. Grace] Pentasodium pentetate; chelating agent for alkaline earth and heavy metal ions; peroxide bleaching; pale straw clear liq.; m.w. 503.3; water-misc.; sp.gr. 1.29–1.31; dens. 10.8 lb/gal; chel. value 80 mg $CaCO_3$/g min.; pH 11–12 (1%); 40.2% min. act.

Hamp-Ex® Acid. [W.R. Grace] Pentetic acid; chelating agent; wh. cryst. powd.; m.w. 393.4; dens. 7.0 lb/gal; chel. value 253 mg $CaCO_3$/g min.; pH 2.1–2.5; 99% min. act.

Hamp-Ex® M. [W.R. Grace] Trisodium magnesium diethylenetriaminepentaacetate; chelating agent; stabilizer for hydrogen peroxide in kier bleaching of textiles; pale straw clear liq.; m.w. 481.7; water-misc.; sp.gr. 1.17–1.20; dens. 9.8 lb/gal; chel. value 100 mg $CuSO_4 \cdot 5H_2O$/g min.; pH 9–10 (1%); 40% act. solids.

Hamp-Ol® 120. [W.R. Grace] Trisodium HEDTA; general purpose chelating agent for control of iron at pH 6.5–9.5 as well as Ca and Mg; pale straw clear liq.; m.w. 344.2; water-misc.; sp.gr. 1.28–1.31; dens. 10.8 lb/gal; chel. value 120 mg $CaCO_3$/g min.; pH 11–12 (1%); 41.3% min. act.

Hamp-Ol® Acid. [W.R. Grace] HEDTA; chelating agent; wh. cryst. powd.; m.w. 278.3; dens. 6.0 lb/gal; chel. value 353 mg $CaCO_3$/g min.; pH 2.1–2.4; 98% min. act.

Hamp-Ol® Crystals. [W.R. Grace] Trisodium HEDTA dihydrate; chelating agent; wh. cryst. powd.; m.w. 389.3; sol. 59% in water; dens. 6.0 lb/gal; chel. value 255 mg $CaCO_3$/g min.; pH 11–12 (1%); 99.0% min. act.

Hampshire® DEG. [W.R. Grace] Sodium dihydroxyethylglycinate; chelating agent used for for control of iron only in alkaline sol'ns.; pale straw clear liq.; m.w. 185.2; water-misc.; sp.gr. 1.19–1.23; dens. 10 lb/gal; chel. value 50 mg Fe/g min. (@ pH 10); pH 11–12 (1%); 41% act.

Hampshire® EDG. [W.R. Grace] Ethanoldiglycine disodium salt; chelating agent used for the control of iron and chelates calcium; pale straw clear liq.; m.w. 221.1; water-misc.; sp.gr. 1.16–1.20; dens. 9.8 lb/gal; chel. value 110 mg $CaCO_3$/g min (@ pH 11.5); pH 11–12 (1%); 27.5% min. act.

Hampshire® NTA 150. [W.R. Grace] Trisodium NTA; chelating agents used in laundry detergents and specialty cleaning prods., water treatment, textiles, metal finishing, pulp and paper processing, and petrol. industry; nonphosphate detergent builder; pale straw clear liq.; m.w. 257.1; water-misc.; sp.gr. 1.30–1.33; dens. 11.0 lb/gal; chel. value 156 mg $CaCO_3$/g min.; pH 11–12 (1%); 40.0% min. act.

Hampshire® NTA Acid. [W.R. Grace] Nitrilotriacetic acid; general purpose chelating agent; wh. cryst. powd.; m.w. 191.2; dens. 7.5 lb/gal; chel. value 525 mg $CaCO_3$/g min.; pH 2–3; 100% min. act.

Hampshire® NTA Na_3 Crystals. [W.R. Grace] Trisodium NTA monohydrate; chelating agent; detergent builder for low or nonphosphate formulations; wh. cryst. powd.; m.w. 275.1; sol. 50% in water; dens. 6.0 lb/gal; chel. value 365 mg $CaCO_3$/g min.; pH 10.5–11.5 (1%); 100% min. act.

Haro® Chem ALMD-2, ALT. [Harcros] Metal soap stabilizer for cosmetics, oils, fats, lubricants, pharmaceuticals, paints, and printing inks; solid.

Haro® Chem BG, BSG, CBHG, CSG. [Harcros] Metal soap stabilizer for PVC; BG grade also for ABS; solid.

Haro® Chem BP-108X, KB-350X, KB-353A. [Harcros] Solid metal soap stabilizer for rigid and plasticized PVC applics.

Haro® Chem CGL, CGN. [Harcros] Metal soap stabilizer for PVC, LDPE, LLDPE, M/HDPE, PP, ABS, rubbers, thermosets, paint, and fertilizers;

solid.

Haro® Chem CHG, KPR, KS. [Harcros] Metal soap stabilizer for PVC; solid.

Haro® Chem CPR-2. [Harcros] Metal soap stabilizer for PVC, LDPE, LLDPE, M/HDPE, PP, ABS, rubbers, thermosets, cosmetics, paper, pharmaceuticals, paint, fertilizers; solid.

Haro® Chem CZ-31, -35, -36, -37, -38, -39. [Harcros] Nontoxic solid stabilizers for rigid and plasticized PVC.

Haro® Chem KB-214SA, KB-219SA, KB-521SA. [Harcros] Solid metal soap stabilizer for rigid PVC incl. extrusion, profile, extruded foam, and calendered film.

Haro® Chem KB-554A, ZZ-019. [Harcros] Solid metal soap stabilizer for plasticized PVC applics.

Haro® Chem LHG. [Harcros] Metal soap stabilizer for thermosets, oils, fats, and lubricants; solid.

Haro® Chem LIG. [Harcros] Metal soap stabilizer for oils, fats, and lubricants, and wire drawing; solid.

Haro® Chem MF-2. [Harcros] Metal soap stabilizer for ABS, thermosets, cosmetics, pharmaceuticals, fertilizers, and wire drawing; solid.

Haro® Chem NG. [Harcros] Metal soap stabilizer for LDPE, cosmetics, oils, fats, and lubricants, pharmaceuticals, paint, and wire drawing; solid.

Haro® Chem P28G. [Harcros] Normal lead stearate; stabilizer for rigid and plasticized PVC applics.; solid.

Haro® Chem P51. [Harcros] Dibasic lead stearate; stabilizer for rigid and plasticized PVC applics. offers heat stability and good elec. properties; solid.

Haro® Chem PC. [Harcros] Basic lead carbonate; stabilizer for plasticized PVC applics. offers heat stability and good elec. properties.

Haro® Chem PDF. [Harcros] Dibasic lead phosphite; stabilizer for rigid and plasticized PVC applics. offers heat and light stability and good elec. properties.

Haro® Chem PDP-E. [Harcros] Dibasic lead phthalate; stabilizer for plasticized PVC applics. offers heat stability and good elec. properties.

Haro® Chem PPCS-X. [Harcros] Polybasic lead sulfate; stabilizer for rigid and plasticized PVC applics. offers heat stability and good elec. properties.

Haro® Chem PTS-E. [Harcros] Tribasic lead sulfate; stabilizer for rigid and plasticized PVC applics. offers heat stability and good elec. properties.

Haro® Chem ZGN. [Harcros] Metal soap stabilizer for PVC, LDPE, M/HDPE, PS, rubbers, thermosets, and paint; solid.

Haro® Chem ZGN-T. [Harcros] Metal soap stabilizer for PVC, PS, expanded PS, HIPS; solid.

Haro® Chem ZPR-2. [Harcros] Metal soap stabilizer for PVC, LDPE, M/HDPE, PS, rubbers, thermosets, cosmetics, paper, pharmaceuticals, paint; solid.

Haro® Chem ZSG. [Harcros] Metal soap stabilizer for PVC and rubbers; solid.

Haroil SCO-65, -7525. [Graden] Sulfated castor oil; anionic; superfatting agent for cosmetic creams, emulsifier for cosmetic formulations, plasticizer for adhesives, antisagging and antisettling agent for paints and stains; amber liq.; pH 7.0 ± 0.5 (SCO-65, 2% aq.); dens. 1.025 (SCO-7525); 50%, 75% act.

Harol D. [Graden] Highly polymerized naphthalene sulfonate; anionic; stabilizer, visc. depressant; dispersant for carbon blk., clay disps., $CaCO_3$ disps.; paper coating compds.; powd.; dens. 47 lb/ft^3; surf. tens. 71 dynes/cm (1%); pH 9 (1%); 83–85% act.

Harol KG. [Graden] Potassium salts of polymerized alkyl naphthalene sulfonic acids; dispersant used in dispersed dyes; in wax emulsions, coatings, other applics. where sodium salts are undesirable; powd.; 87% act.

Harol RG-71. [Graden] Sodium salt of polymerized alkyl naphthalene sulfonic acid; anionic; dye retardant, leveler; dispersant for pitch, carbon blk., clay, pigments, dyestuffs; used in wet milling, cement wet processing, mfg. of gypsum board, pulp and paper processing, syn. rubber polymerization; buff gran.; dens. 35–37 lb/ft^3; surf. tens. 70–72 dynes/cm (1%); pH 7.8–10.4 (1%); 95% act.

Haro® Mix BF-202, LK-218, LK-228, UK-121. [Harcros] One-pack lead heat and light stabilizers for rigid PVC applics.

Haro® Mix BK-105, BK-107. [Harcros] One-pack lead heat and light stabilizer for rigid PVC applics.

Haro® Mix CE-701, CH-205, CK-203, CK-213, CK-711, MH-204, MK-107, MK-620, MK-744. [Harcros] One-pack lead heat stabilizer systems for rigid PVC applics.

Haro® Mix CH-606, FK-102, IH-108, MK-220. [Harcros] One-pack lead heat stabilizer for plasticized PVC applics. (cable); offers good elec. properties.

Haro® Mix IC-217. [Harcros] One-pack lead heat stabilizer for rigid and plasticized PVC inj. moldings.

Haro® Mix IC-238, SK-602. [Harcros] One-pack lead heat stabilizer for rigid PVC inj. moldings.

Haro® Mix UC-213, VC-501. [Harcros] One-pack lead heat and light stabilizer for rigid PVC inj. moldings.

Haro® Mix YC-601, YK-110, YK-113, YK-307. [Harcros] Ba/Cd/Pb; one-pack heat and light stabilizer for rigid PVC applics. (profiles).

Haro® Mix YE-301. [Harcros] Ba/Pb; one-pack heat and light stabilizer system for rigid PVC sheet.

Haro® Mix ZC-028, ZC-029, ZC-030. [Harcros] Ca/Zn; nontoxic one-pack heat and light stabilizer for rigid PVC applics. (profiles).

Haro® Mix ZC-031, ZC-032. [Harcros] Ca/Zn; nontoxic one-pack heat stabilizer for plasticized PVC applics. (cables); offers good elec. properties.

Haro® Mix ZC-036, ZC-902. [Harcros] Ca/Zn; nontoxic one-pack heat stabilizer for rigid PVC blow moldings (bottles).

Haro® Mix ZC-309, ZC-311. [Harcros] Ca/Zn; nontoxic one-pack heat stabilizer for rigid PVC applics. (pipes).

Haro® Mix ZT-025, ZT-508, ZT-514. [Harcros] Sn; nontoxic one-pack heat stabilizer for rigid PVC applics. (pipes).

Haro® Mix ZT-026, ZT-504. [Harcros] Sn; nontoxic

one-pack heat stabilizer for rigid PVC inj. moldings.

Haro® Mix ZT-905. [Harcros] Sn; nontoxic one-pack heat stabilizer for rigvid PVC blow moldings (bottles).

Haro® Wax L-333. [Harcros] Lubricant blend for rigid and plasticized PVC applics., incl. food contact applics.

Haro® Wax L-344, L-443. [Harcros] Lubricant blend for rigid PVC applics., incl. food contact applics.

Haro® Wax L01-56, L03-58, L04-73, L09-00, L18-78, L21-96, L23-58, L24-98. [Harcros] Lubricant for rigid and plasticized PVC; suitable for food contact applics.

Haro® Wax L02-99, L05-51, L06-62, L07-47, L08-57, L10-58, L11-85, L12-43, L13-89, L14-77, L15-75, L16-92, L17-98, L19-99, L22-99, L25-99. [Harcros] Lubricant for rigid PVC applics.; suitable for food contact.

Haro® Wax L20-98. [Harcros] Lubricant for rigid PVC applics.

Harshaw Antimony Oxide KR. [M&T Harshaw] Antimony oxide; flame retardant for plastics (PVC, PP, PE, ABS, PS, polyester, PU), textiles, paper, paint, rubber; opacifier in ceramics; also as pigment, catalyst, intermediate, modifier; wh. fine powd.; 1–1.3μ particle size; odorless; b.p. 1456 C; sp.gr. 5.7; oil absorp. 9–11; m.p. 656 C; 99.2% conc.

Hartaine CB-40. [Hart Prod.] Coco-betaine; detergent for textiles, household, cosmetics, wetting agent for cottons, bleaching, emulsifier for dispersing waxes, pigments, resins, printing; very lt. liq.; water sol.; dens. 8.85 lb/gal; sp.gr. 1.06; 40% act.

Hartamide 9137. [Hart Chem. Ltd.] Oleic DEA; anionic; coupling agent, emulsion stabilizer, lubricant and antistat; liq.; 80% amide.

Hartamide AD. [Hart Chem. Ltd.] Coconut alkanolamide; nonionic; detergent for textiles, household, cosmetics, industrial cleaning, wetting agent for dyeing, emulsifier for oils, waxes, pigments, resins, thickener, foaming agent; amber liq.; fatty odor; water sol.; 100% act.

Hartamide CE 80. [Hart Chem. Ltd.] Lauric-myristic super amide; nonionic; see Hartamide AD; lt. amber liq.; fatty odor; water sol.; 100% act.

Hartamide CE 90. [Hart Chem. Ltd.] High purity coconut fatty acid alkanolamide; nonionic; see Hartamide AD; lt. amber liq.; fatty odor; water sol.; 100% act.

Hartamide LDA 70, 90. [Hart Chem. Ltd.] Lauric DEA; nonionic; detergent, foam stabilizer and visc. regulator for liq. and powd. detergent systems; clear solid; 100% act.

Hartamide LE 90. [Hart Chem. Ltd.] High purity lauric alkanolamide; nonionic; see Hartamide AD; lt. solid; fatty odor; water sol.; 100% act.

Hartamide LM 11D. [Hart Chem. Ltd.] High purity lauric myristic alkanolamide; nonionic; see Hartamide AD; lt. gel/solid; fatty odor; water sol.; 100% act.

Hartamide LMEA-70. [Hart Chem. Ltd.] Lauric myristic MEA; anionic; foam stabilizer and visc. modifier for detergents and personal care prods.; wh. solid; 95% amide.

Hartamide LMEA-90. [Hart Chem. Ltd.] Lauric MEA; anionic; foam stabilizer for detergents and bubble bath preparations; wh. solid; 95% amide.

Hartamide OD. [Hart Chem. Ltd.] Coconut DEA; nonionic; detergent, foam stabilizer and visc. regulator for liq. and powd. detergent systems; yel. liq.; dens. 1.02 lb/gal; visc. 900 cps; 100% act.

Hartamide OS. [Hart Chem. Ltd.] Coco-oleic super amide; nonionic; detergent for textiles, laundering, wetting agent, emulsifier for cutting and sol. oils, dispersing waxes, resins, pigments; lt. amber liq.; fatty odor; sol. in water; 100% act.

Hartenol LAS-30. [Hart Prod.] Sodium lauryl sulfate; anionic; detergent for textiles, skins, leather, laundering, wetting agent, emulsifier for dispersing waxes, pigments, resins; very lt. liq.; 30% act.

Hartex San Q 50. [Hart Chem. Ltd.] Lauryl dimethylbenzyl ammonium chloride; cationic; germicide, sanitizer, disinfectant; amber liq.; sp.gr. 0.980; visc. 60 cps; ref. index 1.4412; 50% act.

Hartex V63, V64. [Hart Chem. Ltd.] Sulfonated castor oil; anionic; penetrant, emulsifier for terpineols, min. and fatty acids, lubricant; amber liq.; dens. 1.02–1.056 lb/gal; visc. 250–520 cps; 70% and 45% act.

Hartolan, Hartolan Super. [Croda] Lanolin alcohols; spreading agent, dispersant, stabilizer, plasticizer, o/w emulsifier and emollient for cosmetic and pharmaceutical systems; brn. solid wax and pale amber solid wax resp.; m.p. 58 and 60 C resp.; sapon. no. 5 mg. max.

Hartolite. [Croda Ltd.] Lanolin alcohols fraction; nonionic; w/o emulsifier, emollient, skin conditioner, moisturizing agent; pale yel. soft waxy solid; 100% act.

Hartolon 368. [Hart Chem. Ltd.] Fatty acid condensate; nonionic; softener and antistat on yarns; amber liq.; pH neutral; 95% act.

Hartolon AL. [Hart Chem. Ltd.] Fatty acid condensate; nonionic; softener and antistat for yarns; wh. paste; pH neutral; 95% act.

Hartolon NA. [Hart Chem. Ltd.] Sat. esters and alkyl condensate; nonionic; softener, lubricant, antistat used in textiles; wh. paste; 25% act.

Hartonyl L531. [Hart Chem. Ltd.] Ethoxylated fatty amine; cationic; leveling aid for dyeing polyamides; migrating leveling agent and dispersant for acid and disperse dye stuffs; increases contrast and clarity between nylon fibers of different affinities; effective in continuous and winch dyeing of carpets; yel. liq.; ref. 1.407; pH alkaline; 45% act.

Hartonyl L535. [Hart Chem. Ltd.] Ethoxylated fatty amine; cationic; dyeing assistant, antiprecipitant, dispersant; dyeing assistant for leveling and dispersing of acid and disperse dye stuffs; promotes contrast; yel. liq.; sp.gr. 1.025; pH acid; 33% act.

Hartonyl L537. [Hart Chem. Ltd.] Ethoxylated fatty amine; cationic; leveling and retarding agent for acid dyes; used for dyeing of dyeable nylon carpet; effective in continuous and winch dyeing; yel. liq.; sp.gr.

0.999; ref. index 1.412; pH alkaline; 60% act.

Hartopol L44. [Hart Chem. Ltd.] Polyoxyalkylene glycol; nonionic; dispersant and demulsifier; liq.; HLB 16.0; cloud pt. 69–73 C; 100% act.

Hartopol L64. [Hart Chem. Ltd.] Polyoxyalkylene glycol; nonionic; dispersant and emulsifier in emulsion polymerization; liq.; HLB 15.0; cloud pt. 58–62 C; 100% act.

Hartopol L81. [Hart Chem. Ltd.] Polyoxyalkylene glycol; nonionic; industrial defoamer; liq.; HLB 2.0; cloud pt. 18–22 C; 100% act.

Hartopol P65, P85. [Hart Chem. Ltd.] Polyoxyalkylene glycol; nonionic; dispersant and demulsifier; paste; HLB 17.0 and 16.0 resp.; cloud pt. 80–84 and 83–87 C resp.; 100% act.

Hartopon FS. [Hart Chem. Ltd.] Modified linear alcohol ethoxylate; nonionic; detergent for soaping off of dyestuffs, emulsifying and suspending agent; clear liq.; dens. 1.04 lb/gal; biodeg.; 75% act.

Hartosoft 75. [Hart Chem. Ltd.] Quat. imidazoline; cationic; softener and antistat; wh. liq.; sp.gr. 0.989; visc. 530 cps; pH acid; 15% act.

Hartosoft 171. [Hart Chem. Ltd.] Esters and quat. amines; cationic; softener; wh. liq.; visc. 1400 cps; pH acid; 15% act.

Hartosoft ORF. [Hart Chem. Ltd.] Fatty nitrogenous compd., modified; cationic; textile softener; wh. paste; pH acid; 100% act.

Hartosoft TAF. [Hart Chem. Ltd.] Polyamine; cationic; softener; cream liq.; pH acid; 18% act.

Hartotrope AXS 40. [Hart Chem. Ltd.] Ammonium xylene sulfonate; anionic; detergent, solubilizer and cloud pt. depressant for lt. duty and built liq. detergent systems; clear liq.; sp.gr. 1.125; 40% act.

Hartotrope KTS 44. [Hart Chem. Ltd.] Potassium toluene sulfonate; anionic; hydrotrope; for use where sodium ion undesirable; liq.; 44% act.

Hartotrope KTS 50. [Hart Chem. Ltd.] Potassium toluene sulfonate; anionic; detergent, solubilizer and cloud pt. depressant for lt. duty and built liq. detergent systems; clear liq.; sp.gr. 1.24; 50% act.

Hartotrope STS 40, Powd. [Hart Chem. Ltd.] Sodium toluene sulfonate; anionic; detergent, solubilizer and cloud pt. depressant for lt. duty and built liq. detergent systems; clear liq., wh. powd. resp.; dens. 1.19 lb/gal (liq.); 40%, 93% act. resp.

Hartotrope SXS 40, Powd. [Hart Chem. Ltd.] Sodium xylene sulfonate; anionic; detergent, solubilizer and cloud pt. depressant for lt. duty and built liq. detergent systems; clear liq., powd. resp.; sp.gr. 1.19 (liq.); 40 and 90% act.

HB-40. [Monsanto] Partially hydrog. terphenyl; plasticizer extender, polymer modifier, resin solvator for vinyl sheeting, films, fabric or paper coatings, vinyl protective coatings, adhesives.

HCl Thickener. [Tomah] Amine deriv.; cationic; thickener, visc. builder for HCl cleaners; amber visc. liq.

HD Eutanol. [Henkel] Oleyl alcohol; ultra pure grade used as as solubilizer for dyes and waxes, emollient; lt. clear oily liq.; bland odor.

HD-Ocenol. [Henkel] Oleyl alcohol; emollient, superfatting agent, carrier for cosmetics; oil-sol.

HD-Ocenol 92/96. [Henkel] Oleyl alcohol; superfatting agent for alcohol preparations and emulsions; clear oily liq.

Herbavert. [Henkel] 3,3,5-Trimethylcyclohexylethyl ether; fragrance raw material esp. for shampoos and foam baths.

Herco®. [Hercules] Pine oil; various grades for household and industrial cleaners; disinfectant, antifoam agent, textile specialties; disp. in water.

Hercoflat® Texturing and Flatting Pigment. [Hercules] Special type of PP; easily dispersed in coating vehicles for textured, nonglare finishes; used in most finishes, maintains original texture in fully cured, baked systems; powd. avail. in various particle sizes.

Hercoflex® 707. [Hercules] Pentaerythritol ester; plasticizer for PVC with outstanding heat resistance; wire insulation for high-temp. service and government specification cable construction and high-quality plastisol formulations; Pt-Co 150 max.; b.p. 143.3; sp.gr. 1.006–1.016; visc. 110 cps; pour pt. –51 C; sapon no. 390; ref. index 1.455; flash pt. 300 C (COC); fire pt. 325 C; vol. resist. 4.9×10^{11} ohm-cm.

Hercoflex® 707A. [Hercules] Pentaerythritol ester; see Hercoflex 707; Pt-Co 600 max.; b.p. 143.3 C; sp.gr. 1.010–1.020; visc. 150 cps; pour pt. –46 C; sapon. no. 390; ref. index 1.455; flash pt. 268 C (COC); fire pt. 325 C.; vol. resist. 4.9×10^{11} ohm-cm.

Hercoflex® 900. [Hercules] Polyester; plasticizer for use with PVAc; used in latex adhesives, paints, emulsion waxes and polishes; rug backings; used in food pkg. adhesives; Gardner 6 max. liq.; sol. in esters, ketones, aromatic hydrocarbons; sp.gr. 1.22; dens. 10.2 lb/gal; visc. (G-H) Z–Z2; flash pt. 238 C (COC).

Hercofloc® Flocculant Polymers. [Hercules] High m.w., syn., water-sol. polymers, anionic or cationic; flocculants and coagulant aids for industrial and municipal liq.-solids separation processes; powd. or visc. liq.

Hercolube® 302, 304. [Hercules] Polyol esters; ready-to-use completely compd. lubricating oils; wide-temp. variance gear oils designed to reduce energy consumption and maintenance in automotive and industrial uses; combine low channel point and superior thermal stability, low volatility at elevated temps. and high flash pt.

Hercolube® 404. [Hercules] Polyol esters; crankcase lubricant that meets API Engine Oil Performance and Engine Service Classifications SE and CC; contains balanced mixed-additive pkg. formulated to meet Detroit diesel and MIL-46152 requirements.

Hercolyn® D. [Hercules] Methyl hydrog. rosinate; resinous plasticizer or tackifier in finished prods. such as lacquers, inks, adhesives, floor tiles, vinyl plastisols, artificial leather, and antifouling paints; fixative and carrier in perfumes and cosmetic preps.; lt. amber visc. liq.; low odor; sol. in ester, ketones, alcohols, ethers, coal tar, petrol. hydrocarbons, and veg. and min. oils; insol. in water; dens. 1.02 kg/l; visc. (G-H) Z2–Z3; b.p. 360–364 C; acid no. 7; sapon. no. 155; ref. index 1.52; flash pt. 183 C (COC).

Hercosett® 57. [Hercules] Reactive polyamide-epichlorohydrin resins; cationic; wool shrinkproofing, antistat finishing; printing pretreatment; insolubilizing BAR fiber; aq. sol'n.

Hercosett® 70. [Hercules] Reactive polyamide-epichlorohydrin resins; see Hercosett 57; aq. sol'n.

Hercosett® 125. [Hercules] Reactive polyamide-epichlorohydrin resins; see Hercosett 57; aq. sol'n.

Hercotac® LA95BHT. [Hercules] Aromatic modified C_5 polymer; resin designed for compatibility in aliphatic- and aromatic-containing formulas; used in hot melt adhesives; tackifier for natural rubber, SBR, NBR, and SIS and SDS block polymer; contains antioxidant; Gardner 8 solid, flake; low odor; sol. in all standard adhesive formulations; dens. 1.015 kg/l; visc. 272 cps (70% solids in toluene); soften. pt. (R&B) 94 C; acid no. < 1; sapon. no. < 1; flash pt. 270 (COC).

Hercules® 137 Defoamer. [Hercules] Silica org.; defoamer used in paper and food pkg. applic.; pulp washing; lt. tan liq.; dens. 0.88 kg/l; visc. 1500–2500 cps.

Hercules® 187 Defoamer. [Hercules] Hydrocarbon oil-based; defoamer for kraft pulpmill brown stock washing systems; used in food pkg.; lt. brn. liq.; dens. 0.92 kg/l; visc. 1500–3000 cps.

Hercules® 248. [Hercules] Dk. resin; dry size used with alum in paperboard and building prods. to produce high level of resistance to water and aq. sol'ns.; brn.; water-sol.; dens. 28 lb/ft^3; 96% total solids.

Hercules® 388 Defoamer. [Hercules] Hydrocarbon oil-based; defoamer in aq. systems; used in mill effluent and waste treatment systems; kraft pulpmill screening and bleaching operations, papermaking systems; food pkg.; tan oily liq.; water-disp.; dens. 0.864 kg/l; 100% conc.

Hercules® 752. [Hercules] Nonfortified dk. paste rosin; see Hercules 248; also primarily for unbleached kraft southern pine pulps; dk. amber paste to visc. liq.; dens. 9.2 lb/gal (140 F); Stormer visc. 1300 cps (140 F); 70% total solids.

Hercules® 831 Defoamer. [Hercules] Hydrocarbon oil based; defoamer used in size press applic. and paper making and coatings; food pkg.; grnsh. brn. oily liq.; dens. 0.91 kg/l; visc. 800–1000 cps; 100% conc.

Hercules® 845 Defoamer. [Hercules] Hydrocarbon oil-based; quick-acting defoamer for coatings in paper processing; food pkg.; lt. gray liq.; dens. 7.7 lb/gal; visc. 750 cps.

Hercules® 1331. [Hercules] Fortified pale dry rosin; see Hercules 248; used primarily in high-brightness, high-color-stability papers, or in photographic papers; lt. tan; water-sol.; dens. 28 lb/ft^3; 96% total solids.

Hercules® 1512 Defoamer. [Hercules] Hydrocarbon oil based; see Hercules 187 Defoamer; gray liq.; dens. 0.92 kg/l; visc. 600 cps.

Hercules® 2051 Defoamer. [Hercules] Hydrocarbon oil based; see Hercules 187 Defoamer; gray liq.; dens. 0.92 kg/l; visc. 4000 cps.

Hercules® 2470 Defoamer. [Hercules] Silica-org.; defoamer for pulp washing; food grade paper, paperboards; lt. tan liq.; dens. 0.88 kg/l; visc. 2500 cps.

Hercules® Eff-101 Defoamer. [Hercules] Hydrocarbon oil based; defoamer for plant effluent and waste treatment systems; kraft pulpmill screening and pulpmill bleaching operations; mfg. of food paper applic.; grnsh. brn. oily liq.; water-disp.; dens. 0.91 kg/l; visc. 800–1000 cps; 100% conc.

Hercules® Cellulose. [Hercules] CMC, food grade; gum used as thickener, stabilizer, binder, rheology modifier for cosmetics, pharmaceuticals, industrial uses; powd.

Hercules® Cellulose Gum. [Hercules] Sodium CMC; food-grade gums used as thickeners, protective colloid, suspending aids, binder/lubricant, extrusion aid, stabilizer used in human and animal food, toothpaste, creams and lotions; adhesive and cohesive agent in denture adhesive; lt. cream to tan powd. in various sizes and visc.; odorless; sol. in cold and hot water; 99.5% min. act.

Hercules® CMC. [Hercules] Sodium CMC; anionic; thickener, binder, stabilizer, protective colloid, suspending agent; used in food system, cosmetics, pharmaceuticals, paper prod., adhesives, ceramics; liq.; water-sol.; sp.gr. 1.0068 (2%); dens. 0.75 g/ml; ref. index 1.3355 (2%); pH 7.5 (2%); 99.5% act.

Hercules® CMC-T. [Hercules] Sodium CMC, tech.; suspending agent for soil particles; prevents redeposition; improves detergency; permits use of more builder and less soap in built soaps; sol. in water and other water-misc. liqs.; avail. in full range of visc.; 70 and 96% conc.

Hercules® CMC-Warp Size. [Hercules] Sodium CMC; thickener; size for fabrics made from polyester and cotton blends, cotton, syn. fibers, cellulose yarns, other blends; sol. in cold and hot water.

Hercules® CMHEC-37L. [Hercules] Carboxymethyl hydroxyethyl cellulose; anionic/nonionic; hydrophilic colloid with good flocculating action on suspended solids and water-binding capacity; possibility for complexing and crosslinking reactions; lt. cream to wh. granular powd.; water-sol.; dens. 0.6 g/ml; visc. 25–85 cps; pH 6.5–8.5 (2%); 96% min. conc.

Hercules® Ester Gum 8BG. [Hercules] Purified glycerol ester of wood rosin, beverage grade; thermoplastic resin gum used in beverage industry as clouding agent; improves stability of citrus oils; USDA Rosin N max. flakes; sol. in aromatic and aliphatic hydrocarbons, terpenes, esters, ketones, and citrus and essential oils; dens. 1.08 kg/l; soften. pt. 90 C; acid no. 6.5.

Hercules® Ester Gum 8D. [Hercules] Deodorized glycerol ester of wood rosin, chewing gum grade; thermoplastic resin used as masticatory ingred. in chewing gums; resin modifier for film-formers, elastomers, and waxes, and adhesives, inks, and protective coatings where min. odor is required; USDA Rosin N max. flakes; sol. in aromatic and aliphatic hydrocarbons, esters, ketones, and CCl_4; sp.gr. 1.08; dens. 1.07 kg/l; soften. pt. 90 C; acid no. 6.5.

Hercules® Ester Gum 8D-SP. [Hercules] Deodorized glycerol ester of tall oil rosin-chewing gum grade; hard thermoplastic resin; used in chewing gums, adhesives, inks, and protective coatings; USDA Rosin Scale: WG solid, flakes; low odor; sol. see Hercules Ester Gum 8D; dens. 1.07 k/gl; soften. pt. (R&B) 80 C; acid no. 6.9.

Hercules® Ester Gum 10D. [Hercules] Glycerol ester of a partially dimerized rosin; thermoplastic resin used as a softener or plasticizer for elastomeric masticatory agents used in chewing gums; modifier for rubbers, film-formers, and waxes in adhesive and protective coating compositions; USDA Rosin M max. flakes; sol. in aromatic, aliphatic, and chlorinated hydrocarbons, esters, ketones; insol. in water; visc. (G-H) H; soften. pt. 116 C; acid no. 7.

Hercules® IAD Dry-Strength Additive. [Hercules] Cellulosic polymer; anionic; increases the efficiency of Kymene wet-strength resins for paper; sol. in water.

Hercules® X Dry Size. [Hercules] Nonfortified pale dry rosin; size; see Hercules 248; lt. tan; water-sol.; dens. 28 lb/ft^3; 96% total solids.

Hetamide LA. [Heterene] Lauric acid; nonionic; foam stabilizer, emulsifier, visc. builder, detergent for car shampoos, cleaning, cosmetics, dispersants, lubricants; liq.; sp.gr. 1.00; acid no. 10–14; pH 9–10 (1% aq.); 100% act.

Hetamide MC. [Heterene] Cocamide DEA; nonionic; see Hetamide LA; liq.; sp.gr. 0.98; pH 9.5–10.5 (1% aq.); 100% act.

Hetamide ML. [Heterene] Lauramide DEA; nonionic; see Hetamide LA; solid; sp.gr. 0.96; m.p. 35 C; 100% act.

Hetamide MMC, OC. [Heterene] Mixed fatty acids; nonionic; foam stabilizer, emulsifier, visc. builder, detergent, for car shampoos, cleaning, cosmetics, dispersants, lubricants; liq.; sp.gr. 1.01; pH 9–10.5 (1% aq.); 100% act.

Hetamide RC. [Heterene] Cocamide DEA; nonionic; see Hetamide LA; liq.; sp.gr. 1.01; pH 9–10 (1% aq.); 100% act.

Hetamine 5 L 25. [Heterene] Stearamidopropyl dimethylamine lactate; antistat, conditioner for hair; Gardner 4 max. stratified liq. to paste; pH 4.0–5.0; 19.0–21.0% solids.

Hetester ISS. [Heterene] Isostearyl stearoyl stearate; emollient for use in stick prods., pigmented emulsion-type systems; binding oil for use in pressed powds.; yel. liq.; sol. @ 5% in most cosmetic oils; partly sol. in 95% ethanol; insol. in water; acid no. 4.0 max.; sapon. no. 115–135; hyd. no. 20 max.

Hetester MS. [Heterene] Myristyl stearate; bodying agent in creams and lotions; in certain anionic systems, imparts pearling effects; replacement for spermaceti; wh. to off-wh. flake; m.p. 43–47 C; acid no. 5.0 max.; iodine no. 1.0 max.; sapon. no. 109–120.

Hetester PCA. [Heterene] PPG-1 ceteth-3 acetate; antichalking agent, emollient used in personal care prods.; colorless clear liq.; sol. @ 5% in water; cloud pt. 15 C; sapon. no. 110–130; pH 6.0–7.0.

Hetester PHA. [Heterene] PPG-1 isoceteth-3 acetate; antichalking agent, emollient used in personal care prods.; wh.-pale yel. clear liq.; self-emulsifying in water; sapon. no. 110–130; pH 6.0–7.0.

Hetester PMA. [Heterene] Propylene glycol myristyl ether acetate; emollient, solv., and plasticizer for anhyd. oil systems, emulsions; colorless clear liq.; sol. @ 5% in most cosmetic oils; insol. in water, glycols, 70% ethanol; acid no. 0.2 max.; sapon. no. 140–160; hyd. no. 5.0 max.; pH 6.0–7.0 (5% in 50/50 IPA/water); cloud pt. ≈ 0 C.

Hetester SSS. [Heterene] Stearyl stearoyl stearate; emollient used in cosmetic stick formulations, emulsion systems; cream to lt. tan cryst. solid; sol. @ 5% and 45 C in most cosmetic oils; partly sol. in 95% ethanol; insol. in water; acid no. 4.0 max.; sapon. no. 130–140; hyd. no. 20 max.

Hetester TICC. [Heterene] Triisocetyl citrate; oily liq. emollient useful in stick and pigmented emulsion-type prods.; pigment dispersing properties; yel. clear liq.; sol. @ 5% in castor oil, min. oil, safflower oil, octyl palmitate, oleyl alcohol, SD-40 ethyl alcohol (95%); insol. in water and propylene glycol; acid no. 4.0 max.; sapon. no. 165–180.

Hetlan AC. [Heterene] Acetylated lanolin alcohols; emollient for creams and lotions; pale yel. liq., faint char. odor; acid no. 1.0 max.; sapon. no. 180–200; hyd. no. 8.0 max.

Hetoxamate FA-5. [Heterene] PEG-4 tallate; nonionic; detergent, emulsifier, lubricant, softener for cosmetics, textiles, leather, metal cleaning; liq.; sol. in IPA, min. oil, disp. in water; HLB 8.4; acid no. 2.0; sapon. no. 100–120.

Hetoxamate FA-20. [Heterene] PEG-20 tallate; nonionic; see Hetoxamate FA-5; semisolid; sol. in water, IPA; disp. min. oil; HLB 14.9; acid no. 2.0; sapon. no. 50–60.

Hetoxamate LA-5. [Heterene] PEG-5 laurate; nonionic; see Hetoxamate FA-5; liq.; sol. in IPA; disp. in water; HLB 9.8; acid no. 2.0; sapon. no. 125–145.

Hetoxamate LA-9. [Heterene] PEG-9 laurate; nonionic; see Hetoxamate FA-5; liq.; sol. in water, IPA; HLB 13.1; acid no. 2.0; sapon. no. 89–96.

Hetoxamate MO-2. [Heterene] PEG-2 oleate; detergent, emulsifier for personal care prods.; softener for leather; Gardner 2 max. liq.; sol. in IPA, min. oil; insol. in water; HLB 5.3; sapon. no. 145–160.

Hetoxamate MO-5. [Heterene] PEG-5 oleate; nonionic; see Hetoxamate FA-5; liq.; sol. in IPA; disp. water; HLB 8.0; acid no. 2.0; sapon. no. 115–125.

Hetoxamate MO-9. [Heterene] PEG-9 oleate; nonionic; see Hetoxamate FA-5; liq.; sol. in IPA; disp. water; HLB 11.4; acid no. 2.0; sapon. no. 80–88.

Hetoxamate MO-15. [Heterene] PEG-15 oleate; nonionic; see Hetoxamate FA-5; liq.; sol. in water, IPA; HLB 13.5; acid no. 2.0; sapon. no. 60–70.

Hetoxamate SA-5. [Heterene] PEG-5 stearate; nonionic; see Hetoxamate FA-5; solid; sol. in IPA; disp. hot water; HLB 8.0; acid no. 2.0; sapon. no. 120–130.

Hetoxamate SA-7. [Heterene] PEG-7 monostearate; see Hetoxamate MO-2; Gardner 1 max. solid; sol. in IPA; disp. in water; HLB 10.5; sapon. no. 90–100.

Hetoxamate SA-9. [Heterene] PEG-9 stearate; nonionic; see Hetoxamate FA-5; solid; sol. in IPA; disp.

hot water; HLB 11.5; acid no. 2.0; sapon. no. 84–93.

Hetoxamate SA-13. [Heterene] PEG-12 stearate; see Hetoxamate MO-2; Gardner 2 max. liq.; sol. in water, IPA; HLB 13.4; sapon. no. 60–70.

Hetoxamate SA-23. [Heterene] PEG-20 stearate; see Hetoxamate MO-2; Gardner 2 max. solid; sol. see Hetoxamate SA-13; HLB 15.6; sapon. no. 39–49.

Hetoxamate SA-35. [Heterene] PEG-35 stearate; nonionic; see Hetoxamate FA-5; solid; sol. in water, IPA; HLB 16.9; acid no. 2.0; sapon. no. 24–34.

Hetoxamate SA-40. [Heterene] PEG-40 stearate; see Hetoxamate MO-2; Gardner 2 max. flake; sol. see Hetoxamate SA-13; HLB 17.2; sapon. no. 24–34.

Hetoxamate SA-90. [Heterene] PEG-90 stearate; nonionic; see Hetoxamate FA-5; solid; sol. in water IPA; HLB 18.6; acid no. 2.0; sapon. no. 10–18.

Hetoxamine C-2. [Heterene] PEG-2 cocamine; cationic; emulsifier, softener, antistat, water repellent, desizing agent in agriculture, waxes, oils, textile/ leather, metal cleaning; liq.; m.w. 285; sol. IPA, min. oil; gel in water.

Hetoxamine C-5. [Heterene] PEG-5 cocamine; cationic; see Hetoxamine C-2; liq.; m.w. 425; sol. water, IPA, min. oil.

Hetoxamine C-15. [Heterene] PEG-15 cocamine; cationic; see Hetoxamine C-2; liq.; m.w. 860; sol. water, IPA.

Hetoxamine O-2. [Heterene] PEG-2 oleamine; nonionic; see Hetoxamine C-2; liq.; m.w. 350; sol. in IPA, min. oil.

Hetoxamine O-5. [Heterene] PEG-5 oleamine; nonionic; see Hetoxamine C-2; liq.; m.w. 492; sol. in IPA, min. oil.

Hetoxamine O-15. [Heterene] PEG-15 oleamine; nonionic; see Hetoxamine C-2; liq.; m.w. 930; sol. in water, IPA.

Hetoxamine S-2. [Heterene] PEG-2 soyamine; cationic; see Hetoxamine C-2; liq.; m.w. 350; sol. in IPA, min. oil.

Hetoxamine S-5. [Heterene] PEG-5 soyamine; cationic; see Hetoxamine C-2; liq.; m.w. 480; sol. in IPA; partly sol. in min. oil; gel in water.

Hetoxamine S-15. [Heterene] PEG-15 soyamine, cationic; desizing agent, antistat; emulsifier in agriculture, waxes and oils, leather processing, and metal cleaning industries; water repellent and wet spinning assistant in textile industries; Gardner 12 max. liq., solid; sol. see Hetoxamate SA-13; 95.0% min. tert. amine.

Hetoxamine ST-2. [Heterene] PEG-2 stearamine; nonionic; see Hetoxamine C-2; solid; m.w. 388; sol. in IPA, min. oil.

Hetoxamine ST-5. [Heterene] PEG-5 stearamine; nonionic; see Hetoxamine C-2; solid; m.w. 520; sol. in IPA, min. oil.

Hetoxamine ST-15. [Heterene] PEG-15 stearamine; see Hetoxamine S-15; Gardner 8 max. solid; sol. see Hetoxamate SA-13; 95.0% min. tert. amine.

Hetoxamine ST-50. [Heterene] PEG-50 stearamine; see Hetoxamine S-15; Gardner 8 max. solid; sol. see Hetoxamate SA-13.

Hetoxamine T-2. [Heterene] PEG-2 tallow amine; cationic; see Hetoxamine C-2; liq.; m.w. 350; sol. in IPA, min. oil.

Hetoxamine T-5. [Heterene] PEG-5 tallow amine; cationic; see Hetoxamine C-2; liq./paste; m.w. 490; sol. in water, IPA, min. oil.

Hetoxamine T-15. [Heterene] PEG-15 tallow amine; cationic; see Hetoxamine C-2; liq./paste; m.w. 925; sol. in water, IPA.

Hetoxamine T-20. [Heterene] PEG-20 tallow amine; cationic; see Hetoxamine C-2; liq./paste; m.w. 1150; sol. in water, IPA.

Hetoxide BN-13. [Heterene] B-naphthol ethoxylate; nonionic; emollient, emulsifier, visc. control agent, lubricant, dispersant, perfume solubilizer, used in cosmetics, household, textile industry, metal treating and plating; intermediate; paste; sol. in water, IPA, min. oil; acid no. 1.0.

Hetoxide BP-3. [Heterene] Modified butanol ethoxylate; nonionic; emollient, emulsifier, used in cosmetics, household, textile industry; liq.; sol. in water, IPA.

Hetoxide BY-3. [Heterene] PEG-3 butynediol; nonionic; emollient, emulsifier used in cosmetics, household, textile industry; liq.; sol. in water, IPA.

Hetoxide C-2. [Heterene] PEG-2 castor oil; nonionic; emollient, emulsifier, used in cosmetics, household, textile industry; liq.; sol. in IPA, min. oil; HLB 4.3; acid no. 1.0; sapon. no. 155–170.

Hetoxide C-9. [Heterene] PEG-9 castor oil; nonionic; see Hetoxide C-2; liq.; sol. in IPA, min. oil; HLB 11.2; acid no. 1.0; sapon. no. 120–136.

Hetoxide C-15. [Heterene] PEG-15 castor oil; nonionic; see Hetoxide C-2; liq.; sol. in IPA, min. oil; HLB 13.6; acid no. 1.0; sapon. no. 95–105.

Hetoxide C-25. [Heterene] PEG-25 castor oil; nonionic; see Hetoxide C-2; liq.; sol. in IPA, min. oil; HLB 15.6; acid no. 1.0; sapon. no. 74–82.

Hetoxide C-30. [Heterene] PEG-30 castor oil; nonionic; perfume solubilizer, emollient, emulsifier, visc. control and scouring agent, lubricant, dispersant for cosmetic formulations; dyeing assistant, dye carrier for textiles; used in household cleaning comps., metal treatment, metal plating, chemical intermediate; Gardner 5 max. liq.; sol. in IPA; disp. in water; HLB 11.7; sapon. no. 65–75.

Hetoxide C-40. [Heterene] PEG-40 castor oil; nonionic; see Hetoxide C-2; paste; sol. in water, IPA; HLB 16.9; acid no. 1.0; sapon. no. 55–65.

Hetoxide C-200. [Heterene] PEG-200 castor oil; nonionic; see Hetoxide C-2; solid; sol. in water, IPA; acid no. 1.0; sapon. no. 16–18.

Hetoxide C-200-50%. [Heterene] PEG-200 castor oil; see Hetoxide C-30; Gardner 4 max. liq.; sol. in water, IPA; sapon no. 7–10; 49.0–51.0% moisture.

Hetoxide DNP-4, DNP-9.6. [Heterene] Nonyl nonoxynol-4, -9.6; nonionic; see Hetoxide C-2; liq.; sol. in IPA, min. oil, disp. in water; HLB 6.6, 10.7; hyd. no. 120–140, 65–75.

Hetoxide G-7. [Heterene] Glycereth-7; nonionic; see Hetoxide C-2; liq.; sol. in IPA, water; acid no. 1.0.

Hetoxide G-26. [Heterene] Glycereth-26; humectant for pressure-sensitive adhesives; Gardner 2 max. liq.;

sol. in water, IPA; acid no. 2.0 max.

Hetoxide HC-16. [Heterene] PEG-16 hydrog. castor oil; nonionic; see Hetoxide C-2; liq.; sol. in IPA, min. oil; HLB 11.3; acid no. 1.0; sapon. no. 85–95.

Hetoxide HC-40. [Heterene] PEG-40 hydrog. castor oil; see Hetoxide C-30; Gardner 4 max. semisolid; sol. see Hetoxamate SA-13; HLB 13.1; sapon. no. 50–60.

Hetoxide HC-60. [Heterene] PEG-60 hydrog. castor oil; see Hetoxide C-30; Gardner 4 max. solid; sol. see Hetoxamate SA-13; HLB 14.8; sapon. no. 41–51.

Hetoxide MPC. [Heterene] m,p-Cresol hydrophobe; solv. for lacquers and coatings; Gardner 5 max. liq.; sol. in IPA, min. oil; disp. in water; hyd. no. 275–295.

Hetoxide MTG. [Heterene] Alcohol ethoxylate; nonionic; see Hetoxide C; liq.; sol. in IPA, min. oil, water; acid no. 1.0.

Hetoxide NP-4. [Heterene] Nonoxynol-4; nonionic; see Hetoxide C; liq.; sol. in IPA, min. oil; HLB 7.6; hyd. no. 135–140.

Hetoxide NP-9. [Heterene] Nonoxynol-9; nonionic; see Hetoxide C; liq.; sol. in water, IPA; HLB 13.0; hyd. no. 85–95.

Hetoxide NP-40. [Heterene] Nonoxynol-40; nonionic; see Hetoxide C; liq.; sol. in water, IPA; HLB 17.0.

Hetoxol 15 CSA. [Heterene] Ceteareth-15; nonionic; detergent, emulsifier, leveling agent, intermediate; used in personal care prods., wax, oil and textiles, scouring agents, dyes, household formulations, silicone emulsification, surfactants; Gardner 1 max. solid; sol. see Hetoxamate SA-13; HLB 14.2.

Hetoxol CA-2, CA-10, CA-20. [Heterene] Ceteth-2, -10, –20; nonionic; detergent, emulsifier, leveling agent, intermediate, used for cosmetics, household formulations, silicone emulsification, textile processing; solid; sol. in IPA; CA-2 sol. in min. oil; CA-10 sol. in water, IPA; CA-20 sol. in water; HLB 5.3, 12.7, 15.7; hyd. no. 160–180 (CA-2), 45–60 (CA-20).

Hetoxol CS-4, CS-5. [Heterene] Ceteareth-4, –5; see Hetoxol 15 CSA; Gardner 1 max. solid; sol. in IPA; disp. in water; HLB 8.2 and 9.2 resp.

Hetoxol CS-9. [Heterene] Ceteareth-9; nonionic; see Hetoxol CA-2; solid; sol. in water, IPA; HLB 12.0; hyd. no. 80–90.

Hetoxol CS-15. [Heterene] Ceteareth-15; see Hetoxol 15 CSA; Gardner 1 max. solid; sol. water, IPA; HLB 14.2.

Hetoxol CS-20. [Heterene] Ceteareth-20; nonionic; see Hetoxol CA-2; solid; sol. in water, IPA; HLB 15.4; hyd. no. 50–70.

Hetoxol CS-30. [Heterene] Ceteareth-30; nonionic; see Hetoxol CA-2; solid; sol. in water, IPA; HLB 16.7; hyd. no. 40–52.

Hetoxol CS-50, CS-50 Special. [Heterene] Ceteareth-50; see Hetoxol 15 CSA; Gardner 2 max. flake; sol. in water, IPA.

Hetoxol L-3N. [Heterene] Laureth-3; nonionic; see Hetoxol CA-2; liq.; sol. in IPA, min. oil; HLB 7.9; hyd. no. 170–176.

Hetoxol L-4N. [Heterene] Laureth-4; nonionic; see Hetoxol CA-2; liq.; sol. in IPA, min. oil; HLB 9.7; hyd. no. 145–165.

Hetoxol L-9N. [Heterene] Laureth-9; nonionic; see Hetoxol CA-2; liq.; sol. in water, IPA; HLB 11.8; hyd. no. 90–110.

Hetoxol L-23N. [Heterene] Laureth-23; nonionic; see Hetoxol CA-2; solid; sol. in water, IPA; HLB 16.9; hyd. no. 40–55.

Hetoxol LS-9. [Heterene] Laureth-9; nonionic; see Hetoxol CA-2; solid; sol. in water, IPA; HLB 13.3; hyd. no. 90–100.

Hetoxol M-3. [Heterene] Myreth-3; emulsifier and pigment dispersant in makeup; sol. in IPA; disp. in water; HLB 7.6; pH 5.5–7.0.

Hetoxol OA-3 Special. [Heterene] Oleth-3; emulsifier and pigment dispersant for cosmetic applics.; pale yel. liq.; sol. in IPA, min. oil; insol. in water; HLB 6.4.

Hetoxol OA-5 Special. [Heterene] Oleth-5; see Hetoxol OA-3 Special; pale yel. liq.; sol. in IPA, min. oil; disp. in water; HLB 9.0.

Hetoxol OA-10 Special. [Heterene] Oleth-10; see Hetoxol OA-3 Special; wh. semisolid; sol. in water, IPA; HLB 12.4.

Hetoxol OA-20 Special. [Heterene] Oleth-20; see Hetoxol OA-3 Special; wh. solid; sol. see Hetoxol OA-10 Special; HLB 15.3.

Hetoxol OL-2. [Heterene] Oleth-2; nonionic; see Hetoxol CA-2; liq.; sol. in IPA, min. oil; HLB 4.9; hyd. no. 152–162.

Hetoxol OL-4. [Heterene] Oleth-4; nonionic; see Hetoxol CA-2; liq.; sol. in IPA, min. oil; HLB 8.0; hyd. no. 120–135.

Hetoxol OL-5. [Heterene] Oleth-5; nonionic; see Hetoxol CA-2; liq.; sol. in water, IPA HLB 9.1; hyd. no. 110–120.

Hetoxol OL-23. [Heterene] Oleth-23; nonionic; see Hetoxol CA-2; semisolid; sol. in water, IPA; HLB 15.9; hyd. no. 47–62.

Hetoxol OL-40. [Heterene] Oleth-40; nonionic; see Hetoxol CA-2; solid; sol. in water, IPA; HLB 17.4; hyd. no. 30–45.

Hetoxol PLA. [Heterene] PPG 30 lanolin ether; oily emollient; liq.; sol. in min. oil, IPA; insol. in water; acid no. 1.0 max.

Hetoxol SP-15. [Heterene] PPG-15 stearyl ether; oily emollient material in cosmetics; ASTM 100 max. liq.; sol. in IPA, min. oil; insol. in water; acid no. 1.5 max.; sapon. no. 2.0 max.; hyd. no. 62–70.

Hetoxol STA-2. [Heterene] Steareth-2; nonionic; see Hetoxol CA-2; liq.; sol. in IPA; HLB 4.9; hyd. no. 145–165

Hetoxol STA-10. [Heterene] Steareth-10; see Hetoxol 15 CSA; Gardner 1 max. solid; sol. in water, IPA; HLB 12.4.

Hetoxol STA-20. [Heterene] Steareth-20; see Hetoxol 15 CSA; Gardner 1 max. solid; sol. in water, IPA; insol. in min. oil; HLB 15.3.

Hetoxol STA-30. [Heterene] Steareth-30; see Hetoxol 15 CSA; Gardner 2 max. flake; sol. see Hetoxol STA-20; HLB 16.6; m.p. 46–50 C.

Hetoxol TD-3. [Heterene] Trideceth-3; nonionic; see

Hetoxol CA-2; liq.; sol. in IPA, min. oil; HLB 7.9; hyd. no. 165–175.

Hetoxol TD-6. [Heterene] Trideceth-6; nonionic; see Hetoxol CA-2; liq.; sol. in IPA, water; HLB 11.3; hyd. no. 115–125.

Hetoxol TD-12. [Heterene] Trideceth-12; nonionic; see Hetoxol CA-2; liq.; sol. in IPA, water; HLB 14.5; hyd. no. 72–87.

Hetsorb L-10. [Heterene] PEG-10 sorbitan laurate; detergent, emulsifier, lubricant for cosmetics; liq.; sol. in water, IPA; HLB 8.4; sapon. no. 66–76.

Hetsorb L-20. [Heterene] Polysorbate 20; see Hetsorb L-10; liq.; sol. in water, IPA; HLB 16.7; sapon. no. 40–50.

Hetsorb O-20. [Heterene] Polysorbate 80; see Hetsorb L-10; liq.; sol. in water, IPA; HLB 15.0; sapon. no. 45–55.

Hetsorb S-20. [Heterene] Polysorbate 60; see Hetsorb L-10; liq.; sol. in water, misc. in IPA; HLB 14.9; sapon. no. 45–55.

Hetsulf 40, 40X. [Heterene] Sodium dodecylbenzene sulfonate; anionic; wetting agent, emulsifier, dispersant, intermediate, detergent, liq. formulation syndet; paste, liq.; pH 7–8.5 (5% aq.); biodeg.; 38% and 35% act.

Hetsulf 50A. [Heterene] Ammonium dodecylbenzene sulfonate; anionic; wetting agent, emulsifier, dispersant, for lt. duty detergent formulations; slurry; pH 6.0–7.0 (5% aq.); biodeg.; 48% act.

Hetsulf 60S. [Heterene] Sodium dodecylbenzene sulfonate; anionic; wetting agent, emulsifier, dispersant, base for formulated prods.; slurry; pH 7.0–8.0 (5% aq.); biodeg.; 57% act.

Hetsulf 60T. [Heterene] Amine dodecylbenzene sulfonate; anionic; wetting agent, emulsifier, dispersant, for cosmetic, bath and shampoos uses; liq.; pH 7.0–7.5 (5% aq.); biodeg.; 52% act.

Hetsulf Acid. [Heterene] Dodecylbenzene sulfonic acid; anionic; wetting agent, emulsifier, dispersant, intermediate, base for neutralized surfactant; liq.; dens. 1.05; biodeg.; 97% act.

Hetsulf IPA. [Heterene] MIPA-dodecylbenzene sulfonate; anionic; wetting agent, emulsifier, dispersant; liq.; pH 5.0–6.5 (5% aq.); biodeg.

Hexaplant Richter. [Henkel] Polyvalent herbal extract in aq. alcohol; emollient for aq. and hydroalcoholic herbal cosmetics, emulsified preparations; dk. brn. liq.; herbal odor.

Hexcel FO 425A. [Hexcel/Rezolin] Petrol. solv.; general purpose cleaning and degreasing solv. for cleaning filled telecommunications cables and tools; noncorrosive; clear; odorless; sp.gr. 0.753; flash pt. 200 F (COC).

Hexetidine. [Angus] Substituted hexa hydropyrimidine; antimicrobial, antifungal agent for oral hygiene prods.; liq.; sol. in glycols, nonpolar solvs.; sparingly sol. in water; 90 and 95% act.

Hexol Q. [GAF] Complex phosphate compd.; sequestrant and complexing agent, water softener, and conditioner; textile aux.; liq.

HF-19. [Chem-Trend] Water-glycol type; fire-resistant hydraulic fluid; also provides protection from rust and corrosion; red liq.; sp.gr. 1.07; visc. 180 ± 10 SUS (100 F); visc. index > 150; pH 9.0.

HFR-201. [M&T Harshaw] Proprietary pentavalent antimony sol, alkalized (aq. colloidal disp.); flame retardant for industrial textile applic.; wh. milky liq., 30–100 mμ primary particle size; odorless; b.p. 212 F; sp.gr. 1.590 ± 0.020; dens. 13.4 lb/gal; visc. < 75 cps; pH 2.8–4.0; 48–50% solids (dried @ 110 C); 30% Sb.

HFR-301. [M&T Harshaw] Proprietary pentavalent antimony oxide, alkalized; flame retardant used in highly pigmented and clear/transparent vinyl coated textiles/fabrics; wh. powd., 30–100 mμ primary particle size; 8–15 mμ aggregate particle size; odorless; dens. 1.35 ± 0.05 g/cc; dens. 4.50 ± 0.10 g/cc; pH 8.5–11 (10% water slurry); 100% solids; 58.5–59.5% Sb.

Hinosan®. [Bayer] Edifenphos; nonmercurial fungicide for rice; clear ylsh. liq.; m.w. 310.4; sol. in acetone and xylene; b.p. 154 C (0.01 mbar).

Hi-Pflex®. [Pfizer] Calcium carbonate, surface-treated; high-performance reinforcing agent for plastics; applics. incl. PVC (pipe, wire and cable), PP (interior and exterior auto parts, toys, pallets, corrugated boxes), HDPE (pipe and other rigid applics., wire and cable and other flexible applics.); very wh. powd.; 3.0 μ avg. particle size; sp.gr. 2.71; surf. area 3.5 m^2/g; oil absorp. 14 lb/ 100 lb; dry brightness 95; 97% $CaCO_3$.

Hi-Pflex® 100. [Pfizer] Calcium carbonate, surface-treated; filler improving impact str. and flex. mod. in polymers; very wh. free-flowing powd.; 3.0 μ avg. particle size; sp.gr. 2.71; surf. area 3.5 m^2/g; oil absorp. 14 lb/ 100 lb; dry brightness 91; 94–97% $CaCO_3$.

Hipochem ADN. [High Point] Quat. ammonium complex; nonionic; dyeing assistant, leveling agent of acid dyes on nylon; amber liq.; mild alcohol odor; sol. in water; dens. 8.73 lb/gal; sp.gr. 1.048; 50% act.

Hipochem C-95. [High Point] Quat. ammonium compd.; cationic; dyeing assistant, retarder for cationic dyes on acrylics; amber liq.; mild alcohol odor; water sol.; dens. 8.29 lb/gal; sp.gr. 0.995; 50% act.

Hipochem CAD. [High Point] Aromatic ester, emulsified; amphoteric; dyeing assistant for textiles; liq.; 100% conc.

Hipochem CDL. [High Point] Fatty alcohol ethoxylate; nonionic; dyeing assistant and scouring agent for textiles; nonionic; liq.; 30% conc.

Hipochem D2. [High Point] Quat. ammonium compd.; cationic; dyeing assistant, retarder for cationic dyes on acrylics; amber liq.; mild alcohol odor; water sol.; dens. 8.08 lb/gal; sp.gr. 0.97; 40% act.

Hipochem Dispersol GTO. [High Point] Sulfated glycerol trioleate; anionic; detergent, wetting agent, emulsifier, dyeing assistant; dk. amber liq.; mild sulfate odor; water sol.; dens. 8.66 lb/gal; sp.gr. 1.04; 55% act.

Hipochem Dispersol SB. [High Point] Sulfated butyl oleate; anionic; see Hipochem Dispersol GTO; dk. amber liq.; mild sulfate odor; water disp.; dens. 8.41 lb/gal; sp.gr. 1.01; 55% act.

Hipochem Dispersol SCO. [High Point] Sulfated castor oil; anionic; see Hipochem Dispersol GTO; dk. amber liq.; mild sulfate odor; water disp.; dens. 8.33 lb/gal; sp.gr. 1.0; 70% act.

Hipochem Dispersol SP. [High Point] Sulfated propyl oleate; anionic; dyeing assistant, detergent, emulsifier used in textiles; liq.; 60% conc.

Hipochem EFK. [High Point] Aliphatic sulfate; detergent, textile lubricant, substantive to cellulosic fibers; wh. paste; water sol.; 40% act.

Hipochem LCA. [High Point] Sulfonated oleyl alcohol; anionic; detergent, lubricant, textile dye leveling agent; wh. paste; mild sulfate odor; disp. in water (10%); 40% act.

Hipochem No. 3. [High Point] Sulfated oil plus soap; anionic; detergent and dye leveling agent used in textiles; liq.; 45% conc.

Hipochem No. 40-L. [High Point] Modified ABS; anionic; detergent, wetting agent, emulsifier, dyeing penetrant; dk. amber liq.; water sol.; dens. 8.33 lb/gal; sp.gr. 1.0; 40% act.

Hipochem No. 641. [High Point] Alkyldiaryl sulfonate; anionic; dyeing assistant for polyamide fibers; lt. br. liq.; water sol.; dens. 9.7 lb/gal; sp.gr. 1.16; 45% act.

Hipochem Retarder CJ. [High Point] Quat. ammonium compd.; cationic; dyeing assistant; retarder for cationic dyes on acrylic fibers; dye retarder for jet machines; amber liq.; mild amine odor; water-sol.; sp.gr. 0.985; dens. 8.21 lb/gal; pH acid; 40% act.

Hi-Point® 90. [Witco/Argus] MEK peroxide in dimethyl phthalate; catalyst/initiator for R.T. cures of polyesters; clear liq.; 9.0% act. oxygen.

Hi-Point® 90 Red. [Witco/Argus] MEK peroxide in dimethyl phthalate; dk. red formulation of High-Point 90 specifically designed as a visual aid; 9.0% act. oxygen.

Hi-Point® PD-1. [Witco/Argus] MEK peroxide in dimethyl phthalate; catalyst/initiator for R.T. cures of polyesters; clear liq.; 5.25% act. oxygen.

Hi-Sil® 132. [PPG Industries] Precipitated silica; reinforcing agent for rubber; thickener and carrier; powd.; 16 nm particle size; dens. 10 lb/ft^3; surf. area (BET) 200 m^2/g; pH 7.

Hi-Sil® 210. [PPG Industries] Amorphous hydrated silica; wh. pigment, reinforcing filler for rubber prods.; diluent, grinding aid, stabilizer, thixotrope; agric. chemicals; promotes adhesion in natural and syn. rubber-based adhesives; anticaking agent in lawn fertilizers, fungicides, grinding wheel abrasives, drain cleaners, laundry sour compds., hexamethylenetetramine for phenolic molding compds.; absorbent carrier and flow conditioner for solids and visc. control in liqs.; pellets; pH 6.5–7.3; 87.0% max. SiO_2 hydrate.

Hi-Sil® 233. [PPG Industries] Amorphous hydrated silica; see Hi-Sil 210; powd., 0.5% max. retained on 325-mesh screen; pH 6.5–7.3; 87.0% max. SiO_2 hydrate.

Hi-Sil® 243LD. [PPG Industries] Precipitated silica; reinforcing agent for rubber; nugget; 19 nm particle size; dens. 17 lb/ft^3; surf. area (BET) 150 m^2/g; pH 7.

Hi-Sil® 250. [PPG Industries] Amorphous hydrated silica; see Hi-Sil 210; powd., 0.5% max. retained on 325-mesh screen; pH 6.5–7.3; 87.0% max. SiO_2 hydrate.

Hi-Sil® 532EP. [PPG Industries] Precipitated silica; reinforcing agent for rubber, aids extrusion and resilience; carrier; powd.; 46 nm particle size; dens. 10 lb/ft^3; surf. area (BET) 55 m^2/g; pH 8.

Hi-Sil® ABS. [PPG Industries] Precipitated silica; carrier to convert ester and liq. resin plasticizers and bonding agents to free-flowing powds. for introduction to rubber compds.; grinding aid, suspension aid, reinforcing filler, and anticaking agent; wh. particulate; 15 μm median particle diam.; sp.gr. 2.0; bulk dens. 8 lb/ft^3; surf. area 150 m^2/g; oil absorp. 305 cc/100 g; pH 7.0 (5% aq. slurry); 97.5% SiO_2.

Hi-Sil® T-600. [PPG Industries] Syn. silica, amorphous; thickener used to increase visc. and provide thixotropic action for liqs.; prevents sagging of paints, sealants, and materials with vertical surf.; stabilizes emulsions of immiscible liqs. acting to prevent phase separation; used in coatings, laminating, and epoxy resins, adhesives, caulks, putties, sealants, pharmaceuticals, cosmetics, toothpaste, printing ink, plasticols, areosols, waxes and greases; antisettling agent for coarse particles in liq.; wh. powd., 0.021 μ particle size; 0.002% 325-mesh wet sieve residue; sp.gr. 2.1; dens. 17.5 lb/gal; dens. 2–4 lb/ft^3; ref. index 1.455; pH 6.5–7.3 (5%); 97.5% SiO_2.

Hi-Sil® T-690. [PPG Industries] Precipitated silica; thixotrope for resins, adhesives; powd.; 1.3 μm median size; dens. 4 lb/ft^3 (tapped); surf. area 170 m^2/g; oil adsorp. 220 linseed oil lb/100 lb; pH 7.

Hitox®. [Hitox] Titanium dioxide; buff-colored pigment developed as alternative to wh. titanium dioxide; used in alkyds, acrylic urethanes, high solids systems, water reducibles, water bases, powd. coatings, inks, and adhesives; buff; 1.5 μ avg. particle size; Hegman grind 6–6.5; sp.gr. 4.1; dens. 45 lb/ft^3 (poured), 76 lb/ft^3 (tapped); surf. area 0.7133 m^2/g; oil absorp. 18–22; pH 6.5–7.0; 95% rutile TiO_2.

Hi-Tri®. [Dow] Trichloroethylene; solv.

Hodag FD Series. [Hodag] Silicones; food and industrial defoamers; liq.

Hodag MR-216. [Hodag] High m.w. polymeric ester; release agent for nonaq. coatings; liq.

Hodag RA-2, RA-7. [Hodag] Fatty esters; nonionic; release agent for paper industry; liq.

Hoe S 2817. [Hoecht Celanese] Alkyl substituted dicarboxylic anhydride; nonionic; hydrotrope/solubilizer for formulations with high electrolyte concs.; liq.; sol. in min. oil, gasolines; insol. in water; 100% act.

Hoechst Wax LP. [Hoecht Celanese] Base crude montan wax; water repellent for textiles.

Hombitan® LOCR-K. [Sachtleben] Surf.-treated, micronized anatase TiO_2; wh. pigment with high tint reduction; for plastics; dens. 3.9 g/ml; pH 8.

Hombitan® LW. [Sachtleben] Micronized, pure wh. anatase TiO_2; wh. pigment with high tint reduction; for plastics; dens. 3.9 g/ml; pH 8.

Hombitan® R 101 D. [Sachtleben] Surf.-treated,

micronized rutile TiO_2; wh. pigment producing clear and brilliant color shades in polymer systems; dens. 4.1 g/ml; pH 8.

Hombitan® R 610 K. [Sachtleben] Stabilized, surf.-treated, micronized rutile TiO_2; wh. pigment with high resistance to weathering; used in PVC compds., fluidized bed and powd. coatings; dens. 4.0 g/ml; pH 8.

Hombitan® R 610 L. [Sachtleben] Stabilized, surf.-treated, micronized rutile TiO_2; wh. pigment with high photochem. stability; used for urea-formaldehyde and melamine-formaldehyde molding materials, esp. laminated plastics; dens. 4.0 g/ml; pH 7.

Homodan. [Grindsted] Special emulsifiers and blends; plasticizer, improves plasticity of margarine for flakier puff pastry, nonionic; flakes, liq., pills, solid; 100% act.

Homogenol L-18. [Kao] Polymeric surfactant; dispersant in oil vehicles, nonpolar solv.; liq.

Homotex PT. [Kao] Monoglyceride, med. chain; nonionic; bacteria control; liq.; HLB 3.2; 50% min.

Honoralin PL. [Takemoto] Sulfated oil; anionic; wetting and penetrating agent; dyeing aux. for cotton and wool; liq.

Horse Head® Standard Zinc Dust 22. [Zinc Corp. of Am.] Zinc dust with unslaked lime; blue-gray pigment which imparts corrosion resistance in metal protective coatings; 8 μ avg. particle size; 98% thru 325 mesh; sp.gr. 7.0; dens. 58.6 lb/solid gal; pkg. dens. 220 lb/ft³; 96.1% metallic zinc.

Horse Head® Standard Zinc Dust 44. [Zinc Corp. of Am.] Zinc dust; blue-gray pigment imparting corrosion resistance in metal protective coatings; used in lubricants, soot removers, smoke screens, brake linings, mfg. of dyestuffs and other chemicals; 8 μ avg. particle size; 98% thru 325 mesh; sp.gr. 7.0; dens. 58.6 lb/solid gal; pkg. dens. 220 lb/ft³; 96.7% metallic zinc.

Horse Head® Standard Zinc Dust 122. [Zinc Corp. of Am.] Zinc dust; blue-gray pigment for use in org. metal-protective paints imparting rust-inhibition and weather resistance; 8.0 μ avg. particle size; 98% thru 325 mesh; sp.gr. 7.0; dens. 58.6 lb/solid gal; pkg. dens. 220 lb/ft³; 97.5% metallic zinc.

Horse Head® Standard Zinc Dust 222. [Zinc Corp. of Am.] Zinc dust with unslaked lime; blue-gray pigment for use in org. metal-protective paints imparting rust-inhibition and weather resistance; 8.0 μ avg. particle size; 98% thru 325 mesh; sp.gr. 7.0; dens. 58.6 lb/solid gal; pkg. dens. 220 lb/ft³; 97.1% metallic zinc.

Horse Head® Standard Zinc Dust 422. [Zinc Corp. of Am.] Zinc dust with unslaked lime; blue-gray pigment for use in org. coatings imparting corrosion resistance to zinc dust primers; 6.0 μ avg. particle size; 99% thru 325 mesh; sp.gr. 7.0; dens. 58.6 lb/solid gal; pkg. dens. 200 lb/ft³; 94.6% metallic zinc.

Horse Head® Standard Zinc Dust 444. [Zinc Corp. of Am.] Zinc dust; blue-gray pigment for use in inorg. coatings imparting corrosion resistance to zinc dust primers; used in the mfg. of sodium hydrosulfite and sodium sulfoxylate formaldehyde; reducing agent for org. compds. during the preparation of dyestuffs, pharmaceuticals, and other fine chemicals; 6.0 μ avg. particle size; 99% thru 325 mesh; sp.gr. 7.0; dens. 58.6 lb/solid gal; pkg. dens. 200 lb/ft³; 95.0% metallic zinc.

Horse Head® XX®-4. [Zinc Corp. of Am.] Amer. process zinc oxide; pigment for the rubber industry; provides uniform activation and reinforcement; also used in rubber adhesives for industrial tapes, in lubricants, abrasive sheets; 0.27 μ mean particle size; 99.97% thru 325 mesh; sp.gr. 5.6; sp.vol. 0.18; pkg. dens. 30 lb/ft³; surf. area 4.0 m²/g; oil absorp. 18 lb oil/100 lb ZnO; 99.0% zinc oxide.

Horse Head® XX®-32. [Zinc Corp. of Am.] Amer. process zinc oxide; pelleted form of Horse Head XX-4; preferred for ceramic and glass applics.; 0.27 μ mean particle size; sp.gr. 5.6; dens. 46.7 lb/solid gal; pkg. dens. 65 lb/ft³; surf. area 4.0 m²/g; oil absorp. 16 lb oil/100 lb pigment; 99.0% zinc oxide.

Horse Head® XX®-78. [Zinc Corp. of Am.] French process zinc oxide; pigment and chemical used in wh. and tinted rubber goods, insulated wire, packings, latex, rubber adhesives, ceramics, rayon, floor coverings, and metal treatment; provides reinforcement, tack retention; 0.31 μ mean particle size; 99.99% thru 325 mesh; sp.gr. 5.6; sp.vol. 0.18; pkg. dens. 40 lb/ft³; surf. area 3.5 m²/g; oil absorp. 12 lb oil/100 lb ZnO; 99.8% zinc oxide.

Horse Head® XX®-85. [Zinc Corp. of Am.] French process zinc oxide; pigment for reactive polymers; yel.-buff; 0.31 μ mean particle size; 99.99% thru 325 mesh; sp.gr. 5.6; sp.vol. 0.18; pkg. dens. 55 lb/ft³; surf. area 2.8 m²/g; oil absorp. 119 lb oil/100 lb ZnO; 99.3% zinc oxide.

Horse Head® XX®-503. [Zinc Corp. of Am.] Amer. process zinc oxide; pigment offering max. tint retention and mildew protection for exterior house paints, glazes in ceramics industry; 0.8 μ mean particle size; 99.95% thru 325 mesh; sp.gr. 5.6; dens. 46.7 lg/solid gal; pkg. dens. 60 lb/ft³; surf. area 1.3 m²/g; oil absorp. 10 lb oil/100 lb pigment; 99.2% zinc oxide.

Horse Head® XX®-600. [Zinc Corp. of Am.] Amer. process zinc oxide; pigment offering max. tint retention and mildew protection in exterior house paints; 0.6 μ mean particle size; 99.95% thru 325 mesh; sp.gr. 5.6; dens. 46.7 lb/solid gal; pkg. dens. 35 lb/ft³; surf. area 1.7 m²/g; oil absorp. 12 lb oil/100 lb pigment; 99.0% zinc oxide.

Horse Head® XX®-601. [Zinc Corp. of Am.] Amer. process zinc oxide; pigment featuring max. tint retention and mildew protection; used in oil-based exterior house paints; 0.27 μ mean particle size; 99.97% thru 325 mesh; sp.gr. 5.6; dens. 46.7 lb/solid gal; pkg. dens. 30 lb/ft³; surf. area 4.0 m²/g; oil absorp. 16 lb oil/100 lb pigment; 99.0% zinc oxide.

Horse Head® XX®-631. [Zinc Corp. of Am.] Amer. process zinc oxide; organophilic surface treatment; pigment featuring max. tint retention and mildew protection; used in exterior house paints; 0.6 μ mean particle size; 99.97% thru 325 mesh; sp.gr. 5.6; dens. 46.7 lb/solid gal; pkg. dens. 35 lb/ft³; surf. area 1.7 m²/g; oil absorp. 12 lb oil/100 lb pigment; 99.0% zinc oxide.

Hostacerin PN 73. [Hoecht Celanese] Sodium salt of an acrylic acid copolymer; thickener for cosmetic use; powd.; water-sol.

Hostacerin T-3. [Hoechst Celanese] Ceteareth-3; nonionic; emulsifier, superfatting agent, base for ointments, creams, liq. emulsions, shampoo additive; wh. soft, waxlike substance; sol. warm in all hydrocarbons, fatty alcohols; sp.gr. 0.905 (50 C); visc. 15 ± 3 cps (50 C); HLB 7–8; cloud pt. 54 C (in butyl diglycol); flash pt. 220 C; sapon. no. 1 max.; 100% act.

Hostacor 2098, 2125, 2732. [Hoecht Celanese] Complex carboxylic acid; corrosion inhibitor in aq. systems; sol. in water.

Hostacor BK. [Hoechst Celanese] Complex carboxylic acid condensation prod.; corrosion inhibitor, emulsifier, and lubricant for formulation of drilling aids and aq. cooling systems; brn. clear visc. liq.; sol. in water, min. oil, glycols; dens. 1.0; pH 9 (1% aq.); 85% act.

Hostacor BM. [Hoechst Celanese] Alkyl sulfonamidocarboxylic acid ester 70%; emulsifier and corrosion inhibitor in aq. systems and coolants; yel., brn. clear liq.; water sol; sp.gr. 1.1; pour pt. 14 F; pH 9 (1%).

Hostacor BS. [Hoechst Celanese] Aryl sulfonamidocarboxylic acid ester 70%; emulsifier and corrosion inhibitor in coolants, syn. cutting fluids, aq. systems; yel., brn. clear liq.; sol. in water, min. oils; sp.gr. 1.1; pour pt. 14 F; flash pt. 302 F; pH 9 (1%); 70% conc.

Hostacor DT. [Hoechst Celanese] Fatty acid alkanolamide; corrosion inhibitor used in metalworking fluids; brn. oily liq.; sol. in water, min. oils; sp.gr. 1.01; flash pt. 210 C (Marcusson); pH 9 (1% aq.).

Hostacor H Liq. [Hoechst Celanese] Arylsulfonamidocarboxylic acid; anionic; chemical intermediate for mfg. of corrosion inhibitors for aq. systems; yel. visc. liq.; sol. in ethanol, glycol, water insol.; sp.gr. 1.17 ± 0.05 (122 F).

Hostacor KS 1. [Hoechst Celanese] Alkanolamine salt of a nitrogen-containing condensation prod.; anionic; corrosion inhibitor, lubricant used in cooling and machining aux. for metalworking; pale brn. clear visc. liq.; dissolves in water; sp.gr. 1.16 ± 0.02; pH 8.0–8.5 (1% aq.); flash pt. > 200 C.

Hostacor R. [Hoechst Celanese] Barium salt of alkylsulfamido carboxylic acid; corrosion inhibitor for preparation of rust-inhibiting oils; dk. brn. clear oily liq.; sol. in min. oils, org. solvs.; insol. in water; sp.gr. 0.99–1.00; flash pt. 140 C.

Hostacor TP 2125. [Hoecht Celanese] Complex carboxylic acid; corrosion inhibitor in aq. systems; liq.

Hostalux KCB. [Hoechst Celanese] Benzoxazole type; optical brightener, whitening agent used in all types of polymer processing; used in plastic films, press molding, and inj. molding material fibers and bristles, paints and lacquers; ylsh.-grn. powd.; sol. (mg/100 ml): 1020 mg in toluene; 950 mg in perchloroethylene; 860 mg in mesitylene; 800 mg in CCl_4; 760 mg in dimethyl formamide; 170 mg in acetone; m.p. 211 C.

Hostapal BV Conc. [Hoechst Celanese] Alkylaryl polyglycol ether sulfate, sodium salt; anionic; dispersant for pigments, wetting agent, detergent for fibers, emulsifier for emulsion polymerization; yel. brn. gelatinous paste; misc. with water; dens. 9.10 lb/gal; 49–51% act.

Hostaphat AW. [Hoecht Celanese] Aromatic phosphate ester; low foaming corrosion inhibitor and lubricant for aq. systems; sol. in water, acetone, ethanol, benzene.

Hostaphat CP. [Hoecht Celanese] Phosphate ester; anionic; high temp. antiwear additive; lubricant, antiseize corrosion inhibitor; solid; oil-sol.

Hostaphat F Brands. [Hoechst Celanese] Phosphoric acid complex org. ester; anionic; antistat finish for syn. fibers; liq.; 95% conc.

Hostaphat L Brands. [Hoechst Celanese] Org. phosphoric acid ester; anionic; antistat for textile industry.

Hostaphat OPS. [Hoechst Celanese] Alkyl phosphonic acid; anionic; hydrotrope used in electrolyte sol'ns.; solubilizer; paste; 99% conc.

Hostastat® HS1. [Hoechst Celanese] Sodium alkyl sulfonate; antistat for thermoplastics incl. PS, HIPS, ABS, PVC, styrenic alloys, copolymers, and acrylic esters; can be incorporated before processing or applied externally to finished surfaces; FDA approved; nearly wh.; avail. in fine grain or masterbatch form; < 5 mm particle size; sol. in water, chlorinated hydrocarbons, IPA; sp.gr. 1.05 (30% aq.); dens. 450 g/l (fine grain); pH 7–8 (2% aq.).

Hostastat® Systems E 3952, E 3953, E 3954. [Hoechst Celanese] Sodium alkyl sulfonate predispersed in PS; antistat systems for extruded or molded PS, HIPS, and styrene copolymer parts; FDA approved; pellets.

Hostastat® System E 5951. [Hoechst Celanese] Laurylamide in a PP carrier; antistat system for polyolefins; esp. suited for electronics applics.; FDA approved; pellets.

Hostastat® System E 6952. [Hoechst Celanese] Laurylamide in a LLDPE carrier; antistat system for polyolefins; esp. suited for electronics applics.; FDA approved; pellets.

Hostavin® ARO 8. [Hoechst Celanese] Benzophenone-12; uv absorber for plastics, esp. LDPE, HDPE, PP, polyisobutylene, cellulosics, PC, EVA copolymers, plasticized PVC; pale yel. cryst. powd.; m.w. 326; sol. (g/100 g): 75 g acetone, 70 g ethyl acetate, 65 g toluene, 58 g chloroform, 25 g petrol. ether; sp.gr. 1.1; m.p. 48 C.

Hostavin® N 20. [Hoechst Celanese] Hindered amine light stabilizer; effective for thin-walled polyolefins, natural or dyed polymers incl. LDPE, HDPE, PP, polyisobutylene, EVA, ABS, PU, PC, and cellulosics; wh. cryst. powd.; sp.gr. 1.06; bulk dens. 0.46 g/cc; m.p. 225–227 C.

Hostavin® VP NiCS 1. [Hoechst Celanese] Nickel-containing light stabilizer, antioxidant for polyolefins, esp. HDPE and LDPE; ideal for polyolefin films for agric. use; grn. powd.; m.w. 747; sol. (g/100 g): 35 g chloroform, 30 g toluene; m.p. 150 C.

H.P.X. [RITA] Lyophilized human placenta proteins

containing essential amino acids; hair conditioners, creams, and lotions; sol. in aq. sol'ns.

HT-400 E 1/8´´. [M&T Harshaw] Cobalt molybdate; catalyst for desulfurization and denitrogenation of refinery prods.; extrusion, 1/8´´ diam.; dens. 50 lb/ft³.

HT-500 E 1/8´´. [M&T Harshaw] Nickel molybdate; catalyst for denitrogenation, desulfurization, and polyaromatic saturation; extrusion, 1/8´´ diam.; dens. 50 lb/ft³.

HT-Proteolytic. [Miles Biotech] Bacterial protease; enzyme for hydrolysis of proteins (baking, brewing, etc.); liq./powd.

HTSA #1. [Hexcel] Oleyl palmitamide; release agent providing slip, antiblocking to thermoplastics incl. PP film, nylon; m.p. 66–80 C.

HTSA #3. [Hexcel] Stearyl erucamide; release agent providing slip, antiblocking to thermoplastics incl. PP film, nylon; sol. in alcohols, ethyl acetate, benzene; m.p. 69–77 C; b.p. 824 F.

Huber ARO 60. [Huber] Carbon black for the rubber industry; 90 nm avg. particle diam.; 0.05% max 325-mesh residue; sp.gr. 1.8; pour dens. 32 lb/ft³.

Huber N110. [Huber] Carbon black for the rubber industry; 20 nm avg. particle diam.; 0.05% max 325-mesh residue; sp.gr. 1.8; pour dens. 22 lb/ft³.

Huber N220. [Huber] Carbon black for the rubber industry; 22 nm avg. particle diam.; 0.05% max 325-mesh residue; sp.gr. 1.8; pour dens. 22 lb/ft³.

Huber N234. [Huber] Carbon black for the rubber industry; 21 nm avg. particle diam.; 0.05% max 325-mesh residue; sp.gr. 1.8; pour dens. 21 lb/ft³.

Huber N299. [Huber] Carbon black for the rubber industry; 23 nm avg. particle diam.; 0.05% max 325-mesh residue; sp.gr. 1.8; pour dens. 21 lb/ft³.

Huber N326. [Huber] Carbon black for the rubber industry; 26 nm avg. particle diam.; 0.05% max 325-mesh residue; sp.gr. 1.8; pour dens. 28 lb/ft³.

Huber N330. [Huber] Carbon black for the rubber industry; 29 nm avg. particle diam.; 0.05% max 325-mesh residue; sp.gr. 1.8; pour dens. 23 lb/ft³.

Huber N339. [Huber] Carbon black for the rubber industry; 25 nm avg. particle diam.; 0.05% max 325-mesh residue; sp.gr. 1.8; pour dens. 21 lb/ft³.

Huber N343. [Huber] Carbon black for the rubber industry; 24 nm avg. particle diam.; 0.05% max 325-mesh residue; sp.gr. 1.8; pour dens. 21 lb/ft³.

Huber N347. [Huber] Carbon black for the rubber industry; 26 nm avg. particle diam.; 0.05% max 325-mesh residue; sp.gr. 1.8; pour dens. 21 lb/ft³.

Huber N351. [Huber] Carbon black for the rubber industry; 28 nm avg. particle diam.; 0.05% max 325-mesh residue; sp.gr. 1.8; pour dens. 21 lb/ft³.

Huber N375. [Huber] Carbon black for the rubber industry; 24 nm avg. particle diam.; 0.05% max 325-mesh residue; sp.gr. 1.8; pour dens. 21 lb/ft³.

Huber N539. [Huber] Carbon black for the rubber industry; 35 nm avg. particle diam.; 0.05% max 325-mesh residue; sp.gr. 1.8; pour dens. 24 lb/ft³.

Huber N550. [Huber] Carbon black for the rubber industry; 35 nm avg. particle diam.; 0.05% max 325-mesh residue; sp.gr. 1.8; pour dens. 22 lb/ft³.

Huber N650. [Huber] Carbon black for the rubber industry; 60 nm avg. particle diam.; 0.05% max 325-mesh residue; sp.gr. 1.8; pour dens. 23 lb/ft³.

Huber N660. [Huber] Carbon black for the rubber industry; 60 nm avg. particle diam.; 0.05% max 325-mesh residue; sp.gr. 1.8; pour dens. 27 lb/ft³.

Huber N683. [Huber] Carbon black for the rubber industry; 60 nm avg. particle diam.; 0.05% max 325-mesh residue; sp.gr. 1.8; pour dens. 21 lb/ft³.

Huber N762. [Huber] Carbon black for the rubber industry; 88 nm avg. particle diam.; 0.05% max 325-mesh residue; sp.gr. 1.8; pour dens. 32 lb/ft³.

Huber N774. [Huber] Carbon black for the rubber industry; 88 nm avg. particle diam.; 0.05% max 325-mesh residue; sp.gr. 1.8; pour dens. 30 lb/ft³.

Huber N787. [Huber] Carbon black for the rubber industry; 62 nm avg. particle diam.; 0.05% max 325-mesh residue; sp.gr. 1.8; pour dens. 28 lb/ft³.

Huber N990. [Huber] Carbon black for the rubber industry; 320 nm avg. particle diam.; 0.05% max 325-mesh residue; sp.gr. 1.8; pour dens. 40 lb/ft³.

Huber S212. [Huber] Carbon black for the rubber industry; 0.05% max 325-mesh residue; sp.gr. 1.8; pour dens. 27 lb/ft³.

Huber S315. [Huber] Carbon black for the rubber industry; 0.05% max 325-mesh residue; sp.gr. 1.8; pour dens. 28 lb/ft³.

Huber SM. [Huber] Surface-treated organo-functional muscovite mica; filler for improved flexural, tensile, and impact properties in polyolefin plastics; see also Huber WG-1; wh. powd., flakes; 82.0–87.0% –325 mesh; negligible odor; negligible sol. in water; sp.gr. 2.80; bulk dens. 11.0–13.0 lb/ft³; m.p. ≈ 1000 C; pH 7.0–8.0 (@ 28% solids); flash pt. none.

Huber WG-1. [Huber] Wet-ground muscovite mica; filler for plastics (reinforcement in many polyolefins, nylons, thermoplastic and thermosetting polyesters, ABS, etc.), rubber, coatings, and pearlescent pigment applics.; off-wh.; 87.0–92.0% –325 mesh; negligible sol. in water; sp.gr. 2.80; bulk dens. 11.0–13.0 lb/ft³; m.p. ≈ 1000 C; pH 7.0–8.0 (@ 28% solids); flash pt. none.

Huber WG-2. [Huber] Wet-ground muscovite mica; see Huber WG-1; off-wh. powd., flake; 82.0–87.0% –325 mesh; negligible odor; negligible sol. in water; sp.gr. 2.80; bulk dens. 11.0–13.0 lb/ft³; m.p. ≈ 1000 C; pH 7.0–8.0 (@ 28% solids); flash pt. none.

Huberfil® 96. [Huber] Silicate; paper filler; 4 μm avg. particle size; surf. area 60 m²/g; oil absorp. 120 cc/100 g; pH 10.5(20%).

Hubersil® 162. [Huber] Silica, precipitated amorphous; reinforcing filler, carrier for rubber applics. incl. hose, belting, molded and extruded parts, tires, footwear, wire and cable, thermolastic elastomers, caulks and sealants; wh. powd.; 12 μ avg. particle size; dens. 2.0 gml; surf. area 160 m²/g; oil absorp. 190 cc/100 g; ref. index 1.44; pH 7.0 (5%).

Hubersorb® 600. [Huber] Silicate; conditioning agent, carrier; 3 μm avg. particle size; surf. area 300 m²/g; oil absorp. 450 cc/100 g; pH 10.0 (20%).

Humectol C, C Highly Conc. [Hoechst Celanese AG] Sulfonated fatty acid amines; anionic; wetting and leveling agent, dispersant for dyeing and printing; liq.

HVP 5-SD. [Hercules] Hydrolyzed veg. protein with salt and caramel color; flavor enhancer; beef-type flavor used in food applics.; tan powd.; pH 5.4; 97% total solids; 46% flavor solids.

HVP-A. [Hercules] Hydrolyzed veg. protein with salt and caramel color; flavor enhancer; smooth, beefy flavor useable in food applics. in highly seasoned prods.; tan powd.; pH 5.4; 97% total solids; 43.5% flavor solids.

HVP-LS. [Hercules] Hydrolyzed veg. protein, low sodium; flavoring for sodium-restricted dietary applics. for meaty flavors in food; brn. powd.; pH 5.2; 97% total solids; 75% flavor solids.

H-White. [Huber] Calcium carbonate; functional filler extender; used in coatings, plastic and rubber fillers, building prods., ceramics flux, paper fillers, adhesives, cleaning compds., polishes; wh. irreg., uniaxial particles; particle size 3.0 μ 99.99% thru 500 mesh; water-sol.; sp.gr. 2.71; dens. 22.6 lb/solid gal, 45 lb/ft^3; oil absorp. 16 lb oil/100 lb; ref. index 1.6; hardness 3 Moh; 97.7% $CaCO_3$.

Hyamine® 10-X. [Lonza] Methylbenzethonium chloride; cationic; germicide, disinfectant, sanitizer in restaurant and pharmaceutical uses; wh. cryst.; dens. 27.5 lb/ft^3; surf. tens. 40 dynes/cm (0.01% aq.); 100% act.

Hyamine® 1622. [Lonza] Benzethonium chloride; cationic; see Hyamine 10-X; also antistat, bacteriostat on fabrics; preservative for starch, glue, casein; cocatalyst for curing polyesters; lt. amber liq.; dens. 8.6 lb/gal; flash pt. 110 F (TOC); pour pt. 25 F; 50% act.

Hyamine® 2389. [Lonza] Dodecylbenzyl trimonium chloride (80%), dodecylxylylditrimonium chloride (20%); cationic; see Hyamine 10-X; also antistat, bacteriostat for textiles, sanitized paper; preservative; lt. amber liq.; dens. 8.3 lb/gal; visc. 360 cps; flash pt. > 200 F (TOC); pour pt. 25 F; surf. tens. 43 dynes/cm (0.01% aq.); 50% act.

Hyamine® 3500. [Lonza] n-Alkyl (50% C_{14}, 40% C_{12}, 10% C_{16}) dimethyl benzyl ammonium chloride; cationic; see Hyamine 10-X; also antistat, bacteriostat on fabrics, sanitized paper, deodorant, preservative; VCS 2–5 liq.; dens. 8.0 lb/gal; visc. 42 cps; flash pt. 105 F (TOC); pour pt. 15 F; surf. tens. 40 dynes/cm (0.01% aq.); 50% act.

Hyamine® 3500-NF. [Lonza] Benzalkonium chloride; bactericide used as disinfectant, sanitizer, deodorant, preservative in liq. and powd. formulations; humectant; pale yel. liq.; mild odor.

Hybase C-300. [Witco] Highly basic oil-sol. calcium sulfonate; detergent, rust inhibitor used in marine or stationary diesels; dk. brn. liq.; sol. in org. solvs.; dens. 9.4 lb/gal; sp.gr. 1.13 (60 F); visc. 800 SUS (210 F); flash pt. 380 F; 29% act.

Hybase M-300, -400. [Witco] Highly basic oil-sol. magnesium sulfonate; lubricant, corrosive inhibitor, used in marine or stationary diesels; dk. br. liq.; petrol. odor; sol. in org. solv.; dens. 9.3–9.6 lb/gal; sp.gr. 1.12–1.15 (60 F); visc. 300 cs (210 F); flash pt. 375–380 F; 27–28% act.

Hydagen DEO. [Henkel] Triethyl citrate and BHT; act. ingred. for deodorant systems and personal care prods.; pale yel. clear oily liq.; sol. in hydroalcoholic sol'ns.; sp.gr. 1.11–1.12; 99.0% act.

Hydagen F. [Henkel] Sodium salt of polyhydroxycarboxylic acid; skin moisture regulator; for moisturizing creams and lotions; wh. powd.

Hydagen P. [Henkel] Diethylene tricaseinamide; softening agent, conditioner for emulsion-type shampoos, hair care preparations; paste.

Hy Dense Calcium Stearate HP Gran., RSN Powd. [Mallinckrodt] Calcium stearate; release agent in PVC processing; gran. and powd. resp.

Hy Dense Zinc Stearate XM Powd. [Mallinckrodt] Zinc stearate; release agent for use where flow and green strength are critical and in intricate die cavities; powd.

Hy Dense Zinc Stearate XM Ultra Fine. [Mallinckrodt] Zinc stearate; release agent used where sintered strength and dimensional stability are important; powd.

Hydrenol DD. [Henkel] Cetyl/stearyl alcohol; emollient, consistency giving agent for skin creams and lotions; wh. flakes.

Hydrex®. [Huber] Syn. sodium aluminum silicate; high brightness filler; 4 micrometer avg. particle size; surf. area 75 m^2/g; oil absorp. 115 cc/100 g; pH 9.7.

Hydrex® R. [Huber] Syn. sodium magnesium aluminosilicate, precipitated amorphous; filler for the rubber industry; wh. powd.; 3 μ avg. particle size; 0.2% max. #325 mesh residue; dens. 2.1 g/ml; surf. area 70 m^2/g; oil absorp. 125 cc/100 g; brightness 98.5%; ref. index 1.55; pH 10.5 (20%).

Hydrine. [Gattefosse Ets.] PEG-2 stearate; stabilizer for ointments, cream lotions; solid.

Hydrokote® 25. [Capital City] Hydrog. veg. oil; specialty base used as replacement for cocoa butter in cosmetic and pharmaceutical applics.; Lovibond R1.5 max.; m.p. 35.6–36.7 C; iodine no. 1–5.

Hydrokote® 27. [Capital City] Hydrog. veg. oil; specialty base used as replacement for cocoa butter in cosmetic and pharmaceutical applics.; Lovibond R1.5 max.; m.p. 38.3–40.0 C; iodine no. 1–5.

Hydrokote® 79. [Capital City] Hydrog. veg. oil; specialty base used as replacement for cocoa butter in cosmetic and pharmaceutical applics.; Lovibond R1.5 max.; m.p. 41.1–42.8 C; iodine no. 1–5.

Hydrokote® 711. [Capital City] Hydrog. veg. oil; specialty base used as replacement for cocoa butter in cosmetic and pharmaceutical applics.; Lovibond R1.5 max.; m.p. 45.6–47.8 C; iodine no. 1–5.

Hydrokote® S-7. [Capital City] Hydrog. veg. oil; specialty base used as replacement for cocoa butter in cosmetic and pharmaceutical applics.; Lovibond R1.5 max.; m.p. 33.3–35.0 C; iodine no. 3–6.

Hydrokote® SP. [Capital City] Hydrog. veg. oil; specialty base used as replacement for cocoa butter in cosmetic and pharmaceutical applics.; Lovibond R1.5 max.; m.p. 32.8–33.3 C; iodine no. 6–8.

Hydrolactin 2500. [Croda] Hydrolyzed milk protein; skin and hair care conditioner; cream colored powd.; slight, char. odor; sol. in water; partly sol. in water/

ethanol; pH 5–7; 76.5–89.5% protein.

Hydromond. [Croda Ltd.] Hydrolyzed almond protein; conditioner for skin and hair care prods.; aq. sol'n.

Hydropalat 535. [Henkel] Amine neutralized polyester; suspending agent for water-reducible paints.

Hydro-Pruf 200. [Sandoz] Silicone emulsion; durable water repellent.

Hydrosan. [Goldschmidt] Flocculating aid for water treatment.

Hydrosoy 2000. [Croda] Hydrolyzed veg. protein; conditioner for cosmetic prods.; liq.

Hydrosoy 2000/SF. [Croda] Hydrolyzed soy protein; conditioner, moisturizer; clear amber liq.; m.w. 4000; 20% act.

Hydrotriticum QL. [Croda] Laurdimonium hydrolyzed whole wheat protein; skin and hair conditioning agent; lt. amber clear liq., mild char. odor; sol. in water, glycerin, propylene glycol, 50/50 water/ethanol; pH 4–5; 28–32% solids.

Hydrotriticum QM. [Croda] Cocodimonium hydrolyzed whole wheat protein; skin and hair conditioning agent; lt. amber clear liq., mild char. odor; sol. in water, glycerin, propylene glycol, 50/50 water/ethanol; pH 4–5; 28–32% solids.

Hydrotriticum QS. [Croda] Steardimonium hydrolyzed whole wheat protein; skin and hair conditioning agent; lt. amber opaque paste, mild char. odor; sol. in water, glycerin, propylene glycol, 25/75 water/ethanol; pH 4–5; 28–32% solids.

Hydroxylan. [Fanning] Hydroxylated lanolin; emulsifier, conditioner.

Hyfil 10. [Solem] Alumina trihydrate and fillers; resin extender used in fiberglass-reinforced polyester applic.; resin extension, lower glass requirements, improved processing, faster mold turnover, and longer mold life; particulate; median particle size 12–15 μ, 85–92% through 325 mesh; sp.gr. 2.5–2.6.

Hyonic NP-40. [Henkel] Nonoxynol-4; nonionic; emulsifier for solv. cleaning compds., waterless hand cleaners, agric. emulsifiable concs., silicone prods., detergent for petrol. oils; visc. reducer for plastisols; intermediate for prod. of anionic surfactants; metal cleaners; food contact applics.; clear liq.; insol. in water; dens. 1.02 g/ml; HLB 9; pour pt. 26 C; pH 7.0 (1% aq.); > 99% act.

Hyonic NP-60. [Henkel] Nonoxynol-6; nonionic; emulsifier, wetting agent, detergent for textile lubricants, agric. emulsifiable concs.; intermediate for textile antistats; color enhancer in latex paint; post-polymerization stabilizer; food applics.; clear liq.; disp. in water; dens. 1.04 g/ml; HLB 12; pour pt. –32 C; cloud pt. < 0 C (1% aq.); pH 7.0 (1% aq.); surf. tens. 30 dynes/cm (0.01%); > 99% act.

Hyonic NP-90. [Henkel] Nonoxynol-9; nonionic; detergent and penetrant for household and industrial cleaning compds., textile processing; degreasing compds.; wetting agent for paper toweling or tissue; emulsifier in agric. applics.; clear liq.; sol. in water; dens. 1.06 g/ml; HLB 13; pour pt. 2 C; cloud pt. 54 C (1% aq.); pH 7.0 (1% aq.); surf. tens. 31 dynes/cm (0.01%); > 99% act.

Hyonic NP-100. [Henkel] Nonoxynol-10; nonionic; surfactant, wetting agent, emulsifier; base for liq. dishwashing detergents, laundry preparations, household and industrial cleaners; agric. applics.; penetrant for corrosion inhibitors; stabilizer for latex; rewetting agent for paper toweling and tissue; food applics.; clear liq.; sol. in water; dens. 1.07 g/ml; HLB 13.2; pour pt. 4 C; cloud pt. 68 C (1% aq.); pH 7.0 (1% aq.); surf. tens. 32 dynes/cm (0.01%); > 99% act.

Hyonic NP-110. [Henkel] Nonoxynol-11; nonionic; see Hyonic NP-100; clear liq.; sol. in water; dens. 1.07 g/ml; HLB 13.6; pour pt. 13 C; cloud pt. 72 C (1% aq.); pH 7.0 (1% aq.); surf. tens. 34 dynes/cm (0.01%); > 99% act.

Hyonic NP-120. [Henkel] Nonoxynol-12; nonionic; detergent, wetting agent, penetrant, coemulsifier; food and agric. applics.; liq., semisolid; sol. in water; dens. 1.07 g/ml; HLB 14; pour pt. 15 C; cloud pt. 91 C (1% aq.); pH 7.0 (1% aq.); surf. tens. 36 dynes/cm (0.01%); > 99% act.

Hyonic NP-407. [Henkel] Nonoxynol-40; nonionic; stabilizer for water dispersions; primary emulsifier in mfg. of vinyl acetate or acrylic latex; wetting agent for min. acid and alkali sol'ns.; coemulsifier for agric. formulations; food applics.; clear liq.; sol. in water; dens. 1.10 g/ml; HLB 17.6; pour pt. –6 C; cloud pt. 100 C (1% aq.); pH 7.0 (1% aq.); 70% act.

Hyonic NP-500. [Henkel] Nonoxynol-50; nonionic; see Hyonic NP-407; clear solid; sol. in water; dens. 1.08 g/ml; HLB 18; pour pt. 24 C; cloud pt. 100 C (1% aq.); pH 7.0 (1% aq.); > 99% act.

Hyonic OP-40. [Henkel] Octoxynol-4; nonionic; emulsifier in solv. cleaning compds., waterless hand cleaners, agric. emulsifiable concs., silicone prods.; metal cleaners; detergent for petrol. oils; visc. reducer for plastisols; intermediate for prod. of anionic surfactants; agric. use; clear liq.; disp. in water; dens. 1.02 g/ml; HLB 8; pour pt. –26 C; cloud pt. 53 C (1% aq.); pH 7.0 (1% aq.); surf. tens. 28 dynes/cm (0.01%); > 99% act.

Hyonic OP-70. [Henkel] Octoxynol-7; nonionic; surfactant, emulsifier, wetting agent, intermediate for textile lubricants, agric. emulsifiable concs.; color development enhancer in latex paint; post-polymerization stabilizer; clear liq.; disp. in water; dens. 1.06 g/ml; HLB 12; pour pt. 11 C; cloud pt. 23 C (1% aq.); pH 7.0 (1% aq.); surf. tens. 27 dynes/cm (0.01%); > 99% act.

Hyonic OP-100. [Henkel] Octoxynol-9; wetting agent, coemulsifier; detergent for household and industrial cleaning compds. and textile processing; penetrant; degreasing compds.; wetting agent for paper toweling/tissue; agric. applics.; clear liq.; sol. in water; dens. 1.07 g/ml; HLB 13; pour pt. 10 C; cloud pt. 65 C (1% aq.); pH 7.0 (1% aq.); surf. tens. 30 dynes/cm (0.01%); > 99% act.

Hyonic OP-407. [Henkel] Octoxynol-40; nonionic; stabilizer for water dispersions; mfg. and stabilize latexes; wetting agent in min. acid and alkali formulations; coemulsifier and stabilizer in agric. formulations; clear liq.; sol. in water; dens. 0.91 g/ml; HLB

18; pour pt. –2 C; cloud pt. 73 C (1% aq.); pH 7.0 (1% aq.); surf. tens. 48 dynes/cm (0.01%); 70% act.

Hyonic OP-705. [Henkel] Octoxynol-70; see Hyonic OP-407; clear liq.; sol. in water; dens. 1.10 g/ml; HLB 18.5; pour pt. –12 C; cloud pt. 91 C (1% aq.); pH 7.0 (1% aq.); surf. tens. 49 dynes/cm (0.01%); 50% act.

Hyonic PE-40. [Henkel] Nonyl phenol ethoxylate; nonionic; emulsifier for solv. cleaning compds., waterless hand cleaners, metal cleaners, surfactant intermediate; clear liq.; dens. 8.5 lb/gal; cloud pt. 0 C (1% aq.); pour pt. –29 C; surf. tens. 28 dynes/cm (0.01%); pH 7.0 (1% aq.); > 99% act.

Hyonic PE-90. [Henkel] Nonyl phenol ethoxylate; nonionic; detergent, wetter, emulsifier, stabilizer for household and industrial cleaning, textile, paper processing, latex adhesives; clear liq.; sp.gr. 1.06; dens. 8.8 lb/gal; cloud pt. 54 C (1% aq.); pour pt. 2 C; surf. tens. 31 dynes/cm (0.01%); pH 7.0 (1% aq.); > 99% act.

Hyonic PE-100. [Henkel] Nonoxynol-10; nonionic; detergent, wetter, emulsifier, base, penetrant for household and industrial cleaners, paper toweling; latex stabilizer; clear liq.; sp.gr. 1.068; dens. 8.9 lb/gal; cloud pt. 68 C (1% aq.); pour pt. 4 C; surf. tens. 32 dynes/cm (0.01%); pH 7.0 (1% aq.); > 99% act.

Hyonic PE-120. [Henkel] Nonyl phenol ethoxylate; nonionic; detergent, penetrant, wetting agent, coemulsifier; liq./semisolid; dens. 8.9 lb/gal; cloud pt. 91 C (1% aq.); pour pt. 15 C; surf. tens. 36 dynes/cm (0.01%); pH 7.0 (1% aq.); > 99% act.

Hy-Pel GP-4. [GAF] Aluminum salt-wax emulsion; nonionic; water-disp. nondurable water repellent; used in woven, knit, or nonwoven textiles and produces a spray rating of 100; gel.

Hypermer Series. [ICI Am.] Polymeric dispersants and emulsifiers for dispersion and fluidization of solid particles in aq. and nonaq. systems; liq. to solid; sp.gr. 0.84–1.7; m.p. 25–60 C.

Hysoft 975. [Lyndal] Difatty quat.; cationic; fabric softener; liq.; 75% act.

Hysoft LC Conc. [Lyndal] Fatty amine/amide; cationic; fabric softener; solid; 81% act.

Hystar® CG. [Lonza] Hydrog. starch hydrolysate; lubricant, humectant for toiletries, dentifrices; clear wh. liq.

Hystar® TPF. [Lonza] Hydrog. starch hydrolysate; lubricant; maintains optimal moisture control in industrial, pet food, tobacco, teat dip, oral hygiene prods.; clear wh. liq.

Hystrene® 1835. [Witco/Humko] Mixt. tallow/coconut acid; chemical intermediate, emulsifier; used for personal care prods., waxes, textile aux., pharmaceuticals; solid. pt. 40 C max.; acid no. 214–222; sapon. no. 211-220.

Hystrene® 3022. [Witco/Humko] 30% behenic and arachidic acid; chemical intermediate, emulsifier; used for personal care prods., waxes, textile aux., pharmaceuticals; solid. pt. 50-54 C; acid no. 193–201; sapon. no. 193–202.

Hystrene® 3675. [Witco/Humko] 75% dimer acid; corrosion inhibitor, intermediate; derivs. used as syn. lube components, corrosion inhibitors for petrol. processing, as extenders and crosslinking agents for high polymeric systems; Gardner 9 max. color; visc. 9000 cSt; acid no. 189–197; sapon. no. 189–199.

Hystrene® 3675C. [Witco/Humko] 75% dimer acid, 3% monomer; corrosion inhibitor, intermediate; derivs. used as syn. lube components, corrosion inhibitors for petrol. processing, as extenders and crosslinking agents for high polymeric systems; Gardner 9 max. color; visc. 7500 cSt; acid no. 189–197; sapon. no. 189–199.

Hystrene® 3680. [Witco/Humko] 80% dimer acid; corrosion inhibitor, intermediate; derivs. used as syn. lube components, corrosion inhibitors for petrol. processing, as extenders and crosslinking agents for high polymeric systems; Gardner 8 max. color; visc. 8500 cSt; acid no. 190–197; sapon. no. 190–199.

Hystrene® 3695. [Witco/Humko] 95% dimer acid; corrosion inhibitor, intermediate; derivs. used as syn. lube components, corrosion inhibitors for petrol. processing, as extenders and crosslinking agents for high polymeric systems; Gardner 6 max. color; visc. 7500 cSt; acid no. 190–196; sapon. no. 190–202.

Hystrene® 4516. [Witco/Humko] 45% palmitic acid; lubricant, textile aux., emulsifier, plasticizer, intermediate, used in cosmetics, shampoos, pharmaceuticals; solid; sapon. no. 207–212; 100% conc.

Hystrene® 5012. [Witco/Humko] Hydrog. coconut acid; chemical intermediate, emulsifier; used for personal care prods., waxes, textile aux., pharmaceuticals; solid. pt. 24–33 C; acid no. 250–266; sapon. no. 250–266.

Hystrene® 5016. [Witco/Humko] Triple pressed stearic acid; see Hystrene 4516; also as stabilizer; solid; sapon. no. 208–211; 100% conc.

Hystrene® 5016 NF FG. [Witco/Humko] Triple pressed stearic acid; food grade acids used as lubricants, release agents, binders, and defoamers, and in components for producing other food grade additives; Lovibond 1.0Y-0.1R; solid pt. 54.5–56.5 C; acid no. 206–210; sapon. no. 206–211.

Hystrene® 5460. [Witco/Humko] 60% trimer acid; corrosion inhibitor, lubricant; liq.; visc. 40,000 cs; acid no. 182–190; sapon. no. 190–198.

Hystrene® 5522. [Witco/Humko] 55% behenic and arachidic acid; chemical intermediate, emulsifier; used for personal care prods., waxes, textile aux., pharmaceuticals; solid. pt. 60–63 C; acid no. 178–185; sapon. no. 179–186.

Hystrene® 7018. [Witco/Humko] 70% stearic acid; see Hystrene 4516; solid; acid no. 199–205; sapon. no. 200–206; 100% conc.

Hystrene® 7018 FG. [Witco/Humko] Stearic acid; see Hystrene 5016 NF FG; Lovibond 1.0Y-0.1R; acid no. 200–205; sapon. no. 200–206.

Hystrene® 7022. [Witco/Humko] 70% arachidic and behenic fatty acid; see Hystrene 4516; solid; acid no. 171–181; sapon. no. 172–182; 100% conc.

Hystrene® 9014. [Witco/Humko] 90% myristic fatty acid; see Hystrene 4516; solid; acid no. 240–243; sapon. no. 241–244; 100% conc.

Hystrene® 9016. [Witco/Humko] 90% palmitic fatty

acid; see Hystrene 4516; solid; acid no. 216–220; sapon. no. 217–244; 100% conc.

Hystrene® 9022. [Witco/Humko] 90% behenic acid; see Hystrene 4516; solid; acid no. 168–178; sapon. no. 169–179; 100% conc.

Hystrene® 9512. [Witco/Humko] 95% lauric fatty acid; see Hystrene 4516; solid; acid no. 276–281; sapon. no. 277–282; 100% conc.

Hystrene® 9514. [Witco/Humko] 95% Myristic acid; chemical intermediate, emulsifier; used for personal care prods., waxes, textile aux., pharmaceuticals; solid. pt. 51.9–53.5 C; acid no. 243–246; sapon. no. 243–247.

Hystrene® 9718. [Witco/Humko] 92% stearic acid; see Hystrene 4516; solid; acid no. 195–200; sapon. no. 196–201; 100% conc.

Hystrene® 9718 NF FG. [Witco/Humko] 92% stearic acid; see Hystrene 5016 NF FG; acid no. 196–201; sapon no. 196–201; 92% conc.

Hystrene® 9912. [Witco/Humko] 99% lauric acid; chemical intermediate, emulsifier; used for personal care prods., waxes, textile aux., pharmaceuticals; solid. pt. 43–44 C; acid no. 277–281; sapon. no. 278–281.

I

I7860. [Hüls] Isopropyldimethylchlorosilane; sterically hindered blocking agent; used in prostaglandin synthesis.

Ibercarrier 300-L. [A. Harrison] Biphenyl and butyl benzoate; carrier for polyesters.

Ibercarrier OPP. [A. Harrison] o-Phenol type; carrier for acrylics.

Iberpenetrant-114. [A. Harrison] Alkylphenol polyethylene sulfonate; anionic; textile wetting agent, dispersant, foaming agent; liq.

Iberquestrene. [A. Harrison] Tetrasodium EDTA; chelating agent, dyeing and finishing assistant; water-sol.

Ice® #2. [Van Den Bergh] Glyceryl stearate and polysorbate 80; stabilizer and emulsifier for the food industry; lubricant for textiles and plastics; fabric softener; HLB 5.2; m.p. 59–63 C; 32–38% alpha monoglyceride.

Ice® #12. [Van Den Bergh] Mono- and diglycerides and polysorbate 65; see Ice #2; HLB 4.5; m.p. 59–63 C; 32–36% alpha monoglyceride.

Ice® #81. [Van Den Bergh] Mono- and diglycerides and fatty acids polyglycerol esters; see Ice #2; HLB 4.8; m.p. 58–63 C; 34–40% alpha monoglyceride.

Icinol H260. [ICI Australia] Polyoxyalkylene glycol ether; syn. lubricant, defoamer; high visc. index solv.; liq.; water-sol.; visc. 260 SUS (100 F).

Icinol H280X. [ICI Australia] Polyoxyalkylene glycol ether; inhibited version of Icinol H260.

Icinol H660. [ICI Australia] Polyoxyalkylene glycol ether; see Icinol H260; liq.; water-sol.; visc. 660 SUS.

Icinol H660YA. [ICI Australia] Polyoxyalkylene glycol ether; inhibited version of Icinol H660.

Icinol H5100. [ICI Australia] Polyoxyalkylene glycol ether; see Icinol H260; liq.; water-sol.; visc. 5100 SUS.

Icinol HT 90000. [ICI Australia] Polyoxyalkylene glycol; high visc. index lubricant, thickener; high-visc. liq.; water-sol.; 90% act.

Icinol L65. [ICI Australia] Polyoxyalkylene glycol ether; see Icinol H260; liq.; water-insol.; visc. 65 SUS (100 F).

Icinol L285. [ICI Australia] Polyoxyalkylene glycol ether; see Icinol H260; liq.; water-insol.; visc. 285 SUS.

Icinol L300X. [ICI Australia] Polyoxyalkylene glycol ether; inhibited version of Icinol L285.

Icinol L385. [ICI Australia] Polyoxyalkylene glycol ether; see Icinol H260; liq.; water-insol.; visc. 385 SUS.

Icinol L625. [ICI Australia] Polyoxyalkylene glycol ether; see Icinol H260; liq.; water-insol.; visc. 625 SUS.

Icinol L650X. [ICI Australia] Polyoxyalkylene glycol ether; inhibited version of Icinol L625.

Icinol L1715. [ICI Australia] Polyoxyalkylene glycol ether; see Icinol H260; liq.; water-insol.; visc. 1715 SUS.

Icodimeen T-30. [BASF] Tallow diamine adduct (30 EO); cationic/nonionic; wetting agent, penetrant, emulsifier, stabilizer, dispersant, antistat, lubricant; wax; water-sol.; m.w. 1488; sp.gr. 1.07; HLB 17.0; m.p. 28 C; pH 8.0–9.5 (5% aq.); 100% act.

Icomeen 18-5. [BASF] PEG-5 stearamine; cationic/nonionic; see Icodimeen T-30; also as solubilizer; Gardner 8 max. cast solid; disp. in water; sp.gr. 0.98; HLB 8.9; pH 9 (5% aq.); 100% act.

Icomeen O-30. [BASF] PEG-30 oleamine; cationic/nonionic; see Icomeen 18-5; Gardner 10 max. paste; m.w. 1590; water-sol.; sp.gr. 1.05; HLB 16.2; pH 8.0–9.5 (5% aq.); 100% act.

Icomeen O-30-80%. [BASF] PEG-30 oleamine; cationic/nonionic; see Icomeen 18-5; Gardner 10 max. liq.; water-sol.; sp.gr. 1.05; HLB 16.2; pH 7.5–9.0 (5% aq.); 80% act.

Icomeen S-5. [BASF] PEG-5 soyamine; cationic/nonionic; see Icomeen 18-5; disp. in water; sp.gr. 0.98; HLB 9.0; pH 9 (5% aq.); 100% act.

Icomeen T-2. [BASF] PEG-2 tallow amine; cationic/nonionic; see Icodimeen T-30; Gardner 8 max. paste; insol. in water; m.w. 350; sp.gr. 0.92; HLB 5.0; pH 10 (5% aq.); 100% act.

Icomeen T-5. [BASF] PEG-5 tallow amine; cationic/nonionic; see Icodimeen T-30; Gardner 8 max. liq., paste; water-sol.; sp.gr. 0.96; HLB 9.2; cloud pt. 45 C (1% aq.); pH 9 (5% aq.); surf. tens. 33 dynes/cm (0.1% aq.); 100% act.

Icomeen T-7. [BASF] PEG-7 tallow amine; cationic/nonionic; see Icodimeen T-30; Gardner 8 max. liq.; water-sol.; m.w. 570; sp.gr. 0.98; HLB 10.8; cloud pt. 85 C (1% aq.); pH 9 (5% aq.); surf. tens. 36 dynes cm (0.1% aq.); 100% act.

Icomeen T-20. [BASF] PEG-20 tallow amine; cationic/nonionic; see Icodimeen T-30; Gardner 8 max. liq.; water-sol.; m.w. 1145; sp.gr. 1.04; HLB 15.4; pH 9 (5% aq.); 100% act.

Icomeen T-25. [BASF] PEG-25 tallow amine; cationic/nonionic; see Icodimeen T-30; Gardner 8 max. liq.; water-sol.; m.w. 1365; sp.gr. 1.05; HLB 16.0; pH 9 (5% aq.); 100% act.

Icomeen T-25 CWS. [BASF] Tallow amine ethoxylate, modified; cationic/nonionic; surfactant for gel formation; see also Icodimeen T-30; Gardner 8 max. liq.; water-sol.; sp.gr. 1.05; HLB 16.0; pH 7.5–9.0 (5% aq.); 100% act.

Icomeen T-40. [BASF] PEG-40 tallow amine; cationic/nonionic; see Icodimeen T-30; Gardner 8 max. cast solid; water-sol.; m.w. 2030; sp.gr. 1.09; HLB 17.2; m.p. 33 C; pH 8.0–9.5 (5% aq.); 100% act.

Icomeen T-40-80%. [BASF] PEG-40 tallow amine; cationic/nonionic; see Icodimeen T-30; Gardner 8 max. liq.; water-sol.; m.w. 2030; sp.gr. 1.08; HLB 17.2; pH 8.0–9.5 (5% aq.); 80% act.

Icomid HT-50. [BASF] PEG-50 hydrog. tallow amide; noninic; emulsifier, foam stabilizer, finishing agent and lubricant; Gardner 8 max. wax. solid; water-sol.; m.w. 1200; sp.gr. 1.14; HLB 17.7; pH 8–10 (5% aq.); surf. tens. 47 dynes/cm (0.1% aq.); 100% act.

Icomid O-5. [BASF] PEG-5 oleamide; nonionic; see Icomid HT-50; Gardner 8 max. liq.; disp. in water; m.w. 510; sp.gr. 1.0; visc. 500 cps; HLB 8.8; pH 5–7 (5% aq.); surf. tens. 35 dynes/cm (0.1% aq.); 100% act.

Iconol 28. [BASF] Coco DEA; nonionic; see Icomid HT-50; Gardner 12 max. liq.; water-sol.; sp.gr. 1.01; visc. 2300 cps; cloud pt. > 100 C; pH 8.0–10.0 (5% aq.); 100% act.

Iconol COA. [BASF] Coco DEA; nonionic; see Icomid HT-50; Gardner 10 max. liq.; water-sol.; sp.gr. 1.02; visc. 1300 cps; cloud pt. > 100 C; pH 8.5–11.0 (5% aq.); 100% act.

Iconol DDP-10. [BASF] Dodoxynol-10; nonionic; emulsifier, wetting and cleaning agent, penetrant, detergent, stabilizer, dispersant, coemulsifier; Gardner 1 max. liq.; water-sol.; m.w. 725; sp.gr. 1.04; visc. 400 cps; HLB 12.6; pour pt. 10 C; cloud pt. 39–44 C (1% aq.); pH 6.0–7.5 (5% aq.); 100% act.

Iconol DNP-8. [BASF] Nonyl nonoxynol-8; nonionic; see Iconol DDP-10; Gardner 4 max. clear liq.; disp. in water; m.w. 700; sp.gr. 1.0; visc. 500 cps; HLB 10.0; pour pt. 9 C; cloud pt. < 25 C (1% aq.); pH 6.0–7.5 (5% aq.); 100% act.

Iconol DNP-24. [BASF] Nonyl nonoxynol-24; nonionic; see Iconol DDP-10; Gardner 1 max. paste; water-sol.; m.w. 1400; HLB 15.0; m.p. 33 C; cloud pt. > 95 C (1% aq.); pH 6.0–7.5 (5% aq.); 100% act.

Iconol DNP-150. [BASF] Nonyl nonoxynol-150; nonionic; see Iconol DDP-10; Gardner 1 max. flakes, cast solid; water-sol.; m.w. 6900; sp.gr. 1.05; HLB 19.0; m.p. 55 C; cloud pt. > 100 C (1% aq.); pH 6.0–7.5 (5% aq.); 100% act.

Iconol NP-7. [BASF] Nonoxynol-7; nonionic; wetting and cleaning agent, penetrant, detergent, emulsifier, stabilizer, dispersant, coemulsifier; used in food; APHA 100 max. liq.; water-sol.; m.w. 523; sp.gr. 1.05; visc. 300 cps; HLB 11.9; pour pt. 5 C; cloud pt. 22–27 C (1.0% aq.); surf. tens. 30 dynes/cm (0.1% aq.); 100% conc.

Iconol NP-12. [BASF] Nonoxynol-12; nonionic; see Iconol NP-7; liq.; HLB 14.0; 100% conc.

Iconol OP-30. [BASF] Octylphenol ethoxylate; nonionic; emulsifier, wetting agent, dispersant, syn. latex stabilizer, and detergent in formulating industrial, institutional, and household cleaning prods.; primary emulsifier for acrylic and vinyl emulsion polymerization and for asphalt emulsion systems; solid; HLB 17; cloud pt. > 100 C (1% aq.); surf. tens. 46.4 dynes/cm (0.1%).

Iconol OP-30-70%. [BASF] Octylphenol ethoxylate; nonionic; see Iconol OP-30; liq.; HLB 17; cloud pt. > 100 C (1% aq.).

Iconol WA-1. [BASF] Alkoxylated phenolic; nonionic; surfactant; pigment dispersant for aq. systems; Gardner 6 clear liq.; sol. in water, alcohols, toluene; sp.gr. 1.02; visc. 300 cps; HLB 13; cloud pt. 53–58 C (1% aq.); pH 6.0–7.5 (5% aq.); 90% act.

Iconol WA-4. [BASF] Alkoxylated phenolic; nonionic; dispersant; liq.

Idet 5L SP NF. [Swastik] Specially formulated alkylaryl sulfonate; anionic; detergent and wetting agent for wet processes in textile industry, dispersing and penetrating agent; yel. clear visc. liq.; pH 6.8 (1%); 25% conc.

Igepal® CA-880. [Rhone-Poulenc Surf.] Octylphenoxypoly ethyleneoxy ethanol; nonionic; emulsifier for fats and oils, vinyl acetate and acrylate polymerization; post-stabilizer for syn. latices; dyeing assistant; solid; HLB 17.4; 100% conc.

Igepal® CA-887. [Rhone-Poulenc Surf.] Octoxynol-30; nonionic; emulsifier, stabilizer for plastics; dyeing assistant; pale yel. liq.; aromatic odor; sp.gr. 1.10; HLB 17.4; cloud pt. > 212 F (1% aq.); flash pt. > 200 F (PMCC); pour pt. 36 ± 2 F; surf. tens. 39 dynes/cm (0.01%); 70% act.

Igepal® CA-890. [Rhone-Poulenc Surf.] Octoxynol-40; nonionic; see Igepal CA-887; off-wh. wax; aromatic odor; sol. in butyl Cellosolve, ethanol, water; sp.gr. 1.08 (50 C); HLB 18.0; cloud pt. > 212 F (1% aq.); flash pt. > 200 F (PMCC); pour pt. 115 ± 2 F; surf. tens. 42 dynes/cm (0.01%); 100% act.

Igepal® CA-897. [Rhone-Poulenc Surf.] Octoxynol-40; nonionic; see Igepal CA-887; pale yel. liq.; aromatic odor; sp.gr. 1.10; HLB 18.0; cloud pt. > 212 F (1% aq.); flash pt. > 200 F (PMCC); pour pt. 25 ± 2 F; surf. tens. 48 dynes/cm (0.01%); 70% act.

Igepal® CA-950. [Rhone-Poulenc Surf.] Octylphenoxypoly ethyleneoxy ethanol; nonionic; see Igepal CA-880; solid; HLB 18.7; 100% conc.

Igepal® CO-210. [Rhone-Poulenc Surf.] Nonoxynol-1.5; nonionic; foam and emulsion stabilizer, detergent, coemulsifier, intermediate, defoamer, dispersant; yel. liq., aromatic odor; sol. in wh. min. oil, kerosene, Stod., naphtha, xylene, butyl Cellosolve, perchloroethylene, ethanol; sp.gr. 0.99; visc. 300–400 cps; HLB 4.6; flash pt. > 200 F (PMCC); pour pt. –3 ± 2 F; 100% act.

Igepal® CO-430. [Rhone-Poulenc Surf.] Nonoxynol-4; nonionic; coemulsifier, plasticizer, stabilizer, antistat, detergent, dispersant; pale yel. liq., aromatic odor; sol. in kerosene, Stod., naphtha, xylene, butyl Cellosolve, perchloroethylene, ethanol; sp.gr. 1.02; visc. 250–325 cps; HLB 8.8; flash pt. > 200 F (PMCC); pour pt. –15 ± 2 F; 100% act.

Igepal® CO-520. [Rhone-Poulenc Surf.] Nonoxynol-5; nonionic; intermediate, emulsifier, coupling agent, dispersant; rust inhibitor in jet aircraft and automotive fuels; pale yel. liq., aromatic odor; sol. in Stod., naphtha, xylene, butyl Cellosolve, perchloroethylene, ethanol; sp.gr. 1.03; visc. 240–300 cps; HLB 10.0; flash pt. > 200 F (PMCC); pour pt. –24 ± 2 F; surf. tens. 30 dynes/cm (0.01%); 100% act.

Igepal® CO-530. [Rhone-Poulenc Surf.] Nonoxynol-6; nonionic; emulsifier, detergent, dispersant for agric., petrol., paper industries; pale yel. liq., aromatic odor; sol. in Stod., naphtha, xylene, butyl Cellosolve, perchloroethylene, ethanol; sp.gr. 1.04; visc. 230–300 cps; HLB 10.8; flash pt. > 200 F (PMCC); pour pt. –26 ± 2 F; surf. tens. 28 dynes/cm (0.01%); 100% act.

Igepal® CO-630. [Rhone-Poulenc Surf.] Nonoxynol-9; nonionic; detergent, wetting and rewetting agent, corrosion inhibitor, penetrant, emulsifier for textile, paper, household/industrial cleaners, paints, metal processing; almost colorless liq., aromatic odor; sol. in naphtha, xylene, butyl Cellosolve, perchloroethylene, ethanol, water; sp.gr. 1.06; visc. 225–300 cps; HLB 13.0; cloud pt. 126–133 F (1%); flash pt. > 200 F (PMCC); pour pt. 31 ± 2 F; surf. tens. 31 dynes/cm (0.01%); 100% act.

Igepal® CO-660. [Rhone-Poulenc Surf.] Nonoxynol-10; nonionic; see Igepal CO-630; pale yel. liq., aromatic odor; sol. see Igepal CO-630; sp.gr. 1.06; visc. 225–275 cps; HLB 13.2; cloud pt. 140–149 F (1%); flash pt. > 200 F (PMCC); pour pt. 46 ± 2 F; surf. tens. 31 dynes/cm (0.01%); 100% act.

Igepal® CO-710. [Rhone-Poulenc Surf.] Nonoxynol-10 (10–11 EO); nonionic; see Igepal CO-630; pale yel. liq., aromatic odor; sol. see Igepal CO-630; sp.gr. 1.06; visc. 240–300 cps; HLB 13.6; cloud pt. 158–165 F (1%); flash pt. > 200 F (PMCC); pour pt. 49 ± 2 F; surf. tens. 32 dynes/cm (0.01%); 100% act.

Igepal® CO-720. [Rhone-Poulenc Surf.] Nonoxynol-12; nonionic; see Igepal CO-630; opaque liq., aromatic odor; sol. see Igepal CO-630; sp.gr. 1.06; visc. 260–340 cps; HLB 14.2; flash pt. > 200 F (PMCC); pour pt. 62 ± 2 F; surf. tens. 34 dynes/cm (0.01%); 100% act.

Igepal® CO-730. [Rhone-Poulenc Surf.] Nonoxynol-15; nonionic; detergent, dispersant, wetting agent, penetrant, emulsifier for industrial cleaners; yel. liq., aromatic odor; sol. see Igepal CO-630; sp.gr. 1.07; visc. 450–550 cps; HLB 15.0; cloud pt. 203–212 F (1%); flash pt. > 200 F (PMCC); pour pt. 71 ± 2 F; surf. tens. 36 dynes/cm (0.01%); 100% act.

Igepal® CO-850. [Rhone-Poulenc Surf.] Nonoxynol-20; nonionic; detergent, wetting agent, dispersant, emulsifier, for industrial cleaners, polyester resins; latex stabilizer; demulsifier for crude petrol. oil emulsions; pale yel. wax; aromatic odor; sol. see Igepal CO-630; sp.gr. 1.08 (50 C); HLB 16.0; cloud pt. > 212 F (1% aq.); flash pt. > 200 F (PMCC); pour pt. 91 ± 2 F; surf. tens. 39 dynes/cm (0.01%); 100% act.

Igepal® CO-880. [Rhone-Poulenc Surf.] Nonoxynol-30; nonionic; see Igepal CO-850; also textile scouring, solubilizer; pale yel. wax; aromatic odor; sol. see Igepal CO-630; sp.gr. 1.08 (50 C); HLB 17.2; cloud pt. > 212 F (1% aq.); flash pt. > 200 F (PMCC); pour pt. 109 ± 2 F; surf. tens. 43 dynes/cm (0.01%); 100% act.

Igepal® CO-887. [Rhone-Poulenc Surf.] Nonoxynol-30; nonionic; see Igepal CO-880; pale yel. liq.; aromatic odor; sp.gr. 1.09; HLB 17.2; flash pt. > 200 F (PMCC); pour pt. 34 ± 2 F; 70% act.

Igepal® CO-890. [Rhone-Poulenc Surf.] Nonoxynol-40; nonionic; emulsifier, stabilizer, wetting agent, dyeing assistant for plastics, floor polishes, etc.; off-wh. wax; aromatic odor; sol. in ethanol, water; sp.gr. 1.09 (50 C); HLB 17.8; cloud pt. > 212 F (1% aq.); flash pt. > 200 F (PMCC); pour pt. 112 ± 2 F; 100% act.

Igepal® CO-897. [Rhone-Poulenc Surf.] Nonoxynol-40; nonionic; see Igepal CO-890; pale yel. liq.; aromatic odor; sp.gr. 1.10; HLB 17.8; flash pt. > 200 F (PMCC); pour pt. 46 ± 2 F; 70% act.

Igepal® CO-970, -977. [Rhone-Poulenc Surf.] Nonoxynol-50; nonionic; see Igepal CO-890; off-wh. wax and pale yel. liq. resp.; aromatic odor; sol. in ethanol, water; sp.gr. 1.10 (50 C); HLB 18.2; flash pt. > 200 F (PMCC); pour pt. 114 ± 2 F and 52 ± 2 F resp.; 100 and 70% act. resp.

Igepal® CO-980. [Rhone-Poulenc Surf.] Nonylphenoxypoly ethyleneoxy ethanol; nonionic; stabilizer and dyeing assistant; polymerization emulsifier for vinyl acetate and acrylic emulsions; solid; water-sol.; HLB 18.7; cloud pt. > 212 F (1% aq.); 100% conc.

Igepal® CO-985, -987. [Rhone-Poulenc Surf.] Nonylphenoxypoly ethyleneoxy ethanol; nonionic; see Igepal CO-980; liq.; water-sol.; HLB 18.7; cloud pt. > 212 F (1% aq., CO-987); 50 and 70% conc. resp.

Igepal® CO-990. [Rhone-Poulenc Surf.] Nonoxynol-100; nonionic; see Igepal CO-890; off-wh. wax; aromatic odor; sol. in ethanol, water; sp.gr. 1.12 (50 C); HLB 19.0; cloud pt. > 212 F (1% aq.); flash pt. > 200 F (PMCC); pour pt. 122 ± 2 F; 100% act.

Igepal® CO-997. [Rhone-Poulenc Surf.] Nonoxynol-100; nonionic; see Igepal CO-890; pale yel. liq.; aromatic odor; sp.gr. 1.11; HLB 19.0; cloud pt. > 212 F (1% aq.); flash pt. > 200 F (PMCC); pour pt. 68 ± 2 F; 70% act. in water.

Igepal® DM-710. [Rhone-Poulenc Surf.] Dinonyl phenyl ethoxylate; nonionic; detergent, emulsifier, antistat for textiles, leather, metal cleaners, paper, latex, pesticides; pale yel. opaque liq.; aromatic odor; sol. in naphtha, xylene, perchloroethylene, ethanol, water; dens. 8.72 lb/gal; sp.gr. 1.045; visc. 19 cps (100 C); HLB 13.0; cloud pt. 48–52 C (1%); flash pt. 500–520 F (COC); pour pt. 18 C; surf. tens. 29 dynes/cm (0.01%); 100% act.

Igepal® DM-880. [Rhone-Poulenc Surf.] Nonyl nonoxynol-49; nonionic; emulsifier, solubilizer for emulsion polymerization, latex stabilization, pesticides, essential oils, cleaners; pale yel. wax; aromatic odor; sol. in xylene, ethanol, water; sp.gr. 1.050 (50 C); visc. 55 cps (100 C); HLB 17.2; cloud pt. > 100 C (1%); flash pt. 500–520 F (COC); pour pt. 47 C; surf. tens. 38 dynes/cm (0.01%); 100% act.

Igepal® DM-970. [Rhone-Poulenc Surf.] Nonyl nonoxynol-150; nonionic; detergent, dispersant, wetter, stabilizer for laundry, household, textile, hard-surface detergents, cosmetics, insecticides, paper, petrol., paints; wh. flakes; sol. in water, aromatic solv., methanol, ethanol; sp.gr. 1.05 (80 C); HLB 19; cloud pt. > 100 C (1%); 100% act.

Igepal® OD-410. [Rhone-Poulenc Surf.] Ethoxylated alkylphenol; nonionic; solv. for vinyl, phenolic, polyester, alkyd, NC, and cellulose acetate resins; ingred. of metal cleaners, paint strippers, and cleaning compds. where solvs. and solv. boosters are required; used as ink vehicle; liq.; 100% act.

Igepal® RC-620. [Rhone-Poulenc Surf.] Dodecylphenoxypoly (ethyleneoxy) ethanol; nonionic; detergent, wetting agent, emulsifier, dispersant for agric., household/industrial cleaners; dedusting agent; APHA 200 max. liq.; HLB 10.0; cloud pt. 38–42 C (1% aq.); 100% act.

Igepal® RC-630. [Rhone-Poulenc Surf.] Dodecylphenoxypoly (ethyleneoxy) ethanol; nonionic; detergent, wetting agent, emulsifier, dispersant for household/industrial cleaners; APHA 200 max. liq.; HLB 12.7; cloud pt. 62–66 C (1% aq.); 100% act.

Igepon AC-78. [Rhone-Poulenc Surf.] Sodium cocoyl isethionate; anionic; surfactant with good dispersing, foaming for detergent bars, dentifrices, shampoos, bubble baths, other cosmetics; powd.; ≥ 83% act.

Igepon CN42. [Rhone-Poulenc Surf.] Sodium-N-cyclohexyl-N-palmitoyl taurate; anionic; low-foaming detergent and dispersant for mechanical dishwasher and industrial cleaners, stabilizer for syn. rubber emulsions; lt. tan, soft firm paste; fatty odor; m.w. 467.2; water sol.; pH 8.0–9.5 (10%); 23% min. act.

Igepon T-33. [Rhone-Poulenc Surf.] Sodium-N-methyl-N-oleoyl taurate; anionic; detergent, wetting agent, dispersant for textiles, industrial detergents, paper industry; pale yel. nearly clear liq., perfumed alcoholic odor; m.w. 425; water sol.; pH 6.5–8.0 (10%); 32% act.

Igepon T-43. [Rhone-Poulenc Surf.] Sodium-N-methyl-N-oleoyl taurate; anionic; see Igepon T-33; wh. visc. liq. slurry, fatty odor; m.w. 425; water sol.; pH 6.5–8.0 (5%); 33% act.

Igepon T-51. [Rhone-Poulenc Surf.] Sodium-N-methyl-N-oleoyl taurate; anionic; see Igepon T-33; also latex stabilizer; lt. amber gel, perfumed odor; m.w. 425; water sol.; pH 6.5–8.0 (10%); 13.9% act.

Igepon T-77. [Rhone-Poulenc Surf.] Sodium-N-methyl-N-oleoyl taurate; anionic; see Igepon T-33; cream flakes, fatty odor; m.w. 425; water sol.; pH 6.5–8.0 (5%); 67% act.

Igepon TC-42. [Rhone-Poulenc Surf.] Sodium methyl cocoyl taurate; anionic; detergent, foamer, dispersant in cosmetics; wh. soft, smooth paste; m.w. 363; pH 7.0–8.5 (10%); 24% act.

Igepon TK-32. [Rhone-Poulenc Surf.] Sodium-N-methyl-N-tall oil acid taurate; anionic; detergent, suspending/dispersing agent, precipitation inhibitor for petrol. industry; VCS 8 max. clear liq.; m.w. 439; water sol.; pH 8–10 (10%); 20% min. act.

Igepon TN-74. [Rhone-Poulenc Surf.] Sodium methyl palmitoyl taurate; anionic; detergent with suspending and dispersing agent; wh. fine powd.; m.w. 406; water sol.; pH 6.5–8.0 (2%); 44% min. act.

Imerol DU. [Sandoz Ltd.] EO adduct; dry cleaning assistant, softener, antistatic agent, nonionic; liq.

Imerol NCP, NCP Liq. [Sandoz Ltd.] Sulfonate containing solvs.; wetting agent, emulsifer, dispersant, used for print paste dyes; sp.gr. 1.09; 57% act.

Imicure® AMI-2. [Pacific Anchor] 2-Methyl imidazole; curing agent for printed circuit board laminates, powd. coatings, adhesives, encapsulation; accelerator for dicyandiamide and anhydrides; wh. powd.; dens. 8.4 lb/gal; m.p. 165 F.

Imicure® EMI-24. [Pacific Anchor] 2-Ethyl, 4-methyl imidazole; curing agent for adhesives, composites, filament winding, solder-resistant inks, potting compds.; accelerator for anhydride curing agents; Gardner 8 color; dens. 8.2 lb/gal; visc. 65 poise.

Imicure® EMI-24S. [Pacific Anchor] 2-Ethyl, 4-methyl imidazole; high purity solid version of Imicure EMI-24; uring agent for adhesives, composites, filament winding, solder-resistant inks, potting compds.; accelerator for anhydride curing agents; designed for elec. applics.; yel. solid; dens. 8.2 lb/gal; m.p. 110–115 F.

Improved Kelmar. [Kelco] Potassium alginate; gellant, emulsifier, and stabilizer in food and indust. applic.; used for water holding in foods and industry; cream granular particles; visc. 400 cps; pH neutral.

Imsil® A-8. [Unimin] Silica; filler; used in finishes, enamels, maintenance paints, plastic film antiblocks, urethane rubber, polishes; wh. particulate; 100% thru 325 mesh; 2.1 μ median particle size; oil absorp. 28 g/100 g; surf. area 2.0 m/g; pH 6.4; brightness 87.7; 99.0% SiO_2.

Imsil® A-10. [Unimin] Amorphous silica; filler/extender in solv.-based or latex paints; permits high pigment loading without increase in visc.; thickener, carrier, free flow agent; powd. coating; paint flatting agent; used in protective and wire and cable coatings, PU elastomers, epoxy, phenolic, buffing, polishing and polyester compds., silicone rubbers, ceramic glazes, PU foams, adhesives, elec. resistors, refractory prods., insulations, antiblock agents, toothpastes, agric. compositions, cosmetic, metal castings, inj. thermoset moldings; particulate; particle size 1.55 (Fisher subsieve), 99.0% < 10 μ; sol. in hydrofluoric acid; sp.gr. 2.65; dens. 22.07 lb/solid gal, 21–23 lb/ft³; m.p. 1722 C; oil absorp. (Gardner Coleman) 29–31; ref. index 1.54–1.55; pH 7; 99.5 ± 0.5% SiO_2.

Imsil® A-15. [Unimin] Amorphous silica; see Imsil A-10; particulate; particle size 1.82 (Fisher subsieve), 99.0% < 15 μ; sol. in hydrofluoric acid; sp.gr. 2.65; dens. 22.07 lb/solid gal, 22–25 lb/ft³; m.p. 1722 C; oil absorp. (Gardner Coleman) 29–31; ref. index 1.54–1.55; pH 7; hardness 6.5 Moh; 99.5 ± 0.5% SiO_2.

Imsil® A-25. [Unimin] Amorphous silica; see Imsil

A-10; particulate; particle size 1.97 (Fisher subsieve), 99.9+% thru 400-mesh screen; sol. see Imsil A-10; sp.gr. 2.65; dens. 22.07 lb/ft³; m.p. 1722 C; oil absorp. (Gardner Coleman) 29–31; ref. index 1.54–1.55; pH 7; hardness 6.5 Moh; 99.5 ± 0.5% SiO_2.

Imsil® A-30. [Unimin] Silica; filler, flatting agent for coatings, urethane rubber compds., buffing compds.; wh. particulate; 5.4 μ median particle size; 99.6% thru 325 mesh; oil absorp. 28 g/100 g; GE brightness 85.9; surf. area 1.1 m/g; pH 6.5; 99% SiO_2.

Imsil® A-75. [Unimin] Silica; filler for nonskid coatings, mastics and adhesives, cements, buffing compds.; wh. particulate; 6.2 μ median particle size; 95% thru 325 mesh; oil absorp. 27 g/100 g; surf. area 1.3 m/g; pH 6.5; 99% SiO_2.

Imsil® A-108. [Unimin] Amorphous silica; see Imsil A-10; particulate; particle size 1.12 (Fisher subsieve), 99.0% < 8 μ; sol. see Imsil A-10; sp.gr. 2.65; dens. 22.07 lb/solid gal, 20–22 lb/ft³; m.p. 1722 C; oil absorp. (Gardner Coleman) 29–31; ref. index 1.54–1.55; pH 7; hardness 6.5 Moh; 99.5 ± 0.5% SiO_2.

Imwitor 191. [Hüls] Glyceryl stearate; coemulsifier, dispersant for personal care prods.; emulsifier in o/w and w/o emulsions; lubricants and binders used in the pharmaceutical industry; suspending agent, stabilizer, thickener; Gardner 4 max. powd.; sol. in oils, molten fats; m.p. 56–61 C; solid. pt. 63–68 C; sapon. no. 155–170; 90% 1-monoglycerides.

Imwitor 308. [Hüls] Glyceryl caprylate; solubilizer for pharmaceutical drugs; carrier/vehicle for drugs in capsules; wh. cryst. mass; m.p. 30 C.

Imwitor 310. [Hüls] Glyceryl caprate; see Imwitor 308; wh. cryst. mass.

Imwitor 312. [Hüls] Glyceryl laurate; coemulsifier, solubilizer, carrier for lipophilic drugs; wh. cryst. mass.

Imwitor 595. [Hüls] Monoglyceride, molecular dist.; nonionic; stabilizer and plasticizer for emulsions; dispersant; powd.; HLB 4.4; 100% conc.

Imwitor 742. [Hüls] Glyceryl mono/dicaprylate/caprate; nonionic; plasticizer for hard fats, solv. for lipophilic ingreds.; emollient; coemulsifier, solubilizer, carrier for lipophilic drugs; paste; HLB 4.0; 100% conc.

Imwitor 900. [Hüls] Glyceryl stearate; coemulsifier, dispersant for personal care prods.; emulsifier in o/w and w/o emulsions; lubricants and binders used in the pharmaceutical industry; suspending agent, stabilizer, thickener; Gardner 4 max. powd.; sol. in fats, oils, waxes; m.p. 56–61 C; solid. pt. 56–61 C; sapon. no. 162–173; 40–50% 1-monoglyceride.

Imwitor 900 K. [Hüls] Glyceryl stearate; nonionic; stabilizer, emulsifier, thickener for cosmetics; wh. waxy solid; slightly fatty odor; sol. in all fats, oils, waxes, chloroform, ether, ethanol (@ 60 C); m.p. 56–60 C; acid no. 3 max.; HLB 3.8; sapon. no. 160–176; 42–48% monoglycerides.

Imwitor 908. [Hüls] Glyceryl mono/dicaprylate; see Imwitor 312; ylsh. oily liq.

Imwitor 914. [Hüls] Glyceryl mono/dimyristate; see Imwitor 312; ylsh. cryst. mass.

Imwitor 940. [Hüls] Palm oil glycerides; nonionic; stabilizer for cosmetics; dispersant for clay systems; sec. emulsifier and stabilizer for asphalt emulsions; flakes, powd.; sol. in fats, oils, waxes; HLB 3.8; 100% conc.

Imwitor 940 K. [Hüls] Glycerol monostearate/palmitate Ph. Eur.; see Imwitor 191; Gardner 4 max. powd.; m.p. 53–57 C; solid. pt. 54–60; sapon. no. 165–178; 42–48% 1-monoglyceride.

Imwitor 945. [Hüls] Palmitic/stearic acid glycerol monodiester; nonionic; stabilizer for cosmetics; powd.; HLB 3.8% 100 conc.

Imwitor 960. [Hüls] Glyceryl stearate SE; emulsion base for cosmetics; suspending agent in creams, lotions, emulsions, cosmetics; Gardner 4 max. powd.; sol. in fats, oils, waxes; m.p. 56–61 C; sapon. no. 155–175; 30–40% 1-monoglyceride.

Imwitor 960 K. [Hüls] Glyceryl stearate SE; anionic; emulsifier and ointment base for pharmaceutical and cosmetic creams of the o/w type; wh. hard waxy solid; typical odor; sol. in all fats, oils, waxes, in chloroform, benzene, ether, ethanol; HLB 12.0; m.p. 56–60 C; sapon. no. 155–170; 30–40% monoester; 100% conc.

Imwitor 965. [Hüls] Palm oil glycerides and potassium stearate; anionic; o/w emulsifier for cosmetics; cream base; suspending agent in creams, lotions, emulsions, cosmetics; flakes, powd.; sol. in fats, oils, waxes; 100% conc.

Imwitor 965 K. [Hüls] Glycerol monostearate/palmitate and potassium stearate; see Imwitor 960; Gardner 4 max. flakes; m.p. 53– 57 C; solid. pt. 56–62 C; sapon. no. 150– 170; 30–42% 1-monoglyceride.

Incrocas 10. [Croda] PEG-10 castor oil; nonionic; emulsifier, solubilizer, emollient, superfatting agent, lubricant for personal care prods., detergents, metalworking fluids, insecticides, herbicides, household prods.; lubricant, antistat, softener, surfactant, and dye leveling agent for textile; pale yel. liq.; disp. in water; sol. @ 10% in ethanol, naphtha, MEK, oleic acid, trichloroethylene; HLB 6.3; cloud pt. 20 C (1% in 10% brine); sapon. no. 120–130; pH 6.0–7.5 (3% aq.); 100% act.

Incrocas 30. [Croda] PEG-30 castor oil; nonionic; see Incrocas 10; also lime soap dispersant in alkaline scouring systems; pale yel. liq.; sol. in water, ethanol, oleyl alcohol, naphtha, MEK, oleic acid, trichloroethylene; HLB 12.5; cloud pt. 35–40 C (1% in 10% brine); sapon. no. 72–82; pH 6.0–7.5 (3% aq.); surf. tens. 41.5 dynes/cm (0.1% DW); 100% act.

Incrocas 40. [Croda] PEG-40 castor oil; nonionic; see Incrocas 10; pale yel. liq.; sol. @ 10% see Incrocas 30; HLB 13.0; cloud pt. 50 C (1% in 10% brine); sapon. no. 60–65; pH 6.0–7.5 (3% aq.); surf. tens. 40.90 dynes/cm (0.1% DW); 100% act.

Incrocas 60. [Croda] PEG-60 castor oil; nonionic; see Incrocas 30; pale yel. soft paste; sol. see Incrocas 30; HLB 14.7; cloud pt. 60 C (1% in 10% brine); sapon. no. 45–50; pH 6.0–7.5 (3% aq.); surf. tens. 43.20 dynes/cm (0.1% DW); 100% act.

Incrocas 100. [Croda] PEG-100 castor oil; nonionic;

see Incrocas 10; also lime soap dispersant in alkaline scouring systems; pale yel. waxy solid; sol. see Incrocas 30; HLB 16.5; cloud pt. 66 C (1% in 10% brine); sapon. no. 25–35; pH 6.0–7.5 (3% aq.); surf. tens. 41.60 dynes/cm (0.1% DW); 100% act.

Incromate ALL. [Croda] Almondamidopropyl dimethylamine lactate; conditioner providing slip for wet comb; visc. amber liq.; 75% act.

Incromate AVL. [Croda] Avocadamidopropyl dimethylamine lactate; conditioner providing slip for wet comb; amber visc. liq.; 75% act.

Incromate BAL. [Croda] Babassamidopropyl dimethylamine lactate; conditioner with good foam for improved dry comb; amber visc. liq.; 68% act.

Incromate CDL. [Croda] Cocamidopropyl dimethylamine lactate; cationic; surfactant for personal care prods., conditioners; base for cationic emulsions; yel. visc. liq.; sol. in water; pH 6–7 (5% sol'n.); 95–97% solids.

Incromate CDP. [Croda] Cocamidopropyl dimethylamine propionate; cationic; see Incromate CDL; Gardner 4 max. liq.; pH 6.0–7.0; 40% act.

Incromate CPML. [Croda] Cocamidopropyl morpholine lactate; cationic; see Incromate CDL; yel. visc. liq.; sol. in water; pH 4–5 (5%); 29–31% act.

Incromate IDL. [Croda] Isostearamidopropyl dimethylamine lactate; cationic; see Incromate CDL; yel. visc. liq.; pH 6–7 (5%); 95–97% act.

Incromate LDL. [Croda] Linoleamidopropyl dimethylamine lactate; cationic; see Incromate CDL; yel. visc. liq.; pH 6–7 (5%); 28–30% act.

Incromate Mink L. [Croda] Minkamidopropyl dimethylamine lactate; conditioner and foamer for hair and skin care prods.; amber visc. liq.; 75% act.

Incromate OLL. [Croda] Olivamidopropyl dimethylamine lactate; conditioner and foamer for hair and skin care prods.; amber visc. liq.; 75% act.

Incromate PPI. [Croda] Stearamidopropyl dimethylamine lactate, ethylene glycol monostearate, propylene glycol emulsion; cationic; conditioner, emulsifier, pearlizing agent, humectant, visc. control agent for creme rinses, conditioning shampoos, hand cleansers and lotions; wh. thick paste; pH 4–5 (5%); 30.5–32.5% solids.

Incromate SDL. [Croda] Stearamidopropyl dimethylamine lactate; cationic; softener for hair shampos, fabrics; raw material; Gardner 3 max. slurry; pH 4–5 (10%); 24–26% conc.

Incromate SEL. [Croda] Sesamidopropyl dimethylamine lactate; conditioner and foamer for good slip, wet combing, and detangling; amber visc. liq.; sol. in alcohol and water; 75% act.

Incromate WGL. [Croda] Wheat germamidopropyl dimethylamine lactate; conditioner with exc. slip, wet comb, and dry feel; amber visc. liq.; sol. in alcohol; 75% act.

Incromectant AMEA-70. [Croda] Acetamide MEA; solv., humectant for cosmetics; clarifying agent for shampoos; conditioner for creme rinses; sol'n.; sp.gr. 1.07–1.17; dens. 9.3 lb/gal; pH 6.0–8.5; 70% min. act.

Incromectant AMEA-100. [Croda] Acetamide MEA; clarifying detangling agent for shampoos, conditioners, cream rinses; humectant in creams and lotions; liq.; 100% act.

Incromectant AQ. [Croda] Acetamidopropyl trimonium chloride; cationic; antistat for shampoos and conditioners; humectant; plasticizer for hair conditioning/setting polymers; liq.; 75% act.

Incromectant LAMEA. [Croda] Acetamide MEA and lactamide MEA; humectant, moisturizing agent for hair and skin care prods.; Gardner 4 max. liq.; acid no. 15 max.; pH 5–8 (10%); 95% min. act.

Incromectant LMEA. [Croda] Lactamide MEA; clarifying detangling agent for shampoos, conditioners, and cream rinses; humectant in creams and lotions; liq.; 100% act.

Incromectant LQ. [Croda] Lactamidopropyl trimonium chloride; cationic; see Incromectant AQ; liq.; 75% act.

Incromide ALD. [Croda] 1:1 Almondamide DEA; nonionic; conditioner, visc. builder, foam stabilizer; clear liq.

Incromide AVD. [Croda] 1:1 Avocadamide DEA; nonionic; visc. builder, foam stabilizer; clear liq.

Incromide BAD. [Croda] 1:1 Babassamide DEA; nonionic; visc. builder, foam stabilizer, emulsifier; clear liq.

Incromide BED. [Croda] 1:1 Behenamide DEA; nonionic; high melting pearling and opacifying agent used in stick preps.; pale yel. wax.

Incromide BEM. [Croda] 1:1 Behenamide MEA; nonionic; structural wax for antiperspirant and other stick prods.; pale yel. flake.

Incromide CA. [Croda] 1:1 Cocamide DEA; nonionic; surfactant, foam stabilizer, emulsifier, and thickener used in household, cosmetic, and industrial formulations; clear liq.; 100% act.

Incromide CAC. [Croda] 1:1 Cocamide DEA and cocoyl sarcosinate; anionic; low irritation foamer, clarifying agent for clear soap and shampoo bar preps.; Gardner 4 max. clear liq., bland odor; water-sol.; pH 8.6–9.2 (3%).

Incromide CAL. [Croda] Cocamide DEA; nonionic; visc. builder and foam enhancer for bath prods. and shampoo systems; Gardner 5 max. liq.; bland odor; 100% act.

Incromide CM. [Croda] 1:1 Cocamide MEA; nonionic; visc. builder, foam stabilizer, and opacifier for cosmetic, industrial, household applics.; esp. for alpha olefin sulfonate formulations; yel. flake; 100% act.

Incromide L-90. [Croda] Lauramide DEA; nonionic; foam stabilizer, thickener, detergent, and foaming agent in household, industrial and institutional cleaning comps., car washes, rug and upholstery cleaners, and personal care prods.; Gardner 2 max. solid; mild fatty odor; sol. in alcohol, chlorinated and aromatic hydrocarbons; disp. in water; 100% act.

Incromide LA. [Croda] Linoleic DEA; nonionic; superfatting agent and thickener for personal care and household prods.; useful in increasing visc. of various sulfate and ether sulfate dilutions; conditioner and lubricant; Gardner 5 max. liq.; bland odor;

100% act.

Incromide LEP. [Croda] Coco DEA; nonionic; see Incromide CA; liq.; 100% act.

Incromide LI. [Croda] Lauric MIPA; nonionic; visc. builder and foam booster; wh. flake; m.p. 54 C; 100% act.

Incromide LL. [Croda] Lauric DEA; nonionic; foam booster, stabilizer, visc. builder, solubilizer, coupler used in personal care prods. and liq. detergent formulations; Gardner 4 max. liq., paste; bland odor; pH 9.7–10.3; 100% conc.

Incromide LLA. [Croda] Lauramide DEA, linoleamide DEA; nonionic; thickener, foam stabilizer, wetting agent, emulsifier, and emulsion stabilizer used in personal care prods.; Gardner 5 max. liq.; 100% act.

Incromide LMD. [Croda] Lauric diglycolamide; nonionic; surfactant, thickener, and foam stabilizer; Gardner 3 max. flake; m.p. 52 C; 100% act.

Incromide LMI. [Croda] Lauric/myristic MIPA; nonionic; visc. builder and foam booster; Gardner 4 max. flake; m.p. 48–50 C; 100% act.

Incromide LMM. [Croda] Lauramide MEA; nonionic; foam enhancer, visc. builder, gelling and wetting agent, thickener, and foam stabilizer used in personal care prods.; Gardner 5 max. flake; slight amine odor; 100% act.

Incromide LR. [Croda] Lauramide DEA; nonionic; thickener, foam stabilizer used in personal care prods., dishwash detergents and industrial cleaners; Gardner 5 max. liq.; bland odor; 100% act.

Incromide Mink D. [Croda] 1:1 Minkamide DEA; nonionic; visc. builder and foam stabilizer; clear liq.

Incromide OD. [Croda] 1:1 Oleamide DEA; nonionic; low color visc. builder with good sol. in anionic surfactants; clear liq.

Incromide OLD. [Croda] 1:1 Olivamide DEA; nonionic; visc. builder, foam stabilizer; clear liq.

Incromide OM. [Croda] Oleic MEA; nonionic; visc. builder for personal care prods.; Gardner 6 max. solid; m.p. 50–52 C; 100% act.

Incromide OPD. [Croda] Oleic DEA; nonionic; visc. builder, conditioner, emulsifier, foam stabilizer, and thickener for personal care prods.; solvent degreasers; yel. paste, liq.; bland odor; 100% act.

Incromide OPM. [Croda] Oleic MEA, modified; nonionic; visc. builder used in personal care prods.; Gardner 5 max. solid; m.p. 49–50 C; 100% act.

Incromide SED. [Croda] 1:1 Sesamide DEA; nonionic; visc. builder and foam stabilizer; clear liq.

Incromide SM. [Croda] Stearamide MEA; nonionic; softener and dye carrier in textile industry; pearlizier, opacifier and thickener in cosmetic preparations; flake; 100% conc.

Incromide UM. [Croda] Undecylenamide MEA; nonionic; foam stabilizer for antidandruff and fungicidal shampoos; Gardner 8 max. flake; m.p. 60 C; 100% act.

Incromide WGD. [Croda] 1:1 Wheat germamide DEA; nonionic; visc. builder and foam stabilizer; clear liq.

Incromine BB. [Croda] Behenamidopropyl dimethylamine; nonionic/cationic; emollient conditioner, lubricant, and moisturizer for personal care prods.; yel. flakes; m.w. 420–450; m.p. 70–72; 100% act.

Incromine CPM. [Croda] Cocamidopropyl morpholine; nonionic/cationic; intermediate used in personal care prods.; wh.-cream soft solid; m.p. 32 C; 100% act.

Incromine IB. [Croda] Isostearamidopropyl dimethylamine; nonionic/cationic; o/w emulsifier and lubricant for personal care prods.; yel. visc. liq., soft paste; m.w. 364–384; 100% act.

Incromine ISM. [Croda] Isostearamidopropyl morpholine; nonionic/cationic; hair conditioner in shampoos; yel. paste, liq.; 100% act.

Incromine L-40. [Croda] Lauramine oxide; nonionic; foam stabilizer, visc. builder, booster, softener and conditioner used in cosmetic and multifunctional applics.; wetting agent in electrolyte sol'ns.; liq.; 40% conc.

Incromine LSB. [Croda] Linoleamidopropyl dimethylamine; nonionic/cationic; see Incromine CPM; yel. visc. liq., soft solid; 100% act.

Incromine OPB. [Croda] Oleamidopropyl dimethylamine; nonionic/cationic; emollient conditioner, lubricant and moisturizer for personal care prods.; yel. soft paste; 100% act.

Incromine OPM. [Croda] Oleamidopropyl dimethylamine; nonionic/cationic; see Incromine BB; soft solid; 100% conc.

Incromine Oxide AL. [Croda] Almondamidopropylamine oxide; nonionic; visc. builder, conditioner for hair care prods.; pale yel. gel; 25% act.

Incromine Oxide AV. [Croda] Avocadamidopropylamine oxide; nonionic; visc. builder, conditioner for hair care prods.; pale yel. gel; 25% act.

Incromine Oxide B-30P. [Croda] Behenamine oxide; nonionic; softener and conditioner for hair care prods.; visc. builder, emulsifier, lubricant, wetting, and foam stabilizer used in personal care prods.; wh. paste; pH 6.5–8.0 (3%); 29–31% act.

Incromine Oxide BA. [Croda] Babassamidopropylamine oxide; nonionic; foamer, foam stabilizer; pale yel. liq.; 30% act.

Incromine Oxide C. [Croda] Cocamidopropyl amine oxide, aq. sol'n.; nonionic; visc. builder, foam stabilizer, conditioner used in personal care prods.; lt. straw liq.; pH 6.5–7.5 (5%); 29.5–31.5% act.

Incromine Oxide I. [Croda] Isostearamidopropyl amine oxide; nonionic; foam stabilizer, thickener, lubricant, and visc. builder used in cosmetic, household, and janitorial prods.; wetting agent for conc. electrolyte sol'ns.; Gardner 1 max. gel; pH 6.5–7.5; 24.5–26.5% act.

Incromine Oxide ISMO. [Croda] Isostearamidopropyl morpholine oxide; nonionic; see Incromine Oxide I; Gardner 2 max. liq.; pH 6.5–7.5; 24.5–26.5% amine oxide.

Incromine Oxide L. [Croda] Lauramine oxide; nonionic; surfactant for dishwashing comps., household prods., hair conditioner used in personal care prods.; colorless clear liq.; pH 7.0–8.0 (5% aq.); 29.0–31.0% amine oxide.

Incromine Oxide L-40. [Croda] Lauramine oxide; nonionic; foaming agent, foam stabilizer, degreaser; for shampoos and lt.-duty liqs.; essentially colorless liq.; 40% act.

Incromine Oxide M. [Croda] Myristamine oxide; nonionic; surfactant, emulsifier, emollient, conditioner used in personal care prods.; visc. liq.; pH 7.0–8.0 (5%); 29.5–30.5% amine oxide.

Incromine Oxide MC. [Croda] Myristamine oxide and cetamine oxide; nonionic; foam stabilizer and visc. builder used in personal, household and industrial applics.; wetting agent in electrolyte sol'ns.; liq.

Incromine Oxide Mink. [Croda] Minkamidopropylamine oxide; nonionic; foaming agent, visc. builder; pale yel. gel; 25% act.

Incromine Oxide O. [Croda] Oleamidopropyl amine oxide; nonionic; see Incromine Oxide I; colorless gel; pH 6.5–7.5; 24.5–26.5% amine oxide.

Incromine Oxide OD-50. [Croda] Oleamine oxide; nonionic; see Incromine Oxide I; Gardner 1 max. liq.; pH 7–8 (10%); 50–52% amine oxide.

Incromine Oxide OL. [Croda] Olivamidopropylamine oxide; nonionic; thickener for clear systems; yel. gel; 25% act.

Incromine Oxide S. [Croda] Stearamine oxide; nonionic; conditioner and cosmetic emulsifier for personal care prods.; wh. paste; 24.5–26.5% amine oxide.

Incromine Oxide SE. [Croda] Sesamidopropylamine oxide; nonionic; visc. builder, conditioner for hair care prods.; pale yel. gel; 25% act.

Incromine Oxide WG. [Croda] Wheat germamidopropylamine oxide; nonionic; visc. builder, conditioner for hair care prods.; yel. gel; 25% act.

Incromine PB. [Croda] Palmitamidopropyl dimethylamine; nonionic/cationic; see Incromine CPM; substantive conditioner and emulsifier used in personal care prods.; off-wh. flakes; m.w. 340–350; m.p. 56–57 C; 100% act.

Incromine SB. [Croda] Stearamidopropyl dimethylamine; nonionic/cationic; see Incromine CPM; cr. waxy flakes; faint char. odor; m.p. 63 C; 100% act.

Incromine TB. [Croda] Tallowamidopropyl dimethylamine; nonionic/cationic; see Incromine CPM; soft solid; 100% act.

Incronam 30. [Croda] Cocamidopropyl betaine; amphoteric; surfactant, emulsifier, coupling agent, visc. builder, foam detergent for personal care prods., chemical specialities, rug and upholstery shampoos, dishwashing compds.; Gardner 5 max. clear liq.; sol. in water; pH 5.5–7.5 (10% aq.); 30% act.

Incronam AL-30. [Croda] Almondamidopropyl betaine; amphoteric; foam booster/stabilizer for skin prods.; yel. liq.; sol. in water, alcohol; 30% act.

Incronam AV-30. [Croda] Avocadamidopropyl betaine; amphoteric; foam booster/stabilizer for skin prods.; pale yel. liq.; sol. in water, alcohol; 30% act.

Incronam B-40. [Croda] Behenyl betaine; amphoteric; foaming surfactant, conditioner, lubricant used in personal care prods.; wh. paste; pH 5.5–7.0 (3%); 40% act.

Incronam BA-30. [Croda] Babassamidopropyl betaine; amphtoeric; foam booster/stabilizer; yel. liq.; sol. in alcohol, water; 30% act.

Incronam I-30. [Croda] Isostearamidopropyl betaine; amphoteric; see Incronam B-40; yel. clear visc. liq.; pH 6–7.5 (5%); 29.5–31% act.

Incronam ISM-30. [Croda] Isostearamidopropyl morpholine betaine; amphoteric; conditioner, visc. builder used in personal care prods.; yel. liq.; pH 5–7 (5%); 29.5–31.0% act.

Incronam Mink 30. [Croda] Minkamidopropyl betaine; amphoteric; conditioner; foam booster/stabilizer, visc. builder; amber liq.; sol. in alcohol, water; 30% act.

Incronam OL-30. [Croda] Olivamidopropyl betaine; amphoteric; foam booster/stabilizer for skin care prods.; amber liq.; sol. in water, alcohol; 30% act.

Incronam OP-30. [Croda] Oleamidopropyl betaine; amphoteric; see Incronam B-40; yel. clear liq.; pH 6.0–7.2 (5%); 30% act.

Incronam P-30. [Croda] Palmitamidopropyl betaine; amphoteric; thickener and conditioner used in personal care prods.; yel. visc. liq., slurry; pH 6–7.2 (5%); 30% min. act.

Incronam SE-30. [Croda] Sesamidopropyl betaine; amphoteric; foam booster/stabilizer, conditioner with good slip; yel. liq.; forms clear sol'ns. in alcohol, clear gels in water; 30% act.

Incronam WG-30. [Croda] Wheat germamidopropyl betaine; amphoteric; foam booster/stabilizer; yel. liq.; forms clear sol'ns. in water and hot alcohol; 30% act.

Incronol ALS. [Croda Ltd.] Ammonium lauryl sulfate; anionic; detergent dispersant, suspending, and foaming agent used in personal care prods.; pH 6.3–6.8 (10%); 27–30% act.

Incropearl Liquid. [Croda Ltd.] Sodium laureth sulfate and hydroxyethyl MIPA; anionic; pearling agent for shampoos, dishwashing, bubble baths, shower cleansers, rug cleaners; creamy, pearly paste, mild char. odor; pH 8.0–8.6; 35–40% solids.

Incropearl Liquid NI. [Croda Ltd.] Hydroxyethyl stearamide MIPA and PPG-5 ceteth-20; nonionic; see Incropearl Liquid; creamy, pearly paste, mild char. odor; pH 7–8; 18–22% solids.

Incropol 233. [Croda] Propoxylated alcohol; nonionic; surfactant used as emulsifier, lubricant, detergent for industrial and general household prods.; coupling agent, antistat, fiber lubricant and solubilizer in personal care prods.; colorless liq.; pH 5.5–7.5 (3%).

Incropol 290. [Croda] Alcohol, ethoxylated; nonionic; see Incropol 233; colorless liq.

Incropol CS-12. [Croda] Ceteareth-12; nonionic; see Incropol 233; Gardner 1 max. solid; pH 5.5–7.5 (3%); 100% act.

Incropol CS-20. [Croda] Ceteareth-20; nonionic; see Incropol 233; Gardner 1 max. solid; pH 5.5–7.5 (3%).

Incropol CS-30. [Croda] Ceteareth-30; nonionic; see Incropol 233; Gardner 1 max. solid; pH 5.5–7.5 (3%); 100% act.

Incropol CS-40. [Croda] Ceteareth-40; nonionic; see Incropol 233; wh. flake; pH 5.5–7.5 (1%).

Incropol CS-50. [Croda] Ceteareth-50; nonionic; see Incropol 233; Gardner 1 max. flake; pH 6–8 (3%); 100% act.
Incropol CS-60. [Croda] Ceteareth-60; nonionic; see Incropol 233; wh. flake; pH 5.5–7.5 (1%).
Incropol L-2, L-7. [Croda] Laureth-2, –7 resp.; nonionic; see Incropol 233; Gardner 1 max. liq.; pH 5.5–7.5 (3%); 100% act.
Incropol L-7-90. [Croda] Laureth-7; nonionic; see Incropol 233; Gardner 1 max. liq.; pH 5.5–7.5 (3%); 90% act.
Incropol L-12. [Croda] Laureth-12; nonionic; see Incropol 233; Gardner 1 max. paste; pH 5.5–7.5 (3%); 100% act.
Incropol L-23. [Croda] Laureth-23; nonionic; see Incropol 233; solid; 100% act.
Incropol L-30. [Croda] Laureth-30; nonionic; see Incropol 233; Gardner 1 max. paste; pH 5.5–7.5 (3%); 100% act.
Incroquat 100. [Croda] Methyl bis (hydrog. tallow amido ethyl) 2-hydroxyethyl ammonium chloride; cationic; conditioner, antistat for hair care prods.; flake or powd.; sol. in hot alcohol; precipitates when cold; insol. in water; 100% act.
Incroquat 248. [Croda] Quaternium-72; cationic; hair conditioner; yel. solid; 90% act.
Incroquat AL-85. [Croda] Almondamidopropalkonium chloride; cationic; conditioner, foamer for hair care prods.; amber liq.; 85% act.
Incroquat AV-85. [Croda] Avocadamidopropalkonium chloride; cationic; conditioner, foamer for hair care prods.; good slip and wet comb; amber liq.; clearly sol. in alcohol and water; 85% act.
Incroquat B65C. [Croda] Behenalkonium chloride, cetyl alcohol; substantive conditioner for hair care prods.; wh. flake; 65% act.
Incroquat BA-85. [Croda] Babassamidopropalkonium chloride; cationic; conditioner, foamer, antistat for hair care prods.; amber liq.; 85% act.
Incroquat BDQ-25P. [Croda] Behenalkonium chloride; cosmetic hair conditioner; paste.
Incroquat Behenyl BDQ/P. [Croda] Behenalkonium chloride; cationic; o/w emulsifier and conditioner; paste; 25% conc.
Incroquat Behenyl TMC/P. [Croda] Behenyl trimethyl ammonium chloride; cationic; o/w emulsifier and conditioner; paste; 25% conc.
Incroquat Behenyl TMS. [Croda] Behenyl trimethyl ammonium methosulfate, cetearyl alcohol; cationic; conditioner, softener, emollient, o/w emulsifier used in personal care prods.; flake; 25% act.
Incroquat BTQ-25. [Croda] Behenyl trimethyl ammonium methosulfate; cosmetic hair conditioner; flake.
Incroquat CR Conc. [Croda] Cetearyl alcohol, PEG-40 castor oil, stearalkonium chloride; cationic; see Incroquat Behenyl TMS; wh. flake; mild fatty odor; pH 6.3 (5% disp.); 96% min. act.
Incroquat CTC-25. [Croda] Cetrimonium chloride; cationic; conditioner in specialty hair preparations; see also Incroquat Behenyl TMS; APHA 100 max. liq.; pH 3.5–4.0 (2% aq.); 24.0–26.0% min. act.
Incroquat CTC-30. [Croda] Cetrimonium chloride; cationic; see Incroquat CTC-25; Gardner 2 max. liq.; pH 3.5–4.0 (5%); 29% act.
Incroquat DBM-90. [Croda] Dibehenyldimonium methosulfate; conditioner with good wetting and slip for hair care prods.; wh. flake; sol. in hot alcohol; insol. in water, cold alcohol; 90% act.
Incroquat I-85. [Croda] Isostearamidopropalkonium chloride; cationic; conditioner with good slip and detangling for hair prods.; yel. liq.; water-sol.; 85% act.
Incroquat Mink-85. [Croda] Minkamidopropalkonium chloride; cationic; conditioner, foamer with exc. slip for hair prods.; amber liq.; 85% act.
Incroquat O-50. [Croda] Olealkonium chloride; cationic; conditioner with good slip for hair care prods.; pale yel. liq.; 50% act.
Incroquat OL-85. [Croda] Olivamidopropalkonium chloride; cationic; conditioner with good foaming and slip for hair prods.; amber liq.; 85% act.
Incroquat S-75 CG. [Croda] Quaternium-27; cationic; conditioner, antistat, and ingred. used in personal care prods., fibers, and synthetics; textile and paper softener; dispersant for pigments and dyestuffs; Gardner 5 max. liq.; pH 5–7 (5%); 74–76% act.
Incroquat S-85. [Croda] Stearalkonium chloride; cationic; conditioner, softener; see also Incroquat Behenyl TMS; wh. waxy flake; m.p. 140 F; pH 5–7 (0.5%); 85.0% min. act.
Incroquat SDQ-25. [Croda] Stearalkonium chloride; cationic; surfactant used as ingred. in personal care prods., textile and paper; dispersant for pigments and dyestuffs; antistat for fibers and synthetics; hair conditioner; paste; m.p. 140 F; pH 3.0–4.0 (1%).
Incroquat SE-85. [Croda] Sesamidopropalkonium chloride; cationic; conditioner, antistat with good foam for hair care prods.; amber liq.; 85% act.
Incroquat WG-85. [Croda] Wheat germamidopropalkonium chloride; cationic; conditioner with good foam and slip for hair prods.; amber liq.; sol. in alcohol and water; 85% act.
Incrosoft 100. [Croda] Methyl bis (hydrog. tallow amido ethyl) 2-hydroxyethyl ammonium chloride; fabric softener with good hand and antistatic properties; pale yel. flake or powd.; 100% act.
Incrosoft 248. [Croda] Quaternium-72; softening agent in dryer sheets; yel. solid; 90% act.
Incrosoft CFI. [Croda] Quaternium-72; cationic; fluid disp. for softening fabrics and tissue or as paper debonding agents; Gardner 7 clear liq.; sp.gr. 0.95; flash pt. (PM) 73 F; pH 5.0–6.5; 74–76% solids.
Incrosoft O-90. [Croda] Complex difatty quat. sulfate; fabric softener for clear softening prods.; used in home and commercial laundry prods.; yel. clear visc. liq.; pH 4–5 (3%); 88–90% act.
Incrosoft S-75. [Croda] Quaternium-27; softener base, lubricant, antistat and rewetting agent for fabrics and syns.; Gardner 5 max. liq.; water-sol.; pH 4–7 (5%); 74–76% act.
Incrosoft S-90. [Croda] Quaternium-27; see Incrosoft S-75; also used in textile maintenance and finish-

ing; yel. semisolid; disp. in water; pH 5.0–7.0 (5% in DW); 85–88% act.

Incrosoft S-90M. [Croda] Quaternium-27; see Incrosoft S-75; Gardner 4 max. liq.; water-sol.; pH 5–7 (5%); 89–90% act.

Incrosoft T-75. [Croda] Quaternium-53; fabric softener, lubricant and antistat for home and commercial laundry prods.; liq.; disp. in water; 75% act.

Incrosoft T-90. [Croda] Quaternium-53; see Incrosoft T-75; Gardner 6 max. visc. liq.; disp. in water; pH 4–7 (3%); 88–90% act.

Incrosoft T-90HV. [Croda] Quaternium-53; softening agent with exc. rewetting; general purpose fabric softener; yel. paste; 90% act.

Incrosorb S-60. [Croda Ltd.] Polysorbate 60; nonionic; o/w emulsifier and solubilizer for alkaline and acid pH systems; used in personal care prods.; pale yel. liq. gel; sapon. no. 45–55.

Incrosul LAFS. [Croda] Disodium laneth-5 sulfosuccinate; anionic; conditioning surfactant used in personal care prods.; yel. slurry/liq.; 40% act.

Incrosul LMA. [Croda] Diammonium lauramido MEA sulfosuccinate; surfactant, conditioner.

Incrosul LMS. [Croda] Disodium lauramido MEA sulfosuccinate; surfactant, conditioner.

Incrosul LS. [Croda] Disodium lauryl sulfosuccinate; surfactant, conditioner.

Incrosul LSA. [Croda] Diaamonium lauryl sulfosuccinate; surfactant, conditioner.

Incrosul OTS. [Croda] Disodium oleth-3 sulfosuccinate; anionic; surfactant, conditioner for mild shampoos and bath prods.; lowers irritation of SLS and SLES; exc. foaming; Gardner 2 max. clear liq.; mild char. odor; sol. in water, water/alcohol, glycerin; pH 6–7; 34–37% solids.

Incrosul TS. [Croda] Disodium tridecyl sulfosuccinate; surfactant, conditioner.

Incrosul UMS, UMS-45. [Croda] Disodium undecylenamido MEA sulfosuccinate; anionic; surfactant used in personal care prods.; lime soap dispersant and fungicidal activity; yel. clear liq.; faint char. odor; pH 6–7 (3%) and 5.25–6.75 (5%) resp.; 49–51 and 43.5–46% act. resp.

Indalca CD30. [Hercules] Deriv. of guar; nonionic; thickener used in carpet printing; lump-free visc. sol'ns. obtained by adding prod. to water with agitation to disp. the particles; off-wh. to tan powd.; water-sol.; water-disp.; visc. 20,000–24,000 cps; pH 7.0–7.5.

Indalca N-110-70. [Hercules] Highly refined, deriv. of natural gum; nonionic; thickener for cationic dyes used in fabric printing, rotary screen machines w/fine mesh, acid and premetallized dyes on nylon, silk, or wool, and disp. dyes on polyester, nylon, acetate, and triacetate; off-wh. to tan powd.; water-sol.; visc. 20,000–24,000 cps; pH 7.0–7.5.

Indalca XCD. [Hercules] Natural gum deriv.; nonionic; thickener for fabric-printing dyes, used in space dye, TAK, and Kuster applic.; wh. to off-wh. powd.; water-sol.; water-disp.; visc. 23,000–27,000 cps; pH 7.0–7.2.

Indalca XCDA/1. [Hercules] Modified natural gum; nonionic; thickener for carpet printing and Kuster and TAK dyeing; off-wh. to tan dustless powd.; water-disp.; dissolves in acid pH; visc. 300–400 cps (5%); 17,000–19,000 cps (1.5%).

Indulin® AS-1. [Westvaco] Amine-based; cationic; antistripping additive for highway paving applics.; asphalt adhesion agent; dk. brn. visc. liq.; sp.gr. 0.970; dens. 8.10 lb/gal; visc. 2450 cps; pour pt. 55 F; flash pt. (COC) 320 F; 100% act.

Indulin® AS-Special. [Westvaco] Amine-based; cationic; antistripping additive for highway paving applics.; asphalt adhesion agent; dk. brn. liq.; sp.gr. 0.985; dens. 8.22 lb/gal; visc. 370 cps; pour pt. 54 F; flash pt. (COC) 280 F; 100% act.

Indulin® AT. [Westvaco] Kraft pine lignin; used in polymeric applics. where dispersant and adsorp. properties are required; brn. powd.; sol. in ethylene glycol, dixoane, monoethanolamine, dimethyl formamide, Cellosolve; sp.gr. 1.3; dens. 26 lb/ft^3; flash pt. 349 F; pH 6.5 (2% aq.); surf. tens. 43 dynes/cm; 97% lignin content

Indulin® C. [Westvaco] Kraft pine lignin, sodium salt; anionic; stabilizer with dispersant properties for asphalt emulsions; primary emulsifier; industrial emulsions used in dispersion of fillers; brn. powd.; dens. 30 lb/ft^3; pH 10.2 (2% aq.); surf. tens. 43 dynes/cm (1% aq.)

Indulin® SAL. [Westvaco] Kraft lignin salt, formulated; anionic; emulsifier and stabilizer for asphalt; liq.; 40% conc.

Indulin® W-1. [Westvaco] Amine deriv. of pine lignin; cationic; emulsifier in asphalt emulsions, retarding agent in cement; brn. powd.; water sol. ≥ pH 8.3 and ≤ pH 4.2; dens. 22 lb/ft^3 (loose); visc. 100 cps @ 20% solids; surf. tens. 47 dynes (1% aq.); pH 10.5 (2% aq.); 100% conc.

Indulin® W-2. [Westvaco] Lignin, modified, amine deriv.; cationic; emulsifier and stabilizer for o/w emulsions; powd.; 100% conc.

Indulin® W-3. [Westvaco] Lignin amine; cationic; emulsifier for asphalt emulsions, slurry seal applics.; primary emulsifier, stabilizer, and retarder in emulsions; brn. liq.; amine odor; water-sol.; sp.gr. 1.090; dens. 9.1 lb/gal; visc. 450 cps; pH 12.0 (1%); 35% solids.

Indu-Sol 238. [Alkaril Canada] Wetting agent, suspending agent, lubricant, rust inhibitor, used in grinding and general machining operations; grnsh.–red, translucent liq.; dens. 11 lb/gal (Imp.); pH 9.2–9.6 (5%); biodeg.

Industrene® 104. [Witco/Humko] Oleic acid, low titer; chemical intermediate; solid. pt. 4 C max.; acid no. 198–204; sapon. no. 199–205.

Industrene® 105. [Witco/Humko] Oleic acid; used in alkyd resins, rubber compding., water repellents, polishes, soaps, abrasives, cutting oils, candles, crayons, emulsifiers; FG grades as lubricant, release agent, binder, defoamer in foods, intermediate for food emulsifiers; Gardner 6 liq.; solid. pt. 145 C max.; acid no. 195–204; sapon. no. 197–205; 100% conc.

Industrene® 106. [Witco/Humko] Oleic acid, low

titer; chemical intermediate; solid. pt. 6 C max.; acid no. 198–204; sapon. no. 199–205.

Industrene® 120. [Witco/Humko] Linseed acid; chemical intermediate; solid. pt. 15-22 C; acid no. 197–202; sapon. no. 197–203.

Industrene® 126. [Witco/Humko] Soya acid; chemical intermediate; solid. pt. 24–29 C; acid no. 196–205; sapon. no. 196–206.

Industrene® 130. [Witco/Humko] Linoleic acid; chemical intermediate; solid. pt. 30–38 C; acid no. 199–206; sapon. no. 199–207.

Industrene® 143. [Witco/Humko] Dist. tallow acid; see Industrene 105; Gardner 5 paste; sapon. no. 203–207; 100% conc.

Industrene® 145. [Witco/Humko] Tallow acid; chemical intermediate; solid. pt. 44–50 C; acid no. 195 min.; sapon. no. 195 min.

Industrene® 205. [Witco/Humko] Oleic acid; see Industrene 105; wh. liq.; sapon. no. 197–205; 100% conc.

Industrene® 206. [Witco/Humko] Oleic acid; see Industrene 105; liq.; solid. pt. 6 C max.; 100% conc.

Industrene® 206LP. [Witco/Humko] Oleic acid, low titer; chemical intermediate; solid. pt. 6 C max.; acid no. 199–205; sapon. no. 200–206.

Industrene® 223. [Witco/Humko] Hydrog. coconut acid; chemical intermediate, emulsifier; used for personal care prods., waxes, textile aux., pharmaceuticals; solid. pt. 23–26 C; acid no. 266–274; sapon. no. 267–276.

Industrene® 224. [Witco/Humko] Oleic-linoleic acid; chemical intermediate; solid. pt. 17–22 C; acid no. 195–204; sapon. no. 195–205.

Industrene® 225. [Witco/Humko] Dist. soya acid; see Industrene 105; Gardner 3–4 liq.; solid. pt. 25 C max.; acid no. 195–201; sapon. no. 197–204; 100% conc.

Industrene® 226. [Witco/Humko] Dist. soya acid; see Industrene 105; Gardner 3–4 liq.; solid. pt. 26 C max.; acid no. 195–203; sapon. no. 197–204; 100% conc.

Industrene® 226 FG. [Witco/Humko] Dist. soya acid; lubricant, release agent, binder, defoaming agent and intermediate for food additives; liq.; Lovibond 25.0Y/2.5R; solid. pt. 26 C max.; acid no. 195–203; sapon. no. 195–204; 100% conc.

Industrene® 325. [Witco/Humko] Dist. coconut acid; see Industrene 105; paste; sapon. no. 260–280; 100% conc.

Industrene® 328. [Witco/Humko] Stripped coconut acid; see Industrene 105; paste; sapon. no. 254–261; 100% conc.

Industrene® 365. [Witco/Humko] Mixt. caprylic/capric acid; see Industrene 105; sapon. no. 345–360; 100% conc.

Industrene® 4516. [Witco/Humko] 45% palmitic acid; see Industrene 105; solid; sapon. no. 206–212; 100% conc.

Industrene® 4518. [Witco/Humko] Single pressed stearic acid; see Industrene 105; 3 Gardner solid; sapon. no. 205–212; 100% conc.

Industrene® 5016. [Witco/Humko] Double pressed stearic acid; see Industrene 105; solid; sapon. no. 208–211; 100% conc.

Industrene® 7018. [Witco/Humko] 70% Stearic acid; see Industrene 105; solid; sapon. no. 201–208; 100% conc.

Industrene® 7018 FG. [Witco/Humko] 70% Stearic acid; see Industrene 226 FG; Lovibond 2.0Y/0.2R solid; solid. pt. 58–62 C; acid no. 200–207; sapon. no. 200–208; 100% conc.

Industrene® 8718 FG. [Witco/Humko] 87% Stearic acid; see Industrene 226 FG; Lovibond 2.0Y-0.2R; solid. pt. 64.5–67.5 C; acid no. 196–201; sapon. no. 196–202; 92% conc.

Industrene® 9018. [Witco/Humko] 90% Stearic acid; see Industrene 105; solid; sapon. no. 197–202; 100% conc.

Industrene® B. [Witco/Humko] Hydrog. stearic acid; see Industrene 105; solid; sapon. no. 200–208; 100% conc.

Industrene® M. [Witco/Humko] Oleic-stearic acid; chemical intermediate; solid. pt. 35 C max.; acid no. 175–190; sapon. no. 175–200.

Industrene® R. [Witco/Humko] Hydrog. rubber grade stearic acid; see Industrene 105; 10 Gardner solid; sapon. no. 194–214; 100% conc.

Inhibiteur I.C. 59. [Henkel] Modified polyoxyalkylene condensate of a heterocyclic amine; corrosion inhibitor for drilling mud; liq.

Inhibitor RT 212. [Zschimmer & Schwarz] Fatty acid alkanolamide; anticorrosion additive; liq.; 100% act.

Inhibitor RT 212. [Zschimmer & Schwarz] Fatty acid alkanolamide; nonionic; anticorrosion additive for metal cleaning and metal working; lubricant in drilling and cutting oils; brn. clear liq.; sp.gr. 1.0; sol. in min. oil, disp. in water; 100% act.

Inipol 002. [Ceca] N-Oleyl propylene diamine dioleate; rust inhibitor; paint additive; liq.

Inipol OT2. [Ceca] N-Oleyl propylene diamine ditallate; rust inhibitor, paint additive; liq.

Inipol S.43. [Ceca] Salt of fatty acid, polyalkylated N-alkyl propylene diamine; lubricant additive, paint additive, rust inhibitor; liq.

Intercide® 2 DIDP. [Akzo] 2% OBPA in diisodecylphthalate; microbicide for PVC compds.

Intercide® ABF. [Azko] 10,10′-Oxybisphenoxyarsine; microbicide for protection of plastics incl. PVC; avail. concs. 1, 2, and 5%.

Intercide® ABF 1 ESBO. [Akzo] 1% OBPA in epoxidized soybean oil; microbicide for PVC compds.

Intercide® ABF 2 DIDP. [Akzo] 10,10′-Oxybisphenoxyarsine in diisodecyl phthalate; fungicide/bactericide for plasticized PVC and other polymers; lt. yel. liq.; dens. 990 kg/m^3; visc. 100 mPa•s; 2% sol'n.

Intercide® ABF 2 ESBO. [Akzo] 10,10′-Oxybisphenoxyarsine in epoxidized soybean oil; fungicide/bactericide for plasticized PVC and other polymers; lt. yel. liq.; dens. 990 kg/m^3; visc. 500 mPa•s; 2% sol'n.

Intercide® FC, T-0. [Akzo] Tributyltin compd.; antimicrobial for control of fungi and bacteria in coated fabrics for exterior uses, sol'n. vinyl coatings; sol. in common org. solvs.

Intercide® N-628, TMP. [Akzo] Tributyltin compd.; antimicrobial for control of fungi and bacteria in coated fabrics for exterior uses.

Interlube A. [Anchor] Blend of zinc fatty acids and lubricants; lubricant for rubber compds.; improves mold and mill roll release; straw paste; sp.gr. 0.81; m.p. 107 C.

Interlube P/DS. [Anchor] Blend of zinc fatty acids and lubricants on inert carrier; see Interlube A; cream to buff dust-suppressed powd.; sp.gr. 1.1.

Intermediate 300. [Witco/Humko] Coco/tallow MEA; nonionic; thickener, foam booster, superfatting agent; solid; 100% conc.

Intermediate 325. [Witco/Humko] Coco MEA; nonionic; see Intermediate 300; solid; 100% conc.

Intermediate 512. [Witco/Humko] Lauric DEA; nonionic; see Intermediate 300; solid; 100% conc.

Interstab® 761-28. [Akzo] Barium-cadmium-zinc; high zinc stabilizer offering heat and lt. stability; nonsulfide staining for plastisol and flexible compds.; liq.

Interstab® 761-28A. [Akzo] Barium-cadmium-zinc; low zinc stabilizer offering nonsulfur staining char.; used for plastisol applics.; liq.

Interstab® BC-100S. [Akzo] Barium-cadmium; zinc-free stabilizer offering heat stability; used for flexible compds., esp. hose and profile, filled systems, film and sheet; liq.

Interstab® BC-103. [Akzo] Barium-cadmium-zinc; low zinc stabilizer offering heat and lt. stability; used for plastisol applics.; liq.

Interstab® BC-103A. [Akzo] Barium-cadmium-zinc; med. zinc stabilizer offering heat and lt. stability; nonsulfur staining; used for palstisol applics. and flexible compds.; liq.

Interstab® BC-103C. [Akzo] Barium-cadmium-zinc; med. zinc stabilizer offering heat and lt. stability; nonsulfur staining; used for plastisol applics., flexible compds. esp. hose, profile, filled systems.

Interstab® BC-103L. [Akzo] Barium-cadmium-zinc; low zinc stabilizer offering heat and lt. stability; used for plastisol applics. and flexible compds.; liq.

Interstab® BC-109. [Akzo] Barium-cadmium-zinc; high zinc stabilizer offering heat and lt. stability, non plate-out, plastisol air release, plastisol visc. control for plastisol applics.; liq.

Interstab® BC-110. [Akzo] Barium-cadmium-zinc; high zinc stabilizer offering heat and lt. stability, non plate-out, plastisol air release, and plastisol visc. control for plastisol applics.; liq.

Interstab® BC-4362. [Akzo] Barium/cadmium/zinc; stabilizer for flexible calendered prods. at reduced usage levels; provides heat stability and plate-out performance, exc. clarity in clear applics.; Gardner 8 max. liq.; sp.gr. 1.010–1.020; visc. Gardner 50 cps max.

Interstab® CA-18-1. [Akzo] Calcium stearate; lubricant, release agent, and processing aid in extrusion, calendering, inj., or blow molding of rigid or flexible PVC compds.; lubricant or metal scavenger in PP prod.; used in plastics, rubber, coatings, cosmetic, and metallurgical fields; food contact applic.; fine particle size.

Interstab® CH-55. [Akzo] Organophosphites; chelating agents to improve clarity in PVC compd.; aux. heat and light stabilizers; food contact and medical applics.

Interstab® CH-55R. [Akzo] Organophosphite; see Interstab CH-55.

Interstab® CZ-10. [Akzo] Calcium-zinc soap; nontoxic stabilizer offering nonstaining; suggested for flexible compds.; liq.

Interstab® CZ-11. [Akzo] Calcium-zinc soap; nontoxic stabilizer offering nonstaining; used for flexible compds.; paste.

Interstab® CZ-11D. [Akzo] Calcium-zinc soap; nontoxic stabilizer offering nonstaining; suggested for flexible compds.; paste.

Interstab® CZ-19A. [Akzo] Calcium-zinc soap; nontoxic stabilizer offering nonstaining; used for flexible compds.; powd.

Interstab® CZ-22. [Akzo] Calcium-zinc soap; nontoxic stabilizer offering nonstaining; used for flexible compds.; powd.

Interstab® CZ-23. [Akzo] Calcium-zinc soap; nontoxic stabilizer offering nonstaining; suggested for flexible compds.; powd.

Interstab® CZ-4359. [Akzo] Complex calcium and zinc soap of alkyl carboxylic acids combined with org. auxs.; stabilizer providing excellent performance in extruded blown film where compatibility and roll or contact clarity are of primary importance; used in food contact applics.; wh. paste; sp.gr. 1.02.

Interstab® CZL-710. [Akzo] Complex calcium and zinc soaps of carboxylic acids combined with org. auxs.; stabilizer primarily for food contact applics. for extruded blown film where compatibility and clarity are needed; stabilizer/activator for foamed PVC cap liners; Gardner 8 max. pourable liq.; sp.gr. 0.990–1.015; visc. 1000–6000 cps.

Interstab® CZL-712. [Akzo] Complex of zinc soaps of carboxylic acids combined with org. auxs.; see Interstab CZL-710; Gardner 7 max. pourable liq.; sp.gr. 0.995–1.020; visc. 1000–6000 cps.

Interstab® CZL-715. [Akzo] Complex of zinc soaps of carboxylic acids combined with org. auxs.; see Interstab CZL-710 Gardner 7 max. pourable liq.; sp.gr. 0.995–1.020; visc. 1000–6000 cps.

Interstab® E-82. [Akzo] Epoxy-modified ether-ester; heat and lt. stabilizer used in areas where epoxy plasticizers may have adverse effect on physical properties; offers high oxirane content with low hydroxyl and iodine values.

Interstab® F-402. [Akzo] Calcium-zinc soap; nontoxic stabilizer offering heat stability, clarity, non-plate-out, and nonstaining; powd.

Interstab® G-140. [Akzo] Fatty ester; syn. lubricant used in rigid calendering, extrusion, blow and inj. molding as internal lubricant; excellent clarity, sparkle, and surf. finish to compds.; food contact applic.

Interstab® G-8257. [Akzo] Ethylene bis-stearamide

wax; lubricant; functions at low levels as internal lubricant in rigid PVC applic.; food contact applic. in PVC and polymers; wax.

Interstab® LF 3623, LF 3653. [Akzo] Lead sulfate based; stabilizer for interior profiles.

Interstab® LF 3626, 3645, 3669, 10898/4, 10898/6. [Akzo] Lead phosphite based; stabilizer for exterior profiles.

Interstab® LF 3631/1. [Akzo] Pb/Ba/Cd; one-pack stabilizer system for PVC window profiles.

Interstab® LF 3631/2. [Akzo] Pb/Ba/Cd; stabilizer for window profiles, esp. for acrylic modified systems.

Interstab® LF 3631/3. [Akzo] Pb/Ba/Cd; one-pack stabilizer system incl. lubricant for PVC window profiles.

Interstab® LF 3634. [Akzo] Lead phosphite/Ba/Cd based; stabilizer for exterior profiles.

Interstab® LF 3638. [Akzo] Lead coprecipitate, Pb phosphite based; one-pack stabilizer for PVC window profiles.

Interstab® LF 3675, LF 3751. [Akzo] Lead phosphite/sulfate based; stabilizer for exterior profiles.

Interstab® LF 10773/25. [Akzo] Lead coprecipitate, Pb phosphite based; one-pack stabilizer for PVC window profiles; high whiteness.

Interstab® LF 11298. [Akzo] Pb/Ba/Cd; one-pack stabilizer for PVC window profiles.

Interstab® LF 11323/1. [Akzo] Pb/Ba/Cd; one-pack stabilizer for PVC window profiles.

Interstab® LF 11359. [Akzo] Lead coprecipitate, Pb phosphite based; one-pack stabilizer for PVC window profiles; improved heat stability.

Interstab® LL 3289. [Akzo] Lead stabilizer for interior profiles based on E-PVC; liq.

Interstab® LP 3103. [Akzo] Tribasic lead sulfate; stabilizer for PVC, profiles; 83% lead content.

Interstab® LP 3104. [Akzo] Tetrabasic lead sulfate; stabilizer for PVC, profiles; 85% lead content.

Interstab® LP 3139. [Akzo] Dibasic lead phosphite; stabilizer for PVC window profiles; 83.5% lead content.

Interstab® LP 3150. [Akzo] Dibasic lead stearate; stabilizer for PVC, profiles; 51% lead content.

Interstab® LP 3153. [Akzo] Dibasic lead phthalate; stabilizer for PVC; 76% lead content.

Interstab® LP 3155. [Akzo] Normal lead stearate; stabilizer for PVC, profiles; 28% lead content.

Interstab® LP 3190. [Akzo] Tetrabasic lead sulfate modified; stabilizer for PVC.

Interstab® LP 3289. [Akzo] Liquid lead complex; stabilizer for PVC; 33% lead content.

Interstab® LP 3631/5. [Akzo] Pb-phosphite-lubricant-antioxidant combination; stabilizer system for use in hot climates

Interstab® LT 4289. [Akzo] Calcium/zinc; stabilizer suitable for vinyl plastisol applics.; resistance to sulfide staining; uv stability; used for plastisol and organosol applics., e.g., fabric and paper coating, slush and rotational molding, chemically blown vinyl, highly filled plastigels, flooring wear-layers; Gardner 3 max. clear liq.; sp.gr. 0.948; visc. 50 cps max.

Interstab® LT 4308. [Akzo] Calcium/zinc; stabilizer for vinyl plastisol applics.; resistance to sulfide staining; uv stability; for vinyl compositions in contact with rubber, neoprene, and other sulfur-containing polymers; at 1.0–2.0 phr in plastisol and organosol applics., e.g., fabric/paper coating, chemically blown vinyl, highly-filled plastigels, and flooring wear-layers; color retention for high-temp. or low-temp.; cadmium-free; Gardner 2 max. clear liq.; sp.gr. 1.016; visc. 35 cps max.

Interstab® LT 11122/10. [Akzo] Lead/barium/cadmium complex; one-pack stabilizer for PVC window profiles.

Interstab® M85. [Akzo] Barium/cadmium complex; stabilizer for general-purpose flexible applics. incl. calendering, extrusion, and inj. molding of flexible PVC compds.; creamy wh. flake; nil sol. in water; sp.gr. 1.2.

Interstab® M341. [Akzo] Barium/cadmium complex; stabilizer for extrusion of rigid PVC compds. for outdoor applics., calendering of semirigid and flexible PVC compds.; creamy wh. flake; nil sol. in water; sp.gr. 1.3.

Interstab® M722, M763, M767. [Akzo] Barium/zinc; stabilizer for flexible PVC.

Interstab® M727. [Akzo] Barium/zinc; stabilizer for rigid PVC.

Interstab® M731. [Akzo] Potassium/zinc; stabilizer for PVC plastisols.

Interstab® M744, M767, M11301. [Akzo] Barium/zinc; stabilizer for PVC plastisols.

Interstab® M803, M11289. [Akzo] Calcium/zinc; stabilizer for flexible PVC.

Interstab® M809. [Akzo] Calcium/zinc; stabilizer for PVC plastisols.

Interstab® M876. [Akzo] Calcium/zinc; stabilizer for rigid PVC.

Interstab® M3187. [Akzo] Barium/cadmium complex; stabilizer similar to Interstab M-85 but provides a more compatible lubrication system for rigid and flexible applics. where surface characteristics are critical; creamy wh. flake; nil sol. in water; sp.gr. 1.2.

Interstab® MF981, MF985. [Akzo] Barium/cadmium; stabilizer for flexible PVC.

Interstab® MP10581/3. [Akzo] Calcium/zinc; stabilizer for rigid PVC.

Interstab® MT981. [Akzo] Barium/cadmium complex; stabilizer for rigid PVC, window profiles; polyol-free version; useful for dark pigmentations; nondusting tablets.

Interstab® MT982. [Akzo] Barium/cadmium complex; polyol stabilizer for PVC window profiles; nondusting tablets.

Interstab® MT11303/1. [Akzo] Calcium/zinc; stabilizer for rigid PVC.

Interstab® R-4048. [Akzo] Cadmium-free, low toxicity stabilizer for heat stability, nonplate-out, nonsulfide stain; for plastisol applics. (slush molding, rotational molding, dipping, spreading, film and sheet).

Interstab® R-4052. [Akzo] Cadmium-free, low toxicity stabilizer for nonplate-out and nonsulfide stain; used for plastisols esp. spreading applics.

Interstab® R-4101. [Akzo] Barium-cadmium-zinc; med. zinc stabilizer for heat and lt. stability; nonsulfide staining; used for flexible compds., hose and profile, filled systems, and film and sheet; liq..

Interstab® R-4109. [Akzo] Barium-cadmium-zinc; low zinc stabilizer for heat stability, lubricity; for flexible compds. esp. hose and profile applics.

Interstab® R-4114. [Akzo] Barium-cadmium-zinc; low zinc stabilizer for heat stability; flexible compds., film and sheet applics.; liq.

Interstab® R-4137. [Akzo] Barium-cadmium-zinc; low zinc stabilizer for heat stability, lubricity, nonplate-out, nonsulfur stain; flexible compds., esp. inj. molding and film and sheet applics.; liq..

Interstab® ZN-18-1. [Akzo] Zinc stearate; lubricant exhibiting internal and external properties used in plastics industry; excellent early color and clarity to clear PVC prods.; efficient color dispersant and lubricant in polyolefins and PS; food contact applic.; fine particle size.

Interwax G 8140. [Akzo] Alpha-olefin copolymer; lubricant for PVC.

Interwax G 8200. [Akzo] Glyceryl monooleate; lubricant for PVC.

Interwax G 8204. [Akzo] Glyceryl monostearate; lubricant for PVC.

Interwax G 8205. [Akzo] Glyceryl fatty acid ester; lubricant for PVC.

Interwax G 8206. [Akzo] C_{16}–C_{18} fatty alcohol; lubricant for PVC.

Interwax G 8207. [Akzo] Stearic acid; lubricant for PVC.

Interwax G 8208. [Akzo] Paraffin wax; lubricant for PVC.

Interwax G 8212. [Akzo] 12-Hydroxy stearic acid; lubricant for PVC.

Interwax G 8213. [Akzo] Hydrocarbon wax; lubricant for PVC.

Interwax G 8252. [Akzo] Polyethylene wax; lubricant for PVC.

Interwax G 8253. [Akzo] Octyl stearate; lubricant for PVC.

Interwax G 8257. [Akzo] Amide wax; lubricant for PVC.

Interwax G 8259. [Akzo] Calcium montanate; lubricant for PVC.

Interwax G 8268. [Akzo] Syn. paraffin wax; lubricant for PVC.

Interwax M 3142. [Akzo] Calcium stearate; lubricant for PVC.

Interwet® 33. [Akzo] Nonionic; emulsifier, penetrant for coatings, adhesives, insecticides, wax dispersions, bitumen emulsions; yel. liq., mild fatty odor; sol. in alcohols, acetones, toluol, disp. in water; sp.gr. 1.01; visc. 85 cps; f.p. 2 C; b.p. 290 C; HLB 11.5; 100% act.

Intralan® Salt HA. [Crompton & Knowles] Neutralized naphthalene condensate; anionic; dispersing and leveling agent for disperse dyes and dyeing of syn. fibers, dyeing aux.; powd.; water sol.

Intralan® Salt N. [Crompton & Knowles] EO condensate; nonionic; dye assistant, leveling agent, penetrant for textiles; yel. liq.; sol. in boiling water.

Intraphasol COP. [Crompton & Knowles] Hydrocarbons, solubilized, aliphatic and sulfonic acid salts; anionic; emulsifier, wetting agent assistant for dyeing of fabrics and carpets; liq.

Intraphasol PC. [Crompton & Knowles] Sulfuric acid ester salt of polyglycol ether compd.; anionic; wetting agent, emulsifier, dispersant used in printing of polyamide and polyester carpets; amber liq.; alcoholic odor; water sol.; dens. 8.7 lb/gal.

Intraphor AC. [Henkel] Polymerized sodium salt of naphthalene sulfonic acids; general purpose nonfoaming dispersant for dyestuffs, pigments, pesticides; lt. powd.

Intrapol 1014. [Henkel] Solv. and alkylpolyglycol ether; nonionic; foaming detergent and dispersant; kier boiling assistant; liq.; 80% conc.

Intratex® A. [Crompton & Knowles] Complex amino condensate; amphoteric; nonfoaming leveling agent for dyeing wool; tan liq.; water sol.; dens. 9 lb/gal.

Intratex® AN. [Crompton & Knowles] EO condensate; nonionic; leveling agent for wool dyeing, dye stripping; liq.; water sol.

Intratex® B. [Crompton & Knowles] Polyglycol ether deriv.; amphoteric; leveling agent, detergent, for dye leveling of wool; tan liq.; water sol.; dens. 8.7 lb/gal.

Intratex® C. [Crompton & Knowles] Complex amino condensate; amphoteric; leveling agent for dyeing wool; clear liq.; dens. 8 lb/gal.

Intratex® OR. [Crompton & Knowles] Low molecular polyamide, nonionic; leveling agent, vat dyeing assistant; yel. brn. liq.; water sol.; dens. 8.6 lb/gal; sp.gr. 1.04.

Intratex® POK. [Crompton & Knowles] Sulfonated benzimidazole deriv.; anionic; dispersant, leveling agent for vat and disperse dyes; yel. brn. liq.; water sol.

Intratex® W New. [Crompton & Knowles] EO condensate; nonionic; leveling agent, penetrant, for wool dyeing; yel. liq.; water sol.; dens. 8.5 lb/gal.

Intravon® JF, JU. [Crompton & Knowles] EO condensate; nonionic; detergent, emulsifier, dispersant, wetting agent, penetrant for textile, household and cosmetic applics.; yel. liq.; water sol.; dens. 8.4–8.5 lb/gal.

Inversol 140, 170, 190. [Keil] Complex polyglycol ester; nonionic; lubricity agent for metalworking liqs.; liq.; 100% conc.

Ionet 300. [Sanyo] POE alkyl ether type surfactant; anionic; scouring agent for fibers; leveling agent; penetrant; dispersant; liq.

Ionet DL-200. [Sanyo] POE dilaurate; nonionic; emulsifier for emulsion polymerization, metal processing lubricant and personal care prods.; liq.; HLB 6.6; 100% conc.

Ionet DO-200, -400, -600, -1000. [Sanyo] POE dioleate; nonionic; see Ionet DL-200; liq. except DO-

1000 (solid); HLB 5.3, 8.4, 10.4, and 12.9 resp.; 100% conc.

Ionet DS-300, -400. [Sanyo] POE distearate; nonionic; see Ionet DL-200; solid; HLB 7.3 and 8.5 resp.; 100% conc.

Ionet S-20. [Sanyo] Sorbitan monolaurate; nonionic; emulsifier for personal care prods.; lubricant, rust inhibitor; pigment dispersant; spreading agent for agric. pesticides; base for textile lubricants; liq.; HLB 8.6; 100% conc.

Ionet S-60 C. [Sanyo] Sorbitan monostearate; nonionic; see Ionet S-20; solid; HLB 4.7; 100% conc.

Ionet S-80. [Sanyo] Sorbitan monooleate; nonionic; see Ionet S-20; liq.; HLB 4.3; 100% conc.

Ionet S-85. [Sanyo] Sorbitan trioleate; nonionic; see Ionet S-20; liq.; HLB 1.8; 100% conc.

Ionet T-20 C. [Sanyo] Polysorbate 20; nonionic; base and emulsifier for personal care prods., metal processing, lubricant and rust inhibitor; pigment dispersant; spreader sticker for agric. pesticides; base for textile lubricants; liq.; HLB 16.7; 100% conc.

Ionet T-60 C. [Sanyo] POE sorbitan monostearate; nonionic; see Ionet T-20 C; liq.; HLB 14.9; 100% conc.

Ionet T-80 C. [Sanyo] POE sorbitan monooleate; nonionic; see Ionet T-20 C; liq.; HLB 15.0; 100% conc.

Ionol. [Shell] 2,6-Di-tert-butyl-4-methylphenol; antioxidant for petrol. prods.; rubber compds.; stabilizer for neoprene and plastic; Pt-Co 45 max.; sol. in wh. oil; solid. pt. 68.8 C min.; 98.0% min. purity.

Ionol CP. [Shell] BHT; antioxidant for rubber, paraffin, and plastic used in food and drug prods.; paper pkg.; Pt-Co 15 max.; solid. pt. 69.4 C min.; 99.0% min. purity.

IPP. [PPG Industries] Diisopropyl peroxydicarbonate; initiator for polymerization of unsat. monomers; reduces time of batch runs; helps control polymerization; wh. solid; m.w. 206.18; sol. 0.04% in water; sp.gr. 1.080 (15.5/4C); m.p. 8–10 C; 98.5% min. conc.; 7.8% act. oxygen.

Ircogel® 900. [Lubrizol] Calcium org.; thixotropic, antisag, and flow control agent for use in coating plastisols and organosols, cloth coating plastisols, PVC sealants, polysulfide sealants, polymercaptan sealants; exhibits wetting or dispersant effect on fillers; lt. br. soft gel; sp.gr. 1.10 (15.6 C); dens. 9.15 lb/gal (15.6 C); flash pt. (COC) 204 C; 100% calcium org. gel.

Ircogel® 901. [Lubrizol] Calcium org. gel in dioctyl phthalate; thixotropic, antisag, and flow control agent for use in coating plastisols and organosols, cloth coating plastisols, PVC sealants; exhibits wetting or dispersant effect on fillers; lt. br. liq.; sp.gr. 1.09 (15.6 C); dens. 9.08 lb/gal (15.6 C); flash pt. (COC) 204 C; 66.7% calcium org. gel; 33.3% dioctyl phthalate.

Ircogel® 903. [Lubrizol] Calcium complex in diisononyl phthalate; thixotrope and visc. control agent for coating plastisols and organosols, fabric coatings, PVC sealants; liq.

Ircogel® 904. [Lubrizol] Calcium org. gel in Rule 66 Stoddard solv.; thixotropic, antisag, and flow control agent for use in coating organosols, PVC sealants; exhibits wetting or dispersant effect on fillers; lt. br. liq.; sp.gr. 0.98 (15.6 C); dens. 8.16 lb/gal (15.6 C); flash pt. (PMCC) 38 C min.; 70% calcium org. gel; 30% Stoddard solv.

Ircogel® 2354. [Lubrizol] Zinc complex; thixotropic, antisag, and flow control agent for acrylic and butyl caulk formulations; lt. colored liq.; sp.gr. 1.385 (15.6 C); visc. (Gardner Holdt) Z_2–Z_3; dens. 11.54 lb/gal (15.6 C); 7.8% zinc.

Irgalube® 53. [Ciba-Geigy] POE alkyl ester and fatty ester; nonionic; dyeing assistant for fabrics; dye bath lubricant; paste

Irganox® 245. [Ciba-Geigy] Triethyleneglycol bis [3- (3′-tert-butyl-4′-hydroxy-5′-methylphenyl) propionate]; antioxidant/stabilizer for use in org. polymers; wh. to sl. ylsh. cryst. powd.; odorless; m.w. 586.8; sol. > 50% in acetone; > 40% in chloroform and methylene chloride; 37% in ethyl acetate; 18% in benzene; 12% in methanol; ≈ 6% in styrene; < 0.01% in water; sp.gr. 1.14; m.p. 76–79 C; decomp. pt. > 220 C; flash pt. (PM) > 302 F.

Irganox® 259. [Ciba-Geigy] 1,6-Hexamethylene bis-(3,5-di-tert-butyl-4-hydroxyhydrocinnamate); antioxidant; stabilizer for polyolefin, elastomer, styrenic, polyacetal, petrol. prods., and org. substrates; off-wh. cryst. powd.; m.w. 639; sol. in benzene, acetone, chloroform, min. oil, hexane, and water; m.p. 103–108 C.

Irganox® 1010. [Ciba-Geigy] Tetrakis [methylene (3,5-di-tert-butyl-4-hydroxyhydrocinnamate)] methane; antioxidant; stabilizer for org. and polymeric materials; polyolefin, elastomer, food pkg. and adhesive applic., petrol. prods.; wh. cryst. powd.; m.w. 1178; sol. in benzene, acetone, chloroform; sp.gr. 1.45; m.p. 110–125 C.

Irganox® 1035. [Ciba-Geigy] Thiodiethylene bis-(3,5-di-t-butyl-4-hydroxy) hydrocinnamate; antioxidant for polymer, org. substrate, and polyolefin; used in elastomer, petrol. prods., food pkg. applic.; wh. cryst. powd.; m.w. 642; sol. in toluene, acetone, methanol, min. oil, and water; sp.gr. 1.19; m.p. 63 C min.

Irganox® 1076. [Ciba-Geigy] Octadecyl 3,5-di-tert-butyl-4-hydroxyhydrocinnamate; antioxidant used in org. and polymeric material; stabilizer for polyolefins, styrenics, elastomers, PVC, urethane, acrylic coatings, adhesive, and petrol. prods.; food pkg. applic.; wh. cryst. powd.; odorless; m.w. 531; sol. in benzene, xylene, ethyl acetate, acetone, hexane, and water; m.p. 50–55 C.

Irganox® 1093. [Ciba-Geigy] O,O-Di-n-octadecyl-3,5-di-tert-butyl-4-hydroxybenzyl phosphonate; antioxidant for org. material; stabilizer for fiber, monofilament, film, PS, styrene copolymers, PC, polyester, PVC, polyethylene, acrylics, and adhesives; wh. cryst. powd.; m.w. 804; sol. in benzene, hexane, acetone, methanol, min. and veg. oils; insol. in water; m.p. 52 C min.

Irganox® 1098. [Ciba-Geigy] N,N′-hexamethylene bis (3,5-di-tert-butyl-4-hydroxy-hydrocinnamamide); antioxidant for polymers, rubber, SBR, polyacetals, linear saturated polyesters, PVC, polyolefins; wh. cryst. powd.; m.w. 637; sol. in chloroform, methanol, acetone, ethyl acetate, water, benzene, and hexane; m.p. 156–161 C.

Irganox® 1520. [Ciba-Geigy] 2,4-Bis [(ocylthio) methyl]-o-cresol; antioxidant for polymers; used in the base stabilization of elastomers and the compd. stabilization of adhesives; effective during elastomer dynamic processing without the use of phosphites; pale yel. low-visc. liq.; m.w. 424.7; sol. in methanol, ethanol, acetone, ethyl acetate, methylene chloride, chloroform, toluene, n-hexane; sol. < 0.01 g/100 g water; dens. 0.9787 g/ml; flash pt. > 200 C.

Irganox® B-215. [Ciba-Geigy] Irganox 1010/Irgafos 168, 1:2 ratio; antioxidant-process stabilizer combination; food pkg. applic.; colorless powd.; m.w. 647; sol. in chloroform, methylene chloride, benzene, hexane, ethyl acetate, acetone, methanol, and water; m.p. 180–185 C.

Irganox® B-225. [Ciba-Geigy] Irganox 1010/Irgafos 168, 1:1 ratio; see Irganox® B-215; m.w. 647; sol. see Irganox B-215; m.p. 180–185 C.

Irganox® MD-1024. [Ciba-Geigy] Hindered phenolic; antioxidant/metal deactivator; used in extending the lifetime of PP, polyethylene, and certain thermoplastic elastomers, EPDM, peroxide, nylon, and polyacetal; wh. cryst. powd.; sol. in THF, acetone, toluene, water, and paraffin oil; sp.gr. 1.12; m.p. 227–229 C.

Irgasan DP300. [Ciba-Geigy] 2,4,4′-Trichloro-2′-hydroxydiphenyl ether; broad spectrum bacteriostat for deodorant prods., e.g., bar soaps, fabric sanitization, institutional fabric softeners; 99% min. act.

Irgasol® N. [Ciba-Geigy] Sodium lignosulfonate; anionic; dye dispersant; liq.

Irgastab® 2002. [Ciba-Geigy] Nickel bis[O-ethyl (3,5-di-tert-butyl-4-hydroxybenzyl)] phosphonate; UV lt. stabilizer and antioxidant for polyolefins; useful in pigmented or opaque fibers and in films; dyesite for nickel chelatable dyes for polyolefins; provides stability during processing of PP; imparts negligible odor to substrates; resistant to extraction; resists discoloration by nitrogen oxides; low volatility; sulfur-free; lt. tan powd.; m.w. 713.5; sol. 100g/100 ml or more in most common hydrocarbon solvs.; m.p. 180 C min.; 8.24% nickel content.

Irgastab® T-265. [Ciba-Geigy] Mixed mono- and dioctyltin (C_{10-16}) alkylthioglycollates; organotin heat stabilizer for food-grade PVC bottle and sheet applics.; for improved start-to-finish color protection; low volatility and extraction resistance in processing; almost colorless clear liq.; sp.gr. 1.04; flash pt. 302 F; pH 3 (@200 g/l water).

Irgastab® T-634. [Ciba-Geigy] Butyltin carboxylate, mercpatide-reinforced; heat and light stabilizer for vinyl systems, incl. rigid and plasticized PVC; slightly ylsh. low-visc. liq.; sol. 40 ppm in water; sp.gr. 1.08; b.p. > 250 C; m.p. < 20 C; ref. index 1.485; pH 4 (2g/10 ml water).

Iscolan. [Croda Ltd.] Lanolin ester; nonionic; emollient, wetting and spreading agent, used for skin, make-up and hair care cosmetics; pale yel., slightly opalescent liq.

Isobutyl Niclate. [Vanderbilt] Nickel diisobutyldithiocarbamate; antioxidant/antiozonant for protection in epichlorohydrin; grnsh. powd. 99.8% min thru 100 mesh; m.w. 467.47; sol. in acetone, chloroform, toluene; pract. insol. in water; dens. 1.27 ± 0.03 mg/m^3; m.p. 173–181 C; 11.5–13.5% nickel content.

Isocreme. [Croda Ltd.] Lanolin-derived base; emollient, moisturizer, emulsifier for cosmetics and pharmaceuticals; soft solid.

Isonate® 125M. [Dow] 4,4′-Diphenyl methane diisocyanate; processing aid, intermediate for the prod. of cast, RIM, and thermoplastic PU elastomers, adhesives, binders, coatings, and sealants; solid; b.p. 200 C (5 mm Hg); dens. 1.18 g/ml (43 C); visc. 5 cps (43 C); decomp. pt. 230 C; isocyanate equiv. 125.5; flash pt. (COC) 199 C; 99% purity; 33.5% NCO content.

Isonate® 143L. [Dow] 4,4′-Diphenyl methane diisocyanate; processing aid for PU industry for use alone or with prepolymers; used in cast and RIM processing, adhesives, binders, coatings, and sealants; liq.; b.p. 200 C (5 mm Hg); dens. 1.21 g/ml; visc. 35 cps; decomp. pt. 230 C; isocyanate equiv. 144.9; flash pt. (COC) 199 C; 29.0% NCO content.

Isonate® 181. [Dow] 4,4′-Diphenyl methane diisocyanate; processing aid for PU industry; polyether quasi-prepolymer for use in formulating cast elastomers and microcellular rubber for shoe soles and related applics., adhesives, binders, coatings, and sealants; liq.; b.p. 200 C (5 mm Hg); dens. 1.21 g/ml; visc. 850 cps; decomp. pt. 230 C; isocyanate equiv. 183; flash pt. (COC) 199 C; 22.9% NCO content.

Isonate® 191. [Dow] 4,4′-Diphenyl methane diisocyanate; processing aid for PU industry for use in automotive energy management foams and high modulus RIM, adhesives, binders, coatings, and sealants; liq.; b.p. 200 C (5 mm Hg); dens. 1.22 g/ml; visc. 40 cps; decomp. pt. 230 C; isocyanate equiv. 139.5; flash pt. (COC) 199 C; 30% NCO content.

Isonate® 240. [Dow] 4,4′-Diphenyl methane diisocyanate; processing aid for PU industry; polyester quasi-prepolymer yielding high physical property elastomers or microcellulars via the cast, low pressure or RIM dispensing process; for structural polymers, dynamic elastomers, adhesives, binders, coatings, and sealants; liq.; b.p. 200 C (5 mm Hg); dens. 1.22 g/ml; visc. 1100 cps; decomp. pt. 230 C; isocyanate equiv. 225; flash pt. (COC) 199 C; 18.7% NCO content.

Isonox 129. [Schenectady] 2,2′-Ethylidenebis (4,5-di-tert.-butylphenol); antioxidant and thermal stabilizer for polymers; food pkg. applic.; used in PP, polyethylene, PVC, PS, ABS, hydrocarbon resins, EVA-modified compds.; wh. cryst. powd.; m.w. 438; b.p. > 550 F; sol. in acetone, toluene, heptane, ethanol, and water; sp.gr. 1.01 ± 0.03; dens. 35 lb/ft^3; m.p. 161–163 C; flash pt. > 193 C (COC); 99%+ purity.

Isonox 132. [Schenectady] 2,6-Di-tert-butyl-4-sec-butylphenol; antioxidant used in polyols and rubber systems; pale to straw yel. clear liq.; m.w. 262; b.p.

275 C @ 760 mm; sp.gr. 0.902; visc. 75 cps; m.p. 18 C; flash pt. > 94 C (FTCC); 95% min.

Isopropylan 33. [Amerchol] Isopropyl palmitate/ lanolin oil; binder in talc and pearl powd. systems; plasticizer, emollient, and moisturizer; lt. yel. clear oily liq.; slight char. odor; sapon. no. 145–165.

I.T. 3X [Vanderbilt] Hydrous magnesium calcium silicate; industrial talc used as filler/extender in NR and syn. rubbers, paints, sealants, mastics, latex compds.; wh. powd.; median particle size 9.3 µm; 98% min. thru 325 mesh; dens. 2.85 ± 0.03 mg/m^3; oil absorp. 29; pH 9.4; 56.0% SiO_2, 30.7% MgO, 7.0% CaO.

I.T. 5X [Vanderbilt] Hydrous magnesium calcium silicate; see I.T. 3X; wh. powd.; median particle size 7.6 µm; 99.25% min. thru 325 mesh; dens. 2.85 ± 0.03 mg/m^3; oil absorp. 30; pH 9.4; 56.0% SiO_2, 30.7% MgO, 7.0% CaO.

I.T. 325 [Vanderbilt] Hydrous magnesium calcium silicate; see I.T. 3X; wh. powd.; median particle size 5.5 µm; 99.9% min. thru 325 mesh; dens. 2.85 ± 0.03 mg/m^3; oil absorp. 29; pH 9.4; 56.0% SiO_2, 30.7% MgO, 7.0% CaO.

I.T. FT [Vanderbilt] Hydrous magnesium calcium silicate; see I.T. 3X; wh. powd.; median particle size 7.0 µm; 99.4% min. thru 325 mesh; dens. 2.85 ± 0.03 mg/m^3; oil absorp. 29; pH 9.4; 56.0% SiO_2, 30.7% MgO, 7.0% CaO.

I.T. X. [Vanderbilt] Industrial talc; see I.T. 3X; particulate; median particle size 10.2 µm; dens. 23.7 lb/ gal; oil absorp. 23; pH 9.4; 56.0% SiO_2, 30.7% MgO, 7.0% CaO.

Ivex® 10. [CasChem] Zinc undecylenate and undecylenic acid; antifungal powds.; wh. powd.; 99.9% act.

J

JAQ Powdered Quat. [Huntington] n-Alkyl dimethyl benzyl ammonium chloride; for formulation of disinfectants, sanitizers, and swimming pool algicides; 95% act.

Jasmacyclat. [Henkel] Methylcyclooctylcarbonate; fragrance raw material for floral notes.

Jeffamine® BuD-2000. [Texaco] Urea condensate of POP polyamine; epoxy modifier; nonreactive additive used in conc. of 5-20 phr to provide enhancement of metal-to-metal adhesion, thermal shock properties; results in increased elongation, higher impact and tensile strength, and lowered modulus, while heat deflection values are only slightly affected; reactive with formaldehyde to produce polymeric materials; lt. yel. visc. liq.; visc. 22,000 cps; flash pt. 471 F (PMCC).

Jeffamine® D-230. [Texaco] POP diamine; epoxy curing agent and modifier used in heat-cured sol'n. coatings, castings, adhesives, and laminates; colorless to slight yel. slightly hazy liq.; m.w. 230; sol. in water, glycol ethers, esters, alcohols, ketones, and hydrocarbons; sp.gr. 0.9480; dens. 7.9 lb/gal; visc. 8.7 cs; flash pt. (COC) 256 F; ref. index 1.466; pH 11.3 (1% aq.); distort. temp 60–75 C (264 psi).

Jeffamine® D-400. [Texaco] POP diamine; epoxy curing agent and modifier used in coatings, castings, adhesives; colorless to pale yel. slightly hazy liq.; m.w. 400; sol. see Jeffamine D-230; sp.gr. 0.9702; dens. 8.1 lb/gal; visc. 22 cs; flash pt. (COC) 347 F; ref. index 1.4482; pH 11.3 (1% aq.); distort. temp. 42–45 C (264 psi).

Jeffamine® D-2000. [Texaco] POP diamine; epoxy curing agent and modifier useable alone or in combination; pale yel. slightly hazy liq.; m.w. 2000; sol. see Jeffamine D-230; slightly sol. in water; sp.gr. 0.9964; dens. 8.3 lb/gal; visc. 265 cs; flash pt. (COC); 460 F; ref. index 1.4514; pH 10.1 (1% aq.).

Jeffamine® D-4000. [Texaco] Polyoxyalkylene diamine; epoxy curing agent; also in polyamide, polyurea, modified urethane resins, in adhesives, elastomers, foam formulas; intermediate for textile and paper treatment chemicals; colorless to lt. yel. hazy liq.; m.w. 4000; sp.gr. 0.9996; visc. 874 cSt; flash pt. 415 F (PMCC); pH 11.8 (5% aq.)

Jeffamine® DU-700. [Texaco] Urea condensate of POP polyamine; epoxy curing agent useable alone or in combination; pale yel. slightly visc. liq.; visc. 1500 cps; distort. temps < 25 C (264 psi).

Jeffamine® DU-1700. [Texaco] Urea condensate of POP polyamine; epoxy curing agent and modifier; yel. visc. liq.; visc. 200,000 cps.

Jeffamine® DU-3000. [Texaco] Urea condensate of POP polyamine; epoxy curing agent and modifier; metal-to-metal adhesion; yel. slightly visc. liq.; visc. 1800 cps.

Jeffamine® ED-600. [Texaco] POE polyamine; epoxy curing agent and modifier; flexibilizer; lt. yel. liq.; water-sol.; visc. 60 cps.

Jeffamine® ED-900. [Texaco] POE polyamine; epoxy curing agent and modifier; lt. yel. liq.; water-sol.; visc. 120 cps.

Jeffamine® ED-2001. [Texaco] POE polyamine; epoxy curing agent and modifier; wh. waxy solid.

Jeffamine® EDR-148. [Texaco] Polyoxyalkylene diamine; see Jeffamine D-4000; colorless to lt. yel. liq.; m.w. 148; sp.gr.1.0154 (25/4 C); visc. 8 cSt; flash pt. 265 F (PMCC).

Jeffamine® T-403. [Texaco] POP triamine; epoxy curing agent and modifier; colorless to pale yel. slightly hazy liq.; m.w. 403; sol. see Jeffamine D-230; sp.gr. 0.9812; visc. 76.5 cs; ref. index 1.4606; pH 11.2 (1% aq.); flash pt. 380 F (COC); distort. temp. 60–70 C (264 psi).

Jeffamine® T-3000. [Texaco] Polyoxyalkylene triamine; see Jeffamine D-4000; colorless to lt. yel. hazy liq.; m.w. 3000; sp.gr.1.1203; visc. 467 cSt; flash pt. 455 F (PMCC).

Jeffamine® T-5000. [Texaco] Polyoxyalkylene triamine; see Jeffamine D-4000; colorless to lt. yel. hazy liq.; m.w. 5000; sp.gr. 0.9967 (25/4 C); visc. 829 cSt; flash pt. 410 F (PMCC).

Jeffox® PPG-400. [Texaco] PEG; lubricant, intermediate for ester synthesis; m.w. 400; water-sol.

Jeffox® PPG-2000. [Texaco] PEG; lubricant, intermediate for ester synthesis; defoamer for detergent sol'ns.; polyol for urethane elastomers; m.w. 2000.

Jeffox® WL. [Texaco] Mixed ethoxylated and propoxylated alcohol; conditioner; sol. in water, alcohol.

Jeffox® WL-660, -1400, -5000. [Texaco] Mixed ethoxylated and propoxylated alcohol or glycol; emollient, stabilizer; WL-660 in solv. applics. to remove fluid by a water rinse; sol. in water, alcohol; visc. 660, 1400, 5000 SUS resp. (100 F).

Jet Amine D-C, D-O, D-T. [Jetco] Diamine; cationic; emulsifier of bitumen, fuel oil and gasoline additive, corrosion inhibitor, min. flotation, bactericide; liq., paste; 100% conc.

Jet Amine DE-13. [Jetco] Ether diamine R-O-$(CH_2)_3$-N-$(CH_2)_3$-NH_2; emulsifier, corrosion inhibitor; liq.

Jet Amine PC, PHT, P-O, P-S, PT. [Jetco] Primary amine; cationic; corrosion inhibitor, ore flotation agent, emulsifier, paint pigment dispersant, mold release agent, lube oil additive, fertilizer anticake; liq., solid; 100% conc.

Jet Amine PE-13, -810, -1215. [Jetco] Primary amine R-O-$(CH_2)_3$-NH_2; ore flotation, emulsifier, corrosion inhibitor, fuel oil additive; liq.

Jet Fil® 100. [Vanderbilt] Platy talc, hydrous magnesium silicate; filler for PP and other resin systems; wh. powd.; 13 μm mean particle size; dens. 53 lb/ft³ (tapped); oil absorp. 27; pH 9.5.

Jet Fil® 200. [Vanderbilt] Platy talc, hydrous magnesium silicate; see Jet Fil 100; wh. powd.; 9 μm mean particle size; dens. 48 lb/ft³ (tapped); oil absorp. 36; pH 9.5.

Jet Fil® 350. [Vanderbilt] Platy talc, hydrous magnesium silicate; see Jet Fil 100; wh. powd.; 6 μm mean particle size; dens. 43 lb/ft³ (tapped); oil absorp. 40; pH 9.5.

Jet Fil® 500. [Vanderbilt] Platy talc, hydrous magnesium silicate; see Jet Fil 100; wh. powd.; 4.5 μm mean particle size; dens. 39 lb/ft³ (tapped); oil absorp. 44; pH 9.5.

Jet Quat 2C-75. [Jetco] Dicoco dimethyl ammonium chloride; cationic; bactericide, textile softener, asphalt emulsifier, petrol. processing; liq.; 75% conc.

Jet Quat 2HT-75. [Jetco] Dihydrog. tallow dimethyl ammonium chloride; cationic; see Jet Quat 2C-75; liq.; 75% conc.

Jet Quat C-50. [Jetco] Coco trimethyl ammonium chloride; cationic; see Jet Quat 2C-75; liq.; 50% conc.

Jet Quat S-50. [Jetco] Soya trimethyl ammonium chloride; cationic; see Jet Quat 2C-75; liq.; 50% conc.

Jet Quat T-2C-50. [Jetco] Tallow dicoco quat. ammonium chloride; cationic; see Jet Quat 2C-75; liq.; 50% conc.

Jet Quat T-50. [Jetco] Tallow trimethyl ammonium chloride; cationic; see Jet Quat 2C-75; liq.; 50% conc.

JF 77. [Tri-K] Drometrizole; uv absorber.

Jorchem PRM-98. [PPG-Mazer] Phosphate ester; anionic; wetting and scouring agent, penetrant for all fibers; liq.; 90% conc.

Jordamide 22. [PPG-Mazer] Modified 2:1 coconut alkanolamide; nonionic; high visc. and low foam for hard surface cleaners; emulsifier; thickener; suggested for waterless hand cleaners; brn. liq.; acid no. 50–56; 18–21% free amine; 100% conc.

Jordamide 29-78. [PPG-Mazer] Modified 2:1 alkanolamide; see Jordamide 22; brn. liq.; acid no. 22–32; 32–36% free amine.

Jordamide 201. [PPG-Mazer] Oleamide DEA (2:1); nonionic; emulsifier and wetting agent in cleaning applics.; lubricant for metal cutting fluids, wire and deep metal drawing, syn. grinding and cutting fluids; brn. liq.; 100% conc.

Jordamide 1214. [PPG-Mazer] Lauramide DEA (2:1); nonionic; emulsifier, wetting agent, foam stabilizer for cleaning applics., personal care prods.; lubricant for metal cutting fluids, wire and deep metal drawing, syn. grinding and cutting fluids; solubilizer for oils and fragrances into detergent systems; brn. liq.; acid no. 10–17; 20–27% free amine; 100% conc.

Jordamide 1281. [PPG-Mazer] Modified 2:1 coconut alkanolamide; nonionic; foam booster and stabilizer, visc. builder, detergent and emulsifier; self-coupling; used for hard surface cleaning formulations; amber liq.; acid no. 50–56; 20–25% free amine; 100% conc.

Jordamide CCO. [PPG-Mazer] Cocamide DEA (2:1); nonionic; emulsifier and wetting agent in cleaning applics.; lubricant for cutting fluids; liq.; 100% conc.

Jordamide CFAM. [PPG-Mazer] Cocamide MEA (1:1); nonionic; foam stabilizer, visc. builder for household, industrial, and cosmetic prods.; lt. yel. flake; 100% conc.

Jordamide CLD. [PPG-Mazer] Cocamide DEA (1:1); nonionic; foam stabilizer, visc. builder for household, industrial, cosmetic prods.; amber liq.; 6–8% free amine; 100% conc.

Jordamide CLM. [PPG-Mazer] Lauramide DEA (1:1); nonionic; see Jordamide CLD; yel. liq.; 8.4–11.3% free amine; 100% conc.

Jordamide CMEA, CMEA Extra. [PPG-Mazer] Cocamide MEA (1:1); foam stabilizer, visc. builder, thickener, and emulsifier for personal and household prods.; lt. yel. flake; solid. pt. 68 C (CMEA Extra); 2% max. free amine; 100% conc.

Jordamide CP. [PPG-Mazer] Cocamide DEA (2:1), modified; nonionic; self-coupling foam booster and stabilizer, visc. builder, detergent, and emulsifier for hard surface cleaners; amber liq.; acid no. 45–55; 15–19% free amine; 100% conc.

Jordamide JR-100. [PPG-Mazer] Cocamide DEA (2:1), modified; emulsifier and wetting agent for hard surface cleaners, waterless hand cleaners; lubricant for cutting fluids; brn. liq.; acid no. 42–53; 18–23% free amine; 100% conc.

Jordamide JT-128. [PPG-Mazer] Cocamide DEA (1:1); nonionic; foam stabilizer, thickener and emulsifier in personal, household, and industrial prods.; yel. liq.; 4.0–8.5% free amine; 100% conc.

Jordamide JT-1286. [PPG-Mazer] Cocamide DEA superamide; see Jordamide 1214; yel. liq.

Jordamide LLD. [PPG-Mazer] Linoleamide DEA (1:1); nonionic; see Jordamide JT-128; amber liq.; 5–7% free amine; 100% conc.

Jordamide OW. [PPG-Mazer] Cocamide DEA (2:1); see Jordamide 201; liq.

Jordamide PCS. [PPG-Mazer] Coconut alkanolamide, modified; nonionic/anionic; foam booster, stabilizer, visc. builder, detergent and emulsifier; liq.; 100% conc.

Jordamide RO. [PPG-Mazer] Modified 2:1 alkanolamide; emulsifier, low foam thickener for hard surface cleaners, waterless hand cleaners; brn. liq.; acid no. 75–85; 33–35% free amine.

Jordamide SCD. [PPG-Mazer] 1:1 Linoleamide/cocamide DEA; foam stabilizer and visc. builder for household, industrial, and cosmetic prods.; amber liq.; 6–7% free amine.

Jordamide TC. [PPG-Mazer] Tallow DEA, modified; nonionic; wool fulling and scouring agent, thickener; liq.; 100% conc.

Jordamide WC Conc. [PPG-Mazer] Cocamide DEA (1:1); nonionic; see Jordamide PCS; yel. liq.; 6.0–8.5% free amine; 100% conc.

Jordamine 1281. [PPG-Mazer] Modified 2:1 alkanolamide; visc. builder for hard surface cleaners; amber liq.; acid no. 50–56.

Jordamine DAPI. [PPG-Mazer] Isostearamidopropyl dimethylamine; cationic; intermediate used in quats., amine oxides; oil and grease additive; emollient, conditioner for skin and hair care prods.; solid; 99% conc.

Jordamine DAPL, DMCAPA. [PPG-Mazer] Cocamidopropyl dimethylamine; cationic; see Jordamine DAPI; solid and liq. resp.; 99 and 90% conc. resp.

Jordamine DAPSA. [PPG-Mazer] Stearamidopropyl dimethylamine; cationic; see Jordamine DAPI; solid; 99% conc.

Jordamine DDBSA. [PPG-Mazer] Dodecylbenzene sulfonic acid; anionic; intermediate for formulation of household and industrial detergents; liq.; 97% conc.

Jordamine S-13. [PPG-Mazer] Stearamidopropyl dimethylamine; cationic; see Jordamine DAPI; liq.; 99% conc.

Jordamine S-13 Lactate. [PPG-Mazer] Stearamidopropyl dimethylamine lactate; hair conditioner for shampoos, conditioners; liq.; 23% aq.

Jordamine SHCFA. [PPG-Mazer] Cocamidopropyl dimethylamine; cationic; see Jordamine DAPI; solid; 99% conc.

Jordamox CAPA. [PPG-Mazer] Cocamidopropylamine oxide; nonionic; wetting and scouring agent, foam booster and stabilizer, conditioner, visc. builder, detergent used in cosmetic, household and industrial formulations; liq.; 29.5–31.5% conc.

Jordamox CDA. [PPG-Mazer] Palmitamine oxide; conditioner for hair and fabrics; liq.; 30% act.

Jordamox CDA-40. [PPG-Mazer] Palmitamine oxide; nonionic; see Jordamox CAPA; liq.; 39–41% conc.

Jordamox LDA. [PPG-Mazer] Lauramine oxide; nonionic; see Jordamox CAPA; also grease emulsifier; liq.; 29–31% conc.

Jordamox MDA. [PPG-Mazer] Myristamine oxide; nonionic; wetting agent, foam booster/stabilizer, visc. builder for cosmetics, liq. detergents; liq.; 29–31% conc.

Jordamox ODA. [PPG-Mazer] Oleamine oxide; conditioner for hair and fabrics; visc. builder for liq. detergents; liq.; 50–52% act.

Jordamox SDA. [PPG-Mazer] Stearamine oxide; nonionic; conditioner, softener for hair and fabrics, emulsifier with low foam, emollient; visc. builder for liq. detergents; paste; 24.5–26.5% conc.

Jordanol NP-95. [PPG-Mazer] Nonylphenol, ethoxylated; nonionic; emulsifier, detergent for formulation of household and industrial detergents; solubilizer of oily materials into water; liq.; 100% conc.

Jordanol SXS. [PPG-Mazer] Sodium xylene sulfonate; anionic; coupler, solubilizer, cloud pt. reducer for detergent formulation, with or without builder; liq.; 40% act.

Jordaphos 151. [PPG-Mazer] Aromatic phosphate ester; anionic; emulsifier for chlorinated solvs. and aerosol propellents; detergent surfactant; lubricant; clear liq.; 100% conc.

Jordaphos 236. [PPG-Mazer] Aliphatic phosphate ester; anionic; emulsifier for aliphatic solvs.; coupling agent, detergent; Gardner 3 max. clear liq.; 100% conc.

Jordaphos DT. [PPG-Mazer] Aliphatic phosphate ester; anionic; surfactant for hard surface cleaning and wetting, textile scouring; lubricant; clear liq.; 100% conc.

Jordaphos FDED. [PPG-Mazer] Phosphate ester blend; low-foaming surfactant, coupling agent, lubricant; clear liq.; 100% act.

Jordaphos FDEO. [PPG-Mazer] Aliphatic phosphate ester; anionic; low foaming surfactant; coupling agent; Gardner 8 max. clear liq.; 100% conc.

Jordaphos JA-60. [PPG-Mazer] Aliphatic phosphate ester; anionic; detergent, surfactant, and coupling agent, lubricant; Gardner 6 max. clear visc. liq.; 100% conc.

Jordaphos JB-40. [PPG-Mazer] Aliphatic phosphate ester; anionic; lubricity additive, rust inhibitor, emulsifier; Gardner 9 max. opaque visc. liq.; oil and water-sol.; 100% conc.

Jordaphos JE-41. [PPG-Mazer] Aromatic phosphate ester; anionic; emulsifier for aromatic solvs.; detergent, solubilizer, coupling agent, lubricant; Gardner 6 max. clear visc. liq.; 100% conc.

Jordaphos JE-61. [PPG-Mazer] Phosphate ester; coupling agent for surfactants in presence of builders and salts; contributes detergency; liq.; 100% act.

Jordaphos JM-51. [PPG-Mazer] Aromatic phosphate ester; emulsifier, coupling agent for aliphatic and chlorinated solvs. and aerosol propellents used in cutting oils and lubricants; hazy liq.; 100% conc.

Jordaphos JP-70. [PPG-Mazer] Aromatic phosphate ester; anionic; nonfoaming lubricant, coupler; Gardner 6 max. clear liq.; water-sol.; 100% conc.

Jordaphos JS-61. [PPG-Mazer] Aliphatic phosphate ester; anionic; lubricant, load bearing additive, cutting oil emulsifier; Gardner 6 max. hazy visc. liq.; 100% conc.

Jordaphos JS-71. [PPG-Mazer] Aliphatic phosphate ester; anionic; detergent, emulsifier, wetting agent, dispersant; Gardner 6 max. opaque visc. liq.; 100% conc.

Jordaphos RA-60. [PPG-Mazer] Aliphatic phosphate ester; anionic; emulsifier for aliphatic solvs.; coupling agent, detergent; Gardner 4 max. clear liq.; 100% conc.

Jordapon® CI. [PPG-Mazer] Sodium cocoyl isethionate; anionic; detergent; lime soap dispersant; powd.; 80% conc.

Jordaquat 40. [PPG-Mazer] Polyquaternium 6; homopolymer quat. for hair and skin prods.; emollient; clear visc. liq.; 39.0–41.0% act.

Jordaquat 41. [PPG-Mazer] Polyquaternium 7; high m.w. copolymer quat. used for hair and skin prods.;

clear visc. liq.; 8.1–9.1% act.

Jordaquat 350, 358. [PPG-Mazer] Benzalkonium chloride; used for prods. requiring bacteriostatic, germicidal, and algicidal activity; also for static elimination at low use levels; liqs.; 50 and 80% act. resp.

Jordaquat 522. [PPG-Mazer] Isostearamidopropyl ethyldimonium ethosulfate; cationic; conditioner for hair conditoners and shampoos; clear visc. liq.; exc. water sol.; 100% act.

Jordaquat 1033. [PPG-Mazer] Soya ethyldimonium ethosulfate; cationic; conditioner for hair conditioners and shampoos; antistat for cleaners, rug shampoos; low foaming; clear yel. liq.; 58% act.

Jordaquat Dimer 12. [PPG-Mazer] Hydroxypropyl bislauryldimonium chloride; cationic; conditioner for skin and hair; low skin and eye irritation; paste; 50% act.

Jordaquat Dimer 16. [PPG-Mazer] Hydroxypropyl biscetyldimonium chloride; cationic; conditioner for skin and hair; low skin and eye irritation; paste; 50% act.

Jordaquat Dimer 18. [PPG-Mazer] Hydroxypropyl bisstearyldimonium chloride; cationic; conditioner for hair and skin, perms, mousses; emulsifier; wh. paste; 50% act.

Jordaquat Dimer 22. [PPG-Mazer] Hydroxypropyl bisbehenyldimonium chloride; cationic; conditioner for skin and hair; low skin and eye irritation; paste; 50% act.

Jordaquat JN. [PPG-Mazer] Ricinolamidopropyl ethyldimonium ethosulfate; cationic; substantive to hair; liq.; water-sol.; 100% act.

Jordaquat JO-50. [PPG-Mazer] Olealkonium chloride; cationic; conditioner, antistat for clear hair rinses; clear yel. liq.; 50% act.

Jordaquat JS-25. [PPG-Mazer] Stearalkonium chloride; cationic; conditioner, softener, emollient for hair; wh. paste; 25% act.

Jordawet DMDS. [PPG-Mazer] Disodium oleamido PEG-2 sulfosuccinate; anionic; wetting and penetrating agent for hair, bath and home prods.; liq.; 30% conc.

Jorgard AMF. [PPG-Mazer] Tributyltin maleate; bacteriostatic and fungistatic protection to carpets; 25% act. sol'n.

Jorphox KCAO. [PPG-Mazer] Potassium salt of phosphated n,n-bis (hydroxyethyl) coco amine oxide; corrosion inhibitor, conditioner for cosmetic to industrial uses; visc. liq.; 33.0% act.

Jorphox KTAO. [PPG-Mazer] Potassium salt of phosphated n,n-bis (hydroxyethyl) tallow amine oxide; see Jorphox KCAO; visc. liq.; 33.0% act.

Jorquest 100. [PPG-Mazer] Tetrasodium EDTA; chelating agent for Ca, Mg, other metal ions; liq.; 40% act. sol'n.

Jortaine C, CAB-35. [PPG-Mazer] Cocamidopropyl betaine; amphoteric; wetting agent, stabilizer, foamer and visc. builder in personal care, household, and industrial prods.; liq.; 34–35% conc.

Jortaine CB-40. [PPG-Mazer] Coco betaine; amphoteric; wetting agent, stabilizer, foam booster, and visc. builder in personal care and industrial prods.; liq.; 40% conc.

Jortaine CFA-35. [PPG-Mazer] Cocoamidopropyl betaine; amphoteric; see Jortaine CB-40; liq.; 35% conc.

Jortaine COSB. [PPG-Mazer] Cocamidopropyl hydroxysultaine; amphoteric; wetting agent, stabilizer, foamer, visc. builder used in detergents and personal care prods.; liq.; 50% conc.

Jortaine CSB, CSB-50. [PPG-Mazer] Cocamidopropyl hydroxysultaine; amphoteric; visc. builder used in personal care prods. and cleaners; liq.; 35 and 50% conc. resp.

Jortaine LMAB. [PPG-Mazer] Lauramidopropyl betaine; amphoteric; see Jortaine COSB; liq.; 35% conc.

Jortaine TM. [PPG-Mazer] Dihydroxyethyl tallow glycinate; thickener, conditioner, foamer; for high or low pH formulations, shampoos; visc. liq. to gel; 39% act.

KA 301. [Kenrich] Diisobutyl (oleyl) aceto acetyl aluminate; coupling agent; lt. yel. liq.; sol. in IPA, xylene, toluene, DOP, min. oil; sp.gr. 0.97; visc. $<$ 1000 cps; flash pt. 72 F; pH 6.0.

KA 322. [Kenrich] Diisopropyl (oleyl) aceto acetyl aluminate; coupling agent; pale grn. liq.; sol. in xylene, toluene, DOP, min. oil; sp.gr. 0.99; visc. $<$ 1000 cps; flash pt. 70 F; pH 6; 95% solids in IPA.

Kadif 50 Flakes. [Witco UK] Sodium dodecylbenzene sulfonate; general purpose wetting and penetrating agent, detergent for industrial, institutional and household uses; lt. flakes; practically odorless; water sol.; pH 8.0 (1% aq.); 50% act.

Kadox®-15. [Zinc Corp. of Am.] French process zinc oxide; pigment providing highest activating power and reinforcement in rubber; also used in insulated wire, transparent rubber, mechanical goods, latex, lubricating oils; improves lubrication and inhibits corrosion; 0.12 μ mean particle size; 99.99% thru 325 mesh; sp.gr. 5.6; sp.vol. 0.18; pkg. dens. 30 lb/ft^3; surf. area 9 m^2/g; 99.6% zinc oxide.

Kadox®-25. [Zinc Corp. of Am.] French process zinc oxide; pigment used in food-can mfg., sealants based on dry rubber or latex compds., coatings on the interior of cans; the high reactivity prevents formation of black tin and iron sulfides that discolor foods; 0.11 μ mean particle size; 99.99% thru 325 mesh; pkg. dens. 30 lb/ft^3; surf. area 10 m^2/g; oil absorp. 12 lb oil/100 lb ZnO; 99.6% zinc oxide.

Kadox®-72. [Zinc Corp. of Am.] French process zinc oxide; pigment used in latex, mechanical goods, insulated wire, footwear, and tires where uniform activation and a degree of reinforcement desired; also in the prod. of resinates, ceramics, textiles, phosphate solutions; 0.18 μ mean particle size; 99.99% thru 325 mesh; sp.gr. 5.6; sp.vol. 0.18; pkg. dens. 35 lb/ft^3; surf. area 6 m^2/g; oil absorp. 13 lb oil/100 lb ZnO; 99.7% zinc oxide.

Kadox®-215. [Zinc Corp. of Am.] French process zinc oxide; pelleted Kadox-15; 0.12 μ mean particle size; sp.gr. 5.6; sp.vol. 0.18; pkg. dens. 60 lb/ft^3; surf. area 9 m^2/g; oil absorp. 14 lb oil/100 lb ZnO; 99.6% zinc oxide.

Kadox®-272. [Zinc Corp. of Am.] French process zinc oxide; pelleted form of Kadox-72; 0.18 μ mean particle size; 99.99% thru 325 mesh; sp.gr. 5.6; sp.vol. 0.18; pkg. dens. 60 lb/ft^3; surf. area 6 m^2/g; oil absorp. 13 lb oil/100 lb ZnO; 99.7% zinc oxide.

Kadox® 930. [Zinc Corp. of Am.] Amer. process zinc oxide; pigment for use as activator for both wh. and colored rubber compds.; 0.36 μ mean particle size; 99.97% thru 325 mesh; sp.gr. 5.6; sp.vol. 0.18; dens. 35 lb/ft^3; surf. area 3.0 m^2/g; oil absorp. 13 lb oil/100 lb ZnO; 98.9% zinc oxide.

Kadox® 720C, 720CP. [Zinc Corp. of Am.] Amer. process zinc oxide treated with propionic acid; pigment for use as activator for both wh. and colored rubber compds.; 0.27 μ mean particle size; 99.97% thru 325 mesh; sp.gr. 5.6; sp.vol. 0.18; dens. 50 and 70 lb/ft^3 resp.; surf. area 4.0 m^2/g; oil absorp. 18 lb oil/100 lb ZnO; 98.8% zinc oxide.

Kalcohl 5-24, 6-24, 7-24. [Kao] Lauryl myristyl alcohol; raw material for sodium lauryl sulfate; metal rolling agent; liq.; 100% conc.

Kalcohl 08H. [Kao] n-Octyl alcohol; raw material for plasticizer; tobacco-offshoots controller; liq.; 100% conc.

Kalcohl 10H. [Kao] n-Decyl alcohol; tobacco-offshoots controller; raw material for antioxidant, alkyl phosphate, ethoxylate, chloride; liq.; 100% conc.

Kalcohl 20. [Kao] Lauryl alcohol; see Kalcohl 10H; liq.; 100% conc.

Kalcohl 40. [Kao] Myristyl alcohol; see Kalcohl 10H; solid; 100% conc.

Kalcohl 60. [Kao] Cetyl alcohol; see Kalcohl 10H; beads or solid; 100% conc.

Kalcohl 68. [Kao] Cetyl stearyl alcohol; base for ointment and cream; beads, solid; 100% conc.

Kalcohl 80. [Kao] Stearyl alcohol; see Kalcohl 68; beads, solid; 100% conc.

Kalcohl 86. [Kao] Stearyl cetyl alcohol; lubricant for plastic processing; beads, solid; 100% conc.

Kalex. [Hart Prod.] Org. amino acid; chelating and sequestrant for alkaline earths and polyvalent metals; ion exchange agent; hard water softener; dyeing assistant for metal senstive dyes; water conditioner in textile applic.; deflocculating agent in photographic developing and fixing sol'ns.; bottle washing and dishwashing prods.; floor cleaning compds.; leather processing; cold rubber polymerization recipes; straw liq.; water-sol.

Kalex 100. [Hart Prod.] Tetrasodium EDTA; chelating agent; misc. in water.

Kalex 220 Crystal. [Hart Prod.] Tetrasodium EDTA; anionic; chelating agent; powd.; 82% conc.

Kalex Acids. [Hart Prod.] EDTA; anionic; sequestrant for preparation of amine or alkali metal salts; used where sodium ion is undesirable; wh. powd.; pH acid; 99% act.

Kalex Conc. [Hart Prod.] Tetrasodium EDTA; chelating agent; misc. with water.

Kalex IR. [Hart Prod.] Iron specific chelating agent

for use in caustic sol'ns.

Kalex Liq. 50%. [Hart Prod.] Tetrasodium EDTA; anionic; general purpose chelating agent; complexes Ca, Mg, other common metals; liq.; 40% conc.

Kalex OH. [Hart Prod.] Trisodium HEDTA; anionic; sequesters Ca, Mg, iron @ pH 8.0–10.5; liq.; 41.5% conc.

Kalex Penta. [Hart Prod.] Pentasodium DTPA; anionic; sequestrant used when slightly higher chelate stability is required and when other strong complexing agents are present; yel. liq.; 40% conc.

Kalex Powd. FC. [Hart Prod.] Tetrasodium EDTA dihydrate; anionic; sequestrant; formulating dry prods.; ylsh. powd.; pH alkaline; 82% conc.

Kalex Regular. [Hart Prod.] Tetrasodium EDTA; chelating agent; misc. in water.

Kantstik Q Powd. [Specialty Prod.] Refined syn. wax ester; internal lubricant for inj. molding; improves plastic flow in hard-to-reach mold areas; valuable in heavily pigmented molding resin mix; used in butyrate, PP, nylon, glass-filled nylon, PC, SAN, styrene, polyethylene, ABS, high-impact PS, PVC; granular powd.

Kaolin RC 32. [Sachtleben] Wh. filler with very fine particle size; semireinforcing properties in elastomers; dens. 2.6 g/ml; pH 5.

Kara Lube ECM. [Lyndal] Fatty condensate; nonionic; lubricant and softener with resistance to yellowing; water-disp.

Kara Sperse DDL. [Lyndal] Modified sulfonate; dispersant, suspending agent for wettable powds.; water-sol.

Kara Sperse DDL-12. [Lyndal] Sodium org. sulfonate; nonionic; pigment dispersant; liq.; 50% act.

Karathane® Liq. Conc., WD. [Rohm & Haas] Dinitrooctylphenyl crotonate; agric. fungicide/miticide for fruit, ornamentals; 35 and 18.25% conc. resp.

Katanol 387. [Rhone-Poulenc Surf.] Alkylated biphenyl; nonionic; dye carrier for pressure dyeing of polyester; liq.

Katapol® OA-860. [GAF] PEG-30 oleamine; hydrophilic emulsifier, leveling agent; textile dyeing assistant; antiprecipitant for dyeing processes; solid; water-sol.; 99% act.

Katapol® OA-910. [GAF] PEG-30 oleamine; cationic; emulsifier and textile dyeing assistant; antiprecipitant to prevent cross-staining; stripping agent and dye leveler for acid dyes; liq.; 98% conc.

Katapol® PN-430. [GAF] PEG-5 tallow amine; cationic; emulsifier for min. oils, acid corrosion inhibitor for ferrous alloys; brn. liq.; sol. in kerosene with slight haze; pH 8.0–10.0 (10% aq. disp.); 99% act.

Katapol® PN-730, PN-810. [GAF] PEG-15, -20 tallow amine; cationic; emulsifier, dye assistant, softener, antistat, lubricant for textiles and leathers; amber clear liq.; water sol.; 99% min. act.

Katapol® VP-532. [GAF] POE alkylamine; cationic; retarder in applic. of cationic dyes to acrylic fibers; antiprecipitant for acid and cationic dyes used in combination in same bath in dyeing acid-dyeable/ cationic-dyeable acrylic and wool/acrylic fibers; liq.

Katapone® VV-328. [GAF] Quat. ammonium chloride; acid corrosion inhibitor for metals; used in cleaners, dairy equipment, leather dyeing, petrol. processing, drilling, and acidizing; liq.; water-sol.; > 75% act.

Katemul IB-70. [Scher] N-(3-Isostearamidopropyl)-N,N-dimethylamino glycolate; imparts substantive hair conditioning, fly-away control, enhanced wet and dry combing and softening to shampoos; liq.; water-sol.

Katemul IG-70. [Scher] Isostearamidopropyl dimethylamine glycolate; cationic; conditioner and softener for hair prods.; dk. yel. slight visc. liq.; mild, typical odor; water-sol.; sp.gr. 0.99; visc. 600; pH 7.0 (as is); 5.5 (5.0%); 70% nonvolatiles; 30% propylene glycol.

Katemul IGU-70. [Scher] Isostearamidopropyl dimethylamine gluconate; see Katemul IG-70; dk. amber visc. liq.; sp.gr. 1.06; visc. 7500; pH 7.0 (as is); 5.5 (5.0%).

Kathon® 886. [Rohm & Haas] 5-Chloro-2-methyl-4-isothiazolin-3-one; preservative for metalworking liqs.; aq. sol'n.

Kathon® 886 MW. [Rohm & Haas] 5-Chloro-2-methyl-4-isothiazolin-3-one; antimicrobial for aq. metalworking fluids; sol. in propylene glycol; sparingly sol. in water.

Kathon® 893. [Rohm & Haas] N-octyl-isothiazolone in propylene glycol; industrial fungicide; 45% conc.

Kathon® 925. [Rohm & Haas] 4,5-Dichloro-N-octyl-isothiazolone; antimicrobial; liq.

Kathon® 4200. [Rohm & Haas] 2-n-Octyl-4-isothiazolin-3-one; fabric mildewcide; 25% act. in propylene glycol.

Kathon® CG. [Rohm & Haas] 1.15% 5-Chloro-2-methyl-4-isothiazolin-3-one and 0.35% 2-methyl-4-isothiazolin-3-one ; antimicrobial, preservative for cosmetics and toiletries; lt. amber clear liq.; mild odor; misc. in water, lower alcohols and glycols; low sol. in hydrocarbons; sp.gr. 1.21; pH 3.5–5.0; 1.5% act. in water and magnesium salts.

Kathon® CG/ICP. [Rohm & Haas] 5-Chloro-2-methyl-4-isothiazolin-3-one; antimicrobial, preservative for in-container prods.; 1.5% sol'n.

Kathon® DP. [Rohm & Haas] Methyl-isothiazolone plus chloromethyl isothiazolone; preservative for domestic prods.; liq.

Kathon® ICP. [Rohm & Haas] Methyl-isothiazolone plus chloromethyl isothiazolone; in-container preservative; liq.

Kathon® LM. [Rohm & Haas] 2-n-Octyl-4-isothiazolin-3-one; fabric mildewcide; 5% sol'n. in propylene glycol.

Kathon® LP. [Rohm & Haas] 2-n-Octyl-4-isothiazolin-3-one; mildewcide for processing hides, leather; 8.1% sol'n. in propylene glycol.

Kathon® LX. [Rohm & Haas] 5-Chloro-2-methyl-4-isothiazolin-3-one; antimicrobial, preservative for polymer emulsions.

Kathon® WT. [Rohm & Haas] 5-Chloro-2-methyl-4-isothiazolin-3-one; antimicrobial for cooling tower water, slimicide for paper mills; water-sol.

Katioran AF. [BASF AG] Fatty acid hydroxylalkylamide and fatty alcohol, ethoxylated; nonionic; emulsifier and thickener for cosmetic preparations; waxy.

Kaydol. [Witco] Wh. min. oil USP; emollient and lubricant in cosmetics and pharmaceuticals; visc. (Saybolt) 345–355 (100 F); pour pt. 0 F.

K-Cop. [Griffin] Copper-ammonium complex; agric. fungicide; liq.; 8% elemental copper.

Kelacid®. [Kelco] Alginic acid; used as gelling agent, emulsifier and stabilizer in food, pharmaceutical, and industrial applics.; stabilizer in paper and textile industry; wh. fibrous particles; sol. in alkaline sol'n.; swells in water; pH 2.9 (1% aq.); surf. tens. 53 dynes/cm; 7% moisture.

Kelate CDS [Tri-K] Calcium disodium EDTA, FCC; chelating agent.

Kelco-Gel®. [Kelco] Sodium alginate; stabilizer for food preparations; gives sol. and clarity in gels, films, and sol'ns.

Kelco-Gel® Gellan Gum. [Kelco] Purified gellan gum; high m.w. anionic polysaccharide; gelling agent for use in foods, pet foods, personal care prods., industrial applics.; cream to wh. dry free-flowing powd.; 100% thru 28 mesh.

Kelco-Gel® HV. [Kelco] Low-calcium sodium alginates; see Kelacid; cream fibrous particles; sp.gr. 1.64; dens. 43.38 lb/ft³; visc. 400 cps; ref. index 1.3342; pH 7.2; 9% moisture.

Kelco-Gel® LV. [Kelco] Low-calcium sodium alginates; see Kelacid; cream fibrous particles; sp.gr. 1.64; dens. 43.38 lb/ft³; visc. 50 cps; ref. index 1.3342; pH 7.2; 9% moisture.

Kelcoloid®. [Kelco] Propylene glycol alginate; gum used as emulsion stabilizer with thixotropic flow properties; for food preparations; sol. in water.

Kelcoloid® D. [Kelco] Propylene glycol alginate; see Kelacid; cream fibrous particles; sp.gr. 1.46; dens. 33.71 lb/ft³; visc. 170 cps; ref. index 1.3343; pH 4.4; surf. tens. 58 dynes/cm; 13% max. moisture

Kelcoloid® DH, DO, DSF. [Kelco] Propylene glycol alginate; see Kelacid; cream agglomerated; sp.gr. 1.46; dens. 33.71 lb/ft³; visc. 400, 25, 20 cps resp.; ref. index 1.3343; pH 4.0, 4.3, 4.0 resp.; surf. tens. 58 dynes/cm.

Kelcoloid® HVF, LVF, O, S. [Kelco] Propylene glycol alginate; see Kelacid; cream fibrous particles; sp.gr. 1.46; dens. 33.71 lb/ft³; visc. 400, 120, 25, 20 cps resp.; ref. index 1.3343; pH 4.0, 4.0, 4.3, 4.0 resp.; surf. tens. 58 dynes/cm.

Kelco-Pac®. [Kelco] Sodium alginate; see Kelacid; ivory gran. particles; water-sol.; sp.gr. 1.64; dens. 43.38 lb/ft³; visc. 55 cps; pH 7.2; surf. tens. 70 dynes/cm; 9% moisture.

Kelcosol®. [Kelco] Sodium alginate; see Kelacid; cream fibrous particles; water-sol.; sp.gr. 1.64; dens. 43.38 lb/ft³; visc. 1300 cps; pH 7.2; surf. tens. 70 dynes/cm; 9% moisture.

Kelene 77. [Lowenstein] Sodium dihydroxyethylglycinate and trisodium hydroxyethyl ethylenediamine triacetic acid; chelating agent; slightly yel. clear liq.; pH 9.5 min.; 41% min. act.

Kelex 100. [Sherex] Chelating agent for solv. extraction; sp.gr. 0.986; dens. 8.20 lb/gal; visc. 15 cps; pour pt. –15 C; flash pt. (PM) > 200 F.

Kelfo®. [Kelco] Xanthan gum/limestone blend; gum for use in animal feed; thickener, stabilizer.

Kelgin®. [Kelco] Sodium alginate; gum for use in industrial preparations and food contact paper; controls penetration and film-forming of surface sizing in paper/paperboard; sol. in water.

Kelgin® F. [Kelco] Sodium alginate, refined; see Kelacid; ivory gran. particles; sp.gr. 1.59; dens. 54.62 lb/ft³; visc. 300 cps; ref. index 1.3343; surf. tens. 62 dynes/cm; 13% moisture.

Kelgin® HV, LV, MV. [Kelco] Sodium alginate; see Kelacid; ivory gran. particles; sp.gr. 1.59; dens. 54.62 lb/ft³; visc. 800, 60, 400 cps resp.; ref. index 1.3343; pH 7.5; surf. tens. 62 dynes/cm; 13% moisture

Kelgin® QH, QL, QM. [Kelco] Treated sodium alginate; see Kelacid; improved disp.; ivory gran. particles; visc. 400, 30, 180 cps resp.; pH neutral.

Kelgin® RL, XL. [Kelco] Refined sodium alginate; see Kelacid; ivory gran. particles; sp.gr. 1.59; dens. 54.62 lb/ft³; visc. 10 and 30 cps resp.; ref. index 1.3343; pH 7.5; surf. tens. 62 dynes/cm; 13% moisture.

Kelig 32. [Daishowa] Lignosulfonate and modified sugar acids; anionic; sequestrant, dispersant for the lubricant removal from zinc phosphate coatings, oil well cement retardant; used for cooling water treatment compds., alkaline cleaners, scale control; powd; 100% conc.

Kelig 100. [Daishowa] Sodium lignosulfonate; anionic; metal cleaning component for industrial cleaning applics.; corrosion inhibitor; powd., liq.; 55% liq., 95% powd.

Kelmar®, Kelmar Improved. [Kelco] Potassium alginate; gellant, emulsifier, and stabilizer in food and indust. applic.; gum, bodying agent for creams and lotions, dental impression materials; used for water holding in foods and industry; cream gran. and fibrous particles resp.; water-sol.; visc. 270 and 400 cps resp.; pH neutral.

Kelset®. [Kelco] Alginate; see Kelacid; self-gelling gum; lt. ivory fibrous particles; water-sol.; pH neutral.

Keltex®, P, S. [Kelco] Industrial sodium alginate; gellant, emulsifier, and stabilizer in food and indust. applic.; print paste thickener in textiles; tan gran. particles; water-sol.; visc. 800, 765, and 1300 cps resp.; pH neutral.

Keltone®. [Kelco] Sodium alginate; gellant, emulsifier, and stabilizer in food and indust. applic.; cream fibrous particles; water-sol.; sp.gr. 1.64; dens. 43.38 lb/ft³; visc. 400 cps; pH 7.2; surf. tens. 70 dynes/cm; 9% moisture.

Keltose®. [Kelco] Calcium alginate and ammonium alginate; gellant, binder, emulsifier, and stabilizer in food and indust. applic.; ivory gran. particles; water-sol.; visc. 250 cps; pH neutral.

Keltrol®, Keltrol F. [Kelco] Food-grade xanthan gum; stabilizer for foods; thickener and emulsion stabilizer in creams and lotions; binder in toothpaste;

suspending agent for fruit pulp; cream dry powd.; 80 and 200 mesh size resp.; water-sol.; swells in glycerin and propylene glycol; sp.gr. 1.5; dens. (bulk) 52.2 lb/ft^3; visc.1400 cps (1% visc. with 1% electrolyte added, LVF, 60 rpm); surf. tens. 75 dynes/cm; pH 7.0; 11% moisture.

Kelvis®. [Kelco] Sodium alginate, refined; gellant, emulsifier, and stabilizer in food and indust. applic.; ivory gran. particles; water-sol.; sp.gr. 1.59; dens. 54.62 lb/ft^3; visc. 760 cps; ref. index 1.3343; pH 7.5; surf. tens. 62 dynes/cm. 13% moisture.

Kelzan®, D, M, XC Polymer. [Kelco] Industrial grade xanthan gum; foam stabilizer, flocculant suspending, gelling agent, rheology modifier, lubricant for industrial applics. incl. abrasives, adhesives, agric. herbicides, fertilizers, ceramics, cleaners, emulsions, gels, mining, thixotropic paints, paper, petrol., pigments; viscosifier for drilling fluids; Kelzan is the std. industrial grade; D grade is for use with galactomannens; M grade is a more refined industrial grade for use in systems which require a lower salt content; XC Polymer is esp. useful as an additive to oil well drilling mud; cream dry powd.; 40 mesh size; water-sol.; sp.gr. 1.6; dens. (bulk) 52.4 lb/ft^3; visc. 850 cps; f.p. O C (1% aq.); surf. tens. 75 dynes/cm; pH 7.0.

Kelzan® D35. [Kelco] Dispersible heteropolysaccharide prod.; gum providing rheological control to aq. systems, esp. in oil field applics; suspends solids slurries; fully disperses in neutral to acidic pH sol'ns.

Kelzan® S. [Kelco] Dispersible xanthan gum prod.; gum providing suspension of solids slurries and rheological control for aq. systems; disperses in water sol'ns. at pH neutral to acidic.

Kemamide® B. [Witco/Humko] Behenamide; lubricant, slip, antiblock, and mold release agent for plastics; crayons, petrol. prods., asphalts, inks, metals, and textiles; mold release agent for thermoplastic resins in inj. molding applic.; defoamer and water repellent component in industrial and household applic.; corrosion inhibitor; pigment grinding aid and dyestuff dispersant in paints, enamels, varnishes, and lacquers; intermediate for textile emulsifiers and softeners and durable water repellents; foam stabilizer in household detergents; Gardner 4 max. waxy solid, powd., and pellet; m.w. 312; sol. (g/100 g solv. @ 60 C) > 10 g in IPA, MEK, 10 g in methanol and toluene; water-insol.; dens. 0.807 g/ml (130 C); visc. 6.5 cP (130 C); m.p. 98–108 C; acid no. 4 max.; flash pt. 257 C (COC); 95% conc.

Kemamide® E. [Witco/Humko] Erucamide; see Kemamide B; Gardner 5 max solid; m.w. 335; sol. (g/100 g solv.) 25 g in chloroform, 8 g in IPA, 4 g in MEK, 3 g in methyl alcohol and toluene; water insol.; m.p. 76–86 C; 100% conc.

Kemamide® E-180. [Witco/Humko] Stearyl erucamide; see Kemamide B; wh. powd.; sol. (g/100 g solv. @ 50 C) < 50 g in chloroform and toluene, 50 g in dichloroethane, 40 g in VM&P naphtha, 30 g in IPA, 25 g in MEK; dens. 0.8074 (110 C); m.p. 72–75 C; flash pt. (CC) 258 C.

Kemamide® E-221. [Witco/Humko] Erucyl erucamide; see Kemamide B; wh. powd.; sol. (g/100 g solv. @ 40 C) < 50 g in dichloroethane and VM&P naphtha, 50 g in IPA, 30 g in MEK; dens. 0.8165 (110 C); m.p. 55–58 C; flash pt. (CC) 260 C.

Kemamide® O. [Witco/Humko] Oleamide, tech.; see Kemamide B; solid; 100% conc.

Kemamide® P-181. [Witco/Humko] Oleyl palmitamide; see Kemamide B; wh. powd.; sol. (g/100 g solv @ 50 C) 50 g in MEK, < 50 g in dichlorethane, IPA, VM&P naphtha, toluene, 5 g in methyl alcohol; dens. 0.8076 g/ml (110 C); m.p. 69–72 C; acid no. 10 max.; flash pt. 262 C (CC).

Kemamide® S. [Witco/Humko] Stearamide; see Kemamide B; Gardner 4 max. waxy solid, powd., and pellet; m.w. 278; sol. (g/100 g solv. @ 50 C) > 10 g in chloroform, 10 g in IPA, 4 g in MEK, 3 g in methyl alcohol, 2 g in toluene; dens. 0.809 g/ml (130 C); visc. 5.8 cP (130 C); m.p. 98–108 C; acid no. 4 max.; flash pt. 246 C (COC); fire pt. 268 C (COC); 95% min. amide.

Kemamide® S-180. [Witco/Humko] Stearyl stearamide; see Kemamide B; wh. powd.; sol. (g/100 g solv. @ 60 C) 50 g in toluene, < 50 g in chloroform, 20 g in dichloroethane, 15 g in IPA, MEK, and VM&P naphtha; dens. 0.8042 g/ml (110 C); m.p. 92–95 C; acid no. 10 max.; flash pt. 246 C (CC).

Kemamide® S-221. [Witco/Humko] Erucyl stearamide; see Kemamide B; wh. powd.; sol. (g/100 g solv. @ 50 C) < 50 g in chloroform and toluene, 20 g in VM&P naphtha, 14 g in dichloroethane and IPA, 7 g in MEK; dens. 0.7877 g/ml (110 C); m.p. 72–75 C; acid no. 10 max.; flash pt. 268 C (CC).

Kemamide® U. [Witco/Humko] Oleamide; see Kemamide B; Gardner 5 max. waxy solid, powd., and pellet; m.w. 275; sol. (g/100 g solv. @ 30 C) > 30 g in IPA, 28 g in methyl alcohol, 25 g in toluene, > 20 g in MEK; dens. 0.823 g/ml (130 C); visc. 5.5 cP (130 C); m.p. 68–78 C; acid no. 4 max.; flash pt. 245 C (COC); 95% min. amide.

Kemamide® W-20. [Witco/Humko] Ethylene dioleamide; see Kemamide B; also useful as internal and external lubricants in ABS, PS, polyethylene, PP, PVC, nylon, cellulose acetate, PVAc, and phenolic resins; defoamer in paper industry blk. liquoring, fabric dyeing, latex systems; metal processing, asphalts; Gardner 6 max. powd. and flake; sol. (g/100 g solv. @ 35 C) > 20 g in toluene, 12 g in IPA, 4 g in dichloroethane, 3 g in MEK; m.p. 120 C; acid no. 10 max.; flash pt. 296 C (COC); fire pt. 315 C (COC).

Kemamide® W-39. [Witco/Humko] Ethylene distearamide; see Kemamide B and W-20; Gardner 18 max. flakes; sol. (g/100 g solv. @ 70 C) 2.0 g toluene, 1.6 g dichloroethane; 1.4 g IPA, 0.9 g MEK; m.p. 140 C; acid no. 10 max.; flash pt. 299 C (COC); fire pt. 315 C (COC).

Kemamide® W-40. [Witco/Humko] Ethylene distearamide; see Kemamide B and W-20; Gardner 3 max. powd. and flake; sol. see Kemamide W-39; m.p. 140 C; acid no. 10 max.; flash pt. 299 C (COC); fire pt. 315 C (COC).

Kemamide® W-40/300. [Witco/Humko] Ethylene distearamide; see Kemamide B; Gardner 3 max.

atomized very fine powd.; m.p. 140 C; acid no. 10 max.

Kemamide® W-40DF. [Witco/Humko] Ethylene distearamide, defoamer grade; see Kemamide B and W-20; Gardner 3 max. powd. and flake; sol. see Kemamide W-39; m.p. 140 C; acid no. 5 max.; flash pt. 299 C (COC); fire pt. 315 C (COC).

Kemamide® W-45. [Witco/Humko] Ethylene distearamide; see Kemamide B and W-20; Gardner 3 max. powd. and flake; sol. see Kemamide W-39; m.p. 145 C; acid no. 10 max.; flash pt. 304 C (COC); fire pt. 322 C.

Kemamine® A650. [Witco/Humko] Coconut amine acetate; cationic; pigment dispersant, fungicide, petrol. additive, collectors in ore flotation and boiler water treatment; solid; 100% conc.

Kemamine® A970. [Witco/Humko] Hydrog. tallow amine acetate; cationic; see Kemamine A650; solid; 100% conc.

Kemamine® A974. [Witco/Humko] Tallow amine acetate; cationic; see Kemamine A650; solid; 100% conc.

Kemamine® A990. [Witco/Humko] Octadecyl amine acetate; cationic; pigment dispersant, fungicide, petrol. additive; solid; 100% conc.

Kemamine® AD 650. [Witco/Humko] N-coco 1,3-propylene diamine diacetate; cationic; see Kemamine A990; solid; 100% conc.

Kemamine® AD 974. [Witco/Humko] N-tallow 1,3 propylene diamine diacetate; cationic; see Kemamine A990; solid; 100% conc.

Kemamine® AS-650. [Witco/Humko] Nitrogen derivs.; cationic; antistat for polyolefins, styrenics, and other plastics, esp. film applics.; lubricity aid, mold release aid, pigment dispersant; FDA approved; pale straw clear liq.; sp.gr. 0.9058; m.p. 4 C; 97% min. tert. amine.

Kemamine® AS-974. [Witco/Humko] Nitrogen derivs.; cationic; see Kemamine AS-650; ambert paste; sp.grt. 0.904; m.p. 25 C; 97% min. tert. amine.

Kemamine® AS-974/1. [Witco/Humko] Nitrogen derivs.; cationic; see Kemamine AS-650; wh. powd.; 60% min. tert. amine.

Kemamine® AS-989. [Witco/Humko] Nitrogen derivs.; cationic; see Kemamine AS-650; pale straw clear liq.; sp.gr. 0.904; m.p. –1 C; 96.8% min. tert. amine.

Kemamine® AS-990. [Witco/Humko] Nitrogen derivs.; cationic; see Kemamine AS-650; wh. powd.; sp.gr. 0.8065 (100 C); m.p. 50–55 C; 97% min. tert. amine.

Kemamine® BQ-2802C. [Witco/Humko] Behenalkonium chloride; cationic; antistat, textile softener, dyeing aid, corrosion inhibitor, emulsifier; used in personal care prods., e.g., creams, lotions, shampoos, hair conditioners; Gardner 4 max.; water-disp. or misc.; m.w. 475; pH 9 max. (5%); 75% min. act.

Kemamine® BQ-9702C. [Witco/Humko] Dimethyl hydrog. tallow benzyl ammonium chloride; cationic; germicide, sanitizer, slimicide, antistat, textile softener, dyeing aid, corrosion inhibitor, emulsifier; also for personal care prods.; Gardner 4 max. liq.; m.w. 420; disp. in water; pH 9 max. (5%); 75% act.

Kemamine® BQ-9742C. [Witco/Humko] Tallow alkonium chloride; cationic; antistat, textile softening agent, dyeing aid, corrosion inhibitor, emulsifier; water-misc.; Gardner 6 max.; m.w. 420; pH 9 (5%); 75% min. act.

Kemamine® D-150, D-190. [Witco/Humko] Arachidyl-behenyl 1,3-propylenediamine; cationic; gasoline detergent, bactericide, corrosion inhibitor in petrol. prod., epoxy hardener; Gardner 5 max. solid; 88% conc.

Kemamine® D-650. [Witco/Humko] N-coconut 1,3-propylenediamine; cationic; see Kemamine D-150; Gardner 6 max. paste; 88% conc.

Kemamine® D-970. [Witco/Humko] N-hydrog. tallow 1,3-propylenediamine; cationic; see Kemamine D-150; Gardner 6 max. solid; 88% conc.

Kemamine® D-974. [Witco/Humko] N-tallow 1,3-propylenediamine; cationic; see Kemamine D-150; Gardner 12 max. solid; 88% conc.

Kemamine® D-989. [Witco/Humko] N-oleyl 1,3-propylenediamine; cationic; see Kemamine D-150; Gardner 7 max. liq.; 88% conc.

Kemamine® D-999. [Witco/Humko] N-Oleic-linoleic 1,3-propylenediamine; see Kemamine D-190; Gardner 8 max. paste; 85% min. act.

Kemamine® DD-3680. [Witco/Humko] Dimer diamine; chemical intermediate, extender, crosslinking agent in polymeric systems; corrosion inhibitor; in epoxy systems; Gardner 14 max. color.

Kemamine® DP-3680, DP-3695. [Witco/Humko] Dimer diprimary amine; chemical intermediate, extender, crosslinking agent in polymeric systems; corrosion inhibitor; in epoxy systems; Gardner 14 max. color.

Kemamine® P-150, P-150D. [Witco/Humko] 50% arachidyl-behenyl primary amine (P-150D—dist.); cationic; emulsifier, flotation agent, dispersing and flushing agent, intermediate, used in metalworking oils, as fuel oil additive; mold release for rubber and plastics; lubricant and spinning aid in metalworking oils; Gardner 3 and 1 resp., solid; sol. in common org. solv.; 93 and 97% conc.

Kemamine® P-190, P-190D. [Witco/Humko] 90% arachidyl-behenyl primary amine (P-190D—dist.); cationic; see Kemamine P-150; Gardner 3 and 1 resp., solid; sol. in common org. solv.; 93, 97% conc.

Kemamine® P-650, P-650D. [Witco/Humko] Coco amine (P-650D—dist.); cationic; see Kemamine P-150; Gardner 3 and 1 resp., liq.; sol. in common org. solv.; 93%, 97% conc.

Kemamine® P-690, P-690D. [Witco/Humko] Lauramine (P-690 D—dist.); cationic; see Kemamine P-190; liq.; 100% conc.

Kemamine® P-790, P-790D. [Witco/Humko] Tetradecyl primary amine (P-790 D—dist.); cationic; see Kemamine P-190; solid; 100% conc.

Kemamine® P-880, P-880D. [Witco/Humko] Palmityl primary amine (tech. and dist. resp.); cationic; see Kemamine P-150; Gardner 3 and 1 resp., solid; sol. in common org. solv.; 93, 97% conc.

Kemamine® P-970, P-970D. [Witco/Humko] Hydrog. tallow amine (tech. and dist. resp.); cationic; see Kemamine P-150; Gardner 3 and 1 resp., solid; sol. in common org. solv.; 93, 97% conc.

Kemamine® P-974, P-974D. [Witco/Humko] Tallow amine (tech. and dist. resp.); cationic; see Kemamine P-150; Gardner 3 and 1 resp., paste; sol. in common org. solv.; 93, 97% conc.

Kemamine® P-989, P-989D. [Witco/Humko] Oleamine (tech. and dist. resp.); cationic; see Kemamine P-150; Gardner 3 and 1 resp., liq.; sol. in common org. solv.; 93, 97% conc.

Kemamine® P-990, P-990D. [Witco/Humko] Stearamine (tech. and dist. resp.); cationic; see Kemamine P-150; Gardner 3 and 1 resp., solid; sol. in common org. solv.; 93, 97% conc.

Kemamine® P-997, P-997D. [Witco/Humko] Soyamine (tech. and dist. resp.); cationic; see Kemamine P-150; Gardner 3 and 2 resp., paste; sol. in common org. solv.; 93, 97% conc.

Kemamine® P-999. [Witco/Humko] Tech. oleic-linoleic amine; see Kemamine P-970; Gardner 5 max.; 93% min. act.

Kemamine® Q-1302C. [Witco/Humko] Dimethyl di-30% arachidyl-behenyl quat. ammonium chloride; cationic; emulsifier, corrosion inhibitor, textile and paper softener, household laundry uses; solid; 75% conc.

Kemamine® Q-1902C. [Witco/Humko] Dibehenyl/diarachidyl dimonium chloride; cationic; germicide, sanitizer, slimicide, antistat, textile softening agent, dyeing aid, corrosion inhibitor, emulsifier; Gardner 4 max. solid; m.w. 680; disp. in water; pH 9 max (5%); 75% act.

Kemamine® Q-2802C. [Witco/Humko] Dibehenyldimonium chloride; see Kemamine BQ-2802C; Gardner 4 max.; m.w. 690; pH 9 max. (5%); 75% min. act.

Kemamine® Q-6502C. [Witco/Humko] Dimethyl dicoconut ammonium chloride; cationic; see Kemamine BQ-9702C; Gardner 4 max. liq.; m.w. 465; disp. in water; pH 9 max. (5%); 75% act.

Kemamine® Q-6503 B. [Witco/Humko] Dicoco dimethyl ammonium chloride; cationic; see Kemamine Q 1902 C; liq.; 50% conc.

Kemamine® Q-9702C. [Witco/Humko] Quaternium-18; cationic; see Kemamine BQ-2802C; Gardner 2 max. solid; disp. in water; m.w. 575; pH 9 max. (5%); 75% min. act.

Kemamine® Q-9743C. [Witco/Humko] Tallow trimonium chloride; cationic; antistat, textile softening agent, dyeing aid, corrosion inhibitor, emulsifier; water-misc.; Gardner 4 max.; m.w. 335–355; pH 9 (5%); 65% min. act.

Kemamine® Q-9743CHGW. [Witco/Humko] Tallow trimonium chloride; cationic; antistat, textile softening agent, dyeing aid, corrosion inhibitor, emulsifier; water-misc.; Gardner 4 max.; m.w. 335–355; pH 9 (5%); 65% min. act.

Kemamine® S-190. [Witco/Humko] Di-90% arachidyl-behenyl sec. amine; cationic; industrial use, grease, corrosion inhibitor; solid; 90% conc.

Kemamine® S-650. [Witco/Humko] Dicoconut sec. amine; cationic; see Kemamine S-190; solid; 100% conc.

Kemamine® S-970. [Witco/Humko] Hydrog. di-tallow amine; cationic; see Kemamine S-190; solid; 100% conc.

Kemamine® T-1902D. [Witco/Humko] Dist. dimethyl-90% arachidyl-behenyl tert. amine; cationic; chemical intermediate for quat. ammonium derivs., acid scavenger in petrol. prods.; epoxy hardener, catalyst in mfg. of flexible PU foams; Gardner 1 max. liq.; 95% conc.

Kemamine® T-6501. [Witco/Humko] Methyl dicoconut tert. amine; cationic; see Kemamine T-1902D; Gardner 3 max. liq.; 95% conc.

Kemamine® T-6502D. [Witco/Humko] Dist. dimethyl cocamine; cationic; see Kemamine T-1902D; Gardner 1 max.; 95% conc.

Kemamine® T-9701. [Witco/Humko] Dihydrog. tallow methylamine; cationic; see Kemamine T-1902D; Gardner 3 max. liq.; 95% conc.

Kemamine® T-9702D. [Witco/Humko] Dist. dimethyl hydrog. tallow amine; cationic; see Kemamine T-1902D; Gardner 1 max. liq.; 95% conc.

Kemamine® T-9742D. [Witco/Humko] Dist. dimethyl tallow amine; cationic; see Kemamine T-1902D; Gardner 1 max. liq.; 95% conc.

Kemamine® T-9892D. [Witco/Humko] Dist. dimethyl oleamine; intermediate for quats. used in cosmetics and textiles; acid scavenger in petrol. prods.; Gardner 1 max. color; 95% tert. amine.

Kemamine® T-9902D. [Witco/Humko] Dist. dimethyl stearamine; cationic; see Kemamine T-1902D; Gardner 1 max. liq.; 95% conc.

Kemamine® T-9972D. [Witco/Humko] Dist. dimethyl soya amine; cationic; see Kemamine T-1902D; Gardner 1 max. liq.; 95% conc.

Kemamine® T-9992D. [Witco/Humko] Dist. dimethyl oleic-linolenic amine; intermediate for quats. used in cosmetics and textiles; acid scavenger in petrol. prods.; Gardner 2 max. color; 95% tert. amine.

Kemester® 104. [Witco/Humko] Methyl oleate; nonionic; emulsifier, emollient for cosmetics; lubricant for leather; carrier for agric. spray prods.; Gardner 2 max. liq.; acid no. 2 max.; sapon. no. 190–200; 100% conc

Kemester® 105. [Witco/Humko] Methyl oleate; intermediate in prod. of superamides, in metalworking lubricants, specialized solv.; opacifier, visc. control agent; Gardner 2 liq.; acid no. 4.0 max.; sapon. no. 194–203; 100% conc.

Kemester® 115. [Witco/Humko] Methyl oleate; see Kemester 105; Gardner 4 max.; acid no. 4.0 max.; sapon. no. 185–205.

Kemester® 143. [Witco/Humko] Methyl tallowate; see Kemester 105; also lubricant, plasticizer for cosmetics, leather, rubber prods.; Gardner 8 max.; acid no. 4.0 max.; sapon. no. 195 min.

Kemester® 205. [Witco/Humko] Methyl oleate; see Kemester 105; Gardner 1 liq.; acid no. 4.0 max.; sapon. no. 194–203; 100% conc.

Kemester® 213. [Witco/Humko] Methyl oleate/linoleate; see Kemester 104; Gardner 3 max.; acid no.

10.0 max.; sapon. no. 194–204.

Kemester® 226. [Witco/Humko] Methyl soyate; see Kemester 105; Gardner 4 max. liq.; acid no. 7.0 max.; sapon. no. 195–200; 100% conc.

Kemester® 325. [Witco/Humko] Methyl cocoate; see Kemester 105; Gardner 1 max. liq.; sapon. no. 246–256; 100% conc.

Kemester® 1000. [Witco/Humko] Triolein; emollient used in cosmetics, textiles, metalworking lubricants; Gardner 6 max. color; m.p. –8 C; acid no. 5 max.; sapon. no. 190–198.

Kemester® 1816. [Witco/Humko] Synthesized polyol ester; lubricant used for crankcase, turbine, compressor, gear, and metalworking lubricant formulations; Gardner 4 max.; visc. 5 cSt (100 C); pour pt. –40 C max.; acid no. 0.1 max.; flash pt. 250 C (COC).

Kemester® 1846. [Witco/Humko] Synthesized polyol ester; see Kemester 1816; Gardner 4 max.; visc. 4 cSt (100 C); pour pt. –60 C max.; acid no. 0.1 max.; flash pt. 250 C (COC).

Kemester® 2000. [Witco/Humko] Glycerol oleate; nonionic; emollient, emulsifier, stabilizer, wetting agent for textiles; Gardner 6 max. liq.; m.p. 25 C; acid no. 3 max.; sapon. no. 160–175; 100% conc.

Kemester® 2050. [Witco/Humko] Methyl eicosenate; intermediate in prod. of alkanolamides, in metalworking lubricants, as specialized solvs.; foam depressant and nutrient in fermentation; Gardner 14 max. color; acid no. 25 max.; sapon. no. 180–190.

Kemester® 3681. [Witco/Humko] Dioctyl dilinoleate; see Kemester 1816; also cosmetic emollient, pearling and bodying agent; Gardner 10 max.; visc. 12.5 cSt (100 C); pour pt. –40 C; acid no. 3.0 max.; flash pt. 310 C (COC).

Kemester® 3684. [Witco/Humko] Ditridecyl dimerate ester; see Kemester 3681; golden liq. (@ ambient temp.); acid no. 3.5.

Kemester® 3689. [Witco/Humko] Synthesized dimerate ester; see Kemester 1816; Gardner 10 max.; visc. 18 cSt (100 C); pour pt. –40 C; acid no. 4.0 max.; flash pt. 305 C (COC).

Kemester® 4000. [Witco/Humko] Butyl oleate; nonionic; emollient, wetting agent; plasticizer for textiles, leathers, elastomers; Gardner 2 max. liq.; m.p. 2 C; acid no. 2 max.; sapon. no. 164–172; 100% conc.

Kemester® 4516. [Witco/Humko] Methyl stearate; intermediate in prod. of alkanolamides, in metalworking lubricants, as specialized solvs.; foam depressant and nutrient in fermentation; Gardner 1 max. color; acid no. 3 max.; sapon. no. 192–202.

Kemester® 5221SE. [Witco/Humko] PEG-2 stearate SE; anionic; emollient, emulsifier, plasticizer, lubricant for cosmetics, rubber, textiles; Gardner 3 max. flake; m.p. 46–56 C; acid no. 95–105; sapon. no. 163–178; 100% conc.

Kemester® 5415. [Witco/Humko] Isobutyl stearate; emollient, lubricant for textiles, metalworking fluids; Gardner < 1 liq.; m.p. 21 C; acid no. 4 max.; sapon. no. 170–176.

Kemester® 5500. [Witco/Humko] Glyceryl stearate; nonionic; emollient, emulsifier; stabilizer, plasticizer, lubricant for cosmetic, paper, textile, and industrial uses; Gardner 3 max. bead, flake; m.p. 56–60 C; acid no. 3 max.; sapon. no. 164–177; 100% conc.

Kemester® 5510. [Witco/Humko] Butyl stearate; emollient for cosmetics; lubricant, plasticizer; Gardner < 1 max. color; m.p. 23 C; acid no. 2 max.; sapon. no. 170–176.

Kemester® 5651. [Witco/Humko] Synthesized adipate ester; see Kemester 1816; Gardner 2 max.; visc. 2.25 cSt (100 C); pour pt. –60 C; acid no. 0.1 max.; flash pt. 191 C (COC).

Kemester® 5653. [Witco/Humko] Synthesized adipate ester; see Kemester 1816; Gardner 2 max.; visc. 3.25 cSt (100 C); pour pt. –60 C; acid no. 0.1 max.; flash pt. 205 C (COC).

Kemester® 5654. [Witco/Humko] Ditridecyl adipate; see Kemester 3681; Gardner 2 liq.; f.p. –50 C; acid no. 0.1.

Kemester® 5721. [Witco/Humko] Tridecyl stearate; emollient, dye carrier, textile lubricant; Gardner 1 max. color; m.p. 8 C; acid no. 2 max.; sapon. no. 117–126.

Kemester® 5822. [Witco/Humko] Isocetyl stearate; emollient, plasticizer for cosmetics; Gardner 1 max. color; m.p. 0 C; acid no. 2 max.; sapon. no. 102–114.

Kemester® 6000. [Witco/Humko] Glyceryl stearate; nonionic; cosmetic and industrial emulsifier, emollient; plasticizer for elastomers; Gardner 3 max. bead; m.p. 57–60 C; acid no. 3 max.; sapon. no. 162–176; 100% conc.

Kemester® 6000E. [Witco/Humko] Glyceryl stearate; nonionic; emulsifier, stabilizer, plasticizer; bulk, beads; 100% conc.

Kemester® 6000SE. [Witco/Humko] Glyceryl stearate SE; anionic; emulsifier, emollient, lubricant, plasticizer for cosmetic, paper and textiles; Gardner 3 max. beads; m.p. 56–61 C; acid no. 10 max.; sapon. no. 140–156; 100% conc.

Kemester® 7018. [Witco/Humko] Methyl stearate; lubricant and plasticizer for cosmetics, textiles, leather prods.

Kemester® 9018. [Witco/Humko] Methyl stearate; see Kemester 105; Gardner 1 max.; acid no. 4.0 max.; sapon. no. 185–192.

Kemester® 9022. [Witco/Humko] Methyl behenate; see Kemester 105; Gardner 2 max.; acid no. 20 max.; sapon. no. 150–160.

Kemester® BE. [Witco/Humko] Behenyl erucate; industrial lubricant, cosmetic emollient; Gardner 1 solid; m.p. 40–44 C; acid no. 1; sapon. no. 80–95.

Kemester® CP. [Witco/Humko] Cetyl palmitate; see Kemester BE; wh. solid; m.p. 46–53 C; acid no. 1.

Kemester® DMP. [Witco/Humko] Dimethyl phthalate; emollient used in cosmetics, textiles, metalworking lubricants; APHA 10 max. color; m.p. 1 C; acid no. 0.1 max.

Kemester® E. [Witco/Humko] Erucamide; release agent for elastomers, plastics; solid.

Kemester® EE. [Witco/Humko] Erucyl erucate; see Kemester 3681; Gardner 2 liq. (@ ambient temp.); acid no. 1.

Kemester® EGDS. [Witco/Humko] Glycol distearate; intermediate in prod. of superamides, in metalworking lubricants, specialized solv.; industrial lubricant; opacifier, pearling additive, thickener for cosmetics and pharmaceuticals; Gardner 2 max. solid; m.p. 60–63 C; acid no. 6 max.; sapon. no. 190–200; 100% conc.

Kemester® EGMS. [Witco/Humko] Glycol stearate; see Kemester EGDS; Gardner 1 solid; m.p. 56–60 C; acid no. 1; sapon. no. 179–195.

Kemester® GMS (Powd.). [Witco/Humko] Glyceryl stearate; industrial lubricant; cosmetic emollient; wh. powd.; m.p. 58–59 C; acid no. 1.

Kemester® HMS. [Witco/Humko] Homosalate; uv absorber; sunscreen agent; used in cosmetic skin preps.; clear liq.; sp.gr. 1.049–1.053; ref. index 1.516–1.519; 99% min. purity.

Kemester® JO. [Witco/Humko] Erucyl arachidate; emollient for cosmetics; liq.

Kemester® MM. [Witco/Humko] Myristyl myristate; see Kemester 3681; wh. m.p. 38–40 C; acid no. 1.

Kemester® THFO. [Witco/Humko] Tetrahydrofurfuryl oleate; emollient used in cosmetics, textiles, metalworking lubricants; Gardner 1 max. color; m.p. < –20 C; acid no. 0.5 max.; sapon. no. 156.

Kemplex 100. [GAF] Tetrasodium EDTA; chelating agent, sequestrant for calcium, magnesium and common metals; liq.

Kempore® 60/14FF. [Olin] Azodicarbonamide-based; blowing agent for dynamic foaming processes, e.g., inj. molding, extrusion, and calendering; pale yel. powd.; sp.gr. 1.52 dens. 30 lb/ft^3; gas yield 200 ml/g.

Kempore® 60E. [Olin] 1,1′-Azobisformamide (azodicarbonamide), modified; self-dispersing blowing agent for PVC plastisol applics.; pale yel. powd.; sp.gr. 1.52; dens. 20 lb/ft^3; decomp.pt. 212 C (in air).

Kempore® 125E. [Olin] 1,1′-Azobisformamide (azodicarbonamide), modified; see Kempore 60E; pale yel. powd.; sp.gr. 1.52; dens. 17 lb/ft^3; decomp.pt. 212 C (in air).

Kempore® 125FF. [Olin] Azodicarbonamide-based; see Kempore 60/14FF; pale yel. powd.; sp.gr. 1.52; dens. 21 lb/ft^3; gas yield 200 ml/g.

Kempore® E. [Olin] Modified azodicarbonamide; self-dispersible blowing agent for easy incorporation into PVC and rubber compositions; powd.

Kempore® FF. [Olin] Modified azodicarbonamide; blowing agent for inj. molding and extrusion of plastics.

Kempore® MC. [Olin] Modified azodicarbonamide; blowing agent for inj. molding and extrusion of plastics; powd.

Kemstrene® 96.0%. [Witco/Humko] Refined glycerin USP; humectant, solv.; used in cosmetics, liq. soaps, confections, inks, and lubricants; intermediate used in polyester and PU formulations; sp.gr. 1.25165 min; 96.0% purity.

Kemstrene® 99.0%. [Witco/Humko] Refined glycerin USP; see Kemstrene 96.0%; sp.gr. 1.25945 min.; 99.0% purity.

Kemstrene® 99.5%. [Witco/Humko] Refined glycerin USP; see Kemstrene 96.0%; sp.gr. 1.26073 min.; 99.5% purity.

Kemstrene® 99.7%. [Witco/Humko] Refined glycerin USP; see Kemstrene 96.0%; sp.gr. 1.26124; 99.7% purity.

Kemstrene® High Gravity. [Witco/Humko] Refined glycerin; see Kemstrene 96.0%; sp.gr. 1.2587.

Ke-Mul 181. [Georgia-Pacific] Lignosulfonate; anionic; asphalt emulsifier and stabilizer; liq.; 45% conc.

Kencure MPP, MPPJ. [Kenrich] Curing agents; sp.gr. 1.19 and 1.13 resp.

Kenflex® A. [Kenrich] Alkylated naphthalene methylene oligomer; nonmigratory plasticizer offering good heat aging for Hypalon and neoprene; improves processing and smoothness of extrudates and calendered stocks, retards neoprene cryst.; lt. amber solid; m.w. 662; sp.gr. 1.08; m.p. 58–74 C; i.b.p. 204 C; flash pt. (COC) 240 ± 20 C; acid no. 0.3; iodine no. 65; 100% act.

Kenflex® N. [Kenrich] Alkylated naphthalene methylene oligomer; see Kenflex A; yel. to lt. brn. liq.; m.w. 472 ± 15; sp.gr. 1.01; i.b.p. 176 C; flash pt. (COC) 190 ± 20 C; pour pt. 2 C.

Ken Kem® CP-45. [Kenrich] Cumyl phenol; modifier for epoxy, furan, and phenolic resins; epoxy cure accelerator; chemical intermediate; Gardner > 18 liq.; sp.gr. 1.07; dist. range 340–730 F; flash pt. (COC) 250 F min.; pour pt. 30 F; hyd. no. 119; 45% act., 50% α-methyl styrene oligomer, 5% light end components.

Ken Kem® CP-99. [Kenrich] Cumyl phenol; see Ken Kem CP-45; sp.gr. 1.08; 99% cumyl phenol.

Kenplast® 3070. [Kenrich] Plasticizer; sp.gr. 1.01.

Kenplast® A-450. [Kenrich] Catalytic reformer petrol. distillate; plasticizer for urethanes, PVC, cellulosics; nonreactive diluent for epoxy; Gardner < 9 color; sp.gr. 1.0; visc. 25 cps; b.p. 410–550 F; flash pt. (TCC) 199 F.

Kenplast® AP-19. [Kenrich] Plasticizer; sp.gr. 0.86.

Kenplast® APK. [Kenrich] Plasticizer; chips and blocks; sp.gr. 0.86.

Kenplast® BG. [Kenrich] Plasticizer; sp.gr. 1.02.

Kenplast® ES-2. [Kenrich] Cumyl-phenyl acetate; plasticizer for urethanes; high flash pt. reactive diluent for epoxy, reducing odor and irritation; ideal for polyamide-cured epoxy floorings and coal tar pipe coatings; comonomer and impact modifier for phenolics; improves adhesion of vinyl plastisols coating metals and polar plastics; Gardner 9 liq.; sp.gr. 1.08; dens. 9.0 lb/gal; visc. 325 cps; dist. range 321–355 C; flash pt. (COC) 160 C; .pour pt. –12 C; acid no. < 1; sapon. no. 154.

Kenplast® ESB. [Kenrich] Cumyl-phenyl benzoate; primary plasticizer for PVC and PVC/nitrile; process aid for semirigid PVC extrudates; solvates conductive polyester and acrylic inks to improve impact; Gardner > 18 liq.; sp.gr. 1.13 dens. 9.4 lb/gal; visc. 15,000 cps; dist. range 250–427 C; flash pt. (COC) 171 C; pour pt. 16 C; acid no. < 1; sapon. no. 124; 70% act. in α-methyl styrene oligomer.

Kenplast® ESI. [Kenrich] Biscumylphenyl trimellitate; plasticizer, process aid for extrusion of PVC, urethanes; lubricant and process aid for filled PS, PC; reactive diluent and flow promoter for epoxy powd. coatings; prevents oxidative depolymerization in nylons; Gardner > 18 powd.; sp.gr. 1.18; dens. 9.8 lb/gal; soften. pt. 80 C; flash pt. (COC) 260 C; acid no. < 1; sapon. no. 172; 90% act. in α-methyl styrene oligomer.

Kenplast® ESN. [Kenrich] Cumylphenyl neodecanoate; primary plasticizer for PVC, PVC/nitrile; process aid for semirigid PVC extrusions; solvates and impact modifies conductive polyester and acrylic inks; Gardner > 18 liq.; sp.gr. 1.05; dens. 8.75 lb/gal; visc. 5800 cps; flash pt. (COC) 196 C; pour pt. 1.66 C; acid no. < 1; sapon. no. 130; 85% act. in α-methyl styrene oligomer.

Kenplast® G. [Kenrich] Mixed alkylated phenanthrenes; nonreactive primary plasticizer for dark PVC, nitrile, and urethane compds.; epoxy diluent; Gardner 9+ color; sp.gr. 1.06; visc. 25 cps; i.b.p. 510 F; flash pt. (COC) 290 F.

Kenplast® LG. [Kenrich] Plasticizer; sp.gr. 1.08.

Kenplast® LT. [Kenrich] Mixed dibasic ester; plasticizer for low temp. polymers incl. PVC, NBR, SBR; offers exc. heat stability and good electricals; coalescing agent, flow improver for latex alkyds and acrylics; Gardner 2 liq.; sp.gr. 0.9; visc. 20 cps; i.b.p. 320 F; flash pt. (COC) 260 F; sapon. no. 260.

Kenplast® PG. [Kenrich] Phenyl glycol ether; plasticizer; nonreactive diluent and flexibilizer for epoxies; Gardner 3 liq.; sp.gr. 1.11; b.p. 450 F; f.p. –50 F; flash pt. (COC) 230 F.

Kenplast® PPE. [Kenrich] Plasticizer; sp.gr. 1.20.

Kenplast® RD. [Kenrich] Plasticizer; sp.gr. 0.992.

Kenplast® RDN. [Kenrich] Plasticizer; sp.gr. 1.00.

Kenplast® RG. [Kenrich] Plasticizer; sp.gr. 1.09.

Ken-React® Series. [Kenrich] Titanate and zirconate coupling agents; monoalkoxy titanate (KR 7, 9S, 12, 26S, 33DS, 38S, 39DS, 44, TTS); chelate titanate (KR 133DS, 134S, 138S, 158FS, 212, 238S, 262ES); quat titanate (KR 138D, 158D, 238A, 238J, 238M, 238T, 262A); coordinate titanate (KR 41B, 46B, 55); cycloheteroatom titanate (KR OPP2, OPPR); coupling agents which sometimes also act as adhesion promoters (esp. KR 12, 44, 38S, 41B, 46B, 55, 138S, 158FS, TTS), antioxidants, antistats, antifoaming agents, accelerators, blowing agent activators (esp. KR 9S, 38S, TTS), catalysts (esp. KR 9S, 41B, TTS), curatives, corrosion inhibitors, disp. aids (esp. KR 9S, 12, 38S, 55, 58FS, 134BS, 134S, 138S, 212, 212 Quats, 238S, TTS), emulsifiers, flame retardants, foaming agents, grinding aids, hardeners, hydrophobes (most titanates), impact modifiers, internal lubes, metal primers, process aids, pigment intensifiers, peroxide activators, release agents, retarders, stabilizers, surfactants, suspension aids, thixotropes, wetting agents; recommended for aging, corrosion, and acid resistance are KR 44, 55, 38S, 138S, 238S, TTS; for conductivity: KR 46B; for thermoplastic polymers: KR 39DS, 134S, 238M, TTS; for polyolefins: KR 9S, 12, 38S, 138S, TTS; for elastomers: KR 7, 39DS, 134S; for thermosets and coatings: various KR grades; various forms.

Ken-React® 7 (KR 7). [Kenrich] Isopropyl dimethacryl isostearoyl titanate (monoalkoxy); see Ken-React Series; dk. red br. liq.; b.p. 250 F (initial); sol. in IPA, xylene, toluene, DOP, min. oil; insol. in water; sp.gr. 1.02 (16 C); visc. 220 cps; pH 5 (sat. sol'n.); flash pt. (TCC) 110 F; 95+% solids in IPA.

Ken-React® 9S (KR 9S). [Kenrich] Isopropyl tridodecylbenzenesulfonyl titanate (monoalkoxy); see Ken-React Series; transparent redsh. br. liq.; b.p. 130 F (initial); sol. in IPA, xylene, toluene, min. oil; reacts slowly in DOP; insol. in water; sp.gr. 1.08 (16 C); visc. 8000 cps; pH 2 (sat. sol'n.); flash pt. (TCC) 97 F; 88+% solids in IPA.

Ken-React® 12 (KR 12). [Kenrich] Isopropyl tri(dioctylphosphato) titanate (monoalkoxy); see Ken-React Series; translucent to translucent off-wh. liq.; b.p. 170 F (initial); sol. in IPA, xylene, toluene, min. oil; reacts slowly in DOP; insol. in water; sp.gr. 1.04 (16 C); visc. 1500 cps; pH 4.5 (sat. sol'n.); flash pt. (TCC) 150 F; 95+% solids in IPA.

Ken-React® 26S (KR 26S). [Kenrich] Isopropyl 4-aminobenzenesulfonyl di(dodecylbenzenesulfonyl) titanate (monoalkoxy); see Ken-React Series; grey liq.; b.p. 250 F (initial); sol. in IPA, xylene, toluene; reacts slowly in DOP; insol. in min. oil, water; sp.gr. 1.12 (16 C); visc. 30,000 cps; pH 6.5 (sat. sol'n.); flash pt. (TCC) 75 F; 95+% solids in IPA.

Ken-React® 33DS (KR 33DS). [Kenrich] Alkoxy trimethacryl titanate (monoalkoxy); see Ken-React Series; tan to red br. liq.; b.p. 170 F (initial); sol. in IPA, xylene, toluene; reacts slowly in DOP; insol. in min. oil, water; sp.gr. 1.11 (16 C); visc. 2500 cps; pH 3.5 (sat. sol'n.); flash pt. (TCC) 120 F; 78+% solids in IPA.

Ken-React® 38S (KR 38S). [Kenrich] Isopropyl tri(dioctylpyrophosphato) titanate (monoalkoxy); see Ken-React Series; yel. to amber liq.; b.p. 170 F (initial); sol. in IPA, xylene, toluene, min. oil; reacts slowly in DOP; insol. in water; sp.gr. 1.09 (16 C); visc. 1500 cps; pH 2 (sat. sol'n.); flash pt. (TCC) 100F; 99+% solids in IPA.

Ken-React® 39DS (KR 39DS). [Kenrich] Alkoxy triacryl titanate (monoalkoxy); see Ken-React Series; tan to red br. liq.; b.p. 220 F (initial); sol. in IPA, xylene, toluene; reacts slowly in DOP; insol. in min. oil, water; sp.gr. 1.07 (16 C); visc. 100 cps; pH 3 (sat. sol'n.); flash pt. (TCC) 180 F; 49+% solids in IPA.

Ken-React® 41B (KR 41B). [Kenrich] Tetraisopropyl di (dioctylphosphito) titanate (coordinate type); see Ken-React Series; yel. liq.; b.p. 160 F (initial); sol. in IPA, xylene, toluene, DOP, min. oil; insol. in water; sp.gr. 0.96 (16 C); visc. 15 cps; pH 6 (sat. sol'n.); flash pt. (TCC) 130 F; 98+% solids in IPA.

Ken-React® 44 (KR 44). [Kenrich] Isopropyl tri(N ethylamino-ethylamino) titanate (monoalkoxy); see Ken-React Series; yel. br. liq.; b.p. 180 F (initial); sol. in IPA; reacts slowly in DOP; insol. in xylene, toluene, min. oil, water; sp.gr. 1.2 (16 C); visc. 36,000 cps; pH 10 (sat. sol'n.); flash pt. (TCC) 130F; 95+% solids in IPA.

Ken-React® 46B (KR 46B). [Kenrich] Tetraoctyloxytitanium di (ditridecylphosphite) (coordinate type); see Ken-React Series; pale yel. liq.; b.p. 160 F (initial); sol. in IPA, xylene, toluene, DOP, min. oil; insol. in water; sp.gr. 0.92 (16 C); visc. 50 cps; pH 6 (sat. sol'n.); flash pt. (TCC) 180 F; 98+% solids in IPA.

Ken-React® 55 (KR 55). [Kenrich] Tetra (2, diallyloxymethyl-1 butoxy titanium di (di-tridecyl) phosphite (coordinate type); see Ken-React Series; yel. liq.; b.p. 120 F (initial); sol. in IPA, xylene, toluene, DOP, min. oil; insol. in water; sp.gr. 0.97 (16 C); visc. 50 cps; pH 5 (sat. sol'n.); flash pt. (TCC) 200 F; 78+% solids in IPA.

Ken-React® 133DS (KR 133DS). [Kenrich] Titanium dimethacrylate oxyacetate (oxyacetate chelate type); see Ken-React Series; amber liq.; b.p. 272 F (initial); limited sol. in IPA, xylene, toluene, min. oil; reacts slowly in DOP; insol. in water; sp.gr. 1.06 (16 C); visc. 100 cps; pH 2.5 (sat. sol'n.); flash pt. (TCC) 150 F; 45+% solids in IPA.

Ken-React® 134S (KR 134S). [Kenrich] Titanium di (cumylphenylate) oxyacetate (oxyacetate chelate type); see Ken-React Series; red liq.; b.p. 180 F (initial); sol. in xylene, toluene, DOP; insol. in IPA, min. oil, water; sp.gr. 1.14 (16 C); visc. 8000 cps; pH 5 (sat. sol'n.); flash pt. (TCC) 195F; 95+% solids in IPA.

Ken-React® 138D (KR 138D). [Kenrich] KR 138S and 2-dimethylamino methyl propanol (quat type); see Ken-React Series; tan br. liq.; b.p. 180 F (initial); sol. in IPA, xylene, toluene, DOP, water; insol. in min. oil; sp.gr. 1.06 (16 C); visc. < 1000 cps; pH 8 (sat. sol'n.); flash pt. (TCC) 95 F; 94+% solids in IPA.

Ken-React® 138S (KR 138S). [Kenrich] Titanium di (dioctylpyrophosphate) oxyacetate (oxyacetate chelate type); see Ken-React Series; pale yel. liq.; b.p. 160 F (initial); sol. in IPA, xylene, toluene, DOP; insol. in min. oil, water; sp.gr. 1.12 (16 C); visc. 1000 cps; pH 3 (sat. sol'n.); flash pt. (TCC) 100 F; 99+% solids in IPA.

Ken-React® 158D (KR 158D). [Kenrich] KR 158FS and 2-dimethylamino methyl propanol (quat type); see Ken-React Series; pale yel. liq.; b.p. 162 F (initial); sol. in IPA, water; limited sol. in xylene, toluene; reacts slowly in DOP; insol. in min. oil; sp.gr. 1.08 (16 C); visc. 400 cps; pH 7.2 (sat. sol'n.); flash pt. (TCC) 85 F; 92% solids in IPA.

Ken-React® 158FS (KR 158FS). [Kenrich] Titanium di (butyl, octyl pyrophosphate) di (dioctyl, hydrogen phosphite) oxyacetate (oxyacetate chelate type); see Ken-React Series; yel.-tan liq.; b.p. 170 F (initial); sol. in IPA, xylene, toluene, DOP; insol. in min. oil, water; sp.gr. 1.16 (16 C); visc. ≤200 cps; pH 3 (sat. sol'n.); flash pt. (TCC) 110; 80+% solids in IPA.

Ken-React® 212 (KR 212). [Kenrich] Di (dioctylphosphato) ethylene titanate (A, B ethylene chelate type); see Ken-React Series; pale orange-red liq.; b.p. 185 F (initial); sol. in IPA, xylene, toluene, DOP, min. oil; insol. in water; sp.gr. 0.98 (16 C); visc. 300 cps; pH 4.8 (sat. sol'n.); flash pt. (TCC) 70 F; 75+% solids in IPA.

Ken-React® 238A (KR 238A). [Kenrich] KR 238S and acrylic functional amine (quat type); see Ken-React Series; yel. liq.; b.p. 170 F (initial); sol. in IPA, water; insol. in xylene, toluene, DOP, min. oil; sp.gr. 1.03 (16 C); visc. < 300 cps; pH 8 (saturated sol'n.); flash pt. (TCC) 95 F; 95+% solids in IPA.

Ken-React® 238J (KR 238J). [Kenrich] KR 238S and methacryl functional amine (quat type); see Ken-React Series; amber/red liq.; b.p. 220 F (initial); sol. in IPA, water; insol. in xylene, toluene, DOP, min. oil; sp.gr. 1.06 (16 C); visc. < 400 cps; pH 7 (saturated sol'n.); flash pt. (TCC) 85 F; 88+% solids in IPA.

Ken-React® 238M (KR 238M). [Kenrich] KR 238S and methacrylic functional amine (quat type); see Ken-React Series; yel. to amber liq.; b.p. 170 F (initial); sol. in IPA, xylene, toluene, DOP; insol. in min. oil, water; sp.gr. 1.04 (16 C); visc. < 400 cps; pH 5.5 (sat. sol'n.); flash pt. (TCC) 95 F; 94+% solids in IPA.

Ken-React® 238S (KR 238S). [Kenrich] Di (dioctylpyrophosphato) ethylene titanate (A, B ethylene chelate type); see Ken-React Series; pale orange-red liq.; b.p. 160 F (initial); sol. in IPA, xylene, toluene, DOP, min. oil; insol. in water; sp.gr. 1.08 (16 C); visc. 800 cps; pH 3 (sat. sol'n.); flash pt. (TCC) 70 F; 78+% solids in IPA.

Ken-React® 238T (KR 238T). [Kenrich] KR 238S and triethylamine (quat type); see Ken-React Series; pale yel. to orange/red liq.; b.p. 170 F (initial); sol. in IPA, water; insol. in xylene, toluene, DOP, min. oil; sp.gr. 1.02 (16 C); visc. < 200 cps; pH 7 (sat. sol'n.); flash pt. (TCC) 100 F; 95+% solids in IPA.

Ken-React® 262A (KR 262A). [Kenrich] KR 262ES and acrylic functional amine (quat type); see Ken-React Series; pale yel. liq.; b.p. 170 F (initial); sol. in IPA, water; limited sol. in xylene, toluene, DOP; insol. in min. oil; sp.gr. 1.11 (16 C); visc. < 5000 cps; pH 6.5 (sat. sol'n.); flash pt. (TCC) 95 F; 88+% solids in IPA.

Ken-React® 262ES (KR 262ES). [Kenrich] Di (butyl, methyl pyrophosphato) ethylene titanate di (dioctyl, hydrogen phosphite) (A, B ethylene chelate type); see Ken-React Series; yel.-tan liq.; b.p. 170 F (initial); sol. in IPA, xylene, toluene, DOP, min. oil; insol. in water; sp.gr. 1.17 (16 C); visc. 100 cps; pH 2 (sat. sol'n.); flash pt. (TCC) 85 F; 82+% solids in IPA.

Ken-React® OPP2 (KR OPP2). [Kenrich] Dicyclo (dioctyl) pyrophosphato, titanate (cycloheteroatom type); see Ken-React Series; ylsh.-br. liq.; b.p. 210 F (initial); sol. in xylene, toluene; limited sol. in min. oil; reacts slowly in IPA, DOP, water; sp.gr. 1.13 (16 C); visc. 7000 cps; pH 5 (sat. sol'n.); flash pt. (TCC) 140 F; 66% solids in IPA.

Ken-React® OPPR (KR OPPR). [Kenrich] Dicyclo (dioctyl) pyrophosphato dioctyl titanate (cycloheteroatom type); see Ken-React Series; ylsh.-br. liq.; b.p. 280 F (initial); sol. in xylene, toluene; limited sol. in min. oil; reacts slowly in IPA, DOP, water; sp.gr. 1.06 (16 C); visc. 110 cps; pH 6 (sat. sol'n.); flash pt. (TCC) 160 F; 66% solids in IPA.

Ken-React® TTS (KR TTS). [Kenrich] Isopropyl

triisostearoyl titanate (monoalkoxy); see Ken-React Series; transparent rdsh.-br. liq.; b.p. 300 F (initial); sol. in IPA, xylene, toluene, DOP, min. oil; insol. in water; sp.gr. 0.95 (16 C); visc. 125 cps; pH 5.5 (sat. sol'n.); flash pt. (TCC) 200+ F; 95+% solids in IPA.

Kerasol. [Croda] Hydrolyzed animal keratin; proteinic conditioner for hair and nail care prods.; slightly hazy amber liq., char. odor; m.w. 125,000; sol. in water; pH 5–7; 15% act. aq. sol'n.

Kera-Tein 1000. [Amerchol] Hydrolyzed animal keratin; conditioner for hair care prods.; water sol'n.

Kera-Tein 1000AS. [Amerchol] Ethyl hydrolyzed animal keratin; used in hair spray resins; imparts gloss and feel; alcoholic sol'n.

Kera-Tein 1000SD. [Amerchol] Hydrolyzed animal keratin; used in hair care prods.; powd.

Kerensim L 10 P. [Henkel] Alkyl polyethoxy esters and ethers; antistatic, humectant fiber treatment; liq.

Kerinol 2012 F. [Henkel] Fatty acid alkanolamide; nonionic; detergent, foam stabilizer and thickener; liq.; 100% conc.

Kerinol AA 62. [Henkel] Fatty acid alkanolamide; nonionic; see Kerinol 2012 F; paste; 100% conc.

Keripon NC. [Sandoz] Fatty acid ester, glycol; nonionic; wetting, rewetting; antistat; leather degreaser and penetrant with fat liquors; used in paper mfg.; liq.; 50% conc.

Kerobit® BPD. [BASF AG] N,N′-di-sec-butyl-p-phenylene diamine; antioxidant for crude oil distillates; liq.; 100% conc.

Kerobit® COM. [BASF AG] N,N-di-sec-butyl-p-phenylene diamine and N,N′-disalicylidene-1,2-propane diamine; antioxidant and metal deactivator for crude oil distillates; liq.; 99% conc.

Keroflux® 5323, 5486, H, M. [BASF AG] Low m.w. wax; cold flow improver for middle distillates; liq. disp. in solv.

Keromet MD 60, MD 80. [BASF AG] N,N′-Disalicylidene-1,2-propane diamine; chelating agent and metal deactivator for fuels and lubricating oils; liq.; 60 and 80% conc. resp.

Keromet MD 100. [BASF AG] N,N′-Disalicylidene-1,2-propane diamine; see Keromet MD 60; solid; 100% conc.

Kerostat® 5009. [BASF AG] Salts of amino carboxylic acid; antistatic additive for fuels; liq.

Kessco® 639. [Stepan] IPP, IPM, isopropyl stearate; nonionic; emollient, lubricant, solv. for cosmetics and toiletries formulations; APHA 30 liq.; f.p. 7 C; b.p. 170 C; sp.gr. 0.849–0.855; dens. 7.1 lb/gal; visc. 5.9 cps; acid no. 1.0 max.; ref. index 1.436; flash pt. 305 F (COC).

Kessco® 653. [Stepan] Cetyl palmitate; nonionic; syn. spermaceti wax, emollient, base material, thickener, visc. booster for pharmaceutical and cosmetic prods.; lubricant for metalworking fluids; wh. flakes; sol. in boiling alcohol, ether, chloroform, other waxes, oils, hydrocarbons; insol. in water; m.p. 51–55 C; sapon. no. 109–117; acid no. 2.0 max.

Kessco® 654. [Stepan] Syn. spermaceti wax; nonionic; thickener for cosmetics.

Kessco® 874. [Stepan] Pentaerythritol tetracaprylate/caprate; heat-stable lubricant for textile and metalworking compds.; clear liq.; f.p. 21 F; flash pt. 500 F; acid no. 1.0; sapon. no. 350; 100% act.

Kessco® 887. [Stepan] Trimethylolpropane tricaprylate/caprate; see Kessco 874; clear liq.; f.p. –60 F; flash pt. 500 F; acid no. 1.0; sapon. no. 310; 100% act.

Kessco® 891. [Stepan] Alkoxy monooleate/stearate; lubricant and emulsifier for textile finishing and metalworking applics.; coupling agent; water- and oil-sol.; yel. liq.; HLB 12.8; acid no. 1.0; sapon. no. 65; 100% act.

Kessco® 894. [Stepan] Alkoxy dioleate/stearate; lubricant and emulsifier for textile and metalworking applics.; coupling agent; yel. liq.; water- and oil-sol.; HLB 8.8; acid no. 1.0; sapon. no. 100; 100% act.

Kessco® 3283. [Stepan] Esters of veg. oil fatty acids; SE, food grade emulsifier, antioxidant solv., defoamer; liq.

Kessco® Acetin. [Stepan] Glycerol ester of fatty acid; nonionic; emulsifier, stabilizer, emollient, opacifying and bodying agent for cosmetics; APHA 75 max. liq.; sp.gr. 1.188–1.198; dens. 9.9 lb/gal; visc. 66.8 cps; f.p. –30 C; flash pt. (COC) 305 F.

Kessco® BS. [Stepan] Butyl stearate; nonionic; emollient, lubricant, solv., plasticizer, solubilizer for cosmetics, syn. fiber spin finishes; APHA 20 and 100 resp. liq.; sp.gr. 0.850–0.860 and 0.853–0.859; dens. 7.1 lb/gal; visc. 7 cps; f.p. 18–19 C; b.p. 200 C (4 mm); flash pt. (COC) 370 F; ref. index 1.442.

Kessco® BSC. [Stepan] Butyl stearate; emollient in cosmetics; skin conditioning agent; plasticizer; lubricant for textile and metalworking compds.

Kessco® Butoxyethyl Oleate. [Stepan] Nonionic; plasticizer for rubber and PVC; APHA 250 liq.; sp.gr. 0.881–0.889; dens. 7.4 lb/gal; visc. 10 cps; f.p. –37 C; b.p. 220 C (4mm); ref. index 1.449.

Kessco® Butoxyethyl Stearate. [Stepan] Nonionic; plasticizer for rubber and PVC; APHA 60 liq.; sp.gr. 0.874–0.880; dens. 7.3 lb/gal; visc. 11 cps; f.p. 12 C; b.p. 218 C (4 mm); flash pt. (COC) 380 F; ref. index 1.444.

Kessco® Butyl Oleate. [Stepan] Nonionic; emollient, lubricant, solv., plasticizer, solubilizer for creams and lotions, aerosol hair sprays, lipsticks, syn. fiber spin finishes, rubber and PVC prods.; APHA 200 liq.; sp.gr. 0.861–0.869; dens. 7.2 lb/gal; visc. 8 cps; f.p. < –15 C; b.p. 185 C (4 mm); flash pt. (COC) 380 F; ref. index 1.450.

Kessco® Diacetin. [Stepan] Glycerol diacetate; see Kessco Acetin; liq.; sp.gr. 1.173–1.183; dens. 9.8 lb/gal; visc. 36 cps; f.p. -35 C; flash pt. (COC) 295 F.

Kessco® Dibutoxyethyl Phthalate. [Stepan] Nonionic; plasticizer for rubber and PVC prods.; APHA 100 liq.; sp.gr. 1.054–1.060; dens. 8.8 lb/gal; visc. 30 cps; f.p. –55 C; flash pt. (COC) 400 F; ref. index 1.482.

Kessco® Diethylene Glycol Monostearate. [Stepan] PEG-2 stearate; nonionic; opacifier and bodying agent in cosmetic creams and lotions; imparts soft velvety feel; flakes; m.p. 44.5–47.5 C; acid no. 5.0 max.; flash pt. 395 F (COC).

Kessco® Diglycol Stearate Neutral. [Stepan] PEG-

2 stearate; see Kessco Diethylene Glycol Monostearate; flakes; m.p. 42–48 C; acid no. 103 max.; flash pt. 360 F (COC).

Kessco® Diglycol Stearate SE. [Stepan] PEG-2 stearate SE, stearic acid; see Kessco Diethylene Glycol Monostearate; flakes; m.p. 48–53 C; acid no. 103 max.; flash pt. 345 F (COC).

Kessco® EGAS. [Stepan] Glycol stearate and stearamide AMP; pearlescent and bodying agent for shampoos, liq. hand soaps; imparts a soft, smooth skin feel to formulations; wh. to cream flakes; m.p. 57 C; acid no. 5.0 max.

Kessco® EGDS. [Stepan] Glycol distearate; nonionic; opacifier in liq. cream shampoos; pearlizer, emollient, emulsifier; suggested for use when no additional visc. response is desired, e.g., high-solids formulations; wh. to cream flake; HLB 1.5; m.p. 61 C; acid no. 10.0 max.; flash pt. 390 F (COC).

Kessco® EGMS. [Stepan] Glycol stearate; pearlescent in shampoos and liq. hand soaps; bodying agent and emulsion stabilizer; wh. to cream flakes; HLB 2.9; m.p. 58 C; acid no. 2.0 max.; flash pt. 390 F (COC).

Kessco® EGMS-70. [Stepan] Glycol stearate; see Kessco Ethylene Glycol Distearate; flakes; m.p. 52–56 C; acid no. 2.0 max.; flash pt. 370 F (COC).

Kessco® GDL. [Stepan] Glyceryl dilaurate; emollient for lotions; wh. solid; HLB 4.0; m.p. 30 C; acid no. 5.0 max.; 100% act.

Kessco® GDS 386F. [Stepan] Glyceryl distearate; emulsifier, opacifier, thickener for creams, lotions, antiperspirants, hair care prods., sunscreens; wh. flake; HLB 2.4; m.p. 56 C; acid no. 5 max.

Kessco® Glycerol Dioleate. [Stepan] Glyceryl dioleate; see Kessco Acetin; yel. to amber liq.; sp.gr. 0.923–0.929; dens. 7.7 lb/gal; visc. 90 cps; f.p. O C; flash pt. (COC) 520 F; acid no. 5.0 max.

Kessco® Glycerol Monostearate 860. [Stepan] Glyceryl stearate; see Kessco Acetin; also as tablet coatings in pharmaceuticals; wh. flakes; m.p. 58.5–61.5 C; flash pt. (COC): 450F; acid no. 3.0 max.

Kessco® Glycerol Monostearate DH-1. [Stepan] Glyceryl stearate; see Kessco Acetin; also as tablet coatings for pharmaceuticals; wh. flakes; m.p. 55–58 C; flash pt. (COC): 410 F; acid no. 5.0 max.

Kessco® Glycerol Monostearate Pure. [Stepan] Glyceryl stearate; see Kessco Acetin; also as tablet coatings for pharmaceuticals; wh. flakes; m.p. 56.5–58.5 C; flash pt. (COC): 410 F; acid no. 3.0 max.

Kessco® GMC-8. [Stepan] Glyceryl caprylate/caprate; solubilizer and emulsifier for vitamins, flavors, and medicaments; lt. yel. liq.; HLB 8.3; acid no. 1.5 max.

Kessco® GML. [Stepan] Glyceryl laurate; primary emulsifier for w/o emulsions; emollient for cosmetic formulations; wh. solid; HLB 4.9; m.p. 46 C; flash pt. (COC) 425 F; acid no. 4.0 max.

Kessco® GMO. [Stepan] Glyceryl oleate; w/o emulsifier; emollient, spreading agent in bath oil; pigment dispersant in makeup; slip agent in moisturizing creams; lubricant for textiles, metalworking compds.; yel. liq.; sp.gr. 0.923–0.953; dens. 7.9 lb/gal; visc. 204 cps; HLB 3.8; m.p. 20 C; flash pt. (COC) 435 F; acid no. 3.0 max.

Kessco® GMS. [Stepan] Glyceryl stearate; emulsifier, opacifier, boyding agent for creams, lotions, antiperspirants, hair care prods., sunscreens; lubricant for textiles, metalworking compds.; wh. flakes; HLB 3.8; m.p. 56 C; acid no. 3 max.; 100% act.

Kessco® GMS SE. [Stepan] Glyceryl stearate; self-emulsifying; see Kessco Acetin; acid stable grade; also as tablet coatings for pharmaceuticals and primary emulsifiers; wh. to cream flakes; m.p. 54–58 C; acid no. 20.0 max.

Kessco® GMS SE AS. [Stepan] Glyceryl stearate and PEG-100 stearate; nonionic; emulsifier, SE cream base, hair and skin conditioner providing good electrolyte stability; wh. to cream flake; HLB 11.2; m.p. 53 C; acid no. 3 max.

Kessco® IBP. [Stepan] Isobutyl palmitate; see Kessco BSC; liq.; 100% act.

Kessco® IBS. [Stepan] Isobutyl stearate; slip agent for min. oil formulations; wetting agent for pigments; used in lipsticks, bath oils, nail polishes and removers, skin cleansers, creams, lotions; lubricant for textiles, metalworking compds.; APHA 35 liq.; m.p. 15 C; acid no. 1.0 max.; 100% act.

Kessco® ICS. [Stepan] Isocetyl stearate; emollient for makeup; APHA 200 max. liq.; m.p. 0 C; acid no. 2 max.; 100% act.

Kessco® IPM. [Stepan] IPM; emollient, blending agent, solubilizer, vehicle for highly pigmented prods.; used in pre- and after-shave prods., liq. and cream makeup, lipsticks, bath oils, creams, lotions, and hair preps.; lubricant for spin finish, coning oils, carding, and dye bath in textile industry; APHA 20 max. liq.; m.p. –3 C; acid no. 1.0; 100% act.

Kessco® IPP. [Stepan] IPP; see Kessco IPM; APHA 20 liq.; m.p. 13 C; acid no. 1 max.; 100% act.

Kessco® Isobutyl Palmitate. [Stepan] Nonionic; emollient, lubricant for syn. fiber spin finishes; APHA 60 liq.; sp.gr. 0.850–0.856; dens. 7.1 lb/gal; visc. 6.7 cps; f.p. 13 C; b.p. 170 C (4 mm); flash pt. (COC) 325 F; ref. index 1.437.

Kessco® Isobutyl Stearate. [Stepan] Isobutyl stearate; emollient, lubricant, and solv. used in creams and lotions; plasticizer in aerosol hair sprays, cosolv. in nonaq. preparations; dye solubilizer in lipstick, and syn. fiber finishes; speciality plasticizer in rubber and PVC prods.; APHA 35 max. liq.; f.p. 15 C; b.p. 200 C; sp.gr. 0.849–0.855; dens. 7.1 lb/gal; visc. 8.5 cps; ref. index 1.441; flash pt. 360 F (COC).

Kessco® Isocetyl Stearate. [Stepan] Nonionic; emollient; APHA 200 liq.; sp.gr. 0.853–0.859; dens. 7.1 lb/gal; visc. 32 cps; f.p. 0 C; flash pt. (COC) 450 F; acid no. 3 max.; ref. index 1.452.

Kessco® Isopropyl Myristate. [Stepan] Nonionic; emollient, lubricant, solv., plasticizer for cosmetics; APHA 20 liq.; sp.gr. 0.849–0.855; dens. 7.1 lb/gal; visc. 4.8 cps; f.p. –3 C; b.p. 160 C (4 mm); flash pt. (COC) 305 F; acid no. 1 max.; ref. index 1.433.

Kessco® Isopropyl Palmitate. [Stepan] Nonionic; emollient, lubricant, solv., plasticizer for cosmetics;

APHA 20 liq.; sp.gr. 0.849–0.855; dens. 7.1 lb/gal; visc. 6.7 cps; f.p. 13 C; b.p. 170 C (4 mm); flash pt. (COC) 325 F; acid no. 1 max.; ref. index 1.437.

Kessco® Myristyl Myristate. [Stepan] Emollient melting at body temp.; waxy solid; 100% act.

Kessco® Octyl Isononoate. [Stepan] Octyl isononanoate; emollient with skin breathing properties for creams, lotions, makeup, lipsticks, antiperspirants; APHA 20 max. liq.; f.p. < –50 C; b.p. 150 C; sp.gr. 0.853–0.859; dens. 7.2 lb/gal; visc. 4.3 cps; acid no. 1.0 max.; ref. index 1.434; flash pt. 265 F (COC).

Kessco® Octyl Oxystearate. [Stepan] Emollient for prevention of skin defatting; liq.; 100% act.

Kessco® OP. [Stepan] Octyl palmitate; emollient for dry, lt., silky skin feel; enhances gloss in stick makeup, hair grooms, suntan prods., bath oils; binder for pressed powds.; lubricant for textile and metalworking compds.; APHA 25 max. liq.; m.p. 0 C; acid no. 1.5 max.; 100% act.

Kessco® PEG Series. [Stepan] Polyethylene glycol esters; nonionic; surfactants used in cosmetic, pharmaceutical, food, agric., plastic, and other industries; thickener for shampoos and cream rinses; solubilizer in bath oils and fragrance compositions; in ointments and nonaq. hair preparations to provide washability; emollient, opacifier, spreading agent, wetting agent, dispersant.

Kessco® PEG 200 Dilaurate. [Stepan] PEG-4 dilaurate; see Kessco PEG series; lt. yel. liq.; f.p. < 9 C; sol. in IPA, acetone, CCl_4, ethyl acetate, toluol, IPM, wh. oil; water-disp.; sp.gr. 0.951; dens. 7.9 lb/gal; HLB 5.9; acid no. 10.0 max.; sapon. no. 176–186; flash pt. (COC) 460 F; fire pt. 510 F.

Kessco® PEG 200 Dioleate. [Stepan] PEG-4 dioleate; lt. amber liq.; f.p. < –15 C; sol. in naptha, kerosene, IPA, acetone, CCl_4, ethyl acetate, toluol, IPM, peanut oil, wh. oil; water-disp.; sp.gr. 0.942; dens. 7.9 lb/gal; HLB 6.0; acid no. 10.0 max; sapon. no. 148–158; pH 5.0 (3%); flash pt. 545 F (COC).

Kessco® PEG 200 Distearate. [Stepan] PEG-4 distearate; see Kessco PEG series; wh. to cream soft solid; sol. in naptha, kerosene, IPA, acetone, CCl_4, ethyl acetate, toluol, IPM, peanut oil, wh. oil; water-disp.; sp.gr. 0.9060 (65 C); HLB 5.0; m.p. 34 C; acid no. 10.0 max.; sapon. no. 153–162; pH 5.0 (3%); flash pt. 475 F (COC).

Kessco® PEG 200 Monolaurate. [Stepan] PEG-4 laurate; see Kessco PEG series; lt. yel. liq.; f.p. < 5 C; sol. in IPA, acetone, CCl_4, ethyl acetate, toluol, IPM, water-disp.; sp.gr. 0.985; dens. 8.2 lb/gal; HLB 9.8; acid no. 5.0 max.; sapon. no. 132–142; pH 4.5; flash pt. 385 F (COC).

Kessco® PEG 200 Monooleate. [Stepan] PEG-4 oleate; see Kessco PEG series; lt. amber liq.; f.p. < –15 C; sol. in IPA, acetone, CCl_4, ethyl acetate, toluol, water-disp.; sp.gr. 0.973; dens. 8.1 lb/gal; HLB 8.0; acid no. 5.0 max.; sapon. no. 115–124; pH 5.0; flash pt. 395 F (COC).

Kessco® PEG 200 Monostearate. [Stepan] PEG-4 stearate; see Kessco PEG series; wh. to cream soft solid; sol. in IPA, acetone, CCl_4, ethyl acetate, toluol, IPM, peanut oil; water-disp.; sp.gr. 0.9360 (65 C); HLB 7.9; m.p. 31 C; acid no. 5.0 max.; sapon. no. 120–129; flash pt. 410 F (COC).

Kessco® PEG 300 Dilaurate. [Stepan] PEG-6 dilaurate; see Kessco PEG series; lt. yel. liq.; f.p. < 13 C; sol. in naptha, IPA, acetone, toluol, IPM; water-disp.; sp.gr. 0.975; dens. 8.1 lb/gal; HLB 9.8; acid no. 10.0 max.; sapon. no. 148–158; flash pt. 475 F (COC).

Kessco® PEG 300 Dioleate. [Stepan] PEG-6 dioleate; see Kessco PEG series; lt. amber liq.; f.p. < –5 C; sol. in IPA, acetone, CCl_4, ethyl acetate, toluol, IPM, peanut oil; water-disp.; sp.gr. 0.962; dens. 8.0 lb/gal; HLB 7.2; acid no. 10.0 max.; sapon. no. 128–137; pH 5.0; flash pt. 510 F (COC).

Kessco® PEG 300 Distearate. [Stepan] PEG-6 distearate; see Kessco PEG series; wh. to cream soft solid; sol. in naptha, kerosene, IPA, acetone, CCl_4, ethyl acetate, toluol, IPM, peanut oil, wh. oil; water-disp.; HLB 6.5; m.p. 32 C; acid no. 10.0 max.; sapon. no. 130–139; pH 5.0.

Kessco® PEG 300 Monolaurate. [Stepan] PEG-6 laurate; see Kessco PEG series; lt. yel. liq.; f.p. < 8C; sol. in IPA, acetone, CCl_4, ethyl acetate, toluol; water-disp.; sp.gr. 1.011; dens. 8.4 lb/gal; HLB 11.4; acid no. 5.0 max.; sapon. no. 104–114; pH 4.5; flash pt. 445 F (COC).

Kessco® PEG 300 Monooleate. [Stepan] PEG-oleate; see Kessco PEG series; lt. amber liq.; f.p. < –5 C; sol. in IPA, acetone, CCl_4, ethyl acetate, toluol; water-disp.; sp.gr. 0.998; dens. 8.3 lb/gal; HLB 9.6; acid no. 5.0 max.; sapon. no. 94–102; pH 5.0 (3% disp.); flash pt. 450 F (COC).

Kessco® PEG 300 Monostearate. [Stepan] PEG-6 stearate; see Kessco PEG series; wh. to cream soft solid; sol. in IPA, acetone, CCl_4, ethyl acetate, toluol, IPM, peanut oil; water-disp.; sp.gr. 0.9660 (65 C); HLB 9.7; m.p. 28 C; acid no. 5.0 max.; sapon. no. 97–105; pH 5.0 (3% disp.); flash pt. 475 F (COC).

Kessco® PEG 400 Dilaurate. [Stepan] PEG-8 dilaurate; see Kessco PEG series; lt. yel. liq.; f.p. 18 C; sol. in naptha, IPA, acetone, CCl_4, ethyl acetate, toluol, IPM, peanut oil; water-disp.; sp.gr. 0.990; dens. 8.3 lb/gal; HLB 9.8; acid no. 10.0 max.; sapon. no. 127–137; flash pt. 480 F (COC).

Kessco® PEG 400 Dioleate. [Stepan] PEG-8 dioleate; see Kessco PEG series; lt. amber liq.; f.p. < 7 C; sol. in naptha, IPA, acetone, CCl_4, ethyl acetate, toluol, IPM, peanut oil; water-disp.; sp.gr. 0.977; dens. 8.1 lb/gal; HLB 8.5; acid no. 10.0 max.; sapon. no. 113–122; pH 5.0; flash pt. 520 F (COC).

Kessco® PEG 400 Distearate. [Stepan] PEG-8 distearate; see Kessco PEG series; wh. to cream soft solid; sol. in naptha, IPA, acetone, CCl_4, ethyl acetate, toluol, IPM, peanut oil, wh. oil; water-disp; sp.gr. 0.9390 (65 C); HLB 8.0; m.p. 36 C; acid no. 10.0 max.; sapon. no. 115–124; pH 5.0 (3% disp.); flash pt. 500 F (COC).

Kessco® PEG 400 Monolaurate. [Stepan] PEG-8 laurate; see Kessco PEG series; lt. yel. liq.; f.p. 12 C; sol. in water, IPA, acetone, CCl_4, ethyl acetate, toluol; sp.gr. 1.028; dens. 8.6 lb/gal; HLB 13.1; acid no. 5.0 max.; sapon. no. 86–96; flash pt. 475 F (COC).

Kessco® PEG 400 Monooleate. [Stepan] PEG-8 oleate; see Kessco PEG series; lt. amber liq.; f.p. < 10 C; sol. in IPA, acetone, CCl_4, ethyl acetate, toluol; water-disp.; sp.gr. 1.013; dens. 8.4 lb/gal; HLB 11.4; acid no. 5.0 max.; sapon. no. 80–89; pH 5.0 (3% disp.); flash pt. 510 F (COC).

Kessco® PEG 400 Monostearate. [Stepan] PEG-8 stearate; see Kessco PEG series; wh. to cream soft solid; sol. in IPA, acetone, CCl_4, ethyl acetate, toluol; water-disp.; sp.gr. 0.9780 (65 C); HLB 11.6; m.p. 32 C; acid no. 5.0 max.; sapon. no. 83–92; pH 5.0 (3% disp.); flash pt. 480 F (COC).

Kessco® PEG 600 Dilaurate. [Stepan] PEG-12 dilaurate; see Kessco PEG series; f.p. 24 C; sol. in IPA, acetone, CCl_4, ethyl acetate, toluol, IPM; water-disp.; sp.gr. 0.9820 (65 C); HLB 12.2; acid no. 10.0 max.; sapon. no. 102–112; flash pt. 465 F (COC).

Kessco® PEG 600 Dioleate. [Stepan] PEG-12 dioleate; see Kessco PEG series; lt. amber liq.; f.p. 19 C; sol. in IPA, acetone, CCl_4, ethyl acetate, toluol, IPM, peanut oil; water-disp.; sp.gr. 1.001; dens. 8.3 lb/gal; HLB 10.5; acid no. 10.0 max.; sapon. no. 92–102; pH 5.0 (3% disp.); flash pt. 495 F (COC).

Kessco® PEG 600 Distearate. [Stepan] PEG-12 distearate; see Kessco PEG series; wh. to cream soft solid; sol. in IPA, acetone, CCl_4, ethyl acetate, toluol, IPM, peanut oil; water-disp.; sp.gr. 0.9670 (65 C); HLB 10.6; m.p. 39 C; acid no. 10.0 max.; sapon. no. 93–102; pH 5.0 (3% disp.); flash pt. 490 F (COC).

Kessco® PEG 600 Monolaurate. [Stepan] PEG-12 laurate; see Kessco PEG series; lt. yel. liq.; f.p. 23 C; sol. in water, Na_2SO_4, IPA, acetone, CCl_4, ethyl acetate, toluol; sp.gr. 1.050; dens. 8.8 lb/gal; HLB 14.9; acid no. 5.0 max.; sapon. no. 64–74; flash pt. 475 F (COC).

Kessco® PEG 600 Monooleate. [Stepan] PEG-12 oleate; see Kessco PEG series; lt. amber liq.; f.p. 23 C; sol. in water, IPA, acetone, CCl_4, ethyl acetate, toluol; sp.gr. 1.037; dens. 8.7 lb/gal; HLB 13.5; acid no. 5.0 max.; sapon. no. 60–69; pH 5.0 (3% disp.); flash pt. 525 F (COC).

Kessco® PEG 600 Monostearate. [Stepan] PEG-12 stearate; see Kessco PEG series; wh. to cream soft solid; sol. in water, IPA, acetone, CCl_4, ethyl acetate, toluol; sp.gr. 1.000 (65 C); HLB 13.6; m.p. 37 C; acid no. 5.0 max.; sapon. no. 61–70; pH 5.0 (3% disp.); flash pt. 480 F (COC).

Kessco® PEG 1000 Dilaurate. [Stepan] PEG-20 dilaurate; see Kessco PEG series; cream soft solid; f.p. 38 C; sol. in water, IPA, acetone, CCl_4, ethyl acetate; toluol, IPM; sp.gr. 1.015 (65 C); HLB 14.5; acid no. 10.0 max.; sapon. no. 68–78; flash pt. 475 (COC).

Kessco® PEG 1000 Dioleate. [Stepan] PEG-20 dioleate; see Kessco PEG series; cream soft solid; f.p. 37 C; sol. in water, IPA, acetone, CCl_4, ethyl acetate, toluol; sp.gr. 1.005 (65 C); HLB 13.1; acid no. 10.0 max.; sapon. no. 64–74; pH 5.0 (3% disp.); flash pt. 505 F (COC).

Kessco® PEG 1000 Distearate. [Stepan] PEG-20 distearate; see Kessco PEG series; cream wax; sol. in water, IPA, acetone, CCl_4, ethyl acetate, toluol; sp.gr. 1.005 (65 C); HLB 12.3; m.p. 40 C; acid no. 10.0 max.; sapon. no. 65–74; pH 5.0 (3% disp.); flash pt. 485 F (COC).

Kessco® PEG 1000 Monolaurate. [Stepan] PEG-20 laurate; see Kessco PEG series; cream soft solid; f.p. 40 C; sol. in water, propylene glycol (hot), Na_2SO_4, IPA, acetone, CCl_4, ethyl acetate, toluol; sp.gr. 1.035 (65 C); HLB 16.5; acid no. 5.0 max.; sapon. no. 41–51; flash pt. 490 (COC).

Kessco® PEG 1000 Monooleate. [Stepan] PEG-20 oleate; see Kessco PEG series; cream soft solid; f.p. 39 C; sol. in water, Na_2SO_4 (5%), IPA, acetone, CCl_4, ethyl acetate, toluol; sp.gr. 1.035 (65 C); HLB 15.4; acid no. 5.0 max.; sapon. no. 40–49; pH 5.0; flash pt. 515 F (COC).

Kessco® PEG 1000 Monostearate. [Stepan] PEG-20 stearate; see Kessco PEG series; cream wax; sol. in water, Na_2SO_4 (5%), IPA, acetone, CCl_4, ethyl acetate, toluol; sp.gr. 1.030 (65 C); HLB 15.6; m.p. 41 C; acid no. 5.0 max.; sapon. no. 40–48; pH 5.0 (3% disp.); flash pt. 475 F (COC).

Kessco® PEG 1540 Dilaurate. [Stepan] PEG-32 dilaurate; see Kessco PEG series; cream wax; f.p. 42 C; sol. in water, Na_2SO_4 (5%); hot in propylene glycol, IPA, acetone, CCl_4, ethyl acetate, toluol; sp.gr. 1.04 (65 C); HLB 15.7; acid no. 10.0 max.; sapon. no. 48–56; pH 4.5 (3% disp.); flash pt. 450 F (COC).

Kessco® PEG 1540 Dioleate. [Stepan] PEG-32 dioleate; see Kessco PEG series; cream wax; f.p. 44 C; sol. in water, propylene glycol, Na_2SO_4; hot in IPA, acetone, CCl_4, ethyl acetate, toluol; sp.gr. 1.025 (65 C); HLB 15.0; acid no. 10.0 max.; sapon. no. 45–55; pH 5.0 (3% disp.); flash pt. 480 F (COC).

Kessco® PEG 1540 Distearate. [Stepan] PEG-32 distearate; see Kessco PEG series; cream wax; sol. in water, IPA, acetone, CCl_4, ethyl acetate, toluol; sp.gr. 1.015 (65 C); HLB 14,8; m.p. 45 C; acid no. 10.0 max.; sapon. no. 49–58; pH 5.0; flash pt. 490 F (COC).

Kessco® PEG 1540 Monolaurate. [Stepan] PEG-32 laurate; see Kessco PEG series; cream wax; f.p. 46 C; sol. in water, Na_2SO_4 (5%); sol. hot in propylene glycol, IPA, acetone, CCl_4, ethyl acetate, toluol; sp.gr. 1.06 (65 C); HLB 17.6; acid no. 5.0 max.; sapon. no. 26–36; pH 4.5 (3% disp.); flash pt. 445 F (COC).

Kessco® PEG 1540 Monooleate. [Stepan] PEG-32 oleate; see Kessco PEG series; cream wax; f.p. 45 C; sol. in water, propylene glycol, Na_2SO_4; sol. hot in IPA, acetone, CCl_4, ethyl acetate, toluol; sp.gr. 1.050 (65 C); HLB 17.0; f.p. 47 C; acid no. 5.0 max.; sapon. no. 28–37; pH 5.0 (3% disp.); flash pt. 520 F (COC).

Kessco® PEG 1540 Monostearate. [Stepan] PEG-32 stearate; see Kessco PEG series; cream wax; sol. in water, Na_2SO_4 (5%), IPA, acetone, CCl_4, ethyl acetate, toluol; sp.gr. 1.050 (65 C); HLB 17.3; m.p. 47 C; acid no. 5.0 max.; sapon. no. 27–36; pH 5.0; flash pt. 495 F (COC).

Kessco® PEG 4000 Dilaurate. [Stepan] PEG-75 dilaurate; see Kessco PEG series; cream wax; f.p. 52 C; sol. in water, Na_2SO_4 (5%); sol. hot in propylene glycol, IPA, acetone, CCl_4, ethyl acetate, toluol; sp.gr. 1.065 (65 C); HLB 17.6; acid no. 5.0 max.; sapon. no. 20–30; pH 4.5 (3% disp.); flash pt. 495 F (COC).

Kessco® PEG 4000 Dioleate. [Stepan] PEG-75 dioleate; see Kessco PEG series; cream wax; f.p. 49 C; sol. in water, propylene glycol, Na_2SO_4; sol. hot in IPA, acetone, CCl_4, ethyl acetate, toluol; sp.gr. 1.060 (65 C); HLB 17.8; acid no. 5.0 max.; sapon. no. 19–27; pH 5.0; flash pt. 500 F (COC).

Kessco® PEG 4000 Distearate. [Stepan] PEG-75 distearate; see Kessco PEG series; cream wax; sol. in water, Na_2SO_4 (5%), IPA, acetone, CCl_4, ethyl acetate, toluol; sp.gr. 1.060 (65 C); HLB 17.3; m.p. 51 C; acid no. 5.0 max.; sapon. no. 19–27; pH 5.0 (3% disp.); flash pt. 515 F (COC).

Kessco® PEG 4000 Monolaurate. [Stepan] PEG-75 laurate; see Kessco PEG series; cream wax; f.p. 55 C; sol. in water, Na_2SO_4 (5%); sol. hot in propylene glycol, IPA, acetone, CCl_4, ethyl acetate, toluol; sp.gr. 1.075 (65 C); HLB 18.8; acid no. 5.0 max.; sapon. no. 9–18; pH 4.5; flash pt. 515 F (COC).

Kessco® PEG 4000 Monooleate. [Stepan] PEG-75 oleate; see Kessco PEG series; cream wax; f.p. 55 C; sol. in water, propylene glycol, Na_2SO_4; sol. hot in IPA, acetone, CCl_4, ethyl acetate, toluol; sp.gr. 1.075 (65 C); HLB 18.3; acid no. 5.0 max.; sapon. no. 10–18; pH 5.0; flash pt. 495 F (COC).

Kessco® PEG 4000 Monostearate. [Stepan] PEG-75 stearate; see Kessco PEG series; cream wax; sol. in water, Na_2SO_4 (5%), IPA, acetone, CCl_4, ethyl acetate, toluol; sp.gr. 1.075 (64 C); HLB 18.6; m.p. 56 C; acid no. 5.0 max.; sapon. no. 10–18; pH 5.0; flash pt. 465 F (COC).

Kessco® PEG 6000 Dilaurate. [Stepan] PEG-150 dilaurate; see Kessco PEG series; cream wax; f.p. 57 C; sol. in water, Na_2SO_4 (5%); sol. hot in propylene glycol, IPA, acetone, CCl_4, ethyl acetate, toluol; sp.gr. 1.077 (65 C); HLB 18.7; m.p. 56 C; acid no. 9.0 max.; sapon. no. 12–20; pH 4.5 (3% disp.); flash pt. 435 F (COC).

Kessco® PEG 6000 Dioleate. [Stepan] PEG-150 dioleate; see Kessco PEG series; cream wax; f.p. 56 C; sol. in water, propylene glycol, Na_2SO_4 (5%); sol. hot in IPA, acetone, CCl_4, ethyl acetate, toluol; sp.gr. 1.070 (65 C); HLB 18.3; acid no. 9.0 max.; sapon. no. 13–21; pH 5.0 (3% disp.); flash pt. 500 F

Kessco® PEG 6000 Distearate. [Stepan] PEG-150 distearate; see Kessco PEG series; cream wax; sol. in propylene glycol, Na_2SO_4 (5%), IPA, acetone, CCl_4, ethyl acetate, toluol; sp.gr. 1.075 (65 C); HLB 18.4; m.p. 55 C; acid no. 9.0 max.; sapon. no. 14–20; pH 5.0 (3% disp.); flash pt. 475 F (COC); 100% act.

Kessco® PEG 6000 Monolaurate. [Stepan] PEG-150 laurate; see Kessco PEG series; cream wax; f.p. 61 C; sol. in water, Na_2SO_4 (5%); sol. hot in propylene glycol, IPA, acetone, CCl_4, ethyl acetate, toluol; sp.gr. 1.085 (65 C); HLB 19.2; acid no. 5.0 max.; sapon. no. 7–13; pH 4.5.

Kessco® PEG 6000 Monooleate. [Stepan] PEG-150 oleate; see Kessco PEG series; cream wax; f.p. 59 C; sol. in water, propylene glycol, Na_2SO_4 (5%); sol. hot in IPA, acetone, CCl_4, ethyl acetate, toluol; sp.gr. 1.085 (65 C); HLB 19.0; acid no. 5.0 max.; sapon. no. 7–13; pH 5.0; flash pt. 470 F

Kessco® PEG 6000 Monostearate. [Stepan] PEG-150 stearate; see Kessco PEG series; cream wax; sol. in water, propylene glycol, Na_2SO_4 (5%), IPA, acetone, CCl_4, ethyl acetate, toluol; sp.gr. 1.080 (65 C); HLB 18.8; m.p. 61 C; acid no. 5.0 max.; sapon. no. 7–13; pH 5.0 (3% disp.); flash pt. 480 F (COC).

Kessco® PGML. [Stepan] Propylene glycol laurate; emollient and aux. emulsifier imparting a soft, velvety skin feel to cosmetics; clear liq.;low odor; HLB 3.2; m.p. 10 C; acid no. 3.0 max.

Kessco® PGMS. [Stepan] Propylene glycol stearate; aux. emulsifier and opacifier in hand care prods., makeup preparations, and solidified perfume compositions; m.p. close to body temp. improving applic. properties; wh. to cream flakes; HLB 3.4; m.p. 35 C; acid no. 3.0 max.; flash pt. 390 F (COC).

Kessco® Synthetic Spermaceti N.F. [Stepan] Used for raising the visc. or m.p. of face creams, cold creams, lipsticks, etc.; wh. flakes; sol. in boiling alcohol, ether, chloroform, other waxes, oils, hydrocarbons; insol. in water; m.p. 43–47 C; acid no. 2 max.; sapon. no. 109–117.

Kessco® Triacetin. [Stepan] Triacetin; see Kessco Acetin; also as solv. for essential oils, food flavors; oily liq.; APHA 50 max; sol. in benzol, ethyl acetate, MEK, IPA, methanol, propylene glycol; partly sol. in castor oil; 6.1% sol. in water; sp.gr. 1.152–1.158; dens. 9.6 lb/gal; visc. 16 cps; f.p. -50 C (gel); flash pt. (COC): 290 F.

Ketjenblack® E.C. [Akzo] Carbon black; elec. conductive carbon black for use as internal antistat for polyolefins, PVC, elastomers.

Ketjenflex® 8. [Akzo] Toluene sulfonamide; plasticizer for resins, coatings, electroplating sol'ns., thermoplastics and thermosets, nitrocellulose lacquers; lt. yel. visc. liq.; dens. 1.2 g/ml; visc. 400 mPa•s; m.p. 0 C; b.p. > 340 C; flash pt. (COC) 224 C.

Ketjenflex® 9. [Akzo] Toluene sulfonamide; plasticizer for difficult resins, e.g., nylon, other polyamides, shellac, cellulose, cellulose-acetate, casein, and other protein materials; fungicide in protective coatings; brightener in electroplating; plasticizer in thermoplastics; component in fluorescent resins; wh. to lt. cream fine cryst. powd.; m.p. 102–107 C; flash pt. (COC) 180 C; 1.5% max. moisture.

Ketjensil® SM 405. [Akzo] Hydrated aluminum silicate, precipitated amorphous; filler in disp. paints for partial replacement of TiO_2; 5.5 μm avg. particle size; dens. 2.1 g/cc; bulk dens. 180 kg/m³; surf. area 70–80 m²/g; pH 10.5 (10% aq. disp.); 82% SiO_2, 9% Al_2O_3.

Ketomax® GI-100. [UOP] Immobilized glucose isomerase derived from Streptomyces; enzyme which converts glucose to fructose; activity 1150 IGIU/g min.; tan gran. solid; dens. 17 lb/ft³ (dry).

Keycide® X-10. [Witco] Tributyltin oxide, stabilized; antimildew additive, antimicrobial for PVAc latex paints for packaged stability and mildew-resistant applied films; Gardner 1 clear liq.; disp. in water; dens. 7.85 lb/gal; 12% act.

Kieralon B, B Highly Conc. [BASF AG] Nonionic/anionic surfactants; detergent, wetting agent, dispersant for textile processing; brnish. liq.; water-sol.; dens. 1.03 and 1.05 g/ml resp.; set pt. –3 and –16 C.

Kilfoam. [Arol] Silicone, emulsifiers, and stabilizer emulsion; defoamer for textile and industrial use; emulsifier for paints; milky wh. liq.; mild odor; disp. in water; sp.gr. 0.97.

Kingsolve SPA. [GAF] Polymeric-organo-metallic compd.; anionic; slip-proofing agent over wide pH range for incorporation into finishing formulas where previously not possible; easily cleanable; emulsion.

Kisuma 5A. [Kyowa] Noble magnesium hydroxide compd.; nontoxic fire retardant; compatibile with plastic to produce high quality composites; used with thermoplastic resins; eliminates toxic gas emissions and reduces smoke emissions; heat stabilizer for resins containing halogen; superfine particles; sp.gr. 0.69 m/100 g; dens. 2.36; ref. index 1.56; hardness 2.5 Mohs; 96.8% $Mg(OH)_2$.

Kisuma 5B. [Kyowa] Noble magnesium hydroxide compd.; see Kisuma 5A; superfine particles; sp.gr. 0.58 m/100 g; dens. 2.36; ref. index 1.56; hardness 2.5 Mohs; 98.1% $Mg(OH)_2$.

Kisuma 5E. [Kyowa] Noble magnesium hydroxide compd.; see Kisuma 5A; superfine particles; sp.gr. 0.56 m/100 g; dens. 2.36; ref. index 1.56; hardness 2.5 Mohs; 98.8% $Mg(OH)_2$.

Kito 40. [Hamblet & Hayes] Tributyltin complex; fungicide for leather; yel. to amber liq.; misc. in water, alcohol, polyamines.

Kito 703. [Hamblet & Hayes] Linear alcohol, ethoxylated polyhydric; nonionic; wetting agent, emulsifier, detergent, defoamer; rewetting in papermaking; colorless liq.; mild odor; f.p. –32 C; sol. in oils and org. media; water-insol.; sp.gr. 0.924; visc. 37 cs; HLB 9.0; pH neutral; flash pt. 370 F; biodeg.; 100% act.

Klearfac AA040. [BASF] Phosphate ester; anionic; solubilizer for nonionics in high electrolyte sol'ns.; for nonionic surfactants and dedusters for powd. alkalies; emulsifier; antistat used with syn. fiber yarns; used in textile processing to improve effectiveness of kier boiling, scouring, bleaching, soaping, and print-washing operations; used in heavy-duty cleaning formulations, pesticides, herbicides, insecticides, liq. fertilizers, and plastics; liq.; sp.gr. 1.112; visc. 240 cps; pour pt. 12.8 C; cloud pt. 2 C (1% aq.); biodeg.; 60% min. act.

Klearfac AA270. [BASF] Phosphate ester; anionic; see Klearfac AA040; liq.; sp.gr. 1.165; visc. 5900 cps; pour pt. –3.9 C; cloud pt. 95 C (1% aq.); 90% min. act.

Klearfac AA420. [BASF] Phosphate ester; anionic; see Klearfac AA040; liq.; sp.gr. 1.118; visc. 2950 cps; pour pt. –6.7 C; cloud pt. 67 C (1% aq.); 90% min. act.

Klearfac AB270. [BASF] Phosphate ester; anionic; see Klearfac AA040; liq.; sp.gr. 1.088; visc. 800 cps; pour pt. –15 C; cloud pt. 88 C (1% aq.); 95% min. act.

Klearol. [Witco] Wh. min. oil NF; emollient in cosmetics; visc. (Saybolt) 50–60 (100 F); pour pt. 0 F.

Klerzyme®. [Int'l. Bio-Synthetics] Pectinase; enzyme for depectinizing fruit juices; liq.

Klucel® Series. [Aqualon] Hydroxypropyl cellulose; nonionic; film-former, thickener, stabilizer, suspending agent, film barrier, thermoplastic, or protective colloid used in coatings, adhesives, extrusions and moldings, paper, paint removers, encapsulations, inks, and other applics.; food grades (denoted by F) used in food, cosmetics, and pharmaceuticals, as emulsifier and whipping aid in toppings, thickener and stabilizer in dips, edible protective coating for candies, and binder for molding and extruding fabricated foods; wh. to off wh. gran. solid; 99% min. thru 20 mesh; sol. in ethanol, methanol, propylene glycol, many polar organic solvs., and in water < 38 C; sp.gr. 1.010 (2% aq., 30 C); dens. (bulk) 0.5 g/ml (varies with type); visc. avail. in wide range of viscs.; soften. pt. 130 C; ref. index 1.337 (2% aq.); surf. tens. 43.6 dynes/cm (0.1%); ref. index 1.337 (2% aq.); pH 5.0–8.5 (aq.); avail. in a wide range of concentrations in both water and anhyd. ethanol.

Klucel® E. [Aqualon] Hydroxypropylcellulose; see Klucel Series; m.w. 80,000; visc. 150–700 cps (10% aq.).

Klucel® EF. [Aqualon] Hydroxypropylcellulose; food grade; see Klucel Series; m.w. 80,000; visc. 200–600 (10% aq.).

Klucel® G, GF. [Aqualon] Hydroxypropylcellulose; see Klucel Series; m.w. 370,000; visc. 150–400 cps (2% aq.).

Klucel® H, HF. [Aqualon] Hydroxypropylcellulose; see Klucel Series; m.w. 1,150,000; visc. 1500–3000 cps (1% aq.).

Klucel® J, JF. [Aqualon] Hydroxypropylcellulose; see Klucel Series; m.w. 140,000; visc. 150–400 cps (5% aq.).

Klucel® L, LF. [Aqualon] Hydroxypropylcellulose; see Klucel Series; m.w. 95,000; visc. 75–150 cps (5% aq.).

Klucel® M, MF. [Aqualon] Hydroxypropylcellulose; see Klucel Series; m.w. 850,000; visc. 4000–6500 cps (2% aq.).

KLX Flakes. [Van Den Bergh] Partially hydrog. veg. oil; icing stabilizer; m.p. 124–130 F.

Knightset M-4. [GAF] Multimetallic salt; accelerant; water-sol. catalyst used in curing selected or modified thermosetting resins of the glyoxal and carbonate types; liq.

Kobate C. [Rhone-Poulenc SpA] Triazine deriv.; high action spectrum bactericide; liq.; 80% conc.

Kocide® 20/20. [Griffin] Copper hydroxide plus nutritional zinc; wettable powd. agric. fungicide, bactericide, and nutritional; 30.7% act. (copper hydroxide).

Kocide® 101. [Griffin] Cupric hydroxide; wettable powd. agric. fungicide, bactericide for many crops; 77% act.

Kocide® 404S. [Griffin] Cupric hydroxide-sulfur flowable agric. fungicide; 27% cupric hydroxide; 15.5% sulfur.

Kocide® 606. [Griffin] Cupric hydroxide; flowable

agric. fungicide for many crops; 37.5% act.

Kocide® SD. [Griffin] Cupric hydroxide sol'n.; agric. fungicide, seed protectant for field crops; 30% act.

Kodaflex® DBP. [Eastman] Dibutyl phthalate; plasticizer used in coatings industry as primary plasticizer-solv. for nitrocellulose lacquers; for rubbers and CAB, ethyl cellulose, PVAc, and syn. resins; solv. for oil-sol. dyes, insecticides, peroxides, and org. compds.; antifoamer and fiber lubricant in textile mfg.; Pt-Co 15 ppm liq.; m.w. 278; f.p. –35 C; b.p. 340 C; sp.gr. 1.048; dens. 1.04 kg/l; visc. 15 cP; ref. index 1.4920; flash pt. 190 C (COC); fire pt. 202 C (COC); vol. resist. 3.0 × 10^9 ohm cm; 99.0% min. assay.

Kodaflex® DEP. [Eastman] Diethyl phthalate; plasticizer; wetting agent in grinding pigments; pigment-disp. medium in cellulose acetate sol'ns. and plastics, and solv. for natural resins and polymers; PVC prods. due to relatively high volatility; Pt-Co 10 ppm liq.; m.w. 222; f.p. < –50 C; b.p. 298 C; sol. 0.12 g/l in water; sp.gr. 1.120; dens. 1.12 kg/l; visc. 9.5 cP; ref. index 1.4990; flash pt. 161 C (COC); fire pt. 171 C (COC); vol. resist. 1.45 × 10^9 ohm cm; 99.0% min. assay.

Kodaflex® DMEP. [Eastman] Dimethoxyethyl phthalate; plasticizer with solv. power; used in cellulose acetate; imparts stability to uv lt.; compatible with NC, CAB, ethyl cellulose, chlorinated rubber, polyvinyl resins, and methyl methacrylate; end uses incl. cellulose acetate cast sheeting and molding compositions, adhesives, laminating cements, and flashbulb lacquers; Pt-Co 20 ppm. liq.; m.w. 282; f.p. –40 C; b.p. 340 C; sol. 0.94 g/l in water; sp.gr. 1.171; dens. 1.17 kg/l; visc. 32 cP; ref. index 1.5020; flash pt. 182 (COC); fire pt. 216 C (COC); vol. resist. 5.7 × 10^7 ohm cm; 99.0% min. assay.

Kodaflex® DMP. [Eastman] Dimethyl phthalate; plasticizer with high solv. power for cellulose acetate extrusion compds.; compatible with ethyl cellulose, CAB, PS, PVAc, polyvinyl butyral, and PVC; used in NC-based printing inks; Pt-Co 5 ppm liq.; m.w. 194; f.p. –1 C; b.p. 284 C; sol. 0.45 g/l in water; sp.gr. 1.192; dens. 1.19 kg/l; visc. 11.0 cP; ref. index 1.513; flash pt. 157 C (COC); fire pt. 168 C (COC); vol. resist. 1.07 × 10^9 ohm cm; 99.0% min. assay.

Kodaflex® DOA. [Eastman] Dioctyl adipate; plasticizer providing flexibility at low temps. to vinyl prods.; used in unfilled garden hose, clear sheeting, elec. insulation; Pt-Co 20 ppm liq.; m.w. 370; f.p. < –70 C; b.p. 417 C; sol. < 0.1 g/l in water; sp.gr. 0.927; dens. 0.924 kg/l; visc. 12 cP; ref. index 1.4472; flash pt. 206 C (COC); fire pt. 229 C (COC); vol. resist. 9.3 × 10^{11} ohm cm; 99.0% min. assay.

Kodaflex® DOP. [Eastman] Dioctyl phthalate; all-purpose plasticizer used with PVC resins incl. film and sheeting for upholstery, clothing, food pkg., paper coatings, molded vinyl prods., elec. wire insulation; compatible with PS, methyl methacrylate, chlorinated rubber, NC, and CAB; low odor, relatively low toxicity, and low volatility; Pt-Co 15 ppm liq.; m.w. 390.57; f.p. –50 C; b.p. 384 C; sol. < 0.1 g/l in water; sp.gr. 0.9852; dens. 0.982 kg/l; visc. 56.5 cP; ref. index 1.4836; flash pt. 216 C (COC); fire pt. 240 C (COC); vol. resist 2.2 × 10^{11} ohm cm.

Kodaflex® DOTP. [Eastman] Dioctyl terephthalate; primary plasticizer used with PVC resins, in PVC plastisols, rubber; applic. incl. wire coatings, automotive and furniture upholstery; compatible with acrylics, CAB, cellulose nitrate, polyvinyl butyral, styrene, oxidizing alkyds, and nitrile rubber; Pt-Co 15 ppm liq.; m.w. 390.57; f.p. 48 C; b.p. 400 C; sol. 0.004 g/l in water; sp.gr. 0.9835; dens. 0.980 kg/l; visc. 63 cP; ref. index 1.4867; flash pt. 238 C (COC); fire pt. 266 C (COC); vol. resist. 3.9 × 10^{12} ohm-cm.

Kodaflex® HS-3. [Eastman] Butyl octyl phthalate; primary plasticizer for polyvinyl homopolymer and copolymer resins, vinyl plastisols, expanded vinyl foams, and rotational molding and dip coating; Pt-Co 35 ppm liq.; f.p. –48 C; b.p. 350 C; sol. < 0.01 g/l in water; sp.gr. 0.988–0.996; dens. 0.989 kg/l; visc. 42 cP; ref. index 1.4848; flash pt. 208 C (COC); fire pt. 224 C (COC).

Kodaflex® HS-4. [Eastman] 60% dioctyl phthalate, 5% dibutyl phthalate; high-solvating primary plasticizer; used in formulating vinyl plastisols having processing chars. required to be mechanically frothed; for PVC formulations, rotational molding, slush molding, dip coating, and filament coating applic.; Pt-Co 35 ppm liq.; f.p. –47 C; b.p. 337 C; sol. < 0.01 g/l in water; sp.gr. 0.993; dens. 0.989 kg/l; visc. 31.8 cP; ref. index 1.4798; flash pt. 182 C (COC); fire pt. 200 C (COC); vol. resist. 2.0 × 10^{14} ohm-cm.

Kodaflex® PA-5. [Eastman] Proprietary mixt. (80% di(2-ethylhexyl phthalate); adhesion-promoting plasticizer for PVC resins and copolymers; used with other plasticizers in vinyl films ≥ 12 mils; improves adhesion to metals; used in automotive and appliance sealants, dip coating, and foamed and unfoamed insulation; Gardner 1 clear liq.; mild odor; insol. in water; sp.gr. 1.000; dens. 0.997 kg/l; visc. 700 cP; pour pt. –22 C; acid no. 2.0; ref. index 1.4890; flash pt. 225 C (COC); fire pt. 241 C (COC).

Kodaflex® TOTM. [Eastman] Trioctyl trimellitate; primary plasticizer used in vinyl film and vinyl-coated fabrics; Pt-Co 100 ppm liq.; m.w. 547; f.p. –38 C; b.p. 414 C; sol. 0.006 g/l in water; sp.gr. 0.989; dens. 0.984 kg/l; visc. 194 cP; ref. index 1.4832; flash pt. 263 C (COC); fire pt. 297 C (COC); 99.0% min. assay.

Kodaflex® Triacetin. [Eastman] Glyceryl triacetate; low-toxicity plasticizer for vinyl compds.; used in adhesives, resinous and polymeric coatings, paper, and paperboard for food contact; water-insol. hydroxyethyl cellulose films; Pt-Co 5 liq.; slight fatty odor; m.w. 218; b.p. 258 C; moderate sol. in water; sp.gr. 1.160; dens. 1.16 kg/l; visc. 17 cP.

Kodaflex® TXIB. [Eastman] 2,2,4-Trimethyl-1,3-pentanediol, diisobutyrate; primary plasticizer used in surf. coatings, vinyl floorings, moldings, and vinyl prods.; compatible with film-forming vehicles used in lacquers for wood, paper, and metals; primary plasticizer for PVC plastisols for rotocasting and slush molding; used in PVC organosols processed by extrusion and inj. molding; clear liq.; slightly fruity odor; m.w. 286.4; f.p. –70 C; b.p. 280 C; sol. 0.42 g/l in water; sp.gr. 0.945; dens. 0.941 kg/l; visc. 9 cP; ref. index 1.4300; flash pt. 143 C (COC); fire pt. 152 C (COC); vol. resist. 1.5×10^{11} ohm-cm.

Komeen®. [Griffin] Copper-ethylenediamine complex; aquatic herbicide for hydrilla control in golf course, ornamental, and fish ponds, potable water reservoirs, fresh water lakes, fish hatcheries; 8% elemental copper.

Koraid PSM. [Kaopolite] Silica, modified; suspension aid for pigments and abrasive particles in water-based systems without increasing visc.; used for paints, polishes, inks. agric. and pharmaceutical formulations; wh. microgran.; disp. in water with good shear; pH 9–10 (5%).

Korantin® BH Liq. [BASF AG] But-2-yne-1,4-diol; corrosion inhibitor in acid pickles and cleaners; water-misc.; 33% conc.

Korantin® BH Solid. [BASF AG] But-2-yne-1,4-diol; see Korantin BH Liq.; water-sol.

Korantin® CD. [BASF AG] Fatty acid DEA condensate; corrosion inhibitor, emulsifier, solubilizer for aq. alkaline systems, e.g., metal treating fluids; liq.; 98% conc.

Korantin® LUB. [BASF AG] Polyether acid phosphate; corrosion inhibitor in weakly alkaline systems and cleaners; lubricant for aq. cutting fluids; liq.; water-disp.; 100% conc.

Korantin® MAT. [BASF AG] Alkanolamine salt of an org. acid containing nitrogen; corrosion inhibitor in neutral and weakly alkaline systems and cleaners; liq.; water-misc.; 100% conc.

Korantin® PA. [BASF AG] Org. acid containing nitrogen; corrosion inhibitor as salts for alkaline aq. systems, e.g., cleaners, cutting fluids; liq.; 80% conc.

Korantin® PAT. [BASF AG] TEA salt of a nitrogen-containing org. acid; corrosion inhibitor for alkaline aq. systems, e.g., cleaners, cutting fluids; liq.; 80% conc.

Korantin® SH. [BASF AG] Fatty acid condensate; anionic; anticorrosive agent for metal treating as the alkanolamine salt; liq.; 100% conc.

Korantin® SMK. [BASF AG] Phosphoric acid monoester; corrosion inhibitor in aq. alkaline systems, cleaners, detergents; liq.; 100% conc.

Korantin® TD. [BASF AG] Fatty acid DEA condensate; corrosion inhibitor and emulsifier for aq. alkaline systems, e.g., cleaners, cutting fluids; liq.; 99% conc.

Korthix. [Kaopolite] Bentonite, refined; thixotropic agent for water-based paints, inks, polishes, adhesives, and for household prods.; wh. fine powd.; 0% retained 325 mesh; disp. in water; visc. 102–508 cps (6%); pH 7.5–8.5 (2% aq.).

Korthix H. [Kaopolite] Bentonite, modified refined; thickening agent for water-based paints, inks, polishes, adhesives, and for household prods.; wh. fine powd.; 0% retained 325 mesh; disp. in water; visc. 254–1516 cps (6%); pH 7.5–8.5 (2% aq.).

Kosmos® 10. [Goldschmidt AG] Stannous octoate; tin-org. catalyst for the mfg. of polyether PU foam; accelerates the gel reaction and intensifies the activation of the blowing reaction; slightly yel. liq.; sp.gr. 1.067 ± 0.02; visc. < 200 mPas; solid. pt. < –60 C; flash pt. (COC) > 170 C; ref. index 1.4887 ± 0.009; 33% sol'n. in dioctyl phthalate; > 9.035% tin content.

Kosmos® 15. [Goldschmidt AG] Stannous octoate; see Kosmos 10; slightly yel. liq.; sp.gr. 1.107 ± 0.02; visc. < 180 mPas; solid. pt. < –60 C; flash pt. (COC) > 157 C; ref. index 1.490 ± 0.009; ; 50% sol'n. in dioctyl phthalate; > 14.0% tin content.

Kosmos® 16. [Goldschmidt AG] Stannous octoate; see Kosmos 10; slightly yel. liq.; sp.gr. 1.024 ± 0.007; visc. ≤ 150 mPas; solid. pt. < –10 C; flash pt. (COC) > 120 C; ref. index 1.48 ± 0.02; 50% sol'n. in min. oil; ≥ 13.9% tin content.

Kosmos® 19. [Goldschmidt AG] Di-n-butyltin-dilaurate; catalyst for PU flexible, rigid, and semi-rigid foams, for cold cure high resiliency PU foams, and as cocatalyst; clear yel. liq.; sol. with polyols, plasticizers, and common org. solvs.; sp.gr. 1.07 ± 0.01; visc. < 80 mPas; solid. pt. < –10 C; flash pt. 200 C; ref. index 1.479 ± 0.009; 18.5 ± 0.5% tin content.

Kosmos® 21. [Goldschmidt AG] Org. tin compd.; catalyst for PU polymerization, for molded PU foams, e.g., microcellular foams, rigid foams, high resilient foams, and RIM; clear slightly yel. liq.; sp.gr. 1.020 ± 0.01; visc. 25 ± 7 mPas; f.p. < –2 C; flash pt. (COC) 260 C; ref. index 1.502 ± 0.007; 18.5 ± 0.5% tin content.

Kosmos® 23. [Goldschmidt AG] Dioctyltinmercaptide; catalyst for the prod. of PU plastics, esp. where extended pot life is required; clear to slightly yel. liq.; sol. in polyols; insol. in water; dens. 1.085 ± 0.01 g/cm²; visc. 100 ± 50 mPas; flash pt. (OC) 130 C; ref. index 1.499 ± 0.008.

Kosmos® 24. [Goldschmidt AG] Organo-tin compd.; catalyst for the prod. of PU plastics; yel. liq.; slight odor; sol. in polyols; insol. in water; dens. 1.085 ± 0.005 g/cm³; visc. 95 ± 10 mPas; flash pt. > 100 C; ref. index 1.499 ± 0.008.

Kosmos® 25. [Goldschmidt AG] Organo-tin (IV) catalyst for the prod. of PU plastics; lt. yel. liq.; slight odor; sol. in polyols; insol. in water; dens. 1.028 g/cm³; visc. 35 mPas; flash pt. (OC) 109 C; ref. index 1.499 ± 0.008.

Kosmos® 29. [Goldschmidt AG] Stannous octoate; catalyst for the gelling reaction during mfg. of polyether PU foams; pale liq.; sol. in polyols and most org. solvs.; insol. in water, alcohol; sp.gr. 1.25 ± 0.02; visc. < 500 mPas; solid. pt. < –25 C; ref. index 1.491 ± 0.008; 28 ± 0.5% tin content.

Kosmos® 64. [Goldschmidt AG] Potassium octoate and polyglycol; catalyst for the prod. of polyisocyanurate foams; clear yel. to slightly brn. liq.; sp.gr. 1.092 ± 0.02; visc. 400 ± 20 mPas; flash pt. (COC) 200 C; hyd. no. 420; 43.3 ± 0.5% potassium octoate content.

KP-140. [FMC] Tributoxyethyl phosphate; leveling agent in floor polish formulations allowing uniform coverage, eliminates high and low spots in gloss, and preventing streaking, crazing, powd., and film contracting; flame retardant for plastics or syn. rubbers of lower flammability; imparts low temp. flexibility to plastics or acrylonitrile rubbers; reduces visc. in plastisols; Pt-Co 75 max. liq.; mild butyl odor; m.w. 398; f.p. < –70 C; b.p. 215–228 C; sol. in org. liqs. and gasoline; sp.gr. 1.018 ± 0.002; dens. 1018 kg/m³; visc. 12.2 cp; pour pt. < –70 C; ref. index 1.434 ± 0.002; flash pt. 224 C; fire pt. 252 C.

K-Pool. [Griffin] Copper-TEA complex; algicide for swimming pools; 8% elemental copper.

Kricinol 35. [Climax Performance] Potassium ricinoleate; mold release; water-sol.

Kronitex® 50. [FMC] Triaryl phosphate; flame retardant used in PVC compositions, flexible PU, cellulosic resins, and syn. rubber; processing aid in engineering resins; syn. equivalent to cresyl diphenyl phosphate (lower sp.gr. and cost); Pt-Co 75 max. clear liq.; odorless; m.w. 375; sp.gr. 1.170–1.180; dens. 1174 kg/m³; visc. 70 cp; pour pt. –26 C; ref. index 1.553; flash pt. 260 C (COC); fire pt. 329 (COC); 8.3% phosphorus.

Kronitex® 100. [FMC] Triaryl phosphate; flame retardant plasticizer for PVC; aids fusion; in plastisols, visc. stability; compatibilizing plasticizer; catalyst carrier and pigment vehicle for PU; processing aid in rubber belting and mech. goods; flame retardant and processing aid in engineering resins; Pt-Co 75 max. clear liq.; m.w. 390; b.p. 220–270 C; sp.gr. 1.150–1.165; dens. 1150 kg/m³; visc. 90 cp; pour pt. –25 C; ref. index 1.552; flash pt. 254 C (COC); fire pt. 332 C (COC); 7.9% phosphorus.

Kronitex® 3600. [FMC] Phosphate ester; plasticizer with improved low temp. flexibility, high flame retardance, and low smoke evolution; ideal for vinyl film and sheeting, wire and cable insulation, coated fabrics, plastisols; APHA 100 max. clear liq.; mild org. odor; sp.gr. 1.030–1.090; dens. 8.94 lb/gal; visc. 28 cp; pour pt. –48 C; flash pt. (PMCC) 175 C; fire pt. (COC) 275 C; 7.4% phosphorus.

Kronitex® PB-460. [FMC] Brominated triaryl phosphate; flame retardant for PPO, PBT, PET, PC, and ABS molding compds., alloys of these resins, EVA wire and cable coatings, PU, PET fibers, and textile coatings; off-wh. solid; sp.gr. 2.3; m.p. 110 C; flash pt. (PMCC) > 228 C; 3.9% phosphorus; 60% bromine.

Kronitex® TCP. [FMC] Tricresyl phosphate; general purpose flame retardant for vinyl compds.; low air, oil, and water loss; processing aid by improving flux rate of compds. containing slow-solvating plasticizers; rapid gelation and fusion rate make it useful in plastisols; plasticizer for NC lacquers and coatings; plasticizer and processing aid for rubbers; flame retardant sheeting; Pt-Co 75 max. clear liq.; slight odor; m.w. 370; b.p. 241–255 C; sp.gr. 1.160–1.175; dens. 1162 kg/m³; visc. 80 cp; pour pt. –28 C; ref. index 1.554; flash pt. 252 C (COC); fire pt. 338 C (COC); 8.4% phosphorus.

Kronitex TXP. [FMC] Trixylenyl phosphate; flame retardant with better milling action in filled PVC compds.; for superior elect. compds. (wire and cable applic.); Pt-Co 100 max. clear liq.; slight odor; m.w. 400; b.p. 248–265 C; sp.gr. 1.130–1.155; dens. 1150 kg/m³; visc. 190 cp; pour pt. –20 C; ref. index 1.553; flash pt. 263 C (COC); fire pt. 343 C (COC); 7.8% phosphorus.

K-Tea. [Griffin] Copper-TEA complex; algicide for use in golf course, ornamental, and fish ponds, potable water reservoirs, fresh water lakes, and fish hatcheries; 8% elemental copper.

Kureton 200. [Kao] Dist. monoglyceride/food additives blend; defoaming agent for soybean curd; bead.

Kureton A. [Kao] Dist. monoglyceride; defoaming agent for soybean curd; bead.

KZ 55. [Kenrich] Tetra (2,2 diallyloxymethyl) butyl, di (ditridecyl) phosphito zirconate; coupling agent; lt. brn. liq.; sol. in IPA, xylene, toluene, DOP, min. oil; sp.gr. 1.00; visc. 100 cps; i.b.p. 380 F; flash pt. (TCC) > 200 F; pH 5.7; 90+% solids in IPA.

KZ OPPR. [Kenrich] Cyclo (dioctyl) pyrophosphato dioctyl zirconate; coupling agent; pale yel. liq.; sol. in xylene, toluene, DOP, min. oil; sp.gr. 1.12; visc. 1000 cps; flash pt. (TCC) 150 F; 90% solids in methyl naphthalene solv.

KZ TPP. [Kenrich] Cyclo [dineopentyl (diallyl)] pyrophosphato dineopentyl (diallyl) zirconate; coupling agent; ylsh. brn. liq.; sol. in IPA, xylene, toluene, DOP, min. oil; sp.gr. 1.18; visc. 3280 cps; i.b.p. 170 F; flash pt. (TCC) > 200 F; 95% solids in IPA.

KZ TPPJ. [Kenrich] Cycloneopentyl, cyclo (dimethylaminoethyl) pyrophosphato zirconate, di mesyl salt; coupling agent; tan liq.; sol. in IPA, toluene, DOP, water; sp.gr. 1.08.

K-Zinc. [Griffin] Zinc oxide formulation; flowable nutrient for rice seed dressing; 25.5% zinc.

L

Labs-100. [Zohar] Linear alkylbenzene sulfonic acid; anionic; detergent intermediate; liq.; 100% conc.
Lactil®. [Goldschmidt] Moisturizers, amino acids, and fission prods. of collagen and urea; nonionic; humectant, moisturing agent for creams and lotions; liq.; 50% conc.
Lactodan. [Grindsted] Monoglyceride lactic acid esters; nonionic; food emulsifier, improves aeration and foam stabilization; solid, flakes; HLB 3.5; 100% act.
Lactozym. [Novo] Lactase; enzyme for hydrolysis of lactose in dairy industry; liq.
Lactozym 1500 L, type GP and 3000 L, type HP. [Novo] Beta-galactosidase (lactase) derived from the yeast Kluyveromyces fragilis; food grade enzymes; 1500 L is general purpose type recommended for applics. like hydrolysis of whey and whey permeate, pretreatment of cheese-milk; 3000 L is high purity type recommended for treatment of milk for direct consumption; activity 1500 LAU/ml (1500 L); 3000 LAU/ml (3000 L); liq.; water-misc.
Lamacit ER. [Grünau] PEG-20 glyceryl ricinoleate and ricinoleamide DEA; nonionic; solubilizer and emulsifier for cosmetics; liq.; 100% conc.
Lamacit GML 12. [Grünau] PEG-12 glyceryl laurate; nonionic; see Lamacit ER; liq.; 100% conc.
Lamacit GML 20. [Henkel KGaA] PEG-20 glyceryl laurate; solubilizer for essential oils in aq./alcoholic systems; liq.
Lamacit GMO 25. [Grünau] PEG-25 glyceryl oleate; nonionic; see Lamacit ER; liq.; 100% conc.
Lamecerin 50-80. [Grünau] POE glycol lanolin derivate; nonionic; refatting agent for personal care prods.; liq.; water-sol.; 50% conc.
Lamefix 680. [Grünau] Fatty acid polyglycol ether and PEG; nonionic; accelerator for HT-fixation of polyester and triacetate; liq.; 98% conc.
Lamegin EE. [Grünau] Acetylated hydrog. tallow glyceride; nonionic; emulsifier and plasticizer for cosmetic, food, and edible coatings; solid; 100% conc.
Lamegin GLP 10, 20. [Grünau] Hydrog. tallow glyceride lactate; nonionic; emulsifier and plasticizer for cosmetics, foods, and drugs; solid; 100% conc.
Lamepon 287 SF. [Grünau] Sulfonic acid; anionic; dispersant, wetting agent, protective colloid for dyestuffs; liq.; 40% conc.
Lamepon A. [Grünau] Protein fatty acid; anionic; dispersant, protective colloid for dyes; stabilizer for peroxide bleaching; liq.; 45% conc.
Lamepon N. [Grünau] Sulfonic acid; anionic; dispersant and protective colloid for dyes; liq.; 40% conc.
Lamepon RE. [Grünau] Sulfonic acid; anionic; dispersant and leveling agent for dyes; liq.; 40% conc.
Lamequat L. [Henkel KGaA] Collagen hydrolysate; cationic; conditioning component for hair and body care preps.; liq.; 35% conc.
Lamesoft LMG. [Grünau] Glycerol monolaurate, dispersed; anionic; refatting agent and thickener for foam bath and shampoos; liq.; 25% conc.
Lamigen ES-30, -60, -100. [Dai-ichi Kogyo] PEG lanolin fatty acid ester; nonionic; softener and antistat for textile; oiling agent for leather; emulsifier for emulsion polymerization; additive to lubricating oil and water sol. paint; paste, solid; HLB 11.0, 13.0, 14.5 resp.; 100% conc.
Lamigen ES-180, ET-20, -70, -90, -180. [Dai-ichi Kogyo] PEG lanolin fatty alcohol ether; nonionic; see Lamigen ES-30; solid, paste; HLB 16.0, 12,0, 14.0, 15.0, 16.0 resp.; 100% conc.
Lanacet® 1705. [Henkel/Emery] Acetylated lanolin USP; nonionic; emollient, superfatting agent; used in personal care prods.; solid; sol. in IPM, min. oil; 100% conc.
Lan-Aqua-Sol. [Fanning] Ethoxylated lanolins; anionic; emulsifier for cosmetic and pharmaceutical emulsions; emollient, superfatting agent, conditioner for skin and hair care prods., household detergents; solubilizer, wetting agent, dispersing aid; yel. gel, solid; trace odor; sol. in water, alcohol; sp.gr. 1.00–1.03; m.p. 45–51 C; 50 and 100% act.
Lan-Aqua-Sol Hydrophilic 50, 100. [Fanning] PEG-40 lanolin; nonionic; primary o/w emulsifier; emollient, wetting agent, dispersing aid, conditioner and superfatting agent; used in personal care prods., cleaners, dishwashing detergents; Gardner 10–11 max. resp.; faint odor; water-sol.; sp.gr. 1.00–1.03 and 1.02–1.06 resp.; m.p. 45–51 C; sapon. no. 5–25; pH 4.0–6.0 and 4.5–6.5 resp.; 50 and 100% act. resp.
Lan-Aqua-Sol Hydrophilic-Plus 50, 100. [Fanning] PEG-50 lanolin; nonionic; see Lan-Aqua-Sol Hydrophilic 50; Gardner 10 and 11 max. resp.; faint odor; sol. in water; sp.gr. 1.00–1.04 and 1.02–1.06 resp. (50/4 C); m.p. 45–51 C; sapon no. 7–10 and 14–20 resp.; pH 5.0–8.0 and 4.0–7.0 resp.; 50 and 100% act. resp.
Lan-Aqua-Sol Super-Hydrophilic 50, 100. [Fanning] PEG-85 lanolin; nonionic; see Lan-Aqua-Sol Hydrophilic 50; Gardner 9 and 11 max. resp.; faint odor; water-sol.; sp.gr. 1.00–1.05 and 1.03–1.07 resp. (50/4 C); m.p. 46–52 C; sapon. no. 5–12 and 14–22 resp.; pH 4.0–6.0; 50 and 100% act. resp.

Lan-Aqua-Sol xtra-Hydrophilic 50, 100. [Fanning] PEG-75 lanolin; nonionic; see Lan-Aqua-Sol Hydrophilic 50; Gardner 10 and 11 max. resp.; faint odor; water-sol.; sp.gr. 1.00–1.05 and 1.03–1.07 resp. (50/4 C); m.p. 46–52 C; sapon. no. 6.5–15 and 14–22 resp.; pH 4.0–6.0; 50 and 100% act. resp.

Lanbritol Wax N21. [Ronsheim & Moore] Cetearyl alcohol, ceteth-12, and oleth-12; nonionic; SE wax, emulsifier for pharmaceuticals, cosmetics and hair care preparations; wh. waxy solid; m.p. 45–53 C; biodeg.; 100% conc.

Lanesta L. [Westbrook Lanolin] Isopropyl lanolate; nonsticky emollient with lubricant properties; liq.

Lanesta S, SA30. [Westbrook Lanolin] Isopropyl lanolate; nonsticky emollient rapidly absorbed by skin; lubricant; soft wax.

Lanethyl. [Croda Ltd.] Alcohol extract of lanolin alcohols; nonionic; plasticizer or film modifier for hair setting resins, w/o emulsifier, conditioner used in alcohol-based systems; amber brittle wax; sol. in ethanol; lipophilic; 100% conc.

Laneto 40. [RITA] PEG-40 lanolin; nonionic; moisturizer, lubricant, and solubilizer surfactant for soap and detergent systems; emollient, resin modifier and solubilizer for personal care prods.; glossing agent for hair; Gardner 12 max.; char. odor; sol. in alcohol; slightly sol. in water; 50 ± 1% water.

Laneto 100. [RITA] PEG-75 lanolin; nonionic; emollient, lubricant, solubilizer, emulsifier, plasticizer for cosmetics and pharmaceuticals; wax; water sol.; 100% conc.

Lanette 14. [Henkel] Myristyl alcohol; nonionic; emollient, consistency agent for cosmetic and pharmaceutical o/w and w/o creams, emulsions, sticks; wh. flakes; 100% conc.

Lanette 16. [Henkel] Cetyl alcohol; nonionic; see Lanette 14; wh. flakes; 100% conc.

Lanette 18. [Henkel] Stearyl alcohol; nonionic; see Lanette 14; wh. flakes; 100% conc.

Lanette 18-22. [Henkel] Fatty alcohol C_{18}-C_{22}; nonionic; see Lanette 14; flakes; 100% conc.

Lanette 22. [Henkel] Behenyl alcohol; nonionic; see Lanette 14; wh. fused flakes; 100% conc.

Lanette O. [Henkel] Cetearyl alcohol; nonionic; emollient, base, consistency factor for ointments, creams, emulsions; wh. gran.; insol. in water; 100% conc.

Lanette SX. [Henkel] 90% cetyl stearyl alcohol and 10% sodium lauryl sulfate; anionic; emulsifier; SE base for mfg. of o/w ointment, creams and emulsions; gran.

Lanette W. [Henkel] Cetearyl alcohol and sodium lauryl sulfate; anionic; SE base for mfg. of ointments, creams and liniments; gran.; 100% conc.

Lanette Wax B.P. [Henkel] SE wax; anionic; solid; biodeg.

Lanette Wax CAT. [Henkel] Cetearyl alcohol, cetrimonium bromide, and laureth-12; cationic; SE wax for prep. of hair conditioners, creams, lotions; waxy solid; biodeg.; 100% conc.

Lanette Wax SX, SXBP. [Henkel] Cetearyl alcohol and sodium C_{12-15} alcohols sulfate (SXBP complies to B.P. specifications); anionic; o/w emulsifier, SE wax for use in toiletry, pharmaceutical preparations, creams, ointments and lotions; cream to pale yel. waxy flakes; faint char. odor; partially sol. in alcohol, almost insol. in water; m.p. 50 C; biodeg.; 100% conc.

Lanex. [Croda Ltd.] Alcohol extract of pure lanolin; nonionic; emollient, plasticizer and conditioner in cosmetics, hair sprays and preshave lotions, superfatting agent, film modifier; amber visc. liq.; sol. in ethanol; lipophilic; 100% conc.

Lanexol. [Croda Ltd.] Alkoxylated liq. lanolin; conditioner, emollient for cosmetic skin and hair care prods. incl. detergents; plasticizer for hair spray resins; lubricant; liq.; sol. in alcohol and water.

Lanexol AWS. [Croda Ltd.] PPG-12-PEG-50 lanolin; nonionic; emollient, conditioner, superfatting agent, foam stabilizer, and lubricant for alcoholic and aq. compositions, plasticizer for hair sprays, o/w emulsifier, solubilizer; amber liq.; sol. in oil, water, ethanol and mixts.; cloud pt. 65–80 C (1% aq.); pour pt. 13 C max.; acid no. 2 max.; sapon. no. 10–20; pH 7.0–7.0 (1% aq.); 97% conc.

Lanfrax®. [Henkel/Emery] Lanolin wax; nonionic; cosmetic emulsion stabilizer, w/o emulsifier and waxing agent, o/w aux. emulsifier; slip reducing agent; for floor finishing prods., polishes; wax; 100% conc.

Lanfrax® 1776. [Henkel/Emery] USP lanolin, wax fraction; nonionic; emulsifier, emollient; waxing agent, o/w aux. emulsifier; slip reducing agent in floor finishing compds.; w/o emulsion stabilizer and thickener; wax; 100% conc.

Lanfrax® 1777 Deodorized. [Henkel/Emery] Specially deodorized lanolin wax fraction; emollient for cosmetics.

Langford® Clay. [Vanderbilt] Soft kaolin; inert filler for all elastomers, nonblack stocks; cream powd.; 99% min. thru 300 mesh; dens. 2.62 ± 0.03 mg/m³.

Lankrocell® D15L. [Harcros] Mechanical foam promoter normally used in conj. with a liq. metal soap stabilizer for PVC.

Lankrocell® KLOP, KLOP/CV. [Harcros] Stabilizer and mechanical foam promoter for foamed PVC plastisols.

Lankroflex® ED6. [Harcros] Epoxidized ester of fatty acids; plasticizer/stabilizer for epoxy.

Lankroflex® GE. [Harcros] Epoxidized soya bean oil; plasticizer/stabilizer for epoxy.

Lankroflex® L. [Harcros] Epoxidized linseed oil; plasticizer/stabilizer for epoxy.

Lankromark® BL277, BM286. [Harcros] Butyl tin carboxylate; transparent heat and light stabilizer for plasticized PVC applics.

Lankromark® BM271, BM400. [Harcros] Butyl tin carboxylate; transparent heat and light stabilizer for rigid and plasticized PVC applics.

Lankromark® BT050, BT105, BT120, BT120A, BT190, BT339. [Harcros] Butyl thiotin; transparent heat stabilizer for rigid and plasticized PVC applics.

Lankromark® DLTDP, DSTDP. [Harcros] Dialkyl thiodipropionate; stabilizer for polyolefins, ABS,

MBS, and SBR.

Lankromark® LC68, LC244, LC299, LC310, LC431, LC486, LC662. [Harcros] Cadmium-based, nonlubricating stabilizers for suspension PVC resins; LC310 grade also for emulsion resins; liqs.

Lankromark® LC90. [Harcros] Cd-Zn; stabilizer/activator for chemically blown PVC; fast action; liq.

Lankromark® LC475, LC541, LC563, LC585, LC629, LC651. [Harcros] Cadmium-based stabilizers for suspension PVC resins; high efficiency, self-lubricating; liqs.

Lankromark® LE65. [Harcros] Triphenyl phosphite; stabilizer for rigid and flexible PVC, PU; epoxy curing agent.

Lankromark® LE76. [Harcros] Alkylaryl phosphite; stabilizer for rigid and flexible PVC.

Lankromark® LE87. [Harcros] Complex phosphite; stabilizer for flexible and nontoxic rigid PVC.

Lankromark® LE98. [Harcros] Alkylaryl phosphite; stabilizer for rigid and flexible PVC; antioxidant for polyolefins.

Lankromark® LE109. [Harcros] Tris (nonyl phenyl) phosphite; stabilizer for nontoxic rigid and flexible PVC; antioxidant for ABS/MBS, polyolefins, SBR.

Lankromark® LE131. [Harcros] Alkylaryl phosphite; stabilizer for rigid and flexible PVC; antioxidant for polyolefins.

Lankromark® LE230. [Harcros] Substituted benzophenone; stabilizer for polyolefins, ABS, MBS, and SBR.

Lankromark® LE274. [Harcros] Benzotriazole derivs.; stabilizer for polyolefins, ABS, MBS, and SBR.

Lankromark® LZ121, LZ1023, LZ1056, LZ1155, LZ1166. [Harcros] Ba-Zn; nonlubricating stabilizer for PVC emulsion resins; liqs.

Lankromark® LZ187. [Harcros] Ba-Zn; stabilizer/activator for chemically blown PVC; slow action; liq.

Lankromark® LZ242. [Harcros] Ba-Zn; nonlubricating stabilizer for PVC suspension resins; suitable for extrusion; liq.

Lankromark® LZ440. [Harcros] Zn; stabilizer/activator for chemically blown PVC; fast action; liq.

Lankromark® LZ495, LZ935. [Harcros] Ca-Zn; nonlubricating stabilizer for PVC suspension and emulsion resins; liqs.

Lankromark® LZ561, LZ1199, LZ1221. [Harcros] K-Zn; stabilizer/activator for chemically blown PVC; fast action; liq.

Lankromark® LZ616, LZ704, LZ770, LZ836, LZ858, LZ968, LZ1067. [Harcros] Ba-Zn; high efficiency self-lubricating stabilizer for PVC suspension resins; LZ616 also for emulsion resins (rotational casting); liqs.

Lankromark® LZ638. [Harcros] Pb-Zn; stabilizer/activator for chemically blown PVC; fast action; liq.

Lankromark® LZ649, LZ1188, LZ1210. [Harcros] Ca-Zn; nonlubricating stabilizer for PVC emulsion resins; liqs.

Lankromark® LZ693, LZ792, LZ1144. [Harcros] Ba-Zn; high efficiency nonlubricating stabilizers for PVC suspension and emulsion resins; liqs.

Lankromark® LZ1045, LZ1177. [Harcros] Ca-Zn; nontoxic stabilizer for PVC suspension and emulsion resins; liqs.

Lankromark® LZ1232. [Harcros] K-Zn; stabilizer/activator for chemically blown PVC; med. action; liq.

Lankromark® OT050, OT250, OT335, OT341, OT650. [Hartcros] Octyl thiotin; transparent heat stabilizer for rigid PVC applics.; suitable for food contact applics.

Lankromark® OT450. [Harcros] Octyl thiotin; transparent heat stabilizer for rigid and plasticized PVC applics.; suitable for food contact applics.

Lankro Mud-Aids. [Harcros UK] Series of general workability aids for drilling muds incl. mud surfactants and defoamers.

Lankromul OSD. [Harcros UK] Emulsifier in aliphatic hydrocarbon solv.; emulsifier and dispersant for oil spills; straw clear liq.; mild, fatty odor; disp. in water; sp.gr. 0.808 (10 C); visc. 3.5 cs; flash pt. 88 C (COC); pour pt. < –20 C.

Lankropearl T. [Harcros UK] Blend of pearlizing agents in an anionic base; anionic/nonionic; surfactant, pearlizing agent for toiletries; liq.; 40% conc.

Lankroplast® L. [Harcros UK] Internal and external lubricant for thermoplastics; solid.

Lankroplast® L542. [Harcros] Tackifier for use in highly filled calendered flexible PVC formulations; liq.

Lankroplast® V2012, V2023, V2067, V2100. [Harcros] Visc. modifier for PVC plastisols.

Lankropol ATE. [Harcros UK] Tetrasodium N-(1,2-dicarboxyethyl)-N-octadecyl sulfosuccinamate; primary emulsifier and mechanical stabilizer for emulsion polymers, aux. foaming agent, solubilizing agent; amber slightly hazy liq.; ethanolic odor; sp.gr. 1.119; visc. 36 cs; flash pt. 73 F (Abel CC); pour pt. 3 C; pH 7.0–8.5 (1% aq.); 35% act.

Lankropol KMA. [Harcros UK] Sodium dihexyl sulfosuccinate; anionic; wetting agent esp. in sol'ns. of electrolytes; solubilizer for soaps, emulsion polymerization aid; pale straw hazy liq.; ethanolic odor; sol. in water; sp.gr. 1.082; visc. 31 cs; flash pt. 91 F (Abel CC); pour pt. < 0 C; pH 6.0–7.5 (1% aq.); 60% act. in ethanol.

Lankropol KO Special. [Harcros UK] Sodium dioctyl sulfosuccinate; anionic; solv. emulsifier, water carrier in dry cleaning formulations, dewatering aid, emulsifier for min. oil with nonionics; pale straw liq.; mild odor; disp. in water; sp.gr. 1.050; visc. 1500 cs; flash pt. > 200 F (COC); pour pt. < 0 C; pH 6–7 (1% aq.); 60% act. in min. oil.

Lankropol OPA. [Harcros UK] Potassium salt of a fatty acid sulfonate; anionic; surfactant used in metal industry and for household use; wetting agent, detergent, dispersant; dk. amber clear liq.; mild fatty odor; sol. in water; sp.gr. 1.125; visc. 270 cs; flash pt. > 200 F (COC); pour pt. < 0 C; pH 5.5–6.0 (1% aq.); biodeg.; 50% act.

Lankropol WA. [Harcros UK] Ammonium salt of a sulfated monoester of fatty acid; anionic; low foaming wetting agent in textile industry, dispersion aid in paint industry; solv. emulsifier in degreasing formulations and herbicides, detergent; dk. amber clear liq.; char. odor; sol. in water; sp.gr. 1.030; visc. 120 cs; flash pt. > 200 F (COC); pour pt. < 0 C; pH 6–7 (1% aq.); biodeg.; 50% act.

Lankropol WN. [Harcros UK] Sodium salt of a sulfated monoester of fatty acid; anionic; see Lankropol WA; dk. amber clear liq.; char. odor; sol. in water; sp.gr. 1.050; visc. 75 cs; flash pt. > 200 F (COC); pour pt. < 0 C; pH 6–7 (1% aq.); biodeg.; 50% act.

Lankrosol SXS-30. [Harcros UK] Sodium xylene sulfonate; anionic; hydrotrope and visc. modifier for high act. liq. detergents, solubilizer for anionic surfactants; pale straw clear liq.; negligible odor; sol. in water; sp.gr. 1.128; visc. 2.7 cs; flash pt. > 200 F (COC); pour pt. < 0 C; pH 7 (1% aq.); 30% act.

Lankrostat® 16, 38, 0600, LA3, NP6. [Harcros] Antistats for plasticized PVC.

Lankrostat® 104, LDN. [Harcros] Antistats for polyolefins, PS.

Lankrostat® CA2. [Harcros] Antistat for polyolefins, PS, ABS.

Lankrostat® LME. [Harcros] Antistat for polyolefins, PS, crystal PS.

Lankrostat® QAT. [Harcros] Antistat for plasticized PVC, PS, and crystal PS.

Lanocerin®. [Amerchol] Lanolin wax; w/o emulsifier, emollient, conditioner used in cosmetics; yel.-tan waxy solid; faint pleasant odor; insol. in water; m.p. 41–51 C; acid no. 2.5 max.; sapon. no. 85–115.

Lan-O-Derm. [Rhone-Poulenc Surf.] Syn. ester; emollient for skin care prods.; sol. in aliphatic and aromatic hydrocarbons, esters,ketones, alcohols.

Lanogel® 21, 31, 41, 61. [Amerchol] PEG-27, –40, -75, -85 lanolin; nonionic; emollient, emulsifier, dispersant, wetting agent, solubilizer, foam stabilizer, used in cosmetics, personal care prods., pharmaceuticals, and facial tissues; ASTM 3 max. gel; disp. to sol. in water; sapon. no. 4–24; 50% act.

Lanogen 1500. [Hoechst-Celanese AG] PEG mixt.; ointment base, thickener.

Lanogene®. [Amerchol] Lanolin oil; emollient, moisturizer, and emulsifier which imparts oil sol. and spreading properties; yel.-amber liq.; slight char. odor; sol. in oils, esters, hydrocarbons, and IPA; acid no. 2 max.; sapon. no. 85–105.

Lanol C. [Seppic] Cetyl alcohol; nonionic; emollient, cosmetic assistant; wax; 100% conc.

Lanol CS. [Seppic] Cetearyl alcohol; nonionic; see Lanol C; wax; 100% conc.

Lanol CT, ST. [Seppic] Fatty alcohol, emulsifiable; nonionic; see Lanol C; wax; 100% conc.

Lanol S. [Seppic] Stearyl alcohol; nonionic; see Lanol C; wax; 100% conc.

Lanolic Acid. [Croda Ltd.] Lanolin acid; emollient for cosmetics; lubricant; soft yel. solid.

Lanolin Alcohols LG. [Van Schuppen] Mixt. of alcohols and sterols derived from lanolin; nonionic; emulsifier, stabilizer, emollient, used in textile, wood and paper industries; tan, soft, bleached wax; sol. in hydrocarbons, chlorinated hydrocarbons, veg. and min. oils, insol. in water; acidr no. 6.0 max.; sapon. no. 20–40; 100% conc.

Lanolin Alcohols LO. [Van Schuppen] Nonionic; see Lanolin Alcohols LG; brn. soft solid; sol. in hydrocarbons, chlorinated hydrocarbons, veg. and min. oils, insol. in water; sapon. no. 20–40; 100% conc.

Lanolin Alcohols THG. [Van Schuppen] Mixt. of alcohols and sterols derived from lanolin; nonionic; emulsifier, stabilizer, emollient, used in cosmetic and tech. applics.; tan bleached solid; sol. in hydrocarbons, chlorinated hydrocarbons, veg. and min. oils, insol. in water; acid no. 6.0 max.; sapon. no. 20 max.; 100% conc.

Lanolin Alcohols THO. [Van Schuppen] Nonionic; see Lanolin Alcohols THG; sol. in hydrocarbons, chlorinated hydrocarbons, veg. and min. oils, insol. in water; acid no. 6.0 max.; sapon. no. 20 max.; 100% conc.

Lanoquat® 1756. [Henkel/Emery] Quaternium-33, ethyl hexanediol; cationic; conditioner; emulsifier for skin moisturizers; substantive to hair; provides lubricating, conditioning, antistatic properties; liq.; sol. in water, ethanol, glycols; 50% conc.

Lanoquat® 1757. [Henkel/Emery] Lanolin quat. in ethyl hexanediol; emollient, conditioner, emulsifier in creams, lotions, makeup, shampoos; Gardner < 12 liq.; sol. in water, IPA, IPM; disp. in triolein, glycerin; dens. 8.2 lb/gal; visc. 226 cSt (100 F); pour pt. < 1 C; flash pt. 270 F.

Lanotein AWS 30. [Fanning] PEG-75 lanolin oil and hydrolyzed animal protein; conditioner, film former, lubricant, humectant, and emollient used in hair prods.; lt. amber liq.; bland odor; water-sol.; sp.gr. 1.05–1.08; pH 5.0–6.0.

Lanotex 730. [Tessilchimica] Lanolin, ethoxylated; skin emollient and conditioner for shampoos, bubble bath, cosmetics; liq.

Lanpol 5. [Croda Ltd.] PEG-5 lanolin acids, dist.; nonionic; o/w emulsifier, solv. for bromo acid dyes in lipsticks, skin care and makeup prods., solubilizer, wetting agent, dispersant; pale yel. pasty solid; sol. in oil, disp. in water; HLB 7.5; sapon. no. 105–120; pH 4–7 (3%); 97% conc.

Lanpol 10. [Croda Ltd.] PEG-10 lanolin acids, dist.; nonionic; see Lanpol 5; pale yel. pasty solid; HLB 10.9; sapon. no. 60–80; pH 4–7 (3%); 97% conc.

Lanpol 20. [Croda Ltd.] PEG-20 lanolin acids, dist.; nonionic; see Lanpol 5; solid; 97% conc.

Lanpol 520 [Croda Ltd.] POE lanolin acid; nonionic; see Lanpol 5; pale yel. pasty solid; HLB 14.1; sapon. no. 45–60; pH 4–7 (3%); 97% conc.

Lanpolamide 5. [Croda] PEG-5 lanolinamide, PEG-5 lanolate; nonionic; w/o emulsifier for hydrocarbon sprays, insecticides, room deodorants; emulsion stabilizer, corrosion inhibitor; soft amber solid; HLB 3.6; 50–100% conc.

Lanquell 206, 217. [Harcros UK] Polyglycol-based, low-toxicity antifoam for food industry, used in mfg. of paper for food pkg.; pale yel. liq.; faint odor; disp. in water; sp.gr. 0.984 and 1.005 resp.; visc. 180 cs and 355 cs resp.; flash pt. > 150 and > 200 F resp. (Abel CC); pour pt. –28 and –13 C resp.; pH 7 (1% aq.); 92

and 98% act.

Lantrol® 1673. [Henkel/Emery] Lanolin oil; emollient and moisturizer for makeup, creams, lotions, hair care prods., bath oils, medicinal preps.; pigment dispersant; Gardner < 9 liq.; sol. in min. oil, triolein, IPM; disp. in IPA; dens. 7.8 lb/gal; visc. 948 cSt (100 F); pour pt. 6 C; cloud pt. 18 C; flash pt. 525 F.

Lantrol® 1674. [Henkel/Emery] Lanolin oil; low odor version of Lantrol 1673; Gardner < 10 liq.; sol. in min. oil, triolein, IPM; disp. in IPA; dens. 7.8 lb/gal; visc. 835 cSt (100 F); pour pt. 9 C; cloud pt. 20 C; flash pt. 525 F.

Lantrol® 1674 Deodorized. [Henkel/Emery] Lanolin oil; dispersant, vehicle for pigment grinds in makups; sol. in min. oil, castor oil.

Lantrol® 1675. [Henkel/Emery] Lanolin oil; premium grade for lt. colored or delicate fragrance applics., lip preps., pigment dispersion; Gardner < 7 liq.; sol. in min. oil, triolein, IPM; disp. in IPA; dens. 7.8 lb/gal; visc. 820 cSt (100 F); pour pt. 9 C; cloud pt. < 10 C; flash pt. 525 F.

Lantrol® AWS. [Henkel/Emery] Ethoxylated lanolin; nonionic; aux. o/w emulsifier for personal care prods., emollient; visc. liq.; sol. in water and alcohol; HLB 18.0; 100% conc.

Lantrol® AWS 1692. [Henkel/Emery] PPG-12-PEG-65 lanolin oil; emollient, plasticizer, solubilizer, and conditioner for hair prods., shaving lotions, antiperspirants, body colognes; Gardner < 11 liq.; sol. in water, IPA; disp. in triolein, glycerin; dens. 8.9 lb/gal; visc. 788 cSt (100 F); HLB 18.0; pour pt. 8 C; cloud pt. 57 C; flash pt. 545 F; 100% conc.

Laponite® 508. [Laporte] Syn. sodium magnesium silicate; carrier for act. materials in horticulture and agriculture; stops settling of high solids systems; clay thickener giving high temp. stability.

Laponite® XLG, XLS. [Laporte] Syn. sodium magnesium silicate; inert base/carrier for act. ingreds.; suspending agent; promotes thixotropy giving stable suspensions; thickens cosmetic, toiletry creams, lotions, toothpaste prods.; XLS also clarification aid, adsorbent; forms colloid sol'ns. and clear gels.

Larosol NRL-40. [PPG-Mazer] Sodium alkyl diphenyl oxide disulfonate; anionic; leveling, transfer and streak coverage; nylon carpet dyeing; emulsifier for dye carriers; liq.; 45% conc.

Larostat 88. [PPG-Mazer] Modified soyadimethylethyl ammonium ethosulfate; cationic; noncorrosive mold release agent; antistat; liq.; 11% act.

Larostat 143. [PPG-Mazer] Oleyldimethylethyl ammonium ethosulfate; cationic; antistat; liq.; 100% act.

Larostat 192. [PPG-Mazer] Modified soyadimethylethyl ammonium diethylphosphate; antistat; for syn. blends with cellulosics; liq.; 92.0% act.

Larostat 192 Anhyd. [PPG-Mazer] Modified soyadimethylethyl ammonium diethylphosphate; antistat; yel. liq.; 100.0% act.

Larostat 264-A. [PPG-Mazer] Modified soyadimethylethyl ammonium ethosulfate; cationic; antistat for syn. fibers, fiberglass, plastic, polyethylene; yel. clear liq.; 34.0–36.0% act.

Larostat 264-A Anhyd. [PPG-Mazer] Modified soyadimethylethyl ammonium ethosulfate; antistat for syn. fibers, fiberglass, plastic, polyethylene; gelwax; 100.0% act.

Larostat 264-A Conc. [PPG-Mazer] Modified soyadimethylethyl ammonium ethosulfate; cationic; see Larostat 88; liq.; 90% act.

Larostat 300. [PPG-Mazer] Potassium alkyl phosphate ester; antistat for syn. fibers; liq.

Larostat 377 DPG. [PPG-Mazer] Lauric myristic dimethylethyl ammonium ethosulfate; see Larostat 88; anhyd. liq.; 80% act.

Larostat 451. [PPG-Mazer] Stearyldimethylethyl ammonium ethosulfate; cationic; noncorrosive release agent forming a hard film; imparts gloss and antistatic properties; liq.; 50% act.

Larostat 491. [PPG-Mazer] Stearyldimethyl ethyl ammonium ethosulfate; antistat for hard surfs.; liq.; 50% act.

Larostat 1443. [PPG-Mazer] Oleyldimethylethyl ammonium ethosulfate; cationic; see Larostat 88; liq.; 100% act.

Lathanol® LAL. [Stepan] Sodium lauryl sulfoacetate; anionic; foaming agent and thickener for powd. bubble baths, shampoos, cleansing creams, cream and paste shampoos, syndet bars; wh. powd.; 65% act.

Lauramide 11, D, ME. [Zohar] Coconut DEA; nonionic; foam booster, thickener, and superfatting agent; liq. (11, D); liq. to paste (ME); 80% conc. (11), 90% conc. (ME).

Lauramide R. [Zohar] Fatty acid DEA; nonionic; see Lauramide 11; liq.; 85% conc.

Laurex® 16/18, 16/18D. [Albright & Wilson] Primary fatty alcohol derived from naturally occurring oils and fats; (16/18D—dist. grade); raw material for mfg. of sulfated derivs. and additives for lubricating oils; wh. solid; m.p. 50–54 C; sapon. no. 1.5 and 0.8 resp.

Laurex® 810. [Albright & Wilson] Octyl-decyl alcohol fraction; raw material in the mfg. of alkyl phthalates, phosphoro diethionates, alkyl methacrylate monomers; defoaming agent in drilling muds and fermentation broths; colorless clear liq.; sp.gr. 0.83; flash pt. 92 C; acid no. 0.15; sapon. no. 1.0 max.

Laurex® 4526. [Albright & Wilson] Primary fatty alcohol blend; lubricant for rigid PVC for inj. molding processes; feedstock for ethoxylation; wh. waxy flake; dens. 0.45 g/cc; m.p. 48–53 C; flash pt. 202 C.

Laurex® ARA. [Albright & Wilson] Blended fatty alcohols; aluminum rolling oil additive; lubricant; liq.

Laurex® CH. [Albright & Wilson] Coconut alcohol; superfatting agent in shampoos, raw material for sulfation, ethoxylation; wh. soft solid; m.p. 18–23 C; sapon. no.1.5.

Laurex® CS. [Albright & Wilson] Cetearyl alcohol; raw material for ethoxylation, sulfation, etc.; stabilizer in emulsion polymerization; lubricant in rigid PVC, also for pharmaceutical creams, hand lotions, bath oils, shaving creams; wh. waxy flake; dens. 0.4 g/cc; m.p. 48–53 C; acid no. 0.5 max.; sapon. no. 2.0

max.; flash pt. 150 C.

Laurex® CS/D. [Albright & Wilson] Cetearyl alcohol; raw material for ethoxylation, sulfation, etc.; also for pharmaceutical creams, hand lotions, bath oils, shaving creams; wh. waxy flake; dens. 0.4 g/cc; m.p. 48–53 C; acid no. 0.28 max.; sapon. no. 2.0 max.; flash pt. 150 C.

Laurex® L1. [Albright & Wilson] Lauryl alcohol; raw material for ethoxylation, sulfation, etc.; wh. soft solid; dens. 0.84 g/cc; m.p. 20–25 C; acid no. 0.2 max.; sapon. no. 1.0 max.; flash pt. 130 C.

Laurex® NC. [Albright & Wilson] Lauryl alcohol; raw material for ethoxylation, sulfation, etc.; stabilizer in emulsion polymerization; foam stabilizer for fire-fighting foams; ; wh. soft solid; dens. 0.84 g/cc; m.p. 20–25 C; acid no. 0.2 max.; sapon. no. 1.0 max.; flash pt. 132 C.

Laurex® PKH. [Albright & Wilson] Palm kernel alcohol; see Laurex CH; wh. soft solid; m.p. 18–23 C; sapon. no. 1.5.

Lauridit® KD, KDG. [Akzo BV] Cocamide DEA; nonionic; detergent, emulsifier for cosmetic and household applics., visc. increasing additive; yel. liq.; water sol. 40 and 800 g/l resp.; sp.gr. 1.02; flash pt. 104 and 114 C resp. (PMCC); pH 9.0–9.5; 90 and 85% act.

Lauridit® KM. [Akzo BV] Cocamide MEA; nonionic; detergent for textile, household and cosmetic applics., foam stabilizer for detergents, shampoos, bubble baths; yel. flakes; poor water sol.; sp.gr. 0.97; m.p. 65–70 C; sapon. no. 14 max.; 93% act.

Lauridit® LM. [Akzo BV] Lauric MEA; nonionic; see Lauridit KM; ylsh. flakes; poor water sol.; sp.gr. 1.01 (80 C); m.p. 80–84 C; sapon. no. 20 max; pH 9; 93% act.

Lauridit ®OD. [Akzo BV] Oleic DEA; nonionic; detergent, emulsifier for cosmetic and household applics., visc. increasing additive; yel. liq.; water sol. 100 g/l; sp.gr. 0.97; f.p. 0 C; flash pt. 100 C (PMCC); pH 9; 85% act.

Lauridit® PD. [Akzo BV] Fatty acid poly DEA/fatty acid DEA 1:2; household detergent, corrosion inhibitor for cutting oils, additive for acid cleaners, emulsifier; br. liq.; water sol.; sp.gr. 1.02; flash pt. 105 C (PMCC); 65% act.

Laurox®. [Akzo] Dilauroyl peroxide; initiator for elevated-temp. polyester cures, for prod. of PVC resins, and acrylates; gran.; 98% act.

Laurox® W-25. [Akzo] Dilauroyl peroxide; initiator; suspension; 25% assay; 1% act. oxygen.

Laurox® W-40. [Akzo] Dilauroyl peroxide; efficient initiator for prod. of PVC resins; pumpable form; liq.; 40% act.

Lebaycid®. [Bayer] Fenthion; insecticide for controlling sucking and biting pests; colorless oily liq. (pure a.i.), ylsh.-brn. liq. (tech. a.i.); m.w. 278.3; sol. (g/1000 ml): > 1000 g in methylene chloride and toluene, < 500 and > 1200 g in IPA; 30–100 g in n-hexane; 0.0063 g in water; sp.gr. 1.246; b.p. 87 C (0.01 mbar).

Lebon 15. [Sanyo] Sodium alkyl diaminoethyl glycine; germicide with detergency; for dairy farming, wide range of applics.; liq.

Lebon GM. [Sanyo] Benzalkonium chloride; germicide; liq.

Lecitase. [Novo] Phospholipase A-2; enzyme for use in food industry; powd.

Lecithin W.D. [Troy] Lecithin prod.; wetting agent, dispersant for pigments in water-based paints; 90% conc.

Lecithin Water Dispersible CLR. [Henkel Canada] Hydrolyzed soya lecithin; emollient for aq. skin and hair care preparations, face cleansers, shampoos; brn. turbid visc. prod.; char. odor.

Lekutherm X 50. [Bayer] Bis-(epoxypropyl)-aniline resin; difunctional thinner and catalyst; produces impregnating agents; insulation of components by encapsulation; dielectric in combination with paper, foil, mica; dens. 1.15; visc. 110–150 cps; flash pt. 180 C; Cured properties: tens. str. 265 kp/cm^2; flex. str. 1166 kp/cm^2; hardness 2370 kp/cm^2.

Lekutherm X 80. [Bayer] Diglycidyl ether of a special bisphenol; plasticizing resin used in insulation of components by encapsulation and for sports equipment; dens. 1.24; visc. 9000–13,000 cps; flash pt. 299 C; Cured properties: tens. str. 640 kp/cm^2; flex. str. 1070 kp/cm^2; impact str. 125 kp cm/cm^2; hardness (Ball indent.) 1260 kp/cm^2; distort. temp. (Martens) 50 C; dielec. str. 212 kV/cm (1 mm, 50 Hz).

Lenetol 9130 AH. [ICI Ltd.] C_9-C_{11} alcohol (3 EO); nonionic; emulsifier, solubilizer, stabilizer for emulsion polymerization; paste; 100% conc.

Lenkanol D-48. [Rohm & Haas] Condensed naphthalene sulfonic acid; syn. tanning agent, bleach, retanning agent for leather; dispersant; water sol'n.; 48% solids.

Leocon 1020B. [Lion] Polyoxyalkylene glycol; nonionic; defoamer for industrial effluents; liq.

Leocon 1070B. [Lion] Polyoxyalkylene glycol; nonionic; defoamer for polymerization processes; liq.

Leocon 2100E. [Lion] Polyalkylene glycol; lubricant for hydraulic fluids; paste.

Leocon PL-71L. [Lion] Polyoxyalkylene glycol; nonionic; defoamer for fermentation processes; liq.

Leoguard G. [Lion] Cellulosic resin; cationic; conditioner for hair care prods.; wet powd.

Leoguard GP. [Lion] Cellulosic resin; cationic; conditioner for hair care prods.; powd.

Leomin OR. [Hoechst-Celanese AG] Fatty acid polyglycol ester; nonionic; dispersant and preparation agent for textile applics.; liq.

Leonil DB Powd. [Hoechst-Celanese AG] Dialkyl naphthalene sulfonate; anionic; wetting agent and dyeing auxiliaries for textiles; powd.

Leophen M. [BASF AG] Neutral phosphoric acid ester with nonionic emulsifiers; wetting agent and detergent in textile industry, antifoam; SE; colorless to weak yel. liq.; misc. in water; dens. 1.0 g/ml.

Leophen U. [BASF AG] Nonionic/anionic surfactant blend; wetting agent, detergent and dispersant used in desizing, treating with alkali and bleaching; yel. paste; water sol.; pH 7.5–8.0 (10 g/l).

Levaform HT. [Bayer] Fatty acid deriv. alkali salt; mold lubricant for goods intended to be welded, bonded, varnished, or painted afterwards; whitish thixotropic; dens. 0.9 g/cm³; 25% conc.

Levaform Si Emulsion. [Bayer] Silicone fluid; mold lubricant with good gloss and pleasant feel; good heat stability and leave molds clean; wh. emulsion; dens. 0.98 g/cm³; 35% conc.

Levaform Si Oil. [Bayer] Silicone fluid; see Levaform Si Emulsion; colorless oil; dens. 0.97 g/cm³; 100% conc.

Levaform SiV. [Bayer] Silicone fluid mixed with highly effective org. release agents; see Levaform Si Emulsion; wh. emulsion; dens. 0.98 g/cm³; 20% conc.

Levelan A0192. [Harcros UK] Fatty alcohol ethoxylate; nonionic; leveling agent for direct dyestuffs; clear liq.; faint odor; sol. in water; sp.gr. 1.018; visc. 13 cs; pour pt. < 0 C; pH 6.0–8.0 (1% aq.); flash pt. > 200 F (COC); cloud pt. > 100 (1% aq.); 20% act.

Levelan NKS. [Harcros UK] Amino carboxylic acid; amphoteric; leveling agent for textiles; paste

Levelan P148. [Harcros UK] Nonyl phenol EO condensate; nonionic; primary emulsifier in emulsion polymer industry; latex stabilizer; clear liq.; faint odor; sol. in water; sp.gr. 1.069; visc. 490 cs; cloud pt. > 100 C (1% aq.); flash pt. (COC) > 200 F; pour pt. < 0 C; 80% act.

Levelan P208. [Harcros UK] Nonyl phenol ethoxylate; nonionic; wetting agent at high temp. and high electrolyte conc. in textile industry; latex stabilizer and emulsifier in emulsion polymer industry; clear liq.; faint odor; sol. in water; sp. gr. 1.086; visc. 460 cs; cloud pt. > 100 C (1% aq.); flash pt. (COC) > 200 F; pour pt. < 0 C; pH 6.0–8.0 (1% aq.); 80% act.

Levelan P357. [Harcros UK] Nonyl phenol EO condensate; nonionic; see Levelan P208; clear liq.; faint odor; sol. in water; sp.gr. 1.090; visc. 960 cs; cloud pt. > 100 C (1% aq.); flash pt. (COC) > 200 F; pour pt. 2 C; pH 6.0–8.0 (1% aq.); 70% act.

Levelan R-15, -200. [Harcros UK] Quat. ammonium compd.; cationic; leveling agent for dyestuffs; liq.

Levelan WS. [Harcros UK] Fatty amine condensate, modified; amphoteric; see Levelan R-15; liq.; 50% conc.

Levelene. [Ciba-Geigy] POE alkyl ether, modified; nonionic; rewetting and leveling agent, penetrant and stripper used in texile applics.; liq.

Levenol A Conc. [Kao] POE alkyl amine; nonionic; retarding and stripping agent for textiles; paste; 100% conc.

Levenol DS-1. [Kao] POE alkyl amine; nonionic; stripping agent for textiles; liq.; 100% conc.

Levenol PW. [Kao] POE alkyl ether; nonionic; leveling agent; flake; 100% conc.

Levenol RK. [Kao] Quat. ammonium halide; cationic; retarding agent for dyeing of acrylics; liq.; 41% conc.

Levenol TD-326. [Kao] Quat. ammonium halide; nonionic; leveling and dispersant for polyester dyeing; liq.

Levenol WX, WZ. [Kao] POE alkyl or alkylaryl ether sulfate; nonionic; leveling agent and dye coagulation preventing agent for dyeing applics.; liq.; 25% conc.

Levilite. [Rhone-Poulenc] Amorphous silica; nonabrasive additive for dentifrice; oil absorbent for talcs and sachets; thickener for lotions; powd.; sol. in HF or strong conc. alkali.

Lexaine C. [Inolex] Cocamidopropyl betaine; amphoteric; visc. builder, foam booster, thickener in conditioners, specialty shampoos, personal care prods.; Gardner 4 max. liq.; sp.gr. 1.044; f.p. 0 C; pH 4.5–5.5; 30% act.

Lexaine CS. [Inolex] Cocamidopropyl betaine; amphoteric; surfactant, foam booster, and visc. builder used in personal care prods.; liq.; 35% conc.

Lexamine 22. [Inolex] Stearamidoethyl diethylamine; cationic; emulsifier used in hair care prods.; intermediate forming o/w emulsifiers; its salts are highly substantive to proteins and cellulosic substrates, imparting antistatic properties; used in cream rinse conditioners, protein conditioners, conditioning shampoo; solid; water-sol. when neutralized with water-sol. acids; 100% conc.

Lexamine B-13. [Inolex] Behenamidopropyl dimethylamine; cationic; see Lexamine 22; flakes; 100% conc.

Lexamine C-13. [Inolex] Cocamidopropyl dimethylamine; cationic; see Lexamine 22; soft solid; 100% conc.

Lexamine L-13. [Inolex] Lauramidopropyl dimethylamine; cationic; see Lexamine 22; solid; 100% conc.

Lexamine O-13. [Inolex] Oleamidopropyl dimethylamine; cationic; see Lexamine 22; liq.; 100% conc.

Lexamine P-13. [Inolex] Palmitamidopropyl dimethylamine; cationic; see Lexamine 22; solid; 100% conc.

Lexamine R-13. [Inolex] Ricinoleamidopropyl dimethylamine and glycerin; cationic; see Lexamine 22; liq.; 100% conc.

Lexamine S-13. [Inolex] Stearamidopropyl dimethylamine; cationic; see Lexamine 22; flakes; 100% conc.

Lexamine S-13 Lactate. [Inolex] Stearamidopropyl dimethylamine lactate; cationic; conditioner for hair care prods.; liq.; 23% conc.

Lexate CL-60. [Inolex] Propylene glycol, oleamidopropyl dimethylamine, cocamidopropyl betaine, and oleamide DEA; hair conditioner conc.; Gardner 7 liq.; mild, fatty odor; sp.gr. 1.01; visc. 150 cps; cloud pt. < 5 C; 75% act.

Lexate CRC. [Inolex] Stearamidopropyl dimethylamine, glycol stearate, and ceteth-2; cationic; cream rinse conc. and conditioner, emulsifier; cream to tan, waxy flake; mild, fatty amino odor; easily melted and dispersed in water; m.p. 49–53 C; 100% act.

Lexate IL. [Inolex] Lanolin and min. oil; nonionic; lanolin additive, absorp. base, conditioner in w/o emulsions, cleansing and conditioning lotions; superfatting agent in soaps; emollient additive in spreading or disp. bath oils; straw paste; sol. in oil; insol. in water; m.p. 35 C; acid no. 1; sapon. no. 66; 100% conc.

Lexate PX. [Inolex] Petrolatum, lanolin, and ozokerite, SE; o/w and aux. emulsifier, lanolin cream base, conditioner in soaps and shaving cream, emollient; faintly yel. semisolid; low odor; sol. in oils, insol. in water; m.p. 41 C; acid no. 0.3; sapon. no. 1.0.

Lexate TA. [Inolex] Glyceryl stearate, IPM, and stearyl stearate; aux. lipophilic emulsifier in w/o emulsions, emollient in cosmetic creams and lotions; provides barrier properties and slip; wh. waxy solid; odorless to bland, fatty odor; sol. in oils, insol. in water; m.p. 45–50 C; acid no. 5–9; sapon. no. 159–169.

Lexate TL. [Inolex] Glyceryl stearate, butyl stearate, and stearyl stearate; superfatting agent in milled bar soap, aux. lipophilic emulsifier in w/o systems, emollient; enhances barrier properties in creams and lotions; wh. waxy solid; mild, fatty odor; sol. in oils, insol. in water; m.p. 50–55 C; acid no. 3 max.; sapon. no. 145–152.

Lexein A200. [Inolex] Myristoyl hydrolyzed animal protein; film-forming collagen protein deriv. for hair sprays, makeups, protection skin lotions and creams; powd.

Lexein A210. [Inolex] Collagen polypeptide-fatty acid condensation prod. in anhyd. ethanol; hair fixative, film-former, hair spray resin plasticizer; used in hair preps., lacquers, aftershaves, bath oils, sunscreen prods.; sol. in anhyd. ethanol and IPA, propylene glycol, hexylene glycol, water/ethanol 30/70.

Lexein A520. [Inolex] TEA-abietoyl hydrolyzed animal protein; sebum control additive; causes delay in refatting of the scalp when used in shampoos; water-sol.

Lexein S620S. [Inolex] TEA-coco-hydrolyzed animal protein and sorbitol; anionic; cleansing agent and humectant used in personal care prods.; water sol'n.; HLB 60.0

Lexein S620TA. [Inolex] TEA-coco-hydrolyzed animal protein; anionic; visc. builder and foam modifier; water sol'n.; HLB 40.0

Lexein SLK. [Inolex] Silk amino acids; used in shampoos, cream rinses, hair conditioners, skin treatment prods.; water sol'n.

Lexein X250. [Inolex] Enzymatic hydrolysate of collagen protein; setting lotions, gels, hair groom aids, shampoos, hair conditioners, wave lotions, cream rinses, skin care prods., skin moisturizers; sol. in aq. alcoholic sol'ns., water.

Lexein X300. [Inolex] Collagen protein hydrolysate; cosmetic protein conditioner; water-sol.

Lexein X350. [Inolex] Collagen protein hydrolysate; shampoos, hair conditioners, skin creams and lotions, hair color; sol. in water and hydroalcoholic sol'n.

Lexein X400. [Inolex] Collagen protein hydrolysate; substantivity agent; water-sol.

Lexemul® 55G. [Inolex] Glyceryl stearate; nonionic; surfactant, emulsifier, opacifier used in cosmetics and topical pharmaceuticals; wh. flakes; mild fatty odor; m.p. 55–59 C; sapon. no. 160–170; 100% conc.

Lexemul® 503, 515. [Inolex] Glyceryl stearate; emulsifier, stabilizer, thickener, opacifier in emulsions or surfactant systems; flakes; HLB 3.2 and 3.8 resp.; 100% conc.

Lexemul® AR. [Inolex] Glyceryl stearate and stearamidoethyl diethylamine; cationic; emulsifier, stabilizer, opacifier, and emollient for cationic systems in personal care prods.; wh. to cream flakes; mild fatty odor; disp. in water (60 C); HLB 4.1; m.p. 60 C; acid no. 25–31; sapon. no. 166–174; 100% conc.

Lexemul® AS. [Inolex] Glyceryl stearate, sodium lauryl sulfate; anionic; SE emulsifier, stabilizer, opacifier and emollient in personal care prods.; wh to cream flakes; HLB 4.9; m.p. 60 C; acid no. 14–18; sapon. no. 153–162; 100% conc.

Lexemul® EGDS. [Inolex] Glycol distearate; nonionic; lubricant, opacifier and pearling agent for cosmetic surfactant systems; flakes; HLB 1.5; 100% conc.

Lexemul® EGMS. [Inolex] Glycol stearate; opacifier and pearling agent for personal care prods.; sec. suspending agent in o/w systems; flakes; sol. in hot min. and veg. oils; water-insol.; HLB 2.3; 100% conc.

Lexemul® PEG-200. [Inolex] PEG-4 dilaurate; nonionic; emulsifier, emollient, and lubricant for bath oils, suppositories, creams, lotions for cosmetics, pharmaceuticals, and industrial applics.; straw to yel. liq., mild fatty odor; sp.gr. 0.954; HLB 5.9; m.p. 2–3 C; acid no. 5 max.; sapon. no. 170–180.

Lexemul® PEG-400DL. [Inolex] PEG-8 dilaurate; nonionic; emulsifier, emollient, and lubricant for bath oils, suppositories, creams, lotions for cosmetics, pharmaceuticals, and industrial applics.; straw to yel. liq., mild fatty odor; sp.gr. 0.988; HLB 9.8; m.p. 10–11 C; acid no. 10 max.; sapon. no. 127–137.

Lexemul® PEG-400ML. [Inolex] PEG-8 laurate; nonionic; emulsifier, emollient, and lubricant for bath oils, suppositories, creams, lotions for cosmetics, pharmaceuticals, and industrial applics.; clear pale yel. liq., mild fatty odor; sp.gr. 1.024; HLB 13.1; m.p. 5–6 C; acid no. 4 max.; sapon. no. 86–96.

Lexemul® T. [Inolex] Glyceryl stearate SE; for use as emulsifier, opacifier, stabilizer, and emollient in alkaline anionic systems; flakes; HLB 5.5; 100% conc.

Lexgard B. [Inolex] Butyl paraben; cosmetics preservative; powd.; sol. in ethanol, propylene glycol; slightly sol. in water.

Lexgard E. [Inolex] Ethyl paraben; cosmetics preservative; powd.; sol. in ethanol, propylene glycol; slightly sol. in water.

Lexgard M. [Inolex] Methyl paraben USP; cosmetics preservative; powd.; sol. in ethanol, propylene glycol; slightly sol. in water.

Lexgard P. [Inolex] Propyl paraben USP; cosmetics preservative; powd.; sol. in ethanol, propylene glycol; slightly sol. in water.

Lexol® 60. [Inolex] IPP, IPM, and isopropyl stearate; nonionic; emollient; bath oils, topical pharmaceuticals; personal care prods.; colorless clear liq.; odorless; f.p. 7 C; sol. in acetone, benzene, CCl_4, castor oil, chloroform, ethanol, heptane, IPA; insol. in water; sp.gr. 0.852; dens. 7.1 lb/gal; sapon. no. 190–198; biodeg. 100% conc.

Lexol® 3975. [Inolex] Mixed isopropyl esters of

myristic, palmitic, and stearic acids; emollient; replacement for IPM; sol. see Lexol 60.

Lexol® DIA. [Inolex] Diisopropyl adipate; low visc. emollient; solv. for aromatic chemicals, perfume oils, eosin dyes; conditioner for skin; sol. in alcohols, min. and veg. oils.

Lexol® EHP. [Inolex] 2-Ethylhexyl palmitate; emollient for nonocclusive creams and lotions, bath oils, antiperspirants, other cosmetic and topical formulations; colorless; odorless; sol. in acetone, alcohol, veg. and min. oils.

Lexol® GT 855, GT 865. [Inolex] Caprylic/capric triglyceride; emollient with nonoily skin feel; moisturizer; for creams, lotions, bath oils, lipstick, makeup; solv. for perfume and flavor ingreds.; vehicle for medicinals, antibiotics, vitamins; solubilizer; oxidative stability; odorless; tasteless.

Lexol® IPM. [Inolex] IPM; emollient, solv., penetrant, cloud pt. depressant; used in personal care prods.; colorless clear liq.; odorless; m.w. 270; f.p. 3 C; b.p. 170 C; sol. see Lexol 60; sp.gr. 0.847–0.854; dens. 7.1 lb/gal; visc. 4.8 cp; sapon. no. 202–212; ref. index 1.433; flash pt. 305 F.

Lexol® IPP. [Inolex] IPP; emollient in conditioning cosmetics; solubilizer for cosmetic and topical pharmaceuticals; colorless clear liq.; odorless; m.w. 298; f.p. 12 C; b.p. 172; sol. see Lexol 60; sp.gr. 0.850–0.855; dens. 7.1 lb/gal; visc. 7 cps; sapon. no. 182–191; ref. index 1.437; flash pt. 335 F.

Lexol® PG 800. [Inolex] Propylene glycol dioctanoate; emollient with nonoily feel, oxidation stability; for creams, lotions, topicals, lipsticks, glossers, makeup bases, bath oils, aftershaves; carrier/vehicle for fragrance; sol. in alcohol, min. oil, acetone.

Lexol® PG 855. [Inolex] Propylene glycol dicaprylate/dicaprate; nonionic; emollient, solubilizer for cosmetic creams and lotions; carrier/vehicle for flavors, fragrances, pigmented cosmetics, antibiotics; liq.; 100% act.

Lexol® PG 865. [Inolex] Propylene glycol dicaprylate/dicaprate; emollient, moisturizer with lubricity and nonoily skin deposition for creams, lotions, makeup, bath oils, pre-electric shave lotions, aerosol systems; vehicle for flavors, fragrances, pigmented cosmetics, vitamins, antibiotics, medicinals; solubilizer; sol. in alcohol, min. and veg. oil, acetone.

Lexol® PG 900. [Inolex] Propylene glycol dinonanoate; emollient for bath oils, preshave lotions, aerosol systems, lipsticks, glosses, makeup bases; carrier for fragrances; colorless liq.; odorless

Lexol® SS. [Inolex] Stearyl stearate; emollient for bath oils, creams, lotions; oil sol.

Lexolube 2J-237. [Inolex] Glycol ester; lubricant for polyester filament yarns; liq.; visc. 36 cSt.

Lexolube 2N-212. [Inolex] Glycol ester; lubricant for nylon and polyester filament; liq.; visc. 69 cSt.

Lexolube 2N-237. [Inolex] Glycol ester; lubricant for syn. fibers, industrial applics.; liq.; visc. 20 cSt.

Lexolube 2T-237. [Inolex] Glycol ester; lubricant for polyester and nylon textile and industrial yarns; softener for syn. rubber and other elastomers; liq.; visc. 18 cSt.

Lexolube 2X-108. [Inolex] Dibasic acid ester; lubricant; base stock for hydraulic, compressor, and crankcase oils; liq.; visc. 24 cSt.

Lexolube 2X-109. [Inolex] Dibasic acid ester; base stock for crankcase and compressor oils; fiber lubricant in yarns; liq.; visc. 50 cSt.

Lexolube 2X-114. [Inolex] Dibasic acid ester; lubricant; low temp. base stock for hydraulic and crankcase oils; liq.; visc. 12 cSt.

Lexolube 2X-130. [Inolex] Dibasic acid ester; see Lexolube 2X-108; liq.; visc. 13 cSt.

Lexolube 3G-310. [Inolex] Polyol ester; lubricant component for hydraulic and metalworking oils; liq.; visc. 87 cSt.

Lexolube 3I-310. [Inolex] Polyol ester; lubricant component for turbine and crankcase oils, specialty greases; liq.; visc. 24 cSt.

Lexolube 3N-309. [Inolex] Polyol ester; high temp. lubricant for textile and industrial yarns; liq.; visc. 32 cSt.

Lexolube 3N-310, 3P-309. [Inolex] Polyol ester; heat-stable lubricant for textile fibers, specialty industrial applics.; liq.

Lexolube 4H-415. [Inolex] Polyol ester; high temp. lubricant for hydraulic and chain oils, mold release; liq.; visc. 124 cSt.

Lexolube 4N-415. [Inolex] Polyol ester; high temp. lubricant for filament yarn prod. and processing, tire cord lubricant; liq.; visc. 56 cSt.

Lexolube 4P-415. [Inolex] Polyol ester; high temp. lubricant for filament yarns; liq.; visc. 60 cSt.

Lexolube B-108. [Inolex] Fatty acid ester; lubricant for syn. fibers, metalworking, and industrial applics.; liq.; visc. 21 cSt.

Lexolube B-109. [Inolex] Tridecyl stearate; drawing and heat setting lubricant for textile/industrial filament yarns, plastic extrusion, magnetic tapes; liq.; visc. 27 cSt.

Lexolube BS-Tech. [Inolex] Butyl stearate; general purpose lubricant for textile syn. fibers, metalworking, coatings, inks, plastics, rubber industries; liq.

Lexolube GT-855. [Inolex] Polyol ester; general lubricant for industrial and textile fiber applics.; liq.

Lexolube GT-855IG. [Inolex] Polyol ester; lubricant for mfg. of syn. fibers; APHA 180; sol. in most solvs.; visc. 25.8 cp; m.p. –8 C; flash pt. (COC) 250 C; acid no. 0.2; sapon. no. 330.

Lexolube N-110. [Inolex] Fatty acid ester; cohesive lubricant for nylon and polyester staple; semisolid.

Lexolube NBS. [Inolex] Butyl stearate; see Lexolube BS-Tech; liq.

Lexolube T-110. [Inolex] Fatty acid ester; lubricant for textile, metalworking, plastics industries; liq.

Lexquat® AMG-M. [Inolex] Myristamidopropyl dimethyl 2,3-dihydroxypropyl ammonium chloride; cationic; conditioner, emulsifier for hair and skin prods.; emollient in bath prods., liq. soaps; amber liq., typ. odor; HLB 11.0; pH 6–8; 33–37% solids, 30% min. act.

Lexquat® AMG-O. [Inolex] Oleamidopropyl dimethyl 2,3-dihydroxypropyl ammonium chloride; cationic; conditioner, emulsifier, emollient for hair and skin prods., bath gels; forms clear dilutable gels with other fatty quats; amber liq., typ. odor; HLB 9.7; pH 6–8; 28–32% solids, 25% min. act.

Lexquat® AMG-WC. [Inolex] Cocamidopropyl dimethyl 2,3-dihydroxypropyl ammonium chloride; cationic; foaming conditioner, emulsifier for hair and skin prods., bath gels; amber liq.; typ. odor; HLB 12.0; pH 6–8; 33–37% solids, 30% min. act.

LICA 01. [Kenrich] Neoalkoxy, trineodecanoyl titanate; coupling agents which sometimes also act as adhesion promoters (esp. LICA 01, 12, 38, 44), antioxidants, antistats, antifoaming agents, accelerators, blowing agent activators (esp. LICA 09, 38), catalysts (esp. LICA 01, 09, 44), curatives, corrosion inhibitors, disp. aids (esp. LICA 09, 12, 38, 38 Quats), emulsifiers, flame retardants (esp. LICA 01, 38), foaming agents, grinding aids, hardeners, hydrophobes (most titanates), impact modifiers, internal lubes, metal primers, process aids, pigment intensifiers, peroxide activators, release agents, retarders, stabilizers, surfactants, suspension aids, thixotropes, wetting agents; recommended for aging, corrosion, and acid resistance are LICA 38, 38 Quats; for conductivity: LICA 38, 44; for thermoplastic polymers: LICA 01, 98, 12, 38, 44; for PVC: LICA 09, 38, 44; for elastomers: LICA 01, 09, 12, 38, 97, and LZ 01, 12, 38; for thermosets and coatings various grades; brnsh. orange liq.; b.p. 320 F (initial); sol. in IPA, xylene, toluene, DOP, min. oil; insol. in water; sp.gr. 1.02 (16 C); visc. 850 cps; pH 5 (sat. sol'n.); flash pt. (TCC) 160 F; 95% solids in IPA.

LICA 09. [Kenrich] Neoalkoxy, dodecylbenzenesulfonyl titanate; see LICA 01; grnsh.-br. liq.; b.p. 170 F (initial); sol. in IPA, xylene, toluene, DOP, min. oil; insol. in water; sp.gr. 1.04 (16 C); visc. 2000 cps; pH 2 (sat. sol'n.); flash pt. (TCC) 180 F; 90% solids in IPA.

LICA 12. [Kenrich] Neoalkoxy, tri (dioctylphosphato) titanate; see LICA 01; orange liq.; b.p. 160 F (initial); sol. in xylene, toluene, DOP, min. oil; limited sol. in IPA; insol. in water; sp.gr. 1.03 (16 C); visc. 300 cps; pH 5 (sat. sol'n.); flash pt. (TCC) 160 F; 95% solids in IPA.

LICA 38. [Kenrich] Neoalkoxy, tri (dioctylpyrophosphato) titanate; see LICA 01; grnsh. br. liq.; b.p. 160 F (initial); sol. in IPA, xylene, toluene, DOP, min. oil; insol. in water; sp.gr. 1.13 (16 C); visc. 5000 cps; pH 3.5 (sat. sol'n.); flash pt. (TCC) 160 F; 95% solids in IPA.

LICA 38A. [Kenrich] Acrylate functional amine adduct; see LICA 01; ylsh.-br. liq.; b.p. 235 F (initial); sol. in IPA; limited sol. in toluene, water; reacts slowly in DOP; insol. in xylene, min. oil; sp.gr. 1.08 (16 C); visc. 1100 cps; pH 7.5 (sat. sol'n.); flash pt. (TCC) 150 F; 95+% solids in IPA.

LICA 38J. [Kenrich] Methacrylate functional amine adduct; see LICA 01; ylsh.-red liq.; b.p. 220 F (initial); sol. in IPA, water; limited sol. in xylene, toluene, DOP; insol. in min. oil; sp.gr. 1.09 (16 C); visc. 5100 cps; pH 7.5 (sat. sol'n.); flash pt. (TCC) 160 F; 95+% solids in IPA.

LICA 44. [Kenrich] Neoalkoxy, tri (N ethylaminoethylamino) titanate; see LICA 01; brnsh. orange liq.; b.p. 250 F (initial); sol. in IPA; reacts slowly in DOP; insol. in xylene, toluene, min. oil, water; sp.gr. 1.17 (16C); visc. 10,000 cps; pH 11 (sat. sol'n.); flash pt. (TCC) 200 F; 95% solids in IPA.

LICA 97. [Kenrich] Neoalkoxy, tri (m-amino) phenyl titanate; see LICA 01; br. liq.; b.p. 180 F (initial); sol. in ether solvents; limited sol. in IPA; insol. in xylene, toluene, DOP, min. oil, water; sp.gr. 1.17 (16 C); visc. 2600 cps; pH 6 (sat. sol'n.); flash pt. (TCC) 160 F; 56% solids in phenyl glycol ether.

LICA 99. [Kenrich] Neopentyl (diallyl)oxy, trihydroxy caproyl titanate; coupling agent; tan liq.; sol. in IPA, DOP; sp.gr. 1.03.

Lignosite®. [Georgia-Pacific] Calcium lignosulfonate; anionic; dispersant, emulsifier and emulsion stabilizer; 50% liq., 100% powd.

Lignosite® 17. [Georgia-Pacific] Ammonium lignosulfonate; anionic; 48% liq., 100% powd.

Lignosite® 260. [Georgia-Pacific] Modified lignosulfonate; anionic; emulsifier, dispersant for pigments, insecticides; o/w emulsifier; tan fine powd.; 100% conc.

Lignosite® 401. [Georgia-Pacific] Calcium lignosulfonate; anionic; 40% liq., 100% powd.

Lignosite® 431. [Georgia-Pacific] Sodium lignosulfonate; anionic; dispersant for pigments, wettable powds., wax emulsions, industrial cleaners; brn. fine powd. or liq.; sol. in water; dens. 10.5 lb/gal; pH 4 (10%); 80% conc. (powd.), 50% conc. (liq.).

Lignosite® 458. [Georgia-Pacific] Sodium lignosulfonate; anionic; dispersant for pigments, mfg. of concrete admixtures, wax emulsions, wettable powds., industrial cleaners; brn. fine powd. or liq.; sol. in hot or cold water; dens. 23 lb/ft^3 (powd.), 10.3 lb/gal (liq.); 80% act. (powd.), 46% solids (liq.).

Lignosite® 704. [Georgia-Pacific] Calcium lignosulfonate; anionic; 40% liq., 100% powd.

Lignosite® 823. [Georgia-Pacific] Sodium lignosulfonate; anionic; dispersant for dyestuffs and pigments, mfg. of wax emulsions, wettable powds.; sol. in cold or hot water; dens. 23 lb/ft^3; pH 7–8 (10% aq.); 100% powd., 47% liq.

Lignosite® 854. [GeorgiaPacific] Sodium lignosulfonate; anionic; dispersant; disp. of pigments; mfg. of concrete admixts. and wax emulsions; formulating wettable powds., water-treating compds., and industrial cleaners; brn. powd. or liq.; dissolves in water; dens. 23 lb/ft^3 (powd.); 10.5 lb/gal (liq.); pH 5.0 (10% aq.); 80% conc. (powd.), 50% conc. (liq.)

Lignosite® 889. [Georgia-Pacific] Sodium lignosulfonate; anionic; 100% powd., 47% liq.

Lignosite® 1840. [Georgia-Pacific] Calcium lignosulfonate; extender for adhesive and binder systems, dispersant for mfg. of wax emulsions; lt. fine powd. or liq.; sol. in hot or cold water; dens. 23 lb/ft^3 (powd.), 10.5 lb/gal (liq.); pH 4.5 (10% aq.); 70% conc. (powd.), 50% conc. (liq.).

Lignosite® AC. [Georgia-Pacific] Lignosulfonate, modified; anionic; 44% liq., 100% powd.

Lignosite® FML. [Georgia-Pacific] Ferromagnetic iron lignosulfonate; ferromagnetic fluid used in separation processes; liq.; dens. 10.1 lb/gal; visc. 60 cps; 35% solids.

Lignosite® KLS. [Georgia-Pacific] Potassium lignosulfonate; anionic; dispersant; liq./powd.; 50–100% act.

Lignosol AXD. [Daishowa] Sodium lignosulfonate; anionic; wetting agent for insecticides, herbicides and fungicides, emulsifier, dispersant; boiler feedwater treatment, industrial cleaners, gypsum wallboard additives; br. powd.; water sol.; dens. 28–32 lb/ft^3; pH 4.9 (27%); 95% act.

Lignosol B. [Daishowa] Calcium lignosulfonate; anionic; wetting agent, emulsifier, dispersant, binder; used in refractories, bricks, construction, insecticides, herbicides, fungicides; soil stabilizer; bk. liq.; sp.gr. 1.25; visc. 200 cps (80 F); f.p. –10 C; b.p. 105 C; 50% act.

Lignosol BD. [Daishowa] Calcium lignosulfonate; anionic; see Lignosol B; yel. powd.; water sol.; dens. 28–32 lb/ft^3; pH 4.5 (27%); 95% act.

Lignosol C 60. [Daishowa] Modified chromium lignosulfonate; dispersant; oil well drilling mud thinner; powd.

Lignosol D-10, D-30. [Daishowa] Sodium lignosulfonate; anionic; wetting agent, dispersant for disperse dyes; br. powd.; water sol.; 95% act.

Lignosol DXD. [Daishowa] Sodium lignosulfonate; anionic; wetting agent and dispersant in industrial cleaners, insecticides, herbicides and fungicides, emulsifier, emulsion stabilizer for asphalt emulsions; bk. powd. or liq.; water sol.; dens. 30–32 lb/ft^3; pH 9.0 (27%); 95% act. (powd.) or 42% act. (liq.).

Lignosol F 30. [Daishowa] Modified ferro-chromium lignosulfonate; dispersant; oil well drilling mud thinner; powd.

Lignosol FTA. [Daishowa] Sodium lignosulfonate; anionic; wetting agent, primary dispersant for disperse dye; br. powd.; water sol.; dens. 30–32 lb/ft^3; pH 9.5 (27%); 95% act.

Lignosol HCX. [Daishowa] Sodium lignosulfonate; anionic; wetting agent, dispersant, emulsifier; used in insecticides, herbicides, fungicides; chelating agent for 24D and 45T amines; br. powd.; water sol.; 95% act.

Lignosol LC. [Daishowa] Calcium lignosulfonate; anionic; dispersant, binder; for use in linoleum cement mfg.; liq.; 50% act.

Lignosol NSX 110. [Daishowa] Sodium lignosulfonate; anionic; primary dyestuff dispersant; powd.; 100% act.

Lignosol NSX 120. [Daishowa] Sodium lignosulfonate, modified; see Lignosol FTA; powd.

Lignosol R30. [Daishowa] Calcium sodium lignosulfonate; slime control agent in mining; powd.

Lignosol SF. [Daishowa] Calcium lignosulfonate; anionic; dispersant for Portland cement and concrete, retanning agent; br. powd.; water sol.; dens. 28–32 lb/ft^3; pH 6.0 (27%); 95% act.

Lignosol SFL. [Daishowa] Calcium lignosulfonate; anionic; dispersant for Portland cement and concrete; bk. liq.; water sol.; sp.gr. 1.24; f.p. –10 C; b.p. 100 C; 48% act.

Lignosol SFS. [Daishowa] Sodium lignosulfonate; anionic; dispersant for dyestuff mfg.; SFX for boiler feed water treatment, dyestuff dispersant and extender; powd.; 100% act.

Lignosol SFX. [Daishowa] Sodium lignosulfonate; anionic; dispersant for concrete admixtures, dyestuffs, sludge conditioner; br. powd.; dens. 28–32 lb/ft^3; pH 6.8 (27%); 95% act.

Lignosol SFX-65. [Daishowa] Sodium lignosulfonate; anionic; dyestuff diluent, dispersant; yel. powd.; dens. 28–32 lb/ft^3; pH 6.0 (27%); 95% act.

Lignosol TS. [Daishowa] Ammonium lignosulfonate; anionic; wetting agent, emulsifier, dispersant, tanning extract, slurry water reducer and grinding aid in cement mfg.; bk. liq.; water sol.; sp.gr. 1.23; f.p. –10C; 47% act.

Lignosol TSD. [Daishowa] Ammonium lignosulfonate; anionic; wetting agent, leather retanning agent, emulsifier, dispersant, slurry water reducer and grinding aid in cement mfg.; yel. powd.; water sol.; dens. 28–32 lb/ft^3; pH 4.5 (27%); 95% act.

Lignosol TSF. [Daishowa] Ammonium lignosulfonate; anionic; dispersant for insecticides, fungicides, herbicides; br. powd.; dens. 28–32 lb/ft^3; pH 4.3 (27%); 95% act.

Lignosol WT. [Daishowa] Sodium lignosulfonate; anionic; wetting agent, dispersant, sludge conditioner; yel. powd.; water sol.; 95% act.

Lignosol X. [Daishowa] Sodium lignosulfonate; anionic; dispersant, wetting agent, emulsifier, tanning and retanning agent, used for dye leveling, water reducer, grinding aid; bk. liq.; water sol.; sp.gr. 1.24; f.p. –5 C; b.p. 105 C; 47% act.

Lignosol XD. [Daishowa] Sodium lignosulfonate; anionic; dispersant, wetting agent, emulsifier for wax emulsions, used for retanning leather, water treatment; yel. powd.; water sol.; dens. 28–32 lb/ft^3; pH 6.5 (27%); 95% act.

Lignosol XD-65. [Daishowa] Sodium lignosulfonate; slurry explosive and flotation aid in mining industry; yel. powd.; dens. 28–32 lb/ft^3; pH 6.8 (27%); 95% act.

Liladox. [Berol Nobel] Dicetyl peroxydicarbonate; initiator for suspension and mass polymerization of vinyl chloride; produces polymers with fewer fisheyes; initiator for copolymerization of vinyl chloride with other monomers, polymerization of acrylates and curing unsat. polyesters; mfg. of plastics for food contact applic.; wh. powd.; sol. in hydrocarbons, chlorinated hydrocarbons, ethers, and ketones; insol. in water; dens. 400 kg/m^3; m.p. 52–54 C; 88% peroxide; 2.5% act. oxygen.

Lilamac 140, 160, 170. [Berol Nobel] Acetate amine salt of Lilamin 140, 160, 170 resp.; cationic; flotation aid for crude oil extraction, soil stabilization, wax emulsions, ore flotation; Gardner 8 max. liq.; m.w. 280–332; m.p. 38–60 C; 95% act.

Lilamac S1. [Berol Nobel] Amine acetate; cationic; flotation agent; flakes.

Lilamin 101 D. [Berol Nobel] Arachidyl behenyl amine (dist.); cationic; emulsifier, anticorrosive, lubricant, bactericidal, pigment dispersants in natural and syn. rubbers, mold release agent for hard rubber, additives for crude oil extraction, lubricants; Gardner 3 max. flakes; m.w. 295; 97% act.

Lilamin 115, 115 D. [Berol Nobel] Soya amine (115 D, dist. grade); cationic; see Lilamin 101 D; Gardner 5 max. flakes; m.w. 270; 95% act.

Lilamin 140, 140 D. [Berol Nobel] Hydrog. tallow amine (140 D, dist. grade); cationic; see Lilamin 101 D; also for coating clays; Gardner 3 max. solid; m.w. 266; sp.gr. 0.795 (60 C); visc. 4.40 cps (60 C); m.p. 46 C; flash pt. 158 (OC); 95% act.

Lilamin 142, 142 D. [Berol Nobel] Stearyl amine (142 D, dist. grade); cationic; see Lilamin 101 D; Gardner 3 max. flakes; m.w. 273; 95% act.

Lilamin 151. [Berol Nobel] Tall amine; cationic; see Lilamin 101 D; Gardner 5 max. flakes; m.w. 270; 92% act.

Lilamin 160, 160 D. [Berol Nobel] Coconut amine (160 D, dist. grade); cationic; see Lilamin 101 D; Gardner 3 max. flakes; m.w. 205; sp.gr. 0.781 (60 C); visc. 2.75 cps (60 C); m.p. 16 C; flash pt. 112 C (OC); 95% act.

Lilamin 163, 163 D. [Berol Nobel] Lauryl amine (163 D, dist. grade); cationic; see Lilamin 101 D; Gardner 3 max. flakes; m.w. 190; sp.gr. 0.777 (60 C); visc. 1.80 cps (60 C); m.p. 25 C; flash pt. 110 C (OC); 95% act.

Lilamin 170, 170 D. [Berol Nobel] Tallow amine, dist. grade (170 D); cationic; see Lilamin 101 D; also for coating clays; emulsifier for creams and beauty prods.; Gardner 3 max. paste; m.w. 266; sp.gr. 0.799 (60 C); visc. 3.35 cps (60 C); m.p. 32 C; flash pt. 155 C (OC); 95% act.

Lilamin 172, 172 D. [Berol Nobel] Oleyl amine, dist. grade (172 D); cationic; see Lilamin 101 D; Gardner 3 max. flakes; m.w. 270; sp.gr. 0.802 (60 C); visc. 3.20 cps (60 C); m.p. 19 C; flash pt. 155 C (OC); 95% act.

Lilamin 308 D. [Berol Nobel] Caprylyl dimethyl tert. amine (dist.); cationic; emulsifier, anticorrosive, lubricant, bactericidal, pigment and filler dispersants in natural and syn. rubbers, additives for crude oil extraction, polymerization catalysts in epoxy resins and polyurethane foams; Gardner 2 max. liq.; m.w. 165; 95% act.

Lilamin 310 D. [Berol Nobel] Capryl dimethyl tert. amine (dist.); cationic; see Lilamin 308 D; Gardner 2 max. liq.; m.w. 195; 95% act.

Lilamin 312 D. [Berol Nobel] Lauryl dimethyl tert. amine (dist.); cationic; see Lilamin 308 D; Gardner 2 max. liq.; m.w. 220; 95% act.

Lilamin 314 D. [Berol Nobel] Myristyl dimethyl tert. amine (dist.); cationic; see Lilamin 308 D; Gardner 2 max. liq.; m.w. 250; 95% act.

Lilamin 316 D. [Berol Nobel] Palmityl dimethyl tert. amine (dist.); cationic; see Lilamin 308 D; Gardner 2 max. liq.; m.w. 280; 95% act.

Lilamin 342 D. [Berol Nobel] Stearyl dimethyl tert. amine (dist.); cationic; see Lilamin 308 D; Gardner 2 max. liq.; m.w. 305; flash pt. 164 C (OC); 95% act.

Lilamin 343. [Berol Nobel] Dihydrog. tallow methyl tert. amine; cationic; see Lilamin 308 D; Gardner 2 max. solid; m.w. 525; m.p. 30 C; 95% act.

Lilamin 345 D. [Berol Nobel] Hydrog. tallow-dimethyl tert. amine (dist.); cationic; see Lilamin 308 D; Gardner 2 max. liq.; m.w. 295; flash pt. 160 C (OC); 95% act.

Lilamin 363. [Berol Nobel] Trilauryl tert. amine; cationic; see Lilamin 308 D; Gardner 2 max. liq.; m.w. 529; 95% act.

Lilamin 364. [Berol Nobel] Tri (caprylyl capryl) tert. amine; cationic; see Lilamin 308 D; Gardner 2 max. liq.; m.w. 390; 95% act.

Lilamin 367 D. [Berol Nobel] Coconut dimethyl tert. amine (dist.); cationic; see Lilamin 308 D; Gardner 2 max. liq.; m.w. 230; flash pt. 109 C (OC); 95% act.

Lilamin 368. [Berol Nobel] Tri (caprylyl capryl/ lauryl) tert. amine; cationic; see Lilamin 308 D; Gardner 2 max. liq.; m.w. 436; 95% act.

Lilamin 369. [Berol Nobel] Dicoco methyl tert. amine; cationic; see Lilamin 308 D; Gardner 2 max. liq.; m.w. 390; m.p. 30 C; 95% act.

Lilamin 372 D. [Berol Nobel] Oleyl dimethyl tert. amine (dist.); cationic; see Lilamin 308 D; Gardner 2 max. liq. m.w. 300; 95% act.

Lilamin 381. [Berol Nobel] Tri isooctyl tert. amine; cationic; see Lilamin 308 D; Gardner 2 max. liq.; m.w. 362; 95% act.

Lilamin 382. [Berol Nobel] Tri isodecyl tert. amine; cationic; see Lilamin 308 D; Gardner 2 max. liq.; m.w. 450; 95% act.

Lilamin 383. [Berol Nobel] Tri tridecyl tert. amine; cationic; see Lilamin 308 D; Gardner 2 max. liq.; m.w. 567; 95% act.

Lilamin 540. [Berol Nobel] Hydrog. tallow diamine; cationic; emulsifier, anticorrosive, lubricant, bactericidal, pigment and filler dispersants in natural and syn. rubbers, additives for crude oil extraction, polymerization catalysts in epoxy resins and PU foams; Gardner 5 max. solid; m.w. 340; 85% act.

Lilamin 560. [Berol Nobel] Coconut diamine; cationic; see Lilamin 540; also pigment wetting in coatings; Gardner 6 max. paste; m.w. 265; sp.gr. 0.812 (60 C); visc. 3.85 cps (60 C); m.p. 25 C; flash pt. 125 C (OC); 85% act.

Lilamin 570. [Berol Nobel] Tallow diamine; cationic; see Lilamin 540; Gardner 5 max. paste; m.w. 340; sp.gr. 0.820 (60 C); visc. 4.05 cps (60 C); m.p. 42 C; flash pt. ± 190 C (OC); 85% act.

Lilamin 570 DO. [Berol Nobel] Dioleate of Lilamin 570; cationic; see Lilamin 540; Gardner 12 max. paste; m.w. 885; 95% act.

Lilamin 572. [Berol Nobel] Oleyl diamine; cationic; see Lilamin 540; Gardner 6 max. liq.; m.w. 350; sp.gr. 0.815 (60 C); visc. 3.95 cps (60 C); m.p. 27 C; flash pt. ± 195 C (OC); 85% act.

Lilamin AC-14, -20, -30, -41L. [Berol Nobel] Amine-based; anticaking, antidusting agents for fertilizers; paste.

Lilamin AC-23 T, -24 T, -25 T, -26. [Berol Nobel] Amine-based; anticaking, antidusting agents for fertilizers; melt applic.; prills.

Lilamin AC-81L, -82L. [Berol Nobel] Amine-based; anticaking and antidusting agents for fertilizers;

paste.

Lilamin EO. [Berol Nobel] Ethoxylated amine derivs.; cationic; emulsifier, corrosion inhibitor, antistat, used for crude oil extraction and drilling aids, in textile industry, in metallurgy.

Lilaminox. [Berol Nobel] Tert. amine oxides; gasoline additives, shampoo base.

Lilamuls BG. [Berol Nobel Belgium] N-tallow propylene diamine; cationic; emulsifier, adhesion agent; paste.

Linamine HC-2. [Napp] Polyunsat. amine deriv.; anionic; conditioner for skin and hair prods.; solv. and leveling agent for hair dyes; liq.; 100% act.

Linex® 4L. [Griffin] Linuron suspension; flowable herbicide for for control of grasses and broadleaf weeds in certain crops and noncrop areas; 40.6% act.

Lipal 3 TD. [Stepan/PVO] Trideceth-3; nonionic; emulsifier used in emollient creams, lotions, etc., solubilizer, wetting agent, dispersant; Gardner 1 liq.; sol. in IPA, propylene glycol, peanut and min. oil, water disp.; HLB 8.0 ± 1; cloud pt. 40–43 C (1%); 100% act.

Lipal 4 LA. [Stepan/PVO] Laureth-4; nonionic; see Lipal 3 TD; Gardner 1 liq.; sol. in IPA, propylene glycol, peanut oil, disp. in water and min. oil; HLB 9.7 ± 1; cloud pt. 52–53 C (1%); 100% act.

Lipal 5 L. [Stepan/PVO] PEG-5 laurate; nonionic; emulsifier, solubilizer, wetting agent, dispersant for emollient creams, lotions, shampoos, makeup, etc.; Gardner 2 liq.; sol. in IPA, disp. in water; HLB 9.8 ± 1; sapon. no. 135–150; 100% act.

Lipal 5 OA. [Stepan/PVO] Oleth-5; nonionic; see Lipal 3 TD; Gardner 3 liq.; sol. in IPA, propylene glycol, peanut and min. oil; HLB 9.0 ± 1; cloud pt. 57–60 C (1%); 100% act.

Lipal 5 S. [Stepan/PVO] PEG-5 stearate; nonionic; see Lipal 5 L; Gardner 4 solid; sol. in IPA, disp. in water; HLB 6.8 ± 1; sapon. no. 105–120; 100% act.

Lipal 6 TD. [Stepan/PVO] Trideceth-6; nonionic; see Lipal 3 TD; Gardner 1 liq.; sol. in water, IPA, propylene glycol, min. oil; HLB 11.0 ± 1; cloud pt. 66–69 C (1%); 100% act.

Lipal 9 C. [Stepan/PVO] PEG-9 castor oil; nonionic; see Lipal 5 L; Gardner 4 liq.; sol. in IPA, disp. in water; HLB 8.0 ± 1; sapon. no. 120–136; 100% act.

Lipal 9 L. [Stepan/PVO] PEG-9 laurate; nonionic; see Lipal 5 L; Gardner 2; sol. in IPA, water; HLB 13.0 ± 1; sapon. no. 89–104; 100% act.

Lipal 9 LA. [Stepan/PVO] Laureth-9; nonionic; see Lipal 3 TD; Gardner 1 liq.; sol. in water, IPA, propylene glycol; HLB 13.6 ± 1; cloud pt. 73–76 C (1%); 100% act.

Lipal 9 N. [Stepan/PVO] Nonoxynol-9; nonionic; see Lipal 3 TD; Gardner 1 liq.; sol. in water, IPA, propylene glycol; HLB 15.0 ± 1; cloud pt. 52–56 C (1%); 100% act.

Lipal 9 OL. [Stepan/PVO] PEG-9 oleate; nonionic; see Lipal 5 L; Gardner 4 liq.; sol. in IPA, disp. in water; HLB 12.0 ± 1; sapon. no. 84–100; 100% act.

Lipal 10 OA. [Stepan/PVO] Oleth-10; nonionic; see Lipal 3 TD; Gardner 3, transitional form; sol. in water, IPA; HLB 12.4 ± 1; cloud pt. 48–52 C (1%); 100% act.

Lipal 10 TD. [Stepan/PVO] Trideceth-10; nonionic; see Lipal 3 TD; Gardner 1 solid; sol. in water, IPA, propylene glycol; HLB 13.8 ± 1; cloud pt. 75–85 C (1%); 100% act.

Lipal 12 LA. [Stepan/PVO] Laureth-12; nonionic; see Lipal 3 TD; Gardner 1 solid; sol. in water, IPA, propylene glycol; HLB 14.8 ± 1; cloud pt. 65–68 C (1%); 100% act.

Lipal 15 CSA. [Stepan/PVO] Ceteareth-15; nonionic; see Lipal 3 TD; Gardner 1 solid; sol. in water, IPA, propylene glycol; HLB 14.4 ± 1; cloud pt. 67–70 C (1%); 100% act.

Lipal 15 T. [Stepan/PVO] PEG-15 tallate; nonionic; see Lipal 5 L; Gardner 12 liq.; sol. in water, IPA; HLB 10.5 ± 1; sapon. no. 25–40; 100% act.

Lipal 20 OA. [Stepan/PVO] Oleth-20; nonionic; see Lipal 3 TD; Gardner 3 solid; sol. in water, IPA; HLB 15.4 ± 1; cloud pt. 67–72 C (1%); 100% act.

Lipal 20 SA. [Stepan/PVO] Steareth-20; nonionic; see Lipal 3 TD; Gardner 1 solid; sol. in water, IPA, propylene glycol; HLB 15.3 ± 1; cloud pt. 70–75 C (1%); 100% act.

Lipal 23 LA. [Stepan/PVO] Laureth-23; nonionic; see Lipal 3 TD; Gardner 1 solid; sol. in water, IPA, propylene glycol; HLB 16.9 ± 1; cloud pt. 75–78 C (1%); 100% act.

Lipal 25 C. [Stepan/PVO] PEG-25 castor oil; nonionic; see Lipal 5 L; Gardner 4 liq.; sol. in water, IPA; HLB 14.5 ± 1; sapon. no. 76–88; 100% act.

Lipal 25 S. [Stepan/PVO] PEG-25 stearate; nonionic; see Lipal 5 L; Gardner 4 solid; sol. in water, IPA, propylene glycol HLB 14.0 ± 1; sapon. no. 35–47; 100% act.

Lipal 30 SA. [Stepan/PVO] Steareth-30; nonionic; see Lipal 3 TD; Gardner 1 solid; sol. in water, IPA, propylene glycol; HLB 17.0 ± 1; cloud pt. 76–78 C (1%); 100% act.

Lipal 39 S. [Stepan/PVO] PEG-40 stearate; nonionic; see Lipal 5 L; Gardner 4 solid; sol. in water, IPA, disp. in propylene glycol, peanut oil; HLB 16.9 ± 1; sapon. no. 23–35; 100% act.

Lipal 50 OA. [Stepan/PVO] Oleth-50; nonionic; see Lipal 3 TD; Gardner 3 solid; sol. in water, IPA, propylene glycol; HLB 18.5 ± 1; cloud pt. 75–78 C (1%); 100% act.

Lipal 50 S. [Stepan/PVO] PEG-50 stearate; nonionic; see Lipal 5 L; Gardner 2 solid; sol. in water, IPA; HLB 17.9 ± 1; sapon. no. 18–30; 100% act.

Lipal 52 C. [Stepan/PVO] PEG-52 castor oil; nonionic; see Lipal 5 L; Gardner 4 solid; sol. in water, IPA, propylene glycol; HLB 16.0 ± 1; sapon. no. 46–58.

Lipal 300 DL. [Stepan/PVO] PEG-6 dilaurate; nonionic; emulsifier, solubilizer, wetting agent, dispersant for emollient creams, lotions, shampoos, makeup, etc.; Gardner 1 liq.; sol. in IPA, peanut oil, min. oil, disp. in water; HLB 6.3 ± 1; sapon. no. 148–162; 100% act.

Lipal 300 W. [Stepan/PVO] PEG-6 monooleate; nonionic; see Lipal 300 DL; Gardner 5 liq.; sol. in IPA, min. oil, disp. in water; HLB 9.2 ± 1; sapon. no.

102–115; 100% act.

Lipal 400 DL. [Stepan/PVO] PEG-8 dilaurate; nonionic; see Lipal 300 DL; Gardner 4 liq.; sol. in IPA, peanut oil, min. oil, disp. in water; HLB 10.4 ± 1; sapon. no. 125–140; 100% act.

Lipal 400 DS. [Stepan/PVO] PEG-8 distearate; nonionic; see Lipal 300 DL; Gardner 4 solid; sol. in IPA, peanut and min. oils, disp. in hot water; HLB 7.2 ± 1; sapon. no. 122–132; 100% act.

Lipal 400 DW. [Stepan/PVO] PEG-8 dioleate; nonionic; emulsifier for w/o and o/w systems, solubilizer, wetting agent, dispersant, used for cosmetic and pharmaceutical preparations; liq.; 100% act.

Lipal 400 OL. [Stepan/PVO] PEG-8 monooleate; nonionic; see Lipal 300 DL; Gardner 13 liq.; sol. in IPA disp. in water; HLB 6.8 ± 1; sapon. no. 83–88; 100% act.

Lipal 400 S. [Stepan/PVO] PEG-8 stearate; nonionic; see Lipal 300 DL; Gardner 1 solid; sol. in IPA, disp. in water; HLB 11.3 ± 1; sapon. no. 82–92; 100% act.

Lipal 400 W. [Stepan/PVO] PEG-8 oleate; nonionic; see Lipal 400 DW; liq.; 100% act.

Lipal 600 S. [Stepan/PVO] PEG-12 stearate; nonionic; see Lipal 400 DW; solid; HLB 13.8; 100% act.

Lipal 600 W. [Stepan/PVO] PEG-12 oleate; nonionic; see Lipal 400 DW; liq.; 100% act.

Lipal 610. [Stepan/PVO] Trideceth-8; nonionic; emulsifier, solubilizer, wetting agent, dispersant used in emollient creams, lotions, etc.; Gardner 1 liq.; sol. in water, IPA, propylene glycol; HLB 12.5 ± 1; cloud pt. 42–46 C (1%); 85% act.

Lipal CE 38. [Stepan/PVO] PEG-5 cocoate; nonionic; emulsifier, solubilizer, wetting agent, coupling agent, dispersant used in emollient creams and lotions, shampoos, etc.; Lovibond 3R, transitional form; sol. in IPA, min. oil, trichloroethylene, alcohol; HLB 7.6 ± 1; sapon. no. 130–150; 100% act.

Lipal CE 55. [Stepan/PVO] PEG-11 cocoate; nonionic; see Lipal CE 38; Lovibond 3R, transitional form; sol. in IPA, min. oil, trichloroethylene, alcohol; HLB 11.0 ± 1; sapon. no. 90–115; 100% act.

Lipal CE 64. [Stepan/PVO] PEG-16 cocoate; nonionic; see Lipal CE 38; Lovibond 3R, transitional form; sol. in water, IPA, min. oil, trichloroethylene, alcohol; HLB 12.8 ± 1; sapon. no. 70–100; 100% act.

Lipal CE 71. [Stepan/PVO] PEG-21 cocoate; nonionic; see Lipal CE 38; Lovibond 3R, transitional form; sol. in water, IPA, min. oil, trichloroethylene, alcohol; HLB 14.2 ± 1; sapon. no. 50–80; 100% act.

Lipal EB. [Aquatec] Butyl stearate; emollient for cosmetics; liq.; 100% conc.

Lipal EGDS. [Aquatec] Glycol distearate; nonionic; emulsifier, pearlescent for lotions and creams; flakes; 100% conc.

Lipal EGMS. [Aquatec] Glycol stearate; emulsifier, pearlescent for lotions and creams; flakes; 100% conc.

Lipal LC. [Aquatec] Cetyl lactate; emollient for cosmetics; solid; 100% conc.

Lipal OE 55. [Stepan/PVO] PEG-12 safflower ester; nonionic; see Lipal CE 38; Lovibond 6R, transitional form; sol. in water, IPA, trichloroethylene, min. oil, alcohol; HLB 11.0 ± 1; sapon. no. 70–90.

Lipal OE 64. [Stepan/PVO] PEG-17 safflower ester; nonionic; see Lipal CE 38; Lovibond 6R, liq.; sol. in water, IPA, trichloroethylene, min. oil, alcohol; HLB 12.8 ± 1; sapon. no. 55–75.

Lipal OE 70. [Stepan/PVO] PEG-23 safflower ester; nonionic; see Lipal CE 38; Lovibond 6R, liq.; sol. in water, IPA, trichloroethylene, min. oil, alcohol; HLB 14.0 ± 1; sapon. no. 42–62.

Lipal ST. [Aquatec] Isopropyl stearate; nonionic; emollient for cosmetics; liq.; 100% conc.

Lipal TE 43. [Stepan/PVO] PEG-8 tallowate; nonionic; see Lipal CE 38; Lovibond 6R, liq.; sol. in IPA, trichloroethylene, min. oil, alcohol; HLB 8.6 ± 1; sapon. no. 90–115; 100% act.

Lipal TE 55. [Stepan/PVO] PEG-12 tallowate; nonionic; see Lipal CE 38; Lovibond 6R, transitional; sol. in water, IPA, trichloroethylene, min. oil, alcohol; HLB 11.0 ± 1; sapon. no. 70–100; 100% act.

Lipal TE 70. [Stepan/PVO] PEG-22 tallowate; nonionic; see Lipal CE 38; Lovibond 6R, soft solid; sol. in water, IPA, trichloroethylene, min. oil, alcohol; HLB 14.0 ± 1; sapon. no. 40–65; 100% act.

Lipal TE 76. [Stepan/PVO] PEG-30 tallowate; nonionic; see Lipal CE 38; Lovibond 6R, soft solid; sol. in water, IPA, trichloroethylene, min. oil, alcohol; HLB 15.2 ± 1; sapon. no. 30–60; 100% act.

Lipamide MEAA. [Lipo] Acetamide MEA; nonionic; lubricating humectant for used in personal care prods.; hair conditioner for shampoos, rinses, conditioners; antistat, foam modifier; nontacky liq.; sol. in water, ethanol, IPA, glycols; 75% conc.

Lipamide S. [Lipo] Stearamide DEA; nonionic; emulsifier, opacifier, thickener, emulsion stabilizer, lubricant, used in skin and hair care prods.; off-wh. wax; mild, waxy odor; misc. hot with common oil phase ingred. and most org. solv., disp. in hot water; m.p. 41–46 C; pH 9–10.5 (1% aq.); 100% act.

Lipo DGLS. [Lipo] PEG-2 laurate SE; nonionic; spreading agent, emulsifier, dispersant, lubricant, opacifier, emulsion stabilizer, emollient, visc. builder used in bath oils, creams, lotions; defoamer for process applics.; yel. liq.; water-disp.; HLB 8.3 ± 1; acid no. 4 max.; sapon. no. 160–170; 100% act.

Lipo DGS-SE. [Lipo] Diethylene glycol monostearate SE; see Lipo DGLS; wh./off wh. beads, flakes; HLB 4.0±1; acid no. 90–110; sapon. no. 160–180.

Lipo Diglycol Laurate. [Lipo] Emulsifier, defoamer; liq.; 100% conc.

Lipo EGDS. [Lipo] Ethylene glycol distearate; see Lipo DGLS; wh./off wh. beads, flakes; HLB 1.0 ± 1; acid no. 7 max.; sapon. no. 190–205.

Lipo EGMS. [Lipo] Glycol stearate; nonionic; see Lipo DGLS; also opacifier, pearlizer for shampoos; wh./off wh. beads, flakes; HLB 2.0 ± 1; acid no. 6 max.; sapon. no. 175–190.

Lipo GMS. [Lipo] Glyceryl stearate; nonionic; emulsifier and defoamer for food industry; solid; HLB 3.7; 100% conc.

Lipo GMS 450. [Lipo] Glyceryl stearate; nonionic; general purpose emulsifier, emollient, opacifier and visc. builder in creams and lotions; wh. bead or flake;

HLB 3.6 ± 1; acid no. 5 max.; sapon. no. 165–182; 100% act.

Lipo GMS 470. [Lipo] Glyceryl stearate SE; nonionic; general purpose emulsifier, emollient, opacifier and visc. builder in creams and lotions; wh. bead or flake; HLB 5.8 ± 1; acid no. 5 max.; sapon. no. 138–152; 100% act.

Lipo PGMS. [Lipo] Propylene glycol stearate; nonionic; see Lipo DGLS; wh. solid wax; HLB 3.0 ± 1; acid no. 6 max.; sapon. no. 180–192.

Lipobee 102. [Lipo] Syn. beeswax; oil-misc.; water-insol.

Lipocol. [Croda Ltd.] Lanolin-derived base; emollient, moisturizer, emulsifier for cosmetic and pharmaceutical preps.; liq.

Lipocol B. [Lipo] PEG-9 stearate, PEG-9 laurate and PEG-2 laurate SE; nonionic; spreading agent, emulsifier, dispersant, opacifier, lubricant, used in bath oils, creams, lotions, nonionic; wh. to off-wh., solid wax; water-disp.; HLB 11.0 ± 1; sapon. no. 94–104; 100% act.

Lipocol C. [Lipo] Cetyl alcohol; emollient, consistency builder for creams, lotions, molded stick prods.; oil-misc.; water-insol.

Lipocol C-2, -10, -20. [Lipo] Ceteth-2, -10, -20; nonionic; emulsifier, defoamer, wetting agent, solubilizer, conditioning agent for personal care prods. and pigment disp.; wh., solid wax; HLB 5.3, 12.9, 15.7 resp.; acid no. 1, 1, and 2 max. resp.; 100% act.

Lipocol L. [Lipo] Lauryl alcohol; nonoily afterfeel emollient for creams, lotions, makeup; oil-misc.; water-insol.

Lipocol L-1, -4, -12, -23. [Lipo] Laureth-1, -4, –12, -23; nonionic; see Lipocol C-2; colorless liq. to wh. solid wax; HLB 3.6, 9.7, 14.5, 16.9 resp.; acid no. 2, 2, 1, and 2 max. resp.; 100% act.

Lipocol M-4. [Lipo] Myreth-4; nonionic; see Lipocol C-2; liq.; HLB 8.8; 100% conc.

Lipocol O. [Lipo] Oleyl alcohol; nontacky afterfeel emollient for bath oils, creams, lotions, makeup; oil-misc.; water-insol.

Lipocol O-2, -10, -20. [Lipo] Oleth-2, -10, -20; nonionic; see Lipocol C-2; yel. liq.; HLB 4.9, 12.4, 15.3 resp.; acid no. 1, 2, and 2 max. resp.; 100% act.

Lipocol S. [Lipo] Stearyl alcohol; waxy afterfeel emollient, consistency builder for creams, lotions, molded sticks; oil-misc.; water-insol.

Lipocol S-2, -10, -20. [Lipo] Steareth-2, -10, -20; nonionic; see Lipocol C-2; wh. solid wax; HLB 4.9, 12.4, 15.3 resp.; acid no. 1 max. (all); 100% act.

Lipocol SC-4, -10, -15, -20. [Lipo] Ceteareth-4, -10, -15, -20; nonionic; see Lipocol C-2; solid; HLB 8.0, 12.5, 14.3, 15.4; 100% conc.

Lipocol TD-3, -6, -12. [Lipo] Trideceth-3, -6, -12; nonionic; emulsifier, wetting and scouring agent, dispersant for essential oils; raw material for sulfation and phosphation; TD-12 also solubilizer; liq., paste (TD-12); HLB 14.6 ± 1 (TD-12); 100% act.

Lipolan. [Lipo] Hydrog. lanolin; nonionic; aux. w/o emulsifier; emollient; conditioner; lubricant; paste; wh. to off-wh. paste; mild char. odor; water-insol.; m.p. 37–45; acid no. 1 max.; sapon. no. 5 max.; 100% conc.

Lipolan 31. [Lipo] PEG-24 hydrog. lanolin; nonionic; o/w emulsifiers, solubilizer, emollient, conditioner used in cosmetics, toiletries and topical pharmaceuticals; cream waxy solid; bland, char. odor; sol. in water and ethanol; acid no. 2 max.; sapon. no. 8; 100% act.

Lipolan 1400. [Lion] alpha-Olefin sulfonate; anionic; emulsifier for cosmetics; emulsifier, dispersant for emulsion polymerization; powd.; 100% conc.

Lipolan R. [Lipo] Lanolin oil; emollient, spreading agent, conditioner, cosolv., plasticizer, and lubricant for personal care prods.; dispersant for pigments; yel., amber liq.; mild char. odor; sol. in min. and veg. oils, isopropyl esters, and anhyd. IPA; insol. in water; sapon. no. 85–110; cloud pt. 18 C max.

Lipolan S. [Lipo] Hydrog. lanolin; emollient, lubricant, and conditioner for cosmetics, toiletries, and topical pharmaceuticals; wh./off-wh. paste; mild char. odor; water-insol.; m.p. 45–53; acid no. 1 max.; sapon. no. 5 max.

Lipolan TE, TE(P). [Lion] Alkyl methyl tauride; anionic; scouring agent for textiles; dispersant for pigment; paste, powd. resp.; 25 and 28% conc. resp.

Lipomin LA. [Lion] Alanine type; external antistat for textiles and plastics; liq.

Lipomulse 165. [Lipo] Glyceryl stearate and PEG-100 stearate; nonionic; general purpose emulsifier, emollient, opacifier and visc. builder in creams and lotions; wh. bead or flake; HLB 11.0 ± 1; acid no. 2 max.; sapon. no. 90–100; 100% act.

Liponate 143M. [Lipo] Myreth-3 myristate; emollient ester for adjusting rub-in and afterfeel of personal care prods.; thickener and visc. controller; wh./yel. liq./paste; acid no. 6 max.; sapon. no. 90–100.

Liponate CL. [Lipo] Cetyl lactate; emollient for desired feel and penetration in personal care prods.; thickener and visc. controller; wh. soft solid to liq.; sol. in min. oil, veg. oil, ethanol; partly sol. in water; acid no. 3 max.; sapon. no. 174–195.

Liponate CRM. [Lipo] Cetyl ricinoleate; glosser, emollient with dry afterfeel; oil-sol., water-disp.; 100% act.

Liponate DPC-6. [Lipo] Dipentaerythritol hexacaprate/hexacaprylate; nontacky emollient for treatment prods.; visc. liq.; 100% act.

Liponate GC. [Lipo] Caprylic/capric triglyceride; see Liponate 143M; colorless liq.; sol. in anhyd. alcohol, min. and veg. oils; insol. in water; acid no. 0.1 max.; sapon. no. 325–355.

Liponate IPM. [Lipo] IPM; see Liponate 143M; colorless liq.; sol. in ethanol, min. and veg. oils; insol. in water; acid no. 2 max.; sapon. no. 202–211.

Liponate IPP. [Lipo] IPP; see Liponate 143M; colorless liq.; sol. see Liponate IPM; acid no. 2 max.; sapon. no. 183–190.

Liponate ML. [Lipo] Myristyl lactate; nongreasy, soft afterfeel emollient for bath/body oils, hydroalcoholic systems, creams, lotions, anhyd. and emulsified makeups; sol. in min. oil, veg. oil, ethanol, propylene glycol; insol. in water.

Liponate MM. [Lipo] Myristyl myristate; see Liponate 143 M; wh. solid wax.; sol. in oils; insol. in water; acid no. 5 max.; sapon. no. 120–135.

Liponate NPGC-2. [Lipo] Neopentyl glycol dicaprate/dicaprylate; dry feel emollient for creams, lotions, cleansers, antiperspirants; liq.; oil-sol.; water-insol.; 100% act.

Liponate PB-4. [Lipo] Pentaerythritol tetrabehenate; see Liponate 143M; off-wh. flakes; oil-misc., water-insol.; acid no. 10 max.; sapon. no. 159–169.

Liponate PC. [Lipo] Propylene glycol dicaprylate/dicaprate; see Liponate 143M; colorless liq.; sol. in min. and veg. oils, ethanol; water-insol.; acid no. 0.1 max.; sapon. no. 315–335.

Liponate PO-4. [Lipo] Pentaerythritol tetraoleate; see Liponate 143M; yel. liq.; sol. in min. and veg. oils, IPA; water-insol.; acid no. 10 max.; sapon. no. 185–195.

Liponate PS-4. [Lipo] Pentaerythritol tetrastearate; see Liponate 143M; wh. flakes; oil-misc., water-insol.; acid no. 10 max.; sapon. no. 183–198.

Liponate SPS. [Lipo] Cetyl esters (syn. spermaceti); see Liponate 143M; cream/wh. flakes; oil-misc., water-insol.; acid no. 5 max.; sapon. no. 109–120.

Liponate SS. [Lipo] Stearyl stearate; see Liponate 143M; off-wh. flakes; oil-misc., water-insol.; acid no. 5 max.; sapon. no. 103–117.

Liponate TDS. [Lipo] Tridecyl stearate; emollient for creams and lotions; liq.; 100% act.

Liponate TDTM. [Lipo] Tridecyl trimellitate; non-tacky emollient for treatment prods., hair; visc. liq.; 100% act.

Liponic 70-NC. [Lipo] Noncryst. sorbitol-type polyol; humectant, plasticizer, softener, and lubricant; adds sweet taste and pleasant mouthfeel to oral hygiene prods. such as dentifrices and mouthwashes; oral dosage pharmaceutical, also for adhesives, leather, and paper coatings; clear colorless sol'n.; water-sol.; sp.gr. 1.29–1.32; ref. index 1.455–1.470; pH neutral.

Liponic 76-NC. [Lipo] Noncryst. sorbitol-type polyol; see Liponic 70-NC; colorless to pale yel. clear syrup; odorless; sp.gr. 1.32–1.35; ref. index. 1.468–1.475; 25% max. water.

Liponic EG-1. [Lipo] Glycereth-26; nonionic; humectant in creams and lotions, lubricant, plasticizer for hair resins, foam stabilizer, pigment dispersant, hair conditioner, foam modifier; used in personal care prods.; colorless, clear to slightly hazy visc. liq.; sol. in water, alcohol, acetone and ethyl acetate; acid no. 0.5 max.; 100% act.

Liponic EG-7. [Lipo] Glycereth-7; humectant, hair conditioner, lubricant, and foam modifier for cosmetics and toiletries; adds lubricity and nongreasy luxurious feel to personal care prods.; colorless to pale yel. liq.; odorless; sol. in water, alcohol; 100% act.

Liponic SO-20. [Lipo] Sorbeth-20; humectant and plasticizer for cosmetics and toiletries; nongreasy rich afterfeel with moderate lubricity; yel. visc. liq.; bland char. odor; sol. in water, alcohol; acid no. 1 max.; 1% max. moisture.

Liponox LCR. [Lion] POE alkyl ether; nonionic; dispersant for pigment; wax; 100% conc.

Liponox N-105. [Lion] POE alkyl ether; nonionic; penetrant; liq.; 100% conc.

Liponox NC 6E, NCG, NCI, NCT. [Lion] POE alkylphenol ether; nonionic; detergent, penetrant, emulsifier, dispersant used in dyeing applic. for textiles, min. oil; wax, liq.; 100% conc.

Liponox OCS. [Lion] POE alkyl ether; nonionic; dye leveling agent; wax; 100% conc.

Lipopeg 2-DL. [Lipo] PEG-4 dilaurate; nonionic; dispersant, emulsifier, spreading agent and lubricant in personal care prods., bath oils; yel. liq.; HLB 6.0 ± 1; acid no. 10 max.; sapon. no. 170–185.

Lipopeg 4-DL. [Lipo] PEG-8 dilaurate; nonionic; see Lipopeg 2-DL; yel. liq.; HLB 10.0 ± 1; acid no. 10 max.; sapon. no. 125–142.

Lipopeg 4-DO. [Lipo] PEG-8 dioleate; see Lipopeg 2-DL; amber liq.; HLB 7.2 ± 1; acid no. 10 max.; sapon. no. 113–128.

Lipopeg 4-DS. [Lipo] PEG-8 distearate; see Lipopeg 2-DL; cream soft wax; HLB 8.0 ± 1; acid no. 10 max.; sapon. no. 113–128.

Lipopeg 4-L. [Lipo] PEG-8 laurate; nonionic; spreading agent, emulsifier, dispersant, lubricant for bath oils, creams and lotions; yel. liq.; HLB 13.0 ± 1; acid no. 5 max.; sapon. no. 90–100; 100% act.

Lipopeg 4-S. [Lipo] PEG-8 stearate; nonionic; see Lipopeg 4-L; cream paste; HLB 11.2 ± 1; acid no. 5 max.; sapon. no. 80–90; 100% act.

Lipopeg 6-L. [Lipo] PEG-12 laurate; nonionic; see Lipopeg 4-L; yel. liq.; HLB 14.6 ± 1; acid no. 5 max.; sapon. no. 65–76; 100% act.

Lipopeg 10-S. [Lipo] PEG-20 stearate; nonionic; spreading agent, emulsifier, dispersant, and lubricant for personal care prods.; solid; HLB 15.2; 100% conc.

Lipopeg 15-S. [Lipo] PEG-6-32 stearate; nonionic; see Lipopeg 10-S; solid; HLB 13.8; 100% conc.

Lipopeg 39-S. [Lipo] PEG-40 stearate; nonionic; see Lipopeg 4-L; wh. solid wax; HLB 16.9 ± 1; acid no. 2 max.; sapon. no. 23–35; 100% act.

Lipopeg 100-S. [Lipo] PEG-100 stearate; nonionic; see Lipopeg 4-L; tan flake or bead; HLB 18.8 ± 1; acid no. 1 max.; sapon. no. 9–20; 100% act.

Lipopeg 6000-DS. [Lipo] PEG-150 distearate; nonionic; see Lipopeg 4-L; off-wh. flake; HLB 18.4 ± 1; acid no. 10 max.; sapon. no. 12–20; 100% act.

Lipophos PE9. [Lipo] Nonylphenol ether phosphate ester acid form; anionic; coupling agent for textile scouring, emulsion polymerization, emulsifier and wetting agent; liq.; water-sol.; 100% conc.

Lipophos PL6. [Lipo] Linear alcohol ether phosphate ester, acid form; anionic; see Lipophos PE9; liq.; water-sol.; 100% conc.

Lipoquat C 25. [Lipo] Quaternized alkyl amine ethoxylate; cationic; antistatic agent, latex stabilizer for use where stability in highly acid or alkaline conditions is required; paste; 100% act.

Lipoquat R. [Lipo] Ricinoleamidopropyl ethyldimonium ethosulfate; cationic; conditioner, antistat, emollient, glosser, softener for personal care prods. and anhyd. systems; amber visc. liq.; char. odor; water-sol.; pH 6.5–7.5 (3% aq.); 95% act.

Liposorb L. [Lipo] Sorbitan laurate; nonionic; emulsifier, thickener, lubricant, antistat, all-purpose lipophilic surfactant used with POE Liposorb series; amber liq.; HLB 8.6 ± 1; sapon. no. 158–170; 100% act.

Liposorb L-10. [Lipo] PEG-10 sorbitan laurate; nonionic; o/w emulsifier, lubricant, antistat, all-purpose hydrophilic surfactant used for solubilizing oils and in conjunction with Liposorb esters; yel. liq.; HLB 14.9 ± 1; sapon. no. 66–76; 100% act.

Liposorb L-20. [Lipo] Polysorbate 20; nonionic; see Liposorb L-10; yel. liq.; HLB 16.7 ± 1; sapon. no. 40–50; 100% act.

Liposorb O. [Lipo] Sorbitan oleate; nonionic; see Liposorb L; yel./amber liq.; HLB 4.3 ± 1; sapon. no. 145–160; 100% act.

Liposorb O-5. [Lipo] Polysorbate 81; nonionic; hydrophilic surfactant used for solubilizing oils; emulsifier, lubricant, antistat; liq.; HLB 10.0; 100% conc.

Liposorb O-20. [Lipo] Polysorbate 80; nonionic; see Liposorb L-10; yel. liq.; HLB 15.0 ± 1; sapon. no. 45–55; 100% act.

Liposorb P. [Lipo] Sorbitan palmitate; nonionic; see Liposorb L; tan beads or flakes; HLB 6.7 ± 1; sapon. no. 139–151; 100% act.

Liposorb P-20. [Lipo] Polysorbate 40; nonionic; see Liposorb L-10; yel. liq.; HLB 15.6 ± 1; sapon. no. 40–53; 100% act.

Liposorb S. [Lipo] Sorbitan stearate; nonionic; see Liposorb L; cream beads or flakes; HLB 4.7 ± 1; sapon. no. 147–157; 100% act.

Liposorb S-20. [Lipo] Polysorbate 60; nonionic; see Liposorb L-10; yel. paste; HLB 14.9 ± 1; sapon. no. 45–55; 100% act.

Liposorb SQO. [Lipo] Sorbitan sesquioleate; nonionic; see Liposorb L; amber liq.; HLB 3.7 ± 1; sapon. no. 149–160; 100% act.

Liposorb TO. [Lipo] Sorbitan trioleate; nonionic; see Liposorb L; amber liq.; HLB 1.8 ± 1; sapon. no. 171–185; 100% act.

Liposorb TO-20. [Lipo] Polysorbate 85; nonionic; see Liposorb L-10; yel. liq.; HLB 11.0 ± 1; sapon. no. 82–95; 100% act.

Liposorb TS. [Lipo] Sorbitan tristearate; nonionic; see Liposorb L; cream flakes or beads; HLB 2.1 ± 1; sapon. no. 175–190; 100% act.

Liposorb TS-20. [Lipo] Polysorbate 65; nonionic; see Liposorb L-10; tan solid wax; HLB 10.5 ± 1; sapon. no. 88–98; 100% act.

Lipotin 100, 100J, SB. [Lucas Meyer] Refined soy lecithin, filtrated, deodorized; wetting and disp. agents preventing sedimentation in paints; visc. 10, 20, and 12 Pas resp.; acid no. 30, 34, and 30 max. resp.**Lipotin A.** [Lucas Meyer] Phosphoamino compd.; emulsifier, wetting and disp. agent, stabilizer for latex and emulsion paints, leather finishes, water-disp. and -reducible air-dry and stoving alkyds, offset printing inks, textile auxs.; lt. brn. visc. liq.; visc. 15 Pas; 99% solids.

Lipovol A. [Lipo] Avocado oil; conditioner, glosser, emollient imparting a lt., nongreasy, silky afterfeel to skin and hair prods.; high film gloss and rapid spread; used in personal care prods.; yel. to green visc. oil.; bland char. odor; sol. in oils; sp.gr. 0.908–0.925; acid no. 3 max.; sapon. no. 177–198; ref. index 1.460–1.470.

Lipovol ALM. [Lipo] Sweet almond oil; emollient; see Lipovol A; pale yel. clear oily liq.; bland, odorless; sol. in min. oil, isopropyl esters, ether, chloroform, benzene, and solv. hexane; water-insol.; acid no. 2 max.; sapon. no. 185–200.

Lipovol ALM-S. [Lipo] Veg. oil, sweet almond oil; see Lipovol A; economical replacement for natural sweet almond oil; yel. clear oily liq.; bland char. odor; sol. in min and veg. oils, isopropyl esters; water-insol.; acid no. 2 max.; sapon. no. 185–200.

Lipovol A-S. [Lipo] Veg. oil, avocado oil; economical replacement for natural avocado oil for cosmetic and toiletry applic.; imparts lt., nongreasy, silky afterfeel; rapid spread in bath, body, eye, and throat oils; bland taste, and high film gloss in lip preparations; yel. oily liq.; bland char. odor; sol. in min. oil, isopropyl esters, and ethanol; water-insol.; acid no. 3 max.; sapon. no. 180–198.

Lipovol CP. [Lipo] Cherry pit oil; see Lipovol A; lt. amber oil; sapon. no. 182–202.

Lipovol G. [Lipo] Grape seed oil; see Lipovol A; yel. amber oil; oil-sol.; acid no. 5 max.; sapon. no. 183–205.

Lipovol J. [Lipo] Jojoba oil, refined; emollient, conditioner, and lubricant for cosmetics and toiletries; rapid spread and soft, nontacky afterfeel; used in skin and personal care prods. and oils, anhyd. and emulsified makeups; yel. oil; char. nut-like odor; sol. in min. and veg. oils; insol. in water; acid no. 5 max.; sapon. no. 85–110.

Lipovol P. [Lipo] Apricot kernel oil; emollient used in cosmetics and pharmaceuticals; soft, nontacky afterfeel and high film gloss; straw oily liq.; bland char. fatty odor; sol. in min. oil and isopropyl esters; insol. in water; acid no. 1 max.; sapon. no. 185–195.

Lipovol P-S. [Lipo] Veg. oil, apricot kernel oil; see Lipovol P; economical replacement for natural apricot kernel oil; straw oily liq.; bland char. fatty odor; acid no. 1 max.; sapon. no. 185–195.

Lipovol SAF. [Lipo] Safflower oil; see Lipovol A; yel. oil; acid no. 2 max.; sapon. no. 182–202.

Lipovol SES. [Lipo] Sesame oil; emollient, solv., and vehicle used in cosmetics, toiletries, and pharmaceuticals; offers lt., nontacky feel and enhances gloss and spread of pigmented sticks and pot prods.; yel. clear liq.; bland char. odor; sol. in isopropyl esters and min. oil; insol. in water; acid no. 0.2 max.; sapon. no. 188–195.

Lipovol SES-S. [Lipo] Veg. oil, sesame oil; see Lipovol SES; economical replacement for natural sesame oil; yel. clear liq.; bland char. odor; acid no. 1.0 max.; sapon. no. 188–195.

Lipovol SO. [Lipo] Hybrid safflower oil; see Lipovol A; yel. oil; oil-sol.; acid no. 1 max.; sapon. no. 184–196.

Lipovol SUN. [Lipo] Sunflower seed oil; emollient imparting a pleasant, nongreasy feel to skin and hair care prods. and makeups; adds conditioning, spread, and sheen to personal care prods.; used in preparation of margarine; lt. yel. oily liq.; bland, char. fatty odor; sol. in min. oil and isopropyl esters; insol. in water; acid no. 2 max.; sapon. no. 185–195; 0.05% max. moisture.

Lipovol W. [Lipo] Walnut oil; emollient for makeup, skin, and hair care prods. where a rich, nontacky, persistent afterfeel is desired; rapid spread; yel./ amber oil; bland char. odor; sol. in min. oil and isopropyl esters; acid no. 0.5 max.; sapon. no. 185–202.

Lipovol WGO. [Lipo] Wheat germ oil; emollient imparting perceptible afterfeel to skin and hair care prods.; gloss to anhyd. and emulsified makeups; brn. oil; char. fatty odor; sol. in oils; insol. in water; acid no. 5 max.; sapon. no. 175–195.

Lipowax. [Lipo] Cetearyl alcohol, sodium lauryl sulfate, cetyl esters, myristyl alcohol; nonionic; wax used for building visc. in personal care prods.; solid; 100% conc.

Lipowax C. [Lipo] N,N′-ethylenebisstearamide; nonionic; internal lubricant in thermoplastic and thermosetting resins; defoamer for paper, textile processing; antitacking and antiblocking agent, antistat used in vinyls, PS, polyethylene, ABS; mold release agent, lubricant, antiblock, and detackifying agent in syn. rubbers; surf.-finishing agent in hard rubber; paints and lacquers to aid pigment suspension and salt spray resistance; used in food pkg. materials; lt. cream prill, powd., or atomized powd.; sol. in toluene, naphtha, kerosene, and turpentine; m.p. 140–145 C; acid no. 10 max.; flash pt. 285 C (COC); 100% act.

Lipowax D. [Lipo] Cetearyl alcohol, ceteareth-20; nonionic; o/w SE wax, used in skin and hair creams and lotions, personal care prods.; wh. to off-wh. waxy solid; bland, char. odor; sol. in alcohol, misc. warm with most oil phase ingredients, disp. in warm water; m.p. 46–55 C; HLB 11; sapon. no. 2; 100% act.

Lipowax G. [Lipo] Cetearyl alcohol and ceteareth-20; nonionic; emulsifier for personal care prods.; solid; water-disp.; 100% conc.

Lipowax NI. [Lipo] Cetearyl alcohol and ceteth-20; nonionic; see Lipowax D; solid; water-disp.; 100% conc.

Lipowax P. [Lipo] Blend of fatty alcohols and ethoxylates; nonionic; o/w self-emulsifying wax for neutral and mildly acidic and alkaline pH systems; creamy, waxy solid; bland, char. odor; sol. in alcohol, misc. warm with most oil phase ingred., disp. in warm water; m.p. 48–52 C; HLB 9; sapon. no. 14; 100% act.

LiquaPar. [Mallinckrodt] Isopropyl, isobutyl and n-butyl esters of p-hydroxybenzoic acid; preservative for cosmetics and topical pharmaceuticals; liq.; sol. in alcohol, propylene glycol; slightly sol. in water.

Liquazinc AQ-90. [Witco] Aq. zinc stearate disp.; antitack agent, lubricant for rubber; release aid and lubricant for abrasive papers; wh. disp.; fineness 99.9% thru 325 mesh; sp.gr. 1.02; visc. 1000 cps; 50% solids.

Liquester. [Robeco] Ester C_{30-46} piscine oil; emollient, lubricant, moisturizer for skin applic.; liq.

Liquid Absorption Base Type A, T. [Croda] Min. oil, lanolin alcohol; emollient for liq. make-up to improve dispersion and applic. properties of pigments; primary oil phase ingred. in o/w emulsions; type T is better solv. for oil-sol. dyes; clear yel. liqs.

Liquilan. [Amerchol] Lanolin oil; emollient; oil-sol.

Lite-R-Cobs®. [Andersons] Corncob meal; inert plastic extender and filler; replaces wood flour in wood particle molding with phenolic resins, in profile and sheet stock prod.; filler in glue, asphalt, caulking compds., and rubber; also used in industrial abrasives, as industrial absorbent, as agric. chemical carriers, livestock feed roughage; tan gran.; essentially odorless; 20.7% sol. in 1% sodium hydroxide, 7.4% in hot water, 4.0% in alcohol, 2.4% in 10% sulfuric acid; 2.1% in acetone; sp.gr. 0.8; bulk dens. 8–15 lb/ft^3; oil absorp. 500%; water absorp. 727.0%; flash pt. (OC) 350 F; hardness (Mohs) 1.0; 35.7% cellulose.

Litharge 33. [Eagle-Picher] High-purity lead oxide; acid acceptor, activator, and vulcanizing agent in rubber compding; 0.55 μ median particle diam.; sp.gr. 9.5; dens. 13–17 g/in.^{3}

Lithopone 30% DS, 60% DS. [Sachtleben] Zinc sulfide/barium sulfate, micronized; wh. pigment for thermosets, glass fiber-reinforced thermosets and thermoplastics, and easily oxidizable thermoplastics; dens. 4.3 and 4.2 g/ml resp.; pH 8 and 7 resp.

Lithopone 30% L, 60% L. [Sachtleben] Zinc sulfide/ barium sulfate; see Lithopone 30% DS; dens. 4.3 and 4.2 g/ml resp.; pH 7.

Lithopone D (Red Seal 30% ZnS). [Sachtleben] Lithopone coated with org. surfactantwh. pigment consisting of zinc sulfide and barium sulfate; used as TiO_2 substitute for thermoplastic masterbatches, in thermosets based on urea, melamine, and polyester, in natural and syn. elastomers, and in paper applics.; dens. 4.3 g/cc; bulk dens. 1.70 l/kg (poured), 0.90 l/ kg (tapped); surf. area 3 m^2/g; oil absorp. 9; pH 8; hardness (Mohs) 3.

Lithopone D (Silver Seal 60% ZnS). [Sachtleben] Lithopone coated with org. surfactantsee Lithopone D (Red Seal 30% ZnS); dens. 4.2 g/cc; bulk dens. 1.80 l/kg (poured), 1.00 l/kg (tapped); surf. area 5 m^2/ g; oil absorp. 10; pH 7; hardness (Mohs) 3.

Lithopone DS (Red Seal 30% ZnS). [Sachtleben] Lithoponesee Lithopone D; very good dispersibility; micronized form; dens. 4.3 g/cc; bulk dens. 1.60 l/kg (poured), 0.80 l/kg (tapped); surf. area 3 m^2/g; oil absorp. 8; pH 8; hardness (Mohs) 3.

Lithopone L (Red Seal 30% ZnS). [Sachtleben] Lithopone; see Lithopone D; dens. 4.3 g/cc; bulk dens. 1.80 l/kg (poured), 0.90 l/kg (tapped); surf. area 3 m^2/g; oil absorp. 9; pH 7; hardness (Mohs) 3.

Lithopone L (Silver Seal 60% ZnS). [Sachtleben] Lithoponesee Lithopone D; dens. 4.2 g/cc; bulk dens. 1.90 l/kg (poured), 1.00 l/kg (tapped); surf. area 5 m^2/ g; oil absorp. 10; pH 7; hardness (Mohs) 3.

Lomar D. [Henkel] Highly polymerized naphthalene sulfonate, sodium salt; anionic; dispersing agent for pigments in cement, ceramics, paper, pesticides, for carbon black pigments in latex systems; clear amber powd.; water-sol.; dens. 0.68 g/cc; pH 9.3 (10%); 84% act.

Lomar DL. [Henkel] Liq. version of Lomar D; amber liq.; sp.gr. 1.20; pH 9.0; 30% act.

Lomar HP. [Henkel] Condensed potassium naphthalene sulfonate; anionic; dispersant; powd.; 80% act.

Lomar LS. [Henkel] Condensed sodium naphthalene sulfonate; anionic; dispersant for emulsion polymerization, dyestuff mfg., agric. formulations; emulsifier; leveling agent for dyeing fibers; tan powd.; sol. in slightly hazy aq. sol'ns.; pH 9.5 (10% aq.); 95% act.

Lomar LS Liq. [Henkel] Liq. version of Lomar LS; water-sol.

Lomar PL. [Henkel] Condensed sodium naphthalene sulfonate; dispersant for pigments, extenders, and fillers in aq. media; used in dyeing syn. and natural fibers; used in ceramics, emulsion polymerization; gypsum board, for pigments, printing, rubber and wet milling; food pkg. applics.; agric. prods.; amber liq.; water-sol.; sp.gr. 1.25; pH 9.5 (20%); 41% act.

Lomar PW. [Henkel] Condensed sodium naphthalene sulfonate; anionic; see Lomar PL; also suspending agent, stabilizer for paint and paper industries; tan powd.; water-sol.; dens. 0.66 g/cc; pH 9.5 (10%); 87% act.

Lomar PWA. [Henkel] Condensed ammonium naphthalene sulfonate; anionic; visc. depressant; for molding and extruding operations in ceramics; dispersant for emulsion paints and emulsion polymerization of syn. elastomers; visc. reducer for pigment slurries; stabilizer; lt. tan powd.; sol. in water; pH 7.2 (10%); 92% act.

Lomar PWM. [Henkel] Condensed naphthalene formaldehyde sulfonate; anionic; dispersant for wettable powds., dry flowables, and aq. suspension pesticides; powd.; 85% act.

Lomar ST. [Henkel] Polymerized condensed sodium naphthalene sulfonate; dispersing and suspending agent; water-sol.

Lonzaine® 12C. [Lonza] Coco betaine; amphoteric; conditioner used in personal care prods. and industrial applics.; liq.; biodeg.; 35% conc.

Lonzaine® C, CO. [Lonza] Cocoamidopropyl betaine; amphoteric; foaming agent, conditioner, visc. booster, wetting agent, used in cosmetics, toiletries, detergents, metal finishing, textile finishing, etc.; Gardner 2 nonvisc. liq.; char. odor; surf. tens. 33.6 and 33.9 dynes/cm resp. (0.1%); pH 4.5–5.5 and 6–8 resp. (10%); biodeg.; 30% act.

Lonzaine® CS. [Lonza] Cocoamidopropyl hydroxysultaine; amphoteric; see Lonzaine 12C; liq.; water-sol.; 50% conc.

Lonzest® 143-S. [Lonza] Myristyl propionate; nonionic; emollient, penetrant, and spreading agent in cosmetics; perfume solv. in personal care prods.; humectant; APHA 5–10 clear liq.; f.p. –5 C; sol. in acetone, CCl_4, castor, lanolin, peanut, silicone, and cottonseed oils, ethanol, ethyl acetate, heptane, IPA, toluene; sp.gr. 0.852.

Lonzest® PEG 4-DO. [Lonza] PEG-8 dioleate; nonionic; detergent, lubricant, dispersant, anticorrosive, emulsifier, defoamer, used in textile, petrol., insecticide industries; yel. liq.; sol. in min. oil, IPA, toluol, disp. in water; sp.gr. 0.98; m.p. 0 C; HLB 9.9; acid no. 5–10; sapon. no. 120–130; 100% conc.

Lonzest® PEG 4-L. [Lonza] PEG-8 laurate; nonionic; dispersant, rewetting agent, stabilizer, used in cosmetics, pharmaceuticals, textile, paper, agric. industries; yel. liq.; sol. in water, IPA, toluol; sp.gr. 1.03; m.p. 10 C; HLB 13.1; acid no. 5–10; sapon. no. 90–100; 100% conc.

Lonzest® SML. [Lonza] Sorbitan laurate; nonionic; emulsifier for o/w systems; fiber lubricant and corrosion inhibitor used in food industry; liq.; 100% conc.

Lonzest® SML-20. [Lonza] Polysorbate 20; nonionic; emulsifier used where o/w and w/o emulsions are required; solubilizer; liq.; 100% conc.

Lonzest® SMO. [Lonza] Sorbitan monooleate; nonionic; see Lonzest SML; liq.; 100% conc.

Lonzest® SMO-20. [Lonza] Polysorbate 80; nonionic; emulsifier, solubilizer and stabilizer used in foods, personal care prods. and industrial applics.; liq.; 100% conc.

Lonzest® SMP. [Lonza] Sorbitan palmitate; nonionic; see Lonzest SML; solid; 100% conc.

Lonzest® SMP-20. [Lonza] Polysorbate 40; nonionic; see Lonzest SMO-20; liq.; 100% conc.

Lonzest® SMS. [Lonza] Sorbitan stearate; nonionic; see Lonzest SML; flake; 100% conc.

Lonzest® SMS-20. [Lonza] Polysorbate 60; nonionic; see Lonzest SMO-20; liq.; 100% conc.

Lonzest® STO. [Lonza] Sorbitan trioleate; nonionic; see Lonzest SML; liq.; 100% conc.

Lonzest® STO-20. [Lonza] Polysorbate 85; nonionic; see Lonzest SMO-20; liq.; 100% conc.

Lonzest® STS. [Lonza] Sorbitan tristearate; nonionic; see Lonzest SML; solid; 100% conc.

Lonzest® STS-20. [Lonza] Polysorbate 65; nonionic; see Lonzest SMO-20; solid; 100% conc.

Lorol CT-M. [Albright & Wilson] Syn. fatty alcohol triethoxysulfate and alkanolamide; anionic; pearlescent hair shampoo conc. used in personal care prods.; wh. pearlescent visc. liq.; dilutable in water; sp.gr. 1.054; visc. 4500 ± 1500 cP; pH 7.0 ± 0.8 (2%); 23.0 ± 1.0% act.

Loropan CME. [Triantaphyllou] Cocamide MEA; nonionic; foam stabilizer and thickening agent for shampoos and detergents; bead; 92–96% conc.

Loropan KD. [Triantaphyllou] Cocamide DEA; nonionic; emulsifier, stabilizer, thickener for shampoos and bubble baths; liq., paste; 85–90% conc.

Loropan KM. [Triantaphyllou] Cocamide MEA; nonionic; see Loropan KD; paste; 30% conc.

Loropan LD. [Triantaphyllou] Lauramide DEA; nonionic; foam stabilizer for shampoos, detergents, fortifier for perfumes in soap; solid; 90% conc.

Loropan LM. [Triantaphyllou] Lauramide MEA; nonionic; foam stabilizer and thickening agent for shampoos and detergents; bead; 92–96% conc.

Loropan LMD. [Triantaphyllou] Lauric/myristic DEA; nonionic; foam stabilizer; superfatting and thickening agent for shampoos; fortifier for perfume in soaps; paste; 90% conc.

Loropan OD. [Triantaphyllou] Oleic DEA; nonionic; foam stabilizer, thickener, and superfatting agent for shampoos and bubble bath prods.; liq.; 90% conc.

Lostat 105D. [Ciba-Geigy Switzerland] Lignosulfonate; anionic; dispersant for dyestuffs; powd.; 100% conc

Lo-Vel® 27. [PPG Industries] Precipitated silica; thickener, flatting agent for topcoat lacquers; powd.; 1.7 μm median size; dens. 4–6 lb/ft³ (tapped); surf. area 170 m²/g; oil adsorp. 220 linseed oil lb/100 lb; pH 7.

Lo-Vel® 28. [PPG Industries] Precipitated silica; thickener, flatting agent for coil coating; powd.; 4.2 μm median size; dens. 7 lb/ft³ (tapped); surf. area 170 m²/g; oil adsorp. 220 linseed oil lb/100 lb; pH 7.

Lo-Vel® 29. [PPG Industries] Precipitated silica; thickener, flatting agent for coil coating; powd.; 4.8 μm median size; dens. 7 lb/ft³ (tapped); surf. area 170 m²/g; oil adsorp. 220 linseed oil lb/100 lb; pH 7.

Lo-Vel® 39. [PPG Industries] Precipitated silica; thickener, flatting agent for coatings, micro texture finish; powd.; 6.7 μm median size; dens. 10 lb/ft³ (tapped); surf. area 170 m²/g; oil adsorp. 220 linseed oil lb/100 lb; pH 7.

Lo-Vel® 275. [PPG Industries] Precipitated silica; thickener, flatting agent for general purpose and coil coatings; powd.; 3.6 μm median size; dens. 7 lb/ft³ (tapped); surf. area 170 m²/g; oil adsorp. 220 linseed oil lb/100 lb; pH 7.

Low Crock #5012. [Polymer Research] Self-crosslinking reactive prod.; nonionic; used as binder for roller printing in water phase and aq. systems; compatibility with water phase pigment disps. and printing clears, fastness to dry, wet and perchlor crocking, good lt. and dry cleaning fastness; milky wh. liq. emulsion; dens. 8.8 lb/gal; visc. 200 cps; 46% latex solids.

Low Crock #5013. [Polymer Research] Self-crosslinking reactive prod.; see Low Crock #5012; exc. soft hand; milky wh. liq. emulsion; sp.gr. 0.99; dens. 8.20 lb/gal; visc. 13 cps; pH 8.5–9.0; 38% solids.

Low Crock #5026. [Polymer Research] Self-crosslinking reactive prod.; see Low Crock #5012; also soft hand, redispersibility, mechanical stability and runability on the screens; milky wh. liq. emulsion; dens. 8.07 lb/gal; visc. 14 cps; pH 8.5–9.0; 29.3% latex solids.

Low Crock #5087. [Polymer Research] Self-crosslinking latex stable emulsion polymer; nonionic; all aq. screen/roller print soft hand low crock binder, room temp. cure; used in pigment printing or as a binder for nonwovens; wash and dry clean fastness; milky wh. liq.; dens. 8.9 lb/gal; visc. 14 cps; pH 8.5–9.0; 35% solids.

Low Crock Formula #750. [Polymer Research] Self-crosslinking, nonionic low crock; flame-resistant low crock for water-phase textile printing; extremely soft binder with exc. anticrocking properties on fiberglass, syns. etc., exc. pigment binding properties, pigment loading, color permanency, and low temp. flexibility; dry cured film exhibits superior washability, dry cleanability, and flame resistance; sp.gr. 1.085; dens. 9.05 lb/gal; pH 7; 30% solids.

Loxiol G 10. [Henkel] Monoglycerol ester of unsat. fatty acid; internal lubricant used in clear or opaque applic.; food applic.; costabilizer in combination with tin; aids flow in pigmented and filled rigid PVC compds.; pale yel. liq.; f.p. –1 to –8 C; dens. 7.9 lb/gal; visc. 200–225 cps; acid no. 3.0 max.; ref. index 1.467 ± 0.005; flash pt. 435 F (CC).

Loxiol G 11. [Henkel] Fatty acid ester of polyol; internal lubricant for rigid and soft PVC, used in rigid PVC film and sheeting; liq.; sp.gr. 0.970–0.985; visc. 500–610 mPa.s; acid no. < 1; ref. index 1.473–1.478; flash pt. > 220 C.

Loxiol G 12. [Henkel] Monoglycerol ester of sat. fatty acid; see Loxiol G 10; wh. beads, 20 mesh particle size; dens. 20–30 lb/ft³; acid no. 2 max.; ref. index 1.441 ± 0.005 (80 C); flash pt. 428 F (OC).

Loxiol G 13. [Henkel] Fatty acid ester of polyol; lubricant for rigid PVC esp. in cable and inj. molding; liq.; sp.gr. 0.905–0.915; visc. 35–40 mPa.s; acid no. < 1; ref. index 1.460–1.466; flash pt. > 190 C.

Loxiol G 15. [Henkel] Ester of hydroxy sat. fatty acids; intermediate lubricant with release and flow properties; improves hot-fill properties; low water sensitivity; pale tan flakes, off-wh. beads, 20 mesh particle size; dens. 20–30 lb/ft³; acid no. 5 max.; ref. index 1.445 ± 0.005 (95 C); flash pt. > 300 C (OC).

Loxiol G 16. [Henkel] Glycerol ester of unsat. fatty acid; internal lubricant for rigid PVC, calendered and rigid sheet, clear, opaque inj. molding and profiles; costabilizer in tin-stabilized rigid, semirigid, or plasticized systems; printing and laminating properties to articles; pale yel. fluid; f.p. < –5 C; dens. 7.3 lb/gal; visc. 90–120 cP; acid no. 1.0 max.; ref. index 1.468 ± 0.005; flash pt. 460 (CC).

Loxiol G 20. [Henkel] Stearic acid; external lubricant for rigid and soft PVC; solid; sp.gr. 0.840–0.850 (80 C); visc. 8–11 mPa.s (80 C); acid no. 207–210; ref. index 1.436–1.437 (60 C); flash pt. > 180 C.

Loxiol G 21. [Henkel] Hydroxystearic acid; external lubricant for rigid and soft PVC; antiplate-out effect; solid; sp.gr. 0.885–0.891 (80 C); visc. 30–35 mPa.s (80 C); acid no. 172–180; ref. index 1.440–1.442 (80 C); flash pt. > 210 C.

Loxiol G 22. [Henkel] Hard paraffin; external lubricant for rigid PVC; solid; visc. 10 mPa.s (120 C); acid no. < 0.1; flash pt. > 280 C.

Loxiol G 30. [Henkel] Simple fatty acid esters; internal and external lubricant used in rigid PVC inj. moldings, calendered or extruded sheet, and complicated interior and exterior profiles; superior flow and offers superior lt., heat, and moisture resistance suitable for outdoor applic.; partial substitute for calcium stearates due to excellent flow properties; resists plate-out; wh. soft flakes; dens. 20–25 lb/ft³; acid no. 6.0 max.; ref. index 1.436 ± 0.005 (70 C); flash pt. > 280 F.

Loxiol G 31. [Henkel] Fatty acid ester; lubricant for PS; aux. for dry-coloring gran.; liq.; sp.gr. 0.853–0.857; visc. 8–11 mPa.s; acid no. < 0.5; ref. index 1.441–1.444; flash pt. > 170 C.

Loxiol G 32. [Henkel] Fatty acid ester; all purpose lubricant for rigid PVC extrusion; PC prod.; solid; sp.gr. 0.816–0.819 (80 C); visc. 6–7 mPa.s (80 C); acid no. < 2; ref. index 1.430–1.435 (80 C); flash pt. > 230 C.

Loxiol G 33. [Henkel] Waxy ester of simple fatty acids and fatty alcohols; internal and external lubricant used in rigid PVC inj. molding, calendered or extruded sheet, and complicated interior and exterior profiles; useful in engineering resins; promotes superior flow and superior lt. and heat stability and resistance to water for outdoor applic.; replaces amide waxes; partial replacement for calcium stearates due to excellent flow properties; wh. beads, 20 mesh particle size; dens. 20–25 lb/ft^3; acid no. 3.0 max.; ref. index 1.454 ± 0.005 (70 C); flash pt. > 285 F.

Loxiol G 40. [Henkel] Simple ester from branched chain fatty alcohol and fatty acid; lubricant for calendering rigid and semirigid sheet, extrusion of exterior profiles and semirigid wire and cable compds.; intermediate lubricant in phenolic inj. molding compds. and nylon resins; clear applic.; internal lubricant for semirigid and plasticized applic.; exterior applic.; pale yel. liq.; f.p. 8 C max.; dens. 7.2 lb/gal; visc. 25–35 cps; acid no. 2.0 max.; ref. index 1.452 ± 0.005; flash pt. 401 F.

Loxiol G 41. [Henkel] Fatty acid ester; compatible lubricant for rigid PVC in glass clear articles; lubricant for duroplasts; solid; sp.gr. 0.820–0.827 (80 C); visc. 5–7 mPa.s (80 C); acid no. < 2; ref. index 1.430–1.435 (80 C); flash pt. > 220 C.

Loxiol G 47. [Henkel] Fatty acid ester; externally act. lubricant for rigid PVC extrusion and PC prod.; solid; sp.gr. 0.816–0.821 (80 C); visc. 7–10 mPa.s (80 C); acid no. < 2; ref. index 1.433–1.437 (80 C); flash pt. > 250 C.

Loxiol G 53. [Henkel] Fatty alcohol; internal lubricant for rigid PVC; inj. molding; solid; sp.gr. 0.790–0.802 (80 C); visc. 4–6 mPa.s (80 C); acid no. < 0.2; ref. index 1.427–1.430 (80 C); flash pt. > 160 C.

Loxiol G 60. [Henkel] Dicarboxylic acid ester of sat. aliphatic alcohols; universally applicable internal lubricant for rigid PVC; solid; sp.gr. 0.878–0.884 (80 C); visc. 10–13 mPa.s (80 C); acid no. < 2; ref. index 1.453–1.458 (80 C); flash pt. > 230 C.

Loxiol G 70. [Henkel] Polymeric complex ester of sat. fatty acids; lubricant with good compatibility; used in clear, rigid PVC applic.; replacement of montan esters on cost performance basis; good heat stability and results in superior clarity; wh. flakes; dens. 25–30 lb/ft^3; acid no. 17.5 max.; ref. index 1.454 ± 0.005 (70 C); flash pt. > 280 F.

Loxiol G 71. [Henkel] Complex ester from unsat. fatty acids; external lubricant for rigid PVC processing; effective in semirigid sheet blown film, wire and cable compd.; prods. articles of sparkling clarity; does not adversely affect HDT and hot-fill temps.; replace monton esters on a cost performance basis; lt. visc. liq.; f.p. –10 to 0 C; dens. D 7.86 lb/gal; visc. 500–700 cP; acid no. 17.5 max.; ref. index 1.469 ± 0.005; flash pt. 530 F.

Loxiol G 72. [Henkel] High-molecular complex esters; compatible lubricant with release properties for rigid PVC; useful for rigid PVC bottles, film, and sheeting; solid; sp.gr. 0.887–0.895 (80 C); visc. 18–27 mPa.s (80 C); acid no. < 7; flash pt. > 240 C.

Loxiol G 73. [Henkel] High-molecular complex esters; compatible lubricant for rigid and soft PVC with release properties; soft PVC articles; internally and externally applicable release agent for PU; liq.; sp.gr. 0.905–0.912; visc. 150–185 mPa.s; acid no. < 6; ref. index 1.461–1.466; flash pt. > 220 C.

Loxiol G 74. [Henkel] High-molecular complex ester; release agent for rigid PVC; solid; sp.gr. 0.913–0.917 (80 C); visc. 60–67 mPa.s (80 C); acid no. < 12; ref. index 1.442–1.445 (80 C); flash pt. > 230 C.

Loxiol G 78. [Henkel] High-molecular complex ester; release agent containing metal soaps for prod. of packing articles from rigid PVC; calendered film and calcium zinc stabilized bottles; solid; acid no. 15 max.; flash pt. > 250 C.

Loxiol GH 4. [Henkel] Mixed glycerol esters; intermediate lubricant system used in clear blow-molded bottles and calendered or extruded PVC sheet; produces articles with excellent clarity; strong synergistic effects with tin stabilizers; internal lubricant with metal release effects; wh. flakes; dens. 15–20 lb/ft^3; acid no. 2.0 max.; ref. index 1.4535 ± 0.005 (70 C); flash pt. > 560 F (OC).

Loxiol HOB 7107. [Henkel] Polymeric ester of sat. fatty acids; external lubricant with good compatibility; used in clear rigid PVC applic.; replacement of montan esters on cost performance basis; used in food contact applic.; good heat stability and resists water blush; wh. beads 20 mesh particle size; dens. 15–20 lb/ft^3; acid no. 17.5 max.; ref. index 1.444 ± 0.005 (95 C); flash pt. 585 F.

Loxiol HOB 7108. [Henkel] Mixed glycerol esters and complex polymeric fatty acid esters; one-pkg. lubricant system used in clear blow-molded bottles and calendered or extruded PVC sheet; excellent water and alcohol blush resistance; produces articles of high clarity; promotes melt flow, reduces melt fracture, and contributes to external metal release effects; very pale tan beads, 20 mesh particle size; dens. 20–25 lb/ft^3; acid no. 17.5 max.; ref. index 1.444 ± 0.005 (95 C); flash pt. > 300 F.

Loxiol HOB 7111. [Henkel] Mixed glycerol esters; intermediate lubricant systems used in clear, blow-molded bottles and calendered or extruded PVC sheet; produces high-clarity articles; strong synergistic effect with stabilizers; internal lubricant with external metal release effects; wh. flakes; dens. 15–20 lb/ft^3; acid no. 5 max.; ref. index 1.438 ± 0.005 (95 C); flash pt. > 285 F.

Loxiol HOB 7112. [Henkel] Mixed glycerol esters; balanced, one-pkg. lubricant system used in clear blow-molded bottles, calendered, and/or extruded PVC sheet where FDA approval is required; produces high-clarity articles; promotes melt flow, reduces melt fracture, and contributes external lubrication metal release effects; wh. flakes; dens. 15–20 lb/ft^3; acid no. 15 max.; ref. index 1.4523 ± 0.005 (95 C); flash pt. 460 F.

Loxiol HOB 7119. [Henkel] Long chain fatty acid ester of polyfunctional alcohol; intermediate lubricant used in rigid PVC, PC, and polyesters; imparts good flow properties and release effects without significant adverse effect on clarity; internal and external lubricant with little effect on HDT; wh. beads, 20 mesh particle size; dens. 15–25 lb/ft^3; acid no. 2.0 max.; ref. index 1.4490 ± 0.005 (95 C); flash pt. 580 F (OC).

Loxiol HOB 7121. [Henkel] Partial fatty acid ester; internal lubricant/costabilizer for rigid PVC processing; superior flow in rigid PVC inj. molding and complicated profiles; used in PVC extrusion films, calendered sheets, and blow-molded bottle applic.; good thermal and lt. stability; effective in high calcium stearate compds.; highly costabilizing in CaZn stabilized and lead stabilized PVC systems; resistant to plate-out; off-wh. beads or flakes, 20 mesh bead; dens. 30–35 lb/ft^3 (bead), 15–25 lb/ft^3 (flake); acid no. 2.0 max.; ref. index 1.450 ± 0.004 (70 C); flash pt. 535 F.

Loxiol HOB 7131. [Henkel] Monoglycerol ester of sat. fatty acid; internal lubricant for clear or opaque applic.; food applic.; mildly costabilizing in tin-stabilized systems; good flow in pigmented and filled rigid PVC compds.; wh. beads, 20 mesh bead; dens. 25–30 lb/ft^3; acid no. 3.0 max.; ref. index 1.445 ± 0.005 (70 C); flash pt. 428 F (CC).

Loxiol HOB 7162. [Henkel] Esters of fatty acids and fatty alcohols; internal and external lubricant used in rigid PVC extrusions of complicated interior and exterior profiles; superior flow, lt. and heat stability, resistance to water absorp., and weatherability; well suited to exterior applics.; wide processing latitude; wh. flakes; dens. 20–30 lb/ft^3; ref. index 1.44; flash pt. 280 C.

Lubracal® 48. [Witco] Calcium stearate-based aq. disp.; lubricant and plasticizer for paper coatings; meets lubricity, stability, visc., and color requirements of the most sophisticated processes; improved flow and leveling of coatings enabling smooth finishes with enhanced gloss and printability; wh. powd., 99.9% thru 325 mesh sieve; sp.gr. 1.00; visc. 110 cps; 50.1% solids.

Lubracal® 53. [Witco] Calcium stearate-based aq. disp.; see Lubracal® 48; wh. powd., 99.9% thru 325 mesh sieve; sp.gr. 1.00; visc. 700 cps; 56.0% solids.

Lubracal® 60. [Witco] Calcium stearate-based aq. disp.; see Lubracal® 48; wh. powd., 99.9% thru 325 mesh sieve; sp.gr. 1.02; visc. 350 cps; 60.0% solids.

Lubrajel®. [Guardian] Polyglyceryl methacrylate and propylene glycol; autoclavable nondrying water-sol. lubricant for medical and surgical use; clear, colorless visc. gel; sp.gr. 1.3 g/ml; visc. 300,000–400,000 cps; pH 5.0–6.0.

Lubrazinc® W, Superfine. [Witco] Zinc stearate; lubricant for fiber-reinforced plastics; dry lubricant for powd. metallurgy applic.; used with iron powd. to yield compacts of high uniform density; wh. powd.; sol. in hot turpentine, benzene, toluene, xylene, CCl_4, veg. and min. oils, waxes; insol. in water; sp.gr. 1.10; soften. pt. 120 C; 14.0% conc.

Lubrimet® P 600, P 900. [BASF AG] PPG; lubricant; solubilizer for dyestuffs and surfactants; liq.; misc. with water and oil.

Lubrisol. [Scher] Natural oil; nonionic; textile lubricant; liq.; 100% act.

Lubrizol 2106. [Lubrizol] Barium phenate; heat stabilizer for PVC coatings, films, and fabricated materials; liq.; 28% Ba.

Lubrizol 2116. [Lubrizol] Barium carboxylate; see Lubrizol 2106; liq.; 34% Ba.

Lubrizol 2117. [Lubrizol] Calcium carboxylate; heat stabilizer for PVC coatings, films, bottles, extruded pipe, and food contact applics.; FDA approved; liq.; 14% Ca.

Lubrizol 2163. [Lubrizol] Calcium sulfonate; dispersant for pigments in thermoplastic and epoxy color concs.; visc. stabilizer and reducer in plastisols at 0.5–1.5 phr; liq.

Lubrizol 2164, 2165. [Lubrizol] Succinimide; pigment wetting and dispersing agent for hard-to-disperse pigments in plastisols and color concs.

Lubrol 12A-9. [ICI PLC] Dodecyl alcohol condensate (9.5 moles); nonionic; antistat for polyethylene; paste; HLB 13.8; 100% conc.

Lubrol 17A-10. [ICI PLC] Oleyl/cetyl alcohol condensate (10 moles); nonionic; antifoaming, emulsifier; used in sea water evaporator; oleines, and paraffin waxes; paste; HLB 12.7; 100% conc.

Lubrol 17A-17. [ICI PLC] Fatty alcohol EO condensate; nonionic; emulsifier used to produce o/w dispersions; antistatic lubricant for polyamide continuous-filament yarn and staple fiber; flakes; 100% conc.

Lubrol 101. [ICI PLC] Proprietary blend of fatty acid esters, plasticizers, and dispersing agents; flow promoter for natural/syn. rubber-based compds.; wh. to cream nondusting powd.; sp.gr. 0.7.

Luchem AS-946. [Atochem] Methyl methacrylate/allyl methacrylate copolymer; antishrink additive.

Luchem AS-946-25. [Atochem] Methyl methacrylate/allyl methacrylate copolymer; antishrink additive; 25% sol'n. in ADC.

Lucidol-70. [Atochem] Benzoyl peroxide; initiator for bulk, sol'n., and suspension polymerization and high-temp. and R.T. cure of polyester resins; used in pharmaceutical applic.; granular wet solid; 70% act.; 4.36–4.52% act. oxygen.

Lucidol-78. [Atochem] Benzoyl peroxide; initiator for bulk, sol'n., and suspension polymerization, and high-temp. and R.T. cure of polyester resins; granular wet solid; 78% act.; 4.95–5.28% act. oxygen.

Lucidol-98. [Atochem] Benzoyl peroxide; see Lucidol-78; also used for curing elastomers, cure of acrylic syrup, and polymer modification, thermoplastic crosslinking; granular solid; 98% act.; 6.5% act. oxygen.

Ludigol 60, F. [GAF] m-Nitrobenzene sulfonic acid, sodium salt; accelerates stripping of nickel from steel, brass, or copper; controls reaction rate in pickling nickel-chromium-iron alloys and for dissolving nickel and copper; additive in phosphate coatings of metals; used in textiles for kier boiling of cotton, discharge printing of vats, and oxidizing agent in printing; flakes and gran. resp.; 56% and 94% resp.

Ludox HS-30. [DuPont] Aq. colloidal silica disp.; binder for ceramic casting; water-wh. opalescent liq.; sp.gr. 1.21; dens. 1.21 kg/l; visc. 5 cP; pH 9.8; 30% min. solid silicon dioxide.

Ludox SM. [DuPont] Aq. colloidal silica disp.; see Ludox HS-30; water-wh. opalescent liq.; sp.gr. 1.22; dens. 1.22 kg/l; visc. 5 cP; pH 10.2; 30% min. solid silicon dioxide.

Luperco 101-P20. [Atochem] 20% disp. of 2,5-dimethyl-2,5-di (t-butylperoxy) hexane on a PP powd. carrier; crosslinking agent for elastomers and thermoplastic resins; wh. free-flowing powd.; m.w. 290.45; bulk dens. 32.8 lb/ft³.

Luperco 101-SIL. [Atochem] 2,5-Dimethyl-2,5-di (t-butylperoxy) hexane on inert filler; initiator for high-temp. cure of polyester resins, curing elastomers, and polymer modification thermoplastic crosslinking; solid; 50% act.; 5.51% act. oxygen.

Luperco 101-XL. [Atochem] 2,5-Dimethyl-2,5-di(t-butylperoxy) hexane on inert filler ($CaCO_3$); initiator for curing elastomers, polymer modification thermoplastic crosslinking, and high-temp. cure of polyester resins; free-flowing powd.; m.w. 290.45; dens. 1.248 g/cc; 45% conc.; 4.96–5.29% act. oxygen.

Luperco 130-KE. [Atochem] 2,5-Dimethyl-2,5-di (t-butylperoxy) hexyne-3 on Burgess clay; initiator for high-temp. cure of polyester resins, curing elastomers, and polymer modification thermoplastic cross-linking; solid; 40% act.; 4.47% act. oxygen.

Luperco 130-XL. [Atochem] 2,5-Dimethyl-2,5-di (t-butylperoxy) hexyne-3 on inert filler ($CaCO_3$); crosslinking agent, initiator for thermoplastic modification, curing elastomers, high temp. cure of polyesters, and cure of acrylic syrup; free-flowing powd.; m.w. 286.42; dens. 1.263 g/cc; 45% conc.; 5.03% act. oxygen.

Luperco 230-XL. [Atochem] n-Butyl-4,4-bis (t-butylperoxy) valerate on inert filler; initiator for curing elastomers, and for high-temp. cure of polyester resins; free-flowing powd.; m.w. 334.4; dens. 0.526 g/ml; ; 40% act.; 3.83% act. oxygen.

Luperco 231-KE. [Atochem] 1,1-Bis (t-butylperoxy) -3,3,5-trimethylcyclohexane on Burgess KE clay; initiator for curing elastomers; crosslinking agent for thermoplastic modification; free-flowing powd.; dens. 0.6632 g/ml; 40% conc.;s 4.13–4.34% act. oxygen.

Luperco 231-SRL. [Atochem] 1,1-Di(t-butylperoxy) 2,2,5-trimethyl cyclohexane; crosslinking agent for thermoplastic modification, curing elastomers such as polybutadiene and EPDM, and high temp. cure of polyesters; wh. powd.; bulk dens. 1.413 g/cc; 40% conc.; 4.23% act. oxygen.

Luperco 231-XL. [Atochem] 1,1-Di (t-butylperoxy) 3,3,5-trimethyl cyclohexane on inert filler ($CaCO_3$); initiator for curing elastomers, crosslinking agent for thermoplastic modification; free-flowing powd.; m.w. 302.5; dens. 1.413 g/ml; 40% act.; 4.23% min. act. oxygen.

Luperco 233-KE. [Atochem] Ethyl-3,3-di (t-butylperoxy) butyrate on Burgess clay; initiator for curing elastomers and for polymer modification thermoplastic cross-linking; solid; 40% act.; 4.39% act. oxygen.

Luperco 233-XL. [Atochem] Ethyl-3,3-di (t-butylperoxy) butyrate on inert filler; initiator for curing elastomers and for polymer modification thermoplastic cross-linking; free-flowing powd.; m.w. 292.4; dens. 0.3551 g/ml; 40% act.; 4.39% act. oxygen.

Luperco 331-XL. [Atochem] 1,1-Di (t-butylperoxy) cyclohexane on inert filler; initiator for high-temp. cure of polyester resins and for curing elastomers; free-flowing powd.; m.w. 260.3; dens. 0.3277 g/ml; 40% act.; 4.92% act. oxygen.

Luperco 500-40C. [Atochem] Dicumyl peroxide on calcium carbonate; see Luperco 101XL; free-flowing powd.; m.w. 270.37; dens. 1.611 g/cc; 40% conc.; 2.34–2.46% act. oxygen.

Luperco 500-40KE. [Atochem] Dicumyl peroxide on Burgess KE clay; see Luperco 101XL; free-flowing powd.; m.w. 270.37; dens. 1.579 g/cc; 40% conc.; 2.34–2.46% act. oxygen.

Luperco 500-SRK. [Atochem] Dicumyl peroxide; crosslinking agent for thermoplastics, curing of elastomers, and high temp. cure of polyester resins; powd.; 40% conc.; 2.37% act. oxygen.

Luperco 801-XL. [Atochem] t-Butyl cumyl peroxide on inert filler; initiator for high-temp. cure of polyester resins, curing elastomers, and polymer modification thermoplastic cross-linking; solid; 40% act.; 3.07% act. oxygen.

Luperco 802-40KE. [Atochem] α–α-Bis (t-butylperoxy) diisopropylbenzene on Burgess clay; crosslinking agent for polymer modification, curing elastomers, high-temp. cure of polyester resins; 40% solids; 3.73–3.92% act. oxygen.

Luperco AA. [Atochem] Benzoyl peroxide blend with wheat starch; initiator for polymer modification, thermoplastic crosslinking, high-temp. and R.T. cure of polyester resins; powd.; 33% act.; 2.11–2.18% act. oxygen.

Luperco ACP. [Atochem] Benzoyl peroxide blend with inorg. phosphates; initiator for high-temp. and R.T. cure of polyester resins; powd.; 35% act.; 2.31% min. act. oxygen.

Luperco AFR-250. [Atochem] Benzoyl peroxide; fire retardant initiator for high-temp. and R.T. cure of polyester resins; paste; 25% act.; 1.58–1.78% act. oxygen.

Luperco AFR-400. [Atochem] Benzoyl peroxide; initiator for high-temp. and room-temp. cures of polyester resins; pourable paste; 40% act.; 2.64% act. oxygen.

Luperco AFR-500. [Atochem] Benzoyl peroxide; see Luperco AFR-250; paste; 50% act.; 3.3% min. act. oxygen.

Luperco AFR-501. [Atochem] Benzoyl peroxide; see Luperco AFR-250; paste; 50% act.; 3.3% min. act. oxygen.

Luperco ANS. [Atochem] Benzoyl peroxide with plasticizer; initiator for polymer modification, thermoplastic crosslinking, high-temp. and R.T. cure of polyester resins, and cure of acrylic syrup; paste; 55% act.; 3.6% min. act. oxygen.

Luperco ANS-P. [Atochem] Benzoyl peroxide with plasticizer; crosslinking agent for high temp. and R.T. cure of polyesters, cure of acrylic syrup; paste; 55% conc.; 3.63% act. oxygen.

Luperco AST. [Atochem] Benzoyl peroxide with silicone oil; initiator for curing elastomers; paste; 50% act.; 3.3% min. act. oxygen.

Luperco ATC. [Atochem] Benzoyl peroxide with tricresyl phosphate; see Luperco AA; paste; 50% act.; 3.30–3.43% act. oxygen.

Luperco CST. [Atochem] 2,4-Dichloro benzoyl peroxide in silicone oil; initiator for bulk, sol'n., and suspension polymerization, curing elastomers, and high-temp. cure of polyester resins; paste; 50% conc.; 2.1% min. act. oxygen.

Luperco EFA-4. [Atochem] t-Butyl hydrazinium chloride; crosslinking agent; pale yel. powd.; sol. > 40% in water; sp.gr. 1.13.

Luperco PMA-25. [Atochem] t-Butylperoxymaleic acid; crosslinking agent for emulsion, bulk, sol'n., and suspension polymerization, thermoplastic modification, cure of acrylic syrup; paste; 25% conc.; 2.06% act. oxygen.

Luperco PMA-40. [Atochem] t-Butylperoxymaleic acid in proprietary formula; initiator for bulk, sol'n., emulsion, and suspension polymerization, polymer modification, thermoplastic crosslinking, and high-temp. cure of polyester resins; paste; 40% conc.; 3.30% min. act. oxygen.

Luperfoam 40. [Atochem] Aq. mixt. of t-butylhydrazinium chloride and cupric chloride; chemical blowing agent for unsat. polyester resins.

Luperfoam 329. [Atochem] Aq. mixt. of t-butylhydrazinium chloride and ferric chloride; chemical blowing agent for R.T. foaming of unsat. polyester resins.

Luperox 2,5-2,5. [Atochem] 2,5-Dihydroperoxy-2,5-dimethylhexane; initiator for bulk, sol'n., emulsion, and suspension polymerization; solid; 70% conc. in water; 11.85–12.57% act. oxygen.

Luperox 118. [Atochem] 2,5-Dimethyl-2,5-bis-(benzoylperoxy)hexane; initiator for vinyl polymerization; solid; m.w. 386; sol. 16–20% in ethylene chloride, 11–15% in tetrahydro furan, 6–10% in benzene, 3–5% in ethyl acetate; insol. in water; dens. 24.5 lb/ft³; m.p. 237 F; 92.5% conc.; 7.66% min. act. oxygen.

Luperox 204. [Atochem] Di(2-phenoxyethyl) peroxydicarbonate; initiator for bulk, sol'n., and suspension polymerization, high-temp. curing of polyester resins, and cure of acrylic syrup; cryst. solid; 95% act.; 4.2% min. act. oxygen.

Luperox 500R. [Atochem] Dicumyl peroxide; initiator for bulk, sol'n., and suspension polymerization, polymer modification thermoplastic crosslinking, curing elastomers, high-temp. cure of polyester resins; cryst. solid; m.w. 270.37; sp.gr. 1.00 (40 C); 99% act.; 5.87% min. act. oxygen.

Luperox 500T. [Atochem] Dicumyl peroxide; see Luperox 500R; semicryst. solid; m.w. 270.37; sp.gr. 0.997–1.009 (40 C); 91–93% act.; 5.40% min. act. oxygen.

Luperox 802. [Atochem] α-α-Bis (t-butylperoxy) diisopropylbenzene; crosslinking agent for thermoplastic modification, curing elastomers, and high temp. cure of polyesters; initiator for vinyl polymerization; semicryst. solid; sp.gr. 0.930; 96% conc.; 9.26% act. oxygen.

Luperox 802-40KE. [Atochem] α-α-Bis (t-butylperoxy) diisopropylbenzene on Burgess clay; crosslinking agent for thermoplastic modification, curing elastomers, and high temp. cure of polyesters; free-flowing powd.; bulk dens. 31 lb/ft³; 40% conc.; 3.82% act. oxygen.

Luperox PMA. [Atochem] t-Butylperoxy maleic acid; initiator for bulk, sol'n., suspension, and emulsion polymerization, for high-temp. cure of polyester resins, and for polymer modification thermoplastic cross-linking; wh. cryst. solid; 98% act.; 8.09% act. oxygen.

Lupersol 10. [Atochem] t-Butyl peroxyneodecanoate; initiator for vinyl polymerizations; liq.; insol. in water; sp.gr. 0.90; flash pt. 1.45 F; m.w. 244; 95% min. assay; 6.2% min. act. oxygen.

Lupersol 10-M75. [Atochem] t-Butyl peroxyneodecanoate in odorless min. spirits; see Lupersol 10; sol'n.; m.w. 244; sol. see Lupersol 10; sp.gr. 0.87; visc. 8.2 cps (32 F); ref. index 1.44; flash pt. > 100 F; 74–76% assay; 4.8–5.0% act. oxygen.

Lupersol 11. [Atochem] t-Butyl peroxypivalate in odorless min. spirits; see Lupersol 10; sol'n.; m.w. 174; f.p. –2 F; sol. see Lupersol 10; sp.gr. 0.85; visc. 2.8 cps (32 F); ref. index 1.41; flash pt. 155 F; 74–76% assay; 6.8–7.0% act. oxygen.

Lupersol 47-M75. [Atochem] alpha-Cumylperoxy pivalate in odorless min. spirits; see Luperox 2,5-2,5; sol'n.; 75% conc.; 2.97–3.05% act. oxygen.

Lupersol 70. [Atochem] t-Butyl peroxyacetate in odorless min. spirits; see Lupersol 10; sol'ns.; m.w. 132; f.p. < –22 F; sol. see Lupersol 10; sp.gr. 0.89; visc. 1.4 cps; ref. index 1.40; flash pt. 140 F; 74–76% assay; 8.9–9.2% act. oxygen.

Lupersol 75-M. [Atochem] t-Butyl peroxyacetate in odorless min. spirits; see Lupersol 10; sol'n.; m.w. 132; f.p. < –15 F; sol. see Lupersol 10; sp.gr. 0.83; visc. 1.4 cps; ref. index 1.40; flash pt. 130 F; 49–51% assay; 5.93–6.18% act. oxygen.

Lupersol 76-M. [Atochem] t-Butylperoxy acetate sol'n. in OMS; initiator for bulk, sol'n., and suspension polymerization, for high-temp. cure of polyester resins, and for cure of acrylic syrup; sol'n.; 75% act.; 7.26% act. oxygen.

Lupersol 80. [Atochem] t-Butyl peroxyisobutyrate in odorless min. spirits; see Lupersol 10; sol'n.; m.w. 160; f.p. < –40 F; sol. see Lupersol 10; sp.gr. 0.87; visc. 1.4 cps; ref. index 1.41; flash pt. 140 F; 74–76% assay; 7.4–7.6% act. oxygen.

Lupersol 99. [Atochem] Di-t-butyl diperoxyazelate in odorless min. spirits; initiator for bulk, sol'n., emulsion, and suspension polymerization, and high-temp. cure of polyester resins; liq.; 75% conc.; 7.2%

act. oxygen.

Lupersol 101. [Atochem] 2,5-Dimethyl-2,5-di(t-butylperoxy) hexane; see Luperox 500R; liq.; m.w. 290.45; sp.gr. 0.8650; b.p. 115 C (10 mm Hg); f.p. < 8 C; flash pt. (Seta) 43 C; ref. index 1.4160; 90% act.; 9.92% min. act. oxygen.

Lupersol 130. [Atochem] 2,5-Dimethyl-2 5-di(t-butylperoxy) hexyne-3; initiator for bulk, sol'n., and suspension polymerization, high-temp. curing of polyester resins, and cure of acrylic syrup; liq.; m.w. 286.42; sp.grt. 0.886–0.890; b.p. 113 C (10 mm Hg); fg.p. < 8 C; flash pt. (Seta) 87 C; ref. index 1.4260–1.4300; 90–95% act.; 10.05–10.61% act. oxygen.

Lupersol 188-M75. [Atochem] alpha-Cumylperoxy neodecanoate in odorless min. spirits; see Luperox 2,5-2,5; sol'n.; 75% conc.; 3.86–3.97% act. oxygen.

Lupersol 215. [Atochem] 1,1,3,3-Tetramethyl butyl hydroperoxide; see Luperox 2,5-2,5; liq.; 85% act.; 9.3% min. act. oxygen.

Lupersol 219-M60. [Atochem] Diisononanoyl peroxide in odorless min. spirits; crosslinking agent for bulk, sol'n., and suspension polymerization; sol'n.; 60% conc.; 3.82% act. oxygen.

Lupersol 219-M75. [Atochem] Diisononanoyl peroxide in odorless min. spirits; see Luperox 2,5-2,5; sol'n.; 75% act.; 3.82–3.92% act. oxygen.

Lupersol 220-D50. [Atochem] 2,2-Di(t-butylperoxy) butane in DOP; initiator for bulk, sol'n., and suspension polymerization, high-temp. curing of polyester resins, and cure of acrylic syrup; liq.; m.w. 234.1; sp.gr. 0.924; f.p. –14 C; flash pt. (Seta) 53 C; 50% act.; 6.80–7.0% act. oxygen.

Lupersol 221. [Atochem] Di(n-propyl) peroxydicarbonate; see Luperox 2,5-2,5; liq.; 99% act.; 7.68% min. act. oxygen.

Lupersol 221-M85. [Atochem]; Di(n-propyl) peroxydicarbonate in odorless min. spirits; see Luperox 2,5-2,5; sol'n.; 85% act.; 6.60–6.75% act. oxygen.

Lupersol 223. [Atochem] Di (2-ethylhexyl) peroxydicarbonate; see Luperox 2,5-2,5; liq.; 97% act.; 4.50% min. act. oxygen.

Lupersol 223-M. [Atochem] Di (2-ethylhexyl) peroxydicarbonate in odorless min. spirits; see Luperox 2,5-2,5; sol'n.; 40% act.; 1.85–1.94% act. oxygen.

Lupersol 223-M40. [Atochem] Di (2-ethylhexyl) peroxydicarbonate sol'n. in OMS; initiator for bulk, sol'n., and suspension polymerization; sol'n.; 40% act.; 1.86% act. oxygen.

Lupersol 223-M75. [Atochem] Di (2-ethylhexyl) peroxydicarbonate in odorless min. spirits; see Luperox 2,5-2,5; sol'n.; 75% act.; 3.41–3.51% act. oxygen.

Lupersol 223-T70. [Atochem] Di (2-ethylhexyl) peroxydicarbonate sol'n. in toluene; initiator for bulk, sol'n., and suspension polymerization; sol'n.; 70% act.; 3.25% act. oxygen.

Lupersol 224. [Atochem] 2,4-Pentanedione peroxide; curing agent for unsat. polyester thermoset resins; sol'n.; water-misc.; sp.gr. 1.10; visc. 7.6 cps; ref. index 1.4260; flash pt. 214 F.

Lupersol 225. [Atochem] Di (sec-butyl) peroxydicarbonate; initiator for bulk, sol'n., and suspension polymerization and cure of acrylic syrup; liq.; 98% act.; 6.69% min. act. oxygen.

Lupersol 225-M. [Atochem] Di (sec-butyl) peroxydicarbonate in odorless min. spirits; see Lupersol 225; sol'n.; 75% act.; 5.12–5.26% act. oxygen.

Lupersol 225-M60. [Atochem] Di (sec-butyl) peroxydicarbonate in odorless min. spirits; see Lupersol 225; sol'n.; 60% act.; 4.03–4.17% act. oxygen.

Lupersol 225-M75. [Atochem] Di (sec-butyl) peroxydicarbonate sol'n. in OMS; initiator for bulk, sol'n., and suspension polymerization; sol'n.; 75% act.; 5.12% act. oxygen.

Lupersol 225-T50. [Atochem] Di (sec-butyl) peroxydicarbonate sol'n. in toluene; initiator for bulk, sol'n., and suspension polymerization; sol'n.; 50% act.; 3.41% act. oxygen.

Lupersol 225-X30. [Atochem] Di (sec-butyl) peroxydicarbonate sol'n. in xylene; initiator for bulk, sol'n., and suspension polymerization; sol'n.; 30% act.; 2.05% act. oxygen.

Lupersol 228-Z. [Atochem] Acetyl cyclohexylsulfonyl peroxide in phthalate plasticizer; see Luperox 2,5-2,5; sol'n.; 30% act.; 2.02–2.16% act. oxygen.

Lupersol 230. [Atochem] n-Butyl-4,4-bis (t-butylperoxy) valerate; initiator for bulk, sol'n., and suspension polymerization, for high-temp. cure of polyester resins, for curing elastomers, and for cure of acrylic syrup; liq.; m.w. 334.4; sp.gr. 0.955; f.p. –21 C; flash pt. (Seta) 64–94 C; 90% act.; 8.61% act. oxygen.

Lupersol 231. [Atochem] 1,1-Di (t-butylperoxy) 3,3,5-trimethyl cyclohexane; initiator for bulk, sol'n., and suspension polymerization, curing elastomers, high-temp. cure of polyester resins, and cure of acrylic syrup; colorless liq.; m.w. 302.5; sp.gr. 0.9050; f.p. –40 C; flash pt. (Seta) 46 C; 92% act.; 9.73% min. act. oxygen.

Lupersol 231-P75. [Atochem] 1,1-Di (t-butylperoxy) 3,3,5-trimethyl cyclohexane sol'n. in dibutyl phthalate; initiator for high-temp. cure of polyester resins and for cure of acrylic syrup; liq.; m.w. 302.5; sp.gr. 0.9345; f.p. < –40 C; flash pt. (Seta) 86 C; 75% act.; 7.94% act. oxygen.

Lupersol 233-M75. [Atochem] Ethyl-3,3-di (t-butylperoxy) butyrate in odorless min. spirits; initiator for bulk, sol'n., and suspension polymerization, high-temp. cure of polyesters and acrylics; liq.; m.w. 292.4; sp.gr. 0.9505; f.p. 9 C; flash pt. (Seta) 61 C; 75% act.; 8.21% min. act. oxygen.

Lupersol 233-M90. [Atochem] Ethyl-3,3-di (t-butylperoxy) butyrate sol'n. in paraffinic oil; initiator for bulk, sol'n., and suspension polymerization, for high-temp. cure of polyester resins, for curing elastomers, for cure of acrylic syrup, and for polymer modification thermoplastic cross-linking; sol'n.; 90% act.; 9.86% act. oxygen.

Lupersol 256. [Atochem] 2,5-Dimethyl-2,5-bis(2-ethylhexanoylperoxy) hexane; see Lupersol 10; liq.; m.w. 431; f.p. < –4 F; sol. see Lupersol 10; sp.gr. 0.93; visc. 50.6 cps; ref. index 1.44; flash pt. 190 F; 90% conc.; 6.7% act. oxygen.

Lupersol 288-M75. [Atochem] α-Cumyl peroxyneoheptanoate in odorless min. spirits; sol'n.; 75% conc.; 4.54% act. oxygen.

Lupersol 331-80B. [Atochem] 1,1-Di (t-butylperoxy) cyclohexane in butylbenzyl phthalate; initiator for bulk, sol'n., and suspension polymerization, high-temp. curing of polyester resins, and cure of acrylic syrup; liq.; m.w. 260.3; sp.gr. 0.975; f.p. –56 C; flash pt. (Seta) 70 C; ref. index 1.4546; 80% conc.; 9.70–9.95% act. oxygen.

Lupersol 431-80B. [Atochem] 1,1-Di-(t-amylperoxy) cyclohexane in benzyl butyl phthalate; see Lupersol 331-80B; liq.; 80% conc.; 11.10% act. oxygen.

Lupersol 531-80B. [Atochem] 1,1-Di-(t-amylperoxy) cyclohexane in butyl benzyl phthalate; initiator for bulk, sol'n., and suspension polymerization, for high-temp. cure of polyester resins, and for cure of acrylic syrup; liq.; m.w. 288.4; sp.gr. 0.959; f.p. < –25 C; flash pt. (Seta) 78 C; 80% act.; 11.10% act. oxygen.

Lupersol 531-80M. [Atochem] 1,1-Di (t-amylperoxy) cyclohexane in odorless min. spirits; crosslinking agent for bulk, sol'n., and suspension polymerization and cure of acrylic syrup; liq.; 80% conc.; 8.88% act. oxygen.

Lupersol 533-M75. [Atochem] Ethyl 3,3-di(t-amylperoxy)butyrate in odorless min. spirits; initiator for curing of acrylic syrup; crosslinking agent for bulk, sol'n., and suspension polymerization, thermoplastic modification; sol'n.; m.w. 320; sp.gr. 0.893; f.p. –55 C; flash pt. (Seta) 42 C; 75% conc.; 7.39–7.59% act. oxygen.

Lupersol 546-M75. [Atochem] t-Amylperoxy neodecanoate sol'n. in OMS; initiator for bulk, sol'n., and suspension polymerization, and cure of acrylic syrup; sol'n.; 75% act.; 4.64% act. oxygen.

Lupersol 553-M75. [Atochem] 2,2-Di (t-amylperoxy) propane in odorless min. spirits; initiator for high temp. cure of polyesters, cure of acrylic syrup; crosslinking agent for bulk, sol'n., and suspension polymerization, thermoplastic modification; sol'n.; m.w. 248; sp.gr. 0.857; f.p. –27 C; flash pt. (Seta) 45 C; 75% conc.; 9.53–9.79% act. oxygen.

Lupersol 554-M50, 554-M75. [Lucidol] t-Amylperoxy pivalate sol'ns. in OMS; initiators for bulk, sol'n., and suspension polymerization and for cure of acrylic syrup; sol'n.; 50 and 75% act. resp.; 4.25 and 6.38% act. oxygen resp.

Lupersol 555-M60. [Atochem] t-Amyl peroxyacetate in odorless min. spirits; crosslinking agent for bulk, sol'n., and suspension polymerization, high temp. cure of polyester, and cure of acrylic syrup; 60% sol'n.; 6.6% act. oxygen.

Lupersol 575. [Atochem] t-Amylperoxy-2-ethylhexanoate; initiator for bulk, sol'n., and suspension polymerization, for high-temp. cure of polyester resins, and for cure of acrylic syrup; liq.; 95% act.; 6.60% act. oxygen.

Lupersol 575-M75, 575-P75. [Lucidol] t-Amylperoxy-2-ethylhexanoate sol'ns. in OMS and plasticizer resp.; initiator for bulk, sol'n., and suspension polymerization, for high-temp. cure of polyester resins, and for cure of acrylic syrup; sol'ns.; 75% act.; 5.21% act. oxygen.

Lupersol 601. [Atochem] t-Butyl peroxy neohexanoate in odorless min. spirits; see Lupersol 225; sol'n.; 75% act.; 6.38% act. oxygen.

Lupersol 665-T50. [Atochem] 1,1-Dimethyl-3-hydroxybutylperoxy-2-ethylhexanoate in toluene; crosslinking agent for emulsion, bulk, sol'n., and suspension polymerization, cure of acrylic syrup; 50% sol'n.; 3.07% act. oxygen.

Lupersol 688-M50, 688-T50. [Atochem] 1,1-Dimethyl-3-hydroxybutylperoxyneoheptanoate in odorless min. spirits and toluene resp.; crosslinking agent for bulk, sol'n., emulsion, and suspension polymerization, and cure of acrylic syrup; sol'n.; 50% conc.; 3.25% act. oxygen.

Lupersol 801. [Atochem] t-Butyl cumyl peroxide; initiator for bulk, sol'n., and suspension polymerization, polymer modification thermoplastic crosslinking, curing elastomers, high-temp. cure of polyester resins, and cure of acrylic syrup; liq.; m.w. 208.29; sp.gr. 0.945; f.p. 0 C; flash pt. (Seta) 79 C; ref. index 1.45819; 90–95% act.; 6.91–7.30% act. oxygen.

Lupersol DDA-30. [Atochem] Ketone peroxide in dimethyl phthalate/diallyl phthalate diluent; curing agent for unsat. polyester thermoset resins; filament winding applic.; sol'n.; sp.gr. 1.1120; visc. 13.0; ref. index 1.4840; flash pt. 123 F.

Lupersol DDM. [Atochem] Ketone peroxide in dimethyl phthalate diluent; see Lupersol DDA-30; sp.gr. 1.1050; visc. 17.5 cps; ref. index 1.4540; flash pt. 130 F.

Lupersol DDM-9. [Atochem] MEK peroxide; polymerization initiator for cure of promoted unsat. polyester resins and vinyl ester resins at ambient temps.; promoter, transition metal salt, activates decomposition of peroxide initiator; flexibility in useful conc. range and pot life; clear liq.; f.p. < –30 C; insol. in water; sp.gr. 1.0815; visc. 14.8 cps; ref. index 1.4615; flash pt. 58 C (SETA); 8.8 ± 0.1% act. oxygen.

Lupersol DDM-30. [Atochem] Ketone peroxide in dimethyl phthalate diluent; see Lupersol DDA-30; sol'n.; sp.gr. 1.490; visc. 11.0 cps; flash pt. 143 F.

Lupersol Delta-X. [Atochem] Ketone peroxide in dimethyl phthalate diluent; curative for unsat. polyester thermoset resins; gel coats, spray-up and nonreinforced filled applic.; sol'n.; sp.gr. 1.1300; ref. index 1.4600; flash pt. 135 F.

Lupersol Delta-X-9. [Atochem] MEK peroxide; see Lupersol DDM-9; also formulated to give faster gel and cure times in resin systems; clear liq.; f.p. < –30 C; insol. in water; sp.gr. 1.1471; visc. 15.8 cps; ref. index 1.4758; flash pt. 68 C (SETA); 8.8 ± 0.1% act. oxygen.

Lupersol DFR. [Atochem] Ketone peroxide; initiator for R.T. cure of polyester resins; 8.7% min. act. oxygen.

Lupersol DHD-9. [Atochem] MEK peroxide; see Lupersol DDM-9; also outstanding performance in certain vinyl ester and isophthalic resins; clear liq.; f.p. < –20 C; insol. in water; sp.gr. 1.1363; visc. 16.0

cps; ref. index 1.4777; flash pt. (Seta) 66 C; 8.8 ± 0.1% act. oxygen.

Lupersol DNF. [Atochem] Ketone peroxide in proprietary diluent; see Lupersol Delta-X; sp.gr. 1.0240; visc. 7.1 cps; ref. index 1.4070; flash pt. 79 F.

Lupersol DSW. [Atochem] Ketone peroxide in proprietary diluent; see Lupersol Delta-X; also used in nonreinforced filled water extended resin applic.; sol'n.; sp.gr. 1.0800; visc. 7.2 cps; ref. index 1.4080; flash pt. 115 F.

Lupersol DSW-9. [Atochem] Mixt. of peroxides and hydroperoxides; initiator for cure of acrylic syrup; 8.7% min. act. oxygen.

Lupersol KDB. [Atochem] Di-t-butyl diperoxyphthalate in DBP; see Lupersol 10; sol'n.; m.w. 310; f.p. 41 F; sol. see Lupersol 10; sp.gr. 1.06; visc. 31 cps; ref. index 1.49; flash pt. 235 F; 50% min. assay; 5.2% min. act. oxygen.

Lupersol P-31, P-33. [Atochem] t-Butyl peroctoate and 1,1-di (t-butylperoxy) cyclohexane blend in min. oil; initiator for high-temp. cure of polyester resins; 7.17–7.49 and 6.70–6.97% act. oxygen resp.

Lupersol PDO. [Atochem] t-Butyl peroxy-2-ethylhexanoate in DOP; see Lupersol 10; sol'n.; m.w. 216; f.p. < –22 F; sol. see Lupersol 10; sp.gr. 0.93; visc. 12.0 cps; ref. index 1.45; flash pt. 190 F; 50% min. assay; 3.7% min. act. oxygen.

Lupersol PMA. [Atochem] t-Butyl peroxymaleic acid; see Lupersol 10; solid; m.w. 188; sol. 31–50% in tetrahydro furan, 21–30% in ethyl alcohol, 16–20% in ethyl acetate, 6–10% in MEK, 1–2% in water; m.p. 237; 98% min. act.; 8.3% min. act. oxygen.

Lupersol PMS. [Atochem] t-Butyl peroxy-2-ethylhexanoate in odorless min. spirits; see Lupersol 10; sol'n.; m.w. 216; f.p. < –22 F; sol. see Lupersol 10; sp.gr. 0.82; visc. 2.2 cps; ref. index 1.42; flash pt. 155 F; 50% min. assay; 3.37% min. act. oxygen.

Lupersol TA-46. [Atochem] t-Amyl peroxyneodecanoate; see Luperox 2,5-2,5; liq.; 95% act.; 5.88% act. oxygen.

Lupersol TA-46M75. [Atochem] t-Amyl peroxyneodecanoate in odorless min. spirits; see Lupersol 225; sol'n.; 75% conc.; 4.64% act. oxygen.

Lupersol TA-54M75. [Atochem] t-Amyl peroxy pivalate in odorless min. spirits; see Lupersol 225; sol'n.; 75% conc.; 6.38% act. oxygen.

Lupersol TAEC. [Atochem] OO-t-amyl O-(2-ethylhexyl) monoperoxycarbonate; crosslinking agent for emulsion, bulk, sol'n., and suspension polymerization, high temp. cure of polyester, and cure of acrylic syrup; liq.; 95% conc.; 5.8% act. oxygen.

Lupersol TBEC. [Atochem] OO-t-butyl O-(2-ethylhexyl) monoperoxycarbonate; see Lupersol 231; liq.; 6.17% act. oxygen.

Lupersol TBIC-M75. [Atochem] OO-t-Butyl O-isopropyl monoperoxycarbonate in odorless min. spirits; see Lupersol 10; liq.; m.w. 176; f.p. –65 F; see Lupersol 10; sp.gr. 0.87; visc. 2.05 cps; ref. index 1.41; flash pt. 140 F; 74–76% assay; 6.72–6.90% act. oxygen.

Lustra-Pearl®. [Mallinckrodt] Titanium dioxide/mica; pearlescent pigment for cosmetic eye, face, lip, and body makeup; powd.

Lusynton A. [BASF AG] Mixt. of org. and inorg. substances; corrosion inhibitor; chlorite stabilizer with resistance to hard water and acid with wetting and detergent effects; yel./wh. powd.; sol. in warm water; pH 9.6 (1% aq.).

Lutensit A-EP. [BASF AG] Fatty alcohol alkoxylate acid phosphate ester; anionic; wetting agent, dispersant, solubilizer, and cleaner used in industrial and household cleaning agents; ylsh. liq.; sol. in water, 5% caustic soda, 5% HCl, IPA, wh. spirit, trichlorethylene, xylene; dens. 1.07 g/ml; surf. tens. 39 dynes/cm; pH 2 (0.1% aq.); biodeg.; 100% act.

Lutensit A-ES. [BASF AG] Alkyl phenol ether sulfate, sodium salt; anionic; wetting and dispersing agents for engineering, industrial, and household cleaners; ylsh. liq.; sol. in water, 5% caustic soda, 5% HCl, IPA; dens. 1.09 g/ml; surf. tens. 31 dynes/cm (1 g/l water); pH 7–9 (0.1% aq.); biodeg.; 40% act.

Lutensit A-FK. [BASF AG] Fatty acid condensation prod., sodium salt; anionic; see Lutensit A-ES; ylsh. liq.; sol. in water, trichlorethylene, xylene; dens. 1.00 g/ml; pour pt. 5 C; surf. tens. 36 dynes/cm (1 g/l water); pH 8–9.5 (0.1% water); biodeg.; 55% act.

Lutensit A-LBA. [BASF AG] Alkylbenzene sulfonate, alkanolamine salt; anionic; see Lutensit A-ES; ylsh. liq.; sol. in water, 5% caustic soda, 5% HCl, IPA, trichlorethylene, xylene; dens. 1.08 g/ml; pour pt. –5 C; surf. tens. 33 dynes/cm (1 g/l water); pH 7–9 (0.1% aq.); biodeg.; 55% act.

Lutensit A-PS. [BASF AG] Alkylsulfonate, sodium salt; anionic; wetting, dispersing and cleansing agents for industrial and household uses; yel. liq.; sol. in water, 5% caustic soda, 5% HCl; dens. 1.2 g/ml; surf. tens. 34 dynes/cm (1 g/l water); pH 7–8 (0.1% aq.); biodeg.; 65% act.

Lutensit K-LC, K-LC 80. [BASF AG] Benzalkonium chloride; cationic; biocidal, wetting and dispersing agents for prod. of disinfectant cleaners for beverages and foodstuffs, trading concerns and household; clear to ylsh. liq.; sol. in water, 5% sodium hydroxide, 5% HCl, 5% sodium chloride, alcohol, chlorinated hydrocarbons, oil, wh. spirit; dens. 0.98 g/cm^3; visc. 80 and 500 mPa•s resp.; cloud pt. –4 and –10 C; surf. tens. 37 mN/m (1 g/l); pH 6 (1% aq.); 80% biodeg.; 50 and 80% act.

Lutensit K-OC. [BASF AG] Benzalkonium chloride; cationic; see Lutensit K-LC; ylsh. liq.; sol. in water, 5% sodium hydroxide, alcohol; dens. 0.98 g/cm^3; visc. 100 mPa•s; cloud pt. –3 C; surf. tens. 31 mN/m; pH 6 (1% aq.); 80% biodeg.; 50% act.

Lutensit K-TI. [BASF AG] Iodine surfactant complex; disinfectant, biocidal for cleaners; liq.

Lutensol® A 8. [BASF AG] PEG-8 $C_{12}C_{14}$ fatty alcohol; nonionic; detergent, wetting, dispersing, and emulsifying agent; clear liq.; sol. in water, 5% caustic soda, 5% HCl, min. oil, gasoline, chlorinated hydrocarbon; cloud pt. 55 C (0.1% aq.); pH 6.0–7.5 (1% aq.); 90% act.

Lutensol® A 80. [BASF AG] Tallow alcohol ethoxylate; nonionic; detergent, wetting, dispersing, and emulsifying agent; wh. or lt. ylsh., waxy powd.; sol. in water, 5% caustic soda, 5% HCl, alcohol, ketones, benzene, xylene, chlorinated hydrocarbon; cloud pt. > 100 C (1% aq.); pH 6.0–7.5 (1% aq.); 100% act.

Lutensol® AO 3, AO 5. [BASF AG] PEG-3 and PEG-5 resp. straight chain syn. fatty alcohol; nonionic; detergent, emulsifying and dispersing agent used in household and industrial detergents; clear, dull liq.; sol. in petrol. fractions, alcohols, aromatic hydrocarbons; slightly sol. water; dens. 0.93 and 0.96 g/cm³; visc. 40 mPa•s; flash pt. 130 and 150 C; pH 7 (1% aq.); > 80% biodeg.; 100% act.

Lutensol® AO 7, AO 8. [BASF AG] PEG-7 and PEG-8 resp. straight chain syn. fatty alcohol; nonionic; see Lutensol AO 3; clear, dull liq.; sol. water, 5% HCl, petrol. fractions, alcohols, aromatic and chlorinated hydrocarbons; dens. 0.98 and 0.96 g/cm³; visc. 120 mPa•s, 30 mPa•s (60 C); cloud pt. 43 and 52 C (1%); flash pt. 190 and 200 C; pH 7 (1% aq.); > 80% biodeg.; 100% act.

Lutensol® AO 10, AO 11, AO 12. [BASF AG] PEG-10, -11, and -12 resp. straight chain syn. fatty alcohol; nonionic; see Lutensol AO 3; wh. soft paste (AO 10, 11), wh. solid paste (AO 12); sol. water, 5% HCl, alcohols, aromatic and chlorinated hydrocarbons; dens. 0.98, 0.99, and 1.00 g/cm³ (60 C); visc. 40 cps (60 C); cloud pt. 80, 86, and 91 C (1%); flash pt. 200 C; pH 7 (1% aq.); > 80% biodeg.; 100% act.

Lutensol® AO 30. [BASF AG] PEG-30 straight chain syn. fatty alcohol; nonionic; see Lutensit A-FK; wax; 100% conc.

Lutensol® AO 109. [BASF AG] PEG-10 straight chain syn. fatty alcohol; nonionic; see Lutensol AO 3; clear, dull liq.; sol. water, 5% HCl, petrol. fractions, alcohols, aromatic and chlorinated hydrocarbons; dens. 1.03 g/cm³; visc. 200 cps; cloud pt. 80 C (1%); flash pt. 190 C; pH 7 (1% aq.); > 80% biodeg.; 90% act.

Lutensol® AP 6. [BASF AG] PEG-6 alkylphenol; nonionic; detergent, wetting, emulsifying and dispersing agent, used in cleaners, detergents, leather, fur, paper, paint and dye industries; ylsh. liq.; sol. in min. oil, alcohols, ketones, aromatic and chlorinated hydrocarbons; sp.gr. 1.05; visc. 400 cps; surf. tens. 31 dynes/cm; 100% act.

Lutensol® AP 7. [BASF AG] PEG-7 alkylphenol; nonionic; see Lutensol AP 6; clear liq.; sol. in min. oil, alcohols, ketones, aromatic and chlorinated hydrocarbons; sp.gr. 1.04; visc. 350 cps; surf. tens. 31 dynes/cm; 100% act.

Lutensol® AP 8, AP 9. [BASF AG] PEG-8 and PEG-9 resp. alkylphenol; nonionic; see Lutensol AP 6; clear liq.; sol. in water, 5% HCl, alcohols, ketones, aromatic and chlorinated hydrocarbons; sp.gr. 1.05; visc. 350 cps; surf. tens. 31 dynes/cm; 100% act.

Lutensol® AP 10. [BASF AG] PEG-10 alkylphenol; nonionic; see Lutensol AP 6; clear liq.; sol. in water, 5% HCl, alcohols, ketones, aromatic and chlorinated hydrocarbons; sp.gr. 1.06; visc. 350 cps; surf. tens. 32 dynes/cm; 100% act.

Lutensol® AP 14. [BASF AG] PEG-14 alkylphenol; nonionic; see Lutensol AP 6; clear liq.; sol. in water, 5% HCl, alcohols, ketones, aromatic and chlorinated hydrocarbons; sp.gr. 1.05 (50 C); visc. 250 cps (30 C); surf. tens. 37 dynes/cm; 100% act.

Lutensol® AP 20. [BASF AG] PEG-20 alkylphenol; nonionic; see Lutensol AP 6; wh. soft wax; sol. in water, 5% HCl, alcohols, ketones, aromatic and chlorinated hydrocarbons; sp.gr. 1.06 (50 C); surf. tens. 42 dynes/cm; 100% act.

Lutensol® AP 30. [BASF AG] PEG-30 alkylphenol; nonionic; see Lutensol AP 6; wax; 100% act.

Lutensol® AT 11. [BASF AG] PEG-11 sat. $C_{16}C_{18}$ alcohol; nonionic; detergent, wetting, dispersing and emulsifying agent used in chemical industry, for household and industrial detergents; wh. to pale ylsh. paste; sol. in dist. water; visc. 30 cps (60 C); cloud pt. 86 C (0.5% aq.); pH practically neutral (1%); 100% act.

Lutensol® AT 18. [BASF AG] PEG-18 sat. $C_{16}C_{18}$ alcohol; nonionic; see Lutensol AT 11; paste; 100% act.

Lutensol® AT 25. [BASF AG] PEG-25 sat. $C_{16}C_{18}$ alcohol; nonionic; see Lutensol AT 11; wh. powd.; sol. in dist. water, benzene, chlorinated hydrocarbons, ethanol, IPA; visc. 70 cps (60 C); cloud pt. > 100 C (0.5% aq.); pH practically neutral (1%); 100% act.

Lutensol® AT 50. [BASF AG] PEG-50 sat. $C_{16}C_{18}$ alcohol; nonionic; see Lutensol AT 11; wh. powd.; sol. in dist. water, benzene, chlorinated hydrocarbons, ethanol; visc. 125 cps (60 C); cloud pt. > 100 C (0.5% aq.); pH practically neutral (1%); 100% act.

Lutensol® ED 140. [BASF AG] EO-PO ethylene diamine compd.; nonionic; antistat, detergent, dispersant, defoamer, wetting agent, gelling agent, solubilizer, emulsifier, demulsifier, lubricant, foam suppressor for household and industrial uses; clear ylsh. liq.; sol. @ 10% in water, 5% HCl, benzene, ethanol, IPA, trichlorethylene; dens. 1.05 g/cm³; visc. 750 cps; cloud pt. 80 C (1% aq.); pour pt. –20 C; surf. tens. 44 mN/m (1 g/l); pH 9 (5% aq.); 100% act.

Lutensol® ED 310. [BASF AG] EO-PO ethylene diamine compd.; nonionic; see Lutensol ED 140; clear to dull ylsh. liq.; sol. @ 10% in benzene, ethanol, IPA; dens. 1.00 g/cm³; visc. 800 cps; cloud pt. 20 C (1% aq.); pour pt. –25 C; surf. tens. 33 mN/m (1 g/l); pH 9 (5% aq.); 100% act.

Lutensol® ED 370. [BASF AG] EO-PO ethylene diamine compd.; nonionic; see Lutensol ED 140; wh. to ylsh. fine powd.; sol. @ 10% in water, 5% HCl, ethanol, IPA, trichlorethylene; dens. 1.05 g/cm³; visc. 600 cps; m.p. 50 C; cloud pt. > 100 C (1% aq.); surf. tens. 44 mN/m (1 g/l); pH 8 (5% aq.); 100% act.

Lutensol® ED 610. [BASF AG] EO-PO ethylene diamine compd.; nonionic; see Lutensol ED 140; colorless to ylsh. dull liq.; sol. @ 10% in benzene, ethanol, IPA; dens. 1.02 g/cm³; visc. 1300 cps; cloud pt. 15 C (1% aq.); pour pt. –25 C; surf. tens. 33 mN/m (1 g/l); pH 8 (5% aq.); 100% act.

Lutensol® FA 12. [BASF AG] Oleyl amine ethoxylate; nonionic; emulsifier, dispersant, wetting and degreasing agent used in heavy-duty and other detergents, shampoos; ylsh. liq.; sol. @ 10% in dist. water,

5% caustic soda, 5% HCl, alcohol; dens. 1.02 g/cm³ (20 C); visc. 200 cps; surf. tens. 35 dynes/cm (1 g/l aq.); pH 8 (1% aq.); > 80% biodeg.; 100% act.

Lutensol® FSA 10. [BASF AG] Oleic acid amide ethoxylate; nonionic; emulsifier, dispersant, detergent, wetting agent, used in fine detergents, hand cleaners, fur dressing; rdsh. liq. with some sediment; sol. see Lutensol FA 12; dens. 1.01 (60 C); visc. 70 cps (60 C); cloud pt. 80 C (1% aq.); surf. tens. 35 dynes/cm (1 g/l aq.); pH 8 (1% aq.); > 80% biodeg.; 100% act.

Lutensol® LF 400. [BASF AG] Alkoxylated straight chain alcohol; nonionic; low-foaming detergent, wetting and dispersing agent, acidic rinse aid, dishwashing, glass cleaners, metal cleaning; colorless clear liq.; sol. @ 10% in water, 5% HCl, wh. spirit, benzene, ethanol, IPA, trichlorethylene; dens. 0.97 g/cm³; visc. 60 cps; cloud pt. 31 C (1% aq.); surf. tens. 30 dynes/cm (0.1% aq.); pH 7 (5% aq.); 100% act.

Lutensol® LF 401. [BASF AG] Alkoxylated straight chain alcohol; nonionic; low-foaming detergent, wetting and dispersing agent for heavy-duty detergents, acid pickling, metal cleaning; colorless clear liq.; sol. @ 10% in water, 5% HCl, wh. spirit, benzene, ethanol, IPA, trichlorethylene; dens. 1.05 g/cm³; visc. 330 cps; cloud pt. 77 C (1% aq.); surf. tens. 40 dynes/cm (0.1% aq.); pH 7 (5% aq.); 100% act.

Lutensol® LF 600. [BASF AG] Alkoxylated straight chain alcohol; nonionic; low-foaming detergent, wetting and dispersing agent for car washing, hard surface cleaners; colorless clear liq.; sol. @ 10% in water, 5% HCl, wh. spirit, benzene, ethanol, IPA, trichlorethylene; dens. 0.97 g/cm³; visc. 110 cps; cloud pt. 52 C (1% aq.); surf. tens. 40 dynes/cm (0.1% aq.); pH 7 (5% aq.); 100% act.

Lutensol® LF 700. [BASF AG] Alkoxylated straight chain alcohol; nonionic; low-foaming detergent, wetting and dispersing agent for dishwashing, rinse aids, dairy cleaners, household and industrial cleaners; slightly ylsh. clear liq.; sol. @ 10% in 5% HCl, wh. spirit, benzene, ethanol, IPA, trichlorethylene; dens. 0.96 g/cm³; visc. 110 cps; cloud pt. 77 C (1% aq.); surf. tens. 34 dynes/cm (0.1% aq.); pH 7 (5% aq.); 100% act.

Lutensol® LF 711. [BASF AG] Alkoxylated straight chain alcohol; nonionic; low-foaming detergent, wetting and dispersing agent for dishwashing machines, rinse aids, metal cleaning; colorless clear liq.; sol. @ 10% in water, 5% HCl, wh. spirit, benzene, ethanol, IPA, trichlorethylene; dens. 1.0 g/cm³; visc. 110 cps; cloud pt. 36 C (1% aq.); surf. tens. 28 dynes/cm (0.1% aq.); pH 7 (5% aq.); 100% act.

Lutensol® LF 1300. [BASF AG] Alkoxylated straight chain alcohol; nonionic; see Lutensol LF 700; slightly ylsh. clear liq.; sol. @ 10% in wh. spirit, benzene, ethanol, IPA, trichlorethylene; dens. 0.97 g/cm³; visc. 140 cps; surf. tens. 35 dynes/cm (0.1% aq.); pH 7 (5% aq.); 100% act.

Lutensol® LSV. [BASF AG] Alkoxylated straight chain alcohol; nonionic; low-foaming detergent, wetting and dispersing agent for bottle washing, lime and scale dispersant; slightly ylsh. clear liq.; sol. @ 10% in benzene, 5% HCl, ethanol, IPA, trichlorethylene; sol. opaque in water; dens. 1.04 g/cm³; visc. 500 cps; cloud pt. 30 C (1% aq.); surf. tens. 42 dynes/cm (0.1% aq.); pH 8–9 (5% aq.); 100% act.

Lutostat 171. [Henkel] Betaine; amphoteric; textile antistat; liq.; 30% conc.

Lutostat MSW 30. [Henkel] Modified alkyl amine; antistat for PS; 100% conc.

Lutostat MSW 88. [Henkel] Modified alkyl amine; antistat for ABS and other plastics; 100% conc.

Lutrol E 300, E 400. [BASF AG] PEG-6 and PEG-8 resp.; emollient, lubricant, and solv. for liq. preps.; liq.; 100% conc.

Lutrol E 1500. [BASF AG] PEG-32; nonionic; binder, solubilizer, resorption promoter for substances insol. or sparingly sol. in water; microbeads; 100% conc.

Lutrol E 4000. [BASF AG] PEG-75; nonionic; see Lutrol E 1500; microbeads; 100% conc.

Lutrol E 6000. [BASF AG] PEG-150; nonionic; see Lutrol E 1500; microbeads; 100% conc.

Lutrol F 127. [BASF AG] EO/PO block copolymer; nonionic; thickening and gelling agent; microbeads; 100% conc.

Luviflex® D 430 I, D 455 I. [BASF AG] Vinylpyrrolidone/vinyl acetate/alkylaminoacrylate terpolymer; film-forming agent in hair fixing lotions; 50% conc. in IPA/water and 50% in IPA resp.

Luviquat® Series. [BASF] Methylvinylimidazolinium chloride/vinylpyrrolidone copolymer; hair fixative for shampoos, rinses, conditioners, liq. soaps, shower gels, related skin prods.

Luviquat® FC 370. [BASF] Methylvinylimidazolium chloride/vinylpyrrolidone copolymer (30:70 ratio) aq. sol'n.; cationic; substantive cationic polymer used as conditioner in prods. for hair and skin care; film former; foam stabilizing and lubricating effects; ylsh. clear visc. liq.; slight, char. odor; sol. in water, ethanol, water/IPA mixts., anionic surfactants; pH 7 ± 1; 40 ± 2% solids.

Luviquat® FC 550. [BASF] Methylvinylimidazolium chloride/vinylpyrrolidone copolymer (50:50 ratio) aq. sol'n.; cationic; see Luviquat FC 370; ylsh. clear visc. liq.; slight, char. odor; sol. in water, ethanol, water/IPA mixts., anionic surfactants; pH 7 ± 1; 40 ± 2% solids..

Luviquat® FC 905. [BASF] Methylvinylimidazolium chloride/vinylpyrrolidone copolymer; hair conditioners, rinses, shampoos, bleaches, dyes, liq. soaps, bath preps., skin care prods.; 40% act. in water.

Luviset CA 66. [BASF] Vinyl acetate/vinylpropionate/crotonic acid terpolymer; hair fixative for aerosols, hair sprays, setting lotions, conditioners; colorless solid; when neutralized sol. in water, water/alcohol and hydrocarbon/alcohol blends; acid no. 66 ± 4.

Luviset CAP. [BASF] Vinyl acetate/vinyl propionate/crotonic acid terpolymer; film-forming agents for hair-sprays and fixing lotions; compatible in aerosols with propane/butane; powd.; acid no. 66 ± 4.

Luviset CAP X. [BASF] Crotonic acid/vinyl acetate/vinyl propionate terpolymer; see Luviset CA 66;

colorless solid; sol. see Luviset CA 66.

Luviskol. [BASF] PVP; hair fixative, film binder, stabilizer, thickener, protective colloid, coating in cosmetic, pharmaceutical applics., e.g., hair sprays, setting lotions, shampoos, protective bandages, shoe polish, metal coatings; sol. in water, most org. solvs.; 50% ethanol sol'n.

Luviskol K12, K17, K30, K60. [BASF] PVP; film-forming agent, hair fixative, thickener, protective colloid, suspending agent, and dispersant for cosmetics industry, tech. applics.; 50% conc. in water, powd. or 50% in water, powd. or 30% in water, and 45% in water resp.

Luviskol K80, K90. [BASF] PVP; see Luviskol K12; powd. or 20% in water.

Luviskol VA 28 E, VA 28 I. [BASF] PVP/VA copolymer (20:80 ratio); film-forming agents for hairsprays and fixing lotions; 50% conc. in ethanol and 50% in IPA resp.

Luviskol VA 37 E, VA 37 I. [BASF] PVP/VA copolymer (30:70 ratio); see Luviskol VA 28 E; 50% in ethanol and 50% in IPA resp.

Luviskol VA 55 E, VA 55 I. [BASF] PVP/VA copolymer (50:50 ratio); see Luviskol VA 28 E; 50% in ethanol and 50% in IPA resp.

Luviskol VA 64, VA 64 I. [BASF] PVP/VA copolymer (60:40 ratio); see Luviskol VA 28 E; powd. and 50% IPA sol'n. resp.

Luviskol VA 73 E. [BASF] PVP/VA copolymer (70:30 ratio); see Luviskol VA 28 E; 50% conc. in ethanol.

Luviskol VAP 343 E, VAP 343 I. [BASF] PVP/VA/vinyl propionate terpolymer (30:40:30 ratio); see Luviskol VA 28 E; compatible in aerosols with propane/butane; 50% in ethanol and 50% in IPA resp.

Luvitol EHO. [BASF AG] Cetearyl octanoate; nonionic; emollient oil component for cosmetics and pharmaceuticals; liq.; 100% conc.

Luvitol HP. [BASF AG] Hydrog. polyisobutene; nonionic; emollient oil component for cosmetics and pharmaceuticals; liq.; 100% conc.

Luxor 1517V. [Hercules] Hydrolyzed veg. protein with salt, MSG, and partially hydrog. veg. oil; premium flavor and flavor enhancer; background flavor for poultry and seafood dishes; lt. yel. powd.; pH 4.0; 96% total solids; 53% flavor solids; 11% MSG.

Luxor 1576. [Hercules] Hydrolyzed veg. protein with salt, caramel color, disodium guanylate, and disodium inosinate; flavor used as replacement for beef extract; meaty flavor; brn. powd.; pH 5.3; 97% total solids; 52% flavor solids.

Luxor 1626. [Hercules] Hydrolyzed veg. protein with salt, caramel color, corn syrup, disodium guanylate, and disodium inosinate; mild beef flavor used in food applics.; dk. brn. powd.; pH 5.2; 96% total solids; 55% flavor solids.

Luxor 1639V. [Hercules] Hydrolyzed veg. protein with salt, partially hydrog. veg. oil, citric acid, disodium guanylate, disodium inosinate, and oleoresin celary; high degree of flavor enhancement; used in soups and gravies for tangy and spicy flavor; lt. tan powd.; pH 6.1; 96% total solids; 37% flavor solids.

Luxor 1658V. [Hercules] Hydrolyzed veg. protein with salt and partially hydrog. veg. oil; background flavor for poultry and seafood dishes; lt. yel. powd.; pH 4.8; 96% total solids; 53% flavor solids.

Luxor E-40V. [Hercules] Hydrolyzed veg. protein with partially hydrog. soybean oil; flavor used as an MSG extender; beef background taste; lt. tan powd.; pH 5.6; 97% total solids; 54% flavor solids.

Luxor E-50V. [Hercules] Hydrolyzed veg. protein with partially hydrog. soybean oil; flavor; MSG extender; contributes mild seafood char. similar to sweet clams; lt. tan powd.; pH 5.7; 97% total solids; 53% flavor solids.

Luxor EB-2V. [Hercules] Hydrolyzed veg. protein with MSG, caramel color, and partially hydrog. veg. oil; beeflike flavor used in canned, frozen, and dehydrated foods requiring hearty meat flavor; tan powd.; pH 5.2; 96% total solids; 57% flavor solids; 12% MSG.

Luxor EB-400V. [Hercules] Hydrolyzed veg. protein with MSG, partially hydrog. veg. oil, disodium guanylate, and disodium inosinate; beeflike flavor used in food applic.; lt. tan powd.; pH 5.1; 96% total solids; 55% flavor solids; 12% MSG.

Luxor GR-100. [Hercules] Hydrolyzed veg. protein; flavor providing distinctive roasted char. to food applic.; dk. brn. powd.; pH 6.4; 97% total solids; 52% flavor solids.

Luxor GR-150. [Hercules] Hydrolyzed veg. protein; flavor providing moderately browned flavor profile where browned notes are required; when less roasted char. is desired; golden brn. powd.; pH 5.6; 97% total solids; 54% flavor solids.

Luxor GR-200. [Hercules] Hydrolyzed veg. protein; strong browned flavor used when roasted note is desired; dk. brn. powd.; pH 5.5; 97% total solids; 53% flavor solids.

Luxor KB-300V. [Hercules] Hydrolyzed veg. protein with caramel color, partially hydrog. soybean oils; beefy flavor used in food applic.; brn. powd.; pH 5.2; 97% total solids; 56% flavor solids.

Luxor KB-320V. [Hercules] Hydrolyzed veg. protein with caramel color, yeast, soy flour, and partially hydrog. soybean oil; beeflike flavor for food applic.; brn. powd.; pH 5.6; 97% total solids; 59% flavor solids.

Luxor KB-330V. [Hercules] Hydrolyzed veg. protein with caramel color and partially hydrog. soybean oil; beef broth flavor for background of spice blends and food applic.; brn. powd.; pH 5.3; 96% total solids; 57% flavor solids.

Luxor KB-350V. [Hercules] Hydrolyzed veg. protein with MSG, caramel color, and partially hydrog. soybean oil; flavor-enhanced vers. of Luxor KB-300V; robust beeflike aroma and strong meaty taste used in food applic.; brn. powd.; pH 5.3; 97% total solids; 59% flavor solids; 12% MSG.

Luxor KB-400V. [Hercules] Hydrolyzed veg. protein with partially hydrog. soybean oil; used in flavor systems where high degree of natural flavor enhancement is desired; chicken and mushroom applic.; tan powd.; pH 5.7; 97% total solids; 53% flavor solids.

Luxor KB-500V. [Hercules] Hydrolyzed veg. protein with partially hydrog. soybean oil and caramel color; flavor used in poultry-type flavor systems; lt. tan powd.; pH 4.8; 97% total solids; 53% flavor solids.

Luxor KB-530V. [Hercules] Hydrolyzed veg. protein with partially hydrog. soybean oil; mild meaty flavor for middle-range meat types such as pork and veal; lt. tan powd.; pH 5.1; 96% total solids; 56% flavor solids.

Luxor KB-600V. [Hercules] Hydrolyzed veg. protein with partially hydrog. soybean oil; bland flavor for use in snackfood and seafood applic.; nutty aroma and slightly salty taste; tan powd.; pH 6.0; 97% total solids; 53% flavor solids.

Luxor MB-40. [Hercules] Hydrolyzed veg. protein; flavor used in petfood industry to improve palatability of animal foods; brn. powd.; 97% total solids; 53% flavor solids.

Luxor MB-110. [Hercules] Hydrolyzed veg. protein; see Luxor MB-40; brn. powd.; 97% total solids; 53% flavor solids.

Luxor MB-120. [Hercules] Hydrolyzed veg. protein; see Luxor MB-40; brn. powd.; 97% total solids; 53% flavor solids.

Luxor R-100V. [Hercules] Hydrolyzed veg. protein with thiamine hydrochloride and partially hydrog. soybean oils; flavor providing stewed/cooked meaty aromatics to all types of meat or meat systems; lt. tan powd.; pH 5.0; 96% total solids; 54% flavor solids.

LV-31. [Solem] Alumina trihydrate; filler with flame retarding and smoke suppressing properties; used as a resin extender in polyester, vinyls, PU, latex, neoprene foam systems, wire and cable insulation, vinyl wall and floor coverings, epoxies; also suitable for high loadings and high visc. systems like rubber compds.; used in polyester cast onyx; particulate, 60–70% thru 325 mesh; sp.gr. 2.42; dens. 1.0 g/cm^3; surf. area 0.50 m^2/g; ref. index 1.57; hardness 2.5–3.5 Moh; 65% Al_2O_3.

LV-336. [Solem] Alumina trihydrate; see LV-31; also for high loadings and moderate rate of visc. build up; wet out and processing chars.; FRP contact molding; particulate; median particle diam. 15 μ, 90–94% thru 325 mesh; sp.gr. 2.42; dens. 0.7–0.8 g/cm^3; surf. area 2–3 m^2/g; oil absorp. 22–24; ref. index 1.57; hardness 2.5–3.5 Moh; 64.9% Al_2O_3.

Lyogen BE Liq. [Sandoz] Dyeing assistant used on nylon; yel. liq.; 55% sol. in water @ 70 F; sp.gr. 1.16; 45% act.

Lyogen DFT Liq. [Sandoz] Detergent, wetting agent, emulsifier, dispersant, dyeing assistant for polyester; yel. liq.; misc. with water; sp.gr. 1.05; 30% act.

Lyogen F Liq. [Sandoz] Leveling agent, dyeing assistant used on wool; brn. liq.; misc. with water; sp.gr. 1.06; 85% act.

Lyogen KF Liq. [Sandoz] Detergent, wetting agent, emulsifier, dispersant, used with vat dyes; yel. liq.; sp.gr. 1.05; 17% act.

Lyogen MS Liq. [Sandoz] Dyeing and leveling agent for wool and nylon; yel. liq.; misc. with water; sp.gr. 1.08; 33% act.

Lyogen NL Liq., P Liq. [Sandoz] Dyeing assistant used on nylon with acid dyes; amber and yel. resp. liq.; misc. with water (NL Liq.); sp.gr. 1.05 and 1.09 resp.; 34 and 35% act. resp.

Lyogen PAA Liq. [Sandoz] Antiprecipitant used in dyeing acrylic fibers and blends with wool and nylon where cationic and anionic dyes are used simultaneously; amber liq.; misc. with water; 43% act.

Lyogen SF Liq. [Sandoz] Dyeing assistant used in wool, silk, nylon; yel. liq.; sp.gr. 1.03; 25% act.

Lyogen SMK-40 Liq., SMK Paste. [Sandoz] Dye leveling agent, antiprecipitating agent for dyes; amber liq., brn. gel, paste; misc. with water; sp.gr. 1.03; 20–49.5% act.

Lyogen V (U) Liq. [Sandoz] Detergent, wetting agent, surfactant and leveling agent for scouring and printing of polyester and nylon; amber liq.; water disp.; sp.gr. 1.04.

Lyogen WD Liq. [Sandoz] Dyeing assistant, leveling agent; amber liq.; misc. with water; sp.gr. 1.05; 40% act.

Lytron 284, 288, 295, 300, 305, 308, 621. [Morton Int'l.] Modified PS latex; opacifier for cosmetics, shampoos, aq. ammonia, dishwash liqs., lt. duty and industrial detergents; misc. with water.

Lytron 614 Latex. [Morton Int'l.] Modified PS latex; anionic; opacifier for liq. detergents; wh. low-visc. liq.; sp.gr. 1.03; visc. 100 cps max.; pH 5.5 ± 0.3; 40% act.

LZ 01. [Kenrich] Neoalkoxy trisneodecanoyl zirconate; coupling agents which sometimes also act as adhesion promoters (esp. LZ 01, 44, 38), antioxidants, antistats, antifoaming agents, accelerators, blowing agent activators, catalysts), curatives, corrosion inhibitors, disp. aids, emulsifiers, flame retardants, foaming agents, grinding aids, hardeners, hydrophobes (most titanates), impact modifiers, internal lubes, metal primers, process aids, pigment intensifiers, peroxide activators, release agents, retarders, stabilizers, surfactants, suspension aids, thixotropes, wetting agents; recommended for elastomers: LZ 01, 12, 38; for thermosets and coatings: various grades; amber liq.; b.p. 188 F (initial); sol. in xylene, toluene, DOP, min. oil; limited sol. in IPA; insol. in water; sp.gr. 0.96 (16 C); visc. 40 cps; pH 8 (sat. sol'n.); flash pt. (TCC) 195 F; 95% solids in IPA.

LZ 09. [Kenrich] Neoalkoxy tris (dodecyl) benzene sulfonyl zirconate; see LZ 01; brn. liq.; b.p. 300 F (initial); sol. in xylene, toluene, DOP; limited sol. in IPA, min. oil; insol. in water; sp.gr. 1.09 (16 C); visc. 190 cps; pH 4 (sat. sol'n.); flash pt. (TCC) 150 F; 95% solids in IPA.

LZ 12. [Kenrich] Neoalkoxy tris (dioctyl) phosphato zirconate; see LZ 01; orange/red liq.; b.p. 220 F (initial); sol. in xylene, toluene, DOP, min. oil; limited sol. in IPA; insol. in water; sp.gr. 1.06 (16 C); visc. 160 cps; pH 6 (sat. sol'n.); flash pt. (TCC) 170 F; 95% solids in IPA.

LZ 38. [Kenrich] Neoalkoxy tris (dioctyl) pyrophosphato zirconate; see LZ 01; reddsh. liq.; b.p. 345 F (initial); sol. in IPA, xylene, toluene, DOP; insol. in min. oil, water; sp.gr. 1.10 (16 C); visc. 360 cps; pH

6 (sat. sol'n.); flash pt. (TCC) 170 F; 95% solids in IPA.

LZ 38J. [Kenrich] Methacrylamide functional amine; adduct of a LZ 38 analog; coupling agent; amber-orange liq.; sol. in IPA, xylene, toluene, DOP, min. oil, water; sp.gr. 1.02 (16 C); visc. 1200 cps; i.b.p. 220 F; flash pt. (TCC) 90 F; pH 9; 85+% solids in IPA.

LZ 44. [Kenrich] Neoalkoxy tris (ethylene diamino) ethyl zirconate; see LZ 01; yel./orange liq.; b.p. 300 F (initial); sol. in IPA; limited sol. in xylene, toluene; reacts slowly in DOP; insol. in min. oil, water; sp.gr. 1.17 (16 C); visc. > 50 cps; pH 11 (sat. sol'n.); flash pt. (TCC) > 220 F; 95% solids in IPA.

LZ 97. [Kenrich] Neoalkoxy tris (m-amino) phenyl zirconate; see LZ 01; brn. liq.; b.p. 300 F (initial); sol. in ether solvents; limited sol. in IPA; reacts slowly in DOP; insol. in xylene, toluene, min. oil, water; sp.gr. 1.20 (16 C); visc. 4400 cps; pH 6 (sat. sol'n.); flash pt. (TCC) 190 F; 67% solids in phenyl glycol ether.

LZ 817. [Lubrizol] Phenolic; antioxidant for industrial lubricants; solid; oil-sol.

LZ 850, 859. [Lubrizol] Carboxy acid deriv.; corrosion inhibitor for industrial lubricants; liq.; oil-sol.

LZ 5022. [Lubrizol] Carbamate; antioxidant for industrial lubricants; liq.; oil-sol.

LZ 5340. [Lubrizol] Sulfurized hydrocarbon; act. EP additive, lubricant for metalworking; liq.; oil-sol.

LZ 5907. [Lubrizol] Hydrocarbon; tackiness agent, lubricant for industrial applics.; liq.; oil-sol.

LZ 5985. [Lubrizol] Used to suspend graphite, silica, other solids in min. oil and solvs.; liq.; oil-sol.

M

M-7 Mold Release®. [TSE] Fluorotelomer disp. in Freon TF solv.; mold release lubricant; wh. translucent fluid; slight ethereal odor; 5 μ particle size; sp.gr. 1.58; dens. 13.15 lb/gal; visc. 5.8 cps; m.p. –35 C; b.p. 47 C; flash pt. none; 2.45% solids.

MA. [PMC] Methyl anthranilate, tech.; industrial deodorant; aromatic; colorless to pale yel. liq.; strong grapelike odor and taste; sol. in ≥ 5 vols. 60% ethanol; f.p. 21 C; 98% purity.

Macaloid®. [Rheox] Magnesium aluminum silicate; suspending agent, thickener, emulsion stabilizer, rheological additive for creams and lotions; lt. cream powd.; water-disp.; sp.gr. 2.65.

Mackalene 116. [McIntyre] Cocamidoamine lactate; cationic; softener and hair conditioner for cream rinses; liq.; 100% act.

Mackalene 117. [McIntyre] Cocamidopropyl dimethylamine propionate; conditioner.

Mackalene 216. [McIntyre] Ricinoleamido amine lactate; cationic; see Mackalene 116; liq.; 100% act.

Mackalene 316. [McIntyre] Stearamido amine lactate; cationic; see Mackalene 116; liq.; 25% act.

Mackalene 326. [McIntyre] Stearamido morpholine lactate; cationic; see Mackalene 116; liq.; 25% act.

Mackalene 416. [McIntyre] Isostearamido amine lactate; cationic; see Mackalene 116; liq.; 100% act.

Mackalene 426. [McIntyre] Isostearamido morpholine lactate; cationic; see Mackalene 116; liq.; 100% act.

Mackalene 716. [McIntyre] Wheat germamido dimethylamine lactate; cationic; hair conditioner.

Mackam BA. [McIntyre] Behenamidopropyl betaine; amphoteric; conditioner, visc.builder; liq.; 25% act.

Mackam CB, CB-35, CB-LS. [McIntyre] Coco betaine; amphoteric; foam booster and visc. builder; liq.; 45, 35, and 32% conc. resp.

Mackam HV. [McIntyre] Oleamidopropyl betaine; amphoteric; hair conditioner, foamer, emollient, visc. builder; liq.; 35% conc.

Mackam ISA. [McIntyre] Isostearylamidopropyl betaine; amphoteric; conditioner, thickener for shampoos; liq.; water-sol.; 35% conc.

Mackam LMB-LS. [McIntyre] Lauramidopropyl betaine; amphoteric; visc. builder and foam booster; amber liq.; 35% conc.

Mackam OB. [McIntyre] Oleyl betaine; amphoteric; hair conditioner foamer, emollient, visc. builder; liq.; 50% conc.

Mackam OB-30. [McIntyre] Oleyl betaine; amphoteric; visc. builder for alkaline cleanser; amber liq.; 30% conc.

Mackam RA. [McIntyre] Ricinoleamidopropyl betaine; conditioner; amber liq.

Mackam TM. [McIntyre] Dihydroxyethyl tallow glycinate; conditioner for shampoos; liq.; water-sol.

Mackam WGB. [McIntyre] Wheat germamidopropyl betaine; conditioner.

Mackamate WGD. [McIntyre] Disodium wheat germamido PEG-2 sulfosuccinate; nonirritating emollient surfactant; yel. liq.

Mackamide 100-A. [McIntyre] Cocamide DEA; nonionic; foam stabilizer and thickener; liq.; 100% conc.

Mackamide AME-75. [McIntyre] Acetamide MEA; nonionic; humectant; liq.; 75% conc.

Mackamide AN55. [McIntyre] Alkylolamide DEA; nonionic; foam enhancer, visc. builder; liq.; 100% conc.

Mackamide C. [McIntyre] Coco DEA; nonionic; foam stabilizer and thickener for shampoos; liq.; 100% conc.

Mackamide CD. [McIntyre] Cocamide DEA and diethanolamine; nonionic; foam enhancer and visc. builder; liq.; 100% conc.

Mackamide CD-8, -25, CDM, CDT. [McIntyre] Cocamide, modified; nonionic; degreaser, rust inhibitor, visc. builder; liq.; 100% conc.

Mackamide CDX. [McIntyre] Amide, modified; nonionic; see Mackamide CD-8; liq.; 100% conc.

Mackamide CMA. [McIntyre] Cocamide MEA; nonionic; foam enhancer and visc. builder; solid; 100% conc.

Mackamide CS. [McIntyre] Cocamide DEA; nonionic; see Mackamide CMA; liq.; 100% conc.

Mackamide CSA. [McIntyre] Cocamide, modified; nonionic; see Mackamide CD-8; liq.; 100% conc.

Mackamide EC. [McIntyre] Cocamide DEA; nonionic; see Mackamide CMA; liq.; 100% conc.

Mackamide ISA. [McIntyre] Isotearamide DEA; nonionic; lubricant; liq.; 100% conc.

Mackamide L-95, LLM, LMD. [McIntyre] Lauramide DEA; nonionic; foam enhancer and visc. builder; solid and liq. resp.; 100% conc.

Mackamide LME. [McIntyre] Lactamide MEA; conditioner for shampoos; liq.; water-sol.

Mackamide LMM. [McIntyre] Lauramide MEA; nonionic; see Mackamide L-95; solid; 100% conc.

Mackamide MC. [McIntyre] Cocamide DEA; nonionic; see Mackamide L-95; liq.; 100% conc.

Mackamide O. [McIntyre] Oleamide DEA; nonionic; solv. degreaser, visc. builder; liq.; 100% conc.

Mackamide PK. [McIntyre] Palm kernel DEA; nonionic; visc. builder; liq.; 100% conc.
Mackamide R. [McIntyre] Ricinoleic DEA; nonionic; emulsifier; softener; liq.; 100% conc.
Mackamide S. [McIntyre] Soyamide DEA; nonionic; thickener, hair conditioner; liq.; 100% conc.
Mackamine BAO. [McIntyre] Behenamido propylamine oxide; nonionic; conditioner and foam stabilizer; paste; 25% conc.
Mackamine CO. [McIntyre] Cocamine oxide; amphoteric; foam booster and stabilizer; liq.; 30% conc.
Mackamine IAO. [McIntyre] Isostearamidopropylamine oxide; nonionic; conditioner and foam stabilizer; liq.; 30% conc.
Mackamine LAO. [McIntyre] Lauramidopropyl dimethylamine oxide; nonionic; foam booster and visc. builder; water-wh. liq.; 30% conc.
Mackamine LO. [McIntyre] Lauramine oxide; nonionic; see Mackamine LAO; water-wh. liq.; 30% conc.
Mackamine O2. [McIntyre] Oleamine oxide; nonionic; see Mackamine LAO; visc. builder; amber liq.; 50% conc.
Mackamine OAO. [McIntyre] Oleamidopropylamine oxide; nonionic; see Mackamine LAO; amber liq.; 50% conc.
Mackamine WGO. [McIntyre] Wheat germamido propyl dimethylamine oxide; conditioner; amber liq.
Mackanate DG30. [McIntyre] Disodium silicone polyol sulfosuccinate; conditioner for hair; liq.
Mackanate DOS-75. [McIntyre] Sodium dioctyl sulfosuccinic; anionic; wetting agent, dispersant, and penetrant; liq.; 75% conc.
Mackanate IM. [McIntyre] Disodium isostearamido MEA sulfosuccinate; anionic; mild conditioner; liq.; 35% act.
Mackanate OD-35. [McIntyre] Disodium oleamide PEG-2 sulfosuccinate; anionic; mild conditioning shampoo; liq.; 35% act.
Mackanate WG. [McIntyre] Disodium wheat germamido MEA sulfosuccinate; anionic; mild conditioner; liq.; 35% act.
Mackanate WGD. [McIntyre] Disodium wheat germamido PEG-2 sulfosuccinate; anionic; emollient surfactant; liq.; 35% conc.
Mackazoline C. [McIntyre] Cocoyl hydroxyethyl imidazoline; nonionic; emulsifier, corrosion inhibitor; amber liq.; 100% conc.
Mackazoline L. [McIntyre] Lauryl hydroxyethyl imidazoline; nonionic; see Mackazoline C; amber liq.; 100% conc.
Mackazoline O. [McIntyre] Oleyl hydroxyethyl imidazoline; nonionic; see Mackazoline C; amber liq.; 100% conc.
Mackernium SDC-25. [McIntyre] Stearalkonium chloride; cream rinse hair conditioner; wh. paste.
Mackernium SDC-50. [McIntyre] Stearalkonium chloride; cationic; cream rinse hair conditioner; paste; 50% act.
Mackernium SDC-85. [McIntyre] Stearalkonium chloride; cream rinse hair conditioner; wh. paste.
Mackester Series. [McIntyre] Glycol mono/diesters; nonionic; emulsifiers, emollient bases for cosmetics; liq.; 100% conc.
Mackine 101. [McIntyre] Cocamido amine; cationic; softener; intermediate convertible to salts for hair conditioners; liq.; 100% act.
Mackine 201. [McIntyre] Ricinoleamido amine; cationic; softener; intermediate convertible to salts for hair conditioners; liq.; 100% act.
Mackine 301. [McIntyre] Stearamido amine; cationic; softener; intermediate convertible to salts for hair conditioners; solid; 100% act.
Mackine 321. [McIntyre] Stearamido morpholine; cationic; softener; intermediate convertible to salts for hair conditioners; solid; 100% act.
Mackine 401. [McIntyre] Isostearamido amine; cationic; softener; intermediate convertible to salts for hair conditioners; liq.; 100% act.
Mackine 421. [McIntyre] Isostearamido morpholine; cationic; softener; intermediate convertible to salts for hair conditioners; liq.; 100% act.
Mackine 501. [McIntyre] Oleamidopropyl dimethylamine; hair conditioner; amber liq.
Mackine 601. [McIntyre] Behenamidopropyl dimethylamine; nonionic; hair conditioner; solid; 100% act.
Mackine 701. [McIntyre] Wheat germamidopropyl dimethylamine; hair conditioner; amber liq.
Mackine 801. [McIntyre] Lauramidopropyl dimethylamine; hair conditioner; amber liq.
Mackine 901. [McIntyre] Soyamidopropyl dimethylamine; hair conditioners; amber liq.
Mackpearl LV. [McIntyre] Anionic; pearl agent for cold blends; liq.; 50% conc.
Mackpro NLP. [McIntyre] Quaternium-79 hydrolyzed animal protein; hair conditioner; liq.
Macol® 1. [PPG-Mazer] Poloxamer 181; syn. lubricant base fluid for metalworking and textile lubricants, chemical intermediates; APHA < 100 liq.; water-sol.; sp.gr. 1.015; dens. 8.5 lb/gal; visc. 285 cps; HLB 3.0; pour pt. –29 C; ref. index 1.4520; flash pt. (PMCC) 455 F; cloud pt. 24 C (1% aq.).
Macol® 2, 4, 8. [PPG-Mazer] Poloxamer 182, 184, 188 resp.; nonionic; defoamer, deduster, demulsifier, detergent, dispersant, dye leveler, gelling agent, antistat; liq., paste, flake; HLB 7.0, 15.0, 29.0 resp.; 100% conc.
Macol® 15. [PPG-Mazer] Meroxapol 105; nonionic; defoamer, pulp and paper additive, dispersant, lubricant, leveling aid, wetting agent; liq.; HLB 15.0; 100% conc.
Macol® 18. [PPG-Mazer] Meroxapol 171; see Macol 1; APHA < 100 liq.; water-sol.; sp.gr. 1.018; dens. 8.5 lb/gal; visc. 285 cps; HLB 6.0; pour pt. –27 C; ref. index 1.4516; flash pt. (PMCC) > 450 F; cloud pt. 32 C (1% aq.).
Macol® 19. [PPG-Mazer] Meroxapol 172; see Macol 1; APHA < 100 liq.; water-sol.; sp.gr. 1.030; dens. 8.6 lb/gal; visc. 450 cps; HLB 8.0; pour pt. –25 C; ref. index 1.4535; flash pt. (PMCC) > 450 F; cloud pt. 35 C (1% aq.).
Macol® 20. [PPG-Mazer] EO-PO block copolymers; nonionic; defoamer, pulp and paper additive, disper-

sant, lubricant, leveling aid, wetting agent; liq.; HLB 3.0; 100% conc.

Macol® 21, 24, 25, 30. [PPG-Mazer] Oxethylated straight chain alcohol, modified; nonionic; detergent, dispersant, wetting agent, rinse additive; liq.; HLB 7.0, 10.0, 10.0, 9.0 resp.; 100% conc.

Macol® 31. [PPG-Mazer] EO-PO block copolymer; see Macol 1; APHA < 100 liq.; water-sol.; sp.gr. 1.025; dens. 8.5 lb/gal; visc. 165 cps; HLB 6.3; pour pt. –32 C; ref. index 1.4515; flash pt. (PMCC) 439 F; cloud pt. 37 C (1% aq.).

Macol® 32. [PPG-Mazer] Meroxapol 251; see Macol 1; APHA < 100 liq.; water-sol.; sp.gr. 1.017; dens. 8.5 lb/gal; visc. 460 cps; HLB 4.0; pour pt. –27 C; ref. index 1.4521; flash pt. (PMCC) < 450 F; cloud pt. 28 C (1% aq.).

Macol® 33. [PPG-Mazer] Meroxapol 311; see Macol 1; APHA < 100 liq.; water-sol.; sp.gr. 1.018; dens. 8.5 lb/gal; visc. 578 cps; HLB 4.0; pour pt. –25 C; ref. index 1.4522; flash pt. (PMCC) < 450 F; cloud pt. 25 C (1% aq.).

Macol® 34. [PPG-Mazer] Meroxapol 254; nonionic; see Macol 15; liq.; HLB 10.0; 100% conc.

Macol® 35. [PPG-Mazer] Poloxamer 105; nonionic; see Macol 1; HLB 8.0; 100% conc.

Macol® 42. [PPG-Mazer] Poloxamer 122; see Macol 1; APHA < 100 liq.; water-sol.; sp.gr. 1.018; dens. 8.5 lb/gal; visc. 250 cps; HLB 12.0; pour pt. –26 C; ref. index 1.4541; flash pt. (PMCC) 450 F; cloud pt. 37 C (1% aq.).

Macol® 44, 46, 72, 77, 85, 108. [PPG-Mazer] Poloxamer 124, 101, 212, 217, 235, and 338 resp.; nonionic; see Macol 1; liq., flake, paste; HLB 3.0, 18.5, 2.0, 24.5, 16.0, 17.5 resp.; 100% conc.

Macol® 81. [PPG-Mazer] EO-PO block copolymer; see Macol 1; APHA < 100 liq.; water-sol.; sp.gr. 1.019; dens. 8.5 lb/gal; visc. 475 cps; pour pt. –37 C; ref. index 1.4526; flash pt. (PMCC) < 450 F; cloud pt. 20 C (1% aq.).

Macol® 101. [PPG-Mazer] Poloxamer 331; see Macol 1; APHA < 100 liq.; water-sol.; sp.gr. 1.020; dens. 8.5 lb/gal; visc. 800 cps; HLB 1.0; pour pt. –23 C; ref. index 1.4524; flash pt. (PMCC) < 450 F; cloud pt. 15 C (1% aq.).

Macol® 121. [PPG-Mazer] EO-PO block copolymer; nonionic; see Macol 1; HLB 5.0; 100% conc.

Macol® 620. [PPG-Mazer] N-Butoxy POE POP glycols; emollient where industrial rinse water is required for applics.; low color liq.; low odor.

Macol® 660. [PPG-Mazer] PPG-12 buteth-16; nonionic; defoamer, rubber lubricant, intermediate; hydraulic, heat transfer, and metal working fluids; mold release agent; APHA 100 max. clear visc. liq.; sol. in acetone, propylene glycol, oleic acid, castor oil, water, toluene, IPA; sp.gr. 1.047; dens. 8.72 lb/gal; visc. 660 s (100 F); pour pt. –40 F; acid no. 0–1; ref. index 1.459; pH 5–7; flash pt. 430 F (COC).

Macol® 3520. [PPG-Mazer] PPG-28 buteth-35; nonionic; see Macol 660; APHA 100 max. clear visc. liq.; sol. see Macol 660; sp.gr. 1.050; dens. 8.75 lb/gal; visc. 3520 s (100 F); pour pt. –20 F; acid no. 0–1; ref. index 1.461; pH 5–7; flash pt. 430 F (COC).

Macol® 5100. [PPG-Mazer] PPG-33 buteth-45; nonionic; see Macol 660; APHA 100 max. clear visc. liq.; sol. see Macol 660; sp.gr. 1.050; dens. 8.75 lb/gal; visc. 5100 s (100 F); pour pt. –20 F; acid no. 0–1; ref. index 1.462; pH 5–7; flash pt. 430 F (COC).

Macol® CSA-2. [PPG-Mazer] Ceteareth-2; nonionic; detergent, wetting agent, emulsifier, dispersant, solubilizer, and coupling agent for cosmetics, textiles, metalworking lubricants, household prods., and industrial applics.; wh. solid; sol. in IPA; insol. in water; HLB 5.1; m.p. 39 C.

Macol® CSA-4. [PPG-Mazer] Ceteareth-4 nonionic; see Macol CSA-2; wh. solid; sol. in IPA; insol. water and min. oil; HLB 7.9; m.p. 38 C.

Macol® CSA-10. [PPG-Mazer] Ceteareth-10; nonionic; see Macol CSA-2; wh. solid; sol. in IPA; disp. in water; HLB 12.6; m.p. 39 C.

Macol® CSA-15. [PPG-Mazer] Ceteareth-15; nonionic; see Macol CSA-2; wh. solid; sol. in water, IPA; insol. min. oil; HLB 14.3; m.p. 40 C.

Macol® CSA-20. [PPG-Mazer] Ceteareth-20; nonionic; see Macol CSA-2; wh. solid; sol. in water, IPA; insol. min. oil; HLB 15.4; m.p. 40 C.

Macol® E-200, -300, -400, -600, -1000, -1450, -3350, -8000. [PPG-Mazer] PEG-4, -6, -8, -12, -20, -32, -75, and -150 resp.; plasticizer, humectant, lubricant for cosmetic and pharmaceutical creams, lotions, ointments; liq. to wh. solid or flake; water-sol.

Macol® LA-4. [PPG-Mazer] Laureth-4; nonionic; detergent, wetting agent, emulsifier, dispersant, solubilizer, stabilizer, coupling agent for cosmetics, textiles, metalworking lubricants, household prods., industrial uses; colorless liq.; sol. in IPA, min. oil; HLB 9.7; 100% conc.

Macol® LA-9. [PPG-Mazer] Laureth-9; nonionic; see Macol LA-4; colorless liq.; sol. in water, IPA; HLB 11.8; 100% conc.

Macol® LA-12, -23. [PPG-Mazer] Laureth-12 and -23 resp.; nonionic; see Macol LA-4; wh. solid; sol. in water, IPA; HLB 14.5 and 16.9 resp.; m.p. 32 C (LA-23); 100% conc.

Macol® LA-790. [PPG-Mazer] Laureth-7; nonionic; see Macol CSA-2; colorless liq.; sol. in water, IPA; insol. min. oil; HLB 11.0; pour pt. 5 C.

Macol® LF-110. [PPG-Mazer] Wetting aid and low surface tension surfactant; syn. lubricant base fluid for metalworking, hard-surface cleaning, and metal cleaning and degreasing; used in cleaners and rinse aids; metalworking base fluid; APHA < 100 liq.; water-sol.; sp.gr. 1.050; dens. 8.4 lb/gal; visc. 175 cps; pour pt. < 0 C; flash pt. (PMCC) 375 F; cloud pt. 12 C (1% aq.); 100% act.

Macol® LF-111. [PPG-Mazer] Polyalkoxylated aliphatic ether; nonionic; see Macol LF-110; APHA < 100 liq.; water-sol.; sp.gr. 1.040; dens. 8.4 lb/gal; visc. 200 cps; pour pt. < 0 C; flash pt. (PMCC) 380 F; cloud pt. 12 C (1% aq.); 100% act.

Macol® LF-111A. [PPG-Mazer] Polyalkoxylated aliphatic ether; nonionic; see Macol LF-110; APHA < 100 liq.; water-sol.; sp.gr. 1.040; dens. 8.4 lb/gal; visc. 225 cps; pour pt. < 0 C; flash pt. (PMCC) 380 F; cloud pt. 10 C (1% aq.); 100% act.

Macol® LF-120. [PPG-Mazer] Polyalkoxylated aliphatic ether; nonionic; see Macol LF-110; APHA 150 liq.; water-sol.; sp.gr. 1.050; dens. 8.4 lb/gal; visc. 75 cps; pour pt. < 0 C; flash pt. (PMCC) 310 F; cloud pt. 18 C (1% aq.); 100% act.
Macol® NP-4, -6. [PPG-Mazer] Nonoxynol-4 and -6 resp.; nonionic; emulsifier, detergent, dispersant, solubilizer, stabilizer, oil surfactant and intermediate; liq.; HLB 8.8 and 10.8; 100% conc.
Macol® NP-9.5. [PPG-Mazer] Nonoxynol-9 (9.5 EO); nonionic; detergent emulsifier, solubilizer, wetting agent, stabilizer; surfactant used in paints, textile and pesticide formulation; liq.; water-sol.; HLB 13.0; 100% conc.
Macol® NP-11, -20. [PPG-Mazer] Nonoxynol–11 and –20 resp.; nonionic; wetting agent, solubilizer, emulsifier, stabilizer, and detergent; liq. and waxy solid resp.; water-sol.; HLB 14.2 and 16.0; 100% conc.
Macol® NP-30, -30(70). [PPG-Mazer] Nonoxynol-30; nonionic; detergent, wetting agent, solubilizer, stabilizer, and emulsifier used in pesticide, textile treatment; HLB 17.2; 100 and 70% conc. resp.
Macol® NP-70, -100. [PPG-Mazer] Nonoxynol–70 and –100 resp.; nonionic; emulsifier, stabilizer, solubilizer; detergent and sanitizer formulations; solid (both), flake (NP-100); HLB 18.4 (NP-70); 100% conc.
Macol® OA-2, -4. [PPG-Mazer] Oleth-2 and -4 resp.; nonionic; see Macol LA-4; colorless liq.; sol. in IPA, min. oil; HLB 4.9 and 8.0; 100% conc.
Macol® OA-5. [PPG-Mazer] Oleth-5; nonionic; see Macol CSA-2; colorless liq.; sol. in IPA, min. oil; insol. in water; HLB 8.7.
Macol® OA-10, -20. [PPG-Mazer] Oleth-10 and -20 resp.; nonionic; see Macol LA-4; cream solid; sol. in water, IPA; disp. min. oil; HLB 12.4 and 15.3; m.p. 16 and 30 C; 100% conc.
Macol® P-500, -1200, -2000, -4000. [PPG-Mazer] PPG-9, -20, -26, -30 resp.; nonionic; defoamer, mold release applics., chemical intermediates for fatty acid esters, components for urethane resins; liq.
Macol® SA-2, -5, -10. [PPG-Mazer] Steareth-2, -5, and –10 resp.; nonionic; see Macol LA-4; wh. solid; sol. in IPA; HLB 4.9, 8.9, and 12.4; m.p. 43 C (SA-2), 38 C (SA-10); 100% conc.
Macol® SA-15, -20, -40. [PPG-Mazer] Steareth-15, -20, and -40 resp.; nonionic; see Macol LA-4; wh. solid; sol. in water, IPA; HLB 14.2, 15.3, and 17.4; m.p. 38, 38, and 40 C; 100% conc.
Macol® TD-3, -4, -6, -8. [PPG-Mazer] Trideceth-3, -4, –6, and -8 resp.; nonionic; see Macol LA-4; liq.; HLB 7.9, 9.3, 11.4, 12.7; 100% conc.
Macol® TD-10. [PPG-Mazer] Trideceth-10; nonionic; see Macol LA-4; semisolid; HLB 13.7; 100% conc.
Macol® TD-610. [PPG-Mazer] Trideceth-6; nonionic; see Macol LA-4; liq.; HLB 12.7; 86% conc.
Macrobase 600. [Sartomer] Blend of methacrylate-terminated PS and isooctyl acrylate; dispersant for hydrophobic pigments; used in uv/eb curable adhesives, uv-curable inks and coatings; APHA 50 liq., mild odor; sp.gr. 0.0960; visc. 2885 cps; 95+% act.
Macrobase 610. [Sartomer] Blend of methacrylate-terminated PS and tripropylene glycol diacrylate; see Macrobase 600; APHA 50 liq., mild odor; sp.gr. 1.060; visc. 100,000–125,000 cps; 95+% act.
Macrobase 620. [Sartomer] Blend of methacrylate-terminated PS, ethoxylated TMPTA, and tripropylene glycol diacrylate; see Macrobase 600; APHA 50 liq., mild odor; sp.gr. 1.090; visc. 250,000–300,000 cps; 95+% act.
Macromer® 13K-PSMA. [Sartomer] Methacrylate-terminated PS; pour pt. depressant; used in inks, coatings, solv.-based sealants, glass reinforcements, uv-cured heat-seal adhesives, graft copolymers; wh. gran. solid; m.w. 13,000; sp.gr. 1.05; 90+% act.
Mafloc® 764. [PPG-Mazer] Polyacrylate polymer; cationic; flocculating power towards suspension of solids; liq.; water-disp.
Mafloc® 900. [PPG-Mazer] Polyacrylate polymer; anionic; see Mafloc 764; powd.; water-disp.
Mafo® 13. [PPG-Mazer] Potassium salt of complex n-stearyl amino acid; amphoteric; emulsifier, wetting agent, corrosion inhibitor, suspending agent, solubilizer for difficult materials, dairy chain and metal-to-metal lubricant, emollient; clear amber liq.; sol. in water and most popular solvs.; sp.gr. 1.015; visc. 140–200 cps; b.p. 230 F; flash pt. 230 F; pour pt. 10 C; pH 10–11 (100%); biodeg.; 70% act.
Mafo® C-12. [PPG-Mazer] Potassium salt of complex n-coco amino acid; amphoteric; solubilizer, suspending agent, surfactant used for hard surface cleaners, agric. emulsions, cosmetics, steam cleaners, dishwashing; lt. yel. clear liq.; sol. @ 5% in 18% KOH, 7% NaOH, 15% HCl, 10% trisodium phosphate, sea water @ pH 4 and 8, disp. at pH 9–10; dens. 8.47 lb./gal; sp.gr. 1.02; flash pt. 212 F (COC); pH 10–11; 29–31% solids.
Mafo® CAB. [PPG-Mazer] Cocamidopropyl betaine; amphoteric; detergent, dispersant, surfactant, conditioner, foam and visc. stabilizer, chelating agent, wetting agent, solubilizer, lubricant, emulsifier; used in personal care, dishwashing, rug and carpet cleaning applics.; liq.; water-sol.; 30% conc.
Mafu®. [Bayer] Dichlorvos; contact, stomach, and breathing poison for control of mosquitoes, flies, cockroaches, mites, etc.; for control of public health and stored-prod. pests; colorless to ylsh. liq., mild aromatic odor; m.w. 220.98; sol. in most org. solvs., 1% in water; sp.gr. 1.4 (20/4 C); b.p. 74 C (1 mm Hg).
Magnabrite F. [Am. Colloid] Magnesium aluminum silicate NF; stabilizing and suspending agent for cosmetics and pharmaceuticals; wh. fine powd.; visc. 150–450 cps; pH 9–10 (5% disp.).
Magnabrite HV. [Am. Colloid] Magnesium aluminum silicate NF; see Magnabrite F; wh. soft flakes.; visc. 800–2200 cps; pH 9–10 (5% disp.).
Magnabrite K. [Am. Colloid] Magnesium aluminum silicate NF, acid-stable; see Magnabrite F; wh. soft flakes.; visc. 100–300 cps; pH 9–10 (5% disp.).
Magnabrite S. [Am. Colloid] Magnesium aluminum silicate NF; see Magnabrite F; wh. soft flakes; visc. 225–600 cps; pH 9–10 (5% disp.).

Magnabrite T. [Am. Colloid] Magnesium aluminum silicate, tech.; stabilizing and suspending agent for household and industrial specialties, e.g., paints and cleaning compds.; wh. soft flakes.; visc. 250 cps min.; pH 9–10 (5% disp.).

Magnamite® AS1.. [Hercules] PAN-based graphite fiber, surf.-treated; reinforcing agent for prepregging; also suitable for filament winding, pultrusion, and molding compds.; avail. in 10K filament-count tows; 8 μ filament diam.; dens. 1.80 g/cc; tens. str. 3103 MPa; elong. 1.32% (ultimate); 92% C.

Magnamite® AS2. [Hercules] PAN-based graphite fiber, surf.-treated, sized; reinforcing agent; for use in weaving, prepregging, filament winding, pultrusion, and molding compds.; avail. in 12K filament-count tows; 8 μ filament diam.; dens. 1.80 g/cc; tens. str. 2759 MPa; elong. 1.53% min.; 94% C.

Magnamite® AS4. [Hercules] PAN-based graphite fiber, surf.-treated; see Magnamite AS2; avail. in 3K, 6K, and 12K filament-count tows; 8 μ filament diam.; dens. 1.80 g/cc; tens. str. 3587 MPa; elong. 1.53% (ultimate); 94% C.

Magnamite® AS6. [Hercules] PAN-based grapite fiber, surf.-treated, sized; reinforcing agent for use in prepregging, filament winding; avail. in 12K filament-count tows; dens. 1.83 g/cc; tens. str. 4137 MPa; elong. 1.65% min.; 94% C.

Magnamite® Chopped. [Hercules] PAN-based chopped graphite fiber, sized; reinforcing agent for polymer-matrix molding compds.; avail. grades: 1800/AS for polyamide, polyacetal, PU; 1805/AS for PC, PVC; 1810/AS for polysulfone, PPS, PPO; 1815/AS for slurry processing; thin flake; 1/4 in. chopped fiber; 8 μ filament diam.; water-disp. (1815/AS); dens. 1.77 g/cc; bulk dens. 280 g/l; tens. str. 2480 MPa; elong. 1.2% (ultimate).

Magnamite® HMS. [Hercules] Graphite fiber, surf.-treated; reinforcing agent for use in prepregging, filament winding; avail. in 10K filament-count tows; 8 μ filament diam.; dens. 1.83 g/cc; tens. str. 2206 MPa; elong. 0.58% (ultimate); 99.7% C.

Magnamite® HMU. [Hercules] PAN-based graphite fiber, sized; reinforcing agent for use in carbon-carbon composite applics.; exc. dimensional stability in extreme temp. ranges; avail. in 1K, 3K, 6K, and 12K filament-count tows; 8 μ filament diam.; dens. 1.84 g/cc; tens. str. 2758 MPa; elong. 0.70% (ultimate); 99.7% C.

Magnamite® IM6. [Hercules] PAN-based graphite fiber, surf.-treated, sized; see Magnamite AS6; avail. in 12K filament-count tows; dens. 1.73 g/cc; tens. str. 4378 MPa; elong. 1.50% (ultimate); 94% C.

Magox® Super Premium. [Premier Refractories] Magnesium oxide; adsorption, absorption agent and scavenger with high surf. area for plastics, rubber industries, chemical neutralization, etc.; 2 μ median particle size; 98% min. thru 325 mesh; bulk dens. 27 lb/ft^3 (loose), 42 lb/ft^3 (tapped); surf. area 120 m^2/g min.

Makon® 8240. [Stepan] PEG-36 castor oil; nonionic; wetting agent, lubricant, coupling agent, defoamer for cutting fluids, drawing compds., corrosion inhibitors; liq.; HLB 13.0; pH 5–7 (1%); 100% act.

Makon® DN 50, DN 70. [Stepan] Dinonylphenol alkoxylates; emulsifier, dispersant; liq.

Makon® NF-5, NF-12. [Stepan] Polyalkoxylated aliphatic base; nonionic; low foaming wetting agent and penetrant for textile applics.; NF-5 useful in scouring and dye baths; lubricant, coupling agent, defoamer for cutting fluids, drawing compds., corrosion inhibitors; yel. liq.; cloud pt. 15 C (NF-12); pH 10.5 and 10.0 resp. (1%); 97 and 100% act. resp.

Makon® NI 10, NI 20, NI 30. [Stepan] Alkylphenol alkoxylates; emulsifier, dispersant; visc. liq.

Makon® NP-40, 40-70. [Stepan] Nonylphenol ethoxylate; nonionic; wetting agent, dispersant, and emulsifier; solid, liq. resp.; 100 and 70% conc. resp.

Makon® TP 60, TP 100. [Stepan] Tristyrylphenol alkoxylates; emulsifier, dispersing agent; visc. liq.

Maltrin® M040. [Grain Processing] Maltodextrin; nonsweet, nutritive polymer useful for wet binding and anticaking; adds sol'n. visc., good mouthfeel; for pharmaceuticals, foods; DE 4–7; bulk dens. 0.51 g/cc (packed); pH 4–5.

Maltrin® M050. [Grain Processing] Maltodextrin; nonsweet, nutritive polymer useful for adding visc. and mouthfeel; for pharmaceuticals, foods; DE 4–7; bulk dens. 0.56 g/cc (packed); pH 4–5.

Maltrin® M100. [Grain Processing] Maltodextrin; nonsweet, nutritive polymer useful as carrier and bulking agent; provides exc. mouthfeel for chewable tablets; functions in binding, coating, and spray drying; for pharmaceuticals, foods; DE 9–12; bulk dens. 0.56 g/cc (packed); pH 4.0–4.7.

Maltrin® M150. [Grain Processing] Maltodextrin; very slightly sweet, nutritive polymer useful for binding properties; directly compressible; for pharmaceuticals, foods; DE 13–17; bulk dens. 0.61 g/cc (packed); pH 4.0–4.7.

Maltrin® M180. [Grain Processing] Maltodextrin; slightly sweet, nutritive polymer; directly compressible; binding properties; for pharmaceuticals, foods; DE 16.5–19.5; bulk dens. 0.63 g/cc (packed); pH 4.0–4.7.

Maltrin® M200. [Grain Processing] Corn syrup solids; dried glucose syrup with very good coating and binding properties; directly compressible; for pharmaceutical use; DE 20–23; bulk dens. 0.64 g/cc (packed); pH 4.0–4.7.

Maltrin® M250. [Grain Processing] Corn syrup solids; dried glucose syrup with good binding properties; directly compressible; for pharmaceutical use; DE 23–27; bulk dens. 0.67 g/cc (packed); pH 4.5–5.5.

Maltrin® M365. [Grain Processing] High maltose corn syrup solids; dried glucose syrup with noticeable sweetness; for pharmaceutical use; DE 34–38; bulk dens. 0.67 g/cc (packed); pH 4.5–5.2.

Maltrin® M440. [Grain Processing] Agglomerated maltodextrin; agglomerated form of Maltrin M040; exhibits exc. dispersibility and dissolution; for pharmaceutical use; free-flowing gran.; DE 4–7; bulk dens. 0.30 g/cc (packed); pH 4.0–5.1.

Maltrin® M500. [Grain Processing] Agglomerated

maltodextrin; agglomerated form of Maltrin M100; exhibits exc. dispersibility and dissolution; directly compressible binder and good carrier; for pharmaceutical use; free-flowing gran.; DE 9–12; bulk dens. 0.34 g/cc (packed); pH 4.0–5.1.

Maltrin® M510. [Grain Processing] Agglomerated maltodextrin; flowable form of Maltrin M100; exhibits exc. dispersibility and dissolution; directly compressible binder and diluent; for pharmaceutical use; free-flowing gran.; DE 9–12; bulk dens. 0.56 g/cc (packed); pH 4.0–4.7.

Maltrin® M550. [Grain Processing] Agglomerated maltodextrin; agglomerated form of Maltrin M150; exhibits exc. dispersibility and dissolution; directly compressible binder and good carrier; for pharmaceutical use; free-flowing gran.; DE 13–17; bulk dens. 0.37 g/cc (packed); pH 4.0–5.1.

Maltrin® M580. [Grain Processing] Maltodextrin; for pharmaceutical use; DE 16.5–19.5; bulk dens. 0.40 g/cc (packed); pH 4.0–5.1.

Maltrin® M600. [Grain Processing] Agglomerated corn syrup solids; agglomerated form of Maltrin M200; directly compresible binder and good carrier; for pharmaceutical use; free-flowing gran.; DE 20–23; bulk dens. 0.40 g/cc (packed); pH 4.0–5.1.

Maltrin® M700. [Grain Processing] Agglomerated maltodextrin; agglomerated form of Maltrin M100; exhibits exc. dissolution; very good carrier when low bulk dens. is required; for pharmaceutical use; free-flowing gran.; DE 9–12; bulk dens. 0.13 g/cc (packed); pH 6.0–7.0

Manex. [Griffin] Maneb and zinc disp.; flowable fungicide for many crops; 37% act.

Manro AO 3OC. [Manro] N-alkyl dimethyl amine oxide; foam booster; detergent thickener for bleaches; liq.

Manro BES 70. [Manro] Sodium alkyl ether sulfate; anionic; foaming agent, emulsifier for chlorinated solvs., lime soap dispersant, intermediate used in cosmetics, household cleaning prods. and fire fighting foams; pale yel., mobile gel; mild odor; sp.gr. 1.05; pH 7–9 (10% aq.); biodeg.; 67.5% min. act.

Manro CD, CDS, CDX. [Manro] Cocamide DEA; nonionic; foam stabilizer, solubilizer used in liq. detergents, shampoos, bubble baths; clear golden yel. visc. liq.; mild odor; sp.gr. 0.982, 0.990, 1.01 resp.; visc. 950, 800, 950 cps; biodeg.; 90, 85, and 70% act.

Manro CMEA. [Manro] Cocamide MEA; nonionic; foam booster, foam stabilizer used in liq. and powd. detergent systems; cream waxy flake; mild odor; biodeg.; 94% min. act.

Manro DPM 2169. [Manro] Amphoterics blend; thickener for phosporic and citric acids; liq.

Manro DS 35. [Manro] Sodium alkyl sulfate; anionic; solubilizer, wetting agent used in industrial cleaners; clear pale yel. liq.; char. odor; sp.gr. 1.07; visc. 50 cps; pH 7–8 (10% aq.); 35% act.

Manro FCM 95LV. [Manro] Acid compd. based on an aromatic sulfonic acid; catalyst in prod. of resin bound sand castings; curing agent in prod. of resins; dk. brnsh. liq.; slight aromatic odor; sp.gr. 1.28; visc. 1100 cps; 91% min. act.

Manro FCM 100. [Manro] Acid compd. based on an aromatic sulfonic acid; see Manro FCM 95LV; lt. brnsh. clear liq.; slight odor; sp.gr. 1.24; visc. 175 cps; 90% act.

Manro KXS 40. [Manro] Potassium xylene sulfonate; solubilizing agent; pale yel. clear liq.; negligible odor; sp.gr. 1.18 (40 C); pH 7–10 (10% aq.); 39% min. act. in water.

Manro ML 33. [Manro] MEA-lauryl sulfate; anionic; used in cosmetics and toiletries; pale yel. clear to slightly hazy visc. liq.; mild odor; sp.gr. 1.03; visc. 7500 cps; pH 6.3–7.3 (10% aq.); biodeg.; 32% act.

Manro PC 35N. [Manro] Anionic/nonionic; pearlized lotion conc. for shampoos, bubble baths; wh., visc. liq.; negligible odor; sp.gr. 1.00; visc. 16,000 cps; pH 6.5–7.0 (10% aq.); 35% act.

Manro PTSA 65 E. [Manro] Toluene sulfonic acid; catalyst in prod. resin bound sand castings and in the mfg. of esters; amber clear liq.; slight odor; sp.gr. 1.24; visc. 10 cps; 65% act.

Manro PTSA 65 H. [Manro] Toluene sulfonic acid; see Manro PTSA 65 E; amber clear liq.; slight odor; sp.gr. 1.24; visc. 10 cps; 62.5% act.

Manro PTSA 65 LS. [Manro] Toluene sulfonic acid; see Manro PTSA 65 E; amber clear liq.; slight odor; sp.gr. 1.24; visc. 10 cps; 65% act.

Manro PTSA Crystals. [Manro] Para toluene sulfonic acid monohydrate; catalyst in prod. of esters; curing agent for coating resins; adhesive systems; mfg. of dyes and pigments; wh. to pink crystals; 97% min. act.

Manro SCS 40. [Manro] Sodium cumenesulfonate; hydrotrope used in liq. and spray-dried powd. formulations; liq.; 40% act.

Manro SCS 90. [Manro] Sodium cumenesulfonate; hydrotrope used in liq. and spray-dried formulations; wh. powd.

Manro STS 40. [Manro] Sodium toluene sulfonate; solubilizing agent; pale yel. clear liq.; negligible odor; sp.gr. 1.19; pH 7–10 (10% aq.); 40% act. in water.

Manro STS 90. [Manro] Sodium toluene sulfonate; hydrotrope for reduction of slurry visc. before spray-drying in heavy-duty detergent formulations; powd.; 90% conc.

Manro SXS 30, 40. [Manro] Sodium xylene sulfonate; solubilizing agent; pale yel. clear liq.; negligible odor; sp.gr. 1.15–1.17 (30 C); pH 7–10 (10% aq.); 30 and 40% act. in water.

Manro SXS 93. [Manro] Sodium xylene sulfonate; solubilizing agent; powd.; 93% conc.

Manroteric AT 1200. [Manro] Betaine; thickener for HCl/phosphoric acid cleaners; visc. liq.; 40% act.

Mapeg® 200 DL. [PPG-Mazer] PEG-4 dilaurate; nonionic; emulsifier, dispersant used in cosmetics, pharmaceuticals, metalworking and fiber lubricants, etc.; yel. clear liq.; sol. in IPA, toluol, soybean and min. oil, water disp.; sp.gr. 0.95; m.p. 10 C; HLB 6.8; acid no. 10 max.; sapon. no. 176–192; 100% conc.

Mapeg® 200 DO. [PPG-Mazer] PEG-4 dioleate; nonionic; see Mapeg 200 DL; yel. clear liq.; sol. in IPA, toluol, soybean and min. oil, water disp.; sp.gr.

0.95; m.p. < –10 C; HLB 6.0; acid no. 10 max.; sapon. no. 148–158; 100% conc.

Mapeg® 200 DOT. [PPG-Mazer] PEG 200 ditallate; surfactant, emulsifier, emollient for hair preps., creams and lotions; solubilizer for bath oils and fragrances; liq.; sol. in IPA, min. spirits, toluene, min. oil; disp. in water; sp.gr. 0.95; HLB 6.0; flash pt. (PMCC) > 350 F; pour pt. –18 C; acid no. 10 max.; sapon. no. 150.

Mapeg® 200 DS. [PPG-Mazer] PEG-4 distearate; nonionic; see Mapeg 200 DL; wh. solid; sol. in IPA, toluol, soybean and min. oil, hot water disp.; m.p. 34 C; HLB 4.7; acid no. 10 max.; sapon. no. 155–165; 100% conc.

Mapeg® 200 ML. [PPG-Mazer] PEG-4 laurate; nonionic; see Mapeg 200 DL; yel. clear liq.; sol. in IPA, toluol, soybean and min. oil, water disp.; sp.gr. 0.99; m.p. 5 C; HLB 9.8; acid no. 5 max.; sapon. no. 139–159; 100% conc.

Mapeg® 200 MO. [PPG-Mazer] PEG-4 oleate; nonionic; see Mapeg 200 DL; yel. clear liq.; sol. in IPA, toluol, soybean oil, water disp.; sp.gr. 0.98; m.p. < –10 C; HLB 8.0; acid no. 5 max.; sapon. no. 115–125; 100% conc.

Mapeg® 200 MOT. [PPG-Mazer] PEG-4 tallate; nonionic; see Mapeg 200 DL; liq.; HLB 8.0; 100% conc.

Mapeg® 200 MS. [PPG-Mazer] PEG-4 stearate; nonionic; see Mapeg 200 DL; wh. solid; sol. in IPA, toluol, soybean oil, hot water disp.; m.p. 33 C; HLB 8.0; acid no. 5 max.; sapon. no. 120–130; 100% conc.

Mapeg® 350 MS. [PPG-Mazer] PEG 350 stearate; nonionic; see Mapeg 200 DL; solid; HLB 11.0; 100% conc.

Mapeg® 400 DL. [PPG-Mazer] PEG-8 dilaurate; nonionic; see Mapeg 200 DL; lt. yel. liq.; sol. in IPA, toluol, soybean oil, water disp.; sp.gr. 0.98; m.p. 18 C; HLB 9.8; acid no. 10 max.; sapon. no. 130–140; 100% conc.

Mapeg® 400 DO. [PPG-Mazer] PEG-8 dioleate; nonionic; see Mapeg 200 DL; yel. liq.; sol. in IPA, toluol, soybean and min. oil, water disp.; sp.gr. 0.98; m.p. < 7 C; HLB 8.5; acid no. 10 max.; sapon. no. 114–122; 100% conc.

Mapeg® 400 DOT. [PPG-Mazer] PEG-8 ditallate; nonionic; see Mapeg 200 DL; liq.; HLB 8.5; 100% conc.

Mapeg® 400 DS. [PPG-Mazer] PEG-8 distearate; nonionic; see Mapeg 200 DL; wh. solid; sol. in IPA, toluol, soybean and min. oil, hot water disp.; m.p. 36 C; HLB 8.1; acid no. 10 max.; sapon. no. 116–125; 100% conc.

Mapeg® 400 DSLM. [PPG-Mazer] PEG 400 distearate; nonionic; surfactant, emulsifier, emollient for hair preps., creams and lotions; solubilizer for bath oils and fragrances; solid; sol. in IPA, min. spirits, toluene, min. oil; disp. hot in water; HLB 8.1; flash pt. (PMCC) > 350 F; pour pt. 36 C; acid no. 10 max.; sapon. no. 124.

Mapeg® 400 ML. [PPG-Mazer] PEG-8 laurate; nonionic; see Mapeg 200 DL; lt. yel., liq.; sol. in IPA, toluol, water; sp.gr. 1.01; m.p. 12 C; HLB 13.1; acid no. 5 max.; sapon. no. 89–96; 100% conc.

Mapeg® 400 MO. [PPG-Mazer] PEG-8 oleate; nonionic; see Mapeg 200 DL; yel. liq.; sol. in IPA, toluol, soybean oil, water disp.; sp.gr. 1.01; m.p. < 10 C; HLB 11.4; acid no. 5 max.; sapon. no. 80–88; 100% conc.

Mapeg® 400 MOT. [PPG-Mazer] PEG-8 tallate; nonionic; see Mapeg 200 DL; liq.; HLB 11.4; 100% conc.

Mapeg® 400 MS. [PPG-Mazer] PEG-8 stearate; nonionic; see Mapeg 200 DL; wh. solid; sol. in IPA, toluol, soybean oil, hot water disp.; m.p. 33 C; HLB 11.5; acid no. 5 max.; sapon. no. 84–93; 100% conc.

Mapeg® 600 DL. [PPG-Mazer] PEG-12 dilaurate; dispersant and emulsifier for metalworking lubricants, fiber lubricants and softeners, solubilizers, defoamers, antistats, cosmetics, pharmaceuticals, and chemical intermediates; lt. yel. semisolid; sol. in IPA, toluol, soybean oil; partly sol. min. oil; disp. water; sp.gr. 0.99; HLB 12.2; m.p. 24 C; acid no. 10 max.; sapon. no. 102–112.

Mapeg® 600 DO. [PPG-Mazer] PEG-12 dioleate; nonionic; see Mapeg 200 DL; yel. liq.; sol. in IPA, toluol, soybean oil, water disp.; sp.gr. 1.00; m.p. 20 C; HLB 10.5; acid no. 10 max.; sapon. no. 92–102; 100% conc.

Mapeg® 600 DOT. [PPG-Mazer] PEG-12 ditallate; see Mapeg 600 DL; amber liq.; sol. IPA, toluol, soybean oil; disp. water; sp.gr. 1.00; HLB 13.2; m.p. 15 C; acid no. 10 max.; sapon. no. 85–95.

Mapeg® 600 DS. [PPG-Mazer] PEG-12 distearate; nonionic; see Mapeg 200 DL; wh. solid or flake; sol. in IPA, toluol, soybean oil, hot water disp.; m.p. 41 C; HLB 10.6; acid no. 10 max.; sapon. no. 94–104; 100% conc.

Mapeg® 600 ML. [PPG-Mazer] PEG-12 laurate; see Mapeg 600 DL; lt. yel. liq.; sol. in water, IPA, toluol; sp.gr. 1.02; HLB 14.9; m.p. 23 C; acid no. 5 max.; sapon. no. 64–74.

Mapeg® 600 MO. [PPG-Mazer] PEG-12 oleate; nonionic; see Mapeg 200 DL; yel. liq.; sol. in IPA, toluol, soybean oil, water; sp.gr. 1.03; m.p. 25 C; HLB 13.5; acid no. 5 max.; sapon. no. 60–70; 100% conc.

Mapeg® 600 MOT. [PPG-Mazer] PEG-12 tallate; see Mapeg 600 DL; amber liq.; sol. in water, IPA, toluol; sp.gr. 1.04; HLB 10.3; m.p. 20 C; sapon. no. 60–70.

Mapeg® 600 MS. [PPG-Mazer] PEG-12 stearate; nonionic; see Mapeg 200 DL; wh. solid; sol. in IPA, toluol, soybean oil, water, propylene glycol, disp. min. oil; m.p. 36 C; HLB 13.6; acid no. 5 max.; sapon. no. 62–70; 100% conc.

Mapeg® 1000 MS. [PPG-Mazer] PEG-20 stearate; nonionic; see Mapeg 200 DL; wh. solid or flake sol. in IPA, toluol, propylene glycol, water; m.p. 42 C; HLB 15.7; acid no. 5; sapon. no. 41–49; 100% conc.

Mapeg® 1500 MS. [PPG-Mazer] PEG-6-32 stearate; nonionic; see Mapeg 200 DL; solid; HLB 16.8; 100% conc.

Mapeg® 1540 DS. [PPG-Mazer] PEG-32 distearate; nonionic; see Mapeg 200 DL; wh. solid or flake; sol.

in IPA, toluol, soybean oil, propylene glycol, water; m.p. 45 C; HLB 14.8; acid no. 10 max.; sapon. no. 49–58; 100% conc.

Mapeg® 4000 MS. [PPG-Mazer] PEG-75 stearate; nonionic; see Mapeg 200 DL; liq.; HLB 18.7; 100% conc.

Mapeg® 6000 DS. [PPG-Mazer] PEG-150 distearate; nonionic; see Mapeg 200 DL; wh. solid or flake; sol. in IPA, toluol, propylene glycol, water; m.p. 55 C; HLB 18.4; acid no. 9 max.; sapon. no. 14–20; 100% conc.

Mapeg® 6000 MS. [PPG-Mazer] PEG-150 stearate; see Mapeg 600 DL; lt. yel. waxy solid, flake; sol. in water, IPA, toluol; HLB 18.8; m.p. 60 C; acid no. 5 max.; sapon. no. 7–14.

Mapeg® CO-16H. [PPG-Mazer] PEG-16 hydrog. castor oil; surfactant, emulsifier, emollient for hair preps., creams and lotions; solubilizer for bath oils and fragrances; liq.; sol. in toluene, min. oil; disp. in water, IPA, min. spirits; sp.gr. 1.010; HLB 8.6; flash pt. (PMCC) > 350 F; pour pt. 7 C; acid no. 2 max.; sapon. no. 105.

Mapeg® CO-25. [PPG-Mazer] PEG-25 castor oil; solubilizer, coupling agent for oils, solvs., waxes; for cosmetic, paper, metalworking fluid, and emulsion polymerization; liq.

Mapeg® CO-25H. [PPG-Mazer] PEG-25 hydrog. castor oil; nonionic; surfactant for formulation of gels; solubilizer, coupling agent; for cosmetic, paper, metalworking fluids, and emulsion polymerization; liq.; HLB 10.8; 100% conc.

Mapeg® CO-30. [PPG-Mazer] PEG-30 castor oil; surfactant, emulsifier, emollient for hair preps., creams and lotions; solubilizer for bath oils and fragrances; liq.; sol. in water, IPA, toluene; sp.gr. 1.046; HLB 11.8; flash pt. (PMCC) > 350 F; pour pt. 9 C; acid no. 2 max.; sapon. no. 75.

Mapeg® CO-36. [PPG-Mazer] PEG-36 castor oil; surfactant, emulsifier, emollient for hair preps., creams and lotions; solubilizer for bath oils and fragrances; liq.; sol. in water, IPA, toluene; sp.gr. 1.055; HLB 12.6; flash pt. (PMCC) > 350 F; pour pt. 12 C; acid no. 2 max.; sapon. no. 73.

Mapeg® CO-200. [PPG-Mazer] PEG-200 castor oil; surfactant, emulsifier, emollient for hair preps., creams and lotions; solubilizer for bath oils and fragrances; solid; sol. in water, IPA; HLB 18.1; flash pt. (PMCC) > 350 F; pour pt. 50 C; acid no. 2 max.; sapon. no. 17.5.

Mapeg® DGLD. [PPG-Mazer] Diethylene glycol laurate; nonionic; see Mapeg CO-25H; also plasticizer, stabilizer, emulsifier, wetting agent; HLB 8.3; 100% conc.

Mapeg® EGDS. [PPG-Mazer] Glycol distearate; nonionic; see Mapeg 200 DL; also thickener, opacifier, pearling additive; wh. solid or flake; sol. in IPA, toluol, soybean and min. oil, m.p. 63 C; HLB 1.4; acid no. 6 max.; sapon. no. 190–199; 100% conc.

Mapeg® EGMS. [PPG-Mazer] Glycol stearate; nonionic; see Mapeg 200 DL; also thickener, opacifier, pearling additive; wh. to cream solid or flake; sol. in IPA, toluol, soybean and min. oil; m.p. 56 C; HLB 2.4; acid no. 4 max.; sapon. no. 180–188; 100% conc.

Mapeg® PGDS. [PPG-Mazer] Propylene glycol distearate; see Mapeg 600 DL; lt. yel. solid; sol. in IPA, toluol, propylene glycol, soybean and min. oils; insol. water; HLB 1.2; m.p. 40 C; acid no. 10 max.; sapon. no. 183–193.

Mapeg® PGMS. [PPG-Mazer] Propylene glycol stearate; see Mapeg 600 DL; lt. yel. solid; sol. in IPA, toluol, propylene glycol, soybean and min. oil; insol. in water; HLB 2.3; m.p. 36 C; acid no. 5 max.; sapon. no. 172–182.

Mapeg® PPG-400. [PPG-Mazer] PPG (400); lubricant, antifoam, binder; intermediate for emulsifiers and dispersing surfactants; water-wh. liq.

Mapeg® PPG-500. [PPG-Mazer] PPG (500); see Mapeg PPG-400; water-wh. liq.

Mapeg® S-40. [PPG-Mazer] PEG-40 stearate; nonionic; see Mapeg 200 DL; wh. solid or flake; sol. in IPA, toluol, propylene glycol, water; m.p. 48 C; HLB 17.4; acid no. 1 max.; sapon. no. 25–35; 100% conc.

Mapeg® S-100. [PPG-Mazer] PEG-100 stearate; surfactant, emulsifier, emollient for hair preps., creams and lotions; solubilizer for bath oils and fragrances; flake; sol. in water, IPA, toluene, propylene glycol; disp. in min. spirits; HLB 18.7; flash pt. (PMCC) > 350 F; pour pt. 50 C; acid no. 1 max.; sapon. no. 16.

Mapeg® S-150. [PPG-Mazer] PEG-150 stearate; surfactant, emulsifier, emollient for hair preps., creams and lotions; solubilizer for bath oils and fragrances; flake; sol. see Mapeg S-100; HLB 19.0; flash pt. (PMCC) > 350 F; pour pt. 51 C; acid no. 1 max.; sapon. no. 9.5.

Mapeg® TAO-10. [PPG-Mazer] PEG-220 tallate; solubilizer, surfactant used in cosmetic, pharmaceutical, and industrial applics.; liq.

Mapeg® TAO-15. [PPG-Mazer] PEG 660 tallate; nonionic; see Mapeg 200 DL; clear amber liq.; sol. in IPA, toluol, water; m.p. 20 C; HLB 13.8; acid no. 5 max.; sapon. no. 55–65; 100% conc.

Maphos® 4. [PPG-Mazer] Complex org. phosphate acid ester; anionic; wetting agent used in detergent formulations, emulsion polymerization, oil and cutting fluids; solubilizer; liq.; oil-sol.; water-disp.; 100% conc.

Maphos® 17. [PPG-Mazer] Aromatic phosphate ester; anionic; emulsifier for emulsion polymerization; solubilizer; yel. clear visc. liq.; sol. @ 5% in water @ pH 2.0 and 9.5; sp.gr. 1.11; pour pt. < 0 C; pH 1.5–2.5 (1% aq.); 68.5–71.5% act.

Maphos® 31. [PPG-Mazer] Aliphatic phosphate ester; lubricant with anticorrosive/antifrictional properties for oil and water-sol. lubricant systems, e.g., greases, syn. cutting oils, drawing compds., chain-belt lubricants, gear oils, rust preventatives; water-wh. clear liq.; sp.gr. 1.450; 67% act.

Maphos® 32. [PPG-Mazer] Aliphatic phosphate ester; see Maphos 31; lt. yel. clear liq.; sp.gr. 1.140; acid no. 210 (to pH 5.2); 386 (to pH 9.5); 99.5% act.

Maphos® 33. [PPG-Mazer] Aliphatic phosphate ester; see Maphos 31; pale yel. paste; sp.gr. 1.156;

acid no. 233 (to pH 5.2); 461 (to pH 9.5); 99.5% act.

Maphos® 41. [PPG-Mazer] Aromatic phosphate ester; see Maphos 31; soft, waxy paste; acid no. 105 (to pH 5.2); 210 (to pH 9.5); 99.5% act.

Maphos® 58. [PPG-Mazer] Aliphatic phosphate ester; see Maphos 31; lt. yel. liq.; sp.gr. 1.265; acid no. 173 (to pH 5.2); 330 (to pH 9.5); 90% act.

Maphos® 60. [PPG-Mazer] Aliphatic phosphate ester; anionic; hydrotrope used in hard surface cleaners, dedusting applics., metalworking, emulsion polymerization; detergent, corrosion inhibitor, emulsifier; antistat fortextile yarn and fibers for dry cleaning; lt. yel. clear visc. liq.; sol. @ 5% in dist. water @ pH 2.0 and 9.5; sp.gr. 1.08; pour pt. < 0 C; pH 1.5–2.5 (1% aq.); 100% act.

Maphos® 60A. [PPG-Mazer] Complex org. phosphate acid ester; anionic; textile wetting, hard surface detergent; lubricant, anticorrosive, dispersant, hydrotrope, solubilizer, emulsifier; yel. clear liq.; water-sol.; sp.gr. 1.160; acid no. 175 (to pH 5.2); 325 (to pH 9.5); 100% conc.

Maphos® 66. [PPG-Mazer] Potassium salt of phosphate ester; see Maphos 60; colorless clear liq.; sol. in dist. water @ pH 9.5 @ 5%; sp.gr. 1.18; pour pt. < 0 C; pH 9–9.5 (1% aq.); 45–47% act.

Maphos® 66H. [PPG-Mazer] Phosphate acid ester; anionic; hard surface cleaning hydrotrope; antistat for solubilization of low foam and conventional surfactants in alkaline liqs.; liq.; 50% conc.

Maphos® 76. [PPG-Mazer] Aromatic phosphate ester; anionic; detergent, antistat, anticorrosive agent, emulsifer, solubilizer, hydrotrope used in drycleaning formulations, hard surface cleaners, emulsion polymerization; yel. clear visc. liq.; sol. in dist. water @ pH 9.5 (@ 5%), in perchloroethylene (@ 10%); sp.gr. 1.09; pour pt. < 0 C; pH 1.5–2.5 (1% aq.); 100% act.

Maphos® 76 NA. [PPG-Mazer] Partial sodium salt of aromatic phosphate ester; anionic; detergent, antistat, anticorrosive agent, solubilizer used in drycleaning formulations; lt. yel. clear liq.; sol. in dist. water @ pH 2.0 and 9.5 (@ 5%), in perchloroethylene (@ 10%); sp.gr. 1.08; pour pt. 8 C; pH 5.0–6.0 (1% aq.); 87–89% act.

Maphos® 77. [PPG-Mazer] Aliphatic phosphate ester; see Maphos 31; lt. yel. clear liq.; sp.gr. 1.010; acid no. 34 (to pH 5.2); 65 (to pH 9.5); 99.5% act.

Maphos® 78. [PPG-Mazer] Aliphatic phosphate ester; anionic; see Maphos 31; also hydrotrope with detergency properties; lt. yel. clear liq.; sp.gr. 1.365; acid no. 370 (to pH 5.2); 650 (to pH 9.5); 99.5% act.

Maphos® 79. [PPG-Mazer] Aliphatic phosphate ester; see Maphos 31; water-wh. clear liq.; sp.gr. 1.130; 99.5% act.

Maphos® 91. [PPG-Mazer] Aromatic phosphate ester; anionic/nonionic; lubricant, anticorrosive, and antifrictional additive for greases, syn. cutting oils (water- and oil-based), drawing compds., chain belt lubricant, gear oils, rust preventatives; solubilizer; yel. clear visc. liq.; sol. in dist. water @ pH 2.0 and 9.5 (@ 5%), perchloroethylene (@ 10%); xylene (@ 10%); clear to boiling in 10% STPP and 5% TSP (@ 1%); clear to 30 C in 16% NaOH (@ 1%); sp.gr. 1.10; pour pt. 2 C; pH 1.5–2.5 (1% aq.); 100% act.

Maphos® 6600. [PPG-Mazer] Complex org. phosphate acid ester; anionic; hydrotrope used in cleaning applics.; liq.; 100% conc.

Maphos® 8078. [PPG-Mazer] Aromatic phosphate ester; anionic; lubricant additive, greases, syn. cutting oils, drawing compds., chain belt lubricants, grease oils, rust preventatives; amber clear liq.; sol. in xylene (@ 10%), clear to boiling in 10% STPP and 5% TSP (@ 1%); sol. in dist. water (@ 5%); sp.gr. 1.11; pour pt. 5 C; pH 1.5–2.5 (1% aq.); 100% act.

Maphos® 8135. [PPG-Mazer] Aromatic phosphate ester; anionic; dispersant, hydrotrope, emulsifier, lubricant additive for greases, syn. cutting oils, drawing compds., and hard surf. cleaners; lt. amber visc. liq.; sol. in dist. water @ pH 2.0 and 9.5 (@ 5%); sp.gr. 1.07; pour pt. 2 C; pH 1.5–2.5 (1% aq.); 100% act.

Maphos® L-4. [PPG-Mazer] Aromatic phosphate ester; anionic/nonionic; lubricant, anticorrosive, and antifrictional additive for greases, syn. cutting oils (water- and oil-based), drawing compds., chain belt lubricant, gear oils, rust preventatives; solubilizer; yel. clear visc. liq.; sol. in perchloroethylene, min. oil, and xylene (@ 10%); sp.gr. 1.09; pour pt. < 0 C; pH 1.5–2.5 (1% aq.); 100% act.

Maphos® L-6. [PPG-Mazer] Aromatic phosphate ester; anionic; detergent, antistat, anticorrosive agent, lubricant, coupling agent for drycleaning and lubricant formulations, hard surf. cleaning, emulsion polymerization; yel. visc. liq.; sol. @ 10% in perchloroethylene, min. oil, and xylene; sp.gr. 1.05; pour pt. 8 C; pH 1.5–2.5 (1% aq.); 100% act.

Maphos® L-10. [PPG-Mazer] Aliphatic phosphate ester; anionic/nonionic; see Maphos 91; lt. yel. clear visc. liq.; sol. in dist. water @ pH 9.5 (@ 5%), and @ 10% in perchloroethylene, min. oil, and xylene; sp.gr. 1.045; pour pt. < 10 C; pH 1.5–2.5 (1% aq.); 100% act.

Maphos® L-12. [PPG-Mazer] Aliphatic phosphate ester; anionic; detergent, lubricant, emulsifier for hard surface cleaners, lubricant formulations, emulsion polymerization; clear lt. yel. visc. liq.; sol. @ 5% in dist. water @ pH 9.5, @ 10% in min. oil and xylene; sp.gr. 1.01; pour pt. 12 C; pH 1.5–2.5 (1% aq.); 100% act.

Maphos® L-13. [PPG-Mazer] Aliphatic phosphate ester; anionic/nonionic; lubricant, anticorrosive, coupling agent for metalworking, dry cleaning, hard surf. cleaning, dedusting, lubrication, emulsion polymerization; lt. yel. amber clear liq.; sol. in dist. water @ pH 9.5 (@ 5%), in perchloroethylene, min. oil, and xylene (@ 10%), clear to boiling in 5% TSP (@ 1%); sp.gr. 1.015; pour pt. 3 C; pH 1.5–2.5 (1% aq.); 100% act.

Maphos® L-22. [PPG-Mazer] Aliphatic phosphate ester; anionic/nonionic; see Maphos 91; lt. yel. clear liq.; sol. see Maphos L-13; sp.gr. 1.03; pour pt. 17 C; pH 1.5–2.5 (1% aq.); 100% act.

Maprofix® LK (USP). [Stepan] Sodium lauryl sulfate; anionic; foaming agent, dispersant used in powdered detergent formulations for household and personal care prods., industrial pigment dispersant; powd.; sp.gr. 0.35; flash pt. > 200 F; 90% act.

Maprolyte 101. [Stepan] DEA lauryl sulfate/amphoteric-2; anionic/amphoteric; conditioner for shampoos and bubble baths; liq.; sp.gr. 1.03; pH 7.5–8.5; flash pt. > 200 F; pH 7.5–8.5; 33–37% solids.

Maprolyte C. [Stepan] Cocoamidopropyl betaine; nonionic; detergent, wetting and foaming agent, stabilizer used in shampoos and bubble baths; liq.; sp.gr. 1.04; flash pt. > 200 F; 33–37% solids.

Maprolyte LX. [Stepan] DEA-lauryl sulfate, DEA-lauraminopropionate and sodium lauraminopropionate; anionic/amphoteric; foaming agent, conditioner, mild conc. for shampoos, body soaps, bubble baths; liq.; sp.gr. 1.03; flash pt. > 200 F; pH 7.5–8.5; 33–37% solids.

Maquat 4450-E. [Mason] Didecyl dimonium chloride; disinfectants, algicides, sanitizers, deodorant; clear to lt. straw liq.; m.w. 361; sol. in water, lower alcohols, ketones, glycols; pH 6–8 (10% aq.); 50% quat.

Maquat 4480-E. [Mason] Didecyldimonium chloride; see Maquat 4450-E; clear to lt. straw liq.; m.w. 361; sol. in water, lower alcohols, ketones, glycols; pH 6–8 (10% aq.); 80% quat.

Maquat DLC-1214. [Mason] n-Alkyl dimethyl dichlorobenzyl ammonium chloride; germicide with good wetting and penetration; APHA 100 max. liq.; m.w. 425; sol. in water and most polar solvs.; dens. 8.0–8.4 lb/gal; pH 7–8 (10% sol'n.); 50 and 80% act. in water.

Maquat LC-12S-50%, LC-12S-80%. [Mason] Benzalkonium chloride; antimicrobial; clear to lt. straw liq.; m.w. 360; sol. in water, lower alcohols, ketones, and glycols; pH 6–8 (10% aq.); 50 and 80% act.

Maquat MC-1412. [Mason] Benzalkonium chloride; germicide with good wetting and penetration; APHA 100 max. liq.; m.w. 358; sol. in water and most polar solvs.; dens. 8.0–8.4 lb/gal; pH 7–8 (10%); 50% act. in water, 80% act. in IPA or ethanol.

Maquat MC-1416. [Mason] Benzalkonium chloride; germicide with good wetting and penetration; APHA 100 max. liq.; m.w. 380; sol. in water and most polar solvs.; dens. 7.8–8.2 lb/gal; pH 7–8 (10% sol'n.); 50% act. in water; 80% act. in IPA or ethanol.

Maquat MC-6025-50%. [Mason] Benzalkonium chloride; germicide with good wetting and penetration; clear to lt. straw liq.; m.w. 365; sol. in water, lower alcohols, ketones, glycols; pH 6–8 (10% aq.); 50% act.

Maquat MQ-2525-50%, MQ-2525-80%. [Mason] 25% Benzalkonium chloride, 25% alkyl dimethyl ethylbenzyl ammonium chloride; antimicrobial with hard water tolerance; clear to lt. straw liq.; m.w. 384; sol. in water, lower alcohols, ketones, glycols; pH 6–8 (10% aq.); 50 and 80% act.

Maquat MQ-2525M-50%, MQ-2525M-80%. [Mason] Benzalkonium chloride (A) and n-alkyl dimethyl ethylbenzyl ammonium chloride (B); germicide with good wetting and penetration; APHA 100 max. liq.; m.w. 384; sol. in water, polar solvs.; dens. 7.8–8.2 lb/gal; pH 7–8 (10% sol'n.); 50% act. (25% A, 25% B) in water or 80% act. (40%A, 40% B) in 20% IPA.

Maquat SC-18. [Mason] Stearalkonium chloride; cationic; germicide with good wetting and penetration; paste, flake, or powd.; 25%, 85% and 94% act. resp.

Maquat SC-1632. [Mason] Stearalkonium chloride, PEG-40 castor oil, cetearyl alcohol; see Maquat SC-18; wh. waxy flakes; mild fatty odor; sol. in water, polar solvs.; pH 5-7 (5% aq.); 96% act. solids.

Maquat TC-76-50%. [Mason] 42% Benzalkonium chloride and 8% n-dialkyl methyl benzyl ammonium chloride; germicide with good wetting and penetration; clear to lt. straw liq.; m.w. 410; sol. in water, lower alcohols, ketones, glycols; pH 6–8 (10% aq.); 50% act.

Maracarb. [Daishowa] Modified lignosulfonate; dispersant, humectant, chelating agent for mfg. of alkaline metal cleaners; water-sol.

Maracarb N-1. [Daishowa] Sodium lignosulfonate; anionic; modifier, dispersant, and humectant in dyestuff pastes; industrial cleaners; plant foliar spray; chelates metal ions; powd. and liq.; pH 8.1–9.0 (powd.); 7.0–8.0 (liq.); 100% powd. or 52% liq.

Maracell XC. [Daishowa] Sodium lignosulfonate; anionic; chelating agent; org. expander in the negative plates in lead acid storage batteries; stabilizer for 2,4-D amine sol'ns. in hard water; powd.; 100% conc.

Maracell XE. [Daishowa] Sodium lignosulfonate; anionic; chelating agent for prevention of scale formation in treatment of boiler water, industrial cleaners; powd.; 100% conc.

Marasperse. [Daishowa] Modified lignosulfonate; dispersing agent, chelating agent for alkaline and acid cleaners; suspends toxicants in agric. formulations or industrial cleaners; water-sol.

Marasperse 41G-3. [Daishowa] Lignosulfonate; anionic; dispersant for inorg. mins. such as cement slurry, gypsum slurry; 45% liq., 100% powd.

Marasperse C-21. [Daishowa] Calcium lignosulfonate; anionic; dispersant, emulsion stabilizer used in agric. chemical formulations, gypsum board additive, industrial cleaners, in mfg. of brick, tile, refractories, pottery and porcelain ware; brn. powd.; water sol.; dens. 35–40 lb/ft^3; surf. tens. 49.4 dynes/cm (1%); pH 7.0–8.2 (3%); 100% conc.

Marasperse CB. [Daishowa] Sodium lignosulfonate; anionic; dispersant, o/w emulsion stabilizer used in textile dyeing, agric. chemical formulations, gypsum board additives, industrial cleaners; blk. powd., brn. powd. resp.; water sol.; dens. 43–47 lb/ft^3; surf. tens. 51.4 dynes/cm (1%); pH 8.5–9.2 (3%).

Marasperse CBOS-3. [Daishowa] Partially desulfonated sodium lignosulfonate; anionic; dispersant for aq. carbon blk. slurries, dyestuffs, pigments, wettable powd. insecticides; powd.; 100% act.

Marasperse CBX-2. [Daishowa] Sodium lignosulfonate; anionic; primary dyestuff dispersant; powd.; 100% act.

Marasperse N-22. [Daishowa] Sodium lignosulfonate; anionic; dispersant, o/w emulsion stabilizer, emulsifier; mfg. of disperse dyes for dyeing acetate and polyesters; dispersant and sequestering agent in cooling water treatments; agric. chemical formulations; gypsum board additive; industrial cleaners; brn. powd.; water-sol.; dens. 35–40 lb/ft^3; surf. tens. 52.8 dynes/cm (1%); pH 7.5–8.5 (3%); 100% conc.

Mar'blend. [Georgia Marble] Calcium carbonate; single-filler system (preproportioned blend of 15M and 40-200) for cultured marble; sp.gr. 2.71; pH 9.0–9.5; hardness (Moh) 3; 95.0% min. total carbonates.

Marchon® C21. [Albright & Wilson] Copper chromium oxide; catalyst in hydrog. of chemical compds. incl. fatty acids and esters, aldehydes, ketones, unsat. alcohols, aromatic nitro compds.; contains barium to inhibit deactivation during use; blk. powd.; 98% thru 75 μ sieve; bulk dens. 1.15 g/cc (loose), 1.35 g/cc (packed); surf. area 35–55 m^2/g; 34% Cu, 30.5% Cr, 6.8% Ba.

Marchon® DC 1202. [Albright & Wilson] Amine blend; brine sol. corrosion inhibitor; liq.

Margel. [Kelco] Calcium alginate, ammonium alginate; gellant, emulsifier, and stabilizer in food and indust. applic.; cream granular particles; pH 9.4.

Mark 152, 1043A, 1178, 1500, OTM. [Witco/Argus] Octyltins, calcium zincs, and phosphite stabilizers; stabilizers for food wrap. pkg. film, beverage tubing, bottles, and vacuum-formed sheet; nontoxic.

Mark 155, 565A, GS, RFD. [Witco/Argus] Ca/Zn and Ba/Zn stabilizers; liqs. and solids.

Mark 158. [Witco/Argus] Phosphite complex; antioxidant for use in stabilization of polyolefin resins; used in extrusion and molding compds., monofilament and fiber applic.; pale yel. liq.; sol. in hexane, MEK, and benzene; f.p. < –20 C; sp.gr. 0.982; Gardner Z1 visc.; ref. index 1.488; flash pt. 169 C (COC).

Mark 202A. [Witco/Argus] Substituted benzophenone; UV absorber for vinyls, polyolefins, and other polymers incl. PC, PU, cellulose acetate/butyrate, ABS, vinylidene chloride, polyesters, epoxies, and PS, require optimum lt. stability; protects PVC, polyethylene, and PP against UV degradation in outdoor weathering for fluorescent lt. exposure; high absorp. throughout the UV spectrum (300–400 mu); does not affect the color of the substrate; exhibits compatibility and permanence, low volatility, and low extraction; off-wh. powd.; sol. 60% by wt. in toluene; 50% by wt. in MEK.

Mark 217. [Witco/Argus] Phosphite; antioxidant for use in stabilization of PP resins; extrusion and molding compd., monofilament, and fiber; pale yel. liq.; sol. see Mark 158; sp.gr. 1.000; Gardner Z visc.; ref. index 1.4910.

Mark 232B, 550, 556. [Witco/Argus] Barium lead stabilizer; stabilizer for DWV pipe, elec., conduit, wire and cable, records, etc.

Mark 260. [Witco/Argus] Phosphite; antioxidant for stabilization of polyolefins; pale yel. visc. liq.; m.w. 1238; sol. see Mark 158; sp.gr. 0.943; ref. index 1.515; flash pt. 138 C.

Mark 281B, 630. [Witco/Argus] One-pkg. stabilizer/antifog systems for meat and produce wrap; stabilizer/activators and cell stabilizers for plastisol and calendered foam; nitrogenous flooring stabilizers, and other special compds.

Mark 366, C. [Witco/Argus] Phosphite stabilizer; for use as aux. stabilizers with Ba/Cd and Ca/Zn systems to improve color and clarity in flexibles and rigids for both general purpose and nontoxic end uses.

Mark 462, 755, LL. [Witco/Argus] Barium/cadmium/phosphite stabilizers; for all clear and pigmented, flexible and semirigid compds., plastisols, organosols, and vinyl sol'ns.; many with zinc to improve sulfide stain resistance and optimize stability; liq.

Mark 522. [Witco/Argus] Phosphite; antioxidant for stabilization of PP compds.; pale yel. visc. liq.; m.w. 1881; sol. see Mark 158; sp.gr. 0.975; ref. index 1.502; flash pt. 193 C. (COC).

Mark 649, 1772A, 1900, 1905. [Witco/Argus] Butyl, methyl, and octyl organotins; stabilizers for high impact and unmodified rigid compds., incl. the super tins, weather-resistant systems, etc., potable water pipe, vinyl siding, and rigid foam; avail. in sulfur and nonsulfur types; solids and liqs.

Mark 684A, 684B, 1092. [Witco/Argus] Stabilizers for asbestos-filled PP compds.; imparts max. heat stability protection to compds. containing either Chrysotile or Anthophyllite asbestos; suggested for use at levels of 1.0–2.0 phr; pale yel. powd.; sp.gr. 1.21, 1.17, and 1.15 resp.

Mark 1178B. [Witco/Argus] Phosphite; antioxidants; chelating agents used with stabilizers for polymers and plastics; food pkg., hot melt formulations; pale yel. visc. liq.; sp.gr. 0.985; ref. index 1.5250; flash pt. 210 C (COC).

Mark 1216. [Witco/Argus] Organo complex; FDA-cleared stabilizer for ABS polymers where the retention of initial color and physical properties are important; effective for stabilizing compds. based on rubber-modified polymers, e.g. ABS; pale yel. liq.; sp.gr. 0.971.

Mark 1220. [Witco/Argus] Organo complex; antioxidant/stabilizer for styrene and ABS polymers; food applic.; stabilizer for ABS latex; pale yel. visc. liq.; sp.gr. 0.978; visc. 4125 cps; ref. index 1.5152; flash pt. 186 C.

Mark 1259A. [Witco/Argus] Phosphite complex; antioxidant for polymers; water-wh. visc. liq.; sp.gr. 1.156; ref. index 1.541; flash pt. 151 C.

Mark 1295. [Witco/Argus] Phosphite complex; see Mark 1259A; pale yel. visc. liq.; sp.gr. 1.1660; ref. index 1.5268; flash pt. 160 C.

Mark 1314, WS, TT. [Witco/Argus] Barium-cadmium stabilizers; for plasticized and rigid compds. for calendering, extrusion, inj. molding, powd. molding, slush and rotational molding, dip-and spread-coating, etc.; sulfide stain protection, plate-out resistance, good weatherability, and high heat distort.; solid

Mark 1409. [Witco/Argus] Phenolic-phosphite; antioxidant/stabilizer for polyolefins; inj. molding and extrusion of film and fiber; food additives; lt. brn.

visc. liq.; sp.gr. 0.989; visc. 20,000; ref. index 1.5291; flash pt. 190 C.

Mark 1413. [Witco/Argus] 2-Hydroxy-4-n-octoxy-benzophenone; uv absorber in vinyls, polyolefins, and other polymers; lt. straw powd.; sp.gr. 1.16; m.p. 46–48 C.

Mark 1490S. [Witco/Argus] Phenolic-phosphite; see Mark 1409; granular solid; sp.gr. 1.18.

Mark 1535. [Witco/Argus] Benzophenone; see Mark 1413; liq.

Mark 1589. [Witco/Argus] Phenolic complex; antioxidant used in polyolefins and elastomers; wh. powd.; sp.gr. 1.17; m.p. 208 C.

Mark 1589B. [Witco/Argus] Phenolic-thio-complex; see Mark 1589; wh. powd.; sp.gr. 1.06.

Mark 1600. [Witco/Argus] Ba/Zn stabilizer; stabilizer for applics. where the absence of cadmium is desired, without loss in dynamic heat stability; liq.

Mark 2100, 2100A. [Witco/Argus] Sulfur-free organotins; heat stabilizers for rigid PVC compds.; additives to boost the heat and lt. stability of std. organotin mercaptide stabilized compds.; stabilizers for halogenated resins other than PVC; wh. to off-wh. powd.

Mark 2112. [Witco/Argus] Substituted triphenyl phosphite; antioxidant for polymer processing; stabilizer for PP, HDPE, and LDPE; wh. powd.; sol. in org. solv.; insol. in water; sp.gr. 0.98; m.p. 186 C.

Mark 2140. [Witco/Argus] Pentaerythrityl hexylthiopropionate; stabilizer for use in polyolefins and other polymeric systems; synergistic with primary antioxidants; color which reduces or eliminates the need for phosphite; used at 0.1–0.3 phr in PP, 0.05–0.10 phr in HDPE, 0.25–0.5 phr in elastomers, and 0.5–1.0 phr in S.B. latex; very pale straw liq.; mild odor; m.w. 825; sol. in common org. solvs.; sp.gr. 1.0570; flash pt. (COC) 274 C.

Mark 2180. [Witco/Argus] Organo-metallic compd.; stabilizer; nucleating agent for PP; reduces haze in PP when used at 0.1–0.3 phr; increases the crystallization temp. without affecting the heat aging of the compd.; wh. powd.; sol. in hot water; sp.gr. 1.24; m.p. decomp. @ 300 C.

Mark 5050, 5060. [Witco/Argus] Distearyl pentaerythritol diphosphite; heat and lt. stabilizer/antioxidant for use in a wide range of polymers; heat/processing additive for molding, extrusion, fibers, and films with polymers; synergistic with hindered phenolic antioxidants, thioesters, UV absorbers, and/or both amine and nickel type lt. stabilizers; 5060 is used where improved hydrolytic stability is required; wh. flakes; sp.gr. 1.0 (both); 7.2–8.0% and 7.0–7.8% phosphorus content resp.

Mark DDHP. [Witco/Argus] Didecyl hydrogen phosphite; stabilizer; pale yel. liq.; m.w. 362; sp.gr. 0.922–0.924; flash pt. (COC) 132 C; 8.56% phosphorus.

Mark DDMPP. [Witco/Argus] Didecyl mono phenyl phosphite; stabilizer; water-wh. liq.; m.w. 438; sp.gr. 0.935–0.965; dens. 7.8 lb/gal; flash pt. (COC) 211 C; 7.1% phosphorus.

Mark DPHP. [Witco/Argus] Diphenyl hydrogen phosphite; stabilizer; sp.gr. 1.20–1.22; dens. 10.1 lb/gal; flash pt. (COC) 148 C; 13.2% phosphorus.

Mark MDDPP. [Witco/Argus] Monodecyl diphenyl phosphite; stabilizer; water-wh. liq.; m.w. 374; sp.gr. 1.02–1.04; dens. 8.6 lb/gal; flash pt. (COC) 195 C; 8.3% phosphorus.

Mark TDP. [Witco/Argus] Tridecyl phosphite; stabilizer; water-wh. liq.; m.w. 502; sp.gr. 0.880–0.9050; flash pt. (COC) 140 C; 6.1% phosphorus.

Mark TNPP. [Witco/Argus] Trisnonyl phenyl phosphite; stabilizer; pale yel. liq.; m.w. 715; sp.gr. 0.980–1.000; dens. 8.25 lb/gal; flash pt. (COC) 185 C; 4.3% phosphorus.

Mark TPP. [Witco/Argus] Triphenyl phosphite; stabilizer; water-wh. liq.; m.w. 310; sp.gr. 1.172–1.186; dens. 9.86 lb/gal; m.p. 25 C; flash pt. (COC) 155 C; 9.5% phosphorus.

Markstat AL-12. [Witco/Argus] Quat. ammonium chloride deriv. of polyalkoxy tert. amines; cationic; antistat additive for PU films and thermoplastics; lt. amber oily liq.; sol. in IPA, acetone, MEK, ethanol, water; sp.gr. 1.01; visc. 1800 cps; 100% act.

Markstat AL-22. [Witco/Argus] Quat. ammonium chloride deriv., modified; antistat used as surf. treatment for plastics; amber clear liq.; water-sol.; sp.gr. 1.042; Gardner V-W visc.; ref. index 1.4726; 100% act.

Markstat Antistats. [Witco/Argus] Antistatic agents providing extra heat and lt. stability in PVC compds., permanence, and low lubricity; for surface treatments of plastics, fabrics, and fibers.

Marlamid D 1218, DF 1218. [Huls AG] Cocamide DEA; nonionic; foam stabilizer, thickener, superfatting agent for liq. detergents; liq.; water-sol; 100% act.

Marlamid D 1885. [Huls AG] Oleamide DEA; nonionic; foam stabilizer in wash liquors, superfatting agent; liq.; water-sol.; 100% act.

Marlamid KL. [Huls AG] Cocamidopropyl lauryl ether; nonionic; pearlescent surfactant, foam stabilizer, thickener, opacifier for liq. and paste detergents, shampoos; solid; insol. in water; 100% conc.

Marlamid KLA. [Huls AG] Fatty acid alkylolamide and fatty alcohol ether sulfate; nonionic; pearlescent surfactant for personal care prods.; liq.; 40% conc.

Marlamid KLP. [Huls AG] Fatty alcohol ether sulfate and pearlescents disp.; pearlescent base for shampoos and bubble baths; fluid paste.

Marlamid M 1218. [Huls AG] Cocamide MEA; nonionic; foam stabilizer, thickener in household, personal, industrial detergents, superfatting agent; solid; water-insol.; 100% act.

Marlamid PG 20. [Huls AG] Pearlescent base for shampoos and bubble baths; fluid paste.

Marlate® 2-MR. [Kincaid Enterprises] Tech. methoxychlor; emulsifiable insecticide for use on certain crops, livestock, and in builidings; 24% act.

Marlate® 50. [Kincaid Enterprises] Methoxychlor; wettable powd.; biodegradable insecticide controlling wide range of pests for vegetable, fruit, and forage crops, grain storage bins, and warm-blooded animals; 50% act.

Marlate® Tech. [Kincaid Enterprises] Methoxychlor; see Marlate 50; lt. yel. flake or chips; slightly fruity odor; m.w. 345.65; sol. in aromatic, chlorinated, and ketone solvs., paraffinic types and veg. oils; insol. in water; sp.gr. 1.41; m.p. 77 C; 100% act.

Marlican. [Huls AG] Straight-chain dodecylbenzene; anionic; detergent intermediate, solubilizer; liq.; 100% act.

Marlipal 24/20, /30, /40, /50, /60, /70, /80, /90, /100, /110, /120, /140, /200, /300. [Huls AG] C_{12-14} alcohol polyglycol ethers; nonionic; dispersant, wetting agent, for washing, cleaning, soil suspending, and homogenizing applics.; liq. to solid; HLB 6.2, 8.1, 9.5, 10.6, 11.5, 12.3, 12.9, 13.4, 13.9, 14.3, 14.6, 15.2, 16.4, and 17.4 resp.; 100% conc.

Marlipal BS. [Huls AG] Fatty acid polyglycol ester; nonionic surfactant, superfatting agent, thickener in detergents, dishwashing and cosmetic preparations; wax; 100% act.

Marlon AS_3. [Huls AG] Dodecylbenzene sulfonic acid; anionic; intermediate for mfg. of anionic surfactants; liq.; 100% act.

Marlophen 83, 84, 85, 86, 86 S, 87, 88, 89, 810, 812, 814, 820, 825, 850. [Huls AG] Nonylphenol polyglycol ethers; detergent, wetting and dispersing agent; sol. in oil and water; liq. to wax; 100% act.

Marlophen DNP 16, 18, 30. [Huls AG] Dinonylphenol polyglycol ethers; nonionic; nonionic; raw material for textile and paper auxiliaries, dispersant; wax; 100% act.

Marlophor ND. [Huls AG] Partial phosphate ester; anionic; wetting agent for textile and paper industries, antistat for natural and syn. fibers used in drycleaning detergents; liq.; water- and oil-sol.; 100% act.

Marlophor ND-Acid. [Huls AG] Org. phosphate ester; anionic; detergent; wetting agent; antistat; component for textile aux. agents and drycleaning formulations; antistat; liq.; 100% conc.

Marlophor ND DEA Salt, ND Na-Salt. [Huls AG] Org. phosphate ester/alkylphenol polyglycol ether; anionic/nonionic; detergent, wetting agent, antistat; component for textile aux. agents and drycleaning formulations; liq.; 100 and 92% conc. resp.

Marlopon CA. [Huls AG] TEA-dodecylbenzene sulfonate, modified; anionic; detergent; superfatting and dishwashing agent; personal care prods.; liq.; 55% conc.

Marlosoft IQ 75, 90. [Huls AG] Imidazolinium methosulfate; cationic; base for fabric softeners; liq.; 75 and 90% conc. resp.

Marlosol 183, 189, 1820, 1825, OL 7, OL 8, OL 10, OL 15, OL 20. [Huls AG] Fatty acid polyglycol esters; nonionic; raw material for finishing agents in the syn. fiber industry; liq. to wax; 100% act.

Marvanol 55% SPO. [Marlowe-Van Loan] Sulfated propyl oleate; anionic; detergent, wetting and leveling agent; amber liq.; 55% act.

Marvanol 60% SBO. [Marlowe-Van Loan] Sulfated butyl oleate; anionic; detergent, wetting and leveling agent, emulsifier; amber liq.; 60% act.

Marvanol 75% SCO. [Marlowe-Van Loan] Sulfated castor oil; detergent, wetting agent, dyeing assistant and lubricant used in finishing operations; amber liq.; 75% act.

Marvanol GC. [Marlowe-Van Loan] Sulfosuccinamide; anionic; textile fiber wetting and dye leveling agent; amber gel; 65% act.

Marvanol Penetrant 35. [Marlowe-Van Loan] Amine neutralized sulfonic acid; detergent, wetting agent, textile dyeing, finishing, leveling and retarding agents for acid dyes; amber liq.; 40% act.

Marvanol RE-1274. [Marlowe-Van Loan] 1:1 coconut fatty amide; cationic; emulsifer, visc. builder, dye leveler; amber liq.; sp.gr. 0.950; 97% act.

Marvanol RE-1281. [Marlowe-Van Loan] 2:1 coconut fatty amide; cationic; emulsifer, visc. builder, dye leveler; amber liq.; sp.gr. 1.02; 98% act.

Marvelin W-50. [Matsumoto] Alkylene oxide addition prods.; nonionic; dye leveling agent of wool; flake; 100% conc.

Masil® 260, 260A. [PPG-Mazer] Specialty silicone fluid; release agent; for high-visc. coatings and air-dry coating systems; water-wh. clear liq.; odorless, tasteless; sp.gr. 0.960 (both); visc. 25 and 20 cSt resp.; pour pt. –65 C (both); ref. index 1.3985 and 1.3973 resp.; flash pt. (CC) 60 C (both).

Masil® 263. [PPG-Mazer] Alkyl methyl polysiloxane; processing aid, lubricant, mold release; for lubricating formulations, personal care prods., textiles and fibers; colorless liq.; sp.gr. 0.88; visc. 1200 cSt; flash pt. (PMCC) 280 F; ref. index 1.454; 98% act. silicone.

Masil® 264. [PPG-Mazer] Alkyl methyl polysiloxane; processing aid, lubricant, mold release; for aerosol pkg., ink/printing industry; colorless liq.; sp.gr. 0.903; visc. 825 cSt; flash pt. (PMCC) 350 F; ref. index 1.448; 100% act. silicone.

Masil® 265. [PPG-Mazer] Mixed alkyl methyl polysiloxane; lubricant, release agent providing release where subsequent finishing or printing is required; clear yel. liq.; odorless; sp.gr. 0.930; visc. 1500 cSt; pour pt. < –20 C; ref. index 1.463; flash pt. (PMCC) 350 F; surf. tens. 28 dynes/cm; 100% act. silicone.

Masil® 265HV. [PPG-Mazer] Aryl methyl polysiloxane; higher visc. homolog of Masil 265; release agent for rubber goods, plastic articles, and metal moldings which require subsequent finishing; clear yel. liq.; odorless; sp.gr. 1.015; visc. 2000 cSt; pour pt. < –20 C; ref. index 1.49; flash pt. (PMCC) 350 F; surf. rtens. 22.5 dynes/cm.

Masil® 266. [PPG-Mazer] Aq. silicone emulsion; nonionic; emulsifier; release aid in molding, extrusion, laminating, and casting for rubber, plastics, and metals; for leather, glass, and vinyl cleaners, polish formulations, textile softeners, and textile and fiber lubricants; used in aq. systems; paintable mold release aid; emulsion; sp.gr. 1.00; visc. 900 cSt; 50% act.

Masil® 270. [PPG-Mazer] Specialty silicone fluid; release agent used during fabrication of urethane and epoxy parts; sp.gr. 0.992; visc. 260 cSt; pour pt. < –20 C; ref. index 1.4055; flash pt. (CC) 66 C; 100% act.

Masil® 271. [PPG-Mazer] Silicone fluid; release agent for molded plastics and elastomers; liq.

Masil® 272. [PPG-Mazer] Specialty silicone fluid; waterproofing agent for fabric and leather; clear liq.; sp.gr. 0.862; visc. 25 cSt; pour pt. < –20 C; ref. index 1.4304; flash pt. (CC) 40 C.

Masil® 280. [PPG-Mazer] Dimethicone copolyol; antistat, wetting agent for personal care prods., etc.; paste; sp.gr. 1.065; cloud pt. 78 C (1% aq.); flash pt. (PMCC) > 300 F; pour pt. –35 C; surf. tens. 28.5 dynes/cm (1% aq.).

Masil® 280LP. [PPG-Mazer] Silicone glycol; see Masil 280; liq.; sp.gr. 1.065; visc. 500 cSt; cloud pt. 76 C (1% aq.); flash pt. (PMCC) > 300 F; pour pt. –16 C; surf. tens. 20.0 dynes/cm (1% aq.).

Masil® 290D. [PPG-Mazer] Specialty silicone fluid; masonry water repellent; used on brick, mortar, sandstone, concrete, stucco, terrazzo, and concrete blocks; easily diluted with min. spirits; sp.gr. 0.884; visc. 10 cSt; pour pt. < –20 C; ref. index 1.4025; flash pt. (CC) 21 C; 60% act. resin conc.

Masil® 290F. [PPG-Mazer] Specialty silicone fluid; see Masil 290D; easily diluted with min. spirits; sp.gr. 0.068; visc. 10 cSt; pour pt. < –20 C; ref. index 1.4020; flash pt. (CC) 54 C; 70% act. resin conc.

Masil® 1066C. [PPG-Mazer] Dimethicone copolyol; lubricant and antistat for plastics, textiles, metal processing; wetting and leveling char.; antifog for glass cleaners; liq.; water-sol.; sp.gr. 1.02; visc. 1800 cSt; cloud pt. 42 C (1% aq.); flash pt. (PMCC) > 300 F; pour pt. –50 C; surf. tens. 32 dynes/cm (1% aq.).

Masil® 1066D. [PPG-Mazer] Dimethicone copolyol; lubricant and antistat for plastics, textiles, and metal processing; liq.; water-sol.; sp.gr. 1.03; visc. 1050 cSt; cloud pt. 37 C (1% aq.); flash pt. (PMCC) > 300 F; pour pt. –33 C; surf. tens. 28.4 dynes/cm (1% aq.).

Masil® 2132, 2133, 2134. [PPG-Mazer] Silicone glycol; antistat and wetting agent for personal care prods., textile, plastics, lubricants and formulations; solv. based coating, dispersant, and antifoam; liq.; sp.gr. 1.03, 1.03, and 1.02 resp.; visc. 2000, 1050, 1800 cSt; cloud pt. 38, 37, 39 C (1% aq.); flash pt. (PMCC) > 300 F; pour pt. –40, –33, –50 C; surf. tens. 26.5, 25.8, 32.0 dynes/cm (1% aq.); 100% conc.

Masil® EM 14. [PPG-Mazer] Dimethylpolysiloxane fluids aq. emulsion; nonionic; see Masil 266; emulsion; sp.gr. 1.00; visc. 350 cSt; 14% act.

Masil® EM 62. [PPG-Mazer] Dimethylpolysiloxane fluids aq. emulsion; nonionic/anionic; see Masil 266; emulsion; sp.gr. 0.99; visc. 25 cSt; 25% act.

Masil® EM 100. [PPG-Mazer] Dimethylpolysiloxane fluids aq. emulsion; nonionic; see Masil 266; recommended for glass hard surf. cleaner applics., and imparts nonsmearing, low gloss, and ease of wipe properties to such formulations; emulsion; sp.gr. 0.99; visc. 100 cSt; 35% act.

Masil® EM 100 Conc. [PPG-Mazer] Dimethylpolysiloxane fluids aq. emulsion; nonionic; see Masil 266; emulsion; sp.gr. 0.99; visc. 100 cSt; 60% act.

Masil® EM 100D. [PPG-Mazer] Dimethylpolysiloxane fluids aq. emulsion; nonionic; see Masil 266; emulsion; sp.gr. 0.99; visc. 100 cSt; 30% act.

Masil® EM 100P. [PPG-Mazer] Dimethylpolysiloxane fluids aq. emulsions; nonionic; see Masil 266; used in automotive and furniture polish formulations; quick breaking emulsion, ease of rub-out and uniform deposition of silicone; emulsion; disp. in water; sp.gr. 0.99; visc. 100 cSt; 60% act.

Masil® EM 250 Conc. [PPG-Mazer] Dimethylpolysiloxane fluids aq. emulsion; nonionic; see Masil 266; recommended for glass hard surf. cleaners imparting nonsmearing, low gloss, and ease of wipe properties to such formulations; emulsion; sp.gr. 0.99; visc. 250 cSt; 60% act.

Masil® EM 266. [PPG-Mazer] Mixed alkyl methyl polysiloxane; aq. emulsion of Masil 265; milky wh. emulsion; sp.gr. 0.95; visc. 100 cSt; flash pt. (PMCC) none; 50% act. silicone.

Masil® EM 266 (35%). [PPG-Mazer] Mixed alkyl methyl polysiloxane emulsion; processing aid, lubricant, mold release; for paintable release market, personal care prods., textiles and fibers; milky wh. emulsion; sp.gr. 0.98; visc. 80 cSt; flash pt. (PMCC) none; 35% act. silicone.

Masil® EM 266HV. [PPG-Mazer] Aryl methyl polysiloxane emulsion; processing aid, lubricant, mold release; for die-cast applics., ink/printing industries; milky wh. emulsion; sp.gr. 0.95; visc. 120 cSt; flash pt. (PMCC) none; 50% act. silicone.

Masil® EM 350. [PPG-Mazer] Dimethylpolysiloxane fluids aq. emulsion; nonionic; see Masil 266; food grade release emulsion used in mfg. of articles in contact with food; emulsion; sp.gr. 0.99; visc. 350 cSt; 35% act.

Masil® EM 350X. [PPG-Mazer] Dimethylpolysiloxane fluids aq. emulsion; nonionic; see Masil 266; emulsion; sp.gr. 0.99; visc. 350 cSt; 35% act.

Masil® EM 350X Conc. [PPG-Mazer] Dimethylpolysiloxane fluids aq. emulsion; nonionic; see Masil 266; recommended for glass, leather, and vinyl cleaners, in textile softening applics.; printing release agent minimizing ink smearing and scuffing at folding bars and ink transfer between freshly printed sheets; emulsion; sp.gr. 0.99; visc. 350 cSt; 60% act.

Masil® EM 1000. [PPG-Mazer] Dimethylpolysiloxane fluids aq. emulsion; nonionic; see Masil 266; emulsion; sp.gr. 0.99; visc. 1000 cSt; 35% act.

Masil® EM 1000 Conc. [PPG-Mazer] Dimethylpolysiloxane fluids aq. emulsion; nonionic; see Masil 266; emulsion; sp.gr. 0.99; visc. 1000 cSt; 60% act.

Masil® EM 1000P. [PPG-Mazer] Dimethylpolysiloxane fluids aq. emulsion; nonionic; see Masil EM-100P; emulsion; sp.gr. 0.99; visc. 1000 cSt; 60% act.

Masil® EM 10,000. [PPG-Mazer] Dimethylpolysiloxane fluids aq. emulsion; see Masil 266; emulsion; sp.gr. 1.09; visc. 10,000 cSt; 35% act.

Masil® EM 10,000 Conc. [PPG-Mazer] Dimethylpolysiloxane fluids aq. emulsion; nonionic; see Masil 266; recommended for fiber and textile lubricants; emulsion; sp.gr. 0.99; visc. 10,000 cSt; 60% act.

Masil® EM 60,000. [PPG-Mazer] Dimethylpolysiloxane fluids aq. emulsion; nonionic; see Masil 266;

emulsion; sp.gr. 0.99; visc. 60,000 cSt; 35% act.

Masil® EM 100,000. [PPG-Mazer] Dimethylpolysiloxane fluids aq. emulsion; nonionic; see Masil 266; emulsion; sp.gr. 1.00; visc. 100,000 cSt; 35% act.

Masil® EM-N. [PPG-Mazer] Dimethylpolysiloxane fluids aq. emulsion; anionic; see Masil 266; emulsion; sp.gr. 0.97; visc. 15 cSt; 25% act.

Masil® SF 5. [PPG-Mazer] Dimethicone; release aid; internal lubricant for plastics, rubber, and metal; higher visc. fluids (> 10,000 cSt) recommended to formulate band-ply lubricants and mold release prods. to mfg. plastics and rubber parts; foam control agent for nonaq. processes, esp. in the petrol., foods, and printing inks industries; also in furniture and auto-wax polishes, household and personal care prods.; textile lubricant; lower visc. fluids recommended for cosmetic applications; water-wh. oily, clear liq.; odorless, tasteless; sp.gr. 0.916; visc. 5 cSt; pour pt. –84 C; ref. index 1.3970; flash pt. (CC) 138 C.

Masil® SF 20. [PPG-Mazer] Dimethicone; see Masil SF 5; water-wh. oily, clear liq.; odorless, tasteless; sp.gr. 0.953; visc. 20 cSt; pour pt. –65 C; ref. index 1.4010; flash pt. (CC) 202 C.

Masil® SF 50. [PPG-Mazer] Dimethicone; see Masil SF 5; water-wh. oily, clear liq.; odorless, tasteless; sp.gr. 0.963; visc. 50 cps; pour pt. –55 C; ref. index 1.4020; flash pt. (CC) 238 C.

Masil® SF 100. [PPG-Mazer] Dimethicone; see Masil SF 5; water-wh. oily, clear liq.; odorless, tasteless; sp.gr. 0.968; visc. 100 cps; pour pt. –55 C; ref. index 1.4030; flash pt. (CC) 238 C.

Masil® SF 200. [PPG-Mazer] Dimethicone; see Masil SF 5; water-wh. oily, clear liq.; odorless, tasteless; sp.gr. 0.972; visc. 200 cSt; pour pt. –50 C; ref. index 1.4031; flash pt. (CC) 238C.

Masil® SF 350. [PPG-Mazer] Dimethicone; see Masil SF 5; water-wh. oily, clear liq.; odorless, tasteless; sp.gr. 0.973; visc. 350 cSt; pour pt. –50 C; ref. index 1.4032; flash pt. (CC) 260 C.

Masil® SF 500. [PPG-Mazer] Dimethicone; see Masil SF 5; water-wh. oily, clear liq.; odorless, tasteless; sp.gr. 0.973; visc. 500 cSt; pour pt. –50 C; ref. index 1.4033; flash pt. (CC) 260 C.

Masil® SF 1000. [PPG-Mazer] Dimethicone; see Masil SF 5; water-wh. oily, clear liq.; odorless, tasteless; sp.gr. 0.974; visc. 1000 cSt; pour pt. –50 C; ref. index 1.4035; flash pt. (CC) 260 C.

Masil® SF 5000. [PPG-Mazer] Dimethicone; see Masil SF 5; water-wh. oily, clear liq.; odorless, tasteless; sp.gr. 0.975; visc. 5000 cSt; pour pt. –49 C; ref. index 1.4035; flash pt. (CC) 260 C.

Masil® SF 10,000. [PPG-Mazer] Dimethicone; see Masil SF 5; water-wh. oily, clear liq.; odorless, tasteless; sp.gr. 0.975; visc. 10,000 cSt; pour pt. –47 C; ref. index 1.4035; flash pt. (CC) 260 C.

Masil® SF 12,500. [PPG-Mazer] Dimethicone; see Masil SF 5; water-wh. oily, clear liq.; odorless, tasteless; sp.gr. 0.975; visc. 12,500 cSt; pour pt. –47 C; ref. index 1.4035; flash pt. (CC) 260 C.

Masil® SF 30,000. [PPG-Mazer] Dimethicone; see Masil SF 5; water-wh. oily, clear liq.; odorless, tasteless; sp.gr. 0.976; visc. 30,000 cSt; pour pt. –46 C; ref. index 1.4035; flash pt. (CC) 260 C.

Masil® SF 60,000. [PPG-Mazer] Dimethicone; see Masil SF 5; water-wh. oily, clear liq.; odorless, tasteless; sp.gr. 0.977; visc. 60,000 cSt; pour pt. –44 C; ref. index 1.4035; flash pt. (CC) 260 C.

Masil® SF 100,000. [PPG-Mazer] Dimethicone; see Masil SF 5; water-wh. oily, clear liq.; odorless, tasteless; sp.gr. 0.978; visc. 100,000 cSt; pour pt. –40 C; ref. index 1.4035; flash pt. (CC) 260 C.

Masil® SF 300,000. [PPG-Mazer] Dimethicone; see Masil SF 5; water-wh. oily, clear liq.; odorless, tasteless; sp.gr. 0.978; visc. 300,000 cSt; pour pt. –40 C; ref. index 1.4035; flash pt. (CC) 260 C.

Masil® SF 600,000. [PPG-Mazer] Dimethicone; see Masil SF 5; water-wh. oily, clear liq.; odorless, tasteless; sp.gr. 0.979; visc. 600,000 cSt; pour pt. –34 C; ref. index 1.4035; flash pt. (CC) 260 C.

Maslip® 500. [PPG-Mazer] Proprietary formula containing no chlorinated or sulfonated compds.; lubricant base for metalworking fluids which require extreme-pressure qualities for heavy-duty work; red-br. liq.; water-sol.; sp.gr. 1.110; dens. 9.2 lb/gal; pour pt. < –12 C; flash pt. (PMCC) > 200 F; 36% water.

Maslip® 501 Base. [PPG-Mazer] Proprietary formula containing no chlorinated or sulfonated compds.; see Maslip 500; red-br. liq.; water-sol.; sp.gr. 1.102; visc. 800 cps; dens. 9.2 lb/gal; pour pt. 0 C; flash pt. (PMCC) > 300 F; 7% water.

Matacil®. [Bayer] 3-Methyl-4-dimethylaminophenyl-N-monomethylcarbamate; insecticide for controlling bollworms and leafworms of cotton and hard-to-kill caterpillars on vegetables; wh. cryst.; odorless; m.w. 208.25; slightly sol. in water, moderately sol. in aromatic solvs., readily sol. in polar org. solvs.; m.p. 93–94 C.

Matexil AA-NS. [ICI Ltd.] Mixed polyglycol fatty acid; anionic; antifoam agent for textile dyeing processes; liq.

Matexil DA-AC. [ICI Ltd.] Naphthalene sulfonic acid/formaldehyde condensate, disodium salt; anionic; dispersant for dyeing applics.; powd.

Matexil DN-VL500. [ICI Ltd.] Fatty alcohol ethoxylate; nonionic; antistat; dye leveling agent; dispersant, emulsifier; flakes

Matexil LC-CWL. [ICI Ltd.] EO condensate; nonionic; level dyeing assistant; liq.

Matexil LC-RA. [ICI Ltd.] Quat. ammonium compd.; cationic; retarding agent for dyeing applics.; liq.

Matexil LN-RD. [ICI Ltd.] Fatty acid ethoxylate; nonionic; level-dyeing assistant for disperse dyes; liq.

Matexil PN-HT. [ICI Ltd.] Fatty acid ethoxylate; nonionic; used in dyeing applics.; liq.

Maxacal® F400. [Int'l. Bio-Synthetics] Protease; enzyme which breaks down protein into water-sol. prods.; used in high alkaline detergent powds.; encapsulated form.

Maxacal® P400,000. [Int'l. Bio-Synthetics] Protease; encapsulated, high alkaline enzyme for use in powd. detergent compositions; 500 μ avg. particle size; water-sol.; dens. 750–950 g/l.

Maxahibit 100. [Climax Performance] Sodium tolyltriazole; corrosion inhibitor for nonferrous metals incl. copper, brass, and bronze; 50% aq. sol'n.

Maxaliq. [Int'l. Bio-Synthetics] Bacterial alpha-amylase, tech.; starch hydrolyzing enzyme for ethanol industry; liq.

Maxamyl®. [Int'l. Bio-Synthetics] Bacterial amylase; enzyme for high temp. liquefaction of starch; liq.

Maxamyl® WL7000. [Int'l. Bio-Synthetics] α–Amylase; enzyme for use in liq. laundry detergents; clear ylsh. liq.

Maxatase® LS400. [Int'l. Bio-Synthetics] Protease; enzyme which breaks down proteins into water-sol. prods.; for liq. detergents; liq.

Maxatase® MP375. [Int'l. Bio-Synthetics] Protease/amylase; enzyme for breakdown of proteins and starch into water-sol. prods.; used in powd. detergents; encapsulated form.

Maxatase® P440. [Int'l. Bio-Synthetics] Protease; see Maxatase MP375; encapsulated form.

Maxazyme® GI-IMMOB. [Int'l. Bio-Synthetics] Glucose isomerase; enzyme which catalyzes the isomeration of glucose to fructose; immobilized.

Maxilact®. [Int'l. Bio-Synthetics] Lactase; enzyme which hydrolyzes lactose to simple sugars; liq.

Maxinvert®. [Int'l. Bio-Synthetics] Invertase; enzyme used in confections and the prod. of invert syrups; powd. or liq.

May-Tein C. [Amerchol] Potassium-coco hydrolyzed animal protein; mild surfactant and conditioner used in shampoos; water sol'n.

May-Tein CT. [Amerchol] TEA-coco hydrolyzed animal protein; mild surfactant and conditioner used in shampoos; water sol'n.

May-Tein SK. [Amerchol] Sodium-coco hydrolyzed animal protein; foaming protein and conditioner used in shampoos and skin cleansers; water sol'n.

Mazamide® 25. [PPG-Mazer] 2:1 coconut sulfonic acid alkanolamide; nonion/anionic; thickeners for liq. detergent systems and shampoos, emulsifiers, foam boosters, rust inhibitors; used in hard surface cleaners, metalworking fluids/syn. coolants, waterless hand cleaners, and automotive specialties; liq.; sp.gr. 1.053; biodeg.

Mazamide® 65, 65CZ, 66. [PPG-Mazer] 2:1 mixed fatty acid DEA; nonionic/anionic; see Mazamide 25; also solubilizer; liq.; sp.gr. 0.99–1.02; pH 9.0–10.0; biodeg.

Mazamide® 70. [PPG-Mazer] 2:1 Cocamide DEA and diethanolamine; nonionic; emulsifier, detergent, solubilizer, thickener used in hard surface cleaners, dishwash liq., metalworking fluids; corrosion inhibitor for sol. oils; liq.; sp.gr. 0.99–1.02; pH 9–10; biodeg.

Mazamide® 80. [PPG-Mazer] 1:1 Cocamide DEA and diethanolamine; nonionic; thickener, foam stabilizer, solubilizer, emulsifier used in cosmetic and toiletry formulations, hard surface wetters and cleaners; liq.; sp.gr. 0.98–1.00; pH 9–10.5; biodeg.

Mazamide® C-5. [PPG-Mazer] PEG-6 cocamide MEA; nonionic; emulsifier, lubricant, rust inhibitor, buffing compd., thickener, foam booster, detergent; liq.; sp.gr. 1.0–1.1; pH 9.5–10.5; biodeg.

Mazamide® CS-148. [PPG-Mazer] 1:1 Cocamide DEA; nonionic; thickener, foam stabilizer, emulsifier, solubilizer used in cosmetic and toiletry formulations; liq.; sp.gr. 0.98–1.00; pH 9.0–10.5; biodeg.

Mazamide® L-5. [PPG-Mazer] PEG-6 lauramide DEA; nonionic; emulsifier, lubricant, rust inhibitor, buffing compd.; liq.; sp.gr. 1.0–1.1; pH 9.5– 10.5; biodeg.

Mazamide® L-298. [PPG-Mazer] 2:1 Lauramide DEA and diethanolamine; nonionic/anionic; foam builder and stabilizer, emulsifier, dispersant, visc. builder for hard surface cleaners, dishwashing, shampoos, metalworking fluids, automotive specialties, fiber and hair conditioners, dry cleaning, agric. sprays, leather/fur preparations, emulsifiable waxes, rust inhibitors, polishes, paint removers, rug shampoos, fuel oil additives, textile detergents; solid; water sol.; sp.gr. 1.00; pH 9.4 (5% aq.); biodeg.

Mazamide® LM-21. [PPG-Mazer] 2:1 Lauramide DEA and diethanolamine; nonionic/anionic; see Mazamide L-298; liq.; water sol.; sp.gr. 1.00; pH 10.0 (5% aq.); biodeg.

Mazamide® LS-173, -196. [PPG-Mazer] 1:1 Lauramide DEA; nonionic; thickener, foam stabilizer, emulsifier used in cosmetic and toiletry formulations; liq.; sp.gr. 0.98–1.00; pH 9.0–10.5; biodeg.

Mazamide® O-10. [PPG-Mazer] 1:1 Oleic alkanolamide; nonionic; see Mazamide L-298; paste; insol. in aq. systems; sp.gr. 0.94 (40 C).

Mazamide® O-20. [PPG-Mazer] 2:1 Oleamide DEA and diethanolamine; nonionic; emulsifier, dispersant, lubricant, thickener, solubilizer, corrosion inhibitor, for metalworking fluids, buffing compds.; liq.; sp.gr. 0.99–1.01; pH 9–10; biodeg.

Mazamide® SS-10. [PPG-Mazer] 1:1 Linoleamide DEA; nonionic; emulsifier, lubricant, thickener, solubilizer, corrosion inhibitor, buffing compd.; liq.; sp.gr. 0.98–1.00; pH 9.0–10.5; biodeg.

Mazamide® SS-20. [PPG-Mazer] 2:1 Linoleamide DEA; nonionic/anionic; see Mazamide L-298; liq.; water sol.; sp.gr. 1.01; pH 9.8 (5% aq.); biodeg.

Mazamide® T-10. [PPG-Mazer] 1:1 Tall oil alkanolamide; nonionic; see Mazamide L-298; liq.; insol. in aq. systems; sp.gr. 0.99; pH 9.5 (5% aq.); biodeg.

Mazamide® T-15, -20. [PPG-Mazer] 2:1 tall oil alkanolamide; nonionic/anionic and nonionic resp.; see Mazamide L-298; solid and liq. resp.; water sol.; sp.gr. 1.00; pH 9.5 (5% aq.); biodeg.

Mazamide® TO-10. [PPG-Mazer] 1:1 Tall oil alkanolamide; nonionic; emulsifier, thickener, foam booster, rust inhibitor, detergent for industrial lubricants; liq.; sp.gr. 0.98–1.00; pH 9.0–10.0; biodeg.

Mazawax® 163R. [PPG-Mazer] Cetearyl alcohol and polysorbate 60; nonionic; emulsifier for pharmaceutical and cosmetic applics.; base; emolliency and thickening properties; wh. flake; 100% conc.

Mazawet® DOSS. [PPG-Mazer] Dioctyl sodium sulfosuccinate; antifog for glass cleaners; liq.

Mazeen® 173. [PPG-Mazer] Tetrahydroxypropyl ethylenediamine; insecticide and herbicide emulsifier, antistat and rewetting agent, grease additive, textile lubricant; emulsifier for lubricants, inks, and cosmetics; Gardner 2 liq.; m.w. 292; sol. in water, benzene, acetone, IPA; sp.gr. 1.01; surf. tens. 51.4 dynes/cm (0.1%).

Mazeen® 174. [PPG-Mazer] Tetrahydroxypropyl ethylenediamine; see Mazeen 173; Gardner < 1 liq.; m.w. 292; sol. in water, acetone, IPA; sp.gr. 1.00; surf. tens. 54.2 dynes/cm (0.1%).

Mazeen® C 2. [PPG-Mazer] PEG-2 cocamine; cationic; emulsifier, rewetting agent, lubricant, coupler used in insecticides and herbicides, grease additives, textile lubricants, water-based inks, cosmetics; plastics antistat; Gardner 11 liq.; m.w. 285; sol. in benzene, acetone, IPA, min. oil, forms gel in water; sp.gr. 0.874; surf. tens. 28 dynes/cm (0.1%); 100% conc.

Mazeen® C 5. [PPG-Mazer] PEG-5 cocamine; cationic; see Mazeen C 2; Gardner 11 liq.; m.w. 425; sol. in water, benzene, acetone, IPA, min. oil; sp.gr. 0.976; surf. tens. 33 dynes/cm (0.1%); 100% conc.

Mazeen® C 10. [PPG-Mazer] PEG-10 cocamine; cationic; see Mazeen C 2; Gardner 11 liq.; m.w. 645; sol. in water, benzene, acetone, IPA; sp.gr. 1.017; surf. tens. 39 dynes/cm (0.1%); 100% conc.

Mazeen® C 15. [PPG-Mazer] PEG-15 cocamine; cationic; see Mazeen C 2; Gardner 9 liq.; m.w. 860; sol. in water, benzene, acetone, IPA; sp.gr. 1.042; surf. tens. 41 dynes/cm (0.1%); 100% conc.

Mazeen® DBA. [PPG-Mazer] Dibutylamino ethanol; see Mazeen 173; Gardner 9 liq.; m.w. 192; sol. in benzene, acetone, IPA, min. oil; insol. in water; sp.gr. 0.860.

Mazeen® S 2. [PPG-Mazer] PEG-2 soyamine; cationic; see Mazeen C 2; Gardner 14 liq.; m.w. 350; sol. in benzene, acetone, IPA, min. oil; sp.gr. 0.911; surf. tens. 26 dynes/cm (0.1%); 100% conc.

Mazeen® S 5. [PPG-Mazer] PEG-5 soyamine; cationic; see Mazeen C 2; Gardner 14 liq.; m.w. 480; sol. in benzene, IPA; partly sol. acetone, min. oil; sp.gr. 0.951; surf. tens. 33 dynes/cm (0.1%); 100% conc.

Mazeen® S 10. [PPG-Mazer] PEG-10 soyamine; cationic; see Mazeen C 2; Gardner 14 liq.; m.w. 710; sol. in water, benzene, acetone, IPA; sp.gr. 1.020; surf. tens. 40 dynes/cm (0.1%); 100% conc.

Mazeen® S 15. [PPG-Mazer] PEG-15 soyamine; cationic; see Mazeen C 2; Gardner 18 liq.; m.w. 930; sol. in water, benzene, acetone, IPA; sp.gr. 1.040; surf. tens. 43 dynes/cm (0.1%); 100% conc.

Mazeen® T 2. [PPG-Mazer] PEG-2 tallow amine; cationic; see Mazeen C 2; Gardner 11 liq.; m.w. 350; sol. in benzene, acetone, IPA, min. oil; sp.gr. 0.916; surf. tens. 29 dynes/cm (0.1%); 100% conc.

Mazeen® T 5. [PPG-Mazer] PEG-5 tallow amine; cationic; see Mazeen C 2; Gardner 12 liq.; m.w. 480; sol. in benzene, acetone, IPA, min. oil; gels in water; sp.gr. 0.966; surf. tens. 34 dynes/cm (0.1%); 100% conc.

Mazeen® T 15. [PPG-Mazer] PEG-15 tallow amine; cationic; see Mazeen C 2; Gardner 18 liq.; m.w. 925; sol. in water, benzene, acetone, IPA; sp.gr. 1.028; surf. tens. 41 dynes/cm (0.1%); 100% conc.

Mazol® 159. [PPG-Mazer] PEG-7 glyceryl cocoate; emulsifier for food prods.; emollient; used in cosmetics, toiletries, pharmaceuticals, lubricants, mold release compds.; plasticizer in syn. fabrics and plastics; Gardner 1 clear liq.; HLB 13.0; acid no. 5 max.; sapon. no. 90.

Mazol® 165C. [PPG-Mazer] Glyceryl stearate and PEG-100 stearate; SE; see Mazol 159; wh. flake; HLB 11.2; acid no. 2 max.; sapon. no. 95.

Mazol® 300. [PPG-Mazer] Glyceryl oleate; nonionic; antifoam agent for sugar and protein processing, coemulsifier; GRAS dispersant for oil or solv. systems; yel. liq.; sol. in soybean oil, propylene glycol; disp. in ethanol; HLB 3.0; acid no. 5 max.; sapon. no. 150–160; 100% conc.

Mazol® 300 K. [PPG-Mazer] Glyceryl oleate; GRAS dispersant for oil or solv. systems; liq.; HLB 3.8; acid no. 2 max.

Mazol® 1400. [PPG-Mazer] Caprylic/capric triglyceride; carrier for flavors, fragrances, vitamins, antibiotics, pigmented cosmetics, medicinals; Gardner 1 liq.; HLB 1.0; acid no. 0.5 max.; sapon. no. 335–360.

Mazol® GDO. [PPG-Mazer] Glyceryl dioleate; see Mazol 159; Gardner 4 max. clear liq.; HLB 2.9; sapon. no. 182.

Mazol® GMO. [PPG-Mazer] Glyceryl oleate; nonionic; GRAS dispersant for oil or solv. systems; antifoam for food processing; base for cosmetic creams, lotions, ointments; w/o emulsifier with emolliency, thickening properties; plasticizer; lubricant; antifog for PVC; yel. liq.; sol. in soybean oil; disp. in ethanol, propylene glycol; HLB 2.4; acid no. 5 max.; sapon. no. 167–177; 100% conc.

Mazol® GMO K. [PPG-Mazer] Glyceryl oleate; GRAS-type antistat for plastic fibers in contact with food; dispersant for oil or solv. systems; Gardner 2 paste; HLB 3.8; acid no. 2 max.; 100% act.

Mazol® GMR. [PPG-Mazer] Glyceryl ricinoleate; base for cosmetic creams, lotions, ointments; w/o emulsifier with emolliency, thickening properties; plasticizer; wh. liq.; HLB 6.0; acid no. 7 max.; sapon. no. 138–145.

Mazol® GMS. [PPG-Mazer] Glyceryl stearate; lubricant, emulsifier, plasticizer, and thickener for foods, drugs, and cosmetics; wh. flake; HLB 3.9; acid no. 5 max.; sapon. no. 172.

Mazol® GMS-D. [PPG-Mazer] Glycerol stearate SE; see Mazol 159; wh. flake; HLB 6.0; acid no. 3.5 max.; sapon. no. 142.

Mazol® GMS-HM. [PPG-Mazer] Glycerol stearate; see Mazol 159; base for cosmetic creams, lotions, ointments; w/o emulsifier with emolliency, thickening properties; plasticizer; wh. solid, flake; HLB 3.3; sapon. no. 165.

Mazol® PG-810. [PPG-Mazer] Propylene glycol dicaprylate/caprate; solubilizer, carrier for flavors, fragrances, vitamins, antibiotics, pigmented cosmetics, medicinals; liq.

Mazol® PGMS. [PPG-Mazer] Propylene glycol stearate; nonionic; coemulsifier for edible oil and shortenings, dispersing aid for nondairy creamers; wh. cream solid; sol. in propylene glycol, disp. in ethanol; m.p. 39–46 C; HLB 3.4; acid no. 5 max.;

sapon. no. 170–180; 100% conc.
Mazol® PGO-104. [PPG-Mazer] Decaglycerol tetraoleate; nonionic; emulsifier for food prods., cosmetics, lubricants, etc.; also food-grade solubilizer and carrier for essential oils and flavors; Gardner 8 max. liq.; water-disp.; HLB 6.2; acid no. 8 max.; sapon. no. 125–150.
Mazol® PGS-61. [PPG-Mazer] Hexaglycerol stearate; see Mazol 159; Gardner 10 max. solid; HLB 8.87; sapon. no. 100.
Mazoline® OA. [PPG-Mazer] Oleyl imidazoline; cationic; emulsifier for preparing w/o emulsions, used in metalworking; antistat for syn. fibers and blends; dispersant for pigments and clays in aq. latex and solv. systems; liq.; disp. freely in nonpolar solvs., e.g., kerosene and min. oils, lipophilic; 100% conc.
Mazoline® OA4. [PPG-Mazer] Fatty imidazoline; corrosion inhibitor for metalworking applics.; liq.
Mazoline® T. [PPG-Mazer] Tall oil substituted imidazoline; cationic; see Mazoline OA; liq.; 100% conc.
Mazon® 1045A, 1086, 1096. [PPG-Mazer] POE sorbitol fatty acid ester; nonionic; emulsifier for pesticide, herbicide, metalworking, die-cast lubricant formulations, and emulsion polymerization; humectant, emollient; liq.; water-sol.; 100% conc.
Mazon® RI-3. [PPG-Mazer] High-performance ferrous corrosion inhibitor for metalworking fluids and coolants, metal treating, and metal cleaning formulations; clear visc. liq.; water-sol.; sp.gr. 1.090; visc. 450 cps; pour pt. 20 C; pH 8.0 (1% aq.).
Mazon® RI-4A. [PPG-Mazer] Amine-based; ferrous corrosion inhibitor developed to replace sodium nitrite in metalworking fluids; lt. yel. clear liq.; water-sol.; sp.gr. 1.180; visc. 2000 cps; pour pt. 20 C; pH 9.5 (1% aq.).
Mazon® RI-7B. [PPG-Mazer] Low-foaming ferrous corrosion inhibitor for metalworking and coolant formulations; lt. yel. clear liq.; water-sol.; sp.gr. 1.140; visc. 940 cps; pour pt. 20 C; pH 8.2 (1% aq.).
Mazon® RI-8B. [PPG-Mazer] Corrosion inhibitor specifically for water-based, syn. metalworking fluids and coolants; contributes lubricity and antiwear properties for lubricant formulations; lt. yel. clear liq.; water-sol.; sp.gr. 1.130; visc. 400 cps; pour pt. 20 C; pH 9.6 (1% aq.).
Mazon® RI-9. [PPG-Mazer] Ferrous corrosion inhibitor with lubricity and antiwear properties; lt. yel. clear liq.; water-sol.; sp.gr. 1.080; visc. 200 cps; pour pt. 10 C; pH 9.2 (1% aq.).
Mazon® RI-10A. [PPG-Mazer] Ferrous corrosion inhibitor for syn. metalworking and recirculating coolant formulations; lt. yel. clear liq.; water-sol.; sp.gr. 1.130; visc. 640 cps; pour pt. –7 C; pH 9.9 (1% aq.).
Mazon® RI-12. [PPG-Mazer] Corrosion inhibitor for ferrous, copper, brass, and aluminum metals; amber visc. liq.; oil-sol.; water-insol.; sp.gr. 0.962; visc. 800 cps; pour pt. 10 C.
Mazon® RI-13. [PPG-Mazer] Blend of alkanolamines and carboxylates; rust inhibitor in aq. systems; esp. water-sol. syn. and semisyn. metalworking fluids and coolants; yel. clear liq.; water-sol.; sp.gr. 1.110; visc. 370 cps; pour pt. –7 C; pH 9.5 (1% aq.).
Mazon® RI-14. [PPG-Mazer] Blend of amines and carboxylates; rust inhibitor for aq. systems, esp. water-sol. syn. and semisyn. metalworking fluids and coolants; yel. clear liq.; water-sol.; sp.gr. 1.060; visc. 1100 cps; pour pt. 20 C; pH 9.3 (1% aq.); 100% act.
Mazon® RI-110. [PPG-Mazer] Corrosion inhibitor for ferrous metals as well as for copper, brass, and aluminum; also imparts lubricity and antiwear properties to metalworking fluids and industrial lubricants; yel. to br. clear liq.; water-sol.; sp.gr. 1.095; visc. 400 cps; pour pt. 20 C; pH 8.1 (1% aq.).
Mazu® DF 100S. [PPG-Mazer] Silicone; antifoamer used in food processing as direct food additive; translucent lt. gray syrupy liq.; sp.gr. 1.01; dens. 8.4 lb/gal; visc. 2000 cSt; 100% conc.
Mazu® DF 110S. [PPG-Mazer] Silicone; nonionic; emulsifier; antifoamer used used in food processing as direct food additive; latex processing, boiler water defoaming, leather finishing, metal working, and waste treatment applic.; wh. lt. cream emulsion; water-disp.; sp.gr. 1.00; dens. 8.3 lb/gal; visc. 450 cS; 10% conc.
Mazu® DF 130S. [PPG-Mazer] Silicone; defoamer used as direct food additive; wh. med. cream emulsion; water-disp.; sp.gr. 1.01; 30% conc.
Mazu® DF 200S. [PPG-Mazer] Silicone; food-grade defoamer; liq.; sp.gr. 1.00; dens. 8.3 lb/gal; visc. 2000 cSt; 100% conc.
Mazu® DF 200SP. [PPG-Mazer] Simethicone; see Mazu DF-100S; also is formulated to meet the specific needs of pharmaceutical industry; liq.; sp.gr. 1.00; dens. 8.3 lb/gal; visc. 1720 cSt; 100% conc.
Mazu® DF 200SX. [PPG-Mazer] Silicone; defoamer for industrial use; used in adhesives, solv.-based inks and paints, insecticides, resin polymerization, and petrol. industry; liq.; sp.gr. 1.00; dens. 8.3 lb/gal; visc. 2000 cSt; 100% conc.
Mazu® DF 200SX Special. [PPG-Mazer] Silicone; see Mazu DF 200SX; sp.gr. 1.00; dens. 8.3 lb/gal; visc. 2500 cSt; 100% conc.
Mazu® DF 210S. [PPG-Mazer] Silicone; nonionic; see Mazu DF 110S; liq.; water-disp.; sp.gr. 1.00; dens. 8.3 lb/gal; visc. 400 cS; 10% conc.
Mazu® DF 210SX. [PPG-Mazer] Silicone; nonionic; emulsifier; defoamer for adhesives mfg., antifreeze, hot aq. systems, water-based inks and paints, insecticides, vinyl latex binders and emulsions, petrol., textile and paper applic.; liq.; sp.gr. 1.00; dens. 8.3 lb/gal; visc. 1900 cSt; 10% conc.
Mazu® DF 210SX Mod 1. [PPG-Mazer] Silicone; nonionic; see Mazu DF 210SX; also used as an antifoam in carpet cleaning applic.; liq.; sp.gr. 1.00; dens. 8.3 lb/gal; visc. 1000 cSt; 3.5% conc.
Mazu® DF 215SX. [PPG-Mazer] Silicone emulsion; industrial grade defoamer for dilution and repkg.; liq.; 15% act.
Mazu® DF 230S. [PPG-Mazer] Silicone; nonionic; emulsifier; defoamer used as direct food additive; latex and food processing, boiler water defoaming, and waste treatment; liq.; sp.gr. 1.00; dens. 8.3 lb/gal; visc. 1500 cSt; 30% conc.

Mazu® DF 230SX. [PPG-Mazer] Silicone; nonionic; emulsifier; defoamer for adhesive, water-based paints; soap mfg., antifreeze, hot aq. systems, insecticides, textile, and paper processing, petrol. applic., vinyl latex binders and emulsions, boiler water defoaming, leather finishing, metal working, and waste treatment; liq.; sp.gr. 1.00; dens. 8.3 lb/gal; visc. 3000 cSt; 30% conc.

Mazu® DF 243. [PPG-Mazer] Silicone; nonionic; see Mazu DF 230SX; liq.; sp.gr. 1.00; dens. 8.3 lb/gal; visc. 3000 cSt; 30% conc.

MBEO 1.8 Adduct. [Air Prod.] Ethoxylated acetylenic alcohol; corrosion inhibitor in min. acid systems; leveler and brightener agent for use in electroplating.

MBT. [Akrochem] 2-Mercaptobenzothiazole; very act., nondiscoloring org. accelerator; cream to lt. yel. powd.; disp. readily; sp.gr. 1.51; m.p. 165–180 C.

MBTS. [Akrochem] Benzothiazyl disulfide; nonstaining org. accelerator; very act. at temps. above 280 F; cream to lt. yel. powd.; disp. readily; sp.gr. 1.50; m.p. 160–170 C.

MBTS Pellets. [Akrochem] Benzothiazyl disulfide; see MBTS; cream to lt. yel. 2-mm pellets; disp. readily; sp.gr. 1.50; m.p. 158 C min.

McNamee® Clay. [Vanderbilt] Kaolin; extender and reinforcing filler for polyester, styrene, epoxies, phenol formaldehyde resins; preferred over calcium carbonate giving substantially better flexural modulus, and improved flow chars.; wh. to cream powd., diam 3.0 μm, fineness 99.7% min. thru 325-mesh screen; dens. 2.62 mg/m^3; oil absorp. 35; 44.46% SiO_2, 39.84% Al_2O_3.

M-C-Thin 45. [Lucas Meyer] Lecithin; nonionic; release aid based on phopholipids; liq.; 30-40% conc.

M-C-Thin AF-1. [Lucas Meyer] Lecithin, natural; amphoteric; emulsifier, wetting agent and stabilizer used in the food industry; liq.; 100% conc.

Mearlin. [Mearl] Titanium dioxide/mica; for pearlescent, metallic, antique, and iridescent effects in powd. coating systems.

Mearlmica MMCF. [Mearl] Mica; extender in loose powds.; increased slip in pressed powds.; binder and reinforcement in lipsticks and other lanolin and wax-based make-ups; 26 μ avg.

Mearlmica MMSV. [Mearl] Mica; see Mearlmica MMCF; 7–8 μ avg.

MEB 6046. [Bayer] 2,2,2-Trichloro-1-(3,4-dichlorophenyl) ethyl acetate; insecticide for control of domestic pests; colorless cryst.; m.w. 336.4; sol. > 60 g/100 g in cyclohexanone, methylene chloride, toluene; m.p. 84.5 C.

Megum® Series. [Chemetall] Bonding agent for bonding rubber to metals and other materials under vulcanizing conditions; used in automotive industry, mechanical engineering, etc.

Mekon® White. [Petrolite] Hard microcryst. wax; release agent; used in hot-melt coatings and adhesives, cup and paper coatings, printing inks, plastic modification (as lubricant and processing aid), lacquers, paints, and varnishes, as binder in ceramics, for potting, filling, and impregnant in elec./electronic components, in investment casting, rubber and elastomers (plasticizer, antisunchecking, antiozonant), as emulsion wax size in papermaking, as fabric softener ingred. in permanent-press fabrics, in emulsion and latex coatings, and in cosmetic hand creams and lipsticks; wax; very low sol. in org. solvs.; sp.gr. 0.93; visc. 11.7 cps (98.9 C); m.p. 92.8 C.

Mercerol GVC-65 Liq. [Sandoz Ltd.] Wetting agent, mercerizing penetrant for efficient wetout in caustic soda saturator; yel. liq.; disp. in water; sp.gr. 1.03; 65% act.

Merix® Anti-Fog. [Merix] Antifog for glass and plastic surfaces; prevents fog, mist, and steam.

Merix®#79 Conc. [Merix] Antistat for hard surfaces, e.g., plastics and paper, solar conductors; conductive coating in electronic and computer applics.; clear, nonflam.

Merix® 79 Special. [Merix] Low sodium, low calcium antistat for hard surfaces and plastics, esp. for sensitive applics. such as microconductors, delicate instruments.

Merix® 79-OL Conc. [Merix] Odorless antistat for soft surfaces, e.g., carpets, textiles, PE; clear, nonflam.

Merkyl MAP. [Vikon] Phenyl mercury ammonium propionate; mildewcide for outdoor textiles, water-based paints; misc. in water; ammonium acetate sol'n.

Merkyl PM-TL. [Vikon] Phenyl mercury triethanol ammonium lactate; antimicrobial, algicide, mildewcide for outdoor textiles; misc. in water.

Merpol C. [DuPont] Long-chain alcohol sulfate; anionic; wetting agent, detergent, penetrant, emulsifier used in textile and paper industries; lt. amber liq.; misc. with water; dens. 7.5 lb/gal; pH 8 (1%).

Merpol DA. [DuPont] EO condensate; nonionic; low-foaming leveling agent and dyeing assistant for textile dyes; lt. yel. clear, slightly visc. liq.; alcoholic odor; sol. in water; > 50% sol. in ethanol, ethylene glycol, cellosolve, acetone, MEK, oleic acid; dens. 8.5 lb/gal; visc. 110 cps; cloud pt. 10 C (lower), > 100 C (upper 1%); clear pt. 12 C; surf. tens. 50 dynes/cm (0.1%); pH 9 (1%); 60% act.

Merpol HCS. [DuPont] EO condensate; nonionic; wetting agent, penetrant, antistat, leveling agent, detergent, dyeing assistant, emulsifier, stabilizer used in textiles, paints, inks, medicinal ointments, antiperspirants, cosmetics; lt. yel. clear, visc. liq.; mild aromatic odor; 40% sol. in water, sol. in org. solvs. that are misc. with water; dens. 8.63 lb/gal; HLB 15.5; cloud pt. 43 F; flash pt. > 235 F; surf. tens. 42.9 dynes/cm (0.1%); pH 6–8 (10%); 60% act.

Merpol HCW. [DuPont] Higher fatty alcohol EO condensate; nonionic; emulsifier, leveling agent, stabilizer, wetting agent, dispersant, detergent, dyebath additive for all fibers; clear, visc. liq.; mild, alcoholic odor; water sol.; dens. 8.32 lb/gal; f.p. < 3 F; cloud pt. > 212 F; flash pt. > 80 F; pH 6–7 (10%); 30% act.

Merpol LF-H. [DuPont] Polyether; nonionic; detergent, nonfoam wetting, scouring and dyeing assistant for textiles, emulsifier, lubricant; pale yel. clear liq.;

mild odor; 40% max. sol. in water below 70 F; dens. 7.96 lb/gal; visc. 28 cps; cloud pt. 78 F (upper 1% aq.); flash pt. > 200 F (TCC); pH 6–8; biodeg.

Merpol OJS. [DuPont] EO condensate; nonionic; wetting and dispersing agent for textiles, emulsifier, antistat, detergent; slightly yel. clear liq.; mild fatty odor; sol. in water, sol. or disp. in alcohol, ketones, aliphatic and aromatic hydrocarbons; dens. 8.04 lb/gal; sp.gr. 0.9648; visc. 46 cps; HLB 12.7; cloud pt. 30 F; flash pt. 167 F (TOC); surf. tens. 33.5 dynes/cm (0.1%); pH 6–8; 60% act.

Merpol SE. [DuPont] Long-chain fatty alcohol EO condensate; nonionic; wetting and rewetting agent, emulsifier, dispersant; colorless to pale yel. clear liq.; 0.1% sol. in water, cloudy disp. at higher conc.; dens. 8.1 lb/gal; HLB 10.5; pH 6.2 (0.1%); 95% act.

Merquat® 100. [Calgon] Polyquaternium-6; cationic; conditioner, antistat for hair shampoos, conditioners, rinses, moisturizing creams, lotions, bath prods., skin care; sol. in water, propylene glycol; limited sol. in alcohol.; 40% act.

Merquat® 280. [Calgon] Polyquaternium-22; cationic; conditioner contributing softness, slip, lubricity to hair and skin prods., baby prods.; improves clarity of bath gels and shampoos; 35% act.

Merquat® 550. [Calgon] Polyquaternium-7; cationic; foam stabilizing film former for pearlized/opaque systems; used in soaps, skin care, and hair care prods.; 8% act.

Merquat® S. [Calgon] Polyquaternium-7; foam stabilizing film formers for hair shampoos, rinses, conditioners, bath and skin care prods.; sol. see Merquat 100; 8% act.

Merrol® 418S. [Merrand] n-Butyl stearate; plasticizer for CR, CPE, Hypalon elastomers, PVAc, cellulosics, PS, ABS, ink systems.

Merrol® 418T. [Merrand] n-Butyl oleate; plasticizer for CR, CPE, Hypalon elastomers, PVAc, cellulosics, PS, ABS, ink systems; lubricant additive.

Merrol® 600TM. [Merrand] Tri n-hexyl trimellitate; plasticizer for CR, CPE elastomers, PVC, engineering resins, polyesters, alloys.

Merrol® 810A. [Merrand] Di (n-octyl, n-decyl) adipate; plasticizer for PVC, PVAc, cellulosics, PS, ABS.

Merrol® 810TM. [Merrand] Tri (n-octyl-n-decyl) trimellitate; plasticizer for CR, CPE elastomers, PVC, engineering resins, polyesters, alloys; very broad low to high temp. range.

Merrol® 818T. [Merrand] Alkyl tallate; plasticizer for CR, CPE, Hypalon elastomers, ink systems, correctable ribbon/tape; lubricant additive.

Merrol® 3810. [Merrand] Triethylene glycol dicaprate-caprylate; plasticizer for NBR, NBR/PVC, CR, CPE elastomers, polyacrylate, fluroelastomers, inks; lubricant additive; fuel resistance.

Merrol® 3900. [Merrand] Triethylene glycol dipelargonate; plasticizer for NBR, NBR/PVC, CR, CPE elastomers, polyacrylate, fluroelastomers; lubricant additive.

Merrol® 4200. [Merrand] Dibutoxyethyl sebacate; low temp. plasticizer for NBR, NBR/PVC, CR, CPE elastomers.

Merrol® 4206. [Merrand] Dibutoxyethyl adipate; plasticizer for NBR, NBR/PVC, CR, CPE elastomers, polyacrylate, fluoroelastomers, PVAc, cellulosics, PS, ABS.

Merrol® 4208. [Merrand] Dibutoxyethyl phthalate; plasticizer for NBR elastomers, PVC, PVAc, cellulosics, PS, ABS, cellulose acetate; good uv resistance.

Merrol® 4218. [Merrand] Butoxyethyl oleate; plasticizer for CR, CPE, Hypalon elastomers, ink systems; ink solv.

Merrol® 4220. [Merrand] Dibutoxyethoxyethyl sebacate; see Merrol 4200.

Merrol® 4221. [Merrand] Dibutoxyethoxyethyl formal; low temp. plasticizer for NBR, NBR/PVC, CR, CPE, Hypalon elastomers, polyacrylate, fluoroelastomers; good antistatic properties.

Merrol® 4226. [Merrand] Dibutoxyethoxyethyl adipate; plasticizer for NBR, NBR/PVC, CR, CPE elastomers, polyacrylate, fluoroelastomers.

Merrol® 4228. [Merrand] Dibutoxyethoxyethyl phthalate; plasticizer for NBR elastomers, PVC, PVAc, cellulosics, PS, ABS.

Merrol® 4295. [Merrand] Glycol ether glutarate; plasticizer for NBR, NBR/PVC elastomers, polyacrylate, fluoroelastomers, PVC, engineering resins, polyesters, alloys, semiconductive PVC.

Merrol® 4425. [Merrand] Glycol ether glutarate; plasticizer for NBR, NBR/PVC elastomerts, polyacrylate, fluoroelastomers; exc. general purpose plasticizer for very polar polymers.

Merrol® 4800. [Merrand] Tetraethylene glycol di 2-ethylhexoate; plasticizer for CR, CPE, Hypalon elastomerts, PVC, PVAc, cellulosics, PS, ABS, engineering resins, polyesters, alloys, low temp. fuel hose, lacquers, and coatings.

Merrol® C-102. [Merrand] Triethyl citrate; plasticizer for PVC, PVAc, cellulosics, PS, ABS; acceptable for many FDA applics.

Merrol® C-103. [Merrand] Acetyl triethyl citrate; plasticizer for PVC, PVAc, PS, ABS; exc. cellulosics plasticizer.

Merrol® C-104. [Merrand] Tri n-butyl citrate; plasticizer for PVC, PVAc, cellulosics, PS, ABS, adhesives, and coatings; low toxicity.

Merrol® C-105. [Merrand] Acetyl tri n-butyl citrate; plasticizer for PVC, PVAc, cellulosics, PS, ABS, food wrap film; low toxicity.

Merrol® DBP. [Merrand] Di n-butyl phthalate; general purpose plasticizer for NBR, NBR/PVC, CR, CPE elastomers, PVC, PVAc, cellulosics, PS, ABS, inks.

Merrol® DBS. [Merrand] Di n-butyl sebacate; low temp. plasticizer for NBR, NBR/PVC, CR elastomers, polyacrylate, fluoroelastomers, PVC, PVAc, cellulosics, PS, ABS; also for ink systems, lubricant additive; ink solv. and dispersant.

Merrol® DIBA. [Merrand] Diisobutyl adipate; plasticizer for PVAc, cellulosics, PS, ABS, coatings; lubricant additive.

Merrol® DIDA. [Merrand] Diisodecyl adipate; plasticizer for NBR, NBR/PVC, CR, CPE elastomers, PVC, PVAc, cellulosics, PS, ABS.

Merrol® DIDN. [Merrand] Diisodecyl nylonate/glutarate; plasticizer for NBR, NBR/PVC, CR, CPE elastomers, PVC, PVAc, cellulosics, PS, ABS.

Merrol® DIOA. [Merrand] Diisooctyl adipate; plasticizer for NBR, NBR/PVC, CR, CPE elastomers, PVC, PVAc, cellulosics, PS, ABS.

Merrol® DOA. [Merrand] Di 2-ethylhexyl adipate; plasticizer for NBR, NBR/PVC, CR, CPE elastomers, PVC, PVAc, cellulosics, PS, ABS; suitable for food wrap film.

Merrol® DOS. [Merrand] Di 2-ethylhexyl sebacate; low temp. plasticizer for NBR, NBR/PVC, CR elastomers, polyacrylate, fluoroelastomers, PVC, PVAc, cellulosics, PS, ABS; also for ink systems, lubricant additive.

Merrol® DOZ. [Merrand] Di 2-ethylhexyl azelate; plasticizer for NBR, NBR/PVC, CR, CPE elastomers, PVC, PVAc, cellulosics, PS, ABS.

Merrol® DTDA. [Merrand] Ditridecyl adipate; plasticizer for PVC, PVAc, cellulosics, PS, ABS, engineering resins, polyesters, alloys; lubricant additive.

Merrol® E-45. [Merrand] Epoxidized alkyl tallate; low temp. plasticizer for PVC, plastisols.

Merrol® E-52. [Merrand] Epoxidized oleate ester; plasticizer for PVC, engineering resins, polyesters, alloys; good lubricity.

Merrol® E-68. [Merrand] Epoxidized soybean oil; plasticizer for NBR, NBR/PVC elastomers, PVC, PVAc, cellulosics, PS, ABS, inks; lubricant additive; PVC stabilizer.

Merrol® E-70. [Merrand] Epoxidized soybean oil; plasticizer for PVC, PVAc, cellulosics, PS, ABS, engineering resins, polyesters, alloys; pigment dispersions.

Merrol® N-301. [Merrand] Substituted benzenesulfonamide; plasticizer for engineering resins, alloys.

Merrol® N-302. [Merrand] n-Ethyl o,p-toluenesulfonamide; plasticizer for polyacrylate, fluoroelastomers, PVC, PVAc, cellulosics, PS, ABS, engineering resins, polyesters, alloys, PVAc adhesives; liq.

Merrol® N-303. [Merrand] n-Butylbenzene sulfonamide; plasticizer for polyacrylate, fluoroelastomers, PVC, PVAc, cellulosics, PS, ABS; nylon processing aid; liq.

Merrol® N-304. [Merrand] p-Toluenesulfonamide; plasticizer for PVC, PVAc, cellulosics, PS, ABS; strong antimicrobial agent; solid.

Merrol® N-305. [Merrand] o,p-Toluenesulfonamide; plasticizer for PVC, PVAc, cellulosics, PS, ABS, engineering resins, polyesters, alloys.

Merrol® N-306. [Merrand] Benzenesulfonamide; plasticizer for PVAc, cellulosics, PS, ABS, engineering resins, polyesters, alloys; solid.

Merrol® N-307. [Merrand] Toluenesulfonamide-formaldehyde; plasticizer for PVAc, cellulosics, PS, ABS, engineering resins, polyesters, alloys, nitrocellulose lacquers; solid.

Merrol® P-1030. [Merrand] Polyester sebacate; plasticizer for NBR, NBR/PVC, CR elastomers, polyacrylate, fluoroelastomers, PVC, PVAc, cellulosics, PS, ABS; also for ink systems.

Merrol® P-1030LV. [Merrand] Polyester azelate; plasticizer for NBR, NBR/PVC elastomers, PVC, PVAc, cellulosics, PS, ABS, engineering resins, polyesters, alloys; low migration.

Merrol® P-5510. [Merrand] Polyester glutarate; plasticizer for PVC, PVAc, cellulosics, PS, ABS; exc. weatherability, low migration.

Merrol® P-5511. [Merrand] Polyester nylonate; plasticizer for NBR, NBR/PVC elastomers, polyacrylate, fluoroelastomers, PVC, PVAc, cellulosics, PS, ABS, engineering resins, polyesters, alloys, PVC sealants; good oil resistance.

Merrol® P-6303. [Merrand] Polyester adipate; plasticizer for NBR, NBR/PVC, CR, CPE elastomers, polyacrylate, fluoroelastomers, PVC, PVAc, cellulosics, PS, ABS.

Merrol® P-6310. [Merrand] Polyester adipate; nonmigrating general purpose polymeric plasticizer for NBR, NBR/PVC, CR, CPE elastomers, polyacrylate, fluoroelastomers, PVC.

Merrol® P-6311. [Merrand] Polyester adipate; general purpose plasticizer for PVC, appliance gaskets.

Merrol® P-6320. [Merrand] Polyester adipate; plasticizer for NBR, NBR/PVC, CR, CPE elastomers, PVC; low temp. flex for elec. tape and wire insulation.

Merrol® P-6403. [Merrand] Polyester adipate; plasticizer for NBR, NBR/PVC, CR, CPE elastomers, PVC, PVAc, cellulosics, PS, ABS.

Merrol® P-6410. [Merrand] Polyester adipate; plasticizer for PVC, esp. for tape.

Merrol® P-6420. [Merrand] Polyester adipate; general purpose plasticizer for PVC.

Merrol® P-6422. [Merrand] Polyester adipate; low visc. plasticizer for PVC, plastisols and fabric coating.

Merrol® P-6424. [Merrand] Polyester adipate; plasticizer for NBR, NBR/PVC, CR, CPE elastomers, polyacrylate, fluoroelastomers, PVC.

Merrol® P-6510. [Merrand] Polyester adipate; plasticizer for PVC, PVAc, cellulosics, PS, ABS, engineering resins, polyester, alloys; outstanding uv and weather resistance.

Merrol® P-8227. [Merrand] Polyester phthalate/adipate; plasticizer for NBR elastomers, PVC, engineering resins, polyesters, alloys.

Merrol® P-8425. [Merrand] Polyester phthalate; plasticizer for NBR elastomers, PVC, PVAc, cellulosics, PS, ABS, engineering resins, polyesters, alloys; exc. light stability.

Merrol® P-9500. [Merrand] Polyester azelate; plasticizer for NBR, NBR/PVC, CR, CPE elastomers, PVC, PVAc, cellulosics, PS, ABS, engineering resins, polyesters, alloys; outstanding extraction resistance, good heat aging; clear amber visc. liq.; mild odor; sp.gr. 1.075; visc. 58,000 cps; acid no. 2 max.

Merrol® PGB-50. [Merrand] Di (propylene/ethylene) glycol dibenzoate; plasticizer for NBR, NBR/PVC, CR, CPE, Hypalon elastomers, PVC, PVAc, cellulosics, PS, ABS.

Merrol® PGB. [Merrand] Dipropylene glycol dibenzoate; plasticizer for PVC, PVAc, cellulosics, PS, ABS, PU; highly solvating in various adhesives.

Merrol® TIOTM. [Merrand] Triisooctyl trimellitate; plasticizer for NBR, NBR/PVC, CR, CPE elastomerts, PVC, PVC wire insulation.

Merrol® TOTM. [Merrand] Tri 2-ethylhexyl trimellitate; plasticizer for NBR, NBR/PVC, CR, CPE elastomerts, PVC, PVC wire insulation; good high temp. permanence.

Mesamoll. [Bayer] Alkyl sulfonic ester of phenol; universal plasticizer used in PVC calendering, extrusion, inj. molding, dip coating, high-pressure foam, rotational and compression moldings, film for linings and food pkg., shower curtains, floorcoverings, imitation leather, tarpaulins, protective clothing, cable insulation and sheathing, structure profiles, tubing, blown film, shoes, tech. articles, toys; also used with PS, joint sealants, natural, S/B, nitrile/butadiene, chlorinated and butyl rubber; cleansing agent for PU processing equipment; Hazen < 450; dens. 1.030–1.070 g/cc; visc. 95–125 mPa.s; acid no. < 0.1; ref. index 1.4970–1.5000; flash pt. 210–240 C (OC); 70:30 PVC:plasticizer properties: tens. str. 22 mPa; tens. elong. 340%; hardness (Shore D) 30.

Mesurol®. [Bayer] Methiocarb; insecticide for control of sucking and biting insect pests, spider mites, slugs, and snails, and for repelling depradating birds; wh. cryst. powd.; m.w. 225.3; sol. g/1000 ml: > 200 g in dichloromethane, 50–100 g in 2-propanol, toluene; m.p. 119 C.

Metasystox® R. [Bayer] Oxydemeton-methyl; systemic insecticide and acaricide for the control of sucking pests like aphids, spider mites, etc.; ylsh. liq., practically odorless; m.w. 246.3; sol. in most org. solvs.; misc. with water; sp.gr. 1.289 (20/4 C); b.p. 106 C (0.01 mm Hg).

Metasystox®(i). [Bayer] Demeton-S-methyl; systemic insecticide and acaricide for the control of sucking pests and sawflies; pale ylsh. oily liq., penetrating odor reminiscent of leeks; m.w. 230.3; sol. in most org. solvs.; sol. 3300 mg/l in water; sp.gr. 1.21 (20/4 C); b.p. 118 C (1 mm Hg).

Methocel® A. [Dow] Methylcellulose; nonionic; plasticizer for ceramic and refractory shapes and furnace linings; imparts visc. stability to latex and emulsion paints; thickener; wh. to off-wh. powd.; odorless; sp.gr. 1.39; dens. 11.6 lb/gal; dens. 0.25–0.70 g/cc; f.p. 0.0 C (2%); pH 7; surf. tens. 47–53 dynes/cm; ref. index 1.336 (2%); 27.5–31.5% methoxyl.

Methocel® A4C, A4M, A15-LV. [Dow] Methylcellulose; food gums used as thickener, stabilizer, emulsifier, adhesive, and gelling agent used in food industry; firm gel structure; water-sol.; visc. 400, 4000, 15 cps. (2% aq.) resp.

Methocel® E. [Dow] Hydroxypropyl methylcellulose; nonionic; see Methocel A; wh. to off-wh. powd.; odorless; sp.gr. 1.39; dens. 11.6 lb/gal; dens. 0.25–0.70 g/cc.

Methocel® E4M. [Dow] Hydroxypropyl methylcellulose or carbohydrate gum; see Methocel A4C; semifirm gel structure; water-sol.; visc. 4000 cps (2% aq.).

Methocel® E5, E15-LV, E50-LV, E4M, F50-LV, F4M, K3, K35, K100-LV, K4M, K15M, K100M. [Dow] Hydroxypropyl methylcellulose or carbohydrate gum; see Methocel A4C; soft, semifirm gel structure; water-sol.; visc. 5, 15, 50, 4000, 50, 4000, 3, 35, 100, 4000, 15,000, and 100,000 cps (2% aq.) resp.

Methocel® F. [Dow] Hydroxypropyl methylcellulose; a stock sol'n. or as a glycol-powd. slurry in latex paints to provide good thickening efficiency; lower gelation temp. of 60–70 C restricts the temp. or types of colorants used for paint batch shading compared to Methocel J or K prods.; wh. powd.; sp.gr. 1.39; dens. 11.6 lb/gal; f.p. 0.0 C (2%); pH 7; surf. tens. 44–50 dynes/cm; ref. index 1.336 (2%); 27–30% methoxyl; 4.0–7.5% hydroxypropoxyl.

Methocel® F50-LV, F4M. [Dow] Hydroxypropyl methylcellulose; see Methocel A4C; semifirm gel structure; water-sol.; visc. 50 and 4000 cps resp. (2% aq.); gel pt. 62–68 C.

Methocel® J. [Dow] Hydroxypropyl methylcellulose; high solids aq. slurry in latex paints, provides uniform visc. development, improve film build and hiding per coat, reduced drying time, increased prod. rates, and greater film thickness per coat in industrial latex paints; wh. powd.; sp.gr. 1.39; dens. 11.6 lb/gal; f.p. 0.0 C (2%); pH 7; surf. tens. 48–52 dynes/cm; 16.5–20.0% methoxyl; 23.0–32.0% hydroxypropoxyl.

Methocel® K. [Dow] Hydroxypropyl methylcellulose; nonionic; see Methocel A; wh. to off-wh. powd.; odorless; sp.gr. 1.39; dens. 11.6 lb/gal; dens. 0.25–0.70 g/cc; f.p. 0.0 C (2%); pH 7; surf. tens. 50–56 dynes/cm; 19.0–24.0% methoxyl; 4.0–12.0% hydroxypropoxyl.

Methocel® K3, K35, K100-LV, K4M, K15M, K100M. [Dow] Hydroxypropyl methylcellulose; see Methocel A4C; soft gel structure; water-sol.; visc. 3, 35, 100, 4000, 15,000, and 100,000 cps resp. (2% aq.); gel pt. 70–90 C.

Methocel® K100-LV. [Dow] Hydroxypropyl methylcellulose or carbohydrate gum; see Methocel A4C; soft gel structure; water-sol.; visc. 100 cps (2% aq.).

Methyl Cumate®. [Vanderbilt] Copper dimethyldithiocarbamate; accelerator for high-speed vulcanization of SBR, IIR, EPDM; dk. br. powd.; also avail. rods 99.9% thru 100 mesh (powd.); m.w. 303.98; sol. in acetone, toluene, chloroform; pract. insol. in water, alcohol, gasoline; dens. 1.75 ± 0.03 mg/m^3; m.p. > 325 C; 18–20% copper content.

Methyl-Ethyl Tuads®. [Vanderbilt] Methyl Tuads/ Ethyl Tuads, ratio 60:40; thiuram ultra accelerator; used in nonblooming heat resistant sulfurless and low sulfur stocks of NR and SBR; also in nonblooming EPDM vulcanizates; wh. to cream rods; sol. in toluene, chloroform; practically insol. in water; dens. 1.34 ± 0.03 mg/m^3; m.p. 62 C min.; 11.9% avail. sulfur.

Methyl Ledate®. [Vanderbilt] Lead dimethyldithiocarbamate; ultra accelerator; NR, SBR, IIR, IR, BR ultra accelerator for high speed, high temp. vulcanization; effective under continuous curing conditions; generally used with thiazole modifiers; wh. powd., 99.9% thru 100 mesh; gray rods; m.w. 447.65; dens. 2.43 ± 0.03 mg/m³ (powd.); m.p. > 310 C (powd.); 300 C min. (rods); 45.5–47.5% lead content (powd.); 41.0–43.0% lead content (rods).

Methyl Niclate. [Vanderbilt] Nickel dimethyldithiocarbamate; antioxidant for epichlorohydrin and peroxide vulcanized elastomers; grnsh. powd.; 99.8% min. thru 100 mesh; m.w. 299.12; insol. in water; dens. 1.77 ± 0.03 mg/m³; m.p. > 290 C.; 18-20% nickel content.

Methyl Selenac®. [Vanderbilt] Selenium dimethyldithiocarbamate; ultra accelerator for NR, SBR, IIR; also vulcanizing agent; effective in low sulfur and sulfurless heat resistant compds.; nondiscoloring in lt. stocks; generally used with thiazoles to balance scorch and curing chars.; yel. powd., 99.9% thru 100 mesh; rods; m.w. 559.78; sol. in toluene, carbon disulfide, chloroform; dens. 1.58 ± 0.03 mg/m³ (powd.); m.p. 140–172 C (powd.); 138 C min. (rods); 13.0–15.0% selenium content (powd.); 12.5–14.0% selenium content (rods).

Methyl Tuads®. [Vanderbilt] Tetramethylthiuram disulfide; ultra accelerator; for NR and syn. rubbers (esp. IIR, CR); accelerator and vulcanizing agent; cure modifier for Neoprene (retards G types; accelerates vulcanization of W types); nondiscoloring in lt. stocks; wh. to cream powd. and thread grade, 99.9% thru 100 mesh; also avail. in wh. to cream and blue rods; m.w. 240.44; sol. in toluene, carbon disulfide, chloroform; dens. 1.42 ± 0.03 mg/m³; m.p. 142–156 C (powd. and thread grade); 140 C min. (rods); 13.3% avail. sulfur (powd., thread grade); 12% (rods).

Methyl Zimate®. [Vanderbilt] Zinc dimethyldithiocarbamate; ultra accelerator for NR and syn. rubbers; latex accelerator; act. over wide temp. range; generally requires thiazole modifier for safe processing and wide curing range; nondiscoloring in lt. stocks; wh. powd., 99.9% thru 100 mesh; wh. and pink rods; 50% act. slurry; m.w. 305.82; practically insol. in water; dens. 1.71 ± 0.03 mg/m³ (powd.); m.p. 242–257 C (powd.); 239 C min. (rods); 19.5–23.0% zinc content (powd.); 19.0–21.0% zinc content (rods).

Metso Beads® 2048. [PQ] Anhyd. sodium metasilicate; corrosion inhibitor for metal surfs.; used in detergents to protect metal, china, glass, ceramic prods.; sol. in water.

Metso Pentabead® 20. [PQ] Sodium metasilicate pentahydrate; see Metso Beads 2048; sol. in water.

Meturon 4L. [Griffin] Fluometuron suspension; flowable herbicide controlling annual grasses and broadleaf weeds in cotton and sugarcane; 41.2% act.

Mg-0601 T 1/8″. [M&T Harshaw] Magnesia; catalyst support; gray-wh. tablet, 1/8″ diam.; dens. 72 lb/ft³; 98% conc.

Mibiron. [Rona] Bismuth oxychloride coated mica; pearlescent for lipsticks, pressed powds., eye makeups, for frosted effects.

Michel XO-24. [M. Michel] Modified amine; antistat additive, integral release agent for linear polyethylene and PVC.

Michel XO-85. [M. Michel] Modified amine; antistat additive, integral release agent for PP.

Michel XO-108. [M. Michel] Modified amine; antistat for PP, polyethylene, polyester film; water-insol.

Michel XO-144. [M. Michel] Hexadecyl alcohol; emollient; liq.

Michel XO-144B. [M. Michel] Hexadecyl alcohol; coupler; cosolv.; misc. in ethanol/water mixts.

Michel XO-146. [M. Michel] Isostearyl alcohol; coupler; cosolv.; misc. in ethanol/water mixts.

Micral® 532. [Solem] Alumina trihydrate; ultrafine flame retardant; ground powd.; 4–6 μ median particle diam.; 99.98% thru 325 mesh; sp.gr. 2.42; bulk dens. 0.55 g/cc (loose), 0.9 g/cc (packed); surf. area 4.5–5.5 m²/g; oil absorp. 30–31; ref. index 1.57.

Micral® 855. [Solem] Alumina trihydrate; ultrafine flame retardant for halogen-free compding. of thermoplastics.

Micral® 916. [Solem] Alumina trihydrate; halogen-free smoke suppressor/flame retardant for wire and cable insulation, inj.-molded polyolefins, coatings, adhesives, rubber goods, paper filler and coating, PVC, EPDM, EPR, ABS, XLPE, and compr.-molded thermosets; powd.; 0.8 μ median particle diam.; 100% through 325 mesh; sp.gr. 2.42 g/cm³; bulk dens. 0.15 g/cm³ (loose), 0.35 g/cm³ (packed); surf. area 13 m²/g; oil absorp. 46 ml/100 g; brightness (Photovolt) 97+; 64.9% Al_2O_3.

Micral® 932. [Solem] Alumina trihydrate; see Micral 916; powd.; 1.5 μ median particle diam.; 100% thru 325 mesh; sp.gr. 2.42 g/cm³; bulk dens. 0.3 g/cm³ (loose), 0.5 g/cm³ (packed); surf. area 7 m²/g; oil absorp. 38 ml/100 g; brightness (Photovolt) 97+; 64.9% Al_2O_3.

Microbloc®. [Pfizer] Platy talc with surf. treatment; antiblocking agent and processing aid for polyolefin films; 2 μ avg. particle size; dry brightness 90; sp.gr. 2.71; bulk dens. 13 lb/ft³ (loose), 37 lb/ft³ (tapped); pH 8.8; 61.2% SiO_2; 31.7% MgO.

Micro-Cel A. [Manville] Syn. calcium silicate; functional filler used as carriers, grinding aids, anticaking agent, and conditioner in agric. chemicals; toxicants; carriers for liq. seed inoculants; inert carriers to convert sticky visc. liq. to dry liqs.; off-wh. particulate, 8.0% 325-mesh residue; sp.gr. 2.33; dens. 7.2 lb/ft³; oil absorp. 375%; ref. index 1.53; pH 9.2 (10% aq.); 53.7% SiO_2, 23.5% CaO.

Micro-Cel B. [Manville] Syn. calcium silicate; see Micro-Cel A; off-wh. particulate, 7.0% 325-mesh residue; sp.gr. 2.20; dens. 12.5 lb/ft³; oil absorp. 220%; ref. index 1.52; pH 9.1 (10% aq.); 53.6% SiO_2, 22.1% CaO.

Micro-Cel C. [Manville] Calcium silicate; functional filler used to convert sticky, visc. liqs. to dry liqs. for use in rubber, agric., chemical, plastics, food processing, animal feed, and pharmaceutical industries; wh. particulate 4.0% 325-mesh residue; sp.gr. 2.26; dens. 8.2 lb/ft³; oil absorp. 380%; ref. index 1.55; pH 9.8 (10% aq.).

Micro-Cel E. [Manville] Calcium silicate; see Micro-Cel A; also used as inert catalyst carriers, vanadium catalyst used in mfg. of sulfuric acid, and phosphoric acid catalyst of petrol. industry; off-wh. powd., 7.0% 325-mesh residue; sp.gr. 2.45; dens. 5.5 lb/ft^3; oil absorp. 490%; ref. index 1.55; pH 8.4 (10% aq.); 58.6% SiO_2, 22% CaO.

Micro-Cel T-21. [Manville] Functional filler; lt. gray; sp.gr. 2.4; dens. 12.8 lb/ft^3; ref. index 1.54; pH 7.6 (10% aq. slurry); 65% silica, 14.9% magnesia.

Micro-Cel T-38. [Manville] Calcium silicate; extender pigment, opacifier contributing good hiding power to emulsion paints; flatting properties for flat wall paints; wh. particulate; dens. 18.8 lb/gal; oil absorp. 310; pH 9.8.

Micro-Cel T-49. [Manville] Functional filler; off-wh.; sp.gr. 2.1; dens. 12.0 lb/ft^3; ref. index 1.53; pH 10.0 (10% aq. slurry); 36% silica, 33% lime.

Micro-Cel T-70. [Manville] Syn. calcium silicate; extender pigment, opacifier contributing hiding power to emulsion paints; good nonsettling properties for use in traffic paints; wh. particulate; dens. 20.07 lb/gal; oil absorp. 205; pH 9.5.

Micro-Chek 11, 11D, 11DIDP, 11S-711. [Ferro] 2-n-Octyl-4-isothiazolin-3-one in plasticizers (epoxidized soybean oil, DOP, diisodecyl phthalate, and mixed dialkyl phthalates resp.); mildewcide, antimicrobial, fungicide, preservative for PVC, PU used for roofing membranes, automotive trim, awnings, pond liners, marine upholstery, outdoor furniture, interior applics; dens. 8.3, 8.21, 8.1, and 8.2 lb/gal resp.; 4% sol'ns.

Microduct®. [Dow Corning/Wickhen] Maltodextrin; carrier for fragrances; emollient oils, bath prods.; food additive; powd.; 100% act.

Micromesh No. 3. [Mearl] Water-ground muscovite mica; filler; nearly wh. platy-shaped; –400 mesh (well below); sp.gr. 2.8–2.9; bulk dens. 7 lb/ft^3; surf. area 6–8 m^2/g; ref. index 1.58; 42–47% SiO_2; 34–38% Al_2O_3; 9–12% K_2O; 0.5–2.5% Fe_2O_3; 4–5% water.

MicroPflex 1200. [Pfizer] Surface-modified microtalc; filler; powd.; 1.5 μ median particle size; nil retained on 325 mesh; sp.gr. 2.70; dens. 11.0 lb/ft^3 (loose), 21.5 lb/ft^3 (tapped); dry brightness 90.0; hardness (Mohs) 1; 60.5% SiO_2; 31.7% MgO; 1.9% acid solubles.

Microtalc MP10-52. [Pfizer] High purity talc; filler; Hegman grind 6.5+; sp.gr. 2.70; bulk dens. 0.10 g/cc (loose), 0.35 g/cc (tapped); oil absorp. 52; dry brightness 91.5; pH 8.8.

Microtalc MP12-50. [Pfizer] High purity talc; filler; Hegman grind 6.0+; sp.gr. 2.70; bulk dens. 0.11 g/cc (loose), 0.36 g/cc (tapped); oil absorp. 50; dry brightness 90.5; pH 8.8.

Microtalc MP15-38. [Pfizer] High purity talc; filler; Hegman grind 6.0; sp.gr. 2.70; bulk dens. 0.17 g/cc (loose), 0.53 g/cc (tapped); oil absorp. 44; dry brightness 91; pH 8.8.

Microtalc MP20-40. [Pfizer] High purity talc; filler; Hegman grind 6–; sp.gr. 2.70; bulk dens. 0.17 g/cc (loose), 0.54 g/cc (tapped); oil absorp. 44; dry brightness 88.5; pH 8.8.

Microtalc MP25-38. [Pfizer] High purity talc; filler; Hegman grind 5.5+; sp.gr. 2.70; bulk dens. 0.21 g/cc (loose), 0.59 g/cc (tapped); oil absorp. 43; dry brightness 91; pH 8.8.

Microtalc MP26-38. [Pfizer] High purity talc; filler; Hegman grind 5.5; sp.gr. 2.70; bulk dens. 0.21 g/cc (loose), 0.59 g/cc (tapped); oil absorp. 43; dry brightness 91; pH 8.8.

Microtalc MP30-36. [Pfizer] High purity talc; filler; Hegman grind 5.0; sp.gr. 2.70; bulk dens. 0.18 g/cc (loose), 0.43 g/cc (tapped); oil absorp. 36; dry brightness 88; pH 8.3.

Microtuff 1000. [Pfizer] Surface-treated talc; filler for polyolefins; platy; 1.5 μ avg. particle size; uncoated properties: sp.gr. 2.70; bulk dens. 13 lb/ft^3 (loose); dry brightness 90.5; oil absorp. 50; pH 8.8; 61.2% SiO_2; 31.7% MgO.

Microtuff F. [Pfizer] Surf.-treated talc; filler for polyolefins for food pkg. such as thermoformed containers, blown film, and inj. molded pkg.; 1.5 μ avg. particle size; sp.gr. 2.70; bulk dens. 7.0 lb/ft^3 (loose), 22.0 lb/ft^3 (tapped); dry brightness 90; pH 8.8.

Miglyol 810. [Hüls] Caprylic/capric triglyceride; nonionic; emollient, dispersant, lubricant, suspending agent, solubilizer; act. ingred. for cosmetics and pharmaceuticals; carrier/vehicle and solv. for inj. prods., topical ointments, creams, lotions, suppositories; dietetic prods.; dispersant; Gardner 3 max. liq.; neutral odor; sol. in diethyl ether, petrol. ether, chloroform, IPA, toluene, alcohol, min. oil, acetone; sp.gr. 0.94–0.95; visc. 27–30 cps; acid no. 0.1 max.; iodine no. 0.5 max.; sapon. no. 340–360; pH neutral; cloud pt. 0 C.

Miglyol 812. [Hüls] Caprylic/capric triglyceride; dispersant, lubricant, anticaking agent, carrier, solv., solubilizer, suspending agent for cosmetics; liq.; sol. in alcohol, min. oil, acetone; 100% conc.

Miglyol 818. [Hüls] Caprylic/capric/linoleic triglyceride; emollient for topicals; liq.

Miglyol 829. [Hüls] Caprylic/capric diglyceryl succinate; emollient, suspending agent for cosmetic and pharmaceutical topicals with dens. above 1.0; visc. liq.

Miglyol 840. [Hüls] Propylene glycol dicaprylate/dicaprate; see Miglyol 810; liq.; sol. in alcohol, min. oil, acetone.

Miglyol 840 Gel. [Hüls] Propylene glycol dicaprylate/dicaprate and stearalkonium hectorite; stabilizer, emollient; anhyd. preps. (ointments, make up, lip gloss); improves consistency and thermal stability in cosmetics and topical pharmaceuticals; emulsion stabilizers; base for w/o creams in combination with suitable emulsifier; cream to lt. brn. paste; pract. odorless; sol. in oils, fats, waxes; acid no. 0.5 max.; sapon no. 290–310; pH neutral.

Miglyol Gel. [Hüls] Caprylic/capric triglyceride and stearalkonium hectorite; stabilizer; improves consistency and thermostability in cosmetic and pharmaceutical creams; emulsion stabilizer; as a base for w/o and o/w creams when combined with suitable emulsifiers; does not melt at even 100 C; cream to lt. brn. paste; nearly odorless; sol. in oils, fats, waxes; acid no. 0.5 max.; sapon no. 295–315; pH neutral; 100% conc.

Migregal 2N. [Nippon Senka] Alkyl sulfate; anionic; level dyeing of nylon piece goods with acid dyes; liq.

Migregal NC-2. [Nippon Senka] Alkyl sulfate and nonionic surfactant; anionic/nonionic; leveling agent for fabrics; liq.

Milco-150. [Milport] Sodium glucoheptonate, alpha and beta isomers; chelating agent for bottle washing, metal cleaning, brew kettle cleaning; food additive grade; lt. yel. sol'n.; 50% act.

Milezyme® AFP. [Miles Biotech] Acid fungal protease; enzyme for protein hydrolysis; powd.

Milezyme® APG, APL. [Miles Biotech] Alkaline protease; enzyme for detergents, prespotters; gran. and liq. resp.

Milezyme® Brand Hemicellulase. [Miles Biotech] Galactomannase/cellulase; enzyme for hydrolysis of galactomannans in coffee extracts, foods, drilling fluids; powd.

Milezyme® Bromelain. [Miles Biotech] Protease; enzyme for protein hydrolysis; powd.

Milezyme® Catalase L. [Miles Biotech] Liver catalase; enzyme for decomposition of hydrogen peroxide; liq.

Milezyme® Cellulase, Cellulase TV Conc. [Miles Biotech] Cellulase; enzyme for breakdown of cellulosic materials for improved processing; powd.

Milezyme® DAL. [Miles Biotech] Alpha-amylase; enzyme for detergent, prespotters; liq.

Milezyme® Fungal Amylase. [Miles Biotech] Fungal alpha-amylase; enzyme for baking; powd.

Milezyme® Fungal Lactase. [Miles Biotech] Lactase; enzyme for hydrolysis of lactose to sugars; powd.

Milezyme® Fungal Protease. [Miles Biotech] Protease; enzyme for baking, protein hydrolysis; powd.

Milezyme® Lipase Powds. [Miles Biotech] Lipase/esterase; enzyme for flavor development; powd.

Milezyme® Pancreatic Lipase. [Miles Biotech] Lipase; enzyme for hydrolysis of fats and fatty acid esters; powd.

Milezyme® Pancreatin 4NF. [Miles Biotech] Amylase/protease/lipase; enzyme for pharmaceuticals; powd.

Milezyme® Papain. [Miles Biotech] Protease; enzyme for protein hydrolysis; powd.

Millathane Adhesive 2000. [TSE] Aromatic diisocyanate; bonding agent; clear to wh. liq., methylene chloride odor; sp.gr. 1.32; b.p. 40 C; flash pt. (CC) 199 C.

Millifoam. [Stepan] Sodium ether sulfate; gypsum board foamer; clear liq.; 50% act.

Millifoam A, ODE-60A. [Stepan] Ammonium ether sulfate; gypsum board foamer; clear liq.; 50 and 60% act. resp.

Millithix 925. [Milliken] Sorbitol acetal; thixotrope and gellant for org. systems; solid wh. powd.

Mineral Jelly No. 5. [Penreco] Petrolatum; preblended base for cosmetic mfg.; odorless; tasteless; misc. with cosmetic ingred. used in oil-base formulations; visc. 38–43 SUS (210 F); pour pt. 75–85 F.

Mineral Jelly No. 10. [Penreco] Petrolatum; preblended base for cosmetic mfg.; odorless; tasteless; misc. with cosmetic ingred. used in oil-base formulations; visc. 40–43 SUS (210 F); melt pt. (Saybolt) 97–105 F; pour pt. 95–105 F.

Mineral Jelly No. 15. [Penreco] Petrolatum; preblended base for cosmetic mfg.; odorless; tasteless; misc. with cosmetic ingred. used in oil-base formulations; visc. 40–44 SUS (210 F); m.p. (Saybolt) 97–108 F; pour pt. 95–105 F.

Mineral Jelly No. 20. [Penreco] Petrolatum; preblended base for cosmetic mfg.; odorless; tasteless; misc. with cosmetic ingred. used in oil-base formulations; visc. 37–40 SUS (210 F); m.p. (Saybolt) 111–116 F; pour pt. 110– 120 F.

Mineral Jelly No. 25. [Penreco] Petrolatum; preblended base for cosmetic mfg.; odorless; tasteless; misc. with cosmetic ingred. used in oil-base formulations; visc. 38–40 SUS (210 F); m.p. (Saybolt) 103–108 F; pour pt. 100–110 F.

Minex 2. [Unimin] Anhyd. sodium potassium aluminosilicate (nepheline syenite); filler for traffic paint, textured coatings, adhesives, caulks, and sealants; wh. particulate; 14.8 μ median particle size; 98.35% thru 325 mesh; oil absorp. 22.4 g/100 g; GE brightness 89.64; surf. area 4.2 m_2/g; pH 10.2; 60.712% SiO_2, 22.92% $Al2O_3$, 10.78% Na_2O.

Minex 3. [Unimin] Anhyd. sodium potassium aluminosilicate (nepheline syenite); see Minex 2; wh. particulate; 10.8 μ median particle size; 98.35% thru 325 mesh; oil absorp. 24.2 g/100 g; GE brightness 90.2; surf. area 6.8 m^2/g; pH 10.2; 60.78% SiO_2, 22.98% $Al2O_3$, 10.42% Na_2O.

Minex 4. [Unimin] Anhyd. sodium potassium aluminosilicate (nepheline syenite); see Minex 2; also for polyester BMC and SMC; wh. particulate; 8.9 μ median particle size; 100% thru 325 mesh; oil absorp. 26.7 g/100 g; GE brightness 89.55; surf. area 7.3 m^2/g; pH 10.3; 60.25% SiO_2, 23.41% Al_2O_3, 10.71% Na_2O.

Minflo Mining Flotation Reagents. [Hercules] Water-sol. polymers; selectively separate gangue materials in min. flotation.

Minugel 200. [Floridin] Colloidal attapulgite clay; suspending agent in suspension fertilizers and other agric. suspensions; coarse ground.

Minugel 400. [Floridin] Colloidal attapulgite clay; thickening, stabilizing and rheology control agent for latex and oil paints; stabilizer for flowable emulsions; suspending agent; very fine ground; lt. tan-gray, fine particle size colloid; fineness 7.5 Hegman; dens. 26 lb/ft^3; visc. 2200 cP (7% clay disp. in water); pH 9.6.

Minugel FG. [Floridin] Colloidal attapulgite clay; min filler in spackles and seam sealers; suspending agent; very fine ground.

Minugel LF. [Floridin] Colloidal attapulgite clay; thickener, stabilizer for liq. animal feeds containing immiscible materials (fat emulsions, vitamins, limestone powd.); suspending agent; coarse ground.

Miramine® CC. [Rhone-Poulenc Surf.] Cocoyl imidazoline; cationic; corrosion inhibitor in cutting oils, lubricant for metal surfaces, used in acid shampoos, with min. spirits to emulsify grease and oil; amber liq.; bland to faintly ammoniacal odor; sol. in many org. solvs., disp. in water; sp.gr. 0.940; biodeg.; 99% act.

Miramine® CPC. [Rhone-Poulenc Surf.] Caprylic fatty acid substituted imidazoline; cationic; see Miramine CC; liq.; 100% conc.

Miramine® DD. [Rhone-Poulenc Surf.] Stearamidopropyl dimethylamine; cationic; softener, emulsifier, and conditioner for hair and skin prods.; solid; 100% conc.

Miramine® GS. [Rhone-Poulenc Surf.] Stearic fatty acid substituted imidazoline; cationic; emulsifier for asphalt, fuel and cutting oils, metal lubricants; corrosion inhibitor; solid; water-disp.; 100% conc.

Miramine® GT. [Rhone-Poulenc Surf.] Tall oil fatty acid substituted imidazoline; cationic; see Miramine GS; liq.; 100% conc.

Miramine® IC. [Rhone-Poulenc Surf.] Isostearic acid substituted imidazoline; cationic; corrosion inhibitor, lubricant, fabric softener; liq.; 100% conc.

Miramine® OC. [Rhone-Poulenc Surf.] Oleyl hydroxyethyl imidazoline; cationic; detergent, thickener of HCl, corrosion inhibitor in metal treating and cutting fluids; emulsifier; amber liq. to pasty solid; ammoniacal odor; oil sol. and water disp.; sp.gr. 0.930; biodeg.; 99% min. act.

Miramine® SC. [Rhone-Poulenc Surf.] Soya imidazoline; cationic; emulsifier and coemulsifier in solv./water systems, corrosion inhibitor in cutting oils, lubricant for metal surfaces; amber liq.; ammoniacal odor; sol. in many org. solvs., disp. in water; sp.gr. 0.930; biodeg.; 99% act.

Miramine® TA-26. [Rhone-Poulenc Surf.] Substituted stearyl ammonium chloride; cationic; hair conditioner, fabric softener; liq.; 26% act.

Miramine® TC. [Rhone-Poulenc Surf.] Tallow fatty acid substituted imidazoline; cationic; see Miramine CC; solid; 100% conc.

Miramine® TOC. [Rhone-Poulenc Surf.] Tall oil hydroxyethyl imidazoline; cationic; see Miramine CC; solid; water-disp.; 100% conc.

Miranate® LEC. [Rhone-Poulenc Surf.] Sodium laureth-13 carboxylate; anionic; aux. detergent for shampoo systems, lime soap dispersant; clear, hazy gel, pH 8.0 (10%); 70% solids.

Miranol® C2M Anhyd. Acid. [Rhone-Poulenc Surf.] Cocoamphodipropionic acid; amphoteric; detergent, wetting agent, high foaming surfactant for disp. on caustic soda and on powd. mixes, etc., leveling agent in tin plating from acid baths; amber stiff gel; sol. in water and alcohol, slightly sol. in nonpolar org. solvs.; pH 4.6 (50%); 100% act.

Miranol® C2M Conc. [Rhone-Poulenc Surf.] Cocoamphodiacetate; amphoteric; detergent used in nonirritating shampoos, skin cleansers, make-up removers, emulsifier, solubilizer and stabilizer in pharmaceuticals, household and industrial uses; clear, visc. liq.; sol. in water; pH 8.0–8.5 (20% aq.); 38% act.

Miranol® C2M-SF Conc. [Rhone-Poulenc Surf.] Cocoamphodipropionate; amphoteric; see Miranol C2M-SF 70%; also coupling agent, solubilizer, wetting agent; clear, amber liq.; sol. in water and alcohol; pH 9.4–9.8; biodeg.; 39% act.

Miranol® CM Conc. [Rhone-Poulenc Surf.] Cocoamphoacetate; amphoteric; detergent, wetting and foaming agent, sequestrant, emulsifier, dispersant, germicidal, visc. builder; lt. amber visc. liq.; pH 9.0–13.0; biodeg.; 37% conc.

Miranol® CM-SF Conc. [Rhone-Poulenc Surf.] Cocoamphopropionate; amphoteric; see Miranol CM Conc.; lt. amber clear liq.; sol. in water and alcohol; pH 9.5–10.5; biodeg.; 36–38% solids.

Miranol® CS Conc. [Rhone-Poulenc Surf.] Cocoamphohydroxypropylsulfonate; amphoteric; detergent, wetting agent, corrosion inhibitor, emulsifier, sequestrant, foaming agent, for household and industrial uses; liq.; water-sol.; biodeg.; 38% conc.

Miranol® DM. [Rhone-Poulenc Surf.] Stearoamphoacetate; amphoteric; antistat, household softener, lubricant, dispersant used in textile industry, hair conditioners; wh. creamy paste, readily pourable @ 60 C; pH 5.4–6.0 (65 C); biodeg.; 25–27% solids.

Miranol® DM Conc. 45. [Rhone-Poulenc Surf.] Stearoamphoacetate; lubricant, softener, conditioner for cosmetics and textiles; paste; water-disp.

Miranol® Ester PO-LM4. [Rhone-Poulenc Surf.] Oligomeric polyester; nonionic; emulsifier, emollient, conditioner, antistat for personal care prods.; visc. liq.; 100% conc.

Miranol® H2M Conc. [Rhone-Poulenc Surf.] Lauroamphodiacetate; amphoteric; similar to Miranol® C2M Conc.; lt. amber visc. liq.; water sol.; pH 8.0–8.5 (20% aq.); biodeg.; 49–51% solids.

Miranol® H2M-SF 70%. [Rhone-Poulenc Surf.] Lauroamphodipropionate; used in place of Miranol C2M-SF Conc. where higher quality is required; for cosmetic formulations; paste at R.T., pours and pumps readily at 50 C; 70% act.

Miranol® H2M-SF Conc. [Rhone-Poulenc Surf.] Lauroamphodipropionate; see Miranol H2M-SF 70%; lt. amber clear liq.; sol. in water, alcohol; pH 9.2–9.8; biodeg.; 38–40% solids.

Miranol® HM Conc. [Rhone-Poulenc Surf.] Lauroamphoacetate; amphoteric; detergent, wetting and foaming agent, sequestrant, emulsifier, dispersant, germicidal; lt. amber, visc. liq.; pH 9.0–9.5; biodeg.; 43–45% solids.

Miranol® HS Conc. [Rhone-Poulenc Surf.] Lauroamphohydroxypropylsulfonate; amphoteric; general surfactant, sequestrant for heavy metal, detergent, wetting and foaming agent, emulsifier; liq.; 38% conc.

Miranol® ISM. [Rhone-Poulenc Surf.] Monocarboxylic isostearic deriv., sodium salt; amphoteric; lubricant, conditioner for cosmetics and industrial prods.; visc. and foam booster for shampoos; visc. liq.; 35% conc.

Miranol® J2M Anhyd. Acid. [Rhone-Poulenc Surf.] Anhydrous acid of Miranol J2M Conc.; amphoteric; wetting agent, penetrant used in bottle washing, wax stripping formulations; stiff amber gel; sol. in water, alcohol, slightly sol. in nonpolar org. solvs.; pH 4.6 (50%); 100% act.

Miranol® JEM Conc. [Rhone-Poulenc Surf.] Sodium salt of a dicarboxylated caprylic and ethylhexoic imidazoline deriv.; amphoteric; see Miranol J2M Anhyd. Acid; clear liq.; surf. tens. 30.5 dynes/cm (0.12% in water); pH 10.0–10.5; biodeg.; 33–35% solids.

Miranol® JS. [Rhone-Poulenc Surf.] Sulfonated caprylic deriv., sodium salt; caustic soda thickener; liq.; water-sol.

Miranol® JS Conc. [Rhone-Poulenc Surf.] Caprylo-amphopropylsulfonate; amphoteric; wetting agent, corrosion inhibitor used in pickling acids; liq.; water-sol.; biodeg.; 40% conc.

Miranol® LM-SF Conc. [Rhone-Poulenc Surf.] Tallamphopropionate; similar to Miranol L2M-SF Conc. but more hydrophobic, suggested as lubricant for ferrous and nonferrous metals; visc. liq. which sets to a pasty gel on aging at R.T., pourable @ 60 C; pH 9.5–10.5; biodeg.; 36–38% solids.

Miranol® OM-SF Conc. [Rhone-Poulenc Surf.] Oleoamphopropionate; amphoteric; wetting and foaming agent, sequestrant, emulsifier, dispersant, germicidal; sol. in water, alcohol; pH 9.8–10.2; biodeg.; 36–38% solids.

Miranol® OS. [Rhone-Poulenc Surf.] Oleoamphohydroxypropylsulfonate; amphoteric; general surfactant, sequestrant for heavy metal, detergent, wetting and foaming agent, emulsifier; paste; 21% conc.

Miranol® S2M Conc. [Rhone-Poulenc Surf.] Caproamphodiacetate; amphoteric; wetting and foaming agent, sequestrant, emulsifier, dispersant, germicidal; clear liq.; pH 8.1–8.3; biodeg.; 50–52% solids.

Miranol® S2M-SF Conc. [Rhone-Poulenc Surf.] Caproamphodipropionate; see Miranol S2M Conc.; liq.; pH 8.8–9.2; biodeg.; 38–40% solids.

Mirapol® 9, 95, 175. [Rhone-Poulenc Surf.] Polyquaternium-27; conditioner, antistat, emollient for hair and skin prods.; improves thickening of Bentonite clays in aq. systems; m.w. 20,000; pH 8.0 (10%); 62% act. in water.

Mirapol® A-15. [Rhone-Poulenc Surf.] Polyquaternium-2; cationic; softening, conditioning, lubricant, antistat, surface modifying agent used in cream rinses, conditioning-type shampoos, textile processing; amber visc. liq.; m.w. 2260; dissolves readily in water; pH 8.5; 64% act. in water.

Mirapol® AD-1. [Rhone-Poulenc Surf.] Polyquaternium-17; cationic; conditioner and antistat for personal care prods.; amber visc. liq.; m.w. 50,000; water-sol.; pH 8.0 (10%); 62% act. in water.

Mirapol® AZ-1. [Rhone-Poulenc Surf.] Polyquaternium-18; cationic; conditioner and antistat for personal care prods.; clear to hazy straw-colored liq.; m.w. 50,000; water-sol.; pH 8.0 (10%); 62% act. in water.

Mirapol® WT. [Rhone-Poulenc Surf.] Polyquat. ammonium chloride; cationic; antiscaling compd. in water systems; textile softener; corrosion inhibitor; liq.; water-sol.; 63% conc.

Mirapon® FM. [Rhone-Poulenc Surf.] Alkylamino acid, potassium salt; amphoteric; detergent, emulsifier, wetting agent and lubricant; liq.; 70% conc.

Mirataine® BB. [Rhone-Poulenc Surf.] Lauramidopropyl betaine; amphoteric; visc. builder and foam booster for shampoos and dishwashing liqs.; liq. 30% conc.

Mirataine® BD. [Rhone-Poulenc Surf.] Cocamidopropyl betaine; amphoteric; detergent and visc. builder for shampoos, bubble baths; liq.; 30% conc.

Mirataine® CBC, CBR. [Rhone-Poulenc Surf.] Cocamidopropyl betaine; amphoteric; visc. builder and foam booster for shampoos and dishwashing liqs.; liq.; 30% conc.

Mirataine® CBS. [Rhone-Poulenc Surf.] Cocamidopropyl hydroxysultaine; amphoteric; surfactant, visc. builder, foam booster for shampoo formulations; liq.; water-sol.; dens. 9.1 lb/gal; cloud pt. ≤ –12 C; pH 8.2; 50% act. in water.

Mirataine® COB. [Rhone-Poulenc Surf.] Coco oleamidopropyl betaine; amphoteric; surfactant and conditioner for use in personal care prods.; oil solubilizer, visc. builder, foam booster, stabilizer; yel. visc. liq.; water-sol.; pH 7.0; 34.0% solids.

Mirataine® H2C. [Rhone-Poulenc Surf.] Disodium lauriminodipropionate; amphoteric; detergent, lubricant, antistat, corrosion inhibitor, and solubilizer used in detergent and industrial formulations, personal care prods., leather, and textile fibers; yel. thin liq.; pH 10.5; flash pt. (PM) 132 C; biodeg.; 30.0% act.

Mirataine® HTS. [Rhone-Poulenc Surf.] Hydrog. tallow hydroxy sultaine; amphoteric; lime soap dispersant; conditioner for hair and skin preparations; gel; 35% conc.

Mirataine® ODMB-35%. [Rhone-Poulenc Surf.] Oleyl betaine; amphoteric; visc. builder and conditioner for personal care prods.; yel. liq.; dens. 8.4 lb/gal; pH 7.0 (10% aq.); 30% conc.

Mirataine® T2C. [Rhone-Poulenc Surf.] Disodium tallowiminodipropionate; amphoteric; detergent, solubilizer, moderate foaming surfactant used in textile, leather, industrial and personal care prods.; liq.; sol. clear (30% solids in 1% NaOH); surf. tens. 31.6 dynes/cm (1% aq.); pH 11.5; biodeg.; 35% act. in water.

Mirataine® TABS. [Rhone-Poulenc Surf.] Tallowamidopropyl hydroxysultaine; amphoteric; lime soap dispersant in industrial and household cleaners; clear to slightly hazy gel; pH 8.3; biodeg.; 41.0% solids.

Mirataine® TM. [Rhone-Poulenc Surf.] Dihydroxyethyl tallow glycinate; amphoteric; conditioner for shampoos; HCl thickener; lt. amber slightly hazy visc. liq.; pH 5.1; 35% conc.

Mixxim® BB/100. [Fairmount] Bis [2-hydroxy-5-t-octyl-3-(benzotriazol-2-yl)phenyl] methane; high m.w. uv lt. absorber for engineering resins, e.g., nylon, PC, PET, PBT, PPO, and other polymers, e.g., polyolefin fibers, PVC, acrylics; wh. flowable powd.; m.w. 658; sol. (g/100 ml): 25 g in xylene, 3 g in ethyl acetate, < 1 g in ethanol, n-hexane, acetone; m.p. 196 C.

Mixxim® HALS 57. [Fairmount] Tetrakis (2,2,6,6-

tetramethyl-4-piperidyl)-1,2,3,4-butane tetracarboxylate; hindered amine lt. and heat stabilizer for polyolefins such as PP, polyethylene, PS, ABS, PVC, PU, and engineering plastics; wh. powd.; sol. (g/100 ml) > 10 g in ethanol, chloroform, toluene, THF; 0.1 g in water; m.p. 130–140 C.

Mixxim® HALS 62. [Fairmount] Hindered amine lt. and heat stabilizer for polymers (polyolefins, styrene, engineering plastics, PVC), paints, and latexes; amber liq.; m.w. ≈ 900; sol. (g/100 g) > 10 g in ethanol, benzene, ethyl acetate, THF, n-hexane; < 0.1 g in water; sp.gr. 0.98; visc. 9500 cp; ref. index 1.476; 100% act.

Mixxim® HALS 63. [Fairmount] [1,2,2,6,6-Pentamethyl-4-piperidyl/β,β,β′,β′-tetramethyl-3,9-(2,4,8,10- tetraoxaspiro (5,5) undecane) diethyl]-1,2,3,4-butane tetracarboxylate; high m.w. hindered amine lt. and heat stabilizer for PP, polyethylene, PS, ABS, engineering plastics, and elastomers, esp. in PP monofilament, tapes, molded and extruded prods., polyethylene blown film; lt. yel. powd.; m.w. ≈ 2000; sol. (g/100 ml) > 10 g in benzene, ethyl acetate, THF, n-hexane; < 0.1 g in methanol and water; m.p. 80–90 C.

Mixxim® HALS 67. [Fairmount] See Mixxim HALS 62; amber liq.; m.w. ≈ 900; sol. (g/100 g) > 10 g in methanol, ethanol, benzene, ethyl acetate, THF, n-hexane; < 0.1 g in water; sp.gr. 0.98; visc. 5000 cp; ref. index 1.471; 100% act.

Mixxim® HALS 68. [Fairmount] [2,2,6,6-Tetramethyl-4-piperidyl/β,β,β′,β′ -tetramethyl-3,9-(2,4,8,10- tetraoxaspiro (5,5) undecane) diethyl]-1,2,3,4-butane tetracarboxylate; see Mixxim HALS 63; lt. yel. powd.; m.w. ≈ 1900; sol. (g/100 ml) > 10 g in methanol, benzene, ethyl acetate, THF; < 0.1 g in water; m.p. 70–80 C.

Mo-1907 T 3/16″. [M&T Harshaw] Iron molybdena; catalyst used for the prod. of formaldehyde; tablet, 3/16″ diam.; dens. 80 lb/ft^3; 55% conc.

M.O.D. [Gattefosse Ets.] Octyldodecyl myristate; emollient; rancidless additive for cosmetics and pharmaceuticals; oily liq.

Modaflow®. [Monsanto] Resin modifier for nonaq. coatings and adhesives; reduces surf. imperfections for powd. coatings; used for food container linings, gel coats, silk screen inks, general inks, coil coatings, primers.

Modaflow® Powd. II. [Monsanto] Resin modifier for nonaq. coatings; reduces surf. imperfections and flow and leveling problems in industrial powd. coatings based on epoxy, polyester, acrylic resin systems.

Modicol L. [Henkel] PEG fatty ester; nonionic; coagulant, wetting and dispersing agent, used in latex and resin emulsions; clear, yel. liq.; sol. in ethanol, acetone, ethyl acetate, toluol, veg. oil, water, min. oil; dens. 8.6 lb/gal; acid no. 1.0; pH 7.5 (2%); 99% act.

Modicol N. [Henkel] Fatty amido condensate; nonionic; stabilizer for natural and syn. latex; liq.; 100% act.

Modicol S. [Henkel] Sulfonated fatty prod.; anionic; stabilizer; in rubber industry to stabilize natural and syn. latexes during compding., storage, and applic; ensures mechanical and chemical stability; prevents premature coagulation during high-speed agitation, acidification, or pigmentation; dk. rdsh.-brn. liq.; fatty odor; sol. in water and sol. alcohols; hazy sol'n. in 4% sodium hydroxide, in toluol, and in min. and veg. oils; sp.gr. 1.1; pH 7.5 (2%), 60% act.

Modicol VD. [Henkel] Aq. sol'n. of a modified sodium polyacrylate; thickener, stabilizer, and protective colloid for natural and syn. rubber latexes, and PVCa and acrylate emulsions; inert to hydrolysis and bacterial action; offers ease of handling; clear visc. pale straw liq.; dilutability in water; dens. 9.0 lb/gal; visc. 75,000 cps; pH 10.5 (2%); 15% solids.

Modulan®. [Amerchol] Acetylated lanolin; conditioner, emollient, softener, lubricant for cosmetic and pharmaceutical prods.; yel.-amber soft solid; faint, pleasant odor; sol. in min. oil; m.p. 30–40 C; acid no. 2.5 max.; sapon. no. 95–120; hyd. no. 10 max.

Moellon R. [Atlas Refinery] Syn. anhyd. moellon from pure refined herring oil; lubricant and softener for pastel-colored leathers; nonyel.; lt. brn. liq.; dens. 8.31 lb/gal; visc. 165 SUS (210 F); iodine no. 72; 10% free fatty acids; 14% oxidized fatty acids; 100% act.

Moldbrite 25. [Disco] Extender/modifying agent for silicone emulsions used as release agents for all rubber compds.

Moldbrite 30. [Disco] Conc. silicone mold release for rubber compds.

Mold-Ease. [Merix] Nonsilicone mold lubricant for plastics, ceramic, and rubber articles.

Mollisan 4NY, 379 Conc. [Lyndal] Fatty ester; nonionic; softener, o/w emulsifier; paste; 100% conc.

Mollisan Conc. [Lyndal] Fatty condensate; nonionic; lubricant, softener, wetting agent; o/w emulsifier; paste; 100% conc.

Molyhibit 100 Series. [Climax Performance] Sodium molybdates; corrosion inhibitor for general ferrous and aluminum corrosion control; solids and 35% aq. sol'n.

Molyhibit 200 Series. [Climax Peformance] Potassium molybdates; corrosion inhibitor.

Molyhibit 300 Series. [Climax Performance] Lithium molybdate; corrosion inhibitor for specialized applics., e.g., inhibiting lithium brines; solid and 40% aq. sol'n.

Molykote®. [Dow Corning] Silicone, MoS_2 lubricant base; environment lubricants.

Molysulfide Suspension, Tech., Tech. Fine. [Amax] Molybdenum disulfide; lubricant for automobile, mining, aerospace, railroad, metalworking, and petrol. industries; blue-gray powd.; 2, 20, and 5 μm mean particle diam. resp.; m.w. 160.06; dens. 4.8–5.0 g/cc; m.p. dissociates at 1600 C; hardness (Mohs) 1; conduct. 0.13 W/m•K (40 C).

Molyvan A. [Vanderbilt] Molybdenum oxysulfide dithiocarbamate; antiwear and EP agent, antioxidant; used in long-life chassis greases for ball joints, steering linkages, and lubricating greases at high temps.; additive for petrol. and syn. lubricants; used in non-petrol. base valve lubricants; yel. powd.; insol. in water; dens. 1.58 mg/m^3; m.p. 251 C.

Molyvan L. [Vanderbilt] Sulfurized oxymolybdenum organophosphorodithioate; antiwear agent, friction reducer, antioxidant, EP agent; used in hypoid and engine oils, metalworking compositions, industrial and automotive lubricating oils, greases, and specialties; additive for break-in oils, forms residual lubricating films on metal parts; reduces friction and operating temps; formulation of single-phase oils containing molybdenum; dk. grn. liq.; sol. in petrol. and syn. lubricant bases; insol. in water; dens. 1.08 mg/m^3; visc. 9.0 cSt (100 C); pour pt. –37 C; flash pt. 166 C (COC).

Moly-White® 101. [PMC] Zinc molybdenum; corrosion inhibiting pigment for coatings; wh.

Moly-White® 212. [PMC] Calcium zinc molybdenum; corrosion inhibiting pigment for coatings; wh.

Moly-White® 331, ZNP. [PMC] Basic zinc molybdate; corrosion inhibiting pigment for coatings; wh.

Mona AT-1200. [Mona] Amphoteric; surfactant, thickener, corrosion inhibitor for acids; used in acid bowl cleaners, metal cleners, pickling baths, petrol. acidizing sol'ns., etc.; biodeg.; Gardner 9 clear visc. liq.; pH 5.5; 40% act.

Monacor 39. [Mona] Amido-ester carboxylic acid deriv.; corrosion inhibitor for slushing oils, rolling oils, lubricating oils; amber visc. liq.; sol. in min. oil, IPA, butyl Carbitol, benzene, perchloroethylene; sp.gr. 0.97; dens. 8.1 lb/gal; acid no. 70; 100% act.

Monacor BE. [Mona] MEA-borate, MIPA-borate; corrosion inhibitor for ferrous metals; used in metalworking lubricants, cooling towers; clear visc. liq.; water-sol.; sp.gr. 1.107; dens. 9.24 lb/gal; pH 9.1 (1%); 100% act.

Monafax 057. [Mona] Aromatic-based phosphate ester; low foaming EP lubricant in metalworking fluids and high performance syn. coolants; corrosion inhibitor; low coeff. of friction, good antiwear and antiweld properties; provides hydrotropic properties for conc. electrolyte systems and low foam for detergent compd.; used in cutting fluids, chain lubricants, hydraulic fluids, rust preventives, metal and maintenance cleaners; lt. yel. clear visc. liq.; sol. (@ 10%) in water, ethanol, butyl Carbitol, aromatic hydrocarbon, chlorinated paraffin, and naphthenic oil (100 SSU); sp.gr. 1.2; acid no. 155; pH < 2.5 (10% aq.); 100% act.

Monafax 060. [Mona] Phosphate ester, aliphatic-based; hydrotrope and solubilizer for nonionics in alkaline systems; used in soak tank, alkaline, and steam cleaners; amber clear visc. liq.; sol. @ 10% in water, ethanol, ethylene glycol monobutyl ether; sp.gr. 1.335; dens. 11.3 lb/gal; pH 2.5 (10%); 100% act.

Monafax 785. [Mona] Mixt. of mono and di-phosphate esters, aromatic-based; anionic; emulsifier, lubricant, antistat, detergent, corrosion inhibitor for agric., industrial use; antisoil redeposition for dry cleaning; clear to slightly hazy visc. liq.; sol. in water, ethanol, perchlorethylene, xylene; dens. 9.2 lb/gal; sp.gr. 1.115; acid no. 70 ± 3; pH < 2.5 (10%); 100% act.

Monafax 786. [Mona] Nonoxynol-6 phosphate; anionic; see Monafax 785; clear to slightly hazy visc. liq.; sol. @ 10% in ethanol, perchloroethylene, Stod., xylene; disp. water; dens. 9.0 lb/gal; sp.gr. 1.085; acid no. 57 ± 3; pH < 2.5 (10%); 100% act.

Monafax 794. [Mona] Mixt. of mono and di-phosphate esters; anionic; emulsifier for polymerization, corrosion inhibitor; soft waxy paste; sol. @ 10% in water, ethanol, perchloroethylene, xylene; dens. 9.8 lb/gal; sp.gr. 1.175; acid no. 98 ± 5; pH < 2.5 (10%); 100% act.

Monafax 831. [Mona] Mixt. of mono and di-phosphate esters; anionic; emulsifier, lubricant, antistat, detergent, corrosion inhibitor, coupler for cleaning and industrial use; clear visc. liq.; sol. @ 10% in water, ethanol, perchloroethylene, min. oil, Stod., xylene; dens. 9.0 lb/gal; sp.gr. 1.08; acid no. 110 ± 3; pH < 2.5 (10%); 100% act.

Monafax 872. [Mona] Org. phosphate ester, potassium salt; anionic; hydrotrope and solubilizer for detergent systems; lt. straw clear liq.; sol. @ 10% in water, disp. in ethanol, ethylene glycol monobutyl ether, cottonseed oil; dens. 11.24 lb/gal; sp.gr. 1.35; acid no. 20 ± 2; pH 7–9 (10%); 50% act.

Monafax 939. [Mona] Alcohol phosphate ester in acid form; emulsifier, lubricant used in metalworking lubricants and syn. cutting fluids, low coeff. of friction, rust inhibition, reduced surf. tension, EP lubricity, antiwear, and anitweld properties; used in water and oil-based formulations; amber clear liq.; m.w. 297; sol. at 1.0% and 10% in ethanol, chlorinated and aromatic hydrocarbons, min. spirits, kerosene, and min. oil; insol. in water; dens. 8.6 lb/gal; acid no. 220 ± 3; pH 2.5 max.; 100% act.

Monafax 1293. [Mona] Org. phosphate ester; hydrotrope, coupling agent for nonionics and other detergents which are only slightly sol. in high electrolyte concs.; surfactant, wetting agent, antistat, dispersant; biodeg.; for household and industrial hard surf. detergents, metal cleaners, agric., textile, paper/pulp, drycleaning, emulsion polymerization; Gardner 3 clear liq.; sol. in polar and nonpolar solvs.; sp.gr. 1.07; dens. 8.9 lb/gal; acid no. 111 (@ pH 4.5), 185 (@ pH 9.4); pH < 2 (10%); 98% act.

Monafax H-15. [Mona] Mixt. of mono and di-phosphate esters; anionic; emulsifier for herbicides and insecticides, lubricant, antistat, detergent, corrosion inhibitor; hazy visc. liq.; sol. in water, ethanol, perchloroethylene, xylene; dens. 8.9 lb/gal; sp.gr. 1.07; acid no. 40 ± 3; pH < 2.5 (10%); 100% act.

Monafax L-10. [Mona] Mixt. of mono and di-phosphate esters; anionic; see Monafax H-15; slightly hazy visc. liq.; sol. @ 10% in ethanol, perchloroethylene, kerosene, Stod., xylene; disp. water and min. oil; dens. 8.8 lb/gal; sp.gr. 1.067; acid no. 50 ± 5; pH < 2.5 (10%); 100% act.

Monalube 29-78. [Mona] Alkanolamide; corrosion inhibitor, lubricant for metalworking lubricants, fiber lubricants, textile specialties, wire and deep metal drawing; dark amber liq.; sol. in water, aromatic hydrocarbons, chlorinated solvs.; acid no. 22–32.

Monalube 780. [Mona] Alkanolamide; nonionic/anionic; lubricant for copper drawing; corrosion inhibitor; amber liq.; disp. in water; dens. 8.2 lb/gal;

acid no. 7.5; 100% act.

Monamid® 7-100. [Mona] 1:1 Cocamide DEA; nonionic; foam booster/stabilizer, emulsifier, detergent, wetting agent, corrosion inhibitor, visc. builder, lubricant, dispersant, used in textiles, cosmetics, metalworking, agric., leather and fur applics.; Gvcs-33 6 max. liq.; sol. @ 10% in ethanol, chlorinated and aromatic hydrocarbons, min. spirits, kerosene, wh. min. oil, natural fats and oils, water disp.; dens. 8.00 lb/gal; sp.gr. 0.96; acid no. 0–2; pH 8–9 (10%); biodeg.; 100% act.

Monamid® 7-153 CS. [Mona] Modified cocamide DEA (1:1); nonionic/anionic; see Monamid 7-100; Gvcs-33 3 max. liq.; sol. @ 10% in ethanol, chlorinated and aromatic hydrocarbons, min. spirits, kerosene, wh. min. oil, disp. in water and natural oils and fats; dens. 8.75 lb/gal; sp.gr. 1.05; acid no. 0–2; pH 8.5–9.5 (10%); biodeg.; 100% act.

Monamid® 15-70W. [Mona] 1:1 Linoleamide DEA; nonionic, see Monamid 7-100; also hair and fiber conditioner; Gvcs-33 11 max. liq.; sol. @ 10% in ethanol, chlorinated and aromatic hydrocarbons, min. spirits, kerosene, natural oils and fats, gels in water; dens. 8.00 lb/gal; sp.gr. 0.96; acid no. 0–1; pH 10–11 (10%); biodeg.; 100% act.

Monamid® 15-MW. [Mona] Alkanolamide; thickener, visc. builder in shampoos, bubble baths, other aq. systems; sol. in aromatic hydrocarbons, chlorinated solvs.; sol to disp. in water.

Monamid® 150-AD, 150-ADD. [Mona] 1:1 Cocamide DEA; nonionic; see Monamid 7-100; Gvcs-33 11 and 4 max. resp., liqs.; sol. @ 10% in water, ethanol, chlorinated and aromatic hydrocarbons, natural oils and fats; dens. 8.25 and 8.30 lb/gal; sp.gr. 0.99 and 1.00; acid no. 0–3; pH 9.8–10.8 and 10–11 (10%); biodeg.; 100% act.

Monamid® 150-ADY. [Mona] 1:1 Linoleamide DEA; nonionic; see Monamid 7-100; Gvcs-33 10 max. liq.; sol. @ 10% in ethanol, chlorinated and aromatic hydrocarbons, min. spirits, kerosene, gels in water; dens. 8.20 lb/gal; sp.gr. 0.97; acid no. 0–1; pH 10–11 (10%); biodeg.; 100% act.

Monamid® 150-CW. [Mona] 1:1 Capramide DEA; nonionic; see Monamid 7-100; Gvcs-33 7 max. liq., crystallizes on aging; sol. @ 10% in water, ethanol, chlorinated and aromatic hydrocarbons, natural oils and fats; dens. 8.25 lb/gal; sp.gr. 0.99; acid no. 2 max.; pH 10.3–11.3 (10%); biodeg.; 100% act.

Monamid® 150-DR. [Mona] 1:1 Cocamide DEA; nonionic; see Monamid 7-100; Gvcs-33 5 max. liq.; sol. @ 10% in ethanol, min. spirits, kerosene, wh. min. oil, natural oils and fats, disp. in water, chlorinated and aromatic hydrocarbons; dens. 8.20 lb/gal; sp.gr. 0.98; acid no. 0–5; pH 8.5–9.5 (10%); biodeg.; 100% act.

Monamid® 150-GLT. [Mona] Modified lauramide DEA; nonionic; thickener and foam stabilizer for cosmetic and toiletry prods.; amber liq.; sol. @ 10% in ethanol, chlorinated and aromatic hydrocarbons, min. spirits, kerosene; gels in water; sp.gr. 0.96; dens. 8.00 lb/gal; acid no. 0–1; pH 10.3–11.3 (10%); 100% act.

Monamid® 150-IS. [Mona] 1:1 Isostearamide DEA; nonionic; see Monamid 7-100; Gvcs-33 6 max. liq.; sol. 10% in ethanol, chlorinated and aromatic hydrocarbons, min. spirits, kerosene, wh. min. oil, natural oils and fats, water disp.; dens. 8.00 lb/gal; sp.gr. 0.96; acid no. 5–10; pH 8.8–9.8 (10%); biodeg.; 100% act.

Monamid® 150-LMW-C. [Mona] Lauramide DEA; nonionic; see Monamid 7-100; Gvcs-33 3 max. solid; sol. @ 10% in ethanol, chlorinated and aromatic hydrocarbons, min. spirits, gels in water; dens. 8.20 lb/gal; sp.gr. 0.98 (40 C); acid no. 0–1; pH 10.2–11.2 (10%); biodeg.; 100% act.

Monamid® 150-LW, 150-LWA. [Mona] 1:1 Lauric DEA; nonionic; see Monamid 7-100; Gvcs-33 3 max. solid; sol. @ 10% in ethanol, chlorinated and aromatic hydrocarbons, disp. in water; dens. 8.20 lb/gal; sp.gr. 0.98 (40 C); acid no. 0–1; pH 10–11 and 9.5–10.5 (10%); biodeg.; 100% act.

Monamid® 150-MW. [Mona] 1:1 Myristamide DEA; nonionic; see Monamid 7-100; Gvcs-33 2 max. solid; sol. @ 10% in ethanol, chlorinated and aromatic hydrocarbons, disp. in water; dens. 8.20 lb/gal; sp.gr. 0.98 (45 C); acid no. 0–3; pH 9.5–10.5 (10%); biodeg.; 100% act.

Monamid® 664. [Mona] Coconut alkanolamide; nonionic; foam booster, stabilizer and thickener for industrial and household detergents; liq.; 100% conc.

Monamid® 716. [Mona] Lauramide DEA; nonionic; foam booster and stabilizer, visc. builder, solubilizer, coupler, wetting agent, used in shampoos, bubble baths, and liq. detergent formulations; lt. amber clear liq.; sol. @ 10% in water, ethanol, chlorinated and aromatic hydrocarbons, natural oils and fats; sp.gr. 0.98; dens. 8.2 lb/gal; cloud pt. < 1 C; acid no. 0–3; pH 10.0–11.0 (10%); biodeg.; 100% act.

Monamid® 718. [Mona] Stearamide DEA; nonionic; see Monamid 7-100; Gvcs-33 4 max. solid; sol. @ 10% in ethanol, aromatic hydrocarbons, gels in water; dens. 8.00 lb/gal; sp.gr. 0.96 (45 C); acid no. 21 ± 3; pH 9.3–10.3 (10%); biodeg.; 100% act.

Monamid® 759. [Mona] Cocamide DEA; foam booster and stabilizer, visc. builder, and conditioner for personal care prods.; amber liq.; alkali no. 35 ± 5; acid no. 1 max.; pH 10.5 ± 0.5 (10%); cloud pt. 0 C; clear pt. 33 C; 100% act.

Monamid® 770. [Mona] Lauramide DEA; nonionic; see Monamid 7-100; Gvcs-33 4 max. liq.; sol. @ 10% in water, ethanol, chlorinated and aromatic hydrocarbons, natural fats and oils; dens. 8.25 lb/gal; sp.gr. 0.99; acid no. 0–1; pH 9.2–10.2 (10%); biodeg.; 85% act.

Monamid® 853. [Mona] Alkanolamide, modified; nonionic/anionic; detergent, wetting emulsifier, foam stabilizer used in shampoos, liq. household detergents, dairy cleaners, wool scouring, waterless hand cleaners; liq.; 100% conc.

Monamid® 982, 1007. [Mona] 1:1 Lauramide DEA and linoleamide DEA; foam booster and stabilizer, visc. builder, used in shampoos, bubble baths, skin cleansers, household and industrial liq. detergents; clear, amber liq.; dens. 8.16 lb/gal; 100% act.

Monamid® 1034. [Mona] Lauramide DEA; nonionic; foam booster and visc. builder for shampoos; liq.; 100% conc.

Monamid® ADY. [Mona] Alkanolamide; thickener, visc. builder in shampoos, bubble baths, aq. systems; sol. in aromatic hydrocarbons, chlorinated solvs.; sol. to disp. in water.

Monamid® CMA. [Mona] 1:1 Cocamide MEA; nonionic; see Monamid 7-100; tan granular; sol. @ 10% in ethanol, disp. in water, chlorinated and aromatic hydrocarbons, min. spirits, kerosene, wh. min. oil, natural oils and fats; acid no. 0–1; pH 9.4–10.8 (10%); biodeg.; 100% act.

Monamid® LIPA. [Mona] 1:1 Lauramide MIPA; nonionic; see Monamid 7-100; tan granular; sol. @ 10% in ethanol, disp. in water, kerosene, wh. min. oil, natural fats and oils; acid no. 0–1; pH 10.3–11.3 (10%); biodeg.; 100% act.

Monamid® LMA. [Mona] 1:1 Lauramide MEA; nonionic; see Monamid 7-100; tan granular; sol. @10 in ethanol, disp. in water, aromatic hydrocarbons, kerosene, wh. min. oil, natural fats and oils; acid no. 0–1; pH 10–11 (10%); biodeg.; 100% act.

Monamid® LMIPA. [Mona] Lauramide MIPA; nonionic; surfactant, thickener used in personal care prods. and detergent systems; off-wh. solid; m.p. 61 ± 2 C; acid no. 1 max.; alkali no. 10–25; pH 10.5 ± 0.5 (10% aq.); 100% act.

Monamid® LMMA. [Mona] 1:1 Lauramide MEA; nonionic; see Monamid 7-100; tan granular; sol. @ 10% in ethanol, disp. in water, aromatic hydrocarbons, kerosene, wh. min. oil, natural oils and fats; acid no. 0–1; pH 9.7–10.7 (10%); biodeg.; 100% act.

Monamid® LMW-C. [Mona] Alkanolamide; thickener, visc. builder in shampoos, bubble baths, aq. systems; sol. in aromatic hydrocarbons, chlorinated solvs.; sol. to disp. in water.

Monamid® R31-42. [Mona] Lauramide DEA and propylene glycol; nonionic; see Monamid 7-100; 2 max (Gvcs-33) liq.; sol. @ 10% in water, ethanol, chlorinated and aromatic hydrocarbons, disp. in natural oils and fats; dens. 8.25 lb/gal; sp.gr. 0.99; acid no. 0–1; pH 10–11 (10%); biodeg.; 80% act.

Monamid® S. [Mona] 1:1 Stearamide MEA; nonionic; see Monamid 7-100; tan granular; disp. @ 10% in water, chlorinated and aromatic hydrocarbons, min. spirits, kerosene, wh. min. oil, natural oils and fats; acid no. 0–1; pH 9.5–11.0 (10%); biodeg.; 100% act.

Monamine Series. [Mona] Fatty acid alkanolamide; foam booster and stabilizer, emulsifier, detergent, wetting agent, corrosion inhibitor, visc. builder, lubricant, dispersant for cosmetics, household and industrial cleaners, textiles, agric. sprays, leather and fur preparations.

Monamine 779. [Mona] DEA-laureth sulfate and cocamide DEA; nonionic/anionic; foaming agent, visc. builder, detergent, soil suspending agent, solubilizer, wetting and penetrating agent used in shampoos, bubble baths, household and industrial cleaners, germicides, uv absorbers; lt. yel. clear visc. liq.; sol. @ 10% in water, polar solvs., alcohol, glycols, fatty acid esters, chlorinated and aromatic hydrocarbons, dens. 8.75 lb/gal; sp.gr. 1.05 ± 0.1; cloud pt. 2 C; surf. tens. 29 dynes/cm (0.1%); pH 9.25 ± 0.5 (10%); biodeg.; 100% act.

Monamine 928. [Mona] Tallowamide DEA and diethanolamine; nonionic/anionic; emulsifier, lubricant, corrosion inhibitor used in syn. coolants and metalworking fluids; cream paste; dens. 8.0 lb/gal (50 C); acid no. 5 max.; 100% act.

Monamine 929. [Mona] Modified long-chain alkanolamide; nonionic/anionic; emulsifier, lubricant, corrosion inhibitor used in syn. coolants and metalworking fluids; cream paste; dens. 8.3 lb/gal; 100% act.

Monamine AA-100. [Mona] 1:2 Dist. cocamide DEA and diethanolamine; nonionic/anionic; see Monamine Series; Gvcs-33 7 max. liq.; sol. @ 10% in water, ethanol, chlorinated and aromatic hydrocarbons; dens. 8.30 lb/gal; sp.gr. 1.00; acid no. 28–32; pH 9.5–10.5 (10%); biodeg.; 100% act.

Monamine AC-100. [Mona] Cocamide DEA and linoleamide DEA; nonionic/anionic; see Monamine Series; Gvcs-33 13 max. liq.; sol. @ 10% in ethanol, chlorinated and aromatic hydrocarbons, gels in water; dens. 8.30 lb/gal; sp.gr. 1.00; acid no. 22–32; pH 9.5–10.5 (10%); biodeg.; 100% act.

Monamine ACO-100. [Mona] Lauramide DEA and diethanolamine; nonionic/anionic; see Monamine Series; Gvcs-33 5 max. paste; sol. @ 10% in water, ethanol, chlorinated and aromatic hydrocarbons; dens. 8.40 lb/gal; sp.gr. 1.01; acid no. 10–14; pH 9.5–10.5 (10%); biodeg.; 100% act.

Monamine AD-100, ADD-100. [Mona] Cocamide DEA and diethanolamine; nonionic; see Monamine Series; Gvcs-33 11 and 3 max. resp., liqs.; sol. @ 10% in water, ethanol, chlorinated hydrocarbons, disp. in aromatic hydrocarbons; dens. 8.50 lb/gal; sp.gr. 1.02; acid no. 2–8 and 2–8; pH 9.5–10.5 (10%); biodeg.; 100% act.

Monamine ADDS-100. [Mona] 1:2 Mixed fatty acid DEA; nonionic/anionic; see Monamine Series; Gvcs-33 7 max. liq.; sol. @ 10% in water, ethanol, chlorinated and aromatic hydrocarbons; dens. 8.30 lb/gal; sp.gr. 1.00; pH 9.0–10.0 (10%); biodeg.; 100% act.

Monamine ADS-100. [Mona] Cocamide DEA; nonionic/anionic; see Monamine Series; Gvcs-33 11 max. liq.; sol. @ 10% in water, ethanol, chlorinated and aromatic hydrocarbons; dens. 8.30 lb/gal; sp.gr. 1.00; acid no. 48–52; pH 9.0–10.0 (10%); biodeg.; 100% act.

Monamine ADY-100. [Mona] Linoleamide DEA and diethanolamine; nonionic/anionic; see Monamine Series; Gvcs-33 9 max. liq.; sol. @ 10% in ethanol, min. spirits, kerosene; disp. in water, chlorinated and aromatic hydrocarbons; dens. 8.20 lb/gal; sp.gr. 0.98; acid no. 0–2; pH 10.5–11.5 (10%); biodeg.; 100% act.

Monamine AF-100. [Mona] 1:2 Mixed fatty acid DEA; nonionic/anionic; see Monamine Series; Gvcs-33 12 max. liq.; sol. @ 10% in water, ethanol, chlorinated and aromatic hydrocarbons, min. spirits;

dens. 8.25 lb/gal; sp.gr. 0.99; pH 9–10 (10%); biodeg.; 100% act.

Monamine ALX-80 SS, ALX-100 S. [Mona] Cocamide DEA and DEA-dodecylbenzene sulfonate; nonionic/anionic; see Monamine Series; Gvcs-33 6 and 8 max. resp. liq.; sol. @ 10% in water, ethanol; dens. 8.75 and 8.85; sp.gr. 1.05 and 1.06; acid no. 52–60 and 62–70; pH 8.5–9.5 (10%); biodeg.; 80 and 100% act. resp.

Monamine ARA-100. [Mona] 1:2 Cocamide DEA; nonionic/anionic; see Monamine Series; Gvcs-33 7 max. liq.; sol. @ 10% in water, ethanol, chlorinated and aromatic hydrocarbons; dens. 8.30 lb/gal; sp.gr. 1.00; pH 9.5–10.5 (10%); biodeg.; 100% act.

Monamine CD-100. [Mona] Linoleamide DEA and diethanolamine; lubricant and emulsifier for syn. cutting and drawing fluids; metal corrosion inhibitor; dk. br. visc. liq.; water-disp.; acid no. 10–15; pH 9.5–10.5 (10% disp.); 100% act.

Monamine CF-100 M. [Mona] Cocamide DEA and diethanolamine; nonionic/anionic; see Monamine Series; Gvcs-33 11 max. liq.; sol. @ 10% in water, ethanol, chlorinated and aromatic hydrocarbons; dens. 8.40 lb/gal; sp.gr. 1.01; acid no. 56–64; pH 8.5–9.5 (10%); biodeg.; 100% act.

Monamine I-76. [Mona] 1:2 Cocamide DEA; nonionic/anionic; see Monamine Series; Gvcs-33 12 max. liq.; sol. @ 10% in water, ethanol, chlorinated and aromatic hydrocarbons; dens. 8.30 lb/gal; sp.gr. 1.00; acid no. 45–55; pH 8.5–9.5 (10%); biodeg.; 100% act.

Monamine LM-100. [Mona] Lauramide DEA and diethanolamine; nonionic/anionic; see Monamine Series; Gvcs-33 4 max. liq.; sol. @ 10% in water, ethanol, chlorinated and aromatic hydrocarbons; dens. 8.50 lb/gal; sp.gr. 1.02; acid no. 18–23; pH 9.5–10.5 (10%); biodeg.; 100% act.

Monamine R8-26. [Mona] Linoleamide DEA and diethanolamine; nonionic/anionic; see Monamine Series; amber liq.; sol. @ 10% in water, ethanol, chlorinated and aromatic hydrocarbons; dens. 8.25 lb./gal; sp.gr. 0.99; acid no. 75–85; pH 9.0–10.0 (10%); biodeg.; 100% act.

Monamine T-100. [Mona] Tallamide DEA and diethanolamine; nonionic/anionic; see Monamine Series; Gvcs-33 10 max. liq.; sol. @ 10% in ethanol, chlorinated and aromatic hydrocarbons, min. spirits, kerosene, disp. in water; dens. 8.10 lb/gal; sp.gr. 0.97; acid no. 10–16; pH 10.0–11.0 (10%); biodeg.; 100% act.

Monamulse 653-C. [Mona] Alkanolamide, modified; anionic/nonionic; emulsifier, solubilizer, degreaser for solvs.; liq.; 100% conc.

Monamulse 748. [Mona] Alkanolamide, modified; emulsifier and dispersant for crude oil; liq.; 100% conc.

Monamulse CI. [Mona] Imidazoline, modified; corrosion inhibitor in greases improving water resistance, oil-based lubricant systems, and on cast iron; improves emulsion stability; penetrant; amber clear liq.; sol. in alcohol, min. spirits, kerosene, wh. min. oil, natural fats and oils, and chlorinated and aromatic hydrocarbons; sp.gr. 0.959 ± 0.005; acid no. 0–2.

Monamulse R10-29M. [Mona] Imidazoline, modified; nonionic/cationic; emulsifier, solubilizer, corrosion inhibitor for fuel oils; liq.; disp. in water; 100% conc.

Monaquat AT-1074. [Mona] Quat. compd.; cationic; thickener with foaming and cleaning properties for HCl sol'ns.; used in acid bowl and pipe line cleaners, rust removers, and oil well sizing applics.; Gvcs-33 3 clear to slightly hazy liq.; sol. in HCl; sp.gr. 1.2; dens. 10 lb/gal; solid pt. < 0 C; pH 5.5 (10% aq.); 30% act.

Monaquat ISIES. [Mona] Isostearyl ethylimidonium ethosulfate; cationic; antistat, lubricant, softener, corrosion inhibitor used in cosmetic industry, in industrial and textile applics.; amber liq.; sol. @ 10% in water, ethanol, butyl Cellosolve, aromatic, chlorinated and fluorinated hydrocarbons; sp.gr. 1.03; pH 6.9 (10%); biodeg.; 100% act.

Monaquat P-TC. [Mona] Cocamidopropyl PEG-dimonium chloride phosphate; cationic; bactericidal, conditioner, antistat, detergent, foamer, emulsifier, solubilizer, dispersant, thickener, and wetting agent used in personal care, household, pharmaceutical, veterinary prods., fire fighting foams, petrol. prod., photographic processes, agric., mining, and textiles; amber clear liq.; sp.gr. 1.10; dens. 9.1 lb/gal; pH 7.0 (10%); surf. tens. 37.7 dynes/cm (1%); 40% act.

Monaquat P-TD. [Mona] Lauramidopropyl PEG-dimonium chloride phosphate; cationic; see Monaquat P-TC; amber clear liq.; sp.gr. 1.05; dens. 8.7 lb/gal; pH 7.5 (10%); surf. tens. 43.1 dynes/cm (1%); 34% act.

Monaquat P-TL. [Mona] Lauramphо PEG-glycinate phosphate; cationic; see Monaquat P-TC; clear to opaque visc. liq.; sp.gr. 1.10; dens. 9.1 lb/gal; pH 7.5 (10%); surf. tens. 33.7 dynes/cm (1%); 30% act.

Monaquat P-TS. [Mona] Stearamidopropyl PG-dimonium chloride phosphate; cationic; thickener, foamer, emulsifier, skin and hair conditioner; used in moisturizers, sunscreens, hair conditioners; lt. yel. paste; sp.gr. 1.01; dens. 8.4 lb/gal; m.p. 40–50 C; pH 7.6; biodeg.; 30% act.

Monaquat P-TZ. [Mona] Cocohydroxyethyl PEG-imidazolinium chloride phosphate; cationic; see Monaquat P-TC; amber clear liq.; sp.gr. 1.07; dens. 8.9 lb/gal; pH 7.0 (10%); surf. tens. 37.6 dynes/cm (1%); 30% act.

Monaquat TG. [Mona] Bishydroxyethyl dihydroxypropyl stearamonium chloride; cationic; surfactant used in personal care prods.; conditioner for hair rinses; antistat and foaming used in fabric laundering and softening prods.; thickener for acid bowl cleaners, naval gels; Gardner 2 clear to slightly hazy liq.; sp.gr. 1.011; dens. 8.39 lb/gal; solid. pt. –10 to –12 C; cloud pt. 3 C; pH 4 (10%); biodeg.; 30% act.

Monastat 1195. [Mona] Cationic antistat/cleaner for glass and plastic, e.g., TV screens, computers, medical diagnostic equip., safety goggles; clear amber liq.; sol. in water, IPA, ethanol, methanol, propylene glycol, ethylene glycol, perchloroethylene; dens. 8.2 lb/gal; pH 6.6 (10%); 80% act.

Monaterge 85. [Mona] Fatty acid amido complex;

nonionic/anionic; detergent, wetting agent, emulsifier, solubilizer for chlorinated solvs., used in cleaners; clear, amber liq.; sol. in water, ethanol, aromatic hydrocarbons, natural oils and fats; dens. 8.5 lb/gal; sp.gr. 1.02; surf. tens. 27.9 dynes/cm (0.05%); pH 8.9 (10%); biodeg.; 85% act.

Monateric 805. [Mona] Cocoamphodiacetate and disodium cocamido MIPA-sulfosuccinate; amphoteric/anionic; emulsifier, wetting agent, foamer, corrosion inhibitor, detergent, solubilizer for industrial and personal care prods.; lt. amber, visc. liq.; dens. 9.16 lb/gal; sp.gr. 1.10; pH 7.8 ± 0.5 (10%); biodeg.; 39–41% act.

Monateric 810-A-50. [Mona] Caprylic/capric monocarboxylic propionate, imidazoline-derived, salt-free; amphoteric; surfactant used in industrial detergents, cleaners, and cosmetics; hydrotrope, coupling agent, and/or solubilizer; corrosion inhibitor in metalworking systems, oil well flooding, and aerosol pkgs.; br. clear liq.; sp.gr. 1.07; dens. 8.9 lb/gal; pH 4.4 (10%); biodeg.; 50% act.

Monateric 811. [Mona] Capryloamphodipropionate; amphoteric; corrosion inhibitor, detergent, wetting agent in noncorrosive cleaners and industrial detergents; amber liq.; water sol.; dens. 8.7 lb/gal; sp.gr. 1.04; pH 10.4 (10%); biodeg.; 50% act.

Monateric 985A. [Mona] Lauroamphodiacetate; amphoteric; see Monateric 810-A-50; amber clear liq.; sp.gr. 1.07; dens. 8.9 lb/gal; pH 9.3 (10%); biodeg.; 36% act.

Monateric 1000. [Mona] Capryloamphopropionate; amphoteric; corrosion inhibitor, detergent, wetting agent for metal cleaning, cutting fluids, syn. lubricants; amber liq.; water sol.; dens. 8.8 lb/gal; sp.gr. 1.05; pH 11.2 (10%); biodeg.; 50% act.

Monateric 1202. [Mona] Dihydroxyethyl tallow glycinate; amphoteric; surfactant used as a conditioner, coupling, and visc. control agent in personal care prods.; clear visc. liq.; sol. in water; pH 5.5; 35.0% act.

Monateric ADA. [Mona] Cocamidopropyl betaine; used in air drilling, foam drilling, foam blanketing, air entraining agent for cement, gypsum, wallboard; clear, amber liq.; pH 7.9; biodeg.; 38% solids.

Monateric CA-35%. [Mona] Cocoamphopropionate; amphoteric; detergent, wetting agent, emulsifier, dispersant used in cosmetic, household, and industrial prods.; coupling agent, solubilizer; amber liq.; m.w. 360; dens. 8.5 lb/gal; sp.gr. 1.02; surf. tens. 29.5 dynes/cm (0.1% conc.); biodeg.; 35% act.

Monateric CAM-40. [Mona] Cocoamphopropionate; amphoteric; see Monateric 810-A-50; amber clear liq.; sp.gr. 1.05; dens. 8.8 lb/gal; pH 9.3 (10%); 40% act.

Monateric CDL. [Mona] Cocoamphodiacetate and sodium lauryl sulfate; amphoteric; see Monateric 810-A-50; yel. clear liq.; sp.gr. 1.11; dens. 9.2 lb/gal; pH 8.5 (10%); 31% act.

Monateric CDS. [Mona] Cocoamphodiacetate and sodium lauryl sulfate; amphoteric; see Monateric 810-A-50; yel. clear liq.; sp.gr. 1.09; dens. 9.1 lb/gal; pH 8.5 (10%); 31% act.

Monateric CDTD. [Mona] Cocoamphodiacetate and sodium trideceth sulfate; amphoteric; see Monateric 810-A-50; yel. clear liq.; sp.gr. 1.11; dens. 9.2 lb/gal; pH 8.5 (10%); 44% act.

Monateric CDX-38 Mod. [Mona] Cocoamphodiacetate; amphoteric; see Monateric 810-A-50; yel. clear liq.; sp.gr. 1.18; dens. 9.8 lb/gal; pH 8.8 (10%); 39% act.

Monateric CEM-38%. [Mona] Cocoamphodipropionate; amphoteric; detergent used in liq. detergent systems, wetting agent, emulsifier, dispersant, solubilizer, hydrotrope; amber, clear to hazy liq.; dens. 8.75 lb/gal; surf. tens. 35 dynes/cm (0.1%); pH 8.5 ± 0.5 (10%); biodeg.; 38% act.

Monateric CEM-38CG. [Mona] Cocoamphopropionate; amphoteric; see Monateric 810-A-50; amber clear to hazy liq.; sp.gr. 1.07; dens. 8.9 lb/gal; pH 9.8 (10%); 38% act.

Monateric CNa-40. [Mona] Coconut monocarboxylic propionate, imidazoline-derived, salt-free; amphoteric; see Monateric 810-A-50; amber clear liq.; sp.gr. 1.09; dens. 9.1 lb/gal; pH 10.9 (10%); 40% act.

Monateric COAB. [Mona] Cocamidopropyl betaine; amphoteric; see Monateric 810-A-50; yel. clear liq.; sp.gr. 1.04; dens. 8.7 lb/gal; pH 7.9 (10%); 32% act.

Monateric CyA-50, CyMM-40. [Mona] Caprylic dicarboxylic propionate, imadazoline-derived, salt-free; amphoteric; see Monateric 810-A-50; dk. br. and amber clear liq. resp.; sp.gr. 1.07 and 1.10 resp.; dens. 8.9 and 9.2 lb/gal resp.; pH 5.6 and 9.8 (10%) resp.; 50 and 40% act. resp.

Monateric CyNa-50%. [Mona] Capryloamphopropionate; amphoteric; detergent, emulsifier, coupling agent, solubilizer, wetting agent, low to moderate foaming surfactant used in conc. electrolyte systems; amber liq.; pH 10.5 ± 0.5 (10%); biodeg.; 50% act.

Monateric ISA-35%. [Mona] Isostearoamphopropionate; amphoteric; surfactant used in cosmetic and industrial prods.; conditioner, lubricant, thickener; amber clear to hazy flowable gel; sol. @ 10% in water, alcohol, glycol ethers and alkanolamines, disp. in min. oil and natural oils and fats; surf. tens. 30 dynes/cm (0.1%); pH 5–6 (10%); biodeg.; 35% act.

Monateric LMAB. [Mona] Lauramidopropyl betaine; amphoteric; see Monateric 810-A-50; lt. yel. clear liq.; sp.gr. 1.04; dens. 8.7 lb/gal; pH 8.3 (10%); 30% act.

Monateric LMM-30. [Mona] Lauroamphoacetate; amphoteric; see Monateric 810-A-50; amber visc. liq.; sp.gr. 1.09; dens. 9.1 lb/gal; pH 9.2 (10%); 30% act.

Monateric MCB. [Mona] Cocamidopropyl betaine; amphoteric; see Monateric 810-A-50; lt. yel. clear liq.; sp.gr. 1.02; dens. 8.5 lb/gal; pH 4.8 (10%); 30% act.

Monateric TA-35. [Mona] Tallamphopropionate; amphoteric; see Monateric 810-A-50; dk. br. gel; sp.gr. 1.02; dens. 8.5 lb/gal; pH 5.2 (10%); 35% act.

Monateric TDB-35. [Mona] Disodium tallowiminodipropionate; amphoteric; hydrotrope and detergent for high electrolyte systems such as heavy-duty liq. cleaners and wax strippers; yel. clear-hazy liq.; sp.gr. 1.03; dens. 8.6 lb/gal; pH 11.25 (1%); 35% act.

Monatrope 1250. [Mona] Sodium alkanoate; anionic; surfactant hydrotrope for formulating alkaline built liq. detergent concs.; coupling agent for nonionic and other surfactants in high concs. of electrolytes; for household and industrial detergents, spray washes, textiles, hypochlorite detergents/sanitizers, dishwash liqs.; Gardner 1 clear liq.; sol. in 10% sodium hydroxide, 20% potassium hydroxide; pH 10; 45% solids.

Monatrope 1296. [Mona] Org. phosphate ester; hydrotrope for formulating highly built liq. detergents for household or industrial use; solubilizer for nonionics in high electrolyte systems; used for spray and soak tank cleaners, liq. laundry detergents, mech. dishwashing; Gardner 2 clear liq.; sp.gr. 1.23; pH 5; 50% solids.

Monawet MB-45. [Mona] Sodium diisobutyl sulfosuccinate; anionic; wetting, dispersing, emulsifying, penetrating and solubilizing agent used in emulsion polymerization of S/B for rug backing; clear, colorless liq.; m.w. 332; sol. in water, fairly sol. in polar and nonpolar solvs.; dens. 9.3 lb/gal; sp.gr. 1.12; cloud pt. 13 C; flash pt. 215 F (PMCC); surf. tens. 54 dynes/cm (0.1%); pH 6 ± 1; 45% act.

Monawet MM-80. [Mona] Sodium dihexyl sulfosuccinate; anionic; see Monawet MB-45; also used in paint and food industries; clear, colorless liq.; m.w. 388; sol. 33 g/100 g water, in polar and nonpolar solvs.; dens. 9.2 lb/gal; sp.gr. 1.10; cloud pt. < 0 C; flash pt. 110 F (PMCC); surf. tens. 46 dynes/cm (0.1%); pH 6 ± 1; 80% act.

Monawet MO-65-150. [Mona] Sodium dioctyl sulfosuccinate, anhyd.; anionic; wetting, penetrating and spreading agent, emulsifier used in oil well cleaning, drycleaning detergents, solv. cleaners and strippers; clear, colorless liq.; dens. 8.75 lb/gal; sp.gr. 1.05; flash pt. 42–43 C (COC); pH 5.5 ± 1.0 (10% sol'n.); 65% act.

Monawet MO-70. [Mona] Sodium dioctyl sulfosuccinate; anionic; wetting, dispersing, emulsifying, penetrating and solubilizing agent used in emulsion and suspension polymerization, textile, fertilizer, cosmetic, food industries; clear, colorless liq.; m.w. 444; sol. in polar and nonpolar solvs.; dens. 9.0 lb/gal; sp.gr. 1.08; cloud pt. < –5 C; flash pt. 325 F (PMCC); surf. tens. 29 dynes/cm (0.1%); pH 6 ± 1; 70% act.

Monawet MO-70-150. [Mona] Sodium dioctyl sulfosuccinate; anionic; see Monawet MO-65-150; liq.; 70% conc.

Monawet MO-70E. [Mona] Sodium dioctyl sulfosuccinate; anionic; see Monawet MO-70; clear, colorless liq.; m.w. 444; sol. in polar and nonpolar solvs.; dens. 9.0 lb/gal; sp.gr. 1.08; cloud pt. < –5 C; flash pt. 82 F (PMCC); surf. tens. 29 dynes/cm (0.1%); pH 6 ± 1; 70% act.

Monawet MO-70R. [Mona] Sodium dioctyl sulfosuccinate; anionic; wetting, dispersing, emulsifying, penetrating and solubilizing agent for textile, printing, agric., cosmetic, food industries; clear, colorless liq.; m.w. 444; sol. in polar and nonpolar solvs.; dens. 8.8 lb/gal; sp.gr. 1.06; cloud pt. < –5 C; flash pt. 280 F (PMCC); surf. tens. 29 dynes/cm (0.1%); pH 6 ± 1; 70% act.

Monawet MO-75E. [Mona] Sodium dioctyl sulfosuccinate; anionic; see Monawet MO-70R; clear, colorless liq.; m.w. 444; sol. in polar and nonpolar solvs.; dens. 9.0 lb/gal; sp.gr. 1.08; cloud pt. < –5 C; flash pt. 80 F (PMCC); surf. tens. 29 dynes/cm (0.1%); pH 6 ± 1; 75% act.

Monawet MO-84R2W. [Mona] Sodium dioctyl sulfosuccinate; anionic; see Monawet MO-70R; lt. yel. visc. liq.; m.w. 444; sol. in polar and nonpolar solvs.; dens. 9.2 lb/gal; sp.gr. 1.10; cloud pt. < –10 C; flash pt. 223 F (PMCC); surf. tens. 29 dynes/cm (0.1%); pH 5.5 ± 1; 83% act.

Monawet MT-70. [Mona] Sodium ditridecyl sulfosuccinate; anionic; wetting, dispersing, emulsifying, penetrating and solubilizing agent used in emulsion and suspension polymerization, indirect food additives; lt. straw clear liq.; m.w. 584; sol. in polar and nonpolar solvs.; dens. 8.5 lb/gal; sp.gr. 1.02; cloud pt. –2 C; flash pt. 230 F (PMCC); surf. tens. 29 dynes/cm (0.1%); pH 6 ± 1; 70% act.

Monawet MT-70E. [Mona] Sodium ditridecyl sulfosuccinate; anionic; see Monawet MT-70; lt. straw clear liq.; m.w. 584; sol. in polar and nonpolar solvs.; dens. 8.4 lb/gal; sp.gr. 1.01; cloud pt. –15 C; flash pt. 86 F (PMCC); surf. tens. 29 dynes/cm (0.1%); pH 6 ± 1; 70% act.

Monawet MT-80H2W. [Mona] Sodium ditridecyl sulfosuccinate; anionic; wetting, dispersing, emulsifying, penetrating and solubilizing agent used in printing inks, indirect food additives; lt. yel. clear liq.; m.w. 584; sol. in polar and nonpolar solvs.; dens. 8.5 lb/gal; sp.gr. 1.02; cloud pt. < 0 C; flash pt. 225 F (PMCC); surf. tens. 29 dynes/cm (0.1%); pH 5.5 ± 1; 80% act.

Monawet SNO-35. [Mona] Tetrasodium dicarboxyethyl stearyl sulfosuccinamate; anionic; wetting agent, solubilizer, emulsifier, dispersant, mild detergent used in textile, cosmetic, agric. prods.; clear, lt. amber liq.; m.w. 653; sol. in water, high electrolyte salt sol'ns.; dens. 9.5 lb/gal; sp.gr. 1.12–1.16; visc. 16–18 s (#2 Zahn cup); f.p. 45 ± 5 F; surf. tens. 43 dynes/cm (0.1%); pH 7–8; biodeg.; 35% solids.

Monazoline Series. [Mona] 1-Hydroxyethyl-2-alkyl imidazolines; cationic; wetting agent, emulsifier for nonpolar liq., detergent, thickener, corrosion inhibitor, antistat, softener, bactericide used in paint and textile industries; liq.; sol. in oil, disp. in water; biodeg.

Monazoline C. [Mona] Cocoyl hydroxyethyl imidazoline; see Monazoline Series, also dispersant for clay and pigments, in acid dairy cleaners, in oil well acidifying and sec. recovery operations; amber liq., may crystallize on aging; m.w. 282; sol. @ 10% in ethanol, chlorinated hydrocarbons, min. oil, toluene, kerosene, min. spirits, veg. oil, disp. in water; dens. 7.75 lb/gal; sp.gr. 0.93; pH 10.5–12.0 (10% disp.); biodeg.; 90% min. imidazoline.

Monazoline CY. [Mona] Capryl hydroxyethyl imidazoline; see Monazoline Series; amber liq., may crystallize on aging; m.w. 212; sol. @ 10% in ethanol, chlorinated hydrocarbons, toluene, veg. oil; dens. 8.25 lb/gal; sp.gr. 0.99; pH 10.5–12.0 (10% disp.); biodeg.; 90% min. imidazoline.

Monazoline IS. [Mona] Isostearyl hydroxyethyl imidazoline; cationic; corrosion inhibitor and lubricant; liq.; 100% conc.

Monazoline O. [Mona] Oleyl hydroxyethyl imidazoline; see Monazoline Series, also dispersant for clays and pigments, in agric. preparations, acid dairy cleaners; amber liq., may crystallize on aging; m.w. 345; sol. @ 10% in ethanol, chlorinated hydrocarbons, min. oil, toluene, kerosene, min. spirits, veg. oil; dens. 7.66 lb/gal; sp.gr. 0.92; pH 10.0–11.5 (10% disp.); biodeg.; 90% min. imidazoline.

Monazoline T. [Mona] Tall oil hydroxyethyl imidazoline; see Monazoline Series; also dispersant for clays and pigments, aids gravel-to-asphalt bonding, rinse aid for automatic car washes, printing ink additive, protective metal coatings; amber liq., may crystallize on aging; m.w. 350; sol. @ 10% in ethanol, chlorinated hydrocarbons, min. oil, toluene, kerosene, min. spirits, veg. oil, disp. in water; dens. 7.75 lb/gal; sp.gr. 0.93; pH 10.0–11.5 (10% disp.); biodeg.; 90% min. imidazoline.

Monceren®. [Bayer] Pencycuron; fungicide with specific action against *Rhizoctonia solani*; colorless cryst. (pure a.i.), powd. (tech. a.i.); m.w. 328.8; sol. (g/1000 ml): 100–1000 g in dichloromethane, 10–30 g in toluene, 1–10 g in 2-propanol.

Mondur CB-75. [Mobay] Adduct of toluene diisocyanate and polyol; cross-linking agent that improves resistance to heat, greases, oils, plasticizers, org. solvs., and the adhesion to many substrates; primarily used in adhesives; pale yel. clear liq.; dens. 1.2 g/cm^3; visc. 20 p; 75% solids in ethyl acetate; 13% NCO content (of sol'n.).

Mondur MR. [Mobay] Polymethylene polyphenyl isocyanate; see Mondur CB-75; dk. br. liq.; dens. 1.2 g/cm^3; visc. 2 p; 100% solids; 31% NCO content.

Mondur MRS. [Mobay] Polymethylene polyphenyl isocyanate; see Mondur CB-75; dk. br. liq.; dens. 1.2 g/cm^3; visc. 2 p; 100% solids; 31.5% NCO content.

Mondur PF. [Mobay] Modified polyisocyanate based on 4,4′-diphenyl methane diisocyanate; see Mondur CB-75; pale yel. liq.; dens. 1.2 g/cm^3; visc. 8 p; 100% solids; 22.5% NCO content.

Monflor 32. [ICI Am.] Fluorochemical surfactant; anionic; leveling and wetting agent, dispersant, emulsifier for ink/paint pigments; liq.; dens. 1.09 g/cc; 30% act.

Monflor 51. [ICI Am.] Fluorochemical surfactant; nonionic; dispersant, emulsifier, foam stabilizer, leveling agent, wetting agent for mold release, industrial cleaning, plastics, suspension/emulsion polymerization, ink/paint pigments; waxy solid; sol. in alcohols, acetone, MEK, trichloroethylene, water, toluene; dens. 1.25 g/cc; 100% act.

Monflor 52. [ICI Am.] Fluorochemical surfactant; nonionic; dispersant, emulsifier, foam stabilizer, leveling agent, wetting agent for mold release, industrial cleaning, plastics, ink/paint pigments; visc. liq.; sol. in alcohols, acetone, MEK, trichloroethylene, toluene, water; dens. 1.32 g/cc; 100% act.

Mono-Coat® 65-RT. [Chem-Trend] Solv.-based release agent system formulated to cure and adhere to the mold surface at R.T.; for use on laminates and/or composite molding; also as primary release for epoxy and polyester resin systems with graphite, aramid, or fiberglass reinforcements; slightly hazy fluid; dens. 8.62 lb/gal; visc. < 100 cP (21 C); flash pt. none.

Mono-Coat® E76. [Chem-Trend] Solv.-based release system for all types of rubber molding and for mold conditioning; colorless water-thin, clear liq.; dens. 10.52 lb/gal; visc. < 10 cP (21 C); flash pt. none.

Mono-Coat® E91. [Chem-Trend] Solv.-based release system for molding operations where high degree of slip is required; used in rubber molding (compr., transfer, and inj. of peroxide-cured fluoroelastomers, "dry" EPDM compds., some epichlorohydrin compds.), and in composite/laminate molding; colorless clear fluid; dens. 10.71 lb/gal; visc. < 10 cP (21 C); flash pt. none.

Monogen. [Dai-ichi Kogyo] Sulfated fatty alcohol, sodium salt; anionic; textile scouring and washing agent; dye dispersant and leveling agent; paste; 60% conc.

Monoglyme. [Ferro] Ethylene glycol dimethyl ether dimethoxyethane (DME); solv. which tends to solvate cations; used in electrochemistry, polymer and boron chemistry, physical processes such as gas absorp., extraction, stabilization, industrial prods. such as fuels, lubricants, textiles, pharmaceuticals, pesticides; colorless; ethereal, nonresidual odor; m.w. 90.12; f.p. –69 C; b.p. 85.2 C; water-sol.; sp.gr. 0.8683; dens. 7.24 lb/gal; visc. 1.1 cP; ref. index 1.3792; pH neutral; flash pt. –6 C (CC); 99.6% min. purity.

Monolan 1206/2. [Harcros UK] EO-PO block polymer; nonionic; antifoam for detergents, wetting agents, syn. lubricants; colorless clear liq., faint odor; water-insol.; sp.gr. 1.003; visc. 445 cs; flash pt. > 400 F (COC); pour pt. < 0 C; pH 6–8 (1% aq.); 100% act.

Monolan 2000 E/12. [Harcros UK] EO-PO block polymer; nonionic; detergent, wetting agent, antifoam, dispersant, rinse aid; colorless clear liq., faint odor; water-insol.; sp.gr. 1.020; visc. 390 cs; flash pt. > 400 F (COC); pour pt. < 0 C; pH 6–8 (1% aq.); 100% act.

Monolan 3000 E/60, 8000 E/80. [Harcros UK] EO-PO block polymer; nonionic; detergent, wetting agent, antifoam for industrial and domestic detergents, emulsion polymerization, metal cleaning, resin plasticizers, latex stabilization, textile processing; wh. solid and wh. flake resp., faint odor; water-sol.; sp.gr. 1.052 and 1.070; visc. 355 cs and 1100 cs (60 C) resp.; flash pt. > 400 F (COC); pour pt. 31 and 49 C; pH 6–8 and 7–8 (1% aq.); 100% act.

Monolan 12,000 E/80. [Harcros UK] EO-PO copolymer; nonionic; latex stabilizer and sec. emulsifier; toilet block component; solid; HLB 16.0; 100% conc.

Monolan P222. [Harcros UK] EO-PO block polymer; nonionic; antifoam for detergents, wetting agents, syn. lubricants; colorless clear liq., faint odor; water-insol.; sp.gr. 1.004; visc. 440 cs; flash pt. > 400 F (COC); pour pt. < 0 C; pH 6–8 (1% aq.); 100% act.

Monolan PB, PC. [Harcros UK] Complex EO–PO copolymer; nonionic; wetting agent, low foam detergent and rinse aid, pigment dispersant; colorless liq.; faint odor; disp. and sol. in water resp.; sp.gr. 1.020 and 1.040; visc. 680 and 800 cs; cloud pt. 19 and 44 C (1% aq.); flash pt. > 400 F (COC); pour pt. < 0 and 8 C; pH 6–8 (1% aq.); 100% act.

Monolan PPG440, PPG1100, PPG2200. [Harcros UK] PPG; nonionic; lubricant, antistat, plasticizer, cosolvs. in dyestuff, ink, resin, rubber industries, cosmetic preparations, intermediates in surfactants and plastic prod.; colorless liq.; faint odor; sol. in water (PPG440), insol. in water (PPG1100, 2200); m.w. 400, 1000, 2000 resp.; sp.gr. 1.010, 1.005, and 1.004; visc. 80, 180, and 450 cs; flash pt. > 450 F (COC); pour pt. < 0 C; 100% act.

Monolan PT. [Harcros UK] Complex EO–PO copolymer; nonionic; low foam detergent and rinse aid, pigment dispersant; colorless liq.; faint odor; water sol.; sp.gr. 1.037; visc. 750 cs; cloud pt. 28 C (1% aq.); flash pt. > 400 F (COC); pour pt. < 0 C; pH 6–8 (1% aq.); 100% act.

Monomuls 60-10. [Grünau] Lard glycerides; nonionic; emulsifier, stabilizer, dispersant, opacifier for cosmetics, foods and drugs; solid; 100% conc.

Monomuls 60-15. [Grünau] Hydrog. lard glycerides; nonionic; see Monomuls 60-10; powd.; 100% conc.

Monomuls 60-20. [Grünau] Tallow glycerides; nonionic; see Monomuls 60-10; solid; 100% conc.

Monomuls 60-25. [Grünau] Hydrog. tallow glycerides; nonionic; see Monomuls 60-10; solid; 100% conc.

Monomuls 60-25/2. [Grünau] Hydrog. tallow glycerides with 2% sodium stearate; nonionic; see Monomuls 60-10; powd.; 100% conc.

Monomuls 60-25/5. [Grünau] Hydrog. tallow glycerides with 5% sodium stearate; nonionic; see Monomuls 60-10; powd.; 100% conc.

Monomuls 60-30. [Grünau] Palm oil glycerides; anionic; see Monomuls 60-10; solid.; 100% conc.

Monomuls 60-35. [Grünau] Hydrog. palm oil glycerides; nonionic; see Monomuls 60-10; powd.; 100% conc.

Monomuls 60-40. [Grünau] Sunflower seed oil glycerides; nonionic; see Monomuls 60-10; solid; 100% conc.

Monomuls 60-45. [Grünau] Hydrog. soybean oil glycerides; nonionic; see Monomuls 60-10; powd.; 100% conc.

Monomuls 90-10. [Grünau] Dist. lard glyceride; nonionic; see Monomuls 60-10; solid; 100% conc.

Monomuls 90-15. [Grünau] Dist. hydrog. lard glyceride; nonionic; see Monomuls 60-10; powd.; 100% conc.

Monomuls 90-20. [Grünau] Dist. tallow glyceride; nonionic; see Monomuls 60-10; solid; 100% conc.

Monomuls 90-25. [Grünau] Dist. hydrog. tallow glyceride; nonionic; see Monomuls 60-10; powd.; 100% conc.

Monomuls 90-25/2, 90-25/5. [Grünau] Dist. hydrog. tallow glyceride with 2% and 5% sodium stearate resp.; anionic; see Monomuls 60-10; powd.; 100% conc.

Monomuls 90-30. [Grünau] Dist. palm oil glyceride; nonionic; see Monomuls 60-10; solid; 100% conc.

Monomuls 90-35. [Grünau] Dist. hydrog. palm oil glyceride; nonionic; see Monomuls 60-10; powd.; 100% conc.

Monomuls 90-40. [Grünau] Dist. sunflower seed oil glyceride; nonionic; see Monomuls 60-10; solid; 100% conc.

Monomuls 90-45. [Grünau] Dist. hydrog. soybean oil glyceride; nonionic; see Monomuls 60-10; powd.; 100% conc.

Monomuls 90-L12. [Grünau] Glyceryl laurate; nonionic; refatting agent and thickener for personal care prods.; solid; 100% conc.

Monomuls 90-O18. [Grünau] Glyceryl oleate; nonionic; emulsifier, stabilizer for cosmetics, food and drugs; solid; 100% conc.

Monoplex® DDA. [C.P. Hall] Diisodecyl adipate; lubricant; APHA 20 clear liq.; m.w. 427; sol. in hexane, toluene, kerosene, ethanol, acetone, min. oil; sp.gr.0.916; visc. 30 cps; f.p. –43 C; flash pt. 227 C; acid no. 0.1; iodine no. nil; sapon. no. 263; ref. index 1.4501.

Monoplex® DIOA. [C.P. Hall] Diisooctyl adipate; lubricant; APHA 50 clear liq.; m.w. 373; sol. in hexane, toluene, kerosene, ethanol, acetone, min. oil; sp.gr. 0.926; f.p. –65 C; flash pt. 193 C; acid no. 0.05; iodine no. < 1; sapon. no. 303; ref. index 1.446.

Monoplex® DOA. [C.P. Hall] Di-2-ethylhexyl adipate; plasticizer for PVC food pkg. film; clear liq.

Monoplex® DOS. [C.P. Hall] Di-2-ethylhexyl sebacate; plasticizer for elec. PVC compds., low temp. sheet, film; clear liq.

Monoplex® NODA. [C.P. Hall] n-Octyl, n-decyl adipate; plasticizer; APHA 20 clear liq.; m.w. 400; sol. in hexane, toluene, kerosene, ethanol, acetone, min. oil; sp.gr. 0.913; visc. 16 cps; f.p. 0 C; flash pt. 212 C; acid no. 0.1; iodine no. nil; sapon. no. 280; ref. index 1.446.

Monoplex® S-73. [C.P. Hall] Epoxidized octyl tallate; PVC stabilizer, plasticizer; clear liq.

Monoplex® S-75. [C.P. Hall] Epoxidized glycol dioleate; PVC stabilizer, plasticizer.

Monopol Oil 75. [GAF] Sulfated castor oil; anionic; wetting agent, penetrant, leveling agent, softener/ lubricant or softener additive for cellulosic fibers, used in dyeing; lt. yel., clear liq.; sp.gr. 1.05; 75% act.

Monosteol. [Gattefosse Ets.] Propylene glycol stearate; nonionic; stabilizer for ointments, cream lotions; solid; HLB 4.0; 100% conc.

Monosulph. [Henkel] Highly sulfonated castor oil; anionic; penetrant, emulsifier; textile dyeing assistant; fat liquor for suede leather; paper coating evener; plasticizer for starch, glues; emulsifier for latex; liq.; 68% conc.

Montaclere®. [Monsanto Europe] Styrenated phenol; nonstaining antioxidant for NR, syns., latexes; liq.

Montegal 150 RG, AP 80, DZ. [Seppic] Alkylpolyethoxy ether; nonionic; textile leveling agent; liq., flakes, wax resp.; 50, 100, and 100% conc. resp.

Montegal OL 50. [Seppic] Castor oil, ethoxylated; nonionic; see Montegal 150 RG; liq.; 100% conc.

Montegal SH 25. [Seppic] Fatty amine, ethoxylated; nonionic; see Montegal 150 RG; liq.; 45% conc.

Monthyle. [Gattefosse Ets.] Glycol stearate; stabilizer for ointments, cream lotions; solid.

Montosol PF-10. [Pulcra SA] Sodium lauryl ether (2.8) sulfate; anionic; mfg. of personal care prods.; solubilizer for perfumes; liq.; 27 ± 1% conc.

Montosol PF-14. [Pulcra SA] Sodium lauryl ether (2) sulfate; anionic; see Montosol PF-10; liq.; 27 ± 1% conc.

Montosol PF-26, PG-10. [Pulcra SA] Sodium lauryl ether (2.4) sulfate; anionic; see Montosol PF-10; liq., gel resp.; 27 ± 1 and 70 ± 2% conc. resp.

Montosol PG-17. [Pulcra SA] MEA lauryl ether (1) sulfate; anionic; dispersant for personal care prods.; liq.; 27 ± 1% conc.

Montosol PL-16. [Pulcra SA] Sodium lauryl ether (2.4) sulfate; anionic; see Montosol PF-10; liq.; 27 ± 1% conc.

Montosol PQ-17. [Pulcra SA] Sodium laureth-3 sulfate; anionic; see Montosol PF-10; liq.; 59±2% conc.

Montovol GJ-12. [Pulcra SA] Sodium tallow alcohol sulfate; anionic; detergent, dispersant; paste; 34±1% conc.

Morestan®. [Bayer] Chinomethionat; organic fungicide with a specific action against powdery mildews; good acaricidal act.; pale yel. cryst., odorless; m.w. 234.3; sol. in hot aromatic hydrocarbons, dioxane, dimethylformamide; slightly sol. in cold org. solvs.; insol. in water; m.p. 169.8–170.0 C.

Morfax®. [Vanderbilt] 4-Morpholinyl-2-benzothiazole disulfide; sulfenamide accelerator for NR and syn. rubbers; provides good curing activity with adequate processing safety; used for tires and mech. goods requiring max. strength and wearing quality; cream to lt. yel. friable powd.; m.w. 284.35; sol. in toluene, chloroform; practically insol. in water; dens. 1.51 ± 0.03 mg/m³; m.p. 116–128 C.

Morlex® DEEA. [Union Carbide] N,N-diethyl ethanolamine; controls corrosion in boiler water and returnline condensate steam systems; water-sol.

Morlex® DMEA. [Union Carbide] N,N-dimethyl ethanolamine; see Morlex DEEA; water-sol.

Morwet D-425. [DeSoto] Sodium sulfonate of naphthalene formaldehyde condensate; anionic; dispersant for pesticides, dyestuff; powd.; 90% act.

Morwet DB. [DeSoto] Sodium di-n-butyl naphthalene sulfonate; anionic; dispersant for pesticides, etc.; powd.; 75% act.

Mould Release Agent N 32. [Chemetall] Solvent-free mold release agent on a plastic base for the rubber and plastics industry.

M-P-A® 14. [Rheox] Organophilic clay; antisettling additive; lt. cream powd.; sp.gr. 2.3.

M-P-A® 60MS. [Rheox] Antisettling agent in 75% min. spirits; translucent wh. paste; dens. 6.79 lb/gal; flash pt. 105 F.

M-P-A® 60T. [Rheox] Antisettling agent in 75% toluene; translucent wh. paste; dens. 7.26 lb/gal; flash pt. 40 F.

M-P-A® 60X. [Rheox] Antisettling agent in 75% xylene; translucent wh. paste; dens. 7.26 lb/gal; flash pt. 74 F.

M-P-A® 1075. [Rheox] Antisettling agent in 55% n-butanol; translucent wh. paste; dens. 7.0 lb/gal; flash pt. (Seta) 100 F.

M-P-A® 1078X. [Rheox] Antisettling agent in 60% xylene; translucent wh. paste; dens. 7.40 lb/gal; flash pt. 74 F.

M-P-A® 2000T. [Rheox] Antisettling agent in 80% toluene; translucent wh. liq.; sp.gr. 0.88; dens. 7.35 lb/gal; flash pt. (Seta) 39 F.

M-P-A® 2000X. [Rheox] Antisettling agent in 80% xylene; translucent wh. liq.; sp.gr. 0.88; dens. 7.34 lb/gal; flash pt. (Seta CC) 81 F.

M-P-A® 3000 MS. [Rheox] Proprietary org.; pourable antisettling agent for aliphatic solv.-based systems, esp. high solids systems; provides sag control in certain systems; translucent wh. liq.; dens. 6.7 lb/gal; visc. 200–400 cps; 20% solids in min. spirits.

M-P-A® MS. [Rheox] Antisettling agent in 60% min. spirits; translucent wh. paste; dens. 6.90 lb/gal; flash pt. 105 F.

M-Pyrol®. [GAF] N-methyl-2-pyrrolidone; solv., stabilizer for vinyls, urethanes, acrylic (co)polymers, difficult sol. resins and waxes, wire insulation enamels; spinning solv. for syn. fibers; used in commercial petrochemical processing, incl. aromatic hydrocarbon extraction, gas purification, butadiene extraction, dehydration of natural gas; used for agric. chemicals; used in removal of strippable coatings and paints; for lube oil processing; reactive solv. for sand core binder; used as nonaq. syn. medium for polyol reactions; used in org. reactions; residue cleaning solv. for PVC and urethane prod. equipment; liq.; f.p. –24.4 C; b.p. 202 C; flash pt. 96 C (OC).

M-Quat® 32. [PPG-Mazer] Octadecyl diethanol methyl ammonium chloride; cationic; emulsifier, defoamer, and coagulents; liq.; 50% conc.

M-Quat® 2475. [PPG-Mazer] Dicocodimonium chloride and isopropanol; cationic; see M-Quat 32; liq.; 75% conc.

MS-122. [Miller-Stephenson] Tetrafluorethylene telomer; release agent, dry lubricant for use on cold molds, esp. for epoxy potting/encapsulating, PU, nylon, acrylics, PP, PC phenolics, PS, foams, rubber molding; solid.

MS-136. [Miller-Stephenson] Tetrafluorethylene telomer; release agent for hot molds, esp. for PU, nylon, ABS, polypropylene oxides, acetals, vinyls, PVC, cellulosics, phenolics, elastomers, TPE, melamines, etc.; solid.

Mulsifan RT 18. [Zschimmer & Schwarz] Alkylaryl polyglycol ether; solubilizer for perfumes and essential oils; liq.; HLB 13; 100% act.

Mulsifan RT 69. [Zschimmer & Schwarz] Triglyceride, ethoxylated; nonionic; emulsifier for fat and oils, solubilizer for perfumes and volatile oils; liq., paste; HLB 13.0; 100% conc.

Mulsifan RT 141. [Zschimmer & Schwarz] Polysorbate 20; nonionic; solubilizer for perfumes and volatile oils; straw-colored visc. liq.; sol. in water, ethanol, IPA, acetone, benzene, toluene, chlorinated hydrocarbons; sp.gr. 1.09; HLB 15–16; pH 5–7 (10%); 100% act.

Mulsifan RT 146. [Zschimmer & Schwarz] Polysorbate 80; nonionic; see Mulsifan RT 141; yel. visc. liq.; sp.gr. 1.09; sol. in water, ethanol, IPA, acetone, benzene, toluene, chlorinated hydrocarbons; HLB 15–16; pH 5–7 (10%); 100% act.

Mulsifan RT 203/80. [Zschimmer & Schwarz] Pareth-25-12; nonionic; see Mulsifan RT 141; colorless liq.; sp.gr. 1.04; sol. in water; HLB 14.0; cloud pt. 89–94 C (1% aq.); pH 5–7 (10%); 80% act.

Mulsifan RT 302. [Zschimmer & Schwarz] Ethoxylated hydrog. castor oil; solubilizer for perfumes and essential oils; wax.

Multiflow. [Monsanto] Resin modifier for nonaq. coatings; improves flow, substrate wetting in industrial powd. coatings.

Multinol C. [Nippon Senka] Special sulfonic acid deriv.; anionic; one-bath scouring, bleaching and dyeing agent; paste

Multiwax® 180-M. [Witco] Microcryst. wax NF; plasticizer or modifier for polymeric coatings and adhesives; hot melt adhesives and coatings, chewing gum base, protective coatings; FDA approved; lt. yel.; misc. with petrol. prods., many essential oils, most animal and veg. fats, oils, and waxes; m.p. 82–88 C; visc. 14.3–18.0 cSt (99 C); flash pt. (COC) 277 C min.

Multiwax® HS. [Witco] Microcrystalline wax; wax used in foil/tissue laminations, heat-seal laminations, glassine laminations; FDA approved; lt. yel.; m.p. 71–77 C; visc. 14.3–18.0 cSt (99 C); flash pt. (COC) 274 C min.

Multiwax® ML-445. [Witco] Microcryst. wax NF; plasticizer or modifier for polymeric coatings and adhesives; laminating agent for paper, film, foil, sealant, rustproofing compds., candles; FDA approved; lt. yel.; misc. with petrol. prods., many essential oils, most animal and veg. fats, oils, and waxes; m.p. 77–82 C; visc. 14.3–18.0 cSt (99 C); flash pt. (COC) 274 C min.

Multiwax® W-445. [Witco] Microcryst. wax NF; plasticizer or modifier for polymeric coatings and adhesives; cheese coating, laminating of cellophane and plastic film, waterproofing, protective lining; used in hair dressings, medicated creams, chewing gum base, dental waxes, lubricants, candles, hot-melt adhesives; FDA approved; wh. wax; misc. with petrol. prods., many essential oils, most animal and veg. fats, oils, and waxes; m.p.77–82 C; visc. 14.3–18.0 cSt (99 C); flash pt. (COC) 274 C min.

Multiwax® W-835. [Witco] Microcryst. wax; plasticizer or modifier for polymeric coatings and adhesives; used in silk screen printing, cold creams, cleansing creams, hair pomades, pharmaceuticals, crayons, paste-up adhesive; FDA approved; wh.; m.p. 74–79 C; visc. 14.3–18.0 cSt (99 C); flash pt. (COC) 246 C min.

Multiwax® X-145A. [Witco] Microcryst. wax NF; plasticizer or modifier for polymeric coatings and adhesives; cheese coating, laminating of cellophane and plastic film, waterproofing, protective lining; FDA approved; lt. yel.; misc. with petrol. prods., many essential oils, most animal and veg. fats, oils, and waxes; m.p. 66–71 C; visc. 14.3–18.0 cSt (99 C); flash pt. (COC) 260 C min.

Musloid 815D, 815M, 815S. [Rhone-Poulenc] Fatty acid ethoxylate; nonionic; dispersant, stabilizer, leveling agent, low-foaming surfactant in emulsion paint systems; amber visc. liq.; sp.gr. 0.95, 1.01, and 0.98 resp.; 100% act.

Myacide SP. [Inolex] 2,4-Dichlorobenzyl alcohol; antifungal agent, preservative; sol. (g/100 ml): 95 g in acetone, 80 g in methanol, 45 g in propylene glycol, 0.1 g in water.

Mycolase®. [Int'l. Bio-Synthetics] Fungal alpha-amylase; enzyme which converts low DE acid or modified starch hydrolysates to high maltose syrups; liq.

Myritol 318. [Henkel] Caprylic/capric triglyceride; nonionic; emollient for pharmaceutical and cosmetic preps. in emulsion form; oily component with solv. capacity; solubilizer; liq.; 100% conc.

Mytab. [Hexcel] Myrtrimonium bromide; surfactant used as emulsifier and antistat in hair rinses; antimicrobial for cosmetics, topicals; wh. powd.; char. odor; sol. in water, alcohols, chloroform; m.w. 336.40; pH 5–8 (1% aq.); 100% act.

Myvacet® 5-07(K). [Eastman] Dist. acetylated monoglyceride from hydrog. veg. oils; emulsifier, emollient; forms films with good moisture vapor barrier properties; waxy solid; sp.gr. 0.94 (80 C); m.p. 41–46 C; 48.5–51.5% acetylation.

Myvacet® 7-00. [Eastman] Dist. acetylated monoglyceride from hydrog. lard; see Myvacet 5-07(K); waxy solid; sp.gr. 0.94 (80 C); m.p. 37–40 C; 66.5–69.5% acetylation.

Myvacet® 7-07(K). [Eastman] Dist. acetylated monoglyceride from hydrog. veg. oil; see Myvacet 5-07(K); waxy solid; sp.gr. 0.94 (80 C); m.p. 37–40 C; 66.5–69.5% acetylation.

Myvacet® 9-08(K). [Eastman] Dist. acetylated monoglyceride from hydrog. coconut oil; emulsifier, emollient; liq.; sp.gr. 0.94 (80 C); m.p. 4–12 C; 96% min. acetylation.

Myvacet® 9-40. [Eastman] Dist. acetylated monoglyceride from edible lard; emulsifier; food-grade lubricant and emollient; deaerator in some systems; liq.; sp.gr. 0.94 (80 C); m.p. 4–12 C; 96% min. acetylation.

Myvacet® 9-45. [Eastman] Dist. acetylated monoglyceride from partially hydrog. soybean oil; see Myvacet 9-40; liq.; sp.gr. 0.94 (80 C); m.p. 4–12 C; 96% min. acetylation.

Myvatem®. [Eastman] Diacetyl tartaric acid esters of dist. monoglycerides; dispersant and emulsifier for baked goods; lt. amber sticky, visc. semisolid, faint acetic acid odor; 70% sol. in soybean oil; 50% sol. in water; sp.gr. 1.056 (60 C); HLB 8; m.p. 30 C; flash pt. (TOC) 208 C; acid no. 62–76; sapon. no. 380–425.

Myvatem® 06(K). [Eastman] Diacetyl tartaric acid

ester of dist. monoglycerides (from hydrog. soybean oil); emulsifier, dispersant for food, pharmaceutical, and cosmetic applics.; solid; m.p. 47 C.

Myvatem® 30. [Eastman] Diacetyl tartaric acid ester of dist. monoglycerides (from edible tallow); see Myvatem 06(K); semisolid; m.p. 33 C.

Myvatem® 35K. [Eastman] Diacetyl tartaric acid ester of dist. monoglycerides (from refined palm oil); see Myvatem 06(K); semisolid; m.p. 26 C.

Myvatem® 92K. [Eastman] Diacetyl tartaric acid ester of dist. monoglycerides (from refined sunflower oil); see Myvatem 06(K); liq.; m.p. < 0 C.

Myvatex Mighty Soft. [Eastman] Dist. monoglyceride prepared from edible veg. oil; softener for food industry; wh. powd.; odorless; water-disp.; sp.gr. 0.94; m.p. 62 C min.; acid no. 3.0 max.; 90% min. act.

Myverol® P-06 (K). [Eastman] Dist. monoester from hydrog. soybean oil and propylene glycol; food emulsifier, stabilizer, aerating agent; small beads; sp.gr. 0.89 (80 C); m.p. 45 C; acid no. 3 max.; iodine no. 5 max.; 90% min. monoester content.

N-30. [R.H. Carlson] Susp. of fluorocarbon telomer in a nonflam., low toxicity halocarbon; mold release agent and lubricant providing release for most casting resins or rubbers from surfs. such as plaster, wood, silicone rubber, metal, or epoxy; little buildup on mold or transfer to part; disp.

Na-0101 T 1/8′′. [M&T Harshaw] Sodium bifluoride; catalyst used for removal of HF; tablet, 1/8′′ diam.; dens. 81 lb/ft³; 99% conc.

Nacap. [Vanderbilt] Sodium 2-mercaptobenzothiazole; metal deactivator; corrosion inhibitor for water, alcohol, and glycol systems; used in antifreeze; chemical intermediate; lt. amber liq.; sol. in water, alcohols, and glycols; dens. 1.25 mg/m³; 50% aq.

Nacconol® 35SL. [Stepan] Sodium dodecylbenzenesulfonate; anionic; detergent, wetting and foaming agent, emulsifier, dispersant, penetrant used in textile, cosmetics, paper, leather, food industries; lt. colored clear liq.; faint odor; water sol.; dens. 9.12 lb/gal; visc. 103 cps; cloud pt. 0 C; surf. tens. 41.8 dynes/cm (0.1%); pH 6.0–7.5; 90% min. biodeg.; 35–38% act.

Nacconol® 40F. [Stepan] Sodium dodecylbenzenesulfonate; anionic; wetting and foaming agent, dispersant, detergent used in cement, textile, cleaning industries; wh. flakes; faint odor; water sol.; dens. 18.7–22.0 lb/ft³ (untamped); surf. tens. 31 dynes/cm (0.1%); pH 6.4–7.6 (1% aq.); 90% min. biodeg.; 40% act.

Nacconol® 40G. [Stepan] Sodium dodecylbenzene sulfonate; anionic; foamer, dispersant, wetting agent, detergent for agric., cement, dyeing, emulsion polymerization, textile, metal cleaning, metalworking, mining, paper industries; cream gran. powd.; sol. 2 g/100 ml in water; dens. 35 lb/ft³; pH 6.4–7.6 (1% aq.); biodeg.; 38–42% act.

Nacconol® 90F. [Stepan] Sodium dodecylbenzenesulfonate; anionic; wetting agent, emulsifier, dispersant, foaming agent, detergent used in agric., cleaners, shampoos, textile dyeing, metal cleaning, mining, paper; lt. cream colored flakes; water sol.; dens. 22.0 lb/ft³ (untamped); surf. tens. 35 dynes/cm (0.1%); pH 6.0–7.5 (10% aq.); 90% min. biodeg.; 90% act.

Nacconol® 90G. [Stepan] Sodium dodecylbenzene sulfonate; see Nacconol 40G; also emulsifier for latex polymerization; lt. cream gran. powd.; sol. 15 g/100 ml in water; dens. 30 lb/ft³; pH 6.0–7.5 (1% aq.); biodeg.; 90% act.

Nacopa®. [PMC] Monosodium-4-chlorophthalate; modifying agent for phthalocyanine pigments to retard crystallization; resin modifier; off-wh. powd.; 65% min. assay.

Na Cumene Sulfonate 40, Sulfonate Powd. [Hüls AG] Sodium cumene sulfonate; anionic; hydrotrope, solubilizer, coupling agent, sol'n. aid for liq. detergents and slurries; 40% act. liq., 96% act. powd. resp.

Nadone®. [Allied-Signal] Cyclohexanone; general solv.; coating resins, printing inks, adhesives, fats, oils, waxes; sol. in acetone, ethanol, benzene, ethylene glycol.

Naftocit® Di 4. [Chemetall] Zinc dimethyldithiocarbamate; vulcanizing agent for rubber and latex processing; powd.

Naftocit® Di 7. [Chemetall] Zinc diethyldithiocarbamate; see Naftocit Di 4; powd.

Naftocit® Di 13. [Chemetall] Zinc dibutyldithiocarbamate; see Naftocit Di 4; powd.

Naftocit® DPG. [Chemetall] N,N′-diphenyl guanidine; see Naftocit Di 4; powd.

Naftocit® DPTT. [Chemetall] Dipentamethylene thiuram tetrasulfide; see Naftocit Di 4; powd.

Naftocit® MBT. [Chemetall] 2-Mercaptobenzothiazole; see Naftocit Di 4; powd.

Naftocit® MBTS. [Chemetall] 2,2′-Dibenzothiazyl disulfide; see Naftocit Di 4; powd.

Naftocit® Mi 12 C. [Chemetall] Ethylene thiourea (2-mercaptoimidazoline); see Naftocit Di 4; powd.

Naftocit® NaDBC. [Chemetall] Sodium dibutyldithiocarbamate; see Naftocit Di 4; powd; liq.

Naftocit® NaDMC. [Chemetall] Sodium dimethyldithiocarbamate; see Naftocit Di 4; powd; liq.

Naftocit® Thiuram 16. [Chemetall] Tetramethylthiuram disulfide; see Naftocit Di 4; powd.

Naftocit® ZBEC. [Chemetall] Zinc dibenzyldithiocarbamate; see Naftocit Di 4; powd.

Naftocit® ZMBT. [Chemetall] Zinc-2-mercaptobenzothiazole; see Naftocit Di 4; powd.

Naftolen® H, NV, ZD, ZD103, ZD105, ZD106, ZM. [Chemetall] Aromatic plasticizer and extender oil for natural and syn. rubber; sp.gr. 0.947, 0.976, 0.961, 0.971, 0.949, 0.981, 0.957 resp.; visc. 39, 260, 880, 135, 1250, 950, 500 mm²/s resp. (40 C).

Naftolen® N400, N401, N402, N403, N404, N405, N406, N407, N408. [Chemetall] Naphthenic plasticizer and extender oil for natural and syn. rubber; sp.gr. 0.874, 0.868, 0.868, 0.863, 0.857, 0.853, 0.848, 0.855, 0.866 resp.; visc. 20, 12.7, 31, 46, 68, 100, 185, 460, 21.7 mm²/s resp. (40 C).

Naftolen® ND, P 603, P 611. [Chemetall] Relatively naphthenic plasticizer and extender oil for natural and syn. rubber; sp.gr. 0.834, 0.827, 0.828 resp.; visc.

110, 24, 24 mm²/s resp. (40 C).

Naftolen® P600, P604, P606, P612, P613, P614, P616. [Chemetall] Paraffinic plasticizer and extender oil for natural and syn. rubber; sp.gr. 0.820, 0.817, 0.814, 0.815, 0.799, 0.806, and 0.806 resp.; visc. 120, 146, 77, 85.5, 570, 89, and 91 mm²/s resp. (40 C).

Naftonox® 2246. [Chemetall] 2,2′-Methylenebis (4-methyl-6-t-butylphenol); antioxidant for prod. and processing of natural and syn. rubbers and latexes, fuels and oils, and adhesives.

Naftonox® BBM. [Chemetall] 4,4′-Butylidene bis (2-t-butyl-5-methylphenol); antioxidant for solid rubber and latex processing, adhesives, plastics.

Naftonox® BHT. [Chemetall] 2,6-Di-t-butyl-4-methylphenol; antioxidant for prod. and processing of natural and syn. rubbers and latexes, thermoplastics, adhesives, fuels and oils, foodstuffs and animal feeds.

Naftonox® IMB . [Chemetall] 2,2′-Isobutylenebis (4,6-dimethylphenol); antioxidant for solid rubber and latex processing, adhesives.

Naftonox® PA. [Chemetall] Sterically hindered high alkylated phenols; antioxidant for syn. rubber and latex prod. and processing.

Naftonox® PS. [Chemetall] Sterically hindered styrenated phenols; antioxidant for syn. rubber and latex prod. and processing.

Naftonox® TMQ. [Chemetall] Polymerized 2,2,4-trimethyl-1,2-dihydroquinoline; antioxidant for solid rubber and latex processing, adhesives, plastics; flakes and powd.

Naftonox® ZMP. [Chemetall] Polymerized sterically hindered polyphenol; antioxidant for solid rubber and latex processing, adhesives, plastics.

Naftopast® Antimontrioxid-CP. [Chemetall] Antimony trioxide; processing aid in rubber compounding; solid disp.; 80% conc.

Naftopast® Antimontrioxid. [Chemetall] Antimony trioxide; processing aid in rubber compounding; solid disp.; 77% conc.

Naftopast® Di7-P. [Chemetall] Zinc diethyldithiocarbamate; processing aid in rubber compounding; solid disp.; 75% conc.

Naftopast® Di13-P. [Chemetall] Zinc dibutyldithiocarbamate; processing aid in rubber compounding; solid disp.; 85% conc.

Naftopast® GMF. [Chemetall] GMF; processing aid in rubber compounding; solid disp.; 60% conc.

Naftopast® Litharge A. [Chemetall] Lead oxide; processing aid in rubber compounding; solid disp.; 90% conc.

Naftopast® MBT-P. [Chemetall] 2-Mercaptobenzothiazole; processing aid in rubber compounding; solid disp.; 70% conc.

Naftopast® MBTS-A. [Chemetall] 2,2′-Dibenzothiazyl disulfide; processing aid in rubber compounding; solid disp.; 70 and 75% concs.

Naftopast® MBTS-P. [Chemetall] 2,2′-Dibenzothiazyl disulfide; processing aid in rubber compounding; solid disp.; 70% conc.

Naftopast® MgO-A. [Chemetall] MgO; processing aid in rubber compounding; solid disp.; 60 and 66.6% conc.

Naftopast® Mi12-P. [Chemetall] Ethylene-thiourea (2-mercaptoimidazoline); processing aid in rubber compounding; solid disp.; 80% conc.

Naftopast® Red Lead A, P. [Chemetall] Pb_3O_4; processing aid in rubber compounding; solid disp.; 90%conc.

Naftopast® Schwefel-P. [Chemetall] Sulfur; processing aid in rubber compounding; solid disp.; 75% conc.

Naftopast® Thiuram 16-P. [Chemetall] Tetramethylthiuram disulfide; processing aid in rubber compounding; solid disp.; 70% conc.

Naftopast® TMTM-P. [Chemetall] TMTM; processing aid in rubber compounding; solid disp.; 70% conc.

Naftopast® ZnO-A. [Chemetall] ZnO; processing aid in rubber compounding; solid disp.; 80% conc.

Naftozin® N, Spezial. [Chemetall] Stearic acid; processing aid for solid rubber processing.; lubricant in PVC processing; flakes and pearls; iodine no. 8.0 and 3.0 max. resp.

Nailsyn. [Rona] Syn. pearl pigments; pearlescent for nail polish.

Nalan GN. [DuPont] Dispersed polymer and waxes; cationic; durable water repellent for applic. to all fabrics; homogeneous disp.; milky wh.; acetic odor; readily disp. in water in all proportions at temps. of 21–82 C; dens. 8.14 lb/gal; visc. readily pourable; nonflam.; pH 3.5–4.5; 30% solids.

Nalan W. [DuPont] Modified resin dispersed in water; cationic; water repellent extender for use with Zepel fabric fluoridizer; end uses incl. rainwear, general apparel fabrics, home furnishings, and fabrics for use as automotive and marine tops, awnings, and on outdoor furniture; mobile liq.; milky wh.; mildly acrylic odor; miscible with water in all proportions; dens. 8.24 lb/gal; visc. 10 cps (26.7 C); nonflam.; pH 3.5–4.5; 25% solids.

Nalco® 70. [Nalco] Acylated polyamide; nonionic; antifoam for highly caustic systems; liq.

Nalco® 131. [Nalco] Glycol-type surfactants; nonionic; antifoam for beet sugar and enzyme operations; liq.; 100% act.

Nalco® 1180. [Nalco] Crosslinked polyacrylic acid salt; absorbent for water or aq. sol'ns. offers moderate capacity, high gel str.; dehydrating agent; used in diapers, bandages, agric. applics., construction materials, fire extinguishing, waste sludge stabilization; gran. solid; slight to no odor; 45–1000 μ particle size; bulk dens. 550–750 g/l; pH 5.5–6.5 (5% aq. gel).

Nalco® 1181. [Nalco] Crosslinked polyacrylic acid salt; see Nalco 1180; wh. to slightly yel. powd.; slight to no odor; < 45 μ particle size; bulk dens. 30–40 lb/ft³; pH 5.5–7.5 (5% aq. gel).

Nalco® 2300. [Nalco] Silicone; defoamer for coatings; liq.

Nalco® 2305. [Nalco] Emulsified silicone; defoamer for water-based coatings and adhesives; liq.; water-disp.

Nalco® 2335. [Nalco] Polyelectrolyte; anionic; dispersant for pigments, slurries, coatings; liq.

Nalco® 2343. [Nalco] Polyglycol and fatty type surfactants; nonionic; antifoam; liq.

Nalkylene® 500. [Vista] C_{10-12} linear undecylbenzene sulfonate; surfactant intermediate for prod. of biodegrad. detergent prods.; Saybolt 30+ liq.; odorless; m.w. 237; sp.grt. 0.8654; dens. 7.209 lb/gal (60 F); visc. 4.5 cSt (100 F); b.p. 536 F; f.p. < –70 F; flash pt. (PM) 280 F; 100% act.

Nalkylene® 515, 550. [Vista] Linear alkylbenzene; see Nalkylene 500; water-wh. oily liq. and Saybolt 30+ liq. resp., odorless; sp.gr. 0.861 and 0.862; m.p. < –90 F; 100% act. (550).

Nalkylene® 550L. [Vista] C_{10-14} linear dodecylbenzene sulfonate; see Nalkylene 500; Saybolt 30+ liq.; m.w. 241; sp.gr. 0.860 (60 F); dens. 7.2 lb/gal (60 F); visc. 4.3 cSt (100 F); b.p. 521 F; flash pt. (PM) 298 F; 100% act.

Nalkylene® 600. [Vista] Linear alkylbenzene; see Nalkylene 500; Saybolt 30+ liq.; odorless; sp.gr. 0.865; visc. 46.9 SSU (100 F); m.p. < –90 F; 100% act.

Nalkylene® 600L. [Vista] C_{12-14} linear tridecylbenzene sulfonate; intermediate to produce surfactants, powd. detergents, industrial and specialty chemicals; liq.; 100% act.

Nalzin 2. [Rheox] Zinc hydroxyphosphite; anticorrosive pigment; wh. powd. sp.gr. 3.9; dens. 33 lb/gal.

Nalzin ZP. [Rheox] Zinc phosphate; anticorrosive pigment; wh. powd.; sp.gr. 3.25.

Nansa® 1042. [Albright & Wilson] Dodecylbenzene sulfonic acid; intermediate used in mfg. of detergents, laundry prods.; dk. br. visc. liq.; sp.gr. 1.05; visc. 1900 cs; 95% act.

Nansa® 1042/P. [Albright & Wilson] Dodecylbenzene sulfonic acid; intermediate for neutralization; sodium salts as emulsifiers in emulsion polymerization of plastics and syn. rubbers; dk. br. visc. liq.; sp.gr. 1.05; visc. 1900 cs; 95% act.

Nansa® BMC. [Albright & Wilson] Anionic/nonionic; air entraining agent for mortar and concrete; clear yel. liq.; m.w. 336; sp.gr. 1.045; visc. 800 cs; cloud pt. 15 C; set pt. –6 C; pH 7.0 (5% aq.); 27 ± 1% act.

Nansa® BXS. [Albright & Wilson] Sodium alkyl naphthalene sulfonate; anionic; wetting and dispersing agent for colloidal systems without detergent properties, used in agric., leather, paper and pulp, textile industries; cream-colored powd.; pH 7.5–9.5 (1%); 38% act.

Nansa® EVM50. [Albright & Wilson] Calcium dodecylbenzene sulfonate in aromatic solv.; emulsifier, dispersant for textiles, surf. coatings, polymerization, leather industries; brn. visc. liq.; flash pt. (CC) > 48 C; pH 5.5–7.5 (3%); 50 ± 1.5% act.

Nansa® EVM62/H. [Albright & Wilson] Calcium dodecylbenzene sulfonate in isobutanol; see Nansa EVM50; brn. visc. liq.; flash pt. (CC) > 28 C; pH 6.5–8.0 (3%).; 62 ± 2% act.

Nansa® EVM70. [Albright & Wilson] Calcium dodecylbenzene sulfonate in isobutanol; see Nansa EVM50; brn. visc. liq.; flash pt. (CC) > 28 C; pH 6.5–8.0 (3%); 68.5 ± 1.5% act.

Nansa® EVM70/B. [Albright & Wilson] Calcium dodecylbenzene sulfonate in hexanol; see Nansa EVM50; brn. visc. liq.; flash pt. (CC) > 28 C; pH 5.5–7.5 (3%); 68.5 ± 1.5% act.

Nansa® EVM70/E. [Albright & Wilson] Calcium dodecylbenzene sulfonate in isobutanol; see Nansa EVM50; brn. visc. liq.; flash pt. (CC) > 57 C; pH 5.5–7.5 (3%); 67 ± 2% act.

Nansa® HS40-AU. [Albright & Wilson] Sodium dodecylbenzene sulfonate; wetting agent, dispersant in scouring powds., industrial applics., automotive, floor, wall, and hard surface cleaners; cream coarse powd.; dens. 0.55 g/cm³; pH 8.5 ± 1.0 (1%); > 80% biodeg.; 38 + 3% act.

Nansa® HS40 Soft. [Albright & Wilson] Sodium dodecylbenzene sulfonate; anionic; wetting and dispersing agent, emulsifier used in detergent compd. for domestic and industrial use; cream-colored, free-flowing coarse powd.; dens. 0.55 g/cm³; pH 8.5 ± 1 (1%); > 80% biodeg.; 38 ± 3% act.

Nansa® HS80-AU. [Albright & Wilson] Sodium alkylbenzene sulfonate; wetting agent, dispersant in laundry, detergent compds., insecticides, metal pickling sol'ns.; cream coarse powd.; dens. 0.55 g/cm³; pH 8.5 ± 1.0 (1%); > 80% biodeg.; 78 + 3% act.

Nansa® HS80P. [Albright & Wilson] Sodium dodecylbenzene sulfonate; anionic; see Nansa HS40 Soft; cream-colored powd.; dens. 0.52 g/cm³; pH 8.5 ± 1 (1%); > 80% biodeg.; 80 ± 3% act.

Nansa® HS80/S. [Albright & Wilson] Sodium dodecylbenzene sulfonate; anionic; formulation of detergents, hard surface and bottle cleaners; metal treatment and paper processing; scouring and wetting agent for textile industry; foamer and mortar plasticizer in building industry; cream powd.; sp.gr. 0.65; pH 9–11 (1%); 80% act.

Nansa® HS80SK. [Albright & Wilson] Sodium dodecylbenzene sulfonate; built detergent, wetting, dispersing agent for dairy, laundry, heavy-duty, hard surface detergents; off-wh. to cream coarse powd.; dens. 0.55 g/cm³; pH 9.0 ± 0.8 (1%); > 80% biodeg.; 75 + 3% act.

Nansa® HS80 Soft. [Albright & Wilson] Sodium dodecylbenzene sulfonate; anionic; see Nansa HS40 Soft; cream-colored coarse powd.; dens. 0.55 g/cm³; pH 8.5 ± 1 (1%); > 80% biodeg.; 78 ± 3% act.

Nansa® HS85/S. [Albright & Wilson] Sodium dodecylbenzene sulfonate; anionic; see Nansa HS80/S; also as rubber/plastics emulsifier; cream flake; dens. 0.5 g/cm³; pH 9–11 (1%); biodeg.; 85 ± 3% act.

Nansa® MA30. [Albright & Wilson] Sodium dodecylbenzene sulfonate and ethoxylated nonionic blend; detergent base for dishwashing and hard surf. cleansers; scouring agent for textiles; mortar plasticizer in mfg. of masonry cement; yel. liq.; sp.gr. 1.05; visc. 400 cs; cloud pt. 0 C; pH 6.2–7.2 (5% aq.); 20% act., 10% nonionic in water.

Nansa® SBA. [Albright & Wilson] Dodecylbenzene sulfonic acid; anionic; detergent, intermediate; dk. br. liq.; visc. 20,000 cs; 96% act.

Nansa® SSA. [Albright & Wilson] Dodecylbenzene sulfonic acid; anionic; detergent, intermediate; dk.

br. liq.; visc. 2000 cs; 96% act.

Nansa® SSA/P. [Albright & Wilson] Dodecylbenzene sulfonic acid; intermediate for neutralization; sodium salts as emulsifiers in emulsion polymerization of plastics and syn. rubbers; dk. br. visc. liq.; sp.gr. 1.05; visc. 1900 cs; 95% act.

Nansa® TS 60. [Albright & Wilson] TEA-dodecylbenzene sulfonate; anionic; detergent, emulsifier, pigment dispersant used in cleaners, desizing agent for syn. fibers; med. br. visc. liq.; sp.gr. 1.09; visc. 9000 cs; cloud pt. < 0 C (50%); pH 6.6–7.0 (2% aq.); 60% act.

Nansa® YS94. [Albright & Wilson] Isopropylamine dodecylbenzene sulfonate; emulsifier for solv.-based hand cleaners, coupling agent for water in charge detergent systems; amber visc. liq.; sp.gr. 1.02; visc. 6000 cs; cloud pt. < 0 C; pH 5–8 (2% aq.); biodeg.; 94% act. in water.

Na-Sul 611. [Vanderbilt] Carbonated basic barium dinonylnaphthalene sulfonate in lt. min. oil; rust and corrosion inhibitor in petrol. and syn. lubricating oils and greases; acid neutralizing and deactivating properties; dk. brn. visc. liq.; sol. in petrol. and syn. lubricant bases; insol. in water; dens. 1.08 mg/m³; visc. 65 cSt (100 C); pour pt. 2 C; flash pt. 165 C (COC).

Na-Sul 707. [Vanderbilt] Lithium dinonylnaphthalene sulfonate in lt. min. oil; see Na-Sul 611; dk. brn. liq.; sol. in petrol. and syn. lubricant bases; slightly sol. in water; dens. 0.97 mg/m³; visc. 65 cSt (100 C); pour pt. 4 C; flash pt. 175 C (COC).

Na-Sul 729. [Vanderbilt] Calcium dinonylnaphthalene sulfonate in lt. min. oil; rust inhibitor; demulsifier in oils; grease penetrant; dk. brn. visc. liq.; sol. see Na-Sul 611; dens. 0.98 mg/m³; visc. 80 cSt (100 C); pour pt. 4 C; flash pt. 175 C (COC).

Na-Sul AS. [Vanderbilt] Ammonium dinonylnaphthalene sulfonate in min. seal oil; acid neutralizer; rust and corrosion inhibitor used in galvanized steel, petrol. fuels and lubricating greases; antistall or carburetor deicing agent in motor gasoline; nonstaining additive for aluminum rolling oils; visc. liq.; sol. see Na-Sul 707; dens. 0.94 mg/m³; visc. 40 cSt (100 C); pour pt. 2 C; flash pt. 135 C (COC).

Na-Sul BSB. [Vanderbilt] Basic barium dinonylnaphthalene sulfonate in lt. min. oil; rust and corrosion inhibitor, acid neutralizer, stabilizer used in metal working lubricants (containing chlorinated paraffins), greases, lubricating oils, and specialty prods.; dk. brn. visc. liq.; sol. see Na-Sul 611; dens. 1.06 mg/m³; visc. 50 cSt (100 C); pour pt. 2 C; flash pt. 175 C.

Na-Sul BSN. [Vanderbilt] Neutral barium dinonylnaphthalene sulfonate in lt. min. oil; corrosion, rust inhibitor, demulsifier used in industrial lubricants; slushing oils for cold-rolled sheet steel; dk. brn. visc. liq.; sol. see Na-Sul 611; dens. 1.02 mg/m³; visc. 55 cSt (100 C); pour pt. 4 C; flash pt. 175 C.

Na-Sul BSN/W765. [Vanderbilt] Neutral barium dinonylnaphthalene sulfonate in microcryst. wax; rust inhibitor in protective coatings; tan flakes; dens. 1.10 mg/m³; m.p. 96 C; 65% act. sodium sulfate.

Na-Sul BSN/W780. [Vanderbilt] Neutral barium dinonylnaphthalene sulfonate in microcryst. wax; see Na-Sul BSN/W765; tan flakes; dens. 1.13 mg/m³; m.p. 96 C; 80% act.

Na-Sul CA-50. [Vanderbilt] Carbonated basic calcium dinonylnaphthalene sulfonate in lt. min. oil; rust inhibitor for petrol. lubricants; dk. brn. liq.; sol. see Na-Sul 611; dens. 0.97 mg/m³; visc. 20 cSt (100 C); flash pt. 175 C (COC).

Na-Sul DTA. [Vanderbilt] Diethylenetriamine dinonylnaphthalene sulfonate in lt. min. oil; ashless rust inhibitor in petrol. and syn. lubricant base; high m.w., purity, solubility; demulsifier; dk. brn. liq.; sol. see Na-Sul 611; dens. 0.94 mg/m³; visc. 30 cSt (100 C); flash pt. 180 C (COC).

Na-Sul EDS. [Vanderbilt] Ethylene diamine dinonylnaphthalene sulfonate in kerosene; ashless rust inhibitor for petrol. fuels, pipelines; fingerprint suppressors and removers, lubricating greases; antistat and deicing agent in petrol. fuels; dk. brn. visc. liq.; sol. see Na-Sul 611; dens. 0.93 mg/m³; visc. 30 cSt (100 C); pour pt. –4 C; flash pt. 77 C (COC).

Na-Sul LP. [Vanderbilt] Ethylene diamine dinonylnaphthalene sulfonate in mixed solv.; rust inhibitor for distillate fuels and LPG; corrosion inhibitor in petrol. prods. pipelines; antistat in petrol. fuels; carburetor deicing additive; dk. brn. liq.; sol. see Na-Sul 611; dens. 0.91 mg/m³; visc. 45 cSt (40 C); pour pt. –34 C; flash pt. 54 C (COC).

Na-Sul LS. [Vanderbilt] Lead dinonylnapthalene sulfonate in lt. min. oil; rust inhibitor, demulsifier; additive for metalworking lubricants; industrial and automotive oils, wire rope lubricants; dk. brn. visc. liq.; sol. see Na-Sul 611; dens. 1.06 mg/m³; visc. 30 cSt (100 C); pour pt. –7 C; flash pt. 171 C (COC).

Na-Sul SS. [Vanderbilt] Sodium dinonylnaphthalene sulfonate in lt. min. oil; rust and corrosion inhibitor for petrol. and syn. lubricants, greases, and oils (slushing, preservative, metalworking); dk. brn. visc. liq.; sol. see Na-Sul 707; dens. 0.97 mg/m³; visc. 50 cSt (100 C); pour pt. –1 C; flash pt. 175 (COC).

Na-Sul ZS. [Vanderbilt] Zinc dinonylnaphthalene sulfonate in lt. min. oil; high m.w.; purity, solubility; rust inhibitor used in industrial and automotive oils, greases, and metal working lubricants; dk. brn. liq.; sol. see Na-Sul 611; dens. 0.96 mg/m³; visc. 35 cSt (100 C); pour pt. 18 C; flash pt. 175 C (COC).

Nasuna B. [Henkel Canada] Vinyl pyrrolidone/vinyl acetate copolymer; basic material for hair setting lotions, sprays; sol. in alcohol, water; powd.

Natac®. [Whitney & Oettler] Resin acids-amine resin soaps blend; tackifier for IIR, NR, SBR; NBR, SBR molding aid; slightly activates thiazole and thiuram accelerators; rdsh. br. solid; shatters at R.T.; practically odorless; sp.gr. 1.075–1.085; soften. pt. (R&B) 145–155 F.

Na Toluene Sulfonate 30, 40. [Hüls AG] Sodium toluene sulfonate; hydrotrope, solubilizer, and coupling agent for lt.-duty detergents and heavy-duty detergent slurries; sol. in water.

Natrosol® 250 Series. [Aqualon] Hydroxyethyl cellulose; nonionic; water-sol. polymer for use as a thickener, protective colloid, binder, stabilizer, and suspending agent in industrial applics. incl. pharmaceuticals, textiles, paper, adhesives, decorative and protective coatings, emulsion polymerization, ceramics, etc.; avail. in ten visc. grades, as R-grades which display fast dispersion without lumping in water, as B-grades with superior biostability, and in three particle sizes (regular, X grind, and W grind); wh. powd.; odorless and tasteless; sol. in water, DMSO; wh. to lt. tan powd.; 10% max. on 40 mesh (regular grind), 0.5% on 60 mesh (X grind), 0.5% on 80 mesh (W grind); sp.gr. 1.0033 (2%); ref. index 1.336 (2%); pH 7.

Natrosol® 250ER. [Aqualon] Hydroxyethylcellulose; see Natrosol 250 Series; visc. 25–105 cps (2%).

Natrosol® 250GR. [Aqualon] Hydroxyethylcellulose; see Natrosol 250 Series; m.w. 300,000; visc. 150–400 cps (2%).

Natrosol® 250H4R. [Aqualon] Hydroxyethylcellulose; see Natrosol 250 Series; m.w. 1.1 × 10^6; visc. 2600–3300 cps (1%).

Natrosol® 250HHR. [Aqualon] Hydroxyethylcellulose; see Natrosol 250 Series; m.w. 1.3 × 10^6; visc. 3400–5000 cps (1%).

Natrosol® 250HR. [Aqualon] Hydroxyethylcellulose; see Natrosol 250 Series; m.w. 1.0 × 10^6; visc.1500–2500 cps (1%).

Natrosol® 250JR. [Aqualon] Hydroxyethylcellulose; see Natrosol 250 Series; visc. 150–400 cps (5%).

Natrosol® 250KR. [Aqualon] Hydroxyethylcellulose; see Natrosol 250 Series; visc. 1500–2500 cps (2%).

Natrosol® 250LR. [Aqualon] Hydroxyethylcellulose; see Natrosol 250 Series; m.w. 90,000; visc. 75–150 cps (5%).

Natrosol® 250MHR. [Aqualon] Hydroxyethylcellulose; see Natrosol 250 Series; visc. 800–1500 cps (1%).

Natrosol® 250MR. [Aqualon] Hydroxyethylcellulose; see Natrosol 250 Series; m.w. 720,000; visc. 4500–6500 cps (2%).

Natrosol® Plus, Grade 330. [Aqualon] Modified hydroxyethylcellulose polymer; thickener offering enhanced rheology control for interior and exterior latex paints; wh. to off-wh. powd.; 10% on 40 mesh; bulk dens. 0.55–0.75 g/ml; visc. 150–500 mPas (1%); pH 6.0–8.5; 5% moisture.

Natrovis®. [Aqualon] Hydroxypropyl hydroxyethylcellulose; nonionic; water-sol. polymer; additive to hydraulic binder and latex-modified building materials, e.g., gypsum plasters, cement stuccos, mortars, masonry cements; controls water balance, workability, adhesion, consistency, tackiness; cream to tan powd.; bulk dens. 0.6 kg/l; visc. varies with grade; surf. tens. 56.5 mN/m; pH 7; 4.5% moisture.

Naturechem CR. [CasChem] Cetyl ricinoleate; emollient for cosmetics; wh. solid, liquefies on skin; 100% act.

Naturechem EGHS. [CasChem] Ethylene glycol hydroxystearate; nonionic; aux. emulsifier, emollient, thickener, opacifier for cosmetics, household prods.; wh. flakes; HLB 2.0; 100% conc.

Naturechem GMHS. [CasChem] Glyceryl hydroxystearate; nonionic; aux. emulsifier, emollient, opacifier, bodying and thickening agent for cosmetics, household prods.; wh. flakes; HLB 3.4; 100% conc.

Naturechem OHS. [CasChem] Octyl hydroxystearate; emollient for cosmetics; refatting additive for soaps, cleansers; lt. yel. liq.; 100% act.

Naturechem PGHS [CasChem] Propylene glycol hydroxystearate; nonionic; aux. emulsifier, dispersant, opacifier, thickener, emollient, stabilizer for cosmetics, household prods.; wh. flakes; HLB 2.6; 100% conc.

Naturechem PGR. [CasChem] Propylene glycol ricinoleate; pigment/dye dispersant providing emolliency, gloss, plasticization to cosmetics, household prods.; pale yel. liq.; 100% act.

Naturechem THS-200. [CasChem] PEG-200 trihydroxystearin; nonionic; emulsifier, emollient, thickener, stabilizer for cosmetics, household prods.; wh. wax-like solid; sol. in water and alcohol; HLB 18; 100% conc.

Naturon. [Rona] Natural guanine derived from fish; pearlescent in nail polish, lotions.

Naugalube® 438. [Uniroyal] Alkylated sec. aromatic amine; antioxidant in automatic transmission fluids, turbine oil, and syn. lubricants used in jet turbine engines; thermal stabilizer for automatic transmission fluid at high temps.; tan solid; sol. in most org. liqs.; insol. in water; 5.5% sol. in petrol. oil; sp.gr. 0.97; m.p. 82–87 C; flash pt. (PM) 192 C.

Naugalube® 438-L. [Uniroyal] Alkylated sec. aromatic amine; liq. form of Naugalube 438; used in petrol.-based greases, motor oils, turbine oils, and industrial oils; br. visc. liq.; sol. in most org. liqs.; insol. in water; misc. in petrol. oil; sp.gr. 0.99; visc. 3900 SUS (38 C), 86 SUS (99 C); flash pt. (PM) 154 C; 100% act.

Naugalube® 438-R. [Uniroyal] Alkylated secondary aromatic amine; refined form of Naugalube 438; used in syn. lubricants that must meet exacting military and commercial requirements; off-wh. solid; sol. in most org. liqs.; insol. in water; 5.0% sol. in petrol. oil; sp.gr. 0.97; m.p. 84–95 C; flash pt.(PM) 210 C.

Naugard® 76. [Uniroyal] Octadecyl 3,5-di-t-butyl-4-hydroxyhydrocinnamate; antioxidant for stabilizing polymeric substances such as polyolefins, styrenics, EPDM, and PVC; provides good thermal and color stability; wh. powd.; m.w. 531; sol. in hexane, acetone, ethyl acetate, and xylene; m.p. 50–55 C.

Naugard® 431. [Uniroyal] Phenolic/bisphenolic; antioxidant; stabilizer used in hot melt adhesives applic.

Naugard® 492. [Uniroyal] Phosphite; see Naugard 431.

Naugard® 524. [Uniroyal] Tris (2,4-di-t-butyl phenyl) phosphite; antioxidant used in thermoplastic and thermoset polymers where color and processing stability are critical; wh. powd.; m.w. 647; sol. in chlorobenzene, hexane, hot ethanol; insol. in water; m.p. 180–186 C.

Naugard® BHT. [Uniroyal] Phenolic/bisphenolic; see Naugard 431.

Naugard® NBC. [Uniroyal] Nickel dibutyldithiocarbamate; nickel chelating uv stabilizer for polyolefins; dk. grn. powd.; m.w. 467.5; sol. (g/100 ml solv.): 101 g benzene, 58 g toluene, ethylene dichloride, 28 g CCl_4, 11 g acetone; sp.gr. 1.26; m.p. 86–90 C; flash pt. 500 F (TOC).

Naugard® P. [Uniroyal] Phenolic phosphite; antioxidant; processing and color stabilizer for substrates; hot melt adhesive applic.

Naugard® PHR. [Uniroyal] Phenolic phosphite; see Naugard P.

Naugard® SP. [Uniroyal] Phenolic/bisphenolic; antioxidant and processing stabilizer; used in EVA and polyamide hot melt adhesives.

Naugard® XL-1. [Uniroyal] 2,2′-Oxamido bis-[ethyl 3-(3,5-di-tert-butyl-4-hydroxyphenyl) propionate]; antioxidant and metal deactivator; used in polymerization, processing, and in end use applics., wire and cable insulation, pipe and inj. parts for automobiles and appliances; processing stabilizer for polyolefins, film, sheet, and blow molded bottles; wh. powd.; m.w. 697; sol. in chloroform, acetone, styrene, methanol, hexane, and water; m.p. 175–177 C; flash pt. 260 C (TOC); f.p. 273 C (TOC).

Naugawhite. [Uniroyal] Phenolic/bisphenolic; antioxidant for thermoplastics; hot melt adhesives.

Naxol®. [Allied-Signal] Cyclohexanol; solv. for fats, oils, resins; coupling agent; misc. with ether, ethanol, aromatic hydrocarbons.

Naxonate® 4AX. [Ruetgers-Nease] Ammonium xylene sulfonate; anionic; hydrotrope, stabilizer, solubilizer used in formulating detergents, inks, electroplating baths, dyestuffs, polymers; Klett 50 max. liq.; dens. 9.50 lb/gal; visc. 13.08 cSt (38 C); pH 6.5–8.5; 40% min. act.

Naxonate® 4KT. [Ruetgers-Nease] Potassium toluene sulfonate; anionic; see Naxonate 4AX; Klett 50 max. liq.; dens. 10.2 lb/gal; pH 7–9; 40% min. act.

Naxonate® 4L. [Ruetgers-Nease] Sodium xylene sulfonate; anionic; see Naxonate 4AX; Klett 50 max. liq.; dens. 9.75 lb/gal; visc. 4.49 cSt (38 C); pH 7–9; 40% min. act.

Naxonate® 4ST. [Ruetgers-Nease] Sodium toluene sulfonate; anionic; see Naxonate 4AX; Klett 50 max. liq.; dens. 9.63 lb/gal; visc. 11.93 cSt (38 C); pH 7–9; 40% min. act.

Naxonate® 5KT. [Ruetgers-Nease] Potassium toluene sulfonate; anionic; see Naxonate 4AX; Klett 50 max. liq.; dens. 9.60 lb/gal; visc. 2.98 cSt (38 C); pH 7–9; 50% min. act.

Naxonate® 5L. [Ruetgers-Nease] Sodium xylene sulfonate; anionic; see Naxonate 4AX; Klett 60 max. liq.; dens. 10 lb/gal; visc. 8.79 cSt (38 C); pH 7–9; 50% min. act.

Naxonate® 6AC. [Ruetgers-Nease] Ammonium cumene sulfonate; anionic; see Naxonate 4AX; Klett 80 max. liq.; pH 7–8; 60% min. act.

Naxonate® 45SC. [Ruetgers-Nease] Sodium cumene sulfonate; anionic; see Naxonate 4AX; Klett 75 max. liq.; dens. 9.83 lb/gal; visc. 2.27 cSt (38 C); pH 7–9; 44% min. act.

Naxonate® AX. [Ruetgers-Nease] Ammonium xylene sulfonate; anionic; see Naxonate 4AX; cream powd.; 95% min. act.

Naxonate® G. [Ruetgers-Nease] Sodium xylene sulfonate; anionic; see Naxonate 4AX; wh. to cream powd.; m.w. 208.2; pH 7–9; 93% min. act.

Naxonate® KT. [Ruetgers-Nease] Potassium toluene sulfonate; anionic; see Naxonate 4AX; wh. to cream powd.; m.w. 210.3; pH 7–9; 93% min. act.

Naxonate® SC. [Ruetgers-Nease] Sodium cumene sulfonate; anionic; see Naxonate 4AX; wh. to cream powd.; m.w. 222.2; pH 7–9; 93% min. act.

Naxonate® ST. [Ruetgers-Nease] Sodium toluene sulfonate; anionic; see Naxonate 4AX; wh. to cream powd.; m.w. 194.2; pH 7–9; 93% min. act.

NB. [Angus] 2-Nitro-1-butanol; chemical and pharmaceutical intermediate, in tire cord adhesives, as formaldehyde release agents, deodorants, antimicrobials; m.w. 119.1; sol. 54 g/100 ml water; m.p. –48 C; b.p. 105 C; flash pt. > 200 F (TCC); pH 4.5 (0.1 M aq. sol'n.).

N-Butyl Acid Phosphate. [Richardson] Org. phosphate; corrosion inhibitor, industrial catalyst and additive, visc. modifier; clear liq.; dens. 10.5 lb/gal; acid no. 430; pH 2.1 (10%); 100% act.

NE. [Angus] Nitroethane; intermediate, stabilizer for halogenated solvs., as fuels, explosives, and solvs. for coatings or industrial processes; m.w. 75.1; dens. 1.05 g/ml; b.p. 114 C; flash pt. 87 F (TCC).

Nekal® BA-77. [GAF] Sodium alkyl naphthalene sulfonate; anionic; wetting agent, dispersant, stabilizer without detergent properties, foaming agent used in textiles, paints, pesticides, latex emulsions; tan powd.; m.w. 314; pH 8–10 (5%); 75% act.

Nekal® BX-78. [GAF] Sodium alkyl naphthalene sulfonate; anionic; emulsifier, wetting and dispersing agent, foamer for dyes, pigments, industrial use; cream to tan powd.; m.w. 326; water-sol.; pH 6–8 (5%); 70% act.

Nekal® BX Conc. Paste, BX Dry. [GAF] Sodium alkyl naphthalene sulfonate; anionic; see Nekal BX-78; ylsh. high visc. paste, pale hygroscopic powd. resp.; water sol.; pH 10–12 and 9–11 (5% aq.); 60 and 65% act.

Nekal® NF. [GAF] Sodium alkyl naphthalene sulfonate; anionic; low-foaming wetting agent, dispersant, leveling agent, penetrant used in textile dyeing; clear, amber liq.; misc. with water; sp.gr. 1.04; pH 6–8.

Nekal® WS-21, 25. [GAF] Sulfonated aliphatic polyester; anionic; wetting, rewetting agent, dispersant for textile processing, latex paints; water-wh. slightly visc. liq.; slightly alcoholic odor; sol. in water; sp.gr. 0.9–1.0 and 1.0–1.1 resp.; pH 6.0–7.5 (10%); 48 and 55–60% conc. resp.

Nekanil 910. [BASF AG] Alkyl phenol adduct (9–10 moles EO); nonionic; detergent, wetting and dispersing agent used in textile industry; 100% act.

Nemacur®. [Bayer] Fenamiphos; systemic nematicide; colorless cryst.; m.w. 303.4; sol. (g/1000 ml): > 1200 g in IPA, methylene chloride, toluene; sp.gr. 1.14 (melt); m.p. 49.2 C.

Neobee® 18. [Stepan] Hybrid safflower oil; emollient

oil for cosmetics and pharmaceuticals, solubilizer, solv.; liq.; bland taste and odor; sol. in alcohol, min. oil, acetone; sp.gr. 0.915; visc. 70 cps; sapon. no. 190–200; surf. tens. 31.6 dynes/cm.

Neobee® 20. [Stepan] Propylene glycol dicaprylate/dicaprate; emollient oil for cosmetic and pharmaceutical applics, e.g., bath oils, creams, lotions, lipsticks, glosses, makeup bases, pre- and after-shave lotions, flavor and fragrance carrier/extender, vehicle for vitamins, antibiotics, medicinals; APHA 50 liq.; m.p. –20 C; acid no. 0.1 max.

Neobee® 62. [Stepan] Tristearin; solubilizer, stabilizer used in food applics.; flakes; sapon. no. 189–195.

Neobee® 1053. [Stepan] Coconut oil derived triglycerides; diluent vehicle for flavoring, medicinals, colorings, clouding agent in beverages, liq.; odorless and tasteless; sol. in min. oil, acetone, alcohol; sp.gr. 0.930–0.960; sapon.no. 325–355.

Neobee® 1054. [Stepan] Propylene glycol diester of coconut fatty acids; cosolv. for cosmetics and pharmaceuticals, carrier for flavors and colors; liq.; sol. in alcohol, min. oil, acetone; sp.gr. 0.910–0.923; sapon. no. 315–335.

Neobee® 1062. [Stepan] Coconut oil-derived triglycerides; cosolv., carrier for fat-sol. vitamins, medicinals, and ointments in food, cosmetic and pharmaceutical industries; liq.; sol. in min. oil, acetone; sp.gr. 0.925–0.945; sapon. no. 295–315.

Neobee® 1223 (Neobee O). [Stepan] Caprylic/capric triglyceride; see Neobee 1062; also spreading agent, penetrant, solubilizer; liq.; sol. in min. oil, acetone; sp.gr. 0.938; visc. 30 cps; sapon. no. 300–315; surf. tens. 32 dynes/cm.

Neobee® M-5. [Stepan] Caprylic/capric triglyceride; emollient, diluent vehicle/carrier, solubilizer, cosolv. for flavoring, medicinals, colorings used in foods, beverages, cosmetic and pharmaceutical prod., emollient; APHA 125 max. liq.; sol. in alcohol, min. oil, acetone; sp.gr. 0.930–0.960; visc. 23 cps; m.p. –5 C; acid no. 0.1 max.sapon. no. 335–360; surf. tens. 32.3 dynes/cm.

Neobee® M-20. [Stepan] Propylene glycol dicaprylate/dicaprate; emollient oil for cosmetics and pharmaceuticals, carrier for flavors and colors in foods; solubilizer, cosolv.; liq.; sol. in alcohol, min. oil, acetone; sp.gr. 0.920; visc. 9 cps; sapon. no. 315–335; surf. tens. 31.0 dynes/cm.

Neobee® O. [Stepan] See Neobee 1223.

Neocation G. [Nikko] Quat. ammonium salt; cationic; surfactant, leveling agent for cationic dyes; liq.; 50% conc.

Neodene® 6. [Shell] Linear C_6 alpha olefin; intermediate for surfactants and specialty industrial chemicals; liq.; 100% conc.

Neodene® 8. [Shell] Linear C_8 alpha olefin; see Neodene 6; liq.; 100% conc.

Neodene® 10. [Shell] Linear C_{10} alpha olefin; see Neodene 6; liq.; 100% conc.

Neodene® 12. [Shell] Linear C_{12} alpha olefin; see Neodene 6; liq.; 100% conc.

Neodene® 14. [Shell] Linear C_{14} alpha olefin; see Neodene 6; liq.; 100% conc.

Neodene® 16. [Shell] Linear C_{16} alpha olefin; see Neodene 6; liq.; 100% conc.

Neodene® 18. [Shell] Linear C_{18} alpha olefin; see Neodene 6; liq.; 100% conc.

Neodene® 20. [Shell] Linear C_{20} alpha olefin; see Neodene 6; wax; 100% conc.

Neodene® 1214. [Shell] C_{12}-C_{14} alpha olefin; see Neodene 6; liq.; 100% conc.

Neodene® 1416. [Shell] C_{14}-C_{16} alpha olefin; see Neodene 6; liq.; 100% conc.

Neodene® 1418. [Shell] C_{14}-C_{18} alpha olefin; see Neodene 6; liq.; 100% conc.

Neodene® 1618. [Shell] C_{16}-C_{18} alpha olefin; see Neodene 6; liq.; 100% conc.

Neodol® 1. [Shell] C_{11} alcohol; detergent intermediate; clear liq.; m.w. 173; sp.gr. 0.831; visc. 11 cSt (100 F); m.p. 42–57 F; pour pt. 52 F; flash pt. (PMCC) 250 F; acid no. < 0.001; 100% act.

Neodol® 1-5. [Shell] Pareth-1-5; nonionic; detergent intermediate, surfactant; colorless liq.; misc. with many hydrocarbon-based formulations; sp.gr. 0.9663; dens. 8.1 lb/gal; m.p. –4 to +12 C; cloud pt. 4 C (1% aq.); flash pt. (PMCC) 289 F; acid no. < 0.001; pH 6.0 (1% aq.).

Neodol® 23. [Shell] C_{12}-C_{13} alcohol; detergent intermediate; liq.; m.w. 194; sp.gr. 0.833; visc. 14 cSt (100 F); m.p. 45–72 F; pour pt. 63 F; flash pt. (PMCC) 279 F; acid no. < 0.001; 100% act.

Neodol® 23-1. [Shell] Pareth-23-1; nonionic; detergent intermediate used in preparation of sulfates for high-foaming liq. detergents; APHA 5–10 color.; m.w. 238; sp.gr. 0.873; visc. 13 cSt (100 F); m.p. 27–48 F; HLB 3.7; cloud pt. 13.6 F (1% aq.); flash pt. 289 F (PMCC); pour pt. 41 F; acid no. < 0.001; pH 10.1 (1%); 100% act.

Neodol® 23-3. [Shell] C_{12}–C_{13} pareth-3; detergent intermediate used in preparation of sulfates for high-foaming liq. detergents; APHA 50 max., clear to slightly hazy liq.; m.w. 310–342; dens. 7.7 lb/gal; sp.gr. 0.925; visc. 19 cs (100 F); m.p. 5–6 C; HLB 8.1; flash pt. 300 F (PMCC); pour pt. 4 C; 100% act.

Neodol® 23-5. [Shell] Pareth-23-5; nonionic; detergent intermediate used in preparation of sulfates for high-foaming liq. detergents; APHA 5–10 color.; m.w. 413; sp.gr. 0.965; visc. 23 cSt (100 F); m.p. 27–61 F; HLB 10.7; flash pt. 315 F (PMCC); pour pt. 45 F; acid no. < 0.001; pH 6.0 (1%); 100% act.

Neodol® 23-6.5. [Shell] C_{12}–C_{13} pareth-7 (6.5 EO); see Neodol 23-3; also wetting agent; practically colorless liq., mild odor; m.w. 484; dens. 8.08 lb/gal (100 F); sp.gr. 0.963; visc. 29.5 cs (100 F); m.p. 11–15 C; HLB 12.0; flash pt. 330 F (PMCC); pour pt. 16 C; 100% act.

Neodol® 23-6.5T. [Shell] Pareth-23-6.5; nonionic; detergent intermediate used in preparation of sulfates for high-foaming liq. detergents; APHA 10–15 color.; m.w. 529; sp.gr. 0.993; visc. 33 cSt (100 F); m.p. 36–66 F; HLB 12.6; cloud pt. 147 F (1% aq.); flash pt. 289 F (PMCC); pour pt. 61 F; acid no. < 0.001; pH 6.5 (1%); 100% act.

Neodol® 23-12. [Shell] Pareth-23-12; nonionic; detergent intermediate used in preparation of sulfates for high-foaming liq. detergents; APHA 10–20 color.; m.w. 719; sp.gr. 1.006 (122/77 F); visc. 53 cSt (100 F); m.p. 63–90 F; HLB 14.6; cloud pt. 177 F (5% aq. NaCl); flash pt. 399 F (PMCC); pour pt. 79 F; acid no. < 0.001; pH 10.1; 100% act.

Neodol® 25. [Shell] C_{12}-C_{15} alcohol; detergent, emulsifier intermediate; liq.; m.w. 203; sp.gr. 0.834; visc. 15 cSt (100 F); m.p. 54–77 F; pour pt. 66 F; flash pt. (PMCC) 286 F; acid no. < 0.001; 100% act.

Neodol® 25-3. [Shell] C_{12}–C_{15} pareth-3; nonionic/anionic; detergent intermediate used in preparation of sulfates for high-foaming liq. detergents; colorless, clear to slightly hazy liq.; mild odor; m.w. 320–350; dens. 7.70 lb/gal; sp.gr. 0.925; visc. 19 cs (100 F); m.p. 5–6 C; HLB 7.9; flash pt. 315 F (PMCC); pour pt. 0 C; 100% act.

Neodol® 25-7. [Shell] Pareth-25-7; nonionic; detergent intermediate used in preparation of sulfates for high-foaming liq. detergents; APHA 5–10 paste-like, mild odor.; m.w. 524; sp.gr. 0.965 (122/77 F); visc. 34 cSt (100 F); m.p. 36–70 F; HLB 12.3; cloud pt. 121 F (1% aq.); flash pt. 367 F (PMCC); pour pt. 66 F; acid no. < 0.001; pH 6.0 (1%); 100% act.

Neodol® 25-9. [Shell] Pareth-25-9; nonionic; detergent intermediate used in preparation of sulfates for high-foaming liq. detergents; APHA 5–10 paste-like.; m.w. 597; sp.gr. 0.982 (122/77 F); visc. 41 cSt (100 F); m.p. 57–77 F; HLB 13.1; cloud pt. 163 F (1% aq.); flash pt. 370 F (PMCC); pour pt. 70 F; acid no. < 0.001; pH 6.0 (1%); 100% act.

Neodol® 25-12. [Shell] Pareth-25-12; nonionic; detergent intermediate used in preparation of sulfates for high-foaming liq. detergents; APHA 5–10 color.; m.w. 729; sp.gr. 0.999 (122/77 F); visc. 53 cSt (100 F); m.p. 68–86 F; HLB 14.4; cloud pt. 173 F (5% aq. NaCl); flash pt. 433 F (PMCC); pour pt. 81 F; acid no. < 0.001; pH 6.0 (1%); 100% act.

Neodol® 45. [Shell] C_{14}-C_{15} alcohol; see Neodol 23; liq.; m.w. 218; sp.gr. 0.820; visc. 18 cSt (100 F); m.p. 59–97 F; pour pt. 84 F; flash pt. (PMCC) 315 F; acid no. < 0.001; 100% act.

Neodol® 45-2.25. [Shell] Pareth-45-2.25; nonionic; detergent intermediate used in preparation of sulfates for high-foaming liq. detergents; APHA 5–10 color.; m.w. 319; sp.gr. 0.903; visc. 19 cSt (100 F); m.p. 48–68 F; HLB 6.3; cloud pt. 21 F (1% aq.); flash pt. 336 F (PMCC); pour pt. 59 F; acid no. < 0.001; pH 6.5 (1%); 100% act.

Neodol® 45-7. [Shell] Pareth-45-7; nonionic; detergent intermediate used in preparation of sulfates for high-foaming liq. detergents; APHA 5–10 paste-like.; m.w. 529; sp.gr. 0.959 (122/77 F); visc. 35 cSt (100 F); m.p. 48–75 F; HLB 11.8; cloud pt. 112 F (1% aq.); flash pt. 365 F (PMCC); pour pt. 66 F; acid no. < 0.001; pH 6.0 (1%); 100% act.

Neodol® 45-7T. [Shell] Pareth-45-7; nonionic; detergent intermediate used in preparation of sulfates for high-foaming liq. detergents; APHA 10–15 color.; m.w. 567; sp.gr. 0.966 (122/77 F); visc. 39 cSt (100 F); m.p. 46–73 F; HLB 12.3; cloud pt. 131 F (1% aq.); flash pt. 441 F (PMCC); pour pt. 66 F; acid no. < 0.001; pH 6.8 (1%); 100% act.

Neodol® 45-13. [Shell] Pareth-45-13; nonionic; detergent intermediate used in preparation of sulfates for high-foaming liq. detergents; APHA 5–10 paste-like.; m.w. 790; sp.gr. 1.003 (122/77 F); visc. 59 cSt (100 F); m.p. 77–93 F; HLB 14.5; cloud pt. 178 F (5% aq. NaCl); flash pt. 480 F (PMCC); pour pt. 86 F; acid no. < 0.001; pH 6.4 (1%); 100% act.

Neodol® 91. [Shell] C_9-C_{11} alcohol; nonionic; see Neodol 23; clear liq.; m.w. 160; sp.gr. 0.829; dens. 6.95 lb/gal; visc. 9 cSt (100 F); m.p. 3–25 F; pour pt. 10 F; flash pt. (PMCC) 228 F; acid no. < 0.001; 100% act.

Neodol® 91-2.5, 91-6, 91-8. [Shell] C_9–C_{11} pareth-3 (2.5 EO), -6, and -8 (8.4 EO) resp.; nonionic; oil-sol. emulsifier and wetting agent, intermediate, dispersant, surfactant; 91-6 and 91-8 as solubilizer; colorless liq.; mild odor; m.w. 270, 424, 529 resp.; sol. in wide range of oxygenated and hydrocarbon solvs., 91-2.5 partly sol. in water, 91-8 readily disp. in water; dens. 7.73, 8.24, 8.42 lb/gal; sp.gr. 0.934, 0.991, 1.002; visc. 12, 23, 30 cs (100 F); HLB 8.1, 12.5, 14.0; m.p. –2 to –31 F, 21–52, and 45–68 F; flash pt. 280, 334, 349 F (PMCC); pour pt. –12, 7, 16 C; acid no. < 0.001; pH 6.0 (1% aq. disp.); biodeg.; 100% act.

Neolyn® 23-75T. [Hercules] Rosin-derived elastomeric resin in toluene; modifier for film-formers, contributing pigment-wetting properties, excellent adhesion, outstanding gloss, and broad compatibility without lowering strength of the film-former; used in prod. of high-quality protective and decorative NC- and vinyl-based lacquers for wood, metal, textiles, concrete, glassine in organosol disp. in latex-type adhesives; sol. in aromatic hydrocarbons, ester, and ketones; dens. 1.07 kg/l; visc. (G-H) V; soften pt. 72 C; acid no. 5; flash pt. 11 C (TCC); 75% solids.

Neolyn® 35D. [Hercules] Rosin-derived alkyd-type resin; resin modifier for polymers used in mfg. of adhesives and coatings, offering grease resistance, gloss, adhesion, and flexibility without detriment to film strength; amber solid, flake; low odor; sol. in acetone, MEK, ethyl acetate, butyl acetate, toluene, CCl_4, and ethylene dichloride; insol. in water; dens. 1.208 kg/l; visc. (G-H) Z3; soften. pt. 87 C; acid no. 8; ref. index 1.5467; flash pt. 217 C (COC).

Neolyn® 40. [Hercules] Rosin-derived alkyd, balsamic resin; used in adhesives, plastisols, organosols, vinyl-type inks, and specialty lacquers; plasticizer for polymeric resins, formulation of vinyl inks, hot-melt adhesives; amber; low odor; sol. in ketones, esters, low m.w. alcohols, and aromatic and chlorinated hydrocarbon solvs.; insol. in water; dens. 1.06 kg/l; visc. 1.8 poise (121 C); acid no. 13; ref. index 1.5629; flash pt. 201 C (COC).

Neoprotex KM. [Nikko] Methylol amide surfactant; anionic; softener and water-repellent agent for resin treatment; liq.; 40% conc.

Neorate NA-30. [Nikko] Alkyl phosphate; anionic; penetrant for polyester fiber; liq.; 65% conc.

Neospinol 264. [Toho] Blend of special anionics, nonionics, and min. oil; dispersant for sulfur in rayon prod.; liq.; 100% conc.

Neospinol 358. [Toho] Special nonionic complex (nitrogen compd.); dispersant for sulfur in rayon prod.; liq.; 100% conc.

Neo-Voronit®. [Bayer] Fuberidazole (0.5%) and sodium dimethyl dithiocarbamidate (30%); nonmercurial liq. seed dressing for cereals.

NEPD. [Angus] 2-Nitro-2-ethyl-1,3-propanediol; chemical and pharmaceutical intermediate, in tire cord adhesives, as formaldehyde release agents, deodorants, antimicrobials; m.w. 149.2; sol. 400 g/100 ml water; m.p. 56 C; b.p. decomposes; pH 5.5 (0.1 M aq. sol'n.).

Neustrene® 045, 053. [Humko] Hydrog. menhaden oil; textile lubricant, pharmaceutical intermediate, emulsifier, mold release agent, buffing compd.; solid; acid no. 5–6; sapon. no. 186–201; 100% conc.

Neustrene® 059, 060. [Humko] Hydrog. tallow glycerides (060—refined); textile lubricant, pharmaceutical intermediate, emulsifier, mold release agent, buffing compd.; Gardner 5 max. (059) solids; acid no. 10 and 3 max. resp.; sapon. no. 193–205; 100% conc.

Neustrene® 064. [Humko] Hydrog. soybean oil; textile lubricant, pharmaceutical intermediate, emulsifier, mold release agent, buffing compd.; solid; acid no. 4 max.; sapon. no. 188–200; 100% conc.

Neutral Degras. [Fanning] Fatty esters of wool grease; lubricant with EP and slip chars.; wire drawing compds., slushing and cutting oils, lubricants; rust preventative with excellent adherence to metal surfs.; plasticizer and lubricant for adhesives; textile lubricant formulations to improve rheological and adhesional properties; inhibits crystallization of wax components; ink formulations; dispersant for other waxes; used in leather in stuffing greases; waterproofing agent; fat-like substance; sol. in chloroform, trichloroethylene, and 100 parts of boiling anhyd. alcohol; insol. in water.

Neutrase® 0.5 L, 1.5 G. [Novo] Bacteria proteinase derived from Bacillus subtilis; neutral enzyme for breakdown of proteinaceous matter for prod. of peptides and amino acids; used in brewing industry to substitute malt proteinases when malt is replaced by barley or other unmalted cereals, and in industries modifying and upgrading proteins of vegs. as well as animal origin; brn. clear liq. and brn. gran. resp.; water-sol.; dens. 1.25 g/ml and 1.0 g/cm^3 resp.

Neu-Tri®. [Dow] Trichloroethylene; solv.

Neutronyx® 600, 656. [Onyx] Nonoxynol-9 (9.5 EO) and -11 resp.; nonionic; detergent, dispersant, wetting agent; liq.; sp.gr. 1.05; cloud pt. 54 and 71 C; flash pt. > 200 F; 100% conc.

Nevastain 21. [Neville] Hindered styrenated phenol; antioxidant used in mastic adhesives, rubber goods such as mech. and molded goods, caulking compds., cement, and antiskinning agents; Gardner 7 visc. liq.; m.w. 300; sol. in alcohols, chlorinated hydrocarbons, esters, ethers, ketones (except acetone), aromatic, naphthenic, and terpene hydrocarbons; sp.gr. 1.075; hyd. no. 190; flash pt. 325 F (COC).

Nevastain 30L. [Neville] Alkylated bisphenol; nonstaining, nondiscoloring antioxidant for syn. and natural rubbers.

Nevastain 76. [Neville] Hindered phenolic compd.; see Nevastain 21; Gardner 18 flakes (50% in toluene); m.w. 635; sol. see Nevastain 21; sp.gr. 1.080; soften. pt. 134 C; hyd. no. 122; flash pt. 350 F (COC).

Nevastain 2170. [Neville] Alkylated phenol; antioxidant for natural, syn., nitrile, polychloroprene rubbers; nonstaining; flakes.

Nevastain A. [Neville] Hindered phenolic compd.; see Nevastain 21; Gardner 16 visc. liq.; sol. in aromatics and aliphatics; m.w. 230; sol. see Nevastain 21; sp.gr. 1.082; hyd. no. 210; flash pt. 275 F (COC).

Nevastain B. [Neville] Hindered phenolic compd.; see Nevastain 21; Gardner 15 flakes; m.w. 490; sol. see Nevastain 21; sp.gr. 1.100; soften. pt. 90 C; hyd. no. 120; flash pt. 300 F (COC).

Nevpene 9500. [Neville] Modified hydrocarbon resin; aromatic producing an exceptional tackifier with modification; used in adhesives, coatings, rubber, and caulking compds.; Gardner 8 (50% in toluene) solid, flakes; m.w. 1040; sol. see Nevtac 80; sp.gr. 1.050; soften. pt. (R&B) 95 C; flash pt. 450 F (COC).

Nevtac 80. [Neville] Syn. polyterpene resin; syn. tackifier used in adhesives, coatings, rubber prods., concrete-curing compds., and caulking compds.; Gardner 5 (50% in toluene) solid; m.w. 1070; sol. in esters, ethers, ketones (except acetone), and chlorinated, aromatic, naphthenic, and terpene hydrocarbons; sp.gr. 0.950; soften. pt. (R&B) 81 C; flash pt. 440 F (COC).

Nevtac 99 Super. [Neville] Syn. polyterpene resin; see Nevtac 80; Gardner 7 (50% in toluene) solid, flakes; m.w. 1040; sol. see Nevtac 80; sp.gr. 0.980; soften pt. (R&B) 99 C; flash pt. 450 F (COC).

Nevtac 100. [Neville] Syn. polyterpene resin; see Nevtac 80; Gardner 6 (50% in toluene) solid, flakes; m.w. 1280; sol. see Nevtac 80; sp.gr. 0.960; soften pt. (R&B) 102 C; flash pt. 450 F (COC).

Nevtac 115. [Neville] Syn. polyterpene resin; see Nevtac 80; Gardner 7 (50% in toluene) solid, flakes; m.w. 1415; sol. see Nevtac 80; sp.gr. 0.970; soften. pt. (R&B) 114 C; flash pt. 460 F (COC).

Nevtac 130. [Neville] Syn. polyterpene resin; see Nevtac 80; Gardner 8 (50% in toluene) solid, flakes; m.w. 1410; sol. see Nevtac 80; sp.gr. 0.980; soften. pt. (R&B) 131 C; flash pt. 475 F (COC).

Newcol 3-80, 3-85. [Nippon Nyukazai] Sorbitan oleate; nonionic; antistat, lubricant, emulsifier, corrosion inhibitor, emulsion solubilizer; liq.; 100% conc.

Newcol 20. [Nippon Nyukazai] Sorbitan laurate; nonionic; antistat, emulsifier, corrosion inhibitor, EP lubricant, emulsion solubilizer; solid; 100% conc.

Newcol 25. [Nippon Nyukazai] POE sorbitan laurate; nonionic; see Newcol 20; liq.; 100% conc.

Newcol 40. [Nippon Nyukazai] Sorbitan palmitate; nonionic; see Newcol 20; solid; 100% conc.

Newcol 45. [Nippon Nyukazai] POE sorbitan palmitate; nonionic; see Newcol 20; solid; 100% conc.

Newcol 60. [Nippon Nyukazai] Sorbitan stearate; nonionic; see Newcol 20; solid; 100% conc.

Newcol 65. [Nippon Nyukazai] POE sorbitan stearate; nonionic; see Newcol 20; solid; 100% conc.

Newcol 80. [Nippon Nyukazai] Sorbitan oleate; nonionic; see Newcol 20; liq.; 100% conc.

Newcol 85. [Nippon Nyukazai] POE sorbitan oleate; nonionic; see Newcol 20; liq.; 100% conc.

Newcol 150. [Nippon Nyukazai] POE laurate; nonionic; emulsifier, dispersant, lubricant; liq.; 100% conc.

Newcol 170. [Nippon Nyukazai] POE oleate; nonionic; see Newcol 150; liq.; 100% conc.

Newcol 180. [Nippon Nyukazai] POE stearate; nonionic; see Newcol 150; solid; 100% conc.

Newcol 290K, 290M, 290P. [Nippon Nyukazai] Sodium alkyl sulfosuccinate; anionic; wetting and dispersing agent, solubilizer, penetrant; liq.; 75, 75, and 30% conc.

Newcol 405, 410, 420. [Nippon Nyukazai] POE lauryl amine; nonionic; corrosion inhibitor, intermediate for textile industry; liq., solid; 100% conc.

Newcol 560. [Nippon Nyukazai] POE nonylphenyl ether; nonionic; wetting agent, penetrant, spreading and dispersing agent, emulsifier, base material for detergents; liq.; 100% conc.

Newcol 560SF. [Nippon Nyukazai] Ammonium POE nonylphenyl ether sulfate; anionic; emulsifier, wetting agent, penetrant, detergent used in emulsion polymerization; liq.; 50% conc.

Newcol 560SN. [Nippon Nyukazai] Sodium POE nonylphenyl ether sulfate; anionic; emulsifier, wetting agent, penetrant, detergent used in emulsion polymerization; liq.; 30% conc.

Newcol 561H, 562, 564, 565. [Nippon Nyukazai] POE nonylphenyl ether; nonionic; antifoaming agent, mold lubricant, wetting agent, penetrant, spreading and dispersing agent, emulsifier, base material for detergents; liq.; 100% conc.

Newcol 566. [Nippon Nyukazai] POE nonylphenyl ether; nonionic; wetting agent, penetrant, spreading and dispersing agent, emulsifier, base material for detergents, agric. chemicals, machine oils; liq.; 100% conc.

Newcol 600, 700 Series. [Nippon Nyukazai] POE alkylaryl ether; nonionic; emulsifier, solubilizer, detergent used in emulsion polymerization; liqs. and solids; 100% conc.

Newcol 861SE. [Nippon Nyukazai] Sodium octylphenoxyethoxyethyl sulfonate; anionic; emulsifier, wetting agent, penetrant, detergent used in emulsion polymerization; liq.; 30% conc.

Newcol 864. [Nippon Nyukazai] POE octylphenyl ether; nonionic; antifoaming agent, mold lubricant; liq.; 100% conc.

Newcol 865. [Nippon Nyukazai] POE octylphenyl ether; nonionic; wetting, penetrating, spreading, and dispersing agent, emulsifier, base for detergents, agric. chemicals, machine oils; liq.; 100% conc.

Newcol 1010, 1020. [Nippon Nyukazai] POE octyl ether; nonionic; stabilizer of latex; solid; 100% conc.

Newcol 1100. [Nippon Nyukazai] POE lauryl ether; nonionic; stabilizer of latex; solid; 100% conc.

Newcol 1105. [Nippon Nyukazai] POE octyl ether; nonionic; emulsifier of min. oil, lubricant, detergent, dyeing additive agent; liq.; 100% conc.

Newcol 1110. [Nippon Nyukazai] POE lauryl ether; nonionic; stabilizer of latex; solid; 100% conc.

Newcol 1120. [Nippon Nyukazai] POE lauryl ether; nonionic; emulsifier, dyeing aid; solid; 100% conc.

Newcol 1200. [Nippon Nyukazai] POE oleyl ether; nonionic; emulsifier of min. oil, dyeing additive agent; solid; 100% conc.

Newcol 1203. [Nippon Nyukazai] POE lauryl ether; nonionic; emulsifier of min. oil, dyeing additive agent; liq. 100% conc.

Newcol 1204, 1208, 1210. [Nippon Nyukazai] POE oleyl ether; nonionic; emulsifier of min. oil, dyeing additive agent; liq. (1204), solid (1208, 1210); 100% conc.

Newcol 1305. [Nippon Nyukazai] POE tridecyl ether; nonionic; emulsifier for silicone, dyeing additive agent, lubricant; solid; 100% conc.

Newcol 1305SN, 1310SN. [Nippon Nyukazai] Sodium POE tridecyl ether sulfate; anionic; emulsifier, wetting agent, penetrant, detergent used in emulsion polymerization; liq.; 30% conc.

Newcol 1310, 1515, 1525, 1545. [Nippon Nyukazai] POE tridecyl ether; nonionic; emulsifier for silicone, dyeing additive agent, lubricant; solid (1310, 1545), liq. (1515, 1525); 100% conc.

Newcol 1610, 1620. [Nippon Nyukazai] POE cetyl ether; nonionic; emulsifier of paraffin wax, latex stabilizer; solid; 100% conc.

Newcol 1807, 1820. [Nippon Nyukazai] POE stearyl ether; nonionic; emulsifier of paraffin wax, latex stabilizer; solid; 100% conc.

Newcol B4, B10, B18. [Nippon Nyukazai] POE naphthyl ether; nonionic; dispersant, suspending and diffusion agent; liqs. (B4, B10), solid (B18); 100% conc.

Newkalgen NX 405 H. [Takemoto] POE alkylphenol sulfoacetate; dispersant for wettable powds.; powd.

Newpol PE-61, -62, -64, -68, -74, -75, -78, -88. [Sanyo] EO/PO block copolymer; nonionic; base material for household and industrial detergents; plasticizer, antistat for phenol resins; emulsifier for agric. pesticides and emulsion polymerization; pigment and pitch dispersant; liq.; solid; HLB 5.8, 6.3, 10.1, 14.9, 10.1, 10.7, 14.8, and 14.6 resp.; 100% conc.

Ni-0104 P. [M&T Harshaw] Nickel, reduced and stabilized; catalyst for slurry applics.; powd.; dens. 37 lb/ft^3; 60% conc.

Ni-0104 T 1/8″. [M&T Harshaw] Nickel, reduced and stabilized; catalyst for hydrogenation of both aromatics and aliphatics; tablet, 1/8″ diam.; dens. 90 lb/ft^3; 58% conc.

Ni-0301 T 1/8″. [M&T Harshaw] Nickel, unreduced; catalyst used for the selective hydrogenation of dialkenes to monoalkenes; tablet, 1/8″ diam.; dens. 68 lb/ft^3.

Ni-0707 T 1/8″. [M&T Harshaw] Nickel, unreduced; catalyst; tablet, 1/8″ diam.; dens. 62 lb/ft^3.

Ni-0901 S 1/2″. [M&T Harshaw] Nickel, unreduced; catalyst for reforming hydrocarbons and ammonia decomposition; spheres, 1/2″ diam.; dens. 75 lb/ft^3.

Ni-1404 T 3/16″. [M&T Harshaw] Nickel, reduced and stabilized; catalyst for hydrogenation, hydrogenolysis, and amination; tablet, 3/16″ diam.; dens. 63 lb/ft^3.

Ni-2002 C 1″. [M&T Harshaw] Nickel, unreduced; gas reforming catalyst; cubes; dens. 37 lb/ft^3.

Ni-3210 T 3/16″. [M&T Harshaw] Nickel, reduced and stabilized; catalyst for hydrogenation, hydrogenolyses, or amination reactions; tablet, 3/16″ diam.; dens. 61 lb/ft^3.

Ni-3250 T 3/16″. [M&T Harshaw] Nickel, reduced and stabilized; catalyst for alkaline hydrogenation; dens. 65 lb/ft^3; 50% conc.

Ni-3266 E 1/16″. [M&T Harshaw] Nickel, reduced and stabilized; catalyst for diffusion controlled reactions; extrusion, 1/16″ diam.; dens. 45 lb/ft^3; 50% conc.

Ni-4301 E 1/16″. [M&T Harshaw] Nickel tungsten; catalyst for saturation of mono and polycyclic aromatic compds., denitrogenation, and desulfurization; extrusion, 1/16″ diam.; dens. 60 lb/ft^3.

Ni-5124 3/16″. [M&T Harshaw] Nickel, reduced and stabilized; catalyst for difficult hydrogenations; tablet, 3/16″ diam.; dens. 76 lb/ft^3; 65% conc.

Ni-5132 P. [M&T Harshaw] Nickel, reduced and stabilized; catalyst for slurry applic.; powd.; dens. 27 lb/ft^3.

Ni-5333 T 3/16″. [M&T Harshaw] Nickel, reduced and stabilized; catalyst for hydrogenation of aromatics; tablet, 3/16″ diam.; dens. 62 lb/ft^3; 33% conc.

Niaproof® Anionic Surfactant 4. [Niacet] Sodium tetradecyl sulfate; penetrant in sol'ns. containing low concs. of dissolved solids; wetting agent; essentially colorless liq.; mild, char. odor; misc. in water; sp.gr. 1.031; dens. 8.58 lb/gal; b.p. 92 C; flash pt. (COC) none; surf. tens. 47 dynes/cm; pH 8.5; 27% act.

Niaproof® Anionic Surfactant 08. [Niacet] Sodium 2-ethylhexyl sulfate; penetrant, wetting agent, coupling agent; useful for applics. involving the penetration of conc. acid, alkali, or salt sol'ns.; essentially colorless liq.; mild, char. odor; misc. in water; sp.gr. 1.109; dens. 9.23 lb/gal; b.p. 95 C; flash pt. (COC) none; surf. tens. 63 dynes/cm; pH 7.3; 39% act.

Niax® Catalyst C-224. [Union Carbide] Catalyst for molded rigid foams incorporating both fluorocarbon and water blown formulations utilizing polymeric MDI isocyanates; sp.gr. 0.8978; dens. 7.47 lb/gal; flash pt. (Tag CC) 107 F.

Niax® PPG 2025. [Union Carbide] PPG; antifoam for detergent/cleaning sol'ns.; urethane elastomer polyol; m.w. 2000; water-insol.

Nicron JS 216. [Montana Talc] Platy talc; low oil absorp. filler for paints and coatings; produces max. loading for applics. requiring low VOC and high solids formulations; 19 μ median particle size; 95% thru 200 mesh; Hegman grind 2.0; sp.gr. 2.8; bulk dens. 43 lb/ft^3 (loose), 78 lb/ft^3 (tapped); oil absorp. 16; dry brightness 78–82; 99% conc.

Nicron JS 322. [Montana Talc] Platy talc; low oil absorp. filler for paints, plastisols, and coatings; produces max. loading and min. oil absorp.; 15 μ median particle size; 99% thru 200 mesh; Hegman grind 0–1; sp.gr. 2.8; bulk dens. 35 lb/ft^3 (loose), 65 lb/ft^3 (tapped); surf. area 9 m^2/g; oil absorp. 22; dry brightness 78–80.

Nicron JS 422. [Montana Talc] Platy talc; filler providing min. oil absorp. for low VOC and high solids applics.; 10 μ median particle size; 100% thru 325 mesh; Hegman grind 4; sp.gr. 2.8; bulk dens. 28 lb/ft^3 (loose), 50 lb/ft^3 (tapped); surf. area 11 m^2/g; oil absorp. 22; dry brightness 80.

Nicron JS 426. [Montana Talc] Platy talc; filler providing controlled oil absorp. for low VOC and high solids applics.; 8 μ median particle size; 100% thru 325 mesh; Hegman grind 4; sp.gr. 2.8; bulk dens. 28 lb/ft^3 (loose), 50 lb/ft^3 (tapped); surf. area 14 m^2/g; oil absorp. 26; dry brightness 82–84.

Nicron JS 634. [Montana Talc] Platy talc; filler providing min. oil absorp. for high solids applics. and wherever low VOC is specified; 2 μ median particle size; 100% thru 325 mesh; Hegman grind 6+; sp.gr. 2.8; bulk dens. 9 lb/ft^3 (loose), 18 lb/ft^3 (tapped); surf. area 9 m^2/g; oil absorp. 34; dry brightness 86–88.

Nikkol Batyl Alcohol 100, EX. [Nikko] Batyl alcohol; nonionic; emulsifier, hydrotrope; cosmetic and pharmaceutical preparations; powd.; 100% conc.

Nikkol BEG-1630. [Nikko] Isoceteth-30; nonionic; emulsifier and solubilizer used for cosmetics and pharmaceuticals; solid; HLB 15.5; 100% conc.

Nikkol Behenyl Alcohol 65, 80. [Nikko] Behenyl alcohol; emollient for cosmetics and pharmaceuticals; sol. in warm ethanol, min. oil, 2-hexyldecanol, IPM; insol. in water.

Nikkol BPS-5. [Nikko] PEG-5 phytosterol; nonionic; emulsifier for o/w and w/o compds., solubilizer, dispersant, emollient, foam stabilizer, visc. modifier, conditioner used in cosmetics, pharmaceuticals; wh. to pale yel. paste; sol. in propylene glycol, ethanol; partly sol. in water; HLB 9.5; acid no. 0.25 max.; pH 4.8 (5%).

Nikkol BPS-10. [Nikko] PEG-10 phytosterol; nonionic; see Nikkol BPS-5; wh. to pale yel. paste or solid; sol. in water, propylene glycol, ethanol; HLB 12.5; acid no. 0.18 max.; pH 5.2 (5%).

Nikkol BPS-15, -20. [Nikko] PEG-15 and -20 resp. phytosterol; see Nikkol BPS-5; wh. to pale yel. wax-like solid; sol. in water, propylene glycol, ethanol; HLB 15.0 and 15.5; acid no. 0.06 and 0.07 max. resp.; pH 5.0 and 5.5 (5%).

Nikkol BPS-25, -30. [Nikko] PEG-25 and -30 resp. phytosterol; nonionic; see Nikkol BPS-5; wh. to pale yel. wax-like solid; sol. in water, propylene glycol, ethanol; HLB 18.0; acid no. 0.11 and 0.09 max. resp.; pH 5.7 (5%).

Nikkol BWA-5. [Nikko] PEG-5 lanolin alcohol; nonionic; emulsifier, solubilizer, and softener for shampoos; paste; HLB 12.5; 100% conc.

Nikkol BWA-10. [Nikko] PEG-10 lanolin alcohol; emollient for cream, milk lotion; solid.

Nikkol BWA-20, -40. [Nikko] PEG-20 and -40 resp. lanolin alcohol; nonionic; see Nikkol BWA-5; solid; HLB 16.0 and 17.0 resp.; 100% conc.

Nikkol Chimyl Alcohol 100. [Nikko] Palmityl glyceryl ether; nonionic; emulsifier, hydrotrope used for cosmetic and pharmaceutical prods.; powd.; 100% conc.

Nikkol CO-3. [Nikko] PEG-3 castor oil; nonionic; hydrotrope, emulsifier used in cosmetic and pharmaceutical preparations; liq.; HLB 3.0; 100% conc.

Nikkol CO-10, -20TX, -40TX. [Nikko] PEG-10, -20, and -40 castor oil; nonionic; see Nikkol CO-3; liq.; HLB 6.5, 10.5, and 12.5; 100% conc.

Nikkol CO-60TX. [Nikko] PEG-60 castor oil; nonionic; see Nikkol CO-3; paste; HLB 14.0; 100% conc.

Nikkol DDP-2. [Nikko] Di-PEG-2 alkyl ether phosphate; anionic; emulsifier, stabilizer, dispersant, anticorrosive agent and detergent used in cosmetics, drugs, agric. chemicals and general industrial use; liq.; HLB 6.5; 100% conc.

Nikkol DDP-4. -6. [Nikko] Di-PEG-4 and -6 alkyl ether phosphate; anionic; see Nikkol DDP-2; liq.; HLB 9.0; 100% conc.

Nikkol DDP-8, -10. [Nikko] Di-PEG-8 and -10 alkyl ether phosphate; anionic; see Nikkol DDP-2; semisolid; HLB 11.5 and 13.5; 100% conc.

Nikkol Decaglyn 1-IS. [Nikko] Decaglycerin monoisostearate; nonionic; emulsifier, solubilizer, dispersant for cosmetics, pharmaceuticals and foods; liq.; HLB 12.5; 100% conc.

Nikkol Decaglyn 1-L. [Nikko] Decaglycerin monolaurate; nonionic; see Nikkol Decaglyn 1-IS; liq.; HLB 17.0; 100% conc.

Nikkol Decaglyn 1-LN. [Nikko] Decaglycerin monolinoleate; nonionic; see Nikkol Decaglyn 1-IS; liq.; HLB 12.0; 100% conc.

Nikkol Decaglyn 1-M. [Nikko] Decaglycerin monomyristate; nonionic; see Nikkol Decaglyn 1-IS; paste; HLB 14.5; 100% conc.

Nikkol Decaglyn 1-O. [Nikko] Decaglycerin monooleate; nonionic; see Nikkol Decaglyn 1-IS; liq.; HLB 13.5; 100% conc.

Nikkol Decaglyn 1-S. [Nikko] Decaglycerin monostearate; nonionic; see Nikkol Decaglyn 1-IS; solid; HLB 12.5; 100% conc.

Nikkol Decaglyn 3-O. [Nikko] Decaglycerin trioleate; nonionic; o/w and w/o emulsifier; gelling agent for hydrocarbon; liq.; HLB 6.5; 100% conc.

Nikkol Decaglyn 3-S. [Nikko] Decaglycerin tristearate; nonionic; see Nikkol Decaglyn 3-O; flake; HLB 6.5; 100% conc.

Nikkol Decaglyn 5-IS. [Nikko] Decaglycerin pentaisostearate; nonionic; emulsifier, lubricant, coating agent and anticrystallization agent; liq.; HLB 3.5; 100% conc.

Nikkol Decaglyn 5-O. [Nikko] Decaglycerin pentaoleate; nonionic; see Nikkol Decaglyn 5-IS; liq.; HLB 4.0; 100% conc.

Nikkol Decaglyn 5-S. [Nikko] Decaglycerin pentastearate; nonionic; see Nikkol Decaglyn 5-IS; solid; HLB 3.5; 100% conc.

Nikkol Decaglyn 7-IS. [Nikko] Decaglycerin heptaisostearate; nonionic; see Nikkol Decaglyn 5-IS; liq.; 100% conc.

Nikkol Decaglyn 7-O. [Nikko] Decaglycerin heptaoleate; nonionic; see Nikkol Decaglyn 5-IS; liq.; 100% conc.

Nikkol Decaglyn 7-S. [Nikko] Decaglycerin heptastearate; nonionic; see Nikkol Decaglyn 5-IS; solid; 100% conc.

Nikkol Decaglyn 10-IS. [Nikko] Decaglycerin decaisostearate; nonionic; see Nikkol Decaglyn 5-IS; liq.; 100% conc.

Nikkol Decaglyn 10-O. [Nikko] Decaglycerin decaoleate; nonionic; see Nikkol Decaglyn 5-IS; liq.; 100% conc.

Nikkol Decaglyn 10-S. [Nikko] Decaglycerin decastearate; nonionic; see Nikkol Decaglyn 5-IS; solid; 100% conc.

Nikkol DGO-80. [Nikko] Glyceryl dioleate; nonionic; emollient, emulsifier; liq.; 100% conc.

Nikkol DGS-80. [Nikko] Glyceryl distearate; nonionic; see Nikkol DGO-80; powd.; 100% conc.

Nikkol DLP-10. [Nikko] Dilaureth-10 phosphate; anionic; emulsifier, dispersant, hydrotrope; surfactant, solubilizer; paste; 100% conc.

Nikkol Estepearl 10, 15. [Nikko] Ethylene glycol distearate; pearlescent for shampoos, rinse and hair conditioners; flake; 100% conc.

Nikkol Estepearl 30, 35. [Nikko] Triethylene glycol distearate; pearlescent for shampoos, rinse and hair conditioners; flake; 100% conc.

Nikkol GM-18IS. [Nikko] Batyl isostearate; nonionic; emollient, emulsion stabilizer used in cosmetics; paste; 100% conc.

Nikkol GM-18S. [Nikko] Batyl stearate; nonionic; see Nikkol GM-18IS; solid; sol. in warm ethanol, castor oil, olive oil, oleyl alcohol; insol. in water; 100% conc.

Nikkol GO-430. [Nikko] PEG-30 sorbitan tetraoleate; nonionic; emulsifier, solubilizer, superfatting agent used in drugs and cosmetics, for emulsion polymerization, agric. chemicals, printing inks; pale yel. liq.; sol. in ethanol, ethyl acetate, xylene; partly sol. in water; sp.gr. 1.048; HLB 11.5; ref. index 1.4727; 100% conc.

Nikkol GO-440. [Nikko] PEG-40 sorbitan tetraoleate; nonionic; see Nikkol GO-430; pale yel. liq.; sol. in ethanol, ethyl acetate, xylene; partly sol. in water, propylene glycol; sp.gr. 1.054; HLB 12.5; 100% conc.

Nikkol GO-460. [Nikko] PEG-60 sorbitan tetraoleate; nonionic; see Nikkol GO-430; pale yel. liq.; sol. in water, ethanol, ethyl acetate, xylene; partly sol. in propylene glycol; sp.gr. 1.060; HLB 14.0; 100% conc.

Nikkol GS-460. [Nikko] PEG-60 sorbitan tetrastearate; nonionic; see Nikkol GO-430; pale yel. liq.; sol. in water, ethanol, ethyl acetate, xylene, olive oil; m.p. 35 C; HLB 13.0; 100% conc.

Nikkol HCO-5, -7.5, -10, -20. [Nikko] PEG-5, -7.5, -10, and -20 hydrog. castor oil; nonionic; hydrotrope, emulsifier used in cosmetics and pharmaceuticals; liq.; HLB 6.0, 6.0, 6.5, and 10.5; 100% conc.

Nikkol HCO-30, -30(FF). [Nikko] PEG-30 hydrog. castor oil; nonionic; see Nikkol HCO-5; liq.; HLB 11.0; 100% conc.

Nikkol HCO-40, -40(FF). [Nikko] PEG-40 hydrog. castor oil; nonionic; see Nikkol HCO-5; liq.; HLB 12.5; 100% conc.

Nikkol HCO-50, -50(FF). [Nikko] PEG-50 hydrog. castor oil; nonionic; see Nikkol HCO-5; paste; HLB 13.5; 100% conc.

Nikkol HCO-60, -60(FF). [Nikko] PEG-60 hydrog. castor oil; nonionic; see Nikkol HCO-5; semi-solid; HLB 14.5; 100% conc.

Nikkol HCO-80. [Nikko] PEG-80 hydrog. castor oil; nonionic; see Nikkol HCO-5; solid; HLB 15.0; 100% conc.

Nikkol HCO-100, -100(FF). [Nikko] PEG-100 hydrog. castor oil; nonionic; see Nikkol HCO-5; solid; HLB 16.5; 100% conc.

Nikkol Hexaglyn 1-S. [Nikko] Hexaglyceryl monostearate; nonionic; emulsifier, dispersant, lubricant; solid; HLB 9.5; 100% conc.

Nikkol ICIS. [Nikko] Isocetyl isostearate; emollient used as oil phase component; low congealing pt.; liq.; 100% conc.

Nikkol ICS-R. [Nikko] Isocetyl stearate; see Nikkol ICIS; liq.; 100% conc.

Nikkol Jojoba Oil N. [Nikko] Jojoba oil; emollient; liq.; sol. in min. and veg. oils.

Nikkol Jojoba Oil S. [Nikko] Jojoba oil; decolored and deodorized emollient; liq.; sol. in min. and veg. oils.

Nikkol MGS-A, -B, -DEX. [Nikko] Glyceryl monostearate; nonionic; emulsifier, stabilizer used in foods and cosmetics; flake; HLB 4.5, 5.0, and 6.5 resp.; 100% conc.

Nikkol MGS-ASE, -BSE. [Nikko] Glyceryl monostearate SE; nonionic; see Nikkol MGS-A; flake; HLB 6.5 and 5.5 resp.; 100% conc.

Nikkol MYO-10. [Nikko] PEG-10 oleate; solubilizer for lotions, toiletries; liq.

Nikkol MYS-1EX. [Nikko] PEG-1 stearate; nonionic; emulsifier or solubilizer for cosmetics and pharmaceuticals; flake; HLB 2.0; 100% conc.

Nikkol MYS-2. [Nikko] PEG-2 stearate; nonionic; see Nikkol MYS-1EX; flake; HLB 4.0; 100% conc.

Nikkol MYS-4. [Nikko] PEG-4 stearate; nonionic; see Nikkol MYS-1EX; solid; HLB 6.5; 100% conc.

Nikkol MYS-10. [Nikko] PEG-10 stearate; nonionic; see Nikkol MYS-1EX; solid; HLB 11.0; 100% conc.

Nikkol MYS-25, MYS-25(FF). [Nikko] PEG-25 stearate; nonionic; see Nikkol MYS-1EX; solid; HLB 15.0; 100% conc.

Nikkol MYS-40, MYS-40(FF). [Nikko] PEG-40 stearate; nonionic; see Nikkol MYS-1EX; solid; HLB 17.5; 100% conc.

Nikkol MYS-45, -55. [Nikko] PEG-45 and -55 stearate; nonionic; see Nikkol MYS-1EX; flake; HLB 18.0; 100% conc.

Nikkol NP-7.5. [Nikko] Nonoxynol-8 (7.5 EO); nonionic; emulsifier, detergent, solubilizer, and wetting agent used in cosmetics, insecticide, and other industrial applics.; liq.; HLB 14.0; 100% conc.

Nikkol NP-10, -15. [Nikko] Nonoxynol-10 and -15; nonionic; see Nikkol NP-7.5; liq.; HLB 16.5 and 18.0 resp.; 100% conc.

Nikkol NP-18TX. [Nikko] Nonoxynol-18; nonionic; see Nikkol NP-7.5; paste; HLB 19.0; 100% conc.

Nikkol N-SP. [Nikko] Cetyl palmitate; syn. spermaceti; emollient; solid; sol. in warm oils; 100% conc.

Nikkol OP-3. [Nikko] Octoxynol-3; nonionic; emulsifier, detergent, solubilizer, and wetting agent used in cosmetics, insecticide and other industrial applics.; liq.; HLB 6.0; 100% conc.

Nikkol OP-10, -30. [Nikko] Octoxynol-10 and -30 resp.; nonionic; see Nikkol OP-3; liq. and paste resp.; HLB 11.5 and 17.0; 100% conc.

Nikkol OTP-100S. [Nikko] Sodium di-2-ethylhexyl sulfosuccinate; anionic; dispersant, wetting agent; sponge-like solid; 98% conc.

Nikkol PBC-31. [Nikko] PPG-4 ceteth-1; nonionic; emulsifier, solubilizer, dispersant used in cosmetic, pharmaceuticals and other industrial applics.; liq.; HLB 9.4; 100% conc.

Nikkol PBC-33. [Nikko] PPG-4 ceteth-10; nonionic; see Nikkol PBC-31; paste; HLB 10.5; 100% conc.

Nikkol PBC-34. [Nikko] POE-POP cetyl ether; nonionic; see Nikkol PBC-31; solid; HLB 16.5; 100% conc.

Nikkol PBC-41. [Nikko] PPG-8 ceteth-1; nonionic; see Nikkol PBC-31; liq.; HLB 9.5; 100% conc.

Nikkol PBC-44. [Nikko] PPG-8 ceteth-20; nonionic; see Nikkol PBC-31; HLB 12.5; 100% conc.

Nikkol PBC-44(FF). [Nikko] POE-POP cetyl ether; nonionic; see Nikkol PBC-31; solid; 100% conc.

Nikkol PEMS. [Nikko] Pentaerythritol stearate; emollient for creams, makeups; flake.

Nikkol PEN-4612. [Nikko] PPG-6-decyltetradeceth-12; nonionic; solubilizer for cosmetic lotions and toiletry prods.; solid; HLB 8.5; 100% conc.

Nikkol PEN-4620. [Nikko] PPG-6-decyltetradeceth-20; nonionic; see Nikkol PEN-4612; solid; HLB 11.0; 100% conc.

Nikkol PEN-4630. [Nikko] PPG-6 decyltetradeceth-30; nonionic; see Nikkol PEN-4612; solid; HLB 12.0; 100% conc.

Nikkol PMS-1C. [Nikko] Propylene glycol monostearate; nonionic; emulsifier, dispersant for cosmetics and pharmaceuticals; solid; HLB 3.5; 100% conc.

Nikkol PMS-1CSE. [Nikko] Propylene glycol monostearate SE; nonionic; see Nikkol PMS-1C; solid; HLB 4.0; 100% conc.

Nikkol Sarcosinate OH. [Nikko] Oleoyl sarcosine; corrosion inhibitor; liq.

Nikkol Syncelane 30. [Nikko] δ-Olefin oligomer; emollient for creams, lotions, makeups, hair preps.; sol. in olive oil, min. oil, ester, 2-octyl dodecanol.

Nikkol TDP-2. [Nikko] C_{12-15} pareth-2 phosphate; anionic; emulsifier and solubilizer for cosmetics, drugs, agric. chemicals, dispersant, anticorrosive agent and detergent for general industrial use; liq.; HLB 7.0; 100% conc.

Nikkol TDP-4, -6, -8. [Nikko] Tri-PEG-4, -6, and -8 resp. alkyl ether phosphate; anionic; see Nikkol TDP-2; liq.; HLB 7.0, 8.0, and 11.5; 100% conc.

Nikkol TDP-10. [Nikko] Tri-PEG-10 alkyl ether phosphate; anionic; see Nikkol TDP-2; semisolid; HLB 14.0; 100% conc.

Nikkol TI-10. [Nikko] PEG-20 sorbitan isostearate; nonionic; emulsifier, solubilizer for cosmetics; liq.; HLB 15.0; 100% conc.

Nikkol Trifat S-308. [Nikko] Glycerin tri-2-ethylhexanoate; emollient for creams, lotions, makeups, hair preps.; sol. in ethanol, olive oil, min. oil.

Nikkol TS-10, TS-10(FF). [Nikko] Polysorbate 60; emulsifier, solubilizer for o/w prods., cosmetics, pharmaceuticals; pale yel. to yel., visc. liq.; sol. in water, ethanol, ethyl acetate, toluene; HLB 14.9; sapon. no. 43–49; pH 5.7–7.7 (5%); 100% conc.

Nikkol TS-30. [Nikko] Polysorbate 65; nonionic; see Nikkol TS-10; liq.; HLB 11.0; 100% conc.

Nikkol TW-10, -20, -30. [Nikko] PEG-10, -20, and -30 lanolin; nonionic; solubilizer or bodying agent for soap, cleanser, shampoos, and rinses; paste; HLB 12.0, 13.0, and 15.0; 100% conc.

Nilofoam 60, M, XC. [Sandoz] Silicone emulsion; anionic, nonionic, nonionic resp.; defoamer used in pressure machines; beam batching operation for knit goods; wh. visc. emulsion; sol. in water; pH 6.0–6.5 (1%)

Nimcolan® 1740. [Henkel/Emery] Absorp. base of lanolin esters, alcohols, and sterols; nonionic; emollient, w/o emulsifier, aux. o/w emulsifier; soft solid; 100% act.

Nimcolan® 1747. [Henkel/Emery] Petrolatum, lanolin and lanolin alcohol; nonionic; emollient w/o emulsifier and aux. o/w emulsifier; soft solid; 100% conc.

Nimlesterol® 1730. [Henkel/Emery] Fractionated blend of cholesterol and related sterols; hypoallergenic skin penetrant, emollient, base; sol. in min. oil, IPM; water-insol.

Nimlesterol® 1732. [Henkel/Emery] Fractionated blend of cholesterol and related sterols; nonionic; hypoallergenic w/o emulsifier, emollient, aux. o/w emulsifier for skin care prods.; base; liq.; 100% act.

Ninate® 411. [Stepan] Alkylamine alkylbenzene sulfonate; anionic; emulsifier, penetrant, dye dispersant, wetting agent for textile applics.; defoamer for metalworking applics.; yel. liq.; oil-sol.; 90% act.

Ninate® ABS-60C. [Stepan] Calcium dodecylbenzene sulfonate; anionic; intermediate for emulsifiers; liq.; 60% conc.

Ninate® DS 70. [Stepan] Dioctyl sulfosuccinate; dispersing, wetting agent, emulsifier; liq.

Ninate® L70, R63, R70. [Stepan] Calcium alkylbenzene sulfonate; base emulsifier, dispersant; visc. liq.

Ninate® PA. [Stepan] Sodium polycarboxylate; dispersing, fluidifying, stabilizing agent; visc. liq.

Ninex TDO 5, 9, 14. [Stepan] PEG tall oil dioleates; coemulsifier of min. oils, fatty acids; lubricant, sanforizer, fiber processing aid and disperse dye carrier in textile industry; component of textile and metalworking lubricants; liq.; oil-sol.; 100% conc.

Ninex TMO 5, 7, 9, 14. [Stepan] Tall oil monooleate; nonionic; textile emulsifier, lubricant, fiber processing aid; liq.; HLB 8.3, 10.3, 11.5 and 13.4 resp.; 100% conc.

Ninol® 11-CM. [Stepan] Cocamide DEA; emulsifier and lubricant with antistatic properties for textiles; corrosion inhibitor in cutting fluids, drawing compds., cleaners; lt. color liq.; 100% act.

Ninol® 30-LL. [Stepan] Lauramide DEA; foam booster, thickener for liq. detergents; stabilizer for shampoos, hand soaps, bath prods.; clear amber liq.; 100% act.

Ninol® 40-CO, 49-CE. [Stepan] Cocamide DEA; nonionic; detergent, foam booster and stabilizer, thickener in detergent formulations, textile applics.; lt. color liq.; 100% act.

Ninol® 41-CO. [Stepan] 1:1 Cocamide DEA; surfactant, visc. modifier for bubble baths, personal care prods.; Gardner 4 liq.

Ninol® 50-LL. [Stepan] Lauramide DEA; foam stabilizer and booster; lt. color liq. to gel; 100% act.

Ninol® 51-LL. [Stepan] Lauramide DEA (1:1); nonionic; surfactant, visc. builder for personal care prods.; Gardner 6 liq.

Ninol® 52-LL. [Stepan] Lauramide DEA (1:1); nonionic; surfactant, foam enhancer, visc. builder for personal care prods.; Gardner 4 liq.

Ninol® 55-LL. [Stepan] Lauramide DEA (1:1); nonionic; see Ninol 52-LL; Gardner 3 liq.; 100% act.

Ninol® 70-SL. [Stepan] Lauramide DEA; nonionic; foam booster, stabilizer, visc. builder for shampoos, hand soaps, bath prods.; solid; 100% act.

Ninol® 71-LL. [Stepan] Lauric/myristic DEA (1:1); nonionic; see Ninol 52-LL; Gardner 2 liq.

Ninol® 96-SL. [Stepan] Lauramide DEA; thickener, foam stabilizer and booster, detergent; lt. color solid; 100% act.

Ninol® 128 Extra. [Stepan] Cocamide DEA; nonionic; foam booster and stabilizer, detergent, thickener used in shampoos, bubble baths, dishwash; yel. liq.; water-disp.; dens. 8.3 lb/gal; 100% conc.

Ninol® 201. [Stepan] Oleamide DEA; nonionic; emulsifier in industrial lubricant systems; thickener for personal care and liq. detergent prods.; lubricant, antistat for textiles; corrosion inhibitor in cutting fluids; amber clear liq.; sol. in oils, disp. in water; dens. 8.23 lb/gal; pH 9.5–10.5; 100% act.

Ninol® 1281. [Stepan] Modified fatty alkanolamide; emulsifier and corrosion inhibitor in cutting fluids, drawing compds., cleaners; liq.

Ninol® 1301. [Stepan] Modified fatty alkanolamide; foam stabilizer and thickener for textile applics.; emulsifier and corrosion inhibitor in cutting fluids, drawing compds., cleaners; lt. tan paste; 1005 act.

Ninol® 2012 Extra. [Stepan] Cocamide DEA; nonionic; foam booster and stabilizer, detergent, thickener used in shampoos, bubble baths, dishwash; yel. liq.; water-disp.; dens. 8.2 lb/gal; 100% conc.

Ninol® 4812. [Stepan] Cocamide; see Ninol 128 Extra; water-sol.

Ninol® 4821. [Stepan] Lauramide DEA; nonionic; foam booster/stabilizer, visc. builder for personal care prods.; clear amber liq.; water-sol.; dens. 8.2 lb/gal; 90% act.

Ninol® 5024. [Stepan] Modified fatty diethanolamide; highly salt tolerant thickener base for hard surf. cleaners; dk. amber liq.; 100% act.

Ninol® A-10MM. [Stepan] Mixed fatty alkanolamide; visc. building base for hard-surf. cleaners; highly salt tolerant; liq.; 100% act.

Ninol® AA-62, AA-62 Extra. [Stepan] Lauramide DEA; nonionic; foam booster and stabilizer, detergent, thickener used in shampoos, bubble baths, dishwash; paste and wh. wax resp.; dens. 8.34 and 8.2 lb/gal; 100% act.

Ninol® AC-201. [Stepan] Oleic fatty acid polydiethanolamide; nonionic; anti-corrosion additive for cutting oils; liq.; 100% conc.

Ninol® CBR. [Stepan] Cocamide MEA; thickener, foam booster, foam stabilizer; wh. wax.

Ninol® CMP. [Stepan] Cocamide MEA; foam booster and thickener for liq. detergents; also sueful in detergent blocks or bars; may be melted and molded; wh. beads; 100% act.

Ninol® CNR. [Stepan] Cocamide MEA; nonionic; thickener, superfatting agent, foam booster; solid; 100% conc.

Ninol® CX. [Stepan Canada] Alkanolamide; nonionic; foam and visc. booster; liq.; 100% conc.

Ninol® DS, 4821 F. [Stepan] Cocamide DEA; thickener, foam booster, foam stabilizer; visc. liq.

Ninol® GR. [Stepan] Cocamide DEA; foam booster, stabilizer, visc. builder for shampoos, hand soaps, bath prods.; liq.; 100% act.

Ninol® L-9. [Stepan] Lauramide DEA; see Ninol GR; solid; 100% act.

Ninol® LD. [Stepan] Cocamide DEA (2:1); nonionic; thickener, detergent builder, foam booster; liq.; 100% conc.

Ninol® LDL 2. [Stepan] Alkyl DEA; thickener, foam booster, foam stabilizer; detergency enhancer; visc. liq.

Ninol® LMP. [Stepan] Lauric/myristic MEA; foam booster, thickener for liq. detergents; also useful in detergent blocks or bars; may be melted and molded; wh. beads; 100% act.

Ninol® P-616. [Stepan] Lauramide DEA; see Ninol AA-62 Extra.

Ninol® P-621. [Stepan] Lauramide DEA; nonionic; foam booster and stabilizer, detergent, thickener used in shampoos, bubble baths, dishwash; wh. paste; water-disp.; dens. 8.2 lb/gal; 100% conc.

Ninol® P-650. [Stepan] Cocamide DEA; see Ninol 2012 Extra.

Ninol® SNR. [Stepan Europe] Stearamide MEA; nonionic; thickener, superfatting agent, pearlescent; solid; 100% conc.

Ninol® SR-100. [Stepan] Oleamide DEA; low-foaming visc. builder for hard-surf. cleaners; emulsifier and corrosion inhibitor in cutting fluids, drrawing compds., cleaners; liq.; 100% act.

Ninox® CA. [Stepan] Cocamidopropylamine oxide; nonionic/cationic; surfactant, foamer, foam stabilizer, conditioner, thickener for lt.-duty detergents, shampoos, etc.; clear straw liq.; 30% act.

Ninox® CS. [Stepan] Alkyl amine oxide; emollient, thickener, mild detergent; liq.

Ninox® FCA. [Stepan] Cocamidopropyl dimethylamine oxide; nonionic; wetting, foaming, foam stabilizer, thickener, antistat in personal care prods. and detergents; liq.; 30% conc.

Ninox® L. [Stepan] Lauramine oxide; see Ninox CA; water-wh. clear liq.; 30% act.

Ninox® M. [Stepan] Myristamine oxide; see Ninox CA; water-wh. clear liq.; 30% act.

Ninox® O. [Stepan] Oleylamine oxide; thickener, mild detergency enhancer; liq.

Ninox® S. [Stepan] Stearylamine oxide; thickener emollient, mild detergent; paste.

Niox AK-40. [Pulcra SA] PEG-200 oleate; spreading agent, antifoamer, kerosene and min. oil emulsifier, dispersant; water-disp.

Niox AK-44. [Pulcra SA] PEG-400 oleate; lubricant, min. and veg. oil emulsifier, dispersant; water-disp.; sol. in min. oil.

Niox EO-10. [Pulcra SA] Cetyl alcohol blend; base for cosmetic creams and lotions; water-insol.

Niox EO-12. [Pulcra SA] Linear fatty alcohol, ethoxylated sat.; nonionic; surface act. agent for detergents; dispersant; soap additive; flakes; HLB 16.0; 100% conc.

Niox EO-14, 23. [Pulcra SA] Linear fatty alcohol ethoxylated sat.; nonionic; foam controller in detergents; dispersant; flakes; HLB 17.8 and 17.4 resp.; 100% conc.

Niox EO-26. [Pulcra SA] Cetearyl alcohol blend; base for cosmetic creams and lotions; water-insol.

Niox EO-32, -35. [Pulcra SA] Linear fatty alcohol ethoxylated sat.; nonionic; see Niox EO-12; flakes; HLB 16.2 and 15.5 resp.; 100% conc.

Niox EO-33. [Pulcra SA] PEG diester; conditioning agent in shampoos, skin creams and lotions; visc. builder; sol. in water, alcohols, glycols.

Niox KF-12. [Pulcra SA] Linear fatty alcohol ethoxylated sat.; nonionic; detergent, wetting agent, dispersant; emulsifier for waxes, oils, and fats; stabilizer for syn. resins; solid; HLB 14.5; 100% conc.

Niox KF-13. [Pulcra SA] Linear fatty alcohol ethoxylate; nonionic; emulsifier for min. oils, aliphatic solvs.; visc. depressant; intermediate for sulfonation; liq.; HLB 7.6; 100% conc.

Niox KF-17. [Pulcra SA] Linear fatty alcohol ethoxylate; nonionic; emulsifier for min. oils; additive for detergents; intermediate for sulfonation; liq.; oil-sol.; HLB 6.1; 100% conc.

Niox KF-26. [Pulcra SA] Linear fatty alcohol ethoxylate; nonionic; see Niox KF-17; liq.; HLB 6.8; 100% conc.

Niox KH Series. [Pulcra SA] Fatty acid, ethoxylated; nonionic; raw material used as emulsifier and coemulsifier for paraffinic waxes and compds., dyeing assistant, lubricant, antistat; emulsifier for the textile industry, min. oils, defoamer, softener, detergent and solubilizer for all industries; 100% conc.

Niox KI Series. [Pulcra SA] Ethoxylated castor oil; nonionic; surface act. agent used as detergent, emulsifier, dispersant, and hydrosolubilizer for conc. pesticides in the metal, leather, cosmetics, toiletries, pharmaceutical and polymer industries; textile assistant; 100% conc.

Niox KI-29. [Pulcra SA] Ethoxylated castor oil; solubilizer for flavors, perfumes, vitamins, oils; water-sol.

Niox KJ-10. [Pulcra SA] Ethoxylated oleyl alcohol; emulsifier, dispersant, fragrance grade w/o emulsifier; lipophilic cosolv.; solubilizer; liq.; oil-misc.

Niox KJ-55. [Pulcra SA] Linear fatty alcohol, ethoxylated sat.; nonionic; surface act. agent for detergents; soap additive; dispersant; solid; HLB 13.2; 100% conc.

Niox KJ-56. [Pulcra SA] Linear fatty alcohol ethoxylated sat.; nonionic; emulsifier intermediate raw material for mfg. of ethoxysulfates used in detergent and industrial specialties; solid; HLB 8.1; 100% conc.

Niox KJ-61. [Pulcra SA] Linear fatty alcohol ethoxylated sat.; nonionic; see Niox KJ-55; solid; HLB 14.1; 100% conc.

Niox KJ-66. [Pulcra SA] Linear fatty alcohol ethoxylated sat.; nonionic; see Niox KJ-56; solid; HLB 10.2; 100% conc.

Niox KL-19. [Pulcra SA] Linear fatty alcohol ethoxylated sat.; nonionic; see Niox KF-17; liq.; HLB 8.0; 100% conc.

Niox KP-68. [Pulcra SA] Isotridecanol, ethoxylated; nonionic; detergent, foamer, solubilizer; scouring and leveling agent for the textile industry; paste; HLB 14.1; 100% conc.

Niox KP-69. [Pulcra SA] Isotridecanol, ethoxylated; nonionic; detergent, foamer, solubilizer used in foaming detergents; emulsifier and wetting agent for textile fiber finishing; liq.; HLB 12.7; 100% conc.

Niox KQ-20. [Pulcra SA] Fatty oxo-alcohol, ethoxylated; nonionic; aux. for wool scouring; emulsifier for hydrocarbon solv.; wetting and dispersant for detergents; paste; HLB 13.0; 100% conc.

Niox KQ-30. [Pulcra SA] Fatty oxo-alcohol, ethoxylated; nonionic; see Niox KF-17; liq.; HLB 7.8; 100% conc.

Niox KQ-32. [Pulcra SA] Fatty oxo-alcohol, ethoxylated; nonionic; see Niox KQ-20; paste; HLB 12.0; 100% conc.

Niox KQ-33. [Pulcra SA] Fatty oxo-alcohol, ethoxylated; nonionic; emulsifier for silicones and hydrocarbon solvs.; dry cleaning assistant; intermediate raw material for sulfonation; liq.; HLB 9.2; 100% conc.

Niox KQ-36. [Pulcra SA] Fatty oxo-alcohol, ethoxylated; nonionic; see Niox KQ-20; paste; HLB 11.0; 100% conc.

Niox KQ-55, -56. [Pulcra SA] Fatty oxo-alcohol, ethoxylated; nonionic; o/w emulsifier for hydrocarbon solvs. and min. oils; detergent; wetting agent and antistat for textile industry; assistant for wool scouring and speciality cleaners; liq., paste resp.; water-sol.; HLB 12.8 and 13.7 resp.; 100% conc.

Niox KQ-70, LQ-13. [Pulcra SA] Fatty oxo-alcohol, ethoxylated; nonionic; see Niox KQ-20; paste, liq. resp.; 100 and 88% conc. resp.

Niox KS Series. [Pulcra SA] PEGs; humectant for food, cosmetic, pharmaceutical prods.; lubricant, solv. for dyestuffs and resins; solv. and ointment base for cosmetics and pharmaceuticals; sol. in water; partly sol. in some org. solvs.

NiPar S-10. [Angus] 1-Nitropropane; intermediate, solvs. for coatings; m.w. 89.1; dens. 1.0 g/ml; b.p. 131 C; flash pt. 96 F (TCC).

NiPar S-20. [Angus] Nitropropane; used in solv. blends for inks and coatings; esp. for NC, chlorinated rubber, vinyl, epoxy, acrylic, PU, polyamide systems; automotive finishes; sol. in many nonaq. systems.

Nissan Amine AB. [Nippon Oils & Fats] Stearyl amine; intermediate for cationic surfactants; anticorrosive agent, germicide, wetting agent, mold release agent, softener, emulsifier, dispersant, intermediate used in textiles, water treatment, concrete, asphalt, agriculture, ceramics; wh. waxy solid or flake; solid. pt. 47–53 C; 98% min. primary amine.

Nissan Amine ABT. [Nippon Oils & Fats] Primary hydrog. tallow amine; see Nissan Amine AB; wh. waxy solid or flake; solid. pt. 40–46 C; 98% min. primary amine.

Nissan Amine ABT_2. [Nippon Oils & Fats] Primary tallow-alkylamine; see Nissan Amine AB; wh. waxy solid; solid. pt. 30–40 C; 98% min. primary amine.

Nissan Amine BB. [Nippon Oils & Fats] Primary dodecylamine; see Nissan Amine AB; APHA 120 max. liq. in summer, solid in winter; solid. pt. 24–28 C; 98% min. primary amine.

Nissan Amine DT. [Nippon Oils & Fats] Tallow-alkyl propylenediamine; asphalt emulsifier, corrosion inhibitor, wetting agent, dispersant used in water treatment, pigment flushing and dispersing, ore flotation, gasoline, grease, fuel oil additive; yel. br. solid; solid. pt. 25–34 C.

Nissan Amine DTH. [Nippon Oils & Fats] Tallow-hydrog.-alkyl propylenediamine; see Nissan Amine DT; ylsh. br. flake.

Nissan Amine FB. [Nippon Oils & Fats] Primary cocoalkylamine; see Nissan Amine AB; APHA 120 max., liq. in summer, solid in winter; solid. pt. 16–22 C; 98% min. primary amine.

Nissan Amine MB. [Nippon Oils & Fats] Primary tetradecylamine; see Nissan Amine AB; wh. waxy solid; solid. pt. 31–35 C; 98% min. primary amine.

Nissan Amine OB. [Nippon Oils & Fats] Primary oleylamine; see Nissan Amine AB; 140 max APHA, liq. in summer, solid in winter; solid. pt. 30 C max.; 98% min. primary amine.

Nissan Amine PB. [Nippon Oils & Fats] Primary hexadecylamine; see Nissan Amine AB; wh. waxy solid; solid. pt. 38–46 C; 98% min. primary amine.

Nissan Amine SB. [Nippon Oils & Fats] Primary soybean alkylamine; see Nissan Amine OB; liq., solid; 100% conc.

Nissan Amine VB. [Nippon Oils & Fats] Behenylamine; see Nissan Amine OB; solid; 100% conc.

Nissan Anon BF. [Nippon Oils & Fats] Dimethyl cocoalkyl betaine; amphoteric; plastic antistatic agent, softener, germicide in foods and cleaning industries, asphalt antistripping agent, textile treatment; lt. yel. liq.; 25% min act.

Nissan Anon BL. [Nippon Oils & Fats] Dimethyl dodecyl betaine; amphoteric; antistatic agent, softener, germicide in foods and cleaning industries used in textile industry, rinse aid in cosmetics, extraction aid in fermentation; lt. yel. liq.; pH 4.5–6.5; 30–33% min. act.

Nissan Anon LG. [Nippon Oils & Fats] Alkyl di-(aminoethyl) glycine; amphoteric; antistatic agent, softener, germicide in foods and cleaning industries used in water treatment, as duckweed killer, extraction aid for fermentation; lt. yel. liq.; pH 8.5–10.5; 30% act.

Nissan Cation AB. [Nippon Oils & Fats] Octadecyl trimethyl ammonium chloride; cationic; germicide in water treatment, petrol., paper, foods and textile industries, antistat in plastics, pulp and paper industries, rinse agent, pigment coating agent, dispersant, coagulant, softener; lt. yel. liq.; 23% min. act.

Nissan Cation ABT-350, 500. [Nippon Oils & Fats] Tallow-hydrog. alkyl trimethyl ammonium chloride; cationic; see Nissan Cation AB; lt. yel. liq.; 33 and 50% act.

Nissan Cation AR-4. [Nippon Oils & Fats] Alkyl imidazoline chloride; cationic; see Nissan Cation AB; yel.-br. liq.; 35% min. act.

Nissan Cation BB. [Nippon Oils & Fats] Dodecyl trimethyl ammonium chloride; cationic; see Nissan Cation AB; lt. yel. liq.; 30% min. act.

Nissan Cation F_2-10R, -20R, –40E, -50. [Nippon Oils & Fats] Coco-alkyl dimethyl benzyl ammonium chloride; cationic; see Nissan Cation AB; colorless to lt. yel. liq.; 9.5, 19–21, 40–44, and 50% act. resp.

Nissan Cation FB, FB-500. [Nippon Oils & Fats] Coco-alkyl trimethyl ammonium chloride; cationic; see Nissan Cation AB; lt. yel. liq.; 30 and 50% min act.

Nissan Cation L-207. [Nippon Oils & Fats] POE dodecyl monomethyl-ammonium chloride; cationic; see Nissan Cation AB; ylsh.-br., visc. liq.; 90% min. act.

Nissan Cation M_2-100. [Nippon Oils & Fats] Tetradecyl dimethyl benzyl ammonium chloride; cationic; see Nissan Cation AB; wh. or lt. yel. cryst. powd.; 85% min. act.

Nissan Cation MA. [Nippon Oils & Fats] Tetradecylamine acetate; cationic; corrosion inhibitor and germicide in water treatment and petrol. industries, duckweed killer, pigment dispersant, antistatic agent used in textile industry; wh. to lt. yel. flake; solid. pt. 60–72 C.

Nissan Cation PB-40, -300. [Nippon Oils & Fats] Hexadecyl trimethyl ammonium chloride; cationic; see Nissan Cation AB; lt. yel. liq.; 40 and 27% act. resp.

Nissan Cation S_2-100. [Nippon Oils & Fats] Octadecyl dimethyl benzyl ammonium chloride; cationic; see Nissan Cation AB; wh. or lt. yel. cryst. powd.; 85% min. act.

Nissan Cation SA. [Nippon Oils & Fats] Octadecylamine acetate; cationic; see Nissan Cation MA; wh. to lt. yel. flake; solid. pt. 72–82 C.

Nissan Diapon T. [Nippon Oils & Fats] N-methyl tallow-alkyl taurate, sodium salt; anionic; dyeing aux. for hair dye detergent, emulsifier; powd.; 30% conc.

Nissan Diapon TO. [Nippon Oils & Fats] N-methyl oleoyl taurate, sodium salt; anionic; detergent, emulsifier; scouring agent for wool; dyeing aux.; powd.; 30% conc.

Nissan Disfoam C Series. [Nippon Oil & Fats] Polyalkylene glycol derivs.; defoamers for fermentation field, pulp, synthetics, etc.; liq.

Nissan Elegan S-100. [Nippon Oils & Fats] Special nonion; antistat for ABS, PS, PP, and PE plastics; solid.

Nissan New Elegan A. [Nippon Oil & Fats] Special cation; antistat for soft PVC; paste.

Nissan New Elegan ASK. [Nippon Oil & Fats] Special cation; antistat for rigid PVC; powd.

Nissan Nonion DS-60HN. [Nippon Oils & Fats] PEG distearate; emulsifier, thickener used in cosmetics, textile, industrial uses, nonionic; solid; sol. in water, methanol, warm in diethylene glycol; m.p. 54–62 C; HLB 19; acid no. 2 max.; 100% conc.

Nissan Nonion E-205, 215, 230. [Nippon Oils & Fats] POE oleyl ether; nonionic; emulsifier, dispersant, detergent, wetting agent used in textile processing, cosmetics, metalworking, agric. preparations, industrial cleaners; APHA 140 max. liq., semisolid, solid resp.; sol. in xylene, kerosene, methanol, disp. in water (E-205); sol. in water, methanol, cloudy in xylene (E-215, -230); HLB 9.0, 14.2 and 16.6; cloud pt. 0, 95 and 100 C resp. (1% aq.); 100% conc.

Nissan Nonion K-202, -203. [Nippon Oils & Fats] POE lauryl ether; nonionic; emulsifier, dispersant, detergent, wetting agent used in textile processing, cosmetics, metalworking, agric. preparations, industrial cleaners; APHA 200 max. liq.; sol. in xylene, ether, methanol; disp. in kerosene; HLB 6.0 and 8.0 resp.; cloud pt. 0 C max. (1% aq.); 100% conc.

Nissan Nonion K-204. [Nippon Oils & Fats] POE lauryl ether; nonionic; see Nissan Nonion K-202; APHA 200 max. semisolid; sol. in xylene, ether, methanol; sol. warm in kerosene; disp. in water; HLB 9.2; cloud pt. 0 C max. (1% aq.); 100% conc.

Nissan Nonion K-207. [Nippon Oils & Fats] POE lauryl ether; nonionic; see Nissan Nonion K-202; APHA 50 max. semisolid; sol. in water, xylene, methanol; sol. warm in ether; HLB 12.1; cloud pt. 55–63 C max. (1% aq.); 100% conc.

Nissan Nonion K-211. [Nippon Oils & Fats] POE lauryl ether; nonionic; see Nissan Nonion K-202; APHA 50 max. semisolid; sol. in water, xylene, methanol; HLB 14.1; cloud pt. 90–98 C max. (1% aq.); 100% conc.

Nissan Nonion K-215. [Nippon Oils & Fats] POE lauryl ether; nonionic; see Nissan Nonion K-202; APHA 140 max. semisolid; sol. in water, xylene, methanol; HLB 15.2; cloud pt. 100 C min. (1% aq.); 100% conc.

Nissan Nonion K-220. [Nippon Oils & Fats] POE lauryl ether; nonionic; see Nissan Nonion K-202; APHA 120 max. solid; sol. in water, xylene, methanol; HLB 16.2; cloud pt. 100 C min. (1% aq.); 100% conc.

Nissan Nonion K-230. [Nippon Oils & Fats] POE lauryl ether; nonionic; see Nissan Nonion K-202; APHA 120 max. semisolid; sol. in water, xylene, methanol; HLB 17.3; cloud pt. 100 C min. (1% aq.); 100% conc.

Nissan Nonion LP-20•R, LP-20•RS. [Nippon Oils & Fats] Sorbitan monolaurate; nonionic; emulsifier for cosmetic, pharmaceutical and food applics., o/w emulsion stabilizer and thickener, fiber lubricant and softener; Gardner 5 max. oily liq.; sol. in methanol, ethanol, acetone, xylene, ethyl ether, kerosene, disp. in water; HLB 8.6; 100% conc.

Nissan Nonion MP-30R. [Nippon Oils & Fats] Sorbitan monomyristate; nonionic; see Nissan Nonion LP-20•R; solid; oil-sol.; HLB 6.6; 100% conc.

Nissan Nonion OP-80•R. [Nippon Oils & Fats] Sorbitan monooleate; nonionic; see Nissan Nonion LP-20•R; Gardner 9 max. oily liq.; oil-sol.; HLB 4.3; 100% conc.

Nissan Nonion OP-83•RAT. [Nippon Oils & Fats] Sorbitan sesquioleate; nonionic; see Nissan Nonion LP-20•R; Gardner 9 max. oily liq.; sol. in ethanol, acetone, xylene, ethyl ether, kerosene, methanol, warm in water; HLB 3.7; 100% conc.

Nissan Nonion OP-85•R. [Nippon Oils & Fats] Sorbitan trioleate; nonionic; see Nissan Nonion LP-20•R; Gardner 9 max. oily liq.; oil-sol.; HLB 1.8; 100% conc.

Nissan Nonion P-6. [Nippon Oils & Fats] PEG monopalmitate; nonionic; emulsifier and thickener for cosmetic applics., textile and other industrial uses; semisolid; HLB 13.8; 100% conc.

Nissan Nonion P-208. [Nippon Oils & Fats] POE cetyl ether; nonionic; emulsifier, dispersant, detergent, wetting agent used in textile processing, cosmetics, metalworking, agric. preparations, industrial cleaners; APHA 120 liq.; sol. in water, xylene, methanol; HLB 11.9; cloud pt. 43–53 C (1% aq.); 100% conc.

Nissan Nonion P-210. [Nippon Oils & Fats] POE cetyl ether; nonionic; see Nissan Nonion P-208; APHA 120 max. semisolid; sol. in water, xylene, methanol; HLB 12.9; cloud pt. 64–74 C (1% aq.); 100% conc.

Nissan Nonion P-213. [Nippon Oils & Fats] POE cetyl ether; nonionic; see Nissan Nonion P-208; APHA 120 max. semisolid; sol. in water, kerosene, ether, methanol; HLB 14.1; cloud pt. 87–97 C (1% aq.); 100% conc.

Nissan Nonion PP-40•R. [Nippon Oils & Fats] Sorbitan monopalmitate; nonionic; see Nissan Nonion LP-20•R; Gardner 7 max. waxy solid; oil-sol.; HLB 6.7; 100% conc.

Nissan Nonion S-2. [Nippon Oils & Fats] PEG monostearate; nonionic; emulsifier, thickener used in cosmetics, textile, industrial uses; Gardner 4 max. semisolid; sol. in methanol, warm in xylene, disp. in water, kerosene, ether; m.p. 33–41 C; HLB 8.0; 100% conc.

Nissan Nonion S-4. [Nippon Oils & Fats] PEG monostearate; nonionic; see Nissan Nonion S-2; Gardner 4 max. semisolid; sol. in kerosene, ether, methanol, warm in xylene, disp. in water; m.p. 30–40 C; HLB 11.6; 100% conc.

Nissan Nonion S-6. [Nippon Oils & Fats] PEG monostearate; nonionic; see Nissan Nonion S-2; Gardner 4 max. semisolid; sol. in water, methanol, warm in xylene, ether, cloudy in kerosene; m.p. 35–41 C; HLB 13.6; 100% conc.

Nissan Nonion S-10. [Nippon Oils & Fats] PEG monostearate; nonionic; see Nissan Nonion S-2; Gardner 4 max. semisolid; sol. in water, xylene, ether, methanol; m.p. 38–44 C; HLB 15.2; 100% conc.

Nissan Nonion S-15. [Nippon Oils & Fats] PEG monostearate; nonionic; see Nissan Nonion S-2; Gardner 4 max. semisolid; sol. in kerosene, ether, methanol, warm in xylene; disp. in water; m.p. 39–45 C; HLB 12.8; 100% conc.

Nissan Nonion S-15.4. [Nippon Oils & Fats] PEG monostearate; nonionic; see Nissan Nonion S-2; Gardner 4 max. semisolid; sol. in water, xylene, ether, methanol; m.p. 42–48 C; HLB 16.7; 100% conc.

Nissan Nonion S-40. [Nippon Oils & Fats] PEG monostearate; nonionic; see Nissan Nonion S-2; Gardner 6 max. semisolid; sol. in water, xylene, ether, methanol; m.p. 49–55 C; HLB 18.2; 100% conc.

Nissan Nonion S-207. [Nippon Oils & Fats] POE stearyl ether; nonionic; see Nissan Nonion E-205; APHA 120 max. semisolid; sol. in xylene, kerosene, liq. paraffin, soybean oil, tetrachloromethan, methanol, diethylene glycol; water-disp.; HLB 10.7; cloud pt. 0 C max. (1% aq.); 100% conc.

Nissan Nonion S-215. [Nippon Oils & Fats] POE stearyl ether; nonionic; see Nissan Nonion E-205; APHA 120 max. semisolid; sol. in water, xylene, kerosene, soybean oil, ether, tetrachloromethan, methanol, diethylene glycol; HLB 14.2; cloud pt. 100 C min. (1% aq.); 100% conc.

Nissan Nonion S-220. [Nippon Oils & Fats] POE stearyl ether; nonionic; see Nissan Nonion E-205; APHA 120 max. semisolid; sol. see Nissan Nonion S-215; HLB 15.3; cloud pt. 100 C min. (1% aq.); 100% conc.

Nissan Nonion SP-60•R. [Nippon Oils & Fats] Sorbitan monostearate; nonionic; see Nissan Nonion LP-20•R; Gardner 5 max. waxy solid; sol. in methanol, ethanol, xylene, kerosene, ethyl ether, disp. in warm water; HLB 4.7; 100% conc.

Nissan Nonion T-15. [Nippon Oils & Fats] PEG monotallow acid ester; nonionic; see Nissan Nonion P-6; liq.; HLB 12.8; 100% conc.

Nissan Nymeen DT-203, -208. [Nippon Oils & Fats] POE tallow-alkylpropylene diamine; nonionic; pigment wetting and dispersing agent, emulsifier and acid cleaner additive for metals, textile antistatic agent and auxiliary; lt. br. liq.; cloud pt. 0 and 100 C resp.

Nissan Nymeen L-201. [Nippon Oils & Fats] Oxyethylene dodecylamine; nonionic; see Nissan Nymeen DT-203; colorless to lt. yel. liq.; cloud pt. 0 C max.

Nissan Nymeen L-202, 207. [Nippon Oils & Fats] POE dodecylamine; nonionic; see Nissan Nymeen DT-203; lt. yel. and lt. br. liq. resp.; cloud pt. 0, 80 C.

Nissan Nymeen S-202, S-204. [Nippon Oils & Fats] POE octadecylamine; nonionic; see Nissan Nymeen DT-203; ylsh.-br. solid; cloud pt. 0 C max.

Nissan Nymeen S-210. [Nippon Oils & Fats] POE octadecylamine; nonionic; see Nissan Nymeen DT-203; lt. br. solid in winter, liq. in summer; cloud pt. 95 C min.

Nissan Nymeen S-215, -220. [Nippon Oils & Fats] POE octadecylamine; nonionic; see Nissan Nymeen DT-203; ylsh.-br. solid; cloud pt. 100 C min.

Nissan Nymeen T_2-206, -210. [Nippon Oils & Fats] POE tallow-alkylamine; nonionic; see Nissan Nymeen DT-203; lt. br. liq.; cloud pt. 95 C.

Nissan Nymeen T_2-230, -260. [Nippon Oils & Fats] POE tallow-alkylamine; nonionic; see Nissan Nymeen DT-203; lt. br. solid; cloud pt. 100 C.

Nissan Nymide MT-215. [Nippon Oils & Fats] POE alkylamide; nonionic; modifier for soaps; solid; 100% conc.

Nissan Panacete 810. [Nippon Oils & Fats] Triglyceride, medium chain; nonionic; diluent for perfumes; raw material for pharmaceuticals and special foods; liq.; 100% conc.

Nissan Persoft EK. [Nippon Oils & Fats] Sulfated fatty alcohol ethoxylate, sodium salt; anionic; emulsifier for vinyl polymerization; textile detergent; dyeing assistant; degreaser; liq.; 30% conc.

Nissan Persoft NK-60, -100. [Nippon Oils & Fats] Fatty alcohols, ethoxylated; nonionic; detergent, emulsifier, wetting agent; textile detergent; dyeing assistant; base for liq. detergent; dispersant for pulp pitch; lubricant for bottle cleaning processing; liq.; 90% conc.

Nissan Persoft SK. [Nippon Oils & Fats] Sulfated fatty alcohol, sodium salt; anionic; detergent, emulsifier, dispersant, wetting agent; base for shampoo and liq. detergent; liq.; 30% conc.

Nissan Plonon 102, 104. [Nippon Oils & Fats] POE-POP ether; nonionic; emulsifier, solubilizer, dispersant, detergent, antifoaming agent, wetting agent used in metal cleaning, emulsion polymerization, fermentation, paper industries; colorless liq.; odorless and tasteless; m.w. 1250 and 1650 resp.; sol. in water, ethanol, acetone, benzol, dioxane; visc. 105 cs (37.8 C); HLB 7 and 15; cloud pt. 22 and 64 C (10% aq.); surf. tens. 41.1 and 43.5 dynes/cm (0.1%); 100% conc.

Nissan Plonon 108. [Nippon Oils & Fats] POE-POP ether; nonionic; see Nissan Plonon 102; m.w. 4000; HLB 30.5; cloud pt. 100 C (10% aq.); 100% conc.

Nissan Plonon 171. [Nippon Oils & Fats] POE-POP ether; nonionic; see Nissan Plonon 102; m.w. 2000; sol. in ethanol, acetone, benzol, dioxane; visc. 160 cs (37.8 C); HLB 3; cloud pt. 24 C (10% aq.); surf. tens. 39.5 dynes/cm (0.1%); 100% conc.

Nissan Plonon 172. [Nippon Oils & Fats] POE-POP ether; nonionic; see Nissan Plonon 102; m.w. 2400; sol. in ethanol, acetone, benzol, dioxane; visc. 210 cs (37.8 C); HLB 7; cloud pt. 28 C (10% aq.); 100% conc.

Nissan Plonon 201. [Nippon Oils & Fats] POE-POP ether; nonionic; see Nissan Plonon 102; colorless liq.; odorless and tasteless; m.w. 2200; sol. in ethanol, acetone, benzol, dioxane; visc. 190 cs (37.8 C); HLB 3; cloud pt. 21 C (10% aq.); surf. tens. 38 dynes/cm (0.1%); 100% conc.

Nissan Plonon 204. [Nippon Oils & Fats] POE-POP ether; nonionic; see Nissan Plonon 102; colorless paste; odorless and tasteless; m.w. 3400; sol. in water, ethanol; visc. 370 cs (37.8 C); HLB 13.5; cloud pt. 64 C (10% aq.); surf. tens. 39.8 dynes/cm (0.1%); 100% conc.

Nissan Plonon 208. [Nippon Oils & Fats] POE-POP ether; nonionic; see Nissan Plonon 102; colorless paste; odorless and tasteless; m.w. 8000; sol. in water; HLB 28; cloud pt. 100 C (10% aq.); surf. tens. 44.1 dynes/cm (0.1%); 100% conc.

Nissan Polystar OM. [Nippon Oil & Fats] Sodium salt of polymeric carboxylic acid; dispersant for dyes, pigments, clays; liq.

Nissan Polystar OMP. [Nippon Oil & Fats] Sodium salt of polymeric carboxylic acid; dispersant for pigment, clay, dyestuff, and agric. chemicals; powd.

Nissan Rapisol B-30, -80, C-70. [Nippon Oils & Fats] Sodium dioctyl sulfosuccinate; anionic; wetting agent, polymerization agent for PVC, dyeing aux.; liq.; 30, 80, and 70% conc. resp.

Nissan Stafoam DF-1, DF-2. [Nippon Oil & Fats] Cocamide DEA; thickener for shampoos, bubble baths, liq. detergents.

Nissan Stafoam DL. [Nippon Oils & Fats] Lauramide DEA; nonionic; foam stabilizer, thickener for liq. detergents; liq.; 100% conc.

Nissan Stafoam DO, DO-S. [Nippon Oils & Fats] Oleamide DEA; nonionic; thickener, foam stabilizer for shampoo and liq. detergents; liq.; 100% conc.

Nissan Stafoam L. [Nippon Oils & Fats] Lauramide DEA (1:2); nonionic; foam stabilizer for liq. detergents; liq.; 100% conc.

Nissan Stafoam MF. [Nippon Oils & Fats] Cocamide MEA; nonionic; foam stabilizer for paste shampoos; flake; 100% conc.

Nissan Sunbase, Powder. [Nippon Oils & Fats] Alpha-sulfonated fatty acid ester, sodium salt; anionic; lime soap dispersant; detergent base; builder for household detergent; emulsifier and dispersant; paste, gran.; 30 and 40% conc. resp.

Nissan Tert. Amine AB. [Nippon Oils & Fats] Octadecyl-dimethylamine; intermediate for various surfactants, visc. index improver for lubricating oil, curing catalyst for epoxy resin, corrosion inhibitor, germicide; lt. yel. liq. or half-solid; 95% min. tert. amine.

Nissan Tert. Amine ABT. [Nippon Oils & Fats] Tallow-hydrog. alkyl dimethylamine; see Nissan Tert. Amine AB; lt. yel. liq. or half-solid; 95% min. tert. amine.

Nissan Tert. Amine BB. [Nippon Oils & Fats] Dodecyl-dimethylamine; see Nissan Tert. Amine AB; lt. yel. liq. 95% min. tert. amine.

Nissan Tert. Amine FB. [Nippon Oils & Fats] Cocoalkyl dimethylamine; see Nissan Tert. Amine AB; lt. yel. liq.; 95% min. tert. amine.

Nissan Tert. Amine MB. [Nippon Oils & Fats]

Tetradecyl dimethylamine; see Nissan Tert. Amine AB; lt. yel. liq. or half-solid; 95% min. tert. amine.

Nissan Tert. Amine PB. [Nippon Oils & Fats] Hexadecyl-dimethylamine; see Nissan Tert. Amine AB; lt. yel. liq. or half-solid; 95% min. tert. amine.

Nissan Trax H-45. [Nippon Oils & Fats] POE alkylaryl ether sulfate; anionic; penetrant, detergent, wetting agent; scouring agent for wool; dyeing aux. and emulsifier for syn. resins; liq.; 30% conc.

Nissan Unisafe A-LE. [Nippon Oils & Fats] Alkyl dihydroxy ethyl amine oxide; nonionic; foam stabilizer, detergent; liq.; 40% conc.

Nissan Unisafe A-LM. [Nippon Oils & Fats] Alkyl dimethyl amine oxide; nonionic; see Nissan Unisafe A-LE; liq.; 35% conc.

Nitrene 100 SD. [Henkel] Linoleamide DEA; nonionic; thickener, foamer, corrosion inhibitor; emulsifier for oils/greases used in industrial and household cleaners; liq.; 100% conc.

Nitrene 11230. [Henkel] Modified cocamide DEA (2:1); detergent, wetting agent, emulsifier, thickener, foam stabilizer for industrial and specialty cleaners, household hard surface cleaners; liq.; sol. in alcohols, glycols, esters; dens. 8.3 lb/gal 100% conc.

Nitrene 13026. [Henkel] Modified cocamide DEA; see Nitrene 11230; liq.; sol. in alcohols, glycols, esters; 100% conc.

Nitrene A-309. [Henkel] Modified cocamide DEA (2:1); see Nitrene 11230; also chain lubricant; liq.; sol. in alcohols, glycols, esters; dens. 8.4 lb/gal; 100% conc.

Nitrene C. [Henkel] Cocamide DEA (1:1); nonionic; emulsifier, dispersant, wetting agent, foam booster, visc. builder for liq. detergents, car shampoos, solv. cleaners, drycleaning; lt. amber liq.; sol. in water, lower alcohols, PEG, acetone, toluene, xylene; dens. 8.41 lb/gal; pH 9–11 (10%); 100% conc.

Nitrene C Extra. [Henkel] Cocamide DEA (1:1); nonionic; emulsifier, dispersant, wetting agent, foam booster, and visc. builder for detergents and cleansers; intermediate for liq.; detergents; foam stabilizer; clarifier for liq. soaps; solv. cleaners, and dry-cleaning prods.; lt. amber liq.; sol. in water, alcohol, PEG, acetone, toluene, xylene; dens. 8.41 lb/gal; pH 9-11 (10%); 100% conc.

Nitrene L-76. [Henkel] Lauric-myristic (70/30) DEA (1:1); nonionic; wetting, foaming, dispersing agent, foam stabilizer, thickener for household, institutional, and industrial cleaners; pigment dispersant; soft solid to lt. amber liq.; sol. in lower alcohols, propylene glycol, PEG; disp. water; dens. 8.1 lb/gal; pH 9–11 (10%); biodeg.; 100% conc.

Nitrene L-90. [Henkel] Lauramide DEA (1:1); nonionic; detergent, thickener, foam stabilizer for liq. detergents; solid; sol. in alcohols, glycols, esters; dens. 8.3 lb/gal; 100% conc.

Nitrene N. [Henkel] Modified cocyl DEA; nonionic; similar to Nitrene C but offers greater compatibility with inorg. builders; liq.; 97% active.

Nitrene NO. [Henkel] Oleoyl DEA; see Nitrene 11230; sol. in alcohols, glycols, esters.

Nitropore® OBSH. [Olin] 4,4′-Oxybis (benzenesulfonhydrazide); nitrogen-releasing blowing agent for elastomers, thermoplastics, rubber-resin blends; applics. incl. cellular pipe insulation, athletic padding, molded gaskets, carpet underlayment, flotation prods., and coaxial cable; wh. powd.; no odor; sol. in acetone, dimethylformamide, and dimethylsulfoxide; insol. in benzene, hexane, and water; sp.gr. 1.54; dens. 27 lb/ft^3; gas yield 125 ml/g; decomp. pt. 165 C.

NM. [Angus] Nitromethane; intermediate, stabilizer for halogenated solvs., as fuels, explosives, and solvs. for coatings or industrial processes; m.w. 61.0; dens. 1.14 g/ml; b.p. 101 C; flash pt. 96 F (TCC).

NMP. [Angus] 2-Nitro-2-methyl-1-propanol; chemical and pharmaceutical intermediate, in tire cord adhesives, as formaldehyde release agents, deodorants, antimicrobials; m.w. 119.1; sol. 350 g/100 ml water; m.p. 90 C; b.p. 94 C; flash pt. 175 F (TCC); pH 5.1 (0.1 M aq. sol'n.).

Noda Rice Wax FCC. [Fanning] Refined veg. wax extracted from rice bran (esters of lignoceric acid and myricyl alcohol); plasticizer and release agent in chewing gums; replaces carnauba wax and candelilla used in foods, cosmetics; base for lipsticks and cosmetics; coating in candy and chocolate in food industry; pale ylsh. flakes; sol. in chloroform and benzene; insol. in water; sp.gr. 0.93–0.95; m.p. 78–82 C; sapon. no. 75–88.

Noda Rice Wax Industrial. [Fanning] Refined veg. wax derived from rice bran; external PVC lubricant for rigid and semirigid calendering, extruded film and sheet, and inj. molded bottles; polishing agent for floor polishing wax, shoe shining cream, glazing materials; pale ylsh. beads; sol. in chloroform and benzene; insol. in water; sp.gr. 0.93–0.95; m.p. 78–82 C; sapon. no. 75–78.

Nofoam CA 3617. [Clough] Oil, amide-based emulsion; foam control for textile applics.; disp. in water.

Nofoam NO 3990. [Clough] Silicone emulsion; foam control for textile applics.; disp. in water.

Noigen ES90, 120, 140, 160. [Dai-ichi Kogyo] PEG oleic acid ester; nonionic; emulsifier for animal, veg. and min. oil; dispersant for paint and pigments; liq., solid; HLB 9.0, 12.0, 14.0, and 16.0 resp.; 100% conc.

Noiox AK-40. [Pulcra SA] PEG oleate; nonionic; spreading agent, defoamer, kerosene and min. oil emulsifier; liq.; 100% conc.

Noiox AK-41. [Pulcra SA] PEG oleate; nonionic; lubricant, solubilizer and emulsifier; wetting agent and pigment dispersant; antistat for plastics and aux. dispersant; liq.; 100% conc.

Noiox AK-43. [Pulcra SA] PEG stearate; nonionic; lubricant, antistat for textile industry; preparer and after treatment for fibers; paste; 90% conc.

Noiox AK-44. [Pulcra SA] PEG 400 oleate; nonionic; lubricant, emulsifier, dispersant for min. and veg. oils; antifoamer for dispersing assistant; liq.; sol. in min. oil; disp. in water; 100% conc.

Noiox EO-33. [Pulcra SA] PEG 6000 distearate; nonionic; thickener, opacifier, conditioner; used to control visc. in shampoos and lotions; flakes; 100% conc.

Noiox KJ-12. [Pulcra SA] Oleth-18; solubilizer for hydroalcoholic cosmetic and toiletries; wax; oil-misc.

Noiox KJ-15. [Pulcra SA] Oleth-10; see Noiox KJ-12; semisolid; oil-misc.

Noiox KS-10, -12, -13, -14, -16. [Pulcra SA] PEG; nonionic; lubricant, solv. for dyestuffs and resins; plasticizer for casein, gelatin compositions and printing inks; used in cosmetic and pharmaceutical items; intermediate prods.; liq., solid, paste; water-sol.; 100% conc.

Nonal 206, 208, 210. [Toho] Nonylphenol ethoxylate; nonionic; detergent, penetrant, emulsifier, scouring agent; pitch dispersant; liq.; HLB 10.9, 12.3, 13.3; 100% conc.

Nonal 310. [Toho] Octyl phenol ethoxylate; nonionic; detergent, scouring agent; penetrant; liq.; HLB 13.7; 100% conc.

Nonex C5E. [Hart Chem. Ltd.] PEG 500 monococoate; lubricant for textile applics.; liq.; water-sol.; HLB 13.0; 100% conc.

Nonex DL-2. [Hart Chem. Ltd.] POE 200 dilaurate; nonionic; coemulsifier and lubricant in industrial and textile oils; liq.; HLB 7.4; 100% act.

Nonex DO-4. [Hart Chem. Ltd.] POE 400 dioleate; nonionic; emulsifier, solubilizer for oils, fats, solvs.; liq.; HLB 7.2; 100% act.

Nonex S3E. [Hart Chem. Ltd.] POE 300 stearate; nonionic; lubricant and softener for textiles; liq.; HLB 9.6; 100% act.

Nonionic 1017-R, 1025-R. [Hodag] EO/PO block polymers; detergent, antifoam, rinse aids, visc. control, dispersant, demulsifier; liq.; 100% conc.

Nonionic 1035-L, 1044-L, 1061-L, 1062-L, 1064-L, 1068-L, 1088-L. [Hodag] EO/PO block copolymers; nonionic; surfactant used as a detergent, antifoam, wetting agent, emulsifier, antistat, demulsifier, visc. modifier, deduster, gelation aid, metal working lubricant, dispersant; liq., flake; 100% conc.

Nonionic 2017-R, 2025-R, 4017-R, 4025-R, 5025-R. [Hodag] EO/PO block copolymer; nonionic; see Nonionic 1017-R; liq., paste; 100% conc.

Nonipol 20, 40, 55. [Sanyo] POE nonyl phenyl ether; nonionic; penetrant, wetting agent, spreader-sticker; base material for detergent; emulsifier for agric. pesticides and emulsion polymerization, org. solv., machine oils, liq. paraffins; liq.; HLB 5.7, 8.9, 10.5; 100% conc.

Nonipol 60, 70, 85, 95, 100, 110, 120, 130, 160, 200, 400. [Sanyo] POE nonyl phenyl ether; detergent and penetrant; liq.; HLB 10.9, 11.7, 12.6, 13.1, 13.3, 13.8, 14.1, 14.5, 15.2, 16.0, and 17.8 resp.; 100% conc.

Nonipol BX. [Sanyo] POE alkylphenyl ether; nonionic; textile scouring agent and penetrant; liq.; 100% conc.

Nonipol D-160. [Sanyo] POE dinonyl phenyl ether; nonionic; see Nonipol 20; solid; HLB 13.3

Nonipol Soft SS-50, -70, -90. [Sanyo] POE higher alcohol ether; nonionic; penetrant, detergent, emulsifier, dispersant, wetting agent, or base material for household liq. detergent; liq.; HLB 10.5-13.2; 100% conc.

Nonisol 100. [Ciba-Geigy] PEG-8 laurate; nonionic; emulsifier, solubilizing and wetting agent, thickener, used in textiles, cosmetics, hand cleaners, spreading agents; yel. liq.; water-sol.; 100% conc.

Nonisol 210. [Ciba-Geigy] PEG-8 dioleate; nonionic; toxicant emulsifier, dispersant of sludge in fuel oil, textile lubricant; lt. amber liq.; sol. in kerosene; 100% conc.

Nonisol 300. [Ciba-Geigy] PEG-8 stearate; nonionic; emulsifier, wetting agent, starch stabilizer, base for ointments, creams, suppositories, hair dressing, liq. makeup; wh. paste; sol. in xylene, disp. in water; m.p. 30 C; 100% conc.

Nonox® WSP. [ICI Am.] 2,2′-Methylenebis [4-methyl-6-(1-methyl-cyclohexyl) phenol]; antioxidant for plastics, esp. polyolefin, PS; cryst. powd.

Nopalcol Series. [Henkel] PEG fatty esters; dispersant, wetting agent, emulsifier, plasticizer, lubricant, binder, thickener for cosmetics, distillation defoamers, dry cleaning agents, leather and paper industries; solv., textiles, wall tile mastic; liq. to solid; 100% conc.

Nopalcol 1-S. [Henkel] Diethylene glycol monostearate; nonionic; emulsifier, plasticizer, lubricant, wetting agent, binding and thickening agent, used in cosmetics, dry cleaning, leather, textile industries; solid; HLB 3.8; 98% conc.

Nopalcol 1-TW. [Henkel] Diethylene glycol monotallowate; nonionic; see Nopalcol 1-S; solid; HLB 4.1; 99% conc.

Nopalcol 4-O. [Henkel] PEG-8 oleate; nonionic; emulsifier; dispersant for leather pigments; paper coating defoamer, plasticizer and leveling agent; liq.; HLB 12.0; 100% conc.

Nopco® 1419-A. [Henkel] Defoamer for SBR, PVAc, acrylic paint and adhesive systems; amber liq.; forms milky emulsion in water; sp.gr. 0.85; dens. 7.1 lb/gal; pH 6.5 (2%); flash pt. 47 C; 99% act.

Nopco® Colorsperse 66-A. [Henkel] Ammonium polyelectrolyte; anionic; dispersant for titanium dioxide and other pigments; effective with alumina trihydrate and reactive pigments; liq.; 30% act.

Nopco® Colorsperse 188-A. [Henkel] Fatty acid, ethoxylated; nonionic; pigment wetting and dispersing; liq.; 100% conc.

Nopco® JMY. [Henkel] Defoamer for PVAc, acrylic paint and adhesive systems; amber liq.; forms milky emulsion in water; sp.gr. 0.88; dens. 7.3 lb/gal; pH 6.5 (2%); flash pt. 116 C; 99% act.

Nopco® NDW. [Henkel] Defoamer for SBR, PVAc, acrylic, water-sol. resins paint and adhesive systems; amber slightly hazy liq.; disp. in all latex and adhesive systems; sp.gr. 0.90; dens. 7.5 lb/gal; pH 6.5 (2%); flash pt. 79 C; 99.9% act.

Nopco® NXZ. [Henkel] Defoamer for syn. latex emulsions, paint and adhesives from SBR, PVAc, acrylic, water-sol. resins; amber hazy liq.; forms milky emulsion in water; sp.gr. 0.91; dens. 7.6 lb/gal; pH 7 (2%); flash pt. 171 C; 99% act.

Nopco® PD#1-D. [Henkel] Defoamer for adhesives, paints, joint compds., plaster; off-wh. powd.; water-wettable; dens. 22–26 lb/ft^3; pH 8.7 (1%); 65% act.

Nopcocastor, L. [Henkel] Castor oil, sulfated; anionic; emulsifier; superfatting agent for cosmetics; ironing aid; liq.; 75 and 50% conc. resp.

Nopcocide® N-40-D. [Henkel] Chlorothalonil; mildewcide, fungicide for trade sales paints; 40% disp. in water.

Nopcocide® N-96. [Henkel] Tetrachloroisophthalonitrile; broad spectrum microbicide for control of fungi in latex exterior and interior emulsion paints, solv.-based paints; gray 3–5 μ powd.; odorless (pure grade), slight pungent odor (tech. grade); m.w. 265.9; sp.gr. 1.8; m.p. 250–251 C; b.p. 350 C; 96% act.

Nopcocide® N-96-S. [Henkel] 2,4,5,6-Tetrachloroisophthalonitrile; antimicrobial for marine antifouling coatings; micromilled powd., 3–5 μ.

Nopcogen 14-S. [Henkel] Stearamide DEA; textile softener; solid; m.p. 48 C; solid; 100% conc.

Nopcogen 22-O. [Henkel] Oleyl imidazoline; cationic; emulsifier for min. oil, kerosene, wetting agent, corrosion inhibitor, used in textile, asphalt, paper industries; amber, visc. liq.; sol. in water, acetic acid, HCl; misc. with pine oil and min. oil; cloud pt. 0 C; 90% act.

Nopcosant®. [Henkel] Sodium salt of condensed naphthalene sulfonate; dispersant for NR, SR latexes, pigments; used for paints, cements, sealants; tan powd.; water-sol.; sp.gr. 1.02; pH 7.5–8.0 (2%).

Nopcosant® K. [Henkel] Anionic polymer; dispersant for latex systems; clear liq.; misc. with water in all proportions; dens. 10.9 lb/gal; pH 12 (5%); 34% act.

Nopcosant® L. [Henkel] Sulfated naphthalene; anionic; NR, SR latex dispersant for use in aq. systems, paints, paper coatings, textiles, and leather; lt. yel. liq., odorless; water sol.; sp.gr. 1.18; dens. 9.8 lb/gal; pH 9.6 (5%); 25% solids.

Nopcosperse® 44. [Henkel] Sodium neutralized polycarboxylate; anionic; dispersant for paint systems; clear lt. amber liq.; sol. in water and glycols; dens. 10.3 lb/gal; pH 7.8; 35% act.

Nopcosperse® 644A. [Henkel] Ammonium neutralized version of Nopcosperse 44; primary pigment dispersant for coatings, printing inks; clear lt. amber liq.

Nopcosperse® 303-SD. [Henkel] Polycarboxylate; anionic; see Nopcosperse AD-6; powd.; 93% act.

Nopcosperse® AD-6. [Henkel] Polyacrylate; anionic; dispersant for dry flowables, wettable powds., aq. suspension pesticide formulations; powd.; 95% act.

Nopcosulf CA-60, -70. [Henkel] Castor oil, sulfated; anionic; softener in finishing starches, gums; plasticizer for starches; furniture base polish; liq.; 68 and 63% conc. resp.

Nopcosulf TA-30, -45V. [Henkel] Sulfated tallow; anionic; softener for cotton goods; paste; 44 and 78% conc. resp.

Nopcote C-104. [Henkel] Calcium stearate disp.; lubricant for paper coatings; water-disp.

Nopcowax 22-DS. [Henkel] Ethylene bisstearamide; high melting syn. wax; binder, thickener for latex formulation, coatings, adhesives; used in powd. metallurgy as internal lubricant; insol. in water.

Noram 2 C. [Ceca SA] N-dicoco amine; cationic; synthesis intermediate, anticaking agent, flotation, antistripping for road making, soil stabilization; auxs. for fuel additives, rust inhibition, paint; chemical intermediate for quats., betaines, amine oxides; solid.

Noram 2 SH. [Ceca SA] NN-dihydrog. tallow amine; cationic; anti-caking agent; solid.

Noram C. [Ceca SA] N-coco amine; cationic; see Noram 2 C; pasty.

Noram DMC. [Ceca SA] N-coco dimethylamine; cationic; see Noram 2 C; liq.

Noram DMS. [Ceca SA] N-tallow dimethylamine; see Noram 2 C; liq.

Noram DMSH. [Ceca SA] N-hydrog. tallow dimethylamine; cationic; see Noram 2 C; pasty.

Noram M2C. [Ceca SA] N-dicoco methylamine; cationic; see Noram 2 C; pasty.

Noram M2 SH. [Ceca SA] N-dihydrog. tallow methylamine; cationic; see Noram 2 C; solid.

Noram O. [Ceca SA] Oleyl amine; synthesis intermediate, lubricant and textile industries; solid.

Noram S. [Ceca SA] N-tallow amine; cationic; synthesis intermediate; oil industry; solid.

Noram SH. [Ceca SA] N-hydrog. tallow amine; cationic; anticaking agent; solid.

Noramac C. [Henkel-Nopco] N-coco amine acetate; cationic; flotation, bactericide, emulsifier, anticaking, soil stabilization, flocculation, corrosion inhibitor; pasty/solid.

Noramac O. [Henkel-Nopco] Oleyl amine acetate; cationic; see Noramac C; pasty/solid.

Noramac S. [Henkel-Nopco] N-tallow amine acetate; see Noramac C; pasty/solid.

Noramium DA.50. [Ceca SA] Coco dimethyl benzyl ammonium chloride; bactericide, fungicide, demulsification of hydrocarbons, cosmetics, latex coagulation, flotation, electrostatic paints; 50% conc. in water.

Noramium M2C. [Ceca SA] Dicoco dimethyl ammonium chloride; cationic; textile softener; hair conditioner; liq.; 75% conc.

Noramium M2SH. [Ceca SA] Dihydrog. tallow dimethyl ammonium chloride; cationic; textile softener; hair conditioner; pasty; 75% conc.

Noramium MC 50. [Ceca SA] Coco trimethyl ammonium chloride; cationic; additive for antibiotics mfg.; liq.; 50% conc.

Noramium MO 50. [Ceca SA] Oleyl trimethyl ammonium chloride; cationic; see Noramium MC 50; liq.

Noramium MSH 50. [Ceca SA] Hydrog. tallow trimethyl ammonium chloride; cationic; see Noramium MC 50; liq.; 75% conc.

Noramium S 75. [Ceca SA] Tallow dimethyl benzyl ammonium chloride; cationic; see Noramium MC 50.

Noramox C2 to 15. [Henkel-Nopco] N-ethoxylated coco amine; cationic; emulsifier, drying assistant, viscose additive, metal treatment, rust inhibitor; antistat for ABS, PS; liq.

Noramox S2 to 11. [Henkel-Nopco] N-ethoxylated tallow amine; cationic; see Noramox C2; pasty.

Norcure 131. [R.H. Carlson] Aliphatic amine; curing agent (used with DER 741) for compatibility and long pot life at R.T.; fast cure; mix ratio: 15 phr with DER 741; water wh.; slight odor after initial ammoniacal odor dissipates; sp.gr. 0.95; visc. 150 cps; flash pt. (COC) 250 F.

Norcure 138-B. [R.H. Carlson] Modified polyamide; curing agent; blue low visc. liq.; lt. pungent odor after initial ammonia odor dissipates upon opening container; sp.gr. 0.95; flash pt. (COC) 155 C; nonflam.

Norfox 1 Polyol. [Norman, Fox] Hydrophobic block polymer; nonionic; defoamer base for surfactant mixtures; liq.; HLB 3.0; 100% conc.

Norfox 4 Polyol. [Norman, Fox] Hydrophilic block polymer; nonionic; control foam solubilizer; liq.; HLB 11.0; 100% conc.

Norfox 243. [Norman, Fox] Silicone emulsion; defoamer for industrial applics.; liq.; 30% act.

Norfox 916. [Norman, Fox] Linear primary alcohol, ethoxylated; nonionic; wetting agent; wet processing aid for paper, textiles and leather; liq.; HLB 12.5; 100% conc.

Norfox B. [Norman, Fox] Sodium stearate; anionic; gelling agent; stabilizer in cosmetics; lubricant; gran.; 96% conc.

Norfox Coco Powder. [Norman, Fox] Sodium cocoate; anionic; soap base; stabilizer; gelling agent; gran.; 92% conc.

Norfox CS. [Norman, Fox] Calcium stearate; release agent, internal lubricant, water repellent; impalpable powd.

Norfox DC. [Norman, Fox] Coconut oil DEA condensate; nonionic; intermediate for detergent mfg., cosmetics, suds stabilizer and dedusting agent for dry prods.; liq.; 100% conc.

Norfox DCO. [Norman, Fox] Oleic acid modified coconut oil DEA condensate; anionic/nonionic; visc. builder; metal weaving, drawing, stamping or cold rolling lubricant; burnishing, heavy duty cleaner for soft water areas; liq.; 100% conc.

Norfox DESA. [Norman, Fox] Cocamide DEA; nonionic; visc. and foam enhancer; liq.; 100% conc.

Norfox DF210SX. [Norman, Fox] Silicone emulsion; defoamer for spas; liq.; 10% act.

Norfox DLSA. [Norman, Fox] Lauramide DEA; nonionic; specialty foam booster and visc. builder for shampoo and bubble baths; waxy solid; 100% conc.

Norfox EGMS. [Norman, Fox] Ethylene glycol stearate; nonionic; pearlescent ingred. in liq. shampoos and detergents; emulsification aid and bodying agent; beads; 100% org.

Norfox EM 350X, EM-350X Conc. [Norman, Fox] Silicone emulsion; base for vinyl conditioners; release agent; liq.; 35 and 60% act. resp.

Norfox GMS. [Norman, Fox] Glyceryl stearate; nonionic; lotion and cream base in cosmetics; opacifier in liq. shampoos and detergents; emulsion stabilizer; beads; 40–80% alpha mono, 100% org.

Norfox GMS-FG. [Norman, Fox] Glycerol monostearate; nonionic; cosmetic opacifier, food grade emulsifier; flake; HLB 3.9; 100% conc.

Norfox KD. [Norman, Fox] Cocamide DEA and diethanolamine; nonionic; visc. enhancer and foam booster; liq.; 100% conc.

Norfox KO. [Norman, Fox] Potassium oleate; anionic; liq. soap for hand cleaners, tire mounting lubricant; emulsifier and corrosion control in paint strippers; liq.; HLB 20.0; 80% conc.

Norfox M2DS. [Norman, Fox] Specialty stearate; wire drawing lubricant; coarse powd.

Norfox MLD. [Norman, Fox] Methyl ester of lard oil; lubricant ingred. in specialty oils, metalworking fluids; mobile yel. liq.

Norfox NBC. [Norman Fox] Mixed propylene glycol methyl ether; coupler; substitute for ethylene glycol monobutyl ether; liq.; water-sol.

Norfox NP-1. [Norman, Fox] Nonoxynol-1; nonionic; emulsion stabilizer and defoamer; liq.; HLB 4.5; oil-sol.; 100% conc.

Norfox NP-4. [Norman, Fox] Nonoxynol-4; nonionic; emulsifier, detergent and dispersant for petrol. based lubricants; intermediate for sulfanic acid sulfonation to produce foaming agent; liq.; oil-sol.; HLB 9.0; 100% conc.

Norfox PEW. [Norman, Fox] Mono and dialkyl acid phosphate; anionic; detergent; emulsifier; coupling agent; corrosion inhibitor; liq.; 100% conc.

Norfox SEHS. [Norman, Fox] Sodium 2-ethylhexyl sulfate; anionic; foamer with wetting and detergency; coupling agent; liq.; 40% conc.

Norfox Sorbo T-20. [Norman, Fox] Polysorbate 20; nonionic; flavor and fragrance solubilizer; liq.; HLB 16.7; 100% act.

Norfox Sorbo T-80. [Norman, Fox] Polysorbate 80; nonionic; solubilizer for fat-sol. actives; emulsifier; liq.; HLB 16.7; 100% act.

Norfox SXS40, SXS96. [Norman Fox] Sodium xylene sulfonate; coupling agent for detergent applics.; liq., flaked solid resp.

Norfox Syn Lub. [Norman, Fox] Specialty amine soap; conveyor and chain lubricant; liq.

Norfox Terne Oil. [Norman, Fox] Hydrog. glyceride oil; lubricant for prod. of Terneplate steel; waxy solid; m.p. 58–60 C.

Norfox V-5759. [Norman, Fox] Polyhydric alcohol; lubricant, heat transfer fluid; clear fluid.

Norfox Vertex Flakes. [Norman, Fox] Sodium oleate; anionic; textile scouring, fulling, dye leveler; flake; HLB 20.0; 90% conc.

Norfox VMO. [Norman, Fox] Veg. methyl oleate; lubricant for mfg. of food containers, for drying raisins; mobile yel. liq.

Norfox ZNS. [Norman, Fox] Zinc stearate; internal lubricant, water repellent in powds.; impalpable powd.

Norlig. [Daishowa] Calcium and sodium lignosulfonate; anionic; binder, dispersant, used to manufacture clay and brick prods., linoleum mastics, soil stabilizer; liq. or powd.; 50–60% conc. (liq.), 100% conc. (powd.).

Norlig 415. [Daishowa] Modified sodium-calcium lignosulfonate; anionic; dispersant and water reducer for gypsum wallboard mfg.; 100% powd., 45% liq.

Norlig A. [Daishowa] Calcium lignosulfonate; anionic; dispersant, binder, soil and dust stabilizer, carbon blk., briquetting and pelletizing of coal and charcoal, ceramic additive; 100% powd., 50 and 58% liqs.

Norlig NH. [Daishowa] Ammonium lignosulfonate; anionic; dispersant, binder, resin extender; powd., liq.

Norust. [Ceca SA] Fatty amines blend; corrosion inhibitor, acid passivator; liq.; water-disp.

No Stik 802. [Ross] Silicone emulsion; mold release for rubber, plastics, urethane; ingred. in polishes; disp. in water.

Nouryflex 520. [Akzo] Phthalate ester; plasticizer for lacquer and coatings, esp. in varnishes based on NC and alkyds; Gardner 1 max. color; s.gr. 0.988–1.000; visc. 50 mPa•s; b.p. 185–198 C; flash pt. 193 C; acid no. 2 max.; ref. index 1.485.

NovaSize, Dark Fortified. [Georgia-Pacific] Tall oil rosin; internal sizing agent to give water resistance to paper and paperboard; USDA N–M color, translucent paste; dens. 9.2 lb/gal (70 C); visc. 2500 cps max. (70 C); acid no. 10–18 (70%), 12–20 (77%); 70 ± 1% and 77 ± 1% solids.

NovaSize, Dark Unfortified. [Georgia-Pacific] Tall oil rosin; internal sizing agent to give water resistance to paper and paperboard; USDA N–M color, translucent paste; dens. 9.2 lb/gal (70 C); visc. 2500 cps max. (70 C); acid no. 10–18 (70%), 12–20 (77%); 70 ± 1% and 77 ± 1% solids.

NovaSize, Pale Fortified. [Georgia-Pacific] Tall oil rosin; internal sizing agent to give water resistance to paper and paperboard; USDA N min. color, translucent paste; dens. 9.2 lb/gal (70 C); visc. 2500 cps max. (70 C); acid no. 10–18 (70%), 12–20 (77%); 70 ± 1% and 77 ± 1% solids.

Novel® 1412-70. [Vista] Ethoxylated alcohol (11.0 EO); surfactant intermediate; wh. solid; HLB 14.0.

Novol. [Croda Ltd.] Super refined oleyl alcohol; emollient, emulsion stabilizer, superfatting agent, pigment suspending aid, used in cosmetics, personal care prods.; Gardner 1 max. liq.; mild odor; misc. with fat, oil and wax mixts.; sp.gr. 0.845–0.855 (15 C); visc. 24–32 cps; acid no. 0.1 max.; sapon. no. 0.3 max.

Novor 924. [Akrochem] Urethane vulcanizing system; accelerator for high-temp. curing of natural rubber; deep golden br. dustless powd.; sp.gr. 1.27; m.p. 220–230 C; 75% act. in naphthenic process oil.

Noxamium C2-15. [Ceca SA] Cocoamine, ethoxylated, ammonium quat. deriv.; cationic; antistat, bactericidal emulsifier; liq.; 50% conc.

Noxamium S2-11. [Ceca SA] Tallow amines, ethoxylated, ammonium quat. deriv.; cationic; see Noxamium C2-15; liq.; 50% conc.

Noxamium S2/50. [Ceca SA] Ammonium quat. deriv. of ethoxylated tallow amines; antistat, bactericide; liq.

NP-10. [Neville] Aromatic plasticizer; chemically inert, nonsaponifiable grade, with low reactivity; used in adhesives (mastic, pressure sensitive), rubber (cements, mechanical and molded goods, tires), and caulking compds.; Gardner 14 liq. (50% in toluene); m.w. 500; sol. in ethers and chlorinated, aromatic, naphthenic, and terpene hydrocarbons; sp.gr. 0.950; soften. pt. (Ring & Ball) 10 C; flash pt. (COC) 400 F.

NP-25. [Neville] Aromatic plasticizer; see NP-10; Gardner 14 liq. (50% in toluene); m.w. 520; sol. see NP-10; sp.gr. 1.005; soften. pt. (Ring & Ball) 25 C; flash pt. (COC) 450 F.

NPP. [PPG Industries] Di-n-propyl peroxydicarbonate; initiator for polymerization of unsat. monomers; reduces time of batch runs; increases prod.; liq.; m.w. 206.18; immisc. with water; sp.gr. 1.1161 (4/4 C); m.p. < –70 C; ref. index 1.4106; 98.5% min. conc.; 7.8% act. oxygen.

N® Sodium Silicate. [PQ] Sodium silicate; corrosion inhibitor for water systems; used for metals in industrial plants, textile mills, laundries, office bldgs., municipalities, oil refineries; syrupy liq.; sol. in water; dens. 11.6 lb/gal; 37.6% solids.

NTA. [Monsanto] Sodium NTA; sequestering agent; 40% sol'n. or solid.

Nuodex PMA 18. [Servo] Org. mercury compd.; fungicide, bactericide for paint industry; preservative for aq. systems; water-sol.

Nutrapon B. [Clough] Sodium lauryl sulfate and ethylene glycol monostearate; anionic/nonionic; shampoo and detergent blend with pearlizing agent for pearlescent formulations; visc. liq., paste; 32-34% conc.

Nutrol 100. [Clough] PEG-9 octyl phenol ether; nonionic; wetting agent and dispersant; used in metal and acid cleaners and pesticide formulations; liq.; HLB 13.0; 99% min. conc.

Nutrol 600. [Clough] PEG-9 nonylphenol ether; nonionic; dispersant, emulsifier; wetting agent, detergent; liq.; HLB 13.0; 99% min. conc.

Nutrol 611. [Clough] PEG-8 nonylphenol ether; nonionic; see Nutrol 600; liq.; HLB 12.2; 99% min. conc.

Nutrol SXS. [Clough] Sodium xylene sulfonate; anionic; hydrotrope, coupling agent, solubilizer; liq.; 39–40% act.

Nyacol®. [Nyacol] Colloidal silica aq. sols; high temp. resistant, inorg. binder for fibrous insulation materials, ceramic investment casting molds, catalyst; paper and textile antislip and antisoil treatment;

Nyacol® 215. [Nyacol] Colloidal silica; binder for fibers; clear liq. disp., particle 3–4 m μ; sp.gr. 1.1; dens. 9.2 lb/gal; visc. 5 cps; pH 11; 15% silica.

Nyacol® 830. [Nyacol] Colloidal silica; binder for investment casting; sp.gr. 1.2; dens. 10 lb/gal; visc. 6 cps; pH 10.7; 30% silica.

Nyacol® 1430. [Nyacol] Colloidal silica; see Nyacol 830; sp.gr. 1.2; dens. 10 lb/gal; visc. 6 cps.; pH 10.4; 30% silica.

Nyacol® 1440. [Nyacol] Colloidal silica; see Nyacol 215; clear liq. disp., particle 14 m μ; sp.gr. 1.3; dens. 10.8 lb/gal; visc. 16 cps; pH 10.4; 40% silica.

Nyacol® 2030 EC. [Nyacol] Colloidal silica; binder for pkg. prods.; antislip agent, retention aid, and coating for paper applic.; disp., particle 20m μ; sp.gr.

1.2; dens. 10.0 lb/gal; visc. 6 cps; pH 10; 30% silica.

Nyacol® 2040 NH4. [Nyacol] Colloidal silica; see Nyacol 215; clear liq. disp., particle 20 m μ; sp.gr. 1.3; dens. 10.8 lb/gal; visc. 25 cps; pH 9; 40% silica.

Nyacol® 2046 EC. [Nyacol] Colloidal silica; see Nyacol 2030 EC; disp., particle 20 m μ; sp.gr. 1.3; dens. 10.8 lb/gal; visc. 10 cps; pH 10; 46% silica.

Nyacol® 9950. [Nyacol] Colloidal silica; see Nyacol 2030 EC; disp., particle 100 m μ; sp.gr. 1.4; dens. 11.7 lb/gal; visc. 15 cps; pH 9; 50% silica.

Nyacol® A-1530. [Nyacol] Colloidal disp. of antimony pentoxide in water; anionic; flame retardant additive to latex emulsions; durable treatment for fabrics, nonwovens, fiberfill, paper, fiberglass, vinyls; suitable for FR adhesives; disp. particle size 15–30 mμ; sp.gr. 1.37; visc. 5 cps; 30% antimony pentoxide.

Nyacol® A-1540N. [Nyacol] Colloidal antimony pentoxide aq. disp.; flame retardant; used in latex compds. used for textile fabrics, nonwovens, adhesives; aq. disp.; 40% oxide

Nyacol® A-1550. [Nyacol] Colloidal disp. of antimony pentoxide in water; anionic; see Nyacol A-1530; disps. particle size 15–30 mμ; sp.gr. 1.81; visc. 10 cps; 50% antimony pentoxide.

Nyacol® A-1588LP. [Nyacol] Colloidal antimony pentoxide powd.; flame retardant for epoxies; powd.; 10–40 μ particle range; 87% oxide.

Nyacol® AB40. [Nyacol] Colloidal antimony pentoxide disp. in polyester polyol; flame retardant for rigid foams, RIM urethane coatings and elastomers; nonaq. disp.; 40% oxide.

Nyacol® AGO-40. [Nyacol] Disp. of colloidal antimony pentoxide in a liq. polyester resin (unsat.); halogen synergist for flame retardancy in polymer applic.; disp. particle size 0.015–0.030μ; sp.gr. 1.62; 40% colloidal antimony oxide; 50% polyester resin.

Nyacol® AP50. [Nyacol] Colloidal antimony pentoxide disp. in nonreactive high m.w. tert. amine; flame retardant for epoxy resins, ketone sol'ns.; nonaq. disp.; 50% oxide.

Nyacol® APE1540. [Nyacol] Colloidal antimony pentoxide disp. in unsat. polyester; flame retardant for halogenated polyester laminates; nonaq. disp.; 40% oxide.

Nyacol® APVC40. [Nyacol] Colloidal antimony pentoxide disp. in plasticizer (polyester and phthalate); flame retardant for clear and mass tone colored vinyl; nonaq. disp.; 40% oxide.

Nyacol® HA-9. [Nyacol] Pentabromo diphenyl oxide, Nyacol AGO-40, ratio 2:1; flame retardant used where max. FR properties are required from all-liq. system; wide range of compatibility; effective at low addition levels with polymers; pourable and pumpable; nonabrasive; nonsettling; liq.; visc. 190,000 cps; 46% bromine; 14% colloidal antimony pentoxide.

Nyacol® HA-15. [Nyacol] Colloidal antimony pentoxide nonaq. disp. containing 39% halogen chlorine and bromine; flame retardant for urethane coatings, rigid foams, RIM, nonhalogenated polyesters and vinyls; nonaq. disp.; 14% oxide; 39% halogen.

Nyacol® N22. [Nyacol] Bromine/colloidal antimony pentoxide composition; flame retardant blendable with latices (i.e. SBR, acrylic); used for textile coating or impregnation; clear film when dried; nonsettling; fluid; pH 8; 68% solids (48% bromine; 22% colloidal antimony pentoxide).

Nyacol® N24. [Nyacol] Colloidal antimony pentoxide aq. disp.; see Nyacol A-1540N; aq. disp.; 11.2% oxide; 33% halogen (chlorine and bromine); 73% solids.

Nyacol® ZTA. [Nyacol] Colloidal antimony pentoxide powd.; flame retardant for vinyls, ABS, HIPS, PP; powd.; 10–40 μ particle range; 80% oxide.

Nykon® 77. [Rheox] Tetra alkyl or trialkyl aryl ammonium smectites with corrosion inhibitor; corrosion resistant gellant for use in bodying lubricants; wh. powd.; sp.gr. 1.7.

Nylomine Assistant DN. [ICI Am.] Blend of surfactants; anionic; dyeing assistant, leveling agent for dyeing fibers; yel. liq.; sol. in water; sp.gr. 1.0; f.p. 0 C; b.p. 100 C; pH 6.0–7.5 (1%); 55% act.

Nyloset Finish. [Scher] Amide-resin; cationic; nylon builder and softener; liq.; 25% act.

Nysel (Ni-3201 F). [M&T Harshaw] Nickel; catalyst for hydrogenating edible oils, kosher; 25% conc.

Nysel CN-14 (Ni-3611 L). [M&T Harshaw] Nickel; catalyst used where low temp. (< 150 C) hydrogenation is required but suspended in coconut oil rather than soybean oil; liq.

Nysel HK-4 (Ni-3609 F). [M&T Harshaw] Nickel; catalyst for hydrogenation of oils; flakes; 25% conc.

Nysel HK-12 (Ni-5708 L). [M&T Harshaw] Nickel; see Nysel CN-14, kosher; liq.; 20% conc.

Nysel SP-7 (Ni-5169 F). [M&T Harshaw] Nickel; catalyst for applic. for preparation of soybean oil coating fats and "hard" margarine base, kosher; flakes.

Nytal® 99. [Vanderbilt] Talc filler used in PVC, vinyl asbestos tile, polyester in match molded articles, body patching compds., PP, nylon, phenol formaldehyde, polyethylene; resin visc. during processing and improves stiffness and increases heat distortion temp. of compds.; extender in vinyl asbestos tile, polyesters, and polyolefins; used in ceramic wall tile and artware; wh. powd. particle size 38 μ, 2.3–4.0% +200-mesh residue; dens. 2.85 mg/m³; oil absorp. 19; 56.90% SiO_2, 28.60% MgO, 8.28% CaO.

Nytal® 100. [Vanderbilt] Hydrous magnesium silicate; see Nytal 99; also partial replacement for fibrous reinforcement; aux. flux in vitreous ceramic bodies in elec. porcelain or sanitaryware; wh. powd. particle size 34 μ, 2.5–3.2% +200-mesh residue; dens. 2.85 mg/m³; oil absorp. 21; 56.20% SiO_2, 28.80% MgO, 8.58% CaO.

Nytal® 100 HR. [Vanderbilt] Magnesium silicate; filler used in art pottery and hobby ceramic casting slips where control of specific resistance is essential in high talc bodies; powd. particle size 32 μ, 2.5–3.2 + 200-mesh residue; 55.20% SiO_2, 30.00% MgO, 8.42% CaO.

Nytal® 200. [Vanderbilt] Hydrous magnesium silicate; see Nytal 99; also resin extender and reinforcing agent to improve the physical properties of finished prod.; wh. powd. particle size 10 μ; dens. 2.85 mg/m³; oil absorp. 24.

Nytal® 300. [Vanderbilt] Hydrous magnesium silicate; see Nytal 99; also reinforcement and extending thermoplastics as well as thermosetting resins; wh. powd. particle size 5 μ; dens. 2.85 mg/m³; oil absorp. 29.

Nytal® 400. [Vanderbilt] Hydrous magnesium silicate; see Nytal 99; also reinforcement in plastics; wh. powd. particle size 2.3 μ; dens. 2.85 mg/m³; oil absorp. 36.

NZ 01. [Kenrich] Neopentyl (diallyl) oxy, trineodecanoyl zirconate; coupling agent; amber liq.; sol. in xylene, toluene, DOP, min. oil; sp.gr. 0.96 (16 C); visc. 40 cps; i.b.p. 188 F; flash pt. (TCC) 195 F; pH 8; 95% solids in IPA.

NZ 90. [Kenrich] Neopentyl (diallyl) oxy, tri (dodecyl)benzene-sulfonyl zirconate; coupling agent; brn. liq.; sol. in xylene, toluene, DOP; sp.gr. 1.09 (16 C); visc. 190 cps; i.b.p. 300 F; flash pt. (TCC) 150 F; pH 4; 95% solids in IPA.

NZ 12. [Kenrich] Neopentyl (diallyl) oxy, tri(dioctyl) phosphato zirconate; coupling agent; orange/red liq.; sol. in xylene, toluene, DOP, min. oil; sp.gr. 1.06 (16 C); visc. 160 cps; i.b.p. 220 F; flash pt. (TCC) 170 F; pH 6; 95% solids in IPA.

NZ 38. [Kenrich] Neopentyl (diallyl) oxy, tri(dioctyl) pyrophosphato zirconate; coupling agent; reddish liq.; sol. in IPA, xylene, toluene, DOP; sp.gr. 1.10 (16 C); visc. 360 cps; i.b.p. 345 F; flash pt. (TCC) 170 F; pH 6; 95% solids in IPA.

NZ 44. [Kenrich] Neopentyl (diallyl) oxy, tri(N-ethylenediamino) ethyl zirconate; coupling agent; yel./orange liq.; sol. in IPA; sp.gr. 1.17 (16 C); visc. > 50 cps; i.b.p. 300 F; flash pt. (TCC) > 220 F; pH 11; 95% solids in IPA.

NZ 97. [Kenrich] Neopentyl (diallyl) oxy, tri (m-amino) phenyl zirconate; coupling agent; brn. liq.; sol. in ether solvs.; limited sol. in IPA; sp.gr. 1.20 (16 C); visc. 4400 cps; i.b.p. 300 F; flash pt. (TCC) 190 F; pH 6; 67% solids in phenyl glycol ether solv.

NZ 33. [Kenrich] Neopentyl (diallyl) oxy, trimethacryl zirconate; coupling agent; yel./orange liq.; sol. in xylene, toluene, DOP; sp.gr. 1.09 (16 C); visc. < 100 cps; i.b.p. > 300 F; flash pt. (TCC) 150 F; pH 4; 95+% solids in IPA.

NZ 39. [Kenrich] Neopentyl (diallyl) oxy, tri(9,10 epoxy stearoyl) zirconate; coupling agent; amber waxy liq.; sol. in xylene, toluene, DOP, min. oil; sp.gr. 1.01 (16 C).

NZ 89. [Kenrich] Neopentyl (diallyl) oxy, trimercapto-phenyl zirconate; coupling agent; brn. waxy liq.; sol. in xylene, toluene; sp.gr. 1.06 (16 C).

OA 502. [Witco/Argus] Alkylated diphenylamine; antioxidant for industrial oil, grease, and fluid; stabilizer in hydraulic brake fluid; lt. golden liq.; odorless; sol. in hydrocarbons; insol. in water; sp.gr. 0.956; visc. 65 SUS (210 F); flash pt. 425 F (COC).

Obanol 516. [Toho] Polyoctyl polyamino ethyl glycine and POE alkylphenol ether; germicide, disinfectant, deodorant, fungicidal cleaning aid; liq.

Obazoline 662Y. [Toho] Imidazoline deriv.; amphoteric; antistat and softener for syn. fibers; base material for shampoos and hair rinse; liq.; 35% conc.

Obazoline CS-65. [Toho] Imidazoline deriv.; base material for hair rinse; softening agent for textiles; paste.

Obazoline LB-40. [Toho] n-Alkyl betaine; amphoteric; antistat for hair prods.; liq.; 40% conc.

OBTS. [Akrochem] N-oxydiethylene-2-benzothiazole sulfenamide; primary accelerator for natural, SBR, nitrile, and other general-purpose rubbers; tan dustless flake; sp.gr. 1.34 ± 0.03; m.p. 76–88 C.

OCI 56. [Olin] Chloroisocyanurate (sodium dichloro-s-triazinetrione dihydrate); swimming pool stabilizer; also used in dry bleaches for laundries, dishwashing compds., scouring powds., prod. of sanitizers, and as intermediates in industrial applics.; wh. cryst. gran.; mild chlorine odor; m.w. 255.96; easily dissolved; 26.2 g/100 ml water; bulk dens. 55–62 lb/ft³ (coarse gran.); 54–60 lb/ft³ (med. gran.); pH 5.8–7.0 (1%); 55% min. avail. chlorine.

OCI 90. [Olin] Chloroisocyanurate (trichloro-s-triazinetrione); see OCI 56; wh. cryst. gran.; mild chlorine odor; m.w. 232.47; sol. 1.2 g/100 ml water; bulk dens. 58–62 lb/ft³ (coarse gran.); 56–60 lb/ft³ (med. gran.); pH 3 (1%); 89% min. avail. chlorine.

OCI 90-I. [Olin] Trichloro-s-triazinetrione; dry chlorinator as sanitizing agent in food and beverage processing and handling operations; avail. in granular, tablet, and stick form; wh. color, mild chem. odor; m.w. 232.47; sol. 1.2 g/100 ml water; pH 3 (1% sol'n.); 99% act.

Octamine®. [Uniroyal] Reaction prod. of diphenylamine and diisobutylene; antioxidant with low discoloring and staining; used in white sidewall tires, footwear, sponge compds., molded soles and heels, mech. goods, wire insulation, automobile inner tubes; FDA approved; lt. brn. powd., flakes; sol. in gasoline, benzol, toluene, ethylene dichloride, n-hexane, acetone; insol. in water; sp.gr. 0.99; m.p. 77–85 C; flash pt. 257 C.

Octoate® Z. [Vanderbilt] Zinc 2-ethylhexoate; activator for natural and syn. rubbers; used in sol. cure systems in natural and polyisoprene; lt. amber liq.; m.w. 351.77; dens. 1.12 ± 0.02 mg/m³; 80% min. act.; 17.0–19.0% zinc content.

Oftanol®. [Bayer] Isofenphos; insecticide effective against soil insects and leaf-eating plants; colorless oil; m.w. 345.4; sol. (g/100 g): > 60 g in cyclohexanone, IPA, methylene chloride, toluene; sp.gr. 1.134 (20/4 C).

Ogtac-85. [Chem-Y GmbH] Glycidyl trimethyl ammonium chloride; cationic; starch modifier, intermediate for paper, textile and cosmetic industry; liq.; 70% conc.

Ogtac-85 V. [Chem-Y GmbH] Glycidyl trimethyl ammonium chloride; cationic; intermediate for surfactants, starch modifier; liq.; 80% conc.

OHlan®. [Amerchol] Hydroxylated lanolin; nonionic; primary w/o emulsifier, aux. emulsifier and stabilizer, pigment wetting and dispersing agent, emollient and conditioner in personal care prods.; yel.-amber to lt. tan waxy solid; misc. with common oil phase ingredients, sol. at low levels in castor oil; oil-misc.; m.p. 39–46 C; HLB 4; acid no. 10 max.; sapon. no. 95–110; 100% conc.

Ointment Base No. 3, 4, 6. [Penreco] White petrolatum USP; ointment base for eye and skin medications; carriers for medical materials; visc. 55–65, 60–70, and 60–70 SUS resp. (210 F); m.p. 118–125, 118–125, and 122–133 F resp.; congeal pt. 104–115, 109–119, and 120–130 F resp.; solid. pt. N.A., 105–115, and 114–124 F resp.

Olicat® C. [Alox] Polyisobutylene in min. oil; tackiness agent, lubricant in lubricating oils and greases; med. brn. visc. liq.; sp.gr. 0.89–0.93; visc. 5500 ± 2000 cs (100 C); flash pt. 170 C; pour pt. –15 C to –1 C; 95% min. oil content.

Omadine® MDS. [Olin] Magnesium sulfate adduct of 2,2′-dithiobis (pyridine-1-oxide); antidandruff agent for nonalkaline hair care prods.; antimicrobial agent for Gram-negative and Gram-positive bacteria; also inhibits the growth of fungi; wh. to off-wh. powd.; 90% < 100 μ; no or slight odor; m.w. 426.7; sol. in water; sp.gr. 1.730; bulk dens. 0.36 g/cm³; m.p. 210 C; pH 5.5–6.9 (1% in neut. dist. water); 90% min. act.

Omadine® TBAO. [Olin] 1,1-Dimethylethanamine adduct of 2-pyridinethiol-1-oxide; broad spectrum microbiostat for liq. hydrocarbon storage tanks; wh. to gray powd.; sol. in liq. hydrocarbons, water; sp.gr. 1.175; dens. 0.27 g/cm³; m.p. 172–173 C; pH 7.0 (1% in neutral dist. water); 7.4 (10% aq.).

OMTS. [Akrochem] N-oxydiethylene-2-benzothiazole-sulfenamide; delayed-action accelerator for SBR, NR, and nitrile rubbers; tan/br. powd.; sol. in chloroform; sp.gr. 1.40 ± 0.03; m.p. 70 C min.

Onamine 12. [Onyx] n-Dodecyl dimethylamine; cationic; intermediate used in synthesis of surfactants, antioxidants, oil and grease additives; liq.; sp.gr. 0.78; flash pt. > 200 F; 100% conc.

Onamine 14. [Onyx] n-Tetradecyl dimethylamine; cationic; see Onamine 12; liq.; sp.gr. 0.80; flash pt. > 200 F; 100% conc.

Onamine 16. [Onyx] n-Hexadecyl dimethylamine; cationic; see Onamine 12; liq.; sp.gr. 0.80; flash pt. > 200 F; 100% conc.

Onamine 18. [Onyx] n-Stearyl dimethylamine; cationic; see Onamine 12; liq.; sp.gr. 0.80; flash pt. > 200 F; 100% conc.

Onamine 65, 835, 1214, 1416. [Onyx] n-Alkyl dimethylamine; cationic; see Onamine 12; liq.; sp.gr. 0.79; flash pt. > 200 F; 100% conc.

Oncor F-31. [Rheox] Basic lead silico chromate; anticorrosive pigment; orange powd.; sp.gr. 3.95; 100% act.

Oncor M-50. [Rheox] Basic lead silico chromate; corrosion inhibiting pigment for metal protective coatings; orange powd.; 7 µ particle size; 99.7% min. thru 325 sieve; sp.gr. 4.1; dens. 34.1 lb/gal; pH 8.3–8.6; 46.0–49.0% lead oxide; 45.5–48.5% silicon dioxide; 5.1–5.7% chromium trioxide.

Onifine-C. [Otsuka] 20% Azodicarbonamide, 75% polymeric carrier, 5% barium-zinc complex stabilizer; blowing agent for prod. of cross-linked PVC foam of closed-cell structures; cushioning and absorbing materials; insulation, water flotation items; sealant in motor and construction industries; lt. yel. powd.

Onifine-CC. [Otsuka] 40% Azodicarbonamide, modified, 50% polymeric carrier, 10% barium-zinc complex stabilizer; see Onifine-C; lt. yel. powd.

Onifine-CE. [Otsuka] 93.5% Azodicarbonamide, modified, 6.5% barium-zinc complex stabilizer; blowing agent for highly cross-linked PVC foaming prods.; esp. designed for extrusion; PVC expanded prods..

Onyxide 75. [Stepan] n-Alkenyl dimethyl ethyl ammonium bromide; cationic; algicide used in recirculating water systems, swimming pools, humidifiers; paste; sp.gr. 0.95; flash pt. 68 F; 75% min. act.

Onyxide 172. [Stepan] Quaternium-8; cationic; org. nonvolatile antifungal agent and preservative for latex emulsions, paints, adhesives, coated fabrics, cutting oils; liq.; sp.gr. 1.02; flash pt. > 200 F; 77–81% act.

Onyxide 200. [Stepan] Hexahydro-1,3,5-tris (2-hydroxethyl)-s-triazine; preservative for sol. cutting fluids and coolants; bactericide for oil field drilling and completion fluids, fracturing fluids, and enhanced oil recovery applics.; liq.; sp.gr. 1.15; flash pt. > 200 F; 79% act.

Onyxide 3300. [Stepan] Myristalkonium saccharinate; cationic; germicide, conditioner, disinfectant, preservative for toiletry and pharmaceutical formulations; powd.; sol. in polar solvs.; sp.gr. 0.38; flash pt. > 200 F; 95% min. act.

Onyxol® 42. [Stepan] Stearamide DEA; nonionic; softener, dye carrier, pearlescent, opacifying and thickening agent, used in textile and cosmetic industries; wax; sp.gr. 0.97; flash pt. > 200 F; acid no. 18–24.

Onyxol® 336. [Stepan] Lauramide DEA; nonionic; foaming, wetting, and dispersing agent, used in personal care prods.; liq.; sp.gr. 1.01; flash pt. > 200 F; acid no. 18–24; 97% conc.

Onyxol® 345. [Stepan] Lauramide DEA; nonionic; foam stabilizer, wetting agent, dispersant, thickener, detergent used in cosmetic formulations, household detergents; liq.; sp.gr. 1.00; flash pt. > 200 F; acid no. 8–13; 97% conc.

Onyxol® 2062. [Stepan] Mixed fatty acid/tall oil fatty acid modified 2:1 DEA; nonionic; surfactant, detergent, emulsifier, and thickener for hard surface cleaners; liq.; sp.gr. 0.99; flash pt. > 200 F; acid no. 40–50.

Onyxol® SD. [Stepan] Lauramide DEA: nonionic; foam stabilizer and thickener for personal care prods. and lt.-duty detergents; liq.; 100% conc.

OP-2000. [BASF] PPG-26 oleate; cosmetic emollient, lubricant, defoamer, visc. control agent, dispersant, spreading agent for personal care prods.; pale yel. liq., mild, pleasant odor; m.w. 2000; sol. in IPM, IPA; insol. water; sp.gr. 0.986; dens. 8.3 lb/gal; visc. 267 cps; pour pt. –20 F; acid no. < 1.0; sapon no. 23; biodeg.; 100% act.

Oropon. [Rohm & Haas] Protease; enzymes for bating hides before tanning.

Ortho Danitol® 2.4 EC Spray. [Chevron] Fenpropathrin; pyrethorid with insecticidal and miticidal activity for use on ornamentals, fruits, cotton, and other field and veg. crops; liq.; m.w. 349.4; sp.gr. 0.9495; flash pt. 90 F (TCC)

Ortho Orthocide®. [Chevron] cis-N-[(Trichloromethyl) thio]-4-cyclohexene-1,2-dicarboximide; fungicide for control of diseases on a variety of fruit and vegetables; wh. to lt. beige cryst. solid; slight pungent odor; m.w. 300.6; sol. (g/100 ml): 6.5 g xylene, 5 g cyclohexanone, 3.3 g acetone; 3.3 ppm water; m.p. 160–170 C; avail. in three grades: Orthocide 92 Micronized (92% act. powd.), 50 Wettable (50% act. wettable powd.), 80 Wettable (80% act. wettable powd.)

Orthophen® 278. [Pennwalt] Amylphenol; intermediate for chemical specialties; also in mfg. of photographic chemicals, oil demulsifiers, phenolic resins, agric. surfactants, antiskinning agents; sp.gr. 0.918–0.924; b.p. 287–309 C; flash pt. (PMCC) 47 C; 91% min. purity.

Ortho Prunit. [Chevron] (E)-1-(p-Chlorophenyl)-4,4-dimethyl-2-(1,2,4-triazol-1-yl)-1-penten-3-ol; plant growth retardant for wide variety of native and ornamental trees; wh. cryst. solid; sol. in acetone, methanol, ethyl acetate, chloroform, DMF; insol. in water; sp.gr. 1.28; m.p. 147–164 C.

Ortho Select. [Chevron] Selective postemergence grass herbicide; clear amber liq.; m.w. 359.92; sol. in most org. solvs.; sp.gr. 1.15.

Ortho Spotless. [Chevron] Diniconazole; systemic,

sterol-inhibiting fungicide for use on peanuts, apples, grapes, small grains, etc.; wh. cryst. solid; m.w. 326.23; sol. 10–20% in acetone, chloroform, and methanol; 4.0 ppm in water; sp.gr. 1.32; m.p. 134–156 C.

Ortho Sumagic. [Chevron] Uniconizole; plant growth retardant for container-grown ornamentals; wh. cryst. solid; sol. in acetone, methanol, ethyl acetate, chloroform, DMF; insol. in water; sp.gr. 1.28; m.p. 147–164 C.

Orzan® A. [ITT Rayonier] Ammonium lignin sulfonate; anionic; binder; dispersant used to reduce slurry visc. for fine mineral prods., used for skins, leather, metallurgy, etc.; br. powd.; 55% sol. in water; sp.gr. 1.21 (50% aq.); visc. 300 cps (50% aq.); 95% conc.

Orzan® AE. [ITT Rayonier] Lignosulfonate base; SS & CSS asphalt emulsifier, dispersant, suspending agent; liq.

Orzan® AL-50. [ITT Rayonier] Ammonium lignosulfonate aq. sol'n.; binder for linoleum paste; dispersant for gypsum boards; also for briquetting, adhesives, resins; liq.; sp.gr. 1.230–1.240; visc. 500–800 cps; pH 5; biodeg.; 65% lignosulfonate.

Orzan® CG. [ITT Rayonier] Lignosulfonate; dispersant for gypsum board mfg.; liq. and powd.

Orzan® LS. [ITT Rayonier] Sodium lignosulfonate; dispersant, suspending agent in wettable pesticides; sequestrant in water treatment; emulsifier, stabilizer in oil and wax emulsions; industrial cleaners; ore flotation; yel. powd.; rapidly sol. in water; bulk dens. 30 lb/ft^3; pH 6.5 (25%); biodeg.; 59% lignosulfonate.

Orzan® LS-50. [ITT Rayonier] Sodium lignosulfonate aq. sol'n.; see Orzan LS; liq.; sp.gr. 1.255–1.265; visc. 100 cps (20 C); pH 7.0 (25%); biodeg.; 47% sol'n., 59% lignosulfonate.

Orzan® S. [ITT Rayonier] Sodium lignin sulfonate; anionic; dispersant used to reduce slurry visc. for fine min. prods., used for skins, leather, metallurgy, boiler water treatment; emulsion stabilizer; br. powd.; 60% sol. in water; sp.gr. 1.264 (50% aq.); visc. 250 cps (50% aq.); 95% conc.

Orzan® SL-50. [ITT Rayonier] Sodium lignosulfonate aq. sol'n.; binder for carbon black pellets, plywood, particle board; dispersant for portland cements, gypsum board, limestone slurry; also for briquetting, adhesives, resins; liq.; sp.gr. 1.235–1.245; visc. 200 cps (20 C); pH 6.5 (25%); biodeg.; 56% lignosulfonate.

Orzol. [Witco] Wh. min. oil USP; emollient, lubricant for food, drug, and cosmetic industry; water-wh.; visc. (Saybolt) 320–330 (100 F); pour pt. 0 F.

Osimol Grunau 109. [Grünau] Alkylaryl sulfonate; anionic; leveling agent for dyeing polyamide; liq.; 60% conc.

Osimol Grunau 110. [Grünau] Alkylaryl sulfonate and ethylene oxide deriv.; anionic; see Osimol Grunau 109; liq.; 50% conc.

Osimol Grunau DP. [Grünau] Sulfonic acid; anionic; dispersant for dyeing polyester; liq.; 40% conc.

Osimol Grunau EFA. [Grünau] Amide amine deriv.; cationic; leveling agent for dyeing acrylic; liq.; 50% conc.

Osimol Grunau MA. [Grünau] Amide amine deriv.; cationic; migrating agent for dyeing acrylic; liq.; 55% conc.

Osimol Grunau PHT. [Grünau] EO deriv.; cationic; dispersant and leveling agent for polyester dyeing; liq.; 100% conc.

Osimol Grunau RAC. [Grünau] Amide amine deriv.; cationic; see Osimol Grunau EFA; liq.; 50% conc.

Osimol Grunau SF. [Grünau] Sulfonic acid; anionic; dispersant and protective colloid for dyeing applics.; liq.; 45% conc.

Ospin Salt ON, TAN. [Tokai Seiyu] Quat. ammonium salt; cationic; migrator, retarder used for dyeing fibers; liq.; 50% conc.

Ottalume 2100. [Ferro] Fluorescent zinc oxide; uv stabilizer in plastics and paints; opacifier in clear plastics where lt. transmittance is needed, and the lt. needs to be highly scattered; wh. daylt., fluorescent lt. blue, micron-sized powd.

Ottasept® Extra, Tech. [Ferro] 4-Chloro-3,5-xylenol; antimicrobial and preservative for industrial, chemical, and cosmetic uses incl. adhesives, shoe polishes, printing inks, cutting fluids, shampoos, medical powds., antiseptics; also disinfectant, sanitizing soap; Extra grade is purified for cosmetic and pharmaceutical use; wh. cryst solid, faint odor and wh. to off-wh. cryst. solid, char. odor resp.; m.w. 156.6; sol. (g/100 ml solv.) 86.6 g in 95% ethanol, 50 g in IPA; dens. 5 lb/gal; m.p. 114 C and 112 C min. resp.; 99.4% and 98.5% min. act resp.

Ovazyme, XX. [Finnsugar Biochemics] Glucose oxidase (fungal); enzyme for food processing; maintains freshness by preventing oxygen or glucose deterioration; desugarization of egg prods.; optimum pH 4.5–7.5; amber clear liq.

OW-1. [Air Prod.] Sec. acetylenic alcohol; corrosion inhibitor used in min. acid systems; oil well acidizing, steel pickling, and electroplating additive.

Oxaban®-A. [Angus] Dimethyl oxazolidine; cosmetic preservative, antimicrobial; liq.; 78% act. in water.

Oxaban®-E. [Angus] 7-Ethyl bicyclooxazolidine; antibacterial for cosmetics and toiletries; low odor; sol. in water.

Oxamin LO. [ICI Australia] Lauryl dimethyl amine oxide; nonionic; detergent, food foamer and foam stabilizer for personal care prods. and detergent formulations; liq.; 30% conc.

Oxetal C 110, D 104, O 108/112, S 125, T 103, T 106/110, 111/118. [Zschimmer & Schwarz] Fatty alcohol polyglycol ether; nonionic; detergent, dispersant, emulsifier, wetting agent used in detergents and cleaners for household and industry; aux. agent for textile, paper and leather industry; paste, liq.; wax; 100% conc.

Oxi-Chek 114. [Ferro] Hindered phenol; primary antioxidant for rubber polymers, polyolefins, acetals; cream-colored powd.

Oxi-Chek 116. [Ferro] Hindered phenol; antioxidant for polyolefins, styrenics, elastomers, acrylics, adhesives, PVC; wh. to off-wh. powd.

Oxi-Chek 414. [Ferro] Hindered phenol; antioxidant for elastomers, latexes, plastics, adhesives; wh. powd.

Oxitol. [Shell] Ethylene glycol monoethyl ether; med. boiling solv. with toluene dilution ratio of 4.9 and exhibiting strong solv. action on NC and alkyd resins; also used for phenolics, Epon resins, and various oils and dyes; used in mfg. of surf. coatings, esp. lacquers and dopes where it improves gloss and flow out, and in printing and dyeing textiles imparting brighter shades and faster colors due to increased dye solubility; used in leather finishing, where it increases penetration and improves flow; coupling agent for soap-hydrocarbon systems; used in varnish removers and cleaning sol'ns.; colorless liq.; mild odor; water-misc.; sp.gr. 0.9256–0.9286; 0.10% max. water.

Oxypon 288. [Zschimmer & Schwarz] Ethoxylated olive oil; solubilizer and refatting agent for cosmetics; liq.

Oxypon 306. [Zschimmer & Schwarz] Ethoxylated mink oil; solubilizer and refatting agent for cosmetics; liq.; water-sol.

Oxypon 2145. [Zschimmer & Schwarz] PEG-15 glyceryl isostearate; nonionic; superfatting agent; liq.; 100% conc.

Oxypruf 6. [Olin] Mixt. of propoxylated hydrazines; corrosion inhibitor for functional fluids incl. brake fluids, polyglycol-based and water-based hydraulic fluids, syn. lubricants and cutting fluids, and fire-resistant fluids; pale yel. visc. liq.; faint odor; b.p. 169 C; sol. in water, alcohols, and polyglycol ethers; sp.gr. 1.04; visc. 2785 cs (38 C); vapor pressure 19.9 mm Hg (75 C); pH 9.3 (1% aq.); flash pt. 176 C.

Oxypruf 12. [Olin] Mixt. of propoxylated hydrazine; see Oxypruf 6; pale yel. liq., faint odor; sol. in water, alcohols, and polyglycol ethers; sp.gr. 1.02; visc. 358.5 cs (100 F); flash pt. 217 C.

Oxypruf 20. [Olin] Mixt. of propoxylated hydrazine; see Oxypruf 6; pale yel. liq., faint odor; b.p. 490 F; sol. in water, alcohols, and polyglycol ethers; more sol. in hydrocarbons than Oxypruf 6; sp.gr. 1.02; visc. 149.4 cs (100 F).

Oxypruf E. [Olin] Derived from dimethylpyrazole and ethylene oxide; corrosion inhibitor in functional fluids, incl. antifreeze formulations and glycol ether-based brake fluids; pale visc. liq.; b.p. 262 C; readily sol. in water; sp.gr. 1.11; visc. 70.5 cs (100 F); pour pt. –17 C; ref. index 1.477; flash pt. 439 F.

Oxypruf P. [Olin] Derived from dimethylpyrazole and propylene oxide; corrosion inhibitor in functional fluids incl. glycol ether-based brake fluids; pale visc. liq.; b.p. 287 C; greater sol. in hydrocarbons than Oxypruf E; sp.gr. 1.01; visc. 62.8 cs (100 F); pour pt. 5 C; ref. index 1.458; flash pt. 397 F.

P

P-10 Acid. [CasChem] Ricinoleic acid; chemical intermediate; liq.

P0160. [Hüls] Phenyldimethylchlorosilane; sterically hindered blocking agent; protecting reagent.

P-0620 T 1/8′′. [M&T Harshaw] Phosphoric acid on alumina; catalyst; tablet, 1/8′′ diam.; dens. 63 lb/ft^3.

PA-57. [Akrochem] Lightly cross-linked natural rubber extended with a lt.-colored, nonstaining oil; processing aid for extrusions, calendering, and open steam curing; crumb form of rubber compressed into bales; 57% cross-linked rubber.

PA-80. [Akrochem] Lightly cross-linked natural rubber extended with a lt.-colored, nonstaining oil; processing aid when blended with natural rubber, SBR, neoprene, or nitrile rubber; for extrusions, calendering, and open steam cure; crumb form of rubber compressed into bales; 57% cross-linked rubber.

Palatase. [Novo] Fungal esterase/lipase; enzyme for hydrolysis of fats, esp. in dairy prods.; liq.

Pamolyn®. [Hercules] Fatty acid (veg. source); detergent intermediate, emulsifier, fiber lubricant, textile processing aid, defoamer, emulsion breaker; alkali-sol.I

Pancogene S. [Gattefosse Ets.] Soluble animal collagen; humectant for cosmetics; aq. sol'n.I

Papain P-100. [Finnsugar] Protease (papain); enzyme for food processing; meat tenderizer; baking; beer chillproofing; protein hydrolysates; tan powd.

Par®. [Akzo] Oil-based prods. containing fatty amides; pulp mill brn. stock washes defoamer and drainage aid; insol. in water.

Parabis. [Dow] Purer para-para grade of bisphenol A; resin intermediate used to produce PC and polysulfone engineering thermoplastic resins with toughness, clarity, and low dens.; high strength glazing, rugged appliance and tool housings, wheels, and high heat lighting enclosures; antioxidants for rubber, soap, and brake fluids, and fungicides, stabilizers, and dyeability enhancers; wh. flakes; m.w. 228; sol. (g/100 g solv.) 85 g acetone, 33 g ether, 210 g methanol; sp.gr. 1.195; dens. 37–41 lb/ft^3; b.p. 220 C; f.p. 156.5 C min.; b.p. 220 C (4 mm Hg); suspended dust in air can be flam.; flash pt. 415 F (COC).

Parabolix® 100. [Merix] Degreaser, destaticizer for electronics, parabolic light fixtures.

Paracol® 403A6. [Hercules] Paraffin wax-pale rosin emulsion; sizing agent for surf. sizing of paper and paperboard where all-wax emulsions might cause excessive slipperiness; improves sizing and finish of writing and printing papers; used in food-pkg. grades of paper/paperboard; wh. 1.0 μ particle size emulsion; dens. 1.0 kg/l; m.p. 49–54 C; 42% total solids.

Paracol® 404A. [Hercules] Paraffin wax emulsions; sizing agent for food-pkg., writing, and printing papers; wh. emulsion particle size 1.5μ; dens. 0.96 kg/l; m.p. 51–53 C; 47% total solids.

Paracol® 404C. [Hercules] Refined microcrystalline wax emulsion; sizing agent for improving surf. properties of paper and paperboard; detackifier; antiblocking agent; lt. amber emulsion, particle size 1.5 μ; dens. 0.96 kg/l; m.p. 68+ C.

Paracol® 404D. [Hercules] Paraffin wax emulsion; sizing agent for paper and paperboard; emulsion, particle size 1.5 μ; dens. 0.96 kg/l; m.p. 53–54 C; 47% total solids.

Paracol® 404G. [Hercules] Fully refined paraffin wax emulsion; see Paracol 404D; wh. emulsion, particle size 1.5 μ; dens. 0.96 kg/l; m.p. 67 C; 47% total solids.

Paracol® 404N. [Hercules] Paraffin wax emulsion; sizing agent for building prods. and paper/paperboard applics.; sizing hardboard and insulation board; lt. amber emulsion, particle size 1.5 μ; dens. 0.96 kg/l; m.p. 52–57 C; 47% total solids.

Paracol® 447K. [Hercules] Paraffin wax emulsion; sizing agent for improving water resistance of building prods. and paper/paperboard; lubrication in making pencil slats; dens. 0.96 kg/l; m.p. 52–57 C; 47% total solids.

Paracol® 505A. [Hercules] Paraffin wax emulsion; acid-breaking-type emulsion used in specialized internal sizing applics. for food-pkg. grades of paper/paperboard; wh. to lt. amber emulsion, particle size 1.2 μ; dens. 0.91 kg/l; m.p. 49–54 C.

Paracol® 505G. [Hercules] Fully refined paraffin wax emulsion; see Paracol 505A; wh. emulsion, particle size 1.2 μ; dens. 0.91 kg/l; m.p. 65–68 C; 49% total solids.

Paracol® 505N. [Hercules] Paraffin-based aq. emulsion; acid-breaking-type emulsion used in specialized internal sizing applics.; used in molded articles from waste news which need inclusion of wax for water resistance; lt. amber emulsion, particle size 1.0 μ; dens. 0.91 kg/l; m.p. 52–57 C; 49% total solids.

Paracol® 800N. [Hercules] Parafffin wax emulsion; sizing particleboard; superior pumping, mechanical, and spraying stability; lt. amber emulsion, particle size 1.5 μ; dens. 0.96 kg/l; m.p. 52–57 C; 47% total solids.

Paracol® 802A. [Hercules] Paraffin wax emulsion; sizing agent for internal applics. in conjunction with rosin size and for surf.-sizing applics. in food-pkg. grades of paper and paperboard; wh. to lt. amber emulsion, particle size 1.5 μ; dens. 0.96 kg/l; m.p. 49–54 C; 50% total solids.

Paracol® 802N. [Hercules] Paraffin wax emulsion; controls water absorp. in insulating board and hardboard and sizing applics. in nonfood-pkg. grades of paper and paperboard; emulsion, particle size 2 μ; dens. 0.96 kg/l; m.p. 52–57 C; 50% total solids.

Paracol® 802NW. [Hercules] Paraffin wax emulsion; sizing agent for insulation board, hardboard, and nonfood-pkg. grades of paper and paperboard; lt. amber emulsion, particle size 2 μ; dens. 0.96 kg/l; m.p. 52–57 C; 47% total solids.

Paracol® 803A6. [Hercules] Paraffin wax-pale rosin emulsion; sizing agent for surf.-sizing applic. to paper and paperboard; wh. to lt. amber emulsion, particle size 1.0 μ; dens. 1.0 kg/l; m.p. 49–54 C; 50% total solids.

Paracol® 803G6. [Hercules] Fully refined paraffin wax-pale resin emulsion; see Paracol 803A6; pale emulsion, particle size 1.0 μ; dens. 1.0 kg/l; m.p. 66 C; 50% total solids.

Paracol® 810N. [Hercules] Paraffin wax emulsion; sizing agent for particleboard, hardboard, and insulation board; mechanical, pumping, and spraying stability for trouble-free operation in metering pumps and spray nozzles; lt. amber emulsion, particle size 1.5μ; dens. 0.96 kg/l; visc. 80 cps; m.p. 52–57 C; 55% total solids.

Paracol® 810NP. [Hercules] Paraffin wax emulsion; see Paracol 810N; lt. amber emulsion, particle size 2 μ; dens. 0.96 kg/l; visc. 375 cps; m.p. 52–57 C; 55% total solids.

Paracol® 815N. [Hercules] Paraffin wax emulsion; see Paracol 810N; lt. amber emulsion, particle size 1.5 μ; dens. 0.96 kg/l; visc. 50–70 cps; m.p. 52–57 C; 50% total solids.

Paracol® 1886. [Hercules] Paraffin-based aq. emulsion; acid-breaking-type emulsion used in specialized internal sizing applics.; lt. amber emulsion, particle size 1.5 μ; dens. 0.91 kg/l; m.p. 52–57 C; 50% total solids.

Paracol OP. [Nippon Nyukazai] Alkyl phosphate; anionic; anticorrosion agent, emulsion breaker; liq.; 100% conc.

Paraplex® 5-B. [C.P. Hall] Maleic alkyd polyester; resin plasticizer which imparts flexibility and high abrasion resistance to NC lacquer films; used in wood and metal lacquers, book cloth formulations; Gardner 3–6 tough visc. material; misc. with oleoresinous varnishes, alkyd-type resins; dens. 8.2 lb/gal; visc. (G-H) Z_1; acid no. 47–60; 80 ± 2% solids in toluol; also avail. in 100% solid form.

Paraplex® G-25. [C.P. Hall] High m.w. polyester sebacate; plasticizer for elec. tapes, high temp. insulation, coaxial cable, upholstery, coated fabric; high visc. liq.

Paraplex® G-30. [C.P. Hall] Mixed dibasic acid polyester; plasticizer for PVC, insulation, upholstery, wall cover, table cloths; resists heat, humidity, outdoor exposure; liq.

Paraplex® G-31. [C.P. Hall] Mixed dibasic acid polyester; plasticizer for PVC, upholstery, coated fabric, insulation, wall cover, shoe liners, babywear, handbags; resists hydrocarbons; liq.

Paraplex® G-40. [C.P. Hall] High m.w. polyester adipate; plasticizer for PVC, gasoline hose, flooring, tapes, aprons; resists migration into rubber, oils, solvs.; high visc. liq.

Paraplex® G-41. [C.P. Hall] High m.w. polyester adipate; plasticizer for PVC, hoses, pkg., tape, liners, gaskets; high visc. liq.

Paraplex® G-50. [C.P. Hall] Polyester adipate; pigment grinding medium; PVC plasticizer; for insulation, upholstery, window channels, liners, gaskets, coated fabrics; high visc. liq.

Paraplex® G-51. [C.P. Hall] Polyester adipate; PVC plasticizer for wall cover, insulation tape, gaskets, apparel; Gardner 1 visc. liq.; m.w. 2200; sol. in toluene; partly sol. in kerosene, ethanol, acetone, min. oil; sp.gr. 1.10; visc. 2800 cps; f.p. –24 C; flash pt. 268 C; acid no. 1.2; sapon. no. 542; ref. index 1.464.

Paraplex® G-54. [C.P. Hall] Polyester adipate; PVC plasticizer for gaskets, insulation, upholstery, wall cover; high visc. liq.

Paraplex G-56. [C.P. Hall] Polyester adipate; plasticizer; Gardner 2 visc. liq.; m.w. 4200; sol. in toluene; partly sol. in kerosene, ethanol, acetone, min. oil; sp.gr. 1.12; visc. 11,000 cps; f.p. –10 C; flash pt. 293 C; acid no. 0.8; iodine no. nil; sapon. no. 573; ref. index. 1.466.

Paraplex G-57. [C.P. Hall] Polyester adipate; plasticizer; APHA 100 visc. liq.; m.w. 3400; sol. in toluene; partly sol. in kerosene, ethanol, acetone, min. oil; sp.gr. 1.10; visc. 6800 cps; f.p. –10 C; flash pt. 277 C; acid no. 1.0; iodine no. nil; sapon. no. 526; ref. index. 1.466.

Paraplex G-59. [C.P. Hall] Polyester adipate; plasticizer for PVC, NC, CAP, chlorinated rubber; Gardner 4 visc. liq.; m.w. 4900; sol. in toluene; partly sol. in kerosene, ethanol, acetone, min. oil; sp.gr. 1.13; dens. 9.39 lb/gal; visc. 20,000 cps; f.p. 7 C; flash pt. 293 C; acid no. 0.7; iodine no. nil; sapon. no. 555; ref. index 1.471.

Paraplex® G-60. [C.P. Hall] Epoxidized soybean oil; polymeric type plasticizer for coating formulations based on PVC and copolymers, NC, chlorinated rubber and paraffin, and related prods.; permanence in surface coating films under severe exposure, high plasticizing efficiency; good flexibility at low temp.; stabilizes materials with acid-producing components; Gardner 2 max. low-visc. material; dens. 8.2 lb/gal; visc. (G-H) J-O; acid no. 1 max.; 100% solids.

Paraplex® G-62. [C.P. Hall] Epoxidized soybean oil; plasticizer and stabilizer in flexible and semiflexible vinyl compds., chlorinated rubbers; for food pkg., general use; APHA 100 clear liq.; low odor and taste; m.w. 1000; f.p. 5 C; sp.gr. 0.993; dens. 8.3 lb/gal; visc. (G-H) Q; acid no. 0.2; iodine no. 0.9; sapon. no. 183; ref. index 1.471; flash pt. (COC) 590 F.

Paraplex® GA-20. [C.P. Hall] Modified polyester; plasticizer for NC coatings for rubber, sealer, caulks.

Paraplex® RG-2. [C.P. Hall] Oil-modified sebacic acid-type plasticizing coating resin; compatible with ethyl cellulose, polyvinyl butyral; used in cable lacquers, artificial leather, coated fabric formulations requiring exterior durability; used on rubber; Gardner 4–7; dens. 8.0 lb/gal; visc. (G-H) U; acid no. 22–35; Film properties (35% Paraplex/65% NC): tens. str. 8420 psi; tens. elong 5.8% (ultimate); 60 ± 2% solids in toluol; also avail. in 100% solid form.

Paraplex® RG-7. [C.P. Hall] Polymeric plasticizer; coatings resin, flexibility, durability, lt. color, and adhesion in NC lacquers for metal used for automotive parts, office equipment, etc.; compatible with most lacquer components; Gardner 4–7; dens. 8.0 lb/gal; visc. (G-H) U; acid no. 35–48; 60±2% solids in toluol.

Paraplex® RG-8. [C.P. Hall] Oil-modified sebacic acid; all-purpose grinding medium for NC lacquers offering pigment wetting and stability chars.; sole plasticizer in NC films or in combination with castor oil; Gardner 4–7; dens. 8.3 lb/gal; visc. (G-H) Z2–Z3; acid no. 0–3.5; 100% solids.

Paraplex® RGA-2. [C.P. Hall] Oil-modified polyester; NC plasticizer; for cable lacquers, artificial leather, fabric coatings, aircraft dopes, solv., inks.

Paraplex® RGA-7. [C.P. Hall] Polyester; plasticizer for NC lacquers for automotive metal parts office furniture, solv., inks.

Paraplex® RGA-8. [C.P. Hall] Oil-modified polyester; plasticizer; grinding medium for NC lacquers, ethylcellulose coatings, solv. inks.

Par® Clay. [Vanderbilt] Hard kaolin; reinforcer and inert filler for elastomers, nonblack compds.; cream powd.; 99.5% min. thru 325 mesh; dens. 2.62 ± 0.03 mg/m^3.

Paricin® 6. [CasChem] Butyl acetoxy stearate; oxidation-stable plasticizer lubricant for vinyls; liq.; 100% act.

Paricin® 8. [CasChem] Glyceryl (triacetoxystearate); lubricant plasticizer with good elec. properties and extrusion lubricity; high visc. liq.; 100% act.

Paricin® 9. [CasChem] Propylene glycol hydroxystearate; wax modifier, emollient; phys. properties near to those of spermaceti wax; m.p. 53 C; 100% act.

Paricin® 13. [CasChem] Glyceryl hydroxystearate; wax modifier, emollient; phys. properties near to those of beeswax; m.p. 69 C; 100% act.

Paricin® 15. [CasChem] Ethylene glycol hydroxystearate; wax. modifier, emollient; phys. properties near to those of candelilla wax; m.p. 66 C; 100% act.

Paricin® 220. [CasChem] N (2 hydroxyethyl) 12 hydroxystearamide; internal mold release agent, lubricant for polyolefins, PVC, styrenics; solid wax; m.p. 104 C.

Paricin® 285. [CasChem] N,N′-ethylene bis 12-hydroxystearamide; internal lubricant, mold release, slip additive for PVC; solid wax; m.p. 140 C.

Parol 70. [Penreco] Wh. min. oil, tech.; used for animal feed dedusting, food pkg, materials, meat pkg., household cleaners and polishes; sp.gr. 0.828–0.849; visc. 10.8–13.6 cSt (40 C); flash pt. 177 C; pour pt. –9 C.

Parol 80. [Penreco] Wh. min. oil, tech.; see Parol 70; sp.gr. 0.830–0.857; visc. 13.2–17.0 cSt (40 C); flash pt. 179 C; pour pt. -9 C.

Parol 100. [Penreco] Wh. min. oil, tech.; see Parol 70; sp.gr. 0.838–0.864; visc. 17.7–20.2 cSt (40 C); flash pt. 182 C; pour pt. -9 C.

Parsol® MCX. [Givaudan] Octyl methoxycinnamate; uv absorber, sunscreening agent in the wavelength range of 2900–3200 A which causes sunburn and skin damage, stimulates tanning; clear, low-visc., pale yel liq.; pract. odorless; m.w. 290.4; sol. (g/l 100 g solv.): sol. > 50 g in ethanol, 99% IPA, IPM, sweet almond oil, min. oil, coconut oil, dipropylene glycol; 1 g in propylene glycol; sp.gr. 1.007–1.017; b.p. 198–200 C (3 mm); f.p. < -25 C; acid no. 1.0 max.; 98% min. purity.

Patco® 3. [Am. Ingredients] Blend of Emplex sodium stearoyl lactylate and Verv® calcium stearoyl-2-lactylate; conditioner/softener; starch and protein complexing agent for use in yeast-leavened bakery prods.; lt. tan powd.; mild caramel odor; acid no. 55–83; ester no. 137–177; 100% act.

Patcote® 315. [Am. Ingredients] Silicone emulsion; defoamer for fermentation, alcohol, yeast, foods; water-disp.

Patcote® 500. [Am. Ingredients] Silicone process defoamer for latex trade sales; wh. opaque liq.; sp.gr. 1.00 ± 0.012; dens. 8.33 ± 0.10 lb/gal; pour pt. 20 F; flash pt. (PMCC) 400 F; 100% act.

Patcote® 512. [Am. Ingredients] Silicone defoamer for use in urethane-modified resins; lt. amber liq.; sp.gr. 0.892 ± 0.012; dens. 7.43 ± 0.1 lb/gal; pour pt. < –30 F; flash pt. (PMCC) 143 F; 100% act.

Patcote® 513. [Am. Ingredients] Silicone defoamer for use in water-reducible acrylic coatings; wh. cloudy liq.; sp.gr. 0.830 ± 0.012; dens. 6.91 ± 0.1 lb/gal; pour pt. < –30 F; flash pt. (PMCC) 145 F; 100% act.

Patcote® 519. [Am. Ingredients] Silicone defoamer for use in trade sales and industrial acrylic lacquers; wh. cloudy liq.; sp.gr. 0.840 ± 0.012; dens. 7.00 ± 0.1 lb/gal; pour pt. < –30 F; flash pt. (PMCC) 148 F; 100% act.

Patcote® 520. [Am. Ingredients] Silicone defoamer for use in water-reducible alkyd resins; wh. cloudy liq.; sp.gr. 0.839 ± 0.012; dens. 6.99 ± 0.1 lb/gal; pour pt. < –30 F; flash pt. (PMCC) 160 F; cloud pt. 77–85 F; 100% act.

Patcote® 525. [Am. Ingredients] Silicone defoamer for use in water-reducible alkyd systems; clear, slightly opalescent liq.; sp.gr. 0.822 ± 0.012; dens. 6.85 ± 0.1 lb/gal; pour pt. < –30 F; flash pt. (PMCC) 158 F; 100% act.

Patcote® 531. [Am. Ingredients] Silicone defoamer for use in water-reducible alkyds; wh. cloudy liq.; sp.gr. 0.838 ± 0.012; dens. 6.98 ± 0.1 lb/gal; pour pt. < –30 F; flash pt. (PMCC) 158 F; 100% act.

Patcote® 532. [Am. Ingredients] Silicone defoamer for use in exterior semigloss acrylic latexes; wh. cloudy liq.; sp.gr. 0.828 ± 0.012; dens. 6.90 ± 0.1 lb/gal; pour pt. < –30 F; flash pt. (PMCC) 151 F; 100% act.

Patcote® 543. [Am. Ingredients] Silicone defoamer for use in PVA, vinyl-acrylic, or combinations of PVA and acrylic-based trade sales coatings; also in semigloss acrylic paints; amber cloudy liq.; sp.gr. 0.885 ± 0.012; dens. 7.37 ± 0.1 lb/gal; pour pt. 5 F; flash pt. (PMCC) > 300 F; 100% act.

Patcote® 550. [Am. Ingredients] Silicone defoamer for use in water-reducible alkyds and acrylics as well as acrylic latices for both industrial and trade sales formulations; wh. slightly cloudy liq.; sp.gr. 0.874 ± 0.012; dens. 7.28 ± 0.1 lb/gal; pour pt. < –30 F; flash pt. (PMCC) 150 F; 100% act.

Patcote® 577. [Am. Ingredients] Silicone defoamer for use in PVA resins for trade sales and water-reducible alkyds for industrial coatings; lt. amber cloudy liq.; sp.gr. 0.826± 0.012; dens. 6.88 ± 0.1 lb/gal; pour pt. < –30 F; flash pt. (PMCC) 140 F; 100% act.

Patcote® 598. [Am. Ingredients] Silicone defoamer for use in oil-modified urethanes and water-reducible alkyds; amber cloudy liq.; sp.gr. 0.810 ± 0.012; dens. 6.75 ± 0.1 lb/gal; pour pt. < –30 F; flash pt. (PMCC) 159 F; 100% act.

Patcote® 801. [Am. Ingredients] Nonsilicone defoamer for use in PVA-acrylic copolymers and terpolymer emulsions for trade sales; wh. opaque liq.; sp.gr. 0.911 ± 0.012; dens. 7.59 ± 0.1 lb/gal; pour pt. –5 F; flash pt. (PMCC) 300 F; 100% act.

Patcote® 802. [Am. Ingredients] Nonsilicone defoamer for use in acrylic and terpolymer emulsions in trade sales; whitish amber cloudy liq.; sp.gr. 0.905 ± 0.012; dens. 7.54 ± 0.1 lb/gal; pour pt. 5 F; flash pt. (PMCC) > 300 F; 100% act.

Patcote® 803. [Am. Ingredients] Nonsilicone defoamer for use in acrylic and terpolymer emulsions for trade sales; wh. opaque liq.; sp.gr. 0.896 ± 0.012; dens. 7.47 ± 0.1 lb/gal; pour pt. 5 F; flash pt. (PMCC) > 300 F; 100% act.

Patcote® 804. [Am. Ingredients] Nonsilicone defoamer for use in trade sales latex emulsions; wh. amber liq.; sp.gr. 0.897 ± 0.012; dens. 7.47 ± 0.1 lb/gal; pour pt. 5 F; flash pt. (PMCC) > 300 F; 100% act.

Patcote® 805. [Am. Ingredients] Nonsilicone defoamer for use in acrylic latexes for both trade sales and industrial applics.; whitish amber cloudy liq.; sp.gr. 0.913 ± 0.012; dens. 7.61 ± 0.1 lb/gal; pour pt. 5 F; flash pt. (PMCC) > 300 F; 100% act.

Patcote® 818. [Am. Ingredients] Nonsilicone defoamer for use in epoxy and urethane modified alkyd industrial coatings, water and solvent-based systems; clear to lt. amber liq.; sp.gr. 0.825 ± 0.012; dens. 6.87 ± 0.1 lb/gal; pour pt. < –30 F; flash pt. (PMCC) 108 F; 100% act.

Patcote® 834. [Am. Ingredients] Nonsilicone defoamer for use in water-reducible alkyd coatings or in conjunction with Patcote 555 for acrylic latex trade sales paints; amber cloudy liq.; sp.gr. 0.875 ± 0.012; dens. 7.29 ± 0.1 lb/gal; pour pt. 5 F; flash pt. (PMCC) 300 F; 100% act.

Patcote® 847. [Am. Ingredients] Nonsilicone defoamer for use in both solvent- and water-based alkyds; clear liq.; sp.gr. 0.864 ± 0.012; dens. 7.20 ± 0.1 lb/gal; pour pt. < –30 F; flash pt. (PMCC) 107 F; 100% act.

Patcote® 883. [Am. Ingredients] Nonsilicone defoamer for use in trade sales acrylic emulsions; amber cloudy liq.; sp.gr. 0.871 ± 0.012; dens. 7.26 ± 0.1 lb/gal; pour pt. 20 F; flash pt. (PMCC) 395 F; 100% act.

Pationic 122A. [RITA] Sodium capryl lactylate; anionic; surfactant with microbial inhibitor properties; amber clear visc. liq.; sol. in propylene glycol, IPM, IPA, hazy in water; visc. 24 cps (2% aq.); HLB 11.3; acid no. 65–85; sapon. no. 235–265; pH 5.25 (2% aq.); surf. tens. 24.72 dynes/cm (0.1%); 100% conc.

Pationic 138C. [RITA] Sodium lauroyl lactylate; anionic; detergent, conditioner, visc. builder, lipophilic emulsifier, cleansing agent used in shampoos; beige waxy solid; sol. in distilled water; visc. 19 cps; HLB 14.4; m.p. 55–59 C; acid no. 50–70; sapon. no. 175–205; pH 6.90 (2% aq.); 100% conc.

Pationic ISL. [RITA] Sodium isostearoyl-2-lactylate; anionic; surfactant for cosmetics; perfume solubilizer; straw, honey clear visc. liq.; sol. in min. oil, propylene glycol, IPM, IPA; disp. in water; HLB 5.9; sapon. no. 205–225; surf. tens. 26.28 dynes/cm; pH 6.30 (2% aq.); 100% act.

Pationic NMF. [RITA] Straight and branched chain aliphatic esters, alcohol; nonionic; emollient, moisturizer, cosmetics prods.; bland clear liq.; oil-sol.; sp.gr. 0.888; dens. 7.3 lb/gal; visc. 40 cps; acid no. 2.0 max.; sapon. no. 170–190; ref. index 1.4456; 100% conc.

Pationic SSL. [RITA] Sodium stearoyl 2-lactylate; anionic; emulsifier, visc. builder, protein complexer for cosmetics prods.; lt. tan powd.; sol. @ 1% conc. and 25 C; disp. in dist. water and min. oil; visc. 175 cps; HLB 6.5; m.p. 41–46; acid no. 60–70; sapon. no. 210–235; surf. tens. 32.34 dynes/cm (0.1% conc.); pH 5.95 (2% aq.); 100% conc.

Patlac IL. [RITA] Isostearyl lactate; nonionic; surfactant, emollient for cosmetics; pale yel. visc. liq.; sol. in ethanol, propylene glycol, min. oil, IPM; sp.gr. 0.92; HLB 10.3 ± 1.0; pH 4.5 (2% aq.); 100% conc.

Patlac LA. [RITA] Lactic acid; moisture binder, humectant; clear liq.

Patlac NAL. [RITA] Sodium lactate; pH buffer, humectant, stabilizer, component of stratum corneium; clear liq.

Pave 100. [Morton Int'l.] Complex amide amine; asphalt additive, wetting and bonding aid; antistripper used in asphalt cement, cutbacks, and emulsions for paving; dk. brnsh. visc. liq.; dens. 8.2 lb/gal (60 F); sp.gr. 0.985 (60 F); visc. 6600 SUS (100 F); flash pt. 450 F (COC); pour pt. 40 F; 100% act.

Pave 192. [Morton Int'l.] Complex amide amine; see Pave 100; dk. brnsh. liq.; dens. 7.8 lb/gal (60 F); sp.gr. 0.934 (60 F); visc. 1104 SUS (100 F); flash pt. 420 F (COC); pour pt. 10 F; 100% act.

Pawn PPE-501. [GAF] Anionic polymeric blend; pigment-padding binder in continuous dyeing of natural and syn. fabrics and blends; emulsion.

PC-1244. [Monsanto] Ethyl acrylate/2-ethylhexylacrylate copolymer; additive used to suppress foaming tendencies in lubricating oil formulations; oil-disp.

PC-1344. [Monsanto] Ethyl acrylate/2-ethylhexyl-acrylate copolymer; defoamer in nonaq. hydrocarbon and solv. systems; oil-disp.

Pd-0803 E 1/16″. [M&T Harshaw] Palladium; catalyst for hydrogenation of alkenes in aromatic streams; extrusion, 1/16″ diam.; dens. 35 lb/ft^3.

PDI Product 3997. [Pigment Disp.] Diisocyanate; reactive neutralizer designed to limit toxicity dangers from spills containing free NCO; contains two industrial absorbents designed to pick up neutralized material; mildly acrid dry granular mix; dens. 7 lb/gal; flash pt. 65 F (TOC).

PE-68. [Great Lakes] Bis (2,3-dibromopropyl ether) of tetrabromobisphenol A; flame retardant used in PP, polyethylene, polybutylenes, and polyolefin copolymers; effective at low loading levels; off-wh. powd.; m.w. 943.6; sol. g/100 g solv. 50 g methylene chloride, 24 g toluene, 10 g acetone, 7 g MEK, 0.1 g in water and methanol; m.p. 90–100 C; 68% bromine.

Pearl-Glo®. [Van Dyk] Bismuth oxychloride; pearlescent pigment powds. and disps. for cosmetic eye, face, lip, and body makeup.

Pearlex 3105. [Clough] Ethylene glycol stearate; opacifier and pearling agent for liq. detergent and shampoo formulations; disp. in detergent sol'ns.; water-insol.

Pectinex 1XL, 2XL, 3XL. [Novo] Purified pectolytic enzyme derived from Aspergillus niger; food grade enzymes used to break down sol. and insol. pectins with varying degrees of esterification, for reduction of visc., clarification, maceration of plant tissue, and depectinization; used in fruit juice, wine, and citrus industries; brn. liq.; odor typ. of fermentation prods.; water-misc.; pH 4.5.

Pectinex 5XL, SCL. [Novo] Pectin-decomposing enzyme for mash and juice treatment; processing aid in wine industry; liq.

Pectinex Ultra SP-L. [Novo] Aspergillus niger group; enzyme for breakdown of sol. and insol. pectin and the degradation of haze-provoking polysaccharides; liq.

Pectinol® 59L. [Genencor] Pectinase; all purpose enzyme for juice, wine industry; liq.

Pectinol® 60G, 80SB. [Genencor] Pectinase; high strength enzyme; liq.

Pectinol® DL. [Genencor] Anthocyanase; enzyme for breakdown of anthocyanin pigments in wines, juices; liq.

Pectinol® R10. [Genencor] Pectinase; enzyme for processing fruit juice, wine; powd.

Peem 122 Conc. [GAF] LDPE emulsion conc.; nonionic; textile softener (more economical); other properties similar to Peem 397; liq.

Peem 397, 410. [GAF] Polyethylene emulsions; nonionic; textile lubricating agents offering superior color stability, and uniform, pliable films; compatible with textile finishing agents; for padding operations; fluid emulsions.

Peerless No. 1. [Vanderbilt] Sec. kaolin clay; filler used in adhesives, wallboard, paint, paper, fertilizer, roofing gran., crayons, powd. soaps, pharmaceuticals, ceramics; sanitaryware, artware, generalware, floor tile, elec. and chemical porcelain, and special refractories; imparts more plasticity to cast piece; wh. particulate; particle size 1.2 μ, 99.8% thru –325-mesh screen; dens. 2.5 mg/m^3, 16 lb/ft^3; oil absorp. 27; pH 4.6 (10%); 44.29% SiO_2, 39.32% Al_2O_3.

Peerless No. 2. [Vanderbilt] Sec. kaolin clay; see Peerless No. 1; lt. cream particulate; particle size 1.2 μ, 99.6% thru –325 mesh; dens. 2.5 mg/m^3, 16 lb/ft^3; oil absorp. 30; pH 4.8; 44.6% SiO_2, 39.5% Al_2O_3.

Peerless No. 3. [Vanderbilt] Sec. kaolin clay; see Peerless No. 1; cream particulate; particle size 1.1 μ, 99.1% thru –325-mesh screen; dens. 2.5 mg/m^3, 16 lb/ft^3; oil absorp. 30; pH 4.9 (10%); 44.8% SiO_2, 39.2% Al_2O_3.

Peerless No. 4. [Vanderbilt] Sec. kaolin clay; see Peerless No. 1; off-wh. particulate; particle size 1.5 μ; dens. 2.5 mg/m^3; dens. 16 lb/ft^3; oil absorp. 31; pH 4.5 (10%).

Pegafac CA 600. [GAF] Phosphate ester; anionic; detergent, dispersant, and wetting agent in textile wet processing; APHA 200 max. clear liq.; sol. in xylene, perchloroethylene; sp.gr. 1.07; pour pt. < 0 C; pH < 2.5 (5%); flash pt. > 93 C; acid no. 100–115; 100% conc.

Pegafac CE 410, 610. [GAF] Phosphate ester; anionic; see Pegafac CA 600; also antistat, lubricant, and emulsifier for fiber and metal lubricants; APHA 200 max. clear liq.; sol. see Pegafac CA 600; sp.gr. 1.10 and 1.11 resp.; pour pt. 18 and < 0 C resp.; pH < 2.5 (5%); flash pt. > 93 C; acid no. 85–100 and 62–72; 100% act.

Pegafac CP 710, CR 719. [GAF] Complex org. phosphate ester; anionic; see Pegafac CA 600; liq.; 100 and 90% conc. resp.

Pegafac CS 410, 610. [GAF] Phosphate ester; anionic; see Pegafac CE 410; APHA 200 max. hazy to cloudy liq.; sol. see Pegafac CA 600; sp.gr. 1.02 and 1.04 resp.; pour pt. < 0 C; pH < 2.5 (5%); flash pt. > 93 C; acid no. 95–115 and 75–85; 100% act.

Pegafac CS 710. [GAF] Phosphate ester; anionic; see Pegafac CE 410; APHA 200 max. cloudy liq.; sol. in xylene, perchloroethylene; sp.gr. 1.05; pour pt. 18 C; pH < 2.5 (5%); flash pt. > 93 C; acid no. 58–70; 100% act.

Pegafac PDA Free Acid. [GAF] Phosphate ester; anionic; antistat, lubricant, and emulsifier in fiber and metal lubricants; APHA 100 max. hazy to cloudy liq.; sol. in perchloroethylene; sp.gr. 1.02; pour pt. < 0 C; pH < 2.5 (5%); flash pt. > 93 C; acid no. 53–63; 100% act.

Pegafac PDA (Neutralized). [GAF] Phosphate ester; anionic; see Pegafac PDA Free Acid; APHA 100 max. clear liq.; sol. in xylene, perchloroethylene; sp.gr. 1.00; pour pt. < 0 C; pH 7.5–8.5 (5%); 90% act.

Pegafac PEH. [GAF] Complex org. phosphate ester; nonionic; see Pegafac CA 600; liq.; 100% conc.

Pegafac PNP 9. [GAF] Phosphate ester; see Pegafac CA 600; APHA 200 max. clear liq.; sol. in xylene, perchloroethylene; sp.gr. 1.10; pour pt. < 0 C; pH < 2.5 (5%); flash pt. > 93 C; acid no. 50–60; 100% act.

Pegameen 02. [GAF] Fatty amine, ethoxylated; cationic/nonionic; wetting agent, emulsifier, corrosion inhibitor, antistat, dispersant and detergent; liq.; HLB 4.9; 100% conc.

Pegameen 030, 30 80%. [GAF] Fatty amine, ethoxylated; cationic/nonionic; see Pegameen 02; solid and liq. resp.; HLB 16.6; 100 and 80% conc.

Pegameen C2, C5, C10, C25. [GAF] Fatty amine, ethoxylated; cationic/nonionic; see Pegameen 02; liq.; HLB 6.0, 10.4, 13.7, 16.6 resp.; 100% conc.

Pegameen S2, S5, S20. [GAF] Fatty amine, ethoxylated; cationic/nonionic; see Pegameen 02; liq.; HLB 4.9, 9.0, 15.3 resp.; 100% conc.

Pegameen T2, T5, T15, T25. [GAF] Fatty amine, ethoxylated; cationic/nonionic; see Pegameen 02; liq.; HLB 5.0, 9.0, 14.2, 16.1 resp.; 100% conc.

Peganate CO 5, 16, 25, 30. [GAF] Castor oil ethoxylate; nonionic; lubricant, softener, antistat, emulsifier and detergent; liq., solid; HLB 3.9, 8.8, 10.8, 11.7 resp.; 100% conc.

Peganate CO 36, 40, 200, 200 50%. [GAF] Castor oil ethoxylate; nonionic; see Peganate CO 5; liq.; HLB 12.5, 12.9, 18.1, 18.1 resp.; 100% conc.

Peganate COH 25. [GAF] Hydrog. castor oil, ethoxylated; nonionic; see Peganate CO 5; liq.; HLB 11.1; 100% conc.

Peganate COH 200, COH 200 50%. [GAF] Hydrog. castor oil, ethoxylated; nonionic; see Peganate CO 5; solid and liq. resp.; HLB 18.1; 100 and 50% conc.

Peganate L 5, 20. [GAF] Sorbitan ester ethoxylate; nonionic; emulsifier, solubilizer, lubricant and detergent; liq.; HLB 13.3 and 16.7 resp.; 100% conc.

Peganate MA 8. [GAF] Fatty acids, ethoxylated; nonionic; see Peganate CO 5; liq.; HLB 10.0; 100% conc.

Peganate MA 15, 15 80%. [GAF] Fatty acids, ethoxylated; nonionic; see Peganate CO 5; solid and liq. resp.; HLB 14.0; 100 and 80% conc. resp.

Peganate ML 5, 9, 14. [GAF] Fatty acids, ethoxylated; nonionic; see Peganate CO 5; liq.; HLB 9.8, 13.1, 14.9 resp.; 100% conc.

Peganate MO 6, 9, 14. [GAF] Fatty acids, ethoxylated; nonionic; see Peganate CO 5; liq.; HLB 9.8, 11.7, 13.6 resp.; 100% conc.

Peganate MS 8, 14, 23. [GAF] Fatty acids, ethoxylated; nonionic; see Peganate CO 5; solid; HLB 11.3, 13.8, 15.6 resp.; 100% conc.

Peganate O 5, 20. [GAF] Fatty acids, ethoxylated; nonionic; emulsifier, solubilizer, lubricant and detergent; solid; HLB 10.0 and 9.6 resp.; 100% conc.

Peganate S 5, 20. [GAF] Fatty acids, ethoxylated; nonionic; see Peganate O 5; solid; HLB 9.6 and 14.9 resp.; 100% conc.

Peganate TO 9, 16. [GAF] Fatty acids, ethoxylated; nonionic; see Peganate CO 5; liq.; HLB 11.9 and 13.5 resp.; 100% conc.

Peganol A 24, 26. [GAF] Alcohol, ethoxylated; nonionic; wetting agent, emulsifier, dispersant and detergent; solid and liq. resp.; HLB 7.3 and 13.2; 100% conc.

Peganol B 26. [GAF] Alcohol, alkoxylated; nonionic; see Peganol A 24; liq.; HLB 14.0; 100% conc.

Peganol CL 214, 225, 240. [GAF] Alcohol alkoxylate; see Peganol A 24; liq.; 100% conc.

Peganol CSS 10, 25. [GAF] Alcohol, ethoxylated; nonionic; see Peganol A 24; solid; HLB 12.5 and 16.1 resp.; 100% conc.

Peganol D 25. [GAF] Alcohol, alkoxylated; nonionic; see Peganol A 24; liq.; HLB 10.0; 100% conc.

Peganol DA 4, 6. [GAF] Alcohol, ethoxylated; nonionic; see Peganol A 24; liq.; HLB 10.6 and 12.5 resp.; 100% conc.

Peganol DA 211, 214. [GAF] Alcohol, alkoxylated; nonionic; see Peganol A 24; liq.; 100% conc.

Peganol DDP 6, 10. [GAF] Dodecyl phenol ethoxylate; nonionic; see Peganol A 24; liq.; HLB 10.1 and 12.5 resp.; 100% conc.

Peganol DNP 8, 24, 150. [GAF] Dinonyl phenol ethoxylate; nonionic; see Peganol A 24; liq., solid, solid resp.; HLB 10.0, 15.0, 19.0; 100% conc.

Peganol LA 4, 9. [GAF] Alcohol, ethoxylated; nonionic; see Peganol A 24; liq.; HLB 8.0 and 13.2 resp.; 100% conc.

Peganol NP 1.5. [GAF] Nonoxynol-1 (1.5 EO); nonionic; surfactant and chemical intermediate for esters; foam stabilizer for detergents; coemulsifier for surfactant blends; detergent defoamer; emulsion stabilizer; detergent and dispersant for lubricants; Gardner 2 max. clear visc. liq.; sol. in oil, xylene, perchloroethylene; insol. in water; sp.gr. 0.98; visc. 8 cs (100 C); HLB 4.6; pour pt. –19 C; pH 6.0–7.5 (5%); flash pt. > 200 C.

Peganol NP 4. [GAF] Nonoxynol-4; nonionic; detergent and dispersant for lubricants; coemulsifier; plasticizer and antistat; stabilizer for latex emulsions; chemical intermediate for sulfate esters; corrosion inhibitor for engine oils; APHA 100 max. clear visc. liq.; sol. in oil, xylene, perchloroethylene; insol. in water; sp.gr. 1.023; visc. 9 cs (100 C); HLB 8.8; pour pt. –25 C; pH 6.0–7.5 (5%); flash pt. > 200 C.

Peganol NP 5. [GAF] Nonoxynol-5; nonionic; dispersant for lubricants; surfactant; chemical intermediate; emulsifier and coupling agent for surfactant blends; rust inhibitor in storage tanks; deicing fluid for gasoline and jet aircraft fuels; APHA 100 max. clear visc. liq.; sol. see Peganol NP 4; sp.gr. 1.031; visc. 10 cs (100 C); HLB 10.0; pour pt. –30 C; pH 6.0–7.5 (5%); flash pt. > 200 C.

Peganol NP 6. [GAF] Nonoxynol-6; nonionic; detergent and dispersant for lubricants; emulsifier for silicones, agric. chemicals, and min. oil; APHA 100 max. clear visc. liq.; sol. see Peganol NP 4; sp.gr. 1.038; visc. 10.5 cs (100 C); HLB 10.8; pour pt. –32 C; pH 6.0–7.5 (5%); flash pt. > 200 C.

Peganol NP 9. [GAF] Nonoxynol-9; nonionic; detergent, wetting agent, penetrant, corrosion inhibitor, emulsifier in metal processing, textile/leather processing, paper industry, household, institutional, and industrial cleaners, paints, emulsion cleaning; APHA 100 max. clear visc. liq.; sol. in water, xylene, perchloroethylene; sp.gr. 1.055; visc. 11 cs (100 C); HLB 13.0; pour pt. –1 C; pH 6.0–7.5 (5%); cloud pt. 52–56 C (1% aq.).

Peganol NP 10. [GAF] Nonoxynol-10; nonionic; see Peganol NP 9; APHA 100 max. clear liq.; sol. see

Peganol NP 9; sp.gr. 1.057; visc. 12 cs (100 C); HLB 13.2; pour pt. 8 C; pH 6.0–7.5 (5%); flash pt. > 200 C; cloud pt. 61–65 C (1% aq.).

Peganol NP 12. [GAF] Nonoxynol-12; nonionic; see Peganol NP 9; also detergent in scouring raw wool; detergent formulations; APHA 100 max. clear liq.; sol. see Peganol NP 9; sp.gr. 1.064; visc. 14.5 cs (100 C); HLB 14.2; pour pt. 17 C; pH 6.0–7.5 (5%); flash pt. > 200 C; cloud pt. 80–83 C.

Peganol NP 15. [GAF] Nonoxynol-15; nonionic; detergent and dispersant; emulsifier for fats, oils, and waxes; wetting agent and penetrant in caustic formulations; APHA 100 max. clear liq.; sol. see Peganol NP 9; sp.gr. 1.074; visc. 17 cs; (100 C); HLB 15.0; pour pt. 23 C; pH 6.0–7.5 (5%); flash pt. > 200 C; cloud pt. 95–100 C (1% aq.).

Peganol NP 20. [GAF] Nonoxynol-20; nonionic; detergent, wetting agent, and dispersant; emulsifier for fats, oils, waxes, solvs. and polyester resins; demulsifier for crude petrol. oil emulsions; syn. latice stabilizer; component of silicone emulsions used in glass mold release agents; APHA 100 max. (50 C) soft wax; sol. see Peganol NP 9; sp.gr. 1.052; visc. 20 cs (100 C); HLB 16.0; pour pt. 33 C; pH 6.0–7.5 (5%); flash pt. > 200 C; cloud pt. > 100 C.

Peganol NP 30, 30 70%. [GAF] Nonoxynol-30; nonionic; wetting agent, emulsifier, detergent, dispersant; solid; HLB 17.2; 100 and 70% conc. resp.

Peganol NP 40. [GAF] Nonoxynol-40; nonionic; wetting agent in conc. electrolyte sol'ns.; syn. latex stabilizer; polymerization emulsifier for vinyl acetate and acrylic emulsions; floor wax and polish emulsifier and stabilizer; dyeing assistant; APHA 100 max. (50 C) wax; sol. in water; sp.gr. 1.087 (50 C); visc. 44 cs (100 C); HLB 17.8; pour pt. 45 C; pH 6.0–7.5 (5%); flash pt. > 200 C; cloud pt. > 100 C.

Peganol NP 40 70%. [GAF] Nonoxynol-40; nonionic; see Peganol NP 30; liq.; HLB 17.7; 70% conc.

Peganol NP 50. [GAF] Nonoxynol-50; nonionic; see Peganol NP 40; APHA 100 max. (50 C); wax; sol. in water; sp.gr. 1.095 (50 C); visc. 55 cs (100 C); HLB 18.2; pour pt. 46 C; pH 6.0–7.5 (5% aq.); flash pt. > 200 C; cloud pt. > 100 C.

Peganol NP 100. [GAF] Nonoxynol-100; nonionic; see Peganol NP 40; APHA 100 max. (50 C) wax; sol. in water; sp.gr. 1.120 (50 C); visc. 120 cs (100 C); HLB 19.0; pour pt. 50 C; pH 6.0–7.5 (5%); flash pt. > 200 C; cloud pt. > 100 C (1% aq.).

Peganol NP 100 70%. [GAF] Nonoxynol-100; nonionic; see Peganol NP 30; liq.; HLB 19.0; 70% conc.

Peganol OAL 20, 23. [GAF] Alcohol, ethoxylated; nonionic; wetting agent, emulsifier, detergent, dispersant; solid; HLB 15.3 and 15.8 resp.; 100% conc.

Peganol OP 3. [GAF] Octoxynol-3; nonionic; wetting agent, emulsifier, detergent, dispersant; liq.; HLB 7.8; 100% conc.

Peganol OP 6. [GAF] Octoxynol-6; nonionic; see Peganol OP 3; liq.; HLB 10.0; 100% conc.

Peganol OP 8. [GAF] Octoxynol-8; nonionic; see Peganol OP 3; liq.; HLB 12.8; 100% conc.

Peganol OP 10. [GAF] Octoxynol-10; nonionic; see Peganol OP 3; liq.; HLB 13.5; 100% conc.

Peganol OP 20. [GAF] Octoxynol-20; nonionic; see Peganol OP 3; solid; HLB 16.2; 100% conc.

Peganol OP 30, 30 70%. [GAF] Octoxynol-30; nonionic; see Peganol OP 3; solid; HLB 17.3; 100 and 70% conc. resp.

Peganol OP 40, 40 70%. [GAF] Octoxynol-40; nonionic; see Peganol OP 3; solid and liq. resp.; HLB 17.9; 100 and 70% conc. resp.

Peganol OP 70, 70 70%. [GAF] Octoxynol-70; nonionic; see Peganol OP 3; solid and liq. resp.; HLB 18.7; 100 and 70% conc. resp.

Peganol SFE. [GAF] Octyl phenol ethoxylate; nonionic; see Peganol OP 3; solid; HLB 16.0; 100% conc.

Peganol TDA 6, 8. [GAF] Octyl phenol ethoxylate; nonionic; see Peganol OP 3; liq.; HLB 11.4 and 12.7 resp.; 100% conc.

Pegeste SML, SMO, SMP, SMS. [GAF] Sorbitan ester; nonionic; emulsifier, stabilizer, lubricant, corrosion inhibitor and antistat; liq. (SML, SMO), solid (SMP, SMS); HLB 8.6, 4.3, 6.7, and 4.7 resp.; 100% conc.

Pegnol C-14, -18, -20. [Toho] POE cetyl ether; nonionic; emulsifier, dispersant; solid; HLB 14.4, 15.3, and 15.7 resp.; 100% conc.

Pegnol L-6. [Toho] POE lauryl ether; nonionic; emulsifier, dispersant; paste; HLB 11.7; 100% conc.

Pegnol L-8, -10, -12, -15, -20. [Toho] POE lauryl ether; nonionic; see Pegnol L-6; solid; HLB 13.1, 14.1, 14.8, 15.6, and 16.5 resp.; 100% conc.

Pegnol O-6, -16. [Toho] POE oleyl ether; nonionic; see Pegnol L-6; solid; HLB 9.6 and 14.5 resp.; 100% conc.

Pegnol OA-400. [Toho] POE alkyl amine; nonionic; antistat for syn. fibers, emulsifier, dispersant, leveling agent for polyester and polyamide fabrics; solid; HLB 17.4; 100% conc.

Pegnol PDS-60. [Toho] Fatty acid ester blended; thickener in printing of fabrics; solid (block); sol. in water.

Pegol 17 R1, 17 R2, 25 R1, 25 R 2, 31 R 1. [Rhone-Poulenc Surf.] Alkoxylated glycol; nonionic; defoamer, dispersant, wetting agent, emulsifier, demulsifier, leveling agent and detergent; liq.; HLB 6.0, 8.0, 4.0, 6.0, and 4.0 resp.; 100% conc.

Pegol E 200, 400, 600, 1000, 4000. [Rhone-Poulenc Surf.] PEG; nonionic; surfactant intermediate; liq. (E 200, 400, 600), solid (E 1000, 4000); 100% conc.

Pegol F 68, 88. [Rhone-Poulenc Surf.] Alkoxylated glycol; nonionic; see Pegol 17 R1; solid; HLB 29.0 and 18.0 resp.; 100% conc.

Pegol L 31, 35, 43, 61. [Rhone-Poulenc Surf.] Alkoxylated glycol (block copolymers); nonionic; see Pegol 17 R1; liq.; HLB 4.5, 18.5, 16.0, 3.0 resp.; 100% conc.

Pegol L 62, 62 LF. [Rhone-Poulenc Surf.] Alkoxylated glycol (block copolymers); nonionic; see Pegol 17 R1; liq.; HLB 7.0 and 6.6 resp.; 100% conc.

Pegol L 64. [Rhone-Poulenc Surf.] Alkoxylated glycol (block copolymers); nonionic; see Pegol 17 R1; liq.; HLB 15.0; 100% conc.

Pegol L 81, 101, 121. [Rhone-Poulenc Surf.] Alkoxylated glycol; nonionic; see Pegol 17 R1; liq.; HLB 2.0, 1.0, 5.0 resp.; 100% conc.

Pegol P 85. [Rhone-Poulenc Surf.] Alkoxylated glycol; nonionic; see Pegol 17 R1; paste; HLB 16.0; 100% conc.

Pegol P 400, 700, 1000, 2000. [Rhone-Poulenc Surf.] POP; nonionic; see Pegol E 200; liq.; 100% conc.

Pegosperse® 50 DS. [Lonza] Glycol distearate; nonionic; emulsifier, stabilizer for suspensions and dispersions; emollient, lubricant and pigment dispersant in pharmaceuticals and cosmetics; thickener, wetting agent and plasticizer in hair prods.; wh. flakes; sol. in ethanol, min. and veg. oil; insol. in water; HLB 1 ± 1; m.p. 58–63 C; sapon. no. 185–200

Pegosperse® 50 MS. [Lonza] Glycol stearate; nonionic; dispersant, emulsifier for o/w emulsions; wh. flakes; sol. in methanol, ethanol, acetone, ethyl acetate, toluol, naphtha, min. and veg. oils; sp.gr 0.96; m.p. 55–60 C; HLB 0.9 ± 0.5; acid no. < 5; sapon. no. 180–187; pH 4.0–6.0 (3% aq.); 100% conc.

Pegosperse® 100 L. [Lonza] PEG-2 laurate; see Pegosperse 50 MS; straw liq.; sol. in ethanol, toluol, naphtha, min. oil; disp. in water; sp.gr. 0.97; HLB 7.4 ± 0.5; solid. pt. < 13 C; acid no. < 4; sapon. no. 160–170; pH 8–10 (5% aq.); 100% conc.

Pegosperse® 100 ML. [Lonza] PEG-2 laurate; see Pegosperse 50 MS; straw liq.; sol. in methanol, ethanol, acetone, ethyl acetate; sp.gr. 0.96; HLB 6.0 ± 0.5; solid. pt. 16–19 C; acid no. < 4; sapon. no. 180–190; pH 3.0–6.0 (5% aq.); 100% conc.

Pegosperse® 100 MR. [Lonza] PEG-2 ricinoleate; nonionic; see Pegosperse 50 MS; amber liq.; sol. see Pegosperse 100 ML; sp.gr. 0.98; HLB 4.8 ± 0.5; solid. pt. < 0 C; acid no. < 10; sapon. no. 140–150; pH 3–5 (5% aq.); 100% conc.

Pegosperse® 100 O. [Lonza] PEG-2 oleate; nonionic; see Pegosperse 50 MS; amber liq.; sol. in methanol, ethanol, toluol, naphtha; disp. in water; sp.gr. 0.93; HLB 3.5 ± 0.5; solid. pt. < 0 C; acid no. 80–95; sapon. no. 160–175; pH 8–9 (5% aq.); 100% conc.

Pegosperse® 100 S. [Lonza] PEG-2 stearate; nonionic; see Pegosperse 50 MS; wh. beads; sol. in methanol, ethanol, toluol, naphtha, min. and veg. oils; disp. in water; sp.gr. 0.96; m.p. 46–54 C; HLB 3.8 ± 0.5; acid no. 95–105; sapon. no. 165–175; pH 6.8–7.5 (3% aq.); 100% conc.

Pegosperse® 200 DL. [Lonza] PEG-4 dilaurate; nonionic; see Pegosperse 50 DS; lt. yel. liq.; sol. in ethanol, min. and veg. oil; disp. in water; sp.gr. 0.96; HLB 7 ± 1; solid. pt. 3 C; sapon. no. 170–185

Pegosperse® 200 ML. [Lonza] PEG-4 laurate; nonionic; see Pegosperse 50 MS; yel. liq.; sol. in methanol, ethanol, acetone, ethyl acetate, toluol; misc. with water, sp.gr. 0.99; HLB 8.6 ± 0.5; solid pt. < 5; acid no. > 5; sapon. no. 149–159; pH 4.0–6.5 (5% aq.); 100% conc.

Pegosperse® 400 DL. [Lonza] PEG-8 dilaurate; nonionic; see Pegosperse 50 MS; yel. liq.; sol. in methanol, ethanol, acetone, ethyl acetate, toluol, naphtha, min. and veg. oils; sp.gr. 0.99; HLB 10.0 ± 0.5; solid. pt. 5–12 C; acid no. < 5; sapon. no. 127–140; pH 3–5 (5% aq.); 100% conc.

Pegosperse® 400 DO. [Lonza] PEG-8 dioleate; nonionic; see Pegosperse 50 MS; amber liq.; sol. in methanol, ethanol, ethyl acetate, toluol, naphtha, min. and veg. oils; disp. in water; sp.gr. 0.97; HLB 7.2 ± 0.5; solid. pt. < 0 C; acid no. < 10; sapon. no. 115–125; pH 4.5–6.5 (5% aq.); 100% conc.

Pegosperse® 400 DS. [Lonza] PEG-8 distearate; nonionic; see Pegosperse 50 MS; cream soft solid; sol. in ethanol, ethyl acetate, toluol, naphtha, min. and veg. oil, methanol; disp. in water; sp.gr. 0.98; m.p. 29–37 C; HLB 7.8 ± 0.5; acid no. < 10; sapon. no. 115–125; pH 4.0–6.5 (5% aq.); 100% conc.

Pegosperse® 400 DTR. [Lonza] PEG-8 ditriricinoleate; nonionic; see Pegosperse 50 MS; amber liq.; sol. in acetone, ethyl acetate, toluol, naphtha, min. and veg. oils; disp. in water; sp.gr. 0.95–0.97; HLB 1.8 ± 0.5; solid. pt. < 10 C; acid no. < 12; sapon. no. 156–166 pH 3.5–5.0 (5% aq.); 100% conc.

Pegosperse® 400 MC. [Lonza] PEG-8 cocoate; nonionic; see Pegosperse 50 MS; yel. liq.; sol. in methanol, ethanol, acetone, ethyl acetate, toluol, veg. and min. oil; misc. with water; sp.gr. 1.03; HLB 13.9 ± 0.5; solid. pt. < 6 C; acid no. < 4; sapon. no. 88–100; pH 4–6 (5% aq.); 100% conc.

Pegosperse® 400 ML. [Lonza] PEG-8 laurate; nonionic; see Pegosperse 50 MS; straw liq.; sol. see Pegosperse 400 MC; sp.gr. 1.03; HLB 13.9 ± 0.5; solid. pt. < 7 C; acid no. < 3; sapon. no. 90–100; pH 4.0–6.0 (5% aq.); 100% conc.

Pegosperse® 400 MO. [Lonza] PEG-8 oleate; nonionic; see Pegosperse 50 MS; amber liq.; sol. in methanol, ethanol, acetone, ethyl acetate, toluol, naphtha; disp. in water; sp.gr. 1.01; HLB 11.0 ± 0.5; solid. pt. < 0 C; acid no. > 5; sapon. no. 80–88; pH 4–6 (5% aq.); 100% conc.

Pegosperse® 400 MOT. [Lonza] PEG-8 tallate; nonionic; see Pegosperse 50 MS; yel. liq.; sol. in methanol, ethanol, acetone, ethyl acetate, toluol, naphtha, min. and veg. oils; disp. in water; sp.gr. 1.02; HLB 11.0 ± 0.5; solid. pt. < 0 C; acid no. < 5–8; sapon. no. 83–89; pH 3.5–5.0 (5% aq.); 100% conc.

Pegosperse® 400 MS. [Lonza] PEG-8 stearate; nonionic; see Pegosperse 50 MS; wh. soft solid; sol. see Pegosperse 400 MOT; sp.gr. 1.0; m.p. 30 C min.; HLB 11.2 ± 0.5; acid no. > 3; sapon. no. 83–94; pH 3.5–6.0 (5% aq.); 100% conc.

Pegosperse® 600 ML. [Lonza] PEG-12 laurate; nonionic; see Pegosperse 50 MS; yel. liq.; sol. in water, methanol, ethanol, acetone, ethyl acetate, naphtha; sp.gr. 1.01; HLB 14.6 ± 0.5; solid. pt. 16–21 C; acid no. < 1; sapon. no. 65–75; pH 6–8 (5% aq.); 100% conc.

Pegosperse® 600 MS. [Lonza] PEG-12 stearate; nonionic; see Pegosperse 50 MS; wh. soft solid; sol. in methanol, ethanol, acetone, ethyl acetate, toluol, veg. oil; disp. in water; sp.gr. 1.01; m.p. 27–32 C; HLB 13.2 ± 0.5; acid no. < 8.5; sapon. no. 72–78; pH 3–5 (5% aq.); 100% conc.

Pegosperse® 700 TO. [Lonza] PEG-14 oleate; nonionic; see Pegosperse 50 MS; amber liq.; sol. in water, methanol, ethanol, acetone, ethyl acetate, toluol, veg. oil; sp.gr. 1.10; HLB 13.5 ± 0.5; solid pt. < 15 C; acid no. < 2; sapon. value 35–45; pH 6–8 (5% aq.); 100% conc.

Pegosperse® 1000 MS. [Lonza] PEG-20 stearate; nonionic; see Pegosperse 50 MS; cream solid; sol. in methanol, ethanol, acetone, ethyl acetate, toluol, naphtha, veg. oil; disp. in water; sp.gr. 1.02; m.p. 37–43 C; HLB 15.2 ± 0.5; acid no. < 3; sapon. no. 45–55; pH 3–5 (5% aq.); 100% conc.

Pegosperse® 1500 DO. [Lonza] PEG-6-32 dioleate; nonionic; see Pegosperse 50 MS; amber soft solid; sol. see Pegosperse 600 MS; sp.gr. 1.05; m.p. 30–38 C; HLB 7.8 ± 0.5; acid no. < 11; sapon. no. 104–112; pH 3.5–5.0 (5% aq.); 100% conc.

Pegosperse® 1500 MS. [Lonza] PEG-6-32 stearate; nonionic; see Pegosperse 50 MS; cream waxy solid; sol. in ethanol, ethyl acetate, toluol, veg. oil, methanol, naphtha; misc. with water; sp.gr. 1.05; m.p. 27–31 C; HLB 13.8 ± 0.5; acid no. < 3.5; sapon. no. 57–67; pH 3–5 (5% aq.); 100% conc.

Pegosperse® 1750 MS. [Lonza] PEG-40 stearate; nonionic; see Pegosperse 50 DS; wh. flakes; sol. in water, ethanol; HLB 18 ± 1; m.p. 46 C; sapon. no. 25–35

Pegosperse® 4000 MS. [Lonza] PEG-75 stearate; nonionic; see Pegosperse 50 MS; cream waxy solid; sol. see Pegosperse 700 TO; sp.gr. 1.10; m.p. 54–61; HLB 18.0 ± 0.5; acid no. < 6; sapon. no. 17–22; pH 3–5 (5% aq.); 100% conc.

Pegosperse® 6000 DS. [Lonza] PEG-150 distearate; nonionic; see Pegosperse 50 DS; sol. in water, ethanol; HLB 18 ± 1; m.p. 55 C; sapon. no. 17

Pegosperse® 9000 CO. [Lonza] PEG 9000 castor oil; nonionic; see Pegosperse 50 MS; tan solid; sol. in water, methanol, ethanol, acetone; sp.gr. 1.08; m.p. 41–44 C; HLB 18.1 ± 0.5; acid no. < 1; sapon. no. 14–20; pH 6.0–7.0 (3% aq.); 100% conc.

Pegosperse® EGMS-70. [Lonza] Glycol stearate; see Pegosperse 50 DS; wh. flakes; sol. see Pegosperse 50 DS; HLB 2 ± 1; m.p. 52–56 C; sapon. no. 180–188

Pegosperse® MFE. [Lonza] Mixed PEG esters; nonionic; see Pegosperse 50 MS; brn. soft solid; sol. in methanol, ethanol, veg. and min. oil, water; sp.gr. 1.05; HLB 13.0 ± 0.5; m.p. 46–54 C; acid no. < 10; sapon. no. 60–70; pH 3–5 (5% aq.); 100% conc.

Pelex NBL, NB Paste. [Kao] Sodium alkyl naphthalene sulfonate; anionic; wetting agent and penetrant; liq. and paste resp.; 35 and 50% conc. resp.

Pelex TA. [Kao] Sodium dialkyl sulfosuccinate; anionic; foam controller for latex; paste; 35% conc.

Peltex. [Daishowa] Ferro-chromium lignosulfate, modified; anionic; wetting agent, emulsifier, dispersant; oil well drilling mud thinner; blk. powd.; sol. in water; dens. 0.5; 95% act.

Penestrol N-160. [Tokai Seiyu] Sec. alcohol ethoxylate; nonionic; penetrant and wetting agent; bleaching assistant; liq.; 60% conc.

Peneteck. [Penreco] Min. oil tech.; emollient; sp.gr. 0.802–0.811; dens. 6.73–6.81 lb/gal; visc. 3.4–4.7 cSt (40 C); flash pt. 265 F; pour pt. 30 F; ref. index 1.4517.

Pennad 0150. [Pennwalt] DEAE; intermediate, emulsifier, catalyst in urethane foams, curing agent, corrosion inhibitor; m.w. 117.2; sp.gr. 0.880–0.890; b.p. 158–163.5 C; flash pt. (CC) 52 C; 99.5% min. purity.

Pennfloat® 3-2277. [Pennwalt] Mercaptan; intermediate; sp.gr. 0.873; flash pt. (TCC) > 95 C.

Pennmax Five. [Pennwalt] Tin stabilizer; PVC heat stabilizer providing initial color inhibition and process stability; for use in exterior building prod. applics. requiring good resistance to weather; used in siding at levels as low as 1.25 phr; sltly. lubricating; wax-like flakes; m.p. 40–50 C.

Pennmax Four. [Pennwalt] Tin stabilizer; PVC heat stabilizer with good initial color inhibition and process stability; for use at low concs. in rigid PVC applics.; used in multiscrew pipe/conduit at levels as low as 0.2 phr and in single-screw pipe/conduit at 0.5 phr; sltly. lubricating; wax-like flakes; m.p. 50–60 C.

Pennmax Four-A. [Pennwalt] Tin stabilizer; PVC heat stabilizer with initial color inhibition and process stability; used in PVC pipe extrusion, multiscrew pipe/conduit at levels as low as 0.25 phr and in single-screw pipe/conduit at 0.6 phr; sltly. lubricating; wax-like flakes; m.p. 40–50 C.

Pennodorant® 1013. [Pennwalt] 99% min. Tetrahydrothiophene; odorant for natural gas to permit detection of leaks; negligible sol. in water; sp.gr. 1.002 (15 C); dens. 8.35 lb/gal; f.p. < –45.5 C; cloud pt. –45.5 C max.; flash pt. (TCC) < –13 C; 37% max. sulfur.

Pennstop® 1866. [Pennwalt] N,N-diethylhydroxylamine; free radical scavenger used by the rubber industry as an emulsion polymerization inhibitor; vapor phase inhibitor for olefin or styrene monomer recovery systems; in-process inhibitor for prod. of styrene, divinyl benzene, butadiene, isoprene; intermediate in the synthesis of silicone rubber and photographic developers; APHA 250 max. color; m.w. 89.1; sp.gr. 0.902; flash pt. (TCC) 46 C; 85% min. purity.

Pennstop® 2049. [Pennwalt] N,N-diethylhydroxylamine; see Pennstop 1866; colorless to lt. straw liq.; m.w. 89.1; sp.gr. 0.902; dens. 7.5 lb/gal; b.p. 125–133 C; f.p. –25 C; flash pt. (TCC) 46 C; 85% min. purity.

Pennstop® 2697. [Pennwalt] N,N-diethylhydroxylamine; see Pennstop 1866; colorless to lt. yel. liq.; m.w. 89.1; sp.gr. 0.865; dens. 7.2 lb/gal; b.p. 125–130 C (decomposes); f.p. –6 C; flash pt. (TCC) 50 C; 98% min. purity.

Pennwalt 4P®. [Pennwalt] Tertiary dodecyl mercaptan; modifier in polymerization reactions, esp. for SBR and NBR; reducing initiator; water-wh.; m.w. 202.4; sp.gr. 0.858–0.860 (15.5 C); dens. 7.16 lb/gal; flash pt. (TCC) 188 F; 98.5% min. act.

Pennwalt n-Dodecyl Mercaptan. [Pennwalt] Modifier in polymerization reactions, esp. for SBR; reducing initiator and chain transfer agent; water-wh.; m.w. 202.4; sp.gr. 0.858–0.860; dens. 7.16 lb/gal; b.p. 227 C min.; flash pt. (COC) 205 F; 98.5% min. act.

Penreco 1520, 3070. [Penreco] Petrolatum, tech.;

used in rubber processing aids, carbon papers, buffing and polishing compds., corrosion preventatives, general purpose lubricants, printing inks, solder pastes; dk. green; visc. 70–115 and 70–95 SUS resp. (210 F); m.p. 115–135 and 125–140 F.

Penreco 2251 Oil. [Penreco] High purity hydrocarbon solv.; processing solv., foam control agent, in waterless hand cleaners, agric. sprays, polishes, fruit and veg. processing, cleaning oils; sp.gr. 0.779–0.797 (60 F); dens. 6.56 lb/gal (60 F); visc. 30.5 SUS (100 F); b.p. 375 F min.; flash pt. (COC) 165 F; pour pt. –40 F.

Penreco 2257 Oil. [Penreco] High purity hydrocarbon solv.; see Penreco 2251 Oil; sp.gr. 0.793–0.806 (60 F); dens. 6.64 lb/gal (60 F); visc. 33.2 SUS (100 F); b.p. 430 F min.; flash pt. (COC) 220 F; pour pt. –10 F.

Penreco 2259 Oil. [Penreco] High purity hydrocarbon solv.; see Penreco 2251 Oil; sp.gr. 0.793–0.811 (60 F); dens. 6.69 lb/gal (60 F); visc. 33.6 SUS (100 F); b.p. 445 F min.; flash pt. (COC) 240 F; pour pt. –10 F.

Penreco 2260 Oil. [Penreco] High purity hydrocarbon solv.; see Penreco 2251 Oil; sp.gr. 0.806–0.825 (60 F); dens. 6.79 lb/gal (60 F); visc. 40.2 SUS (100 F); b.p. 500 F min.; flash pt. (COC) 280 F; pour pt. 25 F.

Penreco 2263 Oil. [Penreco] High purity hydrocarbon solv.; see Penreco 2251 Oil; sp.gr. 0.779–0.797 (60 F); dens. 6.56 lb/gal (60 F); visc. 30.5 SUS (100 F); b.p. 375 F min.; flash pt. (COC) 165 F; pour pt. –40 F.

Penreco Amber. [Penreco] Petrolatum USP; emollient, base for cosmetic and pharmaceutical preparations; waterproofing agent for butcher paper; lubricant, water repellent, moisture barrier for textile and paper; carrier for modeling clays, soldering paste and flux; pigment carrier for carbon paper; binder and conditioner for crayons; visc. 68–82 SUS (210 F); m.p. 122–135 F; congeal pt. 123 F; solid. pt. 122 F.

Penreco Blond. [Penreco] Petrolatum USP; emollient, base for cosmetic and pharmaceutical preparations; lubricant, water repellent, moisture barrier for textile and paper; visc. 68–82 SUS (210 F); m.p. 122–135 F; congeal pt. 123 F; solid. pt. 122 F.

Penreco Cream. [Penreco] Wh. petrolatum USP; emollient, base, and carrier for cosmetic and pharmaceutical preparations; lubricant for textile and paper; visc. 64–75 SUS (210 F); m.p. 122–135 F; congeal pt. 125 F; solid. pt. 122 F.

Penreco Frost. [Penreco] Wh. petrolatum USP; emollient, base, and carrier for cosmetic and pharmaceutical preparations; lubricant for textile and paper; visc. 60–75 SUS (210 F); m.p. 125–135 F; congeal pt. 124 F; solid. pt. 123 F.

Penreco Green. [Penreco] Tech. petrolatum; tech.-grade base; dk. gr.; visc. 60–125 SUS (210 F); m.p. 115–140 F.

Penreco Lily. [Penreco] Wh. petrolatum USP; emollient, base, and carrier for cosmetic and pharmaceutical preparations; lubricant for textile and paper; visc. 64–75 SUS (210 F); m.p. 122–135 F; congeal pt. 124 F; solid. pt. 123 F.

Penreco Red. [Penreco] Tech. petrolatum; tech.-grade base; base and binder for polishes; red; visc. 70–82 SUS (210 F); m.p. 120–135 F.

Penreco Regent. [Penreco] Wh. petrolatum USP; emollient, base, and carrier for cosmetic and pharmaceutical preparations; lubricant for textile and paper; water repellent, moisture barrier for textile and paper; plasticizer and softener for putty; visc. 57–70 SUS (210 F); m.p. 118–130 F; congeal pt. 120 F; solid. pt. 119 F.

Penreco Royal. [Penreco] Petrolatum USP; emollient, base, and carrier for cosmetic and pharmaceutical preparations; lubricant for textile and paper; visc. 57–70 SUS (210 F); m.p. 118–130 F; congeal pt. 118 F; solid. pt. 115 F.

Penreco Snow. [Penreco] Wh. petrolatum USP; emollient, base, and carrier for cosmetic and pharmaceutical preparations; solv. and emollient in hand creams; sanitary lubricant in food prod. machinery; rust-preventive coating in food processing equipment; carrier in adhesive tapes and compds.; lubricant for textile and paper; visc. 64–75 SUS (210 F); m.p. 122–135 F; congeal pt. 123 F; solid. pt. 121 F.

Penreco Super. [Penreco] Wh. petrolatum USP; emollient, base, and carrier for cosmetic and pharmaceutical preparations; lubricant for textile and paper; visc. 60–75 SUS (210 F); m.p. 122–135 F; congeal pt. 125 F; solid. pt. 124 F.

Penreco Ultima. [Penreco] Wh. petrolatum USP; emollient, base, and carrier for cosmetic and pharmaceutical preparations; lubricant for textile and paper; visc. 60–70 SUS (210 F); m.p. 130–140 F; congeal pt. 130 F; solid. pt. 128 F.

Pentachlorophenol DP-2. [Dow] Pentachlorophenol, tech.; antimicrobial for commercial use esp. for preservation of wood prods.; lt. tan block solid; phenolic odor; low sol. in water; sp.gr. 1.97; b.p. 275 C; 85% pentachlorophenol, 10% other chlorophenols.

Pentalan. [Croda Ltd.] Pentaerythritol tetraester; temporary corrosion protectives; grease; oil-sol.

Pentalyn® 225. [Hercules] Rosin-derived complex polyol partial ester; thermoplastic syn. resin as component of solv.- and water-based gravure and flexographic inks; used in polymer-based emulsion floor polishes as leveling agent and gives clarity and brilliance of the film; used in food pkg. and processing operations; Gardner 6 flakes; sol. in ethanol, IPA, diethylene glycol, triethylene glycol, propylene glycol, dipropylene glycol, and aq. alkalies; insol. in water; dens. 1.14 kg/l; visc. (G-H) J; soften. pt. 171 C; acid no. 190.

Pentalyn® 255. [Hercules] Rosin-derived complex polyol partial ester; thermoplastic syn. resin, used as resin component of solv.-and water-based gravure and flexographic inks; also as binder, leveling agent in polymer-based emulsion floor polishes, adhesives, varnishes; Gardner 6 (60% solids in 2B ethanol) flakes; sol. in ethanol, isopropanol, diethylene glycol, triethylene glycol, propylene glycol, dipropylene glycol and water; ammonia disp.; dens. 1.14

kg/l; visc. (G-H) J (60% solids in 2B ethanol); soften. pt. (Hercules Drop) 171 C; acid no. 190.

Pentalyn® 261. [Hercules] Rosin-derived complex polyol partial ester; pale, hard thermoplastic syn. resin, broad sol. and compatibility; used as a leveling agent for films with clarity and depth of gloss in polymer-based emulsion floor polishes; for gravure and flexographic inks; Gardner 4 (40% solids in toluene) flake; sol. in alcohol and aq. ammonia, esters, ketones, and aromatic hydrocarbons; dens. 1.15 kg/l; visc. (G-H) J (40% solids in toluene); soften. pt. (Hercules Drop) 166 C; acid no. 205; pH 8.1 (15% solids in aq. ammonia).

Pentalyn® 269. [Hercules] Rosin-derived complex polyol partial ester; thermoplastic syn. resin suitable as component in solv.- and water-based gravure and flexographic inks, and leveling agent in polymer-based emulsion floor polishes; food pkg. and processing operations; Gardner 7 flake; sol. in ethanol, IPA, diethylene glycol, triethylene glycol, propylene glycol, dipropylene glycol, and aq. alkalies; dens. 1.16 kg/l; visc. (G-H) J; soften. pt. 165 C; acid no. 200; pH 9.0.

Pentalyn® 830. [Hercules] Internally plasticized, modified pentaerythritol ester of rosin; rapid solv. release; modifier resin used with variety of film-formers to impart/improve gloss, adhesion, toughness, and hardness; good alcoholic solubility and sol'n. stability making it useful in flexographic inks, spirit varnishes, conventional lacquers, and specialty coatings and finishes; food contact; Gardner 7 flakes; low odor; sol. in ethanol; sp.gr. 1.14; dens. 9.5 lb/gal; visc. (Gardner-Holdt) E; soften. pt. 118 C; acid no. 78.

Pentalyn® 856. [Hercules] Modified pentaerythritol ester of rosin; thermoplastic resin suitable for modifying compositions based on alcohol- and water-sol. film-formers, which contributes high gloss, block resistance, fast solv. release, and good adhesion; used in printing inks; replacement or extender for shellac, in formulating water-based inks, and paper coatings, adhesives, wax emulsions, and water-reducible paints; food contact applic.; USDA Rosin WG flakes; sol. in aq. sol'ns. of alkalies, ethanol, and IPA, ethers, alcohol-ethers, esters, ketones, aromatic and chlorinated hydrocarbons, and to 25–39% solids in diethylene glycol; sp.gr. 1.10; dens. 1.09 kg/l; soften pt. 131 C; acid no. 140.

Pentalyn® A. [Hercules] Pentaerythrityl rosinate; tackifier; hard, pale, thermoplastic resin used as varnish resin with both soft and hard oils and producing exc. color and alkali resistance; used in enamel applics., linoleum cements, hot-melt adhesives and coatings, and in pressure-sensitive adhesives; contributes adhesion, high aliphatic solubility, gloss, solv. release, stability with reactive pigments, and good pigment wetting to printing inks; food contact use; Gardner 9.5 (60% in min. spirits) solid, flakes; sol. in esters, ketones, aromatic and aliphatic hydrocarbons, and higher alcohols; water-insol.; dens. 1.07 kg/l; visc. (G-H) G (60% in min. spirits); soften pt. 111 C; acid no. 12; ref. index 1.544.

Pentalyn® H. [Hercules] Pentaerythrityl hydrog. rosinate; hard, thermoplastic resin designed as tackifier for various types of polymer- and elastomer-based adhesives, where it contributes improved resistance to oxidation, greater heat stability, better color retention, and greater retention of tack-imparting properties than ordinary rosin ester resin; food-pkg. and processing operations; USDA Rosin N flakes; sol. in aromatic and aliphatic hydrocarbons, esters, and ethers; dens. 1.07 kg/l; visc. (G-H) D (60% in min. spirits; soften. pt. 104 C; acid no. 12.

Pentaphen® 67. [Pennwalt] p-t-Amylphenol; intermediate for chemical specialties; in germicidal formulations; also in mfg. of photographic chemicals, oil demulsifiers, phenolic resins, agric. surfactants, antiskinning agents; m.w. 164.2; sp.gr. 0.92; b.p. 265 C; flash pt. (PMCC) 121 C; 98.5% min. purity.

Pentaquest Extra. [Clough] Pentasodium pentetate; for iron chelation up to pH 11.5; misc. in water.

Pentex 40. [Rhone-Poulenc] Alkylaryl sulfonate; anionic; wetting agent, penetrant, emulsifier; textile and industrial processing; improves dye uniformity; aids rewetting of leather, paper, and paper-mill felts; red-br. liq.; sp.gr. 1.15; pour pt. 0 C; pH 9.5 (5%)

Pentex 99. [Rhone-Poulenc] Dioctyl sodium sulfosuccinate and propylene glycol; anionic; detergent; textile scouring and dispersant for dyes; paper rewetting and felt washing surfactant; wetting agent in cosmetics; detergent additive in dry cleaning fluids; dishwashing compds.; wallpaper removers; emulsion polymerization; water-based paint formulations; antifog; clear liq.; water-sol.; sp.gr. 1.08–1.13 (70 F); visc. 500–1000 cps (70 F); pour pt. 40 F; 75% conc.

PentoXone. [Shell] 4-Methoxy-4-methyl-pentanone-2; high boiling solv. used in preparation of NC, acrylic, vinyl, and cellulose acetate butyrate lacquers, alkyd and epoxy enamels, and urethane and thermosetting acrylic coatings; used in cleaning compds.; colorless liq.; pleasant odor; sp.gr. 0.899–0.909; 0.10% max. water.

Pepol A-0858. [Toho] Oxyethylated fatty alcohol, modified; nonionic; detergent, penetrant; paste; 100% conc.

Peptein 2000®. [Hormel] Hydrolyzed animal protein; conditioner for cosmetic applics.; esp. substantive to hair; Gardner 12 max., bland odor; m.w. 1500–2500; sp.gr. 1.15; visc. 100 cps max.; pH 5.8–6.3; 55% min. solids.

Peptizer 566. [C.P. Hall] Naphthenic oil/sulfonate ester blend; amber clear, oily liq.; sol. in hexane, toluene, kerosene, min. oil; sp.gr. 0.922; visc. 75 cps; f.p. –35 C; flash pt. 179 C; acid no. 1.5; sapon. no. 1; ref. index 1.502.

Peptizer 932. [C.P. Hall] Min. oil/sulfonate blend; amber clear, oily liq.; sol. in hexane, toluene, kerosene, min. oil; sp.gr. 0.843; visc. 43 cps; f.p. –5 C; flash pt. 135 C; acid no. 1.1; sapon. no. 1; ref. index 1.466.

Peptizer 965. [C.P. Hall] Min. oil/sulfonate ester blend; amber clear, oily liq.; sol. in hexane, toluene, kerosene, min. oil; sp.gr. 0.866; visc. 15 cps; f.p. –5

C; flash pt. 135 C; acid no. 0.1; sapon. no. 10; ref. index 1.4655.

Peptizer 7010. [C.P. Hall] Min. oil/sulfonate blend; amber clear, oily liq.; sol. in hexane, toluene, kerosene, min. oil; sp.gr. 0.850; visc. 18 cps; f.p. –5 C; flash pt. 74 C; acid no. 8.1; sapon. no. 8; ref. index 1.465.

Perchem® 44. [Akzo] Modified refined montmorillonite clay; rheological additive for visc. control, to prevent pigment settling, and reduce sag in paints, lubricating greases, printing inks, mastics and sealants; pale cream free-flowing powd.; 95% < 75 μm.

Perchem® 97. [Akzo] Modified refined montmorillonite clay; rheological additive for visc. control and antisettling in med. to high polarity systems, e.g., aromatic hydrocrbons, esters, ketones, syn. oils, plasticizers, epoxies, polyesters; off-wh. free-flowing powd.; 95% < 75 μm.

Perchem® 108. [Akzo] Modified refined montmorillonite clay; see Perchem 44; off-wh. free-flowing powd.; 95% < 75 μm.

Perchem® AMU 60 X, AMU X. [Akzo] Dispersed polymeric hydrocarbon in xylene; thixotropic rheological additive, antisag, anticaking for paints; 60 X grade is softer and more readily dispersed, AMU X offers higher solids; translucent wh. waxy solid; dens. 0.872 and 0.875 g/cc resp.; 24 and 40% NV resp.

Perchem® EAG. [Akzo] Modified montmorillonite clay; rheological additive for visc. control, antisettling, sag control; for all aq. systems, e.g., emulsion paints, water-based coatings; lt. tan free-flowing pod.; 95% < 75 μm.

Perchem® Easigel. [Akzo] Modified refined montmorillonite clay; rheological additive, gelling agent, antisettling agent, visc. control agent for aromatic, aliphatic, or polar systems; off-wh. free-flowing powd.; 95% < 75 μm; dens. 1.43 g/cc.

Perchem® Econogel. [Akzo] Modified montmorillonite clay; rheological additive, antisettling, antisag for paints, inks, sealants, mastics, plastisols, undercar coatings, and foundry mold washes; lt. tan free-flowing powd.; 95% < 75 μm; dens. 1.8–1.9 g/cc; oil absorp. 40–50 ml/100 g.

Perchem® TEA. [Akzo] Modified hectorite clay; rheological additive, viscosifier, sag control agent for all aq. systems, e.g., aq. printing inks, emulsion paints, other water-based coatings; very pale cream free-flowing powd.; 95% < 75 μm.

Peregal® D. [GAF] Complex alkylamine deriv.; leveling agent in dyeing of nylon with acid dyes to eliminate or decrease barre; liq.

Peregal® O. [GAF] POE fatty alcohol; nonionic; wetting agent, penetrant, dye assistant; leveling agent and stripping assistant for dyes; clear yel.-brn. liq.; misc. with water; sp.gr. 1.02.

Peregal® OK. [GAF] Methyl polyethanol quat. amine; cationic; wetting agent, solubilizer, textile aux. used for leveling, retarding, antistripping agents, and improved penetration for dyes; cloudy pale amber liq.; misc. with water; sp.gr. 1.03; pH 7.6.

Peregal® ST. [GAF] PVP polymer; cationic; dispersant, stripping assistant, detergent; used for dyes; used in long liquors, pkg. machines, or jigs; fulling soap additive; rag-stripping assistant in paper industry; detergent and soil-suspending agent for washing yarns; VCS 5 max. liq.; water-sol.; pH 5–7 (5% aq.); 30–32% act.

Perenol EI. [Henkel] Polyvinyl isobutyl ether; defoamer for petrol. solv.-based paints.

Perenol F3. [Henkel] Polyacrylate; anticaking agent; silicone-free additive for coatings to improve flow and leveling in high solids and conventional solv.-based coatings; reduces craters, fish eyes, and insufficient wetting of the substrate; slightly yel. clear visc. liq.; sp.gr. 0.96–0.98; flash pt. 212–224 C.

Perenol S4. [Henkel] Modified polysiloxane; anticaking agent; paint additive to improve slip properties, mar and scratch resistance, imparts better flow, and reduces the tendency of film surfs. to pick up dirt; used in solvent and solventless systems; pale ylsh. liq.; sp.gr. 0.961; flash pt. 61 C; 50% in higher aromatics.

Perenol S5. [Henkel] Modified polysiloxane; anticaking agent; paint additive to improve slip properties of water-reducible coatings, both air dry and baking systems; improves mar and scratch resistance, imparts better flow, and reduces tendency of film to pick up dirt; pale ylsh. liq.; sp.gr. 0.981; flash pt. 65 C; 50% act. in glycol-monobutylether.

Perfecta®. [Witco] Petrolatum; emollient; misc. with all petrol. prods., many essential oils, most animal and veg. fats, oils, waxes; visc. 55–75 (210 F); m.p. 135-140 F.

Perfecta® USP. [Witco] Petrolatum USP; cosmetic/pharmaceutical grade with lightest color, med. consistency, high m.p.; as carrier, lubricant, moisture barrier, protective agent, softener; Lovibond color 0.3Y; visc. 9–14 cSt (100 C); m.p. 57–60 C.

Perk. [Rhone-Poulenc Basic] Perchloroethylene; solv. for cleaning and bleaching, in paper deinking; dry cleaning solv. and carrier solv. in impregnation of fibers and fabrics; cold degreasing mixts.; solv. for oils, waxes, tars, gums, and silicones; in an industrial grade is used as chemical intermediate in org. synthesis and mfg. of fluorocarbons, in metal finishing, drying of metals, vapor degreasing; colorless liq.; f.p. –22.4 C; b.p. 120.4–122.0 C; slightly sol. in water; sp.gr. 1.619; dens. 13.47 lb/gal; 99.97% min. conc.

Perkadox® 14. [Akzo] α,α′-Bis (t-butylperoxy) diisopropyl benzene; low reactivity peroxide useful as a finishing initiator at high temps. for styrenics; synergist for some halogen-containing flame retardants; also for cross-linking of olefin copolymers, EPDM, SBR, Neoprene, Hypalon; liq.; 96% act.

Perkadox® 14/40. [Akzo] α,α′-Bis(t-butylperoxy) diisopropylbenzene with inorg. phlegmatizer; initiator for elevated-temp. polyester cures when long ambient temp. or high ambient temp. storage of molding compds. is necessary; cross-linking of olefin copolymers, EPDM, SBR, Neoprene, Hypalon; powd.; 40% act.

Perkadox® 14/40 Bpd. [Akzo] α,α′-Bis (t-butylperoxy) diisopropyl benzene; initiator where very long or high ambient temp. storage of molding compds. is necessary; also for cross-linking of olefin copolymers, EPDM, SBR, Neoprene, Hypalon; powd.; 40% act. on inorg. filler.

Perkadox® 16. [Akzo] Bis(4-t-butylcyclohexyl) peroxydicarbonate; ultrafast initiator for polyester cure above 180 F; pultrusion, matched die molding; short ambient temp. compd. shelf life.; powd.; 98% act.

Perkadox® 16/35. [Akzo] Bis-(4-t-butyl cyclohexyl) peroxydicarbonate; ultrafast initiator for curing polyester; powd.; 35% on inorg. filler.

Perkadox® 16N. [Akzo] Bis (4-t-butyl cyclohexyl) peroxydicarbonate; initiator for PVC prod., acrylic polymerization; liq.; 98% act.

Perkadox® 16/W40. [Akzo] Bis(4-t-butylcyclohexyl) peroxydicarbonate; see Perkadox 16N; liq.; 40% act. in water.

Perkadox® 16/W70. [Akzo] Bis-(4-t-butylcyclohexyl) peroxydicarbonate; see Perkadox 16/35; 16/W70 also used as initiator for PVCs; powd.; 70% act. in water.

Perkadox® BC. [Akzo] Dicumyl peroxide; high-temp. initiator used as a flame retardant synergist; also as cross-linking agent for a variety of natural and syn. rubbers and olefins; solid; 98% act.

Perkadox® BC-40Bpd, BC-40Kpd. [Akzo] Dicumyl peroxide; cross-linking agent for a variety of natural and syn. rubbers as well as olefins; powd.; 40% act. on calcium carbonate and on silane-treated clay resp.

Perkadox® IPP-AT50. [Akzo] Di(isopropyl) peroxydicarbonate; initiator for use in PVC; liq.; 50% act. in toluene.

Perkare. [PPG Industries] Activated perchlorethylene; dry cleaning solv. with a detergent; antistat; APHA clear, free of suspended material; pleasant odor; sp.gr. 1.614; dens. 13.4 lb/gal; pH close to neutral.

Perlankrol ASC38. [Harcros UK] Fatty alcohol ethoxylate; anionic; toiletry intermediate; liq.; 28% conc.

Perlankrol DGS. [Harcros UK] Sodium primary alcohol sulfate; anionic; solubilizer, textile aux.; detergent base in household and industrial cleaners; foaming agent in syn. latex industry; primary emulsifier in emulsion polymerization; wh. slurry; faint fatty alcohol odor; water-sol.; sp.gr. 1.041; visc. 220 cs (40 C); flash pt. > 200 F (COC); pour pt. 12 C; pH 7.0–8.5 (1% aq.); 28% act.

Perlankrol DSA. [Harcros UK] Syn. primary alcohol sulfate, sodium salt; see Perlankrol DGS; wh. slurry; faint fatty alcohol odor; water-sol.; sp.gr. 1.046; visc. 300 cs (40 C); flash pt. > 200 F (COC); pour pt. 10 C; pH 7.0–8.5 (1% aq.); 28% act.

Perlankrol EAD-60. [Harcros UK] Syn. primary alcohol ether sulfate, ammonium salt; anionic; stabilizer, detergent, industrial and household cleaners; foaming agent for fire fighting foam compds.; dispersant and emulsifier for chlorinated solvs.; clear mobile straw gel; faint alcoholic odor; water-sol.; sp.gr. 1.033; visc. 240 cs; flash pt. 85 F (Abel CC); pour pt. < 0 C; pH 7–9 (1% aq.); 60% act. in ethanol.

Perlankrol ESD. [Harcros UK] Syn. primary alcohol ether sulfate, sodium salt; see Perlankrol EAD-60; water-wh. clear liq.; faint odor; water-sol.; sp.gr. 1.053; visc. 130 cs; flash pt. > 200 F (COC); pour pt. < 0 C; pH 7–9 (1% aq.); 27% act.

Perlankrol ESD-60. [Harcros UK] Syn. primary alcohol ether sulfate, sodium salt; see Perlankrol EAD-60; water-wh. clear to hazy mobile gel; faint alcoholic odor; water-sol.; sp.gr. 1.090; visc. 210 cs; flash pt. 90 F (Abel CC); pour pt. < 0 C; pH 7–9 (1% aq.); 60% act.

Perlankrol ESK-29. [Harcros UK] Syn. primary alcohol ether sulfate, sodium salt; anionic; solubilizer for disinfectants; foam additive for industrial detergents; concrete foaming agent; liq.; 29% conc.

Perlankrol ESS-25. [Harcros UK] Syn. primary alcohol ether sulfate, sodium salt; see Perlankrol EAD-60; water-wh. clear liq.; faint odor; water-sol.; sp.gr. 1.040; visc. 65 cs; flash pt. > 200 F (COC); pour pt. < 0 C; pH 7–9 (1% aq.); 25% act.

Perlankrol FF. [Harcros UK] Alkyl phenol ether sulfate, ammonium salt; anionic; foam booster, stabilizer for industrial detergents; amber hazy visc. liq.; faint ammoniacal odor; water-sol.; sp.gr. 1.115; visc. 4200 cs; flash pt. 84 F (Abel CC); pour pt. 5 C; pH 6.0–8.5 (1% aq.); 90% act.

Perlankrol FN-65. [Harcros UK] Alkyl phenol ether sulfate, sodium salt; anionic; foam stabilizer used in specialty detergent formulations; primary emulsifier in emulsion polymerization; clear straw liq.; faint alcoholic odor; sp.gr. 1.110; visc. 200 cs; flash pt. 85 F (Abel CC); pour pt. < 0 C; pH 6.0–8.5 (1% aq.); 65% act.

Perlankrol N Range. [Harcros UK] Alkylaryl phenol ether sulfates, sodium salt; primary emulsifier for emulsion polymerization; pigment dispersant; agrochemical adjuvants; liq.

Perlankrol O. [Harcros UK] Sodium alkyl sulfate; anionic; clarifying and wetting agent, stabilizer, visc. modifier used in detergent and cleaning formulations; clear lt. amber liq.; char. odor; water-sol.; sp.gr. 1.120; visc. 44 cs; flash pt. > 200 F (COC); pour pt. < 0 C; pH 7.0–8.5 (1% aq.); 40% act.

Perlankrol PA Conc. [Harcros UK] Alkyl phenol ether sulfate, ammonium salt; anionic; foam stabilizer, foaming agent used in liq. detergent formulations; emulsifier for cresylic acid; hazy visc. amber liq.; faint ammoniacal odor; water-sol.; sp.gr. 1.110; visc. 6000 cs; flash pt. 86 F (Abel CC); pour pt. 7 C; pH 6.0–8.5 (1% aq.); 90% act.

Perlankrol S Range. [Harcros UK] Triarylphenol ether sulfates, sodium salts; anionic; see Perlankrol N Range; liq.

Perlankrol SN. [Harcros UK] Alkyl phenol ether sulfate, sodium salt; anionic; foam booster and stabilizer, wetting agent, emulsifier in emulsion polymerization; industrial detergent formulations; water-wh. clear liq.; faint odor; water-sol.; sp.gr. 1.054; visc. 48 cs; flash pt. > 200 F (COC); pour pt. < 0 C; pH 6.0–8.5 (1% aq.); 30% act.

Perlex B.67, B.70. [Rhone-Poulenc] Bismuth compd.; pearlescent for use in lipsticks, emulsion systems; paste; 67 and 70% resp. in castor oil.

Perlex B.67M. [Rhone-Poulenc] Bismuth compd.; see Perlex B.67; 67% in min. oil.

Perlex BU. [Rhone-Poulenc] Bismuth compd.; pearlescent for dry applics. such as face powd., eye shadow or rouge; solid.

Perlex BUA.35. [Rhone-Poulenc] Bismuth compd.; pearlescent for aq.-based emulsions or lotions; 35% disp. in aq. base.

Perlextra B.70. [Rhone-Poulenc] Bismuth compd.; pearlsecent for lipsticks, other wax-based prods., emulsion systems; insol. in water, alcohol; 70% in castor oil.

Perlextra BU. [Rhone-Poulenc] Bismuth compd.; pearlescent for use in eye shadows and powd. prods. or fluid creams and lotions; dry powd.; insol. in water, alcohol.

Perlextra BUA.70. [Rhone-Poulenc] Bismuth compd.; pearlescent for aq. formulations; 70% aq. disp.

Perlglanz-Konzentrat B-30. [Goldschmidt AG] Cocamide DEA; pearlescent for shampoos, bath and shower preps. with superfatting properties; liq.; 30% conc.

Perlglanz-Konzentrat B-48. [Goldschmidt AG] Cocamide DEA; detergent, opacifier, pearlescent; for emulsion shampoos, turbid bath and shower preps. with superfatting properties; semiliq.; 44% conc.

Perlglanzmittel GM 4006. [Zschimmer & Schwarz] Nonionics with fatty alchol ether sulfates; pearlescent for hair shampoos and bath additives; fluid; 30% act.

Perlglanzmittel GM 4055. [Zschimmer & Schwarz] Fatty alcohol ether sulfate and ethylene glycol stearate; opacifier and pearlescent.

Perlglanzmittel GM 4175. [Zschimmer & Schwarz] Fatty alcohol ether sulfate and pearlescents; pearlescent for hair shampoos and bath additives; fluid; 42% act.

Perma Kleer 80. [Lyndal] Trisodium HEDTA; iron chelating agent; liq.; sp.gr. 1.32; flash pt. > 200 F; 41% min. solids.

Perma Kleer 95. [Lyndal] Org. complexing agent; chelates iron in highly alkaline scouring systems in lined kiers; liq.

Perma Kleer 98. [Lyndal] Magnesium complex of DTPA, buffered sol'n.; org. chelating agent which acts as stabilizer in peroxide bleach baths; liq.

Perma Kleer 100. [Lyndal] Tetrasodium EDTA; chelating agent for Ca, Mg, other metal ions; liq.; sp.gr. 1.30; flash pt. > 200 F; 38% min. solids.

Perma Kleer 120. [Lyndal] Trisodium HEDTA; forms chelates with metal ions incl. ferric from pH 1.13; liq.

Perma Kleer 354. [Lyndal] Sodium glucoheptonate; chelates iron in caustic sol'ns.; liq.

Perma Kleer CP Grade. [Lyndal] Tetrasodium EDTA dihydrate; sequestering agent for Ca, iron, other metal ions over wide pH range; powd.

Perma Kleer DI Crystals. [Lyndal] Disodium EDTA; chelating agent for baths or where a prod. of higher alkalinity cannot be tolerated; powd.

Perma Kleer EDTA Acid. [Lyndal] EDTA acid; sequestering agent for polyvalent metal ions; intermediate for several derivs.; powd.

Perma Kleer GM. [Lyndal] Sequestrant for Ca and Mg with high capacity for iron and Cu over wide pH range; liq.; water-sol.

Perma Kleer NTA 150. [Lyndal] Trisodium NTA; general purpose chelating agent; liq.

Perma Kleer NTA Na_3 Crystals. [Lyndal] Trisodium NTA monohydrate; see Perma Kleer NTA 150; powd.

Perma Kleer Tri Crystals. [Lyndal] Trisodium EDTA; chelating agent; powd.

Permalene A-100. [Lyndal] Amine-coconut oil condensation prod.; nonionic; detergent, dispersant, dye leveler and lubricant for rayon, acetate dyeing; liq.; 100% conc.

Permax AW-2. [Yoshimura] Quat. ammonium salt polymer; cationic; antistat for syn. fibers; liq.; 25% conc.

Permax PM 420. [Yoshimura] Quat. ammonium salt polymer; cationic; antistat for polyester textiles; liq.; 20% conc.

Permax PM 503. [Yoshimura] Alkyl phosphate, ethoxylated; anionic; antistat for syn. fibers; liq.; 50% conc.

Permyl® B-100. [Ferro] Proprietary org. lt. stabilizer for polymers; used in fiberglass-polyester corrugated outdoor panels; PVC film for outdoor use; extruded garden hose; cellulosic eyeglass frames; acts as uv absorber; wh. to lt. yel. cryst. powd.; m.w. 214; very sol. in alcohols, ketones, and esters; moderate sol. in aromatic hydrocarbons; low sol. in aliphatic hydrocarbons; sp.gr. 1.33; m.p. 117 C.

Peropal®. [Bayer] Azocyclotin; acaricide for fruits, vegetables, cotton, hops, citrus; colorless powd.; m.w. 436.2; sol. (g/1000 ml): 0–10 g in dichloromethane, n-hexane, 2-propanol, toluene; m.p. 210 C.

Peroxide Stabilizer H. [Hoechst-Celanese AG] Polycarboxylic acid, sodium salt; anionic; stabilizer for hydrogen peroxide bleaching; liq.; 25% conc.

Peroxidol 780. [Reichhold] Epoxidized soya; plasticizer offering heat and lt. stability, migration and extraction properties; does not mar ABS or PS; used in refrigerator gasketing and food wrap; Gardner 2; m.w. 1000; sp.gr. 0.990; dens. 8.2 lb/gal; pour pt. 2 F; ref. index 1.4710.

Peroxidol 781. [Reichhold] Epoxidized tall oil; sec. plasticizer stabilizer; reacts synergistically with metallic stabilizers to give heat and lt. stability; used in chlorinated rubber; Gardner 2; m.w. 420; sp.gr. 0.923; dens. 7.7 lb/gal; pour pt. 2 F; ref. index 1.4579; flash pt. > 500 F.

Pestilizer A. [Stepan] Phosphate ester, acid form; emulsifier for pesticide formulations; dispersant for liq. and powd. formulations; sol. in water, xylene; sp.gr. 1.104; dens. 9.21 lb/gal; visc. 4500 cps; pour pt. –9 C; flash pt. (Seta CC) > 200 F; pH < 2.0.

Pestilizer B. [Stepan] Sulfosuccinate, isopropylamine salt; see Pestilizer A; sol. in water, xylene; sp.gr. 1.088; dens. 9.07 lb/gal; visc. 43,400 cps; pour pt. 9

C; flash pt. (Seta CC) > 200 F; pH 5.5–6.5.

Pestilizer J. [Stepan] Phosphate ester, acid form; see Pestilizer A; sol. in xylene; sp.gr. 1.091; dens. 9.10 lb/gal; visc. 10,750 cps; pour pt. –3 C; flash pt. (Seta CC) > 200 F; pH < 2.0.

Pestilizer M. [Stepan] Phosphate ester, sodium salt; see Pestilizer A; sol. in water, xylene; sp.gr. 1.122; dens. 9.36 lb/gal; visc. 31,250 cps; pour pt. –3 C; flash pt. (Seta CC) > 200 F; pH 6.5–7.0.

Pestilizer N. [Stepan] Phosphate ester, sodium salt; see Pestilizer A; sol. in water, xylene; sp.gr. 1.120; dens. 9.34 lb/gal; visc. 22,250 cps; pour pt. 3 C; flash pt. > 200 F (Seta CC); pH 6.5–7.0.

Petaflex RC-802. [Whittaker] Epoxy curing agent used with Petaflex 55400 to improve the shear and tens. str., chem. and temp. resistance; suitable for food pkg. adhesives; dens. 9.0 lb/gal; visc. 45 cps; 79% solids in ethanol, ethyl acetate.

Petrac® 165. [Syn. Prods.] Petrol. wax; lubricant used in rigid PVC compds. to control external lubrication and fusion; wh. small flakes; dens. 0.92 g/cm^2; flash pt. 510 F (COC).

Petrac® 215. [Syn. Prods.] Oxidized polyethylene wax; lubricant used to control external lubrication of rigid PVC compds. and provide high gloss surf.; off-wh. powd.; soften. pt. 215 F; acid no. 12.

Petrac® 250. [Syn. Prods.] Stearic acid, rubber grade; activator, dispersant, plasticizer and lubricant in rubber compd. processing; thickener and gelling agent for greases; emulsifier in polishing/buffing compds.; FAC 5 color; sapon. no. 199.

Petrac® 270. [Syn. Prods.] Tech. stearic acid; ext. lubricant in flexible PVC processing; chemical intermediate for metallic stearates, esters, etc.; dispersant, plasticizer, activator, lubricant in rubber compounding; thickener for greases; acid no. 205.

Petrac® CP-11. [Syn. Prods.] Calcium stearate; mold release agent, lubricant, pigment suspension aid, and flow control agent for plastics, paints, inks, waterproofing of cement and clay tile, lubrication and glossing in paper coatings; plastics applic.; wh. powd., 97% thru 100 mesh; sp.gr. 1.03; dens. 26 lb/ft^3; m.p. 151 C; acid no. 205; 9.8% CaO.

Petrac® CP-11 LS. [Syn. Prods.] Calcium stearate; see Petrac CP-11; powd., 90% thru 200 mesh; sp.gr. 1.03; dens. 26 lb/ft^3; m.p. 154 C; soften. pt. 150 C; 9.4% CaO.

Petrac® CP-11 LSG. [Syn. Prods.] Calcium stearate; see Petrac CP-11; recommended for applic. where high purity and fine particle size are important; powd., 99.5% thru 325 mesh; sp.gr. 1.03; dens. 24 lb/ft^3; m.p. 154 C; soften. pt. 150 C; 9.4% CaO.

Petrac® CP-12. [Syn. Prods.] Calcium stearate; lubricant and mold release agent for plastics industry, pigment suspension in paints and inks, waterproofing agent for cement and clay tile, lubricant and glossing aid in paper coatings; used in rigid PVC pipe, foundry resins; powd.; 97% thru 100 mesh; sp.gr. 1.03; bulk dens. 25 lb/ft^3 (tapped); soften. pt. 151 C.

Petrac® CP-22G. [Syn. Prods.] Calcium stearate; lubricant, emulsifier; gran.; 100% conc.

Petrac® CZ-81. [Syn. Prods.] Calcium/zinc stearate; lubricant and mold release agent in polyethylene/PP extrusion and molding operations; release agent in bulk and sheet molding compds. of unsat. polyester; particulate, 99% thru 200 mesh; dens. 24 lb/ft^3; m.p. 97–107 C; 0.3% moisture.

Petrac® Eramide®. [Syn. Prods.] Erucamide; slip, release, antitack, and/or internal mold release agent; used in PP for extrusion of sheets, in inj. molding; antistat; in polyvinyls for films and sheeting; in polyethylene it imparts slip and antiblock chars. in film applic.; internal mold release agent in molded prods.; lamination of polyethylene to cellophane and in polyethylene extrusion coatings; withstands high processing temps.; food contact applics.; APHA 10 dry powd., pellets, or flakes; char. mild odor; m.p. 80 C; 99.5% total amide.

Petrac® GMS. [Syn. Prods.] Glycerol stearate; see Petrac CP-12; off-wh. flake; m.p. 140 F; acid no. 2; iodine no. 2.5; sapon. no. 167.

Petrac® MG-20. [Syn. Prods.] Magnesium stearate; anticaking and tableting aid; powd.

Petrac® MG-20 NF. [Syn. Prods.] Magnesium stearate NF; dry lubricant and anticaking agent used in food prods., and filling of pharmaceutical capsules; antistick properties in tableting; improves stability, smoothness, and texture of cosmetic emulsions, creams, ointments; improves texture and water-repellency in baby and medical powds.; wh. fluffy powd., 95% thru 200 mesh; 85% thru 325 mesh; dens. 22 lb/ft^3; soften. pt. 140 C; 7.6% total ash.

Petrac® Slip-Eze. [Syn. Prods.] Oleamide; slip, release, antitack, and/or internal mold release agent; polyethylene film and sheeting; inj. or extruded molded prods.; polyvinyl film; PVC plastisol systems where "quick slip" chars. are required; food pkg. materials; APHA 10 dry powd., pellets, or flakes; char. mild bland odor; m.p. 73C; 99.5% total amide.

Petrac® Slip-Quick®. [Syn. Prods.] Amide blend; slip, antiblock, mold release in PE, PP, PVC film; powd.

Petrac® Vyn-Eze®. [Syn. Prods.] Stearamide; slip, antitack, antiblock, and/or internal mold release agent; polyvinyl molded prods. and PVC plastisol systems; polyethylene film; food pkg. materials; APHA 10 dry powd., pellets, or flakes; char. mild bland odor; m.p. 104 C; 99.5% total amide.

Petrac® ZN-41. [Syn. Prods.] Zinc stearate USP; lubricant and mold release agent in PS, melamine, urea-formaldehyde, phenol-formaldehyde, and polyester molding resins; dusting agent for uncured rubber slabs; suspending and flattening agent in solv.- and water-based paints; anticaking agent used in extinguishers; lubricant in powd. metallurgy; particulate, 99.5% thru 325 mesh; dens. 19 lb/ft^3; m.p. 121 C; pH neutral; 0.3% moisture.

Petrac® ZN-42. [Syn. Prods.] Zinc stearate; release agent in plastics bulk molding or sheet molding compds.; lubricant/stabilizer in vinyl compds.; rubber compd. where high bulk dens. is preferred; particulate, 99% thru 200 mesh; m.p. 121 C; 14% ZnO.

Petrac® ZN-44 HS. [Syn. Prods.] Zinc stearate; mold

release agent and lubricant used in plastic applics. where minimal color development is desired; powd., 99.5% thru 325 mesh; dens. 19 lb/ft^3; m.p. 121 C; 13.5% ZnO.

Petrac® ZW-45. [Syn. Prods.] Zinc stearate in aq. disp.; see Petrac CP-12; wh. disp.; visc. 400 cps; pH 8.7; 45% solids.

Petro® 11. [DeSoto] Sodium alkyl naphthalene sulfonate; hydrotrope, surfactant, germicidal agent, liq. hand soaps, wetting agent in plating and electrolytic baths; 50% act. liq., 95% act. powd.; water-sol.

Petro® AA. [DeSoto] Alkyl naphthalene sulfonate; hydrotrope, detergent; stable in strong acids, alkalies, salts, detergent builders; 50% act. liq., 95% act. powd.

Petro® AG Special. [DeSoto] Sodium polyalkyl naphthalene sulfonate; anticaking agent for fertilizers and other inorg. salts; 50% liq. or 95% powd.

Petro® BA. [DeSoto] Sodium alkyl naphthalene sulfonate; hydrotrope, surfactant for high-solids liq. laundry detergents, electrolytic baths, metal and dairy cleaners; 50% act. liq., 95% act. powd.; water-sol.

Petro® Dispersant 98. [DeSoto] Sodium alkyl naphthalene sulfonate; dispersant for dyestuffs.

Petro® Dispersant 425. [DeSoto] Sodium sulfonated naphthalene-formaldehyde condensate; dispersant for dyestuffs, pigments.

Petro® LBA. [DeSoto] Sodium alkyl naphthalene sulfonate; lighter color form of Petro BA; 50% act. liq., 95% act. powd.

Petro® S. [DeSoto] Alkyl naphthalene sodium sulfonate; anionic; soil conditioner; 50% act. liq.; 95% act. powd.

Petro® Dispersant 98. [DeSoto] Sodium alkyl naphthalene sulfonate; dispersant for dyestuffs.

Petro® Dispersant 425. [DeSoto] Sodium naphthalene-formaldehyde sulfonate; dispersant for dyestuff pigments.

Petrolatum RPB. [Witco] Petrolatum tech.; dk. br.; m.p. 71–77 C.

Petrolite® C-36. [Petrolite] Oxidized microcryst. wax; used in the formulation of polishes; color 1.50 (D1500) wax; m.p. 92.2 C; acid no. 338; sapon. no. 72.

Petrolite® C-400. [Petrolite] Salt of an oxidized polyethylene; wax used as a gelling agent for solvs. in the formulation of paste prods., e.g., shoe or floor polishes; color 1.0 (D1500) wax; m.p. 104.4 C; acid no. 14; sapon. no. 27.

Petrolite® C-700. [Petrolite] Hard microcryst. wax consisting of n-paraffinic, branched paraffinic, and naphthenic hydrocarbons; used in hot-melt coatings and adhesives, cup and paper coatings, printing inks, plastic modification (as lubricant and processing aid), lacquers, paints, and varnishes, as binder in ceramics, for potting, filling, and impregnant in elec./electronic components, in investment casting, rubber and elastomers (plasticizer, antisunchecking, antiozonant), as emulsion wax size in papermaking, as fabric softener ingred. in permanent-press fabrics, in emulsion and latex coatings, and in cosmetic hand creams and lipsticks; color 1.5 (D1500) wax; very low sol. in org. solvs.; sp.gr. 0.93; visc. 12.5 cps (98.9 C); m.p. 91.1 C.

Petrolite® C-1035. [Petrolite] Hard microcryst. wax consisting of n-paraffinic, branched paraffinic, and naphthenic hydrocarbons; see Petrolite C-700; color 0.5 (D1500) wax; very low sol. in org. solvs.; sp.gr. 0.93; visc. 11.7 cps (98.9 C); m.p. 92.8 C.

Petrolite® C-7500. [Petrolite] Oxidized syn. wax; used in the formulation of emulsions having hard films with good gloss and dry-bright polish properties; oil-binding properties make it useful in solv.-based polishes and release agents; color 0.5 (D1500) wax; m.p. 100.0 C; acid no. 15; sapon. no. 31.

Petrolite® C-8500. [Petrolite] Modified oxidized microcryst. wax; used in the formulation of polishes and coatings; produces emulsion polishes with films having good gloss and dry-bright properties; oil-binding properties making it useful in solv.-based polishes and release agents; color 1.5 (D1500) wax; m.p. 96.1 C; acid no. 9; sapon. no. 20.

Petrolite® C-9500. [Petrolite] Oxidized syn. wax; see Petrolite C-7500; color 0.5 (D1500) wax; m.p. 95.6 C; acid no. 32; sapon. no. 52.

Petrolite® CA-11. [Petrolite] Oxidized, modified hydrocarbon wax; used in the formulation of filled investment casting systems; color 5.5 (D1500) wax; m.p. 73.9 C; acid no. 5; sapon. no. 33.

Petrolite® WB-5. [Petrolite] Urethane deriv. of oxidized wax; wax with dye solubility, pigment dispersing, and oil and dye binding properties; used in one-time carbon paper; br. wax; m.p. 88.94 C; acid no. 18; sapon. no. 72.

Petrolite® WB-11. [Petrolite] Urethane deriv. of oxygenated wax; lower melting wax with high functionality for use in highly filled systems, one-time carbon paper; provides dispersed pigment sol'n. without thixotropy in nonvolatile oil systems; br. wax; m.p. 75.0 C; acid no. 6; sapon. no. 46.

Petrolite® WB-16. [Petrolite] Modified hydrocarbon wax; disperses carbon blk. in nonvolatile oil systems without thixotropy; m.p. 181 F.

Petrolite® WB-17. [Petrolite] Urethane deriv. of oxygenated wax; carbon blk. dispersant for filled hot-melt carbon inks; hard br. wax; m.p. 165 F; acid no. 2; sapon. no. 32.

Petromixes. [Witco] Petrol. sulfonate, modified; anionic; emulsifier; base for preparation of cutting oils, solv. degreasers; liq.

Petronate 25C, 25H. [Witco] Calcium petrol. sulfonate; anionic; detergent and rust inhibitor component in lube oil additives; emulsifier for w/o systems; ASTM 3.5 liq.; dens. 8.2 lb/gal; visc. 170 SSU (100 C); flash pt. 360 F; 43–46% calcium sulfonate.

Petronate Basic. [Witco] Barium petrol. sulfonate; anionic; oil additive for lube, industrial oils, fuels, greases; ASTM 7.0 liq.; visc. 250 SSU (100 C); 40–45% barium sulfonate.

Petronate CR. [Witco] Sodium petrol. sulfonate; anionic; wetting agent, emulsifier, dispersant, rust preventative; dk. brn. visc. liq.; petrol. odor; sol. in org. solvs.; dens. 8.5 lb/gal; sp.gr. 1.02 (60 F); visc.

175 s (Furol, 100 C); flash pt. 380 F; 61–63% act.

Petronate HL. [Witco] Sodium petrol. sulfonate; anionic; emulsifier, dispersant, wetting agent; dk. brn. visc. liq.; petrol. odor; sol. in oil., org. solvs.; dens. 8.5 lb/gal; sp.gr. 1.1 (60 F); visc. 135 s (Furol, 100 C); flash pt. 380 F; 61–63% act.

Petronate HMW. [Witco] Calcium petrol. sulfonate; detergent, rust inhibitor; lubricating oil additives, dens. 8.2 lb/gal; visc. 200 SSU (100 C); flash pt. 350 F; 45–46% calcium sulfonate.

Petronate K. [Witco] Sodium petrol. sulfonate; anionic; dispersant, emulsifier and wetting agent in dry cleaning soaps, textile processing; lt. br. visc. liq.; oil-sol.; dens. 8.3 lb/gal; sp.gr. 1.1 (60 F); visc. 85 s (Furol, 100 C); flash pt. 380 F; 61–63% act.

Petronate L. [Witco] Sodium petrol. sulfonate; anionic; dispersant, wetting agent and emulsifier in oil and solv. systems; dr. brn. visc. liq.; petrol. odor; sol. in oil and org. solvs.; dens. 8.5 lb/gal; sp.gr. 1.1 (60 F); visc. 125 s (Furol, 100 C); flash pt. 380 F; 61–63% act.

Petronate Neutral (50-S). [Witco] Barium petrol. sulfonate; anionic; oil additive for lube, industrial and fuel oils, undercoating, and greases; ASTM 3.5 liq.; visc. 100 SSU (100 C); 50–52% barium sulfonate.

Petronate S. [Witco] Sodium didodecylbenzene sulfonate; anionic; see Petronate CR; liq.; 62% conc.

Petronauba® C. [Petrolite] Oxidized microcryst. wax; used in the formulation of polishes and emulsions; carnauba substitute; color 1.5 (D1500) wax; m.p. 92.8 C; acid no. 24; sapon. no. 53.

Petro-Rez 100. [Akrochem] Thermoplastic, aromatic hydrocarbon resin; plasticizer, esp. for colored rubber articles; builds tack; aids processing of min.-filled elastomers; Gardner 10+ (50% in toluol) flake; sol. in benzene, toluol, xylene, hexane, and ketones; insol. in alcohols; sp.gr. 1.07; soften. pt. (B&R) 100 C; acid no. < 1; iodine no. 35.

Petro-Rez 103. [Akrochem] Thermoplastic, aromatic hydrocarbon resin; plasticizer, tack builder; aids processing of min.-filled elastomers; extender in black-loaded compds.; Gardner 11+ (50% in toluol) flake; sol. in benzene, toluol, xylene, hexane, and ketones; insol. in alcohols; sp.gr. 1.07; soften. pt. (B&R) 100 C; acid no. < 1; iodine no. 90.

Petro-Rez 200. [Akrochem] Thermoplastic hydrocarbon resin; plasticizer, tackifier, processing aid; extender in black-loaded compds.; used in rubber compding. applics., e.g., molded mechanical goods, extrusions, shoe soling, elec. insulation, and flooring; Gardner 11+ to 13+ (50% in toluol) flake; sol. in aromatic and aliphatic solvs.; insol. in alcohols; sp.gr. 1.07; soften. pt. (B&R) 97–105 C; acid no. < 1.

Petro-Rez 215. [Akrochem] Aromatic-aliphatic hydrocarbon resin; plasticizer for rubber compding. applics. incl. molding-grade mechanical goods, extrusion compds., shoe soling, elec. insulation, and flooring prods.; reinforcing agent, processing aid; tackifier in adhesive in hot-melt, solv.-type, and emulsion adhesives; Gardner 11+ to 13+ (50% in toluene) flake; sol. in aromatic and aliphatic solvs., incl. low-KB ink oils; insol. in alcohols, esters, and ketones; soften. pt. 112–120 C; acid no. < 1.0; sapon. no. < 1.0.

Petro-Rez PTH. [Akrochem] Low m.w. plasticizer polymerized from mixed petrol. streams; plasticizer/tackifier; modifies other resins and polymers in rubber compding., adhesive systems, and high-solids ink vehicles; Gardner 16–17 (50% in toluene); sp.gr. 1.02; melt visc. 3.6 poises (85 C); soften. pt. (B&R) 25 C; acid no. nil.

Petro-Rez PTL. [Akrochem] Low m.w. plasticizer polymerized from mixed petrol. stream; see Petro-Rez PTH; Gardner 16–17 (50% in toluene); sp.gr. 1.02; melt visc. 1 poise (85 C); soften. pt. (B&R) 10 C; acid no. nil.

Petro-Rez X-95. [Akrochem] Aromatic hydrocarbon resin; low m.w. plasticizer for lt.-colored rubber compds. and vinyl floor tile formulations; modifying resin and processing aid; Gardner 8–10 (50% in xylene); sp.gr. 1.06; visc. (Gardner) M–N (70% in toluene); soften. pt. (B&R) 100 C; acid no. nil; iodine no. (Wijs) 25–30.

Petrosan 102. [Reilly-Whiteman] Methyl palmitate oleate; nonionic; enhances lubricity and wettability of petrol. oils for metals; liq.; 100% conc.

Petroscale 11, 101. [DeSoto] Org. polyphosphonate; chelating agent for industrial water treatment.

Petrosolve® A. [DeSoto] Aromatic solv.; solv. additive for carburetor and soak cleaners.

Petrostep 420, 465. [Stepan] Petrol. sulfonate; anionic; used in enhanced oil recovery; liq.; 50% conc.

Petrostep HMW, LMC, MMW. [Stepan] Petrol. sulfonate; anionic; see Petrostep 420; liq.; 50, 28, and 50% conc. resp.

Petrosul® 545. [Penreco] Sodium petrol. sulfonate; anionic; emulsifier, demulsifier and flotation agent; fuel oil and grease additive; leather, textile and metalworking oils; metal cleaners; ASTM-D-1500 3.5 max. visc. liq.; m.w. 460–475; oil-sol.; dens. 8.3 lb/gal; 50% act.

Petrosul® 550. [Penreco] Sodium petrol. sulfonate; anionic; solubilizer, rust preventative, dispersant; demulsifier, fuel oil and grease additive; lubricating oil additive mfg.; metalworking oils and metal cleaners; ASTM-D-1500 3.5 max. visc. liq.; m.w. 505–520; oil-sol.; dens. 8.3 lb/gal; 50% act.

Petrosul® 742. [Penreco] Sodium petrol. sulfonate; anionic; emulsifier, solubilizer; textile and dry cleaning formulations; fat splitting; demulsifier, flotation agent; fuel oil additive; leather oils; pigment grinding agent; D-1500 3.5 max. visc. liq.; m.w. 415–430; oil-sol.; dens. 8.4 lb/gal; 62% act.

Petrosul® 744 CL. [Penreco] Sodium petrol. sulfonate; anionic; emulsifier, solubilizer, demulsifier, dry cleaning detergent, fat splitting; flotation agent, fuel oil and grease additive; leather, textile and metalworking oils; metal cleaners; pigment grinding agent; D-1500 5 max. visc. liq.; m.w. 440–450; oil-sol.; dens. 8.6 lb/gal; 62% act.

Petrosul® 745. [Penreco] Sodium petrol. sulfonate; anionic; emulsifier; demulsifier, dry cleaning detergent, fat splitting; flotation agent, fuel oil additive; leather and textile oils; pigment grinding aid; D-1500 3.5 max. visc. liq.; m.w. 460–475; oil-sol.; dens. 8.6 lb/gal; 62% act.

Petrosul® 750. [Penreco] Sodium petrol. sulfonate; anionic; dispersant, rust preventative; demulsifier, fuel oil and grease additive; lubricating oil additive mfg.; metalworking oils; metal cleaners; ASTM-D-1500 3.5 max. visc. liq.; m.w. 505–520; oil-sol.; dens. 8.6 lb/gal; 62% act.

Petrosul® H-50, H-60, H-70. [Penreco] Sodium petrol. sulfonate; surfactant and corrosion inhibitor; used as motor and fuel oil additives, rustproofing formulations; also in dry cleaning solvs., leather processing, printing inks, oil well drilling fluids; m.w. 510–550, 510–550, and 510–540 resp.; dens. 8.2, 8.5, and 8.6 lb/gal; 50, 60, 70% min. sulfonate.

Petrosul® HM-62, HM-70. [Penreco] Sodium petrol. sulfonate; surfactant and corrosion inhibitor; used in ore flotation applics., dry cleaning solvs., leather processing, printing inks, oil well drilling fluids; m.w. 485–495; dens. 8.5 and 8.6 lb/gal resp.; 61 and 70% min. sulfonate.

Petrosul® M-50, M-60, M-70. [Penreco] Sodium petrol. sulfonate; surfactant, emulsifier, and corrosion inhibitor for metalworking fluids; also for dry cleaning solvs., leather processing, printing inks, oil well drilling fluids; m.w. 440–475, 440–475, and 450–470 resp.; dens. 8.2, 8.4, 8.6 lb/gal; 50, 60, and 70% min. sulfonate.

Petrosul® Neutral Barium. [Penreco] Barium petrol. sulfonate; anionic; dispersant, fuel and grease additives; rust preventatives; D-1500 5 max. visc. liq.; m.w. 1100–1140; oil-sol.; dens. 8.2 lb/gal; 40% act.

Petrosul® Neutral Calcium. [Penreco] Calcium petrol. sulfonate; anionic; dispersant, rust preventative; fuel oil and grease additive; D-1500 5 max. visc. liq.; m.w. 1000–1040; oil-sol.; dens. 8.0 lb/gal; 40% act.

Petrowet R. [DuPont] Sodium alkyl sulfonate; anionic; wetting agent, detergent, dispersant, penetrant; petrol., metalworking, textile, and paper industries; industrial formulations and cleaning; clear yel. to amber liq.; alcoholic odor; water-sol.; dens. 9 lb/gal; visc. 15 cps (27 C); flash pt. 196 F (OC); surf. tens. 47.1 dynes/cm (0.1%); pH 4.0–5.5 (1% aq.); 22% act.

Pexol® Size. [Hercules] Fortified rosin; fluid size used in bleached papers offering low visc. and ease of handling; food-pkg. grades of paper and paperboard; pale liq.; dens. 1.18 kg/l; visc. 17 poise; 60% solids.

P & G Amide No. 27. [Procter & Gamble] Coconut fatty acid MEA; nonionic; foam stabilizer, thickener; foam booster and visc. builder for detergents and shampoos; flakes; 100% conc.

PH-73. [Great Lakes] 2,4,6-Tribromophenol; reactive intermediate for phenol-based reactions; flame retardant, antifungal agent, or chemical intermediate; lt. cream to tan flakes; sol. g/100 ml 225 g MEK, 84 g methanol, 50 g toluene, 36 g methylene chloride, 0.1 g water; m.p. 92 C; 99.5% assay; 73% bromine.

Pharmasorb (Regular and Colloidal). [Engelhard] Attapulgite clay; adsorbent for pharmaceuticals featuring superior adsorptive properties with effective acid adsorp.; powd., particle size 2.9 and 0.14 μ resp., 0.10 and 0.30% resp.; cream and lt. cream resp.; sp.gr. 2.47 and 2.36 resp.; pH 7.5–9.5 resp.

Phenrez® 130. [Hercules] Petrol. hydrocarbon thermoplastic; used in adhesives and coatings; inks, molding compositions; binders for foundry core sand and fibrous prods.; dk.

Phenyl Sulphonate CA, CAL. [Hoechst-Celanese AG] Calcium alkylaryl sulfonate; anionic; basic material for mfg. of biocide emulsifiers; liq.; 70% conc.

Phosfac 1004, 1006, 1044, 1044FA, 1066, 1066FA, 1068FA. [Lyndal] Phosphate ester; anionic; emulsifier, penetrant, antistat, wetting agent, detergent, lubricant, corrosion inhibitor and dispersant used in liq. detergent formulations; liq.; 99, 99, 99, 90, 90, 99, and 99% conc. resp.

Phosfac 1068 C-FA, 5520. [Lyndal] Org. phosphate ester; controlled foam lubricant for textiles, metalworking fluids, aq. systems; liq.; biodeg.; 100% act.

Phosfac 5513. [Lyndal] Org. phosphate esters; dispersant for metallic pigments in aq. latex systems, carbon blk.; gel, salt; 70% act.

Phosfac 8000, 8608, 9604, 9609. [Lyndal] Phosphate ester; anionic; see Phosfac 1004; liq.; 99% conc.

Phosflex 71B. [Merrand] Butylated triaryl phosphate; plasticizer for NBR elastomers, PVC, PVAc, cellulosics, PS, ABS, engineering resins, polyesters, alloys; PVAc emulsions.

Phosflex 362. [Merrand] 2-Ethylhexyl diphenyl phosphate; plasticizer for PVC, PVAc, cellulosics, PS, ABS, engineering resins, polyesters, alloys; suitable for food film.

Phosflex 370. [Merrand] Alkyl/diaryl phosphate blend; plasticizer for NBR, NBR/PVC, CR, CPE elastomers, polyacrylate, fluoroelastomers, PVC, PVAc, cellulosics, PS, ABS, engineering resins, polyesters, alloys; flame retardant.

Phosflex 390. [Merrand] Isodecyl diphenyl phosphate; plasticizer for NBR elastomers, PVC, PVAc, cellulosics, PS, ABS, engineering resins, polyesters, alloys; esp. for PVC wallcovering.

Phosphate Ester 123. [Hoechst-Celanese] Complex phosphoric acid ester; corrosion inhibitor and lubricant for aq. systems; raw material for formulation of min. oil-free coolants and anticorrosive liq.; lt. yel. clear visc. liq.; partially sol. in water; sp.gr. 1.2 g/ml; pH 2.0–2.4; 100% act.

Phospholan ALF-5. [Harcros UK] Complex phosphate ester from fatty alcohol base; anionic; foam limiter for use with anionic surfactants; off-wh. solid; mild odor; insol. in water; visc. 120 cs (80 C); flash pt. > 300 F (COC); pour pt. 65 C; pH 2.0–3.0 (1% aq.); 100% act.

Phospholan BH-14. [Harcros UK] Complex phosphate ester, free acid; anionic; surfactant for cleaners; coupling agent; lt. yel. clear visc. liq.; dens. 1.27 g/ml; pour pt. < 0 C; pH 2.6 (10%)

Phospholan KPE-4 [Harcros UK] Complex phosphate ester, potassium salt; anionic; hydrotrope;

solubilization of low foaming surfactants into highly built liqs.; industrial and domestic detergents and cleaners; pale straw liq.; mild odor; water-sol.; sp.gr. 1.294; visc. 180 cs; flash pt. 200 F (COC); pour pt. < 0 C; pH 6.0–8.0 (1% aq.); 65% act.

Phospholan PDB-3. [Harcros UK] Fatty alcohol ethoxylate complex phosphate ester; anionic; wetting agent, detergent component, textile and drycleaning applics.; electrolyte metal cleaning; emulsifier for phenols and in emulsion polymerization; pesticides and herbicides; metalworking lubricants with anticorrosive properties; prod. of iodophors; pale straw liq.; mild odor; disp. in water; sp.gr. 1.000; visc. 2400 cs; flash pt. > 300 F (COC); pour pt. 10 C; pH 2.0–3.0 (1% aq.); 100% act.

Phospholan PHB-14. [Harcros UK] Phosphate ester; anionic; hydrotrope and compatibility agent for agrochemical formulations; liq.; 100% conc.

Phospholan PNP-9. [Harcros UK] Nonylphenol ethoxylate complex phosphate ester, acid form; see Phospholan PDB-3; dk. amber visc. liq.; mild odor; water-sol.; sp.gr. 1.200; visc. 9400 cs (40 C); flash pt. > 300 F (COC); pour pt. 15 C; pH 2.0–3.0 (1% aq.); 100% act.

Phospholan PRP-5. [Harcros UK] Complex phosphate ester; dispersant for org. pigments, agrochemical toxicants, latex stabilization; liq.; 100% conc.

Phospholipid EFA. [Mona] Syn. phospholipid with linoleic acid backbone; moisturizer with antimicrobial properties; cationic emulsifier; used as component in skin and hair formulations, esp. hypoallergenic prods., incl. makeup, sunscreens, hair conditioners; clear amber liq.; sp.gr. 1.036; HLB 17–19; pH 7.6 (10% in 50/50 IPA/water); 33% solids.

Phosphoteric® T-C6. [Mona] Substituted carboxylated cocoimidazoline organophosphate; amphoteric; surfactant, hydrotrope; synergizes detergency with ethoxylated nonionics; improves wetting, penetrating, and detergency; clear thin amber liq.; sp.gr. 1.09; dens. 9.1 lb/gal; pH 7.0 (10%); 35% act.

PHT4. [Great Lakes] Tetrabromophthalic anhydride; flame retardant in prod. of unsat. polyester resins and rigid PU polyols; cohardener for epoxy resins; cost efficient additive for latex emulsions; derivs. used as flame retardants in diverse applic. (wire coating, and wool, etc.); off-wh. cryst.; m.w. 463.7; sol. in dimethylformamide, nitrobenzene; insol. in water; dens. 2.87 g/ml (30 C); dens. 86 lb/ft^3; m.p. 276 C; 68.2% bromine.

PHT4-Diol. [Great Lakes] Tetrabromophthalatediol; reactive intermediate used to produce flame retardant rigid urethane foam; can replace chlorinated polyols; for PU elastomers, coatings, adhesives, and fibers; lt. tan visc. liq.; dens. 1.8 g/ml; visc. 135,000 cps; acid no. 0.01; 46% bromine.

Picco® 5070. [Hercules] Aromatic hydrocarbon resin; low m.w. nonpolar thermoplastic resin used as tackifier in various elastomer-based solv., water, and hot-melt sealant and adhesive systems; as processing and reinforcing agent in rubber compd. and extruding applics.; exhibit exc. leafing and solv.-release properties in metallic cold-cut paints; food-pkg. and processing operations; Gardner 11 solid; sol. in aromatic, aliphatic, and chlorinated hydrocarbons, low-KB aliphatic ink oils, benzyl alcohol, cyclohexanol, MEK, butyl Carbitol acetate, and diethyl Carbitol; dens. 1.03 kg/l; visc. (G-H) I (70% in toluene); soften. pt. (R&B) 70 C; acid no. < 1; sapon. no. < 1; flash pt. 221 C (COC).

Picco® 5100. [Hercules] Aromatic hydrocarbon resin; see Picco 5070; Gardner 11 solid, flakes; sol. see Picco 5070; dens. 1.05 kg/l; visc. (G-H) S (70% in toluene); soften. pt. (R&B)100 C; acid no. < 1; sapon. no. < 1; flash pt. 232 C (COC).

Picco® 5110. [Hercules] Aromatic hydrocarbon resin; see Picco 5070; Gardner 11 solid, flakes; sol. see Picco 5070; dens. 1.06 kg/l; visc. (G-H) U (70% in toluene); soften. pt. (R&B)110 C; acid no. < 1; sapon. no. < 1; flash pt. 235 C (COC).

Picco® 6070. [Hercules] Aromatic hydrocarbon resin; low m.w. nonpolar thermoplastic resin used as tackifiers in various elastomer-based solv., water, and hot-melt adhesive systems; as tackifiers, reinforcing agents, and extenders in rubber compd. and extruding applics.; exhibit exc. leafing and solv.-release properties in metallic cold-cut paints; improves drying time and abrasion resistance of lead-free drier systems in oil-based coatings; suitable in heat set letterpress and web offset printing ink vehicles; FDA cleared for use in food-pkg. and processing operations; Gardner 12 (50% in toluene) solid; sol. see Picco 5070; dens. 1.02 kg/l; visc. (G-H) G (70% in toluene); soften. pt. (R&B) 70 C; acid no. < 1; sapon. no. < 1; flash pt. 237 C (COC).

Picco® 6100. [Hercules] Aromatic hydrocarbon resin; see Picco 6070; Gardner 11 solid, flakes; sol. see Picco 5070; dens. 1.06 kg/l; visc. (G-H) T (70% in toluene); soften. pt. (R&B) 100 C; acid no. < 1; sapon. no. < 1; flash pt. 246 C (COC).

Picco® 6110. [Hercules] Aromatic hydrocarbon resin; see Picco 6070; Gardner 11 solid, flakes; sol. see Picco 5070; dens. 1.07 kg/l; visc. (G-H) V (70% in toluene); soften. pt. (R&B) 110 C; acid no. < 1; sapon. no. < 1; flash pt. 254 C (COC).

Picco® 6115. [Hercules] Aromatic hydrocarbon resin; see Picco 6070; Gardner 11 solid, flakes; sol. see Picco 5070; dens. 1.07 kg/l; visc. (G-H) W (70% in toluene); soften. pt. (R&B) 115 C; acid no. < 1; sapon. no. < 1; flash pt. 260 C (COC).

Picco® 6120. [Hercules] Aromatic hydrocarbon resin; see Picco 6070; Gardner 12 solid, flakes; sol. see Picco 5070; dens. 1.07 kg/l; visc. (G-H) X (70% in toluene); soften. pt. (R&B) 120 C; acid no. < 1; sapon. no. < 1; flash pt. 222 C (COC).

Picco® 6130. [Hercules] Aromatic hydrocarbon resin; see Picco 6070; Gardner 11 solid, flakes; sol. see Picco 5070; dens. 1.07 kg/l; visc. 21 stokes (70% in toluene); soften. pt. (R&B) 130 C; acid no. < 1; sapon. no. < 1; flash pt. 274 C (COC).

Picco® 6140. [Hercules] Aromatic hydrocarbon resin; see Picco 6070; Gardner 11 solid, flakes; sol. see Picco 5070; dens. 1.07 kg/l; visc. 38 stokes (70% in toluene); soften. pt. (R&B) 140 C; acid no. < 1; sapon. no. < 1; flash pt. 277 C (COC).

Piccodiene® 2215, 2215SF. [Hercules] Aliphatic hydrocarbon resin produced largely from petrol.-derived dicyclopentadiene monomer; low m.w. heat-reactive, thermoplastic resin used as tackifiers, stiffeners, and processing aids in rubber compds.; tackifiers for fire carcass stocks and in SBR-based mastics and sealants; fluxing resins to aid filler and pigment disp.; aid to flow in molded goods; used in metallic paint vehicles, traffic paints, concrete-curing and form-release compds. for highway and tilt-wall construction and as varnish vehicles for specialty coatings and related prods.; FDA cleared for used in food pkg. and processing operations; Gardner 11 solid, flakes; sol. in aliphatic, aromatic, and chlorinated hydrocarbons; water-insol.; dens. 1.13 kg/l; visc. (G-H) K–L (60% in min. spirits); soften. pt. (R&B) 102 C; acid no. < 1; sapon. no. < 2; flash pt. 224 C (COC).

Piccodiene® 2285, 2285BHT. [Hercules] Aliphatic hydrocarbon resin produced largely from petrol.-derived dicyclopentadiene monomer; see Piccodiene 2215, 2215SF; 2285BHT grade contains an antioxidant; Gardner 10 solid, flakes; sol. see Piccodiene 2215; dens. 1.14 kg/l; visc. (G-H) V (60% in min. spirits); soften. pt. (R&B) 141 C; acid no. < 1; sapon. no. < 2; flash pt. 216 C (COC).

Piccofyn® Resins. [Hercules] Low m.w., nonreactive phenolic-modified terpene hydrocarbon resins; used as tackifiers in adhesives and rubber compding.; in laminating agents, plastics modification, paints, varnishes, and printing inks; lt. solid and flake forms; soften. pt. avail. in grades from 100 to 135 C.

Piccofyn® A100. [Hercules] Highly alkylated phenolic-modified terpene hydrocarbon resin; low m.w. nonreactive thermoplastic resin used as tackifier resins in acrylic-, natural and syn. rubber-, and E/VA-based adhesives; as modifying resins for lacquers, hot-melt polyamide adhesives and coatings, and E/VA-based coatings and sealants, in high-performance pressure sensitives, as tackifying, reinforcing, or processing aids in rubber compds. and as modifier resins in acrylic- and PS-based floor waxes and coatings; food pkg. applics.; Gardner 7 solid, crushed; sol. in aliphatic, aromatic, and chlorinated hydrocarbons, MIBK, and long-chain alcohols; dens. 1.03 kg/l; visc. (G-H) L (70% in toluene); soften. pt. (R&B) 100 C; flash pt. 224 C (COC).

Piccofyn® A115. [Hercules] Highly alkylated, phenolic-modified terpene hydrocarbon resin; see Piccofyn A100; Gardner 7 solid, crushed; sol. see Piccofyn A100; dens. 1.03 kg/l; visc. (G-H) P (70% in toluene); soften. pt. (R&B) 115 C; flash pt. 229 C (COC).

Piccofyn® A135. [Hercules] Highly alkylated phenolic-modified terpene hydrocarbon resin; see Piccofyn A100; Gardner 7 solid, crushed; sol. see Piccofyn A100; dens. 1.03 kg/l; visc. (G-H) V (70% in toluene); soften. pt. (R&B) 135 C; flash pt. 268 C (COC).

Piccofyn® D125. [Hercules] Fully alkylated phenolic-modified terpene hydrocarbon resin; low m.w. nonreactive thermoplastic resin used for heat-sealable hot-melt compositions based on E/VA resins; tackifier in adhesive systems, in coatings and related applics.; FDA cleared for used in food pkg.; Gardner 8 (50% in toluene) solid, flakes; sol. in aromatic and chlorinated hydrocarbons, alcohols, esters, and ethers; water-insol.; dens. 1.04 kg/l; visc. 105 stokes (70% in toluene); soften. pt. (R&B) 125 C; acid no. < 1; sapon. no. < 2.

Piccolastic® A5. [Hercules] Hydrocarbon resin derived from a pure styrene monomer; low m.w. nonpolar thermpolastic resin offering lt. color, and solubility, and wide compatibility; softener and primary plasticizer in hot-melt compositions, adhesives, coatings, and rubber compd.; food-pkg. and processing operations; Gardner 3 liq.; sol. in aromatic, aliphatic, and chlorinated hydrocarbons, ketones, pyridine, carbon bisulfide, ethyl and butyl acetates, and turpentine; insol. in water; dens. 1.04 kg/l; visc. 20 stokes (40 C); soften. pt. 5 C; acid no. < 1; sapon. no. < 1; ref. index 1.57; flash pt. 165 C (COC).

Piccolastic® A25. [Hercules] Hydrocarbon resin derived from a pure styrene monomer; see Piccolastic A5; Gardner 3; sol. see Piccolastic A5; dens. 1.04 kg/l; soften. pt. 25 C; acid no. < 1; sapon. no. < 1; ref. index 1.58; flash pt. 213 C (COC).

Piccolastic® A50. [Hercules] Hydrocarbon resin derived from a pure styrene monomer; see Piccolastic A5; Gardner 2 soft solid; sol. see Piccolastic A5; dens. 1.05 kg/l; soften. pt. 50 C; acid no. < 1; sapon. no. < 1; ref. index 1.58; flash pt. 226 C (COC).

Piccolastic® A75. [Hercules] Hydrocarbon resin derived from a pure styrene monomer; see Piccolastic A5; Gardner 2 hard solid; sol. see Piccolastic A5; dens. 1.06 kg/l; soften. pt. 75 C; acid no. < 1; sapon. no. < 1; ref. index 1.60; flash pt. 260 C (COC).

Piccolastic® BV-50HF. [Hercules] Hydrocarbon resins derived largely from styrene monomer in sol'n.; used as vehicle for leafing metal pigments; esp. with bronze pigments; promotes very bright, complete leafing, does not attack pigments during storage; Gardner 2 sol'n.; dens. 0.95 kg/l; visc. 43 stokes; flash pt. (TCC) 43.3 C; acid no. < 1; 50% solids in proprietary aromatic blends.

Piccolastic® C125. [Hercules] Modified PS hydrocarbon resin offering pale color, uv lt. stability; used in adhesives, coatings, and related applics; plasticizer, reinforcing agent, or tackifier resin in rubber compd. and toner resin; food pkg. and processing operations; Gardner 2 solid, flakes; sol. in aromatic and chlorinated hydrocarbons, ketones, and ethers; dens. 1.05 kg/l; visc. (G-H) X; cloud pt. 138 C; soften. pt. 126 C; acid no. 3; sapon. no. 3; flash pt. 277 C (COC).

Piccolyte® A115. [Hercules] Hydrocarbon resin derived from the terpene monomer alphapinene; low m.w. pale, inert, thermoplastic resin; tackifiers in pressure sensitive and hot-melt systems based on natural and syn. rubbers, esp. SBR, SIS, and related block copolymer thermoplastic elastomers; reinforcing resins in rubber compds.; gloss-promoting agents in paints, varnishes, and coatings; in sealants, caulks, and rubber compositions to improve resistance to moisture, acids, and alkalies; reinforcing modifiers in lost-wax investment castings and in waterproofing agents; tackifier resins in E/VA resin-based hot-melt adhesives and coatings; food contact use; Gardner 4 (50% in toluene) solid; sol. in aliphatic, aromatic, and chlorinated hydrocarbons, ketones, VM&P naphtha, and ethyl acetate; dens. 0.98 kg/l; melt visc. 200 C (1 poise); soften. pt. (R&B) 115 C; ref. index 1.53; flash pt. 216 C (COC).

Piccolyte® A125, A135. [Hercules] Hydrocarbon resin derived from the terpene monomer alpha-pinene; see Piccolyte A115; Gardner 4 and 5 (50% in toluene) resp., solids; sol. see Piccolyte A115; dens. 0.98 kg/l; melt visc. 210 and 220 C resp. (1 poise); soften. pt. (R&B) 125 C and 135 C resp.; ref. index 1.53; flash pt. 219 C and 238 C (COC) resp.

Piccolyte® C100. [Hercules] Terpene hydrocarbon resin produced from d-limonene; pale, neutral, low m.w., highly stable, thermoplastic resin used as tackifiers in rubber-based pressure sensitive and hot-melt adhesives; modifier resins in rubber compds.; in combination with waxes and polybutenes in hot-melt coatings and paper-to-paper laminations; for modifying waxes used in lost-wax investment castings; FDA cleared for used in food pkg. and processing operations; Gardner 4 (50% in toluene) solid, flakes; sol. in aliphatic, aromatic, and chlorinated hydrocarbons, min. oil, VM&P naphtha, turpentine, ether, amyl and butyl acetates, and long-chain aliphatic alcohols; dens. 0.99 kg/l; melt visc. 189 C (1 poise); soften. pt. (R&B) 100 C; flash pt. 230 C (COC).

Piccolyte® C115, C125, C135. [Hercules] Terpene hydrocarbon resin produced from d-limonene; see Piccolyte C100; also cleared under FDA regulations for use as masticatory agents in chewing gum compositions; Gardner 4 (50% in toluene) solid, flakes; sol. see Piccolyte C100; dens. 0.99 kg/l; melt visc. 204 C, 214 C, and 224 C resp. (1 poise); soften. pt. (R&B) 115, 125 C, and 131 C resp.; flash pt. 235 C, 241 C, 260 C (COC).

Piccolyte® D100. [Hercules] Hydrocarbon resin produced primarily from the terpene monomer dipentene; pale, neutral, low m.w., highly stable, thermoplastic resin offering good resistance to aging and exc. balance of tack, adhesive, and cohesive properties; used in adhesive and coatings applics., incl. pressure sensitive systems and hot-melts for laminating, coating, waterproofing, and construction; tackifiers for certain elastomers; reinforcing agents for waxes; modifier for waxes used in lost-wax investment coatings; and in rubber compd.; food pkg. and processing operations; Gardner 4 (50% in toluene) solid, flakes; sol. in aliphatic, aromatic, and chlorinated hydrocarbons, VM&P naphtha, ether, amyl and butyl acetates, and long-chain aliphatic alcohols; dens. 0.98 kg/l; melt visc. 201 C (1 poise); soften. pt. (R&B) 100 C; flash pt. 232 C (COC).

Piccolyte® D115, D135. [Hercules] Hydrocarbon resin produced primarily from the terpene monomer dipentene; see Piccolyte D100; for use in resinous and polymeric coatings, and closures with sealing gaskets for food containers; Gardner 3 and 4 (50% in toluene) resp.; solid, flakes; sol. see Piccolyte D100; dens. 0.98 kg/l; melt visc. 211 C and 232 C resp. (1 poise); soften. pt. (R&B) 115 C and 135 C resp.; flash pt. (COC) 238 C and 254 C resp.

Piccolyte® HM-110 Resin. [Hercules] Modified terpene resin; compatible with EVA copolymers, waxes, SBR and natural rubbers, and SBS and SIS block polymers; used as a universal tackifier in adhesives; esp. useful in adhesives where various polymers are blended to obtain high shear and high tack; food pkg. and processing operations; Gardner 8 (50% in toluene) solid, flakes; sol. in aromatic and aliphatic hydrocarbons; dens. 1.02 kg/l; melt visc. 192 C (1 poise); soften. pt. (R&B) 109 C; acid no. > 1; sapon. no. > 1; flash pt. 525 F; fire pt. 570 F.

Piccolyte® S100. [Hercules] Hydrocarbon resin produced from the monomeric terpene beta-pinene; pale, inert, low m.w. thermoplastic resin; tackifier for natural rubber-based pressure sensitive adhesives, esp. for transparent and friction tapes and in other adhesive systems and in hot-melt compositions; in pale rubber compds., rubber toys, and medical goods; modifier for E/VA resins and waxes for use in hot-melts, esp. roller- and curtain-coating-applied compositions, paper-to-paper laminations, and lost-wax investment castings; stiffening and waterproofing resins in textile sizing and waterproofing applics.; in SBR-based can sealants and in printing inks; food-pkg. and processing operations; Gardner 2 (50% in toluene) solid; sol. in aliphatic, aromatic, and chlorinated hydrocarbons, esters, ethers, long-chain alcohols, and higher ketones; water-insol.; dens. 0.99 kg/l; melt visc. 202 C (1 poise); soften. pt. (R&B) 100 C; flash pt. 232 C (COC).

Piccolyte® S115, S115SF. [Hercules] Hydrocarbon resin produced from the monomeric terpene beta-pinene; see Piccolyte S100; S115SF grade contains an antioxidant; S115 and S115SF offer additional clearances under FDA regulations used in closures with sealing gaskets for food containers, markings on fruits and vegs., chewing gum bases; Gardner 2 solid, flakes; sol. see Piccolyte S100; dens. 0.99 kg/l; melt visc. 220 C; soften pt. 115 C; flash pt. 234 C (COC).

Piccolyte® S125, S125SF. [Hercules] Hydrocarbon resin produced from the monomeric terpene beta-pinene; see Piccolyte S100; S125SF grade contains an antioxidant; S125 and S125SF offers additional clearances under FDA regulations for use in closures with sealing gaskets for food containers, etc.; Gardner 2 (50% in toluene) solid, flakes; sol. see Piccolyte S100; dens. 0.99 kg/l; melt visc. 225 C (1 poise); soften. pt. (R&B) 125 C; flash pt. 238 C (COC).

Piccolyte® S135. [Hercules] Hydrocarbon resin produced from the monomeric terpene betapinene; see Piccolyte S100; S135 offers additional clearances under FDA regulations for use in closures with sealing gaskets for food containers, chewing gum base; Gardner 2 (50% in toluene) solid, flakes; sol. see Piccolyte S100; dens. 0.99 kg/l; melt visc. 232 C (1 poise); soften. pt. (R&B) 135 C; flash pt. 259 C (COC).

Piccomer® 10. [Hercules] Aromatic hydrocarbon resin produced from coal- and petrol.-derived monomers; neutral, thermoplastic resins used as modifier resins for natural and syn. rubbers, in adhesive applic.; in rubber compding.; binder and plasticizer resins in putty, sealants, and caulking compds.; Gardner 1 visc. liq.; sol. in aromatic and chlorinated hydrocarbons, ethyl acetate, ethyl ether, and diethyl Carbitol; insol. in water; dens. 1.04 kg/l; soften. pt. 15 C; acid no. < 1; sapon. no. < 2.

Piccomer® 25. [Hercules] Aromatic hydrocarbon resin produced from coal- and petrol.-derived monomers; see Piccomer 10; Gardner 1 visc. liq.; sol. see Piccomer 10; dens. 1.04 kg/l; soften. pt. 25 C; acid no. < 1; sapon. no. < 2.

Piccomer® 40. [Hercules] Aromatic hydrocarbon resin produced from coal- and petrol.-derived monomers; see Piccomer 10; also suitable for use in blk. and nonblk. printing and industrial processing roll compds. based on NBR and NBR/PVC blends; Gardner 1 soft solid; sol. see Piccomer 10; dens. 1.05 kg/l; soften. pt. 37 C; acid no. < 1; sapon. no. < 2.

Piccomer® 100. [Hercules] Aromatic hydrocarbon resins produced from coal- and petrol.-derived monomers; thermoplastic, inert, low m.w. resins used as vehicles for marine and other metallic paints, varnishes, and related coatings to promote film hardness and high-temp. resistance; modifier resins in rubber and wax compositions; printing ink vehicles contributing hardness, pigment-wetting, and fast-solv.-release properties; food pkg. and processing operations; Gardner 10 solid, flakes; sol. in aromatic and chlorinated hydrocarbons, ethyl acetate, ethyl ether, and diethyl Carbitol; insol. in water; dens. 1.09 kg/l; soften. pt. 102 C; acid no. < 1; sapon. no. < 2; flash pt. 254 C (COC).

Piccomer® 110. [Hercules] Aromatic hydrocarbon resins produced from coal- and petrol.-derived monomers; see Piccomer 100; Gardner 9 flakes, solids; sol. see Piccomer 100; dens. 1.09 kg/l; soften. pt. 110 C; acid no. < 1; sapon. no. < 2; flash pt. 271 C (COC).

Piccomer® 120. [Hercules] Aromatic hydrocarbon resins produced from coal- and petrol.- monomers; see Piccomer 100; Gardner 9 flakes, solids; sol. see Piccomer 100; dens. 1.09 kg/l; soften. pt. 121 C; acid no. < 1; sapon. no. < 2; flash pt. 269 C (COC).

Piccomer® 150. [Hercules] Aromatic hydrocarbon resin produced largely from coal- and petrol.-derived monomers; thermoplastic resin, high gloss and gloss promotability, good solv. release, balance of adhesive and cohesive properties, good pigment-wetting ability and sol. in ink solvs.; in heat set letterpress and web offset inks, conventional "sheet-fed" lithographic and letterpress inks, wax modification, as component in adhesives and specialty protective coatings, and reinforcing agent in rubber compding.; Gardner 11 (50% solids in toluene) solid, crushed or flaked; sol. in aliphatic, aromatic, and chlorinated hydrocarbons, and low-KB ink solvs.; dens. 1.08 kg/l; visc. 2.80 stokes (50% solids in min. spirits); soften. pt. (R&B) 153 C; flash pt. (COC) 276 C; acid no. < 1; sapon. no. < 2.

Piccomer® XX40, XX100. [Hercules] Aromatic hydrocarbon resins; tackifier resins used in adhesives and rubber compd. applics.; promote tensile strength of various elastomer-based adhesives; in rubber compd., improves mold flow, extrusion, and calendering properties; most applic. in blk. stocks and other uses where color is not important; XX40 is low soften. pt. grade and exc. rubber softener and plasticizer; XX100 is an effective reinforcing agent for rubber compds.; Coal Tar 16 max. solid, flakes; sol. in aromatic and chlorinated hydrocarbons; water-insol.; dens. 1.03 and 1.08 kg/l resp.

Picconol® A100, A102. [Hercules] Aliphatic hydrocarbon resin emulsions; anionic; resin emulsions used in combination with various other aq. thermoplastic and/or elastomeric systems to produce coatings, paints, and adhesives; tackifiers for natural and syn. rubber based systems; produces waterproof finishes for paper, textiles, and textile backings; promotes acid-, alkali-, and water-resistance in water-based paints and coatings; A100 esp. suitable for use in high-styrene SBR, E/VA resin, and polyacrylic-based latex paints, and in adhesives for bonding films, fibers, and granular materials; A102 is esp. suitable as tackifier for natural rubber latex-based laminating adhesives; reinforcing agent in SBR-type latices, and in vulcanized latex systems; A100 is used in food-pkg. and processing operations; A102 is FDA approved used in adhesives and rubber articles intended for repeated use; Gardner 9 (50% in toluene) liq.; dens. 0.97 kg/l and 0.99 kg/l resp.; visc. 3000 cps and 35 cps resp.; soften. pt. (R&B) 63 and 93 C resp.; pH 8.5 and 10.5 resp.; 50% aq. disp.

Picconol® A200, A201. [Hercules] Terpene hydrocarbon resin emulsions; anionic; resin emulsions used in combination with other aq. thermoplastic and/or elastomer systems producing exc. adhesives and laminates; tackifiers for natural rubber; tackifier and rubber softener where improved adhesion to substrates, better water repellency, and higher tack and grn.-strength properties are required; FDA cleared for use in food pkg.; Gardner 2 liq.; dens. 0.98 kg/l; visc. 1500 cps and 1000 cps resp.; soften. pt. (R&B) 57 and 80 C resp.; pH 9 and 8.5 resp.; 50% aq. disp.

Picconol® A500, A501. [Hercules] Aromatic hydrocarbon resin emulsions; anionic; resin emulsions with exc. pigment and substrate wetting ability; used in combination with other aq. thermoplastic and/or elastomeric systems to produce exc. adhesives and laminates; tackifiers for rubber latex systems, esp. emulsion adhesives; A500 esp. useful as tackifier, softener, and extender for curing and noncuring latex systems; A501 esp. useful as tackifier and modifier for rubber latices, and for modifying emulsions based on high-soften. pt. resin; FDA cleared for use in food pkg.; Gardner 12–13 (50% in toluene) liq.; dens. 1.00 and 1.01 kg/l resp.; visc. 600 and 70 cps resp.; soften. pt. (R&B) 27 and 49 C resp.; pH 9 and 11.5 resp.

Picconol® A600E. [Hercules] Aromatic hydrocarbon resin-based emulsion; anionic; used with aq. thermoplastic and/or elastomeric systems to make adhesives and laminates; promotes tack in elastomer blend; for use with vinyl and acrylic latices to produce adhesives for polyolefins and hard-to-wet substrates, esp. carpet and upholstery materials; for food pkg.; Gardner 9–10 (base resin, 50% solids in toluene) liq.; dens. 1.03 kg/l; visc. 250 cps; soften. pt. (R&B) 70 C; surf. tens. 36 dynes/cm; pH 10.5; 55% aq. disp.

Piccopale® 70. [Hercules] Aliphatic hydrocarbon resin derived mainly from dienes and other reactive olefin monomers; low m.w. neutral resin with high resistance to moisture, uv stability, tack and tack retention, exc. binding qualities, and good compatibility and solubility; used as tackifier resin in rubber compd. and adhesives applics., esp. in tire carcass stocks, rubber-based pressure sensitive adhesives, and high-ethylene E/VA resin-based hot-melts; saturant and waterproofing agent in various paper, paperboard, and textile prods.; tackifier and binder resin in construction adhesives and other building materials; food-pkg. and processing operations; Gardner 9 (50% in toluene) solid; molten (70SF grade only); sol. in aliphatic, aromatic, and chlorinated hydrocarbons, MIBK, ethyl ether; dens. 0.96 kg/l; soften. pt. (R&B) 70 C; acid no. 1; sapon. no. < 2; flash pt. (COC) 232 C.

Piccopale® 85. [Hercules] Aliphatic hydrocarbon resin derived mainly from dienes and other reactive olefin monomers; see Piccopale 70; avail. in two stabilized grades (85-BHT and 85SF); Gardner 8 (50% solids in toluene) solid; sol.: see Piccopale 70; dens. 0.97 kg/l; melt visc. 205 C (1 poise); soften. pt. (R&B) 85 C; flash pt. (COC) 235 C; acid no. < 1; sapon. no. < 2.

Piccopale® 100, 100BHT, 100SF, 100HM. [Hercules] Aliphatic hydrocarbon resins derived mainly from dienes and other reactive olefin monomers; see Piccopale 70; replacements for gloss oils in paints, varnishes, and coatings; 100BHT, 100 SF, and 100HM are stabilized grades; Gardner 8 (50% solids in toluene) solid, flake (except Piccopale 100); sol.: see Piccopale 70; dens. 0.97 kg/l (all); melt. visc. 230 C (100, 100BHT, and 100SF, @ 1 poise); 225 C (100HM @ 1 poise); soften. pt. (R&B) 100 C (100, 100BHT, and 100SF), 98 C (100 HM); flash pt. (COC) 259 C (all); acid no. < 1 (all); sapon. no. < 2 (all).

Piccotac® 95BHT. [Hercules] Aliphatic hydrocarbon resin; narrow m.w. distribution resin for the adhesives industry; tackifier for hot-melt adhesive applics.; in rubber-based pressure sensitive adhesives and high-ethylene E/VA resin-based hot-melts; as a saturant and waterproofing agent in various paper, paperboard, and textile prods.; in hot-melt road-marking compds.; Gardner 5 (50% solids in toluene) solid, flakes; sol. in aliphatic and aromatic hydrocarbons, MIBK, ethyl ether, and long-chain alcohols; dens. 0.96 kg/l; melt visc. 205 C (1 poise); soften. pt. (R&B) 95 C; flash pt. (COC) 480 F; acid no. < 1; sapon. no. < 2.

Piccotac® B. [Hercules] Aliphatic hydrocarbon resin derived largely from mixed monomers of petrol. origin; low m.w., thermoplastic resin offering lt. color, balance of tack, adhesive, and cohesive properties, heat resistance, and wide compatibility and solubility; tackifier for natural rubber, polyisoprene, and SIS rubbers in pressure sensitive adhesives; as a modifier for waxes used in lost-wax investment casting operations and in EVA resin/wax/low m.w. polyethylene-based hot-melt adhesives; useful in E/VA based hot-melt road-marking compds.; avail. in two stabilized grades (BBHT and BHM); Gardner 5+ (50% solids in toluene) solid, flakes; sol. in aliphatic and aromatic hydrocarbons, long-chain alcohols (ethyl-hexanol), ethylene dichloride, MIBK, and trichloroethylene; dens. 0.97 kg/l; visc. 360 cps (70% solids in toluene); soften. pt. (R&B) 100 C (BBHT) and 98 C (BHM); flash pt. (COC) 249 C; acid no. < 1; sapon. no. < 1.

Piccotac® CBHT. [Hercules] Aliphatic hydrocarbon resin derived largely from modified mixed monomers of petrol. origin; low m.w., thermoplastic resin; tackifier and modifier in natural rubber-based pressure sensitive and hot-melt adhesives and in coatings; in pressure sensitive tape applics. as the sole tackifier resin or with polyterpene resin; as a modifier for other thermoplastic systems, esp. for EVA resin-based hot-melts to promote hot tack, faster speed of set, and reduced open time; Gardner 6+ (50% solids in toluene) solid, flakes; sol. in aliphatic, aromatic, and chlorinated hydrocarbons; dens. 0.96 kg/l; visc. 640 cps (70% solids in toluene); soften. pt. (R&B) 100 C; cloud pt. 155 C max.; flash pt. (COC) 238 C; acid no. < 1; sapon. mo. < 1.

Piccoumaron® 100. [Hercules] Polyindene/related unsat. coal compds. alkylaromatic hydrocarbon resin; binder used in rubber compd., and adhesives; reinforcing agent for SBR in molding and extruding applics.; metallic cold-cut paints; alkyd modifier for coatings; modifier for wax in investment castings; food pkg. and processing operations; Coal Tar 1 solid, flakes; sol. in aliphatic, aromatic, and chlorinated hydrocarbons, ketones, ethyl ether, ethyl acetate, alcohols, glycols, and acetone; insol. in water; dens. 1.12 kg/l; melt visc. 185 C; soften. pt. 100 C; acid no. < 1; sapon. no. < 2; flash pt. 266 C (COC).

Piccoumaron® 110. [Hercules] Polyindene/related unsat. coal compds. alkylaromatic hydrocarbon resin; see Piccoumaron 100; Coal Tar 1 solid, flakes; sol. see Piccoumaron 100; dens. 1.12 kg/l; melt visc. 195 C; soften. pt. 111 C; acid no. < 1; sapon. no. < 2; flash pt. 277 C (COC).

Piccoumaron® 120. [Hercules] Polyindene/related unsat. coal compds. alkylaromatic hydrocarbon resin; see Piccoumaron 100; Coal Tar 1 solid, flakes; sol. see Piccoumaron 100; dens. 1.12 kg/l; melt visc. 206 C; soften. pt. 122 C; acid no. < 1; sapon. no. < 2; flash pt. 282 (COC).

Piccovar® AB165, AB180. [Hercules] Aliphatic hydrocarbon resins from unsat. compds. derived from petrol.; low m.w., thermoplastic resins, high gloss and gloss promotability, good solv. release, balance of adhesive and cohesive properties, good pigment-wetting ability, and wide solubility in common ink solvs.; used in heat set web offset printing inks to provide fast set and hard-inked surfaces; for vehicles for metallic paints and other specialty coatings; as tackifiers in natural and syn. rubber-based adhesive compositions, esp. sealants, mastics, and SBR-based high-temp.-resistant construction adhesives; and as reinforcing agents in rubber compding.; solid, flakes; sol'n. (AB180 only); (Coal Tar) 3; sol. in aliphatic, aromatic, and chlorinated hydrocarbons and in low-KB ink solvs.; dens. 1.05 kg/l (both); visc. 17 and 25 stokes resp. (65% solids in toluene); soften. pt. (R&B) 165 C and 178 C resp.; flash pt. (COC) 299 C and 307 C resp.; acid no. < 1 (both); sapon. no. < 2 (both); AB180 avil. in 48% solids sol'n. in Mentor 28 and 59% solids sol'n. in Magie 470 ink solvs.

Piccovar® AP10. [Hercules] Aromatic hydrocarbon resin; plasticizer, softener and tackifier in rubber compd., pressure sensitive applic., and adhesive systems; food processing operations; Gardner 13 soft solid; sol. in heptane, octane, and min. spirits, aromatic and chlorinated hydrocarbons, ethyl acetate, ethyl ether, and higher m.w. alcohols; insol. in water; dens. 1.02 kg/l; visc. 215 SUS (99 C); soften. pt. 11 C; acid no. < 1; sapon. no. < 2; flash pt. 113 C (COC).

Piccovar® AP25. [Hercules] Aromatic hydrocarbon resin; see Piccovar AP10; Gardner 13 soft solid; sol. see Piccovar AP10; dens. 1.01 kg/l; visc. 100 SFS (99 C); soften. pt. 32 C; acid no. < 1; sapon. no. < 2; flash pt. 221 C (COC).

Piccovar® L30. [Hercules] Dicyclopentadiene alkylaryl hydrocarbon resins; thermoplastic plasticizer and tackifier in solution- and hot-melt applied coatings and adhesives; food pkg.; compatible with petrol. waxes, EVA copolymers, polyethylene, and high m.w. hydrocarbon polymers; Gardner 11 semisolid; sol. in aliphatic, aromatic, chlorinated, and mixed hydrocarbon solvs., ethers, esters, and ketones; insol. in water; dens. 1.05 kg/l; melt visc. 93 C; soften. pt. 30 C; acid no. < 1; sapon. no. < 2; flash pt. 232 C (COC).

Piccovar® L30S. [Hercules] Dicyclopentadiene alkylaryl hydrocarbon resins; see Piccovar L30; Gardner 11 semisolid; sol. see Piccovar L30; dens. 1.05 kg/l; melt visc. 98 C; soften. pt. 31 C; acid no. < 1; sapon. no. < 2; flash pt. 252 C (COC).

Piccovar® L60. [Hercules] Dicylopentadiene alkylaryl hydrocarbon resins; see Piccovar L30; Gardner 11 solid; sol. see Piccovar L30; dens. 1.05 kg/l; melt visc. 129 C; soften. pt. 58 C; acid no. < 1; sapon. no. < 2; flash pt. 243 C (COC).

Pilot SXS-40. [Pilot] Sodium xylene sulfonate; anionic; coupling agt., hydrotrope, solubilizer, solv., stabilizer; used in liq. cleaners, org. polymers and dyestuffs, petrol. industry, pulping, animal glues; lt. yel. clear liq.; dens. 9.9 lb/gal; pH 7.5; 42% total solids.

Pilot SXS-96. [Pilot] Sodium xylene sulfonate; anionic; hydrotrope, visc. reducer, dispersant, extracting aid for detergent formualtions; stabilizes styrene latex opacifier systems for lt.-duty dishwashing liqs.; wh. fine powd.; bulk dens. 220 cc/100 g; pH 7.5 (1%); 96% act.

Pittclor. [PPG Industries] Calcium hypochlorite; algicide, antimicrobial, fungicide used to chlorinate swimming pools and as a disinfectant for cleaning pool parts, locker room floors, etc.; used for municpal water treatment, sewage treatment, in textile and paper mills, tanneries, petrochemical industries to treat waste water effluent; wh. gran. powd.; m.w. 142.994; sol. in water; dens. 61.1 lb/ft^3; 65% act.

PL-2N-204. [Inolex] Glycol ester; low color plasticizer for PVC and syn. rubber; used for neoprene and CAB; APHA 75 max. liq.

PL-2N-204 RG. [Inolex] Glycol ester; plasticizer similar to PL-2N-204 used in applic. where color is less important; Gardner 2 max. liq.

PL-2T-237. [Inolex] Glycol ester; plasticizer providing low temp. flexibility in rubber compd., NC lacquers, and PVC resins; APHA 100 max. liq.

PL-2X-134U. [Inolex] Dibasic acid ester; urethane grade plasticizer for coatings, elastomers, and castings; Gardner 2 max. liq.

PL-4VIP-415. [Inolex] Polyol ester; premium high temp. plasticizer for high quality vinyl coatings, plastisols, and elec. compds.; heat aging properties; APHA 250 max. liq.

PL-26W. [Chem-Trend] Heavy-duty petrol. oil, syn. lubricants, and stabilizers in a water disp; water-based plunger lubricant; brn. opaque fluid; dens. 8.3 lb/gal; visc. < 1000 cP; pH 9–10; flash pt. none.

PL-34. [Chem-Trend] Blend of lubricating oils and high-temp. lubricity additives; oil-based plunger lubricant; clear br. heavy oil; sp.gr. 0.91; dens. 7.6 lb/gal; visc. 40,000 cP (41 C); flash pt. (COC) 221 C.

PL-47. [Chem-Trend] Blend of lubricating oils, EP and temp. additives, and emulsifiers; oil-based plunger lubricant; clear honey-br. liq.; mixes readily with water for cleaning; dens. 8.1 lb/gal; visc. < 6000 cP; flash pt. (COC) 196 C.

PL-48. [Chem-Trend] Blend of lubricating oils, EP and temp. additives, and emulsifiers; see PL-47; clear med.-br. fluid; mixes readily with water for cleaning; dens. 8.3 lb/gal; visc. > 10,000 cP; flash pt. (COC) 227 C.

PL-C. [Chem-Trend] Blend of lubricating oils and high-temp. lubricity additives; see PL-34; clear br. oil; sp.gr. 0.92; dens. 7.7 lb/gal; visc. 10,000 cP (41 C); flash pt. (COC) 254 C.

PL-H. [Chem-Trend] Blend of lubricating oils and high-temp. lubricity additives; see PL-34; clear br. extremely heavy oil; sp.gr. 0.92; dens. 7.7 lb/gal; visc. 1,500,000 cP (41 C); flash pt. (COC) 238 C.

PL-L-35. [Chem-Trend] Blend of lubricating oils and high-temp. lubricity additives; see PL-34; clear br. oil; sp.gr. 0.94; dens. 7.8 lb/gal; visc. 3500 cP (41 C); flash pt. (COC) 249 C.

Plas-Chek 775. [Ferro] Epoxidized soybean oil; stabilizer, plasticizer for PVC compds.; uses incl. food wrap film, refrigerator gaskets, medical tubing and bags, chlorinated paraffin stabilizer, toys, wall coverings, halogen and acid scavenger, PVC foams, infant wear film and sheet, pigment disps., flooring, beverage tubing, ubholstery; Gardner 1 max.; negligible odor; m.w. 1000; sol. in most org. solvs.; sp.gr. 0.992 min.; visc. 5 stokes max.; flash pt. 600 F; pour pt. 25

F; acid pt. 0.5 max.; 7.0% min. oxirane oxygen.

Plas-Chek 795. [Ferro] Epoxidized linseed oil; aux. plasticizer for PVC compds.; sol. in most org. solvs.

Plasdone®. [GAF] PVP; tablet binder and coating agent; promotes dye and pigment disp. in coated and uncoated tablets; cohesive agent, stabilizer, protective colloid; powd.

Plasdone® C-15. [GAF] PVP (Pyrogen-free Povidone USP K-17); low m.w. excipient used primarily for veterinary pharmaceuticals; for parenteral applics.; solubilizer, stabilizer, protective colloid; powd.

Plasdone® C-30. [GAF] PVP; USP grade; for parenteral applics.; solubilizer, stabilizer, protective colloid; minimizes toxic side effects and reduces irritation at site of inj.; accepted blood plasma expander for emergency use; suitable for stock piling; powd.

Plasdone® K-25, K-90. [GAF] Polyvinylpyrrolidone (Povidone USP); pharmaceutical excipient used as tablet binder and coating agent; promotes dye and pigment disp.; as cohesive agent, stabilizer, and protective colloid; detoxicant; drug vehicle and retardant; solubilizer and suspending agent in liq. prods.; for topical applics., liq. pharmaceuticals; film-forming agent in medicinal aerosols; powd.

Plasdone® K-26/28, K-29/32. [GAF] PVP polymer; USP grade used as tablet binder and coating agent; promotes dye and pigment disp. in tablets; used as cohesive agent, stabilizer, and protective colloid, as detoxicant for poisons and irritants, as drug vehicle and retardant; minimizes toxic side effects, prolongs pharmacological action of some drugs; as solubilizer and suspending agent in liq. prods.; as detoxicant and demulcent lubricant in ophthalmic and pharmaceutical topical applics.; as film-forming agent in medicinal aerosols, etc.; powd.

Plasthall® 83SS. [C.P. Hall] Dibutoxyethoxyethyl sebacate substitute; plasticizer for rubber industry; Gardner 18 clear liq.; m.w. 505; f.p. –10 C; sol. in hexane, toluene, kerosene, ethyl alcohol, acetone, min. oil, ASTM oil #1; sp.gr. 0.989; visc. 30 cps; acid no. 1.9; sapon. no. 202; ref. index 1.446; flash pt. 152 C.

Plasthall® 325. [C.P. Hall] Butoxyethyl oleate; Gardner 5 clear oily liq.; m.w. 379; sol. in hexane, toluene, kerosene, ethanol, acetone, min. oil; sp.gr. 0.888; visc. 15 cps; f.p. –35 C; flash pt. 202 C; acid no. 1.0; sapon. no. 148; ref. index 1.42.

Plasthall® 503. [C.P. Hall] Butyl oleate; textile surf. finisher, softener, thread lubricant and antistat; Gardner 11 clear oily liq.; m.w. 388; f.p. –55 C; sol. in hexane, toluene, kerosene, ethyl alcohol, acetone, min. oil, ASTM oil #1; sp.gr. 0.875; visc. 14 cps; acid no. 1.5; iodine no. 127; sapon. no. 165; ref. index 1.457; flash pt. 179 C.

Plasthall® 4141. [C.P. Hall] PEG-3 caprate-caprylate; plasticizer for rubber goods; lubricant for aluminum can industry; Gardner 2 clear liq.; m.w. 430; sol. in hexane, toluene, kerosene, ethanol, acetone, min. oil; sp.gr. 0.968; visc. 25 cps; f.p. –5 C; flash pt. 213 C; acid no. 0.5; sapon. no. 260; ref. index 1.446.

Plasthall® 7006. [C.P. Hall] Alkyl alkylether diester adipate; APHA 125 clear liq.; m.w. 353; sol. in hexane, toluene, kerosene, ethanol, acetone, min. oil; sp.gr. 0.973; visc. 18 cps; f.p. –65 C; flash pt. 188 C; acid no. 0.9; sapon. no. 318; ref. index 1.443.

Plasthall® 7041. [C.P. Hall] Long chain alkyl alkylether diester; rubber plasticizer; transfer aid on correctable ribbon; penetration and tack agent for computer ribbons, carbon paper; yel. liq.; m.w. 492; sol. in hexane, toluene, kerosene, ethanol, acetone, min. oil; sp.gr. 0.925; visc. 15 cps; f.p. –51 C; flash pt. 182 C; acid no. 0.5; sapon. no. 228; ref. index 1.449.

Plasthall® 7049. [C.P. Hall] Alkyl oleate; transfer aid on correctable ribbon; penetration and tack agent for computer ribbons, carbon paper; Gardner 5 clear oily liq.; sol. in hexane, toluene, kerosene, ethanol, acetone, min. oil; sp.gr. 0.874; visc. 25 cps; f.p. –55 C; flash pt. 191 C; acid no. 1.3; sapon. no. 145; ref. index 1.4590.

Plasthall® 7050. [C.P. Hall] Dialkyl diether glutarate; plasticizer for adhesives applic., rubber; yel. liq.; m.w. 450; sol. in toluene, ethyl alcohol, acetone, propylene glycol, min. oil, ASTM oil #1; partly sol. in water; sp.gr. 1.068; visc. 40 cps; acid no. 0.7; sapon. no. 249; ref. index. 1.447; flash pt. 193 C.

Plasthall® BSA. [C.P. Hall] N,N-butyl benzene sulfonamide; plasticizer for emulsion adhesives, pkg., caulk, printing ink, surf. coatings.

Plasthall® DBEA. [C.P. Hall] Dibutoxyethyl adipate; APHA 40 clear liq.; m.w. 346; sol. in hexane, toluene, kerosene, ethanol, acetone; sp.gr. 0.995; visc. 20 cps; f.p. –34 C; flash pt. 188 C; acid no. 0.7; sapon. no. 328; ref. index 1.441.

Plasthall® DBEEA. [C.P. Hall] Dibutoxyethoxyethyl adipate; APHA 100 clear liq.; m.w. 434; sol. in hexane, toluene, kerosene, ethanol, acetone; sp.gr. 1.01; visc. 25 cps; f.p. –25 C; flash pt. 152 C; acid no. 0.5; sapon. no. 232; ref. index 1.4445.

Plasthall® DBEEG. [C.P. Hall] Dibutoxyethoxyethyl glutarate; m.w. 423; sol. in hexane, toluene, kerosene, ethanol, acetone; sp.gr. 1.016; visc. 22 cps; f.p. < –60 C; flash pt. 143 C; acid no. 0.9; sapon. no. 245; ref. index 1.4437.

Plasthall® DBEG. [C.P. Hall] Dibutoxyethyl glutarate; APHA 200 clear liq.; m.w. 335; sol. in hexane, toluene, kerosene, ethanol, acetone; sp.gr. 1.002; visc. 17 cps; f.p. < –60 C; flash pt. 193 C; acid no. 0.8; sapon. no. 338; ref. index 1.440.

Plasthall® DBEP. [C.P. Hall] Dibutoxyethyl phthalate; plasticizer for adhesives; Gardner 2 clear liq.; m.w. 366; sol. in toluene, kerosene, ethanol, acetone; sp.gr. 1.060; visc. 46 cps; f.p. –50 C; flash pt. 266 C; acid no. 0.9; sapon. no. 307; ref. index 1.484.

Plasthall® DBES. [C.P. Hall] Dibutoxyethyl sebacate; APHA 100 clear liq.; m.w. 402; sol. in hexane, toluene, kerosene, ethanol, acetone, min. oil; sp.gr. 0.969; visc. 25 cps; f.p. –20 C; flash pt. 238 C; acid no. 1.0; sapon. no. 280; ref. index 1.445.

Plasthall® DBEZ. [C.P. Hall] Dibutoxyethyl azelate; Gardner 2 clear liq.; m.w. 384; sol. in hexane, toluene, kerosene, ethanol, acetone, min. oil; sp.gr. 0.975; visc. 20 cps; f.p. –25 C; flash pt. 210 C; acid no.

0.6; sapon. no. 292; ref. index 1.4444.

Plasthall® DBZZ. [C.P. Hall] Dibenzyl azelate; yel. liq.; m.w. 372; sol. in toluene, kerosene, ethanol, acetone; sp.gr. 1.072; visc. 38 cps; f.p. 4 C; flash pt. 218 C; acid no. 0.6; sapon. no. 302; ref. index 1.5204.

Plasthall® DDG. [C.P. Hall] Didecyl glutarate; APHA 50 clear liq.; m.w. 411; sol. in hexane, toluene, kerosene, ethanol, acetone, min. oil; sp.gr. 0.918; visc. 32 cps; flash pt. 227 C; acid no. 0.4; sapon. no. 273; ref. index 1.4491.

Plasthall® DIBA. [CP Hall] Diisobutyl adipate; APHA 30 clear liq.; m.w. 259; sol. in hexane, toluene, kerosene, ethanol, acetone, min. oil; sp.gr. 0.950; visc. 13 cps; f.p. –17 C; flash pt. 204 C; acid no. 0.1; sapon. no. 433; ref. index 1.432.

Plasthall® DIBZ. [C.P. Hall] Diisobutyl azelate; amber liq.; m.w. 303; sol. in hexane, toluene, kerosene, ethanol, acetone, min. oil; sp.gr. 0.932; visc. 15 cps; f.p. –30 C; flash pt. 160 C; acid no. 0.6; sapon. no. 369; ref. index 1.4355.

Plasthall® DIDA. [C.P. Hall] Diisodecyl adipate; APHA 50 clear liq.; m.w. 427; sol. in hexane,toluene, kerosene,ethanol, acetone, min. oil; sp.gr. 0.918; visc. 30 cps; f.p. –43 C; flash pt. 227 C; acid no. 0.1; sapon. no. 263; ref. index 1.450.

Plasthall® DIDG. [C.P. Hall] Diisodecyl glutarate; lubricant additive; APHA 50 clear liq.; m.w. 416; f.p. –65 C; sol. see Plasthall 4141; sp.gr. 0.920; visc. 23 cps; acid no. 0.3; sapon. no. 275; ref. index 1.450; flash pt. 204 C.

Plasthall® DIOA. [C.P. Hall] Diisooctyl adipate; APHA 20 clear liq.; m.w. 373; sol. in hexane, toluene, kerosene, ethanol, acetone, min. oil; sp.gr. 0.92; visc. 18 cps; f.p. < –60 C; flash pt. 179 C; acid no. 0.3; sapon. no. 301; ref. index 1.446.

Plasthall® DIODD. [C.P. Hall] Diisooctyl dodecanedioate; lubricant additive; APHA 50 clear liq.; m.w. 454; f.p. –70 C; sol. in hexane, toluene, kerosene, ethyl alcohol, acetone, min. oil, ASTM oil #1; sp.gr. 0.909; acid no. 0.1; sapon. no. 246; ref. index 1.450; flash pt. 238 C.

Plasthall® DOA. [C.P. Hall] Dioctyl adipate; APHA 45 clear liq.; m.w. 373; sol. in hexane, toluene, kerosene, ethanol, acetone, min. oil; sp.gr. 0.925; visc. 18 cps; f.p. –65 C; flash pt. 193 C; acid no. 0.3; sapon. no. 301; ref. index 1.445.

Plasthall® DODD. [C.P. Hall] Dioctyl dodecanedioate dioate; lubricant additive; useful as textile surf. finishes, softeners, thread lubricants and/or antistats; APHA 50 clear liq.; m.w. 454; f.p. –45 C; sol. see Plasthall DIODD; sp.gr. 0.909; acid no. 0.3; sapon. no. 247; ref. index 1.450; flash pt. 222 C.

Plasthall® DOS. [C.P. Hall] Dioctyl sebacate; APHA 25 clear liq.; m.w. 430; sol. in hexane, toluene, kerosene, ethanol, acetone, min. oil; sp.gr. 0.913; visc. 25 cps; f.p. –65 C; flash pt. 216 C; acid no. 0.2; sapon. no. 261; ref. index 1.499.

Plasthall® DOSS. [C.P. Hall] Dioctyl sebacate substitute; plasticizer for rubber industry; amber liq.; m.w. 445; f.p. –55 C; sol. in hexane, toluene, kerosene, ethyl alcohol, acetone, min. oil, ASTM oil #1; sp.gr. 0.918; visc. 51 cps; acid no. 1.5; sapon. no. 264; ref. index 1.449; flash pt. 232 C.

Plasthall® DOZ. [C.P. Hall] Dioctyl azelate; APHA 60 clear liq.; m.w. 412; sol. in hexane, toluene, kerosene, ethanol, acetone, min. oil; sp.gr. 0.915; visc. 25 cps; f.p. –60 C; flash pt. 199 C; acid no. 0.3; sapon. no. 272; ref. index 1.448.

Plasthall® DTDA. [C.P. Hall] Ditridecyl adipate; lubricant additive; APHA 35 clear liq.; m.w. 510; f.p. < –65 C; sol. see Plasthall DIODD; sp.gr. 0.908; visc. 35 cps; acid no. 0.03; sapon. no. 220; ref. index 1.4545; flash pt. 235 C.

Plasthall® HA7A. [C.P. Hall] Polyester adipate; APHA 150 clear visc. liq.; m.w. 3000; sol. in toluene; partly sol. in kerosene, ethanol, acetone, min. oil; sp.gr. 1.153; visc. 22,000 cps; f.p. –10 C; flash pt. 266 C; acid no. 1.5; sapon. no. 589; ref. index 1.4662.

Plasthall® LTM. [C.P. Hall] Linear trimellitate; Gardner 1–2 clear liq.; m.w. 585; sol. in hexane, toluene, kerosene, acetone, min. oil; sp.gr. 0.972; visc. 120 cps; f.p. –55 C; flash pt. 260 C; acid no. 0.08; sapon. no. 288; ref. index 1.482.

Plasthall® NODA. [C.P. Hall] n-Octyl, n-decyl adipate; APHA 25 clear liq.; m.w. 390; sol. in hexane, toluene, kerosene, ethanol, acetone, min. oil; sp.gr. 0.916; visc. 13 cps; f.p. –4 C; flash pt. 198 C; acid no. 0.1; sapon. no. 288; ref. index 1.445.

Plasthall® P-530. [C.P. Hall] Polyester glutarate; plasticizer; Gardner 4 visc. liq.; m.w. 2300; f.p. –15 C; sol. in toluene; sp.gr. 1.10; visc. 2250 cps; acid no. 0.3; sapon. no. 559; ref. index 1.4636; flash pt. 305 C.

Plasthall® P-550. [C.P. Hall] Polyester glutarate; plasticizer for PVC applics.; flexibilizing, permanent, nonmigrating; adhesive for film backing and varieties of tape; APHA 125 visc. liq.; m.w. 2500; f.p. –41 C; sol. sol. in toluene; sp.gr. 1.06; visc. 3700 cps; acid no. 0.7; sapon. no. 480; ref. index 1.463; flash pt. 279 C.

Plasthall® P-554. [C.P. Hall] Polyester glutarate; plasticizer; APHA 150 clear liq.; m.w. 2500; sol. in toluene; partly sol. in kerosene, ethanol, acetone, min. oil; sp.gr. 1.06; visc. 3200 cps; flash pt. 271 C; acid no. 1.0; sapon. no. 490; ref. index 1.464.

Plasthall® P-630. [C.P. Hall] Polyester adipate; plasticizer for PVC applics.; Gardner 1 visc. liq.; m.w. 2200; f.p. –18 C; sol. in toluene; sp.gr. 1.091; visc. 2600 cps; acid no. 1.0; sapon. no. 530; ref. index 1.4644; flash pt. 268 C.

Plasthall® P-640. [C.P. Hall] Polyester adipate; see Plasthall P-550; APHA 100 visc. liq.; m.w. 3400; f.p. –22 C; sol. in toluene; sp.gr. 1.09; visc. 6800 cps; acid no. 1.0; sapon. no. 535; ref. index 1.466; flash pt. 277 C.

Plasthall® P-643. [C.P. Hall] Polyester adipate; see Plasthall P-550; APHA 100 visc. liq.; m.w. 2000; f.p. –50 C; sol. in toluene; sp.gr. 1.08; visc. 2650 cps; acid no. 1.0; sapon. no. 486; ref. index 1.4649; flash pt. 282 C.

Plasthall® P-650. [C.P. Hall] Polyester adipate; APHA 100 visc. liq.; m.w. 1200; sol. in toluene; partly sol. in kerosene, ethanol, acetone, min. oil; sp.gr. 1.05; visc. 3300 cps; f.p. –10 C; flash pt. 260 C; acid no. 0.5; sapon. no. 470; ref. index 1.464.

Plasthall® P-670. [C.P. Hall] Polyester adipate; plasticizer; APHA 100 visc. liq.; m.w. 800; sol. in toluene; partly sol. in kerosene, ethanol, acetone, min. oil; sp.gr. 1.08; visc. 1200 cps; f.p. –45 C; flash pt. 265 C; acid no. 1.0; sapon. no. 505; ref. index 1.464.
Plasthall® P-686. [C.P. Hall] Polyester adipate; APHA 100 visc. liq.; m.w. 2000; sol. in toluene; partly sol. in kerosene, ethanol, acetone, min. oil; sp.gr. 1.08; visc. 2650 cps; f.p. –50 C; flash pt. 282 C; acid no. 1.0; sapon. no. 486; ref. index 1.465.
Plasthall® P-1070. [C.P. Hall] Polyester sebacate; Gardner 3 visc. liq.; m.w. 2000; sol. in hexane, toluene, ethanol, acetone; sp.gr. 1.069; visc. 5000 cps; f.p. –22 C; flash pt. 218 C; acid no. 1.0; sapon. no. 455; ref. index 1.465.
Plasthall® P-7035. [C.P. Hall] Polyester glutarate; plasticizer for flexible PVC; Gardner 7 visc. liq.; m.w. 4500; sol. in toluene; partly sol. in kerosene, ethanol, acetone, min. oil; sp.gr. 1.08; visc. 11,000 cps; f.p. –12 C; flash pt. 260 C; acid no. 1.0; sapon. no. 498; ref. index 1.467.
Plasthall® P-7035M. [C.P. Hall] Polyester glutarate; Gardner 4 visc. liq.; m.w. 3600; sol. in toluene; partly sol. in kerosene, ethanol, acetone, min. oil; sp.gr. 1.089; visc. 6300 cps; f.p. –12 C; flash pt. 260 C; acid no. 1.0; sapon. no. 510; ref. index 1.4653.
Plasthall® P-7046. [C.P. Hall] Polyester glutarate; Gardner 5 visc. liq.; m.w. 4200; sol. in toluene; partly sol. in kerosene, ethanol, acetone, min. oil; sp.gr. 1.11; visc. 12,000 cps; f.p. –25 C; flash pt. 266 C; acid no. 1.0; sapon. no. 530; ref. index 1.466.
Plasthall® P-7092. [C.P. Hall] Polyester glutarate; Gardner 9 visc. liq.; m.w. 5000; sol. in toluene; partly sol. in kerosene, ethanol, acetone, min. oil; sp.gr. 1.11; visc. 24,000 cps; f.p. –20 C; flash pt. 271 C; acid no. 0.8; sapon. no. 520; ref. index 1.467.
Plasthall® P-7092D. [C.P. Hall] Polyester glutarate; Gardner 5 visc. liq.; m.w. 3500; sol. in toluene; partly sol. in kerosene, ethanol, acetone, min. oil; sp.gr. 1.11; visc. 6500 cps; f.p. –25 C; flash pt. 271 C; acid no. 1.0; sapon. no. 530; ref. index 1.465.
Plasthall® R-9. [C.P. Hall] Octyl tallate; transfer aid on correctable ribbon; penetration and tack agent for computer ribbons, carbon paper; yel. to amber liq.; m.w. 398; sol. in hexane, toluene, kerosene, min. oil; sp.gr. 0.873; visc. 17 cps; f.p. –10 C; flash pt. 213 C; acid no. 0.2; sapon. no. 141; ref. index 1.459.
Plasthall® TIOTM. [C.P. Hall] Triisooctyl trimellitate; Gardner 1–2 clear liq.; m.w. 550; sol. in hexane, toluene, kerosene, acetone, min. oil; sp.gr. 0.990; visc. 270 cps; f.p. –52 C; flash pt. 260 C; acid no. 0.08; sapon. no. 306; ref. index 1.484.
Plasthall® TOTM. [C.P. Hall] Trioctyl trimellitate; Gardner 1–2 clear liq.; m.w. 550; sol. in hexane, toluene, kerosene, acetone, min. oil; sp.gr. 0.990; visc. 260 cps; f.p. –54 C; flash pt. 260 C; acid no. 0.08; sapon. no. 306; ref. index 1.484.
Plastigel PG9033. [Plasticolors] Disp. of reactive magnesium oxide in an unsat., nonmonomer-containing polyester vehicle; thickener in the prod. of thickenable molding compd. when a controlled visc. increase is required; liq.; neutral; dens. 12.5 ± 0.5 lb/gal; visc. 12,000–20,000 cps; 37–39% magnesium oxide.
Plastigel PG9037. [Plasticolors] Disp. of reactive magnesium hydroxide in an unsat., nonmonomer-containing polyester vehicle; see Plastigel PG-9033; neutral liq.; dens. 11.4 ± 0.5 lb/gal; visc. 24,000–26,000 cps; 38.4–40.4% magnesium hydroxide.
Plastigel PG9068. [Plasticolors] Disp. of reactive calcium oxide in an unsat., nonmonomer-containing polyester vehicle; see Plastigel PG9033; neutral liq.; dens. 13.8 ± 0.5 lb/gal; visc. 17,000–19,000 cps; 51.0–53.0% calcium oxide.
Plastigel PG9089. [Plasticolors] Disp. of reactive magnesium hydroxide/calcium hydroxide in an unsat. monomer-containing vehicle, see Plastigel PG9033; neutral liq.; visc. 10.500–18.000 cps; 27.5–29.5% magnesium hydroxide plus calcium hydroxide in a 3.75:1 ratio.
Plastigel PG9104. [Plasticolors] Disp. of reactive calcium hydroxide in an unsat., monomer-containing vehicle; see Plastigel PG9033; neutral liq.; dens. 9.6 ± 0.5 lb/gal; visc. 20,000–25,000 cps; 28.5–30.5% calcium hydroxide.
Plastilease 250. [Ram] Silicone; paintable release agent for molding of parts to be post-finished or bonded; effective in overcoming rubber-to-metal friction and vibrational squealing; dens. 6.1 lb/gal; flash pt. 10 F (Seta CC); 98% volatiles.
Plastilease 512-B. [Ram] PVAL resins; film-forming release agent providing fast drying rate and easy cleanup; spray applics. is preferred; deep grn.; dens. 7.7 lb/gal; flash pt. 62 F (Seta CC) 62 F; 92% volatiles.
Plastilease 512-CL. [Ram] PVAL resins; see Plastilease 512-B; clear.
Plastilease 514. [Ram] Polyolefin milky disp.; release agent for processing resins in 200 F range; works well with preform release; textured mold surfs.; dens. 6.7 lb/gal; flash pt. 17 F (Seta CC); 93% volatiles.
Plastoflex® 2307. [Akzo] Epoxidized unsat. soya bean oils; epoxy plasticizer/stabilizer for PVC; aux. plasticizer or stabilizer; uses incl. plastisols, calendered and extruded prods., wetting of pigments, chlorinated paraffins, alkyd resins, neoprene, ethyl/cellulose, PVC/PVA emulsions; Gardner 3 max. clear liq.; dens. 0.99 g/ml; visc. 3–7 poise; acid no. 1.0 max.; ref. index 1.473; 7.0% min. conc.
Plastogen. [Vanderbilt] Plasticizer and softener for elastomers, NR latex films; amber to mahogany liq.; dens. 0.82 ± 0.02 mg/m^3; acid no. 1.0–1.1.
Plastogen R. [Vanderbilt] Plasticizer and processing aid for NR, SBR, IR, IIR, EPDM, CR; for sponge and low durometer stocks; clear amber liq.; dens. 0.85 ± 0.02 mg/m^3; acid no. 8.0–8.2.
Plastolein® 9048. [Henkel/Emery] Dibutyl azelate; low temp. vinyl plasticizer esp. for cellulosic molding compds.; sp.gr. 0.94; dens. 7.8 lb/gal; solid. pt. ≤ –70 F; flash pt. 325 F; acid no. 0.15 max.
Plastolein® 9049. [Henkel/Emery] Low temp. plasticizer; sp.gr. 0.94; dens. 7.8 lb/gal; visc. 8 cSt (100 F); solid. pt. < –70 F; flash pt. 325 F; acid no. 0.15 max.
Plastolein® 9050. [Henkel/Emery] Plasticizer for

use in food pkg. films; color 85/98% min. trans. (440/550 nm); sp.gr. 0.93; dens. 7.7 lb/gal; visc. 8.0 cSt (100 F); solid. pt. –9 F; acid no. 1.0 max.; hyd. no. 4.0 max.; ref. index 1.444; flash pt. 395 F; 45 phr in PVC, 75-mil sheet: tens. str. 3000 psi; elong. 360%; 100% mod. 1250 psi; hardness 84/80 (10 s Durometer A); brittle pt. –62 C.

Plastolein® 9051. [Henkel/Emery] Di-n-hexyl azelate; low temp. plasticizer for food pkg. films; sp.gr. 0.93; dens. 7.7 lb/gal; visc. 8 cSt (100 F); solid. pt. –9 F; flash pt. 395 F; acid no. 0.2 max.; ref. index 1.444.

Plastolein® 9058. [Henkel/Emery] Di-2-ethylhexyl azelate; low temp. vinyl plasticizer imparting exc. hand, drape, and softness; FDA approved; sp.gr. 0.92; dens. 7.6 lb/gal; visc. 10 cSt (100 F); solid. pt. –90 F; flash pt. 415 F; acid no. 1.0 max.; ref. index 1.446.

Plastolein® 9065. [Henkel/Emery] Proprietary; low temp. plasticizer; sp.gr. 0.92; dens. 7.6 lb/gal; visc. 9.4 cSt (100 F); solid. pt. –20 F; flash pt. 400 F; acid no. 1.0 max.; ref. index 1.449.

Plastolein® 9071. [Henkel/Emery] Low-temp. plasticizer for PVC compds.; sp.gr. 0.91; dens. 7.5 lb/gal; visc. 13.5 cSt (100 F); solid. pt. 10 F; acid no. 1.0 max.; hyd. no. 4.0 max.; ref. index 1.451; flash pt. 390 F; 50 phr in PVC, 75-mil sheet: tens. str. 3050 psi; elong. 325%; 100% mod. 1475 psi; hardness 89/86 (10 s Durometer A); brittle pt. –58 C.

Plastolein® 9088. [Henkel/Emery] Proprietary; low temp. plasticizer for PVC; sp.gr. 0.92; dens. 7.6 lb/gal; visc. 9.7 cSt (100 F); solid. pt. –60 F; flash pt. 410 F; acid no. 1.0 max.; ref. index 1.447.

Plastolein® 9091. [Henkel/Emery] Low-temp. plasticizer for PVC; dk. color; sp.gr. 0.92; dens. 7.6 lb/gal; visc. 11 cSt (100 F); solid. pt. –60 F; acid no. 1.0 max.; hyd. no. 4.0 max.; ref. index 1.447; flash pt. 410 F; 45 phr in PVC, 75-mil sheet: tens. str. 3225 psi; elong. 295%; 100% mod. 1575 psi; hardness 89/86 (10 s Durometer A); brittle pt. –57 C.

Plastolein® 9215. [Henkel/Emery] Epoxy plasticizer imparting stability and good low temp. properties to vinyl compositions; Gardner 2 color; sp.gr. 0.95; dens. 7.9 lb/gal; visc. 76 cSt (100 F); solid. pt. 3 F; flash pt. 539 F; acid no. 1.0; iodine no. 3.0; ref. index 1.463.

Plastolein® 9232. [Henkel/Emery] Epoxy soya plasticizer; offers low extraction and low volatility; heat and lt. stability to PVC formulations by acting as HCl scavenger; Gardner 2; All properties 50:50 blend with DOP: sp.gr. 0.99; dens. 8.2 lb/gal; visc. 161 cSt (100 F); solid. pt. 19 F; acid no. 1.0 max.; ref. index 1.470; flash pt. 585 F; fire pt. 649 F; 54 phr in PVC sheet: tens. str. 2650 psi; tens. elong 345%; hardness 86/81 (Durometer A).

Plastolein® 9404. [Henkel/Emery] Triethylene glycol dipelargonate; low temp. plasticizer for syn. elastomers and natural rubbers; sp.gr. 0.965; dens. 8.1 lb/gal; visc. 59 cSt (100 F); solid. pt. < 0 F; flash pt. 410 F; ref. index 1.448.

Plastolein® 9717. [Henkel/Emery] Polymeric plasticizer; low visc. plasticizer providing permanence and low-temp. performance at min. cost; suitable for pigment grinding; Gardner 5 max.; sp.gr. 1.02; dens. 8.5 lb/gal; visc. 255 cSt (100 F); solid. pt. 18 F; acid no. 2.5 max.; ref. index 1.469; flash pt. 450 F; fire pt. 500 F; 56 phr in PVC sheet: tens. str. 3100 psi; tens. elong. 340%; hardness 87/82 (Durometer A).

Plastolein® 9720. [Henkel/Emery] Polymeric plasticizer; easily processed and handled, low volatility and resistance to oil extraction; used in plastisols and calendered fabrics; Gardner 5 max.; sp.gr. 1.03; dens. 8.6 lb/gal; visc. 207 cSt (100 F); solid. pt. 18 F; acid no. 3.0 max.; ref. index 1.462; flash pt. 500 F; fire pt. 560 F; 56 phr in PVC sheet: tens. str. 3125 psi; tens. elong. 355%; hardness 87/80 (Durometer A).

Plastolein® 9730. [Henkel/Emery] Polymeric plasticizer; med. m.w. plasticizer exhibiting low migration, low soapy water extraction, and outstanding humidity aging properties; Gardner 6 max.; sp.gr. 1.06; dens. 8.9 lb/gal; visc. 890 cSt (100 F); solid. pt. 45 F; acid no. 3.0 max.; ref. index 1.483; flash pt. 510 F; fire pt. 560 F; 70 phr in PVC sheet: tens. str. 3000 psi; tens. elong. 350%; hardness 91/85 (Durometer A).

Plastolein® 9731. [Henkel/Emery] Polymeric plasticizer; elec. grade polyester designed to meet UL requirements for 105° wire insulation; Gardner 6 max.; sp.gr. 1.06; dens. 8.9 lb/gal; visc. 880 cSt (100 F); solid. pt. 45 F; acid no. 3.0 max.; ref. index 1.483; flash pt. 510; fire pt. 560 F; 70 phr in PVC sheet: tens. str. 3000 psi; tens. elong. 350%; hardness 91/85 (Durometer A).

Plastolein® 9734. [Henkel/Emery] Polymeric plasticizer; general purpose use with excellent resistance to extraction by soapy water and min. oil, resistance to migration into lacquered surf. and S/B rubber and outstanding humidity aging properties; Gardner 6; sp.gr. 1.04; dens. 8.7 lb/gal; visc. 1616 cSt (100 F); solid. pt. 55 F; acid no. 3.0 max.; ref. index 1.483; flash pt. 500 F; fire pt. 565 F; 70 phr in PVC sheet: tens. str. 2800 psi; tens. elong. 310%; hardness 86/80 (Durometer A).

Plastolein® 9749. [Henkel/Emery] Med. m.w. polymeric general purpose plasticizer offering exc. balance of performance properties incl. resistance to humidity aging; Gardner 5 color; sp.gr. 1.06; dens. 8.9 lb/gal; visc. 857 cSt (100 F); solid. pt. 32 F; flash pt. 535 F; acid no. 3.0 max.; ref. index 1.477.

Plastolein® 9750. [Henkel/Emery] Polymeric plasticizer; med. m.w. general purpose use offering resistance to humidity aging; Gardner 5 max.; sp.gr. 1.06; dens. 8.9 lb/gal; visc. 857 cSt (100 F); solid. pt. 32 F; acid no. 3.0 max.; ref. index 1.477; flash pt. 535 F; fire pt. 600 F; 65 phr in PVC sheet: tens. str. 3000 psi; tens. elong 350%; hardness 86/80 (Durometer A).

Plastolein® 9752. [Henkel/Emery] Low-to-med. visc. polymeric plasticizer with exc. processing properties, good permanence; used for appliance cords, gasketing, wall coverings, plastisols, calendered coated fabrics, automotive constructions; Gardner 3 color; sp.gr. 1.09; dens. 9.1 lb/gal; visc. 1130 cSt (100 F); solid. pt. 40 F; flash pt. 465 F; acid no. 3.0 max.; ref. index 1.496.

Plastolein® 9756. [Henkel/Emery] Polymeric plasticizer; med.-visc. plasticizer used in appliance gasketing, wall coverings, and upholstery; Gardner 4 max.; sp.gr. 1.07; dens. 9.07 lb/gal; visc. 1744 cSt (100 F); solid. pt. 35 F; acid no. 3.0 max.; hyd. no. 20.0 max.; ref. index 1.466; flash pt. 535 F; 62 phr in PVC, 75-mil sheet: tens. str. 2950 psi; elong. 300%; 100% mod. 1425 psi; hardness 84/82 (10 s Durometer A); brittle pt. –17 C.

Plastolein® 9761. [Henkel/Emery] Polymeric plasticizer; suited for PVC compds. in contact with alkyd-type finishes, lacquers, modified and unmodified PS, and ABS; Gardner 2 max.; sp.gr. 1.06; dens. 8.9 lb/gal; visc. 1945 cSt (100 F); solid. pt. 10 F; acid no. 3.0 max.; ref. index 1.469; flash pt. 515 F; fire pt. 600 F; 62 phr in PVC sheet: tens. str. 3050 psi; tens. elong. 340%; hardness 86/80 (Durometer A).

Plastolein® 9762. [Henkel/Emery] Polymeric plasticizer for low-temp. applics.; used in vinyl tapes, elec. insulation, and upholstery; Gardner 2 max.; sp.gr. 1.08; dens. 9.0 lb/gal; visc. 457 cSt (100 F); solid. pt. 5 F; acid no. 3.0 max.; hyd. no. 20.0 max.; ref. index 1.466; flash pt. 560 F; 56 phr in PVC, 75-mil sheet: tens. str. 3080 psi; elong. 300%; 100% mod. 1450 psi; hardness 85/80 (10 s Durometer A); brittle pt. –22 C.

Plastolein® 9765. [Henkel/Emery] Polymeric plasticizer; med.-to-high visc. plasticizer with resistance to migration, extraction media, attack by fungus, and humidity aging; Gardner 6 max.; sp.gr. 1.08; dens. 9.0 lb/gal; visc. 2985 cSt (100 F); solid. pt. 35 F; acid no. 3.0 max.; ref. index 1.479; flash pt. 530 F; fire pt. 595 F; 68 phr in PVC sheet: tens. str. 3025 psi; tens. elong 340%; hardness 86/81 (Durometer A).

Plastolein® 9772. [Henkel/Emery] Polymeric plasticizer; offering permanence, efficiency, migration resistance, and dry-blending ease; vehicle for pigment grinds; Gardner 3 max.; sp.gr. 1.04; dens. 8.7 lb/gal; visc. 375 cSt (100 F); solid. pt. 16 F; acid no. 3.0 max.; ref. index 1.486; flash pt. 520 F; 56 phr in PVC sheet: tens. str. 3300 psi; tens. elong 330%; hardness 87/81 (Durometer A).

Plastolein® 9776. [Henkel/Emery] Polymeric plasticizer; med.-to-high visc. plasticizer used in appliance gasketing, wall coverings, appliance cords, and automotive constructions; Gardner 2 max.; sp.gr. 1.08; dens. 9.0 lb/gal; visc. 2572 cSt (100 F); solid. pt. –4 F; acid no. 2.0 max.; ref. index 1.466; flash pt. 575 F; fire pt. 600 F; 62 phr in PVC sheet: tens. str. 3200 psi; tens. elong. 335%; hardness 88/82 (Durometer A).

Plastolein® 9780. [Henkel/Emery] Polymeric plasticizer; low-temp. performance and migration resistance are critical; recommended for vinyl tapes, elec. insulation, and upholstery; Gardner 7 max.; sp.gr. 1.04; dens. 8.7 lb/gal; visc. 1030 cSt (100 F); solid. pt. –2 F; acid no. 3.0 max.; ref. index 1.466; flash pt. 570 F; fire pt. 640 F; 58 phr in PVC sheet: tens. str. 3125 psi; tens. elong. 355%; hardness 83/78 (Durometer A).

Plastolein® 9781. [Henkel/Emery] Polymeric plasticizer for elec. tape and pressure-sensitive formulations; Gardner 5 max.; sp.gr. 1.09; dens. 9.1 lb/gal; visc. 375–525 cSt (100 F); solid. pt. 30 F; flash pt. 475 F; acid no. 3.0 max.; ref. index 1.465.

Plastolein® 9783. [Henkel/Emery] Polymeric plasticizer; med. m.w. PVC plasticizer used when resistance to migration into pressure-sensitive adhesive backing is required; Gardner 5 max.; sp.gr. 1.08; dens. 9.0 lb/gal; visc. 1121 cSt (100 F); solid. pt. 55 F; acid no. 2.0 max.; ref. index 1.465; flash pt. 540 F; fire pt. 600 F; 62 phr in PVC sheet: tens. str. 3125 psi; tens. elong. 350%; hardness 86/81 (Durometer A).

Plastolein® 9784. [Henkel/Emery] Polymeric plasticizer for pressure-sensitive adhesive applics.; sp.gr. 1.07; dens. 8.9 lb/gal; visc. 1000–1250 cSt (100 F); solid. pt. –5 F; flash pt. 570 F; acid no. 3.0 max.; ref. index 1.469.

Plastolein® 9789. [Henkel/Emery] Polymeric plasticizer; formulations requiring max. permanence; calendering tracking aid and tackifier; Gardner 4 max.; sp.gr. 1.08; dens. 9.0 lb/gal; visc. 16,000 cSt (100 F); solid. pt. –20 F; acid no. 5.0 max.; ref. index 1.460; flash pt. 580 F; fire pt. 635 F; 67 phr in PVC sheet: tens. str. 2800 psi; tens. elong. 325%; hardness 84/79 (Durometer A).

Plastolein® 9790. [Henkel/Emery] Polymeric plasticizer; high m.w. plasticizer offering outstanding performance and permanence; lt. color is of minimal importance; Gardner 7 max.; sp.gr. 1.08; dens. 9.0 lb/gal; visc. 16,000 cSt (100 F); solid. pt. –20 F; acid no. 5.0 max.; ref. index 1.460; flash pt. 580 F; fire pt. 635 F; 67 phr in PVC sheet: tens. str. 2775 psi; tens. elong. 320%; hardness 85/79 (Durometer A).

Plastomag®. [Akrochem] Oil-dispersed magnesium oxide; chemical thickener for polyester resins; anticaking agent; used in syn. rubber compding., adhesives, fuel oil additives, and as acid acceptor for specialty plastics; soft micropellets; 75% thru 40 mesh; 20% on 40 mesh; 5% on 20 mesh; sp.gr. 2.45; bulk dens. 50 lb/ft^3 (aerated); 65% Elastomag 170; 35% naphthenic process oil; 98% MgO.

Plex HT. [Vikon] Chelating agent for textile scouring to remove rust stains; sol. in water, strong caustic sol'n.

Plex PC. [Vikon] Calcium DTPA; chelating agent for peroxide bleach bath; sol. in water.

Plexene D. [Sybron] Pentasodium pentetate; chelating agent in dyeing and peroxide bleaching; 40% act.

Plexene Extra Conc. [Sybron] Tetrasodium EDTA; sequestering agent for Ca and Mg hard water conditions; 39% act.

Plexol® 305. [Rohm & Haas] Octyl phenoxy tetraethoxy-ethanol; component of rust-preventive formulations for syn. lubricants, min. oils, greases.

Pliabrac 519. [Merrand] Propylated triaryl phosphate; flame retardant plasticizer for NBR, NBR/PVC, CR, CPE elastomers, polyacrylate, fluoroelastomers, PVC, PVAc, cellulosics, PS, ABS, engineering resins, polyester, alloys; antifungal PVC.

Pliabrac 521. [Merrand] Propylated triaryl phosphate; flame retardant plasticizer for NBR, NBR/PVC, CR, CPE elastomers, polyacrylate, fluoroelastomers, PVC, PVAc, cellulosics, PS, ABS, engineering resins, polyester, alloys, inks; lubricant additive.

Pliabrac 524. [Merrand] Propylated triaryl phosphate; flame retardant plasticizer for NBR, NBR/PVC, CR, CPE elastomers, polyacrylate, fluoroelastomers, PVC, PVAc, cellulosics, PS, ABS, engineering resins, polyester, alloys; exc. dipping visc. for PVC plastisol.

Pliabrac TBP. [Merrand] Tributyl phosphate; plasticizer for CR, CPE, Hypalon elastomers, PVC, PVAc, cellulosics, PS, ABS, ink systems; lubricant additive; antifoam; ink solv.; low odor.

Pliabrac TCP. [Merrand] Tricresyl phosphate; plasticizer for NBR, NBR/PVC, CR, CPE elastomers, PVC, PVAc, cellulosics, PS, ABS, engineering resins, polyester, alloys; lubricant additive.

Pliabrac TXP. [Merrand] Trixylenyl phosphate; plasticizer for PVC, PVAc, cellulosics, PS, ABS, PVC wire insulation; lubricant additive.

Pluracol® 355. [BASF] Amine-based polyol; cross-linking agent for semiflexible urethane foams, coatings, adhesives, and polymers; APHA 60 max. liq.; sp.gr. 1.03; visc. 3200 cps; pH 11.0.

Pluracol® 364. [BASF] POP deriv. of sucrose; cross-linking agent producing extra strong rigid foams; Gardner 10; sp.gr. 1.09; visc. 22,000 cps; pH 11.0.

Pluracol® 450. [BASF] POP deriv. of pentaerythritol; cross-linking agent for rigid urethane foams; m.w. 405; sp.gr. 1.08; visc. 2200 cps; acid no. 0.06; pH 6.5.

Pluracol® 550. [BASF] POP deriv. of pentaerythritol; see Pluracol 450; m.w. 500; sp.gr. 1.06; visc. 1300 cps; acid no. 0.06; pH 6.5.

Pluracol® 650. [BASF] POP deriv. of pentaerythritol; see Pluracol 450; m.w. 594; sp.gr. 1.05; visc. 1200 cps; acid no. 0.06; pH 6.5.

Pluracol® 669. [BASF] POP deriv. of sucrose; see Pluracol 364; Gardner 10; visc. 44,000 cps; pH 9.8.

Pluracol® E-200. [BASF] PEG-4; intermediate for preparation of nonionic surfactants; binder, base, coating, stabilizer, solv., vehicle, extender, and coupling agent for pharmaceutical, cosmetic, and toiletries; lubricant for metal applics., rubber industry; wood treatment; textile conditioning, antistat, and sizing agent, softener; colorless clear liq.; sol. in water and org. solvs. except aliphatic hydrocarbons; dens. 9.4 lb/gal; sp.gr. 1.12; visc. 4.36 cs (210 F); flash pt. 360 F; surf. tens. 57.2 dynes/cm (1%); pH 6.5 (5% aq.)

Pluracol® E-300. [BASF] PEG-6; see Pluracol E-200; also dispersant in food tablets and preparations; plasticizer; colorless clear liq.; sol. see Pluracol E-200; dens. 9.4 lb/gal; sp.gr. 1.12; visc. 5.75 cs (210 F); flash pt. 410 F; surf. tens. 62.9 dynes/cm (1%); pH 5.7 (5% aq.).

Pluracol® E-400. [BASF] PEG-8; see Pluracol E-300; colorless clear liq.; sol. see Pluracol E-200; dens. 9.4 lb/gal; sp.gr. 1.12; visc. 7.39 cs (210 F); flash pt. 460 F; surf. tens. 66.6 dynes/cm (1%); pH 6.2 (5% aq.).

Pluracol® E-400 NF. [BASF] PEG-8; chemical intermediate, base, coupling agent, thickener, lubricant, mold release agent, defoamer, softener, conditioner, antistat, sizing agent, dispersant for pharmaceutical, cosmetic, and oral care preparations, in metal polishing and cleaning formulations, rubber prods., paper and wood prods., textile processing, ink formulations; liq.; m.w. 400; visc. 7.4 cs (99 C); pour pt. 5 C; flash pt. 182 C (COC).

Pluracol® E-600. [BASF] PEG-12; see Pluracol E-300; colorless clear liq.; sol. see Pluracol E-200; dens. 9.4 lb/gal; sp.gr. 1.12; visc. 10.83 cs (210 F); flash pt. 480 F; surf. tens. 65.2 dynes/cm (1%); pH 5.3 (5% aq.).

Pluracol® E-600 NF. [BASF] PEG-12; see Pluracol E-400 NF; liq.; m.w. 600; visc. 10.8 cs (99 C); pour pt. 20 C; flash pt. 249 C (COC).

Pluracol® E-1000. [BASF] PEG-20; see Pluracol E-400 NF; solid; m.w. 1000; visc. 17.5 cs (99 C); m.p. 38 C; flash pt. 255 C (COC).

Pluracol® E-1450. [BASF] PEG; see Pluracol E-400 NF; solid; m.w. 1450; visc. 28.5 cs (99 C); m.p. 45 C; flash pt. 255 C (COC).

Pluracol® E-1450 NF. [BASF] PEG, NF grade; see Pluracol E-400 NF; solid; m.w. 600; visc. 28.5 cs (99 C); m.p. 45 C; flash pt. 255 C (COC).

Pluracol® E-2000. [BASF] PEG-40; see Pluracol E-400 NF; solid; m.w. 2000; visc. 43.5 cs (99 C); m.p. 52 C; flash pt. > 260 C (COC).

Pluracol® E-4000. [BASF] PEG-75; see Pluracol E-300; wh. waxy solid; sol. see Pluracol E-200; dens. 10.0 lb/gal; sp.gr. 1.20; m.p. 59.5 C; flash pt. > 490 F; surf. tens. 61.9 dynes/cm (1%); pH 6.7 (5% aq.).

Pluracol® E-4000 NF. [BASF] PEG-75; see Pluracol E-400 NF; solid; m.w. 4000; visc. 134 cs (99 C); m.p. 59 C; flash pt. > 260 C (COC).

Pluracol® E-4500. [BASF] PEG; see Pluracol E-400 NF; solid; m.w. 4500; visc. 170 cs (99 C); m.p. 60 C; flash pt. > 260 C (COC).

Pluracol® E-8000. [BASF] PEG; see Pluracol E-400 NF; solid; m.w. 8000; visc. 750 cs (99 C); m.p. 61 C; flash pt. > 260 C (COC).

Pluracol® E-8000 NF. [BASF] PEG, NF grade; see Pluracol E-400 NF; solid; m.w. 8000; visc. 750 cs (99 C); m.p. 61 C; flash pt > 260 C (COC).

Pluracol® P-410. [BASF] PPG-9; chemical intermediate, antifoam agent in fermentation and in paint formulations, antiblooming agent for pentachlorophenol-treated wood; binder and lubricant for ceramics; plasticizer of resin-treated papers; preparation of PU foams; hydraulic and grinding fluids; ore flotation; water-wh. liq.; slight ether-like odor; water-sol.; m.w. 425; dens. 8.36 lb/gal; sp.gr. 1.005; visc. 35 cps (100 F); flash pt. > 400 F (PMCC); pour pt. –35 F; pH 6–7.

Pluracol® P-710. [BASF] PPG-12; see Pluracol P-410; water-wh. liq.; slight ether-like odor; water-sol.; m.w. 775; dens. 8.35 lb/gal; sp.gr. 1.004; visc. 65 cps (100 F); flash pt. > 400 F (PMCC); pour pt. 35 F; pH 6–7

Pluracol® P-1010. [BASF] PPG-17; see Pluracol P-410; water-wh. liq.; slight ether-like odor; insol. in water; m.w. 1050; dens. 8.38 lb/gal; sp.gr. 1.007; visc. 80 cps (100 F); flash pt. > 400 F (PMCC); pour pt. –35 F; pH 6–7.

Pluracol® P-2010. [BASF] PPG-26; see Pluracol P-410; water-wh. liq.; slight ether-like odor; insol. in water; m.w. 2000; dens. 8.34 lb/gal; sp.gr. 1.002; visc. 175 cps (100 F); flash pt. > 400 F (PMCC); pour pt. –35 F; pH 6–7.

Pluracol® P-4010. [BASF] PPG-30; see Pluracol P-410; water-wh. liq.; slight ether-like odor; insol. in water; m.w. 4000; dens. 8.33 lb/gal; sp.gr. 1.00; visc. 550 cps (100 F); flash pt. > 400 (PMCC); pour pt. –20 F; pH 6–7.

Pluracol® V-7. [BASF] Polyol; thickening agent in water-based coolants, lubricants, and metal-working fluids; shear resistance and noncorrosiveness; will not support bacterial growth or form gummy residues; water-sol.; sp.gr. 1.090; visc. 5600 SUS (100 F); flash pt. 510 F; pour pt. 25 F; pH 7–8 (2.5% aq. sol'n.).

Pluracol® V-10. [BASF] Polyoxyalkylene glycol polyol; thickening agent to control the visc. of water-glycol type, fire-resistant hydraulic fluids; resistance to shearing stresses and will not hydrolyze or degrade under use conditions; noncarbonizing and nongumming at high temps.; visc. liq.; water-sol.; sp.gr. 1.089; visc. 210,000 SUS (100 F); flash pt. 510 F.

Pluracol® W-170. [BASF] PPG-5-buteth-7; component in demulsifying and wetting formulations; brake and metalworking fluids; rubber and fiber lubricant; textile applics.; defoamer for hot and cold applics., food and chemical processing; cosmetic formulations; APHA 50 max. visc. liq.; sol. in water, alcohols, ketones, esters, benzene, toluene, glycol ethers, chlorinated solvs.; dens. 8.58 lb/gal; sp.gr. 1.03; visc. 160–180 SUS (100 F); cloud pt. 73 C (1%); flash pt. 360 F (OC); pour pt. –45 F; pH 5.5–7.0 (10% aq.).

Pluracol® W-260. [BASF] Polyalkoxylated polyether; see Pluracol W-170; liq.; m.w. 1000; water-sol.; visc. 260 SUS (37.8 C); pour pt. –40 C; flash pt. (COC) 222 C.

Pluracol® W-660. [BASF] PPG-12-buteth-16; see Pluracol W-170; APHA 50 max. visc. liq.; sol. see Pluracol W-170; dens. 8.79 lb/gal; sp.gr. 1.055; visc. 660 SUS (100 F); cloud pt. 60.5 C (1%); flash pt. 440 F (OC); pour pt. –34 F; pH 5.5–7.5 (10% aq.).

Pluracol® W-2000. [BASF] PPG-20-buteth-30; see Pluracol W-170; APHA 50 max. visc. liq.; sol. see Pluracol W-170; dens. 8.83 lb/gal; sp.gr. 1.06; visc. 2000 SUS (100 F); cloud pt. 57.0 C (1%); flash pt. 440 F (OC); pour pt. –25 F; pH 5.5–7.5 (10% aq.).

Pluracol® W-3520N. [BASF] PPG; see Pluracol W-170; APHA 40 max. visc. liq.; sol. see Pluracol W-170; dens. 8.83 lb/gal; sp.gr. 1.06; visc. 3520 SUS (100 F); cloud pt. 57 C (1%); flash pt. 437 F (OC); pour pt. –20 F; pH 6.0–7.5 (10% aq.).

Pluracol® W-3520N-RL. [BASF] PPG; see Pluracol W-170; APHA 200 max. visc. liq.; sol. see Pluracol W-170; dens. 8.87 lb/gal; sp.gr. 1.065; visc. 1752–2500 SUS (100 F); cloud pt. 55.5 C (1%); flash pt. 440 F (OC); pour pt. 28.4 F; pH 6.0–7.5 (10% aq.).

Pluracol® W-5100N. [BASF] PPG-33-buteth-45; see Pluracol W-170; APHA 40 max. visc. liq.; sol. see Pluracol W-170; dens. 8.83 lb/gal; sp.gr. 1.06; visc. 5100 SUS (100 F); cloud pt. 55 C (1%); flash pt. 437 F (OC); pour pt. –20 F; pH 5.5–7.0 (10% aq.).

Pluracol® WD1400. [BASF] Polyalkoxylated polyether; see Pluracol W-170; liq.; m.w. 2500; water-sol.; visc. 1400 SUS (37.8 C); pour pt. –20 C; flash pt. (COC) 255 C.

Pluradot HA-410. [BASF] Trifunctional polyoxyalkylene glycol; nonionic; emulsifier, solubilizer, deduster, wetting agent, demulsifier, detergent, dispersant; surfactant for hard surface cleaning, machine dishwashing; liq.; m.w. 3200; sp.gr. 1.03; visc. 450 cps; cloud pt. 25 C (1%); pour pt. –15 F; surf. tens. 36 dynes/cm (0.1%); 100% conc.

Pluradot HA-420. [BASF] Trifunctional polyoxyalkylene glycol; see Pluradot HA-410; liq.; m.w. 3600; sp.gr. 1.04; visc. 565 cps; cloud pt. 28 C (1%); pour pt. –10 F; surf. tens. 37 dynes/cm (0.1%); 100% conc.

Pluradot HA-430. [BASF] Trifunctional polyoxyalkylene glycol; see Pluradot HA-410; liq.; m.w. 3900; sp.gr. 1.05; visc. 700 cps; cloud pt. 43 C (1%); pour pt. 20 C; surf. tens. 38 dynes/cm (0.1%); 100% conc.

Pluradot HA-433. [BASF] Trifunctional polyoxyalkylene glycol; see Pluradot HA-410; liq.; m.w. 3900; sp.gr. 1.05; visc. 700 cps; pour pt. 20 F; 100% conc.

Pluradot HA-440. [BASF] Trifunctional polyoxyalkylene glycol; see Pluradot HA-410; liq.; m.w. 4450; sp.gr. 1.05; visc. 750 cps; cloud pt. 54 C (1%); pour pt. 25 C; surf. tens. 39 dynes/cm (0.1%); 100% conc.

Pluradot HA-450. [BASF] Trifunctional polyoxyalkylene glycol; see Pluradot HA-410; paste; m.w. 5700; sp.gr. 1.06; cloud pt. 74 C (1%); pour pt. 60 F; surf. tens. 41 dynes/cm (0.1%)

Pluradot HA-510. [BASF] Trifunctional polyoxyalkylene glycol; see Pluradot HA-410; liq.; m.w. 4600; sp.gr. 1.03; visc. 710 cps; cloud pt. 24 C (1%); pour pt. –15 F; surf. tens. 35 dynes/cm; 100% conc.

Pluradot HA-520. [BASF] Trifunctional polyoxyalkylene glycol; see Pluradot HA-410; liq.; m.w. 5000; sp.gr. 1.03; visc. 760 cps; cloud pt. 27 C (1%); pour pt. 5 F; surf. tens. 36 dynes/cm (0.1%); 100% conc.

Pluradot HA-530. [BASF] Trifunctional polyoxyalkylene glycol; see Pluradot HA-410; liq.; m.w. 5300; sp.gr. 1.04; visc. 850 cps; cloud pt. 42 C (1%); pour pt. 30 F; surf. tens. 37 dynes/cm (0.1%); 100% conc.

Pluradot HA-540. [BASF] Trifunctional polyoxyalkylene glycol; see Pluradot HA-410; liq.; m.w. 6000; sp.gr. 1.05; visc. 1240 cps; cloud pt. 54 C (1%); pour pt. 40 F; surf. tens. 38 dynes/cm (0.1%); 100% conc.

Pluradot HA-550. [BASF] Trifunctional polyoxyalkylene glycol; see Pluradot HA-410; paste; m.w. 7500; sp.gr. 1.06; cloud pt. 77 C (1%); pour pt. 75 F; surf. tens. 39 dynes/cm (0.1%); 100% conc.

Plurafac® RA-20. [BASF] Straight chain primary aliphatic oxyalkylated alcohol; nonionic; detergent, dispersant, wetting agent, emulsifier, defoamer, deduster used in rinse aids and dishwashing prods.; colorless clear liq.; ref. index 1.4530; sol. see Plurafac A-24; dens. 8.3 lb/gal; sp.gr. 0.992; visc. 83 cps;

HLB 10; cloud pt. 45 C (1%); flash pt. 475 F; pour pt. 45 F; surf. tens. 32.8 dynes/cm (0.1%); pH 6–7 (1%); 100% conc.

Plurafac® RA-30. [BASF] Straight chain primary aliphatic oxyalkylated alcohol; see Plurafac RA-20; colorless clear liq.; ref. index 1.4534; sol. see Plurafac RA-20; dens. 8.1 lb/gal; sp.gr. 0.973; visc. 56 cps; HLB 9; cloud pt. 35 C (1%); flash pt. 455 F; pour pt. 50 F; surf. tens. 30.6 dynes/cm (0.1%); pH 6–7 (1%); 100% conc.

Plurafac® RA-40. [BASF] Straight chain primary aliphatic oxyalkylated alcohol; see Plurafac RA-20; colorless clear liq.; ref. index 1.4510; sol. see Plurafac A-24; dens. 8.1 lb/gal; sp.gr. 0.970; visc. 91 cps; HLB 7; cloud pt. 25 C (1%); flash pt. 437 F; pour pt. –15 F; surf. tens. 30.5 dynes/cm (0.1%); pH 6–7 (1%); 100% conc.

Plurafac® RA-43. [BASF] Straight chain primary aliphatic oxyalkylated alcohol; see Plurafac RA-20; wh. opaque liq.; dens. 7.9 lb/gal; sp.gr. 0.950; visc. 120 cps; HLB 7; cloud pt. 24 C (1%); flash pt. 437 F; pour pt. 21 F; surf. tens. 30.5 dynes/cm (0.1%); pH 3–4 (1%); 100% conc.

Pluraflo E4A. [BASF] Nonionic dispersant formulated for use in liq. form; EPA cleared for agric. use; liq.; visc. 820 cps; pour pt. 7 C; flash pt. 69 F (CC); surf. tens. 42.8 dynes/cm (0.1%).

Pluraflo E4B. [BASF] See Pluraflo E4A; liq.; visc. 740 cps; pour pt. 11 C; flash pt. 116 F (CC); surf. tens. 42.3 dynes/cm (0.1%).

Pluraflo E5A. [BASF] See Pluraflo E4A; liq.; visc. 430 cps; pour pt. 5 C; flash pt. 61 F (CC); surf. tens. 42.5 dynes/cm (0.1%).

Pluraflo E5B. [BASF] See Pluraflo E4A; liq.; visc. 520 cps; pour pt. 3 C; flash pt. 110 F (CC); surf. tens. 42.7 dynes/cm (0.1%).

Pluraflo E5BG. [BASF] See Pluraflo E4A; liq.; visc. 250 cps; pour pt. –19 C; flash pt. 138 F (CC); surf. tens. 43.8 dynes/cm (0.1%).

Pluraflo E5G. [BASF] See Pluraflo E4A; liq.; visc. 260 cps; pour pt. –22 C; flash pt. 126 F (CC); surf. tens. 43.9 dynes/cm (0.1%).

Pluraflo N5G. [BASF] See Pluraflo E4A; liq.; visc. 410 cps; pour pt. –21 C; flash pt. 126 F (CC); surf. tens. 46.0 dynes/cm (0.1%).

Pluriol E 200. [BASF AG] PEG; nonionic; solubilizer, impregnating agent, humectant, mold release agent; flow improver, thermal and hydraulic fluid, org. intermediate; detergent and cleaner; dye and pigment dispersant; inks; textile and coatings industry; coloring ceramics; softener in paper industry; plasticizer in adhesives industry and in prod. of cellulose film; ceramics and metalworking lubricant; clear colorless liq.; m.w. 200; water-sol.; dens. 1.12 g/cc; visc. 55–65 cs.; m.p. < –40 C; flash pt. > 150 C; pH 6.0–7.5 (1% aq.); 100% conc.

Pluriol E 300. [BASF AG] PEG; nonionic; see Pluriol E 200; colorless clear liq.; m.w. 300; water-sol.; dens. 1.13 g/cc; visc. 75–95 cs; m.p. –20 to –10 C; flash pt. > 150 C; pH 6.0–7.5 (1% aq.); 100% conc.

Pluriol E 400. [BASF AG] PEG; nonionic; see Pluriol E 200; colorless clear liq.; m.w. 400; water-sol.; dens. 1.13 g/cc; visc. 105–120 cs; m.p. –10 to 10 C; flash pt. > 150 C; pH 6.0–7.5 (1% aq.); 100% conc.

Pluriol E 600. [BASF AG] PEG; nonionic; see Pluriol E 200; semisolid; m.w. 600; water-sol.; dens. 1.15 g/cc; visc. 8–11 cs (99 C); m.p. 10–20 C; flash pt. > 220 C; pH 6.0–7.5 (1% aq.); 100% conc.

Pluriol E 1500. [BASF AG] PEG; nonionic; solubilizer, humectant; binder and hardener in personal care prods.; dispersing dyes and pigments; inks; textile and coating industry; coloring ceramics; paper industry softener; plasticizer in adhesives industry and prod. of cellulose film; ceramics and metalworking lubricants; aux. for copper and nickel electroplating baths; electrolytic polishing of steel; wh. fine powd.; m.w. 1500; water-sol.; bulk dens. 0.6 kg/l; visc. 25–35 cs (99 C); m.p. 45 C; flash pt. > 230 C; pH 6.0–7.5 (1% aq.); 100% conc.

Pluriol E 4000. [BASF AG] PEG; nonionic; see Pluriol E 1500; wh. fine powd.; m.w. 4000; water-sol.; bulk dens. 0.6 kg/l; visc. 80–130 cs (99 C); m.p. 55 C; flash pt. > 240 C; pH 6.0–7.5 (1% aq.); 100% conc.

Pluriol E 6000. [BASF AG] PEG; nonionic; see Pluriol E 1500; wh. fine powd.; m.w. 6000; water-sol.; bulk dens. 0.6 kg/l; visc. 300 cs (99 C); m.p. 60 C; flash pt. > 240 C; pH 6.0–7.5 (1% aq.); 100% conc.

Pluriol E 9000. [BASF AG] PEG; nonionic; see Pluriol E 1500; also used in rubber processing; mold release agent in prod. of latex and rubber; wh. fine powd.; m.w. 9000; water-sol.; bulk dens. 0.6 kg/l; visc. 1900 cs (99 C); m.p. 65 C; flash pt. > 240 C; pH 6.0–7.5 (1% aq.); 100% conc.

Pluriol P 600. [BASF AG] PPG; nonionic; mold release agent, additive for oils and fluids; lubricant and antifoam for rubber; consistency improver and solubilizer; intermediate in industrial applics.; colorless clear liq.; misc. with water and oil; m.w. 600; dens. 1.0 g/cc; visc. 130 cs; flash pt. 216 C; pour pt. –43 C; pH 6.5–7.5 (1% aq.); 100% conc.

Pluriol P 900. [BASF AG] PPG; nonionic; see Pluriol P 600; also in solv.-type cleaners; colorless clear liq.; m.w. 900; dens. 1.0 g/cc; visc. 180 cs; cloud pt. 33 C (1% aq.); flash pt. 220 C; pour pt. –38 C; pH 6.5–7.5 (1% aq.); 100% conc.

Pluriol P 2000. [BASF AG] PPG; nonionic; see Pluriol P 600; colorless clear liq.; m.w. 2000; dens. 1.0 g/cc; visc. 440 cs; flash pt. 222 C; pour pt. –35 C; pH 6.5–7.5 (1% aq.); 100% conc.

Pluriol P 4000. [BASF AG] PPG; defoamer for tech. applics.; liq.

Pluriol PE 3100. [BASF AG] PO/EO block polymer; nonionic; surfactant; dispersant for colorants; defoamer in various applics.; clear colorless liq.; m.w. 1100; ref. index 1.45; sol. in ethanol, 2-propanol, toluene, tetrachloroethylene, min. spirit; dens. 1.02 g/cc; visc. 175 cps: cloud pt. 40 C (1% aq.); pour pt. –20 C; surf. tens. 45 dynes/cm; pH 7 (5% aq.); 100% conc.

Pluriol PE 6100. [BASF AG] PO/EO block polymer; nonionic; defoamer in household dishwashing machines, bottlewashing plants, metal cleaners, boiler feedwater, acid dyebaths; antifreeze; colorless clear liq.; m.w. 2000; ref. index 1.45; sol. in ethanol, 2-propanol, toluene, tetrachloroethylene, min. spirit; dens. 1.02 g/cc; visc. 350 cps; cloud pt. 24 C (1% aq.); pour pt. –20 C; surf. tens. 44 dynes/cm; pH 7 (5% aq.); 100% conc.

Pluriol PE 6101. [BASF AG] PO/EO block polymer; nonionic; defoamer; rinse aid; clear, slightly dull liq.; m.w. 2000; ref. index 1.45; sol. in 10% HCl, ethanol, 2-propanol, toluene, tetrachloroethylene, min. spirit; dens. 1.02 g/cc; visc. 350 cps; cloud pt. 28 C (1% aq.); pour pt. –10 C; surf. tens. 44 dynes/cm; pH 7 (5% aq.); 100% conc.

Pluriol PE 6400. [BASF AG] PO/EO block polymer; nonionic; detergent for dishwashing machines, dairy cleaners; coolant and lubricant for grinding, drilling and cutting; dispersant, emulsifier; emulsion polymerization; slightly dull liq.; m.w. 3000; ref. index 1.45; sol. in Pluriol PE 6200; dens. 1.05 g/cc; visc. 850 cps; cloud pt. 59 C (1% aq.); pour pt. 16 C; surf. tens. 44 dynes/cm; pH 7 (5% aq.); 100% conc.

Pluriol PE 6800. [BASF AG] PO/EO block polymer; nonionic; powd. detergent for household and industry; dust binder, dispersant, solubilizer for cosmetics, emulsifier, prod. of emulsion polymers; wh. fine powd.; m.w. 8500; ref. index 1.45 (70 C); sol. in water, 10% HCl, ethanol, toluene; dens. 1.06 g/cc (70 C); cloud pt. > 100 (1% aq.); surf. tens. 48 dynes/cm; pH 7 (5% aq.); 100% conc.

Pluriol PE 8100. [BASF AG] PO/EO polymer; nonionic; defoamer; surfactant for cleaning processes; breweries and dairies; colorless to clear-slightly dull liq.; m.w. 2600; ref. index 1.45 (70 C); sol. in 2-propanol, toluene, tetrachloroethylene; dens. 1.04 g/cc; visc. 600 cps; cloud pt. 18 C (1% aq.); pour pt. –30 C; surf. tens. 35 dynes/cm; pH 7 (5% aq.); 100% conc.

Pluriol PE 9400. [BASF AG] PO/EO polymer; nonionic; emulsifier in oil industry and prod. of aq. disps., dispersant; dishwashing machines, bottle washing lines and dairy cleaners; wh. waxy; m.w. 4600; ref. index 1.45 (70 C); sol. see Pluriol PE 6200; surf. tens. 40 dynes/cm; pH 7 (5% aq.); 100% conc.

Pluriol PE 10100. [BASF AG] PO/EO block polymer; nonionic; wetting and antifoam agent; cleaners and rinse aids for dishwashing machines; clear to slightly dull liq.; m.w. 3550; ref. index 1.45; sol. in 10% HCl, ethanol, 2-propanol, toluene, tetrachloroethylene, min. spirit; dens. 1.02 g/cc; visc. 1000 cps; cloud pt. 15 C (1% aq.); pour pt. –30 C; surf. tens. 35 dynes/cm; pH 7 (5% aq.); 100% conc.

Pluriol RPE 2540. [BASF AG] PEG-PPG; nonionic surfactant; defoaming and foam controlled detergents; liq.; dens. 1.0 g/cc; visc. 1500 cps; surf. tens. 37 dynes/cm; pH 9 (5% aq.); 100% conc.

Pluriol RPE 3110. [BASF AG] PEG-PPG; nonionic; defoaming and foam controlled detergents; liq.; 100% conc.

Pluronic® 10R5. [BASF] Meroxapol 105; nonionic; emulsifier, wetting agent, binder, stabilizer, plasticizer, lubricant, solubilizer, dispersant, visc. control agent, defoamer, intermediate for hard surface detergents, rinse aids, automatic dishwashing, textile processing; cosmetics; pharmaceuticals, pulp, paper, and petrol. industries, agric. prods., in iodophors, water treating systems, fermentation, cutting and grinding fluids; liq.; m.w. 1970; ref. index. 1.4587; sol. in water, propylene glycol, xylene, IPA, ethyl acetate, perchloroethylene, IPM; sp.gr. 1.058; visc. 400 cps; HLB 21.0; cloud pt. 69 C (1% aq.); flash pt. > 450 F (COC); pour pt. 15 C; surf. tens. 50.9 dynes/cm; 100% act.

Pluronic® 10R8. [BASF] Meroxapol 108; nonionic; see Pluronic 10R5; flakable solid; m.w. 5000; sol. in water, xylene, ethyl acetate; sp.gr. 1.062 (77 C); m.p. 46 C; HLB 33.0; cloud pt. 99 C (1% aq.); flash pt. > 450 F (COC); surf. tens. 54.1 dynes/cm (0.1%); 100% act.

Pluronic® 12R3. [BASF] PO/EO block copolymer; nonionic surfactant for use as defoamer, emulsifier, demulsifier, solubilizer, detergent, dispersant, binder, stabilizer, gelling agent, wetting agent, rinse aid, chemical intermediate in cosmetic, drug, textile, paper, petrol., paint, detergent, and metal cleaning industries; liq.; m.w. 1800; visc. 340 cps; HLB 5.0; pour pt. –20 C; cloud pt. 53 C (1% aq.); surf. tens. 43 dynes/cm (0.1%); 100% act.

Pluronic® 17R1. [BASF] Meroxapol 171; nonionic; see Pluronic 10R5; also for foam control in paper sizing operations; liq.; m.w. 1950; ref. index 1.4516; sol. in xylene, IPA, ethyl acetate, min. oil, perchloroethylene, IPM, trichlorotrifluoroethane; sp.gr. 1.018; visc. 300 cps; HLB 4.0; cloud pt. 32 C (1% aq.); flash pt. > 450 F (COC); pour pt. –27 C; surf. tens. 33.0 dynes/cm (0.1%); 100% act.

Pluronic® 17R2. [BASF] Meroxapol 172; nonionic; see Pluronic 10R5; liq.; m.w. 2100; ref. index 1.4535; sol. in water, xylene, IPA; ethyl acetate, perchloroethylene, IPM, trichlorotrifluoroethane; sp.gr. 1.030; visc. 350 cps; HLB 8.0; cloud pt. 39 C (1% aq.); flash pt. > 450 F (COC); pour pt. –25 C; surf. tens. 41.9 dynes/cm (0.1%); 100% act.

Pluronic® 17R4. [BASF] Meroxapol 174; nonionic; see Pluronic 10R5; liq.; m.w. 2700; ref. index. 1.4572; sol. in water, propylene glycol, xylene, IPA, ethyl acetate, perchloroethylene, IPM, trichlorotrifluoroethane; sp.gr. 1.048; visc. 560 cps; HLB 16.0; cloud pt. 47 C (1% aq.); flash pt. > 450 F (COC); pour pt. 18 C; surf. tens. 44.1 dynes/cm (0.1%); 100% act.

Pluronic® 17R8. [BASF] Meroxapol 178; see Pluronic 10R5; also dry toilet bowl cleaners, dye levelers, solubilizer of drugs, stick type cosmetics; soap bars, dispersant in deinking operations; flakable solid; m.w. 7500; sol. in water, ethyl acetate, IPM; sp.gr. 1.064 (77 C); m.p. 53 C; HLB 32.0; cloud pt. 81 C (1% aq.); flash pt. > 450 F (COC); surf. tens. 47.3 dynes/cm (0.1%); 100% act.

Pluronic® 22R4. [BASF] PO/EO block copolymer; see Pluronic 12R3; liq.; m.w. 3350; visc. 950 cps; HLB 6.3; pour pt. 24 C; cloud pt. 40 C (1% aq.); surf. tens. 43 dynes/cm (0.1%); 100% act.

Pluronic® 25R1. [BASF] Meroxapol 251; nonionic; see Pluronic 10R5; also foam control in paper sizing operations and antifreeze; liq.; m.w. 2800; ref. index 1.4521; sol. in xylene, IPA, ethyl acetate, perchloroethylene, IPM, trichlorotrifluoroethane; sp.gr. 1.017; visc. 460 cps; HLB 2.3; cloud pt. 28 C (1% aq.); flash pt. > 450 F (COC); pour pt. –27 C; surf. tens. 36.3 dynes/cm (0.1%); 100% act.

Pluronic® 25R2. [BASF] Meroxapol 252; nonionic; see Pluronic 10R5; also wetting and rinse aid; lubricant and leveling agent for paper coating; liq.; m.w. 3120; ref. index 1.4541; sol. see Pluronic 25R1; sp.gr. 1.039; visc. 680 cps; HLB 6.3; cloud pt. 33 C (1% aq.); flash pt. > 450 F (COC); pour pt. –5 C; surf. tens. 37.5 dynes/cm (0.1%); 100% act.

Pluronic® 25R4. [BASF] Meroxapol 254; nonionic; see Pluronic 10R5; liq.; m.w. 3800; ref. index 1.4574; sol. in water, propylene glycol, ethyl acetate, perchloroethylene, IPM, trichlorotrifluoroethane; sp.gr. 1.046; visc. 1110 cps; HLB 14.3; cloud pt. 40 C (1% aq.); flash pt. > 450 F (COC); pour pt. 25 C; surf. tens. 40.9 dynes/cm (0.1%); 100% act.

Pluronic® 25R5. [BASF] Meroxapol 255; nonionic; see Pluronic 10R5; also thickener for cosmetic pastes and creams; paste; m.w. 4500; sol. see Pluronic 25R4; sp.gr. 1.036 (60 C); m.p. 33 C; HLB 18.3; cloud pt. 44 C (1% aq.); flash pt. > 450 F (COC); surf. tens. 43.5 dynes/cm (0.1%); 100% act.

Pluronic® 25R8. [BASF] Meroxapol 258; nonionic; see Pluronic 10R5; also dry toilet bowl cleaners, dye levelers for fabrics; solubilizer for drugs; thickener for cosmetics; deinking operations; felt washing operations; flakable solid; m.w. 9000; sol. in water; sp.gr. 1.062; m.p. 56 C; HLB 30.3; cloud pt. 80 C (1% aq.); flash pt. > 450 F (COC); surf. tens. 46.1 dynes/cm (0.1%); 100% act.

Pluronic® 31R1. [BASF] Meroxapol 311; nonionic; see Pluronic 10R5; also floating bath oils; foam control in antifreeze; liq.; m.w. 3200; ref. index 1.4522; sol. Pluronic 25R1; sp.gr. 1.018; visc. 578 cps; HLB 1.7; cloud pt. 25 C (1% aq.); flash pt. > 450 F (COC); pour pt. –25 C; surf. tens. 34.1 dynes/cm (0.1%); 100% act.

Pluronic® 31R2. [BASF] Meroxapol 312; nonionic; see Pluronic 10R5; paper coating color additive; deinking and felt washing operations; liq.; m.w. 3400; ref. index 1.4542; sol. see Pluronic 25R4; sp.gr. 1.030; visc. 818 cps; HLB 5.7; cloud pt. 30 C (1% aq.); flash pt. > 450 F (COC); pour pt. 9 C; surf. tens. 38.9 dynes/cm (0.1%); 100% act.

Pluronic® 31R4. [BASF] Meroxapol 314; nonionic; see Pluronic 10R5; paste; m.w. 4300; sol. see Pluronic 25R1; sp.gr. 1.028 (60 C); m.p. 26 C; HLB 13.7; cloud pt. 31 C (1% aq.); flash pt. > 450 F (COC); pour pt. 26 C; surf. tens. 41.2 dynes/cm (0.1%); 100% act.

Pluronic® F38. [BASF] Poloxamer 108; nonionic; wetting agent, emulsifier, demulsifier, foam and visc. control agent; dispersant, antistat and gelling agent; agric. chemicals; cosmetics and personal care prods.; pharmaceuticals; metal cleaning; pulp and paper industry; rinse aids, textile processing; dyeing assistant and leveler; lubricant; scouring aid; water treatment; prilled; m.w. 5000; sol. in ethanol, propylene glycol, water, toluene, dens. 8.9 lb/gal (77 C); sp.gr. 1.07 (77 C); visc. 260 cps (77 C); m.p. 48 C; HLB 30.5; cloud pt. > 100 C (1% aq.); flash pt. 505 F (COC); surf. tens. 52.2 dynes/cm (0.1% conc.); 100% act.

Pluronic® F68. [BASF] Poloxamer 188; nonionic; see Pluronic F38; prilled; m.w. 8350; sol. in ethanol, water; dens. 8.8 lb/gal (77 C); sp.gr. 1.06 (77 C); visc. 1000 cps (77 C); m.p. 52 C; HLB 29.0; cloud pt. > 100 C (1% aq.); flash pt. 500 F (COC); surf. tens. 50.3 dynes/cm (0.1%); 100% act.

Pluronic® F68LF. [BASF] Poloxamer 108; see Pluronic F38; flake; m.w. 7700; sol. in toluene; sp.gr. 1.06 (77 C); dens. 8.7 lb/gal (77 C); visc. 850 cps (77 C); m.p. 50 C; cloud pt. 32 C (1% aq.).

Pluronic® F77. [BASF] Poloxamer 217; nonionic; see Pluronic F38; prilled; m.w. 6600; sol. in ethanol, water, toluene; dens. 8.7 lb/gal (77 C); sp.gr. 1.04 (77 C); m.p. 48 C; HLB 24.5; cloud pt. > 100 C (1% aq.); flash pt. 485 F (COC); surf. tens. 47.0 dynes/cm (0.1%); 100% act.

Pluronic® F87. [BASF] Poloxamer 237; nonionic; see Pluronic F38; prilled; m.w. 7700; sol. see Pluronic F77; dens. 8.7 lb/gal (77 C); sp.gr. 1.04 (77 C); visc. 700 cps. (77 C); m.p. 49 C; HLB 24.0; cloud pt. > 100 C (1% aq.); flash pt. 472 F (COC); surf. tens. 44.0 dynes/cm (0.1%); 100% act.

Pluronic® F88. [BASF] Poloxamer 238; nonionic; see Pluronic F38; prilled; m.w. 10,800; sol. in ethanol and water; dens. 8.8 lb/gal (77 C); sp.gr. 1.06 (77 C); visc. 2300 cps (77 C); m.p. 54 C; HLB 28.0; cloud pt. > 100 C (1% aq.); surf. tens. 48.5 dynes/cm (0.1%); 100% act.

Pluronic® F98. [BASF] Poloxamer 288; nonionic; see Pluronic F38; prilled; m.w. 13,000; sol. in ethanol, water, perchoroethylene; dens. 8.8 lb/gal (77 C); sp.gr. 1.06 (77 C); visc. 2700 cps (77 C); m.p. 55 C; HLB 27.5; cloud pt. > 100 C (1% aq.); flash pt. 491 F (COC); surf. tens. 43.0 dynes/cm (0.1%); 100% act.

Pluronic® F108. [BASF] Poloxamer 338; nonionic; see Pluronic F38; prilled; m.w. 14,000; sol. in ethanol, water; dens. 8.8 lb/gal (77 C); sp.gr. 1.06 (77 C); visc. 8000 cps (77 C); m.p. 57 C; HLB 27.0; cloud pt. > 100 C (1% aq.); flash pt. 495 F (COC); surf. tens. 41.2 dynes/cm (0.1%); 100% act.

Pluronic® F127. [BASF] Poloxamer 407; nonionic; see Pluronic F38; prilled; m.w. 12,500; sol. in ethanol, water, toluene, perchloroethylene; dens. 8.8 lb/gal (77 C); sp.gr. 1.05 (77 C); visc. 3100 cps (77 C); m.p. 56 C; HLB 22.0; cloud pt. > 100 C (1% aq.); surf. tens. 40.6 dynes/cm (0.1%); 100% act.

Pluronic® L10. [BASF] PO/EO block copolymer; nonionic; see Pluronic 12R3; liq.; m.w. 3200; visc. 660 cps; HLB 14.0; pour pt. –5 C; cloud pt. 32 C (1% aq.); surf. tens. 41 dynes/cm (0.1%); 100% act.

Pluronic® L31. [BASF] Poloxamer 101; nonionic; see Pluronic F38; liq.; m.w. 1100; ref. index 1.4515; sol. in ethanol, propylene glycol, water, toluene, xylene, perchloroethylene; dens. 8.5 lb/gal; sp.gr. 1.02; visc. 165 cps; HLB 4.5; cloud pt. 37 C (1% aq.); flash pt. 439 F (COC); pour pt. –32 C; surf. tens. 46.9 dynes/cm (0.1%); 100% act.

Pluronic® L35. [BASF] Poloxamer 105; nonionic; see Pluronic F38; liq.; m.w. 1900; sol. see Pluronic

L31; dens. 8.8 lb/gal; sp.gr. 1.06; visc. 340 cps; HLB 18.5; cloud pt. 77 C (1% aq.); pour pt. 7 C; surf. tens. 48.8 dynes/cm (0.1%); 100% act.

Pluronic® L42. [BASF] Poloxamer 122; nonionic; see Pluronic F38; liq.; m.w. 1630; ref. index 1.4541; sol. see Pluronic L31; dens. 8.6 lb/gal; sp.gr. 1.03; visc. 250 cps; HLB 8.0; cloud pt. 37 C (1% aq.); flash pt. 450 F (COC); pour pt. –26 C; surf. tens. 46.5 dynes/cm (0.1%); 100% act.

Pluronic® L43. [BASF] Poloxamer 123; nonionic; see Pluronic F38; liq.; m.w. 1850; ref. index 1.4563; sol. see Pluronic L31; dens. 8.7 lb/gal; sp.gr. 1.04; visc. 310 cps; HLB 12.0; cloud pt. 42 C (1% aq.); pour pt. –1 C; surf. tens. 47.3 dynes/cm (0.1%); 100% act.

Pluronic® L44. [BASF] Poloxamer 124; nonionic; see Pluronic F38; liq.; m.w. 2200; ref. index 1.4580; sol. see Pluronic F31; dens. 8.8 lb/gal; sp.gr. 1.05; visc. 440 cps; HLB 16.0; cloud pt. 65 C (1% aq.); flash pt. 464 F (COC); pour pt. 16 C; surf. tens. 45.3 dynes/cm (0.1%); 100% act.

Pluronic® L61. [BASF] Poloxamer 181; nonionic; see Pluronic F38; liq.; m.w. 2000; ref. index 1.4520; sol. in ethanol, toluene, xylene, perchloroethylene; dens. 8.4 lb/gal; sp.gr. 1.01; visc. 285 cps; HLB 3.0; cloud pt. 24 C (1% aq.); flash pt. 455 F (COC); pour pt. –29 C; 100% act.

Pluronic® L62. [BASF] Poloxamer 182; nonionic; see Pluronic F38; liq.; m.w. 2500; sol. see Pluronic L31; dens. 8.6 lb/gal; sp.gr. 1.03; visc. 400 cps; HLB 7.0; cloud pt. 32 C (1% aq.); flash pt. 466 F (COC); pour pt. –4 C; surf. tens. 42.8 dynes/cm (0.1%); 100% act.

Pluronic® L62D. [BASF] Poloxamer 108; see Pluronic F38; liq.; m.w. 2350; sol. see Pluronic L31; sp.gr. 1.04; dens. 8.7 lb/gal; visc. 385 cps; pour pt. –1 C; ref. index 1.4557; cloud pt. 35 C (1% aq.).

Pluronic® L62LF. [BASF] Poloxamer 108; see Pluronic F38; liq.; m.w. 2450; sol. see Pluronic L31; sp.gr. 1.03; dens. 8.6 lb/gal; visc. 400 cps; pour pt. –10 C; ref. index 1.4546; cloud pt. 28 C (1% aq.); 100% act.

Pluronic® L63. [BASF] Poloxamer 183; nonionic; see Pluronic F38; liq.; m.w. 2650; ref. index 1.4562; sol. see Pluronic L31; dens. 8.7 lb/gal; sp.gr. 1.04; visc. 490 cps; HLB 11.0; cloud pt. 34 C (1% aq.); pour pt. 10 C; surf. tens. 43.3 dynes/cm (0.1%); 100% act.

Pluronic® L64. [BASF] Poloxamer 184; nonionic; see Pluronic F38; liq.; m.w. 2900; ref. index 1.4575; sol. see Pluronic L31; dens. 8.8 lb/gal; sp.gr. 1.05; visc. 550 cps; HLB 15.0; cloud pt. 61 C (1% aq.); flash pt. 485 F (COC); pour pt. 16 C; surf. tens. 43.2 dynes/cm (0.1%); 100% act.

Pluronic® L72. [BASF] Poloxamer 212; nonionic; see Pluronic F38; liq.; m.w. 2750; ref. index 1.4542; sol. in ethanol, water, toluene, xylene, perchloroethylene; dens. 8.6 lb/gal; sp.gr. 1.03; visc. 510 cps; HLB 6.5; cloud pt. 25 C (1% aq.); flash pt. 442 F (COC); surf. tens. 39.0 dynes/cm (0.1%); 100% act.

Pluronic® L81. [BASF] Poloxamer 231; nonionic; see Pluronic F38; liq.; m.w. 2750; ref. index 1.4526; sol. in ethanol, toluene, xylene, perchloroethylene; dens. 8.5 lb/gal; sp.gr. 1.02; visc. 475 cps; HLB 2.0; cloud pt. 20 C (1% aq.); pour pt. –37 C; 100% act.

Pluronic® L92. [BASF] Poloxamer 282; nonionic; see Pluronic F38; liq.; m.w. 3650; ref. index 1.4547; sol. see Pluronic L81; dens. 8.6 lb/gal; sp.gr. 1.03; visc. 700 cps; HLB 5.5; cloud pt. 26 C (1% aq.); flash pt. 445 F (COC); pour pt. 7 C; surf. tens. 35.9 dynes/cm (0.1%); 100% act.

Pluronic® L101. [BASF] Poloxamer 331; nonionic; see Pluronic F38; liq.; m.w. 3800; ref. index 1.4524; sol. see Pluronic L81; dens. 8.5 lb/gal; sp.gr. 1.02; visc. 800 cps; HLB 1.0; cloud pt. 15 C (1% aq.); pour pt. –23 C; 100% act.

Pluronic® L121. [BASF] Poloxamer 401; nonionic; see Pluronic F38; liq.; m.w. 4400; ref. index 1.4527; sol. see Pluronic L81; dens. 8.4 lb/gal; sp.gr. 1.01; visc. 1200 cps; HLB 5.0; cloud pt. 14 C (1% aq.); pour pt. 5 C; surf. tens. 33.0 dynes/cm (0.1%); 100% act.

Pluronic® L122. [BASF] Poloxamer 402; nonionic; see Pluronic F38; liq.; m.w. 5000; ref. index 1.4558; sol. see Pluronic L81; dens. 8.6 lb/gal; sp.gr. 1.03; visc. 1750 cps; HLB 4.0; cloud pt. 19 C (1% aq.); flash pt. 490 F (COC); pour pt. 20 C; surf. tens. 33.0 dynes/cm (0.1%); 100% act.

Pluronic® P65. [BASF] Poloxamer 185; nonionic; see Pluronic F38; paste; sol. > 10 g/100 ml in 95% ethanol, propylene glycol, water, toluene, xylene, perchloroethylene; m.w. 3400; dens. 8.8 lb/gal (60 C); sp.gr. 1.06 (60 C); visc. 180 cps (60 C); m.p. 30 C; HLB 17.0; cloud pt. 82 C (1% aq.); pour pt. 27 C; surf. tens. 46.3 dynes/cm (0.1%); 100% act.

Pluronic® P75. [BASF] Poloxamer 215; nonionic; see Pluronic F38; paste; m.w. 4150; sol. see Pluronic F31; dens. 8.8 lb/gal (60 C); sp.gr. 1.06 (60 C); visc. 250 cps (60 C); m.p. 34 C; HLB 16.5; cloud pt. 82 C (1% aq.); pour pt. 27 C; surf. tens. 42.8 dynes/cm (0.1%); 100% act.

Pluronic® P84. [BASF] Poloxamer 234; nonionic; see Pluronic F38; paste; m.w. 4200; sol. see Pluronic F31; dens. 8.6 lb/gal (60 C); sp.gr. 1.03 (60 C); visc. 265 cps (60 C); m.p. 34 C; HLB 14.0; cloud pt. 74 C (1% aq.); flash pt. 442 F (COC); pour pt. 18 C; surf. tens. 42.0 dynes/cm (0.1%); 100% act.

Pluronic® P85. [BASF] Poloxamer 235; nonionic; see Pluronic F38; paste; m.w. 4600; sol. see Pluronic F31; dens. 8.7 lb/gal (60 C); sp.gr. 1.04 (60 C); visc. 310 cps (60 C); m.p. 40 C; HLB 16.0; cloud pt. 85 C (1% aq.); pour pt. 29 C; surf. tens. 42.5 dynes/cm (0.1%); 100% act.

Pluronic® P103. [BASF] Poloxamer 333; nonionic; see Pluronic F38; paste; m.w. 4950; sol. see Pluronic L81; dens. 8.7 lb/gal (60 C); sp.gr. 1.04 (60 C); visc. 285 cps (60 C); m.p. 30 C; HLB 9.0; cloud pt. 86 C (1% aq.); pour pt. 21 C; surf. tens. 34.4 dynes/cm (0.1%); 100% act.

Pluronic® P104. [BASF] Poloxamer 334; nonionic; see Pluronic F38; paste; m.w. 5850; sol. see Pluronic L81; dens. 8.7 lb/gal (60 C); sp.gr. 1.04 (60 C); visc. 550 cps (60 C); m.p. 37.5; HLB 13.0; cloud pt. 81 C (1% aq.); flash pt. 448 F (COC); pour pt. 32 C; surf. tens. 33.1 dynes/cm (0.1%); 100% act.

Pluronic® P105. [BASF] Poloxamer 335; nonionic; see Pluronic F38; paste; m.w. 6500; sol. see Pluronic

L81; dens. 8.8 lb/gal (60 C); sp.gr. 1.05 (60 C); visc. 800 cps (60 C); m.p. 42 C; HLB 15.0; cloud pt. 91 C (1% aq.); pour pt. 35 C; surf. tens. 39.1 dynes/cm (0.1%); 100% act.

Pluronic® P123. [BASF] Poloxamer 403; nonionic; see Pluronic F38; paste; m.w. 5750; see Pluronic L81; dens. 8.5 lb/gal (60 C); sp.gr. 1.02 (60 C); visc. 350 cps (60 C); m.p. 31 C; HLB 8.0; cloud pt. 90 C (1% aq.); pour pt. 31 C; surf. tens. 34.1 dynes/cm (0.1%); 100% act.

PMF® Fiber 204. [Jim Walter] Calcium-alumino silicate; filler-reinforcement; used as a partial replacement for milled and chopped glass fibers in plastics such as nylon, PP, and phenolics; used in combination with particulate fillers, provides friction and wear properties in nonasbestos brake and clutch materials; also used in nonasbestos industrial papers and felts; off-wh. short fibers; 200 μ avg. fiber length; odorless; nil sol. in water; sp.gr. 2.7; 99% min. wool fiber.

PMF® Fiber 204AX. [Jim Walter] Calcium-alumino silicate surface-treated with an organosilane; filler-reinforcement; treated to provide improved wetting and fiber–polymer bonding in a variety of polymers incl. phenolic, epoxy, melamine, nylon, polyimide, PBT, PVC, PC, and urethane.

PMF® Fiber 204BX. [Jim Walter] Calcium-alumino silicate surface treated with an organosilane; see PMF Fiber 204AX.

PMF® Fiber 204CX. [Jim Walter] Calcium-alumino silicate surface treated with a low level of a fatty acid-type material; filler-reinforcement; treated for improved fiber disp. during dry or low-solvent blending.

PMF® Fiber 204EX. [Jim Walter] Calcium-alumino silicate surface treated with an organosilane; filler-reinforcement; treated to provide improved wetting and fiber–polymer bonding in DAP, polybutadiene, urethane, thermoset polyester, and polyolefin materials.

PMP Gluconic Acid 50% Tech. [PMP] D-gluconic acid 50%; sequestrant for iron, calcium, and aluminum; food sanitation and general cleaning fields; lt. yel. sol'n.; sp.gr. 1.23; dens. 10.25 lb/gal; chel. value (calcium) 9.0 g/100 g; pH 1.2; 50% aq.

PMP Liquid Gluconate 60 Tech. [PMP] Buffered sodium acid gluconate; sequestrant; lt. liq.; f.p. 15 F; dens. 11.01 lb/gal; visc. 32 cps; chel. value (calcium) 9.7 g/100 g; pH 3.1–3.3.

PMP Sodium Gluconate Tech. [PMP] Sodium gluconate; sequestrant used in food sanitation and general cleaning fields; wh. powd., gran.; odorless; water-sol.; dens. 56.0 lb/ft^3; chel. value (calcium) > 15.5 g/100 g; pH 6.8–7.2 (10%); 99.5% min. solids.

PO-64P. [Great Lakes] Poly-dibromophenylene oxide; flame retardant which melts into most polymers to optimize physical properites; esp. for cryst. polymers (polyesters, polyamides); enhances flow into thin wall sections; permits higher regrind loading levels; off-wh. powd.; m.w. 20,000; b.p. 750–930 F; sol. 40 g/100 g chloroform; insol. in water; sp.gr. 2.07; soften. pt. 200–230 C; 64% bromine.

PO114. [Hüls] Phenolethylsiloxane; compatibilizer for silicones and hydrocarbons; liq.

Pogol 200. [Hart Chem. Ltd.] PEG; nonionic; solubilizer, antistat, softener, humectant, fiber and metal lubricant, plasticizer, tablet binder; for pharmaceuticals, cosmetics; clear liq.; dens. 1.14; 100% act.

Pogol 300. [Hart Chem. Ltd.] PEG; nonionic; solubilizer, dispersant, emulsifier; herbicides and pesticides; clear liq.; dens. 1.13; 100% act.

Pogol 400. [Hart Chem. Ltd.] PEG; nonionic; emulsifier, solubilizer, dispersant; see Pogol 300; clear liq.; dens. 1.12; visc. 92 cps; 100% act.

Pogol 400 USP. [Hart Chem. Ltd.] PEG, USP grade; nonionic; solubilizer, lubricant; mold release agent and lubricant for both natural and syn. prods.; clear liq.; dens. 1.12; visc; 92 cps; 100% act.

Pogol 600. [Hart Chem. Ltd.] PEG; nonionic; solubilizer, lubricant; used in paper coating mixes; antisticking agents and oil resistors; clear paste; ref. index 1.4699; dens. 1.13; visc. 130 cps; 100% act.

Pogol 1540. [Hart Chem. Ltd.] PEG; nonionic; solubilizer; base for mfg. of fatty acid esters, emulsifiers, and dispersants; wh. solid; 100% act.

Pogol 1570. [Hart Chem. Ltd.] PEG; nonionic; antistat, softener, humectant, fiber and metal lubricant, plasticizer; for formulating gelatin capsules, cosmetics, lotions; base for mfg. of fatty acid esters, emulsifiers, dispersants; liq.; 70% act.

Poiz 530. [Kao] Salt of polymeric carboxylic acid; dispersant for inorg. pigments and fillers; liq.

Polargel. [Am. Colloid] Purified wh. bentonite NF; thickener and suspending agent for cosmetics and pharmaceuticals; wh. fine powd.; 90% finer than 200 mesh (dry); visc. 40–200 cps (5%); pH 9.0–10.0 (5% disp.); brightness (GE) 83–87.

Polargel T. [Am. Colloid] Bentonite tech.; thickener and suspending agent for household and industrial specialties, e.g., paints and cleaning compds.; wh. fine powd.; visc. 200 min cps.

Polarite KB 325. [Am. Colloid] Sodium bentonite; suspending agent and gellant for household and industrial specialties; grd. wh. free-flowing powd.; 75% finer than 200 mesh (dry); visc. 10—50 cps (5% disp.); brightness (GE) 80–84; 85% min. montmorillonite content.

Polawax®. [Croda] Emulsifying wax NF; nonionic; emulsifier, thickener, opacifier, suspending agent; stabilizer for o/w emulsions; powd.; sol. in alcohol; water-disp.

Polawax® A31. [Croda] Emulsifying wax NF; nonionic; used in quick-breaking foams; solid; sol. in alcohol and aerosol propellant; solid; 97% conc.

Polawax® GP200. [Croda] Self-emulsifying wax; nonionic; emulsifier for w/o emulsions; used in cosmetics and pharmaceuticals; solid; 97% conc.

Polawax® NF. [Croda] Self-emulsifying wax; nonionic; solid; 97% conc.

Polectron® 430. [GAF] Styrene/PVP copolymer; binder and adhesive for wood, cotton, paper, glass fiber, flour, concrete; stabilizer and opacifier; laundry processing; stabilizer for detergents; surf., textile, and paper coatings, latex rug backings, floor wax emulsions, and cosmetics; 40% aq. emulsion.

Polefine 51ON. [Takemoto] Alkyl naphthalene sulfonate/formaldehyde condensates; anionic; reducing agent for concrete; liq.

Polidene 33-001. [Scott Bader] Vinylidene chloride copolymer; nonionic/anionic; paper coating; binder or coating for bonding of glass nonwoven; aq. emulsion; particle size 0.25 μ; sp.gr. 1.27; visc. 0.05–0.30 poise; pH 3.0–4.5; 55 ± 1% solids.

Polidene 33-004. [Scott Bader] Vinylidene chloride copolymer; nonionic/anionic; fire retardant coatings, adhesives, and fiber impregnants; pigment binder; aq. emulsion; particle size 0.25 μ; sp.gr. 1.22; visc. 0.05–0.30 poise; pH 3.0–4.5; 50 ± 1% solids.

Polidene 33-021. [Scott Bader] Vinylidene chloride copolymer; anionic; binder for fire-retardant air filters; emulsion; particle size 0.25 μ; sp.gr. 1.22; visc. 0.05–0.30 poise; pH 3.0–4.5; 45 ± 1% solids.

Polidene 33-031. [Scott Bader] Vinylidene chloride copolymer plasticized with a phosphate plasticizer; used in textile fibers; binder for fire-retardant fabrics; impregnant and adhesive for papers; emulsion; particle size 0.3 μ; sp.gr. 1.10; visc. 0.05–0.50 poise; pH 7.0–8.0; 50 ± 1% solids.

Polidene 33-075. [Scott Bader] Vinylidene chloride copolymer; base emulsion and binder for mfg. of paint; specialty surf. coatings; emulsion; particle size 0.25 μ; sp.gr. 1.17; visc. 0.25–1.0 poise; pH 3.0–4.5; 55 ± 1% solids.

Poligen PE. [BASF AG] Polyethylene aq. disp.; drybright floor polish emulsions, release coats; milky liq.; 40% conc.

Polyaldo® DGDO. [Lonza] Polyglyceryl-10 decaoleate; nonionic; w/o and o/w emulsifier, emollient, and lubricant for cosmetics, toiletries, pharmaceuticals, and household speciality prods.; amber clear liq.; sol. in ethanol, min. and veg. oils, insol. in water; HLB 3 ± 1; sapon. no. 155–185.

Polyaldo® DGHO. [Lonza] Polyglyceryl-10 hexaoleate, nonionic; see Polyaldo DGDO; amber clear liq.; sol. see Polyaldo DGDO; HLB 5 ± 1; sapon. no. 130–160.

Polyaldo® HGDS. [Lonza] Polyglyceryl-6 distearate; nonionic; see Polyaldo DGDO; cream beads; sol. in ethanol; disp. in water; HLB 7 ± 1; m.p. 53–57; sapon. no. 120–140.

Polyaldo® TGMS. [Lonza] Polyglyceryl-3 stearate; nonionic; see Polyaldo DGDO; cream beads; sol. see Polyaldo DGDO; HLB 7 ± 1; m.p. 54–58 C; sapon. no. 120–140.

Polycat® 610/50. [Air Prods.] Strong acid salt of Polycat DBU; latent activity catalyst for PU coatings.

Polycat® DBU. [Air Prods.] Tert. amine; catalyst for PU coatings; provides R.T. cure.

Polycat® SA-1. [Air Prods.] Weak acid salt of Polycat DBU; latent activity catalyst for PU coatings.

Polycat® SA-102. [Air Prods.] Moderate acid salt of Polycat DBU; latent activity catalyst for PU coatings.

Polychol 15. [Croda] Laneth-15; nononic; emulsifier, dispersant, solubilizer, emollient and gelling agent for cosmetics and pharmaceuticals; solid; water-sol.; HLB 12.7; acid no. 5 max.; pH 3.5–5.5 (10% aq.); 100% conc.

Polychol 20-40. [Croda] POE lanolin alcohols; dispersant for cosmetic hydrophilic systems; wax; water-sol.

Polyclar® 10. [GAF] Polyvinylpolypyrrolidone; insol. cross-linked polymer; stabilizer for beverage clarification and stabilization; adsorbent in thin-layer and column chromatography; fine powd.

Polyclar® AT. [GAF] Polyvinylpolypyrrolidone polymer; insol., high m.w. polymer for beverage clarification and stabilization; used in cellaring of beer for colloidal and flavor stability; cellaring of wines to prevent browning; used in stabilization of vinegar and fruit juices; adsorbent in thin-layer and column chromatography; powd.

Polyco 2140. [Borden] Carboxylated vinyl acetate homopolymer latex; paper coating emulsion; pigment binder for paper and paperboard coating; fiberglass sizing; emulsion; particle size 0.18 μ; sp.gr. 1.09; visc. 40 cps; surf. tens. 40 dynes/cm; pH 7.0; 47% solids.

Polyco 2142. [Borden] Carboxylated vinyl acetate homopolymer latex; see Polyco 2140; emulsion; particle size 0.15 μ; sp.gr. 1.09; visc. 70 cps; surf. tens. 40 dynes/cm; pH 7.0; 50% solids.

Polyco 2149-C. [Borden] Carboxylated vinyl acetate homopolymer; see Polyco 2140; emulsion; 0.15 μ particle size; sp.gr. 1.08; visc. 60 cps; surf. tens. 39 dynes/cm; pH 7.0; 47.5% solids.

Polyco 2611. [Borden] Vinyl chloride copolymer; nonfilm forming grade at R.T.; used as stiffening and antiblock additive for flexible PVC and other latices; as base for formulating with fire retardant plasticizers; high chlorine content; emulsion; 0.20 μ particle size; sp.gr. 1.18; visc. 30 cps; surf. tens. 40 dynes/cm; pH 9.5; 56% solids.

Polyco 2612. [Borden] Vinyl chloride copolymer; see Polyco 2611; also used in release coatings; emulsion; 0.20 μ particle size; sp.gr. 1.18; visc. 30 cps; surf. tens. 43 dynes/cm; pH 9.5; 56% solids.

Polyco 2617. [Borden] Vinyl chloride copolymer plasticized with phosphate ester; nonwoven binder or saturant for fire retardant fibers; textile finishing operations; emulsion; particle size 0.23 μ; sp.gr. 1.13; visc. 30 cps; surf. tens. 33 dynes/cm; pH 9.5; 56% solids.

Polyco 2618. [Borden] Vinyl chloride copolymer plasticized with DOP; nonwoven binder; textile and paper coating applic.; emulsion; particle size 0.25 μ; sp.gr. 1.13; visc. 30 cps; surf. tens. 33 dynes/cm; pH 9.5; 56% solids.

Polyco 2629. [Borden] Vinyl chloride copolymer plasticized with DOP; see Polyco 2618; emulsion; particle size 0.25 μ; sp.gr. 1.13; visc. 30 cps; surf. tens. 33 dynes/cm; pH 9.5; 56% solids.

Polyco 2638. [Borden] Vinyl chloride copolymer plasticized with DOP; nonwoven saturant; binder; wallpaper coating; emulsion; particle size 0.30 μ; sp.gr. 1.10; visc. 30 cps; surf. tens. 33 dynes/cm; pH 9.5; 58% solids.

Poly-Cone 1000. [Olin] Modified silicone emulsion; mold release agent for rubber; liq.

Polydyol. [Eastern Color] Dye carrier for polyester dyeings; liq.; forms stable emulsions in water.

Polyfac MT-610. [Stepan] PEG-10 monotallate; coemulsifier and emulsifier-solubilizer used in fluids, oils and greases, fiber lubricant, textile applics., preparation of cutting fluids; amber liq.; sol. in IPA, xylene; disp. in water; dens. 8.7 lb/gal; visc. 300 cps; HLB 11.8; pour pt. 41 F; cloud pt. 77 F; sapon. no. 70–80.

Polyfac MT-615. [Stepan] PEG-15 monotallate; see Polyfac MT-610; also emulsifier and coemulsifier in softeners, dye carriers, and detergents; scouring agent and dye bath leveling agent; amber liq. to paste; sol. in water, IPA, xylene; dens. 8.8 lb/gal; visc. 350 cps; HLB 13.3; pour pt. 55 F; cloud pt. 127 F; sapon. no. 55–65.

Polyfac PCW-180. [Stepan] Phosphate ester; surfactant used as lubricant additive for oil-based metalworking and hydraulic fluids; amber visc. liq.; sol. in min. oil; insol. in water; dens. 8.8 lb/gal; visc. 3500 cps; pH 2.1.

Polyfac PCW-181, -182. [Stepan] Complex carboxylated fatty amide salt; additive used as corrosion inhibitor and lubricant in metalworking fluids; amber visc. liq.; visc. 2600 and 3100 cps resp.; pH 8.6 and 8.5 (2% aq.) resp.

Polyfac PN-209. [Stepan] Nonylphenol phosphate, ethoxylated; anionic; surfactant for liq. detergents, textile wet processing, lubricant and corrosion inhibitor; yel. liq.; dens. 9.34 lb/gal; visc. 3600 cps; flash pt. > 300 F (PMCC); pH 2 (1% aq.); 99% act.

Polyfac TDO-9. [Stepan] PEG 400 tall oil dioleate; emulsifier and solubilizer for solvs., fats, and min. oil; emulsifier for kerosene in pesticide formulations, metalworking, and industrial lubricants; amber liq.; sol. see Polyfac TDO-5; dens. 8.2 lb/gal; visc. 130 cps; HLB 8.5; pour pt. 36 F; cloud pt. < 77 F; sapon. no. 110–120.

Polyfac TDO-14. [Stepan] PEG 600 tall oil dioleate; component of defoamers, softeners, and lubricants; amber liq. to paste; disp. in water; sol. in water, IPA, xylene; dens. 8.3 lb/gal; visc. 155 cps; HLB 10.5; pour pt. 64 F; cloud pt. < 77; sapon. no. 90–100.

Polyfac TMO-5. [Stepan] PEG-5 tall oil oleate; emulsifier and solubilizer for solvs., min. and fatty oils; lubricant, emulsifier for textile lubricants, metalworking fluids, and pesticide formulations; amber liq.; sol. in IPA, xylene, min. oil; disp. in water; dens. 8.3 lb/gal; visc. 95 cps; HLB 8.3; pour pt. 23 F; cloud pt. < 77 F; sapon. no. 110–120.

Polyfac TMO-7. [Stepan] PEG-7 tall oil oleate; see Polyfac TMO-5; amber liq.; sol. in IPA, xylene; disp. in water; dens. 8.4 lb/gal; visc. 95 cps; HLB 10.3; pour pt. 32 F; cloud pt. < 77 F; sapon. no. 90–110.

Polyfac TMO-14. [Stepan] PEG-14 tall oil oleate; emulsifier, lubricant, and detergent; textile dye leveling and lubricant formulations; amber liq.; sol. in water, IPA, xylene; dens. 8.7 lb/gal; visc. 195 cps; HLB 13.4; pour pt. 61 F; cloud pt. 126 F; sapon. no. 60–70.

Polyfon F. [Westvaco] Sodium lignosulfonate; dispersant for industrial applics. from agric. to ceramics; br. powd.; dens. 26 lb/ft³; pH 11.0 (2% aq.); surf. tens. 50.4 dynes/cm (1% aq.).

Polyfon H. [Westvaco] Sodium lignosulfonate; dispersant in agric. chemicals and dyestuff formulations; brn. powd.; sol. see Polyfon F; dens. 22–24 lb/ft³; pH 10.0 (2% aq.); surf. tens. 55 dynes/cm (1% aq.).

Polyfon O. [Westvaco] Sodium lignosulfonate; see Polyfon F; brn. powd.; sol. see Polyfon F; dens. 24 lb/ft³; pH 10.6 (2% aq.).

Polyfon OD. [Westvaco] Kraft lignin polymer, sodium salt; see Polyfon F; brn. powd.; sol. see Polyfon F; dens. 24 lb/ft³; pH 8.0 (2% aq.); surf. tens. 40.0 dynes/cm (1% aq.).

Polyfon T. [Westvaco] Sodium lignosulfonate; see Polyfon F; brn. powd.; sol. see Polyfon F; dens. 26 lb/ft³; pH 10.5 (2% aq.).

Poly-G® 76-120. [Olin] EO-capped triol; aids in curing urethane systems; applics. incl. elastomers, coatings, sealants, and foams; m.w. ≈ 1400; sp.gr. 1.044; dens. 8.7 lb/gal; visc. 450 cp; pour pt. –42 C; hyd. no. 120; pH 7.0 (in 10/6 IPA/water); flash pt. (COC) 245 C; 70% min. primary hydroxyl content.

Poly-G® 200. [Olin] PEG; chemical intermediate for prod. of surfactants for cleaners, textiles, paper, cosmetics; carrier for pharmaceuticals; also in cosmetics and personal care prods., textiles, rubber mold releases, printing inks and dyes, metalworking fluids, foods, paints, paper, wood prods., adhesives, agric. prods., ceramics, elec. equipment, petrol. prods., photographic prods., resins; APHA 25 max. liq.; m.w. 200; sol. in water, acetone, ethanol, ethyl acetate, toluene; sp.gr. 1.125; dens. 9.38 lb/gal; visc. 4.3 cs (99 C); flash pt. 171 C (COC).

Poly-G® 300. [Olin] PEG; see Poly-G 200; APHA 25 max. liq.; m.w. 300; sol. in water, acetone, ethanol, ethyl acetate, toluene; sp.gr. 1.125; dens. 9.38 lb/gal; visc. 5.8 cs (99 C); f.p. –15 to –8 C; flash pt. 196 C (COC).

Poly-G® 400. [Olin] PEG; see Poly-G 200; APHA 25 max. liq.; m.w. 400; sol. in water, acetone, ethanol, ethyl acetate, toluene; sp.gr. 1.127; dens. 9.4 lb/gal; visc. 7.3 cs (99 C); f.p. 4–10 C; flash pt. 224 C (COC); pour pt. 3–10 C.

Poly-G® 600. [Olin] PEG; see Poly-G 200; APHA 25 max. liq.; m.w. 600; sol. in water, acetone, ethanol, ethyl acetate, toluene; sp.gr. 1.127; dens. 9.4 lb/gal; visc. 10.5 cs (99 C); f.p. 20–25 C; flash pt. 246 C (COC); pour pt. 19–24 C.

Poly-G® 1000. [Olin] PEG; see Poly-G 200; wh. waxy solid; m.w. 1000; somewhat less sol. in water than liq. glycols; sp.gr. 1.104 (50/20 C); dens. 9.20 lb/gal (50/20 C); visc. 17.4 cs (99 C); f.p. 38–41 C; flash pt. 260 C (COC); pour pt. 40 C.

Poly-G® 1500. [Olin] PEG; see Poly-G 200; wh. waxy solid; m.w. 1500; somewhat less sol. in water than liq. glycols; sp.gr. 1.104 (50/20 C); dens. 9.20 lb/gal (50/20 C); visc. 28 cs (99 C); f.p. 43–46 C; flash pt. 266 C (COC); pour pt. 45 C.

Poly-G® 2000. [Olin] PEG; see Poly-G 200; APHA 25 max. liq.; m.w. 2000; somewhat less sol. in water than lower m.w. glycols; sp.gr. 1.113; dens. 9.26 lb/

gal; visc. 11.7 cs (99 C); f.p. –20 C; 60% aq. sol'n.

Poly-G® B1530. [Olin] PEG; see Poly-G 200; wh. waxy solid; m.w. 900; somewhat less sol. in water than liq. glycols; sp.gr. 1.104 (50/20 C); dens. 9.20 lb/gal (50/20 C); visc. 15 cs (99 C); f.p. 38–41 C; flash pt. 254 C (COC); pour pt. 38 C.

Poly-G® WI 285. [Olin] Polyalkylene glycol derivs.; noncorrosive fluids used in metalworking fluids (cutting, drawing, stamping, rolling, etc.); as hydraulic fluids for boundary lubrication and low-temp. applics.; for corrosion and wear protection; chemical intermediates, defoamers, plasticizers, and solvs. for both inks and dyes; liq.; water-insol.; sp.gr. 0.991; dens. 8.25 lb/gal; visc. 285 SUS (38 C); pour pt. –48 C; flash pt. (COC) 218 C.

Poly-G® WI 625. [Olin] Polyalkylene glycol derivs.; see Poly-G WI 285; liq.; water-insol.; sp.gr. 0.996; dens. 8.29 lb/gal; visc. 616 SUS (38 C); pour pt. –39 C; flash pt. (COC) 227 C.

Poly-G® WI 1715. [Olin] Polyalkylene glycol derivs.; see Poly-G WI 285; liq.; water-insol.; sp.gr. 1.003; dens. 8.35 lb/gal; visc. 1715 SUS (38 C); pour pt. –23 C; flash pt. (COC) 221 C.

Poly-G® WS 100. [Olin] Polyalkylene glycol derivs.; noncorrosive fluids used in metalworking fluids (cutting, drawing, stamping, rolling, etc.); hydraulic fluids for boundary lubrication and low-temp. applics.; for corrosion and wear protection; as chemical intermediates, defoamers, plasticizers, and solvs. for both inks and dyes; liq.; water-sol.; sp.gr. 1.016; dens. 8.46 lb/gal; visc. 100 SUS (38 C); pour pt. –65 C; flash pt. (COC) 166 C.

Poly-G® WS 170. [Olin] Polyalkylene glycol derivs.; see Poly-G WS 100; liq.; water-sol.; sp.gr. 1.035; dens. 8.62 lb/gal; visc. 170 SUS (38 C); pour pt. –61 C; flash pt. (COC) 210 C.

Poly-G® WS 260. [Olin] Polyalkylene glycol derivs.; see Poly-G WS 100; liq.; water-sol.; sp.gr. 1.038; dens. 8.64 lb/gal; visc. 260 SUS (38 C); pour pt. –62 C; flash pt. (COC) 227 C.

Poly-G® WS 280X. [Olin] Heat-transfer fluid and lubricant for plastic pipe extrusion and drawing; annealing and curing medium for thermosetting and thermoplastic resins; Gardner 10 max. clear liq.; sol. no clouding @ 50% by vol. in water; sp.gr. 1.042; visc. 53–63 cs (38 C); pour pt. –37 C; ref. index 1.4584; pH 5.5–7.5 (10% aq.); flash pt (COC) 280 C; sp. heat 0.503 cal/g/C.

Poly-G® WS 660. [Olin] Polyalkylene glycol derivs.; see Poly-G WS 100; liq.; water-sol.; sp.gr. 1.052; dens. 8.75 lb/gal; visc. 660 SUS (38 C); pour pt. –42 C; flash pt. (COC) 227 C.

Poly-G® WS 2000. [Olin] Polyalkylene glycol derivs.; see Poly-G WS 100; liq.; water-sol.; sp.gr. 1.058; dens. 8.80 lb/gal; visc. 2000 SUS (38 C); pour pt. –33 C; flash pt. (COC) 243 C.

Poly-G® WS 3520. [Olin] Polyalkylene glycol derivs.; see Poly-G WS 100; liq.; water-sol.; sp.gr. 1.061; dens. 8.83 lb/gal; visc. 3520 SUS (38 C); pour pt. –28 C; flash pt. (COC) 227 C.

Poly-G® WS 5100. [Olin] Polyalkylene glycol derivs.; see Poly-G WS 100; liq.; water-sol.; sp.gr. 1.063; dens. 8.85 lb/gal; visc. 5100 SUS (38 C); pour pt. –31 C; flash pt. (COC) 227 C.

Poly-G® WT 9150. [Olin] Polyalkylene glycol derivs.; see Poly-G WS 100; liq.; water-thickening; sp.gr. 1.089; dens. 9.06 lb/gal; visc. 9150 SUS (38 C); pour pt. –2 C; flash pt. (COC) 207 C.

Poly-G® WT 90,000. [Olin] Polyalkylene glycol derivs.; see Poly-G WS 100; liq.; water-thickening; sp.gr. 1.095; dens. 9.11 lb/gal; visc. 90,000 SUS (38 C); pour pt. 3 C; flash pt. (COC) 252 C.

Polygard®. [Uniroyal] Tri (mixed mono- and dinonylphenyl) phosphite: nondiscoloring and nonstaining stabilizer for SBR polymers; clear lt. amber liq.; sol. in acetone, alcohol, benzene, CCl_4, naphtha, n-hexane, ligroin; insol. in water but can hydrolyze; sp.gr. 0.99; flash pt. 204 C; fire pt. 268 C.

Polyglycol 15-200. [Dow Europe] PO/EO copolymer; antifoaming agent esp. for high-temp. applics.; liq.

Polyglycol 112-2. [Dow Europe] PO/EO copolymer; lubricant, antifoaming agent in saline water, antifreeze sol'ns., working fluid, surfactant for detergent systems; liq.; visc. 800 cst; surf. tens. 40 dynes/cm (0–1%).

Polyglycol P-1200. [Dow Europe] PPG-20; antifoaming agent for pottery and ceramic industries; liq.

Polyglycol P-2000. [Dow Europe] PPG-26; antifoaming agent for latex and emulsion paints; liq.

Polyglycol P-4000. [Dow Europe] PPG; antifoam agent; liq.; 100% conc.

Polylan®. [Amerchol] Oleyl linoleate, lanolin linoleate; nonionic; emollient, dispersant, conditioner, penetrant, lubricant, cosolv., dye solubilizer used in cosmetics and pharmaceuticals; amber visc. oily liq.; sol. in min. and castor oil, absolute ethyl alcohol, IPA, alkylolamides, ethyl acetate, and D.C. Silicone Fluid #555; insol. in water; sp.gr. 0.890–0.915; visc. 100–120 cps; HLB 8.0; acid no. 5.0 max.; iodine no. 110–130; sapon. no. 90–105; 100% act.

Polylite 31-822. [Reichhold] Monomer-free polyester resin; thermosetting resin used as vehicle for pigment grinding; pigment pastes; visc. 2000–4000 cps.

Polylube #745. [Polymer Research] Hydroxyethylated compds.; dispersant, penetrant, and leveling agent for dyestuffs; misc. with hot water.

Polylube GK. [Hart Chem. Ltd.] Polyoxyalkylene condensate; nonionic; syn. fiber lubricant, antistat; liq.; water-sol.; ref. index 1.467; sp.gr. 1.09; visc. 330 cps; 100% act.

Polylube NPP. [Hart Chem. Ltd.] Polyoxyalkylene and esters; nonionic; lubricant for processing of syn. fiber blends; yel. liq.; sp.gr. 0.99; visc. 230 cps; 87% act.

Polylube SC. [Hart Chem. Ltd.] Fatty acid condensate; nonionic; fiber lubricant for polyester; liq.; water-sol.; ref. index 1.467; sp.gr. 1.02; visc. 110 cps; 87% act.

Polylube WS. [Hart Chem. Ltd.] Polyalkylene glycol ether; nonionic; lubricant for spinning nylon and wool yarns; clear liq.; water-sol.; ref. index 1.4642; sp.gr. 1.086; visc. 500 cps; 100% act.

Polymekon®. [Petrolite] Modified microcryst. wax; used in the formulation of inks and coatings and as binder, antislip and antimar agent; color 2.5 (D1500) wax; m.p. 96.1 C.

Polymekon®. [Goldschmidt] Silicone antifoam emulsion for waste water treatment, metalworking fluids, plastic latexes.

Polymel #7. [Frank B. Ross] Modified polyethylene wax; low m.w. wax incorporated into rubber batches giving exc. mold release; wh.; m.p. 205–215 F; flash pt. 420 F min.; acid no. nil.

Polymer QR-1010. [Rohm & Haas] Acrylate copolymer; kaolin clay dispersant; 44% solids.

Polymeric Acid. [Henkel] Crude polymeric fatty acid; anionic; neutralizer for oil-sol. corrosion inhibitor formulations; liq.; 100% act.

Polymulse 6. [Napp] PEG-6 palmitate; nonionic; emulsifier, dispersant, suspending agent, thickener used in personal care prods.; paste; 100% conc.

Polyox® WSR 205, 301, 1105, N-10, N-12K, N-60K, N-80, N-3000, N-750. [Union Carbide] High m.w. EO polymers; nonionic; thickener; sol. in water, some chlorinated solvs., alcohols, aromatic hydrocarbons, ketones.

Poly-Pale® Ester 10. [Hercules] Glycerol ester of polymerized rosin; pale, thermoplastic resin for lacquers, varnishes, adhesives, and wax modification; tackifying resin in pressure-sensitive rubber-based adhesives, in solv. and emulsion types; in E/VA resin wax hot-melt adhesives and coatings; in varnishes to contribute hardness, rapid drying, and resistance to water and alkali; improves clarity as a wax modifier; USDA Rosin N flakes; sol. in esters, ketones, aromatic and aliphatic hydrocarbons, and chlorinated solvs.; sp.gr. 1.08; dens. 1.08 kg/l; visc. (G-H) F (60% solids in min. spirits); soften. pt. (Hercules Drop) 114 C; acid no. 7.

Polypeg E-400. [Olin] PEG fatty ester; visc. control agent; dispersing and surface active agent; liq.

Polyphos®. [Olin] Sodium hexametaphosphate; municipal and industrial process and potable water treatment chemical which prevents precipitation of mins. and scaling of lines, metal corrosion in the system, and dissolved iron and manganese in the water; ground, powd., plate, lump (walnut and pea-size); 15% thru 100 mesh (ground), 49% thru 100 mesh (powd.); sol. > 150 g/100 g water; dens. 84 lb/ft^3 (ground), 70 lb/ft^3 (powd.); pH 6.8 (1%); 99.0% sodium polymetaphosphate.

Polyplasdone® XL. [GAF] Polyvinylpyrrolidone, pharmaceutical grade; binder, excipient used as tablet disintegrant, complexing agent, detoxifier, and antidiarrhea agent; adsorbent in thin-layer chromatography; powd.

Polyplasdone® XL-10. [GAF] Polyvinylpolypyrrolidone, pharmaceutical grade; dry binder/disintegrant in tablets; suspension aid; adsorbent; finely grd. powd.

Poly Pross. [Disco] Formulated process aid for natural and syn. rubber compds.; off-wh. to ivory small dry flakes; sp.gr. 1.15; soften. pt. 130 F.

Polyquart H. [Henkel] PEG-15 tallow polyamine; cationic; surfactant used in personal care prods.; hair conditioner, antistat; amber clear liq.; sol. in water, alcohol; visc. 1500–4000 cps; pH 5.0–6.0 (1%); 48–51.0% moisture.

Polyquart H 81. [Henkel] Polyglycol-polyamine condensation resin; antistatic and softening agent for shampoos and hair care preps.; liq.; 49–51% conc.

Polyquart H-7102. [Henkel] PEG-15 cocopolyamine and stearalkonium chloride; cationic; see Polyquart H; also hair fixative; Gardner 6.0 max. liq.; sol. in water, alcohol; visc. 1500–4000 cps; pH 5.5–6.5 (1%); 49–52% solids.

Polyrad®. [Hercules] Amine EO adduct; corrosion inhibitor and detergent for petrol. processing equipment; wetting and emulsifying agent; inhibits HCl in industrial and household cleaners; sol. in acid, water.

Polysilicate 48. [DuPont] Polysilicate; binder for refractory, ceramic, metal, inorg. fiber, catalyst support, inorg. paint systems, zinc coatings for marine and industrial applic.; glass surf. modifier; aq. sol'n.; f.p. 0 C; sp.gr. 1.18; dens. 9.8 lb/gal; visc. 12 cps; pH 11; 22.1% total solids.

Polysilicate 85. [DuPont] Polysilicate; see Polysilicate 48; aq. sol'n.; f.p. 0 C; sp.gr. 1.15; dens. 9.6 lb/gal; visc. 8 cps; pH 11; 21.2% total solids.

Polysoft CA. [Scher] Polyethylene; cationic; antistatic agent, lubricant, and softener for the textile industry; liq.; 25% act.

Poly-Solv® DB. [Olin] Diethylene glycol monobutyl ether; solv. for brake fluids, hard-surface cleaners, leather dyeing, paints and coatings, printing inks, textile vat dyeing and printing, adhesives, antifreeze, floor waxes/polishes, insect repellents; solubilize for dyes; plasticizer; extraction and crystallization solv., vinyl chloride dispersant; coupler in cosmetic preparations; APHA 10 max. liq.; mild, char. odor; m.w. 162.22; f.p. –68 C; b.p. 230 C (760 mm); sol. in water; misc. with many org. solvs.; sp.gr. 0.955; dens. 7.95 lb/gal; visc. 6.5 cP; ref. index 1.4316; flash pt. (COC) 116 C; sp. heat 2.283 joules/ g-°C.

Poly-Solv® DE (High Gravity). [Olin] Diethylene glycol monoethyl ether (75%) and ethylene glycol (25%); see Poly-Solv DB; APHA 10 max. liq.; mild, char. odor; f.p. –75 C; b.p. 195 C (760 mm); sol. in water; misc. with many org. solvs.; sp.gr. 1.0253; dens. 8.53 lb/gal; visc. 6.9 cP; ref. index 1.4297; flash pt. (COC) 96 C; sp. heat 2.308 joules/ g-°C.

Poly-Solv® DE (Low Gravity). [Olin] Diethylene glycol monoethyl ether; see Poly-Solv DB; APHA 10 max. liq.; mild, char. odor; m.w. 134.17; f.p. –76 C; b.p. 202 C (760 mm); sol. in water; misc. with many org. solvs.; sp.gr. 0.989; dens. 8.24 lb/gal; visc. 4.59 cP; ref. index 1.4273; flash pt. (TCC) 85 C; sp. heat 2.308 joules/ g-°C; 100% act.

Poly-Solv® DM. [Olin] Diethylene glycol monomethyl ether; see Poly-Solv DB; also in jet fuel system de-icers; APHA 15 max. liq.; mild, char. odor; m.w. 120.15; f.p. –85 C; b.p. 194 C (760 mm); sol. in water; misc. with many org. solvs.; sp.gr. 1.021; dens. 8.51 lb/gal; visc. 3.9 cP; ref. index 1.4263; flash pt. (TCC) 87 C; sp. heat 2.149 joules/ g-°C.

Poly-Solv® DPM. [Olin] Dipropylene glycol monomethyl ether; see Poly-Solv® DB; APHA 15 max. liq.; mild, char. odor; m.w. 148.20; f.p. –83 C; b.p. 187 C (760 mm); misc. with water and many org. solvs.; sp.gr. 0.954; dens. 7.94 lb/gal; visc. 3.9 cP; ref. index 1.419; flash pt. (TCC) 169 C.

Poly-Solv® EB. [Olin] Ethylene glycol monobutyl ether; see Poly-Solv DB; APHA 10 max. liq.; mild, char. odor; m.w. 118.17; f.p. –70 C; b.p. 171 C (760 mm); sol. in water; misc. with many org. solvs.; sp.gr. 0.903; dens. 7.52 lb/gal; visc. 3.4 cP; ref. index 1.4193; flash pt. (TCC) 63 C; sp. heat 2.438 joules/ g-°C.

Poly-Solv® EE. [Olin] Ethylene glycol monoethyl ether; see Poly-Solv DB; APHA 10 max. liq.; mild char. odor; m.w. 90.12; f.p. –70 C; b.p. 135 C (760 mm); sol. in water; misc. with many org. solvs.; sp.gr. 0.931; dens. 7.76 lb/gal; visc. 2.1 cP; ref. index 1.4076; flash pt. (TCC) 45 C; sp. heat 2.321 joules/ g-°C.

Poly-Solv® EM. [Olin] Ethylene glycol monomethyl ether; see Poly-Solv DB; also in jet fuel system de-icers; APHA 10 max. liq.; mild char. odor; m.w. 76.09; f.p. –85 C; b.p. 124 C (760 mm); sol. in water; misc. with many org. solvs.; sp.gr. 0.966; dens. 8.05 lb/gal; visc. 1.7 cP; ref. index 1.4021; flash pt. (TCC) 41 C; sp. heat 2.233 joules/ g-°C.

Poly-Solv® MPM. [Olin] Monopropylene glycol monomethyl ether; see Poly-Solv DB; APHA 10 max. liq.; mild char. odor; m.w. 90.12; f.p. –96 C; b.p. 121 C (760 mm); misc. with water and many org. solvs.; sp.gr. 0.923; dens. 7.68 lb/gal; visc. 1.9 cP; ref. index 1.4036; flash pt. (TCC) 36 C.

Poly-Solv® TB. [Olin] Triethylene glycol monobutyl ether; see Poly-Solv DB; also as additive in de-icing compds.; APHA 50 max. liq.; mild, char. odor; m.w. 206.28; f.p. –41 C; b.p. 188 C (50 mm); misc. with water and many org. solvs.; sp.gr. 0.988; dens. 8.23 lb/gal; visc. 10.9 cP; ref. index 1.4394; flash pt. (COC) 143 C.

Poly-Solv® TE. [Olin] Triethylene glycol monoethyl ether; see Poly-Solv DB; also as additive in de-icing compds.; APHA 50 max. liq.; mild, char. odor; m.w. 178.23; f.p. –21 C; b.p. 256 C (50 mm); misc. with water and many org. solvs.; sp.gr. 1.020; dens. 8.50 lb/gal; visc. 7.8 cP; ref. index 1.4376; flash pt. (COC) 124 C.

Poly-Solv® TM. [Olin] Triethylene glycol monomethyl ether; see Poly-Solv DB; also as additive in de-icing compds.; APHA 50 max. liq.; mild, char. odor; m.w. 164.20; f.p. –55 C; b.p. 249 C (760 mm); misc. with water and many org. solvs.; sp.gr. 1.048; dens. 8.74 lb/gal; visc. 7.5 cP; ref. index 1.4381; flash pt. (COC) 118 C.

Poly-Solv® TPM. [Olin] Tripropylene glycol monomethyl ether; see Poly-Solv DB; APHA 15 max. liq.; mild char. odor; m.w. 206.28; f.p. –78 C; b.p. 242 C (760 mm); misc. with water and many org. solvs.; sp.gr. 0.969; dens. 8.06 lb/gal; visc. 6.1 cP; ref. index 1.428; flash pt. (COC) 121 C.

Polystab RB1483. [Harcros UK] Aromatic alkoxyates blend; solv.; reduces visc. of polyether polyols; gives improved in-mold flowout for PU mfg.; liq.

Polystab RB1523. [Harcros UK] Alkoxylated amino compds. blend; hydrotrope, solubilizer used in PU foam mfg.; liq./paste.

Polystat Agent #5033. [Polymer Research] Phosphoric acid, partial ester; antistat used in plastics.

Polystate C. [Gattefosse] PEG 300 stearate; nonionic; base for cosmetic lotions; solid; 100% conc.

Polystay AA-1. [Goodyear] Anilino-phenyl methacrylamide; antioxidant in emulsion polymers, NBR, SBR, BR, ABS, CR; cryst. powd.

Polystay AA-1R. [Goodyear] Anilino-phenyl methacrylamide; refined, decolorized version of Polystay AA-1; cryst. powd.

Polystep® A-2. [Stepan] Sodium xylene sulfonate; hydrotrope; pale to colorless liq.; 40% act.

Polystep® A-17. [Stepan] Dodecylbenzene sulfonic acid, branched; intermediate for the prod. of sodium salts; neutralized acid as emulsifier; salts as emulsifiers for SBR latex; emulsion polymerization; catalyst in acid catalyzed reactions; visc. liq.; water-sol.; dens. 9.1 lb/gal; sp.gr. 1.09; visc. 9000 cps (30 C); 97.4% act.

Polystep® F-1. [Stepan] Nonoxynol-4; nonionic; pigment dispersant; pale clear liq.; oil-sol.; 100% act.

Polystep® F-2. [Stepan] Nonoxynol-6; nonionic; pigment dispersant; pale clear liq.; oil-sol.; 100% act.

Polystep® F-3. [Stepan] Nonoxynol-8; nonionic; pigment dispersant; pale clear liq.; water-sol.; 100% act.

Polystep® F-5. [Stepan] Nonoxynol-12; nonionic; emulsifier and stabilizer for emulsion polymerization; clear to turbid liq.; 100% act.

Polystep® F-12. [Stepan] Laureth-12; nonionic; for preemulsification and post stabilization in latex systems; wh. paste; 100% act.

Polystep® PN209. [Stepan] Alkylphenol ethoxylate phosphoric ester acid; emulsifier, dispersant, anticorrosion agent; liq.

Polytard®. [Westvaco] Kraft lignin sulfonic acid deriv.; specialty cement additive; powd.

Polytard® MC. [Westvaco] Sulfonic acid deriv. of kraft lignin; additive for masonry cement; imparts improved body and plasticity to mortar; dk. brn. liq, or dk. brn. powd.; sp.gr. 1.170 (liq.); pH 10 (liq.); 35% act. (liq.), 95% min. act. (powd.).

Poly-Tergent® P-17A. [Union Carbide] EO/PO block polymer; nonionic; defoamer, deduster, emulsifier used in dishwashing applics. and laundry detergents; APHA 50 max. clear liq.; mild odor; sol. in water, alcohols, glycol ethers, aromatic and aliphatic hydrocarbons, chlorinated solvs.; sp.gr. 1.01; dens. 8.4 lb/gal; flash pt. 241 C (COC); cloud pt. 28 C (1% aq.); pH 5.0–7.5 (1% aq.); surf. tens. 37 dynes/cm (0.1%); 100% act.

Poly-Tergent® P-17B. [Union Carbide] EO/PO block polymer; nonionic; see Poly-Tergent P-17A; APHA 50 max. clear liq.; mild odor; sol. in water, alcohols, glycol ethers, aromatic hydrocarbons, chlorinated solvs.; sp.gr. 1.04; dens. 8.6 lb/gal; flash pt. 246 C (COC); cloud pt. 31 (1% aq.); pH 5.0–7.5 (1% aq.); surf. tens. 40 dynes/cm (0.1%); 100% act.

Poly-Tergent® P-17D. [Union Carbide] EO/PO

block polymer; nonionic; demulsifier, dedusting dispersant used in rinse aids, dishwashing, metal cleaning and papermaking applics.; liq.; 100% conc.

Poly-Tergent® P-22A. [Union Carbide] EO/PO block polymer; nonionic; see Poly-Tergent P-17D; liq.; 100% conc.

Poly-Tergent® P-32A. [Union Carbide] EO/PO block polymer; nonionic; see Poly-Tergent P-17A; APHA 50 max. clear liq.; mild odor; sol. in alcohols, glycol ethers, aromatic hydrocarbons, chlorinated solvs.; insol. in water; sp.gr. 1.02; dens. 8.5 lb/gal; flash pt. 229 C (COC); cloud pt. 15 C (1% aq.); pH 5.0–7.5 (1% aq.); surf. tens. 43 dynes/cm (0.1%); 100% act.

Poly-Tergent® S-205LF. [Union Carbide] Alkoxylated linear alcohol; nonionic; surfactant for dishwashing applics., metal cleaning baths; foam depressant in hard surface cleaners; cleaners for dairy industry; rewetting agents for paper toweling; APHA 100 max. clear liq.; mild odor; sol. in water, alcohols, glycol ethers, aromatic hydrocarbons, chlorinated solvs.; sp.gr. 1.00; visc. 192 cs; flash pt. 229 C (COC); cloud pt. 10 C (1% aq.); pH 5.5–7.0 (1% aq.); surf. tens. 32 dynes/cm (0.1%); 100% act.

Poly-Tergent® S-305LF. [Union Carbide] Alkoxylated linear alcohol; nonionic; see Poly-Tergent S-205LF; APHA max. clear liq.; mild odor; sol. in alcohols, glycol ethers, aromatic hydrocarbons, chlorinated solvs.; disp. in water; sp.gr. 1.00; dens. 8.3 lb/gal; visc. 171 cs; flash pt. 224 C (COC); cloud pt. 19 C (1% aq.); pH 5.5–7.0 (1% aq.); surf. tens. 31 dynes/cm (0.1%); biodeg.; 100% act.

Poly-Tergent® S-405LF. [Union Carbide] Alkoxylated linear alcohol; nonionic; see Poly-Tergent S-205LF; APHA 100 max. clear liq.; mild odor; sol. see Poly-Tergent S-205LF; sp.gr. 1.01; dens. 8.4 lb/gal; visc. 189 cs; flash pt. 227 C (COC); cloud pt. 28 C (1% aq.); pH 5.5–7.0 (1% aq.); surf. tens. 32 dynes/cm (0.1%); biodeg.; 100% act.

Poly-Tergent® S-505LF. [Union Carbide] Alkoxylated linear alcohol; nonionic; see Poly-Tergent S-205LF; APHA 100 max. clear liq.; mild odor; sol. see Poly-Tergent S-205LF; sp.gr. 1.02; dens. 8.5 lb/gal; visc. 125 cs; flash pt. 232 C (COC); cloud pt. 47 C (1% aq.); pH 5.5–7.0 (1% aq.); surf. tens. 33 dynes/cm (0.1%); biodeg.; 100% act.

Polytex 10. [Napp] Stearamide DIBA-stearate; nonionic; emulsifier, thickener and pearling agent for cosmetic preparations; wax, flakes; 100% conc.

Polytrap® 158. [Dow Corning/Wickhen] Di (2-ethylhexyl) adipate/polymer; entrapped binder/emollient for all powd. and stick systems; powd.; mixes with all powd.; disp. in oil and water systems.

Polytrap® 171. [Dow Corning/Wickhen] Octyl hydroxystearate/acrylates copolymer; binder for emollients in hydrophobic polymer matrix; powd.; disp. in water systems.

Polytrap® 210. [Dow Corning/Wickhen] Petrolatum polymer; see Polytrap 171; powd.

Polytrap® 223. [Dow Corning/Wickhen] Floral fragrance polymer; aromatic; entrapped fragrance in hydrophobic matrix to protect char. and longevity of fragrance; beads; mixes with all powds.; disp. in oil and water systems.

Polytrap® 227. [Dow Corning/Wickhen] Citrus fragrance polymer; see Polytrap 223; micronized powd.; mixes with all powds.; disp. in oil and water systems.

Polytrap® 228. [Dow Corning/Wickhen] Floral fragrance polymer; see Polytrap 223; micronized powd.; mixes with all powds.; disp. in oil and water systems.

Polytrap® 229. [Dow Corning/Wickhen] Min. oil/acrylates copolymer; see Polytrap 171; powd.; mixes with all powds.; disp. in oil and water systems.

Polytrap® 603. [Dow Corning/Wickhen] Acrylates copolymer; cosmetic binder, carrier; sebum adsorp. to prevent buildup of oil on skin; flow additive waste treatment, oil recovery; process aid for syneresis-free formulations; improves pigment disp.; microporous wh. powd.

Polytrap® 801. [Dow Corning/Wickhen] Arachidyl propionate/acrylates copolymer; see Polytrap 171; powd.; disp. in oil and water systems.

Polytrope 1131. [Rheox] Organically modified montmorillonite clay; rheological additive, thickener for unsat. polyester laminating resins.

Polywax® 500. [Petrolite] Polyethylene homopolymer; release agent in carbon ink formulations; modifier for paraffin waxes; component in hot-melt coatings, adhesives, and chewing gum base; lubricant in plastics and rubber processing, elec. insulation, powd. coatings, and printing inks; antiblocking agent; binder; prilled; m.w. 500; melt index > 5000 g/10 min; low sol. in org. solvs., esp. at R.T.; sol. in CCl_4, benzene, xylene, toluene, turpentine; dens. 0.93 g/cc; visc. 3 cps (149 C); m.p. 86 C; soften. pt. 86 C.

Polywax® 655. [Petrolite] Polyethylene homopolymer; see Polywax 500; prilled; m.w. 700; melt index > 5000 g/10 min; sol. see Polywax 500; dens. 0.96 g/cc; visc. 6 cps (149 C); m.p. 102 C; soften. pt. 102 C.

Polywax® 850. [Petrolite] Linear polyethylene; process aid, lubricant for thermoplastic polymers; m.p. 228 F.

Polywax® 1000. [Petrolite] HDPE; see Polywax 500; prilled; m.w. 1000; melt index > 5000 g/10 min; sol. see Polywax 500; dens. 0.96 g/cc; visc. 11 cps (149 C); m.p. 113 C; soften. pt. 113 C.

Polywax® 2000. [Petrolite] HDPE; see Polywax 500; prilled; m.w. 2000; melt index > 5000 g/10 min; sol. see Polywax 500; dens. 0.96 g/cc; visc. 50 cps (149 C); m.p. 125 C; soften. pt. 125 C.

Polywax® E-2020. [Petrolite] Oxidized HDPE; wax producing emulsions for formulating textile lubricants, mold release agents, lubricants, and antiblocking agents; useful in ceramic binders; processing aid in the fabrication of thermoplastic resins; exhibits external and internal lubricating properties; color 0.5 (D1500) wax; m.p. 115.5 C; acid no. 23; sapon. no. 42.

Polywax® OH 425. [Petrolite] Primary, linear polymeric alcohol; functional polymer for modification of PP, PVC, polyethylene, PS, and high-performance engineering resins; acts as antioxidant, heat stabilizer, uv stabilizer, or visc. depressant; promotes emollient protective films onto the skin in cosmetic creams and lotions; used in hot-melt and solv.-based coatings; textile/leather lubricants and finishes; chemical intermediate; off-wh. solid; m.w. 425; sol. in toluene, MIBK, VM&P naphtha, hexane, IPA; visc. 3.2 cps; m.p. 87.8 C; hyd. no. 110.

Polywax® OH 550. [Petrolite] Primary, linear polymeric alcohol; see Polywax OH 425; off-wh. solid; m.w. 550; sol. in toluene, MIBK, VM&P naphtha, hexane; visc. 5.5 cps; m.p. 98.9 C; hyd. no. 85.

Polywax® OH 700. [Petrolite] Primary, linear polymeric alcohol; see Polywax OH 425; off-wh. solid; m.w. 700; sol. in toluene, MIBK, VM&P naphtha; visc. 7.9 cps; m.p. 110 C; hyd. no. 66.

Polywet ND-1. [Uniroyal] Polyfunctional oligomer, sodium salt; anionic; dispersant for mins., pigments and fillers in paints, paper, latex compds., and water treatment applics.; slight yel. liq.; sp.gr. 1.13; visc. 35 cps; pH 6.8–7.8; 25% act.

Polywet ND-2. [Uniroyal] Functionalized oligomer, sodium salt; anionic; dispersant for pigments, mins., extenders, fillers in aq. systems; latex paints and enamels; coatings, adhesives, paper and paperboard; boiler water; colorless to slight yel. liq.; sp.gr. 1.16; visc. 32 cps; pH 6.8–7.8; 25% act.

Polywet Z1766. [Uniroyal] Sodium salt of a polyfunctional oligomer; anionic; dispersant for titanium dioxide, other pigments; liq.; 35% act.

Poly-Zole® AZDN. [Olin] 2,2′-Azobisisobutyronitrile; initiator for addition-polymerization of vinyls, acrylics, and other monomers, in bulk, sol'n., and suspension; wh.or off-wh. powd.; m.w.. 164.2; decomp. pt. 103 C; 99.5% assay.

POM. [Climax Performance] Molybdenum trioxide; corrosion inhibitor for cooling water systems for use in producing molybdate sol'ns. in situ.

Porofor ADC/E. [Bayer] Azodicarbonamide; chemical blowing agent for prod. of plastic foams; used in expanded UPVC pipe and sections; extrusion of foamed polyethylene; orange powd., particle size 23 μm; sol. in dimethyl sulfoxide and dimethylformamide; dens. 1.65 g/cc.

Porofor ADC/F. [Bayer] Azodicarbonamide; chemical blowing agent for prod. of plastic foams; useful in wallpaper and extrusion of thermoplastics; powd., particle size 15μm.

Porofor ADC/K. [Bayer] Azodicarbonamide, activator; blowing agent for rubber goods; yel. powd.; dens. 1.6 g/cm³.

Porofor ADC/M. [Mobay] Azodicarbonamide; chemical blowing agent for prod. of plastic foams and inj. molding; lt. yel. to orange cryst. powd., particle size 5 μm; m.w. 116.1.

Porofor ADC/R. [Mobay] Azodicarbonamide; see Porofor ADC/K; suitable for blowing and expansion processes; fine yel. powd.; dens. 1.6 g/cm³; 99% min. act.

Porofor ADC/S. [Bayer] Azodicarbonamide; chemical blowing agent for prod. of plastic foams; PVC expanded artificial leather and wallpaper; powd., particle size 7 μm.

Porofor B 13/CP 50. [Bayer] Benzene-1,3-disulfonyl hydrazide, chlorinated paraffin, pbw ratio 50:50; see Porofor ADC/K; pale gray crumbly to pasty substance; dens. 1.5 g/cm³.

Porofor BSH Paste. [Mobay] Benzene sulfonyl hydrazide, paraffin oil, pbw ratio 75:25; see Porofor ADC/K; pale gray paste; dens. 1.26 g/cm³.

Porofor BSH Paste M. [Bayer] Benzene sulfonyl hydrazide, paraffin oil, pbw ratio 75:25; see Porofor ADC/K; pale gray paste; dens. 1.26 g/cm³.

Porofor BSH Powder. [Bayer] Benzene sulfonyl hydrazide; see Porofor ADC/K; wh. powd.; dens. 1.48 g/cm³.

Porofor D33. [Bayer] Diphenyl-sulfon-3,3′-disulfohydrazide; blowing agent used in coatings for textiles.

Porofor DNO/F. [Bayer] Dinitrosopentamethylene tetramine, desensitizing agent, pbw ratio 80:20; see Porofor ADC/K; ylsh. powd.; dens. 1.43 g/cm³.

Porofor N. [Bayer] Azoisobutyric dinitrile; blowing agent for prod. of PVC; used in floats, buoys, net floats, bath mats; wh. to pale gray cryst. powd.; sol. in toluene, xylene, dichloroethane, ethyl acetate, chloroform, and most vinyl monomers; insol. in water.

Porofor S 44. [Bayer] Diphenylene-oxide-4,4′-disulfohydrazide; blowing agent for prod. of foamed plastics.

Post-4. [Rheox] Castor oil complex deriv.; thixotrope, gellant; post additive for trade sales and industrial finishes to control sag and pigment settling; no shear required; straw liq.; sp.gr. 1.03; dens. 8.8 lb/gal.

PPG Perchlor. [PPG Industries] Perchlorethylene; solv. avail. in drycleaning grade with unique stabilizer components that are completely removed from garments and leave no residual odor and which are recoverable with solv. after distilling; degreasing and general solv. grade for vapor degreasing, cold cleaning, or drycleaning applics.; heavy duty grade which is more heavily stabilized and intended for use in more rigorous vapor degreasing applics.; water-wh. clear liq.; m.w. 165.85; f.p. –22.3 C; b.p. 121.1 C; sol. 0.015 g sol./100 g water; sp.gr. 1.623–1.628; dens. 13.57 lb/gal; visc. 0.88 cps; ref. index 1.5053.

PPG Trichlor. [PPG Industries] Trichlorethylene; solv. avail. in standard grade, degreasing and general solv. grade, stabilized for general, heavy-duty vapor degreasing and cold cleaning applics.; special grades are avail. for LOX flushing, vapor degreasing, and cleaning electronic components and synthesizing chemicals; water-wh. clear liq.; sweet odor; m.w. 131.40; f.p. –86.4 C; b.p. 86.9 C; sol. 0.11 g sol./100 g water; sp.gr. 1.460–1.470; dens. 12.2 lb/gal; visc. 0.58 cps; ref. index 1.4782; pH 6.7–7.5.

Prapagen WK brands. [Hoechst-Celanese AG] Quat. ammonium salts; textile softener with antistat.

Precirol ATO. [Gattefosse] Glycerol di/tripalmito stearate; nonionic; additive for tablets, binder, lubricant, sustained release; solid; 100% conc.

Precirol WL 2155. [Gattefosse] Glycerol ditristearate; nonionic; additive for tablets mfg.; solid; HLB 2.0; 100% conc.

Prelete® Defluxer. [Dow] Inhibited 1,1,1-trichloroethane with alcohols; solv. for postsolder cleaning; sol. in most org. solvs.; very low water sol.

Premier Maltzymes I and II. [Premier Malt] Bacterial alpha-amylase derived from Bacillus subtilus; enzyme which converts gelatinized starch to dextrins and small amts. of low m.w. carbohydrates; useful for prod. of alcohol from ground corn; liq.

Preventol I. [GAF] Sodium trichlorophenate; biocide, fungicide and bacteriostat for print pastes, finishes, sizes, adhesives, textile and leather; liq.

Primacor 4990 Dispersion. [Dow] Ethylene acrylic acid copolymer disp.; disp. polymer for use as binder for nonwoven fibers incl. PP, polyester, glass, and nylon; provides soft fabrics having exc. antisoil redeposition properties; milky wh, disp.; b.p. 100 C; sol. in water; sp.gr. 0.985 g/cc (liq.), 0.960 g/cc (solids); dens. 8.22 lb/gal; visc. 500 cps; pH 7.75–8.75; surf. tens. 44–46 dynes/cm; 35% solids in ammonia water.

Primafloc® A-10. [Rohm & Haas] Acidic acrylic emulsion; flocculant, oil/water separator; 18% solids.

Primafloc® C-3. [Rohm & Haas] Polyamine; flocculant; 30% solids.

Primapel® C-93. [Rohm & Haas] Acrylic carboxylic copolymer; carpet soil retardant; 25% solids.

Primarol 1006. [Henkel Canada] 2-Hexyl decanol-1; extender and solv. for dyes and fragrance oils; intermediate for prod. of comps. for applics. in lubricants, emulsifiers, metal processing, and textiles; water-wh. clear liq.; sol. in solvs.; flash pt. 154 C (COC); cloud pt. < –20 C; sapon. no. < 5.

Primarol 1107. [Henkel Canada] C_{16}-C_{20} 2-alkyl alcohols; see Primarol 1006; water-wh. clear liq.; cloud pt. < 0 C.

Primarol 1208. [Henkel Canada] 2-Octyl-dodecanol-1; carrier for oil-sol. prods.; pigment dispersant and solubilizer for waxes and fatty alcohols and esters; intermediate for prod. of comps. for applics. in lubricants, emulsifiers, metal processing and textiles; water-wh. clear liq.; cloud pt. < –20 C; sapon. no. 2–6.

Primasol KW. [BASF AG] Mixt. of anionic and nonionic compds., polyfunctional, synergistic; textile aux., wetting agent, detergent, stabilizer; dyeing aux. for continuous dyeing of wool, polyamide, and polyester fibers; used for continuous dyeing and vigoureux printing of slubbing and tow, continuous dyeing of textile floor coverings, and space dyeing; ylsh.-brn.; dissolves readily in cold or warm water; pH 6–6.5.

Primasol SD. [BASF AG] Mixt. of oxyethylation prods.; nonionic; textile aux., stabilizer, wetting agent; aux. for space dyeing of polyamide and acrylic carpet yarns with acid, metal complex, and basic dyes; colorless or lt. ylsh. liq.; 6.5–7.5 (10% sol'n.).

Proaid 9802. [Akrochem] Nonstaining, nondiscoloring processing and dispersing aid for polychloroprene and EPDM; lt br. pellets; sp.gr. 0.91.

Proaid 9810. [Akrochem] Zinc salt of fatty acids; nonstaining, nonblooming processing aid for unsat. polymers; peptizing agent for natural and polyisoprene rubber (and blends); aids disp. of fillers; imparts some mold release benefits; tan pellets; sp.gr. 1.05; m.p. 172 F.

Proaid 9814. [Akrochem] Homogenizing agent and softening resin for use in most elastomers; improves processing; dk. br. flakes; sp.gr. 1.19; soften. pt. 75–85 C.

Proaid 9826. [Akrochem] See Proaid 9814; for lt.-colored compds.; lt. br. pellets; sp.gr. 1.00; soften. pt. 85 C.

Proaid 9831. [Akrochem] Zinc salt of fatty acids; see Proaid 9810; sp.gr. 1.1; m.p. 165–185 F.

Proaid 9904. [Akrochem] Specialty processing aid for use in CM (CPE) and CSM (Hypalon) polymers; gives some mold release properties; stabilizes hot air aging properties; wh. flakes; sp.gr. 1.02.

Procetyl 10. [Croda] PPG-10 cetyl ether; nonionic; emollient, coupler, cosolvent, plasticizer, superfatting, wetting and spreading agent, penetrant; lubricant in cosmetics and personal care prods.; alcoholic and aq. alcoholic compositions; emollient; APHA 150 max. clear liq.; faint, char. sweet odor; sol. in min. oil, acetone, IPM, lanolin oil, alcohol; acid no. 1 max.; pH 6.0–7.5 (3% disp.); 100% act.

Procetyl 20. [Croda] PPG-20 cetyl ether; see Procetyl 10; APHA 150 max. clear liq.; faint, char. sweet odor; pH 6.0–7.5 (3% disp.); 100% act.

Procetyl 30. [Croda] PPG-30 cetyl ether; see Procetyl 10; APHA 150 max. clear liq.; faint, char. sweet odor; pH 6.0–7.5 (3% disp.); 100% act.

Procetyl 50. [Croda] PPG-50 cetyl ether; see Procetyl 10; APHA 150 max. clear liq.; faint, char. sweet odor; sol. in PEG-200, lanolin oil, alcohol; acid no. 1 max.; pH 6–7.5 (3% disp); 100% act.

Procetyl AWS. [Croda] PPG-5 ceteth-20; nonionic; emulsifier, plasticizer, coupler, humectant, dispersant, emollient, and solubilizer in aq. and aq. alcoholic systems, personal care prods., antiperspirants; water-wh. turbid oily liq.; char. sweet odor; water-sol.; sol. in water, alcohol, IPM, PEG-200, glycerin, propylene glycol, lanolin oil; acid no. 1 max.; pH 6.0–7.5 (30% aq.); 100% act.

Prochol. [Croda Ltd.] Propoxylated dist. lanolin alcohols; cosolv.; emollient for cosmetics and skin care prods.; amber liq.; alcohol-sol.

Prochol 30. [Croda Ltd.] Propoxylated lanolin alcohols; superfatting agent for cosmetic use; sol. in water/alcohol.

Prodew 100, 200. [Ajinomoto] L-Pro, PCA-Na Na lactate sorbitol collagen; formulated moisturizer for cosmetics, soaps, hair care prods.; humectant; 50% aq. sol'n.

Product 100. [Henkel KGaA] Fatty alcohol polyglycol ether; dispersant for oil paints and oil-modified emulsion paints; liq.

Product BCO. [DuPont] Cetyl betaine; amphoteric wetting agent, detergent, emulsifier, dispersant, surfactant; dyeing applics.; softener for textiles; leveling and rewetting agent in the paper industry; dyeing assistant and degreaser in the leather industry; antistat on plastic films; brn. clear liq.; mild odor; misc. in water; dens. 8.4 lb/gal; visc. free-flowing; surf. tens. 27.8 dynes/cm (32 C, 0.1%); 33% solids.

Produkt B 2045. [Croda Ltd.] Alkenyl succinic anhydride; corrosion inhibitor additive for nonaq. systems; hardener for cosmetics and pharmaceuticals; forms liq. emulsions; colorless solid; odorless; m.p. 55–63 C; acid no. 1 max.; sapon. no. 2 max.; 100% conc.

Produkt B 3010. [Croda Ltd.] Disodium alkenyl succinate; corrosion inhibitor for aq. systems; visc. liq.; 40% act.

Produkt B 3032. [Croda Ltd.] Trialkanolammonium alkenyl succinate; corrosion inhibitor for aq. systems; visc. liq.; 40% act.

Produkt GM 4055. [Zschimmer & Schwarz] Fatty acid glycol ester and fatty alcohol ether sulfate; opacifier with glossing effect for hair shampoos and bath additives; fluid; 38% act.

Produkt GS 400. [Zschimmer & Schwarz] Boric acid DEA, modified; anticorrosion agent; liq.; 80% act.

Produkt GS 7114/80. [Zschimmer & Schwarz] Boric acid DEA; anticorrosion agent; liq.; 80% act.

Produkt RT 237. [Zschimmer & Schwarz] Alkylaryl polyalkyleneglycol ether; nonionic; spreading aux. for wax; aids gloss and water resistance; ylsh. visc. liq.; sp.gr. 0.99; sol. in alcohol and chlorinated hydrocarbons; insol. in water; pH 4–6 (10%); 100% act.

Produkt RT 245. [Zschimmer & Schwarz] Alkylaryl polyglycol ether; solubilizer for chemo-tech. prods.; liq.; 100% act.

Produkt RT 288. [Zschimmer & Schwarz] Olive oil ethoxylate; solubilizer and emollient for cosmetics; paste; 100% act.

Profan 24 Extra, 128 Extra. [Sanyo] 1:1 Coconut DEA; nonionic; foam stabilizer and thickener for shampoo; liq.; 100% conc.

Profan 2012E. [Sanyo] 1:2 Coconut DEA; nonionic; see Profan 24 Extra; liq.; 100% conc.

Profan AA62. [Sanyo] 1:1 Lauric DEA; nonionic; see Profan 24 Extra; solid; 100% conc.

Profan AB20. [Sanyo] Coconut MEA; nonionic; see Profan 24 Extra; solid; 100% conc.

Profan AD31. [Sanyo] Lauric MIPA; nonionic; see Profan 24 Extra; solid; 100% conc.

Progacyl COS-1. [Lyndal] Guar gum; nonionic; thickener for aq. systems, personal care prods.; water-sol.

Progacyl COS-10. [Lyndal] Carboxymethyl guar; thickener, suspending agent for personal care prods.; water-sol.

Progacyl COS-20, -70. [Lyndal] Hydroxypropyl guar; nonionic; thickener for aq. systems, personal care prods.; sol. in water, alcohol, glycols.

Progacyl CP-7. [Lyndal] Guar; anionic; thickener; water-sol.

Progasol 40. [Lyndal] Alkylaryl sulfonate; anionic; wetting, leveling, and scouring agent; emulsifier used in textiles; liq.; 40% conc.

Progastat 204. [Lyndal] Org. phosphate esters; anionic; antistat for fiber, yarns, fabrics, industrial uses; liq.

Prolase® 300. [Int'l. Bio-Synthetics] Protease (papain); pharmaceutical grade enzyme; protein digestive aid; also for removal of protein from soft contact lenses; powd.

Promulgen® D. [Amerchol] Cetearyl alcohol and ceteareth-20; nonionic; gelling agent; o/w emulsifier, emollient, and stabilizer for cosmetics and pharmaceuticals; wh. waxy solid; odorless; m.p. 47–55 C; acid no. 1 max.; sapon. no. 2 max.; 100% conc.

Promulgen® G. [Amerchol] Stearyl alcohol and ceteareth-20; nonionic; see Promulgen D; yel. liq.; sp.gr. 0.848; visc. 35 cps; m.p. 55–63 C; acid no. 1 max.; sapon. no. 2 max.; 100% act.

Promyr. [Amerchol] IPM; emollient and solv. for cosmetics, toiletries, makeups; nongreasing rub in; water-wh. low visc. liq.; odorless; oil-sol.; acid no. 1.0 max.; sapon. no. 202–211.

Promyristyl PM. [Croda] Propoxylated (1) myristyl alcohol; dry lt. feel emollient; coupling agent, lubricant; sol. in oil, alcohol.

Promyristyl PM-3. [Croda] PPG-3 myristyl ether; low-visc. emollient for clear analgesic, deodorant, and fragrance sticks; Gardner 1 max. clear liq.; sol. in oil, alcohol, lanolin oil; acid no. 1 max.

Pronal 502. [Toho] Polyalkylene glycol ester; defoaming agent for paper, latex, and paint mfg.; liq.

Pronal ST-1. [Toho] POP ether; defoaming agent for food and fermentation industry; liq.

Propacyl CP-2000. [Lyndal] Guar deriv.; nonionic; thickener, suspending agent; water-sol.

Propacyl EM-40. [Lyndal] Guar deriv.; nonionic; thickener for aq. sol'ns.; water-sol.

Propal. [Amerchol] IPP; emollient and solv. for cosmetics, toiletries, makeups; nongreasing rub in; water-wh. liq.; odorless; acid no. 1.0 max.; sapon. no. 182–191.

Propetal 99. [Zschimmer & Schwarz] Fatty alcohol polyalkylene; nonionic; detergent, wetting agent, intermediate for detergents and cleaners; emulsifier, antifoaming additive; liq.; 100% conc.

Propetal 103. [Zschimmer & Schwarz] Fatty alcohol polyalkylene; nonionic; see Propetal 99; liq.; 100% conc.

Propetal 241. [Zschimmer & Schwarz] Fatty alcohol polyalkylene; nonionic; see Propetal 99; liq.; 100% conc.

Propetal 254. [Zschimmer & Schwarz] Fatty alcohol polyalkylene; nonionic; emulsifier, antifoaming additive; liq.; 100% conc.

Proplast 015. [Aquatec] Fatty ester; lubricant for thermoplastics; org. pigments dispersant; water-insol.

Proplast 050, 075. [Aquatec] Glycerol ester; lubricant for thermoplastics; liq.

Proplast 058. [Aquatec] Glycerol ester; antistat for thermoplastics; flakes.

Proplast 060. [Aquatec] Glycerol ester; lubricant for PVC; flakes.

Proplast 290. [Aquatec] Diamide of fatty acid; lubricant for thermoplastics; powd.

Propoxyol® 1695. [Henkel/Emery] PPG-5 lanolin; nonionic; emollient, stabilizer, and pigment dispersant for anhyd. makeups; cosmetic additive; solid; sol. in most cosmetic oils and waxes; 100% conc.

Propyl Zithate. [Vanderbilt] Zinc isopropyl xanthate; accelerator; used in natural and syn. cements and doughs; nondiscoloring in presence of copper or iron; wh. to lt. yel. friable powd.; m.w. 335.83; practically insol. in water; dens. 1.56 ± 0.03 mg/m³; m.p. 140 C min. with decomposition; 19.1–20.5% zinc content.

Prosil®-196. [SCM] γ-Mercaptopropyl trimethoxy silane; coupling agent having both org. and inorg. reactivity; for acrylic, epichlorohydrin, nitrile, polysulfone, PS, PVC, urethane thermoplastics; thermoset acrylic, epoxy, nitrile/phenolic, phenolic, polybutadiene; and elastomerics; lt. straw clear liq.; m.w. 196.3; b.p. 220 C (760 mm Hg); sp.gr. 1.05; ref. index 1.440; flash pt. (TCC) 170 F; 97% purity.

Prosil®-220. [SCM] γ-Aminopropyl triethoxy silane, tech. grade; coupling agent enhancing and promoting chemical bonding between inorg. and org. molecules; for acetal, acrylic, epichlorohydrin, nitrile, NC, polyamide, PC, polyethylene, polyimide, polymethacrylate, PP, polysulfone, PS, PVC, urethane, and vinyl thermoplastics; thermoset acrylic (thermoset and latex), alkyd, epoxy, furan, melamine, nitrile/ phenolic, phenolic, polyester, vinyl butyral/phenolic; and elastomers; lt. straw clear liq.; m.w. 221.3; b.p. 217 C (760 mm Hg); sp.gr. 0.950; flash pt. (COC) 113 F; 85% purity.

Prosil®-221. [SCM] γ-Aminopropyl triethoxy silane; high purity grade; see Prosil-220; Pt-Co APHA 20 liq.; m.w. 221.3; b.p. 217 C (760 mm Hg); sp.gr. 0.946; ref. index 1.420; flash pt. (COC) 182 F; 98% purity.

Prosil®-248. [SCM] γ-Methacryloxypropyl trimethoxy silane; coupling agent having reactive methacrylate and trimethoxysilyl groups; improves adhesion of org. thermoset resins to inorg. materials such as fiberglass, clay, quartz, and other siliceous surfaces; for ABS, acrylic, polyethylene, polyimide, polymethacrylate, PP, PS, silicone, SAN, and urethane thermoplastics; alkyd, DAP, epoxy, polybutadiene, polyester, and crosslinked polyethylene thermosets; and elastomers; Pt-Co APHA 30 clear liq.; m.w. 248.1; b.p. 255 C (760 mm Hg); sp.gr. 1.04; ref. index 1.43; flash pt. (COC) 190 F; 97.5% purity.

Prosol 5. [Hart Chem Ltd] Min. oil and fatty acid condensate; nonionic; min. oil lubricant for processing of wool and wool syn. blends on rigorous mechanical equipment; amber liq.; sp.gr. 0.922; visc. 70 cps; ref. index 1.4913; pH acid.

Prosol 518. [Hart Chem Ltd] Refined min. oil and emulsifiers; nonionic; min. oil lubricant for processing of wool and wool syn. blends; amber liq.; sp.gr. 0.895; visc. 65 cps; ref. index 1.471; pH acid; 100% act.

Prostearyl 15. [Croda Ltd.] PPG-15 stearyl ether; emollient, lubricant for cosmetics; coupler for fragrances; sol. in oil, alcohol.

Protamine-45. [Nat'l. Starch] Amide-amine fatty derivative; cationic; emulsifier, softener and lubricant base; solid; 100% conc.

Protectein. [Hormel] Propyltrimonium hydrolyzed animal protein; substantive quat. reducing irritation potential of surfactants; for skin and hair prods.; amber liq.; bland char. odor; pH 6–7; 55% solids.

Protectol® GDA, GT 50. [BASF AG] Glutaraldehyde; biocide for aq. systems and cleaners; liq.; 50% conc.

Protectol® GL 40. [BASF AG] Glyoxal; see Protectol GDA; liq.; 40% conc.

Protectol® KLC 50, 80. [BASF AG] Benzalkonium chloride; cationic; biocidal surfactant for chemical industry, detergent mfg.; liq.; 50 and 80% conc. resp.

Protectol® TOE. [BASF AG] Thiadiazine deriv.; biocide for use in aq. systems; powd.

Protegin W, WX. [Goldschmidt] Petrolatum, ozokerite, hydrog. castor oil, glyceryl isostearate, polyglyceryl-4 oleate; nonionic; SE w/o emulsifier, emollient, absorp. base for cosmetics and pharmaceuticals; paste; HLB 3.5; 100% conc.

Protegin X. [Goldschmidt] Min. oil, petrolatum, ozokerite, glyceryl oleate, lanolin alcohol; nonionic; paste; see Protegin W; HLB 3.5; 100% conc.

Pro-Tein ES-20. [Amerchol] Ethyl ester of hydrolyzed animal protein; plasticizer for hair resins; improves hair gloss; alcoholic sol'n.; alcohol-sol.

Pro-Tex. [Griffin] Maneb (32.63%) and triphenyltin hydroxide (4.72%) sol'n.; flowable fungicide for potatoes and sugar beets; restricted use.

Protoferm. [Finnsugar] Protease (fungal); enzyme for food processing; meat tenderizer; protein hydrolysates; beer chillproofing; tan powd.

Protol. [Witco] Wh. min. oil USP; emollient, lubricant in cosmetic creams and lotions, pharmaceuticals; visc. 180–190 SUS (100 F); pour pt. 20 F.

Protopet® Alba. [Witco] Petrolatum USP; med. consistency and m.p. petrolatum functioning as carrier, lubricant, emollient, moisture barrier, protective agent, softener for cosmetics, pharmaceutical ointment, industrial applics.; Lovibond color 1.0Y; visc. 10–16 cSt (100 C); m.p. 54–60 C.

Protopet® White 1S. [Witco] Petrolatum USP; see Protopet Alba; Lovibond color 1.5Y; visc. 10–16 cSt (100 C); m.p. 54–60 C.

Protopet® White 2L. [Witco] Petrolatum USP; see Protopet Alba; Lovibond color 8Y0.6R; visc. 10–16 cSt (100 C); m.p. 54–60 C.

Protopet® White 3C. [Witco] Petrolatum USP; see Protopet Alba; Lovibond color 25Y1.0R; visc. 10–16 cSt (100 C); m.p. 54–60 C.

Protopet® Yellow 1E. [Witco] Petrolatum USP; see Protopet Alba; soft solid; m.p. 130–140 F.

Protopet® Yellow 2A. [Witco] Petrolatum USP; see Protopet Alba; Lovibond color 30Y2.5R; visc. 10–16 cSt (100 C); m.p. 54–60 C.

Protowax 90. [Nat'l. Starch] Glyceryl monostearate; nonionic; emulsifier, softener base; solid; 100% conc.

Protowax ES-100. [Nat'l. Starch] Fatty acid, ethoxylated; nonionic; emulsifier, softener, dyeing assistant; solid; 100% conc.

Protowet 3072. [Nat'l. Starch] Alkyl ester, phosphated, free acid; chemical intermediate; antistat in syn. fibers; lt. brn. liq.; insol. in water; pH 6.5 (1%); 100% act.

Protowet 3098. [Nat'l. Starch] Phosphate alcohol free acid; anionic; base for scouring agent, dyeing assistant, emulsifier; liq.; 100% conc.

Protowet C-5. [Nat'l. Starch] Fatty ester sulfates; anionic; bleach stabilizer containing surfactants and lubricants; liq.; 38% conc.

Protowet E-4. [Nat'l. Starch] EO condensate; nonionic; penetrant, dyeing assistant, stabilizer, wetting, emulsifying, scouring and foaming agent used in textiles; liq.; 25% conc.

Protowet FHL. [Nat'l. Starch] Sulfated glycerides, sodium salt; anionic; emulsifier and lubricant in spin finish formulations; scouring agent; emulsifies min. oil, butyl stearate, coconut oil; dye bath lubricant; dyeing assistant; med. amber liq.; pH 5–6 (1%); biodeg.; 75% act.

Protowet MB. [Nat'l. Starch] Ethoxylated alcohol sulfates; anionic; wetting, leveling and dyeing assistant; 68% conc.

Protowet NRW. [Nat'l. Starch] Phosphated alcohol; anionic; penetrant, emulsifier; dyeing assistant in textile industry applics.; lt. yel. clear liq.; pH 7 (1%); 30% act.

Protowet TJ. [Nat'l. Starch] Phosphated alcohol; anionic; emulsifier, dispersant, wetting agent, detergent used in various applics. of the textile industry; pale yel. liq.; water-sol.; sp.gr. 1.10–1.20; 85% act.

Protox®-78. [Zinc Corp. of Am.] French process zinc oxide with propionic acid surface treatment; pigment/reinforcing agent for rubber; 0.31 μ mean particle size; 99.97% thru 325 mesh; sp.gr. 5.6; sp.vol. 0.18; pkg. dens. 50 lb/ft^3; surf. area 3.5 m^2/g; oil absorp. 12 lb oil/100 lb ZnO; 99.3% zinc oxide.

Protox®-166. [Zinc Corp. of Am.] Amer. process zinc oxide with propionic acid; pigment used in the rubber industry; reinforcing agent; 0.27 μ mean particle size; 99.97% thru 325 mesh; sp.gr. 5.6; sp.vol. 0.18; pkg. dens. 50 lb/ft^3; surf. area 4.0 m^2/g; oil absorp. 18 lb oil/100 lb ZnO; 98.8% zinc oxide.

Protox®-167. [Zinc Corp. of Am.] Amer. process zinc oxide with propionic acid and lt. process oil; pigment used in the rubber industry; reinforcing agent; 0.27 μ mean particle size; 99.97% thru 325 mesh; sp.gr. 5.6; sp.vol. 0.18; pkg. dens. 50 lb/ft^3; surf. area 4.0 m^2/g; oil absorp. 18 lb oil/100 lb ZnO; 98.6% zinc oxide.

Protox®-168. [Zinc Corp. of Am.] French process zinc oxide with propionic acid surface treatment; pigment/reinforcing agent for rubber; 0.18 μ mean particle size; 99.97% thru 325 mesh; sp.gr. 5.6; sp.vol. 0.18; pkg. dens. 50 lb/ft^3; surf. area 6 m^2/g; oil absorp. 12 lb oil/100 lb ZnO; 99.3% zinc oxide.

Protox®-169. [Zinc Corp. of Am.] French process zinc oxide with propionic acid surface treatment; pigment for rubber; imparts high reinforcement and activation; 0.12 μ mean particle size; 99.99% thru 325 mesh; sp.gr. 5.6; sp.vol. 0.18; pkg. dens. 40 lb/ft^3; surf. area 9 m^2/g; oil absorp. 12 lb oil/100 lb ZnO; 99.1% zinc oxide.

Provol 10. [Croda] PPG-10 oleyl ether; emollient, superfatting agent, lubricant, pigment dispersant, and coupling agent used in personal care prods.; Gardner 10 max. liq.; sol. in min. oil, IPM, lanolin oil; insol. in water; acid no. 1 max.; pH 5.5–7.5 (3% aq.).

Provol 30. [Croda] PPG-30 oleyl ether; see Provol 10; Gardner 10 max. liq.; sol. in acetone IPM, Fluilan lanolin oil; insol. in water; acid no. 1 max.; pH 5.5–7.5 (3% aq.).

Provol 50. [Croda] PPG-50 oleyl ether; see Provol 10; Gardner 10 max. liq.; sol. see Provol 30; pH 5.5–7.5 (3% aq.).

Prowl®. [Am. Cyanamid/Ag] Pendimethalin; emulsifiable conc. herbicide for control of annual grasses and broadleaf weeds in field crops; 42.3% act.

Proxel CRL. [ICI Am.] 1,2-Benzisothiazolin-3-one sol'n.; microbiostat preservative; dk. brn. liq.; amine-like odor; water-sol.; sp.gr. 1.12; visc. 15 cps.; pH 9.5–11.5; 30–35% act.

PTN® 3.0S. [Novo] Pancreatic trypsins; enzyme for bating in leather industry; powd.

PTZ. [ICI Am.] Phenathiazine; antioxidant; yel.-grn. solid; m.p. 179–185 C.

Pulpex® Thermal Bonding Pulps. [Hercules] Polyolefin pulp; absorbent pad binder; used in feminine hygiene prods., disposable diapers, dressings.

Punctilious® Pure Ethyl Alcohol. [Quantum/USI] Ethanol; solv. and extraction medium; sol. in methanol; misc. in ether, water, chloroform.

Punctilious® Specially Denatured Ethyl Alcohol. [Quantum/USI] Ethanol; solv. or thinner; toilet prods. used externally, cosmetics, pharmaceuticals, detergents; sol. in methanol; misc. in ether, water, chloroform.

Pureco® 76. [Capital City] Coconut oil; Lovibond R1.0 max.; m.p. 76–80 F; iodine no. 12 max.; sapon. no. 248–264.

Pureco® 92. [Capital City] Partially hydrog. coconut oil; Lovibond R1.0 max.; m.p. 100–104 F; iodine no. 4 max.; sapon. no. 248–264.

Pureco® 110. [Capital City] Partially hydrog. coconut and palm oils; Lovibond R1.0 max.; m.p. 112–115 F; iodine no. 4 max.; sapon. no. 246–262.

Pure-Dent® B700. [Grain Processing] Corn starch USP, NF; binder and diluent for granulations and tablets when used wet or dry; disintegrant; for pharmaceutical use; wh. powd.; no odor, bland flavor; pH 5.5–7.0; 9–12.5% moisture.

Pure-Dent® B810. [Grain Processing] Corn starch NF; binder, diluent, absorbent, and disintegrant for pharmaceutical applics.; wh. powd., no odor, bland flavor; pH 4.5–7.0; 8–11% moisture.

Purton CFD. [Zschimmer & Schwarz] Coconut DEA; nonionic; foam stabilizer, thickener and superfatting agent for cosmetics, cleaners; liq.; 100% conc.

Purton SFD. [Zschimmer & Schwarz] Unsat. fatty acids DEA; nonionic; foam stabilizer, thickener and superfatting agent for cosmetics, cleaners; liq. to paste; 100% conc.

Puxol CB-22. [Pulcra SA] TEA dodecylbenzene sulfonate; anionic; used for formulation of liq. detergents, personal care prods., car washing shampoos; emulsifier for emulsion polymerization reactions;

pigment dispersant; liq.; 50 ± 1% conc.

Puxol XB-10. [Pulcra SA] Dodecylbenzene sulfonic acid; anionic; intermediate for mfg. of detergents; liq.; 95% conc.

PVP K-15. [GAF] PVP; binder, stabilizer, protective colloid, carrier; forms films; detoxifier for poison and irritants; visc. modifier for aq. systems; used in cosmetics, adhesives, detergents, coatings, paper, ink, textiles, printing, antifreeze, agric. and specialty formulations; powd.; m.w. 10,000; sol. in water and org. solv.; dens. (bulk) 36 lb/ft^3; 95% min. active.

PVP K-30. [GAF] Polyvinylpyrrolidone polymer; used in cosmetics as film-former, protective colloid, detoxicant in hair rods, cream, lotions, makeup, lipsticks, etc.; off wh. powd.; m.w. 40,000; sol. in water and many organic solvs., incl. alcohols, some chlorinated compds., nitroparaffins, and amines; dens. (bulk) 28 lb/ft^3; 95% min. active.

PVP K-60. [GAF] Polyvinylpyrrolidone polymer; used in adhesives to bond glass, metal, and plastics; as film-former in pressure-sensitive, water-rewettable types; as visc. modifier in polymer based adhesives; m.w. 160,000; dens. 9.3 lb/gal; 45% min. active.

PVP K-90. [GAF] Polyvinylpyrrolidone polymer; used in detergents and soaps as soil-suspending agent, esp. on syns.; in coatings (pigment and dyestuff dispersant, film-leveler); in paper (improves strength, ink receptivity, dyeing and printing chars.); and as pigment dispersant and leveling agent (paper coatings); in inks (dye solubilizer, visc. modifier); in textiles (increases dye receptivity of syn. fibers and latex emulsions); in printing pastes and formulations (dye stripping, delustering, fugitive tinting, sizing, finishing); in printing (with diazo lt. sensitizers in deep-etch plates, as post-etch for lithographic plates); for specialty applics. (agric. formulations, antifreeze sprays); powd. or aq. sol'n.; m.w. 360,000; dens. 20 lb/ft^3 (bulk, powd.); 9.3 lb/gal (sol'n.); 95% min. active (powd.); 20% min. active (sol'n.).

PVP/VA W-735. [GAF] PVP/VA copolymer; film-former used in hairsprays, gels, hair thickeners, tints, and dyes; sol. in most common org. solvs.; 50% aq. sol'n.

Pycal 94. [ICI Am.] POE aryl ether; plasticizer; pale yel. liq.; ref. index 1.502; sol. in water, ether, alcohol, ketone, lower aliphatic ester, and aromatic hydrocarbon solv.; sp.gr. 1.1; visc. approx 50 cps; flash pt. > 300 F; acid no. 2 max.; pH 4–6 (50% aq.).

Pyratex J 1904. [Bayer] Butadiene-styrene-2-vinylpyridine terpolymer; anionic emulsifier; vinyl pyridine latex; improves adhesion of solid rubber to fabric reinforcing materials; for rubber-fabric bonding (tire cord, V-belt, conveyor belting, etc.); pH 11 ± 0.5; 41 ± 1.5% solids.

Pyrax® A. [Vanderbilt] Pyrophyllite; inert filler/extender for NR and syn. rubbers and latexes; off-wh. to tan powd.; 99.5% min. thru 100 mesh, 97% min. thru 200 mesh; dens. 2.80 ± 0.05 mg/m^3.

Pyrax® ABB. [Vanderbilt] Pyrophyllite; diluent or carrier for agric. toxicants; used in cosmetics, pharmaceuticals, water paints, and dry wall plasters; filler for plastics; color additive for drugs and cosmetics; anticaking or blending agent, pelleting aid, or carrier in animal feeds; variable powd., diam. 11μ; pH 6.5 (10% solids).

Pyrax® B. [Vanderbilt] Pyrophyllite; see Pyrax ABB; wh. powd., diam. 10μ; pH 6.9 (10% solids).

Pyrax® WA. [Vanderbilt] Pyrophyllite; see Pyrax ABB; wh. powd., diam. 13μ; pH 6.6 (10% solids).

Pyro-Chek® 60PB. [Ferro] Brominated polystyrene; flame retardant for thermoplastic polyamides and polyesters, other engineering resins; off-wh. powd. (compacted gran.); sp.gr. 1.85; 60% Br.

Pyro-Chek® 68PB. [Ferro] Brominated polystyrene; flame retardant for plastics (esp. engineering plastics); suitable for thermosetting resins and polyolefins; synergist with antimony oxide; off-wh. powd.; also avail. compacted low dust or fine grind; sp.gr. 2.1; soften. pt. 215–225 C; 66% min. bromine.

Pyro-Chek® 77B. [Ferro] Bis (pentabromophenoxy) ethane; flame retardant exhibiting resistance to discoloration on exposure to lt.; used in ABS, HIPS, styrene-maleic anhydride, and polyolefins; synergist with antimony oxide; sp.gr. 2.8; dens. 1.5 cc/g; m.p. 322 C; 77% bromine.

Pyro-Chek® LM. [Ferro] Brominated polystyrene; flame retardant for styrenics; beige/tan powd.; sp.gr. 2.1.

2-Pyrol®. [GAF] 2-Pyrrolidone; monomer for polypyrrolidone; solv. for polymers, insecticides, polyhydroxylic alcohols, sugars, iodine, speciality inks; used in petrol. processing and acrylonitrile mfg.; plasticizer and coalescing agent for acrylic latices and acrylic-styrene copolymers in emulsion type floor coatings; intermediate for N-methylol derivs., alkaloids related to peganine, amino-butyric acid derivs.; liq.; f.p. 25 C; b.p. 245 C; flash pt. 129 C (OC).

Pyronate 40. [Witco] Sodium petrol. sulfonate; anionic; wetting agent, dispersant; oil and froth flotation; dk. liq.; water-sol.; dens. 9.6 lb/gal; sp.gr. 1.18 (60 F); 42% act.

Pyroter CPI-40. [Ajinomoto] PEG-40 hydrog. castor oil monopyroglutamic monoisostearic diester; nonionic; emulsifier, solubilizer and thickener used in personal care prods.; liq.; water-sol.; HLB 9.4; 100% conc.

Pyroter GPI-25. [Ajinomoto] PEG-25 glycerin monopyroglutamic monoisostearic diester; nonionic; moisturizer, emulsifier, solubilizer, dispersant and emollient; liq.; water-sol.; HLB 13.4.

Q-70. [Wacker Silicones] Silicone emulsion; defoamer for aq. foaming systems; disp. in water; 30% act.

Q-93. [Wacker Silicones] Silicone emulsion; defoamer for quick foam knockdown; water dilutable; 10% act.

Q-101. [Wacker Silicones] Silicone compd.; antifoam for nonaq. systems, e.g., cutting oil and agric. foaming systems; disp. in aromatic, aliphatic, and chlorinated hydrocarbons; 100% act.

Q-Cel 200. [PQ] Inorg. silicate microspheres; extender/filler for plastics systems; lower dens. and reduces resin vol. used; fiberglass reinforced plastics, cultured marble, cast polyester furniture and decorative parts, bowling ball cores, cast urethane and epoxy systems, autobody repair fillers, marine putties, PVC plastisol compds., slurry explosives; open mold casting and low pressure forming operations; wh. particulate; particle size 80 μ; closed-cell structure; dens. 5.7 lb/ft^3

Q-Cel 300. [PQ] Inorg. silicate microspheres; see Q-Cel 200; wh. particulate; particle size 75 μ; dens. 7.0 lb/ft^3.

QO® Furan. [QO] Chemical intermediate in the mfg. of herbicides, pharmaceuticals, plastics, and fine chemicals; colorless; sol. in most org. liqs.

QO® Furfural. [QO] 2-Furaldehyde (aldehyde obtained industrially from pentosan-containing agric. residues, e.g., corncobs, bagasse, cottonseed hulls, oat hulls, and rice hulls); chemical intermediate; used in mfg. of derivatives (furan and THF types); solv. for separating sat. from unsat. compds. in petrol. lubricating oil, gas oil, and diesel fuel; extractive distillation of C_4 and C_5 hydrocarbons for the mfg. of syn. rubber; decolorizing agent for wood rosin; solv. and processing aid for anthracene; ingred. in resins; reactive solv. and wetting agent in abrasive wheels and brake linings; colorless when freshly distilled; darkens on contact with air; industrial furfural is lt. yel. to brn. liq.; pungent almond-like odor; m.w. 96.08; dens. 1.1545; visc. 1.49 cps; b.p. 161.7 C (760 mm); f.p. -36.5 C; flash pt. (TCC) 61.7 C; surf. tens. 40.7 dynes/cm (29.9 C); ref. index 1.5235.

QO® Furfuryl Alcohol (FA®). [QO] Used in the prod. of foundry sand binders and corrosion-resistant resins; intermediate for esterification and etherification; impregnating sol'n. and carbon binder; wood adhesive component; solv. and temporary plasticizer for phenolic resins in the mfg. of coldm-molded abrasive wheels; visc. reducer, cure promoter, and carrier in amine-cured epoxy resins; pale yel. liq.; sol. in water and many common org. solvs.

QO® Tetrahydrofuran (THF). [QO] Industrial solv. with high solvency for broad range of materials and low boiling pt.; also used in the prod. of heterocyclic and open-chain compds.; chemical intermediate; colorless liq.; misc. with water; b.p. 66 C.

QO® Tetrahydrofurfuryl Alcohol (THFA®). [QO] High boiling solv. and carrier for pesticides; FDA approved for use in paper processing; chemical intermediate; also in industrial and consumer cleaners, leather and textile dyeing, epoxies, coatings, inks, paints, and adhesives; plasticizer and vinyl stabilizer carrier; colorless liq.; misc. with water; biodeg.

Quad Opacifier #35. [Lonza] Polymer emulsion; opacifier used in household ammonia, laundry and dishwashing detergents; acid bowl cleaner, fabric softener, hair conditioner; dens. 8.5 lb/gal; pH 2–3; 34–36% solids.

Quadrilan AT. [Harcros UK] Quat. fatty amine ethoxylate; cationic; antistat used in PVC and other polymers; amber clear liq.; char. odor; water-sol.; sp.gr. 1.081; visc. 400 cs; flash pt. > 300 F (COC); pour pt. < O C; pH 6–9 (1% aq.); biodeg.; 100% act.

Quadrilan BC. [Harcros UK] Benzalkonium chloride BP grade; cationic; germicide in disinfectants and sanitizers, emulsifier in the dyeing industry; pale yel. liq.; mild char. odor; sol. in water; sp.gr. 0.988; visc. 120 cs; flash pt. > 200 F (COC); pour pt. < O C; pH 7–8 (1% aq.); biodeg.; 50% act.

Quadrilan OD-75. [Harcros UK] Quat. ammonium compd.; cationic; softener, antistat; used in textiles; wh. stiff paste; IPA odor; disp. in water; sp.gr. 0.870 (40 C); visc. 55 cs (50 C); flash pt. 60 F (COC); pour pt. 35–40 C; pH 6.0–9.0 (1% aq.); biodeg.; 75% act.

Quadrilan SK. [Harcros UK] Quat. fatty amine ethoxylate; cationic; antistat used in polymers; liq.; 100% conc.

Quadrol. [BASF] Tetrahydroxypropyl ethylenediamine; polyol; chelating agent; intermediate used in resins, emulsifiers, surfactants, pharmaceuticals, herbicides, fungicides, insecticides, adhesives, and plasticizers; m.w. 292; sol. in water, ethyl alcohol, toluene, ethylene glycol, perchloroethylene; sp.gr. 1.03; ref. index 1.478; 99.2% tert. amine.

Quartamin 24P, 86P Conc. [Kao] Alkyl trimethyl ammonium chloride; cationic; textile softener, sanitizing agent, antistat; liq.; 27% conc.

Quartamin 86P. [Kao] N-Alkyl trimethyl ammonium chlorides; cationic; emulsifier, sanitizing agent, antistat, textile softener; liq.; biodeg.; 50%

conc.

Quartamin 86W. [Kao] Stearyl trimethyl ammonium chloride; cationic; hair rinse base; liq.; 30% conc.

Quartamin DCP. [Kao] Dicoco dimethyl ammonium chloride; cationic; emulsifier and corrosion inhibitor; liq.; 75% conc.

Quat-Pro E, E SD [Amerchol] Triethonium hydrolyzed animal protein ethosulfate; substantive to hair; for oil-free hair conditioners; water sol'n. and spray-dried powd. resp.

Quaternary O. [Ciba-Geigy] Quat. oleyl imidazoline; cationic; detergent, wetting agent, penetrant, antistat used in acids, solvs., germicides, fungistats, polishes, acid cleaners and corrosion inhibitor formulations, ore flotation, asphalt wetting; amber liq.; 100% conc.

Quatrene 7670. [Henkel] Fatty amidoamine quat. compd.; cationic; wetting agent, demulsifier, and corrosion inhibitor for petrol. prod.; liq.; 60% act.

Quatrene C-5-6. [Henkel] Coco benzyl imidazolinium chloride; cationic; corrosion inhibitor for continuous treatment in oil and gas prod., pipelines; red liq.; water-sol.; 60% conc.

Quatrene CA. [Henkel] Coco amido propyl dimethyl benzyl ammonium chloride; cationic; see Quatrene C-5-6; liq.; 80% act.

Quatrene CB. [Henkel] Alkyl dimethyl benzyl ammonium chloride; cationic; see Quatrene C-5-6; liq.; 80% act.

Quatrene CB-50. [Henkel] Alkyl dimethyl benzyl ammonium chloride; cationic; see Quatrene C-5-6; liq.; 50% act.

Quatrene CB-80. [Henkel] Alkyl dimethyl benzyl ammonium chloride; cationic; see Quatrene C-5-6; liq.; 80% act.

Quatrene CE. [Henkel] Coco dialkyl benzyl ammonium chloride; cationic; see Quatrene 7670; liq., paste; 75% act.

Quatrene MB-50. [Henkel] Alkyl dimethyl benzyl ammonium chloride; cationic; see Quatrene C-5-6; liq.; 50% act.

Quatrene MB-80. [Henkel] Alkyl dimethyl benzyl ammonium chloride; cationic; see Quatrene C-5-6; liq.; 80% act.

Quatrisoft Polymer LM-200. [Amerchol Europe] Polyquaternium-24; stabilizer for emulsions; thickener for surfactant systems; substantive conditioner for hair and skin care prods.; tan powd.

Quell-Oil C1. [Harcros UK] Oil spill dispersant; lt. amber clear liq.; mild fatty odor; water-disp.; sp.gr. 0.970; visc. 43 cs; pour pt. < –20 C; pH neutral; flash pt. 101 C (COC).

Querton 14 Br. [Berol Nobel] Tetradecyl trimethyl ammonium bromide; cationic; softening agent; powd.; 96% conc.

Querton 14 Br-40. [Berol Nobel] Tetradecyl trimethyl ammonium bromide; cationic; softening agent; liq.; 40% conc.

Querton 16 CL29. [Berol Nobel] Cetyl trimethyl ammonium chloride; cationic; softener; detergent; liq.; 29% conc.

Querton 16 CL50. [Berol Nobel] Cetyl trimethyl ammonium chloride; cationic; softening agent; liq.; 50% conc.

Querton 210 CL50, 210 CL80. [Berol Nobel] Didecyl dimethyl ammonium chloride; strong germicide used in food processing industry or sanitizers; liq.

Querton 246. [Berol Nobel] Benzalkonium chloride; disinfectant in food processing, medical care, sanitizers; liq.

Querton 442. [Berol Nobel] Di(hydrog. tallow) dimethyl ammonium chloride; cationic; softening agent; paste; 75% conc.

Querton 442-11. [Berol Nobel] Di(hydrog. tallow) dimethyl ammonium chloride; cationic; softening agent; paste; 77% conc.

Querton 442-82. [Berol Nobel] Dimethyl dihydrog. tallow ammonium chloride; cationic; softening agent; solid; 82% conc.

Querton 442S. [Berol Nobel] Distearyl dimethyl ammonium chloride; cationic; softening agent; powd.; 100% conc.

Querton 442-Sx. [Berol Nobel] Distearyl dimethyl ammonium chloride; cationic; softener formulations; powd.

Querton 470. [Berol Nobel] Ditallow dimethyl ammonium chloride; cationic; softening agent; liq.; 75% conc.

Querton 1149. [Berol Nobel] Coco dialkyl benzyl ammonium chloride; algicide in swimming pools; germicide in food processing industry or sanitizers; liq.

Querton BGCL50. [Berol Nobel] Trimethyl tallow ammonium chloride; cationic; emulsifier, dispersant; liq.; 50% conc.

Querton KKB CL50. [Berol Nobel] Coco dimethyl benzyl ammonium chloride; disinfectant in food processing, medical care, sanitizers; liq.

Questal DI. [Clough] Disodium EDTA; chelating agent for use in mildly acidic dry formulations; powd.

Questal FEC. [Clough] Trisodium HEDTA; chelating agent; misc. in water.

Questal Special. [Clough] Tetrasodium EDTA; chelating agent over wide pH range; sol. in water.

Questex 4H. [Rhone-Poulenc Basic] EDTA; chelating agent used in chemical industry, in cosmetic formulations; wh. powd.; sol. < 0.1 g/100 g in water; m.w. 292.25; dens. 54 lb/ft^3; chel. value 340 mg $CaCO_3$/g; pH 2.8; 99% min. act.

Questex 4SW. [Rhone-Poulenc Basic] Tetrasodium EDTA hydrate; general purpose chelating agent; dry cryst.; sol. in water.

Questex 4SW Crystals. [Rhone-Poulenc Basic] Tetrasodium EDTA tetrahydrate, tech. grade; chelating agent; wh. cryst.; m.w. 452.2; sol. 103 g/100 g of water; dens. 45 lb/ft^3; chel. value 219 mg $CaCO_3$/g (@ pH 11); pH 11.0 (1% aq.); 99.0% min. act.

Questric Acid. [Clough] EDTA; chelating agent for use in dry formulations where sodium salt is undesirable; powd.; water-insol.

Quickset® Extra. [Witco/Argus] MEK peroxide;

high purity initiator for R.T. curing of polyester resins; increased reactivity with lower peroxide conc.; liq.; 9.0% act. oxygen.

Quickset® Super. [Witco/Argus] MEK peroxide; high purity initiator for R.T. curing of polyester resins; increased reactivity with lower conc.; liq.; 9.0% act. oxygen.

Quilon C. [DuPont] Chrome complex sol'n. in IPA; water repellent and release agent; used for water-repellent pkg. materials, food pkg.; grease-resistant and release paper when used with PVA, water-repellent nonwoven fabrics, adhesive tapes, release films; blue-grn. liq.; alcoholic odor; b.p. 82 C; water-sol.; dens. 0.97 g/ml; flash pt. 4 C (TOC); 5.7% Cr; 7.8% Cl.

Quilon H. [DuPont] Chrome complex sol'n. in IPA; water repellent and release agent; used on paper and paperboard for water-repellent pkg. materials, food pkg., grease-resistant and release paper when used with PVA; water-repellent coatings and adhesives (for the insolubilization of adhesives, films, protein binders, and inks); water- and stain-resistant leather (shoes, golf bags, etc.), colorfast, water-repellent suede (drycleanable); dk.-grn. liq.; alcoholic odor; b.p. 82 C; water-sol.; dens. 1.04 g/ml; flash pt. –3 C (TOC); 9.2% Cr; 12.6% Cl.

Quilon L. [DuPont] Chrome complex sol'n. in IPA; water repellent and release agent; used on paper and paperboard for water-repellent pkg. materials, food pkg., grease-resistant and release paper when applied with PVA; release and grease-crawl-resistant paper for industrial release sheets for preparation of plastic laminates, food separators and pan liners, separator sheets for pressure-sensitive tapes and labels; water-repellent coatings and adhesives (for the insolubilization of adhesives, films, protein binders, and inks); dk.-grn. liq.; alcoholic odor; b.p. 82 C; water-sol.; dens. 1.03 g/ml; flash pt. –2 C (TOC); 9.2% Cr; 12.7% Cl.

Quilon M. [DuPont] Chrome complex sol'n. in IPA; water repellent; used on resin emulsions and sol'ns. for water-repellent coatings and adhesives (insolubilization of adhesives, films, protein binders, and inks), for water- and stain-resistant leather (golf bags, shoes, etc.), colorfast, water-repellent suede, and water-repellent, odorless feather fillers (pillows, sleeping bags, etc.); dk.-grn. liq.; alcoholic odor; b.p. 82 C; water-sol.; dens. 0.93 g/ml; flash pt. 1 C (TOC); 5.7% Cr; 7.8% Cl.

Quilon S. [DuPont] Chrome complex sol'n. in IPA; water repellent and release agent; used for release and grease-crawl-resistant papers (industrial release sheets for preparation of plastic laminates, food separators and pan liners, separator sheets for pressure-sensitive tapes and labels), water-repellent pkg. materials, food pkg., grease-resistant and release paper when applied with PVA, water-repellent coatings and adhesives (insolubilization of adhesives, films, protein binders, and inks); dk.-grn. liq.; alcoholic odor; b.p. 82 C; water-sol.; dens. 0.95 g/ml; flash pt. 2 C (TOC); 5.7% Cr; 7.8% Cl.

Quimipol DEA OC. [Quimigal-Quimica] Coconut DEA; nonionic; foamer booster, emulsifier, thickener, foam stabilizer; liq.; 85% conc.

Quimipol EA 2507. [Quimigal-Quimica] Syn. primary alcohol ethoxylate; nonionic; wool scouring, antistat, detergent, wetting agent, emulsifier; liq.; HLB 12.2; 100% conc.

Quimipol EA 6803, 6804. [Quimigal-Quimica] Syn. primary alcohol ethoxylate; nonionic; emulsifier and textile lubricant; solid; HLB 6.7 and 8.0 resp.; 100% conc.

Quimipol EA 6806, 6807, 6808. [Quimigal-Quimica] Syn. primary alcohol ethoxylate; nonionic; see Quimipol EA 6803; solid; HLB 10.0, 11.0 and 11.5 resp.; 100% conc.

Quimipol EA 6810. [Quimigal-Quimica] Syn. primary alcohol ethoxylate; nonionic; solubilizer, emulsifier, pigment dispersant, and detergent; solid; HLB 12.5; 100% conc.

Quimipol EA 6812, 6814, 6818, 6820, 6823, 6825, 6850. [Quimigal-Quimica] Syn. primary alcohol ethoxylate; nonionic; see Quimipol EA 6810; solid; HLB 13.5, 14.0, 15.0, 15.5, 15.8, 16.2, 16.7 resp.; 100% conc.

Quimipol ED 2021, 2022. [Quimigal-Quimica] Polypropylene glycol ethoxylate; nonionic; antistat, deduster, defoamer, demulsifier, detergent, dispersant; liq.; HLB 3.0 and 7.0 resp.; 100% conc.

Quimipol ENF 55. [Quimigal-Quimica] Nonylphenol ethoxylate; nonionic; emulsifier for min. oil and aliphatic solvs.; coupling agent; liq.; HLB 10.5; 100% conc.

Quimipol ENF 90. [Quimigal-Quimica] Nonylphenol ethoxylate; nonionic; wool scouring, oil emulsification, antistat, detergent, and wetting agent; liq.; HLB 12.9; 100% conc.

Quimipol ENF 120. [Quimigal-Quimica] Nonylphenol ethoxylate; nonionic; foam stabilizer and suspending agent for liq. detergents; liq.; HLB 14.0; 100% conc.

Quimipol ENF 140, 170. [Quimigal-Quimica] Nonylphenol ethoxylate; nonionic; solubilizer for alkyd aryl sulfonates and essential oils, emulsifier for pesticides and herbicides; paste; HLB 14.5 and 15.4 resp.; 100% conc.

Quso®. [PQ] Precipitated amorphous silica; thickener for pastes, creams, lotions in cosmetics and toiletries; suspending agent; improves free-flowing chars. of fine powds.; microfine; sol. in hot, strong alkaline sol'ns.

Q-White (70% Slurry). [Huber] Calcium carbonate; filler in coatings, rubber, sealants and adhesives, paper, and cleaning compds., and polishing agents; slurry, 0.01% 325-mesh residue; dens. 14.94 lb/gal; visc. 200 cps max.; pH 8.7; 70% solids.

Q-White (Spray-Dried). [Huber] Calcium carbonate; see Q-White (70% Slurry); beads; particle size 0.9 μ, 0.01% 325-mesh residue; sol. 0.0035 g/100 ml water @ 100 C; sp.gr. 2.71; dens. 22.6 lb/solid gal; dens. 52 lb/ft^3; oil absorp. 28%; ref. index 1.6; hardness 3 Moh; 97.7% $CaCO_3$.

R

Racumin®. [Bayer] Coumatetralyl; for control of rats and house mice; ylsh. cryst.; m.w. 292.3; sol. (g/100 ml): 1–5 g in cyclohexanone, 0–1 g in IPA, ligroin, methylene chloride, toluene; m.p. 166–173 C.

Radia 3200. [Oleofina] Hydrog. castor oil; internal lubricant for PVC; lubricant for ABS; flakes or powd.

Radia 7040. [Oleofina] n-Butyl oleate; chemical intermediate, lubricant; chemical synthesis; carbon source in antibiotic culture broths; lubricity improvers in min. oils; formulation of cutting, lamination, and textile oils, rust inhibitors; textile and leather industry; m.w. 327; sol. in most solvs. and min. and veg. oils; sp.gr. 0.854 (37.8 C); visc. 5.30 cps (37.8 C); ref. index. 1.4518; flash pt. 195 C (COC); cloud pt. –15 C.

Radia 7051. [Oleofina] n-Butyl stearate; chemical intermediate, lubricant, plasticizer; chemical synthesis; lubricant in min., cutting, lamination, and textile oils, rust inhibitors; textile and leather applic.; m.w. 325; sol. see Radia 7040; sp.gr. 0.844 (37.8 C); visc. 6.10 cps (37.8 C); ref. index 1.4432; flash pt. 200 C (COC); cloud pt. 22.5 C.

Radia 7060. [Oleofina] Methyl oleate; chemical intermediate, lubricant; chemical synthesis; lubricity improvers in min. oils; formulation of cutting, lamination, and textile oils; rust inhibitors; textile and leather industry; m.w. 286; sol. see Radia 7040; sp.gr. 0.862 (37.8 C); visc. 4.10 cps (37.8 C); ref. index 1.4513; flash pt. 178 C (COC); cloud pt. –11 C.

Radia 7070. [Oleofina] n-Butyl myristate; chemical intermediate, chemical synthesis; lubricant in min., cutting, lamination, textile oils, and rust inhibitors; textile and leather applic.; m.w. 282; sol. see Radia 7040; sp.gr. 0.845 (37.8 C); visc. 4.10 cps (37.8 C); ref. index 1.4388; flash pt. 172 C (COC); cloud pt. 0 C.

Radia 7104. [Oleofina] Glycerol capromyristate; lubricant, chemical intermediate; formulation of cutting, lamination, and textile oils; corrosion inhibitors; chemical synthesis; m.w. 603; sol. in trichlorethylene; sp.gr. 0.881 (98.9 C); visc. 6.25 cps (98.9 C); m.p. 36 C; flash pt. 216 C (COC).

Radia 7110. [Oleofina] Methyl stearate; see Radia 7070; m.w. 285; sol. see Radia 7040; m.p. 31.0; flash pt. 166 C (OC).

Radia 7118. [Oleofina] Methyl laurate; see Radia 7070; m.w. 214; sol. see Radia 7040; sp.gr. 0.855 (37.8 C); visc. 2.30 cps (37.8 C); ref. index 1.4322; flash pt. 132 C (COC); cloud pt. 1.5 C.

Radia 7120. [Oleofina] Methyl palmitate; see Radia 7070; m.w. 267; sol. see Radia 7040; sp.gr. 0.855 (37.8 C); visc. 4.30 cps (37.8 C); m.p. 27.5 C; flash pt. 149 C (COC).

Radia 7127. [Oleofina] Ethylhexyl laurate; see Radia 7070; m.w. 330; ref. index 1.4422; sol. see Radia 7040; sp.gr. 0.846 (37.8 C); visc. 4.50 cps (37.8 C); cloud pt. –25 C; flash pt. 171 C (COC); surf. tens. 30.50 dynes/cm.

Radia 7129. [Oleofina] Ethylhexyl palmitate; see Radia 7070; m.w. 357; sol. see Radia 7040; sp.gr. 0.847 (37.8 C); visc. 7.60 cps (37.8 C); ref. index 1.4467; flash pt. 206 C (COC); cloud pt. –1 C.

Radia 7130. [Oleofina] Isooctyl stearate; see Radia 7070; also plasticizer in paper industry; m.w. 377; sol. see Radia 7040; sp.gr. 0.846 (37.8 C); visc. 9.00 cps (37.8 C); ref. index 1.4479; flash pt. 220 C (COC); cloud pt. 12.5 C.

Radia 7131. [Oleofina] Isooctyl stearate; cosmetics emollient, solv.; plasticizer for PVC; lubricant for PS; liq.

Radia 7161. [Oleofina] Glycerol trioleate; see Radia 7104; also as carbon source in antibiotic culture broths; m.w. 880; sol. in hexane, benzene, IPA, trichlorethylene, min. and veg. oils; sp.gr. 0.908 (37.8 C); visc. 49.20 cps (37.8 C); ref. index 1.4685; flash pt. 274 (COC); cloud pt. –9 C.

Radia 7163. [Oleofina] Glycerol trioleate; see Radia 7161; m.w. 880; sol. see Radia 7161; sp.gr. 0.906 (37.8 C); visc. 42.30 cps (37.8 C); ref. index 1.4688; flash pt. 283 C; cloud pt. –12 C.

Radia 7171. [Oleofina] Pentaerythritol tetraoleate; see Radia 7104; m.w. 1160; sol. (@ 10%) in trichloroethylene, min. and veg. oil, hexane, benzene, IPA; sp.gr. 0.913 (37.8 C); visc. 70.80 cps (37.8 C); ref. index 1.4733; flash pt. 302 C (COC); cloud pt. –9.5 C.

Radia 7176. [Oleofina] Pentaerythritol tetrastearate; see Radia 7104; m.w. 1140; sol. (@ 10%) in trichlorethylene; sp.gr. 0.868 (98.9 C); visc. 14.50 cps (98.9 C); m.p. 64.5 C; flash pt. 290 C.

Radia 7190. [Oleofina] IPM; see Radia 7070; also as emollient, plasticizer, solubilizer of act. components in cosmetics and pharmaceuticals; m.w. 271; sol. see Radia 7040; sp.gr. 0.840 (37.8 C); visc. 3.50 cps (37.8 C); ref. index 1.4345; flash pt. 157 (COC); cloud pt. –3.5 C.

Radia 7194. [Oleofina] Diisopropyl adipate; see Radia 7190; m.w. 230; sol. see Radia 7040; sp.gr. 0.938 (37.8 C); visc. 2.80 cps (37.8 C); ref. index 1.4237; flash pt. 131 C (COC); cloud pt. –14.5 C.

Radia 7195. [Oleofina] Isopropyl stearate; see Radia 7190; m.w. 315; sol. see Radia 7040; sp.gr. 0.842

(37.8 C); visc. 5.10 cps (37.8 C); ref. index 1.4382; flash pt. 175 C (COC); cloud pt. 18.5 C.

Radia 7197. [Oleofina] Diisobutyl adipate; see Radia 7070; m.w. 258; sol. see Radia 7040; sp.gr. 0.939 (37.8 C); visc. 3.90 cps (37.8 C); ref. index 1.4300; flash pt. 153 C (COC); cloud pt. –30 C.

Radia 7200. [Oleofina] IPP; see Radia 7190; m.w. 300; sol. see Radia 7040; sp.gr. 0.843 (37.8 C); visc. 5.05 cps (37.8 C); ref. index 1.4373; flash pt. 167 C (COC); cloud pt. 11.5 C.

Radia 7220. [Oleofina] Isopropyl myristopalmitate; cosmetics emollient, solv.; plasticizer for PVC; lubricant for PS; liq.

Radia 7230. [Oleofina] Isobutyl oleate; see Radia 7040; m.w. 327; ref. index 1.4499; sol. see Radia 7040; sp.gr. 0.854 (37.8 C); visc. 5.60 cps (37.8 C); cloud pt. –17 C; flash pt. 191 C (COC); surf. tens. 31.50 dynes/cm.

Radia 7231. [Oleofina] Isopropyl oleate; see Radia 7060; m.w. 313; sol. see Radia 7040; sp.gr 0.853 (37.8 C); visc. 5.05 cps (37.8 C); ref. index 1.4479; flash pt. 169 C (COC); cloud pt. –17 C.

Radia 7240. [Oleofina] Isobutyl stearate; see Radia 7070; also gloss promoter in special lacquers, and plasticizer in paper industry; m.w. 325; sol. see Radia 7040; sp.gr. 0.844 (37.8 C); visc. 6.20 cps (37.8 C); ref. index 1.4421; flash pt. 181 C (COC); cloud pt. 17 C.

Radia 7241. [Oleofina] Isobutyl stearate; cosmetics emollient, solv.; plasticizer for PVC; lubricant for PS; liq.

Radia 7303. [Oleofina] Glycerol trioleate; see Radia 7161; m.w. 880; sol. see Radia 7161; sp.gr. 0.910 (37.8 C); visc. 45.20 cps (37.8 C); ref. index 1.4709; flash pt. 275 C (COC); cloud pt. 5 C.

Radia 7330. [Oleofina] Isooctyl oleate; see Radia 7060; m.w. 373; sol. see Radia 7040; sp.gr. 0.857 (37.8 C); visc. 7.50 cps (37.8 C); ref. index 1.4542; flash pt. 208 C (COC); cloud pt. –20 C.

Radia 7331. [Oleofina] Ethylhexyl oleate; see Radia 7060; m.w. 373; sol. see Radia 7040; sp.gr. 0.857 (37.8 C); visc. 7.50 cps (37.8 C); ref. index 1.4538; flash pt. 193 C (COC); cloud pt. –23 C.

Radia 7363. [Oleofina] Glycerol trioleate; see Radia 7161; m.w. 880; sol. see Radia 7303; sp.gr. 0.906 (37.8 C); visc. 41.20 cps (37.8 C); ref. index 1.4691; flash pt. 289 C (COC); cloud pt. –13 C.

Radia 7370. [Oleofina] Trimethylpropane trioleate; see Radia 7104; m.w. 900; sol. see Radia 7303; sp.gr. 0.904 (37.8 C); visc. 45.90 cps (37.8 C); ref. index 1.4712; flash pt. 307 (COC); cloud pt. –24 C.

Radia 7500. [Oleofina] Cetyl palmitate; see Radia 7070; also used for wax formulation due to its high m.p.; m.w. 484; sol. (@ 10%) in benzene and trichlorethylene; sp.gr. 0.805 (98.9 C); visc. 4.30 cps (98.9 C); m.p. 49 C; flash pt. 214 C (COC).

Radia 7501. [Oleofina] Cetyl stearate; see Radia 7500; m.w. 534; sol. see Radia 7176; sp.gr. 0.807 (98.9 C); visc. 4.60 cps (98.9 C); m.p. 56 C; flash pt. 216 C (COC).

Radia 7510. [Oleofina] Isononyl stearate; see Radia 7070; m.w. 380; sol. see Radia 7040; sp.gr. 0.851 (37.8 C); visc. 11.30 cps (37.8 C); ref. index 1.4493; flash pt. 201 C (COC); cloud pt. 11 C.

Radia 7514. [Oleofina] Pentaerythritol tetrabehenate; see Radia 7104; m.w. 1300; sol. in trichlorethylene, min. oil; sp.gr. 0.860 (98.9 C); visc. 17.20 cps (98.9 C); m.p. 73.5 C; flash pt. 289 C (COC).

Radiacid 200. [Oleofina] 12-Hydroxystearic acid; internal lubricant for PVC; flakes or powd.

Radiacid 408, 411, 420, 423. [Oleofina] Stearic acid; external lubricants for vinyl plastics; flakes or powd.

Radiacid 1060. [Oleofina] Calcium stearate; internal lubricant for PVC; wh. fluffy powd.

Radiamac 6148, 6149. [Oleofina] Hydrog. tallow amine acetate; cationic; flotation reagent, anticaking aid for fertilizer, corrosion inhibitor, emulsifier; solid; HLB 11; 100% conc.

Radiamac 6169. [Oleofina] Coconut oil amine acetate; cationic; see Radiamac 6148; solid; HLB 11; 100% conc.

Radiamac 6179. [Oleofina] Tallow amine acetate; cationic; see Radiamac 6148; solid; HLB 11; 100% conc.

Radiamine 6140, 6141. [Oleofina] Hydrog. tallow amine (6141—dist.); cationic; min. flotation, corrosion inhibitor, pigment dispersant; cosmetics; lubricant and mold release for hard rubber, textile chemical, chemical synthesis; antistat and antifog additive for plastic foils; solid; 100% conc.

Radiamine 6160, 6161. [Oleofina] Coconut oil amine (6161—dist.); cationic; see Radiamine 6140; liq.; 100% conc.

Radiamine 6163, 6164. [Oleofina] Laurylamine (6164—dist.); cationic; see Radiamine 6140; liq., solid; 100% conc.

Radiamine 6170, 6171. [Oleofina] Tallow amine (6171—dist.); cationic; see Radiamine 6140; paste; 100% conc.

Radiamine 6172, 6173. [Oleofina] Oleyl amine (6173—dist.); cationic; see Radiamine 6140; liq., paste; 100% conc.

Radiamine 6540. [Oleofina] Hydrog. tallow diamine; cationic; corrosion inhibitor, dispersant, emulsifier, intermediate for chemical synthesis; solid; 100% conc.

Radiamine 6560. [Oleofina] Coconut oil diamine; cationic; see Radiamine 6540; paste; 100% conc.

Radiamine 6570. [Oleofina] Tallow diamine; cationic; see Radiamine 6540; paste; 100% conc.

Radiamine 6572. [Oleofina] Oleyl diamine; cationic; see Radiamine 6540; liq.; 100% conc.

Radiamine AC 14, 20, 30, 81, 82, 505, 6144. [Oleofina] Fatty amine compd.; cationic; anticaking and dust inhibitor for fertilizers; paste; 100% conc.

Radiamuls 125. [Oleofina] Sorbitan laurate; nonionic; food emulsifier, dispersant, antistaling agent, spray drying aid, solubilizer, dryness improver, whipping aid; used in food applic.; liq.; HLB 6.8; acid no. 7 max.

Radiamuls 135. [Oleofina] Sorbitan palmitate; see Radiamuls 125; powd. or flake; HLB 5.3; m.p. 50 C; acid no. 7 max.

Radiamuls 137. [Oleofina] Polysorbate 20; see Radiamuls 125; liq.; HLB 16; acid no. 2 max.

Radiamuls 145. [Oleofina] Sorbitan stearate; see Radiamuls 125; powd. or flakes; HLB 5; m.p. 50–60 C; acid no. 7 max.

Radiamuls 147. [Oleofina] Polysorbate 60; see Radiamuls 125; liq.; HLB 15; acid no. 2 max.

Radiamuls 155. [Oleofina] Sorbitan oleate; see Radiamuls 125; liq.; HLB 4.7; acid no. 7 max.

Radiamuls 157. [Oleofina] Polysorbate 80; see Radiamuls 125; liq.; HLB 15.7; acid no. 2 max.

Radiamuls 345. [Oleofina] Sorbitan tristearate; see Radiamuls 125; powd. or flakes; HLB 2.7; m.p. 50–60 C; acid no. 17 max.

Radiamuls MG 2142. [Oleofina] Glyceryl stearate; food emulsifier; lubricant for extruded foods and for food mfg. equip.; powd., flake, liq.; m.p. 56–60 C; HLB 3.0; 40% conc.

Radiamuls PEG. [Oleofina] Fatty acid PEG esters; nonionic; emulsifier and fat dispersant for milk powd. for animal feed; liq., paste, flakes, powd.; 100% conc.

Radiamuls POLY. [Oleofina] Fatty acids polyglycerol esters; nonionic; aerator and stabilizer for food prods.; visc. liq., flakes, powd.; 100% conc.

Radianol 2106. [Oleofina] Glycerol tricaprylate/caprate; lubricant for food mfg. equip.; liq.

Radianol 7106. [Oleofina] Glycerol tricaprate/caprylate; emollient, plasticizer, solubilizer, food additive used in food and feed industry, cosmetics, and pharmaceuticals; low-visc. oil; m.w. 518; sol. @ 10% in hexane, benzene, IPA, trichlorethylene, min. and veg. oils; sp.gr. 0.933 (37.8 C); visc. 15.90 cps (37.8 C); ref. index 1.4402; flash pt. 178 C (COC); cloud pt. < –30 C.

Radianol 7376. [Oleofina] Glycerol triheptanoate; see Radianol 7106; also used as butter tracers; low-visc. oil; m.w. 432; sol. @ 10% in hexane, benzene, IPA, trichlorethylene, min. and veg. oils; sp.gr. 0.952 (37.8 C); visc. 10.10 cps (37.8 C); ref. index 1.4444; flash pt. 207 C (COC); cloud pt. < –30 C.

Radiaquat 6442. [Oleofina] Dihydrog. tallow dimethyl ammonium chloride; cationic; softener; paste; 74% conc.

Radiaquat 6443. [Oleofina] Dialkyl dimethyl ammonium chloride; cationic; softener; liq.; 75% conc.

Radiaquat 6444. [Oleofina] Palmityl trimethyl ammonium chloride; cationic; softener; liq.; 50 or 29% conc.

Radiaquat 6462. [Oleofina] Dicoco dimethyl ammonium chloride; cationic; softener, bactericide for detergents, textiles, fabric softeners; paste; 75% conc.

Radiaquat 6470. [Oleofina] Dialkyl dimethyl ammonium chloride; cationic; softener; liq., paste; 75% conc.

Radiaquat 6471. [Oleofina] Tallow trimethyl ammonium chloride; cationic; softener; liq.; 50% conc.

Radiastar 1060. [Oleofina] Calcium stearate; anticaking agent for powd. food prods.; wh. fluffy powd.

Radiastar 1100. [Oleofina] Magnesium stearate; see Radiastar 1060; wh. fluffy powd.

Radiastar 1208. [Oleofina] Aluminum stearate; see Radiastar 1060; wh. fluffy powd.

Radiasurf 7000. [Oleofina] PEG-20 glycerol stearate; nonionic; chemical intermediate, emulsifier, detergent, lubricant, wetting agent; chemical synthesis; o/w or w/o cosmetic and pharmaceutical emulsions; pearlescent shampoo formulations; detergency and cleaning prods.; formulation of cutting, lamination, and textile oils; rust inhibitors; pigment wetting, faster grinding, and improved gloss in paints, lacquers, and printing inks; lubricant and mold release in plastics; m.w. 1395; sol. (@ 10%) in trichlorethylene, hexane, benzene, IPA, min. and veg. oils; water-disp.; sp.gr. 1.032 (37.8 C); visc. 113.30 cps (37.8 C); HLB 13.0; ref. index 1.4672; flash pt. 244 C (COC); cloud pt. 27.5 C.

Radiasurf 7125. [Oleofina] Sorbitan laurate; nonionic; emulsifier, descouring aid, antistat; anticorrosive agent for pipelines; cleaner for metallic surfaces; superfatting, bodying and antifog aid; pigment dispersant; detergent; emulsion of solv.; cutting oils; textile lubricant additive, concrete and paper additives; leather aux.; cosmetics and pharmaceuticals; plastics; pesticides; dry cleaning formulations; amber liq.; m.w. 450; sol. in hexane, trichlorethylene, min. and veg. oil, benzene, IPA; disp. in water; sp.gr. 1.025; visc. 1193.0 cps (37.8 C); HLB 6.8; acid no. 7 max.; sapon. no. 155–175; ref. index 1.4718; flash pt. 198 C (COC); cloud pt. 23 C.

Radiasurf 7135. [Oleofina] Sorbitan palmitate; nonionic; see Radiasurf 7125; wh. solid; m.w. 500; sol. in IPA, trichlorethylene, min. oil; disp. in water; sp.gr. 0.943 (98.9 C); visc. 51.70 cps (98.9 C); HLB 5.3; m.p. 54.5 C; acid no. 7 max.; sapon. no. 140–160; flash pt. 214 C (COC).

Radiasurf 7136. [Oleofina] Sorbitan-POE-palmitate; nonionic; emulsifier, wetting agent, defoamer, rust inhibitor, pigment grinding, antistat; liq.; HLB 15.5.

Radiasurf 7137. [Oleofina] Polysorbate 20; nonionic; see Radiasurf 7125; amber liq.; m.w. 1340; sol. in water, benzene, IPA, trichlorethylene, min. and veg. oil; sp.gr. 1.095 (37.8 C); visc. 172.3 cps (37.8 C); HLB 16; acid no. 2 max.; sapon. no. 40–55; ref. index 1.4712; flash pt. 248 C (COC); cloud pt. –10 C.

Radiasurf 7140. [Oleofina] Glyceryl stearate; nonionic; see Radiasurf 7136; also internal lubricant for PVC; powd., flakes; HLB 3.0; 40% conc.

Radiasurf 7141. [Oleofina] Glyceryl stearate SE; nonionic; see Radiasurf 7136; powd., flakes.

Radiasurf 7145. [Oleofina] Sorbitan stearate; nonionic; see Radiasurf 7125; wh. solid; m.w. 510; sol. in trichlorethylene; disp. in water; sp.gr. 0.943 (98.9 C); visc. 48.20 cps (98.9 C); HLB 5; m.p. 58 C; acid no. 7 max.; iodine no. 1 max.; sapon. no. 146–158; flash pt. 232 C (COC).

Radiasurf 7146. [Oleofina] Glycerol mono-12-hydroxystearate; nonionic; chemical intermediate, emulsifier, detergent, wetting agent, lubricant; chemical synthesis; cosmetics and pharmaceuticals; pearlescent shampoo formulations; detergency and cleaning prods.; cutting, lamination, and textile oils; rust inhibitor; pigment wetting, grinding, and improved gloss in paints, lacquers, printing inks; paper industry; lubricant and mold release in plastics; textile and leather processing; m.w. 550; sol. (@ 10%); @ 75 C in benzene, IPA, trichlorethylene; sp.gr. 0.925 (98.9 C); visc. 25.80 cps (98.9 C); HLB 1.9; m.p. 70 C; flash pt. 260 C (COC).

Radiasurf 7147. [Oleofina] Polysorbate 60; nonionic; see Radiasurf 7125; amber liq.; m.w. 1400; sol. in trichlorethylene, min. and veg. oil, water, benzene, IPA; sp.gr. 1.068 (37.8 C); visc. 202.6 cps (37.8 C); HLB 15; acid no. 2 max.; sapon. no. 40–56; ref. index 1.4694; flash pt. 260 C (COC).

Radiasurf 7150. [Oleofina] Glycerol oleate; internal lubricant for PVC; liq.

Radiasurf 7151. [Oleofina] Glycerol oleate, SE; nonionic; chemical intermediate, emulsifier, detergent, wetting agent, lubricant; chemical synthesis; cosmetic and pharmaceuticals; detergency and cleaning prods.; cutting, lamination, and textile oils; rust inhibitors; pigment wetting, grinding, and improved gloss in paints, lacquers, and printing inks; paper industry; lubricant and mold release in plastics; textile and leather processing; liq.; m.w. 515; sol. see Radia 7161; sp.gr. 0.947 (37.8 C); visc. 137.70 cps (37.8 C); HLB 3.3; ref. index 1.4696; flash pt. 208 C (COC); cloud pt. 12.0 C.

Radiasurf 7152. [Oleofina] Glycerol oleate; nonionic; see Radiasurf 7151; liq.; m.w. 521; sol. see Radia 7161; sp.gr. 0.934 (37.8 C); visc. 77.10 cps (37.8 C); HLB 2.8; ref. index 1.4697; flash pt. 232 C (COC); cloud pt. 7.5 C.

Radiasurf 7153. [Oleofina] Glycerol ricinoleate; nonionic; see Radiasurf 7151; antistat and antifogging additive in plastics; m.w. 554; sol. (@ 10%); in benzene, IPA, trichlorethylene, min. and veg. oils; sp.gr. 0.982 (37.8 C); visc. 356.20 cps (37.8 C); HLB 2.5; ref. index 1.4774; flash pt. 229 C (COC); cloud pt. –16.0 C; 40% conc.

Radiasurf 7155. [Oleofina] Sorbitan oleate; nonionic; see Radiasurf 7125; amber liq.; m.w. 510; sol. in hexane, IPA, trichlorethylene, min. and veg. oils; sp.gr. 0.987 (37.8 C); visc. 395.2 cps (37.8 C); HLB 4.7; acid no. 7 max.; sapon. no. 145–161; ref. index 1.4778; flash pt. 249 C (COC); cloud pt. 7 C.

Radiasurf 7156. [Oleofina] Pentaerythritol oleate; corrosion inhibitor for lubricating oils and greases; liq.

Radiasurf 7157. [Oleofina] Polysorbate 80; nonionic; see Radiasurf 7125; amber liq.; m.w. 400; sol. in water, benzene, trichlorethylene, IPA, min. and veg. oils; sp.gr. 1.071 (37.8 C); visc. 240.4 cps (37.8 C); HLB 15.7; acid no. 2 max.; sapon. no. 44–56; ref. index 1.4726; flash pt. 287 C (COC); cloud pt. 5.5 C.

Radiasurf 7172. [Oleofina] Trimethylolpropane oleate; corrosion inhibitor for lubricating oils and greases; liq.

Radiasurf 7175. [Oleofina] Pentaerythritol stearate; nonionic; see Radiasurf 7146; solid; m.w. 675; sol. see Radia 7176; sp.gr. 0.886 (98.9 C); visc. 17.60 cps (98.9 C); HLB 2.3; m.p. 53.0 C; flash pt. 259 C (COC).

Radiasurf 7196. [Oleofina] Propylene glycol myristate; nonionic; wetting aid, lubricant, opacifier, antistat, dispersant, w/o emulgent, scouring and detergent aid, defoamer, plasticizer, rust inhibitor; cosmetics and pharmaceuticals, lubricating and cutting oils, textile and leather aids, pigment grinding in paints and printing inks, latex and emulsion paints, plastics, waxes, and maintenance prods., insecticides; wh. to amber liq.; m.w. 378; sol. see Radiasurf 7000; sp.gr. 0.902 (37.8 C); visc. 17.30 cps (37.8 C); HLB 3.9; acid no. 5 max.; sapon. no. 190–205; ref. index 1.4379; flash pt. 145 C (COC).

Radiasurf 7201. [Oleofina] Propylene glycol stearate; nonionic; see Radiasurf 7196; wh. to cream solid; m.w. 457; sol. (@ 10%) in hexane, trichlorethylene, min. oil; sp.gr. 0.849 (98.9 C); visc. 4.50 cps (98.9 C); HLB 1.9; acid no. 2 max.; sapon. no. 177; flash pt. 206 C (COC).

Radiasurf 7206. [Oleofina] Propylene glycol oleate; nonionic; see Radiasurf 7196; amber liq.; m.w. 460; sol. see Radia 7161; sp.gr. 0.900 (37.8 C); visc. 18.80 cps (37.8 C); HLB 2.3; acid no. 2 max.; sapon. no. 172–180; ref. index. 1.4606; flash pt. 157 C (COC); cloud pt. –9 C.

Radiasurf 7269. [Oleofina] Ethylene glycol distearate; nonionic; wh. solid; sol. see Radia 7161; HLB 1.5; m.p. 65 C; acid no. 3 max.; sapon. no. 190–200.

Radiasurf 7270. [Oleofina] Ethylene glycol stearate; nonionic; see Radiasurf 7196; wh. solid; m.w. 445; sol. (@ 10% and 75 C) in benzene, IPA, trichlorethylene, veg. oil; sp.gr. 0.851 (98.9 C); visc. 4.80 cps (98.9 C); HLB 1.8; m.p. 61 C; acid no. 2 max.; sapon. no. 180–190; flash pt. 214 C (COC).

Radiasurf 7345. [Oleofina] Sorbitan tristearate; nonionic; see Radiasurf 7125; solid; m.w. 900; sol. in trichlorethylene, min. and veg. oils; sp.gr. 0.893 (98.9 C); visc. 19.50 cps (98.9 C); HLB 2.7; m.p. 60 C; flash pt. 253 C (COC).

Radiasurf 7372. [Oleofina] Trimethylolpropane oleate; nonionic; see Radiasurf 7151; liq.; m.w. 514; sol. see Radia 7161; sp.gr. 0.935 (37.8 C); visc. 90 cps (37.8 C); HLB 4.4; ref. index 1.4705; flash pt. 224 C (COC); cloud pt. –18.5 C.

Radiasurf 7400. [Oleofina] Diethylene glycol oleate; nonionic; see Radiasurf 7196; amber liq.; m.w. 478; sol. see Radiasurf 7000; sp.gr. 0.929 (37.8 C); visc. 58.50 cps (37.8 C); HLB 3.8; acid no. 4 max.; sapon. no. 155–168; ref. index 1.4642; flash pt. 180 C (COC); cloud pt. –6.5 C.

Radiasurf 7402. [Oleofina] PEG-4 oleate; nonionic; wetting aid, lubricant, opacifier, antistat, dispersant, o/w emulgent, scouring and detergent aid, defoamer, plasticizer, rust inhibitor, visc. modifier, antifog aid; cometics and pharmaceuticals; lubricating and cutting oils; textile and leather aids; pigment grinding aids in paints and printing inks; latex and emulsion paints; plastics; waxes and maintenance prods.; glass fiber; insecticides; silicones; amber liq.; sol. (@ 10%) in benzene, isopropyl, trichlorethylene, hexane, min. and veg. oils; water-disp.; sp.gr. 0.962 (37.8 C); visc. 34.00 cps (37.8 C); HLB 8.4; acid no. < 3; sapon. no. 105–125; ref. index 1.4645; flash pt. 218 C (COC); cloud pt. –10 C.

Radiasurf 7403. [Oleofina] PEG-8 oleate; nonionic; see Radiasurf 7402; amber liq.; m.w. 728; sol. see Radiasurf 7000; sp.gr. 1.007 (37.8 C); visc. 49.50 cps (37.8 C); HLB 11.7; acid no. < 3; sapon. no. 75–90; ref. index 1.4672; flash pt. 261 C (COC); cloud pt. –1 C.

Radiasurf 7404. [Oleofina] PEG-12 oleate; nonionic; see Radiasurf 7402; amber liq.; m.w. 920; sol. (@ 10%) in water, benzene, IPA, trichlorethylene, hexane, min. and veg. oils; sp.gr. 1.030 (37.8 C); visc. 66.30 cps (37.8 C); HLB 13.2; acid no. < 3; sapon. no. 60–75; ref. index 1.4669; flash pt. 254 C (COC); cloud pt. 19 C.

Radiasurf 7410. [Oleofina] Diethylene glycol stearate; nonionic; see Radiasurf 7196; wh. solid; m.w. 475; sol. see Radiasurf 7000; sp.gr. 0.873 (98.9 C); visc. 5.95 cps (98.9 C); HLB 3.5; m.p. 51.5 C; acid no. 3 max.; sapon. no. 160–175; flash pt. 191 C (COC).

Radiasurf 7411. [Oleofina] Diethylene glycol stearate, SE; nonionic; see Radiasurf 7196; wh. solid; m.w. 470; sol. see Radiasurf 7000; sp.gr. 0.878 (98.9 C); visc. 5.90 cps (98.9 C); HLB 4.1; m.p. 52 C; acid no. 3 max.; sapon. no. 160–170; flash pt. 176 C (COC).

Radiasurf 7412. [Oleofina] PEG-4 stearate; nonionic; see Radiasurf 7402; wh. paste; m.w. 522; sol. (@ 10%) in benzene, IPA, trichlorethylene, hexane, min. oil; sp.gr. 0.913 (98.9 C); visc. 6.70 cps (98.9 C); HLB 7.5; m.p. 36 C; acid no. < 3; sapon. no. 120–135; flash pt. 186 C (COC).

Radiasurf 7413. [Oleofina] PEG-8 stearate; nonionic; see Radiasurf 7402; wh. paste; m.w. 722; sol. (@ 10%) in benzene, trichlorethylene, min. oil, hexane; sp.gr. 0.951 (98.9 C); visc. 9.45 cps (98.9 C); HLB 11.9; m.p. 34 C; acid no. < 3; sapon. no. 75–90; flash pt. 248 C (COC).

Radiasurf 7414. [Oleofina] PEG-12 stearate; nonionic; see Radiasurf 7402; wh. paste; m.w. 906; sol. see Radia 7161; sp.gr. 0.981 (98.9 C); visc. 12.60 cps (98.9 C); HLB 13.3; m.p. 38 C; acid no. < 3; sapon. no. 60–75; flash pt. 241 C (COC).

Radiasurf 7417. [Oleofina] PEG-1500 stearate; nonionic; see Radiasurf 7402; wh. solid; m.w. 1812; sol. see Radia 7500; sp.gr. 1.023 (98.9 C); visc. 28.40 cps (98.9 C); HLB 16.5; m.p. 47 C; acid no. < 3; sapon. no. 30–40; flash pt. 248 C (COC).

Radiasurf 7420. [Oleofina] Diethylene glycol laurate; nonionic; see Radiasurf 7196; wh. paste; m.w. 352; sol. see Radia 7161; sp.gr. 0.942 (37.8 C); visc. 15.20 cps (37.8 C); HLB 5.7; acid no. 3 max.; sapon. no. 195–205; flash pt. 164 C (COC); cloud pt. 28.5 C.

Radiasurf 7421. [Oleofina] Diethylene glycol laurate; wh. liq.; m.w. 346; sol. see Radia 7161; sp.gr. 0.901 (98.9 C); visc. 3.60 cps (98.9 C); HLB 6.3; acid no. 3 max.; sapon. no. 183–193; flash pt. 173 C (COC); cloud pt. 29.5 C.

Radiasurf 7422. [Oleofina] PEG-4 laurate; nonionic; see Radiasurf 7402; wh. liq.; m.w. 379; sol. see Radiasurf 7000; sp.gr. 0.983 (37.8 C); visc. 23.40 cps (37.8 C); HLB 9.6; acid no. < 3; sapon. no. 140–150; ref. index 1.4542; flash pt. 188 C (COC); cloud pt. –3 C.

Radiasurf 7423. [Oleofina] PEG-8 laurate; nonionic; see Radiasurf 7402; wh. liq.; m.w. 610; sol. see Radiasurf 7404; sp.gr. 1.023 (37.8 C); visc. 41.70 cps (37.8 C); HLB 13.2; acid no. < 3; sapon. no. 90–100; ref. index 1.4596; flash pt. 238 C (COC); cloud pt. 4 C.

Radiasurf 7431. [Oleofina] PEG-300 oleate; nonionic; see Radiasurf 7402; amber liq.; sol. see Radiasurf 7000; HLB 9.5; acid no. < 3; sapon. no. 95–115; cloud pt. –8 C.

Radiasurf 7432. [Oleofina] PEG-300 stearate; nonionic; see Radiasurf 7402; wh. paste; sol. see Radiasurf 7412; HLB 9.7; m.p. 28–35 C; acid no. < 3; sapon. no. 95–115.

Radiasurf 7443. [Oleofina] PEG-8 dioleate; nonionic; see Radiasurf 7402; amber liq.; m.w. 911; sol. see Radiasurf 7000; sp.gr. 0.962 (37.8 C); visc. 47.00 cps (37.8 C); HLB 7.4; acid no. < 5; sapon. no. 120–130; ref. index 1.4655; flash pt. 248 C (COC); cloud pt. –3 C.

Radiasurf 7453. [Oleofina] PEG-8 distearate; nonionic; see Radiasurf 7402; wh. paste; m.w. 901; sol. (@ 10%) in benzene, IPA, trichlorethylene; sp.gr. 0.920 (98.9 C); visc. 9.85 cps (98.9 C); HLB 7.7; m.p. 39 C; acid no. < 5; sapon. no. 120–130; flash pt. 242 C (COC).

Radiasurf 7454. [Oleofina] PEG-12 distearate; see Radiasurf 7402; wh. solid; m.w. 1101; sol. in benzene, trichlorethylene, hexane, IPA, min. oil; sp.gr. 0.940 (98.9 C); visc. 12.20 cps (98.9 C); HLB 9.7; m.p. 40 C; acid no. < 5; sapon. no. 100–110; flash pt. 249 C.

Radiasurf 7505. [Oleofina] Distearyl phthalate; internal lubricant for PVC; solid.

Radiasurf 7600. [Oleofina] Glycerol stearate; nonionic; see Radiasurf 7125; powd., flakes; HLB 3.3; 60% conc.

Radiasyn 7174. [Oleofina] Pentaerythritol tetra C_5/C_9; syn. ester lubricant; liq.; m.w. 580; sol. see Radia 7161; sp.gr. 0.971 (37.8 C); visc. 29.50 cps (37.8 C); ref. index 1.4543; flash pt. 275 C (COC); cloud pt. –35 C.

Radiasyn 7177. [Oleofina] Pentaerythritol tetra-C_7; syn. ester lubricant; liq.; m.w. 580; sol. see Radia 7161; sp.gr. 0.971 (37.8 C); visc. 24.20 cps (37.8 C); ref. index 1.4526; flash pt. 251 C (COC); cloud pt. –33 C.

Radiasyn 7178. [Oleofina] Pentaerythritol tetra C_8/C_{10}; syn. ester lubricant; liq.; m.w. 690; see Radia 7161; sp.gr. 0.945 (37.8 C); visc. 31.50 cps (37.8 C); ref. index 1.4546; flash pt. 282 C (COC); cloud pt. –2 C.

Radiasyn 7364. [Oleofina] Trimethylolpropane tri C_5/C_9; syn. ester lubricant; liq.; m.w. 465; sol. see Radia 7161; liq.; m.w. 465; sp.gr. 0.951 (37.8 C); visc. 15.40 cps (37.8 C); ref. index. 1.4511; flash pt. 255 C (COC); cloud pt. < –35 C.

Radiasyn 7367. [Oleofina] Trimethylolpropane tri-C_7; syn. ester lubricant; liq.; m.w. 465; sol. see Radia 7161; sp.gr. 0.950 (37.8 C); visc. 14.70 cps (37.8 C); ref. index 1.4514; flash pt. 223 C (COC); cloud pt. < –35 C.

Radiasyn 7368. [Oleofina] Trimethylolpropane tri C_8/C_{10}; syn. ester lubricant; liq.; m.w. 550; sol. see Radia 7161; sp.gr. 0.932 (37.8 C); visc. 22.30 cps (37.8 C); ref. index 1.4527; flash pt. 248 (COC); cloud pt. < –35 C.
Raneoff®. [Eastern Color] Silicone; durable water repellent; water-sol.
Rapidase®. [Int'l. Bio-Synthetics] Bacterial alpha-amylase; enzyme for textile starch desizing; liq.
Rapidblend 1793. [Anchor] Disp. of zinc oxide, magnesium oxide, and liq. alkylated diphenylamine antioxidant (5:4:2 ratio); vulcanizing/antidegradant system for polychloroprene compds.; rods of stiff, thermoplastic putty consistency; sp.gr. 2.0; 100% act.
Ray Krome 340, 400, Fe. [ITT Rayonier] Lignosulfonate; chrome complexed dispersant for drilling mud; powd.
Raylig. [ITT Rayonier] Sodium lignosulfonate; anionic; dispersant, suspending agent, deflocculation and visc. control in water dispersions of solids; brn. liq.; vanilla odor; water-sol.; sp.gr. 1.23; visc. 500 cps. b.p. 100 C; pH 5.5; 50% act.
Raymix. [ITT Rayonier] Desugared sodium lignosulfonate; dispersant for concrete admixtures; liq. and powd.
RC 7. [Releasomers] Fluorocarbon mold release agent and lubricant for silicone rubber molding operations, thermoset plastic molding; solv. sol'n.
Reach® 101, 201, 501. [Reheis] Aluminum chlorohydrate; antiperspirant for increased wetness protection, esp. for aerosols; powd.; 97% thru 325 mesh; pH 4.0–4.4 (15% aq.); 46–48.5, 46–48.5, and 46–48% Al_2O_3 resp., 15.8–17.5, 15.8–17.5, and 15.8–16.8% Cl.
Reax® 15B. [Westvaco] Sodium lignosulfonate; anionic; sequestrant; treatment of industrial water supplies; dispersant for dyestuffs and org. pesticides in wettable powd. formulations; additive in negative plates for lead-acid storage batteries; powd.; pH 10.5; 100% conc.
Reax® 45A. [Westvaco] Modified sodium lignosulfonate; wetting agent and dispersant for pesticides; brn. powd.; water sol.; pH 10 (2% aq.).
Reax® 45L. [Westvaco] Modified sodium lignosulfonate; dispersant providing controlled wetting; suspending agent; sol. in water and alkaline sol'ns.; some sol. in acidic sol'ns.
Reax® 65AE. [Westvaco] Saponified rosin and blended fatty acids; air entrainment aid, plasticizer, masonry and portland cements; liq.; 34% solids.
Reax® 77. [Westvaco] Amine acetate and sulfonated lignin reaction prod.; anionic; dispersant, grinding aid; flow promoter; prod. of portland cements; liq.; sp.gr. 1.200; visc. 40 cps; 70% act.
Reax® 77XF. [Westvaco] Amine acetate and sulfonated polyhydroxy alcohol reaction prod.; anionic; flow promoter and grinding aid for portland cements; liq.; sp.gr. 1.220; visc. 75 cps.
Reax® 80A. [Westvaco] Sodium lignosulfonate; dispersant for dyestuffs, lead acid storage batteries; brn. powd.; dens. 29 lb/ft^3; pH 11.2 (2% aq.); surf. tens. 54.0 dynes/cm (1% aq.).
Reax® 80C. [Westvaco] Sodium lignosulfonate, modified; dispersant, suspending agent for dyestuffs, lead acid storage batteries; brn. powd.; dens. 29 lb/ft^3; pH 9.8 (2% aq.); surf. tens. 54.0 dynes/cm (1% aq.); 100% conc.
Reax® 81A. [Westvaco] Sodium lignosulfonate, modified; dispersant, suspending agent in dye applics. and liq. fertilizer sol'ns.; brn. powd.; dens. 30 lb/ft^3; pH 10.5 (2% aq.); surf. tens. 57.0 dynes/cm (1% aq.); 100% conc.
Reax® 82. [Westvaco] Sodium lignosulfonate, modified; dispersant, suspending agent; brn. powd.; dens. 26 lb/ft^3; pH 11.0 (2% aq.); surf. tens. 45 dynes/cm (1% aq.); 99% conc.
Reax® 83A. [Westvaco] Sodium lignosulfonate, modified; dispersant, suspending agent in dye applics., liq. fertilizer and agric. chemical formulations; brn. powd.; dens. 31 lb/ft^3; pH 10.2 (2% aq.); surf. tens. 61.5 dynes/cm (1% aq.); 100% conc.
Reax® 85A. [Westvaco] Sodium lignosulfonate, modified; see Reax 81A; brn. powd.; dens. 28 lb/ft^3; pH 10.9 (2% aq.); surf. tens. 59.3 dynes/cm (1% aq.); 100% conc.
Reax® 88B. [Westvaco] Sodium lignosulfonate; anionic; dispersant, chelating agent, visc. reducer; used in aq. systems; dye formulations, pesticides, plant foods, fertilizers; visc. 95; pH 11.5; 100% act.
Reax® 90. [Westvaco] Sodium salt of sulfonated hydroxy alkylated Kraft lignin; anionic; dispersant for disperse dyes, esp. diazo type in high pressure applics.; liq.; 33% act.
Reax® 90P. [Westvaco] Sodium lignosulfonate; dispersant in dye systems; brn. powd.; dens. 24 lb/ft^3; pH 10.3 (2% aq.); surf. tens. 42 dynes/cm (1% aq.).
Reax® 92. [Westvaco] Sulfonated kraft lignin; dye dispersant; brn. powd.; dens. 33 lb/ft^3; pH 9.8 (2% aq.); surf. tens. 50.2 dynes/cm (1% aq.).
Reax® 95A. [Westvaco] Sodium lignosulfonate; brn. powd.; dens. 30 lb/ft^3; pH 9.5 (2% aq.); surf. tens. 53.0 dynes/cm (1% aq.).
Reax® 100M. [Westvaco] Sodium kraft lignosulfonate, modified; dispersant for inorg. materials in aq. media; complexing agent for micronutrient formulations; chelating agent; slurry visc. reducing agent; brn. powd.; water-sol.; dens. 24 lb/ft^3; visc. 65 cps; pH 8.9 (2% aq.); surf. tens. 48.9 dynes/cm (1% aq.); 100% act.
Reax® 100M (Liquid). [Westvaco] Kraft lignosulfonate, modified; see Reax 100M; brn. liq.; visc. 20 cps; pH 8.9 (2% aq.); surf. tens. 48.9 dynes/cm (1% aq.); 50% act.
Reax® G-1. [Westvaco] Polyglycol-based liq. with a combination of selected aliphatic hydroxyl functional groups; grinding aid and flow promoter for use in portland cements; processing aid; liq.; readily dilutable in water; sp.gr. 1.125; visc. 50 cp.
Reax® G-2. [Westvaco] Glycol-based reaction prod. with a combination of surface-active groups incl. amines, aliphatic and aromatic hydroxyls, carboxylic acids, and sulfonates; flow promoter and grinding aid for use in portland cements; liq.; sp.gr. 1.110; visc. 30

cp.

Reax® SR-1. [Westvaco] Sodium lignosulfonate; anionic; dyestuff dispersant for textile industry; brn. powd.; dens. 24 lb/ft³; pH 9.9 (2% aq); 100% conc.

Reax® SR-7. [Westvaco] Sodium lignosulfonate; anionic; see Reax SR-1; brn. powd.; pH 6.7 (2% aq.); 100% conc.

Recodan. [Grindsted Denmark] Integrated emulsifiers and stabilizers; nonionic; emulsifier, heat stabilizer; improves fat disp., enhances palatability in sterilized or pasteurized, recombined, filled, imitation, chocolate, and flavored milk prods.; powd.; 100% conc.

Redax. [Vanderbilt] N-Nitrosodiphenylamine; retarder for NR and syn. rubbers; 0.5–1.0% suggested in hot processing compds. as retarder to control scorch; tan to brn. flakes; m.w. 198.23; dens. 1.24 ± 0.03 mg/m³; m.p. 63–66 C.

Refinex. [Floridin] RVM attapulgite; contact adsorbent for reclaiming motor oils; fine powd.

Renacit 7. [Mobay] Pentachlorothiophenol absorbed on clay; peptizing agent facilitating open mill and internal mixer mastication in rubber industry; lt. gray powd.; substantially insol. in common solvs.; sp.gr. 2.3; m.p. partly fusible; 46.5% act.

Renacit 7/WG. [Mobay] Renacit 7 containing stearic acid and paraffin wax; see Renacit 7; lt. gray gran.; dens. 1.1 g/cm³; m.p. partly fusible.

Renacit 8/LG. [Mobay] Activated zinc salts of unsat. fatty acids; see Renacit 7; br. lentil-shaped gran.; dens. 1.1 g/cm³; m.p. ≥ 75 C.

Renex® 36. [ICI Am.] Trideceth-6; nonionic; detergent, wetting agent used in alkaline cleaners; dispersant for solids and paint pigments; dishwashing; cloudy-clear colorless thin oily liq.; sol. in acetone, CCl_4, ethyl acetate, Cellosolve, ethanol, toluene, aniline; disp. in water; sp.gr. 1.0; visc. 80 cps.; HLB 11.4; cloud pt. < 32 F (1% aq.); flash pt. > 300 F; surf. tens. 27 dynes/cm (0.01%); pH 6 (1%); 100% conc.

Renex® 650. [ICI Am.] Nonoxynol-30; nonionic; detergent, emulsifier, stabilizer; surfactant for detergent use; emulsification, and latex stabilizer; wh. solid; sol. in water, propylene glycol, IPA glycol, IPA, xylene; sp.gr. 1.15; HLB 17.1; cloud pt. 212 F; flash pt. > 300 F; pour pt. 91 F; surf. tens. 42 dynes/cm (0.01%); 100% act.

Renex® 751. [ICI Am.] POE octyl phenol ether; nonionic; lubricant, emulsifier, textile processing; liq.; 100% conc.

Rennilase® 11L, 14L, 46L, 150 L type T. [Novo] Proteolytic enzyme derived from the fungus Mucor miehei; milk-clotting enzyme used in cheese-making for coagulation as alternative to calf rennet; liq., yelsh. spray-dried powd.; water-sol.; dens. 1.2–1.25 g/ml.

Rennilase® 14L, Type XL. [Novo] Extra termolabile modification of Rennilase; see Rennilase 11 L; liq.; water-sol.; dens. 1.15 g/ml.

Rennilase® 50L, Type TL. [Novo] Termolabile modification of Rennilase; see Rennilase 11 L; liq.; water-sol.; dens. 1.2 g/ml.

Rennilase® 50TL, 60L. [Novo] Microbial coagulant; enzyme; milk coagulant for cheese prod.; liq.

Reocor 12. [Ciba-Geigy] Carboxylic acid deriv.; corrosion inhibitor for lubricants; clear brn. liq.; oil-sol.

Reocor 190. [Ciba-Geigy] Carboxylic acid; corrosion inhibitor for metalworking fluids, high water-based fluids; off-wh. solid.

Reogen®. [Vanderbilt] High m.w. sulfonic acid with paraffin oil; plasticizer and processing aid in elastomers; clear mahogany liq.; dens. 0.83–0.86 mg/m³; acid no. 8.0–8.2.

Reomet® 39. [Ciba-Geigy] Substituted benzotriazole deriv.; copper corrosion inhibitor for lubricants; liq.; oil-sol.

Reomet® 42. [Ciba-Geigy] Substituted benzotriazole deriv.; copper corrosion inhibitor for metalworking fluids, antifreeze, HWCF's, industrial cooling towers; liq.; sol. in water and glycol.

Repello DC. [Scher] Resin-wax blend; nonionic; fabric water repellent for dry-cleaning industry; paste; 40% act.

Repel-O-Tex 100. [Lyndal] Modified melamine condensate; water repellent, fluorochemical extender; water-disp.

Repel-O-Tex D. [Lyndal] Metallized wax emulsion; anionic; semidurable water repellent for fabric; water-disp.

Repel-O-Tex D-5. [Lyndal] Wax emulsion; imparts nondurable water and spot resistance; hosiery boarding release agent; water-disp.

Reserve Salt Flake. [Ciba-Geigy] Sodium m-nitrobenzene sulfonate; anionic; stabilizer for dyeing of fibers; assistant in discharge printing; flake.

Resin 731D. [Hercules] Modified dehydrogenated (disproportionated) rosin; pale, oxidation-resistant, thermoplastic resin used in hot-melt-applied adhesives and coatings for paper and paperboard substrates, as tackifier and processing aid for rubber-based adhesives and molding compds.; for use in contact with food; USDA Rosin N solid, flakes; sol. in alcohols, esters, ketones, min. spirits, and aromatic hydrocarbons; dens. 1.058 kg/l; soften. pt. (R&B) 73 C; flash pt. (COC) 209 C; acid no. 154; sapon. no. 159.

Resogen® DM. [Crompton & Knowles] Aminoplast precondensate; cationic; fixing agent for dyes on cellulosics; liq.

Resyn® 28-1310, 28-2930. [Nat'l. Starch] Carboxylated PVAc copolymer; hair fixative; for hair sprays, setting lotions, conditioners; when neutralized sol. in water, water/alcohol and hydrocarbon/ethanol blends.

Resyn® 28-3307. [Nat'l. Starch] Carboxylated PVAc; uv absorber for hair and skin prods., protective coatings; when neutralized, sol. in water, alcohol/water mixts.

Retarder A. [Crompton & Knowles] Quat. ammonium deriv.; cationic; retarder for level dyeing on acrylics.

Retarder AK. [Akrochem] Modified phthalic anhydride; nondiscoloring retarding agent to reduce scorching of rubber compds. at processing temps.;

also acts as an activator for certain blowing agents; wh. nondusting powd.; 1.0% max. retained 100 mesh; practically odorless; sol. in alcohol, benzene, and acetone; slightly sol. in water; sp.gr. 1.45–1.51; m.p. 123–132 C.

Retarder BA, BAX. [Akrochem] Predominantly benzoic acid; retarding agent for natural and syn. rubbers and latexes; nonstaining; acts as an activator for certain blowing agents; processing aid with certain cis-polybutadiene rubbers; BAX is oil-treated; wh. flakes; sp.gr. 1.30; m.p. 122 C.

Retarder CA. [Hart Chem. Ltd.] N-alkyl dimethylbenzyl ammonium chloride; cationic; retarder for acrylic fiber dyeing with cationic dyestuff; used at high temps.; clear liq.; sp.gr. 0.98; pH alkaline; 40% act.

Retarder L. [GAF] Lignin complex; nonselective retarder for vat dyes to improve penetration and leveling; slows down strike rate; maintains reduction bath stability and reduces fiber degradation; liq.

Retarder N, N-85. [Hart Prod.] Lauryl dimethyl benzyl ammonium chloride; retarder in dyeing, antistatic agent; liq.; 40 and 85% conc. resp.

Retarder PX. [Akrochem] Phthalic anhydride, oil treated; nondiscoloring retarding agent to reduce scorching of rubber compds. at processing temps.; off-wh. powd.; practically odorless; slightly sol. in water; sol. in alcohol; sp.gr. 1.52; m.p. 129–134 C.

Retarder SAX. [Akrochem] Tech. salicylic acid (90%) and lt. process oil treatment (10%); retarder; vulcanization inhibitor for SBR and natural rubber compds.; also as accelerator for W types of Neoprene; blowing agent activator in sponge rubber compds.; off-wh. cryst. powd.; practically odorless; easily disperses in dry polymers; sp.gr. 1.31.

Reten® 210. [Hercules] Polyquaternium-5; cationic; flocculating agent and retention aid for pulp and paper industry; max retention of fillers; slip agent, thickener, antistat, adhesive, film-former, solids-suspending agent, and/or chemical crosslinking agent; substantive to protein; good specific adhesion to hydrophobic surf.; for paper used in food contact; wh. powd., 30% max. thru 200 mesh; dissolves in water; dens. 42 lb/ft^3; visc. 700 cps; pH 5 (1%); 15% max. volatiles.

Reten® 220. [Hercules] Polyquaternium-5; see Reten 210; wh. powd., 30% max. thru 200 mesh; dens. 42 lb/ft^3; visc. 750 cps (1.0%); pH 5 (1%); 15% max. volatiles conc.

Reten® 300. [Hercules] Polyquaternium-14; high m.w. polymers used as thickeners, solids-suspending agents, flocculants, slip agents, antistats, adhesives, film-formers, chemical cross-linking agents; used in hair-treatment preps. powd. and sol'n. resp.; sol. in water.

Reten® 304. [Hercules] Cationic polymer; retention aid and flocculant for paper industry; functions over a wide pH range in both alum and alum-free systems; liq.; 30% solids.

Reten® 420. [Hercules] Polyacrylamide; nonionic; high m.w. polymers used as thickeners, solids-suspending agents, flocculants, slip agents, antistats, adhesives, film-formers, chemical cross-linking agents; powd.; sol. in water.

Reten® 421, 423, 425. [Hercules] Acrylamide/sodium acrylate copolymer; see Reten 420; powd.; sol. in water.

Reten® 520. [Hercules] Acrylamide-based polymer; nonionic; flocculant, friction reducer, leveling, film-forming, thickener, and solids suspending agent in aq. systems; aq. adhesives to develop "legginess"; improves sheet formation in random-laid processing on long fibers; used in food-grade paper (FDA approved), and fruit processing; off-wh. to wh. powd., 15% max. on 20 mesh screen; dissolves in water; dens. 42 lb/ft^3; visc. 700 cps (1%); pH 8 (1%); 15% max. volatiles.

Reten® 521. [Hercules] Acrylamide-based copolymer; anionic; see Reten 210; wh. powd.; readily dissolves in water; dens. (bulk) 705 kg/m^3; visc. 3300 cps (1%).

Reten® 523. [Hercules] Acrylamide-based copolymer; anionic; see Reten 521; wh. powd.; dissolves in water; dens. 44 lb/ft^3.

Reten® 525. [Hercules] Acrylamide-based copolymer; anionic; see Reten 521; wh. powd.; dissolves in water; dens. 44 lb/ft^3.

Reten® 763. [Hercules] polymer sol'n.; cationic; retention aid for fiber, fines, and fillers and provides drainage and flocculation of wh.-water solids in paper machine and flotation save-all operations; functions over a wide pH range in alum and alum-free systems; used as dry-creping aid, in kraft liquor clarification, and process water treatment; for use in mfg. of paper/paperboard for contact with food (up to 0.2% resin); liq.; dens. 4.24 kg/l; visc. 250–400 cps; f.p. -12 C; pH 4.5; 35% solids.

Reticusol. [Croda] Sol. animal reticulin; moisturizer, conditioner for skin care prods.; pale straw liq.; m.w. 3000; 20% act. in water.

Rewocid® DC 212. [Rewo GmbH] Fatty acid polydiethanolamide; nonionic; lubricating additive, corrosion inhibitor; med. visc. liq.; 100% conc.

Rewocid® DU 185. [Rewo GmbH] Undecylenamide DEA and diethanolamine; nonionic; detergent, emulsifier used as bacteriocide, thickener, foam stabilizer in shampoos; yel. visc. liq.; sol. in water, oil, alcohols, glycols; sp.gr. 1.0; high visc.; 100% act.

Rewocid® SBU 185. [Rewo GmbH] Disodium undecylenamido MEA-sulfosuccinate; anionic; fungicide, detergent, foamer used in personal care prods.; pale yel. powd.; slight odor; sol. in alcohol, glycol; sp.gr. 1.0; pH 6.5–7.5 (5% solids); 50% act.

Rewocid® U 185. [Rewo GmbH] Undecylenamide MEA; nonionic; fungicide, detergent; improves foam quality, stability, superfatting, increases visc.; yel. flakes; strong odor; sol. in alcohol, glycol, disp. in water; m.p. 50 C; 100% act.

Rewocid® UTM 185. [Rewo GmbH] Undecylenic acid propylamido trimethyl ammonium methosulfate; cationic; bactericide, fungicide for toiletries; conditioner for shampoos, antistat; low visc. liq.; 45% conc.

Rewocor AC 28. [Rewo GmbH] Fatty acid alkylolamide; nonionic; corrosion inhibitor for metal working fluids; med. visc. liq.; 100% conc.

Rewocor B 2045. [Rewo GmbH] Alkenyl sulfosuccinic acid anhydride; nonionic; corrosion inhibitor; visc. liq.; 100% conc.

Rewocor B 3010. [Rewo GmbH] Alkenyl succinate, sodium salt; anionic; anticorrosion agent; liq.; 40% conc.

Rewocor B 3032. [Rewo GmbH] Alkenyl succinate, TEA salt; anionic; see Rewocor B 3010; liq.; 40% conc.

Rewocor BAC. [Rewo GmbH] Butyl ammonium caprylate; anionic; see Rewocor B 3010; liq.; water-sol.; 95% conc.

Rewocor RA 178. [Rewo GmbH] Aliphatic/aromatic carboxylic acid, alkanolammonium salt; anionic; anticorrosion agent for aq. systems; med. visc. liq.; 80% conc.

Rewocor RA 280. [Rewo GmbH] Oleic acid dibutylamide; nonionic; corrosion inhibitor for oils; visc. liq.; 100% conc.

Rewocor RA-B 90. [Rewo GmbH] Boric acid ester, modified; anionic; see Rewocor RA 178; visc. liq.; 85% conc.

Rewocor RA-SI. [Rewo GmbH] Alkenyl succinimide; nonionic; corrosion inhibitor for lubricants and hydraulic fluids; visc. liq.; 100% conc.

Rewocor TPAC 100. [Rewo GmbH] Tallow propylene diammonium caprylate; anionic; anticorrosion agent for oils and boiler feed water systems; visc. liq.; 100% conc.

Rewocoros AC 28. [Rewo GmbH] Alkylolamide; anticorrosive for metalworking and lubrication; liq.

Rewocoros B 2045. [Rewo GmbH] Alkenyl sulfosuccinic acid anhydride; rust preventive additive, tar adhesive agent; liq.

Rewocoros B 3010. [Rewo GmbH] Alkenyl succinic acid, disodium salt; rust preventive additive in aq. media; liq.

Rewocoros B 3032. [Rewo GmbH] Alkenyl succinic acid, TEA salt; rust preventive additive in aq. media; liq.

Rewocoros BAC. [Rewo GmbH] Butyl ammonium caprylate; corrosion inhibitor; liq.; water-sol.; 95% conc.

Rewocoros RA 178. [Rewo GmbH] Alkylolammonium salt of an aliphatic/aromatic carboxylic acid; corrosion inhibitor for aq. systems; liq.; 80% conc.

Rewocoros RA 280. [Rewo GmbH] Oleic acid dibutylamide; nonionic; corrosion inhibitor; liq.; 100% conc.

Rewocoros RAB 90. [Rewo GmbH] Modified boric DEA; low foaming corrosion inhibitor for sol. aq. metalworking oils, syn. cold lubricants, water glycol hydraulic fluids, grinding lubricants; liq.

Rewocoros RABE. [Rewo GmbH] Boric acid amine ester; rust preventive additive; liq.; 90% conc.

Rewocoros RASI. [Rewo GmbH] Alkenyl succinimide; nonionic; corrosion inhibitor; liq.; 100% conc.

Rewocoros TPAC 100. [Rewo GmbH] Tallow propylene diammonium caprylate; corrosion inhibitor for oils and boiler feed water systems; liq.; 100% conc.

Rewoderm® ES 90. [Rewo GmbH] Coconut fatty acid monoglyceride polyglycol ether; nonionic; emulsifier for cosmetics, superfatting agent, solubilizer; visc. liq.; 100% conc.

Rewoderm® L 67-75. [Rewo GmbH] Coco monoglyceride polyglycol ether; thickener and superfatting agent for cosmetics; liq.

Rewoderm® LI 48-50, LI S 75. [Rewo GmbH] Tallow monoglyceride polyglycol ether; thickener and superfatting agent for cosmetics; wh. paste.

Rewoderm® LI 63, 67, 67-75. [Rewo GmbH] Coco monoglyceride polyglycol ether; nonionic; see Rewoderm LI 48; also solubilizer and superfatting agent for cosmetics; soft paste, waxy solid, and med.-visc. liq. resp.; 100, 100, and 75% conc.

Rewoderm® LI 420. [Rewo GmbH] Tallow fatty acid monoglyceride polyglycol ether; nonionic; see Rewoderm LI 48; also thickener and superfatting agent for cosmetics; waxy solid; 100% conc.

Rewoderm® LI 420-70. [Rewo GmbH] Tallow fatty acid monoglyceride polyglycol ether; nonionic; mild surfactant, thickener for shampoos, foam baths, baby shampoos; gel; 70% conc.

Rewoderm® LIS 75. [Rewo GmbH] Fatty acid monoglyceride polyglycol ether, modified; mild thickening agent for shampoos, foam baths, baby shampoos, shower gels; gel; 75% conc.

Rewolan® AWS. [Rewo GmbH] Lanolin, ethoxylated; nonionic; superfatting agent for personal care prods.; visc. liq.; 100% conc.

Rewolan® E. [Rewo GmbH] Wool wax alcohol sulfosuccinate; anionic; moisturizing skin-protective agent for creams, lotions, and cleansing formulations; liq., paste; 50% conc.

Rewolan® E 50, 100. [Rewo GmbH] Lanolin polyglycol ether; nonionic; superfatting agent for personal care prods.; visc. liq. and hard wax resp.; 50 and 100% conc. resp.

Rewolan® LP. [Rewo GmbH] Isopropyl lanolate; emollient for skin care, toiletries; liq.; 100% conc.

Rewolub GSM. [Rewo GmbH] Blend of nonionics; lubricant for metalworking fluids; liq.; 100% conc.

Rewolub KSM 14. [Rewo GmbH] Dicarboxylic acid ethoxylate; nonionic; surfactant for syn. cooling oils, lubricant; visc. liq.; 100% conc.

Rewolub TMP 155. [Rewo GmbH] Trimethylolpropane fatty acid ester; lubricant in textile auxs. and metalworking fluids; liq.; 100% conc.

Rewolub TMP 275. [Rewo GmbH] Trimethylolpropane oleic acid ester; see Rewolub TMP 155; liq.; 100% conc.

Rewomat B 2003. [Rewo GmbH] Tetrasodium (1,2-dicarboxyethyl)-N-alkyl sulfosuccinamide; anionic; detergent, foaming agent and stabilizer; emulsion polymerization additive; industrial, metal, household and all-purpose cleaner; solubilizer for detergent raw materials; soldering aid; pigment dispersant; emulsifier for wax and oil; cosmetics; amber clear liq.; sol. in water, alkali and electrolytes; pH 7–9 (10% solids); 35% solids.

Rewomat TMS. [Rewo GmbH] Disodium alkyl sulfosuccinamide; anionic; foaming agent, stabilizer, emulsifier, foam additive for natural and syn. lattices; emulsion polymerization; wetting agent for latex impregnation; soft creamy paste or liq.; pH 9–10 (10% solids); 35% solids.

Rewomid® 203/S. [Rewo GmbH] Lauric DEA; nonionic; detergent, foam stabilizer and visc. modifier for shampoos and detergents; off-wh. solid; slight odor; m.p. 40 C; 100% act.

Rewomid® AC 28. [Rewo GmbH] Groundnut fatty acid polydialkanolamide; nonionic; detergent, emulsifer used as an anticorrosion additive for metalworking lubricants and oils; brn. liq.; dens. 0.98 g/cm^3; sp.gr. 1.0; visc. med.; 100% act.

Rewomid® C 212. [Rewo GmbH] Cocamide MEA; nonionic; detergent used in detergent prods.; stabilizer of emulsions; off-wh. flakes; amidic slight odor; m.p. 73 C; 100% act.

Rewomid® C 220SE. [Rewo GmbH] Cocamide DEA; nonionic; detergent, foam stabilizer, detergent builder; liq.; 100% conc.

Rewomid® CD. [Rewo] Coconut MEA; nonionic; detergent, foam stabilizer, detergent builder; liq.; 100% conc.

Rewomid® DC 212/S. [Rewo GmbH] Cocamide DEA; nonionic; detergent, foam stabilizer and visc. modifier used for shampoos and detergents; lt. yel. visc. liq.; slight odor; dens. 0.99 g/cm^3; 100% act.

Rewomid® DC 212/SE. [Rewo GmbH] Cocamide DEA; nonionic; foam booster, thickener for toiletries; med. visc. liq.; 100% conc.

Rewomid® DC 220/LS. [Rewo GmbH] Cocamide DEA modified; nonionic; thickener and superfatting agent; med. visc. liq.; 100% conc.

Rewomid® DC 220/SE. [Rewo GmbH] Cocamide DEA; nonionic; detergent, thickener, foam stabilizer and superfatting agent; soft paste; 100% act.

Rewomid® DL 203. [Rewo GmbH] Lauramide DEA and diethanolamine; nonionic; detergent; increases visc.; good superfatting; stabilizer of emulsions; fixation of perfumes; wh. wax; dens. 0.97 g/cm^3; sp.gr. 1.0; 100% act.

Rewomid® DL 203/S. [Rewo GmbH] Lauric DEA; nonionic; foam booster and thickener for personal care prods. and general purpose cleaners; wax; 100% conc.

Rewomid® DL 240. [Rewo GmbH] Cocamide DEA (2:1) and diethanolamine; nonionic; detergent added to cosmetic preparations and cleaning and washing agents; stabilizer of emulsions; superfatting; yel. liq.; faint odor; visc. med.; 100% act.

Rewomid® DLMS. [Rewo GmbH] Lauramide DEA; nonionic; foam stabilizer and thickener in detergent systems; wax; 100% conc.

Rewomid® DLM/SE. [Rewo GmbH] Myristamide DEA; nonionic; thickener and superfatting agent for cosmetics; wax; 100% conc.

Rewomid® DO 280/S. [Rewo GmbH] Mixed fatty acid DEA; nonionic; aerosol anticorrosive agent; hair cosmetic specialty; cream bath additive; emulsifier; superfatter; dye and perfume solubilizer; paste; 100% conc.

Rewomid® F. [Rewo GmbH] Linoleamide DEA and diethanolamine; nonionic; superfatting agent and thickener for shampoos; med. visc. liq.; 100% conc.

Rewomid® IPL 203. [Rewo GmbH] Lauric MIPA; nonionic; detergent; additive for solid and paste end prods.; improved washing power; stabilizer of emulsions; wh. flakes; slight odor; m.p. 55 C; 100% act.

Rewomid® IPP 240. [Rewo GmbH] Coconut MIPA; nonionic; see Rewomid IPL 203; wh. flakes; slight odor; m.p. 50 C; 100% act.

Rewomid® L203. [Rewo GmbH] Lauric MEA; nonionic; detergent for detergent preparations; fixation of perfumes; stabilizer of emulsions; wh. flakes; slight odor; m.p. 84 C; 100% act.

Rewomid® OM 101/G. [Rewo] Peanutamide MEA; nonionic; emulsifier, detergent intermediate; solv.; specialty cosmetic prod.; wax; 100% conc.

Rewomid® OM 101/IG. [Rewo] Peanutamide MIPA; nonionic; emulsifier, detergent intermediate; solv.; specialty cosmetic prod.; wax; 100% conc.

Rewomid® OM 101/IG/ER. [Rewo] Mixed fatty acid ethanolamide distillate; nonionic; solv.; specialty cosmetic prod.; wax; 100% conc.

Rewomid® R 280. [Rewo GmbH] Ricinoleic MEA; nonionic; surfactant for soaps; foam stabilizer; wax; 100% conc.

Rewomid® RE. [Rewo GmbH] Blend of anionics and alkylolamides; emulsifier and corrosion inhibitor for metalworking lubricants; liq.; 100% conc.

Rewomid® S 280. [Rewo GmbH] Stearic MEA; nonionic; detergent; superfatting agent; stabilizer of emulsions; wh. flakes; slight odor; m.p. 92 C; 100% act.

Rewomid® U 185. [Rewo GmbH] Undecylenic MEA; nonionic; fungicide; antimycotic; flake; 100% conc.

Rewomine IM-BT. [Rewo GmbH] Imidazoline deriv.; cationic; adhesive additive/agent and emulsifier in acidic bitumen emulsions; rust inhibitor in acidic media; Gardner 13 max. gr.-brn. liq.; amine odor; dens. 0.94 g/cm^3; 100% act.

Rewomine IM-CA. [Rewo GmbH] 1-Hydroxyethyl-2-heptadecenyl imidazoline; cationic; corrosion inhibitor, emulsifier, penetrant, wetting agent; used in leather and metalworking industry, paint and dyes, for carbonization baths; yel. wax; m.w. 285; sol. in nonpolar solvs.; disp. in water; 100% act.

Rewomine IM-OA. [Rewo GmbH] 1-Hydroxyethyl-2-heptadecenyl imidazoline; cationic; coupling agent for rust prevention; emulsifier for the emulsification of oils and bitumens; antistat; raw material for quat. reactions; paint industry; amber liq.; m.w. 345; sol. in polar and nonpolar solvs.; disp. in water; dens. 0.94 g/cm^3; visc. 800 cps; pH 10–11 (1% aq.); 100% act.

Rewominox B 204. [Rewo GmbH] Alkyl amidopropyl dimethylamine oxide; nonionic; foam booster; antistat; hair conditioner; liq.; 35% conc.

Rewominox S 300. [Rewo GmbH] Stearyl dimethylamine oxide; nonionic; foam booster, antistat, conditioner; liq.; 25% conc.

Rewominoxid B 204. [Rewo GmbH] Cocamidopropylamine oxide; nonionic; emulsifier, softener; calcium soap dispersant used in personal care prods.; pale clear liq.; almost odorless; pH 5–7 (10% solids); 35% min. solids.

Rewopal® C 6. [Rewo GmbH] PEG-6 cocamide; nonionic; dispersant, emulsifier, wetting agent for calcium soap, personal care prods.; amber liq.; cloud pt. 80–90 C (2%); pH 8–10 (1% solids); 100% act.

Rewopal® CSF 11. [Rewo GmbH] Ceteareth-11; nonionic; solubilizer; emulsifier for cosmetic and pharmaceutical preparations; flake; 100% conc.

Rewopal® LA 3. [Rewo GmbH] Laureth-3; nonionic; emulsifier, coupler, raw material for the prod. of ether sulfates; shampoos; clear liq.; sol. in min. and org. solvs.; pH 5–7 (1% solids); biodeg.

Rewopal® M 365. [Rewo GmbH] Ricinoleic acid polyglycol ester; nonionic; additive for metal working fluids and oils; med. visc. liq.; 100% conc.

Rewopal® MPG 10. [Rewo GmbH] Phenoxyethanol; nonionic; solv., solubilizer for preservatives; liq.; 100% conc.

Rewopal® MPG 12, 40. [Rewo GmbH] Phenol polyglycol ether; nonionic; solv.; liq.; 100% conc.

Rewopal® MT 2455. [Rewo GmbH] Fatty alcohol EO/PO methyl ether; nonionic; fiber lubricant, detergent; liq.; 100% conc.

Rewopal® MT 5722. [Rewo GmbH] Fatty alcohol EO/PO methyl ether; nonionic; fiber lubricant, detergent; liq.; 100% conc.

Rewopal® O 8. [Rewo GmbH] PEG-9 oleamide; nonionic; detergent; wetting agent, o/w emulsifier, and dispersant for calcium soap; suitable for machine washing formulations; brn. liq.; sol. in water, alcohol, ketone, ester, chlorinated hydrocarbons, benzene, fatty oils; cloud pt. 75–85 C; pH 8–10 (1%); 100% act.

Rewopal® PEG 6000 DS. [Rewo GmbH] PEG-150 distearate; nonionic; thickener for toiletries; flake; 100% conc.

Rewopal® PG 280. [Rewo GmbH] Glycol distearate; anionic; pearlizing agent for cosmetics; flake; 100% conc.

Rewopal® PG 340. [Rewo GmbH] Glycol dibehenate; nonionic; pearlizing agent for cosmetics and detergents; flake; 100% conc.

Rewopal® TA 11. [Rewo GmbH] Talloweth-11; nonionic; wash-act. base, wetting agent, dispersant, emulsifier, and cleaning power; wh. soft wax; sol. in water and org. solvs.; cloud pt. 69–73 C; pH 5–7 (1% solids); biodeg.

Rewopal® TA 25. [Rewo GmbH] Talloweth-25; nonionic; see Rewopal TA 11; wh. powd.; sol. see Rewopal TA 11; cloud pt. 75–90 C (2%); pH 5–7 (1% solids).

Rewopal® TA 25/S. [Rewo GmbH] Talloweth-25; nonionic; detergents surfactant, dispersant; flakes; 100% conc.

Rewophat E 1027. [Rewo GmbH] Alkylphenol polyglycol ether phosphate; anionic; corrosion inhibitor, emulsifier, dispersant, wetting agent; used in pure oil for metals; antistat; yel. liq.; 100% act.

Rewophat EAK 8190. [Rewo GmbH] Lauric fatty alcohol polyglycol ether phosphate; see Rewophat E 1027; yel. liq.; disp. in water; biodeg.; 100% act.

Rewophat NP 90. [Rewo GmbH] Nonylphenol polyglycol ether phosphate; anionic; emulsifier, antistat, raw material for industrial cleaners; liq.; 100% conc.

Rewophat OP 80. [Rewo GmbH] Fatty alcohol polyglycol ether phosphate; anionic; additive for syn. fiber; min. oil emulsifier, antistat; paste; 100% conc.

Rewophat TD 40. [Rewo GmbH] Fatty alcohol polyglycol ether phosphate; anionic; hydrotrope; solubilizer for acid cleaners; liq.; 100% conc.

Rewophat TD 70. [Rewo GmbH] Fatty alcohol polyglycol ether phosphate; anionic; raw material for chemical and acid cleaners; antisoiling finish, flame retarder; liq.; 100% conc.

Rewopol® B 1003. [Rewo GmbH] Tetrasodium dicarboxyethyl stearyl sulfosuccinate; anionic; foaming and antigelling agent for latex foam backings and coatings; emulsion polymerization; flotation agent; paste; 35% conc.

Rewopol® B 2003. [Rewo GmbH] Tetrasodium dicarboxyethyl stearyl sulfosuccinamate; flotation reagent; emulsion polymerization; liq.; 35% conc.

Rewopol® CHT 12. [Rewo GmbH] Coco-EDTA-amide; anionic; sequestering agent, complex building surfactant for detergents; visc. liq.; 40% conc.

Rewopol® NEHS 40. [Rewo GmbH] Sodium dioctyl sulfosuccinate; anionic; wetting agent for alkaline cleaners, mercerizing, electroplating; hydrotrope; liq.; 40% conc.

Rewopol® SBDO 70. [Rewo GmbH] Dioctyl sodium sulfosuccinate; anionic; wetting agent, solubilizer; cosmetic and personal care preparations, household and metal cleaners; used in paper, textile, paint, and dye industries; dry cleaning; colorless liq.; surf. tens. < 30 dynes/cm (0.1%); pH 6.5–7.5 (5% solids); biodeg.; 64% min. act.

Rewopol® SMS 35. [Rewo GmbH] Alkyl disodium sulfosuccinamate; anionic; emulsifier for emulsion polymerization; foaming agent for latex emulsion; antigelling and cleaning agent for paper mill felts; paste; 35% conc.

Rewopol® TMS/F. [Rewo GmbH] Alkyl sulfosuccinamide; anionic; emulsifier for emulsions; foaming agent for latex emulsion; antigelling agent; liq.; 31% conc.

Rewopon® AM-2C. [Rewo GmbH] Coconut-based ampholyte; amphoteric; detergent, foam booster, wetting agent, foam stabilizer used in toiletries; golden-yel. liq.; 49% act.

Rewopon® IM-AN. [Rewo GmbH] Imidazoline deriv.; cationic; water-repellent for car rinses; inhibitor for acid pickling baths; visc. liq., paste; 100% conc.

Rewopon® IM-BT. [Rewo GmbH] Imidazoline deriv.; cationic; improves adhesive properties for coatings; visc. liq., paste; 100% conc.

Rewopon® IM-CA. [Rewo GmbH] 1-Hydroxyethyl-2-alkyl-imidazoline; cationic; corrosion inhibitor, antistat, lubricant adhesive, raw material for quarternization; firm paste; 100% conc.

Rewopon® IM-OA. [Rewo GmbH] 1-Hydroxyethyl-2-alkyl-imidazoline; cationic; corrosion inhibitor, emulsifier, antistat; med. visc. liq.; 100% conc.

Rewopon® IM-OD. [Rewo GmbH] Oleic imidazoline deriv.; anticorrosion additive; imparts hydrophobic effect; liq.

Rewopon® JMBT. [Rewo GmbH] Long-chain imidazoline deriv.; cationic; adhesion agent and emulsifier for acid asphalt; liq.; 100% conc.

Rewopon® JMCA. [Rewo GmbH] Coconut imidazoline deriv.; cationic; emulsifier, anticorrosive agent; firm paste; 100% conc.

Rewopon® JMOA. [Rewo GmbH] Oleic imidazoline deriv.; cationic; rust inhibitor, emulsifier for antistatic preparations; liq.; 100% conc.

Rewoquat B 50. [Rewo GmbH] Benzalkonium chloride; cationic; disinfectant for cleaners, dairy and food industries; algicide; textile dyeing aux.; low visc. liq.; 50% conc.

Rewoquat CPEM. [Rewo GmbH] Coco pentaethoxy methyl ammonium methosulfate; cationic; conditioner for shampoos, emulsifier in emulsion polymerization, antistat; visc. liq.; 100% conc.

Rewoquat CR 3099. [Rewo GmbH] Difatty acid isopropyl ester dimethyl ammonium methylsulfate; cationic; fabric softener, dry cleaning agent; visc. liq.; 95% conc.

Rewoquat DQ 35. [Rewo GmbH] Tallow propylene diamine PEG ammonium methosulfate; cationic; antistat and wetting agent; liq.; 35% conc.

Rewoquat RTM 50. [Rewo GmbH] Ricinoleic acid propylamido trimethyl ammonium methosulfate; cationic; conditioner for personal care prods.; antistat; low visc. liq.; 50% conc.

Rewoquat UTM 185. [Rewo GmbH] Undecylenic quat. ammonium methosulfate; bacteriostat and fungicide for deodorants, shampoos, liq. soaps, antimycobic foot preps.; liq.; 50% conc.

Rewoquat W 222 LM. [Rewo GmbH] Ditallow amidoammonium methosulfate; cationic; fabric softener; liq.; 90% conc.

Rewoquat W 3690, W 3690/PG. [Rewo GmbH] Dioleyl imidazoline methosulfate; cationic; fabric softener; visc. liq.; 75% conc.

Rewoquat W 7500. [Rewo GmbH] Quaternium-27; cationic; fabric softener; visc. liq.; 75% conc.

Rewoquat W 7500 H. [Rewo GmbH] Quat. dialkyl imidazolinium methosulfate, hydrog.; cationic; fabric and hair softener; soft paste; 75% conc.

Rewoquat W 7500/PG. [Rewo GmbH] Dialkyl imidazoline methosulfate, IPA free; cationic; fabric softener; visc. liq.; 75% conc.

Rewoquat W 9000, 9000/PG. [Rewo GmbH] Ditallow imidazolinium methosulfate; cationic; fabric softener; soft paste; 90% conc.

Rewoquat WKH 80. [Rewo GmbH] Quat. imidazoline deriv.; cationic; textile softener; paste; 80% conc.

Rewoquat WP 100. [Rewo GmbH] Quat. imidazoline deriv., modified; cationic; softener; powd.; 100% conc.

Reworyl® ACS 60. [Rewo GmbH] Ammonium cumene sulfonate; anionic; hydrotrope for detergent and cleaning systems; liq.; 60% conc.

Reworyl® B 70. [Rewo GmbH] Benzene sulfonic acid; anionic; catalyst for esterification, polymerization, and polycondensation in foundry resins; straw clear liq.; 70% conc.

Reworyl® C 65. [Rewo GmbH] Cumene sulfonic acid; anionic; see Reworyl B 70; liq.; 65% conc.

Reworyl® NCS 40. [Rewo GmbH] Sodium cumene sulfonate; anionic; see Reworyl ACS 60; liq.; 40% conc.

Reworyl® NTS 40. [Rewo GmbH] Sodium toluene sulfonate; anionic; see Reworyl NCS 40; liq.; 40% conc.

Reworyl® NXS 40. [Rewo GmbH] Sodium xylene sulfonate; anionic; see Reworyl NCS 40; liq.; 40% conc.

Reworyl® T 65. [Rewo GmbH] p-Toluene sulfonic acid; anionic; catalyst for prep. of esters, acid hydrolysis, polymerization reactions; industrial applics.; intermediate; leather tanning agent; toxic crop-dusting agents; prod. of syn. fibers; pale clear liq.; char. odor; 65% act.

Reworyl® TKS 90/F. [Rewo GmbH] Trialkanolammonium dodecylbenzene sulfonate; anionic; detergent raw material for cleansing and defatting agents, car shampoos; chain lubricant; dk. amber visc. liq.; pH 7.0–7.5 (1% solids); 85% min. act.

Reworyl® X 65. [Rewo GmbH] Xylene sulfonic acid; anionic; catalyst for esterification, polymerization, and polycondensation in foundry resins; liq.; 65% conc.

Rewowax CL 1263. [Rewo GmbH] Amide wax; nonionic; gloss emulsion for floors, varnishes, plastics; flake; 100% conc.

Rexfoam 150-A. [Graden] Hydrophobic silica; defoamer for gloss and semigloss latex paints; poorly disp. in water.

Rexfoam B, C. [Graden] Silicone emulsion; defoamer for paint and ink, chemical processing, pulp and paper, adhesive formulations, water treatment; disp. in water; dens. 8.3 lb/gal; 10 and 30% act. resp.

Rexfoam D. [Graden] Silicone emulsion; nonionic; defoamer for paint, ink, pulp and paper mfg., petrol., chemical, and textile processing, resin polymerization, adhesives, metal and water treatment; opaque liq.; sol. in aliphatic, aromatic, and chlorinated solv.; dens. 8.4 lb/gal; 100% conc.

Rexol 25/1. [Hart Chem. Ltd.] Nonoxynol-1; nonionic; defoamer for detergent systems; coemulsifier for surfactant and solv. blends; liq.; HLB 4.6; 100% act.

Rexol 25/4. [Hart Chem. Ltd.] Nonoxynol-4; nonionic; emulsifier, intermediate for anionic sulfonates; clear liq.; ref. index 1.5000; oil-sol.; sp.gr. 1.02; visc. 200 cps; HLB 8.7; 100% act.

Rexol 25/6. [Hart Chem. Ltd.] Nonoxynol-6; nonionic; dispersant in petrol. systems; intermediate for anionic sulfates; antistat plasticizer for plastics; clear liq.; ref. index 1.4909; oil-sol.; sp.gr. 1.04; visc. 250 cps; HLB 10.8; 100% act.

Rexol 25/9. [Hart Chem. Ltd.] Nonoxynol-9; nonionic; detergent, dispersant, emulsifier and wetting agent; used in leather, paint, textile, pulp and paper industries; base for industrial and household detergents; liq.; water-sol.; HLB 13.0; 100% conc.

Rexol 25/10. [Hart Chem. Ltd.] Nonoxynol-10; nonionic; see Rexol 25/9; liq.; water-sol.; HLB 13.4; 100% conc.

Rexol 25/20. [Hart Chem. Ltd.] Nonoxynol-20; nonionic; particle dispersant in aq. systems; solid; HLB 16.0; cloud pt. 72–74 C (1%); 100% act.

Rexol 25/30. [Hart Chem. Ltd.] Nonoxynol-30; nonionic; solubilizer for essential oils and pesticides; solid; HLB 17.2; cloud pt. 74–76 C (1%); 100% act.

Rexol 25/50. [Hart Chem. Ltd.] Nonoxynol-50; nonionic; emulsifier and stabilizer used in floor finishes; solid; HLB 18.2; cloud pt. 74–76 C (1%); 100% act.

Rexol 25/100-70%. [Hart Chem. Ltd.] Nonoxynol-100; nonionic; demulsifier for petrol. oils; pressured textile scouring; emulsifier/stabilizer for latices, asphalt, and floor finishes; liq.; HLB 19.0; cloud pt. 74–76 C (1%); 70% act.

Rexol 25/307. [Hart Chem. Ltd.] Nonoxynol-30; nonionic; see Rexol 25/30; liq.; HLB 17.2; cloud pt. 74–76 C (1%); 70% act.

Rexol 25/407. [Hart Chem. Ltd.] Nonoxynol-40; nonionic; detergent, wetting agent; emulsifier for vinyl acetate and acrylate polymerization; demulsifier of petrol. oils; textile scouring; stabilizer for syn. latices; yel. liq.; water-sol.; sp.gr. 1.10; 70% act.

Rexol 25/507. [Hart Chem. Ltd.] Nonoxynol-50; nonionic; see Rexol 25/50; liq.; HLB 18.2; cloud pt. 74–76 C (1%); 70% act.

Rexol 35/3. [Hart Chem. Ltd.] Alcohol ethoxylate; nonionic; emulsifier for o/w; intermediate for sulfation and phosphation; liq.; HLB 8.0; cloud pt. 49–51 C (1%); 100% act.

Rexol 35/8. [Hart Chem. Ltd.] Alcohol ethoxylate; nonionic; detergent intermediate; wetting agent used in textile applics.; liq.; HLB 12.8; cloud pt. 79–83 C (1%); 85% act.

Rexol 35/100. [Hart Chem. Ltd.] Fatty alcohol, ethoxylated; nonionic; detergent, wetting agent used in scouring of paper machine felts; preparation of industrial detergents; emulsifier, dispersant; clear paste; sp.gr. 1.003; visc. 100 cps; 100% act.

Rexol 45/1. [Hart Chem. Ltd.] Octoxynol-1; nonionic; emulsifier, detergent, dispersant; coemulsifier for surfactant and solv. preparation blends; liq.; oil-sol.; HLB 3.6; 100% act.

Rexol 45/3. [Hart Chem. Ltd.] Octoxynol-3; nonionic; emulsifier, detergent, dispersant, coemulsifier for petrol. oils and solvs.; liq.; HLB 7.8; 100% act.

Rexol 45/5. [Hart Chem. Ltd.] Octoxynol-5; nonionic; emulsifier for fats and oils, detergent, dispersant; liq.; HLB 10.4; cloud pt. 63–66 C (10%); 100% act.

Rexol 45/7. [Hart Chem. Ltd.] Octoxynol-7; nonionic; hard surface detergent; emulsifier, dispersant; liq.; HLB 12.4; cloud pt. 21–23 C (1%); 100% act.

Rexol 45/12. [Hart Chem. Ltd.] Octoxynol-12; nonionic; emulsifier, detergent, dispersant used in metal cleaning, industrial and household liq. detergents and cleaners; liq.; HLB 14.5; cloud pt. 86–90 C (1%); 100% act.

Rexol 45/16. [Hart Chem. Ltd.] Octoxynol-16; nonionic; detergent and wetting agent used in metal cleaning and bottle washing; emulsifier; dispersant; paste; HLB 15.8; cloud pt. 65–69 C (1%); 100% act.

Rexol 45/307. [Hart Chem. Ltd.] Octoxynol-30; nonionic; emulsifier for vinyl acetate and acrylate emulsion polymerization; detergent, dispersant; liq.; HLB 17.4; cloud pt. 73–77 C (1%); 70% act.

Rexol 45/407. [Hart Chem. Ltd.] Octoxynol-40; nonionic; see Rexol 45/307; liq.; HLB 18.0; cloud pt. 73–77 C (1%); 70% act.

Rexol 130. [Hart Chem. Ltd.] Alcohol ethoxylate; nonionic; emulsifier, dyeing assistant; clear liq.; ref. index 1.3850; sp.gr. 1.02; visc. 93 cps; 35% act.

Rexol 2000 HWM. [Hart Chem. Ltd.] Amide ethoxylate, modified; nonionic; antistat emulsifier for mfg. of textile lubricants; amber liq.; ref. index 1.4618; sp.gr. 0.982 visc. 300 cps; 77% act.

Rexol AE-1, AE-2. [Hart Chem. Ltd.] Alcohol ethoxylate; nonionic; intermediate for shampoo and detergent mfg.; liq.; HLB 4.2 and 6.1 resp.; cloud pt. 36–39 and 50–52 C (1%) resp.; 100% act.

Rexol AE-23. [Hart Chem. Ltd.] Alcohol ethoxylate; nonionic; dispersant, emulsifier in cosmetic applics.; solid; HLB 17.0; cloud pt. 90–95 C (1%); 100% act.

Rexol C14. [Hart Chem. Ltd.] Castor oil ethoxylate; nonionic; emulsifier, antistat, demulsifier, leveling agent for dyeing of syn. fibers; solid; 90% conc.

Rexol CCN. [Hart Chem. Ltd.] Castor oil ethoxylate; nonionic; emulsifier, lubricant; syn. fiber mfg.; clear liq.; sp.gr. 0.98; visc. 610 cps; 90% act.

Rexonic 1006. [Hart Chem. Ltd.] Linear alcohol ethoxylate; nonionic; dispersant and leveling assistant for dyeing leather; prescouring aid; clear liq.; 75% act.

Rexonic N-4. [Hart Chem. Ltd.] Linear alcohol ethoxylate; nonionic; intermediate; preparation of liq. detergents; used for wetting, back wetting, sizing, emulsification and dispersant of petrol. oils; clear liq.; oil-sol.; sp.gr. 0.92; visc. 50 cps; HLB 8.0; 100% act.

Rexonic N6.6. [Hart Chem. Ltd.] Linear alcohol ethoxylate; nonionic; detergent, wetting agent, emulsifier; formulation of liq. detergents; foam stabilizer; clear paste; sp.gr. 0.98; HLB 11.9; 100% act.

Rexonic N23-3. [Hart Chem. Ltd.] Linear alcohol ethoxylate; nonionic; detergent, surfactant and chemical intermediate used in solv. emulsion cleaners and dry cleaning detergents; clear liq.; oil-sol.; sp.gr. 0.93; visc. 19 cps; 100% act.

Rexonic N25-3. [Hart Chem. Ltd.] Alcohol ethoxylate; nonionic; emulsifier for oil; backwetting agent, intermediate for detergents; liq.; HLB 7.8; cloud pt. 59–61 C (1%); 100% act.

Rexonic N25-14 85%. [Hart Chem. Ltd.] Linear alcohol ethoxylate; nonionic; detergent for scouring textile fibers; base for mfg. of ether sulfates; dispersant, emulsifier, and wetting agent; used in leather, paint, textile, pulp and paper industries; base surfactant for household and industrial cleaners; liq.; water-sol.; 85% conc.

Rexonic N91-1.6. [Hart Chem. Ltd.] Alcohol ethoxylate; nonionic; intermediate for mfg. of ethoxysulfates; emulsifier, detergent, dispersant; liq.; HLB 6.1; cloud pt. 39–41 C (1%); 100% act.

Rexonic N91-2.5. [Hart Chem. Ltd.] Alcohol ethoxylate; nonionic; emulsifier for oils and solvs.; intermediate for ethoxysulfates; detergent, dispersant; liq.; HLB 8.1; cloud pt. 50–54 C (1%); 100% act.

Rexonic N91-6, N91-8. [Hart Chem. Ltd.] Alcohol ethoxylate; nonionic; detergent for industrial use; emulsifier; shampoo intermediate, dispersant; liq.; water-sol.; HLB 12.5 and 14.1 resp.; cloud pt. 51–53 and 80–82 C resp. (1%); 100% act.

Rexonic P-1. [Hart Chem. Ltd.] Linear alcohol, propoxylated, ethoxylated; nonionic; detergent, wetter, emulsifier, dispersant for natural and syn. fibers; clear liq.; water-sol.; ref. index 1.455; sp.gr. 0.97; visc. 90 cps; 100% act.

Rexopene. [Emkay] Sodium alkylaryl sulfonate; anionic; wetting, leveling, penetrant, scouring assistant; liq.

Rexowet CR. [Emkay] Sulfonated isopropyl oleate and cresylic acid; anionic; wetting agent and penetrant for cotton fabrics; liq.

Rexowet GR. [Emkay] Sulfated diester; anionic; penetrant, wetting agent; liq.

Rexowet MS. [Emkay] Sodium alkyl naphthalene sulfonate; anionic; penetrant for textiles; liq.

Rexowet NF, NFX. [Emkay] Alcohol sulfate salt; anionic; penetrant, wetting agent; textile dyeing assistant; liq.

Rexowet RW. [Emkay] Aliphatic mono and diester, sulfonated; anionic; wetting, rewetting and leveling agent, penetrant; enzyme activator; gel.

Rhenocure AT. [Rhein Chemie] Amine-dialkyldithiophosphate coated with min. oil; accelerator for EPDM rubber, tech. molded and extruded goods; wh. cryst. powd.; sp.gr. 1.0.

Rhenocure CA. [Rhein Chemie] N′N′-diphenylthiourea; accelerator for tech. molded and extruded rubber goods; wh. to ylsh. powd.; sp.gr. 1.3.

Rhenocure CMT. [Rhein Chemie] Accelerator blend; for the cross-linking of EPDM, tech. rubber goods esp. molded articles; ylsh. coated powd.; sp.gr. 1.4.

Rhenocure CMU, EPC. [Rhein Chemie] Accelerator blend; for the cross-linking of EPDM, tech. rubber goods; ylsh. coated powd.; sp.gr. 1.3 (both); 80% act. (EPC).

Rhenocure CUT. [Rhein Chemie] Copper-dialkyldithiophosphate, coated with min. oil; accelerator for EPDM and other diene rubbers, tech. molded and extruded goods; yel.-grn. cryst. powd.; sp.gr. 1.1.

Rhenocure Diuron. [Rhein Chemie] 3-(3,4-Dichlorophenyl)-1,1-dimethylurea bound to ACM; crosslinking agent; for oil-resistant seals based on ACM for automotive applics.; wh.-ylsh. gran.; sp.gr. 1.4.

Rhenocure IS 60. [Rhein Chemie] 60% insol. sulfur, 40% sol. sulfur; curing agent for NR and SR compds. where blooming of sulfur should be avoided; used in tires, conveyor belts; yel. powd.; sp.gr. 1.95.

Rhenocure IS 60-5. [Rhein Chemie] 95% Rhenocure IS 60, 5% oil; see Rhenocure IS 60; yel. powd.; sp.gr. 1.8.

Rhenocure IS 60/G. [Rhein Chemie] 80% Rhenocure IS 60, 20% polymer binder; see Rhenocure IS 60; yel. gran.; sp.gr. 1.6.

Rhenocure IS 90-20. [Rhein Chemie] 80% sulfur, 20% oil; see Rhenocure IS 60; yel. powd.; sp.gr. 1.6.

Rhenocure IS 90-33. [Rhein Chemie] 67% sulfur, 33% oil and inorg. dispersant; see Rhenocure IS 60; yel. powd.; sp.gr. 1.5.

Rhenocure IS 90-40. [Rhein Chemie] 60% sulfur, 40% oil and inorg. dispersant; see Rhenocure IS 60; yel. powd.; sp.gr. 1.4.

Rhenocure IS 90/G. [Rhein Chemie] 70% sulfur, 30% polymer binder; see Rhenocure IS 60; yel. gran.; sp.gr. 1.5.

Rhenocure M. [Rhein Chemie] 4,4′-Dithiodimorpholine; sulfur donor for NR, aging resistant articles based on SBR, NBR, and EPDM, tech. goods; wh. powd.; sp.gr. 1.3.

Rhenocure M/G. [Rhein Chemie] 80% 4,4′-dithiodimorpholine and 20% polymer binder; see Rhenocure M; gray gran.; sp.gr. 1.25.

Rhenocure S/G. [Rhein Chemie] 80% dithiodicaprolactam and 20% polymer binder; sulfur donor for NR and SR, ageing resistant articles based on SBR, NBR, EPDM; brn. gran.; sp.gr. 1.25.

Rhenocure TDD. [Rhein Chemie] Thiadiazole deriv. bound to CM; accelerator for peroxide-free crosslinking of CM and other sat., halogen-containing elastomers; pale gray gran.; sp.gr. 1.4.

Rhenocure TP/G. [Rhein Chemie] 50% zinc dialkyldithiophosphate with 50% polymer binder; accelerator for EPDM and other diene rubbers, tech. molded and extruded goods; beige gran.; sp.gr. 1.25.

Rhenocure TP/S. [Rhein Chemie] Zinc-dialkyldithiophosphate bound to silica; accelerator for EPDM and other diene rubbers, tech. molded and extruded goods; wh. powd.; sp.gr. 1.3.

Rhenocure ZAT. [Rhein Chemie] Zinc-aminedithiophosphate complex, coated with min. oil; accelerator for EPDM and other diene rubbers, tech. molded and extruded goods; wh.-gray powd.; sp.gr. 1.1.

Rhenodiv 20. [Rhein Chemie] Silicone-free mold release agent for rubber; colorless to ylsh. clear liq.; sp.gr. 1.05; pH 7 (10%); 3–10% conc.

Rhenodiv A. [Rhein Chemie] Release agent for rubber compounding, extruding; ylsh. liq.; sp.gr. 1.0; pH 9–10 (10%); 2–5% conc.

Rhenodiv F. [Rhein Chemie] Surfactant–polyvalent alcohols aq. sol'n.; release agent for rubber compounding and extruding; ylsh.-brn. low-visc. liq.; sp.gr. 1.1; pH 10.5–11 (10%); 5–15% conc.

Rhenodiv KS. [Rhein Chemie] Release agent for rubber compounding; yel. to grn. powd.; sp.gr. 1.3; pH 11 (1%); 3–5% conc.

Rhenodiv LE. [Rhein Chemie] Fatty acid salts aq. paste; release agent for rubber compounding and extruding; ylsh.-wh. paste; pH 10–11 (10%); 5–20% conc.

Rhenodiv LL. [Rhein Chemie] Fatty acid salt aq. sol'n.; release agent for rubber compounding and extruding; ylsh.-wh. visc. liq.; sp.gr. 1.0; pH 9.5–10 (10%); 5–20% conc.

Rhenodiv LS. [Rhein Chemie] Fatty acid salts with film-formers bound to silica; release agent for rubber compounding; wh. to yel. powd.; sp.gr. 1.1; pH 9 (1%); 5–20% conc.

Rhenodiv PV. [Rhein Chemie] Release agent for rubber compounding, esp. for EPDM; yel. to grn. powd.; sp.gr. 1.35; pH 11 (1%); 3–5% conc.

Rhenodiv S. [Rhein Chemie] Release agent for rubber compounding, extruding; whitish paste; pH 9.5–10.5 (10%); 3–5% conc.

Rhenodiv ZB. [Rhein Chemie] Zinc stearate aq. disp.; release agent for rubber compounding; wh. low-visc. stable suspension; 3–5% conc.

Rhenofit 1600. [Rhein Chemie] Urea derivs./dispersants in aq. paste form; blow promoter for cellular and microcellular rubber compds.; increases plasticity of uncured compd., accelerates vulcanization; lt.-colored paste; sp.gr. 1.25.

Rhenofit 1987. [Rhein Chemie] Urea/surfactant blend bound to silica; filler activator for rubbers containing lt.-colored reinforcing fillers; wh. fine powd.; sp.gr. 1.5.

Rhenofit 2009. [Rhein Chemie] Urea deriv./surfactant blend bound to silica; filler activator for rubbers containing lt.-colored reinforcing fillers; wh. fine powd.; sp.gr. 1.8.

Rhenofit 2642. [Rhein Chemie] Urea deriv. bound to silica; filler activator for rubbers containing lt.-colored reinforcing fillers; wh. hygroscopic fine powd.; sp.gr. 1.5.

Rhenofit 3555. [Rhein Chemie] Amine derivs. bound to silica; filler activator for rubbers requiring rapid vulcanization and fast-curing molded goods; wh. powd.; sp.gr. 1.4.

Rhenofit B. [Rhein Chemie] Sec. amine; filler activator for colored articles based on NR and syn. rubbers; ylsh. to grn. liq.; sp.gr. 0.9.

Rhenofit BDMA/S. [Rhein Chemie] 70% 1,4-Butanediol dimethacrylate, 30% silica; cross-linking activator for peroxide vulcanization of tech. molded and extruded goods based on EPDM, EPM, NBR, CM, etc.; wh. powd.; sp.gr. 1.2.

Rhenofit CF. [Rhein Chemie] Treated calcium hydroxide; cross-linking activator for fluoro elastomers, tech. rubber goods; wh. powd.; sp.gr. 2.2.

Rhenofit EDMA/S. [Rhein Chemie] 70% Ethyleneglycol dimethacrylate, 30% silica; cross-linking activator for peroxide vulcanization of tech. molded and extruded goods based on EPDM, EPM, NBR, CM, etc.; wh. powd.; sp.gr. 1.25.

Rhenofit NC. [Rhein Chemie] Fatty acid amide-amine; activating accelerator for the cross-linking of CM; ylsh. pellets; sp.gr. .095.

Rhenofit TAC/S. [Rhein Chemie] 70% Triallylcyanurate, 30% silica; cross-linking activator for peroxide vulcanization of tech. molded and extruded goods based on EPDM, EPM, NBR, CM, etc.; wh. fine powd.; sp.gr. 1.25.

Rhenofit TRIM/S. [Rhein Chemie] 70% Trimethylolpropane trimethacrylate, 30% silica; see Rhenofit TAC/S; wh. powd.; sp.gr. 1.25.

Rhenofit UE. [Rhein Chemie] Activators bound to silica; activating accelerator for the cross-linking of CO and ECO, tech. rubber goods; wh., slightly hygroscopic powd.; sp.gr. 1.5.

Rhenomag C2, G3, L3, P2, P3. [Rhein Chemie] Magnesium oxide (G3 in polymer-bound gran. form); acid acceptor and vulcanization activator for rubber goods, tech. molded and extruded articles, adhesives based on CR, erasers, chlorinated paraffins; wh. to lt. gray coated powd., wh.-gray gran., lt. gray paste, wh. powd., and wh. powd. resp.; sp.gr. 2.6, 2.25, 2.0, 3.5, and 3.3 resp.; 97.5% act. (P2), 93% act. (P3).

Rhenopor 1843. [Rhein Chemie] Sodium hydrogen carbonate with dispersants (50:50); inorg. blowing agent for cellular and microcellular rubber articles; wh. powd.; sp.gr. 2.0.

Rhenosin 140. [Rhein Chemie] Thermoplastic copolymeric resin based on selected hydrocarbon fractions; softening agent and homogenizer for dark colored NR and SR rubbers, tires, conveyor belts; dk. brn.-blk. pellets; sp.gr. 1.1; soften. pt. 85 C.

Rhenosin 260. [Rhein Chemie] Thermoplastic aromatic hydrocarbon resin; softening agent and homogenizer for lt. colored NR and SR rubbers, tech. molded and extruded goods; lt. brn. pellets; sp.gr. 1.1; soften. pt. 85 C.

Rhenosorb C. [Rhein Chemie] Coated calcium oxide; desiccant for extruded rubber goods, conveyor belts; gray powd.; sp.gr. 2.8; 90% act.

Rhenosorb C/GW. [Rhein Chemie] Calcium oxide in polymer-bound granular form; desiccant for extruded rubber goods, conveyor belts; pale gray gran.; sp.gr. 2.25; 80% act.

Rhenosorb F. [Rhein Chemie] Calcium oxide; desiccant for seals, molded goods based on fluoro rubber; wh.; sp.gr. 3.0; 96% act.

Rhenovin CBS-70. [Rhein Chemie] 70% N-cyclohexyl-2-benzothiazyl sulfenamide, 30% silica and dispersants; vulcanization accelerator; gray powd.; sp.gr. 1.1.

Rhenovin DDA-70. [Rhein Chemie] 70% Diphenylamine deriv., 30% silica filler; antioxidant; brn. powd.; sp.gr. 1.3.

Rhenovin FH-70. [Rhein Chemie] 70% Aromatic polyether (Vulkanol FH), 30% silica filler; syn. plasticizer; wh. powd.; sp.gr. 1.25.

Rhenovin MBT-70. [Rhein Chemie] 70% 2-Mercaptobenzothiazole, 30% silica and dispersants; vulcanization accelerator; beige powd.; sp.gr. 1.25.

Rhenovin MBTS-70. [Rhein Chemie] 70% Dibenzothiazyl disulfide, 30% silica and dispersants; vulcanization accelerator; yel.-beige powd.; sp.gr. 1.25.

Rhenovin Na-stearat-80. [Rhein Chemie] 80% Sodium stearate, 20% inorg. dispersants; cross-linking activator for ACM; wh. powd.; sp.gr. 1.2.

Rhenovin S-90. [Rhein Chemie] 90% Sulfur, 10% dispersants; vulcanizing agent; yel. powd.; sp.gr. 1.9.

Rhenovin S-stearat-80. [Rhein Chemie] 80% Potassium stearate, 20% inorg. dispersants; cross-linking activator for ACM; wh. powd.; sp.gr. 1.1.

Rhenovin TMTD-70. [Rhein Chemie] 70% Tetramethylthiuram disulfide, 30% silica and dispersants; vulcanization accelerator; whitish powd.; sp.gr. 1.1.

Rhenovin TMTM-70. [Rhein Chemie] 70% Tetramethylthiuram monosulfide, 30% silica and dispersants; vulcanization accelerator; yel. powd.; sp.gr. 1.2.

Rhenovin ZnO-90. [Rhein Chemie] 90% Zinc oxide, 10% dispersants; vulcanization accelerator; yel.-wh. powd.; sp.gr. 4.0.

Rheodol SP-O10. [Kao] Sorbitan monooleate; nonionic; emulsifier and dispersant for printing inks, pastes, and paints; liq.; HLB 4.3; 100% conc.

Rheodol SP-O30. [Kao] Sorbitan trioleate; nonionic; see Rheodol SP-010; liq.; HLB 1.8; 100% conc.

Rheodol SP-S10. [Kao] Sorbitan monostearate; nonionic; see Rheodol SP-010; beads; HLB 4.7; 100% conc.

Rheodol SP-S30. [Kao] Sorbitan tristearate; nonionic; see Rheodol SP-010; beads; HLB 2.1; 100% conc.

Rheodol TW-O106, 0120. [Kao] POE sorbitan monooleate; nonionic; stabilizer in emulsion polymerization; liq.; HLB 10.0 and 15.5 resp.; 100% conc.

Rheodol TW-O320. [Kao] POE sorbitan trioleate; nonionic; see Rheodol TW-0106; liq.; HLB 11.0; 100% conc.

Rheodol TW-P120. [Kao] PEG-20 sorbitan monopalmitate; nonionic; emulsifier for pharmaceuticals, cosmetics; solubilizer for colorants; stabilizer for emulsion polymerization; liq.; HLB 15.6; 100% conc.

Rheodol TW-S106, -S120. [Kao] POE sorbitan monostearate; nonionic; stabilizer in emulsion polymerizaton; solid; HLB 9.6 and 14.9 resp.; 100% conc.

Rheodol TW-S320. [Kao] POE sorbitan tristearate; nonionic; see Rheodol TW-S106; solid; HLB 10.5; 100% conc.

Rheolate 1. [Rheox] Acrylates copolymer; thickener, gellant for water-based trade sales and industrial coatings; used in acrylics, styrene-acrylics, and vinyl-acrylic latexes and water-reducible resins; milky wh. aq. emulsion; sp.gr. 1.07; dens. 8.90 lb/gal; 30% NV.

Rheothik Polymer 80-11. [Henkel] Polysulfonic acid; thickener, suspending agent, slip agent for aq. lubricants; thickens HF and other acids, alkaline systems; gel-like liq.

Rheotol. [Vanderbilt] Polymerized alkyl phosphate; dispersant and wetting agent for pigmented coatings; pale liq.; dens. 0.97 mg/m^3.

Rhodigel 23. [Vanderbilt] Xanthan gum; emulsion stabilizer, suspending agent, thickener for cosmetic and pharmaceutical applics.; water-sol.

Rhodopol [Vanderbilt] Xanthan gum; anionic; emulsion stabilizer, suspending agent, thickener for cosmetic, pharmaceutical, agric., adhesives, binders, coatings, cleaners, explosives, mining, textile applics.; able to be gelled and crosslinked; powd.; 98% thru 50 mesh; sol. in water but insol. in most org. solvs.

Rhodorsil Huiles 47, 633. [Rhone-Poulenc] Organo chlorosilane-silicone oils; excipients for skin protection creams, cosmetics, lubricant, barrier creams.

Rhozyme® 86L. [Genencor] Bacterial alpha amylase; enzyme for starch liquefaction; liq.

Rhozyme® H39. [Genencor] Bacterial alpha amylase; enzyme for starch liquefaction; powd.

Rhozyme® HP-150 Conc. [Genencor] Pentosanase-hexosanase; food grade enzyme for hydrolysis of veg. gums; powd.

Rhozyme® P11. [Genencor] Protease; food grade enzyme for meat tenderization, removing flesh from bones; powd.

Rhozyme® P41. [Genencor] Protease; food grade enzyme for baking operations; powd.

Rhozyme® P53, P64. [Genencor] Protease; food grade enzyme for protein hydrolysis; P64 also for silver recovery from spent film; powd. and liq. resp.

Rhozyme® PF. [Genencor] Protease; industrial grade enzyme for desizing textiles carrying protein sizes; powd.

RIA CS. [Olin] Modified urea; cure accelerator and activator for compds. expanded with Opex blowing agent; powd.

Ridacto®. [Kenrich] Activator; sp.gr. 1.045.

Ridafoam Base 100. [PPG-Mazer] Silicone oil; suppresses foam or prevents its formation; for household, textile, and industrial uses; visc. liq.; 100.0% act.

Ridafoam NS-221. [PPG-Mazer] Nonsilicone-surfactant blend; nonionic; defoamer for household, textile, and industrial uses; opaque liq.; 100.0% act.

Ridafoam S-103-N. [PPG-Mazer] Silicone emulsion; suppresses foam or prevents its formation; for household, textile, and industrial uses; wh. fluid paste; 3.0% act.

Ridafoam S-110-N. [PPG-Mazer] Silicone emulsion; suppresses foam or prevents its formation; for household, textile, and industrial uses; visc. paste; 10.0% act.

Rilanit BS. [Henkel KGaA] n-Butyl stearate; lubricant for metalworking oils, syn. metalworking fluids, rolling, stamping and drawing oils, greases, motor oils, extrusion of ceramics, fatting agents in textile and leather auxs.; liq.

Rilanit CSN. [Henkel KGaA] Cetearyl isononoate; see Rilanit BS; liq.

Rilanit DBA. [Henkel KGaA] Di-n-butyl adipate; see Rilanit BS; liq.

Rilanit DBS. [Henkel KGaA] Dibutyl sebacate; see Rilanit BS; liq.

Rilanit DEHS. [Henkel KGaA] Di-2-ethylhexyl sebacate; see Rilanit BS; liq.

Rilanit DNOP. [Henkel KGaA] Di-n-octyl phthalate; see Rilanit BS; liq.

Rilanit DOA. [Henkel KGaA] Dioctyl adipate; see Rilanit BS; liq.

Rilanit DTP. [Henkel KGaA] Dihydrog. tallow phthalate; see Rilanit BS; solid.

Rilanit EHO. [Henkel KGaA] 2-Ethylhexyl oleate;

see Rilanit BS; liq.
Rilanit EHS. [Henkel KGaA] 2-Ethylhexyl stearate; see Rilanit BS; liq.
Rilanit EHTI. [Henkel KGaA] 2-Ethylhexyl tallowate; see Rilanit BS; liq.
Rilanit G 16. [Henkel KGaA] 1-Hexadecanol; see Rilanit BS; liq.
Rilanit G 20. [Henkel KGaA] 2-Octyldodecanol-1; see Rilanit BS; liq.
Rilanit GDO. [Henkel KGaA] Glyceryl dioleate; see Rilanit BS; liq.
Rilanit GL 401. [Henkel KGaA] Polyglycol ester; see Rilanit BS; liq.
Rilanit GMO. [Henkel KGaA] Glyceryl oleate; see Rilanit BS; liq.
Rilanit GMRO. [Henkel KGaA] Glyceryl ricinoleate; see Rilanit BS; semiliq.
Rilanit GMS. [Henkel KGaA] Glyceryl stearate; see Rilanit BS; solid.
Rilanit GO 401. [Henkel KGaA] Polyglycol ester; see Rilanit BS; liq.
Rilanit GTC. [Henkel KGaA] Glyceryl trifatty acid ester; see Rilanit BS; liq.
Rilanit GTO. [Henkel KGaA] Glyceryl trioleate; see Rilanit BS; liq.
Rilanit GTS. [Henkel KGaA] Glyceryl tristearate; see Rilanit BS; solid.
Rilanit HL. [Henkel KGaA] n-Hexyl laurate; see Rilanit BS; liq.
Rilanit IBO. [Henkel KGaA] Isobutyl oleate; see Rilanit BS; liq.
Rilanit IBS. [Henkel KGaA] Isobutyl stearate; see Rilanit BS; liq.
Rilanit IBTI, IBTIOK. [Henkel KGaA] Isobutyl tallowate; see Rilanit BS; liq.
Rilanit IPM. [Henkel KGaA] IPM; see Rilanit BS; liq.
Rilanit IPP. [Henkel KGaA] IPP; see Rilanit BS; liq.
Rilanit IPS. [Henkel KGaA] Isopropyl stearate; see Rilanit BS; liq.
Rilanit ITS. [Henkel KGaA] Isotridecyl stearate; see Rilanit BS; liq.
Rilanit LTC. [Henkel KGaA] Coconut fatty alcohol caprylic/capric acid ester; see Rilanit BS; liq.
Rilanit MW. [Henkel KGaA] Fatty acid esters and glycerides; artificial wax for bright drying emulsions, carbon paper, paper coatings; flakes.
Rilanit OLO. [Henkel KGaA] Oleyl oleate; see Rilanit BS; liq.
Rilanit PEC 4. [Henkel KGaA] Pentaerythritol tetrafatty acid ester; see Rilanit BS; liq.
Rilanit PES. [Henkel KGaA] Pentaerythritol stearate partial ester; see Rilanit BS; flakes.
Rilanit PMO. [Henkel KGaA] Propylene glycol ester; see Rilanit BS; liq.
Rilanit PPE. [Henkel KGaA] Polyol partial ester; see Rilanit BS; liq.
Rilanit Special. [Henkel Canada] Hydrog. castor oil; thickener for petrol. solv. paint systems.
Rilanit STS-T. [Henkel KGaA] Stearyl stearate; see Rilanit BS; solid.
Rilanit TDO. [Henkel KGaA] TEA dioleate; see Rilanit BS; liq.
Rilanit TMTC. [Henkel KGaA] Trimethylolpropane trifatty acid ester; see Rilanit BS; liq.
Ritacetyl®. [RITA] Acetylated lanolin; nonionic; superfatting agent for soaps, shampoos; film-former for creams and lotions, water resistant films; solid; oil-sol.; 100% conc.
Ritachol®. [RITA] Min. oil and lanolin alcohol; nonionic; used in cosmetics and toiletries; stabilizer for emulsions, dispersions, and suspensions; primary or aux. emulsifier; epidermal moisturizer, lubricant, and emollient; pale straw liq.; faint odor; sp.gr. 0.84–0.86; acid no. 1 max.; sapon. no. 2 max.; 100% act.
Ritaderm®. [RITA] Petrolatum, lanolin, sodium PCA, polysorbate 85; nonionic; emollient, moisturizer, and lubricant used in cosmetics; yel. cream; sol. in oil, alcohol; m.p. 42–48 C; 100% act.
Ritahydrox. [RITA] Hydroxylated lanolin; nonionic; emulsifier, hypoallergenic emollient; paste; HLB 4.0; 100% conc.
Ritalafa®. [RITA] Lanolin acid; nonionic; film-former, emollient; rewetting of makeup preparations; solid; oil-sol.; 100% conc.
Ritalan®. [RITA] Lanolin USP, dewaxed fraction; nonionic; moisturizer, plasticizer, penetrant, emollient; skin lubricant; amber liq.; 100% conc.
Ritalan® AWS. [RITA] Alkoxylated lanolin oil; nonionic; aux. emulsifier, moisturizer, emollient; liq.; sol. in water and alcohol; HLB 18.0; 100% conc.
Ritalan® C. [RITA] IPP and lanolin oil; nonionic; blending agent, emollient, epidermal penetrant, rewetting agent, and solubilizer for waxes and other oil-sol. or disp. materials; personal care prods.; ASTM 2 max. liq.; bland odor; sol. in min. oil; 100% act.
Ritasol. [RITA] Isopropyl lanolate; nonionic; emollient, spreading agent, film former for lip preparations; solid; 100% conc.
Ritawax. [RITA] Lanolin alcohol; nonionic; emulsifier, emollient for skin preparations; solid; sol. in alcohol, IPM, other oils; 100% conc.
Ritawax 5. [RITA] Laneth-5; nonionic; emulsifier, solubilizer, emollient for creams and lotions; solid; 100% conc.
Ritawax 15. [RITA] Laneth-15; nonionic; see Ritawax 5; solid; 100% conc.
Ritawax 40. [RITA] Laneth-40; nonionic; see Ritawax 5; solid; 100% conc.
Ritawax AEO. [RITA] Laneth-10 acetate; nonionic; emollient, lubricant, moisturizer, penetrant, solubilizer, dispersant, plasticizer for personal care prods.; liq.; sol. in water, IPA, castor oil; 100% conc.
Ritawax ALA. [RITA] Cetyl acetate and acetylated lanolin alcohol; nonionic; emollient, lubricant, moisturizer, penetrant; solid; sol. in min. oil, alcohol; 100% conc.
Ritoleth 2. [RITA] Oleth-2; nonionic; o/w emulsifier, solubilizer; liq.; HLB 4.9; 100% conc.
Ritoleth 5. [RITA] Oleth-5; nonionic; see Ritoleth 2; liq.; HLB 9.6; 100% conc.
Ritoleth 10. [RITA] Oleth-10; nonionic; see Ritoleth

2; semisolid; HLB 12.4; 100% conc.

Ritoleth 20. [RITA] Oleth-20; nonionic; see Ritoleth 2; semisolid; HLB 15.1; 100% conc.

RO-40. [Georgia Marble] Ground calcium carbonate; filler for asphalt, putty, ceramic material, foamed compds.; 73 ± 3% min. passing #200 wet screen.

Robane®. [Robeco] Squalane NF; moisturizer, emollient, lubricant, humectant; aids spread of topical agents over the skin, increases skin respiration, prevents insensible water loss, imparts suppleness to skin without greasy feel; cosmetics and pharmaceuticals; colorless liq. oil; odorless.

Robecote. [Robeco] Shark liver oil; emollient for cosmetics, toiletries, dermatologicals; skin protectant.

Robeyl. [Robeco] Squalene, hydrog. shark liver oil; polyunsat. emollient for high color gloss; colorless to faint yel. clear liq.; sp.gr. 0.8375–0.8495; visc. 15–20 cps; ref. index. 1.4750–1.4850.

Robuoy. [Robeco] Pristane; lubricant, buoyant; adjunct to oceanographic research and arctic lubrication; colorless liq.; oil; faint odor; m.w. 268.53; f.p. –60 C; b.p. 290 C; sp.gr. 0.775–0.795; visc. 6.4 cs; acid no. 0–5; sapon. no. 0.5; ref. index 1.435–1.440; 90% min. conc.

Roccal® 50% Technical. [Hilton-Davis] Benzalkonium chloride; cationic; disinfectant and sanitizer used in food or beverage processing plants, food service establishments, industrial plants, and farm use (not for use in hospitals); lt. yel. liq.; mild odor; sol. in water, lower alcohols, ketone, and glycols; dens. 8.2 lb/gal; sp.gr. 0.98; pH 6.8–8.0 (10%); 50% act.

Roccal® II 50%. [Hilton-Davis] Benzalkonium chloride USP; cationic; antimicrobial agent, disinfectant, sanitizer; see Roccal 50% Technical; antistat for fabrics; lt. yel. liq. mild pleasant odor; m.w. 359; sol. in water, lower alcohols, ketone, and glycols; dens. 7.8 lb/gal; sp.gr. 0.94; pH 6.8–8.0 (10%); 50% act.

Roccal® MC-14. [Hilton-Davis] Tetradecyl benzyl dimethyl ammonium chloride dihydrate; cationic; sanitizer, disinfectant, detergent, germicidal laundry prods., swimming pool algicides and slimicides, medicated cosmetics and toiletries; wh. powd.; m.w. 404; sol. in water, ketone, alcohols and glycols; m.p. 60 C min.; pH 7.5 ± 0.5 (10%); 90% act.

Rock Dust. [Georgia Marble] Ground calcium carbonate; med.-ground filler for coal mine dusting; 83 ± 3% min. passing #200 screen.

Rocsol B. [Croda Ltd.] Oxidized hydrocarbon wax; household and industrial polishes; solid wax.

Rocsol C, COB. [Croda Ltd.] Oxidized hydrocarbon wax; household and industrial polishes; flaked wax.

Rocsol Micro 4. [Croda Ltd.] Oxidized hydrocarbon wax; PVC process aid; flaked wax.

Rodea. [Rhone-Poulenc Surf.] Ricinoleamide DEA (2:1); emulsifier for aliphatic hydrocarbons; rust inhibitor; formulation of conveyor chain lubricants; dk. amber visc. liq.; sol. in alcohol, glycol, glycolether, esters, ketone, aliphatic and aromatic hydrocarbons; disp. in water; sp.gr. 1.016; pH 9.0–10.5 (1%); 4% max. free fatty acid.

Rokon. [Vanderbilt] 2-Mercaptobenzothiazole; metal deactivator, copper corrosion inhibitor in fuels, industrial lubricants, automotive chemicals, and industrial cleaners; lt. yel. to lt. buff powd.; sol. 25% in Cellosolve, 20% in butyl Cellosolve, 30% in Carbitol, 1.5% in toluene, and < 5% in highly aromatic oils; slightly sol. in water; dens. 1.50 mg/m³; m.p. 164–175 C.

Ross Thix 700 Series. [Ross] Proprietary; anionic; liq.; thickener for aq. systems; water-disp.

Ross Wax #100. [Frank B. Ross] Fischer-Tropsch wax; mold release in rubber and plastics; external lubricant for PVC; increases m.p., opacity, gloss in wax blends, floor waxes, textiles, coatings, candles, hot melts, paints, inks, asphalt; wh. flakes; sol. hot in most solvs.; visc. 80 SUS (120 C); congeal pt. 200–210 F; acid no. nil; sapon. no. nil.

Ross Wax #140. [Frank B. Ross] Syn. high m.p. wax; used as ingredient, finish or processing aid in adhesives, ammunition, asphalt, explosives, paints, paper, pyrotechnics, lubricants, mold releases, PVC, textile finishes, powd. metallurgy, etc.; wh. flakes, powd., or atomized; sol. in hot naphtha, xylene, Carbitol, IPA, trichlorethylene; sp.gr. 0.97; m.p. 138–140 C; flash pt. 530 F min.; acid no. 3–10; dielec. str. 233 V/mil; vol. resist. 0.99×10^{13} ohm-cm.

Ross Wax #145, 165. [Frank B. Ross] High m.p. fully refined petroleum wax; lubricant for formulating PVC and elec. wire and cable compds.; wh. flakes and chips resp.; sol. in benzene (#145); insol. in acetone and water (#165); sp.gr. 0.930–0.936; m.p. 140–150 and 158–165 F resp.

Ross Wax #160. [Frank B. Ross] Syn. high m.p. wax; used as ingredient, finish or processing aid in adhesives, ammunition, explosives, hot-melt coatings, lubricants, paints, paper, pyrotechnics, release agent, propellants, textile finishes, powd. metallurgy, varnish, plastic film, etc.; tan med. hard wax in lumps, flakes, powd., or atomized; sol. in hot naphtha, xylene, Carbitol, IPA, trichlorethylene, toluene; sp.gr. 1.0232; m.p. 157–162 C; flash pt. 590 F min.; acid no. 10 max.; iodine no. 7.5.

Rotax.® [Vanderbilt] 2-Mercaptobenzothiazole; accelerator; primary accelerator for natural and syn. rubbers; nonstaining and nondiscoloring; used in proofing compds. where lowest odor is desired; pale yel. powd., 99.9% thru 100 mesh; m.w. 167.25; very sol. in dilute caustic, toluene, carbon disulfide, chloroform, alcohol; pract. insol. in water; dens. 1.52 ± 0.03 mg/m³; m.p. 169–180 C.

Rotolan®. [RITA] Proprietary lanolin deriv.; aids in pigment dispersion for use in inks; paste; 100% conc.

RR 5. [Releasomers] Semipermanent mold release agent for most thermosetting rubber and plastic materials; avail. in cold and hot formulations; solv. sol'n.

RR Zinc Oxide-Coated. [Akrochem] Amer. process zinc oxide coated with propionic acid; activator for rubber; powd.; 0.27 μ mean particle size; sp.gr. 5.6; dens. 40 lb/ft³; 98.8% ZnO.

RR Zinc Oxide (Untreated). [Akrochem] Amer.

process zinc oxide; activator for rubber; powd.; 0.27 μ mean particle size; sp.gr. 5.6; dens. 30 lb/ft³; 99.2% ZnO.

RT-3. [Disco] Process aid designed to build tack in elastomers requiring process adhesion; dark, brittle, supercooled liq.; sp.gr. 1.05; m.p. liq. above 135 F.

Rudol®.[Witco] Lt. min. oil NF; white oil functioning as binder, carrier, conditioner, defoamer, dispersant, extender, heat transfer agent, lubricant, moisture barrier, plasticizer, protective agent, and/or softener in adhesives, agric., chemicals, cleaning, cosmetics, food, pkg., plastics, and textiles industries; sp.gr. 0.852–0.870; visc. 28–30 cSt (40 C); flash pt. 188 C; pour pt. –7 C.

Rueterg SA. [Finetex] Dodecylbenzene sulfonic acid; anionic; intermediate for detergents and emulsifiers; liq.; 97% conc.

Runox 1000. [Toho] Salt of condensed naphthalene sulfonic acid; dispersant for dyestuff, pigment; powd.

Rycel 100, 105. [Ryco] Sodium CMC polymers; regular and high visc. additives to drilling muds to influence visc. and fluid loss; filtration control agent and as a viscosifying agent; 1% max. retained on 30 mesh; 5% max. retained on 40 mesh; also avail. in coarse and fine particle grades; powd.; hydrates in water.

Rychem® 17. [Ryco] Polyglycol ester; nonionic; emulsifier for fat and min. oils; lubricant, softener, and rewetting agent; liq.; 90–95% conc.

Rychem® 400-OE. [Ryco] Polymer/alkalies; anionic; dispersant and antifoulant; scale and sludge preventive; used in water treating processing; clear liq.; water-sol.; dens. 9.7 lb/gal.

Rychem® 784. [Ryco] Sodium dimethyldithiocarbamate (15%), nabam (disodium ethylene bisdithiocarbamate) (15%) and 70% inert ingreds.; microbicide for control of fungi and bacteria in water-based drilling muds; yel. grn. to grn.; 30 ± 0.5% act.

Rychem® 808. [Ryco] Polyelectrolyte; dispersant/scale inhibitor used in cooling towers, evaporators, and boiler feed water; pale amber clear liq.; odorless; water-sol.; sp.gr. 1.13.

Rychem® 810. [Ryco] Methylene bis (thiocyanate); antimicrobial for control of fungi, bacteria, slime in pulp and paper mills, recirculating cooling water systems, paper and paperboard for food pkg.; straw liq.; water-sol.; sp.gr. 0.99; dens. 8.30 lb/gal; flash pt. 127 F; 10% act.

Rychem® 830. [Ryco] Sodium dimethyldithiocarbamate (15%), nabam (disodium ethylene bisdithiocarbamate) (15%) in 70% inert ingreds.; antimicrobial, slimicide for control of fungi, bacteria in industrial recirculating cooling towers, air washers, evaporative condensers, paper for food contact use; liq.; water-sol.; sp.gr. 1.150–1.200; dens. 9.58 lb/gal; pH 11.5 ± 1.0; 30% act.

Rychem® CLS. [Ryco] Chrome lignosulfonate; oil well drilling mud conditioner; used in salt water conditions; clay dispersant; brnsh. powd.; water-sol.; dens. 35 lb/ft³; 6 ± 1% moisture content.

Rychem® D941. [Ryco] Nonchromate polymer; oil well drilling mud conditioner; thinner for drilling fluids; dispersant; wh. powd.; m.w. 2000; dens. 30–35 lb/ft³; pH 8.5 ± 0.5.

Rychem® FCL. [Ryco] Ferro chrome lignosulfonate; oil well drilling mud conditioner; used in salt water conditions; brn. powd.; water-sol.; dens. 35 lb/ft³; pH 3.5 ± 0.5; 6 ± 1% moisture content.

Rychem® G-Gard. [Ryco] Fatty amine; corrosion inhibitor, solubilizer, and dispersant for gasoline distribution and storage systems.

Rychem® G-Hib. [Ryco] Fatty amine; see Rychem G-Gard.

Rychem® PD4. [Ryco] Polyeletrolyte; anionic; pitch dispersant; used in pulp and paper mills; amber liq.; sp.gr. 1.034; dens. 8.6 lb/gal.

Ryco® Defoamer 1288. [Ryco] Fatty acid base emulsifiable; anionic; defoamer/antifoam for distillation fermentation; paste; biodeg.

Rycofax® 618. [Ryco; Devan SA] Fatty amidoamine salt; cationic; softener/debonder; release agent for tissue mfg.; liq.; 98% conc.

Rycofax® 3115. [Ryco] Amine deriv.; cationic; softener and antistat for fabrics; liq.; 99% conc.

Rycofax® O. [Ryco] PEG ester; nonionic; wetting and rewetting in pulp/paper and wet process industries such as textile, metal; emulsifier for oils and hydrocarbons; coupler; liq.; 100% conc.

Rycolube® 150 Series. [Ryco] Syn. waxes and proprietary formulations; release agent; prevents adhesion during storage/shipment of rubber, asphalt, plastic in strip, sheet, or tile form; water-sol./disp.

Rycomid® 2120. [Ryco] Coco alkanolamide; nonionic; detergent, thickener, foam stabilizer, wetting agent used for floor, textile, and metal cleaners; liq.; water-sol.; 100% conc.

Rycowax®. [Ryco] Proprietary; syn. extender for natural waxes; components in finishes, coatings, coupling agents.

Rylex NBC. [DuPont] Nickel dibutyldithiocarbamate; stabilizer for PP and polyethylene, esp. HDPE; absorbs strongly in the uv range and functions as an antioxidant; economical for many industrial applics. where its grn. color may be tolerated or masked; synergistic with benzophenone in stabilizing PP and polyethylene; useful for pigmented polyolefin; dk. grn. flakes; mild odor; sol. in most org. solvs. such as acetone, benzene, toluene, and CCl_4; m.p. 88 C; flash pt. 263 C.

Ryoto Sugar Ester LWA-1540. [Ryoto] Sucrose monolaurate; nonionic; o/w and w/o emulsifier, softener, conditioner, and aerating agent in foods; paste; HLB 15.0; 40% conc.

Ryoto Sugar Ester OWA-1570. [Ryoto] Sucrose monooleate; nonionic; see Ryoto Sugar Ester LWA-1540; paste; HLB 15.0; 40% conc.

Ryoto Sugar Ester P-1570, -1670. [Ryoto] Sucrose monopalmitate; nonionic; see Ryoto Sugar Ester LWA-1540; powd.; HLB 15.0 and 16.0 resp.; 100% conc.

Ryoto Sugar Ester S-170. [Ryoto] Sucrose polystearate; nonionic; see Ryoto Sugar Ester LWA-1540; powd.; HLB 1.0; 100% conc.

Ryoto Sugar Ester S-270, -370. [Ryoto] Sucrose ditristearate; nonionic; see Ryoto Sugar Ester LWA-1540; powd.; HLB 2.0 and 2.3 resp.; 100% conc.

Ryoto Sugar Ester S-570, -770. [Ryoto] Sucrose distearate; nonionic; see Ryoto Sugar Ester LWA-1540; powd.; HLB 5.0 and 7.0 resp.; 100% conc.

Ryoto Sugar Ester S-970, -1170. [Ryoto] Sucrose monodistearate; nonionic; see Ryoto Sugar Ester LWA-1540; powd.; HLB 9.0 and 11.0 resp.; 100% conc.

Ryoto Sugar Ester S-1570, -1670. [Ryoto] Sucrose monostearate; nonionic; see Ryoto Sugar Ester LWA-1540; powd.; HLB 15.0 and 16.0 resp.; 100% conc.

S

Sachtolith® HD, HD-S, L. [Sachtleben] Zinc sulfide; wh. pigment used in thermoplastics (HDPE, PP), thermosets (melamine, urea, polyester), flame-resistant plastics, glass-reinforced plastics, pigment concs. and masterbatches, elastomers, textile fibers, apper, sealants, lubricants; opacifier; HD and HD-S grades are coated with a small amt. of org. surfactant and have improved wetting-out and disp. char. in aq. and org. systems; L is the uncoated standard grade; 0.35, 0.30, and 0.35 μm mean particle size; dens. 4.0 g/ml; surf. area 8 m^2/g; oil absorp. 14, 13, and 14; ref. index 2.37; pH 7, 7, and 6; hardness (Mohs) 3; 98% ZnS content.

Safety-Lube® 1776. [Chem-Trend] Blend of parting agents; die lubricant for aluminum castings; tan liq.; mild odor; dilutes easily with water; dens. 8.3 lb/gal; pH 9.5 ± 0.5; flash pt. none.

Safety-Lube® 1976. [Chem-Trend] Blend of parting agents; die lubricant for aluminum castings; tan liq.; mild odor; dilutes easily with water; dens. 8.3 lb/gal; pH 9.5 ± 0.5; flash pt. none.

Safety-Lube® 3220. [Chem-Trend] Blend of parting agents, lubricity agents, EP additives, and corrosion inhibitors; heavy-duty die lubricant for aluminum castings; br. liq.; mild odor; dilutes easily with water; dens. 8.3 lb/gal; pH 9.5 ± 0.5; flash pt. none.

Safety-Lube® 3600. [Chem-Trend] Blend of lubricant base materials and parting agents; die lubricant for aluminum castings; lt. tan liq.; waxy odor; dilutes easily with water; dens. 8.3 lb/gal; pH 9.5 ± 0.5; flash pt. none.

Safety-Lube® 4000. [Chem-Trend] Blend of lubricant base, antiweld chemicals, and EP additives; heavy-duty die lubricant for difficult and complex castings, providing release, protection against pitting and corrosion; creamy liq.; mild odor; dilutes easily with water; dens. 8.3 lb/gal; pH 9.0 ± 0.5; flash pt. none.

Safety-Lube® 6003. [Chem-Trend] Emulsion of wax, antiweld chemicals, and EP additives; wax-based die lubricant for lt.- to med.-duty aluminum die castings; provides release, protection against pitting, corrosion, and staining; tan liq.; mild odor; dilutes easily with water; dens. 8.3 lb/gal; pH 9.5 ± 0.5; flash pt. none.

Safety-Lube® 8155. [Chem-Trend] Emulsion of proprietary lubricity additives, antiweld chemicals, and EP additives; nonstaining heavy-duty aluminum die casting lubricant for complex or large castings; tan liq.; mild odor; dilutes easily with water; dens. 8.3 lb/gal; pH 9.3 ± 0.5; flash pt. none.

Sag® Silicone Antifoam 10. [Union Carbide] Silicone emulsion; antifoam for detergent/cleaning sol'ns., chemical specialty aerosol, textile defoaming; water-disp.; 10% act.

Sag® Silicone Antifoam 30. [Union Carbide] Silicone emulsion; defoamer for detergent/cleaning sol'ns., chemical specialty aerosols; water-disp.; 30% act.

Sag® Silicone Antifoam 100. [Union Carbide] Silica-filled dimethylpolysiloxane compd.; defoamer for detergents, starch processing, insect repellent formulations; used in heavy duty alkaline cleaners, drug extraction, and drug processing; sol. in aromatic, aliphatic solvs.; water-insol.

Sag® Silicone Antifoam 310. [Union Carbide] Silicone emulsion; antifoam for textile and rug shampoo; water-disp.

Sag® Silicone Antifoam 471. [Union Carbide] Organo-modified polysiloxane compd.; leveling aid in floor polishes; defoamer used in coolants, deicers, printing inks, latex paints; water-disp.

Sag® Silicone Antifoam 720. [Union Carbide] Dimethyl polysiloxane emulsion; food grade antifoam; water-disp.; 20% act.

Sag® Silicone Antifoam 4130. [Union Carbide] Organo-modified polysilxoane emulsion; leveling aid in wax emulsions; defoamer for heavy-duty liq. detergents; water-disp.

Sag® Silicone Antifoam 4220. [Union Carbide] Organo-modified polysiloxane emulsion; leveling agent for wax and polish emulsions; defoamer; water-disp.

Sag® Silicone Antifoam 5300. [Union Carbide] Organo-modified polysiloxane fluid; defoamer for hot starch processing used in spray starch/fabric finishes; water sol. at R.T.; water insol. above 90 F.

Sag® Silicone Antifoam 5310. [Union Carbide] Organo-modified polysiloxane fluid; defoamer for chlorinated dry cleaning sol'ns.; sol. in water, many aliphatic/aromtic solvs.; 60% act.

Sag® Silicone Antifoam 5693. [Union Carbide] Silicone/polyalkylene glycol; food grade antifoam; water-disp.; 100% act.

Salabon 50. [Takemoto] Alkyl polyamine ethyl glycine hydrochloride; germicide against wide range of organisms incl. mycobacterium tuberculosis; liq.

Salfax 77. [Chem-Y GmbH] Fatty acid and cationics; cationic; hair conditioner; paste; 22% conc.

Samaron Thickener N. [Hoechst-Celanese AG] High polymeric compd.; thickener for nonionic disp. dyes; wh. paste.

Sanac C. [Capital City] 2-(2-Carboxyethoxy) ethyl 2-[2-(2-hydroxy)] ethoxyethyl methyldodecyl ammonium methyl sulfate, potassium salt; amphoteric; detergent, wetting agent, foamer, solubilizer, heavy duty surface cleaner, metal treatment; straw to amber semivisc. liq.; sol. in water and polar solvs.; dens. 8.7 lb/gal; sp.gr. 1.05; pH 7.5–8.5; biodeg.; 35% act.

Sanac S. [Capital City] [2-(2-Carboxyethoxy) ethyl] [2-[2-(hydroxy) ethoxyethyl] methyloctadecenyl ammonium methyl sulfate, potassium salt; see Sanac C; straw to amber semivisc. liq.; mild typ. odor; water-sol.; dens. 8.7 lb/gal; sp.gr. 1.05; pH 7.5–8.5; biodeg.; 35% act.

Sanac T. [Capital City] Complex carboxylated tallow quat.; amphoteric; detergent, wetting agent and foamer, solubilizer; liq.; 60% conc.

Sandet ALH. [Sanyo] Alkyl diphenyl ether sodium disulfate; anionic; wetting agent, penetrant and base material for making liq. bleaching detergents; liq.; 30% conc.

Sandin EU, VU. [Sandoz] Sulfonated hydrocarbon; antistat for PVC, talc, etc.; visc. liq. and powd. resp.

Sandocorin 8015, 8132, 8132B, 8160. [Sandoz] Complex phosphate ester deriv.; internal corrosion inhibitor for use in internal cooling systems, cooling towers; aq. sol'n.

Sandocorin PP. [Sandoz] Complex phosphate ester deriv. in a polymeric matrix; film-former offering permanent protection to metal surfs.; aq. metal primer; wh. aq. emulsion.

Sandocorin TP. [Sandoz] Complex phosphate ester deriv. in an oil emulsion; film-former offering temporary protection to metal surfs.; tan emulsion.

Sandocorin W. [Sandoz] Complex phosphate ester deriv. in a wax emulsion; film-former offering good protection in marine applics.; tan emulsion.

Sandogen C-CM Liq. [Sandoz] Dye leveling agent; antiprecipitant for dyeing nylon carpet; yel. liq.; misc. with water; sp.gr. 1.04; 31% act.

Sandopan® KD Conc. [Sandoz] Sodium alkyl sulfate, modified; anionic; scouring, milling, wetting agent and dispersant used in textiles; paste.

Sandopan® TFL Liquid. [Sandoz] Fatty acid deriv.; amphoteric; detergent, wetting agent, emulsifier, lubricant, dispersant; dyeing nylon carpets with disperse dyes; lubricant and dispersant in boil-off and dyeing of syn. fabrics; yel. liq.; mild odor; water-misc.; sp.gr. 1.03; pH 9.0; 11.5% act.

Sandostab P-EPQ. [Sandoz] High m.w. phosphonite compd. [tetrakis (2,4-di-tert-butylphenyl) 4,4-biphenylenediphosphonite]; effective processing stabilizer, sec. antioxidant for a broad spectrum of polymers incl. polyolefeins, ABS, PS, polybutylene, polybutylene terephthalate, PEG, nitrile barrier resins; peroxide decomposer for plastics mfg.; prevents polymer yellowing; synergizes uv lt. stability; reduces equipment corrosion in flame-retardant applics.; off-wh. powd.; sol. 100 g/100 ml in ethanol, acetone, ethyl acetate, toluene, benzene; 106 g/100 ml. in cyclohexane @ 50 C; sp.gr. 1.045; dens. (bulk) 38 g/100 ml; m.p. 75 C.

Sandoxylate® 206. [Sandoz] Alkoxylated alcohol; nonionic; wetting agent, dispersant used in cleaning and treatment prods. and processes; liq.; HLB 12–14; 100% conc.

Sandoxylate® 224. [Sandoz] Alkoxylated alcohol; nonionic; see Sandoxylate 206; semisolid; HLB 16–17; 100% conc.

Sandoxylate® 408. [Sandoz] Alkoxylated alcohol; nonionic; see Sandoxylate 206; liq.; HLB 10–12; 100% conc.

Sandoxylate® 412. [Sandoz] Alkoxylated alcohol; nonionic; see Sandoxylate 206; liq.; HLB 12–14; 100% conc.

Sandoxylate® 418. [Sandoz] Alkoxylated alcohol; nonionic; see Sandoxylate 206; liq.; HLB 16.0; 100% conc.

Sandoxylate® 424. [Sandoz] Alkoxylated alcohol; nonionic; see Sandoxylate 206; semisolid; HLB 16.0; 100% conc.

Sandoxylate® AC-9. [Sandoz] Alcohol ethoxylate; nonionic; see Sandoxylate 206; solid; HLB 12.0; 100% conc.

Sandoxylate® AC-24. [Sandoz] Alcohol ethoxylate; nonionic; see Sandoxylate 206; solid, flakes; HLB 15.8; 100% conc.

Sandoxylate® AD-4. [Sandoz] Alcohol ethoxylate; nonionic; see Sandoxylate 206; liq.; 100% conc.

Sandoxylate® AD-6. [Sandoz] Alcohol ethoxylate; nonionic; see Sandoxylate 206; liq.; 100% conc.

Sandoxylate® AD-9. [Sandoz] Alcohol ethoxylate; nonionic; see Sandoxylate 206; liq.; 100% conc.

Sandoxylate® AL-4. [Sandoz] Alcohol ethoxylate; nonionic; see Sandoxylate 206; liq.; 100% conc.

Sandoxylate® AO-12. [Sandoz] Alcohol ethoxylate; nonionic; see Sandoxylate 206; liq.; HLB 12.7; 100% conc.

Sandoxylate® AO-20. [Sandoz] Alcohol ethoxylate; nonionic; see Sandoxylate 206; semisolid; HLB 15.4; 100% conc.

Sandoxylate® AO-60. [Sandoz] Alcohol ethoxylate; nonionic; see Sandoxylate 206; solid; HLB 18.9; 100% conc.

Sandoxylate® AT-6.5. [Sandoz] Alcohol ethoxylate; nonionic; see Sandoxylate 206; liq.; HLB 11.5; 100% conc.

Sandoxylate® AT-12. [Sandoz] Alcohol ethoxylate; nonionic; see Sandoxylate 206; liq.; HLB 14.5; 100% conc.

Sandoxylate® C-10. [Sandoz] Castor oil ethoxylate; nonionic; see Sandoxylate 206; liq.; 100% conc.

Sandoxylate® C-15. [Sandoz] Castor oil ethoxylate; nonionic; see Sandoxylate 206; liq.; HLB 13.5; 100% conc.

Sandoxylate® C-32. [Sandoz] Castor oil ethoxylate; nonionic; see Sandoxylate 206; liq.; HLB 15.0; 100% conc.

Sandoxylate® FO-9. [Sandoz] Fatty acid ethoxylate; nonionic; surfactant, wetting agent, dispersant used in cleaning and treatment prods. and processes; liq.; HLB 11.9; 70% conc.

Sandoxylate® FO-30/70. [Sandoz] Fatty acid ethoxylate; nonionic; see Sandoxylate FO-9; liq.; HLB 13.9; 100% conc.

Sandoxylate® FS-9. [Sandoz] Fatty acid ethoxylate; nonionic; see Sandoxylate FO-9; solid; HLB 11.5; 100% conc.

Sandoxylate® FS-35. [Sandoz] Fatty acid ethoxylate; nonionic; see Sandoxylate FO-9; solid; HLB 16.9; 100% conc.

Sandoxylate® NC-5. [Sandoz] Amine ethoxylate; nonionic; see Sandoxylate FO-9; liq.; 100% conc.

Sandoxylate® NC-15. [Sandoz] Amine ethoxylate; nonionic; see Sandoxylate FO-9; liq.; 100% conc.

Sandoxylate® NSO-30. [Sandoz] Amine ethoxylate; nonionic; see Sandoxylate FO-9; liq.; 100% conc.

Sandoxylate® NT-5. [Sandoz] Amine ethoxylate; nonionic; see Sandoxylate FO-9; liq., paste; 100% conc.

Sandoxylate® NT-15. [Sandoz] Amine ethoxylate; nonionic; see Sandoxylate FO-9; liq., paste; 100% conc.

Sandoxylate® PDN-7. [Sandoz] Amine ethoxylate; nonionic; see Sandoxylate FO-9; liq.; HLB 9.9; 100% conc.

Sandoxylate® PN-6. [Sandoz] Amine ethoxylate; nonionic; see Sandoxylate FO-9; liq.; 100% conc.

Sandoxylate® PN-9. [Sandoz] Alkyl phenol ethoxylate; nonionic; see Sandoxylate FO-9; liq.; HLB 12.9; 100% conc.

Sandoxylate® PN-10.5. [Sandoz] Alkyl phenol ethoxylate; nonionic; see Sandoxylate FO-9; liq.; 100% conc.

Sandoxylate® PO-5. [Sandoz] Alkyl phenol ethoxylate; nonionic; see Sandoxylate FO-9; liq.; HLB 10.4; 100% conc.

Sandoxylate® SX-208. [Sandoz] PO/EO alkoxylate; nonionic; wetting agent for household and industrial applics., textile wet processing; intermediate for anionic surfactants; Gardner < 1 liq.; water-sol.; sp.gr. 0.9625; HLB 11; cloud pt. 22 ± 2 C (1% aq.); pH 6.5 ± 1 (1%); surf. tens. 28.0 dynes/cm; 98 ± 2% act.

Sandoxylate® SX-224. [Sandoz] PO/EO alkoxylate; nonionic; see Sandoxylate SX-208; Gardner < 1 liq.; water-sol.; sp.gr. 1.0262; HLB 15; cloud pt. 93 ± 2 C (1% aq.); pH 6.5 ± 1 (1%); surf. tens. 30.2 dynes/cm; 98 ± 2% act.

Sandoxylate® SX-408. [Sandoz] PO/EO alkoxylate; nonionic; see Sandoxylate SX-208; Gardner < 1 liq.; water-sol.; sp.gr. 0.9650; HLB 11; cloud pt. 22 ± 2 C (1% aq.); pH 6.5 ± 1 (1%); surf. tens. 28.8 dynes/cm; 98 ± 2% act.

Sandoxylate® SX-412. [Sandoz] PO/EO alkoxylate; nonionic; see Sandoxylate SX-208; Gardner < 1 liq.; water-sol.; sp.gr. 0.9822; HLB 13; cloud pt. 38 ± 2 C (1% aq.); pH 6.5 ± 1 (1%); surf. tens. 28.6 dynes/cm; 98 ± 2% act.

Sandoxylate® SX-418. [Sandoz] PO/EO alkoxylate; nonionic; see Sandoxylate SX-208; Gardner < 1 liq.; water-sol.; sp.gr. 1.007; HLB 15; cloud pt. 65 ± 2 C (1% aq.); pH 6.5 ± 1 (1%); surf. tens. 29.4 dynes/cm; 98 ± 2% act.

Sandoxylate® SX-424. [Sandoz] PO/EO alkoxylate; nonionic; see Sandoxylate SX-208; Gardner < 1 solid; water-sol.; sp.gr. 1.0267; HLB 15; cloud pt. 89 ± 2 C (1% aq.); pH 6.5 ± 1 (1%); surf. tens. 30.3 dynes/cm; 98 ± 2% act.

Sandoz Amide CO. [Sandoz] Coconut alkanolamide (2:1); detergent, foamer, wetting and rewetting agent, visc. modifier used in metal cleaning, heavy duty detergents, hard surface cleaning, textile wet processing; amber liq.; 97% act.

Sandoz Amide NP. [Sandoz] Coconut alkanolamide (2:1); see Sandoz Amide CO; amber liq.; 97% act.

Sandoz Amide NT. [Sandoz] Alkanolamide; nonionic; surfactant used as a detergent, emulsifier and visc. aid for household care and cosmetic applics.; liq.; 100% conc.

Sandoz Amide PE. [Sandoz] Lauryl alkanolamide (1:1); detergent, foamer, wetting and rewetting agent, visc. modifier used for liq. dishwashing prods. and personal care prods.; wh. solid; 90–95% act.

Sandoz Amide PL. [Sandoz] Lauryl alkanolamide (2:1); see Sandoz Amide PE; amber liq.; 100% act.

Sandoz Amide PN. [Sandoz] Coconut alkanolamide (2:1); detergent, foamer, wetting and rewetting agent, visc. modifier used for heavy duty cleaning formulations, oil emulsification; amber liq.; 97% act.

Sandoz Amide PO. [Sandoz] Coconut alkanolamide (1:1); detergent, foamer, wetting and rewetting agent, visc. modifier used for personal care prods. and cleaning formulations; lt. yel. visc. liq.; 85% act.

Sandoz Amide PS. [Sandoz] Coconut alkanolamide (2:1); detergent, foamer, wetting and rewetting agent, visc. modifier used in hard surface cleaners, detergents, dishwashing compds.; textile scouring; amber liq.; 97% act.

Sandoz Amine Oxide XA-C. [Sandoz] Cetyl amine oxide; nonionic; surfactant, foam booster, visc. improver for personal care and liq. dishwashing formulations; antistat; textile wet processing; softener and emollient; amber liq.; sol. in water, lower alcohols, glycols; biodeg.; 20% act.

Sandoz Amine Oxide XA-L. [Sandoz] Lauryl amine oxide; nonionic; surfactant, foam booster, visc. improver for personal care and liq. dishwashing formulations; antistat, textile wet processing; amber liq.; sol. in water, lower alcohols, glycols; biodeg.; 30% act.

Sandoz Amine Oxide XA-M. [Sandoz] Myristyl amine oxide; nonionic; surfactant; see Sandoz Amine Oxide XA-L; amber liq.; sol. in water, lower alcohols, glycols; biodeg.; 30% act.

Sandoz Defoamer F. [Sandoz] Silicone emulsion; general purpose defoamer for use in industrial chemical processing; wh. creamy emulsion.

Sandoz Phosphorester 510. [Sandoz] Complex phosphate ester; anionic; detergent, emulsifier in waterless hand cleaners, dry cleaning detergents; antistat, corrosion inhibitor in textile fiber lubricants; primary emulsifier in emulsion polymerization of PVAc, acrylic; clear visc. liq.; water-disp.; sol. in conc. electrolyte systems; pH 1.5–2.5 (10%); 99% solids.

Sandoz Phosphorester 600. [Sandoz] Complex phosphate ester; hydrotrope; solubilizer for nonionic surfactants; wetting agent used in acid and solv.

cleaners; emulsifier for solvs.; clear visc. liq.; sol. in conc. electrolyte systems; pH 1.5–2.5 (10%); 99% solids.

Sandoz Phosphorester 610. [Sandoz] Complex phosphate ester; emulsifier for aromatic and chlorinated solvs.; solubilizer for nonionic surfactants; emulsion polymerization surfactant used in vinyl acetate and acrylic systems; clear visc. liq.; sol. in conc. electrolyte systems; pH 1.5–2.5 (10%); 99% solids.

Sandoz Sulfate 216. [Sandoz] Ammonium laureth sulfate (3 EO); foamer and visc. modifier for personal care prods.; basic ingred. in shampoos and liq. dishwashing prods.; textile scouring; emulsifier and suspending agent in fabrics; pale yel. liq.; visc. 1000 cps max.; cloud pt. 15 C; pH 6.5–7.5 (10%); 58.5–60% act.

Sandoz Sulfate 219. [Sandoz] Sodium laureth sulfate (3 EO); foamer and visc. modifier in personal care prods.; textile scouring; pale yel. liq.; visc. 1000 cps max.; cloud pt. 15 C; pH 7.5–8.5 (15%); 58.5–60% act.

Sandoz Sulfate A. [Sandoz] Ammonium lauryl sulfate; foamer and visc. modifier in personal care prods.; basic ingred. in shampoos and liq. dishwashing prods.; lt. yel. liq.; visc. 6500 cps; cloud pt. 5 C; pH 6.3–7.0 (10%); 28–30% act.

Sandoz Sulfate EP. [Sandoz] DEA-lauryl sulfate; foamer and visc. modifier for personal care prods.; lt. yel. liq.; visc. 225 cps; cloud pt. 6 C; pH 7.5–8.5 (10%); 35–37.5% act.

Sandoz Sulfate ES-3. [Sandoz] Sodium laureth sulfate (3 EO); foamer and visc. modifier for personal care prods.; textile scouring; wetting agent; emulsifier and suspending agent in fabrics; pale yel. liq.; visc. 1000 cps max.; cloud pt. 15 C; pH 7.5–8.5 (10%); 28–30% act.

Sandoz Sulfate K. [Sandoz] Ammonium mixed alkyl sulfate; see Sandoz Sulfate A; brn. liq.; visc. 1000 cps max.; cloud pt. 15 C; pH 6.5–7.0 (10%); 45% act.

Sandoz Sulfate TL. [Sandoz] TEA-lauryl sulfate; see Sandoz Sulfate EP; lt. yel. liq.; visc. 225 cps; cloud pt. 2.5 C; pH 6.4–7.7 (10%); 38.5–42.5% act.

Sandoz Sulfate WA Dry. [Sandoz] Sodium lauryl sulfate; foamer and visc. modifier for personal care, dental, and household prods.; wh. powd.; pH 6.5–7.5 (10%); 90–96% act.

Sandoz Sulfate WAG. [Sandoz] Sodium lauryl sulfate; see Sandoz Sulfate WA Dry; wh. needles; pH 6.5–7.5 (10%); 90–96% act.

Sandoz Sulfate WAS. [Sandoz] Sodium lauryl sulfate; see Sandoz Sulfate WA Special; lt. yel. liq.; visc. 1000–3000 cps; cloud pt. 17–23 C; pH 7.4–8.6 (10%); 27.5–30.5% act.

Sandoz Sulfate WA Special. [Sandoz] Sodium lauryl sulfate; foamer, visc. modifier for personal care, household prods.; wh. paste; visc. 1000–5000 cps; cloud pt. 17–21 C; pH 7.4–8.6 (10%); 27.5–30.5% act.

Sandoz Sulfate WE. [Sandoz] Sodium laureth sulfate (3.5 EO); foamer and visc. modifier for personal care prods.; textile scouring; colorless liq.; visc. 750 cps; cloud pt. 13 C; pH 7.8–8.5 (10%); 27–29% act.

Sandoz Sulfonate 2A1. [Sandoz] Dodecyl diphenyl ether sulfonate, sodium salt; anionic; solubilizer; household detergents, industrial, disinfectant, and metal cleaners, emulsifier in emulsion polymerization; textile aux.; also avail in acid form; yel./brn. liq.; sol. in electrolyte sol'ns., caustic, HCl, TKPP; sp.gr. 1.16; visc. 145 cps; 45% act.

Sandoz Sulfonate 3B2. [Sandoz] Decyl diphenyl ether sulfonate, sodium salt; anionic; wetting agent; agric. formulations; spreader and penetrant for farm usage; also avail. in acid form; amber liq.; sol. see Sandoz Sulfonate 2A1; sp.gr. 1.16; visc. 120 cps; 45% act.

Sanduvor 3206, EPU, VSU. [Sandoz] Oxalanilide deriv.; uv absorber reducing degradation in films and coatings; visc. liq., visc. liq. in solv., and wh. powd. resp.

Sanisol C. [Kao] Benzalkonium chloride; disinfectant, sanitizer for pharmaceuticals; liq.

Sanisol CPR, CR, CR-80%. [Kao] Benzalkonium chloride; cationic; disinfectant, sanitizer for industrial and cooling water treatments, pharmaceuticals; liq., liq., paste resp.; 75, 50, and 80% conc.

Sanisol HTPR. [Kao] Benzalkonium chloride; cationic; see Sanisol CPR; liq./paste; 75% conc.

Sanisol OPR, TPR. [Kao] Benzalkonium chloride; cationic; see Sanisol CPR; liq.; 75% conc.

Sanmorin 11. [Sanyo] POE alkyl ether; nonionic; penetrant and wetting agent for textile processing; liq.

Sanmorin OT 70. [Sanyo] Sodium dioctyl sulfosuccinate; anionic; wetting agent and penetrant for industrial uses; liq.; 70% conc.

Sanstat 2012-A. [Sanyo] Sodium dioctyl sulfosuccinate; cationic; external antistat for plastics incl. polyolefins, PS, polyamide, PVC, ABS; liq.

Santelle-EOM K. [Van Den Bergh] Mono- and diglycerides, ethoxylated; nonionic; food emulsifier; emulsion stabilizer; dispersant aid; paste; m.p. 80–85 F; HLB 13.1; 100% act.

Santicizer 8. [Monsanto] Ethyl toluene-sulfonamide; plasticizes more difficult resins, shellac, cellulose acetate, and protein materials; used in NC lacquers, cellulose acetate compositions, PVAc emulsion adhesives, syn. polyamides; food pkg. applics.; lt. yel. visc. liq.; slight char. odor; m.w. 199; b.p. 196 C; sol. 0.13% in water; sp.gr. 1.190; dens. 9.92 lb/gal; visc. 358 cSt; solid. pt. 0 C; ref. index 1.540; flash pt. 345 F (COC); 15 phr in ethyl cellulose: tens. str. 6900 psi; tens. elong 34%; hardness 70.

Santicizer 9. [Monsanto] o,p-Toluenesulfonamide; plasticizer for thermosetting resins; used in melamine, urea, and phenolic resins; improves wetting action with fillers useful in brake bands, wood-flour filled moldings, and composition wallboard; used in nylon, polyamides, and casein; food pkg. ahesives; wh. to lt. cream fine gran. particles; odorless; m.w. 171; b.p. 214 C; sol. in org. solvs, hot linseed, Chinawood, and castor oil, 1.0% in water @ 34 C; sp.gr. 1.353; pH 4.0 min.; flash pt. 420 F (COC); 15 phr in ethyl cellulose: tens. str. 7325 psi; tens. elong.

40%; hardness 70.

Santicizer 97. [Monsanto] Dialkyl adipate; low temp. flex plasticizer for PVC film, sheeting, and coatings; for film and sheet processing, coated fabrics for rainwear, film for luggage and accessories, coated industrial fabrics exposed to refrigeration; used in PVC plastisols, rubber formulations; APHA 50 max. clear oily liq.; m.w. 370; b.p. 224 C; insol. in water; sp.gr. 0.916–0.924; dens. 7.7 lb/gal; visc. 12.8 cSt; ref. index 1.441–1.447; flash pt. 400 F (COC); fire pt. 450 F (COC).

Santicizer 141. [Monsanto] 2-Ethylhexyl diphenyl phosphate; flame-retardant low-temp. plasticizer for PVC, cellulose nitrate, CAB, ethyl cellulose, polymethyl methacrylate, PS, buna N rubber; used in vinyl film and sheeting, textile coatings, plastisols, organosols, adhesives, and pkg. materials; APHA 40 max. clear oily liq.; odorless; m.w. 362; b.p. 239 C; sol. 0.003% in water; sp.gr. 1.088–1.093; dens. 9.1 lb/gal; visc. 16.4 cSt; pour pt. –54 C; ref. index 1.507–1.510; flash pt. 435 F (COC); fire pt. 460 F (COC); 43 phr in PVC sheet: tens. str. 2930 psi; tens. elong. 320%; hardness (Shore A) 84.

Santicizer 143. [Monsanto] Proprietary, modified triaryl phosphate ester; flame-retardant plasticizer contributing clarity and toughness to end prod. for PVC and vinyl nitrile elastomers, emulsions, cellulosics; APHA 75 max. clear oily liq.; b.p. 232 C; sol. < 0.001% in water; sp.gr. 1.152; dens. 9.58 lb/gal; visc. 44.2 cSt; ref. index 1.542; flash pt. 475 F; fire pt. 525 F; 50 phr in PVC film: tens. str. 3090 psi; tens. elong. 326%; hardness (Shore A) 84.0.

Santicizer 148. [Monsanto] Isodecyl diphenyl phosphate; flame-retardant low-temp. plasticizer for PVC and copolymers, PVAc, acrylics; high solvating; for finished film or coated fabric applics., vinyl plastisols; also for PVC adhesives, ethyl cellulose, NC, SBR and butyl rubbers; clear oily liq.; odorless; m.w. 390; b.p. 245 C; insol. in water; sp.gr. 1.069–1.079; dens. 8.94 lb/gal; visc. 22.5 cSt; pour pt. < –50 C; ref. index 1.503–1.509; flash pt. 465 F (COC); fire pt. 500 F (COC); 43 phr in PVC sheet: tens. str. 2930 psi; tens. elong. 1940 psi; hardness (Shore A) 83.

Santicizer 154. [Monsanto] t-Butylphenyl diphenyl phosphate; flame retardant plasticizer used in PVC, vinyl and vinyl nitrile foams, PVAc emulsions, cellulosic resins; APHA 60 max. clear mobile liq.; odorless; m.w. 368; sol. < 0.001% in water; sp.gr. 1.175–1.185; dens. 9.8 lb/gal; visc. 58 cSt; pour pt. –25 C; ref. index 1.5535–1.5565; flash pt. 505 F (COC); fire pt. 590 F (COC).

Santicizer 160. [Monsanto] Butyl benzyl phthalate; high-solvating, stain-resisting, general-purpose plasticizer used in PVC, NC lacquers and films, PVAc, acrylic coatings, ethyl cellulose; plasticizes difficult resins, chlorinated rubber, cellulose propionate, polyamide; APHA clear oily liq.; slight char. odor; m.w. 312; b.p. 240 C; sol. 0.0003% in water; sp.gr. 1.115–1.123; dens. 9.3 lb/gal; visc. 41.5 ± 1.5 cSt; pour pt. –45 C; ref. index 1.535–1.540; flash pt. 390 F (COC); fire pt. 450 F (COC); 43 phr in PVC sheet: tens. str. 3090 psi; tens. elong. 350%; hardness (Shore A) 86.

Santicizer 261. [Monsanto] Alkyl benzyl phthalate; monomeric plasticizer offering permanence; used in film and sheeting, coated fabrics, plastisols, organosols, vinyl films, and acrylic lacquers, calendering, extrusions and vinyl foams; APHA 75 max. clear oily liq.; slight char. odor; m.w. 368; b.p. 252; sol. < –0.01% in water; sp.gr. 1.065–1.074; dens. 8.9 lb/gal; visc. 53 cSt; pour pt. –45 C; ref. index 1.523–1.529; flash pt. 445 F (COC); hardness (Shore A) 89 (43 phr in PVC).

Santicizer 278. [Monsanto] Benzyl phthalate; high solvating plasticizer offering permanence; used in PVC, paints and acrylic coatings, caulks and sealants based on chlorinated rubber, butyl rubber, polysulfides, or PU; clear oily liq.; slight char. odor; m.w. 455; b.p. 243 C; insol. in water; sp.gr. 1.094–1.101; dens. 9.1 lb/gal; visc. 860 cSt; pour pt. –6.5 C; ref. index 1.517–1.521; flash pt. 440 F (COC); fire pt. 535 F (COC); tens. str. 3180 psi; tens. elong. 258%; hardness (Shore A) 97.

Santicizer 409. [Monsanto] Med. m.w. polymeric plasticizer; used in PVC and rubber formulations; applic. incl. wire and cable coatings, refrigerator applic., film, tape, and coated fabrics; APHA 100 max. clear liq.; faint ester odor; sol. in toluene; insol. in water; sp.gr. 1.080–1.084; dens. 9.0 lb/gal; visc. 29–35 stokes; m.p. 4 C; ref. index 1.4654; pH 2.0; flash pt. 530 F (COC); fire pt. 570 F (COC); 50 phr in PVC sheet: tens. str. 3100 psi; tens. elong. 370%; hardness (Shore A) 92.

Santicizer 412. [Monsanto] Polyester; low visc. plasticizer offering permanence, low temp. flexibility, and fast fusion; used in vinyl compds., formulating plastisols for fabric coatings, slush molding, dipped goods, for dip-made vinyl gloves, baby pants, foamed textile backings, and knife- or roller-coated fabrics; APHA 250 max. clear oily liq.; slight char. odor; insol. in water; sp.gr. 1.030–1.060; dens. 8.7 lb/gal; visc. 2.3–5.0 stokes; m.p –20 C; ref. index 1.453–1.463; pH 4.0; flash pt. 495 F (COC); fire pt. 540 F (COC); hardness (Shore A) 86 (43 phr in PVC).

Santicizer 429. [Monsanto] Med.-high m.w. polyester plasticizer made from glycol reacted with a dibasic acid; used in elec. insulation, refrigerator gasketing, oil-resistant, high-temp. PVC wire and cable compds., in PVC film and sheeting; used for plasticizing ethyl cellulose, NC, acrylic caulking compds., and adhesive systems based on PVAc, S/B, and acrylic latices; APHA 250 max. clear oily liq.; odorless; insol. in water; sp.gr. 1.080–1.110; dens. 9.1 lb/gal; visc. 40–60 stokes; m.p. –18 C; ref. index 1.460–1.470; pH 4.0; flash pt. 550 F (COC); fire pt. 590 F (COC); tens. str. 2910 psi; tens. elong. 350%; hardness (Shore A) 92 (43 phr in PVC).

Santicizer 711. [Monsanto] Dialkyl phthalate; general purpose plasticizer for PVC formulations; compatible in working conc. with vinyl homopolymers and copolymers, chlorinated rubber, SBR, neoprene, and nitrile rubbers, and with common cellulosics except cellulose acetate; useful in plastisol prod.; APHA 25 max. clear oily liq.; slight char. odor; m.w.

414; b.p. 252 C; sol. in org. solvs., < 0.01% in water; sp.gr. 0.965–0.973; dens. 8.06 lb/gal; visc. 41 cSt; pour pt. < –50 C; ref. index 1.480–1.483; flash pt. 440 F (COC); 30% in PVC: tens. str. 2990 psi; tens. elong. 380%; hardness (Shore A) 85.

Santocure®. [Monsanto] N-cyclohexyl-2-benzothiazole sulfenamide; accelerator for sulfur curable elastomers, esp. EPDM; 7 mm pellets or powd.

Santocure® IPS. [Monsanto] N,N-diisopropyl-2-benzothiazole sulfenamide; accelerator for sulfur-curable elastomers; slow cure rate for thick sections; 7 mm pellets.

Santocure® MOR. [Monsanto] 2-(Morpholinothio) benzothiazole; accelerator for sulfur-curable elastomers; millipellet or flakes.

Santocure® NS. [Monsanto] N-tert-butyl-2-benzothiazole sulfenamide; accelerator for sulfur-curable elastomers; fast cure rate; 7 mm pellets or dust-suppressed powd.

Santoflex® 13. [Monsanto] N-(1,3-dimethylbutyl) N′-phenyl-p-phenylenediamine; antiozonant/antiflex cracking agent; SBR stabilizer; monomer polymerization inhibitor; 7 mm pastiles and molten liq.

Santoflex® 77. [Monsanto] N,N′-bis (1,4-dimethylpentyl)-p-phenylenediamine; antiozonant for static applics.; liq.

Santoflex® 134. [Monsanto] Alkylaryl p-phenylenediamine blend; antiozonant/antiflex agent; polymerization inhibitor for monomers; SBR stabilizer; liq.

Santoflex® 715. [Monsanto] Alkylaryl and dialkyl p-phenylenediamine blend; antiozonant for static/dynamic applics.; liq.

Santoflex® AW. [Monsanto] 6-Ethoxy-1,2-dihydro-2,2,4-trimethylquinoline and related prod.; antiozonant and antiflexcracking agent; staining; liq.

Santoflex® IP. [Monsanto] N-Isopropyl-N′-phenyl-p-phenylenediamine; antioxidant/antiozonant for elastomers; deactivator for metal catalyzed degradation; staining; flakes; low sol. in general purpose elastomers.

Santogard® PVI. [Monsanto] N-(cyclohexylthio) phthalimide; prevulcanization inhibitor; delays onset of accelerated sulfur vulcanization; dust-suppressed powd. or wax pellets 80% act.

Santone® 3-1-S. [Van Den Bergh] Polyglyceryl-3 stearate; emulsifier; aeration of lipid systems; solubilizer; color/flavor dispersant in water; cream bead; m.p. 128–130 F; HLB 7.2; sapon. no. 122–139; 27–34% total mono- and diglycerides.

Santone® 3-1-SH. [Van Den Bergh] Polyglyceryl-3 oleate; emulsifier and aerating agent used in food industry, textile and plastic lubricant; color dispersant; HLB 7.2; sapon no. 115–135.

Santone® 8-1-O. [Van Den Bergh] Polyglyceryl-8 oleate; emulsifier for personal care prods.; visc. reducer; emulsion stabilizer; beverage clouding agent; HLB 13.0; sapon. no. 75–85.

Santone® 8-1-S. [Van Den Bergh] Polyglyceryl-8 stearate; emulsifier for personal care prods.; whipping agent; emulsion stabilizer; HLB 13.0; sapon. no. 110–119.

Santone® 10-10-O. [Van Den Bergh] Polyglycerol-10-decaoleate; food emulsifier; solubilizer; emulsion stabilizer; dispersant aid in flavors; liq.; HLB < 1.0; sapon. no. 165–180.

Santoquin. [Monsanto] Ethoxyquin, tech. grade; antioxidant used in fats and rendered prods.; stabilizer, preservative; dk. amber mobile liq.; sp.gr. 1.023–1.024; flash pt. 285–290 F (COC); 90–92% min. assay.

Santoquin Emulsion. [Monsanto] Ethoxyquin, tech.; see Santoquin; milky brn. liq.; 70% conc.

Santoquin Mixture 6. [Monsanto] Ethoxyquin, tech.; see Santoquin; redsh.-brn. gran.; 66.6% conc. on feed grade vermiculite carrier.

Santovar® A. [Monsanto] 2,5-di (tert-amyl) hydroquinone; staining antioxidant in noncuring applics., adhesives; NBR stabilizer; powd.

Santoweb® D, DX. [Monsanto] Treated cellulose fibers; short fiber reinforcing agent for NR, SBR, BR, and CR; bk. fibrous aggregate.

Santoweb® H. [Monsanto] Treated cellulose fibers; short fiber reinforcing agent for EPDM and IIR elastomers; bk. fibrous aggregate.

Santoweb® W. [Monsanto] Treated cellulose fibers; short fiber reinforcing agent for PVC, NBR, and nonblack rubber compds.; cream-colored fibrous aggregate.

Santowhite® Crystals. [Monsanto] 4,4′-Thiobis-(6-tert-butyl-m-cresol); nonstaining thermal antioxidant for NR, syns., and latexes; syn. polymer stabilizer; powd.

Santowhite® PC. [Monsanto] 2,2′-Methylenebis-(4-methyl-6-tert-butyl-phenol); high activity antioxidant for NR, syns., latexes; syn. polymer stabilizer; powd.

Santowhite® Powd. [Monsanto] 4,4′-Butylidenebis-(6-tert-butyl-m-cresol); nonstaining thermal antioxidant and uv dicoloration protectant for NR, syns., latexes; syn. polymer stabilizer; powd.

Sanyo Levelon. [Sanyo] Highly conc. naphthalene sulfonate formaldehyde condensate; dispersant; water reducer for concrete molding fabrications requiring high strength; liq.; sp.gr. 1.2 (15/4 C); pH 9 (1% aq.); 40% solids.

Saret® 500, 515. [Sartomer] Trifunctional crosslinking agents which minimize scorch and offer processing flexibility when used with the peroxide cure of elastomers; suitable for inj., transfer, and compr. molding; Gardner > 10 dk. liq., mild odor; sp.gr. 1.0112; visc. 40–60 cps; b.p. > 200 C (1 mm); compat. with alcohols, ethers, ketones, esters, and aromatic hydrocarbons; dens. 9 lb/gal; flash pt. (PMCC) 212 and 155 F resp.; 95+% act.

Saret® 516. [Sartomer] Difunctional crosslinking agent, coagent; incorporated into peroxide-cured natural and syn. rubber systems; Gardner 5–6 clear liq.; mild odor; sp.gr. 1.0112; visc. 10–15 cps; 98% react. esters.

Saret® 517. [Sartomer] Trifunctional crosslinking agent, coagent; incorporated into peroxide-cured natural and syn. rubber systems; Gardner 5–6 clear liq.; mild odor; sp.gr. 1.0112; visc. 45–60 cps; 99% react. esters.

Saret® 518. [Sartomer] Tetrafunctional crosslinking agent, coagent; incorporated into peroxide-cured natural and syn. rubber systems; Gardner 5–6 clear liq.; mild odor; sp.gr. 1.0112; visc. 130–140 cps; 99+% react. esters.

Sarkosyl® L. [Ciba-Geigy Switzerland] Lauroyl sarcosine; anionic; detergent, corrosion inhibitor, foam booster and stabilizer, wetting agent, lubricant, emulsifier used dentifrices, personal care, and household cleaning prods., pharmaceuticals, metal processing and finishing, metalworking and cutting oils; powd.; m.w. 264–285; sol. in org. solvs.; insol. in water; sp.gr. 0.969; m.p. 35–37 C; 94% min. purity.

Sarkosyl® LC. [Ciba-Geigy Switzerland] Cocoyl sarcosine; see Sarkosyl L; powd.; m.w. 285–300; sol. in org. solvs.; insol. in water; sp.gr. 0.969; m.p. 23–28 C; 94% min. purity.

Sarkosyl® NL-30. [Ciba-Geigy Switzerland] Sodium lauroyl sarcosinate; see Sarkosyl L; dentifrices, shampoos, rug and window cleaners and fine fabric detergents; colorless liq.; pH 7.5–8.5 (1%); 30% act.

Sarkosyl® O. [Ciba-Geigy Switzerland] Oleoyl sarcosine; see Sarkosyl L; powd.; m.w. 340–360; sol. in org. solvs.; insol. in water; sp.gr. 0.948; 94% min. purity.

Sartomer 100. [Sartomer] Aliphatic diol with C_{14}–C_{15} avg. chain length; intermediate for urethanes, polyesters, allyls, and prepolymers; low-cost replacement for 1,6-hexanediol; wh. waxy solid, mild odor; sol. < 0.01 g/100 g water; sp.gr. 0.8961 (60 C); m.p. 38–42 C; flash pt. 370 F; hyd. no. 391.

Sartomer 322. [Sartomer] Triethylene glycol diacetate; plasticizer for films and coatings; colorless clear liq.; mild odor; m.w. 234; 95% min. reactive esters.

Sartomer 640. [Sartomer] Tetrabromo bisphenol A diacrylate; fire retardant for automotive coatings, wire and cable coatings; wh. gran. solid; bromine-like odor; m.w. 741; 96+% reactive esters.

Sartomer 801, 802, 803. [Sartomer] Pigment dispersing aids for wood coatings, wood fillers, solder masks, pigmented coatings; clear liq., mild odor; sp.gr. 1.05, 1,04, 1.03 resp.; visc. 360, 280, and 213 cps.

Sartomer 7001. [Sartomer] Metallic carboxylate; visc. reducing aid for silica-filled rubber compds.; wh. powd.; slightly acidic odor; sp.gr. 1.5; sol. in warm acetic acid; m.p. < 250 C; 98+% reactive esters.

Satexlan. [Croda] PEG-20 hydrog. lanolin; nonionic; o/w emulsifier, emollient, thickener; perfume solubilizer; imparts superfatting properties; Gardner 4 max.; sol. in water and aq./alcoholic sol'ns.; m.p. 42–50 C; sapon no. 7 max.

Satintone® SP-33®. [Engelhard] Calcined aluminum silicate; reinforcing extender for rubber and polymer systems; SP-33 designed as an extender for PVC elec. insulation compds.; very wh. highly pulverized powd.; 1.4 µm avg. particle size; 0.03% max. residue +325 mesh; sp.gr. 2.50 g/cc; bulk dens. 16 lb/ft³ (loose), 30 lb/ft³ (tamped); bulking value 0.048 gal/lb; oil absorp. 50–60 lb/100 lb; ref. index 1.62; pH 5–6; brightness (GE) 84–86; 52.2% SiO_2; 44.3% Al_2O_3.

Satintone® Whitetex®. [Engelhard] Calcined aluminum silicate; see Satintone SP-33; Whitetex is a general-purpose grade used in elastomers, PVC, polyamides; very wh. highly pulverized powd.; 1.4 µm avg. particle size; 0.02% max. residue +325 mesh; sp.gr. 2.63 g/cc; bulk dens. 18 lb/ft³ (loose), 33 lb/ft³ (tamped); bulking value 0.046 gal/lb; oil absorp. 50–60 lb/100 lb; ref. index 1.62; pH 5–6; brightness (GE) 90–92; 52.4% SiO_2; 44.5% Al_2O_3.

Satulan. [Croda] Hydrog. lanolin; nonionic; w/o emulsifier, emollient used in personal care prods.; Gardner 2 max. solid; low odor; HLB 1.0; m.p. 47–54 C; acid no. 1 max.; sapon. no. 7 max.; 100% conc.

Savinase 4.0T. [Novo] Proteinase; enzyme for laundry powd. detergents; off-wh. gran.

Savinase 8.0L. [Novo] Bacterial proteinase; enzyme for breakdown of protein stains; used in liq. detergents; liq.

Savinase 16.0 L, Type W. [Novo] Proteinase; enzyme; defoamer for built liq. detergents; brn. slurry.

Saytex® 102. [Ethyl] Decabromodiphenyl oxide; flame retardant for high-impact PS, thermoset and thermoplastic polyesters, nondrip PP, crosslinked polyethylene, elastomers; off-wh. powd. particle size 3.2µ; insol. in water; m.p. 300–310 C; 83% bromine.

Saytex® 102E. [Ethyl] Decabromodiphenyl oxide; flame retardant; high-purity, elec. grade for wire and cable insulation applic.; off-wh. powd. particle size 3.1µ; insol. in water; m.p. 305 C; 83% bromine.

Saytex® 105. [Ethyl] Pentabromoethylbenzene; flame retardant for thermoset polyester resins, textiles, adhesives, coatings, PU; wh. to cream powd.; sol. g/100 ml 72.5 g in styrene and toluene, 6.8 g in chloroform and ethyl alcohol, 0.1 g in water; m.p. 136–138 C; 79% bromine.

Saytex® 111. [Ethyl] Octabromodiphenyl oxide; flame retardant for ABS, HIPS, polyamides, elastomers, adhesives, and coatings; semiplasticizing additive for styrenic polymers and copolymers such as ABS; lt. tan gran. powd.; somewhat sol. in chlorinated and aromatic solvs.; insol. in water; sp.gr. 2.75; m.p. 70–150 C; 79% bromine.

Saytex® 115. [Ethyl] Pentabromodiphenyl oxide; flame retardant for PU, ABS, unsat. polyesters, epoxies, elastomers, adhesives, and coatings; amber visc. liq.; highly sol. in most org. solvs.; insol. in water; sp.gr. 2.27; 69–72% bromine.

Saytex® 120. [Ethyl] Tetradecabromodiphenoxy benzene; flame retardant for nylon, alloys of styrenics and engineering plastics, engineering thermoplastic polyesters, ABS, cross-linked polyethylene, elastomers, and high-impact PS; wh. powd.; 2.0 µ avg. particle size; insol. in water and common inorgs.; m.p. 370 C min.; 82% bromine.

Saytex® 125. [Ethyl] Blend of 85% Saytex 115 (pentabromodiphenyl oxide) and 15% aromatic phosphate ester; flame retardant for flexible and rigid PU, unsat. polyesters, PVC compds., laminates, adhesives, and coatings; amber low-visc. liq.; sp.gr. 1.95; 59–61% bromine.

Saytex® BCL-462. [Ethyl] Dibromoethyldibromocyclohexane; flame retardant for expandable, cryst.

and high-impact PS, SAN resins, adhesives, coatings, textile treatment, PU; wh. cryst. powd.; sol. g/100 ml 100 g in benzene, 90 g in trichloroethylene, 53 g in ethyl acetate, 48 g in acetone, 6 g in hexane, 5 g in ethanol, 0.06 g in water; dens. 2.27; m.p. 65–80 C; 74% bromine.

Saytex® BN-451. [Ethyl] Ethylenebis dibromonorbornane dicarboximide; thermally stable flame retardant for PP, polyamides 610 and 612, PU elastomers and coatings; off-wh. powd.; m.w. 672.0; sol. ≤0.1% in water, methanol, toluene; sp.gr. 2.05; bulk dens. 777 kg/m^3 (packed); m.p. 297–304 C; acid no. 0.28; 47.6% bromine (theoret.).

Saytex® BT-93. [Ethyl] Ethylene bis-tetrabromophthalimide; flame retardant for high-impact PS, polyethylene, PP, thermoplastic polyesters, nylon, EPDM, rubbers, PC, ethylene copolymers, ionomer resins, textile treatment; lt. yel. powd. particle size 2μ; insol. in water and org. solvs.; sp.gr. 2.66; m.p. 446 C; acid no. 1.0 max.; 66% bromine.

Saytex® BT-93® W. [Ethyl] Ethylenebistetrabromophthalimide; flame retardant for elec. parts, computer housings, film, wire and cable, underhood automotive, and composites for PBT, HIPS, ABS, polyethylene, PP, epoxy, and elastomers; wh. powd.; 99% < 16 μ particle size; m.w. 951.5; sol. < 0.1% in water, acetone, methanol, toluene; sp.gr. 2.77; bulk dens. 67 lb/ft^3 (packed); m.p. 450–455 C); acid no. 1.0 max.; 67.2% Br.

Saytex® FR-1138. [Ethyl] Dibromoneopentyl glycol, bromopentaerythritol, and tribromoneopentyl alcohol; flame retardant for unsat. polyesters, rigid and flexible PU foam, PU elastomers; wh. sm. flakes; m.w. 261.9; sol. 56.4% in methanol, 55.7% in acetone, 42.0% in propylene glycol, 27.2% in ethylene glycol, 2% in water; sp.gr. 2.04; bulk dens. 1028 kg/m^3 (packed); hyd. no. 395–425; 61% bromine (theoret.).

Saytex® HBCD. [Ethyl] Mixed isomers of hexabromocyclododecane; flame retardant for low-dens. expandable and extrudable PS foam; adhesives and coatings intermediate; additive for PU foams; off-wh. cryst. powd.; m.p. 185–195 C; 74% bromine.

Saytex® HBCD-HM. [Ethyl] Hexabromocyclododecane; high melting flame retardant for low-dens. expandable and extruded PS foam, HIPS, adhesives, coatings; wh. powd.; m.w. 641.7; sol. in common solvs.; sol. 10.3% in styrene, 8.6% in acetone, 8.0% in toluene, < 0.01% in water; sp.gr. 2.38; bulk dens. 1520 kg/m^3 (packed); m.p. 187–192 C; 74.7% bromine (theoret.).

Saytex® HBCD-LM. [Ethyl] Hexabromocyclododecane; low melting flame retardant; see Saytex HBCD-HM; wh. powd. or gran.; m.w. 641.7; sol. in common solvs.; sol. 12.3% in styrene, 10.2% in acetone, 9.7% in toluene, < 0.01% in water; sp.gr. 2.36; bulk dens. 1234 kg/m^3 (gran., packed), 1412 kg/m^3 (powd., packed); m.p. 175–183 C; 74.7% bromine (theoret.).

Saytex® RB-34. [Ethyl] Complex polyester, polyether of TBPA; flame retardant for low-visc. urethane and urethane-modified isocyanurate foams, esp. spray and pour-in-place applics.; tan liq.; visc. 2000 cps; 43% bromine.

Saytex® RB-49. [Ethyl] Tetrabromophthalic anhydride; flame retardant; monomer for unsat. polyester; reactive intermediate for preparation of polyols, esters, and imides; m.w. 463.7; sol. in dimethyl formamide, nitrobenzene; m.p. 270 C min; 68% bromine.

Saytex® RB-79. [Ethyl] Complex polyester, polyether of TBPA; flame-retardant for urethane and urethane-modified isocyanurate foams for UL and factory mutual class #1 and 2 specs; slab stock, pour-in-place, and spray foams; also for coatings and elastomers; tan liq.; visc. 100,000 cps; 45% bromine.

Saytex® RB-79 Diol. [Ethyl] Mixed ester of tetrabromo phthalic anhydride with diethylene glycol and propylene glycol; flame retardant for reducing the flame spread of rigid polyurethanes and related polymers; amber liq.; dens. 1.8 g/ml; visc. 100,000 cps; acid no. 0.1; hyd. no. 220; 46% bromine.

Saytex® RB-100. [Ethyl] Tetrabromobisphenol A; reactive or additive source of bromine for flame retardancy; reactive intermediate for preparation of brominated epoxy resins, polycarbonates, and unsat. polyesters; additive for ABS, PS, and phenolic resins; intermediate for other flame retardants; wh. solid; m.w. 543.9; sol. in alcohols and ketones; slightly sol. in aromatic and halogenated solvs.; insol. in water; m.p. 178 C; 58.4% bromine.

Saytex® VBR. [Ethyl] Monomeric vinyl bromide; flame retardant; intermediate in org. synthesis and in the mfg. of flame retardants, polymers, copolymers, pharmaceuticals, fumigants, and other chemicals; also used in textiles, adhesives, coatings, photographic plates and films; off-wh. low-boiling liq.; char. pungent odor; m.w. 106.96; f.p. –139.2 C; b.p. 15.8 C (760 mm Hg); dens. 12.7 lb/gal (20 C); flash pt. (COC) none; 99.8% vinyl bromide; 74.5% bromine.

SB-30. [Solem] Alumina trihydrate; filler featuring flame retarding and smoke suppressing properties; resin extender in polyester, vinyls, PU, latex, neoprene foam systems, wire and cable insulation, vinyl wall and floor coverings, epoxies; used in pharmaceuticals; builds visc. rapidly; exc. suspension properties; used in epoxy elec. laminates, epoxy encapsulating/casting, FRP elec. laminates, SMC/BMC phenolic compr. molding; many applics. are enhanced by surf. treating this grade; also suitable for high loadings and high visc. systems like rubber compds.; wet out and blend easily; used in plastics, rubber, and cellulosics; off-wh. particulate; median particle diam. 100 μ, 2–10% thru 325 mesh; sp.gr. 2.42; dens. 1.2 g/cm^3; surf. area 0.10 m^2/g; ref. index 1.57; hardness 2.5–3.5 Moh; 64.9% Al_2O_3.

SB-31. [Solem] Alumina trihydrate; see SB-30; off-wh. particulate; median particle diam. 35 μ, 35–70% thru 325 mesh; sp.gr. 2.42; dens. 1.0 g/cm^3; surf. area 0.5 m^2/g; 65.0% Al_2O_3.

SB-31C. [Solem] Alumina trihydrate; see SB-30; off-wh. particulate; median particle diam. 90 μ, 3–15% thru 325 mesh; sp.gr. 2.42; dens. 1.2 g/cm^3; surf. area 0.15 m^2/g; 65.0% Al_2O_3.

SB-136. [Solem] Alumina trihydrate; see SB-30, also provides low cost flame retardance and smoke suppression in plastics and rubber applic.; off-wh. particulate; median particle diam. 22 μ, 75–80% thru 325 mesh; sp.gr. 2.42; dens. 0.8 g/cm³; surf. area 2.5 m²/g; ref. index 1.57; hardness 2.5–3.5 Moh; 64.9% Al_2O_3.

SB-331. [Solem] Alumina trihydrate; see SB-30; also builds visc. fast, thus limiting load factor; good wet out, blending, and suspension properties; SMC/BMC compression molding, epoxies, polishes, adhesives, and coatings; particulate; median particle diam. 10–12 μ, 98% thru 325 mesh; sp.gr. 242; dens. 0.6–0.8 g/cm³; surf. area 3–4 m²/g; oil absorp. 27–28; ref. index 1.57; hardness 2.5–3.5 Moh; 65.0% Al_2O_3.

SB-332. [Solem] Alumina trihydrate; see SB-331; particulate; median particle diam. 10–12 μ, 98% thru 325 mesh; sp.gr. 2.42; dens. 0.6–0.8 g/cm³; surf. area 3–4 m²/g; oil absorp. 27–28; ref. index 1.57; hardness 2.5–3.5 Moh; 64.9% Al_2O_3.

SB-335. [Solem] Alumina trihydrate; see SB-30; also flame and smoke suppression; low visc., high filler loading and glass wet-out; for spray-up or land lay-up FRP applic., filament winding, panel prod., resin injection, SMC/BMC/acrylic sheet rigidizing, and cast polyester parts; particulate; median particle diam. 14 μ, 90–94% thru 325 mesh; sp.gr. 2.42; dens. 0.7 g/cm³; surf. area 2–3 m²/g; oil absorp. 25–26 ml/100 g; ref. index 1.57; 65.0% Al_2O_3.

SB-336. [Solem] Alumina trihydrate; see SB-335; particulate; median particle diam. 16 μ, 90–94% thru 325 mesh; sp.gr. 2.42; dens. 0.7 g/cm³; oil absorp. 25–26 ml/100 g; ref. index 1.57; 64.9% Al_2O_3.

SB-431. [Solem] Alumina trihydrate; see SB-30; used for SMC, BMC, and resin inj.; provides optimum in visc., flame, elec., and molding properties; rapid disp. in resin; mold flow and wet-out chars; yield exc. surf. profile, minimal porosity, pigmentation, filler and reinforcement distribution; particulate; wh. median particle diam. 8–10 μ, 99.5% thru 325 mesh; sp.gr. 2.42; dens. 0.6–0.7 g/cm³; surf. area 4–5 m²/g; oil absorp. 27–29; ref. index 1.57; hardness 2.5–3.5 Moh; 64.9% Al_2O_3.

SB-432. [Solem] Alumina trihydrate; see SB-431; off-wh. particulate; median particle diam. 8–10 μ, 99.5% thru 325 mesh; sp.gr. 2.42; dens. 0.6–0.7 g/cm³; surf. area 4–5 m²/g; oil absorp. 27–29; ref. index 1.57; hardness 2.5–3.5 Moh; 64.9% Al_2O_3.

SB-632. [Solem] Alumina trihydrate; see SB-30; also used for plastic and rubber applic.; suitable only for low visc. systems or low levels of loading; particle suspension; poor processing chars.; used in urethane foams, vinyls, wire and cable insulation, SBR belting, EPDM, EPR, adhesives, coatings, ABS, polyethylene, and XPLE; particulate; median particle diam. 3–4 μ, 99.99% thru 325 mesh; sp.gr. 2.42; dens. 0.4–0.5 g/cm³; surf. area 5–6 m²/g; oil absorp. 32; ref. index 1.57; hardness 2.5–3.5 Moh; 64.9% Al_2O_3.

SB-805. [Solem] Alumina trihydrate; superfine flame retardant; 2.2–3 μ median particle diam.; 99.99% thru 325 mesh; sp.gr. 2.42; bulk dens. 0.4 g/cc (loose), 0.7 g/cc (packed); surf. area 5–6 m²/g; oil absorp. 34; ref. index 1.57.

SB-932. [Solem] Alumina trihydrate; see SB-30; used for applic. requiring high surf. area prod.; improved physical and handling properties providing low cost approach to flame retardance and smoke suppression in coatings, adhesives, vinyls, wire and cable insulation, EPDM, EPR, ABS, polyolefins, and XLPE; particulate; median particle diam. 1.5 μ, 100% thru 325 mesh; sp.gr. 2.42; dens. 0.3 g/cm³; surf. area 6.5 m²/g; oil absorp. 38; ref. index 1.57; hardness 2.5–3.5 Moh; 64.9% Al_2O_3.

SBP. [PPG Industries] Di-sec-butyl peroxydicarbonate; initiator for polymerization of unsat. monomers, chiefly with ethylene, vinyl chloride, and vinylidene chloride; efficient for producing med.-dens. polyethylene and polymerization of vinyl chloride; liq.; m.w. 234.2; immisc. with water; sp.gr. 1.067 (4/4 C); m.p. < –80 C; ref. index 1.4132; 98.5% min. assay; 6.83% act. oxygen.

Scav-Ox®. [Olin] Hydrazine aq. sol'n.; corrosion protector in low-, med.-, and high-pressure boilers; oxygen scavenger in feedwater; colorless liq.; b.p. 109.4 C (760 mm Hg); misc. with water; sp.gr. 1.0207; dens. 8.52 lb/gal; visc. 1.0149 cp (30 C); pH 10.7 (1%); flash pt. none; 35% min. hydrazine.

Scav-Ox® II. [Olin] Catalyzed hydrazine; corrosion protector in industrial boilers; lt. pink to lt. orange aq. sol'n.; b.p. 109.4 C (760 mm Hg); misc. in water; pH 10.4 (1%); flash pt. none; 35.0% hydrazine.

Scav-Ox Plus®. [Olin] Catalyzed hydrazine aq. sol'n.; corrosion protector in industrial boilers; pink liq.; b.p. 109.5 C (760 mm Hg); misc. with water; sp.gr. 1.019; dens. 8.504 lb/gal; visc. 1.403 cps; pH 10.4 (1%); flash pt. none; 35% min. hydrazine.

Scepter®. [Am. Cyanamid/Ag] Ammonium salt of imazaquin; herbicide for soybean crops; 17.3% act.

Schercamox C-AA. [Scher] Cocamidopropylamine oxide; nonionic; conditioner, detergent, wetting agent, antistat used in personal care prods. and lt. dishwashing detergents; biodeg.; Gardner 4.0 max. clear to slightly hazy liq.; m.w. 320; sol. in water and hydrophilic solvs.; sp.gr. 0.986 ± 0.01; pH 7.0 ± 0.5; 30% min. act.

Schercamox CMA. [Scher] Dihydroxyethyl cocamine oxide; nonionic; softener, and wetting agent for cosmetics; builds visc. and stabilizes foam in personal care prods.; soft emollient feel on the skin; emolliency, lubricity, and slip to shave creams; conditions and prevents fly-away in hair shampoos; Gardner 3.0 max. clear liq.; mild odor; m.w. 301; sol. in water and hydrophilic solvs.; sp.gr. 0.99 ± 0.05; pH 7.0 ± 0.5 (1.0%); 39% min. conc.

Schercamox DML. [Scher] Lauramine oxide; nonionic; antistat; emulsifier, emulsion stabilizer for used in cosmetics industry; foam booster and visc. builder for shampoos; Gardner 1.0 max. clear liq.; mild odor; m.w. 235; sol. in water and hydrophilic solvs.; sp.gr. 0.99 ± 0.01; pH 7.0 ± 1.0 (1%) 30% min. conc.

Schercamox DMM. [Scher] Myristamine oxide; see Schercamox DML; Gardner 1.0 max. clear liq.; mild odor; m.w. 263; sol. see Schercamox DML; sp.gr.

0.98; pH 7.0 ± 1.0 (1%); 29% min. conc.

Schercamox DMS. [Scher] Stearamine oxide; nonionic; skin emollient, softener, visc. controller, foam stabilizer, and hair conditioner in personal care prods.; wh. to off-wh. paste; typ. odor; m.w. 326; sol. in alcohols, glycols, triols, polyols, and glycol ethers; slightly sol. in water; sp.gr. 0.90; pH 7.0 ± 1.0 (1.0%); 25% min. amine oxide.

Schercamox T-12. [Scher] Dihydroxyethyl tallowamine oxide; nonionic; conditioner, softener, antistat, visc. builder, and foam booster/stabilizer used in personal care prods.; amber clear liq.; mild odor; m.w. 365; sol. in most alcohols, glycols, triols, polyols, glycol ethers, and water; sp.gr. 0.94; pH 7.0 ± 1.0 (1.0% aq.); 49.0% min. act.

Schercemol 185. [Scher] Isostearyl neopentanoate; substantive emollient with low cloud pt.; aids as cloud and freeze point depressant, emulsion and freeze-thaw stabilizer; cosmetic preparations for skin care esp. near eyes; low level of skin and eye irritation; Gardner 2.0 max. clear liq.; slight typ. odor; m.w. 368; sol. in org. solvs.; sp.gr. 0.865; dens. 7.2 lb/gal; acid no. 2.0 max.; sapon. no. 144 ± 8.0; ref. index 1.450 ± 0.002; flash pt. > 180 C (OC); cloud pt. –10 C.

Schercemol 318. [Scher] Isopropyl isostearate; emollient for bath oils, creams, lotions, and lipsticks; lubricity without oiliness; lt. lemon yel. clear liq.; bland odor; m.w. 326; f.p. < –20 C; sp.gr. 0.855 ± 0.01; dens. 7.12 lb/gal; acid no. 1.0 max.; sapon. no. 170 ± 10; ref. index 1.422 ± 0.001; flash pt. > 170 C; 99.0% min. conc.

Schercemol 1688. [Scher] Cetearyl octanoate; emollient; spreads evenly on skin imparting velvety softness; functions as waterproofing agent due to adhesion properties; APHA 100 clear liq.; slight char. odor; m.w. 388; sol. see Schercemol 185; also insol. in water; sp.gr. 0.852; dens. 7.1 lb/gal; acid no. 1.0 max.; sapon. no. 140–150; ref. index 1.4448; flash pt. 170 C (OC); cloud pt. 4.0 C.

Schercemol 1818. [Scher] Isostearyl isostearate; substantive emollient imparting luxurious softness to skin; used in cosmetics imparting slip and lubricity, luster, and sheen; cosolv. and solubilizer in perfumes; yel. clear liq.; slight typ. odor; m.w. 536; f.p. –5 C; sol. see Schercemol 1688; sp.gr. 0.865–0.875; acid no. 2.0 max.; sapon. no. 95–110; ref. index 1.4610; flash pt. > 180 C (OC); cloud pt. 17 C.

Schercemol BE. [Scher] Behenyl erucate; emollient base for lip care cosmetics, skin creams and lotions; chemically comparable to one of main constituents of jojoba oil; melts close to body temp.; nontoxic; cream soft solid; slight typ. odor; m.w. 631; sol. in hydrophobic solvs.; insol. in water; sp.gr. 0.840; dens. 7.0 lb/gal; m.p. 44–48 C; acid no. 2.0 max.; sapon. no. 80–95; flash pt. > 170 C (OC).

Schercemol CL. [Scher] Cetyl lactate; nonoily, soft, waxy emollient; used in creams and hand lotions imparting visc., increased substantivity, and velvety feel to skin, sheen and ease of combing to hair; personal care prods. Gardner 2.0 max. soft solid; slight typ. odor; m.w. 314; sol. see Schercemol 1688; sp.gr. 0.890 ± 0.01 (35 C); dens. 7.4 (35 C); acid no. 3.0 max.; sapon. no. 174–190; flash pt. > 160 C (OC).

Schercemol CM. [Scher] Cetyl myristate; emollient for personal care prods.; imparts visc. and substantivity, velvety feel to skin; melts above body temps.; wh. to pale yel. waxy solid; mild char. odor; m.w. 456; sol. in hydrophobic solvs.; sp.gr. 0.835 (55 C); dens. 7.0 lb/gal (55 C); m.p. 51 C; acid no. 2.0 max.; sapon. no. 110–125; flash pt. > 170 C (OC).

Schercemol CO. [Scher] Cetyl octanoate; solvency properties for use in make-up removers; clear liq.; f.p. 10 C; acid no. 3 max.; iodine no. nil; sapon. no. 140–155.

Schercemol CP. [Scher] Cetyl palmitate; syn. ester, replaces natural spermaceti wax; emollient for creams, lotions, lipsticks; conditioner/opacifier and sheen additive for hair preparations; wh. to pale yel. waxy solid; mild char. odor; m.w. 480; sol. see Schercemol CM; sp.gr. 0.830 (55 C); dens. 7.0 lb/gal (55 C); m.p. 54 C; acid no. 2.0 max.; sapon. no. 110–125; flash pt. > 170 C (OC).

Schercemol CS. [Scher] Cetyl stearate; see Schercemol CP; wh. waxy solid; slight typ. odor; m.w. 494; sol. see Schercemol BE; sp.gr. 0.830 ± 0.01 (55 C); acid no. 2.0 max.; sapon. no. 113 ± 5.0; flash pt. > 200 C (OC).

Schercemol DED. [Scher] Ethyl dimerate; emollient offering lingering effect that is retained on surf. of skin even after washing; used in personal care prods.; lt. yel. clear to slightly hazy liq.; slight char. odor; m.w. 620; f.p. < –11.0 C; sol. see Schercemol 1688; sp.gr. 0.907; dens. 7.55 lb/gal; acid no. 3.0 max.; sapon. no. 175–195; ref. index 1.4448; flash pt. > 170 C (OC).

Schercemol DEIS. [Scher] Decyl isostearate; high m.w. low freeze pt. nonoily lubricant, penetrant, moisturizer for cosmetic and personal care prods.; low iodine value and pigment dispersing properties used in make-up applic.; coupler for castor oil and waxes in lipsticks without the shortcoming of unsaturation; yel. clear liq.; slight char. odor; m.w. 451; f.p. –5.0 C; sol. in most hydrophobic solvs.; insol. in water; sp.gr. 0.87; dens. 7.25 lb/gal; acid no. 3.0 max.; sapon. no. 120–135; ref. index 1.4580; flash pt. > 170 C (OC).

Schercemol DIA. [Scher] Diisopropyl adipate; nonoily penetrating emollient, lubricant, and solv. with mild drying effects used in hydro-alcoholic cosmetic formulations; water wh. clear liq.; faint ester odor; m.w. 230; f.p. –2.0 max.; b.p. 87–89 C; sol. in alcohols, higher glycols, ketones, ester, aromatic, chlorinated, and aliphatic hydrocarbons; insol. in water; sp.gr. 0.960 ± 0.01; dens. 8.0 lb/gal; acid no. 0.5 max.; sapon. no. 487 ± 10; ref. index 1.423 ± 0.001; flash pt. > 170 C.

Schercemol DICA. [Scher] Diisocetyl adipate; nonoily emollient and spreading agent for cosmetics; release and anticlogging additive in aerosols due to strong plasticizing action; film spreader for bath oils; solv. for lipstick colors; Gardner 2.0 max. clear liq.; slight odor; m.w. 594; f.p. < –10.0 C; sol. see Schercemol CM; sp.gr. 0.885 ± 0.005; dens. 7.4 lb/gal; acid no. 2.0 max.; sapon. no. 170–185; ref. index 1.4538 ± 0.001; flash pt. > 180 C (OC).

Schercemol DID. [Scher] Diisopropyl dimerate; nonoily, glossy emollient producing a "cushiony" feel and "body" to skin and makeup preparations; prolongs "play time" and improves disp. and spreading of pigments; binder for pressed powd.; offers sheen, emolliency and prevents "bleeding" and "feathering" in lip preparations; highly substantive; suitable for sun tan preparations requiring some water repellency; Gardner 4.0 max. clear to slightly hazy liq.; slight char. odor; m.w. 650; f.p. < –11.0 C; sol. see Schercemol DED; sp.gr. 0.895–0.905; dens. 7.5 lb/gal; acid no. 3.0 max.; sapon. no. 156–186; ref. index 1.4600–1.4650; flash pt. > 170 C (OC).

Schercemol DIS. [Scher] Diisopropyl sebacate; nonoily emollient, lubricant, solubilizer with mild drying effects used in hydro-alcoholic personal care prods.; solv. and coupling properties; fast spreading action; colorless clear liq.; bland odorless; m.w. 286; f.p. O C; sol. in alcohols, higher glycols, ketones, esters, aromatic, aliphatic, and chlorinated hydrocarbons, min. and natural oils; insol. in water; sp.gr. 0.932 ± 0.01; dens. 7.8 lb/gal; acid no. 1.0 max.; sapon. no. 390± 10; ref. index 1.4320; flash pt. > 170 C; 99% min. ester conc.

Schercemol DISD. [Scher] Diisostearyl dimerate; emollient offering lingering effect retained on skin after washing; used in personal care prods.; yel. to amber clear to slightly hazy liq.; slight char. odor; m.w. 1078; f.p. 5.0 C; sol. see Schercemol 1688; sp.gr. 0.895; acid no. 5.0 max.; sapon. no. 90–110; ref. index 1.4720; flash pt. 170 C max. (OC); 96% conc.

Schercemol DO. [Scher] Decyl oleate; high m.w. low freeze pt. nonoily lubricant, emollient, penetrant, and moisturizer for cosmetic and personal care prods.; Gardner 3.0 max. clear liq.; bland char. odor; m.w. 422; f.p. –13 C; sol. in hydrophobic solvs.; insol. in water; sp.gr. 0.86±0.02; acid no. 5.0 max.; sapon. no. 135 ± 5; ref. index 1.4540; 87.0% conc.

Schercemol EE. [Scher] Erucyl erucate; see Schercemol BE; lt. amber clear to slightly hazy liq.; slight typ. odor; m.w. 624; sol. see Schercemol BE; sp.gr. 0.870; dens. 7.3 lb/gal; acid no. 4.0 max.; sapon. no. 80–100; ref. index 1.4660; flash pt. > 170 C (OC); cloud pt. 22 C.

Schercemol EGMS. [Scher] Glycol stearate; nonionic; emulsifier, opacifier, and pearlescent for cosmetic and personal care prods.; thickener and visc. controller for cosmetic preparations; wh. flakes; m.w. 312; sol. in most org. solvs.; insol. in water; m.p. 55–60 C; acid no. 2.0 max.; sapon. no. 170–190.

Schercemol GMIS. [Scher] Glyceryl isostearate; emulsifier and emollient for creams and lotions; lt. amber clear liq. to soft solid; slight typ. odor; m.w. 385; f.p. 6.0 C; sol. Schercemol 1688; sp.gr. 0.960; dens. 8.0 lb/gal; acid no. 5.0 max.; sapon. no. 150–170; ref. index 1.4715; flash pt. > 170 C (OC).

Schercemol GMS. [Scher] Glyceryl stearate; nonionic; emulsifier, opacifier, stabilizer, and thickener for cosmetic and personal care prods.; wh. to cream flakes; m.w. 342; sol. see Schercemol EGMS; m.p. 55–60 C; acid no. 3.0 max.; sapon. no. 160–176; 40% conc.

Schercemol ICS. [Scher] Isocetyl stearate; nongreasy emollient used in creams; remains liq. even @ low temps.; Gardner 2.0 max. clear liq.; slight typ. odor; m.w. 494; f.p. –5.0; sol. see Schercemol BE; sp.gr. 0.850; dens. 7.1 lb/gal; acid no. 1.0 max.; sapon. no. 112 ± 5.0; flash pt. > 180 C.

Schercemol IPM. [Scher] IPM; emollient and film spreader used in personal care prods.; solv. and solubilizer for additives in cosmetic preparations; nonoily; water wh. clear liq.; odorless; m.w. 270; f.p. –4.0 C; sol. see Schercemol 1688; sp.gr. 0.850; acid no. 1.0 max.; sapon. no. 202–212; ref. index 1.4330–1.4350; flash pt. > 170 C (OC); 99% min. ester conc.

Schercemol IPO. [Scher] Isopropyl oleate; emollient and spreading agent used in cosmetic preparations; lubricant for makeups (exhibits lubricity without much oiliness); Gardner 2.0 max. clear liq.; typ. slight odor; m.w. 324; f.p. < –15.0 C; sol. see Schercemol CM; sp.gr. 0.861 ± 0.005; dens. 7.17 lb/gal; acid no. 2.0 max.; sapon. no. 175–190; ref. index 1.4472 ± 0.0001; flash pt. > 180 C (OC).

Schercemol ISE. [Scher] Isostearyl erucate; nontoxic emollient base for skin creams and lotions; yel. clear to slightly hazy liq.; slight typ. odor; m.w. 589; sol. see Schercemol BE; sp.gr. 0.865; dens. 7.2 lb/gal; acid no. 2.0 max.; sapon. no. 90–105; ref. index 1.4610; flash pt. > 170 C (OC); cloud pt. 16 C.

Schercemol LL. [Scher] Lauryl lactate; cosmetic penetrating emollient; rich velvety emolliency for personal care prods. due to dispersibility in water; antitack agent in roll-on antiperspirants; Gardner 2.0 max. clear liq.; typ. odor; m.w. 275; f.p. –3 C; sol. in org. solvs.; water-disp.; sp.gr. 0.900; dens. 7.6 lb/gal; acid no. 3.0 max.; sapon. no. 200–220; ref. index 1.4430; flash pt. 150 C (OC).

Schercemol MEL-3. [Scher] Myreth-3 laurate; nonoily rich penetrating emollient for cosmetic and personal care prods.; dispersibility and spreadability in bath oils; coupler in hydro-alcoholic systems; emulsifier and solubilizer in lotions; Gardner 3.0 max. clear liq.; mild typ. odor; m.w. 528; f.p. 15 C; sol. in hydrophobic solvs.; water-disp.; sp.gr. 0.907; acid no. 3.0 max.; sapon. no. 100–115; ref. index 1.4510; flash pt. 160 C (OC); 98% ester conc.

Schercemol MEM-3. [Scher] Myreth-3 myristate; see Schercemol MEL-3; wh. to lt. straw clear liq.; slight typ. odor; m.w. 556; f.p. 23 C; sol. see Schercemol MEL-3; sp.gr. 0.901; acid no. 3.0 max.; sapon. no. 95–110; ref. index 1.4525; flash pt. 160 C (OC); 98% conc.

Schercemol MEP-3. [Scher] Myreth-3 palmitate; see Schercemol MEL-3; cream soft wax; mild typ. odor; m.w. 584; sol. see Schercemol MEL-3; sp.gr. 0.890; HLB 4.5; m.p. 26 C; acid no. 3.0 max.; sapon. no. 85–95; flash pt. 160 C (OC); 95% ester conc.

Schercemol ML. [Scher] Myristyl lactate; nonoily soft waxy emollient with low m.p.; conditioner; used in creams and body lotions giving velvety feel to skin;

imparts luster and body to hair in shampoos and cream rinses; reduces tackiness; lubricant and plasticizer for polymer resins in aerosol hair spray formulations and antipersipirants; Gardner 2.0 max.; liq. to soft solid; typ. slight odor; m.w. 286; sol. see Schercemol 1688; sp.gr. 0.90 (30 C); dens. 7.5 lb/gal; m.p. 28–34 C; acid no. 3.0 max.; sapon. no. 170–190; flash pt. > 160 C (OC).

Schercemol MM. [Scher] Myristyl myristate; soft, waxy emollient that melts near body temp.; visc. builder; imparts substantivity to personal care prods.; ease of combing of hair preparations; velvety feel on skin; wh. to pale yel. waxy solid; mild char odor; m.w. 424; sol. see Schercemol CM; sp.gr. 0.839 (45 C); dens. 7.0 lb/gal (45 C); m.p. 36–40 C; acid no. 2.0 max.; sapon. no. 120–135; flash pt. > 170 C (OC).

Schercemol MP. [Scher] Myristyl propionate; emollient for antiperspirants, body oils, creams, and lotions; straw clear liq.; acid no. 2 max.; sapon. no. 190–210.

Schercemol MS. [Scher] Myristyl stearate; waxy emollient for creams and lotions; cream liq.; f.p. 45 C; acid no. 2 max.; sapon. no. 110–120.

Schercemol NGDC. [Scher] Neopentyl glycol dicaprate; solvency properties for use in make-up removers; clear liq.; f.p. 2 C; acid no. 3 max.; iodine no. nil; sapon. no. 255–270.

Schercemol OLO. [Scher] Oleyl oleate; nonoily emollient for cosmetic formulations contributing luster, softness, and high degree of lubricity in skin and hair preparations; cosolv. and solubilizer in perfumes; lubricant for metal working and wire drawing; amber clear liq.; mild oleic odor; m.w. 332; sol. see Schercemol 1818; sp.gr. 0.860 ± 0.01; dens. 7.2 lb/gal; acid no. 1.0 max.; sapon. no. 105±5; ref. index 1.4630 ± 0.001; flash pt. > 180 C (OC); cloud pt. 13 C.

Schercemol OP. [Scher] Octyl palmitate; nonoily emollient ester for cosmetic and personal care prods. giving sheen without greasiness; anticlogging and suspending agent in antiperspirants; soft velvety feel in skin creams, lotions, and aftershaves; Gardner 2.0 max. clear liq.; bland odor; m.w. 368; f.p. 0 C max.; sol. see Schercemol CM; sp.gr. 0.855 ± 0.01; dens. 7.12 lb/gal; acid no. 3.0 max.; sapon. no. 155 ± 8.0; ref. index 1.4460; flash pt. > 170 C (OC).

Schercemol OPG. [Scher] Ethylhexyl pelargonate; dry, nonoily rich penetrating emollient for cosmetic and personal care prods.; anticlogging agent in antiperspirants; soft, luxurious feel in skin creams and aftershaves; Gardner 2.0 max. clear liq.; mild odor; m.w. 270; f.p. < –10. 0 C; sol. see Schercemol CM; sp.gr. 0.857 ± 0.01; dens. 7.13 lb/gal; acid no. 1.0 max.; sapon. no. 207 ± 7; ref. index 1.4363 ± 0.001; flash pt. > 170 C (OC).

Schercemol PGDP. [Scher] Propylene glycol dipelargonate; nonionic; emollient offering low f.p.; cosolv. for perfumed bath oils, creams, and lotions; Gardner 4.0 max. clear liq.; slight typ. odor; m.w. 360; f.p. –25 C; sol. see Schercemol 1688; sp.gr. 0.917; dens. 7.6 lb/gal; acid no. 5.0 max.; sapon. no. 290–310; ref. index 1.440; flash pt. > 170 C (OC).

Schercemol PGML. [Scher] Propylene glycol laurate; nonionic; emollient and solv.; stable base for cosmetics; emulsion stabilizer; solubilizes and couples ingred. such as perfumes, coloring and flavoring agents, sunscreen compds. into natural fatty veg. or min. oils; solv. for org. pesticides for spray applic.; produces sprayable oils which spread well, have good adherence, and are not readily removed by rainfall; plasticizer and stabilizer in vinyl copolymers made from PVAc and PVC; defoaming agent in PVAc emulsions; Gardner 5.0 max. clear liq.; mild odor; m.w. 258 (theoret.); sol. in most org. solvs. such as alcohols, ketones, esters, glycol ethers, veg. oil, min. oil, aliphatic, aromatic, chlorinated hydrocarbons; disp. in glycols, triols, polyols; f.p. 5 C max.; sp.gr. 0.905 ± 0.01; dens. 7.45 lb/gal; flash pt. (OC) 160 C min.; acid no. 5.0 max.; sapon no. 225–240; pH 7.0 (10% disp.).

Schercemol PGMS. [Scher] Propylene glycol stearate; nonionic; see Schercemol EGMS; wh. to cream solid; slight typ. odor; m.w. 326; sol. see Schercemol GMS; m.p. 35 C; acid no. 4.0 max.; sapon. no. 180 ± 5; flash pt. > 170 C (OC).

Schercemol SE. [Scher] Stearyl erucate; see Schercemol ISE; cream soft paste; slight typ. odor; m.w. 589; sol. see Schercemol BE; sp.gr. 0.875; dens. 7.3 lb/gal; m.p. 30–35 C; acid no. 2.0 max.; sapon. no. 90–105; flash pt. > 170 C (OC).

Schercemol TIST. [Scher] Triisostearyl trimerate; emollient; superior gloss and moisturizing chars.; emolliency, shine, visc., and good binding chars.; dk. amber slightly hazy, syrupy liq.; slight typ. odor; m.w. 1656; f.p. –10 C; sol. see Schercemol BE; sp.gr. 0.92; dens. 7.7 lb/gal; acid no. 5.0 max.; sapon. no. 90–120; ref. index 1.4760; flash pt. > 170 C (OC).

Schercemol TT. [Scher] Triisopropyl trilinoleate; binder for pigmented and cosmetic prods.; dk. amber hazy, syrupy liq.; slight typ. odor; m.w. 1026; f.p. –6 C; sol. in hydrophobic solv.; insol. in water; sp.gr. 0.92; dens. 7.7 lb/gal; acid no. 5.0 max.; sapon. no. 160–180; ref. index 1.4670; flash pt. > 170 C (OC).

Schercoat PC-550. [Scher] Substantive poly emulsion; cationic; lubricant for glass containers; liq.; 100% act.

Schercodine B. [Scher] Behenamidopropyl dimethylamine; cationic emulsifier with conditioning properties for hair and skin preparations; tan hard wax; m.w. 394; m.p. 63–68 C; alkali value 135–145; 98% min. amide.

Schercodine C. [Scher] Coco amido alkyl dimethylamine; cationic; emulsifier, intermediate for betaine amphoterics; solid; 100% conc.

Schercodine I. [Scher] Isostearamidopropyl dimethylamine; cationic; versatile liq. o/w emulsifier; lubricant for hair rinses and conditioners; lt. amber liq.; ammoniacal odor; m.w. 368; acid no. 4.0 max.; flash pt. > 160 C (OC); 100% act.

Schercodine L. [Scher] Lauramidopropyl dimethylamine; cationic surfactant, intermediate for betaine amphoterics; lt. tan solid; m.w. 284; m.p. 35–40 C; alkali value 196–206; 98% min. amide.

Schercodine M. [Scher] Myristamidopropyl dimethylamine; cationic o/w emulsifier, conditioner, visc. builder; lt. tan wax; m.w. 312; sol. in org. solvs.; m.p. 45–50 C; alkali value 180–190; 98% min. amide.

Schercodine O. [Scher] Oleamidopropyl dimethylamine; emollient conditioner for hair and skin preparations; amber liq.; m.w. 366; alkali value 150–160; 98% min. amide.

Schercodine P. [Scher] Palmitamidopropyl dimethylamine; substantive conditioner and emulsifier in creams, lotions, rinses; tan hard wax; m.w. 340; sol. in org. solvs.; m.p. 55–60 C; alkali value 160–170; 98% min. amide.

Schercodine S. [Scher] Stearamidopropyl dimethylamine; softener, emulsifier, and conditioner in hair and skin preparations; tan hard wax; m.w. 368; sol. in org. solvs.; m.p. 65–70 C; alkali value 145–155; 98% min. amide.

Schercodine T. [Scher] Tallowamidopropyl dimethylamine; conditioner for cationic emulsions; substantivity and thickening properties; amber liq.; m.w. 366; alkali value 150–160; 98% min. amide.

Scherco Finish 4L. [Scher] Resin disp.; nonionic; lubricant for textiles; liq. 25% act.

Schercolube 707. [Scher] Fatty ester ethoxylate; nonionic; softener for knit goods and lubricant for nylon separator threads in sweater bodies; paste; 25% act.

Schercomid 1-102. [Scher] Cocamide DEA; nonionic; thickener, foam booster, and emulsifier for cosmetic and detergent applics.; exhibits detergency when incorporated into bubble baths and detergent formulations; offers low cloud pt.; clear yel. visc. liq.; mild odor; sol. in water, alcohols, glycols, polyols, glycol ethers, aliphitic (lower members), aromatic, and chlorinated hydrocarbons; disp. in min. oil; sp.gr. 0.99; flash pt. (OC) > 180 C; acid no. 4.0 max.

Schercomid 1214. [Scher] Lauramide DEA and DEA; foam booster/stabilizer for detergent compositions containing alkylaryl sulfonates, alkyl-sulfates, and soaps; good detergency by itself and works synergistically with other surfactants; thickening agent and visc. builder; used in personal care items and cleaners for hard surfaces; clear yel. liq.; slt., typ. odor; sol. in water and in most org. solvs.; disp. in aliphatic hydrocarbons, min. oil, and natural fats; sp.gr. 1.01; dens. 8.4 lb/gal; flash pt. (OC) > 170 C; acid no. 12–16; 100% act.; 60% min. amide content.

Schercomid AME. [Scher] Acetamide MEA; nonionic; solubilizer, humectant, skin and hair conditioner, intermediate, coupling agent, pigment dispersant; Gardner 2.0 max. clear liq.; mild organoleptic odor; sol. in most alcohols, glycols, diols, polyols, glycol ethers, ketones, and water; sp.gr. 1.120; dens. 9.3 lb/gal; acid no. 10.0 max.; alkali no. 15.0 max.; ref. index 1.4700; pH 6.0–8.5 (50% aq.); flash pt. > 180 C (OC).

Schercomid AME-70. [Scher] Acetamide MEA; nonionic; see Schercomid AME; Gardner 2.0 clear liq.; mild organoleptic odor; sol. in most alcohols, glycols, diols, triols, polyols, glycol ethers, and water; sp.gr. 1.100 ± 0.01; dens. 9.2 lb/gal; acid no. 10 max.; alkali no. 15 max.; ref. index 1.4395; pH 6.0–8.5; flash pt. > 180 c (OC); 70% min. act. in water.

Schercomid CCD. [Scher] Cocamide DEA; nonionic/anionic; soil dispersing and suspending agent; for floor cleaners, liq. hand dishwashing, car waxes, liq. hand soaps, all-purpose industrial and household cleaners; dk. amber visc. liq.; sol. in water, alcohols, diols, triols, glycol ethers, polyols, aromatic and chlorinated hydrocarbons; sp.gr. 0.99 ± 0.01; dens. 8.3 lb/gal; acid no. 15–20; flash pt. (OC) > 170 C; pH 9.9 ± 0.5 (10%); 100% act., 60% min. amide.

Schercomid CDA. [Scher] Cocamide DEA and diethanolamine; nonionic/anionic; foam stabilizer, soil suspender, lime soap dispersant, and detergency booster for industrial and household cleaners; lt. amber clear liq.; sol. see Schercomid CCD; sp.gr. 1.01 ± 0.01; dens. 8.33 lb/gal; visc. 1000 cps min.; acid no. 40–50; alkali value 150–170; flash pt. (OC) > 170 C; pH 9.6 ± 0.5 (10%); 100% act.

Schercomid CDA-H. [Scher] Cocamide DEA; nonionic/anionic; thickener; foam booster and stabilizer, detergent, and lime soap dispersant; used in shampoos and other cosmetics; emulsifier for aromatic hydrocarbons and aliphatics if it is first solubilized with sm. amounts of oleic acid; clear yel. liq.; slt. odor; sol. in water, alcohols, glycols, glycol ethers, polyols, aromatic and chlorinated hydrocarbons; disp. in aliphatic hydrocarbons, min. oil, and natural fats; sp.gr. 1.01 ± 0.01; dens. 8.4 lb/gal; visc. 1100 cps min.; flash pt. > 170 C; acid no. 7–14; pH 10 ± 0.5 (10% aq.); 99% active; 60% min. amide content.

Schercomid CDO-Extra. [Scher] Cocoamide DEA and diethanolamine; detergent, wetting agent, foam stabilizer, visc. builder; shampoos, household cleaners, industrial cleaners with builders, liq. dishwashing compds., personal care formulations; visc. yel. liq.; slt. odor; sol. in water, alcohols, diols, triols, glycol ethers, polyols, aromatic and chlorinated hydrocarbons; sp.gr. 1.00 ± 0.01; visc. 950 cps min.; flash pt. (OC) > 170 C; acid no. 4–5; pH 10±0.5 (10% sol'n.); 100% act.

Schercomid CME. [Scher] Cocamide MEA; nonionic; foam booster and visc. builder for shampoos, bubble baths, and powded. detergent compositions; emulsifier for creams and lotions, esp. in cream hair dye formulations; tan wax; ammoniacal odor; sol. in alcohols, glycols, glycol ethers, aliphatic, aromatic, and chlorinated hydrocarbons; disp. in water; m.p. 63 ± 3 C; flash pt. (OC) > 180 C; acid no. 1.0 max.; 100% active; 85% min. amide content.

Schercomid CMI. [Scher] Cocamide MIPA; nonionic; foam booster and stabilizer, antidefatting agent for detergents; lt. tan wax; acid no. 2 max.; alkali value 20 max.; 90% min. amide.

Schercomid HT-60. [Scher] PEG-50 hydrog. tallow amide; nonionic; thickener, detergent, emulsifier, disperant with foam char.; Gardner 7 max. (molten); hard, waxy solid; ammoniacal odor; sol. in alcohols, glycols, triols, polyols, glycol ethers, water, in some aromatic and chlorinated hydrocarbons; sp.gr. 1.064 (60 C); dens. 9.6 lb/gal; m.p. 50–55 C; flash pt. (OC)

> 180 C; pH 9.0–10.0 (10% aq.); 100% active.

Schercomid ID. [Scher] Isostearamide DEA; nonionic; w/o emulsifier, thickening agent, and conditioner for use in cosmetic formulations where lubricity and slip are desired; lt. amber clear liq.; acid no. 2 max.; alkali value 110–130; 70% min. amide.

Schercomid IMI. [Scher] Isostearamide MIPA; nonionic; visc. builder in shampoos; creamy foam and good color stability; in acid media, a soft conditioning effect on hair in shampoos; emulsifier, emollient, and lubricant in night-creams and lotions; lt. yel. soft wax; mild ammoniacal odor; sol. in alcohols, glycols, glycol ethers, aliphatic, aromatic, and chlorinated hydrocarbons, min. and veg. oils; disp. in water; sp.gr. 0.885 ± 0.01 (50 C); dens. 7.4 lb/gal (50 C); flash pt. (OC) > 180 C; acid no. 10.0 max.; pH 10.0 ± 0.7 (10% disp.); 100% active; 85% min. amide content.

Schercomid LD. [Scher] Lauramide DEA and diethanolamine; see Schercomid 1214; also recommended in detergent compositions containing alkyl ether sulfates; yel. visc. liq.; slt., typ. odor; sol. see Schercomid 1214; sp.gr. 1.01 ± 0.01; dens. 8.4 lb/gal; visc. 1000 min.; flash pt. (OC) > 170 C; acid no. 20–26; pH 9.7 ± 0.5 (10% sol'n.); 100% act.; 50% min. amide content.

Schercomid M. [Scher] Isostearamide DEA; nonionic; visc. builder, conditioner, and emulsifier in personal care prods.; clear amber liq., soft solid; 85% min. amide.

Schercomid MD-Extra. [Scher] Myristamide DEA; nonionic; foam stabilizing agent and visc. controller in shampoos, creams and lotions; emulsifier in creams after neutralizing the excess DEA with fatty acids such as stearic acid; cream solid; slt., typ. odor; m.w. 420; sol. in most alcohols, glycols, and glycol ethers; disp. in water; m.p. 47–52 C; m.p. 47–52 C; flash pt. (OC) > 170 C; acid no. 3.0 max.; pH 10.5 ± 1.0 (10% disp.); 100% act.; 70% min. amide content.

Schercomid MME. [Scher] Myristamide MEA; nonionic; emulsifier, pearlescent, thickener, opacifier in cosmetic and pharmaceutical formulations; visc. builder and foam booster in shampoos, liq. dishwashing compds., and powd. detergents; pale wax; slight ammoniacal odor; m.w 271; sol. in most org. solvs.; water-disp.; m.p. 88 ± 4.0 C; acid no. 2.0 max.; flash pt. > 170 C (OC); 100% act.

Schercomid ODA. [Scher] Oleamide DEA and diethanolamine; nonionic/anionic; w/o emulsifier, pigment dispersant, conditioner, corrosion inhibitor, and visc. builder; dispersant for min. clays or pigments; emulsifier for aromatic and aliphitic hydrocarbon solv.; used in shampoo formulations hair conditioning agent; amber liq.; mild, char. odor; sol. in alcohols, glycols, glycol ethers, aliphatic and chlorinated hydrocarbons; water-disp.; sp.gr. 0.950 ± 0.01; dens. 7.9 lb/gal; visc. 1200 cps min.; flash pt. (OC) > 170 C; acid no. 12–16; pH 9.9 ± 0.5 (10% disp.); 100% active; 60% min. amide content.

Schercomid OMI. [Scher] Oleamide MIPA; nonionic; thickener for shampoos, producing a creamy, luxurious foam and stabilizing the foam when used with sodium lauryl ether sulfates; hair conditioning agent; used at concs. < 5% in shampoos; emulsifier for creams and lotions; imparts slip, lubrication, some emolliency, and softening effects upon the skin; amber liq. to soft solid; mild ammoniacal odor; sol. in alcohols, esters, glycol ethers, min. and veg. oils, aliphatic, aromatic, and chlorinated hydrocarbons; sp.gr. 0.90 ± 0.01; dens. 7.5 lb/gal; flash pt. (OC) 180 C; acid no. 10 max.; 100% active; 85% min. amide.

Schercomid SCE. [Scher] Cocamide DEA; nonionic; detergent, visc. builder and foam stabilizer for cosmetic formulations; thickens shampoos, bubble baths, liq. dish wash detergents, and rug shampoos; lt. amber clean, visc. liq.; mild odor; sol. in alcohols, diols, triols, polyols, glycol ethers, aliphatic, aromatic, and chlorinated hydrocarbons; water-disp.; sp.gr. 0.97; dens. 8.1 lb/gal; flash pt. (OC) > 180 C; acid no. 2.0 max.; 100% active; 88% min. amide content.

Schercomid SCO-Extra. [Scher] Cocamide DEA; nonionic; thickener, foam builder, foam stabilizer, and detergency booster for alkylaryl sulfonates and lauryl sulfates; used in shampoos, bubble baths, floor cleaners and liq. and powd. dishwashing detergents; for making self-emulsifiable solv. from aliphatic, aromatic, or chlorinated hydrocarbons, used in waterless hand cleaners, engine shampoos, concrete floor cleaners, wax strippers, and tar removers; clear lt. amber liq.; mild odor; sol. in water, alcohols, glycols, polyols, glycol ethers, aliphatic (lower members), aromatic, chlorinated hydrocarbons, and natural fats; disp. in min. oil; sp.gr. 0.99; dens. 8.25 lb/gal; flash pt. (OC) > 180 C; acid no. 3.0 max.; 100% active; 77% min. amide content.

Schercomid SD-DS. [Scher] Stearamide DEA distearate; nonionic; emulsifier, pearling agent, thickener, and opacifier for creams and shampoos; lt. tan wax; acid no. 5; alkali value 10 max.; 40% min. amide.

Schercomid SI. [Scher] Isostearamide DEA; nonionic; detergent, emulsifier, corrosion inhibitor, thickener, conditioner for personal care prods.; forms w/o emulsions; coupler; yel. clear visc. liq.; mild odor; sol. in most org. solvs., min. and veg. oil; disp. in water; sp.gr. 0.950; dens. 7.9 lb/gal; acid no. 2.0 max.; flash pt. > 180 C; 100% act.

Schercomid SI-M. [Scher] Isostearamide DEA; nonionic; visc. builder in shampoos, conditioning agents, emulsifiers, and low-irritation surfactants; clear amber liq. to soft solid; acid no. 2 max.; alkali value 20–40; 85% min. amide.

Schercomid SLA. [Scher] Lauric/myristic/palmitic DEA; nonionic; visc. builder for shampoos; thick, copious, stable foam useful in shampoos, facial scrubs, bubble baths, hand soaps, and detergent compositions; yel. liq. when fresh (cryst. on aging); mild, fruity odor; sol. in alcohols, glycols, glycol ethers, polyols, aliphatic (lower members), aromatic, and chlorinated hydrocarbons, and natural fats; disp. in water and min. oil; sp.gr. 0.96 ± 0.01; dens. 8.0 lb/gal; visc. 2500 cps min. (10% disp.); flash pt. (OC) >

170 C; acid no. 1.0 max.; pH 10.2 ± 0.5 (10% disp.); 100% active; 88% min. amide content.

Schercomid SLE. [Scher] Linoleamide DEA; nonionic; solubilizer, thickener, w/o emulsifier, conditioner, and emollient for personal care prods.; emulsion stabilizer for o/w emulsions; dk. amber clear liq.; mild, typical odor; sol. see Schercomid SI; sp.gr. 0.965; dens. 8.0 lb/gal; acid no. 1.0 max.; flash pt. > 180 C (OC); 100% act.

Schercomid SL-Extra. [Scher] Lauramide DEA; nonionic; thickener, foam booster/stabilizer for hair shampoos, soaps, syn. detergent formulations; used in gel, paste, and cream shampoos, and beauty parlor conc.; bubble bath applics.; industrial applics. incl. manual dishwashing formulations, liq. heavy-duty laundry detergents, liq. soaps, all-purpose cleaning prods.; emulsifier for aliphatic, aromatic hydrocarbons and oils for o/w emulsions; off-wh. cryst. solid; mild odor; sol. in alcohols, glycols, glycol ethers, polyols, aliphatic (lower members); aromatic, and chlorinated hydrocarbons, and natural fats and oils; disp. in water and min. oils; sp.gr. 0.97 (45 C); dens. 8.1 lb/gal (45 C); m.p. 42 C; flash pt. (OC) > 170 C; acid no. 1.0 max.; 100% active; 88% min. amide content.

Schercomid SLL. [Scher] Lauric/myristic super DEA; nonionic; visc. builder; mild shampoos, bubble baths, aerosol shave creams, facial scrubs, hand soaps, etc.; o/w emulsifier; clear yel. liq.; mild, fruity odor; sol. in alcohols, glycols, glycol ethers, polyols, aliphatic (lower members) and aromatic hydrocarbons, and natural fats; dips. in water and min. oil; sp.gr. 0.963 ± 0.01; dens. 8.02 lb/gal; visc. 1800 cps min. (10% aq.); flash pt. (OC) > 170 C; acid no. 1.0 max.; pH 10.2 ± 0.5 (10% aq.); 100% active; 88% min. amide.

Schercomid SLM. [Scher] Lauramide DEA; nonionic; visc. builder and foaming agent for hair shampoos, aerosol shave creams, etc.; stabilizes foam in bubble baths when used with lauryl sulfates or sulfosuccinates; good o/w emulsifier; lt. yel. liq. when fresh (cryst. on aging); mild odor; sol. in alcohols, glycols, glycol ethers, aliphatic (lower members), aromatic, and chlorinated hydrocarbons, natural fats and oils; disp. in water and min. oil; sp.gr. 0.97 ± 0.01 (45 C); dens. 8.1 lb/gal (45 C); visc. 1800 cps min. (10% disp.); flash pt. (OC) > 170 C; acid no. 20–40; pH 10.2 ± 0.5 (10% disp.); 100% active; 88% min. amide content.

Schercomid SLM-C. [Scher] Lauramide DEA; nonionic; thickener and foam stabilizer with wetting properties; amber clear liq.; acid no. 1 max.; alkali value 30–50; 85% min. amide.

Schercomid SLMC-75. [Scher] Lauramide DEA; visc. builder which generates a thick, copious foam; mild shampoos, bubble baths, aerosol shave creams, and other detergent compositions; liq. at ambient temps. to permit ease of handling; clear yel. liq.; mild odor; sol. in alcohols, glycols, glycol ethers, polyols, aliphatic, aromatic, and chlorinated hydrocarbons, natural fats and oils; disp. in min. oil and water; sp.gr. 0.99; dens. 8.0 lb/gal; flash pt. (OC) > 180 C; acid no. 2.0 max.; 100% active; 80% min. amide content.

Schercomid SL-ML, SLM-LC, SL-ML-LC. [Scher] Lauramide DEA; nonionic; visc. builder which generates thick, copious foam; mild shampoos, bubble baths, aerosol shave creams, and other detergent compositions; emulsifier with balanced hydrophilic-hydrophobic properties so that it functions as both o/w and w/o emulsifier in creams and lotions; clear amber liq., clear yel. liq., clear amber liq., resp.; mild odor; see Schercomid SLMC-75; sp.gr. 0.97, 0.99, 0.98 resp.; dens. 8.0 lb/gal; flash pt. (OC) > 180 C; acid no. 1.5, 2.0, 1.5 max. resp.; 100% act.; 87% min. amide content.

Schercomid SLM-S. [Scher] Lauramide DEA; nonionic; visc. builder for the cosmetic industry; thick, copious foam in shampoos, facial scrubs, bubble baths, hand soaps, etc.; lt. yel. liq. when fresh (cryst. on aging); mild odor; see Schercomid SLMC-75; sp.gr. 0.97 ± 0.01 (45 C); dens. 8.1 lb/gal; flash pt. (OC) > 170 C; acid no. 1.0 max.; 100% act.; 88% min. amide content.

Schercomid SLS. [Scher] Linoleamide DEA; nonionic; conditioner and emollient for personal care prods.; emulsifier for w/o systems and hydrocarbons; dispersant for pigments and min. clays; visc. builder; amber clear liq.; mild fruity odor; sol. see Schercomid SI; sp.gr. 0.980; dens. 8.0 lb/gal; acid no. 2.0 max.; flash pt. > 180 C (OC).

Schercomid SM. [Scher] Myristamide DEA; nonionic; visc. builder and foam booster for hair shampoos, facial cleansers, and body soaps; produces a creamy, rich, tight-knit foam, imparts lubricity to hair, and decreases detergent harshness; synergistic emulsifier for creams and lotions; pale wax; slt. ammoniacal odor; see Schercomid SLMC-75; sp.gr. 0.93 ± 0.01 (60 C); dens. 7.75 lb/gal (60 C); m.p. 48 ± 4 C (freshly made-increases with aging); flash pt. (OC) > 170 C; acid no. 2.0 max.; pH 10.2 ± 0.5 (10% aq.); 100% act.; 87% min. amide content.

Schercomid SME. [Scher] Stearamide MEA; nonionic; detergent, emulsifier, thickener, pearlescent agent for cosmetics, stick deodorants; slower rate of cooling in finished formulation; tan cryst. solid; slight ammoniacal odor; sol. in most org. solvs.; water-disp.; m.p. 98 ± 2.0 C; acid no. 2.0 max.; flash pt. 180 C (OC); 100% act.

Schercomid SME-A. [Scher] Stearamide MEA; nonionic; pearlescent for cosmetic preparations; opacifier, clouding agent, and thickener; pale wax; slight ammoniacal odor; sol. see Schercomid SME; m.p 91 ± 2 C; acid no. 5.0 max.; flash pt. > 180 C (OC); 100% act.

Schercomid SME-M. [Scher] 1:1 stearic acid monoethanolamide; nonionic; pearlescent in cosmetic preparations; opacifier, clouding agent, and thickener in stick deodorants; pale wax; slight ammoniacal odor; sol. see Schercomid SME; m.p. 98 ± 2 C; acid no. 4.0 max.; pH 8.7 ± 0.5 (3% aq.); flash pt. 180 C (OC); 100% act.

Schercomid SME-S. [Scher] Stearamide MEA stearate; nonionic; emulsifier, pearlescent agent, thickener, opacifier in cosmetic and pharmaceutical formulations; pale yel. wax; slight typ. odor; m.w. 565; sol. in most hydrophobic org. solvs.; water-disp.; m.p. 77–82 C; acid no. 10 max.; flash pt. > 180 C (OC); 100% act.

Schercomid SO-A. [Scher] Oleamide DEA; nonionic/anionic; w/o emulsifier, lubricant, conditioner; lt. amber clear liq.; slight typ. oleic odor; sol. in most org. solvs.; water-disp.; sp.gr. 0.95 ± 0.01; dens. 7.9 lb/gal; visc. 450 cps min.; acid no. 8.0 max.; flash pt. > 180 C (OC); 100% act.

Schercomid SO-T. [Scher] Tall oil DEA; anionic/nonionic; w/o emulsifier; amber clear liq.; 100% conc.

Schercomid TO-2. [Scher] Tallowamide DEA and diethanolamine; nonionic; anionic; w/o emulsifier, visc. builder, pigment dispersant; dispersant for min. clays or pigments; emulsifier for aromatic and aliphatic hydrocarbon solv., esp. those of high m.w.; used in shampoos where it generates a creamy, luxurious foam; stabilizes foam when used with surfactants and detergents; hair conditioning agent; clear dk. amber liq.; mild, char. odor; sol. in alcohols, glycols, glycol ethers, aliphatic and chlorinated hydrocarbons; disp. in water; sp.gr. 0.990; dens. 8.25 lb/gal; flash pt. (OC) > 170 C; acid no. 13–17; 100% act.; 60% min. amide content.

Schercomul QW. [Scher] Surfactant blend; nonionic/cationic; w/o and o/w emulsifier, detergent; imparts antistatic properties; liq.; 95% conc.

Schercopearl EA-100. [Scher] Stearamide MEA; nonionic; pearlescing agent for shampoos; solid; 100% act.

Schercophos L. [Scher] Complex phosphate; anionic; detergent builder and water conditioner for the textile industry; liq.; 60% act.

Schercopol CMS-Na. [Scher] Disodium cocamido MEA sulfosuccinate; anionic; solubilizer, softener for personal care prods.; home and industrial detergent cleaning formulations; m.w. 477; sol. in water; partly sol. in most org. solvs.; sp.gr. 1.12; dens. 9.3 lb/gal; visc. 100 cps max.; pH 5–7; cloud pt. 5.0 C; biodeg.; 30% solids.

Schercopol LPS. [Scher] Disodium laureth sulfosuccinate; surfactant; visc. enhancer; yel. clear liq.; pH 5–7; 39% min. dry solids.

Schercopol OMIS-Na. [Scher] Disodium oleamido MIPA sulfosuccinate; foam builder and stabilizer with nonirritating properties; primary surfactant in baby and adult shampoos; clear yel. liq.; mild, char. odor; m.w. 563; sp.gr. 1.1; dens. 9.16 lb/gal; visc. 500 cps max.; cloud pt. 5.0 C max.; pH 5–7; 35% min. solids; diluent: water.

Schercopol OMS-Na. [Scher] Disodium oleamido MEA sulfosuccinate; anionic; solubilizer; surfactant imparting soft, emollient feel on skin and conditioning effect on hair; foamer in toiletries, hand dishwashing detergents, and personal care prods.; yel. clear liq.; mild char. odor; sol. in water; partly sol. in most org. solvs.; sp.gr. 1.10 ± 0.05; dens. 9.16 lb/gal; visc. 1000 cps max.; pH 6.0 ± 1.0; cloud pt. 5.0 C max.; 34% min. solids conc.

Schercoquat CAS. [Scher] Coconut quat.; cationic; w/o and o/w emulsifier, conditioner for hair care prods.; liq.; 100% act.

Schercoquat DAB. [Scher] Quaternium-63; cationic; conditioner for personal care prods.; amber clear to hazy visc. liq.; aromatic odor; m.w. 995; water-sol.; sp.gr. 0.975; dens. 8.1 lb/gal; pH 3–6 (5% aq); 90% min. nonvolatiles; 85% min. quat. assay

Schercoquat DAS. [Scher] Quaternium-61; cationic; see Schercoquat DAB; Gardner 10.0 max.; slight, mild odor; m.w. 1050; water-sol.; sp.gr. 1.01; dens. 8.4 lb/gal; pH 4–7 (5% aq.); flash pt. 90 C (OC); 90% min. nonvolatiles.

Schercoquat FOAS. [Scher] N-safflower-(3-amidopropyl)-N-N-dimethyl-N-ethyl ammonium ethyl sulfate; quat. effective in hair conditioners; good slip, shine, and combability; amber visc. liq.; m.w. 520; water-sol.; 90% min. dry solids.

Schercoquat IAS. [Scher] Isostearamidopropyl ethyldimonium ethosulfate; cationic; conditioner for personal care prods.; Gardner 10 max. clear visc. liq.; slight mild odor; m.w. 522; sp.gr. 0.99 ± 0.01; dens. 8.2 lb/gal; pH 5–7 (5.0% aq.); 90% min. nonvolatiles; 90% min. act.

Schercoquat IAS-LC. [Scher] N-(3-isostearylamidopropyl), N-N dimethyl, N-ethyl ammonium ethyl sulfate; cationic; see Schercoquat IAS; dk. amber clear visc. liq.; slight mild odor; m.w. 550; sp.gr. 0.99 ± 0.01; dens. 8.2 lb/gal; pH 5–7 (5.0% aq.); flash pt. 90 C (OC); 90% min. nonvolatiles; 85% min. act. (assay).

Schercoquat IB. [Scher] Isostearamidopropalkonium chloride; cationic; see Schercoquat IAS; also bactericidal agent and antistat; amber visc. liq.; mild almond odor; m.w. 494; sp.gr. 0.99 ± 0.01 (40 C); dens. 8.2 lb/gal (40 C); dens. 8.2 lb/gal (40 C); pH 4–7 (5.0% aq.); 90% min. nonvolatiles; 85% min. act.

Schercoquat IEP. [Scher] Quaternium-62 (isostearic epoxypropyl chloride); conditioning quat. offering good water-sol. and good compat. with many anionic surfactants; amber visc. liq.; m.w. 486; 80% min. dry solids.

Schercoquat IIB. [Scher] Isostearyl benzylimidonium chloride; cationic; see Schercoquat IAS; dk. amber visc. liq.; aromatic odor; m.w. 505; sp.gr. 0.96; dens. 8.0 lb/gal; pH 4–7 (5.0% aq.); 90% min. nonvolatiles; 75% min. act.

Schercoquat IIS. [Scher] Isostearyl ethylimidonium ethosulfate; cationic; see Schercoquat IAS; Gardner 12.0 max. clear to slighty hazy visc. liq.; slight, mild odor; m.w. 532; sp.gr. 1.036 ± 0.01; dens. 8.6 lb/gal; pH 4–7 (5.0% aq.); 98% min. nonvolatiles; 80% min. quat. (assay).

Schercoquat IIS-R. [Scher] Isostearyl ethyl imidonium ethosulfate; cationic; see Schercoquat IAS; dk. amber clear to slightly hazy visc. liq.; mild amine odor; m.w. 506; sp.gr. 1.020; dens. 8.6 lb/gal; pH 4–7 (5.0% aq.); 98% min. nonvolatiles; 80% min. quat. (assay).

Schercoquat IIS-RD. [Scher] 2-Isoalkyl (C_{14}-C_{20}), 1-hydroxyethyl, 1-ethyl imidazolinium ethyl sulfate; cationic; see Schercoquat IAS; dk. amber clear to slightly hazy visc. liq.; mild amine odor; m.w. 530;

sp.gr. 1.036 ± 0.01; dens. 8.6 lb/gal; pH 4–7 (5.0% aq.); 99% min. nonvolatiles; 80% min. quat. (amine).

Schercoquat MKAS. [Scher] Mink oil quat.; cationic; surfactant imparting softenss and body to hair; liq.; water-sol.

Schercoquat ROAB. [Scher] Rapeseedamidopropyl benzyldimonium chloride; cationic; conditioner and visc. builder for personal care prods.; dk. amber clear to hazy visc. liq.; aromatic odor; m.w. 533; sp.gr. 0.975; dens. 8.1 lb/gal; pH 3–6 (5.0% aq.); 90% min. nonvolatiles; 85% min. quat. (assay).

Schercoquat ROAS. [Scher] Rapeseedamidopropyl ethyldimonium ethosulfate; cationic; see Schercoquat IAS; dk. amber clear to hazy visc. liq.; mild amine odor; m.w. 560; sp.gr. 0.990; dens. 8.2 lb/gal; pH 4–7 (5.0% aq.); flash pt. 90 C (OC); 90% min. nonvolatiles; 80% min. quat.

Schercoquat ROEP. [Scher] N-rapeseed-(3-amidopropyl)-N-N-dimethyl-N-(2,3 epoxypropyl) ammonium chloride; conditioner for conditioning shampoos and hair sprays; dk. amber visc. liq.; m.w. 533; water-sol.; 80% min. dry solids.

Schercoquat SAB. [Scher] Stearamidopropyl benzyldimonium chloride; conditioner for hair rinses; yel. hazy liq.; mild amine odor; m.w. 480; water-sol.; sp.gr. 0.945; dens. 8.0 lb/gal; pH 4–7 (5% aq.); 80% min. dry solids.

Schercoquat SAS. [Scher] Stearamidopropyl ethyldimonium ethosulfate; conditioner for hair rinses; yel. liq.; m.w. 508; water-sol.; 80% min. dry solids.

Schercoquat SOAB. [Scher] Soyamidopropyl benzyldimonium chloride; quat. effective in hair conditioners, good slip, shine, and combability; dk. amber visc. liq.; m.w. 488; water-sol.; 90% min. dry solids.

Schercoquat SOAS. [Scher] Soyamidopropyl ethyldimonium ethosulfate; cationic; see Schercoquat IAS; amber clear visc. liq.; mild amine odor; m.w. 516; water-sol.; sp.gr. 1.04; dens. 8.6 lb/gal; pH 4–7 (5.0% aq.); 90% min. nonvolatiles; 80% min. quat. (assay).

Schercoquat WOAS. [Scher] Wheat germ oil quat.; cationic; surfactant imparting body, bounce, antistatic properties, shine to hair; liq.; 90% act.

Scherco Softener #1. [Scher] Quat.; cationic; substantive softener and finishing agent for orlon and acrilan; liq.; 15% act.

Scherco Softener #366. [Scher] Quat.; cationic; softening agent for acrylics and synthetics; liq.; 15% act.

Schercosol NL. [Scher] Modified coco amide; anionic; lubricant for dry-cleaning industry; liq.; 20% act.

Schercotaine CAB. [Scher] Cocamidopropyl betaine; amphoteric; detergent, wetting agent, foamer, antistat and softener in personal care prods.; pale yel. clear liq.; slight char. odor; m.w. 386; water-sol.; sp.gr. 1.05 ± 0.01; dens. 8.75 lb/gal; visc. 100 cps max.; pH 5.5 ± 1.0; cloud pt. –2.0 C max.; 30% act.

Schercotaine IAB. [Scher] Isostearamidopropyl betaine; conditioner and detergent for shampoos and emollient body treatments; visc. control agent; textile softener; amber visc. liq., soft, opaque gel; slight char. odor; m.w. 477; sol. in aq. alcohol, glycols; sp.gr. 1.05; dens. 8.75 lb/gal; pH 5.5 ± 1.0; 34.0% min. solid.; 30% min. act.

Schercotaine MAB. [Scher] Myristamidopropyl betaine; detergent, thickener, wetting agent with antistatic properties for cosmetic and toiletry preps.; yel. clear to hazy liq.; slt., char. odor; m.w. 393; sp.gr. 1.030; dens. 8.58 lb/gal; visc. 90 cps; cloud pt. -1 C; pH 5.5 ± 1.0 (as is); 30.0% min. solids; 25% min. act.

Schercotaine MKAB. [Scher] Mink oil amido betaine; amphoteric; surfactant, conditioner imparting sheen to hair; softens skin; emulsification properties; liq.; 35% conc.

Schercotaine OAB. [Scher] Oleylamidopropyl betaine; amphoteric; visc. control agent, textile softener and conditioner; liq.; 35% conc.

Schercotaine PAB. [Scher] Palmitamidopropyl betaine; amphoteric; thickening agent, good hair and skin conditioner for lotions and cream rinses; lt. yel. soft gel; sol. in aq. alcohol, glycols; pH 5–7; 35% min. dry solids.

Schercotaine SCAB-KG. [Scher] N-[3-(Cocamido) propyl]-N-(2-hydroxy-3-sulfopropyl)-N-N-dimethyl betaine, potassium salt; visc. stabilizer in natural soap systems; surfactant; lt. yel. clear liq.; typ. odor; m.w. 480; sp.gr. 1.10; dens. 9.1 lb/gal; pH 5–7; 50% min. dry solids.

Schercotaine UAB. [Scher] Bis(undecylenic amidopropyl dimethyl glycinate); amphoteric; surfactant with germicidal/bactericidal activity; for shampoos; liq.; 30% conc.

Schercotaine UAB-Z. [Scher] Zinc-bis(undecylenic amidopropyl dimethyl glycinate); amphoteric; surfactant with some germicidal/bactericidal act.; for shampoos, skin cleansers; amber clear liq.; char. odor; m.w. 717; sp.gr. 1.09; dens. 9.1 lb/gal; pH 4–7; 32% conc.

Schercotarder. [Scher] Fatty amine; cationic; retarding agent for cationic dyes in the textile industry; liq.; 40% act.

Schercoterge 140. [Scher] Fatty amide, ethoxylated, modified; nonionic; detergent, wetting and textile scouring agent, emulsifier, wool fulling; dyeing assistant; post scouring agent; dye bath stabilizer; lemon liq.; mild ammoniacal odor; water-sol.; dens. 8.3 lb/gal; sp.gr. 1.0; pH 9.5 ± 0.5 (1%); 98% act.

Schercoteric STS. [Scher] Stearoamphodiacetate; amphoteric; surfactant and conditioner for personal care prods.; pearly cream paste; 25% min. dry solids.

Schercozoline B. [Scher] Behenyl imidazoline; cationic; antistat, dispersant, wetting agent, emulsifier, microbicide used in acid and emulsion cleaning, cleaners, polishes, surf. treatment, textile and leather processing, agriculture and cosmetic intermediate; tan solid; m.w. 383; m.p. 45–49 C; 90% min. imidazoline.

Schercozoline C. [Scher] Cocoyl imidazoline; cationic; see Schercozoline B; also detergent, intermediate for quat. ammonium compds., and used in primer paints; tan semisolid; m.w. 278; 90% min. imidazoline.

Schercozoline I. [Scher] Isostearyl imidazoline; surfactant, softener, antistat; amber clear liq.; m.w. 378; 90% min. imidazoline.

Schercozoline L. [Scher] Lauryl imidazoline; cationic; see Schercozoline C; also intermediate for quat. ammonium compds.; cream solid; m.w. 268; m.p. 38–42 C; 90% min. imidazoline.

Schercozoline O. [Scher] Oleic imidazoline; cationic; see Schercozoline B; also as w/o emulsifier, corrosion inhibitor, and intermediate for quat. ammonium compds.; dk. amber liq.; m.w. 350; 90% min. imidazoline.

Schercozoline S. [Scher] Stearyl imidazoline; cationic; see Schercozoline I; cream solid; m.w. 336; m.p. 44–48 C; 90% min. imidazoline.

Scheroba Oil. [Scher] Isostearyl/erucyl erucate; similar to jojoba oil; used in cosmetic formulations to provide lubricity, emolliency, penetration, silky after-feel, and moisturization; yel. clear liq.; slight typ. odor; m.w. 600; sol. see Schercemol BE; sp.gr. 0.875; dens. 7.3 lb/gal; acid no. 2.0 max.; sapon. no. 80–100; ref. index 1.4625; flash pt. > 170 C (OC); cloud pt. 15 C.

Schersoftoil P. [Scher] Ester of natural oils; nonionic; textile winding lubricant; liq.; 100% act.

SCMS. [Hilton-Davis] Sodium coconut monoglyceride sulfonate; anionic; wetting agent, dispersant, emulsifier, detergent, foamer; water-base systems; personal care prods.; wh. powd.; mild char. odor; m.w. 366; HLB 40; pH 6–8; 93% min act.

Scourol 700. [Kao] POE alkyl ether; nonionic; textile processing assistant; scouring and desizing agent; liq.

Scripset 500. [Monsanto] Styrene/maleic anhydride, sodium salt; anionic; emulsifier, dispersant, thickener; powd.; 100% conc.

Scripset 520. [Monsanto] Styrene/maleic anhydride copolymer; emulsifier, binder, sizing agent, visc. modifier, stabilizer; starch modifier; pigment dispersant, protective colloid, sizing, coating, water-paint calsomines, adhesives, printing, preparation of emulsifier paints; off-wh. powd.; faint aromatic odor; m.w. 50,000; sol. in alkaline aq. systems.

Scripset 540. [Monsanto] Styrene/maleic anhydride copolymer; see Scripset 520; off-wh. powd.; faint aromatic odor; m.w. 20,000; sol. in alkaline aq. systems, org. solvs.

Scripset 550. [Monsanto] Styrene/maleic anhydride copolymer; see Scripset 520; off-wh. powd.; faint aromatic odor; m.w. 10,000; sol. see Scripset 540.

Scripset 700. [Monsanto] Styrene/maleic anhydride copolymer; see Scripset 520; liq.; faint aromatic odor; m.w. 50,000; water-sol.; dens. 9.5 lb/gal; 30% act.

Scripset 720. [Monsanto] Styrene/maleic anhydride copolymer; see Scripset 520; liq.; faint aromatic odor; m.w. 50,000; water-sol.; 25% act.

Scripset 808. [Monsanto] Styrene/maleic anhydride amide/NH_4OH acid salt; anionic; see Scripset 500; powd.; 100% conc.

SE6920FR. [GE] Flame retardant for Silplus elastomeric systems; translucent; sp.gr. 1.09 ± 0.03.

Sebase. [Westbrook Lanolin] Ethoxylated lanolin plus other derivs.; emollient, lubricant for o/w emulsions; visc. stabilizer for cosmetics; paste.

Secomine TA 02. [Stepan Europe] Alkylamine ethoxylate; lubricating, wetting, antistatic agent, emulsifier; liq./solid.

Secosol® DOS/70. [Stepan Europe] Sodium dioctyl sulfosuccinate; anionic; dispersant and wetting/rewetting agent; textile additive; liq.; 70% conc.

Secoster® 874. [Stepan Europe] Pentaerythritol ester; lubricating agent; liq.

Secoster® 887. [Stepan Europe] Trimethylolpropane ester; lubricating agent; liq.

Secoster® BS. [Stepan Europe] Butoxy ethyl stearate; antistatic and lubricating agent; liq.

Secoster® CL 10. [Stepan Europe] PEG-20 sorbitan laurate; nonionic; emulsifier, dispersant, solubilizer; liq.; 100% conc.

Secoster® CP 10. [Stepan Europe] PEG-20 sorbitan palmitate; nonionic; see Secoster CL 10; liq.; 100% conc.

Secoster® CS 10. [Stepan Europe] PEG-20 sorbitan stearate; nonionic; see Secoster CL 10; liq.; 100% conc.

Secoster® DMS. [Stepan Europe] Glycol distearate; nonionic; emollient, pearlescent, emulsifier for creams and cleansing milks; wh. solid; 100% conc.

Secoster® DO 600. [Stepan Europe] PEG 600 dioleate; nonionic; additive for cutting oils; emulsifier for creams, cleansing milks and pesticides; dispersant; liq.; 100% conc.

Secoster® DS 6000. [Stepan Europe] PEG 6000 distearate; thickener, dispersant; solid.

Secoster® EMS. [Stepan Europe] Glycol stearate; nonionic; see Secoster DMS; wh. solid; 100% conc.

Secoster® KL 10. [Stepan Europe] Sorbitan laurate; nonionic; emulsifier, dispersant, solubilizer; liq.; 100% conc.

Secoster® KP 10. [Stepan Europe] Sorbitan palmitate; nonionic; emulsifier, dispersant, solubilizer; solid; 100% conc.

Secoster® KS 10. [Stepan Europe] Sorbitan stearate; nonionic; emulsifier, dispersant, solubilizer; solid; 100% conc.

Secoster® MA 300. [Stepan Europe] PEG 300 abietate; nonionic; see Secoster DO 600; liq.; 100% conc.

Secoster® ML 300. [Stepan Europe] PEG 300 laurate; nonionic; see Secoster DO 600; liq.; 100% conc.

Secoster® ML 4000. [Stepan Europe] PEG 4000 laurate; nonionic; emulsifier, dispersant; solid; 100% conc.

Secoster® MO 100. [Stepan Europe] Diethylene glycol oleate; nonionic; emulsifier, dispersant; liq.; 100% conc.

Secoster® MO 400. [Stepan Europe] PEG 400 oleate; nonionic; see Secoster DO 600; liq.; 100% conc.

Secosyl. [Stepan Europe] Sodium N-lauroyl sarcosinate; anionic; detergent, foaming agent, base, anticorrosion additive for rug shampoos, for personal care prods.; liq.; 30% conc.

Sedaplant Richter. [Henkel Canada] Polyvalent herbal extract plus anti-irritants in water-alcohol medium; emollient for aq. and hydroalcoholic herbal cosmetics, emulsified preparations; dk. brn. liq.; herbal odor.

Sedefos 75®. [Gattefosse] Glycol stearate, diglycol stearate, and trilaneth-4 phosphate; anionic; SE base for cosmetics and pharmaceuticals; Gardner < 5 waxy solid; weak odor; m.p. 43–48 C; acid no. < 6; iodine no. < 3; sapon. no. 105–120.

Select-A-Sorb. [Vanderbilt] Hydrous magnesium silicate; industrial talc as reinforcing filler for NR and syn. rubbers, wire and cable stocks; wh. powd.; 99.9% min. thru 325 mesh; dens. 2.80 ± 0.03 mg/m^3.

Sellogen 641. [Henkel] Alkyloxyalkyl sodium sulfate; anionic; textile detergent, dispersant and emulsifier; pickling; yel. clear liq.; cloud pt. 5 C; flash pt. 43 C (PMCC); pH 8.5 (2%); biodeg.; 30% act.

Sellogen HR. [Henkel] Sodium dialkyl naphthalene sulfonate; anionic; wetting agent, dispersant; heavy duty household and industrial cleaners; fire control, paint strippers, agric. chemical formulations, emulsion and suspension polymerization aids, latex paints; off-wh. powd.; water-sol.; dens. 0.40 g/cc; surf. tens. 44.1 dynes/cm (0.25%); pH 9.0; 75.5% act.

Sellogen NS-50. [Henkel-Nopco] Sodium alkyl aromatic sulfonate; anionic; anticaking agent in fertilizer and clay; used in phosphate rock acidulation; slurry; 50% conc.

Sellogen W. [Henkel] Sodium dialkyl naphthalene sulfonate; anionic; see Sellogen HR; off-wh. powd.; water-sol.; dens. 0.38 g/cc; surf. tens. 42.1 dynes/cm (0.25%); pH 9.0 (5%); 66.5% act.

Sellogen WL. [Henkel] Sodium alkyl naphthalene sulfonate; anionic; see Sellogen HR; dk. amber liq.; water-sol.; dens. 9.5 lb/gal; sp.gr. 1.140; surf. tens. 48.9 dynes/cm (0.25%); pH 9.5 (10%); 34% act.

SEM-35. [Harcros] Dimethyl silicone fluid-water emulsion; nonionic; emulsifier; lubricant, mold release for rubber and plastics, letterpress and lithographic printing; also for textile softeners, cosmetics; milky wh. liq. emulsion; f.p. 30 F; water-disp.; sp.gr. 0.99; pH 6.0–8.0; 35% act.

SEM-60. [Harcros] Lubricant and mold release for rubber and plastic molding, polishes, leather maintenance; water-disp.; sp.gr. 1.00; 60% act.

SEM-135. [Harcros] Silicone emulsion; lubricant and release agent for rubber, plastics, metalworking; water-disp.; 35% act.

Semacylase. [Novo] Pen-v-acylase; enzyme for mfg. of penicillin intermediates; solid.

Semtol® 40. [Witco] Wh. min. oil, tech.; white oils functioning as binder, carrier, conditioner, defoamer, dispersant, extender, heat transfer agent, lubricant, moisture barrier, plasticizer, protective agent, and/or softener in adhesives, agric., chemicals, cleaning, cosmetics, food, pkg., plastics, and textiles industries; sp.gr. 0.804–0.820; visc. 4–5 cSt (40 C); flash pt. 135 C min.; pour pt. 2 C max.

Semtol® 70. [Witco] Wh. min. oil, tech.; see Semtol 40; sp.gr. 0.837–0.853; visc. 11–14 cSt (40 C); flash pt. 177 C min.; pour pt. –7 C max.

Semtol® 85. [Witco] Wh. min. oil, tech.; see Semtol 40; sp.gr. 0.839–0.855; visc. 14–17 cSt (40 C); flash pt. 179 C min.; pour pt. –7 C max.

Semtol® 100. [Witco] Wh. min. oil, tech.; see Semtol 40; sp.gr. 0.839–0.855; visc. 18–20 cSt (40 C); flash pt. 182 C min.; pour pt. –7 C max.

Semtol® 350. [Witco] Wh. min. oil, tech.; see Semtol 40; sp.gr. 0.850–0.890; visc. 64–90 cSt (40 C); flash pt. 216 C min.; pour pt. –12 C max.

Sencor®. [Bayer] Metribuzin; selective herbicide for weed control in potatoes, tomatoes, soybeans, etc.; colorless crystals; m.w. 214.3; sol. (g/1000 g): > 1000 g in dichloromethane, 100–200 g in toluene, 10–20 g in 2-propanol, 1–20 g in hexane, 1.2 g in water; sp.gr. 1.28; m.p. 125.5–126.5 C; avail. as 35% and 70% act. powd., 70% act. gran., suspension conc.

Senka Antifoam 800. [Nippon Senka] Silicone modified oil emulsion; antifoaming agent; milky wh. liq.

Sepacid® CE 5209. [BASF AG] Quat. ammonium compd.; biocide for oilfield applics.

Sepacid® CE 5265. [BASF AG] Glutaraldehyde deriv.; biocide for oilfield applics.

Sepacid® GA 25, GA 50. [BASF AG] Glutaraldehyde; biocide for oilfield applics.

Sepacorr® HT. [BASF AG] Nitrogen-containing condensation prod.; corrosion inhibitor; liq.

SEQ 80. [Vikon] DTPA; chelating agent for heavy metal ions; water-sol.

SEQ 100. [Vikon] EDTA; see SEQ 80; water-misc.

SEQ 120. [Vikon] HEDTA; see SEQ 80; water-misc.

SEQ NT-15. [Vikon] NTA; see SEQ 80; water-misc.

Seqlene® 190. [Pfanstiehl] Reaction mixt. forming sodium-α-d-glucoheptonate, sodium-β-d-glucoheptonate, aldobionates and other complex carbohydrates; sequestrant which forms nonionic chelates; used in metal applics.; scavenger for antioxidants, bactericides; bottle washing, alkaline cleaning, and textile applics.; brn. amber clear sol'n.; slightly ammoniacal odor; sp.gr. 1.18–1.185; chel. value 190 mg $CaCO_3$/g min. (in 3% caustic soda); 35% solids.

Seqlene® 270. [Pfanstiehl] Reaction mixt. forming sodium-α-d-glucoheptonate, sodium-β-d-glucoheptonate, aldobionates, and other complex carbohydrates; see Seqlene 190; brn. amber clear sol'n.; slightly ammoniacal odor; sp.gr. 1.275–1.28; chel. value 270 mg $CaCO_3$/g min. (3% caustic sol'n.); 50% solids.

Seqlene® 400. [Pfanstiehl] Sodium α-d-glucoheptonate dihydrate with traces of sodium β-d-glucoheptonate; see Seqlene 270; lt. tan to tan cryst. powd.; water-sol.

Seqlene® 540. [Pfanstiehl] Sodium-α-d-glucoheptonate dihydrate; see Seqlene 190; off-wh. to lt. tan cryst.; odorless; water-sol.; chel. value 540 mg $CaCO_3$/g min. (in 3% caustic sol'n.).

Seqlene® ES-40. [Pfanstiehl] Reaction mixt. forming sodium-α-d-glucoheptonate, sodium-β-d-glucoheptonate, aldobionates, and other complex carbohydrates; see Seqlene 190; brn. amber clear sol'n.; slightly ammoniacal odor; water-sol.; sp.gr. 1.21–1.22; chel. value 215 mg $CaCO_3$/g min.; 40% solids.

Seqlene® ES-50. [Pfanstiehl] Reaction mixt. forming sodium-α-d-glucoheptonate, sodium-β-d-glucoheptonate, aldobionates, and other complex carbohydrates; see Seqlene 190; brn. amber clear sol'n.; slightly ammoniacal odor; water-sol.; sp.gr. 1.275–1.28; chel. value 270 mg $CaCO_3$/g min.; 1150 mg ferric hydroxide/g (3% caustic); 50% solids.

Sequestrene® 30A. [Ciba-Geigy] Tetrasodium EDTA; chelating agent used in water softening, liq. soaps, detergents, chemical cleaning, scale removal, beerstone removal, processing of textile, paper, and leather, in metal treatment, and for syn. rubber; straw clear sol'n.; misc. with water; 39% act. in water.

Sequestrene® 220. [Ciba-Geigy] Tetrasodium EDTA dihydrate; chelating agent used in powd. cleaning compds.; scale removal, hair rinses; processing of syn. fibers and textiles, and industrial cleaning preparations; wh. cryst. powd.; water-sol.

Sequestrene® AA. [Ciba-Geigy] EDTA; chelating agent for photographic developer baths, shampoos, cosmetics, electroplating, rare earth separations, metal determinations, liq. soaps, germicides, herbicide sprays; wh. powd.; insol. in water.

Sequestrene® Diammonium. [Ciba-Geigy] Diammonium EDTA; chelating agent used in photographic developer baths, replenishment of blix baths, water treatment, and for boiler cleaning; pale yel. clear aq. sol'n.; misc. with water; 42% act. in water.

Sequestrene® NA2. [Ciba-Geigy] Disodium EDTA; chelating agent for control of trace metal contamination in pharmaceutical and cosmetic prods.; wh. cryst. powd.; odorless; water-sol.; pH 6.0 (5%); 99% act.

Sequestrene® NA2Ca. [Ciba-Geigy] Disodium-calcium EDTA dihydrate; see Sequestrene NA2; wh. cryst. powd.; odorless; 97–102% act.

Sequestrene® NA2Cu. [Ciba-Geigy] Disodium EDTA-Copper; chelating agent used in animal feeds, electroplating; micronutrient in agriculture; colorant in shampoos; 14.2% chelated copper.

Sequestrene® NA3. [Ciba-Geigy] Trisodium EDTA; chelating agent used in personal care prods.; processing of syn. fibers and textiles; stabilizer for resin systems; photographic baths, electrolytic and electroless plating; foam stabilizer, water treatment; powd.; water-sol.; pH 8.5 (10%).

Sequestrene® NA4. [Ciba-Geigy] Tetrasodium EDTA; see Sequestrene NA3; powd.; pH 10.5 (10%).

Sequestrene® NAFe 13% Fe. [Ciba-Geigy] Na(FeEDTA); chelating agent; micronutrient; animal feeds, photographic uses, polymerization catalyst for syn. rubber; powd.

Sequestrene® NH4Fe. [Ciba-Geigy] Ferric ammonium EDTA; chelating agent in photographic baths, blix baths; in fertilizer formulations; red-brn. sol'n.; misc. with water; 40% act.

Sequestrene® Tetraammonium. [Ciba-Geigy] Tetraammonium EDTA; see Sequestrene Diammonium; pale yel. clear aq. sol'n.; misc. with water; 48% act. in water.

Ser-AD FX 1010. [Servo] Nonionic compd.; thickener for aq. systems; primary stabilizer for latex paints; wh. solid.

Serdas GBS, GBU. [Servo] Surfactants in min. oil; defoamer for paint industry; water-disp.

Serdas GE 4010. [Servo] Oil and silicone-free defoamer for aq. paint systems; emulsion.

Serdas GE 4050. [Servo] Oil-free silicone-containing composition; antifoaming agent in aq. paint systems; emulsion.

Serdas GLN. [Servo] Surfactants in min. oil; defoamer for paper, sugar, textile, leather, paint industries; disp. in water.

Serdet DSK 40. [Servo] Sodium 2-ethylhexyl sulfate; anionic; wetting agent; latex stabilizer; spreading, dispersing agent, detergent; emulsion polymerization; liq.; water-sol.; 40% conc.

Serdet Perle Conc. [Servo] Lauryl ether sulfate and alkylolamide; anionic; see Serdet DSK 40; liq.; 34% conc.

Serdolamide POF 61. [Servo] Oleic acid polydiethanolamide; nonionic; foam stabilizer, refatting agent, visc. modifier in detergents and personal care prods.; liq.; 100% conc.

Serdolamide POF 61 C. [Servo] Oleic acid polydiethanolamide; nonionic; component in emulsifiers for cutting oils; corrosion inhibitor; liq.; 100% conc.

Serdolamide PYF 77. [Servo] Fatty acid DEA; nonionic; foam stabilizer, refatting agent, visc. improver in personal care prods.; liq.; 100% conc.

Serdox NDI 100. [Servo] Dinonylphenol polyglycol ether (100 EO); nonionic; intermediate for textile auxs.; HLB 18.5; 100% conc.

Serdox NEL 3. [Servo] C_{12-14} alkyl polyglycol ether (3 EO); nonionic; raw material for sulfation and phosphation; emulsifier, detergent; liq.; HLB 8; biodeg.; 100% conc.

Serdox NEL 12/80. [Servo] C_{12-14} alkyl polyglycol ether (12 EO); nonionic; emulsifier for fats, oils, and waxes; stabilizer for syn. latices; liq.; HLB 14.5; biodeg.; 100% conc.

Serdox NJAD 15. [Servo] Tallow amine polyglycol ether, dist. (15 EO); nonionic; emulsifier and dispersant in acid and neutral formulations, oil and wax emulsifier; wetting agent in metal cleaning compds.; antistat; solid; 100% conc.

Serdox NJAD 20. [Servo] Tallow amine polyglycol ether, dist. (20 EO); nonionic; see Serdox NJAD 15; solid; 100% conc.

Serdox NJAD 30. [Servo] Tallow amine polyglycol ether, dist. (30 EO); nonionic; see Serdox NJAD 15; solid; 100% conc.

Serdox NNP 4. [Servo] Nonoxynol-4; nonionic; detergent; dispersant; emulsifier for insecticides and herbicides; liq.; oil-sol.; HLB 9.0; biodeg.; 100% conc.

Serdox NNP 8.5. [Servo] Nonoxynol-8 (8.5 EO); nonionic; scouring agent for textiles; soaking assistant in the leather industry; detergent for household and industrial purposes; emulsifier for insecticides and herbicides; plasticizer for mortar and concrete; liq.; HLB 12.5; 100% conc.

Serdox NNP 20/70. [Servo] Nonoxynol-20; nonionic; stabilizer for syn. latices; solid; HLB 16.0; 70% conc.

Serdox NNP 30/70. [Servo] Nonoxynol-30; nonionic; dyeing assistant; lime soap dispersant; emulsifier and stabilizer for emulsion polymerization; liq.; HLB 17.0; 70% conc.

Serdox NOP 30/70. [Servo] Octoxynol-30; nonionic; emulsifier and stabilizer for emulsion polymerization; liq.; HLB 17.5; 70% conc.

Serdox NOP 40/70. [Servo] Octoxynol-40; nonionic; see Serdox NOP 30/70; liq.; HLB 18.0; 70% conc.

Serdox NSG 400. [Servo] PEG-9 stearate; nonionic; textile processing and cosmetic emulsions; antistat for plastics; solid; water-disp.; biodeg.; 100% conc.

Serenasol PDN. [Nippon Senka] Dye solubilizer; liq.

Sermul EN 20/70. [Servo] Nonoxynol-20; nonionic; post stabilizer for emulsion polymerization; solid; 70% conc.

Sermul EN 30/70. [Servo] Nonoxynol-30; nonionic; see Sermul EN 20/70; solid; 70% conc.

Sermul EN 145. [Servo] Nonoxynol-30; nonionic; see Sermul EN 20/70; liq.; 70% conc.

Serpol QPA 150. [Servo] Ammonium polyacrylate; pigment dispersant for paints; water-sol.

Serpol QPA 160. [Servo] Sodium polyacrylate; dispersant for paper industry; water-sol.

Servamine KAC 412. [Servo] N-coco-N,N,N-trimethyl ammonium chloride; antistat and lubricant for syn. fibers; bactericide, fungicide, disinfectant, sanitizer; liq.; 50% conc.

Servamine KAC 422. [Servo] N-coco-N,N-dimethyl-N-benzyl ammonium chloride; germicide, disinfectant, sanitizer; liq.; 50% conc.

Servamine KEO 260. [Servo] N (oleyl amido ethyl) N (ethanol amine); cationic; adhesion agent in bitumen and corrosion inhibitor; liq.; 100% conc.

Servamine KEP 4527. [Servo] N (palmityl amido propyl) N,N,N trimethyl ammonium chloride; cationic; emulsifier with bactericide properties; liq.; biodeg.; 50% conc.

Servamine KET 350. [Servo] N (tall oil amido propyl) N,N dimethyl amine; cationic; see Servamine KEO 260; liq.; sol. in IPA, xylene, wh. spirits; 100% conc.

Servamine KOO 330. [Servo] Aminoethyl imidazoline; corrosion inhibitor, intermediate, emulsifier; sol. in oil, xylene, IPA, kerosene; disp. in water.

Servamine KOO 330 B. [Servo] Oleyl amido ethyl oleyl imidazoline; cationic; basic material in mfg. of quat. imidazoline; liq.; biodeg.; 100% conc.

Servamine KOO 360. [Servo] Hydroxyethyl imidazoline; corrosion inhibitor, intermediate; sol. in oil, xylene, IPA, kerosene; disp. in water.

Servamine KOY 4387B. [Servo] Quat. imidazoline; cationic; laundry softener; cloudy visc. liq.; 78% conc.

Servo Brillant Olie B AZ 75. [Servo] Sodium castor oil; anionic; softener and finishing oil; pasting oil for dyestuffs; liq.; 66% conc.

Servo CK 492, CK 494. [Servo] Org. quat. compd.; algicide, fungicide, bactericide for oil industry; sol. in water, IPA, xylene.

Servo CK 601. [Servo] Org. compds.; see Servo CK 492; sol. in water, IPA, xylene.

Servoxyl VLB 1123. [Servo] Sodium monoalkyl polyglycol ether sulfosuccinate; anionic; raw material for personal care prods.; cleaning agent; liq.; biodeg.; 40% conc.

Servoxyl VPI 55. [Servo] Org. phosphate ester; antistat for syn. fibers; sol. in water.

Serwet WH 170. [Servo] Sodium di-2-ethylhexyl sulfosuccinate; anionic; wetting agent; dispersant for pigments; sol. in water and alcohol; 65% conc.

Setsit® 5. [Vanderbilt] Activated dithiocarbamate; accelerator for NR, SBR latexes; primary accelerator for latex; dilute 1:1 with water before adding; sec. accelerator for thiazole activation; redsh. brn. liq.; sol. in alcohol, acetone; dens. 1.01 ± 0.02 mg/m³; flash pt. 55 C.

Setsit® 9. [Vanderbilt] Activated dithiocarbamate blend; accelerator for NR, SBR, and CR latexes; activator for NR, SBR, and NR/SBR latex foam; amber to brn. liq.; mod. sol. in toluene, gasoline, chloroform; misc. with water, alcohol; dens. 1.00 ± 0.02 mg/m³; flash pt. (PM) 55 C.

Setsit® 51. [Vanderbilt] Activated dithiocarbamate; latex accelerator; used alone or as activator for Zetax and other thiazoles; may be added directly to latex without dilution; lt. amber liq.; water-misc.; dens. 1.02 ± 0.02 mg/m³.

Setsit® 104. [Vanderbilt] Activated dithiocarbamate; ultrafast accelerator for NR, SBR, and NR/SBR latex foam and film applics.; little or no precure will occur if zinc oxide is withheld from compd.; should be diluted 1:2 with water; amber liq.; water-misc.; dens. 1.18 ± 0.02 mg/m³.

Seyco Defoamer A-69. [Sherex Polymers] Silicone/nonsilicone emulsion; defoamer for wet processing, agric. sprays; liq.; biodeg.

Seyco Defoamer D-10. [Sherex Polymers] Silicone emulsion; defoamer for wet processing; liq.; biodeg.

Seyco Defoamer G-34, H-61M. [Sherex Polymers] Nonsilicone emulsion; defoamer for pulp and paper industry, wet processing of textiles; liq.; biodeg.

Seyco Defoamer K-20. [Sherex Polymers] Silicone emulsion; defoamer for wet processing; liq.; biodeg.

Seycopon A-95 LCP. [Sherex Polymers] Detergent; foam stabilizer for latex; lt. yel. liq.; sp.gr. 1.02; b.p. 100 C; f.p. 0 C; 30% act.

SF 18. [GE] Polydimethylsiloxane; lubricant, antifoam, mold release; rubber and plastic lubricant; base fluid for grease; mold release for rubber, plastic, and food applic.; antifoam in food applic. and aq. defoaming formulations; water-wh. clear oily liq.; sp.gr. 0.973; visc. 350 cs; pour pt. –58 F; ref. index 1.4030; flash pt. 500 F (CC).

SF 69. [GE] Dimethyl silicone; defoamer, release agent, in cosmetics, polishes, paint additives, and mechanical devices; lubricant in rubber or plastic-to-metal applics.; SF 69 as film modifier and for pigment control in coatings; water-wh. clear, oily fluid; sp.gr. 0.965; visc. 10 cstks; pour pt. –40 F; ref. index 1.4020; flash pt. 145 F; surf. tens. 20.5 dynes/cm; sp. heat 0.36 Btu/lb/°F.

SF 81. [GE] Dimethyl silicone; see SF 69; SF 81 for damping and heat transfer in mechanical/elec. applics.; water-wh. clear, oily fluid; sp.gr. 0.972; visc. 50 cstk; pour pt. –120 F; ref. index 1.4030; flash pt. 460 F; surf. tens. 21.0 dynes/cm; conduct. 0.087 Btu/h-°F ft²/ft; sp. heat 0.36 Btu/lb/°F; dissip. factor 0.0001; dielec. str. 35.0 kV; dielec. const. 2.74; vol. resist. 1×10^{14} ohm-cm.

SF 96. [GE] Polydimethylsiloxane; nonionic; emollient, lubricant for polishes, antifoams, textiles, chemical specialties; plastic and rubber lubrication; dampening or heat transfer fluids; oil defoamer, paint additives; mold release for tires, rubber, plastics; textile softener/modifier; water-wh. liq.; sol. in aliphatic, aromatic, and chlorinated hydrocarbons, alcohols, and ketones, higher hydrocarbons; sp.gr. 0.916–0.974; visc. 5–1000 cs; pour pt. –120 to –58 F; ref. index 1.3970–1.4035; flash pt. 280–500 F (CC); 100% act.

SF 96(5). [GE] Silicone fluid; emollient for hair spray, suntan lotion, antiperspirants; fluid; sol. in lower alcohols, aliphatic, aromatic, and chlorinated hydrocarbons; 100% act.

SF 97. [GE] Dimethyl silicone; see SF 69; SF 97 for transformer in mechanical/elec. applics.; water-wh. clear, oily fluid; sp.gr. 0.0.953–0.974; avail. in 20, 50, 100, 350, 500, and 1000 cstk visc. grades; pour pt. –58 to –85 F; ref. index 1.401–1.4035; flash pt. 395–500 F; surf. tens. 20.8–21.1 dynes/cm; conduct. 0.082–0.092 Btu/h-°F ft²/ft; sp. heat 0.36 Btu/lb/°F; dissip. factor 0.0001; dielec. str. 35.0 kV; dielec. const. 2.68–2.75; vol. resist. 1×10^{14} ohm-cm.

SF 99. [GE] Reactive silicone fluid; forms water-repellent fluids with heat or heat/catalyst; used in textiles, particle treatment, magnesium oxide, and Calrod® units; fluid; sp.gr. 0.970; visc. 10 cstk; ref. index 1.3952; flash pt. 145–345 F; surf. tens. 20.5–21 dynes/cm; sp. heat 0.36 Btu/lb/°F.

SF 1023. [GE] Phenyl containing silicone fluid; flow control agent in polyester coatings; org. finish systems; sp.gr. 1.030; visc. 55 cs; flash pt. 180 F (CC); 100% solids.

SF 1029. [GE] Chlorophenylmethyl silicone fluid; lubricant; API thread lube formulations; hydraulic fluids; base fluid for grease; sliding metal on metal lubrication; excellent high temp. stability, resists gelation for long periods at extreme temps., when in contact with air; liq.; sp.gr. 1.030; visc. 75 cs; pour pt. –100 F; ref. index. 1.4290; flash pt. 425 F (CC).

SF 1066. [GE] Polydimethylsiloxane/EO-PO copolymer; nonionic; lubricant, chemical specialties; cosmetics and toiletries; rubber and plastic mold release; paintable release; textile softeners/modifier; thread and fiber lubes; liq.; sol. in aromatic and chlorinated hydrocarbons, alcohols, higher ketones, water, higher hydrocarbons; sp.gr. 1.04; visc. 800 cs; pour pt. –13 F; ref. index 1.4470; flash pt. 200 F (CC); 100% act.

SF 1080. [GE] Methyl alkyl polysiloxane silicone fluid; nonionic; mold release agent, lubricant; used in rubber, plastic, and metal industries, hair care prods.; internal mold release agent in vinyl slush molding; aluminum die cast mold release agent; lt. yel. liq.; sol. in aliphatic, aromatic, and chlorinated hydrocarbons, higher alcohols, and higher ketones; sp.gr. 1.035; visc. 1500 cs; pour pt. –50 F; ref. index 1.4930; flash pt. 400 F (CC); 100% act.

SF 1091. [GE] Methyl alkyl polysiloxane silicone fluid; paint additive for flow control, mar resistance, and gloss; mold release fluid; high temp. lubricant for soft metals, rubbers, and plastics; thread and fiber lubes; liq.; sp.gr. 0.906; visc. 500 cs; pour pt. –72 F; ref. index 1.4480; flash pt. 400 F (CC).

SF 1093. [GE] Polydimethylsiloxane; mold release agent for high temp. heat transfer applics., aerosol oven sprays, military applics, mech. damping, thermostats; amber liq.; essentially odorless; 50 and 100 cs visc. grades: sp. gr. 0.963 and 0.968 resp.; pour pt. –100 F and –90 F resp.; ref. index. 1.4040; flash pt. (CC) 425 F.

SF 1134. [GE] Methyl alkyl polysiloxane; lubricant; formulated thread and fiber lubricants; sliding metal on metal lubrication; cosmetics and toiletries; compatible with org. materials; wax; sp.gr. 0.93; pour pt. 97 F; flash pt. 399 F (CC).

SF 1147. [GE] Methyl alkyl polysiloxane; lubricant, high temp. hydraulic and lubricating fluid for metal-on-metal; used for impregnating sintered metal bearings for fans and motors, fluid for machine operations with aluminum-drilling, tapping, hobbing, broaching, turning, spinning, drawing, extruding; instrument lubrication; vacuum pumps; liq.; sol. in aliphatic, aromatic, and chlorinated hydrocarbons, higher alcohols, higher ketones; sp.gr. 0.890; visc. 50 cs; pour pt. –60 F; ref. index 1.4433; flash pt. 400 F (CC).

SF 1153. [GE] Polydimethyldiphenyl siloxane; lubricant, dielectric coolant; low temp. damping applics. for aircraft instruments and electronic equipment; heat transfer media; base fluid in silicone greases for ball bearing lubrication from –100 to 400 F; low temp. properties with high temp. performance; liq.; sp.gr. 0.980; visc. 100 cs; pour pt. –100 F; ref. index 1.4219; flash pt. 435 F (CC).

SF 1154. [GE] Polydimethyldiphenyl siloxane; lubricant, dielectric coolant, coupler; high temp. heat transfer applic.; base fluid in high temp. greases; high temp. ultrasonic coupler; high temp. bath and oxide protector for solder baths; outstanding heat resistance; lt. straw clear liq.; sp.gr. 1.050; visc. 190 cs; pour pt. –40 F; ref. index 1.4980; flash pt. 440 F (CC).

SF 1173. [GE] Cyclomethicone; emollient, lubricant used in antiperspirants, skin care prods., sunscreen prods., hair conditioners, facial makeup, particle treatment; liq.; f.p. 17 C; b.p. 175 C; insol. in water; misc. in the lower alcohols and in typ. aliphatic, aromatic, and halogenated hydrocarbon solvs.; sp.gr. 0.954; ref. index 1.394; flash pt. (CC) 55 C.

SF 1188. [GE] Dimethicone copolyol; emollient, lubricant, and release agent for textiles, cosmetics and toiletries, paint, plastic mold release, and rubber lubricants; amber clear fluid; sol. in water below 43 C; sol. in acetone, toluene, lower alcohols, and some hydrocarbons; sp.gr. 1.04; dens. 8.65 lb/gal; visc. 1000 cps; flash pt. (PM) 82 C; surf. tens. 25.5 dynes/cm; 100% silicone.

SF 1201. [GE] Silicone fluid; emollient for antiperspirants, hair care prods.; fluid; sol. in lower alcohols, aliphatic, aromatic, and chlorinated hydrocarbons; 100% act.

SF 1202. [GE] Cyclomethicone; see SF 1173; liq.; f.p. –40 C; b.p. 190–210 C; insol. in water; misc. in the lower alcohols and in typ. aliphatic, aromatic, and halogenated hydrocarbon solvs.; sp.gr. 0.955; ref. index 1.395; flash pt. (CC) 82C.

SF 1204. [GE] Cyclomethicone (85% SF 1173 and 15% SF 1202); see SF 1173; liq.; f.p. 11 C; b.p. 175–210 C; insol. in water; misc. in the lower alcohols and in typ. aliphatic, aromatic, and halogenated hydrocarbon solvs.; sp.gr. 0.954; ref. index 1.394; flash pt. (CC) 55 C.

SF 1214. [GE] Silicone fluid; used in hair and skin cosmetics, personal care prods.; sol. in hydrocarbon solvs., cyclic silicones; 100% act.

SF 1221. [GE] Copolymer of a polydimethylsiloxane and polyoxyalkylene ether; lubricant, paintable mold release agent; fiber and thread lube, hot system defoaming, flow control paint additive; liq.; sol. in water, lower alcohols, and hydrocarbon solvs.; sp.gr. 1.040; visc. 800 cs; pour pt. –13 F; ref. index 1.4470; flash pt. 200 F (CC).

SF 1250. [GE] Chlorophenyl methyl siloxane; lubricant, mold release agent, hydraulic systems; fluid transmission; servomotors and mechanisms; instruments, clocks, and timers; machine tool components; antifriction, rolling, and sliding mechanisms; shock absorbers and damping devices; liq.; sp.gr. 1.050; visc. 70 cs; pour pt. –100 F; ref. index 1.4280; flash pt. 410 F (CC).

SF 1265. [GE] Phenyl silicone; silicone fluid providing higher temp. resistance and improved oxidative properties, increased radiation resistance, improved compat. with org. materials; SF 1265 suggested for use as plastics additive; fluid; sp.gr. 1.10; visc. 450 cstk; ref. index 1.5250; flash pt. 600 F; surf. tens. 25.0 dynes/cm.

SF 1705. [GE] Amine functional polydimethylsiloxane fluid; film modifier for SF 1710; formulated with SF 1710 fluid for polishes having high gloss; also for use in textile treatment, leather protection, and corrosion protection; fluid; sol. or disp. in most aliphatic, aromatic, and chlorinated hydrocarbons; sp.gr. 0.82; visc. 50–500 cps; amine equiv 0.047; flash pt. (CC) 28 C; 50% silicone in min. spirits/IPA.

SF 1706. [GE] Amine functional fluid; durable car polish; liq.; sol. in aromatic, aliphatic solvs, IPA; 100% act.

SFR 100. [GE] Silicone fluid; flame retardant for polyolefins; excellent lubrication during processing; improves impact resistance of PP; easy mold fill and lower temp. processing; colorless clear high visc. liq.; sp.gr. 1.0; dens. 1.04 g/cc; visc. 300,000–900,000 cps; flash pt. 202 C (PM); 100% silicone conc.

Shell Cyclo Sol 53. [Shell] Aromatic hydrocarbon solv. used in dry cleaning, coatings, automotive and chemical specialties; sp.gr. 0.874 (60 F); dens. 7.27 lb/gal (60 F); flash pt. (CC) 116 F; 99.4% aromatics.

Shell Cyclo Sol 63. [Shell] Aromatic hydrocarbon solv. used in dry cleaning, coatings, automotive and chemical specialties; sp.gr. 0.887 (60 F); dens. 7.39 lb/gal (60 F); flash pt. (CC) 142 F; 97.5% aromatics.

Shell Mineral Spirits 135. [Shell] Hydrocarbon solv. for dry cleaning, coatings, automotive, and chemical specialties; sp.gr. 0.783 (60 F); dens. 6.52 lb/gal (60 F); flash pt. (CC) 110 F.

Shell Mineral Spirits 145-EC. [Shell] See Shell Mineral Spirits 135; sp.gr. 0.793 (60 F); dens. 6.60 lb/gal (60 F); flash pt. (CC) 103 F.

Shell Sol 70. [Shell] Hydrocarbon solv. for dry cleaning, coatings, automotive, and chemical specialties; sp.gr. 0.747 (60 F); dens. 6.22 lb/gal (60 F); flash pt. (CC) 104 F.

Shell Sol 71 (H), 71 (WR). [Shell] See Shell Sol 70; sp.gr. 0.759 (60 F); dens. 6.32 lb/gal (60 F); flash pt. (CC) 125 F.

Shell Sol 72. [Shell] See Shell Sol 70; sp.gr. 0.768 (60 F); dens. 6.4 lb/gal (60 F); flash pt. (CC) 137 F.

Shell Sol 140. [Shell] See Shell Sol 70; sp.gr. 0.787 (60 F); dens. 6.56 lb/gal (60 F); flash pt. (CC) 143 F.

Shell Sol 340. [Shell] See Shell Sol 70; sp.gr. 0.765 (60 F); dens. 6.37 lb/gal (60 F); flash pt. (CC) 104 F.

Shell Sol B. [Shell] Aliphatic naphtha solv. for dry cleaning, coatings, automotive, and chemical specialties; low boiling, fast evaporating; sp.gr. 0.670 (60 F); dens. 5.58 lb/gal (60 F); flash pt. (CC) < 0 F.

Shell Super VM&P Naphtha EC. [Shell] Aliphatic naphtha solv. for dry cleaning, coatings, automotive, and chemical specialties; low boiling, fast evaporating; sp.gr. 0.754 (60 F); dens. 6.27 lb/gal (60 F); flash pt. (CC) 55 F.

Shell Toluene. [Shell] Aromatic solv. for dry cleaning, coatings, automotive, and chemical specialties; sp.gr. 0.871 (60 F); dens. 7.25 lb/gal (60 F); flash pt. (CC) 41 F.

Shell Tolu-Sol 5. [Shell] Aliphatic naphtha solv. used in dry cleaning, coatings, automotive, and chemical specialties; low boiling, fast evaporating; sp.gr. 0.702 (60 F); dens. 5.85 lb/gal (60 F); flash pt. (CC) 20 F.

Shell Tolu-Sol 6-EC. [Shell] See Shell Tolu-Sol 5; sp.gr. 0.752 (60 F); dens. 6.26 lb/gal (60 F); flash pt. (CC) 21 F.

Shell Tolu-Sol 19-EC. [Shell] See Shell Tolu-Sol 5; sp.gr. 0.730 (60 F); dens. 6.08 lb/gal (60 F); flash pt. (CC) 24 F.

Shell TS-28. [Shell] Aromatic hydrocarbon solv. for dry cleaning, coatings, automotive, and chemical specialties; sp.gr. 0.852 (60 F); dens. 7.09 lb/gal (60 F); flash pt. (CC) 122 F.

Shell VM&P Naphtha EC. [Shell] See Shell Super VM&P Naphtha EC; sp.gr. 0.750 (60 F); dens. 6.24 lb/gal (60 F); flash pt. (CC) 55 F.

Shellwax. [Shell] Paraffin and microcryst. waxes; binders, coatings, moisture protection, adhesives; m.p. 122–183 F.

Shell Xylene. [Shell] Aromatic solv. for dry cleaning, coatings, automotive, and chemical specialties; sp.gr. 0.871 (60 F); dens. 7.25 lb/gal (60 F); flash pt. (CC) 81 F.

Sherbrite 30. [PMC] Sodium saccharin; industrial grade; brightener in nickel plating; 30% sol'n.

Sherbrite 85. [PMC] Sodium saccharin; industrial grade, not for food use; wh. fine free-flowing gran.

Shur-Coal 159 Promoter. [Sherex] Org. coal flotation reagent; used with a hydrocarbon oil to produce a coalaceous froth which gives high recovery of deep-mined, strip, or stockpiled bituminous coals; yel. liq., low odor; f.p. –15 F; sol. in diesel oil; disp. but insol. in water; sp.gr. 0.92; dens. 7.65 lb/gal; visc. 40 cps (60 F); pH 7.0; flash pt. (PMCC) 220 F.

Shur-Coal 160 Promoter. [Sherex] Org.; see Shur-Coal 159 Promoter; yel. liq.; f.p. –20 F; sol. in diesel oil; disp. but insol. in water; sp.gr. 0.835; dens. 6.95 lb/gal; visc. 35 cps (60 F); flash pt. (PMCC) 157 F.

Shur-Coal 164 Promoter. [Sherex] Org.; see Shur-Coal 159 Promoter; yel. clear liq.; low odor; f.p. < –40 F; sol. in diesel oil; disp. but insol. in water; sp.gr. 0.945; dens. 7.87 lb/gal; visc. 96 cps (60 F); pH 7.0; flash pt. (PMCC) 220 F.

Shur-Coal 166 Promoter. [Sherex] Org.; see Shur-Coal 159 Promoter; lt. yel. clear liq.; low odor; f.p. < –40 F; sol. in diesel oil; disp. but insol. in water; sp.gr. 0.94; dens. 7.83 lb/gal; visc. 85 cps (60 F); pH 7.0; flash pt. (PMCC) 220 F.

Shur-Coal 168 Promoter. [Sherex] Org.; see Shur-Coal 159 Promoter; lt. yel. liq.; low odor; f.p. < –40 F; sol. in diesel oil; disp. but insol. in water; sp.gr. 0.94; dens. 7.8 lb/gal; visc. 65 cps (60 F); pH 7.0; flash pt. (PMCC) 220 F.

Shur-Coal 2000. [Sherex] Cationic flocculant for all types of solid–liq. separations in coal washing operations; used for thickening and filtering coal refuse; straw liq.; f.p. 25 F; infinite sol. in water; visc. 120–250 cps; pH 4–6.

Shur-Coal 2057. [Sherex] Mixt. of higher m.w. alcohols; frother for use in froth flotation circuits; recommended for the flotation of high rank coals; clear liq.; b.p. 125 C min.; 3–5% sol. in water; sp.gr. 0.82; visc. 5–10 cps; flash pt. (TCC) 100 F min.

Sibutol®. [Bayer] Bitertanol; fungicidal seed dressing for cereals; colorless crystals; m.w. 337.4; sol. (g/1000 ml): 200–500 g in dichloromethane, 20–50 g in 2-propanol, 10–20 g in toluene; m.p. 118 C; avail. in 10% act. dry seed dressing, 37.5% act. slurry seed dressing, and various combinations.

Siccatol® Ba-12.5. [Akzo] Based on sat. branched-chain syn. fatty acids; paint drier; Gardner 2 max. liq.; dens. 950–970 kg/m³; visc. 100 mPa•s max.; 12.5 ± 0.3% metal content.

Siccatol® Ca-5, -10, -104, -105. [Akzo] Based on sat. branched-chain syn. fatty acids; paint drier; promotes pigment wetting; Gardner 4, 5, 4, 3, and 3 max. resp.; dens. 880–900, 910–930, 970–990, 830–850, and 850–870 kg/m³ resp.; visc. 300, 500, 30, 10, and 10 mPa•s max. resp.; 4, 5, 10, 4, and 5 ± 0.2% metal content resp.

Siccatol® Ce-6, -10. [Akzo] Based on sat. branched-chain syn. fatty acids; paint drier; yel.-brn.; dens. 860–880 and 930–950 kg/m³ resp.; visc. 200 and 500 mPa•s max. resp.; 6 and 10 ± 0.2% metal content resp.

Siccatol® Co-6, -8, -10, -12. [Akzo] Based on sat. branched-chain syn. fatty acids; paint drier; blue-violet; dens. 880–900, 930–950, 990–1010, and 1100–1150 kg/m³ resp.; visc. 100, 100, 300, and 4000 mPa•s max. resp.; 6, 8, 10, and 12 ± 0.2% metal content; 31–34, 44–48, 56–61, and 68–74% NV resp.

Siccatol® Mn-6, -10. [Akzo] Based on sat. branched-chain syn. fatty acids; paint drier; red-brn.; dens. 880–900 and 980–1000 kg/m³ resp.; visc. 100 and 200 mPa•s max. resp.; 6 and 10 ± 0.2% metal content.

Siccatol® Pb-18, -24, -30, -32, -33. [Akzo] Based on sat. branched-chain syn. fatty acids; paint drier; Gardner 3, 3, 4, 4, and 4 max. resp.; dens. 1000–1020, 1080–1100, 1190–1210, 1260–1290, and 1270–1300 kg/m³ resp.; visc. 50, 100, 300, 500, and 500 mPa•s max. resp.; 18 ± 0.3, 24 ± 0.4, 30 ± 0.4, 32 ± 0.5, and 33 ± 0.5% metal content resp.

Siccatol® Sr-6, -10. [Akzo] Based on sat. branched-chain syn. fatty acids; paint drier; aux. drier catalyst; Gardner 1 max.; dens. 865–885 and 945–965 kg/m³ resp.; visc. 50 and 500 mPa•s max. resp.; 6 and 10 ± 0.2% metal content.

Siccatol® Zr-6, -12, -18. [Akzo] Based on sat. branched-chain syn. fatty acids; paint drier, aux. drier catalyst, good pigment wetting; Gardner 2, 3, and 3 max. resp.; dens. 860–880, 980–1000, and 1115–1135 kg/m³ resp.; visc. 100 mPa•s max. (all); 6 ± 0.2, 12 ± 0.3, and 18 ± 0.3% metal content resp.

Sicolub DSP. [BASF AG] Distearyl phthalate; lubricant for PVC; powd.

Sicolub E. [BASF AG] Montan wax deriv.; lubricant for PVC; powd.

Sicolub EDS. [BASF AG] Ethylene diamine distearyl amide; lubricant for PVC and PS; powd.

Sicolub LB 1, LB 2, LBK, LK 3. [BASF AG] Syn. wax and montan wax-based; lubricant for PVC; powd.

Sicolub OA 2, OA 4. [BASF AG] Oxidized polyethylene wax; lubricant for PVC; powd.

Sicolub OP. [BASF AG] Partly saponified montan wax; lubricant for PVC; powd.

Sicolub TDS. [BASF AG] Isotridecyl stearate; lubricant for PVC; liq.

Silacto. [Kenrich] Activator; sp.gr. 1.10.

Silasorb. [Manville] Syn. calcium silicate; adsorbent; controls free fatty acids; wh. powd.

Silcat® R. [Union Carbide] Vinylsilane; crosslinking agent for modification of polyethylene or its copolymers to yield a moisture crosslinkable system; lt. straw clear liq.; sp.gr. 0.9762; visc. 0.7 cSt; flash pt. (TCC) 20 C.

Silcolapse 430. [ICI Specialty] Silicone; general purpose antifoam; liq.

Silcolapse 431. [ICI Specialty] Silicone; nonaq. antifoam; low visc. liq.

Silcolapse 432. [ICI Specialty] Silicone; nonaq. antifoam; very low visc. liq.

Silcolapse 5000. [ICI Specialty] Silicone; antifoam for foodstuffs applics., pharmaceutical processes; liq.

Silcolapse 5001. [ICI Specialty] Silicone; general purpose antifoam; liq.

Silcolapse 5006. [ICI Specialty] Silicone; antifoam for water-based printing inks; liq.

Silcolapse 5007. [ICI Specialty] Silicone; antifoam

for textile processing; liq.

Silcolapse 5008. [ICI Specialty] Silicone; foodstuffs-grade antifoam for brewery industry; liq.

Silcolapse 5009. [ICI Specialty] Silicone; antifoam conc. for dilution by user; liq.; 30% act.

Silcolapse 5010. [ICI Specialty] Silicone/nat. oil antifoam for nonsurfactant systems; liq.

Silcolapse 5020. [ICI Specialty] Silicone; food-grade antifoam emulsion, general purpose; liq.

Silene® D. [PPG Industries] Amorphous hydrated silica; reinforcing pigment used in syn. and natural rubber compds.; fast, smooth calendering and extrusion; used in soling, footwear, mechanical rubber goods, tire sidewalls; wh. powd.; sp.gr. 1.93; ref. index 1.45; pH 9.6 (5% aq.); 82% SiO_2 hydrate.

Silene® 732D. [PPG Industries] Precipitated silica; reinforcing agent for rubber; aids extrusion and resilience; powd.; 72 nm particle size; dens. 10 lb/ft^3; surf. area (BET) 30 m^2/g; pH 8.5.

Silicate Cluster 102. [Olin] Tris-(tributoxysiloxy) methyl silane; functional fluid for use in specialty defoaming, pressure-sensitive adhesives, heat-transfer fluid, lubricants, mold release, high-performance hydraulics, dielectric coolants; water-wh. to lt. yel. clear liq.; m.w. 833; sol. in acetone, petrol. ether, toluene, Stod., diethylether, and silicones; sp.gr. 0.969 (16/16 C); visc. 39.03 cs (@ 38 C); pour pt. < –79 C; flash pt. (COC) 232 C; surf. tens. 22.05 dynes/cm^2.

Silicone AF-10 FG. [Harcros] Polydimethylsiloxane; nonionic; emulsifier; antifoamer for aq. systems; food applic.; chemical processing (adhesive, ink, and soap mfg., latex and starch processing); textiles and paper, leather finishing, metal working; wh. liq. lt. cream; water-disp.; sp.gr. 1.0; pH 4–5; 10% act.

Silicone AF-10 IND. [Harcros] Polydimethylsiloxane; nonionic; emulsifier; antifoamer for aq. systems; used in water-based industrial processes, coolants, detergents, insecticides, abrasive slurries, cutting oils, adhesive and ink mfg., latex and starch processing, pulp slurries; wh. liq. lt. cream; water disp.; sp.gr. 1.0; pH 4–5.

Silicone AF-30 FG. [Harcros] Polydimethylsiloxane; nonionic; emulsifier; antifoamer for food and latex processing, adhesive, water-base ink, paint and soap mfg., dyeing and sizing, and boiler feed water; wh. creamy liq.; water-disp.; sp.gr. 1.01; pH 4–5; 30% act.

Silicone AF-30 IND. [Harcros] Polydimethylsiloxane; nonionic; emulsifier; antifoamer for antifreeze, cutting oils, industrial soaps, adhesive, and ink mfg.; latex and starch processing, sizes, insecticides, and metal working; wh. creamy liq.; water-disp.; sp.gr. 1.01; pH 4–5; 30% act.

Silicone AF-100 FG. [Harcros] Compd. silicone; antifoamer for aq. and nonaq. systems; used in mfg. of food pkg. materials and direct food additive; adhesive, ink, paint, and corn oil mfg., resin polymerization; translucent syrupy fluid; sol. in aliphatic, aromatic, and chlorinated solvs.; sp.gr. 1.01; flash pt. 316 C min. (OC); 100% act.

Silicone AF-100 IND. [Harcros] Compd. silicone; antifoamer for aq. and nonaq. systems; used in adhesive, paint, and ink mfg., sizes, industrial cooking processes, vacuum distillations, insecticides, deasphalting, and resin polymerization; translucent syrupy fluid; sol. see Silicone AF-100 FG; sp.gr. 1.01; flash pt. 316 C min. (OC); 100% act.

Silicone AF-600M. [Harcros] Polydimethylsiloxane; emulsion antifoamer for aq. systems; pesticides, steam extraction carpet cleaning, cutting oils, ink and adhesive mfg., and cleaning compds.; wh. liq. lt. cream; water-disp.; sp.gr. 0.988; pH 5–6; 10% act.

Silicone C111. [ICI Specialty] Polydimethylsilxoane fluid; emollient for cosmetic and pharmaceutical preps.; liq.; avail. in visc. from 0.65–1000 cst.

Silicone Defoamer #5037. [Polymer Research] Emulsion based on polymeric silicone fluids; defoamer for aq. foaming; used in fermentation applic.; wh. emulsion; water-disp.; sp.gr. 0.980; dens. 8.2 lb/gal; 15% act.

Silicone Emulsion 350, 1M, 10M, 60M. [Akrochem] Formulated emulsions of dimethyl silicone fluids in water; nonionic emulsifier; release agent for most tire and mechanical goods operations, wire and cable, and plastics operations; wh. emulsion; sp.gr. 0.98–1.02; dens. 8.25 lb/gal; visc. in centistokes indicated by number in grade; pH 7.0; 35% silicone content.

Silicone Emulsion 350 Conc. [Akrochem] Formulated emulsion of dimethyl silicone fluid in water; mold release agent; wh. emulsion; sp.gr. 0.96–1.00; dens. 8.3 lb/gal; 60% solids.

Silicone Fluid 350, 1M, 5M, 10M, 30M, 60M. [Akrochem] Dimethylpolysiloxane; release and slip agents for tire and mechanical goods (e.g., fan belts, O rings, floor mats, hose, toys, shoe heels, floor tile, bath mats, wire and cable applics.); additives in rubber and plastics for water-repellent treatments; water-wh. clear fluids; misc. with nonpolar liqs. (hydrocarbons, ethers, etc.); immiscible with water, glycerin, alcohols, and other polar liqs.; sp.gr. 0.95–0.98; dens. 7.8–8.0 lb/gal; visc. 20–60,000 cSt (visc. in centistokes indicated by number in grade).

Silicone M400, M404, M405, M406. [ICI Specialty] Silicone; emulsion mold release agent; liq.

Silicone M407. [ICI Specialty] Silicone; silicone wax coemulsion for the printing industry; liq.

Silicone M409. [ICI Specialty] Silicone grease mold release agent for rubber, plastics, and foundry applics.; liq.

Silicone Release 87-X66. [Ram] Silicone oil release agent for elastomer and resin applics.; dens. 11 lb/gal; nonflam.

Silicone Release Agent #5038. [Polymer Research] Dimethyl silicone fluid emulsion; release agent for rubber and plastics; applied by spraying, brushing, or wiping; milky wh. liq. emulsion; dilutable with water; dens. 8.0 lb/gal; 35% act.

Silk Protein Complex. [Croda Ltd.] Hydrolyzed silk protein; conditioner for hair and skin care prods.; liq.; water-sol.

Sillikolloid P 87. [Hoffmann Min.] Quartz-kaolinite; filler; < 0.1% water sol.; dens. 2.6 g/cm^3; visc. 35 Pas; surf. area 12 m^2/g; oil absorp. 50–55 g/100 g; ref.

index 1.55; pH 7–8; hardness (Mohs) 7/2–2.5; vol. resist. 10^{12} ohm-cm; 67% quartz, 33% kaolinite.

Sillitin N 82. [Hoffmann Min.] Quartz-kaolinite; filler; < 0.1% water sol.; dens. 2.6 g/cm³; visc. 25 Pas; surf. area 13 m²/g; oil absorp. 45–50 g/100 g; ref. index 1.55; pH 7–8; hardness (Mohs) 7/2–2.5; vol. resist. 10^{12} ohm-cm; 78% quartz, 22% kaolinite.

Sillitin N 85. [Hoffmann Min.] Quartz-kaolinite; filler; < 0.1% water sol.; dens. 2.6 g/cm³; visc. 25 Pas; surf. area 12 m²/g; oil absorp. 45–50 g/100 g; ref. index 1.55; pH 7–8; hardness (Mohs) 7/2–2.5; vol. resist. 10^{12} ohm-cm; 80% quartz, 20% kaolinite.

Sillitin V 85. [Hoffmann Min.] Quartz-kaolinite; filler; < 0.1% water sol.; dens. 2.6 g/cm³; visc. 10 Pas; surf. area 8 m²/g; oil absorp. 40–45 g/100 g; ref. index 1.55; pH 7–8; hardness (Mohs) 7/2–2.5; vol. resist. 10^{12} ohm-cm; 86% quartz, 14% kaolinite.

Sillitin Z 86. [Hoffmann Min.] Quartz-kaolinite; filler; < 0.1% water sol.; dens. 2.6 g/cm³; visc. 35 Pas; surf. area 12 m²/g; oil absorp. 45–55 g/100 g; ref. index 1.55; pH 7–8; hardness (Mohs) 7/2–2.5; vol. resist. 10^{12} ohm-cm; 68% quartz, 32% kaolinite.

Sillitin Z 89. [Hoffmann Min.] Quartz-kaolinite; filler; < 0.1% water sol.; dens. 2.6 g/cm³; visc. 35 Pas; surf. area 10 m²/g; oil absorp. 45–50 g/100 g; ref. index 1.55; pH 7–8; hardness (Mohs) 7/2–2.5; vol. resist. 10^{12} ohm-cm; 72% quartz, 28% kaolinite.

Sillum-200. [D.J. Enterprises] Semicalcined silica-alumina; filler and flame retardant for the plastic and epoxy industries; beige; 36.90% 200 mesh; 39.50% 325 mesh; 61.97% silica, 21.25% alumina.

Sillum-200 Q/P. [D.J. Enterprises] Alumina silicate; filler with exc. suspension and dispersion qualities; sp.gr. 2.35; bulk dens. 50–60 lb/ft³; m.p. 2600 F; pH 6–7; 62% SiO_2, 37% Al_2O_3.

Sillum-PL 200. [D.J. Enterprises] Aluminum and silica oxides; flame retardant additive for plastics; sp.gr. 2.6; hardness (Mohs) 3.0; 70.4% SiO_2, 24.8% Al_2O_3.

Silsmooth E, G. [Sandoz] Silicone emulsion; lubricant, softener; emulsifiable.

Silsmooth SR/VEL. [Sandoz] Modified silicone emulsion; softener, lubricant with soil resistance and water wickability; emulsifiable.

Silver Bond 30. [Unimin] Silica; filler, flatting agent for traffic paint, exterior block fillers, mastics and adhesives, buffing compds.; wh. particulate; 8.5 μ median particle size; 99.97% thru 325 mesh; oil absorp. 24 g/100 g; GE brightness 87.8; surf. area 0.7 m/g; pH 6.7; 99.40% SiO_2.

Silver Bond 45. [Unimin] Silica; filler, flatting agent for traffic paint, exterior block fillers, mastics and adhesives, buffing compds., elec. epoxy compds.; wh. particulate; 13.3 μ median particle size; 96.8% thru 325 mesh; oil absorp. 23.4 g/100 g; GE brightness 87.2; surf. area 0.5 m/g; pH 6.9; 99.6% SiO_2.

Silver Bond B. [Unimin] Silica; see Silver Bond 45; wh. particulate; 9.9 μ median particle size; 99.4% thru 325 mesh; oil absorp. 23.7 g/100 g; GE brightness 87.3; surf. area 0.6 m/g; pH 6.7; 99.23% SiO_2.

Silwet® L-77. [Union Carbide] Polyalkylene oxide-modified polymethylsiloxane; surfactant, flow and leveling agent, antistat, dispersant, wetting agent, flotation agent, spreading agent for coatings, printing inks, adhesives, agric., automotive, cleaners, antifogging agent, mining, paper, pharmaceutical applics.; pale amber clear liq.; m.w. 600; sol. in methanol, IPA, acetone, xylene, methylene chloride; disp. in water; sp.gr. 1.007; dens. 8.37 lb/gal; visc. 20 cSt; HLB 5–8; cloud pt. < 10 C (0.1%); flash pt. (PMCC) 240 F; pour pt. 35 F; surf. tens. 20.5 dyne/cm (0.1% aq.); 100% act.

Silwet® L-720. [Union Carbide] Polyalkylene oxide-modified polymethylsiloxane; slip additive for paper; also for pharmaceutical use; colorless clear liq.; m.w. 12,000; sol. in water, methanol, IPA, acetone, xylene, methylene chloride; sp.gr. 1.039; dens. 8.64 lb/gal; visc. 1100 cSt; HLB 9–12; cloud pt. 42 C (1%); flash pt. (PMCC) 205 F; pour pt. –30 F; surf. tens. 29.3 dyne/cm (0.1% aq.); 50% act.

Silwet® L-7001. [Union Carbide] Polyalkylene oxide-modified polymethylsiloxane; surfactant, dispersant, emulsifier, leveling and flow control agent, antifogging agent, lubricant, antiblock, slip additive for adhesives, agric., automotive, coatings, printing inks, textiles, household specialties, cutting fluids, petrol. extraction, paper, personal care prods., plastics and rubber; pale yel. clear liq.; m.w. 20,000; sol. see Silwet L-720; sp.gr. 1.023; dens. 8.50 lb/gal; visc. 1700 cSt; HLB 9–12; cloud pt. 39 C (1%); flash pt. (PMCC) 206 F; pour pt. –55 F; surf. tens. 30.5 dyne/cm (0.1% aq.); 75% act.

Silwet® L-7500. [Union Carbide] Polyalkylene oxide-modified polymethylsiloxane; surfactant, antifoam, dispersant, emulsifier, leveling and flow control agent, lubricant, slip additive for adhesives, automotive, chemical processing, coatings, petrol. extraction, paper, personal care prods., plastics and rubber, pharmaceutical, textile applics.; lt. yel. clear liq.; m.w.3000; sol. in methanol, IPA, acetone, xylene, hexanes, methylene chloride; sp.gr.0.982; dens. 8.16 lb/gal; visc. 140 cSt; HLB 5–8; flash pt. (PMCC) 250 F; pour pt. –45 F; 100% act.

Silwet® L-7600. [Union Carbide] Polyalkylene oxide-modified polymethylsiloxane; surfactant, wetting agent for adhesives, window cleaners, textiles; internal lubricant for plastics and rubber; lt. amber clear liq.; m.w. 4000; sol. in water, methanol, IPA, acetone, xylene, methylene chloride; sp.gr. 1.066; dens. 8.86 lb/gal; visc. 110 cSt; HLB 13–17; cloud pt. 64 C (1%); flash pt. (PMCC) 165 F; pour pt. 35 F; surf. tens. 25.1 dyne/cm; 100% act.

Silwet® L-7602. [Union Carbide] Polyalkylene oxide-modified polymethysiloxane; surfactant, defoamer, dispersant, emulsifier, leveling and flow control agent, gloss agent, lubricant, release agent, antiblock and slip additive, wetting agent for adhesives, agric., automotive specialties, chemical processing, coatings, petrol. extraction, skin care prods., urethane bubble release, pharmaceutical, printing inks, textiles; pale yel. clear liq.; m.w. 3000; sol. in methanol, IPA, acetone, xylene, methylene chloride; disp. in water; sp.gr. 1.027; dens. 8.54 lb/gal; visc. 100 cSt; HLB 5–8; flash pt. (PMCC) 260 F; pour pt.

5 F; surf. tens. 26.6 dyne/cm; 100% act

Silwet® L-7604. [Union Carbide] Polyalkylene oxide-modified polymethylsiloxane; surfactant, dispersant, emulsifier, wetting agent, flotation agent, spreading agent for automotive specialties, coatings, window cleaners, mining, personal care prods., textiles; pale yel. clear liq.; m.w. 4000; sol. in water, methanol, IPA, acetone, xylene, methylene chloride; sp.gr. 1.063; dens. 8.84 lb/gal; visc.420 cSt; HLB 13–17; cloud pt. 50 C (1%); flash pt. (PMCC) 175 F; pour pt. 30 F; surf. tens. 25.4 dyne/cm; 100% act.

Silwet® L-7605. [Union Carbide] Polyalkylene oxide-modified polymethylsiloxane; defoamer, slip additive for chemical processing, coatings; lt. tan waxy solid, melts to clear amber liq. at 32 C; m.w. 6000; sol. in water, methanol, IPA, acetone, xylene, methylene chloride; sp.gr. 1.068 (35/25 C); dens. 8.88 lb/gal; visc. 210 cSt; HLB 13–17; cloud pt. 93 F (1%); flash pt. (PMCC) 280 F; pour pt.90 F; surf. tens. 30.2 dyne/cm; 100% act.

Silwet® L-7607. [Union Carbide] Polyalkylene oxide-modified polymethylsiloxane; surfactant, wetting agent, leveling and flow control agent, grease cleaner, flotation and spreading agent for adhesives, agric., automotive specialties, chemical processing, carpet antistat, mining, metal processing, petrol. extraction, printing inks, textiles; lt. amber clear liq.; m.w. 1000; sol. in water, methanol, IPA, acetone, xylene, methylene chloride; sp.gr. 1.050; dens. 8.73 lb/gal; visc. 50 cSt; HLB 13–17; cloud pt. 58 C (1%); flash pt. (PMCC) 200 F; pour pt. 10 F; surf. tens. 23.4 dyne/cm; 100% act.

Simchin. [RITA] Jojoba oil; moisturizer, emollient, conditioner for skin and hair; clear liq.

Simulsol 1285. [Seppic] Ethoxylated castor oil; solubilizer; liq.

Simulsol 1292. [Seppic] Ethoxylated hydrog. castor oil; solubilizer; liq.

Sipernat® 22, 22 S. [Degussa] Precipitated, spray-dried silica; 22 S is obtained by grinding Sipernat 22; adsorbent, anticaking and free-flow agents; used as aid to convert liqs. into powds.; hydrophilic; wh. fluffy powd., particle size 18 nanometer; dens. 100 and 220 g/l resp.; pH 6.3; 98% SiO_2.

Sipernat® 50, 50S. [Degussa] Silicon dioxide, amorphous; carrier, free-flow agent, anticaking agent.

Sipernat® D 17. [Degussa] Precipitated silica; anticaking and free-flow agent; hydrophobic; wh. fluffy powd., particle size 28 nanometer; dens. 80 g/l; pH 8; 99.5% SiO_2.

Sipex® 280. [Rhone-Poulenc Surf.] Ammonium nonoxynol-4 sulfate; high foaming surfactant, wetting agent, dispersant, emulsifier for shampoos, skin cleansers, lt. duty cleaners, emulsion polymerization; liq.; 58% act.

Sipex® BOS. [Rhone-Poulenc Surf.] Sodium octyl sulfate; anionic; wetting agent, emulsifier, detergent, foamer, rinse aid; post-stabilizer in latex paints; metal treatment; textile and plywood mfg.; visc. control in plywood mfg.; fruit and veg. washing; hard surface cleaners; clear liq.; visc. 50 cps; HLB 42; cloud pt. < 10 C; pH 9.5–10.5 (10%); biodeg.; 39–40% act.

Sipex® CAV. [Rhone-Poulenc Surf.] Sodium isodecyl sulfate; see Sipex BOS, also aid for emulsion polymerization of vinyl, acrylic, SBR, and PVC systems; clear liq.; dens. 8.95 lb/gal; visc. 150 cps; cloud pt. < 10 C: pH 9.5–10.5 (10%); 39–40% act.

Sipex® EC-111. [Rhone-Poulenc Surf.] Sodium cetyl sulfate; anionic; emulsifier, detergent, flotation agent; collector in ore flotation; softener/lubricant for textiles; cosmetics and toiletries; wh. thick paste; pH 8 (10%); 25% act.

Sipex® EST-30. [Rhone-Poulenc Surf.] Sodium trideceth sulfate; anionic; wetting agent, foamer, detergent base; sole emulsifier and post-stabilizer in emulsion polymerization of styrene systems; household and industrial cleaners; scouring of fibers; liq. dishwashing compds.; Gardner 2 max. liq.; cloud pt. 14 C max.; pH 7.5–8.5 (10%); 29–30% act.

Sipex® NB60. [Rhone-Poulenc Surf.] Sodium alkyl ether sulfate; industrial grade foaming agent for dedusting treatments, air drilling, wallboard foaming, and brine water baths; liq.; 60% act.

Sipex® OK. [Rhone-Poulenc Surf.] Alcohol sulfate; anionic; specialty surfactant for use as textile lubricant and spin-finish ingred.; liq.; 65% act.

Sipex® OLS. [Rhone-Poulenc Surf.] Sodium octyl sulfate; anionic; wetting agent; rinse aid; mercerizing agent for cotton goods; surfactant in electrolyte baths for metal cleaning; hard surface cleaning; neoprene dispersant; clear liq.; visc. 100 cps; HLB 42; pH 8 (10%); 33% act.

Sipex® OS. [Rhone-Poulenc Surf.] Sodium oleyl sulfate; anionic; emulsifier, emollient; emulsion polymerization; hair prods. and cosmetics; food pkg.; clear liq.; HLB 42; pH 8 (10%); 26% act.

Sipex® TEA. [Rhone-Poulenc Surf.] TEA-lauryl sulfate; detergent, emulsifier; pigment dispersant and stabilizer; wool scouring; ore flotation; rug shampoos; latex paint formulations; dishwashing detergent; clear liq.; visc. 30 cps; cloud pt. 2 C; pH 7 (10%); 40% act.

Sipon® EA. [Rhone-Poulenc Surf.] Ammonium laureth sulfate; anionic; detergent, wetting agent, foamer, solubilizer, penetrant, personal care prods. and pet shampoos; liq. detergent for fabrics; liq.; mild, pleasant odor; visc. 300 cps; cloud pt. –5 C; pH 6.7 (10%); 27% act.

Sipon® EAY. [Rhone-Poulenc Surf.] Ammonium laureth sulfate; anionic; see Sipon EA; thixotropic product; cloud pt. 0 C; pH 6.7 (10%); 26% act.

Sipon® MLS. [Rhone-Poulenc Surf.] Magnesium lauryl sulfate, pharmaceutical grade; lubricant in drug tableting operations; spray-dried; pH 6.5 (3%); 88% act.

Siponate 330. [Rhone-Poulenc Surf.] Alkylaryl sulfonate; anionic; emulsifier, solubilizer, solv., penetrant; emulsifier for agric. pesticides; degreaser and wax removers; drycleaning; latex paints, pigment dispersant, metal cleaning; amber liq.; sp.gr. 1.03; HLB 11.7; visc. 6500 cps; pH 3–6 (5%); 90% conc.

Siponate A-246 LX. [Rhone-Poulenc Surf.] Sodium C_{14-16} olefin sulfonate; detergent, visc. builder; liq.;

40% act.

Siponate DSB. [Rhone-Poulenc Surf.] Polysulfonate; anionic; surfactant for detergent prods. and emulsion polymerization (SBR, ABS, S/B, PVC, PVAc, acrylic); solubilizer and coupling agent, foaming agent; used in high pressure cleaning, hot caustic paint strippers, bleach formulations; exc. mechanical, thermal, and chemical stability; straw yel. clear liq.; m.w. 569; sol. in water, 20% caustic, 20% HCl; sp.gr. 1.15; visc. 145 cps; pH 7–9 (10%); surf. tens. 32 dynes/cm (1%); 45% act.

Siponic® 25-3. [Rhone-Poulenc Surf.] POE (3) C_{12}–C_{15}; biodeg. detergent, wetting agent, emulsifier for household and industrial cleaners; coupler and solubilizer for perfumes and org. additives; HLB 7.9; 100% act.

Siponic® 25-7. [Rhone-Poulenc Surf.] POE (7) C_{12}–C_{15}; see Siponic 25-3; HLB 12.2; 100% act.

Siponic® 25-9. [Rhone-Poulenc Surf.] POE (9) C_{12}–C_{15}; see Siponic 25-3; HLB 13.3; 100% act.

Siponic® 91-6. [Rhone-Poulenc Surf.] POE (6) C_9–C_{11}; see Siponic 25-3; HLB 12.5; 100% act.

Siponic® E-2. [Rhone-Poulenc Surf.] Ceteareth-4; nonionic; emulsifier, lubricant, emollient, and conditioner for personal care prods.; dyeing assistant, leveling agent, emulsion polymerization; HLB 8.0; 100% act.

Siponic® E-3. [Rhone-Poulenc Surf.] Ceteareth-6; nonionic; see Siponic E-2; HLB 10.1; 100% act.

Siponic® E-5. [Rhone-Poulenc Surf.] Ceteareth-10; nonionic; rubber emulsifier, lubricant, dye and fulling assistant; conditioner and emollient for personal care prods.; emulsion polymerization; floor waxes, paper finishes, pharmaceuticals; wax; HLB 12.4; cloud pt. 73 C (1%); pH 6.5 (1%); 100% act.

Siponic® E-7. [Rhone-Poulenc Surf.] Ceteareth-14; nonionic; see Siponic E-5; also textile lubricant; HLB 14.1; cloud pt. 93 C (1%); pH 6.5 (1%); 100% act.

Siponic® E-10. [Rhone-Poulenc Surf.] Ceteareth-20; nonionic; see Siponic E-2; HLB 15.3; 100% act.

Siponic® E-15. [Rhone-Poulenc Surf.] Ceteareth-30; nonionic; see Siponic E-5; wax; HLB 16.7; cloud pt. > 95 C (1%); pH 6.5 (1%); 100% act.

Siponic® F-90. [Rhone-Poulenc Surf.] Octoxynol-9; perfume solubilizer; liq.

Siponic® F-160. [Rhone-Poulenc Surf.] Octoxynol-16; nonionic; primary surfactant in emulsion polymerization of latices; textile scouring and dye coupling applics.; cosmetic prods., household and industrial cleaners; HLB 15.8; 100% act.

Siponic® F-300. [Rhone-Poulenc Surf.] Octoxynol-30; nonionic; see Siponic F-160; HLB 17.3; 70% act.

Siponic® F-400. [Rhone-Poulenc Surf.] Octoxynol-40; nonionic; see Siponic F-160; HLB 17.9; 70% act.

Siponic® F-707. [Rhone-Poulenc Surf.] Octoxynol-70; nonionic; see Siponic F-160; HLB 18.7; 70% act.

Siponic® L1. [Rhone-Poulenc Surf.] Laureth-1; nonionic; thickener in shampoos and bubble baths; stabilizer for emulsion polymers; emulsifier for monomer systems, floor waxes, paper finishes, rubber; emollient for pharmaceuticals; liq.; HLB 3.7 (1%); pH 6.5 (1%); 100% act.

Siponic® L4. [Rhone-Poulenc Surf.] Laureth-4; nonionic; see Siponic L1; also emulsifier for personal care prods.; liq.; HLB 9.4; pH 6.5 (1%); 100% act.

Siponic® L7. [Rhone-Poulenc Surf.] Laureth-7; nonionic; emulsifier for personal care prods., rubber and monomer systems; stabilizer for emulsion polymers; floor waxes, paper finishes, emollient for pharmaceuticals; wax; HLB 12.5; cloud pt. 56 C (1%); pH 6.5 (1%); 100% act.

Siponic® L7-90. [Rhone-Poulenc Surf.] Laureth-7; nonionic; thickener, emulsifier, visc. control, lubricant, solubilizer for personal care prods.; emulsion polymerization surfactant; textile spin finishes; HLB 12.1; 90% act.

Siponic® L12. [Rhone-Poulenc Surf.] Laureth-12; nonionic; see Siponic L7; wax; HLB 14.6; cloud pt. 89 C (1%); pH 6.5 (1%); 100% act.

Siponic® L16. [Rhone-Poulenc Surf.] Laureth-16; nonionic; see Siponic L7; wax; HLB 15.8; cloud pt. > 95 C (1%); pH 6.5 (1%); 100% act.

Siponic® L25. [Rhone-Poulenc Surf.] Laureth-23; nonionic; see Siponic L7; wax; HLB 16.9; cloud pt. > 95 C (1%); pH 6.5 (1%); 100% act.

Siponic® NP4. [Rhone-Poulenc Surf.] Nonoxynol-4; nonionic; emulsifier; detergent, wetting agent, coupler, dispersant; liq.; oil-sol.; HLB 8.1; 100% conc.

Siponic® NP6. [Rhone-Poulenc Surf.] Nonoxynol-6; nonionic; coemulsifier; intermediate for the synthesis of anionic surfactants; liq.; HLB 10.7; 100% conc.

Siponic® NP7. [Rhone-Poulenc Surf.] Nonoxynol-7; emulsifier, detergent, wetting agent, dispersant, solubilizer, coupler for cosmetic and industrial applics. and in emulsion polymerization; HLB 11.7; 100% act.

Siponic® NP8. [Rhone-Poulenc Surf.] Nonoxynol-8; see Siponic NP-7; HLB 12.3; 100% act.

Siponic® NP9.5. [Rhone-Poulenc Surf.] Nonoxynol-9.5; see Siponic NP-7; HLB 13.2; 100% act.

Siponic® NP10. [Rhone-Poulenc Surf.] Nonoxynol-10; see Siponic NP-7; HLB 13.6; 100% act.

Siponic® NP13. [Rhone-Poulenc Surf.] Nonoxynol-13; see Siponic NP-7; HLB 14.4; 100% act.

Siponic® NP40. [Rhone-Poulenc Surf.] Nonoxynol-40; nonionic; detergent, wetting agent; coemulsifier and vinyl/acrylic latex stabilizer; emulsifier for floor waxes and polishes; liq.; water-sol.; HLB 17.7; 70% conc.

Siponic® NP75. [Rhone-Poulenc Surf.] Nonoxynol-75; nonionic; emulsifier for emulsions; wetting agent and dispersant; liq.; HLB 17.7; 70% conc.

Siponic® NP407. [Rhone-Poulenc Surf.] Nonoxynol-40; see Siponic NP-7; HLB 17.8; 70% act.

Siponic® NP707. [Rhone-Poulenc Surf.] Nonoxynol-70; see Siponic NP-7; HLB 18.6; 70% act.

Siponic® OP1.5. [Rhone-Poulenc Surf.] Octoxynol-1 (1.5 EO); nonionic; emulsifier, detergent, wetting agent, coupler; liq.; oil-sol.; HLB 4.7; 100% conc.

Siponic® OP30. [Rhone-Poulenc Surf.] Octoxynol-30; nonionic; detergent, wetting agent; coemulsifier and vinyl/acrylic latex stabilizer; emulsifier for fats and waxes; liq.; HLB 17.3; 70% conc.

Siponic® OP40. [Rhone-Poulenc Surf.] Octoxynol-40; nonionic; see Siponic OP30; liq.; HLB 17.9; 70% conc.

Siponic® SK. [Rhone-Poulenc Surf.] PEG-8 isolauryl thioether; nonionic; wetting agent, metal cleaning; heavy duty detergent; inhibitor in steel processing; scouring of textiles; agric. formulations; antiskinning agent for paints; hair prods.; wood and paper industry; emulsifier for petrol. oils, chlorinated solvs., silicones, metallic soaps, and emulsion polymers; Gardner 6 max. liq.; sol. in water; dens. 8.5 lb/gal; sp.gr. 1.03; cloud pt. 28 C (1%); surf. tens. 31 dynes/cm (0.05%); 98% act.

Siponic® TD-3. [Rhone-Poulenc Surf.] Trideceth-3; nonionic; emulsifier used in textile scouring, household and metal cleaners and industrial applics.; surfactant in cosmetic emulsions; solubilizer for fragrance and oils; HLB 7.9; 100% act.

Siponic® TD-6. [Rhone-Poulenc Surf.] Trideceth-6; nonionic; emulsifier, solubilizer used in dry cleaning solvs., agric. sprays; degreaser; metal cleaning; clear oily liq.; oil-sol.; dens. 8.2 lb/gal; sp.gr. 0.98; visc. 80 cps; HLB 11.4; flash pt. 355 F; pH 6 (1%); 99% act.

Siponic® TD-9-90. [Rhone-Poulenc Surf.] Trideceth-9; nonionic; emulsifier, emollient and lubricant for skin creams and lotions; solubilizer for fragrance and oils; liq.; water-sol.; HLB 12.9; 90% conc.

Siponic® TD-12. [Rhone-Poulenc Surf.] Trideceth-12; nonionic; see Siponic TD-3; HLB 14.5; 100% act.

Siponic® Y500-70. [Rhone-Poulenc Surf.] Oleth-25; nonionic; wetting agent, emulsifier for polymerization; stabilizer for latices; dispersant in PU foam; textile industry; foamer for latex; floor waxes; pharmaceutical emulsions; metal cleaning; leather industry; food pkg. applics.; liq.; mild odor; visc. 1650 cps; HLB 16.2; cloud pt. 6 C (1%); pH 7.5 (1%); 70% act.

Siponic® Y501. [Rhone-Poulenc Surf.] Oleth-20; nonionic; see Siponic Y500-70; solid; mild odor; sol. in water; m.p. 35 C; HLB 15.4; cloud pt. 65 C (1%); pH 7.5 (1%); 100% act.

Sipothix 1941. [Rhone-Poulenc Surf.] Acrylate/steareth-20 methylacrylate copolymer; pH sensitive thickening agent for liq. detergents, shampoos, and cosmetics; optimum thickening at slightly alkaline pH; opaque liq.; 30% solids.

Sipothix H-65. [Rhone-Poulenc Surf.] Ethylacrylate/methylacrylic acid copolymer; see Sipothix 1941; opaque liq.; 30% solids.

Sitren®. [Goldschmidt] Modified siloxanes; water repellent for treatment of min. fibers insulating materials, plasterboard, etc.

Size 777S. [Hercules] Resin; size used with alum in paperboard and insulation board to provide resistance to water and aq. sol'ns.; amber paste to visc. liq.; dens. 8.5 lb/gal (140 F); visc. 1600 poises (140 F); 70% total solids.

Skane® M-8. [Rohm & Haas] 2-n-Octyl-4-isothiazolin-3-one; mildewcide for paints; sol'n.

Skellite®. [Texaco] Solv. used as stove and lamp fuel; flash pt. –14 F.

SK Fert C13, C14. [KAO SA] Amine-based; cationic; anticaking agent for fertilizers; solid/paste and solid/flakes resp.

SK Fert C24. [KAO SA] Amine-based acetate; cationic; anticaking agent for fertilizers; solid/flakes.

SK Fert F325, F20 A. [KAO SA] Fatty primary amine-based; anticaking agent for fertilizers; paste/liq.

SK Flot 1. [KAO SA] Amine base, acetate salt; cationic; min. flotation reagent; solid, paste; 100% conc.

SK Flot 2. [KAO SA] Amine base, acetate salt; cationic; see SK Flot 1; paste; 100% conc.

SK Flot 3. [KAO SA] Amine base, acetate salt; cationic; see SK Flot 1; solid; 100% conc.

SK Flot 4. [KAO SA] Amine base, acetate salt; cationic; see SK Flot 1; solid, flakes; 100% conc.

Skliro. [Croda] Lanolin acid; emollient; water repellent films; w/o emulsifier; stable emulsions in preparation of waterproof makeup; yel. waxy solid; char. odor; m.w. 45–60 C; acid no. 140–165; sapon. no. 155–185.

Skliro Distilled. [Croda] Lanolin acid; superfatting agent in aerosol shave foams, hand soaps; yel. waxy solid; sol. in oils and esters.

Skyllex. [Aarhus] Dialkyl dimethyl ammonium chloride; fabric softener; Gardner 3; m.p. 50 C; 75% act.

SM 2059. [GE] Cationic emulsion of an amine-functional silicone polymer in water; silicone emulsion cures to a durable, detergent-resistant film for mold release, particle treatment, textile finishes, and polish applics.; off-wh. emulsion; water-disp.; dens. 8.25 lb/gal; visc. 20 cps; 35% silicone.

SM 2061. [GE] Silicone dimethyl o/w emulsion; nonionic; aerosol spray starch, mold release; liq.; 35% act.

SM 2079. [GE] Nonionic emulsion of a high-visc. silicone polymer in water; curable silicone emulsion forms durable, detergent-resistant film for mold release, particle treatment, textile finishes, and polish applics. (requires catalyst); off-wh. emulsion; water-disp.; dens. 8.30 lb/gal; visc. 100–500 cps; 60% silicone.

SM 2086. [GE] Amine-functional silicone polymer aq. emulsion; nonionic; silicone emulsion which cures to a durable, detergent-resistant film as water evaporates; used as polish additive; lubricates wax particles for easy rub-out; contributes water repellency; used in spray-on/wipe-off auto polishes and cream shoe polishes; wh. liq.; dens. 8.30 lb/gal; visc. 300–700 cps max.; 60% silicone.

SM 2109. [GE] Silicone emulsion; for spray bottle auto polishes; disp. in water; 60% act.

SM 2112. [GE] 35% silicone emulsion; used in textiles as softener, hair mousse; disp. in water.

SM 2133, 2135. [GE] Silicone o/w emulsion; nonionic; for furniture, vinyl, auto polishes; liq.; 50% act.

SM 2140. [GE] 10,000 cs dimethyl polysiloxane resin; nonionic; mold release agent used in rubber and plastic prod. operations, in foundry release, aerosol spray starch; good lubricity; wh. liq.; odorless; b.p. 212 F; water-sol.; sp.gr. 1.01; dens. 8.25 lb/gal; visc. 1500 cps; 50% silicone.

SM 2154. [GE] Silicone emulsion; nonionic emulsifier; release agent; useful where rubber, plastic, and metal molded parts must be painted; also for foundry release; wh. to lt. yel. emulsion; easily diluted with water; visc. 1000 cps max.; 50% silicone; 54% total solids.

SM 2155. [GE] Silicone emulsion of dimethyl fluid [SF 96 (1000 cs)]; nonionic; see SM 2140; wh. liq.; dens. 8.25 lb/gal; visc. 1500 cps max.; 50% silicone.

SM 2162. [GE] Silicone dimethyl emulsion; nonionic; see SM 2140; wh. liq.; mild odor; b.p. 212 F; water-sol.; sp.gr. 1.04; dens. 8.25 lb/gal; visc. 1000 cps max.; flash pt. > 575 F; 50% silicone.

SM 2163. [GE] Silicone emulsion; nonionic; lubricant and release agent; used for rubber and plastic release, foundry release, softener/modifier for textiles, in polishes, and printing processes; wh. liq.; water-disp.; dens. 8.25 lb/gal; visc. 2500 cps max.; 60% silicone.

SMA® 1000. [Sartomer] 1:1 Styrene/maleic anhydride copolymer; anionic; soil release agent; used in ammoniacal water sol'n. in carpet shampoos, paints, inks, paper coatings, commercial laundries; in emulsion polymerization, temporary coatings, oven cleaners; as leveling resin for floor polishes; dispersant; powd.; 99% act.

SMA® 1440. [Sartomer] 1:1 Styrene/maleic anhydride copolymer partially esterified with org. alcohol; dispersant for titanium dioxide, other pigments in aq. media; used in paints, inks; flake; sol. in ammoniacal sol'n.

SMA® 1440H. [Sartomer] SMA 1440 in ammoniacal sol'n.; see SMA 1440; liq.; 35% act.

SMA® 2000. [Sartomer] Styrene/maleic anhydride copolymer; anionic; see SMA 1000; powd.; 99% act.

SMA® 2625. [Sartomer] 2:1 Styrene/maleic anhydride copolymer partially esterified with org. alcohol; dispersant; anti-resoil agent for carpet shampoo and carpet mfg.; powd.; sol. in ammoniacal sol'n.

SMA® 3000. [Sartomer] Styrene/maleic anhydride copolymer; anionic; see SMA 1000; powd.; 99% act.

SMA® 17352. [Sartomer] 1:1 Styrene/maleic anhydride half ester; anti-resoil agent for carpet shampoo and carpet mfg.; powd.; sol. in ammoniacal sol'n.; 99% act.

S-Maz® 20. [PPG-Mazer] Sorbitan laurate; nonionic; lubricant, antistat, textile softener, process defoamer, opacifier, coemulsifier, solubilizer, dispersant, suspending agent, coupler; prepares exc. w/o emulsions; with T-Maz Series used as o/w emulsifiers in cosmetics, food formulations, industrial oils, and household prods.; lipophilic; amber liq.; sol. in ethanol, naphtha; water-disp.; sp.gr. 1.0; visc. 4500 cps; HLB 8.6; acid no. 7 max.; sapon. no. 158–170; 100% conc.

S-Maz® 40. [PPG-Mazer] Sorbitan palmitate; nonionic; see S-Maz 20; tan flakes; partly sol. in hot ethanol, acetone, toluol, min. and veg. oils; m.p. 45–46 C; HLB 6.7; acid no. 7.5 max.; sapon. no. 140–150; 100% conc.

S-Maz® 60. [PPG-Mazer] Sorbitan stearate; nonionic; see S-Maz 20; cream flakes; m.p. 50–53 C; HLB 4.7; acid no. 10 max.; sapon. no. 147–157; 100% conc.

S-Maz® 65. [PPG-Mazer] Sorbitan tristearate; nonionic; see S-Maz 20; cream flakes; sp.gr. 1.0; m.p. 53–55 C; HLB 2.1; acid no. 15 max.; sapon. no. 176–188; 100% conc.

S-Maz® 67. [PPG-Mazer] Sorbitan monoisostearate; nonionic; see S-Maz 20; lt. yel. liq.; sol. in min. and veg. oils; water-disp; sp.gr. 1.0; visc. 1000 cps; HLB 4.7; acid no. 10 max.; sapon. no. 146–157; 100% conc.

S-Maz® 80. [PPG-Mazer] Sorbitan oleate; nonionic; see S-Maz 20; amber liq.; sol. see S-Maz 67; sp.gr. 1.0; visc. 1000 cps; HLB 4.3; acid no. 7.5 max.; sapon. no. 149–160; 100% conc.

S-Maz® 83R. [PPG-Mazer] Sorbitan sesquioleate; nonionic; solubilizer, emulsifier and dispersant; liq.; sol. in oils and solvs.; HLB 3.7; 100% conc.

S-Maz® 85. [PPG-Mazer] Sorbitan trioleate; nonionic; see S-Maz 20; amber liq.; sol. in toluol, naphtha, min. and veg. oils; water-disp.; sp.gr. 1.0; visc. 200 cps; HLB 1.8; acid no. 14 max.; sapon. no. 172–186; 100% conc.

S-Maz® 95. [PPG-Mazer] Sorbitan tritallate; nonionic; antistat, textile softener, lubricant, process defoamer, opacifier, coemulsifier; prepares w/o emulsions; together with T-Maz series, as o/w emulsifier in metalworking fluids and coolants, semi-syn. and oil-based metalworking lubricants and coolants, cosmetics, food formulations, industrial oils, and household prods.; amber; sol. in toluol, naphtha, min. oil, veg. oil; misc. with ethanol, acetone; disp. in water; sp.gr. 0.9; visc. 200 cps; HLB 1.7; acid no. 15 max.; sapon. no. 168–182; hyd. no. 46–60.

Snow White 200. [Unimin] 200 mesh mica; filler with exc. whiteness, high refractory char.; used in plastics, microwave cookware; off-wh.; sp.gr. 2.89; dens. 13.7 lb/ft^3; oil absorp. 84.0 g/100 g; ref. index 1.58; pH 5.10; Moh hardness 2.75; 57% SiO_2, 25.8% Al_2O_3.

Snow White 270. [Unimin] 270 mesh mica; filler with exc. whiteness, high refractory char.; used in underbody coatings, joint cement mfg.; off-wh.; sp.gr. 2.89; dens. 14.0 lb/ft^3; oil absorp. 76.0 g/100 g; ref. index 1.58; pH 5.75; Moh hardness 2.75; 56.4% SiO_2, 26.6% Al_2O_3.

Snow White 325. [Unimin] 325 mesh mica; reinforcing filler with exc. whiteness for industrial and trade sale paints, plastics; off-wh.; sp.gr. 2.89; dens. 12.2 lb/ft^3; oil absorp. 81.0 g/100 g; ref. index 1.58; pH 5.7; Moh hardness 2.75; 56.4% SiO_2, 26.6% Al_2O_3.

Socci 30. [Morton Int'l.] Antimicrobial for cordage, rope.

Socci 3500, 3500-WP. [Morton Int'l.] Antimicrobial for textile applics.

Socci 6462. [Morton Int'l.] Antimicrobial; mildew-resistant mold release compd.

Socci 7340. [Morton Int'l.] Antimicrobial for pharmaceutical deodorants.

Sochamine 2662. [Witco SA] Quat. ammonium compd.; catonic; textile softener; liq.; 75% conc.

Sochamine SG 8, 100. [Witco SA] Fatty acid, ethoxylated; nonionic; textile softener; antistat; paste.

Sodium Chlorite Sol'n. 50, Tech. [Olin] Sodium

chlorite; bacterial slimicide in paper mill systems; algicide in cooling towers; used in the electronics industry; as intermediate to generate sodium chloride, an oxidizing agent; wh. slightly cloudy liq.; dens. 11.7 lb/gal (35 C); cryst. pt. 25 C; 37% min. sodium chlorite.

Sodium Chlorite, Tech. [Olin] Sodium chlorite; see Sodium Chlorite Sol'n. 50, Tech.; wh. flakes; dens. 53 lb/ft^3 (loose); 69 lb/ft^3 (packed); 79% min. sodium chlorite.

Sodium Omadine®, 40% Aq. Sol'n. [Olin] Sodium 2-pyridinethiol-1-oxide in water; industrial microbiostat; chelating agent; used in aq. metal coolant and cutting fluids, latex emulsion, inks, fiber lubricants; amber liq.; mild odor; m.w. 149.2 (solid); sol. 53% in neut. water; 19% in 40A ethanol; 17% in dimethylsulfoxide; 13% in propylene glycol; sp.gr. 1.22; dens. 10.2 lb/gal; m.p. 250 C (dec.); pH 8.3 (10% in water); 40% min. act.

Sodium Omadine® Powd. [Olin] Sodium pyridinethiol 1-oxide; see Sodium Omadine 40% Aq. Sol'n.; yel. to tan hygroscopic powd.; m.w. 149.2; sol. see Sodium Omadine 40% Aq. Sol'n.; sp.gr. 1.167; dens. 0.5 g/ml; m.p. 250 C (dec.); pH ≈ 9.8 (10% in neut. dist. water); 90% min. sodium pyrithione.

Sofbon C-1-S. [Takemoto] Quat. ammonium compd.; cationic; softener for polyester jersey; paste.

Sofbon AF Series. [Takemoto] Cationic; softening agents.

Sofnon 105G. [Toho] Imidazoline deriv.; cationic; softener for textiles; wax; 80% conc.

Sofnon GF-2, -5. [Toho] Alkyl amide imidazoline compd. acetate; cationic; softener for glass fibers; liq.

Sof-Plus®. [Am. Ingredients] Hydrated emulsion of high-purity, distilled monoglycerides with sodium stearoyl lactylate and acetic and/or propionic acid; crumb softener in yeast-raised baked goods; wh. semisolid; iodine no. 2 max.; pH 3.5–4.5; 19.2% min. alpha monoester; 77% moisture.

Softal 300. [Tokai Seiyu] Polyamine; cationic; softener and lubricant for syn. fibers; liq.; 15% conc.

Softal A-3. [Tokai Seiyu] Aliphatic amine deriv.; cationic; softener for acrylic fibers; paste; 15% conc.

Softal MR-30. [Tokai Seiyu] Special polyamide deriv.; cationic/nonionic; softener for syn. fibers and blends; liq.

Soft Detergent 95. [Lion] alpha-Olefin sulfonate; anionic; emulsifier and dispersant for emulsion polymerization; powd.; 95% conc.

Softener 64. [Lonza] Textile softener; Gardner 2 color; acid no. 8.

Softener #5039. [Polymer Research] Nonionic softener for use with natural or syn. fabrics; amber liq.; sol. in water; dens. 8.49 lb/gal; pH 7.4 (2%); 100% act.

Softener #FS-18-2. [Polymer Research] Nonionic textile auxiliary imparting softness to fabrics; wh. translucent emulsion; 0.5–1.0 μ particle size; odorless; dilutable in water; dens. 8.3 lb/gal; 40% solids.

Softenol 3100. [Hüls] C_{12-18} fatty acid triglyceride; lubricant for machinery used in mfg. of foodstuffs; additive to textile and glass fiber finishing, sizing agent; block pallets; neutral odor and taste; sol. in MEK, MIBK, ethyl acetate; dens. 0.876 g/cm^3 (80 C); visc. 32.9 mm^2/s (40 C); m.p. 33–35 C; sapon. no. 235–245; biodeg.

Softenol 3107. [Hüls] C_7 fatty acid triglyceride; lubricant for machinery for confectionary and food industries; mold release agent in processing of plastics; additive to cutting oils, lacquer and varnish systems; antiblocking agent for artificial skins; liq.; neutral odor and taste; dens. 0.930 g/cm^3; visc. 20.6 mm^2/s; sapon. no. 385–395; biodeg.

Softenol 3108. [Hüls] C_{8-10} fatty acid triglyceride; lubricant for machinery used in food and confectionary industries and medical/technical equipment; antiblocking agent for artificial skins; oil in mfg. of aluminum foils; mold release agent for plastics; additive to cutting oils and lubricants; hydrophobing agent against moisture; mfg. of textile and glass-fiber sizings; liq.; neutral odor and taste; dens. 0.950 g/cm^3; visc. 31 mm^2/s; sapon. no. 325–345; biodeg.

Softenol 3114. [Hüls] C_{14} fatty acid triglyceride; compressing aid in tableting technical substances; hydrophobing agent against moisture; lubricant for machinery, lacquer and varnish formulations; additive to liq. lubricating systems; powd.; neutral odor and taste; dens. 8.73 g/cm^3 (80 C); visc. 11.2 mm^2/s (80 C); m.p. 52–55 C; sapon. no. 227–235; biodeg.

Softenol 3118. [Hüls] C_{18} fatty acid triglyceride; see Softenol 3114; microfine powd.; neutral odor and taste; dens. 0.863 g/cm^3 (80 C); visc. 17.1 mm^2/s (80 C); m.p. 69–71 C; sapon. no. 188–190; biodeg.

Softenol 3119. [Hüls] Triglyceride; release agent for prod. and processing of PS polymers; microfine powd.

Softenol 3178. [Hüls] C_{8-18} fatty acid triglyceride; emollient, lubricant; paste; neutral odor and taste; dens. 0.878 g/cm^3 (80 C); visc. 31.5 mm^2/s (40 C); m.p. 37–40 C; sapon. no. 240–255.

Softenol 3408. [Hüls] C_{8-10} fatty acid-1,2-propanediol ester; additive in mfg. of textile and glass fibers; lubricant for cutting devices; mold release agent in processing of plastics; liq.; neutral odor and taste; dens. 0.920 g/cm^3; visc. 10.9 mm^2/s; flash pt. 193 C; sapon. no. 320–340; biodeg.

Softenol 3701. [Hüls] Partial glyceride; release agent for prod. and processing of PS polymers; paste.

Softenol 3819. [Hüls] C_{6-10} fatty acid triglyceride; compressing aid in mfg. of dry substances; carrier; lubricant in cutting devices; raw material in preparation of lubricants; emollient; liq.; neutral odor and taste; dens. 0.964 g/cm^3; visc. 22.6 mm^2/s; flash pt. 219 C; sapon. no. 350–450; biodeg.

Softenol 3829. [Hüls] C_{8-10} fatty acids triglyceride, modified; oil in mfg. of aluminum foils; lubricant for machinery, medical/technical equipment, instruments; additive to cutting oils and lubricants; liq.; neutral odor and taste; dens. 1.012 g/cm^3; visc. 230 mm^2/s; flash pt. 265 C; sapon. no. 400–450; biodeg.

Softenol 3900. [Hüls] Glycerol monostearate; nonionic; emulsifier, lubricant, antistat, demolding and antiblocking agent, greaser, plasticizer, transparency and pigment dispersant used in mfg. and processing of plastics, textiles, leather, oils and polishing materials; powd.; neutral odor and taste; dens. 0.893 g/cm³ (80 C); visc. 22.9 mm²/s (80 C); m.p. 58–65 C; sapon. no. 155–175; biodeg.

Softenol 3925, 3945, 3960. [Hüls] Glyceryl stearate; release agent for prod. and processing of PS polymers; powds. except 3945 (microcryst. powd.).

Softenol 3991. [Hüls] Glycerol monostearate; nonionic; see Softenol 3900; liq.; neutral odor and taste; dens. 0.914 g/cm³ (80 C); visc. 34.5 mm²/s (80 C); m.p. 65 C; sapon. no. 155–165; biodeg.; 90% 1-monoester.

Softenol 3995. [Hüls] Glyceryl stearate; see Softenol 3118; powds., 3945, 3995 are microfine powds.

Softex K-304, K-363. [Kao] Polyamide deriv., long chain; cationic; textile softener; liq.

Softigen 701. [Hüls] Glyceryl ricinoleate; nonionic; emollient; refatting and skin protecting agent; emulsifier; personal care prods. and pharmaceuticals; yel.-wh. clear oily liq.; char. odor; sol. in ether, benzene, toluene, xylene, chlorofoam, and dichlorethylene; dens. 0.979–0.981; visc. 640–660 cps; HLB 4.5; sapon. no. 155–170; 100% conc.

Softigen 767. [Hüls] PEG-6 caprylic/capric glycerides; nonionic; refatting and wetting agent, solubilizer, emollient used in cosmetics and pharmaceuticals; detergents; yel. visc. oily liq.; char. odor; sol. in water, acetone, ethyl and butyl acetate; dens. 1.068 g/ml; visc. 110–130 cps; HLB 19; acid no. 1 max.; sapon. no. 90–100; surf. tens. 29.7 dynes/cm (1%); 100% conc.

Softisan 100. [Hüls] Trilaurin; nonionic; emollient, ointment base for personal care prods. and pharmaceutical industry; wh. solid; neutral odor and taste; ref. index 1.4490–1.4510 (40 C); sol. in benzene, ether, xylene, toluene, chloroform, CCl_4, dioxane; sp.gr. 0.950–0.980; m.p. 33–36 C; visc. 30 cps (40 C); solid. pt. 29–33 C; sapon. no. 230–240.

Softisan 133. [Hüls] Hydrog. coco-glycerides; hard wax used in cosmetic and pharmaceutical preparations; wh. blocks; neutral odor and taste; sol. in benzene, toluene, acetone, chloroform, petrol. spirit; insol. in water; m.p. 32–34 C; solid. pt. 29–32 C; sapon. no. 240–255; ref. index 1.445–1.449 (50 C).

Softisan 134. [Hüls] Hydrog. coco-glycerides; see Softisan 133; wh. blocks; neutral odor and taste; sol. see Softisan 133; m.p. 33–36 C; solid. pt. 27–32 C; sapon. no. 225–235; ref. index 1.445–1.449 (50 C).

Softisan 138. [Hüls] Hydrog. coco-glycerides; see Softisan 133; wh. blocks; neutral odor and taste; sol. see Softisan 133; m.p. 37–39 C; solid. pt. 32–36 C; sapon. no. 220–230; ref. index 1.445–1.449 (50 C).

Softisan 142. [Hüls] Hydrog. coco-glycerides; see Softisan 133; wh. blocks; neutral odor and taste; sol. see Softisan 133; m.p. 42–44 C; solid. pt. 37–42 C; sapon. no. 220–230; ref. index 1.445–1.449 (50 C).

Softisan 154. [Hüls] Hydrog. coco-glycerides; see Softisan 133; wh. flakes; neutral odor and taste; sol. see Softisan 133; m.p. 53–58 C; solid. pt. 48–53 C; sapon. no. 200–215; ref. index 1.445–1.460 (50 C).

Softisan 378. [Hüls] Caprylic/capric/stearic triglyceride; nonionic; ointment base, emollient, moisturizer, stabilizer; neutral ointment base with good skin compatibility and resorption chars., for the prep. of nonaq. ointments and creams; waxy; acid no. < 1; sapon no. 240–255; 100% act.

Softisan 601. [Hüls] Glyceryl cocoate, hydrog. coconut oil, and cetereth-25; nonionic; emollient; cosmetic and pharmaceutical base for o/w prods.; wh.-yel. solid; sp.gr. 0.97 (60 C); sapon. no. 120–140; 100% conc.

Softisan 649. [Hüls] Caprylic/capric/isostearic adipic triglycerides; emollient, ointment base for creams, lipsticks, other decorative cosmetics; lanolin substitute; paste; almost odorless; sol. in ether, chloroform; misc. with fats, oils.

Softisan Hard Fats. [Hüls] Glycerol trilaurate/stearate; nonionic; stabilizer for emulsions, sticks; solid, pastille; m.p. 30–55 C; HLB 1.0; 100% conc.

Soi Bact G. [Soitem] Hexahydrotriazine; bactericide for flowable pesticide formulations, metalworking industries; liq.

Soi Mul 60/C. [Soitem] Fatty acid esters; rust inhibitor for metalworking industry; liq.; water-sol.

Soi Mul 440. [Soitem] Sodium petrol. sulfonates; anionic; additive for metal working industries; thick liq.; 60% conc.

Soitem 8 FL/N. [Soitem] Phosphate derivate, amminic salt; anionic; wetting agent and dispersant for pesticide formulations; thick liq.; 100% conc.

Soitem 13 FL/N. [Soitem] Phosphate derivate, amminic salt; anionic; see Soitem 8 FL/N; wax; 100% conc.

Soitem 70. [Soitem] Sodium polycarboxylate; dispersing suspending agent for W.P. pesticides; powd.

Soitem 207. [Soitem] Sodium methane naphthalene sulfonate; dispersing suspending agent for W.P. pesticides; powd.

Soitem 480. [Soitem] Ethoxylated, modified fatty acid and org. sulfate; anionic/nonionic; stabilizer and emulsifier for pesticides; thick liq.; 95% conc.

Soitem SFL/1. [Soitem] Polyglycol ethers; nonionic; wetting agent and dispersant for pesticide formulations; liq.; 100% conc.

Sokalan® CP 2, CP 2 Powd. [BASF AG] Sodium salt of a maleic anhydride/methyl vinylether copolymer; dispersant, anti-incrustation agent; 35% liq. and 92% powd. resp.

Sokalan® CP 5, CP 5 Powd. [BASF AG] Sodium salt of a modified polyacrylic acid; dispersant and anti-incrustation agent; 40% liq. and 92% powd. resp.

Sokalan® CP 7. [BASF AG] Sodium salt of a modified polyacrylic acid; aux. in phosphate-reduced or phosphate-free detergents to improve primary and sec. (antiredeposition and anti-incrustation) washing effects; dispersant; liq.; 40% conc.

Sokalan® CP 8. [BASF AG] Sodium salt of a modified polyacrylic acid; see Sokalan CP 7; liq.; 35% conc.

Sokalan® CP 9. [BASF AG] Sodium salt of maleic acid/olefin copolymer; dispersant for org. and inorg. solids; liq.; 25% conc.

Sokalan® CP 10. [BASF AG] Sodium salt of a modified polyacrylic acid; dispersant for solids and for softening water; liq.; 45% conc.

Sokalan® CP 10S. [BASF AG] Modified polyacrylic acid; dispersant for solids and for softening water; liq.; 50% conc.

Sokalan® CP 11. [BASF AG] Sodium salt of a maleic acid/olefin copolymer; dispersant for org. and inorg. solids; liq.; 40% conc.

Sokalan® CP 12S. [BASF AG] Maleic acid/olefin copolymer; dispersant and anti-incrustation agent; aux. for phosphate-reduced or phosphate-free detergents; liq.; 50% conc.

Sokalan® CP 13S. [BASF AG] Sodium salt of a modified polyacrylic acid; dispersant for org. and inorg. solids; liq.; 35% conc.

Sokalan® CP 45. [BASF AG] Sodium salt of a modified polyacrylic acid, only partly neutralized; dispersant and anti-incrustation agent; liq.; 45% conc.

Sokalan® HP 50. [BASF AG] Polyvinylpyrrolidone; antiredeposition inhibitor for detergents; dispersant; powd.; 95% conc.

Sokalan® PA 15, PA 20. [BASF AG] Sodium salt of polyacrylic acid; aux. for fludification of detergent slurries and the improvement of detergent powd. structure; dispersant for solids and for softening water; liq.; 45% conc.

Sokalan® PA 20 PN. [BASF AG] Sodium salt of polyacrylic acid, only partly neutralized; see Sokalan PA 15; liq.; 45% conc.

Sokalan® PA 25 PN. [BASF AG] Sodium salt of polyacrylic acid, only partly neutralized; see Sokalan PA 15; liq.; 54% conc.

Sokalan® PA 30. [BASF AG] Sodium salt of polyacrylic acid; see Sokalan PA 15; liq.; 45% conc.

Sokalan® PA 40, PA 40 Powd. [BASF AG] Sodium salt of polyacrylic acid; see Sokalan PA 15; 35% liq. and 91% powd. resp.

Sokalan® PA 50. [BASF AG] Sodium salt of polyacrylic acid; dispersant for org. and inorg. solids; liq.; 40% conc.

Sokalan® PA 70 PN. [BASF AG] Sodium salt of polyacrylic acid, only partly neutralized; dispersant for org. and inorg. solids; liq.; 30% conc.

Sokalan® PA 80. [BASF AG] Sodium salt of polyacrylic acid; aux. in phosphate-reduced or phosphate-free detergents to improve primary and sec. (antiredeposition and anti-incrustation) washing effects; dispersant; liq.; 35% conc.

Sokalan® PA 80 S, PA 110 S. [BASF AG] Polyacrylic acid; see Sokalan PA 80.

Solamine AC 20. [Seppic] Quat. ammonium; cationic; softener for cotton; liq.; 30% conc.

Solamine BO5. [Seppic] Imidazoline, modified; cationic; softener; paste; 90% conc.

Solamine HY. [Seppic] Quat. ammonium methosulfate; cationic; hydrophilic softener for toweling; paste; 100% conc.

Solan. [Croda] PEG-60 lanolin; nonionic; hydrophilic emollient, conditioner, thickener, and superfatting agent; foam stabilizer, plasticizer, and humectant used in personal care prods.; solubilizer for incorporating oils and fragrances in aq. and aq./alcoholic systems; o/w emulsifier; Gardner 11 max. wax; sol. in water/alcohol, most polar and nonpolar solvs.; HLB 16.2; m.p 46–54 C; acid no. 2 max.; sapon. no. 8–16; pH 5.5–7.0 (1% aq.); 100% act.

Solan 50. [Croda] PEG-60 lanolin; nonionic; see Solan; Gardner 11 max. liq.; acid no. 2 max.; sapon. no. 8 max.; pH 5.5–7.0 (1% aq.); 50% act.

Solan E. [Croda] POE lanolin; nonionic; emollient and emulsifier; solid; water-sol.; HLB 16.2; 100% conc.

Solan E 50. [Croda] POE lanolin; skin conditioning additive, protective colloid; liq.

Solan X. [Croda] POE lanolin; conditioner for skin and hair care prods.; waxy solid.

Solangel 401. [Croda] PEG-75 lanolin; nonionic; emulsifier, humectant for soap; nonionic; liq.; HLB 16.0; 50% conc.

Solangel 501. [Croda] PEG-50 lanolin; nonionic; see Solangel 401; liq.; HLB 15.8; 50% conc.

Solangel 851. [Croda] PEG-85 lanolin; nonionic; see Solangel 401; liq.; HLB 16.4; 50% conc.

Solanos AD Supra. [Seppic] Modified alkylsulfate; anionic; softener; paste; 60% conc.

Solanos T. [Seppic] Tallow deriv.; anionic; softening agent; paste; 45% conc.

Solar 25. [Guelph] Coconut oil fatty acid amine condensate and amine sulfonate; nonionic/anionic; wetting agent, detergent, air entraining agent in home and industrial prods.; straw visc. liq.; sp.gr. 1.05; pour pt. 0 C; surf. tens. 34.5 dynes/cm (0.1% aq.); pH 8.5 (1%); 100% act.

Solar Air. [Guelph] Coconut fatty acid amine condensate; nonionic; detergent, air-entraining agent used in building industry; yel. liq.; ref. index 1.3427; sp.gr. 1.0015; visc. 5 cps; 25% act.

Solar Air-2. [Guelph] Coconut fatty acid amine condensate; nonionic; see Solar Air; lt. liq.; ref. index 1.3520; sp.gr. 1.0040; visc. 10 cps.

Solar Air-4. [Guelph] Coconut fatty acid amine condensate; nonionic; see Solar Air; lt. liq.; ref. index 1.3695; sp.gr. 1.0090; visc. 22 cps.

Solar CI-387. [Guelph] Imidazoline hydrochloride; cationic; wax-rinse base, corrosion inhibitor used in car cleaning applic.; water repellent film for hard surfs.; amber liq.; mild odor; sp.gr. 0.951; visc. 40 cps; pH 5–7; 52% act.

Solar CO. [Guelph] Coconut oil fatty acid amine condensate; nonionic; stabilizer, detergent, intermediate; cosmetic and detergent formulations incl. liq. dishwashing and other high sudsing liq. where suds stabilization is desired; intermediate in mfg. of modified amides; dedusting and suds stabilization of dry detergents; lt. straw visc. liq.; slt. odor; sp.gr. 1.02; flash pt. 166 C; pour pt. 23 C; pH 9.5 (1% sol'n.); 100% act.

Solar NP. [Guelph] Nonyl phenoxy POE ethanol; nonionic; detergent, emulsifier, wetting agent, dispersant, penetrant; base for cleaning compds.; cosmetics; dairy cleaning; laundry and dry cleaning; leather wetting, defatting, softener; metal cleaning, pickling, plating; paint, plastics, and rubber dispersing; textile scouring, dyeing, desizing, bleaching aid;

agric. wetting and emulsification; brewing industry; visc. clear liq.; odorless; sol. in water, ethylene glycol, ethanol, CCl_4, xylene; sp.gr. 1.06; pour pt. 37 F; pH 7.0 (1%); flash pt. 535–555 F; cloud pt. 125–133 F (1%); 100% act.

Solar Regular. [Guelph] Coconut oil fatty acid amine condensate modified with DEA oleate; nonionic; detergent, wetting agent, corrosion and rust inhibitor; lubricant, thickener; base for cleaning compd.; degreasing, textiles, laundering; making of cutting and sol. oils; amber visc. liq.; sweet odor; water-sol.; sp.gr. 1.07; pour pt. –4 C; pH 9.5 (1%); flash pt. 179 C; 100% act.

Solar TE. [Guelph] Specially modified amide produced primarily from coconut oil; nonionic; detergent; hard surf. cleaners, built formulas, liq. floor scrubbing compds.; liq. steam cleaners; textile fulling and scouring of woolens and raw stock; textile lubricant; couples most alkalis; amber liq.; pour pt. 20 F; pH 9.15 (1%); flash pt. 360 F; fire pt. 370 F.

Solarchem® 0. [CasChem] Octyl dimethyl PABA; selective uv absorber for suncare and cosmetic prods.; clear liq.; odorless.

Solem ATH. [Solem] Alumina trihydrate; filler providing exceptional depth of pattern and translucency in cultured onyx; provides clarity and high filler loading; flame and smoke suppression and better pH stability; wh. powd.

Sole-Onic CDS. [Hodag] Diethylene glycol monolaurate; anionic; defoamer, emulsifier for emulsion paints, industrial oil emulsions; liq.; 100% conc.

Sole-Terge TS-2-S. [Hodag] Sodium 2-ethylhexyl sulfate; anionic; wetting agent, penetrant, industrial penetrant for textiles; liq.; 35% conc.

Solidester. [Robeco] Ester-hydrog. piscine oil; nonionic; emollient, lubricant, moisturizer for skin prods.; solid.

Sollagen®. [Hormel] Soluble animal collagen; natural humectant for cosmetic applics. incl. skin care and body lotions, and hair care prods.; pH 3.6–3.9; 1% sol'n.; 7.5–9.5% total solids.

Solon. [Eastern Color] EDTA and EDTA modified; chelating of all metals, iron specialized; water-sol.

Soloron. [Rona] Rutile titanium dioxide coated mica; pearlescent for pressed powds., eye makeup, frosted pearl effects in cosmetics.

Solricin® 135. [CasChem] Potassium ricinoleate; anionic; emulsifier, mild germicide, glycerized rubber lubricant, foam stabilizer in foamed rubber; making of cutting and sol. oils, household and cosmetic prods.; Gardner 2 liq.; sp.gr. 1.034; dens. 8.6 lb/gal; visc. 0.9 stokes; 32% act.

Sol-Speedi-Dri, Auto-Dri. [Engelhard] Attapulgite clays; min. floor absorbents for promoting safety and cutting floor maintenance costs in metalworking and automotive industries, animal barns, and butcher shops; dens. 29–33 and 33–36 lb/ft^3 resp.; fineness 12/45 and 8/30 mesh resp.

Solulan® 5. [Amerchol] Laneth-5, ceteth-5, oleth-5 and steareth-5; nonionic; emulsifier, wetting agent, dispersant, lubricant, solv., conditioner, plasticizer, emollient used in personal care and dermatology prods.; foam stabilizer for detergent systems; amber semisolid; faint pleasant odor; sol. @ 5% in anhyd. ethanol; disp. in water; acid no. 3 max.; sapon. no. 14 max.; pH 4.5–7.0 (10% aq.); 100% conc.

Solulan® 16. [Amerchol] Laneth-16, ceteth-16, oleth-16 and steareth-16; nonionic; see Solulan 5; lt. tan waxy solid; sol. @ 5% in water, 30% ethanol; HLB 15; cloud pt. 64–70 C; acid no. 3 max.; sapon. no. 8 max.; pH 4.5–7.5 (10% aq.); 100% conc.

Solulan® 25. [Amerchol] Laneth-25, ceteth-25, oleth-25 and steareth-25; nonionic; yel. waxy solid; faint pleasant odor; sol. see Solulan 16; HLB 16; cloud pt. 85–92 C; acid no. 3 max.; sapon. no. 8 max.; pH 4.5–7.5 (10% aq.); 100% conc.

Solulan® 75. [Amerchol] PEG-75 lanolin; nonionic; see Solulan 5; lt. yel.-amber waxy solid; faint pleasant odor; sol. see Solulan 16; HLB 15; cloud pt. 80–87 C; acid no. 3 max.; sapon. no. 10–20; pH 4.5–7.5 (10% aq.); 100% conc.

Solulan® 97. [Amerchol] Polysorbate 80 acetate, cetyl acetate and acetylated lanolin alcohol; nonionic; see Solulan 5; lt. amber visc. liq.; faint pleasant odor; sol. @ 5% in water, ethanol; HLB 15; acid no. 3 max.; sapon. no. 110–130; pH 4.5–7.5 (10% aq.); 100% conc.

Solulan® 98. [Amerchol] Polysorbate 80, acetylated lanolin alcohol, and cetyl acetate; nonionic; see Solulan 5, also aids pearling of stearic acid emulsions; lt. amber visc. liq.; faint pleasant odor; sol. @ 5% in water, ethanol; HLB 13; acid no. 3 max.; sapon. no. 65–75; pH 4.5–7.5 (10% aq.); 100% conc.

Solulan® C-24. [Amerchol] Choleth-24 and ceteth-24; nonionic; see Solulan 5; off-wh. to pale yel. waxy solid; faint pleasant odor; sol. @ 5% in water; HLB 14; cloud pt. 88–95 C; acid no. 1; sapon. no. 3 max.; 100% conc.

Solulan® L-575. [Amerchol] PEG-75 lanolin; nonionic; see Solulan 5; lt. yel.-amber visc. liq.; faint pleasant odor; sol. @ 5% in water, 30% ethanol; HLB 15; cloud pt. 80–87 C; acid no. 1 max.; sapon. no. 10 max.; 50% conc.

Solulan® PB-2. [Amerchol] PPG-2 lanolin ether; nonionic; spreading agent, dispersant, plasticizer; emollient and conditioner for personal care prods., detergents, pharmaceuticals, waxes, polishes, leather treatment; amber semisolid, liq.; sol. in castor oil, IPM, IPP, anhyd. ethanol; HLB 8.0; acid no. 3 max.; sapon. no. 7 max.; 100% conc.

Solulan® PB-5. [Amerchol] PPG-5 lanolin ether; nonionic; see Solulan PB-2; lt. amber clear visc. liq.; sol. see Solulan PB-2; HLB 10.0; acid no. 2 max.; sapon. no. 6 max.; 100% conc.

Solulan® PB-10. [Amerchol] PPG-10 lanolin ether; nonionic; see Solulan PB-2; straw clear heavy-visc. liq.; sol. see Solulan PB-2; HLB 12.0; acid no. 1 max.; sapon. no. 4 max.; 100% conc.

Solulan® PB-20. [Amerchol] PPG-20 lanolin ether; nonionic; see Solulan PB-2; lt. straw clear med-visc. liq.; sol. see Solulan PB-2; HLB 14.0; acid no. 0.75 max.; sapon. no. 3 max.; 100% conc.

Solupret Brands. [Hoechst-Celanese AG] High polymer silicon sol'ns. or emulsions; water repellent finishing for textiles.

Solusoft 350. [Soluol] Silicone emulsion; nonionic; fiber lubricant for finishing of nonwoven fabrics; inclusion in finish formulations to prevent or minimize development of needle cutting or edge melt during fabrication of end prod.; provides lubricity, pleasing hand, improved fabric tear strength, wrinkle recovery, resists retention of chlorine; acts as process aid to provide release of chemical binders from processing equipment; good high temp. resistance; emulsion; water-sol.; 30% conc.

Sol-U-Tein E. [Fanning] Egg protein conc.; powd. binder, protein source for cosmetic, pharmaceutical, and industrial powd. coatings.

Sol-U-Tein EA. [Fanning] Egg albumen; binder, coagulant; used in pharmaceuticals and personal care prods.; dye mordant in textiles, adhesives, veneers, sizing and making papers; gilding leather; bookbinding; food applic.; yel. powd., 100% thru 80 mesh; 90% thru 100 mesh; bland odor; pH 6.5–8.0; 75% ovalbumin, ovoconalbumin, ovomucoid, ovomucin, ovoglobulin, lysozyme, and avidin.

Solvenol®. [Hercules Ltd.] Terpene hydrocarbon solv.; constituent of industrial cleaners, deodorants; liq.

Solvent 80. [W.R. Grace] 80% methyl acetate, 20% methanol; solv. for paint remover compds. and lacquer solv.; clear water-wh.; sp.gr. 0.900–0.910; b.p. 53–57 C.

Solwax C-24. [Van Schuppen] Polyethoxylated cholesterol; nonionic; o/w emulsifier; stabilizer for w/o emulsions; solubilizer for steroids in aq. systems; used in alcoholic and hydrophilic cosmetics; firm waxy solid; faint, pleasant odor; sol. in water and alcohol; m.p. 45–48 C; cloud pt. 85–95 C (1% sol'n. in 5% NaCl aq.); acid no. 1.0 max.; sapon no. 5.0 max.; 100% act.

Solwax L20. [Van Schuppen] Alkoxylated deriv. of pharmaceutical lanolin; nonionic; emulsifier, foam stabilizer, softener; o/w emulsifier; used in hydrophilic cosmetics; in shampoos and shaving compositions, to enhance foam stability; in liq. detergent compositions for softening and skin protection; lt. yel. firm waxy solid; faint, pleasant odor; dissolves in water and alcohol; soften. pt. 45–55 C; cloud pt. 75–80 C (1% sol'n in 5% NaCl aq.); acid no. 3.0 max.; sapon no. 10–30; 100% act.

Solwax LG 35. [Van Schuppen] Alkoxylated deriv. of lanolin alcohols; nonionic; emulsifier, conditioner, emollient, solubilizer in alcoholic and hydrophilic cosmetics; gelling properties; emulsifier for o/w emulsions; yel.-brnsh. firm waxy solid; faint pleasant odor; sol. see Solulan C-24; soften. pt. 40–50 C; acid no. 5.0 max.; sapon. no. 5–15; cloud pt. 75–85 C (1%); 100% act.

Sonojell® Min. Jellies. [Witco] Wh. min. oil, petrolatum, and wax; emollient bases for cleansing creams, cosmetics, pharmaceuticals; m.p. 100–135 F.

Sonojell® No. 4, No. 9. [Witco] Wh. min. oil, petrolatum, paraffin waxes; min. jelly; Lovibond color 0.5Y; visc. 2.6–5.7 Cst (100 C); m.p. 38–52 and 42–49 C resp.

Sopralub ACR 265. [Henkel-Nopco] Polyoxyalkylene fatty glycerides; nonionic; textile softener, plasticizer; liq.; 30% conc.

Sopralub ACR 275. [Henkel-Nopco] Polyoxyalkylene fatty glycerides; nonionic; softener, plasticizer for textile finishes and coatings; fluid cream; 30% conc.

Soprofor 3D33. [Rhone-Poulenc SpA] POE arylphenol phosphate; anionic; dispersing/suspending agent for pesticide flowable formulations; thick liq.; 99% conc.

Soprofor AMC. [Rhone-Poulenc SpA] Ethoxylated polyaryl phenol; nonionic; compatibility/anticaking agent for pesticide wettable powds.; powd.; 60% conc.

Soprofor DO/64. [Rhone-Poulenc SpA] Block polymer dioleate; nonionic; antifoamer; intermediate for industrial applics.; liq.; 100% conc.

Soprofor FL. [Rhone-Poulenc SpA] POE arylphenol phosphate, amine salt; anionic; dispersing/suspending agent for pesticide flowable formulations; thick liq.; 99% conc.

Soprofor NP/20, /30, /100. [Rhone-Poulenc SpA] Nonoxynol-20, -30, -100; nonionic; finishing auxs. used as antifoaming agents, detergents, dispersants, emulsifiers, softeners, and dyeing assistants; solid, solid, powd. resp.; 100% conc.

Soprofor PA/17, PA/19. [Rhone-Poulenc SpA] Alkylpolyethoxyether phosphate; anionic; see Soprofor FL; liq.; 100% conc.

Soprofor PE/220. [Rhone-Poulenc SpA] PEG 200 monolaurate; nonionic; visc. depressant for PVC emulsions; liq.; 100% conc.

Soprofor PE/400. [Rhone-Poulenc SpA] PEG 400 monolaurate; nonionic; see Soprofor PE/220; liq.; 100% conc.

Soprofor PL/Series. [Rhone-Poulenc SpA] POE/POP block polymer; nonionic; detergent, emulsifier, wetting and antistat, defoamer, deduster; 100% conc.

Soprofor PS/17, PS/19. [Rhone-Poulenc SpA] Alkylpolyethoxyether phosphate salt form; anionic; see Soprofor FL; liq.; 80% conc.

Soprofor S/20. [Rhone-Poulenc SpA] Sorbitan monolaurate; nonionic; emulsifier, lubricant and softener for the textile industry; liq.; 100% conc.

Soprofor S/60. [Rhone-Poulenc SpA] Sorbitan monostearate; nonionic; see Soprofor S/20; solid; 100% conc.

Soprofor S/65. [Rhone-Poulenc SpA] Sorbitan tristearate; nonionic; see Soprofor S/20; solid; 100% conc.

Soprofor S/80. [Rhone-Poulenc SpA] Sorbitan monooleate; nonionic; see Soprofor S/20; liq.; 100% conc.

Soprofor S/85. [Rhone-Poulenc SpA] Sorbitan trioleate; nonionic; see Soprofor S/20; liq.; 100% conc.

Soprofor T/20. [Rhone-Poulenc SpA] PEG-20 sorbitan monolaurate; nonionic; emulsifier, solubilizer, antistat and lubricant for textile industry; liq.; 100% conc.

Soprofor T/60. [Rhone-Poulenc SpA] PEG-20 sorbitan monostearate; nonionic; see Soprofor T/20; paste; 100% conc.

Soprofor T/65. [Rhone-Poulenc SpA] Sorbitan tristearate, ethoxylated; nonionic; see Soprofor T/20; solid; 100% conc.

Soprofor T/80. [Rhone-Poulenc SpA] PEG-20 sorbitan monooleate; nonionic; see Soprofor T/20; liq.; 100% conc.

Soprofor T/85. [Rhone-Poulenc SpA] PEG-20 sorbitan trioleate; nonionic; see Soprofor T/20; liq.; 100% conc.

Soprofor VR/50. [Rhone-Poulenc SpA] POE stearic mono diglycerides; nonionic; thickener for shampoo; visc. control; liq.; 50% conc.

Sorba. [Croda Ltd.] Lanolin; emollient for cosmetic and pharmaceutical applics.

Sorban AL. [Witco SA] Sorbitan ester; nonionic; antifoaming; wax; 100% conc.

Sorban AO 1. [Witco SA] Sorbitan ester; nonionic; oil additive, corrosion inhibitor; liq.; 100% conc.

Sorban AST. [Witco SA] Sorbitan ester; nonionic; emulsifier for oils and greases; antifoamer, corrosion inhibitor; cosmetic applics.; flakes; 100% conc.

Sorban CO. [Witco SA] Sorbitan ester; nonionic; antifoamer; corrosion inhibitor; liq.; 100% conc.

Sorbanox AL. [Witco SA] Sorbitan ester, ethoxylated; nonionic; detergent used in veg. cleaning, antistat; liq.; 100% conc.

Sorbanox AOM. [Witco SA] Sorbitan ester, ethoxylated; nonionic; emulsifier; solubilizer of vitamin; liq.; 100% conc.

Sorbax PML-20. [Chemax] Polysorbate 20; nonionic; o/w emulsifier, solubilizer for perfumes and flavors; liq.; water-sol.; HLB 16.7; 100% conc.

Sorbax PMO-5. [Chemax] Polysorbate 81; nonionic; see Sorbax PML-20; liq.; water-sol.; HLB 10.0; 100% conc.

Sorbax PMO-20. [Chemax] Polysorbate 80; nonionic; see Sorbax PML-20; liq.; water-sol.; HLB 15.0; 100% conc.

Sorbax PMP-20. [Chemax] Polysorbate 40; nonionic; see Sorbax PML-20; liq.; water-sol.; HLB 15.6; 100% conc.

Sorbax PMS-20. [Chemax] Polysorbate 60; nonionic; see Sorbax PML-20; liq.; water-sol.; HLB 14.9; 100% conc.

Sorbax PTO-20. [Chemax] Polysorbate 85; nonionic; see Sorbax PML-20; liq., gel; water-sol.; HLB 11.0; 100% conc.

Sorbax PTS-20. [Chemax] Polysorbate 65; nonionic; see Sorbax PML-20; solid; water-sol.; HLB 10.5; 100% conc.

Sorbax SML. [Chemax] Sorbitan monolaurate; nonionic; emulsifier, surfactant for o/w emulsion stabilizers and thickeners used in cosmetic and textile industries; liq.; sol. in oil and org. water systems; HLB 8.6; 100% conc.

Sorbax SMO. [Chemax] Sorbitan monooleate; nonionic; see Sorbax SML; liq.; sol. see Sorbax SML; HLB 4.3; 100% conc.

Sorbax SMP. [Chemax] Sorbitan monopalmitate; nonionic; see Sorbax SML; solid; sol. see Sorbax SML; HLB 6.7; 100% conc.

Sorbax SMS. [Chemax] Sorbitan monostearate; nonionic; see Sorbax SML; solid; sol. see Sorbax SML; HLB 4.7; 100% conc.

Sorbax STO. [Chemax] Sorbitan trioleate; nonionic; see Sorbax SML; liq.; sol. see Sorbax SML; HLB 1.8; 100% conc.

Sorbax STS. [Chemax] Sorbitan tristearate; nonionic; see Sorbax SML; solid; sol. see Sorbax SML; HLB 2.1; 100% conc.

Sorbeth 40HO. [Croda Ltd.] Sorbitol PEG-40 hexaoleate; nonionic; emulsifier/dispersant for agric. chemicals and the oil industry; liq.; HLB 10.5; 100% conc.

Sorbeth 55HO. [Croda Ltd.] Sorbitol PEG-55 hexaoleate; nonionic; emulsifier/dispersant for agric. chemicals and oil industry; liq.; HLB 11.5; 100% conc.

Sorbistat. [Pfizer] Sorbic acid; preservative; broad spectrum against yeast and fungi; sol. in alcohol, propylene glycol; sparingly sol. in water.

Sorbistat-K. [Pfizer] Potassium sorbate; see Sorbistat; extremely water-sol.

Sorbit P. [Ciba-Geigy Switzerland] Alkyl naphthalene sulfonate; anionic; wetting agent and foamer; emulsifier, detergent; germicides, textile leveling; powd.; 65% conc.

Sorbon S-20. [Toho] Sorbitan monolaurate; nonionic; emulsifier, dispersant for w/o emulsion; liq.; 100% conc.

Sorbon S-40. [Toho] Sorbitan monopalmitate; nonionic; used with Sorbon T series; solid; 100% conc.

Sorbon S-60. [Toho] Sorbitan monostearate; nonionic; used with Sorbon T series; solid; 100% conc.

Sorbon S-80. [Toho] Sorbitan monooleate; nonionic; used with Sorbon T series; liq.; 100% conc.

Sorbon T-20. [Toho] PEG-20 sorbitan monolaurate; nonionic; emulsifier, dispersant, solubilizer for o/w emulsion; liq.; HLB 16.7; 100% conc.

Sorbon T-40. [Toho] POE sorbitan monopalmitate; nonionic; used with Sorbon S series; liq.; HLB 15.7; 100% conc.

Sorbon T-60. [Toho] POE sorbitan monostearate; nonionic; used with Sorbon S series; liq.; HLB 15.3; 100% conc.

Sorbon T-80. [Toho] POE sorbitan monooleate; nonionic; used with Sorbon S series; liq. 100% conc.

Sorcinol 35, 40. [Climax Performance] Sodium ricinoleate; mold release; water-sol.

Sorgen 30. [Rilsan] Sorbitan sesquioleate; nonionic; antifoamer and emulsifier for foods, cosmetics, and pharmaceutical prods.; liq.; HLB 3.7; 100% conc.

Sorgen 40. [Rilsan] Sorbitan monooleate; nonionic; see Sorgen 30; liq.; HLB 4.3; 100% conc.

Sorgen 50. [Rilsan] Sorbitan monostearate; nonionic; see Sorgen 30; flake; HLB 4.7; 100% conc.

Sorgen 90. [Rilsan] Sorbitan monolaurate; nonionic; see Sorgen 30; liq.; HLB 8.6; 100% conc.

Sorgen S-30-H. [Rilsan] Sorbitan sesquioleate; nonionic; see Sorgen 30; liq.; HLB 3.7; 100% conc.

Sorgen S-40-H. [Rilsan] Sorbitan monooleate; nonionic; see Sorgen 30; liq.; HLB 4.3; 100% conc.

Soromine AL. [GAF] Complex fatty amido surfactant; amphoteric; softener, lubricant for fibers, leather; finishing agent for processing equip.; wh. paste, essentially odorless; disp. in warm water; pH 6.0–7.5 (10% disp.); 26% act. min.

Soromine AT. [GAF] Complex fatty amido compd.; amphoteric; softener, lubricant, antistat for animal fibers and leather; finishing agent for textiles; paste; > 20% act.

So/San 30M. [Stepan] Softener-sanitizer for fabrics; liq.; 46–51% conc.

So/San 66M. [Stepan] Softener-sanitizer for fabrics; liq.; 64–66% conc.

Sotex 3CW. [Morton Int'l.] Fatty acid ester, long chain; nonionic; dispersant for paints and coatings, ink vehicle; wetting agent for dyes; visc. liq.; sol. in aliphatic and aromatic hydrocarbons; sp.gr. 0.930–0.960; b.p. > 200 C; 100% act.

Sotex COS #2. [Morton Int'l.] Fatty acid ester, long chain; cationic; dispersant for paints and coatings; surface treating pigments; liq.; oil-sol.; sp.gr. 0.900–0.940; b.p. > 200 C; 100% act.

Sotex CW. [Morton Int'l.] Fatty acid ester, long chain; nonionic; visc. depressant and stabilizer in vinyl plastisols; dispersant for paints and coatings; pigments in plasticizers; cosolvent; visc. liq.; sol. in hydrocarbons, ketones; sp.gr. 0.970–1.010; b.p. > 200 C; 100% act.

Sotex N. [Morton Int'l.] Fatty acid ester, long chain; nonionic; dispersant for polar solvs., and inks; visc. liq.; sol. in polar solvs.; sp.gr. 1.02–1.06; b.p. > 200 C; 100% act.

SoyaLec WDF-FG. [Canada Packers] Modified natural lecithin; food grade dispersant, emulsifier; liq.; water-disp.

SP-25. [Schenectady] Alkyl-phenol formaldehyde resin; tackifier for NBR adhesives and sealants; nonheat reactive; yel. to amber lumps; sol. in aromatic and chlorinated hydrocarbons, ketones, esters; sp.gr. 1.07; soften. pt. (R&B) 100–110 C.

SP-553. [Schenectady] Terpene phenolic resin; tackifier for SBR, NBR, NR, and CR compds. and adhesives; lengthens bonding range of Neoprene adhesives; yel. flakes; sol. in aliphatic and aromatic hydrocarbons, ketones, esters, and chlorinated hydrocarbons; sp.gr. 1.00; soften. pt. (R&B) 110–120 C.

SP-560. [Schenectady] Terpene phenolic resin; tackifier for SBR, NR, NBR, CR, and BR compds. and adhesives; lengthens bonding range of Neoprene adhesives with min. effect on heat resistance; lt. amber flakes; sol. in aliphatic and aromatic hydrocarbons, ketones, esters, and chlorinated hydrocarbons; sp.gr. 1.10; soften. pt. (R&B) 146–156 C.

SP-1044. [Schenectady] Alkyl phenol formaldehyde resin; crosslinking agent for butyl rubber, esp. curing bladders for automobile tires; requires addition of Lewis acid (Neoprene W or stannous chloride); lump resin; sol. in aromatic, aliphatic, and chlorinated hydrocarbons, ketones, and esters; sp.gr. 1.03–1.07; m.p. 60–70 C; 7.0–9.5% methylol content.

SP-1045. [Schenectady] Alkyl phenol-formaldehyde resin; crosslinking agent for butyl rubber, esp. curing bladders for automobile tires; requires addition of Lewis acid (Neoprene W or stannous chloride); amber lumps; sp.gr. 1.05; m.p. 60–66 C; 8–11% methylol content.

SP-1055. [Schenectady] Bromomethylated and methylolated alkyl phenol formaldehyde resin; crosslinking agent for unsat. elastomers (butyl or EPDM); used in vulcanizing butyl rubber curing bladders for automobile tires; requires no additional source of labile halogen; rdsh. lumps; sp.gr. 1.10; m.p. 57–66 C; 9.0–12.5% methylol content; 4% max. bromine content.

SP-1056. [Schenectady] Bromomethylated and methylolated alkyl phenol formaldehyde resin; crosslinking agent for unsat. elastomers (butyl, EPDM, and natural rubber); used to partially crosslink elastomer in pressure-sensitive tapes; requires no additional source of labile halogen; rdsh. lumps; sp.gr. 1.10; m.p. 57–66 C; 7.5–11.0% methylol content; 6% min. bromine content.

SP-1068. [Schenectady] Alkyl phenol formaldehyde resin; tackifier for BR, CR, SBR, NBR, NR compds. and in cements; nonheat reactive; lt. amber flakes; sol. in aromatic and aliphatic hydrocarbons, ketones, and esters; sp.gr. 1.02; soften. pt. (R&B) 85–95 C.

SP-1077. [Schenectady] Alkyl phenol formaldehyde resin; tackifier for BR, CR, SBR, NBR, NR, and EPDM elastomers; allows lowest loading of resin component to achieve desired tack level; lt. amber flakes; sol. in aliphatic and aromatic hydrocarbons, ketones, and esters; sp.gr. 1.02; soften. pt. (R&B) 92–103 C.

SP-6700. [Schenectady] Oil-modified phenol formaldehyde resin; resin in NBR, SBR, NR, and CR as plasticizer during processing; acts as hardener with 8% hexa after cure; use where scorching problems preclude use of hexa-containing materials such as SP-6600; brn. flakes; sol. in ketones, alcohols, and esters; sp.gr. 1.16; soften. pt. (R&B) 95 C.

SP-6701. [Schenectady] Oil-modified phenol formaldehyde resin; resin in SBR as plasticizer during processing and hardener after cure with 8% hexa; use where scorch problems preclude use of hexa-containing materials such as SP-6601; amber flakes; sol. in alchols, ketones, and esters; sp.gr. 1.16; soften. pt. (R&B) 95 C.

Span® 20. [ICI Am.] Sorbitan laurate; nonionic; emulsifier, stabilizer, thickener, lubricant, softener, antistatic agent; foods, pharmaceuticals, cosmetics, cleaning compds., textiles; amber liq.; sol. (@ 1%) in IPA, perchloroethylene, xylene, cottonseed oil, min. oil; sol. (hazy) in propylene glycol; visc. 4250 cs; HLB 8.6; 100% act.

Span® 40. [ICI Am.] Sorbitan palmitate; nonionic; see Span 20; tan solid; sol. (@ 1%) in IPA, xylene; sol. (hazy) in perchloroethylene; HLB 6.7; pour pt. 48 C; 100% act.

Span® 60. [ICI Am.] Sorbitan stearate; see Span 20; also dispersant for inorg. pigments in thermoplastics; tan solid; sol. (@1%): sol. in IPA; sol. (hazy) in perchloroethylene, xylene; HLB 4.7; pour pt. 53 C;

100% act.

Span® 65. [ICI Am.] Sorbitan tristearate; nonionic; see Span 20; cream solid; sol. (@ 1%) in IPA, perchloroethylene, xylene; HLB 2.1; pour pt. 53 C; 100% act.

Span® 85. [ICI Am.] Sorbitan trioleate; nonionic; see Span 20; liq.; amber; sol. (@ 1%) in IPA, perchloroethylene, xylene, cottonseed and min. oils; visc. 210 cs; HLB 1.8; 100% act.

Spark-L® HPG. [Miles Biotech] Pectinase; enzyme for depectinization and pulp washing; liq.

Spectra-Pearl®. [Van Dyk] Colored titanium dioxide/mica; pearlescent giving sparkle, luster, and color to cosmetic eye, face, lip, and body makeup.

Spectra-Sorb® UV 9. [Am. Cyanamid] Benzophenone-3; uv lt. absorber; used in sunscreen, cosmetic, and pharmaceutical formulations; pale cream to off-wh. powd.; m.w. 228.25; sol. (g/100 ml solv.): 100 g propylene glycol monostearate, 57 g in ethoxyethyl alcohol, 47 g in diglycol stearate, 15 g in IPM, 11.3 g in soybean oil, 0 g in water; sp.gr. 1.324; m.p. 63.0–64.5 C; 100% assay.

Spectra-Sorb® UV 24. [Am. Cyanamid] Benzophenone-8; uv lt. absorber for cosmetics, sunscreens; stabilizes formulations; yel. free-flowing powd.; m.w. 244.25; sol. (g/100 ml solv.): 38 g in propylene glycol stearate, 34 g in diglycol stearate, 21.8 g in 95% ethanol, 17 g in IPA, 10 g in IPM; sp.gr. 1.382; m.p. 68–70 C; 100% assay.

Spectra-Sorb® UV 5411. [Am. Cyanamid] Octrizole; uv lt. absorber for cosmetics; stabilizes formulations, protects colors; off-wh. free-flowing powd.; m.w. 323.44; sol. (g/100 ml solv.): 65 g in benzene, 27.2 g in ethyl acetate, 0 g in water; sp.gr. 1.18; m.p. 101–105 C.

Spectratech® CM 10540. [Quantum/USI] Dual uv inhibitor for extrusion molding.

Spectratech® CM 10608, 10778, 10779, 11013, 11056, 11513, 11643. [Quantum/USI] Antiblock concs. for film applics.; FDA approved; CM 11513 also slip agent.

Spectratech® CM 11014, 11126, 11172, 11174, 11194. [Quantum/USI] Slip concs. for film applics.

Spectratech® CM 11045, 11638, 77242. [Quantum/USI] Antistat concs. for film and sheet; CM 11638 and 77242 for elec. pkg.; CM 11045 also for inj. and blow molding.

Spectratech® CM 11053, 11489, 11591. [Quantum/USI] Flame retardant concs.; CM 11053 for LDPE film, 11489 for HDPE film, 11591 for LDPE, HDPE, PP film and wire and cable.

Spectratech® CM 11246. [Quantum/USI] UV absorber for pkg., agric. films, molded items.

Spectratech® CM 11340, KM 11264. [Quantum/USI] BHT; antioxidant conc. for HDPE cereal liners.

Spectratech® CM 11357, 11367. [Quantum/USI] UV inhibitor for agric. film, pool coverings, greenhouse film.

Spectratech® CM 11615. [Quantum/USI] UV inhibitor for PP slit tape, HDPE, and PP molded goods.

Spectratech® CM 11616. [Quantum/USI] UV inhibitor and antioxidant for blk. PE and PP mulch film.

Spectratech® CM 11681. [Quantum/USI] UV inhibitor for greenhouse film, PP slit tape, molded goods.

Spectratech® CM 11698. [Quantum/USI] Optical brightener for film applics.

Spectratech® CM 11704, 11720. [Quantum/USI] UV inhibitor for HDPE slit tape, greenhouse and construction film.

Spectratech® CM 11715. [Quantum/USI] UV inhibitor for HDPE slit tape, pkg., and molded goods.

Spectratech® PM 11607, 11725. [Quantum/USI] Processing aid for LLDPE and HDPE extrusion.

Spermafol® 5200. [Sherex] Composed chiefly of esters of sat., long-chain fatty acids and fatty alcohols; wax used as replacement for industrial-grade sperm oil-derived waxes; used in the mfg. of lubricants, textile sizes and finishing aids, leather treatments, drawing compds., and wax specialties; wh. wax-like solid flake; m.w. 487; sp.gr. 0.8324 (60/25 C); m.p. (CT) 48 C; acid no. 3.0 max.; iodine no. 2.0 max.; sapon. no. 105–125; 49% unsaponifiable material.

Spermwax. [Robeco] Cetyl esters wax NF; syn. spermwax used in toiletries, cosmetics, dermatologicals as stiffening agent, slip and visc. aid; solid.

Spezyme AA. [Finnsugar] Alpha-amylase (bacterial); enzyme for textile desizing, starch liquefaction processes; liq.

Spezyme BBA. [Finnsugar] Beta-amylase (barley); enzyme for food processing in high maltose syrups, malt diastatic supplementation; liq.

Spezyme GA. [Finnsugar] Glucoamylase (fungal); enzyme for food processing in saccharification of glucose syrups, low-carbohydrate beer; amber clear liq.

Spezyme IGI. [Finnsugar] Glucose isomerase (Streptomyces); enzyme for food processing; continuous catalytic isomerization of glucose to fructose in deep bed reactors; nonswelling; noncompressible; bisulfite resistant; inert granulate.

Spiroflor. [Henkel] 3-Ethyl-2,4-dioxaspiro (5.5) undec-8-ene; complex, natural odor fragrance raw material.

Spotleak® 1001. [Pennwalt] 80% t-Butyl mercaptan, 20% dimethyl sulfide; odorant for natural gas to permit detection of leaks; negligible sol. in water; sp.gr. 0.816 (15 C); dens. 6.8 lb/gal; f.p. < –56.7 C; cloud pt. –45.5 C max.; flash pt. (TCC) < –18 C; 39% sulfur.

Spotleak® 1003. [Pennwalt] 76.5% t-Butyl mercaptan, 23.5% isopropyl mercaptan; see Spotleak 1001; negligible sol. in water; sp.gr. 0.810 (15 C); dens. 6.75 lb/gal; f.p. < –45.5 C; cloud pt. –45.5 C max.; flash pt. (TCC) < –18 C; 35.8% min. sulfur.

Spotleak® 1007. [Pennwalt] 80% t-Butyl mercaptan, 20% methyl ethyl sulfide; see Spotleak 1001; negligible sol. in water; sp.gr. 0.815 (15 C); dens. 6.79 lb/gal; f.p. < –45.5 C; cloud pt. –45.5 C max.; flash pt. (TCC) < –18 C; 36% sulfur.

Spotleak® 1009. [Pennwalt] 79% t-Butyl mercaptan, 15% isopropyl mercaptan, 6% normal propyl mercaptan; see Spotleak 1001; negligible sol. in water; sp.gr. 0.812 (15 C); dens. 6.76 lb/gal; f.p. < –45.5 C;

cloud pt. –45.5 C max.; flash pt. (TCC) < –18 C; 35% min. sulfur.

Spotleak® 1044. [Pennwalt] 75% Tetrahydrothiophene, 25% t-butyl mercaptan; see Spotleak 1001; negligible sol. in water; sp.gr. 0.948 (15 C); dens. 7.90 lb/gal; f.p. < –45.5 C; cloud pt. –28.8 C max.; flash pt. (TCC) < –18 C; 36% sulfur.

Spotleak® 2323. [Pennwalt] 50% t-Butyl mercaptan, 50% methyl ethyl sulfide; see Spotleak 1001; negligible sol. in water; sp.gr. 0.826 (15 C); dens. 6.90 lb/gal; f.p. < –45.5 C; cloud pt. –45.5 C max.; flash pt. (TCC) < –18 C; 38% sulfur.

Sprayfilm 87-911. [Ram] Film-forming release agent for nongloss parts mfg.; dens. 7.8 lb/gal; flash pt. (Seta CC) 62 F.

Sprayset MEKP. [Witco/Argus] MEK peroxide in anhyd. ethyl acetate; catalyst for spray gun applics. in the curing of polyesters; clear low-visc. liq.; flam.; 5.25% act. oxygen.

Spraywax 60-X5. [Ram] Lt.-duty liq. wax release agent; buffs to high polish; dens. 6.6 lb/gal; flash pt. (Seta CC) 37 F.

Spreading Agent ET0672. [Croda Ltd.] Complex alkyl polyglycol ether; o/w emulsifier; bath oil spreading agent and dispersant; pale straw clear liq.

Squalane. [Croda Ltd.] Squalane; emollient, penetrant used in personal care prods.; water-wh. clear liq.; odorless.

SR-201. [Sartomer] Allyl methacrylate; with 50–185 ppm hydroquinone inhibitor; monomeric acrylic ester capable of polymerizing to hard, infusible resin that is water wh., clear, and glass-like; as comonomer with vinyl-type monomers to produce crosslinked polymers; unsymmetrical crosslinking agent where a two-stage polymerizing or drying action is desired; APHA 50 clear liq.; pungent, disagreeable odor; m.w. 126; sp.gr. 0.929–0.931; dens. 7.7 lb/gal; visc. 12–14 cps; b.p. 150 C; flash pt. (PMCC) 88 F; ref. index 1.4343–1.4350.

SR-204. [Sartomer] Diallyl fumarate; with 60–100 ppm hydroquinone inhibitor; allyl monomer, polymerizes to a hard, infusible resin; crosslinking agent which can raise heat resistance of plastic compositions when properly incorporated; in coatings, polyester additives, casting and hardening formed plastisols, brittle as a homopolymer; lt. to dk. brn. clear liq.; offensive odor; m.w. 196; sol. in alcohols, ethers, ketones, esters, aromatic and aliphatic hydrocarbons; sp.gr. 1.045–1.060; dens. 8.8 lb/gal; visc. 3.0 cps; b.p. 140 C; flash pt. (PMCC) 157 F; sapon. no. 98; ref. index 1.4670.

SR-205. [Sartomer] Triethylene glycol dimethacrylate; with 80 ± 20 ppm hydroquinone inhibitor; noncorrosive, low visc., high boiling crosslinking monomeric ester having polymerization chars. similar to methyl methacrylate; in vinyl plastisols reduces initial visc. and oil extractability, and improves ultimate hardness, heat distort., hot tear strength, and stain resistance; in cast acrylic sheet and cast acrylic rod improves heat distort., machineability, solv., hardness, impact, craze and scratch resistance, etc., e.g., in the button, watch crystal, and contact lens industries, and acrylic teeth producers; in syn. rubber and ion exchange resins; in dental compositions; colorless clear liq.; mild, pleasant odor; m.w. 286; sol. in alcohol, ether, ketones, esters, and low m.w. aliphatic hydrocarbons; sp.gr. 1.072; dens. 8.96 lb/gal; visc. 7.5 cps; flash pt. (PMCC) 146 F; ref. index 1.4595.

SR-206. [Sartomer] Ethylene glycol dimethacrylate; with 40–150 ppm hydroquinone inhibitor; crosslinking agent used in emulsion polymerization, cast acrylic sheet, fiberglass-reinforced polyesters, ion exchange resins, rubber compd.; APHA 50 clear bright liq.; mild musty odor; m.w. 198.21; sol. in alcohols, ethers, ketones, esters, aromatic and aliphatic hydrocarbons; insol. in water; dens. 8.75 lb/gal; visc. 3.4 cps; b.p. 260 C; ref. index 1.4522–1.4532; flash pt. 155 F (PMCC).

SR-209. [Sartomer] Tetraethylene glycol dimethacrylate; with 75 ± 25 ppm hydroquinone inhibitor; crosslinking agent used in castings, plastisols, coatings, fibers, papers, and other fabrications; APHA 150 clear liq.; mild pleasant odor; m.w. 330.37; sol. in alcohols, esters, ethers, ketones, and aromatic hydrocarbons; insol. in water; dens. 9.0 lb/gal; visc. 6.4 cps; b.p. 220 C; acid no. 0.5; ref. index 1.4605–1.4610; flash pt. 161 F (PMCC).

SR-210. [Sartomer] PEG dimethacrylate; with 75 ± 25 ppm hydroquinone inhibitor; see SR-209; colorless clear liq.; mild, pleasant odor; m.w. 330.37; sol. in alcohols, ethers, ketones, esters, and aromatic hydrocarbons; dens. 9.0 lb/gal; visc. 6.4 cps; b.p. > 200 C; acid no. 0.5; ref. index 1.4615; flash pt. 166 F (PMCC).

SR-212. [Sartomer] 1,3-Butylene glycol diacrylate; with 500 ± 20 ppm hydroquinone inhibitor; low visc. monomer; curing agent; polymerizes to hard, insol., infusible, thermoset resin; exothermic polymerization reaction, initiated thermally or by common free radical initiators, i.e., high energy and uv radiation, peroxide compds.; straw clear liq.; musty odor; sol. in aromatics, esters, and ketones; limited sol. in aliphatics; sp.gr. 1.042–1.052; dens. 8.7 lb/gal; visc. 24 cps; flash pt. (PMCC) 133 F; acid no. 1.4; 97% react. esters.

SR-213. [Sartomer] 1,4-Butanediol diacrylate; with 100–150 ppm hydroquinone inhibitor; cross-linking agent used in inks, adhesives, textile prod. modifiers, photoresists, modifiers for castings, polyesters, fiberglass, or radiation cured prods.; APHA 150 clear liq.; pleasant odor; m.w. 198; sp.gr. 1.055–1.105; dens. 8.9 lb/gal; visc. 29 cps; flash pt. 147 F (PMCC).

SR-220. [Sartomer] Cyclohexyl acrylate (monomer); curing agent; APHA 200 clear liq.; sweet, pungent odor; m.w. 154; sp.gr. 0.975; dens. 8.12 lb/gal; flash pt. (PMCC) 162 F; 96% react. esters.

SR-230. [Sartomer] Diethylene glycol diacrylate; with 60–100 ppm hydroquinone inhibitor; crosslinking agent used in radiation curing; used in coatings and adhesives; colorles clear liq.; mild musty odor; m.w. 214; sol. in alcohol, ethers, ketones, aliphatic and aromatic hydrocarbons; insol. in water; sp.gr. 1.1110; dens. 9.25 lb/gal; visc. 24 cps; b.p. >

200 C; ref. index 1.4595; flash pt. 172 F (PMCC).

SR-238. [Sartomer] 1,6-Hexanediol diacrylate; with 100 ppm hydroquinone inhibitor; see SR-213; APHA 250 clear liq.; musty odor; m.w. 226; sp.gr. 1.010; dens. 8.45 lb/gal; visc. 12 cps; flash pt. 152 F (PMCC).

SR-239. [Sartomer] 1,6-Hexanediol methacrylate; with 100 ± 25 ppm hydroquinone inhibitor; cross-linking agent used in casting compds., glass fiber-reinforced plastics, adhesives, coatings, ion-exchange resins, textile prods., plastisols, dental polymers, rubber compding.; APHA 500 clear liq.; bland odor; m.w. 254; sol. in org. solv.; insol. in water; sp.gr. 0.998; dens. 8.31 lb/gal; flash pt. 157 F (PMCC).

SR-247. [Sartomer] Neopentyl glycol diacrylate (monomer); with 225 ± 25 ppm p-methoxyphenol inhibitor; curing agent; diluent for uv irradiation-cured systems (coatings, printing inks, etc.); APHA 150 clear liq.; sharp, pungent odor; m.w. 212; sp.gr. 1.020–1.03; dens. 8.4 lb/gal; visc. 10 cps; ref. index 1.4542; flash pt. (PMCC) 184 F.

SR-259. [Sartomer] PEG 200 diacrylate; with 100–150 ppm hydroquinone inhibitor; cross-linking agent used in radiation-cured coatings, inks, adhesives, textile prods., photoresists; APHA 200 clear liq.; musty odor; m.w. 302; sp.gr. 1.110–1.120; dens. 9.3 lb/gal; visc. 64 cps; flash pt. > 220 F (PMCC).

SR-268. [Sartomer] Tetraethylene glycol diacrylate; with 100–150 ppm hydroquinone inhibitor; see SR-259; APHA 300 clear liq.; musty, acrylic odor; m.w. 302; sp.gr. 1.115–1.125; dens. 9.3 lb/gal; visc. 50 cps; flash pt. 185 F (PMCC).

SR-272. [Sartomer] Triethylene glycol diacrylate (monomer); curing agent; APHA 150 clear liq.; musty odor; m.w. 258; b.p. > 100 (1 mm Hg); sp.gr. 1.099–1.107; dens. 9.16 lb/gal; flash pt. (PMCC) 146 F; 98% react. esters.

SR-285. [Sartomer] Tetrahydrofurfuryl acrylate; with 500 ppm hydroquinone inhibitor; monofunctional monomeric ester of acrylic acid, polymerizes to hard, infusible, insol. thermoset resin; used in uv irradiated coatings, because of low visc. and sensitivity to uv; high boiling with low potential for crosslinking helpful in providing more flexible coating; curing agent; APHA 500 clear liq.; m.w. 156.18; dens. 8.88 lb/gal; flash pt. (PMCC) 148 F; acid no. 0.7.

SR-295. [Sartomer] Pentaerythritol tetraacrylate; with 300–400 ppm MEHQ inhibitor; cross-linking agent in adhesives, coatings, inks, textile prods., photoresists, castings, or as modifiers for polyester or polymers; APHA 250 clear liq.; musty odor; m.w. 352; sp.gr. 1.185–1.200; dens. 9.9 lb/gal; visc. 700 cps; m.p. 15–18 c; flash pt. 158 F (PMCC).

SR-335. [Sartomer] N-lauryl acrylate (monomer); curing agent; APHA 30 clear liq.; bland odor; m.w. 184; sp.gr. 0.870; dens. 7.4 lb/gal; flash pt. (PMCC) 212 F; 95% react. esters.

SR-339. [Sartomer] 2-Phenoxyethyl acrylate; with 100 ppm HQ inhibitor; high boiling monomeric ester of acrylic acid; polymerization initiated by heat, catalysis, and/or radiation; copolymerization with other acrylic-type monomers is easily achieved; curing agent; yel. clear liq.; mild, musty; m.w. 192.2; sol. in aromatic, aliphatic, and chlorinated hydrocarbons, ketones, ethyl ether; sp.gr. 1.090; dens. 9.1 lb/gal; visc. 6 cps; b.p. 110 C; flash pt. (PMCC) 117 F; acid no. 0.06.

SR-340. [Sartomer] Phenoxyethyl methacrylate (monomer); curing agent; APHA 25 clear liq.; sp.gr. 1.0687.

SR-344. [Sartomer] PEG 400 diacrylate (monomer); curing agent; clear liq.; mild odor; water-sol.; sp.gr. 1.1165; visc. 60 cps; flash pt. (PMCC) 51 F; 99% react. esters.

SR-348. [Sartomer] Ethoxylated bisphenol A dimethacrylate; with 350 ± 50 ppm MEHQ inhibitor; cross-linking agent used as additive in peroxide-cured rubber compds., irradiated coatings, and polyesters; adhesives; Gardner 10 clear liq.; musty odor; m.w. 452; sp.gr. 1.110–1.130; dens. 9.2 lb/gal; visc. 910 cps; acid no. < 0.5; flash pt. 168 F (PMCC).

SR-349. [Sartomer] Ethoxylated bisphenol A diacrylate (monomer); curing agent; clear liq.; mild odor; m.w. 424; sp.gr. 1.132; dens. 9.4 lb/gal; visc. 2500 cps; flash pt. (PMCC) > 150 F; 98% react. esters.

SR-350. [Sartomer] Trimethylolpropane trimethacrylate; with 80 ± 20 ppm hydroquinone inhibitor; cross-linking agent; APHA 150 clear liq.; mild odor; m.w. 338.39; sol. in alcohols, ethers, ketones, esters, aliphatic hydrocarbons; insol. in water; dens. 8.8 lb/gal; visc. 85–92 cps; b.p. > 200 C; acid no. 0.5; ref. index 1.4700; flash pt. 149 F (PMCC).

SR-351. [Sartomer] Trimethylolpropane triacrylate; with 100–150 ppm hydroquinone inhibitor; high boiling monomeric ester polymerized by common free radical initiators; curing agent; yel. clear, bright liq.; pungent odor; m.w. 296.31; dens. 9.2 lb/gal; visc. 150 cps; b.p. 200 C; flash pt. (PMCC) 188 F; acid no. 0.7; ref. index 1.4735–1.4750.

SR-355. [Sartomer] Di-trimethylol propane tetraacrylate (monomer); low-toxicity curing agent; APHA 100 clear liq.; mild odor; m.w. 410; sp.gr. 1.1077; visc. 752 cps.

SR-395. [Sartomer] Isodecyl acrylate (monomer); curing agent; APHA 50 clear liq.; mild odor; m.w. 212; b.p. 121 (760 mm); sp.gr. 0.870–0.880; dens. 7.4 lb/gal; visc. 2 cps; ref. index 1.441; flash pt. (PMCC) 185 F; 99% react. esters.

SR-399. [Sartomer] Dipentaerythritol monohydroxypenta acrylate; with 300–500 ppm p-methoxyphenol inhibitor; high m.w., multifunctional acrylate monomer susceptible to polymerization by radiation and uv initiation; curing agent; Gardner 2–5 visc. liq.; mild odor; m.w. 524.51; sol. in esters, ketones; dens. 9.5 lb/gal; visc. 4400 cps; flash pt. (PMCC) 240 F; acid no. 1.4.

SR-440. [Sartomer] Isooctyl acrylate (monomer); curing agent; APHA 50 clear liq.; mild odor; m.w. 184; sp.gr. 0.880; 99% react. esters.

SR-444. [Sartomer] Pentaerythritol triacrylate; with 300–400 ppm MEHQ inhibitor; cross-linking agent used in adhesives, coatings, inks, textile prods., photoresists, castings, modifiers for polyester, fiberglass, or polymers; APHA 250 clear liq.; musty odor; m.w. 298; sp.gr. 1.160–1.190; dens. 9.9 lb/gal; visc. 700 cps; flash pt. 174 F (PMCC).

SR-454. [Sartomer] Ethoxylated trimethylolpropane triacrylate (monomer); curing agent; APHA 250 clear liq.; mild odor; m.w. 428; sp.gr. 1.105–1.115; dens. 9.25 lb/gal; visc. 50–100 cps; ref. index 1.471; flash pt. (PMCC) 188 F.

SR-506. [Sartomer] Isobornyl acrylate (monomer); curing agent; clear liq.; m.w. 208; sp.gr. 0.985; dens. 8.2 lb/gal; visc. 9.5 cps; ref. index 1.4749; flash pt. (PMCC) 240 F; 98% react. esters.

SR-7475. [Firestone] Stereospecific butadiene/styrene copolymer; blendable modifier for thermoplastic resins and asphalt; APHA 5 small pellet; sp.gr. 0.96; melt index 12; ref. index 1.5495; tens. str. 1000 psi; hardness (Shore A) 85; 43% bound styrene.

SRF-1501. [Schenectady] Resorcinol formaldehyde preformed resin sol'n.; bonding agent used in rubber compds. to improve adhesion; red to brnsh. flakes; sol. in alcohols, ketones, esters, and water; sp.gr. 1.37; soften. pt. 100–110 C.

SS 4098. [GE] Solv. sol'n. of silicone polymers and other film-forming components; water repellent; useful for pressurized pkg. to impart water repellency to textiles and leather; may be applied by dipping, swabbing, or mechanical spraying; does not discolor most fabrics; cures without heat by reacting with the moisture in the air; colorless to lt. straw; dilutable in common aliphatic and aromatic hydrocarbons; sp.gr. 0.90; visc. < 10 cps; flash pt. (CC) 70 F; 50% active; solv.: min. spirits.

SS 4129. [GE] Solv. sol'n. of silicone polymers and other film-forming components; water repellent; imparts water repellency to leather goods; cures without heat by reacting with moisture in the air; liq.; lt. yel.; dilutable in common aliphatic and aromatic hydrocarbons; sp.gr. 1.01; visc. 750 cps; flash pt. (CC) 150 F min.; 90% min. active; solv.: min. spirits.

SS 4177. [GE] Silicone polymer blend; release agent for rubber, plastics, and resinous materials, esp. thermoset materials (epoxy resins, polyester resins, urethane elastomers); also used in polishes; clear liq.; dens. 7.9 lb/gal; visc. 20 cps; flash pt. (PMCC) 34 C; 50% silicone in mixed aliphatic/aromatic solvs.

SS 4267. [GE] Silicone fluid; emollient for water-resistant cosmetics and suntan prods.; sol. in aliphatic, aromatic solvs., IPP, IPM, some lt. min. oils; 100% act.

S.S.T.® Sump Saver Tablets. [Angus] Tris (hydroxymethyl) nitromethane; antibacterial agent, preservative for metalworking fluids; 1 oz wafers; water-sol.

Stabilisol S. [Hoechst-Celanese AG] Polysulfide; anionic; stabilizer for hydrogen peroxide bleaching baths; liq.

Stabilizer 9-A. [Givaudan] Mono and diisopropylated meta and para cresols; antioxidant for fatty acids and derivs. used in antistats; tan to brnsh. liq.; phenolic odor; b.p. 90–120 C (5 mm Hg); sol. in org. solv. and alkaline media; sp.gr. 0.950–0.965.

Stabilizer 1097. [Mobay] Org. acid chloride in butyl acetate; stabilizer used to extend the pot life of the bonding agent system/plastisol mixt.; dens. ca 0.95 g/cm³; flash pt. 27 C; 20% solids.

Stabilizer C, I-FF. [Bayer] Diphenyl thiourea and 2-phenylindole resp.; org. stabilizers for PVC; highly effective in concs. as low as 0.3–1.0%; normally restricted to emulsion PVC; used with calcium/zinc compds., suitable for the stabilization of suspension or bulk polymerized PVC food pkg. goods; used in the prod. of unplasticized PVC food pkg. goods e.g., film and bottles; calendered food pkg. film from unplasticized PVC; tarpaulins, industrial gloves, soft floor coverings, unplasticized PVC pipe, sections and sheet; wh. powds.;

Stabilizer T. [Grünau] Protein fatty acid condensate; anionic; stabilizer for bleaching; paste; 50% conc.

Stabland®. [Capital City] Fractionated partially hydrog. soybean oil; specialty oil; Lovibond R1.5 max.; m.p. < 40 F; iodine no. 107–115; sapon. no. 189–195 max.

Staflex 500. [Reichhold] n-Octyl, n-decyl phthalate; plasticizer for outdoor applic.; compatible with NC, PS, neoprene; good low-temp. properties and permanence; APHA 25; m.w. 418; sp.gr. 0.970; dens. 8.1 lb/gal; visc. 140 cSt; pour pt. –40 F; ref. index 1.4820; flash pt. 440 F; cloud pt. –2 F.

Staflex 550. [Reichhold] Butylene glycol polyester; plasticizer for general vinyl applics. and offering permanence, heat stability, low oil extraction; elec. properties and humidity stability in refrigerators and elec. applics.; APHA 175; m.w. 2500; sp.gr. 1.077; dens. 9.0 lb/gal; visc. 2500 cSt; pour pt. 20 F; ref. index 1.4658; flash pt. > 500 F; cloud pt. 20 F.

Staflex 802. [Reichhold] Butylene glycol polyester; plasticizer with elec. properties; useful in wire and cable industry; APHA 100; m.w. 1900; sp.gr. 1.084; dens. 9.0 lb/gal; visc. 3620 cSt; pour pt. 5 F; ref. index 1.4808; flash pt. 485 F; cloud pt. 5 F.

Staflex 803. [Reichhold] Polymeric plasticizer; APHA 100; m.w. 1800; sp.gr. 1.091; dens. 9.1 lb/gal; visc. 4000 cSt; pour pt. 25 F; ref. index 1.4810; flash pt. > 500 F; cloud pt. 25 F.

Staflex 804. [Reichhold] Butylene glycol polyester; plasticizer with low volatility and oil extraction resistance and elec. properties and rubber migration resistance; used in wire and cable industries; APHA 100; m.w. 1900; sp.gr. 1.070; dens. 8.9 lb/gal; visc. 2270 cSt; pour pt. 0 F; ref. index 1.4724; flash pt. > 500 F; cloud pt. 0 F.

Staflex 809. [Reichhold] Butylene glycol polyester; plasticizer offering low odor and odor; offers high resistance to PS migration, humidity stability and elec. properties; used in refrigerator gasketing; APHA 100; m.w. 2100; sp.gr. 1.081; dens. 9.0 lb/gal; visc. 3100 cSt (0 C); pour pt. 10 F; ref. index 1.4748; flash pt. 510 F; cloud pt. 10 F.

Staflex 812. [Reichhold] Polymeric plasticizer; exhibits efficiency, permanence, and low rubber migration in vinyl compds.; outdoor vinyl applic. that require weatherability; APHA 150; m.w. 1100; sp.gr. 1.075; dens. 9.0 lb/gal; visc. 2500 cSt (48 C); pour pt.

5 F; ref. index 1.4645; flash pt. 555 F; cloud pt. 14 F.

Staflex 829. [Reichhold] Polymeric plasticizer; low-temp. properties for PVC wire and cable compds.; APHA 50; m.w. 2000; sp.gr. 1.088; dens. 9.1 lb/gal; visc. 2900 cSt; pour pt. –20 F; ref. index 1.4655; flash pt. 518 F; cloud pt. –20 F.

Staflex BDP. [Reichhold] Phthalate ester; plasticizer useful in vinyl and rubbers; applic. incl. elec. insulation, automotive vinyl and rubber applic., adhesives, roofing, pool and pond liners, consumer goods, pkg., cosmetics, and wall coverings; APHA 25; m.w. 363; sp.gr. 0.985; dens. 8.2 lb/gal; visc. 350 cSt (0 C); pour pt. –60 F; ref. index 1.4850; flash pt. 425 F; cloud pt. < –60 F.

Staflex BOP. [Reichhold] Butyl octyl phthalate; plasticizer; useful where increased solvating action is desired in vinyl; APHA 30; m.w. 368; sp.gr. 0.996; dens. 8.3 lb/gal; visc. 287 cSt (0 C); pour pt. –52 F; ref. index 1.4862; flash pt. 415 F; cloud pt. –22 F.

Staflex DBEA. [Reichhold] Dibutoxy ethyl adipate; plasticizer similar to DOA in most properties with slightly more efficiency and greater heat and lt. stability in vinyls; compatibility and low-temp. properties in rubber.

Staflex DBEP. [Reichhold] Dibutoxy ethyl phthalate; plasticizer; rapid fluxing by high solvating in vinyl; better rolling bank in calendering; better permanence than DBP in NC.

Staflex DBF. [Reichhold] Dibutyl fumarate; plasticizer; paint bases; chemical intermediate; internal plasticizer for copolymerization with vinyl acetate, vinyl chloride, acrylates, and styrene; APHA 50; m.w. 228; sp.gr. 0.986; dens. 8.2 lb/gal; visc. 3.8 cSt (100 F); pour pt. –40 F; ref. index 1.4546; flash pt. 285 F; cloud pt. 0 F.

Staflex DBM. [Reichhold] Dibutyl maleate; comonomer with ester vinyl-type monomers to give an internally plasticized polymer; chemical intermediate due to reactivity of double bond; APHA 50; m.w. 228; sp.gr. 0.993; dens. 8.3 lb/gal; visc. 3.7 cSt (100 F); pour pt. < –76 F; ref. index 1.4545; flash pt. 285 F; cloud pt. < –76 F.

Staflex DBP. [Reichhold] Dibutyl phthalate; general purpose plasticizer for ethylcellulose, rubbers, oil-sol. dyes, printing inks, emulsions; APHA 20; m.w. 278; sp.gr. 1.050; dens. 8.7 lb/gal; visc. 75 cSt (0 C); pour pt. < –80 F; ref. index 1.4910; flash pt. 360 F; cloud pt. < –80 F.

Staflex DIBA. [Reichhold] Adipate ester; plasticizer; aliphatic; often used in blends with phthalate plasticizers for improved low-temp. performance and lower volatility; specialty lubricant applic.; APHA 10; m.w. 258; sp.gr. 0.952; dens. 7.9 lb/gal; visc. 15 cSt (0 C); pour pt. –18 F; ref. index 1.4304; flash pt. 310 F; cloud pt. –18 F.

Staflex DIBCM. [Reichhold] Bis diisobutyl maleate (diisobutyl Carbitol maleate); plasticizer used for special applics. as reactive monomer, formulation of specialty surfactants.

Staflex DIDA. [Reichhold] Diisodecyl adipate; lacquer marring resistance of low temp. plasticizer; heat and lt. stability; APHA 10; m.w. 427; sp.gr. 0.918; dens. 7.6 lb/gal; visc. 50 cSt (0 C); pour pt. < –80 F; ref. index 1.4513; flash pt. 445 F; cloud pt. < –80 F.

Staflex DIDP. [Reichhold] Phthalate ester; see Staflex BDP; APHA 25; m.w. 446; sp.gr. 0.964; dens. 8.0 lb/gal; visc. 700 cSt (0 C); pour pt. –40 F; ref. index 1.4836; flash pt. 465 F; cloud pt. –8 F.

Staflex DINA. [Reichhold] Diisononyl adipate; plasticizer offering heat and lt. stability, low temp. properties, and permanence in vinyls; food pkg. film, upholstery, elec. insulation, gasketing; APHA 10; m.w. 398; sp.gr. 0.919; dens. 7.6 lb/gal; visc. 66 cSt (0 C); pour pt. –59 F; ref. index 1.4502; flash pt. 444 F; cloud < –80 F.

Staflex DINM. [Reichhold] Diisononyl maleate; plasticizer, intermediate used in specialty applic.; lower volatility than DIOM.

Staflex DIOA. [Reichhold] Diisooctyl adipate; plasticizer giving heat and lt. stability, low temp. in vinyls; used in syn. elastomers; provides low initial visc. in plastisols.

Staflex DIOM. [Reichhold] Diisooctyl maleate; stable, high purity plasticizer; comonomer with other monomers such as vinyl acetate; liq.

Staflex DIOP. [Reichhold] Diisooctyl phthalate; plasticizer; provides compatibility in vinyl; low rate of solvation in ethylcellulose and rubbers; APHA 25; m.w. 390; sp.gr. 0.985; dens. 8.2 lb/gal; visc. 280 cSt (0 C); pour pt. –50 F; ref. index 1.4850; flash pt. 425 F; cloud pt. –36 F.

Staflex DMAM. [Reichhold] Dimethyl amyl maleate; plasticizer compatible with syn. and natural resins and rubbers; preparation of corresponding sodium sulfosuccinate as ingred. in detergents; sol. in org. solvs.

Staflex DMP. [Reichhold] Dimethyl phthalate; plasticizer gives molding chars. in molding cellulose acetate; quick tack and low temp. emulsion coalescence in PVA emulsions; peroxide diluent.

Staflex DOA. [Reichhold] Dioctyl adipate; plasticizer used in food wrap; heat and lt. stability in vinyl; low initial visc. in plastisols; APHA 10; m.w. 371; sp.gr. 0.926; dens. 7.7 lb/gal; visc. 30 cSt (0 C); pour pt. < –80 F; ref. index 1.4462; flash pt. 410 F; cloud pt. < –80 F.

Staflex DOF. [Reichhold] Di-2-ethylhexyl fumarate; internal plasticizer for copolymerization with vinyl acetate, vinyl chloride, acrylates, and styrene; exterior paint formulations; APHA 50; m.w. 340; sp.gr. 0.942; dens. 7.8 lb/gal; visc. 16 cSt (100 F); pour pt. < –80 F; ref. index 1.4570; flash pt. 370 F; cloud pt. < –80 F.

Staflex DOM. [Reichhold] Di-2-ethylhexyl maleate; plasticizer; latex paint formulation; internal plasticizer for copolymerization with vinyl acetate, vinyl chloride, acrylates, and styrene; chemical intermediate for prod. of surfactant, sodium dioctyl sulfosuccinate; APHA 50; m.w. 340; sp.gr. 0.943; dens. 7.8 lb/gal; visc. 15 cSt (100 F); pour pt. < –70 F; ref. index 1.4540; flash pt. 370 F; cloud pt. –50 F.

Staflex DOP. [Reichhold] Di (2-ethylhexyl) phthalate; plasticizer for vinyl; uses incl. film sheeting, textile coatings, pkg. materials, adhesives; APHA

35; m.w. 390; sp.gr. 0.985; dens. 8.2 lb/gal; visc. 375 cSt (0 C); pour pt. –60 F; ref. index 1.4851; flash pt. 420 F; cloud pt. < –60 F.

Staflex DOZ. [Reichhold] Dioctyl azelate; plasticizer; in vinyl, provides low volatility, compatibility, low water extraction; resistance to heat and lt.; used in upholstery, raincoats, shower curtains, and outerwear; APHA 45; m.w. 412; sp.gr. 0.920; dens. 7.6 lb/gal; visc. 47 cSt (0 C); pour pt. < –80 F; ref. index 1.4500; flash pt. 455 F; cloud pt. –60 F.

Staflex DTDM. [Reichhold] Ditridecyl maleate; high purity, low odor, and low color plasticizer and intermediate; used in fields of copolymerization and surfactants, the latter via the bisulfite addition process.

Staflex DTDP. [Reichhold] Ditridecyl phthalate; plasticizer with elec. properties, permanence, good compatibility in vinyls; APHA 40; m.w. 531; sp.gr. 0.952; dens. 7.9 lb/gal; visc. 1900 cSt (0 C); pour pt. –48 F; ref. index 1.4833; flash pt. 500 F; cloud pt. < –48 F.

Staflex NODA. [Reichhold] n-Octyl, n-decyl adipate; plasticizer with elec. properties for outdoor wire and cable applic.; heat and lt. stability with high compatibility in vinyls; APHA 10; m.w. 399; sp.gr. 0.918; dens. 7.6 lb/gal; visc. 60 cSt (0 C); pour pt. 26 F; ref. index 1.4458; flash pt. 430 F; cloud pt. < 26 F.

Staflex NONDTM. [Reichhold] n-Octyl, n-decyl trimellitate; plasticizer used in high-temp. wire and cable applics.; low temp. flexibility, permanence, and ease of processing in vinyl applic.; APHA 80; m.w. 585; sp.gr. 0.978; dens. 8.1 lb/gal; visc. 125 cps; pour pt. –30 F; ref. index 1.4821; flash pt. > 500 F; cloud pt. –30 F.

Staflex ODA. [Reichhold] Adipate ester; plasticizer with low temp. flexibility and limited compatibility than that of phthalates; aliphatic in nature; blended with phthalate plasticizers for improved low-temp. performance and lower volatility; specialty lubricant applics.; APHA 10; m.w. 399; sp.gr. 0.922; dens. 7.7 lb/gal; visc. 38 cSt (0 C); pour pt. < –80 F; ref. index 1.4488; flash pt. 430 F; cloud pt. < –80 F.

Staflex ODP. [Reichhold] Octyldecyl phthalate; plasticizer for vinyl applics.; lower in volatility than DOP; APHA 20; m.w. 418; sp.gr. 0.973; dens. 8.1 lb/gal; visc. 500 cSt (0 C); pour pt. –60 F; ref. index 1.4838; flash pt. 440 F; cloud pt. –60 F.

Staflex TIOTM. [Reichhold] Triisooctyl trimellitate; plasticizer used in high-temp. wire and cable applics.; APHA 100; m.w. 546; sp.gr. 0.986; dens. 8.2 lb/gal; visc. 1400 cSt (0 C); pour pt. –40 F; ref. index 1.4830; flash pt. 490 F; cloud pt. –40 F.

Staflex TOTM. [Reichhold] Trioctyl trimellitate; plasticizer used in high-temp. wire and cable applics.; used in vinyls where good low temp. properties must be combined with low volatility; APHA 100; m.w. 546; sp.gr. 0.987; dens. 8.2 lb/gal; visc. 1800 cSt (0 C); pour pt. –30 F; ref. index 1.4838; flash pt. > 500 F; cloud pt. –30 F.

Stafoam. [Nippon Oils & Fats] Fatty acid DEA; detergent, rust inhibitor; liq. detergent for laundry or drycleaning use; textile softener aux.; dyeing aux.; metal cleaning agent; liq. soaps; dirt dispersant; lt. yel. liq.; water-sol.; sp.gr. 0.996 (30 C); visc. 810 cps (30 C); pH 9.5–10.5 (1% aq.); surf. tens. 8 dynes/cm (0.1%).

Stafoam DF-1. [Nippon Oils & Fats] 1:1 Cocamide DEA; detergent, foam booster, softener, antistat; used in shampoo, dishwashing, laundry detergent, drycleaning, textile softener aux., and metal cleaning; lt. yel. liq.; sol. in ethanol, IPA, propylene glycol, diethylene glycol, petrol. ether, tetrachloroethylene, benzene, xylene; sp.gr. 0.981 (30 C); visc. 625 cps (30 C); pH 9.6–10.6 (1% aq.); surf. tens. 14 dynes/cm (0.1%).

Stafoam DF-4. [Nippon Oils & Fats] 1:1 Cocamide DEA; see Stafoam DF-1; lt. yel. liq.; sol. see Stafoam DF-1; sp.gr. 0.974; visc. 675 cps (30 C); m.p. 15 C max.; surf. tens. 16 dynes/cm (0.1%); pH 9.6–10.6 (1% aq.).

Stafoam DL. [Nippon Oils & Fats] 1:1 Lauramide DEA; detergent, foam stabilizer and booster, visc. modifier, softener; shampoos, drycleaning, cosmetics components, textile softening and dyeing auxs., anticlouding agents; wh. solid; sol. > 2% in ethanol, IPA, propylene glycol, diethylene glycol, tetrachloroethylene, benzene, xylene; sp.gr. 0.969 (50 C); visc. 179 cps (50 C); m.p. 42–47 C; surf. tens. 5 dynes/cm (0.1%0.

Stafoam F. [Nippon Oils & Fats] 1:2 Cocamide DEA; detergent, rust inhibitor, foamer, penetrant; liq. detergents, metal cleaning, dirty-dispersal aux., textile processing; dye aux.; lt. yel. liq.; sol. see Stafoam; sp.gr. 0.995 (30 C); visc. 755 cps (30 C); surf. tens. 7 dynes/cm (0.1%); pH 9.5–10.5 (1% aq.).

Stamid HT. [Clough] Cocamide DEA; nonionic; foam stabilizer, emulsifier and thickener used in a variety of household, industrial and cosmetic formulations; liq.; 85% conc.

Stanclere® A-121, A-121 C, A-221. [Akzo] Antimony mercaptide; stabilizer offering heat stability, nonplate-out, plastisol air release, and plastisol visc. control; used in rigid compds. esp. pipe and profile applics.; liq.

Stanclere® T-55. [Akzo] Tin carboxylate; stabilizer for PVC window profiles; outstanding weatherability.

Stanclere® T-57, T-80, T-81. [Akzo] Alkyl-tin-carboxylate; stabilizer for PVC.

Stanclere® T-85, TL. [Akzo] Butyl tin carboxylate; stabilizer offering lt. stability, clarity, nonplate-out, plastisol air release, and plastisol visc. control; used in flexible compds. esp. hose and profile; liq.

Stanclere® T-94 C. [Akzo] Butyl tin mercaptide; stabilizer offering heat stability, clarity, nonplate-out, plastisol air release, and plastisol visc. liq.

Stanclere® T-126. [Akzo] Butyl tin mercaptide; stabilizer offering lubricity, clarity, plastisol air release, and plastisol visc. control, used in rigid compds. esp. blow molding and inj. molding; liq.

Stanclere® T-160. [Akzo] Butyltin stabilizer for profiles.

Stanclere® T-161, T-163, T-164, T-165, T-174, T-182, T-183. [Akzo] Butylmercaptide; stabilizer for PVC.

Stanclere® T-184. [Akzo] Tin mercaptide; stabilizer for PVC window profiles; good light stability.

Stanclere® T-186. [Akzo] Dibutyltin β-mercaptopropionate; stabilizer for PVC.

Stanclere® T-192, T-193, T-194. [Akzo] Butylmercaptide; stabilizer for PVC.

Stanclere® T-197. [Akzo] Ba-tin complex; stabilizer for PVC window profiles; exc. light and weather stability; powd.

Stanclere® T-208. [Akzo] Estertin mercaptide; stabilizer offering heat and lt. stability, clarity, nonplate-out, low odor, plastisol air release, and plastisol visc. control; used for flexible compds. (film, sheet, hose, profile, inj. molding, filled systems) and plastisol applics. (slush molding, rotational molding, dipping, and spreading); liq.

Stanclere® T-222, T-233. [Akzo] Estertin mercaptide; stabilizers offering heat and lt. stability, clarity, nonplate-out, plastisol air release, and plastisol visc. control; used for rigid compds. (pipe, profile, film, sheet, blow molding, inj. molding, siding); liq.

Stanclere® T-250, T-250SD. [Akzo] Estertin mercaptide; stabilizers offering lower processing odor, low migration, and improved lt. stability compared to alkyl-tin mercaptides; T-250 and T-250SD grades feature heat and lt. stability, lubricity, clarity, nonplate-out, plastisol air release, and plastisol visc. control; applics. incl. rigid compds. (pipe, profile, foam profile, siding); liq.

Stanclere® T-470, T-473, T-482, T-483, T-484, T-582. [Akzo] Octylmercaptide; stabilizer for PVC.

Stanclere® T-801. [Akzo] Butyl tin mercaptide; stabilizer offering heat and lt. stability, lubricity, clarity, nonplate-out, plastisol air release, and plastisol visc. control; used in rigid compds., esp. pipe, profile, sheet, film, foam profile, siding; liq.

Stanclere® T-876. [Akzo] Butyl tin carboxylate; stabilizer offering lt. stability, clarity, nonplate-out, plastisol air release, and plastisol visc. control; used in plastisol applics. (slush molding, rotational molding, dipping, spreading); liq.

Stanclere® T-877. [Akzo] Butyl tin carboxylate; stabilizer offering lt. stability, clarity, nonplate-out, plastisol air release, and plastisol visc. control; used in flexible compds., esp. hose and profile applics.; liq.

Stanclere® T-878, T-55. [Akzo] Butyl tin carboxylate; stabilizer offering lt. stability, clarity, nonplate-out; plastisol air release, and plastisol visc. control; used in flexible compds., esp. hose and profile; liq.

Stanclere® T-4356. [Akzo] Sulfur-containing organotin stabilizer; used in filled or pigmented rigid inj. molding applics., siding, and profile market; offers improved uv stability, impact strength, and heat distort. temps. while not sacrificing heat stability chars.; usage levels: 1.0–2.0 phr; Gardner 2 liq.; sp.gr. 1.210; visc. 50 cps.

Stanclere® TM. [Akzo] Dibutyl-tin-maleate; stabilizer for PVC.

Standamid 100. [Henkel] Cocamide MEA; foam stabilizer; superfatting and thickening agent for medicated and cosmetic shampoos, bubble baths, and other detergents; consistency-giving factor for cosmetic preps. in stick form; skin protecting component for industrial ointments; additive to hair treatment preps.; solid; 90% amide content.

Standamid CD. [Henkel] Capramide DEA and diethanolamine; nonionic; foam enhancer for anionic systems; solubilizes fragrances into hydroalcoholic systems; sec. emulsifier in o/w systems, perfume stabilizer; useful in other applics. which take advantage of its hydrophilic properties; amber liq.; sol. in water, propylene glycol, PEG 8, SD 40 alcohol; visc. 675 cps.

Standamid CM. [Henkel] Cocamide MEA; nonionic; foaming agent and stabilizer; powded. or beaded bubble baths; gran.; m.p. 140–148 F; acid no. 8.0–12.0; pH 9.5–10.5 (10% sol'n.); 100% conc.

Standamid CMG. [Henkel] Cocamide MEA; super amide (1:1) derived from whole coconut oil; foam boosters, stabilizer, thickener; shampoos, bubble baths, dishwashing detergents, contains no amide ester; solid, 100% conc.

Standamid ID. [Henkel] Isotearamide DEA; nonionic; visc. builder esp. in sodium lauryl sulfate systems; foam enhancer; hair conditioning agent; lubricant, mold release agent, emulsifier; amber liq.; sol. in SD 40 alcohol; sp.gr. 0.964; dens. 8.04 lb/gal; visc. 1600 cps; pH 8–10 (1.0%); 100% act.

Standamid KD. [Henkel] Cocamide DEA; nonionic; foaming and thickening agent for liq. or gel shampoos and bubble baths; superfatting agent, foam stabilizer, emulsifier, intermediate; visc. amber liq.; m.p. 100 F; acid no. 1.0 max.; pH 9.0–9.5 (10%); 85–90% amide content.

Standamid KDM. [Henkel] 1:2 Cocamide DEA; nonionic; foam builder and stabilizer, thickener, wetting agent, dispersant; shampoos, bubble baths, household detergents, liq. soaps; good lime-soap dispersing and antiredeposition properties; amber visc. liq.; readily sol. in water, alcohol, acetone, and glycols; acid no. 0–5; pH 9.5–10.0 (1%); 100% conc.

Standamid KDO. [Henkel] 1:1 Cocamide DEA; nonionic; visc. builder for anionic systems, good foam enhancement, low odor, good color, broad compatibility, and good handling qualities; used in shampoos, bath prods., liq. hand soaps; amber liq.; sol. in lower alcohols, propylene glycol, and polyethylene glycols; sp.gr. 0.982; acid no. 2.0 max.; pH 9–11 (1.0%).

Standamid KDS. [Henkel] Lauramide DEA (1:1); nonionic; enhances foam dens., lubricity, stability; lt. amber liq.; 80% conc.

Standamid KM. [Henkel] Cocamide MEA; visc. builder for anionic systems; produces lubricious foam for shampoo and bath prods.; flakes, beads; sol. in lower alcohols; sp.gr. 0.884; m.p. 80 C; acid no. 2.0 max.; pH 9–11 (1.0%).

Standamid LD. [Henkel] Lauramide DEA; nonionic; foam booster and stabilizer, superfatting agent, thickening agent, emulsifier for shampoos, bath prods., fortifier for perfumes in soaps; paste; m.p. 105 F; acid no. 0–1.0; pH 9.5–10.5 (10% sol'n.); > 90% amide content.

Standamid LD 80/20. [Henkel] Modified lauric/myristic acid DEA; nonionic; foam booster and stabilizer, detergency builder, emulsifier, thickener, conditioner; shampoos, bubble baths, dishwashing detergents, creams, and lotions; clear amber liq.; cloud pt. 0 C; acid no. 0.5 max.; pH 9.5–10.5 (10% sol'n.); 80% conc.

Standamid LDM. [Henkel] Modified lauramide DEA; nonionic; foam stabilizer, visc. builder; shampoos and bubble baths; clear sltly. visc. liq.; cloud pt. 0 C; acid no. 3.0%; 100% conc.

Standamid LDO. [Henkel] 1:1 Lauramide DEA; nonionic; see Standamid KDO; amber liq.; sol. see Standamid KDO; sp.gr. 0.985; acid no. 2.0 max.; pH 9–11 (1.0%); 100% conc.

Standamid LDS. [Henkel] Lauramide DEA; nonionic; solubilizer for liqs. and oils; used in personal care prods., dishwashing detergents; lt. amber liq.; sol. in lower alcohols, propylene glycol, polyethylene glycol; disp. in water; sp.gr. 0.98; dens. 8.2 lb/gal; pH 9–11 (1.0%); 100% conc.

Standamid LM. [Henkel] Lauramide MEA; see Standamid KM; flakes, beads; sol. in lower alcohols; insol. in water; sp.gr. 0.915; m.p. 80 C; acid no. 2.0 max.; pH 9–11 (1.0 %).

Standamid LP. [Henkel] Lauramide MIPA; nonionic; foam stabilizer, superfatting and thickening agent; shampoos and other detergents; fortifier for perfumes in soaps; solid; acid no. < 2%; > 90% amide content.

Standamid PD. [Henkel] Cocamide DEA (2:1) and diethanolamine; nonionic; more efficient foam builder, less efficient visc. builder than 1:1 superamides; stabilizer; used in shampoos, bubble baths with anionics, nonionics, and amphoterics; amber visc. liq.; 100% conc.

Standamid Resin BC-1283. [Henkel] Nylon; gellant for hydrocarbons and Guerbet alcohols; solid, controlled-release fragrance prods.; lt. amber clear solid.

Standamid SD. [Henkel] Cocamide DEA; nonionic; high foamer and foam stabilizer with pronounced visc. build-up; used in bath gels, liq. bubble baths, shampoos, dishwashing and laundry detergents; derived from whole coconut oil; contains no amide ester; low cost alkanolamide; amber visc. liq.; m.p. 25 F; acid no. 0–1.0; pH 9.5–10.5 (10% sol'n.); 100% conc.

Standamid SDO. [Henkel] Cocamide DEA (1:1); visc. builder and foam enhancer for shampoo, bubble baths, liq. hand soaps, offers excellent handling chars.; residual glycerin to enhance humectancy and hair conditioning effects; amber liq.; sol. in lower alcohols, propylene glycol, polyethylene glycols; sp.gr. 0.991; acid no. 2.0 max.; pH 9–11 (1.0%).

Standamid SM. [Henkel] Cocamide MEA (1:1); see Standamid KM; flakes, beads; sol. in lower alcohols; insol. in water; sp.gr. 0.934; m.p. 62 C; acid no. 1.0 max.; pH 9–11 (1.0%).

Standamid SOD. [Henkel] Linoleic DEA (1:1); nonionic; visc. enhancer, thickener in industrial, institutional and household cleaners; corrosion inhibitor and emulsifier for oils and lubricants; dk. amber liq.; sol. in lower alcohols, glycols, acetone, toluene, xylene; disp. in water; dens. 8.20 lb/gal; pH 9–11 (1%); 100% conc.

Standamid SOMD. [Henkel] Linoleamide DEA; visc. builder with foam enhancement chars.; produces esp. high visc. when used with ethoxylated anionics; used in gel shampoos, bubble baths, liq. hand soaps, and formulation where the amt. of electrolyte must be kept to a min.; low-cost shampoo concentrates; paste/solid; sol. in lower alcohols; sp.gr. 0.935; acid no. 2 max.; pH 9–11 (1.0%).

Standamox C 30. [Henkel] Coco amido amine oxide; wetting agent, foam stabilizer, visc. builder; shampoos, bubble baths, bath oils, dishwash concentrates, and other detergents; liq.; pH 6–7 (as is); 28–30% act.

Standamox CAW. [Henkel] Cocamidopropylamine oxide; foam builder, stabilizer, thickener, emollient for low irritation baby shampoos, bubble baths, skin care preps.; clear liq.; 30% conc.

Standamox LMW-30. [Henkel] Lauryl/myristyl dimethylamine oxide; cationic; foamer and conditioner in household and industrial cleaners; liq.; 30% conc.

Standamox O1. [Henkel] Oleamine oxide; nonionic; lubricant, softener used in industrial applics.; amber clear liq.; 55% conc.

Standamul® 302. [Henkel] Propylene glycol dicaprylate/dicaprate; base, emollient used in pharmaceutical and personal care prods.; clear visc. oily liq.; char. odor; sol. in min. and castor oil, IPM, oleyl alcohol, anhyd. ethanol, and silicone fluid; sp.gr. 0.921; HLB 9; acid no. 0.5 max.; sapon. no. 315–335; 100% conc.

Standamul® 318. [Henkel] Caprylic/capric triglyceride; base, emollient, carrier used in pharmaceutical, personal care and food prods.; oil binder for pressed powders; extender for flavors; clear visc. oily liq.; odorless; sol. in min. and castor oil, IPM, oleyl alcohol and anhyd. ethanol; sp.gr. 0.950; visc. 25 cps; HLB 12; cloud pt. –5 C max.; acid no. 0.5 max.; sapon. no. 340–350; 100% conc.

Standamul® 1414-E. [Henkel] Myreth-3 myristate; carrier, SE base, emulsifier, coupler, emollient and moisturizer used in pharmaceutical and personal care prods.; clear visc. oily liq.; sol. in min. and castor oil, IPM, oleyl alcohol, anhyd. ethanol; HLB 12; cloud pt. 25 C max.; sapon. no. 90–100; 100% conc.

Standamul® 1616. [Henkel] Cetyl palmitate; spermaceti replacement, consistency giving agent used in pharmaceutical and personal care prods.; wh. waxy solid; faint char. odor; sol. in min. and castor oil, IPM, oleyl alcohol; m.p. 45–55 C; HLB 9; sapon. no. 115–130; 100% conc.

Standamul® 7061. [Henkel] Isocetyl stearate; base, emollient used in pharmaceutical and personal care prods.; clear visc. oil; faint char. odor; sol. see Standamul 1616; sp.gr. 0.865; visc. 25 cps; HLB 8; cloud pt. 0 C max.; acid no. 0.5 max.; sapon. no. 105–115; 100% conc.

Standamul® 7063. [Henkel] Octyl dodecyl stearate; see Standamul 7061; clear visc. oil; faint char. odor; sol. see Standamul 1616; sp.gr. 0.865; visc. 100 cps;

HLB 8; cloud pt. 0 C max.; acid no. 0.5 max.; sapon. no. 90–105; 100% conc.

Standamul® 7105. [Henkel] Caprylic/capric triglyceride; base used in pharmaceutical and personal care prods.; clear visc. oil; faint char. odor; sol. see Standamul 318; sp.gr. 1.064; visc. 100 cps max.; HLB 12; cloud pt. –5 C max.; sapon. no. 340–350; 100% conc.

Standamul® B-1. [Henkel] Ceteareth-12; nonionic; emulsifier and solubilizer for personal care and household prods., essential oils; waxy solid; sp.gr. 0.97 (70 C); HLB 12; solid. pt. 35 C; 99.5–100% solids.

Standamul® B-2. [Henkel] Ceteareth-20; nonionic; emulsifier and solubilizer for personal care prods.; additive in the paper and textile industry; household cleaning and leather prods.; waxy solid; sp.gr. 1.0 (70 C); HLB 14; solid. pt. 40 C; 99.7–100% solids.

Standamul® B-3. [Henkel] Ceteareth-30; nonionic; see Standamul B-2; waxy solid; sp.gr. 1.023 (70 C); HLB 15; solid. pt. 43–46 C; 99.5–100% solids.

Standamul® Conc. 1002. [Henkel] Cetearyl alcohol, PEG-40 hydrog. castor oil, stearalkonium chloride; conditioner and softener used in hair care preparations; off-wh. waxy flakes; m.p. 58 C; pH 5.0 (1%); 99.0% min. solids.

Standamul® CTA. [Henkel] Hexyl laurate; emollient, spreading agent, carrier, base, dispersant used in pharmaceutical and personal care prods.; clear visc. oil; odorless; sol. see Standamul 302; sp.gr. 0.85–0.87; HLB 9; cloud pt. 0 C max.; acid no. 0.5 max.; sapon. no. 190–205; 100% conc.

Standamul® CTV. [Henkel] Decyl oleate; see Standamul CTA; clear visc. liq.; faint char. odor; sol. see Standamul 318; sp.gr. 0.86–0.87; HLB 9; cloud pt. 10 C max.; acid no. 0.5 max.; sapon. no. 130–150; 100% conc.

Standamul® G. [Henkel] Octyl dodecanol; carrier for oil-sol. act. ingred.; dissolves dyes; pigment dispersant; coupler for waxes and other materials in personal care prods.; water-wh. clear liq.; odorless; cloud pt. –50 C; acid no. 0.5 max.; sapon. no. 5.0 max.; 100% conc.

Standamul® G-16. [Henkel] Isocetyl alcohol; emollient; extender and solv. for pigments and personal care prods.; binder in cosmetics; water-wh. clear liq.; visc. 100 cps; cloud pt. –50 C; acid no. 0.5 max.; sapon. no. 5.0 max.; 100% conc.

Standamul® G-32/36. [Henkel] Myristyl eicosanol; emollient wax for cosmetic and topical pharmaceuticals; provides structure, stability, and lubricity to cosmetic stick preparations; slip and nontackiness to antiperspirants sticks; opacifier and emollient in specialty bath oils; lt. soft wax; faint char. odor; sol. in min. and castor oil, IPM, oleyl alcohol; HLB 7; m.p. 33–40 C; acid no. 1.0 max.; sapon. no. 65–75.

Standamul® G-32/36 Stearate. [Henkel] Myristyl eicosyl stearate; stabilizer, base, lubricant, opacifier, emollient used in pharmaceutical and personal care prods.; lt. soft wax; faint char. odor; sol. in min. and castor oil, IPM, oleyl alcohol; m.p. 33–40 C; HLB 7; sapon. no. 65–75.

Standamul® GTO-26. [Henkel] Undecylpentadecanol; skin lubricant and pigment wetting agent in cosmetic emulsion prods.; used in personal care prods.; water-wh. clear liq.; insol. in water; visc. 300 cps; cloud pt. –40 C.

Standamul® HE. [Henkel] PEG-7 glyceryl cocoate; nonionic; emollient, refatting agent, solubilizer, coupler, emulsifier used in pharmaceutical and personal care prods.; clear low visc. oil; faint char. odor; sol. in oleyl alcohol, castor oil, anhyd. ethanol, water; sp.gr. 1.050; visc. 200 cps; HLB 16; cloud pt. 0 C; acid no. 5.0 max.; sapon. no. 90–100; 100% conc.

Standamul® LC. [Henkel] Coco caprylate/caprate; dry-feeling penetrating emollient; reduces oiliness in creams, lotions, solid sticks; clear liq.

Standamul® O-5. [Henkel] Oleth-5; nonionic; emulsifier, emollient, bodying and conditioner blended to obtain special emulsification effects; mild to skin; visc. liq.; sol. in alcohols, hydrocarbons, and org. solvs.; 100% conc.

Standamul® O-10. [Henkel] Oleth-10; nonionic; see Standamul O-5; soft waxy solid; sol. see Standamul O-5; 100% conc.

Standamul® O-20. [Henkel] Oleth-20; see Standamul O-5; waxy solid; sol. see Standamul O-5; 100% conc.

Standamul® OXL. [Henkel] PPG-10 ceteareth-20; nonionic; emollient, superfatting agent for personal care prods.; water-wh. liq.; sol. in water; 98% conc.

Standamul® STC-25. [Henkel] Stearalkonium chloride; cationic; antistat, surfactant for personal care prods. and fabrics; wh. slurry; 25% conc.

Standapol® AB-45. [Henkel] Coco betaine; amphoteric; foam and visc. booster, conditioner, antistat, and gelling agent used in personal care prods.; pale yel. liq.; sol. in alcohols, glycols, alkanolamides and water; cloud pt. < 0 C; pH 6.5–7.5; 36.5% min. act.

Standapol® BAW. [Henkel] Cocamidopropyl betaine; amphoteric; conditioner and foamer, antistat; surfactant used in personal care prods.; lt. amber liq.; 34% conc.

Standapol® BC-35. [Henkel] Cocamidopropyl betaine; amphoteric; conditioner, emollient, visc. builder, gelling agent used in personal care prods.; lt. amber liq.; 35% conc.

Standapol® Conc. 7023. [Henkel] Cocamide DEA and DEA-myreth sulfate; anionic; detergent, foamer, solubilizer; conc. shampoos and bath prods. where high foaming is required; gelatin capsules bath beads; solubilizer for fatty alcohols, fatty acid esters, and min. oil; straw-yel. clear visc. liq.; visc. 250 cps max.; pH 8.5–9.5 (1% aq.); cloud pt. 10 C max.; 100% conc.

Standapol® EA-3. [Henkel] Ammonium laureth sulfate; solubilizer used in personal care prods.; water-wh. visc. liq.; water-sol.; dens. 1.020 g/ml; visc. 2500 cps; cloud pt. 5 C; pH 6.0–7.0 (10%); 26–28% act.

Standapol® LF. [Henkel] Sodium octyl sulfate; anionic; wetting agent for metal degreasers, hard surface cleaners, food equipment cleaners, dust control; solubilizer, hydrotrope; liq.; 33% conc.

Standapol® OLB-30. [Henkel] Oleyl betaine; amphoteric; detergent, conditioner and foamer; visc. builder used in personal care prods.; amber visc. liq.; 30% conc.

Standapol® OLB-50. [Henkel] Oleyl betaine; amphoteric; see Standapol OLB-30; lt. amber gel; 50% conc.

Standapol® Pearl Conc. 7130. [Henkel] Glycol distearate, sodium laureth sulfate, propylene glycol, cocamide MEA, laureth-9; pearlescent used in personal care prods.; wh. pearly liq.; misc. in R.T. water; visc. 5000 cps max.; pH 6.5 ± 1.0% (10%); 41.0 ± 3.0% solids.

Standapol® SCO. [Henkel] Sulfated castor oil; emulsifier and dispersant, wetting agent, foam depressant, emollient surfactant and solubilizer used in personal care prods.; amber visc. liq.

Stanno-Plus. [Goldschmidt] Stabilizer for metal salt sol'ns. (electrolytical coloring of aluminum).

Star. [Procter & Gamble] Glycerin USP; humectant.

Starfol® BB. [Sherex] Behenyl behenate; emollient for cosmetic creams and lotions; flakes; sol. in min. oil.

Starfol® CP. [Sherex] Cetyl palmitate; emollient for cosmetic creams and lotions; flakes; sol. in min. oil.

Starfol® IPM. [Sherex] IPM; nonionic; emollient; APHA 30 max. liq.; sapon. no. 202–212; 100% conc.

Starfol® IPP. [Sherex] IPP; nonionic; emollient; APHA 30 max. liq.; sapon. no. 182–192; 100% conc.

Starfol® IS. [Sherex] Isostearyl isostearate; emollient in creams and lotions, solubilizer and solv. for fragrances; lt. amber liq.; 100% solids.

Starfol® ODS. [Sherex] Octododecyl stearate; emollient for cosmetic creams, textile lubricants; liq.; sol. in min. oil.

Starfol® OO. [Sherex] Oleyl oleate; emollient for cosmetic creams, textile lubricants; liq.; sol. in min. oil.

Starfol® OS. [Sherex] Octyldodecyl stearate; emollient for creams and lotions; lt. amber liq.; 100% solids.

Starfol® Wax CG. [Sherex] Cetyl esters; nonionic; syn. spermaceti wax; emollient for cosmetics; APHA 100 max. flake; sol. in min. oil; m.p. 46–49 C; sapon. no. 109–117; 100% conc.

Starfol® Wax CG NF. [Sherex] High m.w. ester of monocarboxylic acids and monohydric alcohols; wax designed as a replacement for spermaceti wax; for cosmetic emulsions, e.g., hand lotions, body lotions, face creams, shaving creams, cold creams, lipsticks; wh. wax-like solid flake; m.p. 48 C; acid no. 1.0; iodine no. 1.0; sapon. no. 114; 48% unsaponifiable material.

Starplex® 70. [Am. Ingredients] Blend of molecularly distilled monoglycerides and 20% fully hydrog. soybean oil; food additive for use where high temp. aq. processing is required; stabilizes emulsions; wh. to cream powd.; m.p. ≈ 58 C; sapon. no. 155–175; 72% min. alpha monoester.

Starplex® 90. [Am. Ingredients] High-purity, molecularly distilled monoglyceride prepared from edible fats or oils and glycerin with BHA and citric acid; provides increased hydration and improved functionality of the monoglyceride; improves softness and extends shelf life in baked goods; emulsifies and stabilizes fat in sauces and gravies; starch complexing agent; wh. to cream powd.; water-disp.; m.p. 66 C; acid no. 4.0 max.; sapon. no. 150–165; 90% min. alpha monoester.

Starwax® 100. [Petrolite] Hard microcryst. wax consisting of n-paraffinic, branched paraffinic, and naphthenic hydrocarbons; wax used in hot-melt coatings and adhesives, cup and paper coatings, printing inks, plastic modification (as lubricant and processing aid), lacquers, paints, and varnishes, as binder in ceramics, for potting, filling, and impregnant in elec./electronic components, in investment casting, rubber and elastomers (plasticizer, antisunchecking, antiozonant), as emulsion wax size in papermaking, as fabric softener ingred. in permanent-press fabrics, in emulsion and latex coatings, and in cosmetic hand creams and lipsticks; color 1.0 (D1500) wax; very low sol. in org. solvs.; sp.gr. 0.92; visc. 11.7 cps (98.9 C); m.p. 86.1C.

Statexan K1. [Mobay] Sulfonated aliphatic hydrocarbon; internal antistat additive for PVC, PS and styrene; also as external additive coating for these materials; lt. wax flakes; sol. in water, methanol, and ethanol.

Staticide®. [ACL] Water-based topical antistat for spray, dip, or wipe-on applic. to any material; biodeg.

Staybelite®. [Hercules] Acidic resin made by partially hydrogenating wood rosin; thermoplastic resin used as plasticizer, tackifier, or processing aid for reclaimed, natural, and syn. rubbers; as component in sealing wax, optical lens pitch, elec. cable sats., waxes for paper coating, syn. resins, metal resinates, soldering fluxes, and ceramic ink vehicles; pale solid; soften. pt. (Drop) 75 C; acid no. 160.

Staybelite® 570. [Hercules] Acidic resin (modified, partially hydrogenated form of rosin); thermoplastic resin related to regular Staybelite resin possessing resistance to oxidation, color and color retention, compatibility, and tack-imparting properties; as tackifier resin in hot-melt coatings and adhesives, incl. hot-melts, solv.-applied, mastics, and other special adhesives (lens-blocking pitches that utilize specific adhesion and narrow softening ranges of acidic resins); used in food-pkg. and -processing operations; USDA Rosin WG_0; sol. in aliphatic, aromatic, and chlorinated hydrocarbons, ketones, esters, and ethers; visc. 0.9 stokes (70% solids in toluene); soften. pt. (R&B) 71 C; acid no. 155.

Staybelite® Ester 5. [Hercules] Glycerol ester of partially hydrogenated wood rosin; thermoplastic syn. resin; resistant to oxidation, discoloration, and aging; as softener/plasticizer for the masticatory agent in chewing gum bases; as resin modifier for film-formers, elastomers, and waxes in adhesive and protective coating compositions where low odor and resistance to aging are important; used in chewing gum bases, and food-pkg. and -processing operations; USDA Rosin solid, flakes; low odor; sol. in esters, ketones, higher alcohols, glycol ethers, and aliphatic, aromatic, and chlorinated hydrocarbons;

dens. 1.06 kg/l; soften. pt. (Hercules Drop) 81 C; acid no. 5.

Staybelite® Ester 10. [Hercules] Glycerol ester of hydrogenated rosin; thermoplastic syn. resin in solv., hot-melt, and emulsion adhesives based on thermoplastic and elastomeric materials (natural and syn. rubbers, block copolymer rubbers, E/VA resins, vinyl resins, and polyamides); used (in emulsion form) as tackifier for natural rubber, SBR, and neoprene latexes in formulation of aq.-based adhesives; as tackifier in industrial tape masses; in lamination of metal, foil, paper, etc.; in hot-melt-applied barrier coatings, chlorinated rubber finishes, etc.; as modifier for paraffin wax; used in food-pkg. and -processing operations; USDA Rosin WG solid, flakes; sol. in esters, ketones, higher alcohols, glycol ethers, and aliphatic, aromatic, and chlorinated hydrocarbons; sp.gr. 1.07; dens. 1.07 kg/l; visc. (G-H) I-M (75% solids in toluene); soften. pt. (Hercules Drop) 83 C; acid no. 8.

Stealim. [Salomon] Talc; anticaking agent for animal feed; powd.

Stearal. [Amerchol] Stearyl alcohol; emollient, aux. emulsifier, texturizer; nonoily, velvety feel; higher visc. in emulsions; wh. waxy solid; mild char. odor; m.p. 55–60 C; acid no. 1.0 max.; sapon. no. 2.0 max.

Stedbac. [Hexcel] Stearalkonium chloride; cationic; hair conditioner, emulsifier imparting softness, antistatic properties to cream rinse formulations; powd.; very sol. in water, alcohol; 100% conc.

Steol CA 460. [Stepan Canada] Ammonium laureth sulfate; anionic; detergent, dispersant, and wetting agent used in shampoos, dishwashers, car washers; liq.; 60% conc.

Steol CS-130, -330, -460. [Stepan Canada] Sodium laureth sulfate; anionic; detergent, dispersant, and wetting agent used in shampoos, dishwashers, car washers; liq.; 25, 25, and 60% conc. resp.

Stepan C25. [Stepan] Methyl caprylate-caprate; intermediate for mfg. of detergents, emulsifiers, wetting agents, stabilizers, lubricants, plasticizers, resins, and textile specialties; water-wh.; m.p. –30 C; acid no. 0.8 max.; sapon. no. 329–335.

Stepan C40, C41. [Stepan] Methyl laurate; see Stepan C25; water-wh.; m.p. 5 and 1 C resp.; acid no. 0.5 max.; sapon. no. 260–264 and 258–264 resp.

Stepan C42, C43. [Stepan] Methyl laurate; see Stepan C25; water-wh.; acid no. 0.6 max.; sapon. no. 254–258 and 252–256 resp.

Stepan C49. [Stepan] Whole coco methyl esters; see Stepan C25; Gardner 1 max.; acid no. 0.5 max.; sapon. no. 250–256.

Stepan C50. [Stepan] Methyl myristate; see Stepan C25; Gardner 1 max.; m.p. 15 C; acid no. 1.5 max.; sapon. no. 230–234.

Stepan C60. [Stepan] Methyl palmitate; see Stepan C25; Gardner 1 max.; m.p. 28 C; acid no. 2.0 max.; sapon. no. 203–210.

Stepan C65. [Stepan] Methyl palmitate-oleate; see Stepan C25; Gardner 1 max.; m.p. 14 C; acid no. 1.0 max.; sapon. no. 197–203.

Stepan C66. [Stepan] Methyl stearate; see Stepan C25; Gardner 1 max.; m.p. 30 C; acid no. 1.5 max.; sapon. no. 195–203.

Stepan C68. [Stepan] Methyl oleate; see Stepan C25; lubricant for metalworking compds.; Gardner 2 max.; m.p. 16 C; acid no. 2.0 max.; sapon. no. 191–197.

Stepan Cationic 10. [Stepan] Dialkyl quat. ammonium methosulfate; cationic; surfactant for household and industrial fabric softeners; antistat; metal cleaners; waterproofing, leather conditioner; textile lubricant; penetrating oils; buffing/polishing compds.; corrosion inhibitor; organoclay thickener; cream fluid; sol. in aromatic solvs.; dens. 8.0 lb/gal; pour pt. 23 C; flash pt. 80 F (CC); pH 3.0 ± 0.5 (5% aq.); surf. tens. 42 dynes/cm (0.1% aq.); 83 ± 3% solids.

Stepan D-50. [Stepan] IPM; emollient for personal care prods.; anticlog agent and dispersant for antiperspirants; water-wh. clear oily liq.; bland, slightly fatty odor; oil-sol.; acid no. 0.22; ref. index 1.4325; 100% act.

Stepan D-70. [Stepan] IPP; see Stepan D-50; water-wh. clear liq.; odorless; oil-sol.; sapon. no. 185–195; ref. index 1.437; 100% act.

Stepan P-621. [Stepan] Lauric-myristic alkylolamide; nonionic; liq. detergent foam stabilizer; lt. solid; water-disp.

Stepanate AM. [Stepan] Ammonium xylene sulfonate; anionic; hydrotrope, solubilizer, coupler in detergent field; heavy duty cleaners, wax strippers, dishwashing detergents; solv. or fluidizer; freeze-thaw stabilizer; additive for high electrolyte or brine systems; colorless liq.; water-sol.; dens. 9.05 lb/gal; pH 6.5–7.2 (1% aq.); 40% act.

Stepanate C-S. [Stepan] Sodium cumene sulfonate; coupler or solubilizer for liq. cleaners; detergent slurries; Klett 70 max.; water-sol.; dens. 9.65 lb/gal; 44–46% act.

Stepanate T. [Stepan] Sodium toluene sulfonate; see Stepanate AM; dens. 10.1 lb/gal; pH 7–9 (1% aq.); 40% act.

Stepanate X. [Stepan] Sodium xylene sulfonate; see Stepanate AM; also as cloud pt. depressant; colorless liq.; water-sol.; dens. 9.83 lb/gal; pH 7–9 (1% aq.); 40% act.

Stepanflote® 24. [Stepan] Sodium alkyl sulfate; flotation reagent for nonmetallic ores; tan paste; 25% act.

Stepanflote® 85L. [Stepan] Sodium alkyl ether sulfate; flotation reagent for molybdenum ore; clear liq.; 30% act.

Stepanflote® 97A. [Stepan] Sodium alkyl ether sulfate; flotation reagent for copper ore; clear liq.; 42% act.

Stepanhold Extra. [Stepan] PVP/ethyl methacrylate/ methacrylic acid copolymer; hair care fixative for super-hold formulations; visc. fluid; 40% solids.

Stepanhold R-1. [Stepan] PVP/ethyl methacrylate/ methacrylic acid polymer; hair fixative for hair sprays, setting lotions, conditioners; 50% alcohol sol'n.

Stepanol AM. [Stepan Canada] Ammonium lauryl

sulfate; anionic; detergent, emulsifier, dispersant for personal care prods.; liq.; 28% conc.

Stepanol WA 100. [Stepan] Sodium lauryl sulfate; anionic; detergent, foamer, wetting and suspending agent; dentrifrice formulations; pharmaceuticals; emulsion polymerization; clay dispersions; wh. powd.; water-sol.; pH 7.5–10 (10%); 98.5% min. act.

Stepanol WAC, WAC P. [Stepan Canada] Sodium lauryl sulfate; anionic; see Stepanol AM; liq.; 30% conc.

Stepanquat HC 80. [Stepan Europe] Alkyl ammonium methoxysulfate; conditioning, antistatic agent for hair prods.; paste.

Stepantan A. [Stepan] Condensed sodium naphthalene-formaldehyde sulfonate; anionic; dispersant, tanning agent; syn. tanning assistant, pitch control agent in papermaking, dye mordant, pigment dispersant; cream powd.; water-sol.; dens. 0.37–0.42 g/cc; pH 7.4 (2% aq.).

Stepantan H-100. [Stepan] Dodecylbenzene sulfonic acid; emulsifier for oils, solvs., waxes and oil field applics.; air entraining agent and foamer for cellular concrete; wetting agent; water-sol.; sp.gr. 1.09; dens. 9.1 lb/gal; visc. 9000 cps.

Stepantan NP 80. [Stepan] Condensed naphthalene sulfonate; dispersant, wetting agent for wettable powds.; powd.

Stepantex® DA 52. [Stepan] Alkyamine alkylaryl sulfonate; detergent, demulsifier, coupler; for treating mops and dust cloths for better dirt pick-up; emulsion degreaser; hand cleaners; emulsifier for dye carriers; dispersant for dyes and pigments; drycleaning detergent; demulsifier in oil recovery; sol. in kerosene, perchlorethylene, Stod. solv., xylene, IPA; water-disp.

Stepantex® DW 85. [Stepan] Fatty quat. methosulfate; softener for textile and home laundry use; conditioner for hair rinses; liq. to paste; 85% conc.

Stepantex® Q 90B. [Stepan] Fatty quat. sulfate; cationic; fabric softener conc. for home and commercial laundry use; pale amber clear liq.; water-disp.; sp.gr. 0.96; dens. 8.03 lb/gal; pour pt. 77 F; 90% act.

Stepantex® Q 185. [Stepan] Fatty quat. sulfate; cationic; softener for textile and home laundry use; liq.; 85% conc.

Stepantex® SO 90. [Stepan] Alkyl ammonium methoxysulfate; hydrophobing, antistatic, lubricant agent; water repellent; liq.

Stepantex® SR 85, VR 80 G, VR 85. [Stepan] Fatty quat. methosulfate; softener for textile and home laundry use; conditioner for hair rinses; liq. to paste; 85% conc.

Stepantex® VG 90, VS 90, VP 85, VRH 80. [Stepan] Dialkyl ammonium methoxysulfates; lubricating, softening, antistatic agent for household and industrial fabric softeners; paste/solid.

Stepfac®. [Stepan] Acid form of complex phosphate ester; coupler for nonionics in high electrolyte systems, esp. caustic sol'ns.; liq. fertilizer sol'ns.; water-sol.

Stepfac® 8170. [Stepan] Anionic, acid form; compatibility agent for liq. fertilizers; pale yel. clear liq.; sol. in xylene, water; 100% act.

Stepfac® 8171, 8172, 8173. [Stepan] Anionic, acid form; compatibility agent for agric. use; pale yel. clear liq.; sol. in xylene.

Stepfac® PN 10. [Stepan] Phosphoric ester; emulsifier, wetting, dispersing agent; visc. liq.

Stepfac® PN 209. [Stepan] Ethoxylated nonylphenol phosphoric ester acid; detergent, wetting, emulsifying, dispersing agent; liq.; 100% conc.

Steramine CD. [Henkel-Nopco] Polyoxyalkylene fatty deriv.; nonionic; softener assistant in textile prods.; fluid paste; 26% conc.

Steramine FPA 197. [Henkel-Nopco] Alkyl amido amine; antistat, lubricant, softener for acrylics and blends; fluid paste; 21% conc.

Steramine PNA 75. [Henkel-Nopco] Alkyl amido amine; antistat, lubricant for acrylics and blends; fluid paste; 20% conc.

Steramine S2. [Henkel-Nopco] Fatty alcohol and acid EO adduct; nonionic; finishing agent for textile prods.; wax; 100% conc.

Steramine TV. [Henkel-Nopco] Polyoxyalkylene fatty deriv.; nonionic; softener assistant for textile prods.; fluid paste; 32% conc.

Stereon® 840. [Firestone] Copolymer rubber of butadiene and styrene; rubber modifier for thermoplastic resins; as blendable modifier for impact PS (imparts extra impact strength), flame-retardant impact PS (restores impact str.), and PP (improves the stress crack resistance of PS exposed to oils and fats); for pkg. applics. involving styrenic materials; for prod. of polymeric alloys with PP from impact PP to flexible olefinic polymers; free-flowing pellets; melt index 3.0; sp.gr. 0.96; visc. (inherent) 0.8; ref. index 1.5495; hardness (Shore A) 85; 43% bound styrene.

Stereon® 840A. [Firestone] Stereospecific copolymer rubber of butadiene and styrene; rubber modifier for thermoplastic resins, esp. for impact PS, flame-retardant impact PS, and PP; pellets; melt index 12; sp.gr. 0.96; inherent visc. 0.8; ref. index 1.5495; tens. str. 1000 psi; hardness (Shore A) 85; 43% bound styrene.

Stereon® 841A. [Firestone] Stereospecific butadiene/styrene copolymer; rubber modifier for thermoplastic resins; blendable modifier for HIPS, polyolefins, asphalt; APHA 5 pellet; sp.gr. 0.96; melt index 12; ref. index 1.5495; tens. str. 1000 psi; hardness (Shore A) 85; 43% bound styrene.

Stereon® 872, 873, 874. [Firestone] Multiblock thermoplastic resins; clear modifier which blends easily with general purpose PS to yield extruded and thermoformed prods. for use in disposable beverage cups, deli lids, food containers, flexibile pkg., printing and decorating applics.; pellets; melt flow 8.0; dens. 1.01 g/cm³; visc. 0.8; m.p. 425–435 F; ref. index 1.5745; hardness (Shore) 67D; 75.0% bound styrene.

Stereon® 875. [Firestone] Multiblock thermoplastic resin; clear modifier specifically designed for deli lids and containers in direct contact with foods containing butter or mayonnaise; pellets; melt flow 4.5 g/10 min; dens. 1.038 g/cm³; m.p. 435 F; tens. str. 2930 psi (break); tens. elong. 60%; flex. str. 6130 psi (yield); impact str. (Izod) 0.45 ft lb/in. notched; hardness (Shore) 77D; soften. pt. (Vicat) 97 C.

Sterling 2XS. [Canada Packers] Sodium xylene sulfonate; hydrotrope, coupler, cloud pt. depressant; used for liq. detergents and general-purpose applics.; liq.; 40% act.

Sterling 66/33. [Canada Packers] Cocamide DEA; nonionic; foam stabilizer, aux. detergent, visc. builder; liq., paste; 100% act.

Sterling Amide 374. [Canada Packers] Cocamide DEA; nonionic; foamer, visc. control agent; base for detergent formulations; oil solubilizer; liq.; 100% act.

Sterling Amide 460. [Canada Packers] Lauramide/myristamide DEA; see Sterling 66/33; liq.; 76% act.

Sterling AOS. [Canada Packers] Sodium C_{14-16} olefin sulfonate; anionic; detergent, foamer and stabilizer used in personal care prods.; general-purpose cleaners; liq. detergents; liq.; 39% act.

Sterling CAB. [Canada Packers] Cocamidopropyl betaine; amphoteric; detergent; foamer, visc. builder used in personal care prods.; liq.; 32% conc.

Sterling DEA. [Canada Packers] Cocamide DEA; nonionic; see Sterling 66/33; liq.; 100% act.

Sterling EDTA. [Canada Packers] Tetrasodium EDTA; chelating agent for detergents, cosmetics, industrial cleaners; liq.; 38% conc.

Sterling Emulsifier #15. [Canada Packers] Pareth-25-15; nonionic; emulsifier and detergent additive; solubilizer; paste; water-sol.; 100% act.

Sterling Granulated Wax. [Canada Packers] Cocamide MEA; detergent, visc. builder, foam stabilizer; flake; 100% act.

Sterling LA Acid-C, Reg. [Canada Packers] Linear dodecylbenzene sulfonic acid; anionic; intermediate for detergent preparations and industrial cleaners; liq.; 96.3% act.

Sterling LA Paste 55%, 60%. [Canada Packers] Sodium linear dodecylbenzene sulfonate; anionic; intermediate for detergent preparations, general purpose and industrial cleaning formulations; paste; 55 and 57% act. resp.

Sterling LDEA-90. [Canada Packers] Lauramide DEA; nonionic; see Sterling 66/33; also used in cosmetics; solid; 100% act.

Sterling LM-70. [Canada Packers] Lauramide/myristamide DEA; see Sterling LDEA-90; solid; 100% act.

Sterling NPX. [Canada Packers] Nonoxynol-9; nonionic; detergent intermediate; household and industrial cleaner formulations; emulsifier; liq.; 100% act.

Sterling Snow White 750. [Canada Packers] Ditallow imidazolinium quat.; softener for home and industrial applics.; liq.; 75% act.

Sterling SXS. [Canada Packers] Sodium xylene sulfonate; anionic; coupling agent, solubilizer, cloud pt. depressant; liq.; 41% act.

Sterotex®. [Capital City] Hydrog. veg. oil; binder and internal lubricant for pressed powds.; powd.

Sterotex® HM & K Veg. Oils. [Capital City] Hydrog. veg. oils; lubricant in pharmaceutical tableting, powd. compression applics; fine wh. powd.

Sterox® DF. [Monsanto] Dodecyl phenol ethoxylate (6 EO); nonionic; wetting agent, emulsifier, paper rewetting, intermediate for metal cleaning compds., o/w agric. emulsifier, dedusting agent, felt washing, latex emulsion stabilizer, paints, cutting and sol. oils; clear yel. visc. liq.; mild odor; misc. with water; sol. in nonpolar solvs. and oils; m.w. 535; ref. index 1.4922; sp.gr. 1.0277; visc. 343 cps; HLB 10.0; flash pt. 495 F; pour pt. 0 F; surf. tens. 29.9 dynes/cm; pH 6.0–7.5 (1%); 99.5% act.

Sterox® ND. [Monsanto] Nonoxynol-4; nonionic; detergent, wetting agent, emulsifier; intermediate for surfactant mfg.; solv. cleaners; additive in petrol. prods.; dedusting; making of cutting and sol. oils; PVAc plasticizer; latex emulsion stabilizer; agric. emulsions; animal skin cleaners; clear yel. visc. liq.; mild odor; misc. with water; sol. in nonpolar solvs.; m.w. 396; ref. index 1.4962; sp.gr. 1.0216; visc. 252 cps; HLB 8.9; flash pt. 435 F; pour pt. –24 F; surf. tens. 29.7 dynes/cm; pH 6.0–7.5 (1%); 99.5% act.

Sterox® NE. [Monsanto] Nonoxynol-5; nonionic; detergent, wetting agent, emulsifier; intermediate for surfactant mfg.; solv. cleaners, additive in petrol. prods.; making of cutting and sol. oils; clear yel. liq.; mild odor; sol. see Sterox ND; ref. index 1.4955; sp.gr. 1.0322; visc. 236 cps; HLB 10.0; flash pt. 445 F; pour pt. –30 F; surf. tens. 28.7 dynes/cm; 99.5% act.

Sterox® NF. [Monsanto] Nonoxynol-6; nonionic; see Sterox NE; also used in emulsified paints; clear yel. liq.; mild odor; sol. see Sterox ND; ref. index 1.4940; sp.gr. 1.0385; visc. 207 cps; flash pt. 480 F; pour pt. –20 F; surf. tens. 28.8 dynes/cm; 99.5% act.

Sterox® NG. [Monsanto] Nonoxynol-6 (6.5 EO); nonionic; see Sterox NE; also additive in solv.-based paints; clear yel. liq.; mild; sol. see Sterox ND; ref. index 1.4933; sp.gr. 1.0426; visc. 229 cps; HLB 11.3; flash pt. 485 F; pour pt. 0 F; surf. tens. 29.1 dynes/cm; 99.5% act.

Stiffener DSC. [Anchor] Amine salt on inert filler; thickener; off-wh. powd.; sp.gr. 2.3; 25% act.

Stoner No E944. [Stoner] Heavy duty rust and corrosion preventive with red tracing dye; leaves film providing long term protection to steel and other metals from moisture and oxidation; lt. red liq.; nonflam.

Stoner No. E922. [Stoner] Rust and corrosion preventive with red tracing dye; leaves film protecting steel and other metals from moisture and oxidation; ideal for molds for plastic and rubber; lt. red liq.; nonflam.

Stoner No. E923. [Stoner] Corrosion inhibitor for steel and other metals; antirust wax leaving heavy film on exposed metal for long-term protection from moisture and oxidation; ideal for molds for plastic and rubber molding; clear to wh. film; nonflam.

Stoner No. E965. [Stoner] Paintable mold release agent for plastic, rubber, wax, and ceramic molding incl. inj., compr., transfer, vacuum form, pour cast, die cast, extrusion, and foam molding operations; antistick agent; allows molded parts to be painted, plated, decorated; clear colorless liq.; nonflam.

Striptron Stripper. [Dow] Inhibited methylene chloride with additives; solv.; stripper for dry film photoresist and screen inks; sol. in most org. solvs.; low water sol.

Strodex MO-100. [Dexter] 2-Ethylhexyl polyphosphoric ester acid anhydride; anionic; emulsifier; pigment dispersant used in oil-based paints and leather coating specialties; liq.; sol. in polar and nonpolar solvs.; insol. in water; 100% conc.

Strodex MOK-70. [Dexter] Strodex MO-100, potassium salt; anionic; dispersant in paints; wetting and rewetting agent; paste; 70% conc.

Strodex MR-100. [Dexter] Polyphosphoric ester acid anhydride; anionic; emulsifier, dispersant for org. pigments in polar and nonpolar solvs.; liq.; 100% conc.

Strodex MRK-98. [Dexter] Strodex MR-100, potassium salt; anionic; emulsifier for grinding and dispersing org. pigments; liq.; 98% conc.

Strodex P-100. [Dexter] Polyphosphoric ester acid anhydride; anionic; emulsifier; dispersant for pigments in polar and nonpolar solvs.; coupler for emulsifiers in aq. and nonaq. systems; used in metal and tile cleaners; pigment grinding aid; liq.; 100% conc.

Strodex PK-80A, PK-95G. [Dexter] Potassium salt of Strodex P-100; pigment dispersant; liq.

Strodex PK-90. [Dexter] Strodex P-100, potassium salt; anionic; emulsifier, dispersant for extender pigments in latex paints, barium sulfate, and iron oxides; wetting agent; oxidation-corrosion inhibitor; used in heavy-duty alkaline cleaners; liq.; 90% conc.

Strodex PSK-28. [Dexter] Phosphate esters, potassium salt; anionic; emulsifier, wetting agent, pigment dispersant; liq.; water-sol.; 58% conc.

Strodex SE-100. [Dexter] Polyphosphoric ester acid anhydride; anionic; dispersant, emulsifier and coemulsifier in aq. and nonaq. systems; liq.; oil-sol.; 100% conc.

Strodex SEK-50. [Dexter] Strodex SE-100, potassium salt; anionic; dispersant, emulsifier, stabilizer for latex emulsions, paints, titanium dioxide and extender pigments; liq.; 50% conc.

Strodex Super V-8. [Dexter] Phosphate ester, potassium salt; anionic; emulsifier, detergent, dispersant, wetting agent used in textile processing applics.; clear liq.; sol. in water, aliphatic, aromatic, and chlorinated solvs.; sp.gr. 1.09–1.11; pH 7–8 (1%); 75–80% act.

Struktol 40 MS. [Struk-Chem] Mixture of rubber-compatible nonhardening resins; resin plasticizer; homogenizing agent for elastomer blend compds.; dk. brn. resinous material.

Struktol 40 MS Flakes. [Struk-Chem] Modified mixture of rubber-compatible nonhardening syn. resins; resin plasticizer; homogenizing agent for elastomer blend compds.; dk. brn. flakes.

Struktol 60 NS. [Struk-Chem] Modified mixture of rubber-compatible nonhardening lt.-colored resins; resin plasticizer; homogenizing agent for lt.-colored elastomer blend compds.; amber transparent resinous material.

Struktol 60 NS Flakes. [Struk-Chem] Mixture of rubber-compatible lt.-colored aliphatic resins; resin plasticizer; homogenizing agent for lt.-colored elastomer blend. compds.; amber flakes.

Struktol TR 016. [Struk-Chem] Blend of metal fatty acid salt and amide with sites avail. for hydrogen bonding; processing agent for thermoplastic elastomers and resins; approved for food and drug contact; pale yel. powd., pellets, or flakes; neutral odor; sp.gr. 1.0; m.p. 99 C; flash pt. > 290 C.

Struktol WB 300. [Struk-Chem] Blend of high molecular aliphatic and aromatic polyesters; plasticizer for nitrile rubber and chloroprene; end uses incl. petrol hoses, printing rollers, oil seals, etc.; processing aid; pale yel. med.-visc. liq.; insol. in aliphatic hydrocarbons, min. oils and greases; dens. 1.13; visc. ≈ 400 (20 C); flash pt. 250 C.

STXS. [Continental] Sodium toluene xylene sulfonate; hydrotrope used in detergent formulations; colorless clear liq.; mild odor; dens. 9.94; sp.gr. 1.18; m.p. 12 F; b.p. 212 F; 42% act.

Stygene Series. [Chemfax] Polynuclear aromatic polymers derived from specially prepared petrol. stream; used in rubber compding., joint cements, plastic compds., protective coatings, fiber board, inks, epoxy potting compds., adhesives, insecticides, sats., briquettes, floor tile, etc.; liqs. and soft solids for plasticizing rubber and as sec. plasticizers; hard grades in rubber as extending and fortifying resin; carrier for insecticidal toxicants; in calendered and extruded goods to produce compds. which process freely and smoothly; in tire tread and camelback stocks, higher melt Stygenes exhibit cut-growth inhibiting properties and provide ease of handling; in heel, sole, and other mechanical goods compds. to improve mold flow, increase resistance to abrasion and tear, etc.; for use in vinyl polymers, natural rubber, Neoprene, etc.; Barrett 22 liq. to hard solids; sol. in aromatics, terpenes, chlorinated hydrocarbons, higher ketones and esters; acid no. 0 (neutral); flam. self-extinguishing when flame source is removed.

Stygene R-2. [Chemfax] Polynuclear aromatic polymer; see Stygene Series; R-2 is plasticizer and softener for natural and syn. rubber, and vinyl polymers; low volatility and good plasticizing action useful in tread and camelback and as a tackifier for carcasses, mechanical goods, and footwear stocks; lowest m.w. member of series; as sat. for wood, paper, fabrics, cord, rope, and felt; liq.; sp.gr. 1.06; dens. 8.9 lb/gal; visc. 90 SSU (210 F); flash pt. (COC) 250 F.

Stygene R-10, R-25. [Chemfax] Polynuclear aromatic polymers; see Stygene Series; visc. liq.; sp.gr. 1.05 and 1.06 resp.; soften. pt. 10 and 25 C resp..

Stygene R-50, R-70, R-85. [Chemfax] Polynuclear aromatic polymer; see Stygene Series; R-50, R-70, R-85 used in low-quality stocks that will not heat up on the mill to the point where Stygene R-100 can be used; solid; sp.gr. 1.09, 1.12, 1.12 resp.; soften. pt. 50 C, 70 C, 85 C resp.

Stygene R-100. [Chemfax] Polynuclear aromatic polymer; see Stygene Series; R-100 is popular grade for general usage; curing agent for tread buffings, ground springs from mold overflow, defective cured and scorched stocks; solid or flaked; sp.gr. 1.16; soften. pt. 100 C.

Stygene R-115, R-150, R-175. [Chemfax] Polynuclear aromatic polymer; see Stygene Series; solid or flakes; sp.gr. 1.21, 1.29, and 1.35 resp.; soften. pt. 115, 150, and 175 C resp.

Sublimed Blue Lead. [Eagle-Picher] Basic blue lead sulfate; lubricating aid and friction aid for the mfg. of brake linings and clutch facings; lubricant additive in oil and high-pressure greases; rust inhibitive pigment for structural steel; 99.9% –325 mesh; sp.gr. 6.17; dens. 51.40 lb/solid gal; bulking value 0.0193 gal/lb; oil absorp. 8.0; 76.0% lead.

Suconox-18. [Hexcel] N-stearoyl-p-amino phenol; processing aid and antioxidant in plastics; m.p. 130–134 C; > 98% assay.

Sufatol LX/B. [Standard Chem.] Sulfated fatty alcohol; anionic; scouring agent, softener, and dyebath aux. for textiles; paste; 25% conc.

Sulfads®. [Vanderbilt] Essentially dipentamethylene thiuram hexasulfide; ultra accelerator for NR and syn. rubbers; vulcanizing agent; lt. yel. to buff powd.; 99.9% thru 100 mesh; rods; m.w. 448.82; sol. in chloroform, toluene, acetone; practically insol. in water; dens. 1.50 ± 0.03 mg/m³ (powd.); m.p. 115 C min.; 35.0% avail. sulfur (powd.); 32.5% avail. sulfur (rods).

Sulfa-Hitech® 0382. [Pennwalt] DMDS-based; solv. used to dissolve sulfur in the prod. of sour gas wells, in sour-gas pipelines, and in refinery and chemical plant flowlines; sp.gr. 1.060; b.p. 107–111 C.

Sulfamin. [Berol Nobel] TEA alkylaryl sulfonate; anionic; detergent, foamer; washing and cleaning agent; textile wetting agent; penetrant; Gardner 5 clear liq.; water-sol.; dens. 1.080 g/cc; visc. 300 cps; clear pt. < 5 C; flash pt. > 100 C; surf. tens. 34 dynes/cm; pH 6.5–7.5 (1% aq.); 50% act.

Sulfasan® R. [Monsanto] 4,4′-Dithiodimorpholine; vulcanizing agent, cross-linking agent for elastomers; sulfur donor for EV and semi-EV cure systems; avail. reg. grind, special grind, thread-grade powd., wax pellets 80% act.

Sulfetal AF. [Zschimmer & Schwarz] Sodium fatty alcohol sulfate (C_{16}–C_{24}); flotation agent, cleansing pastes; paste; 50% act.

Sulfolane W. [Shell] Tetramethylene sulfone; highly polar compd. with outstanding solv. properties and chemical and thermal stability; for most classes of org. compds. and many common polymers; preferred for extraction of benzene, toluene, and other aromatic hydrocarbons from oil refinery streams; used in Sulfinol process to remove acid gases from a variety of sour gas streams and to remove CO_2 from hydrogen and ammonia synthesis gases; used for separation of low boiling alcohols, min. oils, tars, and other fatty acids; plasticizer; polymerization solv.; dielec. in elec. equipment; solv. in surf. coatings; component in hydraulic fluids; sp.gr. 1.2600–1.2615; 2.6–3.0% water.

Sulfonate OA-5. [Tennessee] Sulfonated oleic acid, sodium salt; anionic; wetting agent; antifoam/defoamer, corrosion inhibitor, acid-stable surfactant, solubilizer; lt. brn. liq.; 41% act.

Sulfonate OE-500. [Tennessee] Sulfonated oleic acid, sodium salt, amyl-ester; surfactant base, coupler, solubilizer, emulsifier for dyes and additives used in polyesters.

Sulfonated Castor Oil 50%. [Nat'l. Starch] Fatty glyceride sulfate; anionic; emulsifier, lubricant, dyeing assistant and dispersant; liq.; biodeg.; 50% conc.

Sulfonated Castor Oil 75%. [Nat'l. Starch] Fatty glyceride sulfate; anionic; see Sulfonated Castor Oil 50%; liq.; 70% conc.

Sulfonated Castor Oil GTO. [Nat'l. Starch] Fatty glycerine sulfate; anionic; emulsifier, lubricant, dyeing assistant, dye dispersant and leveler; liq.; biodeg.; 77% conc.

Sulfonated Syn. Sperm Oil 2091. [Clough] Sulfated syn. sperm oil; anionic; emulsifier, lubricant, fat liquor for leather industry; liq.; 59–61% conc.

Sulfonic 800. [Tennessee] Sulfonated oleic acid; intermediate used in textile and industrial cleaning; dk. brn. visc. liq.; water-sol.; sp.gr. 1.09; m.p. < 16 F; b.p. 248–257 F; flash pt.> 230 F (PMCC); pH 0.6; 50–55% act.

Sulfonic 864. [Tennessee] Sulfonic acids; anionic; detergent and penetrant; liq.; biodeg.

Sulfonic Acid LS. [Hart Chem. Ltd.] Linear alkylbenzene sulfonic acid; anionic; intermediate for detergent formulations; amber liq.; dens. 1.055; visc. 1000 cps; 96% act.

Sulfopon WA 1. [Henkel Canada] Sodium alkyl sulfate; anionic; detergent, wetting, emulsifier for SLS; dispersant; solubilizer for emulsion polymers; base for rug and upholstery shampoos; liq.; 30% conc.

Sulfosoft. [Berol Nobel] Linear alkylbenzene sulfonic acid; anionic; intermediate used in prod. of detergents; household, industrial, and washing compds.; emulsifier for solvs. and oils; Hazen 150 visc. liq.; water-sol.; m.w. 320; dens. 1.04 g/cc; visc. 3500 cps; pour pt. 6 C; surf. tens. 33 dynes/cm; pH 2.2 (1% aq.); 97 ± 1.0% act.

Sulfostat KNT. [Zschimmer & Schwarz] Amido alkylamine acetate; antistatic additive for cosmetics; liq.

Sulfotex OA. [Henkel] Sodium octyl sulfate; anionic; wetting agent for animal glue on paper and paperboard, high electrolyte concs., and food processing; detergent and dispersant in industrial cleaners; mercerizing agent in dyeing of fibers; lt. amber; dens. 9.2 lb/gal; pH 8.0–10.0 (10%); biodeg.; 38.5–40.5% act.

Sulfotex PAI. [Henkel] Ammonium alkyl ether sulfate; anionic; foamer and dispersant used in cleaning formulations; pale yel. liq.; dens. 8.31 lb/gal; pH 7.0–7.5; biodeg.; 46–50% act.

Sulfotex PAI-S. [Henkel] Sodium alkyl ether sulfate; anionic; see Sulfotex PAI; pale yel. liq.; dens. 8.69 lb/gal; pH 7.5–8.5; biodeg.; 45–49% act.

Sulfotex SXS. [Henkel] Sodium xylene sulfonate; anionic; hydrotrope, solubilizer for liq. detergents, pine oil in water, inks to prevent gumming; liq.; biodeg.; 40% act.

Sulframin 40. [Witco] Linear alkylaryl sodium sulfonate; anionic; detergent base and wetting agent for household and industrial specialty compds., personal care prods., metal cleaning; lubricant, emulsifier, dye dispersant, penetrant, scouring agent, antistat in textile industry; defoamer for petrol. industry; flakes, gran.; 40% conc.

Sulframin 40DA. [Witco] Linear alkylaryl sodium sulfonate; anionic; detergent for car washing, household and industrial cleaning; wetting agent, lubricant, emulsifier, antistat; metal cleaning; textile surfactant; beads; 40% conc.

Sulframin 40RA. [Witco] Linear alkylaryl sodium sulfonate; anionic; see Sulframin 40DA; also used as base in personal care prods.; beads; 40% conc.

Sulframin 40T. [Witco] Linear alkylaryl TEA sulfonate; anionic; see Sulframin 40DA; liq.; 40% conc.

Sulframin 60T. [Witco] Linear alkylaryl TEA sulfonate; see Sulframin 40T; also for personal care prods.; liq.; 60% conc.

Sulframin 85. [Witco] Linear alkylaryl sodium sulfonate; anionic; detergent, lubricant, emulsifier, antistat, wetting agent for household and industrial specialty compds., metal cleaning, personal care prods., textiles; flakes, powd.; 85% conc.

Sulframin 90. [Witco] Linear alkylaryl sodium sulfonate; anionic; see Sulframin 85; also surfactant for emulsion polymerization and latex stabilization; flakes; 91% conc.

Sulframin 1240, 1245. [Witco] Linear alkylaryl sodium sulfonate; anionic; see Sulframin 85; slurry; 40 and 45% conc. resp.

Sulframin 1250, 1260. [Witco] Linear alkylaryl sodium sulfonate; anionic; see Sulframin 60T; slurry; 51 and 60% conc. resp.

Sulframin 1288. [Witco] Linear alkylaryl sulfonic acid; anionic; penetrant, lubricant, dispersant, detergent, antistat, intermediate for wetting agents, emulsifiers, and detergents for household and industrial specialties, personal care prods.; textile surfactant; liq.; 88% conc.

Sulframin 1298. [Witco] Linear alkylaryl sulfonic acid; anionic; see Sulframin 1288; also used in personal care prods.; surfactant for emulsion polymerization; liq.; 98% conc.

Sulframin 1388. [Witco] Linear alkylaryl sulfonic acid; anionic; see Sulframin 1288; liq.; 88% conc.

Sulframin CSA. [Witco] Cumene sulfonic acid; hydrotrope, coupler, solubilizer, catalyst; used in liq. detergents.

Sulframin TX. [Witco] Toluene sulfonic acid, modified; hydrotrope; coupler, catalyst; solubilizer for liq. detergent.

Sunaptol CA 25, 120, 350, 400, 750. [ICI Ltd.] Castor oil, ethoxylated; nonionic; defoamer, emulsifier, dispersant, softener, leveling agent and antistat for textiles; liq.; 100% conc.

Sunaptol CFA 90, 110. [ICI Ltd.] Fatty acid, ethoxylated; nonionic; emulsifier and antistat; liq.; 100% conc.

Sunaptol DL Conc. [ICI Ltd.] Alkyl phenol, ethoxylated; nonionic; wetting agent and detergent for scouring in textiles; antistat; emulsifier for waxes and greases; liq.; 85% conc.

Sunaptol HB4. [ICI Ltd.] Alkyl phenol, ethoxylated; nonionic; emulsifier, solubilizer and wetting agent used for iodine in Iodophor preparations; liq.; 100% conc.

Sunaptol OA. [ICI Ltd.] Oleic acid, ethoxylated; nonionic; emulsifier, oil slick dispersant, antistat for PVC, ABS; liq., paste.

Sunaptol OA 70, 90, 100, 140. [ICI Ltd.] Oleic acid, ethoxylated; nonionic; see Sunaptol OA; liq., paste.

Sunaptol P Extra Liquid. [ICI Ltd.] Fatty alcohol, ethoxylated; nonionic; dispersant and leveling agent for dyes; dyeing assistant in textiles; emulsifier; liq.; 40% conc.

Sunaptol PP75B. [ICI Ltd.] Coco MEA, ethoxylated; nonionic; foam stabilizer for liq. detergent formulations and personal care prods.; paste; 100% conc.

Sunaptol S 60, 80, 230. [ICI Ltd.] Stearic acid, ethoxylated; nonionic; emulsifier and softener; solid; 100% conc.

Sunaptol TO 28, 85. [ICI Ltd.] Tall oil, ethoxylated; nonionic; defoamer and emulsifier; liq.; 100% conc.

Sunnol NES. [Lion] POE alkylphenyl ether sulfate; anionic; dyeing assistant in textiles; liq.; 30% conc.

Sunolite® 100. [Witco] Petrol.-derived wax; antisunchecking agent and antiozonant used in rubber prods.; sole protectant; lt. yel. flakes; sp.gr. 0.92; m.p. 63–67 C; flash pt. 243 C.

Sunolite® 127. [Witco] Petrol.-derived wax; see Sunolite 100; yel. flakes; sp.gr. 0.92; m.p. 64–68 C; flash pt. 247 C.

Sunolite® 160. [Witco] Refined petrol. wax; external lubricant and processing aid in rigid PVC formulations for extrusion and inj. molding processes; imparts smoothness and high gloss to PVC prods. as pipes and fittings for potable water, natural gas, DWV, sewage, drainage, and irrigation; elec. conduits and profiles for construction industry; food pkg. applic.; wh. flakes or powd.; nil odor and taste; sp.gr. 0.922; visc. 50 SUS; m.p. 70 C; flash pt. 268 C (COC).

Sunolite® 240. [Witco] Petrol.-derived wax; see Sunolite 100; lt. yel. flakes; sp.gr. 0.92; m.p. 66–70 C; flash pt. 248 C.

Sunolite® 666. [Witco] Petrol.-derived wax; antiozonant used in rubber goods; sole protectant; lt. yel. flakes; sp.gr. 0.92; m.p. 63–67 C; flash pt. 249 C.

Sunproofing Wax 1343. [Frank B. Ross] Antiozonant, anticracking and sunchecking wax for the rubber industry; pale yel.; sp.gr. 0.917–0.939; m.p. 70–74 C; acid no. nil; sapon. no. 1.0 max.

SunShade 17-107. [Santech] ZnO/synergist system; wh. uv stabilizer for thermoplastics incl. LDPE, HDPE, PP, and PVC; for cable insulation, pipes, ropes, sheets, tarps, shipping tanks, wraps, drums, and sacks, outdoor furniture, signs, awnings, siding, all-weather carpet, auto trim, safety, sports and pool equipment, $^1/_8$ in. wh. pellets; sp.gr. 1.45; 50% active ingred. in a polymeric base;

SunShade 17-221. [Santech] ZnO/synergist system; see SunShade 17-107; wh. 1/8 in. pellets; sp.gr. 1.82;

76% active ingred. in a polymeric base.

Sunsoflon CK. [Nikko] Alkyl polyamide; cationic; surfactant, softener for textiles; liq.

Sunsoflon K-2. [Nikko] Alkyl polyamide; cationic; see Sunsoflon CK; liq.; 18% conc.

Sunsoflon MT-100. [Nikko] Alkyl polyamide; cationic; see Sunsoflon CK; liq.

Sunsoflon PXP-705. [Nikko] Imidazoline; cationic; softener for textiles; liq.

Sunsoflon PXP-2000. [Nikko] Imidazoline; cationic; softener and lubricant for textiles; soft paste.

Sunsolt RZ-2, -6. [Nikko] Alkylaryl ester; anionic; dispersant and leveling agent for textiles; liq.; 26 and 40% conc. resp.

Super Ad It. [Servo] Org. mercury compd.; fungicide, bactericide for paint industry; oil-sol.

Super Amide GR. [Stepan] Cocamide DEA; nonionic; foam stabilizer, emulsifier, thickener; household, industrial, and cosmetic formulations; max. cost/performance; liq.; sp.gr. 0.99; flash pt. 146 F; acid no. 10 max.; 80% conc.

Super Amide L9, L9C. [Stepan] Lauramide DEA; nonionic; provide thickening and foam stabilizing properties in many cosmetic, industial, and household formulations, e.g. shampoos, bubble baths, lt. duty dishwash detergents, and industrial cleaners; paste; sp.gr. 0.98 and 0.96 resp.; flash pt. 178 and 148 F resp.; acid no. 9 and 2 max. resp.

Super Amide L9A. [Stepan] Lauramide DEA; nonionic; foam stabilizer, emulsifier, thickener used for personal care prods.; paste; 97% conc.

Super Amide LL. [Stepan] Lauramide DEA, modified; nonionic; surfactant used as foam booster, stabilizer, and visc. builder; liq.; sp.gr. 0.98; flash pt. 178 F; 97% conc.

Super Amide LM. [Stepan] Lauramide DEA; nonionic; see Super Amide L9; paste; sp.gr. 0.98; flash pt. 178 F; acid no. 10 max.

Superclear. [Henkel] Natural gum and preservatives; thickener; prevents migration of dyestuffs; protective colloid; water-sol.

Superclear 80-N. [Henkel] Aq. colloidal disp. of a branched polysaccharide macromolecule containing act. carbonyl and hydroxyl groups; nonionic; natural gum for textile dyeing and printing; water modifier controlling mobility of discrete dye particles in aq. systems; hinders migration of dyestuffs during drying of fabrics; dk. rdsh.-brn. semifluid liq.; dens. 10 lb/gal; visc. 11,000 cps; pH 7.0.

Superclear 100-N. [Henkel] Aq. colloidal disp. of a branched polysaccharide macromolecule containing act. carbonyl and hydroxyl groups; nonionic; see Superclear 80-N; dk. rdsh.-brn. semifluid liq.; dens. 10 lb/gal; visc. 25,000 cps; pH 7.0.

Superclear 200-N. [Henkel] Aq. colloidal disp. of a branched hydroxyl groups; nonionic; see Superclear 80-N; dk. rdsh.-brn. semifluid liq.; dens. 10 lb/gal; visc. 60,000 cps; pH 7.0.

Supercol® Guar F. [Aqualon] Galactomannan polysaccharide; edible guar gum for use as thickener, stabilizer, bodying agent for food applics.; wh. to cream; 90% thru 200 mesh.

Supercol® Guar G2. [Aqualon] Galactomannan polysaccharide; edible guar gum with large particle size for use as stabilizer and bodying agent where high visc. and gradual hydration rate are desired; for food applics. white to cream; 90% thru 100 mesh.

Supercol® Guar U. [Aqualon] Galactomannan polysaccharide; edible guar gum used as stabilizer, thickener, bodying agent with rapid hydration for food appalics.; wh. powd.; 80% min. thru 200 mesh; disp. in cold water; 100% conc.

Super Corona. [Croda] Refined anhyd. lanolin; superfatting emollient and emulsifier for cosmetics and pharmaceuticals; improves spreading, penetration, and aesthetic properties of these prods.; plasticizer in aerosol hairsprays; stable w/o emulsions; Gardner 8.5 max. soft solid; odorless; m.p. 36–42 C; acid no. 1.0 max.

Superfine Lanolin Anhydrous USP. [Croda Ltd.] Lanolin; superfatting emollient with some emulsifying properties; yel./amber soft solid; low odor; m.p. 38–44 C.

Super Hartolan. [Croda] Lanolin alcohols; spreading agent, dispersant, stabilizer, plasticizer, o/w emulsifier, and emollient for cosmetics and pharmaceuticals; pale amber solid wax; oil-sol.; m.p. 60 C min.; HLB 1.0; acid no. 1.5 max.; sapon. no. 5 mg max.; 100% conc.

Superloid®. [Kelco] Refined ammonium alginate; gum used as gelling agent, thickener, emulsifier, film-forming agent, suspending agent, and stabilizer in food, pharmaceutical, and industrial applics.; stabilizer in paper and textile industry; tan gran. particles; water-sol.; sp.gr. 1.73; dens. 56.62 lb/ft^3; visc. 1500 cps; ref. index 1.3347; pH 5.5; 13% moisture.

Superol. [Procter & Gamble] Glycerin USP; humectant.

Superox 702. [Reichhold] MEK peroxide in a plasticizer sol'n.; catalyst for R.T. curing of unsaturated polyester and vinyl ester resins; used in gel coats; colorless clear liq.; sol. in polar org. compds. and polyester resins; insol. in water; sp.gr. 1.16; visc. 14 cps; ref. index 1.48; flash pt. > 140 F; f.p. < –22 F.

Superox 732. [Reichhold] MEK peroxide in a low visc. plasticizer sol'n.; See Superox 702; colorless clear liq.; slight odor; f.p. < –22 F; sol. see Superox 702; sp.gr. 1.05; visc. 5 cps; ref. index 1.47; flash pt. > 140 F (Seta).

Superox 744. [Reichhold] Benzoyl peroxide in a nonvolatile plasticizer disp.; polymerization initiator or catalyst for polyester and vinyl ester resins, vinyl monomers; wh. liq. disp.; sol. in org. solvs., unsaturated polyester resins, and monomers; sp.gr. 1.15; dens. 9.6 lb/gal; visc. 5000 cps; flash pt. (Seta) 140 F.

Super-Pflex 100. [Pfizer] Surface-modified precipitated calcium carbonate; filler; provides exc. extrusion processing, impact str. retention, and enhanced color stability on outdoor exposure in rigid PVC siding and profiles; 0.7 micron particulate; sp.gr. 2.7; bulk dens. 38 lb/ft^3 (tapped); surf. area 6.0 m^2/g; oil absorp. 26 g/100 g; 98.4% $CaCO_3$.

Super-Pflex 200. [Pfizer] Surface-modified, precipitated calcium carbonate; reinforcing agent for rigid

PVC; used in elec. conduit, DWV pipe, pressure pipe, PVC fittings, rainwater gutters and downspouts, house siding, and various profiles; powd.; 0.5 μ avg. particle size.

Super-Pro #5A. [Stepan] TEA coco-hydrolyzed animal protein; anionic; detergent, conditioner, emulsifier, moisturizer, foamer used in personal care prods.; Gardner 10 max. liq.; sp.gr. 1.15 min.; visc. 400 cps min.; pH 7.0–7.5; 68% act.

Super Refined Almond Oil NF. [Croda] Almond oil; emollient; provides elegant skin feel and promotes spreading in creams, lotions, bath oils; APHA 40.

Super Refined Apricot Kernel Oil NF. [Croda] Apricot kernel oil; emollient; lubricant and softener in nail oils; conditioner for hair care prods.; APHA 30.

Super Refined Avocado Oil. [Croda] Avocado oil; emollient; provides elegant skin feel and promotes spreading in creams, lotions, bath oils; emollient for hair care prods. APHA 30.

Super Refined Babassu Oil. [Croda] Babassu oil; emollient for sunscreen prods.; APHA 20.

Super Refined Coconut Oil. [Croda] Coconut oil; emollient; improved color and odor for sunscreen prods.; APHA 20.

Super Refined Crossential EPO. [Croda] Evening primrose oil; emollient; contains essential fatty acids vital to health of skin; APHA 20.

Super Refined Grapeseed Oil. [Croda] Grapeseed oil; emollient; APHA 50.

Super Refined Menhaden Oil. [Croda] Menhaden oil; emollient; contains essential fatty acids vital to health of skin; APHA 100.

Super Refined Mink Oil. [Croda] Mink oil; emollient oil for makeup remover systems; APHA 30.

Super Refined Olive Oil. [Croda] Olive oil; cosmetic emollient; lubricant for hair care prods.; APHA 30.

Super Refined Orange Roughy Oil. [Croda] Orange roughy oil; emollient, softener, spreading agent in skin care prods.; APHA 30.

Super Refined Peanut Oil. [Croda] Peanut oil; emollient in skin care prods.; also for nutritional supplements, pharmaceutical delivery systems; APHA 30.

Super Refined Safflower Oil USP. [Croda] Safflower oil; emollient; mobile oil; APHA 30.

Super Refined Sesame Oil. [Croda] Sesame oil; emollient oil used in skin care preps., nutritional supplements, pharmaceutical delivery systems; APHA 30.

Super Refined Shark Oil. [Croda] Shark liver oil; emollient, moisture repellent for skin protection; APHA 40.

Super Refined Soybean Oil. [Croda] Soybean oil; emollient oil used in skin care preps., nutritional supplements, pharmaceutical delivery systems; APHA 60.

Super Refined Wheat Germ Oil. [Croda] Wheat germ oil; emollient; contains essential fatty acids vital to health of skin; used in facial creams; moisturizer in hair care prods.; APHA 250.

Super-Sat. [RITA] Hydrog. lanolin; nonionic; plasticizer, emollient; cosmetics and pharmaceuticals; makeups, night creams, shaving creams; wh. solid; sp.gr. 0.85–0.86; m.p. 48–52 C; 100% act.

Super-Sat AWS-4. [RITA] PEG-20 hydrog. lanolin; nonionic; emollient, emulsifier, plasticizer for emulsion systems; amber solid; slightly water-sol.; 100% conc.

Super Solan Flaked. [Croda] PEG-75 lanolin; emollient, conditioner, superfatting agent, solubilizer; yel. flake; water-sol.

Super Solangel 25. [Croda] PEG-75 lanolin; emollient, conditioner, superfatting agent; yel. liq.; water-sol.

Super Sta-Tac® 80. [Reichhold] Specialty hydrocarbon based on mixed olefins; for pressure-sensitive adhesive, hot-melt, and sealants; tackifier for S-I-S and S-B-S polymers; Gardner color 3–4 (50% N.V. in toluol); soften. pt. (R&B) 80 C.

Super Sterol Ester. [Croda] C_{10-30} cholesterol/lanosterol esters; emollient, lubricant, and moisturizer for dry skin, cosmetics, pharmaceuticals; soft wh. solid; sol. in oil, esters.

Super Sublimed White Lead 41. [Eagle-Picher] Basic sulfate white lead; wh. pigment with exc. color used as the act. ingred. in the mfg. of bank notes and printing inks; also used in brake linings, clutch facings, and high-pressure lubricants; 99.9% –325 mesh; sp.gr. 6.40; apparent dens. 75.0 lb/ft^3; dens. 53.31 lb/solid gal; bulking value 0.0188 gal/lb; oil absorp. 8.0; brightness 83.0%; 83.5% $PbSO_4$; 16.5% PbO.

Super Tin® 4L. [Griffin] Triphenyltin hydroxide sol'n.; flowable fungicide for pecans, potatoes, sugar beets; restricted use; 40% act.

Super Wet. [RITA] Ethoxylated fatty acid and ether; nonionic; wetting agent and emulsifier for pesticides; soil penetrant; liq.; 100% conc.

Supraene®. [Robeco] Purified squalene; natural emollient.

Supragil MNS/90. [Rhone-Poulenc SpA] Sodium methyl naphthalene sulfonate condensate; anionic; dispersing/suspending agent for pesticide wettable powds. and water-disp. gran.; powd.; 90% conc.

Supratol VF. [Hart Chem. Ltd.] Highly sulfonated castor oil; anionic; dispersant, lubricant, penetrant, and conditioner for leather processing, acid and disperse dyes; amber liq.; sp.gr. 1.112; visc. 90 cps; ref. index 1.4022; pH acid; 30% act.

Surco 40-SX. [Stepan] Sodium alkylaryl sulfonate; anionic; detergent base for liq. cleaning systems; hydrotrope coupler; liq.; sp.gr. 1.09; flash pt. > 200 F; pH 7.5–8.5; 37–40% act.

Surco 128T. [Stepan] Cocamide DEA (1:1); nonionic; foam stabilizer, thickener, emulsifier; household, industrial, and cosmetic formulations; liq.; sp.gr. 0.99; flash pt. 146 F; acid no. 2 max.; 100% conc.

Surco CMEA. [Stepan] Cocamide MEA; nonionic; foam stabilizer, thickener, opacifier; industrial, household, and cosmetic formulations; wax or flake; sp.gr. 0.92; flash pt. 146 F; acid no. 2 max.; 100% conc.

Surco Coco Betaine. [Stepan] Cocoamidopropyl betaine; amphoteric; surfactant, foaming agent and stabilizer with wetting properties; used in household, industrial, and dishwashing formulations; sp.gr. 1.04; flash pt. > 200 F; 40.0–44.0% solids.

Surco DC-47. [Stepan] Formulated anionic blend (sodium salt); detergent and foam stabilizer; provides visc. throughout a wide range of dilutions; high active to solid ratio generates max. performance; liq.; sp.gr. 1.07; flash pt. > 200 F; pH 7.0–8.0; 47.0% min. solids.

Surco DC-60. [Stepan] Formulated anionic blend; blended detergent concentrate with a higher solids content than Surco DC-47; provides excellent performance, foam stability, and exhibits good visc. throughout a wide range of use dilutions; liq.; sp.gr. 1.09; flash pt. > 200 F; pH 7.0–7.7; 60.0% min. solids.

Surco DDBSA. [Stepan] Linear dodecylbenzene sulfonic acid; anionic; detergent intermediate; liq.; sp.gr. 1.05; flash pt. > 200 F; 96% min. act.

Surco MA. [Stepan] Coconut fatty acid and mixed fatty acid DDBSA; nonionic; detergent, foamer, visc. builder and degreaser used in hard surface cleaners; liq.; sp.gr. 1.00; flash pt. > 200 F.

Surco SR-200. [Stepan] Coconut fatty acid/DDBSA DEA; anionic; detergent, foamer, visc. builder; hard surface cleaning systems; liq.; sp.gr. 1.03; flash pt. > 200 F; 98% conc.

Surco SXS. [Stepan] Sodium xylene sulfonate; anionic; hydrotrope, coupler, and solubilizer; liq.; sp.gr. 1.17; flash pt. > 200 F; pH 7–10; 39–42% act.

Surco TY. [Stepan] Coconut/mixed fatty acid DEA; nonionic; component in conveyor lubricants for hard surface cleaners; liq.; sp.gr. 0.99; flash pt. > 200 F.

Surco WC Conc. [Stepan] Cocamide DEA; nonionic; detergent, foam booster and stabilizer, thickener used in personal care prods. and dishwash formulations; liq.; sp.gr. 0.99; flash pt. 178 F; acid no. 2 max.; 100% conc.

Surcopur®. [Bayer] Propanil; herbicide for control of weeds in rice crops; m.w. 218; whitish-gray cryst. powd.; slightly sol. in water; sol. in alcohol, chlorobenzene; m.p. 91–92 C.

Surfac® 18-EHP. [Sherex Polymers] Primary amine phosphate salt; cationic; corrosion inhibitor used in metal processing; amber visc. liq.; oil-sol.; dens. 7.9 lb/gal; sp.gr. 0.95; acid no. 130–140; 100% act.

Surfac® P14B. [Sherex Polymers] Primary ether amine acetate; cationic; surfactant, ore flotation agent, corrosion inhibitor; m.w. 268; disp. in water; dens. 7.60 lb/gal; 100% conc.

Surfac® P24M. [Sherex Polymers] Primary ether amine acetate; cationic; see Surfac P14B; m.w. 261; water-disp.; dens. 7.61 lb/gal; 100% conc.

Surfactol® 13. [CasChem] Castor oil, modified; nonionic; wetting agent, wax plasticizer, mold release agent, antifoamer for textiles, leather, paints, household, cosmetics, dyeing, tanning, finishing, sizing, making of cutting and sol. oils, dispersing waxes, pigments, resins, dry powds.; Gardner 4 liq.; faint odor; water-sol.; dens. 8.36 lb/gal; sp.gr. 1.005; visc. 17 stokes; 100% conc.

Surfactol® 318. [CasChem] PEG-5 castor oil; nonionic; emulsifier for oils, waxes; solubilizer for fragrances; emollient; used in textiles, paints, household, cosmetics, dyeing, tanning, finishing, sizing, insecticides, herbicides, fungicides, kier boiling, making of cutting and sol. oils, dispersing waxes, pigments, resins, rewetting dried skins; Gardner 3 liq.; faint odor; sol. in toluene, butyl acetate, MEK; sp.gr. 0.984; visc. 690 cps; HLB 3.6; 100% conc.

Surfactol® 365. [CasChem] PEG-40 castor oil; nonionic; see Surfactol 318; Gardner 3 liq.; sol. in toluene, butyl acetate, MEK, ethanol, and water; sp.gr. 1.054; visc. 500 cps; HLB 13; 100% conc.

Surfactol® 575. [CasChem] PEG-66 trihydroxystearin; see Surfactol 318; wh. solid; water-sol.; alcohol; sp.gr. 1.08; m.p. 39 C; HLB 15.

Surfactol® 590. [CasChem] PEG-200 trihydroxystearin; see Surfactol 318; wh. solid; sol. in water, alcohol; sp.gr. 1.18; m.p. 53 C; HLB 18.

Surfageen S 30. [Chem-Y GmbH] Fatty alcohol ether sulfosuccinate; anionic; raw material for personal care prods.; liq.; 40% conc.

Surfam D89DA. [Sherex Polymers] Alkyl ether diamine; corrosion inhibitor; flotation agent for phosphate and iron ores; lt. liq.; mild ammonia odor; dens. 7.3 lb/gal; pour pt. 13 C.

Surfam P-MEPA. [Sherex Polymers] Methoxyethoxypropylamine; specialty chemical; used in textile dye chemicals, emulsifier intermediate in floor waxes and agric. prods.; Gardner 1 liq.; dens. 7.82 lb/gal; sp.gr. 0.93–0.94; b.p. 210–230 C; flash pt. < 212 F (COC); 97% min. act.

Surfam P5. [Sherex Polymers] Ether amine; cationic; intermediate for ore flotation, corrosion inhibitor, textile aux., lube and fuel additive, water treatment chemical; liq.; 100% conc.

Surfam P5 Dist. [Sherex Polymers] Methoxypropylamine; intermediate; emulsifier in wax emulsions for floor care prods.; textile finishing, paints, and metal working oils; insecticide emulsions; Gardner 2 max. clear liq.; m.w. 90; sol. in water and org. solvs.; dens. 7.29 lb/gal; b.p. 115–123 C; flash pt. 102 F; 98% min. act.

Surfam P5 Tech. [Sherex Polymers] Methoxypropylamine; see Surfam P5 Dist.; Gardner 7 max. clear liq.; sol. see Surfam P5 Dist.; m.w. 91; dens. 7.30 lb/gal; 90% min. act.

Surfam P10. [Sherex Polymers] Primary ether amine; cationic; corrosion inhibitor; emulsifier and lubricant in fiber and fabric processing; raw material in industrial applics.; intermediate in formulation of anticaking agents, specialty surfactant; gasoline and fuel additive; Gardner 4 max. clear liq.; m.w. 169; dens. 7.09 lb/gal; 100% conc.

Surfam P12B. [Sherex Polymers] Primary ether amine; cationic; see Surfam P10; Gardner 4 max. clear liq.; m.w. 200; dens. 7.11 lb/gal; 100% conc.

Surfam P14B. [Sherex Polymers] Primary ether amine; cationic; see Surfam P10; Gardner 4 max. clear liq.; m.w. 229; dens. 7.11 lb/gal; 100% conc.

Surfam P17B. [Sherex Polymers] Primary ether amine; cationic; see Surfam P10; Gardner 4 max.

clear liq.; m.w. 279; dens. 7.30 lb/gal; 100% conc.

Surfam P24M. [Sherex Polymers] Primary ether amine; cationic; see Surfam P10; Gardner 4 max. clear liq.; m.w. 212; dens. 7.10 lb/gal; 100% conc.

Surfam P86M. [Sherex Polymers] Primary ether amine; cationic; see Surfam P10; Gardner 4 max. clear liq.; m.w. 280; dens. 7.05 lb/gal; 100% conc.

Surfam P89MB. [Sherex Polymers] Primary ether amine; cationic; see Surfam P10; Gardner 4 max. clear liq.; m.w. 305; dens. 7.10 lb/gal; 100% conc.

Surfax 250. [E.F. Houghton] Alkyl sulfate sodium salt; dispersant for agric. wettable powds.; powd.

Surfax 585. [E.F. Houghton] Ester of succinic acid; see Surfax 250; powd.

Surfax WO. [E.F. Houghton] Sulfated ester; anionic; penetrant; softener, wetting and rewetting agent; liq.; 64% conc.

Surfine AZI-A. [Finetex] Carboxylate; nonionic; detergent, wetter, and dispersant; liq.; 90% conc.

Surfine WLL. [Finetex] Carboxylate; solubilizer and coupler, dispersant; gel; 70% conc.

Surfine WNG-A. [Finetex] Carboxylate; nonionic; detergent, wetting agent, emulsifier, solubilizer and coupler used in detergents; liq.; 90% conc.

Surfine WNT Conc. [Finetex] Carboxylate; detergent, wetting agent; dispersant for household and cosmetic formulations; liq.; 85% conc.

Surfine WNT Gel. [Finetex] Sodium C_{12-15} pareth-7 carboxylate; see Surfine WNT Conc.; gel; 60% conc.

Surfine WNT LC. [Finetex] Sodium C_{12-15} pareth-7 carboxylate; see Surfine WNT Conc.; liq.; 50% conc.

Surfine WNT LS. [Finetex] Sodium C_{12-15} pareth-7 carboxylate; see Surfine WNT Conc.; liq.; 95% conc.

Surflo®-S24. [CasChem] Alkylphenol and alcohol, oxyethylated; nonionic; dispersant, detergent, wetting and surf. and interfacial tens. reducing agent; used in producing wells and flotation units; liq.; water-sol.; dens. 8.30 lb/gal; sp.gr. 0.996 (68 F); visc. 125 SUS; flash pt. 82 F (PMCC); pour pt. < –10 F; 30% conc.

Surflo®-S32. [CasChem] Sodium alkylaryl sulfonate; anionic; detergent, dispersant, emulsifier; surfactant in oil industry; liq.; water-sol.; dens. 8.16 lb/gal; sp.gr. 0.980 (68 F); visc. 44 SUS; flash pt. 80 F (TOC); pour pt. < –10 F; 20% conc.

Surfonic® HDL. [Texaco] Surfonic N-85 (86%) and triethanolamine (14%); nonionic; emulsifier, wetting agent, dry cleaning detergent, penetrant, solubilizer, lime soap dispersant, antifoamer used in agric. chemicals, cosmetics, industrial cleaners, ceramics and concrete, dust control, wallpaper removal, photographic film developing, fire fighting, emulsion polymerization, indirect food additives, cutting oil emulsifiers; stabilizer for rubber latex and drilling-mud additives; degreaser for leather industry; sol. in water; sp.gr. 1.065; dens. 8.8 lb/gal; visc. 68 SUS (210 F); HLB 17.1; f.p. –7 C; cloud pt. 43 C (1% aq.); flash pt. (PMCC) 425 F; surf. tens. 31.5 dynes/cm (0.1%); ref. index 1.4888; 100% act.

Surfonic® JL-80X. [Texaco] Alkoxypolyalkoxyethanol; nonionic; surfactant used as emulsifier, wetting agent, detergent, penetrant, solubilizer, dispersant for household detergents and industrial prods., agric. sprays, dry cleaning, metal cleaners, ceramics, concrete, textile processing, paper mfg.; clear liq.; sp.gr. 1.0072; dens. 8.4 lb/gal; visc. 159.4 SUS (100 F); f.p. 21.1 F; cloud pt. 56–63 C (1% aq.); flash pt. 375 F (COC); pH 6.0–7.5 (1% aq.); surf. tens. 29.24 dynes/cm (0.1%).

Surfonic® N-10. [Texaco] Nonoxynol-1; nonionic; emulsifier, wetting agent, dry cleaning detergent, penetrant, solubilizer, lime soap dispersant, antifoamer used in agric. chemicals, cosmetics, industrial cleaners, ceramics and concrete, dust control, wallpaper removal, photographic film developing, fire fighting, emulsion polymerization, indirect food additives, cutting oil emulsifiers; stabilizer for rubber latex and drilling-mud additives; degreaser for leather industry; liq.; sol. in acetone, methanol, xylene, CCl_4, min. oil; m.w. 264; ref. index 1.5090; dens. 8.1 lb/gal; sp.gr. 0.98; visc. 675 SUS (100 F); HLB 3.4; flash pt. 355 F (OC); 100% act.

Surfonic® N-31.5. [Texaco] Nonoxynol-3; nonionic; see Surfonic N-10; liq.; sol. see Surfonic N-10; m.w. 358; ref. index 1.4950; dens. 8.4 lb/gal; sp.gr. 1.01; visc. 480 SUS (100 F); HLB 7.7; flash pt. 410 F (OC); 100% act.

Surfonic® N-40. [Texaco] Nonoxynol-4; nonionic; see Surfonic N-10; liq.; sol. in acetone, methanol, xylene, CCl_4, Stoddard; m.w. 396; ref. index 1.4979; dens. 8.5 lb/gal; sp.gr. 1.026; visc. 445 SUS (100 F); HLB 8.9; flash pt. 435 F; surf. tens. 27.5 dynes/cm (0.1%); 100% act.

Surfonic® N-60. [Texaco] Nonoxynol-6; nonionic; see Surfonic N-10; liq.; sol. in acetone, methanol, xylene, CCl_4, Stoddard solv.; disp. in water; m.w. 484; ref. index 1.4938; dens. 8.7 lb/gal; sp.gr. 1.041; visc. 440 SUS (100 F); HLB 10.9; flash pt. 475 F (OC); surf. tens. 28.7 dynes/cm (0.1%); 100% act.

Surfonic® N-85. [Texaco] Nonoxynol-8.5; nonionic; see Surfonic N-10; liq.; sol. in acetone, methanol, xylene, CCl_4, water; m.w. 596; ref. index 1.4923; dens. 8.8 lb/gal; sp.gr. 1.056; visc. 485 SUS (100 F); HLB 12.6; cloud pt. 44 C (1% aq.); flash pt. 500 F (OC); surf. tens. 30.5 dynes/cm (0.1%); 100% act.

Surfonic® N-95. [Texaco] Nonoxynol-10; nonionic; see Surfonic N-10; liq.; sol. see Surfonic N-85; m.w. 632; ref. index 1.4893; dens. 8.8 lb/gal; sp.gr. 1.061; visc. 510 SUS (100 F); HLB 12.9; cloud pt. 54.2 (1% aq.); flash pt. 500 F (OC); surf. tens. 30.8 dynes/cm (0.1%); 100% act.

Surfonic® N-100. [Texaco] Nonoxynol-10; nonionic; see Surfonic N-10; liq.; sol. see Surfonic N-85; m.w. 660; ref. index 1.4888; dens. 8.8 lb/gal; sp.gr. 1.064; visc. 69 SUS (210 F); HLB 13.2; cloud pt. 65 C (1% aq.); flash pt. 510 F (OC); surf. tens. 31.0 dynes/cm (0.1%); 100% act.

Surfonic® N-102. [Texaco] Nonoxynol-10; nonionic; see Surfonic N-10; liq.; sol. see Surfonic N-85; m.w. 668; ref. index 1.4884; dens. 8.8 lb/gal; sp.gr. 1.065; HLB 13.4; cloud pt. 81 C (1% aq.); flash pt. 510 F (OC); surf. tens. 31.2 dynes/cm (0.1%); 100% act.

Surfonic® N-120. [Texaco] Nonoxynol-12; nonionic; see Surfonic N-10; liq.; sol. see Surfonic N-85; m.w. 748; ref. index 1.4869; dens. 8.9 lb/gal; sp.gr. 1.070; visc. 560 SUS (100 F); HLB 14.1; cloud pt. 81 C (1% aq.); flash pt. 525 F (OC); surf. tens. 32.3 dynes/cm (0.1%); 100% act.

Surfonic® N-150. [Texaco] Nonoxynol-15; nonionic; see Surfonic N-10; wh. semisolid; sol. see Surfonic N-85; m.w. 880; ref. index 1.4815 (30 C); dens. 8.95 lb/gal; sp.gr. 1.065 (30/4 C); visc. 89 SUS (210 F); HLB 15.0; cloud pt. 94 C (1% aq.); flash pt. > 500 F (OC); surf. tens. 34.2 dynes/cm (0.1%); 100% act.

Surfonic® N-200. [Texaco] Nonoxynol-20; nonionic; see Surfonic N-10; wh. semisolid; sol. see Surfonic N-85; m.w. 1100; ref. index 1.4720 (50 C); dens. 9.0 lb/gal; visc. 105 SUS (210 F); HLB 15.8; cloud pt. > 100 C (1% aq.); flash pt. > 500 F (OC); 100% act.

Surfonic® N-300. [Texaco] Nonoxynol-30; nonionic; see Surfonic N-10; wh. waxy solid; sol. see Surfonic N-85; m.w. 1540; ref. index 1.4690 (50 C); dens. 9.1 lb/gal; visc. 160 SUS (210 F); HLB 17.1; cloud pt. > 100 C (1% aq.); flash pt. > 500 F (OC); 100% act.

Surfonic® N-400. [Texaco] Nonoxynol-40; nonionic; see Surfonic N-10; wh. waxy solid; sol. see Surfonic N-85; m.w. 1980; dens. 9.1 lb/gal; visc. 210 SUS (210 F); HLB 7.8; cloud pt. > 100 C (1% aq.); flash pt. > 500 F (OC); 100% act.

Surfonic® NB-5. [Texaco] Nonoxynol-30; nonionic; see Surfonic N-10; clear liq.; 70% act.

Surfonic® NB-14. [Texaco] Nonoxynol-40; nonionic; see Surfonic N-10; clear liq.; 70% act.

Surfynol® 82. [Air Prods.] Dimethyl octynediol; nonionic; defoamer, wetting agent used in pesticide concs.; cosmetic ingred.; developer compds.; electroplating baths; wh. cryst. powd.; sp.gr. 0.933; m.p. 51 C; b.p. 222 C; surf. tens. 55.3 (0.1% aq.); 100% conc.

Surfynol® 82S. [Air Prods.] Dimethyl octynediol on amorphous silica; solubilizer and clarifier in shampoos; defoamer/wetting agent in pesticide wettable powds.; low-foam wetting agent in developer compds.; defoamer in electroplating baths; 46% conc.

Surfynol® 104. [Air Prods.] Tetramethyl decynediol; nonionic; defoamer and dye dispersant in paint and ink formulations; surfactant in rinse aids; substrate pigment wetting agent for industrial coatings and adhesives; defoamer in dyestuff formulations; wetting agent for industrial cleaners; visc. reducer for vinyl dispersions; wh. waxy solid; sp.gr. 0.893; m.p. 37 C; b.p. 260 C; surf. tens. 31.6 (0.1% aq.); 100% conc.

Surfynol® 104A. [Air Prods.] Tetramethyl decynediol and 2-ethyl hexanol; nonionic; see Surfynol 104; lt. yel. liq.; sp.gr. 0.869; surf. tens. 33.0 (0.1% aq.); 50% conc.

Surfynol® 104BC. [Air Prods.] Tetramethyl decynediol in 2-butoxyethanol; nonionic; wetting agent, defoamer, dispersant, visc. stabilizer; 50% conc.

Surfynol® 104E. [Air Prods.] Tetramethyl decynediol and ethylene glycol; nonionic; see Surfynol 104; lt. yel. clear liq.; sp.gr. 1.001; surf. tens. 36.2 (0.1% aq.); 50% conc.

Surfynol® 104H. [Air Prods.] Tetramethyl decynediol and ethylene glycol; nonionic; see Surfynol 104; lt. yel. clear liq.; sp.gr. 0.951; surf. tens. 33.8 (0.1% aq.); 75% conc.

Surfynol® 104S. [Air Prods.] Tetramethyl decynediol on amorphous silica; wetting agent, defoamer, dispersant, visc. stabilizer; 46% conc.

Surfynol® 440. [Air Prods.] Tetramethyl decynediol, ethoxylated (3.5 moles); nonionic; water-based industrial finishes; defoamer, rewetting, and leveling agent for paperboard coatings; metal cleaning and plating bath additive; straw clear liq.; sp.gr. 0.982; surf. tens. 31.0 (0.1% aq.); 100% conc.

Surfynol® PC. [Air Prods.] Acetylenic glycol; nonionic; defoamer used in paper coating latices; antishock agent for paper coatings; lt. yel. clear liq.; sp.gr. 0.9585; 100% conc.

Surfynol® SE. [Air Prods.] Acetylenic diol; nonionic; wetting agent, foam control agent in pressure-sensitive adhesives, aq. lubricants, water based paints, inks, dye processing, and paper coatings; liq.

Surpasol E-436. [Climax Performance] Esters in a hydrocarbon base; paper mill defoamer; water emulsifiable.

Sustane® 3. [UOP] Propylene glycol, BHA, propyl gallate, citric acid, ratio 70:20:6:4; preservative and antioxidant used in snack foods, cosmetics, and spices; pale yel. liq.; f.p. –23 C; sp.gr. 1.048; dens. 8.7 lb/gal; visc. 42.3 cS (38 C); ref. index. 1.466.

Sustane® 4A. [UOP] BHA, veg. oil, ratio 30:70; preservative and antioxidant used in nuts, baked goods, and edible oils; pale yel. liq.; sp.gr. 0.951; dens. 7.9 lb/gal; ref. index 1.490.

Sustane® 6. [UOP] BHT, BHA, veg. oil, ratio 22:18:60; preservative and antioxidant used in edible fats, meat prods., vitamins, emulsifiers, and confections; pale yel. liq.; f.p. –26 C; sp.gr. 0.928; dens. 7.8 lb/gal; visc. 37.9 cS (38 C); flash pt. 146 C (OC).

Sustane® 8. [UOP] BHA; citric acid, propylene glycol, ratio 20:20:60; preservative and antioxidant used in animal fats; pale yel. liq.; f.p. –17.8 C; b.p. 188 C (745 mm); sp.gr. 1.088; dens. 9.1 lb/gal; visc. 114.9 cS (38 C); ref. index 1.464.

Sustane® 20. [UOP] TBHQ, citric acid, propylene glycol, ratio 20:10:70; preservative and antioxidant used for edible fats and oils; amber liq.; sp.gr. 1.087; dens. 9.0 lb/gal; ref. index 1.454; flash pt. 99 C (CC).

Sustane® 20-3. [UOP] TBHQ, citric acid, propylene glycol, ratio 20:3:77; preservative and antioxidant for edible oils; yel. liq.; sp.gr. 1.059; dens. 8.8 lb/gal; ref. index 1.457.

Sustane® 20A. [UOP] TBHQ, propyl gellate, citric acid, veg. oil, glycerol monooleate, ratio 20:15:3:30:32; see Sustane 20; yel. liq.; sp.gr. 0.998; dens. 8.1 lb/gal; ref. index 1.483.

Sustane® 31. [UOP] BHA, TBHQ, citric acid, propylene glycol, ratio 20:6:4:70; preservative and antioxidant used in dressing oils, nuts, confections, and flavors; yel. liq.; sp.gr. 1.054; dens. 8.7 lb/gal; ref.

index 1.462.

Sustane® BHA. [UOP] Butylated hydroxyanisole; antioxidant, stabilizer for fats, oils, other foods; wh. tablets, flakes.

Sustane® BHA 1-F. [UOP] Butylated hydroxyanisole; preservative and antioxidant for foods, flavors, cosmetics, vitamins, oils, waxes, essential oils, tallow, sausage, chewing gum base, shortening, lard, food pkg. materials, potatoes, and cereals; inhibits oxidation reaction of oils and fats in presence of air and retards rancidity and off-flavors caused by oxidation; wh. tablets and flakes; m.w. 180.2; b.p. 270 F (@ 5 mm Hg); sol. > 30 g/100 g in soybean oil, cottonseed oil; > 25 g/100 g in acetone, ether; > 10 g/100 g in methanol; neg. sol. in water; visc. 3.3 cS (99 C); m.p. 57 C; flash pt. 130 C (OC); 98.5% min. conc.

Sustane® BHT. [UOP] Butylated hydroxytoluene; preservative and antioxidant for foods, food pkg. materials, tallow, animal feeds, soaps, cosmetics, and plastics; inhibits oxidation of fats and oils and retards rancidity and off-flavors caused by oxidation; wh. cryst.; m.w. 220.3; b.p. 265 C (@ 760 mm Hg); sol. (g/100 solv.); 48 g in lard (50 C); 40 g in benzene; 30 g in min. oil; 28 g in linseed oils; 20 g in methanol; nil in water; sp.gr. 1.048; dens. 37.5 lb/ft³; visc. 3.47 cS (80 C); m.p. 70 C; ref. index 1.486; flash pt. 118 C (CC); 99% min. conc.

Sustane® HW-4. [UOP] BHA, BHT, veg. oil, ratio 20:20:60; preservative and antioxidant used in edible fats, meat prods., vitamins, emulsifiers, and confections; pale yel. liq.; f.p. –23 C; b.p. 258 C; sp.gr. 0.942; dens. 7.8 lb/gal; ref. index 1.492; flash pt. 146 C (OC).

Sustane® P. [UOP] BHA, BHT, ethyl alcohol, ratio 25:25:50; preservative and antioxidant used in dehydrated vegetables; pale yel. liq.; f.p. 12.8 C; sp.gr. 0.884; dens. 7.3 lb/gal; ref. index 1.444; flash pt. 16 C (CC).

Sustane® PA. [UOP] BHA; ethyl alcohol; see Sustane P; pale yel. liq.; f.p. –45.6 C; sp.gr. 0.913; dens. 7.6 lb/gal; ref. index 1.442; flash pt. 16 C (CC).

Sustane® PG. [UOP] Propyl gallate (n-propyl 3,4,5 trihydroxybenzoate); preservative and antioxidant for fats and oils; wh. cryst. powd.; slight odor; m.w. 212; b.p. decomposes; sol. (g/100 g solv.) 170 g in methanol; 121 g in acetone; 103 g in ethanol; 83 g in ethyl ether; 67 g in propylene glycol; m.p. 146–148 C; 100% conc.

Sustane® Q. [UOP] BHA, citric acid, propylene glycol, ratio 40:8:52; preservative and antioxidant used in veg. oils, frozen foods, flavors; pale yel. liq.; sp.gr. 1.066; dens. 8.8 lb/gal; ref. index 1.476.

Sustane® TBHQ. [UOP] Mono-tert.-butyl hydroquinone; preservative and antioxidant for foodstuffs and meat prods.; color stable and useful as substitute for reactive antioxidants that tend to form purple complexes with iron or copper; wh. to lt. tan cryst.; m.w. 166.2; b.p. 295 C; sol. (g/100 g solv.) in methanol; 30 g in propylene glycol; 10 g in corn oil; 5 g in lard (50 C); nil in water; dens. 27 lb/ft³; m.p. 126.5–128.5 C; flash pt. 171 C (CC); 99.0% min conc.

Sustane® TBHQ.4 [UOP] Tert. butyl hydroquinone; see Sustane BHA; wh. to lt. tan cryst.

Sustane® W. [UOP] BHA, BHT, propyl gallate, citric acid, propylene glycol, veg. oil, glycerol monooleate, ratio 10:10:6:6:8:28:32; preservative and antioxidant used in spices, meat prods., baked goods, and snack foods; amber liq.; sp.gr. 0.978; dens. 8.1 lb/gal; visc. 102.4 cS (38 C); ref. index 1.484.

Suttocide® A. [Sutton] Sodium hydroxymethylglycinate; antimicrobial preservative for cosmetics; colorless to lt. yel. clear liq.; mild char. odor; m.w. 127.10; sol. (g/100 g): 60 g in water, 20 g in propylene glycol, 15 g in methanol, 10 g in glycerin; sp.gr. 1.28–1.30; pH 10–12; 49–52% total solids; 5.5–6.1% N.

Swanic 51. [Swastik] Coconut MEA (oil-based); foam booster/stabilizer in detergent powds.; wh. solid paste.

Swanic 52L. [Swastik] Coconut alkylolamide; anionic; foam booster, visc. controller used in shampoos and liq. detergent formulations; yel. liq.; 100% conc.

Swanic 110. [Swastik] Alcohol ethoxylate; nonionic; retarder and dye leveling agent used in textiles; liq.; 30% conc.

Swanic CM. [Swastik] Alkyl phenol POE ether; nonionic; air entraining agent; concrete additive; liq.; 25% conc.

Swanol AM-301. [Nikko] Lauryl betaine; amphoteric; detergent, wetting, dispersant; liq.; 35% conc.

Swanol CA-101. [Nikko] Benzalkonium chloride; antimicrobial; liq.

Sweetzyme® Type A, Type Q. [Novo] Immobilized glucose isomerase derived from Bacillus coagulans; enzymes which catalyze conversion of glucose to fructose; brn. dry extrudate, particle size 0.3–1.0 mm (Type Q), brn. dry powd., particle size 0–0.4 mm (Type A); dens. 1400 g/l, 280–320 kg/m³ (wet bulk).

SXS. [Continental] Sodium xylene sulfonate; hydrotrope used in detergent formulations; colorless liq.; mild odor; dens. 9.80; sp.gr. 1.17; visc. 6.0; m.p. 31 F; b.p. 212 F; 41% act.

Syloid® 244. [Davison] Hydrated silica gels; bonding agent used in adhesives; wh. gel, particle size 4.0μ; dens. 0.15 g/cc; pH 7.6 (5% aq.).

Syloid® 378. [Davison] Ultra-fine amorphous silica gel; see Syloid 244; wh. gel, particle size 4.0μ; dens. 0.26 g/cc; pH 2.5 (5% aq.).

Syncal CAS. [PMC] Calcium saccharin; wh. fine free-flowing powd.; 95% min. assay.

Syncal GS, US. [PMC] Sodium saccharin; syn. sweetening agent for dietetic foods, toothpastes, cosmetics, pharmaceuticals; wh. nondusting, free-flowing gran., sized and unsized resp.

Syncal S. [PMC] Sodium saccharin; see Syncal GS; fine wh. powd.

Syncal SDI. [PMC] Saccharin insol.; wh. powd.; 98–101% assay.

Syncal SDS. [PMC] Sodium saccharin; wh. fine free-flowing particles; 98–101% assay.

Syncrolubes. [Croda Ltd.] Internal and external lubricants and processing aids for polymers; liq./powds.

Syncrowax AWI-C. [Croda] C_{18-36} acid; emulsifier, emollient, opacifier; pale cream waxy flakes, mild waxy odor; oil-sol.; HLB 2.0; m.p. 69–74 C; acid no. 155–170; iodine no. 3 max.; sapon. no. 165–175; 100% conc.

Syncrowax BB4, BB5. [Croda] Syn. beeswax; see Syncrowax AWI-C; also suspending agent for anhyd. systems, aux. w/o emulsifier, thickener for oils and waxes; used in creams and sticks; pale cream solid; oil-sol.; m.p. 60–65 C (BB4).

Syncrowax ERL-C. [Croda] C_{18-36} acid glycol ester; see Syncrowax AWI-C; also lubricant, stabilizer, suspending agent for anhyd. systems, thickener, reducer of bleeding and sweating; gloss improver; sticks, creams; lt. tan solid; mild waxy odor; m.p. 70–75 C; acid no. 10–15; iodine no. 3 max.; sapon. no. 155–160.

Syncrowax HGL-C. [Croda] C_{18-36} acid triglyceride; see Syncrowax AWI-C; also lubricant, suspending agent, strength improver, stabilizer, gloss improver; used in cosmetic makeup; lt. tan solid; mild waxy odor; m.p. 70–75 C; acid no. 6–12; iodine no. 3 max.; sapon. no. 160–175.

Syncrowax HR-C. [Croda] Glyceryl tribehenate; see Syncrowax AWI-C; also suspending agent, thickener, gloss improver used in personal care prods.; off-wh. solid; mild waxy odor; m.p. 60–65 C; acid no. 10 max.; iodine no. 3 max.; sapon. no. 170–175.

Syncrowax HRS-C. [Croda] Glyceryl tribehenate soap, calcium behenate; see Syncrowax AWI-C; also suspending agent for anhyd. systems, aux. w/o emulsifier, gellant, thickener for oils and waxes; pale cream flakes; mild waxy odor; oil-sol.; m.p. 105–115 C; acid no. 8–13; iodine no. 3 max.; sapon. no. 115–125.

Syncrowax PRLC. [Croda] Fatty wax acid propylene glycol ester; reducer of bleeding and sweating; gloss improver; lt. tan solid.

Syn Fac 334. [Milliken] Aryl POE ether; nonionic; emulsifier, dispersant for pigments, insecticides, solvs., cleaning compds. and latex paint; liq. and semiliq. resp.; HLB 11.6 and 12 resp.; 87 and 100% conc. resp.

Syn Fac 905. [Milliken] Nonoxynol-9 (9.5 EO); nonionic; wetting agent and dispersant, emulsifier; liq.; 100% conc.

Syn Fac 8017, 8009. [Milliken] Ethoxylated bisphenol A; nonionic; reactive diluent, dispersant for waterborne coatings and other applics.; liq.; 98% act.

Syn Fac 8026. [Milliken] Alkoxylated BPA; intermediate in epoxy and polyester resins, as urethane cross-linker; liq.; 100% act.

Syn Fac 8031. [Milliken] Alkoxylated bisphenol A; nonionic; reactive diluent, dispersant for solv. coatings and other applics.; liq.; 100% act.

Syn Fac 8210. [Milliken] Polyoxyaryl ether; nonionic; low foaming emulsifier, dispersant for pigments; liq.; 100% act.

Syn Fac 8216. [Milliken] Aryl POE ether; nonionic; pigment dispersant for paints, coatings, and textile print systems; solid; HLB 15.0; 100% conc.

Syn Fac 8337. [Milliken] Phosphated alkoxylated aryl phenol, potassium salt; anionic; dispersant for org. pigments; lt. yel. visc. liq.; dens. 9.0 lb/gal; visc. 1000 cps; HLB 20; flash pt. 310 C; pH 7.0 (5%).

Syn Fac TEA-97. [Milliken] Ethoxylated amine; nonionic; solv. and dispersant in the dye industry; liq.; 100% conc.

Synkad 202. [Keil] Boramide; corrosion inhibitor for syn. cutting fluids; water-sol.

Synkad 204. [Keil] Boramide; corrosion inhibitor for syn. drawing, cutting, and grinding fluids; liq.; water-sol.

Synkad 303. [Keil] Carboxylic acid salt; corrosion inhibitor for syn. grinding fluids; water-sol.

Synkad 404, 500. [Keil] Carboxylic acid salt; rust inhibitor for syn. drawing, cutting, and grinding fluids; liq.; water-sol.

Syn Lube 106 (60%). [Milliken] PEG-200 glyceride ester; nonionic; emulsifier, lubricant; spin finish and viscose additive, tufting aid; liq.; 60% conc.

Syn Lube 107. [Milliken] PEG-25 glyceride ester; nonionic; see Syn Lube 106 (60%); liq.; HLB 15.5; 100% conc.

Synoquart P 50. [Aquatec Quimica] Cetrimonium chloride; cationic; antistatic and conditioning agent for hair care prods., cationic shampoos; emulsifier; liq.

Synotol 119 N. [Aquatec Quimica] Cocamide DEA; nonionic; detergent, wetting agent, thickener, foam booster and stabilizer, lime soap dispersant; used in personal care prods.; liq.; 50% amide.

Synotol CN 20. [Aquatec Quimica] Cocamide DEA; nonionic; foam stabilizer, thickener, superfatting agent for cosmetic and household prods.; liq.; 90% conc.

Synotol CN 60. [Aquatec Quimica] Cocamide DEA; nonionic; see Synotol CN 20; liq.; 50% amide.

Synotol CN 80. [Aquatec Quimica] Cocamide DEA; nonionic; see Synotol CN 20; liq.; 75% amide.

Synotol CN 90. [Aquatec Quimica] Cocamide DEA; see Synotol CN 20; liq.; 85% conc.

Synotol Detergent E. [Aquatec Quimica] Cocamide DEA; see Synotol 119 N; liq.; 50% amide.

Synotol L 60. [Aquatec Quimica] Lauramide DEA; see Synotol 119 N; liq.; 50% amide.

Synotol L 90. [Aquatec Quimica] Lauramide DEA; see Synotol 119 N; solid; 88% conc.

Synotol LM 60. [Aquatec Quimica] Lauramide DEA; see Synotol 119 N; liq.; 50% amide.

Synotol LM 90. [Aquatec Quimica] Lauric/myristic acid DEA; see Synotol 119 N; paste; 88% conc.

Synotol ME-90. [Aquatec Quimica] Coconut fatty acid MEA; nonionic; foam booster, wetting agent, thickener, foam stabilizer, and superfatting agent for cosmetic and household preparations; flakes; 85% conc.

Synox 5LT. [Neville] 2,2′-Methylene bis (4-methyl-6-t-butylphenol); nonstaining antioxidant for natural, S/B, BR, CR, and polyisoprene rubbers; sol. in alcohols, aromatics, esters, ketones.

Synperonic 91/2.5. [ICI PLC] Syn. primary C_{9-11} alcohol ethoxylate; nonionic; detergent and emulsifier; intermediate for sulfation; liq.; HLB 8.2; 100%

conc.

Synperonic 91/5. [ICI PLC] Syn. primary C_{9-11} alcohol ethoxylate; nonionic; wetting and scouring agent for wool textile processing, degreasing hard surface cleaners; coupler; o/w emulsifier; liq.; HLB 11.5; 100% conc.

Synperonic 91/6. [ICI PLC] Syn. primary C_{9-11} alcohol ethoxylate; nonionic; see Synperonic 91/5; also detergent cleaners; liq.; HLB 12.5; 100% conc.

Synperonic 91/8. [ICI PLC] Syn. primary C_{9-11} alcohol ethoxylate; nonionic; see Synperonic 91/6; HLB 14; 100% conc.

Synperonic 91/10. [ICI PLC] Syn. primary C_{9-11} alcohol ethoxylate; nonionic; liq.; HLB 14.6; 100% conc.

Synperonic A2. [ICI PLC] Syn. primary alcohol ethoxylate; nonionic; detergent, emulsifier, coemulsifier; base for emulsifiers, textile antistats, scouring and wetting agents, and specialty cleaners; intermediate for sulfation used in personal care prods.; Hazen 50 max. liq.; sol. in alcohol, glycol ethers, kerosene, and min. oil; dens. 0.897 g/ml; visc. 25 cps; HLB 5.9; pour pt. 2 C; pH 6–8 (1% aq.); 99% min. act.

Synperonic A3. [ICI PLC] Syn. primary alcohol ethoxylate; see Synperonic A2; Hazen 50 max. liq.; sol. see Synperonic A2; dens. 0.918 g/ml; visc. 29 cps; HLB 7.8; pour pt. 5 C; pH 6–8 (1% aq.); 99% min. act.

Synperonic A4. [ICI PLC] Syn. primary alcohol ethoxylate; see Synperonic A2; Hazen 50 max. liq.; sol. in alcohol, glycol ethers, kerosene; sol./disp. in min. oil; dens. 0.937 g/ml; visc. 34 cps; HLB 9.1; pour pt. 9 C; pH 6–8 (1% aq.); 99% min. act.

Synperonic A6. [ICI PLC] Syn. primary alcohol ethoxylate; see Synperonic A2; Hazen 50 max. liq.; sol. in alcohol, glycol ethers; dens. 0.966 g/ml; visc. 46 cps; HLB 11.2; pour pt. 17 C; surf. tens. 29.8 dynes/cm (0.1%); pH 6–8 (1% aq.); 99% min. act.

Synperonic A7. [ICI PLC] Syn. primary alcohol ethoxylate; nonionic; wetting and detergent leveling agent; emulsifier and coemulsifier for oils, solvs., and waxes; Hazen 50 max. visc. liq.; sol. in water, glycol ethers and alcohol; dens. 0.977 g/ml; visc. 60 cps; HLB 12.2; cloud pt. 44–48 C (1% aq.); pour pt. 21 C; surf. tens. 30.2 dynes/cm (0.1%); pH 6–8 (1% aq.); 99% min. act.

Synperonic A9. [ICI PLC] Syn. primary alcohol ethoxylate; see Synperonic A7; wh. paste; sol. see Synperonic A7; dens. 0.984 g/ml (40 C); visc. 35 cps (40 C); HLB 12.5; cloud pt. 62–68 C (1% aq.); pour pt. 24 C; surf. tens. 31.4 dynes/cm (0.1%); pH 6–8 (1% aq.); 99% min. act.

Synperonic A11. [ICI PLC] Syn. primary alcohol ethoxylate; see Synperonic A7; wh. paste; sol. see Synperonic A7; dens. 0.998 g/ml (40 C); visc. 42 cps (40 C); HLB 13.9; cloud pt. 85–89 C (1% aq.); pour pt. 27 C; surf. tens. 33.2 dynes/cm (0.1%); pH 6–8 (1% aq.); 99% min. act.

Synperonic A14. [ICI PLC] Syn. primary alcohol ethoxylate; nonionic; detergent, solubilizer, dispersant, stabilizer; wh. wax; sol. see Synperonic A7; dens. 1.014 g/ml (40 C); visc. 55 cps (40 C); HLB 14.9; cloud pt. > 100 C (1% aq.); pour pt. 30 C; surf. tens. 36.5 dynes/cm (0.1%); pH 6–8 (1% aq.); 99% min. act.

Synperonic A20. [ICI PLC] Syn. primary alcohol ethoxylate; nonionic; see Synperonic A14; wh. hard wax; sol. see Synperonic A7; dens. 1.028 g/ml (50 C); visc. 58 cps (50 C); HLB 16.2; cloud pt. > 100 C (1% aq.); pour pt. 37 C; surf. tens. 40.1 dynes/cm (0.1%); pH 6–8 (1% aq.); 99% min. act.

Synperonic NP4. [ICI PLC] Nonoxynol-4; nonionic; detergent, emulsifier, coemulsifier; intermediate for sulfation in mfg. lubricants and antistats; Hazen 150 max. liq.; sol. see Synperonic A3; dens. 1.022 g/ml; visc. 400 cps; HLB 8.9; pour pt. < 0 C; pH 6–8 (1% aq.); 99% min. act.

Synperonic NP5. [ICI PLC] Nonoxynol-5; nonionic; see Synperonic NP4; Hazen 150 max. liq.; sol. in alcohol, glycol ethers, kerosene, min. oil; insol./disp. in water; dens. 1.035 g/ml; visc. 350 cps; HLB 10.5; pour pt. < 0 C; pH 6–8 (1% aq.); 99% min. act.

Synperonic NP6. [ICI PLC] Nonoxynol-6; nonionic; see Synperonic NP4; Hazen 150 max. liq.; sol. in alcohol, glycol ethers, min. oil; disp. in water; dens. 1.041 g/ml; visc. 355 cps; HLB 10.9; pour pt. < 0 C; pH 6–8 (1% aq.); 99% min. act.

Synperonic NP10. [ICI PLC] Nonoxynol-10; nonionic; detergent, solubilizer, dispersant, stabilizer; Hazen 150 max. liq.; sol. see Synperonic NP9; dens. 1.061 g/ml; visc. 360 cps; HLB 13.3; cloud pt. 62–67 C (1% aq.); pour pt. 5 C; surf. tens. 30.6 dynes/cm; pH 6–8 (1% aq.); 99% min. act.

Synperonic NP12. [ICI PLC] Nonoxynol-12; nonionic; see Synperonic NP10; Hazen 150 max. liq.; sol. Synperonic NP9; dens. 1.062 g/ml; visc. 265 cps; HLB 13.9; cloud pt. 79–84 C (1% aq.); pour pt. 14 C; surf. tens. 35.2 dynes/cm; pH 6–8 (1% aq.); 99% min. act.

Synperonic NP20. [ICI PLC] Nonoxynol-20; nonionic; solubilizer and emulsifier for polar substrates; Hazen 150 max. solid; sol. see Synperonic NP9; dens. 1.073 g/ml (40 C); visc. 168 cps (40 C); HLB 16.0; cloud pt. > 100 C (1% aq.); pour pt. 30 C; surf. tens. 41.7 dynes/cm; pH 6–8 (1% aq.); 99% min. act.

Synperonic NP30. [ICI PLC] Nonoxynol-30; nonionic; see Synperonic NP20; Hazen 150 max. solid; sol. see Synperonic NP9; dens. 1.074 g/ml (50 C); visc. 150 cps (50 C); HLB 7.1; cloud pt. > 100 C (1% aq.); pour pt. 40 C; surf. tens. 42.8 dynes/cm; pH 6–8 (1% aq.); 99% min. act.

Synperonic NP35. [ICI PLC] Nonoxynol-35; nonionic; solubilizer and emulsifier for high polarity materials; solid; HLB 17.4%; 100% conc.

Synperonic PE30/10. [ICI PLC] PO/EO block copolymer; nonionic; defoamer, demulsifier, wetting agent, dispersant, rinse aid for machine dishwashing; liq.; m.w. 1900; sp.gr. 1.017; visc. 325 cs; m.p. < 0 C; HLB 3; cloud pt. 24 C (1% aq.); pH 6–8 (2.5%); 99% min. act.

Synperonic PE30/20. [ICI PLC] PO/EO block copolymer; nonionic; see Synperonic PE30/10; liq.; m.w. 2500; sp.gr. 1.032; visc. 420 cs; m.p. < 0 C; HLB 7; cloud pt. 32 C (1% aq.); surf. tens. 42 dynes/cm; pH 6–8 (2.5%); 99% min. act.

Synperonic PE30/40. [ICI PLC] PO/EO block copolymer; nonionic; see Synperonic PE30/10; visc. liq.; m.w. 2900; sp.gr. 1.05; visc. 600 cs; m.p. < 0 C; HLB 15; cloud pt. 58 C (1% aq.); surf. tens. 41.1 dynes/cm; pH 6–8 (2.5%); 99% act.

Synperonic PE30/80. [ICI PLC] PO/EO block copolymer; nonionic; emulsifier, dispersant, stabilizer; emulsion polymerization; used in pigments in emulsion paints, inks; flake; m.w. 8500; m.p. 50 C; HLB 29; cloud pt. > 100 C (1% aq.); surf. tens. 48 dynes/cm; pH 6–8 (2.5%); 99% min. act.

Synperonic PE39/70. [ICI PLC] PO/EO block copolymer; nonionic; emulsifier, dispersant; wh. flake; m.w. 7500; m.p. 49 C; HLB 24; cloud pt. > 100 C (1% aq.); surf. tens. 45.2 dynes/cm; pH 6–8 (2.5%); 99% min. act.

Synprol Alcohol. [ICI PLC] Syn. primary alcohol, fully sat.; intermediate for prod. of ethoxylates, sulfates and ether sulfates; ref. index. 1.4457; sp.gr. 0.837; visc. 29.1 cps; flash pt. 290 F (PMCC); 99% act.

Synprolam 35. [ICI PLC] C_{13-15} alkyl primary amine; anticaking agent for fertilizer; flotation agent, corrosion inhibitor; metal cleaning formulations, pigment dispersions, rubber processing auxs., bitumen emulsifier; intermediate for prod. of amine salts, ethoxylates and sulfosuccinates; Hazen 50 clear liq.; sp.gr. 0.8; visc. 4.5 cps; flash pt. 127 C (PMCC); 99% primary amine.

Synprolam 35A. [ICI PLC] Syn. C_{13-15} alkyl primary amine; cationic; emulsifier, fertilizer anticaking agent; min. flotation; solid; 100% conc.

Synprolam 35BQC (50). [ICI PLC] Syn. C_{13-15} dimethyl tert. amine benzyl ammonium chloride; cationic; emulsifier, sanitizer, biocide, corrosion inhibitor, textile drying aux., timber preservative; liq.; 50% conc.

Synprolam 35BQC (80). [ICI PLC] Syn. C_{13-15} dimethyl tert. amine benzyl ammonium chloride; cationic; see Synprolam 35BQC (50); liq.; 80% conc.

Synprolam 35DM. [ICI PLC] Syn. tert. fatty amine; chemical intermediate for prod. of quat. salts and amine oxides used in cosmetics; catalyst for PU foam; corrosion inhibitor; Hazen 50 clear liq.; sp.gr. 0.79; visc. 3.9 cps; flash pt. 143 C (PMCC); 97% tert. amine.

Synprolam 35DMA. [ICI PLC] Syn. C_{13-15} dimethyl tert. amine, acetate salt; cationic; emulsifier, biocide, timber preservative; liq.; 100% conc.

Synprolam 35DMBQC. [ICI PLC] C_{13-15} alkyl dimethyl benzyl ammonium chloride; cationic; disinfectant, algicide, fungicide, germicide; general antistat for syn. and nat. fibers; fiber conditioner; also avail. in low foaming grade; liq.; 50 or 80% conc.

Synprolam 35DMO. [ICI PLC] C_{13-15} alkyl dimethyl amine oxide; mildly cationic; foam booster/stabilizer, thickener in personal care prods.; Hazen 50 clear liq.; sp.gr. 0.97; surf. tens. 29 dynes/cm (0.1% aq.); 30% act.

Synprolam 35 Lauramide. [ICI PLC] Difatty amides derived from Synprolam 35; nonionic; low melting lubricant for plastics industry; water repellent; flakes; 100% conc.

Synprolam 35M. [ICI PLC] Syn. (C_{13-15}) alkyl methyl sec. amine; cationic; surfactant intermediate, corrosion inhibitor; liq.; 100% conc.

Synprolam 35MX 1, 35MX 3, 35MX 5. [ICI PLC] PEG-1, -3, and -5 C_{13-15} alkyl methyl amines; cationic; antistat for polyolefin and PVC; liq.; 100% conc.

Synprolam 35MX1/O. [ICI PLC] C_{13-15} alkyl methyl hydroxyethyl amine oxide; nonionic; foam stabilizer, visc. builder for shampoos, bubble baths, dishwashing liqs., detergents; liq.; 30% act.

Synprolam 35MX1QC. [ICI PLC] C_{13-15} alkyl dimethyl hydroxy ethyl ammonium chloride; antistat and conditioner for syn. and nat. fibers; liq.; 30% act.

Synprolam 35N3. [ICI PLC] N-(C_{13-15}) alkyl-1,3-propane diamine; cationic; corrosion inhibitor, bitumen adhesion agent, emulsifier, intermediate; liq.; 100% conc.

Synprolam 35N3 DA. [ICI PLC] C_{13-15} alkyl propane diamine, diacetate salt; general bactericide, algicide, germicide with anticorrosion properties; solid; 100% act.

Synprolam 35N3 Dioleate. [ICI PLC] Dioleate salt of Synprolam 35N3; cationic; rust inhibitor.

Synprolam 35N3X3, 35N3X5, 35N3X10, 35N3X15, 35N3X25. [ICI PLC] N-(C_{13-15}) alkyl-1,3-propane diamine ethoxylate; cationic; emulsifier, corrosion inhibitor, textile processing aids, wetting agent and antistat; liq.; 100% conc.

Synprolam 35 Stearamide. [ICI PLC] Difatty amides derived from Synprolam 35; nonionic; low melting lubricant for plastics industry; water repellent; flakes; 100% conc.

Synprolam 35TMQC. [ICI PLC] (C_{13-15}) alkyl trimethyl ammonium chloride; cationic; biocide, emulsifier, wetting agent and antistat; liq.; 100% conc.

Synprolam 35TMQS. [ICI PLC] C_{13-15} alkyl trimethyl ammonium methosulfate in 1,4-butanediol; plastics antistat; esp. for PU compositions; liq.; 80% act.

Synprolam 35X2, 35X5, 35X10, 35X15, 35X20, 35X25, 35X35, 35X50. [ICI PLC] POE C_{13-15} alkyl amine (2, 5, 10, 15, 20, 25, 35, and 50 EO resp.); cationic; emulsifier, PU catalyst, textile and plastic processing aid, textile dyeing auxs., antistat; liq.; 100% conc.

Synprolam 35X2/O. [ICI PLC] C_{13-15} alkyl bis (2-hydroxyethyl) amine oxide; nonionic; foam stabilizer for personal care prods. and dishwashing detergents; liq.; 30% conc.

Synprolam 35X2QS, 35X5QS, 35X10QS. [ICI PLC] PEG-2, -5, and -10 C_{13-15} alkyl methyl ammonium methosulfate; cationic; antistat for PVC and PU; liq.; 90–100% conc.

Synprolam FS. [ICI PLC] Syn. quat. ammonium compd.; cationic; fabric softener and antistat with rewet properties for textile treatment; liq.; 80% conc.

Synpron 1027. [Syn. Prods.] Antimony mercaptide; PVC heat stabilizer esp. for pipe and conduit applics.; recommended for use with calcium stearate for optimum stabilization; may be used in NSF potable water pipe at 1.0 phr max.; clear straw liq.; sp.gr. 1.2; visc. (Gardner): B.

Synpron 1034. [Syn. Prods.] Antimony mercaptide; see Synpron 1027; clear liq.; visc. (Gardner): B.

Synpron 1066. [Syn. Prods.] Antimony mercaptide; see Synpron 1027; clear straw liq.; sp.gr. 1.0; visc. (Gardner): A.

Syntergent TER-1. [Henkel] Phosphate ester; anionic; wetting agent, textile surfactant, calcium sulfate dispersant; liq.; 82% conc.

Synthionic D 7520, 7525, 9500 [Witco SA] Block polymer; nonionic; wetting agent, detergent, rinse aid, defoamer; liq.

Synthionic P 6040, 8020. [Witco SA] Block polymer; nonionic; see Synthionic D 7520; liq.

Synthrapol N. [ICI Am.] Nonyl phenol ethoxylate; nonionic; detergent, wetting agent, emulsifier, penetrant/dyeing assistant used in textile processing; colorless liq.; mild pleasant odor; sp.gr. 1.0; b.p. 100 C; pH 5.5–7.0; 26% conc.

Syntopon E. [Witco SA] Nonylphenol ethoxylate; nonionic; detergent, emulsifier for metal cleaning; dispersant; wax; 100% conc.

T

T-600. [Olin] PPG triol; plasticizer, chemical intermediate; used in brake fluids and in the mfg. of resins; produces rigid and flexible urethane foams, and nonfoam urethane coatings, adhesives, elastomers, and sealants and caulks; m.w. 600; f.p. –29 C; sp.gr. 1.037; dens. 8.64 lb/gal.

T-1000. [Olin] PPG triol; see T-600; m.w. 1000; f.p. –28 C; sp.gr. 1.025; dens. 8.54 lb/gal.

T1990. [Hüls] 1,1,4,4-Tetramethyl-1,4-bis-(N,N-dimethylamino) disilethylene; difunctional blocking agent for the protection of primary amines.

T2015. [Hüls] 1,1,4,4-Tetramethyl-1,4-dichlorodisilethylene; difunctional blocking agent used to prepare derivatives of primary amines which are stable to strong bases and some oxidizing agents; has been applied to amino acid synthesis.

T2360. [Hüls] (2,4,6-Tri-t-butylphenoxy) dimethylchlorosilane; sterically hindered blocking agent; used for silylenol ethers.

T2520. [Hüls] Triethylchlorosilane; sterically hindered blocking agent.

T2885. [Hüls] Triisopropylchlorosilane; sterically hindered blocking agent; for protection of alcohols; useful with nucleosides and nucleotides.

T2890. [Hüls] Triisopropylsilyl trifluoromethanesulfonate; sterically hindered blocking agent.

T2928. [Hüls] Trimethylbromosilane; blocking agent.

T2950. [Hüls] Trimethylchlorosilane; blocking agent; silylation reagent.

T-3000. [Olin] PPG triol; see T-600; m.w. 3000; f.p. –20 C; sp.gr. 1.009; dens. 8.41 lb/gal.

T3150. [Hüls] 2-Trimethylsiloxypent-2-en-4-one; blocking agent.

T3250. [Hüls] Trimethylsilylacetamide; blocking agent; silylates alcohols; persilylates carbohydrates by fusion.

T3500. [Hüls] Trimethylsilyl-N,N′-diphenylurea; blocking agent; silylates most act. hydrogen-containing compds.; precipitates diphenylurea.

T3550. [Hüls] 2-Trimethylsilylethanol; blocking agent used to prepare trimethylsilylethyl derivatives of carboxylic acids, carbohydrates, amines, and hydroxyl groups.

T3600. [Hüls] N-trimethylsilylimidazole; blocking agent; protecting agent specific for hydroxyl groups.

T3634. [Hüls] Trimethylsilyl methanesulfonate; blocking agent.

T3660. [Hüls] 3-Trimethylsilyl-2-oxazolidinone; blocking agent; for carboxylic acid, amines, alcohols, 1,3-dicarbonyl compds. and thiols; used in penicillin and cephalosporin synthesis.

T3795. [Hüls] Trimethylsilyl trifluoromethanesulfonate; blocking agent; powerful silylation reagent.

T-4000. [Olin] PPG triol; see T-600; m.w. 4000; f.p. –20 C; sp.gr. 1.007; dens. 8.39 lb/gal.

Tactix H31. [Dow] Cycloaliphatic amine blend; epoxy curing agent, hardener; sp.gr. 0.94; visc. 9–10 cps; flash pt. 160 F; 23.78–24.22% N.

Tactix H41. [Dow] Aromatic diamine blend; epoxy curing agent, hardener; sp.gr. 1.009; visc. 80–120 cps; flash pt. 250 F; 16–17% N.

Tagat I, I2. [Goldschmidt] POE glycerol monoisostearate; nonionic; preparation of o/w emulsions; solubilizer for flavors, perfumes, vitamin oils; dispersant and antistat; liq.; HLB 15.6 and 14.2 resp.; 100% conc.

Tagat L. [Goldschmidt] PEG-30 glyceryl laurate; nonionic; see Tagat I; liq.; HLB 17.0; 100% conc.

Tagat L2. [Goldschmidt] PEG-20 glyceryl laurate; nonionic; see Tagat I; liq.; HLB 15.7; 100% conc.

Tagat O. [Goldschmidt] PEG-30 glyceryl oleate; nonionic; see Tagat I; liq.; HLB 16.4; 100% conc.

Tagat O2. [Goldschmidt] PEG-20 glyceryl oleate; nonionic; see Tagat I; liq.; HLB 15.0; 100% conc.

Tagat R 1. [Goldschmidt] PEG-15 glyceryl ricinoleate; nonionic; see Tagat I; liq.; HLB 14.0; 100% conc.

Tagat R40. [Goldschmidt] PEG-40 hydrog. castor oil; see Tagat I; solid; 100% conc.

Tagat R60. [Goldschmidt] PEG-60 hydrog. castor oil; see Tagat I; solid; 100% conc.

Tagat S. [Goldschmidt] PEG-30 glyceryl stearate; nonionic; see Tagat I; liq.; HLB 16.4; 100% conc.

Tagat S2. [Goldschmidt] PEG-20 glyceryl stearate; nonionic; see Tagat I; liq.; HLB 15.0; 100% conc.

Tagat TO. [Goldschmidt] PEG-25 glyceryl trioleate; nonionic; see Tagat I; also refatting agent for hair/bath preps.; liq.; HLB 11.3; 100% conc.

TAHP-50M. [Witco/Argus] Tert.-amyl hydroperoxide sol'n. in min. spirit diluent with sm. amts. of di-tert. amyl peroxide and tert.-amyl alcohol; initiator; liq.; 50% act.; 7.7% act. oxygen.

TAHP-50P. [Witco/Argus] Tert.-amyl hydroperoxide sol'n. in phthalate ester with sm. amts. of di-tert. amyl peroxide, tert.-amyl alcohol; initiator; colorless liq.; 50% act.; 7.7% act. oxygen.

Taka-Sweet®. [Miles Biotech] Glucose isomerase; enzyme for prod. of fructose syrups from glucose; immobilized.

Takatex®. [Miles Biotech] Bacterial alpha-amylase; enzymatic desizing formulation; liq.

Taka-Therm®. [Miles Biotech] Bacterial alpha-amylase; enzyme for starch liquefaction and textile desizing; liq.

Tally® 100 Plus. [Van Den Bergh] Glyceryl stearate, PEG-20 glyceryl stearate, hydrog. soybean oil; emulsifier, dough strengthener and crumb softener used in breads; also as textile lubricant and fabric softener; HLB 6.1; m.p. 55–57 C.

Talofloc®. [Sherex] Flocculant to remove color bodies in cane sugar refining processes.

Tamaron®. [Bayer] Methamidophos; insecticide and acaricide; colorless to slightly ylsh. cryst. mass (tech. a.i.), colorless cryst. (pure a.i.); m.w. 141.1; sol. (g/1000 ml): > 2000 g in water, 1400 g in IPA, 1000 g in methylene chloride; m.p. 46.1 C.

Tamol® 165. [Rohm & Haas] Ammonium salt of carboxylated polyelectrolyte; dispersant for latex paints; lt. yel. aq. sol'n.; dens. 8.8 lb/gal; visc. 125–275 cps; 21% act.

Tamol® 731-25%. [Rohm & Haas] Sodium salt of polymeric carboxylic acid; dispersant for dyes, pigments; lt. yel. aq. sol'n.; dens. 9.2 lb/gal; visc. 200 cps; f.p. 28 F; 25% act.

Tamol® 731-SD. [Rohm & Haas] Sodium salt of polymeric carboxylic acid; dispersant for dyes, pigments; tan powd.; bulk dens. 24 lb/ft^3; flash pt. (TOC) 300 F; surf. tens. 36 dyne/cm (1%).

Tamol® 819. [Rohm & Haas] Sodium salt of condensed naphthalene sulfonic acid; dispersant for cement, pigments, carbon blk.; tan spray-dried solid; bulk dens. 40 lb/ft^3; 93% min. act.

Tamol® 819L-43. [Rohm & Haas] Sodium salt of condensed naphthalene sulfonic acid; dispersant for cement, pigments, carbon blk.; brn. aq. sol'n.; dens. 10.4 lb/gal; visc. 85 cps; f.p. 28 F; 43% act.

Tamol® 850. [Rohm & Haas] Sodium salt of polymeric carboxylic acid; low foaming dispersant for inorg. pigments; lt. yel. aq. sol'n.; dens. 9.9 lb/gal; visc. 250 cps; pour pt. 32 F; surf. tens. 71 dyne/cm (1%).

Tamol® 901. [Rohm & Haas] Ammonium salt of carboxylated polyelectrolyte; pigment dispersant for trade-sales paints; straw aq. sol'n.; dens. 9.4 lb/gal; visc. 75–200 cps; f.p. 32 F; 30% act.

Tamol® 960. [Rohm & Haas] Sodium salt of polymeric carboxylic acid; low foaming dispersant for inorg. pigments; APHA 125 aq. sol'n.; dens. 10.6 lb/gal; visc. 550 cps; pour pt. 32 F; 40% act.

Tamol® 963. [Rohm & Haas] Ammonium salt of carboxylated polyelectrolyte; dispersant.

Tamol® 983. [Rohm & Haas] Potassium salt of carboxylated polyelectrolyte; dispersant.

Tamol® L Conc. [Rohm & Haas] Sodium salt of condensed naphthalene sulfonic acid; anionic; dispersant, tanning agent for leather; amber liq.; dens. 10.4 lb/gal; visc. 20 cps; f.p. 28 F; surf. tens. 71 dyne/cm (1%); 47% act.

Tamol® NH 3901. [BASF AG] Sodium salt of condensed naphthalene sulfonic acid; dispersant for pigments, carbon blk., dyestuffs; stabilizer for dispersions and emulsions; liq.; water-misc.; 38% conc.

Tamol® NH 7519. [BASF AG] Sodium salt of condensed naphthalene sulfonic acid; dispersant for pigments, carbon blk., dyestuffs; stabilizer for dispersions and emulsions; powd.; water-sol.; 75% conc.

Tamol® NH 9103. [BASF AG] Sodium salt of condensed naphthalene sulfonic acid; dispersant for pigments; stabilizer for dispersions and emulsions; plasticizer for concrete, mortar, and gypsum; powd.; water-sol.; 90% conc.

Tamol® NHC 3001, NHC 4001. [BASF AG] Calcium salt of condensed naphthalene sulfonic acid; see Tamol NH 9103; liq.; water-misc.; 30 and 40% conc. resp.

Tamol® NHC 9301. [BASF AG] Calcium salt of condensed naphthalene sulfonic acid; see Tamol NH 9103; powd.; water-sol.; 93% conc.

Tamol® NMC 9301. [BASF AG] Calcium salt of condensed naphthalene sulfonic acid; see Tamol NH 9103; powd.; water-sol.; 95% conc.

Tamol® NN. [BASF AG] 30% aq. sol'n. of Tamol NNOK; anionic; grinding aid; stabilizer for aq. disp. and emulsions; dispersant for pigments; precipitant for basic dyes and cationic compds.; dispersing and plastifying agents for concrete; aux. in metal cleaning; clear pale brn. liq.; pH 9.5–10.5 (10% aq.); 33% act.

Tamol® NN 2406, NN 2901. [BASF AG] Sodium salt of condensed naphthalene sulfonic acid; dispersant for pigments, carbon blk., dyestuffs; stabilizer for dispersions and emulsions; liq.; water-misc.; 24 and 30% conc. resp.

Tamol® NN 4501. [BASF AG] Sodium salt of condensed naphthalene sulfonic acid; see Tamol NN 2406; also dispersant for rubber syntheses, processing of rubber latexes; liq.; water-misc.; 45% conc.

Tamol® NN 7718, NN 8906, NN 9104, NN 9401. [BASF AG] Sodium salt of condensed naphthalene sulfonic acid; see Tamol NN 2406; powd.; water-sol.; 78, 89, 91, and 94% conc. resp.

Tamol® NNa 4109. [BASF AG] Ammonium salt of condensed naphthalene sulfonic acid; dispersant for pigments, carbon blk., dyestuffs; liq.; water-misc.; 41% conc.

Tamol® NNO. [BASF AG] Naphthalenesulfonic acid sodium salt condensation prod.; anionic; see Tamol NN; pale brn. powd.; pH 6.5–7.5 (10% aq.); 95% act.

Tamol® NNOK. [BASF AG] Naphthalenesulfonic acid sodium salt condensation prod.; anionic; see Tamol NN; pale brn. powd.; pH 9.5–10.5 (10% aq.); 95% act.

Tamol® PA Liq. [BASF AG] Polycarboxylic acid sodium salt; anionic; dispersant, solubilizer, protective colloid; chemical industries; water treatment and builder for detergent that act as soil dispersers, dispersants, and complexing agents for calcium and magnesium salts; car shampoos, detergents and clear rinses for glass and ceramics; yel. clear liq.; water-misc.; dens. 1.20–1.25 g/cc; visc. 80–250 cps; pH 6–8; 35 ± 1% solids.

Tamol® PA Powd. [BASF AG] Polycarboxylic acid sodium salt; anionic; see Tamol PA Liq.; wh. to yel.

powd.; water-misc.; dens. 470 g/l; pH 6.5–8.0 (20% aq.); > 91% conc.

Tamol® PP. [BASF AG] Sodium salt of a phenol sulfonic acid condensation prod.; dispersant for the formulation of crop protecting agents, and pigments; powd.; water-sol.; 90% conc.

Tamol® SD. [Rohm & Haas] Sodium salt of condensed naphthalene sulfonic acid; dispersant, tanning agent for leather; solid.

Tamol® SG-1. [Rohm & Haas] Carboxylated polyelectrolyte ammonium salt; anionic; pigment dispersant for latex paints; colorless liq.; f.p. 32 F; dens. 9.6 lb/gal; visc. 260 cps; 35% act.

Tamol® SN. [Rohm & Haas] Sodium salt of condensed naphthalene sulfonic acid; dispersant, tanning agent for leather; tan powd.; bulk dens. 40 lb/ft^3; flash pt. (TOC) 300 F; surf. tens. 71 dyne/cm (1%); 93% min. solids.

Tamsil 8. [Unimin] Silica; filler for finishes, enamels, maintenance paints, plastic film antiblocks, urethane rubber, polishes; wh. particulate; 2.0 μ median particle size; 100% thru 325 mesh; surf. area 2.1 m/g; oil absorp. 28 g/100 g; GE brightness 87.5; pH 6.3; 98.81% SiO_2.

Tamsil 10. [Unimin] Silica; filler for finishes, interior and marine coatings, maintenance paints, exterior coatings, plastic film antiblocks, silicone rubber; wh. particulate; 2.1 μ median particle size; 100% thru 325 mesh; surf. area 2.0 m/g; oil absorp. 28 g/100 g; GE brightness 86.4; pH 6.3; 98.9% SiO_2.

Tamsil 15. [Unimin] Silica; filler for interior coatings, maintenance coatings, marine coatings, exterior coatings, silicone and urethane rubbers; wh. particulate; 2.5 μ median particle size; surf. area 1.6 m/g; oil absorp. 29.9 g/100 g; GE brightness 86; pH 6.3; 98.92% SiO_2.

Tamsil 25. [Unimin] Silica; filler for interior and exterior coatings, maintenance and marine coatings, silicone rubber molding compds.; wh. particulate; 3.5 μ median particle size; surf. area 1.5 m/g; oil absorp. 28.5 g/100 g; GE brightness 85.6; pH 5.9; 99% SiO_2.

Tamsil 30. [Unimin] Silica; filler for interior and exterior coatings, maintenance and marine coatings, urethane rubber, buffing compds.; wh. particulate; 5.8 μ median particle size; 99.6% thru 325 mesh; surf. area 1.0 m/g; oil absorp. 29 g/100 g; GE brightness 85.2; pH 6.2; 99.12% SiO_2.

Tamsil 45. [Unimin] Silica; filler for traffic paint, exterior block fillers, mastics and adhesives, buffing compds.; wh. particulate; 8.4 μ median particle size; 97% thru 325 mesh; surf. area 1.1 m/g; oil absorp. 29 g/100 g; GE brightness 85.2; pH 6.2; 99.07% SiO_2.

Tamsil 75. [Unimin] Silica; filler for nonskid coatings, mastics and adhesives, cements, buffing compds.; wh. particulate; 88.9% thru 325 mesh; oil absorp. 29 g/100 g; pH 6.2; 99.098% SiO_2.

Tamsil 150. [Unimin] Silica; filler for mastics and adhesives, textured coatings, grouts and mortars, buffing compds.; wh. particulate; 15.2 μ median particle size; 79.9% thru 325 mesh; surf. area 0.7 m/g; oil absorp. 32 g/100 g; pH 5.9; 99.17% SiO_2.

Tamsil Gold Bond. [Unimin] Silica; filler for traffic paint, interior/exterior coatings, maintenance and marine coatings, mastics and adhesives, elec. epoxy compds., buffing compds.; wh. particulate; 8.4 μ median particle size; 98.6% thru 325 mesh; surf. area 1.1 m/g; oil absorp. 29 g/100 g; GE brightness 85.2; pH 6.1; 99.075% SiO_2.

Tanapal ME Acid. [Sybron] Phosphate ester; anionic; leveling agent for disperse dyes; liq.; 100% conc.

Tanaquad T-85. [Sybron] Quat. ammonium compd.; cationic; wetting agent, bactericide; sanitizer, dye retarder, antistat; liq.; 85% conc.

Tandem 5K, 8. [ICI Am.] Mono and diglycerides, polysorbate 60 with BHA, citric acid; nonionic; food emulsifier, conditioner/softener for yeast-raised baked goods; cream-wh. plastic solid; sol. in veg. oils; disp. in water; m.p. 128 F; HLB 8.1; flash pt. > 300 F; 100% conc.

Tandem 9. [ICI Am.] Mono- and diglycerides and polysorbate 60 (25%); see Tandem 5K; cream-wh. flakes; bland odor; disp. in water; m.p. 135 F; HLB 6.0; flash pt. > 300 F; 100% conc.

Tandem 11H. [ICI Am.] Hydrated mono- and diglycerides, polysorbate 60 with sodium propionate and lactic acid; see Tandem 5K; cream-wh. soft plastic; disp. in water; HLB 6.3; flash pt. > 300 F; pH 5; 50% conc.

Tandem 552. [ICI Am.] Mono- and diglycerides, polysorbate 60 with BHA, citric acid; see Tandem 5K; golden clear liq.; sol. in veg. oils; disp. in water; m.p. 45 F; HLB 8.1; flash pt. > 300 F; 100% conc.

Tauranol I-78, I-78 Flakes, I-78-3. [Finetex] Sodium cocoyl isethionate; anionic; detergent; foamer and dispersant; conditioner; used in personal care prods.; powd., flakes, and paste resp.; 83, 83, and 27% conc.

TBC. [Croda Ltd.] Tributyl citrate; plasticizer for vinyl and cellulose resins; liq.

TBHP-70. [Witco/Argus] t-Butyl hydroperoxide sol'n. with di-t-butyl peroxide and sm. amts. of t-butyl alcohol and water as diluents; initiator; colorless liq.; 70% act.; 12.4% act. oxygen.

TBP. [FMC] Tributyl phosphate; antifoam agent used in paper coating compds., water adhesives, inks, casein sol'ns., textile sizes, and detergent sol'ns.; plasticizer and solv. for NC, cellulose acetate, chlorinated rubber, and vinyls; solv. for natural gums and syn. resins; component in extraction of rare earth metal salts; Pt-Co 50 max. liq.; char. odor; m.w. 266; f.p. < -80 C; b.p. 137–145 C; sol. in org. liqs., min. oil, and gasoline; sp.gr. 0.978 ± 0.003; dens. 975 kg/m^3; visc. 3.7 cp; pour pt. < -80 C; ref. index 1.423 ± 0.001; flash pt. 166 C; fire pt. 182 C.

TCC. [Monsanto] 3,4,4,'-Trichlorocarbanilide; bacteriostat for bar soaps.

TDP 2000. [Eastman] Thiodipropionate polyester; antioxidant for polyolefins; Gardner 2 paste solid; m.w. 2000; visc. 600–1000 cs (100 C); acid no. 2.5; flash pt. 249 C.

T-Det® BP-1. [Harcros] PO/EO block copolymer; nonionic; emulsifier, detergent, dispersant and polymerization agent; pesticide formulations; solid; HLB 15.7; 100% conc.

T-Det® C-40. [Harcros] PEG-40 castor oil; nonionic; emulsifier, solubilizer, degreaser, lubricant, dispersant, penetrant used in leather, textile and metal processing; paint for dispersing pigment slurries, leveling agent, defoamer, stabilizer; paper mfg.; wax and polish preparations; rubber lubricant; yel. liq.; mild, oily odor; sol. in acetone, butyl Cellosolve, CCl_4, ethanol, ether, ethylene glycol, methanol, veg. oil, water, xylene; dens. 8.7 lb/gal; sp.gr. 1.05; HLB 14.2; flash pt. > 200 F (TOC); pour pt. 60 F; pH 6.0 (1%); 99.5% min. act.

T-Det® EPO-61. [Harcros] EO/PO; anionic; defoamer; crude oil demulsifier; foam control in paper mfg.; water treating compds.; emulsifier and wetting agent in agric. toxicant formulations; clear liq.; sol. in ethanol, toluene, xylene, perchlorethylene; dens. 8.34 lb/gal; sp. gr. 1.01; visc. 271 cps; cloud pt. 23 C (1%); flash pt. > 200 C; pour pt. –28 C; 99.5% min. act.

T-Det® N 1.5. [Harcros] Nonoxynol-1 (1.5 EO); nonionic; surfactant for petrol. oils applics.; stabilizer; coemulsifier; pale yel. liq.; oil-sol.; sp.gr. 0.99 (68 F); dens. 8.2 lb/gal (68 F); HLB 1.5; pour pt. < 0 F; flash pt. (PMCC) 340 F; pH 5–7 (1% aq.); 99.5% min. act.

T-Det® N-4. [Harcros] Nonoxynol-4; nonionic; detergent, wetting agent; agric. toxicant formulations; intermediate for household detergents; dry cleaning soap additive, sludge dispersant additive in fuel oils; gasoline additive; solv. cleaner for metal processing; color development aid, stabilizer and emulsifier for latex paints; pale yel. oily liq.; mild aromatic odor; sol. in butyl Cellosolve, CCl_4, min. and corn oil, ethanol, kerosene, min. spirits, perchloroethylene, Stod., toluene; dens. 8.5 lb/gal; sp.gr. 1.02; visc. 52 SUS (210 F); HLB 8.9; flash pt. > 400 F (OC); pour pt. < O F; pH 7; 99.5% min. act.

T-Det® N-6. [Harcros] Nonoxynol-6; nonionic; emulsifier for silicones, household waxes and polishes; wetting agent; detergent and dispersant for petrol. and fuel oils; agric. toxicant formulations; dry cleaning soap additive; leather and metal processing; solv. cleaner; deinking agent for paper; pale yel. liq.; mild aromatic odor; sol. in butyl Cellosolve, CCl_4, corn oil, diesel fuel, ethanol, kerosene, Stod., xylene; dens. 8.7 lb/gal; sp.gr. 1.04; visc. 58 SUS (210 F); HLB 10.9; flash pt. > 400 F (OC); pour pt. < 0 F; pH 7; 99.5% min. act.

T-Det® N-8. [Harcros] Nonoxynol-8; nonionic; detergent, wetting agent, emulsifier, solubilizer, degreaser; textile, leather and paint processing; agric. toxicant, household and industrial cleaning formulations; pigment dispersant; rewetting, deinking, felt washing, leveling in paper mfg.; oil recovery operations; clear visc. liq.; mild aromatic odor; dens. 8.75 lb/gal; sp.gr. 1.05; visc. 63 SUS (210 F); HLB 12.4; cloud pt. 78 F (1%); flash pt. 500 F (OC); pH 6 (1%); 99.5% min. act.

T-Det® N-9.5. [Harcros] Nonoxynol-9; nonionic; see T-Det N-8; also used as a foamer; clear visc. liq.; mild aromatic odor; sol. in butyl Cellosolve, CCl_4, ethanol, ethylene glycol, HAN, toluene, water, xylene; dens. 8.8 lb/gal; sp.gr. 1.06; visc. 68 SUS (210 F); HLB 13.1; cloud pt. 135 F (1%); flash pt. 500 F (OC); pH 7 (1%); 99.5% min. act.

T-Det® N-10.5. [Harcros] Nonoxynol-10; nonionic; detergent, wetting agent, dispersant, emulsifier, solubilizer; textile processing; household and industrial cleaning prods.; clear visc. liq.; mild aromatic odor; sol. in butyl Cellosolve, ethanol, ethylene glycol, HAN, toluene, water, xylene; dens. 8.8 lb/gal; sp.gr. 1.06; HLB 10.4; cloud pt. 155 F; flash pt. 500 F (OC); pH 7 (1%); 99.5% min. act.

T-Det® N-14. [Harcros] Nonoxynol-14; nonionic; detergent; emulsifier for oils, fats and waxes; wetting agent for metal cleaners; penetrant; corrosion inhibitor; off-wh. semisolid; mild aromatic odor; sol. in butyl Cellosolve, CCl_4, corn oil, ethanol, ethylene glycol, water, xylene; dens. 8.9 lb/gal (68 F); sp.gr. 1.07 (68 F); visc. 89 SUS (210 F); HLB 14.7; cloud pt. 203 F (1%); flash pt. > 400 F (OC); pour pt. 67 F; pH 6 (1%); 99.5% min. act.

T-Det® N-70. [Harcros] Nonoxynol-70; nonionic; detergent, emulsifier for oils, fats and waxes; stabilizer in latex and alkyd emulsions; emulsion polymerization; scouring textile operations; pale yel. waxy solid; mild aromatic odor; dens. 9.0 lb/gal (135 F); sp.gr. 1.08 (135 F); HLB 18.7; cloud pt. 212 F (1%); flash pt. > 500 F; pour pt. 125 F; pH 6.0–7.0 (1%); 99.5% min. act.

T-Det® N-707. [Harcros] Nonoxynol-70; nonionic; see T-Det N-70; clear to hazy liq.; mild aromatic odor; dens. 9.0 lb/gal; sp.gr. 1.08; HLB 18.7; cloud pt. 212 F (1%); flash pt. > 500 F; pour pt. 40 F; pH 5.0–6.5 (1%); 50% min. act.

T-Det® O-6. [Harcros] Octoxynol-6 (6–7 EO); nonionic; surfactant for detergents, textile and leather processing, agric. formulations; dispersant; clear liq.; water-disp.; sp.gr. 1.05 (68 F); dens. 8.7 lb/gal (68 F); HLB 11.6; pour pt. < 0 F; flash pt. (PMCC) > 500 F; pH 6–8 (1%); 99.5% min. act.

T-Det® O-40. [Harcros] Octoxynol-40; nonionic; surfactant for agric. formulations; detergents, textiles, emulsion polymerization; emulsifier, stabilizer; wh. solid; water-sol.; sp.gr. 1.06 (150 F); dens. 8.8 lb/gal (150 F); HLB 17.9; pour pt. 120 F; flash pt. (PMCC) > 500 F; cloud pt. > 212 F (1%); 99.5% min. act.

T-Det® O-407. [Harcros] Octoxynol-40; nonionic; detergent, emulsifier, stabilizer for various applics.; APHA 100 max. liq.; water-sol.; dens. 9.1 lb/gal; visc. 600 cps; cloud pt. > 100 C (1%); flash pt. > 100 C (TOC); pour pt. 30 F; pH 7 (1%); 70 ± 0.5% act.

T-Det® TDA-60. [Harcros] Tridecyl alcohol 60% EO adduct; nonionic; detergent, dispersant, wetting agent; liq.; water-sol.; HLB 12.0; 100% conc.

T-Det® TDA-65. [Harcros] Tridecyl alcohol 65% EO adduct; nonionic; emulsifier, solubilizer; textile processing; hard surface detergent; disperses and suspends solids; wetting and rewetting agent for dust settling, corrosion inhibitor, metal cleaning and pickling; hazy liq.; mild aliphatic odor; water-sol.; dens.

8.20 lb/gal; cloud pt. 130 F (1%); flash pt. > 290 F (PMCC); pour pt. 62 F; pH 5–7; 99.5% min. act.

T-Det® TDA-70. [Harcros] Tridecyl alcohol 70% EO adduct; nonionic; see T-Det TDA-60; paste; water-sol.; HLB 14.0; 100% conc.

Tebol 99. [Arco] High purity tert. butyl alcohol; solv., cosolvent, compatibilizer, coupling agent, processing aid for pharmaceuticals, personal care prods., aq. coatings and adhesives, agric. formulations, polymer processing, cleaners/disinfectants; chlorinated hydrocarbon stabilizer; Pt-Co ≤ 10 color; sol. in water; misc. with most org. solvs.; sp.gr. 0.78; dens. 6.5 lb/gal; visc. 3.3 cps (30 C); i.b.p. 81.5 C (760 mm Hg); f.p. 24.5 C; flash pt. (CC) 11 C; 99.3% act.

Tech Pet F, M. [Witco] Petrolatum tech.; dk. br.; m.p. 57–66 and 51–65 C resp.

Tecquinol®. [Eastman] Hydroquinone, tech.; antioxidant for syn. latexes, fats, oils, monomers, polyester resins.

Tedion V18®. [Duphar B.V.] Tetradifon; acaricide for control of spider mites on almost all crops; m.w. 356.01; sol.: slight to fair in polar org. solvs., poor in apolar solvs., practically insol. in water; m.p. 144 C min. (tech.).

Teepol HB 6. [Shell] Primary alcohol, sulfated, sodium salt; solubilizer, emulsifier; component of hard surface and germicidal cleaners; pale amber liq.; water-sol.; dens. 1.06 kg/l; visc. 20 cs; clear pt. 10 C; pH 8.0 (5% aq.); biodeg.; 34.2% act.

Teepol HB 7. [Shell] C_9-C_{13} primary alcohol sulfate, sodium salt; anionic; solubilizer and emulsifier component of hard surface and germicidal cleaners; liq.; 40% conc.

Tefose® 63. [Gattefosse] PEG-6-32 stearate and glycol stearate; nonionic; SE base for cosmetics, o/w pharmaceutical ointments; Gardner < 5 waxy solid; weak odor; HLB 9–10; m.p. 40–45 C; acid no. < 6; iodine no. < 3; sapon. no. 110–120; 100% conc.

Tefose® 1500. [Gattefosse] PEG-6-32 stearate; nonionic; see Tefose 63; Gardner < 5 pasty solid; weak odor; HLB 10–12; m.p. 39–45 C; acid no. < 6; iodine no. < 3; sapon. no. 75–95.

Tefose® 2000. [Gattefosse] PEG-6 stearate, ceteth-20; nonionic; see Tefose 63; Gardner < 5 pasty solid; weak odor; HLB 10–12; m.p. 32–37 C; acid no. < 6; iodine no. < 3; sapon. no. 65–85.

Tefose® 2561. [Gattefosse] PEG-6 stearate, glyceryl stearate, and ceteth-20; nonionic; see Tefose 63; Gardner < 3 pasty solid; weak odor; HLB 10–12; m.p. 35–40 C; acid no. < 5; iodine no. < 3; sapon.no. 85–105.

Tegamine 18. [Goldschmidt] Stearamidopropyl dimethylamine; cationic; conditioner for hair care and bath prods.; flake; 100% act.

Tegamine P-7. [Goldschmidt] Cocamidopropyl dimethylamine; cationic; conditioner for hair care and bath prods.; flake; 100% act.

Tegamin® Oxide WS-35. [Goldschmidt] Cocamidopropylamine oxide; surfactant, foaming agent, visc. builder for hair and skin cleansing.

Tegiloxan®. [Goldschmidt] Silicone oils; antifoams for min. oils, in rubber and plastics industry; lubricant for tire prod.; additive for polishes.

Tegin® 90. [Goldschmidt] Glyceryl stearate; nonionic; stabilizer for creamy and liq. o/w emulsions in cosmetics, pharmaceuticals; flake; HLB 4.5; 100% conc.

Tegin® 90 NSE. [Goldschmidt] Glyceryl stearate; nonionic; stabilizer for o/w emulsions for cosmetic and pharmaceuticals; powd.; HLB 4.5; 100% conc.

Tegin® 200 L. [Goldschmidt] PEG 200 laurate; nonionic; see Tegin 90; liq.; HLB 7.1; 100% conc.

Tegin® 515. [Goldschmidt] Glyceryl stearate; nonionic; emulsifier, stabilizer for creams and lotions; powd.; HLB 3.8; 100% conc.

Tegin® 515 V.A. [Goldschmidt] Glyceryl stearate; nonionic; see Tegin 515; flake; HLB 3.8; 100% conc.

Tegin® 4100. [Goldschmidt] Glyceryl stearate; nonionic; stabilizer for cosmetics and pharmaceuticals; powd.; HLB 3.8; 100% conc.

Tegin® 4480. [Goldschmidt] Glyceryl laurate; nonionic; emulsifier and stabilizer for cosmetics and pharmaceuticals; semisolid; HLB 3.8; 100% conc.

Tegin® 4600. [Goldschmidt] Glyceryl caprylate/caprate; nonionic; antistat and lubricant for syn. materials; emulsifier for cosmetic and pharmaceutical prods.; liq.; HLB 4.0; 100% conc.

Tegin® 4600 NSE. [Goldschmidt] Glyceryl caprylate; nonionic; antistat for syn. materials; emulsifier for cosmetics and pharmaceuticals; liq.; HLB 3.8; 100% conc.

Tegin® C-1R. [Goldschmidt] Glyceryl stearate SE and hydrog. tallow glyceride citrate; anionic; solubilizer and co-emulsifier for cosmetics; liq.; HLB 17.0; 100% conc.

Tegin® D 1102. [Goldschmidt] Ethylene glycol ester; sheen additive for cosmetics; powd.

Tegin® E-41. [Goldschmidt] Acetylated hydrog. tallow glyceride; nonionic; emulsifier for cosmetic and technical prods.; emollient, plasticizer, food additive; solid; HLB 2.0–3.0; 100% conc.

Tegin® E-41 NSE. [Goldschmidt] Acetylated hydrog. tallow glycerides; nonionic; emulsifier for cosmetic and technical prods.; emollient/plasticizer food additives; semisolid; HLB 2.0–3.0; 100% conc.

Tegin® E-61. [Goldschmidt] Acetylated hydrog. lard glyceride; nonionic; see Tegin E 41; solid; HLB 2.0–3.0; 100% conc.

Tegin® E-61 NSE. [Goldschmidt] Acetylated hydrog. lard glyceride; nonionic; see Tegin E-41 NSE; semisolid; HLB 2.0–3.0; 100% conc.

Tegin® E-66. [Goldschmidt] Acetylated lard glyceride; nonionic; see Tegin E 41; liq.; HLB 2.0–3.0; 100% conc.

Tegin® E-66 NSE. [Goldschmidt] Acetylated lard glyceride; nonionic; see Tegin E-61 NSE; liq.; HLB 2.0–3.0; 100% conc.

Tegin® G. [Goldschmidt] Glycol stearate SE; anionic; sheen additive, opacifier for cosmetics and pharmaceuticals; o/w emulsions; powd.; HLB 12.0; 100% conc.

Tegin® G 1100. [Goldschmidt] Ethylene glycol distearate; sheen additive, pearlescent, opacifier for shampoos; non SE; powd.

Tegin® GO. [Goldschmidt] Ethylene glycol monooleate; anionic; coemulsifier and stabilizer; liq.; HLB 2.8; 100% conc.

Tegin® GRB NSE. [Goldschmidt] Glycerol stearate; nonionic; stabilizer for cosmetic and pharmaceuticals; powd.; HLB 3.8; 100% conc.

Tegin® L 61, L-61 NSE, L 62, L-62 NSE. [Goldschmidt] Hydrog. tallow glyceride lactate; nonionic; emulsifier, emollient, plasticizer for cosmetics; food additive; powd.; HLB 5.0, 5.0, 6.0, and 6.0 resp.; 100% conc.

Tegin® MAV NSE. [Goldschmidt] Glyceryl stearate; anionic; stabilizer for cosmetics and pharmaceuticals; powd.; HLB 3.8; 100% conc.

Tegin® M NSE. [Goldschmidt] Glycerol stearate; nonionic; stabilizer for cosmetics and pharmaceuticals; powd.; HLB 3.8; 100% conc.

Tegin® P. [Goldschmidt] Propylene glycol stearate SE; anionic; emulsifier for o/w lotions and creams; solubilizer for dyestuffs; solid; HLB 4.4; 100% conc.

Tegin® P-411 SE. [Goldschmidt] Propylene glycol stearate SE; o/w creams and lotions; solubilizer for dyestuffs; waxy; HLB 2.8; 100% conc.

Tegin® PL. [Goldschmidt] Propylene glycol laurate; nonionic; see Tegin P-411 SE; liq.; 100% conc.

Tegin® RZ. [Goldschmidt] Glyceryl ricinoleate; nonionic; emulsifier, solubilizer, emollient, plasticizer; paste; HLB 3.0; 100% conc.

Tegin® RZ NSE. [Goldschmidt] Glyceryl ricinoleate; nonionic; o/w emulsifier, solubilizer, emollient, plasticizer; semisolid; HLB 3.0; 100% conc.

Tegin® TGOS. [Goldschmidt] Ethylene triglycol stearate; anionic; sheen additive, opacifier; solid; 100% act.

Tegin® VA. [Goldschmidt] Glyceryl stearate; emulsifier and opacifier for skin care prods.; suitable for nonionic, cationic, and anionic systems.

Tegin® VA 55G. [Goldschmidt] Glycerol stearate; nonionic; emollient for creams and lotions; flake; HLB 3.8; 100% conc.

TegMeR® 703. [C.P. Hall] PEG-3 diheptanoate; lubricant for aluminum can industry; APHA 50 clear liq.; m.w. 388; sol. in hexane, toluene, kerosene, ethanol, acetone, min. oil; sp.gr. 0.990; visc. 28 cps; flash pt. 210 C; acid no. 0.8; sapon. no. 289; ref. index 1.444.

TegMeR® 704. [C.P. Hall] PEG-4 diheptanoate; lubricant for aluminum can industry; APHA 25 clear liq.; m.w. 418; sol. in hexane, toluene, kerosene, ethanol, acetone; sp.gr. 0.997; visc. 26 cps; f.p. –20 C; flash pt. 229 C; acid no. 0.1; sapon. no. 265; ref. index 1.446.

TegMeR® 803. [C.P. Hall] Triethylene glycol di-2-ethylhexoate; plasticizer for adhesives; APHA 50 clear liq.; m.w. 374; sol. in hexane, toluene, kerosene, ethyl alcohol, acetone, min. oil, ASTM oil #1; sp.gr. 0.968; visc. 21 cps; acid no. 0.3; sapon. no. 248; ref. index. 1.4425; flash pt. 204 C.

TegMeR® 804. [C.P. Hall] Tetraethylene glycol di-2-ethylhexoate; plasticizer for rubber industry; textile surf. finishes, softeners, thread lubricants, and/or antistats; Gardner 5 clear liq.; m.w. 449; f.p. –65 C; sol. in hexane, toluene, kerosene, ethyl alcohol, acetone; sp.gr. 0.984; visc. 18 cps; acid no. 0.45; sapon. no. 240; ref. index 1.4445; flash pt. 204 C.

TegMeR® 804 Special. [C.P. Hall] PEG-4 di-2-ethylhexoate; lubricant for aluminum can industry; APHA 60 clear liq.; m.w. 449; sol. in hexane, toluene, kerosene, ethanol, acetone; sp.gr. 0.981; visc. 20 cps; f.p. –62 C; flash pt. 218 C; acid no. 0.4; sapon. no. 242; ref. index 1.4450.

TegMeR® 809. [C.P. Hall] PEG 400 di-2-ethylhexoate; lubricant for aluminum can industry; Gardner 2 clear liq.; m.w. 652; sol. in hexane, toluene, kerosene, ethanol, acetone; sp.gr. 1.02; visc. 50 cps; f.p. –48 C; flash pt. 262 C; acid no. 0.5; sapon. no. 170; ref. index 1.452.

TegMeR® 903. [C.P. Hall] PEG-3 dipelargonate; lubricant for aluminum can industry; APHA 50 clear liq.; m.w. 420; sol. in hexane, toluene, kerosene, ethanol, acetone, min. oil; sp.gr. 0.97; visc. 24 cps; flash pt. 224 C; acid no. 0.2; sapon. no. 254; ref. index 1.447.

Tego® Airex 960. [Tego] Silicone modified org. polymer; deaerator for med. and high polar coating systems, esp. PU paints; also for high solids alkyds, ES spray systems, and urethane-polyesters; clear liq., aromatic odor; insol. in water; sp.gr. 0.91; dens. 7.6 lb/gal; b.p. 282–286 F; flash pt. 25 C; 25% act. in xylene.

Tego® Airex 970. [Tego] Silicone modified org. polymer; deaerator for med.-polar solv.-based systems, esp. NC-containing systems, vinyl paints, and acid-curing baking finishes (alkyd-melamine and urea, acrylic-melamine, epoxy ester, high solids OEM finishes); clear liq., aromatic odor; insol. in water; sp.gr. 0.92; dens. 7.66 lb/gal; b.p. 282–286 F; flash pt. 25 C; 25% act. in xylene.

Tego® Airex 975. [Tego] Silicone modified org. polymer; deaerator for low- to med.-polar solv.-based systems, esp. chlorinated rubber paints, sol'n. vinyls (vehicles for tank linings, pool paints, pond liners); sp.gr. 0.95; dens. 7.9 lb/gal; flash pt. 25 C; 25% act. in xylene.

Tego-Amid D 5040. [Goldschmidt AG] Cocamidopropyl dimethylamine; cationic; emulsifier for creams and lotions; conditioner for hair care prods.; waxy.

Tego-Amid O 18. [Goldschmidt AG] Oleamidopropyl dimethylamine; cationic; see Tego-Amid D 5040; liq.

Tego-Amid S 18. [Goldschmidt AG] Stearamidopropyl dimethylamine; cationic; see Tego-Amid D 5040; flakes.

Tegoamin® 33. [Goldschmidt] Triethylene diamine; catalyst for the mfg. of PU foams and elastomers; sol'n.; sol. in water; sp.gr. 1.033 ± 0.005; solid. pt. < –20 C; b.p. 206–227 C; flash pt. 90 C; 33% sol'n. in dipropylene glycol.

Tegoamin® 100. [Goldschmidt] Triethylene diamine; catalyst for the mfg. of PU foams and elastomers; 100% act.

Tegoamin® BDE. [Goldschmidt] Amine; highly reactive catalyst for the prod. of flexible polyether PU foams; colorless clear liq., typ. amine odor; sol. in water; sp.gr. 0.902; visc. 4.1 mPas; solid. pt. < –80 C; b.p. 186–226 C; flash pt. 75 C; pH 12.8 (33% aq.); 30% dipropylene glycol.

Tegoamin® CPE. [Goldschmidt] Tert. amine; activator for the mfg. of polyester PU foams of reduced odor; sol. in water; sp.gr. 0.958 ± 0.005; solid. pt. –9 C; b.p. 127–324 C; flash pt. 33 C.

Tegoamin® CPU 33. [Goldschmidt] Tert. amine; catalyst for the prod. of conventional flexible PU foam; for slabstock and molded foams; liq.; sol. in water; dens. 0.945 g/cm²; visc. 23 mPas; solid. pt. < –50 C; b.p. 122–231 C; flash pt. 42 C.

Tegoamin® DMEA. [Goldschmidt] Dimethylethanolamine; catalyst for the prod. of flexible PU foams; sp.gr. 0.887; solid. pt. < –20 C; b.p. 133–135 C; flash pt. 40 C.

Tegoamin® EPS. [Goldschmidt] Tert. amine; catalyst in the mfg. of polyester PU slabstock foams; sol. in water; sp.gr. 0.988 ± 0.005; solid. pt. > –31 C; flash pt. 29 C.

Tegoamin® PDD 60. [Goldschmidt] Tert. amine; catalyst used as an activator for the mfg. of flexible PU foam; sol. in water; dens. 0.983 ± 0.005 g/cm³; b.p. 188–230 C; f.p. < –20 C; flash pt. 92 C.

Tegoamin® PMD. [Goldschmidt] Tert. amine; catalyst for the mfg. of flexible PU foams; for slabstock and molded foams; clear colorless liq.; sol. in water; dens. 0.870 ± 0.005 g/cm³; f.p. < –20 C; flash pt. 43 C.

Tegoamin® PTA. [Goldschmidt] Tert. amine; activator for the mfg. of flexible hot-cured and high resilience PU foams; catalyst for the blowing and gelling reactions; sol. in water; sp.gr. 0.962 ± 0.005; solid. pt. < –20 C; b.p. 138–229 C; flash pt. 61 C.

Tegoamin® SMP. [Goldschmidt] Tert. amine; catalyst for the mfg. of conventional flexible PU slabstock and molded foams (hot-cured foam); clear colorless to slightly yel. liq.; sol. in water; sp.gr. 0.941 ± 0.008; visc. 9 ± 2 mPas; solid. pt. < –50 C; b.p. 128–224 C; flash pt. 34 C.

Tego Antiflamm® N. [Goldschmidt] Flame retardant for PU foams.**Tego® Antifoam.** [Goldschmidt] Org. antifoam for waste water treatment.

Tego-Betaine® C. [Goldschmidt] Cocamidopropyl betaine; amphoteric; surfactant used as foam stabilizer and visc. builder in personal care prods., dishwash, liq. soap; liq.; 30% conc.

Tego-Betaine® L-7. [Goldschmidt] Cocamidopropyl betaine; amphoteric; surfactant, foam stabilizer, visc. builder in personal care prods.; liq.; 30% conc.

Tego-Betaine® L-90. [Goldschmidt] Lauramidopropyl betaine; amphoteric; see Tego-Betaine C; liq.; 30% conc.

Tego-Betaine® S. [Goldschmidt] Cocamidopropyl betaine; amphoteric; see Tego-Betaine C; liq.; 30% conc.

Tegochrome® 22. [Goldschmidt] 2,2-Ethylene dithiodiethanol; org. intermediate for producing color developers in the photographic industry.

Tegocoll®. [Goldschmidt] One and two-component PU adhesives; for bonding of glass, metal, and thermoplastics to other substrates, for use with PU foam, in wood processing industry.

Tegocolor®. [Goldschmidt] Pigment disps. in a polyether polyol; heat-stable color pastes for polyether PU foams.

Tegodont®. [Goldschmidt] Chlorinated agent for water treatment.

Tego® EL-HA-CE. [Goldschmidt] Chlorinated agent for water treatment.

Tego IMR® 412 T, 833 T. [Goldschmidt] Internal mold release agent for demolding PU RIM formulations.

Tegomuls® O Spezial. [Goldschmidt] Glyceryl monodioleate; nonionic; defoamer in food industry; semisolid; HLB 3.3; 100% conc.

Tego-Pearl® B-48. [Goldschmidt] Cocamidopropyl betaine, glycol distearate, cocamide DEA, cocamide MEA; pearlescent and opacifier for hair care prods.

Tegopren®. [Goldschmidt] Organo modified siloxanes; surfactants used as antistats, wetting and leveling agents, emulsifiers, dispersants, and for the improvement of lubricity; additive for polishes.

Tegosil®. [Goldschmidt] Silicone-based aerosol for release of inj. molded plastic and rubber moldings.

Tego® Silicone Acrylate RC. [Goldschmidt] Silicone acrylate for release coatings to be cured by electron beam or uv-radiation in the paper and film industries.

Tegosipon®. [Goldschmist] Silicone antifoam for mfg. of stomach and intestinal preps.

Tegosoft 168. [Goldschmidt] Cetyl isooctanoate; emollient for cosmetics; soft solid.

Tegosoft 189. [Goldschmidt] Isooctadecyl isononanoate; emollient, superfatting agent, solv. for lipophilic components; liq.

Tegostab® B 1048. [Goldschmidt] Stabilizer for continuously laminated PU boardstock, refrigeration and pipe insulation, rigid block foams.

Tegostab® B 2219. [Goldschmidt] Stabilizer for continuously laminated PU boardstock, rigid block foams.

Tegostab® B 3752. [Goldschmidt] Foam stabilizer for hot-cured molded PU foam; synergistic with flame retardant.

Tegostab® B 4113. [Goldschmidt] Stabilizer for high resilience cold-cured PU foam.

Tegostab® B 4690. [Goldschmidt] Stabilizer for high resilience cold-cured PU foam; exhibits strong cell regulating effect and weak stabilizing effect.

Tegostab® B 4900. [Goldschmidt] Foam stabilizer with wide processing latitude for hot-cured PU foam.

Tegostab® B 8002, 8021. [Goldschmidt] Foam stabilizer for hot-cured PU foams.

Tegostab® B 8404. [Goldschmidt] Foam stabilizer for continuously laminated PU boardstock, pour-in-place foams, rigid block foams.

Tegostab® B 8406. [Goldschmidt] Stabilizer for PU foams with integral skins, esp. shoe sole systems.

Tegostab® B 8407. [Goldschmidt] Stabilizer for PU refrigeration and pipe insulation, polyisocyanurate foams.

Tegostab® B 8408. [Goldschmidt] Foam stabilizer

for continuously laminated PU boardstock, modified polyisocyanurate foams, spray foams; synergistic with flame retardants.

Tegostab® B 8409. [Goldschmidt] Foam stabilizer for pour-in-place PU foams, refrigeration and pipe insulation.

Tegostab® B 8418, 8444. [Goldschmidt] Foam stabilizer for high-dens. molded PU foams.

Tegostab® B 8418. [Goldschmidt] Foam stabilizer for PU rigid foams.

Tegostab® B 8423. [Goldschmidt] Foam stabilizer and emulsifier for PU formulations with max. demand on flow properties, refrigeration and pipe insulation.

Tegostab® B 8425. [Goldschmidt] Stabilizer for PU refrigeration and pipe insulation.

Tegostab® B 8432. [Goldschmidt] Stabilizer for continuously laminated PU boardstock, spray foams.

Tegostab® B 8680. [Goldschmidt] Stabilizer for high resiliency cold-cured PU foam where high degree of stabilization is required.

Tegostab® B 8681. [Goldschmidt] Stabilizer for high resiliency cold-cured PU foams.

Tegostab® B 8863 T. [Goldschmidt] Silicone-based stabilizer for low-dens. PU foams used in pkg.; also provides open cell structure.

Tegostab® B 8901. [Goldschmidt] Stabilizer for PU foams with integral skins, in shoe sole systems.

Tegostab® B 8906, 8910. [Goldschmidt] Org. stabilizers providing a smooth surf. for PU spray foams.

Tegostab® BF 2370. [Goldschmidt] Foam stabilizer with wide processing latitude for hot-cured PU foam.

Tego® Stannous Oxalate. [Goldschmidt] Tin (II) oxalate; catalyst for stannous esterification.

Tegotrenn® LH 157 A, 525. [Goldschmidt] Release agent for demolding of flexible hot-cured PU foam; solv.-based disp. and water-based emulsion resp.

Tegotrenn® LI 197, 237, 344, 766, 901. [Goldschmidt] Release agent for demolding of integral skinned PU foam, RIM applics.

Tegotrenn® LI 448. [Goldschmidt] Release agent for PU RIM systems.

Tegotrenn® LI 537 W. [Goldschmidt] Release agent suitable for water carrier for PU RIM applics.

Tegotrenn® LI 747, 748. [Goldschmidt] Release agent for PU RIM applics.; silicone-containing and silicone-free resp.

Tegotrenn® LK 104, 260, 498/F, 755, 859, 860. [Goldschmidt] Release agents for demolding of cold cured (high resilience) PU foam.

Tegotrenn® LR 201/F, 236, 309/F, 560. [Goldschmidt] Release agents for demolding or rigid, semirigid, and integral skinned duromeric PU foam.

Tegotrenn® LR 486. [Goldschmidt] Release agent for PU pipe insulation.

Tegotrenn® LS 584. [Goldschmidt] Release agent for PU foams with integral skins.

Tegotrenn® LS 809/F, 814/F, 815, 828, 829/F, 960. [Goldschmidt] Release agent for demolding of PU shoe soles.

Tegotrenn® LS 952. [Goldschmidt] Water-based release agent for demolding microcellular PU foams, RIM parts.

Tegotrenn® S 2100. [Goldschmidt] Silicone conc. release agent for demolding of PU ether shoe soles.

Tektamer® 38. [Calgon] Methyl dibromo glutaronitrile; antimicrobial for preserving cosmetics and personal care prods.

Tek Tan ND. [Hamblet & Hayes] Sodium salt of condensed naphthalene sulfonic acid; tanning, bleaching agent for leather; dispersant; dry form; water-sol.

Tek Tan NL. [Hamblet & Hayes] Sodium salt of condensed naphthalene sulfonic acid; dispersant for applic. in aq. media; aq. sol'n.; misc. with water.

Tek-Wet 951. [Hamblet & Hayes] PEG; nonionic; leather wetting and emulsifying agent; dispersant; sol'n.; biodeg.

Tek-Wet 955. [Hamblet & Hayes] PEG nonylphenol; nonionic; leather wetting and emulsifying; dispersant; sol'n.

Telloy. [Vanderbilt] Tellurium; vulcanizing agent; sec. vulcanizing agent used in NR, IR, SBR; increases state of cure and improves heat aging of sulfurless and low sulfur stocks; reduces curing time in NR hard rubber and allows use of lower sulfur for increase flexibility; steel to dk. gray metallic powd., 100% thru 200 mesh; m.w. 127.61; practically insol. in water; dens. 6.26 ± 0.03 mg/m^3; m.p. > 450 C.

Temasept I. [Hexcel] Dibromo/tribromo salicylanilide; antimicrobial for resins, latex emulsions, plastics.

Temasept IV. [Hexcel] Tribromosalicylanilide; antimicrobial for resins, latex emulsions, plastics.

Tenase®. [Miles Biotech] Bacterial alpha-amylase; enzyme for starch liquefaction; liq./powd.

Tenn-Cop 5E. [Tennessee] Copper salts of fatty and rosin acids; agric. fungicide; lt. grn. powd.

Tenox® BHA. [Eastman] BHA; antioxidant for food applic., food pkg., plastics, rubber; wh. waxy tablets; slight odor; m.w. 180; b.p. 264–270 C (733 mm); sol. in ethanol, diisobutyl adipate, propylene glycol, glyceryl monooleate, soya oil, lard, yel. grease, and paraffin; insol. in water; m.p. 48–55 C.

Tenox® IBP-2. [Eastman] Isobutyric acid, propionic acid, ratio 60:40; grain preservative; colorless liq.; pungent odor; b.p. 146 C; water-sol.; sp.gr. 0.967; flash pt. 55 C (TCC).

Tenox® P. [Eastman] Propionic acid; grain preservative; fungicide; colorless oily liq.; sharp odor; m.w. 74.09; f.p. –22.4 C; b.p. 141.1 C (760 mm); water-sol.; dens. 1.00 kg/l; visc. 1.16 cP; ref. index 1.3850; flash pt. 54 C (TOC); fire pt. 58 C; 99.5% act.

Tenox® PG. [Eastman] Propyl gallate; antioxidant for fats and oils; stabilizer for food applic.; wh. cryst. powd.; slight odor; m.w. 212; b.p. 148 C; sol. in ethanol, propylene glycol, and water; m.p. 146–148 C.

Tenox® TBHQ. [Eastman] t-Butyl hydroquinone; antioxidant and stabilizer used in oils and fats for food applic.; wh. to lt. tan cryst.; slight odor; m.w. 166.22; sol. in ethanol, ethyl acetate; < 1% in water; m.p. 126.5–128.5 C.

Tequat BC. [Tessilchimica] Cetrimonium chloride; softener, after-treatment hair prep.; liq.

Tequat PAN. [Tessilchimica] Lauryl dimethyl benzyl ammonium chloride; germicide, antistat; liq.

Terate® 101, 131. [Hercules] Balsamic resins derived from petrol. aromatic hydrocarbons (high-boiling mixtures of ester-substituted diphenyl and polyphenyl compds.); low unsat. and high resistance to oxidation; highly polar and reactive resins used in rubber compding., adhesives, and sealants, and for plasticizing; dk. brn.

Terate® 202. [Hercules] Aromatic polyester polyols derived from polycarbomethoxy-substituted diphenyls, polyphenyls, and benzyl esters of the toluate family; highly polar resins used as modifier and extender in rigid and semirigid PU foams; plasticizer for polymers and elastomers; dk.; insol. in water.

Terate® 203. [Hercules] Aromatic polyester polyols derived from polycarbomethoxy-substituted diphenyls, polyphenyls, and benzyl esters of the toluate family; see Terate 202; dk.; insol. in water.

Terate® 204. [Hercules] Aromatic polyester polyols derived from polycarbomethoxy-substituted diphenyls, polyphenyls, and benzyl esters of the toluate family; see Terate 202; dk.; insol. in water.

Terate® 303. [Hercules] Polymeric ester of functionally substituted diphenyl and polyphenyl compds. with low unsaturation and high resistance to oxidation; thermoplastic resin which is chemically reactive and highly polar; tackifiers for rubber compding., in hot-melts and rubber-based adhesives, and as binders for wood fiber in formed and pressed paper prods.; soften. pt. (Drop) 120 C.

Terate® HM-AY. [Hercules] Thermoplastic resin used as binder in the prod. of med.-dens. hardboard; contributes to improvements in board strength, moisture resistance, dimensional stability, and weatherability; its resistance to oxidation and its insol. in many solvs. enhance hardboard durability and paintability; offers favorable thermoplastic properties (high melt point and low flow point) which aid in board processing; flakes; sol. limited to certain polar solvs., e.g., acetone, MEK, ethyl acetate, nitroparaffins, and methylene chloride; sp.gr. 1.27; dens. (bulk) 41 lb/ft^3; soften. pt. (Hercules Drop) 128 C; flash pt. (COC) 254 C; acid no. 25; sapon. no. 458.

Terathane® 650. [DuPont] Polytetramethylene ether glycol; functions as a soft segment in PU resins; m.w. 600–700; wh. waxy solid melting to a clear colorless liq. near R.T.; sol. in alcohols, esters, ketones; sp.gr. 0.978; dens. 8.1 lb/gal; ; visc. 100–200 cP (40 C); m.p. 11–19 C; flash pt. (TOC) > 163 C; ref. index 1.462.

Terathane® 1000. [DuPont] Polytetramethylene ether glycol; functions as a soft segment in PU resins; m.w. 950–1050; wh. waxy solid melting to a clear colorless liq. near R.T.; sol. in alcohols, esters, ketones; sp.gr. 0.974 (40 C); dens. 8.1 lb/gal; visc. 260–320 cP (40 C); m.p. 25–33 C; flash pt. (TOC) > 163 C; ref. index 1.463–1.465.

Terathane® 2000. [DuPont] Polytetramethylene ether glycol; functions as a soft segment in PU resins; wh. waxy solid melting to a clear colorless liq. near R.T.; m.w. 1900–2100; sol. in alcohols, esters, ketones; sp.gr. 0.972 (40 C); dens. 8.1 lb/gal; visc. 950–1450 cP (40 C); m.p. 28–40 C; flash pt. (TOC) > 163 C; ref. index 1.464.

Terathane® 2900. [DuPont] Polytetramethylene ether glycol; functions as a soft segment in PU resins; imparts improved elasticity; suitable for castable and thermoplastic PU, flexible coatings, adhesives, syn. leathers; wh. waxy solid melting to a clear colorless liq. near R.T.; m.w. 2825–2975; sol. in alcohols, esters, ketones; sp.gr. 0.97 (40 C); dens. 8.1 lb/gal; visc. 3200–4200 cP (40 C); m.p. 30–43 C; flash pt. (TOC) > 163 C; ref. index 1.464.

Terg-A-Zyme. [Alconox] Alkylaryl sulfonate, lauryl alcohol sulfate, phosphate, carbonate, and protease enzyme; anionic; detergent, wetting agent, sequestering and synergistic agents; used in hospitals, laboratories, dairies; cleaning agent in dairy and pollution processing; powd.; pH 9.0–9.5.

Tergenol G. [Hart Prod.] Sodium n-methyl-n-oleoyl taurate; detergent, wetting agent, emulsifier; used in textiles, skins, leather, laundering, dyeing, household and industrial cleaners, cosmetic prods., dispersing waxes, pigments, resins, dry powd. and printing; corrosion inhibitor in chlorite bleaching; lt. gel; water-sol.; 15% act.

Tergenol S Liquid. [Hart Prod.] Sodium-n-methyl-n-oleoyltaurate; anionic; see Tergenol G; amber liq.; 38% act.

Tergenol Slurry. [Hart Prod.] Sodium-n-methyl-n-oleoyltaurate; anionic; see Tergenol G; wh. paste; water-sol.; 60% act.

Tergitol® 15-S-3. [Union Carbide] C_{11-15} pareth-3; nonionic; detergent, emulsifier, wetting agent used in textile processing applics.; textile lubricant; Pt.Co 75 max.; m.w. 332; dens. 7.68 lb/gal; sp.gr. 0.924; visc. 37 cs; HLB 8.0; cloud pt. < 0 (1% aq.); flash pt. 370 F (COC); pour pt. –32 C; biodeg.; 100% conc.

Tergitol® 15-S-5. [Union Carbide] C_{11-15} pareth-5; nonionic; see Tergitol 15-S-3; Pt.Co 75 max. liq.; m.w. 420; dens. 8.03 lb/gal; sp.gr. 0.965; visc. 51 cs; HLB 10.5; cloud pt. < 0 C (1% aq.); flash pt. 410 F (COC); pour pt. –15 C; pH 6–8 (1%); biodeg.; 100% conc.

Tergitol® 15-S-7. [Union Carbide] C_{11-15} pareth-7; nonionic; see Tergitol 15-S-3; also dyeing aid, dye leveling agent for leather and textile processing; Pt.Co 75 max. liq.; m.w. 508; dens. 8.26 lb/gal; sp.gr. 0.994; visc. 67 cs; HLB 12.1; cloud pt. 37 C (1% aq.); flash pt. 440 F (COC); pour pt. 7 C; surf. tens. 29.1 dynes/cm (0.1%); pH 6–8 (1%); biodeg.; 100% conc.

Tergitol® 15-S-9. [Union Carbide] C_{11-15} pareth-9; nonionic; see Tergitol 15-S-3; also dyeing aid, dye leveling agent for leather and textile processing; Pt.Co 50 max. liq.; m.w. 596; dens. 8.37 lb/gal; sp.gr. 1.006; visc. 78 cs; HLB 13.3; cloud pt. 60 C (1% aq.); flash pt. 470 F (COC); pour pt. 13 C; surf. tens. 30.3 dynes/cm (0.1%); pH 6–8 (1%); biodeg.; 100% conc.

Tergitol® 15-S-12. [Union Carbide] C_{11-15} pareth-12; nonionic; see Tergitol 15-S-3; also dyeing aid, dye leveling agent for leather and textile processing;

Pt.Co 100 max. semisolid; m.w. 728; dens. 8.49 lb/gal; sp.gr. 1.020; visc. 120 cs; HLB 14.5; cloud pt. 90 C (1% aq.); flash pt. 460 F (COC); pour pt. 17 C; surf. tens. 31.5 dynes/cm (0.1%); pH 6–8 (1%); biodeg.; 100% conc.

Tergitol® 15-S-15. [Union Carbide] C_{11-15} pareth-15; nonionic; see Tergitol 15-S-3; Pt.Co. 100 max. solid; m.w. 860; dens. 8.40 (55 C); sp.gr. 1.025; visc. 63 cs (40 C); HLB 15.4; cloud pt. > 100 C (1% aq.); flash pt. 490 F (COC); pour pt. 23 C; pH 6–8 (1% aq.); biodeg.; 100% conc.

Tergitol® 15-S-20. [Union Carbide] C_{11-15} pareth-20; nonionic; see Tergitol 15-S-3; Pt.Co 100 max. solid; m.w. 1080; dens. 8.66 lb/gal (40 C); sp.gr. 1.040; visc. 102 cs (40 C); HLB 16.3; cloud pt. > 100 C (1% aq.); flash pt. 480 F (COC); pour pt. 34 C; pH 6–8 (1%); biodeg.; 100% conc.

Tergitol® 15-S-30. [Union Carbide] C_{11-15} pareth-30; nonionic; see Tergitol 15-S-3; Pt.Co 100 max. solid; m.w. 1520; dens. 8.78 lb/gal (55 C); sp.gr. 1.063; visc. 102 cs (40 C); HLB 17.4; cloud pt. > 100 C (1% aq.); flash pt. 505 F (COC); pour pt. 40 C; pH 6–8 (1%); biodeg.; 100% conc.

Tergitol® 15-S-40. [Union Carbide] C_{11-15} pareth-40; nonionic; see Tergitol 15-S-3; Pt.Co 100 max. solid; m.w. 1960; dens. 8.83 lb/gal (55 C); sp.gr. 1.066; HLB 18.0; cloud pt. > 100 C (1% aq.); flash pt. 515 F (COC); pour pt. 45 C; pH 6–8 (1%); biodeg.; 100% conc.

Tergitol® 24-L-3. [Union Carbide] Primary alcohol PEG ether; nonionic; detergent, emulsifier, surfactant, defoamer, intermediate; liq.; oil-sol.; HLB 7.7; biodeg.; 100% conc.

Tergitol® 25-L-3. [Union Carbide] C_{12-15} pareth-3; nonionic; detergent, wetting agent, emulsifier, textile surfactant and lubricant; Pt-Co 50 max. liq.; m.w. 341; dens. 7.73 lb/gal; sp.gr. 0.929; visc. 36 cs; HLB 7.7; cloud pt. < 0 C (1% aq.); flash pt. 260 F (COC); pour pt. 8 C; pH 5–7; biodeg.; 100% conc.

Tergitol® 25-L-5. [Union Carbide] C_{12-15} pareth-5; nonionic; see Tergitol 25-L-3; Pt.Co 50 max. liq.; m.w. 440; dens. 7.94 lb/gal (30 C); sp.gr. 0.955; visc. 95 cs; HLB 10.4; cloud pt. < 0 C (1% aq.); flash pt. 325 F (COC); pour pt. 16 C; pH 5–7 (1%); biodeg.; 100% conc.

Tergitol® 25-L-7. [Union Carbide] C_{12-15} pareth-7; nonionic; see Tergitol 25-L-3; also in desizing; Pt.Co 50 max. liq.; m.w. 550; dens. 8.19 lb/gal (30 C); sp.gr. 0.985; visc. 51 cps (38 C); HLB 12.4; cloud pt. 50 C (1% aq.); flash pt. 340 F (COC); pour pt. 23 C; surf. tens. 28.7 dynes/cm (0.1%); pH 5–7 (1%); biodeg.; 100% conc.

Tergitol® 25-L-9. [Union Carbide] C_{12-15} pareth-9; nonionic; see Tergitol 25-L-7; Pt.Co 50 max. semisolid; m.w. 585; dens. 8.18 lb/gal (40 C); sp.gr. 0.991; visc. 33 cps (40 C); HLB 12.8; cloud pt. 60 C (1% aq.); flash pt. 285 F (COC); pour pt. 24 C; surf. tens. 29.0 dynes/cm (0.1%); pH 5–7 (1%); biodeg.; 100% conc.

Tergitol® 25-L-12. [Union Carbide] C_{12-15} pareth-12; nonionic; see Tergitol 25-L-7; Pt.Co 100 max. solid; m.w. 730; dens. 8.37 lb/gal (40 C); sp.gr. 1.013; visc. 51 cs (38 C); HLB 14.2; cloud pt. 90 C (1% aq.); flash pt. 410 F (COC); pour pt. 30 C; surf. tens. 32.0 dynes/cm (0.1%); pH 5–7 (1%); biodeg.; 100% conc.

Tergitol® 24-L-45. [Union Carbide] C_{12} and C_{14} primary alcohol ethoxylate (6.3 EO); nonionic; surfactant, detergent, wetting/spreading agent, emulsifier, foaming agent, intermediate, dispersant for household and hard surf. cleaners, textile processing, pulp processing, agric. formulations; APHA 20 color, low, mild char. odor; m.w. 479; misc. with propylene glycol; sp.gr. 0.987; visc. 28 cSt (37.8 C); HLB 11.6; cloud pt. 45 C (1% aq.); flash pt. (PMCC) 182 C; pour pt. 12 C; surf. tens. 29 dynes/cm (1% aq.); pH 6.0 (1% aq.).

Tergitol® 24-L-60. [Union Carbide] C_{12} and C_{14} primary alcohol ethoxylate (7.2 EO); nonionic; see Tergitol 24-L-45; APHA 20 color, mild, low, char. odor; m.w. 519; misc. with propylene glycol, toluene; sp.gr. 0.99; visc. 31 cSt (37.8 C); HLB 12.2; cloud pt. 60 C (1% aq.); flash pt. (PMCC) 174 C; pour pt. 17 C; surf. tens. 29 dynes/cm (1% aq.); pH 6.0 (1% aq.).

Tergitol® 24-L-60N. [Union Carbide] C_{12} and C_{14} primary alcohol ethoxylate (7.0 EO); nonionic; see Tergitol 24-L-45; APHA 20 color, mild, low, char. odor; m.w. 510; misc. with propylene glycol, toluene; sp.gr. 0.99 (30/20 C); visc. 28 cSt (37.8 C); HLB 12.1; cloud pt. 60 C (1% aq.); flash pt. (PMCC) 178 C; pour pt. 16 C; surf. tens. 31 dyne/cm (1% aq.); pH 6.0 (1% aq.).

Tergitol® 24-L-75. [Union Carbide] C_{12} and C_{14} primary alcohol ethoxylate (8.3 EO); nonionic; see Tergitol 24-L-45; APHA 20 hazy liq. to wh. semisolid; low, mild, char. odor; m.w. 567; misc. with ethyl acetate, propylene glycol; sp.gr. 0.99 (30/20 C); visc. 35 cSt (37.8 C); HLB 12.9; cloud pt. 75 C (1% aq.); flash pt. (PMCC) 168 C; pour pt. 21 C; surf. tens. 31 dyne/cm (1% aq.); pH 6.0 (1% aq.).

Tergitol® 24-L-92. [Union Carbide] C_{12} and C_{14} primary alcohol ethoxylate (10.6 EO); nonionic; see Tergitol 24-L-45; wh. solid, low, mild, char. odor; m.w. 668; misc. with propylene glycol, toluene; sp.gr. 1.00 (30 C); visc. 42 cSt (37.8 C); HLB 14.0; cloud pt. 92 C (1% aq.); flash pt. (PMCC) 185 C; pour pt. 29 C; surf. tens. 34 dyne/cm (1% aq.); pH 6.0 (1% aq.).

Tergitol® 24-L-98N. [Union Carbide] C_{12} and C_{14} primary alcohol ethoxylate (11.3 EO); nonionic; see Tergitol 24-L-45; APHA 20 solid; low, mild, char. odor; m.w. 701; misc. with propylene glycol, toluene; sp.gr. 1.01 (30 C); visc. 77 cSt (37.8 C); HLB 14.2; cloud pt. 98 C (1% aq.); flash pt. (PMCC) 204 C; pour pt. 28 C; surf. tens. 36 dyne/cm (1% aq.); pH 6.0 (1% aq.).

Tergitol® 26-L-1.6. [Union Carbide] C_{12}, C_{14}, C_{16} primary alcohol ethoxylate (1.5 EO); nonionic; surfactant, detergent, wetting/spreading agent, emulsifier, foaming agent, intermediate, dispersant, lubricant for household and hard surf. cleaners, textile processing, pulp processing, agric. formulations; APHA 10 clear to slightly hazy liq.; mild char. odor; m.w. 264; misc. with ethyl acetate, ethanol, propylene glycol, IPA, toluene, hexane; sp.gr. 0.895; visc. 14 cSt (37.8 C); HLB 5.0; flash pt. (PMCC) 143 C; pour pt. 10 C; pH 6.5 (1% in 10:6 IPA/water)

Tergitol® 26-L-3. [Union Carbide] C_{12}, C_{14}, C_{16} primary alcohol ethoxylate (3 EO); nonionic; see Tergitol 26-L-1.6; APHA 10 clear to slightly hazy liq.; mild char. odor; m.w. 328; sol. see Tergitol 26-L-1.6; sp.gr. 0.917; visc. 19 cSt (37.8 C); HLB 8.0; flash pt. (PMCC) 157 C; pour pt. 5 C; pH 6.9 (1% in 10:6 IPA/water)

Tergitol® 26-L-5. [Union Carbide] C_{12}, C_{14}, C_{16} primary alcohol ethoxylate (5 EO); nonionic; see Tergitol 26-L-1.6; APHA 10 hazy liq.; mild char. odor; m.w. 419; misc. with propylene glycol, butyl Cellosolve; sp.gr. 0.969; visc. 26 cSt (37.8 C); HLB 10.5; flash pt. (PMCC) 177 C; pour pt. 8 C; pH 6.5 (1% in 10:6 IPA/water)

Tergitol® Min-Foam 1X. [Union Carbide] Nonionic surfactant, foam depressant, wetting agent, detergent for textile processing baths, high pressure metal cleaning; sol. in water, xylene, butyl Cellosolve, anhyd. IPA, perchloroethylene; sp.gr. 0.995; dens. 8.28 lb/gal; visc. 57 cSt; HLB 11.4; solid. pt. –42 C; cloud pt. 42 C (1% aq.); flash pt. (PMCC) 231 F; surf. tens. 30 dynes/cm (0.1% aq.); pH 5–8 (1%); 100% act.

Tergitol® Min-Foam 2X. [Union Carbide] Nonionic surfactant, antifoam, foam depressant, detergent for household and industrial detergents, machine dishwashing, rinse aid, acid metal cleaners; completely rinseable; sol. in xylene, butyl Cellosolve, anhyd. IPA, perchloroethylene; sp.gr. 0.978; dens. 8.14 lb/gal; visc. 66 Cst; HLB 13.8; solid. pt. –47 C; cloud pt. 21 C (1% aq.); flash pt. (PMCC) 311 F; surf. tens. 30 dynes/cm; pH 5–8 (1%); 100% act.

Tergitol® NP-7. [Union Carbide] Nonoxynol-7; nonionic; emulsifier, wetting agent, dispersant used in agric. toxicant formulations, cleaners and santizers; cleaning booster; Pt.Co 75 max. clear liq.; mild char. odor; m.w. 528; sol. in butyl acetate, butyl Cellosolve, corn oil, anhyd. IPA, toluene, water; sp.gr. 1.055; dens. 8.71 lb/gal; visc. 338 cs; HLB 11.7; solid. pt. –6 C; b.p. > 250 C; flash pt. 525 F (COC); cloud pt. 20 C; pH 5–8 (10% aq.); 100% act.

Tergitol® NP-8. [Union Carbide] Nonoxynol-8; nonionic; removal or emulsification of grease or oils; wetting agent, dispersant, emulsifier; Pt.Co 50 max. clear liq.; mild char. odor; m.w. 572; sol. in butyl acetate, butyl Cellosolve, ethylene glycol, anhyd. IPA, toluene, water; sp.gr. 1.056; visc. 325; HLB 12.3; solid. pt. –3 C; b.p. 250 C; flash pt. 400 F (PMCC); cloud pt. 43 C; pH 5.0–8.0 (10% aq.); surf. tens. 30 dynes/cm (0.1% aq.); 100% act.

Tergitol® NP-9. [Union Carbide] Nonoxynol-9; nonionic; detergent, wetting agent, dispersant, emulsifier; yel. clear liq.; mild char. odor; m.w. 616; sol. see Tergitol NP-8; sp.gr. 1.057; dens. 8.80 lb/gal; visc. 318 cs; HLB 12.9; solid. pt. 0 C; b.p. > 250 C; flash pt. 540 F (COC); cloud pt. 54 C; pH 5–8 (10% aq.); surf. tens. 30 dynes/cm (0.1% aq.); 100% act.

Tergitol® NP-10. [Union Carbide] Nonoxynol-10; nonionic; see Tergitol NP-9; yel. clear liq.; mild char. odor; m.w. 682; sol. see Tergitol NP-8; sp.gr. 1.062; dens. 8.84 lb/gal; visc. 327 cs; HLB 13.6; solid. pt. 7 C; b.p. > 250 C; flash pt. 500 F (COC); cloud pt. 63 C (0.5% aq.); 100% act.

Tergitol® NP-14. [Union Carbide] Nonoxynol-4; nonionic; detergent, wetting agent, emulsifier, intermediate; used in textile applics.; Pt.Co 100 max. liq.; m.w. 396; dens. 8.57 lb/gal; sp.gr. 1.031; visc. 448 cs; HLB 8.9; cloud pt. < 0 C (1% aq.); flash pt. 480 F (COC); solid. pt. –40 C; pH 5–8 (10%); 100% conc.

Tergitol® NP-27. [Union Carbide] Nonoxynol-7; nonionic; see Tergitol NP-14; Pt.Co 75 max. liq.; m.w. 528; dens. 8.71 lb/gal; sp.gr. 1.038; visc. 338 cs; HLB 11.7; cloud pt. 22 C (1% aq.); flash pt. 525 F (COC); solid. pt. –6 C; surf. tens. 30.4 dynes/cm (0.1%); pH 5–8 (10%); 100% conc.

Tergitol® NP-44. [Union Carbide] Nonoxynol-40; nonionic; see Tergitol NP-14; Gardner 5 max. solid; m.w. 1980; water-sol.; dens. 8.97 lb/gal (55 C); sp.gr. 1.082; HLB 17.8; cloud pt. > 100 C (1% aq.); flash pt. 525 F (COC); solid. pt. 48 C; pH 4–8 (10%); 100% conc.

Tergitol® NPX. [Union Carbide] Nonoxynol-10 (10.5 EO); see Tergitol NP-14; Pt.Co. 50 max. liq.; m.w. 682; dens. 8.84 lb/gal; sp.gr. 1.062; visc. 318 cs; HLB 13.6; cloud pt. 63 C (0.5% aq.); flash pt. 500 F (COC); solid. pt. 7 C; pH 5–8 (10%); 100% conc.

Tergitol® TMN-3. [Union Carbide] Isolaureth-3; nonionic; wetting agent, penetrant, leveling agent for textile processing; wetting and dispersing agent in wax or resin disps.; spreading agent for emulsions; liq.; m.w. 318; dens. 7.74 lb/gal; sp.gr. 0.936; HLB 8.3; cloud pt. < 0 C (1% aq.); flash pt. 320 F (COC); solid. pt. –40 C; pH 5–8 (10%); 100% conc.

Tergitol® TMN-6. [Union Carbide] Isolaureth-6; nonionic; see Tergitol TMN-3; also for agric. sprays; Pt.Co 150 max. liq.; m.w. 450; dens. 8.40 lb/gal; sp.gr. 1.024; visc. 105 cs; HLB 11.7; cloud pt. 36 C (0.5% aq.); solid. pt. –22 C; pH 5–8 (10%); 90% conc.

Tergitol® TMN-10. [Union Carbide] Isolaureth-10; nonionic; see Tergitol TMN-3; also for agric. sprays; Pt.Co 200 max. liq.; m.w. 626; dens. 8.69 lb/gal; sp.gr. 1.044; visc. 153 cs; HLB 14.1; cloud pt. 76 C (1% aq.); solid. pt. –7.5 C; pH 5–8 (10%); 90% conc.

Tergitol® XD. [Union Carbide] PPG-24-buteth-27; nonionic; dispersant, emulsifier used in dye pigments in textile baths and latex systems; Pt.Co 100 solid; m.w. 2500; dens. 8.66 lb/gal (40 C); sp.gr. 1.041; visc. 428 cs (40 C); cloud pt. 75 C; flash pt. 480 (COC); solid. pt. 33 C; pH 4–7; 100% conc.

Tergitol® XH. [Union Carbide] Polyalkylene glycol ether; see Tergitol XD; Pt.Co 100 solid; m.w. 3500; dens. 8.72 lb/gal (50 C); sp.gr. 1.048; visc. 325 cs (40 C); cloud pt. 95 C (0.5% aq.); solid. pt. 41 C; pH 4–7 (10%); 100% conc.

Tergitol® XJ. [Union Carbide] Polyalkylene glycol ether; see Tergitol XD; Gardner 2 solid; m.w. 2300; dens. 8.51 lb/gal (45 C); sp.gr. 1.027; cloud pt. 50 C (1% aq.); flash pt. 495 F (COC); solid pt. 26 C; pH 5.5–8.0 (1%); 100% conc.

Teric 12A3. [ICI Australia] C_{12-15} pareth-3 (2.6 EO); nonionic; wetting agent and dispersant, coemulsifier, detergent, intermediate used in oil and solv.-based systems; horticultural applics.; w/o emulsions; household and industrial cleaners; mfg. of sulfate and phosphate surfactants; Hazen 100 liq.; sol. in benzene, ethyl acetate, ethyl Icinol, perchlorethylene, ethanol, kerosene, min. and veg. oil, olein; sp.gr. 0.912; visc. 28 cps; m.p. < 0 C; HLB 7.1; pH 6–8 (1% aq.); biodeg.; 100% act.

Teric 12A6. [ICI Australia] C_{12-15} pareth-6; nonionic; wetting agent, dispersant, detergent, chemical intermediate; used in textile processing, hard surface abrasives, metal descaling, detergent for household and industrial use in laundries; metal cleaning, sanitizer; Hazen 100 liq. to paste; sol. in benzene, ethyl acetate, ethyl Icinol, perchlorethylene, ethanol, kerosene, min., veg. and paraffin oil, olein; disp. in water; sp.gr. 0.955 (50 C); visc. 24 cps (50 C); m.p. 10 ± 2 C; HLB 11.2; cloud pt. 40 ± 2 C; surf. tens. 27.8 dynes/cm; pH 6–8 (1% aq.); biodeg.; 100% act.

Teric 12A9. [ICI Australia] C_{12-15} pareth-9; nonionic; see Teric 12A6; Hazen 100 paste; sol. in water, benzene, ethyl acetate, ethyl Icinol, perchlorethylene, ethanol, olein; sp.gr. 0.984 (50 C); visc. 32 cps (50 C); m.p. 20 ± 2 C; HLB 13.2; cloud pt. 82 ± 2 C (1%); surf. tens. 30.2 dynes; pH 6–8 (1% aq.); biodeg.; 100% act.

Teric 12A12. [ICI Australia] C_{12-15} pareth-12 (11.5 EO); see Teric 12A6; Hazen 100 solid; sol. in benzene, ethyl acetate, ethyl Icinol, perchlorethylene, ethanol, olein; sp.gr. 0.993 (50 C); visc. 37 cps (50 C); m.p. 24 ± 2 C; HLB 14.4; cloud pt. 92 ± 2 C (1%); surf. tens. 35.6 dynes/cm; pH 6–8 (1% aq.); biodeg.; 100% act.

Teric 12A16. [ICI Australia] C_{12-15} pareth-16; nonionic; wetting agent, dispersant, emulsifier, detergent; used in degreasing compds., industrial and household hard surface cleaners; sanitizers; Hazen 100 solid; sol. in water, benzene, ethyl acetate, ethyl Icinol, perchlorethylene, ethanol, olein; sp.gr. 1.009 (50 C); visc. 38 cps (50 C); m.p. 27 ± 2 C; HLB 15.3; cloud pt. 98 ± 2 C; surf. tens. 33.7 dynes/cm; pH 6–8 (1% aq.); biodeg.; 100% act.

Teric 12A23. [ICI Australia] C_{12-15} pareth-23; nonionic; see Teric 12A16; Hazen 100 solid; sol. see Teric 12A16; sp.gr. 1.032 (50 C); visc. 75 cps (50 C); m.p. 35 ± 2 C; HLB 16.6; cloud pt. > 100 C; surf. tens. 35.7 dynes/cm; pH 6–8 (1% aq.); biodeg.; 100% act.

Teric 12M2. [ICI Australia] PEG-2 cocamine; nonionic; dispersant, emulsifier, stabilizer; wetting agent of hydrophobic surfaces; used in metal, stone, paper and textile processing; formulation of lubricants and dye bath auxs.; emulsion stabilization; fat liquoring compds.; liq.; sol. in benzene, ethyl acetate, ethyl Icinol, perchlorethylene, ethanol, kerosene, min. and veg. oil; disp. in water; sp.gr. 0.913; visc. 166 cps; m.p. –1 ± 2 C; HLB 11.4; surf. tens. 28.2 dynes/cm; pH 8–10 (1% aq.); 100% act.

Teric 12M5. [ICI Australia] PEG-5 cocamine; nonionic; see Teric 12M2; liq.; sol. in benzene, ethyl acetate, ethyl Icinol, perchlorethylene, ethanol, min., veg. and paraffin oil, olein; disp. in water; sp.gr. 0.971; visc. 311 cps; m.p. –9 ± 2 C; HLB 12.4; surf. tens. 29.2 dynes/cm; pH 8–10 (1% aq.); 100% act.

Teric 12M15. [ICI Australia] PEG-15 cocamine; nonionic; see Teric 12M2; liq.; sol. in water, benzene, ethyl acetate, ethyl Icinol, perchlorethylene, ethanol, veg. oil, olein; sp.gr. 1.038; visc. 343 cps; m.p. –11 ± 2 C; HLB 15.7; cloud pt. > 100 C; surf. tens. 39.6 dynes/cm; pH 8–10 (1% aq.); 100% act.

Teric 13A9. [ICI Australia] Trideceth-9; nonionic; wetting agent and dispersant, emulsifier, detergent used in pigment disps. and agric. and horticultural applics.; solv. emulsion cleaners; Hazen 150 liq.; sol. in water, ethyl acetate, ethyl Icinol, ethanol, olein; sp.gr. 1.018; visc. 142 cps; m.p. < 0 C; HLB 13.3; cloud pt. 63 ± 2 C; surf. tens. 27.9 dynes/cm; pH 6–8 (1% aq.); 90% act.

Teric 16A16. [ICI Australia] Ceteareth-16; nonionic; wetting agent, emulsifier, detergent, solubilizer, dispersant in hydrophobic conditions; dye and pigment carriers; textile applics.; mfg. of wax emulsions and polishes for household and industrial use; antistat; Hazen 100 flakes; sol. in water, benzene, ethyl acetate, ethyl Icinol, perchlorethylene, ethanol, olein; sp.gr. 1.01 (50 C); visc. 65 cps (50 C); m.p. 38 ± 2 C; HLB 14.9; cloud pt. > 100 C; surf. tens. 37.1 dynes/cm; pH 6–8 (1% aq.); biodeg.; 100% act.

Teric 16A22. [ICI Australia] Ceteareth-22; nonionic; see Teric 16A16; Hazen 100 flakes; sol. see Teric 16A16; sp.gr. 1.01 (50 C); visc. 92 cps (50 C); m.p. 40 ± 2 C; HLB 15.8; cloud pt. > 100 C; surf. tens. 41.3 dynes/cm; pH 6–8 (1% aq.); biodeg.; 100% act.

Teric 16A29. [ICI Australia] Ceteareth-29; nonionic; see Teric 16A16; Hazen 100 flakes; sol. see Teric 16A16; sp.gr. 1.05 (50 C); visc. 105 cps (50 C); m.p. 45 ± 2 C; HLB 17.7; cloud pt. > 100 C; surf. tens. 46.0 dynes/cm; pH 6–8 (1% aq.); biodeg.; 100% act.

Teric 16M2. [ICI Australia] PEG-2 soya amine; nonionic; wetting agent, dispersant, emulsifier for waxes and fats; formulation of leather dressing and metal cleaning compds., fiber lubricant applics. in the bldg. industry; liq.; sol. in benzene, ethyl acetate, ethyl Icinol, perchlorethylene, ethanol, kerosene, min., paraffin and veg. oil; olein; sp.gr. 0.906; visc. 191 cps; m.p. –15 ± 2 C; HLB 9.5; surf. tens. 26.6 dynes/cm; pH 8–10 (1% aq.); 100% act.

Teric 16M5. [ICI Australia] PEG-5 soya amine; nonionic; wetting agent and dispersant; used in metal, cleaning, leather dressing and dye leveling compds., agric. sprays; emulsifier for fats and waxes; fiber lubricant; wax emulsion for the bldg. industry; liq.; sol. see Teric 16M2; sp.gr. 0.954; visc. 182 cps; m.p. –16 ± 2 C; HLB 10.5; surf. tens. 29.7 dynes/cm; pH 8–10 (1% aq.); 100% act.

Teric 16M10. [ICI Australia] PEG-10 soya amine; nonionic; wetting agent, dispersant, emulsifier for fats and waxes; used in leather dressing compds., fiber lubricant; wax emulsions for fiber board pkg. and particle board for bldg. industry; liq.; sol. in water, benzene, ethyl acetate, ethyl Icinol, perchlorethylene, ethanol, kerosene, min., veg. and paraffin oil, olein; sp.gr. 0.994; visc. 219 cps; m.p. –22 ± 2 C; HLB 12.1; cloud pt. 89 ± 2 C; surf. tens. 36.1 dynes/cm; pH 8–10 (1% aq.); 100% act.

Teric 16M15. [ICI Australia] PEG-15 soya amine; nonionic; see Teric 16M5; liq.; sol. in water, benzene, ethyl acetate, ethyl Icinol, perchlorethylene, ethanol; sp.gr. 1.021; visc. 266 cps; m.p. –18±2 C; HLB 13.8; cloud pt. > 100 C; surf. tens. 37.4 dynes/cm; pH 8–10 (1% aq.); 100% act.

Teric 17A2. [ICI Australia] Cetoleth-2 nonionic; emulsifier, mfg. of textile lubricants, solv. and waterless hand cleaners; coemulsifier; detergent additive in petrol. oils; intermediate for anionic surfactants; Hazen 150 liq.; sol. in benzene, ethyl acetate, ethyl Icinol, perchlorethylene, ethanol, kerosene, veg., paraffin and min. oils, olein; disp. in water; sp.gr. 0.902; visc. 37 cps; m.p. 11 ± 2 C; HLB 6.0; pH 6–8 (1% aq.); biodeg.; 100% act.

Teric 17A3. [ICI Australia] Cetoleth-3; nonionic; see Teric 17A2; Hazen 150 liq.; sol. see Teric 17A2; sp.gr. 0.918; visc. 41 cps; m.p. 13±2 C; HLB 7.5; pH 6–8 (1% aq.); biodeg.; 100% act.

Teric 17A6. [ICI Australia] Cetoleth-6; nonionic; see Teric 17A2; Hazen 150 liq.; sol. in benzene, ethyl acetate, ethyl Icinol, perchlorethylene, ethanol, veg. oil, olein; disp. in water; sp.gr. 0.957; visc. 60 cps; m.p. 15 ± 2 C; HLB 10.2; pH 6–8 (1% aq.); biodeg.; 100% act.

Teric 17A8. [ICI Australia] Cetoleth-8; nonionic; wetting agent, dispersant, emulsifier; used in processing of yarns and fabrics; textile dyeing; horticultural sprays; removal of oil slicks; Hazen 150 paste; sol. in water, benzene, ethyl acetate, ethyl Icinol, perchlorethylene, ethanol, olein; sp.gr. 0.963 (50 C); visc. 35 cps (50 C); m.p. 17 ± 2 C; HLB 11.6; cloud pt. 41 ± 2 C; surf. tens. 30.8 dynes/cm; pH 6–8 (1% aq.); biodeg.; 100% act.

Teric 17A10. [ICI Australia] Cetoleth-10; nonionic; see Teric 17A8; also as solubilizer for sanitation chemicals; Hazen 150 solid; sol. see Teric 17A8; sp.gr. 0.973 (50 C); visc. 43 cps (50 C); m.p. 24 ± 2 C; HLB 12.7; cloud pt. 72 ± 2 C; surf. tens. 32.9 dynes/cm; pH 6–8 (1% aq.); biodeg.; 100% act.

Teric 17A13. [ICI Australia] Cetoleth-13; nonionic; see Teric 17A10; Hazen 150 solid; sol. see Teric 17A8; sp.gr. 0.986 (50 C); visc. 45 cps (50 C); m.p. 25±2 C; HLB 13.8; cloud pt. 84±2 C; surf. tens. 33.0 dynes/cm; pH 6–8 (1% aq.); biodeg.

Teric 17A25. [ICI Australia] Cetoleth-25; nonionic; see Teric 17A10; Hazen 150 solid; sol. in water, benzene, ethyl acetate, ethyl Icinol, perchlorethylene, ethanol; sp.gr. 1.030 (50 C); visc. 92 cps (50 C); m.p. 37 ± 2 C; HLB 16.2; cloud pt. > 100 C; surf. tens. 42.0 dynes/cm; biodeg.; 100% act.

Teric 17M2. [ICI Australia] PEG-2 tallow amine; nonionic; wetting agent and dispersant; dewatering agent in elec. components and road aggregates; emulsifier for fats, waxes, min. oils and metal working lubricants; corrosion inhibitor; paste; sol. in benzene, ethyl acetate, ethyl Icinol, perchlorethylene, ethanol, kerosene, min. paraffin and veg. oil, olein; sp.gr. 0.896; visc. 56 cps (40 C); m.p. 25 ± 2 C; HLB 9; surf. tens. 26.5 dynes/cm; pH 8–10 (1% aq.); 100% act.

Teric 17M5. [ICI Australia] PEG-5 tallow amine; nonionic; see Teric 17M2; semi-liq.; sol. see Teric 17M2; sp.gr. 0.953; visc. 177 cps; m.p. 11±2 C; HLB 10; surf. tens. 31.5 dynes/cm; pH 8–10 (1% aq.); 100% act.

Teric 17M15. [ICI Australia] PEG-15 tallow amine; nonionic; see Teric 17M2; liq.; sol. in water, benzene, ethyl acetate, ethyl Icinol, perchlorethylene, ethanol, veg. oil., olein; sp.gr. 1.028; visc. 428 cps; m.p. –19 ± 2 C; HLB 13.3; cloud pt. > 100 C; surf. tens. 40.0 dynes/cm; pH 8–10 (1% aq.); 100% act.

Teric 18M2. [ICI Australia] PEG-2 stearamine; nonionic; wetting agent, dispersant; emulsion stabilizer; emulsifier used in agric. toxicants and processing of textiles, paper, leather and bldg. board; corrosion inhibitor in lubricants and greases; solid; sol. in benzene, ethyl acetate, ethyl Icinol, perchlorethylene, ethanol, kerosene, min., paraffin and veg. oil, olein; sp.gr. 0.887 (50 C); visc. 75 cps (50 C); m.p. 50 ± 2 C; HLB 8.5; pH 8–10 (1% aq.); 100% act.

Teric 18M5. [ICI Australia] PEG-5 stearamine; nonionic; see Teric 18M2; also softener and antistat in solv. cleaning of textiles and compd. of plastics; solid; sol. see Teric 18M2; also disp. in water; sp.gr. 0.935 (50 C); visc. 46 cps (50 C); m.p. 35 ± 2 C; HLB 9.5; pH 8–10 (1% aq.); 100% act.

Teric 18M10. [ICI Australia] PEG-10 stearamine; nonionic; see Teric 18M5; liq.; sol. in water, benzene, ethyl acetate, ethyl Icinol, perchlorethylene, ethanol, kerosene, veg. oil, olein; sp.gr. 0.997; visc. 255 cps; m.p. 18±2 C; HLB 11.2; cloud pt. > 100 C; surf. tens. 34.0 dynes/cm; pH 8–10 (1% aq.); 100% act.

Teric 18M20. [ICI Australia] PEG-20 stearamine; nonionic; see Teric 18M2; paste; sol. in water, benzene, ethyl acetate, ethyl Icinol, perchlorethylene, ethanol, olein; sp.gr. 1.025 (50 C); visc. 66 cps (50 C); m.p. 28±2 C; HLB 14.5; cloud pt. > 100 C; surf. tens. 40.3 dynes/cm; pH 8–10 (1% aq.); 100% act.

Teric 18M30. [ICI Australia] PEG-30 stearamine; nonionic; see Teric 18M2; solid; sol. in water, benzene, ethyl acetate, ethyl Icinol, perchlorethylene, ethanol; sp.gr. 1.038 (50 C); visc. 116 cps (50 C); m.p. 28±2 C; HLB 17.8; cloud pt. > 100 C; surf. tens. 43.0 dynes/cm; pH 8–10 (1% aq.); 100% act.

Teric 121. [ICI Australia] Fatty acid EO condensate; nonionic; antifoam agent, emulsifier; coemulsifier of veg. and min oils, paraffinic waxes and solvs.; lubricant for textile processing; foam control agent in paper mfg., industrial fermentation and distillation, textile dyeing, adhesives and waste water treatment; Hazen > 250 liq.; sol. in benzene, ethyl acetate, ethyl Icinol, perchlorethylene, ethanol, veg. oil, olein; sp.gr. 0.987; visc. 353 cps; m.p. < 0 C; HLB 7.4; pH 8 (1% aq.); biodeg.; 100% act.

Teric 124. [ICI Australia] Fatty acid EO condensate; nonionic; see Teric 121; Hazen > 250 liq.; sol. see Teric 121; also sol. in min. oil; sp.gr. 0.966; visc. 118 cps; m.p. < 0 C; HLB 4.5; pH 8 (1% aq.); biodeg.; 100% act.

Teric 127. [ICI Australia] Fatty acid EO condensate; nonionic; see Teric 121; Hazen > 250 liq.; sol. see Teric 124; sp.gr. 0.978; visc. 87 cps; m.p. < 0 C; HLB 7.7; pH 8 (1% aq.); 100% act.

Teric 129. [ICI Australia] Glycerol alkoxylate; defoamer for min. processing and industrial applics.; liq.; 100% conc.

Teric 151. [ICI Australia] Fatty acid EO condensate; nonionic; emulsifier, antistat, lubricant; textile processing aid for fibers; o/w emulsions; Hazen > 250 solid; sol. in benzene, ethyl acetate, ethyl Icinol, perchlorethylene, ethanol, kerosene, veg. oil, olein; disp. in water; sp.gr. 0.994 (50 C); visc. 45 cps (50 C); m.p. 30 ± 2 C; HLB 11.6; pH 7–9 (1% aq.); biodeg.; 100% act.

Teric 154. [ICI Australia] Fatty acid EO condensate; nonionic; antistat and emulsifier; formulation of textile processing aids; veg. and lt. petrol. oils; Hazen > 250 liq.; sol. see Teric 18M10; sp.gr. 1.031; visc. 89 cps; m.p. 7 ± 2 C; HLB 13.3; surf. tens. 30.0 dynes/cm; pH 6–8 (1% aq.); biodeg.; 100% act.

Teric 160. [ICI Australia] Alkylaryl polyoxyalkylene ether; nonionic; wetting agent and dispersant; mfg. of surf. coatings, detergent and sanitizer; Hazen 250 liq.; sol. in water, benzene, ethyl acetate, ethyl Icinol, perchlorethylene, ethanol, veg. oil, olein; sp.gr. 1.032; visc. 312 cps; m.p. < 0 C; HLB 11.3; cloud pt. 29 ± 2 C; surf. tens. 30.4 dynes/cm; pH 6–8 (1% aq.); 100% act.

Teric 161. [ICI Australia] Alkylaryl polyoxyalkylene ether; nonionic; see Teric 160; Hazen 250 liq.; sol. see Teric 160; sp.gr. 1.034; visc. 340 cps; m.p. < 0 C; HLB 12.3; cloud pt. 35 ± 2 C; surf. tens. 33.4 dynes/cm; pH 6–8 (1% aq.); 100% act.

Teric 203. [ICI Australia] Alkylene oxide derivative; nonionic; agric. emulsifier; dispersant for conc. formulations; paste; 100% conc.

Teric 303. [ICI Australia] Anionic; stabilizer for stabilization of iodophor concentrates; corrosion and scale formation inhibitor in feed water systems; Hazen > 250 paste; sol. in water, benzene, ethyl acetate, ethyl Icinol, perchlorethylene, ethanol, olein; sp.gr. 1.015 (50 C); visc. 68 cps (50 C); HLB 14.5; m.p. 28 ± 2 C; cloud pt. > 100 C; surf. tens. 44.8 dynes/cm (0.1%); pH 8-10.

Teric 313. [ICI Australia] Alkylaryl POE; nonionic; solubilizer, emulsifier; liq.; 100% conc.

Teric 350. [ICI Australia] Fatty acid condensate; nonionic; emulsifier and coemulsifier for solvs. and min. oils; formulation of bases for solv. cleaners, degreasing and decarbonizing preparations; lubricant and corrosion inhibitor in lubricant systems used in metal working industry; Hazen > 250 liq.; sol. in benzene, ethyl Icinol, perchlorethylene, ethanol, olein; disp. in water; sp.gr. 1.010; visc. 600 cps; HLB 16.6; pH 8–10 (1% aq.); 100% conc.

Teric 351B. [ICI Australia] Fatty acid condensate; nonionic; see Teric 350; Hazen > 250 liq.; sol. in perchlorethylene, olein; disp. in water; sp.gr. 1.017; visc. 750 cps; HLB 8.2; pH 8–10 (1% aq.).

Teric BL8. [ICI Australia] Syn. alcohol polyalkylene oxide deriv.; detergent; wetting agent and dispersant in printing inks, textiles, wool scouring, dust suppression, household and industrial surf. cleaning; emulsifier of solvs., greases and oils; Hazen 250 liq.; sol. see Teric 18M10; sp.gr. 1.003; visc. 63 cps; m.p. < 0 C; HLB 13.7; cloud pt. 59 ± 2 C; surf. tens. 30.3 dynes/cm; pH 6–8 (1% aq.); biodeg.; 100% act.

Teric BL9. [ICI Australia] Syn. alcohol polyalkylene oxide deriv.; see Teric BL8; Hazen 250 liq.; sol. see Teric 160; sp.gr. 1.018; visc. 100 cps; m.p. < 0 C; HLB 14.5; cloud pt. 67 ± 2 C; surf. tens. 30.0 dynes/cm; pH 6–8 (1% aq.); biodeg.; 98% act.

Teric C12. [ICI Australia] PEG-12 castor oil; nonionic; detergent, emulsifier and coemulsifier in formulation of sol. oils, solv. cleaners and temp. protective coatings; corrosion resistant, used in industrial and institutional cleaning preparations; die lubricant in metal forming operations; mold release agent in plastics; fiber lubricant in textile processing; Hazen > 250 liq.; sol. in water, benzene, ethyl acetate, ethyl Icinol, ethanol, veg. oil, olein; sp.gr. 1.048; visc. 5000 cps.; m.p. 3 ± 2 C; HLB 12.7; cloud pt. 67–69 C; surf. tens. 42.0 dynes/cm; pH 6–8 (1% aq.); 100% act.

Teric CDE. [ICI Australia] Cocamide DEA; nonionic; emulsifier of perchlorethylene, power kerosene; foam booster, thickener and detergent in domestic dishwashing formulations; Hazen > 250 liq.; sol. see Teric 350; sp.gr. 1.010; visc. 600 cps; m.p. < 0 C; HLB 16.6; pH 8–10; biodeg.; 100% act.

Teric CME3. [ICI Australia] PEG-3 cocamide MEA; nonionic; dispersant, detergent; wetting agent for textile applics.; coemulsifier for various oils; emulsion stabilizer; foam booster in detergent blends; Hazen > 250 liq., paste; low odor; sol. in benzene, ethyl acetate, ethyl Icinol, perchlorethylene, ethanol, olein; disp. in water; sp.gr. 0.972 (50 C); visc. 56 cps (50 C); m.p. 18 ± 2 C; HLB 13.8; pH 8–10 (1% aq.); biodeg.; 100% act.

Teric CME7. [ICI Australia] PEG-7 cocamide MEA; nonionic; wetting agent for textile applics.; dispersant, detergent, foam stabilizer; cleaning compds.; Hazen > 250 liq.; low odor; sol. see Teric 18M30; sp.gr. 1.048; visc. 264 cps; m.p. 5 ± 2 C; HLB 15.1; surf. tens. 30.0 dynes/cm; pH 8–10; biodeg.; 100% act.

Teric G9A5. [ICI Australia] C_{9-11} pareth-5; nonionic; emulsifier for solvs.; formulation of hard surf. cleaners, degreasers and dispersants in industrial and domestic applics.; Hazen 250 liq.; sol. in water, veg. oil, olein; sp.gr. 0.976; visc. 52 cps; m.p. < 0 C; HLB 11.6; cloud pt. 36 ± 2 C; surf. tens. 20.1 dynes/cm; pH 5–7 (1% aq.); 100% act.

Teric G9A6. [ICI Australia] C_{9-11} pareth-6; see Teric G9A5; Hazen 250 liq.; sol. in water, benzene, ethyl acetate, ethyl Icinol, perchlorethylene, ethanol, veg. oil, olein; sp.gr. 0.987; visc. 38 cps; m.p. < 0 C; HLB 12.4; cloud pt. 53 ± 2 C; surf. tens. 24.0 dynes/cm; pH 5–7 (1% aq.); 100% act.

Teric G9A8. [ICI Australia] C_{9-11} pareth-8; nonionic; emulsifier, dispersant, wetting agent; used in detergent formulations for domestic use; degreaser in textile applics.; Hazen 250 paste; sol. see Teric G9A6; sp.gr. 0.988 (50 C); visc. 21 cps (50 C); m.p. 13 ± 2 C; HLB 13.7; cloud pt. 78 ± 2 C; surf. tens. 27.0 dynes/cm; pH 5–7 (1% aq.); 100% act.

Teric G9A12. [ICI Australia] C_{9-11} pareth-12; nonionic; see Teric G9A8; Hazen 250 solid; sol. see Teric G9A6; sp.gr. 1.023; visc. 40 cps (50 C); m.p. 24 ± 2 C; HLB 15.4; cloud pt. > 100 C; surf. tens. 34.5 dynes/cm; pH 5–7 (1% aq.); 100% act.

Teric G12A4. [ICI Australia] C_{12-15} pareth-4 (4.5 EO); nonionic; wetting agent and dispersant, detergent; dust suppression; aq. pigment dispersion; coemulsifier for w/o-type emulsion concs.; liq. and powd. detergents for industrial and domestic laundry, dishwashing, metal cleaning, sanitizing, textile processing aids, abrasive cleaners; intermediate in mfg. of sulfate and phosphate surfactants; Hazen 250 liq.; sol. see Teric 18M2; also disp. in water; sp.gr. 0.950; visc. 40 cps; m.p. 3 ± 2 C; HLB 9.8; pH 6–8 (1% aq.); 100% act.

Teric G12A6. [ICI Australia] C_{12-15} pareth-6; nonionic; wetting agent and dispersant, intermediate, textile processing, hard surf. abrasives, metal descaling, domestic and industrial use in laundries, intermediate for metal cleaning, sanitizing, dairy detergents; Hazen 250 liq. to paste; sp.gr. 0.955; visc. 24 cps (50 C); m.p. 10 ± 2 C; HLB 11.2; cloud pt. 40 ± 2 C; surf. tens. 27.8 dynes/cm; pH 6–8 (1% aq.); biodeg.; 100% act.

Teric G12A8. [ICI Australia] C_{12-15} pareth-8 (7.8 EO); nonionic; see Teric G12A6; Hazen 250 paste; sol. see Teric G12A4; sp.gr. 0.971 (50 C); visc. 31 cps (50 C); m.p. 16 ± 2 C; HLB 12.6; cloud pt. 58 ± 2 C; surf. tens. 27.2 dynes/cm; pH 6–8 (1% aq.); biodeg.; 100% act.

Teric G12A12. [ICI Australia] C_{12-15} pareth-12 (11.5 EO); nonionic; see Teric G12A6; Hazen 250 solid; sp.gr. 0.993; visc. 37 cps; m.p. 24 ± 2 C; HLB 14.4; cloud pt. 92 ± 2 C; surf. tens. 35.6 dynes/cm; pH 6–8 (1% aq.); biodeg.; 100% act.

Teric LA4. [ICI Australia] C_{12-15} pareth-4 (4.5 EO); nonionic; see Teric G12A4; liq.; sol. in water, ethyl acetate, ethyl Icinol, ethanol, kerosene, min. oil, olein; sp.gr. 0.955; visc. 45 cps; m.p. 0 ± 2 C; HLB 9.8; pH 6–8 (1% aq.); biodeg.; 88% act.

Teric LA8. [ICI Australia] C_{12-15} pareth-8 (7.8 EO); nonionic; see Teric G12A8; lt. amber liq.; sol. see Teric 18M30; sp.gr. 1.002; visc. 100 cps; m.p. 10 ± 2 C; HLB 12.6; cloud pt. 60 ± 2 C; surf. tens. 27.2 dynes/cm; pH 6–8 (1% aq.); biodeg.; 88% act.

Teric LAN70. [ICI Australia] PEG-70 lanolin; nonionic; emulsifier, dispersant, textile lubricant; yel.-amber soft waxy solid; sol. in water, benzene, ethyl Icinol, perchlorethylene, ethanol, olein; sp.gr. 1.13; visc. 246 cps (50 C); m.p. 42 ± 2 C; HLB 16.3; cloud pt. > 100 C; surf. tens. 48.1 dynes/cm; pH 4.5–7.0 (1% aq.); 100% act.

Teric N2. [ICI Australia] Nonoxynol-2; nonionic; defoamer, wetting agent and dispersant in solv. and oil-based systems; coemulsifier for o/w emulsions; emulsifier for w/o emulsions; used in agric. toxicant prods., industrial solv.-cleaning systems, surf. coating preparations; intermediate for prod. of anionics; Hazen 100 liq.; sol. see Teric 18M2; also disp. in water; sp.gr. 1.001; visc. 620 cps; m.p. < 0 C; HLB 5.7; pH 6–8 (1% aq.); 100% act.

Teric N3. [ICI Australia] Nonoxynol-3 (3.5 EO); nonionic; see Teric N2; Hazen 100 liq.; sol. see Teric N2; sp.gr. 1.017; visc. 395 cps; m.p. < 0 C; HLB 8.2; pH 6–8 (1% aq.); 100% act.

Teric N4. [ICI Australia] Nonoxynol-4; nonionic; see Teric N2; Hazen 100 liq.; sol. see Teric N2; sp.gr. 1.023; visc. 370 cps; m.p. < 0 C; HLB 8.9; pH 6–8 (1% aq.); 100% act.

Teric N5. [ICI Australia] Nonoxynol-5 (5.5 EO); nonionic; see Teric N2; Hazen 100 liq.; sol. in benzene, ethyl acetate, ethyl Icinol, perchlorethylene, ethanol, kerosene, veg. oil, olein; disp. in water; sp.gr. 1.035; visc. 355 cps; m.p. < 0 C; HLB 10.5; pH 6–8 (1% aq.); 100% act.

Teric N8. [ICI Australia] Nonoxynol-8 (8.5 EO); nonionic; wetting agent and dispersant, emulsifier, solubilizer, detergent; used in concrete mfg., agric. sprays, solv. cleaners, paints; detergents; Hazen 100 liq.; sol. see Teric 160; sp.gr. 1.056; visc. 350 cps; m.p. 0 ± 2 C; HLB 12.3; cloud pt. 32 ± 2 C; surf. tens. 29.4 dynes/cm; pH 6–8 (1% aq.); 100% act.

Teric N9. [ICI Australia] Nonoxynol-9; nonionic; see Teric N8; Hazen 100 liq.; sol. in water, benzene, ethyl acetate, ethyl Icinol, perchorethylene, veg. oil, olein; sp.gr. 1.060; visc. 330 cps; m.p. 0 ± 2 C; HLB 12.8; cloud pt. 52 ± 2 C; surf. tens. 30.6 dynes/cm; pH 6–8 (1% aq.); 100% act.

Teric N10. [ICI Australia] Nonoxynol-10; nonionic; see Teric N8; Hazen 100 liq.; sol. see Teric N9; sp.gr. 1.063; visc. 360 cps; m.p. 5 ± 2 C; HLB 13.3; cloud pt. 67 ± 2 C; surf. tens. 30.6 dynes/cm; pH 6–8 (1% aq.); 100% act.

Teric N11. [ICI Australia] Nonoxynol-11; nonionic; see Teric N8; Hazen 100 liq.; sol. in water, benzene, ethyl acetate, ethyl Icinol, perchlorethylene, ethanol, olein; sp.gr. 1.069; visc. 410 cps; m.p. 7 ± 2 C; HLB 13.7; cloud pt. 74 ± 2 C; surf. tens. 34.8 dynes/cm; pH 6–8 (1% aq.); 100% act.

Teric N12. [ICI Australia] Nonoxynol-12; nonionic; see Teric N8; Hazen 100 liq.; sol. see Teric N11; sp.gr. 1.045 (50 C); visc. 67 cps (50 C); m.p. 11 ± 2 C; HLB 13.9; cloud pt. 82 ± 2 C; surf. tens. 35.2 dyens/cm; pH 6–8 (1% aq.); 100% act.

Teric N13. [ICI Australia] Nonoxynol-13; nonionic; see Teric N8; Hazen 100 liq.; sol. see Teric N11; sp.gr. 1.049 (50 C); visc. 75 cps (50 C); m.p. 14+2 C; HLB 14.4; cloud pt. 89 ± 2 C; surf. tens. 34.9 dynes/cm; pH 6–8 (1% aq.); 100% act.

Teric OF4. [ICI Australia] PEG-4 oleate; nonionic; emulsifier used in o/w emulsions of veg. and min. oils, fats, waxes and solvs.; formulation of cutting oils; metal lubricant in metal cleaning formulations and in solv. degreasers; textile and paper finishing applics.; antistat in textile processing and mfg. of syn. fibers; Hazen > 250 liq.; sol. see Teric 151; sp.gr. 0.978; visc. 87 cps; m.p. < 0 C; HLB 7.7; pH 8 (1% aq.); biodeg.; 100% act.

Teric OF6. [ICI Australia] PEG-6 oleate; nonionic; see Teric OF4; Hazen > 250 liq.; sol. see Teric 151; sp.gr. 0.992; visc. 82 cps; m.p. < 0 C; HLB 9.7; surf. tens. 33.9 dynes/cm; pH 7–9 (1% aq.); biodeg.; 100% act.

Teric OF8. [ICI Australia] PEG-8 oleate; nonionic; emulsifier, antistat; aux. product and antistat in textile and paper finishing applics.; Hazen > 250 liq.; sol. see Teric 151; sp.gr. 1.012; visc. 120 cps; m.p. < 0 C; HLB 11.1; pH 8–10 (1% aq.); biodeg.; 100% act.

Teric PE61. [ICI Australia] POP + 5.7 EO; nonionic; demulsifier, intermediate, emulsifier, detergent, wetting agent and dispersant; pigments in latex paints; detergent sanitizer and alkaline cleaner; dewatering aid for treatment of crude oil emulsions; defoamer in paper pulp liquors, starch sizing, glues, and boiler water systems; Hazen 100 liq.; sol. in benzene, ethyl acetate, ethyl Icinol, perchlorethylene, veg. oil, olein; sp.gr. 1.017; visc. 388 cps; m.p. < 0 C; HLB 3; cloud pt. 24 ± 2 C; surf. tens. 39.0 dynes/cm; pH 6 (1% aq.); 100% act.

Teric PE62. [ICI Australia] POP + 17 EO; nonionic; see Teric PE61; also for bloat control in cattle; Hazen 100 liq.; sol. see Teric 160; sp.gr. 1.032; visc. 440 cps; m.p. < 0 C; HLB 7; cloud pt. 32 ± 2 C; surf. tens. 42.1 dynes/cm; pH 6 (1% aq.); 100% act.

Teric PE64. [ICI Australia] POP + 25.5 EO; nonionic; wetting agent, dispersant, emulsifier, detergent, intermediate, demulsifier; used for pigments in latex paints, diary cleaners and sanitizers; bloat control in cattle; oil well applics.; Hazen 100 liq.; sol. see Teric N11; sp.gr. 1.051; visc. 1332 cps; m.p. 8 ± 2 C; HLB 15; cloud pt. 58 ± 2 C; surf. tens. 43.2 dynes/cm; pH 6 (1% aq.); 100% act.

Teric PE68. [ICI Australia] POP + 150 EO; nonionic; wetting agent, dispersant, emulsifier, detergent, intermediate; stabilizer in mfg. of latex paints, paper coatings, diary cleaners and santizers; plasticizer for resins; Hazen 100 flakes; sol. see Teric N11; m.p. 50 ± 2 C; HLB 29; cloud pt. > 100 C; surf. tens. 49.0 dynes/cm; pH 6 (1% aq.); 100% act.

Teric T2. [ICI Australia] PEG-2 tallate; nonionic; wetting agent and dispersant for oil-based systems; emulsifier, detergent, grinding aid for wet milling of pigments and resins; mfg. of lubricants for metal and textile industries, solv. cleaners, and powd. detergent formulations; Hazen > 250 liq.; sol. in benzene, ethyl acetate, ethyl Icinol, perchlorethylene, ethanol, min. and veg. oil, olein; disp. in water; sp.gr. 0.966; visc. 118 cps; m.p. < 0 C; HLB 4.5; pH 6–8 (1% aq.); biodeg.; 100% act.

Teric T5. [ICI Australia] PEG-5 tallate; nonionic; see Teric T2; Hazen > 250 liq.; sol. see Teric T2; sp.gr. 1.018; visc. 196 cps; m.p. < 0 C; HLB 8.2; pH 8–10 (1% aq.); biodeg.; 100% act.

Teric T7. [ICI Australia] PEG-7 tallate; nonionic; see Teric T2; Hazen > 250 liq.; sol. see Teric 121; sp.gr. 1.031; visc. 151 cps; m.p. < 0 C; HLB 10.0; pH 8–10 (1% aq.); biodeg.; 100% act.

Teric T10. [ICI Australia] PEG-10 tallate; nonionic; see Teric T2; Hazen > 250 liq.; sol. see Teric 121; also disp. in water; sp.gr. 1.042; visc. 253 cps; m.p. 5 ± 2 C; HLB 11.7; pH 8–10 (1% aq.); biodeg.; 100% act.

Teric X5. [ICI Australia] Oxtoxynol-5; nonionic; wetting agent and dispersant, detergent used in metal cleaning compds., alkaline and metal cleaners, agric. powds., emulsions, pigment and wax dispersions; textile auxs.; emulsifier in aromatic and aliphatic solvs., emulsions, cleaners and paints; Hazen 100 liq.; sol. see Teric 18M2; sp.gr. 1.046; visc. 465 cps; m.p. < 0 ± 2 C; HLB 10.4; pH 6–8 (1% aq.); 100% act.

Teric X7. [ICI Australia] Octoxynol-7 (7.5 EO); nonionic; see Teric X5; Hazen 100 liq.; sol. see Teric 160; sp.gr. 1.059; visc. 420 cps; m.p. < 0 ± 2 C; HLB 12.3; cloud pt. 28 ± 2 C; surf. tens. 29.5 dynes/cm; pH 6–8 (1% aq.); 100% act.

Teric X8. [ICI Australia] Octoxynol-8 (8.5 EO); nonionic; see Teric X5; Hazen 100 liq.; sol. see Teric 18M20; sp.gr. 1.060; visc. 402 cps; m.p. < 0 ± 2 C; HLB 12.6; cloud pt. 47 ± 2 C; surf. tens. 30.2 dynes/cm; pH 6–8 (1% aq.); 100% act.

Teric X10. [ICI Australia] Octoxynol-10; nonionic; see Teric X5; Hazen 100 liq.; sol. see Teric 18M20; sp.gr. 1.062; visc. 393 cps; m.p. 7 ± 2 C; HLB 13.6; cloud pt. 65 ± 2 C; surf. tens. 31.8 dynes/cm; pH 6–8 (1% aq.); 100% act.

Teric X11. [ICI Australia] Octoxynol-11; nonionic; see Teric X5; Hazen 100 liq.; sol. see Teric 18M20; sp.gr. 1.067; visc. 371 cps; m.p. 8 ± 2 C; HLB 14.0; cloud pt. 80 ± 2 C; surf. tens. 31.0 dynes/cm; pH 6–8 (1% aq.); 100% act.

Teric X13. [ICI Australia] Octoxynol-13; nonionic; see Teric X5; Hazen 100 liq.; sol. see Teric 18M20; sp.gr. 1.052; visc. 70 cps; m.p. 15 ± 2 C; HLB 14.7; cloud pt. 90 ± 2 C; surf. tens. 32.0 dynes/cm; pH 6–8 (1% aq.); 100% act.

Teric X16. [ICI Australia] Octoxynol-16; nonionic; see Teric X5; Hazen 100 semisolid; sol. see Teric 18M2; sp.gr. 1.059 (50 C); visc. 95 cps (50 C); m.p. 24 ± 2 C; HLB 15.2; cloud pt. > 100 C; surf. tens. 35.6 dynes/cm; pH 6–8 (1% aq.); 100% act.

Teric X40. [ICI Australia] Octoxynol-40; nonionic; see Teric X5; Hazen 100 solid; sol. see Teric 18M30; sp.gr. 1.082 (50 C); visc. 245 cps (50 C); m.p. 44 ± 2 C; HLB 17.6; cloud pt. > 100 C; surf. tens. 36.4 dynes/cm; pH 6–8 (1% aq.); 100% act.

Termamyl® 60T. [Novo] Alpha-amylase; enzyme for laundry powd. detergents, dishwashing powds.; off-wh. gran.

Termamyl® 120L. [Novo] Bacterial alpha-amylase; enzyme for liquefaction of gelatinized starch in prod. of dextrose, high fructose, and other syrups; used in alcohol, brewing, and textile industries; liq.

Termamyl® 300T. [Novo] Alpha-amylase; enzyme for nonbuilt liq. detergents, prespotters; lt. brn. liq.

Terracur® P. [Bayer] Fensulfothion; nematocide and insecticide; ylsh.-brn. liq.; m.w. 308.35; sol. in most org. solvs.; sol. 1200 ppm in water; sp.gr. 1.202 (20/4 C); b.p. 138–141 C (0.01 mm Hg); flash pt. (PM) 115 C.

Tetradecene-1. [Ethyl] C_{14} alpha olefins; intermediate for surfactants and industrial chemicals; liq.; 100% conc.

Tetraglyme. [Ferro] Tetraethylene glycol dimethyl ether; solv. which tends to solvate cations; used in electrochemistry, polymer and boron chemistry; physical processes such as gas absorp., extraction, stabilization; used in industrial prods. such as fuels, lubricants, textiles, pharmaceuticals, pesticides; colorless clear; mild, nonresidual odor; m.w. 222.28; f.p. –29.7 C; b.p. 275 C (760 mm Hg); water-sol.;

sp.gr. 1.0132; dens. 8.45 lb/gal; visc. 4.1 cP; pH neutral; flash pt. 141 C (CC); surf. tens. 33.8 dynes/cm; 98.0% min. purity.

Tetralin. [DuPont] 1,2,3,4-Tetrahydronaphthalene; solv. for oils, resins, waxes, rubber, and asphalt; colorless to pale yel. clear liq.; m.w. 132.2; f.p. –31 C; b.p. 207 C (760 mm Hg); negligible sol. in water; sp.gr. 0.970; dens. 8.1 lb/gal; flash pt. 82 C (TCC); 97.0% min. tetrahydronaphthalene.

Tetranol. [Sandoz] Sulfated fatty ester; nonionic; wetting and leveling agent for dyeing processes; assistant for vat dyeing; emulsifier and detergent in scouring processes; water-sol.; pH 5.8–6.0.

Tetronic® 50R1. [BASF] EO/PO ethylene diamine block copolymer; nonionic surfactant series functioning as emulsion stabilizers, solubilizers, dispersants, wetting agents, antistats, penetrants, plasticizers, defoaming agents, demulsifiers in the petrol., paint, paper, cement, ink, cosmetic, drug, plastic, detergent, and metalworking industries; rubber activator; R series for low foaming applics.; liq.; m.w. 2640; visc. 670 cps; HLB 1–7; pour pt. –18 C; cloud pt. 29 C (1% aq.); surf. tens. 40 dynes/cm (0.1%); 100% act.

Tetronic® 50R4. [BASF] EO/PO ethylene diamine block copolymer; see Tetronic 50R1; liq.; m.w. 3740; visc. 1080 cps; HLB 7–12; pour pt. –4 C; cloud pt. 59 C (1% aq.); surf. tens. 46 dynes/cm (0.1%); 100% act.

Tetronic® 50R8. [BASF] EO/PO ethylene diamine block copolymer; see Tetronic 50R1; solid; m.w. 10,200; visc. 650 cps (77 C); HLB 12–18; m.p. 38 C; cloud pt. 88 C (1% aq.); surf. tens. 53 dynes/cm (0.1%); 100% act.

Tetronic® 70R1. [BASF] EO/PO ethylene diamine block copolymer; see Tetronic 50R1; liq.; m.w. 3400; visc. 800 cps; HLB 1–7; pour pt. –17 C; cloud pt. 25 C (1% aq.); surf. tens. 39 dynes/cm (0.1%); 100% act.

Tetronic® 70R2. [BASF] EO/PO ethylene diamine block copolymer; see Tetronic 50R1; liq.; m.w. 3870; visc. 880 cps; HLB 1–7; pour pt. –22 C; cloud pt. 31 C (1% aq.); surf. tens. 40 dynes/cm (0.1%); 100% act.

Tetronic® 70R4. [BASF] EO/PO ethylene diamine block copolymer; see Tetronic 50R1; liq.; m.w. 5230; visc. 2160 cps; HLB 7–12; pour pt. 6 C; cloud pt. 42 C (1% aq.); surf. tens. 43 dynes/cm (0.1%); 100% act.

Tetronic® 90R1. [BASF] EO/PO ethylene diamine block copolymer; see Tetronic 50R1; liq.; m.w. 4580; visc. 940 cps; HLB 1–7; pour pt. –17 C; cloud pt. 21 C (1% aq.); surf. tens. insol. (0.1%); 100% act.

Tetronic® 90R4. [BASF] EO/PO ethylene diamine block copolymer; see Tetronic 50R1; liq.; m.w. 7240; visc. 3870 cps; HLB 1–7; pour pt. 12 C; cloud pt. 43 C (1% aq.); surf. tens. 43 dynes/cm (0.1%); 100% act.

Tetronic® 90R8. [BASF] EO/PO ethylene diamine block copolymer; see Tetronic 50R1; solid; m.w. 18,700; visc. 4000 cps (77 C); HLB 12–18; m.p. 47 C; cloud pt. 88 C (1% aq.); surf. tens. 50 dynes/cm (0.1%); 100% act.

Tetronic® 110R1. [BASF] EO/PO ethylene diamine block copolymer; see Tetronic 50R1; liq.; m.w. 5220; visc. 1000 cps; HLB 1–7; pour pt. –18 C; cloud pt. 21 C (1% aq.); surf. tens. insol. (0.1%); 100% act.

Tetronic® 110R2. [BASF] EO/PO ethylene diamine block copolymer; see Tetronic 50R1; liq.; m.w. 5900; visc. 1320 cps; HLB 1–7; pour pt. –20 C; cloud pt. 27 C (1% aq.); surf. tens. 38 dynes/cm (0.1%); 100% act.

Tetronic® 110R7. [BASF] EO/PO ethylene diamine block copolymer; see Tetronic 50R1; solid; m.w. 13,200; visc. 900 cps (77 C); HLB 7–12; m.p. 47 C; cloud pt. 64 C (1% aq.); surf. tens. 46 dynes/cm (0.1%); 100% act.

Tetronic® 130R1. [BASF] EO/PO ethylene diamine block copolymer; see Tetronic 50R1; liq.; m.w. 6800; visc. 1240 cps; HLB 1–7; pour pt. –18 C; cloud pt. 20 C (1% aq.); surf. tens. insol. (0.1%); 100% act.

Tetronic® 130R2. [BASF] EO/PO ethylene diamine block copolymer; see Tetronic 50R1; liq.; m.w. 7740; visc. 1880 cps; HLB 1–7; pour pt. –3 C; cloud pt. 25 C (1% aq.); surf. tens. 35 dynes/cm (0.1%); 100% act.

Tetronic® 150R1. [BASF] EO/PO ethylene diamine block copolymer; see Tetronic 50R1; liq.; m.w. 8000; visc. 1840 cps; HLB 1–7; pour pt. –17 C; cloud pt. 20 C (1% aq.); surf. tens. insol. (0.1%); 100% act.

Tetronic® 150R4. [BASF] EO/PO ethylene diamine block copolymer; see Tetronic 50R1; paste; m.w. 11,810; visc. 850 cps (60 C); HLB 1–7; pour pt. 34 C; cloud pt. 29 C (1% aq.); surf. tens. 39 dynes/cm (0.1%); 100% act.

Tetronic® 150R8. [BASF] EO/PO ethylene diamine block copolymer; see Tetronic 50R1; solid; m.w. 20,400; visc. 6300 cps (77 C); HLB 7–12; m.p. 53 C; cloud pt. 77 C (1% aq.); surf. tens. 44 dynes/cm (0.1%); 100% act.

Tetronic® 304. [BASF] Poloxamine 304; nonionic; emulsifier, thickener, wetting agent, dispersant, solubilizer, stabilizer, cosmetics and pharmaceuticals; demulsifier in petrol. industry; detergent ingred.; antistat for polyethylene and resin molding powds.; metal treatment; emulsion polymerization; used in latex-based paints, aq.-based syn. cutting fluids and vulcanization of rubber; colorless liq.; m.w. 1650; ref. index 1.4649; water-sol.; sp.gr. 1.06; visc. 450 cps; HLB 16; cloud pt. 94 C (1%); pour pt. –11 C; surf. tens. 53.0 dynes/cm (0.1%); 100% act.

Tetronic® 504. [BASF] Poloxamine 504; nonionic; see Tetronic 304; colorless liq.; m.w. 3400; ref. index 1.4612; water-sol.; sp.gr. 1.04; visc. 800 cps; HLB 15.5; cloud pt. 68 C (1%); pour pt. 7 C; surf. tens. 44.2 dynes/cm (0.1%); 100% act.

Tetronic® 701. [BASF] Poloxamine 701; nonionic; see Tetronic 304; colorless liq.; m.w. 3400; ref. index. 1.4553; sp.gr. 1.02; visc. 575 cps; HLB 3; cloud pt. 22 C (1%); pour pt. –21 C; surf. tens. 36.1 dynes/cm; 100% act.

Tetronic® 702. [BASF] Poloxamine 702; nonionic; see Tetronic 304; colorless liq.; m.w. 4000; ref. index 1.4572; water-sol.; sp.gr. 1.03; visc. 770 cps; HLB 7; cloud pt. 27 C (1%); pour pt. –7 C; surf. tens. 36.8 dynes/cm (0.1%); 100% act.

Tetronic® 704. [BASF] Poloxamine 704; nonionic; see Tetronic 304; colorless liq.; m.w. 5500; ref. index 1.4613; water-sol.; sp.gr. 1.04; visc. 850 cps; HLB 15; cloud pt. 65 C (1%); pour pt. 18 C; surf. tens. 40.3 dynes/cm (0.1%); 100% act.

Tetronic® 707. [BASF] Poloxamine 707; nonionic; see Tetronic 304; flakes; m.w. 12,000; water-sol.; m.p. 49 C; HLB 27; cloud pt. > 100 C (1%); surf. tens. 47.6 dynes/cm (0.1%); 100% act.

Tetronic® 901. [BASF] Poloxamine 901; nonionic; see Tetronic 304; colorless liq.; m.w. 4750; ref. index 1.4545; sp.gr. 1.02; visc. 700 cps; HLB 2.5; cloud pt. 20 C (1%); pour pt. –23 C; 100% act.

Tetronic® 904. [BASF] Poloxamine 904; nonionic; see Tetronic 304; colorless liq.; m.w. 7500; ref. index 1.4604; water-sol.; sp.gr. 1.04; visc. 6000 cps; HLB 14.5; cloud pt. 64 C (1%); pour pt. 29 C; surf. tens. 35.4 dynes; 100% act.

Tetronic® 908. [BASF] Poloxamine 908; nonionic; see Tetronic 304; flakes; m.w. 27,000; water-sol.; m.p. 58 C; HLB 30.5; cloud pt. > 100 C (1%); surf. tens. 45.7 dynes/cm (0.1%); 100% act.

Tetronic® 909. [BASF] EO/PO ethylene diamine block copolymer; see Tetronic 304; solid; m.w. 30,000; visc. 20,000 cps (77 C); HLB > 24; m.p. 59 C; cloud pt. > 100 C (1% aq.); surf. tens. 56 dynes/cm (0.1%); 100% act.

Tetronic® 1101. [BASF] Poloxamine 1101; nonionic; see Tetronic 304; colorless liq.; m.w. 5600; ref. index 1.4540; sp.gr. 1.02; visc. 700 cps; HLB 2; cloud pt. 17 C (1%); pour pt. –15 C; surf. tens. 34.0 dynes/cm; 100% act.

Tetronic® 1102. [BASF] Poloxamine 1102; nonionic; see Tetronic 304; colorless liq.; m.w. 6300; ref. index 1.4557; water-sol.; sp.gr. 1.03; visc. 820 cps; HLB 6; cloud pt. 31 C (1%); pour pt. 7 C; surf. tens. 33.7 dynes/cm (0.1%); 100% act.

Tetronic® 1104. [BASF] Poloxamine 1104; nonionic; see Tetronic 304; paste; m.w. 8300; water-sol.; m.p. 34 C; HLB 14; cloud pt. 72 C (1%); surf. tens. 36.3 dynes/cm (0.1%); 100% act.

Tetronic® 1107. [BASF] Poloxamine 1107; nonionic; see Tetronic 304; solid; m.w. 14,500; water-sol.; m.p. 51 C; HLB 24; cloud pt. > 100 C; surf. tens. 42.9 dynes/cm (0.1%); 100% act.

Tetronic® 1301. [BASF] Poloxamine 1301; nonionic; see Tetronic 304; colorless liq.; m.w. 6800; ref. index 1.4545; sp.gr. 1.02; visc. 1000 cps; HLB 1.5; cloud pt. 16 C (1%); pour pt. –9 C; surf. tens. 33.4 dynes/cm; 100% act.

Tetronic® 1302. [BASF] Poloxamine 1302; nonionic; see Tetronic 304; colorless liq.; m.w. 7800; ref. index 1.4562; water-sol.; sp.gr. 1.03; visc. 1300 cps; HLB 5.5; cloud pt. 20 C (1%) pour pt. 13 C; surf. tens. 34.1 dynes/cm (0.1%); 100% act.

Tetronic® 1304. [BASF] Poloxamine 1304; nonionic; see Tetronic 304; paste; m.w. 10,500; water-sol.; m.p. 36 C; HLB 13.5; cloud pt. 78 C (1%); surf. tens. 35.5 dynes/cm (0.1%); 100% act.

Tetronic® 1307. [BASF] Poloxamine 1307; nonionic; see Tetronic 304; solid; m.w. 18,600; water-sol.; m.p. 54 C; HLB 23.5; cloud pt. > 100 C (1%); surf. tens. 43.8 dynes/cm (0.1%); 100% act.

Tetronic® 1501. [BASF] Poloxamine 1501; nonionic; see Tetronic 304; colorless liq.; m.w. 7900; ref. index 1.4537; sp.gr. 1.02; visc. 1170 cps; HLB 1.0; cloud pt. 15 C (1%); pour pt. –4 C; surf. tens. 33.9 dynes/cm; 100% act.

Tetronic® 1502. [BASF] Poloxamine 1502; nonionic; see Tetronic 304; colorless liq.; m.w. 9000; ref. index 1.4560; water-sol.; sp.gr. 1.03; visc. 1570 cps; HLB 5.0; cloud pt. 70 C (1%); pour pt. 18 C; surf. tens. 34.1 dynes/cm (0.1%); 100% act.

Tetronic® 1504. [BASF] Poloxamine 1504; nonionic; see Tetronic 304; paste; m.w. 12,500; water-sol.; m.p. 41 C; HLB 13.0; cloud pt. 90 C (1%); surf. tens. 37.3 dynes/cm (0.1%); 100% act.

Tetronic® 1508. [BASF] Poloxamine 1508; nonionic; see Tetronic 304; solid; m.w. 27,000; water-sol.; m.p. 60 C; HLB 27; cloud pt. > 100 C (1%); surf. tens. 43.8 dynes/cm (0.1%); 100% act.

Texacar® EC. [Texaco] Ethylene carbonate; solv. for org. and inorg. materials; Rule 66 exempt; also used as reactant and plasticizer in fibers and textiles, plastics and resins, aromatic hydrocarbon extraction, electrolytes, hydraulic brake fluids; wh. cake solid, mild odor; melts to clear liq. > 95 F; m.w. 88.06; sp.gr. 1.3218 (39/4 C); dens. 10.87 lb/gal (40 C); visc. 1.9 cs (40 C); b.p. 248.2 C (760 mm Hg); f.p. 35.5 C min.; flash pt. 305 F; ref. index 1.4158; pH 6.5–7.5 (10% aq.); 99% min. assay.

Texacar® EC-50. [Texaco] Ethylene carbonate/propylene carbonate 50:50 blend; solv., reactant.

Texacar® PC. [Texaco] Propylene carbonate; solv. for org. and inorg. materials; Rule 66 exempt; also used as reactant and plasticizer in fibers and textiles, hydraulic fluids, plastics and resins, gas treating, aromatic hydrocarbon extraction, metal extraction, surf. coatings, foundry sand binders, lubricants, electrolytes, personal care prods.; gellant for clays in greases and cosmetics; clear mobile liq., practically odorless; m.w. 102.09; sp.gr. 1.2057 (20/4 C); dens. 10.05 lb/gal; visc. 2.8 cs; b.p. 241.9 C (760 mm Hg); f.p. –49.2 C; flash pt. 275 F; pour pt. –100 F; ref. index 1.4210; pH 6.5–7.5 (10% aq.); 99.4% min. assay.

Texacat® DD. [Texaco] Catalyst for semiflexible foam urethanes; b.p. 205 C; f.p. < –49 C; flash pt. 170 F (TCC).

Texacat® DM-70. [Texaco] Catalyst for flexible PU foam, polyester and polyether slabstock; sp.gr. 0.99; i.b.p. 150 C; f.p. –32 C; flash pt. (TCC) 102 F.

Texacat® DMDEE. [Texaco] 2,2′-Dimorpholinodiethylether; catalyst for flexible and rigid PU foam, polyester and polyether slabstock, 1-component foam, moisture-cured urethane coatings; sp.gr. 1.1; b.p. 309 C; f.p. –28 C; flash pt. (TCC) 295 F.

Texacat® DME. [Texaco] N,N-Dimethylethanolamine; catalyst for flexible and rigid PU foam, polyether slabstock; acid scavenger in rigid systems with polymeric isocyanates; sp.gr. 0.88; b.p. 130 C; f.p. –59 C; flash pt. (TCC) 105 F.

Texacat® DMP. [Texaco] N,N′-dimethylpiperazine; latent cure catalyst for polyester slabstock flexible foam; sp.gr. 0.85; b.p. 133 C; f.p. –1 C; flash pt. (TCC) 72 F.

Texacat® DPA. [Texaco] N,N-(dimethyl)-N′,N′-diisopropanol-1,3-propanediamine; catalyst for rigid PU foam, pkg. foam; sp.gr. 0.94; f.p. < –26 C; flash pt. (PMCC) 194 F.

Texacat® DPA-50. [Texaco] Texacat DPA and DME (1:1); catalyst for PU pkg. foam; flash pt. (PMCC) 110 F.

Texacat® M-75. [Texaco] Catalyst for flexible PU foam, polyester and polyether slabstock; low odor; sp.gr. 0.90; f.p. < –50 C; flash pt. (PMCC) 126 F; in solv.

Texacat® MM-70. [Texaco] Catalyst for polyester urethane flexible foam slabstock; low odor; sp.gr. 0.94; i.b.p. 157 C; f.p. –80 C; flash pt. 105 F (TCC).

Texacat® NEM. [Texaco] N-ethylmorpholine; catalyst for flexible and rigid PU foam, polyester slabstock; sp.gr. 0.91; b.p. 138 C; f.p. –63 C; flash pt. (TCC) 90 F.

Texacat® NMM. [Texaco] N-methylmorpholine; catalyst for polyester urethane flexible foam, high rise rigid foam panels; sp.gr. 0.92; b.p. 116 C; f.p. –66 C; flash pt. (TCC) 55 F.

Texacat® TD-20. [Texaco] Catalyst for flexible and rigid PU foam; sp.gr. 0.91; i.b.p. 138 C; f.p. < –20 C; flash pt. (PMCC) 108 F.

Texacat® TD-33. [Texaco] 33% Triethylenediamine in propylene glycol; general purpose catalyst for flexible and rigid PU foam, polyether slabstock; sp.gr. 1.0; b.p. 185 C; f.p. –13 C; flash pt. (PMCC) 206 F.

Texacat® TD-33A. [Texaco] 33% triethylenediamine in dipropylene glycol; general purpose catalyst for flexible and rigid PU foam, polyether slabstock; sp.gr. 1.0; f.p. < –24 C; flash pt. (PMCC) 196 F.

Texacat® ZF-10. [Texaco] N,N,N′-trimethyl-N′-hydroxyethyl-bisaminoethylether; catalyst for flexible PU foam, polyether slabstock, pkg. foam; sp.gr. 0.95; b.p. 254 C; flash pt. (PMCC) 245 F.

Texacat® ZF-22. [Texaco] General purpose catalyst for flexible and rigid PU foam, polyether slabstock; sp.gr. 0.90; b.p. 188 C; f.p. < –50 C; flash pt. (TCC) 164 F; 70% in dipropylene glycol.

Texacat® ZF-24. [Texaco] General purpose catalyst for flexible and rigid PU foam, polyether slabstock; sp.gr. 0.98; b.p. 204 C; f.p. –40 C; flash pt. (TCC) 200 F; 23% in dipropylene glycol.

Texacat® ZF-26. [Texaco] General purpose catalyst for flexible and rigid PU foam, polyether slabstock; sp.gr. 1.0; b.p. 216 C; f.p. –30 C; flash pt. (TCC) 210 F; 11% in dipropylene glycol.

Texacat® ZF-51. [Texaco] Blend of Texacat TD-33A, ZF-22, and DME in dipropylene glycol; catalyst for flexible and rigid PU foam, polyether slabstock; sp.gr. 0.98; i.b.p. 152 C; f.p. –40 C; flash pt. (PMCC) 148 F.

Texacat® ZF-52. [Texaco] Blend of Texacat ZF-22, DME, and dipropylene glycol partially neutralized with formic acid; catalyst for polyether slabstock flexible foam; sp.gr. 1.0; i.b.p. 131 C; f.p. –38 C; flash pt. (PMCC) 166 F.

Texacat® ZF-53. [Texaco] Blend of Texacat TD-33A and ZF-22; catalyst for polyether slabstock flexible foam; sp.gr. 1.0; f.p. –39 C; flash pt. (PMCC) 178 F.

Texacat® ZF-54. [Texaco] Texacat ZF-22 partially neutralized with formic acid; catalyst for molded high resilience flexible PU foam; sp.gr. 1.1; b.. 127 C; f.p. –38 C; flash pt. (PMCC) 164 F.

Texacat® ZR-50. [Texaco] N,N-bis-(3-dimethylaminopropyl)-N-isopropanolamine; catalyst for flexible and rigid PU foam, pkg. foam; sp.gr. 0.89 (20/4 C); b.p. 290 C; flash pt. (PMCC) 285 F.

Texacat® ZR-70. [Texaco] 2-(2-Dimethylaminoethoxy) ethanol; catalyst for flexible PU foam, polyether slabstock, pkg. foam; sp.gr. 0.95; b.p. 201 C; f.p. < –50 C; flash pt. (TCC) 199 F.

Texamin AT 1, AT 2, AT 3. [Henkel KGaA] Alkylaminotriazoles in liq. carrier system; corrosion inhibitor for nonferrous metals; liq.

Texamine 84(L). [Zohar] Alkanolamide and ethanolamine alkyl benzene sulfonate; anionic/nonionic; raw material for mfg. of liq. detergents; paste; 84% conc.

Texanol® Ester-Alcohol. [Eastman] 2,2,4-Trimethyl-1,3-pentanediol monoisobutyrate; slow-evaporating solv. used as coalescing agent in latex finishes and water-base ink formulations, PVAc homopolymer and copolymer latices, PVAc-acrylic copolymer latices, and acrylic, EVA, and B/S latices; used as solv. in polyester coil coatings, electrodeposition coatings, disp. lacquer coatings, and printing inks; defoamer in waterborne systems and drilling muds for oil industry; chemical intermediate for prod. of esters. used as stain-resistance plasticizers, lubricant base stocks, solvs., syn. detergents, and herbicides; Pt-Co 20 max. clear liq.; mild char. odor; m.w. 216.3; f.p. –50 C; b.p. 244 C min. (initial, 760 mm); water-insol.; sp.gr. 0.950; dens. 0.95 kg/l; pour pt. –57 C; ref. index 1.4423; flash pt. 120 C (COC); fire pt. 132 C; > 99% volatiles by vol.

Texaphor. [Henkel Canada] Alkyl sulfate; suspending agent for nonpolar solv. paints.

Texaphor 277. [Henkel Canada] Quat. ammonium salt; suspending agent for paints with high chemical and water resistance requirements.

Texaphor Special. [Henkel Canada] Sulfosuccinic acid ester; suspending agent for polar and nonpolar solvs.

Texapon EA-1, -2, -3. [Henkel] Ammonium laureth sulfate; surfactant, visc. builder, solubilizer in personal care prods.; water-wh. visc. liq.; 25–27, 24–26, and 26–28% act. resp.

Texapon ES-1, -2, -3. [Henkel] Sodium laureth sulfate; surfactant and solubilizer for personal care prods.; water-wh. visc. liq.; 24–26, 25–26, and 27.5–29.5% act. resp.

Texapon K-12. [Henkel] Sodium lauryl sulfate; anionic; wetting agent, dispersant and foamer used in personal care prods., dentifrice mfg.; wh. fine powd.; pH 7.5–9.0 (1% aq.); 90% min. act.

Texapon L-100. [Henkel] Sodium lauryl sulfate; wetting and cleansing agent, dispersant used in dentifrice mfg.; wh. powd.; pH 7.0 (1% aq.); 99% act.

Texapon Z Granules. [Henkel] Sodium lauryl sulfate C_{12}-C_{14}; anionic; detergent, wetting agent, dispersant for personal care prods.; gran.; 50% conc.

Texicote 03-060. [Scott Bader] Aq. emulsion of self-crosslinking acrylic/vinyl acetate copolymer; emulsion used as binder for nonwoven fabrics, esp. as component of print bonding adhesives; used as paper impregnant, esp. for wallpaper bases to improve strippability; emulsion; 0.25 μ particle size; sp.gr. 1.08; visc. 3–6 poise; pH 4.0–5.5; 50 ± 1% solids.

Texicote 53-010. [Scott Bader] Styrene homopolymer emulsion; emulsion used as modifier for S/B latices; used to stiffen handle of carpet backing and fabric stiffening compds.; below its high min. film formation temp., dries to wh. powd., but film-forming plasticized versions also avail.; emulsion; 0.5 μ particle size; sp.gr. 1.03; visc. 0.5–2.5 poise; pH 8–9; 60 ± 1% solids.

Texicryl 13-002. [Scott Bader] Acrylic/methacrylic copolymer emulsion; emulsion for surface coating applics., e.g., in self-textured finishes providing good color, stain resistance, and good flexibility, in high-build one-coat finishes, in acrylic wood primers, in paints with exterior durability; added to cement/sand screeds to improve intercoat adhesion, chemical and abrasion resistance; as binder for pigmented leatherboard coatings and fabric backings having wet abrasion resistance and flexibility; emulsion; 0.25 μ particle size; sp.gr. 1.09; visc. 0.75–1.75 poise; pH 9.0–9.5; 55 ± 1% solids.

Texicryl 13-011. [Scott Bader] Acrylic copolymer emulsion; aq. emulsion of alkali-sol. acrylic copolymer used in water-based printing inks, in the mfg. of high gloss and other types of emulsion paint, and as a binder for nonwoven fabrics; emulsion; 0.2 μ particle size; sp.gr. 1.07; visc. 0.05–0.3 poise; pH 2.0–4.0; 41 ± 1% solids.

Texicryl 13-021. [Scott Bader] Self-crosslinking styrene/acrylate copolymer emulsion; emulsion, dries to mod. stiff film; used as binder for nonwoven floor coverings offering good water resistance and heat and lt. stability; good adhesion to many natural and syn. fibers; emulsion; 0.25 μ particle size; sp.gr. 1.05; visc. 0.25–1.25 poise; pH 5.5–6.5; 55 ± 1% solids.

Texicryl 13-022. [Scott Bader] Aq. emulsion of a reactive styrene/acrylic copolymer; emulsion whose particles have high affinity for fillers and extenders, esp. acidic particles such as china clay; increases filler content in paint formulations, etc., while retaining such properties as wet abrasion resistance; upgrades performance; used with starch as clay-coating binder and as binder for wallpaper base-coats to give grounds with wet rub-fastness; good adhesion to many difficult substrates; base for adhesives; emulsion; 0.20 μ particle size; sp.gr. 1.04; visc. 0.5–1.5 poise; pH 9.0–9.5; 50 ± 1% solids.

Texicryl 13-030. [Scott Bader] Styrene acrylic copolymer emulsion; emulsion used as quality binder for water-based paints (varying from gloss finish to high pigment volume concentration low-cost paints), crack fillers, textured coatings, and ceramic tile adhesives; emulsion; 0.2 μ particle size; sp.gr. 1.03; visc. 3.0–5.0 poise; pH 8.0–9.0; 50 ± 1% solids.

Texicryl 13-100, 13-101. [Scott Bader] Acrylic copolymer emulsions containing reactive carboxylic groups; low visc. reactive emulsions with high pigment binding power; form films with adhesion to natural and syn. fibers; 13-101 produces more flexible films than 13-100; 13-100, blended with melamine formaldehyde resin in the ratio of 100:15 used in bonding of polyester high loft waddings and offers wash and dry clean solv. resistance; binder for china clay and calcium carbonate fillers; good compatibility with starch and casein and binder for paper clay coatings which have good lt. stability and low residual odor; 13-101 adheres to textile fibers and confers a soft handle to nonwovens when used as binder; in conjunction with melamine formaldehyde or substituted urea formaldehyde, used in formulation of adhesives for flock coated and printed fabrics; in backing compds. for moquettes, and textile laminating adhesive; used alone as flock adhesive for styrene and ABS thermo-moldable sheet; emulsion; 0.2 μ particle size; sp.gr. 1.07; visc. 0.15–0.75 poise; pH 6.0–7.0; 46 ± 1% solids.

Texicryl 13-203. [Scott Bader] Self-crosslinking acrylic copolymer emulsion; emulsion used as binder in the mfg. of resilient nonwoven fabrics, (dry-laid mfg. of nonwoven fabrics and interlinings); confers a very resilient and lively handle, does not embrittle with age, and has heat and lt. stability; for heat molding processes; used for prod. of wet-laid nonwovens where binder is required to be precipitated at the beater; emulsion; 0.2 μ particle size; sp.gr. 1.04; visc. 0.1–0.4 poise; pH 4–6; 40 ± 1% solids.

Texicryl 13-205. [Scott Bader] Self-crosslinking acrylic copolymer emulsion; emulsion used as binder in the mfg. of spray-bonded syn. fiber waddings; prevents penetration of fibers thru the cover fabric, preserves the high loft of the wadding, and facilitates handling during the making-up process; heat- and lt.-stability, and good wash- and dry-clean solv. resistance; as a high-loft, nonwoven fabric binder; emulsion; 0.2 μ particle size; sp.gr. 1.08; visc. 0.2–0.8 poise; pH 6.0–7.0; 46 ± 1% solids.

Texicryl 13-206. [Scott Bader] Self-crosslinking acrylic copolymer emulsion; emulsion producing dried film, very flexible and when cured is wash-and dry-clean resistant; used as bonding agent for nonwoven fabrics and as component in fabric backings; compatible with Texicryl 13-205; adhesive for bonding precision-cut syn. fiber flock; good heat and lt. stability; emulsion; 0.2 μ particle size; sp.gr. 1.06; visc. 0.5–1.5 poise; pH 6.0–7.0; 50 ± 1% solids.

Texicryl 13-210. [Scott Bader] Self-crosslinking acrylic copolymer emulsion; soft emulsion for the formulation of flocking adhesives, textile laminating adhesives, and used as binder for nonwoven fabrics; adheres to natural and syn. fibers and resistant to washing and dry-cleaning solvs. when cured; emulsion; 0.2 μ particle size; sp.gr. 1.03; visc. 0.2–0.4 poise; pH 6.0–7.0; 45 ± 1% solids.

Texicryl 13-300, 13-302. [Scott Bader] Carboxylated acrylic copolymer aq. emulsions; emulsions used as thickening agents for natural and syn. latices; low visc., high m.w., acidic; 13-302 used where increase in visc. of the water-phase is important in thickening mechanicism, e.g., thickening of emulsion paints;

useful in formulating chemical resistant, high PVC sheen, and semigloss emulsion paints; 13-300 is more effective in bridging across latex particles; for thickening most syn. rubber latices and their compds.; both suitable in thickening acrylic emulsions used as print bonding adhesives in the prod. of nonwoven fabrics; emulsion; 0.2 μ particle size; sp.gr. 1.08; visc. 0.05–0.3 poise; pH 2–4; 40 and 35 ± 1% solids resp.

Texigel 23-002, 23-004, 23-012. [Scott Bader] Aq. polyacrylate gels; thickening agents for natural and syn. latices, e.g., in carpet-backing and similar compds.; pale straw gels; ammoniacal odor (23-002 and 23-004); mildly ammoniacal odor (23-012); sp.gr. 1.06, 1.06, and 1.08 resp.; visc. 300–600, 50–100, and 5–9 poise resp.; pH 10–11.5, 10–11.5, and 7–8 resp.; 13–17% solids (23-002 and 23-004); 16–20% solids (23-012).

Texigel 23-005, 23-018. [Scott Bader] Aq. polyacrylate gels; ammonia-free thickening agents for natural and syn. latices, e.g., in carpet-backing compds.; gels; nonammoniacal odor; sp.gr. 1.06 and 1.07 resp.; visc. 350–850 and 200–400 poise resp.; pH 8–9 and 9–10.5 resp.; 13–17 and 11–14% solids resp.

Texnol IL. [Nippon Nyukazai] Alkyl imidazoline laurate; nonionic; antistat; solid; 100% act.

Texnol R 5. [Nippon Nyukazai] Alkyl benzyl ammonium salt; anionic; emulsifier, fungicide, softener; liq.; 100% act.

Texo LP 528A. [Texo] Ethoxylated alkylaryl; nonionic; mold release agent; liq.; 40% conc.

Texox PPG-400. [Texaco] PPG-400; intermediate yielding esters; useful as lubricants, defoaming agents in rubber and pharmaceuticals, solvs. and humectant modifiers for inks, plasticizers, and functional fluids; water-wh. visc. liq.; faint ether-like odor; m.w. 400; dens. 8.40 lb/gal; visc. 150–175 SUS (100 F); flash pt. 320 F (PMCC); pour pt. –35 F; pH 5–7.

Texox PPG-2000. [Texaco] PPG-2000; see Texox PPG-400; water-wh. visc. liq.; faint ether-like odor; m.w. 2000; dens. 8.37 lb/gal; flash pt. 370 F (PMCC); pour pt. –25 F; pH 5–7.

Texox WL-440. [Texaco] EO/PO derivs.; functional fluids used as lubricants, plasticizers, solvs., coupling agents, frothing agents, heat transfer fluids, intermediates, defoaming agents; used in solder reflow applics., boiler defoaming, ore flotation, inks, dyes; clear liq.; m.w. 488; water-sol.; sp.gr. 1.1515; dens. 9.55 lb/gal; visc. 100 cSt (100 F); flash pt. 580 F (COC); pour pt. –30 F; ref. index 1.4721.

Texox WL-660. [Texaco] EO/PO deriv.; see Texox WL-440; clear liq.; m.w. 1800; water-sol.; sp.gr. 1.0502; dens. 8.75 lb/gal; visc. 143 cSt (100 F); cloud pt. 60 C (1% aq.); flash pt. 475 F (COC); pour pt. –45 F; ref. index 1.4586.

Texox WL-1400. [Texaco] EO/PO deriv.; see Texox WL-440; clear liq.; m.w. 2500; water-sol.; sp.gr. 1.095; dens. 9.15 lb/gal; visc. 306 cSt (100 F); cloud pt. > 99 C (1% aq.); flash pt. 475 F (COC); pour pt. 50 F; ref. index 1.4669.

Texox WL-5000. [Texaco] EO/PO deriv.; see Texox WL-440; clear liq.; m.w. 4365; water-sol.; sp.gr. 1.0557; dens. 8.80 lb/gal; visc. 1080 cSt (100 F); cloud pt. 53 C (1% aq.); flash pt. 460 F (COC); pour pt. –10 F; ref. index 1.4600.

Texsolve B. [Texaco] Highly refined commercial hexane; solv. used for veg. oil and pharmaceutical extraction, compding rubber cements and sealants, and polyolefin prod.; b.p. 4 F; flash pt. –25 F.

Texsolve C. [Texaco] Mixed heptane fraction; solv. used for compding. rubber cements and sealants, extraction of oils and fats, and recrystallization of org. chemicals; flash pt. 13 F (TCC).

Texsolve H. [Texaco] Hexane-heptane fraction; solv. used in rubber tire mfg., rubber cements, sealants, inks, lacquers, and adhesives; flash pt. –20 F.

Texsolve S. [Texaco] Min. spirits; solv. used in paint and protective coatings, dry cleaning, degreasing, wood treating, and charcoal lighter fluid; meets or exceeds standards for min. spirits and Stod.; flash pt. 107 F.

Texsolve S-2. [Texaco] Min. spirits; solv. used in paints, protective coatings, waxes, polishes, cloth impregnation, dry cleaning, and quick-drying applics.; flash pt. 102 F.

Texsolve S-66. [Texaco] Exempt min. spirits with max. aromatic content; solv. used in paint and protective coatings, dry cleaning, degreasing, wood treating, charcoal lighter fluid; meets or exceeds standards for min. spirits and Stod.; flash pt. 107 F.

Texsolve S-LO. [Texaco] Low aromatic min. spirits; solv. used in charcoal lighter fluid, floor finishes, paints, coatings, hand cleaners; low odor; flash pt. 108 F.

Texsolve V. [Texaco] VM&P naphtha; solv. used in paints, coatings, rubber compding., sealants, and chemical absorption; pleasant odor; flash pt. 58 F.

Textamine 05. [Henkel] Fatty imidazoline; cationic; oil-emulsifying agent, corrosion inhibitor and ink dispersant; liq.; 100% act.

Textamine 1839. [Henkel] Fatty-amido tert. amine; cationic; foamer, thickener, ore flotation agent and starting material for quats., betaines, biocides and amine oxides; liq.; 100% act.

Textamine A-5-D. [Henkel] Fatty imidazoline; cationic; corrosion inhibitor, antistripping agent, emulsion breaker, dispersant, and raw material for quat.; liq.; 100% act.

Textamine A-3411. [Henkel] Fatty imidazoline complex; cationic; corrosion inhibitor; liq.; 100% act.

Textamine A-3417. [Henkel] Fatty amido diamine; cationic; corrosion inhibitor, asphalt emulsifier; flotation, wetting, and antistripping agent; intermediate for spray wax emulsifier formulations for car washes; visc. liq.; dens. 7.8 lb/gal; 100% conc.

Textamine A-W-5. [Henkel] Fatty imidazoline; cationic; corrosion inhibitor, emulsion breaker, dispersant; liq.; oil-sol., water-disp.; 100% act.

Textamine O-1. [Henkel] 1-Hydroxyethyl-2-oleyl imidazoline; nonionic; emulsifier for hydrocarbons and nonpolar solvs., agric. sprays; dispersant; lubricant for metalworking compds.; fuel additive for sludge dispersions; used in fungicide, pesticide, and herbicide formulations; thickener and corrosion inhibitor for paints; used in industrial applics.; amber liq.; sol. in alcohol, chlorinated hydrocarbons, oils; disp. in water; dens. 7.66 lb/gal; pH 10.5 (10); 90% imidazoline.

Textamine O-5. [Henkel] Oleo-fatty imidazoline; cationic; corrosion inhibitor for gasoline and diesel fuels; emulsifier, emulsion breaker, dispersant, textile softener and leather working agent; liq.; oil-sol., water-disp.; 100% conc.

Textamine Oxide TA. [Henkel] Tallow amidoalkylamine; nonionic/cationic; foamer, conditioner, and thickener for use in household and industrial cleaners; liq.; 50% act.

Textamine Polymer. [Henkel] Complex resinous fatty amine; cationic; corrosion inhibitor, wetting agent, emulsion breaker, asphalt wetting agent; liq.; 100% act.

Textamine Polymer S. [Henkel] Fatty amido polyamine; cationic; corrosion inhibitor; liq.; 100% act.

Textamine T-1. [Henkel] 1-Hydroxyethyl-2-tall oil imidazoline; nonionic; see Textamine O-1; amber liq.; sol. in alcohol, chlorinated hydrocarbons, aliphatic, and aromatic hydrocarbons, oils; disp. in water; dens. 7.75 lb/gal; pH 11 (10%); 90% min. imidazoline.

Textamine T-5-D. [Henkel] Fatty imidazoline; cationic; emulsifier, oil-sol. corrosion inhibitor, dispersant; liq.; 100% act.

Textamine XO. [Henkel] Aliphatic and heterocyclic mono- and diamines; cationic; steam boiler corrosion and steam volatile inhibitor; liq.; 100% act.

Textamine XO Anhyd. [Henkel] Heterocyclic mono- and diamines; cationic; see Textamine XO; liq.; 100% act.

Textone®. [Olin] Sodium chlorite; non-EPA registered source of chlorine dioxide for oxidizing applics. incl. bleaching of natural foliage, upgrading of fats and oils, stripping dyestuffs from textiles, pulp bleaching, copper etching; wh. flakes; bulk dens. 53 lb/ft^3 (loose), 69 lb/ft^3 (packed); 78–82% sodium chlorite.

Tex-Wet 1002. [Intex] Phosphated alcohol; anionic; dispersant, wetting agent and penetrant for textile wet processing; dyeing assistant for woolen goods; lt. amber liq.; water-sol.; dens. 8.75 lb/gal; pH 7.0±0.5; 55% conc.

Tex-Wet 1004. [Intex] Alkylaryl sulfonate and sulfated nonionic; anionic; detergent, wetting and rewetting agent; wet processing, preparation and/or desizing assistant; dyeing assistant; dens. 8.2 lb/gal; pH 7–8; biodeg.; 35 ± 0.5% solids.

Tex-Wet 1010. [Intex] Long chain phosphated alcohol, modified; anionic; wetting agent, dyeing assistant, detergent, penetrant, boil-off assistant for textiles; liq.; 50% conc.

Tex-Wet 1034. [Intex] Phosphate ester; anionic; wetting agent, detergent, emulsifier, antistat; liq.; 70% conc.

Tex-Wet 1048. [Intex] POE linear alcohol; nonionic; wetting agent, penetrant in textile prods.; liq.; HLB 12.5; 20% conc.

Tex-Wet 1103. [Intex] Phosphated alcohol; anionic; intermediate for preparing penetrants; wetting agent for textile processing; liq.; 100% conc.

Tex-Wet 1131. [Intex] Coconut alkanolamide; nonionic; detergent, emulsifier, foam stabilizer; liq.; 100% conc.

Tex-Wet 1140. [Intex] Sodium xylene sulfonate; anionic; solubilizer, hydrotrope for alkylbenzene sulfonate sol'ns. and detergent formulations; liq.; 40% act.

Tex-Wet 1158. [Intex] Sodium alcohol ether sulfate; anionic; wetting agent, dispersant, emulsifier, detergent, dyeing assistant; fabric scouring agent; cotton processing; stabilizer for dye bath; dyeing of syns.; pale amber liq.; dens. 8.7 lb/gal; 57% min. act.

Tex-Wet 1197. [Intex] Dodecylbenzene sulfonic acid; anionic; detergent intermediate; emulsifier, coemulsifier used in textile dyeing and scouring; dish detergent and cleaning formulations; red-br. liq.; dens. 8.8 lb/gal; 97% act.

Thanecure®. [TSE] Zinc chloride with benzothiozyl disulfide; vulcanization activator for sulfur-curable millable urethane elastomer; cream colored powd.; amine odor; slightly sol. in water; mod. sol. in benzene; sp.gr. 1.85; m.p. 290 ; flash pt. (OC) > 204 C.

Thermact® 1. [Olin] Amine; catalyst developed to allow the use of systems based on Thermolin RF-230 in two-component processing for foam; very sol. in ethanol, acetone, benzene, chloroform; sol. in water; slightly sol. in dioctyl phthalate; insol. in ether; sp.gr. 1.096 (24 C); visc. 207 cp; hyd. no. 520; pH 7.2 (10% aq.); flash pt. (COC) 98–99 C.

Thermacure®. [Freeman] Dicumyl peroxide; vulcanizing agent or polymerizing catalyst in rubber and plastics industries, esp. for unsat. polyester resins, bulk polymerization of vinyl monomers such as styrene; wh. cryst. powd.; m.w. 270.37; sol. in org. solvs., syn. and natural rubbers, polyester resins; insol. in water; sp.gr. 1.012; flash pt. (Seta) 121 C; 99% min. assay; 5.87% act. oxygen.

Thermagel HN. [Polymer Research] Dye migration retardant compatible with most resin emulsions, gums, and resin sol'ns.; water sol'n. of Thermagel HN will remain liq. at R.T. but will solidify when heated to 160 F; when added to pigment paste or pigment emulsion, will prevent dye migration during the heating cycle; improves the hand of the fabric; paste; dissolves easily in cold water mixing.

Thermax® Floform N-990. [Cancarb] Med. thermal carbon blk.; used in rubber prods. with high loadings of carbon blk., in metallurgy, specialty and refractory applics.; soft pellets; mean particle diam. 270 nm; sp.gr. 1.8; pH 9–11; 99.5% min. carbon content.

Thermax® Floform Ultra Pure N-990. [Cancarb] Med. thermal carbon blk. with highest carbon content and extremely low in ash and sulfur; used where low impurities are required; pellets; mean particle diam. 270 nm; sp.gr. 1.8; pH 5–8; 99.6% min. carbon content.

Thermax® Powder N-991. [Cancarb] Med. thermal

carbon blk.; used where disp. and good mixing of dry powd. formulations are critical; also used for its insulting value in high-temp. applics.; powd.; mean particle diam. 270 nm; sp.gr. 1.8; pH 9–11; 99.5% min. carbon content.

Thermax® Powder Ultra Pure N-991. [Cancarb] Med. thermal carbon blk.; powd. equiv. to Ultra Pure N-990; used in metallurgy; powd.; mean particle diam. 270 nm; sp.gr. 1.8; pH 5–8; 99.6% min. carbon content.

Thermax® Stainless Floform N-907. [Cancarb] Med. thermal carbon blk.; nonstaining prod. used in seals and gaskets; soft pellets; mean particle diam. 240 nm; sp.gr. 1.8; pH 9–11; 99.6% min. carbon content.

Thermax® Stainless Powder N-908. [Cancarb] Med. thermal carbon blk.; used for nonstaining end prods. and for its insulating value; powd.; mean particle diam. 240 nm; sp.gr. 1.8; pH 9–11; 99.6% min. carbon content.

Thermax® Stainless Powder Ultra PureN-908. [Cancarb] Med. thermal carbon blk.; used for critical applics. where low oil and high purity are required; mean particle diam. 240 nm; sp.gr. 1.8; bulk dens. 25 lb/ft³; pH 4–8.

Therm-Chek 6-V-6A. [Ferro] Barium-cadmium-zinc lt. and heat stabilizer; primarily for use in plastisols and organisols; in rigidsol disps. containing < 30 phr of plasticizer; provides good clarity and early color retention, good air release and visc. control; recommended for use at 1.0–3.0 phr with 5.0 phr or more of epoxy plasticizer; lt. amber liq.; sp.gr. 1.04.

Therm-Chek 130. [Ferro] Barium-zinc lt. and heat stabilizer; for use in plastisols; useful in organosols and rigidsol disp. containing < 30 phr plasticizer; good air release and visc. control; suggested for use at 1.0-3.0 phr with 5.0 phr or more of epoxy plasticizer for good early color and long-term stability; lt amber liq.; dens. 8.8 lb.gal; visc. (Gardner) A2.

Therm-Chek 344. [Ferro] Calcium-zinc stabilizer; see Therm-Chek 130; lt. yel. liq.; dens. 8.5 lb/gal; visc. (Gardner) A1.

Therm-Chek 707-X. [Ferro] Zinc, epoxy stabilizer; nontoxic stabilizer which produces very clear vinyl prods. having good initial color compared with other nontoxic systems; used at 1.0–3.0 phr with 5–15 phr of an FDA approved epoxy plasticizer; wh. to lt. yel paste.; sp.gr. 0.99.

Therm-Chek 714. [Ferro] Calcium-zinc stabilizer; high efficiency, low odor stabilizer for all types of nontoxic vinyl formulations incl. rigids, but works best in calendered and extruded prods.; recommended for use at 1.0–3.0 phr for all applics.; for optimum stabilization of all plasticized formulations, the addition of 5–10 phr of an epoxy plasticizer is suggested; wh. powd.; dens. (bulk) 2.3 lb/gal.

Therm-Chek 837. [Ferro] Dibutyltin maleate ester; nonsulfur-type heat stabilizer for rigid vinyl resin compds.; provides lt. stability; low level usage will optimize impact properties of a rigid vinyl; will not cross stain during compding. in the presence of lead, cadmium, or other metals that form colored sulfides; does not water blush; suitable for flexible vinyl formulations; for use at 1.0–2.0 phr for rigid extrusions and 0.5–1.0 phr for flexible vinyl systems with 5 phr epoxy plasticizer; Gardner 4 clear pale liq.; dens. 11.2 lb/gal; visc. (Gardner) G; 23.2% tin content.

Therm-Chek 840. [Ferro] Dibutyltin bis isooctyl thioglycolate; heat stabilizer for clear and pigmented rigid vinyl resin extrusions; used for rigid PVC pipe and siding, sheet, bottles, and profiles; suitable for flexible vinyl formulations; combination with epoxy compds. show synergistic heat stability; outstanding processability with no adverse effects of fusion chars.; nonsulfur staining, nonlubricating; will cross stain during compding. in the presence of lead, cadmium, or other metals that form colored sulfides; recommended for use at 2.0–3.0 phr for rigid extrusions and 0.5–1.5 phr with 5 phr epoxy plasticizer for flexible film and sheet; Gardner 3 clear pale liq.; dens. 9.4 lb/gal; visc. (Gardner) A; 18.0% tin; 9.6% sulfur.

Therm-Chek 904. [Ferro] Org. inhibitor; stabilizer; produces a high degree of clarity when used with Therm-Chek 1212A and 1827; improves initial color and heat stability in both clear and pigmented stocks; can be used effectively in formulations containing phosphate plasticizers; recommended at 0.5–1.0 phr with 2.0–3.0 phr Therm-Check 1827 for rigids; 0.5 phr with 1.0–3.0 phr Therm-Check 1212A, 1820, 1840, or 1827 and with 1.0 phr phosphate plasticizer for calendered, molded, and extruded stocks; 0.5 phr with 2.0–4.0 phr Therm-Chek 1212A, 1237, or 1776 and with 1.0 phr phosphate plasticizer for plastisols; wh. to pale yel. liq.; sp.gr. 1.03.

Therm-Chek 1825. [Ferro] Mixed cadmium-barium fatty acid soap with org. inhibitor; stabilizer for rigid formulations and calendered, molded, and extruded plasticized compositions; useful in clear, pigmented, and filled stocks; provides color in calendered expanded vinyls; offers long-term heat stabilization at moderate conc. levels, early color retention in pigmented stocks, compatibility; self-lubricating; slows dissipation of blowing agent in foams; lt. cream powd.; sp.gr. 1.21.

Therm-Chek 5469. [Ferro] Barium-cadmium stabilizer with org. inhibitor; high potency, highly compatible stabilizer for general calendering, extruding, fluid bed, and plastisol applics. requiring max. clarity; produces bright, true colors in pigmented applics.; nonlubricating; recommended in areas where roll or mold plate-out is critical, for use with stearic acid to obtain optimum heat stability and to add lubrication; may be used with an epoxy plasticizer as synergist; recommended for use at 2.0–3.0 phr with 0.25–0.50 phr stearic acid for max. clarity or color retention and at 2.0–4.0 phr for high-clarity plastisols; lt. amber liq.; sp.gr. 1.07.

Thermoguard 210. [M&T] Brominated polymeric epoxy resin; flame retardant additive for various thermosets such as unsat. polyester, phenolics, polyamides, and epoxy resins; lt. straw solid; soften. pt. 60–70 C; epoxy equiv. wt. 510–580 g/eq; 48–50%

bromine.

Thermoguard 212. [M&T] Proprietary formulation based on brominated epoxy resin, halogenated organic phosphates and phosphonates, and a small amt. of styrene; flame retardant additive for thermosets, esp. transparent unsat. polyesters; lt. straw clear liq.; dens. 1.56 g/cm³; visc. 4750 cps; 32.5% bromine.

Thermoguard 213. [M&T] Proprietary formulation based on brominated epoxy resin, antimony oxide, and styrene; efficient flame retardant additive for use in unsat. polyester resins; wh. opaque liq. (stable suspension); dens. 1.7 g/cm³; visc. 3500 cps; 30% bromine; 20% antimony oxide.

Thermoguard 215. [M&T] Brominated epoxy resin and antimony oxide; flame retardant additive for unsat. polyesters, epoxy, phenolic resins, PU, polyamides, and other specialty applics.; wh. powd.; bulk dens. 0.8 g/cc; epoxy equiv. 680 ± 35 g/eq.; 40% bromine; 20% antimony oxide.

Thermoguard 220. [M&T] Brominated polymeric epoxy resin; low m.w. flame retardant additive which can be blended with other resins (solids or liqs.) for use in potting, wet lay-up, pre-preg laminates, adhesives, molding compds., and coatings; Gardner 12 max. semisolid; soften. pt. 51–62 C; epoxy equiv. wt. 350–400 g/eq; 45–48% bromine.

Thermoguard 230. [M&T] Brominated polymeric epoxy resin; flame retardant additive for resin systems esp. nylon; off-wh. powd.; soften. pt. 90–110 C; epoxy equiv. wt. 1700–2100 g/eq; 49–51% bromine.

Thermoguard 240. [M&T] Brominated polymeric epoxy resin; flame retardant additive for thermosets and thermoplastics esp. in engineering resins such as polyester and polyamides; off-wh. solid; soften. pt. 186 C; epoxy equiv. 20,000 g/eq.; 49–51% bromine.

Thermoguard CPA. [M&T] Antimony; flame retardant for use as replacement for antimony oxide in many formulated plastics esp. in flexible PVC applics. such as wire and cable insulation and jacketing; wh. powd.; 99.8% thru 325 mesh (wet sieve); sp.gr. 5.0; bulk dens. 0.5–0.7 g/cc.

Thermoguard FR. [M&T] Antimony-containing compd.; flame retardant for vinyls and other plastics; wh. fine powd.; 99.7% thru 200 mesh (wet sieve); sol. in conc. H_2SO_4, 20% HCl and ammonium salts; insol. in water, acetic acid, and tartaric acid; sp.gr. 4.80; 61.7% antimony.

Thermoguard L. [M&T] Antimony oxide; flame retardant for use with a halogen-containing compd.; flame retardant pigment for PVC; also for use with chlorinated organics for producing flame-retardant polyesters and polyethylene compds.; wh. fine powd.; 3.0 μ avg. particle size; 99.9% thru 325 mesh (wet sieve); insol. in common org. solvs.; very slightly sol. in water, aq. HCl, potassium hydroxide, and tartaric acid; sp.gr. 5.7; m.p. 656 C; ref. index 2.087; 83.0% antimony; 99.5% antimony oxide.

Thermoguard S. [M&T] Antimony oxide; flame retardant for use in plastics, paper, textiles, and paints; flame retardant pigment for PVC; also for use with chlorinated organics for producing flame-retardant polyesters, polyethylene compds., mfg. of flame-retardant films, sheets, textiles, paper, paints; must be used with a halogen-containing compd. for flame retardant effect; wh. superfine powd.; 1.3 μ avg. particle size; 99.9% thru 325 mesh (wet sieve); insol. in common org. solvs.; very slightly sol. in water; sol. in aq. HCl, potassium hydroxide, and tartaric acid; sp.gr. 5.7; m.p. 656 C; ref. index 2.087; 83.0% antimony; 99.5% antimony oxide.

Thermolin® 101. [Olin] Tetrakis (2-chloroethyl) ethylene diphosphate; nonreactive flame retardant additive for use in flexible PU foams for the transportation, bedding, and furniture industries, and in thermoplastic and thermoset resins such as acrylates, polyolefins, PAN, styrene, ABS, polyesters, epoxies, and PET; sol. in many common org. solvs., incl. simple alcohols, ketones, ethers, esters, and aromatic hydrocarbons; immiscible and virtually insol. in dry cleaning solvs.; sol. in TDI and in most polyether polyols for use in urethane foams; sp.gr. 1.45; dens. 12.1 lb/gal; visc. 260 cp; pour pt. –62 C; flash pt. 142 C; 30% chlorine.

Thermolin® RF-230. [Olin] Highly chlorinated polyol preblended with fluorocarbon-11; flame retardant for rigid foam boardstock in the construction industry; dk. amber; sp.gr. 1.5; dens. 12.5 lb/gal; visc. 20,000– 22,0-00 cps; acid no. 1.5 max.; hyd. no. 340.0 ± 10; pH 4.5–7.0 (in 10/6 IPA/water); flash pt. 145 C; 93% act. polyol.

Thiate® E. [Vanderbilt] Trimethylthiourea; accelerator for CR; used in press cured inj. molding and LCM extrusion stocks; up to 1.5 parts gives low compr. set and good scorch resistance; lt. tan flakes; m.w. 118.20; sol. in water and toluene; dens. 1.23 ± 0.03 mg/m³; m.p. 68–78 C.

Thiate® EF-2. [Vanderbilt] Trimethylthiourea; accelerator for CR; used in press cures, inj. molding, LCM extrusion stocks, sponge; wh. cryst. powd.; m.w. 118.20; mod. sol. in toluene, acetone, chloroform; insol. in water; dens. 1.23 ± 0.03 mg/m³; m.p. 80–90 C.

Thiate® H. [Vanderbilt] 1,3-Diethylthiourea; accelerator for CR, EPDM, and chlorobutyl; for fast initial set-up, continuous cured sponge, LCM extrusions, and hot air cures; wh. to lt. yel. flakes; m.w. 132.33; sol. in methanol, chloroform, and acetone; practically insol. in water; dens. 1.11 ± 0.03 mg/m³; m.p. 68–77 C.

Thiate® U. [Vanderbilt] 1,3-Dibutylthiourea; accelerator with same applics. as Thiate H but with slightly slower set-up and better scorch; wh. to lt. tan course powd.; m.w. 188.34; sol. in methanol, chloroform, acetone, toluene; practically insol. in water; dens. 1.03 ± 0.03 mg/m³; m.p. 56–65 C.

Thickener #5004 (All Aq.). [Polymer Research] Carboxyl-containing emulsion; thickener which reacts with neutralized base and becomes very visc.; high efficient thickener for oil-in-water and all aq. systems when maintained on alkaline side of the pH range of 8.0–9.0; can be incorporated in latex, disps., and emulsions in the water media; acts as emulsifier in water systems; milky liq. emulsion; sp.gr. 1.04; dens. 8.69 lb/gal; visc. 30.0 cps; pH 3.0; 34.0% total solids.

Thickener L. [GAF] Modified form of poly (methyl vinyl ether/maleic anhydride); polymer compatible with most latices without adversely affecting dyeing and aging properties for films; forms smooth, free-flowing latices to facilitate uniform applic.; versatile thickener for natural or syn. latices used in upholstery and rug backings; for other latex and emulsion systems, e.g., adhesives, printing pastes, and paints; liq.; pH 9.0 ± 0.5; 15% active.

Thickener LN. [GAF] Modified form of poly (methyl vinyl ether/maleic anhydride); polymer developed specifically as thickening agent for water-based PVC, acrylic, and PVAc emulsion paints; for thickening other emulsion and latex systems as adhesives and water-phase print pastes; as additive for flow-out and leveling improvements of emulsion flat, semi-gloss, gloss, interior and exterior paints; liq.; pH 9.0 ± 0.5; 15% active.

Thimet® 15-G, 20-G. [Am. Cyanamid/Ag.] Phorate; soil and systemic insecticide for veg., field, and forage crops; gran.; restricted use; 15.0 and 20% act. resp.

Thiofide®. [Monsanto] 2,2′-Dithiobis (benzothiazole); primary accelerator for sulfur-curable elastomers; nonstaining, nondiscoloring in wh. stocks; plasticizer/retarder in CR; 2 mm pellets or dust-suppressed powd.

Thiostat® B. [Uniroyal] Sodium dimethyl dithiocarbamate aq. sol'n.; bactericide and fungicide for industrial use; slimicide in the mfg. of paper and paperboard; yel. to amber clear liq.; sol. in water and polar solvents; sp.gr. 1.18; pH 9–12; 40% act.

Thiotax®. [Monsanto] 2-Mercaptobenzothiazole; primary accelerator for sulfur-curable elastomers, e.g., EPDM, IIR, low-temp. NR cures; powd. or dust-suppressed powd.

Thiovanol®. [Evans Chemetics] Thioglycerin; stabilizer for acrylonitrile polymers; cross-linking agent for hard high-gloss coatings; accelerator for epoxy-amine condensation reactions; reducing agent; used in hair waving and straightening, hair dyes, depilatories, textiles, furs, pharmaceuticals, surfactants, foam stabilizing additives for detergents, shampoos, insecticides, pesticides, fungicides, and dessicants; practically clear and colorless sol'n.; mild char. odor; m.w. 108.2; b.p. 118 C (5 mm, anhyd.); misc. in all proportions with water and alcohol; insol. in ether; 90% aq. sol'n.

Thixatrol 1. [Rheox] Rheological additive in 51% min. spirits; translucent off-wh. paste; dens. 6.7 lb/gal; flash pt. 105 F; 75% volatiles.

Thixatrol 53MS. [Rheox] Rheological additive in 58% min. spirits; translucent off-wh. paste; dens. 7.0 lb/gal; flash pt. 105 F; 75% volatiles.

Thixatrol 289. [Rheox] Rheological additive in 50–55% methyl isoamyl ketone; amber liq.; dens. 7.85 lb/gal; b.p. 293 F; flash pt. (Seta) 97 F.

Thixatrol GST. [Rheox] Castor oil inorganically modified deriv.; thixotrope, gellant for solv.-based coatings; used for aliphatic trade sales systems for processing temps. within 57–74 C; aromatic industrial systems for processing temps. within 33–49 C; off-wh. fine powd.; sp.gr. 1.46; dens. 12.16 lb/gal; 100% NV.

Thixatrol SR. [Rheox] Org. thixotrope in 35% petrol. solv. and 35% cyclohexanol; amber liq.; sp.gr. 0.922; dens. 7.70 lb/gal; flash pt. (Seta CC) 121 F.

Thixatrol SR100. [Rheox] Org. thixotrope; off-wh. powd.; sp.gr. 1.0.

Thixatrol ST. [Rheox] Castor oil org. deriv.; see Thixatrol GST; wh. fine powd.; sp.gr. 1.02; dens. 8.50 lb/gal; 100% NV.

Thixcin® E. [Rheox] Glyceryl tris-12-hydroxystearate; stabilizer, thickener, thixotrope, suspending agent for many anhyd. systems; wh. powd.; dens. 8.51 lb/gal.

Thixcin® GR. [Rheox] Castor oil inorganically modified deriv.; thixotrope, gellant for solv.-based coatings; for aliphatic solv. systems where processing temps. fall within 35–54 C; off-wh. fine powd.; sp.gr. 1.477; 100% NV.

Thixcin® R. [Rheox] Trihydroxystearin; see Thixcin GR; wh. fine powd.; sp.gr. 1.02; dens. 8.51 lb/gal; 100% NV.

Thixon® 753. [Whittaker] Vulcanizable bonding agent; one-coat NBR-to-metal water-based adhesive; blk.; sp.gr. 1.09; dens. 9.1 lb/gal; visc. (#1 Zahn) 31 s; flash pt. (Seta) none; 23% NV solids.

Thixseal 1084. [Rheox] Castor oil inorganically modified deriv.; thixotrope, gellant for solv.-based systems; sag and slump control agent for caulks, sealants, and mastics where processing temps. fall within 60–71 C; creamy wh. powd.; sp.gr. 1.36; 100% NV.

Thornel Carbon Fiber T-40 12K. [Union Carbide] Continuous length, high str., high modulus fiber consisting of 12,000 filaments in a one-ply construction; filament diam. 6 μm; dens. 1.81 mg/m^3; surf. area 1 m^2/g; tens. str. 5.65 GPa; tens. elong. 2% (break); 92% carbon assay.

Thornel Carbon Fiber T-500 3K. [Union Carbide] Continuous length, high str., high modulus fiber consisting of 3000 filaments in a one-ply construction; filament diam. 7 μm; dens. 1.79 mg/m^3; surf. area 1 m^2/g; tens. str. 3.65 GPa; tens. elong. 1.5% (break); 92% carbon assay.

Thornel Carbon Fiber T-500 6K. [Union Carbide] Continuous length, high str., high modulus fiber consisting of 6000 filaments in a one-ply construction; filament diam. 7 μm; dens. 1.79 mg/m^3; surf. area 1 m^2/g; tens. str. 3.65 GPa; tens. elong. 1.5% (break); 92% carbon assay.

Thornel Carbon Fiber T-500 12K. [Union Carbide] Continuous length, high str., high modulus fiber consisting of 12,000 filaments in a one-ply construction; filament diam. 7 μm; dens. 1.79 mg/m^3; surf. area 1 m^2/g; tens. str. 3.86 GPa; tens. elong. 1.6% (break); 92% carbon assay.

THQ. [Eastman] Toluhydroquinone; antioxidant for unsat. polyesters.

T-Hydro® Sol'n. [Arco] t-Butyl hydroperoxide; free radical polymerization initiator; co-initiator in polyethylene processes; finishing catalyst in emulsion polymerizations and unsat. polyester resin thermosets; intermediate for peroxygen derivs.; epoxidation reagent; oxidizing reagent; source of act. oxygen for pulp and paper bleaching, metal sulfide oxidations, etc.; clear sol'n., pungent odor; sp.gr. 0.935; visc. 4.1 cps; b.p. 96 C (760 mm Hg); f.p. –2.8 C; flash pt. (TCC) 43 C; ref. index 1.3246; pH 4.3; 69.5% aq. sol'n., 12% act. oxygen.

Ti-0720 T 1/8′′. [M&T Harshaw] Titania; catalyst; tablet, 1/8′′ diam.; dens. 72 lb/ft^3; 100% conc.

Timica. [Mearl] Titanium dioxide/mica; for frosted and iridescent effects in lipsticks, cream makeups, nail enamels, pressed powds.; wh. and colored powds.

Timiron®. [Rona] Titanium dioxide-coated mica; pearlescent for lipsticks, pressed powds., eye makeups, for all frosted effects.

Tinegal® KD. [Ciba-Geigy] EO condensate, modified; amphoteric; dyeing and printing aux. for nylon; liq.

Tinegal® MR-50. [Ciba-Geigy] Quat. ammonium salt; cationic; leveling agent for fibers; liq.

Tinegal® NA. [Ciba-Geigy] Polyoxethylated fatty alcohol; nonionic; dyebath stabilizer, surfactant, emulsifier; liq.

Tinegal® PM. [Ciba-Geigy] Alkylamine polyglycol ether; cationic; leveling agent for nylon; liq.

Tinopal® 5BM Extra Conc. [Ciba-Geigy] Bismethylethanolamino dianilino stilbene disulfonate deriv.; moderately sol. whitener for cellulosics; effective at low wash temps.; used in powd. and liq., anionic and nonionic detergents.

Tinopal® 5BM-GX. [Ciba-Geigy] Stilbene type; whitening agent for powd. and liq. anionic and nonionic detergent; mod. water-sol.

Tinopal® AMS, AMS Pure. [Ciba-Geigy] Dimorpholino dianilino stilbene disulfonate deriv.; slightly sol. whitener for cellulosics; used in nonionic and anionic systems incl. soaps, detergents, dry oxygen bleaches, bluing compds., and laundry breaks.

Tinopal® AMS-GX. [Ciba-Geigy] Stilbene type; whitening agent for powd. anionic and nonionic detergents or laundry soaps; slightly water-sol.

Tinopal® ATS-X. [Ciba-Geigy] Distyrylbiphenyl disulfonate deriv.; blend of lightfast chlorine stable whiteners for cotton and other cellulosics for all temp. applics.; used in commercial laundry detergents.

Tinopal® BLS-X. [Ciba-Geigy] Distyrylbiphenyl disulfonate deriv.; low dusting, lightfast, chlorine stable whitener for cotton and other cellulosics for high wash temp. applics.; commerical laundry detergents.

Tinopal® CBS-X. [Ciba-Geigy] Distyrylbiphenyl disulfonate deriv.; highly soluble, low dusting, lightfast, chlorine stable whitener for cotton and other cellulosics; used in anionic and nonionic laundry detergents, dry bleaches, fabric softeners, commercial laundry prods., toilet bar soaps, liq. prods., and prods. for use at low wash temps.

Tinopal® LPW. [Ciba-Geigy] Bis-diethanolamine dianilino stilbene disulfonate deriv.; highly sol. whitener for cellulosics; effective at low wash temps.; used in liq. detergents.

Tinopal® RA-16. [Ciba-Geigy] Tinopal RBS 200%, Tinopal AMS, ratio 1:6; whitener blend with cellulosic and fine fabric affinity; used in laundry detergents; slightly water-sol.

Tinopal® RBS 200%, RBS WP-S. [Ciba-Geigy] Naphthotriazolstilbene monosulfonate deriv.; chlorine stable whitener for nylon, acetate, and cotton; used in laundry detergents, dry bleaches, and liq. detergents.

Tinopal® SFP. [Ciba-Geigy] Stilbene-2,2′-disulfonic acid type; optical brightening agent for use in photographic prints to enhance the apparent whiteness of low dens. areas incl. that of borders; recommended for addition to the stabilizer or rinse baths during the photo-processing of the photographic paper; recommended conc.: 1–5 g/l of the processing sol'n.; powd.; lt. yel.; m.w. 1305; sol. 150 g/l of deionized water @ 20 C; pH 7.0–8.0 (deionized water).

Tinopal® UNPA. [Ciba-Geigy] Stilbene; whitening agent for liq. detergents, fabric softeners; sol. in systems with pH 7.

Tinuvin® 109. [Ciba-Geigy] Substituted benzotriazole; uv absorber for coatings; liq.

Tinuvin® 144. [Ciba-Geigy] Hindered amine lt. stabilizer; uv lt. stabilizer gives protection against thermal oxidation; resistant to extraction in aq. environments, to volatilization during high-temp. processing; used in coatings systems; off-wh. cryst. powd.; m.w. 685; sol. (g/100 g solv.): sol. 50 g in chloroform; 45 g in methylene chloride; 29 g in benzene; 8 g in ethyl acetate; < 0.01 g in water; sp.gr. 1.07; m.p. 146–150 C.

Tinuvin® 292. [Ciba-Geigy] Sterically hindered tert. amine lt. stabilizer; uv lt. stabilizer, uv absorber for industrial, automotive, and trade sale coatings and inks, incl. latex paints, two-component theromsetting urethane, one and two-component epoxies, moisture-cured urethane, thermoplastic urethane, clear coat/color coat systems, thermosetting and thermoplastic acrylics; low volatility; lt. stability synergism with Tinuvin uv absorbers; clear sltly. yel. liq.; sol. (g/100 g solv. @ 20 C): Sol. > 100 g in hexane, ethyl acetate, MEK, methanol, butyl Carbitol, Texanol, chloroform, benzene, xylene; nil sol. in water; sp.gr. 0.9925; m.p. < 20 C.

Tinuvin® 326. [Ciba-Geigy] Substituted benzotriazole; uv lt. absorber for plastics and coatings; offers greater absorp. of longer wavelengths than Tinuvin P, better compatibility with polyolefins, lower volatility, less likely to discolor with metals, and has less effect on metal driers and metal catalyts; used in PP, polyethylene, polybutylene, and related polymers; exhibits antioxidant properties; lt. yel. cryst. powd.; sol. (g/100 ml solv.): 12.6 g in styrene; 4.9 g in methyl methacrylate; 2.9 g in MEK; 2.5 g in ethyl acetate; 1.8 g in petrol. ether; 0 g in water; m.p. 140–141 C.

Tinuvin® 327. [Ciba-Geigy] 2-(3′,5′-Di-tert-butyl-2′-hydroxyphenyl)-5-chlorobenzotriazole; uv lt.

absorber for plastics and coatings; stability to heat and lt., low volatility, good color, superior wash fastness, and resistance to gas fading make it useful for stabilizing polyolefins, cold-cured polyesters, PU, dyes, and pigments; pale yel. powd.; m.w. 358; sol. (g/100 ml solv.): 16 g in toluene; 15 g in xylene; 14 g in styrene and methyl methacrylate; 8 g in cyclohexane; 7 g in n-butyl acetate; 6 g ethyl acetate; 5 g in min. spirits; 4 g in MEK; 3 g in dioctyl phthalate and hexane; 0.3 g in ethyl alcohol; sp.gr. 1.20; m.p. 154–158 C.

Tinuvin® 328. [Ciba-Geigy] Substituted benzotriazole; uv lt. absorber offering high solubility and desirable balance of properties; used for stabilizing coatings and automative coatings, styrenics, and polyolefins, other substrates requiring uv lt. protection; more solv. sol., better resin compatibility and washfastness, and less volatility off-wh. powd.; sol. (g/100 g solv.): 44 g in xylene; 28 g in n-butyl acetate; 24 g in MEK; 20 g in ethyl acetate; 14 g in min. spirits; nil sol. in water; sp.gr. 0.91; m.p. 81 C.

Tinuvin® 440. [Ciba-Geigy] Hindered amine; lt. stabilizer for acid-cured thermoset coatings; powd.; sol. in most common coating solvs.

Tinuvin® 440L. [Ciba-Geigy] Hindered amine; lt. stabilizer for acid-cured coatings; liq. xylene mixt.

Tinuvin® 622. [Ciba-Geigy] Hindered amine; lt. stabilizer for polyolefin substrates; very low volatility, low migration tendency, no negative influence on polymer color; off-wh. powd.; m.w. > 2000; sol. (g/100 g solv.): > 40 g in chloroform and methylene chloride; > 30 g in benzene; 5 g in ethyl acetate; 2 g in acetone; 0.1 g in methanol; < 0.01 g in hexane and water; m.p. 130–145 C.

Tinuvin® 622LD. [Ciba-Geigy] Polymeric hindered amine; dimethyl succinate polymer with 4-hydroxy-2,2,6,6-tetramethyl-1-piperidine ethanol; lt. and heat stabilizer for polyolefins, PP inj. molded bars, HDPE, linear LDPE plaques, PP tape, LDPE film, PP multifilament, ABS polymer systems, thermoplastic polyolefins, flexible PVC, food pkg.; off-wh. powd.; m.w. > 2500; sol. (g/100 g): > 67 g in chloroform and methylene chloride, 18 g in toluene, 8 g in xylene, 0.01 g hexane and water; sp.gr. 1.16; m.p. 55–70 C.

Tinuvin® 765. [Ciba-Geigy] Hindered amine lt. stabilizer functioning primarily as a free radical scavenger; uv stabilizer for PU; liq.; sol. in most components of urethane polymers.

Tinuvin® 770. [Ciba-Geigy] Sterically hindered amine; lt. stabilizer for polyolefins, incl. natural and pigmented PP multifilament slit film, polyethylene, ethlene-propylene copolymer and terpolymer, PU, ABS, impact PS, and other styrenic polymers and copolymers; synergistic with Tinuvin P in ABS, impact PS, and PU; useful with costabilizers such as high-performance phenolic antioxidants; wh. to sltly. ylsh. cryst. powd.; sol. (g/100 ml): 56 g in methylene chloride; 46 g in benzene; 45 g in chloroform; 38 g in methanol; 19 g in acetone; 5 gin hexane; < 0.01 g in water; m.p. 81–86 C.

Tinuvin® 900. [Ciba-Geigy] 2-[2-Hydroxy-3-5,-di-(1,1-dimethylbenzyl)phenyl]-2H-benzotriazole; uv absorber offering lt. stability in many coating formulations, esp. those systems where high-cure temps. (> 300 F) are used; lowest volatility of all hydroxy phenyl benzotriazoles, a high degree of permanence giving long-term protection; synergistic with Tinuvin hindered amine lt. stabilizers; exhibits high absorbance in the 300–400 nm region and minimal absorbance in the visible region of the spectrum; highly suitable for thermoplastic coatings, coil coatings, powd. coatings, automotive clear coat/color coat systems, automotive high solids systems, extrusion coatings; lt. yel. to yel. powd.; sol. (g/100 ml solv.): sol. 12 g in xylene; 7 g in ethyl acetate; 6 g in 1:1 xylene/butyl cellosolve; 1 g in mek and butyl Carbitol; nil sol. in water; sp.gr. 1.22; m.p. 135–143 C.

Tinuvin® 1130. [Ciba-Geigy] Substituted hydroxyphenyl benzotriazole; uv absorber for coatings, elastomeric coatings, esp. those cured at elevated temps.; yel. to lt. amber clear liq.; misc. > 50 g/100 g in ethyl acetate, ethylene glycol, methyl amyl ketone, MEK, MIBK, n-butanol, toluene, xylene; insol. in water but easily dispersed in waterborne systems; sp.gr. 1.17; visc. 6410 cps (21 C).

Tinuvin® P. [Ciba-Geigy] 2-(2-Hydroxy-5-methylphenyl) benzotriazole; uv lt. absorber (300–400 nm) offering protection to a wide range of plastics; dissipates photochemical energy as thermal energy; high thermal stability; imparts little or no initial color to substrates, does not discolor on lt. or heat aging, antioxidant and thermal stabilizer; used in conventional heat-cured and flame-retardant, halogen-containing polyesters, plasticized and rigid PVC, PVC food pkg., PS homopolymer, impact PS, rubber and other rubber-modified polymers, PC, acrylics, polyolefins, coatings, colorants, adhesives; cryst. powd.; m.w. 225; sol. (g/100 ml): 7.2 g in styrene; 6.0 g in toluene; 5.0 g in methyl methacrylate; 3.5 g in ethyl acetate; 2.5 g in acetone and dioctyl phthalate; 1.7 g in methyl Cellosolve; 1.5 g in min. spirits; nil sol. in water; sp.gr. 1.51; m.p. 128–132 C; b.p. 225 C (10 mm Hg).

Tixosil 311. [Manville] Precipitated silica, amorphous; flatting agents providing gloss and sheen control in coatings, esp. high bake industrial enamels (grades 311, 321); rheology control agent for solv. and water-based systems (grades 331 and 375); for coatings, adhesives, plastics, paper, food, cosmetics, toothpaste, and pharmaceutical applics.; wh. powd.; 4 μ avg. particle size; < 0.1% retained 325 mesh; Hegman grind 6; sp.gr. 2.05; dens. 17.1 lb/gal; surf. area 250 m^2/g; oil absorp. 300%; ref. index 1.45; pH 7; 96% SiO_2.

Tixosil 321. [Manville] Precipitated silica, amorphous; see Tixosil 311; wh. powd.; 3 μ avg. particle size; < 0.1% retained 325 mesh; Hegman grind > 6; dens. 17.1 lb/gal; surf. area 250 m^2/g; oil absorp. 300%; ref. index 1.45; pH 7; 96% SiO_2.

Tixosil 331. [Manville] Precipitated silica, amorphous; see Tixosil 311; wh. powd.; 3 μ avg. particle size; < 0.1% retained 325 mesh; Hegman grind 7; dens. 17.1 lb/gal; surf. area 310 m^2/g; oil absorp.

320%; ref. index 1.45; pH 7; 96% SiO_2.

Tixosil 375. [Manville] Precipitated silica, amorphous; see Tixosil 311; wh. powd.; 2 µ avg. particle size; < 0.1% retained 325 mesh; Hegman grind 7; dens. 17.5 lb/gal; surf. area 180 m^2/g; oil absorp. 290%; ref. index 1.45; pH 6.6; 98% SiO_2.

TL Series. [ICI Advanced Materials] Lubricant produced from fluoropolymer resins; powd. lubricants easily dispersed in a wide variety of liq. or solid media to reduce wear and friction, eliminate stick-slip problems, and increase load bearing and release properties; used as clear, dry film lubricants in place of general purpose solid lubricants (e.g., graphite or MoS2); exhibit low coeff. of friction, broad temp. use range, chemical inertness, and oxidation resistance; finely divided powds.

TL-101. [ICI Advanced Materials] PTFE-based lubricant; general purpose lubricant used for aerosols and solv. disps., and as thickening and EP agent for oils and greases; wh. powd. (spherical particles, 3–8 µ); sp.gr. 2.18–2.28; dens. (bulk) 440 g/l; visc. (melt) 1.5×10^8 poise.

TL-102. [ICI Advanced Materials] PTFE-based lubricant; best film-forming grade lubricant powd.; used in solv. disp. coatings and aerosols because of small, uniform particle size; as thickening and EP agent in oils and greases; provides uniform thickening and retention of consistency in grease formulations; wh. powd. (spherical particles, 3–8 µ); sp.gr. 2.18–2.28; dens. (bulk) 425 g/l; surf. tens. 20.9 dynes/cm.

TL-107. [ICI Advanced Materials] PTFE made from 100% virgin Teflon; lubricant powd. with wide particle size distribution and irregular particle shape; used for friction fighting applic.; internal disp. in solids and liqs.; wh. powd.; sp.gr. 2.18–2.28; dens. 506 g/l.

TL-113. [ICI Advanced Materials] PTFE-based lubricant; lubricant possessing a wide particle size range; film former; used as additive to other media to impart permanent lubricant; used in pastes; not for use with oxidizing agents; powd. (large, irregular particles, 4–35 µ); sp.gr. 2.18–2.28; dens. (bulk) 525 g/l (uncompacted); surf. tens. 20.7 dynes/cm.

TL-115. [ICI Advanced Materials] PTFE; general purpose lubricant powd. used in wide media; excellent bearing chars. when used as additive in polymeric systems; improves PV, static, and dynamic coefficients of friction, and wear factor; dry lubricant additive to thermoset, elastomeric, and rubber compds.; used in rubber industry as release agent when mixed with silicones; belting compds., seals, gaskets; improves tear resistance when added to silicone rubber; powd., 6–25 µ particle size; sp.gr. 2.15–2.25; dens. 475 g/l.

TL-115A, TL-115AH. [ICI Advanced Materials] PTFE-based lubricant; TL-115A is a general-purpose lubricant which impacts exc. bearing chars. when used as an additive in thermoplastic and thermoset polymeric systems; used as thickening and EP agent in oils and greases; TL-115AH contains an inert colloidal additive to prevent agglomeration and facilitate disp. in soft elastomeric compds.; wh. to lt. gray powd. (6–25) 115A: irregular particle shape; sp.gr. 2.18–2.28; dens.(bulk) 630 g/l.

TL-120. [ICI Advanced Materials] Fluorinated ethylene-propylene (FEP)-based lubricant; finest sized TL lubricant powd.; disperses in all liq. and solid media and will not lump or agglomerate on disp.; thickening and EP agent in hydrocarbon and fluorosilicone base oils and greases; imparts uniform thickening, retention of consistency, nonseparation, low leakage, and low C_f; promotes release and antiscuff in printing inks and waxes; used in grease and oil formulations that are utilized in food and pharmaceutical machinery; for food contact use; wh. powd. (1–2 µ); sp.gr. 2.12–2.18; dens. (bulk) 400 g/l; surf. tens. 21.9 dynes/cm.

TL-126, TL-126H. [ICI Advanced Materials] PTFE-based lubricant; TL-126 is a general lubricant whose particles are soft and readily smear to form a dry film; dry lubricant for machinery when improved wear and abrasion resistance are required; can be sprayed or dusted on; TL-126H contains an inert colloidal additive to prevent agglomeration during intensive mixing; as an additive to soft compds. or other media where disp. is difficult; wh. powd. (irregular particles, 6–26 µ); sp.gr. 2.16–2.28; dens. (bulk) 400 g/l; surf. tens. 20.2 dynes/cm.

TL-127. [ICI Advanced Materials] PTFE-based lubricant; lubricant specifically activated to facilitate wetout and disp. in liqs. with surf. tens. as great as water (72.5 dynes/cm); as an additive to latex, ketone, and water volatile systems; not for disp. in other hydrocarbon solvs.; wh. powd. (spherical particles, 3–8 µ); sp.gr. 2.18–2.28.

TL-340. [ICI Advanced Materials] Polytrifluorochloroethylene (PCTFE)-based lubricant; lubricant powd. with unusually high surf. tens. and exhibiting very high oil absorp. value and the greatest thickening action of any fluoropolymer lubricant; as an additive to oils and greases; not for use in high temp. applics. due to relatively low m.p. (390 F); wh. powd. (irregular particles, 6–25 µ); sp.gr. 2.11–2.13; surf. tens. 31.6 dynes/cm.

TL-450. [ICI Advanced Materials] Polyvinylidene fluoride lubricant; lubricant powd. with a higher surf. tens. than the PTFE-based lubricants and exhibits a high oil absorp. value; used as an additive to oils and greases; not for high-temp. applics. due to relatively low melt point (340 F); wh. powd. (4–25 µ particle size); sp.gr. 1.76–1.77; surf. tens. 25.4 dynes/cm.

T-Maz® 20. [PPG-Mazer] Polysorbate 20; nonionic; emulsifier, solubilizer, wetting agent, antistat, stabilizer, dispersant, visc. modifier, suspending agent used in the food, cosmetic, drug, textile and metalworking industries; yel. liq.; sol. in water, ethanol, acetone; sp.gr. 1.1; visc. 400 cps; HLB 16.7; sapon. no. 40–50; 97% act.

T-Maz® 28. [PPG-Mazer] PEG-80 sorbitan monolaurate; emulsifier and solubilizer of essential oils, wetting agent, visc. modifier, antistat, stabilizer and dispersant used in food, cosmetic, drug, textile and metalworking industries; yel. liq.; sol. see T-Maz 20;

sp.gr. 1.0; visc. 1100 cps; acid no. 2 max.; sapon. no. 5–15; 30% max. water.

T-Maz® 40. [PPG-Mazer] Polysorbate 40; see T-Maz 20; yel. liq.; sol. see T-Maz 20; sp.gr. 1.0; visc. 600 cps; HLB 15.6; sapon. no. 43–49; 97% min. act.

T-Maz® 60. [PPG-Mazer] Polysorbate 60; see T-Maz 20; yel. liq.; sol. in water, toluol; sp.gr. 1.1; visc. 550 cps; HLB 14.9; pour pt. 23–25 C; sapon. no. 45–55; 97 min. act.

T-Maz® 60K. [PPG-Mazer] Polysorbate 60; see T-Maz 20; Kosher grade; Gardner 5 gel; sol. @ 5% in water, min. spirits, toluene; HLB 14.9; flash pt. (PMCC) > 350 F; flash pt. (PMCC) > 350 F; sapon. no. 45–55.

T-Maz® 60KHM. [PPG-Mazer] Polysorbate 60; see T-Maz 20; Kosher high melt grade; Gardner 5 gel; sol. @ 5% in water, min. spirits, toluene; HLB 14.9; flash pt. (PMCC) > 350 F; sapon. no. 45–55.

T-Maz® 61. [PPG-Mazer] Polysorbate 61; nonionic; see T-Maz 20; solid; water-sol.; HLB 9.6; 100% conc.

T-Maz® 65. [PPG-Mazer] Polysorbate 65; see T-Maz 20; sol. in ethanol, acetone, naphtha; disp. in water; sp.gr. 1.1; m.p. 30–32 C; HLB 10.5; sapon. no. 88–98; 97% min. act.

T-Maz® 65K. [PPG-Mazer] Polysorbate 65; see T-Maz 20; Kosher grade; Gardner 5 paste; sol. @ 5% in min. oil, min. spirits, veg. oil; disp. in water, toluene; HLB 101.5; flash pt. (PMCC) > 350 F; sapon. no. 88–98.

T-Maz® 67. [PPG-Mazer] PEG-20 sorbitan isostearate; see T-Maz 20; yel. liq.; sol. see T-Maz 60; sp.gr. 1.0; visc. 400 cps; HLB 14.8; sapon. no. 45–55; 97% min. act.

T-Maz® 80. [PPG-Mazer] Polysorbate 80; see T-Maz 20; yel. liq.; sol. in water, ethanol, veg. oil, toluol; sp.gr. 1.0; visc. 400 cps; HLB 15.0; sapon. no. 45–55; 97% min. act.

T-Maz® 80K. [PPG-Mazer] Polysorbate 80, kosher; nonionic; solubilizer and emulsifier for foods; liq.; HLB 15.0; 100% conc.

T-Maz® 80KLM. [PPG-Mazer] Polysorbate 80; see T-Maz 20; Kosher low moisture grade; Gardner 5 liq.; sol. @ 5% in water, veg. oil; disp. in toluene; HLB 15.0; flash pt. (PMCC) > 350 F; sapon. no. 45–55.

T-Maz® 81. [PPG-Mazer] Polysorbate 81; nonionic; see T-Maz 20; liq.; water-sol.; HLB 10.0; 100% conc.

T-Maz® 81K. [PPG-Mazer] Polysorbate 81; see T-Maz 20; Kosher grade; Gardner 6 liq.; sol. @ 5% in min. spirits; disp. in water, min. oil, toluene, veg. oil; HLB 10.0; flash pt. (PMCC) > 350 F; sapon. no. 96–104.

T-Maz® 85. [PPG-Mazer] Polysorbate 85; see T-Maz 20; yel. liq.; sol. in ethanol; disp. in water; sp.gr. 1.0; visc. 300 cps; HLB 11.0; sapon. no. 83–93; 95% min. act.

T-Maz® 85K. [PPG-Mazer] Polysorbate 85; see T-Maz 20; Kosher grade; Gardner 7 liq.; sol. @ 5% in min. spirits; disp. in water, min. oil, toluene, veg. oil; HLB 11.1; flash pt. (PMCC) > 350 F; sapon. no. 83–93.

T-Maz® 90. [PPG-Mazer] PEG-20 sorbitan monotallate; see T-Maz 20; amber liq.; sol. in water, ethanol, toluol, veg. oil; sp.gr. 1.0; visc. 500 cps; HLB 15.0; sapon. no. 45–55; 97% min. act.

T-Maz® 95. [PPG-Mazer] PEG-20 sorbitan tritallate; see T-Maz 28; amber liq.; sol. see T-Maz 85; sp.gr. 1.0; visc. 350 cps; HLB 10.8; acid no. 2 max.; sapon. no. 45–55; 3.0% max. water.

TM/ETD. [Akrochem] 60% Tetramethyl thiuram disulfide/40% tetraethyl thiuram disulfide; very act., sulfur-bearing, nondiscoloring org. accelerator; gives nonblooming cures in EV and semi-EV systems; used in nitrile rubber, SBR and EPDM; off-wh. powd.; sp.gr. 1.35; m.p. 66 C min.; 12.1% avail. sulfur.

TMTM. [Akrochem] Tetramethyl thiuram monosulfide; nonstaining, nondiscoloring, fast-curing accelerator for use alone or in combination in NR, SBR, NBR, butyl rubber, Neoprene, and reclaim rubber; yel. powd.; sp.gr. 1.39; m.p. 105 C min.

T-Mulz® 426. [Harcros] Anionic blend; solv. degreasers, textile lubricant, agric. applics.

T-Mulz® 598. [Harcros] Anionic blend; dry cleaning, general purpose cleaners, agric. formulations, down hole scale inhibitor, emulsifier.

T-Mulz® 734-2. [Harcros] Phosphate ester; anionic; emulsifier for agric. formulations, solvs., detergents, emulsion polymerization, alkaline cleaners; hydrotrope; amber clear liq.; sp.gr. 1.18 (68 F); dens. 9.8 lb/gal (68 F); pour pt. 50 F; flash pt. > 350 F (PMCC); pH 2–3 (1% aq.).

T-Mulz® 800. [Harcros] Anionic blend; alkaline cleaners, textile, hydrotrope, conc. electrolyte sol'ns.

T-Mulz® 844. [Harcros] Anionic blend; alkaline cleaners, textile, hydrotrope.

T-Mulz® 8015. [Harcros] Nonionic blend; spreader sticker for agric.

TOF. [Merrand] Tri 2-ethylhexyl phosphate; low temp. plasticizer for NBR, NBR/PVC, CR, CPE elastomers, PVC, PVAc, cellulosics, PS, ABS; some flame retardance.

Tohol N-220. [Toho] Coconut DEA; nonionic; foam stabilizer and thickener for personal care prods.; liq.; 100% conc.

Tohol N-220X. [Toho] Coconut DEA; nonionic; see Tohol N-220; solid; 100% conc.

Tohol N-230. [Toho] Lauryl DEA; nonionic; see Tohol N-220; liq.; 100% conc.

Tohol N-230X. [Toho] Lauryl DEA; nonionic; see Tohol N-220; solid; 100% conc.

Tokuthion®. [Bayer] Prothiofos; insecticide esp. effective against leaf-eating caterpillars; colorless liq.; m.w. 345.2; misc. in n-hexane, IPA, methylene chloride; sp.gr. 1.3; b.p. 125–128 C (0.13 mbar).

Tolplaz TBEP. [Merrand] Tributoxyethyl phosphate; low temp. plasticizer for NBR, NBR/PVC, CR, CPE elastomers, polyacrylate, fluoroelastomers, PVAc, cellulosics, PS, ABS; lubricant additive; leveling agent in floor polish.

Tomah DA-14. [Tomah] N-decyloxypropyl-1,3-diaminopropane; corrosion inhibitor; used in chemical intermediate areas; lt. amber liq.; m.w. 295; sp.gr.

0.86; pour pt. –50 F; 100% act.; 90% min. diamine.

Tomah DA-17. [Tomah] N-tridecyloxypropyl-1,3-diaminopropane; used in corrosion inhibitor formulations, oilfield chemicals; replacement for coco, soya, or oleyl diamine; amber liq.; m.w. 340; sp.gr. 0.87; pour pt. –30 F; 95% min. amine.

Tomah DA-62. [Tomah] C_{12-14} alkoxypropyl-1,3-diaminopropane; corrosion inhibitor and oilfield chemicals; m.w. 395; sp.gr. 0.85; pour pt. 50 F; 75% min. amine act.

Tomah DA-62 Ditallate. [Tomah] Tall oil fatty salt of Tomah DA-62; corrosion inhibitor in gas pipelines, refinery streams, and primary oil prod.; asphalt antistrip agent; calcium carbonate adhesion aid; pigment wetting aid; sound deadener; caulking compds. and metal working chemicals; amber liq.; sp.gr. 0.91; visc. 3500 cps; pour pt. 60 F.

Tomah E-14-2. [Tomah] Ethoxylated fatty amine; emulsifier, corrosion inhibitor, lubricant used in min. acid inhibition, textile processing; amber liq.; sp.gr. 0.94; HLB 8.3; 95–100% act.

Tomah E-14-5. [Tomah] Ethoxylated fatty amine; see Tomah E-14-2; amber liq.; sp.gr. 0.99; 95–100% act.

Tomah E-14-15. [Tomah] Ethoxylated fatty amine; see Tomah E-14-2; amber liq.; sp.gr. 1.05; 95–100% act.

Tomah E-17-2. [Tomah] Ethoxylated fatty amines; see Tomah E-14-2; amber liq.; sp.gr. 0.93; HLB 5.6; 95–100% act.

Tomah E-18-2. [Tomah] Ethoxylated fatty amines; see Tomah E-14-2; wax; sp.gr. 0.95; HLB 3.0; 95–100% act.

Tomah E-18-5. [Tomah] Ethoxylated fatty amines; see Tomah E-14-2; paste; sp.gr. 0.97; HLB 11.0; 95–100% act.

Tomah E-18-10. [Tomah] Ethoxylated fatty amines; see Tomah E-14-2; amber liq.; sp.gr. 1.02; 95–100% act.

Tomah E-18-15. [Tomah] Ethoxylated fatty amines; see Tomah E-14-2; amber liq.; sp.gr. 1.03; HLB 16.0; 95–100% act.

Tomah E-C-2. [Tomah] Ethoxylated amine; cationic; emulsification; corrosion inhibitor; min. acid inhibition, textile processing and lubricants; liq.; HLB 2.0; 100% conc.

Tomah E-C-5. [Tomah] Ethoxylated amine; cationic; see Tomah E-C-2; liq.; HLB 5.0; 100% conc.

Tomah E-C-15. [Tomah] Ethoxylated amine; cationic; see Tomah E-C-2; liq.; HLB 15.0; 100% conc.

Tomah E-S-2. [Tomah] Ethoxylated fatty soya amines; see Tomah E-14-2; amber liq.; sp.gr. 0.91; HLB 2.0; 95–100% act.

Tomah E-S-5. [Tomah] Ethoxylated fatty soya amines; see Tomah E-14-2; amber liq.; sp.gr. 0.95; HLB 5.0; 95–100% act.

Tomah E-S-10. [Tomah] Ethoxylated fatty soya amines; see Tomah E-14-2; amber liq.; sp.gr. 1.02; 95–100% act.

Tomah E-S-15. [Tomah] Ethoxylated fatty soya amines; see Tomah E-14-2; amber liq.; sp.gr. 1.03; HLB 15.0; 95–100% act.

Tomah PA-10. [Tomah] Hexyloxypropylamine; corrosion inhibitor and antistat; lt. amber liq.; m.w. 165; water-sol.; sp.gr. 0.84; pour pt. –30 F; 95% min. amine.

Tomah PA-14. [Tomah] Decyloxypropylamine; cationic; corrosion inhibitor, antistat, and flotation reagent; emulsification; colorless to lt. amber liq.; m.w. 229; sp.gr. 0.84; pour pt. –50 F; 95% min. amine.

Tomah PA-17. [Tomah] Tridecyloxypropylamine; corrosion inhibitor, cationic emulsification and replacement for oleyl and soya amines; colorless to lt. amber liq.; m.w. 274; sp.gr. 0.85; pour pt. –30 F; 95% min. amine.

Tomah PA-62. [Tomah] C_{12-14} alkyloxypropylamine; cationic; corrosion inhibitor and emulsifier; industrial applic.; replacement for coco amine; liq.; m.w. 340; sp.gr. 0.83; pour pt. 60 F; 70% min. amine.

Tomah PA-1820. [Tomah] C_{14-18} alkyloxypropylamine; corrosion inhibitor; amber liq.

Tomah Q-14-2. [Tomah] Decyloxypropyl dihydroxyethyl methyl ammonium chloride; cationic; quat. used as acid corrosion inhibitor, plastics and textile antistat, and emulsifier; amber liq.; water-sol.; sp.gr. 0.97; pH 6–9 (1% aq.); flam.; 75% conc.

Tomah Q-17-2. [Tomah] Tridecyloxypropyl dihydroxyethyl methyl ammonium chloride; see Tomah Q-14-2; amber liq.; sp.gr. 0.97; pH 6–9 (5% aq.); insol. in water; dens. 8.5 lb/gal; visc. 472 cSt (100 F); HLB 8.9; pour pt. –10 C; flash pt. 430 F; cloud pt. < 25 C; pH 6–9 (5% aq.); 50% in IPA-water solv. blend.

Tomah Q-18-2. [Tomah] Octadecyl dihydroxyethyl methyl ammonium chloride; cationic; see Tomah Q-14-2; paste; sp.gr. 0.92; pH 6–9 (5% aq.); flam.; 75% in IPA-water solv. blend.

Tomah Q-18-2-50%. [Tomah] Quat.; antistat for carpets and textiles; liq.; 50% act.

Tomah Q-18-15. [Tomah] Octadecyl PEG-15 methyl ammonium chloride; cationic; see Tomah Q-14-2; amber liq.; sp.gr. 1.06; pH 6–9 (5% aq.); 100% act.

Tomah Q-62. [Tomah] Complex monofunctional quat.; see Tomah Q-14-2; semisolid; sp.gr. 0.91; pH 6–9 (5% aq.); flam.; 60% act. in IPA-water solv. blend.

Tomah Q-D-62. [Tomah] Complex difunctional quat.; cationic; see Tomah Q-14-2; amber liq.; sp.gr. 0.96; pH 6–9 (5% aq.); 75% act. in isopropanol-water-solv. blend.

Tomah Q-D-T. [Tomah] Tallow dimethyl trimethyl propylene diammonium chloride; cationic; see Tomah Q-14-2; amber liq.; sp.gr. 0.95; pH 6–9 (5% aq.); 50% act. in IPA-water.

Tomah Q-S. [Tomah] Soya trimethyl ammonium chloride; cationic; see Q-14-2; amber liq.; sp.gr. 0.91; pH 6–9 (5% aq.); 50% act. in IPA-water.

Tomah Tallow Amine. [Tomah] Tallow primary amine, tech.; cationic; ore flotation agent, emulsifier, corrosion inhibitor; solid.; 100% conc.

Tomah Tallow Diamine. [Tomah] Tallow diamine; cationic; emulsifier, corrosion inhibitor; solid; 100% conc.

Tonox® 22. [Uniroyal] Aromatic amine; curing agent for epoxy resins; red br. visc. liq.; sp.gr. 1.15;

visc. > 250,000 cps; hyd. equiv. wt. 52.4.

Tonsil (TN). [L.A. Salomon] Activated clay; adsorbent for bleaching oils, fats; powd. or gran.

Topanex 100BT. [ICI Am.] 2-(2´-Hydroxy-5´-methylphenyl) benzotriazole; uv absorber, lt. stabilizer, antioxidant protecting plastics (PVC, styrenics, acrylics, unsat. polyesters), lacquers; effective at 290–380 nm; pale yel. cryst. powd.; m.w. 225; sol. (g/100 ml): 7 g in styrene, 6 g in toluene; m.p. 128–132 C.

Topanol® 205. [ICI Am.] 2,2-Bis[4-(2-(3,5-di-t-butyl-4-hydroxyhydrocinnamoyloxy)) ethoxyphenyl] propane; high m.w. hindered phenolic antioxidant for retardation of thermal and oxidative degradation in polyolefins, styrenics, and engineering polymers; wh. free-flowing powd.; m.w. 836; sol. in common org. solvs.; sp.gr. 1.09; m.p. 100–102 C.

Topanol® CA. [ICI Am.] 1,1,3-Tris (2-methyl-4-hydroxy-5-t-butyl phenyl) butane; antioxidant for polyolefin and styrenic polymers, plasticizers, hot melt adhesives, SBR latexes, polyamides, polyesters; FDA approved; wh. cryst. powd., toluene odor; sol. in acetone, ethanol, diethyl ether, ethyl acetate; dens. 0.5 g/ml; m.p. 181 C min.; flash pt. (Seta CC) 72.8 C.

Topanol® CA-RT. [ICI Am.] 1,1,3-Tris (2-methyl-4-hydroxy-5-t-butylphenyl) butane; antioxidant; wh. powd.; m.p. 181 C.

Topanol® OC. [ICI Am.] BHT; antioxidant; for process stabilization of thermoplastics, heat stabilization of liq. systems, stabilization of foods, fats, oils, animal feeds; wh. powd.

Topanol® OF. [ICI Am.] BHT; see Topanol OC.

Topcithin. [Lucas Meyer] Soya lecithin, highly purified fluid; nonionic; improvement of flowing properties and physical appearance; extends shelf-life; liq.; 100% conc.

Toximul® 80. [Stepan] Alkylaryl polyethoxy ethanol, IPA; nonionic; wetting and spreading agent, penetrant, emulsifier, dispersant, surfactant adjuvant used as spray additive in herbicide formulations; liq.; water-sol.

Toximul® 351. [Stepan] Ammonium alkylaryl ether sulfate, water and alcohol cosolvs.; anionic; emulsifier, dispersant, surfactant ingred. in aq. conc. formulation of herbicides; emulsifier in solv.-based formulations; straw clear liq.; sol. in water and PCP emulsifiable formulations; sp.gr. 1.055; dens. 8.8 lb/gal; flash pt. 79 F (CC); pH 7.0 max. (10% aq.).

Trans-10, -10K. [Trans-Chemco] Silicone emulsion; -10K kosher grade; defoamer for aq. systems incl. food processing; disp. in water; 10% act.

Trans-20, -20K. [Trans-Chemco] Silicone emulsion; -20K kosher grade; see Trans-10; water-disp.; 20% act.

Trans-25, -25K. [Trans-Chemco] Silicone emulsion; -25K kosher grade; see Trans-10; water-disp.; 25% act.

Trans-30, -30K. [Trans-Chemco] Silicone emulsion; -30K kosher grade; see Trans-10; water-disp.; 30% act.

Trans-100. [Trans-Chemco] Silicone compd.; defoamer for detergents, petrol. refining, cutting oils, insect repellents, chemical and food processing; sol. in aliphatic, aromatic, and chlorinated hydrocarbons; 100% act.

Trans-107. [Trans-Chemco] Nonsilicone; defoamer for alkaline systems; slightly disp. in water.

Trans-109. [Trans-Chemco] Nonsilicone; defoamer for paints, coolants, lubricants, cutting oils; slightly disp. in water; 100% act.

Trans-134S. [Trans-Chemco] Emulsion with 11% silicone; defoamer for mfg. of water-based inks; water-disp.

Trans-134S K. [Trans-Chemco] Emulsion with 11% silicone, kosher; defoamer for aq. systems incl. food processing; water-disp.

Trans-135. [Trans-Chemco] Nonsilicone; defoamer for paper/paperboard mfg.; water-disp.

Trans-137. [Trans-Chemco] Nonsilicone; defoamer for alkaline systems, paper and paperboard mfg., wastewater treatment; water-disp.

Trans-138. [Trans-Chemco] Nonsilicone; defoamer for water-based inks, wastewater treatment; disp. in water; sol. in hydrocarbon solvs.; 100% act.

Trans-143. [Trans-Chemco] Nonsilicone; defoamer for mfg. of syn. latex; water-disp.

Trans-149. [Trans-Chemco] Nonsilicone; defoamer for paper/paperboard mfg.; water-disp.

Trans-154, -156. [Trans-Chemco] Nonsilicone; defoamer for starch processing; sol. in hydrocarbon solvs.; slightly disp. in water; 100% act.

Trans-166. [Trans-Chemco] Nonsilicone; defoamer for mfg. of syn. latex, sol. cutting oils; water-disp.

Trans-175, -176. [Trans-Chemco] Nonsilicone; defoamer for mfg. of soya flour and isolate, egg washing; water-disp.

Trans-177S, -177S K. [Trans-Chemco] Emulsion with 5% silicone; -177 S K kosher grade; defoamer for aq. systems incl. food processing, wastewater treatment; water-disp.

Trans-179. [Trans-Chemco] Nonsilicone, water-based; defoamer for mfg. edible phosphates; water-sol.

Trans-198. [Trans-Chemco] Nonsilicone, water-based; defoamer for mfg. of phosphoric acid; water-disp.

Trans-220. [Trans-Chemco] Nonsilicone, water-based; long-lasting antifoam for food processing; water-disp.

Trans-222. [Trans-Chemco] Emulsion with 3% silicone; defoamer/antifoam for aq. systems; water-disp.

Trans-224. [Trans-Chemco] Nonsilicone, water-based; aq. food processing defoamer; water-disp.

Trans-225S. [Trans-Chemco] Silicone emulsion; aq. food processing antifoam; water-disp.; 12.5% act.

Trans-262. [Trans-Chemco] Nonsilicone; defoamer and drainage aid for brn. stock washers; water-insol.; 100% act.

Trans-264. [Trans-Chemco] Surfactants and hydrophobed silica; see Trans-262; insol. in water; 100% act.

Trans-265. [Trans-Chemco] Nonsilicone; water-based ink defoamer; partly sol. in water; 100% act.

Trans-266. [Trans-Chemco] Surfactants and hydrophobed silica; defoamer for paper machine stock, bleach stock, effluent, and red liquor; partly sol. in water; 100% act.

Trans-1030. [Trans-Chemco] Emulsion with 3% silicone; defoamer for sauna baths, whirlpools, rug shampoos; water-disp.

Trans-1030 K. [Trans-Chemco] Emulsion with 3% silicone, kosher; defoamer for aq. systems incl. food processing; water-disp.

Trans-FG2. [Trans-Chemco] Nonsilicone, water-based; defoamer for veg. canning, food processing; water-disp.

Trans-N1. [Trans-Chemco] Nonsilicone, water-based; defoamer for paper/paperboard mfg.; water-disp.

Trans-RC1. [Trans-Chemco] Nonsilicone; defoamer for mfg. of coatings for paper/paperboard, in clay coatings; sol. in hydrocarbon solvs., water-insol.; 100% act.

Translink® 37. [Engelhard] Calcined aluminum silicate with vinyl functional surface treatment; reinforcing extender; used in cross-linked polyethylene and the ethylene-propylene rubbers used in elec. compds.; wh. highly pulverized powd.; 1.4 μm avg. particle size (pretreatment); 0.02% max. residue +325 mesh; sp.gr. 2.63 g/cc; bulk dens. 16 lb/ft^3 (loose), 35 lb/ft^3 (tamped); bulking value 0.046 gal/lb; oil absorp. 45–55 lb/100 lb; ref. index 1.62; brightness (GE) 90–92; 52.4% SiO_2; 44.5% Al_2O_3.

Translink® 77. [Engelhard] Calcined aluminum silicate with vinyl functional surface treatment; reinforcing extender; used in dielectric paper and wire and cable insulation; wh. highly pulverized powd.; 0.8 μm avg. particle size (pretreatment); 0.02% max. residue +325 mesh; sp.gr. 2.63 g/cc; bulk dens. 11 lb/ft^3 (loose), 21 lb/ft^3 (tamped); bulking value 0.046 gal/lb; oil absorp. 80–90 lb/100 lb; ref. index 1.62; brightness (GE) 90–92; 52.4% SiO_2; 44.5% Al_2O_3.

Translink® 445. [Engelhard] Calcined aluminum silicate with amino-silane surface treatment; reinforcing extender for polyamides, polyesters, and other polar polymers; wh. highly pulverized powd.; 1.4 μm avg. particle size (pretreatment); 0.02% max. residue +325 mesh; sp.gr. 2.63 g/cc; bulk dens. 16 lb/ft^3 (loose), 35 lb/ft^3 (tamped); bulking value 0.046 gal/lb; oil absorp. 45–55 lb/100 lb; ref. index 1.62; pH 8.5–9.5; brightness (GE) 90–92; 52.4% SiO_2; 44.5% Al_2O_3.

Translink® 555. [Engelhard] Calcined aluminum silicate with amino-silane surface treatment; reinforcing extender; wh. highly pulverized powd.; 0.8 μm avg. particle size (pretreatment); 0.02% max. residue +325 mesh; sp.gr. 2.63 g/cc; bulk dens. 16 lb/ft^3 (loose), 35 lb/ft^3 (tamped); bulking value 0.046 gal/lb; oil absorp. 80–90 lb/100 lb; ref. index 1.62; pH 8.5–9.5; brightness (GE) 90–92; 52.4% SiO_2; 44.5% Al_2O_3.

Trem-LF-40. [Henkel] Sodium alkylallyl sulfosuccinate; polymerizable surfactant, solubilizer; provides low foaming emulsions with improved water resistance; lt. yel. liq.; water-sol.; 40% act.

Trenbest 500. [Lucas Meyer] Trimethyl hydroxyethyl ammonium ester of a carboxyglycerol phosphoric acid; mold release agent for the plastic and caoutchouc industry; used for molding and extrusion powds., laminated plastics, inj. molding powds., ABS; low-visc. liq.; sp.gr. 0.96; flash pt. < 14 C; 50% solids in isopropyl acetate.

Trennspray Tego®. [Goldschmidt] Silicone-free aerosol for release of inj. molded plastic and rubber moldings.

Triadine® 3. [Olin] Bactericide for metalworking fluids; sol'n.

Triadine® 10. [Olin] Sodium 2-pyridinethiol-1-oxide (6.4%) and hexahydro-1,3,5-tris (2-hydroxyethyl)-s-triazine (63.6%); broad spectrum microbiostat effective against Gram-positive and Gram-negative bacteria and fungi; used in aq.-based metalworking fluids; amber aq. sol'n.; mild amine odor; b.p. 110 C; sp.gr. 1.17; dens. 9.73 lb/gal; pH 10.0–10.5 (10% in neut. dist. water); 70% act. in water.

Tribunil®. [Bayer] Methabenzthiazuron; broad-spectrum herbicide for pre- and post-emergence applic.; colorless cryst.; m.w. 221.3; sol. (g/1000 ml): > 200 g in dichloromethane, 50–100 g in toluene, 20–50 g in 2-propanol; m.p. 119–121 C.

Tricap. [Croda Ltd.] Triethylene glycol dicaprylate/caprate; low temp. plasticizer for plastics and syn. rubbers.

Tri-Ethane. [PPG Industries] Trichloroethane; all-purpose solv. for cold cleaning, vapor degreasing used in resin applics.; for dry-film photoresist processing; adhesive solv.; used in aerosols as solv. and vapor pressure depressant; prod. parts and assemblies can be cleaned by wiping, brushing, dipping, spraying, ultrasonics, and vapor degreasing methods; formulating solv. for printing inks and lubricants; closed textile finishing operations for scouring, desizing, and dyeing to minimize pollution; nonflammable; contains effective stabilizers; water-wh. clear liq.; m.w. 133.42; f.p. –45 C; b.p. 74.1 C; sol. 0.07 g sol./100 g water; sp.gr. 1.304; dens. 10.84 lb/gal; visc. 0.735 cps; ref. index 1.4379; pH 7.0.

Triglyme. [Ferro] Triethylene glycol dimethyl ether; solv. which tends to solvate cations; used in electrochemistry, polymer and boron chemistry; physical processes such as gas absorp., extraction, stabilization; used in industrial prods. such as fuels, lubricants, textiles, pharmaceuticals, pesticides; colorless clear; mild, nonresidual odor; m.w. 178.22; f.p. –45 C; b.p. 216 C (760 mm Hg); water-sol.; sp.gr. 0.9862; dens. 8.23 lb/gal; visc. 3.8 cP; ref. index 1.4224; pH neutral; flash pt. 111 C (CC); surf. tens. 29.4 dynes/cm; 98.0% min. purity.

Trigonal® 14. [Akzo] Benzoin ether; initiator used in u.v. lt. curing of polyester resins; long shelf life in polyester formulation; liq.

Trigonox® 17/40. [Akzo] n-Butyl-4,4-bis(t-butylperoxy) valerate; crosslinking agent for olefin copolymers, EPDM, SBR, Neoprene, Hypalon; powd.; 40% act. with inorg. phlegmatizer.

Trigonox® 17/40 Bpd. [Akzo] 4,4-Di-t-butylperoxy-n-butyl valerate; crosslinking agent for olefin copolymers, esp. chlorinated polyethylene; useful for cross-linking EPDM, SBR, Neoprene, Hypalon; powd.; 40% act. with inorg. phlegmatizer

Trigonox® 21. [Akzo] t-Butyl peroctoate; initiator for acrylates, styrenics, and LDPE polymerization; used where presence of water is objectionable; liq.; 98% act.

Trigonox® 21-C50. [Akzo] t-Butyl peroctoate in min. spirits; used where a diluent can be tolerated; initiator for acrylates; does not require refrigerated storage and transportation; liq.; 50% act.

Trigonox® 21-OP50. [Akzo] t-Butyl-peroctoate in DOP; initiator for elevated-temp. polyester cures; can be substituted where a liq. is desired; used in pultrusion, SMC, and BMC were fast cure is needed; liq.; 50% act.

Trigonox® 22-BB80. [Akzo] 1,1-Di-t-butylperoxy cyclohexane; initiator for SMC and BMC formulations; liq.; 80% in butyl benzyl phthalate.

Trigonox® 23-C75. [Akzo] t-Butylperoxyneodecanoate; highly react. liq. used in polyester pultrusion and continuous sheet curing; initiator for LDPE polymerization, for styrenics, for acrylates, and for PVC polymerization; liq.; 75% act.

Trigonox® 25-C75. [Akzo] t-Butyl perpivalate in odorless min. spirits; initiator for LDPE polymerization, styrenics, and PVC; liq.; 75% act.

Trigonox® 29/40 Bpd. [Akzo] 1,1-Di-t-butylperoxy-3,3,5-trimethylcyclohexane; crosslinking agent for olefin copolymers; useful for cross-linking EPDM, SBR, Neoprene, Hypalon; powd.; 40% act. with inorg. phlegmatizer.

Trigonox® 29-B75. [Akzo] 1,1-Di-t-butylperoxy-3,3,5-trimethylcyclohexane in DBP; initiator for elevated-temp. polyester cures, esp. SMC (conventional and low profile), BMC where longer flow time is required, automotive SMC processes; liq.; 75% act.

Trigonox® 29-C75. [Akzo] 1,1-Di-t-butylperoxy-3,3,5-trimethylcyclohexane; initiator for styrenics; liq.; 75% in odorless min. spirits.

Trigonox® 36-C75. [Akzo] Bis (3,5,5-trimethylhexanoyl) peroxide; initiator for LDPE polymerization, styrenics, acrylates, PVC polymerization; liq.; 75% act.

Trigonox® 40. [Akzo] 2,4-Pentanedione peroxide; initiator for ambient-temp. polyester cures; rapid hardness development in hand and spray lay-up; fire resistant; liq.; 4.0% act. oxygen.

Trigonox® 44P. [Akzo] 2,4-Pentanedione peroxide; initiator for ambient-temp. polyester cures; nonseparating paste form; applic.; incl. autobody repair putties and mine bolt cartridges; paste.

Trigonox® 97-C75. [Akzo] t-Butyl peroxy-2-methylbenzoate; initiator for styrenics; liq.; 75% act. in odorless min. spirits.

Trigonox® 99-C75. [Akzo] α-Cumyl peroxyneodecanoate; low-temp. initiator for all types of PVC prod.; liq.; 98% act.

Trigonox® 101. [Akzo] 2,5-Dimethyl-2,5-di (t-butyl peroxy hexane); initiator for polyester when molding at elevated temps.; also for cross-linking of olefin copolymers, chlorinated polyethylene, EPDM, SBR, and vinyls; initiator for free radical polymerizations of styrene; cross-linking agent for olefin copolymers, chlorinated polyethylene, EPDM, SBR, and vinyls; liq.; 90% act.

Trigonox® 101/45Bpd. [Akzo] 2,5-Dimethyl-2,5-di (t-butyl peroxy hexane); see Trigonox 101; liq.; 45% act. with inorg. phlegmatizer.

Trigonox® 107. [Akzo] Ethyl-o-benzoyl laurohydroximate; high-temp. crosslinking agent; liq.; 90% act.

Trigonox® 123-C75. [Akzo] t-Amyl peroxyneodecanoate; initiator for PVC; liq.; 75% act. in odorless min. spirits.

Trigonox® 125-C75. [Akzo] t-Amyl peroxypivalate; initiator used to provide faster curing of BMC and SMC; initiator for LDPE and PVC polymerization; liq.; 75% in odorless min. spirits.

Trigonox® A-W70. [Akzo] t-Butyl hydroperoxide; initiator for high-temp. acrylate polymerizations; suited for use in redox polymerization of acrylic emulsion polymers; liq.; 70% conc. in water.

Trigonox® ACS-M28. [Akzo] Acetyl cyclohexane sulfonyl peroxide in DMP; reactive initiator avail. for use in producing PVC; liq.; 29% conc.

Trigonox® ADC. [Akzo] Peroxydicarbonate; initiator for low- and med.-density polyethylene; initiator for PVC polymerization; liq.; 98% act.

Trigonox® B. [Akzo] Di-t-butyl peroxide; initiator for LDPE polymerization; lower m.w. at higher temps.; high-temp. peroxide used as finishing initiator for styrenics; crosslinking agent for olefin copolymers; volatile liq. best suited when used in injected extruder applic.; liq.; 99% act.

Trigonox® BPIC. [Akzo] t-Butyl peroxy isopropyl carbonate; initiator for BMC and SMC molding; initiator for free radical polymerizations of styrene and acrylates; cross-linking agent for elastomers incl. SBR, urethanes, EPR, EPDM, and nitrile rubbers; liq.; 75% act. in odorless min. spirits.

Trigonox® C. [Akzo] t-Butyl perbenzoate; initiator for elevated-temp. polyester cures, for LDPE polymerization, and high-temp. polymerization of acrylic emulsion polymers; used in BMC and SMC molding in temp. range 275–325 F; liq.; 98% act.

Trigonox® EHP. [Akzo] Bis(2-ethylhexyl) peroxydicarbonate; versatile, highly reactive initiator for PVC prod.; high m.w. reduces reactor fouling; liq.; 99% act.

Trigonox® EHP-AT70. [Akzo] Bis(2-ethylhexyl) peroxydicarbonate in toluene; see Trigonox EHP; liq.; 70% act.

Trigonox® EHP-C40. [Akzo] Bis(2-ethylhexyl) peroxydicarbonate; highly react. initiator for PVC prod.; liq.; 40% act. in odorless min. spirits.

Trigonox® EHP-C75. [Akzo] Bis(2-ethylhexyl) peroxydicarbonate in odorless min. spirits; see Trigonox EHP; liq.; 75% act.

Trigonox® F-C50. [Akzo] t-Butyl peracetate; initiator for low- and med.-dens. polyethylene and for styrenics; liq.; 50% act. in odorless min. spirits.

Trigonox® KSM. [Akzo] Perester in DBP; SMC/BMC initiator offering good shelf life in compd. and molding temp. range of 230–290 F; polyester cures; liq.; 75% act.

Trigonox® SBP. [Akzo] Di(s-butyl) peroxydicarbonate; initiator for LDPE polymerization; fastest conversion to polymer; used for prod. of PVC; liq.; 99% act.

Trigonox® SBP-AX30. [Akzo] Di(s-butyl) peroxydicarbonate in xylene; efficient initiator give fast cycle times for acrylate polymerization liq.; 30% act.

Trigonox® SBP-C60. [Akzo] Di (s-butyl) peroxydicarbonate); initiator for PVC giving fast cycle times for polymerization; liq.; 60% act. in odorless min. spirits.

Trigonox® SBP-C75. [Akzo] Di(s-butyl)peroxydicarbonate in odorless min. spirits; initiator giving fast cycle times for PVC polymerization; liq.; 75% act.

Trigonox® T. [Akzo] t-Butyl cumyl peroxide; crosslinking agent for EPDM, SBR, Neoprene, Hypalon; low reactivity peroxide useful as a finishing initiator at high temps. for styrenics; a synergist for some halogen-containing flame retardants; liq.; 92% act.

Trilon® A. [BASF] Trisodium NTA; low m.w. general purpose chelate; complexes iron in acid pH range; detergent builder for nonphosphate liqs.; 38% conc. in water.

Trilon® A-92. [BASF] Trisodium NTA; chelating agent; powd.; 92% conc.

Trilon® AS. [BASF] NTA; chelating agent; solid; 100% conc.

Trilon® B. [BASF] Tetrasodium EDTA; general purpose chelate; complexes most common metals over pH range, iron at acidic pH; 87% act. powd. or 40% aq. sol'n.

Trilon® BD. [BASF] Disodium EDTA; chelating agent; solid; 90% conc.

Trilon® B-FA. [BASF] Ferric ammonium EDTA; chelating agent for photographic applics.; liq.; 50% conc.

Trilon® BS. [BASF] EDTA; chelating agent; used where sodium ion is undesirable; solid; water-insol.; 100% conc.

Trilon® C. [BASF] Pentasodium pentetate; chelating agent used when higher metal chelate stability is required; used in peroxide bleach systems; liq.; 40% act. in water.

Trilon® CS. [BASF] Pentetic acid; chelating agent; solid; 100% conc.

Trilon® D. [BASF] Trisodium HEDTA; general purpose chelate for control of iron in pH range of 6.5–9.5, Ca, Mg; liq.; 40% act. in water.

Triple-H. [Hercules] Hydrolyzed veg. protein with salt; flavoring used as background taste in chicken, veal, pork, and seafood applic.; very lt. tan powd.; pH 5.6; 97% total solids; 21% flavor solids.

Tris Amino® (Crystals). [Angus] Tris (hydroxymethyl) aminomethane; pigment dispersant, neutralizing amine, corrosion inhibitor, acid-salt catalyst, pH buffer, chemical and pharmaceutical intermediate, solubilizer; m.w. 121.1; sol. 80 g/100 ml water; m.p. 171 C; b.p. 219 C; pH 10.4 (0.1M aq. sol'n.).

Tris Nitro®. [Angus] Tris (hydroxymethyl) nitromethane; antibacterial agent, preservative for water treatment, metalworking fluids, oil prod., deodorizing; formaldehyde releaser; 100% act. solid or 50% aq. sol'n.

Trisept B. [Tri-K] Butylparaben NF/tech.; preservative.

Trisept E. [Tri-K] Ethylparaben NF/tech.; preservative.

Trisept M. [Tri-K] Methylparaben NF/FCC/tech.; preservative.

Trisept P. [Tri-K] Propylparaben NF/FCC/tech.; preservative.

Trisept SDHA. [Tri-K] Sodium dehydroacetate FCC; preservative.

Tristat K. [Tri-K] Potassium sorbate NF/FCC; preservative; gran. and powd.

Tristat. [Tri-K] Sorbic acid NF/FCC; preservative.

Trisolan 1720. [Henkel/Emery] Lanolin-isopropyl ester; emollient, lubricant for hair sprays, eye shadow creams; sol. in oils, alcohol; insol. in water.

Triton® 770 Conc. [Rohm & Haas] Sodium alkylaryl polyether sulfate; anionic; wetting agent, penetrant used in detergents, cleaners, textile and leather processing; amber liq.; dens. 8.3 lb/gal; visc. 15 cps; flash pt. 78 F (PMCC); pour pt. –20 F; surf. tens. 28 dynes/cm (1%); 30% act.

Triton® CF-10. [Rohm & Haas] Alkylaryl polyether; nonionic; detergent for mechanical dishwashing, rinse aids, laundering, metal and dairy cleaning; nylon dyeing assistant; VCS 4 liq.; dens. 8.9 lb/gal; visc. 250 cps; HLB 14.0; cloud pt. 28 C (1%); flash pt. > 500 F (TOC); pour pt. 60 F; surf. tens. 32 dynes/cm (1%); 100% act.

Triton® CF-32. [Rohm & Haas] Amine polyglycol condensate; nonionic; wetting and antifoamer; detergent for mechanical dishwashing, metal and dairy cleaning; VCS 5 liq.; dens. 8.6 lb/gal; visc. 550 cps; cloud pt. 25 C (1%); flash pt. > 300 F (TOC); pour pt. 15 F; surf. tens. 32 dynes/cm (1%); 95% act.

Triton® CF-54. [Rohm & Haas] Polyethoxy adduct, modified; nonionic; see Triton CF-10; VCS 2 liq.; dens. 8.7 lb/gal; visc. 180 cps; HLB 13.6; cloud pt. 38 C (1%); flash pt. > 200 F (PMCC); pour pt. 35 F; surf. tens. 29 dynes/cm (0.01%); 100% act.

Triton® CS-7. [Rohm & Haas] Alkylaryl polyether alcohol and org. sulfonate; spreading-binding agent; used in agric. concs. with insecticides, fungicides, and miticides; yel. liq.; dens. 8.5 lb/gal; flash pt. 98 F (Setaflash CC); pour pt. –35 F; 60% act.

Triton® DF-12. [Rohm & Haas] Polyethoxylated alcohol, modified; nonionic; see Triton CF-10; lt. straw liq.; dens. 8.6 lb/gal; visc. 60 cps; cloud pt. 17 C (1%); flash pt. 430 F (COC); pour pt. 30 F; surf. tens. 33 dynes/cm (0.01%); 100% act.

Triton® DN-14. [Rohm & Haas] Alkylpolyether alcohol; nonionic; detergent, dye leveling, scouring agent and emulsifier; yel. liq.; dens. 8.8 lb/gal; visc. 400 cps; cloud pt. 90 C (1%); flash pt. > 300 F (TOC); pour pt. 55 F; surf. tens. 39 dynes/cm (1%); biodeg.;

90% act.

Triton® H-55. [Rohm & Haas] Phosphate ester, potassium salt; anionic; hydrotrope/solubilizer used in surfactants for liq. concs.; VCS 3 liq.; dens. 11.2 lb/gal; visc. 40 cps; flash pt. 154 F (PMCC); 50% act.

Triton® H-66. [Rohm & Haas] Phosphate ester, potassium salt; anionic; see Triton H-55; amber liq.; dens. 10.5 lb/gal; visc. 120 cps; flash pt. > 200 F (PMCC); pour pt. –5 F; 50% act.

Triton® N-40. [Rohm & Haas] Nonylphenol polyethoxy ethanol; nonionic; antifogging in plasticized PVC, corrosion inhibitor; dispersant of petrol. oils; liq.

Triton® N-57. [Rohm & Haas] Nonoxynol-5; nonionic; detergent, emulsifier, intermediate; APHA 100 liq.; m.w. 440; oil-sol.; water-disp; dens. 8.5 lb/gal; visc. 240 cps; HLB 10.0; cloud pt. < 0 C (1%); flash pt. > 515 F (COC); pour pt. –25 F; surf. tens. 29 dynes/cm (1%); 100% conc.

Triton® N-60. [Rohm & Haas] Nonoxynol-6; nonionic; see Triton N-57; APHA 125 liq.; m.w. 484; dens. 8.7 lb/gal; visc. 300 cps; HLB 10.9; cloud pt. 53 C (1%); pour pt. –25 F; surf. tens. 28 dynes/cm (1%); 100% act.

Triton® N-101. [Rohm & Haas] Nonoxynol-10; nonionic; wetting agent, detergent, dispersant, solubilizer, emulsifier for pesticides; household and industrial cleaners, textile processing; APHA 100 liq.; m.w. 642; dens. 8.7 lb/gal; visc. 240 cps; HLB 13.4; cloud pt. 54 C (1%); pour pt. 40 C; surf. tens. 29 dynes/cm (1%); 100% act.

Triton® QS-44. [Rohm & Haas] Phosphate surfactant in free acid form; anionic; hydrotrope, detergent; solubilizer for nonionic surfactants in alkaline cleaning baths and for nonionic and anionic surfactants in built concs.; VCS 5 aq. sol'n.; dens. 9.8 lb/gal; visc. 8000 cps; pour pt. 34 F; flash pt. (TOC) 200 F; surf. tens. 34 dyne/cm (0.1%); 80% act.

Triton® RW-10. [Rohm & Haas] Polyethoxy amine; nonionic/cationic; surfactant and detergent used in industrial applications; emulsifier for metalworking for coolants, fiber lubricant; degreaser in annealing nylon fibers; lt. amber clear liq.; amine odor; insol. in water; sp.gr. 0.89; visc. 400 cps; pour pt. –9 C; flash pt. 129 C (SFCC); pH 10.0 (5% aq.).

Triton® RW-20. [Rohm & Haas] Polyethoxy amine; nonionic/cationic; see Triton RW-10; lt. amber clear liq.; amine odor; insol. in water; sp.gr. 0.913; dens. 7.61 lb/gal; visc. 240 cps; HLB 6–8; b.p. 73 C; flash pt. > 167 C (SFCC); pH 9.5–11.0 (5% aq.).

Triton® RW-75. [Rohm & Haas] Polyethoxy amine; nonionic/cationic; see Triton RW 10; lt. amber clear liq.; amine odor; sp.gr. 0.981; dens. 8.17 lb/gal; HLB 14–16; flash pt. > 149 C (CC); cloud pt. 32 C (1% aq.); pH 9.5–11.0 (5% aq.).

Triton® RW-150. [Rohm & Haas] Polyethoxy amine (15 EO); nonionic/cationic; see Triton RW-10; lt. amber clear liq.; amine odor; sp.gr. 1.024; dens. 8.58 lb/gal; visc. 480 cps; HLB > 16; pour pt. –8 F; flash pt. > 149 C (CC); cloud pt. 96 C (1% aq.); pH 9.5–11.0 (5% aq.); surf. tens. 46 dynes/cm (0.01% aq.).

Triton® W-30 Conc. [Rohm & Haas] Sodium alkylaryl ether sulfate; anionic; penetrant, wetter for textiles; emulsifier in emulsion polymerization; amber liq.; dens. 8.2 lb/gal; visc. 9 cps; flash pt. 74 F (PMCC); pour pt. –15 F; surf. tens. 29 dynes/cm (1%); 27% act.

Triton® X-15. [Rohm & Haas] Octoxynol-1; nonionic; coupling agent, emulsifier; APHA 250 liq.; m.w. 250; dens. 8.2 lb/gal; visc. 790 cps; HLB 3.6; flash pt. > 300 F (TOC); pour pt. 15 F; 100% act.

Triton® X-35. [Rohm & Haas] Octoxynol-3; nonionic; see Triton X-15; APHA 125 liq.; m.w. 338; dens. 8.5 lb/gal; visc. 370 cps; HLB 7.8; flash pt. > 300 F (TOC); pour pt. –10 F; surf. tens. 29 dynes/cm (0.01%); 100% act.

Triton® X-45. [Rohm & Haas] Octoxynol-5; nonionic; drycleaning detergent, dispersant, insecticide emulsifier, wetting agent; APHA 100 liq.; m.w. 426; dens. 8.7 lb/gal; visc. 290 cps; HLB 10.4; cloud pt. < 0 C (1%); flash pt. > 300 F (TOC); pour pt. –15 F; surf. tens. 28 dynes/cm (1%); 100% act.

Triton® X-100. [Rohm & Haas] Octoxynol-9 (9–10 EO); nonionic; wetting agent, dispersant, detergent, household and industrial cleaners; textile processing, wool scouring; emulsifier for insecticides and herbicides; solubilizer of perfumes; APHA 100 liq.; m.w. 628; dens. 8.9 lb/gal; visc. 240 cps; HLB 13.5; cloud pt. 65 C (1%); flash pt. > 300 F (TOC); pour pt. 45 F; surf. tens. 30 dynes/cm (1%); 100% act.

Triton® X-102. [Rohm & Haas] Octoxynol-13; nonionic; detergent, wetting agent; metal cleaning, industrial, household liq. detergents and cleaners; solubilizer of anionic detergents; APHA 100 liq.; m.w. 756; dens. 8.9 lb/gal; visc. 330 cps; HLB 14.6; cloud pt. 88 C (1%); flash pt. > 300 F (TOC); pour pt. 60 F; surf. tens. 32 dynes/cm (1%); 100% act.

Triton® X-120. [Rohm & Haas] Octoxynol-9 (9–10 EO); nonionic; wetting agent and dispersant; agric. wettable powds.; wh. solid; m.w. 628; 40% act.

Triton® X-155. [Rohm & Haas] Methylenebisdiamyl phenoxy polyethoxy ethanol; nonionic; surfactant in emulsification of aromatic hydrocarbons; wetting agent and detergent; antistat and lubricant for textile fiber processing; amber clear visc. liq.; mild odor; sol. in water and aromatic hydrocarbons; sp.gr. 1.052; dens. 8.76 lb/gal; visc. 790 cps; HLB 12.5; pour pt. 0 F; flash pt. (TOC) 430 F; cloud pt. < 10 C (1% aq.); pH 7.0–9.0 (1%); surf. tens. 30 dynes/cm (0.1%); 100% act.

Triton® X-155-90%. [Rohm & Haas] Methylenebisdiamyl phenoxy polyethoxy ethanol; nonionic; see Triton X-155; yel.-amber clear liq.; mild odor; sol. see Triton X-155; sp.gr. 1.02; dens. 8.50 lb/gal; visc. 250 cps; HLB 12.5; pour pt. –20 F; flash pt. (Setaflash CC) 53 C; cloud pt. < 10 C (1% aq.); pH 8.50 (5%); surf. tens. 32 dynes/cm (0.01% aq.); 90% act.

Triton® X-200. [Rohm & Haas] Sodium octoxynol-2 ethane sulfonate; anionic; detergent, metal cleaning, pickling and plating baths; household cleaners; polymerization emulsifier and latex post-stabilizer; wetting agent in strong alkaline baths; dispersant for fulling in lime soap; dye leveling agent for acid dyestuffs; wh. liq.; dens. 8.9 lb/gal; visc. 7000 cps; flash pt. > 300 F (TOC); pour pt. 25 F; surf. tens. 29 dynes/cm (1%); 28% act.

Triton® X-202. [Rohm & Haas] Sodium alkylaryl polyether sulfonate; anionic; detergent, wetting agent, emulsifier; metal cleaning, pickling and plating baths; household cleaners; latex post stabilizer; wh. liq.; dens. 8.9 lb/gal; visc. 4800 cps; flash pt. > 212 F (TOC); pour pt. 45 F; surf. tens. 30 dynes/cm (1%); 30% act.

Triton® X-301. [Rohm & Haas] Sodium alkylaryl polyether sulfate; anionic; wetting agent, penetrant; detergent for household and industrial cleaners; cream liq.; dens. 8.8 lb/gal; visc. 4200 cps; flash pt. > 300 F (TOC); pour pt. 30 F; surf. tens. 28 dynes/cm (1%); 20% act.

Triton® X-405. [Rohm & Haas] Octoxynol-40; nonionic; surfactant; coupling agent in naphthol dyeing; APHA 250 liq.; m.w. 1966; dens. 9.2 lb/gal; visc. 490 cps; HLB 17.9; cloud pt. > 100 C (1%); flash pt. > 212 F (TOC); pour pt. 25 F; surf. tens. 44 dynes/cm (1%); 70% act.

Troykyd 42-BA. [Troy] Polymerized oil deriv.; bodying agent; post-additive to increase the visc. of solv.-thinned coatings; air-drying coatings based on long oil and med. oil alkyd resins, and activity with styrene and vinyl toluene modified alkyds, oil-modified PU, epoxy esters, etc.; suitable for clears, primers, flat paints, and gloss and semigloss enamels; used at 0.25–1.0% based on total formulation wt.; visc., amber flowy liq.; sp.gr. 0.91–0.925; dens. 7.6–7.7 lb/gal; visc. (G-H) V–Z; 53–56% N.V.

Troykyd Anti-Float Powd. [Troy] Micromilled, surface-treated calcium carbonate; antifloating, antisilking, and antiflooding agent for solvent-based coatings; improves pigment wetting properties to reduce surf. tens. of the solvent-based system; used in all wh. paints and bases that may be tinted or shaded with colors having a tendency to float and for paints containing pigments with floating tendencies; wh. fine powd., 99.5% min. thru 325 mesh; water-insol.; sp.gr. 2.7; 100% act.

Troykyd Anti-Gel. [Troy] Wetting agent designed to prevent excessive bodying, gellation, and seeding of pigment pastes and pigmented solv.-based coatings; stabilizes paste visc. and permits higher pigment of the finished paint; esp. effective in systems containing reactive pigments such as zinc oxide and red lead, and with high absorptive pigments such as carbon blk., iron blue, org. yels. and org. reds; used in the pigment disp. step of paint mfg. at 0.5–1.0% based on total wt. of paint; clear almost water-wh. liq.; sp.gr. 0.95–0.98; dens. 7.9–8.2 lb/gal; visc. A-3 to A-5; acid no. 1–3.

Troykyd Anti-Settle. [Troy] Silicate compd.; antisettling agent, suspending agent for solv.-based coating systems; wh. fine powd., 98% thru 325 mesh; sp.gr. 2.0–2.2; dens. 16.7–18.4 lb/gal.

Troykyd Anti-Settle Special. [Troy] Antisettling agent, suspending agent for paints, caulks, glazing compds., sealants; off-wh. powd.; sp.gr. 1.18–1.2; dens. 9.8–10.1 lb/gal.

Troykyd Anti-Skin. [Troy] Oxime blend; antiskinning agent used in coating vehicles; lt. mobile liq.; char. mild odor; sp.gr. 0.96–1.00; dens. 8.00–8.33 lb/gal; 100% act.

Troykyd Anti-Skin B. [Troy] Methyl ethyl ketoxime, tech.; antiskinning agent for paint industry; off-wh. mobile liq.; sp.gr. 0.91–0.92 (80 F); dens. 7.6–7.7 lb/gal (80 F); visc. (Gardner) A max.; flash pt. 140 F min. (OC); 98% min. act.

Troykyd Anti-Skin BTO. [Troy] Butyraldoxime, tech.; see Troykyd Anti-Skin; off-wh. mobile liq.; char. odor; sp.gr. 0.902–0.912 (80 F); dens. 7.51–7.60 lb/gal; visc. (Gardner) A-1 max.; flash pt. 115 F min. (OC); 99% min. act.

Troykyd Anti-Skin Odorless Powder. [Troy] Oxime; antiskinning agent for paint industry; off-wh. powd.; b.p. 206 C; sol. in hydrocarbons and alcohols; m.p. 88–90 C; flash pt. 112 C; 97% min. act.

Troykyd Anti-Skin S. [Troy] Phenolic; antiskinning agent and antioxidant used in coating systems; stabilizer used in dip tank coatings; gray-amber liq.; sp.gr. 0.94–0.98; dens. 7.8–8.2 lb/gal; visc. (Gardner) A-3 to A-5; flash pt. 220 F (OC).

Troykyd Anti-Skin Special M. [Troy] Phenolic; antiskinning agent and antioxidant for coating vehicles; lt. amber liq.; sp.gr. 0.94–0.965; dens. 7.8–8.05 lb/gal; visc. A max.

Troykyd Compd. XYZ. [Troy] Proprietary; suspending agent for use in oil systems.

Troykyd Emulso Wet. [Troy] Nonionic surfactant; wetting agent, disp. stabilizer for latex paints when improved stability and compatibility are needed for min. side effects; emulsion stabilizer for any aq. system; improves color acceptance, esp. with red iron oxide and other inorg. pigments; useful in semigloss latex paint systems; should be added in the initial pigment disp. step of paint mfg. at 6 lb/100 gal.; lt. yel. Mobile liq.; sol. in water and most solvs. except aliphatic hydrocarbons; sp.gr. 1.07–1.09; dens. 8.9–9.2 lb/gal; visc. (Gardner): K-P; HLB 15.

Troykyd EX-390. [Troy] Urethane; catalyst; water-wh. to pale amber liq.; sp.gr. 1.150–1.170; visc. (Gardner) F; 10.0 ± 0.2% mercury.

Troykyd Lecithin W.D. [Troy] Lecithin; pigment dispersant for aq. systems; visc. stabilizer; used in paint processing; amber visc. liq.; water-disp.; sp.gr. 1.02–1.05 (80 F); dens. 8.5–8.8 lb/gal (80 F); visc. (Gardner) Z1–Z4; pH 7.3–8.2 (5% disp.); 95% solids.

Troykyd Perma-Dry. [Troy] Cobalt compd. complexed with rare earth and alkaline earth metals; lead-free; feeder drier for paints; blue visc. paste; sp.gr. 1.09–1.12; dens. 9.1–9.3 lb/gal.

Troykyd XYZ. [Troy] Polymerized org. ester; bodying agent for solv.-thinned coatings, sets up a gel structure with the vehicle; in trade sales and industrial coatings, caulks; in formulations based on hydrocarbon solv.; increased visc., imparts high thixotrophy, sag resistance, pigment settling, and penetration; wh. powd.; dens. 8.4–8.6 lb/gal.

Troymax Drier Calcium 4%. [Troy] Calcium drier; used in org. coatings, inks, polyesters; Gardner 4 max. liq.; sp.gr. 0.855–0.890; dens. 7.13–7.43 lb/gal; visc. (Gardner) A max.; flash pt. 104 F (PMCC); 4.0 ± 0.1% metal.

Troymax Drier Calcium 5%. [Troy] Calcium drier; see Troymax Drier Calcium 4%; Gardner 6 max.; sp.gr. 0.941–0.950; dens. 7.8–7.9 lb/gal; visc. (Gardner) B max.; flash pt. 104 F (PMCC); 5.0 ± 0.1% metal.

Troymax Drier Calcium 6%. [Troy] Calcium drier; see Troymax Drier Calcium 4%; Gardner 7 max.; sp.gr. 0.950–0.976; dens. 7.9–8.10 lb/gal; visc. (Gardner) N max.; flash pt. 104 F (PMCC); 6.0 ± 0.1% metal.

Troymax Drier Calcium 8%. [Troy] Calcium drier; see Troymax Drier Calcium 4%; Gardner 5 max.; sp.gr. 0.900–0.960; dens. 7.50–8.00 lb/gal; visc. (Gardner) B max.; flash pt. 104 F (PMCC); 8.0 ± 0.1% metal.

Troymax Drier Calcium 10%. [Troy] Calcium drier; see Troymax Drier Calcium 4%; Gardner 5 max.; sp.gr. 0.950–1.010; dens. 7.91–8.42 lb/gal; visc. (Gardner) B max.; flash pt. 104 F (PMCC); 10.0 ± 0.1% metal.

Troymax Drier Cobalt 6%. [Troy] Cobalt drier; see Troymax Drier Calcium 4%; sp.gr. 0.865–0.915; dens. 7.2–7.6 lb/gal; visc. (Gardner) A max.; flash pt. 104 F (PMCC); 6 ± 0.1% metal.

Troymax Drier Cobalt 12%. [Troy] Cobalt drier; see Troymax Drier Calcium 4%; blue-violet; sp.gr. 1.003–1.055; dens. 8.35–8.79 lb/gal; visc. (Gardner) G max.; flash pt. 104 F (PMCC); 12 ± 0.1% metal.

Troymax Drier Iron 9%. [Troy] Iron drier; see Troymax Drier Calcium 4%; dk. brnsh.; sp.gr. 0.935–9.965; dens. 7.80–8.05 lb/gal; visc. (Gardner) A max.; flash pt. 104 F (PMCC); 9.0 ± 0.1% metal.

Troymax Drier Lead 24%. [Troy] Lead driers; see Troymax Drier Calcium 4%; Gardner 5 max.; sp.gr. 1.08–1.12; dens. 9.0–9.30 lb/gal; visc. (Gardner) A-1 max.; flash pt. 104 F (PMCC); 24.0 ± 0.1% metal.

Troymax Drier Lead 36%. [Troy] Lead driers; see Troymax Drier Calcium 4%; Gardner 7 max.; sp.gr. 1.375–1.425; dens. 11.45–11.87 lb/gal; visc. (Gardner) 3 max.; flash pt. 104 F (PMCC); 36.0 ± 0.2% metal.

Troymax Drier Manganese 6%. [Troy] Manganese driers; see Troymax Drier Calcium 4%; Gardner 18 max.; sp.gr. 0.90–0.92; dens. 7.5–7.7 lb/gal; visc. (Gardner) A max.; flash pt. 104 F (PMCC); 6.0 ± 0.1% metal.

Troymax Drier Manganese 9%. [Troy] Manganese driers; see Troymax Drier Calcium 4%; Gardner 18 max.; sp.gr. 0.998–1.015; dens. 8.3–8.5 lb/gal; visc. (Gardner) C max.; flash pt. 104 F (PMCC); 9.0 ± 0.1% metal.

Troymax Drier Manganese 12%. [Troy] Manganese driers; see Troymax Drier Calcium 4%; dk. brnsh.; sp.gr. 1.045–1.075; dens. 8.70–8.96 lb/gal; visc. (Gardner) R max.; flash pt. 104 F (PMCC); 12.0 ± 0.1% metal.

Troymax Drier Zinc 8%. [Troy] Zinc driers; see Troymax Drier Calcium 4%; Gardner 2 max.; sp.gr. 0.862–0.895; dens. 7.19–7.46 lb/gal; visc. (Gardner) A-3 max.; flash pt. 104 F (PMCC); 8.0 ± 0.1% metal.

Troymax Drier Zinc 16%. [Troy] Zinc driers; see Troymax Drier Calcium 4%; Gardner 4 max.; sp.gr. 1.06–1.08; dens. 8.8–9.0 lb/gal; visc. (Gardner) B max.; flash pt. 104 F (PMCC); 16.0 ± 0.1% metal.

Troymax Drier Zirconium 6%. [Troy] Zirconium driers; see Troymax Drier Calcium 4%; Gardner 5 max.; sp.gr. 0.838–0.875; dens. 6.98–7.30 lb/gal; visc. (Gardner) A-4 max.; flash pt. 104 F (PMCC); 6.0 ± 0.1% metal.

Troymax Drier Zirconium 12%. [Troy] Zirconium driers; see Troymax Drier Calcium 4%; Gardner 5 max.; sp.gr. 0.988–0.999; dens. 8.2–8.3 lb/gal; visc. (Gardner) A-2 max.; flash pt. 104 F (PMCC); 12.0 ± 0.1% metal.

Troymax Drier Zirconium 18%. [Troy] Zirconium driers; see Troymax Drier Calcium 4%; Gardner 6 max.; sp.gr. 1.1–1.2; dens. 9.2–10.0 lb/gal; visc. (Gardner) A max.; flash pt. 104 F (PMCC); 18 ± 0.1% metal.

Troymax Drier Zirconium 24%. [Troy] Zirconium driers; see Troymax Drier Calcium 4%; Gardner 5 max.; sp.gr. 1.235–1.255; dens. 10.30–10.47 lb/gal; visc. (Gardner) J max.; flash pt. 104 F (PMCC); 24 ± 0.1% metal.

Troysan 28. [Troy] Sodium mercaptobenzothiazole; preservative and mildew inhibitor for aq. compositions; paper industry protecting starch and casein coatings for bacterial or fungal activity; textile industry for felt linoleum and rug backings; food pkg. adhesives; used in emulsion polishes, floor waxes; amber liq.; sp.gr. 1.238 (80 F); dens. 10.34 lb/gal (80 F); visc. (Gardner) A-2; pH 10.3–10.4; 50% act.

Troysan 142. [Troy] 3,5-Dimethyltetrahydro-1,3,5,2H-thiadiazine-2-thione; antimicrobial, fungicide; preservative for food pkg. adhesives, dispersed colors, resin emulsions; off-wh. cryst. powd.; slight, distinctive odor; sol. 21% in ethylene dichloride, 19.5% in acetone, 0.12% in water; sp.gr. 1.39; dens. 11.6 lb/gal; m.p. 99.5 C; flash pt. (OC) 280 F; 95% min. act.

Troysan 174. [Troy] 2[(Hydroxymethyl) amino] ethanol; preservative for protection against bacterial spoilage in aq. systems; amber clear liq.; mild odor; water-sol.; sp.gr. 1.144–1.153 (80 F); dens. 9.5–9.6 lb/gal (80 F); visc. (Gardner) L max.; pH 11.0 ± 0.5; 100% act.

Troysan 192. [Troy] 2-[(Hydroxymethyl) amino]-2-methylpropanol; preservative used in aq. systems; pale amber clear liq.; typ. odor; b.p. 99 C; water-sol.; sp.gr. 0.965–0.985 (80 F); dens. 8.00–8.25 lb/gal (80 F); visc. (Gardner) A-3 to A-5; ref. index. 1.42–1.432; pH 11.2 ± 0.5; 100% act.

Troysan 364. [Troy] Troysan 174 (21%), tributyltin oxide (29%); antimicrobial, fungicide, perservative for aq. adhesives; sterilizing agent for latex and paint processing equipment; Gardner 4 max. liq.; water-disp.; sp.gr. 1.011–1.022 (80 F); flash pt. (TOC) 82 F; 50% act.

Troysan Anti-Mildew O. [Troy] N-(trichloromethylthio) phthalimide; mildewcide and fungicide for interior and exterior coatings; off-wh. fine powd.;

mild, char. odor; slightly sol. in aromatic and aliphatic hydrocarbons; insol. in water; sp.gr. 1.75; dens. 14.6 lb/gal; m.p. 329–338 F; 88% act.

Troysan CMP Acetate. [Troy] Chlormethoxypropyl mercuric acetate; preservative for water-based paints; prevents putrefaction, gas formation, or visc. loss when paints are inadequately protected against bacterial and fungal action during shelf life.; colorless clear, mobile liq.; misc. with water; sp.gr. 1.120–1.146 (80 F); dens. 9.34–9.55 lb/gal (80 F); pH 10.0 ± 0.5; 20% act.

Troysan Copper 8. [Troy] Copper naphthenate; preservative for textile, cordage, wood, and cellulosic materials; textile, cordage, and wood dip; sol'n.; sol. in aliphatic and aromatic solvs.; sp.gr. 0.99–1.06 (80 F); dens. 8.2–8.8 lb/gal (80 F); visc. (Gardner) 25 s (10 cm mobilometer); flash pt. 104 F min. (PMCC); 80% max solids.

Troysan PMA-10-SEP. [Troy] Phenyl mercury acetate; preservative and mildew inhibitor for dry formulations where use of a liq. preservative is not feasible; off-wh. fine powd., < 2% retained on 325 mesh sieve; dens. 22 lb/ft^3; 10% conc.

Troysan PMA-30. [Troy] Solubilized phenylmercury acetate; preservative and mildew inhibitor for water-based emulsion for exterior paints; Gardner 4 max. liq.; water-sol.; sp.gr. 1.275–1.320; dens. 10.64–11.00 lb/gal; visc. (Gardner) A max.; pH 7.5–8.0; 30% min. conc.

Troysan PMA-100. [Troy] Phenyl mercury acetate; preservative and mildew inhibitor in emulsion paints; used in water-based paints as pkg. preservative and fungicide for exterior applic.; wh. powd.; 100% act.

Troysan PMDS-10. [Troy] Di (phenyl mercury) dodecenyl succinate; pkg. preservative and mildew inhibitor for aq. latex paints; sol. in org. solvs.; water-insol. but disp. in aq. systems; sp.gr. 0.925–0.947; dens. 7.7–7.9 lb/gal; visc. (G-H) A-2; flash pt. 104 F min. (PMCC); 21% act.

Troysan PMO-30. [Troy] Phenylmercury oleate; mildewcide for paints; Gardner 5 max. liq.; sol. in min. spirits and aromatic solvs.; sp.gr. 0.985–1.020; dens. 8,.24–8.50 lb/gal; visc. (Gardner) A max.; flash pt. (PMCC) 104 F; 30% phenylmercury oleate; 11% metal content.

Troysan Polyphase 17 WD. [Troy] 3-Iodo-2-propynyl butyl carbamate; fungicide, wood preservative; amber liq.; water-disp.; 17% conc.

Troysan Polyphase AF1. [Troy] 3-Iodo-2-propynyl butyl carbamate; fungicide, wood preservative; used in oil-based and latex paints; amber liq.; 40% conc.

Troysan Polyphase Anti-Mildew. [Troy] 3-Iodo-2-propynyl butyl carbamate; antimildew additive for interior and exterior protective coatings; fungicide; amber liq.; char. odor; sol. in aromatics and alcohols; sp.gr. 1.18–1.20; dens. 9.82–10.1 lb/gal; visc. (Gardner) C max.; 40% act.

Troysan Polyphase P-100. [Troy] 3-Iodo-2-propynyl butyl carbamate; fungicide, wood preservative; wh. powd.

Troythix A. [Troy] Polymerized, chemically inert org. ester; bodying agent for trade sales and industrial coatings thinned with hydrocarbons, ketone, and ester solv.; imparts high thixotrophy resistance to sagging, pigment settling, and penetration, wh. powd.; dens. 8.5 lb/gal; 100% active.

Trycol® 5878. [Henkel/Emery] Ethoxylated fatty alcohol ethers; nonionic; emulsifier, lubricant for textiles and detergent shampoo base; solubilizer; personal care prods.; liq.; HLB 9.5; 100% conc.

Trycol® 5882. [Henkel/Emery] Laureth-4; coemulsifier for silicone in polishes, mold release agents; emulsifier in industrial lubricants, textile applics.; Gardner 1 liq.; sol. in butyl stearate, glycerol trioleate, xylene; disp. in water, min. oil; dens. 7.7 lb/gal; visc. 20 cSt (100 F); HLB 9.2; cloud pt. < 25 C; flash pt. 325 F; pour pt. 12 C; biodeg.

Trycol® 5888. [Henkel/Emery] Steareth-20; emulsifier, solubilizer, solv. emulsifier in textile dye carriers; stabilizer in latices; used in dyeing assistants, fruit coatings; Gardner 1 solid; sol. in water, xylene; HLB 15.3; m.p. 40 C; cloud pt. 91 C (5% saline); flash pt. 460 F.

Trycol® 5940. [Henkel/Emery] Trideceth-6; nonionic; emulsifier, dispersant, wetting agent, detergent, foam builder for degreasers, cutting oils, wool scouring; intermediate; Gardner 1 liq.; sol. in butyl stearate, glycerol trioleate, Stod., xylene; disp. in water, min. oil; dens. 8.2 lb/gal; visc. 32 cSt (100 F); HLB 11.4; cloud pt. < 25 C; flash pt. 345 F; pour pt. 12 C; 100% conc.

Trycol® 5946. [Henkel/Emery] Trideceth-18; nonionic; emulsifier, dispersant, detergent, wetting agent used in paper and textile industries; leveling agent, solubilizer, foam stabilizer; intermediate; Gardner 1 solid; sol. in water; dens. 8.5 lb/gal (40 C); HLB 16.0; m.p. 38 C; cloud pt. 83 C (5% saline); flash pt. 475 F; 100% conc.

Trycol® 5949. [Henkel/Emery] Trideceth-8; foam builder, solubilizer, coemulsifier; Gardner 1 liq.; sol. in water, glycerol trioleate, Stod., xylene; disp. in min. oil; dens. 8.3 lb/gal; visc. 39 cSt (100 F); HLB 12.5; cloud pt. 43 C; flash pt. 370 F; pour pt. 8 C.

Trycol® 5950. [Henkel/Emery] Deceth-4; nonionic; wetting agent and penetrant for textile applics.; intermediate; Gardner 1 liq.; sol. in xylene, glycerol trioleate; disp. in water, min. oil; dens. 7.9 lb/gal; visc. 17 cSt (100 F); HLB 10.5; cloud pt. < 25 C; flash pt. 260 F; pour pt. –5 C; 100% conc.

Trycol® 5951. [Henkel/Emery] Deceth-5; nonionic; penetrant, wetting agent and emulsifier for heavy duty cleaners; Gardner 1 liq.; dens. 8.2 lb/gal; visc. 23 cs; HLB 11.6; cloud pt. < 25 C; flash pt. 340 F; pour pt. 8 C; 100% act.

Trycol® 5964. [Henkel/Emery] Laureth-23; nonionic; emulsifier, lubricant used in textiles and detergent shampoo bases; solubilizer in personal care prods.; coemulsifier for silicone in polishes and mold release agents; solv. emulsifier for textile dye carriers; Gardner 1 solid; sol. in water, xylene; dens. 8.6 lb/gal (70 C); HLB 16.7; m.p. 40 C; cloud pt. 93 C (5% saline); flash pt. 440 F; 100% conc.

Trycol® 5968. [Henkel/Emery] Trideceth-8; see

Trycol 5949; Gardner 1 liq.; sol. in water, glycerol trioleate, Stod., xylene; disp. in min. oil; dens. 8.4 lb/gal; visc. 51 cSt (100 F); HLB 12.5; cloud pt. 43 C; pour pt. < –10 C; 90% act. in water.

Trycol® 5971. [Henkel/Emery] Oleth-20; nonionic; emulsifier, dispersant, solubilizer; Gardner 1 solid; sol. in water; dens. 8.5 lb/gal; HLB 15.6; m.p. 39 C; cloud pt. 87 C (5% saline); flash pt. 500 F; 60% conc.

Trycol® 5972. [Henkel/Emery] Oleth-23; nonionic; emulsifier, dispersant, solubilizer, detergent, stabilizer, anticoagulant, dyeing assistant, lubricant used in textiles, cosmetics and processing of animal fibers; Gardner 2 solid; sol. in water, xylene; HLB 15.8; m.p. 47 C; cloud pt. 89 C (5% saline); flash pt. 440 F; 100% conc.

Trycol® 5993. [Henkel/Emery] Trideceth-3; emulsifier, antifoam for textiles; intermediate; Gardner 1 liq.; sol. in butyl stearate, glycerol trioleate, Stod., xylene; disp. in water, min. oil; dens. 7.8 lb/gal; visc. 19 cSt (100 F); HLB 7.9; cloud pt. < 25 C; flash pt. 300 F; pour pt. –15 C.

Trycol® 6720. [Henkel/Emery] Ether; nonionic; low foaming surfactant controlling foam and providing penetration for textile wet processing, metal processing rinses; Gardner 1 liq.; sol. in water, glycerol trioleate, Stod., xylene; dens. 8.2 lb/gal; visc. 39 cSt (100 F); cloud pt. 24 C; flash pt. 360 F.

Trycol® 6802 . [Henkel/Emery] Nonionic; surfactant, dyeing assistant, mild scouring agent; Gardner 1 liq.; sol. in Stod.; disp. in water, min. oil; dens. 8.8 lb/gal; HLB 15.3; cloud pt. 80–90 C (5% saline); flash pt. 485 F; pour pt. 28 C.

Trycol® 6942. [Henkel/Emery] Nonoxynol-100; wetting agent in high electrolyte sol'ns.; stabilizer in syn. latices; Gardner 1 solid; sol. in water; HLB 19.0; m.p. 56 C; cloud pt. 72 C (10% saline); flash pt. 500 F.

Trycol® 6954. [Henkel/Emery] Nonoxynol-150; wetting agent, dispersant; Gardner 1 solid; sol. in water; HLB 19.3; m.p. 60 C; cloud pt. 69 C (10% saline); flash pt. 500 F.

Trycol® 6957. [Henkel/Emery] Nonoxynol-40; detergent, wetting agent, dispersant, coemulsifier, stabilizer; Gardner 1 solid; sol. in water, xylene; HLB 17.8; cloud pt. 90 C (5% saline); flash pt. 560 F.

Trycol® 6960. [Henkel/Emery] Nonoxynol-1; surfactant, coemulsifier, defoamer; Gardner 1 liq.; sol. in min. oil, xylene; dens. 8.3 lb/gal; visc. 150 cSt (100 F); HLB 4.6; cloud pt. < 25 C; flash pt. 385 F; pour pt. –10 C.

Trycol® 6961. [Henkel/Emery] Nonoxynol-4; nonionic; detergent and emulsifier; chemical intermediate; corrosion inhibitor in two-cycle engine oils; Gardner 2 liq.; sol. in Stod., xylene; disp. in min. oil; dens. 8.5 lb/gal; visc. 472 cSt (100 F); HLB 8.9; cloud pt. < 25 C; flash pt. 430 F; pour pt. –10 C; 100% conc.

Trycol® 6962. [Henkel/Emery] Nonoxynol-6; dispersant, wetting agent, coemulsifier in acid cleaning sol'ns., solv. emulsions, detergents; Gardner 2 liq.; sol. in Stod., xylene; disp. in water, min. oil; dens. 8.5 lb/gal; visc. 100 cSt (100 F); HLB 10.9; cloud pt. < 25 C; flash pt. 515 F; pour pt. –10 C.

Trycol® 6963. [Henkel/Emery] Nonoxynol-7; emulsifier, dispersant, wetting agent; Gardner 1 liq.; sol. in Stod., xylene; disp. in water, min. oil; dens. 8.6 lb/gal; visc. 100 cSt (100 F); HLB 11.7; cloud pt. < 25 C; flash pt. 495 F; pour pt. –10 C.

Trycol® 6970. [Henkel/Emery] Nonoxynol-40; see Trycol 6957; Gardner 1 liq.; sol. in water; dens. 9.2 lb/gal; visc. 385 cSt (100 F); HLB 17.8; cloud pt. 90 C (5% saline); pour pt. 7 C; 70% aq.

Trycol® 6971. [Henkel/Emery] Nonoxynol-50; see Trycol 6957; Gardner 1 solid; sol. in water, xylene; HLB 18.2; m.p. 54 C; cloud pt. 76 C (10% saline); flash pt. 520 F.

Trycol® 6972. [Henkel/Emery] Nonoxynol-50; see Trycol 6957; Gardner 1 liq.; sol. in water; dens. 9.0 lb/gal; visc. 440 cSt (100 F); HLB 18.2; cloud pt. 76 C (10% saline); pour pt. –4 C; 70% act.

Trycol® 6974. [Henkel/Emery] Nonoxynol-10; dispersant, emulsifier, wetting agent; sol. in water, Stod., xylene; dens. 8.7 lb/gal; visc. 111 cSt (100 F); HLB 13.2; cloud pt. 62 C; flash pt. 530 F; pour pt. 11 C.

Trycol® 6981. [Henkel/Emery] Nonoxynol-100; see Trycol 6942; Gardner 1 liq.; sol. in water; dens. 9.2 lb/gal; visc. 564 cSt (100 F); HLB 19.0; cloud pt. 72 C (10% saline); pour pt. 18 C; 70% act.

Trycol® 6985. [Henkel/Emery] Nonyl nonoxynol-8; emulsifier in textile dye carrier applics., insecticides, wax emulsions; foam control agent; spreading agent in pigment printing; post-stabilizer in emulsion polymerization; intermediate; Gardner 2 liq.; sol. in xylene, glycerol trioleate; disp. in water, min. oil; dens. 8.4 lb/gal; visc. 173 cSt (100 F); HLB 10.4; cloud pt. < 25 C; flash pt. 525 F; pour pt. 9 C.

Trycol® 6989. [Henkel/Emery] Nonyl nonoxynol-150; emulsifier, dispersant, wetting agent for built detergents, hard surf. cleaners, dairy detergents, pesticides, textiles; leveling agent; Gardner 1 liq.; sol. in water; dens. 9.0 lb/gal; visc. 367 cSt (100 F); HLB 19.0; cloud pt. 85 C (5% saline); pour pt. 23 C; 50% aq.

Trycol® DA-4. [Henkel/Emery] Deceth-4; nonionic; wetting agent, penetrant for yarn and fabrics in dyeing systems and resin pad-bath applics.; intermediate in the prod. of anionics; Gardner < 1 liq.; sol. in glycerol trioleate; disp. in water; dens. 7.9 lb/gal; visc. 30 cs; HLB 10.5; cloud pt. < 25 C; 100% act.

Trycol® DA-6. [Henkel/Emery] Deceth-6; nonionic; wetting agent in clay soils, fire fighting; penetrant and emulsifier in repellent applics.; Gardner 1 liq.; dens. 8.3 lb/gal; visc. 15 cs; HLB 12.5; cloud pt. 40 C; 100% act.

Trycol® DA-69. [Henkel/Emery] Deceth-6; nonionic; see Trycol DA-6; Gardner 1 liq.; dens. 8.3 lb/gal; visc. 28 cSt (100 F); HLB 12.4; pour pt. –15 C; flash pt. 315 F; cloud pt. 42 C; 90% act.

Trycol® DNP-8. [Henkel/Emery] Nonyl nonoxynol-8; foam control agent; emulsifier for polar and nonpolar solvs. and textile jet dye carrier applics.; spreading agent in pigment printing; post-stabilizer in emulsion polymerization; intermediate in mfg. of anionics; surfactant component in acidic cleaners, aerosols, insecticides and wax emulsions; Gardner 2 liq.; disp. in water; dens. 8.4 lb/gal; visc. 410 cs; HLB 10.4; cloud pt. < 25 C; 100% act.

Trycol® DNP-150. [Henkel/Emery] Nonyl nonoxynol-150; dispersant and wetting agent for pesticides; emulsifier, detergent and leveling agent in textile industry; controlled-subs component in detergents and cleaners; Gardner 1 solid, flakes; m.p. 55 C; HLB 19.0; cloud pt. > 100 C; 100% act.

Trycol® DNP-150/50. [Henkel/Emery] Nonyl nonoxynol-150; see Trycol DNP-150; Gardner < 1 liq.; dens. 9.0 lb/gal; HLB 19.0; cloud pt. > 100 C; 50% act.

Trycol® LAL-4. [Henkel/Emery] Laureth-4; nonionic; intermediate; emulsifier for oils and fats in industrial lubricants, textile coning oils; coemulsifier for silicone in cleaner polishes and mold release agents; Gardner 1 liq.; disp. in water; dens. 7.7 lb/gal; visc. 35 cs; HLB 9.2; cloud pt. < 25 C; biodeg.; 100% act.

Trycol® LAL-23. [Henkel/Emery] Laureth-23 nonionic; solubilizer; coemulsifier for silicone in cleaner polishes and mold release agents; solv. emulsifier for textile dye carriers; Gardner < 1 solid; m.p. 46 C; HLB 16.9; cloud pt. > 100 C; 100% act.

Trycol® NP-1. [Henkel/Emery] Nonoxynol-1; nonionic; defoamer; surfactant and coemulsifier; Gardner 1 liq.; sol. in min. oil, xylene; dens. 8.3 lb/gal; visc. 460 cs; HLB 4.6; cloud pt. < 25 C; 100% act.

Trycol® NP-4. [Henkel/Emery] Nonoxynol-4; nonionic; coemulsifier often used as a corrosion inhibitor in two cycle engine oils; Gardner 1 liq.; dens. 8.5 lb/gal; visc. 300 cs; HLB 8.9; cloud pt. < 25 C; 100% act.

Trycol® NP-6. [Henkel/Emery] Nonoxynol-6; dispersant and wetting agent; coemulsifier in acid cleaning sol'ns, solv. emulsions and detergents; Gardner 2 liq.; disp. in water; dens. 8.5 lb/gal; visc. 100 cSt (100 F); HLB 10.9; pour pt. –10 C; flash pt. 515 F; cloud pt. < 25 C.

Trycol® NP-40. [Henkel/Emery] Nonoxynol-40; detergent; wetting agent in sol'ns. of electrolytes; stabilizer for syn. fabrics; coemulsifier for fats, waxes and oils; Gardner 1 solid; HLB 17.8; m.p. 40 C; flash pt. 560 F; cloud pt. 90 C.

Trycol® NP-50. [Henkel/Emery] Nonoxynol-50; see Trycol NP-40; Gardner 1 solid; HLB 18.2; m.p. 54 C; flash pt. 520 F; cloud pt. 76 C.

Trycol® NP-407. [Henkel/Emery] Nonoxynol-40; see Trycol NP-40; Gardner 1 liq.; dens. 9.2 lb/gal; visc. 385 cSt (100 F); HLB 17.8; pour pt. 7 C; cloud pt. 90 C.

Trycol® NP-507. [Henkel/Emery] Nonoxynol-50; see Trycol NP-40; Gardner 1 liq.; dens. 9.0 lb/gal; visc. 440 cSt (100 F); HLB 18.2; pour pt. –4 C; cloud pt. 76 C.

Trycol® NP-1007. [Henkel/Emery] Nonoxynol-100; surfactant used as wetting agent in high electrolyte sol'ns.; stabilizer in syn. latices; Gardner 1 liq.; dens. 9.2 lb/gal; HLB 19.0; pour pt. 18 C; cloud pt. 72 C; 70% act.

Trycol® OAL-23. [Henkel/Emery] Oleth-23; nonionic; dispersant, solubilizer, detergent, stabilizer and anticoagulant for latices and dye pastes; dyeing assistant for wool/acrylic blends; detergent and lubricant for fiber/fabric scouring; emulsifier for waxes used in coating citrus fruits; Gardner 2 solid; m.p. 40 C; HLB 15.9; cloud pt. > 100 C; 100% act.

Trycol® OP-407. [Henkel/Emery] Octoxynol-40; hydrophilic dispersant and wetting agent; primary emulsifier in emulsion of polymerization of acrylic and vinyl monomers; Gardner 1 liq.; dens. 9.0 lb/gal; visc. 220 cSt (100 F); HLB 17.9; pour pt. 13 C; cloud pt. 74 C.

Trycol® SAL-20. [Henkel/Emery] Steareth-20; nonionic; emulsifier and solubilizer used in dyeing assistants and textile dye carriers; stabilizer in latices; wax emulsifier in coatings for citrus fruits; Gardner 1 solid; HLB 15.3; m.p. 40 C; flash pt. 560 F; cloud pt. 91 C.

Trycol® TDA-3. [Henkel/Emery] Trideceth-3; nonionic; emulsifier and antifoam in textile formulations; intermediate for prod. of anionics by sulfation and phosphation; Gardner < 1 liq.; disp. in water; dens. 7.8 lb/gal; visc. 30 cs; HLB 8.0; cloud pt. < 25 C; 100% act.

Trycol® TDA-6. [Henkel/Emery] Trideceth-6; nonionic; dispersant and wetting agent, emulsifier and detergent in degreasers and cutting oils; wool scouring; intermediate for prod. of surfactants by sulfation or phosphation; Gardner < 1 liq.; disp. in water; dens. 8.2 lb/gal; visc. 50 cs; HLB 11.4; cloud pt. < 25 C; 100% act.

Trycol® TDA-8. [Henkel/Emery] Trideceth-8; nonionic; coemulsifier, foam builder and solubilizer for alkylaryl sulfonates, essential oils, aromatic solvs., waxes and fats; Gardner < 1 liq.; dens. 8.3 lb/gal; visc. 65 cs; HLB 12.5; cloud pt. 44 C; 100% act.

Trycol® TDA-18. [Henkel/Emery] Trideceth-18; nonionic; dispersant, solubilizer, emulsifier in cleaners; intermediate in mfg. of anionic surfactants; textile acid dye leveling agent and wet processing detergent; foam stabilizer; Gardner < 1 solid; m.p. 39 C; HLB 16.0; cloud pt. > 100 C; 100% act.

Trycol® TP-2. [Henkel/Emery] PEG-2 tridecyl phenol; nonionic; coemulsifier, foam reducing agent; Gardner 6 liq.; dens. 8.1 lb/gal; visc. 600 cs; HLB 4.6; cloud pt. < 25 C; 100% act.

Trycol® TP-6. [Henkel/Emery] PEG-6 tridecyl phenol; nonionic; dispersant, detergent, emulsifier in metal cleaning, cutting oil and drycleaning formulations, emulsifiable solv. cleaners; Gardner 6 liq.; disp. in water; dens. 8.5 lb/gal; visc. 300 cs; HLB 9.9; cloud pt. < 25 C; 100% act.

Trydet 19. [Henkel/Emery] Ethoxylated mixed rosin and fatty acids, anhyd.; nonionic; see Trydet 20; Gardner 12 liq.; dens. 8.9 lb/gal; visc. 200 cSt (100 F); HLB 13.4; pour pt. 15 C; flash pt. 570 F; cloud pt. 50 C.

Trydet 20. [Henkel/Emery] Ethoxylated mixed esters of rosin and fatty acids; nonionic; detergent and emulsifier; textile leveling agent in dyeing applics.; coemulsifier for xylene, kerosene, trichlorobenzene, o-dichlorobenzene; dens. 9.0 lb/gal; visc. 480 cs; HLB 13.4; cloud pt. > 100 C; 96% act.

Trydet 2636. [Henkel/Emery] PEG-7 stearate; emulsifier, softener, lubricant for textiles, leathers, industrial lubricants; Gardner 1 solid; sol. in xylene; disp. in water; visc. 57 cSt (100 F); HLB 10.1; m.p. 28 C; cloud pt. < 25 C; flash pt. 500 F.

Trydet 2644. [Henkel/Emery] PEG 400 isostearate; surfactant, lubricant for fiber lubricants, textile processing aids, fabric softeners; Gardner 1 liq.; sol. in xylene; disp. in water; dens. 8.5 lb/gal; visc. 70 cSt (100 F); HLB 11.3; cloud pt. < 25 C; flash pt. 450 F; pour pt. 10 C.

Trydet 2670. [Henkel/Emery] PEG-5 stearate; emulsifier, lubricant, softener for textiles, leather, polishes; Gardner 1 liq.; sol. in xylene; disp. in water; dens. 8.6 lb/gal; visc. 44 cSt (100 F); HLB 9.2; pour pt. 25 C; cloud pt. < 25 C; flash pt. 500 F.

Trydet 2671. [Henkel/Emery] PEG-8 stearate; SE lubricant, softener for textile and other industrial applics.; stabilizer in starch sol'ns.; Gardner 1 solid; sol. in xylene; disp. in water; dens. 8.5 lb/gal; HLB 11.4; m.p. 30 C; cloud pt. < 25 C; flash pt. 500 F.

Trydet 2672. [Henkel/Emery] PEG-40 stearate; emulsifier, lubricant used in prod. of textile lubricants and softeners; stabilizer, antigellant for starch sol'ns.; Gardner 1 solid; sol. in water, xylene; HLB 17.3; m.p. 50 C; cloud pt. 75–81 C (5% saline); flash pt. 540 F.

Trydet 2676. [Henkel/Emery] PEG-10 oleate; emulsifier, lubricant for pesticies, metal cleaners, textile detergents and dyeing assistants, leather; rewetting agent for paper; Gardner 2 liq.; sol. in xylene; disp. in water; dens. 8.5 lb/gal; visc. 54 cSt (100 F); HLB 12.2; cloud pt. < 25 C; flash pt. 560 F; pour pt. 14 C.

Trydet 2682. [Henkel/Emery] Ethoxylated mixed rosin and fatty acids; detergent, emulsifier, textile leveling agent; coemulsifier; Gardner 10 liq.; sol. in water; dens. 9.0 lb/gal; visc. 209 cSt (100 F); HLB 13.4; cloud pt. 42–53 C (5% saline); flash pt. 590 F; pour pt. 17 C.

Trydet 2685. [Henkel/Emery] Ethoxylated fatty acid; emulsifier for min. oils and fats; lubricant; Gardner 1 solid; sol. in min. oil, butyl stearate, glycerol trioleate, Stod., xylene; disp. in water; HLB 7.5; m.p. 37 C; cloud pt. < 25 C; flash pt. 350 F.

Trydet DO-9. [Henkel/Emery] PEG-9 dioleate; nonionic; lipophilic emulsifier and solubilizer for min. oils, fats and solvs.; emulsifier for kerosene in agric. and pesticides sprays; emulsification of latex paints, metalworking fluids, solvs., specialty and industrial lubricants; Gardner 3 liq.; disp. in water; dens. 8.1 lb/gal; visc. 45 cSt (100 F); HLB 8.8; pour pt. –6 C; flash pt. 515 F; cloud pt. < 25 C.

Trydet ISA-4. [Henkel/Emery] PEG-4 isostearate; nonionic; lubricant used in textile fiber processing, conc. fabric softeners; Gardner 2 liq.; dens. 8.2 lb/gal; visc. 100 cs; HLB 7.7; cloud pt. < 25 C; 100% act.

Trydet ISA-9. [Henkel/Emery] PEG-9 isostearate; nonionic; see Trydet ISA-4: Gardner 1 liq.; dens. 8.5 lb/gal; visc. 125 cs; HLB 11.7; cloud pt. < 25 C; 100% act.

Trydet LA-5. [Henkel/Emery] PEG-5 laurate; nonionic; emulsifier and coupling agent; defoamer in water base coatings; visc. depressant in vinyl plastic sols.; control additive in hair rinse formulations; paper softener; Gardner liq.; dens. 8.2 lb/gal; visc. 123 cSt (100 F); HLB 9.3; pour pt. 9 C; flash pt. 445 F; cloud pt. < 25 C.

Trydet LA-7. [Henkel/Emery] PEG-7 laurate; emulsifier, nonionic; lubricant component and scrooping agent for textile fibers and yarns; visc. control agent for plastisols; Gardner 2 liq.; dens. 8.4 lb/gal; visc. 55 cs; HLB 11.8; cloud pt. < 25 C; 100% act.

Trydet MP-9. [Henkel/Emery] PEG-9 pelargonic acid; nonionic; surfactant as a base lubricant for syn. fiber finishes; coemulsifier and coupling agent; Gardner 1 liq.; dens. 8.7 lb/gal; visc. 34 cSt (100 F); HLB 14.3; pour pt. 5 C; flash pt. 440 F; cloud pt. 37 C.

Trydet OA-5. [Henkel/Emery] PEG-5 oleate; nonionic; emulsifier for min., cutting, fatty oils and solvs.; solv. emulsifier in metal cleaners and degreasers; w/o emulsifier for pesticide control; textile processing; softener and lubricant leather during tanning; Gardner 3 liq.; disp. in water; dens. 8.1 lb/gal; visc. 55 cs; HLB 8.2; cloud pt. < 25 C; 100% act.

Trydet OA-7. [Henkel/Emery] PEG-7 oleate; nonionic; see Trydet OA-5; Gardner 3 liq.; dens. 8.3 lb/gal; visc. 160 cs; HLB 10.2; cloud pt. < 25 C; 100% act.

Trydet OA-10. [Henkel/Emery] PEG-10 oleate; nonionic; lubricant, emulsifier for solvs. in pesticide carriers, metal cleaners and neatsfoot oil in leather fat liquoring; textile specialty detergent and dyeing assistant; rewetting agent for paper; Gardner 2 liq.; dens. 8.5 lb/gal; visc. 110 cs; HLB 12.1; cloud pt < 25 C; 100% act.

Trydet SA-5. [Henkel/Emery] PEG-5 stearate; nonionic; emulsifier for min. oils and fats; lubricant in leather industry; napping softener in textile industry; min. oil and lard oil emulsions for polish and metal buffing compds.; Gardner 1 solid; m.p. 35 C; HLB 8.7; cloud pt. < 25 C; 100% act.

Trydet SA-7. [Henkel/Emery] PEG-7 stearate; nonionic; waxy emulsifier for oils and fats in industrial lubricants; softener and lubricant to textiles and leather; Gardner < 1 solid; m.p. 30 C; HLB 11.1; cloud pt. < 25 C; 100% act.

Trydet SA-8. [Henkel/Emery] PEG-8 stearate; nonionic; lubricant, softener and scrooping agent for textile and industrial applics.; stabilizer in starch sol'ns.; Gardner < 1 solid; m.p. 30 C; HLB 11.1; cloud pt. < 25 C; 100% act.

Trydet SA-23. [Henkel/Emery] PEG-23 stearate; nonionic; emulsifier for textile lubricants and softeners; emulsifier and thickener in personal care prods. and pharmaceuticals; antigellant in starch sol'ns.; Gardner 1 solid; m.p. 45 C; HLB 15.6; cloud pt. > 100 C; 100% act.

Trydet SA-40. [Henkel/Emery] PEG-40 stearate; nonionic; emulsifier for glycerol monostearate and waxy esters in prod. of conc. textile lubricants and softeners; stabilizer and antigellant for starch sol'ns.; emulsifier for cosmetics and pharmaceuticals; Gardner < 1 solid, flakes; m.p. 50 C; HLB 17.3; cloud pt.

< 25 C; 100% act.

Trydet SA-50/30. [Henkel/Emery] PEG-50 stearate; nonionic; hydrophilic emulsifier for preparing solubilized oils; visc. modifier, softener or plasticizer in acrylic or vinyl resin emulsions; Gardner 1 liq.; dens. 8.5 lb/gal; visc. 671 cSt (100 F); HLB 17.8; pour pt. 0 C; flash pt. 540 F; cloud pt. 81 C.

Tryfac® 525-K. [Henkel/Emery] Phosphate ester; detergent and wetting agent for built liq. detergents; antistat in fiber finishes; Gardner 2 liq.; dens. 8.8 lb/gal; visc. 560 cSt (100 F); pour pt. 18 C; flash pt. 435 F; cloud pt. 76 C.

Tryfac® 610-K. [Henkel/Emery] Phosphate ester potassium salt; anionic; wetting agent, dispersant, emulsifier for chlorinated hydrocarbons in cleaners, detergents and dye carriers; component in metalworking compds.; textile emulsifier and detergent in solv. scouring; scouring agent for cotton goods, antistat in processing oils for fibers; corrosion inhibitor; Gardner < 1 liq.; dens. 9.2 lb/gal; visc. 700 cs; 88% act.

Tryfac® 910-K. [Henkel/Emery] Phosphate ester potassium salt; anionic; detergent and wetting agent in textile desizing; polymer emulsification stabilizer; Gardner 1 liq.; dens. 9.4 lb/gal; visc. 480 cs; 90% act.

Tryfac® 5365. [Henkel/Emery] Phosphate ester; anionic; detergent, wetting and antistat, emulsifier; liq.

Tryfac® 5552. [Henkel/Emery] Phosphate ester, free acid form; anionic; surfactant intermediate; salts used as emulsifiers; Gardner 1 liq.; sol. in water, butyl stearate, Stod., xylene; dens. 8.8 lb/gal; visc. 170 cSt (100 F); cloud pt. 76 C; flash pt. 360 F; pour pt. < –15 C.

Tryfac® 5553. [Henkel/Emery] Phosphate ester (potassium salt of Tryfac 5552); anionic; emulsifier, wetting agent, detergent, antistat; used in heavy duty cleaners, metalworking compds., textile applics.; corrosion inhibitor, dispersant; Gardner 1 liq.; sol. in water, butyl stearate, glycerol trioleate, Stod., xylene; dens. 9.2 lb/gal; visc. 345 cSt (100 F); cloud pt. > 100 C (10% saline); pour pt. < –15 C.

Tryfac® 5554. [Henkel/Emery] Phosphate ester, potassium salt; similar to Tryfac 5553; also stabilizer for polymer emulsification, detergent and wetting agent in starch conversion in textile desizing; Gardner 1 liq.; sol. in water, Stod., xylene; dens. 9.4 lb/gal; visc. 340 cSt (100 F); cloud pt. > 100 C (10% saline); pour pt. –9 C.

Tryfac® 5556. [Henkel/Emery] Complex phosphate ester, free acid form; wetting agent, dispersant, antistat in textile processing; solv. emulsifier for textile scours, detergents, pesticides; drycleaning detergent; also used in emulsion polymerization; Gardner 2 liq.; sol. in water, xylene; dens. 9.3 lb/gal; visc. 1700 cSt (100 F); cloud pt. 68 C (5% saline); flash pt. 450 F; pour pt. 5 C.

Tryfac® 5557. [Henkel/Emery] Phosphate ester, potassium salt; surfactant, wetting agent, dispersant; esp. for textile scours; Gardner 2 liq.; sol. in water; dens. 9.3 lb/gal; pour pt. 5 C; 55% act.

Tryfac® 5559. [Henkel/Emery] Phosphate ester; detergent, wetting agent, oily soil emulsifier in built detergents; antistat in fiber finishes for polyester and PP; Gardner 2 liq.; sol. in water, butyl stearate, glycerol trioleate, Stod., xylene; dens. 8.8 lb/gal; visc. 560 cSt (100 F); cloud pt. 76 C (5% saline); flash pt. 435 F; pour pt. 18 C.

Tryfac® 5560. [Henkel/Emery] Phosphate ester, free acid form; emulsifier, dispersant, antistat; Gardner 2 liq.; sol. in water, butyl stearate, glycerol trioleate, Stod., xylene; dens. 8.8 lb/gal; flash pt. 405 F; pour pt. 5 C.

Tryfac® 5569. [Henkel/Emery] Phosphate ester, free acid form; hydrotrope for nonionic surfactants in alkaline sol'n.; nondiscoloring in contact with solid caustic; Gardner 6 liq.; sol. in water; dens. 10.4 lb/gal; visc. 1180 cSt (100 F); cloud pt. 51 C; flash pt. 425 F; pour pt. 5 C; 100% act.

Tryfac® 5571. [Henkel/Emery] Ethoxylated alcohol phosphate ester, potassium salt; detergent, dispersant and wetting agent used in fabric preparation, hard surface cleaning and antistat protection of fibers; Gardner 1 liq.; sol. in water, butyl stearate, glycerol trioleate, Stod., xylene; dens. 9.2 lb/gal; visc. 650 cSt (100 F); pour pt. < –15 C; cloud pt. 45 C.

Tryfac® 5573. [Henkel/Emery] Phosphate ester, free acid form; mold release agent, antistat, dispersant, emulsifier; Gardner 1 solid; sol. in min. oil, butyl stearate, glycerol trioleate, Sod., xylene; disp. in water; dens. 7.8 lb/gal (40 C); m.p. 35 C; cloud pt. < 25 C; flash pt. 365 F.

Tryfac® 5576. [Henkel/Emery] Ethoxylated alcohol phosphate ester, potassium salt; hydrophilic surfactant used in electrolytes and caustic sol'ns.; antistat, anticorrosive and dispersant; Gardner 1 liq.; sol. in water, butyl stearate, xylene; dens. 9.5 lb/gal; visc. 190 cSt (100 F); pour pt. < –10 C; cloud pt. 49–54 C.

Tryfac® HWD. [Henkel/Emery] Phosphate ester in free acid form; hydrotrope, solubilizer for nonionic surfactants used in built liq. detergents for industrial and institutional cleaners, degreasers, metal, and other hard surf. cleaners; Gardner 3 liq.; sol. in highly alkaline sol'ns., (5%) in water, xylene; dens. 9.2 lb/gal; visc. 306 cSt (100 F); pour pt. 12 C; flash pt. 380 F; cloud pt. 78 C; 100% act.

Trylon® 6733. [Henkel OPG] Long-chain, sat. fatty ester; nonionic; lubricant with heat stability properties; used in textile and industrial processing lubricants; Gardner 2 liq.; sol. 5% in butyl stearate, glycerol trioleate, Stod., xylene; water-disp.; dens. 8.0 lb/gal; visc. 36 cSt (100 F); pour pt. 16 C; flash pt. 515 F; cloud pt. < 25 C.

Trylon® 6751. [Henkel OPG] Fatty ester ethoxylate; coemulsifier for solvs., oils and esters; fiber-to-metal lubricant; Gardner 3 liq.; disp. in water; dens. 8.4 lb/gal; visc. 79 cSt (100 F); pour pt. 10 C; flash pt. 530 F; cloud pt. < 25 C.

Trylon® 6848. [Henkel OPG] Anionic; emulsifier, dye assistant; Gardner 7 liq.; sol. in butyl stearate, Stod., xylene; disp. in water; dens. 8.4 lb/gal; visc. 98 cSt (100 F); flash pt. 180 F; pour pt. < –8 C.

Trylon® EW. [Henkel OPG] Nonionic ether; polymer stabilizer, resin bath penetrant, solv. scour emulsifier, and enzyme bath penetrant in textile applics.; emulsifier for tanning chemicals in the leather industry; used in cold water scours for felted fabrics; Gardner 2 liq.; sol. (5%) in butyl stearate, glycerol trioleate, xylene; disp. in water, Stoddard; dens. 8.6 lb/gal; visc. 126 cSt (100 F); cloud pt. < 25 C; flash pt. 545 F; pour pt. 5 C.

Trylox® 5900. [Henkel/Emery] PEG-5 castor oil; nonionic; emulsifier, dispersant, carrier, foam control agent, lubricant for paints, paper coatings, dye carriers; Gardner 4 liq.; sol. in glycerol trioleate; dens. 8.2 lb/gal; visc. 375 cSt (100 F); HLB 4.0; flash pt. 550 F; pour pt. –3 C; 100% conc.

Trylox® 5902. [Henkel/Emery] PEG-16 castor oil; emulsifier, lubricant for metalworking oils, hydraulic fluids, textiles; Gardner 3 liq.; sol. in xylene; disp. in water; dens. 8.5 lb/gal; visc. 546 cSt (100 F); HLB 8.6; cloud pt. < 25 C; flash pt. 565 F; pour pt. –22 C.

Trylox® 5904. [Henkel/Emery] PEG-25 castor oil; emulsifier, lubricant for formulation of sol. oils, cutting fluids, fiber finishes; Gardner 4 liq.; disp. in water, butyl stearate, glycerol trioleate, Stod., xylene; dens. 8.6 lb/gal; visc. 396 cSt (100 F); HLB 10.8; cloud pt. 66 C (1% saline); flash pt. 565 F; pour pt. –5 C.

Trylox® 5906. [Henkel/Emery] PEG-30 castor oil; emulsifier, pigment dispersant, degreaser, lubricant; visc. and emulsion stabilizer of PVAc and water-based paints; also for paper, textile, and lubricant applics.; Gardner 2 liq.; sol. in water, xylene; dens. 8.8 lb/gal; visc. 309 cSt (100 F); HLB 11.8; cloud pt. 55 C (1% saline); flash pt. 555 F; pour pt. 9 C.

Trylox® 5907. [Henkel/Emery] PEG-36 castor oil; emulsifier for solvs. and oils; lubricant and softener for textiles and leather; coemulsifier for formulating textile dye carriers; Gardner 2 liq.; sol. in water, xylene; dens. 8.8 lb/gal; visc. 363 cSt (100 F); HLB 12.6; cloud pt. 78 C (1% saline); flash pt. 575 F; pour pt. 12 C.

Trylox® 5909. [Henkel/Emery] PEG-40 castor oil; see Trylox 5907; Gardner 2 liq.; sol. in water; dens. 8.8 lb/gal; visc. 313 cSt (100 F); HLB 13.0; cloud pt. 80 C (1% saline); pour pt. 18 C.

Trylox® 5918. [Henkel/Emery] PEG-200 castor oil; lubricant, emulsifier, antistat, humectant for textile fiber processing; Gardner 1 liq.; sol. in water; dens. 8.5 lb/gal; visc. 1015 cSt (100 F); HLB 18.1; cloud pt. 80 C (5% saline); pour pt. 7 C; 50% act.

Trylox® 5921. [Henkel/Emery] PEG-16 hydrog. castor oil; emulsifier, lubricant, softener; used in fabric softeners and aerosol fabric sprays; Gardner 1 liq.; sol. in glycerol trioleate, xylene; disp. in water, min. oil; dens. 8.4 lb/gal; visc. 569 cSt (100 F); HLB 8.6; cloud pt. < 25 C; flash pt. 565 F; pour pt. 7 C.

Trylox® 6753. [Henkel/Emery] PEG-20 sorbitol; humectant, plasticizer; intermediate; used in surfactant sol'ns.; Gardner 1 liq.; sol. in water; dens. 9.7 lb/gal; visc. 200 cSt (100 F); HLB 15.4; cloud pt. > 100 C (10% saline); pour pt. 7 C; flash pt. 435 F.

Trylox® CO-5. [Henkel/Emery] PEG-5 castor oil; emulsifier, nonionic; coemulsifier for chlorinated and aromatic solvs.; dispersant for pigment slurries in water-based paint and clay; carrier for paper coatings; textile foam control agent and emulsifier in dye carriers; Gardner 4 liq.; disp. in water; dens. 8.2 lb/gal; visc. 800 cs; HLB 3.8; cloud pt. < 25 C; 100% act.

Trylox® CO-16. [Henkel/Emery] PEG-16 castor oil; nonionic; emulsifier, lubricant, coemulsifier for metalworking oils, hydraulic fluids, rayon delustrants and fiber lubricants; component in textile leveling and dispersant for vat and naphthol dyes; Gardner 3 liq.; dens. 8.5 lb/gal; visc. 600 cs; HLB 8.6; cloud pt. < 25 C; 100% act.

Trylox® CO-25. [Henkel/Emery] PEG-25 castor oil; nonionic; emulsifier and lubricant in formulation of sol. oils, cutting fluids and fiber finishes; Gardner 4 liq.; dens. 8.6 lb/gal; visc. 396 cSt (100 F); HLB 10.8; pour pt. –5 C; flash pt. 565 F; cloud pt. 66 C.

Trylox® CO-30. [Henkel/Emery] PEG-30 castor oil; nonionic; emulsifier, degreaser, dispersant, visc. stabilizer, lubricant for oils, solvs., waxes, syn. fiber lubricants, pigments, fat liquoring, paints, urethane foams and polyester resins; coemulsifier in fabric softener and dye carrier systems; dyeing assistant; Gardner 2 liq.; dens. 8.8 lb/gal; visc. 720 cs; HLB 11.7; cloud pt. > 100 C; 100% act.

Trylox® CO-36. [Henkel/Emery] PEG-36 castor oil; nonionic; emulsifier for solvs. and oils; lubricant and softener for textiles and leather; coemulsifier with anionic emulsifier in formulating textile dye carriers; Gardner 2 liq.; dens. 8.8 lb/gal; visc. 700 cs; HLB 12.6; cloud pt > 100 C; 100% act.

Trylox® CO-40. [Henkel/Emery] PEG-40 castor oil; see Trylox CO-36; Gardner 2 liq.; dens. 8.8 lb/gal; visc. 650 cs; HLB 12.9; cloud pt. > 100 C; 100% act.

Trylox® CO-80. [Henkel/Emery] PEG-80 castor oil; nonionic; o/w emulsifier and solubilizer; used in fiber finish and textile lubricant formulations; Gardner 2 liq.; dens. 8.9 lb/gal; visc. 322 cSt (100 F); HLB 15.9; pour pt. 20 C; cloud pt. 77 C.

Trylox® CO-200, -200/50. [Henkel/Emery] PEG-200 castor oil; nonionic; lubricant, textile fiber processing as antistat humectant and scrooping agent; emulsifier in blended lubricants for fibers and yarns; Gardner 1 solid and Gardner < 1 liq. resp.; dens. 8.6 lb/gal; m.p. 40 C; HLB 18.1; cloud pt. > 100 C; 100 and 50% act.

Trylox® HCO-5. [Henkel/Emery] PEG-5 hydrog. castor oil; nonionic; emulsifier, coemulsifier for syn. esters used in the textile industry; fiber lubricant and softener; Gardner 1 liq.; disp. in water; dens. 7.9 lb/gal; visc. 1200 cs; HLB 3.8; cloud pt. < 25 C; 100% act.

Trylox® HCO-16. [Henkel/Emery] PEG-16 hydrog. castor oil; nonionic; emulsifier for castor oil; lubricant and softener in fabric softeners and aerosol fabric sprays; Gardner 1 liq.; low odor; dens. 8.4 lb/gal; visc. 2100 cs; HLB 8.2; cloud pt. < 25 C; 100% act.

Trylox® HCO-25. [Henkel/Emery] PEG-25 hydrog. castor oil; nonionic; emulsifier for textile resin finishing and softener lubricant systems; coemulsifier for lanolin and dye carrier systems; Gardner 1 liq.; disp. in water, dens. 8.6 lb/gal; visc. 1100 cs; HLB

10.8; cloud pt. 25 C; 100% act.

Trylox® HCO-200/50. [Henkel/Emery] PEG-200 hydrog. castor oil; nonionic; emulsifier, lubricant; Gardner < 1 clear liq.; dens. 8.8 lb/gal; HLB 18.1; cloud pt. > 100 C; 50% act.

Trylox® SS-20. [Henkel/Emery] PEG-20 sorbitol; nonionic; humectant, plasticizer; intermediate in the synthesis of fatty acid esters; Gardner 1 liq.; dens. 9.7 lb/gal; visc. 400 cs; HLB 16.1; cloud pt. > 100 C; 100% act.

Trymeen® 6601. [Henkel/Emery] PEG-10 coco amine; coemulsifier, antistat for textile applics.; dispersant; emulsifier in industrial lubricants; Gardner 8 liq.; sol. in water, glycerol trioleate, xylene; dens. 8.4 lb/gal; visc. 68 cSt (100 F); HLB 13.6; cloud pt. 87 C; flash pt. 445 F; pour pt. –10 C.

Trymeen® 6602. [Henkel/Emery] PEG coco amine; cationic; textile antistat, emulsifier, lubricant, dispersant, softener; substantive to metals, fibers, and clays; liq.; HLB 15.2; 100% conc.

Trymeen® 6603. [Henkel/Emery] PEG-8 tallow amine; low foaming emulsifier contributing wetting, lubricating, and softening properties; solv. emulsifier for dye carriers; as dye assistant; emulsifier for polyethylene textile softeners, in industrial lubricants; Gardner 8 liq.; sol. in water, butyl stearate, glycerol trioleate, Stod., xylene; dens. 8.2 lb/gal; visc. 87 cSt (100 F); HLB 11.4; cloud pt. 77 C (1% saline); flash pt. 525 F; pour pt. –2 C.

Trymeen® 6606. [Henkel/Emery] PEG-15 tallow amine; cationic; emulsifier, antiprecipitant for dye baths; leveling agent; intermediate for quats.; antistat for syn. fiber processing; Gardner 8 liq.; sol. in water, glycerol trioleate, xylene; dens. 8.5 lb/gal; visc. 96 cSt (100 F); HLB 14.3; cloud pt. 95 C (5% saline); flash pt. 445 F; pour pt. –10 C; 100% conc.

Trymeen® 6607. [Henkel/Emery] PEG-20 tallow amine; antistat, coemulsifier, and lubricant for textile applics.; antiprecipitant, leveling, and migrating agent in dyeing processes; Gardner 6 liq.; sol. in water; dens. 8.7 lb/gal; visc. 119 cSt (100 F); HLB 15.4; cloud pt. 87 C (10% saline); flash pt. 550 F; pour pt. –2 C.

Trymeen® 6609. [Henkel/Emery] PEG-25 tallow amine; see Trymeen 6607; Gardner 6 liq.; sol. in water, xylene; dens. 8.8 lb/gal; visc. 128 cSt (100 F); HLB 16.0; cloud pt. 87 C (10% saline); flash pt. 540 F; pour pt. 16 C.

Trymeen® 6617. [Henkel/Emery] PEG-50 stearyl amine; emulsifier, antistat for metal buffing compds., latex rubber compding.; anticoagulant; lubricant, leveling agent for textile applics.; Gardner 4 solid; sol. in water; HLB 17.8; m.p. 35 C; cloud pt. 82 C (10% saline); flash pt. 540 F.

Trymeen® 6622. [Henkel/Emery] PEG-30 oleamine; see Trymeen 6623; Gardner 6 liq.; sol. in water; dens. 8.9 lb/gal; visc. 445 cSt (100 F); HLB 16.6; cloud pt. 86 C (10% saline); pour pt. 5 C; 80% aq.

Trymeen® 6623. [Henkel/Emery] PEG-30 oleamine; emulsifier, dyeing assistant, antiprecipitant, stripping agent, dye leveler for textile applics.; Gardner 8 solid; sol. in water; HLB 16.6; m.p. 35 C; cloud pt. 86 C (10% saline); flash pt. 470 F.

Trymeen® 6637. [Henkel/Emery] PEG-40 tallow amine; emulsifier, antistat for commercial carpet maintenance; dyeing assistant, stabilizer for latices; Gardner 4 liq.; sol. in water, xylene; dens. 9.1 lb/gal; visc. 150 cSt (100 F); HLB 17.4; cloud pt. 85 C (10% saline); pour pt. 9 C; 80% aq.

Trymeen® 6640. [Henkel/Emery] PEG-15 tallow propylene diamine; cationic; dispersant in acidic sol'ns.; retarder of acidic, disperse and premetalized dyestuffs; Gardner 18 liq.; dens. 8.4 lb/gal; visc. 120 cSt (100 F); HLB 13.1; pour pt. < –10 C; flash pt. 440 F; cloud pt. 86 C.

Trymeen® 6657. [Henkel/Emery] Amidoamine condensate; lubricant/softener; hydrophobic component with min. oils and fatty esters in formulating fabric finishes; maintains good whiteness retention during fabric or yarn processing; Gardner 4 solid; water-disp.; m.p. 75 C; flash pt. 530 F; cloud pt. < 25 C.

Trymeen® CAM-10. [Henkel/Emery] PEG-10 coconut amine; cationic; coemulsifier and antistat for textile processing; dispersant for inorg. salts in viscose spinning; emulsifier for fats and oils in industrial lubricants; Gardner 8 liq.; dens. 8.4 lb/gal; visc. 260 cs; HLB 13.8; cloud pt. > 100 C; 100% act.

Trymeen® CAM-15. [Henkel/Emery] PEG-15 coconut amine; cationic; emulsifier, antistat, softener; substantive to clay, metals, mins., and textiles; additive in the mfg. of rayon fiber; prevents sludge build-up in spin baths; textile processing; industrial lubricant; Gardner 6 liq.; dens. 8.6 lb/gal; visc. 120 cs; HLB 15.4; cloud pt. > 100 C; 100% act.

Trymeen® OAM-30/80. [Henkel/Emery] PEG-30 oleamine; cationic; emulsifier; textile dyeing assistant; antiprecipitant in cross dyeing; stripping agent and dye leveler for acid dyes; Gardner 6 liq.; dens. 8.9 lb/gal; HLB 15.3; cloud pt. > 100 C; 80% act.

Trymeen® SAM-50. [Henkel/Emery] PEG-50 stearyl amine; cationic; antistat, emulsifier in metal buffing compds.; lubricant for fiberglass; leveling agent for dye applics.; Gardner 4 solid; m.p. 50 C; HLB 17.8; cloud pt. > 100 C; 100% act.

Trymeen® TAM-15. [Henkel/Emery] PEG-15 tallow amine; cationic; emulsifier; antiprecipitant for mixed dye baths; leveling agent for acid dyes; migrating agent for dispersed dyes; intermediate for quat. ammonium compds.; antistat for processing syn. fibers; Gardner 8 liq.; dens. 8.5 lb/gal; visc. 175 cs; HLB 14.2; cloud pt. > 100 C; 100% act.

Trymeen® TAM-20. [Henkel/Emery] PEG-20 tallow amine; cationic; textile antistat, coemulsifier and lubricant for wool and syn. fiber processing; antiprecipitant, leveling and migrating agent in dyeing; Gardner 6 liq.; dens. 8.7 lb/gal; visc. 240 cs; HLB 15.4; cloud pt. > 100 C; 100% act.

Trymeen® TAM-25. [Henkel/Emery] PEG-25 tallow amine; see Trymeen TAM-20; Gardner 6 liq.; dens. 8.8 lb/gal; visc. 170 cs; HLB 16.1; cloud pt. > 100 C; 100% act.

Trymeen® TAM-40. [Henkel/Emery] PEG-40 tallow amine; cationic; emulsifier, antistat additive for commercial carpet maintenance; dyeing assistant and stabilizer for latices; Gardner 6 solid; m.p. 40 C; HLB 17.4; cloud pt. > 100 C; 100% act.

TS 100. [Degussa] Silica; flatting agent for top coats on leathercloth, foil lacquers, and PU-textile coatings; thixotropic agent for lacquer systems; 4 μ avg. particle size; sp.gr. 2.2; bulk dens. 50 g/l (tapped); oil absorp. 360; ref. index 1.45; pH 6–7; 99.8–100% SiO_2.

TSE-2000. [TSE] Diphenylmethane diisocyanate in methylene chloride; one-coat adhesive to bond millathane to a variety of substrates during vulcanization; clear to wh.; sp.gr. 1.32; flash pt. none.

TSE Mold Release®. [TSE] Glycol surfactant; mold release esp. for Viton parts; milky wh.; odorless; sp.gr. 1.01; b.p. 100 C; flash pt. none.

T.S.S. [Hart Chem. Ltd.] Trisodium sulfosuccinate; anionic; anticaking agent in powd. detergents; compatibilizer for detergent blends; liq. 42% act.

Turgum® S. [Whitney & Oettler] Turpene-resin acid blend; plasticizer and conditioner for SBR; retarder for high, intermediate, and super abrasion furnace blk./natural rubber stocks; brn. solid; shatters at R.T.; pine odor; sp.gr. 1.06–1.07; soften. pt. (R&B) 135–145 F.

Türkischrotöl 100%, 50%. [Zschimmer & Schwarz] Sodium sulforicinoleate; solubilizer for cosmetics; liq.; 85 and 45% act. resp.

Turpinal 4 NL. [Henkel] Tetrasodium etidronate; chelating agent for heavy metal ions; stabilizer for cosmetic/pharmaceutical hair/skin preps.; liq.; 29–30% conc.

Turpinal SL. [Henkel] Etidronic acid; chelating agent for heavy metal ions; stabilizer, antioxidant for cosmetic/pharmaceutical hair/skin preps.; liq.; 58–61% conc.

Tween® 20. [ICI Am.] Polysorbate 20; nonionic; solubilizer; o/w emulsifier; antistat and fiber lubricant used in textile industry; pale yel. liq.; sol. in water, methanol, ethanol, IPA, propylene glycol, ethylene glycol, cottonseed oil; sp.gr. 1.1; visc. 400 cps; HLB 16.7; flash pt. > 300 F; sapon. no. 40–50; 100% act.

Tween® 20 SD. [ICI Am.] POE sorbitan monolaurate; anionic; see Tween 20; liq.; 100% act.

Tween® 21. [ICI Am.] Polysorbate 21; nonionic; see Tween 20; yel. oily liq.; sol. in corn oil, dioxane, Cellosolve, CCl_4, methanol, ethanol, aniline; disp. in water; sp.gr. 1.1; visc. 500 cps; HLB 13.3; flash pt. > 300 F; sapon. no. 100–115; 100% act.

Tween® 40. [ICI Am.] Polysorbate 40; nonionic; see Tween 20; pale yel. liq.; sol. in water, methanol, ethanol, IPA, ethylene glycol, cottonseed oil; sp.gr. 1.08; HLB 15.6; flash pt. > 300 F; sapon. no. 41–52; 100% act.

Tween® 60. [ICI Am.] Polysorbate 60; nonionic; see Tween 20; pale yel. liq.; sol. in water, IPA, ethyl alcohol; sp.gr. 1.1; visc. 600 cps; HLB 14.9; flash pt. > 300 F; sapon. no. 45–55; 100% act.

Tween® 61. [ICI Am.] Polysorbate 61; nonionic; see Tween 20; tan waxy solid; sol. in methanol, ethanol; disp. in water; sp.gr. 1.06; HLB 9.6; flash pt. > 300 F; pour pt. 100 F; sapon. no. 95–115; 100% act.

Tween® 65. [ICI Am.] Polysorbate 65; nonionic; see Tween 20; tan waxy solid; sol. in IPA, ethyl alcohol, veg. and min. oil; sp.gr. 1.05; HLB 10.5; flash pt. > 300 F; pour pt. 92 F; sapon. no. 88–98; 100% act.

Tween® 80. [ICI Am.] Polysorbate 80; nonionic; see Tween 20; pale yel. liq.; sol. in water, IPA, ethanol; sp.gr. 1.08; visc. 425 cps; HLB 15.0; flash pt. > 300 F; sapon. no. 45–55; 100% act.

Tween® 81. [ICI Am.] Polysorbate 81; nonionic; see Tween 20; amber oily liq.; sol. in min. and corn oil, dioxane, Cellosolve, methanol, ethanol, ethyl acetate, aniline; disp. in water; sp.gr. 1; visc. 450 cps; HLB 10.0; flash pt. > 300 F; sapon. no. 96–104; 100% act.

Tween® 85. [ICI Am.] Polysorbate 85; nonionic; see Tween 20; amber liq.; sol. in veg. oil, Cellosolve, lower alcohols, aromatic solvs., ethyl acetate, min. oils and spirits, acetone, dioxane, CCl_4 and ethylene glycol; disp. in water; sp.gr. 1.0; visc. 300 cps; HLB 11.0; flash pt. > 300 F; sapon. no. 80–95; 100% act.

Twitchell 6805 Oil. [Henkel/Emery] Sulfonated min. oil; anionic; rewetting agent, softener and textile fiber lubricant, deduster; liq.; 100% conc.

Twitchell 6808 Oil. [Henkel/Emery] Sulfated fatty min. deriv.; anionic; lubricant, wetting agent, and dyestuff deduster; liq.; 100% conc.

Twitchell 6809 Oil. [Henkel/Emery] Min. fatty derivs.; nonionic; wetting, lubricant, dyestuff deduster; liq.; 100% conc.

Tylose C 30 L. [Hoechst-Celanese AG] Sodium CMC; binder in textile industry.

Tylose C, CB Series. [Hoechst-Celanese AG] Sodium CMC; anionic; binder in pencil leads; thickener in batteries, rubber industry, cosmetics, foodstuffs, pharmaceuticals, tobacco and textile industry; dispersant, emulsifier for insecticidal, fungicidal and herbicidal prods.; plasticizer in ceramics; surface sizing in paper industry; press aid and lubricant in welding electrodes; gran., powd.

Tylose CR, CBR Series. [Hoechst-Celanese AG] Sodium CMC, tech.; binder for coatings; sedimenting aid in mining; gelling agent/binder/thickener in chemical tech. and rubber industry; sizing for paper and textile industry; plasticizer/filler in soaps and hand cleaning pastes; gran..

Tylose H Series. [Hoechst-Celanese AG] Hydroxyethylcellulose; nonionic; binder, thickener, plasticizer, visc. control agent, protective colloid in ceramics, emulsion polymerization, tobacco and textile industry, cosmetics, soaps, and hand cleaning pastes; water-sol.

Tylose MH, MHB, MB. [Hoechst-Celanese AG] Methyl hydroxyethylcellulose; nonionic; binder, thickener, pigment, foam, and filler stabilizer, dispersant, emulsifier, plasticizer, visc. control and sedimenting aid, and protective colloid used in coatings, paints, resins, mining, batteries, insecticidal, fungicidal, and herbicidal prods., rubber, textile, and leather industry; ceramics, suspension polymerization, and pharmaceuticals; gran.; water-sol.

Tylose MH-K, MH-xp, MHB-y, MHB-yp.

[Hoechst-Celanese AG] Methyl hydroxyethylcellulose; see Tylose MH.

Tylose P, P-x, PS-x, P-Z Series. [Hoechst-Celanese AG] Methyl hydroxycellulose; binder in plasters, adhesive, and troweling compds.; powd.

Tyzor AA. [DuPont] Acetylacetonate chelate; catalyst for esterification and olefin polymerization; used in coatings; resin crosslinking agent for automotive prods., coatings, elastomers, films/paints, graphic arts, plastics; surf. modifier for coatings, cosmetics, electronics, films, glass, graphic arts, metals, petrol. prods., plastics; red liq.; m.w. 364; sol. in IPA, IPA-water, ethyl acetate, IPA; sp.gr. 0.99; visc. 15 cP; flash pt. 12 C (PMCC); 75% in IPA.

Tyzor DC. [DuPont] Ethylacetoacetate chelate; see Tyzor AA; yel. to amber liq.; m.w. 424; sol. in IPA, acetone, ethanol, Freon TF solv.; sp.gr. 1.05; visc. 50 cP; flash pt. 27 C (PMCC); 100% conc.

Tyzor LA. [DuPont] Lactic acid chelate, ammonium salt; see Tyzor AA; lt. amber liq.; m.w. 294 (of solids); water-misc.; sp.gr. 1.21; visc. 8 cP; 50% in water.

Tyzor TBT. [DuPont] Tetrabutyl titanate; see Tyzor AA; pale yel. liq.; m.w. 340 (of solids); sol. in n-butanol, n-heptane; sp.gr. 0.99; visc. 65 cP; flash pt. 47 C (PMCC); 100% conc.

Tyzor TE. [DuPont] TEA chelate; see Tyzor AA; pale yel. liq.; m.w. 462 (of solids); sol. in water, IPA; sp.gr. 1.06; visc. 90 cP; flash pt. 16 C (PMCC); 80% in IPA.

Tyzor TOT. [DuPont] Tetrakis (2-ethylhexyl) titanate; see Tyzor AA; pale yel. liq.; m.w. 565 (of solids); sol. in n-heptane, IPA; sp.gr. 0.91; visc. 140 cP; flash pt. 60 C (PMCC); 100% conc.

Tyzor TPT. [DuPont] Tetraisopropyl titanate; see Tyzor AA; pale yel. liq.; m.w. 284 (of solids); sol. in IPA, n-heptane; sp.gr. 0.95; visc. 3 cP; flash pt. 21 C (PMCC); 100% conc.

Ucar® Additive PN-1. [Union Carbide] N-Phenyl-α-naphthylamine; antioxidant in petrol. prods., syn. lubricants, and rubber prods; avail. as free-flowing grd. powd., solid, and molten; lt. tan color gradually darkens upon exposure to air and light; m.w. 219.29; sol. 2% in petrol. oils; dissolves readily in lubricating oils and greases above 57.2 C; sp.gr. 1.2; dens. 5.5 lb/gal; m.p. 59 C.

Ucar® Cement Grinding Aids. [Union Carbide] Glycol based; grinding aids for cement, gypsum, pigments; dispersant; liq.; water-sol.

Ucarcide® 225. [Union Carbide] Glutaral; preservative, antimicrobial for cosmetic, toiletry, and chemical specialty prods.; Pt-Co 100 color; sol. in water; sp.gr. 1.066; dens. 8.87 lb/gal; visc. 3.2 cps; b.p. 100.5 C (760 mm Hg); f.p. –10 C; flash pt. (TCC) none; surf. tens. 45 dynes/cm; ref. index 1.375; pH 3.1–4.5; 25% act. min.

Ucarcide® 250. [Union Carbide] Glutaral; see Ucarcide 225; Pt-Co 100 color; sol. in water; sp.gr. 1.131; dens. 9.41 lb/gal; visc. 21 cps; b.p. 101.2 C (760 mm Hg); f.p. –21 C; flash pt. (TCC) none; surf. tens. 48 dynes/cm; ref. index 1.421; pH 3.1–4.5; 50% act. min.

Ucare® Polymer JR, LR. [Union Carbide] Modified hydroxyethyl cellulose, ethoxylated cellulose; cationic; hair fixative; for cosmetics, toiletries, hair and skin care prods.; sol. in water.

Ucar® Latex 131. [Union Carbide] PVA homopolymer emulsion; nonionic surfactant; externally plasticized homopolymer for tape joint compds., general mastic-type compds. for trowel and brush; general purpose binder; 0.8 μ particle size; sp.gr. 1.15 (polymer dens.); dens. 9.2 lb/gal; visc. 2600 cP (latex); pH 4.5; 60% solids.

Ucar® Latex 828. [Union Carbide] Self-crosslinking vinyl-acrylic emulsion; anionic/nonionic surfactant; hard, durable polymer for use as a binder in fiberfill, batting, padding, and firm-handed nonwovens; emulsion; 0.3 μ particle size; sp.gr. 1.19 (polymer dens.); dens. 9.0 lb/gal; visc. 300 cP (latex); pH 4.0; 45% solids.

Ucar® Latex 879. [Union Carbide] Self-crosslinking acrylic emulsion; anionic surfactant; mod. stiff, salt-stable polymer with durability to laundering and dry cleaning; fiberfill binder for nonwoven prods., tenting, awning fabrics; emulsion; 0.3 μ particle size; sp.gr. 1.13 (polymer dens.); dens. 8.9 lb/gal; visc. < 100 cP (latex); pH 4.5; 45% solids.

Ucarsan® Sanitizers. [Union Carbide] Glutaraldehyde/detergent; industrial sanitizers, poultry hatchery sanitizers; conc. sol'ns.; water-sol.

Ucarsil® FR-1A, FR-1B. [Union Carbide] Organosilicon chemical; additives allowing the processing of up to 70% alumina trihydrate into polyolefins.

Ucarsil® MD-200. [Union Carbide] Organosilicon; dispersant for magnetic iron oxide; liq.

Ucar® Silicone ALE-56. [Union Carbide] Amino silicone emulsion; detergent-resistant vinyl conditioner; polishes; water-disp.

Ucarsil® Organosilicon Chemical AF-1, AF-2. [Union Carbide] Organosilicon chemical; stabilizer for silicate inhibitors in antifreeze; clear liq.

Ucar® Super Wetter FP. [Union Carbide] Org. silicon; nonionic; wetting agent, penetrant, coemulsifier, lubricant, dispersant; liq.; sol./disp. in water; 100% conc.

Uconex® Antimicrobials. [Union Carbide] Glutaraldehyde; microbicide for metalworking fluids; 15 and 45% sol'ns.

Ucon® Fluid 50-HB-Series. [Union Carbide] Mixed ethoxylated and propoxylated butanol; emollient; for applics. where fluid is to be removed by a water rinse; sol. in water, alcohol; visc. in SUS @ 100 F corresponds to numerical value.

Ucon® Fluid 75-H-1400. [Union Carbide] EO/PO copolymer; see Ucon Fluid 50-HB; sol. in water, alcohol; visc. in SUS @ 100 F corresponds to numerical value.

Ucon® Fluid LB-, -625, -1145, -1715, -3000. [Union Carbide] Propoxylated butanol; emollient; cosolv. for most natural oils in aq. alcoholic compositions; low color and odor; sol. in alcohol; visc. in SUS @ 100 F corresponds to numerical value.

Ucon® Fluid LO-500. [Union Carbide] Propoxylated oleyl alcohol; emollient; cosolv. for min. oil-ethanol systems; sol. in alcohol; visc. in SUS @ 100 F corresponds to numerical value.

Ultra NCS Liquid. [Witco] Ammonium cumene sulfonate; anionic; hydrotrope, coupling agent, solubilizer; cloud pt. depressant for liq. detergents; anti-blocking agent for powd. detergents; liq.; 40% conc.

Ultra NXS Liquid. [Witco] Ammonium xylene sulfonate; see Ultra NCS Liquid; liq.; 45% conc.

Ultra SCS Liquid. [Witco] Sodium cumene sulfonate; see Ultra NCS Liquid; liq.; 40% conc.

Ultra SXS Liq., Powder. [Witco] Sodium xylene sulfonate; see Ultra NCS Liquid; 40% liq.; 90% powd.

Ultra-Cure I-100. [PMC] Chlorothioxanthone; photoinitiator used to decrease drying time of uv-cured inks and coatings; sol. in IPA, MEK, benzene, chlorobenzene; slightly sol. in acetone, methanol, chloroform.

Ultraflex®. [Petrolite] Microcryst. wax; plastic wax offering high ductility, flexibility at very low temps., low surf. tens., solvency for amorphous hydrocarbon resins and syn. elastomers; provides protective barrier properties against moisture vapor and gases; uses incl. hot-melt laminating adhesives for papers, films, and foils; hot-melt coatings; in antisunchecking agents in rubber goods, elec. insulating agents, leather treating agents, water repellents for textiles, solv.-based rustproof coatings, cosmetic ingreds., and as plasticizer for other petrol. waxes used in crayons, dental compds., chewing gum base, and candles; color 1.5 (D1500); also avail. in wh.; dens. 0.92 g/cc; visc. 11 cps (98.8 C); m.p. 64.4 C; flash pt. 293.3 C.

Ultramoll I. [Bayer] Polyadipate; polymeric plasticizer; good gelation, resistance to oils, fats, greases, and normal-grade petrol; used in calendering, extrusion, inj. molding; food pkg., oil tank linings, self-adherent film, crash pad film (antifogging), protective clothing, conveyor belts, tubing for milking machines and oil, shoe soles; used in VC copolymers, NC, cellulose acetobutyrate, and in natural, S/B, nitrile/butadiene, chlorinated, and butyl rubber; Hazen ≤ 150; dens. 1.075–1.090 g/cc; visc. 2000–3000 mPa.s; flash pt. 280–300 C; acid no. ≤ 1.0; sapon. no. 490–510; ref. index 1.472–1.473; 70:30 PVC:plasticizer suspension: tens. str. 23 MPa; tens. elong. 330% (break); hardness (Shore D) 41.

Ultramoll II. [Bayer] Polyadipate; polymeric plasticizer; resistance to oils, fats, greases, bitumen, and normal-grade petrol; used for calendering, coating, extrusion, and inj. molding; pkg., foils, oil tank linings, self-adhesive film, crash pad film (antifogging), protective clothing, tablecloths, insulating sleeves, and strands for the radio and television industry, oil tubes, shoe soles; used in VC copolymers, PVAc, NC, cellulose acetobutyrate, and natural, S/B, nitrile/butadiene, chlorinated, and butyl rubbers; Hazen ≤ 150; dens. 1.100–1.115 g/cc; visc. 2000–3000 mPa.s; flash pt. 280–300 C; acid no. ≤ 1.0; sapon. no. 510–540; ref. index 1.472–1.473; 70:30 PVC:plasticizer suspension: tens. str. 24 MPa; tens. elong. 320% (break); hardness (Shore D) 40.

Ultramoll III. [Bayer] Polyadipate; polymeric plasticizer; resistance to oils, fats, greases, bitumen, and normal-grade petrol; much lower paste visc. than Ultramoll I and II; used in coating, calendering, extrusion, inj. molding, and dip-coating for tarpaulins, protective clothing, imitation and expanded imitaion leather, pkg., foils, self-adhesive film, cable sheathing, sealing profiles, boots, gloves; used in VC copolymers, PVAc, NC, cellulose acetobutyrate, and in natural, S/B, chlorinated, and butyl rubber; Hazen ≤ 150; dens. 1.100–1.110 g/cc; visc. 1000–1300 mPa.s; flash. pt. 270–290 C; acid no. ≤ 1.0; sapon. no. 510–530; ref. index 1.469–1.470; 70:30 PVC: plasticizer suspension: tens. str. 23 MPa; tens. elong. 330% (break); hardness (Shore D) 44.

Ultramoll PP. [Bayer] Polyphthalate; polymeric plasticizer; gelation and resistance to saponification, good elec. properties, lt. stability, good tropical resistance, better resistance to oils, fats, and greases than with monomeric plasticizers; gives pastes of low visc., migration resistant with respect to plastics, surface coatings, and adhesives based on e.g., acrylates; used in calendering, coating, extrusion, inj. molding, dip-coating, and rotational molding for tarpaulins, protective clothing, wallpapers, imitation and expanded imitation leather, self-adhesive film, elec. insulating film, for babies' panties, automotive films, protective foils for buildings, cables with resistance to high temps., profiles, tubing, conveyor belts, shoes, soles, gaskets, technical inj. moldings, industrial gloves, bellows, technical articles; used in VC copolymers; Hazen ≤ 100; dens. 1.035–1.045 g/cc; visc. 1200–1500 mPa.s; flash pt. 230–265 C; acid no. ≤ 0.5; sapon. no. 200–320; ref. index. 1.502–1.504; 70:30 PVC:plasticizer suspension: tens. str. 25 MPa; tens. elong. 245% (break); hardness (Shore D) 38.

Ultramoll PU. [Bayer] PU; polymeric plasticizer; high mechanical values, very good low-temp. stability, high abrasion resistance, lt. stability, resistance to oils, fats, greases, bitumen, and petrol, resistance to extraction by almost all org. solvs. (incl. perchloroethylene, trichloroethylene), nonvolatility; used in melt roll coating, extrusion, inj. molding, calender coatings for the clothing, upholstery, bag, and shoe industries, films (for garments to withstand chemical cleaning, plasters, optical industry, lining split leather and food pkg.), computer cables, cable sheathing, tubing, structural profiles, gaskets, bellows, shoe soles, boots for the oil and fisheries industry; powd.; dens. 1.14 g/cc; 70:30 PVC: plasticizer suspension: tens. str. 24 MPa; tens. elong. 275%; hardness (Shore D) 59.

Ultramoll TGN. [Bayer] Polyphthalate; polymeric plasticizer; good gelation, resistance to oils, fats, greases, bitumen, and normal-grade petrol, comparatively low visc., and good spreading properties; used in pastes for spreading and dip-coating, calendering, extrusion, inj. molding for floor coverings (top layers and inlays), imitation leather, protective clothing and foils, tarpaulins, boots, gloves, self-adhesive films and tapes; used as desensitizing agents for peroxides; Hazen ≤ 100; dens. 1.090–1.100 g/cc; visc. 2000–2500 mPa.s; flash pt. 220–240 C; acid no. ≤ 1.0; sapon. no. 300–320; ref. index 1.503–1.505; 70:30 PVC:plasticizer suspension: tens. str. 30 MPa; tens. elong. 210% (break); hardness (Shore D) 55.

Ultranox® 226. [GE] 2,6 Di-t-butyl-4-methylphenol; antioxidant for use in plastics incl. polyolefins, styrenics, PU, rubber, and food-pkg. materials; antioxidant for oils, greases, and lubricants; stabilizer in fuels, lubricating and min. oils; wh. cryst. solid; m.w. 220; sol. in acetone, benzene, CCl_4, ethyl acetate, toluene; sp.gr. 1.03; dens. 8.6 lb/gal; m.p. 69 C; b.p. 265 C (760 mm Hg); flash pt. (COC) 127 C; 99% min. purity.

Ultranox® 236. [GE] 4,4′-Thio-bis (2-t-butyl-5-methylphenol); antioxidant for use in adhesives, rubber articles for repeated use, polymers incl. polyolefins, PVC, acrylic ethyl cellulose; antioxidant for lubricants, cutting oils, water-sol. oils, hydraulic oils; wh. cryst. solid; m.w. 359; sol. in acetone, methanol, ethanol, ethyl acetate; sp.gr. 0.9–1.0; dens. 7.51–8.35 lb/gal; m.p. 158 C; flash pt. (COC) 205 C min.; 99% min. purity.

Ultranox® 246. [GE] 2,2′-Methylene-bis-(4-methyl-6-t-butylphenol); stabilizer, antioxidant for PP, polyethylene, polyoxymethylene copolymer, polyoxymethylene homopolymer, styrenics, and rubber; wh. cryst. powd.; m.w. 341; sol. in acetone, dioxane, ethyl acetate, ethanol, benzene; sp.gr. 1.07; dens. 8.93 lb/gal; m.p. 124–128 C; flash pt. (COC) 190 C min.

Ultranox® 254. [GE] Polymeric 2,2,4-trimethyl-1,2-dihydroquinoline; antioxidant for natural and syn. rubber and their vulcanizates, e.g., in wire and cable applics; cream powd.; m.w. 173; sol. in acetone, ethyl acetate, methylene chloride, CCl_4, benzene, chloroform, ethanol; sp.gr. 1.08; dens. 9 lb/gal; m.p. 75 C; flash pt. (COC) 235 C min.

Ultranox® 257. [GE] Polymeric sterically hindered phenol; butylated reaction prod. of p-cresol and dicyclopentadiene; antioxidant for wh., transparent, and colored natural and syn. rubber goods; esp. for latex applics. such as carpet backing, dipping articles; also for EVA-based hot melts, thermoplastic rubber, and styrenics; wh. to cream powd.; m.w. 600–700; sol. in acetone, ethyl acetate, ethanol, methylene chloride, and liq. phosphites; sp.gr. 1.10; m.p. 100 C; flash pt. (COC) 225 C min.

Ultranox® 276. [GE] Octadecyl 3-(3,5-di-t-butyl-4-hydroxyphenyl) propionate; high m.w. antioxidant/stabilizer for styrenics, polyolefins, PVC, urethane and acrylic coatings, adhesives, and elastomers; effective replacement for BHT in polyolefins; wh. cryst. powd.; m.w. 531; sol. in most aprotic org. solvs. such as benzene, xylene, and acetone; sp.gr. 1.02; dens. 22–23 lb/ft^3; m.p. 55 C; flash pt. (COC) 273 C.

Ultranox® 624. [GE] Bis (2,4-di-t-butylphenyl) pentaerythritol diphosphite; Ultranox 626 without the triisopropanolamine; wh. cryst. powd.; m.w. 604; sol. see Ultranox 626; dens. 0.43 g/ml; m.p. 160–175 C; flash pt. (PMCC) 168 C; acid no. 2.5 max.; 10% phosphorus.

Ultranox® 626. [GE] Bis (2,4-di-t-butylphenyl) pentaerythritol diphosphite/0.5–1.2% triisopropanolamine; antioxidant for polyolefin, PVC, PET, styrenics, ABS, and PC polymers; stabilizer; wh. cryst. solid; 40% > 60 mesh, 15% 60–200 mesh; m.w. 604; sol. in methylene chloride, THF, toluene; dens. 0.43 g/ml; m.p. 160–175 C; flash pt. (PMCC) 168 C; acid no. 2.5 max.; 10% phosphorus content.

Ultranox® 626A. [GE] Bis (2,4-di-t-butylphenyl) pentaerythritol diphosphite, ≤ 1.0% triisopropanolamine; larger particle size version of Ultranox 626; fewer fines for easier flow in some additive feeding equip.; 95% greater than 60 mesh.

Ultra-Pflex®. [Pfizer] Ultrafine surface-treated PCC; impact modifier in rigid PVC applics.

Ultra Sulfate AE-3. [Witco] Ammonium alcohol ether sulfate; anionic; wetting agent, penetrant, lubricant, emulsifier, dye dispersant, scouring aid, antistat; base for personal care products; detergent for specialty household and industrial cleaners; textile surfactant; liq.; 58% conc.

Ultra Sulfate SE-5. [Witco] Sodium alcohol ether sulfate; see Ultra Sulfate AE-3; liq.; 58% conc.

Ultra Sulfate SL-1. [Witco] Sodium lauryl sulfate; see Ultra Sulfate AE-3; also used as surfactant for emulsion polymerization and latex stabilization; liq.; 30% conc.

Ultrazym 100. [Novo] Pectin-decomposing enzyme for mash and juice treatment; processing aid in wine industry; pulp wash; liq.

Ultrox. [M&T] Zirconium silicate; glaze opacifier; stabilizes color shades; used in wh. and colored glazes for sanitary ware, wall tile, glazed brick, structural tile, stoneware, dinnerware, special porcelains, refractory compositions, epoxy formulations, encapsulating resins; inert, finely milled heat- and chemically resistant filler; 2 μ avg. particle size; ref. index 2; hardness 7.5 (Mohs).

Ultrox 500W. [M&T] Zirconium silicate; see Ultrox; 500W has higher opacity; 1 μ avg. particle size; ref. index 2; hardness 7.5 (Mohs).

Ultrox 1000W. [M&T] Zirconium silicate; see Ultrox; 1000W recommended for glazes requiring max. opacity, whiteness, and covering power; 0.5 μ avg. particle size; ref. index 2; hardness 7.5 (Mohs).

Unamide® C-2. [Lonza] PEG-3 cocamide; nonionic; foam booster/stabilizer, visc. builder, emulsifier; used in personal care, household and industrial prods.; wh. solid; sapon. no. 15 max.; pH 9–10 (5%); 100% act.

Unamide® C-5. [Lonza] PEG-6 cocamide; nonionic; see Unamide C-2; lt. yel. liq.; sapon. no. 12 max.; pH 9–10 (5%); 100% act.

Unamide® C-72-3. [Lonza] Cocamide DEA (2:1); nonionic; visc. builder, foam stabilizer; used for lt. duty liqs., industrial, household hard surface cleaners, surfactant, emulsifier, corrosion inhibitor, lubricant and personal care prods.; amber liq.; acid no. 0–10; pH 9–10.5 (5%).

Unamide® C-7649. [Lonza] Cocamide DEA, modified; anionic/nonionic; detergent base for floor cleaners; foam stabilizer, emulsifier; liq.; 100% conc.

Unamide® CDX. [Lonza] Cocamide DEA; nonionic; foam booster, stabilizer, visc. builder used in personal care prods.; straw liq.; water-sol.; dens. 8.2 lb/gal; sp.gr. 0.99; acid no. 0–3; pH 9–10.5 (5%); biodeg.; 100% act.

Unamide® CMX. [Lonza] 1:1 Cocamide MEA; nonionic; see Unamide CDX; lt. solid; acid no. 0–3; pH 9–10.5 (5%); 100% act.

Unamide® D-10, GC-75. [Lonza] 2:1 Coco DEA, modified; nonionic; see Unamide C-72-3; dk. amber and yel. resp. liqs.; acid no. 55–75 and 80–100; pH 9–10.5 (5%); 100% act.

Unamide® J-56. [Lonza] Lauramide DEA; nonionic; see Unamide C-72-3; amber liq.; acid no. 18–25; pH 9–10.5 (5%); 100% act.

Unamide® JJ-35. [Lonza] Cocamide DEA; nonionic; see Unamide CDX; lt. straw liq.; acid no. 0–10; pH 9–10.5 (5%); 100% act.

Unamide® L-2. [Lonza] PEG-3 lauramide; nonionic; see Unamide C-2; wh. solid; sapon. no. 15 max.; pH 9–10 (5%); 100% act.

Unamide® L-5. [Lonza] PEG-6 lauramide; nonionic; see Unamide C-2; wh. soft wax, sapon. no. 12 max.; pH 9–10 (5%); 100% act.

Unamide® LDL. [Lonza] Cocamide DEA; nonionic; see Unamide C-72-3; lt. straw liq.; dens. 8.2 lb/gal; sp.gr. 0.98; acid no. 9–10.5; pH 9–10.5 (5%); biodeg.; 100% act.

Unamide® LDX. [Lonza] Lauramide DEA; nonionic; see Unamide CDX; also used in lt. duty detergents; wh. solid; water-sol.; dens. 8.2 lb/gal; sp.gr. 0.99; acid no. 0–3; pH 9–10.5 (5%); biodeg.; 100% act.

Unamide® LMDX. [Lonza] Lauramide DEA; nonionic; see Unamide CDX; wh. solid; water-sol.; dens. 8.2 lb/gal; sp.gr. 0.99; acid no. 0–3; pH 9–10.5 (5%); 100% act.

Unamide® N-72-3. [Lonza] Cocamide DEA; nonionic; see Unamide C-72-3; yel. liq.; acid no. 0–10; pH 9–10.5 (5%); 100% act.

Unamide® S. [Lonza] Stearamide DEA; nonionic; thickener and opacifier in personal care prods.; visc. builder, gelling agent, lubricant, industrial and household prods.; textile lubricant and finishing agent; wh. waxy solid; acid no. 11–15; 100% act.

Unamide® SI. [Lonza] 1:1 Stearic (ethyl amino hydroxyethyl) amide; nonionic; visc. builder, gelling agent, opacifier used in household and industrial prods.; textile lubricant and finishing agent; wh. waxy solid.

Unamide® W. [Lonza] 1:1 Stearamide DEA; see Unamide SI; nonionic; amber liq.

Unamine®-C. [Lonza] Fatty imidazoline 1-hydroxyethyl, 2-undecyl imidazoline; nonionic; surfactant, fungicide; textile antistat; leather treating; base for quats.; acid detergent and wetting agent; corrosion inhibitor; liq.

Unamine®-O. [Lonza] Fatty imidazoline 1-hydroxyethyl, 2-heptadecenyl imidazoline; nonionic; emulsifier, demulsifier; antistat for drycleaning fluids; water displacing agent; corrosion inhibitor; base for quats.; liq.; sol. in acidic sol'ns.

Unamine®-S. [Lonza] Fatty imidazoline 1-hydroxyethyl 2-heptadecyl imidazoline; nonionic; softener, antistat for syn. fabrics; surfactant; filming compd.; corrosion inhibitor; solid; oil-sol.

Unamine®-T. [Lonza] Tall oil hydroxyethyl imidazoline; cationic; surfactant; w/o emulsifier, corrosion inhibitor, filming agent, dewatering compd.; flushing agent for pigments; liq.; acid-sol.; 92% conc.

Unden®. [Bayer] Propoxur; broad spectrum insecticide with fast killing action for the control of sucking and biting pests; colorless cryst.; m.w. 209.2; sol. (g/1000 ml): > 1000 g in methylene chloride, 100–1000 g in IPA, toluene, 1.9 g in water; m.p. 90.7 C (pure a.i.).

Uniflex 300. [Union Camp] Polymeric plasticizer; Gardner 3; sp.gr. 1.10; visc. 3300 cSt; acid no. 0.4; ref. index 1.466; flash pt. 293 C (COC); fire pt. 327 C (COC).

Uniflex 312. [Union Camp] Polymeric plasticizer; Gardner 1; sp.gr. 1.08; visc. 970 cSt; acid no. 0.5; ref. index 1.465; flash pt. 290 C (COC); fire pt. 313 C (COC).

Uniflex 314. [Union Camp] Polymeric plasticizer; APHA 125; sp.gr. 1.19; visc. 4900 cSt; acid no. 33; ref. index 1.479; flash pt. 315 C (COC); fire pt. 338 C (COC).

Uniflex 315. [Union Camp] Polymeric plasticizer; APHA 150; sp.gr. 1.11; visc. 6500 cSt; acid no. 1.0; ref. index 1.465; flash pt. 279 C (COC); fire pt. 315 C (COC).

Uniflex 320. [Union Camp] Polymeric plasticizer; Gardner 4; sp.gr. 1.08; visc. 2500 cSt; acid no. 1.0; ref. index 1.466; flash pt. 279 C (COC); fire pt. 307 C (COC).

Uniflex 327. [Union Camp] Polymeric plasticizer; Gardner 3; sp.gr. 1.06; visc. 2800 cSt; acid no. 1.0; ref. index 1.469; flash pt. 263 C (COC); fire pt. 285 C (COC).

Uniflex 330. [Union Camp] Polymeric plasticizer; APHA 100; sp.gr. 1.09; visc. 5300 cSt; acid no. 1.0; ref. index 1.466; flash pt. 290 C (COC); fire pt. 318 C (COC).

Uniflex BYO. [Union Camp] Butyl oleate; monomeric plasticizer for plastics, rubber, and lubricants; APHA 70; sp.gr. 0.868; visc. 8 cSt; m.p. 10 C; acid no. 0.5; sapon. no. 170; ref. index 1.451; flash pt. 199 C (COC); fire pt. 218 C (COC).

Uniflex DBS. [Union Camp] Dibutyl sebacate; monomeric plasticizer for plastics and rubber; APHA 5; f.p. –11 C; sp.gr. 0.937; visc. 8 cSt; acid no. < 0.1; flash pt. 193 C (COC); fire pt. 213 C (COC).

Uniflex DCA. [Union Camp] Dicapryl adipate; monomeric plasticizer for plastics and rubber; APHA 10; f.p. –32 C; sp.gr. 0.914; visc. 13 cSt; acid no. < 0.1; ref. index 1.439; flash pt. 207 C (COC); fire pt. 224 C (COC).

Uniflex DCP. [Union Camp] Dicapryl phthalate; monomeric plasticizer for plastics and rubber; APHA 25; f.p. –60 C; sp.gr. 0.974; visc. 56 cSt; acid no. < 0.1; ref. index 1.479; flash pt. 202 C (COC); fire pt. 232 C (COC).

Uniflex DOA. [Union Camp] Dioctyl adipate; monomeric plasticizer for plastics and rubber; f.p. –70 C; sp.gr. 0.927; visc. 12 cSt; acid no. < 0.1; ref. index 1.447; flash pt. 207 C (COC); fire pt. 230 C (COC).

Uniflex DOS. [Union Camp] Dioctyl sebacate; monomeric plasticizer for plastics, rubber, and lubricants; APHA 35; f.p. –55 C; sp.gr. 0.915; visc. 19 cSt; acid no. < 0.1; ref. index 1.448; flash pt. 227 C (COC); fire pt. 260 C (COC).

Uniflex TCTM. [Union Camp] Tricapryl trimellitate; monomeric plasticizer for plastics and rubber; APHA 80; sp.gr. 0.978; visc. 230 cSt; pour pt. –40 C; acid no. 0.5; ref. index 1.478; flash pt. 246 C (COC); fire pt. 268 C (COC).

Uniflex TOTM. [Union Camp] Trioctyl trimellitate;

monomeric plasticizer for plastics, rubber, and lubricants; APHA 100; sp.gr. 0.990; visc. 215 cSt; pour pt. –45 C; acid no. < 0.1; ref. index 1.485; flash pt. 260 C (COC); fire pt. 285 C (COC).

Unihib® 106. [Lonza] Organo-phosphonate; dispersant, scale inhibitor; liq.

Unihib® 305. [Lonza] Aminotrimethylene phosphonic acid; corrosion and scale deposition inhibitor in aq. systems; sludge conditioner for boiler water treatment; pale yel. clear low visc. liq.; pH < 2 (1%); 50% solids.

Unihib® 314. [Lonza] Sodium salt of organophosphonate; dispersant, scale inhibitor; liq.

Unilin 425. [Petrolite] High m.w. C_{30} avg. linear primary alcohol; defoamer for pulp and paper processing; solid.

Unilin 550. [Petrolite] High m.w. C_{40} avg. linear primary alcohol; defoamer for pulp and paper processing; solid.

Unilin 700. [Petrolite] High m.w. C_{50} avg. linear primary alcohol; defoamer for pulp and paper processing; solid.

Unilink 4100. [UOP] Aromatic diamine; chain extender for PU elastomers; dk. reddish liq.; m.w. 220; sp.gr. 0.94 (60 F); dens. 7.8 lb/gal; visc. 38.6 cps (60 F); flash pt. (PMCC) 240 F.

Unilink 4102. [UOP] Aromatic diamine; chain extender for PU applics.; dk. liq.; sp.gr. 0.94; dens. 7.8 lb/gal; visc. 36 cps (60 F); flash pt. (PMCC) 265 F.

Unilink 4130. [UOP] 70% Unilink 4100, 30% tetrapropoxylated ethylene diamine; curative giving balance of chain extension and crosslinking properties in PU cast elastomers; dk. reddish liq.; sp.gr. 0.97; dens. 8.02 lb/gal; visc. 192 cps (60 F); flash pt. (PMCC) 285 F.

Unilink 4132. [UOP] 70% Unilink 4102, 30% tetrafunctional hydroxyl crosslinker; curative for low durometer elastomers; dk. liq.; sp.gr. 0.96; dens. 8.0 lb/gal visc. 28 cps (100 F); flash pt. (PMCC) 245 F.

Unilink 4200. [UOP] Aromatic diamine; chain extender in PU elastomers; dk. amber liq.; m.w. 310; sp.gr. 0.996 (60 F); dens. 8.3 lb/gal; visc. 115 cps (100 F); flash pt. (PMCC) 300 F.

Unilink 4230. [UOP] 70% Unilink 4200, 30% tetrapropoxylated ethylene diamine; curative giving balance of chain extension and crosslinking properties in PU cast elastomers; dk. amber liq.; sp.gr. 1.01 (60 F); dens. 8.3 lb/gal; visc. 265 cps (100 F); flash pt. (PMCC) 171 F.

Unilink 8100. [UOP] Aromatic diamine; chain extender for PU elastomers and coatings; contains alkylamine group which acts as internal plasticizer producing lower hardness elastomers; dk. red. liq.; sp.gr. 0.90 (60 F); dens. 7.6 lb/gal; visc. 78 cps (60 F); flash pt. (PMCC) 325 F.

Unilink 8130. [UOP] 70% Unilink 8100, 30% tetrafunctional hydroxyl crosslinker; curative; dk. red liq.; sp.gr. 0.93; dens. 7.8 lb/gal; visc. 51 cps (100 F); flash pt. (PMCC) 325 F.

Uni-Malt®. [Am. Ingredients] Blend of enzyme act. cereal malt and sodium chloride; diastatic malt supplement; starch conditioning agent to produce optimum dough characteristics; lt. tan powd. 95% min. thru 40 mesh; malty odor.

Unimate BYS. [Union Camp] n-Butyl stearate; skin emollient and solv. used in lipsticks, creams, lotions; sol. in min. oil.

Unimate DBS. [Union Camp] Di-n-butyl sebacate; skin emollient and pre-electric shave lotions; sol. in water, 85% ethanol.

Unimate DIPS. [Union Camp] Diisopropyl sebacate; skin emollient, solubilizer, coupler, spreading agent for creams, lotions, bath oils, aerosol toilet preparations; sol. in min. oil, 95% ethanol.

Unimate DOS. [Union Camp] Di-2-ethylhexyl sebacate; emollient for hair grooming, pre-electric shave lotions; sol. in min. oil, 95% ethanol.

Unimate EHP. [Union Camp] 2-Ethylhexyl palmitate; skin emollient for lotions and creams; sol. in min. oil, 91% ethanol.

Unimate IPM. [Union Camp] IPM; skin emollient, lubricant, carrier; used in lotions, creams, antiperspirants, bath oils; sol. in 95% ethanol, min. oil.

Unimate IPP. [Union Camp] IPP; see Unimate IPM; sol. in 95% ethanol, min. oil.

Unimate IPPM. [Union Camp] Isopropyl palmitate/myristate; see Unimate IPM; sol. in 95% ethanol, min. oil.

Unimoll 66 M. [Bayer] Dicyclohexyl phthalate; plasticizer for use in formulation of delayed tack heat sealable coatings; wh. powd., 99.9% < 40 μ; b.p. 248 C; sol. in org. solvs.; insol. in water; sp.gr. 1.15; dens. 400 g/l; m.p. 63–65 C; flash pt. 205–215 C.

Unimoll BB. [Bayer] Benzylbutyl phthalate; plasticizer used for PVC articles with resistance to bitumen, coating, calendering, extrusion, inj. molding, imitation leather, carpet backing, floor coverings, film, structural profiles, joint masking profiles; used in VC copolymer, PVAc, PS, NC, ethyl cellulose, CAB, acrylic surf. coatings, alkyd resins, polymethacrylate, natural, S/B, N/B, and chlorinated rubber; Hazen < 40; dens. 1.120–1.125 g/cc; visc. 61–63 mPa.s; acid no. < 0.1; sapon. no. 350–370; ref. index 1.5396–1.5412; flash pt. 205–215 C (OC); tens. str. 21 MPa; tens. elong. 260%; hardness (Shore D) 30.

Unimoll DB. [Bayer] Dibutyl phthalate; plasticizer used only to assist gelling in combination with other plasticizers of lower gelling capacity; used in PVC pastes, calendering, extrusion, inj. molding, VC copolymers, PVAc, surf. coatings, adhesives, rubber, PS, polyacrylates, and alkyd resins; Hazen < 30; dens. 1.043–1.049 g/cc; visc. 19–21 MPa.s; acid no. < 0.1; sapon. no. 400–410; ref. index 1.4915–1.4930; flash pt. 173–177 C (OC); 30% in PVC: tens. str. 17 MPa; tens. elong. 265%; hardness (Shore D) 28.

Union Carbide® A-151. [Union Carbide] Vinyltriethoxysilane; coupler promoting bonding between org. and inorg. components of a system.

Union Carbide® A-172. [Union Carbide] Vinyl tris-(2-methoxyethoxy) silane; see Union Carbide A-151.

Union Carbide® A-174. [Union Carbide] γ-Methacryloxypropyltrimethoxysilane; see Union Carbide A-151.

Union Carbide® A-186. [Union Carbide] β-(3,4-Epoxycyclohexyl) ethyltrimethoxy silane; see Union Carbide A-151.
Union Carbide® A-187. [Union Carbide] γ-Glycidoxypropyltrimethoxy silane; see Union Carbide A-151.
Union Carbide® A-189. [Union Carbide] γ-Mercaptopropyltrimethoxy silane; see Union Carbide A-151.
Union Carbide® A-1100. [Union Carbide] γ-Aminopropyltriethoxy silane; see Union Carbide A-151.
Union Carbide® A-1102. [Union Carbide] γ-Aminopropyltriethoxy silane, tech.; see Union Carbide A-151.
Union Carbide® A-1120. [Union Carbide] N-β-(aminoethyl) γ-aminopropyltrimethoxy silane; see Union Carbide A-151.
Union Carbide® A-1126. [Union Carbide] Aminofunctional silane in sol'n.; see Union Carbide A-151.
Union Carbide® A-1893. [Union Carbide] β-Mercaptoethyltriethoxy silane; see Union Carbide A-151.
Union Carbide® L-45 Series. [Union Carbide] Dimethylpolysiloxane polymers; emollient for creams and lotions; antifoam for nonaq. petrol. processing systems, hydraulic fluids, pigment flotation on paints; heat transfer fluid for drug mfg.; lubricant; also for cosmetic, car polish, lubricant/release applics., aerosol pkg.; sol. in most nonpolar solvs.; visc. 7–100,000 cstk.
Union Carbide® L-720, -721. [Union Carbide] Organosilicone; nonionic; stabilizer, lubricant, antifoamer, mold release and wetting agent, penetrant; stabilizer, lubricant, antifoamer, mold release and wetting agent; penetrant; solid; water-sol.; 100% conc.
Union Carbide® L-722, -727. [Union Carbide] Organosilicone; nonionic; see Union Carbide L-720; liq.; sol. in alcohol and hydrocarbon; 100% conc.
Union Carbide® L-7001, -7002. [Union Carbide] Organosilicone; nonionic; wetting agent, penetrant, coemulsifier, lubricant, dispersant; liq.; water sol. and disp.; 100% conc.
Union Carbide® L-7500. [Union Carbide] Organosilicone; nonionic; penetrant and dispersant for pigments; liq.; oil-sol.; 100% conc.
Union Carbide® L-7600, -7602, –7604. [Union Carbide] Organosilicone; nonionic; see Union Carbide L-7001; liq.; water sol. and disp.; 100% conc.
Union Carbide® L-7605. [Union Carbide] Organosilicone; nonionic; see Union Carbide L-7001; solid; water sol. and disp.; 100% conc.
Union Carbide® L-7607. [Union Carbide] Organosilicone; nonionic; see Union Carbide L-7001; liq.; water sol. and disp. 100% conc.
Union Carbide® LE-45. [Union Carbide] Dimethylpolysiloxane emulsion; nonionic; lubricant in cosmetics; softener, release agent in textile nonwovens, plastics, glass molding; water-disp.
Union Carbide® LE-458HS. [Union Carbide] Dimethylpolysiloxane emulsion; anionic; lubricant in rubber and plastic mold releases; water-disp.
Union Carbide® LE-467, LE-467 HS. [Union Carbide] Dimethylpolysiloxane emulsion; anionic; release; component in spray starch fabric finishes, lubricant/softeners in textiles; water-disp.
Union Carbide® R20. [Union Carbide] Sodium methyl silanolate; water repellent for masonry surfs., paint additive; water-dilutable; 30% solids in water.
Union Carbide® R272. [Union Carbide] Silicone; water repellent for masonry surfs., as paint additive; solv.-dilutable; 75% solids in volatile silicone.
Union Carbide® R274. [Union Carbide] Silicone; water repellent for masonry surfs.; solv.-dilutable; 75% solids plus catalyst in volatile silicone.
Union Carbide® Y-9567. [Union Carbide] Organosilicon chem.; antistat for glass fibers; clear liq.
Union Carbide® Y-9794. [Union Carbide] Organosilicon chemical; stabilizer for silicate inhibitors in antifreeze; clear liq.
Uniperol AC. [BASF AG] Oxyethylated fatty amine; cationic; dyeing aux., leveling agent; used in the textile industry; lt. br. clear visc. liq.; water-sol.
Uniperol AC Highly Conc. [BASF AG] Oxyethylated fatty amine; cationic; see Uniperol AC; br. visc. oil; water-sol.
Uniperol O. [BASF AG] Fatty alcohol-polyglycol ether; nonionic; emulsifier, dyeing aux., leveling and wetting agent, dispersant, detergent used in textile processing; colorless liq.; water-sol.; 20% act.
Uniperol O Micropearl. [BASF AG] Fatty alcohol-polyglycol ether; nonionic; see Uniperol O; off-wh. powd.
Uniquat® CB-50. [Lonza] Coco imidazoline benzyl chloride; cationic; microbicide to control bacteria, fungi, yeast, mold, and algae in recirculating cooling tower waters; corrosion inhibitor; dk. amber clear to turbid liq.; sp.gr. 0.932; dens. 7.76 lb/gal; visc. 56 cps; f.p. 25 C; flash pt. (COC) 89 F; 50% act. in IPA.
Uniquat® PB-750, PB-802. [Lonza] Pyridine quat.; corrosion inhibitor; liq.
Unislip 1753. [Unichema] Erucamide; slip and antiblock agent; Gardner 2 max. color; 85% < 710 μm; bulk dens. 0.60 g/ml; visc. 12 cSt (100 C); m.p. 79–85 C; flash pt. (COC) 230 C; acid no. 1 max.; 98% amide content.
Unislip 1757. [Unichema] Oleamide; lubricant, slip and antiblock agent, lubricant grade; Gardner 8 max. color; 85% < 710 μm; bulk dens. 0.60 g/ml; visc. 11 cSt (100 C); m.p. 66–72 C; flash pt. (COC) 210 C; acid no. 7 max.; 94% amide content.
Unislip 1759. [Unichema] Oleamide; lubricant, slip and antiblock agent; Gardner 2 max. color; 85% < 710 μm; bulk dens. 0.60 g/ml; visc. 11 cSt (100 C); m.p. 70–76 C; flash pt. (COC) 210 C; acid no. 1 max.; 98% amide content.
Unislip® FW. [Goldschmidt] Slip agent for leather industry.
Unisol UA 20, 80, 150, 300, 400. [ICI Ltd.] Ethoxylated tallow amine, nonionic; emulsifier, bitumen dispersant, antistat, dyebath leveling aid; paste, solid; 100% conc.
Unithox 450. [Petrolite] C_{30} linear alcohol ethoxylate; component for water-based PU mold release agents;

metalworking additive; solid; 50% EO.

Uniwax 1760 EBS. [Unichema] Ethylene bis-stearamide; lubricant, slip and antiblock agent, release agent; Gardner 4 max. color; 85% < 710 μm; bulk dens. 0.55 g/ml; visc. 10 cSt (160 C); m.p. 143 C min.; flash pt. (COC) 290 C; acid no. 7 max.

Urease Fermco. [Finnsugar Biochemicals] Urease; enzyme for urea conversion.

Uresoft 150. [Kao] Complex compd.; anticaking agent for urea fertilizer; liq.

Uricase S. [Novo] Bacterial uricase; enzyme for uric acid determination; powd.

USP®-90MD. [Witco/Argus] 80% sol'n. of 1,1-di (t-amyl peroxy) cyclohexane in odorless min. spirits; initiator for heat-cured polyester resin cures and acrylates; 8.9% act. oxygen.

USP®-240. [Witco/Argus] Ketone peroxide sol'n. from acetylacetone; used for rapidly curing polyester resins with low levels of cobalt accelerators; DOT org. peroxide label is not required; 4.0% act. oxygen.

USP®-245. [Witco/Argus] 2,5-Dimethyl-2,5-di(2-ethyl hexanoyl peroxy) hexane; catalyst for heated curing of polyester resin systems; features rapid cures and outstanding surf. finishes; 90% min. purity; 6.7% act. oxygen.

USP®-355M. [Witco/Argus] 75% sol'n. of 3,5,5-trimethyl hexanoyl peroxide in odorless min. spirits; initiator; 3.8% act. oxygen.

USP®-400P. [Witco/Argus] 80% sol'n. of 1,1-di (t-butyl peroxy) cyclohexane in butyl benzyl phthalate; initiator useful in heat-cured polyester resin systems where improved flow and pot life are critical; 9.8% act. oxygen.

USP®-690. [Witco/Argus] t-Butyl peroxy 2-ethyl hexanoate and 1,1-di (t-butyl peroxy) cyclohexane in a phthalate plasticizer; single initiator for elevated temp. curing of unsat. polyester resins and compds. over a broad temp. range; 6.9% act. oxygen.

USP®-800. [Witco/Argus] 70% t-butyl hydroperoxide with water as major diluent; initiator; colorless liq.; 12.4% act. oxygen.

Ustilan®. [Bayer] Ethidimuron; herbicide for total weed control on noncrop land; colorless cryst.; m.w. 264.3; sol. (g/1000 ml): 100–200 g in dichloromethane, 5–10 g in 2-propanol; 3 g in water; m.p. 155.9–156.0 C.

UV-Absorber Bayer 325. [Bayer] Benzophenone deriv.; uv lt. absorber for the protection of plastics and surface coating material, except polyolefins; in cellulose derivs.; useful in plasticized and unplasticized PVC, PS, polymethacrylates, PVDC, unsat. polyester resins, and paint binders; sltly. ylsh powd.; freely sol. in org. solv.; paint binders, and plasticizers.

UV-Absorber Bayer 340. [Bayer] Cyanacrylic ester deriv.; uv lt. absorber for the protection of plastics and surface coating materials; effective in PU and plasticized and unplasticized PVC, PS, polymethacrylates, NC, alkyd resins, and polyester resins; unsuitable for polyolefins; see UV-Absorber Bayer 325.

UV-Absorber KL 3-2500. [Bayer] Benzophenone deriv.; uv lt. absorber for the protection of polyolefins; see UV-Absorber Bayer 325.

Uvasorb 20H. [Tri-K] Benzophenone-1; uv B absorber with max. absorption at wavelengths 290–320; improves lightfastness of dyes used in cosmetics; also protects colored, liq. detergents; pale yel. powd.; m.w. 214; sol. in polar solvs. (> 10% in acetone, 40% in ethanol, 40% in methanol); insol. in water; m.p. 144–146; 99% min. act.

Uvasorb MET. [Tri-K] Benzophenone-3; uv B filter with max. absorp. at wavelengths 280–320; for cosmetic prods., skin creams, sunscreens; pale yel. powd.; m.w. 228; sol. 20% in chloroform, 10% in flax oil, 4% in ethanol, 3% in methanol; insol. in water; m.p. 62–64 C; 99.5% act.

Uvasorb S-5. [Tri-K] Benzophenone-4; uv B filter with max. absorp. at wavelengths 290–320; for sunscreen preps. and prods. with high water or alcohol content (hair gel, lotions, o/w creams); light stabilizer for colored prods.; pale yel. cryst. powd.; m.w. 308; sol. 50% in methanol, 30% in ethanol, 25% in water; pH 1–2 (1% aq.); 97% min. act.

UV-Chek AM-104. [Ferro] Nickel dibutyldithiocarbamate; lt. stabilizer for polymers; used in PP stretched tape, woven pkg. applics., HDPE pipe and pipe wrap, outdoor carpeting and syn. turf; used as heat, uv, or ozone stabilizer in elastomer and rubber uses of EPDM, epichlorohydrin, SBR, NBR, and neoprene; dk. gr. brittle flakes; m.w. 468; very sol. in aromatic or chlorinated solvs.; moderate sol. in ketones; low sol. in aliphatic solvents; insol. in water; sp.gr. 1.22; m.p. 88 C; 12.5% nickel.

UV-Chek AM-105. [Ferro] Nickel complex thiobisphenol; lt. stabilizer for polymers; used in syn. polypropylene turf, indoor-outdoor carpets, in synergistic combination with ZnO for stabilization of polyolefin, syn. elastomers; gr. brittle flakes; m.w. 1140; very sol. in aromatic or chlorinated solvs.; moderate sol. in aliphatic hydrocarbons and ketones; insol. in water; sp.gr. 1.11; m.p. 130 C; 7.7% nickel.

UV-Chek AM-205. [Ferro] Proprietary nickel org.; lt. stabilizer for polymers; used in polyethylene film for agric., construction, pkg. and greenhouse use, syn. PP turf, indoor-outdoor carpets, with ZnO for stabilization of polyolefins and elastomers; gr. brittle flakes; m.w. 1630; very sol. in aromatic or chlorinated solvs.; moderate sol. in aliphatic hydrocarbons and ketones; insol. in water; sp.gr. 1.19; m.p. 140 C; 13.0% nickel.

UV-Chek AM-300/301. [Ferro] 2-Hydroxy-4-n-octyloxybenzophenone; lt. stabilizer for polymers incl. acetals, acrylics, ABS, alkyds, cellulosics, epoxies, PC, thermoplastic and thermoset polyester, polyethylene, PP, PS, SAN, and surface coatings; costabilizer for polyethylene film in agric. or greenhouse applics., PVC flexible and semirigid applics. such as automotive body side molding, for automotive surface coatings; acts as uv absorber; lt. yel. cryst. powd.; m.w. 326; very sol. in aromatic, aliphatic, and ketone solvs.; low sol. in alcohols; insol. in water; sp.gr. 1.17; m.p. 48 C.

UV-Chek AM-340. [Ferro] 2,4-Di-t-butylphenyl

3,5-di-t-butyl-4-hydroxybenzoate; chemical, lt. and process heat stabilizer which functions alone or in combination with other uv stabilizers and antioxidants, as a terminator of photo-initiated free radicals; PP and HDPE stabilizer, effective in thin films and fibers; imparts the least color to host resins of commonly used uv stabilizers; synergistic with a variety of uv stabilizers; may be used in PP pkg. materials which contact dry or aq. foods; used at 0.1–1.0% in PP and HDPE containing antioxidants; wh. cryst. powd.; sol.: good in aromatic hydrocarbons, chlorinated solvs., ketones, esters; dens. (bulk) 4.3 lb/gal; m.p. 196 C.

UV-Chek AM-595. [Ferro] Proprietary org./inorg. complex containing Ba, Na, and P; lt. and heat stabilizer for polymers; used in rigid or flexible PVC, e.g., rigid profiles, vinyl auto roofs, flexible sheeting; single-ply roofing or pond liners of CPE, flame-retarded plastics; wh. fine powd.; 30 μ avg. particle size; sp.gr. 2.12.

Uvi-Nox 1494. [GAF] Hindered phenolic compd.; antioxidant for polyolefins, syn. rugbacking compds., latex paints, rosin and ester gums; as stabilizer and inhibitor of monomers; liq.; sol. in most org. solvs.; 100% conc.

Uvinul® 400. [BASF] Benzophenone-1; uv absorber; used for polyester, acrylics, PS, in outdoor paints and coatings, varnishes, colored liq. toiletries and cleaning agents, filters for photographic color films and prints, and rubber-based adhesives; wh. to off-wh. powd.; m.w. 214; sol. in alcohols, ether-alcohols, cyclic ethers, ketones, and esters; m.p. 140–143 C; > 98% act.

Uvinul® 408. [BASF] 2-Hydroxy-4-n-octoxy-benzophenone; uv absorber and stabilizer for polyethylene, PP, plasticized and rigid PVC, and other polymers; offers good compatibility, max. protection, and min. color and as a low order of toxicity; lt. yel.; m.w. 326; sol. (g/100 g solv. @ 30 C): 333 g in toluene; 332 g in acetone; 205 g in ethyl acetate; 185 g in MEK; 160 g in methyl pyrrolidone; 36 g in hexane; 5.8 g in ethanol; 3.1 g in methanol; sp.gr. 1.064; dens. (bulk) 0.96 l/kg (MEK); m.p. 48–49 C; 98.5% min. active.

Uvinul® 490. [BASF] Benzophenone-11; uv absorber; used in NC lacquer, fluorescent paint, inks, and for protecting furniture woods, colored liq. toiletries and cleaning agents, isocyanate systems, and butyrate metal lacquers; ylsh. powd.; > 98% act.

Uvinul® D-49. [BASF] Benzophenone-6; most economical of the near-uv absorbers; greater heat stability and more sol. (in chlorinated and aromatic solvs.); gives broad protection to plastics, coatings, textiles such as PVC, chlorinated polyesters, epoxies, acrylics, urethances, cellulosics, and oil-based paints and varnishes; ylsh. powd.; m.w. 274; m.p. 129–131 C; > 98% act.

Uvinul® D-50. [BASF] Benzophenone-2; commercial uv absorber with the broadest uv absorp. spectrum; retards fading of pigments and dyestuffs; prolongs the life of polymeric materials; photostabilizes cosmetic formulations and minimizes discoloration of syn. rubber of plastic latices; ylsh. powd.; m.w. 246; m.p. 195 C; > 98% act.

Uvinul® DS-49. [BASF] Benzophenone-9; sulfonated deriv. of Uvinul D-49; uv absorber in cosmetic formulations to prevent fading of colors and visc. changes caused by uv lt.; in textiles and water-based paints; ylsh. powd.; m.w. 478; sol. in water, alcohol; 67% act.

Uvinul® M-40. [BASF] Benzophenone-3; uv absorber; similar to Uvinul 400 except for higher sol. in aromatic solvs.; good weather resistance in resins and plastics; stabilizes PVC and polyesters against uv-lt. degradation; used in NC lacquers, varnishes, and oil-based paints; wh. to off-wh. powd.; m.w. 228; m.p. 60–62 C; > 98% act.

Uvinul® MS-40. [BASF] Benzophenone-4; sulfonated, water-sol. deriv. of Uvinul M-40; uv absorber in sunscreen prods. and in hair sprays and shampoos for dyed and tinted hair; for leather and textile fibers; wh. to off-wh. powd.; m.w. 308; very sol. in water and alcohol; pH 2 (1%); > 98% act.

Uvinul® N-35. [BASF] Etocrylene; noncolor contributing uv absorber; does not contain aromatic hydroxyl groups; effective under varying pH conditions; for NC lacquers and PVC; used in alkaline systems such as urea-formaldehyde and epoxyamine formulations, and in cosmetics; wh. powd.; m.w. 277; m.p. 94–96 C; > 98% act.

Uvinul® N-539. [BASF] Octocrylene; uv absorber; in flexible and rigid PVC; used in NC lacquers, varnishes, vinyl flooring, and oil-based paints; in aerosol and oil-based suntan lotions; nonreative with metallic driers; yel. clear visc. liq.; m.w. 361; sol. in nonpolar plastics; completely misc. with min. spirits; f.p. –10 C; > 98% act.

Uvinul® P 25. [BASF] PEG-25 PABA; nonionic; uv absorber for sunscreen prods.; pale yel., slightly visc. liq. that becomes clear at 30–40 C; m.w. 1265; 100% conc.

Uvinul® T 150. [BASF] Triazine deriv.; uv filter; liq.; 50% conc.

Uvitex® OB. [Ciba-Geigy] Bis (benzoxazolyl) deriv.; fluorescent whitener of high purity for the optical brightening of polymers at all stages of processing; offers brilliant, bluish wh. effects, good lt. fastness, excellent resistance to heat, high chemical stability, and easy sol. in a wide range of org. solvs.; used in thermoplastics, coatings, printing inks, and man-made fibers; may be combined with dyes to produce particularly bright shades; useful in printing inks to facilitate identification of security bonds and as a safeguard against forgeries, in waxes, fats, and oils, in artificial and expanded artificial leather, sheeting, household articles, elec. goods, pkg. material, etc.; cryst. powd.; yel.; sol. in waxes, paraffins, fats, and min. and veg. oils, and may be mixed with these substances in an org. solv.; m.p. 197–203 C.

V-0701 T 1/8´´. [M&T Harshaw] Vanadia catalyst; oxidation; tablet, 1/8´´ diam.; dens. 54 lb/ft^3.

Valdet 4016. [Air Prods./Valchem] Alcohol ethoxylate, straight chain; nonionic; wetting agent; nonrewetting surfactant and emulsion stabilizer used in fluorocarbon finishes; solv. emulsion scouring operations; emulsifier; APHA 60 max. clear to slightly hazy liq.; water-sol.; sp.gr. 1.011; cloud pt. 39–45 C (1%); acid no. 1 max.; pH 7.0 ± 0.5 (10%); biodeg.; 90% act.

Valfor® 100. [PQ] Sodium alumino-silicate; ion exchange and selective absorp./adsorp.; anticaking agent; wh. powd.; 3–6 μ mean particle size.

Value 3706. [Marubishi] POE nonyl phenyl ether; nonionic; dispersant, emulsifier, wetting agent, detergent; liq.; water-disp.; HLB 10.9; 100% conc.

Value MS-1. [Marubishi] Syn. sperm oil sodium sulfate; anionic; softener, emulsifier for veg., animal oil and wax; paste, liq.; 75% conc.

Vanate Acid. [Vanderbilt] EDTA; chelating agent; used where sodium ion undesirable; cryst. wh. powd.

Vanate PSPa. [Vanderbilt] Pentasodium pentetate; chelating agent; useful in peroxide bleach systems and for control of heavy metal ions; pale straw clear liq.

Vanate TS. [Vanderbilt] Tetrasodium EDTA; general purpose chelating agent; complexes water hardness ions about pH 7, iron up to pH 9.5, and most other common metal ions over entire pH range; pale straw clear liq.

Vanate TSD. [Vanderbilt] Tetrasodium EDTA; high act. form of Vanate TS; cryst. wh. powd.

Vanate TSHE. [Vanderbilt] Trisodium HEDTA; general purpose chelating agent for control of iron at pH 6.5–12.0, as well as for Ca, Mg, heavy metal ions; pale straw clear liq.

Vanate TST. [Vanderbilt] Tetrasodium EDTA tetrahydrate; see Vanate TS; cryst. wh. powd.

Vanax® 552. [Vanderbilt] Piperidinium pentamethylene dithiocarbamate; accelerator for NR, SR, cements, and latexes; peptizer for sulfur-modified G-type neoprenes; creamy wh. to lt. yel. powd.; m.w. 246.47; very sol. in chloroform; mod. sol. in acetone, toluene, alcohol; disp. in water; dens. 1.20 ± 0.03 mg/m^3; m.p. 167 C (decomp.).

Vanax® 808. [Vanderbilt] Butyraldehyde-aniline condensation prod.; accelerator for NR, SBR, CR, IIR, and latexes; activator for acidic accelerators; also for reclaims, hard rubber stocks, CR cements contng. litharge; dk. amber liq.; dens. 0.98 ± 0.02 mg/m^3; flash pt. (PM) 135 C.

Vanax® 833. [Vanderbilt] Butyraldehyde-monobutylamine condensation prod.; accelerator for NR, SR, latexes, and reclaim; also in self-curing CR cements; red-amber liq.; dens. 0.86 ± 0.02 mg/m^3; flash pt. (PM) 96 C.

Vanax® A. [Vanderbilt] 4,4´-Dithiomorpholine; vulcanizing agent; sulfur donor for NR and syn. rubbers; functions as primary accelerator for NR, IR, SBR, NBR, IIR elastomers, and as primary and sec. accelerator in EPDM; powd.; wh. to lt. gray; also avail. in thread grade and rods (80% act.); fineness; 99% thru 100 mesh (powd. and thread grade); m.w. 236.36; sol. in toluene, xylene, and CCl_4; dens. 1.35 ± 0.03 mg/m^3 (rods); m.p. 117–127 C (powd. and thread grade); 115–126 C (rods); 26–29% sulfur (powd. and thread grade); 20.5–23% sulfur (rods).

Vanax® CPA. [Vanderbilt] Dimethylammonium hydrogen isophthalate; accelerator for W and T-type neoprenes; used in press cured, inj. molded, and LCM stocks; wh. to lt. ivory powd.; 99.9% thru 100 mesh; m.w. 211.24; very sol. in dimethylformamide, dimethyl sulfoxide; mod. sol. in acetone, chloroform; slightly sol. in alcohol, water; dens. 1.35 ± 0.03 mg/m^3; m.p. 190 C (decomp.).

Vanax® DOTG. [Vanderbilt] N,N´-di-ortho-tolylguanidine; accelerator for NR and SR; sec. accelerator; wh. to pink powd.; 99.9% thru 100 mesh; m.w. 238.34; very sol. in alcohol, acetone; slightly sol. in toluene, water; dens. 1.20 ± 0.03 mg/m^3; m.p. 173–179 C.

Vanax® DPG. [Vanderbilt] N,N´-diphenyl guanidine; accelerator for NR and SR; sec. accelerator; wh. to lt. yel. with pink tinge powd.; 99.9% thru 100 mesh; m.w. 210.28; very sol. in acetone, alcohol; slightly sol. in toluene, water; dens. 1.20 ± 0.03 mg/m^3; m.p. 144–149 C.

Vanax® MBM. [Vanderbilt] m-Phenylenedimaleimide; accelerator; coagent in peroxide-cured polymers; yel. to lt. br. powd., rods; 99.9% min. thru 100 mesh (powd.); sol. in acetone; mod. sol. in chloroform; pract. insol. in water; dens. 1.44 ± 0.03 mg/m^3.

Vanax® NP. [Vanderbilt] Activated thiadiazine; nonthiourea accelerator for T and W type neoprenes; imparts tensile, heat and scorch resistance and compr. set to blk. loaded compds.; wh. powd.; sol. in acetone, chlororform; slightly sol. in water; dens. 1.35 ± 0.03 mg/m^3; m.p. 90–105 C.

Vanax® NS. [Vanderbilt] N-tert-butyl-2-benzothiazolesulfenamide; accelerator; delayed action accelerator for natural and syn. rubbers; lower scorch and faster cure than Durax; used in tires, mechanical and extruded goods; lt. buff to tan powd., 99.9% min. thru 20 mesh; rods; m.w. 238.37; practically insol. in water; dens. 1.28 ± 0.03 mg/m³; m.p. 105 C min.

Vanax® PML. [Vanderbilt] Di-ortho-tolylguanidine salt of dicatechol borate; accelerator for NR and SR stocks, CR, neoprenes for wire and cable and mechanical goods; activator and mild antioxidant in NR and SBR; tan to lt. br. powd.; 99.9% thru 100 mesh; m.w. 467.37; very sol. in acetone, chloroform; mod. sol. in alcohol, toluene; insol. in water; dens. 1.25 ± 0.03 mg/m³; m.p. 165 C min.

Vanax® PY. [Vanderbilt] 23% poly-p-dinitrosobenzene polymer in wax; chemical conditioner for IIR, cements; dk. br. pellets; 99.9% thru 1/2 in. screen; sol. in toluene, chloroform, acetone; mod. sol. in hexane, gasoline; insol. in water; dens. 0.96 ± 0.03 mg/m³.

Vanblack C. [Vanderbilt] Carbon black; reinforcing pigment for elastomers; dk. gray pellets; dens. 2.05 ± 0.03 mg/m³.

Vanchem® HM-50, HM-4346. [Vanderbilt] Aromatic polyisocyanate in monochlorobenzene and toluene resp.; adhesion promoter in adhesives; primer or adhesive additive for adhering elastomeric coatings to syn. fiber fabrics coated with syn. rubbers; crosslinking agent for elastomers and plastics; dk. br. liqs.; dens. 1.16 ± 0.02 and 1.05 ± 0.02 mg/m³ resp.; flash pt. (PM) 30 C and 7 C min. resp.

Vanchem® NATD. [Vanderbilt] Disodium 2,5-dimercaptothiadiazole; corrosion inhibitor and metal deactivator for non-ferrous metals in aq. systems; chemical intermediate; amber to brn. liq.; water-sol.; dens. 1.30 mg/m³; pH 8.9; 40% aq. sol'n.

Vancide® 51. [Vanderbilt] Sodium dimethyldithiocarbamate (27.6%) and sodium 2-mercaptobenzothiazole (2.4%); fungicide for use as a preservative in latex and starch paste; bactericide for sol. cutting fluids and coolants; paper mill slimicide; used in petrol. storage tanks, recirculating cooling towers, paper and paperboard, cotton fabric; pale yel.-grn. liq.; sol. in water, acetone, alcohol; sp.gr. 1.15 ± 0.02; dens. 9.6 lb/gal; 30% in water.

Vancide® 51Z. [Vanderbilt] Zinc dimethyldithiocarbamate (87%) and zinc 2-mercaptobenzothiazole (7.5%); fungicide for use in neoprene compositions; also for preservation of adhesives, for industrial cooling water slime control, in sanitizing cleansing compds., for textile mildew and bacterial growth inhibition, and as mold inhibitor in caulking compds.; wh. to cream powd.; sol. in caustic, ammonia, hydrocarbons, chloroform; very slight sol. in water; sp.gr. 1.72 ± 0.02; m.p. 240–246 C.

Vancide® 51Z Disp. [Vanderbilt] Dispersed form of Vancide 51Z; fungicide for addition to latex or for aq. applics.; misc. with water; sp.gr. 1.26 ± 0.02; 50% act. in water.

Vancide® 89. [Vanderbilt] N-Trichloromethylthio-4-cyclohexene 1,2-dicarboximide; fungicide for natural and syn. rubber compds. containing susceptible plasticizers; industrial preservative for vinyl, polyethylene, paint, lacquer, soap, wallpaper flour paste; lt. tan powd.; m.w. 300.6; sol. in tetrachloroethane, chloroform, xylene, cyclohexanone, ethylene dichloride; sp.gr. 1.69 ± 0.03; m.p. 168–174 C; 90% min. assay.

Vancide® 89 RE. [Vanderbilt] N-Trichloromethylthio-4-cyclohexene-1,2-dicarboximide; antimicrobial and preservative for cosmetics and pharmaceuticals, veterinary prods.; wh. to off-wh. fine powd.; sol. (g/100 g solv.): 7.5 g xylene, 5.8 g cyclohexanone, 5.3 g chloroform, 5.2 g tetrachloroethane, 5.0 g ethyl acetate, 4.2 g acetone; dens. 1.7 mg/m³; m.p. 171–176 C; pH 5–6 (1% disp.); 97% assay.

Vancide® MZ-96. [Vanderbilt] Zinc dimethyldithiocarbamate; antimicrobial, preservative for starch and syn. latex adhesives, food pkg. adhesives; wh. powd.; m.w. 305.82; mod. sol. in dilute caustic, toluene, carbon disulfide, chloroform; water-disp.; dens. 1.71 ± 0.03 mg/m³; m.p. 252–260 C; 96% assay.

Vancide® PA. [Vanderbilt] Trans-1,2-bis (n-propylsulfonyl) ethene; antimicrobial, mold inhibitor for paints; wh. powd.; dens. 1.40 mg/m³.

Vancide® PA Disp. [Vanderbilt] Trans-1,2-bis (n-propylsulfonyl) ethene; see Vancide PA; off-wh. disp.; dens. 1.01 mg/m³; flash pt. 40 C.

Vancide® TH. [Vanderbilt] Hexahydro-1,3,5-triethyl-s-triazine; fungicide for use as industrial preservative for latex, adhesives, cutting fluids, marine lubricants; colorless to lt. yel. liq.; m.w. 171.29; sol. in acetone, ethanol, ether, water; mod. sol. in hydrocarbon solvs.; dens. 0.89 ± 0.02 mg/m³; b.p. 205–210 C; flash pt. (TCC) 66 C.

Vandex. [Vanderbilt] Selenium; sec. vulcanizing agent for NR, IR, SBR; increases state of cure and improves heat aging of sulfurless and low sulfur stocks; reduces curing time in NR hard rubber and allows use of lower sulfur for increased flexibility; dk. gray metallic powd., 99% thru 200 mesh; m.w. 78.96; sol. in carbon disulfide; practically insol. in water; dens. 4.80 ± 0.03 mg/m³; m.p. > 217 C.

Vanfre® DFL. [Vanderbilt] Processing aid for mold lubrication; corrosion inhibitor; clear to straw liq.; dens. 1.08 ± 0.02 mg/m³; visc. 1200 cps min.; pH 7.2–8.5 (10%).

Vanfre® HYP. [Vanderbilt] Processing aid for Hypalon; reduces plasticity, improves mold flow and release; off-wh. to lt. tan friable flake; very sol. in toluene, chloroform; mod. sol. in acetone, hexane, gasoline; insol. water; dens. 0.97 ± 0.03 mg/m³; m.p. 81–97 C.

Vanfre® IL-1. [Vanderbilt] Sodium alkyl sulfates; internal lubricant for CR, NBR, CSM, and NR; improves flow chars. of highly loaded compds.; improves release from mill rolls; provides smoother extrusions; disperses easily; wh. to creamy wh. powd.; dens. 1.16 ± 0.03 mg/m³.

Vanfre® IL-2. [Vanderbilt] Fatty alcohols on inert carrier; internal lubricant; improves flow and molding chars. of highly loaded elastomer compds.; does not reduce tensile strength or appreciably affect water swell of vulcanizates; gray powd., 99.9% min. 100 mesh; mild fatty odor; dens. 1.17 ± 0.03 mg/m³.

Vanfre® M. [Vanderbilt] Processing aid for natural

and syn. rubbers; reduces plasticity, improves mold flow; amber pellets; dens. 0.87 ± 0.03 mg/m³; m.p. 80 C min.

Vanfre® TK. [Vanderbilt] Inorg. acid on inert carrier; dusting agent to prevent surface tack in crosslinking of elastomers by peroxides; wh. to gray powd.; 99.0% min. thru 100 mesh; dens. 1.51 ± 0.03 mg/m³; pH 3.5–4.5 (10%).

Vanfre® UN. [Vanderbilt] Lubricant, processing aid for ethylene/acrylic elastomers to improve release; lt. straw liq.; dens. 0.98 ± 0.02 mg/m³; visc. 160–180 cps; flash pt. (PM) 93 C; pH 2.0–2.5 (1%).

Vanfre® VAM. [Vanderbilt] Org. phosphate ester free acid; processing aid for VAMAC, ethylene/acrylic elastomer, other solv.-resistant polymers; tan to lt. br. waxy solid; dens. 0.97 ± 0.03 mg/m³ (60 C); acid no. 98–110; pH 3.0–4.0 (1%).

Van Gel®. [Vanderbilt] Magnesium aluminum silicate, colloidal; thixotrope for aq. coating systems; emulsion stabilizer, suspending agent, thickener for industrial applics.; resistant to bacterial and enzymatic degradation; off-wh. flakes; Hegman 7 min. (4% disp.); disperses and swells in water; dens. 2.6 mg/m³; visc. 800 ± 400 cps; pH 9.5 ± 0.5 (4% disp.); 12% max. moisture.

Van Gel® B. [Vanderbilt] Magnesium aluminum silicate (smectite); thickener and visc. stabilizer for dispersions; off-wh. small flakes; dens. 2.60 ± 0.03 mg/m³; visc. 300–900 cps (4%).

Vanisol BIS sodico-2. [Rhone-Poulenc SpA] Sodium bistridecyl sulfosuccinate; anionic; visc. depressant, stabilizer, and emulsifier for emulsion polymerization of PVC; latex surf. tens. stabilizer; dispersant for resins, pigments in plastics and org. media; base for rust inhibitors; liq.; oil-sol., water-disp.; 60% conc.

Vanlube 26. [Vanderbilt] Alkylated diphenylamines/zinc dithiocarbamate; antioxidant for greases, metal deactivator; oxidation inhibitor for petrol. oils lubricant; brn. liq.; sol. in petrol. and syn. lubricant bases; insol. in water; dens. 0.99 mg/m³; visc. 17.74 cSt (100 C); pour pt. 7 C; flash pt. 193 (COC) C.

Vanlube 71. [Vanderbilt] Lead diamyldithiocarbamate; EP and antiscuff agent, antioxidant, corrosion inhibitor; multifunctional additive for industrial lubricating oils and greases, crankcase and industrial gear oils; amber liq.; sol. in petrol. and syn. lubricant bases; insol. in water; dens. 1.10 mg/m³; visc. 5.73 cSt (100 C); pour pt. –23 C; flash pt. 171 C (COC).

Vanlube 73. [Vanderbilt] Antimony dialkyldithiocarbamate; EP and antiscuff agent, antioxidant for industrial and automotive lubricants; antiwear additive; bearing corrosion inhibitor in motor, compressor, and gas engine oils, lubricating greases; partial replacement for zinc dithiophosphate in engine oils; dk. amber liq.; sol. see Vanlube 71; dens. 1.04 mg/m³; visc. 11.6 cSt (100 C); pour pt. –23 C; flash pt. 171 C (COC).

Vanlube 81. [Vanderbilt] p,p´-Dioctyldiphenylamine; antioxidant, corrosion for petrol. and syn. lubricants; off-wh. powd.; sol. see Vanlube 26; dens. 1.01 mg/m³; m.p. 95 C.

Vanlube 601. [Vanderbilt] Heterocyclic sulfur nitrogen compd.; metal deactivator, corrosion and rust inhibitor, color stabilizer, extreme pressure agent; used in petrol. fuels and solv.; dk. amber liq.; sol. in petrol. and syn. lubricant bases; insol. in water; dens. 0.96 mg/m³; visc. 12.96 cSt (100 C); pour pt. –1 C; flash pt. 149 C (COC).

Vanlube 622. [Vanderbilt] Antimony dialkylphosphorodithioate; EP, antiwear/antiscuff agent, antioxidant, friction reducer for steel mill and industrial gear oils, petrol. oils and greases, base lubricants; amber liq.; sol. see Vanlube 71; dens. 1.20 mg/m³; visc. 6.04 cSt (100 C); pour pt. –34 C; flash pt. 171 C (COC).

Vanlube 648. [Vanderbilt] Antimony dialkylphosphorodithioate; EP, antiwear/antiscuff agent, corrosion inhibitor, anitoxidant in crankcase oils, lubricating greases, automotive and industrial gear oils, metalworking lubricants; amber liq.; sol. see Vanlube 71; dens. 1.05 mg/m³; visc. 6.33 cSt (100 C); pour pt. –34 C; flash pt. 163 C (COC).

Vanlube 672. [Vanderbilt] Amine phosphate; antiwear and EP agent for industrial lubricants; metal working lubricants such as drawing, stamping, and forming compds.; improves EP performance of conventional EP materials; antiwear additive in syn. lubricants; lt. yel. visc. liq.; sol. in water, petrol., and syn. lubricant bases; dens. 1.05 mg/m³; visc. 133.17 cSt (100 C); pour pt. 27 C; flash pt. 116 C (COC).

Vanlube 692. [Vanderbilt] Aromatic amine-phosphate; EP and antiwear agent, mild antioxidant for nonmetallic industrial gear oils giving high load carrying properties; used for lubricants based on petrol. oils and syns.; enhances EP properties of sulfurized olefins, chlorinated paraffins, dithiocarbamates and phosphorodithioates; dk. amber visc. liq.; sol. see Vanlube 71; dens. 0.99 mg/m³; visc. 40.52 cSt (100 C); flash pt. 154 C (COC).

Vanlube 732. [Vanderbilt] Dithiocarbamate deriv.; antioxidant and antiwear agent used in industrial and automotive oils; antiscuff agent in engine oils; component of additive pkgs.; rdsh.-brn. liq.; sol. see Vanlube 26; dens. 1.08 mg/m³; visc. 41.7 cSt (100 C); flash pt. 121 C (COC).

Vanlube 793. [Vanderbilt] Proprietary 2,5-dimercapto-1,3,4-thiadiazole deriv.; metal deactivator for nonferrous metals; corrosion inhibitor; used in turbine and gear oils, hydraulic and metal working fluids, greases; yel. liq.; sol. see Vanlube 601; dens. 0.99 mg/m³; visc. 5.78 cSt (100 C); flash pt. 157 C (COC); 50% conc. in oil.

Vanlube 7723. [Vanderbilt] 4,4´-Methylene bis (dibutyldithiocarbamate); antioxidant and extreme pressure agent; used in petrol. lubricant; component of additive pkgs.; rdsh. amber liq.; sol. see Vanlube 26; dens. 1.05 mg/m³; visc. 14.5 cSt (100 C); flash pt. 177 C (COC).

Vanlube AZ. [Vanderbilt] Zinc diamyldithiocarbamate; antioxidant, metal deactivator, copper corrosion inhibitor, color stabilizer, antiwear agent used in engine and industrial oils, greases; inhibits oxidation on metal surfaces; straw liq.; sol. see Vanlube 26; dens. 0.98 mg/m³; visc. 9.5 cSt (100 C); pour pt. –23

C; flash pt. 171 C (COC).

Vanlube DND. [Vanderbilt] Dinonyldiphenylamine; antioxidant for industrial oils and greases; dk. amber liq.; sol. see Vanlube 26; dens. 0.95 mg/m³; visc. 19.6 cSt (100 C); flash pt. 227 C (COC).

Vanlube NA. [Vanderbilt] Alkylated diphenylamines; see Vanlube DND; brn. liq.; sol. see Vanlube 26; dens. 0.94 mg/m³; visc. 12.4 cSt (100 C); pour pt. –6.7 C; flash pt. 229 C (COC).

Vanlube PC. [Vanderbilt] 2,6-Di-t-butyl-p-cresol; antioxidant, color stabilizer, synergist; oxidation inhibitor in fuel and industrial oils, petrol. wax, polyolefins, industrial fats and oils; wh. cryst. flakes; sol. see Vanlube 26; dens. 1.04 mg/m³; m.p. 69 C; flash pt. 135 C (COC).

Vanlube RD. [Vanderbilt] Polymerized 1,2-dihydro-2,2,4-trimethylquinoline; antioxidant used in polyglycols, "Ucon" fluid, and diester syn. lubricant; high temp. inhibitor for petrol. and syn. lubricants; amber small pellets; sol. in diesters, polyglycols and Ucon fluids; insol. in water; dens. 1.08 mg/m³; soften. pt. 74 C.

Vanlube SL. [Vanderbilt] Octylated diphenylamines; antioxidant for industrial lubricants, greases and oils; rdsh. brn. liq.; sol. see Vanlube 26; dens. 0.99 mg/m³; visc. 16.28 cSt (100 C); pour pt. 7 C; flash pt. 210 C (COC).

Vanlube SS. [Vanderbilt] Octylated diphenylamines; antioxidant for industrial petrol. and syn. lubricants; corrosion inhibitor in silane and siloxane syn. lubricants; lt. tan powd.; sol. see Vanlube 26; dens. 1.02 mg/m³; m.p. 90 C.

Vanox® 3C. [Vanderbilt] N-Isopropyl-N′-phenyl-p-phenylenediamine; antioxidant, antiozonant for dk. colored NR and SR; purple-bk. flakes; m.w. 226.4; very sol. in alcohol; mod. sol. in toluene; insol. in water; dens. 1.10 ± 0.03 mg/m³; m.p. 70–79 C.

Vanox® 6H. [Vanderbilt] N-Cyclohexyl-N′-phenyl-p-phenylenediamine; antioxidant, antiozonant for dk. colored NR and SR; m.w. 264.4; very sol. in acetone; mod. sol. in toluene; insol. in water; gray-violet powd.; dens. 1.18 ± 0.03 mg/m³; m.p. 110–120 C.

Vanox® 12. [Vanderbilt] p,p′-Dioctyldiphenylamine; antioxidant for elastomers used in adhesives, hot melts; off-wh. gran.; m.w. 393.66; sol. in alcohol, toluene, gasoline; insol. in water; dens. 1.01 ± 0.03 mg/m³; m.p. 94–100 C.

Vanox® 13. [Vanderbilt] Polyalkyl phosphited polyphenols, modified; antioxidant for rubber; lt. amber liq.; sol. in chloroform, toluene, xylene, and petrol. hydrocarbons; pract. insol. in water; dens. 0.93±0.02 mg/m³.

Vanox® 100. [Vanderbilt] Phenolic; antioxidant for rubbers and peroxide crosslinked chlorinated polyethylene; amber liq.; sol. in toluene, acetone, chloroform; pract. insol. in water; dens. 1.03 ± 0.02 mg/m³.

Vanox® 102. [Vanderbilt] Styrenated phenol; antioxidant; lt. straw liq.; sol. in petrol. hydrocarbons, alcohols, esters; pract. insol. in water; dens. 1.08 ± 0.02 mg/m³.

Vanox® 1003. [Vanderbilt] Polybutylated bisphenol A; antioxidant for PU foam; oxidation inhibitor; scorch protector during mfg. and in bun storage; amber liq.; sol. in chloroform, petrol. hydrocarbons; dens. 0.945-0.965 mg/m³.

Vanox® 1004. [Vanderbilt] Polybutylated bisphenol A; antioxidant for polyethylene and PP; intermediate between BHT and Vanox GT; gray to tan powd.; dens. 1.26 ± 0.03 mg/m³; 73% on inert carrier.

Vanox® 1005. [Vanderbilt] Alkylated-arylated bisphenolic phosphite; antioxidant for PS; gel inhibitor during polymerization; polymer stabilizer in syn. rubber mfg.; amber visc. liq.; sol. in benzene, chloroform, gasoline; pract. insol. in water; dens. 0.955–0.975 mg/m³; visc. 200–700 cps (60 C).

Vanox® 1013. [Vanderbilt] Polyalkyl phosphited polyphenols, modified; antioxidant for ABS resins; lt. amber liq.; dens. 0.93 mg/m³.

Vanox® 1030. [Vanderbilt] Phenolic-thio-ester; antioxidant for PP; wh. powd.; dens. 1.04 mg/m³; m.p. 61 C min.

Vanox® 1081. [Vanderbilt] p,p′-Dioctyldiphenylamine; antioxidant for hot melts; off-wh. gran.; sol. in oils, gasoline, acetone, alcohol, and toluene; pract. insol. in water; dens. 1.02 ± 0.03 mg/m³; m.p. 94–100 C.

Vanox® 1290. [Vanderbilt] 2,2′-Ethylidene bis(4,6-di-t-butylphenol); antioxidant; oxidative inhibitor for polymers; process stabilizer for polyolefins; stabilizer for PU and PS; wh. cryst. powd.; sol. in acetone, heptane, toluene, water; dens. 1.01 mg/m³; m.p. 162–163 C; flash pt. +380 F (COC); 99% assay.

Vanox® 1320. [Vanderbilt] 2,6-Di-t-butyl-4-sec-butyl phenol; antioxidant for PU foam; oxidation inhibitor and scorch preventer for mfg. and storage of bun stock; stabilizer for PU foam; straw yel. liq.; sol. in oil, acetone, alcohol, heptane, toluene; insol. in water; dens. 0.93 ± 0.02 mg/m³; visc. 3.30 cSt (40 C); flash pt. 94 C min. (TCC).

Vanox® 2246. [Vanderbilt] 2,2′-Methylene bis 4 methyl-6-t-butyl-phenol; antioxidant for NR, SR, and latexes, esp. wh. and lt. colored stocks; copper inhibitor in NR, CR, and their latexes; wh. to cream powd.; m.w. 340.58; very sol. in acetone, chloroform, toluene; mod. sol. in alcohol; pract. insol. in water; dens. 1.08 ± 0.03 mg/m³; m.p. 125–130 C.

Vanox® 3240. [Vanderbilt] Aromatic sulfonate; min. deactivator; dk. brn. liq.; dens. 0.98 mg/m³; visc. 18 cSt min. (100 C); flash pt. 170 C min. (COC).

Vanox® 3245. [Vanderbilt] Aromatic sulfonate; min. deactivator; brn. flake.

Vanox® AM. [Vanderbilt] Diphenylamine and acetone low-temp. reaction prod.; antioxidant for NR and SR; lt. gr. tan powd.; 99.0% min. thru 100 mesh; very sol. in acetone, toluene; mod. sol. in alcohol; insol. in water; dens. 1.15 ± 0.03 mg/m³; soften. pt. (B&R) 85–97 C.

Vanox® AT. [Vanderbilt] Butyraldehyde-aniline condensation prod.; antioxidant for CR, NR, SBR, EPDM stocks and latexes; activator for thiuram and thiazoles; amber liq.; dens. 1.02 ± 0.02 mg/m³; flash pt. (PM) 80 C.

Vanox® GT. [Vanderbilt] Tris(3,5-di-t-butyl-4-hydroxy benzyl) isocyanurate; antioxidant for PP, polyethylene, PU, and polymers; wh. powd.; m.w. > 500; sol. in toluene, acetone, chloroform, dimethylformamide; pract. insol. in water; dens. 1.03 mg/m³; m.p. 221 C; flash pt. 553 F (COC).

Vanox® MTI. [Vanderbilt] 2-Mercaptotoluimidazole; antioxidant for NR, SR; synergist with Agerite antioxidants; lt. tan powd.; 99.9% min. thru 100 mesh; m.w. 165.25; mod. sol. in acetone, alcohol; insol. in water; dens. 1.33 ± 0.03 mg/m³; m.p. 250 C min.

Vanox® NBC. [Vanderbilt] Nickel di-n-butyldithiocarbamate; antioxidant-antiozonant for SBR, NBR, CR, CSM, and ECO; dk. gr. cryst. powd.; 99.9% thru 20 mesh; m.w. 467.47; very sol. in chloroform, toluene; mod. sol. in acetone; insol. water; dens. 1.26 ± 0.03 mg/m³; m.p. 86–90 C; 11.8–12.8% nickel content.

Vanox® ODP. [Vanderbilt] Dioctylated diphenylamine; general purpose antioxidant for all elastomers; brnsh. gray powd.; m.w. 393.666; dens. 0.99 ± 0.03 mg/m³; m.p. 85–97 C.

Vanox® PC. [Vanderbilt] 2,6-Di-t-butyl-p-cresol; industrial grade antioxidant/stabilizer for animal and veg. oils, color stabilization, oxidation in waxes; wh. cryst. solid.

Vanox® PCX. [Vanderbilt] 2,6-Di-t-butyl-p-cresol; food grade antioxidant/stabilizer for animal and veg. oils; wh. cryst. solid.

Vanox® SKT. [Vanderbilt] 3,5-Di-t-butyl-4-hydroxyhydrocinnamic acid triester of 1,3,5-tris (2-hydroxyethyl)-s-triazine-2,4,6-(1H,3H,5H)-trione; antioxidant; stabilizer for polyolefins; hot melt and food pkg. applics.; wh. cryst.; m.w. 1042; sol. in aromatics; dens. 0.921 mg/m³; m.p. 126–131 C; flash pt. 512 F (COC); f.p. 554 F (COC).

Vanox® SWP. [Vanderbilt] 4,4′-Butylidene-bis-(6-t-butyl-m-cresol); antioxidant for natural and syn. latexes; nonstaining; wh. fine cryst.; m.w. 382.5; very sol. in alcohol; mod. sol. in toluene; insol. in water; dens. 1.03 ± 0.03 mg/m³; m.p. 210–218 C.

Vanox® ZMTI. [Vanderbilt] Zinc 2-mercaptotoluimidazole; antioxidant for NR and SR, EPDM, nitrile stock; synergist with Agerite antioxidants; lt. tan powd.; 99.0% min. thru 100 mesh; m.w. 393.85; sol. in methanol, ethanol; pract. insol. in water; dens. 1.69 ± 0.03 mg/m³; m.p. > 300 C; 15.5% min. zinc content.

Vanox® ZS. [Vanderbilt] Fortified phenol; nonstaining antioxidant for NR, SR and their latexes; gel inhibitor during SBR processing; wh. to pale yel. powd.; 99.0% thru 100 mesh; slightly sol. in most org. solvs.; insol. in water; dens. 1.30 ± 0.03 mg/m³.

Vanplast® 3S. [Vanderbilt] 2,2′-Dibenzamidodiphenyl disulfide on inert carrier; peptizer for natural and syn. rubber; reduces plasticity during mastication; lt. grnsh. gray powd.; 99.9% thru 100 mesh; dens. 1.89 ± 0.03 mg/m³.

Vanplast® 201. [Vanderbilt] Barium dinonylnaphthalene sulfonate; corrosion inhibitor used in automotive protective coatings; ASTM color 6 max. visc. liq.; m.w. 459.77; dens. 1.02 ± 0.02 mg/m³; visc. 475 SUS max. (100 C); flash pt. (COC) 160 C min.; 50% min. assay; 6.5% min. barium.

Vansil® W-9. [Vanderbilt] Wollastonite; extender pigment for solvent-thinned and latex paints; dens. 24.2 lb/solid gal; oil absorp. 17; pH 9.8 (10% slurry).

Vansil® W-10. [Vanderbilt] Wollastonite; see Vansil W-9; dens. 24.2 lb/solid gal; oil absorp. 19; pH 9.8 (10% slurry).

Vansil® W-20. [Vanderbilt] Wollastonite; see Vansil W-9; dens. 24.2 lb/solid gal; oil absorp. 20; pH 9.8.

Vansil® W-30. [Vanderbilt] Wollastonite; see Vansil W-9; dens. 24.2 lb/solid gal; oil absorp. 21; pH 9.8 (10% slurry).

Vansil® W-50. [Vanderbilt] Wollastonite; see Vansil W-9; dens. 24.2 lb/solid gal; oil absorp. 27; pH 9.8 (10% slurry).

Vanstay 137. [Vanderbilt] Zinc and cadmium complex; stabilizer with good lt. and aging stability for chemically blown flexible vinyl, low-dens. foam; liq.

Vanstay 145. [Vanderbilt] Nontoxic PVC stabilizer; offers high-temp. protection; used in surgical gloves, syringes, pkg. film, calendered film for food wrap; fluid paste.

Vanstay 162-B. [Vanderbilt] Barium/cadmium/zinc stabilizer; features low cost, high clarity, minimal plate-out, and good lt. stability; used for clear applics. where moderate temps. are employed, plastisols, and organosols; liq.

Vanstay 246. [Vanderbilt] Barium/cadmium/zinc stabilizer; for broad applic. and offering outstanding low cost; used for clear film and sheeting, shoe soles, plastisols, calendered prods. for semi-rigid and flexible PVC; liq.

Vanstay 3027. [Vanderbilt] Barium/cadmium/zinc stabilizer; offers property retention upon high-temp. aging; used in calendered and extruded prods. for PVC and other copolymers; cream wh. powd.

Vanstay 4017. [Vanderbilt] Barium/cadmium stabilizer; similar to Vanstay HTA with improved plate-out resistance; used for calendered prods.; wh. powd.

Vanstay 4030. [Vanderbilt] Barium/cadmium stabilizer; stabilizer with long-term heat stability and retention of properties after aging; used for wire and cable, and general flexible PVC applics.; wh. powd.

Vanstay 4039. [Vanderbilt] Barium/cadmium stabilizer; low-cost, general-purpose stabilizer for profile extrusions, calendered prods., and wire and cable applics.; wh. powd.

Vanstay 5510. [Vanderbilt] Org. lead stabilizer; for wire and cable applics.; also used in phonograph records and inj. molded prods.; powd.

Vanstay 5515. [Vanderbilt] Organically modified lead stabilizer; low cost, lubricating primary heat stabilizer for low-temp. wire and cable, 60 C, 80 C UL rated insulation, flexible communications wire, expanded vinyl, elec. tape; processing aid for wire and cable; powd.

Vanstay 5563. [Vanderbilt] Blend of barium and lead compds. and complexing agents; general purpose heat and lt. stabilizer for PVC; used for wire and cable compds. where lubrication is desired; powd.

Vanstay 5630. [Vanderbilt] Complexing agents and

organo-metallic compds.; stabilizer offering good heat stability, lt. aging, and dimensional stability; used in vinyl asbestos floor tile; powd.

Vanstay 5698. [Vanderbilt] Low cost, non-lubricating, easy processing stabilizer for vinyl asbestos floor tile; nonwater curling in min. filled vinyl; powd.

Vanstay 6032. [Vanderbilt] Barium/cadmium/zinc stabilizer; minimizes roll plate-out; low lubricating stabilizer with good clarity and edge color; used for clear calendered sheeting; liq.

Vanstay 6040. [Vanderbilt] Barium/cadmium/zinc stabilizer; good visc. stability and air release properties for plastisols, lubricating type; used for plastisols and organosols; liq.

Vanstay 6053. [Vanderbilt] Barium/cadmium/zinc stabilizer; clarity and color retention and good lt. stability; used for inj. molding, calendered sheet, and automotive mats; liq.

Vanstay 6055. [Vanderbilt] Barium/cadmium/zinc stabilizer; low cost, general purpose stabilizer with good long-term stability; used for automotive calendered goods; liq.

Vanstay 6074. [Vanderbilt] Barium/cadmium/zinc stabilizer; low cost, general purpose stabilizer with low plate-out, low sulfur staining, and good clarity; used in automotive calendered goods; liq.

Vanstay 6078. [Vanderbilt] Barium/cadmium/zinc stabilizer; low cost stabilizer noted for clarity; controls plasticizer migration; used in clear and tinted inj. molded automotive goods; liq.

Vanstay 6133. [Vanderbilt] Barium/cadmium/zinc stabilizer; low cost, general purpose stabilizer with good long-term stability; similar to Vanstay 162-B; used in plastisols and organosols, and clear vinyl prods.; liq.

Vanstay 6172. [Vanderbilt] Barium/cadmium/zinc stabilizer; designed for good color; eliminates air entrapment and stabilizes plastisol visc.; used in plastisols and organosols; liq.

Vanstay 6191. [Vanderbilt] Barium/cadmium/zinc stabilizer; high potency stabilizer with long-term stability; used in semirigid, thin gauge film where max. calender speeds are required; liq.

Vanstay 6201. [Vanderbilt] Barium/cadmium/zinc stabilizer; balanced with low plate-out and low sulfur staining; used in rotational molding and pigmented calendering applics.; liq.

Vanstay 7024. [Vanderbilt] Barium/cadmium stabilizer; low-cost, high-potency, general-purpose heat stabilizer for filled and unfilled flexible vinyl; liq.

Vanstay 7025. [Vanderbilt] Barium/cadmium stabilizer; similar in performance to 7024 with improved early color and control of color in vinyl compds.; used in flexible and rigid PVC, calendering, and extrusion; liq.

Vanstay 7032. [Vanderbilt] Barium/cadmium stabilizer; for calendered and extruded prods.; liq.

Vanstay 8073. [Vanderbilt] Nontoxic stabilizer for PVC offering optimum color and clarity, and safety margin for severe processing conditions; used in clear and tinted PVC food wraps for flexible and semirigid PVC; paste.

Vanstay 8120. [Vanderbilt] Mixt. of barium and zinc orgs.; stabilizer for use where resistance to sulfide staining is important; nonplating; used in calendering, plastisols and organosols; liq.

Vanstay 8210. [Vanderbilt] Tris nonylphenol phosphite; general purpose chelator used with nontoxic primary stabilizer for tubing, pkg. film, and calendered prods.; liq.

Vanstay 8300. [Vanderbilt] Lead/zinc org. complex; low cost stabilizer; used for cellular flexible vinyl; liq.

Vanstay 8308. [Vanderbilt] Zinc org. complex; low cost, low temp. actuation, nonstaining stabilizer; for chemically blown flexible vinyl; liq.

Vanstay A, B, C, M. [Vanderbilt] Low-toxicity stabilizers containing no barium, cadmium, or lead; eliminate plate out and sulfur staining while offering heat stability, initial color, and lt. stability; A used for filled and pigmented flame-retardant systems containing phosphate plasticizers; B recommended for formulations containing high levels of $CaCO_3$; C used for automotive upholstery material requiring long exposure at low temps.; clear low-visc. liq.; colorless; very low odor.

Vanstay CE. [Vanderbilt] Complex of barium and zinc compds.; low cost stabilizer with heat and lt. stability and nonsulfur staining properties; used in highly-filled vinyl tile; powd.

Vanstay FA. [Vanderbilt] Nontoxic, self-lubricating stabilizer with good long-term heat and lt. stability; used in clear medical tubing, and as general stabilizer for semirigid and flexible PVC; wh. powd.

Vanstay HA. [Vanderbilt] Barium/cadmium/zinc stabilizer; low cost, high efficiency stabilizer with good early color; prevents color drift in pigmented stock; used in homogeneous vinyl tile; used in conjunction with Vanstay RAZ; wh. powd.

Vanstay HT. [Vanderbilt] Barium/cadmium/zinc stabilizer, offers heat protection, high bulk dens., and low dusting; for general use in flexible and semirigid PVC, calendered prods.; wh. powd.

Vanstay HTA. [Vanderbilt] Barium/cadmium stabilizer; offers high temp. stability, outstanding clarity and color in PVC compds.; for general use in flexible and semirigid PVC; cream wh. powd.

Vanstay HTB. [Vanderbilt] Barium soap of mixed fatty acids; aux. stabilizer for flexible and rigid PVC.

Vanstay HTC. [Vanderbilt] Cadmium soap of mixed fatty acids; aux. stabilizer for flexible and rigid PVC.

Vanstay HTE. [Vanderbilt] Barium/cadmium/zinc stabilizer; lubricating stabilizer for high-temp. use; offers early color and low plate-out; used in calendered and extruded prods.; fine wh. powd.

Vanstay HTF. [Vanderbilt] Barium/cadmium/zinc stabilizer; offers superior sulfur staining resistance and low plate-out; used in homogeneous floor tile and highly filled vinyl prods.; cream wh. powd.

Vanstay L. [Vanderbilt] Phosphate-type; low cost, nontoxic stabilizer providing max. lt. aging; universal uv lt. stabilizer for vinyl; wh. powd.

Vanstay RAZ. [Vanderbilt] Cadmium and zinc org. complex; synergistic heat stabilizer which protects color of pigmented PVC stocks; used with Vanstay HA for filled PVC flooring and PVC tile compds.; liq.

Vanstay RR. [Vanderbilt] Barium/cadmium stabilizer; with stabilizing and lt. aging properties; used for flexible and semirigid PVC, calendered prods.; liq.

Vanstay RRE. [Vanderbilt] Barium/cadmium/zinc stabilizer; high temp. heat and lt. stabilizer with good clarity; used in plastisols and organosols; calendered prods., clear and filled; liq.

Vanstay RRZ. [Vanderbilt] Barium/cadmium/zinc stabilizer; offers best early color in tinted and color pigmented compds.; minimizes color drift while providing long-term stability; used in calendered sheeting; liq.

Vanstay SC. [Vanderbilt] Phosphite org. complex; chelator/stabilizer which significantly improves clarity and early color; can be used with all primary Ba/Cd and Ba/Cd/Zn stabilizers for clear semirigid and plasticized PVC formulations which are calendered and extruded; water-wh. liq.

Vanstay SD. [Vanderbilt] Org. phosphite; low cost chelator/stabilizer/synergist; used with Ba/Cd/Zn and Ba/Cd stabilizers in flexible PVC; water-wh. liq.

Vanstay SE. [Vanderbilt] Org. phosphite; low cost, general purpose chelator/stabilizer/synergist for PVC; used with Ba/Cd/Zn stabilizers for improved long-term, high-temp. aging; water-wh.liq.

Vanstay SF. [Vanderbilt] Org. phosphite; chelator/stabilizer/synergist to improve initial clarity and long-term aging of PVC compds.; used with Ba/Cd and Ba/Cd/Zn primary stabilizers; water-wh. liq.

Vanstay SG. [Vanderbilt] Org. phosphite; chelator/stabilizer/synergist to improve initial long term, low-temp. aging; used with Ba/Cd and Ba/Cd/Zn stabilizers; water-wh. liq.

Vanstay SH. [Vanderbilt] Org. phosphite; low cost, general purpose chelator/stabilizer/synergist for use with all Ba/Cd and Ba/Cd/Zn stabilizers; water-wh. liq.

Vanstay Zinc Stearate. [Vanderbilt] Aux. stabilizer for flexible and rigid PVC; avail. in tech. and dense grades.

Vantalc® 6H. [Vanderbilt] Magnesium silicate, hydrous; pigment for coatings; filler; dens. 22.91 lb/solid gal; oil absorp. 52; pH 9.5; 60.8% SiO_2; 31.9% MgO.

Vantalc® 500. [Vanderbilt] Platy talc, hydrous magnesium silicate; min. filler for polyester body patch compds.; produces low-visc. formulations, improves troweling and sanding, extends shelf life; low binder demand filler for plastics; extender in filled PP; off-wh. powd.; platy particle shape; mean particle size 15 μm, 4% +325 mesh residue; dens. 2.8 mg/m^3, 60 lb/ft^3; oil absorp. 21; pH 9.2; 37.0% SiO_2, 34.0% MgO.

Vantalc® 700. [Vanderbilt] Platy talc, hydrous magnesium silicate; see Vantalc 500; off-wh. powd. platy particle shape; mean particle size 12 μm, 3% +325 mesh residue; dens. 2.8 mg/m^3, 62 lb/ft^3; oil absorp. 22; pH 9.2.

Vantalc® 900. [Vanderbilt] Platy talc, hydrous magnesium silicate; see Vantalc 500; off-wh. powd. platy particle shape; mean particle size 5 μm, trace +325 mesh residue; dens. 40 lb/ft^3; oil absorp. 31; pH 9.2.

Vantalc® 1350. [Vanderbilt] Platy talc, hydrous magnesium silicate; min. filler produced by flotation process and offering high purity and wh. color; superior filler for PP, giving longevity at elevated temp. aging; wh. powd. platy particle shape; mean particle size 5.5 μm, 0.05% +325 mesh residue; dens. 2.7 mg/m^3, 60 lb/ft^3; oil absorp. 40; pH 9.5; 60.8% SiO_2, 31.9% MgO.

Vantalc® 1500. [Vanderbilt] Platy talc, hydrous magnesium silicate; see Vantalc 1350; wh. powd. platy particle shape; mean particle size 4 μm; dens. 2.7 mg/m^3, 62 lb/ft^3; oil absorp. 44; pH 9.5; 60.8% SiO_2, 31.9% MgO.

Vantoc AL. [ICI Specialty] Alkyl trimethyl ammonium bromides; bactericide for food and drink processing plants; liq.

Vantoc CL. [ICI Specialty] Lauryl dimethyl benzyl ammonium chloride; cationic; surfactant and biocide; liq.

Vanwax® H. [Vanderbilt] Protective wax for sunchecking inhibition of elastomerics; cream to lt. yel. flakes; mod. sol. in toluene, chloroform; pract. insol. in alcohol; insol. water; dens. 0.90 ± 0.03 mg/m^3; drop pt.. 70–80 C.

Vanwax® H Special. [Vanderbilt] Petrol. wax; protectant for elastomers exposed to sunlight and/or ozone under static conditions; wh. flakes; mod. sol. in toluene, chloroform; pract. insol. alcohol; insol. water; dens. 0.93 ± 0.03 mg/m^3; m.p. 65 C min.

Vanwax® OZ. [Vanderbilt] Wax blend; sunchecking inhibitor for elastomerics; wh. to lt. cream flakes; sol. in chloroform, toluene, other aromatics; insol. water and alcohol; dens. 0.90 ± 0.03 mg/m^3; drop pt. 58–63 C.

Varamide® A-2. [Sherex] Refined coconut DEA (2:1); nonionic; thickener, foam stabilizer, and detergent for soaps and hand cleaners; enhances anionic and nonionic detergency; used with alkylaryl sulfonates in bubble baths, textile applics. as a scouring and fulling agent, and in rug and floor cleaners to improve detergency; used in soap bars to increase gloss and limesoap dispersibility while tending to reduce cracking; clear yel. liq.; water-disp.; dens. 8.5 lb/gal; pH 10.2 (1%); 100% act.

Varamide® A-7. [Sherex] Oleic DEA (2:1); nonionic; rust inhibitor and base for o/w emulsifiers, detergents, and thickener for waterless hand cleaners; degreasers; amber clear liq.; dens. 8.2 lb/gal; pH 10.4 (1%); 100% act.

Varamide® A-10. [Sherex] Modified coconut DEA (2:1); nonionic; thickener, foam stabilizer, detergent, rust inhibitor for floor cleaners and general-purpose cleaners; reduces necessary rinsing and increases visc., perfume and dye solubility in anionic base cleaners; suitable for chain belt lubricants and metal-lathe working sol'ns.; clear amber liq.; water-sol.; dens. 8.3 lb/gal; pH 8.8 (1%); 100% act.

Varamide® A-12, A-83. [Sherex] Modified coconut DEA (2:1); see Varamide A-10; A-12 is a higher salt tolerance version designed for use at phosphate levels as high as 10%; A-83 is a lower cost version of A-10 used in industrial, institutional, and household cleaners, wh. sidewall cleaners, wax strippers, degreasers, and for textile scouring and fulling; clear amber liq.; dens. 8.2 and 8.3 lb/gal resp.; pH 9 (both, 1%); 100% act.

Varamide® AC-28. [Sherex] Modified mixed DEA; emulsifier, anticorrosive in cutting oils and metalworking liqs.; liq.; sol. in hydrocarbons, emulsified in water.

Varamide® BWA. [Sherex] Emulsifier blend; degreaser and engine cleaner removable with cold water; liq.; 100% solids.

Varamide® C-212. [Sherex] Coconut MEA (1:1); foam booster and stabilizer in aq. systems; flake; 100% solids.

Varamide® Degreaser Conc. [Sherex] Emulsifier/solv.; engine cleaner with solv. for superior oil removal; liq.; 100% solids.

Varamide® DO 280. [Sherex] Modified DEA; anticorrosive lubricant; liq.; 100% conc.

Varamide® DU 185. [Sherex] Undecylenic DEA; nonionic; fungicidal and bacteriostatic additive for shampoos, soaps, creams; liq.; 100% conc.

Varamide® L-1. [Sherex] Lauramide DEA and diethanolamine; nonionic; detergent; thickener and foam booster, stabilizer for shampoos; paste; 100% act.

Varamide® L-203. [Sherex] Lauramide MEA (1:1); foam booster and stabilizer for aq. systems; flake; 100% act.

Varamide® MA-1. [Sherex] Refined cocamide DEA (1:1); nonionic; foam stabilizer and booster, thickener; basic liq. superamide for shampoos, bubble bath, and dishwasher; low cost equivalent to lauric superamide; does not require melting; gives higher visc. and foam stability than conventional 2:1 alkanolamides in liq. detergent systems; Gardner 3+ max. clear liq.; readily disp. in water; dens. 8.2 lb/gal; pH 9.8 (1%).; 100% act.

Varamide® MA-4. [Sherex] Lauramide DEA; nonionic; detergent, foam booster and stabilizer, thickener used in dishwashing detergents and shampoos; liq.; 100% conc.

Varamide® ML-1. [Sherex] Lauric DEA; nonionic; thickener and foam stabilizer for shampoo, bubble bath, and hand laundry detergent; gives the highest visc., foam level, and stability of the superamides in series; wh. wax; water-disp.; dens. 8.1 lb/gal; pH 10.2 (1%).

Varamide® ML-4. [Sherex] Lauramide DEA; nonionic; exhibits similar properties to Varamide ML-1 but gives high formulation visc.; wh. wax; dens. 8.2 lb/gal; pH 10.2 (1%).

Varamide® SL-9. [Sherex] Lauramide DEA and propylene glycol; nonionic; detergent, foam stabilizer and booster used in shampoos, detergents, laundry prods. and hard surface cleaners; Gardner 2 clear liq.; dens. 8.3 lb/gal; solid. pt. 84 F; pH 12.5; surf. tens. 33.0 dynes/cm.

Varamide® SL-910. [Sherex] Lauric-palmitic superamide; see Varamide SL-9; liq. 100% act.

Varamide® U 185. [Sherex] Undecylenic acid MEA; nonionic; see Varamide DU 185; flakes; 100% conc.

Varcum® 564, 4326-WP. [Reichhold] Phenolic resin; curing agent for epoxy resins; powd. and flake resp.; soften. pt. (R&B) 80–90 C.

Varcum® 647. [Reichhold] Thermoplastic phenolic novolac; nonheat reactive tackifier for SBR and butyl rubbers; flake; sol. in aromatics, aliphatics, esters, ketones; partially sol. in alcohols; soften. pt. (R&B) 95–105 C.

Varcum® 1198. [Reichhold] Alkyl phenolic resin; heat-reactive resin; curing agent for butyl rubber; modifier for pressure-sensitive adhesives based on SBR and natural rubber; crushed; sol. (equal parts solv. and resin) in aromatics, aliphatics, esters, and ketones; soften. pt. (R&B) 80–90 C.

Varcum® 2053. [Reichhold] Phenolic resin; bonding agent in rag-filled molding compds.; liq.; visc. 1500–1900 cps; 72–76% solids in water.

Varcum® 2217. [Reichhold] Thermoplastic phenolic novolac; nonheat reactive tackifier for chlorobutyl and other syn. rubbers; flake; sol. in aromatics, esters, ketones; partially sol. in aliphatics, alcohols; soften. pt. (R&B) 125–140 C.

Varcum® 4727. [Reichhold] Phenolic novolac resin; thermoplastic, nonheat reactive tackifier with SBR, butyl, chlorobutyl, and other syn. rubbers; lump; sol. (equal parts solv. and resin) in aromatics, aliphatics, alcohols, esters, and ketones; soften. pt. (R&B) 105–120 C.

Varcum® 6404. [Reichhold] Epoxy resin; powdered epoxy resin for use as modifier to soften phenolic resins; shellac substitute; coarse particles; soften. pt. (Cap.) 60–70 C.

Varcum® 7608. [Reichhold] Phenolic resin; short flow, fast cure grinding wheel resin; used with slower curing resins as modifier to produce quick setting and hasten dimensional stability; powd.; soften. pt. (Cap.) 80–90 C; 7.2–7.5% hexa content.

Varcum® 8169. [Reichhold] Phenolic resin; bonding agent, impregnating agent for impervious carbon; liq.; visc. 800–1200 cps; gel time 26–30 min (121 C); 74.5–78.5% resin solids in water-phenol.

Varcum® 8199. [Reichhold] Phenolic resin; specialty resin for bonding and impregnation of impervious carbon; liq.; visc. 200–275 cps; 70–74% solids in water.

Varcum® 8251. [Reichhold] Furan resin; narrow specification liq. furfuryl alcohol polymer, for impregnation and bonding carbon and graphite; requires the addition of an acid to promote cure; liq.; visc. 160–320 cps; 50–55% solids in furfuryl alcohol.

Varcum® 8261. [Reichhold] Furan resin; liq. furfuryl alcohol polymer for impregnation and bonding carbon and graphite; requires the addition of an acid to promote cure; liq.; visc. 1500–2000 cps; 66–70% solids.

Varcum® 8267. [Reichhold] Furan resin; resin with high solids and low residual moisture; recommended for bonding dense shapes requiring min. volatile evolution; requires the addition of an acid to promote cure; liq.; visc. 8000–35,000 cps (sheared); 77–83%

solids.

Varcum® 8345. [Reichhold] Phenolic resin; FDA-approved and MID-sanctioned resin used in metal coatings; slow cure modifying resin with good vinyl compatibility; improves vinyl adhesion; crushed; sol. (equal parts solv. and resin) in esters, ketones; sp.gr. 1.195–1.205; visc. 150–300 cps (60% sol'n. in 1:1 ethanol: MIBK); soften. pt. (R&B) 90–100 C.

Varcum® 8506. [Reichhold] Phenolic resin; inorg. modification; modifying resin to reduce swelling and slumping in high dens. wheels; powd.; soften. pt. (Cap.) 77–87 C; 6.5–7.5% hexa content.

Varcum® 9008. [Reichhold] Two-step phenolic resin; thermosetting resin for reinforcement of SBR compds.; powd.; sol. (equal parts solv. and resin) in esters, ketones; soften. pt. (Cap.) 62–72 C.

Varcum® C-86. [Reichhold] Two-step phenolic resin; thermosetting resin used as rubber reinforcing agent; powd.; sol. (equal parts solv. and resin) in alcohols, esters, and ketones; soften. pt. (Cap.) 70–80 C.

Varcum® DR-406. [Reichhold] Phenolic novolac resin; thermoplastic, nonheat reactive tackifier and processing aid for SBR, butyl, and chlorobutyl rubber; flaked; sol. (equal parts solv. and resin) in aromatics, aliphatics, esters, and ketones; soften. pt. (R&B) 85–105 C.

Varidri 40. [Sherex] Dioctyl sulfosuccinate; dewatering aid in mining industry; dispersant; sol. in water, polar solvs.

Varine C. [Sherex] Coco hydroxyethyl imidazoline; cationic; emulsifier, anticorrosive, raw material; shampoo base, penetrating oils, antistats, corrosion inhibitors, paints, printing inks, textiles; Gardner 9 liq.; m.w. 274; disp. in water; sol. in polar solvs. and hydrocarbons; m.p. 31 C; 100% act.

Varine O®. [Sherex] Oleyl hydroxyethyl imidazoline; cationic; see Varine C; Gardner 8 liq.; m.w. 352; sol. see Varine C; m.p. –10 C; 100% act.

Varine O Acetate. [Sherex] Oleic imidazoline acetate; anticorrosive; liq.; 100% solids.

Varine T. [Sherex] Tall oil hydroxyethyl imidazoline; see Varine C; Gardner 8; m.w. 350; sol. in polar solvs. and hydrocarbons; disp. in water; m.p. –10 C; 85% tert. amine.

Varion® AM-KSF-40%. [Sherex] Coco amphopropionate; salt-free amphoteric; surfactant for heavy-duty cleaners, steam cleaners, metal cleaners; offers coupling ability with foam and detergency in high-electrolyte systems; liq.; 40% conc.

Varion® AM-V. [Sherex] Caprylic glycinate; amphoteric; coupling agent for use in low-foam wetting, acidic or basic systems; liq.; 35% solids.

Varion® CADG. [Sherex] Cocamidopropyl betaine; amphoteric; detergent, foamer, conditioner, solublizer used in shampoo prods.; liq.; 32% act.

Varion® CADG-HS, CADG-LS. [Sherex] Cocamidopropyl betaine; amphoteric; surfactant; detergent; foam booster and visc. modifier; personal care prods. and heavy-duty detergents; LS is low-salt version; yel. clear liq.; pH 6.5–8.5 and 5.0–7.0 resp.; 35–37% solids.

Varion® CADG-W. [Sherex] Cocamidopropyl betaine; see Varion CADG-HS; liq.; 35% solids.

Varion® CDG. [Sherex] Lauryl betaine; amphoteric; foamer, detergent, solubilizer, conditioner used in shampoos; liq.; 31% act.

Varion® SDG. [Sherex] Stearyl betaine; amphoteric; detergent, softener for fabric and hair applics.; liq.; 50% conc.

Varion® TEG. [Sherex] Tallow dihydroxyethyl betaine; amphoteric; foamer in personal care prods.; thickener in household and industrial detergents; HCl gelling agent used in acid stimulation in the petrol. industry; Gardner 7 clear liq.; flash pt. (CC) > 200 F; pH 4.1–5.3 (5% aq.); 41–44% act.

Varion® TEG-40%. [Sherex] Tallow glycinate; amphoteric; lower visc. version of Varion TEG; for thickening of hydrochloric and phosphoric acid toilet bowl systems; liq.; 40% solids.

Variquat® 50AC. [Sherex] Benzalkonium chloride; IPA solv. version of Variquat 50MC; liq.; 50% solids.

Variquat® 50AE. [Sherex] Benzalkonium chloride; ethanol solv. version of Variquat 50MC; liq.; 50% solids.

Variquat® 50MC. [Sherex] Benzalkonium chloride; cationic; germicide, algicide, disinfectant, sanitizer, deodorant; used in pesticides and mfg. of sanitizers; food processing, dairy, restaurant, industrial and household prods.; Gardner 2 max. liq.; m.w. 358; flash pt. (PM) 120 F; 50% act.

Variquat® 50ME. [Sherex] Dimethyl alkyl (C_{12}–C_{16}) benzyl ammonium chloride (50%), ethyl alcohol (7.5%) in water; specialty quat., germicide used for disinfection and sanitizing for hospitals, beautician instruments, food processing plants; Gardner 2 max. liq.; m.w. 358; sol. in water or alcohol; sp.gr. 0.96; dens. 8.0 lb/gal; flash pt. 120 F (P-M); pH 5–8; 57.5% act.

Variquat® 60LC. [Sherex] Benzalkonium chloride; cationic; swimming pool and water treatment algicide; Gardner 2 max. liq.; m.w. 351; flash pt. (PM) 130 F; 49.5–51.5% quat.

Variquat® 80AC. [Sherex] Benzalkonium chloride; IPA version of Variquat 80MC; liq.; 80% solids.

Variquat® 80AE. [Sherex] Benzalkonium chloride; ethanol version of Variquat 80MC; liq.; 80% solids.

Variquat® 80LC. [Sherex] Benzyl chloride quat.; algicide conc. for swimming pools; liq.; 80% solids.

Variquat® 80MC. [Sherex] Dimethyl alkyl (C_{12-16}) benzyl ammonium chloride; cationic; germicide, disinfectant, deodorant used in pesticides and sanitizers; corrosion inhibitor for water inj. systems; liq.; m.w. 358; flash pt. (PM) 100 F; 80% act.

Variquat® 80ME. [Sherex] Dimethyl alkyl (C_{12}–C_{16}) benzyl ammonium chloride; germicidal conc. for disinfection and sanitization; liq.; Gardner 2 max.; m.w. 358; flash pt. 100 F (P-M); 81.5–82.5% quat.

Variquat® 475. [Sherex] Tallow imidazoline methyl sulfate; corrosion inhibitor for improved water sol. formulations; 75% conc.

Variquat® 638. [Sherex] PEG-2 cocomonium chloride; cationic; antistat, plating bath foam blanket, emulsifier; Gardner 9 max. liq.; m.w. 353; flash pt. 56 F (PM); 74–75% quat.

Variquat® B200. [Sherex] Benzyltrimethyl ammonium chloride; cationic; dispersant, dye leveler and retarder, emulsifier used in textile industry; APHA 50 max. liq.; pH 6.5–8.0; 60% act.

Variquat® E228. [Sherex] Cetrimonium chloride; cationic; softener, conditioner used in hair applics.; Gardner 1 max. liq.; m.w. 320; flash pt. > 200 F (PM); 74–76% solids.

Variquat® K300. [Sherex] Dimethyl dicoco ammonium chloride; cationic; emulsifier, dispersant; Gardner 5 max. liq.; m.w. 439; flash pt. 50 F (PM); 75% conc.

Variquat® LC60. [Sherex] Benzalkonium chloride; algicide for swimming pools; 60% quat.

Variquat® LC80. [Sherex] Dimethyl alkyl (C_{10-18}) benzyl ammonium chloride; cationic; algicide for use in swimming pools and industrial water treatment; corrosion inhibitor for water inj. systems; Gardner 2 max. liq.; m.w. 354; flash pt. (PM) 102 F; 80% act.

Varisoft® 110. [Sherex] Quaternium-18; cationic; fabric softener conc. for home and commercial laundries, textile processing; Gardner 6 max. paste; flash pt. 99 F (PM); 74–76% solids.

Varisoft® 133H, 135H, 136, 242, 369ON-90%. [Sherex] Quat. ammonium blends; fabric softener concs.

Varisoft® 135. [Sherex] Sulfated quat. blend; softener, antistat for fabric treatment in dryers; Gardner 4 max. clear liq.; dens. 7.4 lb/gal (140 F); m.p. 40–50 C; flash pt. 76 F (PMCC); pH 6.5; 91–93% solids.

Varisoft® 137. [Sherex] Dihydrog. tallow dimonium methosulfate; cationic; quat. for home and commercial laundry fabric and tissue softeners, debonding agent and antistat; emulsifier; Gardner 5 solid paste; m.w. 649; disp. in water; dens. 7.4 lb/gal; flash pt. 78 F (PMCC); pH 5–8; 89–91% solids.

Varisoft® 190. [Sherex] Distearyl dimethyl ammonium methyl sulfate; see Varisoft 110; Gardner 4 max. paste; flash pt. 78 F (89–91% solids); 65 F (74–76% solids).

Varisoft® 190-100P. [Sherex] Dimethyl distearyl ammonium methyl sulfate; see Varisoft 110; powd.; 97% solids.

Varisoft® 222. [Sherex] Methyl bis (tallowamidoethyl) 2-hydroxyethyl ammonium methyl sulfate; cationic; see Varisoft 110; Gardner 6 max. liq.; flash pt. 72 F (PM); 74–76% solids.

Varisoft® 222-LM-90. [Sherex] Methylbis (tallow amidoethyl) 2-hydroxy ethyl ammonium methyl sulfate; softener with exc. handling properties for laundry rinse-cycle softeners; liq.; 90% solids.

Varisoft® 222-LT-90. [Sherex] Complex difatty (oleic) quat.; cationic; softener conc. for formulation of liq. detergent-softeners or high solids softener prods.; liq.

Varisoft® 238. [Sherex] Methyl bis (tallow-amidoethyl) 2-hydroxypropyl ammonium methyl sulfate; see Varisoft 110; Gardner 6 max. liq.; flash pt. 60 F (PM); 74–76% solids.

Varisoft® 445. [Sherex] Methyl (1) hydrog. tallow amidoethyl (2) hydrog. tallow imidazolinium methosulfate; see Varisoft 110; liq.; 77% solids.

Varisoft® 472-80%. [Sherex] Blend of dihydrog. tallow ammonium and ditallow imidazolinium methosulfate; softener offering optimum static control; Gardner 5 max. soft paste; slightly hazy liq. at 45 C; sp.gr. 0.94; pH 4.5–7.5 (5% in 1:1 IPA/water); flash pt. (PM) ≈ 65 F; 79–83% solids in IPA.

Varisoft® 475. [Sherex] Quaternium-27; cationic; see Varisoft 110; Gardner 5 max. liq.; flash pt. 72 F (PM); 75–77% solids.

Varisoft® 920. [Sherex] Tallow bis hydroxyethyl methyl ammonium chloride; cationic; softener conc. for anionic/nonionic based detergent/softeners; liq.; 75% act.

Varisoft® 3262. [Sherex] Quaternium-18, IPA, stearyl alcohol, and glycol stearate; cationic; cream rinse conc.; solid; 75% conc.

Varisoft® 3690. [Sherex] Methyl (1) oleylamidoethyl (2) oleyl imidazolinium methosulfate; cationic; see Varisoft 110; Gardner 7 max. liq.; m.w. 723; flash pt. 66 F (PM); 74–76% quat.

Varisoft® 6112. [Sherex] 1-Ethylene bis (2-tallow, 1-methyl, imidazolinium-methyl sulfate); see Varisoft 110; Gardner 7 max. liq.; flash pt. < 70 F (PM); 74–76% solids.

Varisoft® CA-50. [Sherex] Cationic; hair conditioner; flake.

Varisoft® CRC. [Sherex] Cetearyl alcohol, dicetyl dimonium chloride, stearamidopropyl dimethylamine; cream rinse conc. for formulating cream rinses and conditioners; waxy flake; m.p. 49 C; pH 8.8 (5% in 1:1 IPA/water); 99% solids.

Varisoft® DHT. [Sherex] Dimethyl dihydrog. tallow ammonium chloride; cationic; hair conditioner and antistat in cream rinse concs.; Gardner 3 max. soft solid; m.w. 569; disp. in water; dens. 7.12 lb/gal; flash pt. 68 F (PMCC); pH 6–9; 74–77% act. in aq. IPA.

Varisoft® E-228. [Sherex] Cetrimonium chloride; cationic; conditioner, softener; liq.; 28% conc.

Varisoft® E-290. [Sherex] Cetrimonium chloride; base for hair conditioners and creme rinses; liq.; 28–30% solids.

Varisoft® SA-50. [Sherex] Distearyl dimonium chloride, stearyl alcohol; cationic; base for hair and skin conditioners; flake; 100% act.

Varisoft® SDC. [Sherex] Stearalkonium chloride; hair softener and conditioning cream rinse; Gardner 2 max. paste; flash pt. 116 F (PM); 24–26% solids.

Varisoft® SDC-65%. [Sherex] Stearalkonium chloride; cationic; used in hair care prods.; wh. paste; pH 3–5; 65–70% act.

Varisoft® SDC-85S. [Sherex] Stearalkonium chloride and stearyl alcohol; base for hair conditioner, emulsifier for creams and lotions; flakes; 100% solids.

Varisoft® SDC-100P. [Sherex] Stearalkonium chloride; cationic; for hair conditioners; powd.; 100% act.

Varisoft® SDC-W. [Sherex] Stearalkonium chloride; cationic; antistat used in hair care prods.; wh.

paste; flash pt. > 200 F (PM); pH 3.5–4.0; 24–26% solids in water.

Varisoft® TSC. [Sherex] Steartrimonium chloride; base for hair conditioners; liq.; 25% solids in water.

Varonic® 32-E20. [Sherex] Alkoxylated oleyl alcohol; emulsifier, stabilizer; paste; HLB 11.3; 100% solids.

Varonic® 200 Series. [Sherex] 2–30 mol EO adducts on coco, soya, tallow, and hydrog. tallow amines; corrosion inhibitors for the formulation of improved water-sol. inhibitor systems for o/w emulsions, and used to combat HCl acid corrosion and H_2S in gas pipelines; 100% conc.

Varonic® 320. [Sherex] POE oleyl alcohol; nonionic; emulsifier, lubricant.

Varonic® 400MS. [Sherex] PEG 400 monostearate; nonionic; emulsifier, thickener used in emollients, creams and lotions; solid; 100% act.

Varonic® 1000MS. [Sherex] PEG 1000 monostearate; nonionic; solubilizer, humectant, surfactant; solid.

Varonic® 1800MS. [Sherex] PEG 1800 monostearate; nonionic; see Varonic 1000MS; solid.

Varonic® DM55. [Sherex] Methyl capped glycol ether; solv. with low toxicity and exc. grease cutting properties for surf. cleaners; liq.; water-emulsifiable; 100% solids.

Varonic® K202. [Sherex] PEG-2 cocamine; emulsifier; corrosion inhibitor in metal finishing (e.g. as cutting oil additives); detergent, antifouling, antistalling, and deicing agent in gasoline; also in textile lubricants, oil field emulsification; Gardner 2 liq.; sol. in IPA, benzene, Stod., CCl_4, MEK; insol. in water; sp.gr. 0.916; HLB 6.2; 100% solids.

Varonic® K205. [Sherex] PEG-5 cocamine; corrosion inhibitor, antifouling agent in gasoline; antistat, lubricant in wool spinning; emulsifier and leveling agent in textile dyeing; insecticide and herbicide systems; metal finishing; Gardner 7 liq.; sol. in IPA, benzene, Stod., CCl_4, MEK; sp.gr. 0.977; HLB 11.0; 100% solids.

Varonic® K205LC. [Sherex] PEG-5 cocamine; low color dye leveler, coemulsifier; liq.; HLB 11.0; 99% solids.

Varonic® K210. [Sherex] PEG-10 cocamine; dye leveling agent, dispersant in paper industry; wetting and spreading agent and emulsifier in insecticides and herbicides; tanning of leather and furs; wetting and redeposition aids in fur treating; Gardner 7 liq.; sol. in water, IPA, benzene, CCl_4; sp.gr. 0.995; HLB 13.8; 100% solids.

Varonic® K210LC. [Sherex] PEG-10 cocamine; low color dye leveler, coemulsifier; liq.; HLB 13.8; 99% solids.

Varonic® K215. [Sherex] PEG-15 cocamine; wetting agent and redeposition aid in fur treatment; textile additive in spinning bath to reduce jet clogging and disperse sulfur/zinc sulfide particles to prevent cratering; in petrol. industry for corrosion resistance; Gardner 7 liq.; sol. in water, IPA, benzene, CCl_4, MEK; sp.gr. 1.040; HLB 15.4; 100% solids.

Varonic® K215LC. [Sherex] PEG-15 cocamine; low-color version of Varonic K215; wetting agent during liming stage in tanning of leathers and furs; textile aid reducing jet clogging in spinning bath and dispersing sulfur/zinc sulfide particles preventing cratering; Gardner 3 liq.; sol. see Varonic K215; sp.gr. 1.040; HLB 15.4; 100% solids.

Varonic® L202. [Sherex] PEG-2 soyamine; nonionic; antistat, dispersant; emulsifier for latex, dyes, oils; liq.; HLB 4.7; 100% act.

Varonic® L205. [Sherex] PEG-5 soyamine; nonionic; see Varonic K205; Gardner 7 liq.; sol. in IPA, benzene, CCl_4, MEK; sp.gr. 0.985; 100% solids.

Varonic® L205LC. [Sherex] PEG-5 soyamine; nonionic; low-color version of Varonic L205; liq.; HLB 8.4; 100% act.

Varonic® L230, L230-60, L230-80. [Sherex] PEG-30 soyamine; nonionic; see Varonic L 202; liq.; 100, 60, and 80% act. resp.

Varonic® LI-42. [Sherex] PEG-20 glyceryl monotallowate; nonionic; low-irritation detergent, emulsifier, solubilizer for personal care prods.; Gardner 2 paste; HLB 13; m.p. 27 C; surf. tens. 44 dynes/cm (1%); 100% solids.

Varonic® LI-48. [Sherex] PEG-80 glyceryl tallowate; nonionic; emulsifier, solubilizer, antiirritant surfactant used in personal care prods.; Gardner 2 hard solid; sol. in water, methanol, ethanol, IPA, acetone, and butyl Cellosolve; HLB 18 ± 1; m.p. 41 C; pH 7.0; surf. tens. 49.5 dynes/cm.

Varonic® LI-63. [Sherex] PEG-30 glyceryl monococoate; nonionic; low-irritation detergent, emulsifier, solubilizer for personal care prods.; Gardner 2 paste; HLB 15; m.p. 27 C; surf. tens. 40 dynes/cm (1%); 100% solids.

Varonic® LI-67. [Sherex] PEG-78 glyceryl cocoate; nonionic; see Varonic LI-63; Gardner 2 solid; HLB 18; m.p. 42 C; surf. tens. 47 dynes/cm (1%); 100% solids.

Varonic® LI-67-75. [Sherex] PEG-78 glyceryl monococoate; see Varonic LI-63; liq.; HLB 18.0; 75% solids.

Varonic® LI-420-80. [Sherex] PEG-200 glyceryl tallowate; nonionic; see Varonic LI-48; Gardner 2 hard solid; sol. see Varonic LI-48; HLB 19 ± 1; m.p. 53 C; pH 7.0; surf. tens. 54.0 dynes/cm; 80% solids.

Varonic® MT 65. [Sherex] Methyl capped alkoxylated fatty alcohol; nonionic; low-foam detergent, textile spinning lubricant; Gardner 2 clear liq.; sol. in polar and nonpolar solvs.; pH 3.5 (10%); surf. tens. 32 dynes/cm; 100% solids.

Varonic® Q202. [Sherex] PEG-2 oleylamine; anticorrosive emulsifier for metalworking; liq.; HLB 4.5; 100% solids.

Varonic® S202. [Sherex] PEG-2 stearylamine; polymer additive and emulsifier; solid; HLB 5.0; 100% solids.

Varonic® T105. [Sherex] Ethoxylated alcohol; surfactant for drilling aids and emulsifiers; thickener in hydraulic fracturing; defoamer in cementing in the petrol. industry; soft paste; flash pt. > 200 F (CC); 100% conc.

Varonic® T202. [Sherex] PEG-2 tallowamine; nonionic; lubricant, softener and antistat for syn. fibers; used in conjunction with other cationics to give improved visc. in syn. latex paints; emulsifier for latex, dyes, and oils; dispersant; Gardner 2 semiliq.; sol. in IPA, benzene, Stod., CCl_4; insol. in water; sp.gr. 0.915; HLB 5.1; 100% solids.

Varonic® T202 SR. [Sherex] PEG-2 tallowamine; special grade for optimum visc. control in acid cleaners; paste; HLB 5.1; 100% solids.

Varonic® T205. [Sherex] PEG-5 tallowamine; nonionic; lubricant, softener, and antistat for syn. fibers; antistat/lubricant in wool spinning; emulsifier/leveling agent in water/oil dye systems; emulsifier and spreading-wetting agent in insecticides and herbicides; Gardner 7 liq.; sol. in IPA, benzene, Stod., CCl_4, MEK; emul. in water; sp.gr. 0.966; HLB 9.2; 100% solids.

Varonic® T205LC. [Sherex] PEG-5 tallowamine; nonionic; low-color version of Varonic T205; liq.; 100% act.

Varonic® T210. [Sherex] PEG-10 tallowamine; nonionic; lubricant, scouring aid, dye leveler for textile industry; process modifier in polymer industry; raw material for quat. and amphoteric surfactants; liq.; HLB 12.6; 100% solids.

Varonic® T215. [Sherex] PEG-15 tallowamine; nonionic; lubricant, softener, and antistat for syns.; wetting aid/emulsifier in tanning; in petrol. industry used with caustic to break o/w crude oil emulsions and to impart corrosion resistance; Gardner 7 liq.; sol. in water, IPA, benzene, CCl_4, MEK; sp.gr 1.029; HLB 14.4; 100% conc.

Varonic® T215LC. [Sherex] PEG-15 tallowamine; low-color version of Varonic T215; liq.; HLB 14.4; 100% solids.

Varonic® T403. [Sherex] PEG-3 tallow diamine; corrosion inhibitors for the formulation of improved water-sol. inhibitor systems for o/w emulsions, and used to combat HCl acid corrosion and H_2S in gas pipelines; 100% conc.

Varonic® T410. [Sherex] PEG-10 tallow diamine; corrosion inhibitors for the formulation of improved water-sol. inhibitor systems for o/w emulsions, and used to combat HCl acid corrosion and H_2S in gas pipelines; 100% conc.

Varonic® U105. [Sherex] Ethoxylated alcohol; see Varonic T105; hard paste; flash pt. > 200 F (CC); 100% conc.

Varonic® U202. [Sherex] PEG-2 cetyl/stearylamine; antistat for textile dye leveling; stabilizer for asphalt emulsions; corrosion inhibitor and foam prevention in boilers; used in felted paper; Gardner 2 solid; sol. in IPA, benzene, CCl_4; insol. in water; sp.gr. 0.916; HLB 5.0; 100% conc.

Varonic® U205, 205LC. [Sherex] PEG-5 cetyl/stearylamine; nonionic; antistat, emulsifier, dispersant, dye intermediate; semiliq., solid; 100% act.

Varonic® U215. [Sherex] PEG-15 cetyl/stearylamine; see Varonic U202; solid; HLB 14.3; 100% solids.

Varonic® U250. [Sherex] PEG-50 cetyl/stearylamine; nonionic; corrosion inhibitor in metal working; cutting oil additive; emulsifier and spreading and wetting agent in insecticides and herbicides; plasticizer for NC lacquer, resinous PC; leveling agent; Gardner 7 solid; sol. in water, IPA, CCl_4; sp.gr. 0.968; HLB 17.9; 100% conc.

Varonic® U250-50%. [Sherex] PEG-50 cetyl/stearylamine; liq. version of Varonic U250; HLB 17.9; 50% solids.

Varox®. [Vanderbilt] 2,5-Dimethyl-2,5-di(t-butylperoxy) hexane; crosslinking agent used for vulcanization; used in EPDM, EPM, PE, and NBR; water-wh. to lt. yel. liq.; wh. powd., 99.9% thru 100 mesh; m.w. 290.45; dens. 0.87 ± 0.02 mg/m³; 45% min. act. (powd.), 90% min. act. (liq.)

Varox® 130, 130-XL. [Vanderbilt] 2,5-Dimethyl-2,5-di(t-butylperoxy)-3-hexyne; vulcanizing agent for EPDM, EPM, and CPE; crosslinking agent for PE and EVA; lt. yel. liq. and wh. powd. (99% min. thru 100 mesh) resp.; m.w. 286.5; dens. 0.89 ± 0.02 and 1.26 ± 0.03 mg/m³ resp.; 90–95% and 45–55% assay resp.; 10.05–10.65 and 4.52–5.83% act. oxygen resp.

Varox® 185E. [Vanderbilt] Dihydroxyethyl C_{12-15} alkoxypropylamine oxide; foam stabilizer for anionic surfactants used in liq. dishwashing, shampoos and fabric detergents; Gardner 1 liq.; pH 4.5–5.5; surf. tens. 31.5 dynes/cm; 39–43% amine oxide.

Varox® 185PH. [Vanderbilt] Hydroxyethyl hydroxypropyl C_{12-15} alkoxypropylamine; see Varox 185E; Gardner 1 max. liq.; pH 4.5–5.5; surf. tens. 30.9 dynes/cm.

Varox® 188E. [Vanderbilt] Dihydroxyethyl C_{8-10} alkoxypropylamine oxide; see Varox 185E; pH 4.5–5.5; surf. tens. 34.8 dynes/cm.

Varox® 191E. [Vanderbilt] Dihydroxyethyl C_{9-11} alkoxypropylamine oxide; see Varox 185E; Gardner 1 max. liq.; pH 4.5–5.5; surf. tens. 34.0 dynes/cm.

Varox® 230-XL. [Vanderbilt] n-Butyl-4,4-bis (t-butylperoxy) valerate; crosslinking agent for elastomers, e.g., polybutadiene, CPE, EPDM; wh. powd.; 99% min. thru 100 mesh; m.w. 334.46; dens. 0.95 ± 0.03 mg/m³; 39–41% assay; 3.83% min. act. oxygen.

Varox® 231, 231-XL. [Vanderbilt] 1,1-Bis (t-butylperoxy)-3,3,5-trimethylcyclohexane; vulcanizing agent for SBR, NBR, EPDM, EPM, CPE, and silicones; crosslinking agent for PE and EVA; lt. yel. liq. and wh. powd. (99% min. thru 100 mesh) resp.; m.w. 302.46; dens. 0.91 ± 0.02 and 1.41 ± 0.03 mg/m³ resp.; 90% min. and 39–41% assay resp.; 9.52% min. and 3.8% min. act. oxygen. resp.

Varox® 365. [Vanderbilt] Lauryl dimethylamine oxide; foam booster/stabilizer for anionic surfactants for shampoo and detergent systems; Gardner 1 liq.; 30% solids.

Varox® 375. [Vanderbilt] Lauryl dimethylamine oxide; foam booster/stabilizer; liq.; 40% solids.

Varox® 743. [Vanderbilt] Coco dialkoxy amine oxide; nonionic; detergent, foam booster and stabilizer; liq.; 50% conc.

Varox® DBPH, DBPH-50. [Vanderbilt] 2,5-Dimethyl-2,5-di(t-butylperoxy) hexane; DBPH-50 on inert min. carrier; vulcanizing and crosslinking agent for elastomers and polyolefins, e.g., EPDM, EPM, PE, and NBR; water-wh. to lt. yel. liq. and wh. powd. (99.9% thru 100 mesh) resp.; m.w. 290.45; dens. 0.87 ± 0.02 and 1.45 ± 0.03 mg/m³ resp.; 90% min. and 45% min. assay resp.; 9.92% min. and 4.96% min. act. oxygen resp.

Varox® DCP-40C. [Vanderbilt] Dicumyl peroxide on calcium carbonate; crosslinking agent; wh. powd.; dens. 1.60 ± 0.03 mg/m³; 39–41% assay.

Varox® DCP-40KE. [Vanderbilt] Dicumyl peroxide on KE clay; crosslinking agent; wh. powd.; dens. 1.61 ± 0.03 mg/m³; 39–41% assay.

Varox® DCP-R. [Vanderbilt] Dicumyl peroxide; crosslinking agent for PE, EPR, silicone, EVA, and VAE; nonsulfur curing agent for NR, IR, SBR, BR, NBR, EPDM, and CR; pale yel. recryst. cryst. solid; m.w. 270.37; dens. 1.001 mg/m³; 99% min. assay.

Varox® DCP-T. [Vanderbilt] Dicumyl peroxide; see Varox DCP-R; pale yel. semicryst. solid; m.w. 270.37; dens. 0.977–1.009 mg/m³ (40 C); 91% min. assay.

Varox® VC-40KE. [Vanderbilt] α,α′-Bis (t-butylperoxy)-diisopropylbenzene on KE clay; nonsulfur curing agent for NR and SR; crosslinking agent for PE, EPR, silicone EVA and VAE; wh. to cream powd.; 99.0% thru 40 mesh; dens. 1.50 ± 0.03 mg/m³; 39–41% assay; 3.6% min. act. oxygen.

Varstat® 10. [Sherex] Ethoxylated alcohol; non-nitrogen-containing antistat for flexible and rigid PVC; nearly colorless paste; lt.-colored liq. @ 50 C; visc. 90 cps.

Varstat® 55. [Sherex] Quat.; cationic; antistat used in treating fibers and textiles; hydrophilic.

Varstat® 66. [Sherex] Ethyl bis (poly hydroxyethyl) alkyl ammonium ethyl sulfate; cationic; antistat used in treating fibers and textiles; hydrophilic; Gardner 8 max. liq.; m.w. 1110; flash pt. > 200 F (PM); 68–76% conc.

Varstat® K-22. [Sherex] N,N-Bis (2 hydroxyethyl) alkyl amine; internal antistat for polyolefins; aids dispersion where pigments are added; APHA 100 max. clear liq.; pH 10–12; 100% act.

Varstat® T-22. [Sherex] N,N-Bis (2 hydroxyethyl) alkyl amine; internal antistat for polyolefins; aids dispersion where pigments are added; APHA 100 max. liq.-paste; pH 10–12; 100% act.

Varsulf® 5. [Sherex] Lanolin sulfosuccinate; surfactant for hand cleaners; refatting agent; paste; 50% conc.

Varsulf® S1333. [Sherex] Ricinoleic sulfosuccinate; refatting agent for liq. dish detergents and soaps; liq.; 40% solids.

Varsulf® SBDO-70. [Sherex] Dioctyl sulfosuccinate; anionic; wetting agent and solubilizer for textiles, emulsion polymerization, agric. chemicals; dispersant for paints, pigments; liq.; sol. in water, polar solvs.; 70% conc.

Varsulf® SBU-185. [Sherex] Disodium undecylenamido MEA sulfosuccinate; anionic; detergent, fungicide for specialty toiletries; Gardner 4 liq., powd.; pH 7.0; 50% solids.

Vazo 52. [DuPont] 2,2′-Azobis(2,4-dimethylvaleronitrile); initiator for bulk, sol'n., emulsion, and suspension polymerization of vinyl monomers; polymerization of unsat. polyesters and copolymerization of vinyl compds.; source of radicals for vinyl polymerizations and chain reactions; decomposes in solvs. to give free radicals with no evidence of induced chain decomposition; used in pigmented or dyed systems that may be susceptible to oxidative degradation; wh. cryst. solid; m.w. 248.37; sol. in functional compds. and aromatic hydrocarbons; insol. in water; dens. 400 kg/m³; 98% min. act.

Vazo 64. [DuPont] 2,2′-Azobis(isobutyronitrile); see Vazo 52; wh. cryst. solid; m.w. 164.21; sol. see Vazo 52; sp.gr. 1.128; dens. 400 kg/m³; 98% min. act.

Veegum®. [Vanderbilt] Complex colloidal magnesium aluminum silicate derived from natural smectite clays; thickener for emulsions, suspensions, and sol'ns.; visc. modifier for liq., creams, and pastes; imparts thixotropy to cosmetics, toothpaste, pharmaceuticals, paints, and textile finishes; stabilizer in emulsions; suspending agent for powd. and pigments; reduces stringiness and tackiness of org. gum sol'ns.; binder for inorg. powd. and pigments; disintegrating agent in tablets; disperses pigments; improves spreadability of lotions, creams, and ointments; for cosmetic, toiletry, and pharmaceutical formulations; in chemical specialty and industrial applic.; wh. sm. flakes; swells to many times original vol. in water to form colloidal disp; dens. 2.6 mg/m³; visc. 250 cps ± 25% (5% aq. disp.); pH 9 (5% aq disp.); < 8% moisture.

Veegum® F. [Vanderbilt] Magnesium aluminum silicate; see Veegum; microfine grade used in tablets for dry blending of fine powd.; prep. of ointments and pastes where dry incorporation is essential; wh. microfine powd.; sol. & pH see Veegum; visc. 250 cps ± 25% (5.5% disp.); < 8% moisture.

Veegum® HS. [Vanderbilt] Magnesium aluminum silicate; see Veegum; max. electrolyte stability; wh. flakes.; sol.& pH see Veegum; visc. 150 cps ± 50% (5.5% disp.); pH 9 (5% aq. disp.); < 8% moisture.

Veegum® HV. [Vanderbilt] Magnesium aluminum silicate; see Veegum; high visc. at low solids is desired; emulsification and suspension are obtained at low solids; cosmetics and pharmaceuticals; wh. flakes; sol. & pH see Veegum; visc. 250 cps ± 25% (4% disp.); < 8% moisture.

Veegum® K. [Vanderbilt] Magnesium aluminum silicate; see Veegum; for pharmaceutical acid suspensions; has low acid demand and high acid compatibility; wh. flakes; sol. & pH see Veegum; visc. 220 cps ± 25% (5.5% disp.); < 8% moisture.

Veegum®, Neutral. [Vanderbilt] See Veegum; neutral pH and very low acid demand; for pharmaceutical suspensions requiring nonalkaline Veegum; somewhat difficult to disperse in water and has a low visc.; wh. flakes; sol. see Veegum; visc. 100 cps ± 50% (7% disp.); pH 6.5–7.5 (5% aq. disp.); < 8% moisture.

Veegum® S-728. [Vanderbilt] See Veegum; good wh. color, electrolyte stability, and visc. stability on aging; wh. flakes; sol. & pH see Veegum; visc. 250

cps ± 25% (5.0% disp.); < 8% moisture.

Veegum® T. [Vanderbilt] Magnesium aluminum silicate; see Veegum; tech. grade for industrial use; indicated wherever an inorg., thixotropic material is desired to disperse pigments and retard settling, to impart body, or stabilize emulsions; large amounts are used in paint; near wh. flakes; sol. & pH see Veegum; dens. 2.8 mg/m³; visc. 200 cps ± 50% (4% disp.); < 8% moisture.

Veegum® WG. [Vanderbilt] See Veegum; low acid demand; a disintegrator in the dry compding. of tablets; in ointments and pastes where dry incorporation is essential; powd. (50 mesh); wh.; sol. & pH see Veegum; < 8% moisture.

Velsan D8P-3. [Sandoz] Isopropyl PPG-2 isodeceth-7-carboxylate; noncomedogenic emollient that helps to solubilize cosmetic actives; provides emulsion visc., wetting, and softer feel for skin care, sun care, hair care, bath, and pigmented prods.; lt. yel. oil; sol. @ 5% in SDA-40 ethanol, propylene glycol; disp. in dist. water, glycerin; sp.gr. 1.02; HLB 14.

Velsan P 8-3. [Sandoz] Isopropyl C_{12-15} pareth-9-carboxylate; emollient providing emulsion visc., wetting, and softer feel to skin care, sun care, bath, and pigmented prods.; lt. yel. oil; sol. @ 20% in dist. water, SDA-40 ethanol; sp.gr. 1.01; HLB 18.

Velsan P 8-16. [Sandoz] Cetyl C_{12-15} pareth-9-carboxylate; see Velsan P 8-3; yel. semisolid paste; disp. @ 20% in hot min. oil, glycerin; sp.gr. 0.95; HLB 9.

Velvamine HSB. [Lyndal] Alkyl imidazoline deriv.; cationic; softener in hair rinses, for leather, textiles, hair rinses; paste; 45% conc.

Velvetex® 710L. [Henkel] Lauraminopropionic acid; amphoteric; surfactant used in personal care prods.; corrosion inhibition for aerosol pkg.; amber clear liq.; sol. in water, acid and alkali systems and alcohol; sp.gr. 0.994; visc. 2500 cps; pH 5.5–6.5 (1%); 48.0–52.0% solids.

Velvetex® AB-45. [Henkel] Coco betaine; amphoteric; surfactant and conditioner used in cleansing emulsions and personal care prods.; visc. builder, gelling agent; Gardner 2 clear liq.; water-sol.; sp.gr. 1.03; visc. < 100 cps; cloud pt. < 0 C; pH 6.5–8.5 (10%); 43–45% solids.

Velvetex® BC. [Henkel] Coco betaine, amphoteric; see Velvetex AB-45; liq.; dens. 7.7 lb/gal; 59% conc.

Velvetex® BC-35. [Henkel] Cocamidopropyl betaine; amphoteric; surfactant, antistat, foamer, and conditioner for personal care prods.; lt. amber liq.; 35% solids.

Velvetex® BK-35. [Henkel] Cocamidopropyl betaine; surfactant for hair care prods.; visc. builder; yel. clear liq.; water-sol.; sp.gr. 1.039; visc. < 100 cps; pH 4.5–5.5; 34.0–36.0% solids.

Velvetex® OLB-50. [Henkel] Oleyl betaine; amphoteric; surfactant, conditioner used in personal care prods., mfg. of fibers and cutting oils; amber translucent gel; water-sol.; sp.gr. 0.953 (60 C); pH 6.0–8.0 (10%); 48.0–52.0% solids.

Verdoxan. [Henkel] 2,2,5,5-Tetramethyl-4-isopropyl-1,3-dioxane; fragrance raw material for grn. or fruity notes.

Versamag® B-16. [Morton Int'l.] Magnesium hydroxide, coated; see Versamag DC.

Versamag® DC. [Morton Int'l.] Magnesium hydroxide; inorg. filler providing flame retardancy and smoke suppression to elastomers, plastics, and thermosets incl. EPDM, PP, PE, PVC; used in wire and cable compds., conduit/tubing, film and sheet; micropellet; sp.gr. 2.36; bulk dens. 0.40 g/ml; ref. index 1.57; 96% assay.

Versamag® SB. [Morton Int'l.] Magnesium hydroxide, silane treated; see Versamag DC.

Versamag® Tech. [Morton Int'l.] Magnesium hydroxide; see Versamag DC; wh. powd.; 1–2 μ avg. particle size; 0.5% retained on 325 mesh; sp.gr. 2.36; bulk dens. 0.25 g/ml; ref. index 1.57; 96% assay.

Versamag® UF. [Morton Int'l.] Magnesium hydroxide; fire retardant/smoke suppressant filler for plastics and elastomers incl. flexible and rigid PVC; wh. ultra-fine powd.; 70% < 2 μ; 95% min. assay.

Versene®. [Dow] Tetrasodium EDTA; chelating agent for agric., cosmetics, metalworking, water treatment; sp.gr. 1.3; chel. value 100 mg $CaCO_3$/g; 39% act.

Versene® 100. [Dow] Tetrasodium EDTA; chelating agent used in metal ions; micronutrient in agriculture prods.; cosmetics and toiletries; metal working; pulp and paper processing; rubber and polymer chemistry; lt. straw liq.; m.w. 380.2; f.p. –20 C; b.p. 106 C; water-misc.; sp.gr. 1.31; dens. 10.9 lb/gal; chel. value 102 mg $CaCO_3$/g min.; pH 11.0–11.8 (1% aq.); 39% act.

Versene® 220 Crystals. [Dow] Tetrasodium EDTA; see Versene 100; wh. cryst. powd.; m.w. 452.2; sol. 48–50% in water; dens. 45 lb/ft³; chel. value 219 mg $CaCO_3$/g min. (@ pH 11); pH 10.5–11.5 (1%); 99.0% act.

Versene® Acid. [Dow] EDTA; see Versene 100; wh. powd.; m.w. 292.2; sol. < 0.1% in water; dens. 54 lb/ft³; chel. value 339 mg $CaCO_3$/g min. (@ pH 11); pH 2.5–3.0; 99% act.

Versene® AG 100. [Dow] Tetrasodium EDTA; chelating agent used in agric. prods.; lt. straw liq.; f.p. –31 C; b.p. 106 C; water-misc.; sp.gr. 1.31; 39% act.

Versene® CA. [Dow] Calcium disodium EDTA; chelating agent used in foods; wh. solid; sol. 100 g/100 g in water; dens. 43 lb/ft³; 99% act.

Versene® Diammonium EDTA. [Dow] Diammonium EDTA; chelating agent; lt. straw liq.; ammonia odor; f.p. –14 C; b.p. 104 C; water-misc.; sp.gr. 1.21; chel. value 137 mg $CaCO_3$/g min.; pH 4.6–5.2; 40% act.

Versene® NA. [Dow] Disodium EDTA; see Versene CA; wh. cryst. powd.; sol. 10 g/100 g in water; dens. 34 lb/ft³; pH 4.3–4.7 (1%); 99% act.

Versene® Tetraammonium EDTA. [Dow] Tetraammonium EDTA; see Versene Diammonium EDTA; liq.; sp.gr. 1.16–1.19; pH 9.0–9.5; 38.0% act.

Versenex 80. [Dow] Pentasodium pentetate; chelating agent; lt. straw liq.; m.w. 503.1; b.p. 106 C; f.p. < –20 C; water-misc.; sp.gr. 1.30; dens. 11.0 lb/gal; chel. value 80.0 mg $CaCO_3$/g min.; pH 11.0–11.8 (1% aq.); 40.2% act.

Versenol 120. [Dow] Trisodium HEDTA; chelating agent for control of iron; lt. straw liq.; m.w. 344.2; b.p. 107 C; f.p. < –20 C; water-misc.; sp.gr. 1.29; dens. 10.8 lb/gal; chel. value 120 mg $CaCO_3$/g; pH 11.0–11.8 (1% aq.); 41.3% act.

Versilan MX 112. [Henkel-Nopco] Alcohol ethoxylate, modified; nonionic/anionic; solv. used in degreasers; liq.

Verv®. [Am. Ingredients] Calcium stearoyl-2-lactylate; starch and protein complexing agent, softener for use in yeast-leavened bakery prods.; conditioning agent in dehydrated potatoes; cream powd.; mild caramel odor; slightly tart taste; acid no. 50–86; ester no. 125–164; 100% conc.

Victawet 12. [Akzo] EO reaction prod. with 2-ethyl hexanol and P_2O_5; nonionic; wetting agent, penetrant, dispersant, stabilizer; for pkg. dyeing of nylon, acid-type cleaners, emulsion polymerization, starch coatings; APHA 500 max. liq.; sol. in alcohol, acetone, toluene; disp. in water; dens. 9.34 lb/gal; sp.gr. 1.121; visc. 91.2 cs (100 F); b.p. 325 F; flash pt. 250 F (COC); pour pt. –48 F; surf. tens. 29.4 dynes/cm^2 (0.2%); pH 7.0–7.4 (0.5%); 100% act.

Victawet 35B. [Akzo] 2-Ethyl hexanol P_2O_5, NaOH reaction prod.; anionic; wetting agent, penetrant, dispersant, stabilizer; surfactant; lt. tan to wh. soft paste; sol. in water, ethylene glycol; dens. 9.7 lb/gal; sp.gr. 1.17; visc. > 100,000 cps; b.p. 200 F; flash pt. 200 F (COC); surf. tens. 26.2 dynes/cm^2; pH 6.9–7.4 (0.5%); 68–72% act.

Victawet 58B. [Akzo] Capryl alcohol, P_2O_5, NaOH reaction prod.; anionic; wetting agent, penetrant, dispersant, stabilizer, surfactant; lt. tan to wh. soft paste; sol. in water, ethylene glycol; dens. 9.77 lb/gal; sp.gr. 1.121; visc. > 100,000 cps; flash pt. 180 F (COC); surf. tens. 29.0 dynes/cm^2; pH 6.8–7.3 (5%); 68–72% act.

Victawet 85X. [Akzo] Buffered salt of Victawet 58B; anionic; stabilizer for vinyl resins, dispersant; powd.; 94% act.

Victory®. [Petrolite] Microcryst. wax; plastic wax offering high ductility, flexibility at very low temps., low surf. tens., solvency for amorphous hydrocarbon resins and syn. elastomers; provides protective barrier properties against moisture vapor and gases; used in hot-melt adhesives for papers, films, and foils; hot-melt coatings; in antisunchecking agents in rubber goods, elec. insulating agents, leather treating agents, water repellents for textiles, solv.-based rustproof coatings, cosmetic ingreds., and as plasticizer for other petrol. waxes used in crayons, dental compds., chewing gum base, and candles; color 1.5 (D1500); also avail. in wh. and br.; dens. 0.93 g/cc; visc. 12 cps (98.8 C); m.p. 79.4 C; flash pt. 293.3 C.

Vigilan. [Fanning] Lanolin oil; nonionic; emulsifier; skin and hair substantive emollient and moisturizer; solubilizer for immisc. fluids; o/w and/or w/o emulsions; adds elegance, lubricity, and hydration to makeup and facial preparations; spreads rapidly; Gardner 8 max. liq. (Regular), 6 max. (Superfine); acid no. 1.0 max.; sapon. no. 95–110; 100% conc.

Vigilan AWS. [Fanning] PEG-75 lanolin oil; o/w emulsifier; plasticizer, emollient, solubilizer, and wetting agent in personal care prods.; sol. in water and alcohol; disperses to form milky emulsion in water used in bath oils; Gardner 9–11; sol. in water and alcohol; acid no. 4.0 max.; sapon. no. 10–25; 1.5% max. moisture.

Vikol® AF-25. [Vikon] Bis (tri-n-butyltin) oxide; cationic; SE antimicrobial, algicide, bacteriostat, and fungistat for textiles receiving minimal outdoor exposure; misc. in water.

Vikol® LO-25. [Vikon] Bis (tri-n-butyltin) oxide, SE; durable bacteriostat and fungistat for cellulosic fibers; SE in water.

Vikol® PX-15. [Vikon] Bis (tri-n-butyltin) oxide, stabilized; antimicrobial for vinyl to control bacterial growth and mildew; sol. in vinyl plasticizers and in vinyl polymer.

Vikol® RQ. [Vikon] Benzalkonium chloride; synergistic bacterostat and fungistat for textiles in combination with Vikol LO-25; misc. with water.

Vikol® THP. [Vikon] Triclosan; wash and dry cleaning resistant antimicrobial treatment for textiles; SE liq.

Vikosperse KDS. [Vikon] Sulfonated naphthalene formaldehyde condensate; anionic; low dusting, low foaming dispersant for disperse dyes; 95% act.

Vinnapas® A 50. [Wacker-Chemie] VA homopolymer disp.; binder for fabrics, glass fibers; textile auxiliary and coating; paper coatings; disp.; 0.15 μm particle size; visc. 85 ± 30 mPa•s; pH 5; tens. str. 12 N/mm^2; tens. elong. 100% (break); 50 ± 1% solids.

Vinnapas® CEF 10. [Wacker-Chemie] VA/ethylene/vinyl chloride terpolymer disp.; binder for emulsion paints, textured finishes, and thermal insulation systems; disp.; 0.1 μm particle size; visc. 2700 ± 800 mPa•s; pH 6; tens. str. 4 N/mm^2; tens. elong. 550% (break); 50 ± 1% solids.

Vinnapas® CEZ 16. [Wacker-Chemie] VA/ethylene/vinyl chloride terpolymer disp.; binder for emulsion paints, textured finishes, and thermal insulation systems; adhesive for flooring, walls, foam, tiles; disp.; 0.7 μm particle size; visc. 12,000 ± 3000 mPa•s; pH 4.5; tens. str. 5 N/mm^2; tens. elong. 700% (break); 50 ± 1% solids.

Vinnapas® DZN 220. [Wacker-Chemie] VA homopolymer disp. with reactive groups; binder for fabrics, glass fiber; textile auxiliary and coating; disp.; 0.7 μm particle size; visc. 6500 ± 3000 mPa•s; pH 4; 50 ± 1% solids.

Vinnapas® EN 424. [Wacker-Chemie] VA/ethylene copolymer disp. with reactive groups; binder for fabrics, glass fiber; textile auxiliary and coating; disp.; 0.1–0.2 μm particle size; visc. 300 ± 200 mPa•s; pH 4–4.5; crosslinked film prop.: tens. str. 3 N/mm^2; tens. elong. 600% (break); 45 ± 1% solids.

Vinnapas® EN 426. [Wacker-Chemie] VA/ethylene copolymer disp. with reactive groups; see Vinnapas EN 424; disp.; 0.2–0.3 μm particle size; visc. 100 ± 50 mPa•s; pH 4–4.5; tens. str. 4.5 N/mm^2 (crosslinked film); tens. elong. 400% (break); 51 ± 1% solids.

Vinnapas® EN 427. [Wacker-Chemie] VA/ethylene copolymer disp. with reactive groups; see Vinnapas

EN 424; disp.; 0.2–0.3 μm particle size; visc. 100 ± 50 mPa•s; pH 4; tens. str. 4.5 N/mm² (crosslinked film); tens. elong. 400% (break); 51 ± 1% solids.

Vinnapas® EN 428. [Wacker-Chemie] VA/ethylene copolymer disp. with reactive groups; see Vinnapas EN 424; disp.; 0.2–0.3 μm particle size; visc. 200 ± 150 mPa•s; pH 4–4.5; tens. str. 4.5 N/mm² (crosslinked film); tens. elong. 400% (break); 52 ± 1% solids.

Vinnapas® EN 429. [Wacker-Chemie] VA/ethylene copolymer disp. with reactive groups; see Vinnapas EN 424; disp.; 0.2–0.3 μm particle size; visc. 100 ± 50 mPa•s; pH 4; crosslinked film prop.: tens. str. 4.5 N/mm²; tens. elong. 400% (break); 51 ± 1% solids.

Vinnapas® EP 16. [Wacker-Chemie] VA/ethylene copolymer disp.; binder for emulsion paints, textured finishes, and thermal insulation systems; disp.; 0.5–2 μm particle size; visc. 9000 ± 3000 mPa•s; pH 4; tens. str. 3 N/mm²; tens. elong. 800% (break); 50 ± 1% solids.

Vinnapas® EV 25 [Wacker-Chemie] VA/ethylene copolymer disp.; binder for emulsion paints, textured finishes, and thermal insulation systems; disp.; 0.5–1 μm particle size; visc. 11,000 ± 4000 mPa•s; pH 4; tens. str. 0.3 N/mm²; tens. elong. 3000% (break); 50 ± 1% solids.

Vinnapas® EZ 15. [Wacker-Chemie] VA/ethylene copolymer disp.; binder for emulsion paints, textured finishes, and thermal insulation systems; disp.; 1 μm particle size; visc. 2100 ± 800 mPa•s; pH 4.5; tens. str. 2 N/mm²; tens. elong. 1500% (break); 55 ± 1% solids.

Vinnapas® H 54/15 C. [Wacker-Chemie] VA homopolymer disp. with plasticizer; binder for fabrics, glass fibers; additive to hydraulic binders; adhesive for paper, pkg.; disp.; 1–3 μm particle size; visc. 4000 ± 1000 mPa•s; pH 4; tens. str. 8 N/mm²; tens. elong. 650% (break); 54 ± 1% solids.

Vinnapas® H 60. [Wacker-Chemie] VA homopolymer disp.; binder for fabrics, glass fibers; adhesive for wood, plastic, paper, pkg.; textile auxiliary and coating; disp.; 1–3 μm particle size; visc. 40,000 ± 10,000 mPa•s; pH 4; tens. str. 15 N/mm²; 60 ± 1% solids.

Vinnapas® LT 420. [Wacker-Chemie] VA/acrylate copolymer disp. with reactive groups; binder for fabrics, glass fiber; textile auxiliary and coating; disp.; 0.3 μm particle size; visc. 120 ± 80 mPa•s; pH 4.5; tens. str. 25–30 N/mm² (cross-linked film); 45 ± 1% solids.

Vinnapas® M 50/300. [Wacker-Chemie] VA homopolymer disp.; binder for fabrics, glass fibers; adhesive for paper, pkg.; textile auxiliary and coating; disp.; 0.5–2 μm particle size; visc. 1000 ± 300 mPa•s; pH 4; tens. str. 24 N/mm²; 50 ± 1% solids.

Vinnapas® SAF 54. [Wacker-Chemie] Styrene/acrylate disp.; binder for emulsion paints, textured finishes, and thermal insulation systems; adhesive for flooring, walls, foam, tiles; disp.; 0.1 μm particle size; visc. 9000 ± 3000 mPa•s; pH 7; tens. str. 4 N/mm²; tens. elong. 600% (break); 50 ± 1% solids.

Vinnapas® T 50/30 VL. [Wacker-Chemie] VA/vinyl laurate/vinyl chloride terpolymer disp.; binder for emulsion paints, textured finishes, and thermal insulation systems; disp.; 0.5–2 μm particle size; visc. 10,000 ± 3000 mPa•s; pH 4.5; tens. str. 1 N/mm²; tens. elong. 1100% (break); 50 ± 1% solids.

Vinnapas® Z 50. [Wacker-Chemie] VA homopolymer disp.; binder for fabrics, glass fibers; textile auxiliary and coating; paper coatings; sizing agent for glass fibers; disp.; 0.5–2 μm particle size; visc. 5500 ± 2000 mPa•s; pH 4; tens. str. 15 N/mm²; 50 ± 1% solids.

Vinnapas® Z 54/20 C. [Wacker-Chemie] VA homopolymer disp. with plasticizer; binder for fabrics, glass fibers, emulsion paints, textured finishes, and thermal insulation systems; disp.; 1–2 μm particle size; visc. 8500 ± 3000 mPa•s; pH 4; tens. str. 1.5 N/mm²; tens. elong. 1200% (break); 54 ± 1% solids.

Vinnol® 50. [Wacker-Chemie] Vinyl chloride/vinyl acetate copolymer disp.; binder for fabrics, glass fiber; textile auxiliary and coating; paper coatings; disp.; 0.15 μm particle size; visc. 80 ± 60 mPa•s; pH 5; 50 ± 1% solids.

Vinnol® 50/25 C. [Wacker-Chemie] Vinyl chloride/vinyl acetate copolymer disp. with 25% dibutyl phthalate plasticizer; binder for fabrics, glass fiber; textile auxiliary and coating; paper coatings; disp.; 0.15 μm particle size; visc. 100 ± 30 mPa•s; pH 5; tens. str. 4 N/mm²; tens. elong. 300% (break); 50 ± 1% solids.

Vinnol® CE 35. [Wacker-Chemie] Vinyl chloride/vinyl acetate/ethylene terpolymer disp.; binder for fabrics, glass fiber; textile auxiliary and coating; paper coatings; disp.; 0.15 μm particle size; visc. 50 ± 30 mPa•s; pH 6; 50 ± 1% solids.

Vinnol® K 704. [Wacker-Chemie] PVC/polyacrylate graft polymer; impact modifier for rigid PVC; produces flexible, unplasticized PVC components; spherical coarse particles; < 2000 μm particle size; sp.gr. 1.16; bulk dens. 0.64 g/cc; tens. str. 20 N/mm² (ultimate); tens. elong. 200% (break); hardness (Shore A) 95.

Vinsol® Emulsion. [Hercules] Ammonia disp. of Vinsol resin in water; deriv. of Vinsol Resin; extender for latices and liq. phenolic-resin systems; used in adhesives; < 1 μ particle size emulsion; dens. 8.9 lb/gal; visc. 500 cps; pH 8.5; 39.5–41.0% solids.

Vinsol® MM. [Hercules] Vinsol resin and fatty acid sodium soap; water repellent used in masonry cements; provides air entrainment, increases water-repellency of hardened mortar; dk. brn. free-flowing powd.; water-sol.; 96.0% min. solids.

Vinsol® NVX. [Hercules] Sodium soap of Vinsol; powd. form permitting elimination of the neutralization step; functions as an air-entraining agent in air-entraining and masonry cement and concrete, and as an asphalt emulsifier for anionic, slow-setting emulsions; lt. tan powd.; sol. in water; .

Vinsol® Soap Solutions. [Hercules] Aq. sol'ns. of the sodium and potassium soaps of Vinsol resin; used as portland cement air-entraining agents, and as emulsifiers for preparation of asphalt emulsions; 25 to 60% solids.

Vinylube® 36, 38. [Lonza] Ester wax; specialty plastic lubricant; cream bead; m.p. 59 and 60 C resp.; flash pt. 210 and 220 C resp.

Vinyzene® BP-5-2-MEK, BP-5-2-MS. [Morton Int'l.] Antimicrobial for solv. systems in adhesive/latex applics.

Vinyzene® BP-5-2-PG. [Morton Int'l.] Antimicrobial for bedding, carpet backing, and polyether urethane foams.

Vinyzene® BP-5-2-U. [Morton Int'l.] Antimicrobial for polyester urethane.

Vinyzene® BP-505 DIDP. [Morton Int'l.] 5% 10,10′-Oxybisphenoxarsine in diisodecyl phthalate; see Vinyzene BP-505; m.w. 502.2 (OBPA); sp.gr. 0.993; flash pt. (COC) 205 F.

Vinyzene® BP-505 DOP. [Morton Int'l.] 5% 10,10′-Oxybisphenoxarsine in di 2-ethylhexyl phthalate; see Vinyzene BP-505; m.w. 502.2 (OBPA); sp.gr. 1.007; flash pt. (COC) 215 F.

Vinyzene® BP-505 S160. [Morton Int'l.] 5% 10,10′-Oxybisphenoxarsine in butylbenzyl phthalate; see Vinyzene BP-505; m.w. 502.2 (OBPA); sp.gr. 1.013; flash pt. (COC) 215 F.

Vinyzene® BP-505. [Morton Int'l.] 5% 10,10′-Oxybisphenoxarsine in epoxidized soybean oil; antimicrobial, bacteriostat, fungistat for PVC, PU, other plastics, syn. rubber; recommended for film and sheet, extruded profiles, plastisols, molded goods, organosols, fabric coatings; m.w. 502.2 (OBPA); sp.gr. 1.049; flash pt. (COC) 205 F.

Vinyzene® SB-1 EAA, SB-1-PR. [Morton Int'l.] Antimicrobial for polyolefins.

Vinyzene® SB-1 ELV. [Morton Int'l.] Antimicrobial for thermoplastic rubber.

Vinyzene® SB-1-NY. [Morton Int'l.] Antimicrobial for nylon.

Vinyzene® SB-1-PS. [Morton Int'l.] 5% 10,10′-Oxybisphenoxarsine in a polymeric resin carrier; antimicrobial for hot melt adhesives, polyolefins, PU, PS, CPE; clear pellets; $^1/_{16} \times ^1/_{16}$ in. pellet size.

Vinyzene® SB-1. [Morton Int'l.] Antimicrobial for PVC.

Vinyzene® Sil 3. [Morton Int'l.] Antimicrobial for silicone caulking.

Vinyzene® T-129. [Morton Int'l.] Antimicrobial for high tech filtration papers.

Viplex 525. [Crowley] Highly aromatic petrol. fraction; plasticizer/extender for PU, primer coatings; provides visc. reduction permitting higher filler loadings, good wetting of pigments, fillers, fiberglass; aids leveling, penetration; sp.gr. 1.10 (60 F); visc. 123 SSU (100 F); i.b.p. 538 F; flash pt. (COC) 335 F; pour pt. 0 F.

Viplex 885. [Crowley] Highly aromatic (99%) petrol. oil; plasticizer/extender for epoxy systems, coatings, potting/encapsulation; promotes low system visc., better wetting of fillers and pigments, higher loading levels; modifier for polyesters; lt. yel. liq., may form yel. cryst. when cooled or aged, oily aromatic odor; sp.gr. 1.11 (60 F); dens. 9.25 lb/gal; visc. 147 SSU (100 F); i.b.p. 660 F; flash pt. (COC) 365 F; pour pt. 0 F; 97% aromatic content.

Viscasil®. [GE] Dimethicone; nonionic; defoamer, release agent, emollient in cosmetics, polishes, paint additives, and mechanical devices; lubricant in rubber or plastic-to-metal applics.; suggested for auto polish, skin care, hair care, textile softeners, antifoams for petrol. refining, rubber and plastic mold release, film modifier in coatings, damping in mechanical/elec. applics.; water-wh. clear, oily fluid; sp.gr. 0.975–0.980; visc. 5000, 10,000, 12,500, 30,000, 60,000, 100,000, 300,000, and 600,000 cstk grades; pour pt. –58 to –25 F; ref. index 1.4035; flash pt. 500 F; surf. tens. 21.1–21.3 dynes/cm; conduct. 0.090–0.092 Btu/h-°F ft²/ft; sp. heat 0.36 Btu/lb/°F; dissip. factor 0.0001; dielec. str. 35.0 kV; dielec. const. 2.75; vol. resist. 1×10^{14} ohm-cm; 100% act.

Viscon 103. [Akzo] Acrylic acid ester copolymer; for thickening aq. systems; suspending agent; disp. in water.

Vissurf S, 1400. [Kao] Polymeric nonionic surfactant; thickener for syn. resin emulsion; liq.

Vista LPA, LPA-140, LPA-210. [Vista] Mixts. of hydrotreated isoparaffins and naphthenes with very low levels of aromatics; solv. for food applics., pesticides, coatings, household prods., water, paper, and mining chemicals; used as textile lubricants, chemical process solvs., freeze pt. depressants, printing inks solvs.; Saybolt +30, +30, and +15 color resp., clear liqs., mild odor; sp.gr. 0.806, 0.803, 0.825 (60 F); visc. 2.10, 1.94, 3.82 cSt; i.b.p. 370, 370, 470 F; f.p. –75, –98, –46 F; flash pt.(PM) 150, 140, 215 F; pour pt. < –30 F (all).

Vista STXS. [Vista] Sodium toluene/xylene sulfonate; anionic; hydrotrope; solubilizer for heavy- and lt.-duty liq. formulations; liq.; 40% act.

Viton Curative No. 20. [DuPont] 33% organophosphonium salt and 67% fluoroelastomer; used in combination with Viton Curative No. 30 or No. 40 as a curing system for Viton fluoroelastomer; wh. to off-wh. pellets; sp.gr. 1.37.

Viton Curative No. 30. [DuPont] 50% dihydroxy aromatic compd. and 50% fluoroelastomer; used in combination with Viton Curative No. 20 as a curing system for Viton fluoroelastomers; tan to gray brn. pellets; sp.gr. 1.50.

Viton Curative No. 40. [DuPont] 33% benzophenone compd. and 67% fluoroelastomer; used in combination with Viton Curative No. 20 as a curing system for Viton fluoroelastomers; yel. pellets; sp.gr. 1.44.

Vocol®, S-62. [Monsanto] Zinc-O,O-di-n-butylphosphorodithioate; accelerator for EPDM cures; sec. accelerator for thiazoles and sulfenamides; 100% act. liq. and 62% act. powd.

Volan. [DuPont] Methacrylato chromic chloride; bonding agent and coupler for glass fiber-reinforced laminates; dk. grnsh.; alcoholic odor; b.p. 79 C; water-misc.; sp.gr. 1.02; pH 3.1 (1% aq.); flash pt. 6 (TOC); tens. str. 46,400 psi; flex. str. 81,200 psi; 19–21% act.

Volan L. [DuPont] Methacrylato chromic chloride; see Volan; dk. grn.; alcoholic odor; b.p. 77 C; water-misc.; sp.gr. 0.95; pH 3.1 (1% aq.); flash pt. 7 C (TOC); tens. str. 48,400 psi; flex str. 79,800 psi;

17–18% act.

Volatile Silicone 7158, 7207, 7349. [Union Carbide] Cyclic dimethylsiloxane; emollient, lubricant for skin creams and lotions, antiperspirants, bath oils, shaving prods., suntan lotions, colognes, hair care prods.; sol. in alcohols, aerosol propellants, fatty esters, min. oil; insol. in water.

Volaton®. [Bayer] Phoxim; low-mammalian-toxic, foliage- and soil-applied insecticide with a broad spectrum of activity; lt. yel. oily liq. (pure a.i.), reddish-brn. oily liq. (tech. a.i.); m.w. 298.3; sol. > 60 g/100 g in ligroin, IPA, toluene, methylene chloride; sp.gr. 1.176 (20/4 C); m.p. 5–6 C.

Volclay HPM-20. [Am. Colloid] Sodium bentonite, tech.; suspending agent, gellant, binder for household prods., automotive prods., aerosols, paints, enamels; micron-sized particles; 99% finer than 200 mesh (dry); pH 8.5–10.5 (5% susp.).

Volclay NF-BC. [Am. Colloid] Sodium bentonite; suspending agent, gellant, binder for pharmaceuticals, cosmetics, and personal care prods.; particulate; 99% finer than 200 mesh; pH 9.5–10.5 (2% disp.); 63.02% SiO_2, 21.08% Al_2O_3.

Volpo 3. [Croda] Oleth-3; nonionic; emollient, lubricant, emulsifier, solubilizer for cosmetic applic.; emulsifier in astringent creams and lotions; clear gel formation; superfatting in shampoos, foaming bath preparations, and Carbopol gels; used in cold waves, depilatories, and hair straighteners; solubilizer for bromo acids in lipsticks and liq. rouge; spreading agent for bath oils; hazy liq.; sol. in alcohols, glycols, ketones, and chlorinated and aromatic solvs., min. oil, and nonpolar oils; insol. in water; HLB 6.6; acid no. 2.0 max.; pH 5–7 (3% aq.).

Volpo 5. [Croda] Oleth-5; nonionic; see Volpo 3; hazy liq.; sol. see Volpo 3; HLB 8.8; acid no. 2.0 max.; pH 5–7 (3% aq.).

Volpo 10. [Croda] Oleth-10; nonionic; see Volpo 3; also effective in producing solubilizations and clear gels at low ratios of emulsifier to oil; wh. semisolid; sol. in alcohols, glycols, ketones, chlorinated and aromatic solvs., and water; HLB 12.4; acid no. 2.0 max.; pH 5–7 (3% aq.).

Volpo 20. [Croda] Oleth-20; nonionic; see Volpo 3; wh. solid; HLB 15.4; acid no. 2.0 max.; pH 5–7 (3% aq.).

Volpo 25D3. [Croda] PEG-3 C_{12-15} ether; nonionic; wetting, scouring, leveling agent, dispersant, solubilizer, emulsifier, antistat, plasticizer in personal care prods., pharmaceuticals, hard-surface cleaners, textile, agric., paper, and metal industries; water-wh. liq.; sol. in ethanol, min. oil, kerosene, trichloroethylene, oleic acid, oleyl alcohol; HLB 7.8; acid no. 1.0 max.; pH 6.0–7.5 (3%); 97% min. act.

Volpo 25D5. [Croda] PEG-5 C_{12-15} ether; nonionic; see Volpo 25D3; off-wh. opaque liq.; sol. in ethanol, trichloroethylene, oleic acid; HLB 10.3; acid no. 1.0 max.; surf. tens. 29.5 dynes/cm (0.1%); pH 6.0–7.5 (3%); 97% min. act.

Volpo 25D10. [Croda] PEG-10 C_{12-15} ether; see Volpo 25D3; off-wh. soft paste; sol. in water, ethanol, trichloroethylene; HLB 13.6; cloud pt. 83 C (1% aq.); acid no. 1.0 max.; surf. tens. 33.5 dynes/cm (0.1% aq.); pH 6.0–7.5 (3%).

Volpo 25D15. [Croda] PEG-15 C_{12-15} ether; see Volpo 25D3; wh. soft solid; sol. see Volpo 25D10; HLB 15.3; cloud pt. 73 C; acid no. 1.0 max.; surf. 36.5 dynes/cm (0.1% aq.); pH 6.0–7.5 (3%).

Volpo 25D20. [Croda] PEG-25 C_{12-15} ether; see Volpo 25D3; wh. soft solid; sol. see Volpo 25D10; HLB 16.1; cloud pt. 80 C; acid no. 1.0 max.; surf. tens. 38.5 dynes/cm (0.1% aq.); pH 6.0–7.5 (3%).

Volpo CS3. [Croda] Ceteareth-3; nonionic; see Volpo 25D3; off-wh. soft waxy solid; sol. in ethanol, kerosene, trichloroethylene, oleic acid, oleyl alcohol; HLB 7.3; acid no. 1.0 max.; pH 6.0–7.5 (3%).

Volpo CS5. [Croda] Ceteareth-5; nonionic; see Volpo 25D3; off-wh. soft waxy solid; sol. see Volpo 25D5; HLB 9.5; acid no. 1.0 max.; surf. tens. 33.0 dynes/cm (0.1% aq.); pH 6.0–7.5 (3%).

Volpo CS10. [Croda] Ceteareth-10; nonionic; see Volpo 25D3; off-wh. soft waxy solid; sol. see Volpo 25D5; HLB 12.9; cloud pt. 68 C (1% aq.); acid no. 1.0 max.; surf. tens. 34.5 dynes/cm (0.1% aq.); pH 6.0–7.5 (3%).

Volpo CS15. [Croda] Ceteareth-15; nonionic; see Volpo 25D3; off-wh. waxy solid; sol. see Volpo 25D5; HLB 14.6; cloud pt. 94 C (1% aq.); acid no. 1.0 max.; surf. tens. 35.5 dynes/cm (0.1% aq.); pH 6.0–7.5 (3%).

Volpo CS20. [Croda] Ceteareth-20; nonionic; see Volpo 25D3; off-wh. hard waxy solid; sol. in water, ethanol, oleic acid, trichloroethylene; HLB 15.6; cloud pt. 78 C; acid no. 1.0 max.; surf. tens. 41.5 dynes/cm (0.1% aq.); pH 6.0–7.5 (3%).

Volpo L3 Special. [Croda] Laureth-3; dispersant for use in bath oils; liq.

Volpo L4. [Croda] Laureth-4; nonionic; emulsifier and dispersant; liq.; 97% conc.

Volpo L23. [Croda] Laureth-23; nonionic; o/w emulsifier and solubilizer for cosmetics, pharmaceuticals and household prods.; solid; 97% conc.

Volpo N3. [Croda] Oleth-3, dist.; nonionic; solubilizer, emulsifier, dispersant, wetting, scouring and gelling agent, solubilizer; liq.; 97% conc.

Volpo N5. [Croda] Oleth-5, dist.; nonionic; see Volpo N3; liq.; 97% conc.

Volpo N10. [Croda] Oleth-10, dist.; nonionic; see Volpo N3; paste; 97% conc.

Volpo N15. [Croda] Oleth-15, dist.; nonionic; see Volpo N3; paste; 97% conc.

Volpo N20. [Croda] Oleth-20, dist.; nonionic; see Volpo N3; solid; 97% conc.

Volpo O3. [Croda] Oleth-3; nonionic; see Volpo 25D3; pale straw liq.; sol. see Volpo 25D3; HLB 6.7; acid no. 1.0 max.; pH 6.0–7.5 (3%).

Volpo O5. [Croda] Oleth-5; nonionic; see Volpo 25D3; pale straw liq.; sol. see Volpo 25D5; HLB 9.0; acid no. 1.0 max.; surf. tens. 31.0 dynes/cm (0.1% aq.); pH 6.0–7.5 (3%).

Volpo O10. [Croda] Oleth-10; nonionic; see Volpo 25D3; pale straw paste; sol. see Volpo 25D10; HLB 12.4; cloud pt. 57 C (1% aq.); acid no. 1.0 max.; surf. tens. 33 dynes/cm (0.1% aq.); pH 6.0–7.5 (3%).

Volpo O15. [Croda] Oleth-15; nonionic; see Volpo 25D3; pale straw paste; sol. see Volpo 25D10; HLB 14.2; cloud pt. 91 C (1% aq.); acid no. 1.0 max.; surf. tens. 34.5 dynes/cm (0.1% aq.); pH 6.0–7.5 (3%).

Volpo O20. [Croda] Oleth-20; nonionic; see Volpo 25D3; pale straw soft solid; sol. see Volpo 25D10; HLB 15.5; cloud pt. 78 C; acid no. 1.0 max.; surf. tens. 41.0 dynes/cm (0.1% aq.); pH 6.0–7.5 (3%).

Volpo S100. [Croda] Steareth-100; nonionic; solubilizer for creams and lotions; wetting agent in stick systems to improve antiperspirant efficacy; wh. waxy solid flakes; water-sol.; HLB 18.8.

Volpo T3. [Croda] Trideceth-3; nonionic; see Volpo 25D3; water-wh. liq.; sol. see Volpo 25D3; HLB 8.0; acid no. 1.0 max.; pH 6.0–7.5 (3%).

Volpo T5. [Croda] Trideceth-5; nonionic; see Volpo 25D3; water-wh. opaque liq.; sol. in ethanol, kerosene, trichloroethylene, oleic acid; HLB 10.4; acid no. 1.0 max.; surf. tens. 29.0 dynes/cm (0.1% aq.); pH 6.0–7.5 (3%).

Volpo T10. [Croda] Trideceth-10; nonionic; see Volpo 25D3; off-wh. soft paste; sol. see Volpo 25D10; HLB 13.7; cloud pt. 61 C (1% aq.); acid no. 1.0; surf. tens. 29.5 dynes/cm (0.1% aq.); pH 6.0–7.5 (3%).

Volpo T15. [Croda] Trideceth-15; nonionic; see Volpo 25D3; off-wh. soft paste; sol. see Volpo 25D10; HLB 15.5; cloud pt. 94 C (1% aq.); acid no. 1.0 max.; surf. tens. 32.0 dynes/cm (0.1% aq.); pH 6.0–7.5 (3%).

Volpo T20. [Croda] Trideceth-20; nonionic; see Volpo 25D3; off-wh. soft solid; sol. see Volpo 25D10; HLB 16.3; cloud pt. 75C; acid no. 1.0 max.; surf. tens. 36.5 dynes/cm (0.1% aq.); pH 6.0–7.5 (3%).

Voltalef 1 S. [Atochem] Chlorotrifluoroethylene light oil; lubricant, EP additive; used for compressor lubricants, hydraulic fluids, pump fluids, damping fluids, heat transfer fluids, etc.; Hazen 20 max. color; m.w. 500; sol. in aromatic and aliphatic hydrocarbons, chlorinated and fluorinated solvs., alcohols, ketones, esters; insol. in water, min. acids; sp.gr. 1.86; visc. 5 cSt; b.p. 141 C (760 mm Hg); pour pt. < –55 C; acid no. 0.02 max.; surf. tens. 23 dynes/cm; ref. index 1.400.

Voltalef 3 S. [Atochem] Chlorotrifluoroethylene med. oil; see Voltalef 1 S; Hazen 20 max. color; m.w. 650; sol. see Voltalef 1 S; sp.gr. 1.93; visc. 60 cSt; b.p. 256 C (760 mm Hg); pour pt. –45 C; acid no. 0.02 max.; surf. tens. 28 dynes/cm; ref. index. 1.405; dielec. str. 20 kV/mm.

Voltalef 3 X. [Atochem] Chlorotrifluoroethylene med. oil; wider fraction of dist. than 3 S grade; see Voltalef 1 S; Hazen 20 max. color; sol. see Voltalef 1 S; sp.gr. 1.93; visc. 60 cSt; b.p. 141 C (760 mm Hg); pour pt. –50 C; acid no. 0.02 max.; ref. index 1.405.

Voltalef 10 S. [Atochem] Chlorotrifluoroethylene heavy oil; see Voltalef 1 S; Hazen 20 max. color; m.w. 800; sol. see Voltalef 1 S; sp.gr. 1.96; visc. 800 cSt; b.p. 323 C (760 mm Hg); pour pt.0 C; acid no. 0.02 max.; surf. tens. 30 dynes/cm; ref. index 1.410.

Voltalef 90. [Atochem] Grease based on Voltalef oils thickened with inert gelling agent; lubricant; translucent wh.; drop pt. 300 F; bulk dens. 1.5.

Voltalef 901. [Atochem] Voltalef 90 grease plus PTFE lubricant.

V-Pyrol. [GAF] N-vinyl-2-pyrrolidone; stabilized with N,N′-di-sec-butyl-p-phenylenediamine; reaction rate accelerator; copolymerizes readily with most other vinyl monomers; permits control of hydrophilic chars. and modifies other properties in copolymer systems, incl. adhesives, coatings, cosmetics, textile and other finishes, syn. fibers, textile sizes, protective colloids, lube oil additives; low visc., reactive diluent in UV- and electron beam-curable systems; solvency power is used in aiding pigment disp. (inks) and/or introducing other additives into radiation-curable formulations; intermediate for modified phenolic resins of interest as plasticizers, dye intermediates, textile assistants; undergoes addition reactions with hydrogen, amides, phenol, and many phenolic derivs.; liq.; f.p. 13.5 C; b.p. 193 C (400 mm); flash pt. 98 C (OC).

Vueguard 901®. [Panelgraphic] Coating treatment for plastic sheet and molded articles providing resistance to scratching and abrasion; esp. for acrylic and PC substrates; avail. as water-clear (WC) and anti-glare (AG) grades.

Vueguard 911, 912. [Panelgraphic] Coating treatment for plastic sheet and molded articles; provides resistance to uv degradation and abrasion; 911 esp. for PC substrates, 912 esp. for acrylics; avail. as water-clear (WC) grade.

Vueguard 931 AF. [Panelgraphic] Coating treatment providing resistance to fogging for PC substrates.

Vueguard 941. [Panelgraphic] Antistatic coating for PC substrates.

Vulcabond® VP. [Akzo] 25% sol'n. of an isocyanurate-type polymer of TDI in dibutyl phthalate; bonding agent added to PVC plastisols to improve adhesion to a reinforcing substrate, e.g., nylon and polyester textiles; yel. tinged visc., slightly cloudy liq.; sp.gr. 1.05; visc. 20–30 poises; flash pt. (PMCC) 146 C; 3.5–3.8% NCO content.

Vul-Cup 40KE, R. [Hercules] Bisperoxide; 40KE grade supported on Burgess KE clay; efficient, scorch-resistant, low-odor vulcanizing agent and polymerization catalyst for elastomers and plastics; esp. suitable for transfer, inj., and rotational molding applics.; compds. can be extruded at higher rates because of higher extrusion temps. permissible.; wh. to off-wh. free-flowing powd. and tan to yel.-wh. semicryst. solid resp.; m.w. 338; sol. in aliphatic, aromatic, and ketone solvs.; water-insol.; sp.gr. 1.50 and 9.930 resp.; m.p. 45–55 C (R); 39.5–41.5% and 96–100% act. peroxide resp.

Vulcuren 2. [Bayer] 2-(4-Morpholinyldithio) benzothiazole; curing and vulcanizing agent; sulfur donor; heat resistant rubber applic.; yel. to beige powd.; dens. 1.48 g/cm³; m.p. 1.27.

Vulcuren 2/EGC. [Mobay] 2-(4-Morpholinyl-dithio) benzothiazole; sulfur donor or vulcanizing agent with a much delayed onset of cure and high efficiency in the vulcanization of natural and syn.

rubber goods that must withstand heat; used in aircraft and racing car tires, heat-resistant goods, technical goods, esp. seals and conveyor belting, and bulky goods; ylsh. to beige gran.; dens. 1.48 g/cm^3; m.p. 127 C.

Vulkacit 576. [Bayer] Butyraldehyde/aniline condensation prod.; semi-ultra-accelerator for elastomers (NR, SBR, BR, NBR) that are subject to heavy dynamic stresses and also for ebonite; used for tires, buffers, roll covers, mech. goods, goods containing lg. amts. of reclaim or scrap; rdsh. brn. liq.; sol. in benzene, aliphatic hydrocarbons, methylene chloride, CCl_4, ethanol, ethyl acetate, acetone; sp.gr. 0.99.

Vulkacit 1000/C. [Mobay] o-Tolyl-biguanide, min. oil coated; accelerator for NR, IR, BR, SBR, NBR, and acrylic rubbers; used in mech. goods, erasers, repair compds.; wh. to grayish powd.; m.w. 191; sol. in ethanol; sparingly sol. in acetone, ethyl acetate, methylene chloride, and water; sp.gr. 1.2; m.p. 140 C min.; 95% min. act.

Vulkacit CRV/LG. [Mobay] 3-Methyl-thiazolidinethione-2; accelerator for polychloroprene elastomers, halobutyl rubbers esp. chlorobutyl; used for mech. goods, cables, hoses, membranes, fabric proofings, vulcanizing sol'ns.; beige-brn. pellets; m.w. 133; sol. in methylene chloride, acetone, dimethyl formamide, methanol, toluene, ethyl acetate; sp.gr. 1.39; m.p. > 64 C; 98% min. act.

Vulkacit CT-N. [Bayer] Tricrotonylidene tetramine with dispersants; accelerator for ebonite mixes containing lg. amts. of scrap or reclaim; used mainly for press or steam cured technical goods; brn. visc. oil; dens. 1.05 g/cm^3.

Vulkacit CZ/EGC. [Mobay] Benzothiazyl-2-cyclohexyl sulfenamide; fast accelerator giving delayed onset of cure; used for tires, dynamically stressed technical goods, technical moldings and extrudates; coated gran.; dens. 1.30 g/cm^3; m.p. ≥ 96 C.

Vulkacit CZ/MGC. [Mobay] Benzothiazyl-2-cyclohexyl sulfenamide, min. oil coated; accelerator; fast and safe accelerator for NR, IR, BR, SBR, and NBR, giving a much delayed onset of cure; used mainly for tires; suitable for dynamically stressed technical goods, e.g., buffers and conveyor belting, and for technical moldings and extrudates in general, e.g. seals, hoses, and strip, footwear, cable sheathings, and insulations; lt. gray coated microgran.; m.w. 264.1; sol. in benzene, CCl_4, methylene chloride, ethyl acetate, acetone; sp.gr. 1.31; m.p. ≥ 96 C; 95% min. act.

Vulkacit D/C. [Mobay] Diphenyl guanidine; accelerator for NR, IR, BR, SBR, NBR, CR, with very slow onset of cure; used alone for bulky goods, e.g., tires, buffers, roll covers; sec. accelerator with mercaptos for mech. goods, extrudates and moldings, footwear, fabric proofings, cable sheathings and insulation; wh. to beige powd.; m.w. 211.1; sol. in acetone, ethanol, ethyl acetate, methylene chloride, benzene; sp.gr. 1.19; m.p. > 144 C; 96% min. act.

Vulkacit D/EGC. [Mobay] Diphenyl guanidine; accelerator for NR, IR, BR, SBR, NBR, CR, with slow onset of cure; used alone for bulky goods, e.g., tires, buffers, roll covers; sec. accelerator with mercaptos for mech. goods, extrudates and moldings, footwear, fabric proofings, cable sheathings and insulation; wh. to beige gran.; m.w. 211.1; sol. see Vulkacit D/C; sp.gr. 1.19; m.p. ≥ 144 C; 96% min. act.

Vulkacit DM/C. [Mobay] Dibenzothiazyl disulfide, min. oil coated; delayed action, semi-ultra accelerator for NR, IR, BR, SBR, NBR, CR, IIR, chlorosulfonated polyethylene; used for mech. goods, tires, conveyor belts, cables, hoses, rubber footwear, expanded rubber goods, proofed fabrics, intricate molded goods; ylsh. coated powd.; m.w. 332.2; slightly sol. in benzene, CCl_4, methylene chloride, ethanol, acetone; sp.gr. 1.51; m.p. ≥ 169 C; 94% min. act.

Vulkacit DM/MGC. [Mobay] Dibenzothiazyl disulfide, min. oil coated; see Vulkacit DM/C; ylsh. microgran.; m.w. 332.2; sol. see Vulkacit DM/C; sp.gr. 1.34; m.p. > 169 C; 94% min. act.

Vulkacit DM/EGC. [Mobay] Dibenzothiazyl disulfide; see Vulkacit DM/C; coated gran.; dens. 1.51 g/cm^3; m.p. ≥ 169 C.

Vulkacit DOTG/C. [Mobay] Di-o-tolyl guanidine; see Vulkacit D/C; wh. to beige powd.; m.w. 239.1; sol. in acetone, ethanol, ethyl acetate, methylene chloride; sp.gr. 1.18; m.p. > 171 C; 96% min. act.

Vulkacit DOTG/EGC. [Mobay] Di-o-tolyl guanidine; see Vulkacit D/EGC; wh. to beige gran.; m.w. 239.1; sol. see Vulkacit DOTG/C; sp.gr. 1.18; m.p. ≥ 171 C; 96% min. act.

Vulkacit DZ/C. [Mobay] Benzothiazyl-2-dicyclohexyl sulfenamide; see Vulkacit CZ/MGC; coated ground powd.; dens. 1.20 g/cm^3; m.p. ≥ 96 C.

Vulkacit DZ/EGC. [Mobay] Benzothiazyl-2-dicyclohexyl sulfenamide; accelerator with delayed onset of cure for dynamically stressed rubber, e.g., NR, IR, BR, SBR, NBR; used for large tires, conveyor belts, shock absorbers, intricate molded goods with long flow periods; cream to lt. brn. extrusion gran.; m.w. 346.1; sol. in benzene, CCl_4, methylene chloride, ethyl acetate, acetone, ethanol, aliphatic hydrocarbons; sp.gr. 1.20; m.p. ≥ 96 C; 95% min. act.

Vulkacit H 30. [Bayer] Hexamethylene tetramine (97%) with additions to prevent lump formation (3%); slow accelerator for bulky goods; sec. accelerator used with prods. of the mercapto class; mainly for press or steam cured technical goods and ebonite; wh. fine powd.; dens. 1.3 g/cm^3.

Vulkacit HX. [Bayer] Cyclohexyl ethyl amine; sec. accelerator for dipped goods, fabric proofings, and self-curing sol'ns.; colorless to slightly ylsh. liq.; dens. 0.85 g/cm^3; b.p. 165 C.

Vulkacit J. [Mobay] Dimethyl diphenyl thiuram disulfide; delayed action curing agent and accelerator for NR, IR, BR, SBR, and NBR; used for molded goods, conveyor blets, hoses, dipped goods, seals, proofed fabrics; wh. cryst. powd.; m.w. 364.2; sol. in benzene, methylene chloride; sp.gr. 1.33; m.p. ≥ 175 C.

Vulkacit L. [Mobay] Zinc-N-dimethyl dithiocarbamate; very fast accelerator used alone or in conjunction with other accelerators; used for technical, food contact, latex, and dipped goods, footwear and cables, fabric proofings, self-curing mixes and sol'ns., sheets; wh. to ylsh. wh. powd.; dens. 1.7 g/cm³; m.p. 250 C.

Vulkacit LDA. [Mobay] Zinc-N-diethyl dithiocarbamate; see Vulkacit L; wh. to ylsh. wh. powd.; dens. 1.49 g/cm³; m.p. ≥ 175 C.

Vulkacit LDB/C. [Mobay] Zinc-N-dibutyl dithiocarbamate; see Vulkacit L; wh. to ylsh. wh. coated powd.; dens. 1.26 g/cm³; m.p. ≥ 104 C.

Vulkacit LZ/LG. [Mobay] Benzothiazyl-2-diisopropyl sulfenamides; see Vulkacit CZ/EGC; lentil-shaped gran.; dens. 1.20 g/cm³; m.p. ≥ 50 C.

Vulkacit Merkapto/C. [Mobay] 2-Mercaptobenzothiazole, min. oil coated; semi-ultra accelerator for NR, IR, BR, SBR, NBR, IIR, EPDM; used alone in bulky goods or in combination for molded and extruded goods, hoses, conveyor belts, tires, footwear, cables, expanded rubber goods, proofed fabrics, latex articles; ylsh. coated powd.; m.w. 167.1; sol. in ethyl acetate, acetone, methylene chloride, ethanol; sp.gr. 1.52; m.p. ≥ 172 C; 93% min. act.

Vulkacit Merkapto/EGC. [Mobay] 2-Mercaptobenzothiazole; see Vulkacit DM/C; coated gran.; dens. 1.52 g/cm³; m.p. ≥ 172 C.

Vulkacit Merkapto/MGC. [Mobay] Zinc salt of 2-mercaptobenzothiazole, min. oil coated; see Vulkacit Merkapto/C; ylsh. coated microgran.; m.w. 167.1; sol. see Vulkacit Merkapto/C; sp.gr. 1.52; m.p. > 172 C; 93% min. act.

Vulkacit MOZ-90/SG. [Mobay] 90% Benzothiazyl-2-sulfene morpholide, 10% Vulkacit DM; see Vulkacit CZ/EGC; cylindrical gran.; dens. 1.40 g/cm³; m.p. 75–130 C.

Vulkacit MOZ/LG. [Mobay] Benzothiazyl-2-sulfene morpholide; accelerator with pronounced delayed onset of cure; used for dynamically stressed rubber goods, NR, IR, BR, SBR, and NBR; used for tires, conveyor belts, shock absorbers, mountings, seals, hoses, cables, footwear, intricate molded goods; cream to lt. brn. lentil-shaped gran.; m.w. 252.1; sol. in acetone, methylene chloride, CCl_4, benzene, ethyl acetate, methanol; sp.gr. 1.38; m.p. ≥ 75 C; 95% min. act.

Vulkacit MOZ/SG. [Mobay] Benzothiazyl-2-sulfene morpholide; see Vulkacit MOZ/LG; cream to lt. brn. cylindrical gran.; m.w. 252.1; sol. see Vulkacit MOZ/LG; sp.gr. 1.38; m.p. ≥ 75 C; 95% min. act.

Vulkacit NPV/C2. [Bayer] 2-Mercaptoimidazoline (ethylene thiourea); special accelerator for CR (solid rubber and latex); used in conjunction with other accelerators for all other diene rubbers; used mainly for technical goods, insulations, and sheathings, proofings, rubber sol'ns.; wh. coated powd.; dens. 1.42 g/cm³; m.p. ≥ 195 C.

Vulkacit NZ. [Mobay] Benzothiazyl-2-t-butyl sulfenamide; see Vulkacit CZ/MGC; gray to beige small gran.; dens. 1.29 g/cm³; m.p. ≥ 103 C.

Vulkacit NZ/EGC. [Mobay] Benzothiazyl-2-t-butyl sulfenamide, oil treated; accelerator with delayed onset of cure for dynamically stressed rubber goods of NR, IR, BR, SBR, NBR; used for tires, conveyor belts, shock absorbers, mountings, seals, hoses, sleeves, cables, footwear; grayish beige cylindrical gran.; m.w. 238; sol. in aliphatic hydrocarbons, benzene, methylene chloride, CCl_4, ethyl acetate, acetone, ethanol; sp.gr. 1.29; m.p. ≥ 105 C; 95% min. act.

Vulkacit P. [Mobay] Pentamethylene-ammonium-N-pentamethylene dithiocarbamate; very fast accelerator; used for technical goods, footwear and cables, dipped goods, fabric proofings, latex goods, self-curing mixes and solutions, and sheets; ylsh. to wh. cryst. powd.; dens. 1.15 g/cm³; m.p. ≥ 168C.

Vulkacit P Extra N. [Mobay] Zinc-N-ethylnophenyl dithiocarbamate; very fast accelerator; used for technical goods, footwear and cables, dipped goods, fabric proofings, latex goods, self-curing mixes and solutions, and sheets; wh. to ylsh. wh. powd.; dens. 1.50 g/cm³; m.p. ≥ 20 C.

Vulkacit Thiuram, Thiuram/C. [Mobay] Tetramethyl thiuram disulfide; thiuram accelerator; fast accelerator used alone or in conjunction with other accelerators, esp. for efficient or sulfurless vulcanization; used in transparent, wh., and colored rubber goods, esp. technical goods of all types that need to withstand heat; suitable for cable sheathings and insulations, food-contacting, surgical and dipped goods, bathing articles, fabric proofings, ebonite; ylsh. wh. to grysh. powd.; dens. 1.40 g/cm³; m.p. ≥ 140 C.

Vulkacit Thiuram/EGC. [Mobay] Tetramethyl thiuram disulfide; see Vulkacit J; ylsh. to graysh. wh. coated gran.; dens. 1.40 g/cm³; m.p. ≥ 140 C.

Vulkacit Thiuram MS/C. [Mobay] Tetramethyl thiuram monosulfide; see Vulkacit Thiuram; pale yel. coated powd.; dens. 1.39 g/cm³; m.p. ≥ 104 C.

Vulkacit Thiuram MS/EGC. [Mobay] Tetramethyl thiuram monosulfide;nondiscoloring ultra-acclerator for NR, IR, BR, SBR, NBR, CR, EPDM; used alone or in combination for mech. goods, hoses, cables, seals, tires, conveyor belts, proofed textiles, latex goods, ebonite; pale yel. extrusion gran.; m.w. 208.1; sol. in methylene chloride, benzene, acetone, ethyl acetate; sp.gr. 1.39; m.p. ≥ 105 C; 96% min. act.

Vulkacit TR. [Bayer] Polyethylene polyamines; accelerator used in mixes containing acidic or retarding ingreds. or lg. amts. of reclaim or scrap; used for goods that are cured in presses, steam, or hot air and are subjected to dynamic stresses; ylsh. brn. to rdsh. brn. liq.; dens. 0.99 g/cm³.

Vulkacit ZM. [Mobay] Zinc salt of 2-mercaptobenzothiazole; semi-ultra accelerator for NR, IR, BR, SBR, NBR; used for mech. goods, footwear, cables; sec. accelerator for latex goods; sensitizing agent for natural and syn. latex mixes, esp. in the prod. of foam rubber; ylsh. powd.; m.w. 397.7; slightly sol. in benzene, methylene chloride, CCl_4, acetone, ethanol; sp.gr. 1.70; m.p. > 310 C; 17% zinc.

Vulkacit ZP. [Mobay] Zinc-N-pentamethylene dithiocarbamate; see Vulkacit L; wh. to ylsh. wh. powd.; dens. 1.60 g/cm³; m.p. > 220.

Vulkadur A. [Mobay] Phenol-formaldehyde resin plus hardener; reinforcing and hardening agent used mainly for goods based on NBR, e.g., in the mfg. of brake and clutch linings; also used as an ingred. of adhesives; wh. to yel. powd.; dens. 1.3 g/cm³; soften. pt. 75–90 C.

Vulkadur T, 40%. [Bayer] Precondensed resorcinol formaldehyde resin; bonding agent for latex dips for textiles, tire cord, and rubber goods; rdsh. brn. liq.; dens. 1.17 g/cm³.

Vulkalent B/C. [Mobay] Phthalic anhydride; retarder; increasing additions improve safety of mixes undergoing processing and prolong the flow times at curing temps.; not suitable for mixes containing Vulkacit Thiuram and without sulfur; wh. cryst. powd.; dens. 1.51 g/cm³; m.p. 124 C.

Vulkalent E/C. [Mobay] Aromatic sulfonamide deriv. plus 5% calcium carbonate and 1–2% min. oil; vulcanization retarder to control the processing safety and flow time of sulfur-curing natural and syn. rubbers; useful for replasticization of slightly scorched compds.; beige-wh. fine powd.; sol. in ethyl acetate, benzene, methylene chloride, CCl_4; sp.gr. 1.68; m.p. 110 C; 90% min. act.

Vulkalent TM. [Mobay] Blend of 4- and 5- methylbenzotriazole; prevulcanization retarder for sulfur-modified CR and halobutyl rubbers; also effective in NBR when Vulkacit DM/sulfur curing systems are used; applics. incl. conveyor belting, hose, molded goods; lt. beige gran.; m.w. 133.2; sol. in acetone, ethyl acetate, ethanol, MEK, methylene chloride; sp.gr. 1.2; m.p. 83 C; 98% min. act.

Vulkanol 80. [Bayer] Polyglycol ester of a mixt. of monocarboxylic acids; plasticizer to improve low-temp. flexibility and elastic behavior of vulcanizates; bright yel. clear liq.; dens. 0.97–0.99 g/cc.

Vulkanol 81. [Mobay] Mixt. of a thioester and a carboxylic ester; syn. plasticizer used in tech. goods based on natural and syn. rubber, e.g., hoses and seals for vehicles, aircraft and machinery, moldings and extrudates, conveyor and transmission belting, cable sheathings, proofings; yel. clear liq.; b.p. cannot be distilled without decomp.; dens. 0.96–0.99 g/cm³.

Vulkanol 85. [Mobay] Ether thioether; plasticizer used to impart antistat properties to flexible PVC; low-temp. strength in plasticized PVC; applic. incl. conveyor belts, mine air ducts, floorcoverings, shoe soles, protective work clothes, films; ylsh. liq.; unpleasant odor; m.w. 650; sol. in org. solvs.; insol. in water; sp.gr. 1.03–1.06; visc. 40–80 MPa.s; ref. index 1.472–1.473; flash pt. 190 C (OC).

Vulkanol 88. [Bayer] Dibutyl methylene bisthioglycolate; plasticizer to improve low-temp. flexibility and elastic behavior of vulcanizates; used for nitrile and chloroprene rubber; yel. to brn. liq.; b.p. 180–227 C; dens. 1.07–1.10 g/cc.

Vulkanol 90. [Mobay] Di-2-ethylhexyl ester thiodiglycolate; plasticizer for natural and syn. rubber incl. NBR, NBR/PVC blends, CR, SBR, BR, chlorosulfonated polyethylene, and chlorinated polyethylene; used for hoses, seals, shock absorbers, roll covers, conveyor belting; lt. brn. clear liq.; sol. in aliphatic hydrocarbons, toluene, ethanol, acetone; sp.gr. 0.97; visc. 18 cps; b.p. cannot be distilled without decomp.; solid. pt. –60 C; flash pt. (OC) 200 C.

Vulkanol BA. [Bayer] Dibenzyl ether; see Vulkanol 80; colorless liq.; b.p. 174–185 C; dens. 1.04 g/cc.

Vulkanol FH. [Bayer] Xylene/formaldehyde condensation prod.; resinous plasticizer and tackifier for rubber mixes based on NBR, CR, SBR; gives high bldg. tack; pale yel. highly visc. liq.; misc. with aromatic and chlorinated hydrocarbons; partly sol. in ethers, esters, and aliphatic hydrocarbons; sp.gr. 1.08; visc. 12,000 cps (50 C); flash pt. (OC) 185 C; pour pt. 20 C.

Vulkanol KA. [Mobay] Polyglycol ether; syn. plasticizer used in tech. goods based on natural and syn. rubber, e.g., hoses and seals for vehicles, aircraft and machinery, moldings and extrudates, conveyor and transmission belting, cable sheathings, proofings; also offers antistatic properties; brn. visc. liq.; misc. with org. solvs. other than aliphatic hydrocarbons and with water; sp.gr. 1.12; visc. 675 cps; b.p. cannot be distilled without decomp.; flash pt. (OC) 205 C; pour pt. 1 C.

Vulkanol OT. [Bayer] Ether thioether; plasticizer to improve low-temp. flexibility and elastic behavior of vulcanizates; used for nitrile and chloroprene rubber; applics. incl. hoses, seals, mech. goods; pale yel. liq.; misc. with most org. solvs.; sp.gr. 0.97; visc. 40 cps; flash pt. (OC) 200 C; pour pt. –60 C.

Vulkanol SF. [Bayer] Alkyl sulfonic acid alkyl phenyl ester; see Vulkanol 80; ylsh. transparent liq.; b.p. > 350 C; dens. 0.995–1.06 g/cc.

Vulkanox 3100. [Mobay] Blend of diaryl p-phenylene diamine; staining antioxidant and antiflex-cracking agent for most diene rubbers (NR, IR, SBR, NBR); moderate antiozonant properties; used for tires, buffers, engine mounts, conveyor belts, mech. goods; brn. flakes; sol. in acetone, ethyl acetate, toluene, methylene chloride, ethanol, benzene, CCl_4; sp.gr. 1.20; m.p. 90–105 C.

Vulkanox 4010 NA. [Mobay] N-isopropyl-N′-phenyl-p-phenylene diamine; staining and discoloring antioxidant/antiozonant for protection of rubber from ozone attack, oxidation, heat aging, flexcracking, and rubber poisons; suitable for natural and syn. rubbers; used for dynamically stressed goods, tech. goods, spring components, conveyor and transmission belting, seals, insulation; lt. gray to dk. brn. flakes; m.w. 226; sol. in benzene, methylene chloride, CCl_4, ethyl acetate, acetone, ethanol; sp.gr. 1.07; m.p. > 75 C; 95% min. act.

Vulkanox 4020. [Mobay] N-(1,3-dimethylbutyl)-N′-phenyl-p-phenylene diamine; staining stabilizer for cold- and hot-polymerized emulsion and sol'n. SBR and BR; protects vulcanizates against heat, oxidation, and flexcracking; brn. to brn. violet flakes; m.w. 268; sol. in benzene, methylene chloride, CCl_4, ethyl acetate, acetone, ethanol, aliphatic hydrocarbons; sp.gr. 1.1; m.p. 45 C min.; 97% min. act.

Vulkanox 4022. [Mobay] Blend of alkylaryl-p-phenylenediamines; staining and discoloring antioxidant offering protection to natural and syn. rubbers from dynamic stresses, oxidation, heat, ozone, and rubber poisons; used in tires, tech. goods, spring components, conveyer belting, hoses, transmission belting and seals, cable sheathings and insulations, inner tubes, roll covers; also offers antiflexcracking effects; dk. purple liq.; sol. in benzene, methylene chloride, aliphatic hydrocarbons, acetone; sp.gr. 1.01; visc. 80 SUS (75 C).

Vulkanox 4023. [Mobay] Blend of alkylaryl-p-phenylenediamines and dialkyl-p-phenylenediamines; staining and discoloring antiozonant and antioxidant protecting natural and syn. rubbers against ozone attack, oxidation, heat aging, flexcracking, and rubber poisons; used in dynamically stressed goods, tires, conveyor belts, spring components; dk. purple liq.; sol. in benzene, methylene chloride, aliphatic hydrocarbons, acetone; sp.gr. 1.00; visc. 70 SUS (75 C).

Vulkanox 4025. [Mobay] Blend of alkylaryl-p-phenylenediamines and dialkyl-p-phenylenediamines; see Vulkanox 4022; also offers antiflexcracking effects; dk. purple liq.; dens. 0.97 g/cm³.

Vulkanox 4027. [Mobay] Blend of alkylaryl p-phenylene diamines; staining and discoloring antiozonant/antioxidant which protects natural and syn. rubber against ozone attack, oxidation, heat aging, flexcracking, and rubber poisons; used in dynamically stressed goods, tires, conveyor belts, spring components; dk. purple liq.; sol. in benzene, methylene chloride, aliphatic hydrocarbons, acetone; sp.gr. 0.96; visc. 80 SUS (75 C).

Vulkanox 4030. [Mobay] N,N′-di-(1,4-dimethylpentyl)-p-phenylene diamine; staining and discoloring antiozonant/antioxidant protecting natural rubber and syn. rubbers except CR from ozone attack, oxidation, heat aging, and flexcracking; used for dynamically stressed goods, tires, conveyor belts, hoses, spring components, cables and seals; dk. red low-visc. liq.; sol. in aliphatic hydrocarbons, benzene, methylene chloride, CCl_4, ethyl acetate, acetone, ethanol; sp.gr. 0.91; b.p. 237 C (14 Torr); 93% min. act.

Vulkanox AFD. [Mobay] Unsat. aromatic ether; nonstaining antiozonant for NR, SBR, BR, NBR/PVC, and CR; liq.; dens. 1.02 g/cm³.

Vulkanox AFS-50. [Mobay] Unsat. acetal absorbed on inorganic filler; see Vulkanox AFD; gray to beige powd.; dens. 1.34 g/cm³; m.p. 90 C (of act.); 50 ± 2% act.

Vulkanox AFS/LG. [Mobay] Unsat. acetal; see Vulkanox AFD; gray to beige lentil-shaped gran.; dens. 1.06 g/cm³; m.p. 90 C.

Vulkanox BKF. [Mobay] 2,2′-Methylene-bis(4-methyl-6-t-butylphenol); nonstaining antioxidant for natural and syn. rubbers; used in transparent, bathing, surgical tech., and latex goods, fabric proofings; stabilizer for hot- and cold-polymerized emulsion-SBR, BR, and NBR, and sol'n.-BR, IR, and ABS; wh. cryst. powd.; m.w. 340; sol. in ethanol, acetone, ethyl acetate, methylene chloride, CCl_4, benzene; sp.gr. 1.04; m.p. > 125 C.

Vulkanox CS. [Mobay] Sterically hindered bisphenol absorbed on clay; nonstaining antioxidant imparting resistance to oxygen and heat; wh. to beige powd.; dens. 1.91 g/cm³; m.p. ≥ 150 C (a.i.); 25 ± 2% bisphenol.

Vulkanox DDA. [Mobay] Styrenated diphenyl amine; staining and discoloring antioxidant for natural and syn. rubbers, exp. CR, giving protection against oxidation, heat, flexcracking, and rubber poisons; used for inner tubes, sponge rubber, seals/gaskets, soles, latex goods, heat-resistant articles; stabilizer for cold and hot-polymerized NBR and NBR or SBR latexes; orange to brn. visc. liq.; sol. in aliphatic hydrocarbons, benzene, CCl_4, methylene chloride, acetone, ethyl acetate; sp.gr. 1.08; visc. 15,000 cps; b.p. > 300 C.

Vulkanox DS. [Mobay] Alkyl and aralkyl-substituted phenols; antioxidant for latex, transparent, dipped, bathing, and tech. goods, fabric proofings; yel. to rdsh. clear visc. liq.; b.p. 140 C; dens. 0.92 ± 0.05 g/cm³.

Vulkanox DS/F. [Mobay] Alkyl and aralkyl-substituted phenols; see Vulkanox DS; wh. to beige-yel. powd.; b.p. > 140 C; dens. 1.30 ± 0.05 g/cm³.

Vulkanox HS/Pellets. [Mobay] 2,2,4-Trimethyl-1,2-dihydroquinoline, polymerized; weakly staining and discoloring antioxidant for NR, IR, BR, SBR, NBR%, EPDM; used for heat-resistant goods, tire, conveyor belts, hoses, seals, footwear, cables; yel. to tan pellets; m.w. $(172)_n$; sol. in acetone, ethyl acetate, ethanol, benzene, methylene chloride, CCl_4; sp.gr. 1.07; soften. pt. 75 C min.; 95% min. act.

Vulkanox HS/Powd. [Mobay] 2,2,4-Trimethyl-1,2-dihydroquinoline, polymerized; see Vulkanox HS/Pellets; yel. to tan powd.; m.w. $(172)_n$; sol. see Vulkanox HS/Pellets; sp.gr. 1.07; soften. pt. > 75 C.

Vulkanox KB. [Mobay] 2,6-Di-t-butyl-p-cresol; nondiscoloring/nonstaining antioxidant for lt. colored and transparent natural and syn. rubber goods; used in fabric proofings, toys, bathing, latex, and dipped goods; stabilizer in emulsion and sol'n.-polymerized elastomers (SBR, NBR, BR, IR); costabilizer for ABS; colorless cryst.; m.w. 220; sol. in aliphatic hydrocarbons, benzene, CCl_4, methylene chloride, acetone, ethyl acetate, ethanol; sp.gr. 1.03; solid. pt. 69.2 C; 99.8% min. act.

Vulkanox MB-2/MGC. [Mobay] 4- and 5-Methylmercaptobenzimidazole, min. oil coated; nondiscoloring/nonstainaing antioxidant for natural and syn. rubbers; synergist; protects against rubber poisons; sensitizing agent for latex; yel.-wh. microgran.; m.w. 164; sol. in dimethyl formamide, acetone, ethanol; sp.gr. 1.31; 96% min. act., 2% oil.

Vulkanox NKF. [Mobay] 2,2′-Isobutylidene-bis (4,6-dimethyl phenol); nonstaining/nondiscoloring antioxidant for natural and syn. elastomers; used for proofed and dipped goods, foam rubber, hoses, footwear, floor coverings, latex goods, carpet backing, other tech. goods; stabilizer for hot- and cold-polymerized emulsion-SBR, BR, and NBR, and sol'n.-BR, SBR, and IR; also effective in ABS; wh. cryst. powd.; m.w. 298; sol. in acetone, ethyl acetate, ethanol, methylene chloride, CCl_4; sp.gr. 1.12; m.p. > 156 C.

Vulkanox OCD. [Mobay] Octylated diphenyl amine;

see Vulkanox HS/LG; pale gray brn. gran.; dens. 1.0 ± 0.05 g/cm³; m.p. > 88 C.

Vulkanox OCD/SG. [Mobay] Octylated diphenyl amine; weakly staining and discoloring antioxidant for CR, NR, IR, SBr, BR, NBR, protecting against rubber poisons; used for dynamically stressed goods, tires, air hoses, belts, sponge rubber, footwear, profiles; pale grayish-brn. cylindrical gran.; sol. in acetone, ethyl acetate, benzene, methylene chloride, CCl_4, aliphatic hydrocarbons; sp.gr. 1.0; m.p. ≥ 88 C.

Vulkanox PAN. [Mobay] Phenyl-alpha-naphthyl amine; see Vulkanox HS/LG; also antiflexcracking agent on NR and IR; lt. brn. to lt. violet gran.; dens. 1.11 ± 0.03 g/cm³; solid. pt. > 52 C.

Vulkanox SP. [Mobay] Styrenated phenol; nonstaining antioxidant; yel. to brnsh. visc. liq.; dens. 1.08 g/cm³; 25 ± 2% bisphenol.

Vulkanox ZKF. [Mobay] 2,2′-Methylene-bis(4-methyl-6-cyclohexyl phenol); see Vulkanox BKF and NKF; wh. powd.; dens. 1.08 g/cm³; m.p. > 125 C.

Vulkanox ZMB-2/G. [Mobay] 4- and 5-methylmercaptobenzimidazole; see Vulkanox MB-2/MGC; wh. to pale beige coated powd.; dens. 1.75 g/cm³; m.p. > 300 C.

Vulkanox ZMB2/C5. [Mobay] Zinc salt of 4- and 5-methylmercaptobenzimidazole; nonstaining and nondiscoloring antioxidant for natural and syn. rubber; synergist; used mainly for heat-resistant thiuram-cured goods, wh. and colored and transparent goods, and latex foam; sensitizer for latexes; wh. to beige powd.; m.w. 391.5; slightly sol. in aliphatic hydrocarbons, benzene, CCl_4; sp.gr. 1.45; m.p. ≥ 300 C; 15.5% zinc, 5% oil.

Vulkasil A1. [Bayer] Precipitated sodium aluminum silicate with a med. reinforcing effect; filler used in all rubbers except silicone rubber; wh. amorphous powd.; dens. 2.0 g/cm³; surf. area (BET) 65 m²/g; pH 10–12.

Vulkasil C. [Bayer] Precipitated silica with a small amt. of calcium silicate and a med. reinforcing effect; see Vulkasil A1; wh. amorphous powd.; dens. 2.0 g/cm³; surf. area (BET) 60 m²/g; pH 8.5–9.5.

Vulkasil N. [Bayer] Reinforcing precipitated silica; filler for all rubbers except silicone rubber; ingred. of Cohedur or Cofill bonding systems; wh. amorphous powd. and gran.; dens. 2.0 g/cm³; surf. area (BET) 130 m²/g; pH 6.5–7.3.

Vulkasil N/GR-S. [Bayer] Reinforcing precipitated silica; see Vulkasil N; wh. gran.; dens. 2.0 g/cm³; surf. area (BET) 130 m²/g; pH 6.5–7.3.

Vulkasil S. [Bayer] Reinforcing precipitated silica; see Vulkasil N; wh. amorphous powd. and gran.; dens. 2.0 g/cm³; surf. area (BET) 170 m²/g; pH 5.5–7.0.

Vulkazon AFD. [Mobay] Unsat. aromatic ether; nonstaining and nondiscoloring antiozonant for CR, and for NR, BR, SBR in conjunction with ozone protective waxes; used for extruded profiles, hoses, cables, and proofed fabrics; colorless to pale yel. liq.; bitter almond odor; sol. in acetone, ethanol, ethyl acetate, toluene, methylene chloride, aliphatic hydrocarbons; sp.gr. 1.02; b.p. 135 C (4.5 Torr).

Vulkazon AFS-50. [Mobay] Bis- (1,2,3,6-tetrahydrobenzaldehyde) pentaerythritol acetal absorbed on silica, min. oil coated; nonstaining antiozonant for CR, IIR, CIIR, BIIR; used for hoses, extruded profiles, cables, sheeting, proofed goods; beige to gray powd.; m.w. 310; sol. in acetone, ethyl acetate, toluene, methylene chloride, ethanol; sp.gr. 1.34; m.p. 85 C; 50% act.

Vulkazon AFS/LG. [Mobay] Bis (1,2,3,6-tetrahydrobenzaldehyde) pentaerythritol acetal; see Vulkazon AFS-50; beige to lt. gray lentil-shaped pellets; m.w. 310; sol. see Vulkazon AFS-50; sp.gr. 1.06; m.p. 85 C min.

Vultac® 2. [Pennwalt] Alkyl phenol disulfide; NR, NBR, SBR vulcanizer; plasticizer; NBR, SBR tackifier; accelerates SBR; br. to straw tacky, resinous solid; sp.gr. 1.1–1.2; soften. pt. 50–60 C; flash pt. (COC) 235 C; 21.8–23.8% sulfur.

Vultac® 3. [Pennwalt] Alkyl phenol disulfide; vulcanizing agent for rubber and adhesives industries; used in chlorobutyl rubber, tires; brn. solid; sp.gr. 1.20; soften. pt. 78–93 C; flash pt. (COC) 171 C; 27–29% sulfur.

Vultac® 4. [Pennwalt] Alkyl phenol disulfide and stearic acid; NR, NBR, SBR vulcanizer; plasticizer; tackifier; br. tacky, resinous solid; char. odor; sp.gr. 1.05–1.15; soften. pt. 52–60 C; flash pt. (COC) 231 C; 15.3–16.7% sulfur.

Vultac® 5. [Pennwalt] Alkyl phenol disulfide on inert carrier; NR, NBR, SBR vulcanizer; accelerates SBR, NBR cures; br. lt. powd.; sp.gr. 1.435; flash pt. (COC) 171 C; 18.5–21% sulfur.

Vultac® 7. [Pennwalt] Alkyl phenol disulfide; see Vultac 3; dk. brn. flakes; sp.gr. 1.20; soften. pt. 105–125 C; flash pt. (COC) 204 C; 29.35–31.75% sulfur.

Vultac® 710. [Pennwalt] Alkyl phenol disulfide, 10% stearic acid; vulcanizing agent for rubbers, incl. CR, NR, EPDM, chlorinated rubbers, and blends of SBR with reclaim or natural rubber; replacement for sulfur or thiuram; used in tires; dk. brn. flakes; soften. pt. 75–90 C; flash pt. (COC) 204 C; 26.4–28.4% S.

Vultamol®. [BASF AG] Sodium salt of condensed naphthalene sulfonic acid; dispersant for rubber syntheses and processing of rubber latexes; powd.; water-sol.; 91% conc.

Vultamol® SA Liq. [BASF AG] Sodium salt of condensed naphthalene sulfonic acid; dispersant for rubber syntheses and processing of rubber latexes; liq.; water-misc.; 45% conc.

Vybar® 103. [Bareco] Ethylene-derived hydrocarbon polymer; lubricant, anticaking agent, modifier; used in paraffin making it more opaque and harder on solidification without affecting cloud pt. and visc. of the molten blend; used in candles to replace stearic acid and opacifies the candle and imparts resistance to thermal shock; pigment disp. to wet inorg. pigments and fillers at high loading levels; suitable for hot melt inks, mold release compds., plastic lubricants, protective coatings, polishes, slip and antimar additives; 0.5 solid; limited sol. in org. solvs.; dens. 0.92 g/cc; visc. 360 cps (98.9 C); m.p. 72 C.

Vybar® 260. [Bareco] Olefin-derived hydrocarbon polymer; paraffin wax modifier; hard, low-melting polymer with low melt visc; additive producing blends with paraffin which are harder and more opaque; upgrades properties of lower-grade waxes; useful in polishes, industrial coatings, hot-melt inks, mold-release compds., plastic lubricants, and slip and antimar additives; sol. in min. spirits, xylene, and trichloroethylene; dens. 0.9 g/cc; visc. 915 cps (65.6 C); m.p. 51 C; iodine no. 15.

Vybar® 825. [Bareco] Ethylene-derived hydrocarbon polymer; lubricant with good slip chars.; additive in specialty lubricant applic., polishes, cutting oils, and release agents; 0.0 liq.; sol. in org. solvs., min. oil, and selected silicones; dens. 0.86 g/cc; visc. 2800 cps; m.p. 30 F.

Vydax WD. [DuPont] Fluorotelomer disp.; mold lubricant and release agent where applic. to hot pressing surfs. is required; specially compded. to permit water dilution; excellent antistick properties; release coatings, water-based paints and inks; dry lubricant films; 5 μ particle size disp.; dens. 1.56 g/ml; visc. 585 cps.

Vykacet L, T. [Croda Ltd.] Acetylated monoglyceride; nonionic; mold release agent in food industry; liq. and solid resp.; 100% conc.

Vykamol N/E, S/E. [Croda Ltd.] Glyceryl monostearate; nonionic; emulsifier, dispersant, wetting, scouring and gelling agent, solubilizer; solid.

Wax OTO. [Fanning] Hard natural wax derived from Douglas fir tree bark; defoamer; replacement or extender for carnauba, candelilla, and beeswax in cosmetics and toiletries; dk. grn. wax; visc. 208 SUS (210 F); m.p. 138 F; acid no. 80; sapon. no. 145.

Waxenol® 801. [CasChem] Arachidyl propionate ester; binder; emollient, solubilizer, and ingred. in cosmetic and toiletries; lubricant for pressed powds.; metal working lubricant and corrosion inhibitor for metal surf.; sol. in min. oil, IPM, chloroform, anhyd. ethanol @ 45 C; misc. with lanolin @ 45 C.

Waxenol® 810. [CasChem] Myristyl myristate; emollient with unusual afterfeel; sol. in IPM, min. oil, oleyl alcohol; insol. in water.

Waxenol® 815. [CasChem] Cetyl palmitate; emollient additive; internal lubricant and binder in pressed powds.; in metalworking lubricant coatings; powd.

Waxenol® 816. [CasChem] Cetyl palmitate; nonionic; internal lubricant and binder in pressed binder operations; emollient additive; metalworking lubricant and corrosion inhibitor coating for metal surfs.; flakes; 100% act.

Waxenol® 821 S.B. [CasChem] Syn. beeswax; emollient; very fine flakes.

Waxenol® 822. [CasChem] Arachidyl behenate; emollient for sticks, creams, lotions; waxy flakes.

Wayhib®. [Olin] TEA phosphate ester; anionic; detergent builder, sequestration agent, scale inhibitor; liq.; 70% conc.

Wayhib® S. [Olin] Org. polyphosphate ester; anionic; corrosion inhibitor, pipeline scale inhibitor, water circulating systems; for air conditioning, boiler treatment compds.; liq.; 60% act.

Wayplex 45-K. [Olin] Org. polyphosphonate, potassium salt; anionic; scale inhibitor for cooling tower treatment; liq.; 40% act.

Wayplex 55-A. [Olin] Org. polyphosphonate, acid form; anionic; dispersant in detergent applics., coating paper mfg., water treatment; sequestrant for iron, Ca, Mg; liq.; 45% conc.

Wayplex 55-S. [Olin] Org. polyphosphonate, sodium salt; anionic; dispersant in detergent applics., clay and TiO_2 disps.; sequestrant for iron, Ca, Mg; stabilizer for hydrogen peroxide and chlorine bleaches; liq.; 50% conc.

Wayplex 68-K. [Olin] Org. polyphosphonate, potassium salt; anionic; scale inhibitor for bottle washing formulations; liq.; 33% act.

Wayplex HEDP-A. [Olin] Org. polyphosphonate, acid form; anionic; scale and corrosion inhibitor, sequestrant for iron, coupler for herbicides; liq.; 60% act.

Wayplex NTP-A. [Olin] Org. polyphosphonate, acid form; anionic; scale inhibitor; crystal growth modifier; liq.; 50% act.

Wayplex NTP-S. [Olin] Polyphosphonate, sodium salt; anionic; detergent builder, scale inhibitor; liq.; 40% conc.

Weston 63. [GE] Mixt. of DPDP and TPP; stabilizer; APHA 50 max. clear liq.; sp.gr. 1.048–1.054 (25/15.5 C); dens. 1.05 g/ml; visc. 11.0 cps (38 C); flash pt. (COC) 213 C; acid no. 0.50 max.; ref. index 1.5280–1.5310; 8.6% phosphorus.

Weston 310. [GE] Triphenyl phosphite (contains 1.0% wt. triisopropanolamine); stabilizer used in epoxies, hot-melt adhesives, PU, polyesters, SBR, PP; used in molding and extrusion of PP, HDPE, LDPE, high-impact PS, PC, ABS, PVC, polyesters, in calendering of ABS and PVC, in film applics. of PP, polyethylene, PVC, fiber applics. of PP and polyesters, etc.; APHA 50 max. clear liq.; m.w. 310; sp.gr. 1.180–1.186 (25/15.5 C); dens. 1.18 g/ml; visc. 12.0 cps (38 C); flash pt. (COC) 168 C; acid no. 0.50 max.; ref. index 1,5860–1.5890; 10.0% phosphorus.

Weston 399. [GE] Trisnonylphenyl phosphite (contains 0.75% wt. triisopropanolamine); see Weston 310; Gardner 3 max. clear liq.; m.w. 688; sp.gr. 0.980–0.992 (25/15.5 C); dens. 0.98 g/ml; visc. 250.0 cps (60 C); flash pt. (PM) 196 C; acid no. 0.50 max.; ref. index 1.5250–1.5280; 4.3% phosphorus.

Weston 399B. [GE] Trisnonylphenyl phosphite (contains 1.0% wt. triisopropanolamine); see Weston 310; Gardner 3 max. clear liq.; m.w. 688; sp.gr. 0.980–0.992 (25/15.5 C); dens. 0.98 g/ml; visc. 250.0 cps (60 C); flash pt. (PM) 196 C; acid no. 0.50 max.; ref. index 1.5250–1.5280; 4.3% phosphorus.

Weston 430. [GE] Tris(dipropyleneglycol) phosphite; stabilizer for hot-melt adhesives, PU, polyesters; used in molding, extrusion, and film applics. in PP, HDPE, LDPE, PVC, and polyesters; also useful for PP fiber applics. and calendering of PVC; APHA 50 max. clear liq.; m.w. 430; sp.gr. 1.088–1.098 (25/15.5 C); dens. 1.09 g/ml; visc. 42.2 cps (38 C); flash pt. (COC) 132 C; acid no. 0.20 max.; 7.2% phosphorus.

Weston 439. [GE] Poly 4,4′ isopropylidenediphenol neodol 25 alcohol phosphite; chelating agent in conjunction with metallic soaps to enhance the clarity and color of PVC formulations; improves color and light stability and clarity of PVC formulations; FDA approved; APHA 100 max. clear liq.; sol. in most common aprotic org. solvs.; sp.gr. 0.965–0.990

(25/15.5 C); b.p. 165 C (5 mm Hg); flash pt. (COC) 216 C; acid no. 0.2 max.; hyd. no. 20–40; ref. index 1.5040–1.5095; flash pt. (COC) 216 C.

Weston 474. [GE] Tris Neodol-25 phosphite; see Weston 430; APHA 50 max. clear liq.; m.w. 670; sp.gr. 0.875–0.885 (25/15.5 C); dens. 0.88 g/ml; visc. 21.0 cps (38 C); flash pt. (COC) 196 C; acid no. 0.05 max.; ref. index 1.4580–1.4610; 4.6% phosphorus.

Weston 491. [GE] Diphenyl didecyl (2,2,4-trimethyl-1,3-pentanediol) diphosphite; stabilizer for PU, hot melt adhesives; used in molding and extrusion of PP, HDPE, LDPE, PC, ABS, PVC; used in film applics. for HDPE, LDPE, and PP; used in calendering of PVC, molding of PU, and fiber PP applics.; APHA 50 max. clear liq.; m.w. 706; sp.gr. 0.985–0.998 (25/15.5 C); dens. 0.99 g/ml; visc. 11.5 cps (38 C); flash pt. (PM) 168 C; acid no. 0.20 max.; ref. index 1.4895–1.4925; 8.8% phosphorus.

Weston 494. [GE] Diisooctyl octylphenyl phosphite; see Weston 491; APHA 50 max. clear liq.; m.w. 494; sp.gr. 0.920–0.930 (25/15.5 C); dens. 0.92 g/ml; visc. 24.5 cps (38 C); flash pt. (PM) 152 C; acid no. 0.05 max.; ref. index 1.4750–1.4800; 6.3% phosphorus.

Weston 600. [GE] Diisodecyl pentaerythritol diphosphite; see Weston 430; APHA 100 max. clear liq.; m.w. 508; sp.gr. 1.020–1.040 (25/15.5 C); dens. 1.04 g/ml; visc. 110.0 cps (38 C); flash pt. (PM) 152 C; acid no. 0.20 max.; ref. index 1.4710–1.4750; 12.1% phosphorus.

Weston 618. [GE] Distearyl pentaerythritol diphosphite; color and heat stabilizer; improves the processing and thermal stability of PP; synergistic when used with lt. stabilizers such as benzophenones, benzotriazoles, and hindered amines; used in olefin polymers, PS, and rubber-modified PS for food contact and pkg. applics.; wh. solid flake (also avail. in ground version); fineness: 95% between 0.25 and 0.0232 in.; m.w. 732; sp.gr. 0.920–0.935 (60/15.5 C); dens. (bulk): 0.464 g/ml (flake); m.p. 40–70 C; flash pt. (PM) 185 C; acid no. 1.0 max.; ref. index 1.4560–1.4590 (60 C); 7.2–7.8% phosphorus.

Weston 619. [GE] Distearyl pentaerythritol diphosphite (contains 1.0% wt. triisopropanolamine); see Weston 618; wh. flakes (also avail. in ground version); fineness: 95% between 0.25 and 0.0232 in.; m.w. 732; sp.gr. 0.920–0.935 (60/15.5 C); dens. (bulk): 0.397 g/ml (flake); m.p. 40–70 C; flash pt. (PM) 188 C; acid no. 1.0 max.; ref. index 1.4560–1.4590 (60 C); 7.2–7.8% phosphorus.

Weston 732. [GE] Distearyl pentaerythritol diphosphite (contains 5.0% wt. calcium stearate); stabilizer for hot-melt adhesives, PU, polyesters, SBR, ABS; used in molding and extrusion of PP, HDPE, LDPE, high-impact PS, PC, ABS, PVC, polyesters; for PP, HDPE, LDPE film applics.; for PP and polyester fiber applics.; for calendering of PVC, etc.; wh. solid; m.w. 732; sp.gr. 0.920–0.926 (60/15.5 C); dens. (bulk) 0.530 g/ml (flake); m.p. 40–60 C; flash pt. (COC) 249 C; acid no. 4.0 max.; ref. index 1.4540–1.4580 (60 C); 6.8–7.4% phosphorus.

Weston 800. [GE] Blend of (1:1) Weston 618/Irganox 1010; stabilizer; wh. solid flake; fineness 90% > 0.0061 in.; dens. (bulk) 0.460 g/ml (flake); m.p. 40 C; flash pt. (COC) > 260 C; acid no. 1.0 max.; 3.6–3.9% phosphorus.

Weston 801. [GE] Blend of (2:1) Weston 618/Irganox 1010; stabilizer; wh. solid flake; fineness 90% > 0.0661 in.; dens. (bulk) 0.450 g/ml (flake); m.p. 40 C; flash pt. (COC) 196 C; acid no. 1.0 max; 4.8–5.2% phosphorus.

Weston 802. [GE] Blend of (3:1) Weston 618/Irganox 1010; stabilizer; wh. solid; fineness 90% > 0.0661 in.; dens. (bulk) 0.460 g/ml (flake); m.p. 40 C; flash pt. (COC) > 260 C; acid no. 1.0 max.; 5.4–5.8% phosphorus.

Weston DHOP. [GE] Poly (dipropyleneglycol) phenyl phosphite; see Weston 491; APHA 50 max. clear liq.; m.w. 2102; sp.gr. 1.168–1.180 (25/15.5 C); dens. 1.00 g/ml; visc. 200.0 cps (38 C); flash pt. (PM) 179 C; acid no. 0.10 max.; ref. index 1.5340–1.5380; 11.8% phosphorus.

Weston DLP. [GE] Dilauryl phosphite; stabilizer; APHA 50 max. clear liq.; m.w. 418; sp.gr. 0.898–0.906 (25/15.5 C); dens. 0.91 g/ml; visc. 8.5 cps (60 C); flash pt. (COC) 210 C; acid no. 4.00 max.; ref. index 1.4500–1.4530; 7.4% phosphorus.

Weston DOPI. [GE] Diisooctyl phosphite; see Weston 430; APHA 50 max. clear liq.; m.w. 306; sp.gr. 0.925–0.933 (25/15.5 C); dens. 0.93 g/ml; visc. 8.0 cps (38 C); flash pt. (PM) 146 C; acid no. 3.0 max.; ref. index 1.4415–1.4430; 9.8% phosphorus.

Weston DPDP. [GE] Diphenyl isodecyl phosphite; stabilizer for flexible and rigid PVC; reacts principally by chelation of metallic chlorides during PVC compding.; for use in conjunction with primary heat stabilizers; APHA 50 max. clear liq.; m.w. 374; sol. in most common aprotic org. solvs.; sp.gr. 1.022–1.032 (25/15.5 C); dens. 8.6 lb/gal; visc. 10.3 cps (38 C); flash pt. (PM) 154 C; acid no. 0.05 max.; ref. index 1.5160–1.5190; 8.3% phosphorus.

Weston DPP. [GE] Diphenyl phosphite; see Weston 310; APHA 50 max. clear liq.; m.w. 234; sp.gr. 1.200–1.220 (25/15.5 C); flash pt. (PM) 143 C; acid no. 15.0 max.; ref. index 1.5550–1.5575; 13.3% phosphorus.

Weston DSP. [GE] Distearyl phosphite; see Weston 430; lt. yel. waxy solid (@ R.T.); m.w. 586; sp.gr. 0.860–0.880 (60/15.5 C); flash pt. (PM) 185 C; acid no. 3.0 max.; ref. index 1.4450–1.4480 (60 C); 5.3% phosphorus.

Weston DTDP. [GE] Di-tridecyl phosphite; stabilizer; APHA 50 max. clear liq.; m.w. 446; sp.gr. 0.906–0.917 (25/15.5 C); dens. 0.91 g/ml; visc. 27.5 cps (38 C); flash pt. (COC) 224 C; acid no. 3.00 max.; ref. index 1.4530–1.4560; 7.0% phosphorus.

Weston EGTPP. [GE] Triphenyl phosphite (contains 0.5% wt. triisopropanolamine); reactive diluent for epoxy applics. incl. adhesives, coatings, laminates, potting and soldering compds., tooling; visc. reducer; APHA 50 max. clear liq.; m.w. 310; sp.gr. 1.180–1.186 (25/15.5 C); dens. 9.8 lb/gal; visc. 12.0 cps (38 C); m.p. 22–25 C; flash pt. (COC) 187C; acid no. 0.50 max.; ref. index 1.5860–1.5890; 10.0% phosphorus.

Weston ODPP. [GE] Diphenyl isooctyl phosphite;

see Weston 491; APHA 50 max. clear liq.; m.w. 346; sp.gr. 1.040–1.047 (25/15.5 C); dens. 1.04 g/ml; visc. 8.2 cps (38 C); flash pt. (PM) 182 C; acid no. 0.05 max.; ref. index 1.5210–1.5230; 9.0% phosphorus.

Weston PDDP. [GE] Phenyl diisodecyl phosphite; see Weston 491; APHA 50 max. clear liq.; m.w. 438; sp.gr. 0.938–0.947 (25/15.5 C); dens. 0.94 g/ml; visc. 12.0 cps (38 C); flash pt. (PM) 160 C; acid no. 0.05 max.; ref. index 1.4780–1.4810; 7.1% phosphorus.

Weston PNPG. [GE] Phenyl neopentylene glycol phosphite; see Weston 491; APHA 35 max. clear liq.; m.w. 227; sp.gr. 1.130–1.139 (25/15.5 C); dens. 1.14 g/ml; visc. 13.0 cps (38 C); flash pt. (PM) 152 C; acid no. 0.14 max.; ref. index 1.5140–1.5188; 13.7% phosphorus.

Weston PTP. [GE] Heptakis (dipropylene-glycol) triphosphite; see Weston 430; APHA 50 max. clear liq.; m.w. 1022; sp.gr. 1.105–1.120 (25/15.5 C); dens. 1.11 g/ml; visc. 130.0 cps (38 C); flash pt. (COC) 129 C; acid no. 0.20 max.; ref. index 1.4660–1.4710; 9.1% phosphorus.

Weston TDP. [GE] Triisodecyl phosphite; see Weston 430; APHA 50 max. clear liq.; m.w. 502; sp.gr. 0.884–0.904 (25/15.5 C); dens. 0.89 g/ml; visc. 11.6 cps (38 C); flash pt. (PM) 160 C; acid no. 0.05 max.; ref. index 1.4530–1.4610; 6.2% phosphorus.

Weston THOP. [GE] Tetraphenyl dipropyleneglycol diphosphite; see Weston 491; APHA 50 max. clear liq.; m.w. 566; sp.gr. 1.164–1.188 (25/15.5 C); dens. 1.17 g/ml; visc. 39.0 cps (38 C); flash pt. (PM) 350 C; acid no. 0.20 max.; ref. index 1.5570–1.5620; 10.9% phosphorus.

Weston TIOP. [GE] Triisooctyl phosphite; see Weston 430; APHA 50 max. clear liq.; m.w. 418; sp.gr. 0.886–0.900 (25/15.5 C); dens. 0.90 g/ml; visc. 14.1 cps (38 C); flash pt. (COC) 146 C; acid no. 0.05 max.; ref. index 1.4470–1.4520; 7.4% phosphorus.

Weston TLP. [GE] Trilauryl phosphite; see Weston 430; APHA 50 max. clear liq.; m.w. 586; sp.gr. 0.870–0.885 (25/15.5 C); dens. 0.88 g/ml; visc. 16.0 cps (38 C); flash pt. (COC) 232 C; acid no. 0.05 max.; ref. index 1.4545–1.4595; 5.3% phosphorus.

Weston TLTTP. [GE] Trilauryl trithio phosphite; see Weston 430; APHA 50 max. clear liq. (@ 100 C); m.w. 634; sp.gr. 0.900–0.935 (25/15.5 C); flash pt. (PM) 218 C; acid no. 0.50 max.; ref. index 1.4985–1.5025; 4.9% phosphorus.

Weston TNPP. [GE] Trisnonylphenyl phosphite; see Weston 310; Gardner 2 max. clear liq.; m.w. 688; sp.gr. 0.980–0.992 (25/15.5 C); dens. 0.98 g/ml; visc. 250.0 cps (60 C); flash pt. (PM) 207 C; acid no. 0.10 max.; ref. index 1.5255–1.5280; 4.3% phosphorus.

Weston TPP. [GE] Triphenyl phosphite; see Weston 310; APHA 50 max. clear liq.; m.w. 310; sp.gr. 1.180–1.186 (25/15.5 C); dens. 1.18 g/ml; visc. 12.0 cps (38 C); flash pt. (PM) 146 C; acid no. 0.50 max.; ref. index 1.5880–1.5900; 10.0% phosphorus.

Weston TSP. [GE] Tristearyl phosphite; see Weston 430; wh. waxy solid (@ R.T.); m.w. 838; sp.gr. 0.845–0.855 (50/15.5 C); dens. 0.85 g/ml; visc. 15.0 cps (62 C); flash pt. (COC) 246 C; acid no. 0.20 max.; ref. index 1.4490–1.4530 (50 C); 3.7% phosphorus.

Westvaco Diacid® H-240. [Westvaco] Dicarboxylic org. acid (C_{21}); coupler for built detergents, phenolic disinfectants, liq. cleaners; liq.

Westvaco WPD-120. [Westvaco] Fatty derivs. and anionic emulsifiers; defoamer for pulp/paper applics.; milk wh. emulsion; dens. 8.0 ± 0.05 lb/gal; visc. 700 ± 200 cps; pH 9.0 ± 1.0 (1%).

Westvaco WPD-140. [Westvaco] Org. and inorg. derivs., anionic and nonionic emulsifiers; defoamer for pulp/paper applics.; milk wh. emulsion; partially disp. in water; dens. 7.8 lb/gal; visc. 500 ± 350 cps.

Westvaco WPD-145. [Westvaco] Org. and inorg. derivs., anionic and nonionic emulsifiers; defoamer for pulp/paper applics.; milk wh. emulsion; partially disp. in water; dens. 7.8 lb/gal; visc. 500 ± 200 cps.

Westvaco WPD-203. [Westvaco] Water-extended oil-based prod.; defoamer for pulp/paper applics.; yel. wh. emulsion; insol. in water; dens. 7.6 lb/gal; visc. 1150 ± 200 cps.

Westvaco WPD-220. [Westvaco] Water-extended oil-based prod.; defoamer for pulp/paper applics.; yel. wh. emulsion; f.p. 30 F; insol. in water; dens. 7.7 lb/gal; visc. 1200 ± 300 cps.

Westvaco WPD-300. [Westvaco] Fatty derivs. and nonionic emulsifiers; defoamer for paper machines; amber clear liq.; disp. in water; dens. 7.1 lb/gal; flash pt. (COC) 130 F; 100% act.

Westvaco WPD-305. [Westvaco] Oil-based defoamer for pulp/paper applics.; amber opaque liq.; disp. in water; dens. 7.5 lb/gal; visc. 1500 ± 500 cps; flash pt. (COC) 330 F; 100% act.

Westvaco WPD-390. [Westvaco] Oil-based nonionic emulsifiers and fatty derivs.; defoamer for effluent systems in pulp/paper industry; dk. amber clear liq.; disp. in water; dens. 7.4 lb/gal; pour pt. –15 to –10 C; flash pt. (COC) 330 F; 100% act.

Westvaco WPD-700. [Westvaco] Oil-based defoamer and drainage aid for the brn. stock washer system and pulp mill applics. amber opaque liq.; insol. in water; dens. 7.4 lb/gal; visc. 1050 ± 350 cps; flash pt. (COC) 330 F; 100% act.

Westvaco WPD-710. [Westvaco] Water-based paste defoamer composed of fatty derivs. with nonionic and anionic emulsifiers; defoamer in pulp/paper mill applics.; wh. emulsion; f.p. –1 C; disp. in water @ 37 C; dens. 7.7 lb/gal (37 C); pH 7.5 ± 1.0 (1% sol'n.); 100% act.

Wettol® D 1. [BASF AG] Sodium salt of phenolsulfonic acid condensation prod.; dispersant for formulation of wettable powds. for crop protection; powd.; water-sol.; 95% conc.

Wettol® D 2. [BASF AG] Sodium salt of condensed naphthalene sulfonic acid; dispersant for the formulation of wettable powds. for crop protection; powd.; water-sol.; 95% conc.

WFT-78 Microbicide. [Huntington Lab.] Hexahydro-1,3,5 tris (2-hydroxyethyl)-s-triazine; for use as a preservative in sol. cutting fluids and syn. coolants; 78.5% act.

White Swan. [Croda Ltd.] Lanolin BP, anhyd.; conditioner, moisturizer; w/o emulsifier, and emollient for personal care prods., pharmaceuticals; superfat-

ting agent in soap; aids pigment dispersion; yel. unctuous mass.
Wickenol® 101. [CasChem] IPM; emollient, solubilizer, and lubricant for use in cosmetic and toilet preparations; sol. in alcohol, animal, veg., and wh. oils; insol. in water.
Wickenol® 105. [CasChem] IPM, IPP; see Wickenol 101; sol. see Wickenol 101.
Wickenol® 111. [CasChem] IPP; see Wickenol 101; sol. see Wickenol 101.
Wickenol® 127. [CasChem] Isopropyl stearate; emollient and lubricant; sol. in acetone, min. and veg. oils, ethyl acetate, alcohol; insol. in water.
Wickenol® 131. [CasChem] Isopropyl isostearate; lubricant, emollient, solubilizer; sol. in acetone, castor oil, corn oil, ethyl acetate, ethanol, min. oil.
Wickenol® 136. [CasChem] IPM, IPP, and isopropyl stearate; emollient, solubilizer, vehicle, and solv. for cosmetic and toilet preparations; liq.; sol. in alcohols, ketones, aromatic and aliphatic hydrocarbons; insol. in water.
Wickenol® 139. [CasChem] Syn. jojoba oil; emollient, plasticizer, lubricant for hair and skin care prods.; replacement for natural jojoba oil; clear liq. wax; sol. in alcohol, animal and veg. oils, wh. oils.
Wickenol® 141. [CasChem] Butyl myristate; emollient, solubilizer, and lubricant for use in cosmetic and toilet preparations; plasticizer; sol. in alcohol, animal, wh., and veg. oils; insol. in water.
Wickenol® 142. [CasChem] Octyldodecyl myristate; emollient ingred., plasticizer for cosmetic and pharmaceutical preparations; sol. see Wickenol 101.
Wickenol® 143. [CasChem] Oleyl oleate; lubricant used in cosmetic and toilet preparations; replacement for sperm oil in addition to functioning as cosmetic additive; sol. in alcohol, animal, wh. and veg. oils; insol. in water.
Wickenol® 151. [CasChem] Isononyl isononanoate; emollient; silky emolliency and solv. chars.; sol. in alcohol, animal, veg., and min. oil; insol. in water.
Wickenol® 152. [CasChem] Isodecyl isononanoate; silky emollience and solv. char.; sol. in alcohol, animal, veg., and min. oils.
Wickenol® 153. [CasChem] Isotridecyl isononanoate; silky emollience and solv. char.; sol. in alcohol, animal, veg., and min. oils.
Wickenol® 155. [CasChem] Octyl palmitate; emollient; increases water vapor porosity of fatty components used in cosmetic and topical pharmaceutical preparations; sol. in alcohol, animal, veg., and min. oils; insol. in water.
Wickenol® 156. [CasChem] Octyl stearate; see Wickenol 155; sol. in alcohol, animal, veg., and min. oils.
Wickenol® 158. [CasChem] Dioctyl adipate; see Wickenol 155; sol. in alcohol, animal, veg., and min. oils.
Wickenol® 159. [CasChem] Dioctyl succinate; see Wickenol 155; sol. in alcohol, animal, veg., and min. oils.
Wickenol® 160. [CasChem] Octyl pelargonate; see Wickenol 155; sol. in alcohol, animal, veg., and min. oils.
Wickenol® 161. [CasChem] Dioctyl adipate, octyl stearate, octyl palmitate; see Wickenol 155; sol. in alcohol, animal, veg., and min. oils.
Wickenol® 163. [CasChem] Dioctyl adipate, octyl stearate, octyl palmitate; emollient, solubilizer; fragrance enhancer; sol. in alcohol, animal, veg., and min. oils.
Wickenol® 171. [CasChem] Octyl hydroxystearate; see Wickenol 155; also used as refatting agent, solubilizer; sol. in acetone, castor, corn, and min. oil, chloroform, ethyl acetate, ethanol.
Wickenol® 174. [CasChem] Myristyl octanoate; emollient; provides soft satiny skin afterfeel; liq.
Wickenol® 303. [CasChem] Aluminum chlorohydrate; act. ingred. in antiperspirant and deodorant formulations; sol. in water, methanol, ethanol, propylene glycol, glycerin; 50% sol'n.
Wickenol® 308. [CasChem] Aluminum sesquichlorohydrate; act. ingred. in antiperspirant formulations; beads; sol. in water, 190 proof ethanol.
Wickenol® 321. [CasChem] Aluminum chlorohydrate; for antiperspirant formulations requiring fine particle size; beads; sol. in water.
Wickenol® 323. [CasChem] Aluminum chlorohydrate; for antiperspirant formulations requiring dry, discrete particles; impalpable powd.; sol. in water.
Wickenol® 368, 369. [CasChem] Aluminum/zirconium chlorohydrate; act. ingred. for topical antiperspirants such as roll-ons, stick, creams; sol. in water, ethanol.
Wickenol® 370. [CasChem] Aluminum-zirconium glycine; act. ingred. for topical antiperspirants, e.g., roll-ons, stock, creams; super fine powd.; sol. in water.
Wickenol® 372. [CasChem] Aluminum zirconium tetrachlorohydrate; act. ingred. for antiperspirant formulations; liq.; sol. in ethanol, water.
Wickenol® 373. [CasChem] Aluminum zirconium tetrachlorohydrate; see Wickenol 372; impalpable powd.; sol. in water.
Wickenol® 374, 375. [CasChem] Aluminum zirconium tetrachlorohydrex glycine; see Wickenol 372; sol. in water; 50% sol'n. and powd. resp.
Wickenol® 379. [CasChem] Aluminum zirconium trichlorohydrex glycine; act. ingred. for antiperspirants; powd.
Wickenol® 506. [CasChem] Myristyl lactate; emollient with unusual afterfeel; sol. in ethanol, min. oil, IPM, propylene glycol; insol. in water.
Wickenol® 535. [CasChem] Wheat germ glycerides; emollient, emulsifier, skin lubricant, and anticounter irritant; sol. in veg. and min. oils, IPA, acetone, chloroform; insol. in water.
Wickenol® 535 Vita-Cos®. [CasChem] Wheat germ glycerides; hydrophilic/hydrophobic emollient; anti-irritant; liq.; 100% act.
Wickenol® 545. [CasChem] Glucose glutamate; humectant, emulsifier, surfactant, thickener used in personal care prods.; sol. in water, slightly sol. in alcohol.
Wickenol® 550 Microduct®. [CasChem] Malto-

dextrin; absorbent for lipophilic materials for powd. bath applics.; food-grade carrier for flavors; powd.; 100% act.

Wickenol® 707. [CasChem] PPG-30 cetyl ether; all-purpose fluid, nongreasy emollient with hydroalcoholic compatibility; sol. in animal and veg. oils, ethanol; insol. in water.

Wickenol® 727. [CasChem] PPG-30 lanolin ether; nongreasy emollient, solubilizer; also low odor derivative; sol. see Wickenol 707.

Wickenol® CPS® 325. [CasChem] Aluminum chlorohydrate; controlled particle size act. ingred. for aerosol antiperspirants; sol. in water, ethanol.

Wickenol® CPS® 331. [CasChem] Aluminum chlorohydrate; used in suspensoid solid state antiperspirant sticks; super fine powd.; sol. in water.

Wickenol® CPS® 336. [CasChem] Aluminum chlorohydrate; act. ingred. for antiperspirants; extra fine powd.

Wickenol® CPS® 370. [CasChem] Aluminum-zirconium glycine; act. ingred. for topical antiperspirants, e.g., roll-ons, stock, creams; super fine powd.; sol. in water.

Wiltrol P. [Olin] Surface-treated phthalic anhydride; scorch inhibitor for rubber stocks; powd.

Windale White. [Georgia Marble] Calcium carbonate; general use filler; 7.0 μ median particle size; 0.200% retained on #325 wet screen; Hegman grind 3; dry brightness 94.

Wingstay Antioxidants. [Goodyear] Hindered phenols; antioxidants used in rubber compds. to resist deterioration from oxidation in presence of heat and lt.; liq, powd., and flakes.

Wingstay K. [Goodyear] Reaction prod. of p-nonyl phenol, dodecane thiol, and formaldehyde; highly act. self-synergized nonstaining and nondiscoloring antioxidant for raw polymer stabilization, esp. for SBR, NBR, and polybutadiene; water-wh. to amber liq.; sp.gr. 0.90–0.95; visc. 500–600 cP.

Winnofil S, SP. [ICI Am.] Precipitated calcium carbonate; filler and impact modifier for thermoplastics and paints.

Witcamide® 61. [Witco] Oleamide MIPA and isopropanolamine; nonionic; hair conditioner, emulsifier, lubricant; cosmetics and toiletries.

Witcamide® 70. [Witco; Witco SA] Stearamide MEA; nonionic; opacifier, conditioner, lubricant, thickener, gelling agent, mold release agent, binder; cosmetics and toiletries; base for antiperspirant and makeup sticks; wax.

Witcamide® 82. [Witco] Cocamide DEA and diethanolamine; nonionic; foam stabilizer, visc. modifier, lubricant, conditioner, emulsifier, wetting agent, penetrant, dye dispersant, scouring aid, antistat; cosmetics and toiletries, base for scrub soap; thickener; industrial and textile surfactant; metal processing; liq.; 100% conc.

Witcamide® 272. [Witco] Alkanolamide, modified; nonionic; detergent, emulsifier, lubricant, wetting agent, penetrant, dye dispersant, scouring aid, antistat for industrial, textile, drilling fluids, metal processing; liq.; sol. in water, high concs. of electrolytes; 90% conc.

Witcamide® 511. [Witco] Alkanolamide; nonionic; lubricant, wetting agent, penetrant, dye dispersant, scouring aid, antistat, w/o emulsifier for aerosol and industrial applics., drilling muds, metal processing, textiles; liq., paste.

Witcamide® 823/10. [Witco] Isostearic alkanolamide; nonionic; hair conditioner, lubricant used in personal care prods.; lt. amber liq.; sol. in aromatic solv., chlorinated hydrocarbons, min. spirits and oil, kerosene, veg. oil, ethanol; disp. in water; cloud pt. –5 C; acid no. 10 max.; pH 9.5 ± 0.5 (10% disp.); 100% act.

Witcamide® 5130. [Witco] Cocamide DEA; anionic/nonionic; emulsifier, conditioner, visc. modifier, lubricant, dispersant, suspending and wetting agent, penetrant, dye dispersant, scouring aid, antistat, foam stabilizer for aerosol formulations, textiles, industrial use, metal processing; amber clear visc. liq.; sol. (@ 5%) in water, glycols, alcohols, and chlorinated and aromatic hydrocarbons; sp.gr. 1.0; acid no. 43; pH 9.2 (1% aq.); flash pt. > 200 F (PMCC); 98% act.

Witcamide® 5133. [Witco] Diethanolamine; nonionic; emulsifier, foam stabilizer, visc. modifier, lubricant, conditioner, wetting agent, penetrant, dye dispersant, scouring aid, antistat for textiles, industrial use, personal care prods., metal processing; liq.; 100% conc.

Witcamide® 5138. [Witco] Alkanolamide; nonionic; emulsifier, detergent, lubricant, wetting agent, penetrant, dye dispersant, scouring aid, antistat for industrial use, textiles, metal processing, dry cleaning, oleoresinous coatings, petrol. industry; clear to slightly hazy liq.; sol. (5%) in kerosene, xylene, IPA; insol. in water; sp.gr. 0.99; pH 9.5 (5% aq.); flash pt. > 93 C.

Witcamide® 5140. [Witco] Alkanolamide, modified; nonionic; detergent, wetting agent, dispersant; industrial cleaning agent; foam stabilizer in drilling mud surfactants and aq. media; air-entraining agent; liq.; 100% conc.

Witcamide® 5195. [Witco] Lauramide DEA; nonionic; visc. modifier, foam stabilizer, lubricant, conditioner, lubricant, emulsifier, wetting agent, thickener, penetrant, dye dispersant, scouring aid, antistat; cosmetics and toiletries; industrial foamer and stabilizer; metal processing; textile surfactant; aerosol formulations.

Witcamide® 6310. [Witco] Lauramide DEA; nonionic; conditioner, foam stabilizer, gelling agent, lubricant, and visc. modifier for cosmetics and toiletries; industrial detergent; paste; water-sol.

Witcamide® 6445. [Witco] Alkanolamide, modified; anionic/nonionic; industrial detergent, o/w emulsifier, lubricant, visc. modifier, foamer and stabilizer; liq.; water-sol.

Witcamide® AL69-58. [Witco] Alkanolamide; cationic/nonionic; lubricant, wetting agent, penetrant, dye dispersant, scouring aid, antistat; industrial w/o emulsifier; metal processing; textile surfactant; liq.; 100% conc.

Witcamide® CD. [Witco] Alkanolamine condensate; detergent, thickener; liq.; 100% conc.

Witcamide® CDA. [Witco] Alkanolamide, modified; nonionic; see Witcamide 6445; liq.; water-sol.

Witcamide® CDS. [Witco] Alkylolamide; nonionic; additive for cutting oils, corrosion inhibitor; liq.; 100% conc.

Witcamide® LDTS. [Witco] Alkylolamide; nonionic; detergent, foam stabilizer, softener, visc. control for cosmetics; liq.; 100% conc.

Witcamide® LM, LMT. [Witco] Alkylolamide; nonionic; thickener for personal care prods.; flakes; 100% conc.

Witcamide® M-3. [Witco] Cocamide DEA; nonionic; cleansing agent, conditioner, lubricant, and visc. modifier for cosmetics and toiletries; detergent, foamer and stabilizer for industrial detergents; liq.; water-sol.

Witcamide® MAS. [Witco] Stearamide MEA-stearate; opacifier, conditioner, lubricant, gelling agent for personal care, household, and institutional liq. soaps; used as partial or total replacement for veg. waxes in polishes; coating agent for paper and textiles; mold release agent for industrial processing; additive for raising melting pts. of petrol. waxes or glyceride waxes and fats; ingred. of insulating coatings or barriers, and of water-repellent compds.; lt. cream waxy flakes; mild char. odor; sol. in aliphatic, aromatic, and chlorinated hydrocarbons, alcohol, ketones, esters; sp.gr. 0.844; m.p. 81 C; acid no. 5.0.

Witcamide® MM. [Witco] Myristamide MEA; nonionic; see Witcamide MAS; also binder; wax.

Witcamide® NA, NA1. [Witco] Alkylolamide; nonionic; foam stabilizer, softener; liq.; 100% conc.

Witcamide® S-771. [Witco] Alkanolamide, modified; anionic/nonionic; industrial foamer and stabilizer, lubricant; liq.; water-sol.

Witcamide® S-780. [Witco] Alkanolamide; anionic/nonionic; see Witcamide S-771; liq.; water-sol.

Witcamide® STD-HP. [Witco] Lauramide DEA; nonionic; conditioner, foam stabilizer, lubricant for cosmetics and toiletries; solid; water-sol.

Witcamine® 209. [Witco] 1-(2-Aminoethyl)-2-n-alkyl-2-imidazoline; cationic; emulsifier, lubricity and wetting agent; penetrant, dye dispersant, scouring aid, antistat, corrosion inhibitor intermediate in petrol. industry, metal processing, textiles; liq.; sol. in IPA, heavy aromatic naphtha, toluene; insol. in water.

Witcamine® 210. [Witco] Alkyl amidoamine; cationic; emulsifier, lubricant; corrosion inhibitor intermediate in petrol. industry and metal processing; sol. in IPA, heavy aromatic naphtha, toluene; insol. in water.

Witcamine® 211. [Witco] Mixed 1-(2-aminoethyl)-2-n-alkyl-2-imidazoline; see Witcamine 209; sol. in IPA, heavy aromatic naphtha, toluene; insol. in water.

Witcamine® 235. [Witco] 1-Polyaminoethyl-2-n-alkyl-2-imidazoline; see Witcamine 209.

Witcamine® 3164. [Witco] Alkylamidoamine; see Witcamine 210.

Witcamine® AL42-12. [Witco] Fatty imidazoline; corrosion inhibitor for ferrous metals; salts act as antistat, emulsifier, wetting agent, dispersant, and detergent; demulsification of anionic emulsions, ore flotation, leather and textile softening, degreasing, and pickling operations, asphalt bonding; amber clear liq.; faint amine odor; m.w. 350; sp.gr. 0.935; 87.0% tert. amine content.

Witcamine® PA-60B. [Witco] Salt of Witcamine AL42-12; cationic; industrial antistat, corrosion inhibitor, dispersant and wetting agent; liq.; sol. in water.

Witcamine® PA-78B. [Witco] Salt of Witcamine AL42-12; cationic; dispersant aiding grinding in nonpolar resins, and soil in hydrocarbon systems; suspending agent for pigments in oleophilic carrier liqs.; visc. reduction of slurries based on org. carrier liqs.; wetting agent enabling pigment particles and substrates to be coated by paint vehicles; water-displacing surfactant for org. coatings; dk. amber clear liq.; sol. (5%) in kerosene, xylene, IPA; insol. in water; sp.gr. 0.925; acid no. 115; pH 6.5; flash pt. > 93 C (PMCC); 99.5% act.

Witcamine® RAD. [Witco SA] POE rosin amine; corrosion inhibitor for water treatment; liq.

Witcamine® RAD-0500, 0515, –1100, -1110. [Witco] POE amine; cationic/nonionic; corrosion inhibitor intermediate, scouring agent, softener, dye assistant, emulsifier for acid-stable emulsions; liq.; 100% conc.

Witcarb® Columbia® JXC. [Witco] Activated carbon; adsorbent for vapor-phase applics. e.g., in the solvent-recovery field, purification of industrial gases and process-gas streams, gas separation, as catalyst carriers, in air conditioning, etc.; pelletized; dens. 0.48 g/cc min.

Witcarb® Granular Activated Carbon Grade 965. [Witco] Activated carbon; adsorbent for industrial respirator field; gran.; dens. 0.45–0.48 g/cc (mesh size 8 × 16); 0.43–0.45 g/cc (mesh size 12 × 20).

Witco® 97H Acid. [Witco] Alkylaryl sulfonic acid; anionic; industrial detergent, dispersant, o/w emulsifier, wetting agent; liq.; sol. in water and oil.

Witco® 912. [Witco] POE carboxylic acid esters and sulfonates; pigment dispersant for aq. systems; solubilizer for methyl cellulose; paint surfactant.

Witco® 915. [Witco] Imidazoline; cationic; pigment dispersant and suspending agent, wetting and grinding aid for oil-based paint systems; emulsifier for latex paints.

Witco® 916. [Witco] Fatty acid ester; spreading agent for aq. paint systems.

Witco® 918. [Witco] Alkylaryl sulfonate; pigment wetting and grinding aid for paints; pigment dispersant; oil-sol.

Witco® 934. [Witco] Alkanolamide; pigment dispersant and suspending agent, wetting and grinding aid for oil-based paint systems; surfactant for latex paints.

Witco® 1298 Hard Acid. [Witco] Dodecylbenzene sulfonic acid; anionic; see Witco 97H Acid; liq.; sol. in oil and water.

Witco® 1298 Soft Acid. [Witco] Dodecylbenzene

sulfonic acid; anionic; detergent intermediate, o/w emulsifier, solubilizer, wetting agent, and detergent for metal cleaning; emulsion polymerization surfactant for latex stabilization and pigment dispersion; liq.; sol. in oil and water.

Witco® Acid B. [Witco] Dodecylbenzene sulfonic acid; anionic; detergent intermediate; detergent, dispersant, o/w emulsifier, wetting agent for industrial and metal cleaning; solubilizer, emulsion polymerization surfactant for latex stabilization and pigment dispersion; liq.; sol. in oil and water.

Witco® ACS 60. [Witco SA] Ammonium cumene sulfonate; hydrotrope; liq.

Witco® Aluminum Stearate 18. [Witco] Aluminum di/tristearate; thickener for hydrocarbon fluids; sol. hot with gelation on cooling in aliphatic and aromatic solvs., oils.

Witco® Aluminum Stearate 22. [Witco] Aluminum distearate; pigment suspending agent for oil-based paints; thickener for hydrocarbon fluids; lubricant for wire drawing, stamping, metalworking; sol. hot in aromatic and aliphatic solvs. and oils; water-insol.

Witco® Aluminum Stearate Non Gel A. [Witco] Modified aluminum stearate; water repellent in solv. sol'n.; for masonry; sol. hot with gelation on cooling in aromatic solvs. and oils; insol. in water.

Witco® Calcium Stearate A. [Witco] Calcium stearate; lubricant in processing of PP, ABS plastics; sol. in hot aromatic and aliphatic solvs. and oils; insol. in water, alcohol.

Witco® Calcium Stearate EA. [Witco] Calcium stearate; anticaking additive for food spices; sol. in hot aromatic solvs., min. oils, fatty acids.

Witco® Calcium Stearate F. [Witco] Calcium stearate; water repellent additive for concrete mixes; lubricant in processing of PVC; sol. in hot aromatic and aliphatic solvs. and oils; insol. in water.

Witco® D51-29. [Witco] Alkylaryl sulfonic acid; anionic; industrial detergent, dispersant, o/w emulsifier; liq.; water-sol.

Witco® DTA-150, -180. [Witco] Fatty acid polymers; corrosion inhibitor intermediate; petrol. industry.

Witco® DTA-350. [Witco] Dimer-trimer acid; anionic; corrosion inhibitor for petrol. industry; liq.; sol. in IPA, kerosene, xylene.

Witco® Heat Stable Sodium Stearate. [Witco] Sodium stearate; lubricant for impact-modified PS; sol. in hot water, alcohol.

Witco® Magnesium Stearate D. [Witco] Magnesium stearate; lubricant for ABS plastics; sol. in hot aromatic and aliphatic solvs. and oils; insol. in water, alcohol.

Witco® MRC. [Witco] Org. phosphate ester; mold release agent for unsat. polyester plastics.

Witco® SCS 40. [Witco SA] Sodium cumene sulfonate; hydrotrope; liq.

Witco® Sodium Stearate C-1, C-7. [Witco] Sodium stearate; gelling agent for cosmetic and toiletry stick prods.; sol. hot with gelation on cooling in water, methanol, ethanol, certain lower glycols.

Witco® STS 40. [Witco SA] Sodium toluene sulfonate; hydrotrope; liq.

Witco® SXS 40. [Witco SA] Sodium xylene sulfonate; hydrotrope; liq.

Witco® TX Acid. [Witco] Modified toluene sulfonic acid; anionic; hydrotrope, catalyst; coupler and solubilizer for liq. detergents; anticaking aid in dry neutralization; catalyst in org. reactions (e.g., esterification); liq. to 10–15 C; 95% act.

Witco® Zinc Stearate NW. [Witco] Zinc stearate; external lubricant, antitack agent for rubber articles, slabs, belting, hose; internal lubricant in phenolic molding resins, PS; water-insol.

Witco® Zinc Stearate Polymer Grade. [Witco] Zinc stearate; lubricant for PS; sol. in hot aromatic and aliphatic solvs. and oils; water-insol.

Witco® Zinc Stearate REP. [Witco] Zinc stearate; lubricant for fiber-reinforced polyester plastics; sol. in hot aromatic and aliphatic solvs. and oils; water-insol.

Witcolate 58. [Witco] Ammonium alcohol ether sulfate; wetting agent, penetrant, lubricant, emulsifier, dye dispersant, scouring aid, antistat; detergent base; detergent for metal cleaning; textile surfactant.

Witcolate 1247H. [Witco] Alcohol ether sulfate; anionic; industrial detergent, foamer; petrol. industry intermediate; electrolyte tolerant; sol. in IPA and water.

Witcolate 1259. [Witco] Alcohol ether sulfate; anionic; industrial coupler, detergent, foamer, solubilizer, o/w emulsifier, wetting agent for metal cleaning, dry cleaning, petrol. industry; electrolyte tolerant; liq.; sol. in water, naphthenic oils, IPA, perchloroethylene.

Witcolate 1276. [Witco] Alcohol ether sulfate; anionic; wetting agent, penetrant, lubricant, emulsifier, dye dispersant, scouring aid, antistat; foamer for wallboard mfg. and petrol. industry; detergent for metal cleaning; textile surfactant; emulsion polymerization and latex stabilization; liq.

Witcolate 7031. [Witco] Alcohol ether sulfate; anionic; industrial detergent, dispersant, foamer, wetting agent, penetrant; electrolyte tolerance; liq.; water-sol.

Witcolate A. [Witco] Sodium lauryl sulfate; anionic; cleansing agent, detergent base, foamer, emulsifier, solubilizer for cosmetics and toiletries, industrial detergents; liq.; water-sol.

Witcolate D-510. [Witco] Sodium 2-ethylhexyl sulfate; wetting agent and penetrant for industrial use and polymerization reactions; dispersant for bleaching powds.; lime soap and grease dispersant; clear liq.; sol. (5%) in water, IPA; sp.gr. 1.10; pH 10.5 (10% aq.); flash pt. > 93 C; 55% moisture.

Witcolate D51-51. [Witco] Sodium alkylaryl polyether sulfate; anionic; antistat for syn. fibers and polymer prods.; surfactant, wetting agent and dispersant; clear liq.; sol. in water, ethanol; sp.gr. 1.06; visc. 2500; pH 8.0 (5% aq.); flash pt. (PMCC) > 93.3 C; surf. tens. 29.6 dynes/cm (at CMC); 29% act.

Witcolate T. [Witco] TEA-lauryl sulfate; anionic; cleansing agent, and foamer, emulsifier, solubilizer for cosmetics and toiletries; detergent base; wetting agent for industrial detergents; liq.; water-sol.

Witcomul 78. [Witco] Sorbitan monotallate; w/o emulsifier, thickener, emulsion stabilizer for industrial applics.; drilling fluid additive.

Witcomul 1557. [Witco] POE sorbitan monotallate; drilling fluid additive in petrol. industry.

Witcomul 3126. [Witco] Amine ester; drilling fluid additive for petrol. industry.

Witcomul 4016. [Witco] Complex alkylate; see Witcomul 3126.

Witconate 60B. [Witco] Sodium dodecylbenzene sulfonate; anionic; cleansing agent, foamer, solubilizer for cosmetics and toiletries; detergent base, foamer and wetting agent, emulsifier for industrial detergents; liq.; water-sol.

Witconate 60T. [Witco] TEA-dodecylbenzene sulfonate; anionic; see Witconate 60B; liq.; water-sol.

Witconate 93S. [Witco] Amine alkylaryl sulfonate; anionic; detergent, detergent base, emulsifier, foamer, wetting agent, solubilizer for the detergent industry; liq.; oil-sol.

Witconate 605A. [Witco] Calcium alkylaryl sulfonate; anionic; industrial detergent, dispersant, o/w and w/o emulsifier, lubricant, wetting agent; liq.; oil-sol.

Witconate 605T. [Witco] Alkylaryl sulfonate; penetrant, lubricant, scouring aid, antistat, o/w emulsifier, dispersant, oil wetting agent for industrial uses; additive for lube oils; textile surfactant; oil-sol.

Witconate 1075X. [Witco] Amine alkylaryl sulfonate; anionic; industrial detergent, dispersant, o/w emulsifier, foamer, and wetting agent; liq.; water-sol.

Witconate 1250 Slurry. [Witco] Sodium dodecylbenzene sulfonate; anionic; see Witconate 60B; liq.; water-sol.

Witconate 1260 Slurry. [Witco] Sodium dodecylbenzene sulfonate, sodium xylene sulfonate; anionic; see Witconate 60B; also industrial and textile surfactant; liq.; water-sol.

Witconate 1840X. [Witco] Fatty acid sulfonate; detergent, coupler, wetting agent, emulsifier, penetrant, lubricant, dye dispersant, scouring aid, antistat; industrial surfactant for alkaline systems; metal processing; textile surfactant; dk. liq.; sol. in water, IPA; sp.gr. 1.10; flash pt. > 93 C; pH 6–8 (5% aq.).

Witconate 3009-15. [Witco] Sodium alkylaryl ether sulfate; latex emulsifier and stabilizer for polymerizations.

Witconate AOS. [Witco] Sodium C_{14-16} olefin sulfonate; anionic; see Witconate 60B; also textile and emulsion polymerization surfactant; liq.; water-sol.

Witconate NCS. [Witco] Ammonium cumene sulfonate; anionic; antiblocking agent, coupler, solubilizer, cloud pt. depressant, and hydrotrope for detergent industry; liq.; water-sol.

Witconate NXS. [Witco] Ammonium xylene sulfonate; hydrotrope, solubilizer, coupler and processing aid in detergent mfg. and industrial processes; antiblocking and anticaking agent in powd. prods.; formulates shampoos, aerosols, cutting oils, glue; textile finishing; Klett 30 liq.; dens. 1.1 g/cc; pH 8.0; 40% act.

Witconate P-1052N. [Witco Israel] Dodecylbenzene sulfonic acid, amine salt; anionic; o/w emulsifier, fuel oil additive, solubilizer, degreaser for emulsions; liq.

Witconate P-1059. [Witco] Dodecylbenzene sulfonic acid, amine salt; anionic; emulsifier, solubilizer, detergent, and wetting agent for oil-based systems; dispersant in oil and water-based systems; used in dry-cleaning surfactants; hydrotrope for liq. detergents; amber clear liq.; sol. in water, kerosene, xylene, IPA; sp.gr. 1.02; flash pt. > 93 C; pH 4.8 (20% in 25% IPA); 40% act.

Witconate P-1073F. [Witco] Alkylaryl sulfonic acid, amine salt; drilling fluid additive used in the petrol. industry.

Witconate SCS. [Witco] Sodium cumene sulfonate; anionic; see Witconate NXS; Klett 50 liq., powd.; dens. 0.32–1.2 g/cc; pH 7.5–9.0; 45% act. liq., 905 act. powd.

Witconate STS. [Witco] Sodium toluene sulfonate; see Witconate NXS; Klett 150 liq., powd.; dens. 0.32–1.2 g/cc; pH 9.0–10.0; 40% act. liq., 90% act. powd.

Witconate SXS. [Witco] Sodium xylene sulfonate; see Witconate NXS; Klett 40 liq., powd.; dens. 0.32–1.2 g/cc; pH 8.6–9.6; 40% act. liq., 905 act. powd.

Witconate TAB. [Witco] TEA-dodecylbenzene sulfonate; anionic; see Witconate 60B; detergent, foamer, and wetting agent in industrial surfactants, paints; electrolyte tolerant; liq.; water-sol.

Witconate TX Acid. [Witco SA] Modified toluene sulfonic acid; hydrotrope, coupler and solubilizer for liq. detergents; anticaking aid in dry neutralization; catalyst in org. reactions; liq. to 10–15 C.

Witconate YLA. [Witco] Amine alkylaryl sulfonate; anionic; detergent base, emulsifier, solubilizer, foaming and wetting agent for the detergent, metal cleaning, and textile industries; liq.; oil-sol.

Witconol 14. [Witco] Polyglyceryl-4 oleate; nonionic; w/o emulsifier, lubricant, and antifoamer for industrial use; liq; 100% conc.

Witconol 18L. [Witco] Polyglyceryl-4 isostearate; nonionic; see Witconol 14; also used as solubilizer; liq.; 100% conc.

Witconol 171, 172. [Witco] Polyalkylene glycol ether; defoamer for soap and detergent foam control; surfactant.

Witconol 1206. [Witco] Alkyl POE glycol ether; wetting agent for industrial uses; visc. and flow control agent for polymerization reactions.

Witconol APEB. [Witco SA] PPG-26 buteth-26; emollient, superfatting agent, skin lubricant; perfume solubilizer; vehicle for suntan oils and lotions; wax.

Witconol APEM. [Witco] PPG-3-myreth-3; nonionic; lubricant, emulsifier, wetting agent, penetrant, dye dispersant, scouring aid, antistat, solv. coupler; syn. oils for personal care prods.; metal processing; textile surfactant; liq.; sol. in 3A ethanol, min. oil.

Witconol APM. [Witco] PPG-3 myristyl ether; nonionic; see Witconol APEM; also emollient oil for cosmetics and toiletries, solubilizer; liq.

Witconol APS. [Witco] PPG-11 stearyl ether; nonionic; see Witconol APEM; also emollient oil for cosmetics and toiletries, solubilizer; liq.

Witconol CA. [Witco] Glyceryl stearate SE; nonionic; lubricant, bodying agent, emulsifier, opacifier used in industrial, cosmetic and aerosol formulations; wax; 100% conc.

Witconol CAD. [Witco] Diethylene glycol monostearate; anionic; bodying agent, lubricant, and opacifier for cosmetics and toiletries; wax; 100% conc.

Witconol CD-17. [Witco] PPG-34; nonionic; antistat, emulsifier for personal care prods.; syn. lubricant oil and foam modifier for aerosol formulations.

Witconol CD-18. [Witco] Polypropoxylated polyol; nonionic; emollient for personal care prods., cosmetic emulsifier, lubricant, antistat; sol. in lower hydrocarbon solvs., lower alcohols.

Witconol F26-46. [Witco] PPG-36 oleate; nonionic; emollient oil, spreading agent and coupler for cosmetic oil systems; conditioner in hair grooms; surfactant with aux. lubricating, dispersing, and coupling properties; dispersant for industrial use; spreading and anticaking agent for aerosol prods.; visc. and flow control agent; APHA 150 clear liq.; sol. in common alcohols, water/alcohol sol'ns., and hydrocarbons; sp.gr. 0.987; visc. 270 cps; acid no. 1.3; sapon. no. 25.0; flash pt. > 93 C; 0.1% moisture.

Witconol H-31. [Witco] PEG-8 tallate; nonionic; industrial defoamer, dispersant and o/w emulsifier; metal processing emulsifiable oil; liq.; sol. in oil.

Witconol H-31A. [Witco] PEG-8 oleate; nonionic; lubricant, aux. lubricant, and plasticizer in oils and polymers; o/w emulsifier for min. and veg. oils, and solvs.; improves flow and leveling of coatings, increases spreadability of personal care prods.; lt. amber liq.; sol. in alcohol, xylene, kerosene, perchloroethylene, wh. min. oil; partially sol. in water; sp.gr. 0.99; acid no. 7; pH 3.7 (3% aq.); flash pt. > 93 C (PMCC); 100% conc.

Witconol L32-45. [Witco] PEG 150 distearate; thickening agent for aq. systems; stabilizer for o/w emulsions; in hair shampoos, bubble bath formulations, cosmetic and toiletry lotions, water-based paints and lubricants, and protective coatings; aux. agent in corrosion inhibitor formulations; off-wh. waxy flakes; m.p. 57 C; acid no. 7; sapon. no. 18; pH 5.0 (3% aq.).

Witconol MST. [Witco] Glycerol mono- and distearate; nonionic; cosmetic, pharmaceutical, aerosol formulations; internal lubricant, plasticizer, and emulsifier in industrial applics.; flow control agent for polymerization reactions; solid; 100% conc.

Witconol NP-40. [Witco] Alkylaryl polyether alcohol; nonionic; detergent, o/w emulsifier, solubilizer for oils in metal processing; liq.; sol. in naphthenic and paraffinic oil.

Witconol NP-100. [Witco] Alkyl polyether; pigment dispersant, latex stabilizer, and leveling agent for polymerization reactions.

Witconol NP-300. [Witco] Alkylaryl polyether alcohol; nonionic; paint industry emulsifier for emulsion polymerization; pigment wetting and grinding agent for aq. systems; spreading agent; solid, liq.; water-sol.

Witconol NS500K. [Witco] Polyalkoxylated butyl ether; nonionic; industrial antistat, coupler, detergent, dispersant, o/w emulsifier, lubricant, solubilizer, spreading agent; paste; water-sol.

Witconol RHP. [Witco] Propylene glycol monostearate SE; nonionic; emollient, conditioner, emulsifier, lubricant, opacifier and visc. modifier for cosmetics and toiletries; flake; sol. in oil.

Witconol RHT. [Witco] Glyceryl stearate SE; nonionic; lubricant, plasticizer, emulsifier used in industrial applics.; paste; 100% conc.

Witcopaque 11, 12, 25. [Witco] Modified PS latexes; opacifier for dishwash, shampoo, personal care prods.; wh. liq.; 40% solids.

Witcor 3192. [Witco] Complex amine phosphate; cationic; corrosion inhibitor for petrol. industry; liq.; sol. in IPA, kerosene, xylene; disp. in water.

Witcor 3194, 3195. [Witco] Fatty amide; cationic/nonionic; corrosion inhibitor for petrol. industry; liq.; sol. in IPA, kerosene, xylene.

Witcor CI-6. [Witco] Complex cationic; corrosion inhibitor for HCl; sol. in xylene, IPA; insol. in water.

Witcor CI-3117. [Witco] Amide; intermediate/finished corrosion inhibitor; sol. in xylene, IPA, kerosene; insol. in water.

Witcor SI-3065. [Witco] Complex phosphate ester; anionic; scale inhibitor for water treatment; used in petrol. industry; liq.

Witepsol Suppository Bases. [Huls] Hydrog. coco-glycerides; suppository bases for hydrophilic and lipophilic drugs; pellets.

Wytox® 240. [Olin] Tris (2,4-di-tert-butylphenyl) phosphite; sec. process stabilizer for plastics and rubber; powd.

Wytox® 312. [Olin] Tris (nonylphenyl) phospite; antioxidant for PE, PP, PS, PVC, ABS, nylon, food pkg.; sol. in MEK, kerosene, amyl alcohol.

Wytox® 320. [Olin] Alkaryl phosphite; antioxidant for ABS,PVC, elastomeric polymers; sol. in ketones, alcohols, hydrocarbons.

Wytox® 345. [Olin] Polymeric phosphite; antioxidant for emulsion-type SBR polymers and latexes; NBR stabilization; sol. in min. oil, benzene, hexane.

Wytox® 540. [Olin] Polymeric phenol phosphite; antioxidant for SBR, carboxylated SBR latex; sol. in ketones, org. solvs.

Wytox® 604. [Olin] Polymeric phosphited hindered phenol; antioxidant for elastomers, SBR, thermoplastic and thermoset polymers; sol. in acetone, benzene, min. spirits.

Wytox® 604LMS. [Olin] Polymeric phosphited hindered phenol; antioxidant for hot-processed elastomers, hot-melt adhesives, ABS, polyolefins; sol. in acetone, benzene, min. spirits.

Wytox® ADP-F. [Olin] Alkylated diphenyl amine; antioxidant for rubber stocks, soling; sol. in common org. solvs., carbon disulfide.

Wytox® AF Series, ATP Series, LTS Series. [Olin]

Multifunctional antioxidants with synergistic stabilizing properties.

Wytox® HPM. [Olin] High m.w. hindered phenol primary antioxidant; heat aging stabilizer accepted for FDA-regulated polyolefin applics.

Wytox® PAP. [Olin] Polymeric hindered phenol; antioxidant for nat. and syn. rubber, latexes, elastomers, adhesives; sol. in acetone, benzene, toluene, min. spirits.

Wytox® PAP-SE. [Olin] Polymeric hindered phenol; SE in aq. latex systems; sol. in acetone, benzene, toluene, min. spirits.

Wytox® PMW. [Olin] Polymeric hindered phenol; primary antioxidant for latex and rubber stocks; sol. in ketones, org. solvs., alcohols.

XY

X-743. [Neville] Aromatic plasticizer; aromatic, chemically inert, nonsaponifiable plasticizer exhibiting low reactivity; used in adhesives (mastic, pressure sensitive), rubber (cements, mechanical and molded goods, tires), and caulking compds.; dk. brn. liq.; m.w. 450; sol. in ethers and chlorinated, aromatic, naphthenic, and terpene hydrocarbons; sp.gr. 1.075; flash pt. (COC) 235 F.

Xanco-Frac®. [Kelco] Heteropolysaccharide prod.; gum for use as a viscosifier in oil field hydraulic fracturing fluids; suspending agent; water-sol.

Xanflood®. [Kelco] Industrial-grade xanthan gum; foam stabilizer, flocculant, suspending, gelling agent, rheology modifier, lubricant for industrial applics. esp. for sec. and tert. oil recovery.

XCE-89. [Air Prods.] t-Butyltoluenediamine; crosslinker providing urea linkages; chain extender for PU/urea and polyurea elastomers; used for RIM automotive fascia, bumper covers, and body panels, high-solids elastomeric coatings; sp.gr. 1.02; m.p. 38 C; b.p. 188 C (20 mm Hg); flash pt. > 110 C; 97% min. purity.

XK 22. [Releasomers] Nonsilicone semipermanent mold release agent for the composite industry; offers fast cure, high chemical and abrasion resistance, and gives a film with exc. adhesion to the mold; solv. sol'n.

XNS-6066. [Dow] Brominated unsat. polyester alkyd; flame retardant; dissolved in monomers to yield a thermosetting resin; excellent lt. stability, chemical resistance, strength, and elec. chars. in molding compds.; solid; dens. 1.81 g/ml; visc. 300–350 cps (180 C); acid no. 30–35; 45–48% bromine.

XO White. [Georgia Marble] Ground calcium carbonate; screen-controlled filler for neutralization of acids, syn. marble, aggregate for cement finishes, vinyl asbestos tile, welding rods, polyester and epoxy floor tiles; 1% max. retained on #16 screen; 15% min. passing #40 screen; sp.gr. 2.71; pH 9.0–9.5; hardness (Moh) 3; 95.0% min. total carbonates.

XPV1 Reinforcing Whisker. [J.M. Huber] Cobweb whisker occurring as both individual fibers and as intergrown fiber bundles; whiskers are predominantly amorphous silica; reinforcing additive in thermosets, thermoplastics, adhesives, elastomers, ceramic bodies, paints, porcelain enamels, and glasses; tan to olive cobweb whisker-like fiber; 10 nm median diam. individual fiber; fiber bundles are roughly spherical with median diam. of 3000–4000 nm; sp.gr. 2.25 g/cc; bulk dens. 64–80 kg/m^3; surf. area 35–50 m^2/g; 75.6% amorphous silicon dioxide; 20.2% cryst. elemental silicon.

XR 7. [Releasomers] Semipermanent mold release agent; forms a tough, durable film with exc. adhesion to mold surf.; gives easy and multiple releases with most thermosetting rubber and plastic materials; solv. sol'n.

XT 66. [Releasomers] Mold sealer for tool sealing in the composite industry; higher solids allow 1 coat sealing of most porosity and minor surf. imperfections; water-wh. clear sol'n.; sp.gr. 0.88 ± 0.03; flash pt. (TCC) 45 F; 5% solids in toluene.

Yeoman. [Croda Ltd.] Anhyd. lanolin BP; emollient used in personal care prods., pharmaceuticals; superfatting agent for soap; aids dispersion of pigments into anhyd. systems; yel. unctuous mass; lipophilic.

Z

ZB-100. [Union Carbide] Sodium alumino silicate zeolite; detergent builder; chelating agent removing hardness ions by ion exchange.

ZB-130 (TN). [Union Carbide] Sodium alumino silicate zeolite; detergent builder in formulations with high silicate levels; chelating agent.

ZB-300. [Union Carbide] Sodium alumino silicate zeolite; see ZB-100.

Zeeospheres® 200. [Zeelan] Silica-alumina ceramic; strong, hard, inert, thick-walled hollow spheres for use as filler for a variety of plastic resins in inj. molding, extrusion, SMC, BMC, RTM, compression molding, potting/encapsulating, adhesives, tooling, casting, flooring, grouting, sealants, mastics, coatings, films, and other applics.; lt. gray spheres; 1.3 μ avg. particle size; 7+ Hegman grind; 0.01% retained on 325 mesh; sp.gr. 2.3 g/cc (D153); bulking value 19.2 lb/gal; soften. pt. 1200 C; pH 4.0–7.0; compr. str. > 60,000 psi; hardness 7 (Mohs); 99.4% spheres.

Zeeospheres® 400. [Zeelan] Silica-alumina ceramic; see Zeeospheres 200; lt. gray spheres; 1.6 μ avg. particle size; 67+ Hegman grind; 0.015% retained on 325 mesh; sp.gr. 2.2 g/cc; bulking value 18.3 lb/gal; soften. pt. 1200 C; pH 4.0–7.0; compr. str. > 60,000 psi; hardness 7 (Mohs); 99.4% spheres.

Zeeospheres® 600. [Zeelan] Silica-alumina ceramic; see Zeeospheres 200; lt. gray spheres; 1.8 μ avg. particle size; 3+ Hegman grind; 0.15% retained on 200 mesh; sp.gr. 2.1 g/cc; bulking value 17.5 lb/gal; soften. pt. 1200 C; pH 4.0–7.0; compr. str. > 60,000 psi; hardness 7 (Mohs); 99.4% spheres.

Zeeospheres® 800. [Zeelan] Silica-alumina ceramic; see Zeeospheres 200; lt. gray spheres; 3.0 μ avg. particle size; 7.0% retained on 200 mesh; sp.gr. 2.0 g/cc; bulking value 16.7 lb/gal; soften. pt. 1200 C; pH 4.0–7.0; compr. str. > 60,000 psi; hardness 7 (Mohs); 99.4% spheres.

Zeeospheres® 850. [Zeelan] Silica-alumina ceramic; see Zeeospheres 200; lt. gray spheres; 17.0 μ avg. particle size; 18.0% retained on 200 mesh; sp.gr. 2.0 g/cc; bulking value 16.7 lb/gal; soften. pt. 1200 C; pH 4.0–7.0; compr. str. > 60,000 psi; hardness 7 (Mohs); 99.4% spheres.

Zelec NE. [DuPont] Alcohol phosphate, neutralized; anionic; antistat for textiles, plastics, and films; internal mold release agent for unsat. thermosetting polyester resins; recommended where external heat is applied during curing cycle; protects ferrous metal molds from corrosion during storage; improves surf. appearance of molded parts; eliminates adhesion problems during subsequent applic. of finishes, printing inks, or adhesives to molded part; yel. visc. paste; mild fatty alcohol odor; sol. in water, polar solvs., glycerol, Freon TF, IPA, styrene, toluene; sp.gr. 1.10; dens. 1.10 g/ml; visc. > 100,000 cP; pH 7.0–7.5; flash pt. > 99 C (PMCC); 100% act.

Zelec NK. [DuPont] Fatty alcohol phosphate; anionic; antistat for textiles, plastics, and films; lt. yel. visc. paste; mild fatty alcohol odor; sol. in ethanol, IPA, ethylene glycol, benzene, toluene, or Freon TF solv.; disp. in water; sp.gr. 1.05; dens. 8.7 lb/gal; pH 7.0–7.5 (10% disp.); flash pt. > 210 F (PMCC); 100% act.

Zelec TY. [DuPont] Alcohol phosphate; anionic; antistat for polyolefin fibers, plastics, and films; lt. amber liq.; mild odor; water-sol.; dens. 9.8 lb/gal; pH 6.7–7.3 (10%).

Zelec UN. [DuPont] Alcohol phosphate, unneutralized; anionic; antistat; additive for oils to improve lubricity and provide corrosion protection; formulation of textile lubricants and finishes; release agent in plastic molding operations; pale yel. liq.; mild fatty alcohol odor; self-disp. in water; sol. in Freon TF, IPA, styrene, and toluene; sp.gr. 0.98; dens. 0.98 g/ml; visc. 165 cP; pH 2–3; flash pt. > 99 C (PMCC); 100% act.

Zeo® 49. [Huber] Syn. precipitated silica; polishing agent, catalyst support; 9.0 μ avg. particle size; surf. area 250 m^2/g; oil absorp. 92 cc/100 g; pH 7.0.

Zeodent® 113. [Huber] Syn. precipitated silica; polishing agent, antiskid agent; 9.0 μ avg. particle size; surf. area 200 m^2/g; oil absorp. 92 cc/100 g; pH 7.0.

Zeofree® 80. [Huber] Syn. precipitated silica; defoamer, carrier; anticaking and free-flow aid for powd. detergents; 6.0 μ avg. particle size; surf. area 140 m^2/g; oil absorp. 190 cc/100 g; pH 7.0.

Zeofree® 153. [Huber] Syn. precipitated silica; carrier, filler; 7.0 μ avg. particle size; surf. area 120 m^2/g; oil absorp. 165 cc/100 g; pH 7.0.

Zeolex® 7. [Huber] Syn. sodium aluminum silicate; conditioning and anticaking agent; absorbent carrier; 6 μ avg. particle size; surf. area 115 m^2/g; oil absorp. 115 cc/100 g; pH 7.0.

Zeolex® 7A. [Huber] Syn. sodium aluminum silicate; anticaking and free-flow aid for powd. detergents; absorbent carrier; 4 μ avg. particle size; surf. area 200 m^2/g; oil absorp. 145 cc/100 g; pH 7.0.

Zeolex® 23. [Huber] Syn. sodium aluminum silicate; reinforcing filler for rubber; powd.; 5 μ avg. particle size; 0.2% max. 325 mesh residue; dens. 2.1 g/ml; surf. area 73 m^2/g; oil absorp. 115 cc/100 g; ref. index

1.55; pH 10.2 (20%).

Zeolex® 23A. [Huber] Syn. sodium aluminum silicate; conditioning and anticaking agent; absorbent carrier; 5 μ avg. particle size; surf. 73 m²/g; oil absorp. 115 cc/100 g; pH 10.2.

Zeolex® 23P. [Huber] Syn. sodium aluminum silicate; paper filler; 5 μ avg. particle size; surf. area 73 m²/g; oil absorp. 115 cc/100 g; pH 10.2.

Zeolex® 35P. [Huber] Syn. sodium aluminum silicate; used in printing ink; 6 μ avg. particle size; surf. area 35 m²/g; oil absorp. 70 cc/100 g; pH 9.5.

Zeolex® 40. [Huber] Syn. sodium aluminum silicate; paper filler; 4 μ avg. particle size; surf. area 55 m²/g; oil absorp. 115 cc/100 g; pH 9.7.

Zeolex® 80. [Huber] Syn. sodium aluminum silicate; titanium dioxide extender for rubber applics.; powd.; 4 μ avg. particle size; 0.2% max. 325 mesh residue; dens. 2.1 g/ml; surf. area 115 m²/g; oil absorp. 115 cc/100 g; ref. index 1.51; pH 7.0 (20%).

Zeomatt 155. [Huber] Silica; flatting agent; 4 μm avg. particle size; surf. area 150 m²/g; oil absorp. 180 cc/100 g; pH 7.0 (5%).

Zeosyl® 100. [Huber] Syn. precipitated silica; carrier, reinforcing agent; 6.0 μ avg. particle size; surf. area 140 m²/g; oil absorp. 190 cc/100 g; pH 7.0.

Zeosyl® 110SD. [Huber] Syn. precipitated silica; carrier, reinforcing agent; surf. area 140 m²/g; oil absorp. 190 cc/100 g; pH 7.0.

Zeosyl® 200. [Huber] Syn. precipitated silica; carrier, rheology agent, thickener for liq. detergents; adsorbent that converts liqs. to free flowing powds.; 5.0 μ avg. particle size; surf. area 250 m²/g; oil absorp. 200 cc/100 g; pH 7.0.

Zeothix® 95. [Huber] Syn. precipitated silica; flatting agent; 2.3 μ avg. particle size; surf. area 175 m²/g; oil absorp. 210 cc/100 g; pH 7.0.

Zeothix® 175. [Huber] Syn. precipitated silica; defoamer, conditioner; 2.5 μ avg. particle size; surf. area 175 m²/g; oil absorp. 220 cc/100 g; pH 7.0.

Zeothix® 177. [Huber] Syn. precipitated silica; thickener, thixotrope; 1.5 μ avg. particle size; surf. area 175 m²/g; oil absorp. 235 cc/100 g; pH 7.0.

Zeothix® 265. [Huber] Syn. precipitated silica; thickener, thixotrope; anticaking and free-flow agent for hygroscopic powds.; thickener for liq. detergents; adsorbent that converts liqs. to powds.; 1.7 μ avg. particle size; surf. area 250 m²/g; oil absorp. 220 cc/100 g; pH 7.0.

Zepel B, DR. [DuPont] Fluoropolymer disp. in water; cationic; fabric fluoridizer imparting oil and water repellency, and resistance to staining by both aq. and oily materials, to textiles; produces finishes which are durable to laundering and to drycleaning; used for outerwear, rainwear, home furnishings, general apparel fabrics; does not adversely affect the whiteness of tensile strength of fabrics, does not promote yellowing or discoloration on exposure to lt.; heat, or gas fumes; Zepel B gives slightly better repellency esp. at lower concs.; DR is less thermoplastic and is preferred in applics. where superior resistance to build-up on pad rolls or deposits on hot metal rolls in contact with damp facbric is required; conc. range in formulating finishes is 2.0–3.0% on the wt. of the fabric; liq.; bluish-white; mildly acrylic odor; miscible in water in all proportions at R.T.; sp.gr. 1.027; dens. 8.57 lb/gal; visc. 8 cps (27 C); pH 3.5–4.5; 14% active ingred.

Zerogen 11. [Solem] Inorg. halogen-free flame retardant/smoke suppressor for thermoplastics such as EVA, polyethylene, PP, PBT, and nylon; acid scavenger; 2 μ avg. particle size; 0.01% retained on 325 mesh; sp.gr. 2.4; bulk dens. 0.27 g/cc (loose), 0.63 g/cc (packed); surf. area 15 m²/g; oil absorp. 56.

Zerogen 15. [Solem] Halogen-free proprietary composition; flame retardant and smoke suppressor for thermoplastic and elastomeric formulations incl. styrenics, polyamides, polyolefins, and thermoplastic polyesters; powd.; 2.3 μ avg. particle size; 0.01% retained 325 mesh; sp.gr. 2.26 g/cm³; bulk dens. 0.27 g/cm³ (loose), 0.63 g/cm³ (packed); surf. area 15 m²/g; oil absorp. 56 cc/100 g.

Zerogen 33. [Solem] Halogen-free proprietary blend; flame retardant and smoke suppressor for thermoplastics and elastomers, incl. styrenics, polyamides, polyolefins, and thermoplastic polyesters; 1.1 μ avg. particle size; 0.01% retained 325 mesh; sp.gr. 2.38 g/cc; bulk dens. 0.3 g/cc (loose), 0.6 g/cc (packed); surf. area 15 m²/g; oil absorp. 56 cc/100 g.

Zerogen 35. [Solem] Halogen-free proprietary blend; flame retardant and smoke suppressor for thermoplastics and elastomers, incl. styrenics, polyamides, polyolefins, and thermoplastic polyesters; offers improved disp. and flow properties; 1.1 μ avg. particle size; 0.01% retained 325 mesh; sp.gr. 2.26 g/cc; bulk dens. 0.3 g/cc (loose), 0.6 g/cc (packed); surf. area 15 m²/g; oil absorp. 56 cc/100 g.

Zetax®. [Vanderbilt] Zinc 2-mercaptobenzothiazole; sec. accelerator in latex foam curing systems; pale yel. powd., 99.9% thru 100 mesh; m.w. 397.86; practically insol. in water; dens. 1.70 ± 0.03 mg/m³; m.p. > 300 C; 15.0–18.0% zinc content.

Zinc Omadine®, Cosmetic Grade. [Olin] Zinc 2-pyridinethiol 1-oxide; antidandruff agent for shampoos; inhibits growth of Gram-positive and Gram-negative bacteria and fungi; cosmetic preservative; off-wh. aq. dispersion (particle size: 95% < 7μ; 90% < 5μ); mild odor; m.w. 317.7 (zinc pyrithione); insol. in water; dens. 10 lb/gal; m.p. ≈ 240 C (powd. dec.); pH 6.5–8.5 (5% act. slurry in pH 7 water); 48% aq. disp.

Zinc Omadine®, Industrial Grade, 48% Aq. Disp. [Olin] Zinc 2-pyridinethiol 1-oxide; antimicrobial inhibiting growth of Gram-negative and Gram-positive bacterial, fungi, mold, and yeast; used in metal-working fluids and on laundered fabrics; off-wh. disp. (particle size 90% < 5μ); mild odor; m.w. 317.7 (zinc pyrithione); dens. 10 lb/gal; pH 6.5–9.0 (5% active slurry in neut. water); 48–5% act.

Zinc Omadine®, Industrial Grade, Powder. [Olin] Zinc 2-pyridinethiol 1-oxide; antimicrobial inhibiting growth of Gram-negative and Gram-positive bacterial, fungi, mold and yeast; used in aq. metal coolant and cutting fluids, PVC plastics; tan powd. (particle size 90% < 20 mesh); mild odor; m.w. 317.7;

sp.gr. 1.782; dens. 0.35 g/ml; m.p. ≈ 240 C (dec.); pH 7.3 (10% aq.); 95% act.

Zinc Oxide 35. [Akrochem] Precipitated zinc oxide; accelerator activator for rubber articles based on natural and syn. elastomers, and latex applics., e.g., transparent and translucent rubber goods, dynamically stressed articles; crosslinking agent for metal oxide-curable elastomers; reinforcing filler; nonblooming; wh. to ylsh. grn. powd.; sp.gr. ≈ 5.4 g/cm³; 93–96% zinc oxide.

Zinc Oxide No. 185. [Eagle Zinc] Tert. zinc oxide; accelerator activator, pigment, reinforcing agent for rubber and applics. not requiring a high degree of purity; also for prod. of zinc compds.; 99.6% –325 mesh; sp.gr. 5.6; bulk dens. 60 lb/ft³; 96.5% ZnO.

Zinc Oxide No. 318. [Eagle Zinc] Amer. process zinc oxide, lead-free; slow curing fine particle size pigment and reinforcing agent for rubber industry; also for mfg. of various zinc compds.; 0.38 μ median particle diam.; 99.93% –325 mesh; sp.gr. 5.65; bulk dens. 38 lb/ft³; 99.2% ZnO.

Zinc Oxide Transparent. [Mobay] Highly disperse precipitated zinc oxide; vulcanization accelerator activator for transparent rubber goods, in natural and syn. rubbers; acid acceptor in polychloroprene adhesives; lt. colored reinforcing filler; used for vulcanizates needing high elasticity or transparency, vulcanizates cured in hot air or with sulfenamides, vulcanizates cured with metal oxides and without sulfur, in food-contacting goods; used in latex.; wh. to slightly ylsh. grn. powd.; dens. 3.5 g/cm³; pH 9–10; 70–73% ZnO; < 0.02% PbO.

Zinc Pyrion. [Ruetgers-Nease] Zinc salt of 2-mercapto-pyridine-N-oxide; antidandruff agent, preservative, antibacterial, antimicrobial; aq. disp.

Zinc Stearate 11. [Witco] Zinc stearate; lubricant for fiber-reinforced and PS plastics; wh. powd.; 99.9% thru 325 mesh; sol. in hot turpentine, benzene, toluene, xylene, CCl_4, veg. and min. oils, waxes; sp.gr. 1.09; soften. pt. 119 C.

Zinc Stearate 44. [Witco] Zinc stearate; lubricant for PS rubber; wh. powd.; 99.9% thru 325 mesh; sol. see Zinc Stearate 11; sp.gr. 1.09; soften. pt. 120 C.

Zinc Stearate Disperso. [Witco] Zinc stearate; lubricant and antitack agent for rubber; wh. disp.; sol. see Zinc Stearate 11; sp.gr. 1.09; soften. pt. 120 C.

Zinc Stearate Heat-Stable. [Witco] Zinc stearate; lubricant for PS plastics; wh. powd.; 99.9% thru 325 mesh; sol. see Zinc Stearate 11; sp.gr. 1.09; soften. pt. 122 C.

Zinc Stearate LV. [Witco] Zinc stearate; lubricant for fiber-reinforced plastics, SMC, BMC; wh. powd.; 99.7% thru 325 mesh; sol. see Zinc Stearate 11; sp.gr. 1.09; soften. pt. 120 C.

Zinc Stearate NW. [Witco] Zinc stearate; lubricant and antitack agent for rubber; wh. powd.; 99.9% thru 325 mesh; sol. see Zinc Stearate 11; sp.gr. 1.09; soften. pt. 120 C.

Zinc Stearate Polymer Grade. [Witco] Zinc stearate; lubricant for PS plastics; wh. powd.; 99.9% thru 325 mesh; sol. see Zinc Stearate 11; sp.gr. 1.09; soften. pt. 120 C.

Zinc Stearate Regular. [Witco] Zinc stearate; lubricant for urea/formaldehyde, melamine plastics; antitack dustingagent and lubricant for surfs. of uncred rubber slabs; flatting agent for solvent-based paints; wh. powd.; 99.9% thru 325 mesh; sol. see Zinc Stearate 11; sp.gr. 1.09; soften. pt. 120 C.

Zinc Stearate USP. [Witco] Zinc stearate; lubricant, mold release agent, w/o emulsifier for cosmetics, toiletries, pharmaceuticals; wh. powd.; 99.9% thru 325 mesh; sol. see Zinc Stearate 11; sp.gr. 1.09; soften. pt. 120 C.

Zinkoxyd Activ. [Mobay] Highly disperse precipitated zinc oxide; vulcanization accelerator activator for rubber goods based on natural and syn. elastomers and latex applics.; suitable at high levels for dynamically stressed articles, e.g., buffers and rollers, and in low concs. in transparent and translucent goods; lt. colored reinforcing filler; also for food-contacting goods; wh. to slightly ylsh. grn. powd.; dens. 5.5 g/cm³; pH 10–11; 93–96% ZnO; < 1% of water-sol. constituents.

ZMBT. [Akrochem] Zinc salt of 2-mercaptobenzothiazole; semi-ultra accelerator; ylsh. powd., practically odorless, bitter taste; sp.gr.: 1.70 ± 0.03; m.p. 315 C (decomp. occurs).

ZN-7. [Advanced Refractory Tech.] Zirconium diboride, nuclear grade; powd. offering oxidation resistance, exc. mech. properties, corrosion resistance, and high m.p.; used for oxidation-resistant composites, burnable absorber of neutrons, elec. contacts, molten metal crucibles, refractory toughener, cutting tool composites, structural ceramics, wear components, metal matrix composites; gray to blk. hexagonal cryst.; 6–8 μ avg. particle diam.; dens. 6.09 g/cc; m.p. > 3000 C; 79% min. Zr, 18% min. B.

Zn-0312 T 1/4″. [M&T Harshaw] Zinc chromite; catalyst used to synthesize methanol from carbon monoxide and hydrogen; tablet, 1/4″ diam.; dens. 100 lb/ft³.

Zn-0401 E 3/16″. [M&T Harshaw] Zinc oxide; catalyst used to remove sulfur from gas steams; wh. extrusion, 3/16″ diam.; dens. 75 lb/ft³; 100% conc.

Zn-0602 T 1/8″. [M&T Harshaw] Zinc chrome; catalyst; gray tablet, 1/8″ diam.; dens. 68 lb/ft³.

ZO-9. [Disco] Fatty acid salts and petroleum derivs.; plasticizer and lubricant for use in all polymers except Hypalons; opaque beige wax-like solid; practically odorless; sp.gr. 0.92; m.p. 166 ± 4 F; flash pt. 450 F.

Zoharquat 50, 80. [Zohar] Benzalkonium chloride; bacteriocide and algicide; liq.; 50 and 80% conc. resp.

Zoharsoft. [Zohar] Quat. imidazoline deriv.; cationic; fabric softener base; liq. to paste; 75% conc.

Zoharsoft 90. [Zohar] Quat. imidazoline deriv.; cationic; fabric softener base; liq. to paste; 90% conc.

Zoharsoft DAS, DAS-N. [Zohar] Fatty acid deriv.; cationic; softener base for textile industry; flakes.

Zoldine® ZE. [Angus] Oxazolidine; cross-linking agent, catalyst, resin reactant, formaldehyde substitute, corrosion inhibitor, tanning agent, raw material for polymer synthesis; m.w. 143.2; water-sol.; m.p. <

0 C; b.p. 71 C; flash pt. 175 F (TCC); pH 8.9 (0.1M aq. sol'n.).

Zoldine® ZT-55. [Angus] Oxazolidine; cross-linking agent for resorcinol phenol-formaldehyde or protein-based resin systems; used in hair care prods.; m.w. 145.1; water-sol.; m.p. –20 C; b.p. decomposes; flash pt. > 200 F (TCC); pH 7.0 (0.1M aq. sol'n.).

Zonarez 7010. [Arizona] Polyterpene resins (based primarily on dipentene); low m.w., thermoplastic polymer which imparts high levels of tack and adhesion to many elastomeric and polymeric materials; stability in adhesive systems, neutrality, resistance to attack by acids, alkalis, salts, and water, aging properties, and compatibility with numerous org. solvs; used in the mfg. of pressure-sensitive adhesives for tapes and labels, rubber cements, solv.-based adhesives, emulsion adhesives, hot melt adhesives and coatings, can sealants, caulking and general sealant compds., ink, investment casting waxes, paints, concrete waterproofing agents, varnishes, chewing and bubble gum bases, moisture resistant soft gelatin capsules and powds. of ascorbic acid and its salts, etc.; FDA Gardner 2 (1963, 50% heptane) bright, clear; sp.gr. 0.97; soften. pt. (R&B) 10 C; acid no. < 1; sapon. no. < 1.

Zonarez 7025, 7040, 7055. [Arizona] Polyterpene resin; see Zonarez 7010; Gardner 2 (1963, 50% heptane) clear, bright; sp.gr. 0.97, 0.97, 0.98 resp.; soften. pt. (R&B) 25 C, 40 C, 55 C resp.; acid no. < 1; sapon. no. < 1.

Zonarez 7070. [Arizona] Polyterpene resin; see Zonarez 7010; Gardner 3 (1963, 50% in heptane) clear, bright; sp.gr. 0.98; soften. pt. (R&B) 70 C; acid no. < 1 ; sapon. no. < 1.

Zonarez 7085. [Arizona] Polyterpene resin; see Zonarez 7010; Gardner 3 (1963, 50% in heptane) clear, bright; sol. in 2-ethyl hexyl alcohol, aliphatic hydrocarbons, aromatic hydrocarbons, chlorinated hydrocarbons, butyl and amyl acetates, terpenes, ethyl ether, and petrol. ether; sp.gr. 0.97; soften. pt. (R&B) 85 C; acid no. < 1; sapon. no. < 1.

Zonarez 7100. [Arizona] Polyterpene resin; see Zonarez 7010; Gardner 3 (1963, 50% in heptane) clear, bright; sp.gr. 0.97; soften. pt. (R&B) 100 C; acid no. < 1; sapon. no. < 1.

Zonarez 7115, 7125. [Arizona] Polyterpene resin; see Zonarez 7010; Gardner 3 (1963, 50% in heptane) clear, bright; sol. see Zonarez 7085; sp.gr. 0.99; soften. pt. (R&B) 115 C, 125 C resp.; acid no. < 1; sapon. no. < 1.

Zonarez B-10. [Arizona] Polyterpene resin (produced by the polymerization of betapinene); see Zonarez 7010; Gardner 2+ (1963, 50% in heptane) clear, bright; sp.gr. 0.92; soften. pt. (R&B) 10 C; acid no. < 1 apon. no. < 1 .

Zonarez B-25, B-40. [Arizona] Polyterpene resin; see Zonarez 7010; Gardner 2+ (1963, 50% in heptane); clear, bright; sp.gr. 0.93, 0.94 resp.; soften. pt. (R&B) 25 C, 40 C resp.; acid no. < 1; sapon. no. < 1.

Zonarez B-55, B-70. [Arizona] Polyterpene resin; see Zonarez 7010; Gardner 3- (1963, 50% in heptane) clear, bright; sp.gr. 0.96; soften. pt. (R&B) 55 C, 70 C, resp.; acid no. < 1; sapon. no. < 1.

Zonarez B-85, B-115, B-125. [Arizona] Polyterpene resin; see Zonarez 7010; Gardner 3 (1963, 50% in heptane) clear, bright; sol. in 2-ethyl hexyl alcohol, aliphatic hydrocarbons, aromatic hydrocarbons, chlorinated hydrocarbons, ester, terpenes, ethyl ether, and petrol. ether; sp.gr. 0.99, 0.98, 0.97, resp.; soften. pt. (R&B) 85 C, 115 C, 125C, resp.; acid no. < 1; sapon. no. < 1.

Zonester® 55. [Arizona] Glycerol derived ester based on Arizona DR-24 disproportionated tall oil rosin; thermoplastic resin ester offering stability to oxidative attack; tackifier for difficult-to-tackify elastomers, e.g., SBR, and where good flexibility and tackification are required at lower temps.; used in hot melt and pressure sensitive adhesives, contact cements, and latex adhesives; resin; amber; sol. in solvs. usually compded. in pressure sensitive adhesives and contact cements; soften. pt. (R&B) 52 C; acid no. 8; tens. str. 450 psi @ yield, 425 psi (ultimate); tens. elong. 350%.

Zonester® 75. [Arizona] Glycerol derived ester based on Arizona DR-22 disproportionated tall oil rosin; thermoplastic resin ester offering stability to oxidative attack; in hot melt and other applics. where elevated temp. stability is important, and where long term stability under ambient conditions is critical; in emulsion form, tackifies natural, SBR, and neoprene latices; used in hot melt adhesives and coatings, pressure sensitive adhesives, contact cements, and latex adhesives; sol. in most aliphatic and aromatic hydrocarbons, chlorinated hydrocarbons, esters, ketones, ethers, and higher alcohols; soften. pt. (R&B) 78 C; acid no. 8; tens. str. 475 psi (yield), 575 psi (ultimate); tens. elong. 500%.

Zontes PD-101. [Matsumoto] Alkyl polyamide deriv.; cationic; softener for textiles; flake; 100% conc.

Zontes TA-460-15. [Matsumoto] Alkyl polyamide deriv.; cationic; see Zontes PD-101; paste; 15% conc.

Zonyl A. [DuPont] EO-ester condensate; nonionic; emulsifier, lubricant; textile wetting agent for fibers and fabrics during dyeing; leveling agent and antifoam in pigment coatings in paper industry; used in metal processing; clear-pale yel. liq.; mild odor; sol. in water, polar and nonpolar solvs.; sp.gr. 1.07; HLB 6.7; surf. tens. 26 dynes/cm (0.1% aq.); 100% act.

Zonyl FSA. [DuPont] Fluorochemical surfactant; anionic; wetting agent, emulsifier, dispersant, corrosion inhibitor, leveling agent for adhesives, agric., polishes, polymerization, pigment grinding, cleaners, coatings and paints, fire fighting, paper, ink, oil, plastics, and textile industries; liq.; sol > 2% in water, methanol; dens. 9.8 lb/gal; flash pt. 69 F (PMCC); surf. tens. 18 dynes/cm (0.1%); 50% solids in IPA/water.

Zonyl FSB. [DuPont] Flurochemical surfactant; amphoteric; see Zonyl FSA; also foamer, froth flotation agent for solv. cleaners, elastomers; liq.; sol. > 2% in water, IPA, methanol; dens. 8.8 lb/gal; flash pt. 67 F (PMCC); surf. tens. 17 dynes/cm (0.1%); 40% solids in IPA/water.

Zonyl FSC. [DuPont] Fluorochemical surfactant; cationic; see Zonyl FSB; liq.; sol. > 2% in water, IPA, methanol, acetone; dens. 9.7 lb/gal; flash pt. 70 F (PMCC); surf. tens. 18 dynes/cm (0.1%); 50% solids in IPA/water.

Zonyl FSN. [DuPont] Fluorochemical surfactant; nonionic; see Zonyl FSA; liq.; sol. > 2% in water, IPA, methanol, acetone, ethyl acetate, THF; dens. 8.8 lb/gal; flash pt. 72 F (PMCC); surf. tens. 23 dynes/cm (0.1%); 40% solids in IPA/water.

Zonyl FSP. [DuPont] Fluorochemical surfactant; anionic; see Zonyl FSA; liq.; sol. > 2% in water, methanol; dens. 9.6 lb/gal; flash pt. 75 F (PMCC); surf. tens. 19 dynes/cm (0.1%); 35% solids in IPA/water.

Zusolat 1005/85, 1008/85, 1010/85, 1012/85. [Zschimmer & Schwarz] Alkyl polyglycol ether; nonionic; dispersant, emulsifier, wetting agent; liq.; 85% conc.

Zusomin C 108. [Zschimmer & Schwarz] Fatty amine ethoxylate; nonionic; basic material and component for dyeing and textile auxs.; intermediate for quaternization; metal cleaners; liq.; 100% conc.

Zusomin O 102, O 105, O 112. [Zschimmer & Schwarz] Fatty amine ethoxylate; nonionic/cationc; basic material for textile and dyeing auxs., intermediate for quaternization, metal cleaners; liq.; 100% conc.

Zusomin S 110, S 125. [Zschimmer & Schwarz] Fatty amine ethoxylate; nonionic; see Zusomin O 102; liq., paste; 100% conc.

Zotex 319 L. [Hitox] Zinc oxide, French process; see Zotex 319; 0.35 μ mean particle size; 99.7% thru 325 mesh; Hegman grind 5.5; sp.gr. 5.6; dens. 35 lb/ft^3 (poured), 54 lb/ft^3 (tapped); brightness 94; surf. area 3.05 m^2/g; oil absorp. 15–16; 99.5% conc.

Zotex 319. [Hitox] Zinc oxide, French process; pigment/extender for alkyd and latex paints, ceramics, plastics, rubber, and chemical processing; 0.35 μ mean particle size; 99.97% thru 325 mesh; Hegman grind 5.5; sp.gr. 5.6; dens. 35 lb/ft^3 (poured), 54 lb/ft^3 (tapped); brightness 94; surf. area 3.05 m^2/g; oil absorp. 15–16; 99.5% conc.

ZS-7. [Advanced Refractory Tech.] Zirconium diboride, standard grade; powd. offering oxidation resistance, exc. mech. properties, corrosion resistance, and high m.p.; used for oxidation-resistant composites, burnable absorber of neutrons, elec. contacts, molten metal crucibles, refractory toughener, cutting tool composites, structural ceramics, wear components, metal matrix composites; gray to blk. hexagonal cryst.; 5–8 μ avg. particle diam.; dens. 6.09 g/cc; m.p. > 3000 C; 76.5% min. Zr; 18% min. B.

Chemical Component Cross Reference

Acetal resin *Synonyms:* Polyacetal
Vulkanox AFS-50, AFS/LG

Acetamide MEA (CTFA) [CAS #142-26-7] *Synonyms:* N-Acetyl ethanolamine
Incromectant AMEA-70, AMEA-100, LAMEA; Lipamide MEAA; Mackamide AME-75; Schercomid AME, AME-70

Acetamidopropyl trimonium chloride
Incromectant AQ

6-Acetoxy-2,4-dimethyl-m-dioxane
Giv-Gard DXN

Acetylacetonate chelate
Tyzor AA

Acetylated hydrogenated coconut oil glyceride
Myvacet 9-08(K)

Acetylated hydrogenated lard glyceride (CTFA) [CAS #8029-91-2] *Synonyms:* Glycerides, lard mono-, hydrogenated, acetates
Myvacet 7-00; Tegin E 61, E-61 NSE

Acetylated hydrogenated soybean oil glyceride
Myvacet 9-45

Acetylated hydrogenated tallow glyceride [CAS #68990-58-9] *Synonyms:* Glycerides, tallow mono-, hydrogenated, acetates
Lamegon EE; Tegin E 41

Acetylated hydrogenated tallow glycerides [RD #977063-59-4] *Synonyms:* Glycerides, tallow mono-, di- and tri-, hydrogenated, acetates
Tegin E-41 NSE

Acetylated hydrogenated vegetable oil glyceride
Myvacet 5-07(K), 7-07(K)

Acetylated lanolin (CTFA) [CAS #61788-48-5]
Lanacet 1705; Modulan; Ritacetyl

Acetylated lanolin alcohol (CTFA) [CAS #61788-49-6]
Acetol 1706; Acetulan; Acylan; Crodalan AWS, LA; Ethoxyol 1707; Fancol ALA; Hetlan AC; Ritawax ALA; Solulan 97, 98

Acetylated lard glyceride (CTFA) [CAS #8029-92-3] *Synonyms:* Glycerides, lard mono-, acetates
Grindtek AMOS 90; Myvacet 9-40; Tegin E 66, E-66 NSE

Acetylated palm kernel glycerides (CTFA) [RD #977063-60-7]
Grindtek AML 60

Acetylated sucrose distearate (CTFA) [RD #977067-90-5] *Synonyms:* Sucrose distearate, acetates
Crodesta A-10, A-20

Acetyl cyclohexane sulfonyl peroxide
Trigonox ACS-M28

Acetyl cyclohexyl sulfonyl peroxide
Lupersol 228Z

Acetylenic alcohol
OW-1

Acetyl peroxide [CAS #110-22-5] *Synonyms:* Diacetyl peroxide
Acetyl Peroxide-IB25

Acetyl tributyl citrate [CAS #77-90-7] *Synonyms:* 2-(Acetyloxy)-1,2,3-propanetricarboxylic acid, tributyl ester; 1,2,3-Propanetricarboxylic acid, 2-(acetyloxy)- tributyl ester; Acetyl tri n-butyl citrate
Citroflex A-4; Estaflex ATC; Merrol C-105

Acetyl triethyl citrate
Citroflex A-2; Merrol C-103

Acrylamide copolymer
Reten 521

Acrylamide/β-methacryloyloxyethyl trimethyl ammonium methosulfate copolymer. See Polyquaternium-5

Acrylamide/sodium acrylate copolymer (CTFA) [CAS #25085-02-3] *Synonyms:* 2-Propenamide, polymer with 2-propenoic acid, sodium salt
Reten 421, 423, 425

Acrylate monomer
Chemlink 9013, 9015

Acrylates copolymer (CTFA) [RD #977069-05-8]
Acrysol QR-1086; Polymer QR-1010; Polytrap 171, 229, 801; Rheolate 1

Acrylate/steareth-20 methacrylate copolymer
Sipothix 1941

Acrylate terpolymer
Acrysol WTP-1

Acrylic/methacrylic copolymer
Texicryl 13-002

Acrylic resin *Synonyms:* Acrylic fiber; Acrylic polymer
Acralen AFR; Acryloid WR-97; Acrysol ASE-60, ASE-95, ASE-108, ICS-1, TT-615, TT-675; Crystic LS7; Daratak 74L; Duolite ES-132; Emulsion E-1614; Flow Agent WR-100; Forbest VP S7; Texicryl 13-011, 13-100, 13-101, 13-203, 13-205, 13-206, 13-210, 13-300, 13-302; Ucar Latex 879; Viscon 103

Acrylic, styrenated
Acralen ATR

Acrylic/vinyl acetate copolymer
Texicote 03-060

Acrylinoleic acid *Synonyms:* C21-dicarboxylic acid
Diacid 1550

Acrylonitrile-butadiene rubber *Synonyms:* Acrylonitrile rubber; Acrylonitrile butadiene copolymer; Nitrile-butadiene rubber; Nitrile rubber; NBR; Butadiene-acrylonitrile copolymer
Chemigum NBR Polymers; Darex 110L

Acrylonitrile-butadiene-styrene resin *Synonyms:* ABS resin
Acralen BS; Blendex 101, 131, 201, 310, 336, 405, 467, 702; GAF Latex 6000 Series

Acrylonitrile rubber. See Acrylonitrile-butadiene rubber

4-(2-Acryloyloxyethoxy)-2-hydroxybenzophenone
Cyasorb UV 2126

Adipic acid/dimethylaminohydroxypropyl diethylenetriamine copolymer (CTFA) [CAS #61840-27-5; RD #977026-96-2]
Cartaretin F-4, F-8

Albumen (CTFA) [CAS #9006-50-2]
Sol-U-Tein EA

Algin (CTFA) [CAS #9005-38-3] *Synonyms:* Sodium alginate
Kelco-Gel, HV, LV; Kelco-Pac; Kelcosol; Kelgin, F, HV, LV, MV, QH, QL, QM, RL, XL; Keltex, P, S; Keltone; Kelvis

Alginic acid (CTFA) [CAS #9005-32-7]
Kelacid

Alkoxy triacryl titanate
Ken-React 39DS

Alkoxy trimethacryl titanate
Ken-React 33DS

Alkyd resin. See Polyester, thermosetting and alkyd

Alkyl amidopropyl trimethyl ammonium chloride
Empigen CSC

Alkyl amidopropyl trimethyl ammonium methoxysulfate
Catigene CA 30

Alkyl diamido amine acetate
Empigen FKC75K

Alkyl diamido amine lactate
Empigen FKC75L

Alkyl dimethyl benzyl ammonium chloride. See Benzalkonium chloride

Alkyl dimethyl dichlorobenzyl ammonium chloride
Bio-Quat 50-MAC, T-502; Maquat DLC-1214

Alkyl dimethyl ethylbenzyl ammonium chloride
Barquat 4250, 4250Z, 4280, 4280Z; Bio-Quat 50-35, 50-36, 80-35, 80-36; BTC 1326, 2125 80%, 2125 M 80%, 2125 MP-40, 2125 P-40; FMB 3328-5 Quat, 3328-8 Quat, 6075-5 Quat, 6075-8 Quat; Maquat MQ-2525, MW-2525M

Alkyl dimethyl ethylbenzyl ammonium cyclohexyl sulfamate. See Quaternium-8

Alkyl isoquinolinium bromide
Catinal LQ-75

Alkyl methyl polysiloxane
Masil 263, 264, EM 266 (35%)

Alkyl-tin-carboxylate
Stanclere T55, 57, 80, 81

Alkyl trimethyl ammonium bromide
Catinal HTB; HTB-70; Empigen CHB 40; Vantoc AL

Alkyl trimethyl ammonium chloride
Arquad S; Empigen CMC; Genamin KDM-F; Quartamin 24P, 86P, 86P Conc.

Alkyl trimethyl ammonium methoxysulfate
Catigene CT 30/70, LT 45, ST 30/70

Allyl methacrylate
SR-201

Allyltrimethoxy silane [CAS #2551-83-9]
CA0567

Allyltrimethyl silane [CAS #762-72-1]
CA0570

Almond oil
Super Refined Almond Oil NF

Almondamide DEA
Incromide ALD

Almondamidopropalkonium chloride
Incroquat AL-85

Almondamidopropyl betaine
Incronam AL-30

Almondamidopropyl dimethylamine lactate
Incromate ALL

Almondamidopropylamine oxide
Incromine Oxide AL

Aloe vera
Aloe-Moist

Alumina [CAS #1333-84-2 (hydrate); 1344-28-1 (anhydrous)] *Synonyms* Alumina trihydrate; Aluminum oxide
Akrochem Hydrated Alumina; Al Series; Aluminum Oxide C; Catapal Alumina; CM-66; FRE; G-431; H-36; Haltex 300, 310, 313, 320; Hyfil 10; LV-31, LV-336; Micral 532, 855, 916, 932; SB-30, -31, -31C, -136, -331, -332, -335, -336, -431, -432, -632, -805, -932; Sillum-PL 200; Solem ATH

Alumina silicate
Sillum-200 Q/P

Aluminum chlorohydrate (CTFA) [CAS #1327-41-9; 12042-91-0] *Synonyms:* Aluminum chloride hydroxide
Dow Corning ACH-303, -323, -331, ACH7-321; Reach 101, 201, 501; Wickenol 303, 321, 323, CPS 325, CPS 331, CPS 336

Aluminum distearate [CAS #300-92-5]
Aluminum Stearate EA; Witco Aluminum Stearate 22

Aluminum di/tristearate (CTFA)
Witco Aluminum Stearate 18

Aluminum 2-ethyl hexanate
Forbest 5724

Aluminum magnesium silicate. See Magnesium aluminum silicate

Aluminum nitride
AlNel A-100, A-200, AG

Aluminum sesquichlorohydrate (CTFA) [RD #977011-08-7]
Dow Corning ACH7-308; Wickenol 308

Aluminum silicate (CTFA) [CAS #1327-36-2] *Synonyms:* Pyrophyllite
Ketjensil SM 405

Aluminum silicate, calcined
Altowhite LL; Satintone SP-33, Whitetex; Translink 37, 77, 445, 555

Aluminum silicate, hydrous
Catalpo, X-1

Aluminum starch octenyl succinate (CTFA) [CAS #9087-61-0] *Synonyms:* Starch, octenylbutanedioate, aluminum salt
Dry Flo

Aluminum stearate [CAS #7047-84-9; #300-92-5] *Synonyms:* Aluminum, dihydroxy (octadecanoato-o-)
ACuflow AF-1; Dehymuls K; Hostacerin WO; Radiastar 1208; Witco Aluminum Stearate Non Gel A

Aluminum/zirconium chlorohydrate
Dow Corning AZG-368, -369; Wickenol 368, 369

Aluminum/zirconium glycine
Dow Corning AZG-370; Wickenol 370, CPS 370

Aluminum/zirconium tetrachlorohydrate (CTFA) [CAS #57158-29-9]
Wickenol 372, 373

Aluminum/zirconium tetrachlorohydrex glycine (CTFA)
Dow Corning AZG-374; Wickenol 374, 375

Aluminum/zirconium trichlorohydrex glycine
Wickenol 379

Amine dodecylbenzene sulfonate
Hetsulf 60T; Witconate P-1052N, P-1059

Amine lignite
Ardril FCA 154

Amine polycarboxylate
Colloid 111M

2-Amino-1-butanol [CAS #96-20-8]
AB

2-(2-Aminoethoxy) ethanol
Diglycolamine Agent (DGA)

N-(2-Aminoethyl)-3-aminopropyl methyldimethoxy silane [CAS #3069-29-2]
CA0699

N-2-Aminoethyl-3-aminopropyl trimethoxysilane [CAS #1760-24-3] *Synonyms:* N-β-(-Aminoethyl) γ-aminopropyltrimethoxysilane
Dow Corning Z-6020; Dynasylan DAMO, DAMO-P, DAMO-T; Union Carbide A-1120

Aminoethyl imidazoline
Servamine KOO 330

Aminoethylpiperazine [CAS #140-31-8] *Synonyms:* 1-(2-Aminoethyl) piperazine
D.E.H. 39

2-Amino-2-ethyl-1,3-propanediol [CAS #115-70-8] *Synonyms:* Aminoamylene glycol; AEPD
AEPD

2-Amino-2-methyl-1,3-propanediol [CAS #115-69-5] *Synonyms:* Aminobutylene glycol; butanediolamine; AMPD
AMPD

Aminomethyl propanol (CTFA) [CAS #124-68-5] *Synonyms:* Isobutanolamine; AMP
AMP, AMP-95

3-Aminopropylmethyldiethoxy silane [CAS #3179-76-8]
CA0742

Aminopropyltriethoxysilane [CAS #919-30-2] *Synonyms:* γ-Aminopropyl triethoxysilane
Dynasylan AMEO, AMEO-P, AMEO-T; Prosil-220, -221; Union Carbide A-1100

Aminopropyltrimethoxysilane [CAS #13822-56-3]
Dynasylan AMMO

Aminotrimethylene phosphonic acid (CTFA) [CAS #6419-19-8] *Synonyms:* Amino tris (methylene phosphonic acid); Nitrilotris (methylene) triphosphonic acid
Arquest 709; Dequest 2000; Fostex AMP; Unihib 305

Ammonium alginate (CTFA) [CAS #9005-34-9]
Keltose; Margel; Superloid

Ammonium carboxylate
Daxad 32; Tamol 165, 901, SG-1

Ammonium C12-15 pareth sulfate [RD #977064-04-2] *Synonyms:* Ammonium pareth-25 sulfate; POE (1–4) C12-15 alcohol ether sulfate, ammonium salt
Neodol 25-3A

Ammonium cumenesulfonate [CAS #37475-88-0]
Eltesol ACS 60; Naxonate 6AC; Reworyl ACS 60; Ultra NCS Liquid; Witco ACS 60; Witconate NCS

Ammonium dimolybdate
ADM

Ammonium dinonylnaphthalene sulfonate
Na-Sul AS

Ammonium dodecylbenzene sulfonate (CTFA) [CAS #1331-61-9]
Conco AAS-50M; Hetsulf 50A

Ammonium ferric EDTA
Hamp-Ene Photoiron

Ammonium imazaquin *Synonyms:* 2-[4,5-Dihydro-4-methyl-4-(1-methylethyl)-5-oxo-1H-imidazol-2-yl]-3-quinolinecarboxylic acid
Scepter

Ammonium laureth sulfate [CAS #32612-48-9 (generic)] *Synonyms:* Ammonium lauryl ether sulfate
Sandoz Sulfate 216; Sipon EA, EAY; Standapol EA-3; Steol CA 460; Texapon EA-1, -2, -3

Ammonium lauryl sulfate (CTFA) [CAS #2235-54-3]
Avirol 200; Conco Sulfate A; Incronol ALS; Mapeg AEG; Sandoz Sulfate A

Ammonium lignosulfonate
Lignosite 17; Lignosol TS, TSD, TSF; Norlig NH; Orzan A, AL-50

Ammonium naphthalene sulfonate
Lomar PWA; Tamol NNa 4109

Ammonium nonoxynol-4 sulfate (CTFA) [CAS #31691-97-1 (generic); 63351-73-5; RD #977060-75-5]
Alipal CO-436; Sipex 280

Ammonium pareth-25 sulfate. See Ammonium C12-15 pareth sulfate

Ammonium polyacrylate
Acrysol G-110; Alcogum 9639; Alcosperse 249; Colloid 102; Daxad 37LA7; Dispersal 130; Serpol QPA 150

Ammonium polycarboxylate
Colloid 252

Ammonium polymethacrylate
Daxad 32, 32S, 34A9

Ammonium polyphosphate
Amgard MC, PI

Ammonium stearate [CAS #1002-89-7]
Ammonium Stearate 33% Liquid; Denlube 33-B

Ammonium xylenesulfonate (CTFA) [CAS #26447-10-9; RD #977055-97-2]
AXS; Eltesol AX 40; Hartotrope AXS 40; Naxonate 4AX, AX; Stepanate AM; Ultra NXS Liquid; Witconate NXS

AMP isostearoyl hydrolyzed animal protein (CTFA)
Crotein AD, AD Anhyd., ADX

Amylase (CTFA) [CAS #9000-92-4]
Amizyme; BAN 120 L, 240 L, 360 S, 1000 S; Biolase Brands; Brew(n)zyme; Clarase; Dex-Lo; Exsize Ha, Reg., Special, Super; Fermalpha; Fungamyl, 800L, 1600S; Maxaliq; Maxamyl, MP375, WL7000; Milezyme DAL, Fungal Amylase, Pancreatin 4NF; Mycolase; Premier Maltzymes I, II; Rapdiase; Rhozyme 86L, H39; Spezyme AA, BBA; Takatex; Taka-Therm; Tenase; Termamyl 60T, 120L, 300T

OO-t-Amyl O-(2-ethylhexyl) monoperoxycarbonate
Lupersol TAEC

t-Amyl hydroperoxide
TAHP-50M, TAHP-50P

Amyloglucosidase
Aldomax GA-100; AMG 200 L; Amigase

t-Amyl peroxyacetate
Lupersol 555M60

t-Amyl peroxybenzoate [CAS #4511-39-1]
Esperox 5100

t-Amyl peroxy 2-ethyl hexanoate [CAS #686-31-7]
Esperox 570P; Lupersol 575, 575M75, 575P75

t-Amyl peroxyneodecanoate
Esperox 545M; Lupersol 546M75, TA-46, TA-46M75; Trigonox 123-C75

t-Amyl peroxyneoheptanoate
Esperox 747M

t-Amyl peroxypivalate
Esperox 551M; Lupersol 554M50, 554M75, TA-54M75; Trigonox 125-C75

p-t-Amylphenol [CAS #80-46-6]
Orthophen 278; Pentaphen 67

Amyltrichlorosilane [CAS #107-72-2]
CA0900

Anilazine
Dyrene

Anilino-phenyl methacrylamide
Polystay AA-1, AA-1R

Animal collagen amino acids (CTFA) *Synonyms:* Collagen amino acids: Crotein CAA
Anthocyanase; Crotein CAA/SF; Pectinol DL

Animal keratin amino acids
Crotein HKP/SF

Anthranilic acid (CTFA) [CAS #118-92-3] *Synonyms*: o-Aminobenzoic acid; 2-Aminobenzoic acid; 1-Amino-2-carboxybenzene; Carboxyaniline; o-Carboxyaniline; 2-Carboxyaniline; o-Amidobenzoic acid; Vitamin L
AA

Antimony [CAS #7440-36-0]
Thermoguard CPA, FR

Antimony mercaptide
Stanclere A-121, A-121 C, A-221; Synpron 1027, 1034, 1066

Antimony oxide [CAS #1327-33-9] *Synonyms:* Antimony peroxide
Harshaw Antimony Oxide KR; HFR-301; Thermoguard 213, 215, L, S

Antimony pentoxide [CAS #1314-60-9] *Synonyms:* Antimonic anhydride; Antimonic acid; Stibic anhydride
Nyacol A-1540N, A-1588LP, A-1530, A-1550, AB40, AGO-40, AP50, APE1540, APVC40, HA-9, HA-15, N22, N24, ZTA

Antimony trioxide [CAS #1309-64-4] *Synonyms:* Antimony white; Sb_2O_3
Naftopast Antimontrioxid, -CP

Apricot kernel oil (CTFA) [RD #977006-46-4] *Synonyms:* Persic oil
Lipovol P, P-S; Super Refined Apricot Kernel Oil NF

Arachidic acid [CAS #506-30-9] *Synonyms:* Eicosanoic acid (CTFA)
Hystrene 7022

Arachidyl behenate
Waxenol 822

Arachidyl behenyl amine
Arosurf MG-101D; Kemamine P-150, P-150D, P-190, P-190D; Lilamin 101 D

Arachidyl-behenyl 1,3-propylene diamine
Kemamine D-150, D-190

Arachidyl propionate (CTFA) [CAS #65591-14-2] *Synonyms:* Eicosanyl propanoate
Polytrap 801; Waxenol 801

Arnica extract (CTFA) [CAS #8057-65-6]
Arnica Oil CLR

Aryl methyl polysiloxane
Masil EM266HV

Atrazine *Synonyms:* 2-Chloro-4-ethylamino-6-isopropylamino-S-triazine
Atrabute+, Atrabute+ II

Attapulgite (CTFA) [CAS #1337-76-4]
Attacote; Attaflow; Attagel 40, 50, 150, 350; Florco X; Florex; Min-U-Gel 200, 400, FG, LF; Pharmasorb; Refinex; Sol-Speedi-Dri, Auto-Dri

Avocadamide DEA
Avamid 150; Incromide AVD

Avocadamidopropalkonium chloride
Incroquat AV-85

Avocadamidopropylamine oxide
Incromine Oxide AV

Avocadamidopropyl betaine
Incronam AV-30

Avocadamidopropyl dimethylamine lactate
Incromate AVL

Avocado oil (CTFA) [CAS #8024-32-6] *Synonyms:* Alligator pear oil
Avamid 150; Avocado Oil CLR; Lipovol A, A-S; Super Refined Avocado Oil

Azidosilane
Az-Cup MC

Azinphos-ethyl
Gusathion A

Azinphos-methyl [CAS #86-50-0] *Synonyms:* o,o-Dimethyl-S-4-oxo-1,2,3-benzotriazin-3(4H)-yl methyl phosphorodithioate
Gusathion

2,2´-Azobis (2,4-dimethylvaleronitrile)
Vazo 52

2,2´-Azobisisobutyronitrile [CAS #78-67-1] *Synonyms:* Azobisisobutylonitrile; Azodiisobutyronitrile
Poly-zole AZDN; Vazo 64

Azocyclotin
Peropal

Azodicarbonamide [CAS #123-77-3] *Synonyms:* 1,1´-Azobisformamide
Celogen AZ 3990, AZ; Ficel AC2, AC3, AC3F, AC4, AC-SP4, EPA, EPB, EPC, EPD, EPE, LE, SCE; Kempore 60E, 125E, E, FF, MC; Onifine-C, -CC, -CE; Porofor ADC/E, ADC/F, ADC/K, ADC/M, ADC/R, ADC/S

2,2-Azodiisobutyronitrile [CAS #78-67-1] *Synonyms:* Azobisisobutyronitrile
Ficel AZDN-LF, AZDN-LMC

2,2-Azodi (2-methylbutyronitrile)
Ficel AZM

Azoisobutyrodinitrile
Porofor N

Babassamide DEA
Incromide BAD

Babassamidopropalkonium chloride
Incroquat BA-85

Babassamidopropyl betaine
Incronam BA-30

Babassamidopropyl dimethylamine lactate
Incromate BAL

Babassamidoproylamine oxide
Incromine Oxide BA

Babassu oil
Super Refined Babassu Oil

Balsamic resin
Terate 101, 131

Barium-cadmium
Interstab BC-100S, M 341, M 3187, M-85, MF981, MF985, MT981, MT982; Mark 1314, WS, TT; Therm-Chek 5469; Vanstay 4017, 4030, 4039, 7024, 7025, 7032, HTA, RR

Barium-cadmium-lead
Haro Mix YC-601, YK-110, YK-113, YK-307

Barium-cadmium-phosphite
Mark 462, 755, LL

Barium-cadmium-zinc
Ferro 1288; Interstab 761-28, 761-28A, BC-103, BC-103A, BC-103C, BC-103L, BC-109, BC-110, BC-4362, R-4101, R-4109, R-4114, R-4137; Therm-Chek 6-V-6A; Vanstay 162-B, 246, 3027, 6032, 6040, 6053, 6055, 6074, 6078, 6133, 6172, 6191, 6201, HA, HT, HTE, HTF, RRE, RRZ

Barium carbonate sulfonate
Barium Petronate Basic

Barium carboxylate
Lubrizol 2116

Barium dinonylnaphthalene sulfonate [CAS #25619-56-1]
Na-Sul 611, BSB, BSN, BSN/W765, BSN/W780; Vanplast 201

Barium ferrite
Magnacomp A, AP-86, AP-88

Barium-lead
Haro Mix YE-301; Mark 232B, 550, 556; Vanstay 5563

Barium phenate
Lubrizol 2106

Barium sulfate [CAS #7727-43-7] *Synonyms:* Barytes (natural); Blanc fixe (artificial, precipitated); Basofor)
Bartex 80; Blanc fixe F, N, micro; Lithopone 30% DS, 30% L, 60% DS, 60% L

Barium sulfonate
Alox 2291; Barium Petronate 50-S Neutral

Barium-tin *Synonyms:* Barium stannate
Stanclere T 197

Barium-zinc
Baerostab UBZ 630, 632, 791, 793; Interstab M722, 727, 744, 763, 767, M11301; Lankromark Lz187, LZ121, LZ242, LZ616, LZ693, LZ704, LZ770, LZ792, LZ836, LZ858, LZ968, LZ1023, LZ1056, LZ1067, LZ1144, LZ1155, LZ1166; Mark 1600; Therm-Chek 130; Vanstay 8120, CE

Batyl alcohol (CTFA) [CAS #544-62-7] *Synonyms:* Octadecyl ether of glycerin; Stearyl glyceryl ether; 3-(Octadecyloxy)-1,2-propanediol)
Nikkol Batyl Alcohol 100, EX

Batyl isostearate (CTFA)
Nikkol GM-18IS

Beeswax (CTFA) [CAS #8006-40-4 (white), 8012-89-3 (yellow)]
Alcolan 40; Cremba B6; Cutina LM; Cyclochem 326A; Dehymuls K; Dermalcare 326A; Hostacerin WO; Lipobee 102; Waxenol 821 S.B.

Behenalkonium chloride (CTFA) [CAS #16841-14-8] *Synonyms:* Behenyl dimethyl benzyl ammonium chloride; Benzyldocosyldimethylammonium chloride
Incroquat B65C, BDQ-25P, Behenyl BDQ/P; Kemamine BQ-2802C

Behenamide [CAS #3061-75-4] *Synonyms:* Behenic acid amide; Docosanamide
Kemamide B

Behenamide DEA
Incromide BED

Behenamide MEA
Incromide BEM

Behenamidopropylamine oxide
Makcamine BAO

Behenamidopropyl betaine
Mackam BA

Behenamidopropyl dimethylamine (CTFA) [RD #977063-18-5]
Incromine BB; Lexamine B-13; Mackine 601; Schercodine B

Behenamine *Synonyms:* Behenyl amine
Nissan Amine VB

Behenamine oxide
Incromine Oxide B-30P

Behenic acid (CTFA) [CAS #112-85-6] *Synonyms:* Docosanoic acid
Crodacid B; Hystrene 7022, 9022

Behenoxy dimethicone
Abil-Wax 2440

Behenoyl-PG-trimonium chloride (CTFA) [CAS #69537-38-8]
Akypoquat 131

Behentrimonium chloride (CTFA) [CAS #17301-53-0] *Synonyms:* Behenyl trimethyl ammonium chloride
Genamin KDM; Incroquat Behenyl TMC/P

Behentrimonium methosulfate *Synonyms:* Behenyl trimethyl ammonium methyl sulfate
Incroquat Behenyl TMS, BTQ-25

Behenyl alcohol (CTFA) [CAS #661-19-8] *Synonyms:* 1-Docosanol
Adol 60; Dehydag Wax 22 (Lanette); Lanette 22; Nikkol Behenyl Alcohol 65, 80

Behenyl behenate [CAS #17671-27-1]
Starfol BB

Behenyl betaine
Incronam B-40

Behenyl erucate (CTFA) [CAS #18312-32-8] *Synonyms:* Docosyl 13-docosenoate
Crodamol BE; Kemester BE; Schercemol BE

Behenyl hydroxyethyl imidazoline *Synonyms:* Behenyl imidazoline
Schercozoline B

Bentonite (CTFA) [CAS #1302-78-9] *Synonyms:* Soap clay
Korthix, H; Polargel, T

Benzalkonium bromide *Synonyms:* Alkyl dimethyl benzyl ammonium bromide
Bio-Quat 50-MAB; Catigene BR/80B

Benzalkonium chloride (CTFA) [CAS #8001-54-5] *Synonyms:* Alkyl dimethyl benzyl ammonium chloride
Alacsan 7LUF; Alkaquat DMB-451; Arosurf PT 50; Arquad B-50 USP, B-90 USP, B-1000, DMCB, DMMCB-50, DMMCB-75; Bardac 205M; Barquat 1552, 4250, 4250Z, 4280, MB-50, MB-80, OJ-50; Bio-Dac 205; Bio-Quat 50-24, 50-25, 50-28, 50-30, 50-35, 50-36, 50-40, 50-42, 50-60, 50-65, 80-24, 80-28, 80-35, 80-36, 80-40, 80-42, T-501; BTC 50, 50 USP, 65, 65 USP, 100, 776, 835, 885, 1326, 2125 80%, 2125 M 80%, 2125 MP-40; BTC 2125 P-40, 2565, 2568, 8248, 8249, E-8358; Carsoquat 621, 621 80%; Catigene 4513, 4513/80, CS 40, T 50, T 80, T 80 F; Crapol AU-20, AU-21, AU-24; Empigen BAC, BAC50, BAC50/BP, BAC80, BAC90, BCB50, BCF80; FMB 65-15 Quat, 65-28 Quat, 451-5 Quat, 451-8 Quat, 500-15 Quat USP, 504-5 Quat, 1210-5 Quat, 1210-8 Quat, 3328-5 Quat, 3328-8 Quat, 4500-5 Quat, 4500-8 Quat, 6075-5 Quat, 6075-8 Quat; Gardiquat 12H, 1450, 1480, SV 480; Hyamine 3500, 3500-NF; JAQ Powdered Quat; Jordaquat 350, 358; Lebon GM; Lutensit K-LC, K-LC 80, K-OC; Maquat LC-12S, MC-1412, MC-1416, MC-6025-50%, MQ-2525, MQ-2525M, TC-76; Protectol KLC 50, 80; Quadrilan BC; Quatrene CB, CB-50, CB-80, MB-50, MB-80; Querton 246; Retarder CA; Rewoquat B 50; Roccal 50% Tech., II 50%; Sanisol C, CPR, CR, CR-80%, HTPR, OPR, TPR; Swanol CA-101; Synprolam 35DMBQC; Variquat 50AC, 50AE, 50MC, 50ME, 60LC, 80AC, 80AE, 80MC, 80ME, LC60, LC80; Vikol RQ; Zoharquat 50, 80

Benzene (CTFA) [CAS #71-43-2] *Synonyms:* Annulene; Benzol; Benzole; Banzolene; Bicarburet of Hydrogen; Carbon oil; Coal Naphtha; Cyclohexatriene; Mineral naphtha; Motor Benzol; Nitration benzene; Pyrobenzol; Pyrobenzole
Amsco Benzene

Benzene-1,3-disulfonyl hydrazide
Porofor B 13/CP 50

Benzenesulfonamide
Merrol N-306

Benzene sulfonyl hydrazide
Porofor BSH Paste, BSH Paste M, BSH Powd.

Benzethonium chloride [CAS #121-54-0] *Synonyms:* Diisobutyl phenoxy ethoxy ethyl dimethyl benzyl ammonium chloride
Hyamine 1622

1,2-Benzisothiazolin-3-one
Proxel CRL

Benzoic acid (CTFA) [CAS #65-85-0]
Retarder BA, BAX

Benzoin butyl ether
Escacure EB3

Benzoin ether
Trigonal 14

Benzophenone-1 (CTFA) [CAS #131-56-6] *Synonyms:* 2,4-Dihydroxybenzophenone; Benzoresorcinol; 4-Benzoyl resorcinol
Uvasorb 20H; Uvinul 400

Benzophenone-2 [CAS #131-55-5] *Synonyms:* 2,2′,4,4′-Tetrahydroxy benzophenone
Uvinul D-50

Benzophenone-3 (CTFA) [CAS #131-57-7] *Synonyms:* 2-Hydroxy-4-methoxybenzophenone; (2-Hydroxy-4-methoxyphenyl) phenylmethanone; Oxybenzone
Anti UVA; Cyasorb UV 9; Escalol 567; Spectra-Sorb UV 9; Uvasorb MET; Uvinul M-40

Benzophenone-4 [CAS #4065-45-6] *Synonyms:* 2-Hydroxy-4-methoxybenzophenone-5-sulfonic acid; Sulisobenzone
Uvasorb S-5; Uvinul MS-40

Benzophenone-6 [CAS #131-54-4] *Synonyms:* 2,2′-Dihydroxy-4,4′-dimethoxybenzophenone
Uvinul D-49

Benzophenone-8 (CTFA) [CAS #131-53-3] *Synonyms:* 2,2′-Dihydroxy-4-methoxybenzophenone
Cyasorb UV 24; Spectra-Sorb UV 24

Benzophenone-9 [CAS #3121-60-6] *Synonyms:* Disodium 2,2′-dihydroxy-4,4′-dimethoxy-5,5′-disulfobenzophenone
Uvinul DS-49

Benzophenone-11 [CAS #1341-54-4]
Uvinul 490

Benzophenone-12 (CTFA) [CAS #1843-05-6] *Synonyms:* 2-Hydroxy-4-(octyloxy) benzophenone; 2-Hydroxy-4-n-octoxybenzophenone; Octabenzone
Carstab 700, 705; Cyasorb UV 531; Ferro UV-Chek AM-300; Hostavin ARO 8; Mark 1413; UV-Chek AM-300/301; Uvinul 408

Benzothiazyl-2-t-butyl sulfenamide. See N-t-Butyl-2-benzothiazolesulfenamide

Benzothiazyl-2-cyclohexyl sulfenamide. See N-cyclohexyl-2-benzothiazolesulfenamide

Benzothiazyl disulfide [CAS #71-43-2] *Synonyms:* Benzothiazole disulfide (CTFA); Benzothiazolyl disulfide; 2-Benzothiazolyl disulfide; Bis (benzothiazolyl) disulfide); Bis (2-benzothiazyl) disulfide); Di-2-benzothiazolyldisulfide; Dibenzothiazyl disulfide; 2,2′-Dibenzothiazyldisulfide; Dibenzoylthiazyl disulfide; Dibenzthiazyl disulfide; 2,2′-Dithiobis (benzothiazole); 2-Mercaptobenzothiazoledisulfide; 2-Mercaptobenzothiazyldisulfide
Amax No. 1; MBTS, MBTS Pellets

2-Benzothiazyl-N-morpholine disulfide
Accelerator MF

Benzothiazyl-2-sulfene morpholide
Vulkacit MOX-90/SG, MOZ/LG, MOZ/SG

1H-Benzotriazole [CAS #95-14-7] *Synonyms:* Benzisotriazole; 1,2-Aminozophenylene; Benzotriazole
Cobratec 99, PT

Benzotriazole, substituted
Tinuvin 109, 326, 328, 1130

Benzoxazole [CAS #273-53-0] *Synonyms:* 1-Oxa-3-azaindene
Ecco Polyester Optical 525; Ecco White OP

Benzoyl peroxide [CAS #94-36-0] *Synonyms:* Dibenzoyl peroxide; Benzoperoxide
Cadet BPO-70, BPO-70W, BPO-78, BPO-78W; Cadox 40E, BCP, BEP-50, BFF-50, BFF-60W, BP-55, BS, BT-50, BTA; Lucidol-70, -78, -98; Luperco AA, ACP, AFR-250, AFR-400, AFR-500, AFR-501, ANS, ANS-P, AST, ATC; Superox 744

Benzyl butyl phthalate [CAS #85-68-7] *Synonyms:* BBP; 1,2-Benzenedicarboxylic acid, butyl phenylmethyl ester; Butyl benzyl phthalate
Bonding Agent TN/S50; Unimoll BB

o-Benzyl-p-chlorophenol [CAS #120-32-1]
Dowicide OBCP

Benzyldimethyl ketal
Escacure KB1, KB60

Benzyloctyl adipate
Adimoll BO

Benzyl phthalate
Santicizer 278

Benzyltrimonium chloride *Synonyms:* Benzyl trimethyl ammonium chloride
Arosurf PT 20; Variquat B200

Benzyltrimonium hydrolyzed animal protein (CTFA) [RD #977066-14-0]
Crotein BTA

BHA (CTFA) [CAS #25013-16-5] *Synonyms:* Butylated hydroxyanisole
Sustane 3, 4A, 6, 8, 31, BHA, BHA 1-F, HW-4, P, PA, Q, W; Tandem 5K, 8, 552; Tenox BHA

BHT (CTFA) [CAS #128-37-0] *Synonyms:* Butylated hydroxytoluene; 2,6-Di-t-butyl-p-cresol; 2,6-t-Butyl-4-methylphenol
CAO-1, -3, -3 Blend 29; Hydagen DEO; Ionol, Ionol CP; Naftonox BHT; Spectratech CM 11340, KM 11264; Sustane 6, BHT, HW-4, P, W; Topanol OC, OF; Ultranox 226; Vanlube PC; Vanox PC, PCX; Vulkanox KB

Biphenyl [CAS #92-52-4] *Synonyms:* Bibenzene; 1,1′-Biphenyl; Diphenyl; Limonene; Phenadox-X; Phenylbenzene; PHPH; Xenene
Ibercarrier 300-L

Biphenyl, alkylated
Katanol 387

3(N,N-Bisacetoxyethyl) amino acetanilide [CAS #27059-08-1]
Emery 5758

3(N,N-Bisacetoxyethyl) amino benzanilide [CAS #43051-43-0]
Emery 5761

3(N,N-Bisacetoxyethyl) amino-4-methoxy acetanilide [CAS #23128-51-0]
Emery 5753

N,N-Bisacetoxyethyl aniline [CAS #19249-34-4]
Emery 5704

N,N-Bisacetoxyethyl-m-toluidine [CAS #21615-36-1]
Emery 5713

Bisacrylate of POE tetrabromobisphenol A
BA-43

Bis (allyl ether) of tetrabromobisphenol A
BE-51

Bis (4-t-butylcyclohexyl) peroxydicarbonate *Synonyms:* Di-(4-t-butylcyclohexyl) peroxydicarbonate
Perkadox 16, 16/35, 16N, 16-W40, 16/W70

α,α′-Bis (t-butylperoxy) diisopropylbenzene [CAS #25155-25-3]
Luperox 802, 802-40KE; Perkadox 14, 14/40, 14/40 Bpd; Varox VC-40KE

Bis (t-butylperoxy isopropyl) benzene
Akrochem VC-40C, VC-40K; Akroform VC-40 EPMB; Akrosperse VC-40 EPMB

1,1-Bis (t-butylperoxy)-3,3,5-trimethylcyclohexane [CAS #6731-36-8]
Luperco 231-KE; Varox 231, 231-XL

Bis (o-chlorobenzoyl) peroxide
Cadox OS

Bis (p-chlorobenzoyl) peroxide [CAS #94-17-7] *Synonyms:* Di-(4-chlorobenzoyl) peroxide
Cadox PS

Biscumylphenyl trimellitate
Kenplast ESI

Bis (2,3-dibromopropyl ether) of tetrabromobisphenol A
PE-68

2,2-Bis [4-(2-(3,5-di-t-butyl-4-hydroxyhydrocinnamoyloxy) ethoxyphenyl] propane
Topanol 205

Bis (2,4-di-t-butylphenyl) pentaerythritol diphosphite
Ultranox 624. 626, 626A

Bis (2,4-dichlorobenzoyl) peroxide [CAS #80-43-3]
Cadox TS-50

Bis-diethanolamine dianilino stilbene disulfonate deriv.
Tinopal LPW

Bis diisobutyl maleate
Staflex DIBCM

N,N-Bis (3-dimethylaminopropyl)-N-isopropanolamine
Texacat ZR-50

N,N′-Bis (1,4-dimethylpentyl)-p-phenylenediamine [CAS #3081-14-9] *Synonyms:* Diheptyl-p-phenylene-diamine
Santoflex 77

Bis [4-(diphenylsulfonio)phenyl] sulfide-bishexafluorophosphate
Degacure KI 85

Bis (epoxypropyl) aniline resin
Lekutherm X 50

Bis (2-ethylhexyl) peroxydicarbonate *Synonyms:* Dioctyl peroxydicarbonate
Trigonox EHP, EHP-AT70, EHP-C40, EHP-C75

Bis (glycidoxypropyl) tetramethyldisiloxane [CAS #126-80-7]
CB2405

N,N-Bis (2-hydroxyethyl) alkylamine
Chemstat 122, 122/60DC, 172, 182, 182/67DC, 192; Varstat K-22, T-22

Bis (hydroxyethyl) aminopropyltriethoxy silane [CAS #7538-44-5]
CB2408

Bis (2-hydroxyethyl) coco amine [CAS #61791-31-9]
Antistatic Agent 273-C; Armostat 475

Bishydroxyethyl dihydroxypropyl stearaminium chloride (CTFA)
Monaquat TG

N,N-Bis (2-hydroxyethyl)-N-(3-dodecyloxy-2-hydroxypropyl) methyl ammonium methosulfate
Cyastat 609

Bis (2-hydroxyethyl ether) of tetrabromobisphenol A
BA-50, -50P

Bishydroxyethylisodecyloxypropylamine oxide
AO-14-2

Bis (2-hydroxyethyl) octyl methyl ammonium p-toluene sulfonate
Antistatic Agent 106G-90%

Bis-2-hydroxyethyl stearamine
Antistatic Agent 273-E

Bis (2-hydroxyethyl) tallow amine [CAS #61791-44-4]
Armostat 375

Bis [2-hydroxy-5-t-octyl-3-(benzotriazol-2-yl) phenyl] methane [CAS #103597-45-1]
Mixxim BB/100

1,8-Bis (hydroxyphenyl) pentadecane [CAS #68390-52-3]
Cardolite NC-512

Bis-(N-methylbenzamide) ethoxymethyl silane [CAS #16230-35-6]
CB2409.5

Bis-methylethanolamino dianilino stilbene disulfonate deriv.
Tinopal 5BM Extra Conc.

Bismuth compound
Perlex B.67, B.70, B.67M, BU, BUA.35; Perlextra B.70, BU, BUA.70

Bismuth dimethyl dithiocarbamate [CAS #21260-46-8] *Synonyms:* Tris (dimethyldithiocarbamate) bismuth
Bismate; BISMET

Bismuth oxychloride (CTFA) [CAS #7787-59-9] *Synonyms:* Basic bismuth chloride; Bismuthine, chloroxo; Bismuth subchloride
Biju; Bi-Lite; Mibiron; Pearl-Glo

2,4-Bis [(octylthio) methyl]-o-cresol
Irganox 1520

Bis (pentabromophenoxy) ethane
Pyro-Chek 77B

Bisperoxide
Vul-Cup 40KE, R

Bisphenol A [CAS #80-05-7] *Synonyms:* 4,4′-Isopropylidenediphenol (CTFA); 2,2-Bis (4-hydroxyphenol) propane; p,p'-Dihydroxydiphenylpropane
Parabis

Bisphenol A, polybutylated
Vanox 1003, 1004

Bisphenolic phosphite, alkylated-arylated
Vanox 1005

trans-1,2-Bis (n-propylsulfonyl) ethene
Vancide PA, PA Disp.

Bispyrithione (CTFA) [CAS #3696-28-4] *Synonyms:* Pyridine, 2,2′-Dithiobis-, 1,1′-dioxide
Omadine MDS

N,N-Bis-stearoylethylenediamide
Aktiplast AS

Bis (1,2,3,6-tetrahydrobenzaldehyde) pentaerythritol acetal
Vulkazon AFS-50, AFS/LG

N,N′-Bis (2,2,6,6-tetramethyl-4-piperidinyl)-1,6-hexanediamine polymer with 2,4,6-trichloro-1,3,5-triazine and 2,4,4-trimethyl-1,2-pentanamine [CAS #70624-18-9]
Chimassorb 944FL, 944LD

Bis (tribromophenoxy) ethane
FF 680; Firemaster 680

Bis (tri-n-butyltin) oxide [CAS #56-35-9]
Vikol AF-25, LO-25, PX-15

Bis (trichloromethyl) sulfone (CTFA) [CAS #3064-70-8]
Amerstat 294

Bis [3-(triethoxysilyl) propyl] tetrasulfide [CAS #40372-72-3]
CB2494

Bis (trimethoxysilylpropyl) ethylene diamine [CAS #68845-16-9]
CB2493

Bis (3,5,5-trimethylhexanoyl) peroxide
Trigonox 36-C75

Bis (trimethylsilyl) acetamide [CAS #10416-59-8]
B2500; Dynasylan BSA

Bis (trimethylsilyl) carbamate
B2539

Bis (trimethylsilyl) sulfate
B2559.5

Bis (trimethylsilyl) trifluoroacetamide
B2570

Bis (trimethylsilyl) urea [CAS #18297-63-7]
B2595; CB2595

Bis (undecylenic amidopropyl dimethyl glycinate)
Schercotaine UAB

Bitertanol
Baycor; Sibutol

Boramide
Synkad 202, 204

Boric acid DEA
Produkt GS 400, GS 7114/80; Rewocoros RAB 90

Boron nitride
Combat HPC-40, HPF-325, HTP-40, HTP-325, SHP-40, SHP-325, SRG-40, SRG-325

Brominated triaryl phosphate [CAS #2788-11-6]
Kronitex PB-460

Bromine [CAS #7726-95-6]
DD-8133; Nyacol HA-15, N22

Bromoacenaphthylene [CAS #208-96-8D]
Con-BACN

3-Bromochloro-5,5-dimethyl hydantoin [CAS #126-06-7]
GSD 550; Halobrom

5-Bromo-5-nitro-1,3-dioxane (CTFA) [CAS #30007-47-7] *Synonyms:* 1,3-Dioxane, 5-bromo-5-nitro-
Bronidox L, L 5

2-Bromo-2-nitropropane-1,3-diol (CTFA) [CAS #52-51-7] *Synonyms:* 1,3-Propanediol, 2-bromo-2-nitro
Bronopol; Bronopol-Boots

Bromopentaerythritol
Saytex FR-1138

Burgess clay
Di-Cup 40KE; Esperox 10-KXL; Luperco 231-KE; Luperox 802-40KE; Varox DCP-40KE, VC-40KE; Vul-Cup 40KE

Butadiene/acrylonitrile copolymer (CTFA) [CAS #9003-18-3]
Perbunan N Latex 2818

Butadiene-acrylonitrile-organic acid terpolymer
Krynac 110, 211, 211, 231

Butadiene-acrylonitrile rubber. See Acrylonitrile-butadiene rubber

Butadiene rubber. See Polybutadiene rubber

Butadiene-styrene-acrylonitrile. See Acrylonitrile-butadiene-styrene

Butadiene/styrene copolymer *Synonyms:* B/S
SR-7475; Stereon 840, 840A

Butadiene/styrene/vinylidene chloride terpolymer
Darex 529L, 535L

Butadiene-styrene-2-vinylpyridine terpolymer
Pyratex J 1904

1,4-Butanediol
Dabco BDO

1,4-Butanediol diacrylate
SR-213

1,4-Butanediol dimethacrylate
Rhenofit BDMA/S

Butoxy diglycol (CTFA) [CAS #112-34-5] *Synonyms:* PEG-2 butyl ether; Diethylene glycol monobutyl ether; 2-(2-Butoxyethoxy) ethanol; Ethanol, 2-(2-butoxyethoxy)-
Butyl Dioxitol; Dowanol DB; Poly-Solv DB

Butoxyethanol (CTFA) [CAS #111-76-2] *Synonyms:* 2-Butoxyethanol; Ethylene glycol monobutyl ether
Butyl Oxitol; Dowanol EB; Ektasolve EB; Poly-Solv EB

Butoxyethanol acetate
Ektasolve EB Acetate

Butoxyethyl oleate
Kessco Butoxyethyl Oleate; Merrol 4218; Plasthall 325

Butoxyethyl stearate
Kessco Butoxyethyl Stearate; Secoster BS

Butoxy POE-POP glycol
Macol 620, 660, 3520, 5100

Butoxytriglycol [CAS #143-22-6] *Synonyms:* Triethylene glycol monobutyl ether; PEG-3 butyl ether
Poly-Solv TB

Butyl acetate [n- CAS #123-86-4; s- CAS #105-46-4; t- CAS #540-88-5]
Desmodur N-75

Butyl acetoxystearate
Paricin 6

Butyl acetyl ricinoleate (CTFA) [CAS #140-04-5] *Synonyms:* Butyl 12-(acetyloxy)-9-octadecenoate
Flexricin P-6

Butyl alcohol (CTFA) [CAS #71-36-3] *Synonyms: n*-Butyl alcohol, 1-Butanol; *n*-Butanol; Propyl carbinol)
Alfol 4; Tebol 99

Butyl ammonium caprylate
Rewocor BAC; Rewocoros BAC

Butylated d: (dimethylbenzyl) phenol
Wingstay C

Butylated PVP
Ganex P904

Butylated triaryl phosphate
Phosflex 71B

Butylbenzene phthalate
Vinyzene BP-505 S160

N,N-Butyl benzene sulfonamide
Merrol N-303; Plasthall BSA

Butyl benzoate (CTFA) [CAS #136-60-7] *Synonyms:* Benzoic acid-n-butyl ester; n-Butyl benzoate
Dai Cari XBN; Ibercarrier 300-L

N-t-Butyl-2-benzothiazolesulfenamide [CAS #95-31-8] *Synonyms:* Benzothiazyl-2-t-butyl sulfenamide
BBTS; Santocure NS; Vanax NS; Vulkacit NZ

Butyl benzyl phthalate (CTFA) [CAS #85-68-7] *Synonyms:* 1,2-Benzenedicarboxylic acid, butyl phenylmethyl ester
Esperox 570P; Santicizer 160; USP-400P

n-Butyl-4,4-bis (t-butylperoxy) valerate [CAS #995-33-5]
Luperco 230-XL; Lupersol 230; Varox 230-XL

t-Butyl cumyl peroxide
Luperco 801-XL; Lupersol 801; Trigonox T

t-Butyldimethylchloro silane [CAS #18162-48-6]
B2790; CB2790

t-Butyldimethylsililimidazole
B2794

2-(t-Butyldimethylsiloxy) pent-2-en-4-one
B2792

t-Butyldimethylsilyl trifluoromethanesulfonate
B2796

t-Butyldiphenylchlorosilane [CAS #58479-61-1]
B2805; CB2805

1,3-Butylene glycol diacrylate *Synonyms:* 1,3-Butanediol diacrylate
SR-212

1,4-Butylene glycol diacrylate *Synonyms:* 1,4-Butanediol diacrylate
SR-213

OO-t-Butyl O-(2-ethylhexyl) monoperoxycarbonate
Lupersol TBEC

t-Butylhydrazinium chloride
Luperco EFA-4; Luperfoam 40, 329

t-Butyl hydroperoxide [CAS #75-91-2]
TBHP-70; T-Hydro Sol'n.; Trigonox A-W70; USP-800

t-Butyl hydroquinone (CTFA) [CAS #1948-33-0] *Synonyms:* TBHQ; Mono-tert-butyl hydroquinone
Eastman MTBHQ; Sustane 20, 20-3, 20A, 31, TBHQ, TBHQ.4; Tenox TBHQ

4,4´-Butylidenebis (6-t-butyl-m-cresol) [CAS #85-60-9] *Synonyms:* 4,4´-Butylidenebis (3-methyl-6-t-butylphenol)
Santowhite Powd.; Vanox SWP

4,4´-Butylidene bis (2-t-butyl-5-methylphenol)
Naftonox BBM

t-Butyl mercaptan *Synonyms:* 2-Methyl-2-butanethiol
Spotleak 1001, 1003, 1007, 1009, 1044, 2323

Butylmercaptide
Stanclere T160, 161, 163, 164, 165, 174, 182, 183 184, 192, 193, 194

Butyl methacrylate (CTFA) [CAS #97-88-1] *Synonyms:* Butyl 2-methyl-2-propenoate
Acryloid F-10

Butyl myristate (CTFA) [CAS #110-36-1] *Synonyms:* Butyl n-tetradecanoate)
Bumyr; Radia 7070; Wickenol 141

Butyl naphthalene sodium sulfonate
Emkal BNS, BNX Powder

Butyl naphthalene sulfonic acid
Emkal BNS Acid

Butyl octyl phthalate
Kodaflex HS-3; Staflex BOP

Butyl oleate (CTFA) [CAS #142-77-8] *Synonyms:* Butyl 9-octadecenoate
Emerest 2328; Graden Butyl Oleate; Kemester 4000; Kessco Butyl Oleate; Merrol 418T; Plasthall 503; Radia 7040

Butylparaben (CTFA) [CAS #94-26-8] *Synonyms:* Butyl p-hydroxybenzoate
Lexgard B; LiquaPar; Trisept B

t-Butyl peracetate [CAS #107-71-1] *Synonyms:* t-Butyl peroxyacetate
Lupersol 70, 75M, 76-M; Trigonix F-C50

t-Butyl perbenzoate [CAS #614-45-9] *Synonyms:* t-Butyl peroxybenzoate
Esperox-10, -10KXL, -10XL; Trigonox C

t-Butyl peroctoate [CAS #62695-55-0] *Synonyms:* t-Butyl peroxy 2-ethyl hexanoate
Esperox-28, -28MD, -28PD; Lupersol P-31, PDO, PMS; Trigonox 21, 21-C50, 21-OP50; USP-690

t-Butyl peroxyacetate. See t-Butyl peracetate

t-Butyl peroxybenzoate. See Butyl perbenzoate

t-Butyl peroxycrotonate [CAS #23474-91-1]
Esperox 13M

t-Butyl peroxy 2-ethylhexanoate See Butyl peroctoate

t-Butyl peroxy 2-ethylhexyl carbonate [CAS #34443-12-4]
Esperox C-496

t-Butyl peroxyisobutyrate [CAS #109-13-7] *Synonyms:* t-Butyl perisobutyrate
Lupersol 80

t-Butyl peroxy isopropyl carbonate [CAS #2372-21-6] *Synonyms:* BPIC
Lupersol TBIC-M75; Trigonox BPIC

t-Butyl peroxymaleic acid [CAS #1931-62-0] *Synonyms:* t-Butyl permaleic acid
Experox-41-25; Luperco PMA-25, PMA-40; Lupersol PMA; Luperox PMA

t-Butyl peroxy 2-methyl benzoate
Esperox-497M; Trigonox 97-C75

t-Butyl peroxyneodecanoate
Esperox-33M; Lupersol 10, 10-M75; Trigonox 23-C75

t-Butyl peroxyneoheptanoate [CAS #26748-38-9]
Esperox 750M

t-Butyl peroxyneohexanoate
Lupersol 601

t-Butyl peroxypivalate [CAS #927-07-1]
Esperox-31M; Lupersol 11; Trigonox 25-C75

(p-t-Butylphenethyl) dimethylchloro silane
B2809; DB2809

t-Butylphenyl diphenyl phosphate
Santicizer 154

Butyl ricinoleate
Flexricin P-3

Butyl stearate (CTFA) [CAS #123-95-5] *Synonyms:* Butyl octadecanoate
Crodamol BS; Emerest 2321, 2325, 2326; Kemester 5510; Kessco BSC, Butyl Stearate Cosmetic, Distilled; Lexate TL; Lipal EB; Merrol 418S; Radia 7051; Rilanit BS; Unimate BYS

Butyl thiotin
Lankromark BT050, BT105, BT120, BT120A, BT190, BT339

Butyltin carboxylate
Advastab T-52N Conc., T-290; Irgastab T 6343; Lankromark BL277, BM271, BM286, BM400; Stanclere T-55, T-85, T-876, T-877, T-878, TL

Butyltin mercaptide
Advastab TM-180; Dabco T-120; Stanclere T-94 C, T-801

t-Butyl toluenediamine
XCE-89

But-2-yne-1,4-diol [CAS #110-65-6] *Synonyms:* 1,4-Butynediol
Korantin BH Liq., BH Solid

Butynediol hydroxyethyl ethers
Butoxyne 497

Butyraldehyde-aniline condensation product [CAS #34562-31-7]
Vanax 808

Butyraldehyde-aniline condensation product [CAS #68411-20-1]
Vanox AT

Butyraldehyde-monobutylamine condensation prod. [CAS #68411-19-8]
Vanax 833

m-Butyraldehyde oxime [CAS #110-69-0] *Synonyms:* Butyraldoxime
Troykyd Anti-Skin BTO

Butyrolactone [CAS #96-48-0] *Synonyms:* γ-Butyrolactone
BLO

C5 hydrocarbon resin
Wingtack 10, 86, 115, Extra, Plus

C6 alcohol
Epal 6

C6-10 acid triglyceride
Softenol 3819

C6-10 alcohol
Alfol 610, 610ADE; Epal 108, 610

C7 acid triglyceride
Softenol 3107

C8 alcohol
Epal 8

C8-10 acid-1,2-propanediol ester
Softenol 3408

C8-10 acid triglyceride
Softenol 3108, 3829

C8-10 alcohols
Alfol 810, 810FD; Epal 810

C8-10 ether amine
Adogen 188, 588

C8-10 ether amine acetate
Arosurf MG-98A, MG-98A3

C8-10 trimethyl ammonium chloride
Adogen 464

C8-18 acid triglyceride
Softenol 3178

C9-11 alcohols (CTFA) [CAS #68551-08-6]
Dobanol 91; Neodol 91

C9-11 ether amine
Arosurf MG-91

C9-11 ether amine acetate

Arosurf MG-91A3

C9-11 pareth-3 [CAS #68439-46-3 (generic)] *Synonyms:* Pareth-91-3; PEG-3 C9-11 alcohol ether
Dobanol 91-2.5; Lenetol 9130 AH; Neodol 91-2.5

C9-11 pareth-5 *Synonyms:* Pareth-91-5; PEG-5 alkyl C9-11 alcohol
Teric G9A5

C9-11 pareth-6 [CAS #68439-46-3 (generic)] *Synonyms:* Pareth-91-6; PEG-6 C9-11 alcohol ether
Neodol 91-6; Teric G9A6

C9-11 pareth-8 [CAS #68439-46-3 (generic)] *Synonyms:* Pareth-91-8
Neodol 91-8; Teric G9A8

C9-11 pareth-12
Teric G9A12

C10 ether amine acetate
Arosurf MG-70A5

C10-12 alcohols
Alfol 1012; Epal 1012

C10-12 undecylbenzene sulfonate
Nalkylene 500

C10-14 dodecylbenzene sulfonate
Nalkylene 550L

C10-30 cholesterol/lanosterol esters
Super Sterol Ester

C11 alcohol
Neodol 1

C11-15 pareth-3 *Synonyms:* Pareth-15-3
Tergitol 15-S-3

C11-15 pareth-5 *Synonyms:* Pareth-15-5
Tergitol 15-S-5

C11-15 pareth-7 *Synonyms:* Pareth-15-7
Tergitol 15-S-7

C11-15 pareth-9 *Synonyms:* Pareth-15-9
Tergitol 15-S-9

C11-15 pareth-12 *Synonyms:* Pareth-15-12
Tergitol 15-S-12

C11-15 pareth-15 *Synonyms:* Pareth-15-15
Tergitol 15-S-15

C11-15 pareth-20 *Synonyms:* Pareth-15-20
Tergitol 15-S-20

C11-15 pareth-30 *Synonyms:* Pareth-15-30
Tergitol 15-S-30

C11-15 pareth-40 *Synonyms:* Pareth-15-40
Tergitol 15-S-40

C12 benzene
Aristol E

C12 dibasic acid [CAS #693-23-2]
Corfree M2, M3

C12-13 alcohols (CTFA) [RD #977063-82-3]
Dobanol 23; Neodol 23

C12-13 pareth-2 *Synonyms:* Pareth-23-2
Dobanol 23-2

C12-13 pareth-3 [CAS #66455-14-9 (generic)] *Synonyms:* Pareth-23-3; PEG-3 C12-13 fatty alcohol ether
Dobanol 23-3; Neodol 23-3

C12-13 pareth-7 [CAS #66455-14-9 (generic)] *Synonyms:* Pareth-23-7; PEG-7 C12–13 fatty alcohol ether
Dobanol 23-6.5; Neodol 23-6.5

C12-14 alcohols
Alfol 1214, 1214-GC, 1216-CO; Epal 12/70, 12/85

C12-14 alkoxypropyl-1,3-diaminopropane
Tomah DA-62

C12-14 alkoxypropyl-1,3-diaminopropane ditallate
Tomah DA-62 Ditallate

C12-14 pareth-3 *Synonyms:* PEG-3 C12-14 alcohol
Serdox NEL 3

C12-14 pareth-4 hydrogen phosphate *Synonyms:* PEG-4 C12-14 alcohol hydrogen phosphate
Merpiphos PP

C12-14 pareth-8 *Synonyms:* PEG-8 C12-14 fatty alcohol
Lutensol A 8

C12-14 pareth-12 *Synonyms:* PEG-12 C12-14 alkyl ether
Serdox NEL 12/80

C12-14 tridecylbenzene sulfonate
Nalkylene 600L

C12-15 alcohols (CTFA) [RD #977063-83-4]
Dobanol 25; Neodol 25

C12-15 alcohols benzoate [RD #977039-90-9]
Finsolv TN

C12-15 alcohols lactate (CTFA) [RD #977064-17-7]
Ceraphyl 41

C12-15 ether amine
Adogen 185

C12-15 pareth-2 phosphate *Synonyms:* Pareth-25-2 phosphate; PEG-2-C12-15 alcohols phosphate
Crodafos 25 D2 Acid; Nikkol TDP-2

C12-15 pareth-3 [CAS #68131-39-5 (generic); RD #977061-52-1] *Synonyms:* Pareth-25-3; PEG-3 C12–15 fatty alcohol ether
Alkasurf LA-3; ; Dobanol 25-3; Neodol 25-3; Tergitol 25-L-3; Teric 12A3 (2.6 EO); Volpo 25D3

C12-15 pareth-4
Teric G12A4 (4.5 EO), LA4 (4.5 EO)

C12-15 pareth-5 *Synonyms:* PEG-5 C12-15 alkyl ether
Tergitol 25-L-5; Volpo 25D5

C12-15 pareth-5 phosphate
Crodafos 25 D5 Acid

C12-15 pareth-6 *Synonyms:* PEG-6 C12-15 alcohol
Teric 12A6, G12A6

C12-15 pareth-7 [CAS #68131-39-5 (generic); RD #977054-91-3] *Synonyms:* Pareth-25-7; PEG-7 C12–15 fatty alcohol ether
Alkasurf LA-7; Tergitol 25-L-7

C12-15 pareth-8 *Synonyms:* PEG-8 C12-15 alkyl ether
Teric G12A8.(7.8 EO), LA8 (7.8 EO)

C12-15 pareth-9 [CAS #68131-39-5 (generic)] *Synonyms:* Pareth-25-9
Tergitol 25-L-9; Teric 12A9

C12-15 pareth-10
Volpo 25D10

C12-15 pareth-10 phosphate
Crodafos 25 D10 acid

C12-15 pareth-12 *Synonyms:* Pareth 25-12
Tergitol 25-L-12; Teric 12A12 (11.5 EO), G12A12 (11.5 EO)

C12-15 pareth-15
Sterling Emulsifier #15; Volpo 25D15

C12-15 pareth-16
Teric 12A16

C12-15 pareth-23
Teric 12A23

C12-15 pareth-25 *Synonyms:* PEG-25 C12-15 alkyl ether
Volpo 25D20

C12-16 alcohols [CAS #68855-56-1]
Alfol 1216, 1412; Epal 1214

C12-18 acid triglyceride [CAS #67701-26-2]
Softenol 3100

C12-18 alcohols
Alfol 1218; Epal 1218

C13-15 alkyl bis (2-hydroxyethyl) amine oxide
Synprolam 35X2/0

C13-15 alkyl-1,3-propane diamine
Synprolam 35N3

C13-15 alkyl trimonium chloride
Synprolam 35TMQC

C13-15 amine
Synprolam 35, 35A

C13-15 dimethyl amine benzyl ammonium chloride
Synprolam 35BQC (50), 35BQC (80)

C13-15 dimethyl amine oxide
Synprolam 35DMO

C13-15 methyl amine
Synprolam 35M

C14 acid triglyceride
Softenol 3114

C14 ether diamine acetate
Arosurf MG-84A3

C14–15 alcohols (CTFA)
Dobanol 45; Neodol 45

C14-16 alcohols
Epal 1416

C14-18 alcohols
Alfol 1418-DDB; Epal 1418

C14-18 alkyloxypropylamine
Tomah PA-1820

C16-18 alcohols
Alfol 1618; Epal 618, 1618; Interwax G 8206

C16-18 benzene
Aristol B, D

C16-18 pareth-4 hydrogen phosphate *Synonyms:* PEG-4 C16-18 alcohol hydrogen phosphate
Merpiphos TP

C16-18 pareth-8 hydrogen phosphate *Synonyms:* PEG-8 C16-18 alcohol hydrogen phosphate
Merpiphos MP, VP

C16-18 pareth-11 *Synonyms:* PEG-11 sat. C16-18 alcohol
Lutensol AT 11

C16-18 pareth-18
Lutensol AT 18

C16-18 pareth-25
Lutensol AT 25

C16-18 pareth-50
Lutensol AT 50

C16-18 pareth-80
Lutensol AT 80

C16-20 alcohols
Alfol 1618-CG, 1620

C18 acid triglyceride
Softenol 3118

C18 alcohol
Epal 18NF

C18-22 alcohols
Lanette 18-22

C18-36 acid (CTFA) [CAS #112-85-6]
Syncrowax AW1C

C18-36 acid glycol ester (CTFA)
Syncrowax ERL-C

C18-36 acid triglyceride (CTFA)
Syncrowax HGL-C

C24–28 alkyl methicone
Abil-Wax 9810

C30 alcohol
Unilin 425

C40 alcohol
Unilin 550

C50 alcohol
Unilin 700

Cadmium [CAS #7440-43-9] *Synonyms:* Cd
Lankromark LC68, LC244, LC299, LC310, LC431, LC475, LC486, LC541, LC563, LC585, LC629, LC651, LC662

Cadmium diamyldithiocarbamate
Amyl Cadmate

Cadmium diethyldithiocarbamate [CAS #14239-68-0]
Ethyl Cadmate

Cadmium-zinc
Lankromark LC90; Vanstay RAZ

Calcium [CAS #7440-70-2]
Ircogel 900, 901; Troymax Drier Calcium 4%, 5%, 6%, 8%, 10%

Calcium alginate [CAS #9005-35-0]
Keltose; Margel

Calcium alumino silicate
PMF Fiber 204, 204AX, 204BX, 204CX, 204EX

Calcium behenate
Syncrowax HRS-C

Calcium carbonate (CTFA) [CAS #471-34-1] *Synonyms:* Carbonic acid, calcium salt
Calwhite, II; Camel-CAL, Slurry, ST; Camel-CARB; Camel-FIL; Camel-TEX; Camel-WITE, Slurry, ST; D-70; Di-Cup 40C; Esperox-10XL; GamaCo, II; Gama-Fil 55, 90, D-1, D-2; Gama-Plas; Gama-Sperse 5, 80, 140, 255, 6451, 6532, 6532 NSF, CS-11; G-White; Hi-Pflex, 100; H-White; Mar'blend; Q-White; RO-40; Rock Dust; Super-Pflex 100, 200; Troykyd Anti-Float Powd.; Varox DCP-40C; Windale White; Winnofil S, SP; XO White

Calcium carboxylate
Lubrizol 2117

Calcium dinonylnaphthalene sulfonate
Na-Sul 729, CA-50

Calcium disodium EDTA (CTFA) [CAS #62-33-9]
Kelate CDS; Versene CA

Calcium dodecylbenzene sulfonate [CAS #26264-06-2]
Alkasurf CA; Conco AAS-75S; Nansa EVM50, EVM62/H, EVM70, EVM70/B, EVM70/E; Ninate ABS-60C

Calcium DTPA *Synonyms:* Calcium diethylene triamine pentaacetic acid; Calcium pentetate
Plex PC

Calcium hydroxide (CTFA) [CAS #1305-62-0] *Synonyms:* Calcium hydrate; Hydrated lime
Cri-Spersion CRI-ACT-45, -45-1/1, -45-LV; Plastigel PG9089, PG9104; Rhenofit CF

Calcium hypochlorite [CAS #7778-54-3] *Synonyms:* Calcium oxychloride
Pittclor

Calcium lignosulfonate (CTFA) [CAS #8061-52-7]
Lignosite, 401, 704, 1840; Lignosol B, BD, LC, SF, SFL; Marasperse C-21; Norlig, A; PB-92

Calcium montanate
Interwax G 8259

Calcium naphthalene sulfonate
Tamol NHC 3001, NHC 4001, NHC 9301, NMC 9301

Calcium oxide (CTFA) [CAS #1305-78-8] *Synonyms:* Lime
Plastigel PG9068; Rhenosorb C, C/GW, F

Calcium saccharin [CAS #6485-34-3]
Syncal CAS

Calcium silicate (CTFA) [CAS #10101-39-0] *Synonyms:* Silicic acid, calcium salt
Micro-Cel A, B, C, E, T-38, T-70; Silasorb

Calcium sodium lignosulfonate
Lignosol R30

Calcium stearate (CTFA) [CAS # 1592-23-0]
Akrochem Calcium Stearate; Calcium Stearate Regular; Gama-Sperse CS-11; Hy Dense Calcium Stearate HP Gran., RSN

Powd.; Interstab CA-18-1; Interwax M 3142; Lubracal 48, 53, 60; Nopcote C-104; Norfox CS; Petrac CP-11 LS, CP-11 LSG, CP-12, CP-22G; Radiacid 1060; Radiastar 1060; Witco Calcium Stearate A, EA, F

Calcium stearoyl lactylate (CTFA) [CAS #5793-94-2]
Patco 3; Verv

Calcium sulfate (CTFA) [CAS #7778-18-9] *Synonyms:* Anhydrite; Calcium sulfonate; Gypsum; Plaster of Paris
Alox 2292; Hybase C-300; Lubrizol 2163

Calcium-zinc
Haro Mix ZC-028, ZC-029, ZC-030, ZC-031, ZC-032, ZC-036, ZC-902; Lankromark LZ495, LZ649, LZ935, LZ1045, LZ1177, LZ1188, LZ1210; Interstab LT-4289, LT-4308, M803, M809, M876, M11289, MP 10581/3, MT11303/1; Therm-Chek 344, 714

Calcium-zinc-molybdenum
Moly-White 212

Calcium-zinc stearate
Petrac CZ-81

Calendula extract (CTFA) [RD #977025-25-4] *Synonyms:* Extract of calendula; Marigold extract
Calendula Oil CLR

Candelilla wax (CTFA) [CAS #8006-44-8]
Cutina LM

Capramide DEA [CAS #136-26-5] *Synonyms:* Capric acid diethanolamide
Alkamide 1509; Monamid 150-CW; Standamid CD

Caproamphodiacetate (CTFA) [CAS #7075-05-9] *Synonyms:* Caproamphocarboxyglycinate
Miranol S2M Conc.

Caproamphodipropionate (CTFA) [CAS #68815-45-2] *Synonyms:* Caproamphocarboxypropionate
Miranol S2M-SF Conc.

Capryl hydroxyethyl imidazoline [CAS #37478-68-5] *Synonyms:* Caprylic imidazoline
Alkateric EHCIB; Miramine CPC; Monazoline CY

Caprylic acid (CTFA) [CAS # 124-07-2]
Industrene 365

Caprylic alcohol (CTFA) [CAS #111-87-5] *Synonyms:* n-Octyl alcohol; 1-Octanol; n-Octanol
Alfol 8; Kalcohl 08H

Caprylic/capric acid
Industrene 365

Caprylic/capric diglyceryl succinate (CTFA)
Miglyol 829

Caprylic/capric glycerides
Imwitor 742

Caprylic/capric/isostearic adipic triglycerides (CTFA) [CAS #82249-33-0]
Softisan 649

Caprylic/capric/lauric triglyceride [RD #977069-11-6]
Captex 350

Caprylic/capric/linoleic triglyceride (CTFA)
Miglyol 818

Caprylic/capric/stearic triglyceride [RD #977064-06-4]
Softisan 378

Caprylic/capric triglyceride (CTFA) [RD #977059-83-8]
Captex 300, 355; Lexol GT 855, GT 865; Liponate GC; Mazol 1400; Miglyol 810, 812, Gel; Myritol 318; Neobee 1223, 1400 Oil, M-5, O; Standamul 318, 7105

Caprylic glycinate
Varion AM-V

Caprylic triglyceride
Captex 8000

Capryloamphodipropionate (CTFA) [CAS #68815-55-4] *Synonyms:* Capryloamphocarboxypropionate
Monateric 811

Capryloamphohydroxypropylsulfonate (CTFA) [CAS #68610-39-9] *Synonyms:* Capryloamphopropylsulfonate
Miranol JS Conc.

Capryloamphopropionate (CTFA)
Monateric CyNa-50%

Carbodiimide [CAS #622-16-2]
Activator 2013-P

Carbofuran [CAS #1563-66-2] *Synonyms:* 2,3-Dihydro-2,2-dimethyl-7-benzofuranylmethylcarbamate
Curaterr

Carbomer 910 (CTFA)
Carbopol 910

Carbomer 934 (CTFA) [CAS #9007-16-3]
Carbopol 934

Carbomer 934P (CTFA) [CAS #9003-01-4; RD #977007-23-0]
Carbopol 934P

Carbomer 940 (CTFA) [CAS #9003-01-4; 9007-17-4]
Carbopol 940

Carbomer 941 (CTFA) [CAS #9003-01-4; RD #977056-22-6]
Carbopol 941

Carbon, activated *Synonyms:* Active carbon; Activated charcoal
Witcarb Columbia JXC, Gran. Activated Carbon 965

Carbon black [CAS #1333-86-4] *Synonyms:* Furnace black; Thermal black; Channel black
Huber ARO 60, N110, N220, N234, N299, N326, N330, N339, N343, N347, N351, N375, N539, N550, N650, N660, N683, N762, N774, N787, N990, S212, S315; Ketjenblack E.C.; Thermax Floform N-990, Floform Ultra Pure N-990, Powder N-991, Powder Ultra Pure N-991, Stainless Floform N-907, Stainless Powder N-908, Stainless Powder Ultra Pure N-908; Vanblack C

Carbon fiber *Synonyms:* Graphite fiber
Thornel Carbon Fiber T-40 12K, T-500 3K, T-500 6K, T-500 12K; Torlon 7233, 7330

Carboxy benzotriazole [CAS #60932-58-3]
Cobratec CBT

Carboxymethylcellulose. See Cellulose gum

Carboxymethyl guar
Progacyl COS-10

Carboxymethyl hydroxyethyl cellulose (CTFA) [CAS #9004-30-2; 9088-04-4] *Synonyms:* Sodium carboxymethyl hydroxyethyl cellulose
CMHEC-37L; Hercules CMHEC-37L

Carboxymethyl mercaptosuccinic acid [CAS #99-68-3]
Evanacid 3CS

Carboxymethylmethylcellulose
Benecel CM; Culminal CMMC

Cardanol
Cardolite NC-700

Cardanol acetate
Cardolite NC-548

Carnauba (CTFA) [CAS #8015-86-9] *Synonyms:* Brazil wax
BASF Wax MCE; Cutina LM

Carotene (CTFA) [CAS #7235-40-7] *Synonyms:* β-Carotene
Carrot Oil CLR

Carrageenan (CTFA) [CAS #9000-07-1] *Synonyms:* Chondrus; Irish moss extract; 3,6-Anhydro-d-galactan; Carrageen
CC-603; Frimulsion 6G; Genu Carrageenan; Genugel; Genulacta Series; Genuvisco

Carrot extract (CTFA) [RD #977027-46-5] *Synonyms:* Daucus carota sativa extract; Extract of carrot
Carrot Oil CLR

Carrot oil (CTFA) [CAS #801-58-8] *Synonyms:* Carrot seed oil; Oils, carrot seed
Carrot Oil CLR

Casein. See Milk protein

Castor oil (CTFA) [CAS #8001-79-4]
AA Standard, AA USP; Bentone Gel CAO; Cosmetol X; Crystal Crown, O; Cutina LM; Diamond Quality; Incroquat CR Conc.; Surfactol 13

Castor oil, hydrogenated. See Hydrogenated castor oil

Catalase
Fermcolase; Milezyme Catalase L

Cellulase
Celluclast 2.0 L Type X, 200L, 200 L Type N; Celluferm; Celluzyme 2400 T; Genencor Cellulase 150L, 300P; Milezyme

Brand Hemicellulase, Cellulase, Cellulase TV Conc.

Cellulose (CTFA) [CAS #9004-34-6] *Synonyms:* Wood pulp, bleached
Celquat; Elcema G 250, P 100; Santoweb D, DX, H, W

Cellulose acetate *Synonyms:* CA; Cellulose acetate ester
Crystic Release Agent No. 1

Cellulose fibers
CF-42,500-T, Med., CF-70,000-WDK, Ex. Superfine, CFR-72,500-T Superfine

Cellulose gum (CTFA) [CAS #9004-32-4] *Synonyms:* Carboxymethyl cellulose; Cellulose, carboxymethyl ether; Sodium carboxymethylcellulose; CMC; Sodium CMC
Aqualon C, R, Cellulose Gum, CMC-T, CMC-Warp Size; Aquasorb A250; Avicel RC-591; Blanose Cellulose Gum, Refined CMC; CMC-6-DG-L, CMC-6CTL, CMC-T, CMTC-T, CMC Warp Size; Hercules Cellulose; Hercules Cellulose Gum, CMC, CMC-T, CMC-Warp Size; Rycel 100, 105; Tylose C, C 30 L, CB Series, CR, CBR Series

Cellulose propionate. See Cellulose acetate propionate

Cellulosic resin
Leoguard G, GP

Ceresin wax (CTFA) [CAS #8001-75-0]
Dehymuls K

Cetalkonium chloride [CAS #122-18-9] *Synonyms:* Cetyl dimethyl benzyl ammonium chloride; Palmityl dimethyl benzyl ammonium chloride
Ammonyx T

Cetamine oxide
Incromine Oxide MC

Ceteareth-2 (CTFA) *Synonyms:* PEG-2 cetyl/stearyl ether; POE (2) cetyl/stearyl ether; PEG 100 cetyl/stearyl ether
Macol CSA-2

Ceteareth-2 phosphate
Crodafos CS2 Acid

Ceteareth-3 (CTFA) [RD #977054-58-2]
Hostacerin T-3; Volpo CS3

Ceteareth-4 (CTFA) [RD #977054-59-3]
Hetoxol CS-4; Lipocol SC-4; Macol CSA-4; Siponic E-2

Ceteareth-5 (CTFA) [RD #977054-60-6]
Hetoxol CS-5; Volpo CS5

Ceteareth-5 phosphate (CTFA)
Crodafos CS5 Acid

Ceteareth-6 [RD #977054-61-7]
Siponic E-3

Ceteareth-9
Hetoxol CS-9

Ceteareth-10 (CTFA) [RD #977054-63-9]
Lipocol SC-10; Macol CSA-10; Siponic E-5; Volpo CS10

Ceteareth-10 phosphate
Crodafos CS10 Acid

Ceteareth-11 (CTFA)
Rewopal CSF 11

Ceteareth-12 (CTFA) [RD #977054-64-0]
Eumulgin B-1; Incropol CS-12; Standamul B-1

Ceteareth-14
Siponic E-7

Ceteareth-15 (CTFA) [RD #977054-65-1]
Hetoxol 15 CSA, CS-15; Lipal 15 CSA; Lipocol SC-15; Macol CSA-15; Volpo CS15

Ceteareth-16
Teric 16A16

Ceteareth-20 (CTFA) [RD #977054-66-2]
Acconon W-230; Cosmowax J, K, P; Empilan KM 20; Emthox 5885; Eumulgin B-2; Hetoxol CS-20; Incropol CS-20; Lipocol SC-20; Lipowax D, G; Macol CSA-20; Promulgen D, G; Siponic E-10; Standamul B-2; Volpo CS20

Ceteareth-22
Teric 16A22

Ceteareth-25
Softisan 601

Ceteareth-29
Teric 16A29

Ceteareth-30 (CTFA) [RD #977063-72-1]
Eumulgin B-3; Hetoxol CS-30; Incropol CS-30; Siponic E-15; Standamul B-3

Ceteareth-40 (CTFA)
Incropol CS-40; Ceteareth-50; Hetoxol CS-50, CS-50 Special; Incropol CS-50

Ceteareth-60
Incropol CS-60

Cetearyl alcohol (CTFA) [CAS #8005-44-5] *Synonyms:* Cetostearyl alcohol; Cetyl/stearyl alcohol
Adol 640; Amphocerin K; Cetax 50; Cosmowax J, P; Crodacol CS50; Crodex A, C, N; Cutina CBS; Cyclochem EM 560, POL; Dehydag Wax N, O, SX, W; Empiwax SK, SK/BP; Hydrenol DD; Incroquat Behenyl TMS; Incroquat CR Conc.; Kalcohl 68, 86; Lanbritol Wax N21; Lanette O, SX, W, Wax CAT, Wax SX, Wax SXBP; Lanol CS; Laurex CS, CS/D; Lipowax; Lipowax D, G, NI; Maquat SC-1632; Mazawax 163R; Nioix EO-26; Promulgen D; Standamul Conc. 1002; Varisoft CRC

Cetearyl isononanoate (CTFA) *Synonyms:* Isononanoic acid, cetyl/stearyl ether
Cetiol SN; Rilanit CSN

Cetearyl lactate
Crodamol CSL

Cetearyl octanoate (CTFA) [CAS #8005-44-5]
Crodamol CAP; Luvitol EHO; Schercemol 1688

Cetearyl palmitate (CTFA) [RD #977069-12-7] *Synonyms:* Hexadecanoic acid, cetyl/stearyl ether
Crodamol CSP

Cetearyl stearate
Crodamol CSS

Ceteth-2 (CTFA) [CAS #9004-95-9 (generic); RD #977054-67-3] *Synonyms:* PEG-2 cetyl ether; POE (2) cetyl ether; PEG 100 cetyl ether
Ethosperse CA-2; Hetoxol CA-2; Lexate CRC; Lipocol C-2

Ceteth-5 (CTFA) [CAS #9004-95-9 (generic)]
Solulan 5

Ceteth-10 (CTFA) [CAS #9004-95-9 (generic), #9035-85-2; RD #977054-70-8]
Brij 56; Hetoxol CA-10; Lipocol C-10

Ceteth-12 (CTFA) [CAS #9004-95-9 (generic); RD #977048-84-2]
Lanbritol Wax N21

Ceteth-16 (CTFA) [CAS #9004-95-9 (generic); RD #977064-09-7]
Solulan 16

Ceteth-20 (CTFA) [CAS #9004-95-9 (generic); RD #977042-39-9]
Brij 58; Crodex N; Cyclochem POL; Dermalcare C-20; Hetoxol CA-20; Lipocol C-20; Lipowax NI; Tefose 2000, 2561

Ceteth-24 [CAS #9004-95-9 (generic)]
Solulan C-24

Ceteth-25 (CTFA) [CAS #9004-95-9 (generic); RD #977069-13-8]
Solulan 25

Cetethyldimonium bromide (CTFA) [CAS #124-03-8] *Synonyms:* Cetyl dimethyl ethyl ammonium bromide; Quaternium-17
Bretol; Cycloton D256B/99

Cetoleth-2 *Synonyms:* PEG-2 cetyl/oleyl ether; POE (2) cetyl/oleyl ether
Teric 17A2

Cetoleth-3
Teric 17A3

Cetoleth-6
Teric 17A6

Cetoleth-8
Teric 17A8

Cetoleth-10
Lubrol 17A-10; Teric 17A10

Cetoleth-13
Teric 17A13

Cetoleth-25 (CTFA)
Teric 17A25

Cetrimonium bromide (CTFA) [CAS #57-09-0] *Synonyms:* Cetyl trimethyl ammonium bromide
Acetoquat CTAB; Bromat; Cycloton M242B/99; Lanette Wax CAT

Cetrimonium chloride (CTFA) [CAS #112-02-7] *Synonyms:* Cetyl trimethyl ammonium chloride; Palmityl trimethyl ammonium chloride; Hexadecyl trimethyl ammonium chloride
Adogen 444; Ammonyx CETAC, CETAC-30; Arquad 16-29, 16-50; Bio-Quat ASH-29; Carsoquat CT 429, CTM-29, CTM-429; Chemquat 16-50; Cycloton M242C/29; Dehyquart A; Genamin CTAC; Incroquat CTC-25, CTC-30; Nissan Cation PB-40, -300; Querton 16 CL 29, 16 CL50, 24 CL 35; Radiaquat 6444; Synoquart P 50; Tequat BC; Variquat E228; Varisoft E-228, E-290

Cetrimonium methosulfate *Synonyms:* Cetyl trimethyl ammonium methosulfate
Catigene CT 30

Cetrimonium tosylate (CTFA) [CAS #138-32-9] *Synonyms:* Cetyl trimethyl ammonium *p*-toluene sulfonate
Cetats

Cetyl acetate [CAS #629-70-9] *Synonyms:* Hexadecyl acetate
Acetol 1706; Acetulan; Crodalan AWS, LA; Crodamol AC; Ethoxyol 1707; Ritawax ALA; Solulan 97, 98

Cetyl alcohol (CTFA) [CAS #36653-82-4] *Synonyms:* Palmityl alcohol; C16 linear primary alcohol; n-Hexadecyl alcohol; 1-Hexadecanol
Adol 52, 52 NF, 54, 520, 520 NF; Alfol 16; Amerchol 400; Cachalot C-50, C-51; Cetal; Cetax 16; CO-1695; Crodacol C, C70, C95NF; Cutina LM; Dehydag Wax 16; Emery 1787; Epal 16NF; Incroquat B65C; Kalcohl 60; Lanette 16; Lanol C; Lipocol C; Michel XO-144, XO-144B; Niox EO-10; Rilanit G 16

Cetyl betaine (CTFA) [CAS #693-33-4]
Darvan NS; Product BCO

Cetyl C12-15 pareth-9-carboxylate
Velsan P 8-16

Cetyl dimethicone
Abil-Wax 9801

Cetyl dimethicone copolyol
Abil B 9806

Cetyl esters (CTFA) [CAS #8002-23-1; RD #977067-67-6] *Synonyms:* Synthetic spermaceti wax
Cetina, TE; Crodamol SS; Cyclochem EM 560, SPS; Dermalcare SPS; Kessco 654, Synthetic Spermaceti N.F.; Liponate SPS; Lipowax; Spermwax; Starfol Wax CG

Cetyl ethyl... See Cetethyl...

Cetyl isooctanoate
Tegosoft 168

Cetyl lactate (CTFA) [CAS #35274-05-6] *Synonyms:* n-Hexadecyl lactate
Cegesoft C 19; Ceraphyl 28; Crodamol CL; Cyclochem CL; Lipal LC; Liponate CL; Schercemol CL

Cetyl myristate (CTFA) [CAS #2599-01-1] *Synonyms:* Tetradecanoic acid, hexadecyl ester
Schercemol CM

Cetyl octanoate (CTFA) [CAS #59130-69-7; RD #977063-15-2] *Synonyms:* Cetyl 2-ethylhexanoate; n-Hexadecyl 2-ethylhexanoate
Exceparl HO; Schercemol CO

Cetyl palmitate (CTFA) [CAS #540-10-3]
Crodamol CP; Cutina BW, CBS, CP, CP-A; Cyclochem CP; Kemester CP; Kessco 653; Nikkol N-SP; Radia 7500; Schercemol CP; Standamul 1616; Starfol CP; Waxenol 815, 816

Cetylpyridinium bromide
Acetoquat CPB

Cetylpyridinium chloride (CTFA) [CAS #123-03-5] *Synonyms:* Hexadecylpyridinium chloride
Acetoquat CPC; CPC

Cetyl ricinoleate (CTFA) [CAS #10401-55-5] *Synonyms:* Hexadecyl 12-hydroxy-9-octadecenoate
Liponate CRM; Naturechem CR

Cetyl stearate (CTFA) [CAS #1190-63-2] *Synonyms:* n-Hexadecyl stearate
Radia 7501; Schercemol CS

Cetyl/stearyl. See Cetearyl ...

Cetyl trimethyl ammonium... See Cetrimonium...

Cherry pit oil (CTFA) [CAS #0822-29-5] *Synonyms:* Cherry kernel oil
Lipovol CP

Chinomethionat
Morestan

Chlorendic anhydride *Synonyms:* Hexachloroendomethylenetetrahydrophthalic anhydride
CA-57

Chlorhexidiene diacetate (CTFA) [CAS #56-95-1] *Synonyms:* Chlorhexidine acetate; 1,6-Bis (5-(p-chlorophenyl) bi-guanidino) hexane diacetate; 1,6-Di (4′-chlorophenyldiguanidino) hexane diacetate; 1,1′-Hexamethylenebis (5-(p-chlorophenyl) biguanide diacetate
Arlacide A

Chlorhexidine digluconate (CTFA) [CAS #18472-51-0, #14007-07-9] *Synonyms:* Chlorhexidine gluconate; N,N′-Bis (4-chlorophenyl)-3,12-diimino-2,4,11,13-tetraazatetradecane-diimidamide compd. with D-gluconic acid
Arlacide G

Chlorhexidine dihydrochloride (CTFA) [CAS #3697-42-5] *Synonyms:* Chlorhexidine hydrochloride; N,N′-Bis (4-chlorophenyl)-3,12-diimino-2,4,11,13-tetraazatetradecanediimidamide, dihydrochloride
Arlacide H, HM]

Chlorinated paraffin. See Paraffin, chlorinated

Chlorinated rubber. See Rubber, chlorinated

Chlorine [CAS #7782-50-5]
DD-8133

Chlorine dioxide [CAS #10049-04-4]
Anthium Dioxide

1-(3-Chloroallyl)-3,5,7-triaza-1-azoniaadamantane chloride. See Quaternium-15

1,1,1-Chlorodifluoroethane [CAS #75-68-3] *Synonyms:* 1,1,1-Difluorochloroethane; Difluoromonochloroethane
Dymel 142

Chlorodifluoromethane [CAS #74-97-5] *Synonyms:* Monochlorodifluoromethane; Difluorochloromethane; Difluoromonochloromethane; Refrigerant 22
Dymel 22

N-Chloroethyl-N-ethyl aniline [CAS #92-49-9]
Emery 5770

N-Chloroethyl-N-ethyl-m-toluidine [CAS #22564-43-8]
Emery 5771

2-Chloroethylmethyldichloro silane [CAS #7787-85-1]
CC3005

Chloro-2-hydroxypropyl trimonium chloride *Synonyms:* 3-Chloro-2-hydroxypropyltrimethyl ammonium chloride
CHPTA

Chloromethoxypropyl mercuric acetate
Troysan CMP Acetate

Chloromethyldimethylchloro silane [CAS #1719-57-9]
CC3270

Chloromethyl isothiazoline
Kathon DP, ICP

5-Chloro-2-methyl-4-isothiazolin-3-one. See Methylchloroisothiazolinone

Chloromethylmethyldichlorosilane [CAS #1558-33-4]
CC3275

Chloromethyltrimethylsilane [CAS #2344-80-1]
CC3285

m-Chlorophenyl diethanolamine [CAS #92-00-2]
Emery 5715, 5717

(E)-1-(p-Chlorophenyl)-4,4-dimethyl-2-(1,2,4-triazol-1-yl)-1-penten-3-ol
Ortho Prunit

Chlorophenylmethylsiloxane
SF 1029, 1250

Chloroprene/acrylonitrile copolymer
Neoprene Latex 450A

Chloroprene/dichlorobutadiene copolymer
Neoprene Latex 400, 460, 650, 750

Chloroprene rubber. See Polychloroprene

Chloroprene/sulfur copolymer
Neoprene Latex 571, 572

3-Chloropropylmethyldimethoxysilane [CAS #18171-19-2]
CC3290

Chloropropyltrichlorosilane [CAS #2550-06-3]
CC3291

3-Chloropropyltriethoxysilane [CAS #5089-70-3]
Dynasylan CPTEO

3-Chloropropyltrimethoxysilane [CAS #2530-87-2]
Dow Corning Z-6076; Dynasylan CPTMO

Chlorothalonil. See Tetrachloroisophthalonitrile

Chlorothioxanthone
Ultra-Cure I-100

Chlorotrifluoroethylene polymer *Synonyms:* Polytrifluorochloroethylene; PCTFE; Fluorothene
TL-340; Voltalef 1 S, 3 S, 3 X, 10 S

Chloroxylenol (CTFA) [CAS #88-04-0] *Synonyms:* p-Chloro-m-xylenol; 4-Chloro-3,5-dimethylphenol
Ottasept Extra, Tech.

Cholesterol (CTFA) [CAS #57-88-5] *Synonyms:* Cholest-5-en-3-ol (3B)
Dastar; Dermatein GSL; Nimlesterol 1730, 1732

Choleth-24 (CTFA) [CAS #27321-96-6 (generic)] *Synonyms:* PEG-24 cholesteryl ether
Solulan C-24

Chondroitin sulfate
Cromoist CS

Chromium alumina
Cr-0211 T 5/32′′

Chromium lignosulfonate
Lignosol C 60; Rychem CLS

Citric acid (CTFA) [CAS #77-92-9] *Synonyms:* 2-Hydroxy-1,2,3-propanetricarboxylic acid
Citrosol 50, 50E, 50T, 50W; Sustane 3, 8, 20, 20-3, 20A, 31, Q, W; Tandem 5K, 8, 552

Clonitralid
Bayluscide

Cobalt [CAS #7440-48-4] *Synonyms:* Super cobalt
Co-0127, -0138, 0164; Troykyd Perma-Dry; Troymax Drier Cobalt 6%, 12%

Cobalt molybdate [CAS #13762-14-6] *Synonyms:* Cobalt molybdenum oxide
CoMo-0402 T 1/8′′, -0603 T 1/8′′; HT-400 E 1/8′′

Cobalt oxide [Cobaltic oxide CAS #1308-04-9; Cobaltous oxide CAS #1307-96-6]
Co-0301

Cocamide (CTFA) [CAS #61789-19-3] *Synonyms:* Coconut oil amides; Coconut acid amide
Armid C; Ninol 4812; Schercosol NL

Cocamide DEA (CTFA) [CAS #61791-31-9; 68603-42-9] *Synonyms:* Coconut diethanolamide; Cocoyl diethanolamide; N,N-bis (2-hydroxyethyl) coco amides
Accomid C; Alkamide 1423, 2104, 2204, CDE, CDE-Extra, CDO; Aminol HCA, KDE; Calamide C, CW-100; Calsuds CD-6; Carsamide 7644, C-3, CA, SAC; Cedemide DX; Chimipal DCL; Comperlan COD, KD, KDO, KM, LS, PD, SDO; Condensate PA, PC, PO, PS; Cyclomide DC212, DC212M, DC212/S, DC212/SE, KD; Detergyl S; Emid 6514, 6515, 6521, 6529, 6531, 6533, 6534, 6538; Empilan CDE, 2502, CDE, CDEY, CDX, FD, FD20, FE; ESI-Terge 10, B-15, C-5, S-10; Ethylan LD, LDG, LDS; Gafamide CDD-518; Hartamide OD; Hetamide MC, RC; Iconol 28, COA; Incromide CA, CAC, CAL, LEP; Jordamide CLD, CCO, CLD, JT-128, JT-1286, OW, WC Conc.; Lauramide 11, D, ME; Lauridit KD, KDG; Loropan KD; Mackamide 100-A, C, CD, CS, EC, MC; Manro CD, CDS, CDX; Maprofix AEG; Marlamid D 1218, DF 1218; Mazamide 70, 80, CS-148; Monamid 7-100, 7-153 CS, 150-AD, 150-ADD, 150-DR, 759; Monamine 779, AA-100, AC-100, AD-100, ADD-100, ADD-100 LE, ADS-100, ALX-80 SS, ALX-100 S, ARA-100, CF-100 M, I-76; Ninol 40-CO, 41-CO, 49-CE, 128 Extra, 2012 Extra, 4821 F, DS, LD, P-650; Nissan Stafoam DF-1, DF-2; Nitrene 11230, 13026, A-309, C, C Extra, N; Norfox DC, DCO, DESA, KD; Perlglanz-Konzentrat B-30, B-48; Profan 24 Extra, 128 Extra, 2012E; Purton CFD; Quimipol DEA OC; Rewomid C 220SE, DC 212/S, DC 212/SE, DC 220/LS, DC 220/SE, DL 240; Schercomid 1-102, CCD, CDA, CDA-H, CDO-Extra, SCE, SCO-Extra; Stafoam DF-1, DF-4, F; Stamid HT; Standamid KD, KDO, PD, SD, SDO; Standapol Conc. 7023; Steinamid DC 212/S, DC 220/SE, DL 240; Sterling 66/33, Amide 374, DEA; Super Amide GR; Surco 128T, WC Conc.; Synotol 119 N, CN 60, CN 80, CN 90, Detergent E; Tego-Pearl B-48; Teric CDE; Tohol N-220, N-220X; Unamide C-72-3, C-7649, CDX, D-10, GC-75, JJ-35, LDL, N-72-3; Varamide A-2, A-10, A-12, A-83, MA-1, MA-4; Witcamide 82, 5130, M-3

Cocamide MEA (CTFA) [CAS #68140-00-1] *Synonyms:* Coconut monoethanolamide; N-(2-hydroxyethyl) coco fatty acid amide
Alkamide CME, CMO; Aminol CM, CM Flakes, CM-C Flakes, CM-D Flakes; Carsamide CMEA; Chimipal MC; Comperlan SM; Cyclomide C212; Cyclosheen 100; Emid 6500; Empilan CM, CME; Euperlan PK 771, PK 776, PK 789, PK-810, PK 900; Foamole M; Genapol PGM Conc.; Incromide CM; Intermediate 325; Jordamide CFAM, CMEA, CMEA Extra; Lauridit KM; Loropan CME, KM; Mackamide CMA; Manro CMEA; Marlamid M 1218; Monamid CMA; Ninol CBR, CNR; Nissan Stafoam MF; P & G Amide No. 27; Profan AB20; Rewomid C 212, CD; Schercomid CME; Standamid 100, CM, CMG, KM,

SM; Standapol Pearl Conc. 7130; Steinamid C 212; Sterling Granulated Wax; Surco CMEA, MGS-35; Swanic 51; Synotol ME-90; Tego-Pearl B-48; Unamide CMX; Varamide C-212

Cocamide MIPA (CTFA) [CAS #68333-82-4, #8039-67-6] *Synonyms:* Coconut monoisopropanolamide
Empilan CIS; Rewomid IPP 240; Schercomid CMI; Steinamid IPP 240

Cocamido betaine
Alkateric CAB; Amfotex FV-28; Amido Betaine C

Cocamidopropylamine oxide [CAS #68155-09-9] *Synonyms:* Cocamidopropyl dimethylamine oxide
Alcolec 634; Alkamox CAPO; Aminoxid WS 35; Ammonyx CDO; Barlox C; Cyclomox CO; Empigen OS/A; Incromine Oxide C; Jordamox CAPA; Ninox CA, FCA; Rewominoxid B 204; Schercamox C-AA; Standamox C 30, CAW; Tegamin Oxide WS-35

Cocamidopropyl betaine (CTFA) [CAS #61789-40-0] *Synonyms:* CADG; Cocamidopropyl dimethyl glycine
Aerosol 30; Amphosol CA, CG; Carsonam 3, 33-S, 3147; Chimin AX; Cycloteric BET C-30, BET-CB, BET-W; Dehyton K; Emcol DG, NA-30; Emery 5430, 6748; Incronam 30; Jortaine C, CAB-35, CFA-35; Lexaine C, CS; Lexate CL-60; Lonzaine C, CO; Mafo CAB; Maprofix AEG; Maprolyte C; Mirataine BD, CBC, CBR; Monateric ADA, COAB, MCB; Schercotaine CAB; Standapol BAW, BC-35; Sterling CAB; Surco Coco Betaine; Tego-Betaine C, L-7, S; Tego-Pearl B-48; Varion CADG, CADG-W; Velvetex BC-35, BK-35

Cocamidopropyl dimethyl acetyl betaine
Varion CADG-HS, CADG-LS

Cocamidopropyl dimethylamine (CTFA) [CAS #68140-01-2] *Synonyms:* N-[3-Dimethylamino) propyl]coco amides
Cyclomide CODI; Empigen AS; Jordamine DAPL, DMCAPA, SHCFA; Lexamine C-13; Mackine 101; Schercodine C; Tegamine P-7; Tego-Amid D 5040

Cocamidopropyl dimethylamine betaine
Empigen BS, BS/H, BS/P

Cocamidopropyl dimethylamine lactate (CTFA) [CAS #68425-42-3] *Synonyms:* N-[3-(Dimethylamino) propyl] cocamide lactate
Incromate CDL; Mackalene 116

Cocamidopropyl dimethylamine oxide. See Cocamidopropylamine oxide

Cocamidopropyl dimethylamine propionate (CTFA) [CAS #68425-43-4] *Synonyms:* N-[3-(Dimethyl-amino)propyl]coco amides, propionates
Emcol 1655; Incromate CDP; Mackalene 117

Cocamidopropyl dimethyl benzyl ammonium chloride
Aremsan C40; Quatrene CA

Cocamidopropyl dimethyl 2,3-dihydroxypropyl ammonium chloride
Lexquat AMG-WC

Cocamidopropyl hydroxysultaine (CTFA) [CAS #68139-30-0] *Synonyms:* (3-Cocamidopropyl)(2-hy-droxy-3-sulfopropyl)dimethyl quaternary ammonium compounds, hydroxides, inner salt
Jortaine COSB, CSB, CSB-50; Lonzaine CS; Mirataine CBS

Cocamidopropyl lauryl ether (CTFA) [RD #977016-04-8]
Marlamid KL

Cocamidopropyl morpholine
Incromine CPM

Cocamidopropyl morpholine lactate
Incromate CPML

Cocamidopropyl PG-dimonium chloride phosphate [CAS #83682-78-4]
Monaquat P-TC

Cocamidopropyl trimethyl ammonium chloride
Empigen CSC

Cocamine [CAS #61788-46-3] *Synonyms:* Coconut amine
Adogen 160 (D); Amine KK; Armeen C, CD; Arosurf MG-160; Kemamine P-650, P-650D; Lilamin 160, 160 D; Nissan Amine FB; Noram C; Radiamine 6160, 6161

Cocamine acetate
Acetamin 24, C; Armac C; Kemamine A650; Lilamac 160; Noramac C; Radiamac 6169

Cocamine oxide [CAS #61788-90-7] *Synonyms:* Coco dimethylamine oxide; Coconut dimethylamine oxide
Aromox DMC, DMCD, DMC-W; Barlox 12; Conco XA-Y; Cyclomox C; Empigen 5083; Genaminox CS, KC; Mackamine CO

Cocaminobutyric acid (CTFA) [CAS #68649-05-8]
Armeen Z, Z-9

Cocaminopropionic acid [RD #977056-35-1] *Synonyms:* N-coco-2-aminopropionic acid
Cycloteric CAPA; Deriphat 151C

Coco alkyl-2,2′-iminobisethanol [CAS #61791-31-9]
Armostat 410

Cocoamphoacetate [CAS #68334-21-4; 68390-66-9] *Synonyms:* Cocoamphoglycinate
Miranol CM Conc.

Cocoamphodiacetate (CTFA) [CAS #68650-39-5, #68647-53-0] *Synonyms:* Cocoamphocarboxyglycinate
Alkateric 2CIB; Dehyton G; Miranol C2M Conc.; Monateric 805, CDL, CDS, CDTD, CDX-38, CDX-38 Mod

Cocoamphodipropionate (CTFA) [CAS #68919-41-5] *Synonyms:* Cocoamphocarboxypropionate
Amphoterge K2; Miranol C2M-SF Conc.; Monateric CEM-38%

Cocoamphodipropionic acid (CTFA) [CAS #68919-40-4] *Synonyms:* Cocoamphocarboxypropionic acid
Miranol C2M Anhydrous Acid

Cocoamphohydroxypropyl sulfonate *Synonyms:* Cocoamphopropylsulfonate
Miranol CS Conc.

Cocoamphopropionate (CTFA) [CAS #68919-41-5] *Synonyms:* Cocoamphocarboxypropionate
Amphoteric C; Miranol CM-SF Conc.; Monateric CA-35%, CAM-40, CEM-38CG

Coco-betaine (CTFA) [CAS #68424-94-2; RD #977056-36-2] *Synonyms:* Coco dimethyl glycine; Coconut betaine
Accobetaine CL; Alkateric BC; Ampho B11-34; Amphoram CB A30; Carsonam BCW; Cycloteric BET-C41; Dehyton AB-30; Emcol CC-37-18; Hartaine CB-40; Jortaine CB-40; Lonzaine 12C; Mackam CB, CB-35, CB-LS; Standapol AB-45; Velvetex AB-45, BC

Coco caprylate/caprate (CTFA)
Cetiol LC; Rilanit LTC; Standamul LC

Coco dialkyl benzyl ammonium chloride
Quatrene CE; Querton 1149

Coco diamine
Adogen 560

Coco dimethyl amine. See Dimethyl cocamine

Coco dimethyl benzyl ammonium chloride *Synonyms:* Coco dimonium chloride
Arquad B90; Nissan Cation F2-10R, -20R, -40E, -50; Noramium DA.50; Querton KKB CL50; Servamine KAC 422

Cocodimonium hydrolyzed animal keratin
Croquat WKP

Cocodimonium hydrolyzed animal protein
Croquat M

Cocodimonium hydroxyethyl cellulose
Crodacel QM

Coco-EDTA-amide
Rewopol CHT 12

Cocoglycerides *Synonyms:* Glycerides, coconut, mono-, di-, and tri-
Cutina CBS

Cocohydroxyethyl PEG-imidazolinium chloride phosphate
Monaquat P-TZ

Coco imidazoline benzyl chloride
Uniquat CB-50

Coco imidazoline betaine
Carsonam C, C-SF, C-SPCL

Coco imidazoline betaine dicarboxylate
Carsonam DC, DC-SF, DC-70%-SF

Coco imidazoline dicarboxylate
Antaron MC-44

Coco/lauric DEA
Alkamide CL63

Coco morpholine
Armeen N-CMD

Coco nitrile
Arneel C

Coconut acid (CTFA) [CAS #61788-47-4]
Industrene 325, 328

Coconut alcohol (CTFA) [CAS #68425-37-6] *Synonyms:* Coconut fatty alcohol
Laurex CH

Coconut diamine
Lilamin 560; Radiamine 6560

Coconut diethanolamide. See Cocamide DEA

Coconut monoethanolamide. See Cocamide MEA

Coconut monoisopropanolamide. See Cocamide MIPA

Coconut oil (CTFA) [CAS #8001-31-8] *Synonyms:* Copra oil
Cobee 76, 92, 110; Pureco 76; Super Refined Coconut Oil

Coconut oil amides. See Cocamide

Coco/oleamidopropyl betaine (CTFA) [CAS #86438-79-1]
Mirataine COB

Coco/oleic DEA
Ninol SR 100

Coco-pentaethoxy methyl-ammonium methosulfate
Rewoquat CPEM

Coco-1,3-propanediamine *Synonyms:* Coco-1,3 diaminopropane
Duomeen C, CD, C Special

Coco-1,3-propanediamine diacetate *Synonyms:* Coco-1,3-diaminopropane diacetate
Duomac C

Coco 1,3-propylene diamine
Dinoram C; Kemamine D-650

Coco propylene diamine acetate
Dinoramac C

Coco-1,3-propylene diamine diacetate
Kemamine AD 650

Coco/tallow MEA
Intermediate 300

Coco taurine
Amphoram CT 30

Cocotrimonium chloride (CTFA) [CAS #61789-18-2] *Synonyms:* Coconut trimethyl ammonium chloride; Cocoyl trimethyl ammonium chloride
Adogen 461; Arquad C-33, C-33-W, C-50; Jet Quat C-50; Nissan Cation FB, FB-500; Noramium MC 50; Rewopon JMCA; Servamine KAC 412

N^2 Cocoyl-L-arginine ethyl ester DL-pyrrolidone carboxylic acid salt
CAE

Cocoyl hydroxyethyl imidazoline [CAS #61791-38-6] *Synonyms:* Coconut hydroxyethyl imidazoline; Cocoyl imidazoline
Alkaquat C; Alkazine C; Amphotensid 9 M, CT, GB 2009; Crapol AV-10; Mackazoline C; Miramine CC; Monazoline C; Schercozoline C; Varine C

Cocoyl sarcosinate
Incromide CAC

Cocoyl sarcosine (CTFA) [CAS #68411-97-2; RD #977056-38-4] *Synonyms:* N-Cocoyl-N-methyl glycine
Sarkosyl LC

Collagen (CTFA) [CAS #9007-34-5] *Synonyms:* Collagen fiber
Clearcol; Elastosol

Collagen hydrolysates. See Hydrolyzed animal protein

Copolyester-carbonate resin
Lexan 3250

Copper (CTFA) [CAS #7440-50-8]
Cu-0803 T 1/8'', -2501 G 4-10

Copper-ammonium complex
K-Cop

Copper chromite
Cu-0203 T 1/8'', -0211 P, -0223 P, -0233 T 1/8'', -1106 P, -1107 T 1/8'', -1117 T 1/8'', -1129 P, -1132 T 1/8'', -1413 P, -1422 T 1/8'', -1800 P, -1803 P, -1808 T 1/8'', -1809 T 1/8'', -1920 P, -3829 P

Copper chromium oxide
Marchon C21

Copper dialkyldithiophosphate
Rhenocure CUT

Copper dimethyldithiocarbamate (CTFA) [CAS #137-29-1] *Synonyms:* Dimethyldithiocarbamic acid

copper salt; Wolfen
Akrochem Cu.D.D., Cu.D.D.-PM; Methyl Cumate

Copper-ethylenediamine complex
Komeen

Copper hydroxide [CAS #20427-59-2] *Synonyms:* Cupric hydroxide
Kocide 20/20, 101, 404S, 606, SD

Copper naphthenate [CAS #1338-02-9]
Troysan Copper 8

Copper sulfate, tribasic [CAS #7758-98-7] *Synonyms:* Cupric sulfate; Basic copper sulfate
Basicop

Copper-triethanolamine complex
K-Pool; K-Tea

Corn cob meal (CTFA) [RD #977056-40-8]
Grit-O'Cobs; Lite-R-Cobs

Corn starch
Pure-Dent B700, B810

Corn syrup (CTFA) [CAS #8029-43-4; RD #977004-12-8]
ND-201-C Syrup, -305 Syrup

Corn syrup solids [CAS #68131-37-3]
Maltrin M200, M250, M365, M600

Cottonseed glyceride (CTFA) [CAS #8029-44-5] *Synonyms:* Cottonseed oil monoglyceride; Glyceryl mono cottonseed oil
Drewmulse 15; Rexowet CR

Coumatetralyl
Racumin

m,p-Cresol [CAS #108-39-4 (m-cresol); CAS #106-44-5 (p-cresol)]
Hetoxide MPC

Crotonic acid/vinyl acetate/vinyl propionate terpolymer
Luviset CAP X

Cumene hydroperoxide [CAS #80-15-9]
CHP-158

Cumene sulfonic acid
Eltesol CA 65, CA 96; Reworyl C 65; Sulframin CSA

Cumylperoxyneodecanoate
Esperox-939M; Lupersol 188M75; Trigonox 99-C75

α-Cumylperoxyneoheptanoate
Esperox 740M; Lupersol 288M75

α-Cumylperoxypivalate
Lupersol 47M75

Cumyl phenol
Ken Kem CP-45, CP-99

Cumyl-phenyl acetate
Kenplast ES-2

Cumyl-phenyl benzoate
Kenplast ESB

Cumyl-phenyl neodecanoate
Kenplast ESN

Cupric chloride [CAS #1344-67-8] *Synonyms:* Copper chloride
Luperfoam 40

N-Cyanoethyl-N-acetoxyethyl aniline [CAS #22031-33-0]
Emery 5725, 5731

3(N-Cyanoethyl) amino-4-methoxy acetanilide [CAS #26408-28-6]
Emery 5736

N-Cyanoethyl-N-butyl aniline [CAS #61852-40-2]
Emery 5727

3(N-Cyanoethyl-N-ethyl) amino-4-methoxy acetanilide [CAS #19433-94-4]
Emery 5751

N-Cyanoethyl-N-ethyl aniline [CAS #148-87-8]
Emery 5723

N-Cyanoethyl-N-ethyl-m-toluidine [CAS #148-69-6]
Emery 5728

3(N-Cyanoethyl-N-hydroxyethyl) amino-4-methoxy acetanilide [CAS #22588-78-9]
Emery 5737

N-Cyanoethyl-N-hyroxyethyl aniline [CAS #92-64-8]
Emery 5724

N-Cyanoethyl-N-hydroxyethyl-m-toluidine [CAS #119-95-9]
Emery 5730

N-Cyanoethyl-N-methyl aniline [CAS #94-34-8]
Emery 5722

2-Cyanoethyltriethoxysilane [CAS #17932-62-6]
CC3433

2-Cyanopropyltrichloro silane [CAS #1071-22-3]
CC3555

Cycloaliphatic amine
Epilink 148, 149

Cyclo [dineopentyl (diallyl)] pyrophosphato dineopentyl (diallyl) zirconate
KZ TPP

Cyclo (dioctyl) pyrophosphato dioctyl zirconate
KZ OPPR

Cyclohexane (CTFA) [CAS #110-82-7] *Synonyms:* Hexahydrobenzene; Hexamethylene; Hexanaphthene
Amsco Cyclohexane

1,4-Cyclohexane dimethanol dibenzoate
Benzoflex 352

Cyclohexanol [CAS #108-93-0] *Synonyms:* Hexahydrophenol
Naxol

Cyclohexanone [CAS #108-94-1] *Synonyms:* Ketohexamethylene
Nadone

Cyclohexanone peroxide [CAS #12262-58-7] *Synonyms:* 1-Hydroperoxycyclohexyl 1-hyroxycyclohexy peroxide
Cyclonox BT-50

Cyclohexyl acrylate
SR-220

Cyclohexyl amine acetate
Coagulant CHA

N-Cyclohexyl-2-benzothiazolesulfenamide (CTFA) [CAS #95-33-0] *Synonyms:* Benzothiazyl-2-cyclohexyl sulfenamide; N-Cyclohexyl-2-benzothiazyl sulfenamide
CBTS; Rhenovin CBS-70; Santocure; Vulkacit CZ/EGC, CZ/MGC

1,4-Cyclohexylenedimethylene terephthalate/isophthalate copolymer
Kodar A150 Copolyester, PETG 6763

Cyclohexylethyl amine
Vulkacit HX

Cyclohexyl-N′-phenyl-p-phenylenediamine [CAS #101-87-1]
Vanox 67H

N-Cyclohexylthio phthalimide
Santogard PVI

Cyclomethicone (CTFA) [CAS #69430-24-6] *Synonyms:* Cyclic dimethylsiloxane; Cyclic dimethyl polysiloxane with n= 3–6
Arlamol S3, S7; Bentone Gel VS-5; Dow Corning 344 Fluid, 345 Fluid; F-222, -251; SF 1173, 1202, 1204, 1214; Volatile Silicone 7158, 7207, 7349

Cycloneopentyl, cyclo (dimethlaminoethyl) pyrophosphato zirconate, di mesyl salt
KZ TPPJ

Cyfluthrin
Baythroid; Cyfluthrin

DDBSA. See Dodecylbenzene sulfonic acid

DEA-cetyl phosphate (CTFA) [CAS #69331-39-1] *Synonyms:* Cetyl DEA phosphate
Crodafos CDP

DEA-dodecylbenzene sulfonate [CAS #26545-53-9]
Monamine ALX-80 SS, ALX-100 S

DEAE [CAS #100-37-8]
Pennad 0150

DEA-lauraminopropionate [CAS #65104-36-1]
Maprolyte LX

DEA-laureth sulfate (CTFA) [RD #977061-59-8] *Synonyms:* Diethanolamine laureth sulfate
Monamine 779

DEA-lauryl sulfate (CTFA) [CAS #143-00-0, #68585-44-4] *Synonyms:* Diethanolamine lauryl sulfate
Conco Sulfate EP; Condanol DLS 35; Maprolyte LX; Sandoz Sulfate EP

DEA-myreth sulfate
Standapol Conc. 7023

DEA-oleth-3 phosphate [RD #977060-94-8]
Crodafos N3 Neutral

DEA-oleth-10 phosphate [RD #977060-97-1]
Crodafos N10 Neutral

Decabromobiphenyl oxide *Synonyms:* Decabromodiphenyl oxide
DE-83R; FR-300-BA; FR-1210; Saytex 102, 102E

Decaglyceryl... See Polyglyceryl-10...

Decahydronaphthalene [CAS #91-17-8] *Synonyms:* Perhydronaphthalene
Decalin

Decamethyl cyclopentasiloxane (CTFA) [CAS #541-02-6]
Abil B 8839; CD3770

Decamethyltetrasiloxane [CAS #141-62-8]
CD3780

Decanoyl peroxide
Decanox-F

Deceth-3 *Synonyms:* PEG-3 decyl ether; POE (3) decyl ether
Alkasurf DA-3

Deceth-4
Alkasurf DA-4; Chemal DA-4; Trycol 5950, DA-4

Deceth-4 phosphate (CTFA) [CAS #9004-80-2 (generic)] *Synonyms:* PEG-4 decyl ether phosphate
Gafac RA-600

Deceth-5
Trycol 5951

Deceth-6
Alkasurf DA-6; Chemal DA-6; Trycol DA-6, DA-69

Deceth-9
Chemal DA-9

Decyl alcohol (CTFA) [CAS #112-30-1] *Synonyms:* C10 linear primary alcohol
Alfol 10; Epal 10; Kalcohl 10H

Decylamine oxide (CTFA) [CAS #2605-79-0] *Synonyms:* Capric dimethyl amine oxide; Decyl dimethyl amine oxide
Barlox 10S

c-Decyl betaine (CTFA) [CAS #2644-45-3]
Darvan NS

Decyl diphenyloxide disulfonate
Dowfax 3B0

Decyl diphenyloxide disulfonic acid
Conco Sulfate 3B2 Acid

Decyl diphenyl phosphite
Mark MDDPP

Decyl isostearate (CTFA) *Synonyms:* Decyl isooctadecanoate
Schercemol DEIS

Decyl oleate (CTFA) [CAS #3687-46-5]
Ceraphyl 140; Cetiol V; Crodamol DO; Dehymuls K; Schercemol DO; Standamul CTV

Decyloxypropylamine
Tomah PA-14

N-Decyloxypropyl-1,3-diaminopropane
Tomah DA-14

Decyloxypropyl dihydroxyethyl methyl ammonium chloride
Tomah Q-14-2

Dehydroabietylamine
Amine D

Dehydroabietylamine acetate
Amine D Acetate

Demeton-S-methyl
Metasystox (i)

Denatonium benzoate NF [CAS #3734-33-6]
Bitrex

Dextranase
Dextranase Novo 25 L; DN 25L

Dextrose. See Glucose (CTFA)

Diacetylated hydrogenated soybean oil monoglyceride, tartaric acid ester
Myvatem 06(K)

Diacetylated palm oil monoglyceride, tartaric acid ester
Myvatem 35K

Diacetylated sunflower oil monoglyceride, tartaric acid ester
Myvatem 92K

Diacetylated tallow monoglyceride, tartaric acid ester
Myvatem 30

Diacrylate (monomer)
Chemlink 9014, 9040

Diacrylate (monomer), alkoxylated
Chemlink 9024, 9025

Dialkyl dimethyl ammonium chloride *Synonyms:* Quaternium 31
Radiaquat 6443, 6470; Skyllex

Dialkyl dimethyl benzyl ammonium chloride
Adogen 432; Bio-Dac 50-20, 205

Dialkyl methyl benzyl ammonium chloride
BTC 776

Diallyl fumarate
SR-204

Diallyl phthalate [CAS #131-17-9] *Synonyms:* DAP
Cadox M-30, MDA-30

Diammonium EDTA *Synonyms:* Diammonium edetate; Diammonium ethylene diamine tetraacetate
Hamp-Ene Diammonium EDTA; Sequestrene Diammonium; Versene Diammonium EDTA

Diamylhydroquinone (CTFA) [CAS #79-74-3] *Synonyms:* DAHQ; 2,5-Di (t-amyl) hydroquinone
Santovar A

1,1-Di (t-amylperoxy) cyclohexane [CAS #15667-10-4]
Lupersol 431-80B, 531-80M; USP-90MD

2,2-Di (t-amylperoxy) propane [CAS #3052-70-8]
Lupersol 553-M75

Diamyl sodium sulfosuccinate (CTFA) [CAS #922-80-5] *Synonyms:* Sodium diamyl sulfosuccinate
Aerosol AY, AY-65, AY-100

Diarachidyl-behenyl sec. amine
Kemamine S-190

Diaryldisulfides
Aktiplast 6 R

Diaryl p-phenylene diamine
Akrochem Antioxidant MPD-100; Vulkanox 3100; Wingstay 100, 100AZ, 200

Diastase
Aquazym 120L

Diatomaceous earth (CTFA) [CAS #7631-86-9] *Synonyms:* Diatomaceous silica; Diatomite
Celite 21-A, 110, 209, 219, 263, 266, 270, 275, 281, 289, 292, 305, 315, 321, 321A, 350, 388, 400, 408, 410, 499, 1200, CAS-30KR, CS-22R, F.C., HSC, R-625, R-633, R-643, R-685, R-690, Snow Floss, Super Fine Super Floss, Super Floss, White Mist

Diazolidinyl urea [CAS #78491-02-8]
Germaben II, II-3; Germall II

Dibasic acid (C12 and C11) [CAS #72162-23-3]
Corfree M1

Dibehenyl/diarachidyl dimonium chloride (CTFA)
Kemamine Q-1302C, Q-1902C

Dibehenyldimonium chloride (CTFA) [CAS #26597-36-4]
Kemamine Q-2802C

Dibehenyldimonium methosulfate
Incroquat DBM-90

2,2´-Dibenzamidodiphenyl disulfide [CAS #135-57-9]
Vanplast 3S

Dibenzothiazyl disulfide
Naftocit MBTS; Naftopast MBTS-A, MBTS-P; Rhenovin MBTS-70; Vulkacit DM/MGC, DM/C, DM/EGC

Dibenzyl azelate
Plasthall DBZZ

Dibenzylidene sorbitol (CTFA) [CAS #32647-67-9] *Synonyms:* Sorbitol acetal
Millithix 925

1,3-Dibromo-5,5-dimethyl hydantoin
Glybrom Reagent Grade

Dibromoethyldibromocyclohexane
Saytex BCL-462

Dibromoneopentyl glycol [CAS #3296-90-0] *Synonyms:* Dibromopentaerythritol
Emery 9336; FR-521, -522, -1138; Saytex FR-1138

2,2-Dibromo-3-nitrilopropionamide
Amerstat 300; Biobrom C-103L, C-103 Tech.; Biosperse 240; Dow Antimicrobial 7287, 8536

Dibromophenol
Emery 9331; FR-612

Dibromopropanol [CAS #96-13-9] *Synonyms:* 2,3-Dibromo-1 propanol
DBP

Dibromo/tribromo salicylanilide
Temasept I

Di-t-butoxydiacetoxysilane [CAS #13170-23-5]
Dynasylan BDAC

Dibutoxyethoxyethyl adipate
Merrol 4226; Plasthall DBEEA

Dibutoxyethoxyethyl formal
Merrol 4221

Dibutoxyethoxyethyl glutarate
Plasthall DBEEG

Dibutoxyethoxyethyl phthalate
Merrol 4228

Dibutoxyethoxyethyl sebacate
Merrol 4220; Plasthall 83SS

Dibutoxyethyl adipate
Merrol 4206; Plasthall DBEA; Staflex DBEA

Dibutoxyethyl azelate
Plasthall DBEZ

Dibutoxyethyl glutarate
Plasthall DBEG

Dibutoxyethyl phthalate [CAS #117-83-9] *Synonyms:* n-Butyl glycol phthalate
Kessco Dibutoxyethyl Phthalate; Merrol 4208; Plasthall DBEP; Staflex DBEP

Dibutoxyethyl sebacate
Merrol 4200; Plasthall DBES

Dibutyl adipate (CTFA) [CAS #105-99-7] *Synonyms:* Hexanedioic acid, dibutyl ester
Cetiol B; Rilanit DBA

Dibutylamino ethanol [CAS #102-80-8]
Mazeen DBA

Dibutylammonium oleate
Activator 1102

Dibutyl azelate [CAS #2917-73-9]
Plastolein 9048

2,6-Di-t-butyl-4-s-butyl phenol [CAS #17540-75-9]
Isonox 132; Vanox 1320

2,6-Di-t-butyl-p-cresol. See BHT

Di-t-butyldichlorosilane
D4165

2,6-Di-t-butyl-α-dimethylamino-p-cresol
Ethanox 703

Di-t-butyl diperoxyazelate
Lupersol 99

Di-t-butyl diperoxyphthalate [CAS #2155-71-7]
Lupersol KDB

Dibutyl fumarate [CAS #105-76-9] *Synonyms:* Fumaric acid, dibutyl ester
Staflex DBF

Di-t-butylhydroquinone (CTFA) [CAS #88-58-4] *Synonyms:* 2,5-Di-t-butyl hydroquinone; 2,5-Bis (1,1-dimethylethyl)-1,4-benzenediol
Eastman DTBHQ

3,5-Di-t-butyl-4-hydroxybenzoic acid, n-hexadecyl ester
Cyasorb UV 2908

3,5-Di-t-butyl-4-hydroxyhydrocinnamic acid triester of 1,3,5-tris (2-hydroxyethyl)-s-triazine-2,4,6-(1H,3H,5H)-trione [CAS #34137-09-2]
Good-rite 3125; Vanox SKT

2-(3′,5′-Di-t-butyl-2′-hydroxyphenyl)-5-chlorobenzotriazole
Tinuvin 327

Dibutyl maleate [CAS #105-76-0] *Synonyms:* 2-Butenedioic acid, dibutyl ester; DBM
Staflex DBM

Dibutyl methylene bisthioglycolate
Vulkanol 88

2,6-Di-t-butyl-4-methylphenol. See BHT

Di (butyl, methyl pyrophosphato) ethylene titanate di (dioctyl, hydrogen phosphite)
Ken-React 262A, 262ES

Di-t-butyl peroxide [CAS #110-05-4] *Synonyms:* t-Butyl peroxide; DTBP
Trigonox B

2,2-Di (t-butylperoxy) butane [CAS #2167-23-9]
Lupersol 220-D50

4,4-Di-t-butylperoxy-n-butyl valerate *Synonyms:* n-Butyl-4,4-bis(t-butylperoxy) valerate
Trigonox 17/40, 17/40 Bpd

1,1-Di (t-butylperoxy) cyclohexane [CAS #3006-86-8]
Luperco 331-XL; Lupersol 331-80B, P-31; Trigonox 22-BB80; USP-400P, -690

Di-s-butyl peroxydicarbonate [CAS #19910-65-7]
Espercarb 438M-60; Lupersol 225, 225-M, 225-M60, 225-M75, 225-T50, 225-X30; SBP; Trigonox SBP, SBP-AX30, SBP-C60, SBP-C75

1,1-Di (t-butylperoxy) 2,2,5-trimethyl cyclohexane
Luperco 231-SRL, 231-XL

1,1-Di (t-butylperoxy) 3,3,5-trimethyl cyclohexane [CAS #6731-36-8]
Lupersol 231, 231-P75, 231-XL; Trigonox 29/40 Bpd, 29-B75, 29-C75

2,6-Di-t-butylphenol [CAS #128-39-2]
Ethanox 701

2,4-Di-t-butylphenyl 3,5-di-t-butyl-4-hydroxybenzoate
UV-Chek AM-340

N,N′-Di-s-butyl-p-phenylenediamine [CAS #101-92-2]
Kerobit BPD, COM

Dibutyl phthalate (CTFA) [CAS #84-74-2] *Synonyms:* 1,2-Benzenedicarboxylic acid, dibutyl ester; Di n-butyl phthalate
Kodaflex DBP, HS-4; Merrol DBP; Staflex DBP; Unimoll DB; Vulcabond VP

Dibutyl sebacate (CTFA) [CAS #109-43-3] *Synonyms:* Dibutyl decanedioate; Di-n-butyl sebacate
Hallco DBS; Merrol DBS; Rilanit DBS; Uniflex DBS; Unimate DBS

1,3-Dibutylthiourea [CAS #109-46-6]
Thiate U

Dibutyltin bisisooctyl thioglycolate
Therm-Chek 840

Dibutyltin (bis) mercaptide
Dabco T-131

Dibutyltin diacetate [CAS #1067-33-0]
Dabco T-1; Fomrez SUL-3

Dibutyltin dilaurate [CAS #77-58-7]
Cata-Chek 820; Dabco T-12; Fomrez SUL-4; Kosmos 19

Dibutyltin disulfide
Dabco T-5

Dibutyltin maleate [CAS #15535-69-0] *Synonyms:* DBM
Advastab T-340; Stanclere TM; Therm-Chek 837

Dibutyltin β-mercaptopropionate
Stanclere T186

Dicapryl adipate (CTFA) [CAS #105-97-5] *Synonyms:* Didecyl hexanedioate
Uniflex DCA

Dicapryl phthalate *Synonyms:* DCP; Di-(2-octyl) phthalate
Uniflex DCP

Dicetyl dimonium chloride (CTFA) [CAS #1812-53-9] *Synonyms:* Quaternium-31; 1-Hexadecanaminium, N-hexadecyl-N,N-dimethyl-, chloride; N-Hexadecyl-N,N-dimethyl-1-hexadecanaminium chloride; Dicetyl dimethyl ammonium chloride
Adogen 432 CG, 432-ET; Varisoft CRC

Dicetyl peroxydicarbonate
Liladox

Dicetyl thiodipropionate (CTFA) [CAS #3287-12-5] *Synonyms:* Propanoic acid, 3,3´-thiobis-, dihexadecyl ester
Evanstab 16

Dichlofluanid
Euparen

o-Dichlorobenzene [CAS #95-50-1] *Synonyms:* 1,2-Dichlorobenzene
Dizene

2,4-Dichlorobenzoyl peroxide
Cadox TDP; Luperco CST

2,4-Dichlorobenzyl alcohol (CTFA) [CAS #1777-82-8]
Myacide SP

1,3-Dichloro-5,5-dimethyl hydantoin [CAS #118-52-5] *Synonyms:* DDH; Dichlorantin
Dantoin DCDMH; Glychlor Reagent Grade

4,5-Dichloro-N-octyl isothiazoline
Kathon 925

Dichlorophene [CAS #97-23-4] *Synonyms:* 2,2´-Dihydroxy-5,5´-dichlorodiphenylmethane; DDDM
G-4 Pure, Tech.

3-(3,4-Dichlorophenyl)-1,1-dimethylurea
Rhenocure Diuron

1,3-Dichlorotetraisopropyldisiloxane
D4368

Dichlorvos
Dedevap; Mafu

Dicocamine (CTFA) [CAS #61789-76-2] *Synonyms:* Dicoco alkyl amine
Armeen 2C; Kemamine S-650; Noram 2 C

Dicocodimonium chloride (CTFA) [CAS #61789-77-3] *Synonyms:* Dicoco dimethyl ammonium chloride
Accoquat 2C-75, 2C-75H; Adogen 462, R-6; Arquad 2C-75, S-2C-50, T-2C-50; Jet Quat 2C-75; Kemamine Q-6502C, Q-6503B; M-Quat 2475; Noramium M2C; Quartamin DCP; Radiaquat 6462; Variquat K 300

Dicoco methylamine
Kemamine T-6501; Noram M2C

Dicumyl peroxide [CAS #80-43-3] *Synonyms:* Di-α-cumyl peroxide; Cumene peroxide; Cumyl peroxide; Diisopropylbenzene peroxide; Isopropylbenzene peroxide
Akrochem DCP-40C, DCP-40K; Akroform DCP-40 EPMB; Akrosperse DCP-40 EPMB; Di-Cup 40C, 40KE, R, T; Esperal

115RG; Luperco 500-40C, 500-40KE, 500-SRK; Luperox 500R, 500T; Perkadox BC, BC-40Bpd, BC-40Kpd; Thermacure; Varox DCP-40C, DCP-40KE, DCP-R, DCP-T

Dicyandiamide *Synonyms:* Cyanoguanidine
D.E.H. 40

Dicyclo (dioctyl) pyrophosphato dioctyl titanate
Ken-React OPPR

Dicyclo (dioctyl) pyrophosphato titanate
Ken-React OPP2

Dicyclohexylamine nitrite [CAS #3129-91-7] *Synonyms:* Dodecahydrophenylamine nitrate
Dichan 100

Dicyclohexylbenzothiazyl-2-sulfenamide *Synonyms:* Benzothiazyl-2-dicyclohexyl sulfenamide
Vulkacit DZ/C, DZ/EGC

Dicyclohexyl phthalate *Synonyms:* DCHP
Cadox BFF-50, BFF-60W; Edenol DCHP; Unimoll 66 M

Dicyclohexyl sodium sulfosuccinate (CTFA) [CAS #23386-52-9] *Synonyms:* Sodium dicyclohexyl sulfosuccinate
Aerosol A-196

Dicyclopentadiene [CAS #77-73-6]
Piccovar L30, L30S, L60

Didecyldimonium chloride (CTFA) [CAS #173-51-5] *Synonyms:* Didecyl dimethyl ammonium chloride
Arquad 2-10/50; Bardac 205M, 2250; Bio-Dac 50-22; BTC 99, 818, 885, 1010; BTCO 1010; FMB 210-8 Quat, 210-15 Quat, 504-5 Quat, 1210-5 Quat, 1210-8 Quat; Maquat 4450-E; Querton 210 CL50, 210 CL 80

Didecyldimonium methoxysulfate *Synonyms:* Didecyl dimethyl ammonium methoxysulfate
Catigene DE 80

Didecyl glutarate
Plasthall DDG

Didecyl hydrogen phosphite
Mark DDHP

Didecyl phenyl phosphite
Mark DDMPP

Di (dioctylphosphato) ethylene titanate
Ken-React 212

Di (dioctylpyrophosphato) ethylene titanate
Ken-React 238A, 238J, 238M, 238S, 238T

Diene-styrene copolymer
Glissoviscal SB, SG

Diethanolamine (CTFA) [CAS #111-42-2] *Synonyms:* DEA; 2,2´-Iminobisethanol
Aminol COR-2C, N-1918; Calamide CW-100; Condensate PL, PS; Cyclomide DC212, DIN 295/S, DO280; Emid 6531, 6533, 6534, 6541; Empilan LDX; Mackamide CD; Mazamide 70, 80, O-20; Monamine 928, AA-100, ACO-100, AD-100, ADD-100, ADD-100 LE, ADY-100, CD-100, CF-100 M, LM-100, R8-26, T-100; Norfox KD; Rewocid DU 185; Rewomid DL 203, DL 240, F; Schercomid CDA, CDO-Extra, LD, ODA, TO-2; Standamid CD, PD; Varamide L-1; Witcamide 82, 5133

Diethchinalphion
Bayrusil

Diethylaminoethanol [CAS #100-37-8] *Synonyms:* Diethylethanolamine
Morlex DEEA

Diethylaminoethyl stearate (CTFA) [CAS #3179-81-5]
Cerasynt 303

N,N-Diethylaminotrimethyl silane [CAS #996-50-9]
CD4450; D4450

Diethylene glycol... See also PEG-2...

Diethylene glycol diacrylate
SR-230

Diethylene glycol diethyl ether [CAS #112-36-7] *Synonyms:* Diglycol diethyl ether
Ethyl Diglyme

Diethylene glycol dimethyl ether *Synonyms:* Diglycol methyl ether
Diglyme

Diethylene glycol monobutyl ether. See Butoxydiglycol

Diethylene glycol monoethyl ether. See Ethoxydiglycol

Diethylene glycol monomethyl ether. See Methoxydiglycol

Diethylene glycol monopropyl ether
Ektasolve DP

Diethylenetriamine [CAS #111-40-0] *Synonyms:* Aminoethylethandiamine
D.E.H. 20, 52, 58

Diethylene triamine dinonylnaphthalene sulfonate
Na-Sul DTA

Diethylene triamine pentaacetic acid. See Pentetic acid

Diethylene triamine penta (methylene phosphonic acid)
Dequest 2060

Diethylene tricaseinamide (CTFA)
Hydagen P

Diethylethanolamine. See Diethylaminoethanol

Di-2-ethylhexyl... See Dioctyl...

N,N-Diethylhydroxylamine [CAS #3710-84-7]
Pennstop 1866, 2049, 2697

Diethyl phthalate [CAS #84-66-2] *Synonyms:* Ethyl phthalate; Phthalic acid, diethyl ester; DEP
Kodaflex DEP

1,3-Diethylthiourea [CAS #105-55-5]
Akrochem D.E.T.U. Accelerator; Thiate H

Diethyl toluene diamine
Ethacure 100

Difenzoquat methyl sulfate *Synonyms:* 1,2-Dimethyl-3,5-diphenyl-1H-pyrazolium methyl sulfate
Avenge

Diflubenzuron [CAS #35367-38-5] *Synonyms:* 1-(4-Chlorophenyl))-3-(2,6-difluorobenzoyl) urea
Dimilin

1,1-Difluoroethane [CAS #75-37-6] *Synonyms:* Ethylidene fluoride
Dymel 152

Diglyceryl... See Polyglyceryl-2...

Diglyceryl borate sesquistearate
Emulbon S-260

Diglycol... See PEG-2...

Di-n-hexyl azelate [CAS #109-31-9]
Plastolein 9051

Dihexyl sodium sulfosuccinate [CAS #3006-15-3] *Synonyms:* Sodium dihexyl sulfosuccinate
Aerosol MA-80; Empimin MA; Lankropol KMA; Monawet MM-80

Dihydroabietyl alcohol (CTFA) [CAS #26266-77-3] *Synonyms:* Dodecahydro-1,4a-dimethyl-7-(1-methylethyl)-1-phenanthrenemethanol; Hydroabietyl alcohol
Abitol

Dihydrogenated tallow amine
Adogen 240; Noram 2SH

Dihydrogenated tallow ammonium methosulfate
Varisoft 472-80%

Dihydrogenated tallow dimethyl methosulfate
Cycloton D261S/90

Dihydrogenated tallow dimonium chloride. See Quaternium-18

Dihydrogenated tallow dimonium methosulfate *Synonyms:* Dimethyl dihydrogenated tallow ammonium methyl sulfate
Accosoft 748; Varisoft 137

Dihydrogenated tallow methylamine [CAS #61788-63-4]
Kemamine T-9701; Noram M2 SH

Dihydrogenated tallow phthalate (CTFA) *Synonyms:* 1,2-Benzenedicarboxylic acid, dihydrogenated tallow ester
Rilanit DTP

Dihydro-4-methyl-4-(1-methylethyl)-5-oxo-1H-imidazol-2-yl]-3-pyridinecarboxylic acid with 2-propanamine salt
Arsenal

2,5-Dihydroperoxy-2,5-dimethylhexane
Luperox 2,5-2,5

1,2-Dihydro-2,2,4-trimethyl quinoline
Agerite MA, Resin D; Flectol H, Pastilles; Vanlube RD

2,4-Dihydroxybenzophenone. See Benzophenone-1

Dihydroxyethyl C8-10 alkoxypropylamine oxide
Varox 188E

Dihydroxyethyl C9-11 alkoxypropylamine oxide
Varox 191E

Dihydroxyethyl C12-15 alkoxypropylamine oxide (CTFA)
Varox 185E

Dihydroxyethyl cocamine oxide (CTFA) [CAS #61791-47-7] *Synonyms:* N,N (bis (2-hydroxyethyl) cocamine oxide
Alkamox C2-0; Aromox C/12, C/12-W, CD/12; Schercamox CMA

Di-(2-hydroxyethyl)-5,5-dimethyl hydantoin
Dantocol DHE

Di-(2-hydroxyethyl)-5,5-dimethyl hydantoin dilaurate
Dantoest DHE DL

Dihydroxyethyl stearamine oxide [CAS #14048-77-2] *Synonyms:* Bis (2-hydroxyethyl)stearylamine oxide
Aromox 18/12

Dihydroxyethyl tallowamine oxide (CTFA) [CAS #61791-46-6]
Aromox T/12; Schercamox T-12

Dihydroxyethyl tallow glycinate (CTFA) [CAS #61791-25-1] *Synonyms:* Tallow dihydroxyethyl betaine
Cycloteric BET-T2 40; Jortaine TM; Mackam TM; Mirataine TM; Monateric 1202

2,2´-Dihydroxy-4-methoxybenzophenone. See Benzophenone-8

Diiodomethyl p-tolyl sulfone
Amical 48, 50, Flowable

Diisobutyl adipate (CTFA) [CAS #141-04-8] *Synonyms:* DIBA
DIBA; Merrol DIBA; Plasthall DIBA; Radia 7197

Diisobutyl azelate
Plasthall DIBZ

Diisobutyl (oleyl) aceto acetyl aluminate
KA 301

Diisobutyl sodium sulfosuccinate [CAS #127-39-9] *Synonyms:* Sodium diisobutyl sulfosuccinate
Aerosol IB-45; Monawet MB-45

Diisocetyl adipate (CTFA) [CAS #57533-90-1; 58262-41-2; RD #977052-98-4] *Synonyms:* Hexanedioic acid, bis (2-hexyl decyl) ester
Schercemol DICA

Diisodecyl adipate (CTFA) [CAS #27178-16-1] *Synonyms:* DIDA; Hexanedioic acid, diisodecyl ester
Merrol DIDA; Monoplex DDA; Plasthall DIDA; Staflex DIDA

Diisodecyl glutarate
Plasthall DIDG

Diisodecyl nylonate/glutarate
Merrol DIDN

Diisodecyl pentaerythritol diphosphite
Weston 600

Diisodecyl phthalate *Synonyms:* DIDP
Intercide 2 DIDP, ABF 2 DIDP; Vinyzene BP-505 DIDP

Diisononanoyl peroxide
Lupersol 219-M60, 219-M75

Diisononyl adipate
Adimol DN; Staflex DINA

Diisononyl maleate
Staflex DINM

Diisooctyl adipate *Synonyms:* DIOA
Merrol DIOA; Monoplex DIOA; Plasthall DIOA; Staflex DIOA

Diisooctyl dodecanedioate
Plasthall DIODD

Diisooctyl maleate
Staflex DIOM

Diisooctyl octylphenyl phosphite
Weston 494

Diisooctyl phosphite
Weston DOPI

Diisooctyl phthalate [CAS #27554-26-3] *Synonyms:* DIOP
Staflex DIOP

Diisooctyl sodium sulfosuccinate
Condanol SBDO70; Steinapol SB-DO 70

Diisopropyl adipate (CTFA) [CAS #6938-94-9]
Ceraphyl 230; Crodamol DA; Lexol DIA; Radia 7194; Schercemol DIA

N,N-Diisopropylbenzothiazyl-2-sulfenamide *Synonyms:* N,N-Diisopropyl benzothiazole-2-sulfenamide; Benzothiazyl-2-diisopropyl sulfenamide
BIBBS; Santocure IPS; Vulkacit LZ/LG

Diisopropyl dimerate (CTFA) [RD #977012-60-4] *Synonyms:* Bis(1-methylethyl)dimerate; Diester of isopropyl alcohol and dimer acid
Schercemol DID

Diisopropyl (oleyl) aceto acetyl aluminate
KA 322

Diisopropyl perdicarbonate [CAS #105-64-6] *Synonyms:* Diisopropyl peroxydicarbonate; Isopropyl percarbonate; Isopropyl peroxydicarbonate
IPP; Perkadox IPP-AT50

Diisopropyl sebacate (CTFA) [CAS #7491-02-3]
Schercemol DIS; Unimate DIPS

Diisostearyl dilinoleate (CTFA) [RD #977047-44-1] *Synonyms:* Diisostearyl dimerate
Schercemol DISD

Diisotridecyl phthalate
Edenol W 300 S

Dilaureth-10 phosphate (CTFA)
Nikkol DLP-10

Dilauroyl peroxide [CAS #105-74-8]
Laurox, W-25, W-40

Dilauryl phosphite
Weston DLP

Dilauryl thiodipropionate (CTFA) [CAS #123-28-4] *Synonyms:* Didodecyl 3,3′-thiodipropionate; Thiodipropionic acid, dilauryl ester
Carstab DLTDP; Cyanox LTDP; Evanstab 12

Dimelamine phosphate
Amgard ND

Dimer acid (CTFA)
Hystrene 3675, 3675C, 3680, 3695

Dimer amine
Kemamine DP-3680, DP-3695

2,5-Dimercapto-1,3,4-thiadiazole
Vanlube 793

Dimer diamine
Kemamine DD-3680

Dimethicone (CTFA) [CAS #9006-65-9; 9016-00-6; 63148-62-9] *Synonyms:* Dimethyl polysiloxane; Dimethyl silicone; Dimethylsiloxane; Polydimethylsiloxane; Poly [oxy(dimethylsilylene], α-(trimethylsilyl)-ω-methyl-; PDMS
Abil 10-100000; AF-60, AF-66; Akrochem SWS-201; Antifoam Compound SWS-201, SWS-202, SWS-203; Dow Corning 200 Fluid, 1500, 1520, QF1-3593A; FS-1265 Fluid; Foamkill 810F, 830F; Formasil Silicone Emulsion 45; Masil EM 14, 62, 100, 100 Conc., 100D, 100P, 250 Conc., EM 350, 350X, 350X Conc., 1000, 1000 Conc., 1000P, 10,000, 10,000 Conc., 60,000, 100,000, EM-N; Masil SF 5, 20, 50, 100, 200, 350, 500, 1000, 5000, 10,000, 12,500, 30,000, 60,000, 100,000, 300,000, 600,000; Sag Silicone Antifoam 100, 720; SEM-35; SF 18, 69, 81, 96, 97, 1093, 1188, 1214, 1705; Silicone AF-10 FG, AF-10 IND, AF-30 FG, AF-30 IND, AF-600M; Silicone C111; Silicone Emulsion 350, 350 Conc., 1M, 10M, 60M; Silicone Fluid 350, 1M, 5M, 10M, 30M, 60M; SM 2140; Silicone Release Agent #5038; SM 2061, 2155, 2162; Union Carbide L-45 Series, LE-45, 458HS, LE-467, LE-467HS; Viscasil

Dimethicone copolyol (CTFA) [CAS #64365-23-7; RD #977058-72-2] *Synonyms*: Dimethylsiloxane-glycol copolymer
Abil B 8842, B 8843, B 8851, B 8852, B 8863, B 8873, B 88183; Dow Corning 3225C Formulation Aid, FF-400; Masil 280,

280LP, 1066C, 1066D

Dimethicone propyl PG-betaine
Abil B 9950

Dimethicone thiosulfate
Abil S201, S255

Dimethoate [CAS #60-51-5] *Synonyms:* O,O-Dimethyl-S-(N-methylcarbamoylmethyl)phosphorodithioate
Cygon 400

Dimethoxyethyl phthalate [CAS #117-82-8] *Synonyms:* DMEP; 1,2-Benzenedicarboxylic acid bi (2-methoxy ethyl) ester
Kodaflex DMEP

Dimethyl adipate [CAS #627-93-0]
DBE, DBE-2, DBE-2SPG, DBE-3, DBE-9

p-Dimethylamino-benzenediazo sodium sulfonate
Bayer 5072

(N,N-Dimethylamino) (3,3-dimethylbutyl) dimethylsilane
D5115

Dimethylaminoethanol. See Dimethylethanolamine

2-(2-Dimethylaminoethoxy) ethanol
Texacat ZR-70

2-Dimethylamino-2-methyl-1-propanol *Synonyms:* 2-Dimethylaminomethyl propanol
Ken-React 138D, 158D

Dimethylaminopropyl ricinoleamide benzyl chloride
ES-1239

Dimethylaminotrimethylsilane [CAS #2083-91-2]
CD5400; D5400

Dimethylammonium hydrogen isophthalate [CAS #71172-17-3]
Vanax CPA

Dimethyl amyl maleate
Staflex DMAM

Dimethyl arachidyl-behenyl amine
Kemamine T-1902D

Dimethyl azelate
Emery 2914

Dimethyl behenamine (CTFA) [CAS #215-42-9; RD #977042-42-4] *Synonyms:* Behenyl dimethyl amine
Crodamine 3.ABD

2,5-Dimethyl-2,5-bis (benzoylperoxy) hexane
Luperox 118

2,5-Dimethyl-2,5-bis (t-butylperoxy) hexane
Luperco 101-P20

2,5-Dimethyl-2,5-bis (2-ethylhexanoylperoxy) hexane
Lupersol 256

3,3-Dimethylbutyldimethychloro silane
CD5430; D5430

N-(1,3-Dimethylbutyl)-N′-phenyl-p-phenylene diamine
Akrochem Antiozonant PD-2; Antozite 67, 67F; Santoflex 13; Vulkanox 4020

Dimethyl C10–18 alkyl amine
Armeen DMAD

Dimethyl C12–14 alkyl amine
Adma 24

Dimethyl C12–14–16 alkyl amine
Adma 246-451, 246-621

Dimethyl C14–16 alkyl amine
Adma 46

Dimethyl caproamide
Hallcomid M-6

Dimethyl caprylamide-capramide
Hallcomid M 8-10

Dimethyl capryl amine
Lilamin 310 D

Dimethyl caprylyl amine
Lilamin 308 D

Dimethylchlorosilane [CAS #1066-35-9]
CD5470

Dimethyl cocamine (CTFA) [CAS #61788-93-0] *Synonyms:* Coco dimethyl amine
Adogen 367-D; Amine 2MKK; Armeen DMCD, DMMCD; Kemamine T-6502D; Lilamin 367 D, 369; Nissan Tertiary Amine FB; Noram DMC

3,6-Dimethyl-3-cyclohexene-1-carbaldehyde
Cyclovertal

Dimethyl cyclohexyl ammonium dibutyl dithiocarbamate
Akrochem Accelerator CZ-1

Dimethylcyclosiloxane
CD5487

Dimethyl decylamine *Synonyms:* Decyl dimethylamine; C10 alkyl dimethylamine
Adma C10; Armeen DM10D

Dimethyldiacetoxysilane
D5490

2,5-Dimethyl-2,5-di (t-butylperoxy) hexane [CAS #78-63-7]
Luperco 101-SIL, 101-XL; Lupersol 101; Trigonox 101, 101/45Bpd; Varox, DBPH, DBPH-50

2,5-Dimethyl-2,5-di (t-butylperoxy)-3-hexyne [CAS #1068-27-5]
Luperco 130-KE, 130-XL; Lupersol 130; Varox 130, 130-XL

Dimethyldichloro silane [CAS #75-78-5]
CD5550

Dimethyldiethoxy silane [CAS #78-62-6]
CD5600; D5600

2,5-Dimethyl-2,5-di(2-ethyl hexanoyl peroxy) hexane [CAS #13052-09-0]
USP-245

Dimethyl dihydrogenated tallow ammonium chloride. See Quaternium-18

N,N-Dimethyl)-N′,N′-diisopropanol-1,3-propanediamine
Texacat DPA, DPA-50

Dimethyl diphenyl thiuram disulfide
Vulkacit J

Dimethyl erucylamine
Crodamine 3.AED

Dimethylethanolamine [CAS #108-01-0] *Synonyms:* 2-Dimethylaminoethanol; Deanol
Morlex DMEA; Tegoamin DMEA; Texacat DME

Dimethyl glutarate [CAS #1119-40-0]
DBE, DBE-2, DBE-2SPG, DBE-3, DBE-5, DBE-5SPG, DBE-9

N,N′-Di-3(5-methylheptyl)-p-phenylenediamine
Antozite 2

Dimethyl hydantoin. See DM hydantoin

Dimethyl hydantoin formaldehyde. See DMHF

Dimethyl hydrogenated tallow amine (CTFA) [CAS #61788-95-2] *Synonyms:* Hydrogenated tallow dimethyl amine
Adogen 345-D; Amine 2MHBG; Armeen DMHTD; Kemamine T-9702D; Lilamin 343, 345 D; Nissan Tertiary Amine ABT; Noram DMSH

1,1-Dimethyl-3-hydroxybutylperoxy-2-ethylhexanoate
Lupersol 665T50

1,1-Dimethyl-3-hydroxybutylperoxyneoheptanoate
Lupersol 688M50, 677T50

Dimethyl lauramide
Hallcomid M-12

Dimethyl lauramine (CTFA) [CAS #112-18-5] *Synonyms:* Lauryl dimethyl amine; Dodecyldimethylamine
Adma 2; Amine 2M12; Armeen DM12D; Empigen AB; Lilamin 312 D; Nissan Tertiary Amine BB; Onamine 12

Dimethyl methylphosphonate
Antiblaze 19

Dimethyl myristamine [CAS #112-75-4] *Synonyms:* Dimethyl myristylamine; Myristyl dimethylamine; Tetradecyl dimethylamine
Adma 4; Amine 2M14; Armeen DM14D; Lilamin 314 D; Nissan Tertiary Amine MB; Onamine 14

Dimethyloctadecylchloro silane [CAS #18643-08-8]
CD5636

Dimethyl octylamine *Synonyms:* C8 alkyl dimethylamine; Octyl dimethylamine
Adma C8; Armeen DM8D

Dimethyl octynediol (CTFA) [CAS #1321-87-5] *Synonyms:* 3,6-Dimethyl-4-octyne-3,6-diol
Surfynol 82, 82S

Dimethyl oleamide
Hallcomid M 18-OL

Dimethyl oleamine [CAS #28061-69-0] *Synonyms:* Oleyl dimethyl amine
Armeen DMOD; Crodamine 3.AOD; Kemamine T-9892D; Lilamin 372 D

Dimethyl oleic-linolenic amine
Kemamine T-9992D

Dimethyl oxazolidine (CTFA)
Oxaban-A

Dimethyl PABA ethyl cetearyldimonium tosylate
Escalol 537Q

Dimethyl palmitamine [CAS #112-69-6] *Synonyms:* Palmityl dimethylamine; Hexadecyl dimethylamine; Cetyl dimethylamine
Adma 6; Amine 2M16; Armeen DM16D; Crodamine 3.A16D; Lilamin 316 D; Nissan Tertiary Amine PB; Onamine 16

Dimethyl phthalate (CTFA) [CAS #131-11-3] *Synonyms:* Dimethyl 1,2 benzenedicarboxylate; DMP
Cadox M-30, M-50; DMP; Hi-Point 90, 90 Red, PD-1; Kemester DMP; Kodaflex DMP; Staflex DMP

N,N´-Dimethylpiperazine
Texacat DMP

Dimethylpolysiloxane. See Dimethicone

2,2-Dimethylpropane-1,3-diacrylate *Synonyms:* Neopentyl glycol diacrylate
SR-247

(2,3-Dimethylpropyl) dimethylchloro silane [CAS #71864-46-5]
CD5610

N-(1,2-Dimethylpropyl)-N´-ethyl-6-(methylthio)-1,3,5-triazine-2,4-diamine
Belclene 322

Dimethylsiloxane. See Dimethicone

Dimethylsiloxane-glycol copolymer. See Dimethicone copolyol

Dimethyl soyamine (CTFA) [CAS #61788-91-8] *Synonyms:* Soya dimethyl amine; Soyaalkyl dimethylamine
Armeen DMSD; Kemamine T-9972D

Dimethyl stearamide
Hallcomid M-18

Dimethyl stearamine (CTFA) [CAS #124-28-7] *Synonyms:* Stearyl dimethyl amine; Octadecyl dimethyl amine
Adma 8; Adogen 342-D; Armeen DM18D; Crodamine 3.A18D; Kemamine T-9902D; Lilamin 342 D; Nissan Tertiary Amine AB; Onamine 18

Dimethyl succinate [CAS #106-65-0]
DBE, DBE-2, DBE-2SPG, DBE-3, DBE-4, DBE-9

Dimethyl sulfide [CAS #75-18-3] *Synonyms:* Methyl sulfide
Spotleak 1001

Dimethyl tallowamine (CTFA) [CAS #68814-69-7; RD #977056-62-4] *Synonyms:* Tallow dimethylamine
Armeen DMTD; Kemamine T-9742D; Noram DMS

3,5-Dimethyl tetrahydro-2-H,1,3,5-thiadiazone-2-thione [CAS #533-74-4]
Amerstat 233; Troysan 142

Dimorpholino dianilino stilbene disulfonate deriv.
Tinopal AMS, AMS Pure, RA-16

2,2´-Dimorpholino diethylether
Texacat DMDEE

Dimyristyl thiodipropionate (CTFA) [CAS #16545-54-3] *Synonyms:* Ditetradecyl 3,3´-thiobispropanoate; Propanoic acid, 3,3´-thiobis-, ditetradecyl ester
Carstab DMTDP; Cyanox MTDP; Evanstab 14

N,N´-Di-β-naphthyl-p-phenylenediamine *Synonyms:* DNPD; sym-Di-β-naphthyl-p-phenylenediamine
Agerite White; Anchor DNPD

Diniconazole
Ortho Spotless

Dinitrooctylphenyl crotonate
Karathane Liq. Conc., WD

Dinitrosopentamethylene tetramine [CAS #101-25-7] *Synonyms:* DNPT
Porofor DNO/F

Dinonyldiphenylamine
Vanlube DND

Dinonyl phenol [CAS #1323-65-5]
Berol Dinonylphenol

O,O-Di-n-octadecyl-3,5-di-t-butyl-4-hydroxybenzyl phosphonate
Irganox 1093

Dioctyl adipate (CTFA) [CAS #103-23-1] *Synonyms:* Bis (2-ethylhexyl) hexanedioate; Di-(2-ethylhexyl) adipate; Hexanedioic acid, bis (2-ethylhexyl) ester; DOA
Adimoll DO; Crodamol DOA; DOA; Kodaflex DOA; Merrol DOA; Monoplex DOA; Plasthall DOA; Polytrap 158; Rilanit DOA; Staflex DOA; Uniflex DOA; Wickenol 158, 161, 163

Dioctyl azelate [CAS #103-24-2] *Synonyms:* Di(2-ethylhexyl) azelate; DOZ
Merrol DOZ; Plasthall DOZ; Plastolein 9058; Staflex DOZ

Dioctyl cyclohexane
Cetiol S

Di (n-octyl, n-decyl) adipate
Merrol 810A

Dioctyl dilinoleate (CTFA) *Synonyms:* Dioctyl dimerate
Kemester 3681

Dioctyl dimonium chloride *Synonyms:* Dioctyl dimethyl ammonium chloride
Bardac LF, LF80; BTC 818, 885; FMB 504-5 Quat

p,p´-Dioctyldiphenylamine [CAS #101-67-7] *Synonyms:* Dioctylated diphenylamine; Di-n-octyl diphenylamine
Agerite Hipar T, HP-S; Vanlube 81; Vanox 12, 1081, ODP

Dioctyl dodecanedioate dioleate
Plasthall DODD

Dioctyl fumarate [CAS #141-02-6] *Synonyms:* Di (2-ethylhexyl) fumarate
Staflex DOF

Dioctyl maleate [CAS #2915-53-9; #56235-92-8] *Synonyms:* Di-N-octyl maleate
Ceraphyl 45; Staflex DOM

Dioctyl peroxydicarbonate [CAS #16111-62-4] *Synonyms:* Di-2-ethylhexyl peroxydicarbonate; Bis (2-ethylhexyl) ester, peroxydicarbonic acid
EHP; Espercarb 840, 840M, 840M-40, 840M-70; Lupersol 223, 223-M, 223-M40, 223-M75, 223-T70

N,N´-Di (2-octyl)-p-phenylenediamine
Antozite 1

Dioctyl phthalate (CTFA) [CAS #117-81-7] *Synonyms:* DOP; Di(2-ethylhexyl) phthalate; di-s-octyl phthalate
Ircogel 901, 904; Kodaflex DOP, HS-4, PA-5; Rilanit DNOP; Staflex DOP; Vinyzene BP-505 DOP

Dioctyl sebacate (CTFA) [CAS #122-62-3, #2432-87-3] *Synonyms:* Di-2-ethylhexyl sebacate; Bis (2-ethylhexyl) decanedioate
Merrol DOS; Monoplex DOS; Plasthall DOS; Rilanit DEHS; Uniflex DOS; Unimate DOS

Dioctyl sodium sulfosuccinate (CTFA) [CAS #577-11-7] *Synonyms:* Sodium dioctyl sulfosuccinate; Sodium di-(2-ethylhexyl) sulfosuccinate
Aerosol OT-75, OT-100, OT-B; Alconate SBDO; Alkasurf SS-O-75; Alrowet D-65; Avirol SO-70P; Complemix 100; Condanol SBDO60; Cyclopol SBDO; Denwet CM; Emcol 4500, 4560; Empimin OP45, OP70, OT; Gemtex PA-75, -85P, PAX-60, SC-40, -70, -75, -75E, SC Powd.; Geropon SDS; Lankropol KO Special; Mackanate DOS-75; Mazawet DOSS; Monawet MO-65-150, MO-70, MO-70-150, MO-70E, MO-70R, MO-75E, MO-84R2W; Nikkol OTP-100S; Nissan Rapisol B-30, -80, C-70; Pentex 99; Rewopol NEHS 40, SB-DO 70; Sanmorin OT 70; Sanstat 2012-A; Secosol DOS/70; Serwet WH 170; Thorowet G-40, -75

Dioctyl succinate [CAS #2915-57-3] *Synonyms:* Bis (2-ethylhexyl) butanedioate
Wickenol 159

Dioctyl sulfosuccinate
Arylene M40, M60; Atlas WA-100; Ninate DS 70; Varidri 40; Varsulf SBDO-70

Dioctyl terephthalate
Kodaflex DOTP

Dioctyl thiodiglycolate
Vulkanol 90

Dioctyltin mercaptide [CAS #58229-88-2] *Synonyms:* Di-n-octyltin mercaptide
Kosmos 23

Dioleyl imidazoline methosulfate
Rewoquat W 3690, W 3690/PG

Dipentaerythritol hexacaprate/hexacaprylate [CAS #68130-24-5]
Liponate DPC-6

Dipentaerythritol hydroxypentaacrylate
SR-399

Dipentamethylene thiuram hexasulfide [CAS #120-54-7]
DPTT, DPTT-S; Sulfads

Dipentamethylene thiuram tetrasulfide
Naftocit DPTT

Dipentene [CAS #138-86-3] *Synonyms:* Cinene; Limonene, inactive
Piccolyte D100, D115, D135

Di (2-phenoxyethyl) peroxydicarbonate
Luperox 204

Diphenylamine-acetone reaction prod. [CAS #9003-79-6]
Agerite Hipar T, Superflex, Superflex Solid G; Vanox AM

Diphenylamine, alkylated
OA 502

Diphenylcresyl phosphate
Disflamoll DPK

Diphenyldichloro silane [CAS #80-10-4]
CD5950

Diphenyldidecyl (2,2,4-trimethyl-1,3-pentanediol) diphosphite
Weston 491

Diphenyldiethoxy silane [CAS #2553-19-7]
CD6000

Diphenyldimethoxy silane [CAS #6843-66-9]
CD6010

Diphenylene oxide-4,4´-disulfohydrazide
Porofor S 44

Diphenyl guanidine [CAS #102-06-7] *Synonyms:* N,N´-Diphenylguanidine; 1,3-Diphenylguanidine; DPG
Akrochem DPG; Anchor DPG; DPG; Naftocit DPG; Vanax DPG; Vulkacit D/C, D/EGC

Diphenyl hydrogen phosphite
Mark DPHP

Diphenyl isodecyl phosphite
Weston 63, DPDP

Diphenyl isooctyl phosphite
Weston ODPP

Di (phenyl mercury) dodecenyl succinate
Troysan PMDS-10

4,4´-Diphenyl methane diisocyanate [CAS #101-68-8] *Synonyms:* MDI; Methylene di-p-phenylene isocyanate; Methylene bisphenyl isocyanate
Desmodur MP-225; Isonate 125M, 143L, 181, 191, 240; Mondur PF; TSE-2000

2-(Diphenylmethylsily) ethanol
D6102

Diphenylnitrosamine. See Nitrosodiphenylamine

Diphenyloctyl phosphate
Disflamoll DPO

N,N´-Diphenyl-p-phenylene diamine [CAS #74-31-7] *Synonyms:* DPPD
Agerite DPPD, Hipar T, HP-S

Diphenyl phosphite
Weston DPP

Diphenylsilanediol [CAS #947-42-2]
CD6150

Diphenyl-sulfon-3,3´-disulfohydrazide
Porofor D33

N,N´-Diphenyl thiourea [CAS #102-08-9] *Synonyms:* N,N´-Diphenylthiocarbamide; N,N´-Diphenylthiourea; 1,3-Diphenylthiourea; 1,3-Diphenyl-2-thiourea; Sulfocarbanilide; Thiocarbanilide
A-1; Akrochem Accelerator Thio No. 1; Rhenocure CA; Stabilizer C

Di (propylene/ethylene) glycol dibenzoate
Merrol PGB-50

Dipropylene glycol... See also PPG-2...
Dabco 33-LV; Texacat TD-33A

Dipropylene glycol butyl ether [CAS #29911-28-2]
Dowanol DPnB

Dipropylene glycol dibenzoate
Merrol PGB

Dipropylene glycol salicylate. See PPG-2 salicylate

Di-n-propyl peroxydicarbonate [CAS #16066-38-9] *Synonyms:* n-Propyl percarbonate
Lupersol 221, 221-M85; NPP

N,N´-Disalicylidene-1,2-propane diamine *Synonyms:* N,N´-disalicylidene4-1,2-diaminopropane; Disalicylalaminopropane disalicylalpropylenediimine
Kerobit COM; Keromet MD 60, MD 80, MD 100

Disodium C12-15 pareth sulfosuccinate *Synonyms:* Disodium pareth-25 sulfosuccinate
Emcol 4300

Disodium-calcium EDTA
Sequestrene NA2Ca

Disodium cocamido MEA-sulfosuccinate (CTFA) [CAS #61791-66-0; 68784-08-7]
Schercopol CMS-Na

Disodium cocamido MIPA-sulfosuccinate (CTFA) [CAS #68515-65-1; 68909-21-7]
Monateric 805

Disodium cocamido sulfosuccinate
Alconate CPA

Disodium coconut sulfosuccinamate
Condanol CMS

Disodium deceth-4 sulfosuccinate
Alkasurf SS-DA4-HE

Disodium deceth-6 sulfosuccinate (CTFA) [CAS #68311-03-5 (generic); 68630-97-7 (generic)]
Aerosol A-102; Alconate D-6

Disodium 2,5-dimercaptothiadiazole
Vanchem NATD

Disodium EDTA (CTFA) [CAS #139-33-3] *Synonyms:* Disodium edetate
Hamp-Ene Na2, OH Powd.; Perma Kleer DI Crystals; Questal DI; Sequestrene NA2; Trilon BD; Versene NA

Disodium EDTA-copper (CTFA) [CAS #14025-15-1] *Synonyms:* Copper versenate
Sequestrene NA2Cu

Disodium ethanoldiglycine
Hampshire EDG

Disodium ethylene bisdithiocarbamate ethylene diamine
Amerstat 274

Disodium hexamethylene bisthiosulfate
Duralink HTS

Disodium hydrogenated cottonseed glyceride sulfosuccinate (CTFA)
Emcol 4072

Disodium isodecyl sulfosuccinate (CTFA) [CAS #37294-49-8; 55184-70-8]
Aerosol A-268

Disodium isostearamido MEA sulfosuccinate (CTFA)
Mackanate IM

Disodium laneth-5 sulfosuccinate (CTFA) [CAS #977059-29-2; #68890-92-6 (generic)]
Incrosul LAFS; Rewolan 5

Disodium lauramido MEA-sulfosuccinate [CAS #25882-44-4; 55101-78-5]
Alconate LEA; Condanol SBL 203; Steinapol SBL 203

Disodium laureth sulfosuccinate [CAS #39354-45-5 (generic); 40754-59-4; 42016-08-0] *Synonyms:* Disodium lauryl ether sulfosuccinate
Geronol ACR/4; Schercopol LPS; Steinapol SBFA 30 40%

Disodium lauriminodipropionate [CAS #3655-00-3] *Synonyms:* Disodium N-lauryl β-iminodipropionate; Disodium 3,3´-(dodecylimino) dipropionate
Deriphat 160; Mirataine H2C

Disodium lauryl sulfosuccinate [CAS #13192-12-6; 19040-44-9; 26838-05-1]
Condanol SBF 12, SBF 12-Powd.; Emcol 4400-1; Steinapol SBF 12, SBR 12-Powd.

Disodium maleic anhydride/diisobutylene copolymer
Empicryl APD, APD/B

Disodium methylenebis (naphthalene sulfonic acid)
Dispersol T

Disodium myristamido MEA-sulfosuccinate [CAS #37767-42-3]
Emcol 4100M

Disodium naphthalene sulfonic acid/formaldehyde condensate
Matexil DA-AC

Disodium nonoxynol-10 sulfosuccinate (CTFA) [CAS #67999-57-9 (generic); RD #977069-21-8]
Aerosol A-103; Alconate SBN-862

Disodium oleamido MEA-sulfosuccinate [CAS #68479-64-1]
Alconate SBG-280; Schercopol OMS-Na

Disodium oleamido MIPA-sulfosuccinate (CTFA) [CAS #43154-85-4; 67815-88-7] *Synonyms:* Disodium oleoyl isopropanolamide sulfosuccinate
Emcol 4161L; Schercopol OMIS-Na

Disodium oleamido PEG-2 sulfosuccinate (CTFA) [CAS #56388-43-3; 68227-80-5] *Synonyms:* Disodium oleamido diglycol sulfosuccinate
Jordawet DMDS; Mackanate OD-35

Disodium silicone polyol sulfosuccinate
Mackanate DG30

Disodium stearyl sulfosuccinamate (CTFA) [CAS #14481-60-8; 26446-37-7] *Synonyms:* Disodium octadecyl sulfosuccinamate
Aerosol 18; Alconate SBTA-269; Alkasurf SS-TA

Disodium tallowiminodipropionate (CTFA) [CAS #61791-56-8] *Synonyms:* Disodium N-tallow-β iminodipropionate
Deriphat 154; Mirataine T2C; Monateric TDB-35

Disodium undecylenamido MEA-sulfosuccinate [CAS #26650-05-5; 37311-67-4; 40839-40-5] *Synonyms:* Disodium undecylenic monoethanolamide sulfosuccinate
Alconate SBU-185; Cyclopol SBU-185; Incrosul UMS, UMS-45; Rewocid SBU 185; Varsulf SBU-185

Disodium wheat germamido MEA sulfosuccinate (CTFA) *Synonyms:* N-[2-[(Sulfosuccinyl) oxy] ethyl] wheat germ oil amides
Mackanate WG

Disodium wheat germamido PEG-2 sulfosuccinate
Mackamate WGD

Disoya dimethyl ammonium chloride
Arquad 2S-75

Distearyldimonium chloride (CTFA) [CAS # 107-64-2] *Synonyms:* Distearyl dimethyl ammonium chloride; Quaternium-5
Arosurf TA-100, TA-101; Cation DS; Comperlan LD 9; Dehyquart DAM; Genamin DSAC; Querton 442S, 442-Sx; Varisoft SA-50

Distearyldimonium methosulfate
Varisoft 137, 190, 190-100P

Distearyl pentaerythritol diphosphite
Mark 5050, 5060; Weston 618, 619, 732, 800, 801, 802

Distearyl phthalate
Radiasurf 7505; Sicolub DSP

Distearyl thiodipropionate (CTFA) [CAS #693-36-7] *Synonyms:* 3,3′-Thiobispropanoic acid, dioctadecyl ester; 3,3′-Dioctadecyl thiodipropionate
Carstab DSTDP; Cyanox STDP; Evanstab 18

Distyrylbiphenyl disulfonate deriv.
Tinopal ATS-X, BLS-X, CBS-X

Distyryl-phenyl
Ecco White FW-5

Disulfoton
Disyston

Ditallow amidoammonium methosulfate
Rewoquat W 222 LM

Ditallow diamido methosulfate
Incrosoft T-75

Ditallow dimonium chloride [CAS #68783-78-8] *Synonyms:* Ditallow dimethyl ammonium chloride; Quaternium-48
Adogen 470; Armosoft L; Querton 470

Ditallow imidazolinium methosulfate
Rewoquat W 9000, 9000/PG; Varisoft 472-80%

2,2´-Dithiobis (benzothiazole) [CAS #120-78-5] *Synonyms:* 2-Mercaptobenzothiazyl disulfide
Thiofide

Dithiodicaprolactam
Rhenocure S/G

4,4´-Dithiodimorpholine
Akrochem Accelerator R; Rhenocure M, M/G; Sulfasan R

4,4´-Dithiomorpholine [CAS #103-34-4]
Vanax A

Di-o-tolyl guanidine [CAS #97-39-2] *Synonyms:* DOTG; N,N´-Di-o-tolyl guanidine
Akrochem DOTG; Anchor DOTG; Vanax DOTG; Vulkacit DOTG/C, DOTG/EGC

Di-o-tolyl guanidine salt of dicatechol borate [CAS #16971-82-7]
Vanax PML

Ditridecyl adipate (CTFA) [CAS #26401-35-4]
Kemester 5654; Merrol DTDA; Plasthall DTDA

Ditridecylamine [CAS #5910-75-8]
Adogen 283, 2283

Ditridecyl dilinoleate (CTFA) [CAS #16958-92-2] *Synonyms:* Ditridecyl dimerate; Dimer acid, ditridecyl ester
Kemester 3684

Ditridecyl maleate
Staflex DTDM

Ditridecyl phosphite
Weston DTDP

Ditridecyl phthalate [CAS #119-06-2] *Synonyms:* 1,2-Benzenedicarboxylic acid, ditridecyl ester
Staflex DTDP

Ditridecyl sodium sulfosuccinate [CAS #2673-22-5] *Synonyms:* Sodium bistridecyl sulfosuccinate; Sodium ditridecyl sulfosuccinate
Aerosol TR-70; Emcol 4600; Monawet MT-70, MT-70E, MT-80H2W

Ditridecyl thiodipropionate (CTFA) [CAS #10595-72-9] *Synonyms:* 3,3´-Thiobispropanoic acid, ditridecyl ester
Cyanox 711; Evanstab 13

Ditrimethylolpropane tetraacrylate
SR-355

Diuron [CAS #330-54-1] *Synonyms:* 3-(3,4-Dichlorophenyl)-1-1dimethyl-urea
Direx 4L

Divinyltetramethyldisiloxane [CAS #2627-95-4]
CD6210

Dixylyldisulfides
Aktiplast 6 N

DMDM hydantoin (CTFA) [CAS #6440-58-0]
Dantoin DMDMH-55; Glydant

DME
Texacat DPA-50

DMHF (CTFA) [CAS #9065-13-8; RD #977056-71-5] *Synonyms:* Dimethyl hydantoin formaldehyde
Dantoin DMHF, DMHF-75, DMHF Refined

DM hydantoin (CTFA) [CAS #77-71-4] *Synonyms:* 5,5-Dimethylhydantoin
Dantoin DMH

Dodecachloro dodecahydro dimethanodibenzo cyclooctene
Dechlorane Plus 25, 515, 2520

Dodecyl... See Lauryl...

Dodecylbenzene sulfonic acid (CTFA) [CAS #27176-87-0] *Synonyms:* DDBSA
Alkasurf LA Acid; Arsul DDB, LAS, S Acid; Arylan SBC Acid, SC Acid, SP Acid; Bio Soft S-100; Calsoft LAS-99; Carsosulf UL-100 Acid; Conco AAS-98S; Condasol Sulfonic Acid K; DeSonate SA, SA-H; Emulsifier 99; Hetsulf Acid; Jordamine DDBSA; Marlon AS3; Nansa 1042, 1042/P, SBA, SSA, SSA/P; Polystep A-17; Puxol XB-10; Rueterg SA; Steinaryl-Sulfonic Acid K; Stepantan H-100; Sterling LA Acid-C, Reg.; Surco DDBSA; Tex-Wet 1197; Witco 1298 Hard Acid, 1298 Soft Acid, Acid B

Dodecyl benzotriazole [CAS #3663-25-0]
Cobratec DBT-50, DBT-100

Dodecylbenzyl trimonium chloride [CAS #1330-85-4; 19014-05-2] *Synonyms:* Dodecylbenzyl trimethyl ammonium chloride
Hyamine 2389

Dodecyl dimethyl ethylbenzyl ammonium chloride. See Quaternium-14

Dodecyl dimethyl (2-phenoxyethyl) ammonium bromide [CAS #538-71-6] *Synonyms:* Domiphen bromide (CTFA)
Fungitex R

Dodecyl diphenyloxide disulfonic acid
Conco Sulfate 2A1 Acid; Dowfax 2A0

n-Dodecyl mercaptan
Pennwalt n-Dodecyl Mercaptan

t-Dodecyl mercaptan
Pennwalt 4P

4-Dodecyloxy-2-hydroxybenzophenone
Eastman Inhibitor DOBP

Dodecyltrichlorosilane [CAS #4484-72-4] *Synonyms:* Trichlorododecylsilane
CD6220

Dodecylxylylditrimonium chloride [RD #977065-94-3] *Synonyms:* Quaternium-29
Hyamine 2389

Dodine [CAS #2439-10-3] *Synonyms:* N-dodecylguanidine acetate
Cyprex 65-W

Dodoxynol-6 (CTFA) [CAS #9002-92-0 (generic); 9014-92-0 (generic); 26401-47-8 (generic); RD #977064-28-0]
Sterox DF

Dodoxynol-10
Iconol DDP-10

Drometrizole (CTFA) [CAS #2440-22-4] *Synonyms:* 2-(2′-Hydroxy-5′-methylphenyl) benzotriazole
JF 77; Tinuvin P; Topanex 100BT

DTPA. See Pentetic acid

DTPA-H_5
Dissolvine DZ

DTPA-Na FeH
Dissolvine D FE

Edifenphos
Hinosan

EDTA (CTFA) [CAS #60-00-4] *Synonyms:* Edetic acid; Ethylene diamine tetraacetic acid
Cheelox BF Acid; Hamp-Ene Acid; Kalex Acids; Perma Kleer EDTA Acid; Questex 4H; Questric Acid; SEQ 100; Sequestrene AA; Solon; Trilon BS; Vanate Acid; Versene Acid

EDTA-H_4
Dissolvine Z

EDTA KH
Dissolvine K 3

EDTA-Mg_2
Dissolvine MG 2

EDTA Na_2Ca
Dissolvine CA

EDTA NaFe
Dissolvine NAFE

EDTA-Na_2H_2
Dissolvine NA 2

EDTA $(NH_4)_4$
Dissolvine AM 4

EDTA $(NH_4)_2H_2$
Dissolvine AM 2

EDTA $(NH_4)_3H$
Dissolvine AM 3

EDTA NH_4Fe NH_4OH
Dissolvine AMFE

Egg albumen. See Albumen

Egg oil (CTFA) [CAS #8001-17-0]
Emcon E

Egg powder (CTFA) [RD #977054-46-8] *Synonyms:* Egg, dried
Sol-U-Tein E

Elastin hydrolysate. See Hydrolyzed elastin

Emulsifying wax NF
Emerwax 1257

EO/PO block polymer
Atmer 1040, 1041, 1042; Berol 371, 372 Flakes, 374; Genapol B, PN 30; Industrol N2, N3; Lutrol F 127; Macol 1, 18, 19, 31, 32, 33, 42, 81, 101; Pluronic 10R5, 10R8, 12R3, 17R1, 17R2, 17R4, 17R8, 22R4, 25R1, 25R2, 25R4, 25R5, 25R8, 31R1, 31R2, 31R4, F38, F68, F68LF, F77, F87, F88, F98, F108, F127, L10, L31, L35, L42, L43, L44, L61, L62, L62D, L62LF, L63, L64, L72, L81, L92, L101, L121, L122, P65, P75, P84, P85, P103, P104, P105, P123; Polyglycol 15-200, 112-2; Tetronic 50R1, 50R4, 50R8, 70R1, 70R2, 70R4, 90R1, 90R4, 90R8, 110R1, 110R2, 110R7, 130R1, 130R2, 150R1, 150R4, 150R8, 304, 504, 701, 702, 704, 707, 901, 904, 908, 909, 1101, 1102, 1104, 1107, 1301, 1302, 1304, 1307, 1501, 1502, 1504, 1508

EPDM. See Ethylene-propylene-diene terpolymer

EPM. See Ethylene-propylene rubber

Epoxidized butyl esters of linseed oil fatty acids
Epoxol 8-2B

Epoxidized glycol dioleate
Monoplex S-75

Epoxidized linseed oil *Synonyms:* Epoxidized flaxseed oil
Admex ELO; Drapex 10.4; Epoxol 9-5; Lankroflex L; Plas-Chek 795

Epoxidized octyl stearate
Drapex 3.2

Epoxidized octyl tallate
Drapex 4.4; Epoxol 5-2E; Flexol Plasticizer EP-8; Monoplex S-73

Epoxidized oleate
Merrol E-52

Epoxidized soybean oil [CAS #8013-07-8]
Admex 710, 711; Drapex 6.8; Epoxol 7-4; Estabex 138-A, 2307; Flexol Plasticizer EPO; Intercide ABF 1 ESBO, ABF 2 ESBO; Lankroflex GE; Merrol E-68, E-70; Paraplex G-60, G-62; Peroxidol 780; Plas-Chek 775; Plastoflex 2307; Plastolein 9232

Epoxidized tallate
Admex 746; Merrol E-45

Epoxidized tall oil
Peroxidol 781

Epoxy acrylate
Chemlink 3000

Epoxy, brominated
Thermoguard 210, 212, 213, 215, 220, 230, 240

Epoxy coal tar system
Interset Epoxy Coal Tar

2-(3,4-Epoxycyclohexyl) ethyltrimethoxy silane [CAS #3388-04-3] *Synonyms:* β-(3,4-Epoxycyclohexyl) ethyltrimethoxysilane
CD6250; CE6250; Union Carbide A-186

Epoxy mono ester [CAS #141-38-8]
Estabex 2381

Epoxy resin
Barco Bond MB-14X, MB-127, MB-165, MB-175, MB-185; Lekutherm X 50; Varcum 6404

Epoxy-silicone
Eccosil 4712

Erucamide (CTFA) [CAS #112-84-5] *Synonyms:* Erucyl amide; Erucic acid amide; 13-Docosenamide
Adogen 58; Armoslip EXP; Crodamide E, ER; Kemamide E; Kemester E; Petrac Eramide; Unislip 1753

Erucyl arachidate (CTFA) *Synonyms:* Erucyl eicosanoate
Kemester JO

Erucyl erucamide
Kemamide E-221

Erucyl erucate (CTFA) [CAS #27640-89-7]
Kemester EE; Schercemol EE

Erucyl stearamide
Kemamide S-221

Esterase
Milezyme Lipase Powds.

Esterase/lipase
Palatase

Estertin mercaptide
Stanclere T-208, T-222, T-233, T-250, T-250SD

ETFE. See Ethylene/tetrafluoroethylene copolymer

1,1-(1,2-Ethanediyl) bis (3,3,5,5-tetramethylpiperazinone)
Good-rite UV3034

Ethanol. See Ethyl alcohol

2,2′-(1,2-Ethenediyldi-4,1-phenylene) bisbenzoxazole
Eastobrite OB-1

Ether thioether
Vulkanol 85, OT

Ethidimuron
Ustilan

Ethiofencarb
Croneton

N-(p-Ethoxycarbonylphenyl)-N′-ethyl-N′-phenylformamidine
Givsorb UV-2

Ethoxydiglycol (CTFA) [CAS #111-90-0] *Synonyms:* Diethylene glycol monoethyl ether
Dioxotol-High Gravity, -Low Gravity; Dowanol DE; FC-520; Poly-Solv DE (High Gravity), (Low Gravity)

6-Ethoxy-1,2-dihydro-2,2,4-trimethylquinoline [CAS #91-53-2] *Synonyms:* Ethoxyquin
Santoflex AW; Santoquin, Emulsion, Mixture 6

Ethoxyethanol (CTFA) [CAS #110-80-5] *Synonyms:* Ethylene glycol monoethyl ether
Ektasolve EE; Oxitol; Poly-Solv EE

Ethoxyethanol acetate (CTFA) [CAS #111-15-9] *Synonyms:* Ethylene glycol monoethyl ether acetylated
Ektasolve EE Acetate

Ethoxytriglycol [CAS #112-50-5] *Synonyms:* Triethylene glycol monoethyl ether; PEG-3 ethyl ether
Poly-Solv TE

Ethyl acetate [CAS #141-78-6] *Synonyms:* Acetic ether; Acetic ester; Vinegar naphtha
Desmodur RE, RFE

Ethylacetoacetate chelate
Tyzor DC

Ethyl acrylate/2-ethylhexylacrylate copolymer
PC-1244, -1344

Ethylacrylate/methylacrylic acid copolymer
Sipothix H-65

Ethyl alcohol [CAS #64-17-5] *Synonyms:* Ethanol; Undenatured ethanol; Alcohol (CTFA); Grain alcohol; EtOH
Filmex; Punctilious Pure Ethyl Alcohol, Specially Denatured Ethyl Alcohol; Sustane P, PA

Ethyl-o-benzoyl laurohydroximate
Trigonox 107

Ethyl benzylammonium chloride
Catigene 4513, 4513/80

7-Ethyl bicyclooxazxolidine
Oxaban-E

Ethyl bis (polyhydroxyethyl) alkyl ammonium ethosulfate
Varstat 66

Ethyl-2-t-butylcyclohexylcarbonate
Floramat

Ethyl cellulose (CTFA) [CAS #9004-57-3] *Synonyms:* Cellulose, ethyl ether
Ethyl Cellulose NF Grade

Ethyl 3,3-di (t-amylperoxy) butyrate [CAS #67567-23-1]
Lupersol 533-M75

Ethyl-3,3-di (t-butylperoxy) butyrate
Luperco 233-KE, 233-XL; Lupersol 233-M75, 233-M90

Ethyl dihydroxypropyl PABA (CTFA) [CAS #58882-17-0] *Synonyms:* Ethyl dihydroxypropyl-p-aminobenzoate; Ethyl p-aminobenzoate, 2 mole propoxylate
Amerscreen P, P 80/20

S-Ethyl diisobutylthiocarbamate *Synonyms:* Butylate
Atrabute+, Atrabute+ II

Ethyl dimerate
Schercemol DED

3-Ethyl-2,4-dioxaspiro (5.5) undec-8-ene
Spiroflor

Ethylene/acrylate copolymer (CTFA) [RD #977064-63-3] *Synonyms:* Ethylene/acrylic acid copolymer; Ethylene/methacrylic acid copolymer; Ethylene/methacrylate copolymer
A-C Copolymer 540A; A-C Polyethylene 540; PE 2205, 2207, 2255, 2260; Primacor 4990 Dispersion

Ethylenebis dibromonorbornane dicarboximide [CAS #52907-07-0]
Saytex BN-451

2,2′-Ethylene bis (4,6-di-t-butylphenol) [CAS #35958-30-6]
Vanox 1290

N,N′-Ethylenebis 12-hydroxystearamide
Paricin 285

Ethylenebisstearamide. See Ethylene distearamide

1-Ethylene bis (2-tallow, 1-methyl, imidazolinium-methosulfate)
Varisoft 6112

Ethylene bis-tetrabromophthalimide
Saytex BT-93

Ethylene brassylate (CTFA) [CAS #105-95-3] *Synonyms:* 1,4-Dioxacycloheptadecane-5,17-dione; Ethylene undecane dicarboxylate
Emeressence 1150

Ethylene carbonate [CAS #96-49-1] *Synonyms:* Glycol carbonate; Dioxolone-2
Texacar EC, EC-50

Ethylene copolymer [CAS #9010-77-9]
ACuflow AF-1

Ethylene diamine dinonylnaphthalene sulfonate
Na-Sul EDS, LP

Ethylene diamine distearyl amide
Sicolub EDS

Ethylene dioleamide (CTFA) [CAS #110-31-6] *Synonyms:* N,N′-1,2-Ethanediylbis-9-octadecenamide; 9-Octadecenamide, N,N′-1,2-ethanediylbis
Advawax 240; Kemamide W-30

Ethylene distearamide (CTFA) [CAS #110-30-5] *Synonyms:* N,N′-Ethylene bisstearamide
Acrawax C, C DF #1, C DF #2, C SG; Advawax 275, 280, 290; EBS Wax; Alkamide STEDA; Interstab G-8257; Kemamide W-39, W-40, W-40/300, W-40DF, W-45; Lipowax C; Nopcowax 22-DS; Uniwax 1760 EBS

2,2-Ethylene dithiodiethanol
Tegochrome 22

Ethylene dodecanedioate
Emeressence 1151

Ethylene glycol. See Glycol

Ethylene glycol butyl ether. See Butoxyethanol

Ethylene glycol diethyl ether [CAS #629-14-1] *Synonyms:* 1,2-Diethoxyethane
Ethyl Glyme

Ethylene glycol dimethacrylate [CAS #97-90-5] *Synonyms:* 1,2-Ethanediol dimethacrylate
Rhenofit EDMA/S; SR-206

Ethylene glycol dimethyl ether *Synonyms:* Dimethoxyethane; DME; Glycol dimethyl ether
Monoglyme

Ethylene glycol hydroxystearate
Paricin 15

Ethylene glycol monoethyl ether. See Ethoxyethanol

Ethylene glycol monoethyl ether acetate. See Ethoxyethanol acetate

Ethylene glycol monopropyl ether
Ektasolve EP

Ethylene thiourea [CAS #96-45-7]
Akrochem ETU-22 PM

Ethylene thiourea (2-mercaptoimidazoline)
Naftocit Mi 12 C; Naftopast Mi12-P

Ethylene triglycol stearate
Tegin TGOS

Ethylene/vinyl acetate copolymer (CTFA) [CAS #24937-78-8] *Synonyms:* EVA copolymer; Acetic acid, ethenyl ester, polymer with ethene
A-C Copolymer 400A; A-C Polyethylene 400; BASF Wax EVA 1

Ethyl hexanediol (CTFA) [CAS #94-96-2]
Lanoquat 1756, 1757

2-Ethylhexanol
Surfynol 104A

2-Ethyl hexanol and P_2O_5 reaction prod.
Victawet 12

2-Ethyl hexanol, P_2O_5 and NaOH reaction prod.
Victawet 35B

2-Ethylhexyl... See Octyl..., Caprylic...

2-Ethylhexyl p-dimethyl aminobenzoate. See Octyl dimethyl PABA

2-Ethylhexyl diphenyl phosphate
Phosflex 362; Santicizer 141

Ethylhexyl laurate
Radia 7127

2-Ethylhexyl oleate
Radia 7331; Rilanit EHO

2-Ethylhexyl palmitate. See Octyl palmitate

Ethylhexyl pelargonate. See Octyl pelargonate

2-Ethylhexyl tallowate
Rilanit EHTI

2-Ethylhexyl triglyceride
Exceparl TGO

Ethyl hydrolyzed keratin protein [CAS #69430-36-0] *Synonyms:* Ethyl hydrolyzed animal keratin
Kera-Tein 1000AS

N-Ethyl-N-hydroxyethyl-m-toluidine [CAS #91-88-3]
Emery 5714

Ethyl hydroxymethyl oleyl oxazoline (CTFA) [CAS #68140-98-7]
Alkaterge-E

2,2′-Ethylidenebis (4,5-di-t-butylphenol)
Isonox 129

2,2′-Ethylidenebis (4,6-di-t-butylphenyl) fluorophosphonite [CAS #118337-09-0]
Ethanox 398

Ethyl methacrylate (CTFA) [CAS #97-63-2] *Synonyms:* Ethyl 2-methyl-2-propenoate
Acryloid B-72

Ethyl morpholine
Texacat NEM

Ethyl myristate (CTFA) [CAS #124-06-1] *Synonyms:* Ethyl tetradecanoate
Carsemol A-500

Ethyl palmitate (CTFA) [CAS #628-97-2] *Synonyms:* Ethyl hexadecanoate
Carsemol A-500

Ethyl paraben (CTFA) [CAS #120-47-8] *Synonyms:* Ethyl 4-hydroxybenzoate
Lexgard E; Trisept E

Ethyl PEG-15 cocamine sulfate (CTFA) [RD #977069-64-9] *Synonyms:* Quaternium-54; N-ethyl-N-cocoammonium ethoxylate (15) sulfate
Antaron PC-37

Ethyl pelargonate (CTFA) [CAS #123-29-5] *Synonyms:* Ethyl nonanoate
Carsemol A-500

Ethyl phenyl ethanolamine [CAS #92-50-2]
Emery 5707

Ethyl stearate (CTFA) [CAS #111-61-5] *Synonyms:* Ethyl octadecanoate
Carsemol A-500

Ethyl toluene sulfonamide (CTFA) [CAS #80-39-7; 1077-56-1]
Merrol N-302; Santicizer 8

Ethyltriacetoxysilane [CAS #17689-77-9]
CE6345

Ethyltrichlorosilane [CAS #115-21-9] *Synonyms:* Trichloroethylsilane
CE6350

Etidronic acid (CTFA) [CAS #2809-21-4] *Synonyms:* (1-Hydroxyethylidene) bisphosphonic acid
Fostex P; Turpinal SL

Etocrylene (CTFA) [CAS #5232-99-5] *Synonyms:* Ethyl-2-cyano-3,3-diphenylacrylate
Uvinul N-35

EVA copolymer. See Ethylene/vinyl acetate copolymer

Evening primrose oil
Crossential EPO; Super Refined Crossential EPO

Fenamiphos
Nemacur

Fenitrothion
Folithion

Fenpropathrin
Ortho Danitol 2.4 EC Spray

Fensulfothion
Terracur P

Fenthion
Baytex; Lebaycid

FEP resin. See Fluorinated ethylene-propylene

Ferric ammonium EDTA
Sequestrene NF4Fe; Trilon B-FA

Ferric chloride [CAS #7705-08-0] *Synonyms:* Iron chlorides
Luperfoam 329

Ferro-aluminum silicate
Ferrosil 14

Ferro-chromium lignosulfate
Lignosol F 30; Peltex; Rychem FCL

Fischer-Tropsch wax,. See Synthetic wax

Fischer-Tropsch wax, oxidized. See Synthetic wax

Flubenzimine
Cropotex

Flucythrinate
AAstar

Fluometuron
Meturon 4L

Fluorinated ethylene-propylene *Synonyms:* FEP resin
TL-120

Fluoropolymer
Zepel B, DR

Fluorotelomer *Synonyms:* Fluorocarbon telomer
M-7 Mold Release; N-30; Vydax WD

Food starch
Frimulsion Q8

Formaldehyde ethyl cyclo dodecyl acetal
Boisambrene Forte

Formaldehyde methyl cyclododecyl acetal
Boisambrene

Formaldehyde naphthalene/alkylaryl sulfonate copolymer
Lomar BDG

Fuberidazole
Neo-Voronit

Furan polymer *Synonyms:* Furan resin
Cardolite NC-552; QO Furan; Varcum 8251, 8261, 8267

Furfural (CTFA) [CAS #98-01-1] *Synonyms:* 2-Furaldehyde
QO Furfural

Furfuryl alcohol
QO Furfuryl Alcohol; QuarCorr Resin/Catalyst System

Galactomannan
GFS

Galactomannan polysaccharide [CAS #9000-30-0]
Supercol Guar F, G2, U

Galactomannase
Gamanase 1.5L; Milezyme Brand Hemicellulase

β-Galactosidase [CAS #9031-11-2] *Synonyms:* Lactase
Lactozym, 1500 L type GP, 3000 L type HP; Maxilact; Milezyme Fungal Lactase

Gelatin (CTFA) [CAS #9000-70-8]
Crodyne BY19; Frimulsion Q8, RA, RF; Gelrite

Gellan gum
Kelcogel Gellan Gum

Glucanase
Cereflo 200 L; Finizym 200L

Glucoamylase
Diazyme; Spezyme GA

Gluconic acid [CAS #526-95-4] *Synonyms:* Glyconic acid
PMP Gluconic Acid 50% Tech.

Glucose (CTFA) [CAS #50-99-7 (anhydrous), 5996-10-1 (hydrous)] *Synonyms:* Dextrose
CC-603; Frimulsin 6G, Q8, RA

Glucose glutamate (CTFA) [RD #977056-78-2]
Wickenol 545

Glucose isomerase
Ketomax GI-100; Maxazyme GI-IMMOB; Spezyme IGI; Sweetzyme Type A, Q; Taka-Sweet

Glucose oxidase
Fermcozyme 1307, BG, BGXX, CBB, CBBXX, M; Ovazyme XX

Glutaral (CTFA) [CAS #111-30-8] *Synonyms:* Glutaraldehyde; Glutaric dialdehyde; Pentanedial
Aqucar Microbiocides; Protectol GDA, GT 50; Sepacid CE 5265, GA 25, GA 50; Ucarcide 225, 250; Ucarsan Sanitizers; Uconex Antimicrobials

Glycereth-7 (CTFA) [CAS #31694-55-0] *Synonyms:* PEG-7 glyceryl ether; POE (7) glyceryl ether
Hetoxide G-7; Liponic EG-7

Glycereth-26 (CTFA) [CAS #31694-55-0 (generic); RD #977054-72-0]
Acconon ETG; Ethosperse G-26; Hetoxide G-26; Liponic EG-1

Glycerides, acetates. See Acetylated glycerides

Glycerides, hydrogenated. See Hydrogenated glycerides

Glycerin (CTFA) [CAS #56-81-5] *Synonyms:* Glycerol; 1,2,3-Propanetriol
Croderol G7000; Emery 912, 916; Kemstrene 96.0%, 99.0%, 99.5%, 99.7%, High Gravity; Lexamine R-13; Star; Superol

Glyceryl caprate (CTFA) [CAS #26402-22-2]
Imwitor 310

Glyceryl capromyristate
Radia 7104

Glyceryl caprylate (CTFA) [CAS #26402-26-6]
Imwitor 308; Tegin 4600 NSE

Glyceryl caprylate-caprate (CTFA)
Capmul MCM; Drewmulse GMC-8

Glyceryl cocoate (CTFA) [CAS #61789-05-7] *Synonyms:* Glycerides, coconut oil mono-; Glycerol mono coconut oil; Glyceryl coconate
Drewmulse CNO; Softisan 601

Glyceryl diacetate
Kessco Diacetin

Glyceryl dilaurate (CTFA) [CAS #27638-00-2] *Synonyms:* Dilaurin
Cithrol GDL N/E, GDL S/E; Emulsynt GDL; Kessco Glycerol Dilaurate

Glyceryl dioleate (CTFA) [CAS #25637-84-7]
Cithrol GDO N/E, GDO S/E; Kessco Glycerol Dioleate; Mazol GDO; Nikkol DGO-80; Rilanit GDO

Glyceryl distearate (CTFA) [CAS #1323-83-7]
Cithrol GDS N/E, GDS S/E; Nikkol DGS-80

Glyceryl di/tripalmitostearate
Precirol ATO

Glyceryl di/tristearate
Precirol WL 2155

Glyceryl hydrogenated rosinate
Foral 85; Staybelite Ester 5, 10

Glyceryl hydroxystearate (CTFA) [CAS #1323-42-8] *Synonyms:* Glyceryl 12-hydroxystearate
Cutina BW; Naturechem GMHS; Paricin 13; Radiasurf 7146

Glyceryl isostearate (CTFA) [CAS #32057-14-0; 66085-00-5; RD #977064-41-7]
Emerest 2410; Schercemol GMIS

Glyceryl laurate (CTFA) [CAS #142-18-7; 27215-38-9] *Synonyms:* Glyceryl monolaurate
Aldo MLD; Alkamuls GML-45; Cithrol GML N/E, GML S/E; Grindtek ML 90; Imwitor 312; Kessco Glycerol Monolaurate; Lamesoft LMG; Monomuls 90-L12; Tegin 4480

Glyceryl linoleate (CTFA) [CAS #2277-28-3; 26545-74-4] *Synonyms:* Monolinolein
Grindtek MOL 90

Glyceryl mono/dicaprylate
Imwitor 908

Glyceryl mono/dilaurate
Aldo ML

Glyceryl mono/dimyristate
Imwitor 914

Glyceryl mono/dioleate
Aldo MO FG, MOD FG; Tegomuls O Spezial

Glyceryl mono/distearate
Atmer 122, 123; Witconol MST

Glyceryl myristate (CTFA) [CAS #589-68-4; 27214-38-6] *Synonyms:* Monomyristin
Grindtek MM 90

Glyceryl oleate (CTFA) [CAS #111-03-5; 25496-72-4; SE grade RD #977064-71-3] *Synonyms:* Monoolein; Glyceryl monooleate; GMO
Alkamuls GMO-45, GMO-45LG, GMO-55LG; Armostat 810; Atmer 121, 1007, 1010; Capmul GMO; Cithrol GMO N/E, GMO S/E; CPH-31-N, -205-NX, -362-N; Cyclochem GMO; Drewmulse 85, 200, GMO; Dur-Em GMO; Emerest 2421; Interwax G 8200; Kemester 2000; Kessco Glycerol Monooleate, GMO; Mazol 300, 300 K, GMO, GMO K; Monomuls 90-O18; Protegin, W, WX, X; Radiasurf 7150, 7151 (SE), 7152; Rilanit GMO; Sustane 20A, W

Glyceryl PABA (CTFA) [CAS #136-44-7] *Synonyms:* 1-(4-Aminobenzoate)-1,2,3-propanetriol; Glyceryl p-aminobenzoate
Escalol 106

Glyceryl ricinoleate (CTFA) [CAS #141-08-2; 1323-38-2] *Synonyms:* Glyceryl monoricinoleate; Mono-ricinolein
Alkamuls GMR; Cithrol GMR N/E, GMR S/E; Cutina LM; Drewmulse GMRO; Flexricin 13; Mazol GMR; Radiasurf 7153; Rilanit GMRO; Softigen 701; Tegin RZ, RZ NSE

Glyceryl rosinate
Hercules Ester Gum 8BG, 8D, 8D-SP, 10D; Poly-Pale Ester 10; Vinsol Ester Gum; Zonester 55, 75

Glyceryl stearate (CTFA) [CAS # 123-94-4; 11099-07-3; 31566-31-1; 61789-07-9; 56-81-5; 57-11-4; SE grade RD #977053-96-5] *Synonyms:* Glyceryl monostearate; GMS; Monostearin
Ahcovel Base N-15; Aldo HMS, MS LG, PME; Aldosperse O-20 FG; Alkamuls GMS, GMS-45; Armostat 801; Atmer 122, 123, 125, 126, 128, 129; Capmul GMS; Cerasynt 945, GMS, M, MN, SD; Cithrol GMS Acid Stable, GMS A/E ES 0743, GMS N/E, GMS S/E; CPH-53-N; Cutina CBS, GMS, MD; Cyclochem GMS, GMS 21, GMS 165, GMS/SE; Dermalcare GMS,

GMS-165, GMS/SE; Drewmulse 20/200/V, 21, 30/900, 300, 365, 700, 900, 1128, 1129, GMS, HM-100, TP, V, V-SE (SE); Drewsoft 100; Dur-Em 117; Emerest 2400, 2401; Empilan GMS LSE40, GMS LSE80, GMS MSE40, GMS NSE40, GMS NSE90, GMS SE40, GMS SE70; Gelot 64; Grindtek MSP 32-6, 40, 40F, 52, 90; Ice #2; Imwitor 191, 900, 900 K, 960, 960 K; Interwax G 8204; Kemester 5500, 6000, 6000E, 6000SE, GMS (Powd.); Kessco Glycerol Monostearate 860, DH-1, Pure, SE; Lexate TA, TL; Lexemul 55G, 503, 515, AR, AS; Lexemul T (SE); Lipo GMS, GMS 450, GMS 470 (SE); Lipomulse 165; Mazol 165C, GMS, GMS-D (SE), GMS-HM; Nikkol MGS-A, MGS-ASE (SE), MGS-B, MGS-BSE (SE), MGS-DEX; Norfox GMS, GMS-F6; Petrac GMS; Protowax 90; Radiamuls MG 2412; Radiasurf 7140, 7141 (SE), 7600; Rilanit GMS; Schercemol GMS; Softenol 3900, 3925, 3945, 3945, 3960, 3991, 3995; Tally 100 Plus; Tefose 2561; Tegin 90, 90 NSE, 515, 515 V.A., 4100, C-1R (SE), GRB NSE, ISO, M NSE, MAV NSE, VA, VA 55G; Vykamol N/E, S/E; Witconol CA (SE), RHT (SE)

Glyceryl stearate palmitate
Imwitor 940 K, 965 K

Glyceryl stearate SE (CTFA). See under Glyceryl stearate

Glyceryl triacetate. See Triacetin

Glyceryl triacetyl hydroxystearate (CTFA) [CAS #27233-00-7] *Synonyms:* Glyceryl (triacetoxystearate)
Paricin 8

Glyceryl triacetyl ricinoleate (CTFA) [CAS #101-34-8] *Synonyms:* 9-Octadecenoic acid, 12-(acetyloxy)-, 1,2,3-propanetriol ester
Flexricin P-8

Glyceryl tribehenate (CTFA) [CAS #18641-57-1] *Synonyms:* Tribehenin
Syncrowax HR-C, HRS-C

Glyceryl tribenzoate
Benzoflex S-404

Glyceryl tricaprate/caprylate
Radianol 2106, 7106

Glyceryl triheptanoate
Radianol 7376

Glyceryl tri-12-hydroxystearate. See Trihydroxystearin

Glyceryl triisostearate. See Triisostearin

Glyceryl trilaurate. See Trilaurin

Glyceryl trilaurate/stearate
Softisan Hard Fats

Glyceryl trioctanoate (CTFA) [CAS #538-23-8] *Synonyms:* Glycerin tri-2-ethylhexanoate
Nikkol Trifat S-308

Glyceryl trioleate. See Triolein

Glyceryl tristearate. See Tristearin

(3-Glycidoxypropyl)-methyldiethoxy silane [CAS #2897-60-1]
CG6710

3-Glycidoxypropyltrimethoxysilane [CAS #2530-83-8] *Synonyms:* γ-Glycidoxypropyltrimethoxysilane
Dow Corning Z-6040; Dynasylan GLYMO; Union Carbide A-187

Glycidyl trimethyl ammonium choride
Ogtac-85, -85V

Glycol (CTFA) [CAS #107-21-1] *Synonyms:* Ethylene glycol; 1,2-Ethanediol
Dow Corning Q2-7224; Poly-Solv DE (High Gravity); Surfynol 104E, 104H

Glycol, acrylated
Chemlink 2000

Glycol dibehenate (CTFA) [CAS #344-64-8] *Synonyms:* 1,2-Ethanediyl docosanoate
Rewopal PG 340

Glycol dilaurate *Synonyms:* Ethylene glycol dilaurate
Cithrol EGDL N/E, EGDL S/E

Glycol dioleate *Synonyms:* Ethylene glycol dioleate
Cithrol EGDO N/E, EGDO S/E

Glycol distearate (CTFA) [CAS #627-83-8] *Synonyms:* Ethylene glycol distearate; EGDS
Cithrol EGDS N/E, EGDS S/E; Cutina AGS; Cyclochem EGDS; Drewmulse EGDS; EGDS-VA; Elfan L310; Emerest 2355; Euperlan PK 771, PK 776, PK 789; Genapol PGM Conc., PMS; Kemester EGDS; Kessco Ethylene Glycol Distearate; Lexemul EGDS; Lipal EGDS; Lipo EGDS; Mapeg EGDS; Nikkol Estepearl 10, 15; Pegosperse 50 DS; Radiasurf 7269; Rewopal PG 280; Secoster DMS; Standapol Pearl Conc. 7130; Tegin G 1100; Tego-Pearl B-48

Glycol ether glutarate
Merrol 4295, 4425

Glycol hydroxystearate (CTFA) [CAS #33907-46-9] *Synonyms:* Ethylene glycol hydroxystearate
Naturechem EGHS

Glycol laurate *Synonyms:* Ethylene glycol monolaurate
Cithrol EGML N/E, EGML S/E

Glycol oleate (CTFA) [CAS #4500-01-0] *Synonyms:* Ethylene glycol monooleate
Cithrol EGMO N/E, EGMO S/E; Tegin GO

Glycol ricinoleate (CTFA) [CAS #106-17-2] *Synonyms:* Ethylene glycol monoricinoleate
Cithrol EGMR N/E, EGMR S/E; Flexricin 15

Glycol stearate (CTFA) [CAS #111-60-4; SE grade RD #977064-73-5] *Synonyms:* Ethylene glycol monostearate; EGMS
Alkamuls EGMS; Cerasynt IP, LP; Chimipal SGE; Cithrol EGMS N/E, EGMS S/E; CPH-37-NA; Cutina EGMS; Cyclochem EGMS, EGMS/SE, EM 324, POL, SEG; Cyclosheen 100, 202; Dermalcare EGMS/SE; Drewmulse EGMS; EGMS-VA; Emerest 2350; Empilan EGMS; Euperlan PK-810; Incromate PPI; Kemester EGMS; Kessco Ethylene Glycol Monostearate, 70; Lexate CRC; Lexemul EGMS; Lipal EGMS; Lipo EGMS; Mapeg EGMS; Monthyle; Norfox EGMS; Nutrapon B; Pearlex 3105; Pegosperse 50 MS, EGMS-70; Radiasurf 7270; Schercemol EGMS; Secoster EMS; Sedefos 75; Tefose 63; Tegin G (SE); Varisoft 3262

Glyoxal (CTFA) [CAS #107-22-2] *Synonyms:* Biformyl; Ethanedial; Oxalaldehyde
Protectol GL 40

Grapeseed oil (CTFA) [CAS #8024-22-4]
Lipovol G; Super Refined Grapeseed Oil

Guanine (CTFA) [CAS #73-40-5]
Naturon

Guar gum (CTFA) [CAS #9000-30-0] *Synonyms:* Guar flour
CC-603; Cosmedia Guar U; Frimulsion 10, Q8, X5; Galactasol; Genuzan 1038A, 1063X; Guar Gum; Guartex CM, FM, SJM; Indalca CD30; Progacyl COS-1, CP-7; Propacyl CP-2000, EM-40

Guar hydroxypropyltrimonium chloride (CTFA) [CAS #65497-29-2]
Cationic Guar C-261

Hectorite (CTFA) [CAS #12173-47-6; RD #977022-31-3]
Bentone LT; Perchem TEA

HEDTA (CTFA) [CAS #150-39-0] *Synonyms:* Hydroxyethyl ethylenediamine triacetic acid
Hamp-Ol Acid; SEQ 120

Heptakis (dipropylene glycol) triphosphite
Weston PTP

Heptane (CTFA) [CAS #142-82-5] *Synonyms:* n-Heptane
Amsco Heptane; Texsolve C, H

Hexabromocyclododecane [CAS #3194-55-6]
CD-75P; FR-1206; Saytex HBCD, HBCD-HM, HBCD-LM

1-Hexadecanol. See Cetyl alcohol

Hexadecyl... See Cetyl..., Palmityl...

n-Hexadecyl alcohol. See Cetyl alcohol

Hexaethylene glycol. See PEG-6

Hexaglyceryl... See Polyglyceryl-6...

Hexahydropyrimidine
Hexetidine

Hexahydrotriazine [CAS #13980-04-6] *Synonyms:* Hexahydro-1,3,5-s-triazine
Soi Bact G

Hexahydro-1,3,5-triethyl-s-triazine [CAS #7779-27-3]
Vancide TH

Hexahydro-1,3,5-tris (2-hydroxyethyl)-s-triazine
Onyxide 200; Triadine 10; WFT-78 Microbicide

Hexamethoxymethylmelamine
Cymel 303

1,1,3,3,5,5-Hexamethylcyclotrisilazane [CAS #1009-93-4]
CH7250; H7250

Hexamethylcyclotrisiloxane [CAS #541-05-9]
CH7260

Hexamethyldisilane [CAS #1450-14-2]
CH7280

Hexamethyldisilazane [CAS #999-97-3] *Synonyms:* Bis (trimethylsilyl) amine
CHY300EG; Dynasylan HMDS; H7300

Hexamethyldisiloxane (CTFA) [CAS #107-46-0] *Synonyms:* Oxy bis (trimethylsilane)
CH7310

N,N-Hexamethylene bis (3,5-di-t-butyl-4-hydroxyhydrocinnamamide) *Synonyms:* 1,6-Hexamethylene bis (3,5-di-t-butyl-4-hydroxyhydrocinnamate)
Irganox 259

Hexamethylene diamine tetra (methylene phosphonate)
Dequest 2051

Hexamethylene diisocyanate
Desmodur N-75

Hexamethylene tetramine [CAS #100-97-0] *Synonyms:* Methenamine; HMTA
Vulkacit H30

Hexane [CAS #110-54-3] *Synonyms:* n-Hexane
Amsco Hexane; Texsolve B, H

1,6-Hexanediol. See Hexamethylene glycol

1,6-Hexanediol diacrylate
SR-238

1,6-Hexanediol methacrylate
SR-239

Hexapotassium hexamethylene diamine tetra (methylene phosphonate)
Dequest 2054

Hexyl alcohol (CTFA) [CAS #111-27-3] *Synonyms:* Hexanol
Alfol 6; Nansa EVM70/B

Hexyl decanol
Primarol 1006

Hexyl dodecanol
Eutanol G16

Hexylene glycol (CTFA) [CAS #107-41-5] *Synonyms:* 2-Methyl-2,4-pentanediol
Cerasynt LP; Variquat K375

Hexyl laurate (CTFA) [CAS #34316-64-8]
Cetiol A; Dermalcare HL; Rilanit HL; Standamul CTA

Hexyloxypropylamine
Tomah PA-10

Hexyltrichlorosilane [CAS #928-65-4]
CH7332

Homosalate (CTFA) [CAS #118-56-9] *Synonyms:* Homomenthyl salicylate; Metahomomenthyl salicylate
Kemester HMS

Hyaluronic acid (CTFA) [CAS #9004-61-9]
BioCare Polymer HA-24; Cromoist HYA

Hyaluronic acid/protein complex
Cromoist HYA

Hybrid safflower oil (CTFA) [RD #977064-78-0]
Neobee 18

Hydramethylnon
Amdro

Hydrated silica. See Silica, hydrated

Hydrazine [CAS #302-01-2]
Scav-Ox, II, Plus

Hydrocarbon solvent [CAS #64742-46-7]
Penreco 2257 Oil, 2259 Oil, 2260 Oil

Hydrogenated C16-22 amine
Arosurf MG-102

Hydrogenated castor oil (CTFA) [CAS #8001-78-3]
Castorwax MP-70, MP-80; Cenwax G; Ceroxin Special; Cutina HR; Protegin W, WX; Radia 3200; Rilanit Special

Hydrogenated cocoglycerides (CTFA) [CAS #977056-87-3]
Softisan 133, 134, 138, 142, 154; Witepsol Suppository Bases

Hydrogenated coconut acid
Hystrene 5012; Industrene 223

Hydrogenated coconut oil (CTFA) [RD #977056-87-3]
Pureco 92, 110; Softisan 601

Hydrogenated cottonseed oil
Capital 5330

Hydrogenated ditallow amine (CTFA) [CAS #61789-79-5] *Synonyms:* Dihydrogenated tallow amine; Bis(hydrogenated tallow alkyl) amines
Armeen 2HT; Kemamine S-970

Hydrogenated fish glycerides
Neustrene 045, 053

Hydrogenated glyceride oil
Norfox Terne Oil

Hydrogenated lanolin (CTFA) [CAS #8031-44-5]
Dist. Lipolan; Fanchem HL; Fancol HL; Lipolan, S; Satulan; Super-Sat

Hydrogenated lard glyceride (CTFA) [CAS #8040-05-9]
Monomuls 9015

Hydrogenated lard glycerides (CTFA) *Synonyms:* Hydrogenated lard mono-, di-, and tri- glycerides
Monomuls 60-15

Hydrogenated menhaden acid (CTFA)
Hystrene 3022, 5522

Hydrogenated menhaden oil
Neustrene 045, 053

Hydrogenated methyl rosinate. See Methyl hydrogenated rosinate

Hydrogenated palm glycerides (CTFA) *Synonyms:* Hydrogenated palm mono-, di-, and tri-glycerides
Monomuls 60-35

Hydrogenated palm oil
Pureco 110

Hydrogenated palm oil glyceride (CTFA) [CAS #67784-87-6]
Monomuls 90-35

Hydrogenated polyisobutene (CTFA) [CAS #61693-08-1]
Luvitol HP

Hydrogenated rosin
Foral AX

Hydrogenated shark liver oil (CTFA) [RD #977062-72-8]
Robeyl

Hydrogenated soybean oil (CTFA) [CAS #8016-70-4]
Capital 5370; Neustrene 064; Stabland; Tally 100 Plus

Hydrogenated soybean oil glycerides (CTFA) *Synonyms:* Hydrogenated soybean oil mono-, di- and tri-glycerides
Monomuls 60-45

Hydrogenated soy glyceride (CTFA) [CAS #61789-08-0] *Synonyms:* Hydrogenated soybean oil monoglyceride
Monomuls 90-45

Hydrogenated starch hydrolysate (CTFA) *Synonyms:* Hydrogenated corn syrup
Hystar CG, TPF

Hydrogenated tallowalkonium chloride *Synonyms:* Hydrogenated tallow dimethyl benzyl ammonium chloride
Arquad DMHTB-75; Empigen BCM75, BCM75/A; Kemamine BQ-9702C

Hydrogenated tallow amide (CTFA) [CAS #61790-31-6]
Armid HT; Armoslip HT

Hydrogenated tallow amine (CTFA) [CAS #61788-45-2]
Adogen 140 (D); Amine HBG; Armeen HT, HTD; Arosurf MG-140; Crodamine 1.HT; Kemamine P-970, P-970D; Lilamin 140, 140 D; Nissan Amine ABT; Noram SH; Radiamine 6140, 6141

Hydrogenated tallow amine acetate [CAS #61790-59-8]
Acetamin HT; Armac HT; Kemamine A970; Lilamac 140; Radiamac 6148, 6149

Hydrogenated tallow amine oxide (CTFA) [CAS #61788-94-1] *Synonyms:* Hydrogenated tallow dimethyl amine oxide
Aromox DMHT

Hydrogenated tallow diamine
Lilamin 540; Radiamine 6540

Hydrogenated tallow dimethyl amine. See Dimethyl hydrogenated tallow amine

Hydrogenated tallow glyceride (CTFA) [CAS #61789-09-1] *Synonyms:* Hydrogenated tallow monoglyceride
Monomuls 90-25, 90-25/2, 90-25/5

Hydrogenated tallow glyceride citrate (CTFA) [CAS #68990-59-0]
Tegin C-1R

Hydrogenated tallow glyceride lactate (CTFA) [CAS #68990-06-7]
Lamegin GLP 10, GLP 20; Tegin L 61, L-61 NSE, L 62, L-62 NSE

Hydrogenated tallow glycerides (CTFA) [CAS #68308-54-3; RD #977056-92-0] *Synonyms:* Hydrogenated tallow mono-, di- and tri- glycerides
Monomuls 60-25, 60-25/2, 60-25/5; Neustrene 059, 060

Hydrogenated tallow hydroxysultaine
Mirataine HTS

Hydrogenated tallow 1,3-propylene diamine
Dinoram SH; Kemamine D-970; Nissan Amine DTH

Hydrogenated tallowtrimonium chloride (CTFA) [CAS #61788-78-1] *Synonyms:* Hydrogenated tallow trimethyl ammonium chloride
Adogen 441; Nissan Cation ABT-350, -500; Noramium MSH 50

Hydrogenated vegetable glycerides phosphate (CTFA) [RD #977046-65-3]
Emphos F27-85

Hydrogenated vegetable oil
Hydrokote 25, 37, 79, 711, S-7, SP

Hydrolyzed almond protein
Hydromond

Hydrolyzed animal elastin (CTFA) *Synonyms:* Elastin, hydrolyzed; Elastin hydrolysate; Hydrolyzed elastin
Crolastin, 10 Powd., 30 Powd.

Hydrolyzed animal keratin (CTFA) [CAS #69430-36-0] *Synonyms:* Hydrolyzed keratin
Crotein ASK, K, WKP; Kera-Tein 1000, 1000SD; Kerasol

Hydrolyzed animal protein (CTFA) [CAS #9015-54-7] *Synonyms:* Collagen hydrolysates; Hydrolyzed collagen
Collagen Hydrolyzate Cosmetic 55, SD; Crodyne BY-19; Cromoist CS, HYA; Cropepsol; Cropeptone; Crotein A, C, O, SPA, SPA55, SPC, SPO; Lamequat L; Lanotein AWS 30; Lexein X250, X300, X350, X400; Peptein 2000

Hydrolyzed animal protein, ethyl ester
Pro-Tein ES-20

Hydrolyzed egg
Crotein HWE

Hydrolyzed fibronectin
Cronectin, H

Hydrolyzed keratin. See Hydrolyzed Animal Keratin

Hydrolyzed silk (CTFA) [RD #977077-78-3] *Synonyms:* Hydrolyzed silk protein
Crosilk 10,000; Silk Protein Complex

Hydrolyzed soya lecithin
Lecithin Water Dispersible CLR

Hydrolyzed soy protein
Hydrosoy 2000/SF

Hydrolyzed vegetable protein (CTFA) [CAS #977059-33-8] *Synonyms:* Vegetable protein hydrolysate
HVP; HVP 5-SD, HVP-A, HVP-LS; Hydrosoy 2000; Luxor 1517V, 1576, 1616, 1639V, 1658V, E-40V, E-50V, EB-2V, EB-400V, GR-100, GR-150, GR-200, KB-300V, KB-320V, KB-330V, KB-350V, KB-400V, KB-500V, KB-530V, KB-600V, MB-40, MB-110, MB-120, R-100V; Triple-H

Hydrolyzed wheat protein
Hydrotriticum

Hydroquinone (CTFA) [CAS #123-31-9] *Synonyms:* Hydroquinol; p-Dihydroxybenzene; Quinol
Tecquinol

Hydroquinone monomethyl ether [CAS #150-76-5] *Synonyms:* Hydroquinone methyl ether
Eastman HQMME

2-Hydroxy-4-acryloyloxyethoxy benzophenone
Cyasorb UV 2098

Hydroxyamine phosphate ester
Fostex T

Hydroxyamine phosphate ester, amine salt
Fostex TX

Hydroxyamine phosphate ester, sodium salt
Fostex TN

2-[2-Hydroxy-3,5-di-(1,1-dimethylbenzyl)phenyl]-2H-benzotriazole
Tinuvin 900

1-Hydroxyethyl-1 benzyl-2-coco imidazolinium chloride
Bio-Quat IM-50

Hydroxyethyl cellulose (CTFA) [CAS #9004-62-0; 9063-65-4] *Synonyms:* Cellulose, 2-hydroxyethyl ether
Bentone LT; Cellosize Hydroxyethyl; Natrosol 250ER, 250GR, 250H4R, 250HHR, 250HR, 250JR, 250KR, 250LR, 250MHR, 250MR, Plus, Grade 330; Tylose H Series; Ucare Polymer JR, LR

1-Hydroxyethyl-2-heptadecenyl imidazoline
Rewomine IM-CA, IM-OA; Steinamin IM-CA, IM-OA; Unamine-O

1-Hydroxyethyl 2-heptadecyl imidazoline
Unamine-S

Hydroxyethyl hydroxypropyl C12-15 alkoxypropylamine oxide (CTFA)
Varox 185PH

N (2-Hydroxyethyl) 12-hydroxystearamide
Paricin 220

Hydroxyethylidene diphosphonic acid [CAS #2809-21-4] *Synonyms:* 1-Hydroxyethylidene-1,1-diphosphonic acid
Arquest 710; Dequest 2010

Hydroxyethyl stearamide-MIPA (CTFA) [RD #977068-21-5]
Crodapearl Liq., NI Liq.; Incropearl Liquid, Liquid NI

1-Hydroxyethyl 2-undecyl imidazoline
Unamine-C

Hydroxyhexadecyl dimethyl hydroxyethyl ammonium chloride
Dehyquart E

2-Hydroxy-4-isooctoxybenzophenone
Carstab 701, 702

Hydroxylated lanolin (CTFA) [CAS #68424-66-8]
Hydroxylan; OHlan; Ritahydrox

2-Hydroxy-4-methoxybenzophenone. See Benzophenone-3

2-Hydroxy-4-methoxybenzophenone-5-sulfonic acid. See Benzophenone-4

2[(-Hydroxymethyl) amino] ethanol
Troysan 174, 364

2-[(-Hydroxymethyl) amino]-2-methylpropanol
Troysan 192

2-(2′-Hydroxy-5′-methylphenyl) benzotriazole. See Drometrizole

Hydroxyoctacosanyl hydroxystearate (CTFA)
Elfacos C26

2-Hydroxy-4-n-octoxybenzophenone. See Benzophenone-12

2-(2-Hydroxy-5-t-octylphenyl) benzotriazole
Cyasorb UV 5411

Hydroxy phosphonocarboxylic acid
Belcor 575

Hydroxypropyl bisbehenyldimonium chloride
Jordaquat Dimer 22

Hydroxypropyl biscetyldimonium chloride
Jordaquat Dimer 16

Hydroxypropyl bislauryldimonium chloride
Jordaquat Dimer 12

Hydroxypropyl bisstearyldimonium chloride (CTFA)
Jordaquat Dimer 18

Hydroxypropylcellulose (CTFA) [CAS #9004-64-2] *Synonyms:* Cellulose 2-hydroxypropyl ether
Klucel E, EF, G, GF, H, HF, J, JF, L, LF, M, MF

Hydroxypropyl guar (CTFA) [CAS #39421-75-5] *Synonyms:* Hydroxypropyl guar gum
Gelcharg HP4; Progacyl COS-20, -70

Hydroxypropyl hydroxyethylcellulose
Natrovis

Hydroxypropyl methylcellulose (CTFA) [CAS #9004-65-3] *Synonyms:* Methyl hydroxypropyl cellulose; Methyl cellulose, propylene glycol ether
Benecel MP; Culminal MHPC; Methocel E, E4M, E5, E15-LV, E50-LV, F, F4M, F50-LV, F4M, J, K, K3, K35, K100-LV, K4M, K15M, K100M

1-Hydroxy-2-pyridine
Omadine TBAO

Hydroxystearic acid (CTFA) [CAS #106-14-9; 1330-70-7] *Synonyms:* 12-Hydroxyoctadecanoic acid
Interwax G 8212; Loxiol G 21; Radiacid 200

Imidazolidinyl urea (CTFA) [CAS #39236-46-9]
Germall 115

Imidazolinium methosulfate
Marlosoft IQ 75, 90

Invertase
Fermvertase, 10X, XX; Maxinvert

3-Iodo-2-propynyl butyl carbamate
Troysan Polyphase 17WD, AF1, Anti-Mildew, P-100

Iron [CAS #7439-89-6]
Troymax Drier Iron 9%

Iron molybdena
Mo-1907 T 3/16''

2-Isoalkyl (C14-20), 1-hydroxyethyl, 1-ethyl imidazolinium ethyl sulfate
Schercoquat IIS-RD

Isobornyl acrylate
SR-506

Isobutyl alcohol [CAS #78-83-1] *Synonyms:* Isobutanol; Isopropylcarbinol; 2-Methyl-1-propanol
Nansa EVM62/H, EVM70, EVM70/E

Isobutylated lanolin oil [RD #977064-89-3]
Argonol 40

Isobutylene/isoprene copolymer. See Butyl rubber

2,2′-Isobutylidenebis (4,6-dimethylphenol)
Naftonox IMB; Vulkanox NKF

Isobutyl methacrylate polymer
Acryloid B-67, B-67MT

Isobutyl oleate
Radia 7230; Rilanit IBO

Isobutyl palmitate (CTFA) [CAS #110-34-9] *Synonyms:* 2-Methylpropyl hexadecanoate
Kessco IBP, Isobutyl Palmitate

Isobutylparaben (CTFA) [CAS #4247-02-3] *Synonyms:* Isobutyl p-hydroxybenzoate; Benzoic acid, 4-hydroxy-, 2-methylpropyl ester
LiquaPar

Isobutyl stearate (CTFA) [CAS #646-13-9]
Emerest 2324; Kemester 5415; Kessco IBS, Isobutyl Stearate; Radia 7240, 7241; Rilanit IBS

Isobutyl tallowate
Rilanit IBTI, IBTIOK

Isobutyltrimethoxysilane [CAS #18395-30-7]
Dynasylan IBTMO

Isobutyric acid [CAS #79-31-2] *Synonyms:* Dimethylacetic acid
Tenox IBP-2

Isoceteth-20 (CTFA) [RD #977032-91-9] *Synonyms:* PEG-20 isocetyl ether
Arlasolve 200

Isoceteth-30 (CTFA)
Nikkol BEG-1630

Isocetyl alcohol (CTFA) [CAS #36311-34-9] *Synonyms:* Isohexadecanol; Isohexadecyl alcohol
Ceraphyl ICA; Standamul G-16

Isocetyl isostearate
Nikkol ICIS

Isocetyl stearate (CTFA) [CAS #25339-09-7]
Ceraphyl 494; Cetiol G16S; Crodamol ICS; Kemester 5822; Kessco ICS, Isocetyl Stearate; Nikkol ICS-R; Schercemol ICS; Standamul 7061

Isocyanate
Chemiflex 315XA(80); Desmodur N

Isocyanatopropyltriethoxy silane [CAS #24801-88-5]
CI7840

Isodeceth-4 (CTFA) *Synonyms:* PEG-4 isodecyl ether
Emulphogene DA-530

Isodeceth-6 (CTFA)
Emulphogene DA-630

Isodecyl acrylate [CAS #1330-61-6] *Synonyms:* Isodecyl propenoate
SR-395

Isodecyl benzoate
Benzoflex 131

Isodecyl diphenyl phosphate
Phosflex 390; Santicizer 148

Isodecyl ether amine acetate
Arosurf MG-70A3

Isodecyl isononanoate (CTFA) [CAS #41395-89-5; 59231-35-5] *Synonyms:* 3,5,5-Trimethylhexanoic acid, isodecyl ester
Wickenol 152

Isodecyl oleate (CTFA) [CAS #59231-34-4] *Synonyms:* 9-Octadecenoic acid, isodecyl ester
Ceraphyl 140-A

Isododecane [CAS #13475-82-6]
Bentone Gel ISD

Isofenphos
Oftanol

Isolaureth-3 (CTFA) *Synonyms:* PEG-3 isolauryl ether
Tergitol TMN-3

Isolaureth-6 (CTFA)
Dow Corning Q2-7224; Tergitol TMN-6

Isolaureth-10 (CTFA)
Tergitol TMN-10

Isononyl isononanoate (CTFA) [CAS #42131-25-9; 59219-71-5] *Synonyms:* 3,5,5-Trimethylhexanoic acid, trimethylhexyl ester
Wickenol 151

Isononyl stearate
Radia 7510

Isooctadecyl isononanoate (CTFA) [CAS #42131-25-9]
Tegosoft 189

Isooctyl acrylate
Macrobase 600; SR-440

Isooctyl oleate
Radia 7330

Isooctyl stearate
Radia 7130, 7131

Isophorone diisocyanate
Desmodur Z-4370

Isoprenoidal polymer
Chemprene R-10, R-25, R-50, R-70, R-100, R-115

Isopropanolamine (CTFA) [CAS #78-96-6] *Synonyms:* 1-Amino-2-propanol; Monoisopropanolamine; MIPA; 2-Hydroxypropylamine
Witcamide 61

Isopropyl alcohol (CTFA) [CAS #67-63-0] *Synonyms:* Isopropanol; IPA
Aromox C/12, DM16, DMCD, DMHT, T/12; Arquad 2C-75, 2HT-75, 12-33, 12-50, 16-50, 18-50, B-100, C-33, C-50, S-50, T-50; Carsosoft V-90; Emcol 150; Ethoquad 18/12, C/12, O/12; M-Quat 2475; Varisoft 3262

Isopropylamine dodecylbenzenesulfonate (CTFA) [CAS #26264-05-1]
Calimulse PRS; Nansa YS 94

Isopropyl 4-aminobenzenesulfonyl di (dodecylbenzenesulfonyl) titanate
Ken-React 26S

Isopropyl C12-15 pareth-9-carboxylate
Velsan P 8-3

Isopropyl dimethacryl isostearoyl titanate
Ken-React 7

Isopropyldimethylchlorosilane
I7860

4-Isopropyl-5,5-dimethyl-1,3-dioxane
Anthoxan

4,4′-Isopropylidenediphenol. See Bisphenol A

Isopropyl isostearate (CTFA) [CAS #31478-84-9; 68171-33-5]
Emerest 2310; Schercemol 318; Wickenol 131

Isopropyl lanolate (CTFA) [CAS #63393-93-1]
Amerlate P, W; Crodalan IPL; Emerest 1723; Lanesta L, S, SA30; Rewolan LP; Ritasol; Trisolan 1720

Isopropyl linoleate (CTFA) [CAS #22882-95-7] *Synonyms:* 1-Methylethyl-9,12-octadecadienoate
Ceraphyl IPL

Isopropyl mercaptan
Spotleak 1003, 1009

Isopropyl myristate (CTFA) [CAS #110-27-0] *Synonyms:* IPM; 1-Methylethyl tetradecanoate; Tetradecanoic acid, 1-methylethyl ester
Bentone Gel IPM; Crodamol IPM; Deltyl Extra; Emerest 2314; Exceparl IPM; Kessco 639, IPM, Isopropyl Myristate; Lexate TA; Lexol 60, 3975, IPM; Liponate IPM; Promyr; Radia 7190; Rilanit IPM; Schercemol IPM; Starfol IPM; Stepan D-50; Unimate IPM; Wickenol 101, 105, 136

Isopropyl myristopalmitate *Synonyms:* Isopropyl palmitate/myrisate
Radia 7220; Unimate IPPM

Isopropyl oleate (CTFA) [CAS #112-11-8] *Synonyms:* 1-Methylethyl-9-octadecenoate
Radia 7231; Schercemol IPO

Isopropyl palmitate (CTFA) [CAS #142-91-6] *Synonyms:* Isopropyl n-hexadecanoate; IPP
Crodamol IPP; Emerest 2316; Exceparl IPP; Isopropylan 33; Kessco 639, IPP, Isopropyl Palmitate; Lexol 60, 3975, IPP; Liponate IPP; Propal; Radia 7200; Rilanit IPP; Ritalan C; Starfol IPP; Stepan D-70; Unimate IPP; Wickenol 105, 111, 136

Isopropylparaben (CTFA) [CAS #4191-73-5] *Synonyms:* Isopropyl p-hydroxybenzoate; 1-Methylester-4-hydroxybenzoate
LiquaPar

2-Isopropyl-phenyl-N-methylcarbamate
Etrofolan

N-Isopropyl-N′-phenyl-p-phenylenediamine [CAS #101-72-4] *Synonyms:* p-Isopropylaminodiphenylamine
Santoflex IP; Vanox 3C; Vulkanox 4010 NA

Isopropyl PPG-2 isodeceth-7-carboxylate
Velsan D8P-3

Isopropyl stearate (CTFA) [CAS #112-10-7] *Synonyms:* 1-Methylethyl octadecanoate
Kessco 639; Lexol 60, 3975; Lipal ST; Radia 7195; Rilanit IPS; Wickenol 127, 136

Isopropyl tri (dioctylphosphato) titanate
Ken-React 12

Isopropyl tri (dioctylpyrophosphato) titanate
Ken-React 38S

Isopropyl tridodecylbenzenesulfonyl titanate
Ken-React 9S

Isopropyl tri (N ethylamino-ethylamino) titanate
Ken-React 44

Isopropyl triisostearoyl titanate
Ken-React TTS

Isostearamide DEA (CTFA) [CAS #52794-79-3]
Comperlan ID; Mackamide ISA; Monamid 150-IS; Schercomid ID, M, SI, SI-M; Standamid ID

Isostearamide MIPA (CTFA) *Synonyms:* Isostearic acid monoisopropanolamide; Isostearoyl monoisopropanolamide
Schercomid IMI

Isostearamidopropalkonium chloride (CTFA) [CAS #67633-59-4]
Incroquat I-85; Schercoquat IB

Isostearamidopropylamine oxide (CTFA)
Incromine Oxide I; Mackamine IAO

Isostearamidopropyl betaine (CTFA)
Cycloteric BET I-30; Incronam I-30; Mackam ISA; Schercotaine IAB

Isostearamidopropyl dimethylamine (CTFA) [CAS #67799-04-6]
Cyclomide IODI; Incromine IB; Jordamine DAPI; Mackine 401; Schercodine I

Isostearamidopropyl dimethylamine gluconate (CTFA)
Katemul IGU-70

Isostearamidopropyl dimethylamine glycolate (CTFA)
Katemul IB-70, IG-70

Isostearamidopropyl dimethylamine lactate (CTFA) [CAS #55852-15-8]
Emcol 6613; Incromate IDL; Mackalene 416

Isostearamidopropyl ethyldimonium ethosulfate (CTFA) [CAS #67633-63-0]
Jordaquat 522; Schercoquat IAS, IAS-LC

Isostearamidopropyl morpholine (CTFA)
Incromine ISM; Mackine 421

Isostearamidopropyl morpholine betaine
Incronam ISM-30

Isostearamidopropyl morpholine lactate (CTFA) [RD #977010-33-5]
Emcol ISML; Mackalene 426;

Isostearamidopropyl morpholine oxide (CTFA)
Incromine Oxide ISMO

Isosteareth-2 (CTFA) [CAS #52292-17-8 (generic); RD #977064-96-2] *Synonyms:* PEG-2 isostearyl ether; POE (2) isostearyl ether; PEG 100 isostearyl ether
Arosurf 66-E2

Isosteareth-10 (CTFA) [CAS #52292-17-8 (generic); RD #977059-85-0]
Arosurf 66-E10

Isosteareth-20 (CTFA) [CAS #52292-17-8 (generic); RD #977064-97-3]
Arosurf 66-E20

Isostearoamphopropionate (CTFA) [CAS #68630-96-6]
Monateric ISA-35%

Isostearoyl hydrolyzed animal protein (CTFA) [RD #977066-20-8]
Crotein IP, IPX

Isostearyl alcohol (CTFA) [CAS #27458-93-1; 41744-75-6]
Adol 66; Michel XO-146

Isostearyl benzoate (CTFA) [CAS #68411-27-8]
Finsolv SB

Isostearyl benzylimidonium chloride (CTFA)
Schercoquat IIB

Isostearyl erucate (CTFA) [CAS #977079-10-9] *Synonyms:* Isooctadecyl 13-docosenoate
Schercemol ISE

Isostearyl/erucyl erucate
Scheroba Oil

Isostearyl ethylimidonium ethosulfate (CTFA) [RD #977065-96-5] *Synonyms:* Quaternium-32
Monaquat ISIES; Schercoquat IIS, IIS-R

Isostearyl hydroxyethyl imidazoline (CTFA) [CAS #68966-38-1] *Synonyms:* Isostearyl imidazoline
Miramine IC; Monazoline IS; Schercozoline I

Isostearyl isostearate (CTFA) [CAS #41669-30-1]
Schercemol 1818; Starfol IS

Isostearyl lactate (CTFA) [CAS #42131-28-2]
Patlac IL

Isostearyl neopentanoate (CTFA) [CAS #58958-60-4] *Synonyms:* 2,2-Dimethylpropanoic acid, isooctadecyl ester
Ceraphyl 375; Crodamol ISNP; Cyclochem INEO; Schercemol 185

Isostearyl stearoyl stearate (CTFA)
Hetester ISS

Isotridecyl isononanoate (CTFA) [CAS #42131-27-1; 59231-37-7] *Synonyms:* 3,5,5-Trimethylhexanoic acid, isotridecyl ester
Wickenol 153

Isotridecyl stearate
Rilanit ITS; Sicolub TDS

Jojoba oil (CTFA) [RD #977011-04-3]
Lipovol J; Nikkol Jojoba Oil N, S; Simchin; Wickenol 139

Kaolin (CTFA) [CAS #1332-58-7; 12428-46-5] *Synonyms:* Bolus alba; China clay; Kaolin clay; Hydrated aluminum silicate
Bilt-Cote H-1, H-5, S-1, S-5; Continental; Dixie Clay; Langford Clay; McNamee Kaolin; Par Clay; Peerless No. 1, No. 2, No. 3, No. 4

Keratin amino acids (CTFA)
Crotein HKP, HKP Powd.

Ketone peroxide
Cadox F-85; FR-222

Lactamide MEA
Incromectant LAMEA, LMEA-100; Mackamide LME

Lactamidopropyl trimonium chloride
Incromectant LQ

Lactase. See β-Galactosidase

Lactic acid (CTFA) [CAS #50-21-5]
Patlac LA; Tandem 11H

Lactic acid chelate, ammonium salt
Tyzor LA

Lactol spirits
Amsco Lactol Spirits

Lactylic stearate
Myvatex 40-06S

Laneth-5 (CTFA) [CAS #61791-20-6 (generic); RD #977054-73-1] *Synonyms:* PEG-5 lanolin ether; POE (5) lanolin ether
Ritawax 5; Solulan 5

Laneth-10 acetate [RD #977054-76-4] *Synonyms:* Acetylated PEG-10 lanolin ether
Ritawax AEO

Laneth-15 (CTFA) [CAS #61791-20-6 (generic); RD #977054-77-5]
Polychol 15; Ritawax 15

Laneth-16 (CTFA) [CAS #61791-20-6 (generic); RD #977054-78-6]
Solulan 16

Laneth-20 (CTFA) [CAS #61791-20-6 (generic); RD #977054-79-7]
Aqualose W20, W20/50

Laneth-25 (CTFA) [CAS #61791-20-6 (generic); RD #977054-80-0]
Solulan 25

Laneth-40 (CTFA) [CAS #61791-20-6 (generic); RD #977054-81-1]
Ritawax 40

Lanocetyl acetate
Argonol ACE5

Lanolin (CTFA) [CAS #8006-54-0 (anhyd.), 8020-84-6 (hyd.)] *Synonyms:* Anhydrous lanolin; Cosmetic lanolin; Wool fat; Wool wax
Albalan, 36W, 40; Amerchol 400, BL, C, H-9; Aqualose SLT, SLW; Corona Lanolin; Coronet Lanolin; Cremba, B6; Crodapur; Dusoran MD, NG; Emery 1650, 1656, 1660, 1747; Fancor Lanolin; Fancorp Lanolin; Forlan; Golden Dawn Grade 1, 2; Golden Dawn Superfine; Golden Fleece DF, P-80, P-95, RA; Lantrol 1673, 1674; Lexate IL, PX; Nimcolan 1747; Ritaderm; Ritalan; Sorba; Super Corona; Superfine Lanolin Anhydrous USP; White Swan; Yeoman

Lanolin acid (CTFA) [CAS #68424-43-1] *Synonyms:* Lanolic acids; Lanolin fatty acids
Amerlate LFA, WFA; Emery 1781; Fancor LFA; Lanolic Acid; Nimco 1781; Ritalafa; Skliro

Lanolin alcohol (CTFA) [CAS #8027-33-6]
Alcolan; Amerchol 400, BL, C, CAB, H-9, L-99, L-101, L-500, RC; Argobase 125; Argowax Distilled, Standard; Ceralan; Cremba, B6; Emery 1730, 1732, 1740, 1747; Fancol LA, LAO; Hartolan, Hartolan Super; Lanolin Alcohols LG, LO, THG, THO; Liquid Absorption Base Type A, T; Nimco 1780; Nimcolan 1747; Protegin, X; Ritachol; Ritawax; Super Hartolan

Lanolin alcohol sulfosuccinate
Rewolan E; Varsulf S

Lanolin linoleate (CTFA) [RD #977057-13-8]
Polylan

Lanolin oil (CTFA) [CAS #8038-43-5; 70321-63-0]
Argonol 50, 60; Emery 1720; Fluilan; Fluilanol; Isopropylan 33; Lanogene; Lantrol 1673, 1674, 1675; Lipolan R; Liquilan; Ritalan C; Vigilan

Lanolin wax (CTFA) [CAS #68201-49-0]
Lanfrax, 1777 Deodorized; Lanocerin

Lanosterol (CTFA) [CAS #79-63-0]
Lanosterol

Lapyrium chloride (CTFA) [CAS #6272-74-8]
Emcol E-607L

Lard glyceride (CTFA) [CAS #61789-10-4] *Synonyms:* Lard monoglyceride
Dimodan P, PM, PM 300, S; Grindtek MOP 90; Monomuls 90-10

Lard glycerides (CTFA) [CAS #61789-10-4] *Synonyms:* Lard mono, di- and tri-glycerides
Monomuls 60-10

Lauralkonium bromide (CTFA) [CAS #7281-04-1] *Synonyms:* Benzenemethanaminium, D-dodecyl-N,N-dimethyl-, bromide; N-Dodecyl-N,N-dimethyl benzenemethanaminium bromide
Amonyl BR 1244

Lauralkonium chloride (CTFA) [CAS #139-07-1] *Synonyms:* Lauryl dimethyl benzyl ammonium chloride
Catinal CB-50; Dehyquart LDB; Hartex SAN Q 50; Retarder N, N-85; Tequat PAN; Vantoc CL

Lauramide [CAS #1120-16-7] *Synonyms:* Dodecanamide; Lauric acid amide; Lauryl amide
Hostastat System E 5951, E 6952

Lauramide DEA (CTFA) [CAS #120-40-1] *Synonyms:* Lauric diethanolamide
Agent 804; Alkamide 1182, 2124, L9DE; Aminol COR-4, COR-4C, LM-30C Special; Aramide LDS; Bio Soft LD-95; Carsamide SAL-7, SAL-9; Cedemide AX; Chimipal LDA; Comperlan LD, LDO, LDS, LMD; Condensate PE, PL; Crillon LDE; Cyclomide DL203, DL203/S, DL207/S, LE; Emid 6510, 6518, 6519, 6541, 6590; Empilan LDE, LDX; Ethylan MLD; Hartamide LDA 70, 90; Hetamide ML; Incromide L-90, LL, LLA, LR; Intermediate 512; Jordamide 1214, CLM; Loropan LD; Mackamide L-95, LLM, LMD; Mazamide LS-173, LS-196; Monamid 150-GLT, 150-LMW-C, 150-LW, 150-LWA, 716, 770, 982, 1007, 1034, R31-42; Monamine ACO-100, LM-100; Ninol 50-LL, 51-LL, 52-LL, 55-LL, 96-SL, 4821, AA-62, AA-62 Extra, P-616, P-621; Nissan Stafoam DL, L; Nitrene L-90; Norfox DLSA; Onyxol 336, 345, SD; Profan AA62; Rewomid 203/S, DL203, DL 203/S, DLMS; Schercomid 1214, LD, SL-Extra. SLM, SLM-C, SLMC-75, SL-ML, SLM-LC, SLM-S; Stafoam DL; Standamid KDS, LD, LDM, LDO, LDS; Steinamid 203/S; Sterling LDEA-90; Super Amide L-9, L-9A, L-9C, LL, LM; Synotol L 60, L 90, LM 60; Tohol N-230, N-230X; Unamide J-56, LDX, LMDX; Varamide L-1, MA-4, ML-1, ML-4, SL-9; Witcamide 5195, 6310, STD-HP

Lauramide MEA (CTFA) [CAS #142-78-9] *Synonyms:* Lauric monoethanolamide
Alkamide L9ME; Cedemide MX; Comperlan LM; Crillon LME; Cyclomide L203; Empilan LME; Hartamide LMEA-90; Incromide LMM; Lauridit LM; Loropan LM; Mackamide LMM; Monamid LMA; Monamid LMMA; Rewomid L203; Standamid LM; Steinamid L203; Varamide L-203

Lauramide MIPA (CTFA) [CAS # 142-54-1] *Synonyms:* Lauric monoisopropanolamide
Alkamide LIPA; Cyclochem LIPA; Cyclomide LIPA; Cyclosheen 100; Empicol TLP; Empilan LIS; Incromide LI; Monamid LIPA, LMIPA; Profan AD31; Rewomid IPL 203; Standamid LP; Steinamid IPL 203

Lauramidopropyl acetamidodimonium chloride (CTFA) [CAS #68259-01-8; RD #194-03-2]
Bina Qat-43

Lauramidopropylamine oxide (CTFA) [CAS #61792-31-2] *Synonyms:* Lauramidopropyl dimethylamine oxide
Mackamine LAO

Lauramidopropyl betaine (CTFA) [CAS #4292-10-8; 86438-78-0]
Jortaine LMAB; Mackam LMB; Mirataine BB; Monateric LMAB; Tego-Betaine L-90

Lauramidopropyl dimethylamine (CTFA) [CAS #3179-80-4]
Lexamine L-13; Mackine 801; Schercodine L

Lauramidopropyl PEG-dimonium chloride phosphate [CAS #83682-78-4]
Monaquat P-TD

(3-Lauramidopropyl) trimethyl ammonium methyl sulfate
Cyastat LS

Lauramine (CTFA) [CAS #124-22-1] *Synonyms:* Lauryl amine; Dodecanamine; Dodecylamine
Armeen 12D; Kemamine P-690, P-690D; Lilamin 163, 163 D; Nissan Amine BB; Radiamine 6163, 6164

Lauramine oxide (CTFA) [CAS #1643-20-5] *Synonyms:* Lauryl dimethylamine oxide
Alkamox LO; Ammonyx DMCD-40, LO; Aromox DMMC-W; Conco XA-L; Emcol L; Empigen OB; Incromine L-40; Incromine Oxide L, L-40; Jordamox LDA; Mackamine LO; Ninox L; Oxamin LO; Sandoz Amine Oxide XA-L; Schercamox DML; Varox 365, 375

Lauraminopropionic acid (CTFA) [CAS #1462-54-0; 3614-12-8] *Synonyms:* N-dodecyl-β-alanine
Velvetex 710L

Laurdimonium hydrolyzed animal protein
Croquat L

Laurdimonium hydroxyethyl cellulose
Crodacel QL

Laureth-1 (CTFA) [CAS #4536-30-5; 9002-92-0 (generic); RD #977054-82-2] *Synonyms:* PEG-1 lauryl ether; Ethylene glycol monolauryl ether; Ethylene glycol dodecyl ether
Alkasurf LAN-1; Lipocol L-1; Siponic L1

Laureth-2 (CTFA) [CAS #3055-93-4; 9002-92-0 (generic); RD #977062-02-4] *Synonyms:* Diethylene glycol dodecyl ether; PEG-2 lauryl ether; POE (2) lauryl ether
Dehydol LS 2; Empilan KB 2; Incropol L-7

Laureth-3 (CTFA) [CAS #3055-94-5; 9002-92-0 (generic); RD #977064-47-3] *Synonyms:* Triethylene glycol dodecyl ether
Dehydol LS 3; Empilan KB 3, KC 3; Hetoxol L-3N; Volpo L3 Special

Laureth-4 (CTFA) [CAS #5274-68-0; 9002-92-0 (generic); RD #977054-83-3]
Ameroxol LE-4; Chemal LA-4; Dehydol LS 4; Emthox 5882; Hetoxol L-4N; Lipal 4 LA; Lipocol L-4; Macol LA-4; Siponic L4; Trycol 5882, LAL-4; Volpo L4

Laureth-5 carboxylic acid (CTFA) [CAS #68954-89-2]
Akypo RLM 45

Laureth-7 (CTFA) [CAS #3055-97-8; 9002-92-0 (generic); RD #977062-41-1]
Incropol L-7, L-7-90; Macol LA-790; Siponic L7, L-7-90

Laureth-9 (CTFA) [CAS #9002-92-0 (generic); RD #977007-24-1]
Carsonon L-985; Chemal LA-9; Hetoxol L-9N, LS-9; Lipal 9 LA; Lubrol 12A-9 (9.5 EO); Macol LA-9; Standapol Pearl Conc. 7130

Laureth-10 carboxylic acid (CTFA)
Akypo RLM 100

Laureth-12 (CTFA) [CAS #9002-92-0 (generic); RD #977054-84-4]
Chemal LA-12; Comperlan LS; Emthox 5967; Incropol L-12; Lanette Wax CAT; Lipal 12 LA; Lipocol L-12; Macol LA-12; Polystep F-12; Siponic L12

Laureth-16
Siponic L16

Laureth-23 (CTFA) [CAS #9002-92-0 (generic); RD #977054-85-5]
Alkasurf LAN-23; Ameroxol LE-23; Cerasynt 945; Chemal LA-23; Emthox 5964; Ethosperse LA-23; Hetoxol L-23N; Incropol L-23; Lipal 23 LA; Lipocol L-23; Macol LA-23; Siponic L25; Trycol 5964, LAL-23; Volpo L23

Laureth-30 (CTFA) [CAS # 9002-92-0 (generic); RD #977069-31-0]
Incropol L-30

Lauric acid (CTFA) [CAS #143-07-7] *Synonyms:* Dodecanoic acid
Hetamide LA; Hystrene 9512, 9912

Lauric diethanolamide. See Lauramide DEA

Lauric diglycolamide
Incromide LMD

Lauric monoethanolamide. See Lauramide MEA

Lauric monoisopropanolamide. See Lauramide MIPA

Lauric/myristic DEA
Alkamide L7DE, L7DE-BT; Berol 151; Loropan LMD; Ninol 71-LL; Nitrene L-76; Rewomid DLMS; Schercomid SLL; Standamid LD 80/20; Sterling Amide 460; Sterling LM-70; Synotol LM 90

Lauric/myristic dimethylethyl ammonium ethosulfate
Larostat 377 DPG

Lauric/myristic MEA
Alkamide L7ME; Hartamide LMEA-70

Lauric/myristic MIPA
Incromide LMI

Lauric/myristic/palmitic DEA
Schercomid SLA

Lauroamphoacetate (CTFA) *Synonyms:* Lauroamphoglycinate
Miranol HM Conc; Monateric LMM-30

Lauroamphodiacetate (CTFA) [CAS #14350-97-1] *Synonyms:* Lauroamphocarboxyglycinate
Miranol H2M Conc.; Monateric 985A

Lauroamphodipropionate (CTFA) [CAS #68610-43-5] *Synonyms:* Lauroamphocarboxypropionate
Miranol H2M-SF 70%, H2M-SF Conc.

Lauroamphohydroxypropylsulfonate (CTFA) [CAS #70024-82-7] *Synonyms:* Lauroamphopropylsulfonate
Miranol HS Conc.

Lauroampho PEG-glycinate phosphate [CAS #83682-77-3]
Monaquat P-TL

Lauroyl-1-lysine
Amihope LL-11

Lauroyl peroxide [CAS #105-74-8] *Synonyms:* Bis (1-oxododecyl) peroxide; Dodecanoyl peroxide; Dilauroyl peroxide
Alperox-F

Lauroyl sarcosine (CTFA) [CAS #97-78-9] *Synonyms:* N-methyl-N-(1-oxododecyl) glycine
Sarkosyl L

Laurtrimonium chloride (CTFA) [CAS #112-00-5] *Synonyms:* Lauryl trimethyl ammonium chloride; Dodecyl trimethyl ammonium chloride
Arquad 12-33, 12-50; Chemquat 12-33, 12-50; Dehyquart LT; Empigen 5089; Kemamine Q-6903B; Kosti Kat PE; Nissan Cation BB

Lauryl acrylate
SR-335

Lauryl alcohol (CTFA) [CAS #112-53-8] *Synonyms:* 1-Dodecanol; C12 linear primary alcohol; Dodecyl alcohol
Alfol 12; Epal 12; Kalcohl 20; Laurex L1, NC; Lipocol L

Lauryl amine. See Lauramine

Lauryl betaine (CTFA) [CAS #683-10-3] *Synonyms:* Lauryl dimethyl glycine
Cycloteric BET-L31; Empigen BB; Swanol AM-301; Varion CDG

Lauryl dimethylamine. See Dimethyl lauramine

Lauryl dimethylamine oxide. See Lauramine oxide

Lauryl dimethyl benzyl ammonium chloride. See Lauralkonium chloride

Lauryl dimethyl betaine *Synonyms:* Dimethyl dodecyl betaine
Nissan Anon BL

Lauryl hydroxyethyl imidazoline (CTFA) [CAS #136-99-2] *Synonyms:* Lauryl imidazoline
Crapol AV-11; Mackazoline L; Schercozoline L

Lauryl imidazoline betaine
Carsonam L

Lauryl lactate (CTFA) [CAS #6283-92-7] *Synonyms:* Dodecyl 2-hyroxypropanoate
Ceraphyl 31; Crodamol LL; Cyclochem LVL; Dermalcare LVL; Schercemol LL

Lauryl/myristyl alcohol
Kalcohl 5-24, 6-24, 7-24

Lauryl/myristyl dimethylamine oxide
Bio-Surf PBC-460; Carsamine Oxide LM; Etoxi AC-91; Standamox LMW-30

Laurylpyridinium bisulfate
Dehyquart D

Laurylpyridinium chloride (CTFA) [CAS #104-74-5]
Catigene CLP 50; Dehyquart C, C Crystals

Lauryl/stearyl thiodipropionate
Cyanox 1212

Lauryl trimethyl ammonium chloride. See Laurtrimonium chloride

Lead [CAS #7439-92-1]
Troymax Drier Lead 9%, 24%, 36%; Vanstay 5510, 5515

Lead carbonate [CAS #598-63-0] *Synonyms:* Lead carbonate, basic; Dibasic lead carbonate; Lead subcarbonate; White lead; Lead flake
Cyastab 40, 621; Haro Chem PC

Lead diamyldithiocarbamate [CAS #36501-84-5]
Amyl Ledate; Vanlube 71, AZ

Lead dimethyldithiocarbamate [CAS #19010-66-3]
Methyl Ledate

Lead dinonylnaphthalene sulfonate
Na-Sul LS

Lead oxide *Synonyms:* Lead suboxide; Lead oxide, black
Litharge 33; Naftopast Litharge A

Lead phosphate, dibasic *Synonyms:* Lead monohydrogen phosphate; Lead biorthophosphate
Cyastab 712

Lead phosphite
Interstab LF 3626, 3638, 3645, 3669, 10898/4, 10898/6

Lead phosphite/Ba/Cd
Interstab LF 3634, LF 11298

Lead phosphite, dibasic [CAS #1344-40-7]
Haro Chem PDF; Interstab LP 3139

Lead phosphite/sulfate
Interstab LF 3675, LF 3751

Lead phthalate, dibasic
Cyastab 908, 988; Haro Chem PDP-E; Interstab LP 3153

Lead silicate [CAS #10099-76-0] *Synonyms:* Lead metasilicate
BSWL 202

Lead silicochromate, basic [CAS #11113-70-5]
Oncor F-31, M50

Lead silicosulfate, basic
EPIstatic 112

Lead stearate [CAS #1072-35-1] *Synonyms:* Stearic acid, lead salt
Haro Chem P28G; Interstab LP 3155

Lead stearate, dibasic
Haro Chem P51; Interstab LP 3150

Lead sulfate [CAS #7446-14-2]
Interstab LF 3623, LF 3653

Lead sulfate, basic *Synonyms:* White lead, sublimed; White lead sulfate
Cyastab 523, 563, 593, 800; Super Sublimed White Lead 41

Lead sulfate, blue basic *Synonyms:* Sublimed blue lead; Blue lead
Sublimed Blue Lead

Lead sulfate, polybasic
Haro Chem PPCS-X

Lead sulfate, tetrabasic
Interstab LP 3104, LP 3190

Lead sulfate, tribasic
EPIstatic 103; Haro Chem PTS-E; Interstab LP 3103

Lead-zinc
Vanstay 8300

Lecithin (CTFA) [CAS #8002-43-5]
Actiflo 68, 70; Asol; Centrocap 162SS, 162US, 273SS, 273US; Centrol Series, 2FSB, 2FUB, 3FSB, 3FUB, CA; Centrolene Series, A, S; Centrolex Series, C, D, F, G, P, R; Centromix Series, CPS, E; Centrophase Series, 152, C, HR, HR2B, HR2U, HR4B, HR4U, HR6B, NV; Centrophil Series, K, M, W; Chocothin; Clearate LV, WD, Special Extra, WDF; Dugista; Dulectin, Granular, Powdered; EmCon E-5, E-20; Emulthin M-35; Lecithin W.D.; Lipotin 100, 100J, SB; M-C-Thin 45, AF-1; SoyaLec WDF-FG; Topcithin; Troykyd Lecithin W.D.

Limestone [CAS #1317-65-3]
Franklin T-11, T-13, T-14, T-325

Limonene [CAS #138-86-3]
Piccolyte C100, C115, C125, C135

Linoleamide/cocamide DEA
Jordamide SCD

Linoleamide DEA (CTFA) [CAS #56863-02-6] *Synonyms:* Linoleic diethanolamide
Comperlan F, VOD; Cyclomide DIN295, DIN295/S, DOTS; Incromide LA, LLA; Jordamide LLD; Monamid 15-70W, 150-ADY, 982, 1007; Monamine AC-100, ADY-100, CD-100, R8-26; Nitrene 100 SD; Rewomid F; Schercomid SLE, SLS; Standamid SOD, SOMD

Linoleamidopropyl dimethylamine
Incromine LSB

Linoleamidopropyl dimethylamine lactate
Incromate LDL

Linoleic acid (CTFA). See Soy acid

Linolenic acid (CTFA) *Synonyms:* Linseed acid
Industrene 120

Linseed oil, epoxidized. See Epoxidized linseed oil

Linuron [CAS #330-55-3] *Synonyms:* 3-(3,4-Dichlorophenyl)-1-methoxy-1-methylurea
Linex 4L

Lipase [CAS #9001-62-1]
Fermlipase; Milezyme Lipase Powds., Pancreatic Lipase, Pancreatin 4NF

Lithium dinonylnaphthalene sulfonate
Na-Sul 707

Lithium molybdate
Molyhibit 300 Series

Lithopone [CAS #1345-05-7]
Lithopone D, DS, L

Locust bean gum (CTFA) [CAS #9000-40-2]
Frimulsion 6G, 10, Q8, X5

Magnesia. See Magnesium oxide

Magnesium aluminum silicate (CTFA) [CAS #1327-43-1] *Synonyms:* Aluminum magnesium silicate
Bentone EW; Macaloid; Magnabrite F, HV, K, S, T; Van Gel, B; Veegum, F, HS, HV, K, Neutral, S-728, T, WG

Magnesium carbonate
Elastocarb Tech Light, Tech Heavy, UF

Magnesium DTPA
Perma Kleer 98

Magnesium ethylenebisdithiocarbamate. See Maneb

Magnesium hydroxide (CTFA) [CAS #1309-42-8] *Synonyms:* Magnesium hydrate in aq. suspension; Milk of magnesia; Magnesia magma
FR-20; Kisuma 5A, 5B, 5E; Plastigel PG9089; Versamag B-16, DC, SB, Tech., UF

Magnesium lauryl sulfate (CTFA) [CAS #3097-08-3]
Sipon MLS

Magnesium oxide (CTFA) [CAS #1309-48-4] *Synonyms:* Magnesia
Cri-Spersion CRI-ACT-45, -45-1/1, -45-LV; Elastomag 100, 100R, 170, 170 Micropellet, 170 Special; Magox Super Premium; Mg-0601 T 1/8″; Naftopast MgO-A; Plastigel PG9033, PG9037; Plastomag; Rapidblend 1793; Rhenomag C2, G3, L3, P2, P3

Magnesium silicate (CTFA) [CAS #1343-88-0; 13776-74-4] *Synonyms:* Silicic acid, magnesium salt (1:1)
Ceramitalc 10AC, HDT

Magnesium silicate, hydrous. See Talc

Magnesium stearate (CTFA) [CAS #557-04-0]
Hostacerin WO; Petrac MG-20, MG-20 NF; Radiastar 1100; Witco Magnesium Stearate D

Magnesium sulfate (CTFA) [CAS #7587-88-9; 10034-99-8]
Hybase M-300, M-400; Omadine MDS

Magnesium xylene sulfonate
Eltesol MGX

Maize amino acids
Amino Gluten MG

Malathion [CAS #121-75-5] *Synonyms:* S-[1,2-bis (ethoxycarbonyl) ethyl]O,O-dimethyl phosphorodithioate
Cythion, Emulsifiable Conc.

Maleated soybean oil [CAS #68648-66-8] *Synonyms:* Soybean oil, maleated
Ceraphyl GA

Maleic acid/olefin copolymer
Sokalan CP 12S

Malt
Uni-Malt

Maltodextrin (CTFA) [CAS #9050-36-6]
Maltrin M040, M050, M100, M150, M180, M440, M500, M510, M550, M580, M700; Microduct; Wickenol 550 Microduct

Maneb [CAS #12427-38-2] *Synonyms:* Magnesium ethylenebisdithiocarbamate
Manex; Pro-Tex

Maneb/sineb coordinate
Fore WD

Maneb/zinc ion coordinate
Fore

Manganese [CAS #7439-65-5]
Troymax Drier Manganese 6%, 9%, 12%

Mannan depolymerase
Breakerase G

MDI [CAS #101-68-8] *Synonyms:* Methylene di-p-phenylene isocanate; Methylene bisphenyl isocyanate; Diphenylmethane-4,′4-diisocyanate
Isonate 125M, 143L, 181, 191, 240

MDM hydantoin (CTFA) [CAS #116-25-6; RD #977053-13-6] *Synonyms:* 1-(Hydroxymethyl-5,5-dimethyl hydantoin; Monomethylol dimethyl hydantoin
Dantoin MDMH

MEA-borate (CTFA) [CAS #10377-81-8; 68130-12-1] *Synonyms:* Ethanolamine borate
Monacor BE

MEA-laureth sulfate (CTFA) [RD #977067-77-8] *Synonyms:* Monoethanolamine lauryl ether sulfate
Montosol PG-17

MEA-lauryl sulfate (CTFA) [CAS #4722-98-9] *Synonyms:* Monoethanolamine lauryl sulfate
Condanol MLS 35; Manro ML 33

MEK (CTFA) [CAS #78-93-3] *Synonyms:* Methyl ethyl ketone; Ethyl methyl ketone; 2-Butanone
D.E.R. 684-EK40

MEK peroxide. See Methyl ethyl ketone peroxide

Melamine-formaldehyde resin (CTFA) [CAS #9003-08-1] *Synonyms:* Melamine resin
Cymel 303, 370, 373, 380, 1141; Plenco 801 White

Melamine-formladehyde resin, methylated
Cymel 370, 373, 380

Melamine-phenolic
Plenco 713 Grey, 714 Brown, 732 Grey, 755 Brown, 757 Brown

Melamine phosphate
Amgard NH

Melamine resin. See Melamine-formaldehyde resin

Menhaden oil
Super Refined Menhaden Oil

2-Mercaptobenzothiazole [CAS #149-30-4] *Synonyms:* MBT
Captax; Captax-Tuads Blend; Fungicide ZV; MBT; Naftocit MBT; Naftopast MBT-P; Rhenovin MBT-70; Rokon; Rotax; Thiotax; Vulkacit Mercapto/C, Mercapto/EGC

2-Mercaptobenzothiazyl disulfide. See 2,2′-Dithiobis (benzothiazole)

β-Mercaptoethyltriethoxysilane
Union Carbide A-1893

2-Mercaptoimidazoline
Vulkacit NPV/C2

2-Mercapto-4(5)-methyl benzimidazole, zinc salt
Akrochem Antioxidant 58

3-Mercaptopropylmethyldimethoxy silane [CAS #31001-77-1]
CM8450

Mercaptopropyltrimethoxysilane [CAS #4420-74-0] *Synonyms:* γ-Mercaptopropyltrimethoxy silane
Dynasylan MTMO; Prosil-196; Union Carbide A-189

2-Mercaptotoluimidazole [CAS #53988-10-6]
Vanox MTI

Mercury [CAS #7439-97-6]
Nuodex PMA 18

Meroxapol 105 (CTFA) [CAS #9003-11-6 (generic); RD #977066-22-0]
Macol 15; Pluronic 10R5

Meroxapol 108 (CTFA) [CAS #9003-11-6 (generic); RD #977066-23-1]
Pluronic 10R8

Meroxapol 171 (CTFA) [CAS #9003-11-6 (generic); RD #977066-24-2]
Macol 18; Pluronic 17R1

Meroxapol 172 (CTFA) [CAS #9003-11-6 (generic); RD #977066-25-3]
Macol 19; Pluronic 17R2

Meroxapol 174 (CTFA) [CAS #9003-11-6 (generic); RD #977066-26-4]
Pluronic 17R4

Meroxapol 178 (CTFA) [CAS #9003-11-6 (generic); RD #977066-27-5]
Pluronic 17R8

Meroxapol 251 (CTFA) [CAS #9003-11-6 (generic); RD #977066-28-6]
Macol 32; Pluronic 25R1

Meroxapol 252 (CTFA) [CAS #9003-11-6 (generic); RD #977066-29-7]
Pluronic 25R2

Meroxapol 254 (CTFA) [CAS #9003-11-6 (generic); RD #977066-30-0]
Macol 34; Pluronic 25R4

Meroxapol 255 (CTFA) [CAS #9003-11-6 (generic); RD #977066-31-1]
Pluronic 25R5

Meroxapol 258 (CTFA) [CAS #9003-11-6 (generic); RD #977066-32-2]
Pluronic 25R8

Meroxapol 311 (CTFA) [CAS #9003-11-6 (generic); RD #977066-33-3]
Macol 33; Pluronic 31R1

Meroxapol 312 (CTFA) [CAS #9003-11-6 (generic); RD #977066-34-4]
Pluronic 31R2

Meroxapol 314 (CTFA) [CAS #9003-11-6 (generic); RD #977066-35-5]
Pluronic 31R4

Metamitron
Goltix

Methabenzthiazuron
Tribunil

Methacrylate copolymer
Empicryl 6059, 6070, DH122, DH135, DH145, HV125, HV130, HV226, PPT38

Methacrylate polymer
Acryloid 150, 700, 900, 950 Series, 1017, 1019, HF-Series, W-1600; Amphomer; Crylcon 3000, 7000; Empicryl 6047, 6052, 6054, 6058, HV110, PPT144, PPT145, PPT147, PPT148, PPT148/D, PT1334, P1345, PT1397, PT1544, PT1764/DO

Methacrylato chromic chloride
Volan, L

Methacrylic acid [CAS #79-41-4] *Synonyms:* α-Methacrylic acid (monomer); 2-Methylpropenoic acid
Baypren Latex 4 R

Methacrylic acid copolymer
Acrysol A-41

3-Methacryloxypropyltrimethoxysilane [CAS #2530-85-0] *Synonyms:* γ-Methacryloxypropyltrimethoxysilane
Dow Corning Z-6030; Dynasylan MEMO; Prosil-248; Union Carbide A-174

Methacryloyl ethyl betaine/methacrylates copolymer (CTFA) [CAS #87435-35-6]
Amersette

Methamidophos
Tamaron

Methiocarb
Mesurol

Methoxychlor [CAS #72-43-5] *Synonyms:* Methoxy DDT; DMDT; 2,2-bis (p-methoxyphenol)-1,1,1-trichloroethane
Marlate 2-MR, 50, Tech.

Methoxydiglycol (CTFA) [CAS #111-77-3] *Synonyms:* Diethylene glycol monomethyl ehter; Ethanol, 2-(2-methoxyethoxy)-
Ektasolve DM; Poly-Solv DM

Methoxyethanol (CTFA) [CAS #109-86-4] *Synonyms:* Ethylene glycol monomethyl ether; 2-Methoxyethanol
Ektasolve EM; Poly-Solv EM

Methoxyethoxypropylamine
Surfam P-MEPA

4-Methoxy-4-methyl-pentanone-2 [CAS #107-70-0] *Synonyms:* 4-Methoxy-4-methylpentan-2-one
PentoXone

Methoxy PEG-400
Emery 6726

Methoxypropanol (CTFA) [CAS #107-98-2] *Synonyms:* Monopropylene glycol methyl ether
Dowanol PM

Methoxypropylamine *Synonyms:* 3-MPA
Surfam P5 Dist., 95 Dist.

Methyl acetate (CTFA) [CAS #79-20-9] *Synonyms:* Acetic acid, methyl ester
Everflex Solvent 80; Solvent 80

Methyl acetyl ricinoleate (CTFA) [CAS #140-03-4]
Flexricin P-4

Methyl acrylate polymer
Acryloid C-10LV

Methyl alcohol (CTFA) [CAS #67-56-1] *Synonyms:* Methanol
Everflex Solvent 80; Solvent 80

Methyl anthranilate [CAS #134-20-3] *Synonyms:* Methyl-o-aminobenzoate; Neroli oil, artificial
MA

Methyl behenate *Synonyms:* Methyl docosanoate
Kemester 9022

Methylbenzethonium chloride (CTFA) [CAS #25155-18-4] *Synonyms:* Diisobutyl cresoxy ethoxy ethyl dimethyl benzyl ammonium chloride
Hyamine 10-X

4- and 5-Methylbenzotriazole
Vulkalent TM

Methyl bis (hydrogenated tallowamidoethyl) 2-hydroxyethyl ammonium chloride
Incroquat 100; Incrosoft 100

Methyl bis (hydrogenated tallowamidoethyl) 2-hydroxyethyl ammonium methosulfate
Accosoft 440-75

Methyl bis (oleylamidoethyl) 2-hydroxyethyl ammonium methosulfate
Accosoft 750

Methyl bis (tallowamidoethyl) 2-hydroxyethyl ammonium methosulfate
Accosoft 540, 550-90 HHV, 550L-90, 620-90; Varisoft 222, 222LM-90%

Methyl bis (tallowamidoethyl) 2-hydroxypropyl ammonium methosulfate
Varisoft 238

Methyl/butyl methacrylate copolymer
Acryloid B-66

Methyl caprylate/caprate (CTFA)
Emery Methyl Caprylate/Caprate; Stepan C25

Methylcellulose (CTFA) [CAS #9004-67-5] *Synonyms:* Cellulose methyl ether
Benecel M; Combizell AG 070, AG 200, APR 060, APR 200; Culminal MC; Methocel A, A4C, A4M, A15-LV

Methylchloroisothiazolinone (CTFA) [CAS #26172-55-4] *Synonyms:* 5-Chloro-2-methyl-4-isothiazolin-3-one; 4-Isothiazolin-3-one, 5-chloro-2-methyl-
Amerstat 250, 251, 252; Biosperse 250; Kathon 886, 886 MW, CG, CB/ICP, LX, WT

Methyl cocoate (CTFA) [CAS #61788-59-8]
Kemester 325

Methylcyclooctylcarbonate
Jasmacyclat

Methyl dibromo glutaronitrile
Tektamer 38

Methyldichlorosilane [CAS #75-54-7]
CM8750

Methyl dimerate
Emery 2902

3-Methyl-4-dimethylaminophenyl-N-monomethylcarbamate
Matacil

Methyl distearamine
Adogen 343

Methyl eicosenate
Kemester 2050

Methylenebisdiamyl phenoxy polyethoxy ethanol
Triton X-155, X-155-90%

4,4′-Methylenebis (dibutyldithiocarbamate)
Vanlube 7723

4,4′-Methylenebis (2,6-di-t-butylphenol)
Ethanox 702

2,2′-Methylenebis (4-ethyl-6-tert-butylphenol) [CAS #88-24-4] *Synonyms:* Bis (2-hydroxy-3-tert-butyl-5-ethylphenyl) methane; 2,2′-Methylenebis (6-tert-butyl-4-ethylphenol)
Antioxidant 425; Cyanox 425

2,2′-Methylenebis (4-methyl-6-t-butylphenol) [CAS #119-47-1] *Synonyms:* 2,2′-Methylenebis (6-tert-butyl-p-cresol)
Antioxidant 235; CAO-5, -14; Cyanox 2246; Naftonox 2246; Santowhite PC; Synox 5LT; Ultranox 246; Vanox 2246; Vulkanox BKF

2,2′-Methylenebis (4-methyl-6-cyclohexyl phenol)
Vulkanox ZKF

2,2′-Methylenebis [4-methyl-6-(1-methyl-cyclohexyl) phenol]
Nonox WSP

Methylenebis (thiocyanate)
Amerstat 282; Rychem 810

Methylene chloride (CTFA) [CAS #75-09-2] *Synonyms:* Dichloromethane; Methylene dichloride; Methane, dichloro-
Aerothene MM Solv.; Desmodur RF, RU; Striptron Stripper

Methylene sulfonic phenolic resin
Duolite ARC-9359, C-3

Methyl ethyl ketone peroxide [CAS #1338-23-4] *Synonyms:* MEK peroxide; Ethyl methyl ketone peroxide
Cadox L-50, M-30, M-105, MDA-30; Hi-Point 90, 90 Red, PD-1; Lupersol DDM-9, Delta-X-9, DHD-9; Sprayset MEKP; Superox 702, 732

Methyl ethyl ketoxime
Troykyd Anti-Skin B

Methyl ethyl sulfide
Spotleak 1007, 2323

Methyl gluceth-10 (CTFA) [CAS #68239-42-9; RD #977065-07-8] *Synonyms:* PEG-10 methyl glucose ether
Glucam E-10

Methyl gluceth-20 (CTFA) [CAS #68239-43-0; RD #977065-08-9]
Glucam E-20

Methyl gluceth-20 distearate [CAS #98073-10-0]
Glucam E-20 Distearate

Methyl glucose dioleate [CAS #82933-91-3]
Glucate DO

Methyl glucose sesquistearate (CTFA) [CAS #68936-95-8]
Glucate SS

Methyl hydrogenated rosinate (CTFA) [CAS #8050-13-3; 8050-15-5] *Synonyms:* Hydrogenated methyl ester of rosin
Hercolyn D

Methyl (1) hydrogenated tallow amidoethyl (2) hydrogenated tallow imidazolinium methosulfate
Accosoft 808HT-75; Varisoft 445

Methyl hydroxycellulose
Tylose P, P-x, PS-x, P-Z Series

Methyl hydroxyethyl cellulose
Benecel ME; Culminal MHEC; Tylose MH, MHB, MB, MH-K, MH-xp, MHB-y, MHB-yp

Methyl hydroxypropyl cellulose. See Hydroxypropyl methylcellulose

Methyl hydroxystearate (CTFA) [CAS #141-23-1] *Synonyms:* Methyl 12-hydroxyoctadecanoate
Cenwax ME

Methyl isobutyl ketone. See MIBK

Methyl isothiazoline
Kathon DP, ICP

Methylisothiazolinone (CTFA) [CAS #2682-20-4] *Synonyms:* 2-Methyl-4-isothiazolin-3-one; 3(2H)-Isothiazolone, 2-methyl-; 2-Methyl-3(2H)-isothiazolone
Amerstat 250, 251, 252; Biosperse 250; Kathon CG

Methyl lardate
Emery Methyl Lardate; Norfox MLD

Methyl laurate (CTFA) [CAS #111-82-0] *Synonyms:* Methyl dodecanoate
Radia 7118; Stepan C40, C41, C42, C43

4- and 5-Methylmercaptobenzimidazole
Vulkanox MB-2/MGC, ZMB-2/G

Methyl methacrylate/allyl methacrylate copolymer
Luchem AS-946

Methyl methacrylate copolymer
Acryloid B-44, B-48N, B-50, B-82, B-99

Methyl methacrylate polymer
Acryloid A-10, A-11, A-21, A-21LV, A-30, A-101

Methyl methacrylate styrene acrylonitrile copolymer
Blendex 590

Methyl morpholine
Texacat NMM

Methyl myristate (CTFA) [CAS #124-10-7] *Synonyms:* Methyl tetradecanoate
Stepan C50

Methyl oleate (CTFA) [CAS #112-62-9]
Emerest 2301; Emery Methyl Oleate; ; Kemester 104, 105, 115, 205; Norfox VMO; Radia 7060; Stepan C68

Methyl oleate/linoleate
Kemester 213

Methyl oleylamidoethyl oleyl imidazolinium methosulfate
Incrosoft CFI, CFI-75; Varisoft 3690

Methyl palmitate (CTFA) [CAS #112-39-0] *Synonyms:* Methyl hexadecanoate
Radia 7120; Stepan C60

Methyl palmitate oleate
Petrosan 102; Stepan C65

Methylparaben (CTFA) [CAS #99-76-3] *Synonyms:* Methyl 4-hydroxybenzoate
Germaben II, II-3; Lexgard M; Trisept M

2-Methylpentamethylenediamine [CAS #15520-10-2]
Dytek A

Methylphenyldichlorosilane [CAS #149-74-6]
CM8930

Methyl phenyl ethanolamine [CAS #93-90-3]
Emery 5706

Methyl phenyl ketone
Acetophenone-High Purity

N-Methyl-2-pyrrolidone [CAS #872-50-4] *Synonyms:* 1-Methyl-2-pyrrolidinone
M-Pyrol

Methyl ricinoleate (CTFA) [CAS #141-24-2]
Flexricin P-1

Methyl rosinate (CTFA) [CAS #68186-14-1] *Synonyms:* Rosin acid, methyl ester
Abalyn

Methyl soyate
Kemester 226

Methyl stearate (CTFA) [CAS #112-61-8] *Synonyms:* Methyl octadecanoate
Emery Methyl Stearate; Kemester 4516, 7018, 9018; Radia 7110; Stepan C66

α-Methylstyrene [CAS #98-83-9]
Amoco Resin 18-210, 18-240, 18-290; Kenplast ESB, ESI, ESN; Kristalex 1120, 3025, 3070, 3085, 3100

Methyl styrene/α-methylstyrene copolymer
Piccotex 75, 100, 120, LC

1-Methyl-1-tallow amido-ethyl-2-heptadecyl imidazolinium methosulfate
Ammonyx 4080

Methyl tallowate
Kemester 143

3-Methyl-thiazolidinethione-2
Vulkacit CRV/LG

Methyltin mercaptide
Advastab TM-181, TM-181-FS, TM-185, TM-692, TM-2082

Methyltriacetoxysilane [CAS #4253-34-3]
CM8980

Methyl tributyl ammonium chloride
Arosurf PT Bu3

Methyl tri (C8-10) ammonium chloride
Adogen 464; Arosurf PT 64

Methyltrichlorosilane [CAS #75-79-6] *Synonyms:* Trichloromethylsilane
CM9000

Methyltriethoxysilane [CAS #2031-67-6] *Synonyms:* Triethoxymethylsilane
Dynasylan MTES

Methyltrimethoxysilane [CAS #1185-55-3]
Dynasylan MTMS

N-Methyl-N-trimethylsilyltrifluoroacetamide [CAS #24589-78-4]
Dynasylan MSTFA

Methyltris (methylethylketoxime) silane [CAS #22984-54-9]
CM9220

Methyl vinyl ether/maleic anhydride copolymer. See PVM/MA copolymer

Methyl vinyl imidazolium chloride/vinyl pyrrolidone copolymer
Luviquat FC 370, 550, 905

Metribuzin
Sencor

MIBK (CTFA) [CAS #108-10-1] *Synonyms:* Methyl isobutyl ketone; Hexone; Isopropylacetone; 4-Methyl-2-pentanone
D.E.R. 671-MK75, 671-XM75

Mica (CTFA) [CAS #12001-26-2] *Synonyms:* Muscovite mica
Bi-Lite; Huber SM, WG-1, WG-2; Lustra-Pearl; Mearlin; Mearlmica MMCF, MMSV; Mibiron; Micromesh No. 3; Snow White 200, 270, 325; Soloron; Spectra-Pearl; Timica; Timiron

Microcrystalline cellulose (CTFA) [CAS #9004-34-6]
Avicel PH-101, PH-102, PH-103, PH-105, RC-591

Microcrystalline wax (CTFA) [CAS #63231-60-7] *Synonyms:* Petroleum wax, microcrystalline; Waxes, microcrystalline
Antilux 110, 111, 500, 550, 600, 620, 654, 660, 750, L; Be Square 715, 185, 195; Cutina BW, LM; Flexowax C, C Light; Forbest MW 23; Fortex; Mekon White; Multiwax 180-M, HS, ML-445, W-445, W-835, X-145A; Petrolite C-700, C-1035; Polymekon; Shellwax; Starwax 100; Ultraflex; Victory

Microcrystalline wax, oxidized
Cardis 314, 319, 320; Petrolite C-8500; Petronauba C

Milk protein (CTFA) [CAS #9000-71-9] *Synonyms:* Casein
Hydrolactin 2500

Mineral oil (CTFA) [CAS #8012-95-1]
Amerchol BL, L-99, L-101, L-500; Benol; Bentone Gel MIO, Gel MIO A-40; Blandol; Britol; Carnation; Cremba, B6; Cutina LM; Dehymuls K; Drakeol 5, 6, 7, 8, 9, 10, 10B, 13, 15, 19, 21, 32, 34, 35, 50; Emery 1730, 1732, 1740; Ervol; Fancol LAO; Gloria; Hostacerin WO; Kaydol; Klearol; Lexate IL; Liquid Absorption Base Type A, T; Orzol; Parol 70, 80, 100; Peneteck; Polytrap 229; Protegin, X; Protol; Ritachol; Rudol; Semtol 40, 70, 85, 100, 350; Sonojell Min. Jellies, No. 4, 9

Mineral oil/acrylates copolymer
Polytrap 229

Mineral spirits (CTFA) [CAS #8032-32-4] *Synonyms:* Naphtha; Ligroin; Petroleum spirits
Amsco Mineral Spirits 66/3, 75, Odorless Mineral Spirits, Regular Mineral Spirits (Stod.); Bentone Gel S-130; Betaprene BC-100/70MS, BR-100/70MS; Espercarb 438M-60, 840M, 840M-40, 840M-70; Esperox-13M, -28MD, -31M, -33M, -497M, -545M, -551M, 740M, 747M, 750M, -939M; Lupersol 219-M60, 288M75, 531-80M, 533-M75, 553-M75, 555M60, 688M50; Shell Mineral Spirits 135, 145-EC; Texsolve S, S-2, S-66, S-LO; USP-90MD, USP-355M

Minkamide DEA
Incromide Mink D

Minkamidopropalkonium chloride
Incroquat Mink-85

Minkamidopropylamine oxide
Incromine Oxide Mink

Minkamidopropyl betaine
Incronam Mink 30; Schercotaine MKAB

Minkamidopropyl dimethylamine (CTFA) [CAS #68953-11-7]
Foamole B

Minkamidopropyl dimethylamine lactate
Incromate Mink L

Mink oil (CTFA) [RD #977054-30-0] *Synonyms:* Oil of mink
Emulan; Super Refined Mink Oil

MIPA-borate (CTFA) [CAS #68003-13-4] *Synonyms:* Isopropanolamine borate
Monacor BE

MIPA-dodecylbenzene sulfonate (CTFA) [CAS #42504-46-1; 54590-52-2] *Synonyms:* (Mono) isopropanolamine dodecylbenzene sulfonate
Hetsulf IPA

Mixed isopropanolamines lanolate (CTFA)
Tergelan 1790

Mixed isopropanolamines lauryl sulfate (CTFA) [CAS #68877-25-8; RD #977066-38-8]
Carsonol ILS

Mixed isopropanolamines myristate (CTFA) [CAS #10525-14-1]
Tergelan 1790

Mixed tallow/coconut acid (CTFA)
Hystrene 1835

Molybdenum disulfide [CAS #1317-33-5] *Synonyms:* MoS_2; Molybdenum sulfide; Molybdic sulfide
Alphaflex 101, 202; Molykote; Molysulfide Suspension, Tech., Tech. Fine; Thermocomp PFL-4216, PFL-4218, RFL-4216, RFL-4218 RL-4040; Vekton 6PAM

Molybdenum oxysulfide dithiocarbamate
Molyvan A

Molybdenum trioxide [CAS #1313-27-5] *Synonyms:* Molybdenum anhydride; Molybdic oxide; Molybdic acid hydride
POM

Monoethanolamine. See MEA...

Monoisopropanolamine. See MIPA...

Monoolein. See Glyceryl oleate

Monostearin. See Glyceryl stearate

Montan wax (CTFA) [CAS #8002-53-7]
BASF Wax DG, E, ES, L, LCP 2, LCP, LG, LGE, LS, OP, S, SG; Hoechst Wax LP; Sicolub E, OP

Montmorillonite (CTFA) [CAS #1318-93-0]
Baragel 3000; Benathix 1-4-1; Bentone 128, 500, Gel LOI, Gel MIO A-40, Gel S-130, Gel VS-5, SD-1, SD-2, SD-3; Florco; Perchem 44, 97, 108, EAG, Easigel, Econogel; Polytrope 1131

Morpholine (CTFA) [CAS #110-91-8] *Synonyms:* Tetrahydro-1,4-oxazine
Bioban P-1487

2-(Morpholinothio) benzothiazole *Synonyms:* N-oxydiethylene-2-benzothiazolesulfenamide
Santocure MOR

4-Morpholinyl-2-benzothiazole disulfide [CAS #95-32-9]
Morfax

2-(4-Morpholinyldithio) benzothiazole
Vulcuren 2, 2/EGC

Myreth-3 (CTFA) [CAS #27306-79-2 (generic), 63793-60-2; RD #977054-86-6] *Synonyms:* PEG-3 myristyl ether; POE (3) myristyl ether
Hetoxol M-3

Myreth-3 laurate (CTFA) [CAS #977068-97-5]
Schercemol MEL-3

Myreth-3 myristate (CTFA) [RD #977054-88-8] *Synonyms:* PEG-3 myristyl ether myristate
Cetiol 1414E; Liponate 143M; Schercemol MEM-3; Standamul 1414-E

Myreth-3 palmitate (CTFA) *Synonyms:* PEG-3 myristyl ether palmitate
Schercemol MEP-3

Myreth-4 (CTFA) [CAS #27306-79-2 (generic); RD #977054-89-9]
Lipocol M-4

Myristalkonium chloride (CTFA) [CAS #139-08-2] *Synonyms:* Myristyl dimethyl benzyl ammonium chloride; Tetradecyl dimethyl benzyl ammonium chloride
Arquad DM14B-90; Barquat MS-100, MX-50, MX-80; BTC 824, 824 P-100, 2125, 2125 M; Catigene DC 100; Cyncal 80%; Dibactol; Nissan Cation M2-100; Roccal MC-14

Myristalkonium saccharinate (CTFA) [CAS #68989-01-5] *Synonyms:* Myristyl dimethyl benzyl ammonium saccharinate; Quaternium-3
Onyxide 3300

Myristamide DEA (CTFA) [CAS #7545-23-5] *Synonyms:* Myristic diethanolamide
Condensate PM; Monamid 150-MW; Rewomid DLM/SE; Schercomid MD-Extra, MME, SM

Myristamide MEA (CTFA) [CAS #142-58-5; RD #977057-28-5] *Synonyms:* Myristic monoethanolamide; Myristoyl monoethanolamide
Witcamide MM

Myristamidopropyl betaine (CTFA) [CAS #59272-84-3; RD #977011-25-8] *Synonyms:* Myristamidopropyl dimethyl glycine
Schercotaine MAB

Myristamidopropyl dimethylamine (CTFA) [CAS #45267-19-4] *Synonyms:* Dimethylaminopropyl myristamide
Schercodine M

Myristamidopropyl dimethyl 2,3-dihydroxypropyl ammonium chloride
Lexquat AMG-M

Myristamine *Synonyms:* Myristyl amine; Tetradecylamine
Kemamine P-790, P-790D; Nissan Amine MB

Myristamine acetate *Synonyms:* Tetradecylamine acetate
Nissan Cation MA

Myristamine oxide (CTFA) [CAS #3332-27-2] *Synonyms:* Myristyl dimethyl amine oxide; Tetradecyl dimethyl amine oxide
Ammonyx MCO, MO; Barlox 14; Conco XA-M; Emcol M; Empigen OH, OH25; Incromine Oxide M, MC; Jordamox MDA; Ninox M; Sandoz Amine Oxide XA-M; Schercamox DMM

Myristic acid [CAS #544-63-8]
Hystrene 9014, 9514

Myristic diethanolamide. See Myristamide DEA

Myristic monoethanolamide. See Myristamide MEA

Myristoyl hydrolyzed animal protein (CTFA) [RD #977062-07-9]
Lexein A200

Myristyl alcohol [CAS #112-72-1] *Synonyms:* C14 linear primary alcohol; 1-Tetradecanol
Alfol 14, 1412; Cachalot M-43; Cyclochem EM 560; Dehydag Wax 14; Epal 14; Kalcohl 40; Lanette 14; Lipowax

Myristyl amine. See Myristamine

Myristyl/cetyl dimethylamine oxide
Conco XA-MC

Myristyl dimethylamine. See Dimethyl myristamine

Myristyl dimethyl benzyl ammonium chloride. See Myristalkonium chloride

Myristyl dimethyl-naphthylmethyl ammonium chloride
BTC 1100

Myristyl eicosanol (CTFA) *Synonyms:* 1-Eicosanol, 2-tetradecyl
Standamul G-32/36

Myristyl eicosyl stearate (CTFA)
Standamul G-3236 Stearate

Myristyl lactate (CTFA) [CAS #1323-03-1] *Synonyms:* 2-Hydroxypropanoic acid, tetradecyl ester
Cegesoft C 17; Ceraphyl 50, 50-S; Crodamol ML; Cyclochem ML; Liponate ML; Schercemol ML; Wickenol 506

Myristyl myristate (CTFA) [CAS #3234-85-3] *Synonyms:* Tetradecyl tetradecanoate
Ceraphyl 424; Cetiol MM; Crodamol MM; Cyclochem MM/M; Dermalcare MM/M; Kemester MM; Kessco Myristyl Myristate; Liponate MM; Schercemol MM; Waxenol 810

Myristyl octanoate
Wickenol 174

Myristyl propionate (CTFA) [CAS #6221-95-0; RD #977062-14-8] *Synonyms:* 1-Tetradecanol, propanoate
Lonzest 143-S; Schercemol MP

Myristyl stearate (CTFA) [CAS #17661-50-6] *Synonyms:* Octadecanoic acid, tetradecyl ester
Cyclochem MST; Dermalcare MST; Hetester MS; Schercemol MS

Myrtrimonium bromide (CTFA) [CAS #1119-97-7] *Synonyms:* Myristyl trimethyl ammonium bromide
Cycloton M214B/99, M214C/45, M214C/99; Empigen CHB; Mytab; Querton 14 Br, 14 Br-40

Nabam [CAS #142-59-6] *Synonyms:* Disodium ethylene bisdithiocarbamate
Rychem 784, 830

Na(FeEDTA)
Sequestrene NAFe 13% Fe

Naphtha [CAS #8030-30-6]
Amsco Special Naphtholite 66/3, Super High Flash Naphtha; Shell Sol B; Shell Super VM&P Naphtha EC; Shell Tolu-Sol 5, 6-EC, 19-EC; Shell VM&P Naphtha EC; Texsolve V

Naphthalene-formaldehyde sulfonate
Daxad 11; Dyesperse DC; Lomar PWM; Sanyo Levelon

Naphthalene-formaldehyde sulfonate, potassium salt
Daxad 11KLS, 19K

Naphthalene-formaldehyde sulfonate, sodium salt
Daxad 11G, 14B, 15, 16, 17, 19, 19L-33

Naphthalene-formaldehyde sulfonate, sodium/potassium salts
Daxad 14C, 19L-40

2-Naphthalene sulfonic acid [CAS #120-18-3] *Synonyms:* β-Naphthalenesulfonic acid
Demol N, RN; Lenkanol D-48

Naphthenic oil
Cyclolube 62, 85, 120, 132, 210, 213, 270, 413, 2290, 2310, 4053, NN-1, NN-2

Naphthol spirits
Amsco Naphthol Spirits 66/3

Naphthotriazolstilbene monosulfonate deriv.
Tinopal RA-16, RBS 200%, RBS WP-S

Natamycin
Delvocid

Natural rubber. See Polyisoprene

Neoalkoxy dodecylbenzenesulfonyl titanate
LICA 09

Neoalkoxy tri (m-amino) phenyl titanate
LICA 97

Neoalkoxy tri (dioctylphosphato) titanate
LICA 12

Neoalkoxy tri (dioctylpyrophosphato) titanate
LICA 38

Neoalkoxy tri (N-ethylaminoethylamino) titanate
LICA 44

Neoalkoxy trineodecanoyl titanate
LICA 01

Neoalkoxy tris (m-amino) phenyl zirconate
LZ 97

Neoalkoxy tris (dioctyl) phosphato zirconate
LZ 12

Neoalkoxy tris (dioctyl) pyrophosphato zirconate
LZ 38

Neoalkoxy tris (dodecyl) benzene sulfonyl zirconate
LZ 09

Neoalkoxy tris (ethylene diamino) ethyl zirconate
LZ 44

Neoalkoxy trisneodecanoyl zirconate
LZ 01

Neopentyl (diallyl) oxy, triacryl zirconate
NZ 39

Neopentyl (diallyl) oxy, tri (m-amino) phenyl zirconate
NZ 97

Neopentyl (diallyl) oxy, tri (dioctyl) phosphato zirconate
NZ 12

Neopentyl (diallyl) oxy, tri (dioctyl) pyrophosphato zirconate
NZ 38

Neopentyl (diallyl) oxy, tri (dodecyl) benzene-sulfonyl zirconate
NZ 09

Neopentyl (diallyl) oxy, tri (9,10 epoxy stearoyl) zirconate
NZ 49

Neopentyl (diallyl) oxy, tri (N-ethylenediamino) ethyl zirconate
NZ 44

Neopentyl (diallyl) oxy, trihydroxy caproyl titanate
LICA 99

Neopentyl (diallyl) oxy, trimercapto-phenyl zirconate
NZ 89

Neopentyl (diallyl) oxy, trimethacryl zirconate
NZ 33

Neopentyl (diallyl) oxy, trineodecanoyl zirconate
NZ 01

Neopentyl glycol diacrylate
SR-247

Neopentyl glycol dibenzoate
Benzoflex S-312

Neopentyl glycol dicaprate (CTFA)
Schercemol NGDC

Neopentyl glycol dicaprate/dicaprylate [CAS #7063-32-2]
Liponate NPGC-2

Neoprene [CAS #126-99-8] *Synonyms:* Polychloroprene
Neoprene Latex Series

Nickel [CAS #7440-02-0]
Ni-0104 P, T, -0301 T, -0707 T, -0901 S, -1404 T, -2002 C, -3210 T, -3250 T, -3266 E, -5124, -5132 P, -5333 T; Nysel (Ni-3201 F), CN-14 (Ni-3611 L), HK-4 (Ni-3609 F), HK-12 (Ni-5708 L), SP-7 (Ni-5169 F)

Nickel bis[O-ethyl (3,5-di-t-butyl-4-hydroxybenzyl)] phosphonate
Irgastab 2002

Nickel dibutyldithiocarbamate [CAS #13927-77-0]
Naugard NBC; Rylex NBC; UV-Chek AM104; Vanox NBC

Nickel diisobutyldithiocarbamate [CAS #15317-78-9]
Isobutyl Niclate

Nickel dimethyldithiocarbamate [CAS #15521-65-0]
Methyl Niclate

Nickel molybdate
HT-500 E 1/8″

Nickel thiobisphenol
UV-Chek AM-105

Nickel tungsten
Ni-4301 E

Nitrile rubber. See Acrylonitrile-butadiene rubber

Nitrilotriacetic acid [CAS #139-13-9] *Synonyms:* NTA; Triglycine; TGA; Triglycolamic acid
Hampshire NTA Acid; SEQ NT-15; Trilon AS

2-Nitro-1-butanol [CAS #609-31-4]
NB

Nitroethane [CAS #79-24-3]
NE

2-Nitro-2-ethyl-1,3-propanediol [CAS #597-09-1]
NEPD

Nitromethane [CAS #75-52-5]
NM

2-Nitro-2-methyl-1-propanol [CAS #76-39-1]
NMP

p-Nitrophenoxyethanol [CAS #16365-27-8]
Emery 5743

Nitropropane [CAS #108-03-2] *Synonyms:* 1-Nitropropane
NiPar S-10, S-20

Nitrosodiphenylamine [CAS #156-10-5] *Synonyms:* Diphenylnitrosamine; Nitrous diphenylamide
Redax

Nonene [CAS #27214-95-8] *Synonyms:* Nonene (mixed isomers)
Amsco Nonene

Nonoxynol-1 (CTFA) [CAS #9016-45-9 (generic); 26027-38-3 (generic); 27986-36-3; 37205-87-1 (generic); 104-35-8] *Synonyms:* Ethylene glycol nonyl phenyl ether; PEG-1 nonyl phenyl ether; 2-(Nonylphenoxy) ethanol
Alkasurf NP-1; Cedepal CO-210; Chemax NP-1.5 (1.5 EO); DeSonic 1.5N (1.5 EO); Igepal CO-210 (1.5 EO); Norfox NP-1; Peganol NP 1.5 (1.5 EO); Rexol 25/1; Surfonic N-10; T-Det N 1.5 (1.5 EO); Trycol 6960, NP-1

Nonoxynol-2 (CTFA) [CAS #9016-45-9 (generic); 26027-38-3 (generic); 27176-93-8; 37205-87-1 (generic); RD #977057-30-9] *Synonyms:* PEG-2 nonyl phenyl ether; POE (2) nonyl phenyl ether; PEG 100 nonyl phenyl ether
Conco NI-21; Teric N2

Nonoxynol-3
Surfonic N-31.5; Teric N3 (3.5 EO]

Nonoxynol-4 (CTFA) [CAS #7311-27-5; 9016-45-9 (generic); 26027-38-3 (generic); 37205-87-1 (generic)]
Ablunol NP4; Alkasurf NP-4; Carsonon N-4; Cedepal CO-430; Chemax NP-4; Conco NI-43; DeSonic 4N; Hetoxide NP-4; Hyonic NP-40; Igepal CO-430; Macol NP-4; Norfox NP-4; Peganol NP 4; Polystep F-1; Rexol 25/4; Serdox NNP 4; Siponic NP4; Sterox ND; Surfonic N-40; Synperonic NP4; T-Det N-4; Tergitol NP-14; Teric N4; Trycol 6961, NP-4

Nonoxynol-5 (CTFA) [CAS #9016-45-9 (generic); 26027-38-3 (generic); 37205-87-1 (generic); RD #977057-31-0]
Alkasurf NP-5; Igepal CO-520; Peganol NP 5; Sterox NE; Synperonic NP5; Teric N5 (5.5 EO); Triton N-57

Nonoxynol-6 (CTFA) [CAS #9016-45-9 (generic); 26027-38-3 (generic); 37205-87-1 (generic); RD #977057-32-1]
Ablunol NP6; Alkasurf NP-6; Cedepal CO-530; Chemax NP-6; Conco NI-60; Emthox 6962; Hyonic NP-60; Igepal CO-530; Macol NP-6; Peganol NP 6; Polystep F-2; Rexol 25/6; Siponic NP6; Sterox NF, NG (6.5 EO); Surfonic N-60; Synperonic NP6; T-Det N-6; Triton N-60; Trycol 6962, NP-6

Nonoxynol-6 phosphate (CTFA) *Synonyms:* PEG-6 nonyl phenyl ether phosphate
Monafax 786

Nonoxynol-7 (CTFA) [CAS #9016-45-9 (generic); 26027-38-3 (generic); 27177-05-5; 37205-87-1 (generic); RD #977057-33-2]
Iconol NP-7; Siponic NP-7; Tergitol NP-7, NP-27; Trycol 6963

Nonoxynol-8 (CTFA) [CAS #9016-45-9 (generic); 26027-38-3 (generic); 37205-87-1 (generic); RD #977057-34-3]
Carsonon N-8; Nikkol NP-7.5; Nutrol 611; Polystep F-3; Serdox NNP 8.5 (8.5 EO); Siponic NP-8; Surfonic HDL, N-85 (8.5 EO); T-Det N-8; Tergitol NP-8; Teric N8 (8.5 EO)

Nonoxynol-9 (CTFA) [CAS #9016-45-9 (generic); 26027-38-3 (generic); 26571-11-9; 37205-87-1 (generic); RD #977004-65-1]
Alkasurf NP-9; Carsonon N-9; Chemax NP-9; Conco NI-90; Empilan NP9; Emthox 6964; Gradonic N-95 (9.5 EO); Hetoxide NP-9; Hyonic NP-90; Igepal CO-630; Lipal 9 N; Macol NP-9.5 (9.5 EO); Neutronyx 600 (9.5 EO); Nutrol 600; Peganol NP 9; Rexol 25/9; Siponic NP-9.5; Sterling NPX; Syn Fac 905 (9.5 EO); T-Det N-9.5; Tergitol NP-9; Teric N9

Nonoxynol-9 iodine (CTFA) [RD #977059-98-5] *Synonyms:* PEG-9 nonyl phenyl ether iodine complex
Biopal VRO-20

Nonoxynol-9 phosphate (CTFA) [CAS #66197-78-2] *Synonyms:* PEG-9 nonyl phenyl ether phosphate
Gafac RE-610

Nonoxynol-10 (CTFA) [CAS #9016-45-9 (generic); 26027-38-3 (generic); 27177-08-8; 37205-87-1 (generic); RD #977057-35-4]
Alkasurf NP-10; Carsonon N-10; Chemax NP-10; Conco NI-100; Cremophor NP 10; Hyonic NP-100, PE-100; Igepal CO-660, CO-710 (10–11 EO); Nikkol NP-10; Peganol NP 10; Rexol 25/10; Siponic NP-10; Surfonic N-95, N-100, N-102; Synperonic NP10; T-Det N-10.5; Tergitol NP-10, NPX (10.5 EO); Teric N10; Triton N-101; Trycol 6974

Nonoxynol-11 (CTFA) [CAS #9016-45-9 (generic); 26027-38-3 (generic); 37205-87-1 (generic); RD #977065-12-5]
Conco NI-110; Emthox 6965; Hyonic NP-110; Macol NP-11; Neutronyx 656; Teric N11

Nonoxynol-12 (CTFA) [CAS #9016-45-9 (generic); 26027-38-3 (generic); 37205-87-1 (generic); RD #977061-17-8]
Hyonic NP-120; Iconol NP-12; Igepal CO-720; Peganol NP 12; Polystep F-5; Surfonic N-120; Synperonic NP12; Teric N12

Nonoxynol-12 iodine (CTFA) [RD #977014-40-6] *Synonyms:* PEG-12 nonyl phenyl ether iodine complex
Biopal NR-20

Nonoxynol-13 (CTFA) [CAS #9016-45-9 (generic); 26027-38-3 (generic); 37205-87-1 (generic); RD #977065-13-6]
DeSonic 13N; Siponic NP-13; Teric N13

Nonoxynol-14 (CTFA) [CAS #9016-45-9 (generic); 26027-38-3 (generic); 37205-87-1 (generic); RD #977061-55-4]
Cremophor NP 14; T-Det N-14

Nonoxynol-15 (CTFA) [CAS #9016-45-9 (generic); 26027-38-3 (generic); 37205-87-1 (generic); RD #977006-97-5]
Ablunol NP15; Conco NI-150; Igepal CO-730; Nikkol NP-15; Peganol NP 15; Surfonic N-150

Nonoxynol-18 (CTFA) [RD #977065-14-7]
Nikkol NP-18TX

Nonoxynol-20 (CTFA) [CAS #9016-45-9 (generic); 26027-38-3 (generic); 37205-87-1 (generic); RD #977057-36-5]
Alkasurf NP-20, NP-20 70%; Conco NI-185; Ethylan 20; Igepal CO-850; Macol NP-20; Peganol NP 20; Rexol 25/20; Serdox NNP 20/70; Sermul EN 20/70; Soprofor NP/20; Surfonic N-200; Synperonic NP20

Nonoxynol-30 (CTFA) [CAS #9016-45-9 (generic); 26027-38-3 (generic); 37205-87-1 (generic); RD #977006-96-4]
Alkasurf NP-30 70%; Carsonon N-30 70%; Conco NI-187; Igepal CO-880, CO-887; Macol NP-30, NP-30(70); Peganol NP 30, 30 70%; Renex 650; Rexol 25/30, 25/307; Serdox NNP 30/70; Sermul EN 30/70, EN 145; Soprofor NP/30; Surfonic N-300, NB-5; Synperonic NP30

Nonoxynol-35
Alkasurf NP-35 70%; Synperonic NP35

Nonoxynol-40 (CTFA) [CAS #9016-45-9 (generic); 26027-38-3 (generic); 37205-87-1 (generic); RD #977057-37-6]
Alkasurf NP-40, NP-40 70%; Carsonon N-40 70%; Chemax NP-40, NP-40/70; Conco NI-190, NI-197; Emthox 6957; Hetoxide NP-40; Hyonic NP-407; Igepal CO-890, CO-897; Peganol NP 40, 40 70%; Rexol 25/407; Siponic NP40, NP407; Surfonic N-400, NB-14; Tergitol NP-44; Trycol 6957, 6970, NP-40, NP-407

Nonoxynol-50 (CTFA) [CAS #9016-45-9 (generic); 26027-38-3 (generic); 37205-87-1 (generic); RD #977057-38-7]
Alkasurf NP-50 70%; Carsonon N-50 70%; Chemax NP-50, NP-50/70; Conco NI-2000; Emthox 6971, 6972; Hyonic NP-500; Igepal CO-970, CO-977; Peganol NP 50; Rexol 25/50, 25/507; Trycol 6971, 6972, NP-50, NP-507

Nonoxynol-70
Macol NP-70; Siponic NP-707; T-Det N-70

Nonoxynol-75
Siponic NP75

Nonoxynol-100 (CTFA) [CAS #9016-45-9 (generic); 26027-38-3 (generic); 37205-87-1 (generic); RD #977065-11-4]
Carsonon N-100 70%; Chemax NP-100, NP-100/70; Igepal CO-990, CO-997; Macol NP-100; Peganol NP 100, 100 70%; Rexol 25/100-70%; Soprofor NP/100; Trycol 6942, 6981, NP-1007

Nonoxynol-150
Trycol 6954

Nonyl naphthalene sodium sulfonate
Emkal NNS

Nonyl naphthalene sulfonic acid
Emkal NNS Acid

Nonyl nonoxynol-4 *Synonyms:* PEG-4 dinonyl phenyl ether; POE (4) dinonyl phenyl ether
Hetoxide DNP-4

Nonyl nonoxynol-7 phosphate (CTFA) [CAS #66172-78-9; 66172-83-6] *Synonyms:* PEG-7 dinonyl phenyl ether phosphate]
Gafac RM-410

Nonyl nonoxynol-8
Chemax DNP-8; Iconol DNP-8; Trycol 6985, DNP-8

Nonyl nonoxynol-9
Hetoxide DNP-9.6 (9.6 EO)

Nonyl nonoxynol-10 phosphate (CTFA)
Gafac RM-510

Nonyl nonoxynol-15
Chemax DNP-15

Nonyl nonoxynol-15 phosphate (CTFA)
Gafac RM-710

Nonyl nonoxynol-18
Chemax DNP-18

Nonyl nonoxynol-24
Iconol DNP-24

Nonyl nonoxynol-49 (CTFA) [CAS #9014-93-1 (generic); RD #977061-25-8]
Igepal DM-880

Nonyl nonoxynol-100 (CTFA) [CAS #9014-93-1 (generic)]
Serdox NDI 100

Nonyl nonoxynol-150 (CTFA) [CAS #9014-93-1 (generic); RD #977010-96-2, 977069-35-4]
Chemax DNP-150; Iconol DNP-150; Igepal DM-970; Trycol 6989, DNP-150, DNP-150/50

Nonylphenol-dodecanethiol-formaldehyde product
Wingstay K

Novolac resin *Synonyms:* Novolak resin; 2-Stage phenolic resin; Phenol-formaldehyde (novolak) resin
Akrochem P 40, P 49, P 55, P 82, P 86; SP-12, -25, -102, -103, -126, -134, -144, -154, -155, -174, -1044, -1045, -1068, -1077; Varcum 4727, DR-406

NTA. See Nitrilotriacetic acid

NTA H_3
Dissolvine A Z

Nylon (CTFA) [CAS #32131-17-2; #63428-83-1, #9008-75-7] *Synonyms:* Poly [imino (1,6-dioxo-1,6-hexanediyl) imino-1,6-hexanediyl]
Standamid Resin BC-1283

Nylon 6/6
Durethan A 30 S, AKV 30, AKV 30 H, B 25 T, B 30 P, B 30 S, B 31 SK, B 31 F, B 40 E, B 40 F, B 40 SK, B 50 E, BC 30, BC 40, BG 30 X, BKV 30, BKV 30 H, BKV 30N, BKV 30N1, BKV 35, BKV 50 H, C 38 F

Octabromodiphenyl oxide
DE-79; FR-1208; Saytex 111

Octadecyl. See also Stearyl

Octadecyl 3,5-di-t-butyl-4-hydroxyhydrocinnamate
Irganox 1076; Naugard 76

Octadecyl 3-(3,5-di-t-butyl-4-hydroxyphenyl) propionate [CAS #2082-79-3]
Ultranox 276

Octadecyldimethyl [3-(trimethoxysilyl) propyl] ammonium chloride [CAS #27668-52-6]
CO9745

Octadecyltrichlorosilane [CAS #112-04-9] *Synonyms:* Trichlorooctadecylsilane
CO9750

Octadodecanol stearate
Cetiol G20S

Octadodecyl stearate
Starfol ODS

Octamethylcyclotetrasilazane [CAS #1020-84-4]
CO9800

Octamethylcyclotetrasiloxane [CAS #556-67-2; 69430-24-6]
Abil K 4; CO9810

Octamethyltrisiloxane [CAS #105-51-7]
CO9816

Octaphenylcyclotetrasiloxane [CAS #546-56-5]
CO9817

Octasodium diethylene triamine penta (methylene phosphonate)
Dequest 2066

Octocrylene (CTFA) [CAS #6197-30-4] *Synonyms:* Ethylhexyl-2-cyano-3,3-diphenylacrylate)
Uvinul N-539

Octoxynol-1 (CTFA) [CAS #9002-93-1 (generic); 9004-87-9 (generic); 9036-19-5 (generic); 2315-67-5; RD #977054-43-5] *Synonyms:* Ethylene glycol octyl phenyl ether; PEG-1 octyl phenyl ether; POE (1) octyl phenyl ether
Alkasurf OP-1 ; Rexol 45/1; Siponic OP1.5 (1.5 EO); Triton X-15

Octoxynol-3 (CTFA) [CAS #9002-93-1 (generic); 9004-87-9 (generic); 9010-43-9; 9036-19-5 (generic); RD #977057-42-3] *Synonyms:* PEG-3 octyl phenyl ether; POE (3) octyl phenyl ether
Chemax OP-3; Nikkol OP-3; Peganol OP 3; Rexol 45/3; Triton X-35

Octoxynol-4
Hyonic OP-40

Octoxynol-5 (CTFA) [CAS #9002-93-1 (generic); 9004-87-9 (generic); 9036-19-5 (generic); RD #977054-44-6] *Synonyms:* PEG-5 octyl phenyl ether; POE (5) octyl phenyl ether
Alkasurf OP-5; Cedepal CA-520; Chemax OP-5; Rexol 45/5; Teric X5; Triton X-45

Octoxynol-6
Peganol OP 6; T-Det O-6 (6–7 EO)

Octoxynol-7 (CTFA) [CAS #9002-93-1 (generic); 9004-87-9 (generic); 9036-19-5 (generic)]
Hyonic OP-70; Rexol 45/7; Teric X7 (7.5 EO)

Octoxynol-8 (CTFA) [CAS #9036-19-5]
Peganol OP 8; Teric X8 (8.5 EO)

Octoxynol-9 (CTFA) [CAS #9002-93-1 (generic); 9004-87-9 (generic); 9010-43-9; 9036-19-5 (generic); RD #977019-58-1]
Hyonic OP-100; Nutrol 100; Siponic F-90; Triton X-100, X-120 (9–10 EO)

Octoxynol-10 (CTFA) [CAS #9002-93-1 (generic); 9004-87-9 (generic); 9036-19-5 (generic); RD #977067-92-7]
Alkasurf OP-10; Nikkol OP-10; Peganol OP 10; Teric X10

Octoxynol-11 (CTFA) [CAS #9004-87-9 (generic); 9036-19-5 (generic)]
Teric X11

Octoxynol-12
Alkasurf OP-12 (12–13 EO); Rexol 45/12

Octoxynol-13 (CTFA) [CAS #9002-93-1 (generic); 9004-87-9 (generic); 9036-19-5 (generic); RD #977057-43-4]
Teric X13; Triton X-102

Octoxynol-16 (CTFA) [CAS #9004-87-9 (generic); 9036-19-5 (generic); RD #977019-64-9]
Alkasurf OP-16; Rexol 45/16; Siponic F-160; Teric X16

Octoxynol-20
Peganol OP 20

Octoxynol-30 (CTFA) [CAS #9004-87-9 (generic); 9036-19-5 (generic); RD #977019-63-8]
Alkasurf OP-30 70%; Igepal CA-887; Nikkol OP-30; Peganol OP 30, 30 70%; Rexol 45/307; Serdox NOP 30/70; Siponic F-300; Siponic OP30

Octoxynol-40 (CTFA) [CAS #9002-93-1 (generic); 9004-87-9 (generic); 9036-19-5 (generic); RD #977065-20-5]
Alkasurf OP-40 70%; Cedepal CA-897; Chemax OP-40; DeSonic S-405; Dow Corning Q2-7224; Emthox 6984; Hyonic OP-407; Igepal CA-890, CA-897; Peganol OP 40, 40 70%; Rexol 45/407; Serdox NOP 40/70; Siponic F-400; Siponic OP40; T-Det O-40, O-407; Teric X40; Triton X-405; Trycol OP-407

Octoxynol-70 (CTFA) [CAS #9004-87-9 (generic); 9036-19-5 (generic)]
Hyonic OP-705; Peganol OP 70, 70 70%; Siponic F-707

Octrizole
Spectra-Sorb UV 5411

Octyl. See also Caprylic

Octylated diphenylamine
Agerite Stalite, Stalite S; Akrochem Antioxidant S; Anchor ODPA; Flectol ODP;Vanlube SL, SS; Vulkanox OCD, OCD/SG

n-Octyl, n-decyl adipate *Synonyms:* NODA
Monoplex NODA; Plasthall NODA; Staflex NODA

Octyl decyl dimethyl ammonium chloride. See Quaternium-24

n-Octyl, n-decyl phthalate [CAS #119-07-3] *Synonyms:* Octyldecyl phthalate; 1,2-Benzenedicarboxylic acid, decyl octyl ester
Staflex 500; Staflex ODP

n-Octyl, n-decyl trimellitate
Staflex NONDTM

Octyldimethylchlorosilane [CAS #18162-84-0]
CO9819

Octyl dimethyl PABA (CTFA) [CAS #21245-02-3] *Synonyms:* Octyl dimethyl p-aminobenzoate; 4-(Dimethylamino) benzoic acid, 2-ethylhexyl ester; Benzoic acid, 4-(dimethylamino)-2-ethylhexyl ester; 2-Ethylhexyl p-dimethyl aminobenzoate
Arlatone 507; Escalol 507; Solarchem 0

Octyl dodecanol (CTFA) [CAS #5333-42-6] *Synonyms:* 2-Octyldodecanol-1
Amerchol L-500; Cutina LM; Eutanol G; Primarol 1208; Rilanit G 20; Standamul G

Octyldodecyl dimethyl ammonium chloride
BTC 812

Octyldodecyl myristate (CTFA) *Synonyms:* Myristic acid, 2-octyldodecyl ester
Exceparl OD-M; M.O.D.; Wickenol 142

Octyldodecyl oleate
Exceparl OD-OL

Octyldodecyl stearate (CTFA) [CAS #22766-82-1]
Standamul 7063; Starfol OS

Octyldodecyl stearoyl stearate (CTFA)
Ceraphyl 847

Octyl epoxy stearate. See Epoxidized octyl stearate

Octyl epoxy tallate. See Epoxidized octyl tallate

Octyl hydroxystearate (CTFA) [CAS #29383-26-4; 29710-25-6; RD #977057-45-6] *Synonyms:* 2-Ethylhexyl oxystearate
Crodamol OHS; Naturechem OHS; Polytrap 171; Wickenol 171

Octyl isononanoate (CTFA) [RD #977104-46-2] *Synonyms:* 2-Ethylhexyl isononanoate
Kessco Octyl isononanoate

Octyl isothiazoline
Kathon 893

2-n-Octyl-4-isothiazolin-3-one [CAS #26530-20-1]
Kathon 4200, LM, LP; Micro-Chek 11, 11D, 11DIDP, 11S-711; Skane M-8

Octylmercaptide
Stanclere T470, T473, T482, T483, T484, T582

Octyl methoxycinnamate (CTFA) [CAS #5466-77-3] *Synonyms:* 2-Ethylhexyl methoxycinnamate
Escalol 557; Parsol MCX

Octyl oxystearate
Kessco Octyl Oxystearate

Octyl palmitate (CTFA) [CAS #29806-73-3] *Synonyms:* 2-Ethylhexyl palmitate
Cegesoft C 25; Ceraphyl 368; Crodamol OP; Kessco Octyl Palmitate; Lexol EHP; Radia 7129; Unimate EHP; Wickenol 155, 161, 163

Octyl pelargonate (CTFA) [CAS #59587-44-9; 977012-73-9] *Synonyms:* 2-Ethylhexyl pelargonate
Wickenol 160; Schercemol OPG

Octyl phenoxy tetraethoxy-ethanol
Plexol 305

Octyl phosphate. See Trioctyl phosphate

Octyl salicylate [CAS #118-60-5]
Escalol 587

Octyl stearate (CTFA) [CAS #22047-49-0] *Synonyms:* 2-Ethylhexyl stearate
Cetiol 868; Interwax G 8253; Rilanit EHS; Wickenol 156, 161, 163

Octyl tallate
Plasthall R-9

Octyl thiotin
Lankromark OT050, OT250, OT335, OT341, OT450, OT650

Octyltrichlorosilane [CAS #5283-66-9] *Synonyms:* Trichlorooctylsilane
CO9830

Octyltriethoxysilane [CAS #2943-75-1]
Dynasylan OCTEO

Olealkonium chloride (CTFA) [CAS #37139-99-4] *Synonyms:* Oleyl dimethyl benzyl ammonium chloride
Alacsan 7LUF; Ammonyx KP; Cycloton 7LUF; Incroquat O-50; Jordaquat JO-50

Oleamide (CTFA) [CAS #301-02-0] *Synonyms:* 9-Octadecenamide; Oleyl amide
Adogen 73; Armid O; Armoslip CPM; Crodamide O, OR; Kemamide O, U; Petrac Slip-Eze; Unislip 1757, 1759

Oleamide DEA (CTFA) [CAS #93-83-4] *Synonyms:* Oleic diethanolamide; Oleic acid diethanolamide
Aminol OF; Calamide O; Chimipal OLD; Comperlan OD; Cyclomide DO280, DO280/S; Emid 6545; Hartamide 9137; Incromide OD, OPD; Jordamide 201; Lauridit OD; Lexate CL-60; Loropan OD; Mackamide O; Marlamid D 1885; Mazamide O-20; Ninol 201; Ninol AC-201; Nissan Stafoam DO, DO-S; Nitrene NO; Schercomid ODA, SO-A; Serdolamide POF 61, POF 61 C; Steinamid DO 280, DO 280/SE; Varamide A-7

Oleamide MEA (CTFA) [CAS #111-58-0] *Synonyms:* Oleic monoethanolamide
Incromide OM, OPM

Oleamide MIPA (CTFA) [CAS #111-05-7] *Synonyms:* Oleic monoisopropanolamide
Schercomid OMI; Steinamid IPE 280; Witcamide 61

Oleamidopropylamine oxide (CTFA) [CAS #25159-40-4] *Synonyms:* 9-Octadecenamide, N-[3-(dimethylamino) propyl]-, N-oxide; Oleamidopropyl dimethylamine oxide
Incromine Oxide O; Mackamine OAO

Oleamidopropyl betaine (CTFA) [CAS #25054-76-6] *Synonyms:* Oleamidopropyl dimethyl glycine
Cycloteric BET O-30; Incronam OP-30; Mackam HV; Schercotaine OAB

Oleamidopropyl dimethylamine (CTFA) [CAS #109-28-4] *Synonyms:* Dimethylaminopropyl oleamide
Incromine OPM, OPB; Lexamine O-13; Lexate CL-60; Mackine 501; Schercodine O; Tego-Amid O 18

Oleamidopropyl dimethylamine oxide. See Oleamidopropylamine oxide

Oleamidopropyl dimethyl 2,3-dihydroxypropyl ammonium chloride
Lexquat AMG-O

Oleamine (CTFA) [CAS #112-90-3] *Synonyms:* Oleyl amine; 9-Octadcecen-1-amine
Adogen 172 (D); Amine OL; Armeen O, OD; Arosurf MG-172; Crodamine 1.O, 1.OD; Kemamine P-989, P-989D; Lilamin 172, 172 D; Nissan Amine OB; Noram O; Radiamine 6172, 6173

Oleamine acetate
Armac OD; Noramac O

Oleamine oxide (CTFA) [CAS #14351-50-9; 61792-38-9; RD #977057-48-9] *Synonyms:* Oleyl dimethyl amine oxide; Oleylamine oxide
Alkamox ODM; Incromine Oxide OD-50; Jordamox ODA; Mackamine O2; Ninox O; Standamox O1

Oleic acid (CTFA) [CAS #112-80-1] *Synonyms:* cis-9-Octadecanoic acid; Red oil
Industrene 104, 105, 106, 205, 206, 206LP

Oleic acid dibutylamide
Rewocor RA 280; Rewocoros RA 280

Oleic diethanolamide. See Oleamide DEA

Oleic imidazoline acetate
Varine O Acetate

Oleic-linoleic acid
Industrene 224

N-Oleic-linoleic, 1,3-propylene diamine
Kemamine D-999

Oleic-stearic acid
Industrene M

Oleoamphohydroxypropylsulfonate (CTFA) [CAS #68610-38-8] *Synonyms:* Oleoamphopropylsulfonate
Miranol OS

Oleoamphopropionate (CTFA) [CAS #67892-37-9; 70024-80-5]
Miranol OM-SF Conc.

Oleoyl sarcosine (CTFA) [CAS #110-25-8]
Nikkol Sarcosinate OH; Sarkosyl O

Oleth-2 [CAS #9004-98-2 (generic); 25190-05-0 (generic); RD #977057-49-0] *Synonyms:* PEG-2 oleyl ether; POE (2) oleyl ether; PEG 100 oleyl ether
Alkasurf OA-2; Ameroxol OE-2; Ethosperse OA-2; Hetoxol OL-2; Lipocol O-2; Macol OA-2; Ritoleth 2

Oleth-2 phosphate *Synonyms:* PEG-2 oleyl ether phosphate
Crodafos O2 Acid

Oleth-3 (CTFA) [CAS #9004-98-2 (generic); 25190-05-0 (generic); RD #977057-50-3]
Fluilanol; Hetoxol OA-3 Special; Volpo 3, N3, O3

Oleth-3 phosphate (CTFA) [CAS #39464-69-2 (generic); RD #977060-93-7]
Crodafos N3 Acid

Oleth-4 (CTFA) [CAS #9004-98-2 (generic); 25190-05-0 (generic); RD #977065-24-9]
Chemal OA-4; Hetoxol OL-4; Macol OA-4

Oleth-4 phosphate (CTFA) [CAS #39464-69-2 (generic); RD #977065-25-0]
Chemfac PB-184

Oleth-5 (CTFA) [CAS #9004-98-2 (generic); 25190-05-0 (generic); RD #977057-51-4]
Chemal OA-5; Eumulgin M 8, O 5; Hetoxol OA-5 Special, OL-5; Lipal 5 OA; Macol OA-5; Ritoleth 5; Solulan 5; Standamul O-5; Volpo 5, N5, O5

Oleth-5 phosphate
Crodafos N5 Acid, O5 Acid

Oleth-9 (CTFA) [CAS #9004-98-2 (generic); 25190-05-0 (generic)]
Chemal OA-9

Oleth-10 (CTFA) [CAS #9004-98-2 (generic); 25190-05-0 (generic); RD #977057-52-5]
Alkasurf OA-10; Ameroxol OE-10; Brij 96, 97; Eumulgin M8, O10; Hetoxol OA-10 Special; Lipal 10 OA; Lipocol O-10; Macol OA-10; Nioix KJ-15; Ritoleth 10; Standamul O-10; Volpo 10, N10, O10

Oleth-10 phosphate (CTFA) [CAS #39464-69-2 (generic); RD #977060-95-9]
Crodafos N10 Acid, O10 Acid

Oleth-12 [CAS #9004-98-2 (generic); 25190-05-0 (generic)]
Lanbritol Wax N21

Oleth-15 (CTFA) [CAS #9004-98-2 (generic); 25190-05-0 (generic); RD #977065-21-6; 977061-58-7]
Volpo N15, O15

Oleth-16 (CTFA) [CAS #9004-98-2 (generic); 25190-05-0 (generic)]
Solulan 16

Oleth-18
Noiox KJ-12

Oleth-20 (CTFA) [CAS #9004-98-2 (generic); 25190-05-0 (generic); RD #977057-53-6]
Ahco 3998; Alkasurf OA-20; Ameroxol OE-20; Brij 98, 99; Chemal OA-20, OA-20/70; Emulphor ON-870; Hetoxol OA-20 Special; Lipal 20 OA; Lipocol O-20; Macol OA-20; Ritoleth 20; Siponic Y501; Standamul O-20; Trycol 5971; Volpo 20, N20, O20

Oleth-23 (CTFA) [CAS #9004-98-2 (generic); 25190-05-0 (generic); RD #977065-23-8]
Hetoxol OL-23; Trycol 5972, OAL-23

Oleth-25 (CTFA) [CAS #9004-98-2 (generic); 25190-05-0 (generic); RD #977057-54-7]
Siponic Y500-70; Solulan 25

Oleth-40
Hetoxol OL-40

Oleth-50 (CTFA) [CAS #9004-98-2 (generic); 25190-05-0 (generic); RD #977057-55-8]
Lipal 50 OA

Oleyl alcohol (CTFA) [CAS #143-28-2] *Synonyms:* 9-Octadecen-1-ol
Adol 80, 85, 85 NF, 90, 90 NF, 320, 330, 340; Cachalot O-15; Fancol OA 50, OA 70, OA 80, OA 90, OA 95; HD Eutanol; HD-Ocenol, 92/96; Lipocol O; Novol

Oleyl amidoethyl ethanolamine
Servamine KEO 260

Oleyl amidoethyl oleyl imidazoline
Servamine KOO 330 B

Oleyl betaine (CTFA) [CAS #817-37-4] *Synonyms:* Oleyl dimethyl glycine
Cycloteric BET-OB50; Mackam OB, OB-30; Mirataine ODMB-35%; Standapol OLB-30, OLB-50; Velvetex OLB-50

Oleyl diamine
Adogen 572; Lilamin 572; Radiamine 6572

Oleyl dimethyl benzyl ammonium chloride. See Olealkonium chloride

Oleyl dimethylethyl ammonium ethosulfate
Larostat 143, 1443

Oleyl erucate (CTFA) [CAS #17673-56-2] *Synonyms:* 9-Octadecenyl 13-docosenoate; Erucic acid, oleyl ester
Cetiol J 600; Dynacerin 660

Oleyl hydroxyethyl imidazoline [CAS #95-38-5; 21652-27-7; 27136-73-8] *Synonyms:* 1-Hydroxyethyl-2-oleyl imidazoline; Oleic acid imidazoline; Oleyl imidazoline
Alkaquat O; Alkazine O; Amine O; Crodazoline O; Mackazoline O; Mazoline OA; Miramine OC; Monazoline O; Nopcogen 22-O; Quaternary O; Rewopon JMOA; Schercozoline O; Textamine O-1; Varine O

Oleyl imidazoline methosulfate
Empigen FRH75S

Oleyl linoleate (CTFA) [CAS #17673-59-3]
Polylan

Oleyl nitrile
Arneel OD

Oleyl oleate (CTFA) [CAS #3687-45-4] *Synonyms:* 9-Octadecenoic acid, 9-octadecenyl ester
Cetiol; Rilanit OLO; Schercemol OP; Starfol OO; Wickenol 143

Oleyl palmitamide (CTFA) [CAS #16260-09-6]
HTSA #1; Kemamide P-181

N-Oleyl-1,3-propanediamine *Synonyms:* Oleyl-1,3 diaminopropane
Duomeen C

Oleyl 1,3-propylene diamine
Dinoram O; Kemamine D-989

Oleyl propylene diamine acetate
Dinoramac O

Oleyl propylene diamine dioleate
Inipol 002

Oleyl propylene diamine ditallate
Inipol OT2

Oleyl trimethyl ammonium chloride
Arquad S-2C-50; Noramium MO 50

Olivamide DEA
Incromide OLD

Olivamidopropalkonium chloride
Incroquat OL-85

Olivamidopropylamine oxide
Incromine Oxide OL

Olivamidopropyl betaine
Incronam OL-30

Olivamidopropyl dimethylamine lactate
Incromate OLL

Olive oil
Super Refined Olive Oil

Omethoate
Folimat

Orange roughy oil
Super Refined Orange Roughy Oil

Organopolysiloxane. See Polysiloxane

Organosilicon
Union Carbide Y-9567, Y-9794

Organosilicone copolymer
Silwet Surface Active Copolymer L-77, L-720, L-7001, L-7002, L-7004, L-7500, L-7600, L-7604, L-7605, L-7607; Union Carbide L-720, -721, -722, -727, -7001, -7002, -7500, -7600, -7602, -7604, -7605, -7607

Oxalyl bis (benzylidenehydrazide)
Eastman Inhibitor OABH

2,2′-Oxamido bis[ethyl 3-(3,5-di-t-butyl-4-hydroxyphenyl) propionate]
Naugard XL-1

Oxazolidine [CAS #6542-37-6; CAS #7747-35-5]
Amine CS-1135; Bioban CS-1135, N-95; Zoldine ZE, ZT-55

8-α-12-Oxido-13,14,15,16 tetra-norlabdane
Ambroxan

Oxime
Troykyd Anti-Skin, Anti-Skin Odorless Powd.

4,4′-Oxybis (benzenesulfonylhydrazide) [CAS #80-51-3]
Nitropore OBSH

10,10′-Oxybisphenoxyarsine [CAS #58-36-6]; *Synonyms:* OBPA
Intercide 2 DIDP, ABF, ABF 1 ESBO, ABF 2 DIDP, ABF 2 ESBO; Vinyzene BP-5-2-MEK, BP-5-2-MS, BP-505, BP-505 DIDP, BP-505 DOP, BP-505 S160, SB-1-PS

Oxydemeton-methyl
Metasystox R

N-Oxydiethylene 2-benzothiazole sulfenamide [CAS #102-77-2] *Synonyms:* 2-(Morpholinothio) benzothiazole
Amax; Amax No. 1; OBTS; OMTS

Ozokerite (CTFA) [CAS #8021-55-4]
Lexate PX; Protegin, W, WX, X

Palladium [CAS #7440-05-3]
Pd-0803 E 1/16″

Palmamidopropyl betaine (CTFA)
Schercotaine PAB

Palmitamidopropyl betaine (CTFA) [CAS #32954-43-1]
Incronam P-30

Palmitamidopropyl dimethylamine (CTFA) [CAS #39669-97-1]
Incromine PB; Lexamine P-13; Schercodine P

Palmitamidopropyl trimonium chloride
Servamine KEP 4527

Palmitamine (CTFA) [CAS #143-27-1] *Synonyms:* Cetyl amine; Palmityl amine; Hexadecylamine
Armeen 16D; Crodamine 1.16D; Kemamine P-880, P-880D; Nissan Amine PB

Palmitamine oxide (CTFA) [CAS #7128-91-8; RD #977066-47-9] *Synonyms:* Cetamine oxide; Cetyl dimethyl amine oxide; Palmityl dimethylamine oxide; Hexadecyl dimethyl amine oxide
Ammonyx CO; Aromox DM16, DM16D-W; Conco XA-C; Jordamox CDA, CDA-40; Sandoz Amine XA-C

Palmitic acid (CTFA) [CAS #57-10-3] *Synonyms:* Hexadecanoic acid; Cetylic acid
Emersol 143; Hystrene 4516, 9016; Industrene 4516

Palmitic/stearic acid glycerol monodiester
Imwitor 945

Palmityl. See also Cetyl

Palmityl dimethyl amine. See Dimethyl palmitamine

Palmityl glyceryl ether
Nikkol Chimyl Alcohol 100

Palm kernel alcohol (CTFA)
Laurex PKH

Palm kernelamide DEA (CTFA) [RD #977069-44-5] *Synonyms:* Palm kernel oil acid diethanolamide; Diethanolamine palm kernel oil acid amide; N,N-bis (2-hydroxyethyl) palm kernel oil acid amide
Accomid PK; Cyclomide DP240/S; Mackamide PK

Palm kernel glycerides (CTFA)
Grindtek PK 60

Palm oil glyceride (CTFA) [RD #977013-38-9]
Monomuls 90-30

Palm oil glycerides (CTFA)
Imwitor 940, 965; Monomuls 60-30

Paraffin (CTFA) [CAS #8002-74-2] *Synonyms:* Paraffin wax; Paraffin scale; Petroleum wax, crystalline
Antilux 110, 111, 500, 550, 600, 620, 654, 660, L; Cremba B6; Dry Size XL20C; GP-090, -094; Interwax G 8208, G 8268; Jonwax 120; Loxiol G 22; Paracol 404A, 404D, 404G, 404N, 447K, 505A, 505G, 505N, 800N, 802A, 802N, 802NW, 810N, 810NP, 815N, 1886; Shellwax; Sonojell No. 4, 9

Paraffin, brominated
DD-8126

Paraffin, bromochlorinated
DD-8207, -8307; Fyarestor 100

Paraffin, chlorinated
Cereclor 42, 42P, 50LV, 51L, 52P, 70L, AP45, AP52, LP4446, LP4985, S45, S52; Chlorez 700, 700-DF, 760; Clorafin 40, 50; CPF-0001, -0003, -0008, -0019, -0022; Flexchlor 0001, 0002, 0008, 0009, 0010, 0011, 0012, 0018, 0023; Porofor B 13/CP 50

Paraffin oil
Porofor BSH Paste, BSH Paste M

Paraffin wax. See Paraffin

Paraffin-rosin emulsion
Paracol 403A6, 803A6, 803G6

Parathion
Folidol-E605

Parathion-methyl
Folidol M

Pareth 15-.... See C11-15 pareth-...

Pareth 23-.... See C12-13 pareth-...

Pareth 25-.... See C12-15 pareth-...

Pareth 45-.... See C14-15 pareth-...

Pareth 91-.... See C9-11 pareth-...

PCA (CTFA) [CAS #98-79-3; 1698-60-8] *Synonyms:* 2-Pyrrolidone-5-carboxylic acid; 5-Oxo-L-proline; L-Proline, 5-oxo; L-Pyroglutamic acid
Ajidew A-100

PCA 1/2 Na
Ajidew SP-100

PCC
Ultra-Pflex

Peanutamide MEA (CTFA) [RD #977066-48-0] *Synonyms:* N-(2-hydroxyethyl) peanut acid amide
Rewomid OM 101/G

Peanutamide MIPA (CTFA) *Synonyms:* N-(2-Hydroxypropyl) peanut acid amide
Rewomid OM 101/IG

Peanut oil [CAS #8002-03-7] *Synonyms:* Arachis oil
Super Refined Peanut Oil

Pectin (CTFA) [CAS #9000-69-5]
Frimulsion 6G, Q8, RA; Genu 04CG, 04CB, 12CG, 12CB, 18CG, 18CB, 20AS, 20AB, 21AS, 21AB, 102AS, 104AS, 15AB, AA Med.-Rapid Set, BA-KING, BB Rapid Set, DD Extra-Slow Set, DD Extra-Slow Set C, DD Slow Set, JM, Pectin

Pectinase
Calrex L; Extractase L5X, P15X; Klerzyme; Pectinol 59L, 60G, 80SB, R10; Spark-L HPG

PEG-4 (CTFA) [CAS #25322-68-3 (generic); 112-60-7; RD #977007-39-8] *Synonyms:* PEG 200
Alkapol PEG-200; Carbowax PEG 200; Macol E-200; Pluracol E-200

PEG-6 (CTFA) [CAS #25322-68-3 (generic); 2615-15-8; RD #977007-40-1] *Synonyms:* PEG 300
Alkapol PEG-300; Carbowax PEG 300, PEG 540 Blend; Lutrol E 300; Macol E-300; Pluracol E-300

PEG-6-32 (CTFA) [CAS #25322-68-3 (generic); RD #977053-63-6] *Synonyms:* PEG 1500
Alkapol PEG-1500; Lutrol E 1500

PEG-8 (CTFA) [CAS #25322-68-3 (generic); 5117-19-1; RD #977007-41-2] *Synonyms:* PEG 400
Carbowax PEG 400; Emery 6709; Lutrol E 400; Macol E-400; Pluracol E-400, E-400 NF; Texapon SG

PEG-9 (CTFA) [CAS #25322-68-3 (generic)] *Synonyms:* PEG 450
Alkapol PEG-400

PEG-12 (CTFA) [CAS #25322-68-3 (generic); RD #977053-61-4] *Synonyms:* PEG 600
Carbowax PEG 600; Emery 6686; Macol E-600; Pluracol E-600, E-600 NF

PEG-14 (CTFA) [CAS #25322-68-3 (generic); RD #977059-82-7]
Alkapol PEG-600

PEG-20 (CTFA) [CAS #25322-68-3 (generic); RD #977053-62-5] *Synonyms:* PEG 1000
Alkapol PEG-1000; Carbowax PEG 900, 1000; Macol E-1000; Pluracol E-1000

PEG-32 [CAS #25322-68-3 (generic); RD #977007-42-3] *Synonyms:* PEG 1540
Carbowax PEG 540 Blend, PEG 1450; Macol E-1450

PEG-40 (CTFA) [CAS #25322-68-3 (generic); RD #977065-62-5] *Synonyms:* PEG 2000
Incroquat CR Conc.; Pluracol E-2000

PEG-75 (CTFA) [CAS #25322-68-3 (generic); RD #977007-43-4] *Synonyms:* PEG 4000
Carbowax PEG 3350; Lutrol E 4000; Macol E-3350; Pluracol E-4000, E-4000 NF

PEG-100
Carbowax PEG 4600

PEG-150 (CTFA) [CAS #25322-68-3 (generic); RD #977053-64-7; PEG 6000
Alkapol PEG-6000; Carbowax PEG 8000; Lutrol E 6000; Macol E-8000

PEG-200 (CTFA) [CAS #25622-68-3 (generic); RD #977063-53-8] *Synonyms:* Polyethylene glycol 9000
Emery 6773

PEG-350
Carbowax Compound 20M

PEG 100.... See PEG-2...

PEG 200... See PEG-4...

PEG 300... See PEG-6...

PEG 400... See PEG-8...

PEG 450... See PEG-9...

PEG 500... See PEG-10...

PEG 600... See PEG-12...

PEG 800... See PEG-16...

PEG 1000... See PEG-20...

PEG 1500... See PEG-6-32...

PEG 1540... See PEG-32...

PEG 1800... See PEG-36...

PEG 2000... See PEG-40...

PEG 3350
Alkapol PEG-3350

PEG 4000... See PEG-75...

PEG 6000... See PEG-150...

PEG 8000
Alkapol PEG-8000

PEG-6 abietate
Secoster MA 300

PEG-3 aniline [CAS #36356-83-9]
Emery 5702

PEG-100 aniline [CAS #36356-83-9]
Emery 6727, 6749

PEG-200 aniline [CAS #36356-83-9]
Emery 5774

PEG-2 butyl ether. See Butoxy diglycol (CTFA)

PEG-3 butynediol
Hetoxide BY-3

PEG-6.3 C12, C14 alcohol [CAS #68439-50-9]
Tergitol 24-L-45

PEG-7 C12, C14 alcohol
Tergitol 24-L-60N

PEG-7.2 C12, C14 alcohol
Tergitol 24-L-60

PEG-8.3 C12, C14 alcohol
Tergitol 24-L-75

PEG-10.6 C12, C14 alcohol
Tergitol 24-L-92

PEG-11.3 C12, C14 alcohol
Tergitol 24-L-98N

PEG-1.5 C12,C14,C16 alcohol [CAS #68551-12-02]
Tergitol 26-L-1.6

PEG-3 C12,C14,C16 alcohol
Tergitol 26-L-3

PEG-5 C12,C14,C16 alcohol
Tergitol 26-L-5

PEG-2 C13-15 alkyl amine
Synprolam 35X2

PEG-5 C13-15 alkyl amine
Synprolam 35X5

PEG-10 C13-15 alkyl amine
Synprolam 35X10

PEG-15 C13-15 alkyl amine
Synprolam 35X15

PEG-20 C13-15 alkyl amine
Synprolam 35X20

PEG-25 C13-15 alkyl amine
Synprolam 35X25

PEG-35 C13-15 alkyl amine
Synprolam 35X35

PEG-50 C13-15 alkyl amine
Synprolam 35X50

PEG-3 caprylate/caprate
Plasthall 4141

PEG-6 caprylic/capric glycerides (CTFA)
Softigen 767

PEG-2 castor oil (CTFA) *Synonyms:* POE (2) castor oil; PEG 100 castor oil
Hetoxide C-2

PEG-3 castor oil (CTFA) [CAS #61791-12-6 (generic)]
Nikkol CO-3

PEG-5 castor oil (CTFA) [CAS #61791-12-6 (generic)]
Ablunol CO5; Acconon CA-5; Alkasurf CO-5; Chemax CO-5; Surfactol 318; Trylox 5900, CO-5

PEG-6 castor oil
DeSonic 6C

PEG-8 castor oil
Acconon CA-8

PEG-9 castor oil (CTFA) [CAS #61791-12-6 (generic); RD #977065-86-3]
Acconon CA-9; Hetoxide C-9; Lipal 9 C

PEG-10 castor oil (CTFA) [CAS #61791-12-6 (generic)]
Ablunol CO10; Alkasurf CO-10; Etocas 10; Incrocas 10; Nikkol CO-10

PEG-12 castor oil
Teric C12

PEG-15 castor oil
Ablunol CO15; Acconon CA-15; Alkasurf CO-15; Hetoxide C-15

PEG-16 castor oil
Trylox 5902, CO-16

PEG-20 castor oil (CTFA) [CAS #61791-12-6 (generic)]
Alkasurf CO-20; Nikkol CO-20TX

PEG-25 castor oil (CTFA) [CAS #61791-12-6 (generic); RD #977063-56-1]
Acconon CA-25; Alkasurf CO-25; Chemax CO-25; Hetoxide C-25; Lipal 25 C; Mapeg CO-25; Trylox 5904, CO-25

PEG-28 castor oil
Chemax CO-28

PEG-30 castor oil (CTFA) [CAS #61791-12-6 (generic); RD#977059-28-1]
Ablunol CO30; Alkasurf CO-30; Chemax CO-30; DeSonic 30C; Emulphor EL-620; Etocas 30; Hetoxide C-30; Incrocas 30; Mapeg CO-30; Trylox 5906, CO-30

PEG-35 castor oil
Etocas 35

PEG-36 castor oil (CTFA) [CAS #61791-12-6 (generic); RD #977054-99-1]
Chemax CO-36; Cremophor EL; DeSonic 36C; Mapeg CO-36; Trylox 5907, CO-36

PEG-40 castor oil (CTFA) [CAS #61791-12-6 (generic); RD #9777055-00-7]
Alkasurf CO-40, CO-40M; Chemax CO-40; DeSonic 40C; Emulphor EL-719; Etocas 40; Hetoxide C-40; Incrocas 40; Maquat SC-1632; Nikkol CO-40TX; Surfactol 365; T-Det C-40; Trylox 5909, CO-40

PEG-45 castor oil
Ablunol CO45

PEG-52 castor oil
Lipal 52 C

PEG-54 castor oil
DeSonic 54C

PEG-60 castor oil (CTFA) [CAS #61791-12-6 (generic); RD #977065-79-4]
Etocas 60; Incrocas 60; Nikkol CO-60TX

PEG-80 castor oil
Chemax CO-80; Trylox CO-80

PEG-100 castor oil (CTFA) [CAS #61791-12-6 (generic); RD #977063-24-3]
Etocas 100; Incrocas 100

PEG-200 castor oil (CTFA) [CAS #61791-12-6 (generic); RD #977063-54-9]
Alkasurf CO-200; Chemax CO-200/50; Hetoxide C-200, C-200-50%; Mapeg CO-200; Trylox 5918, CO-200, -200/50

PEG 9000 castor oil
Pegosperse 9000 CO

PEG cetyl ethers. See Ceteth Series

PEG cetyl/oleyl ethers. See Cetoleth Series

PEG-2 cetyl/stearylamine
Varonic U202

PEG-5 cetyl/stearylamine
Varonic U205, U205LC

PEG-15 cetyl/stearylamine
Varonic U215

PEG-50 cetyl/stearylamine
Varonic U250, U250-50%

PEG cetyl/stearyl ethers. See Ceteareth Series

PEG-3 cocamide (CTFA) [CAS #61791-08-0 (generic); RD #977061-46-3] *Synonyms:* PEG (5) coconut amide; POE (5) coconut amide; PEG-6 cocoyl amide
Alkamidox CME-2; Unamide C-2

PEG-6 cocamide (CTFA) [CAS #61791-08-0 (generic); RD #977061-47-4]
Alkamidox CME-5; Empilan MAA; Unamide C-5

PEG-6 cocamide MEA
Rewopal C 6; Teric CME3

PEG-7 cocamide MEA
Teric CME7

PEG-2 cocamine (CTFA) [CAS #61791-14-8 (generic); 61788-46-3; RD #977061-61-2] *Synonyms:* PEG 100 coconut amine; POE (2) coconut amine

Accomeen C2; Alkaminox C2; Chemeen C-2; Ethomeen C/12; Hetoxamine C-2; Mazeen C 2; Teric 12M2; Varonic K202

PEG-5 cocamine (CTFA) [CAS #61791-14-8 (generic); RD #977066-78-6]
Accomeen C5; Alkaminox C-5; Chemeen C-5; Ethomeen C/15; Hetoxamine C-5; Mazeen C 5; Teric 12M5; Varonic K205, K205LC

PEG-10 cocamine (CTFA) [CAS #61791-14-8 (generic); RD #977063-21-0]
Accomeen C10; Alkaminox C-10; Chemeen C-10; Ethomeen C/20; Mazeen C 10; Trymeen 6601, CAM-10; Varonic K210, K210LC

PEG-15 cocamine (CTFA) [CAS #8051-52-3; 61791-14-8 (generic)]
Accomeen C15; Chemeen C-25; Hetoxamine C-15; Mazeen C 15; Teric 12M15; Trymeen CAM-15; Varonic K215, K215LC

PEG-75 cocoa butter
Emthox 2730

PEG-5 cocoate (CTFA) [CAS #61791-29-5 (generic); RD #977065-64-7]
Ethofat C/15; Lipal CE 38

PEG-8 cocoate (CTFA) [CAS #61791-29-5 (generic); RD #977055-01-8]
Pegosperse 400 MC

PEG-10 cocoate *Synonyms:* PEG 500 cocoate
Nonex C5E

PEG-11 cocoate
Lipal CE 55

PEG-15 cocoate (CTFA) [CAS #61791-29-5 (generic); RD #977065-34-1; 977059-64-5]
Ethofat C/25

PEG-16 cocoate
Lipal CE 64

PEG-21 cocoate
Lipal CE 71

PEG-2 cocomonium chloride (CTFA) [RD #977066-72-0]
Ethoquad C/12; Variquat 638

PEG-15 cocomonium chloride (CTFA) [RD #977066-75-3]
Ethoquad C/25

PEG-15 cocopolyamine (CTFA)
Polyquart H-7102

PEG decyl ethers. See Deceth Series

PEG-5 DEDM hydantoin (CTFA) *Synonyms:* PEG-5 di-(2-hydroxyethyl)-5,5-dimethyl hydantoin
Dantocol DHE (5)

PEG-10 DEDM hydantoin *Synonyms:* PEG-10 di-(2-hydroxyethyl)-5,5-dimethyl hydantoin
Dantocol DHE (10)

PEG-15 DEDM hydantoin (CTFA) *Synonyms:* PEG-15 di-(2-hydroxyethyl)-5,5-dimethyl hydantoin
Dantocol DHE (15)

PEG-20 DEDM hydantoin *Synonyms:* PEG-20 di-(2-hydroxyethyl)-5,5-dimethyl hydantoin
Dantocol DHE (20)

PEG-5 DEDM hydantoin oleate (CTFA) *Synonyms:* PEG-5 di-(2-hydroxyethyl)-5,5-dimethyl hydantoin oleate
Dantosperse DHE (5) MO

PEG-10 DEDM hydantoin oleate (CTFA) *Synonyms:* PEG-10 di-(2-hydroxyethyl)-5,5-dimethyl hydantoin oleate
Dantosperse DHE (10) MO

PEG-15 DEDM hydantoin oleate (CTFA) *Synonyms:* PEG-15 di-(2-hydroxyethyl)-5,5-dimethyl hydantoin oleate
Dantosperse DHE (15) MO

PEG-20 DEDM hydantoin oleate *Synonyms:* PEG-20 di-(2-hydroxyethyl)-5,5-dimethyl hydantoin oleate
Dantosperse DHE (20) MO

PEG-3 diacetate *Synonyms:* Triethylene glycol diacetate
Sartomer 322

PEG-3 diacrylate
SR-272

PEG-4 diacrylate
SR-259, -268

PEG-8 diacrylate
SR-344

PEG-2 dibenzoate [CAS #120-55-8] *Synonyms:* Diethylene glycol dibenzoate; Benzoic acid, diester with diethylene glycol; Dibenzoyldiethyleneglycol ester
Benzoflex 2-45, 50

PEG-4 dibenzoate
Benzoflex P-200

PEG-2 dibutyl ether *Synonyms:* Diethylene glycol dibutyl ether
Butyl Diglyme

PEG-3 dicaprate-caprylate
Merrol 3810

PEG-3 dicaprylate/caprate
Tricap

PEG-3 di-2-ethylhexoate
TegMeR 803

PEG-4 di-2-ethylhexoate
Merrol 4800; TegMeR 804, 804 Special

PEG-8 di-2-ethylhexoate
TegMeR 809

PEG-3 diheptanoate
TegMeR 703

PEG-4 diheptanoate
TegMeR 704

PEG-2 dilaurate *Synonyms:* Diethylene glycol dilaurate; POE (2) dilaurate; PEG 100 dilaurate
Cithrol DGDL N/E, DGDL S/E

PEG-4 dilaurate (CTFA) [CAS #9005-02-1 (generic); RD #977055-02-9]
Acconon 200-DL; Alkamuls 200-DL; Cithrol 2DL; Cyclochem PEG 200 DL; Emerest 2622, 2704; Ionet DL-200; Kessco PEG 200 Dilaurate; Lexomul PEG 200 DL; Lipopeg 2-DL; Mapeg 200 DL; Nonex DL-2; Pegosperse 200 DL

PEG-6 dilaurate (CTFA) [CAS #9005-02-1 (generic); RD #977055-03-0]
Kessco PEG 300 Dilaurate; Lipal 300 DL

PEG-8 dilaurate (CTFA) [CAS #9005-02-1 (generic); RD #977055-04-1]
Alkamuls 400-DL; Cithrol 4DL; CPH-79-N; Emerest 2652; Kessco PEG 400 Dilaurate; Lexomul PEG 400 DL; Lipal 400 DL; Pipopeg 4-DL; Mapeg 400 DL; Pegosperse 400 DL

PEG-12 dilaurate (CTFA) [CAS #9005-02-1 (generic); RD #977063-27-6]
Alkamuls 600-DL; Cithrol 6DL; Kessco PEG 600 Dilaurate; Mapeg 600 DL

PEG-20 dilaurate (CTFA) [CAS #9005-02-1 (generic); RD #977063-47-0]
Cithrol 10DL; Kessco PEG 1000 Dilaurate

PEG-32 dilaurate (CTFA) [CAS #9005-02-1 (generic); RD #977065-53-4]
Kessco PEG 1540 Dilaurate

PEG-75 dilaurate (CTFA) [CAS #9005-02-1 (generic); RD #977065-82-9]
Kessco PEG 4000 Dilaurate

PEG-150 dilaurate (CTFA) [CAS #9005-02-1 (generic); RD #977055-05-2]
Kessco PEG 6000 Dilaurate

PEG-3 dimethacrylate
SR-205

PEG-4 dimethacrylate
SR-209

PEG-3 dimethyl ether
Triglyme

PEG-4 dimethyl ether
Tetraglyme

PEG dinonyl phenyl ethers. See Nonyl nonoxynol Series

PEG-2 dioleate *Synonyms:* Diethylene glycol dioleate; POE (2) dioleate; PEG 100 dioleate
Alkamuls DEG-DO; Cithrol DGDO N/E, DGDO S/E

PEG-4 dioleate (CTFA) [CAS #9005-07-6 (generic); 52688-97-0 (generic); RD #977065-57-8]
Alkamuls 200-DO; Cithrol 2DO; Ionet DO-200; Kessco PEG 200 Dioleate; Mapeg 200 DO

PEG-6 dioleate (CTFA) [CAS #9005-07-6 (generic); 52688-97-0 (generic); RD #977055-06-3]
Kessco PEG 300 Dioleate

PEG-6-32 dioleate (CTFA) [CAS #9005-07-6 (generic); 52688-97-0 (generic); RD #977065-78-3]
Pegosperse 1500 DO

PEG-8 dioleate (CTFA) [CAS #9005-07-6 (generic); 52688-97-0 (generic); RD #977051-75-4]

Acconon 400-DO; Alkamuls 400-DO; Chemax PEG 400 DO; Cithrol 4DO; CPH-170-N, -211-N; Emerest 2648; Ionet DO-400; Kessco PEG 400 Dioleate; Lexomul PEG 400 DO; Lipal 400 DW; Lipopeg 4-DO; Lonzest PEG-4DO; Mapeg 400 DO; Nonex DO-4; Nonisol 210; Pegosperse 400 DO; Radiasurf 7443

PEG-9 dioleate
Trydet DO-9

PEG-12 dioleate (CTFA) [CAS #9005-07-6 (generic); 52688-97-0 (generic)]
Alkamuls 600-DO; Cithrol 6DO; CPH-213-N; Emerest 2665; Ionet DO-600; Kessco PEG 600 Dioleate; Mapeg 600 DO; Secoster DO 600

PEG-20 dioleate (CTFA) [CAS #9005-07-6 (generic); 52688-97-0 (generic); RD #977063-48-1]
Cithrol 10DO; Ionet DO-1000; Kessco PEG 1000 Dioleate

PEG-32 dioleate (CTFA) [CAS #9005-07-6 (generic); 52688-97-0 (generic); RD #977065-54-5]
Kessco PEG 1540 Dioleate

PEG-75 dioleate (CTFA) [CAS #9005-07-6 (generic); 52688-97-0 (generic); RD #977065-83-0]
Kessco PEG 4000 Dioleate

PEG-150 dioleate (CTFA) [CAS #9005-07-6 (generic); 52688-97-0 (generic); RD #977063-37-8]
Kessco PEG 6000 Dioleate

PEG-3 dipelargonate [CAS #106-06-9] *Synonyms:* Triethylene glycol dipelargonate
Merrol 3900; Plastolein 9404 TGP2; TegMeR 903

PEG-2 distearate (CTFA) [CAS #109-30-8; 9005-08-7 (generic)] *Synonyms:* Diethylene glycol distearate; POE (2) distearate; PEG 100 distearate
Cithrol DGDS N/E, DGDS S/E

PEG-3 distearate (CTFA) [CAS #9005-08-7; RD #977063-57-2] *Synonyms:* Triglycol distearate; Triethylene glycol distearate
Euperlan PK 900; Genapol TS Powd.; Nikkol Estepearl 30, 35

PEG-4 distearate (CTFA) [CAS #9005-08-7 (generic); RD #977065-58-9]
Alkamuls 200-DS; Cithrol 2DS; Kessco PEG 200 Distearate; Mapeg 200 DS

PEG-6 distearate (CTFA) [CAS #9005-08-7 (generic); RD #977059-16-7]
Cithrol 3DS; Ionet DS-300; Kessco PEG 300 Distearate

PEG-8 distearate (CTFA) [CAS #9005-08-7 (generic); RD #977053-29-4]
Alkamuls 400-DS; Cithrol 4DS; Cyclochem PEG 400 DS; Emerest 2642, 2712; Ionet DS-400; Kessco PEG 400 Distearate; Lipal 400 DS; Lipopeg 4-DS; Mapeg 400 DS, 400 DSLM; Pegosperse 400 DS; Radiasurf 7453

PEG-12 distearate (CTFA) [CAS #9005-08-7 (generic); RD #977055-07-4]
Alkamuls 600-DS; Cithrol 6DS; Kessco PEG 600 Distearate; Mapeg 600 DS; Radiasurf 7454

PEG-20 distearate (CTFA) [CAS #9005-08-7 (generic); RD #977055-08-5]
Cithrol 10DS; Kessco PEG 1000 Distearate

PEG-32 distearate (CTFA) [CAS #9005-08-7 (generic); RD #977055-09-6]
Kessco PEG 1540 Distearate; Mapeg 1540 DS

PEG-75 distearate (CTFA) [CAS #9005-08-7 (generic); RD #977055-10-9]
Kessco PEG 4000 Distearate

PEG-150 distearate (CTFA) [CAS #9005-08-7 (generic); RD #977055-11-0]
Ablunol 6000DS; Chimipal DS 6000; Emulvis; Kessco PEG 6000 Distearate; Lipopeg 6000-DS; Mapeg 6000 DS; Pegosperse 6000 DS; Plurest DS 150; Rewopal PEG 6000 DS; Secoster DS 6000; Witconol L32-45

PEG-2 ditallate *Synonyms:* Diethylene glycol ditallate; POE (2) ditallate; PEG 100 ditallate
Alkamuls DEG-DT

PEG-4 ditallate
Mapeg 200 DOT

PEG-8 ditallate (CTFA) [CAS #61791-01-3 (generic); RD #977066-81-1]
Mapeg 400 DOT

PEG-12 ditallate (CTFA) [CAS #61791-01-3 (generic); RD #977063-28-7]
Mapeg 600 DOT

PEG-8 ditriricinoleate (CTFA) [RD #977055-12-1]
Pegosperse 400 DTR

PEG-22/dodecyl glycol copolymer (CTFA)
Elfacos ST 37

PEG-45/dodecyl glycol copolymer (CTFA)
Elfacos ST 9

PEG dodecyl phenyl ethers. See Dodoxynol Series

PEG-25 glycerin pyroglutamic isostearic diester
Pyroter GPI-25

PEG-7 glyceryl cocoate (CTFA) [RD #977064-68-8] *Synonyms:* POE (7) glyceryl monococoate; PEG (7) glyceryl monococoate
Cetiol HE; Mazol 159; Standamul HE

PEG-30 glyceryl cocoate (CTFA)
Varonic LI-63

PEG-78 glyceryl cocoate [CAS #68201-46-7]
Varonic LI-67, LI-67-75

PEG glyceryl ethers. See Glycereth Series

PEG-15 glyceryl isostearate (CTFA)
Oxypon 2145

PEG-12 glyceryl laurate (CTFA) [CAS #59070-56-3 (generic)]
Lamacit GML 12

PEG-15 glyceryl laurate
Glycerox L15

PEG-20 glyceryl laurate (CTFA) [CAS #59070-56-3 (generic); RD #977065-41-0]
Lamacit GML 20; Tagat L-2

PEG-23 glyceryl laurate
Aldosperse ML-23, ML-230

PEG-30 glyceryl laurate (CTFA) [CAS #59070-56-3 (generic); RD #977065-48-7]
Tagat L

PEG-20 glyceryl oleate (CTFA) [CAS #68889-49-6 (generic); RD #977065-42-1]
Tagat O 2

PEG-25 glyceryl oleate (CTFA) [CAS #68889-49-6 (generic)]
Lamacit GMO 25

PEG-30 glyceryl oleate (CTFA) [CAS #68889-49-6 (generic); RD #977065-49-8]
Tagat O

PEG-15 glyceryl ricinoleate (CTFA) [CAS #51142-51-9 (generic); RD #977065-35-2]
Tagat R 1

PEG-20 glyceryl ricinoleate (CTFA) [CAS #51142-51-9 (generic); RD #97065-43-2]
Lamacit CR

PEG-20 glyceryl stearate (CTFA) [RD #977055-13-2] *Synonyms:* PEG 1000 glyceryl stearate; POE (20) glyceryl stearate
Aldo MS-20 FG; Aldosperse MS-20; Durfax EOM; Radiasurf 7000; Tagat S 2; Tally 100 Plus

PEG-30 glyceryl stearate (CTFA) [RD #977065-50-1]
Tagat S

PEG-120 glyceryl stearate (CTFA) [RD #977063-30-1]
Drewmulse 1128

PEG-20 glyceryl tallowate
Varonic LI-42

PEG-80 glyceryl tallowate
Varonic LI-48

PEG-200 glyceryl tallowate (CTFA)
Varonic LI-420

PEG-25 glyceryl trioleate (CTFA) [RD #977014-36-0]
Tagat TO

PEG-5 hydrogenated castor oil (CTFA) [CAS #61788-85-0 (generic); RD #977065-66-9] *Synonyms:* POE (5) hydrogenated castor oil; PEG (5) hydrogenated castor oil
Chemax HCO-5; Nikkol HCO-5; Trylox HCO-5

PEG-7 hydrogenated castor oil (CTFA) [CAS #61788-85-0 (generic)]
Arlacel 989; Nikkol HCO-7.5 (7.5 EO)

PEG-10 hydrogenated castor oil
Croduret 10; Nikkol HCO-10

PEG-16 hydrogenated castor oil (CTFA) [CAS #61788-85-0 (generic); RD #977065-36-3]
Chemax HCO-16; Hetoxide HC-16; Mapeg CO-16H; Trylox 5921, HCO-16

PEG-20 hydrogenated castor oil (CTFA) [CAS #61788-85-0 (generic)]
Nikkol HCO-20

PEG-25 hydrogenated castor oil (CTFA) [CAS #61788-85-0 (generic); RD #977055-14-3]
Arlatone G; Chemax HCO-25; Mapeg CO-25H; Trylox HCO-25

PEG-30 hydrogenated castor oil (CTFA) [CAS #61788-85-0 (generic); RD #977069-47-8]
Croduret 30; Nikkol HCO-30, -30(FF)

PEG-40 hydrogenated castor oil (CTFA) [CAS #61788-85-0 (generic); RD #977055-15-4]
Cremophor RH 40, RH 410, RH 455; Croduret 40; Eumulgin RO-40, HRE 40; Hetoxide HC-40; Nikkol HCO-40, HCO-40(FF); Standamul Conc. 1002; Tagat R40

PEG-50 hydrogenated castor oil (CTFA) [CAS #61788-85-0 (generic)]
Croduret 50 Special; Nikkol HCO-50, HCO-50(FF)

PEG-60 hydrogenated castor oil (CTFA) [CAS #61788-85-0 (generic); RD #977060-08-4]
Cremophor RH 60; Croduret 60; Eumulgin HRE 60; Hetoxide HC-60; Nikkol HCO-60, HCO-60(FF); Tagat R60

PEG-80 hydrogenated castor oil (CTFA) [CAS #61788-85-0 (generic)]
Nikkol HCO-80

PEG-100 hydrogenated castor oil (CTFA) [CAS #61788-85-0 (generic); RD #977063-25-4]
Croduret 100; Nikkol HCO-100, HCO-100(FF)

PEG-200 hydrogenated castor oil (CTFA) [CAS #61788-85-0 (generic); RD #977063-55-0]
Chemax HCO-200/50; Trylox HCO-200/50

PEG-40 hydrogenated castor oil pyroglutamic isostearic diester
Pryoter CPI-40

PEG-20 hydrogenated lanolin (CTFA) [CAS #68648-27-1 (generic); RD #977055-17-6]
Satexlan; Super-Sat AWS-4

PEG-24 hydrogenated lanolin (CTFA) [CAS #68648-27-1 (generic); RD #977055-18-7]
Lipolan 31

PEG-3 hydrogenated tallow propylene diamine
Dinoramox SH 3

PEG isocetyl ethers. See Isoceteth Series

PEG isodecyl ethers. See Isodeceth Series

PEG-8 isolauryl thioether (CTFA)
Siponic SK

PEG-4 isostearate *Synonyms:* POE (4) monoisostearate; PEG 200 monoisostearate
Emerest 2625; Trydet ISA-4

PEG-8 isostearate
Emerest 2644; Trydet 2644

PEG-9 isostearate
Trydet ISA-9

PEG-12 isostearate (CTFA) [CAS #56002-14-3 (generic); RD #977063-29-8]
Emerest 2664

PEG isostearyl ether. See Isosteareth Series

PEG-5 lanolate (CTFA) [CAS #68459-50-7 (generic); RD#977061-92-9] *Synonyms:* PEG-5 lanolin acids
Lanpol 5; Lanpolamide 5

PEG-10 lanolate (CTFA) [CAS #68459-50-7 (generic); RD#977061-91-8]
Lanpol 10

PEG-20 lanolate (CTFA) [CAS #68459-50-7 (generic)]
Lanpol 20

PEG-5 lanolin (CTFA) [CAS #61790-81-6 (generic)] *Synonyms:* PEG (5) lanolin derivative; POE (5) lanolin derivative
Nikkol BWA-5 (PEG-5 lanolin alcohol)

PEG-10 lanolin (CTFA)
Nikkol TW-10

PEG-20 lanolin (CTFA) [CAS #61790-81-6 (generic); RD #977058-73-3]
Nikkol BWA-20 (PEG-20 lanolin alcohol); Nikkol TW-20

PEG-27 lanolin (CTFA) [CAS #61790-81-6]
Lanogel 21

PEG-30 lanolin (CTFA) [CAS #61790-81-6 (generic); RD #977065-51-2]
Aqualose L30; Nikkol TW-30

PEG-40 lanolin (CTFA) [CAS #8051-82-9; 61790-81-6 (generic)]
Lan-Aqua-Sol Hydrophilic 50, 100; Laneto 40; Lanogel 31; Nikkol BWA-40 (PEG-40 lanolin alcohol)

PEG-50 lanolin (CTFA) [CAS #61790-81-6 (generic); RD #977061-54-3]
Lan-Aqua-Sol Hydrophilic-Plus 50, 100; Solangel 501

PEG-60 lanolin (CTFA) [CAS #61790-81-6 (generic); RD #977059-06-5]
Solan, 50

PEG-70 lanolin
Teric LAN70

PEG-75 lanolin (CTFA) [CAS #8039-09-6; 61790-81-6 (generic)]
Aqualose L75, L75/50; Cralane LR-10; Ethoxylan 1685, 1686; Lan-Aqua-Sol xtra-Hydrophilic 50, 100; Laneto 100; Lanogel 41; Solangel 401; Solulan 75, L-575; Super Solan Flaked; Super Solangel 25

PEG-85 lanolin (CTFA) [CAS #61790-81-6 (generic); RD #977055-21-2]
Lan-Aqua-Sol Super-Hydrophilic 50, 100; Lanogel 61; Solangel 851

PEG-5 lanolinamide
Lanpolamide 5

PEG lanolin ethers. See Laneth Series

PEG-75 lanolin oil (CTFA) [CAS #68648-38-4 (generic); RD #977055-20-1]
Lanotein AWS 30; Vigilan AWS

PEG-3 lauramide (CTFA) [CAS #26635-75-6 (generic); RD #977061-49-6] *Synonyms:* POE (3) lauryl amide; PEG (3) lauryl amide
Unamide L-2

PEG-6 lauramide (CTFA) [CAS #26635-75-6 (generic); RD #977061-48-5]
Unamide L-5

PEG-3 lauramine oxide (CTFA) *Synonyms:* POE (3) lauryl dimethyl amine oxide; PEG (3) lauryl dimethyl amine oxide
Empigen OY

PEG-2 laurate (CTFA) [CAS #141-20-8; SE grade RD #977063-42-5] *Synonyms:* Diethylene glycol laurate; Diglycol laurate; PEG 100 monolaurate; POE (2) monolaurate
Alkamuls DEG-ML; Alkasurf L-2; Cithrol DGML N/E, DGML S/E; Lipo DGLS (SE), Diglycol Laurate; Lipocol B (SE); Mapeg DGLD; Pegosperse 100 L, 100 ML; Radiasurf 7420, 7421 (SE); Sole-Onic CDS

PEG-4 laurate (CTFA) [CAS #9004-81-3 (generic); RD #977055-22-3]
Ablunol 200ML; Alkamuls 200-ML; Chemax E-200 ML; Cithrol 2ML; CPH-27-N; Crodet L4; Cyclochem PEG 200 ML; Emerest 2620; Kessco PEG 200 Monolaurate; Mapeg 200 ML; Pegosperse 200 ML; Radiasurf 7422; Soprofor PE/220; Tegin 200 L

PEG-5 laurate
Alkasurf L-5; Hetoxamate LA-5; Lipal 5 L; Trydet LA-5

PEG-6 laurate (CTFA) [CAS #9004-81-3 (generic); RD #977065-75-0]
Emerest 2630; Kessco PEG 300 Monolaurate; Secoster ML 300

PEG-7 laurate
Trydet LA-7

PEG-8 laurate (CTFA) [CAS #9004-81-3 (generic); 35179-86-3; RD #977055-23-4]
Ablunol 400ML; Acconon 400-ML; Alkamuls 400-ML; Chemax E-400 ML; Cithrol 4ML; CPH-30-N; Crodet L8; Cyclochem PEG 400 ML; DyaFac LA9; Emerest 2650; Gradonic 400-ML; Kessco PEG 400 Monolaurate; Lexomul PEG 400 ML; Lipopeg 4-L; Lonzest PEG 4-L; Mapeg 400 ML; Nonisol 100; Pegosperse 400 ML; Radiasurf 7423; Soprofor PE/400

PEG-9 laurate (CTFA) [CAS #106-08-1; 9004-81-3 (generic)]
Alkasurf L-9; Hetoxamate LA-9; Lipal 9 L; Lipocol B

PEG-12 laurate (CTFA) [CAS #9004-81-3 (generic); RD #977055-24-5]
Ablunol 600ML; Alkamuls 600-ML; Cithrol 6ML; CPH-43-N; Crodet L12; Emerest 2661; Kessco PEG 600 Monolaurate; Lipopeg 6-L; Mapeg 600 ML; Pegosperse 600 ML

PEG-14 laurate (CTFA) [CAS #9004-81-3 (generic)]
Alkasurf L-14; Chemax E-600 ML

PEG-20 laurate (CTFA) [CAS #9004-81-3 (generic); RD #977063-49-2]
Cithrol 10ML; Kessco PEG 1000 Monolaurate

PEG-24 laurate
Crodet L24

PEG-32 laurate (CTFA) [CAS #9004-81-3 (generic); RD #977065-55-6]
Kessco PEG 1540 Monolaurate

PEG-40 laurate
Crodet L40

PEG-75 laurate (CTFA) [CAS #9004-81-3 (generic); RD #977065-84-1]
Kessco PEG 4000 Monolaurate; Secoster ML 4000

PEG-100 laurate
Crodet L100

PEG-150 laurate (CTFA) [CAS #9004-81-3 (generic); RD #977063-38-9]
Cithrol 60ML; Kessco PEG 6000 Monolaurate

PEG lauryl ether. See Laureth Series

PEG-6 methyl ether [CAS #9004-74-4 (generic); RD #977065-76-1]
Carbowax MPEG 350

PEG-10 methyl ether
Carbowax MPEG 550

PEG-16 methyl ether
Carbowax MPEG 750

PEG-40 methyl ether
Carbowax MPEG 2000

PEG-100 methyl ether
Carbowax MPEG 5000

PEG-120 methyl glucose dioleate (CTFA)
Glucamate DOE-120

PEG-20 methyl glucose sesquistearate (CTFA) [CAS #68389-70-8]
Glucamate SSE-20

PEG myristyl ethers. See Myreth Series

PEG nonyl phenyl ethers. See Nonoxynol Series

PEG octyl phenyl ethers. See Octoxynol Series

PEG-5 oleamide (CTFA) [RD #977065-67-0] *Synonyms:* POE (5) oleyl amide; PEG (5) oleyl amide
Ethomid O/15; Icomid O-5

PEG-9 oleamide (CTFA)
Rewopal O 8

PEG-2 oleamine (CTFA) [RD #977065-39-6] *Synonyms:* PEG-2 oleyl amine; POE (2) oleyl amine; PEG 100 oleyl amine
Ethomeen O/12; Hetoxamine O-2; Varonic Q202

PEG-5 oleamine (CTFA) [RD #977065-68-1]
Ethomeen O/15; Hetoxamine O-5

PEG-15 oleamine (CTFA) [RD #977063-33-4]
Ethomeen O/25; Hetoxamine O-15

PEG-30 oleamine (CTFA)
Chemeen O-30, O-30/80; Icomeen O-30, O-30-80%; Katapol OA-860, OA-910; Trymeen 6622, 6623, OAM-30/80

PEG-2 oleamonium chloride (CTFA) [RD #977066-73-1]
Ethoquad O/12

PEG-15 oleamonium chloride (CTFA) [RD #977066-76-4]
Ethoquad O/25

PEG-2 oleate (CTFA) [CAS #106-12-7; SE grade RD #977063-43-6] *Synonyms:* Diethylene glycol monooleate; Diglycol oleate; POE (2) monooleate; PEG 100 monooleate
Alkamuls DEG-MO; Alkasurf O-2; Cithrol DGMO N/E, DGMO S/E; Hetoxamate MO-2; Pegosperse 100 O; Radiasurf 7400; Secoster MO 100

PEG-4 oleate (CTFA) [CAS #9004-96-0 (generic); RD #977060-09-5]
Ablunol 200MO; Alkamuls 200-MO; Cithrol 2MO; CPH-39-N; Emerest 2624; Ethylan A2; Kessco PEG 200 Monooleate; Mapeg 200 MO; Radiasurf 7402; Teric OF4

PEG-5 oleate (CTFA) [CAS #9004-96-0 (generic); RD #977065-69-2]
Alkasurf O-5; Chemax E-200 MO; Emulphor VN-430; Ethofat O/15; Hetoxamate MO-5; Trydet OA-5

PEG-6 oleate (CTFA) [CAS #9004-96-0 (generic); RD #977055-25-6]
Acconon 300 MO; Emerest 2632; Ethylan A3; Kessco PEG 300 Monooleate; Lipal 300 W; Radiasurf 7431; Teric OF6

PEG-7 oleate (CTFA) [CAS #9004-96-0 (generic)]
Alkasurf O-7; Trydet OA-7

PEG-8 oleate (CTFA) [CAS #9004-96-0 (generic); RD #977055-26-7]
Ablunol 400MO; Acconon 400 MO; Cithrol 4MO, A; Cyclochem PEG 400 MO; Durpeg 400MO; Emerest 2646; Ethylan A4; Kessco PEG 400 Monooleate; Lexomul PEG 400 MO; Lipal 400 OL, 400 W; Mapeg 400 MO; Nopalcol 4-O; Pegosperse 400 MO; Radiasurf 7403; Secoster MO 400; Teric OF8; Witconol H-31A

PEG-9 oleate (CTFA) [CAS #9004-96-0 (generic); RD #977065-87-4]
Alkamuls 400-MO; Alkasurf O-9; Chemax E-400 MO; Hetoxamate MO-9; Lipal 9 OL

PEG-10 oleate (CTFA) [CAS #9004-96-0 (generic); RD #977063-22-1]
Ethofat O/20; Nikkol MYO-10; Trydet 2676, OA-10

PEG-12 oleate (CTFA) [CAS #9004-96-0 (generic); RD #977055-27-8]
Ablunol 600MO; Alkamuls 600-MO; Cithrol 6MO; CPH-41-N; Emerest 2660; Ethylan A6; Kessco PEG 600 Monooleate; Lipal 600 W; Mapeg 600 MO; Radiasurf 7404

PEG-14 oleate (CTFA) [CAS #9004-96-0 (generic); RD #977063-32-3]
Chemax E-600 MO; Pegosperse 700 TO

PEG-15 oleate
Hetoxamate MO-15

PEG-20 oleate (CTFA) [CAS #9004-96-0 (generic); RD #977063-50-5]
Ablunol 1000MO; Chemax E-1000 MO; Cithrol 10MO; Kessco PEG 1000 Monooleate

PEG-32 oleate (CTFA) [CAS #9004-96-0 (generic); RD #977065-56-7]
Kessco PEG 1540 Monooleate

PEG-75 oleate (CTFA) [CAS #9004-96-0 (generic); RD #977065-85-2]
Cithrol 40MO; Kessco PEG 4000 Monooleate

PEG-150 oleate (CTFA) [CAS #9004-96-0 (generic); RD #977063-39-0]
Cithrol 60MO; Emerest 2617; Kessco PEG 6000 Monooleate

PEG-200 oleate [RD #977043-45-0]
Nioix AK-40

PEG-400 oleate
Nioix AK-44

PEG oleyl ethers. See Oleth Series

PEG-25 PABA [CAS #15716-30-0]
Uvinul P 25

PEG-6 palmitate (CTFA) [CAS #9004-94-8 (generic); RD #977055-28-9] *Synonyms:* POE (6) monopalmitate
Polymulse 6

PEG-8 pelargonate
Emerest 2654

PEG-9 pelargoniate
Trydet MP-9

PEG-5 phytosterol
Nikkol BPS-5

PEG-10 phytosterol
Nikkol BPS-10

PEG-15 phytosterol
Nikkol BPS-15

PEG-20 phytosterol
Nikkol BPS-20

PEG-25 phytosterol
Nikkol BPS-25

PEG-30 phytosterol
Nikkol BPS-30

PEG-PPG-35-9 copolymer (CTFA) [CAS #977066-86-6]
Ucon Fluid 75-H-1400

PEG-20-PPG-10 glyceryl stearate (CTFA)
Acconon TGH

PEG-10 propylene glycol glyceryl laurate
Acconon CON

PEG-120 propylene glycol stearate (CTFA) [RD #977063-31-2]
Drewmulse 1128

PEG-2 ricinoleate (CTFA) [CAS #5401-17-2; 9004-97-1 (generic)] *Synonyms:* Diethylene glycol monoricinoleate; PEG-2 monoricinoleate; POE (2) ricinoleate
Pegosperse 100 MR

PEG-12 safflower ester
Lipal OE 55

PEG-17 safflower ester
Lipal OE 64

PEG-23 safflower ester
Lipal OE 70

PEG-40 sorbitan diisostearate (CTFA)
Emsorb 2726

PEG-40 sorbitan hexaoleate
Sorbeth 40HO

PEG-55 sorbitan hexaoleate
Sorbeth 55HO

PEG-20 sorbitan isostearate (CTFA) [RD #977055-33-6] *Synonyms:* PEG 1000 sorbitan monoisostearate
Crillet 6; Nikkol TI-10; T-Maz 67

PEG-4 sorbitan laurate. See Polysorbate 21

PEG-10 sorbitan laurate (CTFA) [CAS #9005-64-5 (generic); RD #977055-35-8]
Hetsorb L-10; Liposorb L-10

PEG-20 sorbitan laurate. See Polysorbate 20

PEG-80 sorbitan laurate (CTFA) [CAS #9005-64-5 (generic)]
Emsorb 2721; T-Maz 28

PEG-5 sorbitan oleate. See Polysorbate 81

PEG-6 sorbitan oleate (CTFA) [CAS #9005-65-6 (generic)] *Synonyms:* POE (6) sorbitan oleate; PEG 300 sorbitan monooleate)
Nikkol TO-106

PEG-20 sorbitan oleate. See Polysorbate 80

PEG-20 sorbitan palmitate. See Polysorbate 40

PEG-40 sorbitan peroleate (CTFA) [RD #977063-07-2]
Arlatone T

PEG-4 sorbitan stearate. See Polysorbate 61

PEG-20 sorbitan stearate. See Polysorbate 60

PEG-20 sorbitan tallate
T-Maz 90

PEG-30 sorbitan tetraoleate (CTFA)
Nikkol GO-430

PEG-40 sorbitan tetraoleate (CTFA) [CAS #9003-11-6]
Nikkol GO-440

PEG-60 sorbitan tetraoleate (CTFA)
Nikkol GO-460

PEG-60 sorbitan tetrastearate (CTFA)
Nikkol GS-460

PEG-16 sorbitan trioleate
Emsorb 6917

PEG-20 sorbitan trioleate. See Polysorbate 85

PEG-16 sorbitan tristearate
Emsorb 6908

PEG-20 sorbitan tristearate. See Polysorbate 65

PEG-20 sorbitan tritallate
T-Maz 95

PEG sorbitol ethers. See Sorbeth Series

PEG-2 soyamine (CTFA) [CAS #61791-24-0 (generic); RD #977063-44-7] *Synonyms:* POE (2) soya amine; POE (2) soya amine; PEG 100 soya amine
Accomeen S2; Chemeen S-2; Ethomeen S/12; Hetoxamine S-2; Mazeen S 2; Teric 16M2; Varonic L202

PEG-5 soyamine (CTFA) [CAS #61791-24-0 (generic); RD #977055-36-9]
Accomeen S5; Alkaminox SO-5; Chemeen S-5; Ethomeen S/15; Hetoxamine S-5; Icomeen S-5; Mazeen S 5; Teric 16M5; Varonic L 205

PEG-10 soyamine (CTFA) [CAS #61791-24-0 (generic); RD #977063-23-2]
Accomeen S10; Ethomeen S/20; Mazeen S 10; Teric 16M10

PEG-15 soyamine (CTFA) [CAS #61791-24-0 (generic); RD #977063-34-5]
Ethomeen S/25; Hetoxamine S-15; Mazeen S 15; Teric 16M15

PEG-30 soyamine
Chemeen S-30, S-30/80

PEG-5 soya sterol (CTFA) [RD #977065-70-5]
Generol 122 E 5

PEG-10 soya sterol (CTFA) [RD #977065-31-8]
Generol 122 E 10

PEG-16 soya sterol (CTFA) [RD #977065-37-4]
Generol 122 E 16

PEG-25 soya sterol (CTFA) [RD #977065-46-5]
Generol 122 E 25

PEG-2 stearamine (CTFA) [RD #977063-45-8] *Synonyms:* PEG-2 stearyl amine; PEG 100 stearyl amine; POE (2) stearyl amine
Chemeen 18-2; Ethomeen 18/12; Hetoxamine ST-2; Teric 18M2; Varonic S202

PEG-5 stearamine (CTFA) [RD #977065-71-6]
Chemeen 18-5; Ethomeen 18/15; Hetoxamine ST-5; Icomeen 18-5; Teric 18M5

PEG-10 stearamine (CTFA) [RD #977065-32-9]
Ethomeen 18/20; Teric 18M10

PEG-15 stearamine (CTFA) [RD #977063-35-6]
Ethomeen 18/25; Hetoxamine ST-15

PEG-20 stearamine
Teric 18M20

PEG-30 stearamine
Teric 18M30

PEG-50 stearamine (CTFA) [RD #977065-74-9]
Chemeen 18-50; Ethomeen 18/60; Hetoxamine ST-50; Trymeen 6617, SAM-50

PEG-1 stearate
Alkasurf S-1; Nikkol MYS-1EX

PEG-2 stearate (CTFA) [CAS #106-11-6; 9004-99-3 (generic); SE grade RD #977063-17-4] *Synonyms:* Diethylene glycol stearate; Diglycol stearate; PEG 100 monostearate; POE (2) monostearate
Alkamuls DEG-MS; Alkasurf S-2; Cithrol DGMS N/E, DGMS S/E); Cyclochem DGS; Dermalcare GMS-165, SDG; Drewmulse DGMS; Emerest 2717; Hydrine; Kemester 5221SE; Kessco Diethylene Glycol Monostearate, Diglycol Stearate Neutral, Diglycol Stearate SE; Lipo DGS-SE (SE); Nikkol MYS-2; Nopalcol 1-S; Pegosperse 100 S; Radiasurf 7410 (SE), 7411 (SE); Sedefos 75; Witconol CAD

PEG-4 stearate (CTFA) [CAS #106-07-0; 9004-99-3 (generic); RD #977055-37-0]
Ablunol 200MS; Acconon 200-MS; Alkamuls 200-MS; Cithrol 2MS; Crodet S4; Kessco PEG 200 Monostearate; Mapeg 200 MS; Nikkol MYS-4; Radiasurf 7412

PEG-5 stearate (CTFA) [CAS #9004-99-3 (generic); RD #977065-72-7]
Alkasurf S-5; Chemax E-200 MS; Ethofat 60/15; Hetoxamate SA-5; Lipal 5 S; Trydet 2670, SA-5

PEG-6 stearate (CTFA) [CAS #9004-99-3 (generic); RD #977053-32-9]
Cithrol 3MS; Emerest 2636; Kessco PEG 300 Monostearate; Nonex S3E; Polystate C; Radiasurf 7432; Tefose 2000, 2561

PEG-6-32 stearate (CTFA) [CAS #9004-99-3 (generic); RD #977055-38-1] *Synonyms:* PEG 1500 monostearate; POE 1500 monostearate)
Cithrol 15MS; Lipopeg 15-S; Mapeg 1500 MS; Pegosperse 1500 MS; Radiasurf 7417; Tefose 63, 1500

PEG-7 stearate (CTFA) [CAS #9004-99-3 (generic); RD #977065-81-8]
Hetoxamate SA-7; Mapeg 350 MS; Trydet 2636, SA-7

PEG-8 stearate (CTFA) [CAS #9004-99-3 (generic); RD #977055-39-2]
Ablunol 400MS; Acconon 400-MS; Alkamuls 400-MS; Alkasurf S-8; Cithrol 4MS; Crodet S8; Emerest 2640, 2711; Kessco PEG 400 Monostearate; Lipal 400 S; Lipopeg 4-S; Mapeg 400 MS; Nonisol 300; Pegosperse 400 MS; Radiasurf 7413; Trydet 2671, SA-8; Varonic 400MS

PEG-9 stearate (CTFA) [CAS #9004-99-3 (generic); RD #977068-31-7]
Chemax E-400 MS; Cremophor S 9; Emulphor VT-650; Hetoxamate SA-9; Lipocol B; Serdox NSG 400

PEG-10 stearate (CTFA) [CAS #9004-99-3 (generic); RD #977065-33-0]
Ethofat 60/20; Nikkol MYS-10

PEG-12 stearate (CTFA) [CAS #9004-99-3 (generic); RD #977055-40-5]
Ablunol 600MS; Alkamuls 600-MS; Cithrol 6MS; Crodet S12; DyaFac 6-S; Emerest 2662; Hetoxamate SA-13; Kessco PEG 600 Monostearate; Lipal 600 S; Mapeg 600 MS; Pegosperse 600 MS; Radiasurf 7414

PEG-14 stearate (CTFA) [CAS #9004-99-3 (generic)]
Alkasurf S-14

PEG-15 stearate
Ethofat 60/25

PEG-20 stearate (CTFA) [CAS #9004-99-3 (generic); RD #977055-41-6]
Ablunol 1000MS; Cithrol 10MS; Emerest 2610; Hetoxamate SA-23; Kessco PEG 1000 Monostearate; Lipopeg 10-S; Mapeg 1000 MS; Pegosperse 1000 MS; Varonic 1000MS

PEG-23 stearate
Trydet SA-23

PEG-24 stearate
Crodet S24

PEG-25 stearate (CTFA) [CAS #9004-99-3 (generic)]

Lipal 25 S; Nikkol MYS-25, MYS-25(FF)

PEG-32 stearate (CTFA) [CAS #9004-99-3 (generic); RD #977055-43-8] *Synonyms:* PEG 1540 stearate
Kessco PEG 1540 Monostearate

PEG-35 stearate (CTFA) [CAS #9004-99-3 (generic); RD #977068-25-9]
Hetoxamate SA-35

PEG-36 stearate (CTFA) [CAS #9004-99-3 (generic); RD #977068-26-0]
Varonic 1800MS

PEG-40 stearate (CTFA) [CAS #9004-99-3 (generic); RD #977009-12-3]
AF-72, AF-75; Alkasurf S-40; Crodet S40; Drewmulse HM 100; Emerest 2715; Hetoxamate SA-40; Lipal 39 S; Lipopeg 39-S; Mapeg S-40; Nikkol MYS-40, MYS-40(FF); Pegosperse 1750 MS; Trydet 2672, SA-40

PEG-45 stearate (CTFA) [CAS #9004-99-3 (generic); RD #977065-63-6]
Nikkol MYS-45

PEG-50 stearate (CTFA) [CAS #9004-99-3 (generic); RD #977055-45-0]
Emerest 2675; Lipal 50 S; Trydet SA-50/30

PEG-55 stearate
Nikkol MYS-55

PEG-75 stearate (CTFA) [CAS #9004-99-3 (generic); RD #977055-46-1]
Cithrol 40MS; Gelot 64; Kessco PEG 4000 Monostearate; Mapeg 4000 MS; Pegosperse 4000 MS

PEG-90 stearate (CTFA) [CAS #9004-99-3 (generic); RD #977065-88-5]
Drewmulse 1129; Hetoxamate SA-90

PEG-100 stearate (CTFA) [CAS #9004-99-3 (generic); RD #977055-47-2]
Crodet S100; Cyclochem GMS 165; Kessco Glycerol Monostearate SE; Lipomulse 165; Lipopeg 100-S; Mapeg S-100; Mazol 165C

PEG-150 stearate (CTFA) [CAS #9004-99-3 (generic); RD #977055-48-3]
Kessco PEG 6000 Monostearate; Mapeg 6000 MS, S-150

PEG 4400 stearate
Mapeg S-100

PEG-2 stearmonium chloride (CTFA) [RD #977066-74-2]
Ethoquad 18/12

PEG-15 stearmonium chloride (CTFA) [RD #977066-77-5]
Ethoquad 18/25

PEG-5 stearyl ammonium chloride (CTFA) [RD #977065-99-8] *Synonyms:* Quaternium-36
Genamin KS 5

PEG stearyl ethers. See Steareth Series

PEG-2 tallate *Synonyms:* POE (2) monotallate; PEG 100 monotallate; PEG-2 tall oil acid ester
Teric T2

PEG-4 tallate (CTFA) [CAS #61791-00-2 (generic); RD #977065-59-0]
Hetoxamate FA-5; Mapeg 200 MOT

PEG-5 tallate
Teric T5

PEG-7 tallate
Teric T7

PEG-8 tallate (CTFA) [CAS #61791-00-2 (generic); RD #977055-49-4]
Chemax TO-8; Mapeg 400 MOT; Pegosperse 400 MOT; Witconol H-31

PEG-10 tallate (CTFA) [CAS #61791-00-2 (generic)]
Emulphor TO-530; Polyfac MT-610; Teric T10

PEG-12 tallate (CTFA) [CAS #61791-00-2 (generic); RD #977068-27-1]
Actrol 6M25P; Mapeg 600 MOT

PEG-15 tallate
Lipal 15 T; Mapeg TAO-15; Polyfac MT-615

PEG-16 tallate (CTFA) [CAS #61791-00-2 (generic); RD #977055-50-7]
Chemax TO-16; Emulphor TO-639

PEG-20 tallate (CTFA) [CAS #61791-00-2 (generic); RD #977063-51-6]
Hetoxamate FA-20

PEG-220 tallate
Mapeg TAO-10

PEG-8 tall oil dioleate
Polyfac TDO-9

PEG-12 tall oil dioleate
Polyfac TDO-14

PEG-5 tall oil oleate
Polyfac TMO-5

PEG-7 tall oil oleate
Polyfac TMO-7

PEG-14 tall oil oleate
Polyfac TMO-14

PEG-13 tallow amide *Synonyms:* PEG-13 hydrogenated tallow amide; POE (13) hydrogenated tallow amide
Ethomid HT/23 (12.5 EO)

PEG-50 tallow amide (CTFA) [CAS #8051-63-6] *Synonyms:* POE (50) hydrogenated tallow amide
Ethomid HT/60 ; Icomid HT-50; Schercomid HT-60

PEG-2 tallow amine (CTFA) [CAS #61791-26-2 (generic); RD #977063-46-9] *Synonyms:* PEG 100 tallow amine; POE (2) tallow amine; PEG-2 hydrogenated tallow amine
Accomeen T2; Alkaminox T-2; Chemeen HT-2, T-2; DeSomeen TA-2; Ethomeen T/12; Hetoxamine T-2; Icomeen T-2; Mazeen T 2; Teric 17M2; Varonic T202, T202 SR

PEG-5 tallow amine (CTFA) [CAS #61791-26-2 (generic)]
Accomeen T5; Alkaminox T-5; Chemeen HT-5, T-5; DeSomeen TA-5; Ethomeen T/15; Hetoxamine T-5; Icomeen T-5; Katapol PN-430; Mazeen T 5; Teric 17M5; Varonic T205, T205LC

PEG-7 tallow amine
Icomeen T-7

PEG-8 tallow amine (CTFA) [CAS #61791-26-2 (generic); RD #977062-12-6]
Trymeen 6603

PEG-10 tallow amine
Alkaminox T-10; Chemeen T-10; Varonic T210

PEG-12 tallow amine
Alkaminox T-12

PEG-15 tallow amine (CTFA) [CAS # 61791-26-2 (generic); RD #977063-36-7]
Accomeen T15; Alkaminox T-15; Chemeen HT-15, T-15; DeSomeen TA-15; Ethomeen T/25; Hetoxamine T-15; Katapol PN-730; Mazeen T 15; Serdox NJAD 15; Teric 17M15; Trymeen 6606, TAM-15; Varonic T215, T215LC

PEG-20 tallow amine (CTFA) [CAS # 61791-26-2 (generic); RD #977063-52-7]
Atlas G-3780A; Chemeen T-20; DeSomeen TA-20; Hetoxamine T-20; Icomeen T-20; Katapol PN-810; Serdox NJAD 20; Trymeen 6607, TAM-20

PEG-25 tallow amine
Alkaminox T-25; Icomeen T-25; Trymeen 6609, TAM-25

PEG-30 tallow amine
Alkaminox T-30; Serdox NJAD 30

PEG-40 tallow amine (CTFA) [CAS #61791-26-2 (generic); RD #977065-61-4]
Icomeen T-40, T-40-80%; Trymeen 6637, TAM-40

PEG-50 tallow amine (CTFA) [CAS #61791-26-2 (generic); RD #977069-79-6]
Alkaminox T-50; Chemeen HT-50; Ethomeen T/60; Varonic U 250

PEG-3 tallow aminopropylamine (CTFA) [RD #977066-69-5]
Ethoduomeen T/13

PEG-15 tallow aminopropylamine (CTFA) [RD #977066-71-9]
Chemeen DT-15; Ethoduomeen T/25

PEG-2 tallowate *Synonyms:* Diethylene glycol monotallowate; POE (2) tallowate; PEG-2 tallow acid ester
Nopalcol 1-TW

PEG-8 tallowate
Lipal TE 43

PEG-12 tallowate
Lipal TE 55

PEG-22 tallowate
Lipal TE 70

PEG-30 tallowate
Lipal TE 76

PEG-3 tallow diamine
Chemeen DT-3; Varonic T403

PEG-10 tallow diamine
Varonic T410

PEG-30 tallow diamine
Icodimeen T-30

PEG-30 tallow diamine
Chemeen DT-30

PEG tallow ethers. See Talloweth Series

PEG-15 tallow polyamine (CTFA) [RD #977069-49-0]
Polyquart H

PEG-15 N-tallow 1,3-propylene diamine
Trymeen 6640

PEG-3.5 tetramethyl decynediol
Surfynol 440

PEG-100 m-toluidine [CAS #36356-82-8]
Emery 6728, 6730

PEG-200 m-toluidine [CAS #36356-82-8]
Emery 5776, 5778

PEG tridecyl ethers. See Trideceth Series

PEG-2 tridecyl phenol
Trycol TP-2

PEG-6 tridecyl phenol
Trycol TP-6

PEG-3 triethanolamine [CAS #36936-60-4]
Emery 5794

PEG-66 trihydroxystearin (CTFA) [RD #977069-50-3] *Synonyms:* POE (66) trihydroxystearin
Surfactol 575

PEG-200 trihydroxystearin (CTFA) [RD #977069-51-3]
Naturechem THS-200; Surfactol 590

Pencycuron
Monceren

Pendimethalin *Synonyms:* N-(1-ethylpropyl)3-4,-dimethyl-2,6-dinitrobenzenamine
Prowl

Pentabromochlorocyclohexane
FR-651-A

Pentabromodiphenyl oxide
DE-71; FR-1205; Nyacol HA-9; Saytex 115, 125

Pentabromodiphenyl oxide phosphate
DE-60F

Pentabromoethylbenzene
Saytex 105

Pentabromotoluene
FR-705

Pentachlorophenol [CAS #87-86-5] *Synonyms:* PCP
Pentachlorophenol DP-2

Pentachlorothiophenol [CAS #133-49-3] *Synonyms:* Pentachlorobenzenethiol
Renacit 7, 7/WG

3-(n-Penta-8′-decenyl) phenol [CAS #8007-24-7]
Cardolite NC-511

3-(n-Pentadecyl) phenol [CAS #501-24-6, 3158-56-3]
Cardolite NC-507, NC-510

Pentaerythritol hexylthiopropionate
Mark 2140

Pentaerythritol hydrogenated rosinate (CTFA)
Foral 105; Pentalyn H

Pentaerythritol oleate
Alkamuls PEMO; Radiasurf 7156

Pentaerythritol ricinoleate
Flexricin 17

Pentaerythritol rosinate (CTFA) [CAS #8050-26-8; RD #977064-43-9] *Synonyms:* Pentaerythritol ester of rosin
Pentalyn 830, 856, A

Pentaerythritol stearate
Aflux 54; Nikkol PEMS; Radiasurf 7175; Rilanit PES

Pentaerythritol tetraabietate (CTFA) [CAS #127-23-1]
Foral 105

Pentaerythritol tetraacrylate
SR-295

Pentaerythritol tetrabehenate [CAS #61682-73-3] *Synonyms:* Pentaerythrityl tetrabehenate (CTFA)
Liponate PB-4; Radia 7514

Pentaerythritol tetrabenzoate
Benzoflex S-552

Pentaerythritol tetra-C5/C9
Radiasyn 7174

Pentaerythritol tetra-C7
Radiasyn 7177

Pentaerythritol tetra-C8/C10
Radiasyn 7178

Pentaerythritol tetracaprylate/caprate [CAS #68441-68-9]
Crodamol PTC; Kessco 874

Pentaerythritol tetraisostearate
Crodamol PTIS

Pentaerythritol tetraoleate (CTFA) [CAS #19321-40-5] *Synonyms:* Pentaerythrityl tetraoleate
Cyclochem PETO; Liponate PO-4; Radia 7171

Pentaerythritol tetrapelargonate [CAS #14450-05-6]
Emerest 2485, 2486

Pentaerythritol tetrastearate (CTFA) [CAS #115-83-3] *Synonyms:* Pentaerythrityl tetrastearate
Cyclochem PETS; Liponate PS-4; Radia 7176

Pentaerythritol triacrylate [CAS #3524-68-3] *Synonyms:* 2-Propenoic acid-2-(hydroxymethyl)-2-(((1-oxo-2-propenyl) oxy) methyl)-1,3-propanediyl ester
SR-444

Pentamethylene-ammonium-N-pentamethylene dithiocarbamate
Vulkacit P

Pentamethyl-4-piperidyl/β,β,β′,β′-tetramethyl-3,9-(2,4,8,10-tetraoxaspiro (5,5) undecane) diethyl]-[1,2,2,6,6-1,2,3,4-butane tetracarboxylate
Mixxim HALS 63

Pentamethyl tallow propanediammonium dichloride
Duoquad T-50

2,4-Pentanedione peroxide
Lupersol 224; Trigonox 40, 44P

Pentasodium aminotrimethylene phosphonate (CTFA) [CAS #2235-43-0]
Dequest 2006

Pentasodium pentetate (CTFA) [CAS #140-01-2] *Synonyms:* Pentasodium diethylene triamine pentaacetate; Pentasodium DTPA
Cheelox DTPA-14; Chel DTPA-41; Chelon 80; Dissolvine D 40, D 88; Hamp-Ex 80; Kalex Penta; Pentaquest Extra; Plexene D; Trilon C Liq.; Vanate PSPa; Versenex 80

Pentasodium triphosphate [CAS #7758-29-4]
Empiphos STP, STP/1L, STP/6L, STP/6N, STP/AD, STP/D, STP/DMST, STP/E, STP/F, STP/L, STP/L16, STP/M, STP/M16, STP/MST, STP/N

Pentetic acid (CTFA) [CAS #67-43-6] *Synonyms:* DPTA; Diethylene triamine pentaacetic acid
Chel DTPA; Hamp-Ex Acid; SEQ 80; Trilon CS

Pentosanase-hexosanase
Rhozyme HP-150 Conc.

Perchloroethylene [CAS #127-18-4] *Synonyms:* Tetrachloroethylene
Dowper; Perk; Perkare; PPG Perchlor

Peroxydicarbonate
Trigonox ADC

Petrolatum (CTFA) [CAS #8009-03-8 (NF); 8027-32-5 (USP)]
Alcolan, 36W, 40; Amerchol 400, C, CAB, H-9, RC; Cremba; Dehymuls K; Fonoline White, Yellow; Lexate PX; Mineral Jelly No. 5, No. 10, No. 15, No. 20, No. 25; Nimcolan 1747; Ointment Base No. 3, 4, 6; Penreco 1520, 3070, Amber, Blond, Cream, Frost, Green, Lily, Red, Regent, Royal, Snow, Super, Ultima; Perfecta, USP; Petrolatum RPB; Polytrap 210; Protegin,

W, WX, X; Protopet Alba, White 1S, White 2L, White 3C, Yellow 1E, Yellow 2A; Ritaderm; Sonojell Min. Jellies, No. 4, 9; Tech Pet F, M

Petroleum distillate (CTFA) [CAS #8002-32-4; RD# 977057-59-2]
Bentone Gel SS-71; Penreco 2251 Oil, 2263 Oil

Petroleum wax [CAS #8002-74-2; RD #977051-70-9]
Ross Wax #145, #165; Vanwax H Special

Phenalkamine
Cardolite NC-540, -541, -546, -549, -550

1,10-Phenanthroline [CAS #66-71-7] *Synonyms:* 4,5-Diazaphenanthrene; Orthophenanthrodine; 3-Phenanthroline
Activ-8

Phenathiazine
PTZ

2-Phenethyltrichlorosilane [CAS #940-41-0]
CP0110

Phenol disulfide
Vultac 2, 4, 5

Phenolethylsiloxane
PO114

Phenol-formaldehyde (novolak) resin. See Novolac resin

Phenol-formaldehyde resin *Synonyms:* Single-stage phenolic resin
Akrochem P 37, P 87, P-478, P-487; Duolite A-7, S-37, S-761; SP-25, -1044, -1045, -1055, -1056, -1068, -1077, -6700, -6701; Varcum 564, 4326-WP, 8169; Vulkadur A

Phenol-formaldehyde resin, bromomethylated and methylolated
SP-1055, -1056

Phenolic resin. See also Phenol-formaldehyde resin, Novolac resin
Akrochem P-133, P-420, P-486; Amberol ST-149; Nevastain 21, 76, 2170, A, B; Troykyd Anti-Skin S, Special M; Uvi-Nox 1494; Vanox 100, ZS; Varcum 1198, 2053, 8199, 8345, 8506, 9008, C-86; Vulkanox DS, DS/F; Wingstay Antioxidants; Wytox PAP, PAP-SE, PMW

Phenolic/bisphenolic
Naugard 431, BHT, SP; Naugawhite

Phenolic, bromomethyl alkylated
Akrochem P-126

Phenolic, styrenated
Montaclere; Nevastain 21; Vanox 102; Vulkanox SP; Wingstay S

Phenol phosphite
Naugard P, PHR; Wytox 540, 604, 604LMS

Phenol sulfonic acid *Synonyms:* Sulfocarbolic acid
Eltesol PSA 65

Phenol thio ester
Vanox 1030

Phenoxyethanol (CTFA) [CAS #122-99-6] *Synonyms:* 2-Phenoxyethanol; Phenoxytol; Ethylene glycol monophenyl ether
Dowanol EPH; Emeressence 1160 Rose Ether; Emery 6705; Ethylan HB1-TG; Rewopal MPG 10

2-Phenoxyethyl acrylate
SR-339

Phenoxyethyl methacrylate
SR-340

N-Phenylaminopropyltrimethoxy silane [CAS #3068-76-6]
CP0156

Phenyl diethanolamine [CAS #120-07-0]
Emery 5703, 5705

Phenyl diisodecyl phosphite
Weston PDDP

Phenyl dimethicone [CAS #2116-84-9; 63148-58-3] *Synonyms:* Methyl phenyl polysilocane; Polyphenylmethyl siloxane
Dow Corning 556 Fluid

Phenyldimethylchlorosilane [CAS #768-33-2]
CP0160; P0160

m-Phenylenedimaleimide [CAS #3006-93-7]
Vanax MBM

Phenyl ethanolamine [CAS #7568-93-6; #122-98-5]
Emery 5700

2-Phenylindole
Stabilizer I-FF

Phenyl mercuric acetate (CTFA) [CAS #62-38-2] *Synonyms:* Phenylmercury acetate
Troysan PMA-10-SEP, PMA-30, PMA-100

Phenylmercuric oleate *Synonyms:* Phenylmercury oleate
Troysan PMO-30

Phenylmercuritriethanolammonium lactate *Synonyms:* Tris (2-hydroxyethyl) (phenylmercuri) ammonium lactate
Merkyl PM-TL

Phenylmercury ammonium propionate
Merkyl MAP

Phenylmethyl polysiloxane [CAS #68083-14-7]
Abil AV 20-1000, 8853

Phenyl-α-naphthylamine [CAS #90-30-3]
Additin 30; Akrochem Antioxidant PANA; Ucar Additive PN-1; Vulkanox PAN

Phenyl-β-naphthylamine [CAS #135-88-6]
Anchor PBN

Phenyl neopentylene glycol phosphite
Weston PNPG

o-Phenylphenol (CTFA) [CAS #90-43-7] *Synonyms:* 2-Phenyl phenol; o-Hydroxydiphenyl; o-Xenol
Dowicide 1

o-Phenylphenol sodium. See Sodium o-phenylphenate

Phenyl silicone
F 50; SF 1265

5-Phenyltetrazole [CAS #3999-10-8]
Expandex 5PT

Phenyltrichlorosilane [CAS #98-13-5]
CP0280

Phenyltriethoxysilane
CP0320

Phenyltrimethoxysilane [CAS #2996-92-1; #780-69-8]
CP0330

Phorate [CAS #298-02-2] *Synonyms:* o,o-Diethyl-S-ethylmercaptomethyl dithiophosphonate; o,o-Diethyl-s-ethylthiomethyl dithiophosphonate; O,O-Diethyl-s-[(ethylthio) methyl] phosphorodithioate
Aastar; Thimet 15-G, 20-G

Phosphinocarboxylic acid
Belsperse 161

Phospholipase A-2
Lecitase

2-Phosphono-butane-tricarboxylic acid-1,2,4
Bayhibit, Bayhibit-AM

Phosphoric acid (CTFA) [CAS #7664-38-2] *Synonyms:* Orthophosphoric acid
P-0620 T 1/8″

Phosphorus
Amgard CHT, CPC 102, CPC 452, CPC 702

Phoxim
Baythion; Volaton

Phthalic anhydride [CAS #85-44-9] *Synonyms:* 1,3-Isobenzofurandione
Retarder AK; Retarder PX; Vulkalent B/C

α-Pinene resin
Piccolyte A115, A125, A135

β-Pinene resin *Synonyms:* Nopinene
Piccolyte S100, S115, S115SF, S125, S125SF, S135

Pine oil (CTFA) [CAS #8002-09-3]
Herco

Pine resin
Hercosol TP-S

Pinewood resin
Bresin 2 Resin; Vinsol Emulsion

Pinewood resin, sodium soap
Vinsol NVX

Piperidinium pentamethylene dithiocarbamate [CAS #98-77-1]
Accelerator P.P.D.; Vanax 552

POE. See under PEG

POE POP block copolymer. See Poloxamer, EO-PO block copolymer

Poloxamer 101 (CTFA) [CAS #9003-11-6 (generic); RD #977057-62-7]
Macol 46; Pluronic L31

Poloxamer 105 (CTFA) [CAS #9003-11-6 (generic); RD #977057-63-8]
Macol 35; Pluronic L35

Poloxamer 108 (CTFA) [CAS #9003-11-6 (generic); RD #977066-68-4]
Pluronic F38, F68LF, L62D, L62LF

Poloxamer 122 (CTFA) [CAS #9003-11-6 (generic); RD #977057-65-0]
Macol 42; Pluronic L42

Poloxamer 123 (CTFA) [CAS #9003-11-6 (generic); RD #977057-66-1]
Pluronic L43

Poloxamer 124 (CTFA) [CAS #9003-11-6 (generic); RD #977057-67-2]
Macol 44; Pluronic L44

Poloxamer 181 (CTFA) [CAS #9003-11-6 (generic); RD #977057-68-3]
Macol 1; Pluronic L61

Poloxamer 182 (CTFA) [CAS #9003-11-6 (generic); RD #977057-69-4]
Macol 2; Pluronic L62

Poloxamer 183 (CTFA) [CAS #9003-11-6 (generic); RD #977057-70-7]
Pluronic L63

Poloxamer 184 (CTFA) [CAS #9003-11-6 (generic); RD #977054-47-9]
Macol 4; Pluronic L64

Poloxamer 185 (CTFA) [CAS #9003-11-6 (generic); RD #977057-71-8]
Pluronic P65

Poloxamer 188 (CTFA) [CAS #9003-11-6 (generic)]
Macol 8; Pluronic F68

Poloxamer 212 (CTFA) [CAS #9003-11-6 (generic); RD #977057-72-9]
Macol 72; Pluronic L72

Poloxamer 215 (CTFA) [CAS #9003-11-6 (generic); RD #977057-73-0]
Pluronic P75

Poloxamer 217 (CTFA) [CAS #9003-11-6 (generic); RD #977057-74-1]
Macol 77; Pluronic F77

Poloxamer 231 (CTFA) [CAS #9003-11-6 (generic); RD #977057-75-2]
Pluronic L81

Poloxamer 234 (CTFA) [CAS #9003-11-6 (generic); RD #977057-76-3]
Pluronic P84

Poloxamer 235 (CTFA) [CAS #9003-11-6 (generic); RD #977057-77-4]
Macol 85; Pluronic P85

Poloxamer 237 (CTFA) [CAS #9003-11-6 (generic); RD #977057-78-5]
Pluronic F87

Poloxamer 238 (CTFA) [CAS #9003-11-6 (generic); RD #977057-79-6]
Pluronic F88

Poloxamer 282 (CTFA) [CAS #9003-11-6 (generic); RD #977057-80-9]
Pluronic L92

Poloxamer 288 (CTFA) [CAS #9003-11-6 (generic); RD #977057-82-1]
Pluronic F98

Poloxamer 331 (CTFA) [CAS #9003-11-6 (generic); RD #977057-83-2]
Macol 101; Pluronic L101

Poloxamer 333 (CTFA) [CAS #9003-11-6 (generic); RD #977057-84-3]
Pluronic P103

Poloxamer 334 (CTFA) [CAS #9003-11-6 (generic); RD #977057-85-4]
Pluronic P104

Poloxamer 335 (CTFA) [CAS #9003-11-6 (generic); RD #977057-86-5]
Pluronic P105

Poloxamer 338 (CTFA) [CAS #9003-11-6 (generic); RD #977057-87-6]
Macol 108; Pluronic F108

Poloxamer 401 (CTFA) [CAS #9003-11-6 (generic); RD #977057-88-7]
Pluronic L121

Poloxamer 402 (CTFA) [CAS #9003-11-6 (generic); RD #977057-89-8]
Pluronic L122

Poloxamer 403 (CTFA) [CAS #9003-11-6 (generic); RD #977057-90-1]
Pluronic P123

Poloxamer 407 (CTFA) [CAS #9003-11-6 (generic); RD #977057-91-2]
Pluronic F127

Poloxamine 304 (CTFA) [CAS #11111-34-5] (generic); RD #977066-59-3]
Tetronic 304

Poloxamine 504 (CTFA) [CAS #11111-34-5] (generic); RD #977066-60-6]
Tetronic 504

Poloxamine 701 (CTFA) [CAS #11111-34-5] (generic); RD #977066-61-7]
Tetronic 701

Poloxamine 702 (CTFA) [CAS #11111-34-5] (generic); RD #977066-62-8]
Tetronic 702

Poloxamine 704 (CTFA) [CAS #11111-34-5] (generic); RD #977066-63-9]
Tetronic 704

Poloxamine 707 (CTFA) [CAS #11111-34-5] (generic); RD #977066-64-0]
Tetronic 707

Poloxamine 901 (CTFA) [CAS #11111-34-5] (generic); RD #977066-65-1]
Tetronic 901

Poloxamine 904 (CTFA) [CAS #11111-34-5] (generic); RD #977066-66-2]
Tetronic 904

Poloxamine 908 (CTFA) [CAS #11111-34-5] (generic); RD #977066-67-3]
Tetronic 908

Poloxamine 1101 (CTFA) [CAS #11111-34-5] (generic); RD #977066-49-1]
Tetronic 1101

Poloxamine 1102 (CTFA) [CAS #11111-34-5] (generic); RD #977066-50-4]
Tetronic 1102

Poloxamine 1104 (CTFA) [CAS #11111-34-5] (generic); RD #977066-51-5]
Tetronic 1104

Poloxamine 1107
Tetronic 1107

Poloxamine 1301 (CTFA) [CAS #11111-34-5] (generic); RD #977066-52-6]
Tetronic 1301

Poloxamine 1302 (CTFA) [CAS #11111-34-5] (generic); RD #977068-32-8]
Tetronic 1302

Poloxamine 1304 (CTFA) [CAS #11111-34-5] (generic); RD #977066-53-7]
Tetronic 1304

Poloxamine 1307 (CTFA) [CAS #11111-34-5] (generic); RD #977066-54-8]
Tetronic 1307

Poloxamine 1501 (CTFA) [CAS #11111-34-5] (generic); RD #977066-55-9]
Tetronic 1501

Poloxamine 1502 (CTFA) [CAS #11111-34-5] (generic); RD #977066-56-0]
Tetronic 1502

Poloxamine 1504 (CTFA) [CAS #11111-34-5] (generic); RD #977066-57-1]
Tetronic 1504

Poloxamine 1508 (CTFA) [CAS #11111-34-5] (generic); RD #977066-58-2]
Tetronic 1508

Polyacrylamide (CTFA) [CAS #9003-05-8] *Synonyms:* 2-Propenamide, homopolymer
Cyanamer A-370, P-26, P-35, P-70, P-250; Reten 420, 520, 521, 523, 525

Polyacrylate
Alkasperse A-20; Cru Thix 46; Mafloc 764, 900; Nopcosperse AD-6; Perenol F3; Texigel 23-002, 23-004, 23-012, 23-005, 23-018

Polyacrylic acid (CTFA) [CAS #9003-01-4] *Synonyms:* 2-Propenoic acid, homopolymer; Acrylic acid
Acritamer 941; Acrysol A-1, A-3, A-5, ASE-75, LMW-10, LMW-20, LMW-45; Alcogum L; Alcosperse 404; Alkasperse A-2H, A-5H, A-10H; Aquatreat AR4, AR6; Carbopol 615, 616, 617, 907, 1342, 1706, 1720, 1731, 1754; Colloid 117/50, 119/50, 204, 274; Daxad 37L Acid; Good-rite K-722, -732, -752, -7028, -7058, -7200, -7600, Polyacrylates; Sokalan CP 10S, PA 80 S, PA 110 S

Polyamide
Ablusoft ND; Chemiflex 315C; Delsette; GP-2925; Norcure 138-B

Polyamide, acylated
Nalco 70

Polyamide/epichlorohydrin polymer
Hercosett 57, 70, 125

Polyamide/epoxy
Tra-Bond 2143D, 2147

Polyamide polyether
Gaftex LN-110

Polyarylate. See Polyester resin, polyarylate

Polybutadiene, acrylated
Chemlink 5000

Polybutylated bisphenol A
Agerite Superlite, Superlite Emulcon, Superlite Solid

Polybutylene *Synonyms:* Polybutene (CTFA); Polyisobutylene; Polyisobutene
Glissoviscal B; Olicat C

Polybutylene terephthalate (PBT). See Polyester resin, PBT

Polycaprolactone polyurethane. See Polyurethane, polycaprolactone

Polydibromophenylene oxide
PO-64P

Polydimethyldiphenyl siloxane
SF 1153, 1154

Polydimethylsiloxane. See Dimethicone

Polydimethylsiloxane/EO-PPO copolymer
SF 1066

Poly-p-dinitrosobenzene polymer [CAS #9003-34-3]
Vanax PY

Poly (dipropyleneglycol) phenyl phosphite
Weston DHOP

Polyester adipate
Merrol P-6303, P-6310, P-6311, P-6320, P-6403, P-6410, P-6420, P-6422, P-6424, P-6510; Paraplex G-40, G-41, G-50, G-51, G-54, G-56, G-57, G-59; Plasthall HA7A, P-630, P-640, P-643, P-650, P-670, P-686

Polyester azelate
Merrol P-1030LV, P-9500

Polyester, butylene glycol
Staflex 550, 802, 804, 809

Polyester glutarate
Merrol P-5510; Plasthall P-530, P-550, P-554, P-7035, P-7035M, P-7046, P-7092, P-7092D

Polyester nylonate
Merrol P-5511

Polyester phthalate
Merrol P-8425

Polyester phthalate/adipate
Merrol P-8227

Polyester polyol
Terate 202, 203, 204

Polyester resin
Admex 433, 500, 515, 522, 523, 525, 526, 527, 529, 760, 761, 770, 775, 780, 890, 910; Aropol 8101, 8110, 8310, 8319, 8321,

8329, 8329-25, 8339T-12, 8343T-11, 8343T-12, 8343T-21, 8343T-22, 8349-1583449T-09, 8343T-21, 8410-12, 8420, 8420-09, 8449T-15, 8520-14, 8621-16, 8800; Bonding Agent TN, TN/S50; Duowet EHS, EHS Conc.; Hercoflex 900; Hydropalat 535; Paraplex AL-111, G-30, G-31, GA-20, RGA-2, RGA-7, RGA-8; Polylite 31-822; Santicizer 412, 429; Struktol WB 300

Polyester resin, thermoplastic
Magnamite AS4/1655

Polyester resin, thermosetting and alkyd
Cellolyn 95-80T, 98-80T; Neolyn 35D, 40; Paraplex 5-B; XNS-6066

Polyester sebacate
Merrol P-1030; Paraplex G-25; Plasthall P-1070

Polyether
Alkapen EC-101; Gaftex CK-160 Conc., COM-154, COM-154 Conc., DN-159 Conc.; Genapol PS

Polyether glycol. See PEG

Polyether polyol
Colorin 102, 104, 202, 301, 302

Polyethylene (CTFA) [CAS #9002-88-4] *Synonyms:* Ethene, homopolymer
Ablusoft PE; Akrowax PE; Armostat 375; BASF Wax AL 61, FB; Bozemine N 60; Ceranine L; Forbest 410, P 22; Peem 122 Conc., 397, 410; Poligen PE; Polysoft CA; Polywax 500, 655

Polyethylene glycol. See PEG

Polyethylene, high-density *Synonyms:* HDPE
BASF Wax AF 30, AF 31, AF 32, AH 3, AH 6; Denlube 40; Polywax 850, 1000, 2000

Polyethylene, linear low-density *Synonyms:* LLDPE
Hostastat System E 6952

Polyethylene, low-density *Synonyms:* LDPE
BASF Wax A, AF, AL 3, AM 3, AM 6; Peem 122 Conc.

Polyethylene, oxidized [CAS #68441-17-8] *Synonyms:* Oxidized polyethylene (CTFA); Ethene, homopolymer, oxidized
A-C Polyethylene 629A; Cardipol LP, LP 0-25; Petrac 215; Sicolub OA 2, OA 4

Polyethylene polyamines
Vulkacit TR

Polyethylene terephthalate (PET). See Polyester resin, PET

Polyethylene wax
A-C Polyethylene 6, 6A, 8A, 9A, 8, 9; Interwax G 8252; Polymel #7

Polyethylene wax, high-density, oxidized
BASF Wax OA 2, OA 3; Polywax E-2020

Polyethylene wax, low-density, oxidized
BASF Wax OA

Polyglyceryl-2 borate sesquioleate *Synonyms:* Diglyceryl borate sesquioleate
Emulbon S-83

Polyglyceryl-10 decaisostearate *Synonyms:* Decaglycerin decaisostearate
Nikkol Decaglyn 10-IS

Polyglyceryl-10 decaoleate (CTFA) [CAS #11094-60-3; RD #977057-94-5] *Synonyms:* Decaglycerol decaoleate
Caplube 8442; Drewmulse 10-10-O; Drewpol 10-10-O; Nikkol Decaglyn 10-O; Polyaldo DGDO; Santone 10-10-O

Polyglyceryl-10 decastearate (CTFA) [CAS #39529-26-5; RD #977067-09-6] *Synonyms:* Decaglycerol decastearate
Caplube 8448; Caprol 10G10S; Drewmulse 10-10-S; Drewpol 10-10-S; Nikkol Decaglyn 10-S

Polyglyceryl-3 diisostearate (CTFA) [RD #977067-10-9] *Synonyms:* Diglyceryl diisostearate
Emerest 2452

Polyglyceryl-6 distearate (CTFA) [RD #977067-15-4] *Synonyms:* Hexaglycerol distearate
Drewmulse 6-2-S; Drewpol 6-2-S; Polyaldo HGDS

Polyglyceryl-10 heptaisostearate *Synonyms:* Decaglycerin heptaisostearate
Nikkol Decaglyn 7-IS

Polyglyceryl-10 heptaoleate *Synonyms:* Decaglycerin heptaoleate
Nikkol Decaglyn 7-O

Polyglyceryl-10 heptastearate *Synonyms:* Decaglycerin heptastearate
Nikkol Decaglyn 7-S

Polyglyceryl-10 hexaoleate
Polyaldo DGHO

Polyglyceryl-4 isostearate (CTFA) [RD #977068-33-9] *Synonyms:* Tetraglyceryl monoisostearate
Protegin W, WX; Witconol 18L

Polyglyceryl-10 isostearate *Synonyms:* Decaglycerin monoisostearate
Nikkol Decaglyn 1-IS

Polyglyceryl-10 laurate *Synonyms:* Decaglycerin monolaurate
Nikkol Decaglyn 1-L

Polyglyceryl-10 linoleate *Synonyms:* Decaglycerin monolinoleate
Nikkol Decaglyn 1-LN

Polyglyceryl methacrylate
Aloe-Moist; Lubrajel

Polyglyceryl-10 myristate *Synonyms:* Decaglycerin monomyristate
Nikkol Decaglyn 1-M

Polyglyceryl-10 octaoleate *Synonyms:* Decaglycerol octaoleate
Drewmulse 10-8-O; Drewpol 10-8-O

Polyglyceryl-3 oleate (CTFA) [CAS #9007-48-1 (generic); RD #977067-11-0] *Synonyms:* Triglyceryl oleate
Caplube 8410; Caprol 3GO; Drewmulse 3-1-O; Drewpol 3-1-O; Grindtek PGE 25; Santone 3-1-SH

Polyglyceryl-4 oleate (CTFA) [CAS #9007-48-1 (generic); RD #977057-95-6] *Synonyms:* Tetraglyceryl monooleate
Witconol 14

Polyglyceryl-8 oleate (CTFA) [CAS #9007-48-1 (generic)]
Santone 8-1-O

Polyglyceryl-10 oleate *Synonyms:* Decaglycerin monooleate
Nikkol Decaglyn 1-O

Polyglyceryl-10 pentaisostearate *Synonyms:* Decaglycerin pentaiosostearate
Nikkol Decaglyn 5-IS

Polyglyceryl-10 pentaoleate *Synonyms:* Decaglycerin pentaoleate
Nikkol Decaglyn 5-O

Polyglcyeryl-10 pentastearate *Synonyms:* Decaglycerin pentastearate
Nikkol Decaglyn 5-S

Polyglyceryl-3 shortening
Drewpol 3-1-SH

Polyglyceryl-3 stearate (CTFA) [CAS #37349-34-1 (generic); RD #977067-12-1] *Synonyms:* Triglyceryl stearate
Caprol 3GS; Drewmulse 3-1-S; Drewpol 3-1-S; Grindtek PGE 55, 55-6; Polyaldo TGMS; Santone 3-1-S

Polyglyceryl-6 stearate *Synonyms:* Hexaglycerol stearate
Mazol PGS-61; Nikkol Hexaglyn 1-S

Polyglyceryl-8 stearate (CTFA) [CAS #37349-34-1 (generic)]
Santone 8-1-S

Polyglyceryl-10 stearate *Synonyms:* Decaglycerin monostearate
Nikkol Decaglyn 1-S

Polyglyceryl-10 tetracocate *Synonyms:* Decaglycerol tetracocate
Caplube 8445

Polyglyceryl-10 tetraoleate (CTFA) [CAS #34424-98-1] *Synonyms:* Decaglycerol tetraoleate
Caplube 8440; Caprol 10G40; Drewmulse 10-4-O; Drewpol 10-4-O; Mazol PGO-104

Polyglyceryl-2 tetrastearate (CTFA) [RD#977067-17-6] *Synonyms:* Diglycerol tetrastearate
Caprol 2G4S

Polyglyceryl-10 trioleate *Synonyms:* Decaglycerin trioleate
Nikkol Decaglyn 3-O

Polyglyceryl-10 tristearate *Synonyms:* Decaglycerin tristearate
Nikkol Decaglyn 3-S

Polyglycol. See PEG

Polyisobutene. See Polybutylene

Polyisobutylene (PIB). See Polybutylene

Polyisocyanate
Bonding Agent 2001, 2005; Desmodur KA-8267, RC, RU

Polyisoprene (CTFA) [CAS #9006-04-6; 9003-31-0] *Synonyms:* Natural rubber; cis-1,4-Polyisoprene rubber
PA-57, -80

Poly 4,4′ isopropylidenediphenol neodol 25 alcohol phosphite
Weston 439

Polymaleic acid
Belclene 200

Polymaleic acid, sodium salt
Belclene 201

Polymaleic copolymer
Belclene 283

Polymethacrylic acid
Daxad 34, 34S

Polymethylene polyaniline
Curithane 103

Polymethylene polyphenyl isocyanate [CAS #101-68-8] *Synonyms:* Polymer of diphenylmethane 4,4′-diisocyanate
Desmodur VKS-2, VKS-4, VKS-18; Mondur MR, MRS

Polymethylsiloxane
Silwet L-77, L-720, L-7001, L-7500, L-7600, L-7602, L-7604, L-7605, L-7607

Poly-α-methylstyrene
Amoco Resin 18-210, 18-240, 18-290

Poly (α-methylstyrene-styrene acrylonitrile)
Blendex 586, 702

Poly [(6-morpholino-s-triazine-2,4-diyl) [2,2,6,6-tetramethyl-4-piperidyl) imino]-hexamethylene [(2,2,6,6-tetramethyl-4-piperidyl) imino]]
Cyasorb UV 3346 LD

Poly(octadecyl vinyl ether/maleic anhydride)
Gantrez AN-8194

Polyolefin resin. See also Polyethylene, Polypropylene, Polybutylene, Polyisoprene
Epolene C-10, C-13, C-14, C-15, C-16, C-17, C-18, E-10, E-11, E-12, E-14, E-15, E-43, N-10, N-11, N-12, N-14, N-15, N-34; Plastilease 514; Pulpex Thermal Bonding Pulps

Polyol polyborate
Emulbon LB-75

Poly (oxy-1,2-ethanediyl), α-isooctadecyl-ω-hydroxy
Arosurf MSF

Polyoxyethylene (POE). See under PEG

Polyoxypropylene (POP). See under PPG

Poly (pentabromobenzyl) acrylate
FR-1025

Polypropylene [CAS #9003-07-0] *Synonyms:* Propylene polymer; PP
Armostat 475; Hercoflat Texturing and Flatting Pigment; Hostastat System E 5951; Luperco 101-P20

Polypropylene glycol. See PPG

Polyquaternium-2 (CTFA) [CAS #63451-27-4; 68555-36-2]
Mirapol A-15

Polyquaternium-5 (CTFA) [CAS #26006-22-4] *Synonyms:* Quaternium-39; Acrylamide/β-methacrylyloxyethyl trimethyl ammonium methosulfate copolymer
Emcol Q; Reten 210, 220

Polyquaternium-6 (CTFA) [CAS #26062-79-3; 28301-34-0] *Synonyms:* Poly (dimethyl diallyl ammonium chloride); Poly (DMDAAC); Quaternium-40
Jordaquat 40; Merquat 100

Polyquaternium-7 (CTFA) [CAS #26590-05-6] *Synonyms:* N,N-Dimethyl-N-2-propenyl-2-propen-1-aminium chloride, polymer with 2-propenamide; Quaternium-41
Jordaquat 41; Merquat 550, S

Polyquaternium-11 (CTFA) [CAS #53633-54-8] *Synonyms:* Vinylpyrrolidone/dimethylaminoethyl methacrylate copolymer/dimethyl sulfate reaction product
Gafquat 734, 755, 755N

Polyquaternium-14 (CTFA) [CAS #27103-90-8]
Reten 300

Polyquaternium-16 [CAS #29297-22-0] *Synonyms:* Vinylpyrrolidone/vinyl imidazolinium methochloride copolymer
Luviquat Series

Polyquaternium-17
Mirapol AD-1

Polyquaternium-18 [CAS #90624-76-3]
Mirapol AZ-1

Polyquaternium-22
Merquat 280

Polyquaternium-24 [CAS #107987-23-5]
BioCare Polymer HA-24; Quatrisoft Polymer LM-200

Polyquaternium-27
Mirapol 9, 95, 175

Polysiloxane
F-221; Perenol S4, S5; Sag Silicone Antifoam 471, 4130, 4220, 5300, 5310; Silicone Rubber Base B-5, B-9, B-11, B-89, B-100, B-102, B-123, B-124, B-143, B-146, B-147, B-150, B-163, SWS-722, SWS-724, SWS-725, SWS-727, SWS-729, SWS-734; Silicone Rubber Compound C-1564, C-155, C-156, C-549, C-751, C-763, C-780U, C-783, C-914U, C-918U, SWS-06545, SWS-7142U, SWS-7143U, SWS-7152U, SWS-7162U, SWS-7172U, SWS-7182U, SWS-7183U, SWS-7341, SWS-7351, SWS-7355U, SWS-7362, SWS-7371, SWS-7381, SWS-7440U, SWS-7450U, SWS-7460U, SWS-7470U, SWS-7480U, SWS-7532, SWS-7575, SWS-7578, SWS-7580, SWS-7630U, SWS-7645U, SWS-7651U, SWS-7655, SWS-7656U, SWS-7665U, SWS-7666U, SWS-7675U, SWS-7676U, SWS-7677U, SWS-7801U, SWS-7802, SWS-7805, SWS-7810U, SWS-7185U, SWS-7802, SWS-7825U, SWS-7845U, SWS-7865U, SWS-7890U, SWS-9683; SWS-03314

Polysiloxane polydimethyl dimethylammonium acetate copolymer
Abil B 9905, B 9907, B 9908, B 9909

Polysiloxane polyether copolymer [CAS #68937-54-2]
Abil B 8, B 8842, B 8842, B 8843, B 8851, B 8852, B 8863, B 8873

Polysorbate 20 (CTFA) [CAS #9005-64-5 (generic)] *Synonyms:* POE (20) sorbitan monolaurate; PEG-20 sorbitan laurate; Sorbimacrogol laurate 300
Accosperse 20; Ahco 7596T; Alkamuls PSML-20; Armotan PML 20; Capmul POE-L; Crillet 1; Drewmulse POE-SML; Durfax 20; Emasol L-120; Emsorb 2720, 6915; Ethylan GEL-2; Eumulgin SML 20; Hetsorb L-20; Ionet T-20 C; Liposorb L-20; Lonzest SML-20; Mulsifan RT 141; Norfox Sorbo T-20; Radiamuls 137; Radiasurf 7137; Secoster CL 10; Soprofor T/20; Sorbax PML-20; Sorbon T-20; T-Maz 20; Tween 20

Polysorbate 21 (CTFA) [CAS #9005-64-5 (generic); RD #977053-95-4] *Synonyms:* POE (4) sorbitan monolaurate; PEG-4 sorbitan laurate
Ahco 7596D; Alkamuls PSML-4; Crillet 11; Emasol L-106; Ethylan GLE-21; Tween 21

Polysorbate 40 (CTFA) [CAS #9005-66-7] *Synonyms:* POE (20) sorbitan monopalmitate; Sorbimacrogol palmitate 300
Ahco DFP-156; Crillet 2; Emasol P-120; Emsorb 6910; Ethylan GEP-4; Liposorb P-20; Lonzest SMP-20; Rheodol TW-P120; Secoster CP 10; Sorbax PMP-20; T-Maz 40; Tween 40

Polysorbate 60 (CTFA) [CAS #9005-67-8 (generic)] *Synonyms:* POE (20) sorbitan monostearate; PEG-20 sorbitan stearate; Sorbimacrogol stearate 300
Accosperse 60; Ahco DFS-149; Alkamuls PSMS-20; Armotan PMS20; Atmul 700H; Capmul POE-S; Crillet 3; Drewmulse POE-SMS; Durfax 60; Emasol S-120; Emsorb 2728, 6905; Ethylan GES-6; Eumulgin SMS 20; Hetsorb S-20; Incrosorb S-60; Liposorb S-20; Lonzest SMS-20; Mazawax 163R; Nikkol TS-10, TS-10(FF); Radiamuls 147; Radiasurf 7147; Secoster CS 10; Soprofor T/60; Sorbax PMS-20; Tandem 5K, 8, 9, 11H, 552; T-Maz 60, 60K, 60KHM; Tween 60

Polysorbate 61 (CTFA) [CAS #9005-67-8 (generic); RD #977053-94-3] *Synonyms:* POE (4) sorbitan monostearate; PEG-4 sorbitan stearate
Ahco DFS-96; Alkamuls PSMS-4; Crillet 31; Emasol S-106; Emsorb 6906; T-Maz 61; Tween 61

Polysorbate 65 (CTFA) [CAS #9005-71-4; RD #977014-69-9] *Synonyms:* POE (20) sorbitan tristearate; PEG-20 sorbitan tristearate; Sorbimacrogol tristearate 300
Ahco 7166T; Alkamuls PSTS-20; Crillet 35; Drewmulse 365, POE-STS; Durfax 65, 65K; Emasol S-320; Emsorb 2729, 6907; Ice #12; Liposorb TS-20; Lonzest STS-20; Nikkol TS-30; Soprofor T/65; Sorbax PTS-20; T-Maz 65, 65K; Tween 65

Polysorbate 80 (CTFA) [CAS #9005-65-6 (generic)] *Synonyms:* POE (20) sorbitan monooleate; PEG-20 sorbitan oleate; Sorbimacrogol oleate 300
Accosperse 80; Ahco DFO-150; Aldosperse O-20 FG; Amerchol Polysorbate 80; Capmul POE-O; Crillet 4; Crodalan AWS; Drewmulse 700, POE-SMO; Durfax 80; Emasol O-120; Emrite 6120; Emsorb 2722, 6900; Ethoxyol 1707; Ethylan GEO-8; Eumulgin SMO 20; Ice #2; Liposorb O-20; Lonzest SMO-20; Mulsifan RT 146; Norfox Sorbo T-80; Radiamuls 157; Radiasurf 7157; Solulan 98; Soprofor T/80; T-Maz 80, 80K, 80KLM; Tween 80

Polysorbate 80 acetate (CTFA) *Synonyms:* POE (20) sorbitan monooleate acetate
Solulan 97; Sorbax PMO-20

Polysorbate 81 (CTFA) [CAS #9005-65-5 (generic); RD #977053-93-2] *Synonyms:* POE (5) sorbitan monooleate; PEG-5 sorbitan oleate
Ahco DFO-100; Alcolan 40; Alkamuls PSMO-5; Crillet 41; Emasol O-106; Emsorb 6901; Ethylan GOE-21; Liposorb O-5; Sorbax PMO-5; T-Maz 81, 81K; Tween 81

Polysorbate 85 (CTFA) [CAS #9005-70-3] *Synonyms:* POE (20) sorbitan trioleate; PEG-20 sorbitan trioleate; Sorbimacrogol trioleate 300

Ahco DFO-110; Alkamuls PSTO-20; Crillet 45; Emasol O-320; Emsorb 6903; Ethylan GPS-85; Liposorb TO-20; Lonzest STO-20; Ritaderm; Soprofor T/85; Sorbax PTO-20; T-Maz 85, 85K; Tween 85

Polystyrene (CTFA) [CAS #9003-53-6] *Synonyms:* PS; Benzene, ethenyl-homopolymer; Ethenylbenzene, homopolymer
Duolite A-143; Hostastat System E 3952, E 3953, E 3954; Lytron 284, 288, 295, 300, 305, 308, 614, 621; Macrobase 600, 610, 620; Macromer 13K-PSMA; Piccolastic C125; Witcopaque 11, 12, 25

Polystyrene, brominated
Pyro-Chek 60PB, LM

Polysulfide
P-60, -60-S Cold Molding Compounds

Polysulfide/epoxy
Tra-Bond 2133

Polysulfonate
Siponate DSB

Polyterpene resin
Nevtac 80, 99 Super, 100, 115, 130; Wingtack 95; Zonarez 7010, 7025, 7040, 7055, 7070, 7085, 7100, 7115, 7125, B-10, B-25, B-40, B-55, B-70, B-85, B-115, B-125

Polytetrafluoroethylene [CAS #9002-84-0] *Synonyms:* PTFE; TFE; Teflon
Alphaflex 101, 202; FC-60% SS; FC-446CS; FC-477 SM; TL-101, -102, -107, -113, -115, -115A, -115AH, -126, -126H, -127

Polytetramethylene ether glycol
Terathane 650, 1000, 2000, 2900

Polytrifluorochloroethylene. See Chlorotrifluoroethylene polymer

Polyurethane [CAS #9009-54-5]
Graphsize; Specflex; Tegocoll; Troykyd EX-390; Ultramoll PU

Polyurethane-acrylate resin
Chemlink 9503, 9504, 9505

Polyurethane, caprolactone monomer
CAPA Monomer

Polyurethane, polyester
Calthane NF 1300-60A

Polyurethane (polyester)/cotton composite
Hexaform Mesh

Polyurethane, polyether
Adiprene CM; Calthane NF 1500; Rucothane CO-A-5069

Polyvinyl acetate copolymer, carboxylated
Resyn 28-1310, 28-2930, 28-3307

Polyvinyl alcohol (CTFA) [CAS #9002-89-5] *Synonyms:* Ethenol, homopolymer; PVAL; PVOH
Crystic Release Agent No. 1, No. 2; Elvanol 71-30, 75-15, 90-50, T-66; Plastilease 512-B, 512-CL; Ucar Latex 131

Polyvinyl chloride [CAS #9002-86-2] *Synonyms:* PVC
Corvic E7236, S6617, S7106

Polyvinyl chloride/polyacrylate
Vinnol K 704

Polyvinylidene fluoride *Synonyms:* PVDF
TL-450

Polyvinyl isobutyl ether *Synonyms:* PVI; Polyvinyl ether
Perenol EI

Polyvinyl methyl ether (CTFA) [CAS #9003-09-2] *Synonyms:* Methoxyethene, homopolymer; Poly (methylvinyl ether); PVM
Gantrez M-154, -555, -556, -574

Polyvinyl methyl ether/maleic anhydride copolymer. See PVM/MA copolymer

Polyvinylpolypyrrolidone
Polyclar 10, AT

Polyvinylpyrrolidone. See PVP

POP (Polyoxypropylene). See under PPG

Potassium alginate (CTFA) [CAS #9005-36-1] *Synonyms:* Potassium polymannuronate
Kelmar, Kelmar Improved

Potassium 2-(2-carboxyethoxy) ethyl 2-(2-(2-hydroxy) ethoxyethyl methyldodecyl ammonium methosulfate
Sanac C

Potassium 2-(2-carboxyethoxy) ethyl 2-(2-(2-hydroxy) ethoxyethyl methyloctadecenyl ammonium methosulfate
Sanac S

Potassium chloride (CTFA) [CAS #7447-40-7]
Frimulsion 6G

Potassium citrate [CAS #866-84-2]
Frimulsion 6G

Potassium cocamidopropyl hydroxy sulfopropyl dimethyl betaine
Schercotaine SCAB-KG

Potassium coco-hydrolyzed animal protein (CTFA) [CAS #68920-65-0]
May-Tein C

Potassium dichloroisocyanurate [CAS #2244-21-5] *Synonyms:* Dichloroisocyanuric acid potassium salt; Potassium dichloro-s-triazinetrione
ACL 59

Potassium dimethyldithiocarbamate [CAS #128-03-0]
Aquatreat KM

Potassium lauryl phosphate
Crafol AP-31, AP-33

Potassium lignosulfonate
Lignosite KLS

Potassium molybdate [CAS #13446-49-6]
Molyhibit 200 Series

Potassium naphthalene sulfonate
Daxad 11KLS; Harol KG; Lomar HP

Potassium octoate
Kosmos 64

Potassium oleate (CTFA) [CAS #143-18-0] *Synonyms:* Potassium 9-octadecenoate
Atlas KO; Norfox KO

Potassium phosphated bis (hydroxyethyl) coco amine oxide
Jorphox KCAO

Potassium phosphated bis (hydroxyethyl) tallow amine oxide
Jorphox KTAO

Potassium polyacrylate [CAS #25608-12-2]
Daxad 37LK9

Potassium polyphosphonate
Wayplex 45-K, 68-K

Potassium ricinoleate (CTFA) [CAS #7492-30-0]
Kricinol 35; Solricin 135

Potassium rosinate *Synonyms:* Potassium soap of rosin
Dresinate 90, 91, 92, 93, 94, 95

Potassium sorbate (CTFA) [CAS #590-00-1] *Synonyms:* 2,4-Hexadienoic acid, potassium salt
Aflaban KSG; Sorbistat-K; Tristat K

Potassium stearate (CTFA) [CAS #593-29-3]
Imwitor 965, 965 K; Rhenovin S-stearat-80

Potassium tallate *Synonyms:* Tall oil rosin potassium soap
Arizona DRS-40, DRS-42

Potassium toluenesulfonate (CTFA) [CAS #16106-44-8; 30526-22-8]
Eltesol PT 45, PT 93; Hartotrope KTS 44, KTS 50; Naxonate 4KT, 5KT, KT

Potassium xylene sulfonate
Conco PXS; Eltesol PX 40, PX 93; Manro KXS 40

Potassium-zinc
Interstab M731; Lankromark LZ561, LZ1199, LZ1221, LZ1232

PPG-1.... See also under Propylene glycol...

PPG-9 (CTFA) [CAS #253322-69-4 (generic); RD #977055-51-8] *Synonyms:* Polyoxypropylene (9); Polypropylene glycol (9)
Jeffox PPG-400; Macol P-500; Pluracol P-410

PPG-12 (CTFA) [CAS #25322-69-4 (generic); RD #977066-97-9]
Pluracol P-710

PPG-17 (CTFA) [CAS #25322-69-4 (generic); RD #977066-99-1]
Pluracol P-1010

PPG-20 (CTFA) [CAS #25322-69-4 (generic); RD #977097-00-7]
Macol P-1200; Polyglycol P-1200

PPG-26 (CTFA) [CAS #25322-69-4 (generic); RD #977055-52-9] *Synonyms:* Polyoxypropylene (26)
Jeffox PPG-2000; Macol P-2000; Pluracol P-2010; Polyglycol P-2000

PPG-30 (CTFA) [CAS #25322-69-4 (generic); RD #977055-53-0]
Macol P-4000; Pluracol P-4010

PPG-34 (CTFA) [CAS #25322-69-4 (generic)]
Witconol CD-17

PPG 400
Mapeg PPG-400; Texox PPG-400

PPG 425
Alkapol PPG-425

PPG 500
Mapeg PPG-500

PPG 1000
Alkapol PPG 1000

PPG 1200
Alkapol PPG 1200

PPG 2000
Alkapol PPG 2000; Texox PPG-2000

PPG 4000
Alkapol PPG 4000

PPG-5-buteth-7 (CTFA) [CAS #9038-95-3 (generic); RD #977066-93-5] *Synonyms:* POE (7) POP (5) monobutyl ether
Pluracol W-170

PPG-12-buteth-16 (CTFA) [CAS #9038-95-3 (generic); RD #977055-56-3]
Macol 660; Pluracol W-660

PPG-20-buteth-30 (CTFA) [CAS #9038-95-3 (generic); RD #977055-58-5]
Pluracol W-2000

PPG-24-buteth-27 (CTFA) [RD #977066-95-7]
Tergitol XD

PPG-26 buteth-26 (CTFA) [CAS #903-89-5]
Witconol APEB

PPG-28-buteth-35 (CTFA) [CAS #9038-95-3; RD #977055-59-6] *Synonyms:* POE (35) POP (28) monobutyl ether
Macol 3520; Pluracol W3520N

PPG-33-buteth-45 (CTFA) [CAS #9038-95-3 (generic); RD #977055-60-9]
Macol 5100; Pluracol W-5100N

PPG-2-ceteareth-9 *Synonyms:* POP (2) POE (9) cetyl/stearyl ether; POE (9) POP (2) cetyl/stearyl ether
Eumulgin L

PPG-10-ceteareth-20 (CTFA) [RD #977067-21-2]
Standamul OXL

PPG-1 ceteth-3 acetate *Synonyms:* Propylene glycol POE (3) cetyl ether acetate
Hetester PCA

PPG-4 ceteth-1 (CTFA) [CAS #9087-53-0 (generic)] *Synonyms:* POE (1) POP (4) cetyl ether; POP (4) POE (1) cetyl ether
Nikkol PBC-31

PPG-4-ceteth-10 (CTFA) [CAS #9087-53-0 (generic)]
Nikkol PBC-33

PPG-5 ceteth-10 phosphate (CTFA) [RD #977061-68-9] *Synonyms:* POP (5) POE (10) cetyl ether phosphate
Crodafos SG

PPG-5-ceteth-20 (CTFA) [CAS #9087-53-0 (generic); RD #977061-66-7]
Crodamol SS; Crodapearl NI Liq.; Incropearl Liquid NI; Procetyl AWS

PPG-8-ceteth-1 (CTFA) [CAS #9087-53-0 (generic)]
Nikkol PBC-41

PPG-8-ceteth-20 [CAS #9087-53-0 (generic)]
Nikkol PBC-44

PPG-10 cetyl ether (CTFA) [CAS #9035-85-2 (generic); RD #977055-66-5] *Synonyms:* POP (10) cetyl ether; PPG (10) cetyl ether
Procetyl 10

PPG-20 cetyl ether
Procetyl 20

PPG-30 cetyl ether (CTFA) [RD #977055-67-6]
Procetyl 30; Wickenol 707

PPG-50 cetyl ether (CTFA) [CAS #9035-85-2; RD #977060-07-3]
Procetyl 50

PPG-10 cetyl ether phosphate (CTFA) [RD #977060-96-0] *Synonyms:* POP (10) cetyl ether phosphate
Crodafos CAP

PPG-6-decyltetradeceth-12 (CTFA) *Synonyms:* POE (12) POP (6) decyl tetradecyl ether
Nikkol PEN-4612

PPG-6-decyltetradeceth-20 (CTFA)
Nikkol PEN-4620

PPG-6-decyltetradeceth-30 (CTFA)
Nikkol PEN-4630

PPG-3 diacrylate *Synonyms:* Tripropylene glycol diacrylate
Macrobase 610, 620

PPG-2 dibenzoate (CTFA) [CAS #94-51-9] *Synonyms:* Dipropylene glycol dibenzoate; Polypropylene glycol (2) dibenzoate; Polyoxypropylene (2) dibenzoate
Benzoflex 9-88, 9-88 SG, 50

PPG-9 diethylmonium chloride (CTFA) [CAS #9042-76-6] *Synonyms:* POP (9) methyl diethyl ammonium chloride; Quaternium-6
Emcol CC-9

PPG-25 diethylmonium chloride (CTFA) [RD #977062-01-3] *Synonyms:* POP (25) methyl diethyl ammonium chloride; Quaternium-20
Emcol CC-36

PPG-40 diethylmonium chloride (CTFA) [CAS #9076-43-1] *Synonyms:* POP (40) methyl diethyl ammonium chloride; Quaternium-21
Emcol CC-42

PPG-1 isoceteth-3 acetate *Synonyms:* Propylene glycol POE (3) isocetyl ether acetate
Hetester PHA

PPG-3-isosteareth-9 (CTFA) [RD #977068-36-2] *Synonyms:* POE (9) POP (3) isostearyl ether; POP (3) POE (9) isostearyl ether; PEG-9-PPG-3 isostearyl ether
Arosurf 66-PE12

PPG-2 lanolin ether (CTFA) [CAS #68439-53-2 (generic); RD #977067-06-3] *Synonyms:* POP (2) lanolin ether; PPG (2) lanolin ether
Solulan PB-2

PPG-5 lanolin ether (CTFA) [CAS #68439-53-2 (generic); RD #977067-04-1]
Solulan PB-5

PPG-10 lanolin ether (CTFA) [CAS #68439-53-2 (generic); RD #977061-53-2]
Solulan PB-10

PPG-20 lanolin ether (CTFA) [CAS #68439-53-2 (generic); RD #977064-51-9]
Solulan PB-20

PPG-30 lanolin ether (CTFA) [CAS #68439-53-2 (generic); RD #977055-71-2] *Synonyms:* Polyoxypropylene (30) lanolin ether
Hetoxol PLA; Wickenol 727

PPG-2 laurate *Synonyms:* Dipropylene glycol monolaurate; POP (2) laurate; PPG (2) laurate
Cithrol DPGMO N/E, DPGML S/E

PPG-3-laureth-9 (CTFA) *Synonyms:* POE (9) POP (3) lauryl ether; POP (3) POE (9) lauryl ether
Acconon 1300

PPG-2 methyl ether (CTFA) [CAS #13429-07-7; 34590-94-8; 37286-64-9 (generic)] *Synonyms:* Dipropylene glycol monomethyl ether; Methoxy dipropylene glycol; Polypropylene glycol (2) methyl ether
Arcosolv DPM; Dowanol DPM; Poly-Solv DPM

PPG-3 methyl ether (CTFA) [CAS #10213-77-1; 37286-64-9 (generic); RD #977067-03-0] *Synonyms:* Tripropylene glycol monomethyl ether; Polyoxypropylene (3) methyl ether
Arcosolv TPM; Poly-solv TPM

PPG-2 methyl ether acetate
Arcosolv DPM Acetate

PPG-10 methyl glucose ether (CTFA) [CAS #61849-72-7; RD #977067-01-8] *Synonyms:* POP (10) methyl glucose ether; PPG (10) methyl glucose ether
Glucam P-10

PPG-20 methyl glucose ether (CTFA) [CAS #61849-72-7; RD #977067-02-9]
Glucam P-20

PPG-20 methyl glucoside distearate
Glucam P-20 Distearate

PPG-3-myreth-3 (CTFA) [CAS #37311-04-9 (generic); RD #977069-61-6] *Synonyms:* POE (3) POP (3) myristyl ether; POP (3) POE (3) myristyl ether
Witconol APEM

PPG-3 myristyl ether (CTFA) [CAS #63793-60-2 (generic); 78436-05-3 (generic); RD #977068-35-1] *Synonyms:* Tripropylene glycol myristyl ether; POP (3) myristyl ether; PPG (3) myristyl ether
Promyristyl PM-3; Witconol APM

PPG-2 myristyl ether propionate (CTFA) [RD #977078-24-2] *Synonyms:* Polypropylene glycol (2) myristyl ether propionate
Crodamol PMP

PPG-2 oleate *Synonyms:* Dipropylene glycol monooleate; POP (2) monooleate; PPG (2) monooleate
Cithrol DPGMO N/E, DPGMO S/E

PPG-26 oleate (CTFA) [CAS #31394-71-5 (generic); RD #977055-74-5]
OP-2000

PPG-36 oleate (CTFA) [CAS #31394-71-5 (generic); RD #977055-75-6]
Witconol F26-46

PPG-10 oleyl ether (CTFA) [CAS #52581-71-2 (generic); RD #977055-76-7]
Provol 10

PPG-30 oleyl ether (CTFA) [CAS #52581-71-2 (generic); RD #977055-77-8]
Provol 30

PPG-50 oleyl ether (CTFA) [CAS #52581-71-2 (generic); RD #977055-78-9]
Provol 50

PPG-12-PEG-50 lanolin (CTFA) [CAS #684858-58-8 (generic); RD #977062-71-7]
Lanexol AWS

PPG-12-PEG-65 lanolin oil [CAS #68458-58-8 (generic)]
Fluilan AWS, AWS 1692

PPG-40-PEG-60 lanolin oil (CTFA) [RD #977065-80-7]
Aqualose LL100

PPG-2 salicylate (CTFA) [CAS #7491-14-7] *Synonyms:* Dipropylene glycol salicylate; POP (2) monosalicylate
Dipsal

PPG-2 stearate *Synonyms:* Dipropylene glycol monostearate; POP (2) monostearate; PPG (2) monostearate
Cithrol DPGMS N/E, DPGMS S/E

PPG-11 stearyl ether (CTFA) [CAS #25231-21-4 (generic); RD #977069-62-7] *Synonyms:* POP (11) stearyl ether; PPG (11) stearyl ether
Witconol APS

PPG-15 stearyl ether (CTFA) [CAS #25231-21-4 (generic); RD #977061-23-6]
Acconon E; Arlamol E, S3, S7 ; Hetoxol SP-15; Prostearyl 15

PPG-3 trideceth-6 *Synonyms:* POE (6) POP (3) tridecyl ether; POP (3) POE (6) tridecyl ether; PEG-6 PPG-3 trideceyl ether
Hetoxol TDEP-63

PPG-15 trideceth-10
Hetoxol TDEP-15

Pristane [CAS #1921-70-6] *Synonyms:* 2,6,10,14-Tetramethylpentadecane
Robuoy

Prometryn
Cotton-Pro

Propanil
Surcopur

Propineb
Antracol; Fruvit

Propionic acid [CAS #79-09-4] *Synonyms:* Methylacetic acid
Tenox IBP-2, P

Propoxur
Baygon; Unden

Propylated triaryl phosphate
Pliabrac 519, 521, 524

Propylene carbonate (CTFA) [CAS #108-32-7] *Synonyms:* 1,3-Dioxolan-2-one, 4-methyl; 4-Methyl-1,3-dioxolan-2-one
Arconate 1000, HP; Bentone Gel IPM, Gel MIO, Gel SS-71; Texacar EC-50, PC

Propylene glycol (CTFA) [CAS #57-55-6] *Synonyms:* 1,2-Propylene glycol; 1,2-Propanediol; 1,2-Dihydroxypropane; Methylene glycol; Methyl glycol. See also under *PPG-1*
Aloe-Moist; Atmos 2462; Bronidox L, L 5; Ceraphyl 85; Cremophor RH 455; Dowfrost; Emid 6590; Germaben II, II-3; Incromate PPI; Lexate CL-60; Lubrajel; Monamid R31-42; Pentex 99; Solusol 84%; Standapol Pearl Conc. 7130; Sustane 3, 8, 20, 20-3, 31, Q, W; Texacat TD-33; Varamide SL-9

Propylene glycol alginate (CTFA) [CAS #9005-37-2] *Synonyms:* Hydroxypropyl alginate
Kelcoloid D, DH, DO, DSF, HVF, LVF, O, S

Propylene glycol n-butyl ether
Dowanol PnB

Propylene glycol t-butyl ether
Arcosolv PTB

Propylene glycol dibenzoate
Benzoflex 284

Propylene glycol dicaprylate (CTFA) [CAS #7384-98-7; RD #977060-70-0] *Synonyms:* Octanoic acid, 1-methyl-1,2-ethanediyl ester
Crodamol PC

Propylene glycol dicaprylate/dicaprate (CTFA) [RD #977060-55-1]
Aldo DC; Captex 200; Edenol 302; Lexol PG 855, PG 865; Liponate PC; Mazol PG-810; Miglyol 840, 840 Gel; Neobee M-20; Standamul 302

Propylene glycol dioctanoate (CTFA) [CAS #56519-71-2]
Captex 800; Lexol PG 800

Propylene glycol dipelargonate (CTFA) [CAS #41395-83-9] *Synonyms:* Propylene glycol dinonanoate
DPPG; Emerest 2388; Lexol PG 900; Schercemol PGDP

Propylene glycol distearate (CTFA) [CAS #6182-11-2]
Mapeg PGDS

Propylene glycol hydroxystearate (CTFA) [CAS #33907-47-0]
Naturechem PGHS; Paricin 9

Propylene glycol isostearate (CTFA) [CAS #68171-38-0] *Synonyms:* Propylene glycol monoisostearate
Emerest 2384

Propylene glycol laurate (CTFA) [CAS #142-55-2; 27194-74-7]
Cithrol PGML N/E, PGML S/E; Schercemol PGML; Tegin PL

Propylene glycol methyl ether [CAS #107-98-2] *Synonyms:* Polypropylene glycol monomethyl ether
Arcosolv PM; Norfox NBC; Poly-Solv MPM

Propylene glycol methyl ether acetate
Arcosolv PM Acetate; Desmodur Z-4370

Propylene glycol myristate (CTFA) [CAS #29059-24-3]
Radiasurf 7196

Propylene glycol myristyl ether (CTFA) [CAS #54117-34-9] *Synonyms:* Propoxylated (1) myristyl alcohol
Promyristyl PM

Propylene glycol myristyl ether acetate (CTFA)
Hetester PMA

Propylene glycol oleate (CTFA) [CAS # 1330-80-9; RD #977012-90-0]
Cithrol PGMO N/E, PGMO S/E; Radiasurf 7206

Propylene glycol ricinoleate (CTFA) [CAS #26402-31-3]
Cithrol PGMR N/E, PGMR S/E; Flexricin 9; Naturechem PGR

Propylene glycol stearate (CTFA) [CAS #1323-39-3] *Synonyms:* Propylene glycol monostearate
Aldo PMS; Cerasynt PA; Cithrol PGMS N/E, PGMS S/E; CPH-52-SE; Dermalcare PGMS; Drewmulse PGMS; Emerest 2380; Kessco Propylene Glycol Monostearate Pure; Lipo PGMS; Mapeg PGMS; Mazol PGMS; Monosteol; Myvatex 3-50, 40-06S; Nikkol PMS-1C; Radiasurf 7201; Schercemol PGMS; Witconol RHP (SE)

Propylene glycol stearate SE (CTFA) [RD #977067-33-6]
Emerest 2381; Nikkol PMS-1CSE; Tegin P, P-411 SE

Propyl gallate (CTFA) [CAS #121-79-9] *Synonyms:* 3,4,5-Trihydroxybenzoic acid, propyl ester; ; n-Propyl 3,4,5-trihydroxybenzoate
Sustane 3, 20A, PG, W; Tenox PG

Propyl mercaptan
Spotleak 1009

Propyl oleate
Emerest 2302

Propylparaben (CTFA) [CAS #94-13-3] *Synonyms:* Propyl p-hydroxybenzoate
Germaben II, II-3; Lexgard P; Trisept P

n-Propyltrichlorosilane [CAS #141-57-1]
CP0800

n-Propyltrimethoxysilane [CAS #1067-25-0]
CP0810

Propyltrimonium hyrolyzed animal protein
Protectein

Protease
Esperase 4.0T, 8.0L; Gelatinase No. 53; HT-Proteolytic; Maxacal F400, P400,000; Maxatase LS400, MP375, P440; Milezyme AFP, APG, APL, Bromelain, Fungal Protease, Pancreatin 4NF, Papain; Oropon; Papain P-100; Prolase 300; Protoferm; Rhozyme P11, P41, P53, P64, PF

Protease betaglucanase
Brew(n)zyme

Proteinase
Alcalase 2.0T, 2.5L; Esperase 4.0T, 8.0L, 16.0L; Neutrase 0.5 L, 1.5 G; Savinase 4.0T, 8.0L, 16.0 L Type W

Prothiofos
Tokuthion

Pullulanase/amyloglucosidase
Dextrozyme

PVM/MA copolymer (CTFA) [CAS #9011-16-9] *Synonyms:* Methyl vinyl ether/maleic anhydride copolymer; Poly(methyl vinyl ether/maleic anhydride)
Gaftex PT; Gantrez AN-119, -139, -149, -169, -179, S-95, S-97; Thickener L, LN

PVM/MA copolymer, butyl ester [CAS #54018-18-7; 54578-91-5] *Synonyms:* Poly(methylvinyl ether/maleic acid) butyl ester
Gantrez ES-425, -435

PVM/MA copolymer, ethyl ester [CAS #50935-57-4; 54578-90-4]
Gantrez ES-225, SP-215

PVM/MA copolymer, isopropyl ester [CAS #54578-88-0; 56091-51-1] *Synonyms:* Poly(methylvinyl ether/maleic acid) isopropyl ester
Gantrez ES-335

PVM/MA copolymer, sodium salt *Synonyms:* Sodium salt of maleic anhydride/methyl vinyl ether copolymer
Sokalan CP 2, CP 2 Powd.

PVP (CTFA) [CAS #9003-39-8] *Synonyms:* Polyvinylpyrrolidone; Povidone; 1-Ethenyl-2-pyrrolidinone, homopolymer
Ganex P, P904, V, V216, V220, V516; Luviskol, K12, K17, K30, K60, K80, K90; Peregal ST; Plasdone C-15, C-30, K-25, K-26/28, K-29/32, K-90; Polyclar 10, AT; Polyplasdone XL, XL-10; PVP K-15, K-30, K-60, K-90; Sokalan HP 50

PVP/dimethylaminoethyl methacrylate copolymer (CTFA) [RD #977079-80-3] *Synonyms:* Vinylpyrrolidone/dimethylaminoethyl methacrylate copolymer
Copolymer 845, 937, 958

PVP/eicosene copolymer
Ganex V220

PVP/ethyl methacrylate/methacrylic acid polymer (CTFA) [CAS #26589-26-4]
Stepanhold Extra, R-1

PVP/hexadecene copolymer
Ganex V216, V516

PVP/VA copolymer (CTFA) [CAS #25086-89-9] *Synonyms:* Vinylpyrrolidone/vinyl acetate copolymer; Polyvinyl pyrrolidone/vinyl acetate copolymer
Luviskol VA 28 E, VA 28 I, VA 37 E, VA 37 I, VA 55 E, VA 55 I, VA 64, VA 64 I, VA 73 E; Nasuna B; PVP/VA W-735

PVP/VA/vinyl propionate terpolymer
Luviskol VAP 343 E, VAP 343 I

Pyrophyllite [CAS #12269-78-2] *Synonyms:* Hydrated aluminum silicate; Agalmatolite
Pyrax A, ABB, B, WA

2-Pyrrolidone [CAS #616-45-5] *Synonyms:* Butyrolactam
2-Pyrol

Quartz-kaolinite
Sillikolloid P 87; Sillitin N 82, N 85, V 85, Z 86, Z 89

Quaternium-2. See Soyaethyl morpholinium ethosulfate

Quaternium-3. See Myristalkonium saccharinate

Quaternium-6. See PPG-9 diethylmonium chloride

Quaternium-7. See Steapyrium chloride

Quaternium-8 (CTFA) [RD #977066-07-1] *Synonyms:* Alkyl dimethyl ethylbenzyl ammonium cyclohexyl sulfamate
Onyxide 172

Quaternium-9. See Soytrimonium chloride

Quaternium-12
BTC 1010-80

Quaternium-14 (CTFA) [CAS #27479-28-3] *Synonyms:* Dodecyl dimethyl ethylbenzyl ammonium chloride
BTC 2125, 2125 M

Quaternium-15 (CTFA) [CAS #4080-31-3; 51229-78-8] *Synonyms:* 1-(3-Chloroallyl)-3,5,7-triaza-1-azoniaadamantane chloride
Dowicil 75, 200

Quaternium-17. See Cetethyldimonium bromide

Quaternium-18 (CTFA) [CAS #61789-80-8] *Synonyms:* Dimethyl di (hydrogenated tallow) ammonium chloride; Dihydrogenated tallow dimethyl ammonium chloride; Dihydrogenated tallow dimonium chloride
Accosoft 707; Adogen 442, 442-100P, 442 H; Arquad 2HT, 2HT-75, 88; Crapol FU-25; Cycloton D261C/75; Jet Quat 2HT-75; Kemamine Q-9702C; Noramium M2SH; Querton 442, 442-11, 442-82; Radiaquat 6442; Varisoft 110, 3262, DHT

Quaternium-18 bentonite (CTFA) [CAS #68953-58-2] *Synonyms:* Quaternary ammonium compounds, bis (hydrog. tallow alkyl) dimethyl, chlorides, reaction products with bentonite
Bentone 34

Quaternium-18 hectorite (CTFA) [CAS #71011-27-3; RD #977062-10-4] *Synonyms:* Quaternary ammonium compounds, bis (hydrog. tallow alkyl) dimethyl, chlorides, reaction products with hectorite
Bentone 38, Gel MIO, Gel SS-71

Quaterium-18 methosulfate (CTFA) [CAS #61789-81-9] *Synonyms:* Quaternary ammonium compounds, bis (hydrogenated tallow alkyl) dimethyl, methyl sulfates
Accosoft 748; Carsosoft V-90, V-100

Quaternium-21. See PPG-40 diethylmonium chloride

Quaternium-22 (CTFA) [CAS #51812-80-7] *Synonyms:* 1-Propanaminium, 3-(D-gluconoylamino)-N-(2-hydroxyethyl)-N,N-dimethyl-, chloride
Ceraphyl 60

Quaternium-24 (CTFA) [CAS #32426-11-2] *Synonyms:* Decyl dimethyl octyl ammonium chloride; Octyl decyl dimethyl ammonium chloride
Bardac 205M, 2050; BTC 818, 885; FMB 302-8 Quat, 504-5 Quat

Quaternium-25. See Cetethyl morpholinium ethosulfate

Quaternium-26 (CTFA) [CAS #68953-64-0] *Synonyms:* Minkamidopropyl dimethyl 2-hydroxyethyl ammonium chloride
Ceraphyl 65

Quaternium-27 (CTFA) [RD #977065-92-1]
Carsosoft S-75, S-90; Rewoquat W 7500; Varisoft 475

Quaternium-31. See Dicetyl dimonium chloride

Quaternium-32. See Isostearyl ethylimidonium ethosulfate

Quaternium-33 (CTFA) [RD #977077-13-6]
Lanoquat 1756

Quaternium-39. See Polyquaternium-5

Quaternium-52 (CTFA) [CAS #58069-11-7]
Dehyquart SP

Quaternium-53
Incrosoft T-90HV

Quaternium-61 (CTFA)
Schercoquat DAS

Quaternium-62 (CTFA)
Schercoquat IEP

Quaternium-63 (CTFA)
Schercoquat DAB

Quaternium-70 (CTFA) *Synonyms:* Stearamidopropyl dimethyl (myristyl acetate) ammonium chloride
Ceraphyl 70

Quaternium-72
Incroquat 248; Incrosoft 248

Quaternium-75
Finquat CT

Quaternium-79 hydrolyzed animal protein
Mackpro NLP

Quaternium-80
Abil-Quat 3270, 3272

Rapeseedamidopropyl benzyldimonium chloride (CTFA)
Schercoquat ROAB

Rapeseedamidopropyl dimethyl epoxypropyl ammonium chloride
Schercoquat ROEP

Rapeseedamidopropyl ethyldimonium ethosulfate (CTFA)
Schercoquat ROAS

Resorcinol (CTFA) [CAS #108-46-3] *Synonyms:* Resorcin; m-Dihydroxybenzene; 3-Hydroxyphenol; 1,3-Benzenediol
Cohedur RK, RL, RS

Resorcinol-formaldehyde resin
SRF-1501; Vulkadur T, 40%

Resorcinol monobenzoate [CAS #136-36-7]
Eastman Inhibitor RMB

Reticulin, hydrolyzed
Reticusol

Rice bran wax (CTFA) [CAS #8016-60-2]
Noda Rice Wax FCC, Industrial

Ricinoleamide DEA (CTFA) [CAS #40716-42-5] *Synonyms:* Ricinoleic diethanolamide
Alkamide RDO; Aminol CA-2; Cyclomide RODEA; Lamacit ER; Mackamide R; Rodea

Ricinoleamide MEA (CTFA) [CAS #106-16-1] *Synonyms:* Ricinoleoyl monoethanolamide
Rewomid R 280

Ricinoleamidopropyl betaine (CTFA) *Synonyms:* Ricinoleamidopropyl dimethyl glycine
Mackam RA

Ricinoleamidopropyl dimethylamine (CTFA) [CAS #20457-75-4, #977010-66-4]
Lexamine R13; Mackine 201

Ricinoleamidopropyl dimethylamine lactate (CTFA) [CAS #977012-91-1] *Synonyms:* N-[3-(Dimethylamino) propyl] ricinoleamide lactate
Mackalene 216

Ricinoleamidopropyl ethyldimonium ethosulfate (CTFA) [CAS #112324-16-0]
Jordaquat JN; Lipoquat R

Ricinoleamidopropyl trimonium methosulfate
Rewoquat RTM 50

Ricinoleic acid (CTFA) [CAS #141-22-0] *Synonyms:* 12-Hydroxy-9-octadecenoic acid
P-10 Acid

Ricinoleic sulfosuccinate
Varsulf S1333

Rosin (CTFA) [CAS #8050-09-7] *Synonyms:* Colophony; Gum rosin; Rosin gum; WW wood rosin; Tall oil rosin
Dresinol 40, 42; Dry Pexol 200, 243; Dry Size XL20C; Hercules 752, 1331, X Dry Size; NovaSize; Pexol Size; Resin 731D

Rosin, hydrogenated
Staybelite, 570

Saccharin, insoluble [CAS #81-07-2]
Syncal SDI

Saccharose. See Sucrose

Safflowleramidopropyl dimethylethyl ammonium ethyl sulfate
Schercoquat FOAS

Safflower oil (CTFA) [CAS #8001-23-8]
Lipovol SAF; Super Refined Safflower Oil USP

Safflower oil, hybrid (CTFA) [RD #977064-78-0]
Lipovol SO

Salicylic acid (CTFA) [CAS #69-72-7] *Synonyms:* 2-Hydroxybenzoic acid
Retarder SAX

SD Alcohol 40 (CTFA) [RD #977021-81-0]
Bentone Gel MIO A-40

SD Alcohol 40-B (CTFA) [RD #977021-83-2]
Equex AEM

Sebacic acid (CTFA) [CAS #111-20-6] *Synonyms:* Decanedioic acid
Paraplex RG-8

Selenium [CAS #7782-49-2] *Synonyms:* Colloidal selenium
Vandex

Selenium diethyldithiocarbamate [CAS #21559-14-8]
Ethyl Selenac

Selenium dimethyldithiocarbamate [CAS #144-34-4]
Methyl Selenac

Serum albumin (CTFA) [CAS #9048-46-8]
Bovinol 30

Sesame oil (CTFA) [CAS #8008-74-0] *Synonyms:* Gingilli oil
Lipovol SES, SES-S; Super Refined Sesame Oil

Sesamide DEA
Incromide SED

Sesamidopropalkonium chloride
Incroquat SE-85

Sesamidopropylamine oxide
Incromine Oxide SE

Sesamidopropyl betaine
Incronam SE-30

Sesamidopropyl dimethylamine lactate
Incromate SEL

Shark liver oil (CTFA) [CAS #68990-63-6]
Robecote; Super Refined Shark Liver Oil

Shea butter (CTFA) [RD #977026-99-5]
Cetiol SB 45

Silica (CTFA) [CAS #7631-86-9] *Synonyms:* Silicone dioxide, fumed; Silicon dioxide
Aerosil R-972, R-974, R-976; AF-72, AF-75; Cab-O-Sil; Cab-O-Sperse; Crystic Pregel 17, 27; Hercules Defoamer 137, 2470; Hi-Sil 132, 243, 532EP, ABS, T-690; Hubersil 162; Imsil A-8, A-30, A-75; Koraid PSM; Rexfoam 150-A; Rhenocure TP/S; Rhenofit BDMA/S, EDMA/S, TAC/S, TRIM/S; Rhenovin CBS-70, MBT-70, MBTS-70, TMTD-70, TMTM-70; Sillum-PL 200; Silver Bond 30, 45, B; Tamsil 8, 10, 15, 25, 30, 45, 75, 150, Gold Bond; Tixosil 311, 321, 331, 375; TS 100; Vulkazon AFS-50; XPV1 Reinforcing Whisker; Zeo 49; Zeodent 113; Zeofree 80, 153; Zeomatt 155; Zeosyl 100, 110SD, 200; Zeothix 95, 175, 177, 265

Silica-alumina
Sillum-200; Zeeospheres 200, 400, 600, 800, 850

Silica, amorphous. See Silica, hydrated

Silica, colloidal
Ludox HS-30, SM; Nyacol, 215, 830, 1430, 1440, 2030 EC, 2040 NH4, 2046 EC, 9950

Silica, diatomaceous. See Diatomaceous earth

Silica, hydrated (CTFA) [CAS #1343-98-2 (silicic acid); 10279-57-9; 7699-41-4] *Synonyms:* Precipitated silica; Silica gel; Silica hydrate; Silicic acid; Amorphous silica
Britesorb; Flo-Gard AG 110, 130, 150, CC 120, 140, 160, FF 310, 320, 330, 350, 370, 390; Hi-Sil 210, 233, 250, T-600; Imsil A-10, A-15, A-25, A-108; Levilite; Lo-Vel 27, 28, 29, 39, 275; Quso; Silene 732D, D; Sipernat 22, 22 S, 50, 50S, D 17; Syloid 378; Vulksasil, N, N/GR-S, S

Silica, precipitated. See Silica, hydrated

Silicone acrylate
Tego Silicone Acrylate RC

Silicone emulsions and compounds. See also Dimethicone, Cyclomethicone, Polysiloxane *Synonyms:* Organosiloxane
AF 6050, 8805, 8805 FG, 8810, 8810 FG, 8820, 8820 FG, 8830, 8830 FG; Bozemine N 609; Cru Fluid 350; DF 1040; Defoamer S-10, S-100; Dehydran 150; E-170; F-755, -789; Dow Corning 7 Compound, 236 Dispersion, 34, 290, 346, 347, HV-490 Emulsion, 203, 233A Fluid, 1-2531 Release Coating, Q2-7119 Release/Parting Agent; Foamaster DRY, DS, FLD; Foam Burst 2, 5, 10, 87, 100, 105, 150, 151; Foamex AD-50, J-10; Foamkill 8B A, 8G, 80J Series, 400A, 679, 830, 836A, 852, FPF, GCP Series, MS-1, MS Conc.; Hodag FD Series; Hydro-Pruf 200; Levaform Si Emulsion, Si Oil, SiV; Masil 260, 260A, 265, 265HV, 266, 270, 271, 272, 290D, 290F, 1066C, 1066D, 2132, 2133, 2134, EM 266; Mazu DF 100S, 110S, 130S, 200S, 200SX, 200SX Special, 210S, 210SX, 210SX Mod 1, 215SX, 230S, 230SX, 243; Molykote; Nalco 2300, 2305; Nofoam NO 3990; Norfox 243, DF210SX, EM 350X, EM-350X Conc.; No Stik 802; Patcote 315, 500, 512, 513, 519, 520, 525, 531, 532, 543, 550, 577, 598; Plastilease 250; Polymekon; Q-70, -93, -101; Raneoff; Rexfoam B, C, D; Ridafoam Base 100, S-103-N, S-110-N; Sag Silicone Antifoam 10, 30, 310; Sandoz Defoamer F; Senka Antifoam 800; Seyco Defoamer D-10, K-20, 96(5), 99; SF 1201; SFR 100; Silcolapse 430, 431, 432, 5000, 5001, 5006, 5007, 5008, 5009, 5010, 5020; Silicone AF-100 IND; Silicone Defoamer #5037; Silicone M400, M404, M405, M406, M407, M409; Silicone Release 87-X66; SM 2059, 2079, 2086; SS 4098, 4129, 4177, 4267; Silsmooth E, G, SR/VEL; SM 2059, 2079, 2086, 2109, 2112, 2133, 2135, 2154, 2163; Solusoft 350; Tegiloxan; Tego Airex 960, 970, 975; Tegosil; Tegosipon; Tegotrenn S 2100; Trans-10, -10K, -20, -20K, -25, -25K, -30, -30K, -100, -134S, -134S K, -177S, -177S K, -222, -225S, -264, -266, -1030, -1030 K; Union Carbide R272, R274; Viscasil Fluids

Silk amino acids (CTFA) [CAS #977077-71-6]
Amino Silk; Crosilk Liq.; Lexein SLK

Silk powder
Crosilk Powd.

Silver (CTFA) [CAS #7440-22-4] *Synonyms:* Silver, colloidal
Algaesil

Simazine [CAS #122-34-9] *Synonyms:* 2-Chloro-4,6-bis(ethylamine)s-triazine
Belclene 313

Simethicone [CAS #8050-81-5]
Dow Corning Antifoam A, AF, C; Mazu DF 200SP

Smectite clay
Bentone 14, SA-38

Sodium acid gluconate
PMP Liquid Gluconate 60 Tech.

Sodium acrylate
Acrysol 51

Sodium alginate. See Algin

Sodium alkyl diaminoethyl glycine
Lebon 15

Sodium aluminum silicate *Synonyms:* Sodium alumino-silicate
Hydrex; Valfor 100; Vulkasil A1; ZB-100, -130 (TN), -300; Zeolex 7, 7A, 23, 23A, 23P, 35P, 40, 80

Sodium-amine dodecylbenzenesulfonate
Ardet A-25

Sodium bentonite
Polarite KB 325; Volclay HPM-20, NF-BC

Sodium benzoate (CTFA) [CAS #532-32-1]
Aerosol OT-B; Solusol 85%

Sodium bifluoride [CAS #1333-83-1] *Synonyms:* Sodium acid fluoride
Na-0101 T/18″

Sodium bis(naphthalene) sulfonate
Disrol SH

Sodium bistridecyl sulfosuccinate
Vanisol BIS sodico-2

Sodium bromate (CTFA) [CAS #7789-38-0] *Synonyms:* Bromic acid, sodium salt
Dyetone

Sodium C4-12 olefin/maleic acid copolymer (CTFA) [RD #977067-42-7]
Demol EP

Sodium C9-13 alcohols sulfate
Teepol HB 7

Sodium C12-13 pareth sulfate (CTFA) *Synonyms:* Sodium pareth-23 sulfate
Akyposal DS 56

Sodium C12-15 alcohols sulfate (CTFA) [RD #977067-40-5]
Lanette Wax SX, SXBP

Sodium C12-15 alkoxypropyl iminodipropionate
Amphoteric N

Sodium C12-15 pareth-7 carboxylate (CTFA) *Synonyms:* Sodium pareth-25-7 carboxylate
Surfine WNT Gel, WNT LC, WNT LS

Sodium C14-16 olefin sulfonate (CTFA) [CAS #68439-57-6]
Siponate A-246 LX; Sterling AOS; Witconate AOS

Sodium C16-24 alcohol sulfate
Sulfetal AF

Sodium-calcium lignosulfonate
Norlig 415

Sodium capryl lactylate (CTFA) [CAS #29051-57-8; RD #977067-37-0]
Pationic 122A

Sodium carboxylate
Darvan No. 31; Nissan Polystar OM, OMP; Tamol 731-25%, 731-SD, 850, 960

Sodium carboxymethylcellulose. See Cellulose gum

Sodium carboxymethyl hydroxyethyl cellulose. See Carboxymethyl hydroxyethylcellulose

Sodium castorate (CTFA) [CAS #8013-06-7] *Synonyms:* Castor oil, sodium salt
Servo Brillaint Olie B AZ 75

Sodium cetearyl sulfate (CTFA) [CAS #59186-41-3] *Synonyms:* Sodium cetostearyl sulfate; Sodium cetyl/stearyl sulfate
Dehydag Wax E, N

Sodium cetyl diphenyl ether disulfonate *Synonyms:* Sodium hexadecyl diphenyl ether disulfonate
Conco Sulfate 4C3

Sodium cetyl/oleyl sulfate
Gardinol CX

Sodium cetyl sulfate (CTFA) [CAS #1120-01-0]
Berol 472; Sipex EC-111

Sodium chloride (CTFA) [CAS #7647-14-5]
Uni-Malt

Sodium chlorite [CAS #7758-19-2]
Adox 3125; C2 Sodium Chlorite; Sodium Chlorite Solution 50, Tech.; Textone

Sodium-4-chlorophthalate [CAS #56047-23-5]
Nacopa

Sodium CMC. See Cellulose gum

Sodium cocamido DEA sulfosuccinate
Alkasurf SS-CDE

Sodium cocaminobutyrate *Synonyms:* Sodium N-coco-beta-amino butyrate
Armeen SZ

Sodium cocoate (CTFA) [CAS #61789-31-9] *Synonyms:* Coconut oil fatty acid, sodium salt
Norfox Coco Powder

Sodium coco-hydrolyzed animal protein (CTFA) [CAS #68188-38-5]
May-Tein SK

Sodium cocomonoglyceride sulfonate (CTFA)
SCMS

Sodium cocoyl glutamate (CTFA) [CAS #68187-32-6] *Synonyms:* Monosodium N-cocoyl-L-glutamate
Acylglutamate CS-11, GS-11

Sodium cocoyl isethionate (CTFA) [CAS #61789-32-0]
Igepon AC-78; Jordapon CI; Tauranol I-78, I-78 Flakes, I-78-3

Sodium cumenesulfonate (CTFA) [CAS #32073-22-6]
Eltesol SCS 40; Manro SCS 40, 90; Na-Cumene Sulfonate 40, Powd.; Naxonate 45SC, SC; Reworyl NCS 40; Stepanate C-S; Ultra SCS Liquid; Witco SCS 40; Witconate SCS

Sodium cyclohexyl palmitoyl taurate
Igepon CN42

Sodium deceth-3 sulfosuccinate
Alkasurf SS-DA-3

Sodium deceth-6 sulfosuccinate
Alkasurf SS-DA-6

Sodium decyl diphenyl ether disulfonate
Conco 3B2

Sodium decyl diphenyl ether sulfonate [CAS #25167-32-2]
Sandoz Sulfonate 3B2

Sodium decyl diphenyloxide disulfonate
Calfax 10L-45; Conco Sulfate 3B2; Dowfax 3B1, 3B2

Sodium decyl sulfate (CTFA) [CAS #142-87-0]
Empimin SDS

Sodium dehydroacetate
Trisept SDHA

Sodium diamyl sulfosuccinate. See Diamyl sodium sulfosuccinate

Sodium di-n-butyl dithiocarbamate [CAS #136-30-1] *Synonyms:* Sodium dibutyldithiocarbamate; Dibutyldithiocarbamic acid sodium salt
Butyl Namate; Naftocit NaDBC

Sodium dibutyl naphthalene sulfonate
Morwet DB

Sodium dichlorocyanurate [CAS #2893-78-9] *Synonyms:* Sodium dichloroisocyanurate; Dichloroisocyanuric acid sodium salt
ACL 56, 60

Sodium dichloroisocyanurate [CAS #2244-21-5] *Synonyms:* Sodium dichloro-s-triazine-2,4,6-trione
OCI 56

Sodium dicyclohexyl sulfosuccinate. See Dicyclohexyl sodium sulfosuccinate

Sodium didodecylbenzene sulfonate
Petronate S

Sodium dihexyl sulfosuccinate. See Dihexyl sodium sulfosuccinate

Sodium dihydroxyethylglycinate (CTFA) [CAS #139-41-3] *Synonyms:* Sodium N,N-bis-2-hydroxyethyl glycinate
Hampshire DEG; Kelene 77

Sodium diisobutyl sulfosuccinate. See Diisobutyl sodium sulfosuccinate

Sodium diisooctyl sulfosuccinate. See Diisooctyl sodium sulfosuccinate

Sodium diisopropyl naphthalene sulfonate
Geropon IN

Sodium dimethyldithiocarbamate [CAS #148-18-5] *Synonyms:* Sodium N,N-dimethyl dithiocarbamate; Dimethyldithiocarbamic acid, sodium salt; Methyl namate; SDDC
Amerstat 272, 274; Aquatreat SDM; Naftocit NaDMC; Neo-Voronit; Rychem 784, 830; Thiostat B; Vancide 51

Sodium dinaphthyl methane sulfonate *Synonyms:* Dinaphthalene-methane sulfonate sodium salt
Geropon RM/77, RM/77-D

Sodium dinonylnaphthalene sulfonate
Na-Sul SS

Sodium dioctyl sulfosuccinate. See Dioctyl sodium sulfosuccinate

Sodium diphenyl methane sulfonate
Geropon FMS

Sodium ditridecyl sulfosuccinate. See Ditridecyl sodium sulfosuccinate

Sodium dodecylbenzenesulfonate (CTFA) [CAS #25155-30-0] *Synonyms:* Sodium lauryl benzene sulfonate
Calsoft L-40, L-60; Hetsulf 40, 40X, 60S; Kadif 50 Flakes; Nacconol 35SL, 40F, 40G, 90F, 90G; Nansa HS40-AU, HS40 Soft, HS80P, HS80/S, HS80SK, HS80 Soft, HS85/S, MA30; Sterling LA Paste 55%, 60%; Witconate 60B, 1250 Slurry, 1260 Slurry

Sodium dodecyl diphenyl ether disulfonate [CAS #40795-56-0]
Conco 2A1

Sodium dodecyl diphenyl ether sulfonate
Sandoz Sulfonate 2A1

Sodium dodecyl diphenyloxide disulfonate
Conco Sulfate 2A1; Dowfax 2A1, 2EP

Sodium EDTA. See Tetrasodium EDTA

Sodium 2-ethylhexyl sulfate. See Sodium octyl sulfate

Sodium ferric EDTA
Hamp-Ene NaFe Purified Grade

Sodium glucoheptonate
Milco-150; Perma Kleer 354; Seqlene 190, 270, 400, 540, ES-40, ES-50

Sodium gluconate (CTFA) [CAS #527-07-1] *Synonyms:* D-Gluconic acid, monosodium salt
PMP Sodium Gluconate Tech.

Sodium glyceryl oleate phosphate (CTFA) [RD #977067-44-9]
Emphos D70-30C

Sodium hexadecyl diphenyl ether disulfonate
Conco 4C3

Sodium hexadecyl diphenyloxide disulfonate
Dowfax 8390

Sodium hexametaphosphate (CTFA) [CAS #10124-56-8] *Synonyms:* Metaphosphoric acid, hexasodium salt
Polyphos

Sodium hexyl diphenyloxide disulfonate
Dowfax XDS 8292.00

Sodium hydrogenated tallow glutamate (CTFA) [RD #977067-45-0] *Synonyms:* Sodium hydrogenated tallowyl glutamate
Acylglutamate GS-11, HS-11

Sodium hydrogen carbonate
Rhenopor 1843

Sodium hydroxymethylglycinate (CTFA) [CAS #70161-44-3]
Suttocide A

Sodium isethionate (CTFA) [CAS #1562-00-1] *Synonyms:* Sodium 2-hydroxyethanesulfonic acid
Emery 5440

Sodium isodecyl sulfate
Sipex CAV

Sodium isopropyl naphthalene sulfonate
Aerosol OS

Sodium isostearoyl lactylate (CTFA) [CAS #66988-04-3]
Pationic ISL

Sodium lactate (CTFA) [CAS #72-17-3] *Synonyms:* 2-Hydroxypropanoic acid, monosodium salt
Patlac NAL

Sodium laneth-5 sulfosuccinate
Cyclopol SB-5

Sodium lauramido DEA-sulfosuccinate
Alkasurf SS-L7DE

Sodium lauramido MEA-sulfosuccinate
Condanol SBL

Sodium lauraminopropionate (CTFA) [CAS #3546-96-1] *Synonyms:* N-dodecyl-β-alanine, monosodium salt
Maprolyte LX

Sodium laureth-13 carboxylate (CTFA)
Miranate LEC

Sodium laureth MIPA-sulfosuccinate
Condanol SBIL/3, SBZ

Sodium laureth phosphate
Forlanit P; Forlanon

Sodium laureth sulfate (CTFA) [CAS #1335-72-4; 3088-31-1; 9004-82-4 (generic); 13150-00-0; 15826-16-1] *Synonyms:* Sodium lauryl ether sulfate (n=1–4)
Agent 804; Bio Soft LD-95; Calfoam ES-30; Cerasynt LP; Condanol NL2, NL3, NL3 28%; Crodapearl Liq.; Empicol ESB3, ESB3GA, ESB30, ESB50, ESB70, ESB70-AU, ESC70; Empigen XDR121, XDR123, XDR302; Euperlan PK 771, PK 776, PK 789, PK-810, PK 900; Genapol LRO Liq., Paste; Genapol PGM Conc., ZRO Brands; Incropearl Liquid; Montosol PF-10, PF-14, PF-26, PG-10, PL-16, PQ-17; Sandoz Sulfate 219, ES-3, WE; Standapol Pearl Conc. 7130; Steinapol NL 3 28%; Texapon ES-1, -2, -3

Sodium laureth sulfosuccinate
Condanol SBA/1, SBFA/3, SBFA 30 40%

Sodium lauriminodipropionate (CTFA) [CAS #14960-06-6; 26256-79-1] *Synonyms:* Sodium N-lauryl-β-iminodipropionate
Cycloteric SLIP; Deriphat 160C

Sodium lauroyl glutamate (CTFA) [CAS #29923-31-7 (L-form); 29923-34-0 (DL-form); 42926-22-7 (L-form)
Acylglutamate LS-11

Sodium lauroyl lactylate (CTFA) [CAS #13557-75-0; RD #977067-51-8]
Pationic 138C

Sodium lauroyl sarcosinate (CTFA) [CAS #137-16-6] *Synonyms:* N-methyl-N-(1-oxod-decyl) glycine, sodium salt
Crodasinic LS30, LS35; Sarkosyl NL-30

Sodium lauryl/oleyl sulfate
Duponol D Paste

Sodium lauryl sulfate (CTFA) [CAS #151-21-3]
Alcolec 638; Avirol SL-2010; Conco Sulfate WA, WAG, WAS, WA Special; Condanol NLS 30, NLS 90; Crodex A; Cyclochem EM 560; Dehydag Wax SX, W; Duponol C, WA Dry; Empicol 0185, LS30, LX, LX28, LXV, LSV/D, LY28/S, LZ, LZ/D, LZ/E, LZP, LZV, LZV/D, LZV/E; Empiwax SK, SK/BP; Hartenol LAS-30; Lanette SX, W; Lexemul AS; Lipowax; Maprofix LK (USP); Monateric CDL, CDS; Nutrapon B; Sandoz Sulfate WA Dry, WAG, WAS, WA Special; Steinapol NLS 28, NLS 90-Powder; Stepanol WA-100; Texapon K-12, L-100, Z Granules; Ultra Sulfate SL-1; Witcolate A

Sodium lignosulfonate (CTFA) [CAS #8061-51-6] *Synonyms:* Sodium polignate
Darvan No. 2; Dynasperse A, B; Dyqex; Irgasol N; Kelig 100; Lignosite 431, 458, 823, 854, 889; Lignosol AXD, D-10, D-30, DXD, FTA, HCX, NSX 110, NSX 120, SFS, SFX, SFX-65, WT, X, XD, XD-65; Maracarb N-1; Maracell XC, XE; Marasperse CB, CBOS-3, CBX-2, N-22; Norlig; Orzan LS, LS-50, S, SL-50; Polyfon F, H, O, T; Raylig; Raymix; Reax 15B, 45A, 45L, 80A, 80C, 81A, 82, 83A, 85A, 88B, 90, 90P, 95A, SR-1, SR-7

Sodium magnesium aluminosilicate
Hydrex R

Sodium magnesium silicate (CTFA) [CAS #53320-86-8; RD #977068-03-3]
Laponite 508, XLG, XLS

Sodium maleic acid/olefin copolymer
Sokalan CP 9, CP 11

Sodium maleic anhydride/diisobutylene copolymer
Empicryl APD90, APD B90

Sodium MBT [CAS #2492-26-4] *Synonyms:* Sodium 2-mercaptobenzothiazole
Nacap; Troysan 28; Vancide 51

Sodium 2-mercaptobenzothiazole. See Sodium MBT

Sodium metasilicate (CTFA) [CAS #6834-92-0] *Synonyms:* Silicic acid, disodium salt
Crystamet 1020, 2040, 3080; Drymet 59, Fines; Metso Beads 2048, Pentabead 20

Sodium methane napthalene sulfonate
Soitem 207

Sodium methyl cocoyl taurate (CTFA) [CAS #12765-39-8; 61791-42-2] *Synonyms:* Sodium cocoyl methyl taurate
Igepon TC-42

Sodium methylene bisdithiocarbamate
Amerstat 272

Sodium methyl naphthalene sulfonate (CTFA) [CAS #26264-58-4; RD #977052-10-0]
Supragil MNS/90

Sodium methyl oleate sulfate
Darvan SMO

Sodium methyl oleoyl taurate (CTFA) [CAS #137-20-2] *Synonyms:* Sodium N-oleoyl-N-methyl taurate; Oleyl methyl tauride
Igepon T-33, T-43, T-51, T-77; Nissan Diapon TO; Tergenol G, S Liquid, Slurry

Sodium methyl palmitoyl taurate *Synonyms:* Sodium palmitoyl methyl taurate
Igepon TN-74

Sodium methyl silanolate
Union Carbide R20

Sodium methyl tall oil acid taurate
Igepon TK-32

Sodium methyl tallow taurate
Nissan Diapon T

Sodium molybdate
Molyhibit 100 Series

Sodium myristoyl glutamate (CTFA) [CAS #38517-37-2; 71368-20-2]
Acylglutamate MS-11

Sodium myristyl sulfate [CAS #139-88-8]; *Synonyms:* Sodium tetradecyl sulfate
Niaproof Anionic Surfactant 4

Sodium naphthalene formaldehyde sulfonate. See Sodium polynaphthalene sulfonate

Sodium naphthalene sulfonate. See Sodium polynaphthalene sulfonate

Sodium m-nitrobenzenesulfonate (CTFA) [CAS #127-68-4]
Ludigol 60, F; Reserve Salt Flake

Sodium nonoxynol-6 phosphate (CTFA) [CAS #12068-19-8] *Synonyms:* Sodium POE (6) nonyl phenyl ether phosphate
Gafac LO-529

Sodium nonoxynol-9 phosphate (CTFA) [RD #977058-21-1]
Emphos CS-1361

Sodium nonoxynol-4 sulfate (CTFA) [CAS #9014-90-8 (generic); RD #977069-67-2]
Alipal CO-433, EO-526

Sodium nonyl benzene sulfonate
Emkal NOBS

Sodium NTA [CAS #5064-31-3] *Synonyms:* Sodium nitrilotriacetate
NTA

Sodium octoxynol-2 ethane sulfonate (CTFA) [CAS #2917-94-4]
Triton X-200

Sodium octylphenoxyethoxyethyl sulfonate
Newcol 861SE

Sodium octyl sulfate (CTFA) [CAS #126-92-1; #142-31-4] *Synonyms:* Sodium 2-ethylhexyl sulfate; Sulfuric acid, mono (2-ethylhexyl) ester, sodium salt
Avirol SA-4106; Bio Terge PAS 8; Carsonol SHS; Duponol 80; Niaproof Anionic Surfactant 08; Norfox SEHS; Serdet DSK 40; Sipex BOS, OLS; Sole-Terge TS-2-S; Standapol LF; Sulfotex OA; Witcolate D-510

Sodium oleate (CTFA) [CAS #143-19-1] *Synonyms:* Sodium 9-octadecenoate
Norfox Vertex Flakes

Sodium oleic acid sulfate
Dyasulf 2031

Sodium oleic acid sulfonate
Sul-fon-ate OA-5

Sodium oleic acid sulfonate, amyl ester
Sul-fon-ate OE-500

Sodium N-oleoyl sarcosinate
Crodasinic OS35

Sodium oleyl sulfate (CTFA) [CAS #1847-55-8]
Duponol LS Paste; Sipex OS

Sodium PCA (CTFA) [CAS #28874-51-3; 54571-67-4] *Synonyms:* PCA Soda; 5-oxo-DL-proline, sodium salt; Sodium pyroglutamate; Sodium DL-2-pyrrolidone-5-carboxylate
Ajidew N-50; Ritaderm

Sodium petroleum sulfonate [CAS #68608-26-4]
Petrosul H-50, H-60, H-70, HM-62, HM-70, M-50, M-60, M-70

Sodium phenolsulfonate (CTFA) [CAS #825-90-1; 1300-51-2] *Synonyms:* Sodium sulfocarbolate; Hydroxybenzenesulfonic acid, monosodium salt
Tamol PP; Wettol D 1

Sodium o-phenylphenate (CTFA) [CAS #132-27-4] *Synonyms:* o-Phenyl phenol sodium salt; o-Phenylphenol sodium; Sodium o-phenylphenolate
Dowicide A

Sodium polyacrylate [CAS #9003-04-7]
Acrysol 51, GS, HV-1, LMW-10N, LMW-20N, LMW-45N, LMW-100N, LMW-400N, LMW-600N; Alcogum; Alcosperse 107, 124, 149, 157, 602-N; Aquatreat AR602; Colloid 202, 207, 208, 211, 218D, 223, 223D, 230, 233, 245D, 350; Daxad 37LN7, 37LN10, 37NS; Densol 1010; Dispersal 140; Drewsperse 611; Good-rite K-7028, -7058, -7200, -7600; Hostacerin PN 73; Serpol QPA 160; Sokalan CP 5, CP 5 Powd., CP 7, CP 8, CP 10, CP 13S, CP 45, PA 15, PA 20, PA 20 PN, PA 25 PN, PA 30, PA 40, PA 40 Powd., PA 50, PA 70 PN, PA 80

Sodium polycarboxylate
Colloid 111, 225, 226/35, 231; Daxad 30, 30-30, 30S, 31, 35, 37LN10; Ninate PA; Soitem 70; Tamol PA Liq., PA Powd.

Sodium poly-L-glutamate [CAS #28680-04-8]
Ajicoat SPG

Sodium polyisobutylene maleic anhydride copolymer
Daxad 31, 31S

Sodium polymethacrylate (CTFA) [CAS #25086-62-8; 54193-36-1] *Synonyms:* 2-Propenoic acid, 2-methyl–, homopolymer, sodium salt

Aquatreat AR231, AR232; Darvan No. 7; Daxad 30, 30-30, 30S, 34N10, 35, 41

Sodium polynaphthalene-methane sulfonate
Geropon RM/210

Sodium polynaphthalene sulfonate (CTFA) [CAS #9084-06-4] *Synonyms:* Sodium naphthalene-formaldehyde sulfonate
Ablusol ML, NL; Anchoid; Arylan SNS; Blancol, N; Darvan No. 1, 6; Daxad 11, 11G, 13, 17, 19; Intraphor AC; Lomar DL, LS, PL, PW, ST; Morwet D-425; Nopcosant; Petro Dispersant 425; Protodye QNF; Stepantan A; Tamol 819, 819L-43, L Conc., NH 3901, NH 7519, NH 9103, NN, NN 2406, NN 2901, NN 4501, NN 7718, NN 8906, NN 9104, NN 9401, NNO, NNOK, SD, SN; Tek Tan ND, NL; Vultamol, SA Liq.; Wettol D 2

Sodium polyphosphonate
Wayplex 55-S, NTP-S

Sodium potassium aluminosilicate *Synonyms:* Nepheline syenite
Minex 2, 3, 4

Sodium propionate (CTFA) [CAS #137-40-6] *Synonyms:* Propanoic acid, sodium salt
Tandem 11H

Sodium pyrithione (CTFA) [CAS #3811-73-2] *Synonyms:* Sodium 2-pyridinethiol-1-oxide; Sodium (2-pyridylthio)-N-oxide
Sodium Omadine; Triadine 10

Sodium ricinoleate (CTFA) [CAS #5323-95-5] *Synonyms:* 12-Hydroxy-9-octadecenoic acid, sodium salt
Sorcinol 35

Sodium rosinate *Synonyms:* Sodium soap of pale rosin
Dresinate 81

Sodium saccharin [CAS #128-44-9]
Sherbrite 30, 85; Syncal GS, S, SDS, US

Sodium silicate (CTFA) [CAS #1344-09-8] *Synonyms:* Silicic acid, sodium salt; Water glass
D Sodium Silicate; N Sodium Silicate

Sodium stearate (CTFA) [CAS #822-16-2] *Synonyms:* Sodium octadecanoate
Monomuls 60-25/2, 60-25/5, 90-25/2, 90-25/5; Norfox B; Rhenovin Na-stearat-80; Witco Heat Stable Sodium Stearate; Witco Sodium Stearate C-1, C-7

Sodium stearoyl lactylate (CTFA) [CAS #25383-99-7] *Synonyms:* Sodium stearyl-2-lactylate
Patco 3; Pationic SSL

Sodium stearyl sulfosuccinamate *Synonyms:* Sodium octadecyl sulfosuccinamate
Alkasurf SS-TA 125

Sodium styrene maleic anhydride
Scripset 500

Sodium 2-sulfoethyl myristate
Igepon AM-78

Sodium sulforicinoleate
Türkischrotöl 100%, 50%

Sodium tallate *Synonyms:* Tall oil rosin sodium salt
Arizona DRS-43, DRS-44; Dresinate TX

Sodium tallow sulfate (CTFA) [CAS #8052-50-4]
Empicol TAS90; Montovol GJ-12

Sodium tetraborate decahydrate
Borax

Sodium tetrahydronaphthalene sulfonate
Alkanol S

Sodium toluenesulfonate (CTFA) [CAS #12068-03-0; 657-84-1] *Synonyms:* Methylbenzenesulfonic acid, sodium salt
Eltesol ST 34, ST 40, ST 90, ST Pellets; Hartotrope STS 40, Powder; Manro STS 40, STS 90; Marlon ARL; Na Toluene Sulfonate 30, 40; Naxonate 4ST, ST; Reworyl NTS 40; Stepanate T; Witco STS 40; Witconate STS

Sodium toluene xylene sulfonate
STXS; Vista STXS

Sodium tolyltriazole
Cobratec TT-50-S, TT-85; Maxahibit 100

Sodium-2,4,5-trichlorophenate
Preventol I

Sodium trideceth sulfate (CTFA) [CAS #25446-78-0 (n=3); 66161-58-8 (n=4); 3026-63-9] *Synonyms:* Sodium tridecyl ether sulfate; Sodium POE tridecyl sulfate
Monateric CDTD; Sipex EST-30

Sodium undecylenamide-MEA sulfosuccinate
Rewocid SBU 185; Steinazid SBU 185

Sodium xylenesulfonate (CTFA) [CAS #1300-72-7] *Synonyms:* Dimethylbenzene sulfonic acid, sodium salt
Alkatrope SX-40; Carsosulf SXS-Liq.; Conco SXS; Eltesol SX 30, SX 93, SX Pellets; ESI-Terge SXS; Hartotrope SXS 40; Jordanol SXS; Lankrosol SXS-30; Manro SXS 30, 40, 93; Naxonate 4L, 5L, G; Norfox SXS40, SXS96; Nutrol SXS; Pilot SXS-40, SXS-96; Polystep A-2; Reworyl NXS 40; Stepanate X; Sterling 2XS, SXS; Sulfotex SXS; Surco SXS; SXS; Tex-Wet 1140; Ultra SXS Liquid, Powder; Witco SXS 40; Witconate 1260 Slurry, SXS

Soluble collagen (CTFA) *Synonyms:* Soluble animal collagen
Clearcol; Collasol; Elastosol; Pancogene S; Sollagen

Soluble elastin
Elastosol

Soluble keratin
Kerasol

Soluble reticulin *Synonyms:* Soluble animal reticulin
Reticusol

Sorbeth-20 (CTFA) [RD #977058-26-6] *Synonyms:* PEG-20 sorbitol ether; POE (20) sorbitol ether; PEG 1000 sorbitol ether
Ethosperse SL-20; Liponic SO-20; Trylox 6753, SS-20

Sorbic acid (CTFA) [CAS #110-44-1; 22500-92-1] *Synonyms:* 2,4-Hexadienoic acid
Aflaban DF; Sorbistat; Tristat

Sorbitan distearate
Emasol S-20

Sorbitan isostearate (CTFA) [CAS #1338-39-2; 5959-89-7; RD #977010-89-1] *Synonyms:* Anhydrosorbitol monoisostearate
Crill 6; S-Maz 67

Sorbitan laurate (CTFA) [CAS #1338-39-2; 5959-89-7] *Synonyms:* Sorbitan monolaurate; Anhydrosorbitol monolaurate; Sorbitan monododecanoate
Ablunol S-20; Ahco 759; Alkamuls SML; Atmer 100, 110; Crill 1; Drewmulse SML; Durtan 20; Emsorb 2515; Ethylan GL-20; Ionet S-20; Liposorb L; Lonzest SML; Newcol 20; Nissan Nonion LP-20•R, LP-20•RS; Radiamuls 125; Radiasurf 7125; Secoster KL 10; S-Maz 20; Soprofor S/20; Sorbax SML; Sorbon S-20; Sorgen 90; Span 20

Sorbitan myristate
Nissan Nonion MP-30R

Sorbitan oleate (CTFA) [CAS #1338-43-8; 5938-38-5] *Synonyms:* Sorbitan monooleate; SMO; Anhydrosorbitol monooleate; Sorbitan mono-9-octadecenoate
Ablunol S-80; Ahco 832, 944; Alkamuls SMO; Atmer 105; Atpet 80, 100, 200; Capmul O; Crill 4, 50; DeSonic SMO; DeSotan SMO; Drewmulse SMO; Durtan 80; Emasol O-10; Emsorb 2500; Ethylan GO-80; Ionet S-80; Liposorb O; Lonzest SMO; Newcol 3-80, 3-85, 80; Nissan Nonion OP-80•R; Radiamuls 155; Radiasurf 7155; Rheodol SP-O10; S-Maz 80, 80 K; Soprofor S/80; Sorbax SMO; Sorbon S-80; Sorgen 40, S-40-H

Sorbitan palmitate (CTFA) [CAS #26266-57-9] *Synonyms:* Sorbitan monopalmitate
Ahco FP-67; Crill 2; Emasol P-10; Emsorb 2510; Ethylan GP-40; Liposorb P; Lonzest SMP; Newcol 40; Nissan Nonion PP-40•R; Radiamuls 135; Radiasurf 7135; Secoster KP 10; S-Maz 40; Sorbax SMP; Sorbon S-40; Span 40

Sorbitan sesquioleate (CTFA) [CAS #8007-43-0] *Synonyms:* Anhydrosorbitol sesquioleate; Anhydrohexitol sesquioleate
Alcolan, 36W, 40; Crill 43; Dehymuls K; Emasol O-15; Emsorb 2502; Liposorb SQO; Nissan Nonion OP-83•RAT; S-Maz 83R; Sorgen 30, S-30-H

Sorbitan sesquitallate
S-Maz 93 R

Sorbitan stearate (CTFA) [CAS #1338-41-6] *Synonyms:* Sorbitan monostearate; SMS; Anhydrosorbitol monostearate; Sorbitan monooctadecanoate
Ablunol S-60; AF-72, AF-75; Ahco 909; Alkamuls SMS; Crill 3; Drewmulse SMS; Durtan 60; Emasol S-10; Emsorb 2505; Ethylan GS-60; Ionet S-60 C; Liposorb S; Lonzest SMS; Newcol 60; Nissan Nonion SP-60•R; Radiamuls 145; Radiasurf 7145; Rheodol SP-S10; Secoster KS 10; S-Maz 60, 60 K, 60 KHM; Soprofor S/60; Sorbax SMS; Sorbon S-60; Sorgen 50; Span 60

Sorbitan tallate
DeSonic SMT; DeSotan SMT; S-Maz 90; Witcomul 78

Sorbitan trioleate (CTFA) [CAS #26266-58-0] *Synonyms:* Anhydrosorbitol trioleate; Sorbitan tri-9-octadecenoate
Ahco FO-18; Alkamuls STO; Atmer 106; Crill 45; Emasol O-30; Emsorb 2503; Ethylan GT-85; Ionet S-85; Liposorb TO; Lonzest STO; Nissan Nonion OP-85•R; Rheodol SP-O30; S-Maz 85, 85 K; Soprofor S/85; Sorbax STO; Span 85

Sorbitan tristearate (CTFA) [CAS #26658-19-5] *Synonyms:* Anhydrosorbitol tristearate; Sorbitan trioctadecanoate

Ahco FS-21; Alkamuls STS; Crill 35; Drewmulse STS; Emasol S-30; Emsorb 2507; Liposorb TS; Lonzest STS; Radiamuls 345; Radiasurf 7345; Rheodol SP-S30; S-Maz 65, 65 K; Soprofor S/65; Sorbax STS; Span 65

Sorbitan tritallate
S-Maz 95

Sorbitol (CTFA) [CAS #50-70-4] *Synonyms:* D-Glucitol
Lexein S620S

Soy acid (CTFA) [CAS #68308-53-2]
Industrene 225, 226, 226 FG

N-Soya-1,3-diaminopropane
Duomeen S

Soyadimethylethyl ammonium diethylphosphate
Larostat 192, 192 Anhyd.

Soyaethyldimonium ethosulfate [CAS #68308-67-8] *Synonyms:* Soya dimethyl ethyl ammonium ethosulfate
Jordaquat 1033; Larostat 88, 264-A, 264-A Anhyd., 264-A Conc.

Soya hydroxyethyl imidazoline (CTFA) *Synonyms:* Soya imidazoline
Alkaquat S; Miramine SC

Soyamide DEA (CTFA) [CAS #68425-47-8] *Synonyms:* Soya diethanolamide; N,N-bis(hydroxyethyl) soya amides
Empilan 2125-AU; Mackamide S

Soyamidopropyl benzyldimonium chloride (CTFA)
Schercoquat SOAB

Soyamidopropyl dimethylamine (CTFA) [CAS #68188-30-7] *Synonyms:* N-[3-(Dimethylamino) propyl] soya amides
Mackine 901

Soyamidopropyl ethyldimonium ethosulfate (CTFA)
Schercoquat SOAS

Soyamine (CTFA) [CAS #61790-18-9] *Synonyms:* Soya primary amine
Adogen 115 (D); Armeen SD; Kemamine P-997, P-997D; Lilamin 115, 115 D; Nissan Amine SB

Soya oil glyceride *Synonyms:* Glyceryl mono soya oil; Glyceryl monosoyate
Drewmulse 10; Radiamuls 602, MG 2602

Soybean oil (CTFA) [CAS #8001-22-7] *Synonyms:* Oils, soybean; Soya oil
Arnica Oil CLR; Calendula Oil CLR; Carrot Oil CLR; Super Refined Soybean Oil

Soybean oil, epoxidized. See Epoxidized soybean oil

Soybean oil, maleated. See Maleated soybean oil

Soy sterol (CTFA) [RD #977066-08-2]
Generol 122

Soytrimonium chloride (CTFA) [CAS #61790-41-8] *Synonyms:* Soya trimethyl ammonium chloride; Quaternium-9
Adogen 415; Arquad S-50; Jet Quat S-50; Tomah Q-S

Spermaceti, synthetic. See Cetyl esters

Squalane (CTFA) [CAS #111-01-3] *Synonyms:* Dodecahydrosqualene
Robane

Squalene (CTFA) [CAS #111-02-4] *Synonyms:* 2,6,10,15,19,23-Hexamethyl-2,6,10,14,18,22-tetracosahexaene
Robeyl; Supraene

Stannous octoate *Synonyms:* Stannous-2-ethylhexoate; Tin octotate
Fomrez C-2, C-4; Kosmos 10, 15, 16, 29

Steapyrium chloride [CAS #1341-08-8; 14492-68-3; 42566-92-7] *Synonyms:* Quaternium-7; N-stearoyl colamino formyl methyl pyridinium chloride
Emcol E-607S

Stearalkonium chloride (CTFA) [CAS #122-19-0] *Synonyms:* Stearyl dimethyl benzyl ammonium chloride; Octadecyl dimethyl benzyl ammonium chloride
Ablumine 280; Ammonyx 4, 4B, 4-IPA, 485, 490, 4002, CA-Special; Amyx A-25S; Arquad DM18B-90; Carsoquat 816-C, SDQ-25, SDQ-85; Catinal OB-80E; Crapol AU-23; Cycloton M270C/85; Dehyquart STC-25; Empigen BCP/25; Incroquat CR Conc., S-85, SDQ-25; Jordaquat JS-25; Mackernium SDC-25, -50, -85; Maquat SC-18, SC-1632; Nissan Cation S2-100; Polyquart H-7102; Standamul Conc. 1002, STC-25; Stedbac; Varisoft 6112, SDC, SDC-65%, SDC-85S, SDC-100P, SDC-W

Stearalkonium hectorite (CTFA) [RD #977067-62-1; 977061-78-1]
Bentone 27, Gel CAO, Gel IPM; Miglyol 840 Gel, Gel

Stearamide (CTFA) [CAS #124-26-5] *Synonyms:* Octadecanamide; Stearic acid amide
Armid 18; Armoslip 18; Crodamide S, SR; Kemamide S; Petrac Vyn-Eze

Stearamide DEA (CTFA) [CAS #93-82-3] *Synonyms:* Stearic acid diethanolamide; Stearoyl diethanolamide; N,N-bis(2-hydroxyethyl) octadecanamide
Alkamide HTDE; Aminol N-1918; Cetina, TE; Cyclomide DS280, DS280/S; Lipamide S; Monamid 718; Nopcogen 14-S; Onyxol 42; Unamide S, W

Stearamide DEA-distearate (CTFA) *Synonyms:* Stearic diethanolamide distearate
Schercomid SD-DS

Stearamide DIBA-stearate (CTFA) [RD #977058-30-2]
Polytex 10

Stearamide MEA (CTFA) [CAS #111-57-9] *Synonyms:* Stearic acid monoethanolamide; Stearoyl monoethanolamide; N-(2-hydroxyethyl) octadecanamide; Monoethanolamine stearic acid amide
Alkamide HTME; Comperlan HS; Cyclomide S280; Incromide SM; Monamid S; Ninol SNR; Rewomid S 280; Schercomid SME, SME-A, SME-M; Schercopearl EA-100; Steinamid S 280; Witcamide 70

Stearamide MEA-stearate (CTFA) [CAS #14351-40-7]
Cerasynt D; Schercomid SME-S; Witcamide MAS

Stearamidoethyl diethylamine (CTFA) [CAS #16889-14-8] *Synonyms:* Diethylaminoethyl stearamide
Chemical Base 6532; Lexamine 22; Lexemul AR

Stearamidoethyl ethanolamine (CTFA) [CAS #141-21-9] *Synonyms:* Ethanolaminoethyl stearamide
Chemical 39 Base

Stearamidopropalkonium chloride [CAS #65694-10-2] *Synonyms:* N,N-dimethyl-N-[3-[(1-oxooctadecyl)amino] propyl]benzenemethanaminium chloride; Stearamidopropyl dimethyl benzyl ammonium chloride; Stearamidopropyl benzyldimonium chloride
Schercoquat SAB

Stearamidopropylamine oxide (CTFA) [CAS #25066-20-0]
Cyclomox SO

Stearamidopropyl cetearyl dimonium tosylate (CTFA)
Ceraphyl 85

Stearamidopropyl diethylamine
Cyclomide SODI

Stearamidopropyl dimethylamine (CTFA) [CAS #7651-02-7] *Synonyms:* Dimethylaminopropyl stearamide
Incromine SB; Jordamine DAPSA, S-13; Lexamine S-13; Lexate CRC; Mackine 301; Miramine DD; Schercodine S; Tegamine 18; Tego-Amid S 18; Varisoft CRC

Stearamidopropyl dimethylamine lactate (CTFA) [CAS #55819-53-9]
Emcol 3780; Hetamine 5 L 25; Incromate PPI, SDL; Jordamine S-13 Lactate; Lexamine S-13 Lactate; Mackalene 316

Stearamidopropyl dimethyl-B-hydroxyethyl ammonium dihydrogen phosphate
Cyastat SP

Stearamidopropyl dimethyl-B-hydroxyethyl ammonium nitrate
Cyastat SN

Stearamidopropyl ethyl dimonium ethyl sulfate [CAS #67846-16-6] *Synonyms:* Stearamidopropyl ethyldimonium ethosulfate; Stearamidopropyl dimethyl ethyl ammonium ethyl sulfate
Schercoquat SAS

Stearamidopropyl morpholine (CTFA) [CAS #55852-13-6]
Mackine 321

Stearamidopropyl morpholine lactate (CTFA) [CAS #55852-14-7]
Mackalene 326

Stearamidopropyl PG-dimonium chloride phosphate
Monaquat P-TS

Stearamine (CTFA) [CAS #124-30-1] *Synonyms:* Stearyl amine; Octadecylamine; 1-Octadecanamine
Adogen 142 (D); Armeen 18, 18D; Crodamine 1.18D; Kemamine P-990, P-990D; Lilamin 142, 142 D; Nissan Amine AB

Stearamine acetate *Synonyms:* Stearyl amine acetate; Octadecylamine acetate
Acetamin 86; Armac 18D; Kemamine A990; Nissan Cation SA

Stearamine oxide (CTFA) [CAS #2571-88-2] *Synonyms:* Stearyl dimethylamine oxide; Octadecyl dimethylamine oxide
Ammonyx SO, SQ; Aromox DM18D-W; Conco XA-S; Incromine Oxide S; Jordamox SDA; Ninox S; Rewominox S 300; Schercamox DMS

Steardimonium hydrolyzed animal protein
Croquat S

Steardimonium hydroxyethyl cellulose
Crodacel QS

Steareth-2 (CTFA) [CAS #9005-00-9 (generic); RD #977055-79-0] *Synonyms:* PEG-2 stearyl ether; PEG 100 stearyl ether; POE (2) stearyl ether
Alkasurf SA-2; Hetoxol STA-2; Lipocol S-2; Macol SA-2

Steareth-5 (CTFA)
Macol SA-5; Solulan 5

Steareth-10 (CTFA) [CAS #9005-00-9 (generic); RD #977055-80-3]
Alkasurf SA-10; Brij 76; Cosmowax; Hetoxol STA-10; Lipocol S-10; Macol SA-10

Steareth-11 (CTFA) [CAS #9005-00-9 (generic)]
Empilan KM 11

Steareth-15 (CTFA) [CAS #9005-00-9 (generic); RD #977068-89-5]
Macol SA-15

Steareth-16 (CTFA) [CAS #9005-00-9 (generic)]
Solulan 16

Steareth-20 (CTFA) [CAS #9005-00-9 (generic); RD #977053-41-0]
Alkasurf SA-20; Brij 78; Cosmowax; Hetoxol STA-20; Lipal 20 SA; Lipocol S-20; Macol SA-20; Trycol 5888, SAL-20

Steareth-25 (CTFA) [CAS #9005-00-9 (generic)]
Solulan 25

Steareth-30 (CTFA) [CAS #9005-00-9 (generic); RD #977055-81-4]
Hetoxol STA-30; Lipal 30 SA

Steareth-40 (CTFA) [CAS #9005-900-9 (generic)]
Macol SA-40

Steareth-50 (CTFA) [CAS #9005-00-9 (generic)]
Empilan KM 50

Steareth-100 (CTFA) [CAS #9005-99-9 (generic)]
Brij 700; Volpo S100

Stearic acid (CTFA) [CAS #57-11-4] *Synonyms:* n-Octadecanoic acid
Cohedur RS; Crosterene; Dar Chem-11, -12 -13, -14; Emersol 110, 120, 132; Hystrene 5016, 5016 NF FG, 7018, 7018 FG, 9718, 9718 NF FG; Industrene 4518, 5016, 7018, 7018 FG, 8718 FG, 9018, B, R; Interwax G 8207; Kessco Diglycol Stearate SE; Loxiol G 20; Naftozin N, Spezial; Petrac 250, 270; Radiacid 408, 411, 420, 423; Vultac 4

Stearic acid ester amide
CPH-380-N

Stearic diethanolamide. See Stearamide DEA

Stearic (ethyl amino hydroxyethyl) amide
Unamide SI

Stearic monoethanolamide. See Stearamide MEA

Stearoamphoacetate (CTFA) [CAS #68608-63-9] *Synonyms:* Stearoamphoglycinate
Amphoterge S; Miranol DM, DM conc. 45

Stearoamphohydroxypropylsulfonate (CTFA) *Synonyms:* Stearamphopropylsulfonate)
Miranol DS

Stearone (CTFA) [CAS #504-53-0] *Synonyms:* Diheptadecyl ketone
Amerchol 400, RC

Stearoxy dimethicone
Abil-Wax 2434

N-Stearoyl-p-amino phenol
Suconox-18

Stearoyl stearoyl stearate
Hetester SSS

Steartrimonium chloride (CTFA) [CAS #112-03-8] *Synonyms:* Stearyl trimethyl ammonium chloride; Octadecyl trimethyl ammonium chloride
Arquad 18-50; Nissan Cation AB; Quartamin 86W; Varisoft TSC

Steartrimonium hydrolyzed animal protein (CTFA) [CAS #11174-62-0; RD #977012-61-5] *Synonyms:* Trimethyloctadecylammonium hydrolyzed animal protein
Crotein Q

Steartrimonium methosulfate *Synonyms:* Stearyl trimethyl ammonium methosulfate
Catigene ST 30

Stearyl alcohol (CTFA) [CAS #112-92-5] *Synonyms:* n-Octadecanol
Adol 61 NF, 62, 62 NF, 63, 64, 620; Alfol 18; Amerchol RC; Cachalot S-54, S-56; Cetax 18; Cosmowax, K; Crodacol S, S70,

S95NF; Dehydag Wax 18; Emcol 150; Kalcohl 80; Lanette 18; Lanol S; Lipocol S; Promulgen G; Stearal; Varisoft 3262, SA-50, SDC-85S

Stearyl amine. See Stearamine

Stearyl betaine (CTFA) [CAS #820-66-6] *Synonyms:* Stearyl dimethyl glycine
Alkateric STB; Chimin BX; Varion SDG

Stearyl diethanol methyl ammonium chloride
M-Quat 32

Stearyl dihydroxyethyl methyl ammonium chloride
Tomah Q-18-2

Stearyl dimethicone
Abil-Wax 9800

Stearyl dimethyl amine. See Dimethyl stearamine

Stearyl dimethyl benzyl ammonium chloride. See Stearalkonium chloride

Stearyl dimethyl ethyl ammonium ethosulfate
Larostat 451, 491

Stearyl erucamide (CTFA) [CAS #10094-45-8] *Synonyms:* N-Octadecyl-13-docosenamide
HTSA #3; Kemamide E-180

Stearyl erucate (CTFA) *Synonyms:* Octadecyl 13-docosenoate
Schercemol SE

Stearyl heptanoate (CTFA) [RD #977063-12-9] *Synonyms:* Heptanoic acid, octadecyl ester
Crodamol W

Stearyl hydroxyethyl imidazoline (CTFA) [CAS #95-19-2] *Synonyms:* Stearyl imidazoline
Alkazine ST; Amine S; Miramine GS; Schercozoline S

Stearyl methicone
Abil-Wax 9809

Stearyl nitrile
Arneel 18 D

Stearyl PEG-15 methyl ammonium chloride
Tomah Q-18-15

Stearyl stearamide
Kemamide S-180

Stearyl stearate (CTFA) [CAS #2778-96-3]
Cyclochem SS; Dermalcare SS; Lexate TA, TL; Lexol SS; Liponate SS; Rilanit STS-T

Stilbene [CAS #103-30-0] *Synonyms:* Toluelene; trans form of a a,b-diphenylethylene
Eccobrite RB; Tinopal UNPA

Stilbene-2,2′-disulfonic acid type
Tinopal SFP

Stoddard solvent [CAS #8052-41-3]
Ircogel 904

Styrenated diphenyl amine
Vulkanox DDA; Wingstay 29

Styrene/acrylate copolymer (CTFA) [RD #977063-04-9]
Texicryl 13-021, 13-022, 13-030; Vinnapas SAF 54

Styrene-butadiene polymer [CAS #9003-55-8] *Synonyms:* Styrene polymer with 1,3 butadiene (CTFA); Butadiene-styrene resin; 1,3-Butadiene-styrene copolymer; Styrene-butadiene rubber; S/B copolymer; SBR copolymer
GAF Latex 7000 Series, 8000 Series

Styrene-butadiene-styrene block copolymer
C-Flex R70-058 CLHR70, R70-067 CLHR40, R70-068 CLHR50

Styrene homopolymer
Texicote 53-010

Styrene hydrocarbon resin
Piccolastic A5, A25, A50, A75, BV-50HF

Styrene/maleic anhydride [CAS #9011-13-6 (copolymer)] *Synonyms:* SMA
Scripset 520, 540, 550, 700, 720; SMA 1000, 1440, 1440H, 2000, 2625, 3000

Styrene maleic anhydride amide/ammonium hydroxide acid salt
Scripset 808

Styrene polymer with 1,3-butadiene. See Styrene/butadiene polymer

Styrene/PVP copolymer (CTFA) [CAS #25086-29-7]
Polectron 430

3-(N-Styrylmethyl-2-aminoethylamino) propyltrimethoxy silane hydrochloride [CAS #52783-38-7]
CS1590

Sucrose (CTFA) [CAS #57-50-1] *Synonyms:* Table sugar; Saccharose
Frimulsion 6G

Sucrose cocoate (CTFA) [CAS #25339-99-5]
Crodesta SL-40; Grilloten LSE87K

Sucrose distearate (CTFA) [CAS #27195-16-0]
Crodesta F-10, F-50, F-110; Ryoto Sugar Ester S-570, -770

Sucrose di/tristearate
Ryoto Sugar Ester S-170, -270, -370

Sucrose laurate (CTFA) [CAS #25339-99-5]
Grilloten LSE87; Ryoto Sugar Ester LWA-1540

Sucrose mono/distearate
Ryoto Sugar Ester S-970, -1170

Sucrose oleate
Ryoto Sugar Ester OWA-1570

Sucrose palmitate
Ryoto Sugar Ester P-1570, -1670

Sucrose ricinoleate
Grilloten ZT12, ZT40, ZT80

Sucrose stearate (CTFA) [CAS #25168-73-4]
Crodesta F-110, F-160; Grilloten PSE141G; Ryoto Sugar Ester S-1570, -1670

Sulfated butyl oleate
Chemax SBO; Hipochem Dispersol SB; Maravanol 60% SBO

Sulfated butyl oleate, sodium salt
Actrasol SBO

Sulfated castor oil (CTFA) [CAS #8002-33-3] *Synonyms:* Castor oil sulfated; Sulfonated castor oil; Turkey-red oil
Actrasol 167A; Ahco AJ-110; Castor Oil Sulfated 50%, 68%; Chemax SCO; Consos Castor Oil; Cordon NU 890/75; Densol 6920; Dyasulf 1761-A; Eureka 102; Haroil SCO-65, SCO-7525; Hartex V63, V64; Hipochem Dispersol SCO; Marvanol 75% SCO; Monopol Oil 75; Monosulph; Nopcocastor, L; Nopcosulf CA-60, -70; Standapol SCO; Sulfonated Castor Oil 50%, 75%, GTO; Supratol VF

Sulfated castor oil, sodium salt
Actrasol C-50, C-75, C-85

Sulfated glyceryl trioleate
Hipochem Dispersol GTO

Sulfated glyceryl trioleate, sodium salt
Actrasol EO

Sulfated methyl soyate, sodium salt
Actrasol MY-75

Sulfated oleic acid, ammonium salt
Actrasol SR75

Sulfated oleic acid, potassium salt
Actrasol SRK 75

Sulfated propyl oleate
Hipochem Dispersol SP; Marvanol 55% SPO

Sulfated rapeseed oil
Actrasol 6092

Sulfated ricinoleic acid
Actrasol PSR

Sulfated soybean oil, sodium salt
Actrasol CS-75

Sulfated sperm oil
Dymsol S; Eureka 400; Sulfonated Syn. Sperm Oil 2091

Sulfated tall oil
Cordon PB-870

Sulfated tall oil, potassium salt
Actrasol SP-175K

Sulfated tall oil, sodium salt
Actrasol SP

Sulfated tallow
Nopcosulf TA-30, -45V

Sulfolane [CAS #126-33-0] *Synonyms:* Dapsone; Tetramethylene sulfone
Sulfolane W

Sulfonated isopropyl oleate
Rexowet CR

Sulfonated maleic anhydride copolymer
Cyanamer P-80

Sulfonated oleic acid
Sulfonic 800

Sulfonated oleyl alcohol
Hipochem LCA

Sulfonated polystyrene, sodium salt
Flexan 130

Sulfonated tallow
Densulf TA-75

Sulfonic acid
Demol SN-B, SS Paste, SSL, VP

Sulfotepp
Bladafum

Sulfur
Chlorpren-Faktis A; Faktis Asolvan, Asolvan T, Badenia C, Badenia T, DK 10, DK 14, DK 17, HF Braun, Para extra weich, RC 110, RC 111, RC 140, RC 141, RC 144, T-hart, ZD; Naftopast Schwefel-P; Rhenocure IS 60, IS 60-5, IS 60/G, IS 90-22, IS 90-33, IS 90-40, IS 90/G; Rhenovin S-90

Sulfur chloride [CAS #10025-67-9] *Synonyms:* Sulfur subchloride; Sulfur monochloride
Chloropren-Faktis NW; Faktis Rheinau H, Rheinau W, R Spezial, Weib MB

Sulfuric acid [CAS #7664-93-9] *Synonyms:* Hydrogen sulfate; Battery acid; Electrolyte acid
Eltesol TA/E, TA/F, TA/H, TA/K

Sulfurized oxymolybdenum organophosphorodithioate
Molyvan L

Sunflower seed oil (CTFA) [CAS #8001-21-6]
Lipovol SUN

Sunflower seed oil glyceride (CTFA)
Monomuls 90-40

Sunflower seed oil glycerides (CTFA) *Synonyms:* Sunflower seed mono-, di- and tri-glycerides
Monomuls 60-40

Sweet almond oil (CTFA) [CAS #8007-69-0]
Lipovol ALM, ALM-S

Synthetic spermaceti. See Cetyl esters

Synthetic wax (CTFA) [CAS #8002-74-2; RD #977017-52-9] *Synonyms:* Fischer-Tropsch wax; Fischer-Tropsch wax, oxidized
Ross Wax # 100

Talc (CTFA) [CAS #14807-96-6] *Synonyms:* Hydrous magnesium silicate; Industrial talc; Cosmetic talc; Platy talc; Talcum; Soapstone; Steatite
ABT-2500; Ceramitalc 1, 10A; I.T. 3X, 5X, 325, F.T., X; Jet Fil 100, 200, 350, 500; Microbloc; MicroPlfex 1200; Microtalc MP2-40, MP10-52, MP12-50, MP15-38, MP25-38, MP26-38, MP30-36; Microtuff 1000, F; Nicron JS 216, 322, 422, 426, 634; Nytal 99, 100, 100 HR, 200, 300, 400; Select-A-Sorb; Stealim; Vantalc 6H, 500, 700, 900, 1350, 1500

Tallamide DEA (CTFA) [CAS #68155-20-4; RD #977068-70-4] *Synonyms:* Diethanolamine tall oil acid amide; Tall oil acid diethanolamide; Tall oil diethanolamide
Monamine T-100; Schercomid TO-2

Tallamidopropyl dimethylamine
Schercodine T; Servamine KET 350

Tallamphopropionate (CTFA) [CAS #68991-88-8]
Miranol LM-SF Conc.; Monateric TA-35

Tall oil acid (CTFA) [CAS #61790-12-3]
Acofor

Tall oil amine
Adogen 151; Lilamin 151

Tall oil diamine
Adogen 551

Tall oil diethanolamide. See Tallamide DEA

Tall oil dioleate
Ninex TDO 5, 9, 14

Tall oil hydroxyethyl imidazoline (CTFA) [CAS #61791-39-7] *Synonyms:* 1-Hydroxyethyl-2-tall oil imidazoline; Tall oil imidazoline
Alkazine TO; Amine T; Mazoline T; Miramine GT; Miramine TOC; Monazoline T; Textamine T-1; Unamine-T; Varine T

Tall oil oleate
Ninate TMO 5, 7, 9, 14

Tall oil rosin. See Rosin

Tallow (CTFA) [CAS #61789-97-7] *Synonyms:* Beef tallow; Mutton tallow
Dar Chem-030; Foamaster Soap L

Tallow acid [CAS #61790-37-2] *Synonyms:* Fatty acids, tallow
Industrene 143, 145

Tallowalkonium chloride (CTFA) *Synonyms:* Tallow dimethyl benzyl ammonium chloride
Ammonyx 856; Kemamine BQ-9742C; Noramium S 75

Tallow amide (CTFA) [RD #977058-38-0]
Kemamide S-65

Tallowamide DEA (CTFA) [CAS #68140-08-9] *Synonyms:* Tallow diethanolamide
Jordamide TC; Monamine 928; Schercomid SO-T, TO-2

Tallowamidopropylamine oxide (CTFA) [CAS #68647-77-8]
Ammonyx TDO

Tallowamidopropyl dimethylamine (CTFA) [CAS #68425-50-3; RD #977059-03-2] *Synonyms:* Dimethylaminopropyl tallow amide
Incromine TB; Schercodine T

Tallowamidopropyl hydroxysultaine (CTFA)
Mirataine TABS

Tallow amine (CTFA) [CAS #61790-33-8] *Synonyms:* Amines, tallow alkyl
Adogen 170 (D); Amine BG; Armeen TD; Arosurf MG-170; Crodamine 1.T; Kemamine P-974, P-974D; Lilamin 170, 170 D; Noram S; Radiamine 6170, 6171; Tomah Tallow Amine

Tallow amine acetate
Acetamin T; Armac T; Kemamine A974; Lilamac 170; Noramac S; Radiamac 6179

Tallow amine oxide (CTFA) [RD #977058-39-1] *Synonyms:* Tallow dimethylamine oxide
Conco XA-T

Tallowaminopropylamine [CAS #68439-73-6] *Synonyms:* Tallow trimethylene diamine
Kemamine D-974

Tallow betaine
Alkateric TB

Tallow bis hydroxyethyl methyl ammonium chloride
Varisoft 920

Tallow diamine
Adogen 570-S; Arosurf MG-570; Lilamin 570; Radiamine 6570; Tomah Tallow Diamine

Tallow diamine dioleate
Adogen 570-DO

Tallow dihydroxyethyl betaine
Varion TEG

Tallow dimethyl trimethyl propylene diammonium chloride
Tomah Q-D-T

Tallowdimonium propyltrimonium dichloride (CTFA) [CAS #68607-29-4] *Synonyms:* N,N,N′,N′,N′-Pentamethyl-N-tallow alkyl-1,3-propanediammonium dichloride; N-Tallow pentamethyl propane diammonium dichloride
Adogen 477

Talloweth-20
Alkasurf TA-20

Talloweth-25
Rewopal TA 25/S

Talloweth-30
Alkasurf TA-30

Talloweth-40
Alkasurf TA-40

Talloweth-50
Alkasurf TA-50

Tallow glyceride (CTFA) [CAS #61789-13-7] *Synonyms:* Tallow monoglyceride
Monomuls 90-20

Tallow glycerides (CTFA) [RD #977011-09-8] *Synonyms:* Tallow mono, di and tri glycerides
Monomuls 60-20

Tallow glycinate
Varion TEG-40%

Tallow hydroxyethyl imidazoline (CTFA) *Synonyms:* Tallow imidazoline
Accosoft 808; Miramine TC

Tallow imidazolinium methosulfate Synonyms: Tallowalkyl imidazoline methosulfate
Empigen FRB75S, FRC75S, FRC90S, FRG75S; Incroquat S-75 CG; Incrosoft S-75, S-90, S-90M; Variquat 475

Tallow nitrile
Arneel T

N-Tallow-1,3-propanediamine *Synonyms:* Tallow-1,3-diaminopropane
Duomeen T, TX

N-Tallow-1,3-propanediamine diacetate
Duomac T

N-Tallow-1,3-propanediamine dioleate
Duomeen TDO

Tallow propylene diamine
Lilamuls BG; Nissan Amine DT

Tallow propylene diamine acetate
Dinoramac S

N-Tallow 1,3-propylene diamine diacetate
Kemamine AD 974

Tallow propylene diamine PEG ammonium methosulfate
Rewoquat DQ 35

Tallow propylene diammonium caprylate
Rewocor TPAC 100; Rewocoros TPAC 100

N-Tallow trimethylene diamine dioleate
Duomeen TDO-IHF

Tallowtrimonium chloride (CTFA) [CAS #8030-78-2; 7491-05-2] *Synonyms:* Tallow trimethyl ammonium chloride
Adogen 471, R-6; Arquad T-2C-50, T-27W, T-50; Jet Quat T-50; Kemamine Q-9743C, Q-9743CHGW; Quartamin TPR; Querton BGCL50; Radiaquat 6471

Tallow trimonium methosulfate *Synonyms:* Tallow trimethyl ammonium methosulfate
Empigen CM

Tannic acid [CAS #1401-55-4] *Synonyms:* Glycerite; Tannin; Gallotannic acid; Gallotannin
Actan SP; Lexein A520

TEA-coco-hydrolyzed animal protein (CTFA) [CAS #68952-16-9]
Lexein S620S, S620TA; May-Tein CT; Super Pro #5A

TEA-cocoyl-glutamate (CTFA) [CAS #68187-29-1; RD #977068-61-3]
Acylglutamate CT-12

TEA dioleate
Rilanit TDO

TEA-dodecylbenzenesulfonate (CTFA) [CAS #27323-41-7] *Synonyms:* Triethanolamine dodecylbenzene sulfonate
Arylan TE/C; Conco AAS-60S; ESI-Terge T-60; Marlopon CA; Nansa TS 60; Puxol CB-22; Reworyl TKS 90/F; Steinaryl TKS 90/F; Witconate 60T, TAB

TEA hydrochloride (CTFA) [CAS #637-39-8] *Synonyms:* Triethanolamine hydrochloride
Hamp-Ene OH Powd.

TEA-lauroyl animal collagen amino acids (CTFA) [CAS #68952-16-9]; Synonyms: TEA-lauroyl collagen amino acids
Aminofoam C

TEA-lauroyl animal keratin amino acids *Synonyms:* TEA-lauroyl keratin amino acids
Aminofoam K

TEA-lauryl sulfate (CTFA) [CAS #139-96-8] *Synonyms:* Triethanolamine lauryl sulfate; Triethanolammonium lauryl sulfate
Conco Sulfate TL; Condanol TLS 40; Drewpon 40, 40LS; Empicol TLP; Sandoz Sulfate TL; Sipex TEA; Witcolate T

TEA-PEG-3 cocamide sulfate (CTFA)
Genapol AMS

TEA phosphate ester
Wayhib

Tellurium [CAS #13494-80-9]
Telloy

Tellurium diethyl dithiocarbamate
Akrochem TDEC; Ethyl Tellurac

Temephos [CAS #3383-96-8] *Synonyms:* Tetramethyl-o,o´-thiodi-p-phenylene phosphorothioate
Abate

Terbufos [CAS #13071-79-9] *Synonyms:* O,O-Diethyl-s-(t-butyl)methyl phosphorodithioate
Counter

Terbuthylazine [CAS #5915-41-3] *Synonyms:* 4-t-butylamino-2-chloro-6-ethylamino-s-triazine; 2-t-Butylamino-4-chloro-6-ethylamino-s-triazine
Belclene 329

Terpene phenolic resin
SP-553, -560

Terpene resin. See also Dipentene, Pinene
Piccofyn A100, A115, A135, D125, HM-110 Resin; Piccolyte A115, A125, A135, C100, C115, C125, C135, D100, D115, D135, HM-110, S100, S115, S115SF, S125, S125SF, S135; Picconol A200, A201; Solvenol

Tetraacrylate (monomer), alkoxylated
Chemlink 9022

Tetraammonium EDTA
Sequestrene Tetraammonium; Versene Tetraammonium EDTA

Tetrabromobisphenol A
BA-59, -59P; Emery 9350; FR-1524; Saytex RB-100

Tetrabromobisphenol A diacrylate
Sartomer 640

Tetrabromobisphenol A di-2-hydroxyethyl ether
Emery 9353

Tetrabromophthalic anhydride
Saytex RB-49

Tetrabromoxylene [CAS #23488-38-2 (Tetrabromo-p-xylene); CAS #35323-28-1 (a,a,a´,a´-Tetrabromo-m-xylene)]
Emery 9345

Tetrabromophthalate diol
PHT4-Diol

Tetrabromophthalic anhydride
PHT4

Tetrabutyl ammonium bromide [CAS #1643-19-2] *Synonyms:* Tetra-N-butylammonium bromide
Arosurf PT Bu4

Tetrabutyl thiuram disulfide [CAS #1634-02-2] *Synonyms:* Bis (dibutylthiocarbamoyl) disulfide
Butyl Tuads

Tetrabutyl titanate [CAS #5593-70-4] *Synonyms:* TBT; Butyl titanate
Tyzor TBT

Tetrachlorobutyl alcohol
AA-81C

Tetrachloroethylene. See Perchloroethylene

Tetrachloroisophthalonitrile [CAS #1897-45-6] *Synonyms:* Chlorothalonil; Chloroalonil; 1,3-Dicyanotetrachlorobenzene
Bravo W-75; Daconil 2787; Groutcide 75; Nopcocide N-40-D, N-96, N-96-S

Tetrachlorosilane [CAS #10026-04-7]
CT1800

Tetradecabromodiphenoxy benzene
Saytex 120

Tetradecyl. See Myristyl

Tetra (2,2 diallyloxymethyl-1 butoxy titanium di (ditridecyl) phosphite
Ken-React 55

Tetra (2,2 diallyloxymethyl) butyl, di (ditridecyl) phosphito zirconate
KZ 55

Tetradifon
Tedion V18

Tetraethoxysilane [CAS #78-10-4]
Dynasil A

Tetraethylene glycol... See PEG-4...

Tetraethylenepentamine [CAS #112-57-2] *Synonyms:* 1,4,7,10,13-Pentaazatridecane
D.E.H. 26

Tetraethylthiuram disulfide [CAS #97-77-8]
Ethyl Tuads; Methyl-Ethyl Tuads; TM/ETD

Tetrafluoroethylene telomer
MS-122, -136

Tetrahyrofuran [CAS #109-99-9] *Synonyms:* THF
QO Tetrahydrofuran

Tetrahydrofurfuryl acrylate (CTFA)
SR-285

Tetrahydrofurfuryl alcohol [CAS #97-99-4] *Synonyms:* Tetrahydrofuryl carbinol; THFA; Tetrahydro-2-furanmethanol
QO Tetrahydrofurfuryl Alcohol

Tetrahydrofurfuryl oleate
Kemester THFO

Tetrahydronaphthalene [CAS #119-64-2] *Synonyms:* 1,2,3,4-Tetrahydronaphthalene
Tetralin

Tetrahydrothiophene [CAS #110-01-0] *Synonyms:* Thiophane
Pennodorant 1013; Spotleak 1044

Tetrahydroxypropyl ethylenediamine (CTFA) [CAS #102-60-3]
Mazeen 173, 174-75; Quadrol

Tetraisopropyl di (dioctylphosphito) titanate
Ken-React 41B

Tetraisopropyl titanate *Synonyms:* Titanium isopropylate; Isopropyl titanate
Tyzor TPT

Tetrakis (2-chloroethyl) ethylene diphosphate
Thermolin 101

Tetrakis (2,4-di-t-butylphenyl) 4,4-biphenylenediphosphonite
Sandostab P-EPQ

Tetrakis (2-ethoxyethoxy) silane [CAS #18407-94-8]
Dynasil CA

Tetrakis (2-ethylhexyl) titanate
Tyzor TOT

Tetrakis (2-methoxyethoxy) silane [CAS #2157-45-1]
Dynasil CM

Tetrakis [methylene (3,5-di-t-butyl-4-hydroxyhydrocinnamate)]
Irganox 1010, B-215, B-225; Weston 800, 801, 802

Tetrakis (2,2,6,6-tetramethyl-4-piperidyl)-1,2,3,4-butane tetracarboxylate
Mixxim HALS 57

Tetramethoxysilane [CAS #681-84-5]
Dynasil M

1,1,4,4-Tetramethyl-1,4-bis-(N,N-dimethylamino) disilethylene
T1990

1,1,3,3-Tetramethyl butyl hydroperoxide
Lupersol 215

Tetramethyl decynediol [CAS #126-86-3]
Surfynol 104, 104A, 104BC, 104E, 104H, 104S

1,1,4,4-Tetramethyldichlorodisilethylene [CAS #13528-93-3]
CT2015; T2015

Tetramethyldisiloxane [CAS #3277-26-7]
CT2030

Tetramethylene sulfone. See Sulfolane

2,2,5,5-Tetramethyl-4-isopropyl-1,3-dioxane
Verdoxan

[2,2,6,6-Tetramethyl-4-piperidyl/β,β,β′,β′-tetramethyl-3,9-(2,4,8,10-tetraoxaspiro (5,5) undecane) diethyl]-1,2,3,4-butane tetracarboxylate
Mixxim HALS 68

Tetramethylsilane [CAS #75-76-3]
CT2050

Tetramethylthiuram disulfide [CAS #137-26-8] *Synonyms:* Thiram; Thiuram; Bis-(di-methylthiocarbamyl) disulfide
Akrochem TMTD, TMTD Pellet; Akrosperse D-177; Captax-Tuads Blend; Methyl Tuads; Methyl-Ethyl Tuads; Naftocit Thiuram 16; Naftopast Thiuram 16-P; Rhenovin TMTD-70; TM/ETD; Vulkacit Thiuram, Thiuram/C, Thiuram/EGC

Tetramethylthiuram monosulfide Synonyms: TMTM
Naftopast TMTM-P; Rhenovin TMTM-70; TMTM; Vulkacit Thiuram MS/C, MS/EGC

Tetraoctyloxytitanium di (ditridecylphosphite)
Ken-React 46B

Tetraphenyl dipropyleneglcyol diphosphite
Weston THOP

Tetrapotassium pyrophosphate (CTFA) [CAS #7320-34-5] *Synonyms:* Diphosphoric acid, tetrapotassium salt
Empiphos 4KP

Tetrapropoxysilane [CAS #682-01-9]
CT2090

Tetrasodium dicarboxyethyl stearyl sulfosuccinamate (CTFA) [CAS #3401-73-8; 37767-39-8] *Synonyms:* Tetrasodium dicarboxyethyl octadecyl sulfosuccinamate
Aerosol 22, 22N; Alconate 2CSA; Lankropol ATE; Monawet SNO-35; Rewopol B 1003, B 2003

Tetrasodium EDTA (CTFA) [CAS #64-02-8] *Synonyms:* Edetate sodium; Tetrasodium edetate; Ethylene diamine tetraacetic acid, sodium salt; EDTA Na4; Sodium EDTA
Alkaquest EDTA; Aquamolin Brands; Cheelox B-13, BF-12, BF-13, BF-78; Chelon 100; Complexion CA-10; Conco SEQ; Dissolvine E 39, NA; Hamp-Ene 100, 100S, 220, Na4; Iberquestrene; Jorquest 100; Kalex 100, 220 Crystal, Conc., Liq. 50%, Powd. FC, Reg.; Kemplex 100; Perma Kleer 100, CP Grade; Plexene Extra Conc.; Questal Special; Questex 4SW, 4SW Crystals; Sequestrene 30A, 220, NA4; Sterling EDTA; Trilon B; Vanate TS, TSD, TST; Versene, 100, 220 Crystals, AG 100

Tetrasodium etidronate (CTFA) [CAS #3794-83-0] *Synonyms:* Tetrasodium 1-hydroxyethane-1, 1-diphosphonate
Turpinal 4 NL

4,4′-Thiobis-6-(t-butyl-m-cresol) [CAS #96-69-5] *Synonyms:* 6-t-butyl-m-cresol
Santowhite Crystals

4,4′-Thiobis (6-t-butyl-o-cresol) [CAS #96-66-2] *Synonyms:* 4,4′-Thiobis[2-(1,1-dimethyl ethyl)-6-methylphenol
Ethanox 736

4,4′-Thiobis (2-t-butyl-5-methylphenol)
Ultranox 236

2,2′-Thiobis (4-t-octylphenolato)-n-butylamine nickel II
Cyasorb UV 1084

Thiocarbamyl sulfenamide
Cure-rite 18, Accelerator

Thiodiethylene bis (3,5-di-t-butyl-4-hydroxy) hydrocinnamate
Irganox 1035

Thiodiglycol (CTFA) [CAS #111-48-8] *Synonyms:* Thiodiethylene glycol; b-Bis-hydroxyethyl sulfide; Dihydroxyethyl sulfide; 2,2′-Thiodiethanol
Glyecine A

Thiodipropionate polyester
TDP 2000

3,3′-Thiodipropionic acid (CTFA) [CAS #111-17-1] *Synonyms:* Propanoic acid, 3,3′-thiobis
Evanstab A

Thiodipropionic dilauryl ester
Aflux 32

Thioglycerin (CTFA) [CAS #96-27-5] *Synonyms:* 3-Mercapto-1,2-propanediol monothioglycerol
Thiovanol

Thiuram disulfide [CAS #137-26-8] *Synonyms:* Thiram disulfide; Tetramethyl thiuram disulfide; Bis (dimethylthiocarbamyl) disulfide; TMTD)
Baypren 321, 321 GR, 331

Tin [CAS #7440-31-5] *Synonyms:* Stannum
Haro Mix ZT-025, ZT-026, ZT-504, ZT-508, ZT-514, ZT-905

Tin carboxylate
Stanclere T 55

Tin mercaptide
Stanclere T 184

Tin (II) oxalate
Tego Stannous Oxalate

TIPA. See Triisopropanolamine

Titania. See Titanium dioxide

Titanium di (butyl, octyl pyrophosphate) di (dioctyl, hydrogen phosphite) oxyacetate
Ken-React 158FS, 158D

Titanium di (cumylphenylate) oxyacetate
Ken-React 134S

Titanium di (dioctylpyrophosphate) oxyacetate
Ken-React 138D, 138S

Titanium dimethacrylate oxyacetate
Ken-React 133DS

Titanium dioxide (CTFA) [CAS #13463-67-7] *Synonyms:* Titanic anhydride; Titanic acid anhydride; Titanic oxide; Titanium white; Titania
Hitox; Hombitan LOCR-K, LW, R 101 D, F 610 K, R 610 L; Lustra-Pearl; Mearlin; Soloron; Spectra-Pearl; Ti-0720 T 1/8″; Timic, Timiron

Tocopherol (CTFA) [CAS #1406-18-4; 59-02-9 (d-α), 10191-41-0 (dl-α)] *Synonyms:* Vitamin E; D-α tocopherol; DL-α tocopherol
Arnica Oil CLR; Calendula Oil CLR; Carrot Oil CLR; Coviox T-50, T-70, F-350M, F-600, F-1000

Tocopheryl acetate (CTFA) [CAS #1406-70-8; 7695-91-2] *Synonyms:* d-α Tocopheryl acetate)
Covitol 544, 700C, 1100, 1360

Tocopheryl succinate (CTFA) [CAS #4345-03-3; 17407-37-3]
Covitol 1185, 1210

Toluene (CTFA) [CAS #108-88-3] *Synonyms:* Methylbenzene; Phenylmethane; Toluol
Amsco Toluene; D.E.R. 671-T75; Lupersol 665T50, 688T50; Shell Toluene

Toluenesulfonamide
Ketjenflex 8, 9

o-Toluenesulfonamide [CAS #88-19-7]
Merrol N-305; Santicizer 9

p-Toluenesulfonamide [CAS #70-55-3]
Merrol N-304, N-305; Santicizer 9

Toluenesulfonamide-formaldehyde resin (CTFA) [CAS #25035-71-6] *Synonyms:* Benzenesulfonamide, 4-methyl-, polymer with formaldehyde
Merrol N-307

Toluene sulfonic acid (CTFA) [CAS #104-15-4] *Synonyms:* 4-Methylbenzenesulfonic acid
Condasol T 65; Eltesol TA, TA 65, TA 96, TA/E, TA/F, TA/H, TA/K; Manro PTSA 65 E, 65 H, 65 LS, PTSA Crystals; Reworyl T 65; Steinaryl T 65; Sulframin TX; Witco TX Acid; Witconate TX Acid

Toluhydroquinone
THQ

o-Tolyl biguanide
Vulkacit 1000/C

m-Tolyl diethanolamine [CAS #91-99-6]
Emery 5709

o-Tolyl diethanolamine [CAS #28005-74-5]
Emery 5712

p-Tolyl diethanolamine [CAS #3077-12-1]
Emery 5710

o-Tolyl ethanolamine [CAS #136-80-1]
Emery 5711

Tolylfluanid
Euparen M

Tolyltriazole
Cobratec TT-35-A, TT-50-A, TT-100

Triacetin (CTFA) [CAS #102-76-1] *Synonyms:* Glyceryl triacetate; 1,2,3-Propanetriol triacetate; Triacetyl glycerol
Kessco Triacetin; Kodaflex Triacetin

Triacrylate ester (monomer)
Chemlink 9008, 9012

Triacrylate (monomer), alkoxylated
Chemlink 9003, 9020, 9021

Triadimefon
Bayleton

Triadimenol
Bayfidan; Baytan

Triallylcyanurate
Rhenofit TAC/S

Triammonium dodecylbenzene sulfonate
Condasol TKS 90/F

Triaryl phosphate [CAS #68937-41-7]
Kronitex 50, 100

Tribromoneopentyl alcohol [CAS #3296-90-0]
FR-513; Saytex FR-1138

2,4,6-Tribromophenol [CAS #118-79-6] *Synonyms:* Tribromophenol; Bromol
Emery 9332; FR-613; PH-73

Tribromophenyl allyl ether
FR-913

Tribromosalicylanilide [CAS #87-10-5] *Synonyms:* 3,4′,5-Tribromosalicylanilide
Temasept IV

Tributoxyethyl phosphate [CAS #78-51-3]
KP-140; Tolplaz TBEP

Tributyl citrate (CTFA) [CAS #77-94-1] Synonyms: Tri n-butyl citrate
Citroflex C-4; Merrol C-104; TBC

(2,4,6-Tri-t-butylphenoxy) dimethylchlorosilane
T2360

Tributyl phosphate [CAS #126-73-8] *Synonyms:* Tri-n-butyl phosphate
Pliabrac TBP; TBP

Tributyltin maleate
Jorgard AMF

Tributyltin oxide [CAS #56-35-9]
Keycide X-10; Troysan 364

Tri C8-10 amine
Adogen 364; Adogen 2364

Tri C8-10 methyl ammonium chloride
Adogen 2464

Tri C8-10-12 amine
Adogen 368, 2368

Tricaprin (CTFA) [CAS #621-71-6] *Synonyms:* 1,2,3-Propanol tridecanoate
Dynasan 110

Tri (caprylyl/capryl/lauryl) tert. amine
Lilamin 368

Tri (caprylyl/capryl) tert. amine
Lilamin 364

Tricaprylyl methyl ammonium chloride
Aliquat 336-PTC

Trichlorfon
Dipterex

3,4,4′-Trichlorocarbanilide [CAS #101-20-2]
TCC

2,2,2-Trichloro-1-(3,4-dichlorophenyl) ethyl acetate
MEB 6046

Trichloroethane (CTFA) [CAS #71-55-6] *Synonyms:* 1,1,1-Trichloroethane; Ethane, 1,1,1-trichloro-; Methylchloroform
Aerothene TT Solv.; Chlorothene SM Solv.; Dowclene LS; Prelete Defluxer; Tri-Ethane

Trichloroethylene [CAS #79-01-6]
Hi-Tri; Neu-Tri; PPG Trichlor

Trichloroethyl phosphate
Disflamoll TCA

2,4,4′-Trichloro-2′-hydroxydiphenyl ether. See Triclosan

Trichloroisocyanuric acid [CAS #87-90-1] *Synonyms:* Isocyanuric chloride; Trichlorocyanuric acid; 1,3,5-tricloro-s-triazine-2,4,6-trione)
ACL 85; OCI 90, OCI 90-I

N-Trichloromethylthio-4-cyclohexene 1,2-dicarboximide [CAS #133-06-2]
Ortho Orthocide; Vancide 89, 89 RE

N-(Trichloromethylthio) phthalimide [CAS #133-07-3]
Troysan Anti-Mildew O

Trichlorosilane [CAS #10025-78-2]
Dynasylan TCS

Trichloro tetra monopotassium dichloro pentaisocyanurate
ACL 66

Triclosan [CAS #3380-34-5] *Synonyms:* 2,4,4′-Trichloro-2′-hydroxydiphenyl ether
Irgasan DP300; Vikol THP

Tricresyl phosphate (CTFA) [CAS #1330-78-5; 78-30-8] *Synonyms:* TCP; Tritolyl phosphate
Disflamoll TKP; Kronitex TCP; Luperco ATC; Pliabrac TCP

Tricrotonylidene tetramine
Vulkacit CT-N

Trideceth-2 phosphate *Synonyms:* PEG-2 tridecyl ether phosphate
Crodafos T2 Acid

Trideceth-3 (CTFA) [CAS #4403-12-7; 24938-91-8 (generic); RD #977058-48-2] *Synonyms:* PEG-3 tridecyl ether; PEG (3) tridecyl ether; POE (3) tridecyl ether
Chemal TDA-3; Emthox 5993; Emulphogene BC-420; Hetoxol TD-3; Lipocol TD-3; Macol TD-3; Siponic TD-3; Trycol 5993, TDA-3; Volpo T3

Trideceth-4
Macol TD-4

Trideceth-5
Alkasurf TDA-5; Volpo T5

Trideceth-5 phosphate
Crodafos T5 Acid

Trideceth-6 (CTFA) [CAS #24938-91-8 (generic); RD #977058-49-3]
Ahcowet DQ-114; Alkasurf TDA-6; Chemal TDA-6; Emthox 5940; Hetoxol TD-6; Lipal 6 TD; Lipocol TD-6; Macol TD-6, TD-610; Renex 36; Siponic TD-6; Trycol 5940, TDA-6

Trideceth-6 phosphate (CTFA) [CAS #9046-01-9 (generic); RD #977014-26-8]
Gafac RS-610

Trideceth-7
Alkasurf TDA-7; Alkasurf TDA-7.5 (7.5 EO)

Trideceth-7 carboxylic acid (CTFA) [CAS #56388-96-6 (generic); 66161-61-3; RD #977061-70-3]
Carsodet TD-7C

Trideceth-8
Alkasurf TDA-8.5 (8.5 EO); Lipal 610; Macol TD-8, TD-610; Trycol 5949, 5968, TDA-8

Trideceth-9 (CTFA) [CAS #24938-91-8 (generic); RD #977062-57-9]
Chemal TDA-9; Emthox 5941; Siponic TD990; Teric 13A9; Trycol 5941

Trideceth-10 (CTFA) [CAS #24938-91-8 (generic); RD #977058-50-6]
Carsonon TD-10; Emulphogene BC-720; Lipal 10 TD; Macol TD-10; Volpo T10

Trideceth-10 phosphate
Crodafos T10 Acid

Trideceth-11 (CTFA) [CAS #24938-91-8 (generic); RD #977067-68-7]
Emthox 5942

Trideceth-12 (CTFA) [CAS #24938-91-8 (generic); RD #977058-51-7]
Ahcowet DQ-145; Alkasurf TDA-12; Chemal TDA-12; Hetoxol TD-12; Lipocol TD-12; Siponic TD-12

Trideceth-15 (CTFA) [CAS #24938-91-8 (generic); RD #977067-69-8]
Alkasurf TDA-15; Chemal TDA-15; Emulphogene BC-840; Volpo T15

Trideceth-18
Chemal TDA-18; Trycol TDA-18

Trideceth-20
Volpo T20

Trideceth-100
Macol TD-100

Tridecylbenzene sulfonic acid (CTFA) [CAS #25496-01-9]
Conco ATR-98S

Tridecyl ether diamine acetate
Arosurf MG-83A

Tridecyl neopentanoate [CAS #106436-39-9]
Ceraphyl 55

Tridecyloxypropylamine
Tomah PA-17

N-Tridecyloxypropyl-1,3-diaminopropane
Tomah DA-17

Tridecyloxypropyl dihydroxyethyl methyl ammonium chloride
Tomah Q-17-2

Tridecyl phosphite
Mark TDP

Tridecyl stearate (CTFA) [CAS #31556-45-3]
Emerest 2308; Kemester 5721; Liponate TDS

Tridecyl trimellitate [CAS #70225-05-7]
Liponate TDTM

Triethanolamine [CAS #102-71-6] *Synonyms:* 2,2′,2′′-Nitrilotris (ethanol); TEA; Trolamine
Surfonic HDL

Triethanolamine chelate
Tyzor TE

Triethonium hydrolyzed animal protein ethosulfate (CTFA) [CAS #111174-64-2]
Quat-Pro E, E SD

Triethoxysilane [CAS #998-30-1]
CT2500

Triethoxysilyl modified poly (1,2-butadiene)
CPS 078.5

N-[3-(Triethoxysilyl)-propyl] 4,5-dihydroimidazole [CAS #58068-97-6]
Dynasylan IMEO

Triethylchlorosilane [CAS #994-30-9]
CT2520; T2520

Triethyl citrate (CTFA) [CAS #77-93-0] *Synonyms:* Ethyl citrate
Citroflex C-2; Hydagen DEO; Merrol C-102

Triethylene diamine [CAS #280-57-9] *Synonyms:* 1,4-Diazobicyclo [2,2,2] octane
Dabco 33-LV, Crystalline; Tegoamin 33; Texacat TD-33, TD-33A

Triethylene glycol... See PEG-3...

Triethylene glycol bis [3-(3′-t-butyl-4′-hydroxy-5′-methylphenyl) propionate]
Irganox 245

Triethylene glycol monobutyl ether. See Butoxytriglycol

Triethylene glycol monoethyl ether. See Ethoxytriglycol

Triethylene glycol monomethyl ether. See Triglycol monomethyl ether

Triethylenetetramine [CAS #112-24-3] *Synonyms:* Trientine; Trien
D.E.H. 24

Triethylsilane [CAS #617-86-7]
CT2523

Trifluoromethanesulfonic acid [CAS #1493-13-6] *Synonyms:* Triflic acid
FC-24

Trifluoromethanesulfonic acid amine salt
FC-520

Triglyceryl. See Polyglyceryl-3...

Triglycol monomethyl ether [CAS #112-35-6] *Synonyms:* PEG-3 methyl ether; Triethylene glycol monomethyl ether
Poly-Solv TM

Tri n-hexyl trimellitate
Merrol 600TM

Trihydroxy methoxystearin
Cetiol R

Trihydroxystearin (CTFA) [CAS #139-44-6] *Synonyms:* Glyceryl tris-12-hydroxystearate; 12-Hydroxyoctadecanoic acid, 1,2,3-propanetriol ester
Cutina BW; Thixcin E, R

Triisocetyl citrate (CTFA)
Hetester TICC

Triisodecyl amine
Adogen 382; Lilamin 382

Triisodecyl phosphite
Weston TDP

Triisooctyl amine [CAS #2757-28-0] *Synonyms:* 6,6′,6′′-Trimethyltriheptylamine
Adogen 381, 2381; Lilamin 381

Triisooctyl phosphite [CAS #25103-12-2]
Weston TIOP

Triisooctyl trimellitate
Merrol TIOTM; Plasthall TIOTM; Staflex TIOTM

Triisopropanolamine (CTFA) [CAS #122-20-3] *Synonyms:* TIPA; 1,1′,1′′-Nitrilotris-2-propanol
Ultranox 626

Triisopropylchlorosilane
T2885

Triisopropylsilyl trifluoromethane sulfonate
T2890

Triisopropyl trilinoleate (CTFA) [RD #977079-86-9] *Synonyms:* Triisopropyl trimerate
Schercemol TT

Triisostearin (CTFA) [RD #977042-88-8] *Synonyms:* Glyceryl triisostearate
Cyclochem GTIS; Dermalcare GTIS

Triisostearyl trilinoleate (CTFA) [RD #977076-81-5] *Synonyms:* Triisostearyl trimerate
Schercemol TIST

Trilaneth-4 phosphate (CTFA) [RD #977058-52-8] *Synonyms:* PEG-4 lanolin ether triphosphate
Sedefos 75

Trilaurin (CTFA) [CAS #538-24-9] *Synonyms:* Glyceryl trilaurate
Cyclochem GTL; Softisan 100

Trilaurylamine (CTFA) [CAS #102-87-4] *Synonyms:* Trilauramine; Tridodecyl amine
Adogen 363, 2363; Lilamin 363

Trilauryl phosphite
Weston TLP

Trilauryl trithiophosphite
Weston TLTTP

Trimer acid (CTFA) [CAS #7049-66-3; 68937-90-6] *Synonyms:* Fatty acids, C18, unsaturated, trimers; 9,12-Octadecadienoic acid, trimer; Trilinoleic acid
Hystrene 5460

1-Trimethoxysilyl-2-(chloromethyl) phenylethane [CAS #68128-25-6]
CT2902

Trimethoxysilylpropyldiethylene triamine [CAS #35141-30-1]
Dynasylan TRIAMO

(N-Trimethoxysilylpropyl)-polyethylenimine
CPS 076

N-Trimethoxysilylpropyl, N,N,N-trimethyl ammonium chloride
CT2925

Trimethylbromosilane
T2928

Trimethylchlorosilane [CAS #75-77-4]
CT2950; T2950

3,3,5-Trimethylcyclohexylethyl ether
Herbavert

2,2,4-Trimethyl-1,2-dihydroquinoline *Synonyms:* TDQP
Akrochem Antioxidant DQ; Naftonox TMQ; Ultranox 254; Vulkanox HS/Powd.

Trimethylethoxysilane [CAS #1825-62-3]
CT2970

3,5,5-Trimethyl hexanoyl peroxide [CAS #3851-87-4]
USP-355M

N,N,N´-Trimethyl-N´-hydroxyethyl-bisaminoethylether
Texacat ZF-10

Trimethylolpropane oleate
Radiasurf 7172, 7372

Trimethylolpropane triacrylate
SR-351

Trimethylolpropane tri-C5/C9
Radiasyn 7364

Trimethylolpropane tri-C7
Radiasyn 7367

Trimethylolpropane tri-C8/C10
Radiasyn 7368

Trimethylolpropane tricaprylate/caprate [CAS #68956-08-1]
Kessco 887

Trimethylolpropane trimethacrylate
Rhenofit TRIM/S; SR-350

Trimethylolpropane trioleate
Radia 7370

Trimethyl-1,3-pentanediol, 2,2,4- diisobutyrate [CAS #6846-50-0]
Kodaflex TXIB

2,2,4-Trimethyl-1,3-pentanediol monoisobutyrate
Texanol Ester-Alcohol

2-Trimethylsiloxypent-2-en-4-one
T3150

Trimethylsilyl acetamide [CAS #13434-12-6]
CT3250; T3250

Trimethylsilyl acetate [CAS #2754-27-0]
CT3254

Trimethylsilylamodimethicone
Dow Corning Q2-7224

Trimethylsilyl-N,N´-diphenylurea
T3500

2-Trimethylsilylethanol
T3550

Trimethylsilyl imidazole [CAS #18156-74-6]
CT3600; T3600

Trimethylsilyl iodide [CAS #16029-98-4]
CT3610

Trimethylsilyl methane sulfonate
T3634

3-Trimethylsilyl-2-oxazolidinone
T3660

Trimethylsilyl trifluoromethane sulfonate [CAS #27607-77-8]
CT3795; T3795

Trimethylthiourea [CAS #2489-77-2]
Thiate E, EF-2

1,3,5-Trimethyl-2,4,6-tris (3,5-di-t-butyl-4-hydroxybenzyl) benzene
Ethanox 330

Trimyristin (CTFA) [CAS #555-45-3] *Synonyms:* Glyceryl trimyristate
Dynasan 114

Tri (n-octyl, n-decyl) trimellitate
Merrol 810TM

Trioctyl phosphate *Synonyms:* Octyl phosphate
Servoxyl VPTZ 100; TOF

Trioctyl trimellitate
Kodaflex TOTM; Merrol TOTM; Plasthall TOTM; Staflex TOTM; Uniflex TOTM

Triolein (CTFA) [CAS #122-32-7] *Synonyms:* Glyceryl trioleate; Olein
Acconon GTO; Alkamuls GTO; Cyclochem GTO; Emerest 2423; Kemester 1000; Radia 7161, 7163, 7303, 7363; Rilanit GTO

Tripalmitin (CTFA) [CAS #555-44-2] *Synonyms:* Glyceryl tripalmitate; Palmitin
Dynasan 116

Tripentaerythritol
XP-1668

Triphenyl methane 4,4′,4′′-triisocyanate
Desmodur RE

Triphenyl phosphate [CAS #115-86-6] *Synonyms:* Phosphoric acid, triphenyl ester; TPP
Disflamoll TP; Weston 63

Triphenyl phosphite [CAS #101-02-0]
Lankromark LE65; Mark 2112, TPP; Weston 310, EGTPP, TPP

Triphenyltin hydroxide [CAS #76-87-9]
Du-Ter; Pro-Tex; Super Tin 4L

Tripropylene glycol... See PPG-3...

Tripropylene glycol n-butyl ether
Dowanol TPnB

1,3,5-Tris (4-t-butyl-3-hydroxy-2,6-dimethylbenzyl)-1,3,5-triazine-2,4,6-(1H,3H,5H)-trione
Cyanox 1790, 2777

Tris (3,5-di-t-butyl-4-hydroxy benzyl) isocyanurate [CAS #27676-62-6]
Good-rite 3114; Vanox GT

Tris (2,4-di-t-butylphenyl) phosphite
Cyanox 2777; Naugard 524; Wytox 240

2,4,6-Tris-(N-1,4-dimethylpentyl-p-phenylenediamino)-1,3,5-triazine [CAS #121246-28-4]
Durazone 37

Tris (dipropyleneglycol) phosphite
Weston 430

Tris (hydroxymethyl) aminomethane [CAS #77-86-1]
Tris Amino (Crystals)

Tris (hydroxymethyl) nitromethane (CTFA) [CAS #126-11-4] *Synonyms:* (2-Hydroxymethyl) 2-nitro-1,3-propanediol; Trimethylolnitromethane
S.S.T. Sump Saver Tablets; Tris Nitro

Tris (p-isocyanato-phenyl) thiophosphate
Desmodur RFE

1,1,3-Tris (2-methyl-4-hydroxy-5-t-butylphenyl) butane
Topanol CA, CA-RT

Tris Neodol-25 phosphite
Weston 474

Trisnonylphenyl phosphite (CTFA)
Lankromark LE109; Mark TNPP; Vanstay 8210; Weston 399, 399B, TNPP; Wytox 312

Trisodium EDTA (CTFA) [CAS #150-38-9] *Synonyms:* Edetate trisodium; Trisodium ethylenediamine tetraacetate; Ethylene diamine tetraacetic acid, trisodium salt
Hamp-Ene Na3 Liq., Na3T; Perma Kleer Tri Crystals; Sequestrene NA3

Trisodium HEDTA (CTFA) [CAS #139-89-9] *Synonyms:* Trisodium hydroxyethyl ethylene diamine triacetate
Cheelox B-13, HE-24; Chel DM-41; Chelon 120; Dissolvine H 40, H 84; Hamp-Ol 120, Crystals; Kalex OH; Kelene 77; Perma Kleer 80, 120; Questal FEC; Trilon D Liq.; Vanate TSHE; Versenol 120

Trisodium magnesium diethylenetriamine pentaacetate
Hamp-Ex M

Trisodium NTA (CTFA) [CAS #5064-31-3] *Synonyms:* Trisodium nitrilotriacetate; Nitrilo triacetic acid sodium salt
Cheelox NTA-14, NTA-Na3; Dissolvine A 40, A 92; Hampshire NTA 150, NTA Na3 Crystals; Perma Kleer NTA 150, NTA Na_3 Crystals; Trilon, A 92, A Liq.Trisodium nitriloacetate

Trisodium sulfosuccinate
T.S.S.

Tristearin (CTFA) [CAS #555-43-1] *Synonyms:* Glyceryl tristearate
Capital 5380; Dynasan 118; Neobee 62; Rilanit GTS

Tristearyl phosphite
Weston TSP

Tris-(tributoxysiloxy) methyl silane (CTFA)
Silicate Cluster 102

Tritridecyl amine
Adogen 383, 2382; Lilamin 383

Trixylenyl phosphate *Synonyms:* Tri (dimethylphenyl) phosphite
Kronitex TXP; Pliabrac TXP

Trypsin [CAS #9002-07-7]
PTN 3.0S

Undecylenamide DEA (CTFA) [CAS #25377-64-1; 60239-68-1] *Synonyms:* Undecylenoyl diethanolamide
Rewocid DU 185; ; Varamide DU 185

Undecylenamide MEA (CTFA) [CAS #20545-92-0; 25377-63-3] *Synonyms:* Undecylenoyl monoethanolamide
Comperlan UDM; Incromide UM; Rewocid U 185; Rewomid U 185; Steinazid U 185; Varamide U185

Undecylenamidopropyl trimonium methosulfate *Synonyms:* Undecylenic acid propylamido trimethyl ammonium methosulfate
Rewocid UTM 185

Undecylenic acid (CTFA) [CAS #112-38-9] *Synonyms:* 10-Undecenoic acid
Ivex 10

Undecylenic quat. ammonium methosulfate
Rewoquat UTM 185

Undecylpentadecanol (CTFA) [CAS #68444-33-7] *Synonyms:* Isohexacosanol
Standamul GTO-26

Uniconizole
Ortho Sumagic

Urea (CTFA) [CAS #57-13-6]
Bio Soft LD-95

Urease (CTFA) [CAS #9002-13-5]
Urease Fermco

Urethane. See Polyurethane

Uricase
Uricase S

Vanadium [CAS #7440-62-2]
V-0701 T 1/8″

Vegetable oil (CTFA) [CAS #68956-68-3]
Lipovol ALM-S, A-S, P-S, SES-S; Sustane 4A, 6, 20A, HW-4, W

Vegetable oil glyceride *Synonyms:* Vegetable oil monoglyceride
D-9

Vegetable oil mono/diglycerides
Emuldan HV 40 K, HVF 52 K

Vinyl acetate [CAS #108-05-4]
Chemplex 3050; Daratak 83L, SP1030; Everflex 81L; Polyco 2140, 2142, 2149-C; Vinnapas A 50, DZN 220, H54/15 C, H 60, M 50/300, Z 50, Z 54/20 C

Vinyl acetate/acrylate
Vinnapas LT 420

Vinyl acetate/crotonic acid/vinyl propionate copolymer [CAS #25035-26-1]
Luviset CA 66, CAP

Vinyl acetate/ethylene copolymer
Vinnapas EN 424, EN 426, EN 427, EN 428, EN 429, EP 16, EV 25, EZ 15

Vinyl acetate/ethylene/vinyl chloride terpolymer
Vinnapas CEF 10, CEZ 16

Vinyl acetate/vinyl laurate/vinyl chloride terpolymer
Vinnapas T 50/30 VL

Vinyl acrylic
Everflex T, TMF; Ucar Latex 828

N[2-Vinylbenzylamino) ethyl]-3-aminopropyl trimethoxysilane
Dow Corning Z-6032

Vinyl bromide [CAS #593-60-2]
Saytex VBR

Vinylcaprolactam/vinyl pyrrolidone/dimethylaminoethyl methacrylate terpolymer
Gaffix VC-713

Vinyl chloride copolymer
Polyco 2611, 2612, 2617, 2618, 2629, 2638

Vinyl chloride/vinyl acetate copolymer
Vinnol 50, 50/25 C

Vinyl chloride/vinyl acetate/ethylene terpolymer
Vinnol CE 35

Vinyl compounds and polymers. See also Polyvinyl chloride, Vinyl chloride, Vinyl acetate
Darex SP-1566; Everflex 515L

Vinyl ether polymer
BASF Wax V

Vinylidene chloride [CAS #75-35-4]
Daran SL143; Polidene 33-001, 33-004, 33-021, 33-031, 33-075

Vinyl isobutyl ether [CAS #109-53-5] *Synonyms:* Isobutyl vinyl ether; IVE; Poly(vinyl isobutyl ether)
Gantrez B-773

Vinylmethyldichlorosilane [CAS #124-70-9]
CV4772

Vinyl polymers. See Vinyl compounds and polymers

N-Vinyl-2-pyrrolidone [CAS #88-12-0]. See also PVP
V-Pyrol

Vinylpyrrolidone/dimethylaminoethyl methacrylate copolymer. See PVP/dimethylaminoethyl methacrylate copolymer

Vinylpyrrolidone/styrene copolymer
Polectron 430

Vinylpyrrolidone/vinyl acetate/alkylaminoacrylate terpolymer
Luviflex D 430 I, D 455 I

Vinylpyrrolidone/vinyl acetate copolymer. See PVP/VA copolymer

Vinylpyrrolidone/vinyl imidazolinium methochloride copolymer. See Polyquaternium-16

Vinyl silane
Silcat R

Vinyltoluene alkyd copolymer
Chempol 13-2211

Vinyltriacetoxy silane [CAS #4130-08-9]
CV4800; Dow Corning Z-6075

Vinyltrichlorosilane [CAS #75-94-5]
Dynasylan VTC

Vinyltriethoxysilane [CAS #78-08-0]
Dynasylan VTEO; Union Carbide A-151

Vinyltrimethoxysilane [CAS #2768-02-7]
Dynasylan VTMO

Vinyl tris (methoxyethoxy) silane [CAS #1067-53-4]
Dynasylan VTMOEO; Union Carbide A-172

Vinyl tris (1-methoxy-2-propoxy) silane
Dynasylan VTPMO

Vinyl tris (methylethylketoxime) silane [CAS #2224-33-1]
CV5050

Vinyl tris (trimethylsiloxy) silane [CAS #5356-84-3]
CV5100

Walnut oil (CTFA) [CAS #8024-09-7]
Lipovol W

Wheat germamide DEA
Incromide WGD

Wheat germamidopropalkonium chloride
Incroquat WG-85

Wheat germamidopropylamine oxide (CTFA)
Incromine Oxide WG; Mackamine WGO

Wheat germamidopropyl betaine (CTFA) *Synonyms:* Wheat germ oil amido betaine
Incronam WG-30; Mackam WGB

Wheat germamidopropyl dimethylamine (CTFA)
Mackine 701

Wheat germamidopropyl dimethylamine lactate (CTFA)
Incromate WGL; Mackalene 716

Wheat germ glycerides (CTFA) [CAS #68990-07-8; #8046-25-1] *Synonyms:* Wheat germ oil mono-, di-, and triglycerides
Wickenol 535, 535 Vita-Cos

Wheat germ oil (CTFA) [CAS #8006-95-9]
Emcon W; Lipovol WGO; Super Refined Wheat Germ Oil

Wollastonite [CAS #13983-17-0]
Vansil W-9, W-10, W-20, W-30, W-50

Xanthan gum (CTFA) [CAS #11138-66-2] *Synonyms:* Corn sugar gum; Xanthan
Biozan, SPX 5423; Genuzan 1038A, 1063X; GFS; Kelfo; Keltrol, F; Kelzan, D, M, S, XC Polymer; Rodigel 23; Xanflood

Xylene (CTFA) [CAS #1330-20-7] *Synonyms:* Benzene, dimethyl-; Dimethylbenzene
Amsco Xylene; D.E.R. 671-X75, 671-XM75; Desmodur N-75, Z-4370; Shell Xylene

Xylene sulfonic acid (CTFA) [CAS #25321-41-9]
Eltesol 4009, 4018, XA, XA 65, XA 90, XA/M65; Reworyl X 65

Zeolite synthetic [CAS #68989-22-0]
EZA

Zinc [CAS #7440-66-6]
Ficel AF100; Ircogel 2354; Lankromark LZ440; Manex; Troymax Drier Zinc 8%, 16%; Vanstay 8308

Zinc-bis (undecylenic amidopropyl dimethyl glycinate)
Schercotaine UAB-Z

Zinc borate (CTFA) [CAS #1332-07-6; 120007-67-9] *Synonyms:* Boric acid, zinc salt
Firebrake ZB

Zinc-cadmium
Vanstay 137

Zinc chloride with benzothiozyl disulfide [CAS #14239-75-9]
Thanecure

Zinc chromite
Zn-0312 T 1/4″; Zn-0602 T 1/8″

Zinc cyclopentamethylene dithiocarbamate
Accelerator Z.P.D.

Zinc dialkyl dithiophosphate
Rhenocure TP/G, TP/S

Zinc diamyldithiocarbamate [CAS #15337-18-5]
Amyl Zimate

Zinc dibenzyl dithiocarbamate
Accelerator Z.B.E.D.; Naftocit ZBEC

Zinc dibutyl dithiocarbamate [CAS #136-23-2]
Accelerator BZ; Butasan; Butyl Zimate; Naftocit Di 13; Naftopast Di 13-P; Vulkacit LDB/C

Zinc dibutyl dithiophosphorate *Synonyms:* Zinc O,O-di-n-butyl-phosphorodithioate
Akrochem Accelerator VS; Vocol, S-62

Zinc diethyldithiocarbamate [CAS #14323-55-1]
Accelerator EZ; Anchor ZDBC, ZDEC; Ethyl Zimate; Naftocit Di 7; Naftopast Di7-P; Vulkacit LDA

Zinc dimethyldithiocarbamate [CAS #137-30-4] *Synonyms:* Ziram
Accelerator MZ; Fungicide ZV; Methyl Zimate; Naftocit Di 4; Vancide 51Z, 51Z Disp., MZ-96; Vulkacit L

Zinc dinonylnaphthalene sulfonate
Na-Sul ZS

Zinc dithiocarbamate
Vanlube 26

Zinc dust
Horse Head Standard Zinc Dust 22, 44, 122, 222, 422, 444

Zinc-epoxy
Therm-Chek 707-X

Zinc 2-ethylhexoate [CAS #136-53-8] *Synonyms:* Zinc octoate
Octoate Z

Zinc-N-ethylnophenyl dithiocarbamate
Vulkacit P Extra N

Zinc hydroxyphosphite [CAS #55799-16-1]
Nalzin 2

Zinc isopropyl xanthate [CAS #1000-90-4]
Propyl Zithate

Zinc 2-mercaptobenzothiazole [CAS #155-04-4]
Naftocit ZMBT; Vancide 51Z, 51Z Disp.; Vulkacit Mercapto/MGC; Vulkacit ZM; Zetax; ZMBT

Zinc 2-mercaptopyridine-N-oxide
Zinc Pyrion

Zinc 2-mercaptotoluimidazole [CAS #61617-00-3]
Vanox ZMTI

Zinc 4- and 5-methylmercaptobenzimidazole
Vulkanox ZMB2/C5

Zinc molybdate
Moly-White 101, 331, ZNP

Zinc oxide (CTFA) [CAS #1314-13-2] *Synonyms:* Chinese white; Zinc white
Akrochem 9930; AZO-33, -55, -55TT, -66, -66TT, -77, -77TT; Azodox-55, -55TT; Horse Head XX-4, -32, -78, -85, -503, -600, -601, -631; Kadox-15, -25, -72, -215, -272, -720C, -720CP, -930; K-Zinc; Naftopast ZnO-A; Ottalume 2100; Protox-78, -166, -167, -168, -169; Rapidblend 1793; Rhenovin ZnO-90; RR Zinc Oxide-Coated; RR Zinc Oxide (Untreated); Zinc Oxide 35; Zinc Oxide No. 185, 318; Zinc Oxide Transparent; Zinkoxyd Activ; Zn-0401 E 3/16″; Zotex 319, 319L

Zinc-N-pentamethylene dithiocarbamate
Vulkacit ZP

Zinc phosphate [CAS #7779-90-0]
Nalzin ZP

Zinc 2-pyridinethiol-1-oxide [CAS #13463-41-7]
Zinc Omadine

Zinc ricinoleate (CTFA) [CAS #13040-19-2] *Synonyms:* 12-Hydroxy-9-octadecenoic acid, zinc salt
Grillocin HY-77

Zinc stearate (CTFA) [CAS #557-05-1] *Synonyms:* Zinc octadecanoate; Octadecanoic acid, zinc salt
Akrochem Zinc Stearate; Hy Dense Zinc Stearate XM Powd., XM Ultra Fine; Interstab ZN-18-1; Liquazinc AQ-90; Lubrazine W, Superfine; Norfox ZNS; Petrac ZN-41, ZN-42, ZN-44 HS, ZW-45; Rhenodiv ZB; Witco Zinc Stearate NW, Polymer Grade, REP; Zinc Stearate 11, 44, Disperso, Heat-Stable, LV, NW, Polymer Grade, Regular, USP

Zinc sulfide [CAS #1314-98-3]
Lithopone 30% DS, 30% L, 60% DS, 60% L; Sachtolith HD, HD-S, L

Zinc undecylenate (CTFA) [CAS #557-08-4] *Synonyms:* 10-Undecenoic acid, zinc salt
Ivex 10

Zirconium [CAS #7440-67-7] *Synonyms:* Zircat
Troymax Zirconium 6%, 12%, 18%, 24%

Zirconium diboride *Synonyms:* Zirconium boride
ZN-7; ZS-7

Zirconium silicate (CTFA) [CAS #10101-52-7; 14940-68-2] *Synonyms:* Silicic acid, zirconium salt (1:1); Zircon
Ultrox, 500W, 1000W

Functional Classification of Tradenames

ABSORBENTS

Amerchol BL, C, H-9; Aqualon, C, R; Aquasorb A250; Avicel PH-101, PH-102, PH-103, PH-105; Dispal; Duolite CS-100; Flo-Gard AG110, AG 130, AG 150, CC 120, CC 140, CC 160, FF 310, FF 320, FF 330, FF 350, FF 370, FF 390; Florco; Florco X; Florex Granular Grades; Grillocin HY-77; Grit-O'Cobs; Levilite; Lite-R-Cobs; Magox Super Premium; Nalco 1180, 1181; Pharmasorb (Regular and Colloidal); Polyclar 10, AT; Polyplasdone XL; Polytrap 603; Pure-Dent B810; Refinex; Silasorb; Sipernat 22, 22 S; Sol-Speedi-Dri, Auto-Dri; SorbaSet B, S; Tonsil (TN); Wickenol 550

ACCELERATORS

A-1; Accelerator 399, BZ, D, EZ, MF, MZ, P.P.D., Z.B.E.D., Z.P.D; Actafoam F-2; Activ-8, Activ-8 in Hexylene Glycol; Activator 1102; Activator STAG; Aktiplast, F, PP, T; Altax; Amax; Amax No. 1; Amyl Cadmate, Amyl Ledate, Amyl Zimate; Akrochem 9930 Zinc Oxide Transparent, Accelerator 40, CZ-1, R, Accelerator Thio No. 1, Accelerator VS, Cu.D.D., Cu.D.D.-PM, DCP-40C, DCP-40K, D.E.T.U. Accelerator, DOTG, DPG, ETU-22 PM, TDEC, TMTD, VC-40C, VC-40K, ZIPPAC; Akroform DCP-40 EPMB, VC-40 EPMB; Akrosperse DCP-40 EPMB, VC-40 EPMB; Anchor DBD, DOTG, DPG, ZDBC, ZDEC; AZO-33, AZO-77, AZO-77TT; BBTS; BIBBS; Bismate; Bismet; Butasan; Butyl Eight; Butyl Namate; Butyl Tuads; Butyl Zimate; Butynediol; Captax; Captax-Tuads Blend; Cardolite NC-548, NC-700; CBTS; Cumate; Cure-Rite 18; Cure-rite Accelerator; Desmorapid PP; DPG; DPTT, DPTT-S; Dynamar FX 5166; Emery 5714; Ethyl Cadmate; Ethyl Selenac; Ethyl Tellurac, Ethyl Tuads, Ethyl Zimate; Ken Kem CP-45, CP-99; Ken-React Series, 7 (KR 7), 9S (KR 9S), 12 (KR 12), 26S (KR 26S), 33DS, (KR 33DS), 38S (KR 38S), 39DS (KR 39DS), 41B (KR 41B), 44 (KR 44), 46B (KR 46B), 55 (KR 55), 133DS (KR 133DS), 134S (KR 134S), 138D (KR 138D), 138S (KR 138S), 158 (KR 158D), 158FS (KR 158FS), 212 (KR 212), 238A (KR 238A), 238J (KR 238J), 238M (KR 238M), 238S (KR 238S), 238T (KR 238T), 262A (KR 262A), 262ES (KR 262ES), OPP2 (KR OPP2), OPPR (KR OPPR), TTS (KR TTS); Knightset M-4; Kosmos 10, 15, 16; Lamefix 680; Leguval K 25 R, K 26; LICA 01, 09, 12, 38, 38A, 38J, 44, 97; Ludigol 60, F; LZ 01, 09, 12, 38, 44, 97; MBT; MBTS, MBTS Pellets; Methyl Cumate; Methyl-Ethyl Tuads; Methyl Ledate; Methyl Selenac; Methyl Tuads; Methyl Zimate; Morfax; Natac; Novor 924; OBTS; OMTS; Propyl Zithate; Retarder SAX; Rhenocure AT, CA, CUT, TDD, TP/G, TP/S, ZAT; Rhenofit 1600, CBS-70, MBT-70, MBTS-70, NC, TMTD-70, TMTM-70, UE; Rhenovin ZnO-90; RIA CS; Rotax; Santocure, IPS, MOR, NS; Setsit 5, 9, 51, 104; Sulfads; Thiate E, EF-2, H, U; Thiofide; Thiotax; Thiovanol; TM/ETD; Vanax 552, 808, 833, A, CPA, DOTG, DPG, MBM, NP, NS, PML; Vocol, S-62; V-Pyrol; Vulkacit 576, 1000/C, CRV/LG, CT-N, CZ/EGC, CZ/MGC, D/C, D/EGC, DM, DM/C, DM/EGC, DM/MGC, DOTG/C, DOTG/EGC, DZ/C, DZ/EGC, H 30, HX, J, L, LDA, LDB/C, LZ/LG, Mercapto/C, Mercapto/EGC, Mercapto/MGC, MOZ-90/SG, MOZ/LG, MOZ/SG, NPV/C2, NZ, NZ/EGC, P, P Extra N, Thiuram, Thiuram/C, Thiuram/EGC, Thiuram MS/C, Thiuram MS/EGC, TR, ZM, ZP; Vultac 2, 5; Zetax; Zinc Oxide 35; Zinc Oxide No. 185; Zinc Oxide Transparent; Zinkoxyd Aktiv; ZMBT

ACTIVATORS

Actafoam F-2, R-3, R-10; Activator 1102, 2013-P, STAG; Akrochem 9930 Zinc Oxide Transparent, Accelerator Thio No. 1, DPG, TMTD; AZO-33, AZO-55, AZO-55TT, AZO-66, AZO-66TT, AZO-77, AZO-77TT; Azodox-55, -55TT; Butac; Cri-Spersion CRI-ACT-45, -45/1/1, 45-LV; Eltesol TA 65, TA 96, TA/F, XA65, XA/M65; Interstab CZL-710, CZL-712, CZL-715; Ken-React Series, 7 (KR 7), 9S (KR 9S), 12 (KR 12), 26S (KR 26S), 33DS, (KR 33DS), 38S (KR 38S), 39DS (KR 39DS), 41B (KR 41B), 44 (KR 44), 46B (KR 46B), 55 (KR 55), 133DS (KR 133DS), 134S (KR 134S), 138D (KR 138D), 138S (KR 138S), 158 (KR 158D), 158FS (KR 158FS), 212 (KR 212), 238A (KR 238A), 238J (KR 238J), 238M (KR 238M), 238S (KR 238S), 238T (KR 238T), 262A (KR 262A), 262ES (KR 262ES), OPP2 (KR OPP2), OPPR (KR OPPR), TTS (KR TTS); Lankromark LC90, LZ187, LZ440, LZ561, LZ638, LZ1199, LZ1221, LZ1232; LCA-4, LCA-4LV; LICA 01, 09, 12, 38, 38A, 38J, 44, 97; Litharge 33; LZ 01, 09, 12, 38, 44, 97; Mark 281B, 630; Petrac 250, 270; Porofor ADC/K, ADC/R, B 13/CP 50,, BSH Paste, BSH Paste M, BSH Powder, DNO/F; Retarder BA, BAX, SAX; Rexowet RW; RR Zinc Oxide-Coated; Rhenofit 1987, 2009, 2642, 3555, B, BDMA/S, CF, EDMA/S, NC, TAC/S, TRIM/S, UE; Rhenomag C2, G3, L3, P2, P3; Rhenovin Na-stearat-80, S-stearat-80; RIA CS; Ridacto; RR Zinc Oxide (Untreated); Setsit 9, 51; Silacto; Tegoamin CPE, PTA; Thanecure; Vanax PML; Vanox AT; Zinc Oxide 35, No. 185, Transparent; Zinkoxyd Act.

ADSORBENTS

Duolite A-7, A-143, S-37, S-761; Laponite XLG, XLS; Magox Super Premium; Witcarb Columbia JXC, Granular Activated Carbon Grade 965; Zeosyl 200; Zeothix 265

AERATING AGENTS

Admul 1411; Aldo PGHMS; Cetodan; Cetodan 50-00A; Cetodan 70-00A; Cetodan 90-40

AIR-ENTRAINING AGENTS. SEE ENTRAINMENT AID

ALGICIDES. SEE ANTIMICROBIALS

ANTIBLOCKING AGENTS

ABT-2500; Adogen 58, 73; Alkamide RDO; Armid 18, C, HT; Armoslip 18, CPM, EXP; Celite Super Fine Super Floss, Super Floss; CPH-31-N; Crill 1, 2, 50; Crodamide E, ER, O, OR, S, SR; Denlube 1025-KA; Doittol 891, K21; EBS Wax; Emerest 2650; Emulvis; Kemamide B, E, E-180, E-221, O, P-181, S, S-65, S-180, S-221, U, W-10, W-20, W-39, W-40, W-40/300, W-45; Lipowax C; Microbloc; Paracol 404C; Petrac Eramide; Petrac Slip-Quick; Petrac Vyn-Eze; Polyco 2611, 2612; Polywax 500, 655, 1000, 2000; Silwet L-7001, L-7602; Softenol 3107, 3108, 3900, 3991; Spectratech CM 10608, 10778, 10779, 11013, 11056, 11513, 11643; Ultra NXS Liquid, SCS Liquid, SXS Liquid Powder; Unislip 1753, 1757, 1759, 1760 EBS; Witconate NCS, NXS, SCS, STS, SXS

ANTICAKING AGENTS

Acetamin 24, C, Aerosil 200, Aerosil Colloidal Silicas, Aerosil R 972, R-974, R-976; Bar-Dust 105, 110; Bilt-Cote H-1, H-5, S-1, S-5; Aluminum Oxide C; Aluminum Stearate EA; Anticake 17; Armeen 18, 18D, HT, HTD; Armoflow 48, 65, 66; Attacote; Cab-O-Sil; Dinoramac C, O, S; Drysperse 401, 908H; Elastomag 100, 100R, 170, 170 Micropellet; Elceme 250, P 100; Emcol CC-422; EZA; Farmin 20, 60, 68, 80, 86, AB, C, HT, O, S, T, R 24H, R 86H; Flo-Gard FF 310, FF 320, FF 330, FF 350, FF 370, FF 390; Flotigam Brands; Flow Agent WR-100; Fluidiram; Genamin CC, CS, OL, SH, TA Brands; Good-rite K-7600; Hi-Sil 210, 215, 233, 250, 260, 262, ABS; Jet Amine PC, PHT, P-O, P-S, PT; Lilamin AC-14, AC-20, AC-30, AC-41L, AC-23 T, AC-24 T, AC-25 T, AC-26, AC-81L, AC-82L; Maltrin M040; Micro-Cel A, B, E; Miglyol 812, 840; Noram 2 C, 2 SH, C, DMC, DMS, DMSH, M2 SH, M2C, SH; Noramac C, O, S; Perchem AMU 60 X, AMU X; Perenol F3, S4, S5; Petrac MG-20, MG-20 NF; Petrac ZN-41; Petro AG Special; Plastomag; Pyrax ABB, WA; Radiamac 6148, 6149, 6169, 6179; Radiamine AC 14, 20, 30, 81, 82, 505, 6144; Radiastar 1060, 1100, 1208; Sellogen NS-50; Sipernat 22, 22 S, 50, 50 S, D 17; SK Fert C13, C14, C24, F325, F20 A; Soprofor AMC; Stealim; Surfam P10, P12B, P14B, P17B, P24M, P86M, P89MB; Synprolam 35, 35A; T.S.S.; Uresoft 150; Valfor 100; Vybar 103; Witco Calcium Stearate EA, TX Acid; Witconate NXS, SCS, STS, SXS, TX Acid, F26-46; Zeofree 80; Zeolex 7, 7A, 23A; Zeothix 265

ANTICRACKING AGENTS

Akrowax 130, 145; Diazopon SS-837

ANTIDUSTING AGENTS

Armoflow 48, 65; Lilamin AC-14, AC-20, AC-30, AC-41L, AC-23 T, AC-24 T, AC-25 T, AC-26, AC-81L, AC-82L

ANTIFOAMERS. SEE DEFOAMERS

ANTIFOG AGENTS

Alkamuls GMO-55LG; Armeen Z; Atmer 100, 101, 102, 103, 104, 105, 106, 107, 110, 111, 112, 113, 114, 115, 116, 117, 118, 121, 122, 123, 124, 125, 126, 127, 128, 129, 132, 184, 502, 645, 646, 647, 649, 650, 654, 655, 656, 657, 658, 1007, 1010, 1012, 1020, 7100, 7101, 7107, 7108, 7114, 7115; Durfax 80; Durtan 80; Emery 2421; Emrite 6007, 6120; Gemtex SC-75; Glycolube 810; Glycosperse O-20; Grindtek ML 90, MM 90, MOL 90, MOP 90, MSP 32-6, MSP 40, MSP 40F, MSP 52, MSP 90, PGE 25, PGE 55, PGE 55-6, PGE-DSO, PK 60; Masil 1066C; Mazawet DOSS; Mazol GMO; Merix Anti-Fog; Radiamine 6140, 6141, 6161, 6163, 6164, 6170, 6171, 6172, 6173; Radiasurf 7125, 7135, 7137, 7145, 7147, 7600; Silwet L-7001; Triton N-40

ANTIFOULANTS

Rychem 400-OE; Surflow-S40

ANTIGELLING AGENTS

Aerosol AY; Cosmopon BN; Crodafos N3 Acid, N3 Neutral, N5 Acid, N10 Acid, N10 Neutral, SG, T2 Acid, T5 Acid, T10 Acid; Emerest 2610, 2715; Rewopol B 1003, SMS 35, TMS/F

ANTIMICROBIALS

(INCLUDING BACTERICIDES, GERMICIDES, ALGICIDES FUNGICIDES, DISINFECTANTS, SANITIZERS, DEODORIZERS)

Ablumine 230; Acetamin C; Acetoquat CPB, CPC; CTAB; ACL 56, 59, 60, 66, 85; A-C Polyethylene 629A; Acrawax-C; Actrasol SR 75, SRK 75; Acylglutamate CS-11, CT-12, GS-11, HS-11, LS-11, MS-11; Adox 3125; Aerosol 30, MA-80, OT-75%, OT-100%, OT-B; Agrilan MC-90; Akro-Gel; Akypogene Jod MB 1918; Alconate SBU-185; Algaesil; Algaetrol 76; Alkamide RDO, STEDA; Alkamuls GML-45, GMO-45; Alkapol PEG 3350, 6000; Alkaquat DMB-451; Alkazine C; Amerstat 233, 250, 251, 252, 272, 274, 282, 294, 300; Amical 48, 50, Flowable, WP; Ammonyx T; Amonyl BR 1244; Amphoram CB A30, CT 30; Antracol; Aquatreat DNM-9, DNM-30, DMN-360, KM, SDM; Aqucar Microbiocides; Aremsan C40; Arlacide A, G, H, HM; Armeen 18, SZ; Armoslip 18, CPM, EXP; Arosurf MSF; Arquad 2-10/50, 2C, 2HT-75, B-50 USP, B-90 USP, B-100, DM14B-90, DMCB, DMHTB-75, DMMCB-75, NF-50, S-2C-50, S; AZO-33; Bardac 205M, 2050, LF, LF80; Barquat 1552, 4250, 4250Z, 4280, 4280Z, MB-50, MB-80, MS-100, MX-50, MX-80, OJ-50; Basicop; Baycor, 5072; Bayfidan; Bayleton; Baytan; Belclene 313, 322, 329; Berol Fintex 573; Bioban BNPD, CS-1246, CT, GK; Biobrom C-103L, C-103 Tech.; Bio-Dac 50-20, -22, 205; Biopal VRO-20; Biopen 302, 315, 319, 350; Bio-Quat 50-24, 50-25, 50-30, 50-35, 50-36, 50-40, 50-42, 50-60, 50-65, 50-MAB, 50-MAC, 80-24, 80-28, 80-35, 80-36, 80-40, 80-42, IM-50, T-501, T-502; Bio-Pruf; Biosperse 240, 250; Bio-Surf I-20, I-12LF, I-21LF; Bravo W-75; Bretol; Bromat; Bronopol; Bronopol-Boots; BTC 50, 50 USP, 65, 65 USP, 99, 100, 776, 812, 818, 824, 824 P-100, 885, 1010-80, 1100, 1326, 2125, 2125 80%, 2125 P-40, 2125 M, 2125 80%, 2125MP-40, 2565, 2568, 8248, 8249, E-8358; BTCO 1010

C2 Sodium Chlorite; Carsonam C, C-SF, C-SPCL, DC-70%-SF, DC-SF, DC, L; Catigene 4513, 4513/80, BR/80B, DC 100, DE 80, LT 45, T 50, T 80, T 80 F; Catinal CB 50, HTB-70; Cetats; Comperlan UDM; Conco 2A1; Cralane KR-13, KR-14, LR-10, LR-11; Crapol AU-20, AU-24; Crodasinic LS30, LS35, OS35; Crodex C; Cunilate 2174-NO, 2419, 2419-75; Cuniphen 2713, 2721, 2721-C, 2778-1; Cyclopol SBU-185; Cyncal 80%; Cyprex 65-W; Daconil 2787; Dehyquart A, C, C Crystals, D, LDB, LT; Dibactol; Dinoramac C; Dodigen 226; Dodigen Brands; Dow Antimicrobial 7287, 8536; Dow Corning ACH-323, ACH-3331, ACH7-308, ACH7-321, AZG-368, AZG-369, AZG-370, AZG-374; Dowcide 1, A, OPCP; Dowicil 75, 200; Dry Flo; Du-Ter; Duomac C; Durotex 7603, GPM; Dyrene; Ecco MP; Elcema; Emcol E-607L, E-607S, E-607S; Empigen 5089, BAC, BAC 50, BAC50/BP, BAC 80, BAC 90, BCB50, BCF 80, BCM75, BCM75/A, CJB. CHB 40; Euparen, M; Fekta RT; FMB 302-8 Quat, 451-5 Quat, 451-8 Quat, 4500-5 Quat, 45000-8 Quat, 6075-5 Quat, 6075-8 Quat; Fore, Fore WD; Fruvit; Fungicide ZV; Fungitex R; Fungitrol 11; G-4 Pure, Tech; Gardiquat 12H, 1450, 1480, SV 480; Germaben II, II-E; Germall 115; Giv-Gard DXN; Glydant; Groutcide, 75; Halobrom; Hartex San Q 50; Herco; Hexetidine; Hinosan; Hyamine 10-X, 1622, 2389, 3500, 3500-NF; Incrosul UMS, UMS-45; Intercide 2 DIDP, ABF, ABF 1 ESBO, ABF 2 DIDP, ABF 2 ESBO, FC, T-0, N-628, TMP; Irgasan DP30; Ivex 10; JAQ Powdered Quat; Jet Amine D-C, D-O, D-T; Jet Quat 2C-75, 2HT-75, C-50, S-50, T-2C-50, T-50; Jorgard AMF; Karathane Liq. Conc., WD; Kathon 886 MW, 893, 925, 4200, CG, CG/ICP, LM, LP, LX, WT; K-Cop; Kemamine A650, A970, A974, A990, AD 650, AD 974, BQ-9702C, BQ-9742C, D-150, D-190, D-650, D-970, D-974, D989, D-999, Q-1902C, Q-6502C, Q-6503 B; Ketjenflex 9; Keycide X-10; Kito 40; Kobate C; Kocide 20/20, 101, 404S, 606, SD; K-Pool; K-Tea

Lebon 15, GM; Lilamin 101 D, 115, 115 D, 140, 140 D, 142, 142 D, 151, 160, 160 D, 163, 163 D, 170, 170 D, 172, 172 D, 308 D, 310 D, 312 D, 314 D, 316 D, 342 D, 343, 345 D, 363, 364, 367 D, 368, 369, 372 D, 381, 382, 383; Lutensit K-TI; MA; Macol NP-30, NP-30(70), NP-70, NP-100; Manex; Maquat 4450-E, DLC-1214, LC-12S, MC-1412, MC-1416, MC-6025-50%, MQ-2525, MQ-2525M, SC-18, SC-1632, TC-76; Maypon UD; Merkyl MAP, PM-TL; Merrol N-304; Micro-Chek 11, 11D, 11DIDP, 11S-711; Miranol CM Conc., CM-SF Conc., HM Conc., OM-SF Conc., S2M Conc., S2M-SF Conc.; Monaquat P-TC, P-TD, P-TL, P-TZ; Monazoline Series, C, CY, O, T; Monceren; Morestan; Myacide SP; Mytab; NB; NEPD; Ninol 1301; Nissan Amine AB, ABT, ABT_2, BB, FB, MB, OB, PB, SB, VB; Nissan Anon BF, BL, LG; Nissan Cation AB, ABT-350, 500, AR-4, BB, F_2-10R, -20R, -40E, -50, FB, FB-500, L-207, M_2-100, MA, SA, PB-40, -300, S_2-100; Nissan Tertiary Amine AB, ABT, BB, FB, MB, PB; NMP; Nopcocide N-40-D, N-96, N-96-S; Noramac C, O, S; Noramium DA.50; Norfox NP-9, NP-11; Noxamium S2/50; Nuodex PMA 18; Obanol 51; OCI 90-I; Omadine MDS; Onyxide 75, 172, 3300; Ortho Orthocide; Ortho Spotless; Ottasept; Ottasept Extra, Tech; Oxaban-A, -E; Pationic 122A; Pentachlorophenol DP-2; Petro 11; PH-73; Pinamine K; Preventol I; Pro-Tex; Protectol GDA, GT 50, GL 40, KLC 50, 80, TOE; Quadrilan BC; Quartamin 24P,

86P Conc., 86P; Querton 210 CL50, 210 CL80, 246, 1149, KKB CL50; Radiasurf 7000; Renex 690, 698; Rewocid DU 185, SBU 185, U 185, UTM 185; Rewomid U 185; Rewopon B 50; Rewoquat UTM 185; Rewoteric QAM 50; Roccal 50% Technical; Roccal II 50%; Roccal MC-14; Rychem 810, 830

Salabon 50; Sanisol C, CPR, CR, CR-80%, HTPR, OPR, TPR; Scepter; Schercoquat IB; Schercotaine UAB, UAB-Z; Schercozoline B, C, L, O; Sepacid CE 5209, CE 5265, GA 25, GA 50; Servamine KAC 412, KAC 422, KEP 4527; Servo CK 492, CK 494, CK 601; Sibutol; Skane M-8; Socci 30, 3500, 3500-WP, 6462, 7340; Sodium Chlorite Solution 50 Tech, Tech; Sodium Omadine, 40% Aqueous Solution, Powder; Sol Bact G; Solricin 135; Sorbit P; So/San 30M, 66M; S.S.T. Sump Saver Tablets; Super AD It.; Super Tin 4L; Swanol CA-101; Synprolam 35BQC (50), 35BQC (80), 35DMA, 35DMBQC, 35N3 DA, 35TMQC; Tanaquad T-85; TCC; T-Det N-12; Tektamer 38; Temasept I, IV; Tenox P; Tequat PAN; Teric 12A6, 12A9, 12A12, 12A16, 12A23, 160, 161, 163, 164, G12A4, G12A6, G12A8, G12A12, LA4, LA8; Texnol R 5; Thiostat B; Triadine 3; Tris Nitro; Troysan 142, 364, Anti-Mildew O, PMA-10-SEP, PMA-30, PMO-30; Troysan Polyphase 17 WD, Anti-Mildew, P-100; Ucarcide 225, 250; Ucarsan Sanitizers; Unamine C; Uniquat CB-50; Vancide 51, 51Z, 51Z Disp., 89, 89 RE, MZ-96, PA, PA Disp, TH; Vantoc AL, CL; Varamide DU 185, U 185; Variquat 50AC, 50AE, 50MC, 50ME, 60LC, 80AC, 80AE, 80LC, 80MC, 80ME, LC60, LC80; Varsulf SBU-185; Vikol AF-25, LO-25, PX-15, RQ, THP; Vinyzene BP-5-2-PG, BP-5-2MEK, BP-5-2-MS, BP-5-2-U, BP-505, BP-505 DIDP, BP-505 DOP, BP-505 S160, SB-1, SB-1 EAA, SB-1 ELV, SB-1-NY, SB-1-PR, SB-1-PS, Sil 3, T-129; Zinc Omadine, Cosmetic Grade, Industrial Grade, 48% Aq. Disp., Industrial Grade, Powd.; Zinc Pyrion; Zoharquat 50, 80

ANTIOXIDANTS

AA, Accelerator BZ, Additin 30; Agerite DPPD, Gel, Geltrol, GT, Hipar T, HP-S, MA, NEPA, Stalite, Superflex, Superflex Solid G, Superlite, Superlite Emulcon, Superlite Solid, White; Akrochem Antioxidant 12, 16, 33, 36, 58, PANA, S; Akrochem Antiozonant MPD-100; Alkaterge-E; Anchor DNPD, HDPA, HDPA/SE, ODPA, PBN; Antioxidant 235, 425; Brij 96, 97, 98, 99; Butasan; Butyl Zimate; CAO-1; CAO-3; CAO-3 Blend 29; CAO-5; CAO-14; CAO-41; CAO-42; Carstab DLTDP, DMTDP, DSTDP; Chimassorb 944LD; Cosmetol X; Coviox T-50, T-70; Covitol 544, 700C, 1100, 1185, 1210, 1360, F-350M, F-600F-1000; Cyanox 425, 711, 1212, 1735, 1790, 2246, 2777, LTDP, MTDP, STDP; Cyasorb UV 2908, 3346 LD; Durazone 37; Eastman DTBHQ, HQMME, MTBHQ, TDP 2000; Emulsogen STH; Ethanox 330, 398, 701, 702, 703, 736; Ethanox Antioxidant 376; Evanstab 12, 13, 14, 16, 18, A; Fermcozyme; Flectol H, ODP, Pastilles; Givsorb UV-2; Good-rite 3114, 3125; Hostavin VP NiCS 1; Ionol, Ionol CP; Irganox 245, 259, 1010, 1035, 1076, 1093, 1098, 1520, B-215, B-225, MD-1024; Irgastab 2002; Isobutyl Niclate; Isonox 129, 132; Ken-React Series, 7 (KR 7), 9S (KR 9S), 12 (KR 12), 26S (KR 26S), 33DS, (KR 33DS), 38S (KR 38S), 39DS (KR 39DS), 41B (KR 41B), 44 (KR 44), 46B (KR 46B), 55 (KR 55), 133DS (KR 133DS), 134S (KR 134S), 138D (KR 138D), 138S (KR 138S), 158 (KR 158D), 158FS (KR 158FS), 212 (KR 212), 238A (KR 238A), 238J (KR 238J), 238M (KR 238M), 238S (KR 238S), 238T (KR 238T), 262A (KR 262A), 262ES (KR 262ES), OPP2 (KR OPP2), OPPR (KR OPPR), TTS (KR TTS); Kerobit BPD, COM; Kessco 3283

Lankromark LE98, LE109, LE131; LICA 01, 09, 12, 38, 38A, 38J, 44, 97; LZ 01, 09, 12, 38, 44, 97, 817, 5022; Mark 158, 217, 260, 522, 1178B, 1220, 1259A, 1295, 1409, 1490S, 1589, 1589B, 2112, 2140; Methyl Niclate; Molyvan A, L; Montaclere; Naftonox 2246, BBM, BHT, IMB, PA, PS, TMQ, ZMP; Naugalube 438, 438-L, 438-R; Naugard 76, 431, 492, 524, BHT, P, PHR, SP, XL-1; Naugawhite; Nevastain 21, 30L, 76, 2170, A, B; Na-Sul EDS, LP; Newcol 3-80, 3-85, 20, 25, 40, 45, 60, 65, 80, 85; Newpol PE-61, -62, -64, -68, -74, -75, -78, -88; Ninox FCA; Nonox WSP; OA 502; Oxi-Chek 116, 414; Parabis; Polystay AA-1, AA-1R; Polywax OH 425, 550, 700; PTZ; Rapidblend 1793; Rhenovin DDA-70; Rylex NBC; Sandostab P-EPQ; Santoflex IP; Santoquin, Emulsion, Mixture 6; Santovar A; Santowhite Crystals; Santowhite PC; Santowhite Powder; Spectratech CM 11340, CM 11616, KM 11264; Stabilizer 9-A; Suconox-18; Sustane 3, 4A, 6, 8, 20, 20-3, 20A, 31, BHA, BHA 1-F, BHT, HW-4, P, PA, PG, Q, TBHQ, TBHQ.4, W; Synox 5LT; TDP 2000; Tecquinol; Tenox BHA, PG, TBHQ; THQ; Topanex 100BT; Topanol 205, CA, CA-RT, OC; Troykyd Anti-Skin S, Special M; Turpinal SL; Ucar Additive PN-1; Ultranox 236, 246, 254, 256, 276, 624, 626, 626A; Uvi-Nox 1494; Vanax PML; Vanlube 26, 71, 73, 81, 622, 648, 692, 732, 7723, AZ, DND, NA, PC, RD, SL, SS; Vanox 3C, 6H, 12, 13, 100, 102, 1003, 1004, 1005, 1013, 1030, 1081, 1290, 1320, 2246, AM, AT, GT, MTI, NBC, ODP, PC, PCX, SKT, SWP, ZMTI, ZS; Vulkanox 3100, 4010 NA, 4020, 4022, 4023, 4025, 4027, 4030, BKF, CS, DDA, DS, DS/F, HS/LG, HS/Powd., KB, MB-2/MGC, NKF, OCD, OCD/SG, PAN, SP, ZKF, ZMB-2/G, ZMB2/C5; Wingstay 100, 100AZ, EP, K, L, S, SN-1; Wytox 312, 320, 345, 540, 604, 604LMS, ADP-F, AF Series, ATP Series, HPM, LTS Series, PAP, PMW

ANTIOZONANTS

Akrochem Antioxidant 12, DQ, S; Akrochem Antiozonant MPD-100, PD-2; Akrowax 130, 145, 5050, Micro 23; Antilux 110, 111, 500, 550, 600, 620, 654, 660, 750, L; Antiozonant AFS/LG; Antisun; Antozite 1; Be Square 185; Durazone 37; Fortex; Isobutyl Niclate; Mekon White; Petrolite C-700, C-1035; Santoflex 13, 77, 134, 715, AX, IP; Sunolite 100, 127, 240, 666; Sunproofing Wax 1343; Vanox 3C, 6H, NBC; Vulkanox

3100, 4010 NA, 4027, AFD, AFS-50, AFS/LG; Vulkazon AFD, AFS-50, AFS/LG; Wingstay 100, 100AZ, SN-1

ANTIPERSPIRANTS

Hydagen DEO; Liponate NPGC-2; Reach 101, 201, 501; Wickenol 303, 308, 321, 323, 368, 370, 372, 373, 374, 375, 379, CPS 325, CPS 331, CPS 336, CPS 370

ANTISAG AGENTS

Perchem 44, 108, AMU 60 X, AMU X, EAG, Econogel, TEA

ANTISCALING AGENTS. SEE SCALE INHIBITORS

ANTISETTLING AGENTS

Forbest VP S7, WP; 14, 60MS, 60T, 60X, 1075, 1078X, 2000%, 2000X, 3000MS, MS; Perchem 44, 97, 108, EAG, Easigel, Econogel

ANTISHRINK AGENTS

Luchem AS-946

ANTISKINNING AGENTS

Isol R; Siponic SK; Troykyd Anti-Skin, B, Odorless Powd., S, Special M

ANTISLIP AGENTS

Nyacol, 2030 EC, 2046 EC, 9950; Polymekon

ANTISOILANTS

Alkaril SRP-A, SRP-C, SRP-N; Carpet Guard; Ceranine HCA Gran., PNS Gran.; Crafol AP-50; Good-rite K-7028, K-7058, K-7200, K-7600; Jorgard 10%; Laponite RD, RDS; Nyacol

ANTISTATS

Abil-Quat 3270, 3272; Ablumine 280; Ablunol CO5, CO10, CO15, CO30, CO45; Abluphat LP; Ablusoft ND; Accomeen C2, C5, C10, C15, S2, S5, S10, T2, T5, T15; Accosoft 550, 620, 707, 748, 808, 870, A-155; Adogen 415, 462, 471; Advawax 240; Aerosol C-61, OT-75%, OT-B; Ahco AY 2200, C330; Ahcovel Base N-64; Akypogene KTS; Akypoquat 40; Alipal CO-433, CO-436, EO-526; Alkamide STEDA; Alkaminox C-2, C-5, HT-12, HT-25, HT-30, T-2; Alkamox CAPO, LO; Alkamuls GML-45, GMO-45, GMO-55LG, GMS-45, PSML-20, PSMO-5, SML; Alkaphos 3, 6, 10, B6-56A; Alkapol PEG 200, 300, 3350; Alkaquat C, DAET, DAPT, DMB-ST 25%, O, S, ST, T; Alkasurf 075-5, CO-15, CO-20, NP-4, NP-6; Alkatronic EDP 8-4, EDP 28-1, EDP 28-7, EDP 38-1, EDP 38-4, EDP 38-8, EDP-28-2; Alkazine C, O; Alrowet D-65; Alubrasoft Super 15, Super 100; Alubrasol FE; Amiet 102, 105, 110, 115, 202, 205, 210, 215, 302, 305, 310, 315, 402, 405, 410, 415, 502, 505, 510, 515; Amine O, S, T; Ammmonyx KP; Anstex AK-25; Antistat 680; Antistatic Agent 106G-90%, 273-C; Antistaticum RC 100; Aqualon C, R; Arlatone T; Armosoft WA-201, WA-203, WA-204; Armostat 375, 410, 450, 475, 550, 575, 801, 810; Aromox 18/12, C/12, C/12-W, CD/12, DM16, DMC, DMC-W, DMHT, DMMC-W, DMMCD-W, T/12; Arosurf TA-101; Arquad 2C-75, 2HT-75, 16-29, 16-50, 18-50, DMMCB-50, S-50, T-2C-50, T-50; Aston 123, AP Conc., OI; Atlas G-3780A; Atmer 100, 101, 105, 107, 110, 113, 122, 123, 124, 125, 126, 127, 128, 129, 138, 139, 151, 154, 160, 163, 164, 172, 175, 176, 177, 178, 190, 191, 1001, 1002, 1004, 1005, 1006, 1008, 1009, 1021, 1024, 1025, 1027, 7000, 7001, 7002, 7003, 7004, 7005, 7006, 7007, 7008, 7009, 7010, 7101, 7102, 7103, 7104, 7105, 7106, 7109, 7110, 7111, 7112, 7200, 7201, 7202, 7203; Avitex NA, R

Babinar-801C; Berol 371, 372 Flakes, 808, 809; Berol Fintex 10, 42, 573, 577; Bio-Quat ASH-29; Bromat; CAE; Caplube 8440, 8445; Capstat; Carbowax PEG 200, 300, 400, 600, 900, 1000, 1450, 3350, 4600, 8000; Carsoquat CT 429; Carsosoft T-90; Cassastat WF; Catigene 4513, 4513/80, CA 30, CLP 50, CS 40, CT 30, CT 30/70, DC 100, DE 80, LT 45, ST 30, ST 30/70, T 50, T 80, T 80 F; Catinal CB 50, HTB-70; Catisol AO 100; Ceranine AT, AW, HC; Ceraphyl 60, 65, 70, 85; Chemax E-200 MS, E-400 MS; Chemazine 18, C, O, TO; Chemcol CO-40, T-10, T-15, T-20, T-25; Chemeen 18-2, 18-5, 18-50, C-2, C-5, C-10, C-15, HT-2, HT-5, HT-15, HT-50, S-2, S-5, S-30, S-30/80, T-2, T-5, T-10, T-15, T-20; Chemfac PA-080, PB-082, PB-104, PB-106, PB-109, PB-133, PB-135, PB-184, PB-264, PB-804, PC-188, PD-600, PD-990; Chemistat 6120, 6300, 6300 H; Chemquat 12-33, 12-50, 16-50; Chemstat 122, 122/60DC, 182, 182/67DC, 192; Chimin BX,

P45, P50; Cirrasol AEN-XB, AEN-XF, AEN-XZ, ALN-FP, ALN-GM; Cithrol 2DS, 2ML, 3DS, 4DS, 4ML, 6DS, 6ML, 10DS, 10ML, 60ML, A, DGDL N/E, DGDL S/E, DGDO N/E, DGDO S/E, DGDS N/E, DGDS S/E, DGML N/E, DGML S/E, DGMO N/E, DGMO S/E, DGMS N/E, DGMS S/E, DPGML N/E, DPGML S/E, DPGMO N/E, DPGMO S/E, DPGMS N/E, DPGMS S/E, EGDL N/E, EGDL S/E, EGDO N/E, EGDO S/E, EGDS N/E, EGDS S/E, EGML N/E, EGML S/E, EGMO N/E, EGMO S/E, EGMR N/E, EGMR S/E, EGMS N/E, EGMS S/E, GDL N/E, GDL S/E, GDO N/E, GDO S/E, GDS N/E, GDS S/E, GML N/E, GML S/E, GMO N/E, GMO S/E, GMR N/E, GMR S/E, GMS Acid Stable, PGML GMS N/E, GMS S/E, N/E, PGMO N/E, PGMO S/E, PGMR N/E, PGMR S/E, PGMS N/E, PGMS S/E

Consostat DKM; Cordex DJ; CPH-380-N; Crafol AP-31, AP-33, AP-34; Crapol AU-23, AU-31, AU-40, FU-25; Crestomul T; Crill 1, 2, 3, 35, 50; Crillet 2, 3, 4, 31, 35, 41, 45; Crillon LDE; Crodafos 25 D Series, 25 D2 Acid, 25 D5 Acid, 25 D10 Acid, CAP, CS Series, ID Series, N3 Acid, N3 Neutral, N5 Acid, N10 Acid, N10 Neutral, O Series, SG, T2 Acid, T5 Acid, T10 Acid; Crodamet; Crodasinic LS35, OS35; Cyastat 609, LS, SN, SP; Cyclomide CODI; Cyclophos PL3; Cycloton 75C, D261C/70, M214B/99, M242B/99, M270C/85; Cyncal 80%; Dacospin 9212, PE-146; Deatron N; Dehyquart A, DAM, E, LT, SP, STC-25; Delion 624, 662, A-016; Delion A-160; DeSomeen TA-2, TA-5, TA-15, TA-20; DeSonic 6C, 30C, 36C, 40C, 54C; DeSonic TA-2, TA-15, TA-25CWS; DeSophos 4 CP, 6 DP, 6 MPNa, 6 NPNa, 7 OPNa; DeSotan SMO-20, SMT-20; Doittol P690; Dow Corning FF-400, FF-412, FF-414; Druspin LO-VIS, Supreme; Durfax 20; Durtan 20; Eccospin PPF; Eccostat; Effcol TS-214; Elec AC, QN, RC, TS-5, TS-6, 2; Elimina-254; Emcol ISML, L, M, NA-30, Q, 150, 1655, 3780, CC-9, CC-42, CC-55, CC-57, CC-422; Emerstat 6660; Emkalon AVR; Emkastat PC; Empigen 5107, BS, BS/H, BS/P; Emphos CS-136, CS-1361, D70-30C, PS-121, PS-220, PS-236, PS-400; Empicol 0216; Empigen CM, FRC75S; Empilan BQ 100, CDX, DL 40, DL 100, GMS NSE40, GMS NSE90; Emulgade CRC; Emulphor EL-620, EL-719, EL-980, EL-985; Ethoduoquad T/20; Ethoquad 18/12, 18/25, C/12, C/25, O/12, O/25; Ethoxamine C5, SF11; Ethylan A4, A6, GD, GL-20, GO-80, GP-40, GS-60, GT-85, LD, LDS, LM, MLD, TC, TF, TH-2, TN-10, TT-15; ; Etilenox KM-53; Etocas 10, 35, 40, 60, 100

Farmin 2C, D86; Flexan 130; Forbest 8209; Gafac GB-520, LO-529, MC-470; RB-400, RD-510, RE-410, RE-610, RE-870, RE-877, RE-960, RK-500, RS-410, RS-610, RS-710; Gafstat S, S-100; Gantrez AN, AN-119, AN-139, AN-169, AN-179; Genamin KDM, KDM-F; Geropon PL/68, PL/87; Glytex EL 176, EL 882, EL 905, L 154, L 203; Grindtek ML 90, MM 90, MOL 90, MOP 90, MSP 32-6, MSP 40, MSP 40F, MSP 52, MSP 90, PK 60; Hallco Antistat C-1047; Hallco C-7065; Hartamide 9137; Hartolon 368, AL, NA; Hartosoft 75; Hercosett 57, 70, 125; Hetamine 5 L 25; Hetoxamine C-2, C-5, C-15, O-2, O-5, O-15, S-2, S-5, S-15, ST-2, ST-5, ST-15, ST-50, T-2, T-5, T-15, T-20; Hostaphat F Brands, L Brands; Hostastat HS1, Systems E 3952, E 3953, E 3954, E 5951, E 6952; Hyamine 2389, 3500; Icodimeen T-30; Icomeen 18-5, O-30, O-30-80%, S-5, T-2, T-5, T-7, T-20, T-25, T-25 CWS, T-40, T-40-80%; Igepal CO-430, DM-710; Imerol DU; Incrocas 10, 30, 40, 60, 100; Incromectant AQ, LQ; Incropol 233, 290, CS-12, CS-20, CS-30, CS-40, CS-50, CS-60, L-2, L-7, L-7-90, L-12, L-23, L-30; Incroquat 100, BA-85, S-75 CG, SDQ-25, SE-85; Incrosoft 100, S-75, S-90, S-90M, T-75, T-90; Jordaquat 1033, JO-50; Katapol PN-730, PN-810; Kemamine AS-650, AS-974, AS-974/1, AS-989, AS-990, BQ-2802C, BQ-9702C, BQ-9742C, Q-1902C, Q-2802C, Q-6502C, Q-6503 B, Q-9702C, Q-9743C, Q-9743CHGW; Ken-React Series, 7 (KR 7), 9S (KR 9S), 12 (KR 12), 26S (KR 26S), 33DS, (KR 33DS), 38S (KR 38S), 39DS (KR 39DS), 41B (KR 41B), 44 (KR 44), 46B (KR 46B), 55 (KR 55), 133DS (KR 133DS), 134S (KR 134S), 138D (KR 138D), 138S (KR 138S), 158 (KR 158D), 158FS (KR 158FS), 212 (KR 212), 238A (KR 238A), 238J (KR 238J), 238M (KR 238M), 238S (KR 238S), 238T (KR 238T), 262A (KR 262A), 262ES (KR 262ES), OPP2 (KR OPP2), OPPR (KR OPPR), TTS (KR TTS); Kerensim L 10 P; Keripon NC; Kerostat 5009; Ketjenblack E.C.; Klearfac AA040, AA270, AA420, AB270

Lamigen ES-30, ES-60, ES-100, ES-180, ET-20, ET-70, ET-90, ET-180; Lankrostat 16, 38, 104, 0600, CA2, LA3, LDN, LME, NP6, QAT; Lanoquat 1756; Larostat 88, 143, 192, 192 Anhyd., 264-A, 264-A Anhyd., 264-A Conc., 300, 377 DPG, 491, 1443; Lexamine 22, B-13, C-13, L-13, O-13, P-13, R-13, S-13; LICA 01, 09, 12, 38, 38A, 38J, 44, 97; Lilamin EO; Lipamide MEAA; Lipomin LA; Lipoquat C 25, R; Liposorb L, L-10, L-20, O, O-5, O-20, P, P-20, S, S-20, SQO, TO, TO-20, TS, TS-20; Lipowax C; Lubrol 12A-9; Lutensol ED 140, ED 310, ED 370, ED 610; Lutostat 171, MSW 30, MSW 88; LZ 01, 09, 12, 38, 44, 97; Macol 2, 4, 5, 8, 10; Mapeg 600 DL, 600 DOT, 600 ML, 600 MOT, 6000 MS, PGDS, PGMS; Maphos 60, 66, 66H, 76, 76 NA, L-6; Markstat AL-12, AL-22, Antistats; Marlophor ND, ND-Acid, ND DEA Salt, ND Na-Salt; Masil 280, 280LP, 1066C, 1066D, 2132, 2133, 2134; Mazeen 173, 174-75, C 2, C 5, C 10, C 15, DBA, S 2, S 5, S 10, S 15, T 2, T 5, T 15; Mazoline OA; Mazol GMO K; Merix #79 Conc., 79-OL Conc., 79 Special; Merpol HCS, OJS; Merquat 100; Merrol 4221; Miranol Ester PO-LM4; Mirapol 9, 95, 175, AD-1, AZ-1; Mirataine H2C; Monafax 785, 786, 831, 1293, H-15, L-10; Monaquat ISIES, P-TC, P-TD, P-TL, P-TZ, TG; Monostat 1195; Monazoline Series, C, CY, O, T; Monolan PPG440, PPG1100, PPG2200; Mytab; Niox KH Series, KQ-55, -56; Nissan Anon BF, BL, LG; Nissan Cation AB, MA, SA; Nissan Elegan S-100; Nissan New Elegan A, ASK; Nissan Nymeen DT-203, -208, L-210, -202, -207, S-202, -204, S-210, -215, -220, T_2-206, -210, -230, -260; Noiox AK-41, AK-43; Nonionic 1035-L, 1044-L, 1061-L, 1064-L, 1068-L, 1088-L; Noramax C2 to 15, S2 to 11; Noxamium C2-15, S2-11, S2/50; Obazoline 662Y, LB-40

ANTISTATS (CONT'D.)

Parabolix 100; Pegafac CE 410, 610, CS 410, 610, 710, PDA (Neutralized), PDA Free Acid; Pegameen 02, 030, 30 80%, C2, C5, C10, C25, S2, S5, S20, T2, T5, T15, T20; Peganate CO 5, 16, 25, 30, 36, 40, 200, 200 50%, COH 25, COH 200, COH 200 50%, MA 8, 15 80%, ML 5, 9, 14, MO 6, 9, 14, MS 8, 14, 23, TO 9, 16; Peganol NP 4; Pegeste SML, SMO, SMP, SMS; Pegnol OA-400; Perkare; Permax AW-2, PM 420, PM 503; Petrac Eramide; Phosfac 1004, 1006, 1044, 1044FA, 1066, 1066FA, 1068FA; Plasthall 503; Pluracol E-200, E-300, E-400, E-600, E-1500, E-4000, E-6000, E-400 NF, E-600 NF, E-1000, E-1450, E-1450 NF, E-2000, E-4000 NF, E-4500, E-8000, E-8000 NF; Pluronic F38, F68, F68LF, F77, F87, F98, F108, F127, L31, L35, L42, L43, L44, L61, L62, L62D, L62LF, L63, L64, L72, L81, L92, L101, L121, L122, P65, P75, P84, P85, P94, P103, P104, P105, P123; Pogol 200, 1570; Polylube GK; Polyquart H, H 81, H-7102; Polysoft CA; Polystat Agent #5033; Product BCO; Progastat 204; Proplast 060; Protowet 3072; Quadrilan AT, OD-75, SK; Quartamin 24P, 86P Conc., 86P; Quarternary O; Quimipol EA 2507, ED 2021, 2022; Quimipol ENF 90; Radiamine 6140, 6141, 6161, 6163, 6164, 6170, 6171, 6172, 6173; Radiasurf 7196, 7201, 7206, 7270, 7400, 7402, 7403, 7404, 7410, 7411, 7412, 7413, 7414, 7417, 7420, 7421, 7422, 7423, 7125, 7431, 7432, 7135, 7136, 7137, 7140, 7141, 7443, 7145, 7147, 7453, 7454, 7600; Reten 210, 220, 300, 420, 421, 423, 425, 521, 523, 525; Rewocid UTM 185; Rewomine IM-OA; Rewominox B 204, S 300; Rewophat E 1027, EAK 8190, NP 90, OP 80; Rewopon IM-CA; Rewoquat CPEM, DQ 35, RTM 50; Rexol 25/6, 2000 HWM, C14; Roccal II 50%; Rycofax 3115

Sandin EU, VU; Sandoz Amine Oxide XA-C, XA-L, XA-M; Sandoz Phosphorester 510; Sanstat 2012-A; Schercamox C-AA, DML, DMM, T-12; Schercomul QW; Schercoquat IB, WOAS; Schercotaine CAB, MAB; Schercozoline B, C, I, L, O, S; Secomine TA 02; Secoster BS; Serdox NJAD 15, NJAD 20, NJAD 30, NSG 400; Servamine KAC 412; Servoxyl VPI 55; Silwet L-77; Sipenol IC2, IC10, IO12, IS05, IT11, IT50; S-Maz 20, 40, 60, 60 K, 60 KHM, 65, 65 K, 67, 80, 80 K, 85, 85 K, 90, 93 R, 95; Sochamine SG 8, 100; Softenol 3900, 3991; Soprofor PL/Series, T/20, T/60, T/65, T/80, T/85; Sorbanox AL; Soromine AT; Span 20, 40, 65, 85; Spectratech CM 11045, 11638, 77242; Stafoam DF-1, DF-4; Standamul STC-25; Standapol AB-45, BAW; Statexan K1; Staticide; Stedbac; Stepan Cationic 10; Stepanquat HC 80; Stepantex SO 90, VG 90, VS 90, VP 85, VRH 80; Steramine FPA 197, PNA 75; Sulfostat KNT; Sulframin 40, 40DA, 40RA, 40T, 60T, 85, 90, 1240, 1245, 1250, 1260, 1288, 1298, 1388; Sunaptol CA 25, 120, 350, 400, 750, CFA 90, 110, DL Conc., OA, OA 70, 90, 100, 140; Synoquart P 50; Synprolam 35DMBQC, 35MX 1, 35MX 3, 35MX 5, 35MX1QC, 35N3X3, 35N3X5, 35N3X10, 35N3X15, 35N3X25, 35TMQC, 35TMQS, 35X2, 35X5, 35X10, 35X15, 35X20, 35X25, 35X35, 35X50, 35X2QS, 35X5QS, 35X10QS, FS; Tegin 4600, 4600 NSE

Tegopren; Tequat PAN; Teric 16A16, 16A22, 16A29, 151, 154, OF4, OF6, OF8; Tetronic 50R1, 50R4, 50R8, 70R1, 70R2, 70R4, 90R1, 90R4, 90R8, 110R1, 110R2, 110R7, 130R1, 130R2, 150R1, 150R4, 150R8, 304, 504, 701, 702, 704, 707, 904, 908, 1101, 1102, 1107, 1301, 1302, 1304, 1307, 1501, 1502, 1504, 1508; Texnol IL; Tex-Wet 1034; T-Maz 20, 28, 40, 60, 60K, 60KHM, 61, 65, 65K, 67, 80, 80KLM, 81, 81K, 85, 85K, 90, 95; Tomah PA-10, PA-14, Q-14-2, Q-17-2, Q-18-2, Q-18-2-50%, Q-18-15, Q-D-62, Q-D-T, Q-S; Triton X-155, X-155-90%; Tryfac 525-K, 610-K, 5365, 5553, 5554, 5556, 5559, 5560, 5571, 5573, 5576; Trylox 5918, CO-200, -200/50; Trymeen 6601, 6602, 6606, 6607, 6609, 6617, 6637, CAM-10, CAM-15, SAM-50, TAM-15, TAM-20, TAM-25, TAM-40; Tween 20, 20 SD, 21, 40, 60, 61, 65, 80, 81, 85; Ultra Sulfate AE-3, SE-5, SL-1; Unamine C, O, S; Union Carbide Y-9567; Unisol UA 20, 80, 150, 300, 400; Varine C, O, T; Variquat 638; Varisoft 110, 137, 190, 190-100P, 222, 238, 445, 475, 3690, 6112, DHT, SDC-W; Varonic K205, L202, L205, L205LC, L230, L230-60, L230-80, T202, T205, T205LC, T215, T215LC, U202, U205, U205LC, U215; Varstat 10, 55, 66, K-22, T-22; Velvetex BC-35; Volpo 25D3, 25D5, 25D10, 25D20, CS3, CS5, CS10, CS15, CS20, O5, O10, O15, O20, T3, T5, T10, T15, T20; Vueguard 941; Witamide 82, 272, 511, 5130, 5133, 5138, 5195, AL69-58; Witcamine 209, 211, 235, AL42-12, PA-60B; Witcolate 58, 1276, D51-51; Witconate 605T, 1840X; Witconol APEM, APM, APS, CD-18, NS500K; Zelec NE, NK, TY, UN

ANTISTRIPPING AGENTS:

Acra-500, Agent 765; Base 3059 E; Berol 26; Blendex 600; Diamin HT, O, S, T; Dinoram S; Indulin AS-1, AS-Special; Nissan Anon BF; Noram 2 C, C, DMC, DMS, DMSH, M2 SH, M2C; Pave 100, 192; Polyram S

ANTITACKIFIERS/DETACIFIERS

Ardet SEA; Ceraphyl 31, 41, 368, 375; Cremophor S 9; Crodamol DA, ML; Cyclochem LVL; Dee-Tac; Lipowax C; Paracol 404C; Petrac Eramide; Petrac Slip-Eze; Petrac Vyn-Eze; Schercemol LL; Witco Zinc Stearate NW; Zinc Stearate Disperso, NW, Regular

AROMATICS/ODORANTS

Emeressence 1166 Styrate, 1170 Chamol, 1171 Pseudojasmone, 1172 Privetone, 1173 Parajasmone, 1174 Fir

Balsam; Pennodorant 1013; Spotleak 1001, 1003, 1007, 1009, 1044, 2323

AVERSIVE AGENTS

Bitrex

BASE MATERIALS/RAW MATERIALS

Adogen 432-ET, 442-100P; Ahcovel OB; Alkamide 2104, 2110, 2112, 2204; Alkapol PEG 8000; Alkateric BC; Alox 502, 604, 904, 940, 2019A; Amphocerin K; Argobase EU, LI; Carbowax PEG 540 Blend, 900, 1000; Cardolite NC-507, NC-510, NC-511; Carsosoft S-75, S-90, S-90-M; Catinal HTB-70, OB-80E; Conco Sulfate TL, WA Dry, WAG, WAS, WA Special; Consamine CA; Corona Lanolin; Coronet Lanolin; Croda Absorption Base; Cutina GMS, LM, MD; Cyclomide DC212/M, IODI; Cycloton D261C/75, M270C/85; Dehydag Wax W; Dehymuls K; DeSonate SA; Dobanol 23, 25, 45, 91, 235; Drakeol 6, 7, 9, 13; Duomeen TX; Eltesol CA 65, CA 96, PSA 65, TA, TA/F, XA, XA90; Emal 20C, E-25C, E-70C, NC-35; Emcol ISML; Emerest 2301, 2302, 2485; Emphos PS-21A, PS-236

Empicol ESB3, ESB30, ESB50, ESB70, ESB70-AU, ESC70, Empigen BAC, BAC 90, BCB50, BCF 80, BCM75, BCM75/A, FKC75K, FKC75L, FRC75S; Empilan CIS, CME, LIS, LME; Emulsogen H; ESI-Terge AH 20; Euperlan PK 771, PK 776, PK 810, PK 900; Foam Burst 2; Forlan; Gardiquat CQ; Gelot 64; Genamin CC Brand, CS Brand, CTAC, KDM-F, KS 5, KSE, O Brands, S Brands, T Brands; Genamine C-020, C-050, C-080, C-100, C-150, C-200, C-250, O-020, O-050, O-080, O-100, O-150, O-200, O-250, S-050, S-080, S-100, S-200, S-250; Genapol C-020, C-050, C-080, C-100, C-150, C-200, C-250, CRT 40, O-020, O-050, O-080, O-100, O-100, O-120, O-150, O-200, O-230, O-250, PAF Brands, PF Brands, PL Brands, PN Brands, S-020, S-050, S-080, S-100, S-150, S-200, S-250; Geronol ACR/4, ACR/9; Golden Dawn Grade 1,2, Superfine; Golden Fleece DF, P-80, P-95, RA; Herbavert; Hetsulf 60S, Acid; Hostacerin T-3; Hydrokote 25, 27, 79, 711, S-7, SP; Hyonic NP-100, NP-110, PE-100

Imwitor 960, 960K, 965, 965K; Incromate CDL, CDP, CPML, IDL, LDL, SDL; Incrosoft S-75, S-90, S-90M; Ionet S-20, S-60 C, S-80, S-85, T-20 C, T-60 C, T-80 C; Isomul Extra; Jasmacyclat; Kalcohl 5-24, 6-24, 7-24, 08H, 10H, 20, 40, 60, 68, 80; Lanette O, SX, W; Lanogen 1500; Laponite XLG, XLS; Laurex 16/18, 16/18D, 810, CH, CS, CS/D, L1, NC, PKH; Lexate IL, PX; Lexolube 2X-108, 2X-109, 2X-114, 2X-130; Lilaminox; Lipocol TD-3, TD-6, TD-12; Mackester Series; Macol 1, 18, 19, 31, 32, 33, 35, 42, 44, 46, 72, 77, 81, 85, 101, 108, 121, LF-110, LF-111, LF-111A, LF-120; Marlamid KLP, PG 20; Marlophen DNP 16, 18, 30; Marlosoft IQ 75, 90; Marlosol 183, 189, 1820, 1825, OL 7, OL 8, OL 10, OL 15, OL 20; Maslip 500, 501 Base; Mazawax 163R; Mazol GMO, GMR; Miglyol 840 Gel, Gel; Mineral Jelly No. 5, 10, 15, 20, 25; Molykote

Nimlesterol 1730, 1732; Niox EO-26, KS Series, EO-10; Niox KH Series, KJ-56, KJ-66; Nissan Panacete 810; Nissan Persoft NK-60, -100, SK; Nissan Sunbase, Powder; Noda Rice Wax FCC; Nonipol 20, 40, 55, D-160, Soft SS-50, -70, -90; Nonisol 300; Norfox 1 Polyol; Norfox Coco Powder, EM 350X, EM-350X Conc., GMS; Obazoline 662Y, CS-65; Ointment Base No. 3, 4, 6; Penreco Amber; Penreco Blond; Penreco Cream, Frost, Green, Lily, Red, Regent, Royal, Snow, Super, Ultima; Perlankrol DGS, DSA; Petromixes; Phenyl Sulphonate CA, CAL; Phosphate Ester 123; Pluracol E-200, E-300, E-400, E-400 NF, E-600, E-600 NF, E-1000, E-1450, E-1450 NF, E-2000, E-4000, E-4000 NF, E-4500, E-8000, E-8000 NF; Pogol 1540, 1570; Polyco 2611, 2612; Polystate C; Protamine-45; Protegin W, WX, X; Protowax 90; Protowet 3098; Quartamin 86W; Rewomine IM-OA; Rewopal LA 3, TA 11, 25; Rewophat NP 90, TD 70; Rewopon IM-CA; Rexol 25/9, 25/10; Rexonic N25-14 85%

Sandet ALH; Schercemol BE, EE, ISE, PGML, SE; Secosyl; Sedefos 75; Serdet NZ 60; Serdox NEL 3; Servamine KOO 330 B; Servoxyl VLB 1123; Sipex EST-30; Softenol 3819; Softisan 100, 378, 601, 649; Solar NP, Regular; Sonojell Min. Jellies; Standamul 302, 318, 1414-E, 7061, 7063, 7105, CTA, CTV, G-32/36 Stearate; Sulfonate OE-500; Sulfopon WA 1; Sulframin 40; Surco 40-SX; Surfageen S 30; Tefose 63, 1500, 2000, 2561; Texamine 84(L); Textamine A-5-D; Ultra Sulfate AE-3, SE-5, SL-1; Unamide C-7649; Unamine C, O; Vanisol BIS sodico-2; Varamide A-7; Varine C, O, T; Varisoft E-290, SA-50, SDC-85S, TSC; Verdoxan; Witcamide 70, 82; Witcolate 58, A; Witconate 60B, 60T, 93S, 1250 Slurry, 1260 Slurry, TAB, YLA; Witepsol Suppository Bases; Zoharsoft, 90, DAS, DAS-N

BINDERS

Acetulan; Acralen AFR, ATR, BS; Acrysol ASE-60; Adol 62 NF; Adtac L100BHT; Alkapol PEG 200, 300, 600, 6000, 8000; Aluminum Stearate EA; Ambergum 3021; Avicel PH-101, PH-102, PH-103, PH-105; Benecel CM, M, ME, MP; Be Square 185; Blanose Cellulose Gum; Blanose Refined CMC; Britol; Carbowax Compound 20M, PEG 3350, PEG 4600, PEG 8000; Catapal Alumina; CD-75P; Ceraphyl 28, 31, 45, 50-S, 55, 140A, 368, 375, 847, ICA; Chemal BP-261, BP-262, BP-262LF, BP-263, BP-264, BP-268, BP-268/50; CMC-T; Combizell AG 070, AG 200, APR 060, APR 200; Crosterene; CT-88 Aerosol; Culminal CMMC, MC, MHEC, MHPC; Cyanamer A-370, P-26, P-250; Darex 529L, 535L, SP-1566; Degacure K 126;

BINDERS (CONT'D.)

Dynasan 114; Elfacos ST 37; Elvanol 71-30, 75-15; Emerest 2310; Emery 6686, 6709; Emsorb 2728; Everflex 81L, 515L, T, TMF; Finsolv TN; Fonoline White, Yellow; Forbest VP 13, VP S7; Fortex; GAF Latex 6000 Series, 7000 Series, 8000 Series; Gantrez M-154, M-555, M-556, M-574; Genugel; Genulacta Series; Genuvisco; Glucam P-20 Distearate; Golden Dawn Grade 1,2, Superfine; Golden Fleece DF, P-80, P-95, RA; G-White; Hercules Cellulose, Cellulose Gum, CMC; Hystrene 5016 NF FG 7018 FG, 9718 NF FG; Imwitor 191, 900, 940K; Industrene 105, 138, 143, 154-BG, 205, 206, 225, 225 FG, 226, 232, 239, 325, 328, 333, 365, 3022, 4022, 4516, 4518, 5016, 5016 FG, 5022, 7018, 7018 FG, 8518, 8718 FG, 9018, B, R; Isopropylan 33; Keltose; Keltrol, F; Klucel; Lignosol B, BD, LC; Low Crock #5012, #5013, #5026, #5087, Formula #750; Ludox HS-30, SM; Lutrol E 1500, E 4000, E 6000; Luviskol

Maltrin M040, M100, M150, M180, M200, M250, M500, M510, M550, M600; Mapeg PPG-400, PPG-500; Mearlmica MMCF, MMSV; Mekon White; Methocel A4C, A4M, A15-LV, E4M, E5, E15-LV, E50-LV, E4M, F, F50-LV, F4M, K3, K35, K100-LV, K4M, K15M, K100M; Natrosol 250, 250 ER, GR, H4R, HHR, HR, JR, 250 KR, LR, MHR, MR; Nopalcol Series, 1-S, 1-TW; Nopcowax 22-DS; Norlig, A, NH; Nyacol, 215, 830, 1430, 1440, 2030 EC, 2040 NH4, 2046 EC, 9950; Orzan A, AL-50, SL-50; Papi; Patlac LA; Pawn PPE-501; Penreco Amber, Red; Petrolite C-700, C-1035, C-7500, C-9500, WB-5; Phenrez 13; Piccomer 10, 25, 40; Piccopale 70, 85, 100, 100BHT, 100HM; Piccoumaron 100, 110, 120; Plasdone, K-25, K-26/28, K-29/32, K-90; Pluracol E-200, E-300, E-400, E-600, E-1500, E-4000, E-6000, P-410, P-710, P-1010, P-2010, P-3020, P-4010; Pluriol E 1500, E 4000, E 6000, E 9000, PE 6800; Pluronic 10R5, 10R8, 12R3, 17R1, 17R2, 17R4, 17R8, 22R4, 25R1, 25R2, 25R4, 25R5, 31R2, 31R4, L10; Pogol 200; Polectron 430; Polidene 33-001, 33-004, 33-021, 33-031, 33-038, 33-075; Polyco 2140, 2142, 2149-C, 2617, 2618, 2629, 2838; Polymekon; Polyplasdone XL, XL-10; Polytrap 158, 171, 210, 229, 603, 801; Polywax 500, 655, 1000, 2000; Precirol ATO; Primacor 4990 Dispersion; PT White Wax 602; Pulpex Thermal Bonding Pulps; Pure-Dent B700, B810; Purelast 169; QO Furfuryl Alcohol; Ricon 100, 109, 150, 157; Rudol; Schercemol DID, TIST, TT; Scripset 520, 540, 550, 720; Semtol Series; Shellwax; Sol-U-Tein E, EA; Standamul 318, G-16; Starwax 100; Terate 303, HM-AY; Texicote 03-020, 03-021, 03-060; Texicryl 13-002, 13-011, 13—21, 13-022, 13-030, 13-100, 13-101, 13-203, 13-205, 13-210; Triton CS-7; Tylose C 30 L, C, CB Series, CR, CBR Series, H Series, MH, MHB, MB, MH-K, MH-xp, MHB-y, MHB-yp, P, P-x, PS-x, P-Z Series; Ucar Latex 131, 828, 879; Veegum, F, HS, HV, K, Neutral, S-728, WG; Versaflex 6; Vinnapas A 50, CEF 10, CEZ 16, DZN 220, EN 424, EN 426, EN 427, EN 428, EN 429, EP 16, EV 25, EZ 15, H 54/15 C, H 60, LT 420, M 50/300, SAF 54, T 50/30 VL, Z 50, Z 54/20 C; Vinnol 50, 50/25 C, CE 35; Vinsalyn Resins; Volclay HPM-20, NF-BC; Waxenol 801, 815, 816; Witamide 70

BLOCKING AGENTS

Acrawax-C; B2500, B2539, B2559.5, B2570, B2595, B2790, B2792, B2794, B2796, B2805, B2809; Barco Bond MB-14X, MB-127; Bonding Agent 2001, TN, TN/S50; CA0567; CA0570; CA0699; CA0742; CA0900; CB2408; CB2409.5; CB2493; CB2494; CB2495; CB2790; CB2805; CB2909; CC3005; CC3270; CC3275; CC3285; CC3290; CC3291; CC3433; CC3555; CD3770; CD3780; CD4450; CD5400; CD5430; CD5470; CD5487; CD5550; CD5600; CD5610; CD5636; CD5950; CD6000; CD6010; CD6150; CD6210; CD6220; CD6250; CE6250; CE6345; CE6350; CH7250; CH7260; CH7280; CH7310; CH7332; CI7840; CM8450; CM8750; CM8930; CM8980; CM9000; CM9220; CO9745; CO9750; CO9000; CO9810; CO9816; CO9817; CO9819; CO9830; Coviox T-50, T-70; CP0110; CP0156; CP0160; CP0280; CP0320; CP0330; CP0800; CP0810; CS1590; CT1800; CT2015; CT2030; CT2050; CT2090; CT2500; CT2520; CT2523; CT2902; CT2925; CT2950; CT2970; CT3250; CT3254; CT3600; CT3610; CT3795; CV4772; CV4800; CV5050; CV5100; D4165; D4368; D4450; D5115; D5400; D5430; D5490; D5600; D6102; Dynasil A, CA, CM, M; Dynasylan AMEO, AMMO, BDAC, BSA, CPTEO, CPTMO, DAMO, GLYMO, HMDS, IBTMO, IMEO, MEMO, MSTFA, MTES, MTMO, MTMS, OCTEO, TCS, TRIAMO, VTC, VTEO, VTMO, VTPMO; H7250, H7300; I7860; P0160; T1990, T2015, T2520, T2885, T2890, T2928, T2950, T3150, T3250, T3500, T3550, T3600, T3634, T3660, T3795

BLOWING AGENTS

Alconate SBTA-269; Alkasurf SS-TA 125, SS-TA; Celogen AZ, AZ 3990; Expandex 5PT; Ficel AC, AC2, AC3, AC3F, AC4, AC-SP4, AF100, EP Series, EPA, EPB, EPC, EPD, EPE, LE, SCE; Kempore 60/14FF, 60E, 125E, 125FF, E, FF, MC; Ken-React Series, 7 (KR 7), 9S (KR 9S), 12 (KR 12), 26S (KR 26S), 33DS, (KR 33DS), 38S (KR 38S), 39DS (KR 39DS), 41B (KR 41B), 44 (KR 44), 46B (KR 46B), 55 (KR 55), 133DS (KR 133DS), 134S (KR 134S), 138D (KR 138D), 138S (KR 138S), 158 (KR 158D), 158FS (KR 158FS), 212 (KR 212), 238A (KR 238A), 238J (KR 238J), 238M (KR 238M), 238S (KR 238S), 238T (KR 238T), 262A (KR 262A), 262ES (KR 262ES), OPP2 (KR OPP2), OPPR (KR OPPR), TTS (KR TTS); LICA 01, 09, 12, 38, 38A, 38J, 44, 97; Luperfoam 40, 329; LZ 01, 09, 12, 38, 44, 97; Nitropore OBSH; Onifine-C, CC, CE; Porofor ADC/E, ADC/F, ADC/K, ADC/M, ADC/R, ADC/S, B 13/CP 50, BSH Paste, BSH

Powd., D33, DNO/F, N, S 44; Rhenofit 1600; Rhenopor 1843; Sipex NB60; Texacat ZF-54; Texox WL-440, WL-660, WL-1400, WL-5000

BODYING AGENTS/CONSISTENCY GIVING AGENTS. SEE THICKENERS

BONDING AGENTS

Chemlok 205, 210, 212, 218, 220, 222, 233, 234B, 235, 236A, 238, 246, 250, 252, 253, 402, 459, 607, 608, 610, 810, AP-133, BN, KP-1001, Y-1520A, Y-1530, Y-1540, Y-4310, Y-5323; Cohedur A, A Solid, AS, AS Powder, RK, RL, RS; Crylcon 3000 Series, 7000 Series; Desmodur TT; Dow Corning X1-6106; Dresinol 40, 42; Duralink HTS; Dynasylan AMEO-40, AMEO-P, AMEO-T, DAMO-M; Ganex P904, V216, V220, V516; Gantrez B-773; Glitter Adhesive #5048; Hercules Cellulose Gum; Hi-Sil 210, 215, 233, 250, 260, 262; Kodaflex PA-5; Megum Series; Millathane Adhesive 2000; Natrosol 250; Naugard P, PHR; Naugawhite; Neolyn 35D

Pave 100, 192; Picconol A100, A102, A500, A501; Piccotac B, CBHT; Piccovar AB165, AB180; Plasdone, K-25, K-26/28, K-29/32, K-90; PVP K-60; Pyratex J 1904; Reten 210, 220, 300, 420, 421, 423, 425, 521, 523, 525; Rewomine IM-BT; Rewopon IM-BT, IM-CA, JMBT; Santolite MHP, MS-80%; Servamine KEO 260, KET 350; SRF-1501; Syloid 244, 378; Synprolam 35N3; Tegocoll; Texicryl 13-206, 13-300, 13-302; Thixon 753; Tri-Ethane; TSE-2000; Union Carbide A-151, A-172, A-174, A-186, A-187, A-189, A-1100, A1102, 1120, A-1126, A-1893; Vanchem HM-50, HM-4346; Varcum 8169, 8199, 8251, 8261, 8267, 8345; Volan, L; Vulcabond VP; Vulkadur T, 40%

CARRIERS/VEHICLES

Alkapol PPG-425, 1000; Britol; Captex 200, 350, 355, 800, 810B, 8000; Carbowax PEG 300, 400, 3350, 4600, 8000; Carbowax Sentry; Celite 21-A, 209, 219, 400, 408, 410, 1200, CAS-30KR, CS-22R, F.C., R-625, R-633, R-643, R-685, R-690; Cetiol A, V; Cyclogol Cetyl Alcohol NF; Dai Cari XBN; DEP; Diluex; Drakeol 7, 9; Emerest 2421; Emery Methyl Lardate, Methyl Oleate; Emulsifying Oil C; Emulsion E-1614; Eureka 392; Fillite M-32; Flo-Gard AG110, AG 130, AG 150, CC 120, CC 140, CC 160, FF 310, FF 320, FF 330, FF 350, FF 370, FF 390; Florex Granular Grades; Fonoline White, Yellow; Grit-O'Cobs; HD-Ocenol; Hercolyn D; Hetoxide C-30, C-200-50%, HC-40, HC-60; Hi-Sil 132, 210, 215, 233, 250, 260, 262, 532EP, ABS; Hubersil 162; Hubersorb 600; Ibercarrier 300-L, OPP; Imsil A-10, A-15, A-25, A-108; Imwitor 308, 310, 312, 742, 908, 914; Incromide SM; Kemester 104, 213, 5721; Kronitex 100; Lantrol 1674 Deodorized; Laponite 508; Lexol GT 855, GT 865, PG 800, PG 855, PG 865, PG 900; Lipovol SES, SES-S; Lite-R-Cobs; Maltrin M100, M500, M550, M600, M700; Mazol 1400, PG-810, PGO-104; Micro-Cel A, B, E; Microduct; Miglyol 810, 812, 840; Manro FCM 95LV, FCM 100, PTSA 65 E, PTSA 65 H, PTSA 65 LS, PTSA Crystals; Neobee 1053, 1054, 1062, 1223, O, 1400 Oil, M-5, M-20; Ointment Base No. 3, 4, 6; Paraplex G-50, RG-8; Penreco Amber Cream, Frost, Lily, Regent, Royal, Snow, Super, Ultima; Perfecta USP; Piccolastic BV-50HF; Piccomer 100, 110, 120; Plasdone K-25, K-26/28, K-29/32, K-90; Pluracol E-200, E-300, E-400, E-600, E-4000; Poly-G 200, 300, 400, 600, 1000, 1500, 2000, B1530; Polylite 31-822; Polytrap 603; Protopet Alba, White 2L, 3C, Yellow 1E, 2A; PVP K-15; Pyrax ABB, WA; QO Furfuryl Alcohol, Tetrahydrofurfuryl Alcohol; Rudol; Semtol Series; Sipernat 50, 50 S; Softenol 3819; Sotex 3CW; Standamul 318, CTA, CTV, G; Stygene Series, R-2, R-10, R-50, R-70, R-85, R-100, R-115, R-150, R-175; Sustrelle; Trylox 5900, CO-5; Unimate IPM, IPP, IPPM; Wickenol 136, 550; Witconol APEB; Zeofree 80, 153; Zeolex 7, 7A, 23A; Zeosyl 100, 110SD, 200

CATALYSTS

AB; Advastab LS-202, T-52N Conc.; AEPD; Al-0104 T 3/16″, Al-0109 P, Al-1401 P(MS), Al-1404 T 3/16″, Al-1602 T 1/8″, Al-1605 P, Al-1609 P, Al-3438 T 1/8″, Al-3916 P, Al-3945 E 1/16″, Al-3970 P, Al-3980 T 5/32″, Al-3996 R 3.5 X 1.5 mm, Al-4028 T 3/16″, Al-4126 E 1/16″; Aliquat 336-PTC; Amerzine 35; AMG 200 L; AMP, AMP-95; AMPD; Armeen N-CMD; Arosurf PT 64, , PT 71, PT Bu3, PT Bu4; Arosurf PT Series, PT 20, PT 40, PT 41, PT 50, PT 61, PT 62; Co-0127 T 3/16″; Co-0138 E 1/8″; Co-0164 T 3/16″; Co-0301 E 3/16″; CoMo-0402 T 1/8″; CoMo-0603 T 1/8″; Cr-0211 T 5/32″; Cu-0203 T 1/8″; Cu-0211 P; Cu-0223 P; Cu-0233 T 1/8″; Cu-0803 T 1/8″; Cu-1106 P; Cu-1107 T 1/8″; Cu-1117 T 1/8″; Cu-1129 P; Cu-132 T 1/8″; Cu-1413 P; Cu-1422 T 1/8″; Cu-1800 P; Cu-1803 P; Cu-1808 T 1/8″; Cu-1809 T 1/8″; Cu-1920 P; Cu-2501 G 4-10; Cu-3829 P; Dabco 33-LV, Crystalline, DC-2, T-1, T-5, T-12, T-120, T-125, T-131; Desmorapid LA, PP; Di-Cup 40C, 40KE, R, T; DMAMP, DMAMP-80; Duolite ARC-9351; Eltesol 4009, 4018, CA 65, CA 96, PSA 65, TA, TA 65, TA 96, TA/E, TA/F, TA/H, TA/K, XA, XA65, XA90, XA/M65; Empigen 5089, AB, AH, AM, AY, BAC50, BAC50/BP, BAC80, CHB; FC-24, FC-520; Fomrez C-2, C-4, SUL-3, SUL-4, UL-2, UL-6, UL-8, UL-22, UL-28, UL-29, UL-32; Forbest 780; Harshaw Antimony Oxide KR; Hi-Point 90, PD-1; HT-400 E 1/8″, HT-500 E 1/8″; Hyamine 1622

Kemamine T-1902D, T-6501, T-6502D, T-9701, T-9702D, T-9742D, T-9902D, T-9972D; Ken-React Series,

CATALYSTS (CONT'D.)

7 (KR 7), 9S (KR 9S), 12 (KR 12), 26S (KR 26S), 33DS, (KR 33DS), 38S (KR 38S), 39DS (KR 39DS), 41B (KR 41B), 44 (KR 44), 46B (KR 46B), 55 (KR 55), 133DS (KR 133DS), 134S (KR 134S), 138D (KR 138D), 138S (KR 138S), 158 (KR 158D), 158FS (KR 158FS), 212 (KR 212), 238A (KR 238A), 238J (KR 238J), 238M (KR 238M), 238S (KR 238S), 238T (KR 238T), 262A (KR 262A), 262ES (KR 262ES), OPP2 (KR OPP2), OPPR (KR OPPR), TTS (KR TTS); Knightset M-4; Kosmos 10, 15, 16, 19, 21, 23, 24, 25, 29, 64; Lekutherm X 50; Lilamin 308 D, 310 D, 312 D, 314 D, 316 D, 342 D, 343, 345 D, 363, 364, 367 D, 368, 369, 372 D, 381, 382, 383, 540, 560, 570, 570 DO, 572; LZ 01, 09, 12, 38, 44, 97; Marchon C21; Micro-Cel E; Mo-1907 T 3/16´´; N-Butyl Acid Phosphate; Na-0101 T 1/8´´; Ni-0104 P; Ni-0104 T 1/8´´; Ni-0301 T 1/8´´; Ni-0707 T 1/8´´; Ni-0901 S 1/2´´; Ni-0104 T 3/16´´; Ni-2002 C 1´´; Ni-3210 T 3/16´´; Ni-3210 T 3/16´´; Ni-3250 T 3/16´´; Ni-3266 E 1/16´´; Ni-4301 E 1/16´´; Ni-5124 3/16´´; Ni-5132 P; Ni-5333 T 3/16´´; Niax Catalyst C-224; Nissan Tertiary Amine AB, ABT, BB, FB, MB, PB; Nyacol; Nysel (Ni-3201 F); Nysel CN-14 (Ni-3611 L), HK-4, HK-12 (Ni-5708 L); Nysel SP-7 (Ni-5169 F); P-0620 T 1/8´´; Pd-0803 E 1/16´´; Pennad 0150; Polycat 610/50, DBU, SA-1, SA-102; Quickset 90; Reworyl B 70, C 65, T 65, X 65; Sequestrene NAFe 13% Fe; Sprayset MEKP; Sulframin CSA, TX; Superox 702, 732, 744; Sweetzyme Type A, Type Q; Synprolam 35DM, 35X2, 35X5, 35X10, 35X15, 35X20, 35X25, 35X35, 35X50; Tegoamin 33, 100, BDE, CPU 33, DMEA, EPS, PDD 60, PMD, PTA, SMP; Tego Stannous Oxalate; Texacat DD, DM-70, DMDEE, DME, DMP, DPA, DPA-50, M-75, MM-70, NEM, NMM, TD-20, TD-33, TD-33A, ZF-10, ZF-22, ZF-24, ZF-26, ZF-51, ZF-52, ZF-53, ZF-54, ZR-50, ZR-70; Thermact 1; Thermacure; T-Hydro Sol'n.; Ti-0720 T 1/8´´; Tris Amino; Troykyd EX-300; Tyzor AA, DC, TBT, TE, TOT, TPT; USP-245; V-0701 T 1/8´´; Vul-Cup 40KE, R; Witco TX Acid; Witconate TX Acid; Zn-0312 T 1/4´´, Zn-0401 E 3/16´´, Zn-0602 T-1/8´´

CHAIN EXTENDERS

Dabco BDO; Unilink 4100, 4102, 4200, 8100, 8130; XCE-89

CHELATING AGENTS/SEQUESTRANTS

AA; Alkaquest EDTA; Aquamolin Brands; Bayhibit; Bayhibit-AM; Carsonam C, C-SF, C-SPCL, DC-70%-SF, DC-SF, DC, L; Cheelox B-13, BF-12, BF-78, DTPA-14, FE-12, HE-24, NTA-14, NTA-Na3; Chel DM-41, DTPA, DTPA-41; Chelon 80, 100, 120; Citrosol 50, 50E, 50T, 50W; Cobratec 99, TT-50-S; Complexion CA-10; Conco SEQ; D FE, DZ, E 39, H 40, H 84, K 3, MG 2, NA, NA 2, NAFE, Z; Dehyquart C, C Crystals; Dissolvine A 40, A 92, A Z, AM 2, AM 3, AM 4, AMFE, CA, D 40, D 88; Duolite CS-346; Empiphos STP, STP/1L, STP/6L, STP/6N, STP/AD, STP/D, STP/DMST, STP/E, STP/F, STP/L, STP/L16, STP/M, STP/M16, STP/MST, STP/N; Evanacid 3CS; Fostex AMP, E, P, U;chelat Gantrez S-95, S-97; Gemtex 18, 19, 162; Geropon TA/K; Good-rite K-722, K-732, K-752, K-7058; Hamp-Ene 100, 220, Acid, Diammonium EDTA, Fe-62, Na_2, Na_3, Na_3T, Na_4, NaFe Purified Grade, OH-1, OH Powd, Photoiron; Hamp-Ex 80, Acid, M; Hamp-Ol 120, Acid, Crystals; Hampshire DEG, EDG, NTA 150, NTA Acid, NTA Na_3 Crystals; Iberquestrene; Interstab CH-55, CH-55R; Jorquest 100; Kalex; Kalex 100, Acids, Conc., IR, Liq., OH, Penta, Powd. FC, Regular; Kelate CDS; Kelene 77; Kelex 100; Kelig 32; Kemplex 100; Keromet MD 60, MD 80, MD 100; Lankromark; Lignosol HCX; Mafo CAB; Maracarb; Maracell XC, XE; Marasperse, Marasperse N-22; Mark 1178B; Milco-150; Miranol CM Conc., CM-SF Conc., CS Conc., HM Conc., HS Conc., OM-SF Conc., OS, S2M Conc., S2M-SF Conc.; Naugard NBC; NTA; Orzan LS, LS-50; Pentaquest Extra; Perma Kleer 80, 95, 98, 100, 120, 354, CP Grade, DI Crystals, EDTA Acid, NTA 150, NTA Na_3, Tri Crystals; Petroscale 11, 101; Plex HT, PC; Plexene D; Plexene Extra Conc.; PMP Gluconic Acid 50% Tech; PMP Liquid Gluconate 60 Tech; PMP Sodium Gluconate Tech; Quadrol; Questal DI, FEC; Questal Special; Questex 4H, 4SW, 4SW Crystals; Questric Acid; Reax 15B, 88B, 100M, 100M (Liquid), Rewopol CHT 12; SEQ 80, 100, 120, NT-15; Seqlene 190, 270, 540, ES-40, ES-50, 400; Sequestrene 30A, 220, AA, Diammonium, NA2, NA2Ca, NA2Cu, NA3, NA4, NAFe 13% Fe, NH4Fe, Tetraammonium; Sodium Omadine, 40% Aqueous Solution, Powder; Solon; Sterling EDTA; Terg-A-Zyme; Trilon A-92, AS, B-FA, B, BD, BS, C, CS, D; Turpinal 4 NL, SL; Tyzor AA, DC, TBT, TE, TOT, TPT; Vanate Acid, PSPa, TS, TSD, TSHE, TST; Vanstay 8210, SC, SD, SE, SF, SG, SH; Versene, Versene 100, 220 Crystals, Acid, AG 100, CA, Diammonium, NA, Tetraammonium EDTA; Versenex 80; Versenol 120; Weston 439; ZB-100, ZB-130, ZB-300

CHLORINATING AGENTS

Tego EL-HA-CE; Tegodont

COAGULATING AGENTS

Amerfloc Plus 5270, 5495; Carsoquat CTM-29, CTM-429; Coagulant CHA, WS; Daxad CP-2; Densol 284;

Hercofloc Flocculant Polymers; M-Quat 32, 2475; Modicol L; Nissan Cation AB, ABT-350, 500, AR-4, BB; Sol-U-Tein EA

CONDITIONERS/SOFTENERS

Abil 10-100000, AV 20-100, AV 8853, B 8, B 8839, B 8842, B 8843, B 8851, B 8852, B 8873, B 9905, B 9907, B 9908, B 9909, B 9950, K 4, S201, S255; Abil-Quat 3270, 3272; Ablumine 280; Ablusoft A, ND, SN, SNC; Accosoft 440-75, 540, 550, 550-90 HHV, 550L-90, 620, 620-90, 707, 748, 750, 808, 808HT-75, 870, A-155, M 1154; Acetamin 24; Acryloft, Conc.; Admul CSL, Adogen 432, 432 CG, 432-ET, 442-100P; Aerosol C-61; Ahcovel Base 500, Base 700, Base N-15, N-64; Akrofax 9844 Black, 9851 Light, 9906; Akypoquat 40, 129; Alacsan 7LUF; Alcolec 4135; Alconate SBG-280; Aldo PMS; Aldosperse 40/60, ML 230, O 20, TS 20, TS 40; Algonina AC; Alkaminox C-2, C-5, HT-12, HT-25, HT-30, T-2; Alkamox ODM; Alkamuls 200-DL, 200-DO, 200-DS, 200-MO, 200-MS, 400-DS, 600-DS, GMS-45, PSTS-20, SMO; Alkaquat 1068, C, DAET, DAPT, DHTS, DMB-ST, 25%, O, S, ST, T; Alkasurf CO-20, CO-40M, O-5, S-5, S-8, S65-8, S65-40; Alkateric BC, CAB, STB, TB; Alkazine C; Aloe-Moist, /A; Alphadim 90AB; Alubrasoft SJ-2, SJ-Wax, Super 15, Super 25, Super 100; Amerchol BL, C, L-99, L-101, L-500, RC; Amerlate P, W; Amfotex FV-16, FV-28; Amidan; Amiet 102, 105, 110, 115, 202, 205, 210, 215, 302, 305, 310, 315, 402, 405, 410, 415, 502, 510, 515; Amino Gluten MG; Amino Silk; Aminodermin CLR; Aminofoam C; Aminol CA-2, N-1918; Aminoxid WS 35; Ammonyx 4, 4B, 4-IPA, 485, 490, 4002, 4080, CA-Special, CDO, CETAC, CETAC-30, CO, KP, LO, SO, SQ, TDO; Amphomer; Amphoterge S; Amphoteric C; Amyx A-25S; Aqualose L75, L75/50; Aquatreat AR4, AR6, AR231, AR232; Arlacel 989; Armeen SZ, Z; Armoflo 66; Armosoft L, WA-201, WA-203, WA-204; Arnica Oil CLR; Arosurf TA-100, TA-101; Arquad 2C-75, 2S-75, 2HT, 2HT-75, 12-33, 12-50, 16-29, 16-50, 18-50, 88, DM18B-90, DMMCB-50, S-50, T-2C-50, T-50; Aston AP Conc.; Atlasoft-C; Atmul 700H; Avitex NA, R; Avocado Oil CLR

Babinar-715, -801C; Barre Common Degras; Berol 808, 809; Berol Fintex 10, 42, 572, 577; Bilt-Cote H-1, H-5, S-1, S-5; Bina QAT-43; Biocare Polymer HA-24; Biosulphur Fluid, Powder; Bovinol 30; Britol; Bromat; Cachalot C-51, O-15, S-54; Calendula Oil CLR; Capmul EMG; Carbopol 1342; Carrot Oil CLR; Carsamine Oxide LM; Carsonam BCW; Carsoquat CT 429, CTM-29, CTM-429, SDQ 25, SDQ 85; Carsosoft T-75, T-90; Cartaretin F-4, F-8, F-23; Catigene CA 30, CLP 50, CS 40, CT 30, CT 30/70, LT 45, ST 30, ST 30/70; Cation DS, SF-10, Softener X; Celite 209, 400; Celquat; Centrol Series; Ceranine AT, AW, HC, HCA, HCA Gran., HCL, HCS, L, PAP, PNA, PNL, PNS, PNS Gran., RW, RWN, SG, SIT, SWP, VR, VS, WN; Ceraphyl 60, 65, 85, GA, ICA, IPL; Charlab Leveler AT Special; Charlab Leveler DSL; Cheelox BF-12, BF-13; Chemax E-200 MS, E-400 MS, HCO-5, HCO-16, HCO-2, PEG 600 DO, SBO, SCO; Chemazine 18, C, O, TO; Chemeen T-2, T-5, T-10, T-15, T-20; Chemical 39 Base; Chemical Base 6532; Chemprene R-10, R-25, R-50, R-70, R-100, R-115; Chimin AX; Chimipal DS 6000; Cirrasol LC-HK, LC-PQ; Cithrol 2ML, 2MO, 4ML, 4MO, 6ML, 6MO, 10ML, 10MO, 40MO, 60ML, 60MO; Clearcol; Collagen Hydrolyzate Cosmetic 55, N-55, SD; Collasol; Colsol; Comperlan COD, ID, KD; Compound 9264B; Conco Sulfate EP; Concosoft 90; Consoft CP-50, CPN; Copolymer 845, 937, 958; Corona Lanolin; Cosmopon BL; CPC

Crafol AP-11; Cralane KR-13, KR-14, LR-10, LR-11; Crapol AU-23, AU-31, FU-25; Crestalans; Croda Absorption Base; Crodacel QL, QM, QS; Crodafos N3 Acid, N3 Neutral, N5 Acid, N10 Acid, N10 Neutral, O2 Acid, O5 Acid, O10 Acid, SG, T2 Acid, T5 Acid, T10 Acid; Crodalan AWS, LA; Croderol G7000; Crodyne BY19; Crolastin, 10 Powder, 30 Powder; Cromoist HYA; Cronectin H; Cropepsol; Cropeptone; Croquat K, L, M, S, Soya, WKP; Crosilk 10,000, Liquid, Powder; Crosilkquat; Crotein A, AD Anhyd., AD, ADX, ASC, ASK, BTA, C, CAA, CAA/SF, HKP Powd., HKP/SF, K, O, Q, SPA, SPA55, SPC, SPO, WKP; Cutless 50; Cyclochem GTIS; Cyclomide CODI, DIN 295/S, DS280/2, SODI; Cyclomox CO, SO; Cyclosheen 202; Cycloteric BET C-30, BET-C41, BET I-30, BET-L31, BET O-30, BET-T2 40, BET-W, CAPA, SLIP; Cycloton 7LUF, 75C, D256B/99, D261C/70, D261S/90, M214B/99, M214C/99, M242C/29; Dastar; Daxad 41; Dehyquart A, C Crystals, DAM, E, LT, SP, STC-25; Dehyton K; Densulf TA-75; Dermalcare EGMS/SE, GMS, GMS-165, GMS/SE, GTIS, LVL, MM/M, MST, PGMS, SDG, SS; Dermatein GSL; DeSomeen TA-2, TA-5, TA-15, TA-20; DeSonic 6C, 30C, 36C, 40C, 54C, SMO, SMT, TA-2, TA-15, TA-25CWS; DeSophos 4 CP, 7 OPNa; DeSotan SMT; Dilasoft RS, RW, TF; Dioxitol-High Gravity; Dioxitol-Low Gravity; Dist Lipolan; Double Soft; Dousoft BK 5078; Dow Corning Q2-7224; Dresinate TX, X, XX; Drewpone 60, 65, 80; Drewsoft 100; Duolite C-20-C, C-433; Duponal 80; Durfax EOM, EOM K; Dusoran MD; Dusoran NG, NO; Elacid CLR; Elastosol; Emcol 150, 1655, 3780, 4072, 6613, CC-9, CC-42, CC-55, CC-57, E-607L, E-607S, ISML, NA-30, Q; EmCon W; Emerest 2400, 2423, 2624, 2636, 2652, 2662, 2675, 11723; Emery 912, 1650, 1656, 1660, 1732, 1740, 1747, 5430, 6657, 6748; Emkalon AVR, CL, CNW; Empicol 0216, XDB; Empigen 5083, BCP/25, CM, CMC, CSC, FRB75S, FRC75S, FRC90S, FRG75S, FRH75S, OH25, OS/A; Empiphos STP, STP/1L, STP/6L, STP/6N, STP/AD, STP/D, STP/DMST, STP/E, STP/F, STP/L, STP/L16, STP/M, STP/M16, STP/MST, STP/N; Emplex; Emsorb 2500, 2503, 2507, 6907, 6908; Emulgade CRC; Emulsifier K-700; Epolene E-10, E-12; Espesilor AC-43; Etocas 10, 30, 35, 40, 60, 100; Eureka 400, E-2

Fancol ALA, EOA, LA, LAO, OA 50, OA 70, OA 80, OA 90, OA 95; Fancor Lanwax; Farmin 2C, D86; Finquat

CONDITIONERS/SOFTENERS (CONT'D.)

CT; Finsoft HCM-100; Fluilan; Foamole A, B; Fonoline White, Yellow; Fortex; Gafac GB-520; Gafquat 734, 755, 755N; Gafsoft 325; Genamin KDM, KDM-F, KS 5; Generol 122E5; Glucam E-10, E-20, E-20 Distearate, P-10, P-20, P-20 Distearate; Glucate DO; Golden Dawn Grade 1,2, Superfine; Golden Fleece DF, P-80, P-95, RA; Good-rite Polyacrylates; Grocor 5500; Grotex 1120; H.P.X.; Hubersorb 600; Hydagen P; Hydrolactin 2500; Hydromond; Hydrosoy 2000; Hydrotriticum; Hydroxylan; Hallcomid M-6; Hartolite; Hartolon 368, AL, NA; Hartosoft 75, 171, ORF, TAF; Hercules Ester Gum 10D; Hetamine 5 L 25; Hetoxamate FA-5, FA-20, LA-5, LA-9, MO-2, MO-5, MO-9, MO-15, SA-5, SA-7, SA-9, SA-13, SA-23, SA-35, SA-40, SA-90; Hetoxamine C-2, C-5, C-15, O-2, O-5, O-15, S-2, S-5, S-15, ST-2, ST-5, ST-15, ST-50, T-2, T-5, T-15, T-20; Hi-Sil 210, 215, 233, 250, 260, 262; Hydrosoy 2000/SF; Hysoft 975, LC Conc.; Ice#2, #12, #81; Imerol DU; Incrocas 10, 30, 40, 60, 100; Incromate ALL, AVL, BAL, Mink L, OLL, PPI, SDL, SEL, WGL; Incromectant AMEA-70, AMEA-100, LAMEA; Incromide ALD, LA, OPD, SM; Incromine BB, ISM, L-40, OPB, OPM, Oxide AL, Oxide AV, Oxide C, Oxide L, Oxide M, Oxide S, Oxide SE, Oxide WG; Incronam B-40, I-30, ISM-30, Mink 30, OP-30, P-30, SE-30; Incroquat 100, 248, AL-85, AV-85, BA-85, B65C, BDQ-25P, Behenyl BDQ/P, Behenyl TMC/P, Behenyl TMS, BTQ-25, CR Conc., CTC-25, CTC-30, DBM-90, I-85, Mink-85, O-50, OL-85, S-75 CG, S-75 CG, S-85, SDQ-25, SE-85, WG-85; Incrosoft 100, 248, O-90, T-75, T-90, T-90HV; Isocreme; Isopropylan 3

Jet Quat 2C-75, 2HT-75, C-50, S-2C-50, S-50, T-2C-50, T-50; Jordamine DAPI, DAPL, DMCAPA, DAPSA, S-13, S-13 Lactate, SHCFA; Jordamox CAPA, CDA, CDA-40, LDA, ODA, SDA; Jortaine TM; Jordaquat 522, 1033, Dimer 12, Dimer 16, Dimer 18, Dimer 22, JO-50, JS-25; Jorphox KCAO, KTAO; Kalex; Kara-Lube ECM; Katapol PN-730, PN-810; Katemul IB-70, IG-70, IGU-70; Kemamine BQ-2802C, BQ-9702C, BQ-9742C, Q-1302C, Q-1902C, Q-2802C, Q-6502C, Q-6503 B, Q-9702C; Kerasol; Kera-Tein 1000; Kessco BSC, IBP, ICS, IPM, IPP; Lamequat L; Lamigen ES-30, ES-60, ES-100, ES-180, ET-20, ET-70, ET-90, ET-180; Lan-Aqua-Sol, Lan-Aqua-Sol Hydrophillic 50, 100, Lan-Aqua-Sol Hydrophillic-Plus 50, 100, Lan-Aqua-Sol Super-Hydrophillic 50, 100, Lan-Aqua-Sol xtra-Hydrophillic 50, 100; Lanethyl; Laneto 40, 50, 60; Lanex; Lanexol, Lanexol AWS; Lanocerin; Lanogene; Lanoquat 1756, 1757; Lanotein AWS 30; Lanotex 730; Lantrol 1673, 1674, AWS 1692; Leoguard G, GP; Lexamine S-13 Lactate; Lexate CL-60, CRC, IL, PX; Lexein A220, CP-125, MP, SLK, X250, X300, X350; Lexol DIA, GT 855, GT 865, PG 865; Lexolube 2T-237; Lexquat AMG-M, AMG-O, AMG-WC; Lignosol SFX, WT; Linamine HC-2; Lipamide MEAA; Lipocol, Lipocol C-2, C-10, C-20, L-1, L-4, L-12, L-23, M-4, O-2, O-10, O-20, S-2, S-10, S-20, SC-4, SC-10, SC-15, SC-20; Lipolan, Lipolan 31, R, S; Liponic 70-NC, 76-NC, EG-1, EG-7; Lipoquat R; Lipovol A, ALM, ALM-S, CP, G, J, SAF, SO, SUN; Liquester; Liquid Base; Lonzaine 12C, C, CO, CS; Luviquat FC 370, FC 550, FC 905

Mackalene 116, 117, 216, 316, 326, 416, 426, 716; Mackam BA, HV, ISA, OB, RA, TM, WGB, LME, R, S; Mackamine BAO, IAO, WGO; Mackanate DG30, IM, WG; Mackernium SDC-25, SDC-50, SDC-85; Mackine 101, 201, 301, 321, 401, 421, 501, 601, 701, 801, 901; Mackpro NLP; Mafo CAB; Mapeg 600 DL, 600 DOT, 600 ML, 600 MOT, 6000 MS, PGDS, PGMS; Maprolyte 101, LX; May-Tein C, CT, SK; Mekon White; Merquat 100, 280, S; Micro-Cel A, B, E; Miramine DD, IC, TA-26; Miranol DM, DM Conc. 45, Ester PO-LM4, H2M Conc., ISM; Mirapol 9, 95, 175, A-15, AD-1, AZ-1, WT; Mirataine COB, HTS, ODMB-35%; Modulan; Mollisan 4NY, 379 Conc., Mollisan Conc.; Monamid 759; Monaquat ISIES, P-TC, P-TD, P-TL, P-TS, P-TZ, TG, 1202, ISA-35%; Monazoline Series, C, CY, O, T; Monopol Oil 75; Myvatex Mighty Soft; Neoprotex KM; Nikkol BPS-5, BPS-10, BPS-15, -20, -25, -30, BWA-5, BWA-20, -40; Ninox CA, L, M; Niox EO-33, KH Series; Nissan Amine AB, ABT, ABT_2, BB, FB, MB, OB, PB, SB, VB; Nissan Anon BF, BL, LG; Nissan Cation AB, ABT-350, 500, AR-4, BB; F_2-10R, -20R, -40E, -50, FB, FB-500, L-207, M_2-100; Nissan Cation PB-40, -300, S_2-100; Nissan Nonion LP-20R, LP-20RS, MP-30R, OP-80R, OP-83RAT, OP-85R, PP-40R, SP-60R; Noiox EO-33; Nonex S3E; Nopcogen 14-S; Noramium M2C, M2SH; Nyloset Finish; Obazoline 662Y, CS-65; OHlan; Onyxide 3300; Onyxol 42

Patco 3; Pationic 138C, NMF; Peem 122 Conc.; Peganate CO 5, 16, 25, 30, 36, 40, 200, 200 50%, COH 25, COH 200, 200 50%, MA 8, 15, 15 80%, ML 5, 9, 14, MO 6, 9, 14, MS 8, 14, 23, TO 9, 16; Penreco Amber, Regent; Peptein 2000; Petro S; Petrolite C-700, C-1035; Piccolastic A5, A25, A50, A75; Piccomer XX40, XX100; Picconol A200, A201; Piccovar AP10, AP25; Plasthall 503; Plasthall DODD; Pluracol E-200, E-300, E-400, E-400 NF, E-600, E-600 NF, E-1000, E-1450, E-1450 NF, E-1500, E-2000 E-4000, E-4000 NF, E-4500, E-6000, E-8000, E-8000 NF; Pluriol E 200, E 300, E 400, E 600, E 1500, E 4000, E 6000, E 9000; Pogol 200, 1570; Polylan; Polyqyart H, H 81, H-7102; Polysoft CA; Prapagen WK brands; Proaid 9814, 9826; Prodew 100, 200; Product BCO; Protamine-45; Protopet Alba; Protopet White 2L, 3C; Protopet Yellow 1E, 2A; Protowax ES-100; Pyroter GPI-25; Quadrilan OD-75; Quartamin 24P, 86P, 86P Conc.; Querton 14 BR., 14 BR-40,16 CL29, 14 CL50, 442, 442-11, 442-82, 442S, 442-Sx, 470; Radiaquat 6442, 6443, 6444, 6462, 6470, 6471, Reduce-150, Resyn 28-1310, 2802930, Reticusol; Rewocid UTM 185; Rewoderm 5, SB/LAN 5; Rewominox B 204, S 300; Rewominoxid B 204; Rewoquat CPEM, CR 3099, RTM 50, W 222 LM, W 3690, W 3690/PG, W 7500, W 7500 H, W 7500/PG, W 9000, W 9000/PG, WKH 80, WP 100

Rhenosin 140, 260; Ritaderm; Rudol; Rychem 17, D941, FCL; Rycofax 618, 3115; Ryoto Sugar Ester LWA-

1540, OWA-1570, P-1570, -1670, S-170, -270, -370, -570, -770, -970, -1170, -1570, -1670; Salfax 77; Sandoz Amine Oxide XA-C; Schercamox C-AA, CMA, DMS, T-12; Schercemol CP, CS, ML; Schercodine B, M, O, P, S, T; Schercolube 707; Schercomid AME, AME-70, ID, IMI, M, ODA, OMI, SI, SI-M, SLE, SLS, SO-A, TO-2; Schercophos L; Schercopol CMS-Na, OMS-Na; Schercoquat CAS, DAB, DAS, FOAS, IAS, IAS-LC, IB, IEP, IIB, IIS, IIS-R, IIS-RD, MKAS, ROAB, ROAS, ROEP, SAB, SAS, SOAB, SOAS; Scherco Softener #1, #366; Schercotaine CAB, IAB, MKAB, OAB, PAB; Schercoteric STS; Schercozoline I, S; Semtol Series; Servamine KOY 4387B; Servo Brillant Olie B AZ 75; SF 96, 1066; Silk Protein Complex; Silsmooth E, G, SR/VEL; Simchin; Sipex EC-111; Siponic E-2, E-3, E-5, E-7, E-10, E-15; Skyllex; SM 2112, 2163; S-Maz 20, 40, 60, 60 K, 60 KHM, 65, 65 K, 67, 80, 80 K, 85, 85 K, 90, 93 R, 95; Sochamine 2662, SG 8, 100; Sofbon C-1-S; Sofbon AF Series; Sofnon 105G, GF-2, -5; Sof-Plus; Softal 300, A-3, MR-30; Softener 64; Softener #5039, #FS-18-2; Softex K-304, K-363; Sokalan CP 10S, PA 15, PA 20, PA 20 PN, PA 25PN, PA 30, PA 40, PA 40 Powd.; Solamine AC 20, BO 5, HY; Solan, 50, E 50, X; Solanos AD Supra; Solar NP; Solulan 5, 16, 25, 75, 97, 98, C-24, L-575, PB-2, PB-5, PB-10, PB-20; Solwax L20, LG 35; Sopralub ACR 265, 275; Soprofor NO/20, /30, /100; Soromine AL, AT, So/San 30M, 66M; Span 20, 40, 65, 85; Stafoam, DF-1, DF-4, DL; Standamid ID, LD 80/20; Standamox 01, LMW-30; Standamul Conc. 1002, O-5, O-10, O-20; Standapol AB-45, BAW, BC-35,OLB-30, OLB-50; Starplex 90; Staybelite Ester 5; Stedbac

Stepan Cationic 10; Stepanhold R-1; Stepanquat HC 80; Stepantex DW 85, Q 90B, Q 185, SR 85, VR 80 G, VR 85, VG 90, VS 90, VP 85, VRH 80; Steramine CD, FPA 197, TV; Sterling Snow White 750; Stygene R-2; Sufatol LX/B; Sunaptol CA 25, 120, 350, 400, 750, S 60, 80, 230; Sunsoflon CK, K-2, MT-100, PXP-705, PXP-2000; Super-Pro #5A; Super Refined Apricot Kernel Oil NF, Orange Roughy Oil, Wheat Germ Oil; Super Solan Flaked; Super Solangel 25; Supratol VF; Surfax WO; Synoquart P 50; Synprolam 35DMBQC, 35MX1QC, FS; Tandem 5K, 8, 9, 11H, 552; Tauranol I-78, I-78 Flakes, I-78-3; Tegamine 18, P-7; Tequat BC; Texnol R 5; Textamine O-5; Textamine Oxide TA; Trydet 2636, 2644, 2670, 2671, LA-5, OA-5, OA-7, SA-5, SA-7, SA-8, SA-50/30; Trylox 5907, 5909, 5921, CO-36, CO-40; Trymeen 6602, 6603, 6657, CAM-15; Turgum S; Twitchell 6805 Oil; Unamine S; Uni-Malt; Unihib 305; Union Carbide LE-45; Value MS-1; Vanax PY; Varion CADG, CDG, SDG; Variquat E228; Varisoft 110, 133H, 135H, 136, 190, 190-100P, 222, 222-LM-90%, 222-LT-90, 238, 242, 369-ON-90%, 445, 472-80%, 475, 920, 3690, 6112, CA-50, CRC, DHT, E228, E-290, SDC, SDC-85S, SDC-100P, TSC; Varonic T202, T202 SR, T205, T205LC, T215, T215LC; Velvetex AB-45, BC-35, OLB-50; Verv; White Swan; Witamide 61, 70, 82, 823/10, 5130, 5133, 5195, LDTS, M-3, MAS, MM, NA, NA1, STD-HP; Witcamine RAD-0500, RAD-0515, RAD-1100, RAD-1110; Witconol F26-46, RHP; Zeolex 7, 23A; Zeothix 175; Zontes PD-101

CORROSION INHIBITORS

AA; AB; Acid Thickener; Acihib A9; Acrysol LMW Series; Actan SP; Acto 450, 500, 630, 632, 636, 639; Actrabase PS-470; Actracor 129, 401, 800, 856, 800, 1987, M, T; Actrafos 110, 110A, SN-315, SP-407; Actralube 153, Syn-153; Actramide 202, 5264; Actrasol SS; ADM; Adogen 115 (D), 140 (D), 142 (D), 151, 160, 170, 172, 185, 188, 240, 342-D, 345-D, 367-D, 551, 560, 570-DO, 570-S, 572, 588; AEPD; Aerosol OT-75%, OT-B, TR-70; Agent #5032; Agent 765; Agerite White; Akolidine 10,11,12; Akrolease E-9410, E-9491; Akypogene KTS; Alcophor AC; Alkamide 2204, SDO; Alkaminox C-2, C-5, SO-5, T-5; Alkamox C2-O; Alkamuls GMO-45, PEMO, SML, SMO; Alkaterge-T, T-IV; Alkazine C, O, ST, TO; Alox 111, 164, 165, 318F, 319F, 350, 436A, 502, 502A, 575, 600, 604, 606, 606-55, 606-70, 707, 904, 940A, 1680, 1727, 1727D, 1843, 1846, 2000, 2019A, 2028, 2028L, 2116, 2116A, 2162, 2200, 2201, 2201E, 2211Y, 2213A, 2213A, 2248C, 2248C-50, 2263, 2276A, 2278, 2283, 2284A, 2284AB, 2287, 2289, 2290, 2290A, 2291, 2292; Amerzine 35; Amfotex FV-15; Amine BG, CS-1135, HBG, KK, OL, PMT; Aminox Series; AMP, AMP-95; AMPD; Amphoteric 400, C, N; Ancor 120; Antara LB-400, LK-500; Aqualox 225-100, 225A-100, 232, 2268, 2295; Arcor Series; Arkomon A Conc., SO; Armac 18D, C, HT, OD, T; Armeen 2C, 2HT, 12D, 16D, 18, 18D, C, CD, HT, HTD, O, OD, SD, SZ, TD, Z; Armoflo 66; Armogard F-201, G-101; Armohib 28, 31, F-401; Arneel 18 D, C, OD, T; Arnnate Series; Arnox 500, 515, 1100, 1110; Arquad 2C-75, 2HT-75, 2S-75, 12-33, 12-50, 16-29, 16-50, 18-50, DMMCB-50, S-2C-50, S-50, T-2C-50, T-50; Arquest 709; Astramyl SB; Atpet 80, 100, 200

Barium Petronate 50-S Neutral; Barre Common Degras; Base 11-T; Base 3059 E; Bayhibit; Bayhibit-AM; Baypren Latex 4 R; Belclene 500; Belcor 575; Berol 516, 521, 594, 786, Nonylphenol; Bohrmittel Hoechst; Bubble Breaker 1840X; Butoxyne 497; Butynediol; CA-57; Cachalot O-15; Calamide C; Calcium Petronate; Carsoquat 621, 621 80%; Catisol AO 100; Cavco Mod A, APG; Centrolene Series; Chemazine 18, C, O, TO; Chemeen DT-3, DT-15, DT-30; Chemquat 12-33, 12-50, 16-50; Clink PR-46; Cobratec 99, CBT, DBT-50, DBT-100, PT, TT-35-A, TT-50-A, TT-50-S; Colorol Rust Binder; Corfree M1, M2, M3; Corrosion Inhibitor 500, 600; CPH-205-NX; CPH-353-SE; CPH-362-N; Crafol AP-10, AP-51; Crapol AV-10, AV-11; Crill 1, 2, 50; Crodafos Series; Crodafos 25 D2 Acid, 25 D5 Acid, 25 D10 Acid, 25D Series, CS Series, ID Series, N3 Acid, N3 Neutral, N5 Acid, N10 Acid, N10 Neutral, O Series, SG, T2 Acid, T5 Acid, T10 Acid; Crodasinic LS30, LS35, OS35; Crodinhib RT70, RT70S; Crystamet 1020, 2040, 3080;

CORROSION INHIBITORS (CONT'D.)

Cyclomide AC 28, DO280; Cycloteric MV-SF; Darvan L; Dehyquart C Crystals, SP; Dequest 2000, 2006, 2010, 2051, 2054, 2060, 2066; Deriphat 151C; DeSophos DNP, 6 MPNa, 6 NPNa; Diamiet 503, 508, 520, AB, C; Diamin HT, O, S, T; Diamine B11, BG, KKP; Dichan 100; Dinoramac C, O, S; Dinoramox S 3 to 12, SH 3; Dion Cor-Res 6693FR, 6693HV, 6694, 6694HV, 7000; Dion FR 6308, FR 6604T; Dion-Iso 6246, 6314, 6315HV, 6631T, 8101, 8200; Dodicor Brands; Dodigen Brands; D Sodium Silicate; Duomac C, T; Duomeen C, C Special, L-11, L-15, O, S, T, T Special, TDO; Duoquad T-50; Duroxon B-120; Durtan 80; Drymet 59, Fines; Dytek A; Eccoro; Emcol 14, CC-42, CC-55, D 75-13, 75-33; Emphos PS-236, TS-230; Empigen AB, AD, AF, AG, AH, AM, AS, AT, AY; Emulan FM; Emulsogen B2M, E, KW, S, STH; Estabex 2307, 2307 DEOD; Ethoduomeen T/25; Ethoxamine C5, SF11; Fancol OA 50, OA 70, OA 80, OA 90, OA 95; Fancor Lanwax, LFA; Farmin 20, 60, 68, 80, 86, AB, C, HT, O, S, T, R 24H, R 86H; Finazoline Series; Findet Series; Forbest 600, 850; Fosterge LF Acid, RD, W Acid; Fostex T, TX; Gafac LO-529, RK-500; Gantrez AN, AN-119, AN-139, AN-169, AN-179; Hallcomid M-8-10, M-12, M-18; Horse Head Standard Zinc Dust 22, 44, 122, 222, 422, 444; Hostacor 2098, 2125, 2732, BK, BM, BS, DT, KS 1, R, TP 2125; Hostaphat AW, CP; Hybase C-300, M-300, M-400; Hystrene 3675, 3675C, 3675CS, 3675X, 3680, 3695, 5460

Igepal CO-520, CO-630, CO-660, CO-710, CO-720; Indu-Sol 238; Inhibiteur I.C. 59; Inhibitor RT 212; Inipol 002, OT2, S.43; Ionet S-20, S-60 C, S-80, S-85, T-20 C, T-60 C, T-80 C; Jet Amine D-C, D-O, D-T, DE-13, PC, PE-13, PE-810, PE-1215, PHT, P-O, P-S, PT; Jordaphos JB-40, RA-60; Jorphox KCAO, KTAO; Katapol PN-430; Katapone VV-328; Kelig 100; Kemamide B, E, E-180, E-221, O, P-181, S, S-65, S-180, S-221, U, W-20, W-39, W-40, W-40/300, W-45; Kemamine BQ-2802C, BQ-9702C, BQ-9742C, D-150, D-190, D-650, D-970, D-974, D-989, D-999, DD-3680, DP-3680, DP-3695, Q-1302C, Q-1902C, Q-2802C, Q-6502C, Q-6503 B, Q-9702C, Q-9743C, Q-9743CHGW, S-190, S-650, S-970; Korantin BH Liq., BH Solid, CD, LUB, MAT, PA, PAT, SH, SMK, TD; Lanpolamide 5; Lauridit PD; Levasint; LICA 01, 09, 12, 38, 38A, 38J, 44, 97; Lilamin 101 D, 115, 115 D, 140, 140 D, 142, 142 D, 151, 160, 160 D, 163, 163 D, 170, 170 D, 172, 172 D, 308 D, 310 D, 312 D, 314 D, 316 D, 342 D, 343, 345 D, 363, 364, 367 D, 368, 369, 372 D, 381, 382, 383, 540, 560, 570, 570 DO, 572, EO; Lonzest PEG-4DO, SML, SMO, SMP, SMS, STO, STS; Lusynton A; LZ 01, 09, 12, 38, 44, 97, 850, 859

Mackamide CD-8, CD-25, CDM, CDT, CDX, CSA; Mackazoline C, L, O; Mafo 13; Maphos 60, 60A, 66, 76, 76 NA, 91, L-4, L-6, L-10, L-13; Marchon DC 1202; Maxahibit 100; Mazamide 25, 65, 65CZ, 66, 70, C-5, L-5, O-20, SS-10, TO-10; Mazoline OA 4; Mazon RI-3, RI-4A, RI-7B, RI-8B, RI-9, RI-10A, RI-12, RI-13, RI-14, RI-110; MBEO 1.8 Adduct; Metso Beads 2048; Metso Pentabead 20; Miramine CC, CPC, GS, GT, IC, OC, SC, TC, TOC; Miranol CS Conc., JS Conc.; Mirapol WT; Mirataine H2C; Molyhibit 100, 200, 300; Moly-White 101, 212, 331, ZNP; Monacor 39, BE; Monafax 057, 785, 786, 794, 831, H-15, L-10; Monalube 29-78, 780; Monamid 7-100, 7-153 CS, 15-70W, 150-AD, 150-ADD, 150-ADY, 150-CW, 150-DR, 150-IS, 150-LMW-C, 150-LW, 150-LWA, 150-MW, 718, 770, CMA, LIPA, LMA, LMMA, R31-42, S; Monamine Series, 928, 929, AA-100, AC-100. ACO-100, AD-100, ADD-100, ADDS-100, ADS-100, ADY-100, AF-100, ALX-80 SS, ALX-100 S, ARA-100, CD-100, CF-100 M, I-76, LM-100, R8-26, T-100; Monamulse CI, R10-29M; Monaquat ISIES; Monateric 805, 810-A-50, 811, 985A, 1000, CAM-40, CDL, CDS, CDTD, CDX-38 Mod., CEM-38CG, CNa-40, COAB, CyA-50, CyMM-40, LMAB, LMM-30, MCB, TA-35; Monazoline Series, C, CY, IS, O, T; Morlex DEEA, DMEA

Nalzin 2, ZP; Nissan Tertiary Amine AB, ABT, BB, FB, MB, PB; Nitrene 100 SD; Nopcogen 22-O; Noramac C, O, S; Noramax C2 to 15, S2 to 11; Norfox KO, PEW; Norust; Noryl N300; Noxamium; NP-55-300; Nykon; N-butyl Acid Phosphate; Na-Sul 611, 707, 729, AS, BSB, BSN, BSN/W765, BSN/W780, CA-50, DTA; Na-Sul EDS, LP, LS, SS, ZS; Nacap; Neutral Degras; Newcol 3-80, 3-85, 20, 25, 40, 45, 60, 65, 80, 85, 405, 410, 420; Nikkol DDP-2, -4, -6, -8, -10; Nikkol Sarcosinate OH, TDP-2, -4, -6, -8, -10; Ninol AC-201; Nissan Amine DT, DTH; Nissan Cation MA, SA; N Sodium Silicate; Nykon 77; Oncor F-31, M50; OW-1; Oxypruf 6, 12, 20, E, P; Paracol OP; Pegameen 02; Pegameen 030, 30 80%; Pegameen C2, C5, C10, C25, S2, S5, S20, T2, T5, T15, T25; Peganol NP 4, NP 5, NP 9, 10, 12; Pegeste SML, SMO, SMP, SMS; Pennad 0150; Penreco Snow; Pentalyn 225; Petronate 25C, 25H, CR, HMW, S; Petrosul 550, 750, H-50, H-60, H-70, HM-62, HM-70, M-50, M-60, M-70, Neutral Barium, Neutral Calcium; Phosfac 1004, 1006, 1044, 1044FA, 1066, 1066FA, 1068FA; Phosphate Ester 123; Polyfac PCW-181, -182, PN-209; Polyphos; Polyrad; Polystep PN209; POM; Product WRS 1-66; Produkt B 2045, B 3010, B 3032, GS 400, GS 7114/80; Quartamin DCP; Quatrene 7670, C-5-6, CA, CB, CB-50, CB-80, CE, MB-50, MB-80

Radiamac 6148, 6149, 6169, 6179; Radiamine 6140, 6141, 6160, 6161, 6163, 6164, 6170, 6171, 6172, 6173, 6540, 6560, 6570, 6572; Radiasurf 7000, 7125, 7135, 7136, 7137, 7140, 7141, 7145, 7146, 7147, 7151, 7152, 7153, 7372, 7175, 7196, 7201, 7206, 7270, 7400, 7402, 7403, 7404, 7410, 7411, 7412, 7413, 7414, 7417, 7420, 7421, 7422, 7423, 7431, 7432, 7443, 7453, 7454, 7600; Reocor 12, 190; Reomet 39, 42; Rewocid DC 212; Rewocor AC 28, B 2045, B 3010, B 3032, BAC, RA 178, RA 280, RA-B 90, RA-SI, TPAC 100; Rewocoros AC 28; Rewocoros B 2045, B 3010, B 3032, BAC, RA 178, RA 280, RAB 90, RABE, RASI, TPAC 100; Rewomid AC 28, DO 280/S, RE; Rewomine IM-BT, IM-CA; Rewophat E 1027, EAK 8190;

Rewopon IM-CA, IM-OD, JMCA, JMOA; Rodea; Rokon; Rychem G-Gard, G-Hib; Safety-Lube 4000, 6003; Sandocorin 8015, 8132, 8132B, 8160; Sandostab P-EPQ; Sandoz Phosphorester 510; Sarkosyl L, LC, NL-30, O; Scav-Ox; Scav-Ox II; Scav-Ox Plus; Schercomid ODA, SI; Schercozoline O; Secosyl; Sepacorr HT; Serdolamide POF 61 C; Servamine KEO 260, KET 350, KOO 330, KOO 360; SF 1705; Sol Mul 60/C; Solar CI-387, Regular; Sorban AO 1, AST, CO; Stafoam, F; Standamid SOD; Stepan Cationic 10; Stoner No. E922, E923, E944; Strodex PK-90; Sublimed Blue Lead; Sulfonate OA-5; Surfac 18-EHP, P14B, P24M; Surfam D89DA, P5, P10, P12B, P14B, P17B, P24M, P86M, P89MB; Synkad 202, 204, 303, 404, 500; Synprolam 35, 35BQC (50), 35BQC (80), 35DM, 35M, 35N3, 35N3 DA, 35N3 Dioleate, 35N3X3, 35N3X5, 35N3X10, 35N3X15, 35N3X25

TDA-65; T-Det N-14, Tergenol G; Tergenol S Liquid; Tergenol Slurry; Teric 17M2, 17M5, 17M15, 18M2, 18M5, 18M10, 18M20, 18M30, 303, 350, 351B; Texamin AT 1, AT 2, AT 3; Textamine 05, A-5-D, A-3411, A-3417, A-W-5, O-1, O-5, Polymer, Polymer S, T-1, T-5-D, XO, XO Anhyd; Tomah DA-14, DA-17, DA-62, DA-62 Ditallate, E-14-2, E-14-5, E-14-15, E-17-2, E-18-2, E-18-5, E-18-10, E-18-15, E-C-2, E-C-5, E-C-15, E-S-2, E-S-5, E-S-10, E-S-15, PA-10, PA-14, PA-17, PA-62, PA-1820, Q-14-2, Q-17-2, Q-18-2, Q-18-15, Q-D-62, Q-D-T, Q-S; Tomah Tallow Amine, Diamine; Tris Amino; Triton N-40; Trycol 6961, NP-4; Tryfac 610-K, 5576; Ucar Vehicle 442; Unamide C-72-3, D-10, GC-75, J-56, LDL, N-72-3; Unamine C, O, S, T; Unihib 305; Uniquat PB-750, PB-802; Vanchem NATD; Vanfre DFL; Vanlube 71, 73, 81, 601, 648, 793, SS; Vanox 2246; Vanplast 201; Varamide A-7, A-10, A-12, A-83, AC-28; Varine C, O, O Acetate, T; Variquat 80AC, 80AE, 80MC, 475, LC80, U202; Varonic 200 Series, K202, K205, K215, K215LC, L205, L205LC, T215, T215LC, T403, T410, U215, U250, U250-50%; Velvetex 710L; Waxenol 801, 816; Wayhib S; Wayplex HEDP-A; Witcamide CDS; Witcamine 209, 210, 211, 235, 3164, PA-60B, RAD, RAD-0500, RAD-0515, RAD-1100, RAD-1110; Witco DTA-150, DTA-180, DTA-350; Witcor 3192, 3194, CI-6, CI-3117; Zelec NE, UN; Zonyl FSA, FSB, FSC, FSN, FS

COSOLVENTS. SEE SOLVENTS

COUPLING AGENTS/HYDROTROPES

Ablunol NP6, Accoquat 2C-75, 2C-75H, Actrafos 104, 109, 216, 306, 407, Adol 90 NF, Advawax 240; Agriwet CA, C91, C92; Aldo DC, ML; Alkali Surfactant; Alkamuls 200-ML, GMS, PSML-4, SMO, SMS, STO; Alkaphos R9-47A; Alkapol PEG 400; Alkasurf OP-1, OP-12, OP-16; Alkateric 2CIB, BC, CAB, EHCIB; Alkatrope SX-40; Amphoteric 400, C, N; Antaron MC-44; Arcosolv PTB; Aremul V-91, X-91; Arlatone 285, 289, 970; Arophos DI-403-S, DI-404; Arosurf 66-E2, 66-E10, 66-E20, 66-PE12; Arphos MH-200, MH-202, MK-203; Arsperse Series; Atlox 1096, 1285; Atplus 300F, 555, 989; Az-Cup MC; Butyl Oxitol; CA0567; CA0570; CA0699; CA0742; CA0900; Caplube 8440; Capmul MCM; Captex 200; Carbopol 615, 616, 617; Cardolite NC-510; Carsonam 3, 33-S, 3147; Carsosulf SXS-Liq.; Catinex LB-12; CB2408; CB2409.5; CB2493; CB2494; CB2495; CB2790; CB2805; CB2909; CC3005; CC3270; CC3275; CC3285; CC3290; CC3291; CC3433; CC3555; CD3770; CD3780; CD4450; CD5400; CD5430; CD5470; CD5487; CD5550; CD5600; CD5610; CD5636; CD5950; CD6000; CD6010; CD6150; CD6210; CD6220; CD6250; CE6250; CE6345; CE6350; Ceraphyl 230; CH7250; CH7260; CH7280; CH7310; CH7332; Chemax NP-1.5, NP-4, NP-6, NP-9, NP-10

Chemfac PB-082, PB-104, PB-106, PB-109, PB-133, PB-135; CI7840; CM8450; CM8750; CM8930; CM8980; CM9000; CM9220; CO9745; CO9750; CO9000; CO9810; CO9816; CO9817; CO9819; CO9830; Conco 2A1, 3B2, 4C3; Conco Sulfate 3B2; CP0110; CP0156; CP0160; CP0280; CP0320; CP0330; CP0800; CP0810; CPS 076; CPS 078.5; Crafol AP-60; Crill 4; Crodafos Series; Crodafos 25 D2 Acid, 25 D5 Acid, 25 D10 Acid, 25D Series, CS Series, ID Series, O Series; Crodamol DA, PMP; CS1590; CT1800; CT2015; CT2030; CT2050; CT2090; CT2500; CT2520; CT2523; CT2902; CT2925; CT2950; CT2970; CT3250; CT3254; CT3600; CT3610; CT3795; CV4772; CV4800; CV5050; CV5100; Cyclomide Pinemulse; Denwet RG-7; DeSophos 5AP, 5 BMP, 5 BP; Diacid 1550; Dioxitol-High Gravity; Dioxitol-Low Gravity; Dow Corning X1-6100, X1-6106, Z-6020, Z-6030, Z-6032, Z-6040, Z-6075, Z-6076; Dowfax 3B1, 8390, XDS 8292.00; Dowanol DPM, DPnB, PnB, TPnB; Drewmulse GMC-8; Dymsol S; Dynasil A, CA, CM, M; Dynasylan AMEO, AMMO, BDAC, BSA, CPTEO, CPTMO, DAMO, DAMO-P, DAMO-T, GLYMO, HMDS, IBTMO, IMEO, MEMO, MSTFA, MTES, MTMO, MTMS, OCTEO, TCS, TRIAMO, VTC, VTEO, VTMO, VTPMO; Ektasolve DP, EE, EP; Eltesol ACS 60, AX 40, PT 93, PX 40, PX 93, ST 34, ST 40, ST Pellets, SX 30, SX 93, SX Pellets

Emerest 2620, 2652, 2654; Emery 5703, 5704, 5705, 5706, 5707, 5709, 5711, 5713, 5714, 5715, 5717, 5722, 5723, 5724, 5725, 5727, 5728, 5730, 5731, 5736, 5737, 5751, 5753, 5758, 5761, 5770; Emphos CS-147, CS-733, PS-236, PS-415M, PS-440; Empigen 5107, BS, BS/H, BS/P; Emsorb 2500, 2502, 2503, 2505, 2507, 2510, 2515; Ethylan HB4; Fancol OA 50, OA 70, OA 80, OA 90, OA 95; Findet Series; Fomrez B-306, B-308, B-320; Fosterge W Acid; Gafac RA-600; Gantrez AN, AN-119, AN-139, AN-169, AN-179; Hartamide 9137; Igepal CO-520; Incromide LL; Incronam 30; Incropol 233, 290, CS-12, CS-20, CS-30, CS-40, CS-50, CS-60, L-2, L-7, L-7-90, L-12, L-23, L-30; Jordanol SXS; Jordaphos 236, FDED, FDEO, JA-60, JE-41,

COUPLING AGENTS/HYDROTROPES (CONT'D.)

JE-61, JM-51, JP-70; KA 301, 322; Ken-React Series, 7 (KR 7), 9S (KR 9S), 12 (KR 12), 26S (KR 26S), 33DS, (KR 33DS), 38S (KR 38S), 39DS (KR 39DS), 41B (KR 41B), 44 (KR 44), 46B (KR 46B), 55 (KR 55), 133DS (KR 133DS), 134S (KR 134S), 138D (KR 138D), 138S (KR 138S), 158 (KR 158D), 158FS (KR 158FS), 212 (KR 212), 238A (KR 238A), 238J (KR 238J), 238M (KR 238M), 238S (KR 238S), 238T (KR 238T), 262A (KR 262A), 262ES (KR 262ES), OPP2 (KR OPP2), OPPR (KR OPPR), TTS (KR TTS); KZ 55, OPPR, TPP, TPPJ; LICA 01, 09, 12, 38, 38A, 38J, 44, 97, 99; Lipal CE 38, CE 55, CE 64, CE 71, OE 55, OE 64, OE 70, TE 43, TE 55, TE 70, TE 76; Lipophos PE9, PL6; LZ 01, 09, 12, 38, 38J, 44, 97

Macol CSA-2, CSA-4, CSA-10, CSA-15, CSA-20, LA-4, LA-12, LA-23, LA-790, OA-2, OA-4, OA-5, OA-10, OA-20, SA-2, SA-5, SA-10, SA-15, SA-20, SA-40, TD-3, TD-4, TD-6, TD-8, TD-10, TD-610; Mapeg CO-25, CO-25H, DGLD; Maphos L-6, L-13; Mazeen C 2, C 5, C 10, C 15, S 2, S 5, S 10, S 15, T 2, T 5, T 15; Michel XO-144B, XO-146; Miranol C2M-SF Conc., H2M-SF 70%, H2M-SF Conc., J2M-SF Conc.; Monafax 831, 1293; Monamid 716; Monateric 810-A-50, 985A, 1202, CA-35%, CAM-40, CDL, CDS, CDTD, CDX-38 Mod., CEM-38CG, CNa-40, COAB, CyA-50, CyMM-40, LMAB, LMM-30, MCB, TA-35; Monatrope 1250, 1296; Na Cumene Sulfonate 40; Sulfonate Powd.; Nansa YS 94; Naxol; Niaproof Anionic Surfactant 08; Norfox NBC, PEW, SEHS, SXS40, SXS96; Nutrol SXS; NZ 01, 90, 12, 33, 38, 39, 44, 49, 89, 97; Oxitol; Peganol NP 5; Phospholan BH-14; Pilot SXS-40; Pluracol E-200, E-300, E-400, E-400 NF, E-600, E-600 NF, E-1000, E-1450, E-1450 NF, E-1500, E-2000, E-4000, E-4000 NF, E-4500, E-6000, E-8000, E-8000 NF; Poly-Solv DB, DE (High Gravity), DE (Low Gravity), DM, DPM, EB, EE, EM, MPM, TE, TM, TPM; Procetyl 10, 20, 30, 50, AWS; Promyristyl PM; Prosil-196, -220, -248; Prostearyl 15; Provol 10, 30, 50; Quimipol ENF 55; Rewomine IM-OA; Rewopal LA 3; Rycofax O; Schercemol DEIS, DIS, MEL-3, PGML; Schercomid AME, AME-70, SI; SF 1154; Siponate DSB; Siponic 25-3, 25-7, 25-9, 91-6, F-160, F-300, F-400, F-707, NP4, NP-7, NP-9.5, NP-10, NP-13, NP-407, NP-707, OP1.5; S-Maz 20, 40, 60, 60 K, 60 KHM, 65, 65 K, 67, 80, 80 K, 85, 85 K, 90, 93 R; Solar TE; Standamul 1414-E, G, HE; Stepanate AM, C-S, T, X; Stepantex DA 52; Stepfac; Sterling 2XS, SXS; Strodex P-100; Sulfonate OE-500; Sulframin CSA, TX; Surco SXS; Surfine WLL, WNG-A; Synperonic 91/5, 91/6, 91/8; Tebol 99; Texox WL-440, WL-660, WL-1400, WL-5000; T-Mulz 844; Triton X-15, X-35, X-405; Trydet LA-5; Ultra NCS Liquid, NXS Liquid, SCS Liquid, SXS Liquid Powder; Unimate DIPS; Union Carbide A-151, A-172, A-174, A-186, A-187, A-189, A-1100, A-1102, 1120, A-1126, A-1893; Varion AM-KSF, AM-V; Volan, L; Westvaco Diacid H 240; Witco TX Acid; Witcolate 1259; Witconate 1840X, NCS, NXS, SCS, STS, SXS, TX Acid; Witconol APEM, APM, APS, F26-46, NS500K

CROSSLINKING AGENTS

Ancamine 2049; BA-43; Butynediol; Cadox BS, OS, PS, TS-50; Cyanamer P-26, P-250; Cymel 303, 370, 373, 380, 1411; Dabco CL-485; Desmodur KA-8267, MP-225, N-75, R, RC, RE, RF, RFE, RU, VKS-2, VKS-4, VKS-18, Z-4370; Di-Cup 40C, 40KE, R, T; Duolite A-561; Esperal 115RG; Espercarb 438M-60; Esperox-10XL; Kemamine DD-3680, DP-3680, DP-3695; Luperco 101-P20, 130-XL, 231-SRL, 231-XL, 500-SRK, ANS-P, EFA-4, PMA-25; Luperox 802, 802-40KE; Lupersol 219-M60, 288M75, 531-80M, 533-M75, 553-M75, 555M60, 665T50, 688M50, 688T50, TAEC; Mondur CB-75, MR, MRS, PF; Perkadox 14, 14/40, 14/40 Bpd., BC, BC-40Bpd, BC-40Kpd.; Pluracol 355, 364, 450, 550, 650, 669; Reten 210, 220, 300, 420, 421, 423, 425, 521, 523, 525; Rhenocure CMT, CMU, EPC, Diuron; Rhenofit BDMA/S, CF, EDMA/S, NC, TAC/S, TRIM/S, UE; Rhenovin Na-stearat-80, S-stearat-80; Saret 500, 515, 516, 517, 518; Silcat R; SP-1044, SP-1045, SP-1055, SP-1056; SR-201, SR-204, SR-205, SR-206, SR-209, SR-210, SR-213, SR-230, SR-238, SR-239, SR-259, SR-268, SR-295, SR-348, SR-350, SR-444; Sulfamin R; Syn Fac 8026; Thiovanol; Trigonox 17/40, 17/40 Bpd., 29/40 Bpd., 101, 101/45Bpd., 107, BPIC, T; Tyzor AA, DC, TBT, TE, TOT, TPT; Unilink 4132, 8130; Vanchem HM-50, HM-4346; Varox, 130, 130-XL, 230-XL, 231, 231-XL, DBPH, DBPH-50, DCP-40C, DCP-40KE, DCP-R, DCP-T, VC-40KE; Versaflex 6; Vestowax AO-1699; XCE-89; Zinc Oxide 35; Zoldine ZE, ZT-55

CURING AGENTS

Apilink 148, 149, 353, 354, 355, 356, 360, 361, 370, 375, 380, 381; Armocure 100; C-Cure 111; Cadet BPO-70, BPO-70W, BPO-78, BPO-78W; Cadox BFF-50, BFF-60W; Cardolite NC-540, NC-541, NC-546, NC-549, NC-550; Chemlink 2000, 2100, 3000, 5000, 9000, 9001, 9003, 9008, 9012, 9013, 9014, 9015, 9020, 9021, 9022, 9023, 9024, 9025, 9040, 9503, 9504, 9505; Curithane 103; DC 2900 HW; D.E.H. 20, 24, 26, 29, 39, 40, 52, 58; Dytek A; Eltesol CA 65, CA 96, PSA 65, TA, TA/F, XA, XA90; Emery 5702, 5703, 5707, 5709, 5710; Empigen AB, AD, AF, AG, AH, AM, AY; Ethacure 100; Jeffamine D-230, D-400, D-2000, D-4000, DU-700, DU-1700, DU-3000, ED-600, ED-900, ED-2001, EDR-148, T-403, T-3000, T-5000; Kencure MPP, MPPJ; Ken-React Series, 7 (KR 7), 9S (KR 9S), 12 (KR 12), 26S (KR 26S), 33DS, (KR 33DS), 38S (KR 38S), 39DS (KR 39DS), 41B (KR 41B), 44 (KR 44), 46B (KR 46B), 55 (KR 55), 133DS (KR 133DS), 134S (KR 134S), 138D (KR 138D), 138S (KR 138S), 158 (KR 158D), 158FS (KR 158FS),

212 (KR 212), 238A (KR 238A), 238J (KR 238J), 238M (KR 238M), 238S (KR 238S), 238T (KR 238T), 262A (KR 262A), 262ES (KR 262ES), OPP2 (KR OPP2), OPPR (KR OPPR), TTS (KR TTS); Lankromark LE65; LICA 01, 09, 12, 38, 38A, 38J, 44, 97; Luazo 55, 70, 79, 82, 96; Lucidol-98; Luperox 204; Lupersol 130, 220-D50, 224, 331-80B, 431-80B, DDA-30, DDM, DDM-30, Delta-X, DNF, DSW; LZ 01, 09, 12, 38, 44, 97; Manro FCM 95LV, FCM 100; Norcure 131, 138-B; Pennad 0150; Petaflex RC-802; Poly-G 76-120; QO Furfuryl Alcohol; Rhenocure IS 60, IS 60-5, IS 60/G, IS 90-20, IS 90-33, IS 90-40, IS 90/G; SR-220, SR-247, SR-272, SR-306, SR-335, SR-339, SR-340, SR-344, SR-349, SR-351, SR-355, SR-395, SR-399, SR-440, SR-454, SR-506; Stygene R-100; Tactic H31, H41; Tonox 22; Unilink 4130, 4132, 4230, 8130; Uni-Rez 2100-P75, 2115-C70, 2115-I75, 2115-P80, 2180-B75, 2341, 2355, 2400, 2401, 2415, 2800, 2802, 2810, 2828-D, 2850; USP-240; Varcum 564, 1198, 4326-WP; Varox DCP-R, DCP-T, VC-40KE; Vulcuren 2

DEACTIVATORS

Evanacid 3CS; Irganox MD-1024; Kerobit COM; Keromet MD 60, MD 80, MD 100; Na-Sul 611, 707; Nacap; Naugard XL-1; Rokon; Vanlube 601, 793, AZ; Vanox 3240, 3245

DEAERATORS

Tego Airex 960, 970, 975

DEFLOCCULANTS

Empiphos STP, STP/1L, STP/6L, STP/6N, STP/AD, STP/D, STP/DMST, STP/E, STP/F, STP/L, STP/L16, STP/M, STP/M16, STP/MST, STP/N

DEFOAMERS/ANTIFOAMERS

Ablupol SAE; Actrafoam A, B, C, S; Actrasol MY-75; Acrawax C DF #1, #2; Actrasol SP; Adol 60, 85, 90, 640; Advawax 275, 280, 290; AF 10 FG, 10 IND, 10 TF, 30 FG, 30 IND, 60, 70, 72, 75, 93, 100 FG, 100 IND, 112, 6035, 6050, 8805, 8805 FG, 8810, 8810 FG, 8820, 8820 FG, 8830, 8830 FG, 9000, 9020, CM Conc., GN-11-P, GN-23, GN-46, HL-21, HL-23, HL-26, HL-27, HL-28, HL-40, HL-52; Akrochem Antifoam A, SWS-201; Aldo MO FG, MOD FG, MS-20 FG; Alfol 20, C22; Alkamide STEDA; Alkamuls 200-DL, 200-DO, 600-ML, 600-MS, DEG-DO, DEG-ML; Alkapol PPG-1200, PPG-2000, PPG-4000; Alkasurf CO-5, CO-10, CO-15, L-2, L-9, L-14, NP-1, OP-12, OP-16, S-14; Alkaterge-E; Alkatronic EDP 8-4, EDP 28-1, EDP 28-7, EDP 38-1, EDP 38-4, EDP 38-3, EDP-28-2, EGE 17-1, 17-2, 17-4, 17-8, 25-1, 25-2, 25-5, EGE 25-8, 25-4, PGP 33-1; Alphoxat O 110, O 115; Amphoteric 400; Antifoam 124, Base 263, C-196, CM Conc., Compound SWS-210, SWS-202, SWS-203, CTN, E-20, Emulsion Q-75, Emulsion SWS-211, SWS-213, SWS-214, G, Q-41; Armid HT; Arnox 1003; Axel P 100; Berol 730, 740; Britol; Bubble Breaker 259, 260, 613-M, 622, 730, 737, 746, 748, 776, 900, 913, 917, 1840X, 3009-F, 3017A, 3056-A, 3073-7, D, PR; Capmul GMO; Caprol 3GO; Catinex KB-10; Cedepal CO-210; Cetodan; Chemal BP-261, BP-262, BP-262LF, BP-263, BP-264, BP-268, BP-268/50; Chemax CS-10, CS-30, CS-100, E-200 ML, E-400 ML, PEG 600 DO; Chimin P20; Colloid 1010; Colorin 102, 104, 202, 301, 302; Contraspum 210; CPH-39-N

Crill 1, 2, 4, 50; Cru Fluid 350; Cru Release 900 Series; Defoamer 44, 167, 357, 388, 831, A 50, CK-35, CK-55, CK-75, MJ, S, S-10, S-100, TIP; Defoamer/Drainage Aid CK-25; Degressal SD 20, 21, 23, 30, 40, SNC; Dehydran 150, 240, 241, 420, 610, 630; Dehypon LT 104; Dehyzan Z 4904, 7225; Densol BP-61, BP-62; Densulf TA-75; DeSonic 1.5N, 3K, 4N, 13N, 81-2, 315-3; Dow Corning 1500 Silicone Antifoam, 1520 Silicone Antifoam, Antifoam A; Dowfax 63N10, 63N20; Drysperse 902; EBS Wax; Ecco Defoamer, Defoamer S; Emargol KL; Emcol 14; Emerest 2301, 2302, 2620, 2650; Emery Methyl Lardate, Methyl Oleate; Emka Defoam SD; Emphos PS-415M, TS-230; Empilan BQ 100; Empimin 3093, 3095; Emsorb 2515; Esperase 16.0L; Ethoquad 18/12, 18/25, C/12, C/25, O/12, O/25; Ethylan A2, A3, A4, NP 1; Excel 300; Fancol OA 50, OA 70, OA 80, OA 90, OA 95; Flexricin 13; Foamaster 206-A, 267-A, 335-A, 1407-50, 1719-A, 8034, B, DD-72, DF-122NS, DF-177-F, DF-178, DF-198-L, DNH-1, DR-187, DRY, DS, FGA, FLD, G, JMY, KF-99, NDW, NS-20, NXZ, P, PD-1, Soap L, TBD-1, TDP, V, VL

Foam Burst 5, 10, 87, 100, 105, 150, 151; Foamex AD-50, J-10; Foamgard 73, 161, 200; Foamkill 8BA, 8G, 8J Series, 30 Series, 30C, 30-HP, 80J Series, 400A, 608 Series, 608, 608M, 608G, 618 Series, 618, 618C, 618F, 618J, 627, 628A, 634 Series, 634B-HP, 634C, 634D-HP, 634F-HP, 639 Series, 639 Conc., 639J, 639J-F, 639L, 639P, 644 Series, 644E, 645, 649 Series, 652 Series, 652H, 652-HF, 652L, 660, 660F, 660-HP, 663J, 679, 684 Series, 684A, 684P, 700, 700 Conc., 810F, 830, 830F, 836A, 852, 1001 Series, FPF, GCP Series, MS-1, MS Conc., MSC Series, NSP-1, NSP-3, NSP-4, NSP-5, RP; Forbest 1000B, 1500W, 2000C; Genapol PAF Brands, PF Brands, PL Brands, PN Brands; Grocor 5410, 5510; Hartopol L81; Herco; Hercules 187 Defoamer, 388 Defoamer, 752, 845 Defoamer; Hercules Defoamer 137, 831, 1512, 2051, 2470, Eff-101; Hodag FD Series; Hystrene 5016 NF FG, 7018 FG, 9718 NF FG

Icinol H260, H660YA, H5100, L65, L285, L300X, L385, L625, L1715; Igepal CO-210; Industrene 105, 138,

DEFOAMERS/ANTIFOAMERS (CONT'D.)

143, 154-BG, 205, 206, 225, 225 FG, 226, 232, 239, 325, 328, 333, 365, 3022, 4022, 4516, 4518, 5016, 5016 FG, 5022, 7018, 7018 FG, 8518, 8718 FG, 9018, B, R; Jeffox PPG-2000; Kemamide B, E, E-180, E-221, O, P-181, S, S-65, S-180, S-221, U, W-10, W-20, W-39, W-40, W-40/300, W-45; Ken-React Series, 7 (KR 7), 9S (KR 9S), 12 (KR 12), 26S (KR 26S), 33DS, (KR 33DS), 38S (KR 38S), 39DS (KR 39DS), 41B (KR 41B), 44 (KR 44), 46B (KR 46B), 55 (KR 55), 133DS (KR 133DS), 134S (KR 134S), 138D (KR 138D), 138S (KR 138S), 158 (KR 158D), 158FS (KR 158FS), 212 (KR 212), 238A (KR 238A), 238J (KR 238J), 238M (KR 238M), 238S (KR 238S), 238T (KR 238T), 262A (KR 262A), 262ES (KR 262ES), OPP2 (KR OPP2), OPPR (KR OPPR), TTS (KR TTS); Kessco 3283; Kilfoam; Kito 703; Kodaflex DBP; Kureton 200, A; Lankro Mud-Aids; Lanquell 206, 217; Leocon 1020B, 1070B, PL-71L; Leophen M; LICA 01, 09, 12, 38, 38A, 38J, 44, 97; Lipo DGLS, DGS-SE, Diglycol Laurate, EGDS, EGMS, GMS, PGMS; Lipocol C-2, C-10, C-20, L-1, L-4, L-12, L-23, M-4, O-2, O-10, O-20, S-2, S-10, S-20, SC-4, SC-10, SC-15, SC-20; Lipowax C; Lonzest PEG-4DO; Lubrol 17A-10; Lutensol ED 140, ED 310, ED 370, ED 610; LZ 01, 09, 12, 38, 44, 97

Macol 2, 4, 5, 8, 10, 15, 20, 34, 660, 3520, 5100, P Series; Mapeg 600 DL, 600 DOT, 600 ML, 600 MOT, 6000 MS, PGDS, PGMS, PPG-400, PPG-500; Masil 2132, 2133, 2134; Matexil AA-NS; Mazol 300, GMO; Mazu DF 100S, 110S, 130S, 200S, 200SP, 200SX, 200SX Special, 210S, 210SX, 210SX Mod 1, 215SX, 230S, 230SX, 243; M-Quat 32, 2475; Monolan 1206/2, 2000 E/12, 3000 E/60, 8000 E/80, P222; Nalco 70, 131, 2300, 2305, 2343; Newcol 561H, 562, 565, 864; Niax PPG 2025; Nilofoam 60, M, XC; Nioix AK-40; Niox KH Series; Nissan Disfoam C Series; Nissan Plonon 102, 104, 108, 171, 172, 201, 204, 208; Noiox AK-40, AK-44; Nonionic 1017-R, 1035-L, 1044-L, 1061-L, 1062-L, 1064-L, 1068-L, 1088-L, 1025-R, 2017-R, 2025-R, 4017-R, 4025-R, 5025-R; Nopalcol Series, 4-O; Nopco 1419-A, JMY, NDW, NXZ, PD#1-D; Norfox 243, DF210SX, NP-1; OP-2000; Pamolyn; PAR Series; Patcote 315, 500, 512, 513, 519, 520, 525, 531, 532, 543, 550, 577, 598, 801, 802, 803, 804, 805, 818, 834, 847, 883; Pegol 17 R1, 17 R2, 25 R1, 25 R2, 31 R1, L 31, L 35, L 43, L 61, L 62, L 62 LF, L 64, L 81, L 101, L 121, P 85; Perenol EI; Pliabrac TBP; Pluracol E-400 NF, E-600 NF, E-1000, E-1450, E-1450 NF, E-2000, E-4000 NF, E-4500, E-8000, E-8000 NF, P-410, P-710, P-1010, P-2010, P-3020, P-4010, W-170, W-260, W-660, W-2000, W-3520N, W-3520N-RL, W-5100, W-5100N, WD1400

Plurafac RA-20, RA-40, RA-43; Pluriol P 600, P 900, P 2000, PE 3100, PE 6100, PE 6101, PE 8100, PE 10100, RPE 2540, RPE 3110; Pluronic 10R5, 10R8, 12R3, 17R1, 17R2, 17R4, 17R8, 22R4, 25R1, 25R2, 25R4, 25R5, 31R2, 31R4, L10; Poly-G WI 285, 625, 1715; Poly-G WS 100, 170, 260, 660, 2000, 3520, 5100, WT 9150, 90,000; Polyglycol 15-200, 112-2, P-1200, P-2000, P-4000; Polymekon; Poly-Tergent P-17A, -17B, -32A; Pronal 502, ST-1; Propetal 99, 103, 241, 254; Q-70, Q-93, Q-101; Quimipol ED 2021, 2022; Radiasurf 7136, 7140, 7141, 7196, 7201, , 7206, 7270, 7400, 7402, 7403, 7404, 7410, 7411 , 7412, 7413, 7414, 7417, 7420, 7421, 7422, 7423, 7431, 7432, 7443, 7453, 7454; Rexfoam 150-A, B, C, D; Rexol 25/1; Ridafoam Base 100; Ridafoam NS-221, S-103-N, S-110-N; Rudol; Ryco Defoamer 1288; Sag Silicone Antifoam 10, 30, 100, 310, 471, 720, 4130, 4220, 5300, 5310, 5693; Sandoz Defoamer F; Savinase 16.0 L, Type W; Schercemol PGML; Semtol Series; Senka Antifoam 800; Serdas GBS, GBU, GE 4010, GE 4050, GLN; Seyco Defoamer A-69, D-10, G-34, H-61M, K-20; SF 18, SF 69, SF 97; Silcolapse 430, 431, 432, 5000, 5001, 5006, 5007, 5008, 5009, 5010, 5020; Silicone AF-10 FG, AF-10 IND, AF-30 FG, AF-30 IND, AF-100 FG, AF-100 IND, AF-600M; Silicone Defoamer #5037; Silwet L-7500, L-7602, L-7605; S-Maz 20, 40, 60, 60 K, 60 KHM, 65, 65 K, 67, 80, 80 K, 85, 85 K, 90, 93 R, 95; Sole-Onic CDS; Soprofor DO/64, NP/20, /30, /100, PL/Series; Sorban AL, AST, CO; Sorgen 30, 40, 50, 90, S-30-H, S-40-H; Sulfonate OA-5; Sulframin 40; Sunaptol CA 25, 120, 350, 400, 750, TO 28, 85; Surfactol 13; Surfonic HDL, N-10, N-31.5, N-40, N-60, N-85, N-95, N-100, N-102, N-120, N-150, N-200, N-300, N-400, NB-5, NB-14; Surfynol 82, 82S, 104, 104A, 104BC, 104E, 104H, 104S, 440, PC; Surpasol E-436; Synperonic PE30/10, 30/20, 30/40; Synthionic D 7520, 7525, 9500, P 6040, 8020

TBP; T-Det C-40, EPO-61; Tegiloxan; Tego Antifoam; Tegomuls O Special; Tegosipon; Tergitol 24-L-3, Min-Foam 1X, 2X; Teric 121, 124, 127, 129, N2, N3, N4, N5, PE61, PE62; Tetronic 50R1, 50R4, 50R8, 70R1, 70R2, 70R4, 90R1, 90R4, 90R8, 110R1, 110R2, 110R7, 130R1, 130R2, 150R1, 150R4, 150R8; Texox PPG-400, PPG-2000, WL-440, WL-660, WL-1400, WL-5000; Trans-10, -10K, -20, -20K, -25, -25K, -30, -30K, -100, -107, -109, -134S, -134S K, -135, -137, -138, -143, -154, -156, -166, -175, -176, -177S, -177S K, -179, -198, -220, -222, -224, -225S, -262, -264, -265, -266, -1030, -1030 K, -FG2, N1, RC1; Triton CF-32; Trycol 5993, 6960, NP-1, TDA-3; Trydet LA-5; Unilin 425, 550, 700; Union Carbide L-45 Series, L-720, -721, -722, -727; Varonic T105, U105, U215; Viscasil; Wax OTO; Westvaco WPD-120, WPD-140, WPD-145, WPD-203, WPD-220, WPD-300, WPD-305, WPD-390, WPD-700, WPD-710; Witconol 14, 18L, 171, 172, H-31; Zeofree 80; Zeothix 175; Zonyl A

DEODORIZERS. SEE ANTIMICROBIALS

DEPRESSANTS

Akypopress DB; Chemquat 12-33, 12-50, 16-50; Cyclopol SBDO; Eltesol AX 40, PT 93, PX 93, ST 34, SX 30, SX 93; Ganex P, P904, V, V216, V220, V516; Glucam E-10, E-20, P-10, P-20; Guartex SJM; Harol D; Hartotrope AXS 40, KTS 50, STS 40, Powd, SXS 40, Powd.; Lomar PWA; Niox KF-13; Poly-Tergent S-205LF, S-305LF, S-405LF, S-505LF, SL-42, SL-62, SL-92, SLF-18; Polywax OH 425, 550, 700; Soprofor PE/220, PE/400; Sotex CW; Standapol SCO

DESICCANTS

Rhenosorb C, C/GW, F

DESIZING ASSISTANTS

Alkaminox HT-12, -25, -30; DeSomeen TA-2, TA-15, TA-20; DeSonic TA-2, TA-15, TA-25CWS; Excize HA, Regular Excize, Special Excize, Super Excize; Hetoxamine C-2, C-5, C-15, O-2, O-5, O-15, S-2, S-5, S-15, ST-2, ST-5, ST-15, ST-50, T-2, T-5, T-15, T-20; Hipochem EK-18; Nansa TS 60; Scourol 700; Tergitol 25-L-7, 25-L-9, 25-L-12; Valdet 561

DILUENTS. SEE EXTENDERS

DISINFECTANTS. SEE ANTIMICROBIALS

DISPERSANTS

AB; Ablunol 200 ML, MO, MS, 400ML, MO, MS, 600ML, MO, MS, 1000MO, MS; Ablusol NL; Accomeen T2, T5, T15; Acconon 200-DL, 200-MS, 400-DO, 400-ML, 400-MO, 400-MS, 1300, CA-5, CA-8, CA-9, CA-15, CA-25, CON, E, TGH, W-230; A-C Copolymer 400A, 540A; Acetamin 24; Acryloid 900 Series, WR-97; Acrysol LMW Series, WTP-1; Actiflo 68, 70; Adogen 415, 444, 461, 462, 471, 477; Advawax 240, 275, 280, 290; AEPD; Aerosol 18, 22, 22N, 200, 413, 501, A-102, A-196, A-196-40, AY, AY-65, AY-100, C-61, IB-45, MA-80, OS, OT-75%, OT-B, TR-70; Aflux 32, 42, 54, R, S; Agent 613-95, X-601-52N, X-646-12; Agrilan F491, F502, F513, WP178; Ahco A-117, AJ-110; Ahcowet DQ-114; Ajicoat SPG; Aktiplast, AS, F, PP, T; Akypo RLM 25, RLM 45, RLM Q38; Alconate 2CSA, D-6, LEA, SBDO, SBN-862; Alcosperse 107, 124, 149, 157, 160, 175, 249, 404, 602-N, ML 23, MS 20; Alfonic 1012-60, 1412-40; Alipal CO-433, CO-436, CO-526, EO-526; Alkamide 2112, STEDA; Alkaminox HT-12, HT-25, HT-30, T-2, T-5, T-10, T-12, T-15; Alkamox C2-O; Alkamuls 200-DL, 200-DO, 400-DL, 400-ML, 400-MO, 400-MS, 600-DL, 600-DO, 600-ML, 600-MO, 600-MS, DEG-DO, DEG-DT, DEG-ML; Alkanol A-CN, S, XC; Alkasperse A-2H, A-20, DM-5, MM-2; Alkasurf CA, CO-5, CO-10, CO-15, CO-20, L-2, L-9, L-14, LA-1, LA-7, LA-EP15, LA-EP16, NP-1, NP-4, NP-5, NP-6, NP-9, NP-10, NP-20, NP-40, OA-20, OP-1, OP-5, OP-10, S-14, SS-DA-3, SS-DA4-HE, SS-TA, SS-TA 125, TDA-5, TDA-6; Alkaterge-T, T-IV; Alkatronic EDP 8-4, EDP 28-1, EDP 28-7, EDP 38-1, EDP 38-4, EDP 38-3, EDP-28-2, PGP 18-4, PGP 23-7, 23-8, PGP 33-8, PGP 40-7; Alkazine O, ST, TO; Alphoxat MS 130

Ambergum 3021; Amerlate LFA, P, W, WFA; Ameroxol OE-2; Amiet 102, 105, 110, 115, 202, 205, 210, 215, 302, 305, 310, 315, 402, 405, 410, 415, 502, 510, 515; Amihope LL-11; Amiladin C-1802; Amine C, O, S, T; Ammonyx 856; AMP, AMP-95; AMPD; Amphosol CG; Anchold; Antara LB-400; Antaron FC-34; Antarox G-200; Aquatreat AR4, AR6, AR231, AR232, AR602; Aristonate 430, 460, 500; Armeen O, OD, Z, Z-9; Armogard F-201; Arnox 910, TGE-05, 16, 36, 40, 80; Arofos 200 Conc.; Arsperse Series; Arylan SNS; Atlox 4853 B, 4875; Atolex LDA/40; Avirol SL-2010; Barisol Super BRM; Bayhibit; Bayhibit-AM; Belsperse 161; Bentone SD-1, SD-2; Berol 07, 08 Powd., 09, 049, 081 Flakes, 081 Powd., 263, 472, 475, 594, EGA 07, EGA 192; Berol Fintex 755; Berol Lanco; Berol Nonylphenol; Berol WASC; Bio-Surf DC-730; Blancol; Blancol N; Borax; Britol; ; Cab-O-Sil; Cadox BP-55; Calfax 10L-45; Calfoam NEL-60; Calimulse PRS; Caplube 8385; Capmul GMO, O; Caprol 10G10O; Caprol 10G100; Carbonox; Carboset 514A, 514H, 515, 525, 531, XL-11, XL-19; Carbowax PEG; Carrybon B, L-400; Carryol MD-112

Carsamide CA, SAC; Carsonon L-985, N-4, N-8, N-9, N-10; Catinex KB-15, KB-16, KB-18, KB-19, KB-20, KB-22, KB-23, KB-24, KB-25, KB-26, KB-27, KB-31, KB-32, KB-40, KB-42, KB-49, KB-51; CC-16; Cedepal CA-520, CO-210, CO-430, CO-436, CO-500, CO-530; Cellopal 100; Cellosize Hydroxyethyl; Centrolene Series; Centrolex Series; Centromix Series; Centrophil Series; Ceraphyl 140, 847; Cerasynt 303; Chemal BP-261, BP-262, BP-262LF, BP-263, BP-264, BP-268, BP-268/50, OA-4, OA-9, OA-20, TDA-3, TDA-6, TDA-9, TDA-12, TDA-15, TDA-18; Chemax CO-5, CO-25, CO-28, CO-30, CO-36, CO-40, CO-80, CO-200/50, DNP-8, DNP-15, DNP-150, E-200 ML, E-200 MO, E-400 ML, E-400 MO, E-600 ML, E-1000 MO, HCO-5, HCO-16, NP-1.5, NP-4, NP-6, NP-9, NP-10, OP-3, OP-5, OP-40; Chemeen C-2, C-5, C-10, C-15, T-2, T-5, T-10, T-15, T-20; Chimin BX; Chimipal APG 400, PE 300, PE 302, PE 400; Cirrasol AEN-XZ; Cithrol 2DL, 2DO, 2ML, 2MO, 4DL, 4DO, 4MO, 6DL, 6DO, 6ML, 6MO, 10DL, 10DO, 10ML,

DISPERSANTS (CONT'D.)

10MO, 40MO, 60ML, 60MO, A, DGDL N/E, DGDL S/E, DGDO N/E, DGDO S/E, DGDS N/E, DGDS S/E, DGML N/E, DGML S/E, DGMO N/E, DGMO S/E, DGMS N/E, DGMS S/E, DPGML N/E, DPGML S/E, DPGMO N/E , DPGMO S/E, DPGMS N/E, DPGMS S/E, EGDL N/E, EGDL S/E, EGDO N/E, EGDO S/E, EGDS N/E, EGDS S/E, EGML N/E, EGML S/E, EGMO N/E, EGMO S/E, EGMR N/E, EGMR S/E, EGMS N/E, EGMS S/E, G Range, GDL N/E, GDL S/E, S Range

Clearate LV, WD, Special Extra; Colloid 102, 111, 111M, 117/50, 119/50, 202, 204, 207, 208, 211, 218D, 223, 223D, 225, 226/35, 230, 231, 233, 245D, 252, 274, 350; Colorol 20, 70, E, F, Standard; Complemix 100; Compound 8-S; Conco 2A1, 4C3, NI-21, NI-43, NI-60, NI-90, NI-100, NI-110, NI-150, NI-185, NI-187, NI-190, NI-197, NI-2000; Conco Sulfate 2A1, 3B2, RA, TL, Sulfate WA, WAG, WAS, WA Special, WB-45, WBS-45; Condensate PA, PC, PL, PM, PO; Consos Castor Oil; Corona Lanolin; Coronet Lanolin; CPH-43-N; CPH-79-N; Crapol AV-10, AV-11; Crill 1, 2, 4, 6, 35, 45, 50; Crillet 1, 2, 3, 4, 6, 11, 31, 35, 41, 45; Crodalan AWS; Crodamine 1.0, 1.0D, 1.T, 1HT, 1.16D, 1.18D, 3.A16D, 3.A18D, 3.ABD, 3.AED, 3.AOD; Crodesta A-10, A-20, F-10, F-50, F-110, F-160, SL-40; Crodet L4, L8, L12, L24, L40, L100, S4, S8, S12, S24, S40, S100; Cru Thix 46; Crystic 57; Cyanamer A-370, P-26, P-35, P-70, P-80; Cyastat SN, SP; Cyclochem LVL, ML; Cyclomide DOTS

Darathane WB-4000; Darvan 404, C, L, No. 1, No. 2, No. 3, No. 4, No. 6, No. 7, No. 31, SMO; Daxad 11, 11G, 1KLS, 13, 14B, 14C, 15, 16, 17, 19, 19K, 19L-33, 19L-40, 21, 23, 27, 30, 30-30, 30S, 31, 31S, 32, 32S, 34, 34A9, 34N10, 34S, 35, 37LA7, 37L Acid, 37LK9, 37LN7, 37LN10, 37NS, 41, CP-2; Dehypon Conc.; Dehypon LS 24, LS 36, LS 45, LS 54; Demol AS, C, EP, MS, N, P, RN, SN-B, SS Paste, SSL, VP; Densol 6920, BP-61, BP-62; Denwet CM, RG-7; Depasol AS-27, CM-41; Dequest 2006; D.E.R. 671-EE75, 671-EEA75, 671-EEK75, 671-MK75, 671-T75, 671-X75, 671-XM75; Deriphat BC, BCW; DeSomeen TA-2, TA-5, TA-15, TA-20; DeSonic 6C, 12-3, 30C, 36C, 40C, 54C, TA-2, TA-15, TA-25CWS; DeSophos 7 DP; Detergent CR; Dextrol OC-15, OC-20, OC-50, OC-70; Diamiet 503, 508, 520, AB, C; Diamin DO; Dianol; Dianol 300; Diazopon SS-837; Dinoramox S 3 to 12, SH 31; Dipex 280; Dispatex G; Dispersal 130, 140; Dispersogen ASN; Dispersol T; Disrol SH; Dovanox 23H, 23M, 25N, 231; Dowfax 9N; DRA-1500; Drakeol 9; Dresinate 214, 515, 731, TX, X, XX; Drewfax 0007; Drewmulse 3-1-O, 3-1-S, 6-2-S, 10, 10-4-O, 10-8-O, 10-10-O, 10-10-S, 15, 20/200/V, 30/900, 50, 55, 85, 200, 300, 365, 700, 900, POE-SML, POE-SML, POE-SMO, POE-SMS, POE-STS, SML, SMO, SMS, STS; Drewpon 40, 40LS; Drewsperse 611; Drysperse 401, 801, 902, 908H; Dulectin; Duofol T; Duomeen T, T Special, TDO, TDO-IHF; Duponal 80, C, D Paste, RA; Dur-Em 117; Durfax 65; Durtan 60, 80; Dyasulf 1761-A; Dyesperse DC; Dynasperse A, B

Eccolene OW; Ecco Spersant PDQ; Eccotex P Conc.; Eccowet LF Conc.; Ekaline F; Elimina-254; Emasol O-10, O-30, P-10, S-10, S-30; Emcol 4100M, 4161L, 4300, 4500, 4560, 4560, 4600, CC-9, CC-36, CC-42, CC-55, CC-57, CC-59, CC-422, CS-136, CS-143, CS-151, CS-165, K-8300, P-1045, TS-230; Emerest 1723, 2650, 2704, 11723; Emery 1656, 1660; Emkafix RXC; Emkal BNS, BNS Acid, BNX Powd., NNS, NNS Acid, NOBS; Emkatex DX, PX29; Emphos CS-1361, D70-31, F27-85, PS-21A, PS-220, PS-400, PS-415M; Empicol 0185, LS30, LZ/E, LZV/E; Empicryl APD, APD90, APD/B, APD/B90, DH122, DH135, DH145; Empilan DL 40, DL 100, KA3, KA5, KA590, KA880, KA1080; Empimin AQ60, MA, OP45, OP70, OT, SDS, SQ25, SQ70; Empiphos 4KP, OMP, OSP; Emsorb 2721, 2722, 6900; Emthox 5882, 5940, 5941, 5942, 5964, 5993, 6957, 6962, 6971, 6972, 6984; Emulan OG, OSN, OU; Emulbon S-83, T-80; Emulgane E, O; Emulgen A-60, A-90, A-500; Emulphogene BC-420, TB-970; Emulphor EL-620, EL-719, ON-870, ON-877; Emulthin M-35; Ethofat 60/15, 60/20, 60/25, 142/20, 242/25, C/15, C/25, O/15, O-20; Ethomeen 18/12, 18/15, 18/20, 18/25, 18/60, C/12, C/15, C/20, C/25, O/12, O/15, O/25, S/12, S/15, S/20, S/25, T/12, T/15, T/25, T/60, TD/15, TD/25; Ethomid HT/15, HT/23, HT/60, O/15; Ethoxamine C5, SF11; Ethoxylan 1685, 1686; Ethylan A3, A4, A6, BK 1130, C 160; Eumulgin B-1, B-2, B-3, O 5, O 10; Fancol LA, OA 50, OA 70, OA 80, OA 90, OA 95; Fluilan; Fluilanol; Forbest 610, 8209, VP 13, VP 18, VP 20, VP 33, VP S7, WP; Forlan C-24; Fosterge LFS; Fostex S, SN, TN, TX

Gafac RA-600, RE-410, RE-870, RE-877, RE-960; Gafterge AW-123; Gaftex CD-169, COM-154, COM-154 Conc.; Ganex P; Ganex P904, V, V216, V220, V516; Gantrez AN, AN-119, AN-139, AN-169, AN-179; Gardilene IPA, S25L; Gardinol CX; Gardisperse AC; Gemtex SC-75; Genapol AMS, LRO Liq., LRO Paste, ZRO Brands; Generol 122E25; Geronol ACR/4, ACR/9, TZ/14, TZ/A, TZ/B; Geropon FMS, IN, RM/77, RM/77-D, SC/211, SC/213, SDS, T/36-DF, TA/72, TA/72/S, TA/764, TA/K, TX/99; Glucate DO; Good-rite K-722, K-732, K-752, K-7028, Polycrylates; Gradol 250A, 300, N-95; Grocor 5500; Hallcomid M-18-OL; Harol D, KG, RG-71; Hartolan; Hartolan Super; Hartonyl L531, L 535; Hartopol L44, L64, P65, P85; Hetamide LA, MC, ML, RC; Hetester TICC; Hetoxide BN-13, C-30, C-200-50%, HC-40, HC-60; Hetoxol M-3, OA-3 Special, OA-5 Special, OA-10 Special, OA-20 Special; Hetsulf 40, 40X, 50A, 60S, 60T, IPA; Homotex PS-90; Hostapal BV Conc.; Humectol C, C Highly Conc.; Hypermer

Iberpenetrant-114; Icodimeen T-30; Icomeen 18-5, O-30, O-30-80%, S-5, T-2, T-5, T-7, T-20, T-25, T-25 CWS, T-40, T-40-80%; Iconol DDP-10, DNP-8, DNP-24, DNP-150, NP-7, NP-12, OP-30, OP-30-70%, WA-1; Idet 5LP, 5L SP NF; Igepal CO-210, CO-430, CO-520, CO-530, CO-730, CO-850, CO-880, CO-887, DM-970, RC-620, RC-630; Igepon CN42, T-33, T-43, T-51, T-77, TC-42, TK-32; Imerol NCP, NCP Liq.;

Imwitor 191, 595, 900, 940, 940K; Incrocas 30; Incronol ALS; Incroquat S-75 CG, SDQ-25; Incrosoft CFI; Incrosul UMS, UMS-45; Interstab ZN-18-1; Intralan Salt HA; Intraphasol PC; Intraphor AC, 1014; Intratex POK; Intravon JF, JU; Ionet S-20, S-60 C, S-80, S-85, T-20 C, T-60 C, T-80 C; Jet Amine PC, PHT, P-O, P-S, PT; Jordaphos JS-71; Jordapon CI; Kara Sperse DDL, DDL-12; Kelig 32; Kelzan D35; Kemamide B, E, E-180, E-221, O, P-181, S, S-65, S-180, S-221, U, W-20, W-39, W-40, W-40/300, W-45; Kemamine A650, A970, A974, A990, AD 650, AD 974, AS-650, AS-974, AS-974/1, AS-989, AS-990, P-150, P-150D, P-190, P-190D, P-650, P-650D, P-690, P-690D, P-790, P-790D, P-880, P-880D, P-970, P-970D, P-974, P-974D, P-989, P-989D, P-990, P-990D, P-997, P-997D, P-999; Ken-React Series, 7 (KR 7), 9S (KR 9S), 12 (KR 12), 26S (KR 26S), 33DS, (KR 33DS), 38S (KR 38S), 39DS (KR 39DS), 41B (KR 41B), 44 (KR 44), 46B (KR 46B), 55 (KR 55), 133DS (KR 133DS), 134S (KR 134S), 138D (KR 138D), 138S (KR 138S), 158 (KR 158D), 158FS (KR 158FS), 212 (KR 212), 238A (KR 238A), 238J (KR 238J), 238M (KR 238M), 238S (KR 238S), 238T (KR 238T), 262A (KR 262A), 262ES (KR 262ES), OPP2 (KR OPP2), OPPR (KR OPPR), TTS (KR TTS); Kieralon B, B Highly Conc.

Lamepon 287 SF, A, N, RE; Lan-Aqua-Sol, Lan-Aqua-Sol Hydrophillic 50, 100, Lan-Aqua-Sol Hydrophillic-Plus 50, 100, Lan-Aqua-Sol Super-Hydrophillic 50, 100, Lan-Aqua-Sol xtra-Hydrophillic 50, 100; Lankromul OSD; Lankropol OPA, WA, WN; Lanogel 21, 31, 41, 61; Lanpol 5, 10, 20, 520; Lantrol 1674 Deodorized, 1675; Lecithin W.D.; Lenkanol D-48; Leomin OR; Leophen U; Levenol TD-326; LICA 01, 09, 12, 38, 38A, 38J, 44, 97; Lignosite, Lignosite 260, 431, 458, 823, 854, 1840, KLS; Lignosol AXD, B, BD, C 60, D-10, D-30, DXD, F 30, FTA, HCX, LC, NSX 110, NSX 120, SF, SFL, SFS, SFX, SFX-65, TS, TSD, TSF, WT, X, XD; Lilamin 101 D, 115, 115 D, 140, 140 D, 142, 142 D, 151, 160, 160 D, 163, 163 D, 170, 170 D, 172, 172 D, 308 D, 310 D, 312 D, 314 D, 316 D, 342 D, 343, 345 D, 363, 364, 367 D, 368, 369, 372 D, 381, 382, 383, 540, 560, 570, 570 DO, 572; Lipal 3 TD, 4 LA, 5 L, 5 OA, 5 S, 6 TD, 9LA, 9 C, 9 L, 9N, 9 OL, 10 OA, 10 TD, 12 LA, 15 CSA, 15 T, 20 OA, 20 SA, 23 LA, 25 C, 25 S, 30 SA, 39 S, 50 OA, 50 S, 52 C, 300 DL, 300 W, 400 DL, 400 DS, 400 DW, 400 OL, 400 S, 400 W, 600 S, 600 W, 610, CE 38, CE 55, CE 64, CE 71, OE 55, OE 64, OE 70, TE 43, TE 55, TE 70, TE 76; Lipo DGLS, DGS-SE, Diglycol Laurate, EGDS, EGMS, PGMS; Lipocol B, TD-3, TD-6, TD-12; Lipolan TE, TE(P); Liponic EG-1; Liponox LCR, NC 6E, NCG, NCI, NCT; Lipopeg 2-DL, 4-DL, 4-DO, 4-DS, 4-L, 4-S, 6-L, 10-S, 15-S, 39-S, 100-S, 6000-DS; Lipotin 100, 100J, A, SB; Lomar D, DL, HP, LS, LS Liq., PL, PW, PWA, PWM, ST; Lonzest PEG-4DO, PEG 4-L

Lostat 105D; Lubrizol 2163, 2164, 2165; Lubrol 101; Lutensit A-EP, A-ES, A-FK, A-LBA, A-PS, K-LC, K-LC 80, K-OC; Lutensol A 8, A 80, AO 3, AO 5, AO 7, AO 8, AO 10, AO 11, AO 12, AO 30, AO 109, AP 6, AP 7, AP 8, AP 9, AP 10, AP 14, AP 20, AP 30, AT 11, AT 18, AT 25, AT 50, AT 80, ED 140, ED 310, ED 370, ED 610, FA 12, FSA 10, LF 400, LF 401, LF 600, LF 700, LF 711, LF 1300, LSV, ON 30, ON 50, ON 60, ON 70, ON 80, ON 110; Luviskol K12, K17, K30, K60, K80, K90; Lyogen DFT Liq., KF Liq.; LZ 01, 09, 12, 38, 44, 97; Mackanate DOS-75; Macol 2, 4, 5, 8, 10, 15, 20, 21, 24, 25, 30, 34, CSA-2, CSA-4, CSA-10, CSA-15, CSA-20, LA-4, LA-12, LA-23, LA-790, NP-4, NP-6, OA-2, OA-4, OA-5, OA-10, OA-20, SA-2, SA-5, SA-10, SA-15, SA-20, SA-40, TD-3, TD-4, TD-6, TD-8, TD-10, TD-610; Macrobase 600, 610, 620; Mafo CAB; Makon DN 50, DN 70, NI 10, NI 20, NI 30, NP-40, NP-40-70, TP 60, TP 100; Mapeg 200 DL, 200 DO, 200 DS, 200 ML, 200 MO, 200 MOT, 200 MS, 350 MS, 400 DL, 400 DO, 400 DOT, 400 DS, 400 ML, 400 MO, 400 MOT, 400 MS, 600 DL, 600 DO, 600 DOT, 600 DS, 600 ML, 600 MO, 600 MOT, 600 MS, 1000 MS, 1500 MS, 1540 DS, 4000 MS, 6000 DS, 6000 MS, EGDS, EGMS, PGDS, PGMS, S-40, TAO-15; Maphos 60A, 8135; Maprofix LK (USP)

Maracarb, N-1; Marasperse, 41G-3, C-21, CB, CBOS-3, CBX-2, N-22; Marlophen 83, 84, 85, 86, 86 S, 87, 88, 89, 810, 814, 820, 825, 850; Masil 2132, 2133, 2134; Matexil DA-AC, DN-VL500; Mazamide L-298, LM-21, O-10, O-20, SS-20, T-10, T-15, T-20; Mazoline OA, T; Mazol 300, 300K, GMO, GMO K, PGMS; Merpol HCW, OJS, SE; Merrol DBS; Miglyol 810, 812, 840; Miranate LEC; Miranol CM Conc., CM-SF Conc., DM, HM Conc., OM-SF Conc., S2M Conc., S2M-SF Conc.; Mirataine HTS, TABS; Modicol L; Monafax 1293; Monamid 7-100, 7-153 CS, 15-70W, 150-AD, 150-ADD, 150-ADY, 150-CW, 150-DR, 150-IS, 150-LMW-C, 150-LW, 150-LWA, 150-MW, 718, 770, CMA, LIPA, LMA, LMMA, R31-42, S; Monamine Series, AA-100, AC-100. ACO-100, AD-100, ADD-100, ADDS-100, ADS-100, ADY-100, AF-100, ALX-80 SS, ALX-100 S, ARA-100, CF-100 M, I-76, LM-100, R8-26, T-100; Monamulse 748; Monaquat P-TC, P-TD, P-TL, P-TZ; Monateric CA-35%, CEM-38%; Monawet MB-45, MM-80, MO-70, MO-70E, MO-70R, MO-75E, MO-84R2W, MT-70, MT-70E, MT-80H2W, SNO-35; Monazoline O, T; Monflor 32, 51, 52; Monolan 2000 E/12, PB, PC, PT; Monomer C18; Monomuls 60-10, 60-15, 60-20, 60-25, 60-25/2, 60-25/5, 60-30, 60-35, 60-40, 60-45, 90-10, 90-15, 90-20, 90-25, 90-25/2, 90-25/5, 90-30, 90-35, 90-40, 90-45; Montosol PG-17, GJ-12; Morwet D-425, DB; Musloid 815D, 815M, 815S; Myvatem 06(K), 30, 35K, 92K

Nacconol 35SL, 40F, 40G, 90F, 90G; Nalco 2335; Nansa BXS, EVM50, EVM62/H, EVM70, EVM70/B, EVM70/E, HS40-AU, HS40 Soft, HS80 Soft, HS80-AU, HS80P, HS80SK, TS 60; Naturechem PGHS, PGR; Nekal BA-77, BX-78, NF, WS-21, 25; Nekanil 910; Neodol 91-2.5, -6, -8; Neospinol 264, 358; Neutral Degras; Neutronyx 600, 656; Newcol 150, 170, 180, 290K, 290M, 290P, 560, 561H, 562, 564, 565, 566, 865, B4, B10, B18; Newkalgen NX 405 H; Newpol PE-61, -62, -64, -68, -74, -75, -78, -88; Nikkol BPS-5, BPS-

DISPERSANTS (CONT'D.)

10, BPS-15, -20, -25, -30, DDP-2, -4, -6, -8, -10, Nikkol Decaglyn 1-IS, 1-L, 1-LN, 1-M, 1-O, 1-S, DLP-10, Hexaglyn 1-S, OTP-100S, PBC-31, PBC-33, PBC-34, PBC-41, PBC-44, PBC-44(FF), PMS-1C, PMS-1CSE, TDP-2, -4, -6, -8, -10; Ninate DS 70, L70, R63, R70, PA; Nioix AK-40, AK-44, KJ-10; Niox EO-12, EO-14, 23, 32, 35, KF-12, KI Series, KJ-55, KJ-61, KQ-20, KQ-32, KQ-36, KQ-70, LQ-13; Nissan Amine AB, ABT, ABT_2, BB, DT, DT, FB, MB, OB, PB, SB, VB; Nissan Cation AB, ABT-350, 500, AR-4, BB, F_2-10R, -20R, -40E, -50, FB, FB-500, L-207, M_2-100, MA, SA, PB-40, -300, S_2-100; Nissan Nonion E-205, 215, 230, K-202, -203, -204, -207, -211, -215, -220, -230, P-208, P-210, P-213, S-207, -215, -220; Nissan Nymeen DT-203, -208, L-201, -202, -207, S-202, -204, -210, -215, -220, T_2-206, -210, -230, -260; Nissan Persoft NK-60, -100, SK; Nissan Plonon 102, 104, 108, 171, 172, 201, 204, 208; Nissan Polystar OM, OMP; Nissan Sunbase, Powd.; Nitrene C, C Extra, L-76, N; Noigen ES90, 120, 140, 160; Noiox AK-41, AK-44; Nonal 206, 208, 210; Nonionic 1017-R, 1025-R, 1035-L, 1044-L, 1061-L, 1068-L, 1088-L, 2017-R, 2025-R, 4017-R, 4025-R, 5025-R ; Nonipol Soft SS-50, -70, -90; Nonisol 210; Nopalcol Series, 4-O

Nopco Colorsperse 66-A, 188-A; Nopco NDW; Nopcosant; Nopcosant K, L; Nopcosperse 44, 303-SD, AD-6; Norfox NP-4; Norlig; Norlig 415, A, NH; Nutrol 100, 600, 611; OHlan; Onyxol 336, 345; OP-2000; Orzan A, AE, AL-50, CG, LS, LS-50, S, SL-50; Osimol Grunau DP, PHT, SF; Oxetal C110, D104, O108/112, S125, T103, T106/110, 111/118; Pegafac PB-92, CA 600, CE 410, 610, CP 710, CR 719, CS 410, 610, CS 710, PEH, PNP 9; Pegameen 02, 030 30 80%, C2, C5, C10, C25, S2, S5, S20, T2, T5, T15, T20; Peganol A 24, 26, B 26, CL 214, 225, 240, CSS 10, 25, D 25, DA 211, 214, DDP 6, 10, DNP 8, 24, 150, LA 4, 9, NP 1.5, NP 4, NP 5, NP 6, NP 15, 20, 30, 30 70%, 40 70 %, 100 70%, OAL 20, 23, OP 3, 6, 8, 10, 20, 30, 30 70%, 40, 40 70%, OP 70, 70 70%, SFE, TDA 6, 8; Pegnol C-14, -18, -20, L-6, L-8, -10, -12, -15, -20, O-6, -16, OA-400; Pegol 17 R1, 17 R2, 25 R1, 25 R2, 31 R1, L 31, L 35, L 43, L 61, L 62, L 62 LF, L 64, L 81, L 101, L 121, P 85; Pegosperse 50 DS, 50 MS, 100 L, 100 ML, 100 MR, 100 O, 100 S, 200 DL, 200 ML, 400 DL, 400 DO, 400 DS, 400 DTR, 400 MC, 400 ML, 400 MO, 400 MOT, 400 MS, 600 ML, 600 MS, 700 TO, 1000 MS, 1500 DO, 1500 MS, 1750 MS, 4000 MS, 6000 DS, 9000 CO, EGMS-70, MFE; Peltex; Pentex 99; Peregal ST; Perlankrol EAD-60, ESD, ESD-60, ESS-25, N Range, S Range; Permalene A-100; Pestilizer A, B, J, M, N; Petrac 250, 270; Petro Dispersant 98, 425; Petrolite WB-5, WB-16, WB-17; Petronate CR, HL, K, L, S; Petrosul 550, 750, Neutral Barium, Neutral Calcium; Petrowet R; Phosfac 1004, 1006, 1044, 1044FA, 1066, 1066FA, 1068FA, 5513; Phospholan PRP-5; Pilot SXS-96; Plasdone K-25, K-26/28, K-29/32, K-90; Plastiflow CW-2; Pluracol E-300, E-400, E-400 NF, E-600, E-600 NF, E-1000, E-1450, E-1450 NF, E-1500, E-2000, E-4000, E-4000 NF, E-4500, E-6000, E-8000, E-8000 NF

Pluradot HA-410, -420, -430, -433, -440, -450, -510, -520, -530, -540, -550; Plurafac RA-20, RA-40, RA-43; Pluraflor E4A, E4B, E5A, E5BG, E5G, N5G; Pluriol E 200, E 300, E 400, E 600, E 1500, E 4000, E 6000, E 9000, PE 3100, PE 6400, PE 6800, PE 9400; Pluronic 10R5, 10R8, 12R3, 17R1, 17R2, 17R4, 17R8, 22R4, 25R1, 25R2, 25R4, 25R5, 31R2, 31R4, F38, F68, F68LF, F77, F87, F88, F98, F108, F127, L10, L31, L35, L42, L43, L44, L61, L62, L62D, L62LF, L63, L64, L72, L81, L92, L101, L121, L122, P65, P75, P84, P85, P94, P103, P104, P105, P123; Pogol 300, 400, 1540, 1570; Poiz 530; Poly-Tergent P-17D, -22A; Polychol 15, 20-40; Polyfon F, H, O, OD, T; Polylan; Polylube #745; Polymer QR-1010; Polymulse 6; Polypeg E-400; Polyphos; Polystep F-1, -2, -3, PN-209; Polywet ND-1, -2, Z1766; Primarol 1208; Proaid 9802, 9810, 9831; Procetyl AWS; Product 100;Product BCO; Proplast 015; Propoxyl 1695; Protodyne QNF; Protowet TJ; Provol 10, 30, 50; Puxol CB-22; PVP K-90; Pyronate 40; Pyroter GPI-25; Radiamine 6140, 6141, 6160, 6161, 6163, 6164, 6170, 6171, 6172, 6173, 6540, 6560, 6570, 6572; Radiamuls 125, 135, 137, 145, 147, 155, 157, 345, PEG; Radiasurf 7125, 7135, 7137, 7145, 7147, 7196, 7201, 7206, 7270, 7400, 7402, 7403, 7404, 7410, 7411, 7412, 7413, 7414, 7417, 7420, 7421, 7422, 7423, 7431, 7432, 7443, 7453, 7454, 7600; Ray Krome 340, 400, FE; Raylig; Raymix; Reax 15B, 45A, 45L, 77, 81A, 82, 83A, 85A, 88B, 90, 90B, 92, 100M, 100M (Liquid), SR-1, SR-7; Renex 36; Rewominoxid B 204; Rewopal C 6, O 8, TA 11, 25, TA 25/S; Rewophat E 1027, EAK 8190; Rexol 25/6, 25/9, 25/10, 25/20, 35/100, 45/1, 45/3, 45/5, 45/7, 45/12, 45/16, 45/307, 45/407, 130, AE-23; Rexonic 1006, N-4, N25-14 85%, N91-1.6, N91-2.5, N91-6, N91-8, P-1; Rheodol SP-010, SP-030; Rheotol; Ritawax AEO; Rudol; Runox 1000; Rychem 400-OE, 808, CLS, D941, Rychem-G-Gard, G-Hib

Sandopan KD Conc., TFL Liquid; Sandoxylate 206, 224, 408, 412, 418, 424, AC-9, AD-4, AD-6, AD-9, AL-4, AO-12, AO-20, AO-60, AT-6.5, AT-12, C-10, C-15, C-32, FO-9, FO-30/70, FS-9, FS-35, NC-5, NC-15, NSO-30, NT-5, NT-15, PDN-7, PN-6, PN-9, PN-10.5, PO-5; Santelle-EOM K; Santone 3-1-S, 10-10-S; Sanyo Levelon; Sartomer 801, 802, 803; Schercemol DEIS; Schercomid AME, AME-70, CCD, CDA, CDA-H, HT-60, ODA, SLS, TO-2; Schercozoline B, C, L, O; SCMS; Scripset 500, 520, 540, 550, 720, 808; Secosol DOS/70; Secoster CL 10, CP 10, CS 10, DO 600, DS 6000, KL 10, KP 10, KS 10, MA 300, ML 300, ML 4000, MO 100, MO 400; Sellogen 641, HR, W, WL; Semtol Series; Serdet DSK 40; Serdet Perle Conc.; Serdox NJAD 15, NJAD 20, NJAD 30, NNP 4, NNP 30/70; Serpol QPA 150, 160; Serwet WH 170; Silwet L-77, L-7001, L-7500, L-7602, L-7604; Sipex OLS, TEA; Siponate 330; Siponic NP-4, NP-7, NP-9.5, NP-10, NP-13, NP-75, NP-407, NP-707, Y500-70, Y501; SMA 1000, 1440, 1440H, 2000, 2625, 3000; S-Maz 20, 40, 60, 60 K, 60 KHM, 65, 65 K, 67, 80, 80 K, 83R, 85, 85 K, 90, 93 R; Soft Detergent 95; Softenol

3900, 3991; Sokalan CP 2, CP 2 Powd., CP 7, CP 8, CP 9, CP 10, CP 10S, CP 11, CP 12S, CP 13S, CP 45, HP 50, PA 15, PA 20, PA 20 PN, PA 25 PN, PA 30, PA 40, PA 40 Powd., PA 50, PA 70, PA 80, PA 80 S, PA 110 S; Solar NP; Soltem 8 FL/N, 70, 207, SFL/1; Solulan 5, 16, 25, 75, 97, 98, C-24, L-575, PB-2, PB-5, PB-10, PB-20; Soprofor FL, NO/20, /30, /100, PA/17, PA/19, PS/17, PS/19; Sorbeth 40HO, 55HO; Sorbon S-20, T-20, TR 814, TR 843; Sotex 3CW, COS #2, CW, N; SoyaLee WDF-FG; Stafoam; Standamid KDM; Standamul CTA, CTV, G; Standapol SCO; Stepan D-50, D-70; Stepantan A, NP 80; Stepantex DA 52; Stepfac PN 209; Strodex MO-100, MOK-70, MR-100, P-100, PK-80A, PK-95G, PK-90, PSK-28, SE-100, SEK-50, Super V-8; Sulfonated Castor Oil 50%, 75%, GTO; Sulfopon WA 1; Sulfotex OA, OT, PAI, PAI-S; Sulframin 40, 1288, 1298, 1388; Sunaptol CA 25, 120, 350, 400, 750, OA, OA 70, 90, 100, 140, P Extra Liquid; Sunsolt RZ-2, -6; Super Hartolan; Supragil MNS/90; Supratol VF; Surfactol 13, 318, 365, 575, 590; Surfax 250; Surfine AZI-A, WLL, WNT Conc., WNT Gel, WNT LC, WNT LS; Surflow-S24, S32, S32; Surfonic HDL, JL-80X, N-10, N-31.5, N-40, N-60, N-85, N-95, N-100, N-102, N-120, N-150, N-200, N-300, N-400, NB-5, NB-14; Surfynol 104, 104A, 104BC, 104E, 104H, 104S; Swanol AM-301; Syn Fac 334, 905, 8017, 8009, 8031, 8210, 8216, 8337, TEA-97; Synperonic A14, A20, NP10, NP12, PE30/10, 30/20, 30/40, 30/80, 39/70; Syntergent TER-1

Tagat I, I2, L, 2, O, O2, R1, R40, R60, S, S2, TO; Tamol 165, 731-25%, 731-SD, 819, 819L-43, 850, 901, 960, 963, 983, L Conc., NH 3901, 7519, NH 9103, NHC 3001, NHC 4001, NHC 9301, NMC 9301, NN, NN 2406, NN 2901, NN 4501, NN 7718, NN 8906, NN 9104, NN 9401, NNa 4109, NNO, NNOK, PA Liq., PA Powd., PP, SD, SG-1, SN; Tauranol I-78, I-78 Flakes, I-78-3; T-Det BP-1, C-40, N-4, N-6, N-8, N-9.5, N-10.5, O-6, TDA-60, TDA-65, TDA-70; Tegopren; Tek Tan NL; Tek-Wet 951, 955; Temsperse S 002, S 003; Tergitol 24-L-45, 24-L-60, 24-L-60N, 24-L-75, 24-L-92, 24-L-98N, 26-L-1.6, 26-L-3, 26-L-5, NP-7, NP-8, NP-9, NP-10, TMN-3, XD; Teric 12A3, 12A6, 12A9, 12A12, 12A16, 12A23, 12M2, 12M5, 12M15, 13A9, 16A16, 16A22, 16A29, 16M2, 16M5, 16M10, 16M15, 17A8, 17A10, 17A13, 17A25, 17M2, 17M5, 17M15, 18M2, 18M5, 18M10, 18M20, 18M30, 160, 161, 203, BL8, BL9, CME3, CME7, G9A5, G9A6, G9A8, G9A12, G12A4, G12A6, G12A8, G12A12, LA4, LA8, LAN70, N2, N3, N4, N5, N8, N9, N10, N11, N12, N13, PE61, PE62, PE64, PE68, T2, T5, T7, T10, X5, X7, X8, X10, X11, X13, X16, X40; Tetronic 50R1, 50R4, 50R8, 70R1, 70R2, 70R4, 90R1, 90R4, 90R8, 110R1, 110R2, 110R7, 130R1, 130R2, 150R1, 150R4, 150R8, 304, 504, 701, 702, 704, 707, 904, 908, 1101, 1102, 1107, 1301, 1302, 1304, 1307, 1501, 1502, 1504, 1508; Texapon K-12, L-100, Z Granules; Textamine 05, A-5-D, A-W-5, O-1, O-5, T-1, T-5-D; Tex-Wet 1002, 1158; T-Maz 20, 28, 40, 60, 60K, 60KHM, 61, 65, 65K, 67, 80, 80KLM, 81, 81K, 85, 85K, 90, 95; Toximul 80, 351; Tris Amino; Triton N-40, N-101, X-45, X-100, X-120; Trycol 5940, 5946, 5971, 5972, 6954, 6957, 6962, 6963, 6970, 6971, 6972, 6974, 6984, 6989, DNP-150, DNP-150/50, NP-6, OAL-23, OP-407, TDA-6, TDA-18, TP-6; Tryfac 610-K, 5556, 5571, 5573, 5576, 5557, 5560; Trylox 5900, 5906, CO-5, CO-16, CO-30; Trymeen 6601, 6602, 6640, CAM-10; Tylose C, CB Series, MH, MHB, MB, MH-K, MH-xp, MHB-y, MHB-yp

Ucar Cement Grinding Aids, Super Wetter FP; Ucarsil MD-200; Ultra Sulfate AE-3, SE-5, SL-1; Unihib 106, 314; Union Carbide L-7001, -7002, -7500, -7600, -7602, -7604, -7605, -7607; Uniperol O; Unisol UA 20, 80, 150, 300, 400; Value 3706; Vanisol BIS sodico-2; Varidri 40; Variquat B200, K300; Varonic K210, K215, K215LC, L202, L230, L230-60, L230-80, T202, T202 SR, U205, U205LC; Varstat K-22, T-22; Varsulf SBDO-70; Victawet 12, 35B, 58B, 85X; Vikosperse KDS; Vinsalyn Resins; Volpo 25D3, 25D5, 25D10, 25D20, CS3, CS5, CS10, CS15, CS20, L3 Special, L4, N3, N5, N10, N15, N20, O5, O10, O15, O20, T3, T5, T10, T15, T20; V-Pyrol; Vultamol, SA Liq.; Vybar 103; Vykamol N/E, S/E; Wayplex 55-A, 55-S; Wettol D 1, D 2; White Swan; Witamide 82, 272, 511, 5130, 5133, 5138, 5140, 5195, AL69-58; Witcamine 209, 211, 235, AL42-12, PA-60B, PA-78B; Witco 97H Acid, 912, 915, 918, 934, 1298 Hard Acid, Acid B, D51-29; Witcolate 58, 1276, 7031, D-510, D51-51; Witconate 605A, 605T, 1075X, 1840X, P-1059; Witconol APEM, APM, APS, F26-46, H-31, NP-100, NS500K; Yeoman; Zonyl FSA, FSB, FSC, FSN, FSP; Zusolat 1005/85, 1008/85, 1010/85, 1012/85

DRYING AGENTS

Al-0104 T 3/16″, Al-1404 T 3/16″; Siccatol Ba-12.5, Ca-5, -10, -104, -105, Ce-6, -10, Co-6, -8, -10, -12, Mn-6, -10, Pb-18, -24, -30, -32, -33, Sr-6, -10, Zr-6, -12, -18; Troymax Drier Calcium 4%, 5%, 6%, 8%, 10%, Cobalt 6%, Iron 9%, Lead 36%, Manganese 6%, 9%, 12%, Zinc 8%, 16%, Zirconium 12%, 18%, 24%

DYEING ASSISTANTS

Ahco AJ-110; Airrol CT-1; Alkaminox C-2, C-5; Alkanol A-CN, CNR, ND, WXN; Alkaquat C; Alkasurf CO-15, O-9, OA-20, TA-20, TA-30, TA-40, TA-50, TDA-6; Alkatronic 25-8, 25-4, EGE 17-1, 17-2, 17-4, 17-8, 25-1, 25-5; Amiet 102, 105, 110, 115, 202, 205, 210, 215, 302, 305, 310, 315, 402, 405, 410, 415, 502, 510, 515; Arquad 2C-75, 16-29, 16-50, 18-50, S-50, T-2C-50, T-50; Atlas WA-108; Blancol; Blancol N; Castor Oil Sulfated 50%, 68%; Catinal HTB; Catinex KB-45; Cenegen NWA; Cenekol Liq., NCS Liq.; Chemax E-200 M, E-400 MS, TO-8, TO-16; Chemeen DT-3, DT-15, DT-30, O-30, O-30/80; Cithrol 2ML, 4ML, 6ML, 10ML, 60ML; Conco XA-C, XA-L, XA-M, XA-MC, XA-S, XA-T, XA-Y; Cordon NU 890/

DYEING ASSISTANTS (CONT'D.)

75; Crestomul T; Crotein A, C, O; DeSomeen TA-2, TA-5, TA-15, TA-20; DeSonic 6C, 30C, 54C, S-405; Diazital O; Doittol APS Conc.; Dowfax 2A1, 2EP, 3B0, XDS 8292.00; Dyetone; Eccoterge EO, SCH; Emkal BNS, BNS Acid, BNX Powd., NNS, NNS Acid, NOBS; Emkane Acid, HAD, HAL, HAX; Emkapon Jel BS; Emkatex AES, PX29; Empilan DL 40, DL 100, KA3, KA5, KA590, KA880, KA1080; Emulgane O; Emulphor EL-980, TO-530, TO-639; Ethoduoquad T/20; Ethoquad 18/12, 18/25, C/12, C/25, O/12, O/25; Examide-DA

Gaftex CK-160, KF; Ganex P, P904, V, V216, V220, V516; Gardinol CX; Gemtex PA-75, PA-85P, PAX-60, SC-40, SC-70, SC-75, SC-75E, SC Powd.; Geropon MLS/A; Grocor 5721; Hartonyl L535; Hetoxide C-30, C-200-50%, HC-40, HC-60; Hipochem ADN, C-95, CAD, CDL, D2, Dispersol GTO, Dispersol SB, Dispersol SCO, Dispersol SP, LCA, No. 40-L, No. 641, Retarder CJ; Iberquestrene; Igepal CA-880, CA-887, CA-890, CA-897, CA-950, CO-890, CO-897, CO-970, CO-977, CO-980, CO-985, CO-987, CO-990, CO-997; Intralan Salt N; Intratex OR; Irgalube 53; Irgastab 2002; Kalex; Katanol 387; Katapol OA-860, OA-910, PN-730, PN-810; Kemamine BQ-2802C, BQ-9702C, BQ-9742C, Q-1902C, Q-2802C, Q-6502C, Q-6503 B, Q-9702C, Q-9743C, Q-9743CHGW, ; Leonil DB Powd.; Liponox OCS; Lyogen BE Liq., DFT Liq., F Liq., MS Liq., NL Liq., P Liq., PAA Liq., SF Liq., SMK-40 Liq., SMK Paste, WD Liq.; Marvanol 75% SCO, DC, GD, Penetrant 35, RE-1274, RE-1281; Marvelin W-50; Matexil DN-VL500, LC-CWL, LN-PT, LN-RD, PN-HT; Merpol DA, HCS, LF-H; Monogen; Monosulph; Multinol C; Newcol 1105, 1120, 1200, 1203, 1204, 1208, 1210, 1305, 1310, 1515, 1525, 1545; Niox KH Series; Nissan Diapon T, TO; Peregal O; Permalene A-100; Protopan; Protosan EP, ES-100; Protowet 3098, E-4, MB, NRW

DYES AND PIGMENTS

Abluton CMN, CTP, Acrilev AM, AM-Special; Bartex 80; Hitox; Hombitan LOCR-K, LW, R 101 D, R 610 K, R 610 L; Huber ARO 60, N110, N220, N234, N299, N326, N330, N339, N343, N347, N351, N375, N539, N550, N650, N660, N683, N762, N774, N787, N990, S212, S315; Lithopone 30% DS, 30% L, 60% DS, 60% L, D, DS, L; Nalzin 2, ZP; Ninex TDO 5, 9, 14; Nissan Persoft EK, NK-60, -100; Nissan Trax H-45; Norfox Vertex Flakes; Nylomine Assistant DN; Oncor F-31; Onyxol 42; Peganol NP 40, 50, 100; Sachtolith HD, HD-S, L; Tegocolor; Thermax Floform N-990, Floform Ultra Pure N-990, Powder N-991, Powder Ultra Pure N-991, Stainless Floform N-907, Stainless Powder N-908, Stainless Powder Ultra Pure N-908; Zinc Oxide No. 185, 318; Zotex 319

EMOLLIENTS

AA Standard; AA USP; Abil B 8842, B 8843, B 8847, B 8851, B 8852, B 8873, B 9800, B 9801, B 9806, K 4, ZP 2434; Abil-Wax 2434, 2440, 9800, 9801; Acetol 1706; Acetulan; Acylan; Acylglutamate CS-11, CT-12, GS-11, HS-11, LS-11, MS-11; Adol 42, 52, 52 NF, 54, 60, 61 NF, 62, 62 NF, 63, 64, 66, 80, 85, 85 NF, 90, 90 NF, 330, 340, 520, 520 NF, 620, 620 NF, 630, 640; Aethoxal B; Albalan; Alcolan, 36W; Aldo HMS, PMS; Alfol 16, 18, 20; Alkamox C2-O; Alkamuls GMS; Alkapol PEG 1000; Amerchol 400, BL, C, CAB, H-9, L-99, L-101, L-500, RC; Amerlate LFA, P, W; Ammonyx 4, 4B, 4-IPA, 485, 490, 4002, CA-Special, CETAC, CETAC-30; Amphocerin K; Amyx A-25S; Antaron FC-34; Aqualose L 30, L 75, L75/50, LL 100, SLT, SLW, W 20/50; Argobase 125, EU, LI, MS-5, SI; Argonol 40, 50 Pharmaceutical, 50 Super, 60, ACE5; Arlamol E, GM, S3, S7; Arnica Oil CLR; Aromox DM16D-W, DM18D-W; Arosurf 66-E2, 66-E10, 66-E20, 66-PE12; Avocado Oil CLR; Barlox 10S, 12, 14; Barre Common Degras; Berol 371, 372 Flakes, 374; Biocare Polymer HA-24; Bio-Surf PBC-460; Bumyr

Cachalot C-50, M-43, S-56; Calamide C, O; Calendula Oil CLR; Captex 200, 300, 350, 810B; Carbopol 1706, 1720, 1731, 1754; Carbowax Methoxy PEG; Carbowax Sentry; Carnation; Carrot Oil CLR; Carsodet TD-7C; Cegesoft C 17, C 19, C 25; Centrol Series; Centrolene Series; Centrolex Series; Ceralan; Ceraphyl 41, 45, 50, 50-S, 55, 60, 65, 70, 85, 140, 230, 424, 494, GA, ICA, IPL; Cetal; Cetax 16, 18, 50, DR; Cetina; Cetiol, 868, 1414E, B, G16S, G20S, HE, J 600. LC, MM, R, S, SB, SN, V; Chemical Base 6532; Chimipal FV; CO-1695; CO-1895; Cobee 76, 92, 110; Comperlan OD, VOD; Corona Lanolin; Coronet Lanolin; Cosmetol X; Cralane KR-13, KR-14, LR-10, LR-11; Cremba, B6; Crestalans; Crodacol CS50; Crodalan 0477, AWS, IPL, LA; Crodamol AC, BE, BS, CAP, CL, CP, CSL, CSS, DA, DO, DOA, ICS, IPM, IPP, ISNP, LL, ML, MM, OHS, PC, PMP, PTC, PTIS, SS, W; Croduret 10, 30, 40, 50 Special, 60, 100; Crossential EPO; Crotein AD, ADX; Crystal Crown; Crystal O; Cutina CP-A; Cyclochem CL, CP, EM 560, GMO, GMS 21, GMS 165, GTIS, GTL, GTO, INEO, LVL, ML, MM/M, MST, PEG 200 DL, PEG 200 ML, PEG 400 DS, PEG 400 ML, PEG 400 MO, PETO, SS; Cyclomide DL203/S, DL207/S, DO280/S; Cyclomox CO, SO; Cycloton M242C/29

Dar Chem-030; Deltyl Extra; Dermalcare EGMS/SE, GMS, GMS-165, GMS/SE, GTIS, HL, LVL, MM/M, MST, PGMS, SDG, SS; Diamond Quality; Dist. Lipolan; Dow Corning 556 Fluid, QF1-3593A; DPPG; Drakeol; Dusoran MD; Dynacerin 660; Dynasan 110; Edenol 302; Elfacos ST 9, ST 37; Emcol 1655, 3780, 4072, E-607L, E-607S, ISML, L, M; EmCon E, E-5, E-20, E-L; Emerest 1723, 2310, 2314, 2316, 2325,

2388, 2410, 2452, 2486, 11723; Emersol Stearic Acid; Emery 1650, 1656, 1660, 1720, 1730, 1732, 1740, 1747, 1780, 1781, 1787; Emid 6515; Emphos D70-30C; Empilan 2125-AU, EGMS, KB 2, KB 3, KC 3; Emsorb 2729; Emthox 2730, 5964, 5967; Emulan; Emulsynt GDL; Ervol; Espesilor AC Series; Ethosperse OA-2; Ethoxyol 1707; Etocas 10, 30, 35, 40, 60, 100; Eutanol G, G16; Exceparl HO, IPM, IPP, OD-M, OD-OL, TGO; F-221, F-222, F-251, F-755; Falco S-14S, S-18S; Fanchem HL; Fancol ALA, HL, LA, LAO; Fancor Lanwax, LFA; Finsolv SB, TN, P-8; Fluilan, AWS; Fluilanol; Foamole B; Forlan; Forlan C-24, L

Generol 122, 122E5, 122E10, 122E16, 122E25; Gloria; Glucam E-10, E-20, E-20 Distearate, P-10, P-20, P-20 Distearate; Glucate DO; Glycerox; Golden Dawn Grade 1,2, Lanolin; Golden Fleece DF, Lanolin, P-80, P-95, RA; Grocor 5820 G; Haroil SCO-65, SCO-7525; Hartolan; Hartolan Super; Hartolite; HD Eutanol, HD-Ocenol, HD-Ocenol 92/96; Hetester ISS, PCA, PHA, PMA, SSS, TICC; Hetlan AC; Hetoxide BN-13, BP-3, BY-3, C-2, C-9, C-15, C-25, C-30, C-40, C-200, C-200-50%, DNP-4, DNP-9.6, G-7, HC-16, HC 40, HC-60, MTG, NP-4, NP-9, NP-40; Hetoxol PLA, SP-15; Hexaplant Richter; Hexol Q; Hostacerin T-3; Hydrenol DD; Iscolan; Isocreme; Isopropylan 33; Imwitor 742; Incrocas 10, 30, 40, 60, 100; Incromide LA; Incromine BB, OPM, Oxide M; Incroquat Behenyl TMS, CR Conc., CTC-25, CTC-30, S-85; Intermediate 300, 325, 512; Jeffox WL, WL-660, WL-1400, WL-5000; Jordamine DAPI, DAPL, DMCAPA, DAPSA, S-13, SHCFA; Jordamox SDA; Jordaquat 40, JS-25; Kaydol; Kemester 104, 213, 1000, 3681, 4000, 5221 SE, 5415, 5500, 5510, 5654, 5721, 5822, 6000, 6000 SE, BE, CP, DMP, EE, GMS (Powd.), JO, MM, THFO; Kessco 639, Acetin, BSC, Butyl Oleate, Butyl Stearate Cosmetic, Dist., Diacetin, Glycerol Dilaurate, Glycerol Monolaurate, Glycerol Monooleate, Glycerol Monostearate 860, Glycerol Monostearate DH-1, Glycerol Monostearate Pure, Glycerol Monostearate SE, IBP, ICS, IPM, IPP, Isobutyl Palmitate, Isobutyl Stearate, Isocetyl Stearate, Isopropyl Myristate, Isopropyl Palmitate, Myristyl Myristate, Octyl Isononoate, Octyl Oxystearate, Octyl Palmitate, Triacetin; Klearol

Lanacet 1705; Lan-Aqua-Sol, Hydrophillic 50, 100, Hydrophillic-Plus 50, 100, Super-Hydrophillic 50, 100, xtra-Hydrophillic 50, 100; Lanesta L, S, SA30; Laneto 40, 50, 60, 100; Lanette 14, 16, 18, 18-22, 22, O; Lanex; Lanexol, Lanexol AWS; Lanfarax 1776, 1777 Deoderized; Lanocerin; Lan-O-Derm; Lanogel 21, 31, 41, 61; Lanol C, CS, CT, ST, S; Lanolic Acid; Lanolin Alcohols LG, LO, THG, THO; Lanoquat 1757; Lanotein AWS 30; Lanotex 730; Lantrol, 1673, 1674, AWS, AWS 1692; Lauramide 11, D, ME, R, S; Laurex CH, CS, L1, NC, PKH; Lecithin Water Dispersible CLR; Lexate IL, PX, TA, TL; Lexemul AR, AS, T; Lexol 60, 3975, DIA, EHP, GT 855, GT 865, IPM, IPP, PG 800, PG 855, PG 865, PG 900, SS; Lexquat AMG-M, AMG-O; Lipal EB, ST; Lipo DGLS, DGS-SE, Diglycol Laurate, EGDS, EGMS, GMS 450, GMS 470; Lipocol, Lipocol C, L, O, S; Lipolan, Lipolan 31, S; Lipomulse 165; Liponate 143M, CL, CRM, DPC-6, CG, IPM, IPP, ML, MM, NPGC-2, PB-4, PC, PO-4, PS-4, SPS, SS, TDS, TDTM; Lipoquat R; Lipovol A, ALM, ALM-S, CP, G, J, P, P-S, SAF, SES, SES-S, SO, SUN, W, WGO; Liquester; Liquid Absorption Base Type A, T; Liquilan; Lonzest 143-S; Lutrol E 300, E 400; Luvitol EHO, HP; Mackam HV, OB; Mackanate WGD; Macol 620; Mafo 13; Mapeg 200 DOT, 400 DSLM, CO-16H, CO-30, CO-36, CO-200, S-100, S-150; Marlamid D 1218, DF 1218, D 1885, M 1218; Marlipal BS; Marlopon CA; Mazawax 163R; Mazol 159, 165C, GDO, GMO, GMR, GMS-D, GMS-HM, PGS-61; Mazon 1045A, 1086, 1096; Merpinamid KD 11, KD 12, KM, LD/E, KM, LMIPA, LSD, LSD/E, LSM, OD; Michel XO-144; Miglyol 810, 818, 829, 840, 840 Gel; Miranol Ester PO-LM4, H2M Conc.; Mirapol 9, 95, 175; M.O.D.; Modulan; Monomuls 90-L12; Myritol 318; Myvacet 5-07(K), 7-00, 7-07(K), 9-08(K), 9-40, 9-45

Naturechem CR, EGHS, GMHS, OHS, PGHS, THS-200, Neobee 18, M-20; Nikkol Behenyl Alcohol 65, 80, BPS-5, BPS-10, BPS-15, -20, -25, -30, BWA-10, DGO-80, DGS-80, GM-18IS, 18S, GO-430, GO-440, GO-460, GS-460, ICIS, ICS-R, Jojoba Oil N, Oil S, N-SP, PEMS, PMS-1C, PMS-1CSE, Syncelane 30, Trifat S-308; Nimco 1780, 1781; Nimcolan 1740, 1747; Nimlesterol 1730, 1732; Ninol CNR, SNR; Ninox CS, S; Nissan Cation PB-40, -300, S_2-100; Nissan Diapon T, TO; Nissan Dispanol 16A, TOC; Nissan Monogly L, M; Nissan Nonion CP 08R, DN-202, -203, -209, DS-60HN, E-205, 215, 230, HS-204.5, -206, -208, -210, -215, -220, -240, -270, K-202, -203, -204, -207, -211, -215, -220, -230, L-2, -4, LP-20-R, LP-20-RS, LT-221, NS-202, NS-204.5, NS-206, NS-208.5, NS-210, NS-212, NS-215, NS-220, NS-230, NS-240, -270, O-2, O-3, O-4, O-6, OT-221, P-6, P-208, P-210, P-213, PP-40R, PT-221, S-2, S-4, S-6, S-10, S-15, S-15.4, S-40, S-207, -215, -220, S-207, S-215, S-220; Nopcocastor, L; Novol; OHlan; OP-2000; Orzol; Oxypon 288, 306, 2145; Paracin 9, 13, 15; Pationic NMF; Pegosperse 50 DS, 200 DL, 1750 MS, 6000 DS, EGMS-70; Peneteck; Penreco Amber, Blond, Cream, Frost, Lily, Regent, Royal, Snow, Super, Ultima; Perfecta; Perlganz-Konzentrat B-30, B-48 deter; Polyaldo DGDO, DGHO, HGDS, TGMS; Polychol 15; Polylan; Polytrap 158; Procetyl 10, 20, 30, 50, AWS; Prochol, 30; Produkt RT 288; Promulgen D, G; Promyr; Promyristyl PM, PM-3; Propoxyl 1695; Prostearyl 15; Protamine-45; Protectein; Protegin W, WX, X; Protol; Protopet Alba; Protopet White 2L, 3C; Protopet Yellow 1E, 2A; Provol 10, 30, 50; Purton CFD, SFD; Pyroter GPI-25

Radia 7131; Radianol 7106, 7376; Radiasurf 7125, 7135, 7137, 7145, 7147, 7600; Rewocid U 185; Rewoderm ES 90, L 67-75, LI 48-50, LI S 75; Rewolan AWS; Rewolan E 50, 100; Rewolan LP, SPS; Relowub KSM 14, 80; Rewomid DC 220/LS, DC 220/SE, DL 203, DL 240, DLM/SE, DO 280/S, F, S 280; Ritacetyl; Ritachol; Ritaderm; Ritahydrox; Ritalafa; Ritalan, Ritalan AWS; Ritalan C; Ritasol; Ritawax; Ritawax 5, 15, 40, AEO, ALA; Robane; Robecote; Robeyl; Sandoz Amine Oxide XA-C; Satexlan; Satulan; Scher-

EMOLLIENTS (CONT'D.)

camox CMA, DMS; Schercemol 185, 318, 1688, 1818, CL, CM, CP, CS, DED, DIA, DICA, DID, DIS, DISD, DO, GMIS, ICS, IPM, IPO, LL, MEL-3, MEM-3, MEP-3, ML, MM, MP, MS, OLO, OP, OPG, PGDP, PGML, TIST; Schercodine O; Schercomid IMI, OMI, SLE, SLS; Schercopol OMS-Na; Scheroba Oil; Sebase; Secoster DMS, EMS; Sedaplant Richter; SF 96, 96(5), 1173, 1188, 1201, 1202, 1204; Silicone C111; Simchin; Sipex OS; Siponic E-2, E-3, E-5, E-7, E-10, E-15, L1, L4, L7, L12, L16, L25, TD-9-90; Skliro; Softenol 3178, 3819; Softigen 701, 767; Softisan 100, 378, 601, 649; Solan, 50, E; Solidester; Solulan 5, 16, 25, 75, 97, 98, C-24, L-575, PB-2, PB-5, PB-10, PB-20; Solwax LG 35; Sontex 19, 21, 35, 55, 70, 75, 75 T, 85, 85 T, 95 T, 100, 150; Sorba; Squalane; SS 4267; Standamid 100, KD, LD, LP, PD; Standamox CAW; Standamul 302, 318, 1414-E, 7061, 7063, CTA, CTV, G-16, G-32/36, G-32/36 Stearate, HE, LC, O-5, O-10, O-20, OXL; Standapol BC-35, SCO; Starfol BB, CP, IPM, IPP, IS, ODS, OS, Wax CG, Wax CG NF; Stearal; Stepan D-50, D-70; Super Corona; Super Hartolan; Super Refined Almond Oil NF, Apricot Kernel Oil NF, Avocado Oil, Babassu Oil, Coconut Oil, Crossential EPO, Grapeseed Oil, Menhaden Oil, Mink Oil, Olive Oil, Orange Roughy Oil, Peanut Oil, Safflower Oil USP, Sesame Oil, Shark Oil, Soybean Oil, Wheat Germ Oil; Super-Sat, AWS-4; Super Solan Flaked; Super Solangel 25; Super Sterol Ester; Supraene; Surfactol 318, 365, 575, 590; Syncrowax AWI-C, BB4, BB5, ERL-C, HGL-C, HR-C, HRS-C; Synotol CN, ME-90

Tegin E-41, E-41 NSE, E-61, E-61 NSE, E-66, E-66 NSE, L 61, L-61 NSE, L 62, L-62 NSE, RZ, RZNSE, VA 55G; Tegosoft 168, 189; Trisolan 1720; Ucon Fluid 50-HB-Series, 75-H-1400, LB-, -625, -1145, -1715, -3000, LO-500; Unimate BYS, DBS, DIPS, DOS, EHP, IPM, IPP, IPPM; Union Carbide L-45 Series; Velsan D8P-3, P 8-3, P 8-16; Vigilan, AWS; Viscasil Fluids; Volatile Silicone 7158, 7207, 7349; Volpo 3, 5, 10, 20; Waxenol 801, 810, 816, 821, 822; Wayhib; Wayplex 55-A, 55-S, HEDP-A; White Swan; Wickenol 101, 105, 111, 127, 131, 136, 139, 141, 151, 152, 153, 155, 156, 158, 159, 160, 161, 163, 171, 174, 506, 535, 535 Vita-Cos, 707, 727; Witconol APEB, APEM, APM, APS, CD-18, F26-46, RHP; Yeoman

ENTRAINMENT AIDS/ENTRAINING AGENTS

Air-Plas 200H, 200L; Alipal CD-128; Alkamide 2112; Empimin BMA; Monateric ADA; Nansa BMC

ENZYMES

AA-10; Abex Series; Alcalase 0.6 L, 2.0 T, 2.5 L; Aldomax GA-100; AMG 200 L; Amigase; Amizyme; Amylase-AO; Aquazym 120L; BAN 120 L, 240L, 360 S, 1000 S; Biolase Brands; Breakerase G; Brew(n)zyme; Celluclast 2.0 L Type X, 200L, 200 L Type N; Cellufenn; Celluzyme 2400 T; Cereflo 200 L; Clarase; Clarex L; Dex-Lo; Dextranase; Dextrozyme; Diazyme; Dihydroxy-acetone; DN 25L; Esperase 4.0T, 8.0L, 16.0L; Excize HA, Regular Excize, Special Excize, Super Excize; Extractase L5X, P15X; Fermalpha; Fermcolase; Fermcozyme; Fermlipase; Fermvertase 10X, XX; Finizym 200L; Fungamyl; Fungamyl 800L, 1600S; Gelatinase No. 53; Genencor Cellulase 150L, 300P; Gemanase 1.5L; HT-Proteolytic; Ketomax GI-100; Klerzyme; Lactozym, 1500 L, type GP and 3000 L, type HP; Lecitase; Maxacal F400, P400,000; Maxamyl WL7000; Maxaliq; Maxamyl; Maxatase LS400, MP375, P440; Maxazyme GI-IMMOB; Maxilact; Maxinvert; Milezyme AFP, APG, APL, Brand Hemicellulase, Bromelain, Catalase L, Cellulase, Cellulase TV Conc., DAL, Fungal Amylase, Fungal Lactase, Fungal Protease, Lipase Powds., Pancreatic Lipase, Pancreatin 4NF, Papain; Mycolase; Neutrase 0.5, L, 1.5, G; Oropon; Ovazyme XX; Palatase; Papain P-200; Pectinex 1XL, 2XL, 3XL, 5XL, SCL; Pectinex Ultra SP-L; Pectinol 59L, 60G, 80SB, DL, R10; Premier Maltzymes I and II; Prolase 300; Protoferm; PTN 3.OS; Rapidase; Rennilase 11L, 14L, 46L, 150 L type T, 14L, Type XL, 50 L, Type TL, 50TL, 60L; Rhozyme 86L, H39; Rhozyme HP-150 Conc.; Rhozyme P11, P41, P53, P64, PF; Savinase 4.0T, 8.0L, 16.0 L, Type W; Semacylase; Spark-L HPG; Spezyme AA, BBA, GA, IGI; Sweetzyme Type A, Type Q; Swinease; Taka-Sweet; Takatex; Taka-Therm; Tenase; Terg-A-Zyme; Termamyl 60T, 120L, 300T; Ultrazym 100; Urease Fermco; Uricase S

EXTENDERS/DILUENTS

A-Fax 800; Akrofax 9844 Black, 9906; Altowhite LL; Bartex 80; Bearflex LAO; Benzoflex S-552; Britol; Buca; Califlux 90, 510, GP, LP, LV, SP, TT; Calwhite II; Captex 350, 810B; Cardolite NC-513, NC-548, NC-1307, NC-552, NC-700, NC-1307; Catalpo, X-1; Celite 110, 266, 281, 289, 499, Super Fine Super Floss, Super Floss, White Mist; Cellolyn 95-80T; Cereclor 42, 42P, 50LV, 51L, 52P, 70L, AP45, AP52, LP4446, LP4985, S45, S52; Dixie Clay; Dyqex; Empimin OT; Fancor Lanwax; Ferrosil 14; G-431; G-White; GamaCo, GamaCo II; Grit-O'Cobs; H-White; HB-40; Hi-Sil 532EP; Hyfil 10; I.T. 3X, 5X, 325, FT, X; Imsil A-10, A-15, A-25, A-108; Kemamine DD-3680, DP-3680, DP-3695; Kenplast A-450, ES-2, ESI, G, PG; ; Lignosite 1840; Lignosol SFS; Lite-R-Cobs; Luxor E-40V, E-50V; LV-31, LV-336; Maltrin M510; McNamee Clay; Mearlmica MMCF, MMSV; Moldbrite 25; Naftolen H, N400, N401, N402, N403, N404, N405, N406, N407, N408, ND, NV, P600, P 603, P604, P606, P 611, P612, P613, P614, P616, ZD, ZD103,

ZD105, ZD106, ZM; Nalan W; Norlig N; Nytal 99, 100, 200, 300, 400; Pentalyn 856; Petro-Rez 103, 200; Picco 6070, 6110, 6115, 6120, 6130, 6140; Pluracol E-200, E-300, E-400, E-600, E-1500, E-4000, E-6000; Pure-Dent B700, B810; Pyrax A; Q-Cel 200, 300; Repel-O-Tex 100; Rudol; Rycowax; Satintone SP-33, Whitetex; SB-30, -31, -31C, -136, -331, -332, -335, -336, -431, -432, -632, -932; Semtol Series; Standamul 318, G-16; Terate 202, 203, 204; Translink 37, 77, 445, 555; Vansil W-9, W-10, W-20, W-30, W-50; Viplex 525, 885; Vinsalyn Resins; Wax OTO; Zeolex 80; Zotex 319, 319L

EXTRACTION REAGENTS

Adogen 283, 363, 364, 368, 381, 382, 383, 2283, 2363, 2364, 2368, 2381, 2382, 2464; Nissan Anon BL, LG; Pilot SXS-96

FILLERS

Akrochem Hydrated Alumina; AlNel A-100, A-200, AG; Bartex 80; Blanc fixe F, N, micro; Calwhite; Camel-CAL; Camel-CAL Slurry; Camel-CAL ST; Camel-CARB; Camel-FIL; Camel-TEX; Camel-WITE; Camel-WITE Slurry; Camel-WITE ST; Celite 21-A, 209, 219, 263, 270, 292, 305, 315, 321, 321A, 350, 388, 400, 408, 410, 1200, CAS-30KR, CS-22R, F.C., HSC, R-625, R-633, R-643, R-685, R-690, Snow Floss; Ceramitalc 10AC; CM-66; Continental; Crystic Stopper; D-70; Dilasoft TF; Dixie Clay; DND Compound; Elastocarb Tech Light, Tech Heavy, UF; Empimin OT; Ferrosil 14; Flo-Gard CC 120, CC 140, CC 160; Franklin T-11, T-13, T-14, T-325; FRE; G-431; GamaCo II; Gama-Fil 55, 90, D-1, D-2; Gama-Plas; Gama-Sperse 5, 80, 140, 255, 6451, 6532, 6532 NSF, CS-11; Grit-O'Cobs; H-36; Hi-Pflex 100; Hi-Sil 404, ABS; Huber SM, WG-1, WG-2; Huberfil 96; Hubersil 162; Hydrex, R; I.T. 3X, 5X, 325, FT, X; Imsil A-8, A-10, A-15, A-25, A-30, A-75, A-108; Jet Fil 100, 200, 350, 500; Kaolin RC 32; Ketjensil SM 405; Langford Clay; Lectro 125 M, 125 XLP; Lite-R-Cobs; LV-336; Mar'blend; McNamee Clay; Micro-Cel A, B, C, E, T-21, T-49; Micromesh No. 3; MicroPflex 1200; Microtalc MP10-52, MP12-50, MP15-38, MP20-40, MP25-38, MP26-38, MP30-36; Microtuff F; Microtuff 1000; Minex 2, 3, 4; Minugel FG; Nicron JS 216, 322, 422, 426, 634; Nytal 99, 100, 100HR, 200, 300, 400; Par Clay; Peerless No. 1; PMF Fiber 204, 204AX, 204BX, 204CX, 204EX; PT White Wax 602; Pyrax A, ABB, WA; Q-Cel 200, 300; Q-White (70% Slurry); Q-White (Spray-Dried); Rhenofit 1987, 2009, 2642, 3555, B; RO-40; SB-30, -31, -31C, -136, -331, -332, -335, -336, -431, -432, -632, -932; Select-A-Sorb; Sillikolloid P 87; Sillitin N 82, N 85, V 85, Z 86, Z 89; Sillum-200, -200 Q/P; Silver Bond 30, 45, B; Snow White 200, 270, 325; Solegal W Conc.; Stiffener DSC; Tamsil 8, 10, 15, 25, 30, 45, 75, 150, Gold Bond; Vantalc 6H, 500, 700, 900, 1350; Versamag B-16, DC, SB, Tech., UF; Vulkasil A1, C, N, N/GR-S, S; Windale White; Winnofil S, SP; Zeeospheres 200, 400, 600, 800, 850; Zeolex 23, 40; ZMBT

FLAME RETARDANTS

AA-81C; BA-43; BA-50; BA-50P; BA-59; BA-59P; Bayblend MD-6500, MH-6500; BE-51; Bromoklor 50, 70; CA-57; CD-75P; Cereclor 42, 42P, 50LV, 51L, 52P, 70L, AP45, AP52, LP4446, LP4985, S45, S52; Chlorez 700-DF; Con-BACN; DBP; DD-8126; DD-8133; DD-8207; DD-8307; DE-60F; DE-71; DE-79; DE-83R; Dechlorane Plus 25, 515, 2520; Delvet 68, 70; Dion Cor-Res 6693FR, 6693HV; Dion FR 6114, 6124, 6125, 6127, 6308, 6399, 6604T; 6655T, 6657, 6692T, 830; Disflamoll DPK, DPO, TCA, TPK, TP; Elastocarb Tech Light, Tech Heavy, UF; Emery 9331, 9332, 9336, 9345, 9350, 9353; Epotuf Resin 37-200; FF 680; Fire Retardant RCA; Firebrake ZB; Firemaster 680; Flamtard H, S; Flexchlor 0023; FPC 5000B, 5002, 5006A, 5009; FR-20, -300-BA, -513, -521, -522, -612, -613, -651A, -705, -910, -913, -1025, -1205, -1206, -1208, -1210, -1215, -1524, -1525, -2124, -D; FRE; Fyarestor 100; Fyrid KS1; Grilon A28 NX, A28 NY, A28 NZ; Halofree 22; Haltex 300, 310, 313, 320; Harshaw Antimony Oxide KR; Hetron 32A, 197AT, 470P, 27196; HFR-201, HFR-301; Ken-React Series, 7 (KR 7), 9S (KR 9S), 12 (KR 12), 26S (KR 26S), 33DS, (KR 33DS), 38S (KR 38S), 39DS (KR 39DS), 41B (KR 41B), 44 (KR 44), 46B (KR 46B), 55 (KR 55), 133DS (KR 133DS), 134S (KR 134S), 138D (KR 138D), 138S (KR 138S), 158 (KR 158D), 158FS (KR 158FS), 212 (KR 212), 238A (KR 238A), 238J (KR 238J), 238M (KR 238M), 238S (KR 238S), 238T (KR 238T), 262A (KR 262A), 262ES (KR 262ES), OPP2 (KR OPP2), OPPR (KR OPPR), TTS (KR TTS); Kisuma 5A, 5B, 5E; KP-140; Kronitex 50, 100, 3600, B-460, TCP, TXP; LICA 01, 09, 12, 38, 38A, 38J, 44, 97; Lomod BO100, BO150, BO200, BO220, BO250, BO320, BO520, BO800, BO852; LSF-54; LZ 01, 09, 12, 38, 44, 97; Micral 532, 855, 916, 932; Nyacol A-1530, A-1540N, A-1550, A-1588LP, AB40, AP50, APE1540, APVC40, HA-9, HA-15, N22, N24, ZTA; PE-68; PH-73; Phosflex 370; PHT4; Pliabrac 519, 521, 524; PO-64P; Pyro-Chek 60PB, 68PB, 77, LM; Rewophat TD 70; Sartomer 640; Saytex 102, 102E, 105, 111, 115, 120, 125, BCL-462, BN-451, BT-93, FR-1138, HBCD, HBCD-HM, HBCD-LM, RB-34, RB-49, RB-79, RB-79 Diol., RB-100, VBR; SB-30, -31, -31C, -136, -331, -332, -335, -336, -431, -432, -632, -805, -932; SE6920FR; SFR 100; Sillum-200, -PL 200; Spectratech CM 11053, 11489, 11591; Sylgard 170; Tego Antiflamm N; Thermoguard 210, 212, 213, 215, 220, 230, 240, CPA, FR, L, S; Thermolin 101, RF-230; Versamag B-16, DC, SB, UF; Zerogen 11, 15, 33, 35

FLATTING AGENTS

Colorol 46; Imsil A-30; Silver Bond 30, 45, B; Tixosil 311, 321; TS 100; Zeomatt 155

FLAVOR ENHANCERS

EmCon W; Emery 912; HVP 5-SD, HVP-A, HVP-LS; Luxor 1517V, 1576, 1626, 1639V, 1658V, E-40V, E-50V, EB-2V, EB-400V, GR-100, GR-150, GR-200, KB-300V, KB-320V, KB-330V, KB-350V, KB-400V, KB-500V, KB-530V, KB-600V, MB-40, MB-110, MB-120, R-100V

FLOCCULANTS

Adogen 462; Agent 613-95; Alkafloc A Series, CE Series, N Series; Alkazine O; Amerfloc Plus 5270, 5495; Arklear Series; Arquad DMMCB-50; Catisol AO 100; CMHEC-37L; Cyanamer A-370, P-26, P-250; Daxad 37LN7, 37LN10, 37NS, CP-2; Defloc 50; Dinoramac C, O, S; Drewfloc 3, 4; Duomac T; Galactasol Series; Generol 122E25; Guartex FM, SJM; Hercofloc Flocculant Polymers; Hercules CMHEC-37L; Hydrosan; Kalex; Kelzan, D, M, XC Polymer; Mafloc 764, 900; Noramac C, O, S; Primafloc A-10, C-3; Reten 210, 220, 300, 304, 420, 421, 423, 425, 520, 521, 523, 525; Shur-Coal 2000; Talofloc; Xanflood

FLOTATION REAGENTS

Adogen 464; Amine D, D Acetate; Armac 18D, C, HT, OD, T; Armeen 2C, 2HT, 12D, 16D, 18, 18D, C, CD, HT, HTD; Armeen SD, TD; Arosurf MG-70, MG-70A3, MG70A5, MG-83A, MG-84A3, MG-91, MG-91A3, MG-91A5, MG-98, MG-98A, MG-98A3, MG-101D, MG-102, MG-140, MG-148, MG-160, MG-170, MG-172, MG-570, MG-583, MG-584; Catisol AO 100; Dinoramac C, O, S; Duomac T; Duomeen C, C Special, L-11, L-15, O, S, T, T Special, TDO; Emerest 2301, 2302; Empigen AB, AH, AM, AY, BCM75, BCM75/A; Farmin 20, 60, 68, 80, 86, AB, C, HT, O, S, T; Guartec CM; Jet Amine D-C, D-O, D-T, PC, PE-13, PE-810, PE-1215, PHT, P-O, P-S, PT; Kemamine P-150, P-150D, P-190, P-190D, P-650, P-650D, P-690, P-690D, P-790, P-790D, P-880, P-880D, P-970, P-970D, P-974, P-974D, P-989, P-989D, P-990, P-990D, P-997, P-997D, P-999; Lignosol XD-65; Lilamac 140, 160, 170, S1; Noram 2 C, C, DMC, DMS, DMSH, M2 SH, M2C; Noramac C, O, S; Petrosul 545, 742, 744 CL, 745; Radiamac 6148, 6149, 6169, 6179; Radiamine 6140, 6141, 6160, 6161, 6163, 6164, 6170, 6171, 6172, 6173; Rewopol B 1003, 2003; Shur-Coal 159, 160, 164, 166, 168 Promoter; Silwet L-77, L-7604, L-7607; Sipex EC-111; SK Flot 1, 3, 4; Stepanflote 24, 85L, 97A; Sulfetal AF; Surfac P14B, P24M; Surfam D89DA; Synprolam 35; Textamine 1839; Tomah Tallow Amine

FLOW PROMOTERS/FLOW CONTROL AGENTS

Acrawax C; Avicel PH-101, PH-102, PH-103, PH-105; Emcol TS-230; Flow Agent WR-100; Foamkill MS-1; Ircogel 900, 901, 904, 2354; Kenplast ESI, LT; Keroflux 5323, 5486, H, M; Licowet F 1; Loxiol G 10, G 12; Reax 77, 77XF, G-1, G-2; Silwet L-77, L-7001, L-7500, L-7602, L-7607

FOAM STABILIZERS

Accomid PK; Alkamide 2124, CDE, CDE-Extra, CDO, CL63, L7DE, L7DE-BT, L9DE, L9ME, LIPA; Alkamidox C-2, C-5, CME-2, CME-5, L-2, L-5; Alkamox CAPO, LO; Alkasurf SS-CDE, TDA-7, TDA-8.5; Aminol HCA, KDA, LM-30C Special; Aminoxid WS 35; Amphoteric L; AO-14-2; Aramide CDM-4, CDS, CDX, LDS, LMDS; Aromox DMMC-W, DMMCD-W, T/12; Atmer 152; Avamid 150; Barlox C; Berol 151, 259, 305, 518, 752; Calamide C, CW-100, O; Calfoam ES-30; Calfoam NEL-60; Carsamide SAC, SAL-7, SAL-9; Cedepal FA-406; Cetodan 50-00A, 70-00A, 90-40; Chemal TDA-3, TDA-6, TDA-9, TDA-12, TDA-15, TDA-18; Chimipal DCL, LDA, MC, PE 300, PE 302; Conco XA-C, XA-L, XA-M, XA-MC, XA-S, XA-T, XA-Y; Condensate PA, PE, PL, PM, PO; Crodafos 25D Series, CS Series, ID Series, O Series, 25 D2 Acid, 25 D5 Acid, 25 D10 Acid; Crotein O, SPA, SPC, SPO; Cyclomide DC212/SE, DOTS, DP240/S, KD, L203, LE, LIPA; Cycloteric BET O-30; Denlube 33-B; Deriphat BC, BCW

Ecconol 628; Emal 20C, E-25C, E-70C, NC-35; Emcol L, M, NA-30; Emid 6500, 6534, 6538, 6541, 6545; Emka Defoam BC, BCK, CPT, DP, NC, PKM, SMM; Empicol 0216, 0627, ESB3, ESB3GA, ESB30, ESB50, ESB70, ESB70-AU; Empigen 5083, 5107, BB, BB-AU, BS, BS/H, BS/P, OH25, OS/A; Empilan CDX, CIS, CM, CME, KM 11, KM 20, KM 50, LDE, LDX, LIS, LME, LP2, MMA; Emulphogene BC-720; ESI-Terge 10, S-10, T-60; Ethoxylan 1685, 1686; Ethylan CH, DP, LM, LM2, MLD; Foamole M; Gafamide CDD-518; Genapol PGM; Genu DD Extra-Slow Set, 150 Grade; Geronol ACR/4, ACR/9; Hetamide LA, MC, ML, MMC, OC, RC; Incromide ALD, AVD, CA, CM, L-90, LEP, LMD, LMM, LR, Mink D, OLD, OPD, SED, UM, WGD; Incromine L-40, Oxide B-30, Oxide BA, Oxide C, Oxide I, Oxide ISMO, Oxide L-40, Oxide MC, Oxide O, Oxide OD-50; Incronam AL-30, AV-30, BA-30, Mink 30, OL-30, SE-30, WG-30; Jordamide 1281, CFAM, CLD, CLM, CMEA, CMEA Extra, CP, JT-128, JT-1286, LLD, PCS, SCD, WC Conc.; Jordamox CAPA, CDA-40, LDA, MDA; Kemamide W-40/300; Kerinol 2012 F, AA 62; Lanexol AWS;

Lanogel 21, 31, 41, 61; Lauridit KM, LM; Luviquat FC 370, FC 550; Mackamine BAO, IAO; Manro CD, CDS, CDX, D Paste; Marlamid D 1218, DF 1218, D 1885, KL, M 1218; Mazamide 80, CS-148, L-298, LM-21, LS-173, LS-196, O-10, SS-20, T-10, T-15, T-20; Merquat 550

Monamid 7-100, 7-153 CS, 15-70W, 150-AD, 150-ADD, 150-ADY, 150-CW, 150-DR, 150-GLT, 150-IS, 150-LMW-C, 150-LW, 150-LWA, 150-MW, 664, 716, 718, 759, 770, 853, 982, 1007, CMA, LIPA, LMA, LMMA, R31-42, S; Monamine Series, AA-100, AC-100. ACO-100, AD-100, ADD-100, ADDS-100, ADS-100, ADY-100, AF-100, ALX-80 SS, ALX-100 S, ARA-100, CF-100 M, I-76, LM-100, R8-26, T-100; Monflor 51, 52; Nikkol BPS-5, BPS-10, BPS-15, -20, -25, -30; Ninol 40-CO, 49-CE, 50-LL, 96-SL, 2012 Extra, 4821, 4821 F, AA-62, AA-62 Extra, CBR, DS, LDL 2, P-616, P-621; Ninox CA, FCA, L, M; Nissan Stafoam DL, DO, DO-S, L, MF; Nissan Unisafe A-LE, A-LM; Nitrene 11230, 13026, A-309, C Extra, L-76, L-90, NO; Nofoam CA 3617, NO 3990; Onyxol 345, SD; Oxamin LO; P & G Amide No. 27; Penreco 2251 Oil, 2257 Oil, 2259 Oil, 2260 Oil, 2263 Oil; Perlankrol FN-65, PA Conc.; Pluronic 17R1, 25R1, F38, F68, F68LF, F77, F87, F88, F98, F108, F127, L31, L35, L42, L43, L44, L61, L62, L62D, L62LF, L63, L64, L72, L81, L92, L101, L121, L122, P65, P75, P84, P85, P94, P103, P104, P105, P123; Profan 24 Extra, 128 Extra, 2012E, AA62, AB20, AD31; Purton CFD, SFD; Quimipol ENF 120; Rewomid C 220SE, CD, DC 212/S, DC 220/SE, DLMS, R 280; Rewopon AM-2C; Rexonic N6.6; Rycomid 2120

Schercamox CMA, DMS, T-12; Schercomid 1214, CDA, CDO-Extra, MD-Extra, SCE, SCO-Extra, SL-Extra, SLM-C; Serdolamide POF 61, PYF 77; Seycopon A-95 LCP; Solan, 50; Solricin 135; Solulan 5, 16, 25, 75, 97, 98, C-24, L-575; Solwax L20; Stafoam DL; Stamid HT; Standamid 100, KD, LP, PD, SD; Standamox C 30; Stepan P-621; Sterling 66/33, Amide 460, Granulated Wax, LDEA-90, LM-70; Sunaptol PP75B; Super Amide GR,, LM L9, L9C, L9A; Surco 128T, CMEA, DC-47, DC-60; Surfynol SE; Swanic 51; Synotol 119 N, CN 60, CN 80, CN 90, L 60, L 90, LM 60, LM 90, ME-90; Synprolam 35DMO, 35MX1/0, 35X2/0; T-Det EPO-61; Tego-Betaine C, L-7, L-90, S; Teric 121, 124, 127, CME7; Tex-Wet 1131; Tohol N-220, N-220X, N-230, N-230X; Trycol 5946; Unamide C-2, C-5, C-72-3, C-7649, CDX, CMX, D-10, GC-75, J-56, JJ-35, L-2, L-5, LDL, LDX, LMDX, N-72-3; Varamide A-2, A-10, A-12, A-83, C-212, L-1, L-203, MA-1, MA-4, ML-1, ML-4, SL-9, SL-910; Varion CADG-HS, CADG-W; Varox 185E, 185PH, 188E, 191E, 365, 375, 743; Witamide 82, 5140, 5195, 6310, 6445, CDA, LDTS, M-3, NA, NA1, STD-HP; Xanflood

FROTHERS

Denlube 33-B; Deriphat BC, BCW; Empicol LXV/D; Empimin BMA; Lankrocell D15L, KLOP, KLOP/CV; Millifoam, A, ODE-60A

GELLING AGENTS/THIXOTROPIC AGENTS/RHEOLOGY AGENTS

Aerosil Colloidal Silicas; Aerosol 22N; Agerite Geltrol; Alkatronic EDP 8-4, EDP 28-1, EDP 28-7, EDP 38-1, EDP 38-4, EDP 38-3, EDP-28-2, PGP 23-7, 23-8, PGP 33-8, PGP 40-7; Argowax Distilled, Standard; Baragel 24; Benathix; Bentone 27, 34, 38, EW, Gel LOI, Gel LT, Gel OMS, Gel SS-71LR, Gel TN, Gel VS-5, VS-5 PC; Carbopol 1342; Cerasynt 945; Cirrasol ALN-FP; Crodacid B; Crodafos N3 Acid, N3 Neutral, N5 Acid, N10 Acid, N10 Neutral, SG, T2 Acid, T5 Acid, T10 Acid; Cycloteric BET-OB50; Dermalcare EGMS/SE, GMS, GMS-165, GMS/SE, GTIS, LVL, MM/M, MST, PGMS, SDG, SS; Disflamoll TP; Durlac 100W; Elfapur N 70, N 90, N 120, N 150; Flo-Gard CC 120, CC 140, CC 160; Forbest 5724; Frimulsion 6G; Gantrez AN-119, AN-139, AN-169, AN-179; Gelrite; Genu 04CG or 04CB, 12CG or 12CB, 85-100 Grade, 18CG or 18CB, 70-90 Grade, 20 AS or 20AB, 100 Grade, 21AS or 21 AB, 102AS Buffered, 100 Grade, 104AS or 15AB, 100-120 Grade, AA Medium-Rapid Set, 150 Grade, BA-KING, 150 Grade, BB Rapid Set, 150 Grade, Carrageenan, DD Extra-Slow Set, 150 Grade, DD Extra-Slow Set C, 150 Grade, DD Slow Set, 150 Grade, Pectins; Genugel; Genulacta Series; Genuvisco; Hi-Jell #1; Hi-Sil 210, 215, 233, 250, 260, 262, T-690; Improved Kelmar; Incromide LMM; Ircogel 901, 903, 904, 2354

Kelacid; Kelcogel Gellan Gum; Kelco-Gel HV; Kelcoloid, D, DH, DO, DSF, HVF, LVF, O, S; Kelco-Pac; Kelcosol; Kelgin F, HV, LV, MV, QH, QL, QM, RL, XL; Kelmar, Improved; Kelset; Keltex, P; Keltone; Keltose; Kelvis; Kelzan, D, M, XC Polymer; Ken-React Series, 7 (KR 7), 9S (KR 9S), 12 (KR 12), 26S (KR 26S), 33DS, (KR 33DS), 38S (KR 38S), 39DS (KR 39DS), 41B (KR 41B), 44 (KR 44), 46B (KR 46B), 55 (KR 55), 133DS (KR 133DS), 134S (KR 134S), 138D (KR 138D), 138S (KR 138S), 158 (KR 158D), 158FS (KR 158FS), 212 (KR 212), 238A (KR 238A), 238J (KR 238J), 238M (KR 238M), 238S (KR 238S), 238T (KR 238T), 262A (KR 262A), 262ES (KR 262ES), OPP2 (KR OPP2), OPPR (KR OPPR), TTS (KR TTS); Korthix; Laponite XLG, XLS; LICA 01, 09, 12, 38, 38A, 38J, 44, 97; Lutensol ED 140, ED 310, ED 370, ED 610; Lutrol F 127; LZ 01, 09, 12, 38, 44, 97; Macol 2, 4, 5, 8, 10; Margel; Methocel A4C, A4M, A15-LV, E4M, E5, E15-LV, E50-LV, E4M, F, F50-LV, F4M, K3, K35, K100-LV, K4M, K15M, K100M; Millithix 925; Min-U-Gel 400; Natrovis; Nikkol Decaglyn 3-O, 3-S; Nonionic 1035-L, 1044-L, 1061-L, 1064-L, 1068-L, 1088-L; Norcast 5509-M; Norcast 5615-10/Norcure 178-M; Norfox B, Coco Powder; Nykon 77; Perchem 44, 97, 108, AMU 60 X, AMU X, EAG, Easigel, Econogel, TEA; Petrac 250; Petrolite C-400; Pluronic 12R3, 22R4, F38, F68, F68LF, F77, F87, F98, F108, F127, L10, L31, L35, L42, L43, L44,

GELLING AGENTS (CONT'D.)

L61, L62, L62D, L62LF, L63, L64, L72, L81, L92, L101, L121, L122, P65, P75, P84, P85, P94, P103, P104, P105, P123; Polarite KB 325; Polychol 15; Polytrope 1131; Post-4; Promulgen D, G; Rheolate 1; Standamid Resin BC-1283; Standapol AB-45, BC-35; Superloid; Syncrowax HRS-C; Texacar PC; Thixatrol 1, 53MS, 289, SR, SR100; Thixcin GR, R; Thixseal 1084; Thixotrol GST, ST; Tixosil 331, 375; TS 100; Tylose CR, CBR Series; Unamide S, SI, W; Van Gel; Varion TEG; Velvetex AB-45; Volclay HPM-20, NF-BC; Volpo N3, N5, N10, N15, N20; Vykamol N/E, S/E; Witamide 70, 6310, MAS, MM; Witco Sodium Stearate C-1, C-7; Xanflood; Zeosyl 200; Zeothix 177, 265

GUMS

Avicel RC-591; Biozan SPX 5423; CMTC-T; Cosmedia Guar U; Galactasol Series; Genuzan 1038A, 1063X; GFS; Hercules Cellulose, Cellulose Gum, Ester Gum 8D, Ester Gum 8D-SP, Ester Gum 10D; Kelcoloid; Kelfo; Kelgin; Kelmar, Improved; Kelset; Kelzan D35, S; Methocel A4C, A4M, A15-LV, E4M, E5, E15-LV, E50-LV, E4M, F, F50-LV, F4M, K3, K35, K100-LV, K4M, K15M, K100M

HAIR FIXATIVES

Copolymer 937, 958; Flexan 130; Gaffix VC-713; Gantrez SP-215; Jordaquat JN; Lexein A210; Luviquat Series; Luviset CA 66, CAP X; Luviskol K12, K17, K30, K60, K80, K90; PVP/VA W-735; Stepanhold Extra; Vigilan

HERBICIDES

Aastar; Abate 1-SG, 2-CG, 4-E, 5CG; Cotton-Pro; Crodamine 1.0, 1.0D, 1.T, 1HT, 1.16D, 1.18D, 3.A16D, 3.A18D, 3.ABD, 3.AED, 3.AOD; Direx 4L; Goltix; Komeen; Linex 4L; Mazeen 173, 174-75, DBA; Meturon 4L; Ortho Select; Prowl; Sencor; Surcopur; Tribunil; Ustilan

HUMECTANTS

Acconon ETG; Ajidew A-100, N-50, SP-100; Alkapol PEG 200, 300; Alkasurf CO-200; Amino Gluten MG; Biocare Polymer HA-24; Caprol 10G20; Carbowax PEG 200, 300, 400, 600; Ceraphyl 60; Collasol; Croderol G7000; Crodyne BY-19; Cronectin H; Crosilk Liquid; EmCon E, E-5, E-20; Emery 912, 6686, 6709; Emthox 2730; Ethosperse G-26, SL-20; Etocas 30, 100; Fancol HL, LA, LAO; Fancor Lanwax; Geropon IN; Glucam E-10, E-20, P-10, P-20; Hetoxide G-26; Hyamine 3500-NF; Hystar CG; Incromate PPI; Incromectant AMEA-70, AMEA-100, AQ, LMEA-100, LQ; Kemstrene 96.0%, 99.0%, 99.5%, 99.7%, High Gravity; Kerensim L 10 P; Lactil; Lanotein AWS 30; Lexein S620S; Lipamide MEAA; Liponic 70-NC, 76-NC, EG-1, EG-7, SO-20; Lonzest 143-S; Mackamide AME-75; Macol E; Maracarb, Maracarb N-1; Mazon 1045A, 1086, 1096; Niox KS Series; Pancogene S; Patlac LA, NAL; Pluriol E 200, E 300, E 400, E 600, E 1500, E 4000, E 6000, E 9000; Pogol 200, 1570; Procetyl AWS; Robane; Schercomid AME, AME-70; Solan, 50; Solangel 401, 501, 851; Sollagen; Star; Superol; Trylox 5918, 6753, CO-200, -200/50, SS-20; Varonic 1000MS, 1800MS; Wickenol 545

INHIBITORS

Pennstop 1866, 2049, 2697; T-Mulz 598

INITIATORS

Acetyl Peroxide-IB25; Alperox-F; Cadet BPO-70, BPO-70W, BPO-78, BPO-78W; Cadox 40E, BCP, BEP-50, BFF-50, BFF-60W, BP-55, BT-50, BTA, F-85, L-50, M-30, M-50, M-105, MDA-30, TDP; CHP-158; Cyclonox BT-50; Decanox-F; Degacure KI 85; EPH; Escacure EB3, KB1, KB60; Espercarb 438M-60, 740M, 747M, 750M; Esperfoam FR; Esperox-10, -10KXL, -10XL, -13M, -28, -28MD, -28PD, -31M, -33M, -41-25, -570P (USP-508P), -497M (Esperox-11), -545M, -551M, -939M, 5100, C-496; Ficel AZDN-LF, AZDN-LMC, AZM; FR-222; Hi-Point 90, 90 Red, PD-1; Laurox, W-25, W-40; Liladox; Lucemul 223E60; Lucidol-70, -78, -98; Luperco 101-SIL, 101-XL, 130-KE, 230-XL, 231-KE, 233-KE, 233-XL, 331-XL, 500-40C, 500-40KE, 801-XL, AA, ACP, AFR-250, AFR-400, AFR-500, AFR-501, ANS, AST, ATC, CST, PMA-40; Luperox 2, 5-2, 5, 118, 204, 500R, 500T, PMA; Lupersol 10, 10-M75, 11, 47M75, 70, 75M, 76-M, 80, 99, 101, 130, 130-XL, 188M75, 215, 219-M75, 220-D50, 221, 221-M85, 223, 223-M, 233-M40, 233-M75, 223-T70, 225, 225-M, 2225-M60, 225-M75, 225-T50, 225-X30, 228Z, 230, 231, 231-P75, 231-XL, 233-M75, 233-M90, 256, 331-80B, 431-80B, 531-80B, 533-M75, 546M75, 553-M75, 554M50, 554M75, 575, 575M75, 575P75, 601, 801, DDM-9, Delta-X-9, DFR, DHD-9, DSW-9, KDB, P-31, PDO, PMA, PMS, TA-46, TA-46M75, TA-54M75, TBEC, TBIC-M75; NPP; Pennwalt 4P, n-Dodecyl Mercaptan; Perkadox 14, 14/40, 14/40 Bpd., 16, 16/35, 16N, 16-W40, 16/W70, BC, IPP-AT50; Poly-Zole AXDN; Quickset Extra;

Quickset Super; SBP; Superox 744; TAHP-50M, -50P; TBHP-70; T-Hydro Sol'n.; Trigonal 14; Trigonox 21, 21-C50, 21-OP50, 22-BB80, 23-C75, 25-C75, 29-B75, 29-C75, 36-C75, 40, 44P, 99-C75, 101, 10/45Bpd., 123-C75, 125-C75, A-W70, ACS-M28, ADC, B, BPIC, C, EHP, EHP-AT70, EHP-C75, F-C50, KSM, SBP, SBP-AX30, SBP-C60, SBP-C75, T; Ultra-Cure I-100; USP-90MD, -355M, -400P, -690, -800; Vazo 52, 64

INSECTICIDES/PESTICIDES/MITICIDES/MOLLUSCICIDES

Amdro; Baygon; Bayluscide; Bayrusil; Baytex; Baythion; Baythroid; Bladafum; Counter; Croneton; Cropotex; Curaterr; Cyfluthrin; Cygon 400; Cythion, Emulsifiable Conc.; Dedevap; Dimilin; Dipterex; Disyston; Dowfax 3B2; Etrofolan; Folidol-E605, M; Folimat; Folithion; Gusathion, A; Lebaycid; Mafu; Marlate 2-MR, 50, Tech.; Matacil; Mazeen 173, 174-75, DBA; MEB 6046; Mesurol; Metasystox (i), R; Nemacur; Oftanol; Ortho Danitol 2.4 EC Spray; Peropal; Phosphoteric T-C6; Racumin; Servoxyl VPGZ 7/100, VPIZ 100, VPQZ 9/100, VPTZ 3/100, VPTZ 100, VPUZ, VPYZ 500; Tamaron; Tedion V18; Terracur P; Thimet 15-G, 20-G; Tokuthion; Triton X-45; Unden; Volaton

INTERMEDIATES

AB; Abitol; Ablunol NP4, NP6; A.B.S. 87%; Acconon 300 MO, 400 MO; Adma 2, 4, 6, 8, 24, 46, 246-451, 246-621, C8, C10; Adogen 342-D, 343, 345-D, 367-D; Adol 42, 52 NF, 54, 61 NF, 62 NF, 63, 64, 66, 80, 85 NF, 320, 330, 340, 520 NF, 620 NF, 630; AEPD; Alfol 4, 6, 8, 10, 12, 12, 14, 16, 18, 610, 610-50R, 610ADE, 810, 810-40, 810-60, 810FD, 1012, 1012-40, 1214-GC, 1214-GC-30, 1214-GC-40, 1216, 1216-22, 1216-CO, 1218, 1412, 1418-DDB, 1618, 1618-CG, 1620; Alfonic 1012-60, 1412-40; Alkapol PEG 200, 300, 400, 600, 6000, PPG-425, 1000, 1200; Alkasurf LA-1, LA Acid, LA23-3, LAN-1, NP-4, NP-6, TDA-8.5; Alkawet NP-6; Alkylate 215, 225, 230, A215, A225, A230, H230H, H230L; Alox 600; Amine 2M12, 2M16, 2MHBG, 2MKK; AMPD; Arcosolv DPM, PM, PTB, TPM; Armeen DM16D, DM18D, DMAD, DMCD, DMHTD, DMMCD, DMOD, DMSD, DMTD, HT, HTD, O, OD; Arnox 1003, 1007; Arsul DDB, LAS; Arylan HE Acid, S Acid, SBC Acid, SC Acid, SP Acid; ASA; Berol Dinonylphenol; BLO; Bio Soft S-100; CA0567; CA0570; CA0699; CA0742; CA0900; Calsoft LAS-99; Carbowax MPEG 350, MPEG 550, MPEG 750, MPEG 2000, MPEG 5000; Carbowax PEG 200, 300, 400, 600, 900, 1000, 1450, 3350, 4600, 8000; Cardolite NC-513; Carsamide C-3; Carsonon N-4; Catinex KB-10, KB-11; CB2408; CB2409.5; CB2493; CB2494; CB2495; CB2790; CB2805; CB2909; CC3005; CC3270; CC3275; CC3285; CC3290; CC3291; CC3433; CC3555; CD3770; CD3780; CD4450; CD5400; CD5430; CD5470; CD5487; CD5550; CD5600; CD5610; CD5636; CD5950; CD6000; CD6010; CD6150; CD6210; CD6220; CD6250; CE6250; CE6345; CE6350

Cedepal CO-210, CO-430, CO-500, CO-530; CH7250; CH7260; CH7280; CH7310; CH7332; Cheelox BF Acid; Chemal BP-261, BP-262, BP-262LF, BP-263, BP-264, BP-268, BP-268/50; CI7840; CM8450; CM8750; CM8930; CM8980; CM9000; CM9220; CO9745; CO9750; CO9000; CO9810; CO9816; CO9817; CO9819; CO9830; Conco AAS-98S, ATR-98S; ConcoSulfate 2A1 Acid, 3B2 Acid, 4C3; Consamine CA; CP0110; CP0156; CP0160; CP0280; CP0320; CP0330; CP0800; CP0810; CS1590; CT1800; CT2015; CT2030; CT2050; CT2090; CT2500; CT2520; CT2523; CT2902; CT2925; CT2950; CT2970; CT3250; CT3254; CT3600; CT3610; CT3795; Curithane 103; CV4772; CV4800; CV5050; CV5100; Dantocol DHE, DHE (5), DHE (10), DHE (15), DHE (20); Dantoest DHE DL; Dantoin DCDMH, DMDMH-55, DMH, DMHF, DMHF-75, MDMH; Dantosperse DHE (5) MO, (10) MO, (15) MO, (20) MO; DBE, DBE-2, DBE-2SPG, DBE-3, DBE-4, DBE-5, DBE-5SPG, DBE-9; DBP; DDBS-100; DeSonic 12-3, 81-2, 315-3; DeSonate SA, SA-H; Diacid 1550; Dinoram C, O, SH; Dioxitol-High Gravity; Dioxitol-Low Gravity; Dobanol 23-2, 23-3, 23-6.5, 25-3, 45, 91-2.5, 235-3; Dowfax 2A0, 9N; Duomeen C, C Special, CD, CX, L-11, L-15, O, OX, S, T, T Special, TDO; Dynasil A, CA, CM, M; Dynasylan AMEO, AMMO, BDAC, BSA, CPTEO, CPTMO, DAMO, GLYMO, HMDS, IBTMO, IMEO, MEMO, MSTFA, MTES, MTMO, MTMS, OCTEO, TCS, TRIAMO, VTC, VTEO, VTMO, VTPMO

Eltesol TA, TA 65, TA 96, TA/F, XA65, XA/M65; Emcol P-1020 BU; Emery 5440, 5700, 5703, 5707, 5710, 5717, 5736, 5743, 5771, 5774, 5776, 5778, 6726, 6727, 6728, 6730, 6749, 6773; Empigen AD, AF, AG, AS, AT, CM; Emulphogene BC-420, DA-530, DA-630; Emulsifier 99; Epal 6, 8, 10, 12, 12/70, 12/85, 14, 16NF, 18NF, 108, 610, 618, 810, 1012, 1214, 1218, 1416, 1418, 1618; Ethylan BBC31; Fancol HL, LA, OA 50, OA 70, OA 80, OA 90, OA 95; Farmin 2C, D86, DM20, DM40, DM60, DM80, DM86, DMC, M2C, M2R86; Flexricin 15, 100; Fluowet Brands; Fosterge BA-14 Acid, LF Acid, R Acid; Fostex T; FR-612, FR-613, -1138, -1524, -1525; Glybrom Reagent Grade; Glychlor Reagent Grade; Grocor 5820 G; GSD 550; Harshaw Antimony Oxide KR; Hetoxide BN-13, C-30, C-200-50%, HC-40, HC-60; Hetoxol 15 CSA, CA-2, CA-10, CA-20, CS-4, CS-9, CS-15, CS-20, CS-30, CS-50, CS-50 Special, L-3N, L-4N, L-9N, L-23N, LS-9, OL-2, OL-4, OL-5, STA-2, STA-10, STA-20, STA-30, TD-3, TD-6, TD-12; Hetsulf 40, 40X, Acid; Hostacor H Liq.; Hyonic NP-40, NP-60, OP-40, PE-40; Hystrene 1835, 3022, 3675, 3675C, 3680, 3695, 4516, 5012, 5016, 5522, 7018, 7022, 8016, 8018, 9008, 9010, 9014, 9016, 9018, 9022, 9512, 9514, 9718, 9912; Igepal CO-210, CO-520; Incromine CPM, LSB, PB, SB, TB; Industrene 104, 105, 106, 120, 126, 130, 138, 143,

INTERMEDIATES (CONT'D.)

145, 154-BG, 205, 206, 206LP, 223, 224, 225, 225 FG, 226, 232, 239, 325, 328, 333, 365, 3022, 4022, 4516, 4518, 5016, 5016 FG, 5022, 7018, 7018 FG, 8518, 8718 FG, 9018, B, M, R; Isonate 125M; Jeffox PPG-400, PPG-2000; Jordamine DAPI, DAPL, DMCAPA, DAPSA, S-13, SHCFA

Kemamide B, E, E-180, E-221, O, P-181, S, S-65, S-180, S-221, U, W-20, W-39, W-40, W-40/300, W-45; Kemamine DD-3680, DP-3680, DP-3695, P-150, P-150D, P-190, P-190D, P-650, P-650D, P-690, P-690D, P-790, P-790D, P-880, P-880D, P-970, P-970D, P-974, P-974D, P-989, P-989D, P-990, P-990D, P-997, P-997D, P-999, T-1902D, T-6501, T-6502D, T-9701, T-9702D, T-9742D, T-9892D, T-9902D, T-9972D, T-9992D; Kemester 105, 115, 143, 205, 226, 325, 2050, 4516, 9018, 9022, EGDS, EGMS; Kemstrene 96.0%, 99.0%, 99.5%; Ken Kem CP-45, CP-99; Labs-100; Lexamine 22, B-13, C-13, L-13, O-13, P-13, R-13, S-13; Mackine 101, 201, 301, 321, 401, 421; Macol 1, 18, 19, 31, 32, 33, 35, 42, 44, 46, 72, 77, 81, 85, 101, 108, 121, 660, 3520, 5100, NP-4, NP-6, P Series; Mapeg 600 DL, 600 DOT, 600 ML, 600 MOT, 6000 MS, PGDS, PGMS, PPG-400, PPG-500; Marlican; Marlon AS_3; Monolan PPG440, PPG1100, PPG2200

Nacap; Nalkylene 500, 515, 550, 550L, 600, 600L; Nansa SBA, SSA; NB; NE; Neodene 6, 8, 10, 12, 14, 16, 18, 20, 1214, 1416, 1418, 1618; Neodol 1, 1-5, 23, 23-1, 23-3, 23-5, 23-6.5, 23-6.5T, 23-12, 25, 25-3, 25-7, 25-9, 25-12, 45, 45-2.25, 45-7, 45-7T, 45-13, 91, 91-2.5, 91-6, 91-8; NEPD; Neustrene 045, 053, 059, 060, 064; Newcol 405, 410, 420; Ninate ABS-60C; Niox KF-13, KF-17, KF-26, KL-19, KQ-30; NiPar S-10; Nissan Amine AB, ABT, ABT_2, BB, FB, MB, OB, PB, SB, VB; Nissan Cation F_2-10R, -20R, -40E, -50, FB, FB-500, L-207, M_2-100, PB-40, -300, S_2-100; Nissan Tertiary Amine AB, ABT, BB, FB, MB, PB; Nitrene C Extra; NM; NMP; Noiox KS-10, -12, -13, -14, -16; Noram 2 C, C, DMC, DMS, DMSH, M2 SH, M2C, O, S; Norfox DC, NP-4; Novel 1412-70; OCI 56; Ogtac-85, 85 V; Onamine 12, 14, 16, 18, 65, 835, 1214, 1416; Orthophen 278; Pennad 0150; Pennfloat 3-2277; Pennstop 1866, 2049, 2697; Pentaphen 67; Pluracol E-200, E-300, E-400, E-400 NF, E-600, E-600 NF, E-1000, E-1450, E-1450 NF, E-1500, E-2000, E-4000, E-4000 NF, E-4500, E-6000, E-8000, E-8000 NF, P-410, P-710, P-1010, P-2010, P-3020, P-4010; Pluriol E 200, E 300, E 400, E 600, P 600, P 900, P 2000; Pluronic 10R5, 10R8, 12R3, 17R1, 17R2, 17R4, 17R8, 22R4, 25R1, 25R2, 25R4, 25R5, 31R2, 31R4, L10; Poly-G 200, 300, 400, 600, 1000, 1500, 2000, B1530, WI 285, 625, 1715, WS 100, 170, 260, 660, 2000, 3520, 5100, WT 9150, 90,000; Polywax OH 425, 550, 700; Primarol 1208; Propetal 99, 103, 241; Protowet 3072; Puxol XB-10; P-10; Pamolyn; Parabis; Peganol NP 1.5, NP 4, NP 5; Pegol E 200, 400, 600, 1000, 4000, P 400, 700, 1000, 2000; Perk; Perlankrol ASC38; Petrac 270; PH-73; PHT4-Diol; 2-Pyrol

QO Furan, Furfural, Tetrahydrofuran, Tetrahydrofurfuryl Alcohol; Quadrol; Radia 7040, 7051, 7060, 7070, 7104, 7110, 7118, 7120, 7127, 7129, 7130, 7161, 7163, 7171, 7176, 7190, 7194, 7195, 7197, 7200, 7230, 7231, 7240, 7303, 7330, 7331, 7363, 7500, 7501, 7510, 7514; Radiamine 6540, 6560, 6570, 6572; Radiasurf 7000, 7151, 7152, 7153, 7372; Rewomid OM 101/G, OM 101/IG; Reworyl T 65; Rexol 25/4, 25/6, 35/3, 35/8, AE-1, AE-2; Rexonic N-4, N23-3, N25-3, N91-1.6, N91-2.5, N91-6, N91-8; Rueterg SA; Sandoxylate SX-208, SX-224, SX-408, SX-412, SX-418, SX-424; Sartomer 100; Saytex HBCD, RB-49, RB-100, VBR; Schercodine C, L; Schercomid AME, AME-70; Schercozoline B, C, L, O; Serdox NDI 100; Servamine KOO 330, KOO 360; Sipenol IC2, IC10, IO12, IS05, IT11, IT50; Siponic NP6, NP10.5, NP15; Sodium Chlorite Solution 50 Tech, Tech; Solar CO; Soprofor DO/64; Staflex DBF, DBM, DBM, DINM, DOM, DTDM; Standamid KD; Stepan C25, C40, C41, C42, C43, C49, C50, C60, C65, C66, C68; Sterling LA Acid-C, Reg., LA Paste 55%, 60%, NPX; Sterox DF, ND, NE, NF, NG; Sulfonic 800; Sulfonic Acid LS; Sulfosoft; Sulframin 1288, 1298, 1388; Surco DDBSA; Surfam P5, P5 Dist., P5 Tech, P10, P12B, P14B, P17B, P24M, P86M, P89MB; Syn Fac 8026; Synperonic 91/2.5, A2, A3, A4, A6, NP4, NP5, NP6; Synprol Alcohol; Synprolam 35, 35DM, 35M, 35N3

T-600; T-Det N-4; Tegochrome 22; Terathane 650, 1000, 2000, 2900; Tergitol 24-L-3, 24-L-45, 24-L-60, 24-L-60N, 24-L-75, 24-L-92, 24-L-98N, 26-L-1.6, 26-L-3, 26-L-5, NP-14, -44, NPX; Teric 12A3, 12A6, 12A9, 12A12, 17A2, 17A3, 17A6, G12A4, G12A6, G12A8, G12A12, LA4, LA8, PE61, PE62, PE64, PE68; Tetradecene-1; Texanol Ester-Alcohol; Texox PPG-400, PPG-2000, WL-440, WL-660, WL-1400, WL-5000; Textamine A-3417; Tex-Wet 1103, 1197; T-Hydro Sol'n.; Tris Amino; Triton N-57, N-60; Trycol 5946, 5950, 5968, 5993, 6961, 6985, LAL-4, TDA-3, TDA-6, TDA-18; Tryfac 5552; Trylox 6753, SS-20; Trymeen 6606, TAM-16; Turgum S; Vanchem NATD; Varonic U205, U205LC; V-Pyrol; Witcamine 209, 210, 211, 235, 3164, AL42-12, RAD-0500, RAD-0515, RAD-1100, RAD-1110; Witco 1298 Soft Acid, Acid B, DTA-150, DTA-180; Witcolate 247H; Witcor CI-3117; Zusomin C 108, O 102, O 105, O 112, S 110, S 125

LEVELING AGENTS

Ablunol 200ML, MO, MS, 400ML, MO, MS, 600ML, MO, MS, 1000MO, MS; Ahco 7166T, 7596D, 7596T; Ahcowet DQ-145; Alkaminox T-10, T-12, T-15, T-25, T-30, T-50; Alkamuls 400-ML, 600-ML, 600-MS; Alkasurf L-9, L-14, S-14, TDA-15; Alphenate TFC-76; Amiladin; Arbyl 18/50; Armeen SZ; Arodet AA-350; Arofos 200 Conc.; Atlas WA-108; Atolene RW; Atolex LDA/40; Berol EGA 07; Berol Fintex 755; Blancol; Blancol N; Carboset 514A, 514H, 515, 525, 531, XL-11, XL-19; Cenegen 7; Charlab Leveler AT

Special; Charlab Leveler DSL; Chemcol CO-40, JL-60, JL-120, JL-180; Chemeen C-2, C-5, C-10, C-15, T-2, T-5, T-10, T-15, T-20; Chimipal NH, OS 2; Cibaphasol AS; Conco Sulfate RA; Consoft CP-50; Consos Castor Oil; Consoscour 47; Cordon AES-65, N-400; Crapol AU-21, AU-25; Crillet 2, 3, 31, 35; Crosilk Powd.; Crotein SPA, SPC, SPO; DeSonic 9T, 12T, 15T; Detergent CR; Doittol 14; Dowfax 2A1, 2EP, 3B0, XDS 8292.00; Duponol RA; Ekaline F; Ektasolve EM; Emerest 2650, 2660; Emkatex DX, LS, WNX; Emthox 5940; Emulphor TO-530, TO-639; Ethomid HT/23; Ethoquad 18/12, 18/25, C/12, C/25, O/12, O/25; Ethylan OE, R; Etocas 10, 35, 40, 60, 100; Examide-DA; Fluorad FC-120, FC-128, FC-129, FC-134, FC-135, FC-170-C, FC-171, FC-430, FC-431; Forbest 150, 620, 680, G23; Gaftex 288, CK-160, DN-159 Conc., KF, LN-110, PC-150, ZP; Gantrez B-773; Harol RG-71; Hartonyl L531, L537; Hetoxol 15 CSA, CA-2, CA-10, CA-20, CS-4, CS-9, CS-15, CS-20, CS-30, CS-50, CS-50 Special, L-3N, L-4N, L-9N, L-23N, LS-9, OL-2, OL-4, OL-5, STA-2, STA-10, STA-20, STA-30, TD-3, TD-6, TD-12; Hipochem ADN, No. 3; Humectol C, C Highly Conc.

Incrocas 10, 30, 40, 60, 100; Intralan Salt HA, N; Intratex A, AN, B, C, OR, POK, W New; Katapol OA-860, OA-910; KP-140; Lamepon RE; Larosol NRL-40; Levelan A0192, NKS, R-15, R-200, WS; Levelene; Levenol PW, TD-326, WX, WZ; Linamine HC-2; Lomar LS, LS Liq.; Lyogen F Liq., MS Liq., SMK-40 Liq., SMK Paste, V (U) Liq., WD Liq.; Macol 2, 4, 5, 8, 10, 15, 20, 34; Marvanol 55% SPO, 60% SBO, Penetrant 35; MBEO 1.8 Adduct; Merpol DA; Merpol HCS, HCW; Migregal 2N, NC-2; Miranol C2M Anhyd. Acid; Monflor 32, 51, 52; Monogen; Monopol Oil 75; Montegal 150 RG, AP 80, DZ, OL 50, SH 25; Musloid 815D, 815M, 815S; Nekal NF; Neocation G; Niox KP-68; Nopalcol 4-O; Nylomine Assistant DN; Osimol Grunau 109, EFA, PHT, RAC; Pegnol OA-400; Pegol 17 R1, 17 R2, 25 R1, 25 R2, 31 R1, L 31, L 35, L 43, L 61, L 62, L 62 LF, L 64, L 81, L 101, L 121, P 85; Pentalyn 225, 261, 269; Peregal D, O; Pluronic 25R2, 25R8, F38, F68, F68LF, F77, F87, F98, F108, F127, L31, L35, L42, L43, L44, L61, L62, L62D, L62LF, L63, L64, L72, L81, L92, L101, L121, L122, P65, P75, P84, P85, P94, P103, P104, P105, P123; Polyfac MT-615; Polylev #745; Product BCO; PVP K-90; Reten 520; Rexol C14; Rexonic 1006; Rexopene; Rexowet RW; Sag Silicone Antifoam 471, 4130, 4220; Sandogen C-CM Liq.; Silwet L-77, L-7001, L-7500, L-7602, L-7607; Siponic E-2, E-3, E-7, E-10; SMA 1000, 2000, 3000; Sorbit P; Sulfonated Castor Oil GTO; Sunaptol CA 25, 120, 350, 400, 750, P Extra Liquid; Sunsolt RZ-2, -6; Surfynol 440; Swanic 110; Synperonic A7, A9, A11

Tanapal ME Acid; T-Det C-40, N-8, N-9.5; Tegopren; Tergitol TMN-3; Tetranol; Tinegal MR-50, PM; Tolplaz TBEP; Triton DN-14, X-200; Trycol 5946, 6989, DNP-150, DNP-150/50, TDA-18; Trydet 19, 20, 2682; Trymeen 6606, 6607, 6609, 6617, 6622, 6623, OAM-30/80, SAM-50, TAM-15, TAM-20, TAM-25; Uniperol AC, AC Highly Conc., O; Unisol UA 20, 80, 150, 300, 400; Variquat B200; Varonic K205, K205LC, K210, K210LC, L205, L205LC, T210; Viplex 525; Volpo 25D3, 25D5, 25D10, 25D20, CS3, CS5, CS10, CS15, CS20, O5, O10, O15, O20, T3, T5, T10, T15, T20; Witconol H-31A, NP-100; Zonyl A, FSA, FSB, FSC, FSN, FSP

LUBRICANTS

Abil B 88183; Ablunol 200ML, 200MO, 200MS, 400ML, 400MO, 400MS, 600ML, 600MO, 600MS, 1000MO, 1000MS, CO5, CO10, CO15, CO30, CO45, S-60; Acconon 300 MO, 400 MO, CA9, CA15, CA25, ETG, GTO; A-C Copolymer 540A; Acetulan; A-C Polyethylene 6, 6A, 8A, 9A, 8, 9, 629A; Acrawax C, C DF #1, #2, C SG; Acryloid 950 Series, 1017, HF-Series, W-1600; Activator 1102; Actrafos 104, 109, 110, 110A, 152A, 161, 315, 186, 208, SP-407, T; Actralube Syn-153; Actramide 5264; Actrasol CS-75, SS; Actrol 4DP, 4MP, 628; Adogen 570-DO; Adol 42, 52 NF, 54, 60, 61 NF, 63, 64, 66, 80, 85, 85 NF, 90, 90 NF, 320, 330, 340, 520 NF, 620 NF, 630, 640; Advawax 140, 240, 275, 280, 290; AE-7, AE-1214/6; Aflux 32, 42, 54, R, S; Agriwet CA, C91, C92; Ahco AJ-110, AY 2200, DHS-111; Ahcovel Base N-64; Akro-Gel; Akrolease E-9491; Aldo PMS; Alfol 20, 1216, C22; Alkalube DTDA; Alkamide 2112, RDO, STEDA; Alkamuls 200-DL, 200-DO, 200-DS, 200-MO, 200-MS, 400-DO, 400-DS, 400-MO, 400-MS, 600-DS, DEG-DO, DEG-ML, DEG-MO, DETDA, GML-45, GMO-45, GMO-55LG, GMS-45, GMO, PSML-20, PSMO-5, PSMS-4, PSMS-20, PSTO-20, PSTS-20, SMO, SMS, STS; Alkaphos 6, B6-56A; Alkapol PEG 200, 300, 400, 600, 1000, 1500, 8000, PPG-425, 1000; Alkasurf 075-5, CO-20, CO-25, CO-30, CO-40, CO-40M, CO-200, L-2, O-2, S-5, S-8, S-40, S65-8, S65-40, SA-20, TA-20, TA-30, TA-40, TA-50; Alox 100, 100D, 102, 152, 162, 164, 165, 350, 436A, 488, 502A, 601, 940A, 1689, 2028L, 2138F, 2200, 2213C, 2283, 2284A, 2284AB; Alphaflex 202; Alrosperse 11P Flake; Alubrasoft Super 15, Super 25, Super 100; Alubraspin HS-100; Amerchol 400; Amerlate P, W; Amphoteric 400; Antara HR-719, LB-400, LE-500, LE-700, LF-200, LK-500, LM-400, LM-600, LP-700; Aqualox 225-100, 225A-100, 262, 2268; Aramide CDM-4, LDS; Argonol 50 Pharmaceutical, 50 Super, 60; Aristol A, B, D, E; Arlatone T; Armeen 12D; Armid 18, C, HT; Armoslip 18; Armosoft L; Arnox TGE-05, 16, 36, 40, 80; Arphos MH-200; AT-20GF, AT-30GF, AT-40GF; Atmer 180, 1021, 1025, 1026, 1030, 1040, 1041, 1042; Atolene RW; Avamid 150; Axol, E61, E66

Barium Petronate Basic; BASF Wax AH 3, AL 3, E, OA 2, OP; Be Square 185; Benol; Berol 17, 730, 740; Biocare Polymer HA-24; Blandol; Bohrmittel Hoechst; Borax; Britol; ; CA0567; CA0570; CA0699; CA0742; CA0900; Cachalot C-51, O-15, S-54; Calamide C, CW-100, O; Calcium Stearate, Regular;

LUBRICANTS (CONT'D.)

Caplube 8385, 8410, 8440, 8442, 8445, 8448, 8508; Capmul GMS; Caprol 10G10S, 10G20, 10G100; Captex 355, 800, 8000; Carbopol 934; Carbowax Compound 20M, MPEG 350, MPEG 550, MPEG 750, MPEG 2000, MPEG 5000; Carbowax PEG 200, 300, 400, 600, 900, 1000, 1450, 3350, 4600, 8000; Cardolite NC-507, NC-511; Carnation; Cation Softener X; Catisol AO 100; CB2408; CB2409.5; CB2493; CB2494; CB2495; CB2790; CB2805; CB2909; CC3005; CC3270; CC3275; CC3285; CC3290; CC3291; CC3433; CC3555; CD3770; CD3780; CD4450; CD5400; CD5430; CD5470; CD5487; CD5550; CD5600; CD5610; CD5636; CD5950; CD6000; CD6010; CD6150; CD6210; CD6220; CD6250; CE6250; CE6345; CE6350; Centrol Series; Centrolene Series; Centrophase HR; Cenwax G, ME; Ceraphyl 50, IPL, Cereclor 50LV, 51L, 70L; Cetina; Cetiol G16S, G20S; Cetodan; CH7250; CH7260; CH7280; CH7310; CH7332; Charlab Leveler AT Special, DSL; Chemal LA-4, LA-9, LA-12, LA-23, OA-5, OA-20, OA-20/70; Chemax CO-5, E-200 ML, E-200 MO, E-200 MS, E-200 MS, E-400 ML, E-400 MO, E-400 MS, E-400 MS, E-600 MO, E-1000 MO, HCO-5, HCO-16, HCO-25, HCO-200/50, PEG 600 DO, TO-8, TO-16; Chemeen 18-2, 18-5, 18-50, C-2, C-5, C-10, C-15, HT-2, HT-5, HT-15, HT-50, S-2, S-5, S-30, S-30/80, T-2, T-5, T-10, T-15, T-20; Chemfac PA-080, PB-184, PB-264, PB-804, PC-188, PD-600, PD-990; Chemical 39 Base; Chimin KSP; CI7840; Cindet C-12; Cirrasol ALN-WF, ALN-WY, LC-HK, LC-PQ, LN-GS; Cithrol 2ML, 2MO, 4MO, 6ML, 6MO, 10ML, 10MO, 40MO, 60ML, 60MO, GDO N/E, GDO S/E, GDS N/E, GDS S/E, GML N/E, GML S/E, GMO N/E, GMO S/E, GMR N/E, GMR S/E, GMS Acid Stable, PGML GMS N/E, GMS S/E, N/E, PGMO N/E, PGMO S/E, PGMR N/E, PGMR S/E, PGMS N/E, PGMS S/E

CM8450; CM8750; CM8930; CM8980; CM9000; CM9220; CO9745; CO9750; CO9000; CO9810; CO9816; CO9817; CO9819; CO9830; Colsol; Comboloob; Condensate 640; Consoft CP-50; Consos Castor Oil; Corax; Cordon AES-65, NU 890/75, PB-870; CP0110; CP0156; CP0160; CP0280; CP0320; CP0330; CP0800; CP0810; CPH-211-N; CPH-213-N; Crafol AP-10, AP-51, AP-60; Cremophor S 9; Crill 1, 2, 3, 35, 50; Crillet 2, 3, 4, 31, 35, 41, 45; Crillon LME; Crodacol CS50; Crodafos 25D Series, 25 D2 Acid, 25 D5 Acid, 25 D10 Acid, CS Series, ID Series, O Series; Crodamol DA, DO, ICS, OP, PMP; Crodasinic L&O, LS35, OS35; Crodazoline O; Crosterene; CS1590; CT-88 Aerosol; CT1800; CT2015; CT2030; CT2050; CT2090; CT2500; CT2520; CT2523; CT2902; CT2925; CT2950; CT2970; CT3250; CT3254; CT3600; CT3610; CT3795; Cutina HR; Cutless 50; CV4772; CV4800; CV5050; CV5100; Cyanamer A-370, P-26, P-250; Cyclochem CP, EM 324, EM 560, GMO, INEO, PETS, POL; D-400; D-1000; D-1200; D-1300; Cyclomide DIN295, DS280, RODEA; D-2000; D-3000; D-4000; Dacospin 1735-A, HT-5, HT-50, LS 4100, POE(25)HRG; Dar Chem-11, -12, -13, -14, -030; Darvan L, ME, WAQ; Delion 624; Denlube 40, 1025-KA; Dermalcare EGMS/SE, GMS, GMS-165, GMS/SE, GTIS, HL, LVL, MM/M, MST, PGMS, SDG, SS; DeSonic 6C, 30C, 36C, 40C, 54C, SMO, SMT; DeSophos 7 OPNa; Dist. Lipolan; Dodilube Brands; Dow Corning 7 Compound, 24 Emulsion, 203 Fluid, 344 Fluid, 345 Fluid, 556 Fluid, FF-400, FF-412, FF-414; Drakeol 9, 10, 19; Drewfax R-122; Druspin 1223, 1400, LO-VIS, Supreme; Duomeen TDO; Dur-Em 117, GMO; Durfax 60, EOM; Durtan 20; DyaFac 6-S, 1926-B; Dynasan 110, 114, 116, 118; Dynasil A, CA, CM, M; Dynasylan AMEO, AMMO, BDAC, BSA, CPTEO, CPTMO, DAMO, GLYMO, HMDS, IBTMO, IMEO, MEMO, MSTFA, MTES, MTMO, MTMS, OCTEO, TCS, TRIAMO, VTC, VTEO, VTMO, VTPMO; E-155; Ease Release 500 Series; EBS Wax; Eccospin PPF; EM-90; EM-9400; Emasol O-15 R, S-20, S-106, S-120, S-320; Emcol 150, CC-9, CC-42, CC-57, L, M, TS-230

Emerest 2301, 2302, 2308, 2310, 2314, 2316, 2321, 2324, 2325, 2326, 2328, 2388, 2400, 2410, 2421, 2423, 2617, 2619, 2622, 2624, 2630, 2632, 2636, 2640, 2648, 2652, 2654, 2661, 2662, 2664, 2665, 2704, 2715; Emerlube 5919; Emery 880, 1650, 1787, 2809, 2811, 2831, 2879, 2880, 2890, 2900, 2902, 2905, 2908, 2911, 2913, 2914, 2918, 2924, 2929, 2936, 2939, 2943, 2957, 2958, 2960, 2970, 2971, 2976, 2984, 2990, 2995, 3002, 3004, 3006, 3008, 3010, 6657, 6717, 6724, PAO 2947, PAO 2948, PAO 2950; Emgard 2802; Emkalon CNW; Emkatex AES, WNX; Emphos CS-121, CS-136, CS-141, CS-147, CS-151, CS-330, CS-733, D70-30C, D70-31, F27-85, PS-21A, PS-121, PS-220, PS-236, PS-415M, PS-440, PS-810, PS-900, TS-230; Empicol LX, LX28, LXV, LXV/D, LZ, LZ/D, LZ/E, LZP, LZV, LZV/D, LZV/E, FRC75S; Empilan 0004, GMS LSE40, GMS LSE80, GMS MSE40, GMS NSE40, GMS NSE90, GMS SE40, GMS SE70; Emsorb 2500, 2503, 2505, 2507, 2510, 2729, 6901, 6903, 6905, 6906, 6907, 6908, 6910, 6917; Emulbon LB-75, LB-78, S-260; Emulphor EL-620, EL-719, EL-980, EL-985, VT-650; Emulsogen B2M; Emulthin M-35; Epolene E-10, N-14; Ervol; Espesilor AC-52; Ethomid O/15; Ethylan CF71, GD, GL-20, GO-80, GP-40, GS-60, GT-85, VPK; Etocas 10, 35, 40, 60, 100; Eureka 400, E-2; Eutanol G; Evanstab 12; EZ Mold Lubricant; F-755, F-789; Fancol HL, LA, OA 50, OA 70, OA 80, OA 90, OA 95; Fancor Lanwax; Fancorp Lanolin; FC-60%, FC-446CS, FC-477 SM; Finsolv SB; Flexricin 100, P-1, P-3, P-4; Foamkill 627, 649 Series, 639N, 649P, 663J, 684P; Fonoline White, Yellow; Fortex; Frekote No. 1; Frigid-Go 2815, 2816; Gafac MC-470; RB-400, RD-510, RE-410, RE-610, RE-870, RE-877, RE-960, RK-500, RS-410, RS-610, RS-710; Gaftex COM-154, COM-154 Conc., DN-159 Conc.; GC-44-14; Genagen CD; Genapol B, GEV, PN 30, PS; Getren; Gloria; Glucam E-20 Distearate; Glucate DO; Glycolube P, PG, VL; Glycox PEMS; Glytex EL 176, EL 882, EL 905, L 154, L 203; Grindtek AML 60, AMOS 90, ML 90, MM 90, MOL 90, MOP 90, MSP 32-6, MSP 40, MSP 40F, MSP 52, MSP 90, PK 60

Haro Wax L01-56, L02-99, L03-58, L04-73, L05-51, L06-62, L07-47, L08-57, L09-00, L10-58, L11-85, L12-43, L13-89, L14-77, L15-75, L16-92, L17-98, L18-78, L19-99, L20-98, L21-96, L22-99, L23-58, L24-98, L25-99, L-333, L-344, L-443; Hartamide 9137; Hartex V63, V64; Hartolon NA; Hercolube 404; Hercules Cellulose Gum; Hetamide LA, MC, ML, RC; Hetoxamate FA-5, FA-20, LA-5, LA-9, MO-5, MO-9, MO-15, SA-5, SA-9, SA-35, SA-90; Hetoxide BN-13, C-30, C-200-50%, HC-40, HC-60; Hetsorb L-10, L-20, O-20, S-20; Hipochem EFK, LCA; Hostacor BK, KS 1; Hostaphat AW, CP; Hybase M-300, M-400; Hystar CG, TPF; Hystrene 3675C, 3675CS, 3675X, 4516, 5016, 5016 NF FG, 5460, 7018, 7018 FG, 7022, 8016, 8018, 9008, 9010, 9014, 9016, 9718 NF FG, 9018, 9022, 9512, 9718; Ice#2, #12, #81; Icinol H260, H660, H660YA, H5100, L65, L285, L300X, L385, L625, L1715; Icodimeen T-30; Icomeen 18-5, O-30, O-30-80%, S-5, T-2, T-5, T-7, T-20, T-25, T-25 CWS, T-40, T-40-80%; Icomid HT-50, O-5; Iconol 28, COA; Igralube 53; Imwitor 191, 900, 940K; Incrocas 10, 30, 40, 60, 100; Incromide LA; Incromine BB, IB, OPB, OPM, Oxide B-30, Oxide I, Oxide ISMO, Oxide O, Oxide OD-50; Incronam B-40, I-30, OP-30; Incropol 233, 290, CS-12, CS-20, CS-30, CS-40, CS-50, CS-60, L-2, L-7, L-7-90, L-12, L-23, L-30; Incrosoft S-75, S-90, S-90M, T-75, T-90; Indu-Sol 238; Industrene 105, 138, 143, 154-BG, 205, 206, 225, 225 FG, 226, 232, 239, 325, 328, 333, 365, 3022, 4022, 4516, 4518, 5016, 5016 FG, 5022, 7018, 7018 FG, 8518, 8718 FG, 9018, B, R; Inhibitor RT 212; Inipol S.43; Interlube A, P/DS; Interstab CA-18-1, G-140, G-8257, ZN-18-1; Interwax G 8200, 8204, 8205, 8206, 8207, 8208, 8140, 8212, 8213, 8252, 8253, 8257, 8259, 8268; Interwax M 3142; Inversol 140, 170, 190; Ionet DL-200, DO-200, DO-400, DO-600, DO-1000, DS-300, DS-400, S-20, S-60 C, S-80, S-85, T-20 C, T-60 C, T-80 C; Jeffox PPG-400, PPG-2000; Jordamide 201, 1214, CCO, JR-100, JT-1286, OW; Jordaphos 151, DT, FDED, JA-60, JB-40, JE-41, JP-70, JS-61; Kalcohl 86; Kantstik Q Powd.; Kara-Lube ECM; Katapol PN-730, PN-810; Kaydol; Kelzan, Kelzan D, M, XC Polymer; Kemamide B, E, E-180, E-221, O, P-181, S, S-65, S-180, S-221, U, W-20, W-39, W-40, W-40/300, W-45; Kemamine AS-650, AS-974, AS-974/1, AS-989, AS-990, P-150, P-150D, P-190, P-190D, P-650, P-650D, P-690, P-690D, P-790, P-790D, P-880, P-880D, P-970, P-970D, P-974, P-974D, P-989, P-989D, P-990, P-990D, P-997, P-997D, P-999

Kemester 104, 143, 213, 1816, 1846, 3681, 3684, 3689, 5221SE, 5415, 5500, 5510, 5651, 5653, 5654, 5721, 6000SE, 7018, BE, CP, EE, EGDS, EGMS, GMS (Powd.), MM; Kenplast ESI; Kessco 639, 874, 887, BSC, Butyl Oleate, Butyl Stearate Cosmetic, Dist., GMO, GMS, IBP, IBS, ICS, IPM, IPP, Isobutyl Palmitate, Isobutyl Stearate, Isopropyl Myristate, Isopropyl Palmitate, Octyl Isononoate, Octyl Palmitate; Kodaflex DBP; Korantin LUB; Lamigen ES-30, ES-60, ES-100, ES-180, ET-20, ET-70, ET-90, ET-180; LAN-401; Lanesta S, SA30; Laneto 40, 50, 60, 100; Lanexol, AWS; Lankroplast L; Lanogene; Lanoquat 1756; Lanotein AWS 30; Laurex 4526, ARA; Leadstar; Leocon 2100E; Levaform HT, Si Emulsion, Si Oil, SIV; Lexemul EGDS; Lexolube 2J-237, 2N-212, 2N-237, 2T-237, 2X-108, 2X-109, 2X-114, 2X-130, 3G-310, 3I-310, 3N-309, 3N-310, 3P-309, 4H-415, 4N-415, 4P-415, B-108, B-109, BS-Tech, GT-855, N-110, NBS, T-110; Lexomul PEG 200 DL, 400 DL; LICA 01, 09, 12, 38, 38A, 38J, 44, 97; Lilamin 101 D, 115, 115 D, 140, 140 D, 142, 142 D, 151, 160, 160 D, 163, 163 D, 170, 170 D, 172, 172 D, 308 D, 310 D, 312 D, 314 D, 316 D, 342 D, 343, 345 D, 363, 364, 367 D, 368, 369, 372 D, 381, 382, 383, 540, 560, 570, 570 DO, 572; Lipamide S; Lipo DGLS, DGS-SE, Diglycol Laurate, EGDS, EGMS, PGMS; Lipocol B; Lipolan, Lipolan R, S; Liponic 70-NC, 76-NC, EG-1, EG-7; Lipopeg 2-DL, 4-DL, 4-DO, 4-DS, 4-L, 4-S, 6-L, 10-S, 15-S, 39-S, 100-S, 6000-DS; Liposorb L, L-10, L-20, O, O-5, O-20, P, P-20, S, S-20, SQO, TO, TO-20, TS, TS-20; Lipovol J; Lipowax C; Liquazinc AQ-90; Liquester; Lonzest PEG-4DO, SML, SMO, SMP, SMS, STO, STS; Loobwax 0605, 0638, 0651, 0682, 0750; Loxiol G 10, G 11, G 12, G 13, G 15, G 16, G 20, G 21, G 22, G 30, G 31, G 32, G 33, G 40, G 41, G 47, G 53, G 60, G 70, G 71, G 72, G 73, GH 4, HOB 7107, HOB 7108, HOB 7111, HOB 7112, HOB 7119, HOB 7121, HOB 7131, HOB 7162; Lubracel 48, 53, 60; Lubrajel; Lubrazine W Superfine; Lubrimet P 600, P 900; Lubrisol; Lubrol 17A-17; Lutensol ED 140, ED 310, ED 370, ED 610; Lutrol E 300, E 400; Luviquat FC 370, FC 550; LZ 01, 09, 12, 38, 44, 97, 5340

M-7 Mold Release; ISA; Macol 15, 20, 34, 660, 3520, 5100, E; Mafo 13, CAB; Mapeg 600 DL, 600 DOT, 600 ML, 600 MOT, 6000 MS, PGDS, PGMS, PPG-400, PPG-500; Maphos 31, 32, 33, 41, 58, 60A, 77, 78, 79, 91, 8078, 8135, L-4, L-6, L-10, L-12, L-13, L-22; Marvanol 75% SCO; Masil 263, 264, 265, 1066C, 1066D, EM 266 (35%), EM 266HV, SF 5, SF 20, SF 50, SF 100, SF 200, SF 350, SF 500, SF 1000, SF 5000, SF 10,000, SF 12,500, SF 30,000, SF 60,000, SF 100,000, SF 300,000, SF 600,000; Mazamide C-5, L-5, O-20, SS-10; Mazeen 173, 174-75, C 2, C 5, C 10, C 15, DBA, S 2, S 5, S 10, S 15, T 2, T 5, T 15; Mazol GMO, GMS; Mazon RI-110; Mekon White; Merpol LF-H; Merquat 280; Miglyol 810, 812, 840; Migralube QFL-4536; Miramine CC, CPC, IC, SC, TC, TOC; Miranol DM, DM Conc. 45, ISM, LM-SF Conc.; Mirapol A-15; Mirapon FM; Mirataine H2C; Modulan; Mold-Ease; Mollisan Conc.; Molykote; Molysulfide Suspension, Tech., Tech. Fine; Molyvan A, L; Monafax 057, 785, 786, 831, 939, H-15, L-10; Monalube 29-78, 780; Monamid 7-100, 7-153 CS, 15-70W, 150-AD, 150-ADD, 150-ADY, 150-CW, 150-DR, 150-IS, 150-LMW-C, 150-LW, 150-LWA, 150-MW, 718, 770, CMA, LIPA, LMA, LMMA, R31-42, S; Monamine Series, 928, 929, AA-100, AC-100. ACO-100, AD-100, ADD-100, ADDS-100, ADS-100, ADY-100, AF-100, ALX-80 SS, ALX-100 S, ARA-100, CD-100, CF-100 M, I-76, LM-100, R8-26, T-100; Monaquat ISIES; Monateric ISA-35%; Monolan PPG440, PPG1100, PPG2200; Monoplex DDA, DIOA; MS-122; Myvacet 9-40, 9-45; N-30; Naftozin N, Spezial; Naugalube 438-R; Neustrene 045, 053, 059, 060, 064; Neutral

LUBRICANTS (CONT'D.)

Degras; Newcol 3-80, 3-85, 20, 25, 40, 45, 60, 65, 80, 85, 150, 170, 180, 561H, 562, 564, 565, 864, 1105, 1305, 1310, 1515, 1525, 1545; Nikkol Decaglyn 5-IS, 5-O, 5-S, 7-IS, 7-O, 7-S, 10-IS, 10-O, 10-S, Hexaglyn 1-S;Ninex TDO 5, 9, 14, TMO 5, 7, 9, 14; Nioix AK-44, KS Series; Niox KH Series; Nissan Nonion LP-20R, LP-20RS, MP-30R, OP-80R, OP-83RAT, OP-85R, PP-40R, SP-60R; Nissan Persoft NK-60, -100; Nitrene A-309, OE; Noda Rice Wax Industrial

Noiox AK-41, AK-43, AK-44, KS-10, -12, -13, -14, -16; Nonex C5E, DL-2, S3E; Nonionic 1035-L, 1044-L, 1061-L, 1064-L, 1068-L, 1088-L; Nonisol 210; Nopalcol Series, 1-S, 1-TW; Nopcote C-104; Nopcowax 22-DS; Noram O; Norfox CS, DCO, KO, M2DS, Syn Lab, Terne Oil, V-5759, VMO, ZNS; Olicat C; OP-2000; Orzol; Pamolyn; Paracin 220, 285; Paricin 6, 8; Peem 397, 410; Pegafac CE 410, 610, CS 410, 610, 710, PDA (Neutralized), PDA Free Acid; Peganate CO 5, 16, 25, 30, 36, 40, 200, 200 50%, COH 25, COH 200, 200 50%, L 5, 20, MA 8, 15, 15 80%, ML 5, 9, 14, MO 6, 9, 14, MS 8, 14, 23, O 5, 20, S 5, 20, TO 9, 16; Pegeste SML, SMO, SMP, SMS; Pegosperse 50 DS, 200 DL, 1750 MS, 6000 DS, EGMS-70; Amber, Blond, Cream, Frost, Lily, Regent, Royal, Snow, Super, Ultima; Perfecta USP; Permalene A-100; Petrac 165, 215, 245, 250, 270, CP-11, CP-11 LS, CP-11-LSG, CP-12, CP-22G, CZ-81, MG-20 NF, ZN-41, ZN-42, ZN-44 HS, ZW-45; Petrolite C-700, C-1035; Petronate HMW; Petrosan 102; Petrosul 550; Phosfac 1004, 1006, 1044, 1044FA, 1066, 1066FA, 1068FA, 1068 C-FA, 5520; Phosphate Ester 123; Phospholan PDB-3, PNP-9; PL-26W; PL-34; PL-47; PL-48; PL-C; PL-H; PL-L-35; Plasthall 503, 4141, DIDG, DIODD, DODD, DTDA; Plastilease 250; Pluracol E-200, E-300, E-400, E-400 NF, E-600, E-600 NF, E-1000, E-1450, E-1450 NF, E-2000, E-1500, E-4000, E-4000 NF, E-4500, E-6000, E-8000, E-8000 NF, P-410, P-710, P-1010, P-2010, P-3020, P-4010, W-170, W-260, W-660, W-2000, W-3520N, W-3520N-RL, W-5100, W-5100N, WD1400; Pluriol E 200, E 300, E 400, E 600, P 600, P 900, P 2000, PE 6400; Pluronic 10R5, 10R8, 17R1, 17R2, 17R4, 17R8, 25R1, 25R2, 25R4, 25R5, 31R2, 31R4, F38, F68, F68LF, F77, F87, F98, F108, F127, L31, L35, L42, L43, L44, L61, L62, L62D, L62LF, L63, L64, L72, L81, L92, L101, L121, L122, P65, P75, P84, P85, P94, P103, P104, P105, P123; Pogol 200, 400 USP, 600, 1570

Polyaldo DGDO, DGHO, HGDS, TGMS; Polyfac PCW-180, PCW-181, -182, PN-209, TMO-5, TMO-5, TMO-7, TMO-14; Poly-G WS 280X; Polyglycol 112-2; Polylan; Polylube GK, NPP, SC, WS; Polysoft CA;Polywax 500, 655, 850, 1000, 2000, E-2020, OH 425, 550, 700; Precirol ATO; Procetyl 10, 20, 30, 50; Promyristyl PM; Proplast 050, 075, 060, 290; Prosol 5, 518; Prostearyl 15; Protol; Protopet Alba, White 2L, 3C, Yellow 1E, 2A, FHL; Provol 10, 30, 50; Quimipol EA 6803, 6804, 6806, 6807, 6808; Radia 3200, 7040, 7051, 7060, 7070, 7104, 7110, 7118, 7120, 7127, 7129, 7130, 7131, 7161, 7163, 7171, 7176, 7190, 7194, 7195, 7197, 7200, 7230, 7231, 7240, 7303, 7330, 7331, 7363, 7500, 7501, 7510, 7514; Radiacid 200, 408, 411, 420, 423, 1060; Radiamine 6140, 6141, 6160, 6161, 6163, 6164, 6170, 6171, 6172, 6173; Radiamuls MG 2142; Radianol 2106; Radiasurf 7000, 7146, 7150, 7151, 7152, 7153, 7372, 7175,7196, 7201, 7402, 7403, 7404, 7206, 7270, 7400, 7410, 7411, 7412, 7413, 7414, 7417, 7420, 7421, 7422, 7423, 7431,7432, 7443, 7453, 7454, 7505; Radiasyn 7174, 7177, 7178, 7364, 7367, 7368; RC 7; Renex 751; Rewocid DC212; Rewolub GSM, TMP 155, 275; Rewopal MT 2455, 5722; Rewopon IM-CA; Reworyl TKS 90/F; Rexol CCN; Rhodorsil Huiles 47, 633; Rilanit BS, CSN, DBA, DBS, DEHS, DNOP, DOA, DTP, EHO, EHS, EHTI, G 16; Rilanit G 20, GDO, GL 401, GMO, GMRO, GMS, GO 401, GTC, GTO, GTS, HL, IBO, IBS, IBTI, IBTIOK, IPM, IPP, IPS, ITS, LTC, OLO, PEC 4, PES, PMO, PPE, STS-T, TDO, TMTC; Ritachol; Ritaderm; Ritalan; Ritawax AEO, ALA; Robane; Robuoy; Ross Wax #100, 140, 145, 160, 165; Rudol; Rychem 17; Safety-Lube 1776, 1976, 3220, 3600, 4000, 6003, 8155; Sandopan TFL Liquid; Santone 3-1-SH; Sarkosyl L, LC, NL-30, O; Schercamox CMA; Schercemol 318, 1818, DEIS, DIA, DIS, DO, IPO, ML, OLO; Schercoat PC-550; Schercodine I; Scherco Finish 4L; Schercolube 707; Schercomid ID, IMI, IMO, SM, SO-A; Schercosol NL; Scheroba Oil; Schersoftoll P

Sebase; Secomine TA 02; Secosol 874, 887; Secoster BS; SEM-35, -60, -135; Semtol Series; Servamine KAC 412; SF 18, SF 69, SF 96, SF 97, SF 1029, SF 1066, SF 1080, SF 1091, SF 1134, SF 1147, SF 1153, SF 1154, SF 1173, SF 1188, SF 1202, SF 1204 SF 1221, SF 1250; Sicolub DSP, E, EDS, LB 1, LB 2, LBK, LK 3, OA 2, OA 4, OP, TDS; Silsmooth E, G; Silwet Surface Active Copolymer L-77, L-720, L-7001, L-7002, L-7004, L-7500, L-7600, L-7602, L-7605, L-7607; Sipenol IC2, IC10, IO12, IS05, IT11, IT50; Sipex EC-111; Sipon MLS; Siponic E-2, E-3, E-5, E7, E-10, E-15, L7-90, TD-9-90; SM 2086, 2140, 2155, 2162, 2163; S-Max 20, 40, 60, 60 K, 60 KHM, 65, 65 K, 67, 80, 80 K, 85, 85 K, 90, 93 R, 95; Softal 300; Softenol 3100, 3107, 3108, 3114, 3118, 3178, 3408, 3819, 3900, 3991, 3995; Solar Regular, TE; Solidester; Solricin 135; Solulan 5, 16, 25, 75, 97, 98, C-24, L-575; Solusoft 350; Sontex 19, 21, 35, 55, 70, 75, 75 T, 85, 85 T, 95 T, 100, 150; Soprofor S/20, S/60, S/65, S/80, S/85, T/20, T/60, T/65, T/80, T/85; Soromine AL, AT; Span 20, 40, 65, 85; Staflex DIBA, ODA; Stanclere T-126, T-250, T-250SD, T-801; Standamid ID, KDS; Standamox 01; Standamul G-32/36, G-32/36 Stearate, GTO-26; Starfol OO; Starwax 100; Stepan C68; Stepan Cationic 10; Stepantex SO 90, VG 90, VS 90, VP 85, VRH 80; Steramine FPA 197, PNA 75; Sterotex, HM & K Veg. Oils; Sublimed Blue Lead; Sulfonated Castor Oil 50%, 75%, GTO; Sulfonated Syn. Sperm Oil 2091; Sulframin 40, 40DA, 40RA, 40T, 60T, 85, 90, 1240, 1245, 1250, 1260, 1288, 1298, 1388; Sunolite 160; Sunsoflon PXP-2000; Super Refined Apricot Kernel Oil NF, Olive Oil; Super Sterol Ester; Supratol VF;

Surfam P10, P12B, P14B, P17B, P24M, P86M, P89MB; Syncrolubes; Syncrowax ERL-C, HGL-C; Syn Lube 106 (60%), 107; Synprolam 35 Lauramide, 35 Stearamide

T-Det C-40; Tegiloxan; Tegin 4600; TegMeR 703, 704, 804 Special, 809, 903; Tegopren; Tergitol 15-S-3, 15-S-5, 15-S-7, 15-S-9, 15-S-12, 15-S-20, 15-S-30, 15-S-40, 25-L-3, 26-L-1.6, 26-L-3, 26-L-5; Teric 16M5, 16M10, 16M15, 121, 124, 127, 151, 350, 351B, C12, DD5, DD9, LAN70, OF4, OF6; Texox PPG-400, PPG-2000, WL-440, WL-660, WL-1400, WL-5000; Textamine O-1, T-1; TL Series, -101, -102, -107, -113, -115, -115A, -115AH, -120, -126, -126-H, -127, -340, -450; T-Mulz 426; Tomah E-14-2, E-14-5, E-14-15, E-17-2, E-18-2, E-18-5, E-18-10, E-18-15, E-S-2, E-S-5, E-S-10, E-S-15; Trisolan 1720; Triton RW-10, RW-20, RW-75, RW-150, X-155, X-155-90%; Trycol 5878, 5964, 5972, OAL-23; Trydet 2636, 2644, 2670, 2671, 2672, 2676, 2685, ISA-4, MP-9, OA-10, SA-7, SA-8; Trylon 6733; Trylox 5900, 5902, 5904, 5906, 5907, 5909, 5918, 5921, CO-16, CO-25, CO-30, CO-36, CO-40, CO-200, -200/50, HCO-5, HCO-16, HCO-200/50; Trymeen 6602, 6603, 6607, 6609, 6617, 6657, SAM-50, TAM-20, TAM-25; Tween 20, 20 SD, 21, 40, 60, 61, 65, 80, 81, 85; Twitchell 6805 Oil, 6808 Oil, 6809 Oil; Tylose C, CB Series; Ucar Super Wetter FP; Ultra Sulfate AE-3, SE-5, SL-1; Unamide C-72-3, D-10, GC-75, J-56, LDL, N-72-3, S, SI, W; Unimate IPM, IPP, IPPM; Union Carbide L-45 Series, L-720, -721, -722, -727, -7001, -7002, -7600, -7602, -7604, L-7605, -7607, LE-45, LE-458HS; Unislip 1757, 1759, 1760 EBS; Vanfre IL-1, IL-2; Vanlube 622, 672, 692, AZ; Vanstay 5515, 5563, 6032, 6040, HTE; Varamide DO 280; Varonic 320, K205, L205, L205LC, MT 65, T202, T202 SR, T205, T205LC, T210, T215, T215LC; Vigilan; Vinylube 36, 38; Viscasil; Volatile Silicone 7158, 7207, 7349; Volpo 3, 5, 10, 20; Voltalef 1 S, 3 S, 3X, 10 S, 90, 901; Vybar 103, 825; Vydax WD

Waxenol 801, 815, 816; Wickenol 101, 105, 111, 127, 131, 139, 141, 143, 535; Witcamide 61, 70, 82, 272, 511, 823/10, 5130, 5133, 5138, 5195, 6310, 6445, AL69-58, CDA, M-3, MAS, MM, S-771, S-780, STD-HP; Witcamine 209, 210, 211, 235, 3164; Witco Aluminum Stearate 22, Calcium Stearate A, Calcium Stearate F, Heat Stable Sodium Stearate, Magnesium Stearate D, Zinc Stearate NW, Zinc Stearate Polymer Grade, Zinc Stearate REP; Witcolate 58, 1276; Witconate 605A, 605T, 1840X; Witconol 14, 18L, APEB, APEM, APM, APS, CA, CAD, CD-17, CD-18, F26-46, H-31A, MST, NS500K, RHP, RHT; Xanflood; Zelec UN; Zinc Stearate 11, 44, Disperso, Heat-Stable, LV, NW, Polymer Grade, Regular; ZO-9; Zonyl A

MITICIDES. SEE INSECTICIDES

MODIFIERS

A-C Polyethylene 6, 6A, 8A, 9A, 8, 9; Acryloid HF-Series; Acrysol ASE-60; Admul WOL 1403; Advawax 240, 275, 280, 290; Ajicoat SPG; Akrochem P 87, P-487; Aldo PMS; Amihope LL-11; Amoco Resin 18-210, 18-240, 18-290; Avitex R; Baragel, 3000; Bentone 500; Benzoflex 352, S-312, S-404, S-552; Blendex 101, 131, 201, 310, 336, 338, 405, 420, 424, 467, 586, 590, 702, 703, 27920; Calsuds CD-6; Caprol 2G4S, 10G10S; Cardis 314; Cardolite NC-507, NC-510, NC-511; Cenwax G; CHPTA; CM-88; Combat HPC-40, HPF-325, HTP-40, HTP-325, SHP-40, SHP-325, SRG-40, SRG-325; Complemix 100; Crodamol BE; Crosilk Powd.; Crotein ASC, ASK; Cumate; Darvan NS; Dehydag Wax O; Dispersol T; Dresinol 40, 42; Drewfax 1980; DSX-1514; Eltesol ACS 60, PX 40, ST 40, ST Pellets, SX Pellets; Emargol KL; Emcol 4600, CC-37-18, L, M, NA-30; Emerest 2485, 2662, 2675; Emery 912, 1720, 6686, 6709; Emid 6518, 6519, 6590; Empigen OB, OH, OY; Empilan 2125-AU, CIS, CME, LME; Epolene C-10, C-13, C-14, C-15, C-17, N-15; Espesilor AC-50; Ethyl Tuads; Fluilan AWS; Generol 122, 122E5, 122E10, 122E16; Glucam E-10, E-20, P-10, P-20; Harshaw Antimony Oxide KR; Hartamide LMEA-70; HB-40; Hercules Cellulose, Ester Gum 8D; Jeffamine BuD-2000, D-230, D-400, D-2000, DU-1700, DU-3000, ED-600, ED-900, ED-2001, T-403; Kelzan, D, M, XC Polymer; Ken Kem CP-45, CP-99; Kenplast ES-2, ESN

Ken-React Series, 7 (KR 7), 9S (KR 9S), 12 (KR 12), 26S (KR 26S), 33DS, (KR 33DS), 38S (KR 38S), 39DS (KR 39DS), 41B (KR 41B), 44 (KR 44), 46B (KR 46B), 55 (KR 55), 133DS (KR 133DS), 134S (KR 134S), 138D (KR 138D), 138S (KR 138S), 158D (KR 158D), 158FS (KR 158FS), 212 (KR 212), 238A (KR 238A), 238J (KR 238J), 238M (KR 238M), 238S (KR 238S), 238T (KR 238T), 262A (KR 262A), 262ES (KR 262ES), OPP2 (KR OPP2), OPPR (KR OPPR), TTS (KR TTS); Lanethyl; Laneto 40; Lanex; Lankrosol SXS-30; Lexein S620TA; LICA 01, 09, 12, 38, 38A, 38J, 44, 97; Lipamide MEAA; Liponic EG-1, EG-7; LZ 01, 09, 12, 38, 44, 97; Maracarb N-1; Methyl Tuads; Mirapol A-15; Modaflow Powd. II; Moldbrite 25; Multiwax 180-M, ML-445, W-445, W-835, X-145A; Nacopa; Nikkol BPS-5, BPS-10, BPS-15, -20, -25, -30; Ninol 41-CO; Nissan Nymide MT-215; Nonionic 1035-L, 1044-L, 1061-L, 1064-L, 1068-L, 1088-L; Ogtac-85, 85 V; Palatinol DBP; Paricin 9, 13, 15; Pennwalt 4P, n-Dodecyl Mercaptan; Pentalyn 830, 856; Perlankrol O; Petro-Rez PTH, PTL, X-95; Piccofyn A100, A115, A135; Piccolyte A115, A125, A135, C115, C100, C125, C135, D100, D115, D135, S100, S115, S115SF, S125, S125SF, S135; Piccomer 10, 25, 40, 100, 110, 120, 150; Picconol A500, A501; Piccotac B, CBHT; Piccoumaron 100, 110, 120; Poly-Pale Ester 10; Polysilicate 48; Polywax 500, 655, 1000, 2000; PVP K-15, K-60, K-90; Rewomid DC 212/S; Sandoz Amide CO, NP, PE, PL, PN, PO, PS; Sandoz Sulfate 216, 219, A, EP, ES-3, K, TL, WA Dry, WAG, WAS, WA Special, WE; Scripset 520, 540, 550, 720; SF 69, SF 96, SF 97, SF 1066, SF 1705; SM 2163; SR-213, SR-295, SR-444, SR-7475; Staybelite Ester 5, Ester 10; Stereon 840, 840A, 841A, 872, 873, 874, 875;

MODIFIERS (CONT'D.)

Superclear 80-N, 100-N, 200-N; Terate 202, 203, 204; Texox PPG-400, PPG-2000; T-Maz 20, 28, 40, 60, 61, 65, 67, 80, 81, 85, 90, 95; Trydet SA-50/30; Tyzor AA, DC, TBT, TE, TOT, TPT; Ultra-Pflex; Varcum 1198, 6404, 7608; Vinnol K 704; Viplex 885; Viscasil; V-Pyrol; Vybar 103, 260; Wayplex NTP-A; Winnofil S, SP; Witcamide 82, 5130, 5133, 6445, CDA

MOISTURE BARRIERS. SEE WATER REPELLENTS

NEUTRALIZERS

Emery 5794; Na-Sul AS, BSB; PDI Product 3997; Polymeric Acid

NUTRITIVE ADDITIVES

Maltrin M040, M050, M100, M150, M180, M440, M500, M550

ODORANTS. SEE AROMATICS

OPACIFERS & PEARLESCENTS

Alkamide CME, HTME; Alkamuls 200-DS, 400-DS, 600-DS, DEG-MS, EGMS, GMR; Alkasurf S-1, S-2; Amerlate P; AMS-33; Biju; Bi-Lite; Caplube 8448; Caprol 2G4S, 10G10S; Celite 263, 305, 321, 321A, 388; Cerasynt D, IP, LP, M, MN, PA, SD; Chimipal SGE; Cithrol GMS A/S ES 0743; Crodacol C70, S70; Crodamol BE; Crodapearl Liq., NI Liquid; Cromul 0685; Crosterene; Cutina AGS, EGMS, TS; Cyclochem DGS, EGDS, EGMS, EGMS/SE, GMO, SEG; Cyclosheen 100, 202; Drewmulse 1128, 1129, CNO, DGMS, EGDS, EGMS, GMC-8, GMO, GMRO, GMS, HM-100, PGMS, TP, V, V-SE; EGDS-VA; EGMS-VA; Elfan L 310; Emerest 2350, 2355, 2380, 2381, 2712; Emersol 110, 120, 132, 143; Emery 1787; Empicol 0627; Empilan EGMS; ESI-CRYl 11, 12, 25; Genapol PGM, PGM Conc., PMS, TSM, TS Powd.; Grocor 5221 SE, 5500; Harshaw Antimony Oxide KR; Hetester MS; Incromate PPI; Incromide BED, CM, LEP, SM; Incropearl Liquid, Liquid NI; Kemester 105, 115, 143, 205, 226, 325, 3681, 3684, 5654, 9018, 9022, EE, EGDS, EGMS, MM; Kessco Acetin, Diacetin, Diethylene Glycol Monostearate, Diglycol Stearate Neutral, Diglycol Stearate SE, EGDS, EGMS, Glycerol Monostearate 860, Glycerol Monostearate DH-1, Glycerol Monostearate Pure, Glycerol Monostearate SE, Propylene Glycol Monostearate Pure, Triacetin; Lankropearl T; Lexemul 55G, 503, 515, AR, AS, EGDS, EGMS, T; Lipal EGDS, EGMS; Lipamide S; Lipo DGLS, DGS-SE, Diglycol Laurate, EGDS, EGMS, GMS 450, GMS 470, PGMS; Lipocol B; Lipomulse 165; Lorol CT-M; Lytron 284, 288, 295, 300, 305, 308, 621, 614 Latex; Mackpearl LV; Manro PC 35N; Mapeg EGDS, EGMS; Marlamid KL, KLA; Mearlin; Mibiron; Micro-Cel T-38, T-70; Monomuls 60-10, 60-15, 60-20, 60-25, 60-25/2, 60-25/5, 60-30, 60-35, 60-40, 60-45, 90-10, 90-15, 90-20, 90-25, 90-25/2, 90-25/5, 90-30, 90-35, 90-40, 90-45; Nailsyn; Naturechem EGHS, GMHS, PGHS; Naturon; Nikkol Estepearl 10, 15, 30, 35; Ninol SNR; Noiox EO-33; Norfox EGMS, GMS, GMS-FG; Nuopaque; Nutrapon B; Onyxol 42; Ottalume 2100; Pearl-Glo; Pearlex 3105; Perlex B.67, B.70, B.67M, BU, BUA.35; Perlextra B.70, BU, BUA.70; Perlglanz-Konzentrat B-30, B-48; Perlglanzmittel GM 4006, 4175, 4055; Polawax; Polectron 430; Produkt GM 4055; Quad Opacifier #35; Radiasurf 7402, 7403, 7404, 7412, 7413, 7414, 7417, 7421, 7422, 7423, 7431, 7432, 7443, 7453, 7454, 7196, 7201, 7206, 7270, 7400, 7410, 7411, 7420; Rewopal PG 280, 340; Sachtolith HD, HD-S, L; Schercemol CP, CS, EGMS, FMS; Schercomid MME, SD-DS, SME, SME-A, SME-M, SME-S; Schercopearl EA-100; Secoster DMS, EMS; Sipoteric 1442, 1444, 1450; S-Maz 20, 40, 60, 65, 67, 80, 85, 95; S-Maz 60 K, 60 KHM, 65 K, 80 K, 85 K, 90, 93 R; Soloron; Spectra-Pearl; Standamul G-32/36, G-32/36 Stearate; Standapol Pearl Conc. 7130; Surco CMEA; Syncrowax AWI-C, BB4, BB5, ERL-C, HGL-C, HR-C, HRS-C; Tegin G, G 1100, TGOS, VA; Tego-Pearl B-48; Timiron; Ultrox, 500W, 1000W; Unamide S, SI, W; Vestorwax FT-150, FT-150P, FT-200, FT-300; Vybar 103; Witamide 70, MAS, MM; Witconol CA, CAD, RHP; Witcopaque 11, 12

OPTICAL BRIGHTENERS. SEE WHITENING AGENTS

OXIDIZING AGENTS

ACL 59, 60; T-Hydro Sol'n.

PENETRANTS

Abil-Wax 9800; Acetol 1706; Acetulan; Adsee 775, 799; Alkanol 189-S, XC; Alkasurf DA-3, DA-4, DA-6, LA Acid, NP-15 80%, NP-20, NP-40; Alox 1680, 2000, 2213C; Amerchol H-9, L-101; Amerlate P; Apex

Pentrapex #1923; Armid O; Centromix Series; Chemal DA-4, DA-6, DA-9, TDA-3, TDA-6, TDA-9, TDA-12, TDA-15, TDA-18; Chimin P10; Cibaphasol AS; Cithrol 2ML, 2MO, 4ML, 4MO, 6ML, 6MO, 10ML, 10MO, 40MO, 60ML, 60MO; Conco NI-21, NI-43, NI-60, NI-90, NI-100, NI-110, NI-150, NI-185, NI-187, NI-190, NI-197, NI-2000; Conco Sulfate RA; Consamine PA; Crillet 2, 3; Crodalan LA; Crodamol DO; Crodasinic LS35, OS35; Cyclochem CL, INEO, LVL, ML; Darex 3333; Denwet CM, RG-7; Depasol AS-27, CM-41; DeSophos 6 DP; Detergent CR; Detergyl S; Dovanox 23H, 23M, 25N, 231; Dresinol 40, 42; Drewfax 0007; Duponal 80, RA, WA Dry, WN; Dyasulf 2031; Eccolene OW; Eccotex P Conc.; Eccowet W-50; Ektasolve EM; Emcol CC-9; Emery 1730; Emphos PS-236, PS-440; Emthox 5940, 6964, 6965; Ethylan HB4, HB30; Fancol LAO; Fluilan; Fosterge RD; Gafterge AW-123; Gaftex 288, CD-169, WSP; Hallcomid M-6; Hartex V63, V64; Honoralin PL; Hyonic NP-90, NP-100, NP-110, NP-120, OP-100, PE-100, PE-120; Icodimeen T-30; Icomeen 18-5, O-30, O-30-80%, S-5, T-2, T-5, T-7, T-20, T-25, T-25 CWS, T-40, T-40-80%; Iconol DDP-10, DNP-8, DNP-24, DNP-150, NP-7, NP-12; Idet 5L SP NF; Igepal CO-630, CO-660, CO-710, CO-720, CO-730; Interwet 33; Intralan Salt N; Intratex W New; Intravon JF, JU; Jorchem PRM-98; Jordawet DMDS; Kadif 50 Flakes; Keripon NC

Lantrol 1673, 1674; Levelene; Lexol IPM; Liponox N-105, NC 6E, NCG, NCI, NCT; Liquid Base; Lonzest 143-S; Mackanate DOS-75; Mercerol GVC-65; Merpol C, HCS; Miranol J2M Anhy. Acid, J2M Conc., JEM Conc.; Monamine 779; Monamulse CI; Monawet MB-45, MM-80, MO-65-150, MO-70, MO-70-150, MO-70E, MO-70R, MO-75E, MO-84R2W, MT-70, MT-70E, MT-80H2W; Monopol Oil 75; Monosulph; Nacconol 35SL; Nekal NF; Neobee 1223 (Neobee O), O; Neorate NA-30; Newcol 290K, 290M, 290P, 560, 560SF, 560SN, 561H, 562, 564, 565, 566, 861SE, 865, 1305SN, 1310SN; Niaproof Anionic Surfactant 4, 08; Nimlesterol 1730; Nissan Trax H-45; Nonal 206, 208, 210, 310; Nonipol 20, 40, 55, 60, 70, 85, 95, 100, 110, 130, 160, 200, 400, BX, D-160, Soft SS-50, -70, -90; Oxitol; Peganol NP 9, 10, 12, 15; Pelex NBL, NB Paste; Penestrol N-160; Pentex 40; Pepol A-0858; Peregal O, OK; Petrowet R; Phosfac 1004, 1006, 1044, 1044FA, 1066, 1066FA, 1068FA; Phosphoteric T-C6; Plasthall 7041, 7049, R-9; Polylan; Polylev #745; Procetyl 10, 20, 30, 50; Protowet E-4; Quarternary O; Rewomine IM-CA; Rexopene; Rexowet CR, GR, MS, NF, NFX, RW; Ritalan, Ritalan C; Ritawax AEO, ALA; Sandet ALH; Sandoz Sulfonate 3B2; Sanmorin 11, OT 70; Schercemol DEIS, DO; Scheroba Oil; Seachem 105; Sipon EA, EAY; Siponate 330; Solar NP; Sole-Terge TS-2-S; Squalane; Stafoam F; Sulfamin; Sulfonate 3B2, AAS 35S, AAS 40FS, AAS 45S, AAS 50MS, AAS 60S, AAS 70S, AAS 75S, AAS 90; Sulfonic 864; Sulframin 40, 1288, 1298, 1388; Super Corona; Super Wet; Supratol VF; Surfax WO; Surfonic HDL, JL-80X, N-10, N-31.5, N-40, N-60, N-85, N-95, N-100, N-102, N-120, N-150, N-200, N-300, N-400, NB-5, NB-14; Synthrapol N; T-Det C-40, N-14; Tergitol TMN-3; Tetronic 50R1, 50R4, 50R8, 70R1, 70R2, 70R4, 90R1, 90R4, 90R8, 110R1, 110R2, 110R7, 130R1, 130R2, 150R1, 150R4, 150R8; Tex-Wet 1002, 1010, 1048; Toximul 80; Triton 770 Conc., W-30 Conc., X-301; Trycol 5950, 5951, DA-4; Trylon EW; Ucar Super Wetter FP; Ultra Sulfate AE-3, SE-5, SL-1; Union Carbide L-720, -721, -722, -727, -7001, -7002, -7500, L-7600, -7602, -7604, -7605, -7607; Varine C, O, T; Victawet 12, 35B, 58B; Viplex 525; Witamide 82, 272, 511, 5130, 5133, 5138, 5195, AL69-58; Witcamine 209, 211, 235; Witcolate 58, 1276, 7031, D-510; Witconate 605T, 1840X; Witconol APEM, APM, APS

PEPTIZING AGENTS

Aktiplast, AS, F, PP, T; Blancol; Blancol N; Crotein SPA, SPC, SPO; Empiphos STP, STP/1L, STP/6L, STP/6N, STP/AD, STP/D, STP/DMST, STP/E, STP/F, STP/L, STP/L16, STP/M, STP/M16, STP/MST, STP/N; Proaid 9802

PESTICIDES. SEE INSECTICIDES

PIGMENTS. SEE DYES & PIGMENTS

PLANT GROWTH RETARDANTS

Ortho Prunit, Sumagic

PLASTICIZERS

Abitol; Ablusol ML; Acetulan; Acrawax C; Adimoll BO, DN, DO; Admex 433, 500, 515, 522, 523, 525, 526, 527, 529, 710, 711, 746, 752, 760, 761, 770, 775, 780, 890, 910, ELO; Adol 42, 52 NF, 54, 60, 61 NF, 62 NF, 63, 64, 66, 80, 85, 85 NF, 90, 90 NF, 320, 330, 340, 520 NF, 620 NF, 630, 640; Akrochem P 37, P 49, P 55, P 82, P 86, P 87, P-478, P-486, P-487, LN; Aldo PMS; Alkamide STEDA; Alkamuls DEG-MS; Alkapol PEG 200, 300, 400, 600; Alkasurf NP-4, NP-6; Altax; Amerchol CAB, H-9, L-101; Amerlate LFA, P, WFA; Antistatic KA; Antistaticum RC 100; Aqualose L 30, L 75, L75/50, LL 100, W 20; Arneel 18 D; Arnox TGE-05, 16, 36, 40, 80; Aromatic Oil 745; Arubren CP; Atmer 153, 154, 508; Axol E61, E66; Benzoflex 2-45, 9-88, 9-88 SG, 50, 131, 284, 352, 400, P-200, S-312, S-552; Berol Dinonylphenol; Berol Nonylphenol; Be

PLASTICIZERS (CONT'D.)

Square 175, 185; Bozemine N 60, N 609; Britol; Calcium Stearate, Regular; Captex 300; CB-4-34; Carbowax PEG 200, 300, 400, 600, 900, 1000; Cardolite NC-513, NC-700, NC-1307; Cellolyn 98-80T; Cereclor 42, 42P, 50LV, 51L, 52P, 70L, AP45, AP52, LP4446, LP4985, S45, S52; Cetodan; Citroflex A-2, A-4, A-6, B-6, C-2, C-4; Clorafin 40, 50; Cordon NU 890/75; Corona Lanolin; Coronet Lanolin; CPF-0001; CPF-0003; CPF-0008; CPF-0019; CPF-0022; Cralane KR-13, KR-14, LR-10, LR-11; Crestalans; Crodalan AWS, LA; Crodamol BS, DA; Crodapur; Croderol G7000; Crodet L4, L8, L12, L24, L40, L100; Cyclochem INEO; Dacospin 1735-A, LS 4100, POE(25)HRG; DBE, DBE-2, DBE-2SPG, DBE-3, DBE-4, DBE-5, DBE-5SPG, DBE-9; Densulf TA-75; DEP; D.E.R. 732; DIBA; Disflamoll DPK, DPO, TP; DMP; DOA; Drakeol 9, 10, 19, 21, 32, 34, 35, Drapex 3.2, 4.4, 6.8, 10.4; Dresinate 81, 90, 91, 92, 93, 94, 95; Emcol CC-36; Emerest 2301, 2302, 2675; Empicryl 6047; Emulbon LB-78; Epoxol 5-2E, 7-4, 8-2B, 9-5; Estabex 138-A, 2307, 2307 DEOD, 2381; Estaflex ATC; Ethoxylan 1685, 1686; Ethylan ENTX; Eureka 102; Fancol HL, LA, LAO, OA 50, OA 70, OA 80, OA 90, OA 95; Fancor Lanolin; Fancor Lanwax; Finsolv SB; Flexchlor 0001, 0002, 0008, 0009, 0010, 0011, 0012, 0018, 0023; Flexol Plasticizer 4GO, EP-8, EPO; Flexricin 9, 13, 15, 17, P-1, P-3, P-4, P-6, P-8; Fluilan, AWS; Fonoline White, Yellow; Forlan C-24; Fortex; Fyarestor 100

Gantrez B-773, M-154, M-555, M-556, M-574; Geropon RM/210; Glucam P-20 Distearate; Glucate DO; Graden Butyl Oleate; Gradonic 400-ML; Grindtek AML 60, AMOS 90; Hallco DBS; Haroil SCO-65, SCO-7525; Hartolan; Hartolan Super; Hercoflex 707, 707A, 900; Hercolyn D; Hercules Ester Gum 10D; Hetester PMA; Homodan; Hycar 1312, 1422, 1422X8, 1452P50; Hystrene 4516, 5016, 7018, 7022, 8016, 8018, 9008, 9010, 9014, 9016, 9018, 9022, 9512, 9718; Igepal CO-430; Imwitor 595, 742; Incromectant AQ, LQ; Isopropylan 33; Jeffamine ED-600; Kemester 143, 4000, 5221SE, 5500, 5510, 5822, 6000, 6000E, 6000SE, 7018; Kenflex A, N; Kenplast 3070, A-450, AP-19, APK, BG, ES-2, ESB, ESI, ESN, G, LG, LT, PG, PPE, RD, RDN, RG; Kessco BSC, Butoxyethyl Oleate, Butoxyethyl Stearate, Butyl Oleate, Butyl Stearate Cosmetic, Dist., Dibutoxyethyl Phthalate, Octyl Isononanoate, IBP, ICS, IPM, IPP, Isobutyl Stearate, Isopropyl Myristate, Isopropyl Palmitate; Ketjenflex 8, 9; Kodaflex DBP, DEP, DMEP, DMP, DOA, DOP, DOTP, HS-3, HS-4, PA-5, TOTM, Triacetin, TXIB; Kronitex 100, 3600, TCP; Krynac 810; Lamegin EE, GLP 10, GLP 20; Lanethyl; Laneto 100; Lanex; Lanexol, Lanexol AWS; Lankroflex ED6, GE, L; Lantrol 1673, 1674, AWS 1692; Lexein A210; Lipolan R; Liponic 70-NC, 76-NC, EG-1, SO-20; Lubracel 48, 53, 60; Lubrol 101; Macol E; Mapeg DGLD; Mazol 159, 165C, GDO, GMO, GMR, GMS, GMS-D, GMS-HM, PGS-61; Mekon White; Merrol 418S, 418T, 600TM, 810A, 810TM, 818T, 3810, 3900, 4200, 4206, 4208, 4218, 4220, 4221, 4226, 4228, 4295, 4425, 4800, C-102, C-103, C-104, C-105, DBP, DBS, DIBA, DIDA, DIDN, DIOA, DOA, DOS, DOZ, DTDA, E-45, E-52, E-68, E-70, N-301, N-302, N-303, N-304, N-305, N-306, N-307, P-1030, P-1030LV, P-5510, P-5511, P-6303, P-6310, P-6311, P-6320, P-6403, P-6410, P-6420, P-6422, P-6424, P-6510, P-8227, P-8425, P-9500, PGB, PGB-50, TIOTM, TOTM; Mesamoll; Methocel A, E, K; Monolan PPG440, PPG1100, PPG2200; Monoplex DOA, DOS, NODA, S-73, S-75; Multiwax 180-M, ML-445, W-445, W-835, X-145A; Naftolen H, N400, N401, N402, N403, N404, N405, N406, N407, N408, ND, NV, P600, P603, P604, P606, P611, P612, P613, P614, P616, ZD, ZD103, ZD105, ZD106, ZM; Nansa HS80/S, HS85S, MA30; Neolyn 40; Neutral Degras; Newpol PE-61, -62, -64, -68, -74, -75, -78, -88; Noda Rice Wax FCC; Noiox KS-10, -12, -13, -14, -16; Nopalcol Series, 1-S, 1-TW, 4-O; Nopcosulf CA-60, -70; NP-10, NP-25; Nouryflex 520

Paraplex 5-B, G-25, G-30, G-31, G-40, G-41, G-50, G-51, G-54, G-56, G-57, G-59, G-60, G-62, GA-20, RG-2, RG-7, RG-8, RG-10, RGA-2, RGA-7, RGA-8; Paricin 6, 8; Peerless No. 1, No. 2, No. 3, No. 4; Peganol NP 4; Pegosperse 50 DS, 200 DL, 1750 MS, 6000 DS, EGMS-70; Penreco Regent; Peroxidol 780, 781; Petrac 250, 270; Petro-Rez 100, 103, 200, 215, PTH, PTL, X-95; Petrolite C-700, C-1035; PGE-400-DS; Phosflex, 71B, 362, 370, 390; Piccolastic A5, A25, A50, A75, C125; Piccomer 10, 25, 40, XX40, XX100; Piccovar AP10, AP25, L30, L30S, L60; PL-2N-204; PL-2N-204 RG; PL-2T-237; PL-2X-134U; PL-4VIP-415; Plas-Chek 775, 795; Plasthall 8-10TM, 83SS, 100, 325, 503, 4141, 7006, 7041, 7049, 7050, BSA, DBEA, DBEEA, DBEEG, DBEG, DBEP, DBES, DBEZ, DBZZ, DDG, DIBA, DIBZ, DIDA, DIDG, DIOA, DIODD, DOA, DODD, DOS, DOSS, DOZ, DTDA, HA7A, LTM, NODA, P-530, P-550, P-554, P-630, P-640, P-643, P-650, P-670, P-686, P-1070, P-7035, P-7035M, P-7046, P-7092, P-7092D, R-9, TIOTM, TOTM; Plastoflex 2307; Plastogen R; Plastolein 9048, 9049, 9050, 9051, 9058, 9065, 9071, 9088, 9091, 9215, 9232, 9404, 9717, 9730, 9731, 9734, 9749, 9750, 9752, 9756, 9761, 9765, 9772, 9776, 9780, 9781, 9783, 9784, 9789, 9790; Pliabrac 519, 521, 524, TBP, TCP, TXP; Pluracol E-300, E-400, E-600, E-1500, E-4000, E-6000; P-410, P-710, P-1010, P-2010, P-3020, P-4010; Pluriol E 200, E 300, E 400, E 600, E 1500, E 4000, E 6000, E 9000

Pluronic 10R5, 10R8, 17R1, 17R2, 17R4, 17R8, 25R1, 25R2, 25R4, 25R5, 31R2, 31R4; Pogol 200, 1570; Poly-G WI 285, 625, 1715, Poly-G WS 100, 170, 260, 660, 2000, 3520, 5100, WT 9150, 90,000; Poly-Solv DB, DE (High Gravity), DE (Low Gravity), DM, DPM, EB, EE, EM, MPM, TE, TM, TPM; Polytard MC; Procetyl 10, 20, 30, 50, AWS; Purelast 166, 176, 186, 190, 195; Pycal 94; 2-Pyrol; QO Furfuryl Alcohol; Radia 7051, 7131; Radianol 7106, 7376; Radiasurf 7196, 7201, 7206, 7270, 7400, 7402, 7403, 7404, 7410,

7411, 7412, 7413, 7414, 7417, 7420, 7421, 7422, 7423, 7431, 7432, 7443, 7453, 7454; Reax 77; Reogen; Rexol 25/6; Rhenofit 1600; Rhenovin FH-70; Ritalan; Ritawax AEO; Rudol; Santicizer 8, 9, 97, 141, 143, 148, 154, 160, 261, 278, 409, 412, 429, 711

Sartomer 322; Schercemol DICA, PGML, ML; Semtol Series; Serdox NNP 8.5; Softenol 3900, 3991; Solan, 50; Solulan 5, 16, 25, 75, 97, 98, C-24, L-575, PB-2, PB-5, PB-10, PB-20; Sontex 19, 21, 35, 55, 70, 75, 75 T, 85, 85 T, 95 T, 100, 150; Sopralub ACR 265, 275; SP-6700, SP-6701; Staflex 500, 550, 802, 803, 804, 809, 812, 829, BDP, BOP, DBEA, DBEP, DBF, DBP, DIBA, DIBCM, DIDA, DIDP, DINA, DINM, DIOA, DIOM, DIOP, DMAM, DMP, DOA, DOF, DOM, DOP, DOZ, DTDM, DTDP, NODA, NONDTM, ODA, ODP, TIOTM, TOTM; Staybelite, Ester 5; Sterox ND; Struktol 40 MS, 40 MS Flakes, 60 NS, 60 NS Flakes, WB 300; Stygene R-2; Super Corona; Super Hartolan; Super-Sat, AWS-4; Surfactol 13; T-600; Tamol NH 9103, NHC 3001, NHC 4001, NHC 9301, NMC 9301, NN, NNO, NNOK; TBP; Tegin E-41, E-41 NSE, E-61, E-61 NSE, E-66, E-66 NSE, L 61, L-61 NSE, L 62, L-62 NSE, RZ, RZ NSE; TegMeR 803, 804; Terate 101, 131, 202, 203, 204; Teric PE68; Tetronic 50R1, 50R4, 50R8, 70R1, 70R2, 70R4, 90R1, 90R4, 90R8, 110R1, 110R2, 110R7, 130R1, 130R2, 150R1, 150R4, 150R8; Texacar EC, PC; Texanol Ester-Alcohol; Texox WL-440, WL-660, WL-1400, WL-5000; Thiofide; TOF; Tolplaz TBEP; Tricap; Trydet SA-50/30; Trylox 6753, SS-20; Turgum S; Tylose C, CB Series, H Series, MH, MHB, MB, MH-K, MH-xp, MHB-y, MHB-yp; Ultraflex; Ultramoll I, II, III, PP, PU, TGN; Uniflex 300, 312, 314, 315, 320, 327, 330, BYO, DBS, DCA, DCP, DOA, DOS, TCTM, TOTM; Unilink 8100; Unimoll 66 M, BB, DB; Varonic U250, U250-50%; Victory; Vigilan AWS; Viplex 525, 885; Volpo 25D3, 25D5, 25D10, 25D20, CS3, CS5, CS10, CS15, CS20, O5, O10, O15, O20, T3, T5, T10, T15, T20; Vulkanol 80, 81, 85, 88, 90, BA, FH, KA, OT; Vultac 2, 4; Wickenol 139, 141, 142; Wingtack 10; Witconol H-31A, MST, RHT; X-743; ZO-9

PRESERVATIVES

Admul CSL 2007, CSL 2008; Aflaban DF; Alkaquat DMB-451; Amerstat 251, 252, 282, 300; Amical 50, Flowable; Arlacide A, G, H, HM; Bioban CS-1135, N-95, P-1487; Britesorb; Britesorb A 100; Bronidox L, L 5; CAE; Dow Corning 7 Compound; Dowicil 75; Emercide 1199; Emeressence 1160 Rose Ether; Empigen 5089, BAC50, BAC50/BP, BAC80, CHB; Ethylan HB1-TG; Evanstab 12; FMB 500-15 Quat U.S.P.; Fuelsaver; Fungicide FX; Germaben II, II-3; Germall 115, II; Giv-Gard DXN; Glydant; Hyamine 1622, 2389, 3500; Kathon 886, CG, CG/ICP, DP, ICP, LX; Lexgard B, E, M, P; LiquaPar; Micro-Chek 11, 11D, 11DIDP, 11S-711; Myacide SP; Nuodex PMA 18; Onyxide 172, 200, 3300; Ottasept Extra, Tech; Proxel CRL; Radiamuls 125, 135, 137, 145, 147, 155, 157, 345; Santoquin, Emulsion, Mixture 6; Sorbistat, -K; Starplex 90; Sustane 3, 4A, 6, 8, 20, 20-3, 20A, 31, BHA 1-F, BHT, HW-4, P, PA, PG, Q, TBHQ, TBHQ.4, W; Suttocide A; Synprolam 35BQC (50), 35BQC (80), 35DMA; Tenox P; Trisept B, E, M, P; Tris Nitro; Tristat, K; Troysan 28, 142, 174, 192, 364, CMP Acetate, Copper 8, PMA-10-SEP, PMA- 30, PMA-100, PMDS-10; Ucarcide 225, 250; Vancide 51Z, 89, 89 RE, MZ-96; WFT-78 Microbicide; Zinc Omadine, Cosmetic Grade; Zinc Pyrion

PROCESSING AIDS

Acrawax C; A-C Copolymer 540A; ACuflow AF-1; Akrofax 9844 Black, 9851 Light, 9906; Akrowax PE; Alkamuls GMO-45, GMO-55LG; Amberlig; Amoco Resin 18-210, 18-240, 18-290; Benzoflex S-312, S-404, S-552; Be Square 185; Blendex 590; Cachalot S-53T Series; Celite 270, 292, 350, Super Floss; Chloropren-Faktis A, NW; Cirrasol AEN-XB, AEN-XF, AEN-XZ, ALN-FP, ALN-GM, ALN-TF, ALN-TS, ALN-WY, EN-MB, EN-MP, LAN-SF, LC-HK, LC-PQ, LN-GS; Cycloryl 21SP; Denlube 1025-KA; Durpeg 400MO; Emcol 4600; Emerest 2664; Emsorb 2503, 2505, 2507; Emulphogene TB-970; Estabex 2307, 2307 DEOD; Faktis Asolvan, Asolvan T Badenia C, T DK 10, DK 14, DK 17 HF Braun Para extra wiech R Spezial, Weib MB RC 110, 111, 140, 141, 144 Rheinau H, W T-hart ZD; Fortex; Good-rite 3114, K-7058, K-7200; Interstab CA-18-1; Isonate 125M, 143L, 181, 191, 240; Kenplast ESB, ESI, ESN; Ken-React Series, 7 (KR 7), 9S (KR 9S), 12 (KR 12), 26S (KR 26S), 33DS, (KR 33DS), 38S (KR 38S), 39DS (KR 39DS), 41B (KR 41B), 44 (KR 44), 46B (KR 46B), 55 (KR 55), 133DS (KR 133DS), 134S (KR 134S), 138D (KR 138D), 138S (KR 138S), 158 (KR 158D), 158FS (KR 158FS), 212 (KR 212), 238A (KR 238A), 238J (KR 238J), 238M (KR 238M), 238S (KR 238S), 238T (KR 238T), 262A (KR 262A), 262ES (KR 262ES), OPP2 (KR OPP2), OPPR (KR OPPR), TTS (KR TTS); Kronitex 50, 100, TCP; LICA 01, 09, 12, 38, 38A, 38J, 44, 97; Masil 263, 264, EM 266 (35%), EM 266HV; Mekon White; Microbloc; Naftopast Antimontrioxid, Antimontrioxid-CP, Di7-P, Di13-P, GMF, Litharge A, MBT-P, MBTS-A, MBTS-P, MgO-A, Mi12-P, Red Lead A, Red Lead P, Schwefel-P, Thiuram 16-P, TMTM-P, ZnO-A; Naftozin N, Spezial; Ninex TDO 5, 9, 14, TMO 5, 7, 9, 14; Nordel 2522; Norfox 916; PA-57, 80; Palatinol DBP; Pamolyn; Pectinex 5XL, SCL; Petro-Rez 100, 103, 200, 215, X-95; Petrolite C-700, C-1035; Picco 5070; Piccodiene 2215, 2215SF, 2285, 2285BHT; Piccofyn A100, A115, A135; Plastogen R; Poly Pross; Polywax 850, E-2020; Proaid 9904; Proald 9802, 9810, 9814, 9826, 9831; QO Furfural; Reax G-1; Reogen, Resin 731D; Retarder BA, BAX; Ricon 153, 154, P30/Disperson; Rocsol Micro 4; Ross Wax #140, 160; RT-3; Solusoft 350; Spectratech PM 11607, 11725; Starwax 100; Staybelite; Struktol TR 016; Suconox-18; Sunolite 160;

Syncrolubes; Synprolam 35N3X3, 35N3X5, 35N3X10, 35N3X15, 35N3X25, 35X2, 35X5, 35X10, 35X15, 35X20, 35X25, 35X35, 35X50; Tebol 99; Teric 151, DD5, DD9, G12A4, LA4; Trydet 2644; Ultrazym 100; Vanfre DFL, M, UN; Varcum DR-406; Viton LM; Wingtack 95; Witconate NXS, SCS, STS, SXS

PROPELLANTS

Dymel 22, A, 142, 152

REDUCING AGENTS

CA0567; CA0570; CA0699; CA0742; CA0900; CB2408; CB2409.5; CB2493; CB2494; CB2495; CB2790; CB2805; CB2909; CC3005; CC3270; CC3275; CC3285; CC3290; CC3291; CC3433; CC3555; CD3770; CD3780; CD4450; CD5400; CD5430; CD5470; CD5487; CD5550; CD5600; CD5610; CD5636; CD5950; CD6000; CD6010; CD6150; CD6210; CD6220; CD6250; CE6250; CE6345; CE6350; CH7250; CH7260; CH7280; CH7310; CH7332; Chempol 19-4836; CI7840; CM8450; CM8750; CM8930; CM8980; CM9000; CM9220; CO9745; CO9750; CO9000; CO9810; CO9816; CO9817; CO9819; CO9830; CP0110; CP0156; CP0160; CP0280; CP0320; CP0330; CP0800; CP0810; Crill 1, 2, 50; CS1590; CT1800; CT2015; CT2030; CT2050; CT2090; CT2500; CT2520; CT2523; CT2902; CT2925; CT2950; CT2970; CT3250; CT3254; CT3600; CT3610; CT3795; CV4772; CV4800; CV5050; CV5100; Daxad 19L-40; DD-8126; Drewfax 0007; Dynasil A, CA, CM, M; Dynasylan AMEO, AMMO, BDAC, BSA, CPTEO, CPTMO, DAMO, GLYMO, HMDS, IBTMO, IMEO, MEMO, MSTFA, MTES, MTMO, MTMS, OCTEO, TCS, TRIAMO, VTC, VTEO, VTMO, VTPMO; Lanfarax, 1776; Pilot SXS-96; Thiovanol

REINFORCING AGENTS

Adtac L100BHT; Akrochem P 40, P 55, P 82, P-486; Alphaflex 101; Amoco Resin 18-210, 18-240, 18-290; ASP NC; AZO-33, AZO-55, AZO-55TT; CF-42,500, 70,000, 72,500; Chemprene R-10, R-25, R-50, R-70, R-100, R-115; Emtal 41, 42, 43, 44, 500, 549, 599, 4190; Hi-Pflex; Hi-Sil 132, 210, 215, 233, 243LD, 250, 260, 262, ABS; Huber ARO 60, N110, N220, N234, N299, N326, N330, N339, N343, N347, N351, N375, N539, N550, N650, N660, N683, N762, N774, N787, N990, S212, S315; Hubersil 162; Kaolin RC 32; Petro-Rez 215; Picco 5070, 6070, 6110, 6115, 6120, 6130, 6140; Piccofyn A100, A115, A135; Piccolastic C125; Piccolyte D100, D115, D135; Piccomer 150, XX40, XX100; Picconol A100, A102; Piccotoner 1200; Piccoumaron 100, 110, 120; Protox-78, -166, -167, -168; Santoweb D, DX, H, W; Silene D, 732D; Snow White 325; Super-Pflex 200; Thermax Floform N-990, Floform Ultra Pure N-990, Powder N-991, Powder Ultra Pure N-991, Stainless Floform N-907, Stainless Powder N-908, Stainless Powder Ultra Pure N-908; Thornel Carbon Fiber; Varcum 9413, C-86; Vulkadur A; Vulkasil C; Zecorez Series, 700, 701, 702, 703, 704, 711, 711Z and 712, 730, 731, 732, 733, 734, 760, 761, 762, 763, 764, 765; Zeosyl 100, 110SD; Zinc Oxide 35, No. 318, Transparent; Zinkoxyd Activ

RELEASE AGENTS

Actrafos T; Advawax 240, 275, 280, 290; Akrochem Calcium Stearate, Zinc Stearate; Akrolease E-9410, E-9491; Alkapol PPG-425, 1000; Alkatronic EDP 8-4, EDP 28-1, EDP 28-7, EDP 38-1, EDP 38-4, EDP 38-8, EDP-28-2; Alox 111, 436A; Armid O; Asol; Atmer 7113; Berol 730, 740; BL 3; CA0567; CA0570; CA0699; CA0742; CA0900; Cachalot S-53T Series; Carbowax PEG 200, 300, 400, 600, 900, 1000, 1450, 3350, 4600, 8000; CB2408; CB2409.5; CB2493; CB2494; CB2495; CB2790; CB2805; CB2909; CC3005; CC3270; CC3275; CC3285; CC3290; CC3291; CC3433; CC3555; CD3770; CD3780; CD4450; CD5400; CD5430; CD5470; CD5487; CD5550; CD5600; CD5610; CD5636; CD5950; CD6000; CD6010; CD6150; CD6210; CD6220; CD6250; CE6250; CE6345; CE6350; Centrolene Series; Centrophase HR; Centrophil Series; Ceraphyl 847; CH7250; CH7260; CH7280; CH7310; CH7332; Chemax PEG 200 DO; CI7840; CM8450; CM8750; CM8930; CM8980; CM9000; CM9220; CO9745; CO9750; CO9000; CO9810; CO9816; CO9817; CO9819; CO9830; Coconad RK; Combat HPC-40, HPF-325, HTP-40, HTP-325, SHP-40, SHP-325, SRG-40, SRG-325; CP0110; CP0156; CP0160; CP0280; CP0320; CP0330; CP0800; CP0810; Crill 1, 2, 50; Crodamide E, ER, O, OR, S, SR; Crystic Release Agent No. 1, No. 2; CS1590; CT-88 Aerosol; CV4772; CV4800; CV5050; CV5100; Doittol K21; Dow Corning 1-2531 Release Coating, 7 Compound, 203 Fluid, 233A Fluid, 236 Disp., 290 Emulsion, 346, 347, HV-490 Emulsion, Q2-7119 Release/Parting Agent; Dynasil A, CA, CM, M; Dynasylan AMEO, AMMO, BDAC, BSA, CPTEO, CPTMO, DAMO, GLYMO, HMDS, IBTMO, IMEO, MEMO, MSTFA, MTES, MTMO, MTMS, OCTEO, TCS, TRIAMO, VTC, VTEO, VTMO, VTPMO; E-130; E-131; E-133; E-155; E-2068; Ease Release 200, 300, 400, 500, 2040 Series; EBS Wax; EmCon E-5, E-20; Emerest 2301, 2302, 2652; Emphos D70-31; Epolene C-15, N-12; EZ Mold Lubricant; Fancol HL, LA, OA 50, OA 70, OA 80, OA 90, OA 95; Finazoline Q-4; Flexricin 13; Fluorel 2181; Formasil Silicone Emulsion 45; Garalease 915; Getren; Glycolube VL; GP-0098; Hardwax E-DM; Hodag MR-216, RA-2, RA-7; HTSA #1, #3; Hy Dense Calcium Stearate HP Gran., Calcium Stearate RSN Powd., Zinc Stearate XM Powd., Zinc Stearate XM Ultra Fine; Hystrene 5016 NF

FG, 7018 FG, 9718 NF FG; Industrene 105, 138, 143, 154-BG, 205, 206, 225, 225 FG, 226, 232, 239, 325, 328, 333, 365, 3022, 4022, 4516, 4518, 5016, 5016 FG, 5022, 7018, 7018 FG, 8518, 8718 FG, 9018, B, R; Interstab CA-18-1; Jet Amine PC, PHT, P-O, P-S, PT; Jonwax 120

Kemamide B, E, E-180, E-221, O, P-181, S, S-65, S-180, S-221, U, W-10, W-20, W-39, W-40, W-40/300, W-45; Kemamine AS-650, AS-974, AS-974/1, AS-989, AS-990, P-150, P-150D, P-190, P-190D, P-650, P-650D, P-690, P-690D, P-790, P-790D, P-880, P-880D, P-970, P-970D, P-974, P-974D, P-989, P-989D, P-990, P-990D, P-997, P-997D, P-999; Kemester E; Ken-React Series, 7 (KR 7), 9S (KR 9S), 12 (KR 12), 26S (KR 26S), 33DS, (KR 33DS), 38S (KR 38S), 39DS (KR 39DS), 41B (KR 41B), 44 (KR 44), 46B (KR 46B), 55 (KR 55), 133DS (KR 133DS), 134S (KR 134S), 138D (KR 138D), 138S (KR 138S), 158 (KR 158D), 158FS (KR 158FS), 212 (KR 212), 238A (KR 238A), 238J (KR 238J), 238M (KR 238M), 238S (KR 238S), 238T (KR 238T), 262A (KR 262A), 262ES (KR 262ES), OPP2 (KR OPP2), OPPR (KR OPPR), TTS (KR TTS); Kricinol 35; Lacer Max; Larostat 88, 264-A Conc., 377 DPG, 451, 1443; Lexolube 4H-415; Lexomul PEG 400 DO; LICA 01, 09, 12, 38, 38A, 38J, 44, 97; Lilamin 101 D, 115, 115 D, 140, 140 D, 142, 142 D, 151, 160, 160 D, 163, 163 D, 170, 170 D, 172, 172 D, 308 D, 310 D, 312 D, 314 D, 316 D, 342 D, 343, 345 D, 363, 364, 367 D, 368, 369, 372 D, 381, 382, 383; Lipowax C; Liquazinc AQ-90; Loxiol G 73, G 74, G 78; LZ 01, 09, 12, 38, 44, 97; M-7 Mold Release; Macol 660, 3520, 5100, P Series; Masil 260, 260A, 263, 264, 265, 265HV, 266, 270, 271, EM 14, EM 62, EM 100, EM 100 Conc., EM 100P, EM 250 Conc., EM 266, EM 266 (35%), EM 266HV, EM 350X Conc., EM 350, EM 350X, EM 350X Conc., EM 1000, EM 1000 Conc., EM 1000P, EM 10,000, EM 10,000 Conc., EM 60,000, EM 100,000, EM-N, SF 5, SF 20, SF 50, SF 100, SF 200, SF 350, SF 500, SF 1000, SF 5000, SF 10,000, SF 12,500, SF 30,000, SF 60,000, SF 100,000, SF 300,000, SF 600,000; M-C-Thin 45; Mekon White; Michel XO 24, XO-85; Moldbrite 30; Mono-Coat 65-RT, E76, E91; Mould Release Agent N32; MS-122, -136; N-30; NB; NEPD

Neustrene 045, 053, 059, 060, 064; Nissan Amine AB, ABT, ABT_2, BB, FB, MB, OB, PB, SB, VB; Nissan Cation F_2-10R, -20R, -40E, -50, FB, FB-500, L-207, M_2-100; Nissan Cation PB-40, -300, S_2-100; Nissan Dispanol TOC; NMP; No Stik 802; Noda Rice Wax FCC; Norfox CS, EM 350X, 350X Conc.; Paracin 220, 285; Petrac CP-11, CP-11 LS, CP-11-LSG, CP-12, CZ-81, Eramide, Slip-Eze; Petrac Slip-Quick, Vyn-Eze, ZN-41, ZN-42, ZN-44 HS, ZW-45; Plastilease 250, 512-B, 512-CL, 514; Pluracol E-400 NF, E-600 NF, E-1000, E-1450, E-1450 NF, E-2000; E-4000 NF, E-4500, E-8000, E-8000 NF; Pluriol E 200, E 300, E 400, E 600, P 600, P 900, P 2000; Pogol 400 USP; Poly-Cone 1000; Polymel #7; Polyvinyl Alcohol Solution; Polywax 500, 655, 1000, 2000; Precirol ATO; Proaid 9810, 9831, 9904; Quilon C, H, L, S; Radiamine 6140, 6141, 6160, 6161, 6163, 6164, 6170, 6171, 6172, 6173; Radiasurf 7146, 7151, 7152, 7153, 7372, 7175; RC 7; Repel-O-Tex D-5; Repel-O-Tex D; Rhenodiv 20, A, F, KS, LE, LL, LS, PV, S, ZB; Ross Wax #100, 140, 160; RR 5; Rycofax 618; Safety-Lube 4000, 6003; Schercemol DICA; SEM-35, -60; SF 18, SF 69, SF 96, SF 97, 1066, SF 1080, SF 1091, SF 1093, SF 1188, SF 1221, SF 1250; Silicone Emulsion 350, 1M, 10M, 60M, 350 Conc.; Silicone Fluid 350, 1M, 5M, 10M, 30M, 60M; Silicone M400, M404, M405, M406, M409; Silicone Mold Release SEM-35; Silicone Release Agent #5038; Silicone Release 87-X66; Silwet L-7602; SM 2059, 2061, 2079, 2140, 2154, 2155, 2162, 2163; SMA 1000, 2000, 3000; Socci 6462; Softenol 3107, 3119, 3408, 3701, 3925, 3945, 3960; Sontex 19, 21, 35, 55, 70, 75, 75 T, 85, 85 T, 95 T, 100, 150; Sorcinol 35; SS 4177; Sprayfilm 87-911; Spraywax 60-X5; Standamid ID; Stoner No. E965; Surfactol 13; Tego IMR 412 T, 833 T; Tegosil; Tego Silicone Acrylate RC; Tegotrenn LH 157 A, LH 525, LI 197, LI 237, LI 344, LI 448, LI 537 W, LI 747, LI 748, LI 766, LI 901, LK 104, LK 260, LK 498/F, LK 755, LK 859, LK 860, LR 201/F, LR 236, LR 309/F, LR 486, LR 560, LS 584, LS 809/F, LS 814/F, LS 815, LS 828, LS 928/F, LS 952, LS 960, S 2100; Teric C12; Texo LP 528A; Trenbest 500; Trennspray Tego; Tryfac 5573; TSE Mold Release; Union Carbide L-720, -721, -722, -727, LE-45, LE-458HS, LE-467, LE-467 HS; Unithox 450; Vanfre HYP, IL-1, UN, VAM; Vanstay 6040; Viscasil; Vydax WD; Vykacet L, T; Witamide 70, MAS, MM; Witco MRC; XK 22; XR 7; XT 66; Zelec NE, UN; Zinc Stearate USP

RETARDERS/RETARDANTS

Ablumine PN; Agerite Stalite, Stalite S; Alkasurf NP-1, OP-1; Altax; Cenegen 7; Consos Castor Oil; Consotard No. 895; Crapol AU-21; Crestomul T; Crystal Inhibitor #5; Du Pont Retarder LAN; Ektasolve EB, EB Acetate, EE, EP, PM Acetate; Emkatex LS; Empigen 5089, BAC50, BAC50/BP, BAC80, BCB50, BCY40, CHB, CM; Emulvis; Harol RG-71; Hartonyl L537; Hipochem C-95, D2, Retarder CJ; Indulin W-1, W-3; Katapol VP-532; Ken-React Series, 7 (KR 7), 9S (KR 9S), 12 (KR 12), 26S (KR 26S), 33DS, (KR 33DS), 38S (KR 38S), 39DS (KR 39DS), 41B (KR 41B), 44 (KR 44), 46B (KR 46B), 55 (KR 55), 133DS (KR 133DS), 134S (KR 134S), 138D (KR 138D), 138S (KR 138S), 158 (KR 158D), 158FS (KR 158FS), 212 (KR 212), 238A (KR 238A), 238J (KR 238J), 238M (KR 238M), 238S (KR 238S), 238T (KR 238T), 262A (KR 262A), 262ES (KR 262ES), OPP2 (KR OPP2), OPPR (KR OPPR), TTS (KR TTS); Levenol A Conc., RK; LICA 01, 09, 12, 38, 38A, 38J, 44, 97; LZ 01, 09, 12, 38, 44, 97; Marvanol Penetrant 35; Matexil LC-RA; Ospin Salt ON, TAN; Redax; Retarder A, AK, BA, BAX, L, N, N-85, PX, SAX; Schercotarder; Swanic 110; Thermagel HN; Thiofide; Trymeen 6640; Turgum S; Variquat B200; Vulkalent B/C, E/C, TM

RHEOLOGY AGENTS. SEE GELLING AGENTS

SANITIZERS. SEE ANTIMICROBIALS

SCALE INHIBITORS

Acrysol A-41, LMW-10, LMW-10N, LMW-20, LMW-20N, LMW-45, LMW-45N, QR-1086; Alkasperse A-2H; Aquatreat AR602; Arquest 710; Bayhibit; Belclene 200, 201, 283, 400; Calnox Polymers; Colloid 119/50; Cyanamer P-70, P-80; Darex 41; Daxad 30-30, 37L Acid, 41; Dequest 2000, 2006, 2010, 2051, 2054, 2060, 2066; Dytek A; Emcol CS-1361; Fostex AMP, E, P, S, SN, TN, TX, U; Good-rite K-7028, K-7058, K-7200, K-7600, Polycrylates; Kelig 32; Mirapol WT; Michel XO-24, XO-85, XO-108; Rychem 808; Unihib 305

SEQUESTERING AGENTS. SEE CHELATING AGENTS

SIZING AGENTS

Cyanamer P-250; Elvanol T-25, T-66; Empicol TAS90; Ganex P, V; GP-2925; Hercules CMC-Warp Size; NovaSize, Dark Fortified, Dark Unfortified, Pale Fortified; Paracol 403A6, 404A, 404C, 404D, 404G, 404N, 447K, 800N, 802A, 802N, 802NW, 803A6, 803G6, 810N, 810NP, 815N, 1886; Petrex 7-75T; Pexol Size; Pluracol E-200, E-300, E-400, E-400 NF, E-600, E-600 NF, E-1000, E-1450, E-1450 NF, E-1500, E-2000, E-4000, E-4000 NF, E-4500, E-6000, E-8000, E-8000 NF; Rexonic N-4; Scripset 520, 540, 550, 720; Softenol 3100

SLIP AGENTS

Abil B 88183; Adogen 58, 73; Armid 18, C, HT; Forbest 70, 260, CP, GLW, MW 23, P 22, PAM; Incromate ALL, AVL, SEL, WGL; Incronam SE-30; Incroquat AV-85, I-85, Mink-85, O-50, OL-85, WG-85; Kemamide B, E, E-180, E-221, O, P-181, S, S-65, S-180, S-221, U, W-20, W-39, W-40, W-40/300, W-45; Kingsolve SPA; Lexate TA; Mearlmica MMCF, MMSV; Mono-Coat E91; Reten 210, 220, 521, 523, 525; Rheothik Polymer 80-11; Silwet L-720 L-7001 L-7500 L-7602 L-7605; Spectrataech CM 11014, 11126, 11172, 11174, 11194, 11513; Unislip 1753, 1757, 1759, 1760 EBS, FW

SOFTENERS. SEE CONDITIONERS

SOLUBILIZERS

AB; Abil-Wax 2440, 9801; Accobetaine CL; Acconon 200-DL, 200-MS, 300-MO, 400-DO, 400-ML, 400-MO, 400-MS, 1300, CA-5, CA-8, CA-9, CA-15, CON, E, TGH, W-230; Accosperse 20, 60, 80; Acetulan; Actrafos S-104, S-109; AEPD; Aerosol 18, 22, 22N, A-102, A-103, A-268, MA-80; Ahco 759, 832, 909, 944, DFO-100, DFO-110, DFO-150, DFP-156, DFS-96, DFS-149, EO-102, EO-114, FO-18, FP-67, FS-21; Akyporox CO 400, 600, RC 200; Alconate D-6; Aldo MO FG, MOD FG, MS-20 FG, PMS; Aldosperse ML 23, MS 20, O 20 FG; Alipal CD-128; Alkali Surfactant; Alkamuls 200-ML, 400-DL, 400-DO, 400-MS, 600-DL, 600-DO, PSML-20, PSMO-5; Alkanol A-CN, S; Alkaphos 3, 10, K 380, L6-36S, QS; Alkasurf LAN-23, NP-30 70%, OA-2, OA-10, OA-20, OP-12, OP-16, SA-2, SA-10, SS-DA-6, SS-NO, SS-TA, SS-TA 125, TDA-7, TDA-7.5, TDA-8.5, TDA-12, TDA-15; Alkateric 2CIB, CAB, EHCIB; Alkatrope SX-40; Amerchol Polysorbate 80; Ameroxol LE-4, LE-23, OE-2, OE-10, OE-20; Amfotex FV-28; AMPD; Ampho B11-34; Antaron MC-44, PC-37; Antil 141; Aqualose L 30, L 75, L75/50, LL 100, W 20; Aramide CDX; Arlasolve 200; Arlatone 285, 289, 827, 970, 975, G, T; Armotan PML 20, PMS 20; Arphos MK-203; Berol 521, 522, 525, 733; Bio Terge PAS 8; Brij 56, 58, 76, 78, 96, 97, 98, 99, 721; Calamide C, CW-100; Calfax 10L-45; Calimulse PRS; Capmul POE-L, POE-O, POE-S, 10G10O; Carsonon TD-10; Carsosulf SXS-Liq.; Catinex KB-18, KB-19, KB-20, KB-22, KB-42, KB-43, KB-48; Cedepal ON-877; Cegesoft C 17, C 19, C 25; Centrol Series; Ceraphyl 140-A; Chemal OA-4, OA-5, OA-9, OA-20, OA-20/70, TDA-3, TDA-6, TDA-9, TDA-12, TDA-15, TDA-18; Chemax CO-5, DNP-18, PEG 400 DO; Chemfac PX-322; Cithrol 2ML, 2MO, 2MS, 3MS, 4MO, 4MS, 6ML, 6MO, 6MS, 10ML, 10MO, 10MS, 15MS, 40MO, 40MS, 60ML, 60MO; Complemix 100; Conco Sulfate 3B2; Conco SXS; CPH-27-N; CPH-30-N; CPH-41-N; CPH-43-N; CPH-376-N; Cralane KR-13, KR-14, LR-10, LR-11; Cremophor EL, MP-10, NP-10, NP 14, RH 40, RH 60, RH 410, RH 455; Crillet 1, 4, 6, 11, 41, 45; Crodalan AWS; Crodesta A-10, A-20, F-10, F-50, F-110, F-160, SL-40, SL-40PX; Crodet L4, L8, L12, L24, L40, L100, S4, S8, S12, S24, S40, S100

Croduret 10, 30, 40, 50 Special, 60, 100; Cyanamer A-370, P-26; Cyclophos PE120, PL61; Cyclopol SBDO; Cycloryl DCA; Dacospin 12-R; Dehydol D 3; Denwet CM, RG-7; Deriphat 154, 160, 160C; Dianol; Dianol 300; Diazopon SS-837; DMAMP, DMAMP-80; Dowfax 2A1, 2EP, 3B0, 3B2, XDS 8292.00; Drewmulse 3-1-O, 3-1-S, 6-2-S, 10-4-O, 10-8-O, 10-10-O, 10-10-S, POE-SML, POE-SMO, POE-SMS, POE-STS, SML, SMO, SMS, STS; Drewpol 3-1-O, 3-1-S, 3-1-SH, 6-2-S, 10-4-O, 10-8-O, 10-10-O, 10-10-S; DyaFac LA9; Dymsol S; Eltesol ACS 60, AX 40, PT 93, PX 40, PX 93, SCS 40, ST 34, ST 40, ST Pellets, SX 30,

SX 93, SX Pellets; EM-980, -985; Emasol L-106, O-105 R, O-106, O-120, O-320, P-120; Emcol 4560, 4580PG; CS-136, CS-143, CS-151, CS-165, NA-30; Emerest 2325, 2384, 2648, 2650, 2665; Emery 912; Emphos CS-121, CS-147, PS-236, PS-415M, PS-440; Empigen 5083, OH25; Empilan CDE, CDX, LDE, LDX, LP2, MAA, NP9; Emsorb 2720, 2722, 2726, 6900, 6915; Emthox 5885, 5964; Emulphogene BC-720, BC-840; Emulphor EL-620, EL-719, ON-870, ON-877; Emulsynt 1055; ESI-Terge SXS; Espesilor AC-43, AC-50; Ethoxyol 1707; Ethylan 20, BV, C 160, CD913, CD916, CD919, CD9112, D259, D2512, DP, GEL-2, GEO-8, GEP-4, GES-6, GLE-21, GOE-21, GPS-85, HB4, KEO, PQ; Etocas 10, 30, 35, 40, 60, 100; Etoxi AC-91, KC-10, KC-11, MC-12; Eumulgin HRE 40, 60, L, M 8, O 5, RO 40, SML 20, SMO 20, SMS 20; Finsolv SB, TN; Fluilan AWS; Forlan C-24; Gafac RD-510, RE-877, RM-510, RM-710; Ganex P, P904, V, V216, V220, V516; Gantrez AN; Generol 122, 122E10, 122E16, 122E25; Glucam E-10, E-20, P-10, P-20; Glucamate DOE-120, SSE-20; Glycerox L, L15; Grilloten LSE87, LSE87K, PSE141G, ZT12, ZT40, ZT80; Hallcomid M-6, M-12, M-18, M-18-OL; Hartotrope AXS 40, KTS 50, STS 40, Powd, SXS 40, Powd.; HD Eutanol; Hetoxide BN-13, C-30, C-200-50%, HC-40, HC-60; Hoe S 2817; Hostaphat OPS

Icomeen 18-5, O-30, O-30-80%, S-5; Igepal CO-880, CO-887, DM-880; Imwitor 308, 310, 312, 742, 908, 914; Incrocas 10, 30, 40, 60, 100; Incromide LL; Incropol 233, 290, CS-12, CS-20, CS-30, CS-40, CS-50, CS-60, L-2, L-7, L-7-90, L-12, L-23, L-30; Incrosorb S-60; Jordamide 1214, JT-1286; Jordanol NP-95, SXS; Jordaphos JE-41; Kessco Butyl Oleate, Butyl Stearate Cosmetic, Dist., Isobutyl Stearate, Octyl Isononoate, PEG Series, 200 Dilaurate, 200 Dioleate, 200 Distearate, 200 Monolaurate, 200 Monooleate, 200 Monostearate, 300 Dilaurate, 300 Dioleate, 300 Distearate, 300 Monolaurate, 300 Monooleate, 300 Monostearate, 400 Dilaurate, 400 Dioleate, 400 Distearate, 400 Monolaurate, 400 Monooleate, 400 Monostearate, 600 Dilaurate, 600 Dioleate, 600 Distearate, 600 Monolaurate, 600 Monooleate, 600 Monostearate, 1000 Dilaurate, 1000 Dioleate, 1000 Distearate, 1000 Monolaurate, 1000 Monooleate, 1000 Monostearate, 1540 Dilaurate, 1540 Dioleate, 1540 Distearate, 1540 Monolaurate, 1540 Monooleate, 1540 Monostearate, 4000 Dilaurate, 4000 Dioleate, 4000 Distearate, 4000 Monolaurate, 4000 Monooleate, 4000 Monostearate, 6000 Dilaurate, 6000 Dioleate, 6000 Distearate, 6000 Monolaurate, 6000 Monooleate, 6000 Monostearate; Klearfac AA040, AA270, AA420, AB270; Korantin CD; Lamacit ER, GML 12, GML 20, GMO 25; Lan-Aqua-Sol; Laneto 40, 50, 60, 100; Lanexol AWS; Lankropol ATE, KMA; Lankrosol SXS-30; Lanogel 21, 31, 41, 61; Lanpol 5, 10, 20, 520; Lantrol AWS 1692; Lenetol 9130 AH; Lexol GT 855, GT 865, IPP, PG 855, PG 865; Lipal 3 TD, 4 LA, 5 L, 5 OA, 5 S, 6 TD, 9LA, 9 C, 9 L, 9N, 9 OL, 10 OA, 10 TD, 12 LA, 15 CSA, 15 T, 20 OA, 20 SA, 23 LA, 25 C, 25 S, 30 SA, 39 S, 50 OA, 50 S, 52 C, 300 DL, 300 W, 400 DL, 400 DS, 400 DW, 400 OL, 400 S, 400 W, 600 S, 600 W, 610, CE 38, CE 55, CE 64, CE 71, OE 55, OE 64, OE 70, TE 43, TE 55, TE 70, TE 76; Lipocol C-2, C-10, C-20, L-1, L-4, L-12, L-23, M-4, O-2, O-10, O-20, S-2, S-10, S-20, SC-4, SC-10, SC-15, SC-20, TD-3, TD-6, TD-12; Lipolan 31; Lonzest SML-20, SMO-20, SMP-20, SMS-20, STO-20, STS-20; Lubrimet P 600, P 900; Lutensit A-EP; Lutensol ED 140, ED 310, ED 370, ED 610

Macol CSA-2, CSA-4, CSA-10, CSA-15, CSA-20, LA-4, LA-12, LA-23, LA-790, NP-4, NP-6, NP-9.5, NP-11, NP-20, NP-30, NP-30(70), NP-70, NP-100, OA-2, OA-4, OA-5, OA-10, OA-20, SA-2, SA-5, SA-10, SA-15, SA-20, SA-40, TD-3, TD-4, TD-6, TD-8, TD-10, TD-610; Mafo 13, C-12; Mafo CAB; Manro CD, CDS, CDX, DS 35, MA 35, SCS 40, SCS 90, STS 40, STS 90, SXS 30, SXS 40, SXS 93; Mapeg 200 DOT, 400 DSLM, 600 DL, 600 DOT, 600 ML, 600 MOT, 6000 MS, CO-16H, CO-25, CO-25H, CO-30, CO-36, CO-200, DGLD, PGDS, PGMS, S-100, S-150, TAO-10; Maphos 4, 17, 60, 60A, 66, 66H, 76, 76 NA, 91, 6600, 8135, L-4, L-10, L-22; Marlican; Mazamide 65, 65CZ, 66, 70, 80, CS-148, O-20, SS-10; Mazol PG-810, PGO-104; Miglyol 810, 812, 840; Miranol C2M Conc., C2M-SF Conc., H2M Conc., H2M-SF 70%, H2M-SF Conc., J2M-SF Conc.; Mirataine COB, H2C, T2C; Monafax 060, 872; Monamid 716; Monamine 779; Monamulse 653-C, R10-29M; Monaquat P-TC, P-TD, P-TL, P-TZ; Monaterge 85; Monateric 805, 810-A-50, 985A, CA-35%, CAM-40, CDL, CDS, CDTD, CDX-38 Mod., CEM-38%, CEM-38CG, CNa-40, COAB, CyA-50, CyMM-40, LMAB, LMM-30, MCB, TA-35; Monatrope 1296; Monawet MB-45, MM-80, MO-70, MO-70E, MO-70R, MO-75E, MO-84R2W, MT-70, MT-70E, MT-80H2W, SNO-35; Montosol PF-10, PF-14, PF-26, PG-10, PL-16, PQ-17; Mulsifan RT 18, RT 69, RT 141, RT 146, RT 203/80, RT 302; Myritol 318; Na Cumene Sulfonate 40; Sulfonate Powd.; Na Toluene Sulfonate 30, 40; Naxonate 4AX, 4KT, 4L, 4ST, 5KT, 5L, 6AC, 45SC, AX, G, KT. SC, ST; Neobee 18, 62, 1223, O, 1400 Oil, M-5, M-20; Newcol 3-80, 3-85, 20, 25, 40, 45, 60, 65, 80, 85, 290K, 290M, 290P, 600, 700 Series; Nikkol Batyl Alcohol 100, EX, BEG-1630, BPS-5, BPS-10, BPS-15, -20, -25, -30, BWA-5, BWA-20, -40, Chimyl Alcohol 100, CO-3, CO-10, -20TX, -40TX, -60TX, Decaglyn 1-IS, 1-L, 1-LN, 1-M, 1-O, 1-S, DLP-10, GO-430, GO-440, GO-460, GS-460, HCO-5, -7.5, -10, -20, -30, -30(FF), -40, -40(FF), HCO-50, -50(FF), -60, -60(FF), -80, -100, -100(FF), MYO-6, MYO-10, MYS-1EX, MYS-2, MYS-4, MYS-10, MYS-25, -25(FF), MYS-40, MYS-40(FF), MYS-45, -55, NP-7.5, NP-10, -15, -18TX, OP-3, OP-10, -30, PBC-31, PBC-33, PBC-34, PBC-41, PBC-44, PBC-44(FF), PEN-4612, -4620, -4630, TDP-2, -4, -6, -8, -10, TI-10, TO-10, TO-10(FF), TO-30, TO-106, TS-10, TS-10(FF), TS-30, TW-10, -20, -30; Niox KH Series, KI Series, KI-29, KJ-10, KP-68, KP-69

Nissan Plonon 102, 104, 108, 171, 172, 201, 204, 208; Noiox AK-41, KJ-12, KJ-15; Nonex DO-4; Nonisol 100; Norfox 4 Polyol; Norfox Sorbo T-20, T-80; Nutrol SXS; Oxypon 288, 306; Pationic ISL; Pegante L 5, 20,

SOLUBILIZERS (CONT'D.)

O 5, 20, S 5, 20; Peregal O; Perlankrol DGS, DSA, ESK-29; Petro 11, AA, BA, LBA; Petrosul 742, 744 CL; Phospholan KPE-4, PHB-14; Pilot SXS-40, SXS-96; Plasdone C-15, C-30, K-25, K-26/28, K-29/32, K-90; Pluradot HA-410, -420, -430, -433, -440, -450, -510, -520, -530, -540, -550; Pluriol E 200, E 300, E 400, E 600, E 1500, E 4000, E 6000, E 9000, P 600, P 900, P 2000, PE 6800; Pluronic 10R5, 10R8, 12R3, 17R1, 17R2, 17R4, 17R8, 22R4, 25R1, 25R2, 25R4, 25R5, 25R8, 31R2, 31R4, L10; Pogol 200, 300, 400 USP, 400, 600, 1540; Poly-Solv DB, DE (High Gravity), DE (Low Gravity), DM, DPM, EB, EE, EM, MPM, TE, TM, TPM; Polychol 15; Polyfac MT-610, MT-615, TDO-9, TMO-5, TMO-7; Polylan; Polystab RB1523; Polystep A-2; Primarol 1208; Procetyl AWS; Product WRS 1-66; Produkt RT 245, RT 288; PVP K-90; Pyroter CPI-40, GPI-25; Quimipol EA 6810, 6812, 6814, 6818, 6820, 6825, 6850;Quimipol ENF 140, 170; Radiamuls 125, 135, 137, 145, 147, 155, 157, 345; Radianol 7106, 7376; Rewoderm ES 90; Rewopal CSF 11, MPG 10; Rewophat TD 40; Rewopol NEHS 40, SBDO 70; Reworyl ACS 60, NCS 40, NTS 40, NXS 40; Rexol 25/30, 25/307; Rheodol TW-P120; Ritalan C; Ritawax 5, 15, 40; Ritawax AEO; Ritoleth 2, 5, 10, 20; Rychem-G-Gard, G-Hib; Sanac C, S, T; Sandoz Phosphorester 600, 610; Sandoz Sulfonate 2A1; Santone 3-1-S, 10-10-O; Satexlan

Schercemol 1818, DIS, IPM, MEL-3, MEM-3, MEP-3, OLO, PGML; Schercomid AME, AME-70, SLE; Schercopol CMS-Na, OMS-Na; Secoster CL 10, CP 10, CS 10, KL 10, KP 10, KS 10; Simulsol 1285, 1292; Sipon EA, EAY; Siponate 330, DSB; Siponic 25-3, 25-7, 25-9, 91-6, F-90, L7-90, NP-7, NP-9.5, NP-10, NP-13, NP-407, NP-707, TD-3, TD-6, TD-9-90, TD-12; S-Maz 20, 40, 60, 60 K, 60 KHM, 65, 65 K, 67, 80, 80 K, 83R, 85, 85 K, 90, 93 R; Softigen 767; Solan, 50; Solwax C-24, LG 35; Soprofor T/20, T/60, T/65, T/80, T/85; Sorbanox AOM; Sorbax PML-20, PMO-5, PMO-20, PMP-20, PML-20, PTO-20, PTS-20; Sorbon T-20; Standamid CD, LDS; Standamul B-1, B-2, B-3, HE; Standapol Conc. 7023, EA-3, LF, SCO; Starfol IS; Stepanate AM, C-S, T, X; Sterling Emulsifier #15, SXS; STXS; Sulfonate OA-5, OE-500; Sulfopon WA 1; Sulfotex; Sulframin CSA, TX; Sunaptol HB4; Super Solan Flaked; Super Solangel 25; Surco SXS; Surfactol 318, 365, 575, 590; Surfine WLL, WNG-A; Surfonic HDL, JL-80X, N-10, N-31.5, N-40, N-60, N-85, N-95, N-100, N-102, N-120, N-150, N-200, N-300, N-400, NB-5, NB-14; Surfynol 82S; SXS; Synperonic A14, A20, NP10, NP12, NP20, NP30, NP35; Tamol PA Liq., PA Powd.; T-Det C-40, N-8, N-10.5, TDA-65; Teepol HB 6, 7; Tegin C-1R, P, P-411 SE, PL, RZ, RZ NSE; Teric 16A16, 16A22, 16A29, 17A10, 17A13, 17A25, 313, N8, N9, N10, N11, N12, N13; Tetronic 50R1, 50R4, 50R8, 70R1, 70R2, 70R4, 90R1, 90R4, 90R8, 110R1, 110R2, 110R7, 130R1, 130R2, 150R1, 150R4, 150R8, 304, 504, 701, 702, 704, 707, 904, 908, 1101, 1102, 1107, 1301, 1302, 1304, 1307, 1501, 1502, 1504, 1508; Texapon EA-1, -2, -3, ES-1, -2, -3; Tex-Wet 1140; T-Maz 20, 28, 40, 60, 60K, 60KHM, 61, 65, 65K, 67, 80, 80K, 80KLM, 81, 81K, 85, 85K, 90, 95; T-Mulz 734-2; Trem-LF-40; Tris Amino; Triton H-55, H-66, N-101, QS-44, X-100, X-102; Trycol 5878, 5888, 5946, 5949, 5964, 5968, 5971, 5972, LAL-23, OAL-23, SAL-20, TDA-8, TDA-18; Trydet DO-9; Tryfac 5569, HWD; Trylox CO-80; Türkischrotöl 100%, 50%; Tween 20, 20 SD, 21, 40, 60, 61, 65, 80, 81, 85

Ultra NCS Liquid, NXS Liquid, SCS Liquid, SXS Liquid Powder; Unimate DIPS; Varion CADG, CDG; Varonic 1000MS, 1800MS, LI-42, LI-48, LI-67, LI-67-75, LI-420-80; Varsulf SBDO-70; Velsan D8P-3; Vigilan, AWS; Vista STXS; Volpo 3, 5, 10, 20, 25D3, 25D5, 25D10, 25D20, CS3, CS5, CS10, CS15, CS20, L23, N3, N5, N10, N15, N20, O5, O10, O15, O20, S100, T3, T5, T10, T15, T20; Vykamol N/E, S/E; Waxenol 801; Wickenol 101, 105, 111, 131, 136, 141, 163, 171, 727; Witco 912, 1298 Soft Acid, Acid B, TX Acid; Witcolate 1259, A, T; Witconate 60B, 60T, 93S, 1250 Slurry, 1260 Slurry, AOS, NCS, NXS, P-1052N, P-1059, SCS, STS, SXS, TAB, TX Acid, YLA; Witconol APEB, NP-40, NS500K

SOLVENTS/COSOLVENTS

Acetol 1706; Acetulan; Adogen 283; Adol 42, 52 NF, 54, 60, 61 NF, 62 NF, 63, 64, 66, 80, 85, 85 NF, 90, 320, 330, 340, 520 NF, 620 NF, 630, 640; Aerothene MM Solvent, TT Solvent; Agrisol PX401, 413; Alkapol PEG 200, 300, 400, 600, 1000, 1500, PPG-425, 1000; Ameroxol OE-2; Amsco Lactol Spirits, Min. Spirits, Min. Spirits 66/3, Min. Spirits 75, Naphthol Spirits 66/3, Nonene, Odorless Min. Spirits, Regular Min. Spirits, Rubber Solvent, Solv D, Solv. F, Solv. G, Special Naphtholite 66/3, Super High Flash Naphtha, Tetramer, Textile Spirits, Toluene, Xylene; Arconate 1000, HP; Arcosolv DPM Acetate, PM, PTB, TPM; Arlamol E; Berol 372 Flakes; BLO; Butyl Dioxitol; Butyl Oxitol; Capmul MCM; Captex 200, 350, 810B; Cardolite NC-507, NC-510, NC-511; Cithrol 2DL, 2DO, 4DL, 4DO, 6DL, 6DO, 10DL, 10DO; Corona Lanolin; Coronet Lanolin; Crill 1, 2, 50; Crodamol DA, IPM, PMP; DBE, DBE-2, DBE-2SPG, DBE-3, DBE-4, DBE-5, DBE-5SPG, DBE-9; Decalin; Diglycolamine Agent; Dionil 11, 113; Dioxitol-High Gravity; Dioxitol-Low Gravity; Dizene; Dowanol DB, DE, DPM, EB, EPH, PM; Dowclene EC Solvent, LS; Dowfrost; Dowper; Edenol 302; Ektapro EEP Solvent; Ektasolve DM, DP, EB, EB Acetate, EE, EE Acetate, EM, EP; Emcol CC-42, CC-55, CC-57; Emerest 2452, 2624, 2648, 2665; Emery 912, 6686, 6705, 6709; Emulbon LB-78; Ethylan HB Series, HB4; Everflex Solvent 80; Fancol OA 50, OA 70, OA 80, OA 90, OA 95; Filmex; Flexricin 9; Foam Burst 2; Fosterge LF Acid; Frigen 113 TR-N; Genesolv; Glucam E-10, E-20, P-10, P-20; Glycine A; Grindtek AML 60, AMOS 90; Hallcomid M-6, M-8-10, M-18-OL; Hetester PMA; Hetoxide

MPC; Hexcel FO 425A; Hi-Tri; Icinol H260, H280X, H660, H660YA, H5100, L65, L285, L300X, L385, L625, L1715; Igepal OD-410; Kemester 2050, 4516, EGDS, EGMS; Kemstreme 96.0%, 99.0%, 99.5%, 99.7%, High Gravity; Kessco 639, 3283, Butyl Stearate Cosmetic, Dist., Isobutyl Stearate, Isopropyl Myristate, Isopropyl Palmitate, Octyl Isononoate, Triacetin; Kodaflex DBP, DEP; Kromfax; Lanpol 5, 10, 20, 520; Lewisol 7; Lexol GT 855, GT 865, IPM; Linamine HC-2; Lipolan R; Lipovol SES, SES-S; Lonzest 143-S; Lutrol E 300, E 400, E 1500, E 4000, E 6000; Merrol 4218, DBS; Michel XO-144B, XO-146; Miglyol 810, 812, 840; Monoglyme; Monolan PPG440, PPG1100, PPG2200; Marlowet SWN; Marvanol Solvent Scour 34; Marvanscour KW; Masil 2132, 2133, 2134; M-Pyrol; Nadone; Naxol; NE; Neobee 18, 1054, 1062, 1223 (Neobee O), O, 1400 Oil (Neobee M-5), M-5, M-20; Neu-Tri; Nioix KJ-10, KS Series; Niox KF-13; NiPar S-10; NM; Noiox KS-10, -12, -13, -14, -16; Nonipol 20, 40, 55, D-160; Pliabrac TBP; Pluracol E-200, E-300, E-400, E-600, E-1500, E-4000, E-6000; Poly-G WI 285, 625, 1715, Poly-G WS 100, 170, 260, 660, 2000, 3520, 5100, WT 9150, 90,000; Polylan; PPG Perchlor; Prelete Defluxer; Procetyl 10, 20, 30, 50; Prochol; Promyr; Punctilious Specially Denatured Ethyl Alcohol; Palatinol DBP; Penreco Snow; Perk; Petrosolve A; 2-Pyrol; QO Furfural, Furfuryl Alcohol, Tetrahydrofuran, Tetrahydrofurfuryl Alcohol; Radia 7131; Rewomid OM 101/G, OM 101/IG, 101/IG/ER; Rewopal MPG 10, 12, 40; Rewopol-Emulsifier BWA; Rilanit Special; Schercemol 1818, CO, DIA, DICA, DIS, IPM, NGDC, OLO, PGML; Shell Cyclo Sol 53, 63; Shell Mineral Spirits 135, 145-EC; Shell Sol 70, 71(H), 71(WR), 72, 140, 340, B; Shell Super VM&P Naphtha EC; Shell Toluene; Shell Tolu-Sol 5, 6-EC, 19-EC; Shell TS-28; Shell VM&P Naphtha EC; Shell Xylene; Siponate 330; Skellite; Solulan 5, 16, 25, 75, 97, 98, C-24, L-575; Solvent 80; Sotex CW; Standamul G-16; Starfol IS; Sulfa-Hitech 0382; Sulfolane W; Syn Fac TEA-97; TBP; T-Det O-9; Tebol 99; Tegosoft 189; Tetraglyme; Tetralin; Texacar EC, EC-50, PC; Texanol Ester-Alcohol; Texox PPG-400, PPG-2000, WL-440, WL-660, WL-1400, WL-5000; Texsolve B, C, H, S, S-2, S-66, S-LO, V; Triglyme; Trymeen TAM-8; Ucon Fluid LB-, -625, -1145, -1715, -3000, LO-500; Versilan MX 112; Vista LPA, LPA-140, LPA-210; Wickenol 136; Witconol APEM, APM, APS

SPREADING AGENTS

Abil B 9800, B 9801; Abil-Wax 2434, 2440, 9800; Acetulan; CPH-39-N; Crodamol ICS, IPM, LL, PC; Dow Corning 344 Fluid, 345 Fluid; Emcol NA-30; Emerest 2381; Ethosperse OA-2; Fancol OA 50, OA 70, OA 80, OA 90, OA 95; Fluorad FC-128, FC-134, FC-135, FC-170-C, FC-171, FC-430, FC-431; Hartolan; Hartolan Super; Ionet S-20, S-60 C, S-80, S-85; Iscolan; Lipo DGLS, DGS-SE, Diglycol Laurate, EGDS, EGMS, PGMS; Lipocol B; Lipolan R; Lipopeg 2-DL, 4-DL, 4-DO, 4-DS, 4-L, 4-S, 6-L, 10-S, 15-S, 39-S, 100-S, 6000-DS; Lonzest 143-S; Monawet MO-65-150, MO-70-150; Neobee 1223 (Neobee O), O; Newcol 560, 561H, 562, 564, 565, 566; Newcol 865; Nioix AK-40,AK-40; Procetyl 10, 20, 30, 50; Silwet Surface Active Copolymer L-77, L-720, L-7001, L-7002, L-7004, L-7500, L-7600, L-7604, L-7605, L-7607; Super Refined Almond Oil NF, Avocado Oil, Orange Roughy Oil; Tergitol 24-L-45, 24-L-60, 24-L-60N, 24-L-75, 24-L-92, 24-L-98N, 26-L-1.6, 26-L-3, 26-L-5, TMN-3; T-Mulz 8015; Varonic K210

STABILIZERS

AA; Accelerator BZ; Acetulan; Acofor; A-C Polyethylene 540; Acrysol ASE-108, G-110, QR-1086; Actafoam F-2, R-3, R-10; Activ-8, Activ-8 in Hexylene Glycol; Additin 30; Admex 710, 711, 746, ELO; Adol 42, 52 NF, 54, 61 NF, 62 NF, 63, 64, 66, 80, 85 NF, 90 NF, 320, 330, 340, 520 NF, 620 NF, 630; Advastab LS-202, LS-203, T-52N Conc., T-290, T-340, TM-180, TM-181, TM-185, TM-692, TM-694; Aerosol 22N; Acamate FA 82; Acamizer I, II, III; Advastab TM-181-FS, TM-181 S, TM-697, TM-2082; Aerosol A-196, A-196-40; Agerite Geltrol; Agrilan FS101, FS112; Akrochem Antioxidant 12, Antiozonant PD-2; Aldo HMS, PMS; Alipal EP-110; Alkamide 1182, 1423, 1509, 2124, CDE, CDE-Extra, SDO; Alkamox C2-O, CAPO, LO; Alkamuls GMR; Alkanol S; Alkapol PEG 600; Alkasurf LAN-23, NP-4, NP-20, NP-35 70%, NP-40, NP-40 70%, NP-50 70%; Alkaterge-E; Alkatronic PGP 18-8, PGP 18-8LF; Alphadim 90AB; Alrosol Conc.; Ambergum 3021; Amerchol 400, BL, C, CAB, H-9, L-99, L-101, L-500, RC; Amerlate LFA, P, WFA; Ameroxol OE-2, OE-10, OE-20; Aminol CM, CM Flakes, CM-C Flakes, CM-D Flakes; Amiter LGS-5; Ammonium Stearate 33% Liquid; Ammonyx CDO, CO, DMCD-40, LO, MCO, MO, SO, TDO; AMP. AMP-95; Antioxidant 425; AO728 Special; Aqualon CMC-T; Argobase EU, LI, MS-5, SI; Arlatone 985; Armeen SZ, Z; Armid 18, C, HT; Aromox 18/12, C/12, C/12-W, CD/12, DM16, DMC, DMC-W, DMHT; Arosurf 66-E2, 66-E10, 66-E20, 66-PE12; Arylan DA 36; AST-1001; Atplus 535; Attagel 40, 50, 150, 350; Avirol SA-4106; Baerostab UBZ 630, 632, 791, 793; Bayhibit; Bayhibit-AM; Benecel CM, M, ME, MP; Berol 081 Flakes, 081 Powd., 259; Berol Dinonylphenol; Berol Nonylphenol; Berol PEG 400, 4000; Blanose Cellulose Gum; Blanose Refined CMC; Britesorb; Britesorb A 100; BSWL 202; Butyl Zimate

CA-57; Calsoft L-40, L-60; CAO-1; CAO-3; CAO-42; Capbopol 907, 910, 934, 940, 941; Capmul GMS; Caprol 3GS, 10G4O; Carbopol 1342; Carsamide CMEA; Carsamine Oxide LM; Carsonon N-30 70%, N-40 70%, N-50 70%, N-100 70%; Carstab 700, 701, 702, 705, DLTDP, DSTDP; Cata-Chek 820; Catinex KB-15, KB-32, KB-44, KB-45, KB-50; Cationic Guar C-261; CC-603; Cedemide AX; Cedepal CA-897, CO-210, CO-

STABILIZERS (CONT'D.)

430, CO-500, CO-890, CO-977, CO-990, CO-997; Chelon 80; Chemadox C, L, M; Chemal BP-261, BP-262, BP-262LF, BP-263, BP-264, BP-268, BP-268/50, OA-4, OA-9, OA-20; Chemax NP-1.5, NP-4, NP-6, NP-9, NP-10, NP-40, NP-40/70, NP-50, NP-50/70, NP-100, NP-100/70, OP-3, OP-5, OP-40; Ches 500; Chimassorb 944FL, 944LD; Cirrasol EN-MB; Cithrol G Range, GDO N/E, GDO S/E, GDS N/E, GDS S/E, GML N/E, GML S/E, GMO N/E, GMO S/E, GMR N/E, GMR S/E, GMS Acid Stable, PGML GMS N/E, GMS S/E, N/E, PGMO N/E, PGMO S/E, PGMR N/E, PGMR S/E, PGMS N/E, PGMS S/E; Comperlan F, KM, LD, LDO, LDS, LS, OD, PD; Conco XA-C, XA-L, XA-M, XA-MC, XA-S, XA-T, XA-Y; Condensate PS; Cosmedia Guar C-261, U; Cosmowax; Cosmowax J, K; Crillon LDE; Crodafos CDP, CS2 Acid, CS5 Acid, CS10 Acid; Crodamol BE; Crotein A, C; Cutina CBS, MD; Cyanamer A-370; Cyanox 425, 2777; Cyasorb UV 9, 24, 531, 1084, 2098, 2126, 2908, 3346, 3346 LD, 5411; Cyastab 40, 523, 593, 621, 712, 800, 908, 948, 988, LIPA; Cyclomide C212, DL203; Cyclophos PE30

Dacospin POE(25)HRG; Dariloid Series; Darvan ME, NS, WAQ; Decalin; Densol 284, 1010, BP-61, BP-62; Deriphat 160, 160C; Dermalcare C-20, SPS; DeSophos 30 NP; Diazopon SS-837; Dinoramac C, O, S; Dow Corning FF-400, FF-412, FF-414; Drapex 3.2, 4.4, 6.8, 10.4, Dresinate 81, 91, 92, 93, 94, 95, TX, X, XX; Drewmulse 1128, 1129, CNO, GMC-8, GMO, HM-100, TP, V, V-SE; Drynol E/20, E/30, E/40; DS-207; Dulectin; Duolite C-25D, C-26; Duralink HTS; Dur-Em 127, 127K, 207; Dusoran MD, NG, NO; Dymsol 31-P, L, N; Eastman Inhibitor DOBP, OABH, RMB; Eastman TDP 2000; Elfacos C26, ST 9; Elvanol 75-15; Emasol L-120, P-120; Emcol 4400-1, 4930, CC-37-18, DG; Emerest 2617, 2619, 2640, 2646, 2715; Emery 1730, 1732, 1780, 1787, 5700, 6752; Emid 6510, 6514, 6515, 6518, 6519, 6521, 6529, 6531; Emkatex DX; Empicol TLP, TLR; Empicryl APD, APD/B; Empigen OB, OH, OY; Empilan 2020, 2502, CDE, CIS, CME, DL 40, DL 100, EGMS, GMS NSE40, GMS NSE90, LME; Empiphos 4KP; Emulan OC, OG, OSN, OU; Emuldan HA 40, HA 52, HV 40 K, HV 52 K, HVF 52 K; Emulphogene BC-840; Emulphor ON-870, ON-877; Emulvin S, W; EPIstatic 103; Epolene C-16, C-18; Epoxol 7-4, 8-2B; Espesilor AC-50; Estabex 2307, 2307 DEOD, 2381; Ethanox 330; Ethosperse CA-2, LA-23, OA-2; Ethyl Zimate; Ethylan 20, BV, CL, CRS, HA, HP, KEO, LD, LDG, LDS, N30, N50, N92, OE, R, TCO; Evanstab 12; Fancol LA, LAO, OA 50, OA 70, OA 80, OA 90, OA 95; Fancor LFA; Ferro 1288; Flexricin 9, P-8; Flo-Gard CC 120, CC 140, CC 160; Foamer; Forbest 410; Forlan C-24, L; Frimulsion 6G, 10, Q8, RA, RF, X5

Galactasol Series; Gantrez AN, AN-119, AN-139, AN-169, AN-179; Gantrez ES-225, ES-335, ES-425, ES-435; Gelrite; Genaminox CS, KC; Generol 122, 122E5, 122E10, 122E16; Genu BB Rapid Set, 150 Grade, Carrageenan, JM, Pectins; Genugel; Genulacta Series; Genuvisco; Genuzan 1038A, 1063X; Geropon IN, RM/77-D; GFS; Givsorb UV-2; Good-rite 3114, 3125, Antioxidants, UV3034; Gradonic 400-ML; Hamp-Ex M; Haro Chem ALMD-2, ALT, BG, BP-108X, BSG, CBHG, CGL, CGN, CHG, CPR-2, CSG, CZ-31, CZ-35, CZ-36, CZ-37, CZ-38, CZ-39, KB-214SA, KB-219SA, KB-350X, KB-353A, KB-521SA, KB-554A, KPR, KS, LHG, LIG, MF-2, NG, P28G, P51, PC, PDF, PDP-E, PPCS-X, PTS-E, ZGN, ZGN-T, ZPR-2, ZSG, ZZ-019; Harol D; Haro Mix BF-202, BK-105, BK-107, CE-701, CH-205, CH-606, CK-203, CK-213, CK-711, FK-102, IC-217, IC-238, IH-108, LK-218, LK-228, MH-204, MK-107, MK-220, MK-620, MK-744, SK-602, UC-213, UK-121, VC-501, YC-601, YE-301, YK-110, YK-113, YK-307, ZC-028, ZC-029, ZC-030, ZC-031, ZC-032, ZC-036, ZC-309, ZC-311, ZC-902, ZT-025, ZT-026, ZT-504, ZT-508, ZT-514, ZT-905; Hartamide 9137, LDA 70, LDA 90, LMEA-70, LMEA-90, OD; Hartolan; Hartolan Super; Hercules Cellulose, CMC, Ester Gum 8BG; Hi-Sil 210, 215, 233, 250, 260, 262, T-600; Hostavin N 20, VP NiCS 1; Hydrine; Hyonic NP-60, NP-100, NP-110, NP-407, NP-500, OP-70, OP-407, OP-705, PE-90; Hystrene 5016; Ice#2, #12, #81; Icodimeen T-30; Icomeen 18-5, O-30, O-30-80%, S-5, T-2, T-5, T-7, T-20, T-25, T-25 CWS, T-40, T-40-80%; Iconol DDP-10, DNP-8, DNP-24, DNP-150, NP-7, NP-12, OP-30, OP-30-70%; Igepal CA-880, CA-887, CA-890, CA-897, CA-950, CO-210, CO-430, CO-890, CO-897, CO-970, CO-977, CO-980, CO-985, CO-987, CO-990, CO-997, DM-970; Igepon CN42, T-51; Improved Kelmar; Imwitor 191, 595, 900, 900K, 940, 940K, 945; Incromide LL, LLA; Indulin C, SAL, W-2, W-3; Interstab 761-28, 761-28A, BC-100S, BC-103, BC-103A, BC-103C, BC-103L, BC-109, BC-110, BC-4362, CH-55, CH-55R, CZ-10, CZ-11, CZ-11D, CZ-19A, CZ-22, CZ-23, CZ-4359, CZL-710, CZL-712, CZL-715, E-82, F-402, LF 3623, LF 3626, LF 3631/1, LF 3631/2, LF 3631/3, LF 3634, LF 3638, LF 3645, LF 3653, LF 3669, LF 3675, LF 3751, LF 10773/25, LF 10898/4, LF 10898/6, LF 11298, LF 11323/1, LF 11359, LL 3289, LP 3103, LP 3104, LP 3139, LP 3150, LP 3153, LP 3155, LP 3190, LP 3289, LP 3631/5, LT-4289, LT-4308, LT 11122/10, M-85, M-341, M-722, M-727, M-731, M-744, M-763, M-767, M-803, M-809, M-876, M-3187, M-11289, M-11301, MF-981, MF-985, MP 10581/3, MT981, MT982, MT 11303/1, R-4048, R-4052, R-4101, R-4109, R-4114, R-4137; Ionol; Irganox 245, 259, 1010, 1076, 1093, 1520, B-215, B-225; Irgastab 2002, T-265, T-634

Jeffox WL, WL-660, WL-1400, WL-5000; Jordamide 1214; Jortaine C, CAB-35, CB-40, CFA-35, COSB, LMAB; Kelacid; Kelco-Gel, HV; Kelcoloid, D, DH, DO, DSF, HVF, LVF, O, S; Kelco-Pac; Kelcosol; Kelfo; Kelgin F, HV, LV, MV, QH, QL, QM, RL, XL; Kelmar, Improved; Kelset; Keltex, P; Keltone; Keltose; Keltrol, F; Kelvis; Kelzan, D, M, XC Polymer; Kemamide B, E, E-180, E-221, O, P-181, S, S-65, S-180, S-221, U, W-20, W-39, W-40, W-45; Kemester 5500, 6000E; Ke-Mul 181; Ken-React Series, 7 (KR

7), 9S (KR 9S), 12 (KR 12), 26S (KR 26S), 33DS, (KR 33DS), 38S (KR 38S), 39DS (KR 39DS), 41B (KR 41B), 44 (KR 44), 46B (KR 46B), 55 (KR 55), 133DS (KR 133DS), 134S (KR 134S), 138D (KR 138D), 138S (KR 138S), 158 (KR 158D), 158FS (KR 158FS), 212 (KR 212), 238A (KR 238A), 238J (KR 238J), 238M (KR 238M), 238S (KR 238S), 238T (KR 238T), 262A (KR 262A), 262ES (KR 262ES), OPP2 (KR OPP2), OPPR (KR OPPR), TTS (KR TTS); Kessco Acetin, Diacetin, Glycerol Monolaurate, Glycerol Monooleate, Glycerol Monostearate 860, Glycerol Monostearate DH-1, Glycerol Monostearate Pure, Glycerol Monostearate SE, Triacetin; Kisuma 5A, 5B, 5E; Klucel; KLX Flakes; Kodaflex DMEP; Lactodan; Lamepon A; Lanfarax, 1776; Lankrocell KLOP, KLOP/CV; Lankroflex ED6, GE, L; Lankromark BL277, BM271, BM286, BM400, BT050, BT105, BT120, BT120A, BT190, BT339, DLTDP, DSTDP, LC68, LC90, LC244, LC299, LC310, LC431, LC475, LC486, LC541, LC563, LC585, LC629, LC651, LC662, LE65, LE76, LE87, LE98, LE109, LE131, LE230, LE274, LZ121, LZ187, LZ242, LZ440, LZ495, LZ561, LZ616, LZ638, LZ649, LZ693, LZ704, LZ770, LZ792, LZ836, LZ858, LZ935, LZ968, LZ1023, LZ1045, LZ1056, LZ1067, LZ1144, LZ1155, LZ1166, LZ1177, LZ1188, LZ1199, LZ1210, LZ1221, LZ1232, OT050, OT250, OT335, OT341, OT450, OT650; Lankropol ATE; Lanolin Alcohols LG, LO, THG, THO; Lanpolamide 5; Lenetol 9130 AH; Levelan P148, P208, P357; Lexemul 503, 515, AR, AS, T; Lexol PG 800; LICA 01, 09, 12, 38, 38A, 38J, 44, 97; Lignosite; Lignosol B, BD, DXD; Lipamide S; Lipo DGLS, DGS-SE, Diglycol Laurate, EGDS, EGMS, PGMS; Liponic EG-1; Lipoquat C 25; Lipotin A; Liquid Base

Lomar PW, PWA; Lonzest PEG 4-L, SMO-20, SMP-20, SMS-20, STO-20; Loropan CME, KD, KM, LD, LM, LMD, OD; Loxiol G 10, G 12, G 16, HOB 7121, HOB 7131; Lubrizol 2106, 2116, 2117, 2163; Lusynton A; Luviskol; LZ 01, 09, 12, 38, 44, 97; Macaloid; Mackamine CO, OAO; Macol LA-4, LA-12, LA-23, NP-4, NP-6, NP-9.5, NP-11, NP-20, NP-30, NP-30(70), NP-70, NP-100, OA-2, OA-4, OA-10, OA-20, SA-2, SA-5, SA-10, SA-15, SA-20, SA-40, TD-3, TD-4, TD-6, TD-8, TD-10, TD-610; Mafo CAB; Magnabrite F, HV, K, S, T; Maprolyte C; Maracell XC; Marasperse C-21, CB; Margel; Mark 152, 155, 232B, 281B, 366, 462, 550, 556, 565A, 630, 649, 684, 684B, 755, 1043A, 1092, 1178, 1216, 1220, 1314, 1409, 1490S, 1500, 1600, 1772A, 1900, 1905, 2100, 2100A, 2112, 2140, 2180, 5050, 5060, C, DDHP, DDMPP, DPHP, GS, LL, MDDPP, OTM, RFD, TDP, TNPP, TPP, TT, WS; Markstat Antistats; M-C-Thin AF-1; Merpol HCS, HCW; Merrol E-68; Methocel A, A4C, A4M, A15-LV, E, E4M, F, E5, E15-LV, E50-LV, E4M, F50-LV, F4M, K, K3, K35, K100-LV, K4M, K15M, K100M; Miglyol 840 Gel, Gel; Minugel 400, LF; Miranol C2M Conc., H2M Conc.; Mirataine COB; Mixxim HALS 57, 62, 63, 67, 68; Modicol N, S, VD; Monolan 12,000 E/80; Monomuls 60-10, 60-15, 60-20, 60-25, 60-25/2, 60-25/5, 60-30, 60-35, 60-40, 60-45, 90-10, 90-15, 90-20, 90-25, 90-25/2, 90-25/5, 90-30, 90-35, 90-40, 90-45; Monoplex S-73, S-75; Monosteol; Monthyle; M-Pyrol; Musloid 815D, 815M, 815S; Myvatex Peanut Butter Stabilizer; Myverol P-06

Na-Sul BSB; Natrosol 250; Naturechem PGHS, THS-200; Naugalube 438, 438-L, 438-R; Naugard 76, 431, 492, BHT, NBC, P, PHR, SP, XL-1; Naxonate 4AX, 4KT, 4L, 4ST, 5KT, 5L, 6AC, 45SC, AX, G, KT, SC, ST; NE; Nekal BA-77; Neobee 62; Newcol 1010, 1020, 1100, 1110, 1610, 1620, 1807, 1820; Nikkol DDP-2, -4, -6, -8, -10, GM-18IS, 18S, GO-440, GO-460, GS-460, MGS-A, -B, -DEX, MGS-ASE, -BSE; Nimco 1780; Ninate PA; Ninol 128, AA-62, AA-62 Extra; Ninol P-616, P-621, P-650; Niox KF-12; Nissan Nonion LP-20R, LP-20RS, MP-30R, OP-80R, OP-83RAT, OP-85R, PP-40R, NM; Nonion SP-60R; Nonisol 300; Noramac C, O, S; Norfox B, Coco Powder, DC, GMS, NP-1; Norlig; Norlig A; Novol; OA 502; OCI 56; OHlan; Orzan LS, SL-50, S; Ottalume 2100; Paraplex G-62; Patlac NAL; PDI #0199; Peganol NP 1.5, NP 4, NP 20, 40, 50, 100; Pegeste SML, SMO, SMP, SMS; Pegosperse 50 DS, 200 DL, 1750 MS, 6000 DS, EGMS-70; Pennmax Five, Four, Four-A; Perlankrol EAD-60, ESD, ESD-60, ESS-25, FF, SN, Perlankrol O; Peroxide Stabilizer H; Peroxidol 780, 781; Petrac MG-20 NF, ZN-42; Plas-Chek 775; Plasdone, C-15, C-30, K-25, K-26/28, K-29/32, K-90; Plastiflow LPC; Plastoflex 2307; Plastolein 9232; Pluronic 10R5, 10R8, 12R3, 17R1, 17R2, 17R4, 17R8, 22R4, 25R1, 25R2, 25R4, 25R5, 31R2, 31R4, L10; Polawax; Polectron 430; Polyclar 10, AT; Polysar Bromobutyl 2030, X2; Polysar Butyl 100, 200, 301, 402, XL-20; Polysar Chlorobutyl 1240, 1255; Polystep F-5, -12; Polywax OH 425, 550, 700; Primasol KW, SD; Proaid 9904; Promulgen D, G; Propoxyl 1695; Protowet C-5, E-4; Quatrisoft Polymer LM-200; Quimipol DEA OC

Radiamuls POLY; Recodan; Renex 650; Reserve Salt Flake; Rewocid DU 185, U 185; Rewomat B 2003, TMS; Rewomid C 212, DL 203, DL 240, IPL 203, IPP 240, L203, S 280; Rexol 25/50, 25/100-70%, 25/507; Rheodol TW-0106, 0120, 0320, TW-P120, TW-S106, -S120, -S320; Rhodigel 23; Rhodopol 23; Ritachol; Rylex NBC; Sandostab P-EPQ; Santelle-EOM K; Santoflex 13; Santone 8-1-O, 8-1-S, 10-10-S; Santoquin, Emulsion, Mixture 6; Santovar A; Santowhite Crystals, PC, Powder; Sarkosyl L, LC, NL-30, O; Schercamox DML, DMM; Schercemol 185, GMS; Schercemol PGML; Schercomid CDA-H, CMI, SLM, TO-2; Schercopol OMIS-Na; Schercotaine SCAB-KG; Schercoterge 140; Scripset 520, 540, 550, 720; Seachem 55; Sebase; Ser-AD FX 1010; Serdet DSK 40, Perle Conc.; Serdox NEL 12/80, NNP 20/70, NNP 30/70, NOP 30/70, 40/70; Sermul EN 20/70, 30/70, 145; Sipex BOS, CAV, EST-30, TEA; Siponic L1, L4, L7, L12, L16, L25, NP40, OP30, OP40, Y500-70, Y501; Softisan 378, Hard Fats; Solar CO; Soltem 480; Solwax C-24; Sorbax SML, SMO, SMP, SMS, STO, STS; Sotex CW; Span 20, 40, 65, 85; Stabilizer 1097; Stabilizer C, I-FF, T; Staflex Stabilizers; Stanclere A-121 C, A-221, T-55, T-85, T-126, T-160, T-161, T-163, T-164, T-165, T-174, T-182, T-183, T-184, T-186, T-192, T-193, T-194, T-197, T-208, T-222, T-233, T-250, T-250SD, T-470, T-473, T-482, T-483, T-484, T-582, T-801, T-876, T-877, T-878, T-55, T-4356, TL, TM;

STABILIZERS (CONT'D.)

Standamid CD, CM, CMG, KDM, KDS, LD, LD 80/20, PD; Standamox CAW; Standamul G-32/36 Stearate; Stanno-Plus; Starplex 70, 90; Stepanate AM, T, X; Sterling AOS; Sterox DF, ND; Strodex SEK-50; Super Amide LL; Supercol Supercol Guar F, G2, U; Super Hartolan; Superloid; Surco Coco Betaine, WC Conc.; Surfonic HDL, N-10, N-31.5, N-40, N-60, N-85, N-95, N-100, N-102, N-120, N-150, N-200, N-300, N-400, NB-5, NB-14; Surfynol 104BC, 104E, 104H, 104S; Sustane BHA; Sustrelle; Swanic 52L; Syncrowax ERL-C, HGL-C; Synperonic A14, A20, NP10, NP12, PE30/80; Synpron 1027, 1034, 1066

T55, 57, 80, 81; Tamol NH 7519, NH 9103, NHC 3001, NHC 4001, NHC 9301, NMC 9301, NN, NN 2406, NN 2901, NN 4501, NN 7718, NN 8906, NN 9104, NN 9401, NNO, NNOK; T-Det C-40, N 1.5, N-4, N-70, N-705, O-40, O-407; Tebol 99; Tegin 90, 90 NSE, 515, 515 V.A., 4100, 4480, GO, GRB NSE, MAV NSE, M NSE; Tegostab B 1048, B 2219, B 3752, B 4413, B 4690, B 4900, B 8002, B 8021, B 8404, B 8406, B 8407, B 8408, B 8409, B 8418, B 8423, B 8425, B 8432, B 8444, B 8680, B 8681, B 8863 T, B 8901, B 8906, B 8910, BF 2370; Tenox PG, TBHQ; Teric 12M2, 12M5, 12M15, 16M15, 18M2, 18M5, 18M10, 18M20, 18M30, 303, CME3, PE68; Tetronic 50R1, 50R4, 50R8, 70R1, 70R2, 70R4, 90R1, 90R4, 90R8, 110R1, 110R2, 110R7, 130R1, 130R2, 150R1, 150R4, 150R8, 304, 504, 701, 702, 704, 707, 904, 908, 1101, 1102, 1107, 1301, 1302, 1304, 1307, 1501, 1502, 1504, 1508; Tex-Wet 1158; Therm-Chek 6-V-6A, 130, 344, 707-X, 714, 837, 840, 904, 1825, 5469; Thiovanol; Thixcin E; Tinegal NA; Tinuvin 144, 292, 326, 327, 328, 440, 440L, 622, 622LD, 765, 770; T-Maz 20, 28, 40, 60, 60K, 60KHM, 61, 65, 65K, 67, 80, 80KLM, 81, 81K, 85, 85K, 90, 95; Topanex 100BT; Tribase, E, E XL, E Special, EXL Special, XL; Triton X-200; Troykyd Anti-Gel, Anti-Skin S, Emulso Wet, Lecithin W.D.; Trycol 5888, 5972, 6942, 6957, 6970, 6971, 6972, 6985, DNP-8, NP-40, NP-50, NO-407, NP-507, NP-1007, OAL-23, SAL-20; Trydet 2671, 2672, SA-8, SA-40; Tryfac 910-K, 5554; Trylon EW; Trylox 5906, CO-30; Trymeen 6637, TAM-40; Turpinal 4NL, SL; Tylose MH, MHB, MB, MH-K, MH-xp, MHB-y, MHB-yp; Ucarsil Organosilicon Chemical AF-1, AF-2; Ultranox 226, 276, 626, 626A; Ultrox, 500W, 1000W; Union Carbide L-720, -721, -722, -727, Y-9794; UV-Chek AM-104, AM-105, AM-205, AM-300/301, AM-340, AM-595; Uvi-Nox 1494; Uvinul 408; Valdet 4016; Van Gel, Gel B; Vanisol BIS sodico-2; Vanlube 601, AZ, PC; Vanox 1005, 1290, 1320, PCX, SKT

Vanstay 137, 145, 162-B, 246, 3027, 4017, 4030, 4039, 5510, 5515, 5563, 5630, 5698, 6032, 6040, 6053, 6055, 6074, 6078, 6133, 6172, 6191, 6201, 7024, 7025, 7032, 8073, 8120, 8300, 8308, A, B, C, CE, FA, HA, HT, HTA, HTB, HTC, HTE, HTF, L, M, RAZ, RR, RRE, RRZ, SC, SD, SE, SF, SG, SH, Zinc Stearate; Varonic 32-E20, U202, U215; Veegum, F, HS, HV, K, Neutral, S-728, WG; Victawet 12, 35B, 58B, 85X; Vulkanox 4020, 4022, BKF, DDA, KB, NKF; Wayplex 55-S; Weston 63, 310, 399, 399B, 430, 474, 491, 494, 600, 618, 619, 732, 800, 801, 802, DHOP, DLP, DOPI, DPDP, DPP, DSP, DTDP, EGTPP, ODPP, PDDP, PNPG, PTP, TDP, THOP, TIOP, TLP, TLTTP, TNPP, TPP, TSP; Wingstay 29, 100, 200, C, K, T; Witcamide 5130, 5133, S-771, S-780; Witcomul 78; Witconate 3009-15, TDB; Witconol L32-45; Witconol NP-100; Wytox 240, AF Series, ATP Series, HPM, LTS Series

STRIPPING AGENTS

Armeen CD; Du Pont Retarder LAN; Katapol OA-910; Levelene; Levenol A Conc., DS-1; OL-55-F-10; Peregal O, ST

SUPERFATTING AGENTS

Abil B 88183; Abil-Quat 3270, 3272; Empicol 0627; Skliro Distilled; Super Solan Flaked; Super Solangel 25; Varsulf S1333

SUSPENDING AGENTS

Acrysol ASE-60; Aerosil Colloidal Silicas, R-974, R-976; Agrilan F524, F535; Aldosperse ML 23, MS 20; Ammonyx 856; Aqualon Cellulose Gum, CMC-T; Ardril FCA 154, FCA 2924; Atlox 4862, 4868 B, 4870 B, 4995, 4996 B; Attagel 40, 50; Avirol 200; Benecel CM, M, ME, MP; Bentone EW, Gel CAO, Gel IPM, Gel LOI, Gel MIO, Gel MIO A-40, Gel S-130, Gel SS-71, Gel VS-5, SD-1, SD-2, SD-3; Biozan, SPX 5423; Blanose Cellulose Gum; Blanose Refined CMC; Calcium Stearate, Regular; Carbopol 615, 616, 617, 907, 910, 934P, 940; Castorwax MP-80; CMC-6-DG-L; CMC-T; Colloid 207; Colorol Series; Conco Sulfate RA; Condensate 640; Cotton-Pro; Cremophor S 9; Crodesta F-110, F-160; Cyanamer A-370, P-26, P-70, P-80, P-250; D.E.R. 671-EE75, 671-EEA75, 671-EEK75, 671-MK75, 671-T75, 671-X75, 671-XM75; Daxad 27; Dirsol SH; Emcol CC-9, CC-36, CC-42, K-8300; Ethofat 60/15, 60/20, 60/25, 142/20, 242/25, C/15, C/25, O/15, O-20; Flo-Gard FF 310, FF 320, FF 330, FF 350, FF 370, FF 390; Galactasol Series; Ganex P, P904, V, V216, V220, V516; Genugel; Genulacta Series; Genuvisco; Geronol TZ/14, TZ/A, TZ/B; Geropon CET/50/P, FMS, K/65, K/202, RM/77, SC/211, SC/213, TA/72, TA/72/S, TA/764; GFS; Grocor 6000, 6000 E, 6000 SE; Grotex 1120; Hartopon FS; Hercules Cellulose Gum, CMC, CMC-T; Hi-Sil ABS; Hydropalat 535; Imwitor 191, 900, 940K, 960, 965; Incronol ALS; Indu-Sol 238; Kara Sperse DDL; Keltrol, F; Kelzan, D, D35, M, S, XC Polymer; Ken-React Series, 7 (KR 7), 9S (KR 9S), 12 (KR 12), 26S (KR 26S), 33DS, (KR

33DS), 38S (KR 38S), 39DS (KR 39DS), 41B (KR 41B), 44 (KR 44), 46B (KR 46B), 55 (KR 55), 133DS (KR 133DS), 134S (KR 134S), 138D (KR 138D), 138S (KR 138S), 158 (KR 158D), 158FS (KR 158FS), 212 (KR 212), 238A (KR 238A), 238J (KR 238J), 238M (KR 238M), 238S (KR 238S), 238T (KR 238T), 262A (KR 262A), 262ES (KR 262ES), OPP2 (KR OPP2), OPPR (KR OPPR), TTS (KR TTS)

Klucel, 6, J; Koraid PSM; Laponite XLG, XLS; Lexemul EGMS; LICA 01, 09, 12, 38, 38A, 38J, 44, 97; Lomar PW; Luviskol K12, K17, K30, K60, K80, K90; LV-31, LV-336; LZ 01, 09, 12, 38, 44, 97, 5985; Macaloid; Mafo 13, C-12; Magnabrite F, HV, K, S, T; Marasperse; Miglyol 810, 812, 829, 840; Minugel 200, FG, LF; Monamine 779; Monomer C18; Natrosol, 250, 250 ER, GR, H4R, HHR, HR, JR, 250 KR, LR, MHR, MR; Newcol B4, B10, B18; Novol; Orzan AE, LS, LS-50; Petrac CP-11, CP-11-LS, CP-11-LSG, ZN-41; Plasdone K-25, K-26/28, K-29/32, K-90; Polargel, T; Polarite KB 325; Polawax; Polymulse 6; Polyplasdone XL-10; Propacyl CP-2000; PVP K-90; Quimipol ENF 120; Quso; Raylig; Reax 45L, 80C,, 81A, 82, 83A, 85A; Rheothik Polymer 80-11; Reten 210, 220, 300, 420, 421, 423, 425, 521, 523, 525; Rhodigel 23; Rhodopol 23; Sandoz Sulfate 216, ES-3; Schercemol OP; Schercomid CCD, CDA; Sellogen HR, W, WL; S-Maz 20, 40, 60, 60 K, 60 KHM, 65, 65 K, 67, 80, 80 K, 85, 85 K, 90, 93 R; Soltem 70, 207; Soprofor FL, PA/17, PA/19, PS/17, PS/19; Stepanol WA 100; Superloid; Supragil MNS/90; Syncrowax BB4, BB5, ERL-C, HGL-C, HR-C, HRS-C; T-Det TDA-65; Texaphor; Texaphor 277; Texaphor Special; Thixcin E; T-Maz 20, 40, 60, 60K, 60KHM, 61, 65, 65K, 67, 80, 80KLM, 81, 81K, 85, 85K, 90; Troykyd Anti-Settle, Anti-Settle Special; Troykyd Compd. XYZ; Van Gel; Veegum, F, HS, HV, K, Neutral, S-728, WG; Viscon 103; Volclay HPM-20, NF-BC; Witcamide 5130; Witcamine PA-78B; Witco 915, 934, Aluminum Stearate 22; Xanflood

TACKIFIERS

Abalyn; Abitol; Acrawax C; Adtac B10BHT, B25BHT, L100BHT; Akrochem P 37, P 40, P 49, P 55, P 86, P 87, P-90, P-133, P-420; Amberol ST-149; Amerlate LFA, WFA; Butac; Chemprene R-10, R-25, R-50, R-70, R-100, R-115; Epolene C-16, C-18; Fancol OA 50, OA 70, OA 80, OA 90, OA 95; Foral 85, 105, AX; Gantrez M-154, M-555, M-556, M-574; Hercolyn D; Hercotac LA95BHT; Lankroplast L542; LZ 5907; Natac; Nevpene 9500; Nevtac 80, 99 Super, 100, 115, 130; Olicat C; Pentalyn A, H; Petro-Rez 100, 103, 200, 215, PTH, PTL; Picco 5070, 6070, 6110, 6115, 6120, 6130, 6140; Piccodiene 2215, 2215SF, 2285, 2285BHT; Piccofyn Resins, A100, A115, A135, D125; Piccolastic C125; Piccolyte A115, A125, A135, C100, C115, C125, C135; Piccolyte D100, D115, D135, HM-110 Resin, S100, S115, S115SF, S125, S125SF, S135; Piccomer X40, XX100; Picconol A100, A102, A200, A201, A500, A501, A600E; Piccopale 70, 85, 100, 100 BHT, 100HM; Piccotac 95BHT, B, CBHT; Piccotex 75, 100, 120, LC; Piccovar AB165, AB180, AP10, AP25; Piccovar L30, L30S, L60; Plasthall 7041, 7049, R-9; Plastolein 9789; Poly-Pale Ester 10; Resin 731D; RT-3; SP-25, SP-553, SP-560; SP-1068, SP-1077; Staybelite, 750, Ester 10; Stygene R-2; Super Sta-Tac 80; Terate 303; Varcum 647, 2217, 4727, DR-406; Vultac 2, 4; Wingtack 10, 86, 95, 115, Extra, Plus; Zonarez 7010, 7025, 7040, 7055, 7070, 7085, 7100, 7115, 7125, B-10, B-25, B-40, B-55, B-70, B-85, B-115, B-125; Zonester 55, 75

THICKENERS

Abil-Wax 9810; Ablunol 6000DS; Accomid C, PK; Acid Thickener; A-C Polyethylene 6, 6A, 8A, 9A, 8, 9 UV Absorbers, 400; Acritamer 941; Acryloid 1017, 1019, W-1600; Acrysol 51, A-1, A-3, A-5, ASE-60, ASE-77, ASE-95, ASE-108, G-110, GS, HV-1ICS-1, TT-615, TT-675; Actramide 840; Adogen 442 H; Aerosil 200, R 972; Aerosol 30; Alcogum, 9639, L; Alconate SBG-280; Aldo PMS; Alkamide 1182, 1423, 2124, CDE, CDE-Extra, CDO, CL63, CME, CMO, HTDE, HTME, L7DE, L7DE-BT, L9DE, L9ME, LIPA, SDO; Alkamidox C-2, C-5, L-5; Alkamox C2-O; Alkamuls 400-DS, GMS; Alkateric CAB; Alkatronic PGP 10-1, PGP 23-7, 23-8, PGP 33-8, PGP 40-7; Ambergum 3021; Amfotex FV-28; Aminol CM, CM Flakes, CM-C Flakes, CM-D Flakes, COR-4, 4C, OF; Ammonyx CO, LO, SO; Amphosal CA, CG; Amphoteric C, N; Antil 141; Aramide CDM-4, CDR, LDS; Arnox 910; Aromox DMCD; Attagel 40, 50, 150, 350; Avamid 150; Barlox 10S, 12, 14, C; Benecel CM, M, ME, MP; Bentone 27, 34, 38, 128, 500, EW, LT, SD-1, SD-2, SD-3; Biozan, SPX 5423; Blanose Cellulose Gum; Blanose Refined CMC; Cab-O-Sperse; Calamide C, CW-100, O; Calcium Stearate, Regular; Caprol 2G4S, 10G4O, 10G10S; Carbopol Resins; Carbopol 615, 616, 617, 907, 910, 934, 934P, 940, 941, 1342; Carsamide 7644, CMEA, SAC, SAL-7, SAL-9; Cationic Guar C-261; Cellosize Hydroxyethyl; Cerasynt 303, 945, D, GMS, M, SD; Ceroxin Special; Cetal; Cetiol SB 45; Chemax E-400 ML; Chimipal DCL, DS 6000, LDA, MC, OLD; Cithrol 2ML, 2MO, 2MS, 3MS, 4ML 4MO, 4MS, 6ML, 6MO, 6MS, 10ML, 10MO, 10MS, 15MS, 40MO, 40MS, 60ML, 60MO, G Range; CMC-T

Combizell AG 070, AG 200, APR 060, APR 200; Comperlan COD, F, KD, KDO, LD, LD 9, LDO, LDS, LM, LMD, LS, OD, PD, SDO, SM, VOD; Condensate PA, PC, PE, PL, PM, PO; Consamine CA; Cosmedia Guar C-261; Cosmowax; CPH-53-N; Cremophor S 9; Crester PR; Crodacol C, C70, C95 NF, S, S70, S95NF; Crodafos Series; Crodafos 25 D2 Acid, 25 D5 Acid; Crodafos 25 D10 Acid, 25D Series, CS Series, ID Series, O Series; Crodamol BE, ICS, MM, SS; Crodesta F-110; Crodyne BY-19; Cru Thix 46; Culminal CMMC,

THICKENERS (CONT'D.)

MC, MHEC, MHPC; Cutina GMS, HR; Cyanamer A-370, P-26, P-250; Cyclochem DGS, EGDS, LIPA, NI; Cyclomide C212, CODI, DC212, DC212/M, DC212/S, DC212/SE, DIN 295, DIN 295/S, DIN 295/S, DL207/S, DO280/S, DOTS, DP240/S, DS280, DS280/S, KD, LIPA, RODEA; Cyclomox C, SO; Cyclopol SBG280; Cycloryl DCA; Cyclosheen 202; Cycloteric BET C-30, CB, I-30, O-30, OB50, T2 40, W; Dapral GT Series; Dariloid Series; Dehydag Wax 14, 16, 18, 22; Densol 1010; Dermalcare EGMS/SE, GMS, GMS-165, GMS/SE, GTIS, LVL, MM/M, MST, PGMS, SDG, SS; Detergent CR; Drewmulse 1128, 1129, CNO, GMC-8, GMO, HM-100, POE-SML, TP; DSX-1514; Dymsol L; Ektasolve EP; Elastomag 100, 100R, 170, 170 Micropellet, 170 Special; Elfacos C26, ST 37; Emanon 3199, 3299R, 4110; Emerest 2452, 2610, 2620, 2640, 2642, 2711, 2712, 2717; Emery 1780, 1787, 5430, 6748, 6752; Emid 6500, 6510, 6514, 6515, 6521, 6529, 6531, 6533, 6534, 6538; Emphos D70-31; Empicol TLP, TLR; Empicryl 6045, 6052, 6054, 6058, 6059, 6070, DH122, DH135, DH145, HV110, HV125, HV130, HV226, PT1334, PT1345, PT1397, PT1544, PT1764/DO; Empigen 5083, BB, BB-AU, OH25, OS/A; Empilan CDEY, CDX, FD, FD20, FE, LDX, LIS; Emsorb 2720, 6915; Emthox 5964, 5967; Emulsynt GDL; Emulvis; ESI-Terge 10, 320, AH 20, B-15, C-5, S-10; Espesilor AC Series, AC-43, EC Series; Ethosperse CA-2, LA-23, OA-2; Etoxi AC-91; Fancol LA; Foamole M; Forbest 172; G-431; Gafamide CDD-518; Gaftex PT; Galactasol Series; Gantrez AN, AN-119, AN-139, AN-169, AN-179; Gelcharg HP4; Gelrite; Genaminox CS, KC; Genu 12CG or 12CB, 85-100 Grade, 20 AS or 20AB, 100 Grade, 104AS or 15AB, 100-120 Grade, Carrageenan, Pectins; Genugel; Genulacta Series; Genuvisco; Genuzan 1038A, 1063X; GFS; Glissoviscal B, SB, SG; Glucamate DOE-120

Glucate SS; Good-rite Polyacrylates; Hartamide AD, CE 80, CE 90, LE-90, LM 11D; HCI Thickener; Hercules Cellulose, Cellulose Gum, CMC, CMC-Warp Size; Hetamide LA, MC, ML, MMC, OC, RC; Hetester MS; Hetoxide BN-13, C-30, C-200-50%, HC-40, HC-60; Hi-Sil 132; Hi-Sil 210, 215, 233, 250, 260, 262, T-600, PN 73; Hyonic NP-40, OP-40; Icinol HT90000; Imsil A-10, A-15, A-25, A-108; Imwitor 191, 900, 900K, 940K; Incromate PPI; Incromide ALD, AVD, BEM, CA, CAL, CM, L-90, LA, LEP, LI, LL, LLA, L-90, LMD, LMI, LMM, LR, Mink D, OD, OLD, OM, OPD, OPM, SED, SM, WGD; Incromine L-40, Oxide AL, Oxide AV, Oxide B-30, Oxide C, Oxide I, Oxide ISMO, Oxide Mink, Oxide MC, Oxide O, Oxide OD-50, Oxide OL, Oxide SE, Oxide WG; Incronam 30, ISM-30, P-30; Indalca CD30, N-110-70, XCD, XCDA/1; Intermediate 300, 325, 512; Ircogel 903; Jordamide 22, 29-78, 1281, AXS, CFAM, CLD, CLM, CMEA, CMEA Extra, CP, J-10, JT-128, LLD, PCS, RO, SCD, TC, WC Conc.; Jordamine 1281; Jordamox CAPA, CDA-40, LDA, MDA, ODA, SDA; Jortaine C, CAB-35, CB-40, CFA-35, COSB, CSB, CSB-50, LMAB, TM; Katioran AF; Kelfo; Kelmar, Improved; Keltex, P; Keltrol, F; Kelzan, D, M, XC Polymer; Kemester 105, 115, 143, 205, 226, 325, 3681, 3684, 5654, 9018, 9022, EE, EGDS, EGMS, MM; Kerinol 2012 F, AA 62; Kessco 653, 654, Acetin, Diacetin, Diethylene Glycol Monostearate, Diglycol Stearate Neutral, Diglycol Stearate SE, Glycerol Monolaurate, Glycerol Monooleate, Glycerol Monostearate 860, Glycerol Monostearate DH-1, Glycerol Monostearate Pure, Glycerol Monostearate SE, PEG Series, 200 Dilaurate, 200 Dioleate, 200 Distearate, 200 Monolaurate, 200 Monooleate, 200 Monostearate, 300 Dilaurate, 300 Dioleate, 300 Distearate, 300 Monolaurate, 300 Monooleate, 300 Monostearate, 400 Dilaurate, 400 Dioleate, 400 Distearate, 400 Monolaurate, 400 Monooleate, 400 Monostearate, 600 Dilaurate, 600 Dioleate, 600 Distearate, 600 Monolaurate, 600 Monooleate, 600 Monostearate, 1000 Dilaurate, 1000 Dioleate, 1000 Distearate, 1000 Monolaurate, 1000 Monooleate, 1000 Monostearate, 1540 Dilaurate, 1540 Dioleate, 1540 Distearate, 1540 Monolaurate, 1540 Monooleate, 1540 Monostearate, 4000 Dilaurate, 4000 Dioleate, 4000 Distearate, 4000 Monolaurate, 4000 Monooleate, 4000 Monostearate, 6000 Dilaurate, 6000 Dioleate, 6000 Distearate, 6000 Monolaurate, 6000 Monooleate, 6000 Monostearate, Triacetin

Klucel, 6, EF, GF, HF, J, LF, MF; Korthix H; Lamesoft LMG; Lanfarax 1776; Lankroplast V2012, V2023, V2067, V2100; Lanogen 1500; Laponite 508, XLG, XLS; Lauramide 11, D, ME, R, S; Lauridit KD, KDG, OD; Levilite; Lexaine C, CS; Lexein S620TA; Lexemul 503, 515; Lipamide S; Lipo DGLS, DGS-SE, Diglycol Laurate, EGDS, EGMS, GMS 450, GMS 470, PGMS; Lipomulse 165; Liponate 143M, CG, IPM, IPP, MM, PB-4, PC, PO-4, PS-4, SPS, SS, TDS, TDTM; Liposorb L, O, P, S, SQO, TO, TS; LNP Reprocessed TFE F-2080, Grade 3, OX-60; Lonzaine C, CO; Loropan CME, KD, KM, LM, LMD, OD; Lo-Vel 27, 28, 29, 39, 275; Lubrizol 2163; Lutrol F 127; Luviskol, Luviskol K12, K17, K30, K60, K80, K90; Macaloid; Mackam HV, ISA, LMB-LS, OB, OB-30; Mackamide 100-A, AN55, C, CD, CD-8, CD-25, CDM, CDT, CDX, CMA, CS, CSA, EC, L-95, LLM, LMD, LMM, MC, O, PK, S; Mackamine LAO, LO, O2; Manro AO 3OC, DPM 2169; Manroteric AT 1200; Mapeg EGDS, EGMS; Marlamid D 1218, DF 1218, KL, M 1218; Marlipal BS; Marvanol RE-1274, RE-1281; Mazamide 25, 65, 65CZ, 66, 70, 80, C-5, CS-148, L-298, LM-21, LS-173, LS-196, O-10, O-20, SS-10, SS-20, T-10, T-15, T-20, TO-10; Mazawax 163R; Mazol GMO, GMR, GMS, GMS-HM; Methocel A, A4C, A4M, A15-LV, E, E4M, E5, E15-LV, E50-LV, E4M, F, F50-LV, F4M, J, K, K3, K35, K100-LV, K4M, K15M, K100M; Min-U-Gel 400, LF; Miramine OC; Miranol CM Conc., CM-SF Conc.; Miranol ISM, JS; Mirataine BB, BD, CBC, CBR, CBS, COB, ODMB-35%, TM; Modicol VD; Mona AT-1200; Monamid 7-100, 7-153 CS, 15-70W, 15-MW, 150-AD, 150-ADD, 150-ADY, 150-CW, 150-DR, 150-GLT, 150-IS, 150-LMW-C, 150-LW, 150-LWA, 150-MW, 664, 716,

718, 759, 770, 982, 1007, 1034, ADY, CMA, LIPA, LMA, LMIPA, LMMA, LMW-C, R31-42, S; Monamine Series, 779, AA-100, AC-100, ACO-100, AD-100, ADD-100, ADDS-100, ADS-100, ADY-100, AF-100, ALX-80 SS, ALX-100 S, ARA-100, CF-100 M, I-76, LM-100, R8-26, T-100

Monaquat AT-1074, P-TC, P-TS, P-TZ, TG; Monateric 1202, ISA-35%; Monazoline Series, C, CY, O, T; Monomuls 90-L12; Natrosol 250, 250 ER, GR, H4R, HHR, HR, JR, 250 KR, LR, MHR, MR; Naturechem EGHS, GMHS, PGHS, THS-200; Ninol 40-CO, 49-CE, 51-LL, 52-LL, 55-LL, 96-SL, 128 Extra, 201, 2012 Extra, 4812, 4821, A-10MM, AA-62, AA-62 Extra, CBR, CNR, DS, LD, P-616, P-621, P-650, SNR, SR 100; Ninox CA, CS, FCA, L, M, O, S; Niox EO-33; Nissan Nonion DS-60HN, LP-20R, LP-20RS, MP-30R, OP-80R, OP-83RAT, OP-85R, P-6, PP-40R, S-2, S-4, S-6, S-10, S-15, S-15.4, S-40, SP-60R, T-15; Nissan Stafoam DF-1, DF-2, DL, DO, DO-S; Nitrene 100 SD, 11230, 13026, A-309, C, C Extra, L-76, L-90, N, NO; Noiox EO-33; Nonionic 1017-R, 1025-R, 2017-R, 2025-R, 4017-R, 4025-R, 5025-R; Nonisol 100; Nopalcol Series, 1-S, 1-TW; Nopcowax 22-DS; Norfox DCO, DLSA, EGMS, KD; Onyxol 42, 345, 2062, SD; OP-2000; P & G Amide No. 27; Pationic 138C, SSL; Pegnol PDS-60; Pegosperse 50 DS, 200 DL, 1750 MS, 6000 DS, EGMS-70; Perchem EAG, Easigel, TEA; Petrac 250; Plastigel PG9033, PG9037, PG9068, PG9089, PG9104; Plastomag; Pluracol E-400 NF, E-600 NF, E-1000, E-1450, E-1450 NF, E-2000; E-4000 NF, E-4500, E-8000, E-8000 NF; V-7, V-10; Pluronic 10R5, 10R8, 17R1, 17R2, 17R4, 17R8, 25R1, 25R2, 25R4, 25R5, 25R8, 31R2, 31R4, F38, F68, F68LF, F77, F87, F88, F98, F108, F127, L31, L35, L42, L43, L44, L61, L62, L62D, L62LF, L63, L64, L72, L81, L92, L101, L121, L122, P65, P75, P84, P85, P94, P103, P104, P105, P123; Polargel, T; Polawax; Polymulse 6; Polyox WSR 205, 301, 1105, N-10, N-12K, N-60K, N-80, N-3000, N-750; Polytrope 1131; Profan 24 Extra, 128 Extra; Profan 2012E, AA62, AB20, AD31; Progacyl COS-1, COS-10, COS-20, -70, CP-7, CP-2000, EM-40; Purton CFD, SFD; Pyroter CPI-40; QO Furfuryl Alcohol; Quatrisoft Polymer LM-200; Quimipol DEA OC; Quso; Radiasurf 7402, 7403, 7404, 7412, 7413, 7414, 7417, 7421, 7422, 7423, 7431, 7432, 7443, 7453, 7454, Raylig; Reten 210, 220, 300, 520, 521, 523, 525; Rewocid DU 185, U 185; Rewoderm L 67-75, LI 48-50, LI S 75, LI 420-70, LIS 75

Rewomid DC 212/SE, 220/LS, DC 220/SE, DL 203, DL 203/S, DLMS, DLM/SE, F; Rewopal PEG 6000 DS; Rheolate 1; Rheothik Polymer 80-11; Rhodigel 23; Rhodopol 23; Rilanit Special; Ross Thix 700 Series; Rycel 100, 105; Rycomid 2120; Samaron Thickener N; Sandoz Amide NT; Sandoz Amine Oxide XA-C, XA-L, XA-M; Sartomer 7001; Satexlan; SB-30, -31, -31C, -136, -331, -332, -335, -336, -431, -432, -632, -932; Schercamox CMA, DMS, T-12, DML, DMM; Schercemol EGMS, GMS, MM, PGMS; Schercodine M, T; Schercomid 1-102, 1214, CDA-H, CDO-Extra, CME, HT-60, ID, IMI, LD, ME, ODA, IMI, SCE, SCO-Extra, SD-DS, SI, SI-M, SLA, SLE, SL-Extra, SLL, SLM, SLM-C, SLMC-75, SL-ML, SLM-LC, SL-ML-LC, SLM-S, SM, SME, SME-A, SME-M, SME-S, TO-2; Schercopol LPS; Schercoquat ROAB; Schercotaine MAB, OAB, PAB; Scripset 500, 808; Secoster DS 6000; Ser-AD FX 1010; Serdolamide PYF 77; Shur-Coal 2000; Sipex BOS, CAV; Siponate A-246 LX; Siponic L1, L4, L7-90; Sipoteric CB, COB; Sipothix 1941, H-65; Solan, 50; Soprofor VR/50; Sorbax SML, SMO, SMP, SMS, STO, STS; Span 20, 40, 65, 85; Spermwax; Stamid HT; Standamid 100, CMG, ID, KD, KDM, KDO, LD, LD 80/20, LDM, LDO, LP, SD, SDO, SOD, SOMD; Standamox C 30, CAW; Standapol AB-45, BC-35, OLB-30, OLB-50; Stepan Cationic 10; Sterling 66/33, Amide 460, CAB, DEA, DLEA-90, LM-70, Granulated Wax; Stiffener DSC; Super Amide GR, L9, L9C, L9A, LL; LM; Superclear; Supercol Guar F, G2, U; Superloid; Surco 128T, CMEA, MA, SR-200, WC Conc.; Syncrowax BB4, BB5, HR-C, HRS-C; Synotol 119 N, CN 60, CN 80, CN 90, L 60, L 90, LM 60, LM 90, ME-90; Synprolam 35DMO, 35MX1/0; Tegamin Oxide WS-35; Tego-Betaine C, L-7, L-90, S; Teric CDE; Tetronic 304, 504, 701, 702, 704, 707, 904, 908, 1101, 1102, 1104, 1107, 1301, 1302, 1304, 1307, 1501, 1502, 1504, 1508; Texapon EA-1, -2, -3; Texicryl 13-300, 13-302, 23-002, 23-004, 23-012, 23-005, 23-018; Textamine 1839, O-1, T-1; Textamine Oxide TA; Thickener #5004 (All Aq.); Thickener L, LN

Thixcin E; T-Maz 60K, 60KHM, 65K, 80KLM, 81K, 85K; Tohol N-220, N-220X, N-230, N-230X; Trydet SA-23; Tylose C, CB Series, CR, CBR Series, H Series, MH, MHB, MB, MH-K, MH-xp, MHB-y, MHB-yp; Unamide C-2, C-5, C-72-3, CDX, CMS, D-10, GC-75, J-56, JJ-35, L-2, L-5, LDL, LDX, LMDX, N-72-3, S, SI, W; Van Gel, Gel B; Varamide A-2, A-7, A-10, A-12, A-83, L-1, MA-1, MA-4, ML-1, ML-4; Varion CADG-HS, CADG-W, TEG, TEG-40%; Varonic 400MS, T105, T202, T202 SR, U105; Veegum, F, HS, HV, K, Neutral, S-728, WG; Velvetex AB-45, BK-35; Viscon 103; Vissurf S, 1400; Wickenol 545; Witamide 70, 5195, 6310, CD, LDTS, LM, LMT, M-3; Witco Aluminum Stearate 18, 22; Witcomul 78; Witconol L32-45, RHP; Xanco-Frac; Zeosyl 200; Zeothix 177, 265

THIXOTROPIC AGENTS. SEE GELLING AGENTS

UV ABSORBERS

Aerosol 22N; Alconate 2CSA; Amerscreen P, P 80/20; Anti UVA; Arlatone 507; Butacite; Cyasorb UV 9, 24, 531, 2098, 2126, 5411; Eastman Inhibitor DOBP, RMB; Epoxol 8-2B; Escalol 106, 507, 557, 567, 587; Ferro Permyl B 100; Ferro UV-Chek AM-101, AM-105, AM-205, AM-300, AM-340, AM-595; Finsolv TN; Givsorb UV-2; Good-rite Antioxidants; Hostavin ARO 8; Irgastab 2002; JF 77; Kemester HMS; Lankro-

UV ABSORBERS (CONT'D.)

mark LE; Lusantan 25; Mark 202A, 1413, 1535, 5050, 5060; Mixxim BB/100; Monamine 779; Parsol MCX; Permyl B-100; Resyn 28-3307; Sandovor 3206, EPU, VSU; Solarchem 0; Spectra-Sorb UV 9, 24, 5411; Spectratech CM 10540, 11246, 11357, 11367, 11615, 11616, 11681, 11704, 11720, 11715; Tinuvin 109, 900, 1130, P; Topanex 100BT; UV-Absorber Bayer 325, 340; UV-Absorber KL 3-2500; Uvasorb 20H, MET, S-5; Uvinul 400, 408, 490, D-49, D-50, DS-49, M-40, MS-40, N-35, N-539, P 25

VEHICLES. SEE CARRIERS

VISCOSITY MODIFIERS. SEE THICKENERS

VULCANIZING AGENTS

Di-Cup 40C, 40KE, R, T; DPTT, DPTT-S; Dynamar FC 5157; Ethyl Selenac; Ethyl Tuads; Litharge 33; Methyl Selenac; Methyl Tuads; Naftocit Di 4, Di 7, Di 13, DPG, DPTT, MBT, MBTS, Mi 12 C, NaDBC, NaDMC, Thiuram 16, ZBEC, ZMBT; Rapidblend 1973; Rhenovin S-90; Sulfads; Sulfamin R; Telloy; Thermacure; Vanax A; Vandex; Varox 130, 130-XL, 231, 231-XL, DBPH, DBPH-50; Vul-Cup 40KE, R; Vulcuren 2, 2/EGC; Vultac 2, 3, 4, 5, 7, 710

WATER REPELLENTS/MOISTURE BARRIERS

Abil-Wax 2434, 9809, 9810; Adtec L100BHT; Alkazine C; Aquagard; Aquasan 542; Armid 18, C, HT; Autopoon NI; Autopur WK 4121; Britol; Calcium Stearate, Regular; Ceranine VS; Cerol A Liq., C, EWL, M, ZN; Cyanamer P-35; Cyclophos PE5; Daxad 19L-40; Denphos P-610; DF 1040; Drakeol 35; Dynakoll; Eccopel; Eltesol AX 40, MGX, PT 45, PT 93, PX 93, SCS 40, ST 34, SX 30, SX 93, TA, TA/F; Elvace 1873; Emphos CS-141, CS-151, D70-30C, F27-85; Fancor Lanwax; Flexchem CS; Fungicide M; Fyarestor 100; Graphsize; Hoechst Wax LP; Hetoxamine C-2, C-5, C-15, O-2, O-5, O-15, S-2, S-5, S-15, ST-2, ST-5, ST-15, ST-50, T-2, T-5, T-15, T-20; Hydagen F; Hydro-Pruf 200; Hy-Pel GP-4; Kemamide B, E, E-180, E-221, O, P-181, S, S-65, S-180, S-221, U, W-20, W-39, W-40, W-45; Ken-React Series, 7 (KR 7), 9S (KR 9S), 12 (KR 12), 26S (KR 26S), 33DS, (KR 33DS), 38S (KR 38S), 39DS (KR 39DS), 41B (KR 41B), 44 (KR 44), 46B (KR 46B), 55 (KR 55), 133DS (KR 133DS), 134S (KR 134S), 138D (KR 138D), 138S (KR 138S), 158 (KR 158D), 158FS (KR 158FS), 212 (KR 212), 238A (KR 238A), 238J (KR 238J), 238M (KR 238M), 238S (KR 238S), 238T (KR 238T), 262A (KR 262A), 262ES (KR 262ES), OPP2 (KR OPP2), OPPR (KR OPPR), TTS (KR TTS); LICA 01, 09, 12, 38, 38A, 38J, 44, 97; Masil 272, 290D, 290F; Merpiphos PP, RP, SP, TP; Monafax 060m 872; Monateric 810-A-50, 985A, CAM-40, CDL, CDS, CDTD, CDX-38 Mod., CEM-38%, CEM-38CG, CNa-40, COAB, CyA-50, CyMM-40, LMAB, LMM-30, MCB, TA-35, TDB-35; MS-55-F-4; Nalan GN, W; Neoprotex KM; Neutral Degras; Norfox CS, ZNS

Nissan Cation PB-40, -300, S_2-100; Nissan Nonion E-205, E-215, E-230, K-202, K-203, K-204, K-207, K-211, K-215, K-220, K-230, P-208, P-210, P-213, S-207, S-215, S-220, S-207, S-215, S-220; Nissan Nymeen DT-203, DT-208, L-201, L-202, L-207, S-202, S-204, S-210, S-215, S-220, T_2-206, -210, -230, -260; Nissan Persoft NK-60, -100, SK; Nissan Plonon 102, 104, 108, 171, 172, 201, 204, 208; Nissan Rapisol B-30, -80, C-70; Nissan Trax H-45; Nitrene 11230, 13026, A-309, C, C Extra, L-76, N, NO; Noiox AK-41; Nonarox 575, 730; Nonionic 1035-L, 1044-L, 1061-L, 1062-L, 1064-L, 1068-L, 1088-L, E-4, -5, -6, -7, -10, -12, –20, -30; Nonipol 20, 40, 55, D-160, Soft SS-50, -70, -90; Nonisol 100, 300; Nopalcol Series, 1-S, 1-TW; Nopco 1186A, 2272-R; Nopco Colorsperse 188-A; Nopcogen 14-L, 22-O; Norfox 916, DCS, IM-38, NP-6, PEA-N, SLES-02, SLS, T-60; Novel 1412-70 Ethoxylate; NP-55-40, NP-55-50, NP-55-60, NP-55-80, NP-55-85, NP-55-90, NP-55-95, NP-55-120, NP-55-150, NP-55-300; Nutrol 100, 600, 611, S 60; Penreco Amber, Blond, Regent; Perfecta USP; Petrac MG-20 NF; Picconol A200, A201; Piccopale 70, 85, 100, 100BHT, 100HM; Piccotac 95BHT; Piccotoner 1200; Poly-Pale Ester 10; Purelast 220, 221, 226, 228, 234, 235, 240H, 241, 241H, 245, 245H; Purelast 247, 249, 251, 252, 255, 254; Quilon C, H, L, M, S; Rancoff; Repello DC; Repel-O-Tex 100; Rewopon IM-AN, IM-OD; Rudol; Schercemol 1688, DID; Semtol Series; SF99; Silicone Fluid 350, 1M, 5M, 10M, 30M, 60M; Sipernat D 17; Sitren; Skliro; SM 2086; Softenol 3108, 3114, 3118, 3995; Solar CI-387; Solupret Brands; SS 4098, 4129; Stepan Cationic 10; Stepantex SO 90; Sterling 2XS; Super Refined Shark Oil; Ultraflex; Union Carbide R20; Victory; Vinsol MM; Vista STXS; Witco ACS 60, Aluminum Stearate Non Gel A, Calcium Stearate F, SCS 40, STS 40, SXS 40, TX Acid; Witconate NCS, NXS, P-1059, SCS, STS, SXS, TX Acid; Zepel B, DR

WAXES

Abil-Wax 2434, 2440, 9800, 9801, 9809, 9810; Advawax 275, 280, 290; Aerosol OT-75%, OT-B; Akrowax 130, 145, 5050; Akrowax Micro 23; BASF Wax A, AF, AF 30, AF 31, AF 32, AH 3, AH 6, AL 3, AL 61, AM 3, AM 6, DG, E, ES, EVA 1, FB, L, LCP, LCP 2, LG, LGE, LS, MCE, OA, OA 2, OA 3, OP, S, SG;

Be Square 175, 185, 195; Caprol 10G10S; Cardipol LP, LP 0-25; Cardis 314, 319, 320; Carnapol 77X; Celite 275; Ceraloid 356; Ceramid; Cetina TE; Cetiol MM; Corax; Cosmowax P; Crodamide E, ER, O, OR, S, SR; Crodex A, C, N; Cutina BW, EGMS; Cyclochem PETS; EBS Wax; Emerwax 1251, 1253, 1254, 1257, 1266; Epolene C-10, C-13, C-14, C-15, C-16, C-17, C-18, E-10, E-11, E-12, E-14, E-15, E-43, N-10, N-11, N-12, N-14, N-34; Flexowax C, C Light; Forbest MF II, MW 23, PAM; Glyconol; Glycowax S 932

Kessco 653; Lanbritol Wax N21; Lanette Wax B.P., CAT, SX, SXBP; Lanfrax 1776; Lipowax D, NI, P; Multiwax 180-M, HS, ML-445, W-445, W-835, X-145A; Paracol 505A, 505G; Paraffin Wax Emulsifier CB0674, CB0680; Petrolite C-36, C-400, C-700, C-1035, C-7500, C-8500, C-9500, CA-11, WB-5, WB-7, WB-11, WB-16, WB-17; Petronauba C; Piccomer 100, 110, 120, 150; Polawax A 31 NF, GP200, NF; Polymel #7; Polywax E-2020; Rewowax CL 1263; Rilanit MW; Rocsol B, C, COB; Schercemol CP, CS

Shellwax; Sicolub LB 1, LB 2, LBK, LK 3, OA 2, OA 4, OP, TDS; Silicone M407; Softisan 133, 134, 138, 142, 154; Spermafol 5200; Spermwax; Standamul G-32/36; Starfol Wax CG; Starwax 100; Sunolite 100, 127, 160, 240, 666; Ultraflex; Vanwax H, H Special, OZ; Wax OTO

WHIPPING AGENTS

Aldo PGHMS; PMS-33; Radiamuls 125, 135, 137, 145, 147, 155, 157, 345

WHITENING AGENTS/OPTICAL BRIGHTENERS

Eastobrite OB-1; Eccobrite RB; Ecco Polyester Optical 525; Ecco White FW-5, OP; Formula #633, #733; Hostalux; Ketjenflex 9; Lenkanol D-48; MBEO 1.8 Adduct; MDAC; Spectratech CM 11698; Tinopal 5BM-GX; Tinopal AMS, AMS Pure, AMS-GX, BLS-, Tinopal CBS-X, LPW, RA-16, RBS 200%, RBS WP-S; Uvitex OB

Chemical Manufacturers' Directory

Aarhus Oliefabrik A/S, DK-1800, Aarhus C. Denmark
Skyllex

Aceto Chemical Co., Inc., 126-02 Northern Blvd., Flushing, NY 11368 (Tel 718-898-2300)
Acetoquat; Anti UVA

ACL Inc., 1960 E. Devon Ave., Elk Grove Village, IL 60007 (Tel 312-981-9212; 800-782-8420; FAX 312-981-9278)
Staticide®

Advanced Refractory Technologies, Inc., 699 Hertel Ave., Buffalo, NY 14207 (Tel 716-875-4091; FAX 716-875-0106)
AlNel; ZN-; ZS-

Air Products and Chemicals, Inc., Polyurethane Chemicals Div., 7201 Hamilton Blvd., Allentown, PA 18195-1501 (Tel 800-345-3148; FAX 215-481-5900)
Air Products Nederland B.V., Herculesplein 359, PO Box 85075, NL 3508 AB Utrecht, The Netherlands (Tel 31-30-511828)
Air Products Pacific, Inc., Sakurabashi Yachiyo Bldg., 5-6, Umeda 2-Chome, Kita-Ku, Osaka 530 Japan (FAX 6-341-8680)
Ancamine®; Ancor; Dabco®; MBEO; OW-1; Polycat®; Surfynol®; XCE-89

Air Products, Valchem Div., 403 Carline Rd., Langley, SC 29834 (Tel 803-593-4461)
Valdet

Ajinomoto USA, Inc., Glen Pointe Centre West, 500 Frank West Burr Blvd., Teaneck, NJ 07666 (Tel 201-488-1212; FAX 201-488-6472)
Ajinomoto, Inc., 5-8 Kyobashi 1-chome, Chuo-ku, Tokyo 104, Japan
Acylglutamate; Ajicoat; Ajidew; Amihope; Amiter; CAE; Prodew; Pyroter

Akron Chemical Co. See Akrochem

Akrochem, 255 Fountain St., Akron, OH 44304 (Tel 216-535-2108)
Accelerator; Activator; Akro-Gel; Akrochem®; Akrofax; Akroform®; Akrolease®; Akrosperse®; Akrowax; Antioxidant; BBTS; BIBBS; Bismet; CBTS; Cure-Rite; DPTT; Elastomag®; MBT; MBTS; Novor; OBTS; OMTS; PA-; Plastomag®; Petro-Rez; Proaid; Retarder; RR Zinc Oxide; Silicone Emulsion, Fluid; TM/ETD; TMTM; Zinc Oxide; ZMBT

Akzo Chemicals Inc., 300 S. Riverside Plaza, Chicago, IL 60606 (Tel 312-708-7500; FAX 312-708-7680)
Akzo Chemicals S.A., 13, Ave. Marnix, 1050 Bruxelles, Belgium (Tel 02-5180411)
Akzo Chemicals BV, Stationsstraat 48, PO Box 247, 2800 AE Amersfoort, Netherlands (Tel 033-643911)
Akzo Chemicals Ltd., 1-5 Queens Rd., Hersham, Walton-on-Thames, Surrey KT12 5NL, UK (Tel 0932-247891; FAX 0932-231204)
Akzo Chemicals GmbH, Phillippstrasse 27, PO Box 100132, 5160 Dueren, Germany
Akzo Chemicals Ltd., PO Box 80, Parramatta N.S.W. 2150, Australia
Akzo Chemicals Ltd., 100 University Ave., Suite 906, Toronto, Ontario MSJ IV6, Canada
Akzo Japan Ltd., Godo Kaikan Bldg., 3-27 Kioi-cho, Chiyoda-Ku, Tokyo 102, Japan
Akzo Chemie Italia SpA, Via Vismara, 20020 Arese, Milano, Italy
Armac®; Armeen®; Armid®; Armocure®; Armoflo®; Armogard®; Armohib®; Armoslip®; Armosoft®; Armostat®; Armotan®; Arneel®; Aromox®; Arquad®;

Cadet®; Cadox®; Cyclonox®; Dapral®; Dissolvine®; Duomac®; Duomeen®; Duoquad®; Dynakoll; Elfacos; Elfan®; Epilink®; Estabex®; Estaflex®; Ethoduomeen®; Ethoduoquad®; Ethofat®; Ethomeen®; Ethomid®; Ethoquad®; grapHsize®; Intercide®; Interstab®; Interwax; Interwet®; Ketjenblack®; Ketjenflex®; Ketjensil®; Lauridit®; Laurox®; Nouryflex; Par®; Perkadox®; Perchem®; Plastoflex®; Siccatol®; Stanclere®; Trigonal®; Trigonox®; Victawet®; Viscon; Vulcabond®

Albright & Wilson Ltd., A Tenneco Co., PO Box 3, 210-222 Hagley Rd. West, Oldbury, Warley, West Midlands B68 0NN UK (Tel 021-429-4942; FAX 021-420-5151

Albright & Wilson Inc., PO Box 26229, Richmond, VA 23260 (Tel 804-550-4300; FAX 804-550-4385)

Albright & Wilson Americas, 2 Gibbs Rd., Islington, Ontario M9B 1R1 Canada (Tel 416-234-7000)

Albright & Wilson (Australia) Ltd., 610 St. Kilda Rd., PO Box 4544, Melbourne 3001, Australia

Marchon Espanola SA, Carretera Montblanc Km 2, 4, Alcover (Tarragona), Spain

Marchon France SA, BP 19, F-55300, St. Mihiel, France

Marchon Italiana SpA, Casella Postale No. 30, 1-46043, Castiglione delle Stiviere, Italy

Albrite®; Amgard®; Antiblaze®; Eltesol®; Empicol®; Empicryl®; Empigen®; Empilan; Empimin®; Empiphos; Empiwax; Gardilene; Gardinol; Gardiquat; Gardisperse; Laurex®; Lorol; Marchon®; Nansa®

Alcan Chemicals, Div. of Alcan Aluminum Corp., 3690 Orange Place, Suite 400, Cleveland, OH 44122-4438 (Tel 216-765-2550; 800-321-3864; FAX 216-765-2570)

British Alcan Aluminium PLC, Chalfont Park, Gerrards Cross, Buckinghamshire SL9 0QB, UK (Tel 0753-887373)

Flamtard

Alco Chemical, Div. of National Starch and Chemical Co., 909 Mueller Dr., PO Box 5401, Chattanooga, TN 37406 (Tel 615-629-1405; FAX 615-698-8723)

Alcogum; Alcosperse; Aquatreat

Alcolac, Inc. See Rhone-Poulenc Surfactant and Specialty Div.

Alconox Inc., 215 Park Ave. So., New York, NY 10003 (Tel 212-473-1300; FAX 212-353-1342)

Alconox®; Alcotabs®; Terg-A-Zyme®

Alkaril Chemicals Inc. See Rhone-Poulenc Surfactant and Specialty Div.

Allied-Signal, Inc., PO Box 2332R, Columbia Rd. and Park Ave., Morristown, NJ 07960 (Tel 201-455-2000; FAX 201-445-4807)

Allied Corporation International NV-SA, Haasrode Research Park, B-3030 Heverlee, Belgium

Allied Corporation International NV-SA, International House, Bickenhill Lane, Birmingham B37 7HQ, UK

Allied Chemical International Corporation, P.O. Box 99067, Tsimshatsui Post Office, Hong Kong

A-C®; ACuflow; Nadone®; Naxol®

Alox Corp., 3943 Buffalo Ave., PO Box 517, Niagara Falls, NY 14302 (Tel 716-282-1295)

Alox®; Aqualox®; Olicat

Alphaflex Industries, Inc., PO Box 501049, 6805 Hillsdale Court, Indianapolis, IN 48250-1049

Alphaflex

Amax Mineral Sales, PO Box 1700, Greenwich, CT 06836 (Tel 203-629-6525)

Climax Molybdenum Co., Ltd., 29 Gresham St., London EC2V 7DA, UK

Climax Molybdenum Asia, Ltd., Akasaka Twin Towert, Main Bldg., 1-22 Akasaka, 2-chome, Minato-ku, Tokyo 107, Japan
Molysulfide Susp.

Amerchol Corp., Unit of CPC International Inc., PO Box 4051, 136 Talmadge Road, Edison, NJ 08818 (Tel 201-248-6000; FAX 201-287-4186)
Amerchol Europe, Havenstraat 84, B-1800, Vilvoorde, Belgium
Amerchol, D.F. Anstead, Ltd., Victoria House, Radford Way, Billericay, Essex, CM-12-ODE, UK
Amerchol, Ikeda Corporation, New Tokyo Bldg., No. 3-1, Marunouchi 3-Chome, Chiyoda-Ku, Tokyo 100, Japan
Acetulan®; Alcolan®; Amerchol®; Amerlate®; Ameroxol®; Amerscreen®; Amersette®; Amino Silk; BioCare; Bumyr; Ceralan®; Cetal; Collagen Hydrolyzate; Glucam®; Glucamate®; Glucate®; Isopropylan; Kera-Tein; Lanocerin®; Lanogel®; Lanogene®; Liquilan; May-Tein; Modulan®; OHlan®; Polylan®; Promulgen®; Pro-myr; Propal; Pro-Tein; Quat-Pro; Quatrisoft; Solulan®; Stearal

American Chemical Service, PO Box 190, Griffith, IN 46319 (Tel 219-924-4370)
Epoxol

American Colloid Co., 1500 W. Shure Dr., Arlington Hts., IL 60004-1434 (Tel 312-392-4600; FAX 312-506-6199)
Magnabrite; Polargel; Polarite; Volclay

American Cyanamid Co., Chemicals Group, One Cyanamid Plaza, Wayne, NJ 07470 (Tel 201-831-4111; FAX 201-839-8847)
Polymer Additives Dept., One Cyanamid Plaza, Wayne, NJ 07470 (Tel 201-831-2000)
Cyanamid B.V., Postbus 1523, 3000 BM, Rotterdam, The Netherlands
Cyanamid India Ltd., Nyloc House, 254-D2 Dr. Annie Besant Rd., Bombay 400 025 India
Cyanamid Quimica do Brasil Ltda., Av. Imperatriz Leopoldina, 86, Sao Paulo, Brazil
Cyanamid Taiwan Corp., 8/F Union Commercial Bldg., 137, Nanking E. Rd., Sec. 2, Taipei, Taiwan, R.O.C.
Aerosol®; Antioxidant; Complemix®; Cyanamer; Cyanox®; Cyasorb®; Cyastab®; Cyastat®; Cymel; Spectra-Sorb®

American Cyanamid Co./Agricultural Div., One Cyanamid Plaza, Wayne, NJ 07470
Aastar; Abate®; Amdro®; Arsenal®; Avenge®; Cygon®; Cyprex®; Cythion®; Prowl®; Scepter®; Thimet®

American Ingredients Co., 3947 Broadway, Kansas City, MO 64111 (Tel 800-821-2250)
Double Soft®; Patco®; Patcote®; Sof-Plus®; Starplex®; Uni-Malt®; Verv®

Amoco Chemicals Corp., 200 East Randolph Dr., PO Box 8640A, Chicago, IL 60680 (Tel 312-856-3200; 800-621-4590)
Amoco Chemical Europe S.A., 15, Rue Rothschild, CH-1211 Geneva 21, Switzerland
Amoco Performance Products, Japan Ltd., 10th Floor, Tonichi Building, 2-31 Roppongi 6-Chome, Minato Ku, Tokyo 106, Japan
Amoco®

Anchor Chemical Ltd., Manchester M11 4SR UK (Tel 061-223-2461)
Acclerator; Activator; Anchoid; Anchor; Interlube; Rapidblend; Stiffener

The Andersons, PO Box 119, Illinois Ave., Maumee, OH 43537
The Andersons/Cob Division, 1200 Dussel Dr., Maumee, OH 43537
Grit-O'Cobs®; Lite-R-Cobs®

Angus Chemical Co., 2211 Sanders Rd., Northbrook, IL 60062 (Tel 312-498-6700; 800-323-6209)
Angus Chemie GmbH, Huyssenallee 5, 4300 Essen 1, Germany (Tel 0201-233531)
Angus Chemie GmbH, Le Bonaparte, Centre d'Affaires Paris-Nord, F-93153 Le Blanc Mesnil, Paris, France

Angus Chemie GmbH, 19, Moorgate St., Rotherham, S60 2DA England
Angus Chemical Co., 101 Cecil St., #14-07 Tong Eng Bldg., Singapore 0106
AB; AEPD®; Alkaterge®; Amical®; Amine; Amine PMT®; AMP; AMPD; Bioban®; Bonopol; Comsol; Fuelsaver; Hexetidine; NB; NE; NEPD; NiPar; NM; NMP; Oxaban®; S.S.T.®; Tris Amino®; Tris Nitro®; Zoldine®

Aqualon Co., A Hercules Inc. Co., 1313 North Market St., Wilmington, DE 19894 (Tel 302-594-7600; 800-334-8426; FAX 302-946-2296)
Aqualon Canada Inc., 5407 Eglinton Ave. West, Suite 103, Etobicoke, Ontario M9C 5K6 Canada (Tel 1-416-620-5400)
Aqualon (UK) Ltd., Genesis Centre, Garrett Field, Birchwood, Warrington, Cheshire WA3 7BH, UK (Tel 44-925-830077)
Aqualon France, 44, Ave. de Chatou, F-92508 Rueil Malmaison Cedex, France (Tel 1-4751-2919)
Aqualon GmbH, PO Box 130125, Paul Thomas Strasse 58, D-4000 Düsseldorf 13, Germany (Tel 49-211-7491-0)
Ambergum®; Aqualon®; Aquasorb; Benecel; Blanose; Combizell; Culminal; Klucel®; Natrosol®; Natrovis®; Supercol®

Aquaness Chemicals, 3920 Essex Lane, Houston, TX 77227 (Tel 713-599-7400; FAX 713-599-7460)
Ardril

Aquatec Quimica S/A, Rua Sampaio Viana, 425, CX. Postal 4885, 04004 Sao Paulo Sp., Brazil
Antarol; Axel; Biopen; Cetax; Dispersal; Drewfax; Drewpon; Proplast; Synoquart; Synotol

Arco Chemical, Headquarters, Research & Engineering Center, 3801 West Chester Pike, Newtown Sq., PA 19073 (215-359-2000; 800-321-7000)
Arco Chemical Asia/Pacific Ltd., Toranomon 37 Mori Bldg, 5th Flr, 5-1 Toranomon 3-chome, Minato-Ku, Tokyo 105, Japan
Arco Chemical Euorpe, Inc., Bridge Ave., Maidenhead SL6 1YP, UK (Telex 847436)
Arco Chemical Canada, Inc., 100 Consilium Pl., Suite 306, Scarborough, Ontario, M1H 3E3 Canada
Arco Chemical Pan American, Inc., Paseo de la Reforma, 390 Decimo Piso, 06600 Mexico City, Mexico
Arconate®; Arcosolv®; Tebol; T-Hydro®

Argus Chemical Corp. See Witco Chemical Corp.

Arizona Chemical Co., Div. of International Paper Co., 1001 E. Business 98, Panama City, FL 32401 (Tel 904-785-6700)
Zonarez; Zonester®

Arjay Inc., PO Box 45045, Houston, TX 77045 (Tel 713-599-7400; FAX 713-599-7460)
Aminox; Aramide®; Arcor; Arklear; Amnate; Arphos; Arquest; Arsil; Arsperse; Arsul; Arzoline; Calnox

Arol Chemical Products Co., 649 Ferry St., Newark, NJ 07105 (Tel 201-344-1510)
Detergent CR; Kilfoam

Asarco Tech., 3422 S. 700 West, Salt Lake City, UT 84119 (Tel 801-262-2459)
AZO; Azodox

Astro Chemical Co., Inc., 1205 Godfrey Lane, Schenectady, NY 12309
Barco Bond

Atlas Refinery, Inc., 142 Lockwood St., Newark, NJ 07105
Eureka

Atochem, Groupe elf aquitaine, 4, cours Michelet, La Défense 10-Cedex 42, 92091 Paris La Défense, France (Tel 1-49008080)
Atochem Inc./Polymers Div., 266 Harristown Rd., POB 607, Glen Rock, NJ 07452

(Tel 201-447-3300; FAX 201-447-6797)
Rilsan Corp./Subsid. of Atochem, France, 139 Harristown Rd., Glen Rock, NJ 07452
Sorgen; Voltalef

Atochem North America Inc., Organic Peroxides Div., 1740 Military Rd., PO Box 1048, Buffalo, NY 14240 (Tel 716-877-1740; 800-558-5575; FAX 716-877-1541)
Atochem GmbH, Postfach 1354, 887 Gunzburg/Donau, Germany (Tel 0-8221-98-0)
Atochem Yoshitomi, Ltd., 6-9 Hiranomachi, 2-Chome, Chuo-Ku, Osaka 541, Japan (Tel 201-1161)
Acetyl Peroxide-IB25; Alperox; Decanox; Luchem; Lucidol; Luperco; Luperfoam; Luperox; Lupersol

Bareco. See Petrolite Corp.

Otto Bärlocher GmbH, 8000 München 50, Postfach 500108, Germany (Tel 089-1488-0)
Distributed by R.T. Vanderbilt Co., Inc. (Tel 203-853-1400)
Baerostab®

Baroid Corp., PO Box 1675, 3000 Southwest Freeway, Houston, TX 77251 (Tel 713-987-4000)
Carbonox; CC-

BASF Corp., 100 Cherry Hill Rd., Parsippany, NJ 07054 (Tel 201-316-3000; FAX 201-402-1832)
BASF Canada Ltd., PO Box 430, Montreal, Quebec H4L 4V8, Canada
BASF (UK) Ltd., PO Box 4, Earl Rd., Cheadle Hulme, Cheadle, Cheshire 5K8 60QG, UK
BASF Belgium S.A., avenue Hamoir-Iaan 14, B-1180 Bruxelles/Brussel, Belgium
BASF AG, ESA/WA-H 201, D-6700 Ludwigshafen, West Germany
BASF Espanola S.A., Apartado 762, Barcelona 8, Spain
BASF S.A., Compagnie Francaise, MC-NT, 140, Rue Jules Guesde, 92303 Levallois-Perret, France
BASF India, Ltd., Maybaker House, S.K. Ahire Marg., PO Box 19108, Bombay 400 025 India
BASF Japan Ltd., C.P.O. Box 1757, Toyko 100-91, Japan
BASF; Basopon; Cremophor®; Degressal; Emulan®; Glissofluid®; Glissolube®; Glossoviscal; Icodimeen; Icomeen; Icomid; Iconol; Katioran; Kerobit®; Keroflux®; Keromet; Kerostat®; Kieralon; Klearfac; Korantin®; Leophen; Lubrimet®; Lusynton; Lutensit; Lutensol®; Lutrol; Luviflex®; Luvitol; Luviquat®; Luviset; Luviskol; Nekanil; OP-2000; Pluracol®; Pluradot; Plurafac®; Pluraflo; Pluriol; Pluronic®; Primasol; Protectol®; Quadrol; Sepacid®; Sepacorr®; Sicolub; Sokalan®; Tamol®; Tetronic®; Trilon®; Uniperol; Uvinul®; Vultamol®; Wettol®

Bayer AG, Sitz der Gesellschaft, Leverkusen, Eintragung, Amtsgericht, Leverkusen HRB 1122, Germany
Bayer UK Ltd., Bayer House, Strawberry Hill, Newbury, Berkshire RG13 1JA, UK
Acralen; Adimoll; Antistatic Plasticizer; Antracol®; Arubren; Baycor®; Bayer; Bayfidan®; Baygon®; Bayleton®; Bayluscide®; Bayrusil®; Baytan®; Baytex®; Baythion®; Baythroid®; Bladafum®; Bonding Agent; Coagulant; Cohedur; Croneton®; Cropotex®; Curaterr®; Cyfluthrin; Dedevap®; Desmodur®; Desmorapid; Dipterex®; Disyston®; Dyrene®; Etrofolan®; Euparen®; Folidol®; Folimat®; Folithion®; Fruvit®; Goltix®; Gusathion®; Hinosan®; Lebaycid®; Lekutherm; Levaform; Mafu®; Matacil®; MEB; Mesamoll; Mesurol®; Metasystox®; Monceren®; Morestan®; Nemacur®; Neo-Voronit®; Oftanol®; Peropal®; Porofor; Pyratex; Racumin®; Sencor®; Sibutol®; Stabilizer; Surcopur®; Tamaron®; Terracur®; Tokuthion®; Tribunil®; Ultramoll; Unden®; Unimoll; Ustilan®; UV-Absorber; Volaton®; Vulcuren; Vulkacit; Vulkadur; Vulkanol; Vulkasil

Berol Nobel AB, Nobel Industries Sweden, S-444 85 Stenungsund, Sweden
Berol Nobel, Nobel Industries Sweden, Merritt 8 Corporate Park, 99 Hawley Lane, Stratford, CT 06497 (Tel 203-378-0500; FAX 203-378-5960)
Berol Nobel, Rue Gachard 88, Rte 9, B-1050 Bruxelles, Belgium
Amine; Berol; Diamine; Liladox; Lilamac; Lilamin; Lilaminox; Lilamuls; Querton; Sulfamin; Sulfosoft

Borden Chemical Div./Borden Inc., 180 E. Broad St., Columbus, OH 43215 (Tel 614-225-4000)
Polyco

Cabot Corp., PO Box 188, Tuscola, IL 61953 (Tel 217-253-3370)
Cab-O-Sil®; Cab-O-Sperse®

Calgon Carbon Corp., PO Box 717, Pittsburgh, PA 15230-0717 (Tel 412-787-6700)
Represented by: Goldschmidt Chemicals
Merquat®; Tektamer®

Canada Packers Ltd./Chemical Div., 5100 Timberlea Blvd., Mississauga, Ontario L4W 2S5, Canada (Tel 416-624-7000)
Canada Packers Inc./Edible Oils Div., 2200 St. Clair Ave. West, Toronto, Ontario M6N 1K4, Canada
Gelrite; SoyaLec; Sterling

Cancarb Ltd., 1702 Brier Park Cresc. N.W., PO Box 310, Medicine Hat, Alberta T1A 7G1, Canada (Tel 403-527-1121; FAX 403-529-6093)
Distributed by: R.T. Vanderbilt, 30 Winfield St., Norwalk, CT 06855
Thermax®

Capital City Products Co., PO Box 1759, Janesville, WI 53547 (Tel 608-752-9007; FAX 608-755-9842)
Capital City Products Co., POB 569, Columbus, OH 43216-0569 (Tel 614-299-3131; 800-848-1340; FAX 614-299-8279)
Accobetaine; Accomeen; Accomid; Acconon; Accoquat; Accosoft; Accosperse; Ampho; Capital®; Caplube; Capmul®; Caprol®; Capstat; Captex®; Hydrokote®; Pureco®; Sanac; Stabland®; Sterotex®

The Carborundum Co., Electronic Ceramics Div., PO Box 664, Niagara Falls, NY 14302 (Tel 716-278-3937; FAX 716-278-2900)
Combat®

Cardolite Corp., 500 Doremus Ave., Newark, NJ 07105 (Tel 201-344-5015; FAX 201-344-1197)
Cardolite®

R.H. Carlson Co., 41 Chestnut St., Greenwich, CT 06830 (Tel 203-531-5500)
C-Cure; N-30; Norcure

CasChem Inc., 40 Avenue A, Bayonne, NJ 07002 (Tel 201-858-7900)
AA; Castorwax®; Ches®; Cosmetol; Crystal®; Diamond Quality; ES-1239; Flexricin®; Ivex®; Naturechem; P-10 Acid; Paricin®; Solarchem®; Solricin®; Surfactol®; Surflo®; Waxenol®; Wickenol®

Catawba-Charlab, Inc., 5046 Pineville Rd., PO Box 240497, Charlotte, NC 28224 (Tel 704-523-4242)
Charlab

Ceca SA, 22 Place des Vosges, Cedex 54 92062, LaDefense 5 Paris, France
Amphoram; Dinoram; Dinoramac; Dinoramox; Fluidiram; Inipol; Noram; Noramium; Norust; Noxamium

Central Soya Co. of America, 1300 Ft. Wayne National Bank Bldg., POB 1400, Fort Wayne, IN 46801 (Tel 219-425-5100; 800-348-0960; FAX 219-425-5485)
Central Soya, PO Box 5063, 3008 AB Rotterdam, Netherlands (Tel 31-10-42-39-600; FAX 31-10-42-30-897)

Actiflo®; Centrocap; Centrol®; Centrolene®; Centrolex®; Centromix®; Centrophase®; Centrophil®

Chemax, Inc., PO Box 6067, Highway 25 South, Greenville, SC 29606 (Tel 803-277-7000; 800-334-6234; FAX 803-277-7807)

Chemal; Chemax; Chemazine; Chemeen; Chemfac; Chemquat; Chemstat; Sorbax

Chemetall GmbH, Member of The Metallgesellschaft Group, Reuterweg 14, Postfach 101501, D-6000 Frankfurt am Main 1, Germany (Tel 069-159-0)

The Ore & Chemical Corp., 520 Madison Ave., New York, NY 10022 (Tel 212-715-5262; FAX 212-715-5291)

Megum®; Mould Release Agent N 32; Naftocit®; Naftolen®; Naftonox®; Naftopast®; Naftozin®

Chemfax Inc., 3 Rivers Road, Gulfport, MS 39502 (Tel 601-863-6511)

Chemprene; Stygene

Chem-Trend Inc., 1445 W. McPherson Park Dr., PO Box 860, Howell, MI 48844-0860 (Tel 517-546-4520)

CT-; HF-; Mono-Coat®; PL-; Safety-Lube®

Chem-Y GmbH, Kupferstrasse 1, D4240 Emmerich, West Germany

Akypo; Akypogene; Akypomine; Akypopress; Akypoquat; Akyporox; CHPTA; Defoamer; Ogtac; Salfax; Surfageen

Chevron Chem. Co., Agricultural Chemicals Div., 6750 Poplar, Suite 300, Memphis, TN 38138 (Tel 901-754-3400)

Ortho Danitol®; Ortho Orthocide®; Ortho Prunit; Ortho Select; Ortho Spotless; Ortho Sumagic

Ciba-Geigy Corp., 444 Saw Mill River Rd., Ardsley, NY 10502 (914-478-3131; 800-431-1874)

Ciba-Geigy Corp., Additives Dept., Three Skyline Dr., Hawthorne, NY 10532 (914-347-4700; 800-431-1900)

Ciba-Geigy Corp., CH-4002, Basle, Switzerland

Ciba-Geigy PLC, 30 Buckingham Gate, London SW1E 6LH, UK

Ciba-Geigy Dyestuffs & Chemicals, Ashton New Road, Clayton, Manchester M11 4AR, UK

Alrosol; Alrosperse; Alrowet®; Amine; Belclene®; Belcor; Belsperse; Bina; Chel; Chimassorb®; Cibaphasol®; Fungitex; Irgalube®; Irganox®; Irgasan; Irgasol®; Irgastab®; Levelene; Lostat; Nonisol; Quaternary; Reocor; Reomet®; Reserve Salt Flake; Sarkosyl®; Sequestrene®; Sorbit; Tinegal®; Tinopal®; Tinuvin®; Uvitex®

W.A. Cleary Chemical Corp., 1049 Somerset St., Somerset, NJ 08873 (Tel 908-247-8000)

Clearate

Climax Performance Materials Corp., 7666 W. 63rd St., Summit, IL 60501 (Tel 708-458-8450; FAX 708-458-0286)

Actrabase; Actracor; Actrafoam; Actrafos; Actralube; Actramide; Actrasol; Actrol; ADM; Kricinol; Maxahibit; Molyhibit; POM; Sorcinol; Surpasol

Clough Chemical Co., Ltd., 178 St. Pierre, St.-Jean, Quebec J3B 7B5, Canada (Tel 514-346-6848; FAX 514-346-7263)

Amyx; Castor Oil Sulfated; Dousoft; Nofoam; Nutrapon; Nutrol; Pearlex; Pentaquest; Questal; Questric Acid; Stamid; Sulfonated Syn. Sperm Oil

Colloids Inc. See Rhone-Poulenc

Conap Inc., 1405 Buffalo St., Olean, NY 14760 (Tel 716-372-9650)

Conacure®

Consos, Inc., PO Box 34186, Charlotte, NC 28234 (Tel 704-596-2813)

Consamine; Consoft; Consonyl; Consos; Consoscour; Consostat; Consotard; Defoamer

Continental Chemical Co., 270 Clifton Blvd., Clifton, NJ 07015 (Tel 201-472-5000; FAX 201-472-5221)
Antifoam; Conco®; Concofac; Concosoft; Condensate; STXS; SXS

Cri-Tech, Inc., 150 Recreation Park Dr., Hingham, MA 02043 (Tel 617-740-2020; FAX 617-740-2289)
Cri-Spersion

Croda Chemicals Ltd., Cowick Hall, Snaith Goole, North Humberside DN149AA, UK (Tel 0405-8605551; FAX 0405-860205)
Cithrol; Corona; Coronet; Crestalans; Crester; Crillon; Croda; Crodamet; Crodamide; Crodamine; Crodapur; Crodasinic; Crodax; Crodazoline; Croderol; Crodet; Crodex; Crodinhib; Croduret; Cromul; Cropepsol; Cropeptone; Crosterene; Dastar; DIBA; Etocas; Fluilanol; Glycerox; Hartolite; Hydromond; Incronol; Incropearl; Incrosorb; Iscolan; Isocreme; Lanethyl; Lanex; Lanolic Acid; Lanpol; Lipocol; Pentalan; Prochol; Produkt; Rocsol; Silk Protein; Sorba; Sorbeth; Spreading Agent; Squalane; Superfine Lanolin; Syncrolube; TBC; Tricap; Vykacet; Vykamol; White Swan; Yeoman

Croda, Inc., Div. of Croda International PLC, 180 Northfield Ave., Edison, NJ 08837 (Tel 201-417-0800)
Croda Canada Ltd., 78 Tisdale Ave., Toronto, Canada M4A 1Y7 (Tel 416-751-3571; FAX 416-751-9611)
Croda Italiana Srl, Via Grocco, N917 27036, Mortara (PV), Italay
Croda Chemicals Group Pty. Ltd., PO Box 1012, Richmond, North Victoria 3121, Australia
Croda Japan KK, Aceman Bldg. 5F 3 7, Tokuicho 1-Chome Highashi-ku, Osaka 540, Japan (Tel 6-942-1791)
Croda do Brazil Ltda., Rua Croda 230 Distrito Industrial, CEP 13.053 Campinas/SP-C.P. 1098, Brazil
Acylan; Aminofoam; Amino Gluten; Clearcol; Collasol; Cosmowax; Cremba; Crill; Crillet; Crodacel; Crodacid; Crodacol; Crodafos; Crodalan; Crodamol; Crodapearl; Crodesta; Crodyne; Crolastin; Cromoist; Cronectin; Croquat; Crosilk; Crosilkquat; Crossential; Crotein; Elastosol; Fluilan; Hartolan; Hydrolactin; Hydrosoy; Hydro-triticum; Incrocas; Incromate; Incromectant; Incromide; Incromine; Incronam; Incropol; Incroquat; Incrosoft; Incrosul; Kerasol; Lanexol; Lanpolamide; Liquid Absorption Base; Novol; Polawax®; Polychol; Procetyl; Promyristyl; Prostearyl; Provol; Reticusol; Satexlan; Satulan; Skliro; Solan; Solangel; Super Corona; Super Hartolan; Super Re-fined; Super Solan; Super Solangel; Super Sterol Ester; Syncrowax; Volpo;

Crompton & Knowles Corp., PO Box 33188, Charlotte, NC 28233 (Tel 704-372-5890; FAX 704-372-1522)
Cenegen; Cenekol®; Defoamer; Defoamer/Drainage Aid; Intralan®; Intraphasol; In-tratex®; Intravon®; Resogen®; Retarder A

Crosfield Chemicals, Inc., 101 Ingalls Ave., Joliet, IL 60435 (Tel 815-727-3651; FAX 815-727-5312)
Crystamet; Drymet

Crowley Chemical Co., 261 Madison Ave., New York, NY 10016 (Tel 212-682-1200; 800-424-9300)
Viplex

Crucible Chemical Co., POB 6786, Donaldson Center, Greenville, SC 29606 (Tel 803-277-1284)
Cru Fluid; Cru Release; Cru Thix; Foamkill

Custom Fibers Int'l., 28130 Ave. Crocker, #311, Valencia, CA 91355 (Tel 805-295-0990; 800-321-5324; FAX 805-295-5148)
CF-

Dai-ichi Kogyo Seiyaku Co., Ltd., Miki Building, 3-12-1, Nihombashi, Chuo-ku,

Tokyo, 103 Japan
Amiladin; Dianol; DK Ester; Lamigen; Monogen; Noigen

Daishowa Chemicals Inc., 81 Holly Hill Lane, Greenwich, CT 06830 (Tel 203-625-0701)
Amberlig; Dynasperse; Kelig; Lignosol; Maracarb; Maracell; Marasperse; Norlig; Peltex

Davison. See W.R. Grace & Co.

Dead Sea Bromine Co. Ltd., Bromine Compounds Ltd., Makleff House, POB 180, Beer-Sheva 84101, Israel (Tel 00972-57-667222)
Biobrom; FR-; Halobrom

Degussa Corp., Aerosil and Imported Pigment Prods. Div., 65 Challenger Rd., Ridgefield Park, NJ 07660 (Tel 201-641-6100; FAX 201-807-3200)
Aerosil®; Aluminum Oxide C; Degacure®; Elcema®; Sipernat®; TS

DeSoto, Inc., 3950 Fossil Creek Blvd., Suite 200, Ft. Worth, TX 76137 (817-847-4400; 800-433-2149; FAX 817-847-4444)
Corporate Headquarters, 1700 S. Mt. Prospect Rd., Box 5030, Des Plaines, IL 60017
Arnox; DeSomeen; DeSonic®; DeSophos; DeSotan; Morwet; Petro®; Petroscale; Petrosolve®

Dexter Chemical Corp., 845 Edgewater Rd., Bronx, NY 10474 (Tel 212-542-7700; FAX 212-991-7684)
Atolene; Barisol; Dextrol; Strodex

Disco, Inc., Associated Polymer Products Div., 1010 Greenwood Lake Tpk., Ringwood, NJ 07456 (Tel 201-728-7731)
Moldbrite; Poly Pross; RT-3; ZO-9

D.J. Enterprises, PO Box 31366, Cleveland, OH 44131
Sillum

Dover Chemical Corp., Subsid. of ICC Industries, Inc., Davis at West 15th St., PO Box 40, Dover, OH 44622 (Tel 216-343-7711; FAX 216-364-1579)
AA-81C; Chlorez; DD-

Dow Chemical U.S.A., 2020 W.H. Dow Center, Midland, MI 48674 (Tel 517-636-1000; 800-441-4DOW)
Dow Plastics, 2040 W.H. Dow Center, Midland, MI 48674
Dow Chemical Canada Inc., 1086 Modeland Rd., PO Box 1012, Sarnia, Ontario, Canada N7T 7K7 (Tel 519-339-3131; 800-363-6250)
Dow Chemical Co. Ltd., Stana Place, Fairfield Ave., Staines, Middlesex TW18 4SX, UK
Dow Chemical Europe S.A., Bachtobelstrasse 3, CH-8810, Horgen, Switzerland (Tel 41-1-728-2111)
Dow Chemical Pacific Ltd., 39th Fl., Sun Hung Kai Centre, 30 Harbour Rd., Wanchai, PO Box 711, Hong Kong
Dow Quimica S.A., Sao Paulo, Brazil
Aerothene; Chlorothene; D.E.H.; Dow; Dowanol®; Dowclene®; Dowfax; Dowfrost®; Dowicide®; Dowicil®; Dowper®; FR-; Hi-Tri®; Isonate®; Methocel®; Neu-Tri®; Parabis; Pentachlorophenol DP-2; Polyglycol; Prelete®; Primacor; Striptron; Tactix; Versene®; Versenol; XNS

Dow Corning Corp., Box 0994, Midland, MI 48686-0994 (Tel 517-496-4000)
Dow Corning Ltd., Reading Bridge House, Reading, Berkshire RG1 8PW, UK
Dow Corning®; Molykote®

Dow Corning/Wickhen, East Coast Service Center, Bracken Rd., PO Box 384, Montgomery, NY 12549 (Tel 914-457-9631)
Microduct®; Polytrap®

Drew Chemical Corp., Drew Ind. Div., Ashland Chem. Co., One Drew Plaza, Boonton,

NJ 07005 (Tel 201-263-7600)
Amerfloc; Amerstat®; Amerzine; Biosperse; Drewfax; Drewfloc; Drewsperse

Duphar B.V., PO Box 900, 1380 DA Weesp, The Netherlands
Dimilin®; Dulectin; Dusoran; Tedion®

E.I. duPont de Nemours & Co., Inc., Wilmington, DE 19898 (800-441-9442)
Petrochemicals Dept., Wilmington, DE 19898 (Tel 800-231-0998; 302-999-5053)
DuPont Canada Inc., Box 660, Montreal S., P.Q. H3C 2V1 (Tel 514-861-3861)
DuPont (UK) Ltd., Wedgwood Way, Stevenage, Herts SG1 4QN, UK
DuPont (UK) Ltd./Polymer Prods. Dept., Maylands Ave., Hemel Hempstead, Herts HP2 7DP, UK
DuPont de Nemours International S.A., PO Box, CH-1211, Geneva 24, Switzerland (Tel 022-378111)
DuPont de Nemours (France) S.A., 9, Rue de Vienne, 75008-Paris, France
DuPont Far East Inc., Kowa Bldg. No. 2, 11-39 Akasaka 1-Chome, Minato-Ku, Tokyo 107, Japan (Tel 585-5511)
DuPont S.A. de C.V., Homero 206, Col. Polanco, Mesico 5, D.F. Mexico
Alkanol; Avitex; Compound 8-S; Corfree; Crylcon; DBE; Decalin; Duponol; Du Pont; Dymel®; Dytek; Elvanol; Ludox; Merpol; Nalan; Petrowet; Polysilicate; Product; Quilon; Rylex; Terathane®; Tetralin; Tyzor; Vazo; Viton; Volan; Vydax; Zelec; Zepel; Zonyl

Eagle-Picher Industries, Inc., C & Porter Streets, Joplin, MO 64802 (Tel 417-623-8000)
BSWL; EPIstatic®; EPIthal; Litharge; Sublimed Blue Lead; Super Sublimed White Lead

Eagle Zinc Co., Div. of T.L. Diamond & Co., 30 Rockefeller Plaza, New York, NY 10112 (Tel 212-582-0420; FAX 212-582-3412)
Zinc Oxide

Eastern Color and Chemical Co., 35 Livingston St., Providence, RI 02904 (Tel 401-331-9000; FAX 401-331-2155)
E.B. Golden Glitter, Neutral Glitter; Ecco; Eccobrite; Eccolene; Eccopel; Eccoro®; Ecco Spersant; Eccospin; Eccostat; Eccoterge; Eccotex; Eccowet®; Ecco White®; Polydyol; Raneoff®; Solon

Eastman Chemical Products, Inc., Subsid. of Eastman Kodak Co., PO Box 431, Kingsport, TN 37662 (Tel 615-229-2000)
Eastman Chemical International A.G., Hemel Hempstead, P.O. Box 66, Kodak House, Station Road, Herts, HP1 1JU England
Kodak Mexicana, Calzada de Tlalpan No. 2980, Mexico D.F. 04870, Mexico
Eastman Japan Ltd., Nishi-Shinbashi Mitsui Bldg., 1-24-14 Nishi-Shinbashi, Minato-Ku, Tokyo 105, Japan
Aflaban; Eastman®; Ektapro; Ektasolve®; Epolene; Kodaflex®; Myvacet®; Myvatem®; Myvatex; Myverol®; TDP; Tecquinol®; Tenox®; Texanol®; THQ

ECC International, Suite 200G, 5775 Peachtree-Dunwoody Rd. NE, Atlanta, GA 30342 (Tel 800-334-0122)
Altowhite; Hydrite

Emkay Chemical Co., 319-325 Second St., PO Box 6537, Elizabeth, NJ 07206 (Tel 908-352-7053)
Emka; Emkafix; Emkal; Emkalon; Emkane; Emkapon; Emkastat; Emkatex; Rexopene; Rexowet

Emulan Inc., 3726 Roosevelt Rd., PO Box 582, Kenosha, WI 53141 (Tel 414-654-0734); FAX 414-654-3410
Emulan®

Emulsion Systems Inc., PO Box 307, Lemont, IL 600439 (Tel 312-257-3052; 800-762-7676)
Ashland Chemicals, 2620 Royal Windsor Dr., Mississauga, Ontario L5J 4E7 Canada

Valchem (Australia) Pty. Ltd., Div. of ICI, 20 Joseph St., Blackburn, Victoria 3130 Australia
ESI-Cryl; ESI-Terge

Engelhard Minerals & Chem. Corp., 101 Wood Ave., Iselin, NJ 08830-0770 (Tel 908-321-5000)
Attacote; Attaflow; Attagel; Catalpo®; Pharmasorb; Satintone®; Sol-Speedi-Dri; Trans-link®

Ethyl Corp., 451 Florida Blvd., Baton Rouge, LA 70801 (Tel 504-388-7040; 800-535-3030; FAX 504-388-7686)
Ethyl Corp., Goldlay House, 114 Parkway, Chelmsford, Essex CM2 7PP, England (Tel 245-287-577)
Ethyl S.A., Industrial Chemical Division, 523 Avenue Louise, Boite 19, B-1050 Bruxelles, Belgium (Tel 32-2-642-4411)
Adma®; ASA; Epal®; Ethacure®; Ethanox®; EZA®; Saytex®; Tetradecene

Evans Chemetics. See W.R. Grace & Co.

Exxon Chemical Co., PO Box 2180, Houston, TX 77001
Tomah Prods., 1012 Terra Dr. (POB 388), Milton, WI 53563 (Tel 608-868-6811; FAX 608-868-6810)
Exxon Chemical Ltd., Arundel Towers, Portland Terrace, Southampton SO9 2GW, UK
Acid Thickener; Acto; Alkali Surfactant; Amphoteric; Anticake; AO; Defoamer; HCl Thickener; Tomah

Fairmount Chemical Co., Inc., 117 Blanchard St., Newark, NJ 07105 (Tel 201-344-5790; FAX 201-690-5298)
Mixxim®

The Fanning Corp., 1775 W. Diversity, Chicago, IL 60614-1009 (Tel 312-248-5700)
EmCon; Fanchem; Fancol; Fancor; Fancorp; Hydroxylan; Lan-Aqua-Sol; Lanotein; Neutral Degras; Noda Rice Wax; Sol-U-Tein; Vigilan; Wax OTO

Fermco Biochemics Inc. See Finnsugar Biochemicals, Inc.

Ferro Corp., Ferro Chemicals Group, 7050 Krick Rd., Bedford, OH 44146 (Tel 216-641-8580)
Bromoklor; Butyl Diglyme; Cata-Chek; Diglyme; Ethyl Diglyme; Ethyl Glyme; Ferro; Monoglyme; Ottalume; Ottasept®; Oxi-Chek; Plas-Chek; Permyl®; Pyro-Chek®; Tetraglyme; Therm-Chek; Triglyme; UV-Chek

Finetex Inc., 418 Falmouth Ave., Elmwood Park, NJ 07407 (Tel 201-797-4686; FAX 201-797-6558)
Represented by: Pennine Chemical Ltd., Kent Works, Thomas St., Congltͦon, Cheshire CW12 1QZ, UK
Acrilev; Aminol; Cordex; Cordon; Finazoline; Findet; Finquat; Finsoft; Finsolv®; Gemtex; Rueterg; Surfine; Tauranol

Finnsugar Biochemicals Inc., 1400 N. Meacham Rd., Schaumburg, IL 60173 (Tel 312-843-3200; FAX 312-843-3368)
Celluferm; Extractase; Fermalpha; Fermcolase®; Fermcozyme®; Fermlipase; Fermvertase; Ovazyme; Papain; Protoferm; Spezyme; Urease

Firestone Synthetic Rubber & Latex Co., PO Box 26611, Akron, OH 44319-0006 (Tel 800-282-0222)
SR; Stereon®

Floridin Co., PO Box 510, Quincy, FL 32351 (Tel 800-228-1131)
Diluex; Florco; Florex; Minugel; Refinex

FMC Corp., Chemical Products Group, 2000 Market St., Philadelphia, PA 19103 (Tel 215-299-6000)

FMC Corp., 11475 Cote de Liesse Rd., Dorval, Quebec H9P 1B3, Canada
FMC Corp., Ave. Louise 523, Box 1, B 1050 Brussels, Belgium
Avicel; FR-D; KP-; Kronitex®; TBP

Franklin Limestone Co., 612 Tenth Ave. North, Nashville, TN 37203 (Tel 615-259-4222)
Franklin

Freeman Chemical Corp., A Subsid. of Georgia Gulf Corp., PO Box 996, Pt. Washington, WI 53074-0996 (Tel 414-284-5541)
Thermacure®

GAF Corp., 1361 Alps Rd., Wayne, NJ 07470 (Tel 201-628-3000)
GAF Europe, 40 Alan Turing Rd., Surrey Research Park, Guildford, Surrey, UK
GAF Great Britain Ltd., Tilson Rd., Roundthrone, Wythenshawe, Manchester M23 9PH, UK
Butynediol; Cheelox®; Copolymer; Cutless; Diazopon; Fire Retardant; GAF Latex; Gaftex; Ganex®; Gantrez®; Glyecine; Hexol; Hy-Pel; Katapol®; Katapone®; Kemplex; Kingsolve; Knightset; Ludigol; Monopol; M-Pyrol®; Nekal®; Pawn; Peem; Pegafac; Pegameen; Peganate; Peganol; Pegeste; Peregal®; Plasdone®; Polectron®; Polyclar®; Polyplasdone®; Preventol; PVP K-; PVP/VA; 2-Pyrol®; Retarder; Soromine; Thickener; Uvi-Nox; V-Pyrol

Gattefosse Etablissements, 36 Chemin de Genas, 69800 Saint Priest, France
Gattefosse Corp., 3 Westchester Plaza, Elmsford, NY 10523 (Tel 914-592-8844)
Represented by: Alfa Chemicals Ltd., Broadway House, 7-9 Shute End, Workingham, Berkshire RG11 1BH, UK
DPPG; Gelot 64®; Hydrine; M.O.D.; Monosteol; Monthyle; Pancogene; Polystate; Precirol; Sedefos; Tefose®

GE. See General Electric Co.

General Electric Co.
General Electric Co./Plastics Div., One Plastics Ave., Pittsfield, MA 01201 (Tel 800-845-0600)
General Electric Co./Silicone Prod. Div., Mechanicsville Rd., Waterford, NY 12188
General Electric Plastics B.V., Plasticslaan 1, P.O. Box 117, 4600 AC Bergen op Zoom, The Netherlands
GE Plastics Ltd., Birchwood Park, Risley, Warrington, Cheshire WA3 6DA, England
General Electric (USA) Plastics Japan Ltd., No. 35 Kowa Building 5F, 14-14 Akasaka 1-chome, Minato-ku, Tokyo 107, Japan
GE Europe, Cyprusweg 2, 1044 AA Amsterdam, The Netherlands (Tel 31-20-5806911)
GE Hong Kong, 15/F Convention Center, No. 1 Harbor Rd., Wanchai, Hong Kong (Tel 852-5-8105616)
AF; DF; F; SE; SF; SFR; SM; SS; Viscasil®

GE Specialty Chemicals, 5th and Avery Sts., Parkersburg, WV 26102 (Tel 304-424-5411)
Blendex; EBS Wax; Ultranox®; Weston

Genencor Inc., 180 Kimball Way, S. San Francisco, CA 94080 (Tel 415-742-7500; 800-847-5311; FAX 415-583-8269)
Genencor®; Pectinol®; Rhozyme®

Genstar Stone Products Co., Executive Plaza IV, Hunt Valley, MD 21031 (Tel 301-527-4000; FAX 301-527-4535)
Camel-CAL®; Camel-CARB®; Camel-FIL; Camel-TEX®; Camel-WITE®

Georgia Marble Co., PO Box 356, Tate, GA 30177 (Tel 404-893-2511)
Calwhite®; D-70; Gamaco®; Gama-Fil; Gama-Plas®; Gama-Sperse®; Mar'blend; RO-40; Rock Dust; Windale; XO

Georgia-Pacific Corp., 133 Peachtree St. N.E., PO Box 105605, Atlanta, GA 30348 (Tel 404-521-4711)

CM-; Dyqex®; GP-; Ke-Mul; Lignosite®; NovaSize

Givaudan Corp., 100 Delawanna Ave., Clifton, NJ 07014 (Tel 201-365-8266)

L. Givaudan & Cie SA, CH-1214 Vernier/Geneva, Switzerland

Deltyl®; G-4; Giv-Gard; Givsorb®; Parsol®; Stabilizer

TH. Goldschmidt AG, Goldschmidtstrasse 100, D-4300 Essen 1, Germany (Tel 0201-1731; FAX 0201-224916)

Goldschmidt Chemical Corp., 914 E. Randolph Rd., PO Box 1299, Hopewell, VA 23860 (Tel 804-541-8658; 800-446-1809; FAX 804-541-2783)

Th. Goldschmidt Ltd., Tego House, Victoria Rd., Ruislip, Middx. UK HA4 OYL (Tel 01-4227788; FAX 01-8648159)

Th. Goldschmidt Japan KK, Rm. 1113, Shuwa Kioi-cho TBR Bldg. No. 7, 5-Chome, Koji-machi, Chiyoda-ku, Tokyo 102, Japan

Abil®; Aminoxid; Antil®; Axol®; EGDS; EGMS; Fekta®; Getren®; Hydrosan; Kosmos®; Lactil®; Perlglanz-Konzentrat; Polymekon®; Protegin; Sitren®; Stanno-Plus; Tagat; Tegamine; Tegiloxan®; Tegin®; Tego®; Tego-Amid; Tegoamin®; Tego Antiflamm®; Tego-Betaine®; Tegochrome®; Tegocoll®; Tegocolor®; Tegodont®; Tego IMR®; Tegomuls®; Tego-Pearl®; Tegopren®; Tegosil®; Tegosipon®; Tegosoft; Tegostab®; Tegotrenn®; Trennspray; Unislip®

The BFGoodrich Company, 6100 Oak Tree Blvd., Cleveland, OH 44131 (Tel 216-447-6000; 800-331-1144)

BFGoodrich Chemical (U.K.) Ltd., The Lawn, 100 Lampton Road, Hounslow, Middlesex TW3 4EB, England

BFGoodrich Chemical (Deutschland) GmbH, Goerlitzer Str. 1, 4040 Neuss 1, Germany

Carbopol®; Cure-rite; Good-rite®

Goodyear Tire & Rubber Co., 1485 E. Archwood Ave., Akron, OH 44316 (Tel 216-796-8812)

Goodyear Canada Inc., 45 Raynes Ave., Bowmanville, Ontario, Canada

Polystay; Wingstay

W.R. Grace & Co., Organic Chemicals Div., 55 Hayden Ave., Lexington, MA 02173 (Tel 617-861-6600; FAX 617-862-3869)

W.R. Grace Ltd., Northdale House, North Circular Rd., London NW10 7UH, UK

Darex; Daxad®; Everflex®; Solvent 80

Davison Chemical/Div. W.R. Grace & Co., PO Box 2117, Baltimore, MD 21203

Syloid®

Evans Chemetics/Div. W.R. Grace & Co., 55 Hayden Ave., Lexington, MA 02173

Emulsifier; Evanacid®; Evanstab®; Thiovanol®

Hampshire and Polymer Prods./Div. W.R. Grace & Co., 55 Hayden Ave., Lexington, MA 02173

Hamp-Ene®; Hamp-Ex®; Hamp-Ol®; Hampshire®;

Graden Chemical Co., Inc., 426 Bryan St., Havertown, PA 19083 (Tel 215-449-3808)

Denlube; Denphos; Densol; Densulf; Denwet; Graden; Gradol; Gradonic; Haroil; Harol; Rexfoam

Grain Processing Corp., 1600 Oregon St., Muscatine, Iowa 52761 (Tel 319-264-4265; FAX 319-264-4289)

Maltrin®; Pure-Dent®

Great Lakes Chemical Corp., PO Box 2200, W. Lafayette, IN 47906 (Tel 317-497-6100)

BA; BE; CA; CD; DBP; DE; FF; PE; PH; PHT; PO

Griffin Corp., PO Box 1847, Rocky Ford Rd., Valdosta, GA 31601 (Tel 912-242-8635)

Atrabute; Basicop; Cotton-Pro; Direx®; Du-Ter®; Griffex; K-Cop; Kocide®; Komeen®; K-Pool; K-Tea; K-Zinc; Linex®; Manex; Meturon; Milo-Pro; Pro-Tex; Super Tin®

Grindsted Products Inc., 201 Industrial Pkwy., POB 26, Industrial Airport, KS 66031 (Tel 913-764-8100)

Grindsted do Brazil, Ind. Ecom Ltda., Rodovia Regisé Bitten Court, KM 275, 5, Cx. Postal 172, 06800 Embú S.P. Brazil

Grindsted France S.A.R.L., Parc D'Activités de Tissaloup, Ave. Jean D'Alembert, F-78190 Trappes, France

Grindsted Products A/S, Edwin Rahrs Vej 38, DK-8220 Brabrand, Denmark

Grindsted Products Ltd., Northern Way, Bury St. Edmunds, Suffolk, IP32 6NP, UK (Tel 284769631)

Grindstedvaerket GmbH, Roberts-Bosch Strabe, D-2085 Quickborn, Deutschland, Germany

Cetodan; Emuldan; Grindtek; Homodan; Lactodan; Recodan

Grünau, Chemische Fabrick Grünau GmbH, Robert-Hansen Str. 1, Postfach 1063, D-7918 Jllertissen, Bavaria, West Germany

Arby; Lamacit; Lamecerin; Lamefix; Lamegin; Lamepon; Lamesoft; Monomuls; Osimol Grunau; Stabilizer

Guardian Chemical, Div. of United-Guardian, Inc., PO Box 2500, Smithtown, NY 11787

Lubrajel®

Guelph Soap Co. Inc., 21 Surrey St. W., Guelph, Ontario, Canada N1H 3R3 (Tel 519-824-2581)

Flexichem; Solar

C.P. Hall Co., 7300 South Central Ave., Chicago, IL 60638 (Tel 312-767-4600)

CPH; Emulvis®; Hallco®; Hallcomid®; Monoplex®; Paraplex®; Peptizer; Plasthall®; TegMeR®

Hamblet & Hayes Co., PO Box 730, Colonial Rd., Salem, MA 01970 (Tel 508-745-7250)

Kito; Tek Tan; Tek-Wet

Harcros Chemicals Group, Div. of Harrisons & Crosfield PLC

Harcros Chemicals UK Ltd., Lankro House, POBox 1, Eccles, Manchester M30 0BH, UK (Tel 061-789-7300; FAX 061-788-7886)

Harcros Chemicals Inc., 5200 Speaker Rd., PO Box 2383, Kansas City, KS 66106-1095 (913-321-3131; FAX 913-621-7718)

Harcros Chemicals BV, Haagen House, PO Box 44, 6040 AA Roermond, The Netherlands

Harcros Chemicals France S.a.r.l., BP 40, 441220 St. Laurent, Nouan, France

Lankro Chemicals, 1408 Chia Hsin Bldg., 96 Chung Shan N. Rd., Sec. 2, Taipei 10449, Rep. of China

AF; Agrilan; Algaetrol; Ampholan; Antifoam; Arylan; Crystal Inhibitor; Ethylan®; Haro®; Lankrocell®; Lankroflex®; Lankromark®; Lankro Mud-Aids; Lankromul; Lankropearl; Lankroplast®; Lankropol; Lankrosol; Lankrostat®; Lanquell; Levelan; Monolan; Perlankrol; Phospholan; Polystab; Quadrilan; Quell-Oil; SEM; Silicone AF; T-Det®; T-Mulz®

A. Harrison & Co. Inc., PO Box 494, Pawtucket, RI 02862 (Tel 401-725-7450)

Defoamer; Ibercarrier; Iberpenetrant; Iberquestrene

Hart Chemicals Ltd., 256 Victoria Rd. South, Guelph, Ontario N1H 6K8 Canada (Tel 519-824-3280; FAX 519-824-0755)

Duofol; Dyesperse; Foamer; Hartamide; Hartex; Hartolon; Hartonyl; Hartopol; Hartopon; Hartosoft; Hartotrope; Nonex; Pogol; Polylube; Prosol; Retarder; Rexol; Rexonic; Sulfonic Acid; Supratol; T.S.S.

Hart Products Corp., 173 Sussex St., Jersey City, NJ 07302 (Tel 201-433-6632)
Defoamer; Hartaine; Hartenol; Kalex; Retarder; Tergenol

Henkel Corp., 300 Brookside Ave., Ambler, PA 19022 (Tel 215-628-1000; 800-531-0815; FAX 215-628-1200)
Henkel Canada Ltd., 9550 Ray Lawson Blvd., Ville d'Anjou, Quebec H1J 1L3, Canada (Tel 514-353-7550)
Henkel Chemicals Ltd., Organic Products Division, Merit House, The Hyde, Edgeware Rd., London NW9 5AB, UK
Henkel Argentina S.A., Avda. E. Madero Piso 14, 1106 Capital Federal, Argentina
Henkel KGaA, Postfach 1100 D-4000, Dusseldorf 1, Germany
Henkel-Nopco SA, Process Chem. Div., 185 Ave. de Fontainebleau, 77310 St. Fargeau, Pontheirry, France
Aethoxal; Agriwet; Alcophor; Aliquat; Alphenate; Ambroxan; Aminodermin; Amphocerin; Anthoxan; Arnica Oil CLR; Astramyl; Avirol; Avocado Oil CLR; Base; Biosulphur Fluid; Boisambrene; Bravo; Bronidox; Calendula Oil CLR; Carrot Oil CLR; Cationic Guar C-261; Cegesoft; Ceroxin; Cetiol; Comperlan; Corax; Cosmedia Guar; Coviox; Covitol; Cutina; Cyclovertal; Daconil; Dacospin; Dehydag Wax; Dehydol; Dehydran; Dehymuls; Dehypon; Dehyquart; Dehysan; Dehyton; Delvet; Deriphat; Diazital; Doittol; DSX; Duolite; DyaFac; Dyasulf; Dymsol; Edenol; Elacid; Emulgade; Emulgane; Eumulgin; Euperlan®; Eutanol; Floramat; Flow Agent; Foamaster; Forlanit; Forlanon; Fosterge; Fostex; Galactasol; Generol®; Groutcide; Guartec; HD Eutanol; HD-Ocenol; Herbavert; Hexaplant Richter; Hydagen; Hydrenol; Hydropalat; Hyonic; Inhibiteur; Intraphor; Intrapol; Jasmacyclat; Kerensim; Kerinol; Lamequat; Lanette; Lecithin; Lomar; Loxiol; Lutostat; Modicol; Monosulph; Myritol; Nasuna; Nitrene; Nopalcol; Nopco®; Nopcocaster; Nopcocide®; Nopcogen; Nopcosant; Nopcosperse; Nopcosulf; Nopcote; Nopcowax; Noramac; Noramox; Perenol; Polyquart; Primarol; Product; Quatrene; Rheothik; Rilanit; Sedaplant Richter; Sellogen; Sopralub; Spiroflor; Standamid; Standamox; Standamul®; Standapol®; Steramine; Sulfotex; Sulfopon; Superclear; Syntergent; Texamin; Texaphor; Texapon; Textamine; Trem; Turpinal; Velvetex®; Verdoxan; Versilan

Henkel Corp./Emery Group, 11501 Northlake Dr., Cincinnati, OH 45249 (Tel 513-530-7300; FAX 513-530-7443)
Henkel Corp., Emery Group, 1301 Jefferson St., Hoboken, NJ 07030 (Tel 800-234-4365)
Henkel OPG, 3300 Westinghouse Blvd., Charlotte, NC 28217 (Tel 800-634-2436)
Henkel Corp., Emery Chemicals Ltd., 365 Evans Ave., Toronto, Ontario M8Z 1K2, Canada (Tel 416-259-3751)
Henkel Corp., Emery Group, PO Box 191, World Trade Center, 2-4-1 Hamamatsu-cho, Minato-Ku, Tokyo 105, Japan (Tel 4355611-2)
Acetol®; Clearlan®; Emercide®; Emeressence®; Emerest®; Emerez; Emerlube®; Emersal®; Emersol®; Emerstat®; Emerwax®; Emery®; Emgard®; Emid®; Emrite®; Emsorb®; Emthox®; Ethoxylan®; Frigid-Go®; Lanacet®; Lanfrax®; Lanoquat®; Lantrol®; Nimco®; Nimcolan®; Nimlesterol®; Plastolein®; Propoxyol; Sparkelan; Trisolan; Trycol®; Trydet; Tryfac®; Trylon®; Trylox®; Trymeen®; Twitchell

Hercules Inc., Hercules Plaza, Wilmington, DE 19894 (Tel 302-594-6500)
Hercules Inc./Carbon Fiber Materials, PO Box 98, Magna, UT 84044
Hercules Inc./PFW Div., 33 Sprague Ave., Middletown, NY 10940
Hercules B.V., 8 Veraartlaan, PO Box 5822, 2280 HV Rijswijk, The Netherlands
Hercules Ltd., Hercules European Hdqts., 20 Red Lion St., London WC1R 4PB UK
Abitol®; Adtac®; Amine D®; Az-Cup; Biozan; Cellolyn®; Clorafin®; CMC-; CMHEC-; CMTC-; Defoamer; Delfloc; Delsette; Di-Cup; Dresinate®; Dresinol; Dry Pexol; Dry Size; Foral®; Frimulsion; Gelcharg; Genu; Genugel; Genulacta; Genuvisco; Genuzan; Herco®; Hercoflat®; Hercoflex®; Hercofloc®; Hercolube®; Hercolyn®; Hercosett®; Hercotac®; Hercules®; HVP; Indalca; Luxor; Magnamite®; Minflo; Neolyn®; Pamolyn®; Paracol®; Pentalyn®; Pexol®; Phenrez®; Picco®; Piccodiene®;

Piccofyn®; Piccolastic®; Piccolyte®; Piccomer®; Picconol®; Piccopale®; Piccotac®; Piccoumaron®; Piccovar®; Poly-Pale®; Polyrad®; Pulpex®; Resin; Reten®; Size; Solvenol®; Staybelite®; Terate®; Triple-H; Vinsol®; Vul-Cup

Heterene Chemical Co., Inc., POB 247, 792 21 Ave., Paterson, NJ 07513 (Tel 201-278-2000; FAX 201-278-7512)

Hetamide; Hetester; Hetlan; Hetoxamate; Hetoxamine; Hetoxide; Hetoxol; Hetsorb; Hetsulf

Hexcel Corp., Chemical Products Div., 215 N. Centennial St., Zeeland, MI 49464 (Tel 616-772-2193)

Hexcel Corp./Rezolin Div., 20701 Nordhoff St., Chatsworth, CA 91311

Antistatic Agent; Bretol; Broma; Cetats; CPC; Dibactol; Hexcel; HTSA; Mytab; Stedbac; Suconox; Temasept

High Point Chemical Corp., 601 Taylor St., PO Box 2316, High Point, NC 27261 (Tel 919-884-2214)

Hipochem

Hilton-Davis Chemical Co., Div. of Sterling Drug, Inc., 2235 Langdon Farm Rd., Cincinnati, OH 45237 (Tel 513-841-4000)

Represented by: D.F. Censtead, Ltd., Victoria House, Radford Way, Billericay Essex CM12 0DE, UK

Cyncal®; Roccal®; SCMS

Hitox Corp. of America, PO Box 2544, Corpus Christi, TX 78403 (Tel 512-882-5175; FAX 512-882-6948)

Bartex®; Haltex; Hitox®; Zotex

Hodag Chemical Corp., 7247 N. Central Park Ave., Skokie, IL 60076 (Tel 312-675-3950)

Hodag; Nonionic; Sole-Onic; Sole-Terge

Hoechst Celanese Corp., Int'l. Headquarters, 26 Main St., Chatham, NJ 07928 (Tel 201-635-2600)

Celanese France Sarl, Z.I. Le Broteau 69540, Irigny, France
Hoechst Celanese Plastics Ltd., 78-80 St. Albans Rd., Watford, Herts, UK WD2 4AP
Celanese GmbH, Hordenbachstrasse #40, D-5600-Wuppertal 21, Germany
Hoechst Canada Inc., Plastics Prods. Dept., 4045 Cote Vertu Blvd., Montreal, Quebec H4R 1R6, Canada (Tel 514-333-3500)
Celanese Mexicana S.A., Av. Revolucion No. 1425, Mexico 20, D.F.
Hoechst Japan, 10-33, 4-Chome-Akasaka, Minato-ku, Tokyo, Japan
Hoechst Celanese Far East Ltd., 801 Hong Kong Club Bldg., 3A Chater Rd., Central, Hong Kong

Aquamolin; Arkomon; Biolase; Bohrmittel Hoechst; Bozemine; Cassastat; Defoamer; Dional; Dispersogen; Dodicor; Dodigen; Dodilube; Emulsogen; Flotigam; Frigen; Genagen; Genamin; Genamine; Genaminox; Genapol; Hoe; Hoechst; Hostacerin; Hostacor; Hostalux; Hostapal; Hostaphat; Hostastat®; Hostavin®; Humectol; Lanogen; Leomin; Leonil; Peroxide Stabilizer H; Phenyl Sulphonate; Phosphate Ester 123; Propagen; Samaron; Solupret; Stabilisol; Tylose

Hoffmann Mineral, Franz Hoffmann & Sôhne KG, Postfach 1460, D-8858 Neuburg/Donau, Germany

Sillikolloid; Sillitin

Geo. A. Hormel & Co., 1816 4th St. NE, Austin, MN 55912 (Tel 507-437-5608; FAX 507-437-5489)

Dermatein; Peptein 2000®; Protectein; Sollagen®

E.F. Houghton & Co., Madison & Van Buren Aves., Valley Forge, PA 19482

Surfax

J.M. Huber Corp., PO Box 310, Havre de Grace, MD 21078 (Tel 301-939-3500; FAX

301-939-0394)

J.M. Huber Corp., Carbon Black Div., PO Box 2831, Borger, TX 79008-2831 (Tel 806-274-6331; 800-631-6331)

G-White; Huber; Huberfil®; Hubersil®; Hubersorb®; H-White; Hydrex®; Q-White; XPV1; Zeo®; Zeodent®; Zeofree®; Zeolex®; Zeomatt; Zeosyl®; Zeothix®

Hüls America Inc., PO Box 456, 80 Centennial Ave., Piscataway, NJ 08855-0456 (Tel 201-980-6800; 800-631-5275; FAX 201-980-6970)

Hüls AG, Postfach 1261, Troisdorf, Germany (Tel 49-2241-85-850)Huls (UK) Ltd., Cedars House, Farnborough Common, Orpington, Kent BR6 7TE, UK

Ammonium Cumene Sulfonate 60; B Series; CA; CB; CC; CD; CE; CG; CH; CHY; CI; CM; CO; CP; CPS; CS; CT; CV; D; DEP; DMP; Dynacerin; Dynasan; Dynasil; Dynasylan®; H; I; Imwitor; Marlamid; Marlican; Marlipal; Marlon; Marlophen; Marlophor; Marlopon; Marlosoft; Marlosol; Miglyol; Na Cumene Sulfonate; Na Toluene Sulfonate; P; PO; Softenol; Softigen; Softisan; T; Witepsol

Humko. See Witco

Huntington Laboratories, Inc., 970 East Tipton, Huntington, IN 47650 (Tel 219-356-8100)

FMB; JAQ; WFT

ICI Advanced Materials, 475 Creamery Way, Exton, PA 19341 (Tel 800-854-8774)

FC-; TL-

ICI Americas Inc., Subsid. of Imperial Chemical Industries PLC, Wilmington, DE 19897 (302-575-3000; 800-441-7757; 800-456-3669; FAX 302-575-2459)

ICI Specialty Chemicals, Cherry Lane, PO Box 231, New Castle, DE 19720 (Tel 302-575-4700; 800-456-3669; FAX 302-652-7035)

ICI Europa Ltd., Everslaan 45, B-3078 Kortenberg, Belgium (Tel 02-758-9211; FAX 02-759-5501)

ICI Japan Ltd., Osaka Green Bldg., 1,3-Chome Kitahama, igashi-Ku, J-Osaka 541, Japan

ICI Specialty Chemicals, Cleeve Rd., Leatherhead, GB-Surrey KT22 7SW UK (Tel 0372-376122; FAX 0372-386245)

Atkemix Inc., 70 Market St., PO Box 1085, Brantford, Ontario, Canada

ICI Japan Ltd., Osaka Green Building, 1, 3-Chome Kitahama, Higashi-Ku, J-Osaka 541, Japan

Ahco; Ahcovel; Ahcowet; Arlacel®; Arlacide; Arlamol®; Arlasolve; Arlatone; Atlas; Atlox; Atmer; Atpet; Atplus; Brij®; Cereclor; Cirrasol; Hypermer; Monflor; Nonox®; Nylomine; Proxel; PTZ; Pycal; Renex®; Silcolapse; Silicone; Span®; Synthrapol; Tandem; Topanex; Topanol®; Tween®; Vantoc; Winnofil

ICI Australia Operations PTY Ltd., ICI House, 1 Nicholson St., Melbourne 300, Australia (Tel 03-665-7111)

Acihib; Icinol; Oxamin; Teric

ICI Ltd./Organics Div., Chemical Auxiliaries Business Group, Smith's Rd., Bolton, Lancs, BL3 2QJ, UK

Detergyl; Dispersol; Lenetol; Matexil; Sunaptol; Unisol

ICI PLC/Petrochemicals & Plastics Div., PO Box 90, Wilton Middlesbrough, Cleveland TS6 8JE, UK

Lubrol; Synperonic; Synprol Alcohol; Synprolam

Inolex Chemical Co., Jackson & Swanson Sts., Philadelphia, PA 19148-3497 (Tel 215-271-0800; 800-521-9891)

Represented by: Stanley Black, Ltd., 30/31 Islington Green, London N1 8DU, UK

Lexaine; Lexamine; Lexate; Lexein; Lexemul®; Lexgard; Lexol; Lexolube; Lexomul; Lexquat®; Myacide; PL-

International Bio-Synthetics Inc., PO Box 241068, Charlotte, NC 28224 (Tel 704-527-9000)
Ward Blenkinsop & Co. Ltd., Halebank Widnes, Cheshire WA8 8NS, UK
Amigase®; Breakerase; Delvocid; Dex-Lo®; Klerzyme®; Maxacal®; Maxaliq; Maxamyl®; Maxatase®; Maxazyme®; Maxilact®; Maxinvert®; Mycolase®; Prolase®; Rapidase®

International Dioxide, Inc., 136 Central Ave., Clark, NJ 07066 (Tel 201-499-9660; FAX 201-388-3648)
Adox; Anthium Dioxide

Intex Products, Inc., PO Box 6648, Greenville, SC 29606 (Tel 803-242-6152; FAX 803-299-3475)
Tex-Wet

ITT Rayonier Inc., 1177 Summer St., Stamford, CT 06905 (Tel 203-348-7000)
Orzan®; Ray Krome; Raylig; Raymix

Jetco Chemicals, Inc., PO Box 1898, Corsicana, TX 75110 (Tel 903-872-3011)
Jet Amine; Jet Quat

Kao Corp., International Chemicals Dept., 14-10 Nihonbashi, Kayabacho 1-Chome, Chuo-ku, Tokyo 103, Japan
Kao Corp., Edible Fat and Oil Div., 14-10 Nihonbashi, Kayabacho 1-Chome, Chuo-ku, Tokyo 103, Japan
Acetamin; Antifoam; Demol; Elec; Emal; Emanon; Emasol; Emulgen; Excel; Exceparl; Homogenol; Homotex; Kalcohl; Kureton; Levenol; Pelex; Poiz; Quartamin; Rheodol; Sanisol; Scourol; Softex; Uresoft; Vissurf

KAO Corp. S.A., Puig dels Tudons, 10, 08210 Barbers Del Valles, Barcelona, Spain
Acetamin; Amiet; Diamiet; Diamin; Farmin; SK Fert; SK Flot

Kaopolite, Inc., Subsid. of Combustion Engineering, Office Park, Rte. 22E, Box 3110, Union, NJ 07083 (Tel 201-789-0609)
Ferrosil; Koraid; Korthix

Keil Chemical Div./Ferro Corp., 3000 Sheffield Ave., Hammond, IN 46320 (Tel 219-931-2630; FAX 219-931-6318)
EM-; Inversol; Synkad

Kelco, Div. of Merck & Co., Inc., 8355 Aero Dr., San Diego, CA 92123 (Tel 619-292-4900; FAX 619-569-3436)
Kelco, Westminster Tower, 3 Albert Embankment, London SE1 7RZ, UK (Tel 01-735-0333)
Dariloid®; GFS®; Kelacid®; Kelco-Gel®; Kelcoloid®; Kelco-Pac®; Kelcosol®; Kelfo®; Kelgin®; Kelmar®; Kelset®; Keltex®; Keltone®; Keltose®; Keltrol®; Kelvis®; Kelzan®; Margel; Superloid®; Xanco-Frac®; Xanflood®

Kenobel SA. See Berol Nobel

Kenrich Petrochemicals, Inc., 140 E. 22nd St., PO Box 32, Bayonne, NJ 07002-0032 (Tel 201-823-9000; 800-LICAKPI; FAX 201-823-0691)
KA; Kencure; Kenflex®; Ken Kem®; Kenplast®; Ken-React®; KR; KZ; LICA; LZ; NZ; Ridacto®; Silacto

Kincaid Enterprises, Inc., Box 671, Nitro, WV 25143 (Tel 304-755-3377)
Marlate®

Kyowa America Corp., 2500-T, S.E. Main St., Irvine, CA 92714 (Tel 714-641-0411)
Alcamizer; Kisuma

Laporte Inc., Waverly Mineral Prod. Co., 555 City Line Ave., Bala Cynwyd, PA 19004 (Tel 215-668-2808; 800-829-1829; FAX 215-668-2819)
Laponite®

Laurel Products Corp., Div. of Reilly-Whiteman, 2600 E. Tioga St., Philadelphia, PA 19134 (Tel 215-423-5300; FAX 215-739-6136)
Antistat; Aquasan

Lion Corp., 2-22,1-Chome Yokoami, Sumida-ku, Tokyo, Japan
Dovanox; Leocon; Leoguard; Lipolan; Lipomin; Liponox; Soft Detergent; Sunnol

Lipo Chemicals, Inc., 207 19th Ave., Paterson, NJ 07504 (Tel 201-345-8600)
Represented by: Blagden Campbell & Campbell Ltd., A.M.P. House, Dingwall Rd., Croydon CR9 3QU, UK
Lipamide; Lipo; Lipobee; Lipocol; Lipolan; Lipomulse; Liponate; Liponic; Lipopeg; Lipophos; Lipoquat; Liposorb; Lipovol; Lipowax

Lonza Inc., 17-17 Route 208, Fair Lawn, NJ 07410 (Tel 201-794-2400)
Lonza Ltd., Munchensteinerstrasse 38, POB CH-4002, Basle, Switzerland
Acrawax®; Akolidine; Aldo®; Aldosperse®; Alkawet®; Amphoterge®; Bardac®; Barlox®; Barquat®; Bio-Dac; Bio-Quat; Bio-Surf; Carsamide®; Carsamine; Carsemol; Carsodet; Carsofos; Carsonam; Carsonol®; Carsonon®; Carsoquat®; Carsosoft®; Carsosulf; Ceramid; Dantocol®; Dantoest®; Dantoin®; Dantosperse®; Ethosperse®; Flexowax; Glybrom; Glychlor®; Glycolube®; Glyconol®; Glycowax®; Glycox®; Glydant®; Glytex®; GSD 550; Hyamine®; Hystar®; Lonzaine®; Lonzest®; Pegosperse®; Polyaldo®; Quad Opacifier; Softener; Unamide®; Unamine®; Unihib®; Uniquat®; Vinylube®

Lord Corp., Elastomer Prods., 2000 W. Grandview Blvd., PO Box 10038, Erie, PA 16514-0038 (Tel 814-868-3611; FAX 814-864-3452)
Qyadra Chemical Limited, 2121 Argentia Road, Suite 303, Mississauga, Ontario, Canada L5N 2X4
Chemlok®

Lowenstein Dyes & Cosmetics, Inc., 420 Morgan Ave., Brooklyn, NY 11222
Represented by: Siber Hegner (UK) Ltd., 221-241 Beckenham Rd. Makenzie House, Beckenham, Kent BR3 4UF, UK
Kelene

Lubrizol Corp., 29400 Lakeland Blvd., Wickliffe, OH 44092 (Tel 216-943-4200; FAX 216-943-5337)
Lubrizol Ltd., Waldron House, 57-63 Old Church St., London SW3 5BS, UK (Tel 351-3311-20; FAX 351-3310)
Lubrizol GmbH, Bogenallee 10, 2000 Hamburg 13, Germany
Lubrizol France, Tour Europe, 92400 Courbevoie, France
Lubrizol Japan Ltd., No. 23 Mori Bldg., 5th flr, 23-7 Toranomon, 1-chome, Minato-ku, Tokyo 105, Japan
Angamol; Ircogel®; Lubrizol

Lucidol Div. See Atochem

Lyndal Chemical Co., PO Box 1740, Dalton, GA 30720 (Tel 404-259-4831; FAX 404-259-5979)
Aston; Foamex; Foamgard; Hysoft; Kara Lube; Kara Sperse; Mollisan; Perma Kleer; Permalene; Phosfac; Progacyl; Progasol; Progastat; Propacyl; Repel-O-Tex; Velvamine

3M, Performance Polymers and Additives, 3M Center Bldg., St. Paul, MN 55144-1000 (Tel 612-736-9700)
3M Australia Pty. Ltd., 950 Pacific Hwy., PO Box 99, Pymble, N.S.W. 2073, Australia
3M Canada Inc., PO Box 5757 Terminal A, 1840 Oxford St. East, London, Ontario, Canada N6A 4T1
3M Belgium N.V./S.A., Canadastraat 11, 2730 Zwijndrecht
3M Deutschland GmbH, PO Box 100422, D4040 Neuss 1, Germany
3M United Kingdom PLC, 3M House, PO Box 1, Bracknell, Berkshire RG 12 1JU, UK
Sumitomo/3M Ltd., Central PO Box 490, 33-1 Tamagawadai 2-chome, Setagaya-ku, Tokyo 158, Japan

Dynamar; FC-; Fluorad

M&T Chemicals Inc., Woodbridge Rd., PO Box 1104, Rahway, NJ 07065 (Tel 201-499-0200)

M & T Chemicals Ltd., 670 Strathearne Ave. North, Hamilton, Ontario L8H 7N7 Canada (Tel 416-547-1471)

M & T Chimie S.A., La Defense-Cedex 54, 92062 Paris la Defense 49-04-1010 France

Thermoguard; Ultrox

M&T Harshaw, 1000 Harvard Ave., Cleveland, OH 44109 (Tel 216-349-7300; Int'l. 201-321-5000)

Al-; Co-; CoMo-; Cr-; Cu-; Harshaw; HFR-; HT-; Mg; Mo-; Na-; Ni-; Nysel; P-; Pd-; Ti–; V-; Zn-

Macfarlan Smith Ltd., Wheatfield Rd., Edinburgh, Scotland EH11 2QA

Distributed by Robeco Chemicals, Inc., 99 Park Ave., New York, NY 10016 (212-986-6410; FAX 212-986-6419)

Bitrex

Mallinckrodt Specialty Chemicals, 3200 N. Second St., St. Louis, MO 63147 (Tel 314-895-2000; 800-325-7155)

Van Dyk & Co., Inc./Div. of Mallinckrodt. See under Van Dyk

Mallinckrodt GmbH, Josef-Dietzgen-Strasse 1, PO Box 1462, D-5202, Hennef/sieg 1, Germany

Hy Dense; LiquaPar; Lustra-Pearl®

George Mann & Co., Inc., 123 Georgia Ave., PO Box 9066, Providence, RI 02940 (Tel 401-781-5600)

Aqualease; Ease Release

Manro Products Ltd., Bridge St., Stalybridge, Cheshire SK15 1PH, UK (Tel 061-338-5511; FAX 061-303-2991)

Manro

Manville Filtration & Minerals, PO Box 519, Lompoc, CA 93438-0519 (Tel 805-735-7791; FAX 805-736-7856)

Celite®; Micro-Cel; Silasorb; Tixosil

Marlowe-Van Loan Corp., PO Box 1851, High Point, NC 27261 (Tel 919-886-7126)

Marvanol

Marubishi Oil Chemical Co., Ltd., 3-7-12 Tomobuchi-cho, Miyakojima-ku, Osaka, Japan

Babinar; Value

Mason Chemical Co., 5253 West Belmont Ave., Chicago, IL 60641 (Tel 312-282-0200; 800-362-1855; FAX 312-282-0821)

Algaesil; Maquat

Matsumoto Yushi-Seiyaku Co., Ltd., 1-3,2-Chome, Shibukawa-cho, Yao City, Osaka, Japan

Effcol; Elimina; Marvelin; Zontes

McIntyre Chemical Co., Ltd., 4851 S. St. Louis Ave., Chicago, IL 60632 (Tel 312-927-2401)

Mackalene; Mackam; Mackamate; Mackamide; Mackamine; Mackanate; Mackazoline; Mackernium; Mackester; Mackine; Mackpearl

Mearl Corp., 41 E. 42nd St., New York, NY 10017 (Tel 212-573-8500; FAX 212-557-0742)

Represented by: Cornelius Chemical Co. Ltd., St. James House, 27-43 Eastern Rd., Romford, Essex RM1 3NN, London EE3, UK

Biju; Mearlin; Mearlmica; Micromesh; Timica

Merix Chemical Co., 2234 E. 75 St., Chicago, IL 60649 (Tel 312-221-8242)

AST; Merix®; Mold-Ease; Parabolix®

Merrand Int'l. Corp., 1236 Weathervane Lane, Suite #203, Akron, OH 44313 (Tel 216-869-5800; FAX 216-869-9629)
Merrol®; Phosflex; Pliabrac; TOF; Tolplaz

Lucas Meyer GmbH & Co., Ausschlager Elbdeich 62, D-2000 Hamburg 26, Germany (Tel 040-78955-0; FAX 040-7898329)
Asol; Colorol; Emulthin; Forbest; Lipotin; M-C-Thin; Topcithin; Trenbest

M. Michel & Co., Inc., 90 Broad St., New York, NY 10004 (Tel 212-344-3878; FAX 212-344-3880)
Cachalot; Michel

Miles Laboratories Inc./Biotech Products Div., PO Box 932, 1127 Myrtle St., Elkhart, IN 46515-0952 (Tel 219-264-8111)
Brew(n)zyme®; Diazyme; Gelatinase; HT-Proteolytic; Milezyme®; Spark-L®; Taka-Sweet®; Takatex®; Taka-Therm®; Tenase®

Miller-Stephenson Chem. Co., Inc., George Washington Hwy., Danbury, CT 06810 (Tel 203-743-4447)
MS-

Milliken Chemical, Div. of Milliken & Co., PO Box 1927, Spartanburg, SC 29304 (Tel 803-573-2200; FAX 803-573-2430)
Millithix; Syn Fac; Syn Lube

Millmaster-Onyx International, PO Box 1045, Fairfield, NJ 07007 (Tel 201-227-3995)
Millmaster-Onyx UK, Marlborough House, 30-2 Yarm Rd., Stockton-on-Tees, Cleveland, UK
Cation Softener; Neutronyx®; Onamine; Onyxide; Onyxol®

Milport Chemical Co., 2829 S. 5th Court, Milwaukee, WI 53207 (Tel 414-769-7350)
Milco

Miranol Inc. See Rhone-Poulenc Surfactants and Specialty Chemicals

Mobay Corp., A Bayer USA Inc. Co., Penn Lincoln Parkway West, Pittsburgh, PA 15205-9741 (Tel 412-777-2000)
Additin; Antiozonant; Bayhibit; Bonding Agent; Coagulant; Cohedur; Desmodur®; Desmorapid; Disflamoll; Emulvin; Mondur; Porofor; Renacit; Stabilizer; Statexan; Vulkacit; Vulkadur; Vulkalent; Vulkanol; Vulkanox; Vulkazon; Zinc Oxide; Zinkoxyd Activ

Mona Industries Inc., PO Box 425, 76 E. 24th St., Paterson, NJ 07544 (Tel 201-345-8220)
Avamid; Mona; Monacor; Monafax; Monalube; Monamid®; Monamine; Monamulse; Monaquat; Monastat; Monaterge; Monateric; Monatrope; Monawet; Monazoline; Phosphoteric®

Monsanto Chemical Co., 800 N. Lindbergh Blvd., St. Louis, MO 63167 (Tel 314-694-1000)
Monsanto PLC, Monsanto House, Chineham Court, Chineham, Basingstoke, Hants RG24 OUL, UK (Tel 44256-572-88)
Monsanto Europe S.A., Ave. de Tervuren 270-272, B-1150 Brussels, Belgium (Tel 761-41-11)
A-1; ACL; Alkylate; Butasan®; Dequest®; DOA; DPG; Duralink; EMA; Flectol®; Gelvatol; HB-40; Methasan®; Modaflow®; MonoThiurad®; Montaclere®; Multiflow; NTA; PC-; Phosgard; Santicizer; Santocure®; Santoflex®; Santolite; Santoquin; Santovar®; Santoweb®; Santowhite®; Scripset; Sterox®; Sulfasan®; Tackine®; TCC; Thiofide®; Thiotax®; Thiurad®; Vocol®

Montana Talc Co., 28769 Sappington Rd., Three Forks, MT 59752 (Tel 406-285-3286; FAX 406-285-3530)
Nicron

Morflex, Inc., 2110 High Point Rd., Greensboro, NC 27403 (Tel 919-292-1781)
Citroflex

Morton International, Inc., Specialty Chemicals Group, 2000 West St., Cincinnati, OH 45215-3431 (Tel 513-733-2100)
Morton International, Inc., 150 Andover St., Danvers, MA 01923 (Tel 508-774-3100; 800-621-2847; FAX 508-777-5672)
Morton International Ltd., Specialty Chemicals Group, Industrial Chemicals and Additives, 7900-A Taschereau Blvd., Suite 106, Brossard, Quebec J4X 1C2, Canada (Tel 514-466-7764; FAX 514-466-7771)
N.V. Morton International S.A., Chaussee de la Hulpe 130, Boite 5, B-1050 Brussels, Belgium (Tel 32-2-6602909; FAX 32-2-6604702)
Morton International Inc., Room 2501 Dominion Centre, 37-59A Queen's Road East, Wanchai, Hong Kong
Advastab®; Advawax; Bio-Pruf®; Carstab; Cunilate; Cuniphen; Durotex; Elastocarb; Lytron; Pave; Socci; Sotex; Versamag®; Vinyzene®

Nalco Chemical Co., One Nalco Center, Naperville, IL 60566-1024 (Tel 312-961-9500)
Nalco®

National Starch and Chemical Corp., 10 Finderne Ave., Bridgewater, NJ 08807 (Tel 201-685-5000)
Represented by: National Adhesives & Resins Ltd., Braunston Daventry, Northante NN11 7JL, UK
Amphomer®; Celquat; Condensate; Dry Flo®; Emulsifying Oil; Flexan®; Protamine; Protowax; Protowet; Resyn®; Sulfonated Castor Oil

Neville Chemical Co., 2800 Neville Rd., Pittsburgh, PA 15225 (Tel 412-331-4200)
Aromatic Oil 745; CB-; Nevastain; Nevpene; Nevtac; NP-; Synox; X-743

New Jersey Zinc Co. See Zinc Corp. of America

Niacet Corp., Niagara Falls Blvd. & 47th St., Niagara Falls, NY 14304 (Tel 716-285-1474; 800-828-1207)
Niaproof®

Nikko Chemicals Co., Ltd., 1-4-8 Nihonbashi-Bakurocho, Chuo-ku, Tokyo 103, Japan
Deatron; Dispatex; Neocation; Neoprotex; Neorate; Nikkol; Sunsoflon; Sunsolt; Swanol

Nippon Nyukazai Co., Ltd., 19-9 Ginza 3-Chome, Chuo-ku, Tokyo 104, Japan
Disrol; Newcol; Texnol

Nippon Oils & Fats Co., Ltd., 10-1, Yaraku-Cho, 1-Chome, Chiyoda-Ku, Tokyo 100, Japan
Nissan; Stafoam

Nippon Senka Chemical Industries, Ltd., 17-34 Hanaten-Higashi, 1-chome, Tsurami-ku, Osaka, Japan
Chromosol; Migregal; Multinol; Senka; Serenasol

Norman, Fox & Co., 5511 S. Boyle Ave., PO Box 58727, Vernon, CA 90058 (Tel 213-583-0016)
Norfox

Novo-Nordisk A/S, Novo BioLabs, DK-2880 Bagsvaerd, Denmark (Tel 45-44-490033)
Novo BioLabs, 33 Turner Rd., Danbury, CT 06810-5101 (Tel 800-344-6686)
Novo Laboratories Ltd., 1755 Steels Ave. West, Willowdale, Ontario M2R 3T4 Canada
Novo BioLabs Ltd., Downham House, Downham's Lane, Milton Rd., Cambridge, Cambs CB4 1XG, UK
Novo Yakuhin KK, Sumitomo Higashi Shimbashi, Bldg. No. 2, 12-7, Higashi Shimbashi, 2-Chome, Minato-ku, Tokyo 105, Japan
Alcalase®; AMG; Aquazym®; BAN; Celluclast; Celluzyme; Cereflo®; Dextranase Novo; Dextrozyme; DN 25L; Esperase®; Finizym; Fungamyl®; Gamanase; Lactozym;

Lecitase; Neutrase®; Palatase; Pectinex; PTN®; Rennilase®; Savinase; Semacylase; Sweetzyme®; Termamyl®; Uricase

Occidental Chemical Corp., 360 Rainbow Blvd. South, PO Box 728, Niagara Falls, NY 14302 (Tel 716-286-3000)

Dechlorane®

Oleofina, 4, Rue Jacques de Lalaing, B-1040 Brussels, Belgium (Tel 02-233-91-11; FAX 233-32-99)

Oleofina UK, Petrofina House, 1 Ashley Ave., Epsom, Surrey KT18 5AD, UK (Tel 44-03727-26226; FAX 44-03727-45821

Oleofina Far East, 138 Cecil St. #17-01, Cecil Court, RS Singapore 0106

Radia; Radiacid; Radiamac; Radiamine; Radiamuls; Radianol; Radiaquat; Radiastar; Radiasurf; Radiasyn

Olin Chemicals, 120 Long Ridge Rd., PO Box 1355, Stamford, CT 06904 (Tel 203-356-2602; 800-462-OLIN)

Olin Australia Ltd., 1-3 Atchison St., PO Box 141, St. Leonards 2065, N.S.W., Australia

Olin Brasil Limitada, Rua Galeno de Castro, 165, Jurubatuba, Santo Amaro, 04696 Sao Paulo, SP, Brazil

Olin U.K. Ltd., Suite 7, Kidderminster Road, Cutnall Green, Worcestershire, UK WR9 0NS

Olin Europe S.A., 108-110 Blvd. Haussmann, 75008 Paris, France (Tel 33-1-293-3210)

Olin Japan Inc., Shiozaki Bldg., 7-1 Hirakawa-Cho 2-Chome, Chiyoda-ku, Tokyo 102, Japan (Tel 81-3-263-4615)

Actafoam; C2® Sodium Chlorite; D-; Dee-Tac; Dichan; Dyetone®; Expandex®; Kempore®; Nitropore®; OCI; Omadine®; Oxypruf; Poly-Cone; Poly-G®; Polypeg; Polyphos®; Poly-Solv®; Poly-Zole®; RIA; Scav-Ox®; Silicate Cluster; Sodium Chlorite; Sodium Omadine®; T-; Textone®; Thermact®; Thermolin®; Triadine®; Wayhib®; Wayplex; Wiltrol; Wytox®; Zinc Omadine®

Onyx Chemical Co. See Millmaster Onyx Group

Otsuka Chemical Co., Ltd., 747 Third Ave., 26th Fl., New York, NY 10017 (Tel 212-826-4374; FAX 212-826-5094)

Onifine

Pacific Anchor Chemical Corp., A Company of Air Products and Chemicals, Inc., 5701 S. Eastern Ave., Suite 530, Los Angeles, CA 90040 (Tel 800-423-4391; 213-725-1800; FAX 213-725-1915)

Anchor Chemical Group PLC, Clayton Lane, Clayton, Manchester M11 4SR, UK (Tel 061-223-2451; FAX 061-223-5488)

Curezol®; Dicyanex®; Epodil; Epoxy Modifier ML; Imicure®

Panelgraphic Corp., 10 Henderson Dr., West Caldwell, NJ 07006 (Tel 201-227-1500; 800-222-0618; FAX 201-227-7750)

Vueguard

C.J. Patterson Co., Patco Products Div. See American Ingredient Co.

Pennwalt Corp., Pennwalt Bldg., Three Parkway, Philadelphia, PA 19102-1389 (Tel 215-941-0344; 800-223-3844)

Pennwalt, Inc. Canada, 700 Third Line, PO Box 278, Oakville, Ontario, Canada

Pennwalt S.A. de C.V., Rio San Javier No. 10, Fracc. Vivertos Del Rio, Tlainepantla. Edo. De Mexico CP54060, Mexico

Pennwalt Japan, Ltd., Sumitomo Jimbo-Cho Bldg., 3-25 Kanda Jimbo-Cho, Chiyoda-Ku, Tokyo 101, Japan

Pennwalt Industrial Chemicals Ltd., 12 Parsons Mead, Croydon, Surrey CRO 3SL, UK

Pennwalt Holland B.V., Westblaak 90, 3012 KM Rotterdam, The Netherlands

Pennwalt Chemie GmbH, Liesegangstrasse 18, D-4000 Duesseldorf 1, Germany
Pennvond SARL, 28 Ave. de la Porte de Choisy, F-75013 Paris, France
Orthophen®; Pennad; Pennfloat®; Pennmax; Pennodorant®; Pennstop®; Pennwalt®; Pentaphen®; Spotleak®; Sulfa-Hitech®; Vultac®

Penreco, Div. of Pennzoil Products Co., 106 S. Main St., Butler, PA 16001 (Tel 412-283-5600; 800-245-3952; FAX 412-283-1630)
Drakeol; Draketex; Mineral Jelly; Ointment Base; Parol; Peneteck; Penreco; Petrosul®

Petrolite Corp., Specialty Polymers Group, 6910 E. 14th St., Tulsa, OK 74112 (Tel 800-331-5516; FAX 918-834-9718)
Petrolite Corp./Bareco Div., 6910 E. 14th St., PO Drawer K, Tulsa, OK 74112
Petrolite Ltd., 137 Finchley Rd., London NW3 6JE, UK
Agri-safe; Bar-Dust®; Be Square®; Cardipol®; Cardis®; Fortex®; Mekon®; Petrolite®; Petronauba®; Polymekon®; Polywax®; Starwax®; Ultraflex®; Unilin; Unithox; Victory®; Vybar®

Pfanstiehl Laboratories, Inc., 1219 Glen Rock Ave., PO Box 439, Waukegan, IL 60085 (Tel 312-623-0370; FAX 312-623-9173)
Seqlene®

Pfizer Inc., 235 E. 42nd St., New York, NY 10017 (800-336-9008)
Pfizer Ltd., Chemical Division, 10 Dover Rd., Sandwich, Kent CT13 0BN, UK
ABT; Citrosol; Hi-Pflex®; Microbloc®; MicroPflex; Microtalc; Microtuff; Sorbistat; Super-Pflex; Ultra-Pflex®

Pierce & Stevens Corp., A Pratt & Lambert Co., PO Box 1092, Buffalo, NY 14240-9990 (Tel 716-856-4910)
Dualite

Pigment Dispersions, Inc., 54 Kellog Court, Edison, NJ 08817 (Tel 908-985-7300)
PDI

Pilot Chemical Co., 11756 Burke St., Santa Fe Springs, CA 90670 (Tel 213-723-0036; FAX 213-945-1877)
Aristol; Aristonate; Calamide; Calfax; Calfoam; Calimulse; Calsoft; Calsuds; Emulsifier 99; Pilot

Plasticolors Inc., 2600 Michigan Ave., Ashtabula, OH 44004 (Tel 216-997-5137)
Plastigel

PMC, Inc., Specialties Group, 501 Murray Rd., Cincinnati, OH 45217 (Tel 513-242-3300; 800-543-2466)
AA; CAO®; Cobratec®; MA; Moly-White®; Nacopa®; Sherbrite; Syncal; Ultra-Cure

PMP Fermentation Products, Inc., 7670 N. Port Washington Rd., Milwaukee, WI 53217 (Tel 414-352-3001; FAX 414-352-7252)
PMP

Polymer Research Corp. of America, 2186 Mill Ave., Brooklyn, NY 11234 (Tel 718-444-4300)
Agent; Formula #; Glitter Adhesive; Low Crock; Polylube; Polystat; Silicone Defoamer; Silicone Release Agent; Softener; Thermagel; Thickener

PPG Industries, Inc., One PPG Place, Pittsburgh, PA 15272 (800-243-6745)
Dizene; EHP; Flo-Gard; Hi-Sil®; Lo-Vel®; NPP; Perkare; Pittclor; PPG; SBP; Silene®; Tri-Ethane

PPG-Mazer, Div. of PPG Chemicals Group, 3938 Porett Dr., Gurnee, IL 60031 (312-244-3410; FAX 312-244-9633)
Mazer Chemicals UK Ltd., Carrington Business Park, Carrington, Urmston, Manchester M31 4DD, UK
Mazer de Mexico S.A. de C.V., Londres 226-4 Piso, Colonia Juarez, Mexico D.F. 06600

Mazer Chemicals (Canada), Mississauga, Ontario L4Z 1H8, Canada
PPG Industrial do Brazil Ltda., Edificio Grande Avenida, Paulista Ave. 1754, Suite 153, Sao Paulo, Brazil 01310
Alubrasoft; Alubrasol; Alubraspin; Jorchem; Jordamide; Jordamine; Jordamox; Jordanol; Jordaphos; Jordaquat; Jordawet; Jorgard; Jorphox; Jorquest; Jortaine; Larosol; Larostat; Macol®; Mafloc®; Mafo®; Mapeg®; Maphos®; Masil®; Maslip®; Mazamide®; Mazawax®; Mazawet®; Mazeen®; Mazol®; Mazoline®; Mazon®; Mazu®; M-Quat®; Ridafoam; S-Maz®; T-Maz®

The PQ Corp., PO Box 840, Valley Forge, PA 19482 (Tel 215-293-7200; FAX 215-293-7456)
Britesorb®; D Sodium Silicate; Metso Beads®; Metso Pentabead®; N® Sodium Silicate; Q-Cel; Quso®; Valfor®

Premier Malt Products, Inc., PO Box 36359, Grosse Pointe, MI 48236 (Tel 800-521-1057)
Amizyme; Exsize; Premier

Premier Refractories and Chemicals, Inc., 7887 Hub Pkwy., Valley View, OH 44125 (Tel 216-328-0200)
Magox®

Procter and Gamble Co., Industrial Chemicals Div., 301 E. 6th St., PO Box 599, Cincinnati, OH 45201 (Tel 513-983-5607)
AE-; CO-; P & G; Star; Superol

Pulcra S.A., Avda De Roma, 157, Apartado De Correos 37011, Barcelona 08011, Spain
Amfotex; Catinex; Complexion; Crafol; Cralane; Crapol; Depasol; Espesilor; Etilenox; Etoxi; Montosol; Montovol; Niox; Nioix; Puxol

QO Chemicals, Inc., Subsid. of Great Lakes Chem. Corp., 823 Commerce Dr., Oak Brook, IL 60521 (Tel 312-572-2308)
QO Chemicals (Australia) Inc., Suite #4, 21 Drummond Place, Carlton, Victoria, Australia 3053
QO Chemicals Inc. Belgium, Industriedok 391, B-2030 Antwerp, Belgium
QO Chemicals GmbH, Hermannstrasse 54, D-6078 Neu Isenburg, Germany
Kao-Quaker Co., Ltd., 14-10 Nihonbashi Kayabacho, 1-Chome, Chuo-ku, Tokyo 103, Japan
QO Chemicals Inc.-U.K., Spring House, 231 Glossop Rd., Sheffield S10 2GW, UK (Tel 0742-767842)
QO®

Quantum Chemical Corp., USI Div., 11500 Northlake Dr., PO Box 429550, Cincinnati, OH 45249 (Tel 513-530-6500)
Filmex®; Punctilious®; Spectratech®

Quest International, Lindtsedijk 8, 3336 le Zwijndrecht, The Netherlands
Admul

Quimigal-Quimica de Portugal E.P., Av. Infante Santo No. 2, 1300 Lisboa, Portugal
Quimipol

Ram Chemicals/Div. Lilly Industry, 210 E. Alondra Blvd., Gardena, CA 90248 (Tel 213-321-0710)
Garalease; Plastilease; Silicone Release; Sprayfilm; Spraywax

Reheis Inc., 235 Snyder Ave., Berkeley Heights, NJ 07922 (Tel 201-464-1500; FAX 201-464-7726)
Reheis (Ireland), Kilbarrack Rd., Dublin 5, Ireland
Reach®

Reichhold Chemicals, Inc., RCI Bldg., 525 North Broadway, White Plains, NY 10603 (Tel 914-682-5700)
Acofor; Peroxidol; Polylite; Staflex; Superox; Super Sta-Tac®; Varcum®

Reilly-Whiteman Inc., Washington & Righter Sts., Conshohocken, PA 19428 (Tel 215-828-3800; 800-533-4514; FAX 215-834-7855)
Ryco Inc./Div. Reilly-Whiteman Inc., 801 Washington St., Conshohocken, PA 19428
Carnapol; Petrosan; Rycel; Rychem®; Ryco®; Rycofax®; Rycolube®; Rycomid®; Rycowax®

Releasomers Inc., PO Box 82, Bradfordwoods, PA 15015 (Tel 412-452-4474; FAX 412-452-1965)
BL; RC 7; RR 5; XK 22; XR 7; XT 66

Rhein Chemie, 1008 Whitehead Rd. Extension, Trenton, NJ 08638 (Tel 216-784-1029)
Rheinau GmbH, Muelheimerstrasse, 21-28, D-6800 Mannheim 81, Germany
Aflux; Aktiplast; Antilux; Antistaticum; Chloropren-Faktis; Faktis; Rhenocure; Rhenodiv; Rhenofit; Rhenomag; Rhenopor; Rhenosin; Rhenosorb; Rhenovin

Rheox, Inc., PO Box 700, Hightstown, NJ 08520 (Tel 609-443-2500; FAX 609-443-2446)
Rheox, Inc., Rue de l'Hôpital 31, B 1000 Brussels, Belgium (Tel 02-512-0048)
Baragel®; Benathix; Bentone®; Macaloid®; M-P-A®; Nalzin; Nykon®; Oncor; Polytrope; Post; Rheolate; Thixatrol; Thixcin®; Thixseal

Rhone-Poulenc Basic Chemical Co., One Corporate Dr., Shelton, CT 06484 (Tel 800-642-4200)
Chelon; Perk; Questex

Rhone-Poulenc Performance Resins and Coatings, Industrial Chemicals, 1525 Church St. Ext., Marietta, GA 30060 (Tel 404-422-1250; FAX 404-427-0874)
Rhone-Poulenc Industries, 22 Ave. Montaigne, 75360 Paris Cedex 08, France
Rhone-Poulenc Chimie de Base, Division Specialties Chimiques, Dept. Biochimie, 18, Avenue d'Alsace - F - 92400 Courbevoie, Paris, France
Ceraloid; Colloid; Levilite; Musloid; Pentex; Perlex; Perlextra; Rhodorsil

Rhone-Poulenc Surfactants and Specialty Chemicals, PO Box 425, Cranberry, NJ 08512 (Tel 609-395-4644)
Acryloft; Alcamate; Alconate®; Alipal®; Alkafloc; Alkalube; Alkamide; Alkamidox; Alkaminox; Alkamox; Alkamuls; Alkaphos; Alkapol; Alkaquat; Alkaquest; Alkaril; Alkasperse; Alkasurf; Alkateric; Alkatronic; Alkatrope; Alkazine; Antara®; Antaron; Antarox®; Antifoam; Biopal®; Blancol®; BLO®; Butoxyne; Corrosion Inhibitor; Cyclochem; Cyclomide; Cyclomox; Cyclopol; Cycloryl; Cyclosheen; Cycloteric; Cycloton; Dermalcare; Duowet; Emulphogene®; Emulphor®; Gafac®; Gafamide®; Gaffix®; Gafquat®; Gafsoft; Gafstat; Gafterge; Igepal®; Igepon; Indu-Sol; Katanol; Lan-O-Derm; Miramine®; Miranol®; Mirapol®; Mirapon®; Mirataine®; Pegol; Rodea; Sipex®; Sipon®; Siponate; Siponic®; Sipothix

Rhone-Poulenc SpA, Via Milano 78, 20021 Ospiate Di Bollate, Milano, Italy (Tel (02-3503212; FAX 02-3501770)
Geronol; Geropon; Kobate; Soprofor; Supragil; Vanisol

Rewo. See Sherex

Rilsan. See Atochem

RITA Corp., 1725 Kilkenny Court, PO Box 585, Woodstock, IL 60098 (Tel 815-337-2500; FAX 815-337-2522)
Represented by: Maprecos, 4, Rue des Passe-Loups, 7770 Fontaine, Le Port, France
Acritamer; Barre®; Bovinol; Forlan; Grillocin; Grilloten; H.P.X.; Laneto; Pationic; Patlac; Ritacetyl®; Ritachol®; Ritaderm®; Ritahydrox; Ritalafa®; Ritalan®; Ritasol; Ritawax; Ritoleth; Rotolan®; Simchin; Super-Sat; Super Wet

Robeco Chemicals Inc., 99 Park Ave., New York, NY 10016 (Tel 212-986-6410; FAX 212-986-6419)
Cetina; Liquester; Robane®; Robecote; Robeyl; Robuoy; Solidester; Spermwax; Supraene®

Rohm & Haas Co., Independence Mall West, Philadelphia, PA 19105 (Tel 215-592-3000)
Rohm & Haas Canada Inc., 2 Manse Rd., West Hill, Ontario M1E 3T9 Canada
Rohm & Haas (Australia) Pty. Ltd., 969 Burke Rd., PO Box 115 Camberwell, Victoria 3124 Australia
Rohm and Haas (UK) Ltd., Lennig House, 2 Mason's Avenue, Croydon CR9-3NB, England
Rohm & Haas Co. European Operations, Chesterfield House, Bloomsbury Way, London WC1A 2TP, UK
Rohm and Haas Asia, Ltd., Kaisei Building, 8-10 Azabudai 1-Chome, Minato-ku, Tokyo 106 Japan
Acryloid®; Acrysol®; Amberol; Emulsion E-1614; Fore®; Karathane®; Kathon®; Lenkanol; Oropon; Paraplex®; Plexol®; Polymer QR; Primafloc®; Primapel®; Skane®; Tamol®; Triton®

Rona, Div. of EM Industries, Inc., 5 Skyline Dr., Hawthorne, NY 10532 (Tel 914-592-4660)
Represented by: S. Black (Import & Export) Ltd., The Colonnade, High St., Chesunt Hents EN8 0DJ, UK
Mibiron; Nailsyn; Naturon; Soloron; Timiron®

Ronsheim & Moore Ltd., Div. of Hickson & Welch Ltd., Ings Lane, Castleford, Yorkshire WF10 2JT, UK
Aremsan; Lanbritol

Ross Chemical, Inc., 303 Dale Drive, Fountain Inn, SC 29644 (Tel 803-862-4474)
Foam Bust; No Stik; Ross

Frank B. Ross Co., Inc., PO Box 4085, Jersey City, NJ 07304-0085 (Tel 201-433-4512)
Represented by: Harrisons & Crosfield (Canada) Ltd., St. Laurent, Quebec, Canada H4M 2N6 (Tel 514-748-7911)
Antisun®; Polymel; Ross Wax; Sunproofing Wax 1343

Ruetgers-Nease Chemical Co., Inc., Subsid. of Rütgerswerke AG, 201 Struble Rd., State College, PA 16801 (Tel 814-238-2424; FAX 814-238-1567)
Naxonate®; Zinc Pyrion

Ryco Inc. See Reilly-Whiteman Inc.

Ryoto Co., Ltd., Sun Bldg., 13-12 Ginza 5-Chome, Chuo-ku, Tokyo 105, Japan
Ryoto

Sachtleben Chemie GmbH, Pestalozzistrasse 4, D-4100 Duisburg-Homberg, Germany (Tel 02136-22-0; FAX 021-36-22-660)
Blanc fixe; Hombitan®; Kaolin; Lithopone; Sachtolith®

L.A. Salomon Inc., A Sud-Chemie AG Company, 150 River Rd., Suite L-3B, Montville, NJ 07045 (Tel 201-335-8300; FAX 201-335-1236)
Stealim; Tonsil

Sandoz Chemicals Corp., 4000 Monroe Rd., Charlotte, NC 28205 (Tel 704-331-7038)
Sandoz Chemicals Corporation, Dorva, Quebec H9R 4P5, Canada
Sandoz Ltd./Chemicals Div., Lichtstrasse 35, CH-4002 Basel, Switzerland
Sandoz Ltd., Calverley Lane, Horsforth, Leeds LS18 4RP, UK
Cartaretin®; Ceranine; Cerol; Chemical Base; Dilasoft; Ekaline; Fungicide; Hydro-Pruf; Imerol; Keripon; Lyogen; Mercerol; Nilofoam; Sandin; Sandocorin; Sandogen; Sandopan®; Sandostab; Sandoxylate®; Sandoz; Sanduvor; Silsmooth; Tetranol; Velsan

Santech Inc., 150 Norseman St., Toronto, Ontario M8Z 5M4 Canada (Tel 416-233-1234)
SunShade

Sanyo Chemical Industries, Ltd., 11-1 Ikkyo Nomoto-cho Higashiyama-ku, Kyoto 605, Japan

Carrybon; Carryol; Cation; Chemiflex; Chemistat; Colorin; Ionet; Lebon; Newpol; Nonipol; Profan; Sandet; Sanmorin; Sanstat; Sanyo

Sartomer Co., Oaklands Corp. Center., 468 Thomas Jones Way, Exton, PA 19341 (Tel 215-363-4100)

Sartomer International, Inc., Kingswick House, Sunninghill, Berkshire SL5 7BH, UK (Tel 44-99023491; FAX 44-99026176)

Sartomer International, Inc., Bankastraat 131d, 2585 El Den Haag, The Netherlands

Sartomer Int'l. Deutschland GmbH, Konigsallee 60F, 4000 Dusseldorf 1, Germany

Sartomer International, Inc., 7500A Beach Rd., Unit No. 14-313 The Plaza, Singapore

Chemlink®; Escacure®; Macrobase; Macromer®; Saret®; Sartomer; SMA®; SR-

Schenectady Chemicals, Inc., PO Box 1046, Schenectady, NY 12301 (Tel 518-370-4200)

Isonox; SP-; SRF-

Scher Chemicals Inc., Industrial West & Styertowne Rd., Clifton, NJ 07012 (Tel 201-471-1300; FAX 201-471-3783)

Represented by: Chesham Chemicals Ltd., Cunningham House, Bessborough Rd., Harrow HA1 3DU, UK

Colsol; Dipsal; Katemul; Lubrisol; Nyloset; Polysoft; Repello; Schercamox; Schercemol; Schercoat; Schercodine; Scherco Finish; Schercolube; Schercomid; Schercomul; Schercopearl; Schercophos; Schercopol; Schercoquat; Scherco Softener; Serhcosol; Schercotaine; Schercotarder; Schercoterge; Schercoteric; Schercozoline; Scheroba; Schersoftoil

SCM Chemicals, 7 St. Paul St., Suite 1010, Baltimore, MD 21202 (Tel 301-783-1120; 800-638-3234)

SCM Chemicals Ltd., PO Box 26, Grimsby, South Humberside DN37 8DP, UK (Tel 44-469-571000)

SCM Chemicals Ltd., PO Box 465, Auburn, New South Wales 2144, Australia

SCM Chemicals Ltd., Hop Hing Centre, 22nd Floor, 8-12 Hennessy Rd., Wanchai, Hong Kong

Prosil®

Scott Bader Co. Ltd., Wollaston, Wellingborough, Northamptonshire NN9 7RL, UK

Crystic; Polidene; Texicote; Texicryl; Texigel

Seppic, 75 Quai d'Orsay, 75321 Paris Cedex 07, France

Amonyl; Lanol; Montegal; Simulsol; Solamine; Solanos

Servo Chemische Fabriek B.V., PO Box 1, 7490 AA, Delden, Holland

Fungitrol; Nuodex; Ser-AD; Serdas; Serdet; Serdolamide; Serdox; Sermul; Serpol; Servamine; Servo; Servoxyl; Serwet; Super Ad It

Shell Chemical Co., 2001 Kirby Dr., Houston, TX 77019 (Tel 713-526-4631)

For International Sales: Pecten Chemicals, Inc., One Shell Plaza, Houston, TX 77252-9932 (Tel 713-241-6161)

Shell Chemicals UK Ltd., 1 Northumberland Av., London WC2N 5LA, UK (Tel 71-934-1234)

Shell Chimie France, 27 Rue de Berri, 7539, Paris Cedex 08 France

Shell International Chemical Co., Ltd./Agrochemicals Div. (CSAA), Shell Centre, London SE1 7PG, UK

Shell Nederland Chemie B.V., PO Box 187, 2501 CD, The Hague, Netherlands

Butyl Dioxitol; Butyl Oxitol; Dioxitol; Dobanol; Ionol; Neodene®; Neodol®; Oxitol; PentoXone; Shell; Shellwax; Sulfolane; Teepol

Sherex Chemical Co., Inc., Subsid. of Schering AG, Box 646, Dublin, OH 43017 (Tel 614-764-6500; 800-848-7370)

Nihon Schering KK, 6-64 Nishimiyahara, 2-Chome, Yodogawa Ku, Osaka 532, Japan

Admex; Adogen®; Adol®; Arosurf®; Ficel®; Kelex; Shur-Coal; Spermafol®; Starfol®; Talofloc®; Varamide®; Varidri; Varine; Varion®; Variquat®; Varisoft®; Varonic®;

Varstat®; Varsulf®

Rewo Chemische Werke GmbH, Postfach 1160, D-6497 Steinau, Germany (Tel 011-49-6663-540)

Rewo Chemical Ltd., Gorsey Lane, Widnes, Cheshire WA8 0HE, England

Rewocid®; Rewocor; Rewocoros; Rewoderm®; Rewolan®; Rewolub; Rewomat; Rewomid®; Rewomine; Rewomox; Rewopal®; Rewophat; Rewopol®; Rewopon®; Rewoquat; Reworyl®; Rewowax

Sherex Polymers, Inc., 2525 S. Combee Rd., Lakeland, FL 33801 (Tel 813-665-6226)

Seycopon; Surfac®; Surfam

Soitem-Societa Italiana Emulsionanti Srl, Via R. Cozzi, 34, 20125 Milano, Italy

Soi Bact; Soi Mul; Soitem

Solem Industries, Inc., Subsid. of J.M. Huber Corp., 4940 Peachtree Industrial Blvd., Norcross, GA 30071 (Tel 404-441-1301; FAX 404-368-9908)

CM-; FRE; G-431; H-36; Halofree; Hyfil; LV; Micral®; SB-; Solem; Zerogen

Soluol Chemical Co., Green Hill & Market Sts., Box 112, W. Warwick, RI 02893 (Tel 401-821-8100; FAX 401-823-6673)

Antifoam; Aquagard; Examide; Solusoft

Specialty Products Co., 75 Montgomery St., Box 306, Jersey City, NJ 07303 (Tel 201-434-4700)

Kantstik

Standard Chemical Co., Mill Lane, Cheadle, Chesire SK8 2NX, UK

Atolex; Sufatol

Stepan Co., 22 West Frontage Rd., Northfield, IL 60093 (Tel 312-446-7500)

Stepan Europe, BP127, 38340 Voreppe, France (Tel 01133-7650-8133)

Agent; Ammonyx; Amphosol; Bio Soft®; Bio Terge; BTC®; Catigene; Catisol; Kessco; Lathanol®; Makon®; Maprofix®; Maprolyte; Millifoam; Nacconol®; Neobee®; Ninate®; Ninex; Ninol®; Ninox®; Pestilizer; Petrostep; Polyfac; Polystep®; Secosol®; Secoster®; Secosyl; So/San; Steol®; Stepan; Stepanate®; Stepanflote®; Stepanhold®; Stepan-Mild®; Stepanol®; Stepanquat®; Stepantan®; Stepantex®; Stepfac®; Super Amide; Super-Pro; Surco; Toximul®; Wecobee®

Stepan Co./PVO Dept., 100 West Hunter Ave., Maywood, NJ 07607 (Tel 201-845-3030)

CC-; Cobee; Compound 9264B; Drewmulse; Drewpol; Drewsoft; Druspin; GC-44-14; Lipal

Stepan Canada, Inc., 90 Matheson Blvd. West, Suite 201, Mississauga, Ontario L5R 3P3, Canada (Tel 416-507-1631; FAX 416-507-1633)

Cedepal; Emulphor; Ninol; Steol; Stepanol

Stoner Inc., 1070 Robert Fulton Hwy., PO Box 65, Quarryville, PA 17566 (Tel 717-786-7355; 800-227-5538; FAX 717-786-9088)

Stoner

Struk-Chem, 201 E. Steels Corners Rd., Stow, OH 44224 (Tel 216-928-5188; 800-327-8649; FAX 216-928-8726)

Struktol

Sutton Laboratories, Inc., 116 Summit Ave., PO Box 837, Chatham, NJ 07928-0837 (Tel 201-635-1551; FAX 201-635-4964)

Represented by: Blagden Campbell Chemicals Ltd., A.M.P. House, Dengwall Rd., Croydon, Surrey CR9 3QU, UK

Germaben®; Germall®; Suttocide®

Swastik Household & Industrial Products Ltd., Shahibag House, 13 Walchand Hirachand Marg, Ballard Estate, Bombay 400 038, India

Idet; Swanic

Sybron Chemicals Inc., PO Box 125, Wellford, SC 29385 (Tel 803-439-6333; 800-845-7574; FAX 803-439-1612)
Cellopal; Plexene; Tanapal; Tanaquad

Synthetic Products Co., 301 Barnum Ave. Cutoff, Stratford, CT 06497 (Tel 203-377-5550)
Petrac®; Synpron

Taiwan Surfactant Corp., 8-1 Floor, No. 106, Sec. 2, Changan East Road, Taipei, Taiwan, R.O.C.
Ablumine; Ablunol; Abluphat; Ablupol; Ablusoft; Ablusol; Abluton

Takemoto Oil & Fat Co., Ltd., No. 5, Sec. 2, Minato-Machi, Gamagori, Aichi 443, Japan
Chupol; Delion; Honoralin; Newkalgen; Polefine; Salabon; Sofbon

Tego Chemie Service USA, Div. of Goldschmidt Chemical Corp., PO Box 1299, 914 E. Randolph Rd., Hopewell, VA 23860 (Tel 804-541-8658; 800-446-1809; FAX 804-541-2783)
Tego® Airex

Tennessee Corp., Suite 401, 3400 Peachtree Rd. N.E., Atlanta, GA 30326 (Tel 404-239-6700; FAX 404-239-6701)
Sulfonate; Sulfonic; Tenn-Cop

Terry Laboratories, 3270 Pineda Ave., Melbourne, FL 32940 (Tel 407-259-1630; FAX 407-242-0625)
Aloe-Moist

Tessilchimica SpA, La, Via Baertsch 1, 24100 Bergamo, Italy
Algonina; Chimin; Chimipal; Cosmopon; Lanotex; Tequat

Texaco Chemical Co., PO Box 27707, Houston, TX 77227-7707 (713-235-6304; 800-231-3107)
S.A. Texaco Belgium N.V., Int'l. Congress Center, Citadel Park, B-900 Ghent, Belgium (Tel 011-32-91-41-5920)
Texaco Chemical Deutschland GmbH, Baumwall 5, 2000 Hamburg 11, Germany (Tel 011-49-40-36-3737)
Texaco Ltd., 195 Knightsbridge, London SW7 1RU, UK (Tel 011-411-584-5000)
Texaco France S.A., 5, rue Bellini, Tour Arago, F-92806 Puteaux Cedex, France (Tel 011-33-1-47-78-1655)
Sanseki-Texaco Chemicals Co., Ltd., Kasumigaseki Bldg., PO Box 90, 2,5-, 3-Chome Kasumigaseki, Chiyoda-Ku, Tokyo 100, Japan (Tel 03-580-3611)
Accelerator; Diglycolamine® Agent (DGA®); Jeffamine®; Jeffox®; Skellite®; Surfonic®; Texacar®; Texacat®; Texox; Texsolve

Texo Corp., 2801 Highland Ave., Cincinnati, OH 45212 (Tel 513-731-3400)
Texo

Toho Chemical Industry Co., Ltd., 1-14-9 Kakigara-cho, Nihonbashi, Chuo-ku, Tokyo 103, Japan
Airrol; Anstex; Catinal; Demulfer; DRA; Emulbon; Neospinol; Nonal; Obanol; Obazoline; Pegnol; Pepol; Pronal; Runox; Sofnon; Sorbon; Tohol

Tokai Seiyu Ind. Co. Ltd., 67, 2-Chome, Yamadahigashimachi, Higashi-ku, Nagoya, Japan
Ospin; Penestrol; Softal

Tomah Products, Inc. See Exxon Chemical Americas

Tosoh Corp., 1-7-7, Akasaka, Minato-ku, Tokyo 107, Japan (Tel 03-585-6707; FAX 03-582-7846)
Tosoh USA Inc., 1700 Water Place, Suite 204, Atlanta, GA 30339 (Tel 404-9956-1100; FAX 404-956-7368)
Tosoh Canada Ltd., 1200 Sheppard Ave. East, Suite 511, Willowdale, Ontario M2K

2S5, Canada (Tel 416-756-2226)
Tosoh Europe B.V., World Trade Center Amsterdam, Tower C-Floor 13, Strawinskylaan 1351, 1077 XX Amsterdam, The Netherlands (Tel 020-6644026; FAX 020-6623412)
Con-BACN

Trans-Chemco, Inc., 19235 84th St., Bristol, WI 53104 (Tel 414-857-2363)
Trans

Thomas Triantaphyllou S.A., 405 Tatoiou Av., TK 136 71, Acharnes, Athens, Greece
A.B.S.; Loropan

Tri-K Industries, Inc., 466 Old Hook Rd., PO Box 312, Emerson, NJ 07630 (201-261-2800; FAX 201-261-1432)
Abiol; JF 77; Kelate; Trisept; Tristat; Uvasorb

Troy Chemical Corp., Inc., One Avenue L, Newark, NJ 07105 (Tel 201-589-2500; FAX 201-589-0490)
Actan; Lecithin; Troykyd; Troymax; Troysan; Troythix

TSE Industries, Inc., Millathane Div., 5260 113th Ave. North, POB 17225, Clearwater, FL 33520-7225 (Tel 813-576-5643; 800-621-4141; FAX 813-572-0487)
Activator; EZ Mold Lubricant; M-7 Mold Release®; Millathane; Thanecure®; TSE; TSE Mold Release®

Unical, 1201 W. 5th St., Los Angeles, CA 90017 (Tel 213-977-7181)
Amsco

Unichema Chemicals, Inc., 4650 S. Racine Ave., Chicago, IL 60609 (Tel 312-376-9000; 800-833-2864; FAX 312-376-0095)Unichema International, Postfach 1280, D-4240 Emmerich, Germany
Dar Chem; Unislip; Uniwax

Unimin Specialty Minerals Inc., 258 Elm St., New Canaan, CT 06840 (Tel 203-966-8880)
Imsil®; Minex; Silver Bond; Snow White; Tamsil

Union Camp Corp., 1600 Valley Rd., Wayne, NJ 07470 (Tel 201-628-2000)
Cenwax; Uniflex; Unimate

Union Carbide Corp., Specialty Polymers & Composites Div., 39 Old Ridgebury Rd., Danbury, CT 06817-0001 (Tel 203-794-2000; 800-621-1972)
Union Carbide Agricultural Products Co., Inc., PO Box 12014, Research Triangle Park, NC 27709
Union Carbide Canada Ltd., 10455 Metropolitan E, Montreal East, Quebec, Canada H1B 1A1 (Tel 514-493-2610)
Union Carbide Japan KK, Toranomon 45 Mori Bldg., 1-5 Toranomon, 5-Chome Minato-Ku, Tokyo 105, Japan (Tel 813431-7281)
Union Carbide U.K. Limited, Rickmansworth WD 3 1RB Herts, England
Union Carbide Europe S.A., 11, Ave. Choiseul, CH-1290, Versoix (Geneva), Switzerland (Tel 022-55-47-42)Union Carbide Brazil, Rua Dr. Eduardo De Souza Aranha, 153, Sao Paulo 04530, Brazil
Aqucar; Carbowax®; Cellosize®; Flexol®; Formasil®; Morlex®; Niax®; Polyox®; Poly-Tergent®; Sag®; Silcat®; Silwet®; Tergitol®; Thornel; Ucar®; Ucarcide®; Ucare®; Ucarsan®; Ucarsil®; Uconex®; Ucon®; Union Carbide®; Volatile Silicone; ZB

Uniroyal Chemical Co., Inc., Middlebury, CT 06749 (Tel 203-573-3880; 800-243-3024; FAX 203-573-3393)
Uniroyal Chemical, Brooklands Farm, Cheltenham Rd., Evesham, Worcestershire WR116LW, UK (Tel 011-44-386-47251)
Uniroyal Chimica, S.P.A., Corso Vinzaglio 35, 10121 Torino, Italy
Uniroyal Chemical, Cathay Bldg., 11, Dhoby Ghaut, Suite #14-05, Singapore 0922

Uniroyal Quimica S.A., Avenida Morumbi, 7.029 (CEP 05650) Caixa Postal 30.380, 0100 Sao Paolo, SP- Brazil
Celogen®; Durazone; Naugalube®; Naugard®; Naugawhite; Polywet; Thiostat®; Tonox®

United States Borax & Chemical Corp., 3075 Wilshire Blvd., Los Angeles, CA 90010 (Tel 213-251-5400)
Borax; Firebrake®

UOP, Inc., 25 East Algonquin Rd., Box 5017, Des Plaines, IL 60017-5017 (Tel 312-391-3300; 800-348-0832; FAX 312-391-2758)
Aldomax; Ketomax®; Sustane®; Unilink

Upjohn Co., PO Box 685, La Porte, TX 77571 (Tel 713-473-8161)
Curithane

Van Den Bergh Foods Co., 2200 Cabot Dr., Lisle, IL 60532 (Tel 708-505-5300)
Dur-Em®; Durfax®; Durlac®; Durpeg®; Durtan®; Ice®; KLX; Santelle; Santone®; Tally®

R.T. Vanderbilt Co., Inc., 30 Winfield St., Norwalk, CT 06855 (Tel 203-853-1400)
Activ; Agerite®; Altax®; Amax®; Amyl Cadmate®; Amyl Ledate®; Amyl Zimate®; Antozite®; Baerostab®; Bilt-Cote; Bismate®; Butyl Eight; Butyl Namate®; Butyl Tuads®; Butyl Zimate®; Captax®; Ceramitalc; Continental; Cumate®; Darvan®; Dixie® Clay; Ethyl Cadmate®; Ethyl Selenac®; Ethyl Tellurac®; Ethyl Tuads®; Ethyl Zimate®; Isobutyl Niclate; I.T.; Jet Fil®; Langford®; McNamee®; Methyl Cumate®; Methyl-Ethyl Tuads®; Methyl Ledate®; Methyl Niclate; Methyl Selenac®; Methyl Tuads®; Methyl Zimate®; Molyvan; Morfax®; Nacap; Na-Sul; Nytal®; Octoate®; Par® Clay; Peerless; Plastogen; Propyl Zithate®; Pyrax®; Redax; Reogen®; Rheotol; Rhodigel; Rhodopol; Rokon; Rotax®; Select-A-Sorb; Setsit®; Sulfads®; Telloy; Thiate®; Vanate; Vanax®; Vanblack; Vanchem®; Vancide®; Vandex; Vanfre®; Van Gel®; Vanlube; Vanox®; Vanplast®; Vansil®; Vanstay; Vantalc®; Vanwax®; Varox®; Veegum®; Zetax®

Van Dyk, Div. of Mallinckrodt, Inc., 11 William St., Belleville, NJ 07109 (Tel 201-759-3225; FAX 201-759-5279)
Bi-Lite®; Ceraphyl; Cerasynt; Emulsynt; Escalol®; Foamole; Pearl-Glo®; Spectra-Pearl®

Van Schuppen Chemie, Nieuweweg, Veenendaal, Netherlands
Dusoran; Lanolin Alcohols; Solwax

Velsicol Chemical Corp., 5600 N. River Rd., Rosemont, IL 60018-5119 (Tel 800-843-7759; IL 800-225-4208; FAX 708-698-9714)
Benzoflex; Firemaster

Vikon Chemical Co., Inc., PO Box 1520, Burlington, NC 27215 (Tel 919-226-6331)
Dai Cari; Merkyl; Plex; SEQ; Vikol®; Vikosperse

Vista Chemical Co., 15990 N. Barker's Landing Rd., POB 19029, Houston, TX 77224 (Tel 713-531-3200; FAX 713-531-3236)
Vista Chemical Europe, Hilton Tower, Blvd. de Waterloo #39, 81000 Brussels, Belgium (Tel 32-2-513-7490)
Vista Chemical Far East Inc., Kasumigaseki Bldg., 25th floor, POB 110, Tokyo, Japan 100
Alfol®; Alfonic®; AXS; Catapal®; Nalkylene®; Novel®; Vista

Wacker Silicones Corp., 3301 Sutton Rd., Adrian, MI 49221 (Tel 517-264-8500; FAX 517-263-5711)
Antifoam Compd.; Antifoam Emulsion; E-; F-; Q-

Wacker-Chemie GmbH, Div. L, Prinzregentenstrasse 22, D-8000, Munchen 22, Germany (Tel 089-2109-0; FAX 089-2109-1772)
Wacker Chemicals Ltd., The Clock Tower, Mount Felix, Bridge St., Walton-on-

Thames, Surrey, KT12 1AS, UK (Tel 932-246111; FAX 932-240141)
Wacker-Chemie Danmark A/S, Park Alle 380 A, Postboks 170, DK-2625 Vallensbaek, Denmark
Wacker Quimica Ibérica, S.A., Córcega, 303-2° 3a, E-08008 Barcelona, Spain
Vinnapas®; Vinnol®

Jim Walter Resources, Inc., PO Box 5327, Birmingham, AL 35207 (Tel 205-254-7481)
PMF®

Westbrook Lanolin Co., Argonaut Works, Laisterdyke, Bradford, Yorks BD4 8AU, UK
Albalan; Aqualose; Argobase; Argonol; Argowax; Golden Dawn; Golden Fleece; Lanesta; Sebase

Westvaco Chemical Div., PO Box 70848, Charleston Hts., SC 29415 (Tel 803-740-2300)
Air-Plas®; Diacid®; Indulin®; Polyfon; Polytard®; Reax®; Westvaco

Whitney & Oettler, PO Box 8024, Savannah, GA 31402 (Tel 912-232-7166)
Butac®; Natac®; Turgum®

Whittaker, Dayton Div., 10 S. Electric St., West Alexandria, OH 45381 (Tel 513-839-4612; FAX 513-839-5615)
Petaflex; Thixon®

Witco Corp., 1000 Convery Blvd., Perth Amboy, NJ 08862-1932 (Tel 201-826-7777; 800-231-1542)
Witco Canada Ltd., 2 Lansing Sq./Suite 1200, Willowdale, Ontario M2J 4Z4 Canada (Tel 416-497-9991)
Witco B.V., PO Box 5, Koogaan de Zaan, the Netherlands (Tel 31-75-283854)
Witco Chemical Ltd. (UK), Union Lane, Droitwich, Worcester WR9 9BB, UK
Witco S.A., Rue Gravetel, Saint Pierre les Elbeuf 76320, France
Witco S.A./Cyclo Div., 10, Rue Cambaceres, 75008 Paris, France
Witco Ltd., PO Box 10245, 26112 Haifa Bay, Israel
Adsee®; Agent; Barium Petronate; Bubble Breaker®; Calcium Petronate; Cyclol; Drysperse; Emargol®; Emcol®; Emphos; Ethoxamine; Fomrez; Hybase; Kadif; Keycide®; Petromixes; Petronate; Pyronate; Sochamine; Sorban; Sorbanox; Sulframin; Sunolite®; Synthionic; Syntopon; Ultra; Ultra Sulfate; Witcamide®; Witcamine®; Witcarb®; Witco®; Witcolate; Witcomul; Witconate; Witconol; Witcopaque; Witcor

Witco Corp./Argus Chem. Div., 520 Madison Ave., New York, NY 10022 (Tel 212-605-3600)
CHP-; CPF-; Drapex; Esperal®; Espercarb®; Esperfoam®; Esperox®; Flexchlor; Fyarestor; Hi-Point®; Mark; Markstat; OA 502; Quickset®; Sprayset; TBHP; USP®

Witco Corp./Golden Bear Div., 10100 Santa Monica Blvd., Los Angeles, CA 90067-4183 (Tel 213-277-4511; FAX 213-201-0383)
Bearflex®; Califlux®; Cyclolube®

Witco Corp./Humko Chem. Div., PO Box 125, Memphis, TN 38101-0125 (Tel 901-320-5800)
Hystrene®; Industrene®; Intermediate; Kemamide®; Kemamine®; Kemester®; Kemstrene®; Neustrene®

Witco Corp./Organics Div., 1000 Convery Blvd., Perth Amboy, NJ 08862-1932 (Tel 201-826-7777; 800-231-1542)
Aluminum Stearate EA; Calcium Stearate; Liquazinc; Lubracal®; Lubrazinc®; Zinc Stearate

Witco Corp./Sonneborn Div., 520 Madison Ave., New York, NY 10022-4236 (Tel 212-605-3912)
Benol®; Blandol®; Britol®; Carnation®; Ervol®; Fonoline®; Gloria®; Kaydol®; Klearol®; Multiwax®; Orzol®; Perfecta®; Petrolatum RPB; Protol®; Protopet®; Rudol®; Semtol®; Sonojell®; Tech Pet

Yoshimura Oil Chemical Co., Ltd., Minami 5-chrome-1-1 Honan-cho, Toyonaka-shi, Osaka 561, Japan

Clink; Permax

Zeelan Industries Inc., 220 Endicott Bldg., PO Box 1578, St. Paul, MN 55101 (Tel 612-292-9271)

Zeeospheres®

Zinc Corp. of America, Fourth & Delaware, Palmerton, PA 18071 (Tel 412-773-2295; 800-962-7500)

Horse Head®; Kadox®; Protox®

Zohar Detergent Factory, Kibbutz Dalia, Israel 18920

DDBS 100; Labs; Lauramide; Texamine; Zoharquat; Zoharsoft

Zschimmer & Schwarz GmbH & Co., Max-Schwarz-Strabe 4–5, Postfach 2179, D-5420 Lahnstein, Germany

Zschimmer & Schwarz-France S.a.r.l., 10 rue Saint-Marc, F-75002 Paris, France

Zschimmer & Schwarz-Italiana SpA, Casella Postale N. 1, I-13038 Tricerro (Vc), Italy

Zschimmer & Schwarz-Argentina S.A., Bdo. de Irigoyen 566-5° B, Buenos Aires, Argentina

Alphoxat; Autopoon; Autopur; Contraspum; Inhibitor; Mulsifan; Oxetal; Oxypon; Perlglanzmittel; Produkt; Propetal; Purton; Sulfetal; Sulfostat; Türkischrotöl; Zusolat; Zusomin